METRIC PREFIXES

PREFIX	ABBREVIATION	MEANING	POWER OF TEN
atto	a	quintillionth	10^{-18}
femto	f	quadrillionth	10^{-15}
pico	p	trillionth	10^{-12}
nano	n	billionth	10^{-9}
micro	μ	millionth	10^{-6}
milli	m	thousandth	10^{-3}
centi	c	hundredth	10^{-2}
deci	d	tenth	10^{-1}
deka	da	ten	10
hecto	h	hundred	10^{2}
kilo	k	thousand	10^{3}
mega	M	million	10^{6}
giga	G	billion	10^{9}
tera	T	trillion	10^{12}
peta	P	quadrillion	10^{15}
exa	E	quintillion	10^{18}

COMMON CONVERSIONS

(to three significant digits)

WEIGHT

1 kg = 2.20 lb
1 g = 0.0353 oz
1 oz = 28.3 g
1 lb = 0.454 kg
1 lb = 16 oz

CAPACITY

1 L = 0.264 gal
1 L = 1.06 qt
1 qt = 0.946 L
1 gal = 3.79 L
1 gal = 4 qt

LENGTH

1 cm = 0.394 in
1 m = 3.28 ft
1 km = 0.621 mi
1 in = 2.54 cm
1 ft = 0.305 m
1 mi = 1.61 km
1 mi = 5280 ft

FORCE

1 N = 0.225 lb
1 lb = 4.45 n

PRESSURE

1 kPa = 0.145 psi
1 psi = 6.90 kPa

WORK

1 J = 0.738 ft-lb
1 ft-lb = 1.36 J

TECHNICAL MATHEMATICS WITH CALCULUS

ELEANOR H. NINESTEIN
Fayetteville Technical Community College

HarperCollins*Publishers*

To Edward Elder House

Sponsoring Senior Editor: Anne Kelly
Developmental Editor: Louise Howe
Project Editor: Ann-Marie Buesing
Assistant Art Director: Julie Anderson
Text Design: Lucy Lesiak Design/Barbara Bauer
Cover Design: Lucy Lesiak Design/Lucy Lesiak
Cover Illustration/Photo: Pictor/Uniphoto, Inc.
Photo Researcher: Rosemary Hunter
Production Assistant: Linda Murray
Compositor: Progressive Typographers
Printer and Binder: R. R. Donnelley & Sons Company
Cover Printer: Lehigh Press Lithographers

Technical Mathematics with Calculus

Library of Congress Cataloging-in-Publication Data

Ninestein, Eleanor H.
Technical mathematics with calculus / Eleanor H. Ninestein.
p. cm.
Includes index.
ISBN 0-673-18748-9
1. Mathematics. I. Title.
QA37.2.N56 1992 91-14964
510--dc20 CIP

91 92 93 94 9 8 7 6 5 4 3 2 1

PREFACE

Technical Mathematics with Calculus provides a foundation in mathematics for students in engineering technology or engineering-related technical programs. It was developed to balance the needs of instructors and students in today's classrooms. Instructors, who must reach a large number of students with diverse backgrounds, abilities, and interests, will find that the organization of the text is a valuable feature. It was designed to be truly flexible, not just adaptable, to accommodate a variety of course outlines. The extended applications that appear throughout the text can be used as motivational material, and coding of the applied problems by field makes it unusually easy to choose appropriate examples and assignments for a particular class. Students, who may be trying to juggle school, job, and family responsibilities, will appreciate the book's readability, detailed explanations, and summary boxes. These features and others are described below.

KEY FEATURES

STUDENT-ORIENTED STYLE

The style is direct and accessible. Description and analogy are used instead of definition and proof for difficult concepts, but theory is introduced when it can be done at a level that is consistent with the style. In most cases, concepts are developed through a detailed explanation of a specific example. The method is then generalized and illustrated by additional examples.

VERSATILE ORGANIZATION

The organization allows unusual flexibility in topic selection and order of topics. The chapters of this text are grouped into units. Unit 1 contains the groundwork for the course and should be covered first. Additional prerequisite topics for any given chapter are treated in earlier chapters of the unit that contains it. The only restriction on ordering the chapters after Unit 1 is that before covering any chapter, prior chapters within the same unit should be covered. Before completing a unit, it is possible to branch to another unit (or units), then return to the original unit. This system gives the instructor great freedom, and allows mathematical concepts to be introduced in the sequence that best complements related topics in other courses. Suggested course outlines for various fields are given in the instructor's manual, but ideally, the math instructor and the instructors who teach the technical courses should work together to develop a course outline that best fits the needs of their students.

ABUNDANT EXAMPLES AND EXERCISES

The examples are numerous, carefully chosen, and detailed. Because students often find special cases to be confusing, these are identified and treated in the text. Examples progress steadily from easy to more difficult, and the exercises match the examples in level of difficulty. In most exercise sets, the last problem requires more insight or ingenuity than the routine problems. Frequently, these problems involve a little abstraction and can be used by the instructor as a springboard into more theoretical topics.

FIELD-SPECIFIC APPLICATIONS

A wide range of engineering fields are represented in the applied problems. In particular, applications are drawn from architectural technology, biomedical engineering technology, chemical engineering technology, civil engineering technology, computer engineering technology, electrical/electronics engineering technology, industrial engineering technology, mechanical engineering technology, and mechanical drafting and design technology. A panel of instructors from the various technical fields have carefully screened the applied examples and problems to ensure that the types of problems their students need are included. The members of this panel also supplied problems from their fields to supplement those of the author.

Applications are coded by a number bar like the one shown here. 2 4 8 | 1 5 6 7 9 Each number corresponds to one of nine fields of study, as indicated by the key at the bottom of this page. (The same key appears on text pages where application problems occur.) The problems involve approximately twenty topics, such as statics, dynamics, hydrology, and electrical circuits, that are relevant to one or more of the technologies. If a problem involves a topic considered essential to a particular field, the number corresponding to that field appears in the colored portion of the bar. If the topic is related but nonessential, the number appears in the white portion of the bar. A problem is coded as relevant to a field if the *topic* is relevant, whether or not the specific situation described is from that field. Since engineers do not necessarily agree on what is essential or related knowledge, the coding is a subjective matter. It is intended primarily to provide guidelines for instructors who are not thoroughly familiar with all of the technical fields represented.

EXTENDED APPLICATIONS

Approximately half of the chapters close with an extended application. These features contain complete descriptions of the settings for the problems, which range from banking the curve of a highway to calibrating blood flow. While the examples and exercises show that mathematics is needed to solve technical problems, the extended applications give enough background to explain why someone might need to solve the problem.

PEDAGOGICAL DEVICES

Each section contains one or more boxes summarizing major points. In many cases, the summary is an algorithm for solving a particular type of problem. This feature is useful to students as they learn a procedure for the first time, and as they review it. It also makes the book a valuable reference after the course is completed. A **NOTE ⇨⇨** symbol appears in the margin to highlight commonly made errors, learning hints, or other points that warrant special attention. A second color is used for emphasis, and the end of each example is marked with a ■ for clarity.

EXTENSIVE CHAPTER REVIEWS

The last section of each chapter is a multidimensional review. Key terms, symbols, rules and formulas, and geometric concepts used in the chapter are listed. Each chapter contains approximately eighty-five review exercises, subdivided into four categories. In the *Sequential Review,* the exercises appear in the same order in which the concepts tested appear in the text, and the section in which the concepts are explained is identified. The *Random Review* is a shorter selection of exercises, ordered without regard to sequence. The student cannot depend on position in the exercises to tell him or her how to solve the problem. The *Applied Problems* provide approximately twenty technical applications per chapter. A short *Chapter Test* gives the student one more chance to practice solving a variety of problems, including applied ones, using the concepts presented in the chapter. Answers to odd-numbered problems from all sections are given in the back of the book. Additionally, answers to all of the problems in *Random Reviews* and *Chapter Tests* are provided. The answers have been meticulously checked for both mathematical and typographical errors.

INCLUSION OF GEOMETRY

Topics from geometry are integrated throughout the text. Basic geometrical concepts are explained in Unit 1, and the following chapters use them freely. Additional geometry arises naturally in the discussion of new concepts. Diagrams and graphs are used extensively to emphasize the role of geometry in problem solving.

UP-TO-DATE CALCULUS CONTENT

The text incorporates some of the recommendations stemming from the recent emphasis on reform of calculus instruction. For example, the definite integral is presented before the indefinite integral. Centroids, moments of inertia, and techniques of integration have been de-emphasized but not eliminated. As in the first twenty chapters, the stress is on an intuitive understanding of general principles.

ILLUSTRATION OF CALCULATOR SOLUTIONS

Sample keystroke sequences show how to use the function keys of a scientific calculator with algebraic logic. In many cases, a particular keystroke sequence is recommended for accuracy and efficiency. For more difficult topics, the strategy is explained before the suggested keystroke sequence is given.

GLOSSARY OF APPLIED TERMS

At the end of the text is a glossary of technical terms that are used in the exercises and examples. It is possible to solve most of the problems without knowing what the technical terms mean, and technological background is given in the text when it is needed. Students generally take a course in technical mathematics before they have taken many of their curriculum courses, so the glossary will help the student who wants a better understanding of the technical aspects of the problems.

SUPPLEMENTS

The ***Instructor's Resource Manual,*** prepared by Chris Diorietes, Dwight House, and the author, includes six chapter tests (four open-response and two multiple-choice) for each chapter of the text; suggested course outlines for each of the nine engineering fields; application field coding symbols for the examples that appear in the text; a series of brief articles on the history of mathematics for use in lectures; and computer programs in BASIC, which the instructor can incorporate as circumstances permit.

The ***HarperCollins Test Generator for Mathematics*** enables instructors to select questions for any section in the text, or to use a ready-made test for each chapter. Instructors may generate tests in multiple-choice or open-response formats, scramble the order of questions while printing, and produce twenty-five versions of each test (IBM and Macintosh). The system features printed graphics and accurate mathematical symbols. The program also allows instructors to choose problems randomly from a section or problem type or to choose questions manually while viewing them on the screen, with the option to regenerate variables if desired. The editing feature (IBM and Macintosh) allows instructors to customize the chapter data disks by adding their own problems.

The ***Instructor's Solutions Manual,*** prepared by Susan Ritter and Beverly Hall, contains worked-out solutions to all of the even-numbered exercises that appear in the text.

The ***Student's Solutions Manual,*** prepared by Susan Ritter and Beverly Hall, contains worked-out solutions to all of the odd-numbered exercises that appear in the text. These solutions are given with more detail than those in the ***Instructor's Solutions Manual.***

Interactive Tutorial Software is available in both Apple and IBM versions. It offers interactive modular units for reinforcement of selected topics. The tutorials are self-paced and provide unlimited opportunities to review lessons and to practice problem solving. When a student gives an erroneous answer, he or she can request to see the problem worked out. The program is menu-driven for ease of use, and on-screen help can be obtained at any time with a single keystroke. Students' scores and elapsed time are automatically recorded and can be printed for a permanent record.

USING THE TEXT

Students in engineering programs usually have had some previous algebra. If most of Chapter 1, all of Chapter 2, and parts of Chapter 3 are treated as a review, each section can be covered comfortably in one class period. The explanations are detailed enough, however, that by slowing down the pace, instructors can use the book for courses without an algebra prerequisite. Covering the entire book would require four semesters, but the text contains many topics that are optional for the majority of technical programs. The portions of the text that are essential to most programs can be covered adequately in two or three semesters, or three or four quarters.

ACKNOWLEDGMENTS

It is generally agreed that a book on technical mathematics should contain the mathematics that technicians use in their work. A large number of reviewers helped ensure that this text contains the topics most frequently taught to technicians by the methods most commonly used. If all suggestions to elaborate and to add new material had been followed, however, the book would have grown to gargantuan size. Nevertheless, the reviewing process has led to significant improvements in the final product. I am indebted to the following reviewers for their insight and assistance at various stages of the reviewing process.

Neil Aiken	Milwaukee Area Technical College
A. David Allen	Ricks College
Richard Armstrong	St. Louis Community College–Florissant Valley
Weston H. Beale	Camden County College
Sandra Beken	Horry–Georgetown Technical College
Irving Chaimowitz	Broward Community College
Cheryl Cleaves	State Technical Institute of Memphis
Jan S. Collins	Embry–Riddle Aeronautical University
David M. Crystal	Rochester Institute of Technology
Betty Doversberger	Indiana Vocational Technical College
Rita M. Fischbach	Illinois Central College
John Gill	University of Southern Michigan
Thomas Hinson	Forsyth Technical College
Glenn Jacobs	Greenville Technical College
Judy Ann Jones	Madison Area Technical College
Joseph Jordan	John Tyler Community College
Maryann E. Justinger	Erie Community College–South Campus
Robert E. Lawson	San Antonio College
David Legg	Indiana University–Purdue University at Ft. Wayne
Lynn G. Mack	Piedmont Technical College
Paul Maini	Suffolk County Community College–Selden Campus
Mark Manchester	State University of New York–Morrisville
Robert J. Michaels	San Diego Mesa College
Barbara A. Miller	Lorain County Community College
Wesley J. Orser	Clark College
Douglas Peterson	Green River Community College
Catherine H. Pirri	Northern Essex Community College

Linda Rapp	Randolph Technical College
Doris Schraeder	Texas State Technical Institute
Ric Sexton	Wake Technical Community College
Mary Jane Smith	Tidewater Community College – Portsmouth Campus
Randall C. Sowell	Central Virginia Community College
Thomas J. Stark	Cincinnati Technical College
Susan E. Stokley	Spartanburg Technical College
Kenneth E. Stoll	University of Louisville
Richard Watkins	Tidewater Community College – Virginia Beach
Dr. Walter Weber	Catonsville Community College
Kelly Wyatt	Umpqua Community College
Anne L. Zeigler	Florence – Darlington Technical College

I am also grateful for the help of the following people in reviewing the applied problems.

William Casolara	Tompkins – Cortland Community College
C. Richard Coulson	Marion Technical College
Leon W. Heselton	Mohawk Valley Community College
Richard A. Honeycutt	Davidson County Community College
Daniel L. Kelly	Southwestern Oregon Community College
Daniel Landiss	Forest Park Community College
John McManus	Southern Technical College
Hari Mirchandani	Indiana Vocational Technical College
Ronald S. Nichols	Alfred State College

These individuals also submitted a number of applications that are included in the text. My special thanks go to Larry Oliver, Columbus State Community College, who supplied most of the applied problems in the review sections.

Special thanks also go to my department chairman, Michael D. McLaurin, for scheduling my classes to accommodate my writing and to Tom Cochran and Michael Schneider of Belleville Area College, who are to be commended for their meticulous checking of the answers.

I appreciate the hard work of the staff at HarperCollins Publishers: Anne Kelly, Louise Howe, Laurie Golson, Julie Anderson, Janet Tilden, and Ann Buesing, with whom I have worked closely, as well as many dedicated people I have not met.

My husband, Dr. V. Dwight House, of the department of mathematics and computer science at Fayetteville State University, has read portions of the manuscript and readily put aside his own work on many occasions to give me advice and support. I am also grateful for his cheerfulness each time he was confronted with yet another frozen pizza for dinner. And finally, hugs to my tiny son Edward, whose dimpled smiles remind me to smile.

Eleanor H. Ninestein

TO THE STUDENT

The author has made every effort to make this text one that is easy to read and understand. It is also important for you, the student, to realize that reading mathematics is inherently different from reading a novel. The following list of hints should help you to acquire good study habits.

1. **Attend class.** You'll learn more efficiently and effectively by combining your instructor's explanations with the explanations in the text than from relying on either one alone.

2. **Take notes.** Don't, however, try to write down everything your instructor says. It is important to focus your concentration on understanding rather than on writing. Write down the main points, but leave out things that can be filled in later. After class, go over your notes, and fill in the part you left out while the material is still fresh in your mind.

3. **Read the text.** Read it slowly and carefully. If there is a sentence you don't understand, read it again. If you still don't understand it, read on a little farther. There may be something in the next paragraph that will clear up the confusion. Work through the examples using pencil, paper, and calculator to be sure that you understand the procedure used. If you understand the steps, you will remember them, and memorization can be kept to a minimum. You should realize, too, that different calculators may give answers that are different in the sixth or seventh digit, depending on how they are programmed to do the calculations. You should not be alarmed if some of the answers you get are *slightly* different from the calculator displays shown in the text.

4. **Do your homework *every day.*** The key to learning mathematics is to master the assignments each day. If you get behind, it's hard to catch up.

5. **Check your progress.** Try to do the problems without looking back at your notes or your book. Check your answers with those in the back of the book after every three or four problems. If your answer is different, check to see if you copied the problem correctly. If you did, work it again from scratch instead of reading through your work. If you still disagree, ask your instructor about it.

6. **Look over the next day's lesson.** If you don't have time to read it thoroughly, look at the definition and summary boxes, and read through the examples. Then you will know what is important. When you take notes in class, you won't have to write down what you know is in the text. You can concentrate on what your instructor says and fill in the gaps in your notes after class.

7. **Don't be afraid of the unfamiliar.** Some of the applied problems may involve situations that are entirely new to you, but the problems are stated so that only

mathematics, and not necessarily technology, must be understood to work them. You should not dismiss as unimportant problems that deal with technical areas other than your own. You will still benefit from the practice of analyzing the problems. Pay attention to the units in your answer, especially if they are unfamiliar to you. When a measurement has more than four digits and is used with a metric unit, the digits are written in groups of three, without commas. If the measurement has more than four digits and is used with a U.S. customary unit, commas are used. For example, twelve thousand meters is written 12 000 m, but twelve thousand miles is written 12,000 miles.

Ultimately, a lot of your success will depend on both the quantity and the quality of your studying. The suggestions listed here should maximize the quality of your efforts.

TABLE OF CONTENTS

UNIT 1

UNIT 7

CHAPTER 21 DERIVATIVES 830

CHAPTER 22 APPLICATIONS OF THE DERIVATIVE 888

CHAPTER 23 INTEGRALS 938

CHAPTER 27 INFINITE SERIES 1095

CHAPTER 28 DIFFERENTIAL EQUATIONS 1130

CHAPTER 1 NUMBERS

Computers and calculators are now used routinely to perform much of the arithmetic that was once done by technicians. Nevertheless, an understanding of the principles underlying computation is essential for problem solving in modern technology. In this chapter, we relate number properties and operations to the use of a calculator.

1.1 TYPES OF NUMBERS

natural numbers

The human species' first need for numbers was probably to count. We use the symbols 1, 2, 3, . . . to represent the *counting* or *natural numbers.* The three dots at the end of the list indicate that the pattern continues infinitely. Later, as commerce and trade became important, a need to keep track of debits developed. Negative numbers were viewed with skepticism by many, but their use finally became accepted about A.D. 1600.

integers

The symbols . . . , $-3, -2, -1, 0, 1, 2, 3,$ . . . are used to represent the *integers.* The integers form an extension of the natural numbers. That is, we do not discard the natural numbers and the facts we know about operations on them. Rather, we use what we know about natural numbers to develop additional rules to handle the new situations that arise when we allow negative numbers.

We say that 1, 2, 3, . . . are positive integers and . . . , $-3, -2, -1$ are negative integers. Zero is considered to be neither positive nor negative. Thus, the numbers 0, 1, 2, 3, . . . are called nonnegative integers, and . . . , $-3, -2, -1, 0$ are called nonpositive integers.

rational numbers

NOTE ⇨⇨

When humans became sophisticated enough to count partial objects, the concept of fractions emerged. *Rational numbers* form an extension of the integers. A **rational number** is any number that can be written in the form a/b, where a and b are integers, but b is not 0. ***This definition does not say the numbers* must be *written in the form a/b, but rather that they* can be.** Thus, the integers are included in this number system, since we can write $a/1$ for each integer a. The system also includes numbers like 0.5 or 0.17, which can be written as 1/2 and 17/100, respectively. Furthermore, decimal numbers that have a repeating pattern are included. For instance, 0.33333 . . . or 0.272727 . . . can be written as 1/3 and 3/11, respectively.

irrational numbers

NOTE ⇨⇨

Eventually, it was discovered that decimal numbers that do not form a repeating pattern, such as 0.01001000100001 . . . , do not fit the definition for rational numbers. These numbers are called *irrational numbers.* In 1761 Johann Heinrich Lambert proved that π is irrational. The number π has been computed to over one billion decimal places, but often 3.14 is used as an approximate value. Recall that $\sqrt{x}$ is the positive number that when multiplied by itself produces x. ***A square root of a positive integer x is either an integer or an irrational number.*** For example, $\sqrt{25}$ is the integer 5, but $\sqrt{23}$ is irrational.

real numbers

When the irrational numbers are used to form an extension of the rational numbers, the new system is called the *real number system.* The real numbers can be represented visually by locating each number on a line. We place the positive integers to the right of zero, marking off the line in equal intervals. In the same way, we place negative integers to the left of zero, as shown in Figure 1.1.

Other rational numbers can be located between the integers. A few of these are shown in Figure 1.2. Once we visualize all the rational numbers in place, the holes on the number line represent the irrational numbers.

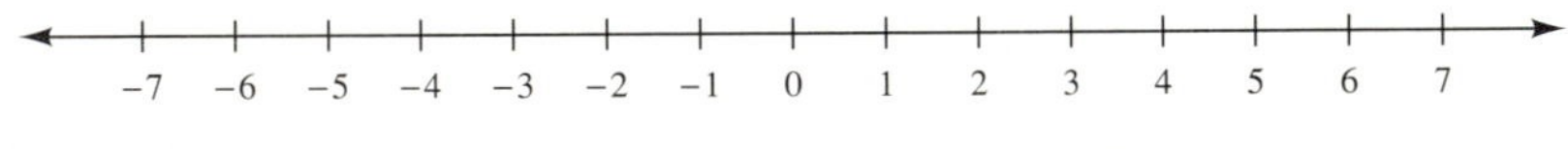

FIGURE 1.1

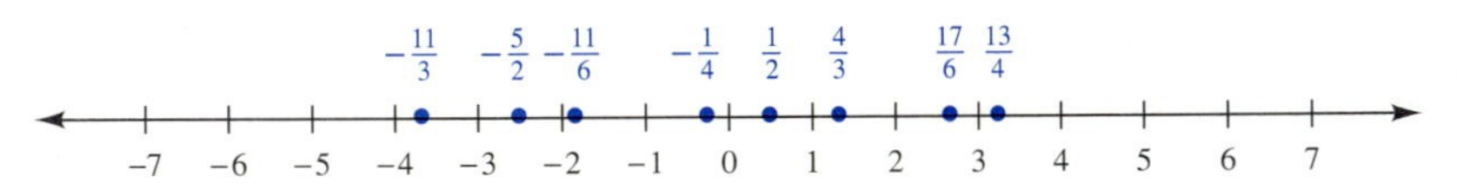

FIGURE 1.2

In order to discuss and compare numbers, we use the concept of *absolute value.* The absolute value of a real number x is denoted $|x|$. A precise definition says that $|x|$ is x if x is greater than or equal to 0, and $|x|$ is $-x$ if x is less than 0. The following definition, however, is more intuitive and will be sufficient until we are ready to work with variable expressions as well as numerical expressions.

DEFINITION The **absolute value** of a nonzero real number x, denoted $|x|$, is the positive number that is the same distance from 0 on the number line as x. The absolute value of 0 is 0.

EXAMPLE 1 Simplify each expression.

(a) $\left|\frac{2}{3}\right|$ **(b)** $|-3|$ **(c)** $|\sqrt{3}|$ **(d)** $-|-2|$

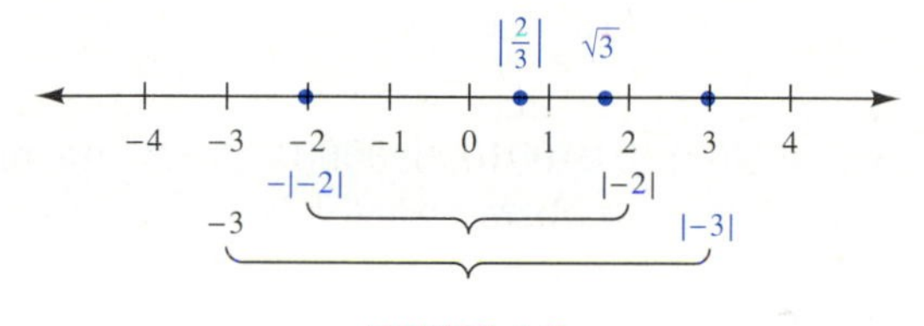

FIGURE 1.3

Solution Consider the number line shown in Figure 1.3.

(a) The number 2/3 is 2/3 of a unit from 0, so $|2/3| = 2/3$.

(b) The number -3 is the same distance from 0 as 3, so $|-3| = 3$.

(c) The number $\sqrt{3}$ is $\sqrt{3}$ units from 0, so $|\sqrt{3}| = \sqrt{3}$.

(d) The number -2 is the same distance from 0 as 2, so $|-2| = 2$, and $-|-2| = -2$. ■

To compare real numbers:

Consider their positions on the number line.

(a) The larger number is to the right of the smaller number.
(b) The smaller number is to the left of the larger number.

EXAMPLE 2 List the following numbers from smallest to largest.

$$2.2, \quad -8, \quad -\pi, \quad \sqrt{3}, \quad -|-6|, \quad |-4|, \quad \frac{17}{3}$$

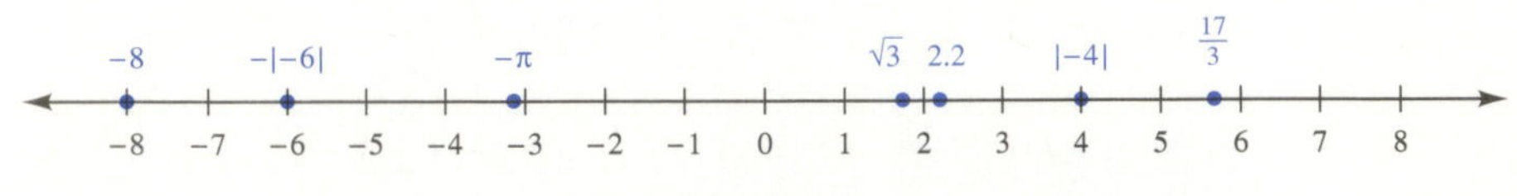

FIGURE 1.4

Solution Note that $-|-6| = -6$, $|-4| = 4$, and $17/3 = 5\ 2/3$. Also, $-\pi$ is about -3.14, and $\sqrt{3}$ is about 1.732. Consider the positions on the number line in Figure 1.4. In order from smallest to largest, the numbers are

$$-8, \quad -|-6|, \quad -\pi, \quad \sqrt{3}, \quad 2.2, \quad |-4|, \quad \frac{17}{3}. \quad ■$$

The symbols used to compare unequal numbers are as follows.

INEQUALITY SYMBOLS

Symbol	Meaning
$>$	is greater than
$\geq$	is greater than or equal to
$<$	is less than
$\leq$	is less than or equal to

EXAMPLE 3 Determine whether each statement is true or false.

(a) $2 < 3$ **(b)** $3 \geq 3$ **(c)** $-2 \leq -3$ **(d)** $3 > 3$

Solution **(a)** $2 < 3$ is true, since 2 is to the left of 3 on a number line.

(b) $3 \geq 3$ is true, since 3 is to the right of or in the same place as 3.

(c) $-2 \leq -3$ is false, since -2 is to the **right** of -3.

(d) $3 > 3$ is false, since 3 is **not** to the right of itself. ■

pure imaginary numbers

The real numbers are sufficient for solving many types of problems. Between 1500 and 1800, however, mathematicians began to recognize the need for an extension of the real number system. In the real number system, it is impossible to find the square root of a negative number. If we allow this type of number, called a *pure imaginary number,* the extended system is called the *complex number system.* That

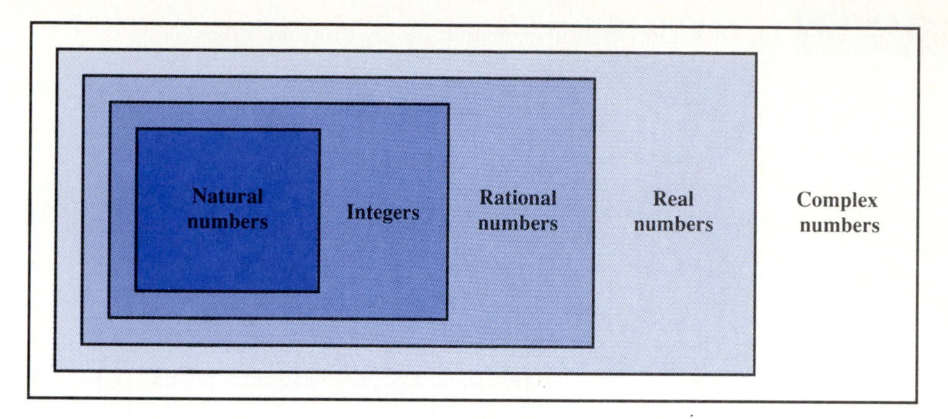

FIGURE 1.5

complex numbers

is, a **complex number** is a number that can be written in the form $a + b\sqrt{-1}$, where a and b are real numbers. Since $a = a + 0\sqrt{-1}$, for each real number a, the real numbers are those complex numbers in which b is 0. The pure imaginary numbers are those complex numbers in which a is 0, but b is not.

The relationship of the different number types is shown in Figure 1.5. ***It is important to realize that a number may belong to more than one number system, but we usually classify a number according to the first system to which it belongs.***

A summary of descriptions of the different types of numbers follows.

TYPES OF NUMBERS

Natural numbers	1, 2, 3, 4, . . .
Integers	. . . −3, −2, −1, 0, 1, 2, 3, . . .
Rational numbers	Numbers that can be put in the form a/b, where a and b are integers, but b is not 0
Irrational numbers	Numbers that do not show a repeating pattern of digits when they are written in decimal form
Real numbers	Numbers that are either rational or irrational
Pure imaginary numbers	Numbers that can be put in the form $b\sqrt{-1}$ where b is a real number, but b is not 0
Complex numbers	Numbers that can be put in the form $a + b\sqrt{-1}$, where a and b are real numbers

EXAMPLE 4 Identify each system to which the number belongs.

$$3, \quad -4, \quad \frac{2}{3}, \quad \sqrt{9}, \quad \sqrt{7}, \quad 4+3\sqrt{-1}, \quad \frac{\sqrt{3}}{2}$$

Solution

Number	*Natural*	*Integers*	*Rational*	*Real*	*Complex*
3	x	x	x	x	x
-4		x	x	x	x
$2/3$			x	x	x
$\sqrt{9}$	x	x	x	x	x
$\sqrt{7}$				x	x
$4+3\sqrt{-1}$					x
$\sqrt{3}/2$				x	x

NOTE ***Notice that $\sqrt{3}/2$ is not a rational number, because the numerator is not an integer.*** ■

Not every ratio represents a rational number. The word *fraction* refers to a number that is written in the form a/b, $b \neq 0$, where a and b are not necessarily integers.

EXAMPLE 5 Which of the following fractions are rational numbers?

$$\frac{0}{3}, \quad \frac{-4}{5}, \quad \frac{\sqrt{2}}{7}, \quad \frac{3\sqrt{-1}}{11}$$

Solution The fraction $\frac{0}{3}$ is rational, since 0 and 3 are integers, and $\frac{-4}{5}$ is rational, since -4 and 5 are integers. The fraction $\frac{\sqrt{2}}{7}$ is not rational, since $\sqrt{2}$ is not an integer, and $\frac{3\sqrt{-1}}{11}$ is not rational, since $3\sqrt{-1}$ is not an integer. The rational numbers, then, are $\frac{0}{3}$ and $\frac{-4}{5}$. ■

When two numbers of the same type are combined using one of the arithmetic operations, the result may be a number of a different type. For example, when the integer 6 is divided by the integer 3, the result is the integer 2. But when 3 is divided by 6, the result is the rational number $1/2$.

EXAMPLE 6 The formula $t = \sqrt{2s/g}$ gives the time of travel t for an object falling from a height of s with an acceleration of g (due to gravity). If s and g are positive rational numbers, what type of number is t? Give the most specific answer that is correct in every case.

Solution If s is a rational number, then so is $2s$. The quotient of two rational numbers is a rational number, so $2s/g$ is rational. However, the square root of a rational number might be an irrational number, so the answer that is correct in every case is that t is a real number. ■

EXERCISES 1.1

In 1–10, simplify each expression.

1. **(a)** $\left|\frac{-1}{4}\right|$ **(b)** $\left|\frac{1}{4}\right|$

2. **(a)** $\left|\frac{\pi}{6}\right|$ **(b)** $\left|\frac{-\pi}{6}\right|$

3. **(a)** $-|-7|$ **(b)** $-|7|$

4. **(a)** $-|\sqrt{11}|$ **(b)** $-|-\sqrt{11}|$

5. **(a)** $-|3.14|$ **(b)** $-|-3.14|$

6. **(a)** $-|-12|$ **(b)** $-|12|$

7. **(a)** $|-\sqrt{5}|$ **(b)** $-|\sqrt{5}|$

8. **(a)** $\left|\frac{-\sqrt{3}}{2}\right|$ **(b)** $-\left|\frac{\sqrt{3}}{2}\right|$

9. **(a)** $-|-2.18|$ **(b)** $|-2.18|$

10. **(a)** $-|0|$ **(b)** $|0|$

In 11–18, write each list of numbers in order from smallest to largest.

11. $\sqrt{5}, 19, 3, |-5|, -9, \pi$

12. $18, |-\pi|, -6, \sqrt{4}, -8, 0$

13. $|\pi|, \sqrt{7}, -7, 21, |-3|, 5$

14. $2, \sqrt{2}, -|-2|, 2\pi, -2\pi, \left|\frac{\pi}{2}\right|, \frac{-\pi}{2}$

15. $\frac{1}{2}, \frac{1}{3}, \frac{1}{4}, \frac{-1}{2}, \frac{-1}{3}, \frac{-1}{4}$

16. $\frac{3}{5}, \frac{2}{5}, \frac{1}{5}, \frac{-1}{5}, \frac{-2}{5}, \frac{-3}{5}$

17. $0.3, -0.3, |0.03|, -|0.33|, \frac{1}{3}, -\left|\frac{-1}{3}\right|, \sqrt{3}$

18. $0.11, -0.11, 1.1, \frac{1}{11}, -|-1.1|, \sqrt{11}, -\sqrt{11}$

In 19–26, determine whether each statement is true or false.

19. **(a)** $4 < 7$ **(b)** $4 \le 7$

20. **(a)** $-4 < -7$ **(b)** $-4 \le -7$

21. **(a)** $5 < 5$ **(b)** $5 \le 5$

22. **(a)** $-3 < 6$ **(b)** $-3 \le 6$

23. **(a)** $8 \ge 2$ **(b)** $8 > 2$

24. **(a)** $9 \ge 9$ **(b)** $9 > 9$

25. **(a)** $-7 \ge 4$ **(b)** $-7 > 4$

26. **(a)** $-1 \ge -2$ **(b)** $-1 > -2$

In 27–42, check each number system to which the number belongs.

	Number	Natural	Integers	Rational	Real	Complex
27.	0					
28.	3					
29.	−4					

	Number	Natural	Integers	Rational	Real	Complex
30.	$3/4$					
31.	$\sqrt{6}$					
32.	$\sqrt{16}$					
33.	$2\ 5/6$					
34.	3.14					
35.	$6 - 3\sqrt{-1}$					
36.	$-2 + \sqrt{-1}$					
37.	$3 + 2\sqrt{-1}$					
38.	$\sqrt{8}/3$					
39.	$\pi/2$					
40.	$7/5$					
41.	-4.71					
42.	$-3\ 3/5$					

In 43–47, consider the possibilities and give the most specific answer that is correct in every case for the given problem.

43. The formula $F = m \cdot a$ gives the force F acting on a body with mass m and acceleration a. If m and a are integers, what type of number is F?
4 7 8 1 2 3 5 6 9

44. The formula $V = D/T$ gives the velocity V for a given distance D and time T. If D and T are integers, what type of number is V?
4 7 8 1 2 3 5 6 9

45. The formula $F = \frac{9}{5}C + 32$ gives the Fahrenheit temperature F for a given Celsius temperature C. If C is a positive integer, what type of number is F? 6 8 1 2 3 5 7 9

46. The formula $I = \sqrt{P/R}$ gives the current I in a device with resistance R and power P. If P and R are positive integers, what type of number is I?
3 5 9 1 2 4 6 7 8

47. The formula $V = \sqrt{2W/C}$ gives the voltage V necessary to charge a capacitor if W is the potential energy stored in the capacitor and C is the capacitance. If W and C are positive integers, what type of number is V? 3 5 9 4 7 8

48. The formula $f = c/\lambda$ gives the frequency f of a light wave of length λ if c is the speed of light. Which has a higher frequency—light with a small wavelength or light with a larger one?
9 1 2 3 5 6 7

49. The formula $A = lw$ gives the area A of a rectangle with length l and width w. How is the area affected if

(a) length and width are both doubled?
(b) length is doubled and width is halved?
(c) length and width are both halved?
1 2 4 7 8 3 5

50. **(a)** If x is a positive number, determine the sign of each number.

$$-x, \quad |x|, \quad -|x|, \quad |-x|$$

(b) If x is a negative number, determine the sign of each number.

$$-x, \quad |x|, \quad -|x|, \quad -|-x|$$

---

1. Architectural Technology
2. Civil Engineering Technology
3. Computer Engineering Technology
4. Mechanical Drafting Technology
5. Electrical/Electronics Engineering Technology
6. Chemical Engineering Technology
7. Industrial Engineering Technology
8. Mechanical Engineering Technology
9. Biomedical Engineering Technology

1.2 ARITHMETIC AND ORDER OF OPERATIONS

sum
difference
product
quotient
dividend
divisor

To perform arithmetical calculations in the technologies, one must know how to add, subtract, multiply, and divide signed numbers. It is also important to know the rules for determining the order in which the calculations are to be done. We begin with a review of the terminology of arithmetic problems. The numbers to be combined in an arithmetic problem are operands. In an addition problem, the answer is called the **sum.** In a subtraction problem, the answer is called the **difference.** In a multiplication problem, the answer is called the **product**. The two numbers that are multiplied are factors. In a division problem, the answer is called the **quotient**. The **dividend** is divided by the **divisor.**

Now we can consider arithmetic with signed numbers.

To add two signed numbers:

(a) If the signs of the numbers are alike, add their absolute values. The sum has the same sign as the numbers.
(b) If the signs of the numbers are different, subtract the smaller absolute value from the larger absolute value. The difference has the same sign as the number with the larger absolute value.

EXAMPLE 1 Perform the indicated additions.

(a) $3 + 5$ **(b)** $-3 + 5$ **(c)** $3 + (-5)$ **(d)** $-3 + (-5)$

Solution

(a) $3 + 5 = 8$ The signs are both positive, so the sum is positive.

(b) $-3 + 5 = 2$ $5 - 3 = 2$; $|5| > |-3|$, so the sum is positive, like 5.

(c) $3 + (-5) = -2$ $5 - 3 = 2$; $|-5| > |3|$, so the sum is negative, like -5.

(d) $-3 + (-5) -8$ $3 + 5 = 8$; both signs are negative, so the sum is negative. ■

EXAMPLE 2 In Fairbanks, Alaska, the temperature was $-10°$ F one day. The next day, it was $5°$ warmer. What was the temperature on the warmer day?

Solution To find the warmer temperature, $5°$ is added to the colder temperature.

$$-10° + 5° = -5°$$ ■

To subtract two signed numbers:

Change the sign of the number being subtracted and add to the other number using the addition rule for signed numbers.

EXAMPLE 3 Perform the indicated subtractions.

(a) $3 - 5$ **(b)** $-3 - 5$ **(c)** $3 - (-5)$ **(d)** $-3 - (-5)$

Solution

(a) $3 - 5 = 3 + (-5) = -2$ Subtracting 5 is equivalent to adding -5.

(b) $-3 - 5 = -3 + (-5) = -8$ Subtracting 5 is equivalent to adding -5.

(c) $3 - (-5) = 3 + 5 = 8$ Subtracting -5 is equivalent to adding 5.

(d) $-3 - (-5) = -3 + 5 = 2$ Subtracting -5 is equivalent to adding 5. ■

To multiply two signed numbers:

(a) If the signs of the factors are alike, multiply their absolute values. The product is positive.
(b) If the signs of the factors are different, multiply their absolute values. The product is negative.

Multiplication may be indicated by placing one or both factors in parentheses.

EXAMPLE 4 Perform the indicated multiplications.

(a) $8(4)$ **(b)** $8(-4)$ **(c)** $-8(4)$ **(d)** $-8(-4)$

Solution

(a) $8(4) = 32$ The signs are alike, so the product is positive.

(b) $8(-4) = -32$ The signs are different, so the product is negative.

(c) $-8(4) = -32$ The signs are different, so the product is negative.

(d) $-8(-4) = 32$ The signs are alike, so the product is positive. ■

A repeated multiplication is sometimes indicated by an exponent. The exponent is sometimes called a power. It indicates the number of times that a particular factor occurs in a multiplication problem. For example, 3^2 has 2 as an exponent, which means 3 is used as a factor twice. That is, 3^2 means (3)(3) or 9. Likewise, 2^3 means that 2 is used as a factor 3 times. That is, 2^3 means (2)(2)(2) or 8.

To divide two signed numbers:

(a) If the signs of the two numbers are alike, divide their absolute values. The quotient is positive.
(b) If the signs of the two numbers are different, divide their absolute values. The quotient is negative.

EXAMPLE 5 Perform the indicated divisions.

(a) $8 \div 4$ **(b)** $-4 \div 8$ **(c)** $4 \div (-8)$ **(d)** $-8 \div (-4)$

Solution **(a)** $8 \div 4 = 2$ The signs are alike; the quotient is positive.

(b) $-4 \div 8 = -\frac{1}{2}$ The signs are different; the quotient is negative.

(c) $4 \div (-8) = -\frac{1}{2}$ The signs are different; the quotient is negative.

(d) $-8 \div (-4) = 2$ The signs are alike; the quotient is positive. ■

More than one operation may be required in a single problem. Consider the problem $2 + 3 \times 5$. Some people might say the answer is 25, because $2 + 3$ is 5, and 5×5 is 25. Others might say the answer is 17, because 3×5 is 15, and $2 + 15$ is 17. So that there will not be two or more different answers to such problems, mathematicians agree on the order in which the operations are to be performed. Knowing the correct order is just as important as knowing how to add and multiply.

Grouping symbols such as parentheses ( ), brackets [ ], or braces { } indicate which operations are to be done first. The fraction bar, absolute value bars | |, and the radical sign $\sqrt{\ }$ may also serve as grouping symbols. If grouping symbols are nested, one inside the other, the arithmetic within the innermost pair is done first.

ORDER OF OPERATIONS

1. Locate the innermost grouping symbols. Within grouping symbols or in the absence of grouping symbols, follow steps 2–4.
2. Perform the multiplications indicated by exponents.
3. Perform multiplications and divisions, working from left to right.
4. Perform additions and subtractions, working from left to right.
5. If there are additional grouping symbols, repeat steps 1–4.

EXAMPLE 6 Evaluate $1 + \{2 - [3 + 2(4 + 5)]\}$.

Solution $1 + \{2 - [3 + 2(4 + 5)]\}$

$= 1 + \{2 - [3 + 2\mathbf{(4 + 5)}]\}$	Locate the innermost grouping symbols.
$= 1 + \{2 - [3 + 2\mathbf{(9)}]\}$	Perform the operation within parentheses.
$= 1 + \{2 - \mathbf{[3 + 2(9)]}\}$	Identify the innermost grouping (brackets).
$= 1 + \{2 - [3 + \mathbf{18}]\}$	Perform the multiplication within brackets.
$= 1 + \{2 - \mathbf{[21]}\}$	Perform the addition within brackets.
$= 1 + \mathbf{\{2 - [21]\}}$	Identify the innermost grouping (braces).
$= 1 + \mathbf{\{-19\}}$	Perform the operation within braces.
$= \mathbf{-18}$	Perform the operation outside the grouping. ■

NOTE ***Multiplication and division are both listed at step 3 in the order of operations because they are performed from left to right as they occur in the expression.*** For example, $8 \div 4 \times 2$ is 2×2 or 4, but $8 \div (4 \times 2)$ is $8 \div 8$ or 1, because the parentheses call for the multiplication to be done first. Addition and subtraction are listed at the same level in the order of operations, so they, too, are performed from left to right as they occur in the expression.

EXAMPLE 7 Evaluate $72 \div 3 \cdot 2^2$.

Solution There are no grouping symbols.

$72 \div 3 \cdot \mathbf{2^2} = 72 \div 3 \cdot \mathbf{4}$	Perform the exponentiation: $2^2 = 4$.
$= \mathbf{24} \cdot 4$	Perform the division: $72 \div 3 = 24$.
$= \mathbf{96}$	Perform the multiplication. ■

EXAMPLE 8 Evaluate $2(4 + 3)^2$.

Solution

$2(4 + 3)^2 = 2\mathbf{(4 + 3)}^2$	Locate the innermost grouping (parentheses).
$= 2\mathbf{(7)}^2$	Perform the operation within parentheses.
$= 2\mathbf{(49)}$	Apply the exponent to the factor 7.
$= \mathbf{98}$	Perform the multiplication. ■

EXAMPLE 9 Evaluate $-1\left[\dfrac{10}{3 + 2} - 5\right]$.

Solution

$$-1\left[\frac{10}{3+2}-5\right] = -1\left[\frac{10}{3+2}-5\right]$$ Locate the grouping symbol (fraction bar).

$$= -1\left[\frac{10}{5}-5\right]$$ Perform the operation under the bar.

$$= -1[2-5]$$ Perform the division within brackets.

$$= -1[-3]$$ Perform the subtraction within brackets.

$$= 3$$ Perform the multiplication. ■

EXAMPLE 10 Evaluate $\sqrt{25-3^2}+|4-7|$.

Solution

$$\sqrt{25-3^2}+|4-7| = \sqrt{25-3^2}+|4-7|$$ Locate the leftmost grouping (square root).

$$= \sqrt{25-9}+|4-7|$$ Perform the exponentiation under the radical sign.

$$= \sqrt{16}+|4-7|$$ Perform the subtraction under the radical sign.

$$= 4+|4-7|$$ Evaluate the square root.

$$= 4+|4-7|$$ Locate the next grouping (absolute value).

$$= 4+|-3|$$ Perform the subtraction within the absolute value bars.

$$= 4+3$$ Evaluate the absolute value.

$$= 7$$ Perform the addition. ■

EXAMPLE 11 The expansion I (in meters) of a section of highway is given by $I = (l/k)(T - t)$. In the formula, l is the length of the highway (in meters), k is a constant that depends on the construction of the highway, t is the temperature (in degrees Celsius) at which it was built, and T is the temperature (in degrees Celsius) when the expansion occurs. Find the expansion if $l = 1200$ m, $k = 150\,000$, $t = 20°$ C, and $T = 0°$ C.

Solution Replacing each letter by the appropriate value, we have

$$I = (1200/150\,000)(0-20) = (0.008)(-20) = -0.16.$$

An expansion of -0.16 m indicates a contraction of 0.16 m. ■

Sometimes, instead of using different letters to represent different quantities, the same letter is used with different *subscripts.* A subscript is placed to the right of and just below the letter it goes with and is usually written in a smaller size. Just as x and y represent different quantities, so do x_1 and x_2. When letters are used to represent a product, the multiplication symbol is often omitted. For example, xy means $x(y)$.

EXAMPLE 12 The equivalent resistance R (in ohms) of two resistors connected in parallel is given by $R = \frac{R_1 R_2}{R_1 + R_2}$, where R_1 and R_2 are the resistances of the two resistors (in ohms). Find the value of R if $R_1 = 15\,000\ \Omega$ and $R_2 = 5000\ \Omega$.

Solution $R = \frac{(15\,000)(5000)}{15\,000 + 5000} = \frac{75\,000\,000}{20\,000} = 3750\ \Omega.$ ■

The rules for the order of operations are ambiguous for a problem like -3^2. Mathematicians treat the negative sign as the sign of the answer. That is, $-3^2 = -9$. Some computers, however, treat the negative sign as part of the number to be squared. Thus, on some computers, $-3^2 = (-3)(-3)$ or 9. Parentheses should be used to avoid confusion. Clearly, $(-3)^2$ means $+9$, and $-(3^2)$ means -9.

NOTE ⇨⇨ Special care must be taken when 0 is one of the operands in an expression. ***It is a common mistake to confuse*** $0/x$ ***(or*** $0 \div x$***) with*** $x/0$ ***(or*** $x \div 0$***).*** Suppose you have \$0. You could divide it with your best friend so that each of you gets \$0 ($0/2$ or $0 \div 2$). But if you have \$2, it is impossible to split it zero ways ($2/0$ or $2 \div 0$). Thus, we say that $x/0$ is undefined. Operations with zero are summarized as follows.

OPERATIONS WITH ZERO

If x is a real number, then

$0 + x = x$ and $x + 0 = x$

$0 - x = -x$ and $x - 0 = x$

$0(x) = 0$ and $x(0) = 0$

$0/x$ (or $0 \div x$) $= 0$ but $x/0$ (or $x \div 0$) is undefined

EXAMPLE 13 Evaluate each expression.

(a) $|-3| + (4 - 4)$ **(b)** $\frac{7-7}{7}$ **(c)** $(2 \cdot 3) - (3^2 - 9)$

(d) $\frac{7 + 3^2}{6 - 6}$ **(e)** $|5 - 5|(6 + 6)$ **(f)** $\frac{5 - (4 + 1)}{-4 + (2 \cdot 2)}$

Solution **(a)** $|-3| + (4 - 4) = 3 + 0 = 3$ **(b)** $\frac{7-7}{7} = \frac{0}{7} = 0$

(c) $(2 \cdot 3) - (3^2 - 9) = 6 - 0 = 6$ **(d)** $\frac{7 + 3^2}{6 - 6} = \frac{16}{0}$, which is undefined

(e) $|5 - 5|(6 + 6) = 0(12) = 0$ **(f)** $\frac{5 - (4 + 1)}{-4 + (2 \cdot 2)} = \frac{0}{0}$, which is undefined ■

EXERCISES 1.2

In 1–18, perform the indicated operation.

1. $7 + 13$
2. $-8 + 12$
3. $6 + (-11)$
4. $(-9) + (-7)$
5. $-11 + 7$
6. $7 - 13$
7. $-8 - 12$
8. $6 - (-9)$
9. $-9 - (-7)$
10. $-10 - 9$
11. $6(13)$
12. $5(-10)$
13. $-10(-12)$
14. $6(-12)$
15. $24 \div (-6)$
16. $-26 \div (-13)$
17. $-27 \div 9$
18. $-33 \div 3$

In 19–42, evaluate each expression.

19. $4 + 2(5)$
20. $(4 + 2)(5)$
21. $8 - 2(3)$
22. $(8 - 2)(3)$
23. $(5 + 3^2)(7)$
24. $(5 + 3)^2(7)$
25. $-8 \div 4 \div 2$
26. $-8 \div (4 \div 2)$
27. $-8 \div 4(2)$
28. $-8 \div (4 \cdot 2)$
29. $4[(5 + 3) - 2^2]$
30. $4[(5 + 3) - 2]^2$
31. $6[7 - (1 - 3)]^2$
32. $6[7 - (1 - 3)^2]$
33. $2[5 + 3(7 - 4^2) - 0]$
34. $2[0 + 3(7 - 4)^2 - 6]$
35. $\dfrac{0(5 + 3)}{2}$
36. $\dfrac{(4 + 1)^2 + 0(3 - 5) - 8}{2 + 3}$
37. $(4 + 1)^2 + \dfrac{2(3 - 5) - 8}{1 + 2}$
38. $|13 - 14| + \sqrt{9 - 5}$
39. $-|15 - 24| - \sqrt{3 + 6}$
40. $|7 - 12| + \sqrt{4(3) + 13}$
41. $|2^2 - 5| - \sqrt{3^2 + 4^2}$
42. $3\{1 + 2[4 + 5(6 - 7) + 8] - 9\}$

---

43. The dry weight of soil S (in pounds) in a wet soil sample is given by $S = \dfrac{W}{1 + 0.01w}$, where W is the weight (in pounds) of the wet soil and w is the moisture content as a percent of the dry weight. Evaluate the formula for S if $W = 1.75$ lb and $w = 8$ percent. **1 2 4 6 8** 5 7 9

44. The power P (in watts) dissipated as heat in a device is given by $P = I^2R$, where I is the current (in amperes) and R is the resistance (in ohms). Evaluate the formula for P if
(a) $I = 0.2$ A and $R = 400\ \Omega$.
(b) $I = 0.2$ A and $R = 150\ \Omega$.
3 5 9 1 2 4 6 7 8

45. After t seconds, the velocity v (in meters/second) of an object dropped with an initial velocity of v_0 (in meters/second) is given by $v = v_0 + 9.8t$. Evaluate the formula for v if
(a) $v_0 = 16$ m/s and $t = 2$ s.
(b) $v_0 = 12$ m/s and $t = 4$ s.
4 7 8 1 2 3 5 6 9

46. The focal length f of a concave spherical mirror is given by $1/f = 1/o + 1/i$ where o and i are the distances from the vertex of the mirror to the object and image, respectively. Is it possible to have a focal length of zero? 3 5 6 9

---

1. Architectural Technology
2. Civil Engineering Technology
3. Computer Engineering Technology
4. Mechanical Drafting Technology
5. Electrical/Electronics Engineering Technology
6. Chemical Engineering Technology
7. Industrial Engineering Technology
8. Mechanical Engineering Technology
9. Biomedical Engineering Technology

47. The volume V of a frustum of a pyramid or cone is given by $V = (1/3)h(A_1 + A_2 + \sqrt{A_1 A_2})$, if A_1 and A_2 are the areas of the bases and h is the altitude. Find the volume (in in^3) of the frustum of a cone if the bases have areas of 9 in^2 and 4 in^2 and the height is 6 in. 1 2 4 7 8 3 5

48. The decibel (dB) is a unit for measuring power levels within electronic equipment. To find the total decibel gain of a complex system, the decibel gains of each section are simply added together. A certain audio system consists of a preamplifier with a power gain of 25 dB, an attenuator with a power gain of -8 dB, and an amplifier with a power gain of 60 dB. What is the total gain? 3 5 9 4 7 8

49. A copper wire has a resistance of 1.00 Ω at 20.0° C. The resistance is 0.88 Ω when the temperature falls to -10.0° C.
(a) How much greater is the original resistance than the final resistance?
(b) How much greater is the original temperature than the final temperature?
(c) What is the change in resistance for each degree change in temperature?
3 5 9 1 2 4 6 7 8

50. In physics, when an object is thrown upward, the force may be indicated as a positive quantity. Explain why the force of gravity is indicated as a negative quantity in such problems.

1. Architectural Technology
2. Civil Engineering Technology
3. Computer Engineering Technology
4. Mechanical Drafting Technology
5. Electrical/Electronics Engineering Technology
6. Chemical Engineering Technology
7. Industrial Engineering Technology
8. Mechanical Engineering Technology
9. Biomedical Engineering Technology

1.3 APPROXIMATE NUMBERS

Numbers are used not only for counting, but also for measuring. The correctness of a measurement is limited by the quality of the measuring device and by the skill of the user. Measurement, by its nature, involves approximations. Most measurements fall between two marks on the scale of the measuring device, and the user must round off the measurement to the nearest mark or estimate how far it goes beyond one of the marks. Thus, there is always some question about the last digit specified in a measurement. For example, suppose measurements are given to the nearest whole number but could, in fact, be measured exactly to tenths. Any measurement in the range of 30.5 to 31.4 would be rounded to 31. When we see a measurement of 31, then, we cannot be sure exactly what the "1" represents.

Consider the sum of 31, 28.3, and 32.69. When the numbers are simply added, the sum is 91.99. This sum appears to represent a number that has been rounded to hundredths, so it could be anything from 91.985 to 91.994. Now we examine the problem again, taking into account that each individual number has been rounded.

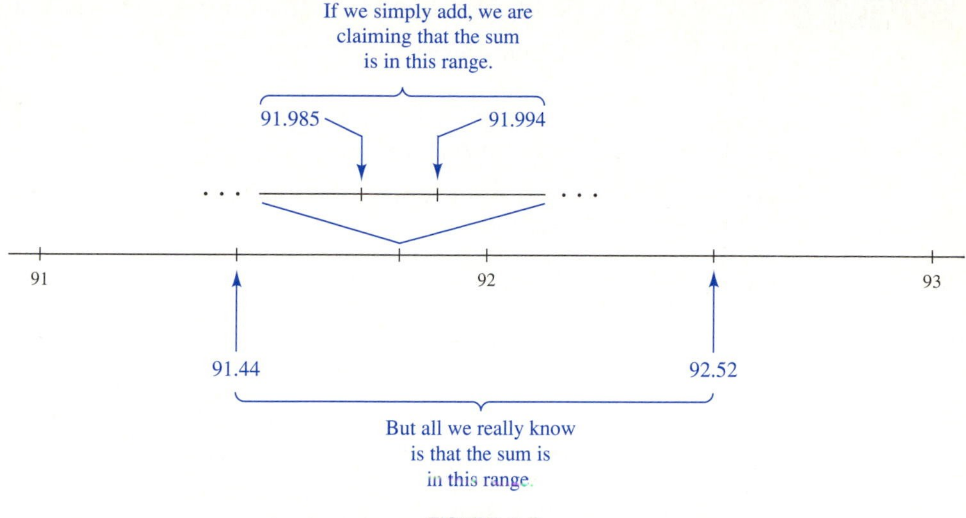

FIGURE 1.6

Suppose the measurements could be made *exactly* to hundredths. A measurement of 31 represents the range from 30.50 to 31.49. A measurement of 28.3 represents the range from 28.25 to 28.34. A measurement of 32.69 is exact. The sum will fall somewhere between 91.44 (30.50 + 28.25 + 32.69) and 92.52 (31.49 + 28.34 + 32.69). Notice the discrepancy shown in Figure 1.6.

NOTE ⇨⇨

The precision of a measurement refers to the number of decimal places shown. ***When approximate numbers are added or subtracted, the result is no more precise than the least precise data used in the computation.*** In our example, the least precise measurement was given to the nearest whole number, 31. Rounding the sum to the nearest whole number gives 92. When we write 92, it is understood that the number is somewhere between 91.5 and 92.4. You can see from the example that it is possible for the sum to fall outside this range, but 92 gives a more realistic range of values than 91.99 does.

Because computers and calculators often display 8 or 10 decimal places, the result of a computation may appear to be more precise than it actually is. Therefore, it is important to know how to round off results obtained through the use of approximate numbers. For most purposes, the following rule works quite well.

To add or subtract approximate numbers:

1. Perform the operation using all of the digits given.
2. Round off the answer to the same number of decimal places as the operand with the fewest decimal places.

EXAMPLE 1 Add or subtract the following approximate numbers, and round off appropriately.

(a) $43.6 + 4.36 + 0.436$ **(b)** $1.42 + 72.3 - 0.51$ **(c)** $632 - 85.4 + 2.31$

Solution **(a)** $43.6 + 4.36 + 0.436 = 48.396$, which is rounded off to one decimal place (43.6 has only one place) to become 48.4.

(b) $1.42 + 72.3 - 0.51 = 73.21$, which is rounded off to one decimal place (72.3 has only one place) to become 73.2.

(c) $632 - 85.4 + 2.31 = 548.91$, which is rounded off to the nearest whole number (632 is a whole number) to become 549. ■

EXAMPLE 2 A tolerance study is an analysis of the range of values that might result if some variation is allowed for specified dimensions. Determine the maximum and minimum values for the length of the object whose front view is shown in Figure 1.7.

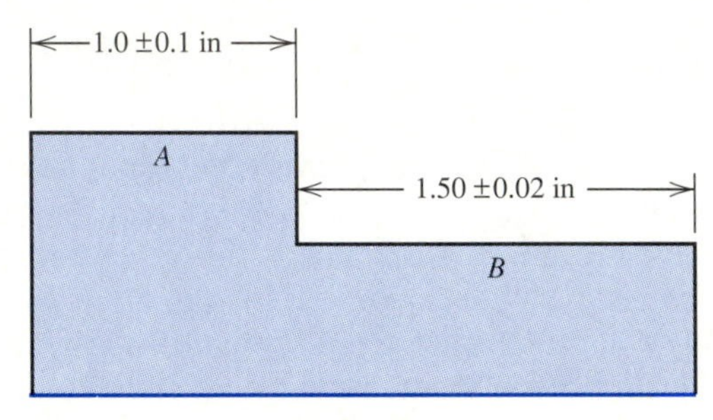

FIGURE 1.7

Solution A measurement of 1.0 ± 0.1 means that the measurement might range from $1.0 - 0.1$ to $1.0 + 0.1$. The length of A, then, could range from 0.9 to 1.1. The range for B is from $1.50 - 0.02$ to $1.50 + 0.02$, or from 1.48 to 1.52. Thus, the total length could be as small as $0.9 + 1.48$, which is 2.4 inches or as large as $1.1 + 1.52$, which is 2.6 inches. ■

Now consider multiplication. When we multiply 493 and 160, for instance, we get 78,880. But if the numbers are approximate, 493 could represent anything from 492.5 to 493.4. Likewise, 160 could represent anything from 159.5 to 160.4. Thus, the product could actually be as low as 492.5×159.5 or as high as 493.4×160.4. That is, the product lies between 78,553.75 and 79,141.36. The product 78,880 (obtained from 493×160) represents a number between 78,879.5 and 78,880.4. Figure 1.8 shows that this range is too narrow.

For multiplication and division, *significant digits* are used to round off results.

DEFINITION A digit is **significant** if

(a) it is a *nonzero* digit,

(b) it is a zero *between* significant digits, or

(c) it is a zero to the *right* of a decimal point, and there are no nonzero digits to the right of it. That is, it is a *final* zero to the *right* of a decimal point.

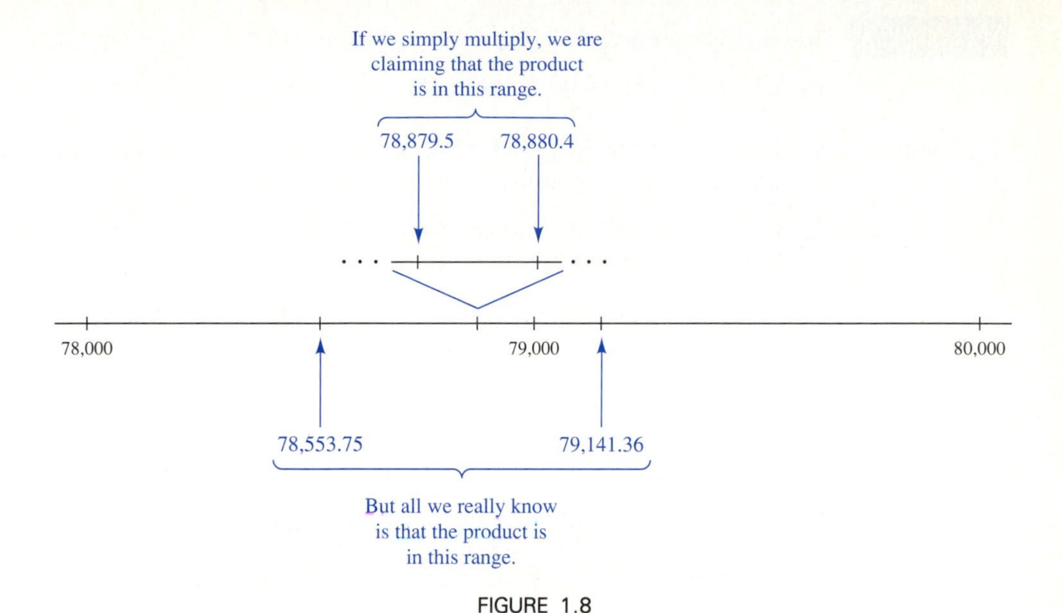

FIGURE 1.8

It might help to see which digits are *not* significant. A zero is not significant, unless specified otherwise, if there are no nonzero digits to the left of it, or if it is to the *left* of a decimal point and there are no nonzero digits to the right of it. That is, a zero is not significant if it is a *beginning* zero or a *final* zero to the *left* of a decimal point.

We say that final zeros before the decimal point are not significant unless there is some additional information to indicate that they are. For example, suppose that two people each buy a computer, one for $788.39 and the other for $800.00. The first might say, "I bought a computer for $800." The two zeros in 800 are not significant, because they indicate rounding to the nearest hundred. The second person might say, "My computer cost $800 to the dollar." The phrase "to the dollar" indicates that the two zeros are significant in this number. The number is correct to the nearest whole number.

EXAMPLE 3 How many significant digits are in each number of each group?

(a) 493, 4.93, 49.3 **(b)** 507, 5.07, 50.7 **(c)** 26.0, 2.60, 0.260

(d) 160, 1600, 16,000 **(e)** 0.18, 0.018, 0.0018

Solution **(a)** There are 3; each digit is a nonzero number.

(b) There are 3; the zero is *between* significant digits.

(c) There are 3; the zeros are *final* zeros *after* a decimal point.

(d) There are 2; the zeros are *final* zeros *before* a decimal point.

(e) There are 2; the zeros are *beginning* zeros. ■

EXAMPLE 4 How many significant digits are in each number?

(a) 0.0020 **(b)** 10.00 **(c)** 10

Solution **(a)** There are 2. The first three zeros are *beginning* zeros; the last zero is a *final* zero *after* the decimal point.

(b) There are 4. The last two zeros are *final after* the decimal point; the first occurs *between* significant digits.

(c) There is only 1. The zero is a *final* zero *before* the decimal point. ■

NOTE ⇨⇨ The accuracy of a measurement refers to the number of significant digits indicated. ***A product or quotient using approximate numbers is no more accurate than the least accurate data used in the computation.***

To multiply or divide approximate numbers:

1. Perform the calculation using all of the digits given.
2. Round off the answer to the same number of significant digits as the operand with the fewest significant digits.

EXAMPLE 5 Perform the multiplication or division using the following approximate numbers, and round off the answer appropriately.

(a) 493×160 **(b)** $5.07 \div 10.00$ **(c)** 5.07×10

Solution **(a)** 493 has 3 significant digits, and 160 has only 2. Since 493×160 is 78,880, the answer is rounded to 2 significant digits: 79,000.

(b) 5.07 has 3 significant digits, and 10.00 has 4. Since $5.07 \div 10.00$ is 0.5070, the answer is rounded to 3 significant digits: 0.507.

(c) 5.07 has 3 significant digits, and 10 has only 1. Since 5.07×10 is 50.7, the answer is rounded to 1 significant digit: 50. ■

EXAMPLE 6 When a block slides with constant velocity on a surface, the coefficient of friction is given by $\mu = F/N$. Calculate μ for the given materials. (F is the magnitude of the force of friction and N is the magnitude of the force exerted by the surface on the block.)

(a) wood on wood: $F = 81$ newtons, $N = 243$ newtons

(b) metal on metal: $F = 48.6$ pounds, $N = 306$ pounds

Solution **(a)** $\mu = 81/243 = 0.33$, to two significant digits.

(b) $\mu = 48.6/306 = 0.159$, to three significant digits. ■

NOTE ⇨⇨ ***It is important to realize that these rules for rounding apply only when using approximate numbers, such as measurements. They do not apply when the numbers are known to be exact values.*** For example, if one voltage regulator cost \$0.59, we would say that three cost $3 \times \$0.59$ or \$1.77, because there are exactly 3 voltage regulators and each one costs exactly \$0.59. In purely mathematical examples, integers are often used. Such numbers are intended to be exact values. When two integers are divided, the answer may have to be rounded. As a general rule, such answers are given to three significant digits.

scientific notation

Scientific notation is often used to write very large or very small numbers. For example, the speed of light is approximately 186,000 miles per second. The number 186,000 can be written as 1.86×10^5. That is, when you multiply 1.86 by 10^5, the decimal point is moved over 5 places to the right. This result is reasonable if you remember, for instance, that 10^2 is 100, and multiplication by 100 moves the decimal point two places to the right. A negative exponent is used to indicate a power of ten in the denominator of a fraction. For instance, 10^{-2} is $1/10^2$. Multiplication by 10^{-2} is equivalent to division by 10^2 and thus moves the decimal point two places to the left. To write a number in scientific notation, express it in the form $n \times 10^k$ where $1 \leq |n| < 10$ and k is an integer.

To express a number in scientific notation:

1. Place the decimal point so that there is exactly one nonzero digit to the left of it.
2. **(a)** Multiply by 10^k if the decimal point should be moved k places to the right.
   (b) Multiply by 10^{-k} if the decimal point should be moved k places to the left.

EXAMPLE 7 Express each number in scientific notation.

(a) 4900 **(b)** 49,000 **(c)** 58.23

Solution **(a)** $4900 = 4.9 \times 10^3$ **(b)** $49{,}000 = 4.9 \times 10^4$ **(c)** $58.23 = 5.823 \times 10^1$ ■

EXAMPLE 8 Express each number in scientific notation.

(a) 0.231 **(b)** 0.0056 **(c)** 5823

Solution **(a)** $0.231 = 2.31 \times 10^{-1}$ **(b)** $0.0056 = 5.6 \times 10^{-3}$ **(c)** $5823 = 5.823 \times 10^3$ ■

NOTE ⇨⇨ ***In Example 8(c), it is a common mistake to write 58.23×10^2. While it is true that $5823 = 58.23 \times 10^2$, this expression is not scientific notation.*** The number 58.23 is not between 1 and 10.

Scientific notation is sometimes used to clarify the number of significant digits in a number. For example, 2×10^2 has only one significant digit, but 2.00×10^2 has three significant digits. Writing "2.00" indicates that the two zeros are significant.

EXAMPLE 9 Rubik's cube, which was developed as an assignment for architecture and design students, is said to have about 43 quintillion different configurations. The exact number is 43,252,003,274,489,856,000. Express this number in scientific notation to two significant digits.

Solution $43{,}252{,}003{,}274{,}489{,}856{,}000 = 4.3 \times 10^{19}$, to two significant digits. ■

Sometimes it is helpful to change a number that is in scientific notation to decimal form.

EXAMPLE 10 In the metric system, multiples and subdivisions of units are designated by prefixes that indicate a power of 10 as in the following table.

Prefix	*Power of Ten*	*Abbreviation*	*Prefix*	*Power of Ten*	*Abbreviation*
pico	10^{-12}	p	kilo	10^{3}	k
nano	10^{-9}	n	mega	10^{6}	M
micro	10^{-6}	μ	giga	10^{9}	G
milli	10^{-3}	m			

Use scientific notation to change each unit of measurement as indicated.

(a) 7 kg to grams (g) **(b)** 9.8 ns to seconds (s) **(c)** 3.3 μm to meters (m)

Solution **(a)** $7 \text{ kg} = 7 \times 10^{3} \text{ g} = 7000 \text{ g}$

(b) $9.8 \text{ ns} = 9.8 \times 10^{-9} \text{ s} = 0.0000000098 \text{ s}$

(c) $3.3\ \mu\text{m} = 3.3 \times 10^{-6} \text{ m} = 0.0000033 \text{ m}$ ■

EXERCISES 1.3

In 1–7, assume that the problems involve approximate numbers. Add or subtract and round appropriately.

1. $6.92 + 14.348$
2. $25.1 + 6.32 - 0.547$
3. $7.53 - 12.214$
4. $34.9 - 7.05 + 0.268$
5. $82 + 53.4 - 6.93$
6. $51.4 - 3.27 + 0.815$
7. $142.3 + 3416 - 29.24$

In 8–19, state the number of significant digits in each number.

8. 25.6
9. 3.14
10. 1.732
11. 7083
12. 80
13. 9000
14. 200
15. 71.30
16. 2.500
17. 6.0
18. 0.046
19. 0.0032

In 20–26, assume that the problems involve approximate numbers. Multiply or divide and round appropriately.

20. $6.92 \div 1.4$ **21.** $0.213 \times 25 \div 4000$ **22.** $0.005 \times 12 \times 2.400$ **23.** 5.32×2.7
24. $31.2 \times 5 \div 400$ **25.** $4.23 \div 0.007$ **26.** $5178 \div 2.6$

In 27–34, write each number in scientific notation.

27. 45,000 **28.** 580,000 **29.** 231,000,000 **30.** 409,000,000
31. 60,000,000 **32.** 0.0008732 **33.** 0.0000271 **34.** 0.000000345

In 35–42, write each number in decimal form.

35. 6.7×10^3 **36.** 2.34×10^2 **37.** 8.35×10^{-2} **38.** 5.08×10^{-1}
39. 9.1×10^{-3} **40.** 8.2×10^2 **41.** 6.204×10^1 **42.** 3.452×10^3

In 43 and 44, use scientific notation and the table given in Example 10 to change each unit of measurement as indicated.

43. **(a)** 2.7 pF to farads (F) (The farad measures capacitance.)
(b) 3.8 kV to volts (V) (The volt measures potential difference.) 3 5 9 1 2 4 6 7 8

44. **(a)** 5.6 mH to henries (H) (The henry measures inductance.)
(b) 2.5 ns to seconds (s) (The second measures time.) 3 5 9 1 2 4 6 7 8

In 45 and 46, write the measurements in scientific notation.

45. **(a)** 1 slug = 14 590 grams (The slug and gram measure mass.)
(b) 1 pound = 445 000 dynes (The pound and dyne measure force.) 4 7 8 1 2 3 5 6 9

46. **(a)** 1 joule = 0.239 calories (The joule and calorie measure energy.)
(b) 1 kilogram = 0.0685 slugs (The kilogram and slug measure mass.) 4 7 8 1 2 3 5 6 9

47. In the computer language BASIC, certain types of variables can store only seven significant digits. When asked to store a number with more than seven significant digits, the computer will round off to seven significant digits and use scientific notation. If the value 312.5476 is stored in A and 0.9786453 is stored in B, what value is stored for the product of A and B? 3 5 9

48. Find the area of the figure shown. 1 2 4 7 8 3 5

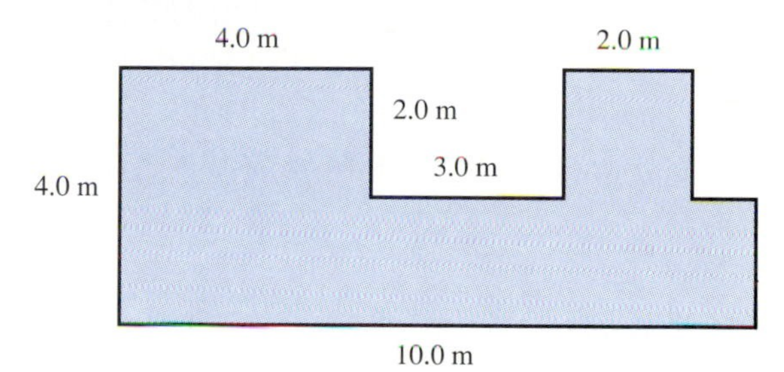

49. The formula $F = ma$ gives the force F exerted on a body of mass m with acceleration a. Calculate F for the given values and round off appropriately.
(a) $m = 4.31$ kilogram, $a = 0.83$ m/s^2 (F will be in newtons)
(b) $m = 2.01$ slugs, $a = 5.5$ ft/s^2 (F will be in pounds) 4 7 8 1 2 3 5 6 9

50. Which is larger, $10^{(10^{10})}$ or $(10^{10})^{10}$?

---

1. Architectural Technology
2. Civil Engineering Technology
3. Computer Engineering Technology
4. Mechanical Drafting Technology
5. Electrical/Electronics Engineering Technology
6. Chemical Engineering Technology
7. Industrial Engineering Technology
8. Mechanical Engineering Technology
9. Biomedical Engineering Technology

1.4 PROPERTIES OF REAL NUMBERS

There are some problems for which it is desirable to do the operations in an order that is different from the order indicated in the original problem. For instance, to perform the addition

$$(17 + 64) + 6,$$

it is not necessary to add 17 and 64 before adding 6. The arithmetic is easier if the problem is done as

$$17 + (64 + 6) = 17 + 70 = 87.$$

This example illustrates the *associative* property. There are three properties of real numbers that permit us to do certain problems in more than one way.

PROPERTIES OF REAL NUMBERS

For real numbers a, b, and c,

Commutative properties: $a + b = b + a$

$ab = ba$

Associative properties: $(a + b) + c = a + (b + c)$

$(ab)c = a(bc)$

Distributive property: $a(b + c) = ab + ac$

EXAMPLE 1 Use the associative property to simplify each computation.

(a) $(47 + 7) + 13$ **(b)** $4[(25)(37)]$ **(c)** $(1/2)(8 \cdot 13)$

Solution **(a)** $(47 + 7) + 13 = 47 + (7 + 13) = 47 + 20 = 67$

(b) $4[(25)(37)] = [4(25)](37) = 100(37) = 3700$

(c) $(1/2)(8 \cdot 13) = [(1/2)(8)](13) = 4(13) = 52$ ■

EXAMPLE 2 **(a)** Calculate the volume of a rectangular building that has dimensions $41' \times 45' \times 11'$.

(b) Use the associative property to find the volume in a different way.

(c) Find the amount of air in the building if a building that has dimensions $21' \times 41' \times 11'$ contains 737 lb of air.

Solution **(a)** Volume is given by $V = lwh$.
$V = (41)(45)(11) = (1845)(11) = 20{,}295 \text{ ft}^3$

(b) $[(41)(45)](11) = (41)[(45)(11)] = (41)(495) = 20{,}295 \text{ ft}^3$

(c) A building with dimensions $21' \times 41' \times 11'$ has a volume of 9471 ft^3. If 9471 ft^3 of air weighs 737 lb, then 1 ft^3 weighs 737/9471 or 0.0778165 lb. It follows that $20{,}295 \text{ ft}^3$ of air would weigh 20,295(0.0778165) or 1600 lb, to two significant digits. ■

The associative property is helpful in avoiding excessive round-off errors. Since

$$\frac{a}{b}(c) = \left(\frac{1}{b}(a)\right)c = \frac{1}{b}(ac) = \frac{ac}{b},$$

the division may be done last.

EXAMPLE 3 Use the formula $V = \frac{4}{3}\pi r^3$ to find the volume of a sphere with radius 10.0 inches. Use $\pi = 3.14$.

Solution $V = \frac{4}{3}\pi r^3 = \frac{4\pi r^3}{3} = \frac{4(3.14)(10.0)^3}{3} = \frac{12.56(1000)}{3} = \frac{12560}{3}$, which is 4190 in^3, to 3 significant digits. ■

Notice that in Example 3, if 4/3 is evaluated first, we have $\frac{4}{3}\pi r^3 = 1.33(3.14)(1000) = 4180$ to three significant digits. There is a small round-off error introduced by using 1.33 as an approximation for the exact value of $4 \div 3$. That error is compounded or magnified when multiplied by 1000. The round-off error is small when the division by 3 is done as the last step.

The distributive property allows us to perform the operations without doing the addition or subtraction in parentheses first.

EXAMPLE 4 Use the distributive property to simplify each computation.

(a) $12(1/2 + 1/3)$ **(b)** $15(2/3 + 3/5)$ **(c)** $7(99)$

Solution **(a)** $12(1/2 + 1/3) = 12(1/2) + 12(1/3) = 6 + 4 = 10$

(b) $15(2/3 + 3/5) = 15(2/3) + 15(3/5) = 10 + 9 = 19$

(c) $7(99) = 7(100 - 1) = 7(100) - 7(1) = 700 - 7 = 693$ ■

The distributive property may be used with formulas also.

EXAMPLE 5 In a series circuit with two resistors, the voltage V is given by $V = IR_1 + IR_2$. Use the distributive property to rewrite the formula.

Solution $V = IR_1 + IR_2 = I(R_1 + R_2)$. ■

NOTE The commutative and associative properties are often confused. ***The commutative properties tell us that in an addition or multiplication problem, the* order *of the operands may be changed without affecting the answer. The associative properties say that the* grouping *of the operands may be changed without affecting the answer to an addition or multiplication problem.***

EXAMPLE 6 Identify the property or properties illustrated by each example.

(a) $3 + (4 + 5) = 3 + (5 + 4)$

(b) $3(4 \cdot 5) = (4 \cdot 5)3$

(c) $3(4 + 5) = 3(4) + 3(5)$

Solution **(a)** The commutative property was used to change the order of the numbers from $3 + (4 + 5)$ to $3 + (5 + 4)$.

(b) The commutative property was used to change the order of the factors from $3(4 \cdot 5)$ to $(4 \cdot 5)3$.

(c) The distributive property was used. ■

EXAMPLE 7 Identify the property or properties illustrated by each example.

(a) $3(4 \cdot 5) = (3 \cdot 4)5$

(b) $3(4 \cdot 5) = (4 \cdot 3)5$

(c) $3 + (4 + 5) = 4 + (3 + 5)$

Solution **(a)** The associative property was used to group 3 and 4 together instead of 4 and 5.

(b) Both the commutative and associative properties were used, since both the order and the grouping of the operands were changed.

(c) Both the commutative and associative properties were used. ■

Subtraction is not a commutative operation, since $a - b$ is not equal to $b - a$ in general. However, since $a - b = a + (-b)$, we have $a - b = (-b) + a$. To add a list of positive and negative numbers, you may want to use the commutative and associative properties to reorder and regroup the numbers. It is sometimes easier to add if the positive numbers are together and the negative numbers are together.

EXAMPLE 8 Perform the indicated operation.

(a) $2 - 3 + 4 - 1 + 7$ **(b)** $1.4 + 3.21 - 4.1 + 0.2 - 0.79$

Solution **(a)** $2 - 3 + 4 - 1 + 7 = (2 + 4 + 7) + [-3 + (-1)] = 13 + (-4) = 9$

(b) $1.4 + 3.21 - 4.1 + 0.2 - 0.79 = (1.4 + 3.21 + 0.2) + [-4.1 + (-0.79)]$
$= 4.81 + (-4.89)$
$= -0.08$ or -0.1 (rounded) ■

EXERCISES 1.4

In 1–8, use the associative property to simplify each computation.

1. $(58 + 17) + 3$ **2.** $23 + (7 + 82)$ **3.** $36 + (4 + 27)$ **4.** $(14 + 38) + 2$
5. $2(5 \cdot 29)$ **6.** $(47 \cdot 4)(25)$ **7.** $(1/3)(9 \cdot 25)$ **8.** $(1/4)(16 \cdot 12)$

In 9–16, use the distributive property to simplify each computation.

9. $14(1/2 + 2/7)$ **10.** $45(3/5 + 5/9)$ **11.** $35(2/5 + 3/7)$ **12.** $21(2/3 + 3/7)$
13. $68(101)$ **14.** $54(99)$ **15.** $15(98)$ **16.** $25(102)$

In 17–32, identify the property or properties illustrated by each example.

17. $2 + (7 + 9) = (2 + 7) + 9$ **18.** $3 + (6 + 8) = 3 + (8 + 6)$
19. $1 + (8 + 5) = (8 + 5) + 1$ **20.** $7 + (4 + 2) = (4 + 7) + 2$
21. $6 + (5 + 3) = (3 + 5) + 6$ **22.** $2(3 + 9) = 2(9 + 3)$ **23.** $8(4 + 7) = 8(4) + 8(7)$
24. $4[6(5)] = [4(6)](5)$ **25.** $2[5(7)] = 2[7(5)]$ **26.** $2[9(6)] = [9(6)](2)$
27. $6[5(3)] = [5(6)](3)$ **28.** $5[6(2)] = [2(6)](5)$ **29.** $3(4 + 8) = (4 + 8)(3)$
30. $9(3 + 8) = 9(3) + 9(8)$ **31.** $3(7 + 1) = (1 + 7)(3)$ **32.** $4(5 + 3) = 4(5) + 4(3)$

In 33–39, use the properties of real numbers to perform the operations so that divisions are last, if possible.

33. The resistance R_2 of a wire is given by $R_2 = \left(\frac{D_1^2}{D_2^2}\right)R_1$. Find R_2 (in ohms) if $D_1 = 0.10$ inch, $D_2 = 0.04$ inch, and $R_1 = 0.50\ \Omega$. In the formula, R_2 is the resistance when the diameter is D_2, and R_1 is the resistance when the diameter is D_1.
3 5 9 | 1 2 4 6 7 8

34. The force F of attraction between two bodies is given by $F = \left(\frac{g}{r^2}\right)(M_1 M_2)$. Find F (in newtons) if $M_1 = 68$ kg, $M_2 = 45$ kg, $r = 0.50$ m, and g is the gravitational constant 9.8 m/s^2. In the formula, M_1 and M_2 are the masses of the bodies, and r is the distance between them.
4 7 8 | 1 2 3 5 6 9

---

1. Architectural Technology
2. Civil Engineering Technology
3. Computer Engineering Technology
4. Mechanical Drafting Technology
5. Electrical/Electronics Engineering Technology
6. Chemical Engineering Technology
7. Industrial Engineering Technology
8. Mechanical Engineering Technology
9. Biomedical Engineering Technology

35. The force F_2 on one side of a lever balanced at the fulcrum is given by $F_2 = \left(\frac{D_1}{D_2}\right)(F_1)$. Find F_2 (in newtons) if $F_1 = 5.40$ N, $D_1 = 2.33$ meters, and $D_2 = 1.50$ meters. In the formula, D_2 is the distance of the force from the fulcrum, and F_1 and D_1 are the force and distance, respectively, on the opposite side of the fulcrum. 4 7 8 1 2 3 5 6 9

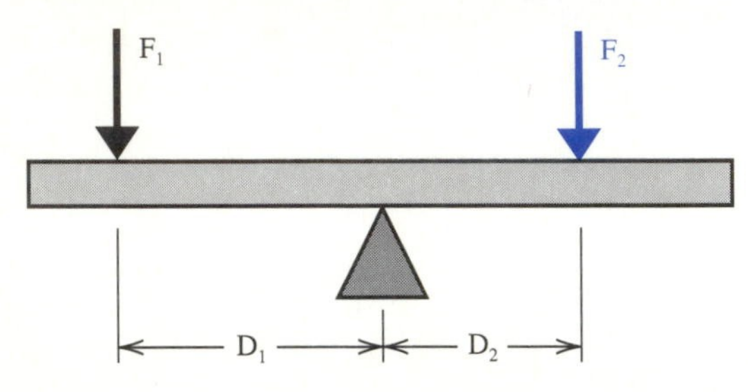

36. The current I in a circuit is given by $I = \frac{V_{emf}}{(R + r)}$. Find I (in amperes) for a dry cell with $V_{emf} = 1.5$ volts and $r = 1.0\ \Omega$ connected to a flashlight bulb with $R = 2.0\ \Omega$. 3 5 9 1 2 4 6 7 8

37. The focal length f of a concave spherical mirror is given by $f = \frac{oi}{(o + i)}$. Find f (in cm) if $o = 8.0$ cm and $i = 3.0$ cm. In the formula, o and i are the distances from the vertex of the mirror to the object and image, respectively. 3 5 7 1 2 4 6 8 9

38. The volume V of a cone is given by $V = (1/3)(\pi r^2 h)$. Find the volume (in cm³) of a cone if $r = 2.5$ cm and $h = 3.0$ cm. Use $\pi = 3.14$. 1 2 4 7 8 3 5

39. The equivalent capacitance C of two capacitors in series (end to end) is given by $C = \frac{(C_1 C_2)}{(C_1 + C_2)}$. Find C (in μF) if $C_1 = 2.0\ \mu$F and $C_2 = 3.0\ \mu$F. In the formula, C_1 and C_2 are the capacitances of the two capacitors. 9 1 2 3 5 6 7

40. Under what conditions is exponentiation commutative? That is, for what values of a and b is $a^b = b^a$ a true statement?

1. Architectural Technology
2. Civil Engineering Technology
3. Computer Engineering Technology
4. Mechanical Drafting Technology
5. Electrical/Electronics Engineering Technology
6. Chemical Engineering Technology
7. Industrial Engineering Technology
8. Mechanical Engineering Technology
9. Biomedical Engineering Technology

1.5 THE SCIENTIFIC CALCULATOR

Problem solving is an important part of every technological field of study. A scientific calculator is an indispensable tool for obtaining accurate answers to a wide range of problems. A scientific calculator is distinguished from other types by the functions that it has. Although not all scientific calculators have the same features, almost any calculator that has trigonometric and logarithmic functions will also have the other functions necessary for scientific and technological work.

Since it is impossible to cover every variation available on the many brands and models of scientific calculators, it is important for you to become thoroughly familiar with your calculator. As new concepts are introduced in the text, the use of a calculator to solve problems related to those concepts will be covered. If you have questions about your particular calculator, you should read the owner's manual or instruction booklet that came with it.

Some of the variations among calculators are minor. The display may show a maximum of only 6 digits or it may show as many as 10 or 12. Answers with more digits than the display will show may be rounded or truncated. A number is truncated if the extra digits are dropped, no matter how large they are. Thus, 9.76 is rounded to 9.8, but truncated to 9.7. Neither the labeling of keys nor the error message is standard. When an invalid operation (such as division by zero) is performed, the display might show ERROR, ERR, E, EO, or something similar.

There is a major variation among calculators in what is called *internal logic.* Scientific calculators generally have either Reverse Polish Notation or Algebraic Logic. This text discusses only calculators that use algebraic logic. These calculators interpret the order of operations just as they are interpreted in the algebra of the real number system (as explained in Section 1.2).

The keys explained in Table 1.1 are the ones you should become familiar with first. If your calculator does not have all of these keys, it is possible that the keys that perform these operations are labeled differently. Sometimes a double keystroke is required. That is, you may have to press a key labeled INV, 2nd, or F followed by a second key, instead of pressing a single key. For example, some calculators don't have a square root key. To find square roots on such calculators, it is necessary to use the INV key followed by the x^2 key.

TABLE 1.1 Calculator Keys

Key	*Meaning*	*Comment*
=	Equals	Completes previously entered arithmetic
+	Add	Completes previously entered arithmetic and instructs calculator to add next entry to displayed number
−	Subtract	Completes previously entered arithmetic and instructs calculator to subtract next entry from displayed number
×	Multiply	Completes previously entered multiplication and division and instructs calculator to multiply displayed number by next entry
÷	Divide	Completes previously entered multiplication and division and instructs calculator to divide displayed number by next entry
y^x	Raise to a Power	Raises the displayed value (y) to the next entry
+/−	Change the Sign	Changes the sign of the displayed number
$1/x$	Reciprocal	Divides displayed number into 1
x^2	Square	Computes square of displayed number
$\sqrt{x}$	Square Root	Computes square root of displayed number
( )	Parentheses	Used as grouping symbols
EE	Enter Exponent	Instructs calculator to accept next entry as exponent of the number 10
STO	Store in Memory	Copies displayed number into memory, leaving number on display also
RCL	Recall Memory	Copies number in memory onto display, leaving number in memory also

The keys +, −, ×, ÷, and y^x (or x^y) perform binary operations. That is, two numbers are required for the operation to be done. The +/−, $1/x$, x^2 and $\sqrt{x}$ keys perform unary operations. That is, only one number is required for the operation to be done. Binary operations require the = key to be used to signal the end of data entry, but unary operations are performed immediately on the displayed number. Consider the examples that follow. A calculator operation will be indicated with a box around it. You should enter the data on your calculator to verify that your calculator follows algebraic logic.

KEYS ON A CALCULATOR WITH ALGEBRAIC LOGIC

(a) The keys [+], [−], [×], [÷], and [y^x] are used between numerical entries and the last number must be followed by [=] to complete the operation.

(b) The keys [+/−], [1/x], [x^2], and [$\sqrt{x}$] will perform the operation immediately on the displayed number.

EXAMPLE 1 Perform the given operations on a calculator.

(a) $1 + 9$ **(b)** $7 - 3$ **(c)** 6×4 **(d)** $8 \div 2$ **(e)** 2^3

Solution

(a) 1 [+] 9 [=] **10**

(b) 7 [−] 3 [=] **4**

(c) 6 [×] 4 [=] **24**

(d) 8 [÷] 2 [=] **4**

(e) 2 [y^x] 3 [=] **8** ■

EXAMPLE 2 Perform the given operations on a calculator.

(a) -9 **(b)** $1/9$ **(c)** 9^2 **(d)** $\sqrt{9}$

Solution

(a) 9 [+/−] **−9**

(b) 9 [1/x] **0.1111111**

(c) 9 [x^2] **81**

(d) 9 [$\sqrt{x}$] **3** ■

EXAMPLE 3 Perform the given operations on a calculator.

(a) $16 - (-4)$ **(b)** $3^2 \times -\sqrt{9}$ **(c)** $\dfrac{1}{2+3}$

Solution (a) 16 [−] 4 [+/−] [=] **20**

(b) 3 [x^2] [×] 9 [$\sqrt{x}$] [+/−] [=] **−27**

(c) 2 [+] 3 [=] [1/x] **0.2**

NOTE ⇨⇨ *Notice in (c) that if you forgot to use the* [=] *key, the sequence* 2 [+] 3 [1/x] *would display only the reciprocal of* 3 *or* 0.3333333. ■

EXAMPLE 4 Perform the given operations on a calculator.

(a) $2(-14) - 3(-12)$

(b) $\sqrt{3^2 + 4^2}$

Solution (a) 2 [×] 14 [+/−] [−] 3 [×] 12 [+/−] [=] **8**

(b) [(] 3 [x^2] [+] 4 [x^2] [)] [$\sqrt{x}$] or

3 [x^2] [+] 4 [x^2] [=] [$\sqrt{x}$] **5**

What would happen if you forgot to press the [=] key in (b)? ■

Notice that Example 4(b) could be done in two ways. More complicated problems can often be worked more than one way.

EXAMPLE 5 The formula $R = \dfrac{R_1R_2}{R_1 + R_2}$ gives the equivalent resistance R of two resistors connected in parallel. R_1 and R_2 are the resistances of the two resistors. Find R if $R_1 = 33\ \Omega$ and $R_2 = 55\ \Omega$.

Solution The problem is to evaluate $\dfrac{(33)(55)}{33 + 55}$. Since the fraction bar serves as a grouping symbol, parentheses may be used to indicate the grouping in the denominator.

33 [×] 55 [÷] [(] 33 [+] 55 [)] [=] **20.625**

To two significant digits, the answer would be 21 Ω. Another way to handle the problem is to use memory to store the numerator while the denominator is simplified. We use [STO] and [RCL] for "store" and "recall." The keys that perform these operations on your calculator may be labeled differently. Other common labels are [M] for "memory" and [MR] for "memory recall." If we recall the numerator from calculator memory and divide, the problem will be "upside down." That is, it will be 88 ÷ 1815, instead of 1815 ÷ 88. The reciprocal key solves the problem.

33 [×] 55 [=] [STO] 33 [+] 55 [=] [÷] [RCL] [=] [1/x] **20.625** ■

To direct your attention to the sequence of keystrokes used in calculator problems, whole numbers have been used in the examples. A calculator, however, is especially useful when the problems involve decimal fractions. ***It is important to remember that a calculator does not consider the accuracy or precision of the data, so the answer is often given with more precision or accuracy than is appropriate.*** You must round off the final answer yourself.

NOTE ⇨⇨

You should try to perform the operations so that rounding is not required until the last step. Judicious use of calculator memory will help. ***If it is necessary to record a result for use in a later calculation, you should retain at least one more digit than required for the accuracy or precision of the answer.*** This precaution will help to minimize round-off errors. When several operations occur in the same problem, it is necessary to decide how many decimal places or significant digits would result for each operation. The final answer should be rounded accordingly. Square roots sometimes occur in technical problems. The following rule is used.

NOTE ⇨⇨

ACCURACY OF A SQUARE ROOT

The accuracy of the square root of a number is considered to be the same number of significant digits as the accuracy of the number itself.

EXAMPLE 6 Perform the calculations on a calculator. Round off appropriately.
$\sqrt{46.7(32.16) + 8.33(1.3)}$

Solution The product 46.7(32.16) has three significant digits (1500), and it is too large for any of the digits to the right of the decimal point to be significant. To two significant digits, the product 8.33(1.3) is 11. Because the leftmost significant digit of 11 is added to the third significant digit of 1500, the sum is precise to tens. It has three significant digits, as does the square root.

46.7 [×] 32.16 [+] 8.33 [×] 1.3 [=] [√x] **38.893457**

Answer: 38.9 to three significant digits ■

All of the detail in Example 6 is a bit cumbersome. It is necessary to work the problem out step by step. Quite often, the results are adequate if the rounding is simply done according to the number of significant digits, even if additions and subtractions are involved. Using this convention, rounding the final answer to two significant digits would have given 39.

EXAMPLE 7 Perform the calculations on a calculator. Round off appropriately.

(a) $4.3(-5.01) + (4.5)^2/0.0013$ **(b)** $(0.4450 + 0.3210 - 0.1051)(-5.7781)$

Solution **(a)** 4.3 [×] 5.01 [+/−] [+] 4.5 [x^2] [÷] 0.0013 [=] **15555.38**

Answer: 16,000 to two significant digits

(b) 0.445 [+] 0.321 [−] 0.1051 [=] [×] 5.7781 [+/−] [=] **−3.8187463**

Answer: −3.819 to four significant digits ■

When an answer has more digits than a calculator can display, it is probably shown in scientific notation. The calculator will not show the number 10, but will show a decimal number followed by a two-digit number. The two-digit number is to be considered an exponent for 10.

When a calculation requires operands that have more digits than the calculator can display, the number may be entered in scientific notation. An [EE] key accepts the next entry as the exponent for 10. If your calculator has a [MODE] key, you may have to use it to enter scientific notation.

EXAMPLE 8 The distance from the earth to the sun is approximately 93,000,000 miles. Light travels 1 mile in 0.0000054 seconds. How long does it take for light to travel from the sun to the earth?

Solution To find the number of seconds it takes for light to travel 93,000,000 miles, we find $93{,}000{,}000 \times 0.0000054$.

$$93{,}000{,}000 \times 0.0000054 = (9.3 \times 10^7) \times (5.4 \times 10^{-6})$$

9.3 [EE] 7 [×] 5.4 [EE] 6 [+/−] [=] **5.022 02**

(Notice that the [+/−] key changed only the sign of the exponent.) The answer, then, is 5.022×10^2 or about 500 seconds. To express the answer in minutes, we divide by 60 to get 8.4 minutes. ■

EXERCISES 1.5

In 1–20, perform the operations on a calculator. Consider the operands to be exact numbers, and round answers to three significant digits, if necessary.

1. $-21 + 17$ **2.** $-18 + (-34)$ **3.** $54 - (-11)$ **4.** $48 - (-13)$

5. $12^2(3)$ **6.** $13^2(9)$ **7.** $4\sqrt{17}$ **8.** $5\sqrt{19}$

9. $\dfrac{1}{7+8}$ **10.** $\dfrac{1}{9+6}$ **11.** $3(-13) - 5(-16)$ **12.** $-2(-2) + 4(17)$

13. $\dfrac{23(5)}{8+6}$ **14.** $\dfrac{27(7)}{13-2}$ **15.** $\dfrac{34(6)}{15-6}$ **16.** $\dfrac{46(7)}{19+7}$

17. $\sqrt{5^2 + 12^2}$ **18.** $\sqrt{17^2 - 8^2}$ **19.** $\sqrt{61^2 - 60^2}$ **20.** $\sqrt{42^2 + 40^2}$

In 21–42, perform the calculations on a calculator. Round off each answer appropriately.

21. $5.2(-3.61) + 121.4$ **22.** $-16.3 + 5.11 \div 0.479$ **23.** $816.0 - 33.4 \times 9.24$

24. $7.5\,(4.11) - 3.2(6.48)$ **25.** $(5.3)^2 \div 0.32$ **26.** $(0.047)^2(-1.1)$

27. $61.4 \div (1.33)^2$ **28.** $-\sqrt{1.48} + 6.33$ **29.** $5.91 + \sqrt{29.0}$

30. $\sqrt{8.61(79.3)}$ **31.** $\dfrac{3.51(-20.1) + 0.0625}{7.33}$ **32.** $\dfrac{13.4(-3.09) - 0.00731}{19.2}$

33. $3.51(-20.1)+\frac{0.0625}{7.33}$

34. $13.4(-3.09)-\frac{0.00731}{19.2}$

35. $\frac{(1.7)^2+9.6}{8.8}$

36. $\frac{0.26+9.6}{8.8-5.1}$

37. $0.26+\frac{9.6}{8.1-5.1}$

38. $\frac{5.0(3.7)}{\sqrt{9.4}}$

39. $\frac{(2.9)^3+8.5}{7.3}$

40. $\frac{6.31(7.62)}{(10.3)^3}$

41. $(8.9\times10^9)\times(6.31\times10^8)$

42. $(9.83\times10^{14})\div(4.5\times10^{-6})$

Use a calculator to solve the following problems.

43. The resistance R (in ohms) of a wire is given by $R=\rho l/A$. Find R if $\rho=0.0175\ \Omega\cdot\text{m}$, $l=60.0$ m, and $A=6.5\ \text{mm}^2$. In the formula, l is the length of the wire (in m), A is its cross-sectional area (in mm^2), and ρ is the specific resistance (in $\Omega\cdot\text{m}$). **3 5 9** 1 2 4 6 7 8

44. The formula $d=\left(\frac{0.080w+0.130h-1.03}{18.3}\right)(a)$ is sometimes used to compute a child's dosage d (in milligrams) of medicine. Compute the dosage for a child with $h=42$ in and $w=41$ lb if $a=425$ mg. In the formula, w is the weight of the child (in pounds), h is the height of the child (in inches), and a is the normal adult dosage (in milligrams). **6 9** 2 4 5 7 8

45. The magnitude f_1 of the soil pressure (in kN/m^2) of a certain type of retaining wall is given by $f_1=\frac{R}{A}\left(1+\frac{6e}{d}\right)$. Compute f_1 if $R=164$ kN, $A=4.40\ \text{m}^2$, $e=0.065$ m, and $d=4.40$ m. **2 4 8** 1 5 6 7 9

46. The tension T (in lb) in the cable at the end of a suspension bridge is given by $T=\frac{1}{2}wa\sqrt{1+\frac{a^2}{16d^2}}$. Compute the tension (in lb) when $w=750.0$ lb/ft, $a=500.0$ ft, and $d=35.0$ ft. In the formula, w is the load (in lb/ft), a is the length of the span (in ft), and d is the sag (in ft). **2 4 8** 1 5 6 7 9

47. The pressure P of a gas (in atmospheres) is given by $P=\frac{0.082n(273+T)}{V}$. Calculate the pressure of a gas when $V=32$ l, $n=1.7$, and $T=25°$ C. In the formula, V is the volume (in liters), n is the number of moles, and T is the temperature (in degrees Celsius). **6 8** 1 2 3 5 7 9

48. The potential difference V_2 (in volts) across the secondary windings of a transformer is given by $V_2=\frac{N_2V_1}{N_1}$. Find V_2 if $V_1=120$ volts, $N_1=250$ and $N_2=50$. In the formula, N_1 and N_2 are the number of turns in the primary and secondary windings, respectively, and V_1 is the potential difference across the primary windings. **3 5 9** 1 2 4 6 7 8

49. The capacitance C for three capacitors in series is given by $\frac{1}{C}=\frac{1}{C_1}+\frac{1}{C_2}+\frac{1}{C_3}$. If the individual capacitances C_1, C_2, and C_3 are measured in nF, then C will be in nF. If $C_1=2400$ pF, $C_2=2.4$ nF, and $C_3=0.0012\ \mu\text{F}$, change all measurements to nF, and solve for C. **3 5 9** 4 7 8

50. The number 22/7 is often used as an approximation for π.

(a) To how many significant digits does 22/7 agree with the value of π as given by the [π] key on your calculator?

(b) To how many significant digits does 355/113 agree with the value of π as given by the [π] key on your calculator?

1. Architectural Technology
2. Civil Engineering Technology
3. Computer Engineering Technology
4. Mechanical Drafting Technology
5. Electrical/Electronics Engineering Technology
6. Chemical Engineering Technology
7. Industrial Engineering Technology
8. Mechanical Engineering Technology
9. Biomedical Engineering Technology

1.6 CHAPTER REVIEW

KEY TERMS

1.1 natural numbers
integers
rational numbers
irrational numbers
real numbers
absolute value
pure imaginary numbers
complex numbers

1.2 sum
difference
product
quotient
dividend
divisor

1.3 significant digit
scientific notation

SYMBOLS

$>$	greater than	$\leq$	less than or equal to
$\geq$	greater than or equal to	$\|x\|$	absolute value of x
$<$	less than	( ), [ ], { }	grouping symbols

RULES AND FORMULAS

Order of Operations:

1. Locate the innermost grouping symbols. Within grouping symbols or in the absence of grouping symbols, follow steps 2–4.
2. Perform the multiplications indicated by exponents.
3. Perform multiplications and divisions, working from left to right.
4. Perform additions and subtractions, working from left to right.
5. If there are additional grouping symbols, repeat steps 1–4.

Commutative properties:

If a and b are real numbers, then $a + b = b + a$.

If a and b are real numbers, then $ab = ba$

Associative properties:

If a, b, and c are real numbers, then $(a + b) + c = a + (b + c)$.

If a, b, and c are real numbers, then $(ab)c = a(bc)$.

Distributive property:

If a, b, and c are real numbers, then $a(b + c) = ab + ac$.

GEOMETRY CONCEPTS

$A = lw$ (area of a rectangle)

$V = lwh$ (volume of a rectangular solid)

$V = \frac{4}{3}\pi r^3$ (volume of a sphere)

$V = \frac{1}{3}\pi r^2 h$ (volume of a cone)

SEQUENTIAL REVIEW

(Section 1.1) *In 1 and 2, simplify each expression.*

1. $|-5.91|$

2. $-\left|\frac{-\pi}{3}\right|$

3. Write the list of numbers in order from smallest to largest. $\sqrt{7}, -11, 5, -\pi, 17, |-3|$

In 4–7, determine whether each statement is true or false.

4. $7 < -9$ **5.** $-7 < -9$ **6.** $6 > 6$ **7.** $6 \geq 6$

In 8–11, check each number system to which the number belongs.

	Number	Natural	Integers	Rational	Real	Complex
8.	2.3					
9.	$\sqrt{5}$					
10.	$-3/5$					
11.	$\pi + 3\sqrt{-1}$					

(Section 1.2) *In 12–15, perform the indicated operation. Consider the numbers to be exact.*

12. $-17 + (-25)$ **13.** $-7 - 19$

14. $5(-13)$ **15.** $-48 \div (-16)$

In 16–19, evaluate each expression. Consider the numbers to be exact.

16. $5[4 - (2 - 5)^2]$ **17.** $-100 \div 20 \div 5$

18. $3 - |2^2 - 7| \div \sqrt{81}$ **19.** $\dfrac{(5 - 1)^2 + 7(2^3 - 8)}{3 + 1}$

(Section 1.3) *In 20 and 21, perform the indicated operation and round off appropriately.*

20. $11.63 + 2.5 - 163.702$ **21.** $3.2(0.61) \div 16$

In 22–25, state the number of significant digits in each number.

22. 92.51 **23.** 73,000 **24.** 0.0086 **25.** 520.00

In 26–28, write each number in scientific notation.

26. 65.98 **27.** 0.00753 **28.** 892,400

In 29–30, write each number in decimal notation.

29. 6.20×10^4 **30.** 7.3×10^{-3}

(Section 1.4) *In 31–34, use the commutative, associative, or distributive property to simplify each computation.*

31. $(64 + 12) + 6$ **32.** $30(1/3 + 2/5)$ **33.** $79(101)$ **34.** $(1/5)(35 \cdot 11)$

In 35–38, identify the property or properties illustrated by each statement.

35. $(5 + 9) + 3 = 5 + (9 + 3)$

36. $(5 + 9) + 3 = 3 + (5 + 9)$

37. $3(5 + 9) = 3(5) + 3(9)$

38. $3(5 + 9) = (9 + 5)3$

(Section 1.5) *In 39–41, perform the calculations on a calculator. Consider the operands to be exact numbers.*

39. $\dfrac{2(-14) - 5(3)}{7}$

40. $\sqrt{41^2 - 40^2}$

41. $\dfrac{38(591)}{29 - 6}$

In 42–45, perform the following calculations on a calculator. Round off each answer to the appropriate number of significant digits.

42. $(8.63 \times 10^5) \div (7.62 \times 10^{11})$

43. $\dfrac{(3.71)^5 + 14.32}{8.8}$

44. $6.33 + \sqrt{8.91}$

45. $(7.33 \times 10^{-2}) \div (1.21 \times 10^{-3})$

APPLIED PROBLEMS

1. The period t (in seconds) of a pendulum is given by $t = 2\pi\sqrt{l/g}$. In the formula l is the length of the pendulum, and g is the acceleration due to gravity. If l and g are both positive rational numbers, what type of number is t? Give the most specific answer that is correct in every case.
4 7 8 1 2 3 5 6 9

2. The bending stress f (in lb/in²) in a structural beam is given by $f = M/S$. If M and S are positive integers, what type of number is f? Give the most specific answer that is correct in every case. In the formula, M is the applied bending moment (in in·lb) and S is a property of the beam's cross-sectional shape (in in³). 1 2 4 8 7 9

3. Change 2.2 Megohms (MΩ) to ohms (Ω).
3 5 9 1 2 4 6 7 8

4. Change 4.7 kilowatts (kW) to watts (W).
3 5 9 1 2 4 6 7 8

5. If the temperature drops from 5° to −3°, by how many degrees does it change?
6 8 1 2 3 5 7 9

6. The formula $R_1 = (l_1/l_2)R_2$ gives the resistance (in ohms) for a wire. Find R_1 if $l_1 = 25.0$ ft, $R_2 = 13.0$ ohms, and $l_2 = 20.0$ ft. In the formula, R_1 is the resistance for a length of l_1 and R_2 is the resistance for a length of l_2. 3 5 9 1 2 4 6 7 8

7. The area of a triangle with sides of length a, b, and c is given by $A = \sqrt{s(s-a)(s-b)(s-c)}$ where $s = (1/2)(a + b + c)$. Find the area (in cm²) of a triangle with sides of lengths $a = 10.0$ cm, $b = 6.0$ cm, and $c = 5.0$ cm. 1 2 4 7 8 3 5

8. A digital multimeter (abbreviated DMM) is used to measure the voltage difference across three resistors in series. The total voltage V (in volts) is given by $V = V_1 + V_2 + V_3$, where V_1, V_2, and V_3 are the individual voltages (in volts). On a 4-1/2 digit DMM $V_1 = 4.123$ volts, $V_2 = 76.45$ volts, and $V_3 = 30.00$ volts. Find the value of V, and round off appropriately.
3 5 9 1 2 4 6 7 8

9. The average velocity v (in cm/s) of blood flowing in a blood vessel is given by $v = \dfrac{1000\,f}{60\,\pi r^2}$. In the formula, f is the flow (in liters/minute), and r is the radius of the blood vessel (in cm). Find v if $f = 4.0$ liters/minute, and $r = 1.5$ cm.
4 7 8 1 2 3 5 6 9

10. The number of threads per inch on a screw is given by $N = \dfrac{6.6}{A + 0.02}$, if A is the diameter (in inches) of the screw. Compute the number of threads per inch for a screw of diameter 0.398 inches. 1 2 4 8 7 9

1. Architectural Technology
2. Civil Engineering Technology
3. Computer Engineering Technology
4. Mechanical Drafting Technology
5. Electrical/Electronics Engineering Technology
6. Chemical Engineering Technology
7. Industrial Engineering Technology
8. Mechanical Engineering Technology
9. Biomedical Engineering Technology

11. A certain type of laser diode has a peak emission wavelength of 675 nanometers. How many micrometers is this? 3 5 6 9

12. The voltage applied to the gate of a particular transistor changed from -2V to -3.6V. By how much did it change? **3 5 9** 4 7 8

13. The moment of inertia I (in kg·m²), of a solid cylinder of radius r (in m), length S (in m), and mass m (in kg) is given by the formula $I = \frac{1}{4} m \left(r^2 + \frac{S^2}{3} \right)$. Find the moment of inertia for a cylinder of mass 3 kg with radius 0.7 m and length 4m. **1 2 4** 8 7 9

14. The speed V (in m/s) of a gas molecule is related to the density D (in kg/m³) and pressure P (in N/m²) of the gas by the formula $V = \sqrt{\frac{3P}{D}}$. What speed would a nitrogen molecule have if if $P = 10^5$ and $D = 1.25$? **4 7 8** 1 2 3 5 6 9

15. The volume of a hollow cylinder used in some microwave equipment is given by the formula $V = \frac{\pi L}{4} [D^2 - d^2]$. Where D and d are its outer and inner diameters respectively, and L is its length. Find V if $L = 27$ cm, $D = 3.5$ cm, and $d = 2.7$ cm. Use $\pi = 3.14$. **3 5 9** 1 2 4 6 7 8

16. The kinetic energy K (in joules) of a moving bullet is given by $K = \frac{1}{2} mv^2$ where m is its mass (in kg) and v its velocity (in m/s). What energy would a 0.0080 kg bullet have moving at 300.0 m/s? **4 7 8** 1 2 3 5 6 9

17. The temperature C of an object in degrees Celsius is given by $C = \frac{5}{9}(F - 32)$ where F is its temperature in Fahrenheit degrees. Find the Celsius temperature corresponding to 95° F. **6 8** 1 2 3 5 7 9

18. The final velocity v (in m/s) of a certain accelerating mechanism is given by $v = \sqrt{2as + v_0^2}$ where v_0 is its initial velocity (in m/s), s the distance it travels (in m), and a its acceleration (in m/s²). Find its velocity if $a = 14.7$ m/s², $s = 0.43$ m, and $v_0 = 2.6$ m/s. **4 7 8** 1 2 3 5 6 9

19. The formula $NA = \sqrt{n_1^2 - n_2^2}$ is used in the study of fiber optics. In the formula n_1 is the index of refraction of the core and n_2 the index of the cladding. Find NA if $n_1 = 1.31$ and $n_2 = 1.22$. 3 5 6 9

20. The output voltage V (in volts) of a certain type of diode is given by the formula $V = 2.73 + 0.000455T$, where T is its temperature (in degrees Celsius). Find V if $T = 35°$ C. **3 5 9** 4 7 8

1. Architectural Technology
2. Civil Engineering Technology
3. Computer Engineering Technology
4. Mechanical Drafting Technology
5. Electrical/Electronics Engineering Technology
6. Chemical Engineering Technology
7. Industrial Engineering Technology
8. Mechanical Engineering Technology
9. Biomedical Engineering Technology

RANDOM REVIEW

1. Identify the property or properties illustrated by each statement.
   (a) $(4 \cdot 3) \cdot 2 = 4 \cdot (3 \cdot 2)$ **(b)** $5 + (2 + 1) = 1 + (2 + 5)$
2. Determine whether each statement is true or false.
   (a) $5 < 5$ **(b)** $-9 < -4$
3. State the number of significant digits in each number.
   (a) 23,500 **(b)** 0.017
4. Write the list of numbers from smallest to largest.

$$\sqrt{5}, \quad -8, \quad 3, \quad \pi, \quad 15/7, \quad |-2|$$

5. Perform the following calculation on a calculator. Round the answer to the appropriate number of significant digits.

$$(7.72 \times 10^4) \div (6.83 \times 10^2)$$

6. Perform the indicated operation. Consider the numbers to be exact.
   (a) $-8-21$ **(b)** $49 \div (-7)$
7. Perform the indicated operation. Consider the numbers to be approximate values and round appropriately.

$$7.734 + 2.17 - 105.9$$

8. Simplify each expression.
   (a) $-|-6.33|$ **(b)** $|12.55|$
9. Evaluate each expression. Consider the numbers to be exact.
   (a) $7 + \sqrt{6 + 2 \cdot 5}$ **(b)** $\dfrac{(9 \div 3) - 2}{|-4-3|}$
10. Evaluate each expression. Consider the numbers to be exact.
   (a) $2 + 5 \cdot 3^2 \div (7-2)$
   (b) $(2+5)(3^2) \div 7 - 2$
11. Write each number in scientific notation.
   (a) 8,436,200 **(b)** 0.000731
12. Check each number system to which the number belongs.

	Number	Natural	Integers	Rational	Real	Complex
(a)	-5.7					
(b)	$\sqrt{14}$					

13. Perform each calculation on a calculator. Consider the numbers to be exact.
   (a) $\dfrac{15(12)}{\sqrt{13^2 - 5^2}}$ **(b)** $\dfrac{4\sqrt{3}}{1+\sqrt{2}}$
14. Write each number in decimal notation.
   (a) 6.415×10^{-4} **(b)** 7.321×10^5
15. Perform the following calculation on a calculator. Consider the numbers to be approximate and round appropriately.

$$\frac{(12.3)^2 - \sqrt{88.4}}{4.7}$$

CHAPTER TEST

1. Simplify each expression and determine whether the statement is true or false.
   (a) $|5-3| < |5| - |-3|$
   (b) $-2 + 7 > -3 - (-4)$
2. Check each number system to which the number belongs.

	Number	Natural	Integers	Rational	Real	Complex
(a)	$4 - 5\sqrt{-1}$					
(b)	1.732					

3. Evaluate each expression. Consider the numbers to be exact.
   (a) $-64 \div 8 \div 4$ **(b)** $\dfrac{4[3-(5-7)^2]}{4+12}$

4. State the number of significant digits in each number and then write the number in scientific notation.
   (a) 0.00072 (b) 600
5. Change each unit of measurement as indicated.
   (a) 3 kc to c (kilocalories to calories)
   (b) 7.2 ml to l (milliliters to liters)
6. Perform the indicated operation and round appropriately.
   (a) $0.411(5.2) \div 3$
   (b) $3.2(7.1) + 6.5$
7. Identify the property or properties illustrated by each statement.
   (a) $4(7 + 2) = 4(7) + 4(2)$
   (b) $(5 + 3) + 2 = 2 + (3 + 5)$
8. Perform the following calculations on a calculator. Consider integers to be exact and decimals to be approximate. Round each answer appropriately.
   (a) $\dfrac{\sqrt{61^2 - 60^2}}{33}$ (b) $\dfrac{(46.2)^3 - 25.43}{9.7}$
9. The current I (in amperes) carried by an inductor is given by $I = \sqrt{\dfrac{2W}{L}}$. If W and L are positive integers, what type of number is I? Give the most specific answer that is correct in every case. In the formula, W is the potential energy (in joules) and L is the inductance (in henries).
10. If a ball of lead with radius r (in cm) is pounded into a circular sheet with thickness t (in cm), the diameter of the sheet is given by $d = \dfrac{4r^3}{3t}$. If a lead ball with radius 2.54 cm is pounded into a circular sheet 0.12 cm thick, how large is the diameter of the sheet?

ALGEBRA CHAPTER 2

Algebra is sometimes described as "generalized arithmetic." That is, addition, subtraction, multiplication, and division are used with letters as well as numbers. In this chapter, algebraic operations are developed and applied to formulas, equations, and problems of the type that arise in technological settings.

2.1 ALGEBRAIC OPERATIONS: ADDITION AND SUBTRACTION

algebraic expression

Letters, or *variables,* are used in algebra to represent numbers with values that may be unknown. When letters and numbers are combined using the arithmetic operations (addition, subtraction, multiplication, division, and finding *n*th roots), the result is an **algebraic expression.** We begin a review of algebra by examining exponents.

An exponent is a symbol used to indicate a repeated multiplication. The expression x^p means that p x's are to be multiplied together. The x in this expression is called the *base*, and p is called the *exponent*. Sometimes we say "x is raised to the power p." Powers of two are often used in computer engineering to specify memory capacity. For example, 2^5 means (2)(2)(2)(2)(2), or 32. In the following definition, the phrase "x is a real number" is used. When this phase is used, it is understood that the "x" referred to can be either a number or a variable that represents a number.

base
exponent

DEFINITION The expression $\boldsymbol{x^p}$ means that x is to be used as a factor p times if x is a real number, and p is a positive integer.

In this chapter, the words *term* and *factor* are used often. If we think of the subtraction of x as the addition of $-x$, we may say that the operands are called *terms* in an expression that contains addition. The operands are called *factors* in an expression that contains multiplication. Thus, $x + y$ has two terms, but xy has two factors.

The distinction between terms and factors is important in an expression such as $2x + 3y$, which contains both addition and multiplication. The rules for order of operations dictate that the last operation performed is addition for this problem, so this expression has two terms. Each term by itself has two factors. In the expression $(x + 2)(x - 1)$, the rules for order of operations dictate that the last operation performed is multiplication, so this expression has two factors. Each factor has two terms.

DEFINITION When an algebraic expression consists of quantities that are added, each quantity along with its sign, is called a **term.** When algebraic expressions consist of quantities that are multiplied, each quantity is called a **factor.**

EXAMPLE 1 Determine the number of terms or factors in each expression.

(a) $2xy - 3$ **(b)** $x^2 - 4x + 1$ **(c)** $x(x - 1)(x - 2)$

Solution **(a)** $2xy - 3$ has two terms. They are $2xy$ and -3. (Subtracting 3 is equivalent to adding -3, so think of the expression as $2xy + (-3)$).

(b) $x^2 - 4x + 1$ has three terms. They are x^2, $-4x$, and 1.

(c) $x(x - 1)(x - 2)$ has three factors. They are x, $(x - 1)$, and $(x - 2)$. ■

We often refer to *like terms* in an algebraic expression. Like terms differ at most by a numerical factor. For example, in the expression $3x^2 + 4y - x^2 + 5$, $3x^2$ and $-x^2$ are like terms, because both are x^2 terms.

DEFINITION **Like terms** of an algebraic expression are terms in which the variable factors are identical in both base and exponent.

EXAMPLE 2 In each expression, name the like terms.

(a) $3x + 4y - 3y - 1$

(b) $3a^2b + 3ab^2 - a^2b$

Solution **(a)** $4y$ and $-3y$ are like terms, because both are y terms.

(b) $3a^2b$ and $-a^2b$ are like terms, because both are a^2b terms. ■

The distributive property may be used to simplify an algebraic expression that contains like terms. For example, $3x + 4x$ is $(3 + 4)x$ or $7x$. As a short cut, we say that $3x$ and $4x$ are combined to yield $7x$.

EXAMPLE 3 Simplify each algebraic expression.

(a) $3x^2 + x - 1 + 4x$ **(b)** $4ab + 2bc + 7ab$ **(c)** $3a^2b - 3ab - a^2b + 2ab$

Solution **(a)** $3x^2 + \mathbf{x} - 1 + \mathbf{4x} = 3x^2 + \mathbf{x + 4x} - 1 = 3x^2 + 5x - 1$

(b) $\mathbf{4ab} + 2bc + \mathbf{7ab} = \mathbf{4ab + 7ab} + 2bc = 11ab + 2bc$

(c) $\mathbf{3a^2b} - 3ab - \mathbf{a^2b} + 2ab = \mathbf{3a^2b - a^2b} - 3ab + 2ab = 2a^2b - ab$ ■

Addition and subtraction problems often involve grouping symbols. We follow the standard order of operations, performing the operations within grouping symbols first *if possible.* In algebraic expressions, however, unless the terms enclosed by grouping symbols are like terms, they cannot be combined. We may have to reorder and regroup, as in Example 3, to put like terms together. In a subtraction problem, the distributive property may be used before like terms are combined. That is, an expression like $-(x + 3)$ can be written as $-1(x + 3)$ and is therefore equal to $-x - 3$.

NOTE ⇨⇨ ***When a grouping in an algebraic expression is preceded by a negative sign, it is very important to change the sign of every term within the grouping when the grouping symbols are removed.*** The distributive property also applies to problems of the form $a(b - c)$, since $b - c$ may be written as $b + (-c)$. That is, $a(b - c) = ab - ac$. With this background, it is possible to add or subtract algebraic expressions.

To add or subtract algebraic expressions:

1. Write the expressions without parentheses, using the distributive property, if necessary.
2. Add like terms.

EXAMPLE 4 Perform the indicated operation.

(a) $(x^2 + x - 1) + (2x^2 + 3x - 4)$

(b) $(x^2 + x - 1) - (2x^2 + 3x + 4)$

(c) $(2x + y - 1) - (3y - 4z + 2)$

Solution **(a)** The expressions within parentheses cannot be combined, since there are no like terms. Thus, we reorder and regroup to put like terms together. This step is often done mentally.

$$\begin{aligned}(x^2 + x - 1) &+ (2x^2 + 3x - 4) \\ &= x^2 + x - 1 + 2x^2 + 3x - 4 && \text{Remove parentheses.} \\ &= (x^2 + 2x^2) + (x + 3x) + (-1 - 4) && \text{Group like terms.} \\ &= 3x^2 + 4x - 5 && \text{Combine like terms.}\end{aligned}$$

(b) The expressions within parentheses cannot be combined, since there are no like terms. Thus, we use the distributive property.

$$\begin{aligned}(x^2 + x - 1) &- \mathbf{(2x^2 + 3x + 4)} \\ &= x^2 + x - 1 \mathbf{- 2x^2 - 3x - 4} && \text{Note: } -(2x^2 + 3x + 4) = -2x^2 - 3x - 4 \\ &= -x^2 - 2x - 5 && \text{Combine like terms: } x^2 - 2x^2 = -x^2,\ x - 3x = -2x, \\ & && \text{and } -1 - 4 = -5\end{aligned}$$

(c)

$$\begin{aligned}(2x + y - 1) &- \mathbf{(3y - 4z + 2)} \\ &= 2x + y - 1 \mathbf{- 3y + 4z - 2} && \text{Note: } -(3y - 4z + 2) = -3y + 4z - 2 \\ &= 2x - 2y + 4z - 3 && \text{Combine like terms: } y - 3y = -2y \\ & && \text{and } -1 - 2 = -3 \quad ■\end{aligned}$$

Algebraic expressions sometimes contain nested grouping symbols. As with arithmetical expressions, the innermost grouping is simplified first.

EXAMPLE 5 Simplify each expression.

(a) $x + [2x + (3x + 2)]$

(b) $a - [3a - (2a - 1) + 5]$

Solution **(a)** The expressions within parentheses cannot be combined, since they are not like terms.

$$\begin{aligned}x + [2x + \mathbf{(3x + 2)}] &= x + [2x \mathbf{+ 3x + 2}] && \text{Remove parentheses.} \\ &= x + [5x + 2] && \text{Combine like terms in brackets.} \\ &= x + 5x + 2 && \text{Remove brackets.} \\ &= 6x + 2 && \text{Combine like terms.}\end{aligned}$$

(b) Once again, the expressions within parentheses cannot be combined. Thus, we begin by removing the parentheses. Notice that the negative sign in front of $(2a - 1)$ affects both terms. That is, $-(2a - 1) = -2a + 1$.

$a - [3a - (2a - 1) + 5]$

$= a - [3a - 2a + 1 + 5]$ Remove parentheses: $-(2a - 1) = -2a + 1$.

$= a - [a + 6]$ Combine like terms: $3a - 2a = a$ and $1 + 5 = 6$.

$= a - a - 6$ Remove brackets: $-[a + 6] = -a - 6$.

$= -6$ Combine like terms: $a - a = 0$. ■

EXAMPLE 6 Simplify each expression.

(a) $by - \{b + [y + by - (b - y)]\}$

(b) $x^2 + \{x - [x^2 - 2x + (x - 1)]\}$

Solution **(a)** $by - \{b + [y + by - (b - y)]\}$

$= by - \{b + [y + by - b + y]\}$ Remove parentheses: $-(b - y) = -b + y$.

$= by - \{b + [2y + by - b]\}$ Combine like terms: $y + y = 2y$.

$= by - \{b + 2y + by - b\}$ Remove brackets.

$= by - \{2y + by\}$ Combine like terms: $b - b = 0$.

$= by - 2y - by$ Remove braces: $-\{2y + by\} = -2y - by$.

$= -2y$ Combine like terms: $by - by = 0$.

(b) $x^2 + \{x - [x^2 - 2x + (x - 1)]\}$

$= x^2 + \{x - [x^2 - 2x + x - 1]\}$ Remove parentheses.

$= x^2 + \{x - [x^2 - x - 1]\}$ Combine like terms.

$= x^2 + \{x - x^2 + x + 1\}$ Remove brackets.

$= x^2 + \{-x^2 + 2x + 1\}$ Combine like terms.

$= x^2 - x^2 + 2x + 1$ Remove braces.

$= 2x + 1$ Combine like terms. ■

Calculators and computers are becoming more sophisticated so that some of them can handle brackets and braces as well as parentheses. Many of these devices, however, only allow the use of parentheses, so one must use extreme care in identifying the innermost grouping. Example 7 illustrates this type of problem.

EXAMPLE 7 Simplify each expression.

(a) $2 + (x - (2x - 3))$ **(b)** $3 - (x + (x - 2) + 3)$

Solution **(a)** $2 + (x - (2x - 3))$

$= 2 + (x - 2x + 3)$ Remove inner parentheses.

$= 2 + (-x + 3)$ Combine like terms.

$= 2 - x + 3$ Remove outer parentheses.

$= 5 - x$ Combine like terms.

(b) $3 - (x + (x - 2) + 3)$

$= 3 - (x + x - 2 + 3)$ Remove inner parentheses.

$= 3 - (2x + 1)$ Combine like terms.

$= 3 - 2x - 1$ Remove outer parentheses.

$= 2 - 2x$ Combine like terms. ■

EXERCISES 2.1

In 1–10, determine the number of terms or factors in each expression.

1. $(x + 1)(x - 3)$
2. $4(y - 5)$
3. $2x + 3y$
4. $4a + 3$
5. $7x^2 + 3x + 2$
6. $(5a - 2)(7a^2 + 3a + 2)$
7. $(4y + 3)(y^2 - y - 5)$
8. $(b + 1)(b - 2)(b - 3)$
9. $(x - 1)(x + 2) - (x + 1)(x - 2)$
10. $x(2x - 3) + x(3x + 1)$

In 11–20, perform the indicated operation.

11. $(2x + 3y) + (x - y)$
12. $(3a + 2b) - (a - b)$
13. $(3x - 1) - (2 - y)$
14. $(7a + 3) + (b - 2)$
15. $(x^2 + 4x - 1) + (3x^2 - 5x + 2)$
16. $(3a^2 + 7a - 4) - (2a - 1)$
17. $(3x + 4y - z) + (2x - y + w)$
18. $(3z^2 + 2z - 1) - (z^2 - z - 4)$
19. $(2x^2 + 3x - 2) - (x^2 - x + 3)$
20. $(7x + 2y - 1) + (x - y + 2)$

In 21–35, simplify each expression.

21. $y + [y + (2y - 1)]$
22. $ax + [3ax + (ax + 2)]$
23. $6m + [3m + (2m - 1) + 2]$
24. $4b + [b + (2b + 6) - 5]$
25. $x + [2x - (x - 3)]$
26. $4x^2 + [x^2 - (x - 2) + 1]$
27. $y^2 + [y - (y^2 + y) - 4]$
28. $x^2 - [x^2 + (x - 3) - 1]$
29. $3z^2 - [z^2 - (z + 3) + z)]$
30. $mn + [mn - (2mn - 1) - 3]$
31. $y - [2y - (3y + 2)]$
32. $az - [3az - (2az + 5)]$
33. $x^2 + \{x - [x^2 + (x - 2) + 1]\}$
34. $2x^2 - \{x^2 + [x - (x + 3)] + 2\}$
35. $3ax + \{b - [ax - (b - 1) + b] + ax\}$

In 36–45, simplify each expression.

36. $3 + (x - (x - 2))$
37. $4 + (y - (2y + 1))$
38. $y - (3y - (y - 1))$
39. $z - (2z + (z - 3))$
40. $x - (x + (x - 2) + 3)$
41. $x + (2x - (3x + 2) - 1)$
42. $z - (3z + (z - 3) + 4)$
43. $x - (2x + (3x - 1) - 1)$
44. $x - (2x + (5x + (7x - 2) + 6)))$
45. $y + (3y - (y - (4y + 5) - 2))$

2.2 ALGEBRAIC OPERATIONS: MULTIPLICATION AND DIVISION

In the previous section, algebraic expressions that had terms were simplified. In this section, we will simplify algebraic expressions that have factors. In the examples that follow, some short cuts are developed so that algebraic expressions may be simplified efficiently.

EXAMPLE 1 Simplify $7^2(7^3)$.

Solution Because the first expression represents two factors of 7, and the second represents three factors of 7, there are a total of five factors of 7 in the product: $7^2(7^3) = (7)(7)(7)(7)(7) = 7^5$. As a short cut, the exponents are added. $7^2(7^3) = 7^{2+3} = 7^5$. ■

If we generalize the concept involved in the previous example, we have one of the rules for working with exponents.

PRODUCT RULE FOR EXPONENTS

If x is a real number and m and n are positive integers,

$$x^m(x^n) = x^{m+n} \quad \text{(2-1)}$$

The product rule requires that both factors have the same base. The product also has this base. If the two factors have different bases, the rule does not apply. For example, $2^3(3^2)$ is written as $8(9)$ or 72.

EXAMPLE 2 Simplify $(3y^2)(4y^5)$.

Solution The commutative and associative properties allow us to reorder and regroup the factors.

$$(3y^2)(4y^5) = (3)(4)(y^2)(y^5) = 12y^7 \quad ■$$

When at least one of the factors of an algebraic expression has two or more terms, the multiplication is based on the distributive property. When both of the factors have two or more terms, each term of one factor distributes over the other factor. For example,

$$\begin{aligned}(x+2)(x-3) &= x(x-3) + 2(x-3)\\ &= x^2 - 3x + 2x - 6\\ &= x^2 - x - 6.\end{aligned}$$

In long problems, like terms may be collected as the multiplication is done.

To multiply two algebraic expressions:

1. Use the distributive property to multiply each term of the first factor times each term of the second factor. If either factor has more than two terms, it is a good idea to collect like terms in columns as the multiplication is done.
2. Combine like terms.

EXAMPLE 3 Multiply $(2x - 1)(3x + 1)$.

Multiply $2x(3x + 1)$. Multiply $-1(3x + 1)$.

Solution
$$(2x - 1)(3x + 1) = \underbrace{6x^2 + 2x}_{} \underbrace{- 3x - 1}_{}$$
$$= 6x^2 - x - 1$$ Combine like terms. ■

EXAMPLE 4 Multiply $(x + 2)(x^2 - x + 1)$.

Solution $(x + 2)(x^2 - x + 1)$

$$= x^3 - x^2 + x$$ Multiply $x(x^2 - x + 1)$.
$$+ 2x^2 - 2x + 2$$ Multiply $2(x^2 - x + 1)$.
$$x^3 + x^2 - x + 2$$ Combine like terms.

x^2's are in 1 column — x's are in 1 column ■

EXAMPLE 5 Multiply $(x^2 + x - 1)(x^2 - x + 1)$.

Solution $(x^2 + x - 1)(x^2 - x + 1)$

$$= x^4 - x^3 + x^2$$ Multiply $x^2(x^2 - x + 1)$.
$$+ x^3 - x^2 + x$$ Multiply $x(x^2 - x + 1)$.
$$- x^2 + x - 1$$ Multiply $-1(x^2 - x + 1)$.
$$x^4 - x^2 + 2x - 1$$ Combine like terms. ■

For longer problems, we follow the standard order of operations.

EXAMPLE 6 Simplify each expression.

(a) $x(2x + 3) - (3x - 1)$ **(b)** $(2x + 3)(3x - 1) + 1$

Solution **(a)** $x(2x + 3) - (3x - 1)$

$$= \mathbf{2x^2 + 3x - 3x + 1}$$ Use the distributive property, since terms within parentheses cannot be combined.
$$= 2x^2 + (3x - 3x) + 1$$ Group like terms, perhaps mentally.
$$= 2x^2 + 1$$ Combine like terms.

(b) $(2x + 3)(3x - 1) + 1$

$= \mathbf{6x^2 - 2x + 9x - 3} + 1$ — Multiply $(2x + 3)(3x - 1)$.

$= 6x^2 + (-2x + 9x) + (-3 + 1)$ — Group like terms, perhaps mentally.

$= 6x^2 + 7x - 2$ — Combine like terms. ■

Example 7 illustrates how multiplication of algebraic expressions may be used to solve problems.

EXAMPLE 7 Find an expression for the area of the rectangle in Figure 2.1.

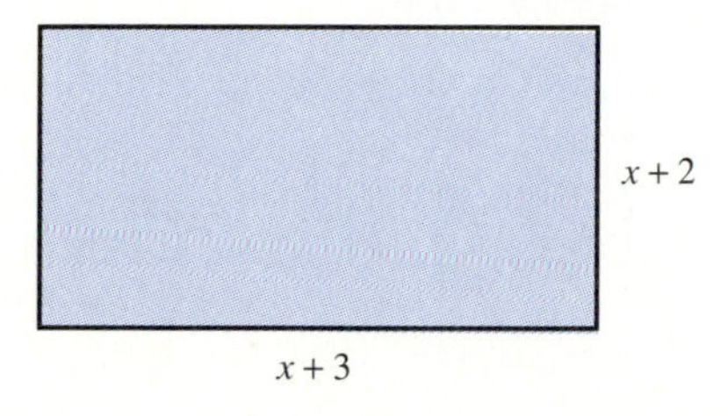

FIGURE 2.1

Solution Since the area of a rectangle is given by the product of length and width, we have the following equation.

$$A = (x + 3)(x + 2) = x^2 + 2x + 3x + 6 = x^2 + 5x + 6 \quad ■$$

We begin a study of division by considering the simplest case.

EXAMPLE 8 Simplify x^5/x^3.

Solution There are five factors of x in the numerator and three factors of x in the denominator. The x's in the denominator divide three of the x's in the numerator, leaving two factors of x, or x^2. As a short cut, the exponents are subtracted. That is, $x^5/x^3 = x^{5-3} = x^2$. ■

The concept involved in Example 8 is generalized as follows.

QUOTIENT RULE FOR EXPONENTS

If x is a nonzero real number, and m and n are positive integers,

$$\frac{x^m}{x^n} = x^{m-n} \text{ if } m > n \qquad \frac{x^m}{x^n} = \frac{1}{x^{n-m}} \text{ if } n > m \qquad \frac{x^m}{x^n} = 1 \text{ if } m = n \tag{2-2}$$

EXAMPLE 9 Simplify each expression.

(a) $\dfrac{3x}{6x^4}$ **(b)** $\dfrac{12a^5}{8a^2}$ **(c)** $\dfrac{45m^2n^3}{9m^4n}$ **(d)** $\dfrac{8xy^7}{56x^5y^3}$

Solution The numerical factors reduce as ordinary fractions.

(a) $\frac{3x}{6x^4} = \frac{1}{2x^3}$ (b) $\frac{12a^5}{8a^2} = \frac{3a^3}{2}$ (c) $\frac{45m^2n^3}{9m^4n} = \frac{5n^2}{m^2}$ (d) $\frac{8xy^7}{56x^5y^3} = \frac{y^4}{7x^4}$ ■

If the numerator contains more than one term, the denominator must be divided into each term of the numerator.

To divide an algebraic expression by an algebraic expression consisting of a single term:

1. Consider the fractions formed by dividing the denominator into each term of the numerator.
2. Simplify each fraction.

EXAMPLE 10 Perform the indicated operation $\frac{9m^2n^2 - 6mn + 3m}{6mn}$.

Solution
$$\frac{9m^2n^2 - 6mn + 3m}{6mn} = \frac{9m^2n^2}{6mn} - \frac{6mn}{6mn} + \frac{3m}{6mn} = \frac{3mn}{2} - 1 + \frac{1}{2n}$$
You may be able to do the intermediate step mentally, writing
$$\frac{9m^2n^2 - 6mn + 3m}{6mn} = \frac{3mn}{2} - 1 + \frac{1}{2n}.$$ ■

polynomial

The divisor, as well as the dividend, of a division problem may contain more than one term. We are now ready to examine division of *polynomials*. A **polynomial** is an algebraic expression that can be obtained using a finite number of additions and subtractions of real numbers and products formed by real numbers and variables. If only these operations are allowed, a variable may not appear in a denominator or under a radical sign. An expression like $2x^2 + \sqrt{3}x + 2/3$ is a polynomial, since $2x^2$ is the product $2 \cdot x \cdot x$, $\sqrt{3}\,x$ is the product of $\sqrt{3}$ and x, and $2/3$ is a real number.

In particular, a polynomial often contains a single variable. It might help to have a mental image of such a polynomial.

DEFINITION

A **polynomial in x** is an expression of the form
$$a_nx^n + a_{n-1}x^{n-1} + a_{n-2}x^{n-2} + \cdots + a_0,$$
where $a_n, a_{n-1}, a_{n-2}, \ldots, a_0$ are real numbers and n is a nonnegative integer.

We say that n is the *degree* of the polynomial and the values a_n, a_{n-1}, a_{n-2}, . . . a_0 are called the *coefficients*. The number a_n is called the leading coefficient.

When both numerator and denominator are polynomials with more than one term, the procedure for division is similar to long division. Consider the similarity between the two problems that follow.

$$\begin{array}{r} 21 \\ 32\overline{)672} \\ -\underline{64} \\ 32 \\ -\underline{32} \\ 0 \end{array}$$

Since $6 \div 32$ is 0, we begin with 67/32. Use 67/30 or 2 to estimate the quotient 67/32. The first digit of the quotient goes above the 7. Multiply: $2(32) = 64$. Subtract 64 from 67. Bring down the next digit. Divide: $32/32 = 1$. The second digit goes above 2. Multiply: $1(32) = 32$. Subtract from 32.

$$\begin{array}{r} 2x + 1 \\ 3x + 2\overline{)6x^2 + 7x + 2} \\ \underline{6x^2 + 4x} \\ 3x + 2 \\ \underline{3x + 2} \\ 0 \end{array}$$

We begin with $(6x^2 + 7x) \div (3x + 2)$. Use $6x^2/3x$ or $2x$ to estimate the quotient $(6x^2 + 7x) \div (3x + 2)$. The first term of the quotient goes above $7x$. Multiply: $2x(3x + 2) = 6x^2 + 4x$. Subtract $6x^2 + 4x$ from $6x^2 + 7x$. Bring down the next term. Estimate $3x + 2$ into $3x + 2$ by dividing: $3x/3x = 1$. The second term goes above the 2. Multiply: $1(3x + 2) = 3x + 2$. Subtract from $3x + 2$.

The procedure is generalized as follows.

To divide two polynomials:

1. Write the problem in long division form, so that the exponents of the variable decrease from left to right in both divisor and dividend.
2. Find the first term of the quotient by dividing the first term of the dividend by the first term of the divisor. Put the quotient over its like term.
3. Multiply quotient times divisor.
4. Subtract each term of the product (in other words, change the sign and add) from the like term in the dividend and bring down the next term.
5. Repeat steps 2–4, using the difference as a new dividend each time, until the remainder is zero or a polynomial of smaller degree than the divisor.

EXAMPLE 11 Perform the indicated operation $\dfrac{6x^2 + 5x + 1}{2x + 1}$.

Solution

$$\begin{array}{rl} 3x + 1 & 6x^2 \div (2x) = 3x \\ 2x + 1\overline{)6x^2 + 5x + 1} & \\ \underline{6x^2 + 3x} & 3x(2x + 1) \\ 2x + 1 & \text{Subtract and bring down 1.} \\ \underline{2x + 1} & 1(2x + 1) \\ 0 & \text{Subtract.} \end{array}$$

Thus, $\dfrac{6x^2 + 5x + 1}{2x + 1} = 3x + 1$. ■

If there are powers that are missing in the dividend, use a coefficient of 0 with them, or leave a blank space where they would be.

EXAMPLE 12 Perform the indicated operation $\dfrac{x^3 - 1}{x - 1}$.

Solution

$$\begin{array}{r|l} & x^2 + x + 1 \\ \hline x - 1 & x^3 + 0x^2 + 0x - 1 \\ & \underline{x^3 - x^2} \\ & x^2 + 0x \\ & \underline{x^2 - x} \\ & x - 1 \\ & \underline{x - 1} \\ & 0 \end{array} \quad \text{or} \quad \begin{array}{r|l} & x^2 + x + 1 \\ \hline x - 1 & x^3 \qquad - 1 \\ & \underline{x^3 - x^2} \\ & x^2 \\ & \underline{x^2 - x} \\ & x - 1 \\ & \underline{x - 1} \\ & 0 \end{array}$$

Thus, $\dfrac{x^3 - 1}{x - 1} = x^2 + x + 1$. ■

EXAMPLE 13 Perform the indicated operation $\dfrac{2x^3 + 4x - 5}{x^2 + 3}$.

Solution

$$\begin{array}{r|l} & 2x \\ \hline x^2 + 3 & 2x^3 + 0x^2 + 4x - 5 \\ & \underline{2x^3 \qquad + 6x} \\ & -2x - 5 \end{array}$$

(Remainder has lower degree than divisor.)

Thus, $\dfrac{2x^3 + 4x - 5}{x^2 + 3} = 2x$ with a remainder of $-2x - 5$. ■

EXERCISES 2.2

In 1–8, simplify each expression.

1. $2x^3(3x^2)$ **2.** $(3^4)(3^7)$ **3.** $2y^2(y^4)(7y)$ **4.** $(3a^2b)(4ab^2)$
5. $6x^3(8x^6)$ **6.** $7m^2n^3(-6mn^2)$ **7.** $5(9xy)(x^2y^2)$ **8.** $3w(3w^2)(3w^3)$

In 9–24, multiply, as indicated.

9. $2x(x - 1)$ **10.** $3a(2a + 5)$ **11.** $3x(x^2 - x + 4)$
12. $y^3(2y^2 + y - 3)$ **13.** $(x + 1)(2x - 3)$ **14.** $(2a + 3b)(4a + 3b)$
15. $(3y - 4)(5y - 1)$ **16.** $(x + 1)(x - 1)$ **17.** $(4a + b)(4a + b)$
18. $(2x - 3)(3x^2 - 2x + 1)$ **19.** $(z - 1)(z^2 + z + 1)$ **20.** $(x + 3)(x^2 - 3x + 9)$
21. $(2y + 1)(4y^2 + 4y + 1)$ **22.** $(x^2 + x + 1)(x^2 - x + 1)$ **23.** $(y^2 - 2y + 1)(y^2 + 3y - 2)$
24. $(2b^2 - b + 3)(b^2 + 3b - 2)$

In 25–32, simplify each expression.

25. $x(x-2)+(3x-1)$

26. $(2x+3)-3(x-2)$

27. $(x^2+2x-1)-3(x+2)$

28. $(x^2+2x-1)-x(x+2)$

29. $(x+1)(x-1)-3$

30. $(x+2)(x+1)-x$

31. $(2x+1)(3x-2)+(x-4)$

32. $(2x-1)(2x-3)-x^2$

In 33–46, perform the indicated operation.

33. $\frac{7^5}{7^3}$

34. $\frac{6^5}{6^7}$

35. $\frac{18x^3y}{9x^2y^3}$

36. $\frac{3ab^2}{15a^3b}$

37. $\frac{27mn}{18m^2n^3}$

38. $\frac{14x^7y^4}{21x^3y^5}$

39. $\frac{24a^4b^3}{16a^3b^4}$

40. $\frac{9x^2y+3xy-6xy^2}{3x^2y^2}$

41. $\frac{8x^2y-12xy^2+2xy}{4xy^2}$

42. $\frac{15a^2-30ab+5b^2}{15ab}$

43. $\frac{18a^4-27a^3+9a^2}{9a^2}$

44. $\frac{12b^5+8b^4-2b^2}{4b^3}$

45. $\frac{8m^2n^2+6mn-2}{2mn}$

46. $\frac{21x^5-14x^4+28x^2}{7x^2}$

In 47–54, perform each division.

47. $\frac{6x^2-x-2}{2x+1}$

48. $\frac{3x^2+5x-2}{3x-1}$

49. $\frac{4x^2-4x+1}{2x-1}$

50. $\frac{3x^2+5x+2}{x+1}$

51. $\frac{x^3+3x^2+3x+1}{x+1}$

52. $\frac{x^3+2x^2-x-2}{x+2}$

53. $\frac{2x^3-3x^2+2x-3}{2x-3}$

54. $\frac{x^3+6x^2+12x+8}{x+1}$

---

55. The formula $T=(100t-h)/100$ is used to estimate the temperature T at a height h, if $h<12\,000$ and t is the temperature at ground level. Find an equivalent formula such that the expression on the right is written as two separate terms. Heights are measured in meters and temperatures in degrees Celsius. 6 8 1 2 3 5 7 9

56. The formula $y=(Px^3-3Pax^2)/(6EI)$ gives the deflection y (in mm) of a cantilever beam at a distance of x from the support. Find an equivalent formula such that the expression on the right is written as two separate terms. In the formula, $x \leq a$, P is the load, a is the length from the support to the load, E is the modulus of elasticity, and I is the moment of inertia. 1 2 4 8 7 9

57. Find the area of the figure shown. 1 2 4 7 8 3 5

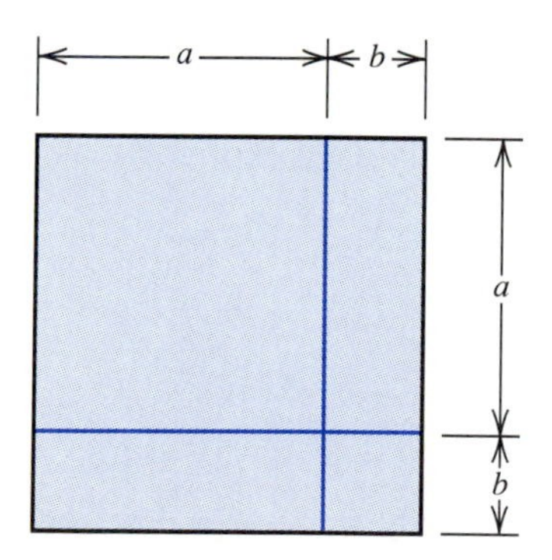

---

1. Architectural Technology
2. Civil Engineering Technology
3. Computer Engineering Technology
4. Mechanical Drafting Technology
5. Electrical/Electronics Engineering Technology
6. Chemical Engineering Technology
7. Industrial Engineering Technology
8. Mechanical Engineering Technology
9. Biomedical Engineering Technology

58. The load P (in kg) that is suspended from a coil spring is given by $P = Gd^4y/(8ND^3)$. Calculate P if $G = 4000$ kg/mm^2, $d = 15$ mm, $N = 12$, $D = 5$ mm, and $y = 12$ mm. In the formula, d is the diameter (in mm) of the wire from which the spring is made, G is the shear modulus (in kg/mm^2), y is the deflection of the spring (in mm), N is the number of coils, and D is the diameter of the coils (in mm). 1 2 4 8 7 9

59. The longitudinal stress s (in megapascals) on a thin-walled vessel under pressure is given by $s = pd/(2t)$. Calculate s if $d = 2$ m, $p = 2$ MPa, and $t = 0.20$ m. In the formula, p is the internal pressure (in MPa), d is the inside diameter (in m), and t is the wall thickness (in m). 1 2 4 8 7 9

1. Architectural Technology
2. Civil Engineering Technology
3. Computer Engineering Technology
4. Mechanical Drafting Technology
5. Electrical/Electronics Engineering Technology
6. Chemical Engineering Technology
7. Industrial Engineering Technology
8. Mechanical Engineering Technology
9. Biomedical Engineering Technology

2.3 ALGEBRAIC OPERATIONS: POWERS AND ROOTS

When an algebraic expression is raised to a power, one must be careful to distinguish between terms and factors. An exponent outside parentheses applies to each separate *factor,* but not to each separate *term.*

EXPONENT RULE

If x and y are real numbers and n is a positive integer,

$$(xy)^n = x^n y^n$$

NOTE ⇨⇨ ***It is a common mistake to try to apply the rule to an expression with two terms, but in general, $(x + y)^n \neq x^n + y^n$.*** If the expression inside parentheses consists of terms, you should apply the multiplication process that you learned in the previous section.

EXAMPLE 1 Simplify each expression.

(a) $(2xy)^3$ **(b)** $(2x + y)^3$ **(c)** $(3ab)^2$ **(d)** $(3a + b)^2$

Solution **(a)** $(2xy)^3 = 2^3x^3y^3$
$= 8x^3y^3$

(b) $(2x + y)^3 = (2x + y)(2x + y)(2x + y)$

$$= (2x + y)(4x^2 + 4xy + y^2) = \begin{array}{r} 8x^3 + 8x^2y + 2xy^2 \\ + 4x^2y + 4xy^2 + y^3 \\ \hline 8x^3 + 12x^2y + 6xy^2 + y^3 \end{array}$$

(c) $(3ab)^2 = 9a^2b^2$

(d) $(3a + b)^2 = 9a^2 + 6ab + b^2$ ■

The exponent rules may be used to solve technical problems as well.

EXAMPLE 2 The landscaping plan for a new home includes a garden area with the shape shown in Figure 2.2. Find the area of the garden if $R = 2r$.

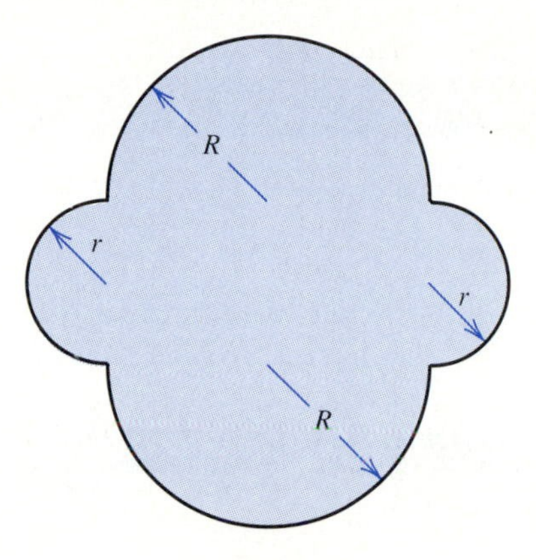

FIGURE 2.2

Solution The combined area of the two large semicircles is πR^2. The combined area of the two small semicircles is πr^2. The area of the rectangle is $4rR$. Thus, $A = \pi R^2 + \pi r^2 + 4rR$. Substituting $2r$ for R, we have $A = \pi(2r)^2 + \pi r^2 + 4r(2r)$, or $4\pi r^2 + \pi r^2 + 8r^2$, which is $5\pi r^2 + 8r^2$. ■

nth root

The ***n*th root** of a number x is a number y such that y^n is x. We say that 2 and -2 are both square roots of 4, because 2^2 is 4 and $(-2)^2$ is 4. But 2 is the only real number that is a cube root of 8. We have $(2)^3 = 8$, but -2 is not a cube root of 8, since $(-2)^3$ is $(-2)(-2)(-2)$ or -8, not 8. A number specified by a radical sign $\sqrt[n]{\ }$ is called a *radical*. The symbol $\sqrt{\ }$ alone means $\sqrt[2]{\ }$. ***The radial sign always restricts attention to a single root.*** When x is a positive number, $\sqrt[n]{x}$ specifies a positive number for both even and odd values of n. When x is negative, $\sqrt[n]{x}$ specifies a negative number for odd values of n, but is not a real number for even values of n. This definition is stated as follows.

radical

NOTE

DEFINITION If n is even, $\sqrt[n]{x} = y$ where $y^n = x$ and $y \geq 0$. If n is odd, $\sqrt[n]{x} = y$ where $y^n = x$.

EXAMPLE 3 Simplify each expression.

(a) $\sqrt{64}$ **(b)** $\sqrt[3]{64}$ **(c)** $\sqrt[3]{-64}$ **(d)** $\sqrt{-64}$

Solution

(a) $\sqrt{64} = 8$, since $8^2 = 64$ and $8 > 0$ **(b)** $\sqrt[3]{64} = 4$, since $(4)^3 = 64$

(c) $\sqrt[3]{-64} = -4$, since $(-4)^3 = -64$ **(d)** $\sqrt{-64}$ is not a real number ■

In mathematics, exact answers rather than approximations are often specified. That is, the answer is given as a radical rather than as a decimal number that has been rounded. When an expression occurs under a radical sign, one must be careful to distinguish between terms and factors. The radical sign applies to each separate *factor,* but not to each separate *term.* This rule is expressed as follows.

RADICAL RULE

If x and y are real numbers and n is a positive integer,

$$\sqrt[n]{xy} = (\sqrt[n]{x})(\sqrt[n]{y})$$

when both of the radicals represent real numbers.

NOTE ⇨⇨ ***It is a common mistake to try to apply the rule to a radical that contains two terms, but in general,*** $\sqrt[n]{x+y} \neq \sqrt[n]{x} + \sqrt[n]{y}$.

EXAMPLE 4 Simplify each radical.

(a) $\sqrt{54}$ **(b)** $\sqrt[3]{54}$ **(c)** $\sqrt{16+9}$ **(d)** $\sqrt{169-25}$

Solution **(a)** $\sqrt{54} = \sqrt{9}\,\sqrt{6} = 3\sqrt{6}$ **(b)** $\sqrt[3]{54} = \sqrt[3]{27}\,\sqrt[3]{2} = 3\sqrt[3]{2}$

(c) $\sqrt{16+9} = \sqrt{25} = 5$ **(d)** $\sqrt{169-25} = \sqrt{144} = 12$ ■

A calculator will give approximately the same answer for both the original radical and its simplified form. You should not be surprised, however, if the last digit or two of the answers are different.

EXAMPLE 5 The effective area of a chimney is given by $E = A - 0.06\sqrt{A}$, where A is the numerical value of the actual area. Find the effective area of a chimney with actual area of 75 in^2.

Solution The problem is to evaluate $75 - 0.06\sqrt{75}$.
Evaluating directly, we have the following keystroke sequence.

75 [−] 0.06 [×] 75 [√x] [=] **74.480385**

Since $\sqrt{75} = \sqrt{25}\,\sqrt{3} = 5\sqrt{3}$, we could also evaluate $75 - (0.06)(5)\sqrt{3}$.

75 [−] 0.06 [×] 5 [×] 3 [√x] [=] **74.480385**

In either case, the answer would be rounded to 74 in^2. ■

Fractions such as $1/\sqrt{3}$ are fairly common in mathematics. When a square root appears in the denominator of a fraction, the quotient may be approximated by using a calculator to do the division. When an exact answer is desired, it is traditional to change the form of the fraction so that the denominator is rational rather than irrational. The process is called *rationalizing the denominator.* When the denominator of a fraction is rationalized, a factor must be chosen so that when both numerator and denominator are multiplied by it, the resulting denominator is a rational number. When the denominator is a square root, the chosen factor is often (but not always) the same as the denominator.

EXAMPLE 6 The formula $I = M/\sqrt{2}$ gives the effective current I (in amps) in an ac circuit if M is the maximum current. Rationalize the denominator.

Solution We need to multiply the irrational denominator by $\sqrt{2}$, since $\sqrt{2}\,\sqrt{2}$ is $\sqrt{4}$, and $\sqrt{4}$ is the rational number 2.

$$I = \frac{M}{\sqrt{2}}\left(\frac{\sqrt{2}}{\sqrt{2}}\right) = \frac{M\sqrt{2}}{\sqrt{4}} = \frac{M\sqrt{2}}{2} \quad \blacksquare$$

The procedure used in Example 6 is generalized as follows.

To rationalize the denominator of a fraction in the form $a/\sqrt{b}$:

1. Find a number c such that $\sqrt{bc}$ is a rational number.
2. Multiply both numerator and denominator by $\sqrt{c}$.
3. Simplify the resulting fraction.

EXAMPLE 7 Simplify each radical, if possible. Find the answer on a calculator for both the original and simplified forms.

(a) $\frac{3}{\sqrt{8}}$ **(b)** $\frac{1}{\sqrt{3}}$

Solution **(a)** Since $(\sqrt{8})(\sqrt{2})$ is $\sqrt{16}$, or 4, which is a rational number, we multiply by $\frac{\sqrt{2}}{\sqrt{2}}$.

$$\left(\frac{3}{\sqrt{8}}\right)\left(\frac{\sqrt{2}}{\sqrt{2}}\right) = \frac{3\sqrt{2}}{\sqrt{16}} = \frac{3\sqrt{2}}{4}.$$

3 [÷] 8 [√x] [=] **1.0606602**

3 [×] 2 [√x] [÷] 4 [=] **1.0606602**

(b) Since $(\sqrt{3})(\sqrt{3})$ is $\sqrt{9}$, or 3, which is a rational number, we multiply by $\frac{\sqrt{3}}{\sqrt{3}}$.

$$\left(\frac{1}{\sqrt{3}}\right)\left(\frac{\sqrt{3}}{\sqrt{3}}\right) = \frac{\sqrt{3}}{\sqrt{9}} = \frac{\sqrt{3}}{3}.$$

1 [÷] 3 [√x] [=] **0.5773503**

3 [√x] [÷] 3 [=] **0.5773503** ■

Finally, when a fraction appears under a radical sign, the radical sign applies to both numerator and denominator. The denominator of the resulting fraction may then be rationalized.

RADICAL RULE

If x and y are real numbers such that $y \neq 0$, and n is a positive integer, $\sqrt[n]{\frac{x}{y}} = \frac{\sqrt[n]{x}}{\sqrt[n]{y}}$ when both of the radicals represent real numbers.

EXAMPLE 8 Simplify each radical, if possible.

(a) $\sqrt{\frac{1}{5}}$ **(b)** $\sqrt{\frac{7}{12}}$

Solution **(a)** $\sqrt{\frac{1}{5}} = \frac{\sqrt{1}}{\sqrt{5}}$ Apply the radical sign to both numerator and denominator.

$= \frac{\sqrt{1}(\sqrt{5})}{\sqrt{5}(\sqrt{5})}$ Multiply both numerator and denominator by $\sqrt{5}$, since $(\sqrt{5})(\sqrt{5})$ is $\sqrt{25}$, or 5.

$= \frac{\sqrt{5}}{5}$ Replace the denominator with the rational number 5.

(b) $\sqrt{\frac{7}{12}} = \frac{\sqrt{7}}{\sqrt{12}}$ Apply the radical sign to both numerator and denominator.

$= \frac{\sqrt{7}(\sqrt{3})}{\sqrt{12}(\sqrt{3})}$ Multiply both numerator and denominator by $\sqrt{3}$, since $(\sqrt{12})(\sqrt{3})$ is $\sqrt{36}$, or 6.

$= \frac{\sqrt{21}}{6}$ Replace the denominator with the rational number 6. ■

EXERCISES 2.3

In 1–12, simplify each expression.

1. $(xy)^3$

2. $(2x)^2$

3. $(3y)^2$

4. $(3a)^3$

5. $(4ab)^2$

6. $(10ax)^3$

7. $(x + y)^3$ compare with 1

8. $(x + 2)^2$ compare with 2

9. $(y + 3)^2$ compare with 3

10. $(a + 3)^3$ compare with 4

11. $(4a + b)^2$ compare with 5

12. $(10a + x)^3$ compare with 6

In 13–22, simplify each radical, if possible.

13. $\sqrt{81}$

14. $\sqrt{25}$

15. $\sqrt[3]{27}$

16. $\sqrt[3]{8}$

17. $-\sqrt{16}$

18. $\sqrt{-81}$ compare with 13

19. $\sqrt{-25}$ compare with 14

20. $\sqrt[3]{-27}$ compare with 15

21. $\sqrt[3]{-8}$ compare with 16

22. $\sqrt{-16}$ compare with 17

In 23–29, simplify each radical, if possible. For each square root, find the answer to three significant digits on a calculator for both the original radical and the simplified form.

23. $\sqrt{50}$

24. $\sqrt{32}$

25. $\sqrt{40}$

26. $\sqrt[3]{32}$

27. $\sqrt[3]{40}$

28. $\sqrt{24}$

29. $\sqrt{250}$

In 30–40, simplify each expression.

30. $1/\sqrt{2}$

31. $2/\sqrt{3}$

32. $3/\sqrt{5}$

33. $1/\sqrt{27}$

34. $1/\sqrt{32}$

35. $\sqrt{36+64}$

36. $\sqrt{144+25}$

37. $\sqrt{169-144}$

38. $\sqrt{25-16}$

39. $\sqrt{289-225}$

40. $\sqrt{225+64}$

---

41. The approximate maximum distance d (in m) that is visible from the top of a building or tower of height h (in m) is given by $d = 0.01\sqrt{37\,455\,h}$

(a) Use the fact that 37 455 = 55(681) to simplify the expression when $h = 55$.

(b) Use a calculator to do the arithmetic for $h = 55$. 1 2 4 7 8 3 5

42. The velocity c of sound in a liquid is given by $c = \sqrt{\beta}/\sqrt{\rho}$. If $\rho = 1.03$ g/cm^3 for water, rationalize the denominator of this expression. In the formula, β is the bulk modulus and ρ is the density of the liquid. 1 2 4 6 8 5 7 9

43. The area of a circle is given by $A = \pi r^2$. If the radius is increased by 2, the new area is given by $A = \pi(r+2)^2$. Simplify this formula. 1 2 4 7 8 3 5

44. The formula $P = I^2R$ gives the power dissipated through a resistance of R when the current is I. If I is increased by 1, the power is given by $P = (I+1)^2R$. Simplify this formula. 3 5 9 1 2 4 6 7 8

45. The power generated by a windmill is given by $P = kV^3$, if V is the velocity of the wind and k is a constant that depends on the windmill. If the velocity is doubled, the power is given by $P = k(2V)^3$. Simplify this formula. 4 7 8 1 2 3 5 6 9

46. The formula $V = s^3$ gives the volume of a cube with each side of length s. If s is halved, the volume is given by $V = (s/2)^3$. Simplify this formula. 1 2 4 7 8 3 5

47. **(a)** Perform the multiplication indicated by $(x+y)^3$.

(b) How many terms are there?

(c) What pattern do the exponents of x follow?

(d) What pattern do the exponents of y follow?

(e) What pattern do the coefficients of the terms follow?

(f) Based on your answers to (b)–(e), what is $(2a+3b)^3$?

---

1. Architectural Technology
2. Civil Engineering Technology
3. Computer Engineering Technology
4. Mechanical Drafting Technology
5. Electrical/Electronics Engineering Technology
6. Chemical Engineering Technology
7. Industrial Engineering Technology
8. Mechanical Engineering Technology
9. Biomedical Engineering Technology

2.4 EXPONENTS

You have seen how multiplication and division of algebraic expressions with exponents are handled. Up to this point, all of the exponents have been positive integers. In this section, the rules are extended to cover exponents that are nonpositive integers. Thus, we will be able to say that x^m/x^n is x^{m-n}, regardless of whether $m \geq n$ or $n > m$. The rules from Section 2.2 are stated for review.

EXPONENT RULES

If x is a real number and m and n are positive integers,

$$x^m(x^n) = x^{m+n}$$ **Product Rule (2-1)**

If x is a nonzero real number and m and n are positive integers,

$$\frac{x^m}{x^n} = x^{m-n}$$ **Quotient Rule (2-2)**

EXAMPLE 1 Simplify x^4/x^4.

Solution If the quotient rule is to be valid when m is equal to n, then x^4/x^4 must be x^0. It is also true that if x is not 0, then x^4/x^4 must be 1. So that there will not be two different answers to the same problem, we define x^0 to be 1. The argument holds for any nonzero value of x and any exponent n that is a positive integer. ■

EXAMPLE 2 Simplify $7^2/7^5$.

Solution By the quotient rule, $7^2/7^5 = 7^{2-5} = 7^{-3}$.
But we also know that $7^2/7^5$ is $1/7^3$. So that there will not be two different answers to the same problem, we say $7^{-3} = 1/7^3$. Since the argument can be generalized, we define x^{-n} to be $1/x^n$ for all nonzero real numbers x and positive integers n. ■

Thus, it is possible to state two more rules.

EXPONENT RULES

If x is a nonzero real number, by definition

$$x^0 = 1, \text{ and}$$ **(2-3)**

$$x^{-n} = 1/x^n \text{ if } n \text{ is a positive integer.}$$ **(2-4)**

These rules are particularly helpful when metric prefixes are used in technical applications.

EXAMPLE 3 Ohm's law is stated as $I = V/R$. In the formula, I is the current (in A), V is the voltage (in volts), and R is the resistance (in Ω).

(a) Find I when $V = 9.0$ volts, and $R = 2.2$ kΩ.

(b) Find I when $V = 9.0$ volts, and $R = 1.0$ MΩ.

Solution Unless you have studied electricity, the terminology might be unfamiliar, but the algebra can be done without knowing the technology.

(a) The formula calls for the resistance in ohms, but it is given in kilohms. Since kilo means 1000, 2.2 kΩ is 2.2×10^3 Ω. Substituting, the numbers for the variables, we have

$$I = \frac{9}{2.2 \times 10^3}$$

But 10^3 in the denominator is equivalent to 10^{-3} in the numerator. Thus,

$$I = \frac{9}{2.2 \times 10^3} = \frac{9 \times 10^{-3}}{2.2} = 4.1 \times 10^{-3} \text{ A.}$$

The formula gives the current in amperes, but 10^{-3} can be indicated by the prefix milli, so 4.1×10^{-3} A $= 4.1$ mA.

(b) The resistance is given in megohms. Since mega means 1 000 000, 1.0 MΩ is 1.0×10^6 Ω. Substituting the numbers for the variables, we have

$$I = \frac{9}{1.0 \times 10^6} = 9 \times 10^{-6} \text{ A.}$$

The formula gives the current in amperes, but 10^{-6} can be indicated by the prefix micro, so 9×10^{-6} A $= 9$ μA. ■

NOTE ⇨⇨ Since a negative exponent can occur in either the numerator or denominator of a fraction, you should know how to simplify either case. ***No zero or negative exponents should appear in the simplified form of a fraction.***

EXAMPLE 4 Simplify x^{-2}/y^{-3}.

Solution $x^{-2} = 1/x^2$ and $y^{-3} = 1/y^3$.
Therefore $x^{-2}/y^{-3} = 1/x^2 \div 1/y^3 = (1/x^2)(y^3/1) = y^3/x^2$. ■

Generalizing this result, we can say that if x and y are nonzero real numbers and m and n are positive integers,

$$\frac{x^{-m}}{y^{-n}} = \frac{y^n}{x^m}$$

We can say that any *factor* with a negative exponent crosses the fraction bar, and the exponent becomes positive. Note that the statement is true for factors, but not for terms.

EXAMPLE 5 Simplify $\dfrac{a^{-2}+b^{-3}}{x^{-2}y^{-3}}$.

Solution x^{-2} and y^{-3} are factors, but a^{-2} and b^{-3} are terms.

$$\frac{a^{-2}+b^{-3}}{x^{-2}y^{-3}}=x^2y^3(a^{-2}+b^{-3})=x^2y^3\left(\frac{1}{a^2}+\frac{1}{b^3}\right)=\frac{x^2y^3}{a^2}+\frac{x^2y^3}{b^3}$$ ■

NOTE Exponent rules 2-1 through 2-4 were stated for exponents that are positive integers. ***All of the rules, however, may be used when the exponents are zero or negative integers as well as when they are positive integers.***

EXAMPLE 6 Simplify each expression.

(a) $3x^3(4x^{-5})$ **(b)** $\dfrac{6y^4}{3y}$ **(c)** 2^0

Solution **(a)** We use the commutative and associative properties to reorder and regroup the factors: $(3)(4)(x^3)(x^{-5}) = 12x^{3-5} = 12x^{-2}$, by the product rule. Since $x^{-n}=\dfrac{1}{x^n}$, we have $\dfrac{12}{x^2}$ as the final answer.

(b) $\dfrac{6}{3}=2$ and $\dfrac{y^4}{y}=y^3$, since $\dfrac{x^m}{x^n}=x^{m-n}$. Therefore, $\dfrac{6y^4}{3y}=2y^3$.

(c) $2^0=1$, since $x^0=1$ for any nonzero x. ■

EXAMPLE 7 Simplify each expression.

(a) $\dfrac{8a^{-3}b^4}{4x^2y^{-2}}$ **(b)** $\dfrac{x^4}{x^{-3}}$

Solution **(a)** We know that $\dfrac{8}{4}$ is 2. The factor a^{-3} crosses the fraction bar to become a^3, and y^{-2} crosses the bar to become y^2. Thus, we have $\dfrac{2b^4y^2}{a^3x^2}$.

(b) We can use the quotient rule: $x^{4-(-3)}=x^7$. Or we can move the factor with the negative exponent across the fraction bar: $x^4(x^3)=x^7$. ■

Example 7(b) shows that there may be more than one correct way to work a problem. We arrive at the same answer, however, regardless of the method we choose. In each case, we could justify the reasoning by citing one of the exponent rules.

Recall from the discussion of order of operations in Section 1.2 that parentheses are used for clarity when a negative number is raised to a power. For example $(-3)^2$ is $(-3)(-3)$ or 9, but -3^2 is $-(3^2)$ or $-(3)(3)$, which is -9. Parentheses are used to clarify algebraic expressions. We now develop exponent rules for use with expressions that have parentheses.

EXAMPLE 8 Simplify each expression.

(a) $(x^2)^3$ **(b)** $(xy)^2$

Solution **(a)** $(x^2)^3 = (x^2)(x^2)(x^2)$. By the product rule, we have x^6. To use a short cut, we multiply exponents. That is, $(x^2)^3 = x^{2(3)} = x^6$.

(b) $(xy)^2 = (xy)(xy)$. Using the commutative and associative properties, we have $(x)(x)(y)(y) = x^2y^2$. To use a short cut, we apply the exponent to each *factor* individually. That is, $(xy)^2 = x^2y^2$. ■

EXAMPLE 9 Simplify $\left(\frac{x}{y}\right)^2$.

Solution $\left(\frac{x}{y}\right)^2 = \left(\frac{x}{y}\right)\left(\frac{x}{y}\right) = \frac{x^2}{y^2}$. To use a short cut, we apply the exponent to both numerator and denominator. ■

The concepts involved in the previous examples can be generalized to give the following exponent rules.

EXPONENT RULES

If x and y are real numbers and m and n are integers

$$(x^m)^n = x^{mn} \qquad \textbf{Power Rule (2-5)}$$

$$(xy)^n = x^ny^n \qquad \textbf{(2-6)}$$

$$(x/y)^n = x^n/y^n \text{ where } y \neq 0 \qquad \textbf{(2-7)}$$

NOTE ⇨⇨ The **power rule** (Rule 2-5), is often confused with the product rule. The product rule $[x^m(x^n) = x^{m+n}]$ applies when two quantities with the same base, each having its own exponent, are multiplied. ***The power rule*** $[(x^m)^n = x^{mn}]$ ***applies when there is one quantity raised to a power, then raised to a power again.*** The following examples review the concepts of this section.

EXAMPLE 10 Simplify each expression.

(a) $(3^2)^4$ **(b)** $(2xy)^3$ **(c)** $(2/3)^2$

Solution **(a)** $(3^2)^4 = 3^8$ by the power rule.

(b) $(2xy)^3 = 2^3x^3y^3 = 8x^3y^3$, since the exponent outside parentheses applies to all factors inside.

(c) $(2/3)^2 = 2^2/3^2 = 4/9$, since the exponent outside the parentheses applies to both numerator and denominator. ■

EXAMPLE 11 Simplify each expression.

(a) $(3a^2b)^3$ **(b)** $4(x^{-2}y)^{-1}$ **(c)** $\left(\frac{2a^{-1}b^2}{3^2xy^{-2}}\right)^3$

Solution **(a)** $(3a^2b)^3 = 3^3(a^2)^3b^3$ Use the rule $(xy)^n = x^ny^n$.

$= 27a^6b^3$ Use the power rule with $(a^2)^3$.

(b) $4(x^{-2}y)^{-1} = 4(x^{-2})^{-1}y^{-1}$ Use the rule $(xy)^n = x^ny^n$ with $(x^{-2}y)^{-1}$.

$= 4x^2y^{-1}$ Use the power rule with $(x^{-2})^{-1}$.

$= 4x^2/y$ Move the factor y across the fraction bar.

(c) $\left(\frac{2a^{-1}b^2}{3^2xy^{-2}}\right)^3 = \frac{2^3(a^{-1})^3(b^2)^3}{(3^2)^3x^3(y^{-2})^3}$ Apply the exponent outside parentheses to each factor in both numerator and denominator.

$= \frac{8a^{-3}b^6}{3^6x^3y^{-6}}$ Use the power rule four times.

$= \frac{8b^6y^6}{729a^3x^3}$ Move factors with negative exponents across the fraction bar. ■

In Example 11(c), it was easy to replace 2^3 with 8. For larger bases and/or higher powers, a calculator may be used or the exponents may be left in the answer. Example 12 shows how the exponent rules may be employed in technical work.

EXAMPLE 12 Consider the formulas $W = 2^{-1}CV^2$ and $V = Q/C$. Write a formula that gives W in terms of Q and V by substituting Q/C for V in the first equation. In these formulas, W is the potential energy of a capacitor, C is the capacitance, V is the potential difference between the plates of the capacitor, and Q is the charge on either of them, but it is not necessary to know the technical terms.

Solution

$W = 2^{-1}CV^2$

$W = 2^{-1}C(Q/C)^2$ Replace V with Q/C.

$= \frac{1}{2}C\left(\frac{Q^2}{C^2}\right)$ Apply the power outside parentheses to both numerator and denominator and use $x^{-n} = 1/x^n$.

$= \frac{1}{2}\left(\frac{Q^2}{C}\right)$ Use the quotient rule.

$= \frac{1}{2}Q\left(\frac{Q}{C}\right)$ Replace Q^2 with $Q(Q)$.

$= \frac{1}{2}QV$ Substitute V for Q/C. ■

As you apply the exponent rules, don't forget to distinguish between terms and factors.

EXAMPLE 13 Simplify each expression.

(a) $\dfrac{(a+b)^{-1}}{ab^{-3}}$ **(b)** $\dfrac{a^{-2}+b^{-3}}{(ab)^{-3}}$

Solution **(a)** The expressions a, b^{-3}, and $(a+b)^{-1}$ are factors. It is important to recognize that the power -3 applies only to the factor b and not to the factor a. Each *factor* with a negative exponent crosses the fraction bar.

$$\frac{(a+b)^{-1}}{ab^{-3}}=\frac{b^3}{a(a+b)}$$

(b) The numerator consists of terms a^{-2} and b^{-3}. The parentheses in the denominator indicate that the expression $(ab)^{-3}$ may be considered as a single factor. The denominator, then, consists of factors, but the numerator consists of terms. Factors cross the fraction bar; terms do not.

$$\frac{a^{-2}+b^{-3}}{(ab)^{-3}}=(ab)^3(a^{-2}+b^{-3})$$ Move the denominator across fraction bar.

$$=a^3b^3(a^{-2}+b^{-3})$$ Apply $(xy)^n=x^ny^n$ to $(ab)^3$.

$$=a^3b^3\left(\frac{1}{a^2}+\frac{1}{b^3}\right)$$ Move factors with negative exponents across the fraction bar in each separate term.

$$=\frac{a^3b^3}{a^2}+\frac{a^3b^3}{b^3}$$ Use the distributive property.

$$=ab^3+a^3$$ Use the quotient rule. ■

Scientific notation, combined with the exponent rules, allows us to work many problems quickly and easily, as illustrated by the next example.

EXAMPLE 14 The formula $C=Q/V$ gives the capacitance C of a capacitor (in farads) if Q is the charge (in coulombs) and V is the voltage (in volts). Find C if $Q=0.0004$ coulombs and $V=200$ volts.

Solution $$\frac{0.0004}{200}=\frac{4\times10^{-4}}{2\times10^{2}}=2\times10^{-6}\text{ F}=2\ \mu\text{F}.$$ ■

EXERCISES 2.4

In 1–43, simplify each expression. There should be no zero or negative exponents in the final answer.

1. 10^0 **2.** 2^{-4} **3.** 10^{-3}

4. $5x^4(3x^{-6})$ **5.** $7x^{-3}(8x^2)$ **6.** $4a^5b^{-4}(5a^{-3}b^2)$

7. $3a^{-2}b^4(6a^{-3}b^2c)$

8. $\dfrac{10a^2bc}{5ab^2c}$

9. $\dfrac{12xy^2}{3xz}$

10. $\dfrac{2^{-3}a}{3b^{-2}}$

11. $\dfrac{5a^{-2}b}{10a^2b}$

12. $\dfrac{10^2b^{-1}}{10^3b}$

13. $\dfrac{10^{-3}y}{10y^{-2}}$

14. $\dfrac{48x^{-3}(10)}{12x(10^{-2})}$

15. $\dfrac{3^2ab^{-1}}{2^{-3}x^{-2}y}$

16. $(a^2)^3$

17. $(3^4)^5$

18. $(x^2)^4$

19. $(xy)^3$

20. $(3a)^2$

21. $(5xy)^3$

22. $\left(\dfrac{2}{x}\right)^3$

23. $\left(\dfrac{3}{4}\right)^2$

24. $\left(\dfrac{y}{6}\right)^5$

25. $\left(\dfrac{2}{3}\right)^3$

26. $(2a^5b)^3$

27. $3(x^2y^{-1})^3$

28. $2(a^5b)^3$

29. $(5xy)^{-1}$

30. $5(xy)^{-1}$

31. $\left(\dfrac{3b}{4}\right)^2$

32. $\left(\dfrac{2x}{y}\right)^3$

33. $\left(\dfrac{x}{y^3}\right)^{-1}$

34. $\dfrac{3(ab)^{-2}}{5}$

35. $\dfrac{3ab^{-2}}{5}$

36. $\left(\dfrac{5ab^{-1}}{c^2}\right)^3$

37. $\left(\dfrac{2ab^{-1}}{c^2}\right)^{-3}$

38. $\left(\dfrac{7xy^{-1}}{3^{-1}ab}\right)^{-2}$

39. $\left(\dfrac{7xy^{-1}}{3^{-1}x^2y^0}\right)^0$

40. $\dfrac{(x+y)^{-2}}{3x^{-3}}$

41. $\dfrac{4a^{-2}b^{-3}}{a(b+3)^{-1}}$

42. $\dfrac{x^{-3}+2^{-1}}{(3x)^{-2}}$

43. $\dfrac{x^{-2}+y^{-1}}{xy^{-3}}$

In 44–46, use scientific notation and the rules of exponents to solve each problem. You may need to use the following prefixes.

Prefix	Power of Ten	Abbreviation	Prefix	Power of Ten	Abbreviation
pico	10^{-12}	p	kilo	10^3	k
nano	10^{-9}	n	mega	10^6	M
micro	10^{-6}	μ	giga	10^9	G
milli	10^{-3}	m			

44. The potential difference V (in volts) across a capacitor with capacitance C (in Farads) and a charge of Q (in coulombs) is given by $V = Q/C$.
(a) Find V if $Q = 5 \times 10^{-4}$ coulombs and $C = 10\ \mu$F.
(b) Find V if $Q = 3 \times 10^{-8}$ coulombs and $C = 300$ pF.
(c) Find V if $Q = 2 \times 10^{-14}$ coulombs and $C = 2\ \mu$F. **3 5 9** 4 7 8

45. The energy W (in joules) used during time t (in seconds) by a device with power input P (in watts) is given by $W = Pt$. Find the energy used by each appliance in the given time.
(a) a disposal with $P = 500$ watts, $t = 1$ minute, 40 seconds
(b) a dishwasher with $P = 1000$ watts, $t = 3$ minutes, 20 seconds
(c) a coffee maker with $P = 1400$ watts, $t = 5$ minutes **3 5 9** 1 2 4 6 7 8

46. The safe load L for a beam is given by $L = kwd^2l^{-1}$. Rewrite the formula without using negative exponents. In the formula, w is the width of the beam, d is its depth, l is the distance between supports, and k is a constant that depends on the material. **1 2 4** 8 7 9

---

1. Architectural Technology
2. Civil Engineering Technology
3. Computer Engineering Technology
4. Mechanical Drafting Technology
5. Electrical/Electronics Engineering Technology
6. Chemical Engineering Technology
7. Industrial Engineering Technology
8. Mechanical Engineering Technology
9. Biomedical Engineering Technology

47. Consider the formulas $P = IV$ and $V = IR$. Write a formula that gives P in terms of I and R by substituting IR for V in the first equation. P is the power (in watts) of an electric current, I is the current (in amps), V is the potential difference across the ends of the circuit, and R is the resistance of the conductor. 3 5 9 1 2 4 6 7 8

48. The length l of a wire is given by $l = AV\rho^{-1}I^{-1}$. Rewrite the formula without using negative exponents. In the formula, A is the cross-sectional area (in cmils), V is the potential difference (in volts), ρ is the specific resistance, and I is the current (in amps). 3 5 9 1 2 4 6 7 8

49. The maximum load P (in lb) that a long column can support is given by $P = \pi^2 EIl^{-2}$. Write the formula without using negative exponents. 1 2 4 8 7 9

50. The cross-sectional area A of the wall of a hollow metal pipe is given by $A = \left(\frac{\pi}{4} D^2 - \frac{\pi}{4} d^2\right)$. Simplify the expression on the right if $d = \frac{1}{4} D$. 1 2 4 7 8 3 5

51. Let ^ represent the operation of exponentiation. That is, $a\hat{}b = a^b$. Is the operation of exponentiation associative?

2.5 EQUATIONS AND FORMULAS

To solve an equation means to find the values of the variables that make the equation true

> **DEFINITION**
>
> A **linear equation in the variable x** is an equation that can be put in the form $ax + b = 0$ where a and b are real numbers, and a is not 0.

The values of the variable that makes the equation true are said to *satisfy* the equation, and are called the *solutions* of the equation. At this time we consider only equations that have a single solution. Two equations with the same solution are said to be *equivalent.*

We will explore two ideas that are important in solving an equation. First, inverse operations are two operations such that one operation "undoes" the other, as do addition and subtraction. Suppose we begin with 2. If we first add 3, then subtract 3, we have $2 + 3 = 5$, and $5 - 3 = 2$. Thus, we end where we started. Multiplication and division are also inverse operations.

Second, if we add or subtract the same number on both sides of the equation, or if we multiply or divide both sides of the equation by the same nonzero number, an equivalent equation is obtained. That is, the solution of the new equation is the same as the solution of the original equation. We are now ready to solve linear equations.

The simplest examples are like the following.

$x + 2 = 6$	$x - 2 = 6$	$2x = 6$	$\frac{x}{2} = 6$
$x + 2 - 2 = 6 - 2$	$x - 2 + 2 = 6 + 2$	$\frac{2x}{2} = \frac{6}{2}$	$\frac{x}{2}(2) = 6(2)$
$x = 4$	$x = 8$	$x = 3$	$x = 12$

Usually, however, more than one step is required. Example 1 shows how to put the steps together.

EXAMPLE 1 Solve $2(x + 3) - 5 = 6$.

Solution **1.** Simplify each side. Use the distributive property, if necessary, and combine like terms.

$$2x + 6 - 5 = 6$$
$$2x + 1 = 6$$

2. Remove "extra" terms: We want to isolate the x on one side of the equation. In this example, "1" is a term that we do not want. Since 1 has been added to $2x$, we can undo the addition with a subtraction. Subtracting 1 from both sides of the equation gives

$$2x + 1 - \mathbf{1} = 6 - \mathbf{1}$$
$$2x = 5$$

3. Remove "extra" factors or divisors. We only want x, not $2x$. We need to remove the factor 2. Since multiplication can be "undone" by division, we divide both sides of the equation by 2.

$$\frac{2x}{\mathbf{2}} = \frac{5}{\mathbf{2}}$$
$$x = 2\ 1/2$$

This equation states the value of x, and the equation is equivalent to the original equation. Thus $x = 2\ 1/2$ is the solution of the original equation. It is a good idea to check an equation. ***Any answer should be checked in the original problem rather than at an intermediate step.***

NOTE

Check:

$$2(5/2 + 3) - 5 = 6$$
$$2(5/2 + 6/2) - 5 = 6$$
$$2(11/2) - 5 = 6$$
$$11 - 5 = 6, \text{ which is true.}$$ ■

The steps used in Example 1 are summarized as follows.

To solve a linear equation in one variable:

1. Simplify each side.
   (a) Use the distributive property, if necessary.
   (b) Combine like terms.
2. Remove extra terms by using inverse operations. (Add or subtract the same number on both sides of the equation.)
3. Remove extra factors or divisors by using inverse operations. (Multiply or divide both sides of the equation by the same nonzero number.)

EXAMPLE 2 Solve $2y - 1 = 3y + 2$.

Solution **1.** Neither side can be simplified.

2. The final answer should have only one y-term. Therefore we begin by removing the extra y-term. To remove $2y$, we subtract $2y$ from both sides. (The equation could also be solved by subtracting $3y$ from both sides.)

$$2y - 1 - 2y = 3y + 2 - 2y$$
$$-1 = y + 2$$

Now we remove the 2 by subtracting 2 from both sides.

$$-1 - 2 = y + 2 - 2$$

3. There are no extra factors or divisors, so the answer is $y = -3$.

Check:
$$2(-3) - 1 = 3(-3) + 2$$
$$-6 - 1 = -9 + 2$$
$$-7 = -7, \text{ which is true.}$$ ■

If an equation contains fractions, it is possible to multiply both sides of the equation by the lowest common denominator of the fractions to find an equivalent equation that does not contain fractions.

EXAMPLE 3 Solve $\frac{z}{3} - \frac{z+1}{6} = \frac{z-1}{2} + \frac{2}{3}$.

Solution The common denominator of the four fractions is 6.

Multiply both sides by 6:
$$6\left(\frac{z}{3} - \frac{z+1}{6}\right) = 6\left(\frac{z-1}{2} + \frac{2}{3}\right)$$

1. Simplify each side:
$$2z - (z + 1) = 3(z - 1) + 4$$
$$2z - z - 1 = 3z - 3 + 4$$
$$z - 1 = 3z + 1$$

2. Remove extra terms:
$$z - 1 - z = 3z + 1 - z$$
$$-1 = 2z + 1$$
$$-1 - 1 = 2z + 1 - 1$$
$$-2 = 2z$$

3. Remove extra factors or divisors:
$$\frac{-2}{2} = \frac{2z}{2}$$
$$-1 = z$$

Check:
$$-1/3 - 0/6 = -2/2 + 2/3$$
$$-1/3 = -3/3 + 2/3$$
$$-1/3 = -1/3, \text{ which is true.}$$ ■

NOTE ⇨⇨ ***When a calculator is used to solve an equation, it is usually best to manipulate the equation algebraically until the variable is isolated.*** The arithmetic can be done in such a way as to minimize round-off errors.

EXAMPLE 4 Solve $1.43x = 1.21(x + 4.56)$.

Solution

$$1.43x = 1.21x + 1.21(4.56)$$ Use the distributive property.

$$1.43x - 1.21x = 1.21(4.56)$$ Subtract $1.21x$ from both sides.

$$x(1.43 - 1.21) = 1.21(4.56)$$ Use the distributive property.

$$= \frac{1.21(4.56)}{1.43 - 1.21}$$ Divide both sides by the factor $(1.43 - 1.21)$.

$$= 25.1 \text{ (rounded)}$$

When an answer is rounded, a check will probably produce a slight discrepancy.

Check:

$$1.43x = 1.21(x + 4.56)$$
$$1.43(25.1) = 1.21(25.1 + 4.56)$$
$$35.893 = 1.21(29.66) \text{ or } 35.8886 \quad ■$$

formula

A **formula** is an expression relating two or more variables by means of an equation. One variable usually appears by itself while the other variables are all on the other side of the equation. A formula is said to be *solved for* the variable that appears alone on one side of the equation. For example, $P = I^2R$ is a formula relating P, I, and R, variables that represent the power, current, and resistance in an electrical device. The formula is solved for P. Sometimes it is necessary to rewrite a formula so that it is solved for a different one of the variables. We could write $R = P/I^2$ and the formula would be solved for R.

Solving a formula is just like solving an equation in one variable, if one has a clear idea of which variable is to be isolated. To leave the desired variable on one side of the equation by itself, we simplify both sides of the equation, remove extra terms, then remove extra factors or divisors.

EXAMPLE 5 The formula $v = v_0 + at$ relates the velocity v of an object to its initial velocity v_0, acceleration a, and time of travel t. Solve for t.

Solution

1. Each side is in simplest form.

2. Since we are solving for t, we subtract v_0 from both sides, leaving only those terms that contain t.

$$v - \mathbf{v_0} = v_0 + at - \mathbf{v_0}$$
$$v - v_0 = at$$

3. Since we are solving for t, we remove the factor a by dividing both sides of the equation by a.

$$\frac{v - v_0}{\mathbf{a}} = \frac{at}{\mathbf{a}}$$

$$\frac{v - v_0}{a} = t \quad ■$$

EXAMPLE 6 The formula $Fr^2 = kQ_1Q_2$ relates the force F of repulsion between two particles to their charges Q_1 and Q_2 and distance r by which they are separated. The letter k represents a constant. Solve for k.

Solution **1.** Each side is in simplest form.

2. There are no extra terms.

3. We remove the extra factors of Q_1 and Q_2 by dividing both sides of the equation by Q_1Q_2.

$$\frac{Fr^2}{Q_1Q_2} = \frac{kQ_1Q_2}{Q_1Q_2}$$

$$\frac{Fr^2}{Q_1Q_2} = k \quad \blacksquare$$

EXAMPLE 7 The formula $A = \frac{h}{2}(B + b)$ relates the area of a trapezoid, shown in Figure 2.3, to the lengths B and b of the bases and the height h. Solve for B.

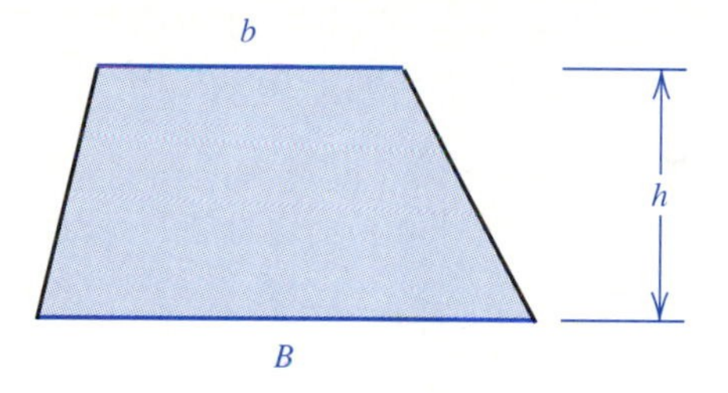

FIGURE 2.3

Solution Multiply both sides by 2, the lowest common denominator.

$$2(A) = 2\left(\frac{h}{2}\right)(B + b)$$

$$2A = h(B + b)$$

1. Simplify each side: $\quad 2A = hB + hb$

2. Remove extra terms: $\quad 2A - hb = hB + hb - hb$

$$2A - hb = hB$$

3. Remove extra factors: $\quad \dfrac{2A - hb}{h} = \dfrac{hB}{h}$

$$\frac{2A - hb}{h} = B \quad \blacksquare$$

Raising a number to the nth power and finding the nth root are inverse operations. That is, the nth root and the nth power "undo" each other. Suppose we begin with 8. If we first find the cube root, then cube the result, we have $\sqrt[3]{8} = 2$, and $2^3 = 8$. Thus, we end where we started. This idea will be used to solve equations with powers and roots when the original equation is defined for real numbers.

To solve $x^n = b$ or $\sqrt[n]{x} = c$ when the equation has real number solutions:

(a) If $x^n = b$, find the nth root (or roots) of b:
$x = \sqrt[n]{b}$ if n is odd, or $x = \pm\sqrt[n]{b}$ if n is even and $b \geq 0$.

(b) If $\sqrt[n]{x} = c$, raise c to the nth power:
$x = c^n$ if n is odd, or if n is even and $c \geq 0$.

EXAMPLE 8 Solve each equation.

(a) $x^3 = 8$ **(b)** $x^3 = -8$ **(c)** $\sqrt[3]{x} = 3$ **(d)** $\sqrt[3]{x} = -3$

Solution **(a)** $x = \sqrt[3]{8} = 2$ **(b)** $x = \sqrt[3]{-8} = -2$

(c) $x = 3^3 = 27$ **(d)** $x = (-3)^3 = -27$ ■

EXAMPLE 9 Solve each equation.

(a) $x^2 = 9$ **(b)** $\sqrt{x} = 9$ **(c)** $x^2 = -9$ **(d)** $\sqrt{x} = -9$

Solution **(a)** The square root sign does not appear in the original equation, so there is no reason for the value of x to be restricted to the positive square root. $x = \pm\sqrt{9} = \pm 3$

(b) $x = 9^2 = 81$

(c) $x^2 = -9$ is not defined for real numbers x, so the equation cannot be solved using real numbers.

(d) $\sqrt{x} = -9$ is not defined for real numbers x, so the equation cannot be solved using real numbers. ■

EXERCISES 2.5

In 1–18, solve each equation.

1. $5x + 3 = 7$ **2.** $4y + 7 = 2$ **3.** $6z + 1 = -3$

4. $\frac{x}{5} + 2 = 7$ **5.** $\frac{3y}{2} - 1 = 3$ **6.** $\frac{m}{3} - 2 = 1$

7. $\frac{p}{4} + 3 = -1$ **8.** $3m + 2 = 5m - 1$ **9.** $5x + 1 = 1 - x$

10. $4x + 3 = 2 - 5x$ **11.** $3(x + 1) = 2x + 3$ **12.** $4(y - 2) = y - 1$

13. $5(1 - p) = 3p - 2$ **14.** $-(x - 2) + 2x = 0$

15. $2(n - 1) + 4 = 3n + 2(n - 1)$ **16.** $-3(x + 2) - 1 = 2x - 4(x + 3)$

17. $-1(y - 3) - 4 = 7y - (y - 7)$ **18.** $2(2z + 3) - 5 = 2z - (3z - 11)$

In 19–23, use a calculator to solve each equation.

19. $(2.3)(x + 1.9) = 4.5x - 7.8$

20. $3.41x - 4.52 = 6.40x + 12.21$

21. $81.44x + 33.66 = 4.32(x - 1.00)$

22. $0.001(x - 0.002) = 0.012x + 3.991$

23. $14.54x + 7.22 = 13.21x - 12.04$

In 24–28, solve each equation that has real number solutions.

24. **(a)** $x^3 = 27$ **(b)** $\sqrt[3]{x} = 2$

25. **(a)** $x^5 = 32$ **(b)** $\sqrt[5]{x} = 10$

26. **(a)** $x^2 = 81$ **(b)** $x^2 = -81$

27. **(a)** $x^2 = 4$ **(b)** $\sqrt{x} = 4$

28. **(a)** $x^4 = 16$ **(b)** $\sqrt[4]{x} = -3$

Problems 29 and 30 use the following information. The work W *done by a force of magnitude* F *acting through a distance of* d *is given by* W = Fd. *Solve for the indicated variable.*

29. Solve for F. 4 7 8 1 2 3 5 6 9

30. Solve for d. 4 7 8 1 2 3 5 6 9

Problems 31 and 32 use the following information. Ohm's law says that the potential difference V *across a conductor with current* I *and resistance* R *is given by* V = IR. *Solve for the indicated variable.*

31. Solve for I. 3 5 9 1 2 4 6 7 8

32. Solve for R. 3 5 9 1 2 4 6 7 8

Problems 33 and 34 use the following information. The inductance L *of a solenoid is given by* $L = \mu N^2 A/l$. *In the formula,* μ *is the magnetic permeability of the core material,* N *is the number of turns of wire,* A *is the cross-sectional area, and* l *is the length. Solve for the indicated variable.*

33. Solve for l. 3 5 7 1 2 4 6 8 9

34. Solve for μ. 3 5 7 1 2 4 6 8 9

35. The illumination I_1 at a distance d_1 is given by $I_1 = (d_2{}^2 I_2)/d_1{}^2$ if the illumination is I_2 at a distance d_2. Solve for I_2. 9 1 2 3 5 6 7

Problems 36 and 37 use the following information. The formula PV = kNT *relates the pressure* P, *volume* V, *number of molecules* N, *temperature* T, *and constant* k *for a gas. Solve for the indicated variable.*

36. Solve for k. 6 8 1 2 3 5 7 9

37. Solve for N. 6 8 1 2 3 5 7 9

Problems 38 and 39 use the following information. The internal resistance r *of a generator with electromotive force* V_{emf} *through which current* I *flows and that is connected to an external resistance* R is given by $r = V_{emf}/I - R$. *Solve for the indicated variable.*

38. Solve for R. 3 5 9 1 2 4 6 7 8

39. Solve for V_{emf}. 3 5 9 1 2 4 6 7 8

40. In a computer program, a group of commands called a loop is executed repeatedly. When one loop includes another loop, the total execution time T is given by $T = M(y + Nx)$. Solve for x. In the formula, M and N are the number of repetitions of the outer and inner loops, respectively, x is the execution time of the inner loop, and y is the additional execution time of the outer loop. 3 5 9

41. The formula $I = bh^3/12$ gives the moment of inertia I of a beam with a rectangular cross section of length h and width b. Solve for h. 1 2 4 8 7 9

---

1. Architectural Technology
2. Civil Engineering Technology
3. Computer Engineering Technology
4. Mechanical Drafting Technology
5. Electrical/Electronics Engineering Technology
6. Chemical Engineering Technology
7. Industrial Engineering Technology
8. Mechanical Engineering Technology
9. Biomedical Engineering Technology

42. The surface area A of a cylinder is given by $A = 2\pi rh + 2\pi r^2$ when r is the radius of the base and h is the height of the cylinder. Solve for π.
1 2 4 7 8 | 3 5

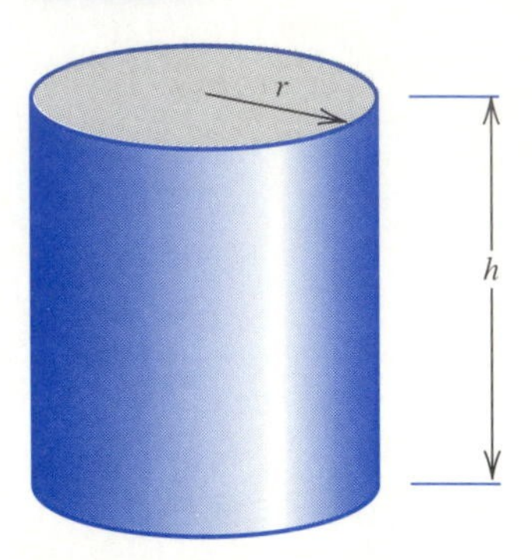

43. The formula $d = v^2/(30f)$ can be used to estimate the distance d that a car traveling at velocity v will skid if f is the coefficient of friction between the car and the road. Solve for v. 4 7 8 | 1 2 3 5 6 9

44. The potential energy w stored by an inductor is given by $w = (1/2)LI^2$ when the inductance is L and the current carried is I. Solve for I.
3 5 7 | 1 2 4 6 8 9

45. The potential difference V across the ends of a circuit is given by $V = \sqrt{PR}$ if P is the power consumed and R is the resistance of the conductor. Solve for P. 3 5 9 | 1 2 4 6 7 8

46. The formula $V = \sqrt[3]{P/K}$ is used to estimate the velocity V of the wind when P is the power generated by a windmill. The value of K depends on the construction of the windmill. Solve for P.
4 7 8 | 1 2 3 5 6 9

47. Show that the formula used in Example 7 (see p. 71) is correct. (Hint: Divide the trapezoid into two triangles as shown in the accompanying diagram.)

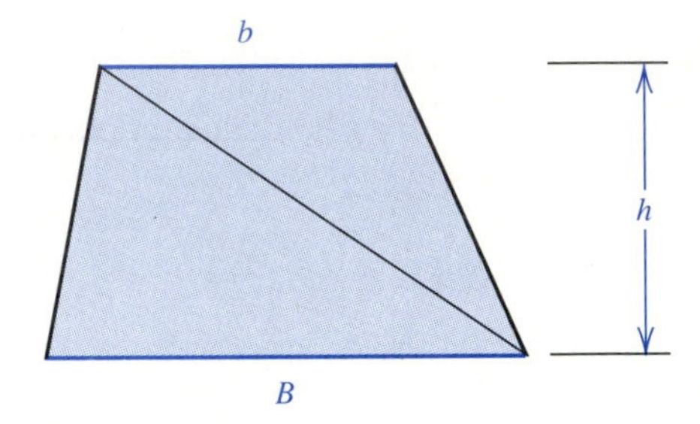

1. Architectural Technology
2. Civil Engineering Technology
3. Computer Engineering Technology
4. Mechanical Drafting Technology
5. Electrical/Electronics Engineering Technology
6. Chemical Engineering Technology
7. Industrial Engineering Technology
8. Mechanical Engineering Technology
9. Biomedical Engineering Technology

2.6 RATIO AND PROPORTION

ratio

denominate numbers

One way to compare two quantities is with a *ratio*. The **ratio** of a to b is the fraction a/b. Since ratios are often comparisons of measurements, the quantities a and b are often numbers with units of measurement, such as feet or pounds, associated. Numbers of this type are called *denominate numbers*. Since ratios are fractions, they are reduced to lowest terms by dividing the numerator and denominator by the same nonzero number.

To reduce a ratio that compares two denominate numbers:

1. Express the numerator and denominator as quantities that have the same units of measurement, if possible.
2. Reduce to lowest terms.

EXAMPLE 1 Reduce each ratio to lowest terms.

(a) $\frac{3\text{ ft}}{24\text{ in}}$ (b) $\frac{1.3\text{ cm}}{5\text{ mm}}$ (c) $\frac{110\text{ mi}}{2\text{ hr}}$ (d) $\frac{2\text{ V}}{2000\ \Omega}$

Solution (a) The two quantities may be expressed in either feet or inches.

$$\frac{3\text{ ft}}{24\text{ in}} = \frac{3\not{\text{ft}}}{2\not{\text{ft}}} = \frac{3}{2} \text{ or } \frac{3\text{ ft}}{24\text{ in}} = \frac{36\not{\text{in}}}{24\not{\text{in}}} = \frac{3}{2}.$$

In either case, the reduced ratio is a dimensionless quantity. That is, the units are divided out, just as numbers are.

(b) $\frac{1.3\text{ cm}}{5\text{ mm}} = \frac{13\not{\text{mm}}}{5\not{\text{mm}}} = \frac{13}{5}$ or $\frac{1.3\text{ cm}}{5\text{ mm}} = \frac{1.3\not{\text{cm}}}{0.5\not{\text{cm}}} = \frac{13}{5}$

(c) $\frac{110\text{ mi}}{2\text{ hr}} = \frac{55\text{ mi}}{1\text{ hr}}$ or 55 mi/hr

The units remain as part of the ratio.

(d) $\frac{2\text{ V}}{2000\ \Omega} = \frac{1\text{ V}}{1000\ \Omega}$ ■

When a ratio compares two quanities with different units, it is called a *rate*. Rates are usually written with a denominator of 1. Thus, the ratio in Example 1(d) could be written as the rate of 0.001 V/Ω.

Ratios may be used to convert a measurement given in one unit to an equivalent measurement given in a different unit. A ratio used in this manner is called a *conversion factor*.

conversion factor

EXAMPLE 2 Convert 5.50 inches to centimeters.

Solution Since we have 1 in = 2.54 cm, the numerator and denominator of the ratio $\frac{2.54\text{ cm}}{1\text{ in}}$ are equal. The ratio is therefore equal to 1. Multiplying 5.50 inches by 1 produces a quantity that is equal to 5.50 inches.

$5.50\not{\text{in}}\left(\frac{2.54\text{ cm}}{1\not{\text{in}}}\right) = 14.0$ cm, to three significant digits. ■

To convert from one unit of measurement to another:

1. Find an equation relating the original unit and the desired unit.
2. Find the conversion factor that is the ratio of the two measurements from step 1. The numerator has the desired unit while the denominator has the original unit.
3. Multiply the original measurement by the conversion factor. Units common to both numerator and denominator will divide or "cancel."

EXAMPLE 3 Convert each measurement to a measurement with the indicated unit.

(a) 16.7 centimeters to inches **(b)** 55 miles to kilometers

Solution **(a)** Since we have 1 in = 2.54 cm, the conversion factor is $\frac{1 \text{ in}}{2.54 \text{ cm}}$.

$$16.7 \cancel{\text{cm}} \left(\frac{1 \text{ in}}{2.54 \cancel{\text{cm}}}\right) = 6.57 \text{ in (Notice that cm divide, leaving in.)}$$

(b) Since we have 1 mi = 1.6 km, the conversion factor is $\frac{1.6 \text{ km}}{1 \text{ mi}}$.

$$55 \cancel{\text{mi}} \left(\frac{1.6 \text{ km}}{1 \cancel{\text{mi}}}\right) = 88 \text{ km (Notice that mi divide, leaving km.)}$$ ■

It may be necessary to use more than one conversion factor.

EXAMPLE 4 Convert 60 mi/hr to ft/s.

Solution 1 mi = 5280 ft, 1 hr = 60 min, and 1 min = 60 s

$$\left(\cancel{60} \frac{\cancel{\text{mi}}}{\cancel{\text{hr}}}\right)\left(\frac{5280 \text{ ft}}{1 \cancel{\text{mi}}}\right)\left(\frac{1 \cancel{\text{hr}}}{\cancel{60} \cancel{\text{min}}}\right)\left(\frac{1 \cancel{\text{min}}}{60 \text{ s}}\right) = 88 \frac{\text{ft}}{\text{s}}$$ ■

proportion

A **proportion** is a statement that two ratios are equal. Since the proportion $\frac{a}{b} = \frac{c}{d}$ is an equation, both sides of it can be multiplied by bd. Thus, $bd\left(\frac{a}{b}\right) = bd\left(\frac{c}{d}\right)$ gives $ad = bc$. The quantities ad and bc are sometimes called *cross-products,* because the following diagram can be used to indicate $ad = bc$.

$$\frac{a}{b} \times \frac{c}{d}$$

Thus, as a short cut, we can solve for a variable in a proportion by equating the cross-products.

To solve for a variable in a proportion:

1. Set the cross-products equal to each other.
2. Solve the resulting equation.

EXAMPLE 5 Solve the following proportions.

(a) $\frac{x}{3} = \frac{7}{10}$ **(b)** $\frac{6}{5} = \frac{2}{y}$

Solution (a) $10x = 21$, so $x = \frac{21}{10}$ or $2\frac{1}{10}$

(b) $6y = 10$, so $y = \frac{10}{6} = 1\frac{4}{6}$ or $1\frac{2}{3}$ ■

Many applied problems can be solved using proportions.

EXAMPLE 6 If a pole of height 2.0 m casts a shadow 1.5 m long, how tall is a building that casts a shadow 11.25 m long?

Solution First, we draw a diagram, as in Figure 2.4.

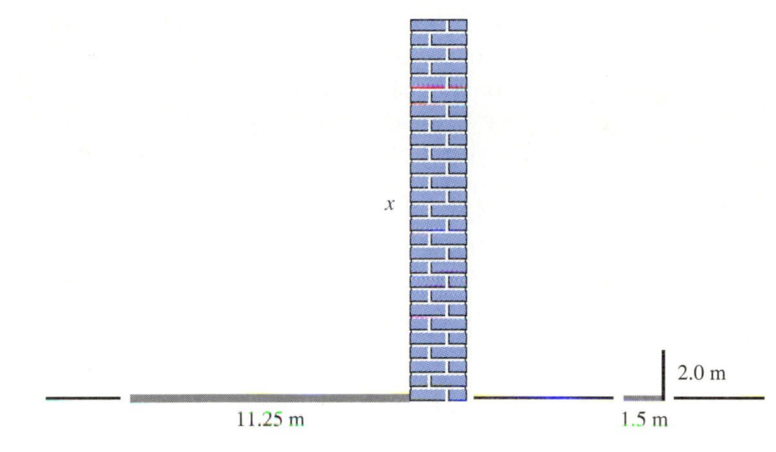

FIGURE 2.4

Let x = height of the building. We compare height to shadow length.

$$\underbrace{\frac{2.0 \text{ m}}{1.5 \text{ m}}}_{\text{POLE}} = \underbrace{\frac{x}{11.25 \text{ m}}}_{\text{BUILDING}}$$

When the ratio on the left is written as a dimensionless number, we have $2.0(11.25 \text{ m}) = 1.5x$. Thus $x = \frac{2.0(11.25) \text{ m}}{1.5}$ or 15 m. ■

In Example 6, all of the measurements were in meters, so the answer was in meters also. Remember that when a ratio has different units in the numerator and denominator, the units are not dropped. If a proportion has a variable for one of the denominators and is written so that both numerators are expressed with the same units, then the denominators have like units. Likewise, if the proportion has a variable for one of the numerators and is written so that both denominators are expressed with the same units, then the numerators have like units. Thus, it is not necessary to carry the units as the arithmetic is done.

EXAMPLE 7 A blueprint uses the scale 1/4″ represents 1 ft. If a measurement is 2 1/8″ on the drawing, what is the actual length?

Solution We want to compare drawing length to actual length. We let $x =$ actual length and set up a proportion.

$$\frac{1/4''}{1\text{ ft}} = \frac{2\ 1/8''}{x}$$

$$\frac{1}{4}(x) = \frac{17}{8}, \text{ so } x = \frac{17(4)}{8} = \frac{17}{2} \text{ or } 8\frac{1}{2}\text{ ft}$$

The answer must be 8 1/2 *feet,* since the denominator of the first ratio was in feet. ■

similar triangles

A common use of proportions is to compare dimensions of objects that have the same shape. Such objects are said to be *similar.* In the study of geometry and trigonometry, similar triangles are important. Two triangles are similar if they have equal angles. Angles are measured in degrees. If a circle is divided by radii into 360 equal parts, the angle between two consecutive radii is 1 degree, denoted 1°. Two sides of a triangle are *corresponding sides* if they occupy the same positions relative to equal angles in the two triangles. For similar triangles, we may set up a proportion with ratios of corresponding sides.

EXAMPLE 8 Find side x of triangle ABC.

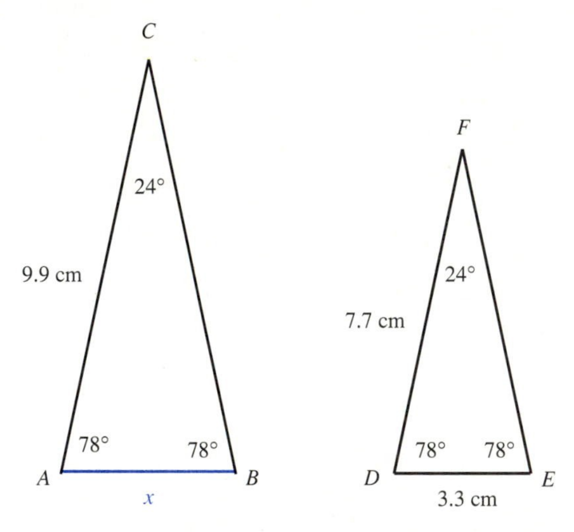

FIGURE 2.5

Solution The 9.9 cm side of triangle ABC is between the 24° and 78° angles. The 7.7 cm side of triangle DEF is between the 24° and 78° angles, so they are corresponding sides. Thus $9.9\text{ cm}/7.7\text{ cm} = x/3.3\text{ cm}$ or $9.9(3.3\text{ cm}) = 7.7x$, so $x = 9.9(3.3\text{ cm})/7.7 = 4.2\text{ cm}$. ■

The triangles in Example 8 are isosceles triangles. An isosceles triangle has two sides of equal length, but can also be recognized by the fact that two of the angles are equal. Therefore we actually know the lengths of all three sides of both triangles.

When working with similar triangles, we may also set up a proportion between the ratio of two sides of one triangle and the ratio of corresponding sides of the other triangle. Example 8 could have been solved by writing 9.9 cm/x = 7.7 cm/3.3 cm. The cross-products are the same as in Example 8. Any proportion may be written in several ways.

EQUIVALENT PROPORTIONS

Proportions	$\frac{a}{b}=\frac{c}{d}$	$\frac{b}{a}=\frac{d}{c}$	$\frac{a}{c}=\frac{b}{d}$	$\frac{c}{a}=\frac{d}{b}$
Cross-products	$ad = bc$	$bc = ad$	$ad = cb$	$cb = ad$

Notice that all four proportions have the same cross-products.

EXERCISES 2.6

In 1–10, write each comparison as a ratio and reduce to lowest terms.

1. 16 m to 24 m

2. 36 ft to 14 ft

3. 2 hr to 45 min

4. 30 s to 3 min

5. 18 in to 4 ft

6. 145 lbs to 5 in^2 (1 lb/in^2 = 1 psi)

7. 24 nautical miles to 2 hours (1 nautical mile/hour = 1 knot)

8. 45 C to 15 s (1 coulomb/second = 1 ampere)

9. 66 kg·m^2 to 11 s^2 (1 kg·m^2/s^2 = 1 joule)

10. 66 kg·m to 11 s^2 (1 kg·m/s^2 = 1 newton)

In 11–20, solve each proportion. Use a calculator and round appropriately.

11. $\frac{x}{3.05}=\frac{7.61}{5.33}$

12. $\frac{7.21}{y}=\frac{4.16}{5.26}$

13. $\frac{12.3}{7.11}=\frac{x}{5.26}$

14. $\frac{9.3}{10.1}=\frac{11.5}{z}$

15. $\frac{7.1}{0.066}=\frac{12.3}{x}$

16. $\frac{19.2}{6.3}=\frac{y}{17.0}$

17. $\frac{z}{7.6}=\frac{0.92}{0.015}$

18. $\frac{12.3}{x}=\frac{0.014}{32.1}$

19. $\frac{400}{3.1}=\frac{y}{0.72}$

20. $\frac{400.0}{3.100}=\frac{y}{0.7200}$

In 21–30, convert each measurement to a measurement with the given unit.

21. 99 pounds to kilograms

22. 2.1 meters to feet

23. 66 kilometers to miles

24. 3.4 ounces to grams

25. 13.2 gallons to liters

26. 4.2 liters to quarts

27. 25 miles/gallon to kilometers/liter

28. 7523 watts to kilowatts

29. 500 picofarads to farads

30. 2741 milliamperes to amperes

Problems 31 and 32 use the following information. A Wheatstone bridge is a device with two fixed resistors of resistance R_1 *and* R_2 *and a calibrated resistor which is set at* R_3*, so that an unknown resistance* R_4 *can be determined. The proportion* $R_1/R_2 = R_3/R_4$ *is used.*

31. If $R_1 = 10.00\ \Omega$, $R_2 = 100.0\ \Omega$, and $R_3 = 24.00\ \Omega$, find R_4. 3 5 9 1 2 4 6 7 8

32. If $R_1 = 10\ \Omega$, $R_2 = 1000\ \Omega$, and $R_3 = 30\ \Omega$, find R_4. 3 5 9 1 2 4 6 7 8

Problems 33 and 34 use the following information. A transformer consists of two coils of wire. The ratio of the number of turns to voltage is the same for both coils.

33. If a transformer has 200 turns in its primary winding and the voltage is 120 V, what is the voltage for the second winding, if it has 400 turns? 3 5 7 1 2 4 6 8 9

34. If a transformer has 200 turns in its primary winding and the voltage is 120 V, how many turns are there in the second coil, if the voltage is 30 V? 3 5 7 1 2 4 6 8 9

Problems 35 and 36 use the following information. On a map 1 inch represents 10 miles.

35. How many miles does a distance of 2 3/4 inches represent? 1 2 4 7 8 3 5

36. How many inches would represent 22 miles? 1 2 4 7 8 3 5

For 37 and 38, consider the similar triangles in the accompanying figure.

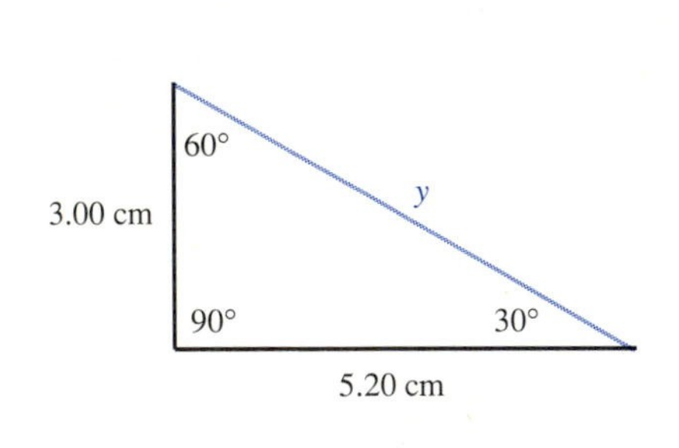

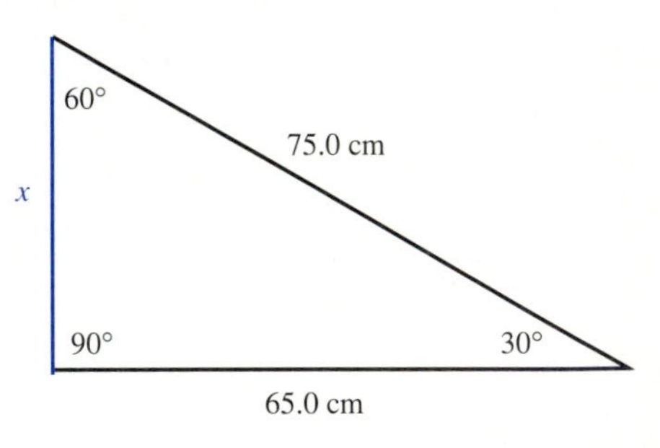

37. Find the length of x. 1 2 4 7 8 3 5

38. Find the length of y. 1 2 4 7 8 3 5

---

1. Architectural Technology
2. Civil Engineering Technology
3. Computer Engineering Technology
4. Mechanical Drafting Technology
5. Electrical/Electronics Engineering Technology
6. Chemical Engineering Technology
7. Industrial Engineering Technology
8. Mechanical Engineering Technology
9. Biomedical Engineering Technology

Problems 39 and 40 use the following information. A surveyor can determine the distance across a river by taking measurements along the bank as shown in the following figure, because the triangles are similar. (Why?)

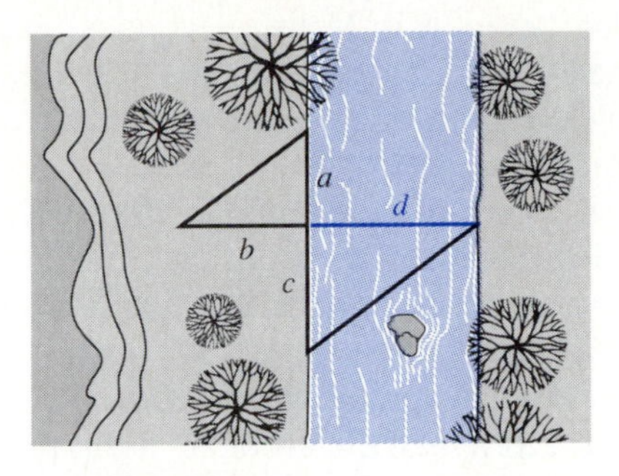

39. Find d if $a = 9.00$ m, $b = 12.0$ m, and $c = 21.0$ m.
1 2 4 7 8 3 5

40. Find d if $a = 12.0$ m, $b = 15.0$ m, and $c = 32.0$ m.
1 2 4 7 8 3 5

41. Many people estimate the distance (in miles) to a storm by dividing the number of seconds between a flash of lightning and the thunder by 5. This rule is based on the fact that sound travels approximately 1 mile in 5 seconds. How far does it travel in 7 seconds? 4 7 8 1 2 3 5 6 9

42. If 1 mile is approximately 1.6 km, formulate a rule similar to the one in Problem 41 to estimate the distance to a storm in kilometers.
4 7 8 1 2 3 5 6 9

43. If the water usage for a city of 60,000 is about 96,000,000 gallons per day, what would be the expected usage for a city of 25,000?
4 8 1 2 6 7 9

44. If the water usage for a city of 60,000 is about 96,000,000 gallons per day, how many people could a plant capable of processing 28,800,000 gallons per day serve? 4 8 1 2 6 7 9

45. If a person has 5×10^6 red blood cells in 0.001 cm^3 of blood, how many red blood cells would be in 1 liter of blood? (1 liter = 1000 cm^3)
6 9 2 4 5 7 8

46. Find the length of cf and bg in the structural frame shown in the figure. 1 2 4 7 8 3 5

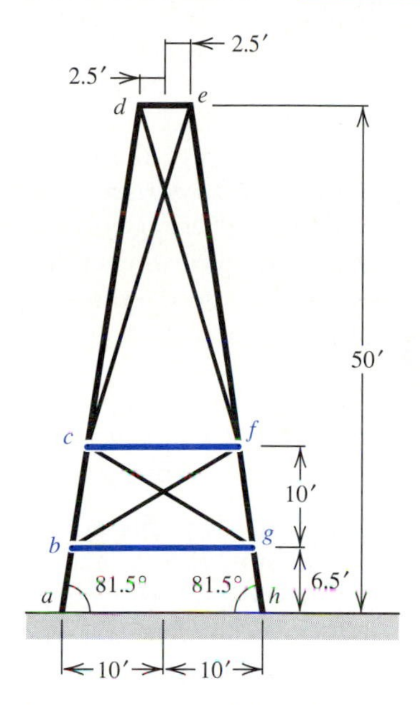

47. Under what conditions is it true that adding a nonzero number to both the numerator and denominator of a fraction produces a fraction equal to the original fraction?

2.7 APPLICATIONS

In solving verbal problems there are no hard and fast "rules." Each problem may be different from the preceding one. There are, however, a few guidelines that may help to get you started.

Read the problem quickly. Don't worry yet about the numbers in the problem. If it says, "A bank teller has $250 in tens and twenties," think to yourself, "A bank teller has *some money* in two denominations." Try to picture the situation. Watch for key words like *sum* and *product,* which indicate that addition and multiplication are involved.

Look for the question. Let the variable represent the quantity you're trying to find. This isn't always the best thing to do, but it *usually* is. If the question is "How long did it take?" write, "Let t = time." If the question is "How far did he travel?" write, "Let d = distance traveled." If the problem asks for more than one thing, write down expressions to represent all of the quantities.

Write down and label relevant information. You will need to reread the problem more carefully at this stage. It helps to have all the bits and pieces of the problem in one place and written in a concise form. Draw a picture or write down a formula.

Write an equation. It is tempting to think that the first step in solving a word problem is to write an equation. If you take the time to analyze the problem, writing the equation is not hard. You will probably want to reread the problem at this stage.

NOTE ⇨⇨ Finally, solve the equation and check your answer. ***In any type of problem, however, it is important to check your answer in the* original *problem.*** If, for example, you write the wrong equation, you may solve that equation correctly, but have the wrong answer for the word problem. Checking the equation (instead of the original problem) would not show you that you had an error. Reread the problem, substituting your answer to see if it meets the conditions described. A summary of these steps follows.

To solve a word problem:

1. Read the problem for understanding.
2. Identify the variable (usually the quantity sought).
3. Write down relevant information about all of the quantities involved in the problem. Draw a picture or write a formula.
4. Write the equation. (It may be necessary to reread the problem to see how the various quantities are related.)
5. Solve the equation.
6. Check the answer by verifying that it meets the conditions of the *original* problem.

EXAMPLE 1 A plane leaves the airport flying west at 300 mph. One hour later another plane leaves the same airport flying east at 350 mph. For how long will each plane travel before they are 1730 miles apart?

Solution

1. Visualize the scene. Don't worry about numbers yet. One plane takes off. Later, another plane takes off headed in the opposite direction. It is going faster than the first plane.

2. Identify the variables. The question is how long will *each* travel, so let the variable represent the time that one plane travels. Write, "Let t = time of the first plane." But the time that the second plane travels should be specified, since we were asked to find that time as well. It took off one hour later, so its travel time is one hour *less.* Write, "$t - 1$ = time of the second plane."

3. Write down other relevant information. We are given the rates of the two planes and the distance between them. We summarize this information.

$$\begin{aligned} \text{Let } t &= \text{time of first plane} \\ t - 1 &= \text{time of second plane} \\ 300 \text{ mph} &= \text{rate of first plane} \\ 350 \text{ mph} &= \text{rate of second plane} \\ 1730 \text{ miles} &= \text{distance between planes} \end{aligned}$$

We might need to use the formula $D = RT$.
A diagram, such as Figure 2.6, of the plane routes might also be useful.

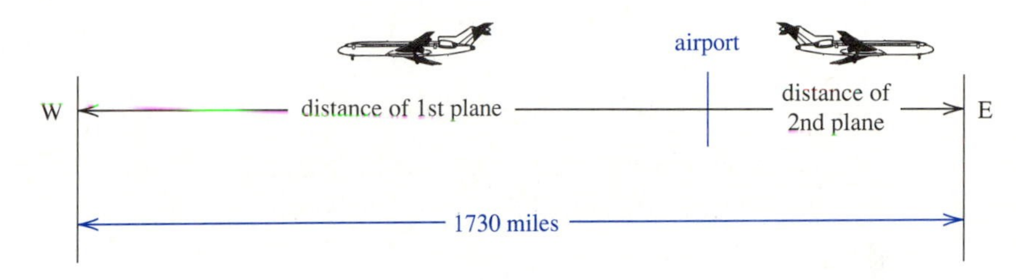

FIGURE 2.6

4. Now we are ready to write the equation. By looking at the information we have summarized in step 3, we can see how to use the formula $D = RT$ to find each individual distance. The distance of the first plane would be its rate multiplied by its time or $300t$. The distance of the second plane would be its rate multiplied by its time or $350(t - 1)$. The diagram shows that the distance 1730 is obtained by adding the two individual distances, so $300t + 350(t - 1) = 1730$.

5. Solve the equation.

$$\begin{aligned} 300t + 350t - 350 &= 1730 && \text{Use the distributive property.} \\ 650t - 350 &= 1730 && \text{Combine like terms.} \\ 650t - 350 + 350 &= 1730 + 350 && \text{Add 350 to both sides.} \\ 650t &= 2080 \\ \frac{650t}{650} &= \frac{2080}{650} && \text{Divide both sides by 650.} \\ t &= 3.2 \end{aligned}$$

Since t was the time of the first plane, and $t - 1$ was the time of the second, we have 3.2 hours (or 3 hours and 12 minutes) for the first, and 2.2 hours (or 2 hours and 12 minutes) for the second.

6. Check the answer. The first plane travels for 3.2 hours at 300 mph, which means that it travels 960 miles. The other plane travels for 2.2 hours at 350 mph, which means that it travels 770 miles. Since the planes traveled in opposite directions, they are a total of $960 + 770$ or 1730 miles apart. Thus, the conditions of the original problem are met. ■

EXAMPLE 2 A chemical company has 220 gallons of a solution that has been diluted to 40% strength, but needs to be stronger. How much of the chemical should be added to bring the solution up to 45% strength?

Solution

1. Try to visualize a vat of this solution. Part of the solution is the chemical; part is a diluting agent. The chemist needs to pour some more of the chemical into a vat to make the mixture stronger.

2. Since the question is, how much chemical should be added, let $c =$ amount of chemical to be added.

3. Write down other relevant information. Summarizing, we have:

$$\begin{aligned} \text{Let } c &= \text{amount of chemical to be added (100\% chemical)} \\ 220 &= \text{original amount of solution (40\% chemical)} \\ c + 220 &= \text{total amount of mixture (45\% chemical)} \end{aligned}$$

4. The amount of chemical is the same whether it is in two containers, (as it was at the start of the problem) or whether it is all mixed together. We equate these amounts. 100% of c is chemical; 40% of 220 gallons is chemical; and 45% of $(c + 220)$ gallons is chemical. A diagram such as Figure 2.7 might be helpful.

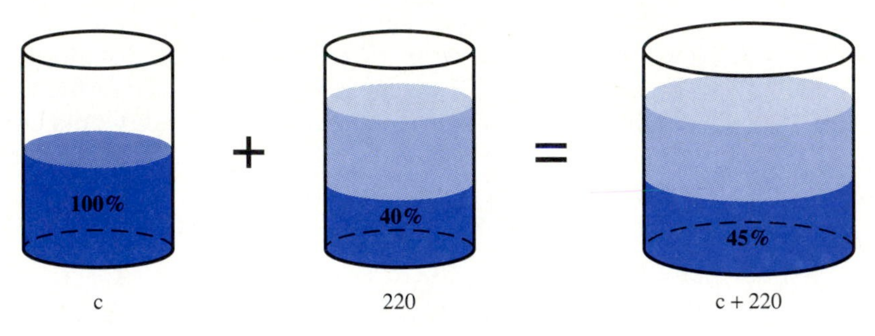

FIGURE 2.7

5. $c + 0.40(220) = 0.45(c + 220)$

$$\begin{aligned} c + 88 &= 0.45c + 99 && \text{Simplify both sides.} \\ 0.55c + 88 &= 99 && \text{Subtract } 0.45c \text{ from both sides.} \\ 0.55c &= 11 && \text{Subtract 88 from both sides.} \\ c &= 20 && \text{Divide both sides by 0.55} \end{aligned}$$

Thus, 20 gallons of the chemical should be added.

6. Check the answer. A chemical company has 220 gallons of a solution that has been diluted to 40% strength, (so 40% of 220 gallons, or 88 gallons would be chemical). Twenty gallons of chemical are added (so there would be 220 + 20 or 240 gallons of solution, and 88 + 20 or 108 gallons would be chemical). The strength would be $108/240 = 45\%$. ■

EXAMPLE 3 The three angles of a triangle are such that the largest is 5 times the smallest. The other angle is 4° less than twice the smallest. Find the measure of all three angles.

Solution **1.** Draw a triangle, such as the one in Figure 2.8.

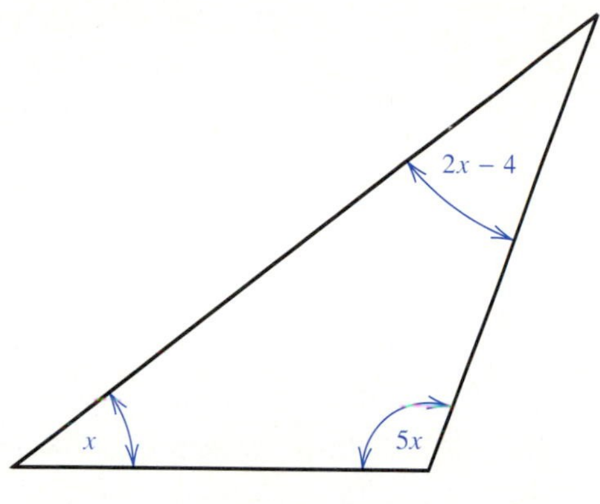

FIGURE 2.8

2. We must find the measures of all three angles of the triangle. We need to know the measure of the smallest to find the other two, so we seek it first.

3.
$$\text{Let } x = \text{measure of smallest angle}$$
$$5x = \text{measure of largest angle}$$
$$2x - 4 = \text{measure of third angle}$$

NOTE ⇨⇨ ***You have probably learned that the sum of the angles of any triangle is 180°.***

4. $x + 5x + (2x - 4) = 180$

5. $8x - 4 = 180$ Simplify the left-hand side.

$8x = 184$ Add 4 to both sides of the equation.

$x = 23$ Divide both sides of the equation by 8.

The angles, then, have measurements of 23°, 115°, and 42°.

6. We have $23° + 115° + 42° = 180°$ and the angles satisfy the conditions described. ■

EXAMPLE 4 An exhibit area for an art museum is to be rectangular with a perimeter of 106 feet. If the length is to be 11 feet longer than the width, find the dimensions that should be used.

Solution **1.** Since we must find both length and width, we may begin with either.

2. Let w = width (We could let l = length.)

3. $w + 11 =$ length (If $l =$ length, then $l - 11 =$ width.) Since the perimeter is the distance around a figure, $2l + 2w$ gives the perimeter of a rectangle. Refer to Figure 2.9.

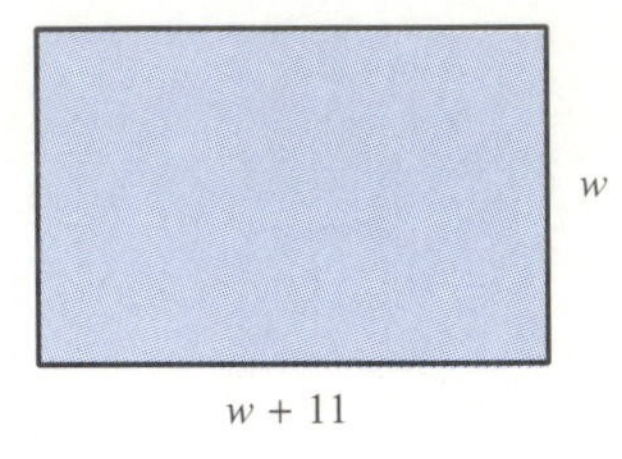

FIGURE 2.9

4. $2(w) + 2(w + 11) = 106$

5. $2w + 2w + 22 = 106$ Use the distributive property.

$4w + 22 = 106$ Simplify the left-hand side.

$4w = 84$ Subtract 22 from both sides.

$w = 21$ Divide both sides by 4.

The width should be 21 ft and the length should be 21 + 11 or 32 ft.

6. Check the answer. A length of 32 ft is 11 ft more than a width of 21 ft. The perimeter of a room 21 ft × 32 ft would be 42 ft + 64 ft or 106 ft. ■

EXERCISES 2.7

In 1–31 use an equation to solve each problem.

1. A student has grades of 85, 89, 92, and 90. What grade must be made on the fifth test in order to have an average of 91? 1 2 6 7 8 3 5 9

2. If the average power required to charge a 12-V storage battery is shown to be 280 watts for a sample of 100 batteries, how many batteries requiring only 250 watts would have to be included to bring the average down to 260 watts? 3 5 9 1 2 4 6 7 8

3. Fourteen-karat gold is 14/24 gold. How much pure gold should be added to 18 grams of 10-karat gold (10/24 gold) to make it 14-karat? 6 9 2 4 5 7 8

4. Ten liters of an acid and water solution are 90% water. How much acid should be added to make it 75% water? 6 9 2 4 5 7 8

5. Two liters of a salt and water solution are 70% water. How much water should be added to make it 80% water? 6 9 2 4 5 7 8

6. Two cars leave point A at the same time, traveling the same direction. The faster car travels at 55.0 mph while the slower car travels at 48.0 mph. How long will it take for the faster car to get ahead by 15.4 miles? 4 7 8 1 2 3 5 6 9

1. Architectural Technology
2. Civil Engineering Technology
3. Computer Engineering Technology
4. Mechanical Drafting Technology
5. Electrical/Electronics Engineering Technology
6. Chemical Engineering Technology
7. Industrial Engineering Technology
8. Mechanical Engineering Technology
9. Biomedical Engineering Technology

7. Two trains leave from the same place heading in opposite directions. The faster train travels at 110 mph while the slower train travels at 96 mph. How long will it take for the trains to be 309 miles apart? 4 7 8 1 2 3 5 6 9

8. Two people begin walking toward each other from points A and B, which are 13.0 miles apart. One person walks at a rate of 0.50 mph faster than the other. If they meet after two hours, how fast was each walking? 4 7 8 1 2 3 5 6 9

9. A plane leaves Chicago flying nonstop to Atlanta at the same time that a plane leaves Atlanta flying nonstop to Chicago. The distance between the two cities is 606 miles. If one plane flies at a rate of 350 mph while the other plane flies at a rate of 340 mph, how far apart will they be after 1/2 hour? after 1 hour? 4 7 8 1 2 3 5 6 9

10. Four water pipes, each 6.0 inches in diameter, are to be replaced by one pipe. What diameter should it have in order for the cross-sectional area to be the same as the four six-inch pipes combined? 4 8 1 2 6 7 9

11. How many water pipes, each 3 inches in diameter, would be required to replace a single pipe 6 inches in diameter, if the total cross-sectional area is to remain the same? 4 8 1 2 6 7 9

12. What is the length of a side of a square window that could be insulated by the same amount of weatherstripping as a window 1.0 m $\times$ 2.0 m? 1 2 4 7 8 3 5

13. What is the diameter of a semicircular window that could be insulated by 3.19 m of weatherstripping? 1 2 4 7 8 3 5

14. The angles of a triangle are such that the ratio of the smallest to the largest is 1/3 and the other angle is 30° less than the largest. What are the measures of the three angles? 1 2 4 7 8 3 5

15. A computer monitor can light up small rectangles called pixels. If the ratio of the number of pixels in a row to the number of pixels in a column is 7/4, and there are to be 120 more pixels in a row than in a column, how many pixels should there be in each row and each column? 3 5 9

16. An easy approximation for changing Fahrenheit temperature F to Celsius temperature C is to subtract 30 and take half. For what temperature does this rule give the same value as the standard formula $C = (5/9)(F - 32)$? 6 8 1 2 3 5 7 9

17. The formula for converting Celsius temperature C to Fahrenheit temperature F is $F = (9/5)C + 32$. At what temperature are the Celsius and Fahrenheit temperatures numerically equal? 6 8 1 2 3 5 7 9

18. The formula for converting Fahrenheit temperature F to Celsius temperature C is $C = (5/9)(F - 32)$. At what Celsius temperature is the Fahrenheit temperature numerically equal to twice the Celsius temperature? 6 8 1 2 3 5 7 9

19. For a horizontal beam, the product of the mass of a safe uniformly distributed load L and the distance between supports d is constant for beams of identical construction. If a beam can support a mass of 582 kilograms when the distance is 2.10 m, what distance should be used to increase the mass of the load to 679 kilograms? 1 2 4 8 7 9

20. The formula Btu = area $\times$ exposure factor $\times$ climate factor is used for determining the Btu rating of an air conditioner. An air conditioner with a rating of 1520.0 Btu is used with an exposure factor of 20.0 and a climate factor of 0.950. How large a room can be cooled? 6 8 1 2 3 5 7 9

21. The power P of an electric current I when the potential difference across the ends of the circuit is V is given by $P = IV$. A 100 watt lightbulb carries a current of 0.83 amperes at 120 volts. What current would it carry at 240 volts? 3 5 9 1 2 4 6 7 8

22. A voltage of 118.25 V is measured at the end of a power line. It is known that the power line has a drop of 1.75%. What is the supply voltage? 3 5 9 1 2 4 6 7 8

23. A generator is known to have an efficiency (output/input) of 80.0% when the input is 8950 watts. If the efficiency could be increased to 85.0%, what input would be necessary to maintain the same output? 3 5 7 1 2 4 6 8 9

24. A microwave oven that leaks radiation at a rate of 1.25 mW/cm^2 at two inches has a rate of leakage that exceeds the level permitted by federal standards. If the leakage were reduced by 40%, the rate would be 75% of the permitted level. What is the permitted level? 6 9 5 8

25. An elevator has a capacity of 1000 kg. Eleven adults of average weight and five children of average weight can ride together. Find the weight of an average adult, if an average child weighs 40 kg less than an adult. **1 2 4** 8 7 9

26. The formula $V = 12.16\sqrt{P}$ gives the velocity V (in ft/s) of water discharged when the pressure is P (in lb/in^2). Find the pressure when the velocity is 86 ft/s. **4 8** 1 2 6 7 9

27. When the adult dose of a drug is 200 mg to three significant digits, the dose for a two-year-old child would be 28.5 mg, based on age. Some physicians multiply the adult dose by a fraction obtained by dividing the child's weight in pounds by 150. For what weight child would the rule give 28.5 mg? **6 9** 2 4 5 7 8

28. A circular duct is to be bent in order to fit through a rectangular hole. The length of the rectangle is to be twice the width. What dimensions would be necessary for a duct of radius 3.00 inches to just fit? **4 8** 1 2 6 7 9

29. A technician needs copies of a diagram to be reproduced with exact dimensions. The copy machine is known to produce copies that are 5.3% smaller than the original. How much should the technician enlarge the dimensions on the original so that they will be correct on the copies? **1 2 4 7 8** 3 5

30. If each joist is 14 feet long and can support a load of 3928 pounds, how far apart should their centers be placed to support a uniformly distributed load of 140 lb/ft^2 of floor surface? **1 2 4** 8 7 9

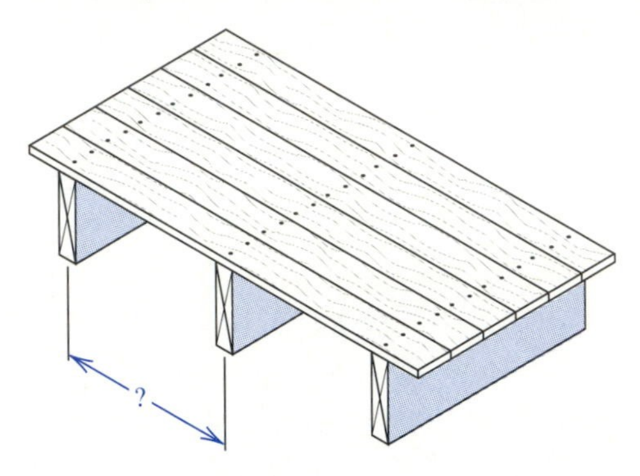

31. The floor of a room is covered with 196 tiles that are 12 inches square. If the tiles are to be replaced with tiles that are 9 inches square, how many tiles are needed? **1 2 4 7 8** 3 5

1. Architectural Technology
2. Civil Engineering Technology
3. Computer Engineering Technology
4. Mechanical Drafting Technology
5. Electrical/Electronics Engineering Technology
6. Chemical Engineering Technology
7. Industrial Engineering Technology
8. Mechanical Engineering Technology
9. Biomedical Engineering Technology

2.8 CHAPTER REVIEW

KEY TERMS

2.1 algebraic expression
base
exponent
terms
factors
like terms

2.2 polynomial
polynomial in x

2.3 nth root
radical

2.5 linear equation in x
formula

2.6 ratio
denominate number
conversion factor
proportion
similar triangles

SYMBOLS

x^p x raised to the power p

$\sqrt[n]{x}$ nth root of x

RULES AND FORMULAS

If x and y are real numbers and n is a positive integer, $\sqrt[n]{xy} = (\sqrt[n]{x})(\sqrt[n]{y})$ when both of the radicals represent real numbers, and $\sqrt[n]{\frac{x}{y}} = \frac{\sqrt[n]{x}}{\sqrt[n]{y}}$ when both of the radicals represent real numbers and $y \neq 0$.

If x is a real number and m and n are integers,

$$x^m(x^n) = x^{m+n}$$

If x is a nonzero real number and m and n are integers,

$$\frac{x^m}{x^n} = x^{m-n}$$

If x and y are real numbers and m and n are integers,

$$(x^m)^n = x^{mn}, \quad (xy)^n = x^n y^n, \quad \text{and} \quad \left(\frac{x}{y}\right)^n = \frac{x^n}{y^n} \text{ if } y \neq 0$$

For any nonzero number x and integer n, $x^0 = 1$ and $x^{-n} = 1/x^n$.

GEOMETRY CONCEPTS

$A = \pi r^2$ (area of a circle)

$A = 2\pi rh$ (surface area of a cylinder, excluding the circular ends)

$A = (h/2)(B + b)$ (area of a trapezoid)

$V = s^3$ (volume of a cube)

$A + B + C = 180°$ if A, B, and C are the angles of any triangle.

SEQUENTIAL REVIEW

(Section 2.1) *In 1–3, determine the number of terms or factors in each expression.*

1. $5x - 2y$ **2.** $(3x + 1)(2y - 5)$ **3.** $5a^2 + 3a - 1$

In 4–7, perform the indicated operations and simplify the results.

4. $(2m^2 - 5m + 3) - (4m - 2)$ **5.** $x + [x - (3x + 2)]$

6. $2x^2 - \{x^2 + [3x - (2x + 1) + 5]\}$ **7.** $2z - (z + (3z - 5) - 1)$

(Section 2.2) *In 8–14, perform the indicated operations and simplify the results.*

8. $2x(3x^2y)(xy^2)$ **9.** $(z^2 + 1)(2z^2 + 3z - 1)$ **10.** $2x(x - 5) - 3(x - 2)$

11. $(x - 2)(x - 3) - x(x + 1)$ **12.** $\frac{12x^3y + 9x^2y^2 + 3xy^3}{6x^2y}$ **13.** $\frac{x^3 + 2x^2 + 3x - 4}{x + 1}$

14. $\frac{x^3 + 9x^2 + 27x + 27}{x + 3}$

(Section 2.3) In 15–17, simplify each expression.

15. $(3xy)^2$

16. $4(xy)^3$

17. $(2a + 3)^3$

In 18 and 19 simplify each radical. For each square root, also find the answer to three significant digits.

18. $\sqrt[3]{48}$

19. $\sqrt{48}$

In 20 and 21, rationalize each denominator.

20. $3/\sqrt{2}$

21. $2/\sqrt{7}$

(Section 2.4) In 22–28, simplify each expression. There should be no zero or negative exponents in your answer.

22. 5^0

23. $4x^{-3}y(3xy^{-2})$

24. $\dfrac{25a^2bc^3}{15ab^2c}$

25. $3(xy^{-2})^2$

26. $\left(\dfrac{3ab^{-1}}{c}\right)^2$

27. $\left(\dfrac{5x^4y}{3xy^{-1}}\right)^{-2}$

28. $\dfrac{x^{-2} + y^{-1}}{x^3y^{-1}}$

(Section 2.5) In 29–32, solve each equation.

29. $2(m + 2) = 4m - 1$

30. $5y/3 + 2 = 7$

31. $3(x + 1) + 2 = 2x - (x - 1)$

32. Use a calculator to solve $3.2(x - 9.1) = 5.4x - 8.7$

33. Solve for w: $V = lwh$

In 34 and 35, solve each equation that has real number solutions.

34. $y^3 = 64$

35. $\sqrt[5]{y} = 3$

(Section 2.6) In 36 and 37, write each comparison as a ratio and reduce to lowest terms.

36. 4 yards to 2 feet

37. 27 inches to 3 feet

In 38 and 39, solve each proportion and round appropriately.

38. $x/7.1 = 8.3/9.4$

39. $13.2/y = 0.0411/2.31$

In 40–43, convert each measurement to the indicated unit.

40. 68.2 kg to lb

41. 5.5 miles to km

42. 3.0 liters to quarts

43. 9.8 newtons to pounds

(Section 2.7) In 44 and 45, solve each problem.

44. How many liters of a 1% solution should be added to 12 liters of an acid and water solution that is 4% acid in order to make a 3% solution? 6 9 2 4 5 7 8

45. Two cars 210.0 km apart are headed toward each other and meet after 2.0 hours. If it is known that one car travels 5.0 km/hr faster than the other, find their rates. 4 7 8 1 2 3 5 6 9

---

1. Architectural Technology
2. Civil Engineering Technology
3. Computer Engineering Technology
4. Mechanical Drafting Technology
5. Electrical/Electronics Engineering Technology
6. Chemical Engineering Technology
7. Industrial Engineering Technology
8. Mechanical Engineering Technology
9. Biomedical Engineering Technology

APPLIED PROBLEMS

1. Metric prefixes are borrowed for use in computer work, but the meanings are slightly different. When used in computer jargon, the prefix k means 1024, which is 2^{10}. Since 256 is 2^8, a memory capacity of 256 k means $(2^8)(2^{10})$ memory cells. Use the exponent product rule to simplify the expression. 3 5 9

2. The formula $C = C_1C_2/(C_1 + C_2)$ gives the equivalent capacitance of two capacitors in series (end-to-end). Solve for C_1. 3 5 9 4 7 8

3. Use the similar triangles shown in the accompanying figure to find the image height of an object viewed through a magnifier. The image distance is 72 cm while the object height is 5.0 cm and the object distance is 12.0 cm. 3 5 6 9

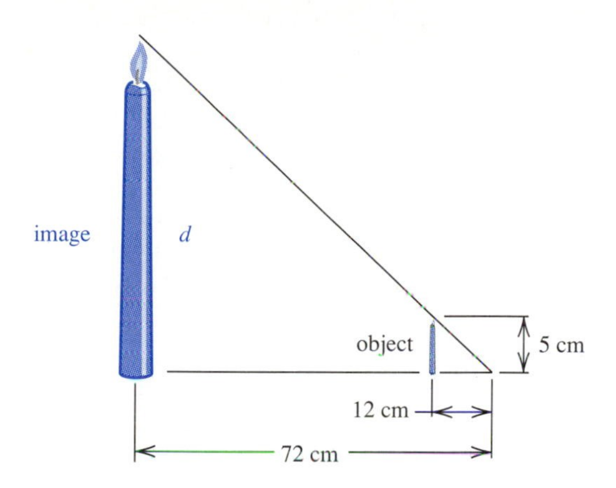

4. A circuit with two resistors connected in parallel (side-by-side) has an equivalent resistance R given by $1/R = 1/R_1 + 1/R_2$. Find R_1 and R_2 if R_1 is twice R_2 and the equivalent resistance is 30.0 Ω. 3 5 9 1 2 4 6 7 8

5. If two pulleys are connected by a belt, the ratio of the speed of the large pulley to the speed of the small one is equal to the ratio of the diameter of the small pulley to the diameter of the large one. Find the speed of the large pulley when the speed of the small pulley is 140 rpm. 4 7 8 1 2 3 5 6 9

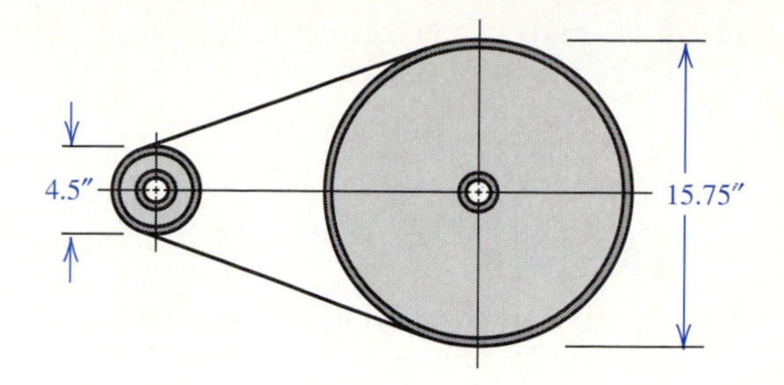

6. The velocity of sound in air at 0° C and 76 cm of pressure is 1086.7 ft/s and increases 1/546 ft/s for each degree the temperature rises. Find the temperature of air in which the velocity of sound is 1086.8 ft/s. 1 3 4 5 6 8 9

7. One cubic foot of water weighs 62.5 pounds. One cubic foot of mercury weighs 850 pounds. What is the ratio of the weight of mercury to the weight of water? 1 2 4 6 8 5 7 9

8. In computer programming, nested parentheses are used to make evaluations of formulas more efficient. Consider, for example, the expression $x(x(x + 4) + 2) + 3$. It requires only two multiplications. Simplify the expression and count the number of multiplications required. (Hint: Remember that an exponent indicates a repeated multiplication. For instance, x^3 represents two multiplications.) 3 5 9

9. Solve the following equation for P.

$$S_v = \frac{P}{NA_v}$$

In the formula, S_v is the shear stress in a riveted boiler joint, P is the load causing the shear, A_v is the cross-sectional area of the rivet sheared, and N is the number of shear planes. 1 2 4 8 7 9

10. The formula $H = \dfrac{4\pi NI}{l}$ gives the magnetic field intensity H (in oersteds) developed by a solenoid. In the formula, N is the number of turns, l is the length (in centimeters), and I is the current (in amperes). Use the laws of exponents to evaluate this expression for $N = 10^3$, $I = 10^{-6}$, and $l = 10^2$. Use 3.14 for π. 3 5 9 4 7 8

11. In the study of conductance in transmission line theory, the following formula arises.

$$G = M^{-1}L^{-3}T^{1}Q^{2}$$

Evaluate this expression for $M = 10^{-15}$, $L = 10^{-6}$, $T = 10^{-3}$, and $Q = 10^{-12}$.
3 5 9 4 7 8

12. If a and b are the focal lengths of two thin lenses separated by a distance d, the focal length of the system is $F = \dfrac{ab}{a+b-d}$. All distances have the same units. If $a = 2$ cm and $b = 3$ cm, for what value of d would F be 6 cm? 3 5 6 9

13. When two sources of light power produce equal illumination on a screen, $\dfrac{P_1}{P_2} = \dfrac{r_1^2}{r_2^2}$. In the formula, P_1 and P_2 represent power (in watts) while r_1 and r_2 represent distances with the same units. At what distance from a 100.0 watt source would the illumination be equal to that of a 50.0 watt source at 20.0 feet? 3 5 6 9

14. The space s passed over by an accelerating airplane in the nth second of its flight is given by $s = v_0 + \dfrac{1}{2}a(2n-1)$ where v_0 is its initial velocity, a its acceleration, and n the nth second. Find s for a plane traveling at 600 ft/s during the 4th second if $a = 450$ ft/s^2. Give your answer to 3 significant digits. **4 7 8** 1 2 3 5 6 9

15. The velocity v (in m/s) of sound in air at temperature T (in degrees Celsius) is given by $v = v_0\sqrt{1 + \dfrac{T}{273}}$ where v_0 is its velocity at 0° C. If a sound wave travels at 330 m/s at 0° C, at what temperature would it travel at 343 m/s? Give your answer to the nearest tenth of a degree.
1 3 4 5 6 8 9

16. In electronics a circuit with two resistors in parallel has an equivalent resistance R_T (in ohms) given by $R_T = \dfrac{R_1R_2}{R_1+R_2}$. In the formula, R_1 and R_2 are the individual resistances (in ohms). What resistance value connected in parallel with a 6 ohm resistor would make the total resistance $R_T = 2\Omega$? **3 5 9** 1 2 4 6 7 8

17. The leakage current of a certain diode is 15.0 nanoamperes at a room temperature of 20.0° C and is known to increase 0.4 nanoamperes for each degree the temperature rises. At what temperature would the leakage double?
3 5 9 4 7 8

18. Computers and programmable calculators can find roots of polynomial equations much faster if the equations are first written using nested parentheses. Rewrite the equation $V = 3t^5 - 6t^4 + 2t^3 + t^2 + 6$ to minimize the number of multiplications. **3 5** 9

19. A piece of glass used to make a fiber optic transmission line starts out as a cylinder 1 meter long with a diameter of 2.5 cm. The cylinder is then heated and drawn out to form the fiber. How many meters of fiber, 100 microns (micrometers) in diameter, could be drawn from this blank? Give your answer to 3 significant digits.
3 5 6 9

20. Five hundred ball bearings with a 5% defect rate are mixed with a batch of 1000 identical bearings with a defect rate of 10%. What defect rate will the mixture have? (Hint: The number of defective bearings is given by the product of defect rate and total number of bearings.) **1 2 6 7 8** 3 5 9

1. Architectural Technology
2. Civil Engineering Technology
3. Computer Engineering Technology
4. Mechanical Drafting Technology
5. Electrical/Electronics Engineering Technology
6. Chemical Engineering Technology
7. Industrial Engineering Technology
8. Mechanical Engineering Technology
9. Biomedical Engineering Technology

RANDOM REVIEW

1. Solve each proportion.
(a) $7.32/x = 5.51/3.66$ **(b)** $1.4/9.6 = y/7.2$

2. Simplify each expression.
(a) $(3x^4y)^2$ **(b)** $(3x^4 + y)^2$

3. A 1-3/8-inch iron chain weighs 17.5 lb/ft and has a breaking strength of 88,301 pounds. What weight added to the chain's own weight would cause a 100-foot long chain to break?

4. Perform the indicated operations and simplify the results.
(a) $(2x^2 + 3x) - (x + 1)$ **(b)** $2 - (x + (3x - 1) + 3)$

5. Simplify each expression. There should be no zero or negative exponents in the final answer.

(a) $(3x^{-2})(3x^2)$ **(b)** $\dfrac{49x^2y}{14xy^2}$

6. Perform the indicated operations and simplify the results.

(a) $3x(x - 1) - x(x + 2)$ **(b)** $\dfrac{x^2 - 7x + 12}{x - 3}$

7. Solve each equation that has real number solutions.
(a) $\sqrt[4]{x} = 3$ **(b)** $x^4 = -16$

8. Simplify each expression. There should be no zero or negative exponents in the final answer.

(a) $\left(\dfrac{3xy^{-1}}{4xy^2}\right)^0$ **(b)** $\dfrac{x^2y}{(x + y)^{-1}}$

9. Simplify each radical.
(a) $\sqrt{128}$ **(b)** $5/\sqrt{3}$

10. Solve each equation.
(a) $4x + 7 = 12$ **(b)** $x + 2y = z$, solve for y

CHAPTER TEST

1. Simplify each expression.
(a) $(x^2 + 5x - 7) - (8x + 4)$ **(b)** $x^2 + [2x - (3x^2 - 1) + 3]$

2. Perform the indicated operations and simplify the results.
(a) $3x(4xy^2)(2x^2y)$ **(b)** $(m^2 + 2)(m^2 - m + 2) - (m - 4)$

3. Perform the indicated operations and simplify the results.

(a) $\dfrac{x^3 + x^2 + x + 6}{x + 2}$ **(b)** $\dfrac{10x^4y^2 - 15x^3y^3}{5xy^2}$

4. Simplify each expression. There should be no zero or negative exponents in the final answer.

(a) $\left(\dfrac{4xy^{-2}}{z}\right)^{-3}$ **(b)** $\dfrac{xy^3}{(x+y)^{-2}}$

In 5 and 6, solve each equation.

5. $5(x-2)+3=7x+(x-3)$

6. $1.1(x+8.2)=6.3x-7.8$ (Use a calculator.)

7. Solve the following proportion: $z/8.12=9.11/4.32$

8. Convert each measurement to the indicated unit.

(a) 7.5 in to cm (1 in = 2.54 cm) **(b)** 165 lb to kg (1 kg = 2.2 lb)

9. A cube of copper that is 1.0 cm on each edge is drawn into a wire with diameter 0.10 cm. Find the length of the wire.

10. A pump delivers oil at the rate of 231 gal/min. How long will it take to pump 15,750 gal of oil?

CHAPTER 3 FUNCTIONS

In technological work, a change in one quantity may produce a change in another quantity. For example, a change in speed produces a change in distance traveled for a given period of time. Or, if the length of a given beam is changed, there is a change in the load that it will bear. It is often important to know how these changes are related. Functions and graphs, presented in this chapter, are mathematical concepts that are used to quantify relationships among variables.

3.1 BASIC CONCEPTS

In the formula for the circumference of a circle, $C = \pi d$, the circumference depends on the diameter of the circle. We say that the circumference is a *function* of the diameter. Functional relationships are common in mathematics and the technologies. We begin our study of functions with a definition.

DEFINITION A **function** is a correspondence between the values of two variables such that there is associated with each value of the first variable exactly one value of the second variable. The set of permissible values of the first variable is called the **domain,** and the set of permissible values of the second variable is called the **range.**

One way to indicate a correspondence is to set up a table. For a function, the values in the domain are listed in the top row, and the corresponding values in the range are listed in the bottom row. This procedure is effective when there are just a few values to display.

EXAMPLE 1 The table shows a correspondence between the temperature T (in degrees Celsius) of a copper wire and its resistance R (in ohms). Determine whether it is a function.

T	0	20	40	60	80
R	9.2	10.0	10.8	11.6	12.4

Solution The correspondence is a function. For each value of the variable T, there is exactly one value of R that corresponds to it. ■

EXAMPLE 2 The table shows the correspondence between the average number n of rainy days in San Antonio and the average precipitation p (in inches) for four different months. Determine whether it is a function.

n	7	7	8	6
p	1.25	2.75	3.50	3.00

Solution The table does *not* show a function, because if we choose the value 7 for the first variable, both 1.25 and 2.75 are associated with it. ***The definition of a function requires that* each *value of the first variable correspond to* exactly *one value of the second variable.*** ■

NOTE ⇨⇨

EXAMPLE 3 The table shows the correspondence between the elapsed time t (in seconds) and the horizontal distance d (in meters) from the rest position of a swinging pendulum bob. Determine whether it is a function.

t	0.25	0.50	0.75	1.00	1.25	1.50	1.75
d	0.26	0.00	0.26	0.50	0.26	0.00	0.26

Solution The correspondence is a function, because for each value of t, there is exactly one value of d associated with it. For example, if we examine the value 0.25 the only value associated with it is 0.26. If we examine 0.50, the only number associated with it is 0.00, and so on. ■

Sometimes there are so many values for each variable that it is not convenient, or even possible, to list them all. In such cases an equation is used. For example, the formula $V = (4/3)\pi r^3$ shows a relationship between the volume V of a sphere and its radius r. Given any value for r, the value for V can be determined from the formula. When r is 1, for example, we have $V = (4/3)\pi$. A table like those in Examples 1–3 may be constructed to display a few corresponding values.

argument of a function

NOTE

A function may also be indicated using *function notation.* The symbol $f(x)$ is read "f of x." The letter f represents the function in this example, but other letters (such as g or k) may also be used. The number in parentheses is called the *argument* of the function. ***Beginning algebra students sometimes think that $f(x)$ is a variable f multiplied by a variable x, but it should be clear from the context of a problem whether $f(x)$ is a multiplication problem or a function value.***

independent variable

dependent variable

Since x can take on any value in the domain, x is often called the *independent variable* of the function f. The value of $f(x)$ depends on the value of x, so $f(x)$ is called the *dependent variable.* The symbol $f(x)$ denotes the value in the range associated with the value x in the domain by the function f. For example, $f(2)$ denotes the value of the dependent variable when the value of the independent variable is 2.

To find $f(a)$ for a real number a:

1. Substitute the number a for x in the function equation.
2. Evaluate the resulting expression.

EXAMPLE 4 Let $f(x) = x^2$ and $g(x) = 2x$. Evaluate each expression.

(a) $f(3)$ **(b)** $g(3)$ **(c)** $f(-3)$

Solution **(a)** $f(3) = 3^2 = 9$ **(b)** $g(3) = 2(3) = 6$ **(c)** $f(-3) = (-3)^2 = 9$ ■

EXAMPLE 5 Let $f(x) = (x-1)^2$ and $g(x) = x^2 - 1$. Evaluate each expression.

(a) $f(-2)$ **(b)** $-f(2)$ **(c)** $g(-2)$

Solution **(a)** Substitute -2 for x in the equation $f(x) = (x-1)^2$.
$f(-2) = (-2-1)^2 = (-3)^2 = 9$

(b) Substitute 2 for x in the equation $f(x) = (x-1)^2$.
$f(2) = (2-1)^2 = 1^2 = 1$
Since $f(2) = 1$, $-f(2) = -1$.

(c) Substitute -2 for x in the equation $g(x) = x^2 - 1$.
$g(-2) = (-2)^2 - 1 = 4 - 1 = 3$ ■

EXAMPLE 6 Let $f(x) = 2x + 3$ and $g(x) = 3x + 2$. Evaluate each expression.

(a) $f(2) + g(2)$ **(b)** $f(2) + f(3)$ **(c)** $f(2) + g(3)$

Solution **(a)** $f(2) = 2(2) + 3 = 4 + 3 = 7$, and $g(2) = 3(2) + 2 = 6 + 2 = 8$
Therefore $f(2) + g(2) = 7 + 8 = 15$.

(b) $f(2) = 7$ and $f(3) = 2(3) + 3 = 6 + 3 = 9$
Therefore $f(2) + f(3) = 7 + 9 = 16$.

(c) $f(2) = 7$ and $g(3) = 3(3) + 2 = 9 + 2 = 11$
Therefore $f(2) + g(3) = 7 + 11 = 18$. ■

It is sometimes important to know the domain of a function. When the domain is not specified, we assume that the domain is the set of all real numbers for which the function is defined. Recall that an expression is not defined in the real number system if it contains zero in the denominator or contains the square root of a negative number.

THE DOMAIN OF A FUNCTION

The domain is the set of real numbers for which the function is defined. Start with the real numbers.

(a) If the function has a denominator, eliminate values for which the denominator is zero.
(b) If the function has a square root (or other even root), eliminate values for which the expression under the radical sign is negative.
(c) If the problem is an applied problem, eliminate values that are physically impossible in the context of the problem.

EXAMPLE 7 Determine whether the equation $y = 1/(2x + 3)$ represents a function, and if it does, state the domain.

Solution The equation $y = 1/(2x + 3)$ represents a function, because each time we substitute a value for x, the equation yields a single value for y. The domain is the set of real numbers, with the exception of any number for which the value of the denominator is 0. Because $2x + 3$ has a value of 0 when x is $-3/2$, the function is undefined when x is $-3/2$. Thus, we say the domain is the set of real numbers, excluding $-3/2$. ■

EXAMPLE 8 When a right triangle is constructed so that the length l of the longest side is one more than the next longest side, the formula $s = \sqrt{2l - 1}$ is used to compute the length s of the shortest side. Determine whether the equation represents a function and if it does, state the domain.

Solution The formula $s = \sqrt{2l - 1}$ is a function, because each time we substitute a value for l, the equation yields a single value for s. The domain is the set of real numbers, with the exception of any number for which the value of the expression under the radical is negative. Because $2l - 1 = 0$ when l is 1/2, the function is undefined for values less than 1/2. Furthermore, a value of exactly 1/2 for l would produce 0 for s, but it does not make sense to have 0 for the length of a side of a triangle. Thus, we say the domain is the set of real numbers greater than 1/2. ■

EXAMPLE 9 Determine whether the equation $A = \pi r^2$, which is the formula for the area of a circle, represents a function. If it does, state the domain.

Solution The relationship between the area of a circle and its radius is given by $A = \pi r^2$, which is a function. Each time we substitute a value for r, the equation yields a single value for A. The domain is the set of positive real numbers, since it does not make sense to talk about a circle with zero or negative radius r. ■

EXAMPLE 10 Determine whether the equation $f(x) = 3$ represents a function, and if it does, state the domain.

Solution The equation $f(x) = 3$ represents a function, because each time we choose a value for x and substitute it into the equation, a single value is returned. For instance, when x is 2, the equation says $f(2) = 3$. When x is -4, the equation is $f(-4) = 3$. The same value (in this case, 3) is associated with each x-value, but nevertheless, it is still true that each x-value has a single y-value associated with it. ■

The function in Example 10 is called a *constant function.* Any function of the form $f(x) = c$, where c is a constant is a **constant function.**

EXERCISES 3.1

In 1–10, determine whether the correspondence in each table is a function.

1.

x	-2	3	-4	5
y	6	-9	12	-15

2.

x	2	2	4	4
y	3	1	5	3

3.

x	-2	-3	-4	-5
y	3	3	3	3

4.

x	2	3	-1	0
y	2	3	-1	0

5.

x	2	3	4	5
y	-2	-3	-4	-5

6.

x	1	2	3
y	1	4	9

7.

x	1	4	9
y	1	2	3

8.

x	4	4	9	9
y	2	-2	3	-3

9.

x	1	1	2	3
y	4	3	2	1

10.

x	4	3	2	1
y	1	1	2	3

In 11–25, evaluate each expression if $f(x) = x^2 - 1$ *and* $g(x) = 3x + 2$.

11. $f(1)$
12. $f(0)$
13. $f(-1)$
14. $f(-2)$
15. $g(1)$
16. $g(0)$
17. $g(-1)$
18. $g(-2)$
19. $f(3) + g(3)$
20. $f(4) - g(4)$
21. $f(-3) + g(-3)$
22. $f(5) - f(3)$
23. $g(4) - g(2)$
24. $f(1) + g(-1)$
25. $f(-1) + g(1)$

In 26–36, determine whether the correspondence is a function. If it is a function, specify the domain.

26. $y = 2x + 1$

27. $y = 3x \pm 2$

28. $y = x^2 - 3$

29. $y = \dfrac{x+4}{x-1}$

30. $y = \dfrac{x-3}{2x+1}$

31. $y = \sqrt{x-4}$

32. $y = \sqrt{2x-1}$

33. $y = x^2$

34. $y = x^3$

35. $y = x^4$

36. $y = \dfrac{\pm\sqrt{x-3}}{x^2+1}$

In 37–42, use a calculator to evaluate each function. Round the answer to the accuracy of the independent variable.

37. Find $f(4.1)$ if $f(x) = 3.72x + 14.3$.

38. Find $g(1.33)$ if $g(x) = 2.61x^2 + 4.71$.

39. Find $h(1.5)$ if $h(x) = \dfrac{2.1x + 3.0}{3.2x - 1.0}$.

40. Find $f(7.92)$ if $f(x) = \sqrt{34.1x}$.

41. Find $g(14.5)$ if $g(x) = \dfrac{x - 3.11}{x + 9.27}$.

42. Find $h(1.55)$ if $h(x) = 0.009x^2$.

In 43–45, determine whether the correspondence in each table is a function.

43. The table shows capacitive reactance X_c (in ohms) of a 10 μF capacitor for various frequencies f (in hertz).

f	30.0	40.0	50.0	60.0	70.0	80.0
X_c	531	398	318	265	227	199

3 5 9 4 7 8

44. The table shows the distance d (in cm) that a spring is stretched by various forces F (in newtons).

d	1.25	2.50	3.75	5.00
F	25.0	50.0	75.0	100.0

4 7 8 1 2 3 5 6 9

---

1. Architectural Technology
2. Civil Engineering Technology
3. Computer Engineering Technology
4. Mechanical Drafting Technology
5. Electrical/Electronics Engineering Technology
6. Chemical Engineering Technology
7. Industrial Engineering Technology
8. Mechanical Engineering Technology
9. Biomedical Engineering Technology

45. The table shows the pressure p (in kg/cm^2) exerted by a liquid on a point at various depths d (in m) below the surface.

d	0.0	5.0	10.0	15.0	20.0
p	0.0	17.5	35.0	52.5	70.0

2 4 8 1 5 6 7 9

46. The period T of a pendulum is approximated by $T = 2.0\sqrt{l}$ for a pendulum of length l. What is the domain of the function? 4 7 8 1 2 3 5 6 9

47. The speed s (in mph) of a car that skids on a wet concrete road is approximated by $s = 3.46\sqrt{d}$, where d is the distance (in ft) of the skid. What is the domain of the function?
4 7 8 1 2 3 5 6 9

48. The distance s (in ft) that an object falls from rest is given by $s = 4.9t^2$ where t is the time of the fall in seconds. What is the domain of the function?
4 7 8 1 2 3 5 6 9

49. The equivalent resistance R of a 10 Ω resistor and a resistor with resistance R_1 is given by $R = 10R_1/(10 + R_1)$ when the resistors are in parallel. What is the domain of the function?
3 5 9 1 2 4 6 7 8

50. A function $f(x)$ is said to be increasing over an interval if $f(b) \geq f(a)$ whenever $b \geq a$, for any two points a and b in the interval. Which of the following functions are increasing for $x \geq 0$? for all x?
(a) $f(x) = x$ **(b)** $g(x) = x^2$
(c) $h(x) = 1/x$ **(d)** $k(x) = x^3$

3.2 RECTANGULAR COORDINATES

x-axis
y-axis
origin

The relationship between the independent and dependent variables x and $f(x)$ is often apparent from a graph. A horizontal line, called the *x-axis*, is used to locate the independent variable. A vertical line, called the *y-axis*, is used to locate the dependent variable. Figure 3.1 shows an x-axis and a y-axis. The point of intersection of the two axes is called the *origin*. For graphing, the dependent variable is frequently represented by y rather than by $f(x)$. For example, instead of writing $f(x) = x^2 - 1$, we write $y = x^2 - 1$.

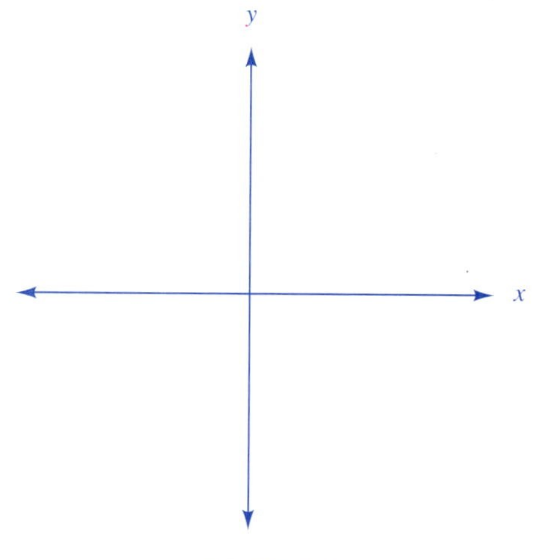

FIGURE 3.1

quadrant

The two lines divide the plane into four regions called *quadrants*. The quadrants are usually numbered with Roman numerals, starting at the upper right with I and proceeding counterclockwise, as shown in Figure 3.2. Such a system is called a

rectangular coordinates

rectangular coordinate system or a *Cartesian coordinate system* in honor of Rene Descartes (1596–1650). Each line is marked off in equal segments to indicate the integers, with 0 on each axis at the origin. On the horizontal axis, positive numbers are to the right of zero, while negative numbers are to the left of zero. On the vertical axis, positive numbers are above zero, while negative numbers are below zero. The same scale is often used on both axes, and the scale may be indicated by a grid as in Figure 3.3.

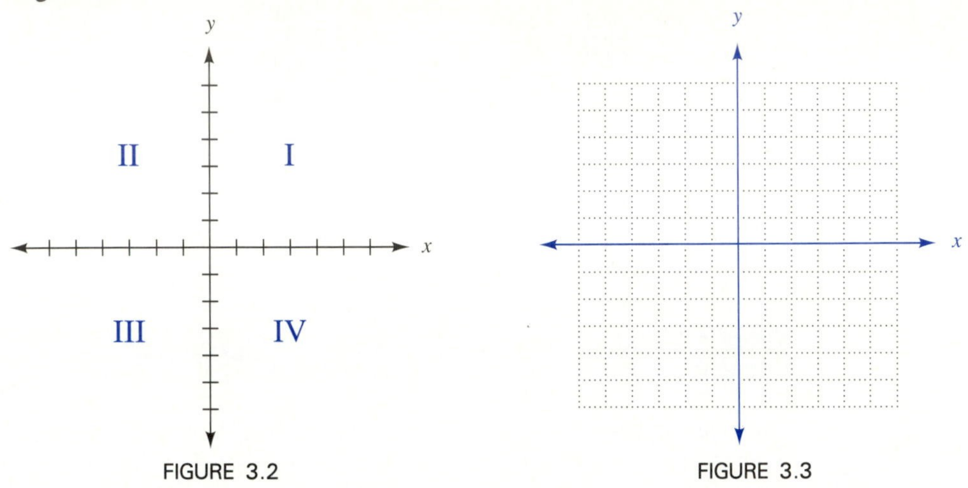

FIGURE 3.2 FIGURE 3.3

For a function, each value of x in the domain has associated with it a unique value of y in the range. For example, $f(2) = 3$ means that when the value of x is 2, the value of y is 3. Two values that are related by the function are often written within parentheses as (x, y). Such a pair of values is called an *ordered pair*. ***The value of the independent variable is always listed first.*** These values determine a unique point on the plane and are called the *coordinates* of the point. For example, the point (2, 3) is located by counting from the origin over two units horizontally and then from that position three units vertically. To *plot* a point on a plane means to indicate its location by a dot. The point (2, 3) is shown in Figure 3.4.

ordered pair

NOTE ⇨⇨

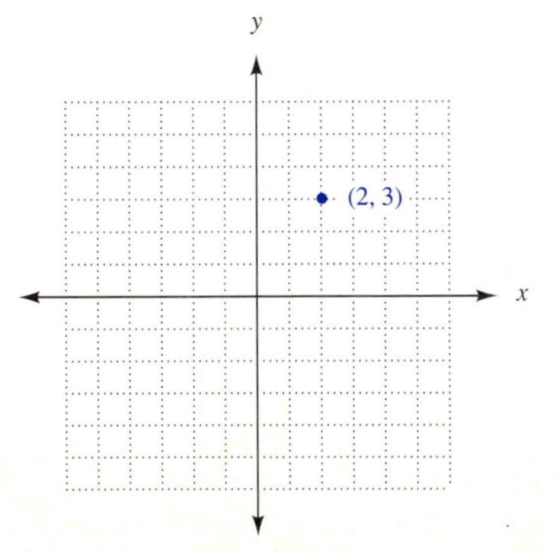

FIGURE 3.4

To graph a function, we plot the ordered pairs (x, y) where x belongs to the domain, and y is the value in the range associated with x.

To sketch the graph of a function given by $y = f(x)$:

1. Sketch and label the x-axis and y-axis.
2. Indicate the scale on each axis. If a grid is used, 1 box = 1 unit unless specified otherwise.
3. Plot all of the pairs (x, y) on the same coordinate system.

EXAMPLE 1 Graph the function indicated by the following table.

x	2	4	-6
y	3	5	6

Solution The graph is shown in Figure 3.5. ■

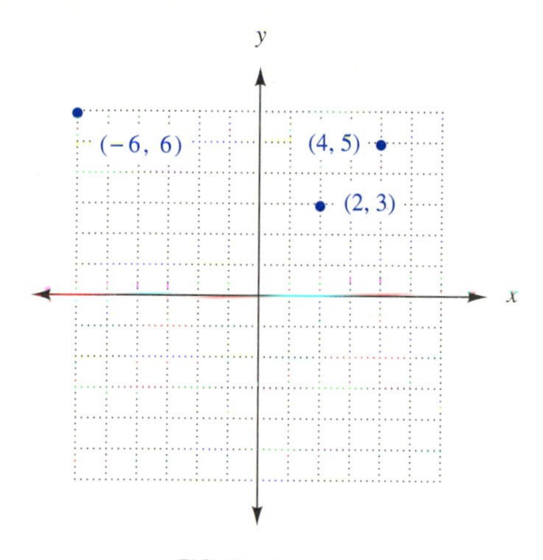

FIGURE 3.5

NOTE We have used the letters x and y, but a function of any variable can be represented on a graph. ***The horizontal axis is used for the independent variable and the vertical axis is used for the dependent variable.***

EXAMPLE 2 The following table gives the resistance R (in ohms) for various lengths l (in meters) of 2.5 mm^2 copper wire. Graph the function.

l	20	30	40	50
R	0.14	0.21	0.28	0.35

Solution The scale must be chosen to accommodate the data. The horizontal axis is used for l, and it is convenient to mark it in units of 10. The vertical axis is used for R, and it is marked in intervals of 0.10. The graph is shown in Figure 3.6. ■

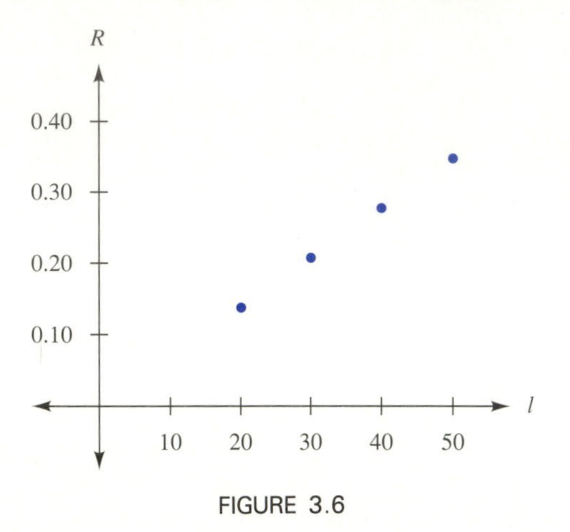

FIGURE 3.6

In Examples 1 and 2, we have only a few values for the variables, so the graphs consist of a few isolated points. If the domain is the set of real numbers, then each value of x on the x-axis will have a y-value associated with it, and the graph may be a straight line or curve, as in Figure 3.7 or Figure 3.8. Eventually we will consider both types of functions.

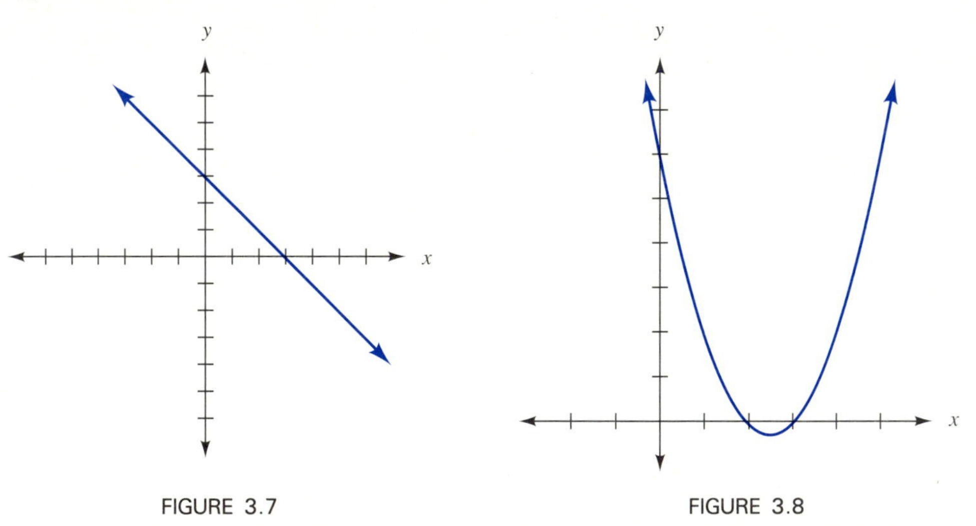

FIGURE 3.7 FIGURE 3.8

Not every graph, however, is the graph of a function. Since the definition of a function requires that only one value of y is to be associated with each value of x, *any* vertical line on the plane will intersect the graph of a function only once, as shown in Figure 3.9. If the graph is not the graph of a function, it is possible to place a vertical line somewhere on the plane so that the line intersects the graph at more than one point, as shown in Figure 3.10.

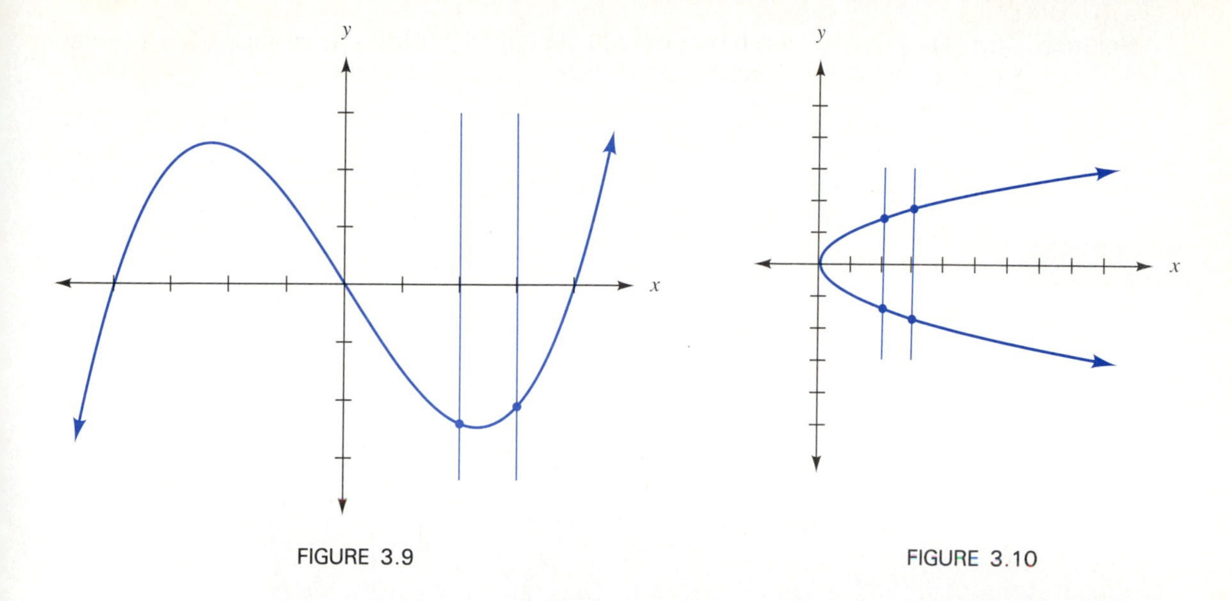

FIGURE 3.9

FIGURE 3.10

CRITERIA FOR DETERMINING WHETHER A GRAPH REPRESENTS A FUNCTION

(a) If it is impossible to draw a vertical line on the plane that intersects the graph more than once, it does represent a function.
(b) If it is possible to draw a vertical line that intersects the graph more than once, it does not represent a function.

EXAMPLE 3 Determine whether each graph represents a function.

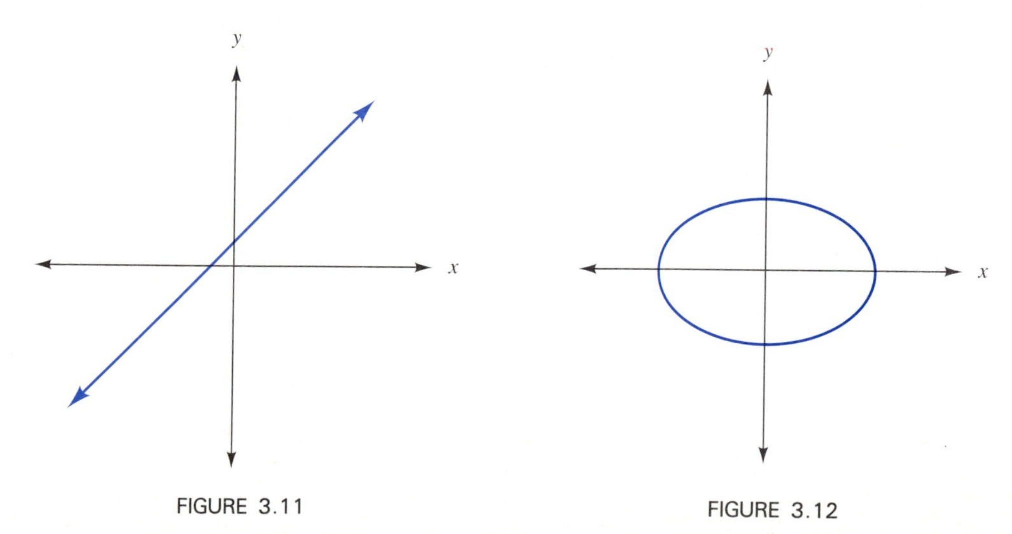

FIGURE 3.11

FIGURE 3.12

Solution **(a)** The graph shown in Figure 3.11 represents a function, because a vertical line drawn anywhere on the plane intersects the graph only once.

(b) The graph shown in Figure 3.12 does *not* represent a function, because a vertical line can be drawn to intersect the graph at more than one point. The y-axis, for example, intersects the graph in two points. ■

EXAMPLE 4 Determine whether each graph represents a function.

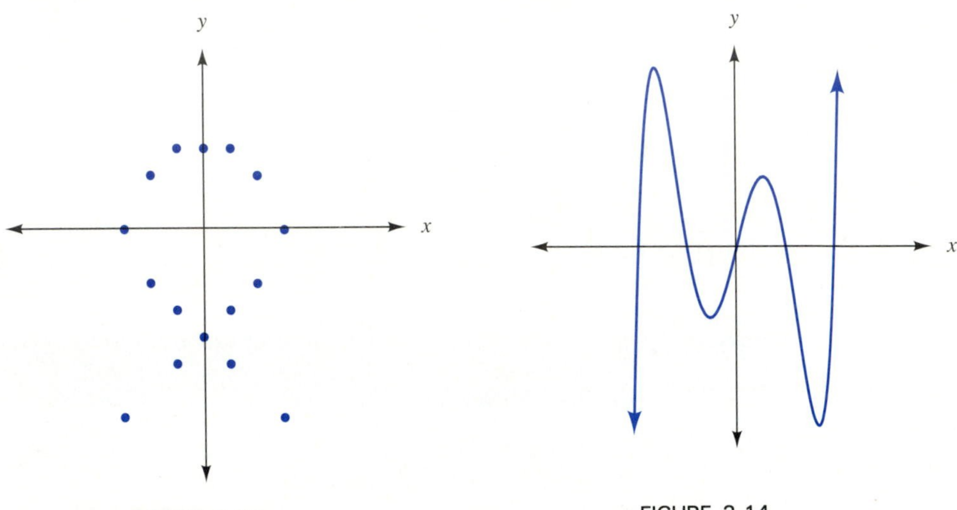

FIGURE 3.13

FIGURE 3.14

Solution **(a)** The graph shown in Figure 3.13 does *not* represent a function, because a vertical line, such as the *y-axis,* can be drawn so that it intersects the graph at more than one point.

(b) The graph in Figure 3.14 represents a function, because anywhere a vertical line is drawn through the graph, it intersects the graph only once. ■

EXERCISES 3.2

In 1–14, graph each function on a separate coordinate system.

1.

x	-2	3	-4
y	6	-9	12

2.

x	1	2	3
y	1	4	9

3.

x	1	4	9
y	1	2	3

4.

x	1	0	-1
y	1	0	-1

5.

x	-2	0	2
y	4	-1	4

6.

x	-1	0	1
y	-1	1	3

7.

x	-2	-3	-4	-5
y	3	3	3	3

8.

x	2	3	-1	0
y	2	3	-1	0

9.

x	2	3	4	5
y	-2	-3	-4	-5

10.

x	-2	-1	0	1
y	7	0	-1	0

11.

x	-2	-1	1	2
y	$-1/2$	-1	1	$1/2$

12.

x	1	2	3	4
y	-1	-1	-1	-1

13.

x	0	1	2	3
y	$1/2$	0	$-1/2$	-1

14.

x	$-1/2$	1	$3/2$	2
y	$1/2$	0	$-1/2$	-1

In 15–30, determine whether each graph represents a function.

15.

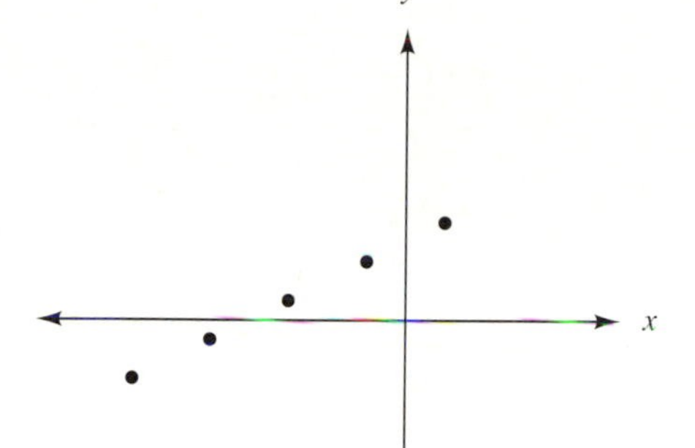

16.

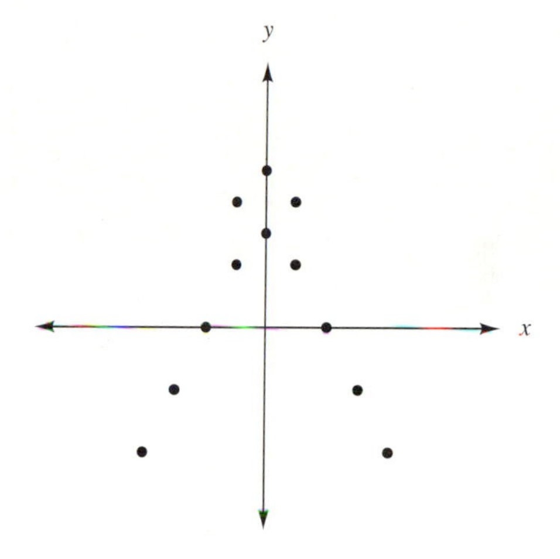

17.

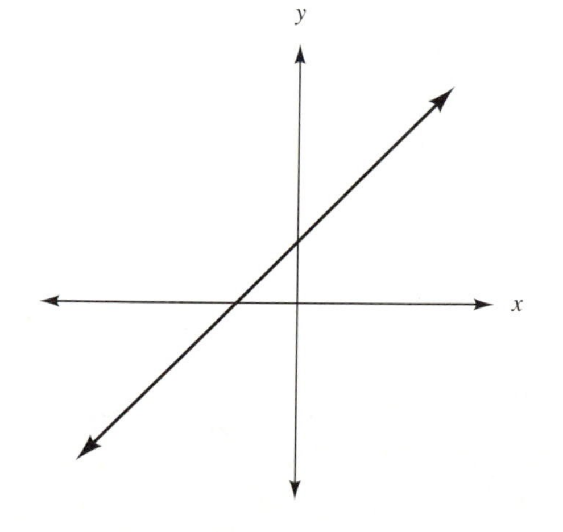

18.

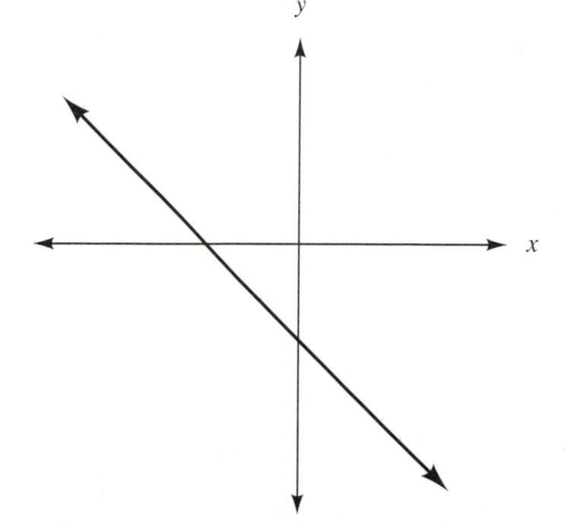

19.

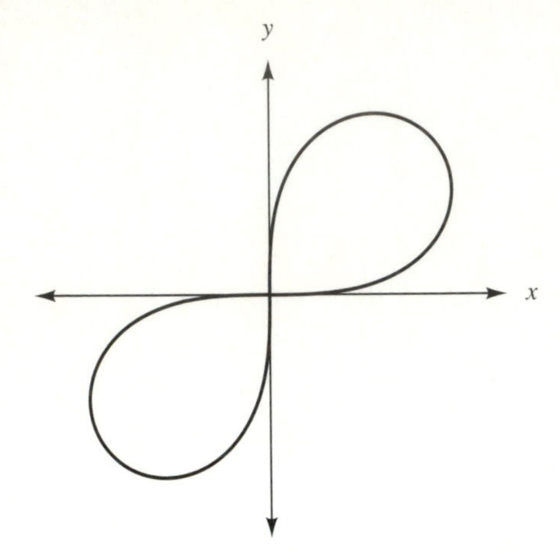

20.

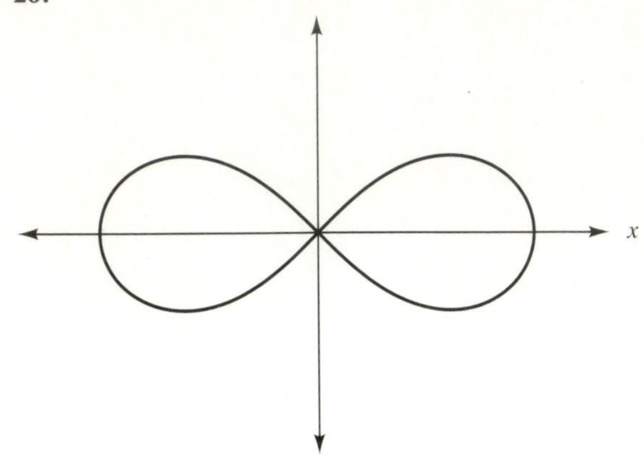

21.

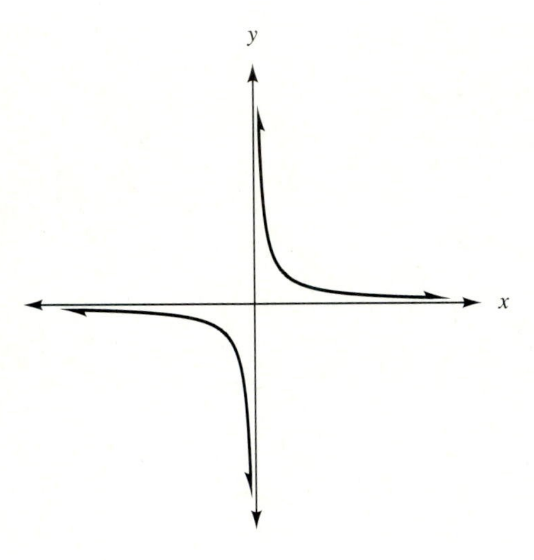

22.

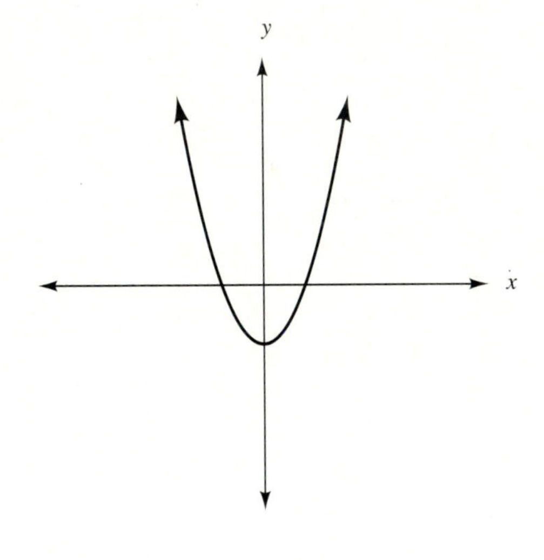

23.

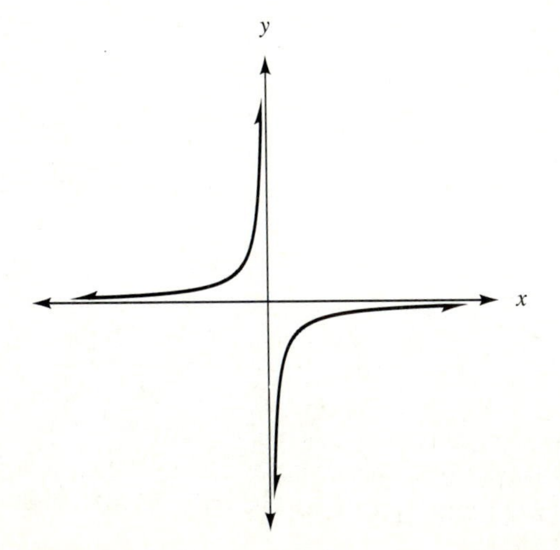

24.

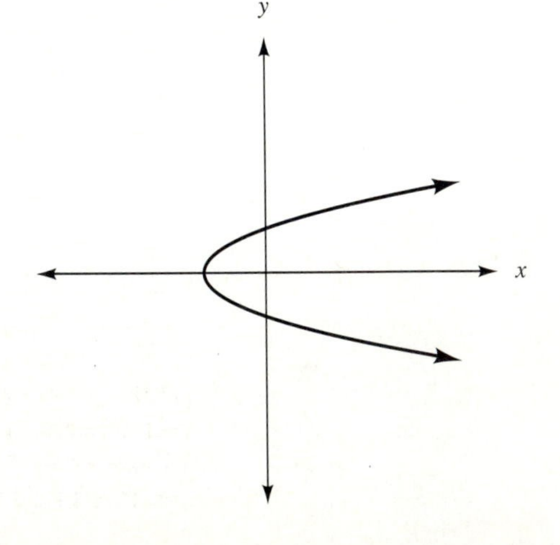

25.

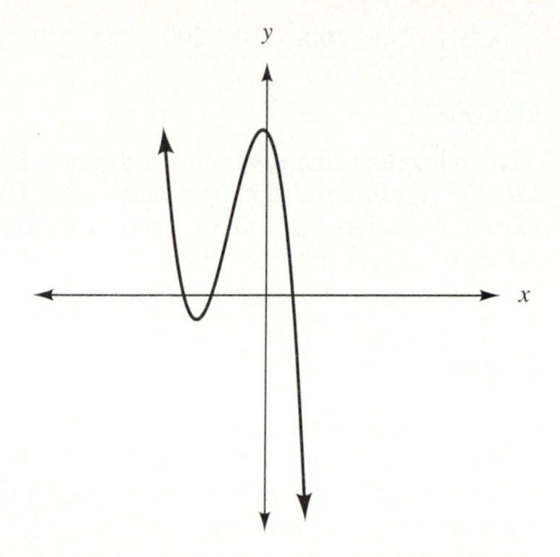

26.

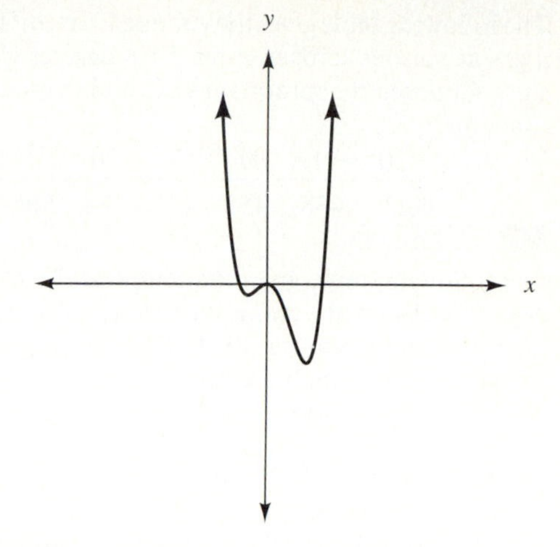

27.

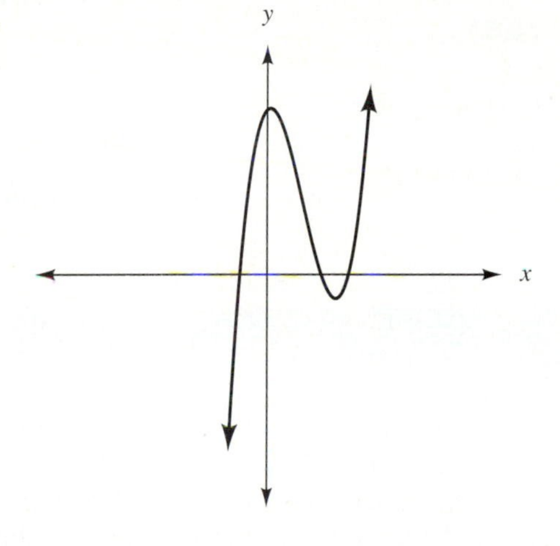

28.

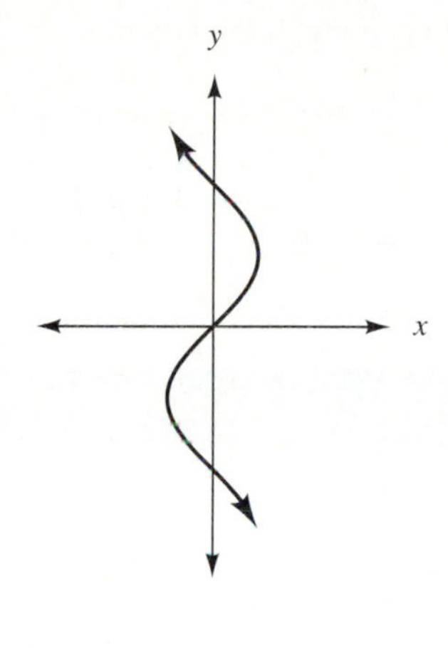

29.

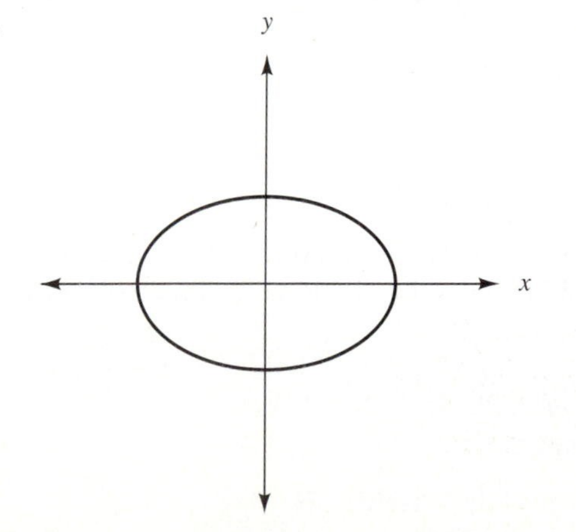

30.

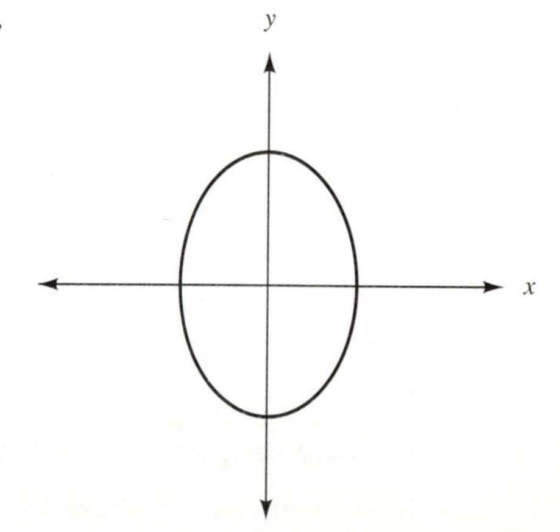

31. The following table gives the volume V (in cm^3) of a gas at various temperatures T (in degrees Celsius). Choose an appropriate scale and graph the function.

T	-20	-10	0	10	20	30
V	422	438	455	472	488	505

6 8 1 2 3 5 7 9

32. The following table gives the temperature T (in degrees Celsius) at various elevations h (in m) if the ground temperature is 0° C. Choose an appropriate scale and graph the function.

h	100	500	1000	1500	2000
T	-1	-5	-10	-15	-20

6 8 1 2 3 5 7 9

33. The following table gives the approximate distance d (in km) that can be seen from a building or tower at various heights h (in km). Choose an appropriate scale and graph the function.

h	50	100	150	200	250	300
d	24	35	43	50	56	61

1 2 4 7 8 3 5

34. The following table gives the resistance R (in Ω/1000 ft) of annealed copper wire at 20° C for various diameters d (in mils). Choose an appropriate scale and graph the function.

d	460.0	324.9	229.4	162.0
R	0.0490	0.0983	0.1970	0.3951

3 5 9 1 2 4 6 7 8

35. The following table gives the pressure P (in kg/cm^2) on an object submerged in sea water at various depths d (in m). Choose an appropriate scale and graph the function.

d	0.00	1.00	2.00	3.00	4.00	5.00
p	1.03	1.13	1.24	1.34	1.44	1.54

2 4 8 1 5 6 7 9

---

1. Architectural Technology
2. Civil Engineering Technology
3. Computer Engineering Technology
4. Mechanical Drafting Technology
5. Electrical/Electronics Engineering Technology
6. Chemical Engineering Technology
7. Industrial Engineering Technology
8. Mechanical Engineering Technology
9. Biomedical Engineering Technology

3.3 LINEAR FUNCTIONS

You have seen that when there are just a few values in the domain of a function, we list them in a table, then plot each ordered pair as a point on the graph. If the domain has an infinite number of values, it is impossible to list every ordered pair. Such functions are usually specified by an equation. When we examine a sample of values, a pattern emerges. Thus, we determine what the graph looks like between the few points that we examine.

EXAMPLE 1 Sketch the graph of $y = 3x - 1$.

Solution

1. We choose four values for x, say -1, 0, 1, and 2.
2. The corresponding values for y are $3(-1) - 1$ or -4; $3(0) - 1$ or -1; $3(1) - 1$ or 2; and $3(2) - 1$ or 5. That is, $(-1, -4)$, $(0, -1)$, $(1, 2)$, and $(2, 5)$ are points on the graph.
3. We plot these points on the graph in Figure 3.15.
4. The points appear to fall along a straight line.
5. We connect the points with a line from left to right. ■

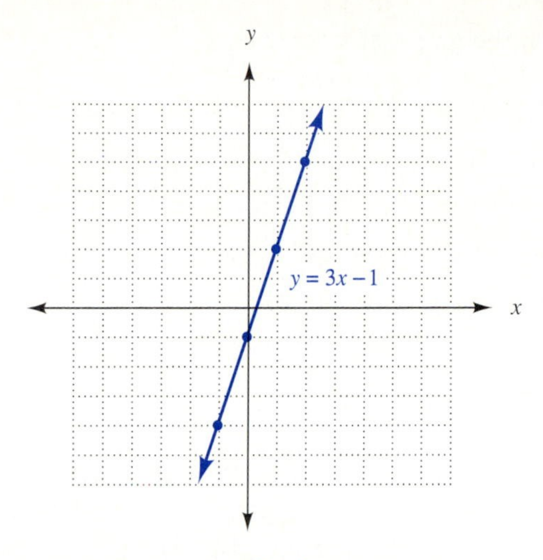

FIGURE 3.15

Any function that fits the pattern $f(x) = mx + b$ (or $y = mx + b$), where m and b are real numbers, is called a **linear function,** because the graph is a straight line. Since a straight line is determined by two points, it is not necessary to plot a lot of points in order to sketch the graph. ***It is a good idea, however, to plot at least three points.*** If you know the function is linear, but the points don't fall on a straight line, you know you have made a mistake and you can find and correct it.

linear function

The numbers m and b in the equation $y = mx + b$ reveal a lot about the graph of this function. When the value of x is 0, the value of y is b. The line crosses the y-axis at the point $(0, b)$, called the *y-intercept.* Although the y-intercept is actually a point, it is traditional to call the value b the y-intercept. The value of m is called the *slope.* It is a measure of the steepness of the line. ***The larger the absolute value of the slope, the more nearly vertical the line is; the closer the slope is to zero, the more nearly horizontal the line is.***

y-intercept

slope

When the slope is written as a fraction, the numerator tells how many units y changes when x changes by the number of units in the denominator. The y-intercept is one point on the line, so the slope and y-intercept can be used to sketch the graph of a linear function.

To sketch the graph of a linear function $y = mx + b$:

1. Identify the slope (m) and y-intercept (b). Write the slope as a fraction.
2. Plot the y-intercept at $(0, b)$. Starting there,
   (a) Use the numerator of the slope to count up (if it is positive) or down (if it is negative).
   (b) Use the denominator of the slope to count over to the right (if it is positive) or to the left (if it is negative).
3. Plot a second point at the position found in step 2.
4. Connect the two points to graph the remaining values of the function.

EXAMPLE 2 Sketch the graph of $y = -x - 2$.

Solution

1. The slope is -1 or $-1/1$ and the y-intercept is -2.
2. The y-intercept is at $(0, -2)$. From that point we count down 1 and over 1 to the right to get $(1, -3)$.
3. Plot the point $(1, -3)$.
4. Connect the two points.

The graph is shown in Figure 3.16.

Alternate Solution

1. The slope is -1 and the y-intercept is -2.
2. The negative sign in the slope may go with either numerator or denominator. If we put it with the denominator $(1/-1)$, we would count (from the y-intercept) up 1 and to the left 1 to get $(-1, -1)$.
3. Plot the point $(-1, -1)$.
4. Connect the two points. ■

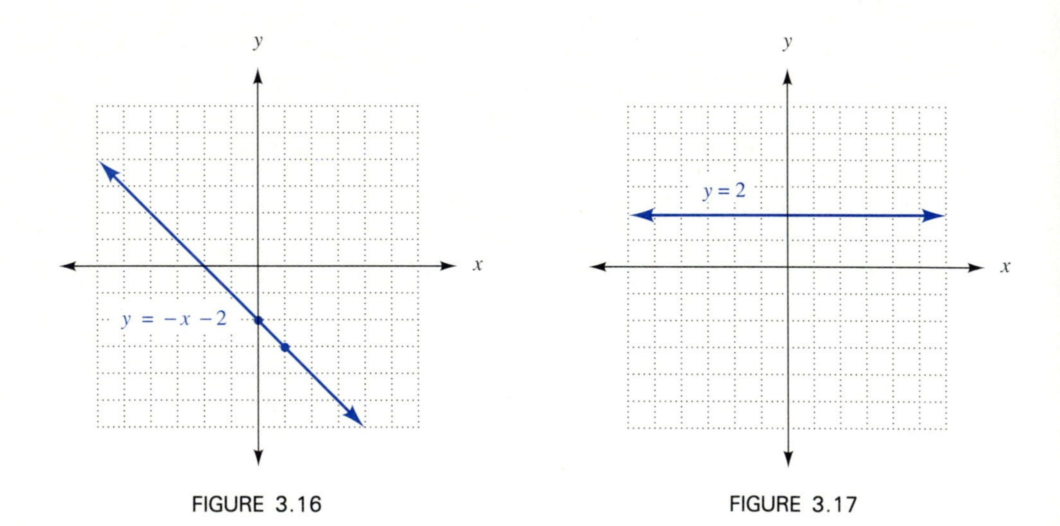

FIGURE 3.16

FIGURE 3.17

EXAMPLE 3 Sketch the graph of $y = 2$.

Solution Since the "x" term is missing, we must have $m = 0$ (i.e., $y = 0x + 2$). A slope of 0 indicates a horizontal line. The y-intercept is 2. In fact, for any value of x, the y-value will be 2. The graph is shown in Figure 3.17. ■

EXAMPLE 4 Sketch the graph of $x = 2$.

Solution Although we may write $x + 0y = 2$, the equation cannot be written in the form $y = mx + b$. It is impossible to change the coefficient of y from "0" to "1." We say the slope is "undefined." The graph will be a vertical line. Since an x-value of 2 has an infinite number of y-values associated with it, the equation is not a function. When x is 2, the entire set of real numbers is associated with x. The graph is shown in Figure 3.18. ■

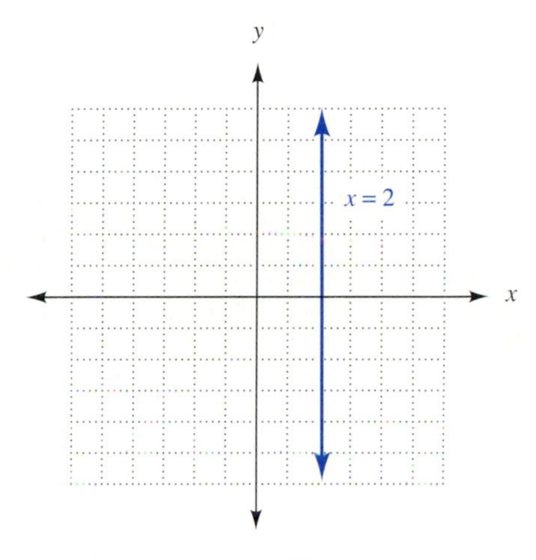

FIGURE 3.18

In technical applications, the variables usually are not x and y. Since the horizontal axis is used for the independent variable, and the vertical axis for the dependent variable, we say that the graph shows the dependent variable versus the independent variable. It is often necessary to choose appropriate scales for the axes.

EXAMPLE 5 The equation $V = 0.707\ V_m$ gives the effective voltage across a power line if V_m is the maximum voltage. Graph V versus V_m for $120 \leq V_m \leq 240$.

Solution The independent variable is V_m. Since V_m takes on values between 120 and 240, we choose to let each box on the horizontal axis represent 10. The dependent variable will range from 0.707(120) or 84.8 to 0.707(240) or about 170. We let each box on the vertical axis represent 10 also. Since b is 0 in the equation $V = 0.707\ V_m$, the graph crosses the vertical axis at the origin. The scale was not chosen to show the origin, however, so we need some other points on the graph. We have already determined that when the value of V_m is 120, the value of V is 84.8, and when the value of V_m is 240, the value of V is 170. We plot the points (120, 84.8) and (240, 170) and connect them. As a check, we can use the slope to find additional points on the graph. It is difficult to count up 0.707 and over 1 with the scale we are using, but 0.707 is approximately 7/10. Thus, a point that is up approximately 7 and over 10 from (120, 84.8) should be on the graph. The graph is shown in Figure 3.19. ■

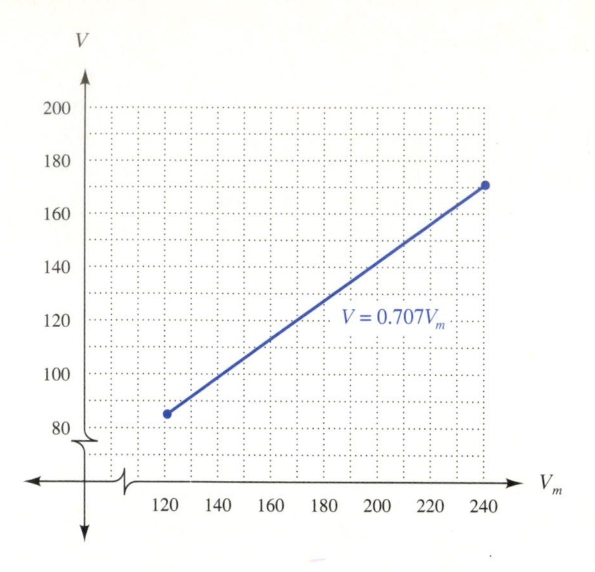

FIGURE 3.19

Linear functions are not always written in the form $y = mx + b$ as the examples have been, but they can be put in this form.

EXAMPLE 6 Sketch the graph of $3y - 2x = 1$.

Solution Solve the equation for y.

$$3y = 2x + 1$$

$$y = \frac{2x}{3} + \frac{1}{3}$$

1. $m = 2/3$ and $b = 1/3$.
2. Plot the y-intercept at (0, 1/3).
3. Count up two units and over three units from (0, 1/3) to (3, 2 1/3).
4. Connect the points.

The graph is shown in Figure 3.20. ■

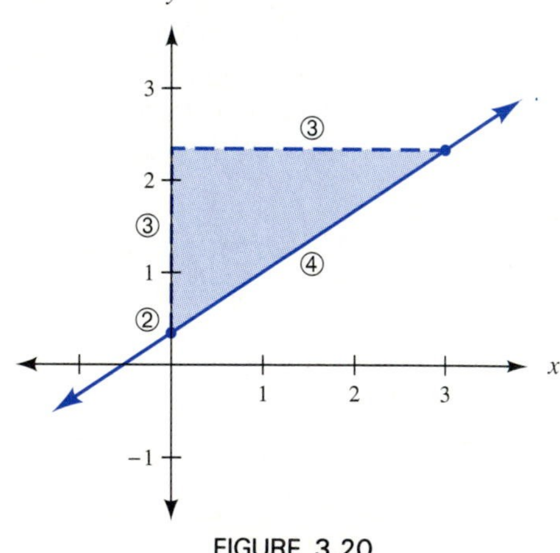

FIGURE 3.20

Another technique for graphing a linear equation is useful when the equation is in the form $ax + by = c$. That technique is to use the x-intercept and y-intercept as two points that determine the line. NOTE ***When x is 0, the line crosses the y-axis, giving the y-intercept. When y is 0, the line crosses the x-axis, giving the x-intercept.***

EXAMPLE 7 Sketch the graph of $2x + 3y = 6$.

Solution When x is 0, $2(0) + 3y = 6$ When y is 0, $2x + 3(0) = 6$

$$3y = 6 \qquad 2x = 6$$
$$y = 2 \qquad x = 3$$

Thus, (0, 2) and (3, 0) are two points on the line. A third arbitrary point may be used as a check. If x is 1, we have

$$2(1) + 3y = 6$$
$$2 + 3y = 6$$
$$3y = 4$$
$$y = \frac{4}{3}$$

The graph is shown in Figure 3.21. ■

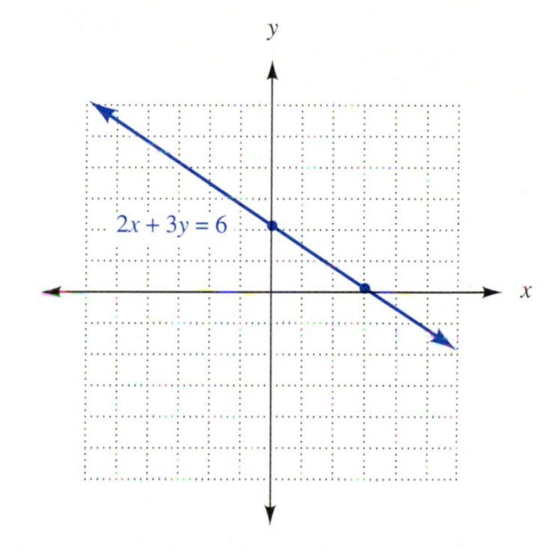

FIGURE 3.21

EXERCISES 3.3

In 1–14, sketch the graph of each equation.

1. $y = 2x + 3$
2. $y = -x - 2$
3. $y = (1/2)x + 3$
4. $y = (2/3)x - 1$
5. $y = (-3/5)x + 4$
6. $y = 1$
7. $y = -1$
8. $x = -1$
9. $x = 1$
10. $f(x) = 2x$
11. $f(x) = -x$
12. $f(x) = 3x - 2$
13. $f(x) = x + 3$
14. $f(x) = -2x + 5$

In 15–24, identify the slope and y-intercept of each function.

15. $2x - y = 4$
16. $2x + 3y = 6$
17. $x = 2y + 1$
18. $3y = 2x - 6$
19. $3x - y = 7$
20. $4x - 5y = 3$
21. $x = 2y$
22. $y = 3$
23. $y = 4$
24. $x = -y$

In 25–34, graph each equation.

25. $x - y = 3$
26. $2x - 3y = 9$
27. $2y = 3x + 4$
28. $2y - x = 3$
29. $x - 2y = 4$
30. $2x - y = 3$
31. $x + y = 5$
32. $3x + y = -1$
33. $2x = 3$
34. $2y = 3$

---

35. The formula $T = 20 - h/100$ gives an approximation for the temperature T (in degrees Celsius) at an elevation h (in m) if the ground temperature is 20° C and $h \leq 12\ 000$ m. Sketch the graph of T versus h. If the height changes 300 m, how much does the temperature change?
6 8 | 1 2 3 5 7 9

36. The formula $C = (5/9)(F - 32)$ gives the Celsius temperature when F is the Fahrenheit temperature. Sketch the graph of C versus F.
6 8 | 1 2 3 5 7 9

37. The formula $v = 88 + t$ gives the velocity v (in ft/s) after time t (in seconds) of a car accelerating at the rate of 1 ft/s² from a speed of 88 ft/s. Sketch the graph of v versus t.
4 7 8 | 1 2 3 5 6 9

38. If a 3-ohm resistor and a variable resistor R are connected in series, and a current of 2 A flows through the circuit, the voltage (in volts) is given by $V = 2(R + 3)$. Sketch the graph of R versus V.
3 5 9 | 1 2 4 6 7 8

39. Suppose a straight line is to be sketched by plotting three points, but an error is made in computing the y-coordinate of one of them. How many different straight lines could be drawn by connecting the points two at a time?

---

1. Architectural Technology
2. Civil Engineering Technology
3. Computer Engineering Technology
4. Mechanical Drafting Technology
5. Electrical/Electronics Engineering Technology
6. Chemical Engineering Technology
7. Industrial Engineering Technology
8. Mechanical Engineering Technology
9. Biomedical Engineering Technology

3.4 GRAPHS OF FUNCTIONS AND THEIR INVERSES

Nonlinear equations may be graphed by finding the values of the dependent variable associated with a few values of the independent variable. The ordered pairs are then plotted and the rest of the graph can be sketched by connecting the points. Example 1 illustrates the technique.

EXAMPLE 1 Sketch the graph of $f(x) = x^2 - 5x + 6$.

Solution Choose 5 values for x: $-2, -1, 0, 1$, and 2. Find the corresponding values of $f(x)$.

$$f(-2) = (-2)^2 - 5(-2) + 6 = 4 + 10 + 6 = 20$$
$$f(-1) = (-1)^2 - 5(-1) + 6 = 1 + 5 + 6 = 12$$
$$f(0) = 0^2 - 5(0) + 6 = 0 - 0 + 6 = 6$$
$$f(1) = 1^2 - 5(1) + 6 = 1 - 5 + 6 = 2$$
$$f(2) = 2^2 - 5(2) + 6 = 4 - 10 + 6 = 0$$

These points are plotted on the graph in Figure 3.22.

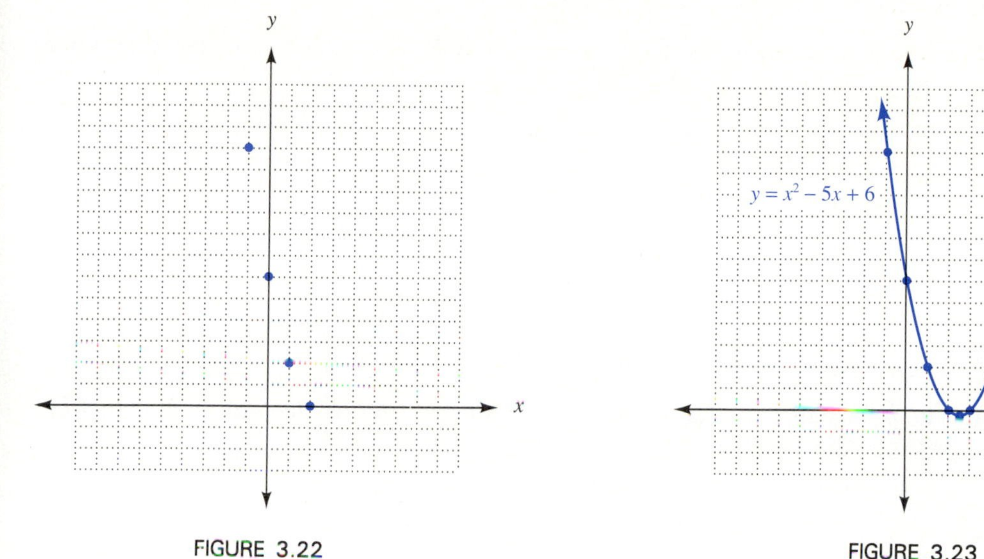

FIGURE 3.22

FIGURE 3.23

The graph appears to be bending, so we examine a couple of points with x-values to the right of 2. We have $f(3) = 0$, and $f(4) = 2$. Since $f(2)$ is 0 and $f(3)$ is 0, we examine $f(2.5)$ in order to determine whether the graph should go straight between these points or dip below the axis and gradually curve back up. Since $f(2.5)$ is -0.25, the graph curves gradually. We connect the points with a smooth curve from left to right as in Figure 3.23. ■

To sketch the graph of a function given by $y = f(x)$:

1. Choose four or five values for the independent variable.
2. Substitute these values into the equation $y = f(x)$ to find the corresponding values of the dependent variable.
3. Plot the points determined in steps 1 and 2.
4. Imagine connecting the points with a smooth curve from left to right.
   (a) If the function is undefined for some value of the independent variable, leave a gap in the graph at that point.
   (b) If the shape of the graph seems to change suddenly, evaluate the function for some value of x between the two points in question.
   (c) If the shape of the graph is not clear, repeat steps 1 – 3, using new values.
5. Draw a smooth curve connecting the points from left to right.

EXAMPLE 2 Sketch the graph of $y = x^3 - x^2 - 2x + 2$.

Solution

1. We choose $-2, -1, 0, 1$, and 2 for x.
2. For convenience, we list these values in a table, and calculate the corresponding function values.

x	$x^3 - x^2 - 2x + 2$	y
-2	$(-2)^3 - (-2)^2 - 2(-2) + 2$	$-8 - 4 + 4 + 2 = -6$
-1	$(-1)^3 - (-1)^2 - 2(-1) + 2$	$-1 - 1 + 2 + 2 = 2$
0	$0^3 - 0^2 - 2(0) + 2$	$0 - 0 + 0 + 2 = 2$
1	$1^3 - 1^2 - 2(1) + 2$	$1 - 1 - 2 + 2 = 0$
2	$2^3 - 2^2 - 2(2) + 2$	$8 - 4 - 4 + 2 = 2$

3. These points are plotted on the graph in Figure 3.24.
4. The points appear to fall along a smooth curve, but we need more information between the x-values of -1 and 0 and between 1 and 2.

$f(-1/2) = (-1/2)^3 - (-1/2)^2 - 2(-1/2) + 2 = -1/8 - (1/4) + 1 + 2 = 2\ 5/8$

$f(3/2) = (3/2)^3 - (3/2)^2 - 2(3/2) + 2 = 27/8 - 9/4 - 3 + 2 = 1/8$

5. We draw the curve from left to right. ■

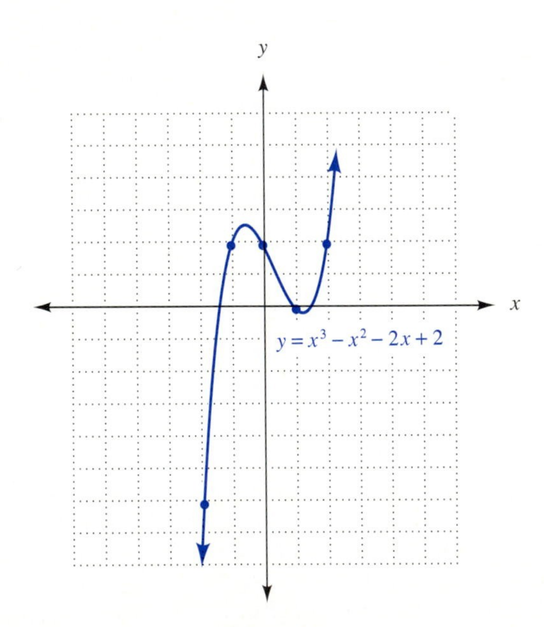

FIGURE 3.24

EXAMPLE 3 Sketch the graph of $y = \sqrt{x+2}$.

Solution

1. Since $x + 2 = 0$ when x is -2, a value smaller than -2 would lead to a negative number under the radical. Thus, the domain is $x \geq -2$. We choose -2, 0, 2, 5, and 7 for x.
2. We list these values in a table and calculate the corresponding function values.

x	$\sqrt{x+2}$	y
-2	$\sqrt{-2+2}$	$\sqrt{0} = 0$
0	$\sqrt{0+2}$	$\sqrt{2}$
2	$\sqrt{2+2}$	$\sqrt{4} = 2$
5	$\sqrt{5+2}$	$\sqrt{7}$
7	$\sqrt{7+2}$	$\sqrt{9} = 3$

3. These points are plotted on the graph in Figure 3.25.
4. The points seem to fall along a smooth curve.
5. We connect the points with a smooth curve from left to right. ■

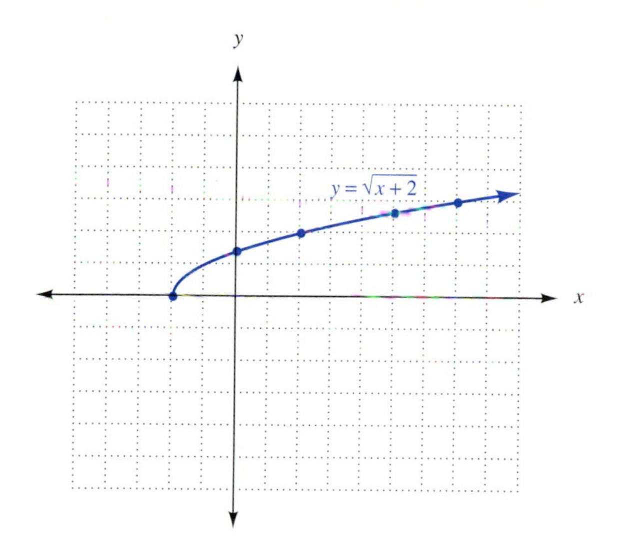

FIGURE 3.25

When working applied problems, there may be values that are excluded from the domain because they are physically impossible. Also, it may be necessary to use different scales on the horizontal and vertical axes in order to accommodate the data.

EXAMPLE 4 The equation $s = -4.9t^2 + 100$ gives the distance s (in meters) above the ground after t seconds of an object released from a height of 100 m. Sketch the graph of the equation.

Solution The elapsed time must be nonnegative. We evaluate the function for several values of t. When t is 5, s is -22.5; so the object hits the ground (that is, $s = 0$) between 4 and 5 seconds after its release.

t	0	1	2	3	4	5	
s	100	95.1	80.4	55.9	21.6	-22.5	(impossible)

We choose an appropriate scale for each axis and sketch the graph, as shown in Figure 3.26. ■

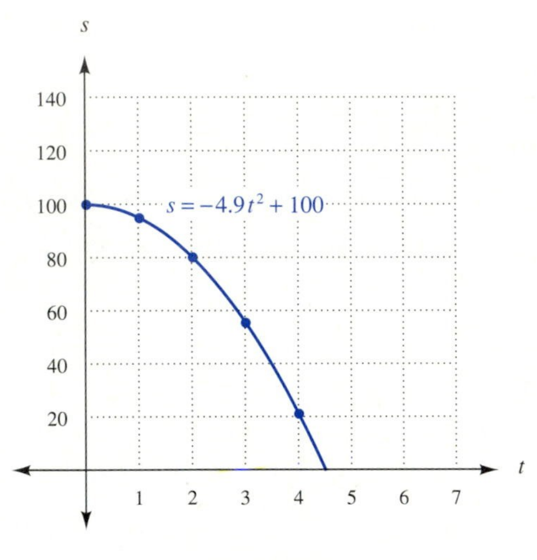

FIGURE 3.26

continuous variable

discrete variable

Variables that are allowed to take on any value between two given values are called **continuous variables.** Variables that may change only by certain increments are called **discrete variables.** When the variables are continuous, the points of the graph are connected with a smooth curve. When the independent variable is discrete, the graph will consist of a series of isolated points. (The points are sometimes connected by straight line segments to make the graph easier to read.) ***When graphing a series of measurements obtained by observation rather than from an equation, it is important to consider whether the variables are continuous or discrete.*** If the variables are continuous, it may be possible to determine what the graph looks like between the given points. That is, the points may be connected by a smooth curve.

NOTE ⇨⇨

EXAMPLE 5 The following table shows the correspondence between the number n of 3 μF capacitors in series and the equivalent capacitance C (in μF). Sketch the graph.

n	1	2	3	4	5	6
C	3.0	1.5	1.0	0.75	0.60	0.50

Solution The independent variable (n) can only take on values that are nonnegative integers. An appropriate scale must be used for each axis. Both axes are marked in intervals of 1, as in Figure 3.27. ■

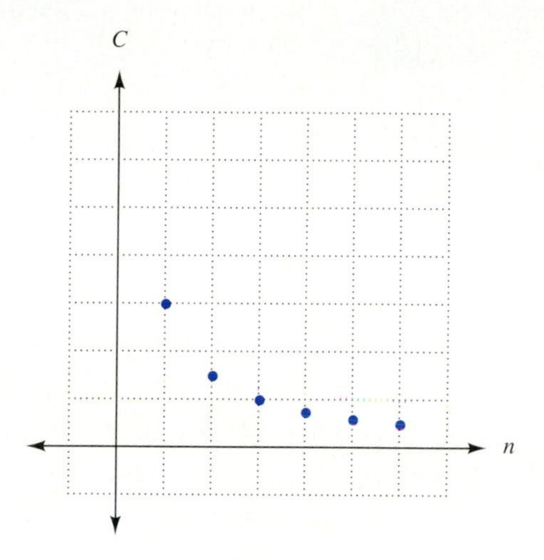

FIGURE 3.27

Recall that operations such as addition and subtraction are called inverse operations because one operation "undoes" what the other "does." Likewise, inverse functions have the property that one function undoes what the other does. In order to fully understand this idea, it is necessary to review function notation. The argument of a function need not be a constant or single variable. It may be an algebraic expression, as in the following examples.

EXAMPLE 6 **(a)** If $g(x) = x - 1$, find $g(x^2)$.

(b) If $f(x) = 2x^2 + 3x$, find $f(x + 1)$.

(c) If $g(x) = (3x - 3)^2$, find $g(2x + 1)$.

Solution **(a)** $g(\mathbf{x^2}) = \mathbf{x^2} - 1$

(b)
$$\begin{aligned} f(\mathbf{x+1}) &= 2(\mathbf{x+1})^2 + 3(\mathbf{x+1}) \\ &= 2(x^2 + 2x + 1) + 3x + 3 \\ &= 2x^2 + 4x + 2 + 3x + 3 \\ &= 2x^2 + 7x + 5 \end{aligned}$$

(c)
$$\begin{aligned} g(\mathbf{2x+1}) &= [3(\mathbf{2x+1}) - 3]^2 \\ &= [6x + 3 - 3]^2 \\ &= [6x]^2 \\ &= 36x^2 \end{aligned}$$
■

When the argument of a function is itself a function, we have a *composition of functions.*

DEFINITION The **composition of functions f and g,** denoted $f \circ g(x)$, is defined by

$$f \circ g(x) = f[g(x)]$$

for all x in the domain of g such that $g(x)$ is in the domain of f.

EXAMPLE 7 If $f(x) = \sqrt[3]{x+2}$, and $g(x) = x^3 - 2$, simplify the following expressions.

(a) $f \circ g(x)$

(b) $g \circ f(x)$

Solution **(a)** $f \circ g(x) = f[g(x)]$. That is, $g(x)$ or $x^3 - 2$, is the argument of the function f. Thus, $x^3 - 2$ replaces x in the equation $f(x) = \sqrt[3]{x+2}$. Then $f(x^3 - 2) = \sqrt[3]{(x^3 - 2) + 2} = \sqrt[3]{x^3} = x$.

(b) $g \circ f(x) = g[f(x)]$. That is, $f(x)$ or $\sqrt[3]{x+2}$ is the argument of the function g. Thus, $\sqrt[3]{x+2}$ replaces x in the equation $g(x) = x^3 - 2$. Then $g(\sqrt[3]{x+2}) = (\sqrt[3]{x+2})^3 - 2 = (x+2) - 2 = x$. ■

In Example 7, notice that we have $f \circ g(x) = x$. That is, when we find the value associated with x by g, then find the value associated with $g(x)$ by f, we get x again. Whatever is done to x by g is undone by f. Likewise $g \circ f(x) = x$.

DEFINITION Two functions f and g are said to be **inverse functions** if

$f \circ g(x) = x$ for all x in the domain of g, and

$g \circ f(x) = x$ for all x in the domain of f.

EXAMPLE 8 Determine whether the functions f and g are inverse functions.

(a) $f(x) = 2x$ and $g(x) = (1/2)x$

(b) $f(x) = 2x + 3$ and $g(x) = 3x - 2$.

Solution **(a)** $f \circ g(x) = f[g(x)] = f[(1/2)x] = 2[(1/2)x] = x$

$g \circ f(x) = g[f(x)] = g(2x) = (1/2)(2x) = x$

Since $f \circ g(x) = x$, and $g \circ f(x) = x$, the functions are inverse functions.

(b) $f \circ g(x) = f[g(x)] = f(3x - 2) = 2(3x - 2) + 3 = 6x - 4 + 3 = 6x - 1$. Since $f \circ g(x) \neq x$, the two functions cannot be inverse functions. ■

When we compare a function f with its inverse function g, we see that the roles of the dependent and independent variables are interchanged. When we evaluate $f \circ g(x)$, x is the independent variable for g. Then $g(x)$, which was the dependent variable for g, becomes the independent variable for f. ***When the graphs of two inverse functions are plotted on the same coordinate plane, it is easy to see that the independent and dependent variables have been interchanged, for one graph is the mirror image of the other about the line $y = x$.*** Often, the mirror image of the graph of a function does not represent a function. Such a graph is said to represent the *inverse of the function,* even though it cannot be called an *inverse function.*

NOTE

EXAMPLE 9 Sketch the graphs of $f(x) = 2x - 1$ and $g(x) = (1/2)(x + 1)$.

Solution The graphs are shown in Figure 3.28. Since $f \circ g(x) = x$, and $g \circ f(x) = x$, the functions are inverse functions. ■

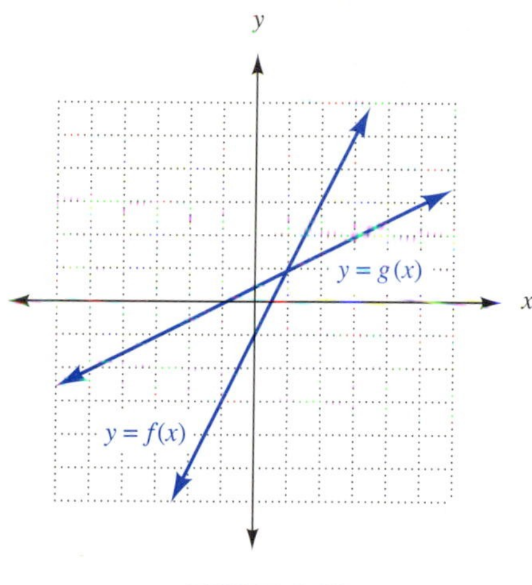

FIGURE 3.28

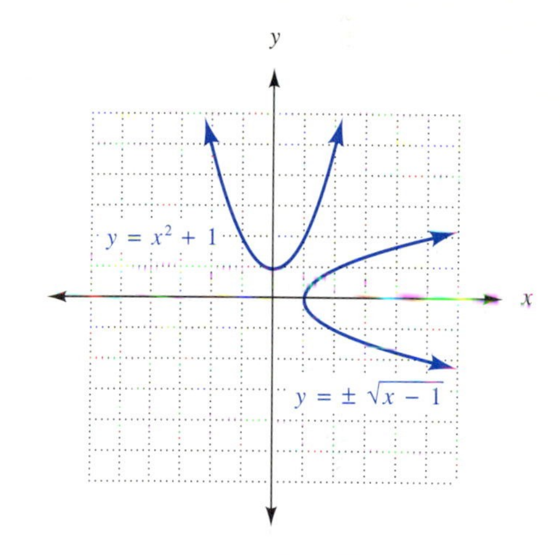

FIGURE 3.29

EXAMPLE 10 Sketch the graphs of $y = x^2 + 1$ and $y = \pm\sqrt{x - 1}$ on the same coordinate plane.

Solution The graphs are shown in Figure 3.29. Notice that $y = \pm\sqrt{x - 1}$ does not define a function. It is correct to say that $y = \pm\sqrt{x - 1}$ defines the *inverse of the function* given by $y = x^2 + 1$, but it does not give the *inverse function* of $y = x^2 + 1$. ■

EXERCISES 3.4

In 1–22, sketch the graph of each equation.

1. $y = x^2 + 7x + 12$
2. $y = -x^2 + 2x + 8$
3. $y = x^2 - 1$
4. $y = x^2 - 9$
5. $y = -x^2 + 4x$
6. $f(x) = x^2 - 6x$
7. $f(x) = 4x^2 - 4x + 1$
8. $f(x) = 4x^2 - 4x - 3$
9. $f(x) = -2x^2 + 3x + 1$
10. $f(x) = -2x^2 - 3x - 1$
11. $y = x^3 + x^2 - 9x - 9$
12. $y = x^3 - 2x^2 - 5x + 6$
13. $y = x^3 - 2x + 1$
14. $f(x) = x^4 - 10x^2 + 9$
15. $f(x) = 9x^4 - 13x^2 + 4$
16. $f(x) = x^4 - 5x^3 + 5x^2 + 5x - 6$
17. $y = \sqrt{x + 4}$
18. $f(x) = \sqrt{x - 4}$
19. $y = \sqrt{-2x + 1}$
20. $y = \sqrt{2x + 1}$
21. $f(x) = \sqrt{3x - 4}$
22. $y = \sqrt{x^2 + 1}$

In 23–30, determine whether f and g are inverse functions.

23. $f(x) = 2x$ and $g(x) = \frac{1}{2}x$
24. $f(x) = x + 3$ and $g(x) = x - 3$
25. $f(x) = 2x - 1$ and $g(x) = 2x + 1$
26. $f(x) = 3x$ and $g(x) = -3x$
27. $f(x) = 3x + 2$ and $g(x) = \frac{1}{3}(x - 2)$
28. $f(x) = \frac{1}{4}x - 1$ and $g(x) = 4(x + 1)$
29. $f(x) = \sqrt{x}$ and $g(x) = x^2$ for $x \geq 0$
30. $f(x) = \sqrt{x - 1}$ and $g(x) = \sqrt{x + 1}$ for $x \geq 1$

In 31–38, sketch the graphs of f and g on the same coordinate plane and determine from the graph whether the functions are inverse functions.

31. $f(x) = 2x$ and $g(x) = \frac{1}{2}x$
32. $f(x) = x + 3$ and $g(x) = x - 3$
33. $f(x) = 2x - 1$ and $g(x) = 2x + 1$
34. $f(x) = 3x$ and $g(x) = -3x$
35. $f(x) = 3x + 2$ and $g(x) = \frac{1}{3}(x - 2)$
36. $f(x) = \frac{1}{4}x - 1$ and $g(x) = 4(x + 1)$
37. $f(x) = \sqrt{x}$ and $g(x) = x^2$, $x \geq 0$
38. $f(x) = \sqrt{x - 1}$ and $g(x) = \sqrt{x + 1}$, $x \geq 1$

---

39. The equation $s = 0.05v^2 + v$ gives the approximate distance s (in feet) required to stop a car traveling at speed v (in mph). Graph s as a function of v for $30 \leq v \leq 60$. The variables are continuous. 4 7 8 1 2 3 5 6 9

40. The equation $f = 0.92\sqrt{k}$ gives the frequency of oscillation of various springs, each supporting a mass of 30 g. Graph f as a function of k for $6 \leq k \leq 12$. The variables are continuous. 4 7 8 1 2 3 5 6 9

---

1. Architectural Technology
2. Civil Engineering Technology
3. Computer Engineering Technology
4. Mechanical Drafting Technology
5. Electrical/Electronics Engineering Technology
6. Chemical Engineering Technology
7. Industrial Engineering Technology
8. Mechanical Engineering Technology
9. Biomedical Engineering Technology

41. The equation $W = (2000/3)(bd^2/l)$ gives the safe load W (in lb) for a white pine beam supported at both ends and loaded in the middle. Graph W as a function of d, if $b = 2$ in, $l = 8$ ft, and $1.5 \le d \le 4.0$. In the formula, b is the width (in inches), d is the depth (in inches), and l is the length (in ft) of the beam. 1 2 4 8 7 9

42. The table shows the correspondence between the number n of 10-ohm resistors connected in parallel and the equivalent resistance R. Graph R as a function of n.

n	1	2	3	4	5
R	10.0	5.00	3.33	2.50	2.00

3 5 9 1 2 4 6 7 8

43. If T is the tensile strength (in tons/in^2) of steel, and X is the fraction of carbon, as shown in the table, graph T as a function of X.

X	0.46	0.66	0.80	0.96
T	33.8	40.0	45.9	52.7

1 2 4 8 7 9

44. The total execution time T for a computer program consisting of two nested loops can be expressed as $T = M(y + Nx)$. In the formula, M and N are the number of repetitions of the outer and inner loops, respectively, x is the execution time of the inner loop, and y is the additional execution time within the outer loop. Assuming that partial loops are permitted (so that the variables are continuous), sketch the graph of N versus M, using $T = 150\ \mu s$, $y = 16\ \mu s$, and $x = 12\ \mu s$.
3 5 9

45. The greatest integer function is denoted $y = [x]$. For each real value of the independent variable, the value of the dependent variable is the greatest integer that is less than or equal to x. Consider the following examples. $[3.2] = 3$, $[7.9] = 7$, and $[-4.1] = -5$. Sketch the graph of $y = [x]$. Is x a continuous variable? Is y a continuous variable?

3.5 VARIATION

You have seen that a function is a relationship between two variables. A change in the independent variable may be accompanied by a change in the dependent variable (unless the function is a constant function). For certain types of functions, we say that one variable "varies as" or "is proportional to" the other. Mathematically, we express these relationships as follows.

EQUATIONS FOR DIRECT AND INVERSE VARIATION

If y is directly proportional to x or y varies directly as x, then

$$y = kx \text{ where } k \text{ is a constant.}$$

If y is inversely proportional to x or y varies inversely as x, then

$$y = k/x \text{ where } k \text{ is a constant.}$$

direct variation

inverse variation

Notice that for *direct variation,* the two quantities change in such a way that one of them increases in absolute value as the other increases in absolute value. For *inverse variation,* one quantity increases in absolute value as the other decreases in absolute value.

EXAMPLE 1 Express each relationship mathematically.

(a) The circumference of a circle varies directly as its diameter.

(b) The area of a circle is directly proportional to the square of its radius.

(c) The resistance of a wire of fixed length varies inversely as its cross-sectional area.

Solution **(a)** $C = kd$ (The constant is π for this example, so we have $C = \pi d$.)

(b) $A = kr^2$ (The constant is π for this example, so we have $A = \pi r^2$.)

(c) $R = k/A$ ■

joint variation

Joint variation occurs when one quantity varies directly as the product of two or more quantities. Since the value of one variable depends on the values of two or more variables, we could say that the first variable is a function of the others. For example, the magnitude of the collision impact I for an automobile varies jointly as its mass m and the square of its speed s. The equation that relates these quantities is $I = kms^2$.

EQUATION FOR JOINT VARIATION

If y varies jointly as x and z, then

$\boldsymbol{y = kxz}$ where k is a constant.

Often joint and inverse variation occur together. Variation problems usually specify a set of values that occur together and require that you find what happens to one quantity as the others change.

To solve a variation problem:

1. Express the relationship mathematically.
2. Substitute a set of values that occur together into the equation.
3. Solve for the constant.
4. Rewrite the equation from step 1 using the value of the constant from step 3.
5. Find the value of the unknown quantity by substituting another set of values that occur together into the equation from step 4.

EXAMPLE 2 Given that y is directly proportional to x, and y is 6 when x is 2, find the value of y when x is 3.

Solution **1.** The equation is $y = kx$.

2. Substituting 6 for y and 2 for x, we have $6 = k(2)$.

3. We solve for k: $k = 3$.
4. The equation from step 1 becomes $y = 3x$.
5. Thus $y = 3(3) = 9$. ■

EXAMPLE 3 Given that s varies inversely as t, and $s = 5$ when $t = 2$, find t when $s = 2$.

Solution

1. The equation is $s = k/t$.
2. We have $5 = k/2$.
3. Thus, $k = 10$.
4. The equation is $s = 10/t$.
5. We have $2 = 10/t$ or $2t = 10$, so $t = 5$. ■

EXAMPLE 4 The maximum safe load of a horizontal beam supported at the ends varies jointly as the width and square of the height and inversely as the length. If an ash beam 12.0 feet long, 4.0 inches wide, and 2.0 inches high can support 180.0 pounds, what is the maximum load that can be supported by a beam 10.0 feet long, 2.0 inches wide, and 4.0 inches high?

Solution The equation is $L = kwh^2/l$.

$$180.0 = \frac{k(4.0)(2.0^2)}{12.0}, \text{ so } k = \frac{180.0(12.0)}{(2.0)^2(4.0)} \text{ or } 135$$

Then $L = \dfrac{135(2.0)(4.0)^2}{10.0}$ or 430 lb, to two significant digits. ■

Once the value of the constant k is found, it is possible to treat the variation equation as a function. Substituting values for one or more independent variables will produce the corresponding values for the dependent variable. The value for k should be rounded to the appropriate accuracy if the equation will be used in this way. If, however, you are only going to find one value of the dependent variable and you are using a calculator, the value of k should not be rounded. That is, after you find the value of k, leave it on the display for use in the remaining calculation. Rounding is done last. Example 5 illustrates this procedure.

EXAMPLE 5 The rate of heat loss through a window varies jointly as the area of the window and the temperature difference between inside and outside. If heat is lost at the rate of 256 Btu (British thermal unit) per hour through a window of area 15.0 ft^2 when the temperature difference is 15.0 Fahrenheit degrees, find the rate of heat loss through a window of area 4.00 ft^2 when the temperature difference is 4.00 Fahrenheit degrees.

Solution The equation is $H = kAT$. We have $256 = k(15.0)(15.0)$ so $k = 256/(15.0)(15.0)$.

The calculator keystroke sequence is

256 [÷] [(] 15 [×] 15 [)] [=] **1.1377778**

The new value of H is found by evaluating $H = kAT$ for $A = 4.00$ and $T = 4.00$. Leaving the value of k on the display, the calculator sequence is

[×] 4 [×] 4 [=] **18.204444**

Rounding the answer, we have 18.2 Btu. ■

EXERCISES 3.5

In 1–15, give the equation relating the variables, and solve the problem for the conditions given. Do not round off answers.

1. Suppose y varies directly as x. When $x = 2$, $y = 3$. Find y when $x = 5$.

2. Suppose s varies directly as t. When $t = 3$, $s = 4$. Find t when $s = 6$.

3. Suppose z is directly proportional to w. If $z = -10$ when $w = 12$, find w when $z = 7$.

4. Suppose w is directly proportional to the square of r. If $w = -27$ when $r = 3$, find r when $w = -12$.

5. Suppose v varies directly as the square root of u. When $u = 4$, $v = -6$. Find v when $u = 9$.

6. Suppose y varies inversely as x. When $x = 2$, $y = 3$. Find y when $x = 5$.

7. Suppose s varies inversely as t. When $t = 3$, $s = 4$. Find t when $s = 6$.

8. Suppose z is inversely proportional to w. If $z = -10$ when $w = 12$, find w when $z = 7$.

9. Suppose w is inversely proportional to the square of r. If $w = -27$ when $r = 3$, find w when $r = 2$.

10. Suppose v varies inversely as the square root of u. When $u = 4$, $v = -6$. Find v when $u = 9$.

11. Suppose z varies jointly as x and y. When $x = 2$ and $y = 3$, $z = 18$. Find z when $x = 1$ and $y = 4$.

12. Suppose r varies jointly as p and q. If $r = 6$ when $p = 3$ and $q = 4$, find r when $p = 4$ and $q = 5$.

13. Suppose w varies jointly as u and v. If $w = -10$ when $u = 20$ and $v = 30$, find w when $u = 10$ and $v = 40$.

14. Suppose s varies jointly as v and the square of t. When $t = 1$ and $v = 12$, $s = 36$. Find v when $t = 2$ and $s = 24$.

15. Suppose t varies jointly as f and the square root of g. If $t = 27$ when $f = 9$ and $g = 4$, find f when $t = 4$ and $g = 9$.

In 16–20, use a calculator to solve each problem for the given conditions. Round appropriately.

16. Suppose y varies directly as x. If $y = 17.34$ when $x = 10.03$, find y when $x = 22.68$.

17. Suppose p is inversely proportional to q. If $p = 3.88$ when $q = 2.14$, find q when $p = 4.77$.

18. Suppose w varies jointly as u and v. If $w = 7.39$ when $u = 5.27$ and $v = 3.15$, find w when $u = 9.73$ and $v = 5.31$.

19. Suppose z varies jointly as x and y and inversely as w. If $z = 7.2$ when $x = 3.1$, $y = 4.2$, and $w = 2.3$, find z when $x = 4.0$, $y = 5.1$, and $w = 2.2$.

20. Suppose s varies jointly as p and q and inversely as r. If $s = 12.6$ when $p = 4.12$, $q = 3.27$, and $r = 4.00$, find s when $p = 2.61$, $q = 2.73$, and $r = 2.00$.

---

21. The resistance of a wire is directly proportional to its length. If an 18.5 ft length of No. 38 copper wire has a resistance of 12.0 Ω, what is the resistance of a 50.0 ft length? **3 5 9** 1 2 4 6 7 8

22. The resistance of a wire of fixed length varies inversely as the square of its diameter. If a wire of diameter 2.0 mm has a resistance of 8.0 Ω, what would the resistance be if the diameter were 3.0 mm? **3 5 9** 1 2 4 6 7 8

23. When an object is dropped from rest, the vertical distance that it falls varies directly as the square of the time it falls. If an object falls 64.0 feet in 2.00 seconds, how far will it fall in 4.00 seconds? **4 7 8** 1 2 3 5 6 9

24. The volume of an ideal gas varies directly as the temperature (in degrees Kelvin) and inversely as the pressure. If the volume is 50.0 cm^3 when the temperature is 350.0° K and the pressure is 140.0 kPa, find the volume when the pressure is 70.0 kPa and the temperature is 375.0° K. **6 8** 1 2 3 5 7 9

25. The illumination of a light source is inversely proportional to the square of the distance from the source. If the illumination of a light source at 10.0 m is 17.5 lumens/m^2, what is the illumination at 5.00 m? **9** 1 2 3 5 6 7

26. The maximum load a cylindrical column can hold varies directly as the fourth power of the diameter and inversely as the square of the height. If a column of diameter 0.50 m and height 3.0 m can hold a load of 4.5 metric tons, what load can be held by a column of diameter 1.0 m and height 6.0 m? **1 2 4** 8 7 9

27. The resonant frequency of a circuit varies inversely as the square root of the product of inductance and capacitance. If the frequency is 318 Hz for an inductance of 5.00 mH and capacitance 50.0 μF, find the frequency for an inductance of 3.00 mH and capacitance 80.0 μF. **3 5 9** 4 7 8

28. The force of gravitational attraction between two objects varies directly as the product of their masses and inversely as the square of the distance between them. If two objects, each of mass 1.0 kg and separated by a distance of 1.0 m, attract each other with a force of 9.8 newtons, find the force of attraction between two objects, each of mass 2.0 kg and separated by a distance of 2.0 m. **4 7 8** 1 2 3 5 6 9

29. The area of contact of an automobile tire with the ground varies directly as the weight of the automobile and inversely as the tire pressure. If an automobile that weighs 2590 lb has a tire pressure of 26.0 psi (lb/in^2) and an area of contact equal to 24.9 in^2, find the weight of an automobile that has a tire pressure of 40.0 psi and an area of contact of 20.1 in^2. **2 4 8** 1 5 6 7 9

30. The weight of a body inside the earth is directly proportional to the distance from the center of the earth. If an object weighs 25 lb at 1000 miles from the center, how far from the earth's center would it have to be in order to weigh 75 lb? **1 2 4 6 8** 5 7 9

31. The number of vibrations made by a pendulum varies inversely as the square root of its length. A pendulum 39.1 inches long makes 1 vibration per second. How long would it have to be to make 2 vibrations per second? **4 7 8** 1 2 3 5 6 9

32. Suppose y varies directly as x, $y = y_1$ when $x = x_1$, and $y = y_2$ when $x = x_2$. Show that $y_1/y_2 = x_1/x_2$.

---

1. Architectural Technology
2. Civil Engineering Technology
3. Computer Engineering Technology
4. Mechanical Drafting Technology
5. Electrical/Electronics Engineering Technology
6. Chemical Engineering Technology
7. Industrial Engineering Technology
8. Mechanical Engineering Technology
9. Biomedical Engineering Technology

3.6 CHAPTER REVIEW

KEY TERMS

3.1 function
domain of a function
range of a function
argument of a function
independent variable
dependent variable

3.2 x-axis
y-axis
origin
quadrant
rectangular coordinates
ordered pair

3.3 linear function
y-intercept
slope

3.4 continuous variable
discrete variable
composition of functions
inverse functions

3.5 direct variation
inverse variation
joint variation

SYMBOLS

$f(x)$ the function value associated with x by the function f

$f \circ g(x)$ the composition of functions $f \circ g(x) = f[g(x)]$

RULES AND FORMULAS

The equation for a line with slope m and y-intercept b is given by

$$y = mx + b.$$

If y is directly proportional to x or y varies directly as x, then

$$y = kx \text{ where } k \text{ is a constant.}$$

If y is inversely proportional to x or y varies inversely as x, then

$$y = k/x \text{ where } k \text{ is a constant.}$$

If y varies jointly as x and z, then

$$y = kxz \text{ where } k \text{ is a constant.}$$

SEQUENTIAL REVIEW

(Section 3.1) *In 1 and 2, determine whether the correspondence in each table is a function.*

1.

x	1	−3	5	−7
y	−2	4	−6	8

2.

x	1	1	3	−3
y	2	−2	−3	3

3. The following table shows the correspondence between the diameter d (in cm) of a circle and its area A (in cm^2). Is the correspondence a function?

d	1.02	2.13	3.24	4.35	5.46
A	0.817	3.56	8.24	14.9	23.4

In 4–6, evaluate each expression if $f(x) = 2x^2$ and $g(x) = 3x + 1$.

4. $f(2)$

5. $g(-3)$

6. $f(3) - g(3)$

In 7 and 8, determine whether the correspondence is a function.

7. $y = \pm\sqrt{x - 3}$

8. $V = (4/3)\pi r^3$

9. Use a calculator to evaluate $f(3.61)$ if $f(x) = \sqrt{2.43\,x}$. Round to the accuracy of the independent variable.

(Section 3.2) *In 10–13, graph each function on a single coordinate plane.*

10.

x	−3	−2	−1
y	1	2	3

11.

x	−3	0	3	5
y	2	0	−2	−5

12.

x	2	4	6
y	5	−5	0

13.

x	−2	4	6	8
y	8	7	7	6

In 14–18, determine whether each graph represents a function.

14.

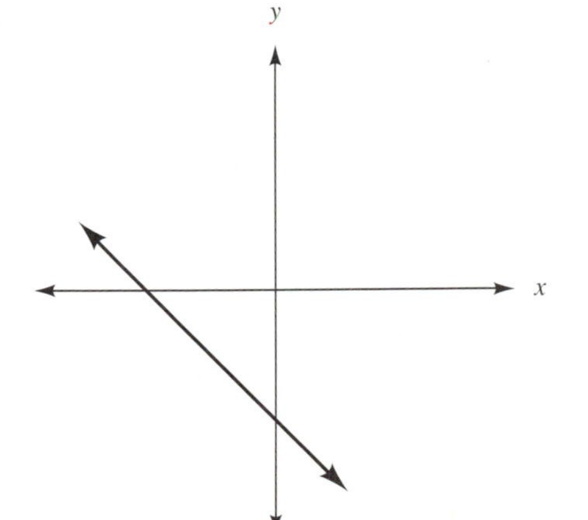

15.

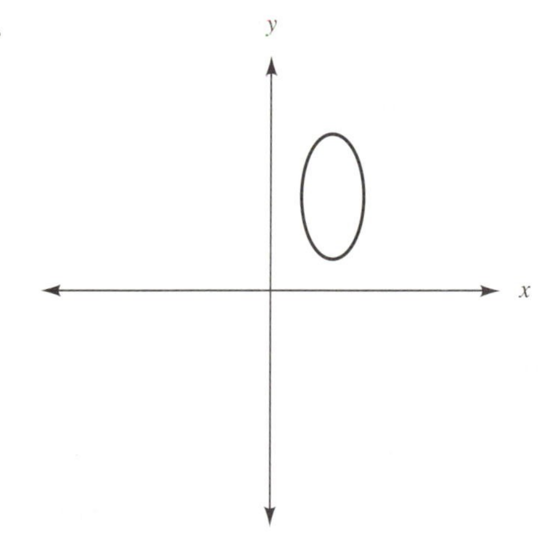

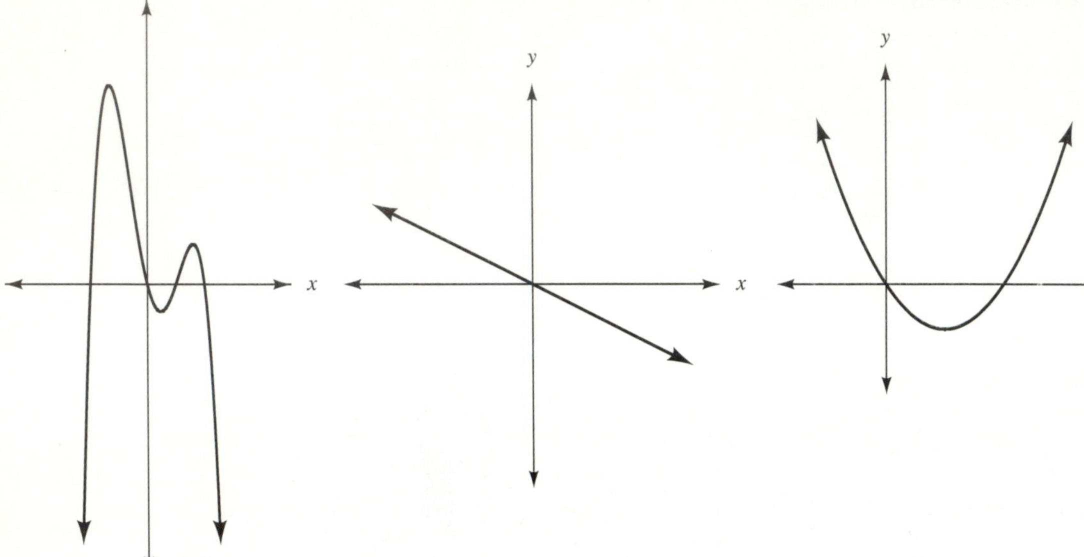

(Section 3.3) *In 19–23, sketch the graph of each equation.*

19. $y = 4x - 3$ **20.** $x + 3y = 10$ **21.** $f(x) = -3x + 1$

22. $y = 3x$ **23.** $x = 3$

In 24–27, identify the slope and y-intercept for each linear function.

24. $y = 2x + 5$ **25.** $x = 3y$ **26.** $y = -1$ **27.** $3x - 2y = 7$

(Section 3.4) *In 28 and 29, sketch the graph of each function.*

28. $y = -x^2 + 2x + 3$ **29.** $y = \sqrt{x + 3}$

In 30–33, determine whether f and g are inverse functions.

30. $f(x) = 2x$ and $g(x) = (1/2)x$ **31.** $f(x) = 3x - 1$ and $g(x) = (1/3)(x + 1)$

32. $f(x) = 5(x + 2)$ and $g(x) = (1/5)(x - 2)$ **33.** $f(x) = \sqrt{x - 3}$ and $g(x) = x^2 + 3$ for $x \geq 3$

34. Sketch the graph of $f(x) = 2x + 5$ and its inverse $g(x) = (1/2)(x - 5)$ on the same coordinate plane.

In 35 and 36, sketch the graphs of f and g on the same coordinate plane and determine from the graph whether f and g are inverse functions.

35. $f(x) = 3x - 1$ and $g(x) = (1/3)x + 1$ **36.** $f(x) = 3x + 2$ and $g(x) = 3x - 2$

(Section 3.5) *In 37–45, give the equation relating the variables and solve the problem for the conditions given. Do not round your answer.*

37. Suppose y varies directly as x. If $y = 1.6$ when $x = 0.2$, find y when $x = 0.3$.

38. Suppose u varies directly as v. If $u = 5$ when $v = 3$, find v when $u = 2$.

39. Suppose y is directly proportional to x. If $y = 1$ when $x = 3$, find y when $x = 6$.

40. Suppose z varies inversely as y. If $z = 1$ when $y = 2$, find y when $z = 2$.

41. Suppose w is inversely proportional to the square of z. If $z = 4$ when $w = 2$, find w when $z = 3$.

42. Suppose s is inversely proportional to t. If $s = 2$ when $t = 3$, find t when $s = 1$.

43. Suppose z varies jointly as y and the square of x. If $z = 16$ when $x = 2$ and $y = 4$, find z when $x = 3$ and $y = 6$.

44. Suppose w varies jointly as u and v. If $w = 4.5$ when $u = v = 1$, find u when $v = 1$ and $w = 13.5$.

45. Use a calculator to solve the problem: s varies jointly as t and the square of v. If $s = 9.2$ when $t = 2.3$ and $v = 3.4$, find s when $t = 1.8$ and $v = 4.7$.

APPLIED PROBLEMS

1. The rubbing velocity V for a continuously rotating shaft is given by $V = \pi DN/1000$, where D is the diameter (in mm) of the shaft and N is the angular velocity (in revolutions per second) of the shaft. If $D = 2.5$ mm, sketch the graph of V versus N for $1 \leq N \leq 10$. **4 7 8** 1 2 3 5 6 9

2. A computer's processing speed is sometimes tested by a program to compute Fibonacci numbers. The nth Fibonacci number is given by $F(n) = F(n - 1) + F(n - 2)$. If $F(1) = 1$ and $F(2) = 1$, find $F(3)$ and $F(4)$. **3 5** 9

3. The frequency (in hertz) of a radar wave varies inversely as the wavelength. If a wave length of 3.2 cm has a frequency of 9.4 gigahertz, find the frequency of a wave length of 3.6 cm. **5** 3 4 7 8 9

4. The weight of the air in a room varies jointly as the length, width, and height of the room. The air in a room that is $10.0' \times 12.0' \times 8.00'$ weighs 74.7 pounds. Find the weight of the air in a room that is $12.0' \times 14.0' \times 9.00'$. **1 2 4 6 8** 5 7 9

5. The oscilloscope display shown in the accompanying graph indicates voltage $v(t)$ over a period of time t (in milliseconds). Determine $v(1)$, $v(9)$, and the equation for voltage $v(t)$ as a function of time from the graph. $\left(\text{Hint: } v(0) = \dfrac{27}{4}\right)$ **3 5 9** 1 2 4 6 7 8

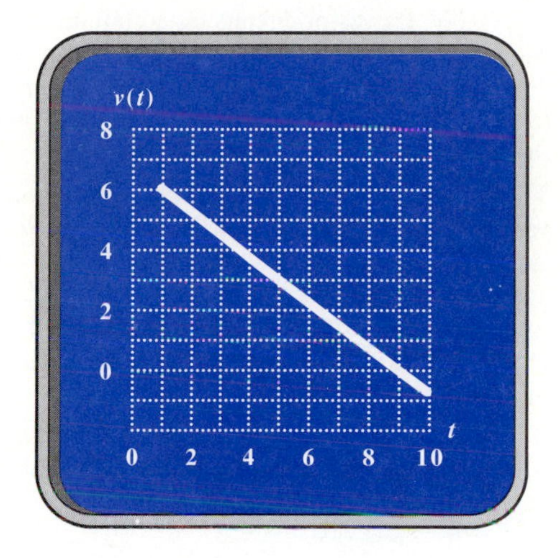

6. If 3 people can paint 3 rooms in 3 days, how long will it take 2 people to paint 2 rooms? (Hint: The number of rooms painted varies jointly as the number of people painting and the number of days spent painting.) **1 2 6 7 8** 3 5 9

7. Capacitive reactance X_c (in ohms) varies inversely as the frequency F of applied voltage. Sketch a graph of X_c versus f. **3 5 9** 4 7 8

8. A body of volume V (in ft^3) immersed in a fluid of density D (in lb/ft^3) will be buoyed up with a force F (in lb) that varies jointly as V and D. A 4 ft^3 sphere immersed in water of density 62.43 lb/ft^3 is

1. Architectural Technology
2. Civil Engineering Technology
3. Computer Engineering Technology
4. Mechanical Drafting Technology
5. Electrical/Electronics Engineering Technology
6. Chemical Engineering Technology
7. Industrial Engineering Technology
8. Mechanical Engineering Technology
9. Biomedical Engineering Technology

buoyed up with a force of 250 lb. How much force would be felt by a 3 ft^3 sphere submerged in mercury with a density of 849 lb/ft^3? Give your answer to three significant digits. 4 8 1 2 6 7 9

9. Neglecting air resistance, for bodies projected vertically upward, the greatest height reached is proportional to the square of the speed of the projection. If a bullet fired at 700 ft/s reaches a height of 7610 ft, how far would one reach if fired at 1400 ft/s? Give your answer to three significant digits. 4 7 8 1 2 3 5 6 9

10. The failure rate for transistor circuits varies jointly as temperature and humidity. If 50 of every 1000 transistor circuits tested fail on a day when the temperature is 70° and the humidity 36%, how many of every 1000 would be expected to fail on a day when the temperature is 90° and the humidity 75%? 1 2 6 7 8 3 5 9

11. The heat H (in calories) developed in a resistor varies jointly as the current I (in amps), the voltage V (in volts), and time t (in seconds). If a voltage of 24 V causes a current of 3.0 amps to flow for 16 seconds and delivers 275 calories, how many calories would 56 volts and 12 amps deliver in 48 seconds? 3 5 9 1 2 4 6 7 8

12. If 4 production lines can assemble 250 personal computers in 5 days, how many lines would be needed to assemble 2000 computers in 15 days? 1 2 6 7 8 3 5 9

13. The force of the head wind on a truck's front surface is proportional to the cube of its speed. At 55 mph a typical truck has to overcome about 700 lb of force just to maintain motion. What force would have to be overcome at 65 mph? Give your answer to two significant digits. 4 7 8 1 2 3 5 6 9

14. The horsepower H, needed for a truck to move a certain load on level ground varies jointly as the distance D and the cube of its speed v, and inversely as the time t taken. If a truck moving at 55 mph for 1 hour needs 100 Hp to move a load, how much will it take to move the same load at 65 mph for 1 hour? Give your answer to the nearest whole number. 4 7 8 1 2 3 5 6 9

15. In deep water the velocity v of an ocean wave is proportional to the square root of its wavelength λ. If a small disturbance causes waves with $\lambda = 1$ ft and sends these waves out at 1.5 mph, how fast would a large disturbance creating waves with $\lambda = 5000$ ft travel? Give your answer to three significant digits. 4 7 8 1 2 3 5 6 9

16. The sound energy contained in a volume of air in a loudspeaker varies jointly as the air's density, the frequency of the sound, and its pressure amplitude. By what percent will this energy change if the density doubles, the frequency is reduced 25%, and the amplitude stays the same? 1 3 4 5 6 8 9

17. In ac circuit theory the quality Q of a resonant circuit varies directly as the square root of the inductance to capacitance ratio $\frac{L}{C}$ and inversely as the resistance R. How much will Q change if R increases 40%, L decreases by 30%, and C increases 15%? 3 5 9 4 7 8

18. The electrical resistance of a wire is directly proportional to its length and inversely proportional to its cross-sectional area. If 900.0 meters of 2.0 mm diameter conductor have a resistance of 39 ohms, how many meters of 3.0 mm wire will have a resistance of 75 Ω? 3 5 9 1 2 4 6 7 8

19. The rungs of a ladder are 1.5 ft apart. When vertical, the shadow cast by this ladder has rungs that are 8 inches apart. The shadow of the entire ladder is 10.7 ft long. How tall is the ladder? (Hint: Use feet for all measurements.) 1 2 4 7 8 3 5

20. The time t it takes for a company's training representatives to cover the country is directly proportional to the number of stops n they must make and inversely proportional to the number of representatives R employed and the speed v at which they drive from city to city. If 5 people driving 55 mph can make 425 stops in 17 days, how many people would be needed to make 1000 stops in 10 days traveling at 65 mph? 1 2 6 7 8 3 5 9

1. Architectural Technology
2. Civil Engineering Technology
3. Computer Engineering Technology
4. Mechanical Drafting Technology
5. Electrical/Electronics Engineering Technology
6. Chemical Engineering Technology
7. Industrial Engineering Technology
8. Mechanical Engineering Technology
9. Biomedical Engineering Technology

RANDOM REVIEW

1. If $f(x) = 2x + 1$ and $g(x) = 3x - 1$, find the following function values.
(a) $f(1) + g(1)$ **(b)** $f(3) - g(2)$

2. If y varies inversely as $\sqrt{x}$, and $y = 5$ when $x = 4$, find x when $y = 10$.

3. Determine whether each correspondence is a function.

(a)

x	2	8	5	7
y	2	5	1	4

(b)

x	7	0	8	4
y	5	2	6	6

4. Sketch the graph of $x = 4$.

5. If s varies jointly as p and q, and $s = 8$ when $p = 2$ and $q = 1$, find s when $p = 3$ and $q = 4$.

6. Graph the following function.

x	5	8	9
y	2	1	4

7. Consider the equation $2y - 3x = 3$.
(a) Identify the slope and y-intercept. **(b)** Sketch the graph.

8. Determine whether f and g are inverse functions.

$$f(x) = \sqrt{2x + 1} \text{ and } g(x) = (1/2)(x^2 - 1) \text{ for } x \geq 0$$

9. Determine whether each correspondence represents a function.
(a) $y = \sqrt{x + 1}$ **(b)** $y = \pm 1/x^2$

10. Sketch the graph of $y = x^3 + 2x^2 - 1$

CHAPTER TEST

1. Determine whether each correspondence represents a function.

(a)

x	9	1	3	6
y	6	1	7	5

(b)

x	3	4	2	3
y	1	9	4	8

2. Determine whether each correspondence represents a function.
(a) $y = x^3 + 3x^2$ **(b)** $y = x^2 \pm 3$

In 3–6, sketch the graph of each function.

3. $3x - 4y = 12$

4. $2x + 5y = 7$

5. $y = 3$

6. $y = \sqrt{2x + 1}$

7. Determine whether f and g are inverse functions, for $x \geq 0$.
   (a) $f(x) = 4x - 3$ and $g(x) = (1/4)x + 3$
   (b)

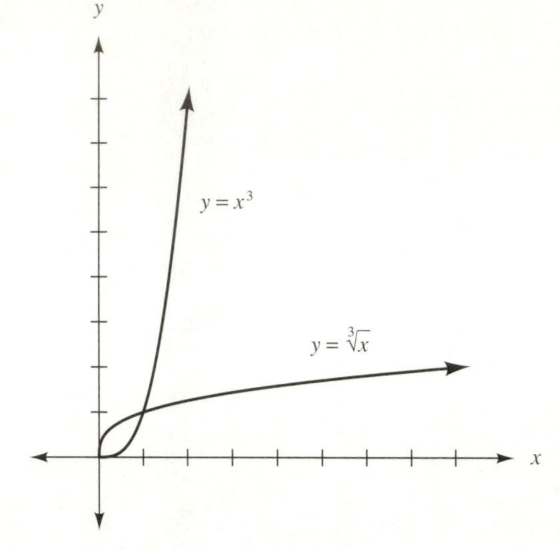

8. (a) If u varies directly as v, and $u = 8$ when $v = 3$, find u when $v = 2$.
   (b) If z varies jointly as x and y, and $z = 10$ when $x = 2$ and $y = 3$, find x when $z = 15$ and $y = 3$.

9. The angular velocity of a circular saw is inversely proportional to its diameter. If a saw has a diameter of 24.0 inches, it makes 1476 revolutions per minute. If the saw is to make 2214 revolutions per minute, what diameter should it have?

10. The amount of oxygen that a person consumes while walking varies jointly as the period of time walked and the amount of oxygen consumed per hour while at rest. A person who consumes 0.05 lb/hr while at rest walks for 2 hours and consumes 0.30 lb while walking. How much will a person who consumes 0.6 lb/hr while at rest consume while walking for 1 1/2 hours?

INTRODUCTION TO TRIGONOMETRY CHAPTER 4

The word trigonometry is derived from trigon (three-sided figure) and metre (to measure). Although it encompasses a larger body of knowledge, trigonometry includes the study of methods for determining the lengths of sides of triangles. One common application for trigonometry is to calculate distances that are difficult to measure directly by using more accessible measurements. The height of a building, for example, can be calculated from measurements made on the ground. In this chapter, we develop the techniques necessary to solve such problems.

4.1 RIGHT TRIANGLE DEFINITIONS

A triangle has six parts: three sides and three angles. If one side and two other parts are known, it is possible to find the other three parts. To solve a triangle means to find these three parts. Angles are usually labeled with upper case letters such as A, B, and C, and the side opposite each angle is labeled with the same letter, but in lower case. A right triangle

acute angle

hypotenuse legs

triangle that has a 90° angle, such as the triangle in Figure 4.1, is called a **right triangle.** For a right triangle, it is traditional to use C for the right angle and the letters A and B for the *acute* angles. An **acute angle** measures less than 90°. The side opposite the right angle, which is labeled c, is called the *hypotenuse.* The perpendicular sides are called *legs.*

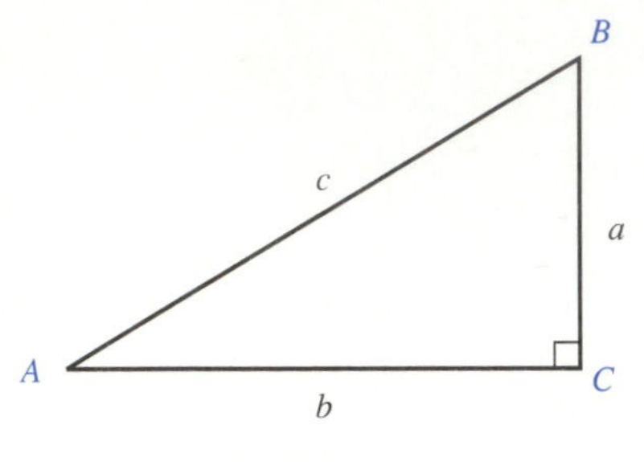

FIGURE 4.1

To solve a right triangle, it may be necessary to use the Pythagorean theorem.

THE PYTHAGOREAN THEOREM

For a right triangle with sides of length a, b, and c where c is the hypotenuse,

$$a^2 + b^2 = c^2$$

EXAMPLE 1 If a right triangle has legs with lengths 5 cm and 12 cm, find the length of the hypotenuse.

Solution We let c = length of the hypotenuse. It does not matter which leg is labeled a and which is labeled b. We arbitrarily choose $a = 5$ cm and $b = 12$ cm. Since $a^2 + b^2 = c^2$, we have

$$5^2 + 12^2 = c^2 \quad \text{or} \quad 25 + 144 = c^2$$
$$169 = c^2 \quad \text{or} \quad 13 = c.$$

(Because the length cannot be negative, $c \neq -13$.) The length is 13 cm. ■

EXAMPLE 2 If a right triangle has hypotenuse of length 50.5 in, side b of length 37.7 in, and $A = 41.7°$, solve the triangle.

Solution The sum of the angles must be 180°.

$$A + B + C = 180°$$
$$41.7° + B + 90° = 180°$$
$$131.7° + B = 180°$$
$$B = 48.3°$$

The Pythagorean theorem gives the remaining side.

$$50.5^2 = a^2 + 37.7^2 \qquad \text{or} \qquad 50.5^2 - 37.7^2 = a^2$$

Using a calculator to do the arithmetic, and rounding the answer to the same accuracy as the given data, we have 33.6 in $= a$. ■

Recall from Section 2.6 that two triangles are similar if they have equal angles, and that it is possible to set up a proportion comparing the lengths of the sides of similar triangles. The ratio comparing the lengths of two sides of one triangle is equal to the ratio comparing the lengths of the corresponding sides of the other triangle.

NOTE ⇨⇨ ***Thus, a ratio comparing two sides of a triangle does not depend on the length of the sides, but rather on the size of the angles.*** Figure 4.2 shows a pair of similar triangles.

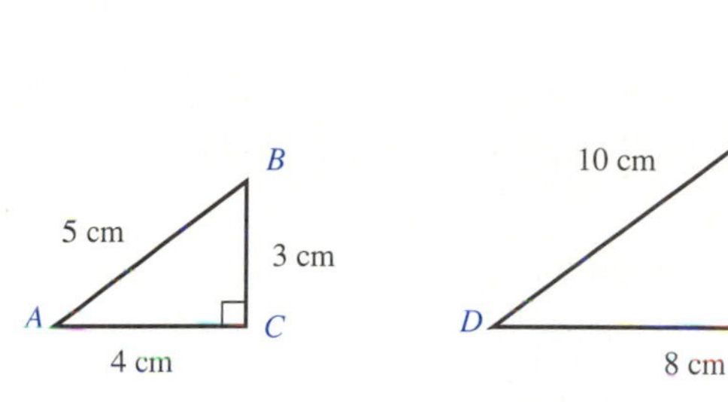

FIGURE 4.2

The hypotenuse is easily identifiable as the side opposite the 90° angle. Choosing one of the acute angles for reference, we refer to the side of the triangle across from the angle as opposite that angle and the side next to the angle as adjacent to it. There are six ratios associated with that angle. If we call the angle θ, each of these associations is, in fact, a function of θ. These six functions are called *trigonometric functions*. They are given special names instead of function labels like f and g. They are defined as follows.

trigonometric functions

THE TRIGONOMETRIC FUNCTIONS

$$\text{sine } \theta = \frac{\text{length of opposite side}}{\text{length of hypotenuse}} \qquad \text{cotangent } \theta = \frac{\text{length of adjacent side}}{\text{length of opposite side}}$$

$$\text{cosine } \theta = \frac{\text{length of adjacent side}}{\text{length of hypotenuse}} \qquad \text{secant } \theta = \frac{\text{length of hypotenuse}}{\text{length of adjacent side}}$$

$$\text{tangent } \theta = \frac{\text{length of opposite side}}{\text{length of adjacent side}} \qquad \text{cosecant } \theta = \frac{\text{length of hypotenuse}}{\text{length of opposite side}}$$

The domain for each function is a set of angles. As long as we are considering right triangles only, the argument of the function must be such that $0 < \theta < 90$. ***The names of the functions are usually abbreviated as sin, cos, tan, cot, sec, and csc, but it does not make sense to use any of these function names with an equals sign unless the***

NOTE ⇨⇨

function has an argument. Tan = 3, for example, is meaningless. The correct notation is $\tan \theta = 3$.

EXAMPLE 3 Find the six trigonometric functions for angle A in the triangle shown in Figure 4.3.

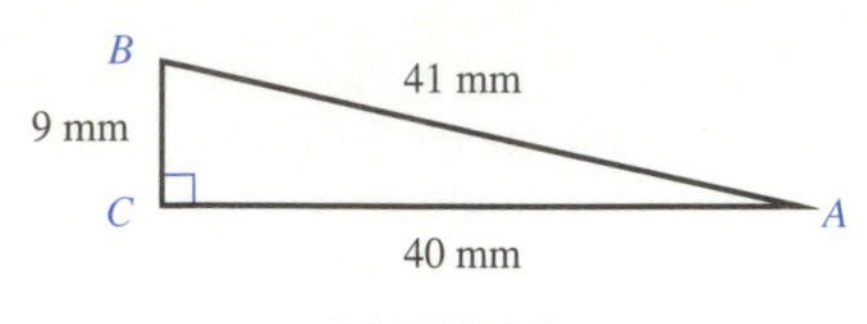

FIGURE 4.3

Solution The hypotenuse is 41 mm in length. The side of length 9 mm is opposite angle A, and the side of length 40 mm is adjacent to A.

$$\sin A = 9/41 \qquad \tan A = 9/40 \qquad \sec A = 41/40$$
$$\cos A = 40/41 \qquad \cot A = 40/9 \qquad \csc A = 41/9$$ ■

EXAMPLE 4 Find the six trigonometric functions for angle B in the triangle of Example 3.

Solution The hypotenuse is 41 mm in length. The side of length 40 mm is opposite angle B, and the side of length 9 mm is adjacent to B.

$$\sin B = 40/41 \qquad \tan B = 40/9 \qquad \sec B = 41/9$$
$$\cos B = 9/41 \qquad \cot B = 9/40 \qquad \csc B = 41/40$$ ■

complementary angles

Two angles whose sum is 90° are said to be *complementary.* Because the sum of the angles of a triangle is 180°, the two acute angles must have a sum of 90°. Thus, angles A and B in Examples 3 and 4 are complementary. Notice from those examples that we have the following relationships:

$$\sin A = \cos B \qquad \tan A = \cot B \qquad \sec A = \csc B$$
$$\cos A = \sin B \qquad \cot A = \tan B \qquad \csc A = \sec B$$

Sine and cosine are called *cofunctions,* as are tangent and cotangent, or secant and cosecant. Cofunctions of complementary angles are equal. That is, if A and B are complementary, a function of A is the cofunction of B and vice-versa.

NOTE ⇨⇨

Figure 4.2 illustrated how the ratio comparing lengths of two sides of a triangle depends on the size of the angle and not the lengths of the sides. ***Since these ratios are real numbers, a trigonometric function is a correspondence between angles and real numbers.*** Thus, it is possible to work with the functions without a particular triangle in mind. The values of the domain and range of a trigonometric function may be obtained from a calculator, or they may be listed in table form. A calculator usually displays an angle as a decimal fraction, but the angles in tables are usually given in degrees and minutes. One degree is subdivided into 60 minutes (denoted ′).

To convert between minutes and decimal form of degree measure:

Use the equation $1° = 60'$ to find the conversion factor.

(a) Multiply a measurement in minutes by $1°/60'$ and round the result to obtain a measurement in decimal form.
(b) Multiply a measurement in decimal form by $60'/1°$ and round the result to obtain a measurement in minutes.

EXAMPLE 5 Convert each measurement to the nearest tenth of a degree.

(a) $14°12'$ **(b)** $68°47'$

Solution **(a)** $12' = 12'(1°/60') = (12/60)° = 0.2°$ so $14°\ 12' = 14.2°$

(b) $47' = 47'(1°/60') = (47/60)° = 0.8°$ so $68°47' = 68.8°$ ■

EXAMPLE 6 For an astronomical telescope, the magnification is given by $m = \alpha/\beta$, where β is the angular separation between two stars and α is the angular separation of their images. If α is $63°26'$ and β is $2°52'$, find m.

Solution Since $m = \alpha/\beta$, $m = 63°26'/2°52'$. Before the division is performed, the measurements are written to the nearest tenth of a degree.

$$26' = 26'(1°/60') = (26/60)° = 0.4°, \text{ so } 63°26' = 63.4°$$
$$52' = 52'(1°/60') = (52/60)° = 0.9°, \text{ so } 2°52' = 2.9°$$

Thus, $m = 63.4°/2.9° = 22$ to the nearest whole number. ■

EXAMPLE 7 Convert each measurement to degrees and minutes.

(a) $44.6°$ **(b)** $87.1°$

Solution **(a)** $0.6° = 0.6°(60'/1°) = 36'$, so $44.6° = 44°36'$

(b) $0.1° = 0.1°(60'/1°) = 6'$, so $87.1° = 87°6'$ ■

A scientific calculator has sin, cos, and tan keys. These functions are like the unary operations in that they accept the number on the display as the argument of the function. It is not necessary to use the (=) key to obtain the function value. To use the functions cot, sec, and csc, it may be necessary to use the following relationships, obtained from the definitions.

RECIPROCAL RELATIONSHIPS

$$\cot \theta = 1/(\tan \theta) \text{ if } \tan \theta \neq 0$$
$$\sec \theta = 1/(\cos \theta) \text{ if } \cos \theta \neq 0$$
$$\csc \theta = 1/(\sin \theta) \text{ if } \sin \theta \neq 0$$

If your calculator doesn't have a [sec] key, for example, you may have to use the sequence [cos] [1/x].

EXAMPLE 8 Find the following function values (to four decimal places) on a calculator.

(a) cos 32.3° **(b)** tan 12° 6′ **(c)** cot 86°

Solution
(a) 32.3 [cos] **0.8452618** Answer: 0.8453
(b) 12 [+] 6 [÷] 60 [=] [tan] **0.2143814** Answer: 0.2144
(c) 86 [tan] [1/x] **0.0699268** Answer: 0.0699 ■

When we begin to solve triangles, it will be necessary to find the angle given the function value. That is, the function value will be the independent variable and the angle will be the dependent variable. Thus, we will be using inverse functions. A scientific calculator usually requires a double keystroke to obtain the inverse functions. You will probably have to use, for example, the sequence [INV] [sin] or [arc] [sin]. (Sometimes the inverse functions are labeled arcsin, arccos, and arctan or $\sin^{-1}$, $\cos^{-1}$, and $\tan^{-1}$.)

EXAMPLE 9 Find the angle (to the nearest hundredth of a degree) associated with each function value.

(a) A if $\sin A = 0.5000$ **(b)** A if $\tan A = 3.2259$ **(c)** B if $\sec B = 5.9774$

Solution
(a) 0.5000 [INV] [sin] **30** Answer: 30°
(b) 3.2259 [INV] [tan] **72.777034** Answer: 72.78°
(c) 5.9774 [1/x] [INV] [cos] **80.369313** Answer: 80.37°

For part (c), remember that $\sec \theta = 1/(\cos \theta)$, so $\cos \theta = 1/(\sec \theta)$. Since 5.9774 is a secant value but the calculator does *not* have a [sec] key, we use [1/x] to get a cosine value on the display. Then the [INV] [cos] sequence can be used. ■

EXAMPLE 10 The equation $\mu = \tan\theta$ gives the coefficient of friction μ between a plane inclined at an angle of θ with the horizontal and an object resting on the plane. If μ is 0.532, find θ.

Solution We have $0.532 = \tan\theta$. Using the INV tan sequence on a calculator, we have $\theta = 28.0°$. ■

EXERCISES 4.1

In 1–10, use the Pythagorean theorem to find the unknown side of each right triangle.

1. $a = 2.8$ cm, $b = 4.5$ cm
2. $a = 5.6$ cm, $b = 3.3$ cm
3. $a = 28.5$ in, $c = 29.3$ in
4. $a = 2.07$ m, $c = 3.05$ m
5. $a = 3.0$ mm, $b = 5.0$ mm
6. $a = 14.0$ in, $b = 14.0$ in
7. $b = 3.6$ km, $c = 8.5$ km
8. $b = 1.17$ cm, $c = 1.25$ cm
9. $a = 11.9$ m, $c = 17.1$ m
10. $a = 21.8$ mm, $c = 23.2$ mm

In 11–20, write the given trigonometric function value as a ratio of the lengths of two of the sides of the triangle described.

11. $\sin A$, $\cos B$, $\tan A$ for $a = 3$ in, $b = 4$ in, $c = 5$ in
12. $\cot A$, $\sec B$, $\csc A$ for $a = 8$ ft, $b = 15$ ft, $c = 17$ ft
13. $\cot B$, $\sec A$, $\csc B$ for $a = 24$ cm, $b = 7$ cm, $c = 25$ cm
14. $\sin B$, $\cos A$, $\tan B$ for $a = 12$ m, $b = 35$ m, $c = 37$ m
15. $\sin A$, $\cos A$, $\tan A$ for $a = 60$ mm, $b = 11$ mm, $c = 61$ mm
16. $\cot B$, $\sec B$, $\csc B$ for $a = 13$ km, $b = 84$ km, $c = 85$ km
17. $\cot A$, $\sec A$, $\csc A$ for $a = 36$ cm, $b = 77$ cm, $c = 85$ cm
18. $\sin B$, $\cos B$, $\tan B$ for $a = 16$ mm, $b = 63$ mm, $c = 65$ mm
19. $\sin A$, $\cos A$, $\tan B$ for $a = 33$ in, $b = 56$ in, $c = 65$ in
20. $\sin B$, $\cos B$, $\tan A$ for $a = 20$ m, $b = 21$ m, $c = 29$ m

In 21–31, use a calculator to find each function value to four decimal places.

21. $\sin 45.7°$
22. $\sin 36°40'$
23. $\cos 27°15'$
24. $\cos 18.4°$
25. $\tan 70.9°$
26. $\tan 89°24'$
27. $\cot 8°35'$
28. $\cot 57.6°$
29. $\sec 63.5°$
30. $\sec 9°45'$
31. $\csc 10°50'$

In 32–42, use a calculator to find an acute angle (to the nearest tenth of a degree) associated with each function value. Then convert the measurement to degrees and minutes.

32. A if $\sin A = 0.4258$
33. B if $\sin B = 0.0175$
34. B if $\cos B = 0.2198$
35. A if $\cos A = 0.9532$
36. A if $\tan A = 7.495$
37. B if $\tan B = 0.9556$
38. B if $\cot B = 0.7813$
39. A if $\cot A = 5.976$
40. A if $\sec A = 1.218$
41. B if $\sec B = 1.352$
42. B if $\csc B = 2.803$

43. The equation $P = IV\cos\theta$ gives the power P (in watts) absorbed in an ac circuit. If $P = 110$ W when $V = 120$ Volts and $I = 1.2$ A, find θ. (In the formula, I is the current (in amperes), V is the voltage (in volts), and θ is the phase angle.) 3 5 9 1 2 4 6 7 8

44. The equation $\sin c = 1/n$ gives the angle of refraction c for which all of the energy of a monochromatic beam goes into the reflected beam. If the index of refraction n is 1.33 for yellow light in water, find c. 3 5 6 9

45. Stakes have been placed at the corners of a proposed building that is to be a rectangle that measures 20 ft by 32 ft as shown in the diagram. In order to determine if the corners are actually laid out at a 90 degree angle, the work crew must check the diagonal distances AC and BD. What should this distance be? 1 2 4 7 8 3 5

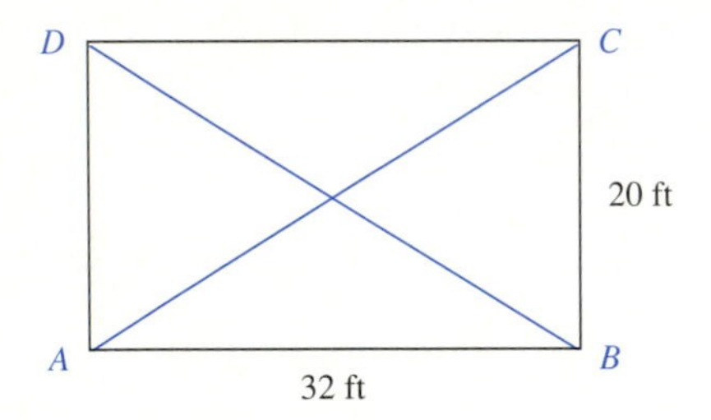

46. The equation $\sin\theta = c/v$ gives the angle θ between the line of motion of a boat traveling at speed v and the front of the bow wave traveling at speed c. Find c if $v = 40.0$ mph and $\theta = 43.1°$. 4 7 8 1 2 3 5 6 9

47. For a cantilever beam under a distributed load, the angle of deflection is given by $\tan\theta = wb^3/(6EI)$. Find θ if $w = 0.02$ kg/mm, $b = 40$ mm, $E = 4000$ kg/mm^2, and $I = 5$ mm^4. In the formula, w is the load per unit length (in kg/mm), b is the length (in mm) of the load, E is the modulus of elasticity (in kg/mm^2), and I is the moment of inertia (in mm^4) of the beam. 1 2 4 8 7 9

48. The horizontal distance d of a pendulum bob from the rest position is given by $d = l\sin\theta$, where l is the length of the pendulum and θ is the angle that the pendulum makes with the vertical. If a clock that is 0.305 m wide has a pendulum 1.00 m long, what is the maximum angle through which it can swing? 4 7 8 1 2 3 5 6 9

49. The diagram shows the layout of a hydraulic cylinder. Determine the length of the cylinder stroke (in inches) that will permit the cylinder to move from point A to point B. 4 7 8 1 2 3 5 6 9

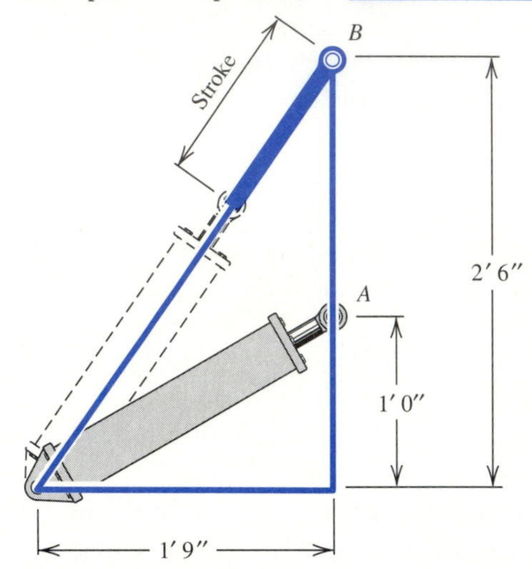

50. A square hopper is constructed so that each of its four sides is a trapezoid, as shown in the accompanying diagram. Find the height of the trapezoid that must be used if the vertical distance between the top and bottom of the hopper is to be 25 inches. 1 2 4 7 8 3 5

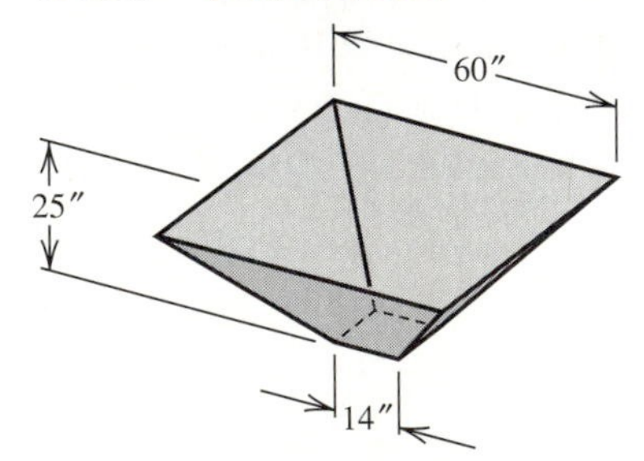

51. Express all six trigonometric functions in terms of $\sin\theta$ and $\cos\theta$. For example, from the reciprocal relationships, it is known that $\sec\theta = 1/\cos\theta$.

4.2 SOLVING RIGHT TRIANGLES

If the length of one side of a triangle and any other two parts are known, it is possible to find the unknown parts. To solve a triangle means to find the unknown parts. The following equations are helpful.

EQUATIONS FOR SOLVING RIGHT TRIANGLES

For a right triangle with acute angles A and B, and sides a, b, and c:

$$a^2 + b^2 = c^2$$

$$A + B = 90°$$

$$\sin \theta = \frac{\text{length of opposite side}}{\text{length of hypotenuse}} \qquad \cos \theta = \frac{\text{length of adjacent side}}{\text{length of hypotenuse}}$$

$$\tan \theta = \frac{\text{length of opposite side}}{\text{length of adjacent side}}$$

EXAMPLE 1 Solve the triangle for which $a = 14.0$ cm, $b = 17.1$ cm, and $C = 90°$.

Solution Figure 4.4 shows a sketch of the triangle. The Pythagorean theorem gives c.

$$c^2 = a^2 + b^2$$
$$c^2 = (14.0)^2 + (17.1)^2$$
$$c = \sqrt{196 + 292.41}$$
$$c = 22.1 \text{ (to three significant digits)}$$

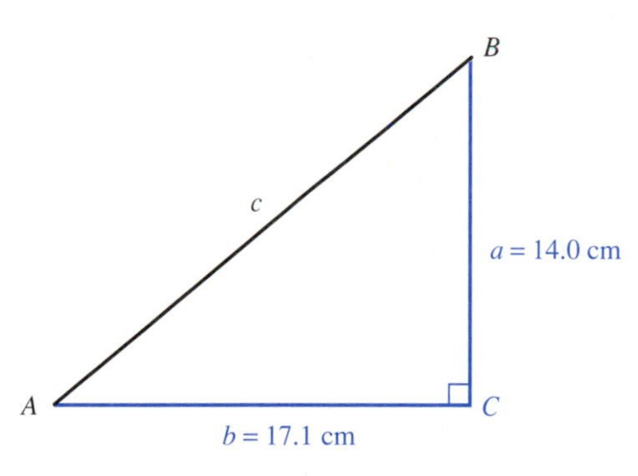

FIGURE 4.4

Now that all three sides are known, we must find either angle A or B. We choose to find A. Any one of the following trigonometric functions could be used.

$$\sin A = 14.0/22.1 \qquad \cos A = 17.1/22.1 \qquad \tan A = 14.0/17.1$$

NOTE ➪➪ ***The best choice is tan*** $A = 14.0/17.1$***, because it depends on information that was given in the problem, rather than on the value 22.1, which was just computed.*** If there was an error in that calculation, it would cause the answer for A to be wrong if sin A or cos A were used. Thus, we have tan $A = 14.0/17.1 = 0.819$ to three significant digits. However, we leave all the digits shown on the calculator display in order to determine the angle A that has this number as its tangent. The inverse tangent is 39.3° to three significant digits. Tan B is given by 17.1/14.0, and $B = 50.7°$. As a check, we could use the following calculations.

$$A + B = 90°$$
$$39.3° + B = 90°$$
$$B = 50.7°$$

All six parts of the triangle are now known. ■

Several important concepts were covered in Example 1. They are summarized here.

HINTS FOR SOLVING RIGHT TRIANGLES

1. Use information that is given in the problem rather than information that has been computed, whenever possible.
2. Round answers in the following manner.
   - **(a)** The square root of a number should be rounded to the accuracy of the number itself.
   - **(b)** If sides are given to one or two significant digits, angles should be rounded to the nearest degree, and vice-versa.
   - **(c)** If sides are given to three significant digits, angles should be rounded to the nearest tenth of a degree, and vice-versa.
   - **(d)** If sides are given to four significant digits, angles should be rounded to the nearest hundredth of a degree, and vice-versa.
   - **(e)** If a trigonometric function value is sought, it should be rounded to the same accuracy as the sides of the triangle.

EXAMPLE 2 Solve the right triangle for which $a = 18.0$ mm, $b = 29.9$ mm, and $c = 34.9$ mm.

Solution We know that $C = 90°$, since the triangle is a right triangle and this value is exact, not rounded. We must use a trigonometric function to determine A or B. We choose to solve for A.

$$\sin A = 18.0/34.9 \text{ (We could use } \tan A = 18.0/29.9 \text{ instead.)}$$
$$\sin A = 0.516 \text{ (to three significant digits—the accuracy of the sides)}$$
$$A = 31.0° \text{ (to the nearest tenth of a degree)}$$

We also find B using information that was given.

$$\sin B = 29.9/34.9$$
$$\sin B = 0.857 \text{ (to three significant digits)}$$
$$B = 59.0°$$

All six parts of the triangle are now known. ■

EXAMPLE 3 Solve the right triangle for which $A = 51.1°$, $C = 90°$, and $b = 2.04$.

Solution Figure 4.5 shows a sketch of the triangle. We find B.

$$A + B = 90°$$
$$51.1° + B = 90°$$
$$B = 38.9°$$

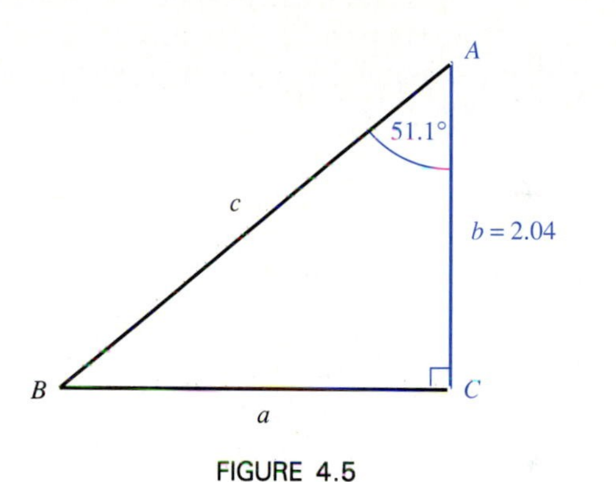

FIGURE 4.5

Since A was given and B was computed, we choose to work with A in order to find a and c.

$$\cos 51.1° = 2.04/c \qquad \text{and} \qquad \tan 51.1° = a/2.04$$

We solve these equations algebraically before using a calculator to do the arithmetic.

$$c(\cos 51.1°) = 2.04 \qquad \text{and} \qquad 2.04(\tan 51.1°) = a$$
$$c = 2.04/(\cos 51.1°)$$

The calculator keystroke sequences are as follows.

2.04 [÷] 51.1 [cos] [=]   2.04 [×] 51.1 [tan] [=]

$c = 3.25$ mm 2.53 mm $= a$

All six parts of the triangle are now known. Since we used the trigonometric functions to obtain the answers, it would be useful to check using the Pythagorean theorem.

$$(3.25)^2 = (2.53)^2 + (2.04)^2$$
$$10.5625 = 6.4009 + 4.1616 \text{ or } 10.5625 \quad ■$$

A triangle with angles of 30°, 60°, and 90° exhibits a special relationship among the sides, as does a triangle with angles of 45°, 45°, and 90°. Examples 4 and 5 illustrate these relationships.

EXAMPLE 4 Consider a triangle with angles $A = 30°$, $B = 60°$, and $C = 90°$. Express the trigonometric functions of 30° without using a calculator or tables.

Solution We start with an equilateral triangle, as in Figure 4.6, with sides of length c. For an equilateral triangle, the three angles are all equal and therefore they are 60° angles. We draw a line segment, called an altitude, from one angle perpendicular to the opposite side. This line segment splits the angle into two 30° angles, and it splits the opposite side into two equal parts, each of length $(1/2)c$. The equilateral triangle has been split into two right triangles. We label the angles of one of them as A, B, and C. Thus, $a = (1/2)c$.

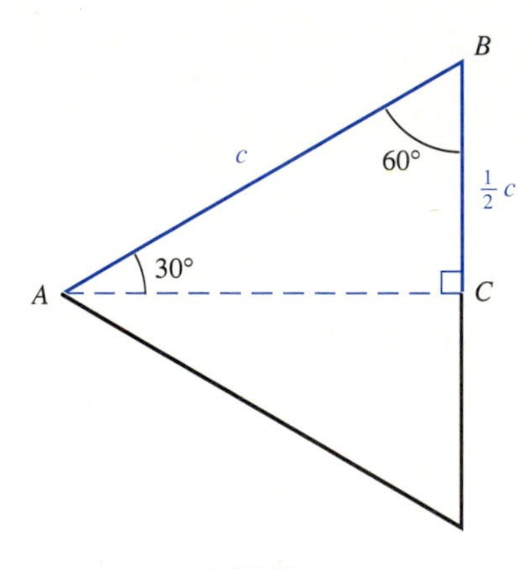

FIGURE 4.6

The Pythagorean theorem is used to find b.

$$[(1/2)c]^2 + b^2 = c^2$$
$$(1/4)\,c^2 + b^2 = c^2$$
$$b^2 = 3/4\,c^2$$
$$b = \sqrt{(3/4)c^2}$$
$$b = \sqrt{(3)c^2}/\sqrt{4}, \text{ which is } (\sqrt{3}/2)c \text{ or } (1/2)c\sqrt{3}$$

The trigonometric functions of 30° are as follows.

$$\sin 30° = \frac{(1/2)c}{c} = \frac{1}{2} \qquad \cot 30° = \frac{(1/2)c\sqrt{3}}{(1/2)c} = \sqrt{3}$$

$$\cos 30° = \frac{(1/2)c\sqrt{3}}{c} = \frac{\sqrt{3}}{2} \qquad \sec 30° = \frac{c}{(1/2)c\sqrt{3}} = \frac{2}{\sqrt{3}} = \frac{2\sqrt{3}}{3}$$

$$\tan 30° = \frac{(1/2)c}{(1/2)c\sqrt{3}} = \frac{1}{\sqrt{3}} = \frac{\sqrt{3}}{3} \qquad \csc 30° = \frac{c}{(1/2)c} = 2$$

Notice that the denominators have been rationalized. ■

The trigonometric functions of 60° could also be determined from Figure 4.6.

EXAMPLE 5 Consider a triangle with angles $A = 45°$, $B = 45°$, and $C = 90°$. Find the trigonometric functions of 45° without using a calculator or tables.

Solution We start with an isosceles triangle with legs of length a, as in Figure 4.7. We know that $C = 90°$. The Pythagorean theorem is used.

$$a^2 + a^2 = c^2 \qquad \text{or} \qquad 2a^2 = c^2$$

Thus, $c = \sqrt{2a^2}$, which is $a\sqrt{2}$.

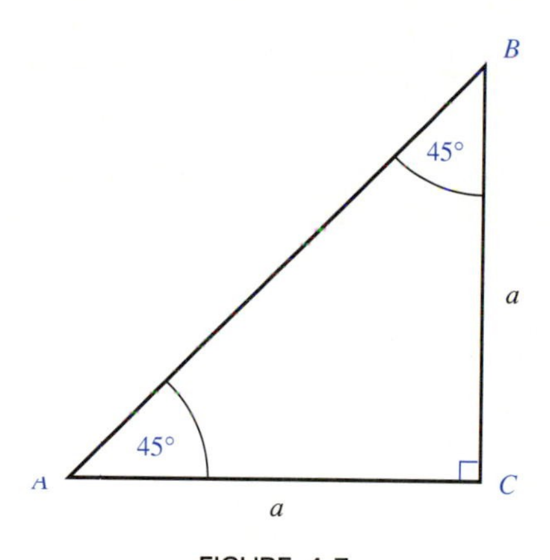

FIGURE 4.7

The trigonometric functions are as follows.

$$\sin 45° = \frac{a}{a\sqrt{2}} = \frac{1}{\sqrt{2}} = \frac{\sqrt{2}}{2} \qquad \cot 45° = \frac{a}{a} = 1$$

$$\cos 45° = \frac{a}{a\sqrt{2}} = \frac{1}{\sqrt{2}} = \frac{\sqrt{2}}{2} \qquad \sec 45° = \frac{a\sqrt{2}}{a} = \sqrt{2}$$

$$\tan 45° = \frac{a}{a} = 1 \qquad \csc 45° = \frac{a\sqrt{2}}{a} = \sqrt{2}$$ ■

The information from Examples 4 and 5 is summarized as follows.

SPECIAL ANGLES

For a 30°-60°-90° triangle:

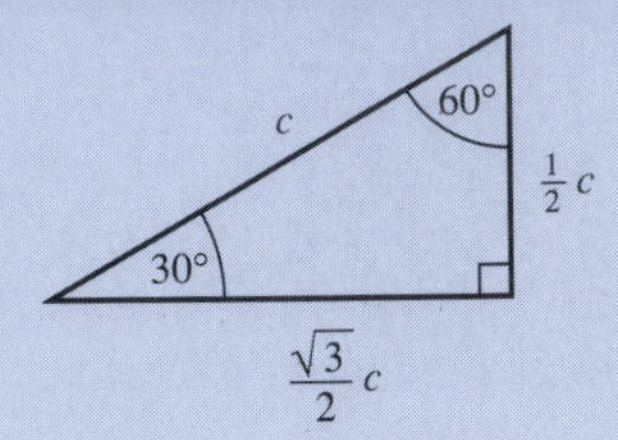

The side opposite the 30° angle is 1/2 the length of the hypotenuse.
The side opposite the 60° angle is $\sqrt{3}/2$ times the length of the hypotenuse.

For a 45°-45°-90° triangle:

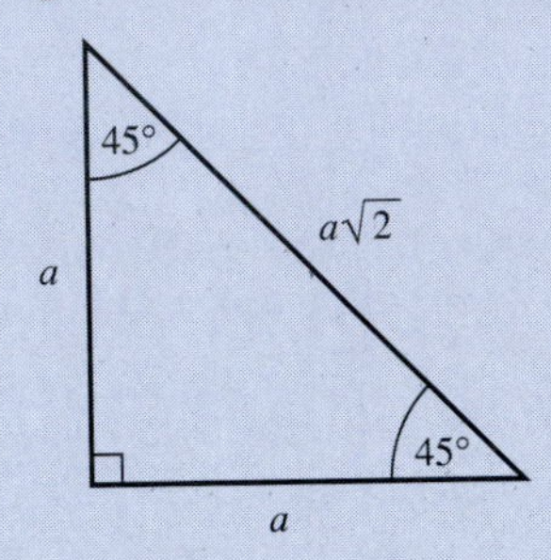

The hypotenuse is $\sqrt{2}$ times the length of a leg.

NOTE ***You should remember how these relationships were derived, since you will frequently have to use them.***

EXERCISES 4.2

In 1–10, solve each right triangle shown.

1.

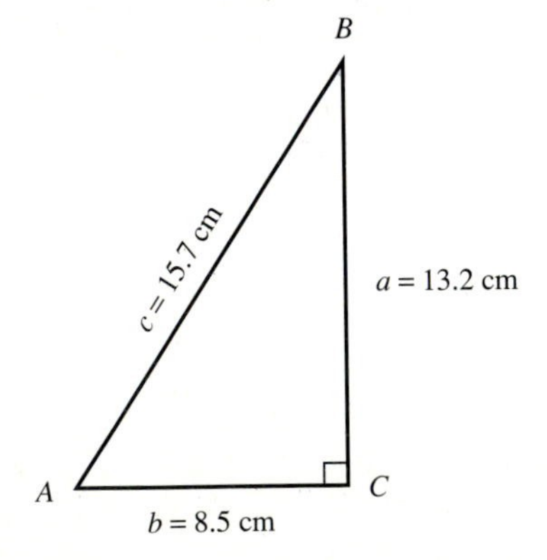

2.

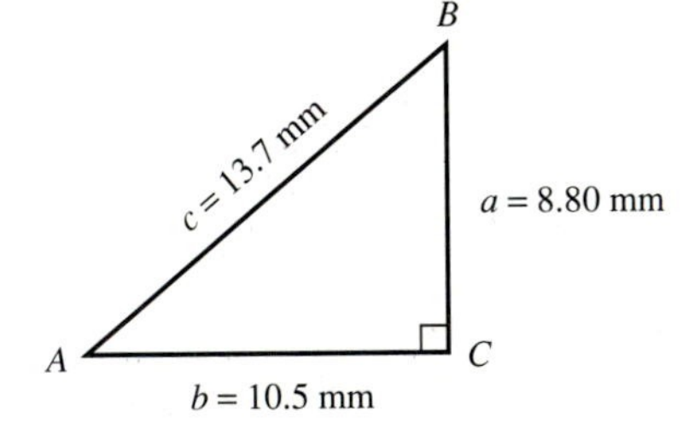

3.

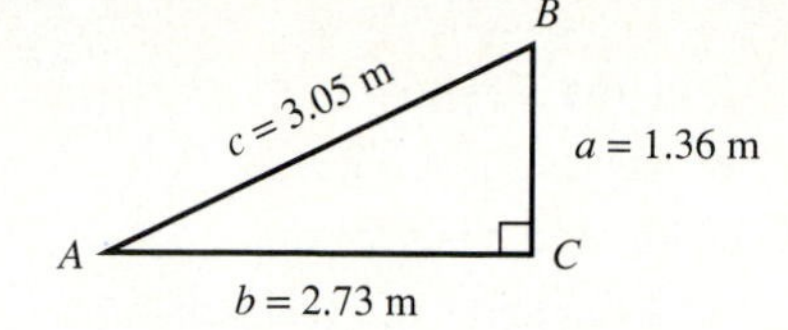

4.

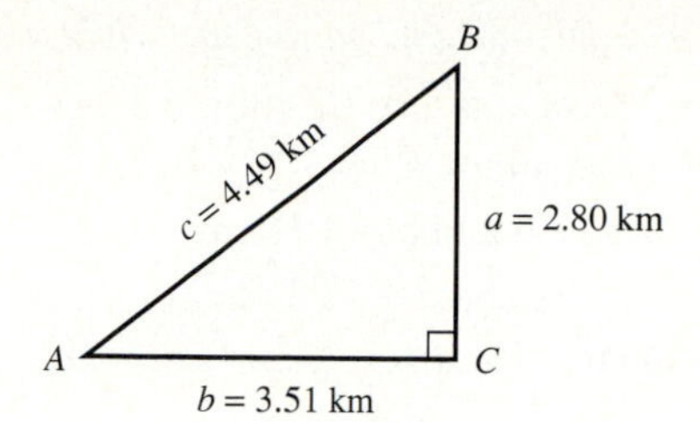

5.

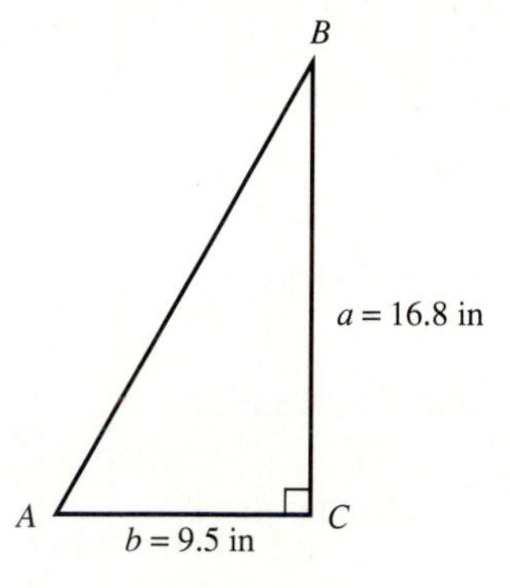

6.

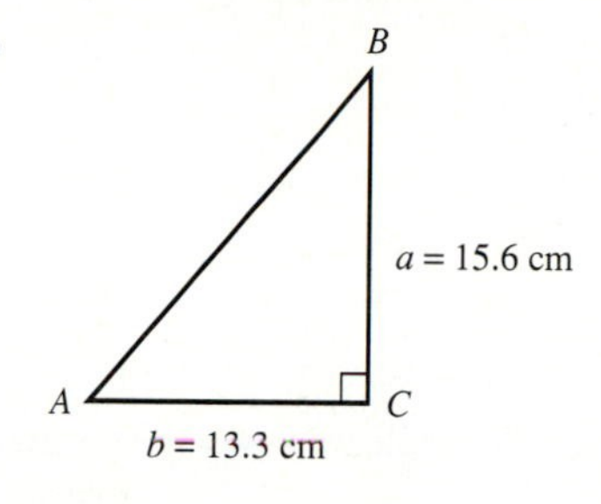

7.

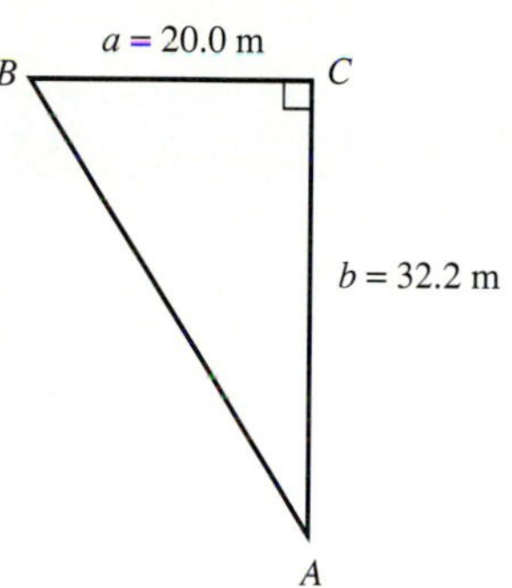

8.

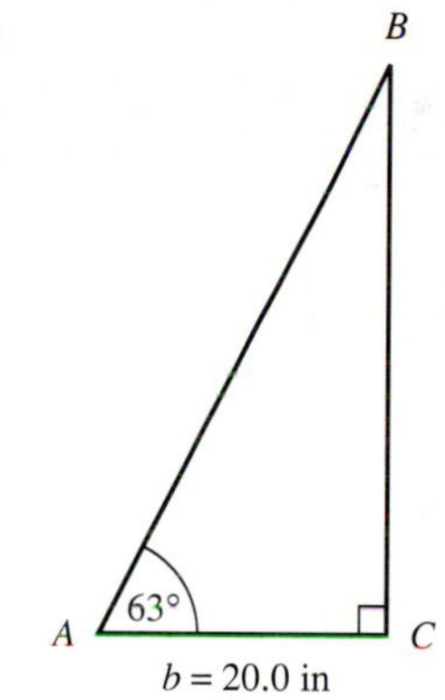

9.

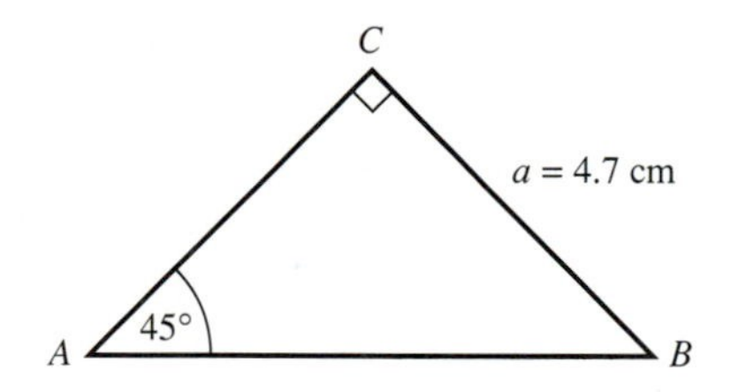

10.

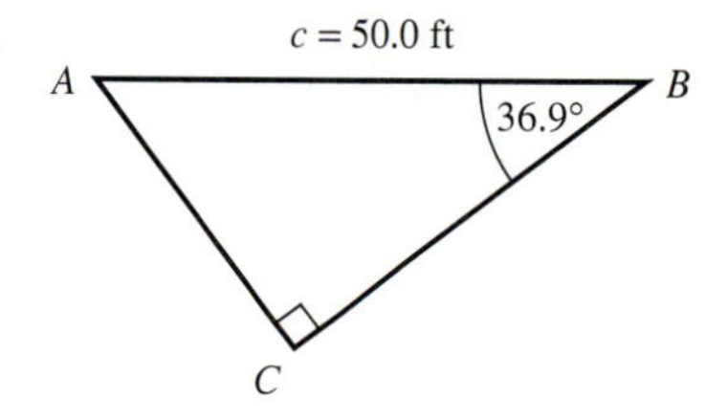

In 11–20, solve each right triangle.

11. $a = 2.08$ m, $b = 1.05$ m, $C = 90°$

12. $a = 2.40$ km, $b = 1.61$ km, $C = 90°$

13. $a = 28.8$ in, $B = 58.4°$, $C = 90°$

14. $b = 22.5$ mm, $A = 50.4°$, $C = 90°$

15. $c = 3.97$ m, $B = 54.9°$, $C = 90°$

16. $c = 5.09$ km, $B = 64.4°$, $C = 90°$

17. $a = 4.3$ in, $b = 2.1$ in, $C = 90°$

18. $a = 3.7$ cm, $A = 70°$, $C = 90°$

19. $b = 5.2$ mm, $B = 40°$, $C = 90°$

20. $a = 6.1$ ft, $A = 10°$, $C = 90°$

In 21–30, find the indicated part of each right triangle.

21. a if $b = 7.3$ cm and $A = 32°$

22. a if $c = 9.2$ m and $B = 15°$

23. c if $a = 4.6$ in and $A = 76°$

24. c if $b = 5.81$ km and $A = 41.3°$

25. A if $a = 1.90$ cm and $b = 2.71$ cm

26. A if $a = 2.83$ m and $c = 5.92$ m

27. B if $b = 13.7$ in and $a = 21.2$ in

28. B if $a = 26.54$ km and $c = 51.41$ km

29. b if $a = 35.62$ m and $B = 37.21°$

30. b if $c = 44.78$ in and $A = 62.43°$

In 31–40, solve each right triangle without using a calculator or tables. Give exact, rather than rounded, answers.

31. $A = 30°$, $a = 6$ ft, $C = 90°$

32. $A = 30°$, $c = 8$ in, $C = 90°$

33. $A = 60°$, $b = 10$ m, $C = 90°$

34. $A = 60°$, $c = 12$ cm, $C = 90°$

35. $B = 30°$, $c = 4$ mm, $C = 90°$

36. $A = 45°$, $C = 8\sqrt{2}$ in, $C = 90°$

37. $A = 45°$, $a = 6$ ft, $C = 90°$

38. $B = 45°$, $a = 4$ cm, $C = 90°$

39. $B = 45°$, $c = 10\sqrt{2}$ mm, $C = 90°$

40. $A = 45°$, $b = 12$ km, $C = 90°$

41. (a) Given a right triangle with A, C (90°), and c known, solve for a, b, and B in terms of A and c.

(b) Given a right triangle with C (90°), b, and c known, solve for A, B, and a in terms of b and c.

4.3 APPLICATIONS

Many applied problems in technological areas require solving a right triangle. For review, the hints that were given in Chapter 2 for solving word problems are repeated.

To solve a word problem:

1. Read the problem for understanding.
2. Identify the variable.
3. Write down relevant information. Draw a picture.
4. Write the equation.
5. Solve the equation and check that the answer satisfies the conditions of the original problem.

NOTE ⇨⇨ ***Drawing a picture is a particularly useful step when a right triangle is involved.*** Geometrical concepts are sometimes needed to draw a picture from the verbal description given.

transversal

When two parallel lines are crossed by a third line, called a *transversal,* eight angles are formed as shown in Figure 4.8.

The symbol for angle is $\angle$. Nonadjacent angles that share a vertex, such as $\angle 1$ and $\angle 4$ are called *vertical angles.* Angles that occupy the same position relative to the transversal on each line, as do $\angle 1$ and $\angle 5$, are called *corresponding angles.* We say that $\angle 1$, $\angle 2$, $\angle 7$, and $\angle 8$ are *exterior angles,* while $\angle 3$, $\angle 4$, $\angle 5$, and $\angle 6$ are called *interior angles.* Interior angles on opposite sides of the transversal, such as $\angle 4$ and $\angle 5$, are *alternate* interior angles. It is important that certain pairs of angles are equal.

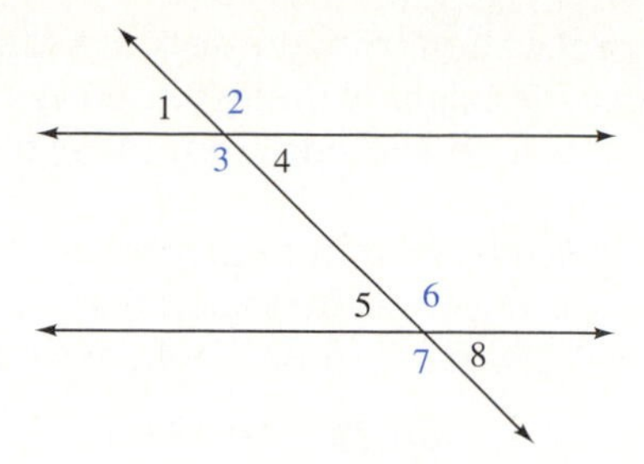

FIGURE 4.8

> **EQUAL ANGLES FORMED WHEN TWO PARALLEL LINES ARE CUT BY A TRANSVERSAL**
>
> **(a)** Alternate interior angles are equal.
> **(b)** Corresponding angles are equal.
> **(c)** Vertical angles are equal.

In Figure 4.8, $\angle 3 = \angle 6$ and $\angle 4 = \angle 5$, because alternate interior angles are equal. Also, $\angle 1 = \angle 5$, $\angle 2 = \angle 6$, $\angle 3 = \angle 7$, and $\angle 4 = \angle 8$, because corresponding angles are equal. Finally, $\angle 1 = \angle 4$, $\angle 2 = \angle 3$, $\angle 5 = \angle 8$, and $\angle 6 = \angle 7$, because vertical angles are equal.

angle of elevation

angle of depression

Problems sometimes use the phrase *angle of elevation* or the phrase *angle of depression.* If a transversal cuts a horizontal line, the angle of elevation is measured from the horizontal *up* toward the transversal. The angle of depression is measured *down* toward the transversal. Imagine two observers, one standing at point A on a cliff, the other at point B at the foot of the cliff, as in Figure 4.9. The transversal through A and B defines the angle of depression α for the observer at A and the angle of elevation β for the observer at B. We also know that $\alpha = \beta$. (Why?)

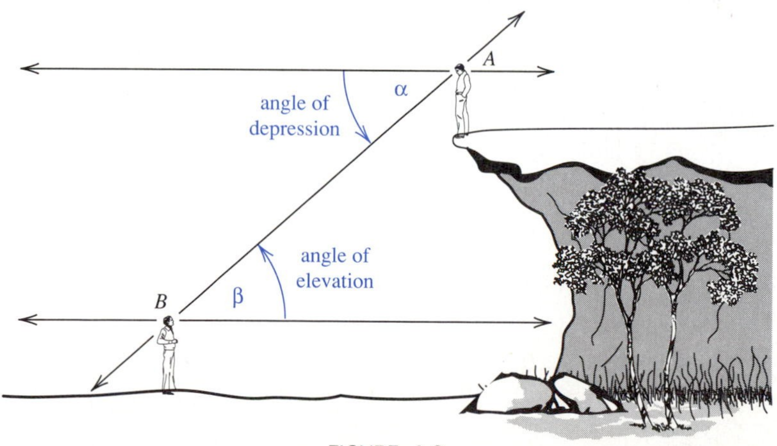

FIGURE 4.9

EXAMPLE 1 The angle of elevation from the deck of a ship to the light in a lighthouse is 29.1°. If it is known that the height of the lighthouse is 198 ft and the deck of the ship is 9 ft above the water, how far is the ship from the lighthouse?

Solution Let $x =$ distance from ship to lighthouse. Figure 4.10 indicates the given information. Since the angle was measured from the deck of the ship, the side opposite the 29.1° angle is 198 ft − 9 ft or 189 ft. We have

$$\tan 29.1^\circ = 189/x$$
$$x \tan 29.1^\circ = 189$$
$$x = 189/\tan 29.1^\circ$$
$$x = 340 \text{ ft (to three significant digits)}$$

Thus, the ship is 340 ft from the lighthouse. ■

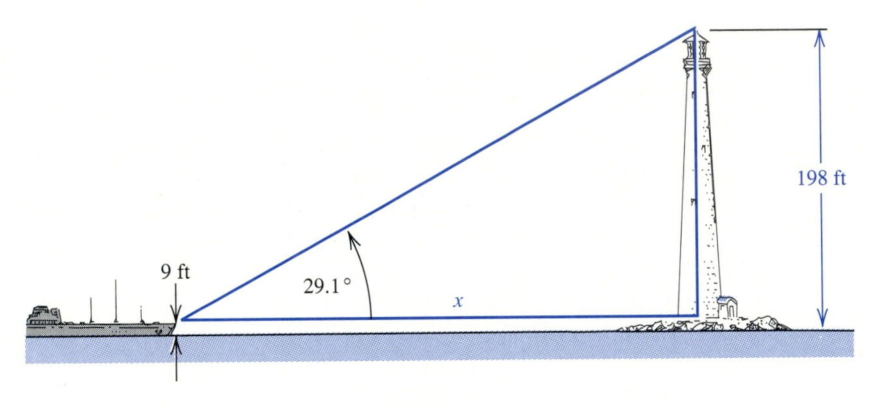

FIGURE 4.10

In Example 1, the curvature of the earth was ignored. For distances under 200 m, it has a negligible effect, but for simplicity, we will continue to ignore it in all problems.

EXAMPLE 2 At the scene of a minor automobile accident, a motorist claimed that the sun's glare had a blinding effect. The motorist who was hit had someone measure her shadow at the time. It was 2 feet long. If she was 5 feet, 6 inches tall, what was the angle of elevation of the sun?

Solution Let $A =$ the angle of elevation of the sun as shown in Figure 4.11.

$$5'6'' = 5.5 \text{ feet}$$
$$\tan A = 5.5/2 = 2.75$$
$$A = 70^\circ \text{ (approximately)}$$

Thus, the angle of elevation to the sun was about 70°. ■

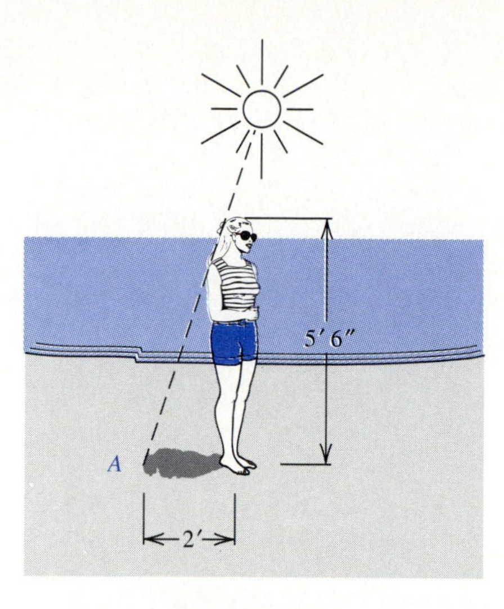

FIGURE 4.11

Navigation problems make extensive use of right triangles. Directions are specified using *bearings*. The letter N or S is given, followed by an acute angle, then the letter E or W. The bearing indicates that the acute angle is measured *from* the north or south, *toward* the east or west. Figure 4.12 shows that N 40° E indicates a bearing from the north 40° toward the east. Figure 4.13 shows that S 30° W indicates a bearing from the south 30° toward the west.

bearings

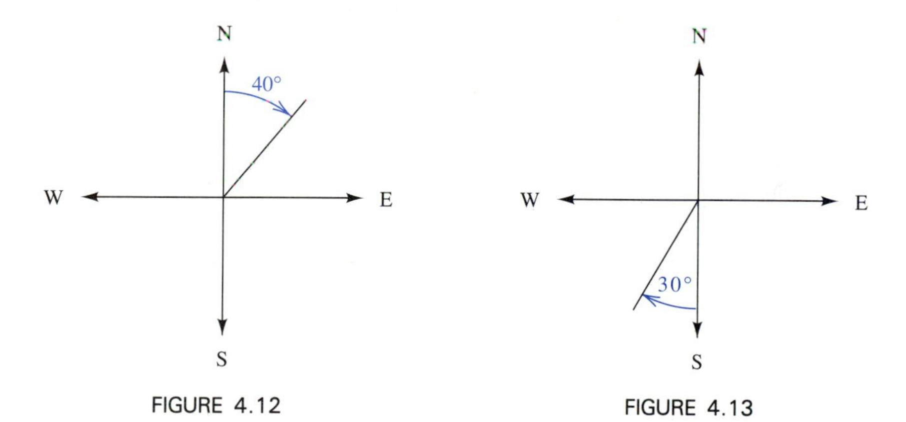

FIGURE 4.12

FIGURE 4.13

EXAMPLE 3 An airplane leaving Phoenix heads N 9.15° W, flying at 314.5 mph for two hours. How far west has the plane traveled?

Solution Let $x =$ distance traveled west. In Figure 4.14, ABC is a right triangle.

$$A = 90° - 9.15° = 80.85°$$

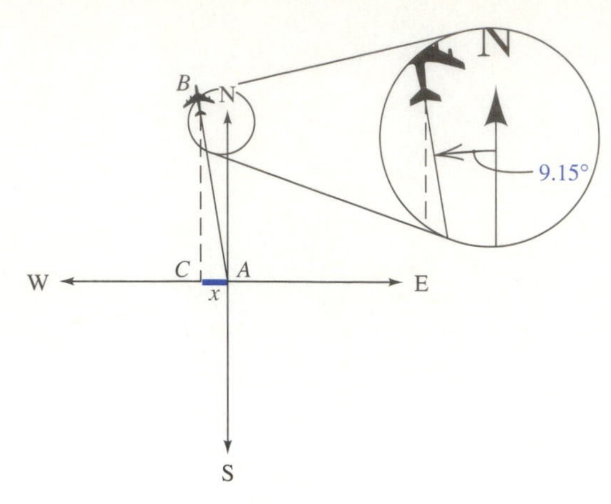

FIGURE 4.14

The plane traveled 2(314.5) or 629.0 miles.

$$\cos 80.85° = x/629.0$$
$$629.0(\cos 80.85°) = x$$
$$100.0 \text{ mi} = x$$

The plane, then, traveled 100.0 miles west. ■

Sometimes more than one triangle is involved in a problem. Example 4 illustrates this type of problem.

EXAMPLE 4 From a fire tower 57.0 m tall, the angle of depression to a forest fire is 17.9°. The angle of depression to a campsite in the same direction is 36.9°. How far is the fire from the campsite?

Solution Let x = distance between fire and camp. In Figure 4.15, there are two right triangles, ACB and BCD. Since A is equal to the angle of depression to the fire, A is 17.9°. The length y of side AC is found as follows.

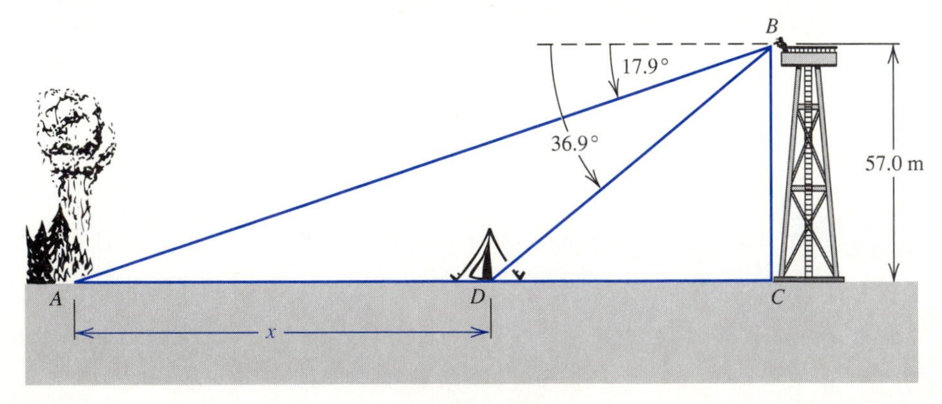

FIGURE 4.15

$$\tan 17.9° = 57.0/y$$
$$y \tan 17.9° = 57.0$$
$$y = 57.0/\tan 17.9°$$
$$y = 176 \text{ m (to three significant digits)}$$

Angle D is equal to the angle of depression to the campsite, so D is 36.9°. The length z of side DC is found as follows.

$$\tan 36.9° = 57.0/z$$
$$z \tan 36.9° = 57.0$$
$$z = 57.0/\tan 36.9°$$
$$z = 75.9 \text{ m (to three significant digits)}$$

The distance x is given by $176 - 75.9$ or 100 m (to three significant digits). ■

EXERCISES 4.3

1. A store wants to fly a large balloon as part of a promotion. There is a building located 12.0 m away. If the angle of elevation to the top of the building is 49.3°, how long a cord would be necessary to ensure that the balloon clears the building when the wind blows it in that direction? See the accompanying figure. 1 2 4 7 8 3 5

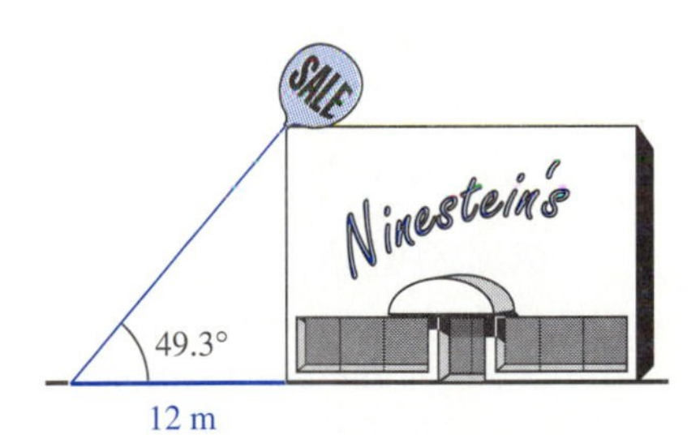

2. A rectangular computer monitor is to have a 13″ diagonal screen. If the screen is to be divided into square pixels with 22 rows and 32 columns, find the dimensions of the screen. (Hint: Let x be the width (in inches) of a pixel. Then the width of the screen is $22x$.) 3 5 9

3. The Leaning Tower of Pisa would be 55 m high if it did not lean. The top is 4.9 m off the vertical. At what angle with respect to the vertical does it lean? 1 2 4 7 8 3 5

4. An architect is designing a circular planetarium that is 20.0 feet in diameter. So that the signs of the zodiac can be displayed, there are to be 12 columns equally spaced around the circumference. What should the straight-line distance between columns be? 1 2 4 7 8 3 5

5. A carousel is to have 10 seats on the outside. If each seat is placed at a distance of 12 feet from the center, what should the distance between seats be? 1 2 4 7 8 3 5

6. An overhang will prevent direct sunlight from shining in a window during the hottest part of the day. If the bottom of the window is 4.0 feet above ground level and the overhang is 12 feet above ground level, how wide should the overhang be if the window is to be shaded when the angle of

elevation of the sun is 75° or more? See the accompanying figure. 1 2 4 7 8 3 5

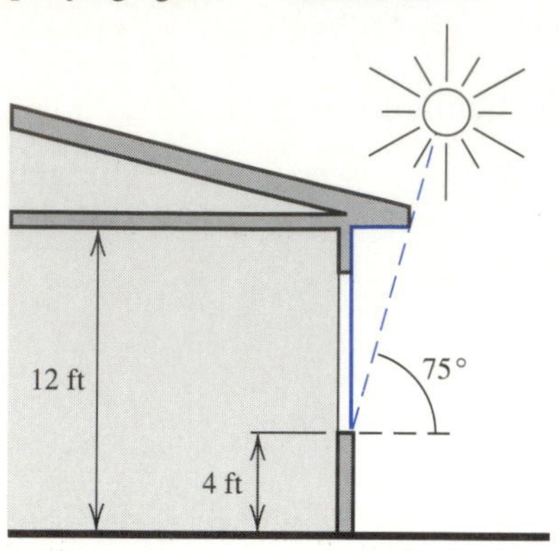

7. The ladder of a firetruck can be extended to 24.0 feet. If it is impossible to place the base of the ladder closer than 6.00 feet to the building, at what angle of elevation should the ladder be inclined to reach its maximum vertical height? 1 2 4 7 8 3 5

8. The ladder on a firetruck is 14.0 feet long. If the maximum angle at which it can be inclined is 75.0°, find the maximum vertical height that can be reached. 1 2 4 7 8 3 5

9. A garden sprinkler is designed so that it can be set to swivel through different angles, as shown in the accompanying figure. The water sprays a distance of 6.3 m. If the sprinkler is set in the middle of a garden that is 8.0 m wide, at what angle should it be set to just miss the walkways on either side? 4 7 8 1 2 3 5 6 9

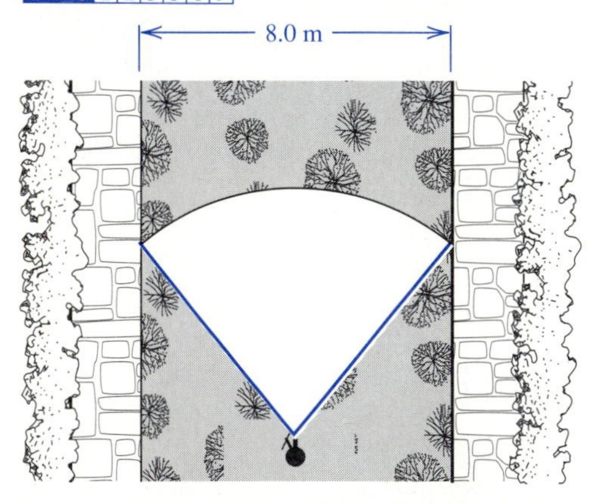

10. From a ship sailing in the Graveyard of the Atlantic, the Cape Hatteras lighthouse can be seen due north and the Ocracoke lighthouse due west. It is known that the Cape Hatteras lighthouse is 30.0 miles N 66.7° E of the Ocracoke lighthouse. How far is it to the Ocracoke lighthouse from the ship? 1 2 4 7 8 3 5

11. From a ship sailing in the Graveyard of the Atlantic, the Cape Hatteras lighthouse can be seen due north and the Ocracoke lighthouse is spotted N 23.3° W. It is known that the Cape Hatteras lighthouse is 30.0 miles N 66.7° E of the Ocracoke lighthouse. How far is it to the Cape Hatteras lighthouse from the ship? 1 2 4 7 8 3 5

12. From a fire tower the bearing to a forest fire is S 40.1° E. From another tower 14.2 miles due south, the bearing is N 49.9° E. Find the distance of the fire from the first tower. 1 2 4 7 8 3 5

13. From a helicopter at a height of 56.0 m, the angle of depression to a craft in distress is 60.8°. The angle of depression to a nearby vessel is 2.7°. How far is the distressed ship from the nearby vessel? 1 2 4 7 8 3 5

14. An inclined ramp is to be constructed in such a way that the height is variable from 3.00 feet to 5.50 feet. If the ramp is 12.0 feet long, what should the minimum and maximum angle settings be? 1 2 4 7 8 3 5

15. The height of a cloud can be measured in the following way. A light is aimed vertically at the cloud. An observer stationed 1000.0 m from the light measures the angle of elevation to the spot on the cloud where the light shines. If the angle of elevation is 52.0°, find the height of the cloud. 1 2 4 7 8 3 5

16. The speed of an airplane can be estimated from the ground by sighting the plane straight ahead and estimating the horizontal angle θ to the spot from which the sound of the plane is coming, as shown in the accompanying figure. Knowing that sound travels at about 700 mph, and assuming that light travels instantaneously, find a formula to estimate the speed of the plane. What is the speed of the plane if the angle is 30.0°? 1 3 4 5 6 8 9

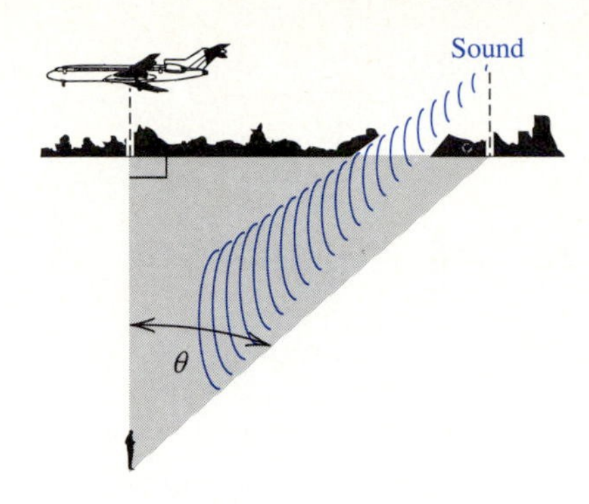

17. A bolt of lightning is observed at an angle of elevation of 27.3°. Five seconds later, the observer hears a clap of thunder. If sound travels at 333 m/s and it is assumed that light travels instantaneously, find the height at which the lightning occurred. 1 3 4 5 6 8 9

18. A vein of copper ore is discovered on the face of a cliff at the top. The heaviest concentration of the ore occurs at the water table, so a vertical shaft is to be sunk so that it intersects the vein, as shown in the accompanying figure. If the angle of depression of the vein is 32.6° and the water table is 21.4 m below the surface, how far from the cliff should the shaft be sunk? 1 2 4 6 8 5 7 9

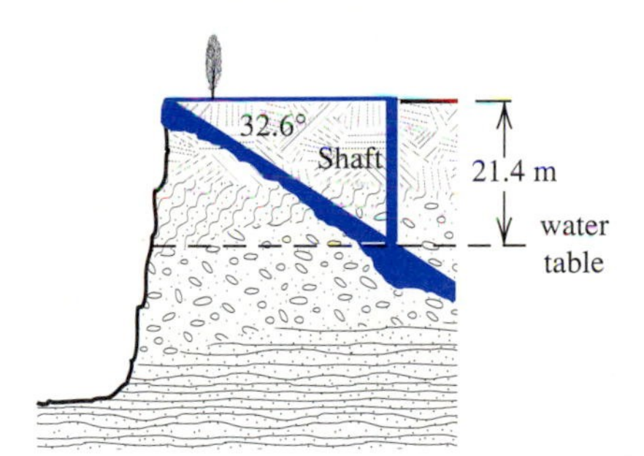

19. A submarine at a depth of 67.3 m detects a ship floating on the ocean surface at an angle of elevation of 25.3°. What is the horizontal distance from the submarine to the ship?
1 2 4 7 8 3 5

20. To calibrate the effective range of VHF ship-to-shore radio communications, it is necessary to know the distance from a point at a height h above the earth to the horizon. If the radius of the earth is 6 378 000 m and the Empire State Building is 380.0 m tall, find the angle of depression from the top of the building to the horizon. See the accompanying figure (which is not drawn to scale).
1 3 4 5 6 8 9

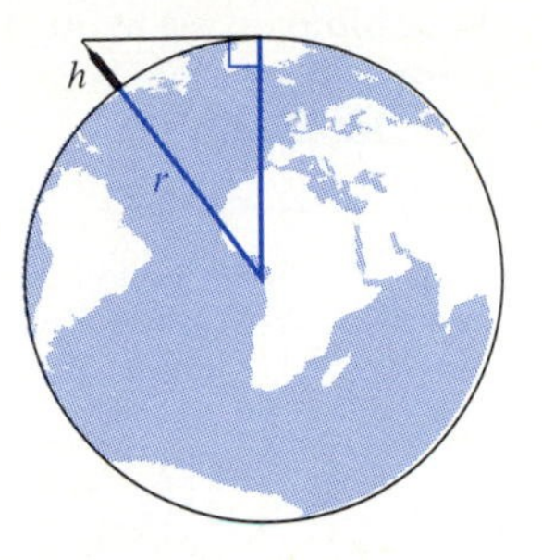

21. Find the height h of the roof frame shown in the accompanying figure. 1 2 4 7 8 3 5

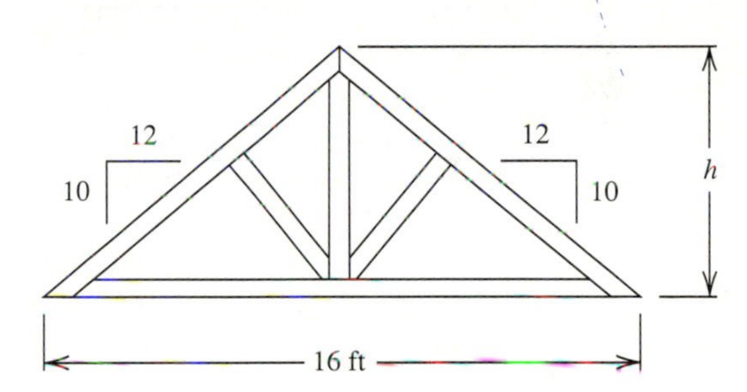

22. The accompanying diagram shows a conveyor belt. The large pulley has a diameter of 18 inches. Find the angle θ of inclination of the belt and the vertical distance from the floor to the top of the large pulley. 1 2 4 7 8 3 5

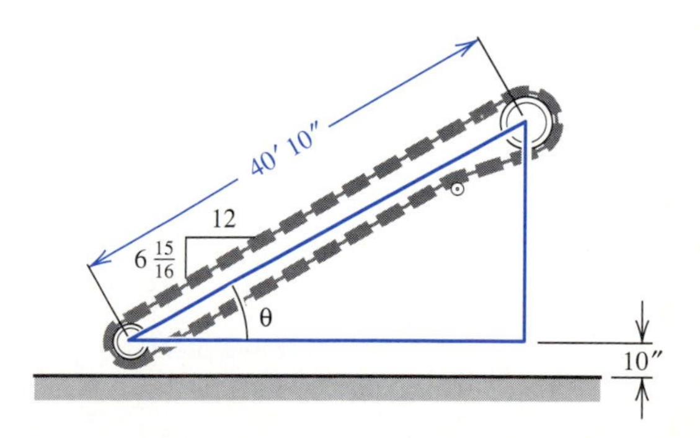

23. A right triangle must be set out at a construction project. One side of the triangle must be 68.43 ft long, and the hypotenuse must make an angle of 26°32′ with that side of the triangle. The triangle can be set without measuring angles in the field if arcs of the proper length are swung from points A and B, as shown in the accompanying figure. Determine the radii of the arcs that should be swung from points A and B. 1 2 4 7 8 3 5

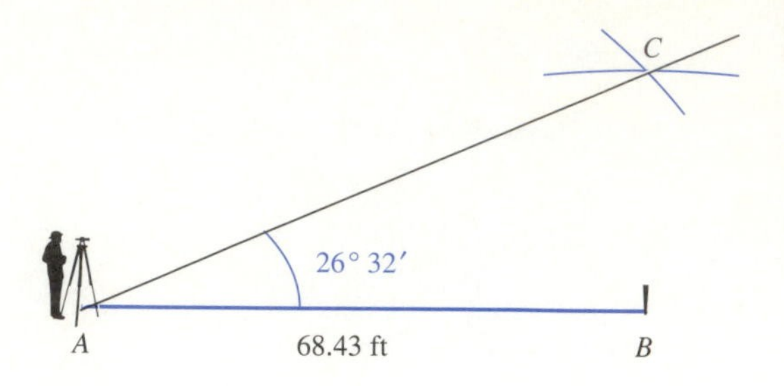

1. Architectural Technology
2. Civil Engineering Technology
3. Computer Engineering Technology
4. Mechanical Drafting Technology
5. Electrical/Electronics Engineering Technology
6. Chemical Engineering Technology
7. Industrial Engineering Technology
8. Mechanical Engineering Technology
9. Biomedical Engineering Technology

4.4 TRIGONOMETRIC FUNCTIONS OF ANY ANGLE

We have introduced trigonometry in the context of triangles, but the domain of the functions can be extended to include all angles, not just those less than 90°.

initial side of an angle

terminal side of an angle

standard position

A line segment of length r, with one end-point at the origin rotating like a clock hand about the origin, sweeps through an angle as shown in Figure 4.16. If it rotates a full 360°, it sweeps out a circle of radius r. The initial position of the line segment locates the *initial side* of the angle and the ending position locates the *terminal side.* Angles measured in the counterclockwise direction are considered positive, while angles measured in the clockwise direction are considered negative. When the initial side is on the positive x-axis, the angle is said to be in *standard position.*

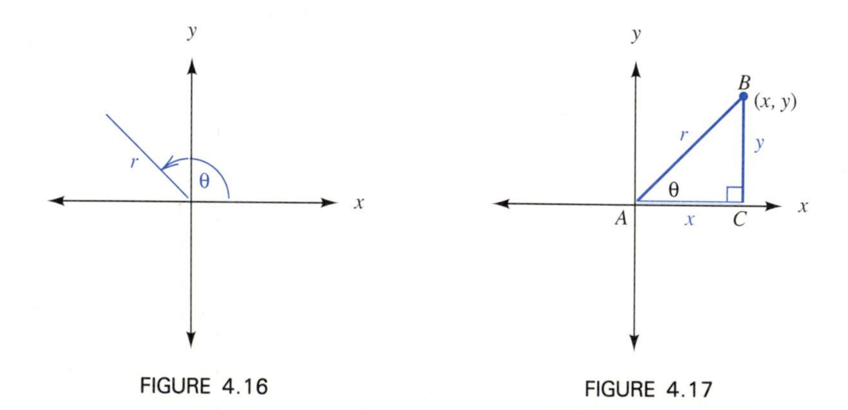

FIGURE 4.16

FIGURE 4.17

Suppose we place a right triangle on a coordinate plane so that one of the acute angles is in standard position as shown in Figure 4.17. Then point B can be specified with coordinates (x, y). The length of AC is x, and the length of BC is y. By the Pythagorean theorem, the hypotenuse has length $r = \sqrt{x^2 + y^2}$.

Using the definitions from Section 4.1, the trigonometric functions can be expressed as follows.

TRIGONOMETRIC FUNCTIONS

If (x, y) is a point on the terminal side of angle θ in standard position, and $r = \sqrt{x^2 + y^2}$ then

$$\sin\theta = y/r \qquad \cot\theta = x/y$$
$$\cos\theta = x/r \qquad \sec\theta = r/x$$
$$\tan\theta = y/x \qquad \csc\theta = r/y$$

NOTE ⇨⇨ Any angle can be placed in standard position. ***If the line segment of length r along the terminal side of the angle ends at the point (x, y), the definitions developed in this section apply to any angle, provided that the definition does not require division by zero.*** If division by zero is required for an angle, the function is undefined for that angle.

EXAMPLE 1 Find the trigonometric function values for θ if $x = 3$ and $y = 4$.

Solution $r = \sqrt{x^2 + y^2} = \sqrt{3^2 + 4^2} = \sqrt{25} = 5$

$$\sin\theta = 4/5 \qquad \cot\theta = 3/4$$
$$\cos\theta = 3/5 \qquad \sec\theta = 5/3$$
$$\tan\theta = 4/3 \qquad \csc\theta = 5/4$$

■

EXAMPLE 2 Find the trigonometric function values for θ if $x = -5$ and $y = 12$.

Solution $r = \sqrt{x^2 + y^2} = \sqrt{(-5)^2 + 12^2} = \sqrt{169} = 13$

$$\sin\theta = 12/13 \qquad \cot\theta = -(5/12)$$
$$\cos\theta = -(5/13) \qquad \sec\theta = -(13/5)$$
$$\tan\theta = -(12/5) \qquad \csc\theta = 13/12$$

■

coterminal angle

The definitions allow two or more angles to have the same trigonometric function values for all six functions. This will happen for angles that have the same initial and terminal sides. Such angles are said to be *coterminal.* One complete rotation of the terminal side through 360° will bring the terminal side back to the same position. Thus θ and $\theta + 360°$ are coterminal. Also, it is possible to measure in the negative direction to find an angle coterminal with a positive angle.

COTERMINAL ANGLES

If θ is an angle in standard position, coterminal angles are given by

$$\theta + n(360°) \text{ and } \theta - n(360°) \text{ for } n = 1, 2, 3, 4, \ldots$$

EXAMPLE 3 If $\theta = 27°$, find an angle coterminal with θ that meets the specified conditions.

(a) an angle greater than 360° **(b)** a negative angle

Solution **(a)** $27° + 360° = 387°$ **(b)** $27° - 360° = -333°$ ■

Since $r = \sqrt{x^2 + y^2}$, the value of r will always be positive. The values of x and y, however, may be positive or negative, depending on the quadrant in which (x, y) lies. If x is 0 or y is 0, the point (x, y) will lie *on* one of the axes. Such points are not *in* any quadrant. The angles associated with points on the axes (0°, 90°, 180°, 270°, and angles coterminal with these angles) are called *quadrantal angles.*

quadrantal angle

Figure 4.18 summarizes information about x, y, and r in the quadrants, and Figure 4.19 summarizes information about x, y, and r for the quadrantal angles.

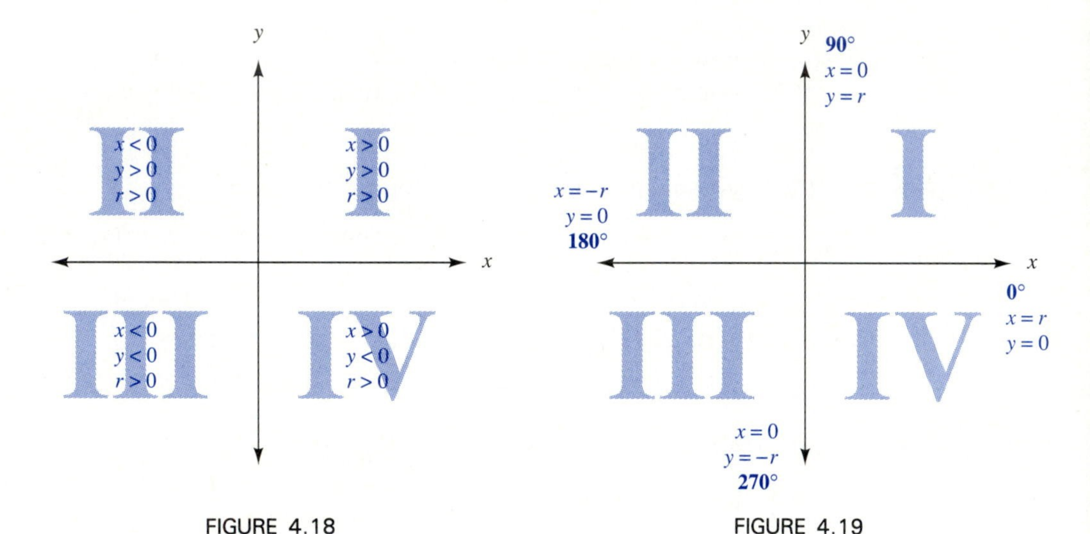

FIGURE 4.18

FIGURE 4.19

NOTE ⇨⇨ ***When the terminal side of an angle lies in a quadrant, we say the angle is in that quadrant, even though it is actually the terminal side of the angle that is in the quadrant.*** Table 4.1 summarizes some general statements about the signs of the trigonometric functions for angles in each quadrant and for quadrantal angles.

TABLE 4.1 Signs of the Trigonometric Functions

	Quadrant				*Quadrantal Angles*			
	I	*II*	*III*	*IV*	*0°*	*90°*	*180°*	*270°*
$\sin\theta = y/r$	+	+	−	−	0	1	0	−1
$\cos\theta = x/r$	+	−	−	+	1	0	−1	0
$\tan\theta = y/x$	+	−	+	−	0	*	0	*
$\cot\theta = x/y$	+	−	+	−	*	0	*	0
$\sec\theta = r/x$	+	−	−	+	1	*	−1	*
$\csc\theta = r/y$	+	+	−	−	*	1	*	−1

* indicates that the function is undefined.

EXAMPLE 4 Determine the sign of the given function.

(a) $\sin 350°$ **(b)** $\cos 225°$

Solution **(a)** Figure 4.20 shows an angle of 350°. The angle is in quadrant IV and $\sin \theta = y/r$. Since $y < 0$ and $r > 0$, $\sin \theta$ is negative.

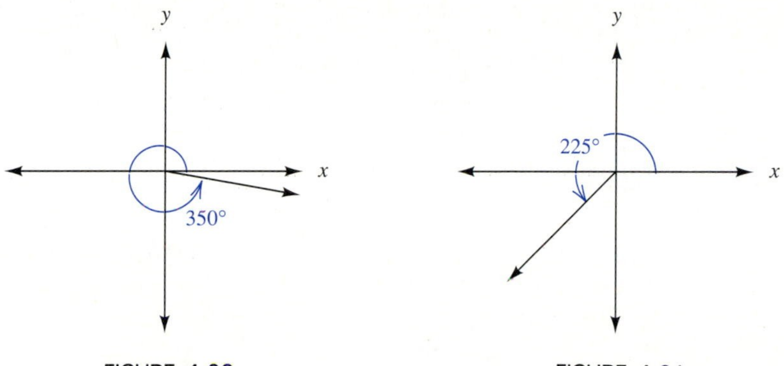

FIGURE 4.20 FIGURE 4.21

(b) Figure 4.21 shows an angle of 225°. The angle is in quadrant III and $\cos \theta = x/r$. Since $x < 0$ and $r > 0$, $\cos \theta$ is negative. ■

EXAMPLE 5 Determine the sign of the given function.

(a) $\tan 450°$ **(b)** $\sec(-200°)$

Solution **(a)** Figure 4.22 shows an angle of 450°. The angle is coterminal with $450° - 360°$ or $90°$ and $\tan \theta = y/x$. Since $y > 0$ and $x = 0$, $\tan \theta$ is undefined for both 90° and 450°.

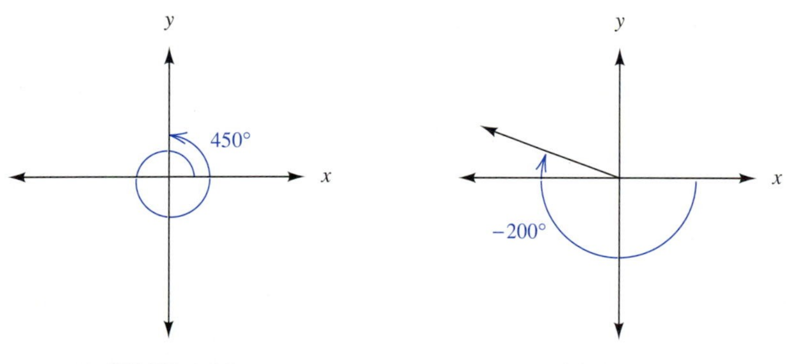

FIGURE 4.22 FIGURE 4.23

(b) Figure 4.23 shows an angle of −200°. The angle is in quadrant II and $\sec \theta = r/x$. Since $x < 0$ and $r > 0$, $\sec \theta$ is negative. ■

EXAMPLE 6 Determine the quadrant in which θ lies under the conditions given.

(a) $\sec\theta < 0$ and $\csc\theta > 0$ **(b)** $\tan\theta < 0$ and $y < 0$

Solution **(a)** Since $\sec\theta = r/x$, $\sec\theta < 0$ when $x < 0$ (quadrants II and III). Since $\csc\theta = r/y$, $\csc\theta > 0$ when $y > 0$ (quadrants I and II). Both conditions are met only if the angle is in the second quadrant.

(b) Since $\tan\theta = y/x$, $\tan\theta < 0$ when x and y have different signs (quadrants II and IV). Since $y < 0$ in quadrants III and IV, θ is in the fourth quadrant. ■

EXERCISES 4.4

In 1–14, find the trigonometric function values for θ, if θ is the angle in standard position with the terminal side determined by the given point.

1. $(12, -5)$ **2.** $(8, -15)$ **3.** $(-24, 7)$ **4.** $(-20, 21)$
5. $(-12, -35)$ **6.** $(-40, -9)$ **7.** $(28, 45)$ **8.** $(16, 63)$
9. $(60, -11)$ **10.** $(-56, 33)$ **11.** $(-48, -55)$ **12.** $(36, 77)$
13. $(-20, 99)$ **14.** $(-84, -13)$

In 15–42, determine the sign of the given function value.

15. $\sin 105°$ **16.** $\cos 120°$ **17.** $\tan 195°$ **18.** $\cot 210°$
19. $\sec 285°$ **20.** $\csc 300°$ **21.** $\csc 135°$ **22.** $\sec 150°$
23. $\cot 225°$ **24.** $\tan 240°$ **25.** $\cos 315°$ **26.** $\sin 330°$
27. $\sin 15°$ **28.** $\csc 30°$ **29.** $\cos(-15°)$ **30.** $\sin(-30°)$
31. $\cot(-165°)$ **32.** $\tan(-150°)$ **33.** $\csc(-255°)$ **34.** $\sec(-240°)$
35. $\sec(-345°)$ **36.** $\csc(-330°)$ **37.** $\tan 465°$ **38.** $\cot 555°$
39. $\sin 645°$ **40.** $\cos 375°$ **41.** $\cos 750°$ **42.** $\sec(-750°)$

In 43–56, determine the quadrant in which the terminal side of θ lies under the given conditions.

43. $\sin\theta > 0$, $\tan\theta < 0$ **44.** $\sin\theta > 0$, $\sec\theta < 0$ **45.** $\cos\theta < 0$, $\tan\theta > 0$
46. $\cos\theta < 0$, $\csc\theta < 0$ **47.** $\tan\theta < 0$, $\sec\theta > 0$ **48.** $\tan\theta < 0$, $\csc\theta < 0$
49. $\sin\theta > 0$, $x < 0$ **50.** $\sin\theta > 0$, $x > 0$ **51.** $\cos\theta < 0$, $y > 0$
52. $\tan\theta < 0$, $y > 0$ **53.** $\tan\theta < 0$, $x < 0$ **54.** $\cot\theta > 0$, $y < 0$
55. $\sec\theta > 0$, $y < 0$ **56.** $\csc\theta < 0$, $x > 0$

In 57–60, specify the angle as a positive angle in standard position for each bearing.

57. N 40° E 1 2 4 7 8 3 5 **58.** S 30° W 1 2 4 7 8 3 5
59. S 20° E 1 2 4 7 8 3 5 **60.** N 50° W 1 2 4 7 8 3 5

1. Architectural Technology
2. Civil Engineering Technology
3. Computer Engineering Technology
4. Mechanical Drafting Technology
5. Electrical/Electronics Engineering Technology
6. Chemical Engineering Technology
7. Industrial Engineering Technology
8. Mechanical Engineering Technology
9. Biomedical Engineering Technology

61. The hand of a clock moves from 3 through an angle of $-330°$. What angle between $0°$ and $360°$ is coterminal with $-330°$? To what number does the clock hand point? 4 7 8 1 2 3 5 6 9

62. Produce scales have a circular dial and will measure up to 10 pounds. Through how many degrees does the needle move to register each pound? If an item causes the needle to move from 0 through an angle of $396°$, how many pounds does the item weigh? 4 7 8 1 2 3 5 6 9

63. For a particular ac circuit, the voltage and current do not peak together. The difference is described by saying that the phase angle is $-45°$. What angle between $0°$ and $360°$ is coterminal with $-45°$? 3 5 9 1 2 4 6 7 8

64. A test driver for a car company reports traveling through an angle of $1800°$ around a circular track. What angle between $0°$ and $360°$ is coterminal with $1800°$? 4 7 8 1 2 3 5 6 9

65. When a projectile is given an initial velocity v_0 from a height h at an angle of θ with the horizontal, its horizontal distance after time t is given by $x = v_0 t \cos\theta$ and its vertical distance by $y = v_0 t \sin\theta - 4.9t^2 + h$. Explain the significance of a negative angle θ for both x and y. 4 7 8 1 2 3 5 6 9

66. For a cantilever beam under a concentrated load, the angle of deflection θ is given by $\tan\theta = Px(x - 2a)/(2EI)$ when $x \le a$. Explain the significance of the sign of $\tan\theta$. (In the formula, P is the load (in kg), x is the distance (in mm) from the support to the point where the deflection occurs, a is the distance (in mm) of the load from the support, E is the modulus of elasticity (in kg/mm^2), and I is the moment of inertia (in mm^4) of the beam.) 1 2 4 8 7 9

67. The equation $\cos\theta = I_R/I$ specifies the phase angle θ between current $I = \sqrt{I_R{}^2 + (I_C - I_L)^2}$ and voltage V for an ac circuit containing a resistor, an inductor, and a capacitor in parallel. Find $\cos\theta$ if $I_R = 1.08$ A, $I_C = 0.93$ A, and $I_L = 0.64$ A. (I_R, I_C, and I_L are the currents (in amps) in the resistor, capacitor, and inductor respectively, and θ is considered positive if $I_C > I_L$ but negative if $I_L > I_C$.) 3 5 9 4 7 8

68. The equation $\tan\theta = (V_L - V_C)/V$ specifies the phase angle θ between voltage $V = \sqrt{V_R{}^2 + (V_L - V_C)^2}$ and current I for an ac circuit containing a resistor, an inductor, and a capacitor in series. Find $\tan\theta$ if $V_R = 7.2$ volts, $V_C = 250$ volts, and $V_L = 240$ volts. (V_R, V_C, and V_L are the voltages in the resistor, capacitor, and inductor respectively, and θ is considered positive if $V_L > V_C$ but negative if $V_C > V_L$.) 3 5 9 4 7 8

69. A mathematician has a bumper sticker that says, "Think in the First Quadrant." What does it mean?

4.5 REFERENCE ANGLES

The trigonometric functions, as defined in the last section, apply to angles larger than $90°$ as well as to acute angles. For any angle, there is a nonnegative angle less than $360°$ coterminal with it, so in this section it is assumed that all angles lie between $0°$ and $360°$.

When a calculator is used to find a trigonometric function value, the angle may be entered directly, even if it is larger than $90°$.

To use a calculator to determine a trigonometric function value:

1. Enter the angle.
2. Press the appropriate trigonometric function key.

EXAMPLE 1 Determine the specified function value directly using a calculator.

(a) sin 110.0° **(b)** cos 200.0° **(c)** cot 290.0°

Solution **(a)** 110 [sin] ***0.9396926***

Answer: 0.940

(b) 200 [cos] ***−0.9396926***

Answer: −0.940

(c) 290 [tan] [1/x] ***−0.3639702***

Answer: −0.364

In part (c), [1/x] is used because $\cot \theta = 1/\tan \theta$. ■

NOTE ⇨⇨

When a function value is known and it is necessary to find the angle associated with it, extreme care must be used. Remember that coterminal angles have the same function values, so there are many angles associated with any function value. ***Furthermore, in most cases, there are two nonnegative angles less than*** **360°** ***associated with a function value. A calculator will return only one of these, and it may not be the correct one for the problem in question.*** It is necessary to know the quadrant in which the angle lies. We avoid confusion by the use of *reference* angles.

> **DEFINITION** The **reference angle** of an angle θ in standard position is the acute angle formed by the x-axis and the terminal side of θ.

Figures 4.24a–4.24d show the relationship of the reference angle to a given angle in any quadrant. The reference angle is shaded. In each case, it can be located by moving from the terminal side of θ to the positive x-axis or negative x-axis, whichever is closer.

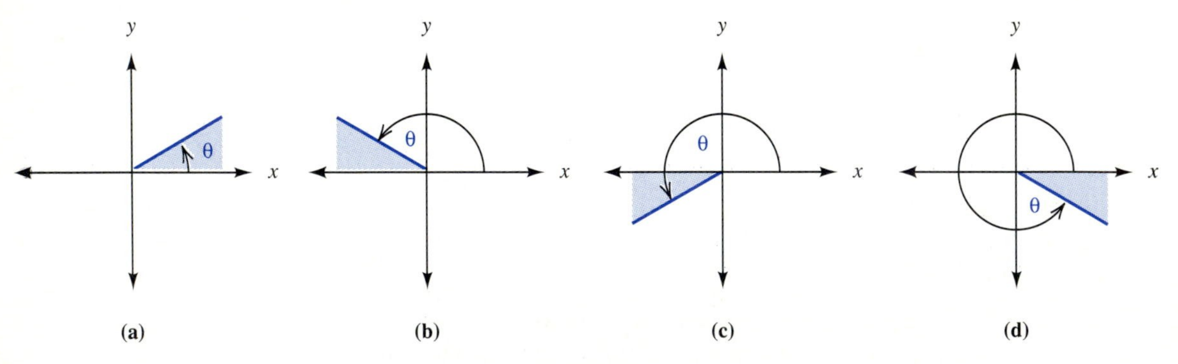

FIGURE 4.24

From the figures, it is possible to see how to determine reference angles (see Table 4.2).

TABLE 4.2 Reference Angles for θ (in degrees)

	Quadrant I	*Quadrant II*	*Quadrant III*	*Quadrant IV*
θ	$0°$ to $90°$	$90°$ to $180°$	$180°$ to $270°$	$270°$ to $360°$
Reference Angle	θ	$180° - \theta$	$\theta - 180°$	$360° - \theta$

The trigonometric function values for an angle are equal in absolute value to the trigonometric function values for its reference angle. The sign can be determined from the quadrant in which the angle lies, just as it was in the previous section. Example 2 shows how reference angles may be used to find trigonometric function values for angles larger than 90°. Although these problems may be done directly on a calculator, the purpose here is to help you understand the relationship between an angle and its reference angle.

EXAMPLE 2 Determine the specified function value.

(a) sin 110.0° **(b)** cos 200.0° **(c)** cot 290.0°

Solution **(a)** Sin 110.0° is positive, since 110.0° is a second quadrant angle. The reference angle θ_R is $180° - 110.0°$ or 70.0°, as in Figure 4.24b. Sin $70.0° = 0.940$, so sin $110.0° = 0.940$ to three significant digits.

(b) Cos 200.0° is negative, since 200.0° is a third quadrant angle. The reference angle θ_R is $200.0° - 180°$ or 20.0°, as in Figure 4.24c. Cos $20.0° = 0.940$, so cos $200.0 = -0.940$ to three significant digits.

(c) Cot 290.0° is negative, since 290.0° is a fourth quadrant angle. The reference angle θ_R is $360° - 290.0°$ or 70.0°, as in Figure 4.24d. Cot $70.0° = 0.364$, so cot $290.0° = -0.364$ to three significant digits. ■

Reference angles are particularly important when it is necessary to find an angle rather than a function value.

To use a calculator to find the angle associated with a given function value:

1. Enter the absolute value of the function value.
2. Use the keystroke sequence that gives the inverse function to obtain the reference angle θ_R on the display.
3. Determine the quadrant in which the angle θ lies.
4. Use the reference angle θ_R from step 2 and the quadrant from step 3 to determine the correct angle θ.

EXAMPLE 3 Use a calculator to find a nonnegative angle less than 360° associated with each function value.

(a) $\sin\theta = 0.7660$, θ is in the second quadrant

(b) $\cos\theta = -0.7660$, θ is in the third quadrant

Solution **(a)** .7660 [INV] [sin] **49.996038**

$\theta_R = 50.00°$

Since θ is in the second quadrant, $\theta = 180° - 50.00°$ or 130.00°.

(b) .7660 [INV] [cos] **40.003962**

$\theta_R = 40.00°$

Since θ is in the third quadrant, $\theta = 180° + 40.00°$ or 220.00°. ■

EXAMPLE 4 Use a calculator to find a nonnegative angle less than 360° associated with each function value.

(a) $\tan\theta = -1.1918$, θ is in the fourth quadrant

(b) $\cot\theta = 1.1918$, θ is in the third quadrant

Solution **(a)** 1.1918 [INV] [tan] **50.001099**

$\theta_R = 50.00°$

Since θ is in the fourth quadrant, $\theta = 360° - 50.00°$ or 310.00°.

(b) 1.1918 [1/x] [INV] [tan] **39.998902**

$\theta_R = 40.00°$

Since θ is in the third quadrant, $\theta = 180° + 40.00°$ or 220.00°. ■

NOTE ⇨⇨ ***The absolute value of a function value is entered into the calculator, so that a first quadrant angle is returned. The first quadrant angle is less than 90° and can be used as the reference angle.*** If a negative function value is entered directly for $\sin\theta$ or $\tan\theta$, a negative acute angle is returned. If a negative function value is entered directly for $\cos\theta$, an angle between 90° and 180° is returned. That angle may not be the desired angle. Using reference angles is generally less confusing.

EXAMPLE 5 The cross section of a road is banked or sloped to help prevent the vehicle from skidding off the road as it rounds a curve. The amount of slope is called the superelevation, and is defined to be the ratio of rise (in ft) to the horizontal distance (in ft) across the road width, as indicated in Figure 4.25. The desired rate of superelevation e is given by $e = \dfrac{V^2}{15R} - f$. In the formula, V is the velocity of the vehicle (in mph), R is the radius of curvature of the curve, and f is the coefficient of friction between the tires and the pavement. Determine the angle of superelevation for a curve with a radius of 1200 ft if the vehicles are expected to travel 60.0 mph, and f is 0.010.

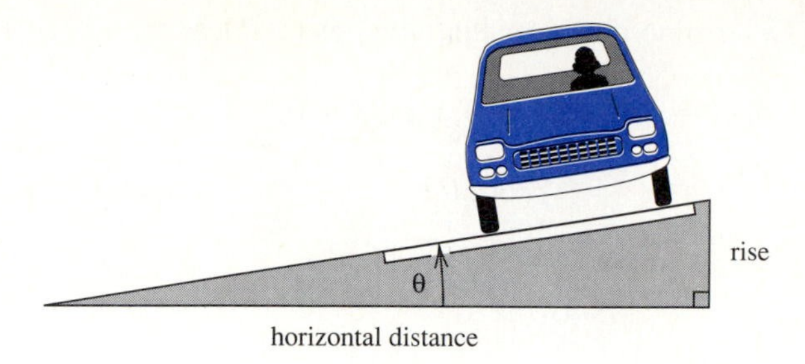

FIGURE 4.25

Solution Figure 4.25 shows that $e = \tan\theta$. But e can be determined from the formula.

$$e = \frac{60.0^2}{15(1200)} - 0.010 = 0.19$$

Hence, $\tan\theta = 0.19$. To solve for θ, we use a calculator.

0.19 [INV] [tan] **10.757967**

Since θ is clearly an acute angle, θ should be about 11°. ■

When the function value is 0, 1, or -1, the associated angle can be determined without the use of a calculator or tables.

EXAMPLE 6 Determine the nonnegative angle(s) θ less than 360° for which $\sin\theta = -1$.

Solution $\sin\theta = y/r$. If $y/r = -1$, then $y = -r$.
Since $x^2 + y^2 = r^2$, we have $x^2 + r^2 = r^2$. Thus, $x^2 = 0$, so $x = 0$. The point $(0, -r)$ is on the terminal side of the angle, so $\theta = 270°$. ■

The steps used in Example 6 are generalized as follows.

To find the angle(s) associated with a function value of 0, 1, −1, or undefined:

1. Choose the appropriate definition y/r, x/r, or y/x for the function.
2. Assign a value of 0, r, or $-r$ to x and y in order to obtain the given function value.
3. Use the values of x, y, and r from step 2 to determine the position of the angle on the coordinate plane.

EXAMPLE 7 Determine the nonnegative angle(s) θ less than $360°$ for which $\cos\theta = 0$.

Solution $\cos\theta = x/r$. If $x/r = 0$, then $x = 0$.

$$\begin{aligned} x^2 + y^2 &= r^2, \\ 0^2 + y^2 &= r^2 \\ y^2 &= r^2 \\ y &= \pm r \end{aligned}$$

The point $(0, r)$ or $(0, -r)$ is on the terminal side of the angle, so $\theta = 90°$ or $\theta = 270°$. ■

Not all of the trigonometric functions are defined for every possible angle. Recall, for example, that $\tan 90°$ is undefined (see Example 5 of Section 4.4). Because $\tan 90°$ is undefined, trying to find $\tan 90°$ on a calculator will produce an error message. Likewise, trying to find an angle associated with certain function values will produce an error message. For the sine and cosine functions, there are no angles associated with values larger than 1 in absolute value. (Why?) Trying to find A if $\sin A = 2$, for example, will produce an error message.

EXERCISES 4.5

In 1–14, determine the specified function value using a calculator.

1. $\sin 130.0°$
2. $\cos 220.0°$
3. $\tan 310.0°$
4. $\cot 330.0°$
5. $\sec 240.0°$
6. $\csc 150.0°$
7. $\csc(-10.0°)$
8. $\sec(-100.0°)$
9. $\cot(-190.0°)$
10. $\tan(-280.0°)$
11. $\cos(370.0°)$
12. $\sin(460.0°)$
13. $\sin(550.0°)$
14. $\cos(640.0°)$

In 15–28, use a calculator to find a nonnegative angle less than 360° associated with each function value.

15. $\sin\theta = 0.4932$, quadrant II
16. $\cos\theta = -0.1405$, quadrant II
17. $\tan\theta = 4.2314$, quadrant III
18. $\cot\theta = 3.3223$, quadrant III
19. $\sec\theta = -2.4132$, quadrant III
20. $\csc\theta = -1.5041$, quadrant IV
21. $\csc\theta = -1.6950$, quadrant III
22. $\sec\theta = 5.0394$, quadrant IV
23. $\cot\theta = -6.9485$, quadrant II
24. $\tan\theta = -7.8576$, quadrant IV
25. $\cos\theta = 0.6950$, quadrant IV
26. $\sin\theta = 0.7869$, quadrant II
27. $\sin\theta = -0.8778$, quadrant III
28. $\cos\theta = 0.9687$, quadrant IV

In 29–42, determine the nonnegative angle(s) less than $360°$ *associated with each function value.*

29. $\sin\theta = 1$

30. $\cos\theta = 1$

31. $\tan\theta = 0$

32. $\cot\theta = 0$

33. $\sec\theta = 1$

34. $\csc\theta = 1$

35. $\csc\theta = -1$

36. $\sec\theta = -1$

37. $\cot\theta$ is undefined

38. $\tan\theta$ is undefined

39. $\cos\theta = 0$

40. $\sin\theta = 0$

41. $\cos\theta = -1$

42. $\sec\theta$ is undefined

43. What is the reference angle for a bearing of N 30° W? 1 2 4 7 8 3 5

44. What is the reference angle for a bearing of S 20° E? 1 2 4 7 8 3 5

45. What is the reference angle for a bearing of S 40° W? 1 2 4 7 8 3 5

46. The formula $\tan\theta = -Pa^2/(2EI)$ gives the angle of deflection θ for a cantilever beam under a concentrated load. Find θ if $P = 2$ kg, $E = 4000$ kg/mm^2, $I = 5$ mm^4, and $a = 30$ mm. In the formula P is the load (in kg), a is the distance (in mm) of the load from the support, E is the modulus of elasticity (in kg/mm^2), and I is the moment of inertia (in mm^4). 1 2 4 8 7 9

47. The equation $\cos\theta = I_R/I$ specifies the phase angle θ between current $I = \sqrt{I_R^2 + (I_C - I_L)^2}$ and voltage V for an ac circuit that contains a resistor, an inductor, and a capacitor in parallel. (In the formula, I_R, I_C, and I_L are the currents (in amps) in the resistor, capacitor, and inductor respectively, and θ is considered positive if $I_C > I_L$ but negative if $I_L > I_C$.) Find θ if $I_R = 1.05$ A, $I_C = 0.97$ A, and $I_L = 0.68$ A. 3 5 9 4 7 8

48. When a projectile is given an initial velocity v_0 from a height h at an angle of θ with the horizontal, its vertical distance after time t is given by $y = v_0 t \sin\theta - 16t^2 + h$. If a projectile is launched with a velocity of 700 ft/s from a height of 200 feet, at what angle should it be launched to hit a target at a height of 100 feet after 2 seconds? 4 7 8 1 2 3 5 6 9

1. Architectural Technology
2. Civil Engineering Technology
3. Computer Engineering Technology
4. Mechanical Drafting Technology
5. Electrical/Electronics Engineering Technology
6. Chemical Engineering Technology
7. Industrial Engineering Technology
8. Mechanical Engineering Technology
9. Biomedical Engineering Technology

4.6 RADIAN MEASURE

Consider a circle of radius 1, called a unit circle, with center at the origin. Imagine wrapping a number line around that circle, just as you might wrap thread around a spool. If 0 is placed at the point (1, 0), distance can be measured along the circumference. With the distance to a point on the circle, there is associated an angle in standard position determined by that point. Since the circumference of a circle is given by the formula $C = 2\pi r$, the circumference of the unit circle is 2π. The distances around this circle of circumference 2π are associated with angles as shown in Figure 4.26. For instance, the distance π is half of the circumference and is associated with 180°, which is halfway around the circle. The distance $\pi/4$ is 1/8 of the circumference and is associated with 45°, which is 1/8 of the way around the circle.

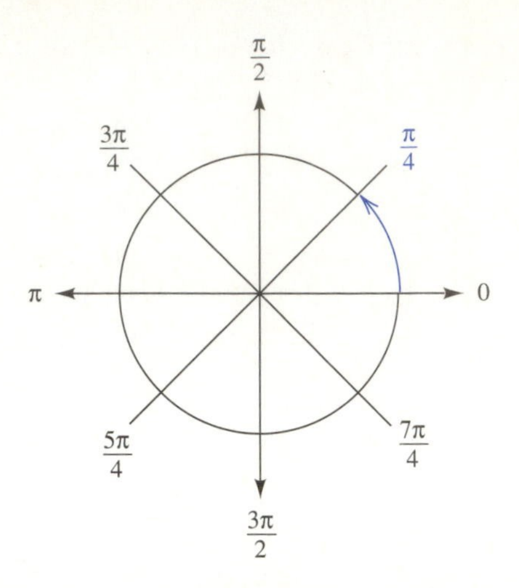

FIGURE 4.26

radian

Angles are sometimes measured in *radians.* One **radian** is the measure of the angle associated with a length of 1 on the circumference of the unit circle. Since the circumference of the unit circle is 2π, there are 2π radians in a full circle, just as there are $360°$ in a full circle. This relationship provides the basis for conversions between degrees and radians. Although we use the ° symbol for degrees, radians are usually specified without any symbol attached, for reasons that will be explained shortly.

To convert between radians and degrees:

Use the equation $360° = 2\pi$ to find the conversion factor.

(a) Multiply a measurement in degrees by $\dfrac{2\pi}{360°}$ or $\dfrac{\pi}{180°}$ to obtain a measurement in radians.

(b) Multiply a measurement in radians by $\dfrac{360°}{2\pi}$ or $\dfrac{180°}{\pi}$ to obtain a measurement in degrees.

EXAMPLE 1 Convert each measurement from radians to degrees.

(a) $\pi/3$ **(b)** $7\pi/6$

Solution **(a)** $(\pi/3)(180°/\pi) = 60°$ **(b)** $(7\pi/6)(180°/\pi) = 210°$ ■

EXAMPLE 2 Convert each measurement from degrees to radians.

(a) 15° **(b)** 300°

Solution **(a)** $15°(\pi/180°) = \pi/12$ **(b)** $300°(\pi/180°) = 5\pi/3$ ■

Just as it is possible to measure an angle larger than 360° or a negative angle in degrees, it is possible to measure an angle larger than 2π or a negative angle in radians.

EXAMPLE 3 Convert each measurement from radians to degrees.

(a) 3π **(b)** $-\pi/4$

Solution **(a)** $(3\pi)(180°/\pi) = 540°$ **(b)** $(-\pi/4)(180°/\pi) = -45°$ ■

The trigonometric functions can be used with angles in radian measure. Because one radian corresponds to a length of 1 along the unit circle, the measure of any angle in radians is numerically equal to the length associated with it on the circumference of the unit circle. Therefore, when angles are measured in radians, a trigonometric function may be regarded as a function with real numbers in the domain. An angle measured in radians is usually specified as a dimensionless quantity, so that it may be treated as an angle or as a real number.

NOTE ⇨⇨ ***When an angle is measured in radians, the variable x is often used instead of θ to emphasize that the argument is to be treated as a real number rather than as an angle.*** If x is a real number, the graph of $y = \sin x$ appears as the graph in Figure 4.27. The trigonometric functions are an important class of functions called *periodic functions.* A periodic function has values that recur at regular intervals. The sine function is sometimes described as an oscillating function, because its values fluctuate between a maximum and minimum value.

periodic function

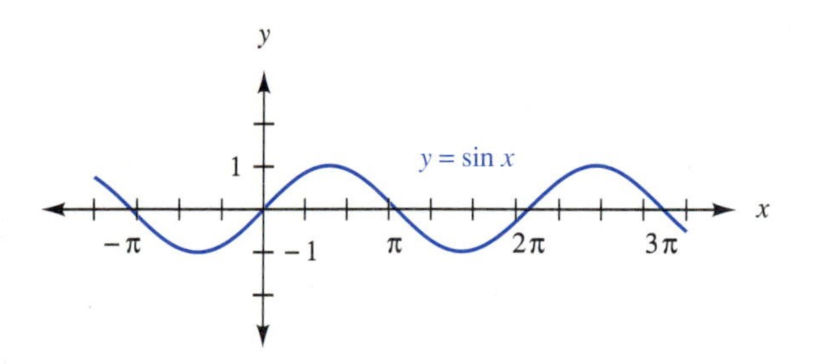

FIGURE 4.27

Special angles, such as 30° and 45°, have function values that can be calculated quickly without the use of tables or a calculator and are often specified as fractions of π. The values for the sine, cosine, and tangent functions for some of the special angles are listed in Table 4.3. Since the cosecant, secant, and cotangent, respectively, are their reciprocals, these function values are not listed.

TABLE 4.3 Trigonometric Functions of Special Angles

Radian	*Degree*	*sin θ*	*cos θ*	*tan θ*
0	0°	0	1	0
$\pi/6$	30°	$1/2$	$\sqrt{3}/2$	$\sqrt{3}/3$
$\pi/4$	45°	$\sqrt{2}/2$	$\sqrt{2}/2$	1
$\pi/3$	60°	$\sqrt{3}/2$	$1/2$	$\sqrt{3}$
$\pi/2$	90°	1	0	*

* indicates that the function is undefined

Calculators will accept angles entered in either degrees or radians. To use radian measure, the calculator must be in *radian mode.* The mode may be set by a special [MODE] key, or there may be a key marked [DRG] that will change the mode.

NOTE ⇨⇨ ***When a calculator is in radian mode, it accepts the argument of a trigonometric function as radian measure and the inverse function keys return an angle in radian measure.*** The display will usually show a decimal number rather than a fraction of π.

Reference angles are required with the inverse functions just as they are when degree mode is used. For a nonnegative angle x such that $0 \leq x < 2\pi$, the reference angles are found as in Table 4.4.

TABLE 4.4 Reference Angles for x (in radians)

	Quadrant I	*Quadrant II*	*Quadrant III*	*Quadrant IV*
x	0 to 1.571	1.571 to 3.142	3.142 to 4.712	4.712 to 6.283
Reference Angle	x	$\pi - x$	$x - \pi$	$2\pi - x$

EXAMPLE 4 Use a calculator to find the following trigonometric function values to four significant digits.

(a) $\sin(2\pi/3)$ **(b)** $\cos(2.766)$ **(c)** $\cot(-1.444)$

Solution Be sure that the calculator is in radian mode.

(a) 2 [×] π [÷] 3 [=] [sin] ***0.8660254***

Answer: 0.8660

(b) 2.766 [cos] ***−0.9302902***

Answer: −0.9303

(c) 1.444 [+/−] [tan] [1/x] ***−0.1274802***

Answer: −0.1275 ■

EXAMPLE 5 Use a calculator to find the angle in radians (to four significant digits) associated with each trigonometric function value.

(a) $\sin x = 0.7670$, x is in the first quadrant

(b) $\cos x = -0.3224$, x is in the second quadrant

(c) $\cot x = 5.684$, x is in the third quadrant

Solution Be sure that the calculator is in radian mode.

(a) In the first quadrant, the reference angle is the angle.

0.7670 [INV] [sin] **0.8741525**

Answer: = 0.8742

(b) Find the reference angle.

0.3224 [INV] [cos] **1.2425325**

Since the angle is in the second quadrant, we subtract the reference angle from π. Leaving the display, we use the following sequence.

[STO] π [−] [RCL] [=] **1.8990602**

Answer: 1.899

or [−] π [=] [+/−] **1.8990602**

Answer: 1.899

(c) Find the reference angle.

5.684 [1/x] [INV] [tan] **0.1741503**

Since the angle is in the third quadrant, we add the reference angle to π. Leaving the display, we use the following sequence.

[+] π [=] **3.3157429**

Answer: 3.316 ■

EXERCISES 4.6

In 1–10, convert each measurement from radians to degrees.

1. $\pi/2$
2. $7\pi/6$
3. $11\pi/6$
4. $\pi/3$
5. $\pi/6$
6. $-2\pi/3$
7. $-\pi/2$
8. $-5\pi/4$
9. $-5\pi/6$
10. $-\pi/3$

In 11–20, convert each measurement from degrees to radians. Express each answer as a fraction of π.

11. 150°
12. 210°
13. 330°
14. 135°
15. 180°
16. 600°
17. 460°
18. −120°
19. −270°
20. 225°

In 21–30, convert each measurement from degrees to radians. Round each answer to three significant digits.

21. 64.0°
22. 55.3°
23. 44.6°
24. 37.5°
25. 112.0°
26. 302.0°
27. 402.5°
28. −36.7°
29. −85.4°
30. 27.0°

In 31–41, use a calculator to find each trigonometric function value. Round each answer to four significant digits.

31. sin 1.549
32. cos 4.276
33. tan 3.367
34. cot 2.458
35. sec 0.450
36. csc 1.369
37. csc −2.071
38. sec −3.187
39. cot −0.962
40. tan −2.278
41. cos −1.763

In 42–52, use a calculator to find the angle in radians associated with each trigonometric function value.

42. $\sin\theta = 0.2356$, quadrant I
43. $\cos\theta = 0.6492$, quadrant I
44. $\tan\theta = 3.7441$, quadrant I
45. $\cot\theta = 2.3837$, quadrant I
46. $\sec\theta = 4.4780$, quadrant IV
47. $\sin\theta = 0.4748$, quadrant II
48. $\cos\theta = 0.9651$, quadrant IV
49. $\cot\theta = 0.5837$, quadrant III
50. $\tan\theta = -0.2478$, quadrant II
51. $\cos\theta = -0.4926$, quadrant II
52. $\sin\theta = -0.3387$, quadrant III

53. Explain the pattern in Table 4.5, and show that the values are equivalent to those in Table 4.3.

TABLE 4.5 Trigonometric Functions of Special Angles

Radian	*Degree*	*sin θ*	*cos θ*	*tan θ*
0	0°	$\sqrt{0}/2$	$\sqrt{4}/2$	$\sqrt{0}/\sqrt{4}$
$\pi/6$	30°	$\sqrt{1}/2$	$\sqrt{3}/2$	$\sqrt{1}/\sqrt{3}$
$\pi/4$	45°	$\sqrt{2}/2$	$\sqrt{2}/2$	$\sqrt{2}/\sqrt{2}$
$\pi/3$	60°	$\sqrt{3}/2$	$\sqrt{1}/2$	$\sqrt{3}/\sqrt{1}$
$\pi/2$	90°	$\sqrt{4}/2$	$\sqrt{0}/2$	*

* indicates that the function is undefined

4.7 ARCS, SECTORS, AND VELOCITY

arc

angle subtended by an arc

Radian measure is useful for certain applications. For example, radian measure is used to compute *arc length* for a circle. A part of a curve is called an **arc.** For a circle, we say that the angle formed by the radii having end points at the end points of the arc is **subtended** by the arc, as shown in Figure 4.28.

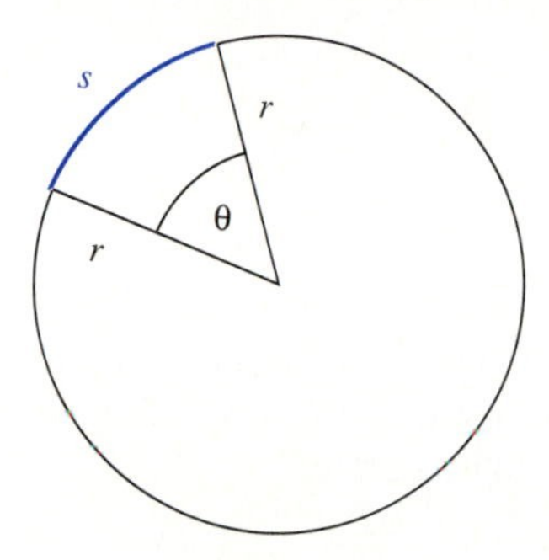

FIGURE 4.28

If a circle has a radius of r, we find a formula for the length of an arc that subtends angle θ, if θ is measured in radians, as follows. Let s = length of the arc. The arc length is the same fraction of the circumference as the angle is of 2π. Therefore, the ratio of the length of the arc to the entire circumference is equal to the ratio of the angle subtended to 2π.

$$\frac{s}{2\pi r} = \frac{\theta}{2\pi} \qquad \text{so} \qquad s = \frac{\theta}{2\pi}(2\pi r)$$

Thus,

$$s = r\theta. \tag{4-1}$$

EXAMPLE 1 Consider a circle with a diameter of 8 inches. Find the length of the arc that subtends an angle of $\pi/3$.

Solution The formula $s = r\theta$ calls for radius. If the diameter is 8 inches, then the radius is 4 inches. Thus,

$$s = 4(\pi/3) \qquad \text{or} \qquad 4\pi/3, \text{ which is about 4.19 inches.} \ ■$$

sector of a circle

Radian measure is also used to find the area of a *sector* of a circle. A **sector of a circle** is the interior part of a circle bounded by two radii and an arc. Thus, a sector of a circle is shaped like a piece of pie, as shown in Figure 4.29.

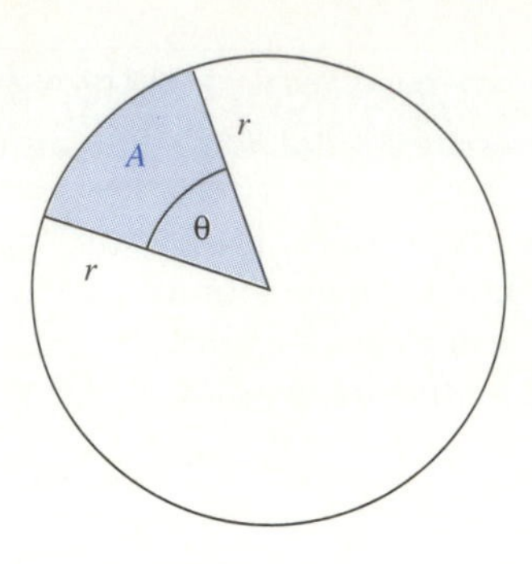

FIGURE 4.29

If a circle has a radius of r, we find a formula for the area of a sector whose arc subtends angle θ, if θ is measured in radians, as follows. Let $A =$ the area of the sector. The area of the sector is the same fraction of the area of the circle as the angle is of 2π. Therefore, the ratio of the area of the sector to the entire area is equal to the ratio of the angle subtended to 2π.

$$\frac{A}{\pi r^2} = \frac{\theta}{2\pi} \qquad \text{so} \qquad A = \frac{\theta}{2\pi}(\pi r^2)$$

Thus,

$$A = \frac{\theta r^2}{2} \qquad \text{or} \qquad (1/2)\theta r^2. \tag{4-2}$$

EXAMPLE 2 Consider a circle with a radius of 6 cm. Find the area of a sector if the arc of the sector subtends an angle of 30°.

Solution The formula $A = (1/2)\theta r^2$ calls for the angle in *radians.*

$$30° = 30(\pi/180) \qquad \text{or} \qquad \pi/6$$

$$A = \frac{1}{2}\left(\frac{\pi}{6}\right)6^2 \text{ or } 3\pi \text{, which is about } 9.42 \text{ cm}^2. \quad ■$$

Equations 4-1 and 4-2 may also be used to find the radius or the angle subtended under certain circumstances. Example 3 illustrates this concept.

EXAMPLE 3 Find the radius of a circle if the area of a sector whose arc subtends an angle of 45° is 6.28 ft^2.

Solution Once again, to use the formula $A = (1/2)\theta r^2$, the angle must be in radians.

$$45° = 45(\pi/180) \qquad \text{or} \qquad \pi/4$$

Substitute the values of A and θ into Equation 4-2.

$$6.28 = \frac{1}{2}\left(\frac{\pi}{4}\right)(r^2)$$

$$\frac{8(6.28)}{\pi} = r^2$$

$$4.00 = r$$

The radius is 4.00 ft to three significant digits. ■

Finally, we find a formula for linear velocity of an object moving around a circle of radius r with angular velocity ω, where ω is in radians per unit time.

Let v = the linear velocity. Linear velocity is given by the formula $v = d/t$. The distance traveled around the circle is given by Equation 4-1 as $s = \theta r$. Thus, we have $v = \theta r/t$. But θ/t is the angular velocity ω in radians per unit time. Thus,

$$v = \omega r. \tag{4-3}$$

EXAMPLE 4 Find the linear velocity of an object moving with an angular velocity of 3 rad/min around a circle whose diameter is 10 in.

Solution Since the diameter of the circle is 10″, the radius is 5″.

$$v = \omega r$$

$$v = (3)(5) \quad \text{or} \quad 15 \text{ inches/minute}$$ ■

NOTE ⇨⇨ ***It is very important that θ is measured in radians when the formulas of this section are used.*** These formulas are summarized as follows.

ARCS, SECTORS, AND VELOCITY

For a circle of radius r, and arc that subtends an angle θ,
$s = r\theta$ where s is the length of the arc and θ is measured in radians.
$A = (1/2)\theta r^2$ where A is the area of the sector and θ is measured in radians.
$v = \omega r$ where v is the linear velocity of an object moving around the circle, and ω is the angular velocity measured in radians per unit time.

In technical applications, angular velocity is often given in revolutions per minute (rpm). Since there are 2π radians in one revolution, rpm's can be converted to radians per minute using a conversion factor of 2π.

EXAMPLE 5 A searchlight on a ship sweeps through an angle of 80.0°. If the range of the light is 121 yards, what area can be searched?

Solution The searchlight sweeps out a sector of a circle as shown in Figure 4.30. The formula for the area of a sector of a circle is given by Equation 4-2 as $A = \dfrac{\theta r^2}{2}$, if θ is in radians. Converting 80.0° to radians, we have $80.0° = 80.0° (\pi/180°) = 1.40$.

$$A = 1.40(121)^2/2$$

The area that can be searched, then, is 10,200 square yards, to three significant digits. ■

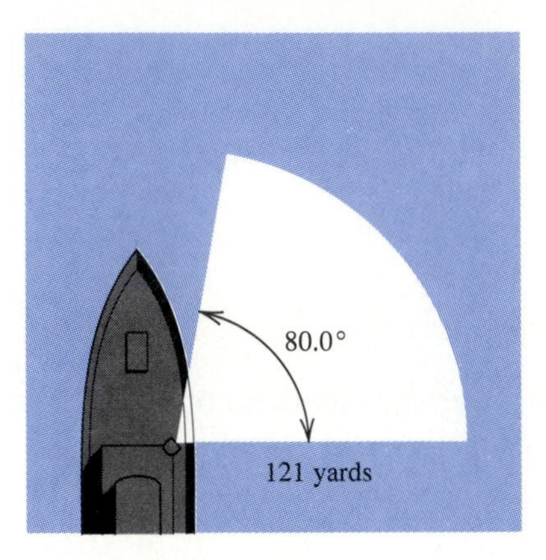

FIGURE 4.30

EXERCISES 4.7

In 1–8, consider a circle of radius r (or diameter d), and find the length of the arc that subtends an angle of θ. Give the answer to three significant digits.

1. $r = 7$ cm, $\theta = \pi/6$ **2.** $r = 5$ in, $\theta = 2\pi/3$ **3.** $d = 10$ ft, $\theta = 3\pi/4$ **4.** $d = 18$ m, $\theta = \pi/2$

5. $r = 3$ m, $\theta = 270°$ **6.** $r = 11$ in, $\theta = 15°$ **7.** $r = 24$ ft, $\theta = 20°$ **8.** $r = 14$ cm, $\theta = 210°$

9. If the length of an arc of a circle is 12.3 m, and the angle subtended is 1.22, find the radius of the circle.

10. If the length of an arc of a circle is 34.8 mm, and the angle subtended is 5.96, find the radius of the circle.

11. If the length of an arc of a circle is 4.8 in, and the radius of the circle is 1.49 in, find the angle subtended by the arc.

12. If the length of an arc of a circle is 7.5 ft, and the radius of the circle is 2.37 ft, find the angle subtended by the arc.

In 13–20, consider a circle of radius r (or diameter d), and find the area of a sector whose arc subtends an angle of θ. Give the answer to three significant digits.

13. $r = 7$ cm, $\theta = \pi/6$ **14.** $r = 5$ in, $\theta = 2\pi/3$ **15.** $d = 10$ ft, $\theta = 3\pi/4$ **16.** $d = 18$ m, $\theta = \pi/2$

17. $r = 3$ m, $\theta = 270°$ **18.** $r = 11$ in, $\theta = 15°$ **19.** $r = 24$ ft, $\theta = 20°$ **20.** $r = 14$ cm, $\theta = 210°$

21. If the area of a sector of a circle is 8.22 m^2, and the angle subtended by the arc of the sector is 1.22, find the radius of the circle.

22. If the area of a sector of a circle is 7.91 cm^2, and the angle subtended by the arc of the sector is 2.38, find the radius of the circle.

23. If the area of a sector of a circle is 8.41 in^2, and the radius of the circle is 2.17 in, find the angle subtended by the arc.

24. If the area of a sector of a circle is 12.7 ft^2, and the radius of the circle is 3.48 ft, find the angle subtended by the arc.

In 25–32, find the linear velocity of an object moving around a circle of radius r (or diameter d) with angular velocity of ω.

25. $r = 9.41$ m, $\omega = 2.65$/s

26. $r = 18.2$ cm, $\omega = 1.43$/s

27. $d = 4.0$ ft, $\omega = 2.0$/min

28. $d = 2.0$ in, $\omega = 1.5$/min

29. $r = 24.3$ m, $\omega = 2.11$ rpm

30. $r = 8.4$ cm, $\omega = 14$ rpm

31. $d = 12$ ft, $\omega = 5.0$ rpm

32. $d = 44$ in, $\omega = 11$ rpm

33. Find the angular velocity of an object moving with linear velocity of 3 ft/s around a circle of diameter 10 ft.

34. Find the angular velocity of an object moving with a linear velocity of 1.5 m/s around a circle of diameter 7.5 m.

35. Find the radius of a circle if an object moves around it with an angular velocity of 2.3/s and a linear velocity of 8.8 m/s.

36. Find the radius of a circle if an object moves around it with an angular velocity of 1.19/s and a linear velocity of 3.72 ft/s.

---

37. A record with a diameter of 12″ is played at 33.3 rpm for 7.1 minutes. How far does a point on the rim travel in that time? 4 7 8 1 2 3 5 6 9

38. An automobile is parked so that there is a nail 7″ behind the rear wheel. If the tire is 26″ in diameter, through what angle will the wheel turn before reaching the nail? 4 7 8 1 2 3 5 6 9

39. A clock of radius 3.2 cm is designed so that the hour marks divide the face into sectors. Adjoining sectors are to be painted different colors. If a small jar of paint contains 15 ml of paint and each ml covers 95 cm^2, how many sectors could be painted with one jar of paint? 1 2 4 7 8 3 5

40. A driveway is constructed so that it is the outer portion of a sector with a central angle of 122°, and the outer edge has a length of 12.8 m. If the driveway is 3.15 m wide, what is its area? 1 2 4 7 8 3 5

41. A flywheel will rotate at 31.2 radians/s. If the linear velocity of a point on the rim is to be 73.71 in/s, what should the radius of the flywheel be? 4 7 8 1 2 3 5 6 9

42. A carousel is designed so that there are two seats side-by-side with 4.00 feet between seats. If the outer seat moves with an angular velocity of 3.89 rpm and a linear velocity of 440.0 ft/min, what is the linear velocity of an inner seat? 4 7 8 1 2 3 5 6 9

43. A circular arc is being laid out as part of a driveway for a new home. Plans show that when measured to the centerline of the driveway, the radius of curvature is 350 ft and the central angle of the arc is 37°25′. Find the length of the driveway along the centerline. If the driveway is 14 ft wide, find its length along the outer edge. 1 2 4 7 8 3 5

---

1. Architectural Technology
2. Civil Engineering Technology
3. Computer Engineering Technology
4. Mechanical Drafting Technology
5. Electrical/Electronics Engineering Technology
6. Chemical Engineering Technology
7. Industrial Engineering Technology
8. Mechanical Engineering Technology
9. Biomedical Engineering Technology

4.8 CHAPTER REVIEW

KEY TERMS

4.1 right triangle
acute angle
hypotenuse
legs
trigonometric functions
complementary angles

4.3 transversal
angle of elevation
angle of depression
bearings

4.4 initial side of an angle
terminal side of an angle
standard position
coterminal angle
quadrantal angle

4.5 reference angle

4.6 radian
periodic function

4.7 arc
angle subtended by an arc
sector of a circle

SYMBOLS

$\angle$ angle
$^\circ$ degree

RULES AND FORMULAS

For a right triangle with sides of length a, b, and c where c is the hypotenuse, $a^2 + b^2 = c^2$.

$\sin\theta = \dfrac{\text{length of opposite side}}{\text{length of hypotenuse}}$ $\cos\theta = \dfrac{\text{length of adjacent side}}{\text{length of hypotenuse}}$ $\tan\theta = \dfrac{\text{length of opposite side}}{\text{length of adjacent side}}$

$\cot\theta = 1/\tan\theta$ $\sec\theta = 1/\cos\theta$ $\csc\theta = 1/\sin\theta$

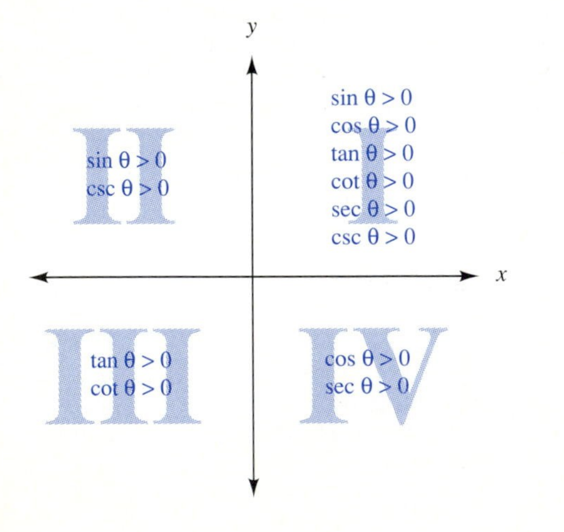

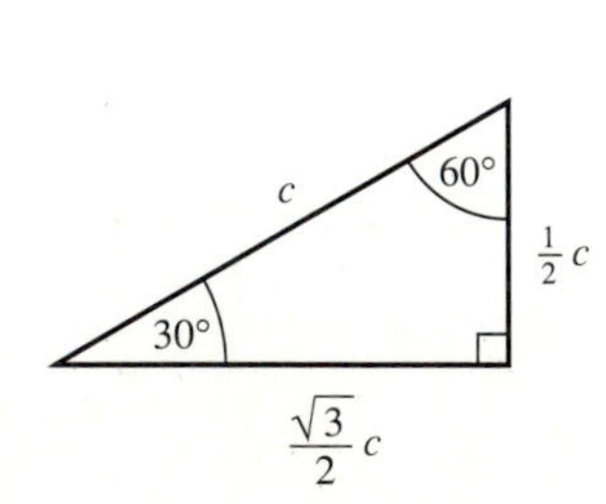

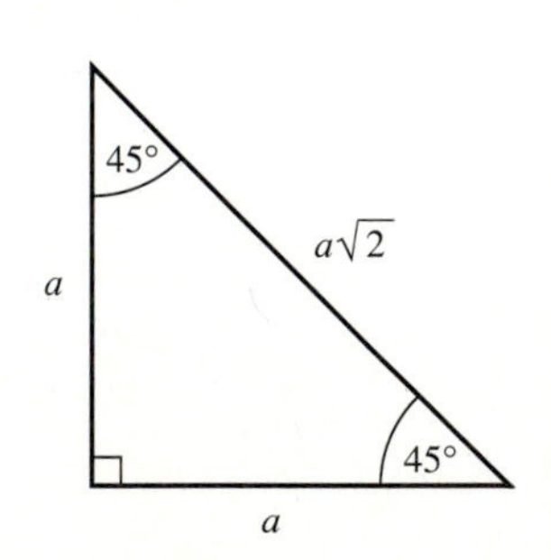

For a circle of radius r, and arc that subtends angle θ,

$s = r\theta$ where s is the length of the arc and θ is measured in radians.

$A = (1/2)\theta r^2$ were A is the area of the sector and θ is measured in radians.

$v = \omega r$ where v is the linear velocity of an object moving around the circle and ω is the angular velocity measured in radians per unit time.

GEOMETRY CONCEPTS

$C = \pi d$ (circumference of a circle)

The angles of an equilateral triangle are equal.

In the accompanying figure, the following pairs of angles are equal.

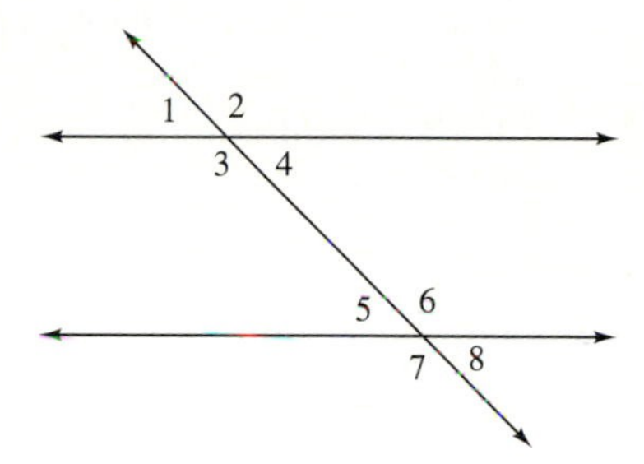

vertical angles: $\angle 1$ and $\angle 4$, $\angle 2$ and $\angle 3$, $\angle 5$ and $\angle 8$, $\angle 6$ and $\angle 7$

corresponding angles: $\angle 1$ and $\angle 5$, $\angle 2$ and $\angle 6$, $\angle 3$ and $\angle 7$, $\angle 4$ and $\angle 8$

alternate interior angles: $\angle 3$ and $\angle 6$, $\angle 4$ and $\angle 5$

SEQUENTIAL REVIEW

(Section 4.1) *In 1 and 2, use the Pythagorean theorem to find the unknown side of each triangle.*

1. $a = 7.4$, $b = 12.3$

2. $b = 7.4$, $c = 9.46$

In 3–5, write the given trigonometric function as a ratio of the lengths of two of the sides of the triangle with a = 28 *mm,* b = 45 *mm, and* c = 53 *mm.*

3. $\sin A$

4. $\csc B$

5. $\sec A$

In 6 and 7, use a calculator to find each function value to four decimal places.

6. $\cos 73.2°$

7. $\cot 14.7°$

In 8 and 9, use a calculator to find the first quadrant angle (to the nearest tenth of a degree) associated with each function value.

8. $\tan \theta = 5.4029$

9. $\sec \theta = 4.6011$

(Section 4.2)

10. Solve the right triangle for which $a = 12.0$ in, $A = 17.1°$, and $C = 90°$.

11. Solve the right triangle for which $a = 7.3$ cm, $b = 4.1$ cm, and $C = 90°$.

In 12–14, find the indicated part of each triangle.

12. a if $B = 12°$ and $c = 7.5$ cm

13. B if $b = 3.4$ m and $a = 4.3$ m

14. c if $A = 15.3°$ and $a = 4.29$ ft

15.

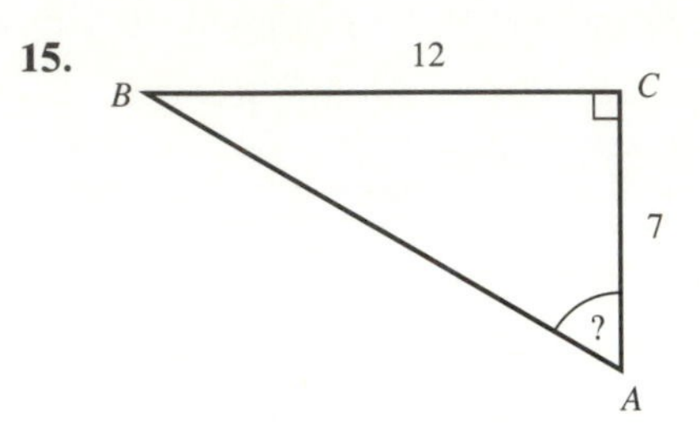

In 16 and 17, give an exact value (i.e., use radicals rather than decimal approximations) for each trigonometric function.

16. sin 30°

17. cos 45°

(Section 4.3)

18. During a solar eclipse, an image of the sun is projected from a pinpoint mirror image to a wall 10.0 feet away, as shown in the accompanying diagram. If the image on the wall has a diameter of 1.00 in, find the angle subtended by the full sun. 3 5 6 9

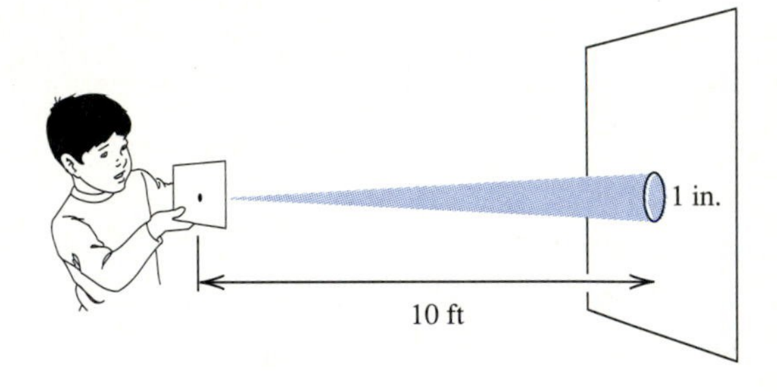

19. If a searchlight is aimed at a target located at a horizontal distance of 110 m with an angle of elevation of 31.5°, but the angle is incorrectly measured as 35.1°, by how far will the light miss the target? 1 2 4 7 8 3 5

(Section 4.4) In 20–22, find the trigonometric function values for θ, if θ is the angle in standard position whose terminal side is determined by the given point.

20. $(-3, -4)$

21. $(12, -35)$

22. $(-7, -24)$

In 23–25, determine the sign of the given function value.

23. sin 110°

24. sec (−25°)

25. tan 215°

In 26–28, determine the quadrant in which θ lies.

26. $\cos \theta > 0$, $y < 0$

27. $\tan \theta < 0$, $x < 0$

28. $\sin \theta < 0$, $\sec \theta > 0$

(Section 4.5) In 29–31, use a calculator to find the given function value to four decimal places.

29. tan (−340.0°)

30. cos 115.0°

31. csc (−35.0°)

1. Architectural Technology
2. Civil Engineering Technology
3. Computer Engineering Technology
4. Mechanical Drafting Technology
5. Electrical/Electronics Engineering Technology
6. Chemical Engineering Technology
7. Industrial Engineering Technology
8. Mechanical Engineering Technology
9. Biomedical Engineering Technology

32. Use a calculator to find θ in the second quadrant if $\cos\theta = -0.5327$.

33. Use a calculator to find θ in the third quadrant if $\tan\theta = 2.1132$.

34. Use a calculator to find θ in the fourth quadrant if $\sec\theta = 1.1132$.

35. Determine the nonnegative angles(s) θ less than 360° for which $\csc\theta$ is undefined.

36. Determine the nonnegative angle(s) θ less than 360° for which $\sin\theta = -1$.

(Section 4.6)

37. Convert $11\pi/3$ to degrees.

38. Convert 47.3° to radians.

39. Use a calculator to find $\cos(-2.4597)$ to four significant digits.

40. Use a calculator to find a second quadrant angle in radians associated with $\sin\theta = 0.5186$.

(Section 4.7)

41. Consider a circle with radius 2.00 cm. Find the length of the arc that subtends an angle of $7\pi/12$.

42. Consider a circle with radius 8 in. Find the length of the arc that subtends an angle of 10°.

43. Consider a circle with diameter 6 m. Find the area of a sector whose arc subtends an angle of $\pi/3$.

44. If the area of a sector of a circle is 18.4 ft^2, and the arc of the sector subtends an angle of $\pi/6$, find the radius of the circle.

45. Find the linear velocity of an object moving around a circle of radius 1.00 cm if the angular velocity is 3.72/s.

APPLIED PROBLEMS

1. The pendulum of a clock travels through an angle of 19.8°. What distance is covered by the pendulum bob if the length of the pendulum is 24.8 cm? 4 7 8 1 2 3 5 6 9

2. A "flat" roof on a building actually slopes downward toward a roof drain at the rate of 1/4 inch per foot of horizontal distance. Find the angle of inclination of the roof. 1 2 4 7 8 3 5

3. A technician must change the light on a microwave tower. The shadow of the tower is 24 feet long, and the angle of elevation from the end of the shadow to the light is 65°. How high is the light? 1 2 4 7 8 3 5

4. A reconnaissance plane flying at a height of 5000 ft sights a submarine breaking the surface ahead of the plane. Directly beyond it is an aircraft carrier. If the angle of depression from the plane to the submarine is 58° and to the aircraft carrier is 28°, find the horizontal distance between the submarine and the aircraft carrier. 1 2 4 7 8 3 5

5. A skydiver jumps from an altitude of 3000 ft and is expected to land at a certain spot. An observer at that spot determines the angle of elevation to the skydiver to be 70°. Through what horizontal distance must the skydiver travel? 4 7 8 1 2 3 5 6 9

6. A carousel moves with a constant angular velocity and completes 3.98 revolutions in 1000 seconds. Find the linear velocity of a point on the edge of the carousel if the radius of the carousel is 12 ft. 4 7 8 1 2 3 5 6 9

7. If an airplane engine rotates too fast, the tip of its propeller blade will break the sound barrier. What rpm would cause a 3.00-ft propeller blade to reach 1090 ft/s (Mach I)? 1 3 4 5 6 8 9

8. The internal critical angle C in a fiber optic transmission line is given by $\sin C = \dfrac{\sqrt{n_1^2 - n_2^2}}{n_1}$ where n_1 and n_2 are the indexes of refraction of the core and cladding, respectively. Find C for a fiber with $n_1 = 1.57$ and $n_2 = 1.43$. 3 5 6 9

9. When a stream of photons in air enters a fiber optic cable at an angle, it changes direction inside the glass. The angle of incidence θ_1 and angle of refraction θ_2 are related by the equation $\frac{\sin \theta_1}{\sin \theta_2} = n$ where n is the index of refraction of the glass. What value of θ_2 would result from a photon striking the glass at a 20° angle? Use $n = 1.5$. 3 5 6 9

10. A high speed centrifuge 1.1 inches in diameter rotates at 20,000 revolutions per second. What is the speed in mph of a point on its outside edge? Give your answer to four significant digits. **4 7 8** 1 2 3 5 6 9

11. Find the speed (in ft/s) with which you would have to pull wire off a full spool 15 inches in diameter to keep up with the feed-out motor if it turns the spool at 75 rpm? **4 7 8** 1 2 3 5 6 9

12. A new radial tire is advertised to have a 45,000 mile tread wear out. If the tire's diameter is 26 in, estimate the number of revolutions the tire will make before wearing out. (Hint: First determine how far the tire travels in one revolution.) **4 7 8** 1 2 3 5 6 9

13. Estimate the angle in degrees through which a person needs to shift his/her eyes to scan the moon from edge to edge. The moon's diameter is 3476 km, and it is 38×10^4 km from the earth. Give your answer to the nearest tenth of a degree. 3 5 6 9

14. Consider a laser rangefinder that focuses a red dot on two objects 5.00 miles away. If there is 1.00 ft between the objects, find the angle between them at the rangefinder. 3 5 6 9

15. A radio transmitter beams power outward through an angle of 26°. If the range of the transmitter is 40.0 miles and the population density is 75 people per square mile, how many people can be served? (Hint: First find the area of the sector served.) 1 3 4 5 6 8 9

16. In ac circuit theory it is common to express some angles as the sum of two angles, one of which is expressed in radians and the other in degrees. Unless one is very careful to do the proper conversion, wrong answers can be obtained. Solve the following equation for v when $t = 0.007$ seconds.

$$v = 622 \sin (377 \cdot t^R + 32°)$$

Give your answer to three significant digits. **3 5 9** 1 2 4 6 7 8

17. Approximately 5900 data bits can be stored per inch along an arc of the circumference of a double density 5-1/4-inch diameter floppy disk. Estimate the number of bits that could be detected per second by a read/write head as it scans the disk near its outside edge. The disk turns at approximately 300 rpm. (Hint: First determine the number of radians traveled in one second.) **3 5** 9

18. The impedance Z (in ohms) of an ac circuit consisting of a resistance R (in ohms) in series with a reactance X (in ohms) is the hypotenuse of a right triangle with sides R and X. The phase angle θ is the angle which has a tangent X/R. A reactance of 100 ohms is placed in series with a resistance of 27 Ω. If R drifts upward 10%, by what percent will Z change? **3 5 9** 1 2 4 6 7 8

19. When a capacitor with X ohms of reactance is placed in series with a resistor with R ohms of resistance, the power factor F of the resulting circuit is given by the equation $F = \cos \theta$ where $\tan \theta = \frac{X}{R}$. Find the power factor of a circuit with $R = 26$ ohms and $X = 45$ ohms. **3 5 9** 4 7 8

20. The phase angle for a circuit containing a resistor, an inductor, and a capacitor in series is given by $\sec \theta = \sqrt{1 + \frac{L}{R^2C}}$. Find θ if $L = 0.33$ henry, $R = 412$ ohms, and $C = 0.47\ \mu$F. In the formula, L is the inductance, R is the resistance, and C is the capacitance. **3 5 9** 1 2 4 6 7 8

---

1. Architectural Technology
2. Civil Engineering Technology
3. Computer Engineering Technology
4. Mechanical Drafting Technology
5. Electrical/Electronics Engineering Technology
6. Chemical Engineering Technology
7. Industrial Engineering Technology
8. Mechanical Engineering Technology
9. Biomedical Engineering Technology

RANDOM REVIEW

1. If the length of an arc of a circle is 4.88 m, and the radius of the circle is 0.53 m, find the angle subtended by the arc.

2. Use a calculator to find sec 86.1° to four decimal places.

3. Use a calculator to find sin (−235°) to four decimal places.

4. Consider a right triangle for which $a = 5.1$ ft and $b = 4.3$ ft. Find A.

5. Determine the sign of csc (315°).

6. Determine the quadrant in which θ lies if $\cos\theta < 0$ and $y > 0$.

7. Give an exact value for cos 60°.

8. Determine the nonnegative angles less than 360° for which $\tan\theta$ is undefined.

9. Use a calculator to find the first quadrant angle θ for which $\csc\theta = 3.7902$. Give your answer to the nearest tenth of a degree.

10. Find the linear velocity of an object moving around a circle of radius 6 in, if the angular velocity is 2 rpm.

CHAPTER TEST

1. Write the given trigonometric function as a ratio of the lengths of two sides of the triangle shown in the accompanying figure.

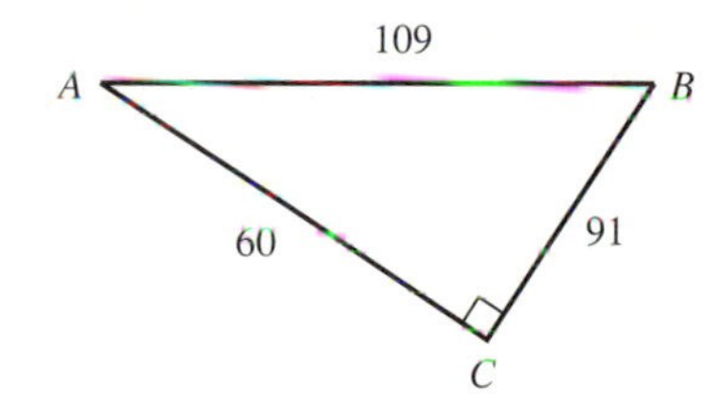

(a) $\sin A$ **(b)** $\tan B$

2. Solve the right triangle for which $a = 6.3$ cm, $B = 24.7°$, and $C = 90°$.

3. Solve the right triangle for which $a = 9.10$ in and $b = 15.8$ in.

4. Determine the quadrant in which θ lies if the given conditions hold.
(a) $\cos\theta < 0, y < 0$ **(b)** $\tan\theta < 0, x > 0$

5. Use a calculator to find the fourth quadrant angle associated with $\tan\theta = -0.8234$.
(a) Give the answer in degrees. **(b)** Give the answer in radians.

6. Use a calculator to find the fourth quadrant angle associated with $\sin\theta = -0.1961$.
(a) Give the answer in degrees. **(b)** Give the answer in radians.

7. Determine the nonnegative angles less than 360° for which the given function is undefined.
(a) $\sec\theta$ **(b)** $\cot\theta$

8. Consider a circle with radius 6 inches and an arc that subtends an angle of $\pi/6$.
(a) Find the length of the arc to three significant digits.
(b) Find the area of the sector bounded by the arc to three significant digits.

9. The "Golden Ratio" is used in architectural and artistic designs. A rectangle that is constructed so that its sides have this ratio is considered to be pleasing to the eye. If the rectangle shown in the accompanying diagram is a golden rectangle, find θ.

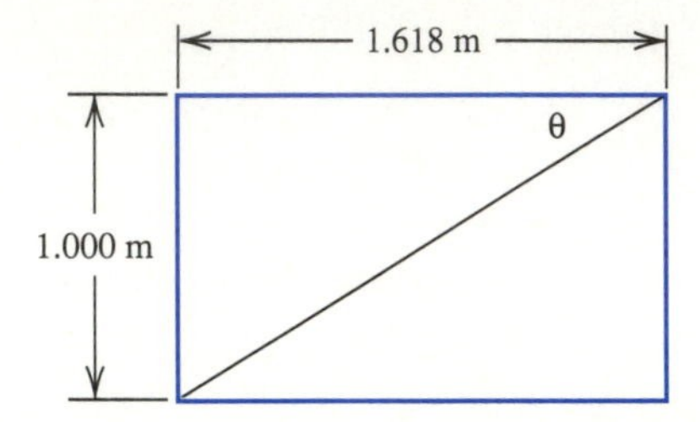

10. When an electric circuit contains a resistance R and a capacitive reactance X_C, the impedance Z is related to X_C and R by the following diagram. Find X_C if $R = 19.0\ \text{k}\Omega$ and $\theta = 71.2°$.

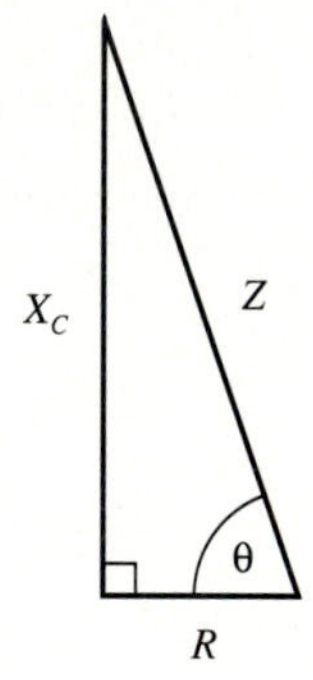

EXTENDED APPLICATION: ARCHITECTURAL TECHNOLOGY

DESIGNING A TRAPEZOIDAL WINDOW

An architectural firm frequently incorporates a trapezoidal shaped thermopane window into the design of a sunroom, as shown in Figure 4.31. To order a custom-made window to fit the opening, the lengths of three of the four sides must be specified. If the window glass is cut exactly the same size as the opening, however, thermal expansion or settling of the building could cause stress within the glass that would cause it to crack or break. Thus, it is a common practice to allow a small margin of 1/8 inch to 1/4 inch between the window and the frame. The margin should be the same around the entire perimeter of the window, because discrepancies may be revealed by a metal bar used to space the two pieces of glass of the thermopane window.

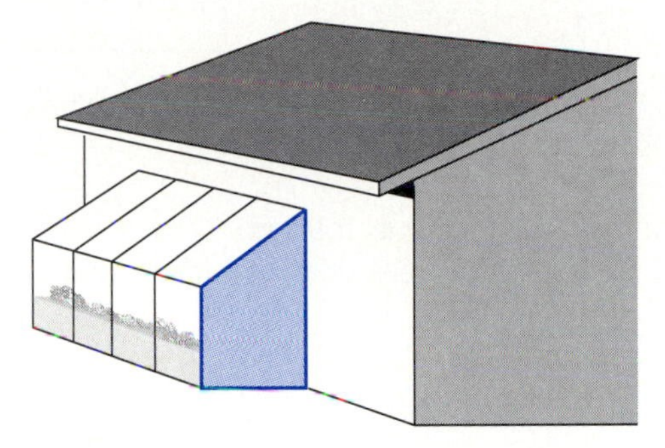

FIGURE 4.31

If the dimensions shown in Figure 4.32 are known, it is possible to use trigonometry to determine the dimensions of the glass needed. The margin x in Figure 4.32 is exaggerated to make the diagram clearer. Figure 4.33 shows the window with additional dimensions labeled.

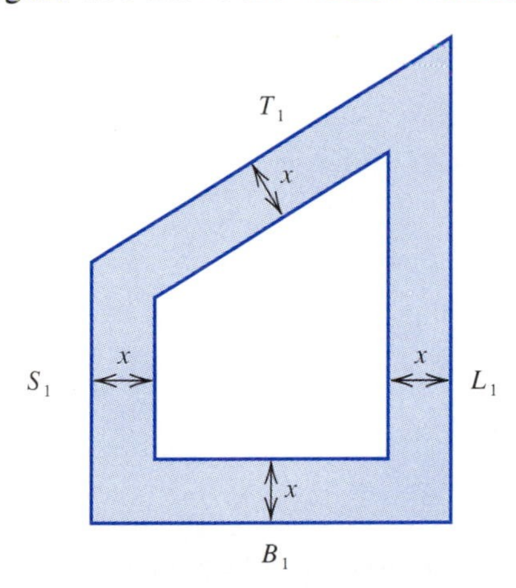

FIGURE 4.32

EXTENDED APPLICATION, continued

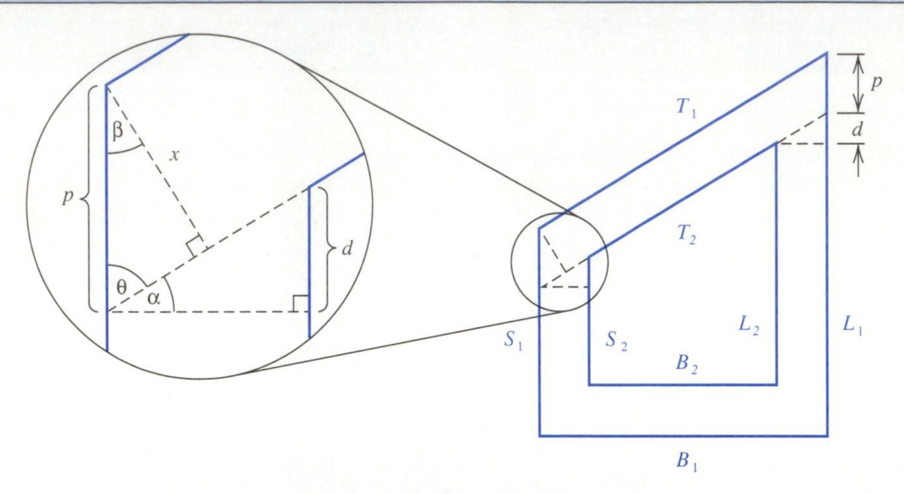

FIGURE 4.33

The lengths B_2, S_2, and L_2 are found as follows.
Since there is a margin of x on both vertical sides, it is clear that

$$B_2 = B_1 - 2x.$$

The angle α can be determined from the "pitch" or slope of the roof. Also, $\alpha = 90^\circ - \theta$ and $\beta = 90^\circ - \theta$, so $\alpha = \beta$.
Thus, we have

$$\sec \alpha = p/x, \text{ so } p = x \sec \alpha, \text{ and}$$
$$\tan \alpha = d/x, \text{ so } d = x \tan \alpha.$$

Therefore,

$$\begin{aligned} S_2 &= S_1 - x - (p - d) \\ &= S_1 - x - p + d \\ &= S_1 - x - x \sec \alpha + x \tan \alpha. \end{aligned}$$

Likewise, we have

$$\begin{aligned} L_2 &= L_1 - x - (p + d) \\ &= L_1 - x - p - d \\ &= L_1 - x - x \sec \alpha - x \tan \alpha. \end{aligned}$$

Thus, the three dimensions B_2, S_2, and L_2 can be calculated from measurements taken from the trapezoidal opening.

■ From "The Dimensions of Window Glass for a Trapezoidal Opening" by Gerard Garland in *The UMAP Journal,* Volume 7, Number 1, 1986, pp. 15–18. Reprinted by permission of COMAP, Inc.

FACTORING, RATIONAL EXPRESSIONS, AND EQUATIONS

CHAPTER 5

Formulas required for scientific and technical applications sometimes contain expressions that have numerators and denominators. To simplify such expressions, additional algebraic techniques are necessary. In this chapter, we examine these techniques and further develop the concepts involved in solving equations. The procedures of this chapter are used to analyze problems involving such diverse topics as beams, medicine, resistance, and time.

5.1 FACTORING AND QUADRATIC EQUATIONS

factoring

If we begin with polynomials and rewrite them as expressions containing factors, the process is called *factoring.* In this chapter, all of the polynomials will have integral coefficients, and the factors must have integral coefficients also. If there is a factor common to *all* of the terms of the polynomial, we use the distributive property. This step is called "factoring out" the common factor.

EXAMPLE 1 Factor the polynomials.

(a) $2x + 6$ **(b)** $3x + 9$ **(c)** $2x^2 + 4x + 4$ **(d)** $6x^3 + 3x^2 - 9x$

Solution **(a)** Since $6 = 2(3)$, the factor 2 is common to both terms. That is, $2x + 6 = 2(x) + 2(3) = 2(x + 3)$ by the distributive property.

(b) $3x + 9 = 3(x) + 3(3) = 3(x + 3)$

(c) $2x^2 + 4x + 4 = 2(x^2) + 2(2x) + 2(2) = 2(x^2 + 2x + 2)$

(d) $6x^3 + 3x^2 - 9x = 3x(2x^2 + x - 3)$ ■

Formulas used in technological settings can sometimes be factored.

EXAMPLE 2 When a rigid beam is uniformly loaded, the bending moment m at a point x feet from the end of the beam is given by

$$m = (1/2)wxl - (1/2)wx^2$$

where w is the weight per unit length and l is the total length of the beam. Factor the expression on the right.

Solution $(1/2)wx$ is the common factor.

$$m = (1/2)wx \cdot l - (1/2)wx \cdot x$$
$$m = (1/2)wx(l - x) \quad ■$$

In Examples 1 and 2 we are, in effect, reversing the procedure used to multiply algebraic expressions in Section 2.2. In that section, however, the polynomial products were often the result of more than one application of the distributive property. In particular, trinomials, which are polynomials having three terms, resulted from two applications of the distributive property. The factors were binomials, which are polynomials with two terms. For instance, $(x + 3)(x - 2) = x^2 + x - 6$. Example 3 shows how we reverse the process to factor $x^2 + x - 6$.

EXAMPLE 3 Factor $x^2 + x - 6$.

Solution The terms have no common factor, so we factor the polynomial as a product of two binomials. We might factor the first term, x^2, as $x(x)$ and the last term, -6, as $3(-2)$. Therefore we "guess" that $(x + 3)$ and $(x - 2)$ are the two factors. We have not yet considered the middle term, but a check shows that it is $3x - 2x$, or x, so the factors $(x + 3)$ and $(x - 2)$ do produce the correct middle term. ■

To factor a trinomial:

1. Factor out the common factor, if there is one.
2. Write two sets of parentheses to represent the factors of the remaining trinomial: ( )( )
3. Factor the first term of the trinomial to get the first term of each factor.
4. Factor the last term of the trinomial to get the last term of each factor.
5. Insert a sign between the terms of each factor.
6. Check the middle term. If it is not correct, try again!

The process of factoring trinomials is often called "trial and error" factorization. As you gain experience with this type of problem, you will find that you make fewer false starts and often guess the correct factors on the first or second try. Refer to the following hints as you study the examples in this section.

HINTS FOR FACTORING TRINOMIALS

(a) Be systematic. If you aren't sure which factors to use at step 3 or 4, choose the two with the smallest difference. If that combination does not work, try the two with the next smallest difference, and so on.

(b) Do not put two terms with a common factor in the same set of parentheses. If there is a common factor, it should have been factored out as the first step.

(c) If the last term of the trinomial is positive, the two signs between terms of the binomial factors are alike and are the same as the sign of the middle term of the trinomial.

(d) If the last term of the trinomial is negative, the two signs between terms of the binomial factors are different.

(e) If you check your guess and the middle term is wrong only in sign, change both signs between terms of the binomial factors.

We use these hints in the examples that follow.

EXAMPLE 4 Factor the polynomials completely.

(a) $y^2 - 3y + 2$ **(b)** $2x^2 + 9x + 4$

Solution **(a)** Since $y^2 = (y)(y)$, and $2 = (2)(1)$, we have $(y \quad 2)(y \quad 1)$. The signs are both negative (see hint c), so our first guess is $(y - 2)(y - 1)$, which is correct, since the middle term is $-y - 2y$ or $-3y$.

(b) We factor $2x^2$ as $(2x)(x)$ and 4 as $(1)(4)$. Although the last term has factors of 2 and 2, with a difference of 0, we know that $(2x + 2)(x + 2)$ will not work, so we do not have to try it (see hint b). Our first guess is $(2x + 1)(x + 4)$, which is correct. ■

EXAMPLE 5 Factor the polynomials completely.

(a) $8a^2 + 22a - 6$ **(b)** $6m^2 + 35m - 6$

Solution **(a)** We start with $2(4a^2 + 11a - 3)$. We factor $4a^2$ as $(2a)(2a)$ and 3 as $(3)(1)$. Our first guess is $2(2a - 3)(2a + 1)$, but the middle term is $2a - 6a$, or $-4a$, which is incorrect. There are no other factors of the last term to try. We factor $4a^2$ as $(a)(4a)$ and try $2(a + 3)(4a - 1)$. The middle term is $-a + 12a$, or $11a$, which is correct.

(b) We factor $6m^2$ as $(2m)(3m)$ and 6 as $(3)(2)$, using factors with the smallest difference. Our first guess is $(2m + 3)(3m - 2)$, but the middle term is incorrect. It is not necessary to factor the last term as $1(6)$, since 6 cannot go in parentheses with either $2m$ or $3m$ (see hint b). Our second guess is $(6m + 1)(m - 6)$, but the middle term is $-36m + m$, or $-35m$, which has the wrong sign. The correct answer is $(6m - 1)(m + 6)$ (see hint e). ■

EXAMPLE 6 The equilibrium constant K of a particular solution is given by

$$K = \frac{x^2}{x^2 - 3x + 2}.$$

Factor the denominator of the expression on the right-hand side of the equation. In the formula, x is the fraction of the chemical that has been dissociated when equilibrium is reached.

Solution

$$K = \frac{x^2}{x^2 - 3x + 2} = \frac{x^2}{(x - 2)(x - 1)}$$ ■

In a subsequent section of the chapter, you will learn the advantages of factoring expressions like the one in Example 6.

Some binomials are treated as trinomials with zero as the middle term. Example 7 illustrates this type of problem.

EXAMPLE 7 Factor the polynomials completely.

(a) $4y^2 - 25$ **(b)** $4p^2 - 100$.

Solution **(a)** Think of $4y^2 - 25$ as $4y^2 + 0y - 25$. The first guess is $(2y + 5)(2y - 5)$, which is correct.

(b) Think of $4p^2 - 100$ as $4p^2 + 0p - 100$. Start with $4(p^2 + 0p - 25)$. Try $4(p - 5)(p + 5)$, which is correct. ■

NOTE ⇨⇨ A binomial like those in Example 7 is called a *difference of squares*. ***Any binomial that is a difference of squares will follow the pattern $a^2 - b^2 = (a - b)(a + b)$.***

There are some polynomials that cannot be factored using integers as coefficients.

EXAMPLE 8 Factor each polynomial completely, if possible.

(a) $x^2 - x + 1$ **(b)** $3x^2 + 2x - 6$

Solution **(a)** We try $(x-1)(x-1)$, but the middle term is incorrect. There are no other possibilities to try.

(b) All possibilities are listed systematically, as follows.
$(3x-2)(x+3)$, but the middle term is $7x$. (It is not necessary to try $(3x+2)(x-3)$. See hint e.)
$(3x-1)(x+6)$, but the middle term is $17x$. (It is not necessary to try $(3x+1)(x-6)$. See hint e.)
We conclude that $3x^2 + 2x - 6$ cannot be factored. ■

Trinomials often occur in equations, and the factored form of the trinomials may be helpful in solving such equations.

DEFINITION A **quadratic equation in *x*** is an equation that can be put in the form $ax^2 + bx + c = 0$ where a, b, and c are real numbers, and $a \neq 0$.

The form $ax^2 + bx + c = 0$ is called the *standard form.* One side of the equation must be zero and the terms must appear with their exponents in descending order from left to right. Since a quadratic equation in standard form has terms, we factor the trinomial so that the following special property of zero, which requires factors, can be used to solve the equation.

ZERO-FACTOR PROPERTY

If a and b are real numbers such that $ab = 0$, then $a = 0$ or $b = 0$ (or both).

NOTE ***This property says that if the product of two numbers is zero, then at least one of the two numbers must be zero.*** Zero is the only number that has this property.

EXAMPLE 9 Solve $6x^2 + x - 2 = 0$.

Solution In order for the equation to follow the pattern $ax^2 + bx + c = 0$, we must include the negative sign as part of the number c. Thus, the equation is in standard form with $a = 6$, $b = 1$, and $c = -2$. We factor the expression on the left-hand side to obtain $(2x-1)(3x+2) = 0$. Since the product of the two factors is zero, one of the factors must be zero. That is, either $(2x-1) = 0$ or $(3x+2) = 0$. But now we have two *linear* equations instead of one *quadratic* equation. If $2x - 1 = 0$, we have $x = 1/2$. If $3x + 2 = 0$, we have $x = -2/3$. Thus, the solution of the equation $6x^2 + x - 2 = 0$ is $x = 1/2$ or $x = -2/3$.

A check shows both answers to be correct.

When x is 1/2, $6(1/2)^2 + 1/2 - 2 = 6(1/4) + 1/2 - 2 = 3/2 + 1/2 - 4/2 = 0$.

When x is $-2/3$, $6\left(\frac{-2}{3}\right)^2 + \left(\frac{-2}{3}\right) - 2 = 6\left(\frac{4}{9}\right) - \frac{2}{3} - \frac{6}{3} = \frac{8}{3} - \frac{2}{3} - \frac{6}{3} = 0$. ■

To solve a quadratic equation:

1. If the equation is not in standard form, write an equivalent equation that is in standard form, $ax^2 + bx + c = 0$. That is, add or subtract the same term(s) on both sides of the equation to make one side equal zero, and arrange the terms in descending order of their exponents.
2. Factor the polynomial.
3. Use the zero-factor property to set each factor equal to zero.
4. Solve the resulting linear equations.

EXAMPLE 10 Solve $y^2 - 5y = 0$.

Solution

$$y^2 - 5y = 0$$

$$y(y - 5) = 0 \qquad \text{Factor } y^2 - 5y.$$

$$y = 0 \quad \text{or} \quad y - 5 = 0 \qquad \text{Set each factor equal to zero.}$$

$$y = 0 \quad \text{or} \quad y = 5 \qquad \text{Solve each linear equation.}$$ ■

EXAMPLE 11 Solve $z^2 + 6 = 5z$.

Solution

$$z^2 + 6 = 5z$$

$$z^2 - \mathbf{5z} + 6 = 5z - \mathbf{5z} \qquad \text{Subtract } 5z \text{ from both sides.}$$

$$z^2 - 5z + 6 = 0$$

$$(z - 2)(z - 3) = 0 \qquad \text{Factor } z^2 - 5z + 6.$$

$$z - 2 = 0 \quad \text{or} \quad z - 3 = 0 \qquad \text{Set each factor equal to zero.}$$

$$z = 2 \quad \text{or} \quad z = 3 \qquad \text{Solve each linear equation.}$$ ■

EXAMPLE 12 Solve $6x^2 - 15x + 9 = 0$.

Solution

$$6x^2 - 15x + 9 = 0$$

$$3(2x^2 - 5x + 3) = 0$$

$$3(2x - 3)(x - 1) = 0$$

One of the factors must be zero, but it is impossible to have $3 = 0$.

$$2x - 3 = 0 \quad \text{or} \quad x - 1 = 0$$

$$2x = 3 \quad \text{or} \quad x = 1$$

$$x = 3/2 \quad \text{or} \quad x = 1$$ ■

EXERCISES 5.1

In 1–21, factor each polynomial completely, if possible.

1. $2x^2 + 2x$
2. $3m^3 - 3m^2 + 6m$
3. $x^2 + 5x + 6$
4. $z^2 - z - 6$
5. $p^2 + 10p + 24$
6. $y^2 - 11y + 24$
7. $6z^2 - 13z + 6$
8. $5x^2 + 26x + 5$
9. $12p^2 + 7p - 12$
10. $3x^2 - 12x + 12$
11. $z^2 + 5z - 2$
12. $4m^2 - 1$
13. $25p^2 + 36$
14. $4y^3 + 8y^2$
15. $p^2 + 9p + 20$
16. $m^2 - 8m + 15$
17. $2a^2 - 24a + 4$
18. $6a^2 - 7a - 5$
19. $2m^2 - 3m - 9$
20. $m^2 + 3m + 1$
21. $9x^2 - 4$

In 22–42, solve each quadratic equation.

22. $x^2 - 5x + 4 = 0$
23. $y^2 + 7y + 12 = 0$
24. $x^2 - 10x + 24 = 0$
25. $m^2 + 2m - 24 = 0$
26. $x^2 - x = 2$
27. $y^2 - 2y = 0$
28. $6m^2 - 13m + 6 = 0$
29. $4x^2 + 4x = 3$
30. $6y^2 + 5y + 1 = 0$
31. $6x^2 - 7x = 3$
32. $2m^2 + 2m = 0$
33. $x^2 - 1 = 0$
34. $s^2 = 25$
35. $25x^2 = 36$
36. $6z^2 - 13z = 5$
37. $2x^2 + 5x = 3$
38. $m^2 = 11m - 24$
39. $r^2 - 10 = 3r$
40. $2y^2 - 7y = 15$
41. $3x^2 + 17x + 10 = 0$
42. $2y^2 = 8$

43. The formula $y = \frac{1}{24EI}(wxl^3 - wx^4 - 2wx^3l)$ gives the deflection y of a simple beam under a distributed load. Factor the expression on the right-hand side of the equation. In the formula, x is the distance from the support at which the deflection is measured, l is the length of the beam, w is the weight of the load per unit length, E is the modulus of elasticity, and I is the moment of inertia. 1 2 4 8 7 9

44. When a circuit contains an inductor and a resistor in series, and the current is switched on, the current does not rise immediately to its steady-state value. It rises gradually, so that the current (in amperes) is given by $I = I_0 - I_0e^{-t/\tau}$. Factor the expression on the right-hand side of the equation. In the formula, I_0 is the steady-state current, e is a constant, $\tau = L/R$ (inductance/resistance), and t is the time elapsed. 3 5 9 1 2 4 6 7 8

45. The formula $f = \frac{1.44}{ac + 2bc}$ gives the frequency of oscillation for a stepping motor (used in data processing) if c is the capacitance and a and b are resistances. Factor the denominator of the expression on the right-hand side of the equation. 3 5 9

46. The formula $V = \pi r_o^2 h - \pi r_i^2 h$ gives the volume V of a circular ring, such as the one in the accompanying diagram. Factor the expression on the

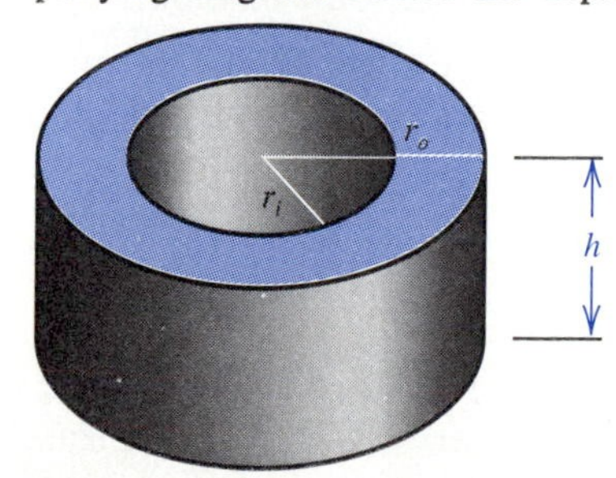

1. Architectural Technology
2. Civil Engineering Technology
3. Computer Engineering Technology
4. Mechanical Drafting Technology
5. Electrical/Electronics Engineering Technology
6. Chemical Engineering Technology
7. Industrial Engineering Technology
8. Mechanical Engineering Technology
9. Biomedical Engineering Technology

right-hand side of the equation. In the formula, r_o is the radius of the outside of the ring, r_i is the radius of the inside of the ring, and h is the height of the ring. 1 2 4 7 8 3 5

47. The formula $w = (1/l^3)(pl - pa)(l^2 + al - 2a^2)$ gives the shearing load w for a beam fixed at both ends under a concentrated load. Factor the expression on the right-hand side. In the formula, p is the weight of the load, a is the distance of the load from a support, and l is the length of the beam. 1 2 4 8 7 9

48. The formula $s = v_0 t + (1/2)at^2$ gives the distance s that an object travels in time t if its initial velocity is v_0 and it has a constant acceleration a. Find the time required for a car to travel 219 feet if it is decelerating ($a < 0$ for deceleration) at a rate of 10 ft/s², and its initial velocity was 60 mph (88 ft/s). 4 7 8 1 2 3 5 6 9

49. The formula $s = T^2 + 517T$ gives the approximate elevation s (in ft) above sea level if T is the number of Fahrenheit degrees below 212 at which water boils at that elevation. Find the temperature at which water boils on the top floor of a skyscraper 1038 feet tall. 6 8 1 2 3 5 7 9

50. The formula $s =$

$$\frac{100(V^4 - 15V^3 + 2045V^2 + 200V)}{V^4 - 15V^3 + 2400V^2 - 31{,}100V + 2{,}400{,}000}$$

gives the saturated oxygen content of blood (as a percent) if V is a number derived from various quantities in a person's blood.

(a) Write both numerator and denominator using grouping symbols within grouping symbols, such as $\{x[x(x + 1) - 2] + 3\}$.

(b) Compare the number of multiplications required to evaluate the expression in the original form and in nested form. An exponent indicates repeated multiplication.

1. Architectural Technology
2. Civil Engineering Technology
3. Computer Engineering Technology
4. Mechanical Drafting Technology
5. Electrical/Electronics Engineering Technology
6. Chemical Engineering Technology
7. Industrial Engineering Technology
8. Mechanical Engineering Technology
9. Biomedical Engineering Technology

5.2 SPECIAL PRODUCTS AND FACTORING

In the previous section, trinomials were factored by trial and error. In that section, we also saw that a difference of squares may be factored without relying on trial and error. It is important to recognize the pattern such problems follow. There are other types of polynomials that may be factored quickly by pattern recognition. We now examine some of them.

EXAMPLE 1 Factor $9x^2 + 12x + 4$ completely.

Solution Using trial and error with the hints in Section 5.1, the first guess would be $(3x + 2)(3x + 2)$ or $(3x + 2)^2$. It is correct. ■

perfect square trinomial

A trinomial like the one in Example 1 is called a *perfect square trinomial.* Trial and error factorization should yield the correct factorization of a perfect square trinomial on the first guess. If you recognize the pattern, however, you will be able to save time by immediately writing $9x^2 + 12x + 4 = (3x + 2)^2$. ***A perfect square trinomial has a square as the first term, a square as the last term, and a middle term that is twice the product of square roots of the end terms.*** The polynomial $9x^2 + 12x + 4$ follows this pattern. The first term is the square of $3x$, the last term is the square of 2,

NOTE ⇨⇨

and the middle term is twice the product of $3x$ and 2. The factored form of a perfect square trinomial is the square of the binomial that contains these square roots as terms.

EXAMPLE 2 Factor $y^2 - 10y + 25$ completely.

Solution The first term is the square of y.
The last term is the square of -5.
Twice the product of y and -5 is $2(-5)(y)$, or $-10y$, the middle term. Thus $y^2 - 10y + 25 = (y - 5)^2$. ■

EXAMPLE 3 Factor $9x^2 + 15x + 4$ completely.

Solution The first term is the square of $3x$.
The last term is the square of 2.
Twice the product of $3x$ and 2 is $12x$, not $15x$.
Thus, $9x^2 + 15x + 4$ is *not* a perfect square trinomial. It can, however, be factored by trial and error. Since 2 and 3 are not the correct factors of 4, we try 4 and 1.
$9x^2 + 15x + 4 = (3x + 4)(3x + 1)$, which is correct. ■

EXAMPLE 4 In an ac series circuit, $Z^2 = R^2 + X^2$, where Z is the magnitude of the impedance (in ohms), R is the resistance (in ohms), and X is the magnitude of the reactance (in ohms). If R is 2 ohms, factor the expression on the right-hand side of the equation, if possible.

Solution When R is 2, $R^2 + X^2 = 2^2 + X^2$. The expression $4 + X^2$ is *not* a difference of squares. If we think of it as the trinomial $4 + 0X + X^2$, we see that it is not a perfect square trinomial. While 4 is the square of 2 (or -2), and X^2 is the square of X (or $-X$), the middle term is not twice the product $2X$ (or $-2X$). Systematically trying all possible factors leads us to the conclusion that $4 + X^2$ cannot be factored. ■

Recall that the first step in any factoring problem is to factor out the common factor. It may be possible to factor a sum of squares by factoring out a common factor.

NOTE ⇨⇨ ***We can say that a polynomial that is the sum of squares* cannot *be factored as the product of two binomials, while a polynomial that is the difference of squares* can *be factored as the product of binomials.***

Both the sum and difference of cubes are factorable, but not as products of binomial factors.

EXAMPLE 5 Factor $x^3 - 8$ completely.

Solution The difference of squares $x^2 - 4$ has as a factor $(x - 2)$, which contains square roots of x^2 and 4 as terms. It is reasonable, then, to guess that the difference of cubes $x^3 - 8$ has as a factor $(x - 2)$, which contains cube roots of x^3 and 8 as terms. The other factor may be obtained by long division.

$$\begin{array}{r} x^2 + 2x + 4 \\ x - 2\overline{)x^3 \qquad\qquad - 8} \\ \underline{x^3 - 2x^2} \qquad\quad \\ 2x^2 \qquad\quad \\ \underline{2x^2 - 4x} \quad \\ 4x - 8 \\ \underline{4x - 8} \end{array}$$

Thus, $x^3 - 8 = (x - 2)(x^2 + 2x + 4)$. ■

Notice the pattern in the factors of $x^3 - 8$ in Example 5. The first factor contains the cube roots of x^3 and -8 as terms. Once these cube roots, x and -2, are identified, they can be used to obtain the second factor. Its end terms are the squares of x and -2. That is x^2 and 4 are the end terms. The middle term is the product of the cube roots, but with opposite sign. That is, the middle term is $-(x)(-2)$ or $2x$.

EXAMPLE 6 Factor $8x^3 + 27$ completely.

Solution Although this polynomial is the *sum* of cubes, it too, follows the pattern outlined in Example 5. The first factor contains the cube roots of $8x^3$ and 27. It is $(2x + 3)$. These cube roots are used to obtain the second factor. Its end terms are $(2x)^2$ and 3^2 or $4x^2$ and 9. The middle term is the product of $2x$ and 3, but with opposite sign. It is $-6x$. Thus,

$$8x^3 + 27 = (2x + 3)(4x^2 - 6x + 9).$$ ■

SPECIAL PRODUCTS

$a^2 + b^2$ cannot be factored as the product of two binomials

$a^2 - b^2 = (a - b)(a + b)$

$a^3 + b^3 = (a + b)(a^2 - ab + b^2)$

$a^3 - b^3 = (a - b)(a^2 + ab + b^2)$

EXAMPLE 7 Factor $8x^3 + 64$ completely.

Solution Always begin by factoring out the common factor.

$$8x^3 + 64 = 8(x^3 + 8)$$

The polynomial $x^3 + 8$ is the sum of cubes. Thus, $8x^3 + 64 = 8(x^3 + 8) = 8(x + 2)(x^2 - 2x + 4)$. ■

EXAMPLE 8 Factor $3a^6 + 3b^6$ completely.

Solution $3a^6 + 3b^6 = 3(a^6 + b^6)$
Since a^6 is the square of a^3 and also the cube of a^2, the polynomial in parentheses fits both the patterns of a sum of squares and a sum of cubes. As the sum of squares, it cannot be factored using *binomial factors.* But we do not know how to factor it using other types of factors. As the sum of cubes, however, it can be factored as follows.

$$3a^6 + 3b^6 = 3(a^2 + b^2)(a^4 - a^2b^2 + b^4) \quad ■$$

EXAMPLE 9 Factor $a^6 - b^6$ completely.

Solution Once again we have a polynomial that fits both the patterns of a difference of squares and a difference of cubes. As the difference of squares, we have

$$\begin{aligned} a^6 - b^6 &= (a^3 - b^3)(a^3 + b^3) \\ &= (a - b)(a^2 + ab + b^2)(a + b)(a^2 - ab + b^2). \end{aligned} \tag{5-1}$$

As the difference of cubes, we have

$$\begin{aligned} a^6 - b^6 &= (a^2 - b^2)(a^4 + a^2b^2 + b^4) \\ &= (a - b)(a + b)(a^4 + a^2b^2 + b^4). \end{aligned} \tag{5-2}$$

However, $(a^2 + ab + b^2)(a^2 - ab + b^2) = a^4 + a^2b^2 + b^4$.
Thus, the polynomial in Equation 5-1 is factored more completely than the polynomial in Equation 5-2. The correct answer, then, is

$$a^6 - b^6 = (a - b)(a^2 + ab + b^2)(a + b)(a^2 - ab + b^2). \quad ■$$

NOTE ⇨⇨ ***Examples 8 and 9 illustrate the principle that if a polynomial may be considered as both a sum of cubes and a sum of squares, then it should be treated as a sum of cubes. On the other hand, if a polynomial may be considered as both a difference of cubes and a difference of squares, then it should be treated as a difference of squares.*** Example 10 illustrates how the difference of cubes might arise in a technical setting.

EXAMPLE 10 Show that the volume of the figure shown in Figure 5.1 is given by

$$V = (2/3)\pi(r_1 - r_2)(r_1{}^2 + r_1r_2 + r_2{}^2).$$

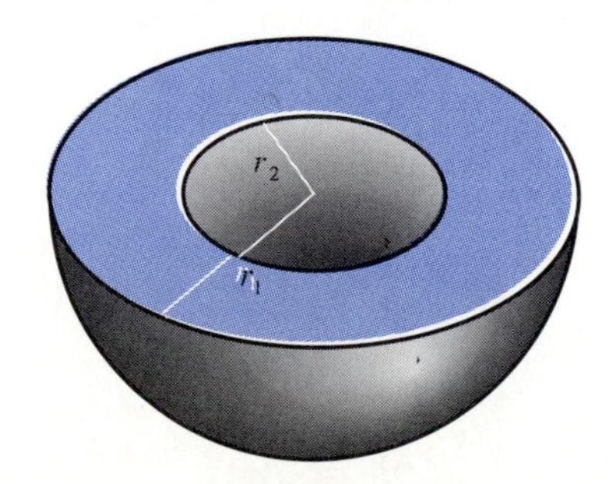

FIGURE 5.1

Solution The figure is a hemisphere with a smaller hemisphere removed from its center. The volume of a sphere is given by $V = (4/3)\pi r^3$, so the volume of a hemisphere is $V = (1/2)(4/3)\pi r^3 = (2/3)\pi r^3$. The volume of the larger hemisphere is $V_1 = (2/3)\pi r_1^3$, and the volume of the smaller hemisphere is $V_2 = (2/3)\pi r_2^3$. The volume of the figure, then, is

$$\begin{aligned} V_1 - V_2 &= (2/3)\pi r_1^3 - (2/3)\pi r_2^3 \\ &= (2/3)\pi(r_1^3 - r_2^3) \\ &= (2/3)\pi(r_1 - r_2)(r_1^2 + r_1 r_2 + r_2^2). \end{aligned}$$ ■

EXERCISES 5.2

In 1–42, factor each polynomial completely, if possible.

1. $4x^2 - 4x + 1$
2. $4y^2 - 12y + 9$
3. $9z^2 - 4$
4. $4m^2 + 25$
5. $4p^2 + 100$
6. $4y^2 - 16$
7. $4z^2 + 20z + 25$
8. $9a^2 - 12a + 4$
9. $4x^2 + 5x + 1$
10. $4b^2 + 13b + 9$
11. $27m^3 - 8$
12. $2p^3 + 16$
13. $2z^3 + 128$
14. $3a^3 - 81$
15. $9b^2 + 16$
16. $16m^2 + 81$
17. $49p^2 - 25$
18. $49y^2 - 28y + 4$
19. $25x^2 + 30x + 9$
20. $x^2 - 17x + 16$
21. $y^2 - 10y + 9$
22. $4z^2 + 34z + 16$
23. $9m^2 - 81m + 72$
24. $25p^2 + 36$
25. $16a^2 + 49$
26. $49b^2 - 16$
27. $64x^2 - 36$
28. $3y^3 - 24$
29. $z^3 - 64$
30. $m^3 + 64$
31. $64p^2 + 100$
32. $4a^2 - 16$
33. $27b^3 - 125$
34. $125x^3 - 64$
35. $8y^3 - 125$
36. $8z^3 + 125$
37. $a^6 + 64$
38. $b^6 - 64$
39. $m^6 - 729$
40. $p^6 + 729$
41. $64x^6 + 729$
42. $64y^6 - 729$

43. Since the length of each side of the square in the accompanying diagram is $(a + b)$, the area of the square is $(a + b)^2$. Find the area of the square by adding the areas of the four regions that compose it. 1 2 4 7 8 3 5

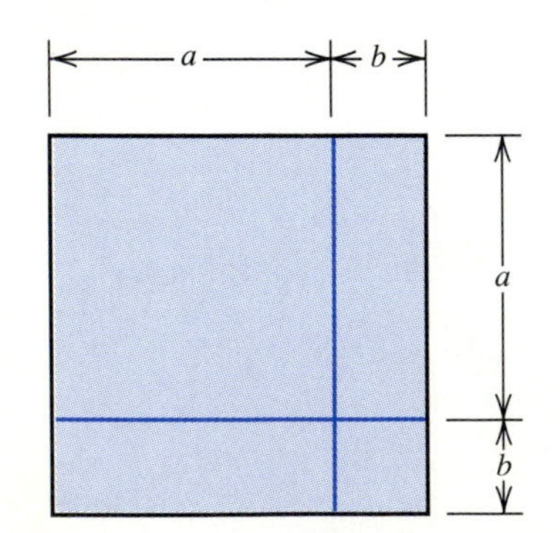

44. The formula $I = kms_2^2 - kms_1^2$ gives the increase I in collision impact when a car of mass m increases its speed from s_1 to s_2. The number k is a constant that depends on the circumstances of the collision. Factor the expression on the right. 4 7 8 1 2 3 5 6 9

45. The formula $I = kv_1^3 - kv_2^3$ gives the increase I in power generated by a windmill when the velocity of the wind increases from v_1 to v_2. The number k is a constant that depends on the characteristics of the windmill. Factor the expression on the right. 4 7 8 1 2 3 5 6 9

46. Show that the volume of the material used to construct the figure shown in the accompanying diagram (a cube with a cube removed) is given by

$$V = (s_2 - s_1)(s_2^2 + s_1 s_2 + s_1^2).$$

1 2 4 7 8 | 3 5

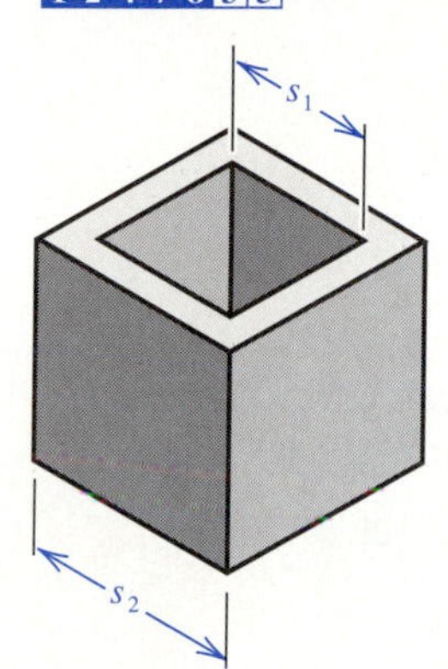

47. When the maximum deflections of two similar cantilever beams bearing the load at the ends are compared, the difference in deflection is given by

$$d = \left(\frac{1}{3EI}\right)(Pl_1^3 - Pl_2^3).$$

Factor the expression on the right-hand side of the equation. The lengths of the beams are l_1 and l_2, E is the modulus of elasticity, I is the moment of inertia, and P is the weight of the load at the end of the beam. 1 2 4 | 8 7 9

48. When the maximum deflections of two similar cantilever beams bearing a distributed load are compared, the difference in deflection is given by

$$d = \left(\frac{1}{8EI}\right)(wl_1^4 - wl_2^4).$$

Factor the expression on the right-hand side of the equation. The lengths of the beams are l_1 and l_2, E is the modulus of elasticity, I is the moment of inertia, and w is the weight per unit length of the load distributed along the beam.
1 2 4 | 8 7 9

49. The formula $R = (1/P)(V^2)$ gives the resistance R (in ohms) when the potential difference is V (in volts) and the power dissipated is P (in watts). If two resistors with potential difference V_1 and V_2 are in series, the resistance is given by $R = (1/P)(V_1^2 + 2V_1V_2 + V_2^2)$. Use this information to show that $V = V_1 + V_2$. 3 5 9 | 1 2 4 6 7 8

50. We stated that a polynomial that is the sum of squares cannot be factored *as the product of two binomials.* Consider $x^4 + 4$, which is the sum of squares. Explain each step of the following factorization.

$$\begin{aligned} x^4 + 4 &= x^4 + 4x^2 + 4 - 4x^2 \\ &= (x^2 + 2)^2 - 4x^2 \\ &= [(x^2 + 2) - 2x][(x^2 + 2) + 2x] \quad \text{or} \\ &\quad (x^2 - 2x + 2)(x^2 + 2x + 2) \end{aligned}$$

1. Architectural Technology
2. Civil Engineering Technology
3. Computer Engineering Technology
4. Mechanical Drafting Technology
5. Electrical/Electronics Engineering Technology
6. Chemical Engineering Technology
7. Industrial Engineering Technology
8. Mechanical Engineering Technology
9. Biomedical Engineering Technology

5.3 FACTORING BY GROUPING

From Sections 5.1 and 5.2, you know that the first step of any factoring problem is to factor out any factors that are common to all terms. The next step depends on the number of terms in the polynomial. If there are only two terms, you should look for a pattern such as the difference of squares, or the sum or difference of cubes. If there are three terms in the polynomial, you should try to factor it as a product of two binomials. If the polynomial has more than three terms, it may be possible to factor it by a technique called **factoring by grouping.**

factoring by grouping

As the name implies, the factoring is done by grouping together two or three terms that form a factorable expression. The key to the technique is to recognize that a factor composed of terms, such as $(x + 1)$, can be treated just like a factor consisting of a single symbol, such as x. Examples 1 and 2 illustrate this concept.

EXAMPLE 1 Factor out the common factor.

(a) $cx + 3c$ **(b)** $(a + b)x + 3(a + b)$

Solution **(a)** $cx + 3c = c(x + 3)$

(b) The expression $(a + b)x + 3(a + b)$ consists of two terms, each having two factors. The factor $(a + b)$ is common to both terms, and can be treated just like the factor c in Example 1(a). The distributive property allows us to write

$$\mathbf{(a + b)}x + 3\mathbf{(a + b)} = (a + b)(x + 3). \quad ■$$

EXAMPLE 2 Factor out the common factor.

(a) $(x + 1)(x - 3) - 2(x + 1)$ **(b)** $(3x + 2)(2x + 3) + (2x + 3)$

Solution **(a)** The expression consists of two terms, each having two factors. The factor $(x + 1)$ is common to both terms.

$$\begin{aligned} \mathbf{(x + 1)}(x - 3) - 2\mathbf{(x + 1)} &= (x + 1)[(x - 3) - 2] \\ &= (x + 1)(x - 5) \end{aligned}$$

(b) The expression consists of two terms. The factor $(2x + 3)$ is common to both. It is important to recognize that the second term can be thought of as $1(2x + 3)$.

$$\begin{aligned} (3x + 2)\mathbf{(2x + 3)} + 1\mathbf{(2x + 3)} &= (2x + 3)[(3x + 2) + 1] \\ &= (2x + 3)(3x + 3) \end{aligned}$$

The expression is now factored, but it is not completely factored.

$$(2x + 3)(3x + 3) = 3(2x + 3)(x + 1) \quad ■$$

EXAMPLE 3 The formula

$$E = 1000(100 - T) + 580(100 - T)^2$$

gives the approximate elevation E (in m) above sea level at which water boils at temperature T (in degrees Celsius). Factor the expression on the right.

Solution

$$\begin{aligned} E &= 1000(100 - T) + 580(100 - T)^2 \\ &= 20(100 - T)[50 + 29(100 - T)] \\ &= 20(100 - T)[50 + 2900 - 29T] \\ &= 20(100 - T)(2950 - 29T) \quad ■ \end{aligned}$$

Example 4 illustrates how the difference of squares pattern may be used when terms are grouped.

EXAMPLE 4 A resistor is deposited on a substrate for an integrated circuit in the form of a ring, as shown in Figure 5.2. The resistance of the ring is inversely proportional to the area of the ring. Assume that the split is only a few microns wide, so that the area A of the circular ring is given by the formula

$$A = \pi(r + d)^2 - \pi r^2.$$

Factor the expression on the right.

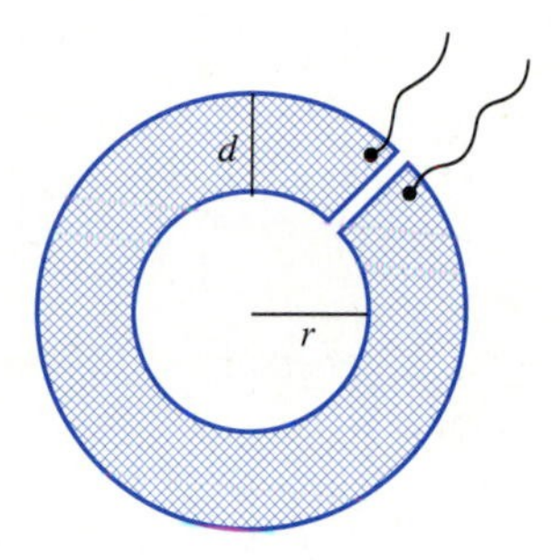

FIGURE 5.2

Solution

$$\begin{aligned} A &= \pi(r + d)^2 - \pi r^2 \\ &= \pi[(r + d)^2 - r^2] \\ &= \pi[(r + d) - r][(r + d) + r] \\ &= \pi d(2r + d) \quad ■ \end{aligned}$$

Notice that in Example 4, if $(r + d)^2$ is expanded, the result will be the same.

$$\begin{aligned} A &= \pi(r + d)^2 - \pi r^2 \\ &= \pi(r^2 + 2rd + d^2) - \pi r^2 \\ &= \pi r^2 + 2\pi rd + \pi d^2 - \pi r^2 \\ &= 2\pi rd + \pi d^2 \\ &= \pi d(2r + d) \end{aligned}$$

When an expression contains more than three terms, it may be possible to group two or three terms to form a factorable expression.

EXAMPLE 5 Factor $x^3 + x^2 + x + 1$.

Solution Since the expression contains four terms, we try breaking it into two expressions, each consisting of two terms.

$$x^3 + x^2 + x + 1 = (x^3 + x^2) + (x + 1)$$

Now, at least $x^3 + x^2$ can be factored.

$$(x^3 + x^2) + (x + 1) = x^2(x + 1) + (x + 1)$$

NOTE ⇨⇨ ***It is a common mistake to stop here, but the expression is not factored. It still contains terms rather than factors.*** The factor $(x + 1)$ is common to both terms.

$$x^2(x + 1) + 1(x + 1) = (x + 1)(x^2 + 1)$$

Thus, $x^3 + x^2 + x + 1 = (x + 1)(x^2 + 1)$. ■

EXAMPLE 6 Factor $3x^3 + x^2 - 12x - 4$.

Solution

$$\begin{aligned} 3x^3 + x^2 - 12x - 4 &= (3x^3 + x^2) + (-12x - 4) \\ &= x^2(3x + 1) - 4(3x + 1) \\ &= (3x + 1)(x^2 - 4) \\ &= (3x + 1)(x - 2)(x + 2) \end{aligned}$$ ■

Sometimes it is necessary to group three factors together. Special products are particularly useful.

EXAMPLE 7 Factor $x^2 + 2x + 1 - y^2$.

Solution

$$\begin{aligned} x^2 + 2x + 1 - y^2 &= (x^2 + 2x + 1) - y^2 \\ &= (x + 1)^2 - y^2 \end{aligned}$$

This expression is the difference of squares.

$$\begin{aligned} (x + 1)^2 - y^2 &= [(x + 1) - y][(x + 1) + y] \\ &= (x + 1 - y)(x + 1 + y) \end{aligned}$$

Thus, $x^2 + 2x + 1 - y^2 = (x + 1 - y)(x + 1 + y)$. ■

It takes practice to recognize the terms that should be grouped together. Grouping the first two terms of the expression in Example 7 does not work.

$$\begin{aligned} x^2 + 2x + 1 - y^2 &= (x^2 + 2x) + (1 - y^2) \\ &= x(x + 2) + (1 - y)(1 + y) \end{aligned}$$

The expression now contains two terms that have no common factor, so it cannot be factored in this form.

Factoring by grouping can be used to factor trinomials. The middle term is split into two terms so that the expression contains four terms instead of three.

EXAMPLE 8 Factor $x^2 + 5x + 6$.

Solution

$$\begin{aligned} x^2 + 5x + 6 &= x^2 + 3x + 2x + 6 \\ &= x(x + 3) + 2(x + 3) \\ &= (x + 3)(x + 2) \end{aligned}$$ ■

To use the technique of factoring by grouping on a trinomial:

1. Multiply the first and last terms of the trinomial.
2. Find two factors of this product that have a sum equal to the middle term of the trinomial.
3. Replace the middle term of the trinomial with an expression that is the sum of the two factors found in step 2.
4. Factor by grouping.

EXAMPLE 9 Factor $6x^2 + 5x - 6$.

Solution The product of the first and last terms of the trinomial is $-36x^2$. Factors of $36x^2$ include $(1x)(36x)$, $(2x)(18x)$, $(3x)(12x)$, $(4x)(9x)$, and $(6x)(6x)$.
Factors of $-36x^2$ have different signs. The two factors $9x$ and $-4x$ have a sum of $5x$, the middle term of the trinomial.

$$\begin{aligned} 6x^2 + 5x - 6 &= 6x^2 + 9x - 4x - 6 \\ &= (6x^2 + 9x) + (-4x - 6) \\ &= 3x(2x + 3) - 2(2x + 3) \\ &= (2x + 3)(3x - 2) \quad \blacksquare \end{aligned}$$

It may be possible to group the terms in more than one way to obtain the correct factorization. In Example 9,

$$\begin{aligned} 6x^2 + 9x - 4x - 6 &= (6x^2 - 4x) + (9x - 6) \\ &= 2x(3x - 2) + 3(3x - 2) \\ &= (3x - 2)(2x + 3). \end{aligned}$$

EXERCISES 5.3

In 1–13, factor by factoring out the common factor.

1. $x(x + 1) + 2(x + 1)$
2. $3y(2y - 3) + (2y - 3)$
3. $z(4z - 2) + (4z - 2)$
4. $x(2x + 3) - 3(2x + 3)$
5. $2y(y + 4) - 1(y + 4)$
6. $3z(z - 5) - (z - 5)$
7. $p(2p + 1) + (2p + 1)(p - 1)$
8. $m(m - 4) + (m - 4)(2m + 3)$
9. $(b - 1)(b + 3) + (b - 1)(b + 2)$
10. $(x + 3)(x + 2) + (x - 3)(x + 2)$
11. $(2p - 1)(2p + 3) - (2p - 1)(3p - 2)$
12. $(3m - 4)(2m + 3) - (3m - 4)(m + 5)$
13. $y^2(3y + 4) - 4(3y + 4)$

In 14–29, factor by grouping.

14. $2x^3 - 6x^2 + x - 3$
15. $y^3 + 2y^2 - y - 2$
16. $z^3 + 3z^2 - 3z - 9$
17. $p^3 - p^2 + 4p - 4$
18. $m^3 + 3m^2 - 4m - 12$
19. $2x^3 + 3x^2 - 8x - 12$
20. $2a^3 + 3a^2 - 8a - 12$
21. $4b^3 + 8b^2 - b - 2$
22. $3x^3 + 2x^2 - 3x - 2$
23. $x^2 + 6x + 9 - y^2$
24. $a^2 - 2a + 1 - 9b^2$
25. $p^2 - 2pq + q^2 - 16$
26. $a^2 + 2ab + b^2 - 25$
27. $x^2 - y^2 + 6x + 9$
28. $x^2 - 9y^2 + 4x + 4$
29. $x^2 - 4y^2 + 12x + 36$

In 30–42 use the technique of factoring by grouping to factor each trinomial.

30. $x^2 + x - 2$
31. $y^2 + y - 6$
32. $x^2 + 11x + 10$
33. $y^2 - 8y + 12$
34. $8x^2 + 2x - 1$
35. $3m^2 - 7m - 6$
36. $5p^2 - 26p + 5$
37. $6m^2 + 13m + 6$
38. $12x^2 - 7x - 12$
39. $2x^2 + 3x - 9$
40. $2m^2 + 24m + 40$
41. $3y^2 + 12y + 12$
42. $6p^2 + 7p - 5$

---

43. The formula $A = \sqrt{(s^2 - sa)(s^2 - sc - sb + cb)}$ gives the area A of a triangle having sides of length a, b, and c with $s = (1/2)(a + b + c)$. The expression under the radical is partially factored. Factor completely. 1 2 4 7 8 3 5

44. The formula $f^{-1} = nr_1^{-1} + nr_2^{-1} - r_1^{-1} - r_2^{-1}$ gives the focal length f of a thin glass lens. Factor the expression on the right. In the formula, n is the index of refraction of glass, and r_1 and r_2 are the radii of curvature of the surfaces of the lens. 3 5 6 9

45. The formula $I_R = \sqrt{I^2 - (I_C - I_L)^2}$ gives the current I_R in a resistor connected in parallel with a capacitor and an inductor across an ac source. Factor the expression under the radical. In the formula, I_C is the current in the capacitor, I_L is the current in the inductor, and I is the effective current for the circuit. 3 5 9 4 7 8

46. When an object is dropped from a height, the distance s that it falls between time t and $t + 1$ is given by

$$s = (1/2)g[(t + 1)^2 - t^2]$$

where g is the acceleration due to gravity. Factor the expression on the right. 4 7 8 1 2 3 5 6 9

47. When the current through a resistor of resistance R is increased from I to $I + 1$, the increase in power P dissipated is given by

$$P = (I + 1)^2R - I^2R.$$

Factor the expression on the right. 3 5 9 1 2 4 6 7 8

48. When the radius of a circle is increased from r to $r + 1$, the increase in area A is given by

$$A = \pi(r + 1)^2 - \pi r^2.$$

Factor the expression on the right. 1 2 4 7 8 3 5

49. A thermistor is sometimes added in series with a transistor to stabilize the current. Assume that the resistance R_1 of the transistor is given by $R_1 = -2t - 2$, and the resistance R_2 of the thermistor is given by $R_2 = (t + 1)^2$. Use the fact that resistances in series are added to find the total resistance. 3 5 9 4 7 8

50. The following method of factoring trinomials is similar to factoring by grouping. It is essential to factor out the common factor from all terms before using it. Work problems 37, 39, and 41 by this method.

1. Use the coefficient of the leading term as the leading coefficient of *each* binomial factor.

---

1. Architectural Technology
2. Civil Engineering Technology
3. Computer Engineering Technology
4. Mechanical Drafting Technology
5. Electrical/Electronics Engineering Technology
6. Chemical Engineering Technology
7. Industrial Engineering Technology
8. Mechanical Engineering Technology
9. Biomedical Engineering Technology

2. Multiply the last term of the trinomial by the leading coefficient.
3. Find two factors of this product that have a sum equal to the coefficient of the middle term of the trinomial.
4. Use these two factors as the final terms of the two binomial factors.
5. Factor out and discard any factor that is common to both terms of either binomial.
Example: $4x^2 + 11x - 3 = (4x \quad)(4x \quad)$
Factors of $4(-3)$ or -12 that have a sum of 11 would be 12 and -1. $(4x + 12)(4x - 1)$ is not correct. Since $4x + 12 = 4(x + 3)$, discard the common factor of 4.

$$4x^2 + 11x - 3 = (x + 3)(4x - 1)$$

5.4 RATIONAL EXPRESSIONS: REDUCTION, MULTIPLICATION, AND DIVISION

In Chapter 1, a *rational number* was defined as any number that can be written as the ratio of two integers, as long as the denominator is not zero. A *rational expression* is defined similarly.

DEFINITION A **rational expression** is a ratio of two polynomials, such that the denominator is not equal to zero.

Rational expressions have much in common with rational numbers in fraction form. To reduce a fraction (or rational expression), we divide both numerator and denominator by all the factors they have in common. When a fraction (or rational expression) cannot be reduced, we say it is in *lowest terms*.

lowest terms for a fraction

First we examine these concepts as they apply to fractions. One way to reduce a fraction is to write both numerator and denominator as products of prime numbers. A *prime number* is an integer greater than 1 such that the only positive integers that divide it evenly are 1 and the number itself. When the numbers involved are relatively small, it is probably unnecessary to go through these steps. But when the numbers are large, the method is a convenient means of finding all of the factors that divide both numerator and denominator.

prime number

EXAMPLE 1 Reduce $\frac{143}{195}$.

Solution Factor the numerator and denominator:
$143 = 11(13)$ and $195 = 3(65)$, but 65 is not a prime number.
Replacing 65 with $5(13)$, we have $195 = 3(5)(13)$.
We divide both numerator and denominator by 13.

$$\frac{143}{195} = \frac{11 \cdot 13}{3 \cdot 65} = \frac{11 \cdot \overset{1}{\cancel{13}}}{3 \cdot 5 \cdot \underset{1}{\cancel{13}}} = \frac{11}{15}$$ ■

EXAMPLE 2 Reduce $\frac{240}{200}$.

Solution

$$\frac{240}{200} = \frac{10 \cdot 24}{10 \cdot 20} = \frac{2 \cdot 5 \cdot 3 \cdot 8}{2 \cdot 5 \cdot 4 \cdot 5} = \frac{2 \cdot 5 \cdot 3 \cdot 2 \cdot 4}{2 \cdot 5 \cdot 2 \cdot 2 \cdot 5} = \frac{\overset{1}{\cancel{2}} \cdot \overset{1}{\cancel{5}} \cdot 3 \cdot \overset{1}{\cancel{2}} \cdot \overset{1}{\cancel{2}} \cdot 2}{\underset{1}{\cancel{2}} \cdot \underset{1}{\cancel{5}} \cdot \underset{1}{\cancel{2}} \cdot \underset{1}{\cancel{2}} \cdot 5} = \frac{6}{5}$$ ■

The process, sometimes called *cancellation,* used in the preceding examples is based on the fact that a/a is 1 if a is a real number, and a is not 0. Since 1 is the identity for multiplication, it has no effect on the remaining factors. That is, by dividing out identical factors in the numerator and denominator, we are, in effect, removing a factor of 1 from the fraction. It is essential that you cancel *factors* of the numerator and denominator rather than *terms,* since 1 is *not* the identity for addition. The same method is used to reduce rational expressions to lowest terms.

To reduce a rational expression:

1. Factor the numerator and denominator completely.
2. Divide out *factors* that are identical in both numerator and denominator.
3. Multiply the remaining factors of the numerator to get the numerator of the reduced expression. Multiply the remaining factors of the denominator to get the denominator of the reduced expression.
4. Assume that the denominator is not equal to zero for rational expressions in this section.

EXAMPLE 3 Reduce $\frac{x^2 - 6x + 9}{x^2 - 9}$.

Solution

1. $\frac{x^2 - 6x + 9}{x^2 - 9} = \frac{(x-3)(x-3)}{(x-3)(x+3)}$ Factor the numerator and denominator.

2. $= \frac{\overset{1}{\cancel{(x-3)}}(x-3)}{\underset{1}{\cancel{(x-3)}}(x+3)}$ Divide out identical factors.

3. $= \frac{(x-3)}{(x+3)}$

4. Assume that $x + 3$ is not zero. ■

EXAMPLE 4 Reduce $\frac{y^2 - 2y}{2y - 4}$.

Solution

$$\frac{y^2 - 2y}{2y - 4} = \frac{y(y-2)}{2(y-2)}$$ Factor the numerator and denominator.

$$= \frac{y\overset{1}{\cancel{(y-2)}}}{2\underset{1}{\cancel{(y-2)}}}$$ Divide out identical factors.

$$= \frac{y}{2}$$ ■

EXAMPLE 5 Reduce $\dfrac{z^2 + 6}{z + 3}$.

Solution Neither $z^2 + 6$ nor $z + 3$ can be factored.
Since the numerator and denominator each consist of two *terms,* we cannot reduce.
That is, $\dfrac{z^2 + 6}{z + 3}$ is in lowest terms. ■

EXAMPLE 6 The equivalent resistance R of three resistors in parallel is given by $R = \dfrac{R_1 R_2 R_3}{R_1 R_2 + R_1 R_3 + R_2 R_3}$. In the formula, R_1, R_2, and R_3 are the resistances of the three resistors. If $R_3 = 3R_2$, reduce the fraction on the right-hand side of the equation.

Solution $R_3 = 3R_2$
Working with numerator and denominator separately, gives the following.

$$\begin{aligned} R_1 R_2 R_3 &= (R_1)(R_2)(3R_2) \\ &= 3R_1 R_2{}^2 \\ R_1 R_2 + R_1 R_3 + R_2 R_3 &= R_1 R_2 + (R_1)(3R_2) + (R_2)(3R_2) \\ &= R_1 R_2 + 3R_1 R_2 + 3R_2{}^2 \\ &= 4R_1 R_2 + 3R_2{}^2 \end{aligned}$$

Thus,

$$\begin{aligned} R &= \frac{3R_1 R_2{}^2}{4R_1 R_2 + 3R_2{}^2} \\ &= \frac{3R_1 R_2{}^2}{R_2(4R_1 + 3R_2)} \\ &= \frac{3R_1 R_2}{4R_1 + 3R_2}. \end{aligned}$$ ■

When multiplying rational numbers, you may multiply and then reduce; or you may reduce and then multiply. The second procedure is usually easier, because the numbers are smaller.

EXAMPLE 7 Find $\left(\frac{3}{4}\right)\left(\frac{8}{9}\right)$.

Solution If we multiply, then reduce, we have $\frac{24}{36} = \frac{2}{3}$.

If we reduce, then multiply, we have $\left(\frac{\overset{1}{\cancel{3}}}{\underset{1}{\cancel{4}}}\right)\left(\frac{\overset{2}{\cancel{8}}}{\underset{3}{\cancel{9}}}\right) = \frac{2}{3}$. ■

EXAMPLE 8 Find the volume of a rectangular box that measures $\frac{3''}{4} \times 1\frac{3''}{5} \times 1\frac{7''}{8}$.

Solution $V = lwh$, so

$$V = \left(\frac{3}{4}\right)\left(\frac{8}{5}\right)\left(\frac{15}{8}\right).$$

If we multiply, then reduce, we have

$$\frac{3}{4}\left(\frac{8}{5}\right)\left(\frac{15}{8}\right) = \frac{360}{160} = \frac{9}{4}.$$

If we reduce, then multiply, we have

$$\frac{3}{4}\left(\frac{\overset{1}{\cancel{8}}}{\underset{1}{\cancel{5}}}\right)\left(\frac{\overset{3}{\cancel{15}}}{\underset{1}{\cancel{8}}}\right) = \frac{9}{4}.$$

The volume of the box is 2 1/4 in^3. ■

When rational expressions are multiplied, the difference in difficulty between the two procedures is even more pronounced. It is much easier to factor both numerator and denominator completely, reduce, and then multiply.

To multiply rational expressions:

1. Factor all the numerators and denominators completely.
2. Divide out any *factors* that are identical in both a numerator and a denominator. (They need not be in the same fraction.)
3. Multiply the remaining factors of the numerators to get the numerator of the product; multiply the remaining factors of the denominators to get the denominator of the product.

EXAMPLE 9 Simplify $\left(\frac{x+1}{x-2}\right)\left(\frac{x^2-4}{x^2+2x+1}\right)$.

Solution

1. $\left(\frac{x+1}{x-2}\right)\left(\frac{x^2-4}{x^2+2x+1}\right) = \frac{(x+1)}{(x-2)} \cdot \frac{(x-2)(x+2)}{(x+1)(x+1)}$ — Factor numerators and denominators.

2. $= \frac{\overset{1}{\cancel{(x+1)}}}{\underset{1}{\cancel{(x-2)}}} \cdot \frac{\overset{1}{\cancel{(x-2)}}(x+2)}{\underset{1}{\cancel{(x+1)}}(x+1)}$ — Divide out common factors.

3. $= \frac{x+2}{x+1}$ — Multiply remaining factors. ■

EXAMPLE 10 Simplify $\left(\frac{m^2-m}{2}\right)\left(\frac{m+1}{m}\right)$.

Solution

$$\left(\frac{m^2-m}{2}\right)\left(\frac{m+1}{m}\right) = \frac{m(m-1)}{2} \cdot \frac{(m+1)}{m}$$

$$= \frac{\overset{1}{\cancel{m}}(m-1)}{2} \cdot \frac{(m+1)}{\underset{1}{\cancel{m}}}$$

$$= \frac{(m-1)(m+1)}{2} \quad ■$$

EXAMPLE 11 Simplify $\left(\frac{6a^2}{4a^2-1}\right)\left(\frac{2a-1}{3a}\right)$.

Solution

$$\left(\frac{6a^2}{4a^2-1}\right)\left(\frac{2a-1}{3a}\right) = \frac{2a(3a)}{(2a-1)(2a+1)} \cdot \frac{(2a-1)}{3a}$$

$$= \frac{2a(\overset{1}{\cancel{3a}})}{\underset{1}{\cancel{(2a-1)}}(2a+1)} \cdot \frac{\overset{1}{\cancel{(2a-1)}}}{\underset{1}{\cancel{3a}}}$$

$$= \frac{2a}{(2a+1)} \quad ■$$

When we divide rational *numbers,* we use the rule, "invert the divisor and multiply." The rule also works for rational expressions.

To divide rational expressions:

Invert the divisor and multiply as rational expressions.

EXAMPLE 12 Simplify $\dfrac{1}{3-2x} \div \dfrac{5}{4x^2-9}$.

Solution Invert the divisor and multiply.

$$\frac{1}{3-2x} \div \frac{5}{4x^2-9} = \left(\frac{1}{3-2x}\right)\left(\frac{4x^2-9}{5}\right)$$
$$= \frac{1}{(3-2x)} \cdot \frac{(2x-3)(2x+3)}{5}$$

$(3-2x)$ and $(2x-3)$ are not identical. However, they differ only in sign. That is, $3-2x = -2x+3 = -1(2x-3)$.

$$\left(\frac{1}{3-2x}\right)\left(\frac{4x^2-9}{5}\right) = \frac{1}{-1(2x-3)} \cdot \frac{(2x-3)(2x+3)}{5}$$
$$= \frac{1}{-1\cancel{(2x-3)}_{1}} \cdot \frac{\overset{1}{\cancel{(2x-3)}}(2x+3)}{5}$$
$$= \frac{(2x+3)}{-5} \quad \text{or} \quad -\frac{2x+3}{5}$$ ■

EXAMPLE 13 Simplify $\dfrac{2p-3}{3p+2} \div \dfrac{2p+3}{3p-2}$.

Solution Invert the divisor and multiply.

$$\frac{2p-3}{3p+2} \div \frac{2p+3}{3p-2} = \left(\frac{2p-3}{3p+2}\right)\left(\frac{3p-2}{2p+3}\right)$$

The numerators and denominators are completely factored, and there are no factors that are identical in both a numerator and a denominator. The fraction cannot be reduced. Answers are generally given in factored form.

$$\frac{2p-3}{3p+2} \div \frac{2p+3}{3p-2} = \frac{(2p-3)(3p-2)}{(3p+2)(2p+3)}$$ ■

EXAMPLE 14 Use the formulas $P = I^2R$ and $I = P/V$ to show that $R = V^2/P$. P is power (in watts) dissipated, I is current (in amps), R is resistance (in ohms), and V is potential difference (in volts).

Solution Only the equation $P = I^2R$ contains the variable R, so we solve that equation for R.

$$R = P/I^2$$

The equation $I = P/V$ can be used to replace I in the equation $R = P/I^2$. First, we find I^2 by squaring both sides of the equation $I = P/V$.

$$I^2 = (P/V)^2 = P^2/V^2$$

Since $R = P/I^2$ means $R = P \div I^2$,

$$\begin{aligned} R &= P \div (P^2/V^2) \\ &= P(V^2/P^2) \\ &= V^2/P. \end{aligned}$$ ■

EXERCISES 5.4

In 1–7, reduce to lowest terms.

1. $\dfrac{140}{210}$
2. $\dfrac{252}{315}$
3. $\dfrac{60}{225}$
4. $\dfrac{80}{96}$
5. $\dfrac{108}{324}$
6. $\dfrac{144}{216}$
7. $\dfrac{252}{420}$

In 8–16, reduce to lowest terms.

8. $\dfrac{x^2+x-2}{x^2+2x-3}$
9. $\dfrac{y^2+5y+6}{y^2-y-6}$
10. $\dfrac{m^2-m}{m^2+m}$
11. $\dfrac{x^3-4x}{x^3+x^2-6x}$
12. $\dfrac{p^2-8p+7}{p^2-49}$
13. $\dfrac{2z^2-5z-12}{2z^2-7z-15}$
14. $\dfrac{2x^2-3x-2}{2x^2-5x-3}$
15. $\dfrac{3y^2+4y-4}{3y^2+y-2}$
16. $\dfrac{a^2-1}{a^2+1}$

In 17–40, multiply or divide, as indicated.

17. $\left(\dfrac{7}{16}\right)\left(\dfrac{8}{21}\right)$
18. $\left(\dfrac{9}{20}\right)\left(\dfrac{25}{27}\right)$
19. $\left(\dfrac{15}{14}\right)\left(\dfrac{12}{35}\right)$
20. $\left(\dfrac{15}{14}\right)\left(\dfrac{26}{33}\right)$
21. $\left(\dfrac{3}{7}\right)\left(\dfrac{5}{9}\right)\left(\dfrac{7}{10}\right)$
22. $\left(\dfrac{2}{15}\right)\left(\dfrac{27}{14}\right)\left(\dfrac{35}{54}\right)$
23. $\left(\dfrac{6}{35}\right)\left(\dfrac{7}{36}\right)(30)$
24. $\left(\dfrac{x-1}{x+2}\right)\left(\dfrac{x^2-4}{x^2-1}\right)$
25. $\left(\dfrac{y^2-y-2}{y^2+2y-3}\right)\left(\dfrac{y^2+y-2}{y^2+3y+2}\right)$
26. $\left(\dfrac{z-5}{4z^2-1}\right)\left(\dfrac{2z+1}{2z^2-10z}\right)$
27. $\left(\dfrac{m+2}{m^2-m-6}\right)\left(\dfrac{m^2+6m+5}{m+1}\right)$
28. $\left(\dfrac{x^2-5x+6}{x}\right)\left(\dfrac{x}{x^2-3x+2}\right)$
29. $\left(\dfrac{a^2-3a}{a-2}\right)\left(\dfrac{a^2-3a+2}{a^3+4a^2+3a}\right)$
30. $\left(\dfrac{x^2+5x+6}{x^2+5x+4}\right)\left(\dfrac{x+1}{x+2}\right)$
31. $\left(\dfrac{p+3}{p+1}\right)\left(\dfrac{p^2-2p-3}{p^2-9}\right)$
32. $\left(\dfrac{4y^2-1}{y+2}\right)\left(\dfrac{y^2+3y+2}{4y^2+4y+1}\right)$
33. $\left(\dfrac{2x^2-5x-3}{x+3}\right)\left(\dfrac{x+3}{2x^2+9x+4}\right)$
34. $\dfrac{x+1}{x^2-49} \div \dfrac{x^2-1}{x-7}$

35. $\frac{2z+1}{3} \div \frac{8z+4}{9}$

36. $\frac{9y^2-3y}{2y+2} \div \frac{18y-6}{y^2+2y}$

37. $\frac{m^2-2m}{m+1} \div \frac{m-2}{m^2+m}$

38. $\frac{a}{a^2-4} \div \frac{a+2}{a^2-4a+4}$

39. $\frac{x-2}{x+3} \div \frac{x^2-4x+4}{x^2+4x+3}$

40. $\frac{x+1}{x-2} \div \frac{x^2-1}{x^2-3x+2}$

---

41. Use the formulas $V = IR$ and $P = I^2R$ to show that $P = V^2/R$. V is potential difference (in volts), I is current (in amps), P is power (in watts) dissipated, and R is resistance (in ohms).
3 5 9 1 2 4 6 7 8

42. Use the formulas $W = amx$, $x = (1/2)at^2$, and $v = at$ to show that $W = (1/2)mv^2$. In the formulas, W is work done (in joules) by a force, m is the mass (in kg) of the object acted on, x is the distance (in m) over which the force acts, a is the acceleration (in m/s^2) produced by the force, t is the time (in seconds) over which the force acts, and v is the velocity (in m/s) of the object.
4 7 8 1 2 3 5 6 9

43. The charge on a capacitor is given by $Q = CV$. The energy stored by the capacitor is given by $W = (1/2)CV^2$. Find the ratio of charge to energy ($Q/W = Q \div W$). **3 5 9** 4 7 8

44. The surface area of a cube of side s is given by $A = 6s^2$. The volume is given by $V = s^3$. Find the ratio of the area to the volume.
1 2 4 7 8 3 5

45. The area of a sphere of radius r is given by $A = 4\pi r^2$. The volume is given by $V = (4/3)\pi r^3$. Find the ratio of the volume to the area ($V/A = V \div A$). **1 2 4 7 8** 3 5

46. The surface area of a cylinder of radius r and height h that is closed on both ends is given by $A = 2\pi r^2 + 2\pi rh$. The volume is given by $V = \pi r^2 h$. Find the ratio of the area to the volume. **1 2 4 7 8** 3 5

47. Consider a rectangular solid with length l, width w, and height h. If the height is equal to the width, find the ratio of the surface area to the volume.
1 2 4 7 8 3 5

48. When a particle has an initial velocity of v_0 and a constant acceleration a, which eventually produces a velocity of v, the time is given by $t = \frac{v - v_0}{a}$, and the distance traveled is given by $s = \frac{v^2 - v_0{}^2}{2a}$. Find the ratio of distance to time.
4 7 8 1 2 3 5 6 9

49. In the study of elasticity, the following formulas are used.

$$\lambda = \frac{\sigma E}{1 - \sigma - 2\sigma^2} \quad \text{and} \quad k = \frac{E}{3 - 6\sigma}$$

Find the ratio of λ to k. In these formulas, λ is Lame's constant, E is Young's modulus, σ is the Poisson ratio, and k is the bulk modulus.
1 2 4 8 7 9

50. Consider the fraction 64/16. A student who does not understand the process of cancellation writes $\not{6}4/1\not{6} = 4$, which is the correct answer.
(a) Give two examples that illustrate that this method does not, in general give the correct answer.
(b) Find another example for which it does give the correct answer.

---

1. Architectural Technology
2. Civil Engineering Technology
3. Computer Engineering Technology
4. Mechanical Drafting Technology
5. Electrical/Electronics Engineering Technology
6. Chemical Engineering Technology
7. Industrial Engineering Technology
8. Mechanical Engineering Technology
9. Biomedical Engineering Technology

5.5 RATIONAL EXPRESSIONS: ADDITION AND SUBTRACTION

Before looking at addition and subtraction of rational expressions, we will review addition and subtraction of rational numbers. We will find the lowest common denominator of two or more fractions in order to add or subtract them.

DEFINITION The **lowest common denominator** (abbreviated LCD) of two or more fractions is the smallest positive number that is evenly divisible by each of their denominators.

For fractions with small denominators, you can probably determine the LCD by simply asking, "What number will both denominators go into evenly?" For $\frac{2}{3}$ and $\frac{3}{4}$, for instance, the LCD is 12. We have

$$\frac{2}{3}+\frac{3}{4}=\frac{2(4)}{3(4)}+\frac{3(3)}{4(3)}=\frac{8}{12}+\frac{9}{12}=\frac{17}{12}.$$

When the numbers involved are larger, however, it is not as easy to see immediately what the LCD should be. Once again, factoring will help, and this procedure is the one we will use with rational expressions as well. In general, to find the LCD of two or more fractions, consider all of the different factors in the denominators of the fractions. The LCD is the product of these different factors, each with an exponent that is the highest exponent to which it is raised in any single denominator.

Consider the problem $\frac{5}{6}+\frac{2}{15}+\frac{4}{35}$. If we factor the denominators, we have $\frac{5}{2\cdot 3}+\frac{2}{3\cdot 5}+\frac{4}{5\cdot 7}$. The factors of the denominators are 2, 3, 5, and 7. Since 1 is the highest exponent of 2 that appears in any one denominator, 2^1 is a factor of the LCD. Since 1 is the highest exponent of 3 that appears in any one denominator, 3^1 is a factor of the LCD. Likewise, 5^1 and 7^1 are factors of the LCD. The LCD, then, is $2\cdot 3\cdot 5\cdot 7$ or 210.

The next step is to change each of the fractions to a fraction with 210 as the denominator. We have

$$\frac{5}{2\cdot 3}+\frac{2}{3\cdot 5}+\frac{4}{5\cdot 7}=\frac{?}{2\cdot 3\cdot 5\cdot 7}+\frac{?}{2\cdot 3\cdot 5\cdot 7}+\frac{?}{2\cdot 3\cdot 5\cdot 7}.$$

It is convenient to work with the factored form, because it is immediately obvious what numbers are needed as factors in order to change the form of each fraction.

In the first fraction, we want to change the denominator from $2\cdot 3$ to $2\cdot 3\cdot \mathbf{5\cdot 7}$. Therefore we multiply both numerator and denominator by $\mathbf{5\cdot 7}$. We have $\frac{5\mathbf{(5\cdot 7)}}{2\cdot 3\mathbf{(5\cdot 7)}}=\frac{175}{2\cdot 3\cdot 5\cdot 7}$.

In the second fraction, we want to change the denominator from $3 \cdot 5$ to $2 \cdot 3 \cdot 5 \cdot 7$. Therefore we multiply both numerator and denominator by $2 \cdot 7$. We have $\frac{2(2 \cdot 7)}{3 \cdot 5(2 \cdot 7)} = \frac{28}{2 \cdot 3 \cdot 5 \cdot 7}$.

In the third fraction, we want to change the denominator from $5 \cdot 7$ to $2 \cdot 3 \cdot 5 \cdot 7$. Therefore we multiply both numerator and denominator by $2 \cdot 3$. We have $\frac{4(2 \cdot 3)}{5 \cdot 7(2 \cdot 3)} = \frac{24}{2 \cdot 3 \cdot 5 \cdot 7}$.

Thus, $\frac{5}{6} + \frac{2}{15} + \frac{4}{35} = \frac{175}{210} + \frac{28}{210} + \frac{24}{210} = \frac{227}{210}$. Only one task remains: to reduce to lowest terms, if possible. The factored form is again useful, because we know that if the fraction can be reduced, it is because one or more of the factors 2, 3, 5, and 7 will divide out. We need only ask if 227 is divisible by any of these factors. It is not. The answer, then, is $\frac{227}{210}$ or $1\,\frac{17}{210}$.

You should examine the next example to be sure that you understand the principles of adding rational *numbers* before we consider adding rational *expressions.*

EXAMPLE 1 Find $\frac{1}{27} - \frac{5}{21} + \frac{3}{49}$.

Solution First, we factor the denominators.

$$\frac{1}{27} - \frac{5}{21} + \frac{3}{49} = \frac{1}{3^3} - \frac{5}{3 \cdot 7} + \frac{3}{7^2}$$

The only factors of the denominators are 3 and 7. Since 3 appears to the third power in the first fraction and to the first power in the second fraction, we choose the higher power. Thus, 3^3 is a factor. Since 7 appears to the first power in the second fraction and to the second power in the third fraction, we choose the higher power. Thus, 7^2 is a factor. The LCD is $3^3 \cdot 7^2$ or 1323.

Change the fractions so that they all have 1323 as the denominator.

$$\begin{aligned}\frac{1}{3 \cdot 3 \cdot 3} - \frac{5}{3 \cdot 7} + \frac{3}{7 \cdot 7} &= \frac{?}{3 \cdot 3 \cdot 3 \cdot 7 \cdot 7} - \frac{?}{3 \cdot 3 \cdot 3 \cdot 7 \cdot 7} + \frac{?}{3 \cdot 3 \cdot 3 \cdot 7 \cdot 7} \\ &= \frac{1(7 \cdot 7)}{3 \cdot 3 \cdot 3(7 \cdot 7)} - \frac{5(3 \cdot 3 \cdot 7)}{3 \cdot 7(3 \cdot 3 \cdot 7)} + \frac{3(3 \cdot 3 \cdot 3)}{7 \cdot 7(3 \cdot 3 \cdot 3)} \\ &= \frac{49 - 315 + 81}{3 \cdot 3 \cdot 3 \cdot 7 \cdot 7} = \frac{-185}{1323}\end{aligned}$$

Since 185 is not divisible by 3 or 7, the fraction is in lowest terms. ■

We now apply the same principles to rational expressions.

To add or subtract rational expressions:

1. Factor each denominator.
2. Find the LCD.
   (a) List all of the different factors from the denominators.
   (b) Raise each factor to the highest power to which it is raised in any single denominator.
3. Multiply the numerator and denominator of each fraction by the factors of the LCD that were not originally in its denominator.
4. Write the numerator as the sum (or difference) of the numerators. Write the denominator as the LCD.
5. Factor the numerator and reduce, if possible.

EXAMPLE 2 The equation $\frac{p}{dg} + \frac{v^2}{2g} + h = k$ is known as Bernoulli's equation. Add the fractions on the left-hand side of the equation. The formula is used to describe fluid flow if d is the density of the fluid, p is the absolute pressure, v is the velocity, g is the acceleration due to gravity, h is the height above an arbitrary reference level, and k is a constant.

Solution

1. Each denominator already contains factors.
2. The LCD is $2dg$.
3. $\frac{p}{dg} + \frac{v^2}{2g} + h = \frac{p(2)}{dg(2)} + \frac{v^2(d)}{2g(d)} + \frac{h(2dg)}{1(2dg)}$ Multiply to get the LCD.
4. $= \frac{2p + dv^2 + 2dgh}{2dg}$ Combine the fractions.
5. The numerator cannot be factored, so the fraction is in lowest terms. ■

EXAMPLE 3 Simplify $\frac{x+2}{x-3} + \frac{x+3}{x-2}$.

Solution The denominators cannot be factored. The LCD is $(x-3)(x-2)$.

$$\begin{aligned}\frac{x+2}{x-3} + \frac{x+3}{x-2} &= \frac{(x+2)(x-2)}{(x-3)(x-2)} + \frac{(x+3)(x-3)}{(x-2)(x-3)} \\ &= \frac{x^2-4}{(x-3)(x-2)} + \frac{x^2-9}{(x-2)(x-3)} \\ &= \frac{2x^2-13}{(x-3)(x-2)}\end{aligned}$$

Since $2x^2 - 13$ cannot be factored, the fraction is in lowest terms. ■

EXAMPLE 4 Simplify $\dfrac{1}{x^2-9}-\dfrac{x}{x^2+6x+9}$.

Solution

$$\frac{1}{x^2-9}-\frac{x}{x^2+6x+9}=\frac{1}{(x-3)(x+3)}-\frac{x}{(x+3)^2}$$ Factor the denominators.

The only factors are $(x-3)$ and $(x+3)$. Since $(x+3)$ is raised to the second power in the second fraction, the LCD is $(x-3)(x+3)^2$.

$$\frac{1}{x^2-9}-\frac{x}{x^2+6x+9}=\frac{1(x+3)}{(x-3)(x+3)(x+3)}-\frac{x(x-3)}{(x+3)(x+3)(x-3)}$$
$$=\frac{x+3-x^2+3x}{(x-3)(x+3)(x+3)}$$
$$=\frac{-x^2+4x+3}{(x-3)(x+3)(x+3)} \quad \text{or} \quad \frac{-x^2+4x+3}{(x-3)(x+3)^2}$$

The numerator cannot be factored, so the fraction is in lowest terms. ■

EXAMPLE 5 Simplify $\dfrac{1}{y+1}+\dfrac{1}{y^2}+\dfrac{1}{y^2+y}$.

Solution

$$\frac{1}{y+1}+\frac{1}{y^2}+\frac{1}{y^2+y}=\frac{1}{y+1}+\frac{1}{y^2}+\frac{1}{y(y+1)}$$ Factor the denominators.

The LCD is $(y+1)(y^2)$ or $y\cdot y(y+1)$.

$$\frac{1}{y+1}+\frac{1}{y^2}+\frac{1}{y^2+y}=\frac{1(y^2)}{(y+1)(y^2)}+\frac{1(y+1)}{y\cdot y(y+1)}+\frac{1(y)}{y(y+1)(y)}$$
$$=\frac{y^2}{y^2(y+1)}+\frac{(y+1)}{y^2(y+1)}+\frac{y}{y^2(y+1)}$$
$$=\frac{y^2+y+1+y}{y^2(y+1)}$$ Combine the fractions.
$$=\frac{y^2+2y+1}{y^2(y+1)}$$ Simplify the numerator.
$$=\frac{(y+1)(y+1)}{y^2(y+1)}$$ Factor the numerator.
$$=\frac{y+1}{y^2}$$ Reduce the fraction. ■

complex fraction

In the technologies, rational expressions often occur in the form of *complex fractions.* A **complex fraction** is a fraction that contains one or more fractions in the numerator or denominator (or both).

EXAMPLE 6 The formula $D_L = \dfrac{1}{\dfrac{1}{D_M} + \dfrac{1}{\theta V_C}}$ gives the relationship between the diffusing capacity D_M of the alveolar membrane and the rate θ at which carbon monoxide combines with hemoglobin in the alveolar capillaries. Simplify the fraction on the right-hand side of the equation. In the formula, D_L is the diffusing capacity for the lung, and V_C is the volume of blood in the capillaries.

Solution The fraction bar serves as a grouping symbol, so the numerator and denominator are treated individually. The LCD of the two fractions in the denominator is $D_M\theta V_C$.

$$D_L = \frac{1}{\dfrac{1}{D_M} + \dfrac{1}{\theta V_C}}$$

$$= \frac{1}{\dfrac{\theta V_C}{D_M\theta V_C} + \dfrac{D_M}{D_M\theta V_C}}$$

$$= \frac{1}{\dfrac{\theta V_C + D_M}{D_M\theta V_C}} = 1 \div \frac{\theta V_C + D_M}{D_M\theta V_C}$$

$$= 1\left(\frac{D_M\theta V_C}{\theta V_C + D_M}\right) = \frac{D_M\theta V_C}{\theta V_C + D_M} \quad \blacksquare$$

EXAMPLE 7 Simplify $\dfrac{\dfrac{x+1}{x-1} + 1}{1 + \dfrac{x-1}{x+1}}$.

Solution The fraction bar serves as a grouping symbol, so the numerator and denominator are treated individually. The LCD of the two fractions in the numerator is $x - 1$, and the LCD of the two fractions in the denominator is $x + 1$.

$$\frac{\dfrac{x+1}{x-1} + 1}{1 + \dfrac{x-1}{x+1}} = \frac{\dfrac{x+1}{x-1} + \dfrac{1(x-1)}{x-1}}{\dfrac{1(x+1)}{x+1} + \dfrac{(x-1)}{x+1}}$$

$$= \frac{\dfrac{2x}{x-1}}{\dfrac{2x}{x+1}} = \frac{2x}{x-1} \div \frac{2x}{x+1}$$

$$= \left(\frac{2x}{x-1}\right)\left(\frac{x+1}{2x}\right) = \frac{x+1}{x-1} \quad \blacksquare$$

EXERCISES 5.5

In 1–28, add or subtract as indicated.

1. $\frac{a}{b}+\frac{b}{a}$

2. $\frac{c}{d}+\frac{1}{c}$

3. $\frac{p}{q}-\frac{q}{p^2}$

4. $\frac{x}{y}+\frac{y}{x^2}$

5. $\frac{a^2}{b}+\frac{a}{b^2}$

6. $\frac{1}{cd^2}+\frac{1}{c^2d}$

7. $\frac{2}{pq}-\frac{1}{p^2q}$

8. $\frac{x}{y^2}+\frac{y}{x^2}-\frac{1}{xy}$

9. $\frac{3}{wz^2}+\frac{z}{w^2z}-\frac{1}{wz}$

10. $\frac{x}{x+1}+\frac{2x}{x-1}$

11. $\frac{y}{y^2-1}+\frac{1}{y^2-2y+1}$

12. $\frac{z}{z^2-3z+2}+\frac{3}{z^2-4z+3}$

13. $\frac{3}{2x^2-3x}-\frac{x}{4x^2-9}$

14. $\frac{7}{6a^2-7a-3}-\frac{2}{3a^2+a}$

15. $\frac{y-1}{y+1}+\frac{y+1}{y-1}$

16. $\frac{2z+3}{z-3}+\frac{2z-3}{z+3}$

17. $\frac{p}{2p^2-2p}-\frac{1}{2p^2+2p}$

18. $\frac{7}{x^2-6x+9}+\frac{x}{x^2-x-6}$

19. $\frac{5m}{m^2-4}+\frac{3}{m-2}$

20. $\frac{5}{t^2-25}-\frac{t}{t^2+10t+25}$

21. $\frac{2x}{9x^2+6x+1}-\frac{1}{3x+1}$

22. $\frac{2}{y}+\frac{3}{y^2}-\frac{1}{2}$

23. $\frac{5}{z}-\frac{7}{z+1}+\frac{z}{z^2+z}$

24. $\frac{2}{m}+\frac{3}{2m+2}-\frac{m}{2m^2+2m}$

25. $\frac{a}{a+1}+\frac{2}{a-1}+\frac{3a}{a^2-1}$

26. $\frac{x}{x-3}+\frac{1}{x+1}-\frac{2x}{x^2-2x-3}$

27. $\frac{5}{m-4}-\frac{m}{m+2}+\frac{3}{m^2-2m-8}$

28. $\frac{y+2}{2y-1}+\frac{y-2}{2y+1}-\frac{5}{4y^2-1}$

In 29–39, simplify each complex fraction.

29. $\dfrac{\frac{3x}{y}}{\frac{6xz}{y^2}}$

30. $\dfrac{\frac{2m^2}{9p}}{\frac{4m}{18p}}$

31. $\dfrac{\frac{1-x}{y}}{\frac{1+x}{y^2}}$

32. $\dfrac{\frac{x-y}{4}}{\frac{x+y}{2}}$

33. $\dfrac{x+\frac{1}{x}}{x^2}$

34. $\dfrac{1}{1-\frac{1}{x}}$

35. $\dfrac{\frac{1}{x}}{1+\frac{1}{x}}$

36. $\dfrac{y-\frac{1}{x}}{xy}$

37. $\dfrac{\frac{1}{x}+\frac{1}{y}}{\frac{1}{x}-\frac{1}{y}}$

38. $\dfrac{\frac{1}{x+1}-1}{\frac{1}{x-1}+1}$

39. $\dfrac{\frac{x}{x+1}-1}{\frac{1}{x+1}-x}$

40. The formula $f_2 = \frac{R}{A} - \frac{6eR}{Ad}$ is used to compute the magnitude f_2 of the soil pressure (in kN/m^2) of a certain type of retaining wall. Write the expression on the right as a single fraction.
1 2 4 6 8 5 7 9

41. If s is the speed of a sound wave with frequency f, the frequency f' of the sound wave received by an observer moving toward the stationary source at speed v is given by $f' = f + \frac{fv}{s}$. Write the expression on the right as a single fraction.
1 3 4 5 6 8 9

42. When a projectile is launched at an angle θ with velocity v_0, the maximum height attained is given by

$$h = v_0(\sin\theta)\left(\frac{v_0 \sin\theta}{g}\right) - \frac{1}{2}g\left(\frac{v_0 \sin\theta}{g}\right)^2$$

where g is the acceleration due to gravity. Show that

$$h = \frac{v_0^2 \sin^2\theta}{2g}.$$

(It is important to know that $\sin^2\theta = (\sin\theta)^2$.)
4 7 8 1 2 3 5 6 9

43. When three capacitors of capacitances C_1, C_2, and C_3 are in series, the formula $\frac{1}{C} = \frac{1}{C_1} + \frac{1}{C_2} + \frac{1}{C_3}$ is used to compute the equivalent capacitance C for the combination. Write the expression on the right as a single fraction. Then solve for C by inverting both sides.
3 5 9 4 7 8

44. For a cantilever beam under a distributed load, the deflection at a distance x from the support is given by

$$y = \frac{wb^4}{24EI} - \frac{wb^3x}{6EI} \quad \text{when} \quad x > b.$$

Write the expression on the right as a single fraction. In the formula, w is the load (in kg/mm), b is the length of the load (in mm), E is the modulus of the elasticity, and I is the moment of inertia.
1 2 4 8 7 9

45. Van der Waal's equation, $P = \frac{RT}{V-b} - \frac{a}{V^2}$, is used to find the pressure P of a gas. Simplify the expression on the right-hand side of the equation. In the formula, R is the gas constant, V is the volume per mole, T is the absolute temperature, and a and b are constants that pertain to a particular substance.
6 8 1 2 3 5 7 9

46. For two resistors in parallel, the equivalent resistance R is given by $R = \frac{1}{\frac{1}{R_1} + \frac{1}{R_2}}$, where R_1 and R_2 are the individual resistances. Show that the formula is equivalent to the usual notation $\frac{1}{R} = \frac{1}{R_1} + \frac{1}{R_2}$.
3 5 9 1 2 4 6 7 8

1. Architectural Technology
2. Civil Engineering Technology
3. Computer Engineering Technology
4. Mechanical Drafting Technology
5. Electrical/Electronics Engineering Technology
6. Chemical Engineering Technology
7. Industrial Engineering Technology
8. Mechanical Engineering Technology
9. Biomedical Engineering Technology

5.6 EQUATIONS WITH RATIONAL EXPRESSIONS

In technological work, it is sometimes necessary to work with equations that contain rational expressions. If both sides of such an equation are multiplied by the LCD of all the rational expressions in the equations, the denominators will divide out.

EXAMPLE 1 The formula $\frac{1}{f_1} + \frac{1}{f_2} = \frac{1}{f}$ relates the focal length f of a system of lenses to the focal lengths f_1 and f_2 of the individual lenses. Solve for f.

Solution Multiply both sides of the equation by ff_1f_2, which is the LCD of the three fractions in the equation.

$$ff_1f_2\left(\frac{1}{f_1}+\frac{1}{f_2}\right)=ff_1f_2\left(\frac{1}{f}\right)$$

$$ff_2+ff_1=f_1f_2$$

Both terms on the left-hand side contain the variable f, which is the one we are to isolate. We cannot combine these terms, however, because they are not like terms. But if we factor, the equation can be written so that the variable appears only once.

$$f(f_2+f_1)=f_1f_2$$

Since we are solving for f, we want to remove the factor f_2+f_1.

$$f=\frac{f_1f_2}{f_2+f_1}$$ ■

EXAMPLE 2 Solve $\dfrac{2}{2x-1}=\dfrac{x}{x+1}$.

Solution Multiply both sides of the equation by $(2x-1)(x+1)$, the LCD.

$$(2x-1)(x+1)\left(\frac{2}{2x-1}\right)=(2x-1)(x+1)\left(\frac{x}{x+1}\right)$$

The denominators divide out, leaving

$$\begin{aligned}(x+1)(2)&=x(2x-1)\\2x+2&=2x^2-x.\end{aligned}$$

Writing the equation in standard form, we have

$$\begin{aligned}0&=2x^2-3x-2\\&=(2x+1)(x-2).\end{aligned}$$

Thus,

$$\begin{aligned}2x+1&=0 &\quad\text{or}\quad x-2&=0\\x&=-1/2 &\quad\text{or}\quad x&=2.\end{aligned}$$

To check this solution, we first replace x with $-1/2$.

$$\frac{2}{-1-1}=\frac{-1/2}{1/2}$$

$$\frac{2}{-2}=-1$$

Then we replace x with 2.

$$\frac{2}{4-1} = \frac{2}{2+1}$$
$$2/3 = 2/3 \quad ■$$

Up to this point, when denominators have included variables, we have said that you may assume there are no zero denominators. We no longer make that assumption. ***Solving the equation that results from eliminating denominators containing variables may produce a value for which the denominator of one of the fractions in the original equation is zero.*** Such a value is not a solution of the original equation.

NOTE

To solve an equation with rational expressions:

1. Multiply both sides of the equation by the LCD of all the rational expressions in the equation.
2. Solve the resulting equation.
3. Eliminate from the solution any value that produces a zero denominator in the original equation.

EXAMPLE 3 Solve for x. $\dfrac{x}{x-3} = \dfrac{3}{x-3} + 3$

Solution 1. Multiply both sides of the equation by $(x-3)$.

$$(x-3)\left(\frac{x}{x-3}\right) = (x-3)\left(\frac{3}{x-3} + 3\right)$$
$$\cancel{(x-3)}\left(\frac{x}{\cancel{x-3}}\right) = \cancel{(x-3)}\left(\frac{3}{\cancel{x-3}}\right) + (x-3)(3)$$
$$x = 3 + (x-3)(3)$$

2. Solve the resulting equation.

$$x = 3 + 3x - 9 \qquad \text{Use the distributive property.}$$
$$x = -6 + 3x \qquad \text{Simplify the right-hand side.}$$
$$-2x = -6 \qquad \text{Subtract } 3x \text{ from both sides.}$$
$$x = 3 \qquad \text{Divide both sides by } -2.$$

3. If x is 3, both denominators in the original equation are zero. Hence, there is no solution to this problem. ■

Example 3 illustrates the importance of checking in the original problem.

EXAMPLE 4 Solve $\frac{1}{3} + \frac{1}{x+1} = \frac{-2}{(x+1)(x-3)}$.

Solution Multiply both sides by $3(x+1)(x-3)$, the LCD.

$$3(x+1)(x-3)\left(\frac{1}{3} + \frac{1}{x+1}\right) = 3(x+1)(x-3)\left(\frac{-2}{(x+1)(x-3)}\right)$$

$$\cancel{3}(x+1)(x-3)\frac{1}{\cancel{3}} + 3\cancel{(x+1)}(x-3)\frac{1}{\cancel{x+1}} = 3\cancel{(x+1)}\cancel{(x-3)}\frac{-2}{\cancel{(x+1)}\cancel{(x-3)}}$$

$$(x+1)(x-3) + 3(x-3) = 3(-2) \quad \text{Divide out the denominators.}$$

$$x^2 - 3x + x - 3 + 3x - 9 = -6 \quad \text{Use the distributive property.}$$

$$x^2 + x - 12 = -6 \quad \text{Simplify the left-hand side.}$$

$$x^2 + x - 6 = 0 \quad \text{Add 6 to both sides.}$$

$$(x+3)(x-2) = 0 \quad \text{Factor the left-hand side.}$$

$$x + 3 = 0 \quad \text{or} \quad x - 2 = 0 \quad \text{Set each factor equal to zero.}$$

$$x = -3 \quad \text{or} \quad x = 2 \quad \text{Solve each linear equation.}$$

Both answers are valid solutions. When x is -3, we have

$$\frac{1}{3} + \frac{1}{-2} = \frac{-2}{(-2)(-6)}$$

$$\frac{2}{6} - \frac{3}{6} = \frac{-2}{12}$$

$$-1/6 = -1/6.$$

When x is 2,

$$\frac{1}{3} + \frac{1}{3} = \frac{-2}{3(-1)} \quad \text{or} \quad \frac{2}{3} = \frac{2}{3}. \quad ■$$

EXAMPLE 5 Solve $\frac{1}{x+1} + \frac{x}{x-2} = \frac{7x-8}{x^2-x-2}$.

Solution Factor $x^2 - x - 2$ as $(x+1)(x-2)$.
Multiply both sides of the equation by the LCD, $(x+1)(x-2)$.

$$(x+1)(x-2)\left(\frac{1}{x+1} + \frac{x}{x-2}\right) = (x+1)(x-2)\left(\frac{7x-8}{(x+1)(x-2)}\right)$$

$$\cancel{(x+1)}(x-2)\frac{1}{\cancel{x+1}} + (x+1)\cancel{(x-2)}\frac{x}{\cancel{x-2}} = \cancel{(x+1)}\cancel{(x-2)}\frac{7x-8}{\cancel{(x+1)}\cancel{(x-2)}}$$

$$x - 2 + x(x+1) = 7x - 8 \quad \text{Divide out the denominators.}$$

$$x - 2 + x^2 + x = 7x - 8 \quad \text{Use the distributive property.}$$

$$x^2 + 2x - 2 = 7x - 8 \quad \text{Simplify the left-hand side.}$$

$$x^2 - 5x + 6 = 0 \qquad \text{Subtract } 7x \text{ and add 8 to both sides.}$$
$$(x-2)(x-3) = 0 \qquad \text{Factor the left-hand side.}$$

Thus $x - 2 = 0$ or $x - 3 = 0$.
We have $x = 2$ or $x = 3$.
But if x is 2, one of the denominators would be zero in the original problem. We conclude that 3 is the only solution. When x is 3, we have

$$\frac{1}{4} + 3 = \frac{21-8}{9-3-2}$$
$$\frac{1}{4} + \frac{12}{4} = \frac{13}{4}. \quad ■$$

EXAMPLE 6 Solve $\dfrac{1}{x-1} + \dfrac{x}{x-3} = \dfrac{4x-6}{x^2-4x+3}$.

Solution Factor $x^2 - 4x + 3$ as $(x-1)(x-3)$.
Multiply both sides of the equation by the LCD, $(x-1)(x-3)$.

$$(x-1)(x-3)\left(\frac{1}{x-1} + \frac{x}{x-3}\right) = (x-1)(x-3)\left(\frac{4x-6}{(x-1)(x-3)}\right)$$
$$\cancel{(x-1)}(x-3)\frac{1}{\cancel{x-1}} + (x-1)\cancel{(x-3)}\frac{x}{\cancel{x-3}} = \cancel{(x-1)}\cancel{(x-3)}\frac{4x-6}{\cancel{(x-1)}\cancel{(x-3)}}$$
$$x - 3 + (x-1)x = 4x - 6$$
$$x - 3 + x^2 - x = 4x - 6$$
$$x^2 - 3 = 4x - 6$$
$$x^2 - 4x + 3 = 0$$
$$(x-3)(x-1) = 0$$
$$x - 3 = 0 \quad \text{or} \quad x - 1 = 0$$
$$x = 3 \quad \text{or} \quad x = 1.$$

Both values would produce zero denominators in the original equation. We conclude that this problem has no solution. ■

From Examples 4 through 6, you can see that eliminating denominators may lead to a quadratic equation. It is possible that of the two solutions, both, only one, or neither is a solution of the original equation involving rational expressions. Examples 7 and 8 illustrate applications of rational equations.

EXAMPLE 7 A boat travels at a rate of 12 mph in still water. It can travel 12 miles upstream in the same length of time that it can travel 20 miles downstream. Find the rate of the current.

Solution Let $x =$ rate of the current

We consider the trip upstream and downstream separately.

I. Trip Upstream

$12 - x =$ rate boat travels upstream (The current slows down the boat.)

$12 =$ distance boat travels upstream (This information was given.)

$\frac{12}{12 - x} =$ time boat travels upstream ($T = D/R$)

II. Trip Downstream

$12 + x =$ rate boat travels downstream (The current speeds up the boat.)

$20 =$ distance boat travels downstream (This information was given.)

$\frac{20}{12 + x} =$ time boat travels upstream ($T = D/R$)

Now we consider the trips together. Since the time for the trip upstream is equal to the time of the trip downstream, we have the following equation.

$$\frac{12}{12 - x} = \frac{20}{12 + x}$$

Since the LCD is $(12 - x)(12 + x)$, both sides of the equation are multiplied by these factors.

$$\cancel{(12 - x)}(12 + x)\frac{12}{\cancel{12 - x}} = (12 - x)\cancel{(12 + x)}\frac{20}{\cancel{12 + x}}$$

$$(12 + x)(12) = 20(12 - x)$$

$144 + 12x = 240 - 20x$ Use the distributive property.

$144 + 32x = 240$ Add $20x$ to both sides.

$32x = 96$ Subtract 144 from both sides.

$x = 3$ mph Divide both sides by 32. ■

EXAMPLE 8 Two computers, working at the same time, can process a certain amount of data in 10 minutes. If the first computer works alone, the same job takes 15 minutes. If the second computer works alone, how long will it take?

Solution Let $x =$ amount of time for the second computer

$\frac{1}{15} =$ fraction of the job done by the first computer in one minute

$\frac{1}{x} =$ fraction of the job done by the second computer in one minute

Assuming that both computers work for 10 minutes, the fraction of the job done by the first computer is $10\left(\frac{1}{15}\right)$ or $\frac{2}{3}$ and the fraction of the job done by the second computer is $10\left(\frac{1}{x}\right)$. Since the two together can finish the entire job, the fraction of the job done by the two together is 1. Thus,

$$\frac{2}{3} + \frac{10}{x} = 1$$

To solve this equation, multiply both sides by $3x$.

$$\mathbf{3x}\left(\frac{2}{3} + \frac{10}{x}\right) = \mathbf{3x}(1)$$

$$2x + 30 = 3x$$

$$30 = x$$

That is, it would take the second computer 30 minutes. ■

EXERCISES 5.6

In 1–34, solve each equation.

1. $\frac{x}{3} + \frac{4}{5} = \frac{1}{2}$
2. $\frac{y-1}{4} + \frac{5}{3} = \frac{2y+7}{6}$
3. $\frac{2x+5}{3x} + \frac{7}{x} = 5$
4. $\frac{m+3}{m} + \frac{5}{2m} = 4$
5. $\frac{p}{4} + \frac{p-2}{2} = \frac{p}{8}$
6. $\frac{y}{4} + \frac{y-1}{6} = \frac{y+2}{3}$
7. $\frac{3z-2}{z} + \frac{7}{2z} = 9$
8. $\frac{x+5}{2x} - 3 = \frac{10}{x}$
9. $\frac{2a+3}{3a} + \frac{5}{3} = \frac{1}{a}$
10. $\frac{y+3}{2y} + 2 = \frac{5}{y}$
11. $\frac{3m+1}{2m} + \frac{3}{2} = \frac{1}{m}$
12. $\frac{p+3}{3p} - 2 = \frac{3}{p}$
13. $\frac{x+2}{2x} + \frac{5}{2} = \frac{1}{x}$
14. $\frac{2z+1}{z} + \frac{5}{3z} = 3$
15. $\frac{3}{x+1} + \frac{2}{3} = \frac{1}{5x+5}$
16. $\frac{3}{m+2} + \frac{2}{3} = \frac{1}{3m+6}$
17. $\frac{2}{y+3} + \frac{3}{2} = \frac{1}{2y+6}$
18. $\frac{2}{y-1} + \frac{1}{3} = \frac{5}{3y-3}$
19. $\frac{2}{x-1} + \frac{3}{5} = \frac{5x-1}{4x-4}$
20. $\frac{4}{z} + \frac{2}{3} = \frac{2z}{3}$
21. $\frac{x}{5} - \frac{5}{x} = \frac{2x-1}{30}$
22. $\frac{2y}{y+2} - \frac{3y}{y-2} = \frac{25}{y^2-4}$
23. $\frac{a+1}{a-3} - \frac{2a+2}{a-2} = \frac{6}{a^2-5a+6}$
24. $\frac{1}{x-1} + \frac{1}{2} = \frac{2}{x^2-1}$
25. $\frac{2}{y-1} + \frac{3y}{y+2} = \frac{y+5}{y^2+y-2}$
26. $\frac{2z}{z-3} + \frac{1}{z+3} = \frac{2z}{z^2-9}$
27. $\frac{a-1}{a+1} - \frac{a-2}{a+2} = \frac{a^2+2a-1}{a^2+3a+2}$
28. $\frac{m+1}{m-2} - \frac{m+2}{m-3} = \frac{-m^2+4}{m^2-5m+6}$
29. $\frac{x}{2x+1} - \frac{1}{x} = \frac{-x^2-3x-1}{2x^2+x}$

30. $\frac{3y}{3y-1} - \frac{2}{y+2} = \frac{-5y+4}{3y^2+5y-2}$

31. $\frac{1}{4p-2} + \frac{p}{p+3} = \frac{p^2-3p+3}{2p^2+5p-3}$

32. $\frac{a-2}{a+3} + \frac{a}{a-1} = \frac{a^2-2a+5}{a^2+2a-3}$

33. $\frac{x-1}{x} + \frac{x}{x+1} = \frac{-3x-1}{2x^2+2x}$

34. $\frac{y-2}{y} + \frac{y}{2y-1} = \frac{y^2-4y+2}{2y^2-y}$

---

35. One technician can run a test on a series of samples in 4 hours. Another technician can run the test on the same number of samples in 3 hours. How long will it take them if they work together? 1 2 6 7 8 3 5 9

36. One pipe can fill a tank in 6 hours. Another pipe can empty the tank in 9 hours. If both pipes are left open, how long will it take to fill the tank? 4 8 1 2 6 7 9

37. The equivalent capacitance C of two capacitors with capacitances of C_1 and C_2 in series is given by $1/C = 1/C_1 + 1/C_2$.
 (a) What capacitance should be used with a 2 μF capacitor to produce an equivalent capacitance of 1.2 μF?
 (b) Solve the formula for C_1. 3 5 9 4 7 8

38. The equivalent resistance R of two resistors with resistances of R_1 and R_2 in parallel is given by $1/R = 1/R_1 + 1/R_2$.
 (a) Two resistors with resistances that differ by 2 Ω produce an equivalent resistance of 1 Ω. Find the resistances.
 (b) Solve the formula for R_2.
 3 5 9 1 2 4 6 7 8

39. One car travels 7 mph faster than another. The slower car leaves point A one hour later than the faster car. When the faster car has traveled 165 miles, the slower car has traveled 96 miles. Find the time that each car travels.
 4 7 8 1 2 3 5 6 9

40. The distance between Baltimore and Dayton is 410 miles. An airplane traveling from Baltimore to Dayton has a rate of 325 mph in still air and a tail wind of 25 mph. How far will the plane have traveled when it reaches the point at which it will take just as long to reach Dayton as it would to turn around and return to Baltimore? (Hint: Don't forget that the wind will slow the plane down on the return trip.) 4 7 8 1 2 3 5 6 9

41. If a car traveled 5 mph faster, it would take one hour less to travel 550 miles. Find the speed of the car. 4 7 8 1 2 3 5 6 9

---

1. Architectural Technology
2. Civil Engineering Technology
3. Computer Engineering Technology
4. Mechanical Drafting Technology
5. Electrical/Electronics Engineering Technology
6. Chemical Engineering Technology
7. Industrial Engineering Technology
8. Mechanical Engineering Technology
9. Biomedical Engineering Technology

5.7 CHAPTER REVIEW

KEY TERMS

5.1 factoring
 quadratic equation in x
5.2 perfect square trinomial
5.3 factoring by grouping
5.4 rational expression
 lowest terms for a fraction
 prime number
5.5 lowest common denominator or LCD
 complex fraction

RULES AND FORMULAS

Zero-factor Property: If a and b are real numbers such that $ab = 0$, then $a = 0$ or $b = 0$ (or both).

Special Products:

$a^2 + b^2$ cannot be factored as the product of two binomials

$a^2 - b^2 = (a - b)(a + b)$

$a^3 + b^3 = (a + b)(a^2 - ab + b^2)$

$a^3 - b^3 = (a - b)(a^2 + ab + b^2)$

GEOMETRY CONCEPTS

$V = (4/3)\pi r^3$ (volume of a sphere)

$V = s^3$ (volume of a cube)

$V = lwh$ (volume of a rectangular solid)

$V = \pi r^2 h$ (volume of a cylinder)

$A = 4\pi r^2$ (area of a circle)

SEQUENTIAL REVIEW

(Section 5.1) In 1–5, factor each polynomial completely, if possible.

1. $a^2 + 11a + 28$ **2.** $2x^2 + 7x + 3$ **3.** $5x^2 + 10x$

4. $y^2 + 4y + 5$ **5.** $4m^2 + 20m - 12$

In 6–8, solve each equation.

6. $2x^2 + 7x + 5 = 0$ **7.** $3y^2 - 27 = 0$ **8.** $3y^2 - 27y = 0$

(Section 5.2) In 9–16, factor each polynomial completely, if possible.

9. $9x^2 + 6x + 1$ **10.** $x^6 - 1$ **11.** $4x^2 + 100$ **12.** $8m^3 - 27$

13. $z^2 - 81$ **14.** $4x^2 - 20x + 25$ **15.** $27z^3 + 1$ **16.** $9x^2 - 36$

(Section 5.3) In 17–23, factor each expression completely, if possible.

17. $x(3x - 2) + 5(3x - 2)$ **18.** $y(7y + 3) - (7y + 3)$

19. $2m^3 + 4m^2 + m + 2$ **20.** $7x^3 - 7x^2 + x - 1$

21. $8x^3 - 16x^2 - x + 2$ **22.** $x^2 + 2xy + y^2 - 4$

23. $m^2 + 2mn + n^2 - 9$

(Section 5.4) In 24–26, reduce to lowest terms.

24. $\frac{70}{105}$ **25.** $\frac{2x^2 - x - 1}{x^2 + x - 2}$ **26.** $\frac{3x^2 + x - 2}{3x^2 - 5x + 2}$

In 27–30, perform the indicated operation.

27. $\frac{26}{35}\left(\frac{14}{39}\right)$ **28.** $\left(\frac{m^2 - 5m + 6}{m^2 - 4}\right)\left(\frac{m^2 + 3m + 2}{m^2 - 9}\right)$

29. $\left(\frac{2y^2 - 5y - 3}{y + 3}\right)\left(\frac{y^2 + 2y - 3}{2y + 1}\right)$ **30.** $\frac{2p - 5}{p + 3} \div \frac{2p^2 - 11p + 15}{p^2 + 4p + 3}$

(Section 5.5) *In 31–38, perform the indicated operation.*

31. $\frac{x}{y}+\frac{y}{x^2}$

32. $\frac{1}{mn}+\frac{2}{m^2n}+\frac{3}{mn^2}$

33. $\frac{z}{x^2}+\frac{x}{z^2}+\frac{1}{x^2z^2}$

34. $\frac{a}{a+2}+\frac{2a}{a-2}$

35. $\frac{z}{2z+3}-\frac{1}{2z^2+z-3}$

36. $\frac{1}{p^2-3p}+\frac{p}{p+2}$

37. $\dfrac{\frac{5x}{y^2}}{\frac{10xz}{y}}$

38. $\dfrac{x}{\frac{1}{x}-1}$

(Section 5.6) *In 39–45, solve each equation.*

39. $\frac{1}{x}+\frac{2}{3x}=1$

40. $\frac{2}{a}+\frac{3}{2}=\frac{5}{6}$

41. $\frac{1}{x+1}+\frac{2}{3}=\frac{4x}{3x+3}$

42. $\frac{y}{y-2}+\frac{3}{5}=\frac{y}{5y-10}$

43. $\frac{1}{x-2}+\frac{2}{x+3}=\frac{2x+1}{x^2+x-6}$

44. $\frac{1}{x-2}-\frac{2}{x+3}=\frac{x^2+3x-5}{x^2+x-6}$

45. $\frac{1}{x+3}+\frac{2}{x-3}=\frac{x^2+3x-6}{x^2-9}$

APPLIED PROBLEMS

1. When a circuit contains a resistor and a capacitor in series, the voltage across the capacitor rises gradually so that the voltage V_C (in volts) is given by $V_C = V - Ve^{-t/\tau}$. Factor the expression on the right-hand side of the equation. In the formula, V is the voltage across the source, e is a constant, t is the time elapsed, and $\tau = RC$ (resistance × capacitance). **3 5 9** 4 7 8

2. The vertical shear V_x at any section of a cantilever beam with a uniformly increasing load is given by

$$V_x = -W + \frac{2Wx}{l} - \frac{Wx^2}{l^2}.$$

Write the expression on the right-hand side as a single fraction. In the formula, W is the total uniform load on the beam, x is the distance from the support to the section being analyzed, and l is the length of the beam. **1 2 4** 8 7 9

3. The focal length f of a concave spherical mirror is given by

$$\frac{1}{f}=\frac{1}{o}+\frac{1}{i},$$

where o and i are the distances from the vertex of the mirror to the object and image, respectively. Write the expression on the right as a single fraction. 3 5 6 9

4. The equivalent resistance R of two resistors with resistances R_1 and R_2 is given by $\frac{1}{R}=\frac{1}{R_1}+\frac{1}{R_2}$.

(a) What resistance should be used with a resistance of 20 Ω to produce an equivalent resistance of 12 Ω?

(b) Solve the equation for R.

3 5 9 1 2 4 6 7 8

1. Architectural Technology
2. Civil Engineering Technology
3. Computer Engineering Technology
4. Mechanical Drafting Technology
5. Electrical/Electronics Engineering Technology
6. Chemical Engineering Technology
7. Industrial Engineering Technology
8. Mechanical Engineering Technology
9. Biomedical Engineering Technology

5. The formula $M_x = \frac{3lwx}{8} - \frac{wx^2}{2}$ gives the bending moment M_x of a beam that is supported at one end, fixed at the other, and bearing a uniformly distributed load. Solve the equation for w. In the formula, l is the length of the beam, w is the uniform load per unit length, and x is the distance from the support to the section under consideration. **1 2 4** 8 7 9

6. The dosage of medicine for a child is often determined by taking a fractional part of the adult dosage. To obtain the fraction, different rules are used for different age groups. One rule says to divide the child's age in months by 150. A rule for a different age group says divide the child's age in years by 12 more than the age. At what age do the two rules agree? **6 9** 2 4 5 7 8

7. When one compares the Celsius and Fahrenheit temperature scales, it can be seen that $\frac{100}{C} = \frac{180}{F - 32}$. Solve this equation for C. **6 8** 1 2 3 5 7 9

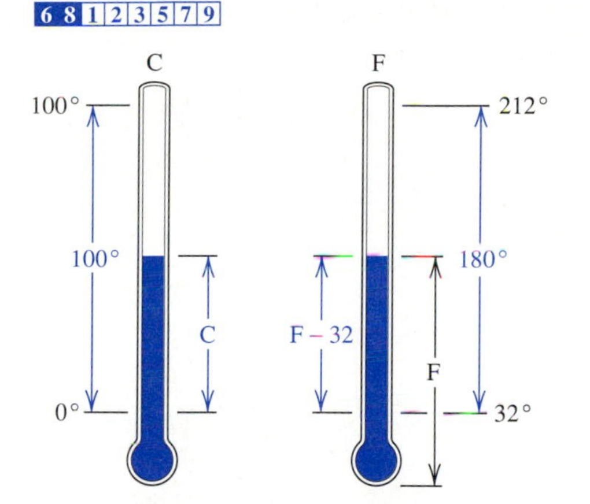

8. A plate for an adjustable capacitor is to be made by punching out and discarding a disk of radius R from the center of a larger aluminum disk of radius nR where n is some positive number. Express the area of the remaining ring in factored form. **3 5 9** 4 7 8

9. From a square copper plate of width 10 cm, a round hole of diameter $\left(\frac{10}{m}\right)$ cm is punched out where m is some number between 1 and 10. Express the area of the remaining copper in factored form. **1 2 4 7 8** 3 5

10. In solving for the peak and valley voltage of a certain type of transistor an equation of the form $I^2 - 1.1I + 0.28 = 0$ arises. Solve for I. **3 5 9** 4 7 8

11. In electronic filter theory a polynomial fraction of the type $F = \frac{P^2 - 17P + 60}{P^3 - 9P^2 + 26P - 24}$ arises. Find the factors of this fraction's numerator. **3 5 9** 1 2 4 6 7 8

12. The formula $R_T = \frac{R}{2}$ gives the equivalent resistance R_T of two identical parallel resistors, each having resistance R. If each resistance changes by an amount NR, the new value of R_T is given by $R_T = \frac{(R + NR)(R + NR)}{(R + NR) + (R + NR)}$. Simplify the fraction on the right. **3 5 9** 1 2 4 6 7 8

13. For three identical resistors in parallel, if each is of resistance R (in ohms) and if each is increased by the same amount NR the new total is given by

$$R_T = [(R + NR)(R + NR)(R + NR)] \div [(R + NR)(R + NR) + (R + NR)(R + NR) + (R + NR)(R + NR)].$$

Simplify the fraction on the right. **3 5 9** 1 2 4 6 7 8

14. The first page of a manuscript investigating volume changes in solids due to thermal expansion has been lost. The second page gives an expanded formula for the new volume of a cylinder when its radius r expands by a and its height h by b as:

$$V = \pi r^2 h + \pi r^2 b + 2\pi arh + 2\pi arb + \pi a^2 h + \pi a^2 b.$$

Factor the expression on the right to find the equation as it would be on the lost page. (Hint: You will need to use factoring by grouping.) **6 8** 1 2 3 5 7 9

15. When three inductors of values L_1, L_2, and L_3 (in henries) are connected in parallel the resulting total inductance L can be found from the equation

$$\frac{1}{L} = \frac{1}{L_1} + \frac{1}{L_2} + \frac{1}{L_3}.$$

Solve this equation for L by combining the fractions on the right. **3 5 9** 4 7 8

16. In electronics it is sometimes possible to approximate the mathematically complicated current/voltage relationships of a device by using simple first or second degree equations. Of course these approximate equations have a limited domain and range of usefulness but are nevertheless useful. Suppose an LED's (light emitting diode) current I (in ma) is approximated by $I = 3V - 6$ when V is given in volts, and a closer approximation is given by $I = (V - 2)^2$. For what values of V do these equations yield the same I?
3 5 9 1 2 4 6 7 8

17. A formula useful for converting degrees Fahrenheit to degrees Celsius is $C = (F + 40)\frac{5}{9} - 40$. Solve this equation for F and factor the result so the new formula has the same form as the old one.
6 8 1 2 3 5 7 9

18. The yield of a certain chemical process was given by the formula $V = QR^2S^3$. Changing each of the variables Q, R, and S gave a new yield different from the original amount by

$$D = R^2S^3 + 2QRS^3 + 3QR^2S^2.$$

Factor out the original yield QR^2S^3 from the expression on the right. 6 9 2 4 5 7 8

---

1. Architectural Technology
2. Civil Engineering Technology
3. Computer Engineering Technology
4. Mechanical Drafting Technology
5. Electrical/Electronics Engineering Technology
6. Chemical Engineering Technology
7. Industrial Engineering Technology
8. Mechanical Engineering Technology
9. Biomedical Engineering Technology

RANDOM REVIEW

1. Reduce to lowest terms. $\dfrac{3x^2 + 8x - 3}{3x^2 - 7x + 2}$
2. Solve for a. $2a^2 + 7a - 15 = 0$
3. Factor completely, if possible. $5x(2x + 3) - (2x + 3)$
4. Perform the indicated operation. $\dfrac{3}{2x + 6} + \dfrac{5}{3x + 9}$
5. Factor completely, if possible. $27x^3 - 8$
6. Solve for x. $\dfrac{3}{x} + \dfrac{2}{3} = \dfrac{1}{9}$
7. Factor completely, if possible. $4x^2 - 81$
8. Factor completely, if possible. $m^2 + 3m - 54$
9. Simplify. $\dfrac{\dfrac{1 + x}{x}}{\dfrac{1}{x} - x}$
10. Perform the indicated operation. $\dfrac{x^3 - 1}{x^2} \div \dfrac{x^2 - 1}{x}$

CHAPTER TEST

In 1–3, factor completely, if possible.

1. $m^2 + 2m - 63$

2. $x^6 + 64$

3. $x^2 + 2xy + y^2 - 81z^2$

In 4 and 5, solve each equation.

4. $6x^2 - x = 1$

5. $\frac{2}{x+3} + \frac{x}{x-1} = \frac{4}{x^2+2x-3}$

In 6 and 7, perform the indicated operation.

6. $\left(\frac{y^2-4}{2y^2+4y}\right)\left(\frac{3y^2+5y-2}{3y-1}\right)$

7. $\frac{x}{y^2} - \frac{y}{x}$

8. Simplify. $\frac{\frac{x-1}{x+1}}{\frac{x+1}{x-1}}$

9. When a fixed beam is under a uniform load, the bending moment (M_x) is found from the following equation.

$$M_x = \frac{-wl^2}{12} + \frac{wxl^2}{2} - \frac{wl^2x^2}{2l^2}$$

Write the expression on the right as a single fraction.

10. Suppose that the total resistance R of a parallel circuit is known, as is R_1. If R_2 has overheated and its color code cannot be determined, it can be found from the following formula.

$$R = \frac{R_1 R_2}{R_1 + R_2}$$

Solve for R_2.

CHAPTER 6 QUADRATIC EQUATIONS AND FUNCTIONS

In Chapter 5, we encountered quadratic equations that could be solved by factoring. The quadratic equations that arise in technical applications, however, are rarely factorable. In this chapter, we introduce other techniques for solving quadratic equations and examine the relationship between quadratic equations and quadratic functions.

6.1 SOLUTIONS BY COMPLETING THE SQUARE

Completing the square is a process that allows us to use the special properties of a perfect square trinomial with other types of polynomials. Recall from Section 5.2 that

$$x^2 + 2ax + a^2 = (x + a)^2 \tag{6-1}$$

The key is to notice that in Equation 6-1, the constant term a^2 of the trinomial is the square of one-half the coefficient of the linear term. $[(1/2)(2a) = a]$

EXAMPLE 1 Add the appropriate constant term to each binomial to make it a perfect square trinomial.

(a) $x^2 + 4x$ **(b)** $x^2 - 8x$ **(c)** $x^2 + 3x$ **(d)** $x^2 - (7/2)x$

Solution **(a)** Since the coefficient of the linear term is 4, one-half of it is 2. The square of 2 is 4, so add 4.

$$x^2 + 4x + 4 = (x + 2)^2$$

(b) Since the coefficient of the linear term is -8, one-half of it is -4. The square of -4 is 16, so add 16.

$$x^2 - 8x + 16 = (x - 4)^2$$

(c) Since the coefficient of the linear term is 3, one-half of it is 3/2. The square of 3/2 is 9/4, so add 9/4.

$$x^2 + 3x + 9/4 = (x + 3/2)^2$$

(d) Since the coefficient of the linear term is $-7/2$, one-half of it is $-7/4$. The square of $-7/4$ is 49/16, so add 49/16.

$$x^2 - (7/2)x + 49/16 = (x - 7/4)^2$$ ■

Completing the square is done in the context of working with equations. Whatever is added to one side of the equation must be added to the other side as well. Example 2 illustrates how to solve a quadratic equation by completing the square.

EXAMPLE 2 Solve $x^2 + 4x + 3 = 0$.

Solution If 3 is subtracted from both sides of the equation, the expression on the left-hand side looks like the binomial of Example 1(a).

$$x^2 + 4x = -3$$

To complete the square on the left-hand side of the equation, we add 4, because that is the square of one-half the coefficient of the linear term. But since $x^2 + 4x = -3$ is an equation, whatever is added to one side must be added to the other side.

$$x^2 + 4x + 4 = -3 + 4$$

Since the expression on the left-hand side is a perfect square trinomial, it can be written as $(x + 2)^2$. Thus, the equation becomes

$$(x + 2)^2 = 1.$$

From Section 2.5, however, we know that an equation of the form $x^2 = b$ has as its solution $x = \pm\sqrt{b}$. Applying this principle, we have

$$(x + 2) = \pm\sqrt{1} \qquad \text{or} \qquad x + 2 = \pm 1.$$

To solve for x, subtract 2 from both sides of the equation.

$$x = -2 \pm 1$$

The $\pm$ sign indicates that there are two solutions. One solution contains the plus sign, and the other contains the minus sign. The solutions, then, are

$$x = -2 + 1, \text{ which is } -1 \text{ or } x = -2 - 1, \text{ which is } -3. \quad \blacksquare$$

Notice in Example 2, that when 2 was subtracted from both sides of the equation, -2 was written to the left of the $\pm$ sign. The term added to or subtracted from the $\pm$ term is usually written in this position to avoid misinterpretation.

The equation in Example 2 could have been solved by factoring.

$$x^2 + 4x + 3 = 0$$
$$(x + 3)(x + 1) = 0$$
$$x + 3 = 0 \quad \text{or} \quad x + 1 = 0$$
$$x = -3 \quad \text{or} \quad x = -1$$

Completing the square, however, works even when factoring does not, as shown in Example 3.

EXAMPLE 3 Solve $x^2 + 3x + 1 = 0$.

Solution

$$x^2 + 3x + 1 = 0$$
$$x^2 + 3x = -1 \qquad \text{Subtract 1 from both sides.}$$

Notice that the expression on the left is like the binomial in Example 1(c). To complete the square, add 9/4 to both sides of the equation.

$$x^2 + 3x + 9/4 = 9/4 - 1$$

$$(x + 3/2)^2 = 5/4 \qquad \text{Replace } x^2 + 3x + 9/4 \text{ with } (x + 3/2)^2 \text{ and } 9/4 - 1 \text{ with } 9/4 - 4/4 \text{ or } 5/4.$$

$$x + \frac{3}{2} = \pm\sqrt{\frac{5}{4}} \qquad \text{Recall that } x^2 = b \text{ means } x = \pm\sqrt{b}.$$

$$x + \frac{3}{2} = \frac{\pm\sqrt{5}}{\sqrt{4}} \qquad \text{Recall that } \sqrt[n]{\frac{x}{y}} = \frac{\sqrt[n]{x}}{\sqrt[n]{y}}.$$

$$x + \frac{3}{2} = \frac{\pm\sqrt{5}}{2} \qquad \text{Replace } \sqrt{4} \text{ with 2.}$$

$$x = -\frac{3}{2} \pm \frac{\sqrt{5}}{2} \qquad \text{Subtract } \frac{3}{2} \text{ from both sides.}$$

$$x = \frac{-3 \pm \sqrt{5}}{2} \qquad \text{Combine the two fractions.} \quad \blacksquare$$

The observation that adding the square of one-half the coefficient of the linear term to a binomial produces a perfect square trinomial holds when the binomial is of the form $x^2 + bx$. If a quadratic equation has a leading coefficient other than 1, it is possible to divide both sides of the equation by the leading coefficient. The entire procedure is summarized as follows.

To solve a quadratic equation by completing the square:

1. Write an equivalent equation that has only the square and linear terms on one side of the equation.
2. Divide both sides of the equation by the leading coefficient, leaving the square term with a coefficient of 1.
3. Add the square of one-half the coefficient of the linear term to both sides of the equation.
4. Factor the trinomial as a perfect square and simplify the expression on the other side of the equation.
5. Solve for x.

EXAMPLE 4 Solve $2x^2 + 12x + 3 = 0$.

Solution

1. $2x^2 + 12x = -3$ — Subtract 3 from both sides.

2. $x^2 + 6x = \frac{-3}{2}$ — Divide both sides by 2.

3. $x^2 + 6x + 9 = \frac{-3}{2} + 9$ — Add $[(1/2)(6)]^2$ or 9 to both sides.

4. $(x + 3)^2 = \frac{15}{2}$ — Replace $x^2 + 6x + 9$ with $(x + 3)^2$ and $\frac{-3}{2} + 9$ with $\frac{-3}{2} + \frac{18}{2}$ or $\frac{15}{2}$.

5. $x + 3 = \pm\sqrt{\frac{15}{2}}$ — Recall that $x^2 = b$ means $x = \pm\sqrt{b}$.

$x + 3 = \frac{\pm\sqrt{15}}{\sqrt{2}}$ — Replace $\pm\sqrt{\frac{15}{2}}$ with $\frac{\pm\sqrt{15}}{\sqrt{2}}$.

$x + 3 = \frac{\pm\sqrt{30}}{2}$ — Rationalize the denominator.

$x = -3 \pm \frac{\sqrt{30}}{2}$ — Subtract 3 from both sides.

$x = \frac{-6 \pm \sqrt{30}}{2}$ — Replace -3 with $-6/2$ and combine terms. ■

Even though the square root that is introduced in completing the square may not be an integer, it may be possible to write the square root in a simplified form. Because $\sqrt{xy} = \sqrt{x}\sqrt{y}$, it may be possible to factor the number under the radical sign so that the square root of one of the factors is an integer. For instance, $\sqrt{18} = \sqrt{9(2)} = \sqrt{9}\sqrt{2} = 3\sqrt{2}$. Example 5 illustrates this point.

EXAMPLE 5 Solve $3y^2 + 6y - 21 = 0$.

Solution

$3y^2 + 6y = 21$	Add 21 to both sides.
$y^2 + 2y = 7$	Divide both sides by 3.
$y^2 + 2y + 1 = 7 + 1$	Add 1 to both sides.
$(y + 1)^2 = 8$	Rewrite both sides.
$y + 1 = \pm\sqrt{8}$	Recall that $x^2 = b$ means $x = \pm\sqrt{b}$.
$y = -1 \pm \sqrt{8}$	Subtract 1 from both sides.
$y = -1 \pm 2\sqrt{2}$	Replace $\sqrt{8}$ with $\sqrt{4(2)}$ or $2\sqrt{2}$. ■

EXERCISES 6.1

Solve each quadratic equation by completing the square.

1. $p^2 + 10p + 24 = 0$
2. $m^2 - 8m + 15 = 0$
3. $z^2 - 4z + 2 = 0$
4. $x^2 - 6x - 4 = 0$
5. $x^2 - x - 6 = 0$
6. $y^2 - 9y - 20 = 0$
7. $p^2 + 5p + 4 = 0$
8. $m^2 - 7m - 8 = 0$
9. $4x^2 + 8x + 3 = 0$
10. $2y^2 + 12y + 11 = 0$
11. $3m^2 - 6m = 2$
12. $5p^2 - 20p = -3$
13. $2m^2 + 8m = 3$
14. $3m^2 + 6m = -1$
15. $3x^2 - 1 = 18x$
16. $4x^2 - 3 = 8x$
17. $2x^2 + 3 = 10x$
18. $3x^2 + 2 = 9x$
19. $4y^2 + 4y - 1 = 0$
20. $2m^2 + 6m - 3 = 0$
21. $5x^2 - x - 1 = 0$
22. $3x^2 - x - 2 = 0$
23. $2x^2 - 9x + 10 = 0$
24. $3y^2 - 5y - 2 = 0$
25. $5m^2 + 3m - 2 = 0$

6.2 THE QUADRATIC FORMULA

Recall that a quadratic equation in x is an equation that can be put in the form $ax^2 + bx + c = 0$, where a, b, and c are real numbers, and a is not zero. The form $ax^2 + bx + c = 0$ is called the *standard form.* We say that a is the coefficient of the quadratic term, b is the coefficient of the linear term, and c is the constant term. A particular equation in this form differs from every other quadratic equation only in the values of a, b, and c. Hence the solution of the equation depends on the values of a, b, and c. The formula known as the "quadratic formula" gives the solution of a quadratic equation in standard form. So that you can become familiar with the formula, its derivation is deferred until the end of the section. In this chapter, only quadratic equations with real number solutions are considered.

THE QUADRATIC FORMULA

For a quadratic equation in standard form $ax^2 + bx + c = 0$, with $a \neq 0$, the solutions are given by

$$x = \frac{-b \pm \sqrt{b^2 - 4ac}}{2a}.$$

In Example 9 of Section 5.1, we solved the equation $6x^2 + x - 2 = 0$ by factoring. Here, we do the same example, using the quadratic formula.

EXAMPLE 1 Solve $6x^2 + x - 2 = 0$.

Solution Identify the coefficients: $a = 6$, $b = 1$, and $c = -2$. Substituting these values into the formula yields the following equation.

$$x = \frac{\overbrace{-1}^{-b} \pm \sqrt{\overbrace{(1)^2 - 4(6)(-2)}^{b^2 - 4ac}}}{\underbrace{2(6)}_{2a}}$$

$$= \frac{-1 \pm \sqrt{1 + 48}}{12}$$

$$= \frac{-1 \pm \sqrt{49}}{12}$$

$$= \frac{-1 \pm 7}{12}$$

One solution contains the plus sign and the other contains the minus sign.

$$x = \frac{-1 + 7}{12} = \frac{6}{12} = \frac{1}{2} \quad \text{or} \quad x = \frac{-1 - 7}{12} = \frac{-8}{12} = \frac{-2}{3}$$ ■

Using the quadratic formula is not as efficient as factoring for many problems. But its advantage is that it may be applied to a quadratic equation that cannot be factored, as well as to one that can be.

discriminant

The expression $b^2 - 4ac$, which appears under the radical sign in the formula, is called the **discriminant.** ***If $b^2 - 4ac$ is negative, you need not complete the computation, for the square root of a negative number is not a real number.*** The steps for solving a quadratic equation using the formula are summarized as follows.

To solve a quadratic equation using the quadratic formula:

1. If the equation is not in standard form $ax^2 + bx + c = 0$, write an equivalent equation that is in standard form. (That is, add or subtract the same term(s) on both sides of the equation to make one side equal zero, and arrange the terms so that the exponents are in descending order.)
2. Identify the values of a, b, and c.
3. Substitute the values of a, b, and c into the formula $x = \dfrac{-b \pm \sqrt{b^2 - 4ac}}{2a}$.
4. Simplify the expression.
   (a) Use the "+" sign between $-b$ and the radical to obtain one solution.
   (b) Use the "−" sign between $-b$ and the radical to obtain the other solution.

EXAMPLE 2 Use the quadratic formula to solve $x^2 - 3x + 1 = 0$.

Solution

1. The equation is in standard form.
2. $a = 1$, $b = -3$, and $c = 1$ — Identify the values of a, b, and c.
3. $x = \dfrac{-(-3) \pm \sqrt{(-3)^2 - 4(1)(1)}}{2(1)}$ — Substitute the values into the formula.
4. $x = \dfrac{3 \pm \sqrt{9 - 4}}{2}$ — Simplify the expression.

$$x = \frac{3 + \sqrt{5}}{2} \quad \text{or} \quad x = \frac{3 - \sqrt{5}}{2} \quad ■$$

EXAMPLE 3 Use the quadratic formula to solve $2x^2 + 3x = -1$.

Solution

1. $2x^2 + 3x + 1 = 0$ — Add 1 to both sides to put the equation in standard form.
2. $a = 2$, $b = 3$, and $c = 1$ — Identify the values of a, b, and c.
3. $x = \dfrac{-3 \pm \sqrt{3^2 - 4(2)(1)}}{2(2)}$ — Substitute the values into the formula.
4. $x = \dfrac{-3 \pm \sqrt{9 - 8}}{4} = \dfrac{-3 \pm \sqrt{1}}{4} = \dfrac{-3 \pm 1}{4}$ — Simplify the expression.

Considering the two solutions, we have

$$x = \frac{-3 + 1}{4} = \frac{-2}{4} = \frac{-1}{2} \quad \text{or} \quad x = \frac{-3 - 1}{4} = \frac{-4}{4} = -1. \quad ■$$

EXAMPLE 4 Use the quadratic formula to solve $x^2 - 5x = 0$.

Solution There are only two terms on the left, but we may consider the equation to be in standard form with $a = 1$, $b = -5$, and $c = 0$.

$$x = \frac{-(-5) \pm \sqrt{(-5)^2 - 4(1)(0)}}{2(1)} = \frac{5 \pm \sqrt{25 - 0}}{2}$$

$$= \frac{5 \pm \sqrt{25}}{2} = \frac{5 \pm 5}{2}$$

$$x = 5 \quad \text{or} \quad x = 0 \quad ■$$

As in the previous section, it may be possible to write the square root in a simplified form. Examples 5 and 6 illustrate this type of problem.

EXAMPLE 5 Use the quadratic formula to solve $x^2 - 5 = 0$.

Solution The equation is in standard form with $a = 1$, $b = 0$, and $c = -5$.

$$x = \frac{-0 \pm \sqrt{0^2 - 4(1)(-5)}}{2(1)} = \frac{-0 \pm \sqrt{0 + 20}}{2} = \frac{\pm\sqrt{20}}{2}$$

$$= \frac{\pm\sqrt{4(5)}}{2} = \frac{\pm\sqrt{4}\sqrt{5}}{2}$$

$$= \frac{\pm 2\sqrt{5}}{2} = \pm\sqrt{5} \quad ■$$

EXAMPLE 6 Use the quadratic formula to solve $2x^2 + 8x + 3 = 0$.

Solution The equation is in standard form with $a = 2$, $b = 8$, and $c = 3$.

$$x = \frac{-8 \pm \sqrt{8^2 - 4(2)(3)}}{2(2)} = \frac{-8 \pm \sqrt{64 - 24}}{4} = \frac{-8 \pm \sqrt{40}}{4}$$

$$= \frac{-8 \pm \sqrt{4(10)}}{4} = \frac{-8 \pm 2\sqrt{10}}{4}$$

At this stage, it is very important to reduce the answer correctly. Factors that are common to both numerator and denominator divide out.

$$x = \frac{-8 \pm 2\sqrt{10}}{4} = \frac{2(-4 \pm \sqrt{10})}{4} = \frac{-4 \pm \sqrt{10}}{2} \quad ■$$

NOTE ⇨⇨ ***Be careful when you reduce fractions like the one in Example 6. Remember that it is factors, not terms, that divide out.*** If a calculator is used to find a decimal approxi-

NOTE ⇨⇨ mation for the solution of a quadratic equation, ***it should not matter whether the radical is simplified before it is evaluated or not.*** For instance, $\sqrt{40}$ and $2\sqrt{10}$, its simplified form, will both be given as 6.3245553. In technical applications, the square root is usually given as a decimal approximation.

EXAMPLE 7 Paper towels 11.0 inches wide are to be packaged in rolls. Since they are to be wrapped in plastic, it is necessary to know the surface area occupied by the sides and two ends of a roll. Packaging considerations dictate that 182 in^2 of plastic can be used for one roll. Determine the radius that a roll of towels should have.

Solution Let $r =$ radius of one roll. Then the ends will each have an area of πr^2 for a total of $2\pi r^2$. The sides can be covered by a rectangular piece of plastic. Its width is 11.0 inches and its length is the circumference of the roll, $2\pi r$, as shown in Figure 6.1.

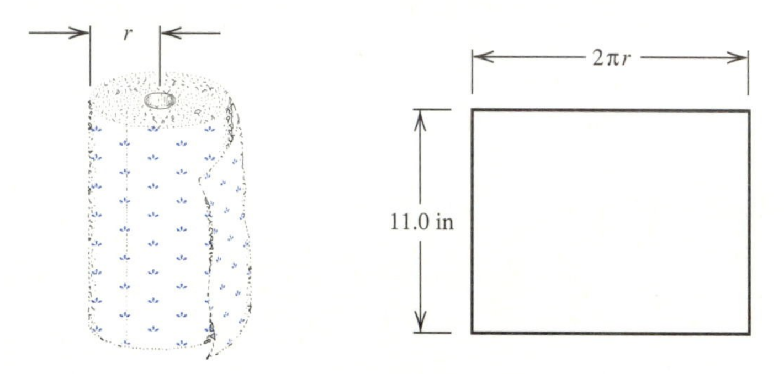

FIGURE 6.1

The rectangular area is $2\pi r(11.0)$. The total area, then, is given by

$$2\pi r^2 + 22.0\pi r = 182.$$

Writing the equation in standard form, we have

$$2\pi r^2 + 22.0\pi r - 182 = 0.$$

Since 2π is a factor of two of the three terms, the equation can be simplified with a division by 2π before applying the quadratic formula.

$$\frac{2\pi}{2\pi} r^2 + \frac{22.0\pi}{2\pi} r - \frac{182}{2\pi} = 0 \qquad \text{or} \qquad r^2 + 11.0r - 28.97 = 0$$

The coefficients of the quadratic equation are given by: $a = 1$, $b = 11.0$, and $c = -28.97$. Thus,

$$r = \frac{-11 \pm \sqrt{11^2 - 4(-28.97)}}{2}.$$

Using a calculator, we evaluate the discriminant first.

11 [x²] [−] 4 [×] 28.97 [+/−] [=] [√x] [STO]

We use the [STO] key to put the value in memory so that it can be recalled for use first with a positive sign and then with a negative sign. Since the value of the discriminant is now on the display, we add it to -11 by subtracting 11.

[−] 11 [=] [÷] 2 [=] **2.1954532**

To find the second solution, we recall the discriminant from calculator memory and change the sign before we subtract 11.

RCL ± − 11 = ÷ 2 = **−13.195453**

We have $r = -13.2$ or $r = 2.20$, to three significant digits. A negative radius, however, does not make sense, so the radius of each roll should be 2.20 inches. ■

EXAMPLE 8 The formula $8M_x = 3lwx - 4wx^2$ gives the moment M_x of a beam, supported at one end and fixed at the other, if the load is uniformly distributed. In the formula, l is the length of the beam (in inches), w is the load (in lb/in), and x is the distance from the support at which the moment is measured. Find the distance x along a 122.0 inch beam bearing a load of 3.71 lb/in, if the bending moment is 203.0 in·lb.

Solution With $l = 122.0$, $w = 3.71$, and $M_x = 203.0$, the equation is

$$8(203.0) = 3(122.0)(3.71)x - 4(3.71)x^2$$
$$0 = 3(122.0)(3.71)x - 4(3.71)x^2 - 8(203.0)$$
$$0 = -4(3.71)x^2 + 3(122.0)(3.71)x - 8(203.0)$$
$$0 = -14.84\,x^2 + 1357.86\,x - 1624$$
$$x = \frac{-1357.85 \pm \sqrt{(1357.86)^2 - 4(-14.84)(-1624)}}{2(-14.84)}$$
$$x = 1.21 \text{ inches} \quad \text{or} \quad x = 90.3 \text{ inches.}$$ ■

Now that you know how to use the formula, you might wonder how we know it will yield the correct solutions. We begin with the general equation $ax^2 + bx + c = 0$. Our objective is to solve this equation for x by completing the square. Subtract c from both sides of the equation.

$$ax^2 + bx = -c$$

Divide both sides of the equation by a.

$$x^2 + \frac{bx}{a} = \frac{-c}{a}$$

Add the square of $\frac{1}{2}\left(\frac{b}{a}\right)$, which is $\left(\frac{b}{2a}\right)^2$ or $\frac{b^2}{4a^2}$, to both sides of the equation.

$$x^2 + \frac{bx}{a} + \frac{b^2}{4a^2} = \frac{b^2}{4a^2} - \frac{c}{a}$$

Factor the expression on the left-hand side of the equation, and simplify the expression on the right-hand side.

$$\left(x + \frac{b}{2a}\right)^2 = \frac{b^2 - 4ac}{4a^2}$$

Since we are solving for x, we need to remove the square on the left-hand side for the next step. Recall from Section 2.5 that the solution of an equation of the form $x^2 = b$ is $x = \pm\sqrt{b}$.

$$x + \frac{b}{2a} = \pm\sqrt{\frac{b^2 - 4ac}{4a^2}}$$

The next step is to remove the term $\frac{b}{2a}$ on the left by subtracting it from both sides.

$$x = \frac{-b}{2a} \pm \sqrt{\frac{b^2 - 4ac}{4a^2}}$$

If $a > 0$, $\sqrt{4a^2} = 2a$. (If $a < 0$, the argument can be modified.)

$$x = \frac{-b}{2a} \pm \frac{\sqrt{b^2 - 4ac}}{2a}$$

With a common denominator, the formula may be written as

$$x = \frac{-b \pm \sqrt{b^2 - 4ac}}{2a}.$$

Since the preceding steps could be done regardless of the values of a, b, and c ($a \neq 0$), it is not necessary to repeat them each time a quadratic equation is solved. We simply use the formula since the solution will follow that pattern.

EXERCISES 6.2

In 1–24, use the quadratic formula to solve each equation.

1. $x^2 - 4x + 2 = 0$ **2.** $y^2 - 3y + 1 = 0$ **3.** $z^2 - 5z + 3 = 0$ **4.** $2x^2 - 3x - 1 = 0$

5. $2y^2 + 5y = 1$ **6.** $2p^2 + 3p = 7$ **7.** $q^2 - 21 = 0$ **8.** $r^2 - 43 = 0$

9. $2x^2 - 7x = 0$ **10.** $3y^2 + 2y - 3 = 0$ **11.** $3z^2 - z - 1 = 0$ **12.** $x^2 - 5x = 7$

13. $p^2 - 6p = 7$ **14.** $10q^2 + q - 1 = 0$ **15.** $r^2 + 8r + 1 = 0$ **16.** $x^2 + 3x + 1 = 0$

17. $2y^2 + 4y = 1$ **18.** $3z^2 + 4z = 1$ **19.** $x^2 + 5x + 2 = 0$ **20.** $y^2 - 5y = 2$

21. $3z^2 + z = 1$ **22.** $x^2 = 3$ **23.** $y^2 = 3y$ **24.** $z^2 - 7z = 8$

---

25. If the paper towels in Example 7 are only 20.0 cm wide, and 1082 cm^2 of plastic can be used, determine the radius of the roll.
1 2 4 7 8 3 5

26. When the sides and top of an 8 ft tall cylindrical tank are painted, it is determined from the amount of paint used, that the area is 251 ft^2. Find the radius of the tank. 1 2 4 7 8 3 5

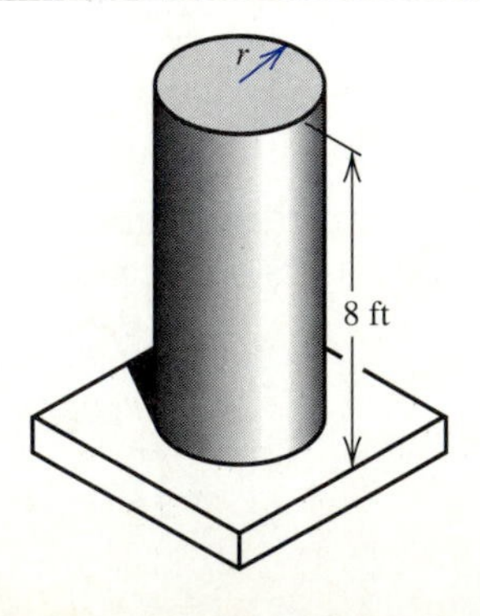

27. The formula $p^2 = \frac{2wp}{A} + \frac{2whE}{AL}$ gives the stress p (in lb/ft^2) exerted upon a bar by a falling object. In the formula, L is the length (in ft) of the bar, w is the weight (in lb) of the object, A is the cross-sectional area (in ft^2) of the bar, h is the height (in ft) from which the object falls, and E is the weight (in lb) of the bar. Find the stress exerted upon a 4.10-ft long bar that weighs 1230 lb and has a cross-sectional area of $0.642\ ft^2$, when a weight of 40.0 lb falls on it from a height of 2.00 ft.
1 2 4 8 7 9

28. The bending moment m (in ft·lb) of a uniformly loaded rigid beam is given by $m = \frac{wx(l-x)}{2}$. In the formula, w is the weight (in lb/ft), l is the length (in ft) of the beam, and x is the distance (in ft) from the end of the beam at which the moment is measured. Find the distance x at which a beam that is 14.21 ft long and weighs 184.7 lb/ft has a bending moment of 2134 ft·lbs.
1 2 4 8 7 9

29. To calculate the temperature T (in degrees C) at which water boils at an elevation E (in m) above sea level, the equation $580x^2 + 1000x - E = 0$ is solved for x. Then $T = 100 - x$ for $x \geq 0$. At what temperature does water boil at the top of the Empire State Building ($E = 380$ m)?
6 8 1 2 3 5 7 9

30. The formula $h^2 + 2rh = d^2$ can be used to calculate the height h of a building or tower. In the formula, r is the radius of the earth (6378 km), and d is the distance from the top of the structure to the horizon. The values r, h, and d must have the same units. Radio signals of a certain type have a maximum range of 51 km. Find the height of a structure from which the signal could be sent to the horizon. 1 3 4 5 6 8 9

31. The accompanying figure shows a potentiometer connected in parallel with a resistor. This arrangement allows for an adjustable value of resistance ranging from 0 to R. The equation $r(x) = \frac{2R(L-x)}{(2L-x)}$ gives the combined resistance $r(x)$ as a function of the position x of the contact point. Find the position of the contact point for which $r(x) = Rx/L$. 3 5 9 1 2 4 6 7 8

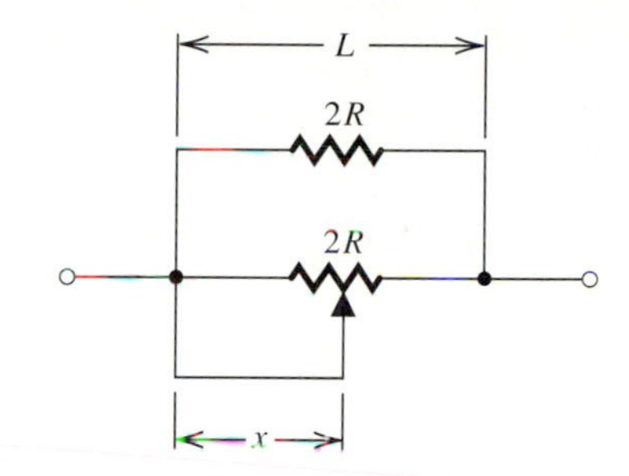

1. Architectural Technology
2. Civil Engineering Technology
3. Computer Engineering Technology
4. Mechanical Drafting Technology
5. Electrical/Electronics Engineering Technology
6. Chemical Engineering Technology
7. Industrial Engineering Technology
8. Mechanical Engineering Technology
9. Biomedical Engineering Technology

6.3 GRAPHING QUADRATIC FUNCTIONS

In the previous two sections, we examined quadratic equations. In this section, we examine quadratic functions.

DEFINITION

A **quadratic function** is a function that can be written in the form $f(x) = ax^2 + bx + c$ (or $y = ax^2 + bx + c$), where a, b, and c are real numbers, and $a \neq 0$.

In Section 3.4, we plotted graphs of functions by choosing values for x and finding the associated value of y for each of those values. In Example 1 of that section, we saw that the graph of $y = x^2 - 5x + 6$ is like the graph shown in Figure 6.2. The shape of this graph is called a *parabola*.

parabola

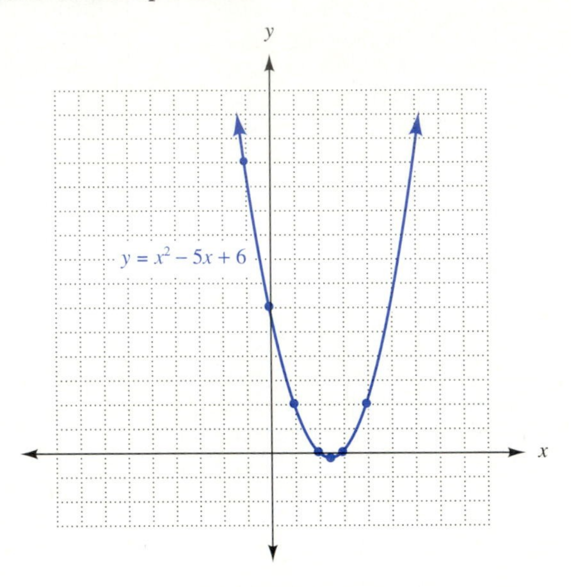

FIGURE 6.2

axis of symmetry

The line given by $x = 5/2$ is called the *axis of symmetry*. We say that the parabola is *symmetric* with respect to this line, because the left-hand side is a mirror image of the right-hand side.

vertex

If we know the coordinates of the **vertex,** which is the highest or lowest point of the parabola, and several other points, we can obtain a reasonably good sketch of the graph by connecting the points to complete a parabola. The x-coordinate of the vertex is given by $-b/(2a)$. By substituting this value for x in the equation that defines the function, we can calculate the y-coordinate.

To see why $-b/(2a)$ is the x-coordinate of the vertex of the parabola, notice that when the graph crosses the x-axis, the y-coordinate is zero. That is, we have $ax^2 + bx + c = 0$. But this equation is the quadratic equation that we saw in the previous section. We consider the case in which it has two real solutions and generalize from the result. The solutions are $\dfrac{-b + \sqrt{b^2 - 4ac}}{2a}$ and $\dfrac{-b - \sqrt{b^2 - 4ac}}{2a}$. Employing the symmetry of the parabola, we see that the x-coordinate of the vertex is halfway between these two values, as shown in Figure 6.3. We average them to find the midpoint.

$$\frac{1}{2}\left(\frac{-b + \sqrt{b^2 - 4ac}}{2a} + \frac{-b - \sqrt{b^2 - 4ac}}{2a}\right)$$

$$= \frac{1}{2}\left(\frac{-2b}{2a}\right) = \frac{-b}{2a}$$

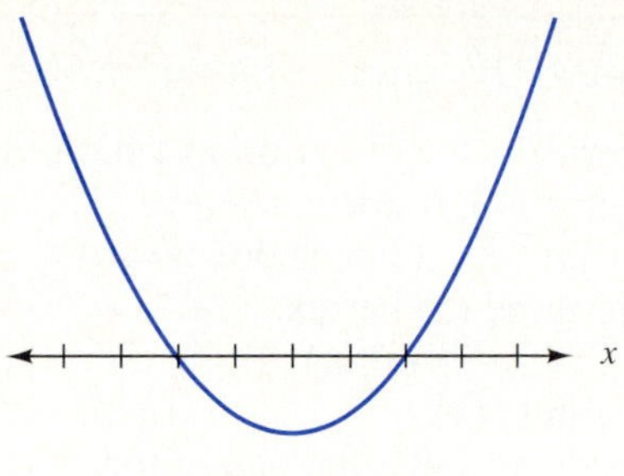

FIGURE 6.3

Once we know where the vertex is, we can find other points on either side of it.

EXAMPLE 1 Sketch the graph of $y = x^2 - 4x + 3$.

Solution For this equation, a is 1, b is -4, and c is 3. The x-coordinate of the vertex is 4/2, which is 2. When x is 2,

$$y = 2^2 - 4(2) + 3 = 4 - 8 + 3 = -1.$$

The vertex, then, is at $(2, -1)$. Consider the x-values $-1, 0, 1, 3, 4$, and 5. Calculate y for each x.

x	$y = x^2 - 4x + 3$
-1	$(-1)^2 - 4(-1) + 3 = 8$
0	$0^2 - 4(0) + 3 = 3$
1	$1^2 - 4(1) + 3 = 0$
3	$3^2 - 4(3) + 3 = 0$
4	$4^2 - 4(4) + 3 = 3$
5	$5^2 - 4(5) + 3 = 8$

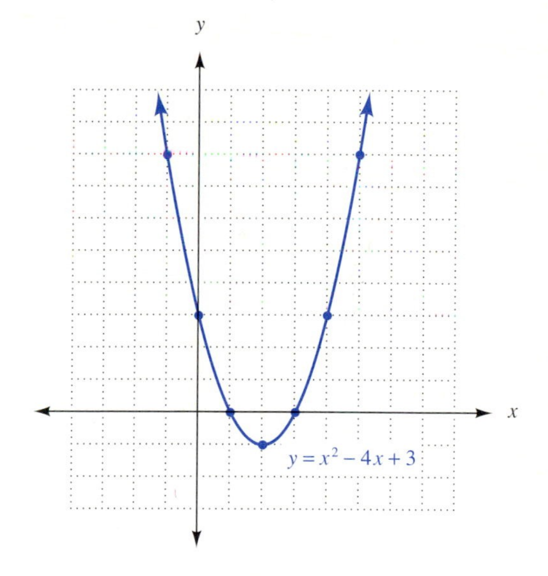

FIGURE 6.4

Figure 6.4 shows the graph. ■

In Example 1, it is interesting to note that the parabola opens upward. In general, when the value of a is positive in the equation $y = ax^2 + bx + c$, the parabola opens upward. When a is negative, the parabola opens downward. We generalize the procedure of Example 1 as follows.

To sketch the graph of a quadratic function $f(x) = ax^2 + bx + c$:

1. Determine the direction in which the parabola opens.
   (a) If $a > 0$, it opens upward.
   (b) If $a < 0$, it opens downward.
2. Determine the vertex.
   (a) Let $x = -b/(2a)$.
   (b) Find $f(x)$.
3. Choose several other values for x (some on each side of the vertex) and find the corresponding function values.
4. Plot these points and connect them to complete the parabola.

EXAMPLE 2 Sketch the graph of $y = -x^2 + 4$.

Solution

1. Since $a = -1$, the parabola opens downward.
2. Since $a = -1$ and $b = 0$, the x-coordinate of the vertex is at $-0/-2$, which is 0. When x is 0, y is 4, and the vertex is at (0, 4).
3. When x is 1, y is $-1 + 4$ or 3, so plot (1, 3).
   When x is 2, y is $-4 + 4$ or 0, so plot (2, 0).
   When x is 3, y is $-9 + 4$ or -5, so plot $(3, -5)$.
   In a similar manner, $(-1, 3)$, $(-2, 0)$, and $(-3, -5)$ are plotted to the left of the vertex.
4. These points are connected to complete the parabola, as shown in Figure 6.5. ■

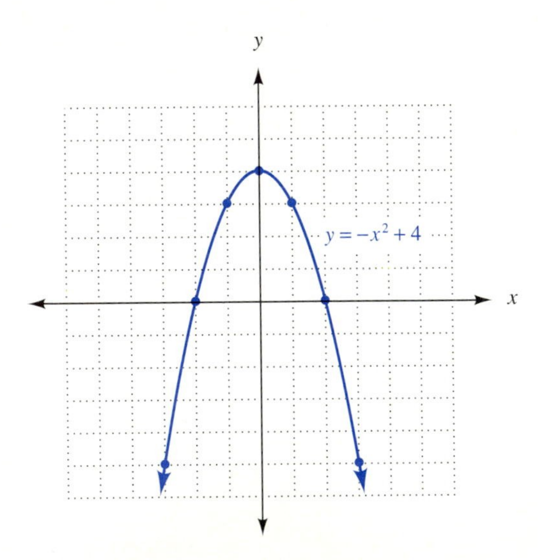

FIGURE 6.5

EXAMPLE 3 Sketch the graph of $y = 2x^2 + 2x + 1$.

Solution

1. Since $a = 2$, the parabola opens upward.
2. The x-coordinate of the vertex is $-2/(2 \cdot 2)$, which is $-1/2$. When x is $-1/2$,

$$y = 2(1/4) + 2(-1/2) + 1 = 1/2 - 1 + 1 = 1/2$$

   Thus, the vertex is at $(-1/2, 1/2)$.

3. The following table of values shows other values for x and the corresponding y-values.

x	$y = 2x^2 + 2x + 1$
$1/2$	$2(1/4) + 2(1/2) + 1 = 5/2$ or $2\ 1/2$
$3/2$	$2(9/4) + 2(3/2) + 1 = 17/2$ or $8\ 1/2$
$-3/2$	$2(9/4) + 2(-3/2) + 1 = 5/2$ or $2\ 1/2$
$-5/2$	$2(25/4) + 2(-5/2) + 1 = 17/2$ or $8\ 1/2$

4. The graph is shown in Figure 6.6. ■

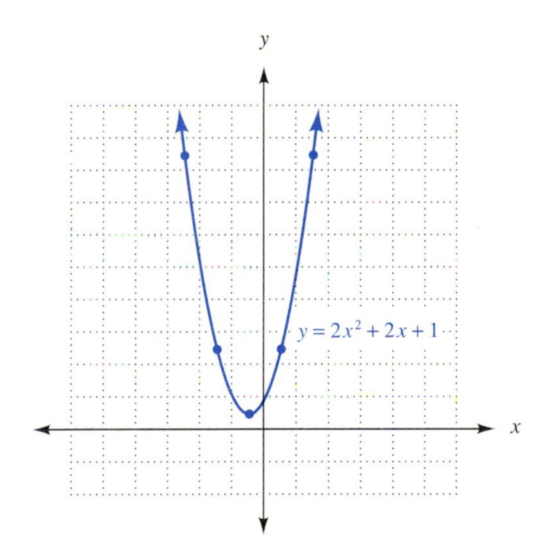

FIGURE 6.6

EXAMPLE 4 The formula $d = 0.05x^2 + x$ gives the approximate distance (in ft) required to bring a car traveling at speed x (in mph) to a complete stop. Let x be the independent variable and d the dependent variable. Sketch the graph of d versus x.

Solution Since a is 0.05, the parabola opens upward. The x-coordinate of the vertex is $-1/0.1$ or -10. When x is -10,

$$d = 0.05(-10)^2 - 10 = 5 - 10 = -5.$$

The vertex is at $(-10, -5)$. A calculator may be used to find additional points. If the value for x is put into calculator memory for each point, the following keystroke sequence may be used.

RCL x^2 × 0.05 + RCL

Points on the parabola are obtained as shown in the following table.

x	-9	-8	-7	-6	-5	-4	-3	-2	-1	0
y	-4.95	-4.80	-4.55	-4.20	-3.75	-3.20	-2.55	-1.80	-0.95	0

Although the vertex of the parabola has negative coordinates, only nonnegative values of x and d make sense for this problem, since the variables represent physical quantities that cannot be negative. Figure 6.7 shows the graph extended into the first quadrant. ■

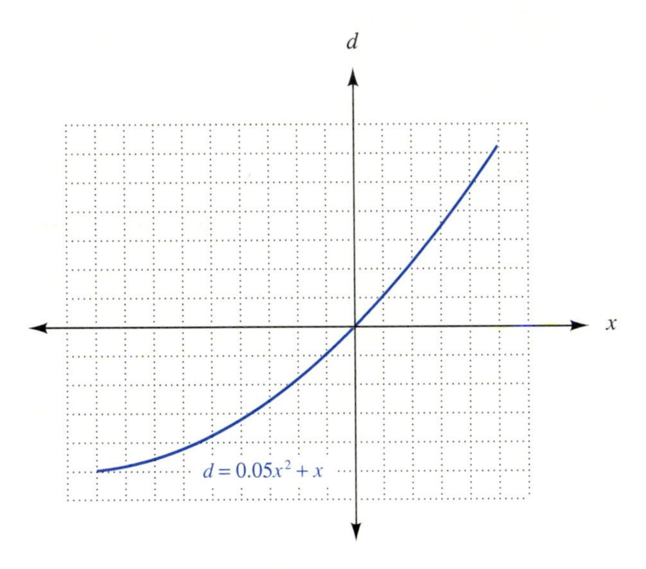

FIGURE 6.7

EXERCISES 6.3

In 1–20, find the vertex and sketch the graph of each parabola.

1. $y = 2x^2 + 8x + 12$
2. $y = -x^2 + 2x + 8$
3. $y = x^2 - 4x - 4$
4. $y = x^2 + 6x + 5$
5. $y = x^2 - 1$
6. $y = 2x^2 - 9$
7. $y = -2x^2 + 4x$
8. $y = 2x^2 - 6x$
9. $y = 4x^2 - 4x + 1$
10. $y = 4x^2 - 4x - 3$
11. $y = x^2 + 2x + 3$
12. $y = -x^2 + 2x + 3$
13. $y = x^2 - 2x + 3$
14. $y = -x^2 - 2x + 3$
15. $y = 2x^2 + 3x + 1$
16. $y = -2x^2 - 3x - 1$
17. $y = -2x^2 + 3x + 1$
18. $y = -x^2 - x + 1$
19. $y = -x^2 + x - 1$
20. $y = x^2 + x - 1$

Problems 21–23 use the following information. The distance s of an object from a fixed point is given by

$$s = s_0 + vt + (1/2)at^2.$$

In the formula, s_0 is the initial distance of the object from the fixed point, v is the initial velocity, a is a constant acceleration, and t is the time during which the object moves. Let t be the independent variable and s the dependent variable.

21. An object is thrown upward from a height of 4 ft with an initial velocity of 12 ft/s. The acceleration due to gravity is -32 ft/s^2.
 (a) Sketch the graph of s versus t.
 (b) Find the length of time until the object strikes the ground. 4 7 8 1 2 3 5 6 9

22. An object is released from a cliff 19.6 m tall. The acceleration due to gravity is -9.8 m/s^2.
 (a) Sketch the graph of s versus t.
 (b) Find the length of time until the object strikes the ground. 4 7 8 1 2 3 5 6 9

23. A bullet is fired vertically from ground level with an initial vertical velocity of 294 m/s. The acceleration due to gravity is -9.8 m/s^2. Find the length of time until the bullet is at a height of 800.0 m. 4 7 8 1 2 3 5 6 9

24. The volume V of a cone is given by

$$V = \frac{\pi}{12} d^2 h.$$

In the formula, d is the diameter of the cone, and h is its height.
 (a) Sketch the graph of volume versus diameter for 12-inch high cones. Use only positive values of d.
 (b) Determine the volume of such a cone having a diameter of 4.60 inches.
1 2 4 7 8 3 5

25. In road construction, a dip in a road is called sag. Under certain circumstances, the elevation Y (in ft) of the road at a horizontal distance X (in ft) from the beginning of the sag is given by

$$Y = \frac{1}{16{,}000} X^2 - \frac{1}{50} X + 250.$$

Sketch the graph of Y versus X.
1 2 4 7 8 3 5

26. Acceleration due to gravity on the moon is 1/6 the acceleration due to gravity on the earth. How much higher will an object thrown upward with an initial velocity of v go on the moon than on earth? Justify your answer by finding the ratio of maximum earth height s_e to maximum moon height s_m.

---

1. Architectural Technology
2. Civil Engineering Technology
3. Computer Engineering Technology
4. Mechanical Drafting Technology
5. Electrical/Electronics Engineering Technology
6. Chemical Engineering Technology
7. Industrial Engineering Technology
8. Mechanical Engineering Technology
9. Biomedical Engineering Technology

6.4 EQUATIONS IN QUADRATIC FORM

Sometimes an equation does not actually fit the definition of a quadratic equation, but is similar to a quadratic equation in the pattern of exponents and coefficients.

> **DEFINITION**
> An equation is said to be in **quadratic form** if it is in the form $ax^{2p} + bx^p + c = 0$ where a, b, c, and p are real numbers, $a \neq 0$, and x is a variable or variable expression.

The key to working with such equations is to recognize that the exponent of one term is twice the exponent of another. The term with the smaller exponent may be treated as the linear term of a quadratic equation.

EXAMPLE 1 Solve $x^4 - 13x^2 + 36 = 0$.

Solution The exponent of the first term is four and is twice the exponent of the second term. The equation may be treated as a quadratic in the following manner.

$$x^4 - 13x^2 + 36 = 0$$

Let $u = x^2$. Then $u^2 = (x^2)^2 = x^4$.

$u^2 - 13u + 36 = 0$ — Replace x^4 with u^2 and x^2 with u.

$(u - 4)(u - 9) = 0$ — Factor the polynomial.

$u - 4 = 0$ or $u - 9 = 0$ — Set each factor equal to zero.

There are now two linear equations in u. The original equation requires a solution for x, not u. Recall that we began by letting $u = x^2$. Replacing u with x^2, we have

$x^2 - 4 = 0$ or $x^2 - 9 = 0$.

$(x - 2)(x + 2) = 0$, or $(x - 3)(x + 3) = 0$ — Factor each polynomial.

$x - 2 = 0, x + 2 = 0$, or $x - 3 = 0, x + 3 = 0$ — Set each factor equal to zero.

$x = 2, x = -2, x = 3$, or $x = -3$ — Solve each equation. ■

EXAMPLE 2 Solve $x^{-4} - 5x^{-2} + 4 = 0$.

Solution $x^{-4} - 5x^{-2} + 4 = 0$

Let $u = x^{-2}$. Then $u^2 = (x^{-2})^2 = x^{-4}$.

$u^2 - 5u + 4 = 0$ — Replace x^{-4} with u^2 and x^{-2} with u.

$(u - 4)(u - 1) = 0$ — Factor the polynomial.

$u - 4 = 0$ or $u - 1 = 0$ — Set each factor equal to zero.

$x^{-2} - 4 = 0$ or $x^{-2} - 1 = 0$ — Replace u with x^{-2}.

$\frac{1}{x^2} - 4 = 0$ or $\frac{1}{x^2} - 1 = 0$ — Recall that $x^{-2} = \frac{1}{x^2}$ if $x \neq 0$.

$1 - 4x^2 = 0$ or $1 - x^2 = 0$ — Multiply both sides by x^2.

$(1 - 2x)(1 + 2x) = 0$ or $(1 - x)(1 + x) = 0$ — Factor each polynomial.

$1 - 2x = 0, 1 + 2x = 0$ or $1 - x = 0, 1 + x = 0$ — Set each factor equal to zero.

$1 = 2x, 2x = -1$ or $1 = x, x = -1$

$x = 1/2, x = -1/2, x = 1$, or $x = -1$ ■

EXAMPLE 3 Solve $(x-1)^2-(x-1)-2=0$.

Solution $(x-1)^2-(x-1)-2=0$

Let $u=(x-1)$. Then $u^2=(x-1)^2$.

$u^2-u-2=0$

$(u-2)(u+1)=0$

$u-2=0$ or $u+1=0$

$(x-1)-2=0$ or $(x-1)+1=0$

$x-3=0$ or $x=0$

$x=3$ or $x=0$ ■

Example 4 illustrates how equations in quadratic form may arise in the technologies.

EXAMPLE 4 When a quadratic equation has coefficients such that $|b|$ is very much larger than $|a|$ and $|c|$, a computer program using the quadratic formula may lead to severe round-off error. In such cases, an alternate formula may be used to find one of the roots. To find this alternate formula, consider the equation that results from multiplying both sides of $Ax^2+Bx+C=0$ by $\frac{1}{x^2}$. The equation is

$$A+B\left(\frac{1}{x}\right)+C\left(\frac{1}{x^2}\right)=0 \qquad \text{or}$$

$$C\left(\frac{1}{x^2}\right)+B\left(\frac{1}{x}\right)+A=0.$$

Solve this equation for x.

Solution The equation is in quadratic form. Let $u=1/x$, then $u^2=1/x^2$. We have $Cu^2+Bu+A=0$. This equation may be solved by the quadratic formula with $a=C$, $b=B$, and $c=A$. Thus,

$$u=\frac{-B\pm\sqrt{B^2-4AC}}{2C}$$

or

$$\frac{1}{x}=\frac{-B\pm\sqrt{B^2-4AC}}{2C}.$$

Hence, $x=\dfrac{2C}{-B\pm\sqrt{B^2-4AC}}$. ■

Sometimes a common factor may be factored from one side of an equation, leaving a quadratic polynomial or a polynomial in quadratic form.

EXAMPLE 5 Solve $x^3 + 5x^2 + 6x = 0$.

Solution

$x^3 + 5x^2 + 6x = 0$

$x(x^2 + 5x + 6) = 0$ Use the distributive property.

$x(x + 2)(x + 3) = 0$ Factor the polynomial in parentheses.

$x = 0,\ x + 2 = 0, \text{ or } x + 3 = 0$ Set each factor equal to zero.

$x = 0,\ x = -2, \text{ or } x = -3.$ Solve each equation. ■

EXAMPLE 6 Solve $x^3 + 3x^2 + x = 0$.

Solution

$$x^3 + 3x^2 + x = 0$$
$$x(x^2 + 3x + 1) = 0$$

The expression in parentheses cannot be factored. Thus,

$$x = 0 \text{ or } x^2 + 3x + 1 = 0$$

Using the quadratic formula to solve the second equation, we have

$$x = \frac{-3 \pm \sqrt{3^2 - 4(1)(1)}}{2(1)} = \frac{-3 \pm \sqrt{5}}{2}.$$

The solution, then, is $x = 0$, $x = \dfrac{-3 + \sqrt{5}}{2}$, or $x = \dfrac{-3 - \sqrt{5}}{2}$. ■

EXERCISES 6.4

In 1–30, solve each equation.

1. $x^4 - 10x^2 + 9 = 0$
2. $36y^4 - 13y^2 + 1 = 0$
3. $4z^4 - 5z^2 + 1 = 0$
4. $x^4 - 17x^2 + 16 = 0$
5. $y^4 - 2y^2 + 1 = 0$
6. $z^4 - 8z^2 + 16 = 0$
7. $x^{-2} - 7x^{-1} + 12 = 0$
8. $y^{-2} - 10y^{-1} + 24 = 0$
9. $z^{-2} - 4z^{-1} + 3 = 0$
10. $2x^{-2} - 3x^{-1} - 2 = 0$
11. $2/y^2 - 5/y + 2 = 0$
12. $2/z^2 + 3/z + 1 = 0$
13. $2/x^2 + 2/x = 0$
14. $2/y^2 + 5/y - 3 = 0$
15. $(x - 2)^2 + 2(x - 2) - 24 = 0$
16. $(x - 3)^2 - (x - 3) - 2 = 0$
17. $(y - 1)^2 - 2(y - 1) = 0$
18. $4(2y + 3)^2 + 4(2y + 3) - 3 = 0$
19. $6(2z - 1)^2 + 5(2z - 1) + 1 = 0$
20. $6(3z - 2)^2 - 7(3z - 2) - 3 = 0$
21. $x^3 - 5x^2 + 6x = 0$
22. $x^3 - 6x^2 + 9x = 0$
23. $2y^3 - 3y^2 + y = 0$
24. $4x^5 - 9x^3 = 0$
25. $6x^4 - 13x^3 + 6x^2 = 0$
26. $3y^3 + 2y^2 = 3y$
27. $3z^3 = z^2 + z$
28. $x^4 = 7x^2 + 5x^3$
29. $y^5 + 81y = 18y^3$
30. $x^6 + 25x^2 = 26x^4$

31. A tank is to be made in the shape of a cylinder of height 10.0 m and radius r capped by a hemisphere on each end as shown in the accompanying diagram.

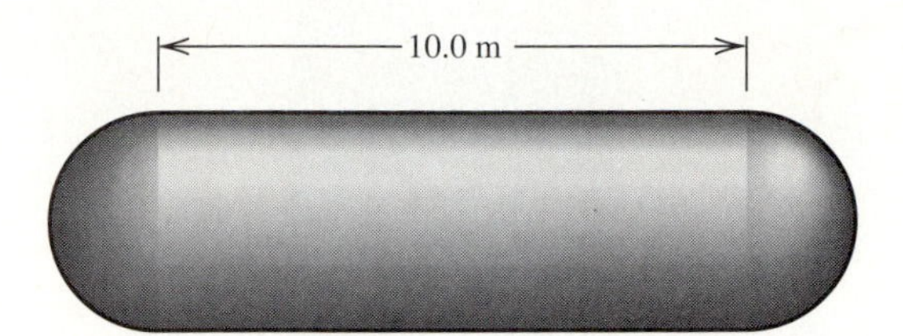

(a) Give a formula for the volume of the tank.
(b) Find a value for r that would make the volume numerically equal to the radius.
1 2 4 7 8 3 5

32. The surface area of a cylinder that is closed on both ends, such as the one in the accompanying diagram, is given by

$$A = 2\pi r^2 + 2\pi rh.$$

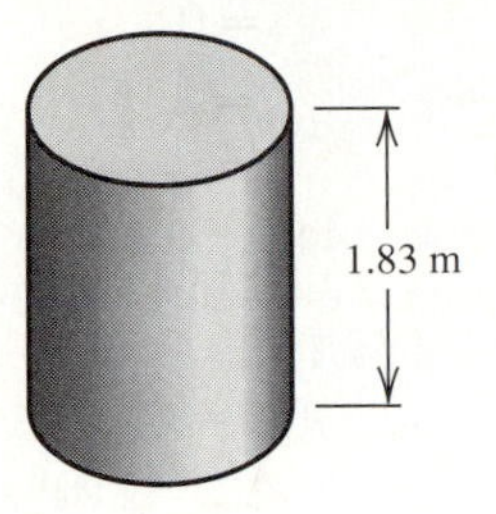

The radius of a cylinder 1.83 m high is increased by an amount denoted Δr, so that the new radius is $r + \Delta r$. If the area of the larger cylinder is 9.54 m^2, solve the equation

$$9.54 = 2\pi(r + \Delta r)^2 + 2\pi(1.83)(r + \Delta r)$$

for $r + \Delta r$. 1 2 4 7 8 3 5

Problems 33 and 34 use the following information. To calculate the temperature (in degrees Celsius) at which water boils at an elevation E (in m) above sea level, the equation

$$E = 580(100 - T)^2 + 1000(100 - T)$$

is used.

33. At what temperature does water boil at an elevation of 200 m? 6 8 1 2 3 5 7 9

34. At what temperature does water boil at an elevation of 50 m? 6 8 1 2 3 5 7 9

35. The maximum height h attained by a projectile launched with an initial velocity v_0 at an angle of θ is given by $h = \dfrac{v_0^2 \sin^2 \theta}{2g}$ and its horizontal distance d is given by $d = \dfrac{2v_0^2 \sin\theta \cos\theta}{g}$. In the formula g is the acceleration due to gravity. If d is 2000 m larger than h, write an equation to express the relationship. Multiply both sides of the equation by the LCD. Show that the resulting equation can be treated as an equation in quadratic form (with $\sin\theta$ as the variable) by identifying the coefficients a, b, and c. 4 7 8 1 2 3 5 6 9

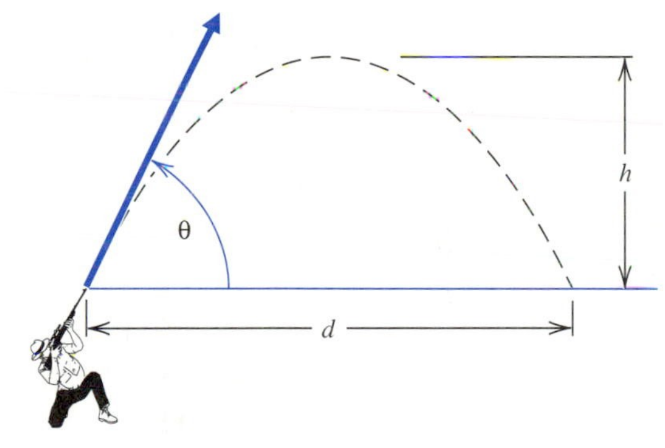

36. To construct a rectangular swimming pool with a fixed area A_1 at one end as shown, the following formula can be used to determine the area needed at the other end.

$$\left[\left(\frac{3V}{L} - A_1\right) - A_2\right]^2 = A_1 A_2$$

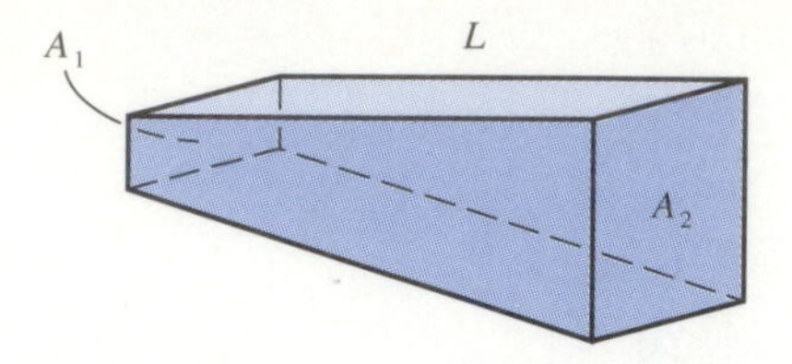

In the formula, V is the volume, and L is the length. Write the formula in quadratic form with A_2 as the variable. 1 2 4 7 8 3 5

37. A cubical box that measures $3' \times 3' \times 3'$ is pushed against a wall. A ladder 15 ft long is placed so that it rests against the edge of the box and the wall. How far is the foot of the ladder from the box?

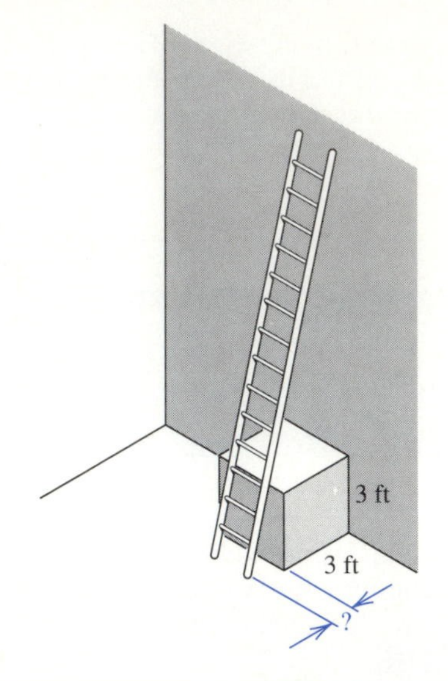

1. Architectural Technology
2. Civil Engineering Technology
3. Computer Engineering Technology
4. Mechanical Drafting Technology
5. Electrical/Electronics Engineering Technology
6. Chemical Engineering Technology
7. Industrial Engineering Technology
8. Mechanical Engineering Technology
9. Biomedical Engineering Technology

6.5 SOLUTIONS BY THE BISECTION METHOD

In this section, we will examine still another method for solving quadratic equations. The method, called the *bisection method,* illustrates the overall nature of an *iterative* process. The approach is to take a "guess," use that guess to obtain a better one, and so on, until we have the answer to the desired degree of precision.

EXAMPLE 1 Solve $x^2 - x - 1 = 0$, given that x is positive. Round your answer to hundredths.

Solution Instead of thinking of the problem as an equation, we might say that we are trying to find the value for x that makes the polynomial $x^2 - x - 1$ equal to 0. Figure 6.8 shows a sketch of $y = x^2 - x - 1$. The graph shows that as x is allowed to take on values between 1 and 2, y takes on values between -1 and 1.

Since zero lies between these two y-values, we know that the x-value associated with a y-value of 0 lies between the x-values associated with y-values of -1 and 1. That is, x lies in the interval between 1 and 2. For our first "guess," we bisect this interval and guess 1.5, which is the average of 1 and 2. When x is 1.5, we have

$$y = (1.5)^2 - 1.5 - 1 = -0.25.$$

Since zero lies between y-values of -0.25 and 1 (that is, between a negative value and a positive value), the x-value we seek must be between 1.5 (which produced the y-value of -0.25) and 2.0 (which produced the y-value of 1), as shown in Figure 6.9. We bisect the interval from 1.5 to 2 and guess 1.75, which is the average of 1.5 and 2, for our second guess.

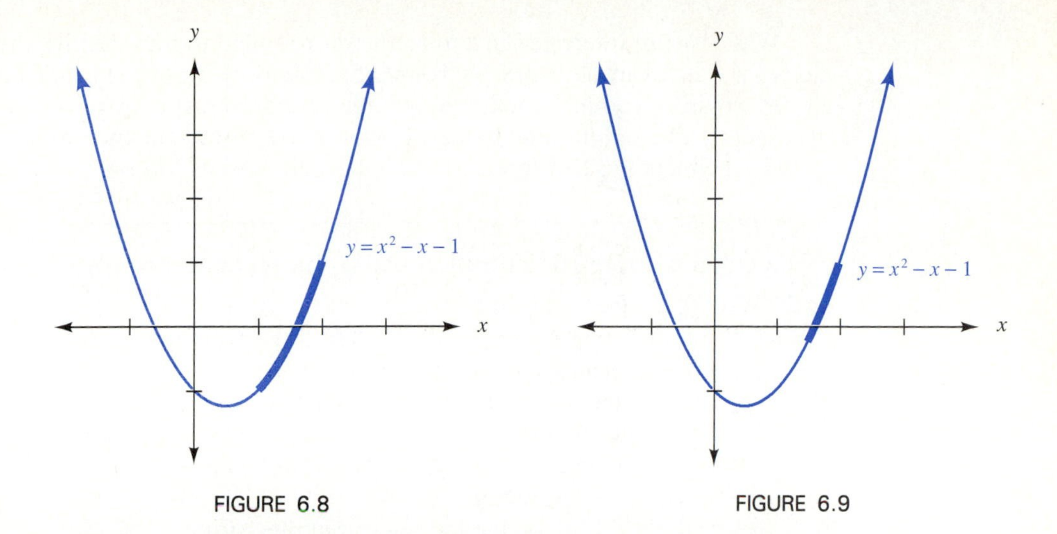

FIGURE 6.8

FIGURE 6.9

We must continue this process until, when x is rounded to the nearest hundredth, we get the same "guess" twice in a row. The repetition is a signal that further computation is unlikely to produce a better guess. (Computer programs may be written to terminate execution when the difference between two x-values is less than some predetermined amount, such as 0.005.) It is easier to see the steps involved if we put the results in table form. The first 3 1/2 lines of Table 6.1 summarize the steps completed up to this point.

TABLE 6.1

Iteration Number	*x*	$x^2 - x - 1$	*Rounded Value of x*
0	1.00000	−1.00	1.00
0	2.00000	1.00	2.00
1	1.50000	−0.25	1.50
2	1.75000	0.31	1.75
3	1.62500	0.02	1.63
4	1.5625	−0.12	1.56
5	1.59375	−0.05	1.59
6	1.609375	−0.02	1.60
7	1.6171875	−0.00	1.62
8	1.6210938	0.01	1.62

Each "guess" comes from realizing that $x^2 - x - 1$ will take on both positive and negative values, but zero must be somewhere between them. Each time we add a new row to the table, we find the most recent entry in the column under $x^2 - x - 1$ that has the *opposite* sign of the new entry. Taking the average of the two corresponding unrounded x-values gives the next x-value. The brackets in the table indicate which entries have been averaged. Notice that as the x-values approach the solution 1.62, the values of $x^2 - x - 1$ approach 0. ■

While we are interested in a final answer rounded to hundredths, the computations for x in Example 1 are not rounded. Rounding all of the calculations to two places would introduce unacceptably large round-off error; therefore the rounding step is last. The polynomial values, however, are rounded, since we are interested only in their signs. The procedure can be generalized as follows.

To solve a quadratic equation using the bisection method:

1. Write the equation in the form $y = ax^2 + bx + c$.
2. Use a sketch of $y = ax^2 + bx + c$ to find the vicinity of the solution.
   (a) Find a value for x that makes y positive.
   (b) find a value for x that makes y negative.
3. For the first iteration
   (a) use the average of the two x-values from step 1.
   (b) calculate the value of y.
   (c) round off the x-value to the desired precision.
4. For the next iteration
   (a) find the most recent iteration for which the y-value is opposite in sign from the y-value calculated in the previous step.
   (b) use the average of the corresponding x-values for the next approximation. Use as many decimal places as your calculator will allow.
   (c) calculate the value of y.
   (d) round off the x-value to the desired precision.
5. Repeat step 4 until the same x-value appears twice consecutively in step 4(d) as the rounded value for x.

EXAMPLE 2 Solve $x^2 - x - 1 = 0$, given that x is positive. Round your answer to thousandths.

Solution **1.** Consider $y = x^2 - x - 1$.

2. Figure 6.10 shows that the positive solution falls between 1 and 2.

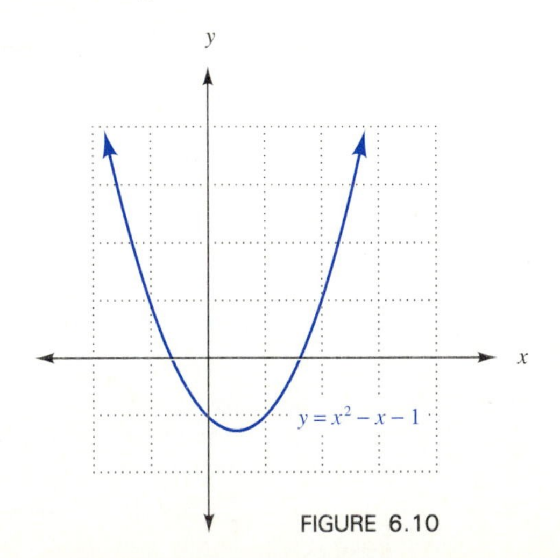

FIGURE 6.10

3–5. Table 6.2 gives a table of values used in the solution. Each x-value is shown rounded to thousandths, although *more digits were used in the calculations.*

TABLE 6.2

Iteration Number	*Rounded Value of x*	*Sign of $x^2 - x - 1$*
0	1.000	−
0	2.000	+
1	1.500	−
2	1.750	+
3	1.625	+
4	1.563	−
5	1.594	−
6	1.609	−
7	1.617	−
8	1.621	+
9	1.619	+
10	1.618	+
11	1.618	−

After the eleventh iteration, the rounded value of x appears as 1.618 twice, so that number is the solution to thousandths. Notice that increasing the desired precision from hundredths to thousandths increased the number of iterations required to find a solution. ■

Example 3 illustrates how a technician might compute a square root using the bisection method with a calculator that does not have a square root key. In fact, computers and calculators do not store tables of trigonometric functions and square roots in memory. Instead, they are programmed to calculate these values when they are needed, and they use iterative methods.

EXAMPLE 3 Compute $\sqrt{3}$ to hundredths.

Solution If $x = \sqrt{3}$, then $x^2 = 3$, and $x^2 - 3 = 0$. The solution falls between 1 and 2.

Iteration	*x-value*	*Keystroke Sequence for y*	*y*	*Keystroke Sequence for Next x*
0	1	1 [×] 1 [−] 3 [=]	−	
0	2	2 [×] 2 [−] 3 [=]	+	2 [+] 1 [=] [÷] 2 [=] [STO]
1	1.5	[×] [RCL] [−] 3 [=]	−	[RCL] [+] 2 [=] [÷] 2 [=] [STO]
2	1.75	[×] [RCL] [−] 3 [=]	+	[RCL] [+] 1.5 [=] [÷] 2 [=] [STO]
3	1.625	[×] [RCL] [−] 3 [=]	−	[RCL] [+] 1.75 [=] [÷] 2 [=] [STO]
4	1.6875	[×] [RCL] [−] 3 [=]	−	[RCL] [+] 1.75 [=] [÷] 2 [=] [STO]
5	1.71875	[×] [RCL] [−] 3 [=]	−	[RCL] [+] 1.75 [=] [÷] 2 [=] [STO]
6	1.734375	[×] [RCL] [−] 3 [=]	+	[RCL] [+] 1.71875 [=] [÷] 2 [=] [STO]
7	1.7265625			

Since $x = 1.73$ for two consecutive iterations, that value is the solution sought. ■

NOTE ⇨⇨ ***Although the bisection method is less efficient than factoring or using the quadratic formula, it has the advantage that it can be used on any equation that can be written in the form $f(x) = 0$, where f represents a function.*** The steps are the same as in the summary box preceding Example 2, if we replace $y = ax^2 + bx + c$ with $y = f(x)$ in steps 1 and 2. In the next chapter, we will use the method to solve higher degree polynomial equations.

EXERCISES 6.5

In 1–5, use the bisection method to solve for x *in the given interval. Round your answer to tenths.*

1. $x^2 + x - 1 = 0$; x is between 0 and 1

2. $2x^2 + x - 2 = 0$; x is between 0 and 1

3. $2x^2 + 3x - 1 = 0$; x is between 0 and 1

4. $2x^2 - 4x - 3 = 0$; x is between 2 and 3

5. $x^2 - x - 5 = 0$; x is between 2 and 3

In 6–10, use the bisection method to solve for x *in the given interval. Round your answer to hundredths.*

6. $3x^2 + 2x - 3 = 0$; x is between 0 and 1

7. $3x^2 - 4x - 6 = 0$; x is between 2 and 3

8. $3x^2 - 4x - 1 = 0$; x is between 1 and 2

9. $3x^2 - 3x - 2 = 0$; x is between 1 and 2

10. $3x^2 - 5x - 1 = 0$; x is between 1 and 2

---

11. A civil engineer on a job needs to know the value of $\sqrt{5}$, but the batteries in her calculator are dead. Show how she can use the bisection method to compute $\sqrt{5}$. Round your answer to hundredths. 1 2 6 7 8 3 5 9

12. A computer engineer needs to program a computer to find square roots. Show how the bisection method can be used to compute $\sqrt{17}$. Round your answer to hundredths. 3 5 9

13. The formula $s = (1/2)at^2 + v_0 t$ gives the distance s that an object travels in time t if its initial velocity is v_0 and it has a constant acceleration a. Use the bisection method to find the time required for a car to travel 300 feet if it is decelerating ($a < 0$ for deceleration) at a rate of 10 ft/s², and its initital velocity was 60 mph (88 ft/s). The answer is between 4 and 5 and should be rounded to tenths. 4 7 8 1 2 3 5 6 9

14. The formula $s = t^2 + 517t$ gives the approximate elevation s (in ft) above sea level if t is the number of Fahrenheit degrees below 212 at which water boils at that elevation. Find the temperature at which water boils on the top floor of a building 1000 ft tall. The value of t is between 1 and 2 and should be rounded to tenths. 6 8 1 2 3 5 7 9

15. When a footing is designed to support a structural column, the area of the soil base must be calculated. Under certain conditions, the area is given by $A = (W + 2h \tan \phi)^2$. In the formula, A is the area (in ft²) of the soil base under the footing, W is the width (in ft) of the soil base, ϕ is the angle (in °) of the sides of the soil base, and h is the depth (in ft) of the soil base beneath the footing. Find h if $A = 32.0$ ft², $\phi = 40.0°$, and $W = 3.00$ ft. Round your answer to hundredths. 1 2 4 6 8 5 7 9

---

1. Architectural Technology
2. Civil Engineering Technology
3. Computer Engineering Technology
4. Mechanical Drafting Technology
5. Electrical/Electronics Engineering Technology
6. Chemical Engineering Technology
7. Industrial Engineering Technology
8. Mechanical Engineering Technology
9. Biomedical Engineering Technology

6.6 CHAPTER REVIEW

KEY TERMS

6.2 discriminant
parabola
vertex
6.3 quadratic function
axis of symmetry
6.4 quadratic form

SYMBOLS

$\pm$ plus or minus

RULES AND FORMULAS

Quadratic formula: For a quadratic equation in standard form $ax^2 + bx + c = 0$, the solutions are given by

$$x = \frac{-b \pm \sqrt{b^2 - 4ac}}{2a}.$$

Vertex of a parabola: If $f(x) = ax^2 + bx + c$, then the x-coordinate of the vertex is given by $x = \frac{-b}{2a}$.

GEOMETRY CONCEPTS

$V = \frac{\pi d^2 h}{12}$ (volume of a cone)

$V = \frac{4}{3}\pi r^3$ (volume of a sphere)

SEQUENTIAL REVIEW

(Section 6.1) In 1–9, solve each equation by completing the square.

1. $x^2 - 8x + 3 = 0$
2. $y^2 + 10y - 1 = 0$
3. $m^2 + 7m - 8 = 0$
4. $p^2 - 5p + 4 = 0$
5. $2x^2 + 14x = 1$
6. $3y^2 - 12y = -2$
7. $2x^2 - 18x = -1$
8. $2x^2 + 3x - 1 = 0$
9. $3x^2 - 5x + 1 = 0$

(Section 6.2) In 10–18, use the quadratic formula to solve each equation.

10. $x^2 - 9x - 6 = 0$
11. $z^2 - 7z = 0$
12. $3m^2 + m - 4 = 0$
13. $2y^2 + 3y - 1 = 0$
14. $x^2 - 6x + 3 = 0$
15. $3y^2 + y - 3 = 0$
16. $x^2 - 5x = 5$
17. $z^2 - 7 = 0$
18. $2r^2 + 2r = 3$

(Section 6.3) In 19–24, find the vertex of each parabola.

19. $y = 2x^2 - 6x$
20. $y = -x^2 - 10x + 3$
21. $y = 3x^2 + 9x - 1$
22. $y = x^2 - 5x + 1$
23. $y = 2x^2 - 3x + 5$
24. $y = -2x^2 - 4x + 1$

\+ *In 25–30, sketch the graph for each parabola.*

25. $y = 3x^2 + 2$

26. $y = 2x^2 - 8x$

27. $y = -2x^2 + 8x + 3$

28. $y = -x^2 - 8x - 3$

29. $y = x^2 + 2x - 3$

30. $y = 2x^2 + 4x - 3$

(Section 6.4) *In 31–40, solve each equation.*

31. $x^4 - 18x^2 + 81 = 0$

32. $y^{-2} - y^{-1} - 6 = 0$

33. $2/z^2 + 3/z - 2 = 0$

34. $(x - 1)^2 + (x - 1) - 12 = 0$

35. $(x + 2)^2 - 2(x + 2) - 8 = 0$

36. $2x^{-2} + x^{-1} - 1 = 0$

37. $4x^4 - 37x^2 + 9 = 0$

38. $1/x^4 - 5/x^2 + 6 = 0$

39. $6x^4 - 7x^3 - 3x^2 = 0$

40. $3x^3 - 14x^2 - 5x = 0$

(Section 6.5) *In 41–44, use the bisection method to solve each equation. Round the answer to tenths.*

41. $3x^2 - 5x - 5 = 0$; x is between 2 and 3

42. $x^2 + x - 3 = 0$; x is between 1 and 2

43. $2x^2 - x - 2 = 0$; x is between 1 and 2

44. $x^2 - 2x - 5 = 0$; x is between 3 and 4

45. Compute $\sqrt{7}$ to hundredths, using the bisection method.

APPLIED PROBLEMS

1. The dosage of medicine for a child is often determined by taking a fractional part of the adult dosage. To obtain the fraction, different rules are used for different age groups. If A is the child's age in years, one rule calls for the fraction $\dfrac{A}{A + 12}$. Another calls for $\dfrac{A + 1}{24}$. For approximately what age (to the nearest year) are the two rules equivalent? 6 9 2 4 5 7 8

2. A window is to be made in the shape of a rectangle topped by a semicircle as shown in the accompanying diagram. Because energy conservation dictates that the window area is to be a certain percentage of the wall area, each window is to be 21.5 ft². The rectangular portion should be 6.00 ft tall. How wide should the window be? 1 3 4 5 6 8 9

6.00 ft

3. A particular electronic circuit produces a voltage v (in volts) that increases as a function of time t (in seconds) according to the equation $v(t) = 2t^2 + 5t + 1$. Find the time for which $v(t)$ is 12 volts. 3 5 9 1 2 4 6 7 8

4. If 0.001 mole of pure acetic acid is mixed with enough water to make 1 liter of solution, then $k = \dfrac{\alpha^2}{10^{-3} - \alpha}$. In the formula, k is the equilibrium constant and has a value of 1.85×10^{-5}, and α is the equilibrium concentration (in moles) of H_3O^+. Solve for α. 6 9 2 4 5 7 8

5. The formula $h^2 + 2rh = d^2$ can be used to calculate the height of a building or tower. In the formula, r is the radius of the earth (6378 km), and d is the distance from the top of the structure to the horizon. The values of r, h, and d must have the same units. Radio signals of a certain type have a maximum range of 71.0 km. Find the height of a structure from which the signal could be sent to the horizon. 1 2 4 7 8 3 5

6. A computer monitor can light up small rectangles called pixels. How many rows and how many columns should there be if the number of columns is to exceed the number of rows by 10 and there are to be a total of 704 pixels? 3 5 9

7. An electric current of I amps flows from a 48-volt battery through a 10-ohm resistor to a load dissipating 8 watts. The current's possible values are given in the equation, $48I = 10I^2 + 8$. Find I to three significant digits. 3 5 9 | 1 2 4 6 7 8

8. In an electronic circuit, the value of an adjustable resistor R (in ohms) needed to assure dissipation of 10 watts of power from a 36-volt source is given in the equation $\left(\frac{36}{10+R}\right)^2 R = 10$. Find R to three significant digits. 3 5 9 | 1 2 4 6 7 8

9. The power output P (in microwatts) of a certain LED (light emitting diode) is given by the equation, $P = -\lambda^2 + 1280\lambda - 409{,}200$ where λ is wavelength (in nanometers). For what values of λ is the power essentially zero? 9 | 1 2 3 5 6 7

10. Shockley's current equation for a certain transistor is given as $I = 25 + 12.49V + 1.56V^2$ when V is measured in volts and I in amperes. If V is allowed to vary, what value would it have (called the cut-off voltage) to reduce this current to zero amps? 3 5 9 | 4 7 8

11. A 36.0-inch wide sheet of aluminum is formed into a cable runway by bending up a certain amount of aluminum from each end. How much should be bent up if the runway must carry 100 round cables each with a cross section of 1.00 square inch? Ignore any space between cables. 1 2 4 7 8 | 3 5

12. A 36.0-inch wide sheet of aluminum is formed into a cable runway by creasing it in the middle and bending the sides up to form a triangular cross section. How high should the ends be raised from the floor to make a runway capable of holding 100 round cables, each with a cross section of 1.00 square inch? Ignore any space between cables. 1 2 4 7 8 | 3 5

13. A circular hole is to be bored into the ground for water storage. The hole is to be 12.0 feet deep and enough waterproofing material to coat 700.0 sq ft of the bottom and side of the hole is available. Find the diameter that should be used for the hole. 1 2 4 7 8 | 3 5

14. The total energy (in ft·lb) of a rocket traveling upward from the earth's surface is given by the equation $E = \frac{wv^2}{64.4} + wvt$, where w is the rocket's weight (in lbs), v is its velocity (in ft/s), and t is the time of flight (in seconds). What velocity would a 5000 lb rocket have to maintain so that after 10 seconds of flight it would have an energy of 44 million ft·lbs? Give your answer to three significant digits. 4 7 8 | 1 2 3 5 6 9

15. A styrofoam model for a propane tank is made by attaching a hemisphere to each end of a 3.0-ft long cylinder. Find the diameter of the cylinder needed if the model must be covered with 6.0 oz of paint advertised to cover 5.0 sq ft/oz. 1 2 4 7 8 | 3 5

16. A ballistics projectile enters a 3-ft long block of gelatin at a speed of 1200 ft/s, where it meets a retarding force that slows it down at a constant rate of $-200{,}000$ ft/s². How long will it be in the gelatin? Distance traveled s is given by $s = v_0t + \frac{1}{2}at^2$ where v_0 is initial velocity, t is time traveled, and a is the acceleration. 4 7 8 | 1 2 3 5 6 9

17. From 250.0 sq ft of canvas, seven perfectly square pieces can be cut and enough is left over for a piece 10.0 ft long and as wide as one of the squares. How big is one of the squares? 1 2 4 7 8 | 3 5

18. When a lens is used to focus a transparency onto a screen, a relationship exists between the focal length of the lens f and the distance from the lens to the object o, and the lens to the screen i.

$$\frac{1}{o} + \frac{1}{i} = \frac{1}{f}$$

To determine how far from a transparency a lens with focal length 26.0 cm should be placed to focus it on a wall 213 cm away, the following equation arises.

$$o^2 - 213o + 5538 = 0$$

Solve for o. 3 5 6 9

1. Architectural Technology
2. Civil Engineering Technology
3. Computer Engineering Technology
4. Mechanical Drafting Technology
5. Electrical/Electronics Engineering Technology
6. Chemical Engineering Technology
7. Industrial Engineering Technology
8. Mechanical Engineering Technology
9. Biomedical Engineering Technology

19. A technician measures the total resistance R_T of two resistors in series and finds 100 ohms. Putting the resistors in parallel she measures 16.5 ohms. Using the fact that in series $R_T = R_1 + R_2$ and in parallel $R_T = \dfrac{R_1 R_2}{R_1 + R_2}$, it can be shown that $R_1^2 - 100R_1 + 1650 = 0$. Solve this equation for R_1. Give the answer to three significant digits.
3 5 9 1 2 4 6 7 8

1. Architectural Technology
2. Civil Engineering Technology
3. Computer Engineering Technology
4. Mechanical Drafting Technology
5. Electrical/Electronics Engineering Technology
6. Chemical Engineering Technology
7. Industrial Engineering Technology
8. Mechanical Engineering Technology
9. Biomedical Engineering Technology

RANDOM REVIEW

1. Sketch the graph of $y = x^2 + 6x + 13$.
2. Use the bisection method to compute $\sqrt{15}$ to hundredths.
3. Solve $0 = x^2 + 7x - 4$ by completing the square.
4. Identify the vertex of the graph given by $f(x) = 2x^2 - 3x$.
5. Solve $(x - 3)^2 + 4(x - 3) + 4 = 0$.
6. Solve $3y^2 - 6y + 1 = 0$ by the quadratic formula.
7. Sketch the graph of $y = 3x^2 - 2$.
8. Solve $6x^3 + 2x^2 = 6x$.
9. Use the bisection method to solve $x^2 + 5x - 3 = 0$, given that x is between 0 and 1. Round the answer to tenths.
10. Solve $2x^2 - 5x + 2 = 0$ by completing the square.

CHAPTER TEST

1. State the quadratic formula.
2. Solve $x^2 - 7x + 4 = 0$ by completing the square.
3. Solve $2y^2 + 1 = 5y$ by the quadratic formula.
4. Identify the vertex of the graph given by $f(x) = 3x^2 + 12x + 12$.
5. Sketch the graph of $y = 2x^2 - 4x$.
6. Solve $1/x^6 - 9/x^3 + 8 = 0$
7. Solve $x^2 - 5x + 3 = 0$, given that x is between 4 and 5. Round the answer to tenths.
8. Use the bisection method to compute $\sqrt{13}$. Round your answer to hundredths.
9. When an object is released from a building with an initial downward velocity of 3.0 ft/s and an acceleration of 32 ft/s^2, the distance s (in ft) fallen after time t (in seconds) is given by $s = 16t^2 + 3t$. If the time is increased by an amount Δt, we have

$$s = 16(t + \Delta t)^2 + 3(t + \Delta t).$$

   Solve the equation for $t + \Delta t$ if $s = 21$ ft.
10. When dc currents flow in the same direction through a resistance R (in ohms), the power P (in watts) dissipated is given as $P = R(I_1 + I_2)^2$, where I_1 and I_2 are the currents (in amperes). Solve for I_2 if $P = 100$ W, $R = 1\ \Omega$, and $I_1 = 6$ A.

EXTENDED APPLICATION: MECHANICAL ENGINEERING TECHNOLOGY

TESTING THE HARDNESS OF MATERIAL

The hardness of material is sometimes measured by the depth of the depression left when a small steel ball resting on the material is subjected to a known pressure. It is, however, easier to measure the diameter of the circular depression than to measure its depth. The diameter may then be used to calculate the depth in the following manner.

Consider the quantities shown in Figure 6.11.

$$r_1 = \text{radius of the ball}$$
$$r_2 = \text{radius of the depression}$$
$$h = \text{depth of the depression}$$

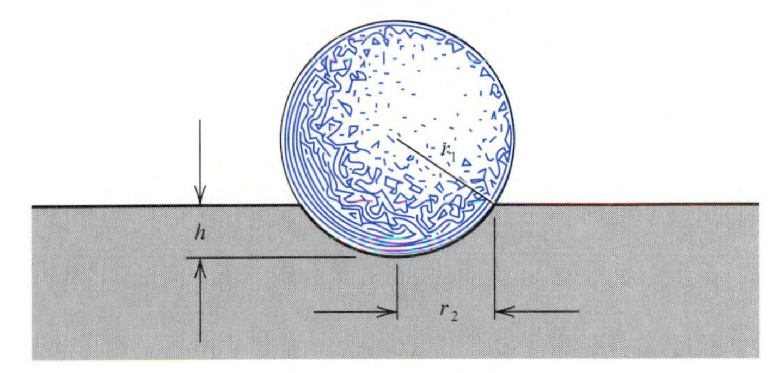

FIGURE 6.11

The triangle shown in Figure 6.12 is a right triangle. From the Pythagorean theorem, it follows that

$$r_1^2 = (r_1 - h)^2 + r_2^2.$$
$$r_1^2 = r_1^2 - 2r_1 h + h^2 + r_2^2$$

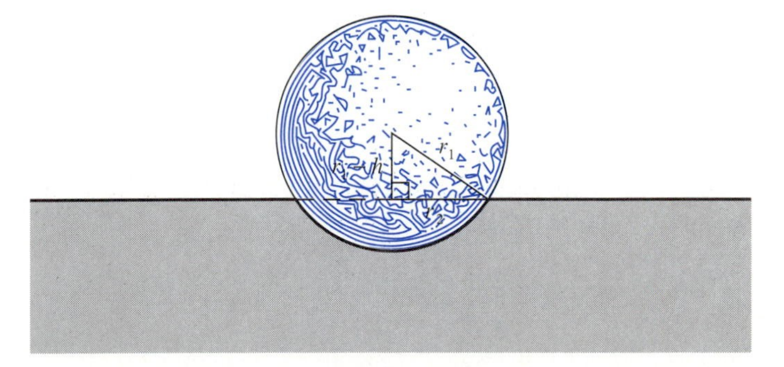

FIGURE 6.12

EXTENDED APPLICATION, continued

Subtracting r_1^2 from both sides, we have

$$0 = h^2 - 2r_1 h + r_2^2.$$

If h is the variable, the equation is a quadratic equation in standard form with $a = 1, b = -2r_1$, and $c = r_2^2$. Substituting these values into the quadratic formula, we obtain the following equation.

$$\begin{aligned} h &= \frac{2r_1 \pm \sqrt{4r_1^2 - 4r_2^2}}{2} \\ &= \frac{2r_1 \pm \sqrt{4(r_1^2 - r_2^2)}}{2} \\ &= \frac{2r_1 \pm 2\sqrt{r_1^2 - r_2^2}}{2} \\ &= r_1 \pm \sqrt{r_1^2 - r_2^2} \end{aligned}$$

The method only works if $h < r_1$, because if the ball is impressed to a depth greater than its radius, as in Figure 6.13, the radius of the impression is r_1, regardless of the depth.

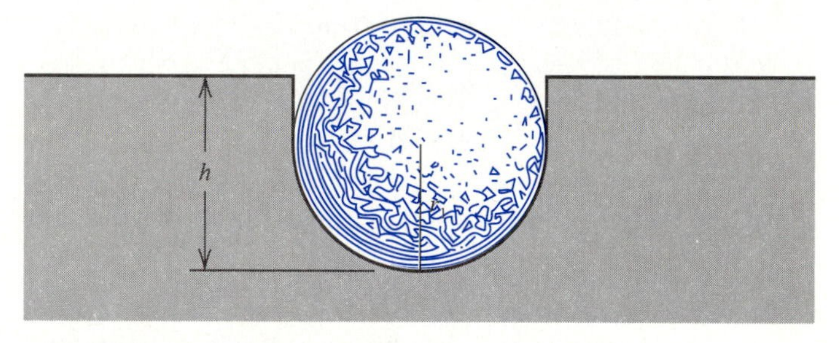

FIGURE 6.13

Assuming, then, that $h < r_1$, we discard the solution $h = r_1 + \sqrt{r_1^2 - r_2^2}$. The relationship among r_1, r_2, and h is given by

$$h = r_1 - \sqrt{r_1^2 - r_2^2}.$$

For example, a steel ball with a radius of 5.0 mm leaves a depression with a radius of 2.0 mm in a piece of white pine. Since r_1 is 5.0, and r_2 is 2.0, we have

$$\begin{aligned} h &= 5.0 - \sqrt{5.0^2 - 2.0^2} \\ &= 5.0 - \sqrt{21} \\ &= 0.4 \text{ mm}. \end{aligned}$$

HIGHER-DEGREE EQUATIONS CHAPTER 7

Some of the equations that arise in technical applications have the form $f(x) = 0$ where $f(x)$ is a polynomial. In previous chapters, we have covered general techniques for solving such equations if $f(x)$ has degree 1 or 2. In Section 6.4 we considered some special cases of higher-degree equations. Our goal in this chapter is to examine more generally the case in which $f(x)$ has degree 3 or more.

7.1 SYNTHETIC DIVISION

In Section 2.3, we considered division problems in which dividend and divisor were polynomials. When the divisor has the form $x - b$, it is easy to perform the division using only the coefficients of the polynomials. Not having to write down all of the powers of the variables saves both time and space. Consider the following division with and without the variables included.

$$
\begin{array}{rl}
 & x^2 + 2x - 8 \\
x - 3 \,\big)\!\! & \overline{x^3 - x^2 - 14x + 20} \\
 & \underline{-x^3 + 3x^2} \\
 & \quad 2x^2 - 14x \\
 & \quad \underline{-2x^2 + 6x} \\
 & \qquad - 8x + 20 \\
 & \qquad \underline{8x - 24} \\
 & \qquad\quad - 4
\end{array}
\qquad\qquad
\begin{array}{rl}
 & 1 + 2 - 8 \\
1 - 3 \,\big)\!\! & \overline{1 - 1 - 14 + 20} \\
 & \underline{-1 + 3} \\
 & \quad 2 - 14 \\
 & \quad \underline{-2 + 6} \\
 & \qquad - 8 + 20 \\
 & \qquad \underline{8 - 24} \\
 & \qquad\quad - 4
\end{array}
$$

Notice that when the variables are omitted, the numbers 1, 2, and 8 each occur three times. Likewise, the numbers 14 and 20 each occur twice. By making three changes in the format, we can delete the repetitions and develop a pattern among the numbers. The first change is to write the divisor as 3, rather than as $1 - 3$. All divisors will be of the form $1x - b$, so it is only necessary to know the value of b. Using b instead of $-b$ also allows us to replace subtraction with addition. The second change is to use less space by omitting repetitions and plus signs while moving the numbers up vertically to fill in the gaps. That is, we write

$$
\begin{array}{r|rrrr}
3 & 1 & -1 & -14 & 20 \\
 & & 3 & 6 & -24 \\
\hline
 & & 2 & -8 & -4.
\end{array}
$$

The third change is to copy the first 1 in the dividend on the bottom line.

The problem appears as

$$
\begin{array}{r|rrrr}
3 & 1 & -1 & -14 & 20 \\
 & & 3 & 6 & -24 \\
\hline
 & 1 & 2 & -8 & -4.
\end{array}
$$

The pattern now emerges.

$$
\begin{array}{r|rrrr}
3 & 1 & -1 & -14 & 20 \\
\hline
\end{array}
$$

Begin with the problem in this form.

$$
\begin{array}{r|rrrr}
3 & 1 & -1 & -14 & 20 \\
 & & 3 & & \\
\hline
 & 1 & & &
\end{array}
$$

Copy the 1 on the bottom line. $3 \times 1 = 3$

$$
\begin{array}{r|rrrr}
3 & 1 & -1 & -14 & 20 \\
 & & 3 & 6 & \\
\hline
 & 1 & 2 & &
\end{array}
$$

$-1 + 3 = 2$ and $3 \times 2 = 6$

$$
\begin{array}{r|rrrr}
3 & 1 & -1 & -14 & 20 \\
 & & 3 & 6 & -24 \\
\hline
 & 1 & 2 & -8 &
\end{array}
$$

$-14 + 6 = -8$ and $3 \times (-8) = -24$

$$
\begin{array}{r|rrrr}
3 & 1 & -1 & -14 & 20 \\
 & & 3 & 6 & -24 \\
\hline
 & 1 & 2 & -8 & -4
\end{array}
$$

$20 + (-24) = -4$

The right-most number on the bottom line is interpreted as the remainder in the original division problem. The remaining numbers on the bottom line are interpreted as the coefficients of the quotient. From *right to left* the exponents of x are 0, 1, and 2, respectively. That is $(x^3 - x^2 - 14x + 20) \div (x - 3) = x^2 + 2x - 8$, remainder -4. The process of doing a division problem in this manner is called **synthetic division** and is generalized as follows.

To divide a polynomial by $x - b$ using synthetic division:

1. Write the coefficients of the dividend when powers of the variable are arranged in descending order. This line will be the top line. Be sure that any missing term is represented by 0.
2. Write b to the left of the polynomial.
3. Skip a line (the middle line) and copy the leading coefficient of the dividend below its original position. This line will be the bottom line.
4. Multiply the number written in the previous step by b. Write the product on the middle line below the next coefficient.
5. Add these two numbers and write the sum on the bottom line.
6. Repeat steps 4 and 5 until each number in the dividend has a number written below it on the bottom line.
7. Interpret the right-most number on the bottom line as the remainder. Interpret the other numbers as the coefficients of the quotient, where the degree is one less than the degree of the dividend.

EXAMPLE 1 If possible, use synthetic division to find the quotient and remainder when $x^3 + x^2 + 4x + 3$ is divided by $x + 1$.

Solution Since the divisor must be in the form $x - b$, think of $x + 1$ as $x - (-1)$. Then $b = -1$.

1–3.
$$\begin{array}{r|rrrr} -1 & 1 & 1 & 4 & 3 \\ \hline & 1 & & & \end{array}$$

4–5.
$$\begin{array}{r|rrrr} -1 & 1 & 1 & 4 & 3 \\ & & -1 & & \\ \hline & 1 & 0 & & \end{array}$$

6.
$$\begin{array}{r|rrrr} -1 & 1 & 1 & 4 & 3 \\ & & -1 & 0 & -4 \\ \hline & 1 & 0 & 4 & -1 \end{array}$$

7. The remainder is -1 and the quotient is $x^2 + 0x + 4$ or $x^2 + 4$. ■

EXAMPLE 2 If possible, use synthetic division to find the quotient and remainder when $x^4 + 3x^2 - x + 2$ is divided by $x^3 + 1$.

Solution The synthetic division process, as described, does not apply unless the divisor is linear (that is, of the form $x - b$). ■

EXAMPLE 3 If possible, use synthetic division to divide $x^3 - 8$ by $x - 2$.

Solution Since there is no x^2 term and no x term, we write 1 0 0 −8 to represent $x^3 + 0x^2 + 0x - 8$.

$$\begin{array}{r|rrrr} 2 & 1 & 0 & 0 & -8 \\ & & 2 & 4 & 8 \\ \hline & 1 & 2 & 4 & 0 \end{array}$$

Hence, $(x^3 - 8) \div (x - 2) = x^2 + 2x + 4$. ■

EXAMPLE 4 Determine whether $x + 3$ is a factor of $x^5 + 2x^4 - 3x^2 + 4x - 5$.

Solution If $x + 3$ is a factor, it will divide $x^5 + 2x^4 - 3x^2 + 4x - 5$ evenly. That is, the remainder will be zero. By synthetic division, we have

$$\begin{array}{r|rrrrrr} -3 & 1 & 2 & 0 & -3 & 4 & -5 \\ & & -3 & 3 & -9 & 36 & -120 \\ \hline & 1 & -1 & 3 & -12 & 40 & -125 \end{array}$$

Since the remainder is -125, $x + 3$ is not a factor of $x^5 + 2x^4 - 3x^2 + 4x - 5$. ■

Example 4 shows that one use of synthetic division is to test whether a factor of the form $x - b$ will divide a polynomial evenly. Even though the synthetic division process cannot be used (without modification) when the divisor is of the form $ax - b$, it is possible to test whether $ax - b$ is a factor of another polynomial. Since $ax - b = a(x - b/a)$ if a is not 0, $ax - b$ will divide a polynomial if both a and $x - b/a$ divide it evenly. In Chapter 5, we considered only polynomials and factors with integral coefficients. The synthetic division process, however, can be used when the coefficients are real (or even complex) numbers, so we can use it to divide by $x - b/a$.

To determine whether a polynomial of the form $ax - b$, where a and b are integers, is a factor of a given polynomial with integral coefficients:

1. Factor $ax - b$ as $a(x - b/a)$.
2. Divide the polynomial by $x - b/a$ using synthetic division.
3. (a) If the remainder is not zero, then $ax - b$ is not a factor.
   (b) If the remainder is zero and a is a factor of the quotient, then $ax - b$ is a factor.

EXAMPLE 5 Determine whether $3x + 2$ is a factor of $3x^3 + 2x^2 - 3x - 2$.

Solution 1. $3x + 2 = 3(x + 2/3)$.

2.
$$\begin{array}{r|rrrr} -2/3 & 3 & 2 & -3 & -2 \\ & & -2 & 0 & 2 \\ \hline & 3 & 0 & -3 & 0 \end{array}$$

3. Since the quotient is $3x^2 - 3 = 3(x^2 - 1)$, it has 3 as a factor. Thus, $3x + 2$ is a factor of $3x^3 + 2x^2 - 3x - 2$. ■

EXAMPLE 6 Determine whether $2x - 1$ is a factor of $6x^2 + x - 3$.

Solution 1. $2x - 1 = 2(x - 1/2)$

2. $$\begin{array}{r|rrr} 1/2 & 6 & 1 & -3 \\ & & 3 & 2 \\ \hline & 6 & 4 & -1 \end{array}$$

3. Since the remainder is -1, $x - 1/2$ does not divide $6x^2 + x - 3$ evenly, so $2x - 1$ does not divide it evenly, either. ■

EXERCISES 7.1

In 1–21, use synthetic division, if possible, to find the quotient and remainder for each division problem.

1. $(x^4 + 3x^3 + 2x^2 + x - 1) \div (x^2 + 3)$
2. $(x^4 - 4x^3 + 3x^2 - 2x + 1) \div (x - 1)$
3. $(x^4 - 4x^3 + 3x^2 - 2x + 1) \div (x + 1)$
4. $(x^3 + 2x^2 - 5x - 6) \div (x + 3)$
5. $(x^3 + 2x^2 - 5x - 6) \div (x - 2)$
6. $(x^3 - 2x^2 - 5x + 6) \div (x - 2)$
7. $(6x^3 + 7x^2 - x - 2) \div (x + 1)$
8. $(6x^3 + 7x^2 - x - 2) \div (2x - 1)$
9. $(6x^3 + 7x^2 - x - 2) \div (3x + 2)$
10. $(2x^5 + 6x^3 - x^2 - 3) \div (x - 3)$
11. $(2x^5 - 6x^3 + x^2 - 3) \div (x - 3)$
12. $(2x^5 - 6x^3 + x^2 - 3) \div (x + 3)$
13. $(x^3 + 8) \div (x + 2)$
14. $(x^3 + 1) \div (x + 1)$
15. $(x^3 - 1) \div (x - 1)$
16. $(x^2 + 1) \div (x + 1)$
17. $(x^2 - 1) \div (x + 1)$
18. $(2x^5 + 3x^3 - 2x^2 - 3) \div (x - 1)$
19. $(3x^5 - 12x^3 + x^2 - 4) \div (x - 2)$
20. $(3x^5 - 12x^3 + x^2 - 4) \div (x + 2)$
21. $(4x^4 - 35x^2 - 9) \div (x^2 + 2)$

In 22–42, determine whether the first polynomial is a factor of the second.

22. $x + 1$; $x^4 + 3x^2 - 4$
23. $x - 1$; $x^4 + 3x^2 - 4$
24. $x - 2$; $x^4 + 3x^2 - 4$
25. $x + 2$; $x^4 + 3x^2 - 4$
26. $x + 3$; $4x^4 + 37x^2 + 9$
27. $x - 3$; $4x^4 + 37x^2 + 9$
28. $2x - 1$; $4x^4 + 37x^2 + 9$
29. $2x + 1$; $4x^4 + 37x^2 + 9$
30. $3x + 2$; $9x^3 + 21x^2 + 16x + 4$
31. $3x - 2$; $9x^3 + 21x^2 + 16x + 4$
32. $2x - 3$; $4x^3 + 8x^2 - 3x - 9$
33. $2x + 3$; $4x^3 + 8x^2 - 3x - 9$
34. $3x + 1$; $9x^5 - x^3 + 9x^2 - 1$
35. $3x - 1$; $9x^5 - x^3 + 9x^2 - 1$
36. $3x - 4$; $3x^3 + 4x^2 - 3x - 4$
37. $3x + 4$; $3x^3 - 4x^2 - 3x + 4$
38. $4x + 3$; $4x^2 - x - 3$
39. $4x - 3$; $4x^2 - x - 3$
40. $4x - 1$; $4x^2 - 7x - 2$
41. $4x + 1$; $4x^2 - 7x - 2$
42. $2x + 5$; $2x^4 + 5x^3 - 2x - 5$
43. If a and b are integers that have no common factor (other than 1) and $x - b/a$ divides a polynomial (with a remainder of 0), do you think that a will always be a factor of the quotient?

    Hint: Answer the following two questions.
    (a) Is $4x - 6$ a factor of $2x^2 + 5x - 12$?
    (b) Is $2x - 3$ a factor of $2x^2 + 5x - 12$?

7.2 THE REMAINDER THEOREM AND THE FACTOR THEOREM

Division problems involving polynomials may be checked in the same manner as long division problems in arithmetic. That is, multiply quotient and divisor and add the remainder to the product. If the result is equal to the dividend, the division is correct. Function notation provides a convenient way to express this statement symbolically. If $f(x)$ is a polynomial, then

$$f(x) = (x - b)[q(x)] + R \tag{7-1}$$

where $q(x)$ is the quotient when $f(x)$ is divided by $(x - b)$, and R is the remainder.

If, in Equation 7-1, we let $x = b$, then

$$\begin{aligned} f(b) &= (b - b)[q(b)] + R \\ &= 0[q(b)] + R \\ &= 0 + R \\ &= R. \end{aligned}$$

This observation leads to a statement known as the **remainder theorem.**

THE REMAINDER THEOREM

If $f(x)$ and $q(x)$ are polynomials and R is a real number such that

$$f(x) = (x - b)[q(x)] + R,$$

then $f(b) = R$.

NOTE The remainder theorem may be stated informally: ***When a polynomial $f(x)$ is divided by $x - b$, the remainder term R, which is a constant, has the same value as $f(b)$.*** It is sometimes easier to find the remainder using synthetic division than to evaluate the function directly.

EXAMPLE 1 $f(x) = x^3 + 4x^2 + x - 6$

(a) Find $f(3)$ by the remainder theorem.

(b) Find $f(3)$ by direct evaluation of the polynomial.

Solution **(a)** Divide $f(x)$ by $x - 3$.

$$\begin{array}{r|rrrr} 3 & 1 & 4 & 1 & -6 \\ & & 3 & 21 & 66 \\ \hline & 1 & 7 & 22 & 60 \end{array}$$

Since $R = 60$, $f(3) = 60$.

(b) $f(3) = 3^3 + 4(3^2) + 3 - 6$
$= 27 + 36 + 3 - 6$
$= 60$ ■

EXAMPLE 2 $g(x) = x^5 + 4x^3 - 3x + 1$

(a) Find $g(-2)$ using the remainder theorem.

(b) Find $g(-2)$ by direct evaluation of the polynomial.

Solution **(a)** Divide $g(x)$ by $x + 2$.

-2	1	0	4	0	-3	1
		-2	4	-16	32	-58
	1	-2	8	-16	29	-57

Since $R = -57$, $g(-2) = -57$.

(b) $g(-2) = (-2)^5 + 4(-2)^3 - 3(-2) + 1$
$= -32 + 4(-8) + 6 + 1$
$= -32 - 32 + 6 + 1$
$= -57$ ■

The remainder theorem has been illustrated using small integers so that you could concentrate on the principles involved rather than on the arithmetic. In practice, however, real world problems usually involve approximate numbers with several significant digits. The remainder theorem is well suited for these problems when a calculator is used. Each time that an operation is performed, the result is used in the next calculation.

EXAMPLE 3 If $f(x) = -x^4 + 160x^3 - 9600x^2 + 48{,}000$, evaluate $f(2.27)$.

Solution

2.27	-1	160	-9600	0	48,000

The following calculator keystroke sequence is used.

1 [+/-] [×] 2.27 [+] 160 [=] [×] 2.27 [−] 9600 [=] [×] 2.27 [×] 2.27 [+] 48000 [=] **377.1411**

The intermediate results will be shown here, although the beauty of the method is that the calculations may be done in one continuous keystroke sequence, without recording the results of each operation.

2.27	-1	160	-9600	0	48,000
		-2.27	358.0471	$-20{,}979.233$	$-47{,}622.859$
	-1	157.73	-9241.9529	$-20{,}979.233$	377.1411 ■

EXAMPLE 4 The volume V of a portion of a sphere between two parallel planes is given by

$$V = (1/6)\pi h^3 + (1/2)h\pi(r_1^2 + r_2^2).$$

In the formula, h is the distance between planes, and r_1 and r_2 are the radii of the circular surfaces. A fountain in a city park is designed to have a basin that is hemispherical in shape with a radius at 10.0 ft as shown in Figure 7.1. There is enough water in the pool so that the radius of the water surface is 8.0 ft. Water is added to increase the radius of the water surface to 10.0 ft. Find the volume of water added.

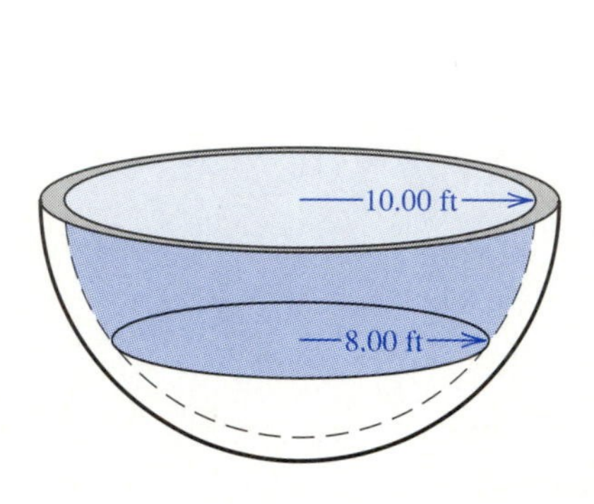

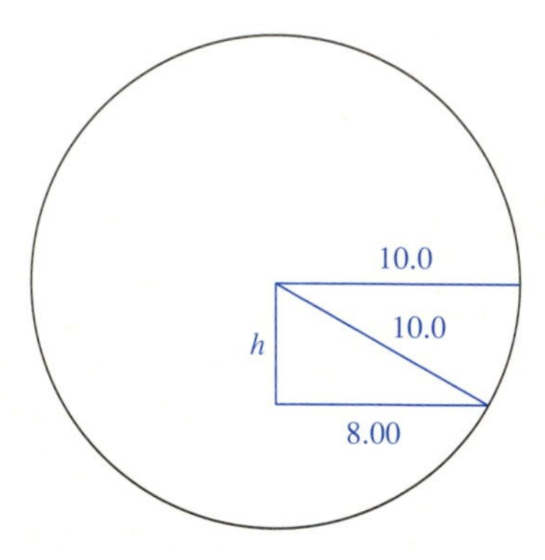

FIGURE 7.1

Solution With the given data, the equation is

$$\begin{aligned} V &= (1/6)(\pi)h^3 + (1/2)(h)(\pi)(8.0^2 + 10.0^2). \\ &= (1/6)(\pi)h^3 + (1/2)(h)(\pi)(64 + 100) \\ &= 0.5236\ h^3 + 257.6\ h \end{aligned}$$

We can find the value of h needed from the Pythagorean theorem.

$$h^2 = 10.0^2 - 8.0^2 = 100 - 64 = 36, \text{ so } h = 6.0 \text{ ft}$$

$$\begin{array}{r|rrrr} 6.0 & 0.5236 & 0 & 257.6 & 0 \\ & & 3.1416 & 18.8496 & 1658.6976 \\ \hline & 0.5236 & 3.1416 & 276.4496 & 1658.6976 \end{array}$$

The volume added, then, is approximately 1700 ft^3. ■

In Example 4 of the previous section, we made the observation that if R is 0 when $f(x)$ is divided by $x - b$, then $x - b$ is a factor of $f(x)$. The remainder theorem says $R = 0$ is equivalent to the condition $f(b) = 0$. These two facts lead to a principle called the *factor theorem.*

THE FACTOR THEOREM

Let $f(x)$ be a polynomial.
If $f(b) = 0$, then $x - b$ is a factor of $f(x)$.
If $f(b) \neq 0$, then $x - b$ is not a factor of $f(x)$.

EXAMPLE 5 Is $x - 1$ a factor of $f(x) = x^3 + 4x^2 + x - 6$?

Solution If $f(1) = 0$, then $x - 1$ is a factor.

$$\begin{aligned} f(1) &= 1^3 + 4(1^2) + 1 - 6 \\ &= 1 + 4 + 1 - 6 \\ &= 0 \end{aligned}$$

Therefore, $x - 1$ is a factor of $x^3 + 4x^2 + x - 6$. ■

EXAMPLE 6 Is $x + 2$ a factor of $g(x) = x^3 - 7x + 6$?

Solution If $g(-2) = 0$, then $x + 2$ is a factor.

$$\begin{aligned} g(-2) &= (-2)^3 - 7(-2) + 6 \\ &= -8 + 14 + 6 \\ &= 12 \end{aligned}$$

Therefore, $x + 2$ is not a factor of $x^3 - 7x + 6$. ■

We now turn our attention to polynomial equations. A **polynomial equation** is an equation of the form $f(x) = 0$, where $f(x)$ is a polynomial. Recall that when the product of two or more factors is zero, one of the factors must equal zero. This concept allows us to relate finding the factors of a polynomial to solving a polynomial equation. In 1799, Carl Friedrich Gauss proved that every polynomial equation has at least one solution or *root*. This statement is known as the **fundamental theorem of algebra.** Once the fundamental theorem was established, it was possible to prove that any nth degree polynomial has n linear factors, and thus, any nth degree polynomial equation has n solutions. Not all of the solutions, however, are required to be real numbers.

root

The basic idea behind the proof is that if $f(x) = 0$, where $f(x)$ is a polynomial of degree n and there is at least one root (r_1), then there is a polynomial $f_1(x)$ such that

$$f(x) = (x - r_1)f_1(x) = 0.$$

Since $f_1(x)$ is found by dividing $f(x)$ by $x - r_1$, we have assumed $x \neq r_1$. If $x - r_1 \neq 0$, then $f_1(x) = 0$. Because $f_1(x)$ must have at least one root (r_2), there is a polynomial $f_2(x)$ such that

$$f_1(x) = (x - r_2)f_2(x) = 0 \text{ and therefore } f(x) = (x - r_1)(x - r_2)f_2(x).$$

The process is repeated until there are n linear factors and

$$f(x) = (x - r_1)(x - r_2)(x - r_3) \ . \ . \ . \ (x - r_n) = 0. \qquad \textbf{(7-2)}$$

To solve such an equation, notice that if one or more of the solutions r_1, r_2, r_3, . . . r_n of Equation 7-2 are known, the corresponding factors can easily be divided out of Equation 7-2 using synthetic division. ***If enough factors are divided out so that the remaining polynomial is quadratic, then the last two solutions may be found by the quadratic formula (or possibly by factoring).*** For the present, you will be given one or more solutions, but the subject of the next section is how to find one or more solutions to start the division process.

NOTE ⇨⇨

To solve a polynomial equation $f(x) = 0$:

1. Assume that k roots r_1, r_2, r_3, . . . r_k are known.
2. Divide $f(x)$ by $x - r_1$, using synthetic division.
3. Let r_i represent the next root. Divide the quotient from the previous step by $(x - r_i)$.
4. Repeat step 3 until the quotient $q(x)$ is a quadratic polynomial.
5. Solve the quadratic equation $q(x) = 0$.

EXAMPLE 7 Given that -2 is one solution, solve the equation

$$x^3 - x^2 - 5x + 2 = 0.$$

Solution **1.** We know that -2 is a solution of the equation $x^3 - x^2 - 5x + 2 = 0$, and therefore, $x + 2$ is a factor of the polynomial $x^3 - x^2 - 5x + 2$.

2.
$$\begin{array}{r|rrrr} -2 & 1 & -1 & -5 & 2 \\ & & -2 & 6 & -2 \\ \hline & 1 & -3 & 1 & \end{array}$$

3–4. Since $x^3 - x^2 - 5x + 2 = (x + 2)(x^2 - 3x + 1)$, further division is not necessary.

5. We must solve $x^2 - 3x + 1 = 0$.

By the quadratic formula, $x = \dfrac{3 \pm \sqrt{9 - 4}}{2} = \dfrac{3 \pm \sqrt{5}}{2}$. ■

When a polynomial contains a factor $(x - b)^n$, the number b is said to be a *multiple root*. The equation in Example 8 has a multiple root.

multiple root

EXAMPLE 8 Given that 2 is a double root, and -2 is also a root, solve $x^5 + x^4 - 13x^3 + 2x^2 + 36x - 24 = 0$.

Solution If 2 is a double root, and -2 is also a root, then $(x-2)^2$ and $(x+2)$ are factors of $x^5+x^4-13x^3+2x^2+36x-24$.

$$\begin{array}{r|rrrrrr} 2 & 1 & 1 & -13 & 2 & 36 & -24 \\ & & 2 & 6 & -14 & -24 & 24 \\ \hline \end{array}$$

Divide $x^5+x^4-13x^3+2x^2+36x-24$ by $x-2$.

$$\begin{array}{r|rrrrr} 2 & 1 & 3 & -7 & -12 & 12 \\ & & 2 & 10 & 6 & -12 \\ \hline \end{array}$$

Divide $x^4+3x^3-7x^2-12x+12$ by $x-2$.

$$\begin{array}{r|rrrr} -2 & 1 & 5 & 3 & -6 \\ & & -2 & -6 & +6 \\ \hline & 1 & 3 & -3 & \end{array}$$

Divide x^3+5x^2+3x-6 by $x+2$.

Notice that x^2+3x-3 is quadratic.

Thus, $x^2+3x-3=0$, and

$$x=\frac{-3\pm\sqrt{9+12}}{2}=\frac{-3\pm\sqrt{21}}{2}. \quad ■$$

EXERCISES 7.2

In 1–14, find the indicated value for each polynomial using (a) the remainder theorem and (b) direct evaluation.

1. $f(x)=x^4+3x^2-2x+1$; find $f(3)$

2. $f(x)=x^4+3x^2-2x+1$; find $f(-3)$

3. $g(x)=2x^3-x^2+3x-4$; find $g(-3)$

4. $g(x)=2x^3-x^2+3x-4$; find $g(3)$

5. $g(x)=3x^3+2x^2-x+5$; find $g(-3)$

6. $g(x)=3x^3+2x^2-x+5$; find $g(3)$

7. $f(x)=x^5-2x^3+3x-1$; find $f(2)$

8. $f(x)=x^5-2x^3+3x-1$; find $f(-2)$

9. $f(x)=2x^5+3x^4-4x^3+5x-6$; find $f(-2)$

10. $f(x)=2x^5+3x^4-4x^3+5x-6$; find $f(2)$

11. $g(x)=3x^5-2x^2+4x-5$; find $g(-2)$

12. $g(x)=3x^5-2x^2+4x-5$; find $g(2)$

13. $g(x)=7x^4-3x^3+2x-6$; find $g(-1)$

14. $g(x)=7x^4-3x^3+2x-6$; find $g(1)$

In 15–28, determine whether the first polynomial is a factor of the second.

15. $x+1$; $x^5-x^3-2x^2+2$

16. $x-1$; $x^5-x^3-2x^2+2$

17. $x-1$; $x^5+2x^3-3x^2-6$

18. $x+1$; $x^5+2x^3-3x^2-6$

19. $x+2$; $x^4-x^3+5x^2-4x-4$

20. $x-2$; $x^4-x^3+5x^2-4x-4$

21. $x-2$; $x^4+2x^3-5x^2-8x+4$

22. $x+2$; $x^4+2x^3-5x^2-8x+4$

23. $x+3$; $2x^3+3x^2-18x-27$

24. $x-3$; $2x^3+3x^2-18x-27$

25. $x-3$; $3x^4-25x^2-18$

26. $x+3$; $3x^4-25x^2-18$

27. $x+1$; $2x^4+2x^3-x^2+x-1$

28. $x-1$; $2x^4+2x^3-x^2+x-1$

In 29–42, find the remaining solutions for each equation, given the solution or solutions indicated.

29. $x=1$; $x^3-2x+1=0$

30. $x=2$; $2x^3-3x^2-5x+6=0$

31. $x=-2$; $x^3+4x^2+x-6=0$

32. $x=-1$; $x^3+4x^2+x-2=0$

33. $x = 3$; $x^3 - 10x + 3 = 0$

34. $x = -3$; $2x^3 + 7x^2 + 2x - 3 = 0$

35. $x = 1, x = -1$; $2x^4 - 3x^2 + 1 = 0$

36. $x = 2, x = -2$; $2x^4 + x^3 - 10x^2 - 4x + 8 = 0$

37. $x = 3, x = -3$; $2x^4 + 3x^3 - 19x^2 - 27x + 9 = 0$

38. $x = 1, x = -2$; $x^4 + 2x^3 - 3x^2 - 4x + 4 = 0$

39. $x = -1, x = 2$; $3x^4 - x^3 - 9x^2 - 3x + 2 = 0$

40. $x = 2, x = 3$; $3x^4 - 15x^3 + 17x^2 + 5x - 6 = 0$

41. $x = -2, x = -3$; $2x^4 + 10x^3 + 9x^2 - 15x - 18 = 0$

42. $x = -2, x = 3$; $3x^4 - 3x^3 - 20x^2 + 2x + 12 = 0$

Problems 43–45 use the following information. The saturated oxygen content (as a decimal fraction) of blood is given by

$$s = \frac{(VP)^4 - 15(VP)^3 + 2045(VP)^2 + 2000(VP)}{(VP)^4 - 15(VP)^3 + 2400(VP)^2 - 31{,}100(VP) + 2{,}400{,}000}.$$

43. Evaluate the numerator in the formula if $VP = 60$
(a) using the remainder theorem.
(b) using direct evaluation. 6 9 2 4 5 7 8

44. Evaluate the denominator in the formula if $VP = 60$
(a) using the remainder theorem.
(b) using direct evaluation. 6 9 2 4 5 7 8

45. If $VP = 70$, find s as a percent. 6 9 2 4 5 7 8

46. To solve a system of n equations in n variables, the number of multiplications and divisions required by the method of Gaussian elimination is given by

$$M = (1/3)n^3 + (3/2)n^2 + (1/6)n.$$

Find M if $n = 12$
(a) using the remainder theorem.
(b) using direct evaluation. 3 5 9

47. For a simple beam under a distributed load, the deflection y (in mm) at a distance x from a support is given by

$$y = \frac{wxl^3}{24EI} - \frac{wx^3l}{12EI} + \frac{wx^4}{24EI}.$$

Assume that $w = 0.05$ kg/mm (the load), $l = 40$ mm (the length of the beam), $E = 4000$ kg/mm² (the modulus of elasticity), and $I = 5$ mm⁴ (the moment of inertia). With the given values, the equation becomes

$$y = \frac{x^4}{9{,}600{,}000} - \frac{x^3}{120{,}000} + \frac{x}{150}.$$

Find y if $x = 20$. 1 2 4 8 7 9

---

1. Architectural Technology
2. Civil Engineering Technology
3. Computer Engineering Technology
4. Mechanical Drafting Technology
5. Electrical/Electronics Engineering Technology
6. Chemical Engineering Technology
7. Industrial Engineering Technology
8. Mechanical Engineering Technology
9. Biomedical Engineering Technology

7.3 THE RATIONAL ROOTS THEOREM

In the previous section, we solved polynomial equations by systematically dividing out factors until a quadratic polynomial remained. It was necessary, however, to know one or more solutions in order to start the process. There are two principles that are often used to simplify the process of finding an initial solution. The rational roots theorem makes it possible to compile a list of rational numbers from which the rational solutions may be determined by trial and error. Descartes' rule of signs

variation in sign

allows us to determine the maximum number of positive solutions and the maximum number of negative solutions. It is based on the number of *variations in sign*. A **variation in sign** occurs when two adjacent terms have different signs.

DESCARTES' RULE OF SIGNS

(a) If the terms of a polynomial equation $f(x) = 0$ are listed so that the exponents are in descending order, the number of positive roots of $f(x) = 0$ is either equal to the number of variations in sign occurring in the coefficients of $f(x)$ or is less than that number by an even number.
(b) If the terms of a polynomial equation $f(-x) = 0$ are listed so that the exponents are in descending order, the number of negative roots of $f(x) = 0$ is either equal to the number of variations in sign occurring in the coefficients of $f(-x)$ or is less than that number by an even number.

EXAMPLE 1 Determine the maximum number of positive and negative roots of $x^3 - x^2 - 5x + 2 = 0$.

Solution Comparing each pair of terms, we see that a variation in sign occurs between the first and second terms; no variation occurs between the second and third terms; and a variation occurs between the third and fourth terms. Since there are two variations in sign, the number of positive roots is either two or is less than two by an even number. That is, it is possible that there are no positive roots. Substituting $-x$ in place of x, we have

$$(-x)^3 - (-x)^2 - 5(-x) + 2 = 0$$
$$-x^3 - x^2 + 5x + 2 = 0.$$

The only variation in sign occurs between the second and third terms. Thus, the maximum number of negative roots is one. Because it is impossible for the number of negative roots to be less than one by an *even* number (that is, we could not have -1 roots), we know that, in fact, there is exactly one negative root. ■

In Example 7 of the previous section, we found the solutions of the equation in this example to be -2, $(3 + \sqrt{5})/2$ or about 2.62, and $(3 - \sqrt{5})/2$ or about 0.382.

EXAMPLE 2 Find the maximum number of positive and negative roots of

$$x^4 + x^3 - x - 1 = 0.$$

Solution

$$x^4 \underbrace{+ x^3 - x} - 1 = 0$$

There is a variation in sign here.

It is impossible for the number of positive roots to be less than one, so we conclude that there is exactly one postive root.

When $-x$ replaces x, we have

$$(-x)^4 + (-x)^3 - (-x) - 1 = 0$$

$$\underbrace{\mathbf{x^4 - x^3}} + x - 1 = 0$$

There is a variation in sign here.

$$x^4 \underbrace{\mathbf{- x^3 + x}} - 1 = 0$$

There is a variation in sign here.

$$x^4 - x^3 \underbrace{\mathbf{+ x - 1}} = 0$$

There is a variation in sign here.

Since there are three variations in sign, the number of negative roots is either three or one. ■

Given that 1 and -1 are solutions of the equation in Example 2, synthetic division is used to divide out the factors $x - 1$ and $x + 1$.

$$\begin{array}{r|rrrrr} 1 & 1 & 1 & 0 & -1 & -1 \\ & & 1 & 2 & 2 & 1 \\ \hline -1 & 1 & 2 & 2 & 1 & \\ & & -1 & -1 & -1 & \\ \hline & 1 & 1 & 1 & & \end{array}$$

Thus, we have $x^2 + x + 1 = 0$. The quadratic formula shows the remaining solutions to be nonreal complex numbers. That is, $x = (-1 \pm \sqrt{-3})/2$. Nonreal complex numbers are not classified as positive or negative. In this chapter, we are seeking only solutions that are real numbers. Descartes' rule of signs only told us that the number of negative solutions is three or one. Because the two nonreal complex solutions are hidden in the equation, we were not able to establish from the rule that there is only one negative solution.

The second principle we will use to solve polynomial equations is the rational roots theorem. First, we will examine *how* it works; later we will see *why* it works.

THE RATIONAL ROOTS THEOREM

In the polynomial equation $a_n x^n + a_{n-1} x^{n-1} + \cdots + a_0 = 0$ where a_n, a_{n-1}, . . ., a_0 are integers, and $a_n \neq 0$, each rational solution can be written as a fraction in lowest terms p/q where p is a factor of a_0 and q is a factor of a_n.

NOTE ⇨⇨ ***The rational roots theorem merely allows us to compile a list of possible solutions. It does not tell us that all of the numbers in the list are, in fact, solutions.***

EXAMPLE 3 Use the rational roots theorem to compile a list of possible rational roots of the equation $x^3 + 2x^2 - 4x - 8 = 0$.

Solution The constant term, a_0, is -8. Its factors are ± 1, ± 2, ± 4, and ± 8. The leading coefficient, a_n, is 1. Its only factors are 1 and -1. Since p/q and $-p/-q$ are equivalent fractions, and $-p/q$ and $p/-q$ are equivalent fractions, it is not necessary to use the $\pm$ sign in both numerator and denominator. The list $\pm 1/1$, $\pm 2/1$, $\pm 4/1$, and $\pm 8/1$ contains all fractions with numerators that are factors of a_0 and denominators that are factors of a_n. That is, the rational roots are found in the list ± 1, ± 2, ± 4, and ± 8. ■

EXAMPLE 4 Use the rational roots theorem to compile a list of possible rational roots of the equation $2x^3 + 6x^2 - x - 3 = 0$.

Solution The factors of the constant term are ± 1 and ± 3. The factors of the leading coefficient are 1 and 2. First, we list ± 1 and ± 3 with the denominator 1; then we list ± 1 and ± 3 with denominator 2. The rational roots are found in the list $\pm 1/1$, $\pm 3/1$, $\pm 1/2$, and $\pm 3/2$. ■

While the rational roots theorem does not tell us which numbers are rational solutions, it does give us a finite list of possibilities. Combined with Descartes' rule of signs, synthetic division allows us to find the rational roots by trial and error. Remember that a number may be a multiple root, as in Example 8 of the previous section, so when a number is found to be a root, it may be necessary to test the same number again.

To find the rational roots of a polynomial equation $f(x) = 0$:

1. Apply Descartes' rule of signs to determine the maximum number of positive and negative solutions.
2. Use the rational roots theorem to list possible solutions.
3. Use synthetic division to test the numbers in the list obtained in step 2.
   - **(a)** If there is only one root of a particular sign, try to find that root first.
   - **(b)** If a remainder of 0 is obtained, the number tested is a root. The quotient represents a new and simpler polynomial equation to which the rational roots theorem applies.
4. Repeat steps 1–3 until the quotient is a quadratic polynomial, which can be solved by the formula, or all possible rational roots have been tested.

EXAMPLE 5 Find all rational solutions of $x^3 + 2x^2 - 4x - 8 = 0$.

Solution

1. There is one positive root. Since replacing x with $-x$ leads to $(-x)^3 + 2(-x)^2 - 4(-x) - 8 = 0$ or $-x^3 + 2x^2 + 4x - 8 = 0$, the number of negative roots is either two or zero.

2. The possible rational solutions were obtained in Example 3. They are $\pm 1, \pm 2, \pm 4$, and ± 8.

3. Since we know there is only one positive solution, we seek it first.

$$\begin{array}{r|rrrr} 1 & 1 & 2 & -4 & -8 \\ & & 1 & 3 & -1 \\ \hline & 1 & 3 & -1 & -9 \end{array} \qquad \begin{array}{r|rrrr} 2 & 1 & 2 & -4 & -8 \\ & & 2 & 8 & 8 \\ \hline & 1 & 4 & 4 & 0 \end{array}$$

A remainder of 0 tells us that 2 is a solution. It is not necessary to seek the negative solutions by synthetic division, since $x^2 + 4x + 4$ is quadratic.

4. We have

$$x^2 + 4x + 4 = 0$$
$$(x + 2)(x + 2) = 0$$
$$x + 2 = 0, \text{ so } x = -2.$$

The solution is a double root, because $x + 2$ occurs twice as a factor. ■

EXAMPLE 6 Find all rational solutions of $2x^4 - 7x^3 - 5x^2 + 28x - 12 = 0$.

Solution 1. The number of positive roots is either three or one. To find the number of negative roots, we examine

$$2(-x)^4 - 7(-x)^3 - 5(-x)^2 + 28(-x) - 12 = 0$$
$$2x^4 + \mathbf{7x^3 - 5x^2} - 28x - 12 = 0.$$

There is only one variation in sign, so there is exactly one negative root.

2. The factors of -12 are $\pm 1, \pm 2, \pm 3, \pm 4, \pm 6$, and ± 12. The factors of 2 are 1 and 2. Thus, the rational solutions are found in the following list:

$\pm 1/1, \pm 2/1, \pm 3/1, \pm 4/1, \pm 6/1, \pm 12/1, \pm 1/2, \pm 2/2, \pm 3/2, \pm 4/2, \pm 6/2, \pm 12/2.$

Eliminating duplicates, we have

$$\pm 1, \pm 2, \pm 3, \pm 4, \pm 6, \pm 12, \pm 1/2, \pm 3/2.$$

3. Since there is only one negative root, we seek it first.

$$\begin{array}{r|rrrrr} -1 & 2 & -7 & -5 & 28 & -12 \\ & & -2 & 9 & -4 & -24 \\ \hline & 2 & -9 & 4 & 24 & -36 \end{array} \qquad \begin{array}{r|rrrrr} -2 & 2 & -7 & -5 & 28 & -12 \\ & & -4 & 22 & -34 & 12 \\ \hline & 2 & -11 & 17 & -6 & 0 \end{array}$$

Thus, we know that -2 is a solution.

4. Consider $2x^3 - 11x^2 + 17x - 6 = 0$. The number of positive roots is either three or one. There are no negative roots, since any negative root of this equation would be a root of $2x^4 - 7x^3 - 5x^2 + 28x - 12 = 0$, and we have found its only negative root to be -2. Factors of 6 are 1, 2, 3, and 6. Factors of 2 are 1 and 2. The rational roots are found in the following list:

$$1/1,\ 2/1,\ 3/1,\ 6/1,\ 1/2,\ 2/2,\ 3/2,\ 6/2 \text{ or } 1,\ 2,\ 3,\ 6,\ 1/2,\ 3/2.$$

$$\begin{array}{r|rrrr} 1 & 2 & -11 & 17 & -6 \\ & & 2 & -9 & 8 \\ \hline & 2 & -9 & 8 & 2 \end{array} \qquad \begin{array}{r|rrrr} 2 & 2 & 11 & 17 & -6 \\ & & 4 & -14 & 6 \\ \hline & 2 & -7 & 3 & 0 \end{array}$$

Thus, 2 is a root, and the remaining roots are found by solving $2x^2 - 7x + 3 = 0$. Factoring or the quadratic formula may be used. In either case, $x = 3$ or $x = 1/2$. ■

Now that you know how to use the rational roots theorem to solve a polynomial equation, let us see why it works.

Consider the equation

$$a_n x^n + a_{n-1} x^{n-1} + \ldots + a_1 x + a_0 = 0.$$

Let r_1 be a rational root written in lowest terms as p/q. Then

$$a_n(p/q)^n + a_{n-1}(p/q)^{n-1} + \ldots + a_1(p/q) + a_0 = 0, \text{ and thus}$$
$$a_n(p^n/q^n) + a_{n-1}(p^{n-1}/q^{n-1}) + \ldots + a_1(p/q) + a_0 = 0.$$

If both sides of the equation are multiplied by q^n, we have

$$a_n p^n + a_{n-1}(p^{n-1}q) + \ldots + a_1(pq^{n-1}) + a_0 q^n = 0. \qquad \textbf{(7-3)}$$

The term $a_0 q^n$ is subtracted from both sides of Equation 7-3 to obtain Equation 7-4.

$$a_n p^n + a_{n-1}(p^{n-1}q) + \ldots + a_1(pq^{n-1}) = -a_0 q^n \qquad \textbf{(7-4)}$$

The term $a_n p^n$ is subtracted from both sides of Equation 7-3 to obtain Equation 7-5.

$$a_{n-1}(p^{n-1}q) + \ldots + a_1(pq^{n-1}) + a_0 q^n = -a_n p^n \qquad \textbf{(7-5)}$$

But p is a factor of each term on the left-hand side of Equation 7-4, so p is a factor of the expression on the right-hand side also. Since we made the assumption that p/q was in lowest terms, p and q have no common factor. Thus, p cannot divide q^n, so p must be a factor of a_0. Similarly, q is a factor of each term on the left-hand side of Equation 7-5, so q is a factor of the expression on the right-hand side also. Since p and q have no common factor, q cannot divide p^n, so q must be a factor of a_n.

EXERCISES 7.3

In 1–14, determine the maximum number of positive and negative roots of each equation.

1. $x^4 + 2x^3 - 3x^2 + x - 1 = 0$

2. $x^4 - 4x^3 + 3x^2 + 2x - 1 = 0$

3. $-x^4 + 3x^3 + 2x^2 + x - 3 = 0$

4. $2x^3 - 2x^2 - x - 7 = 0$

5. $3x^3 + x^2 + 3x - 4 = 0$

6. $-x^3 + 2x^2 + 3 = 0$

7. $-2x^4 - x^3 + x^2 - 1 = 0$

8. $3x^4 + 2x^2 - x - 4 = 0$

9. $x^5 - 2x^4 + x^3 - 2x^2 + x - 1 = 0$

10. $-2x^5 + 3x^4 - 4x^3 + x^2 - 5x + 1 = 0$

11. $3x^4 + 2x^2 + 3 = 0$

12. $-2x^4 - 3x^3 - x = 0$

13. $2x^3 + 2x^2 - x - 3 = 0$

14. $3x^3 - 2x^2 - x + 3 = 0$

In 15–28, use the rational roots theorem to compile a list of possible rational roots of each equation.

15. $x^3 + 2x^2 - 3x + 2 = 0$

16. $x^3 - 3x^2 + 2x - 3 = 0$

17. $x^4 + x^3 - 2x^2 + 5x - 4 = 0$

18. $x^4 - 2x^3 + x^2 + 3x - 6 = 0$

19. $2x^3 + x^2 - x + 3 = 0$

20. $2x^3 - x^2 + 3x - 1 = 0$

21. $2x^4 + 3x^2 - 6 = 0$

22. $2x^4 - 5x^2 + 15 = 0$

23. $3x^3 - x + 3 = 0$

24. $5x^3 + 3x^2 - 5 = 0$

25. $4x^3 - 5x + 10 = 0$

26. $9x^3 + x - 6 = 0$

27. $6x^4 + x^2 - 15 = 0$

28. $15x^4 - x^3 + 35 = 0$

In 29–42, find all rational roots of each equation.

29. $2x^3 - 9x^2 + 12x - 4 = 0$

30. $3x^3 - 4x^2 - x + 2 = 0$

31. $2x^3 - 3x^2 - 2x + 3 = 0$

32. $x^4 - 4x^3 - x^2 + 16x - 12 = 0$

33. $x^4 - 5x^3 + 5x^2 + 5x - 6 = 0$

34. $x^4 - x^3 - 11x^2 + 9x + 18 = 0$

35. $2x^5 - 5x^4 - 3x^3 - 2x^2 + 5x + 3 = 0$

36. $3x^5 + 5x^4 - 2x^3 + 3x^2 + 5x - 2 = 0$

37. $x^5 + x^3 + 3x^2 + 3 = 0$

38. $2x^5 + x^3 + 2x^2 + 1 = 0$

39. $x^4 + x^3 - 7x^2 - 4x + 12 = 0$

40. $4x^4 + 4x^3 - 5x^2 - x + 1 = 0$

41. $2x^4 + x^3 - x^2 + 8x - 4 = 0$

42. $3x^4 + x^3 - 3x - 1 = 0$

7.4 IRRATIONAL ROOTS

The process of finding roots by the rational roots theorem and trial and error can be long and tedious. The following observation can help cut down on the work required.

To narrow the list of possible rational roots of $f(x) = 0$:

1. If all numbers in the bottom row of the synthetic division are greater than or equal to zero when $f(x)$ is divided by $x - c$ and $c > 0$, then there is no root larger than c.
2. If the numbers in the bottom row of the synthetic division alternate in sign (with zero treated as positive or negative as needed), when $f(x)$ is divided by $x - c$ and $c < 0$, then there is no root smaller than c.

EXAMPLE 1 Find all rational roots of $x^4 + x^3 + 5x^2 + 4x + 4 = 0$.

Solution There are no variations in sign, so by Descartes' rule of signs, we know that there are no positive roots. When x is replaced with $-x$, we have

$$(-x)^4 + (-x)^3 + 5(-x)^2 + 4(-x) + 4 = 0$$
$$x^4 - x^3 + 5x^2 - 4x + 4 = 0.$$

Since there are four variations in sign, the number of negative roots is either four, two, or zero.

The constant term has factors of $\pm 1, \pm 2$, and ± 4, and the leading coefficient has only 1 as a factor. Since there is no positive root, the rational roots must appear in the list $-1, -2, -4$.

We use synthetic division to divide the polynomial by $x + 1$.

$$\begin{array}{r|rrrrr} -1 & 1 & 1 & 5 & 4 & 4 \\ & & -1 & 0 & -5 & 1 \\ \hline & 1 & 0 & 5 & -1 & 5 \end{array}$$

We have divided by $x - c$ where c is -1. If we allow 0 to be -0, then the signs on the bottom row alternate. Thus, we know that there is no root smaller than -1. That is, it is not necessary to try -2 and -4, since they are both less than -1. There are no rational roots. ■

Example 1 illustrates the limitations of the rational roots theorem. There must be four solutions to this equation; yet they are not rational. In fact, the roots for this equation must be nonreal complex numbers, since they are neither positive nor negative. In other cases, when the roots are not rational, they may be irrational. If they are irrational, the bisection method of Section 6.5 may be used to obtain an approximate solution. To use the bisection method, however, it is necessary to bracket the root. That is, we must find values x_1 and x_2 such that $f(x_1)$ and $f(x_2)$ have different signs. A graph is often useful.

For higher degree polynomial functions, there are some short cuts similar to those we used to graph linear and quadratic functions, but they involve calculus. Without using calculus, the best thing to do is plot several points and connect them. We make the following observation. A linear function has degree one and zero "bends." A quadratic function has degree two and one "bend." A cubic function has degree three and two "bends." When the degree of the polynomial function is higher than three, however, all we can say is that the number of bends is *at most* one less than the degree. When we have found both bends in the graph of a cubic function, we know that we have a reasonably good graph. Most cubic functions will either look like the graph in Figure 7.2, or the graph in Figure 7.3.

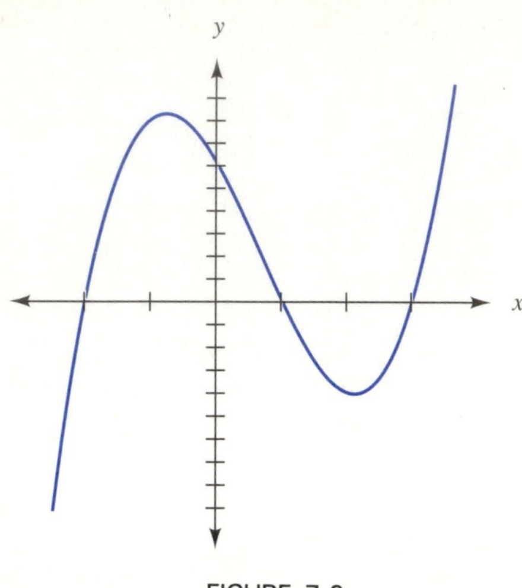

FIGURE 7.2

FIGURE 7.3

EXAMPLE 2 Sketch the graph of $y = x^3 + 3$.

Solution We choose several values for x, say -2, -1, 0, 1, and 2. We list these in table form, and calculate the corresponding function values.

x	$x^3 + 3$	y
-2	$(-2)^3 + 3$	$-8 + 3 = -5$
-1	$(-1)^3 + 3$	$-1 + 3 = 2$
0	$0^3 + 3$	$0 + 3 = 3$
1	$1^3 + 3$	$1 + 3 = 4$
2	$2^3 + 3$	$8 + 3 = 11$

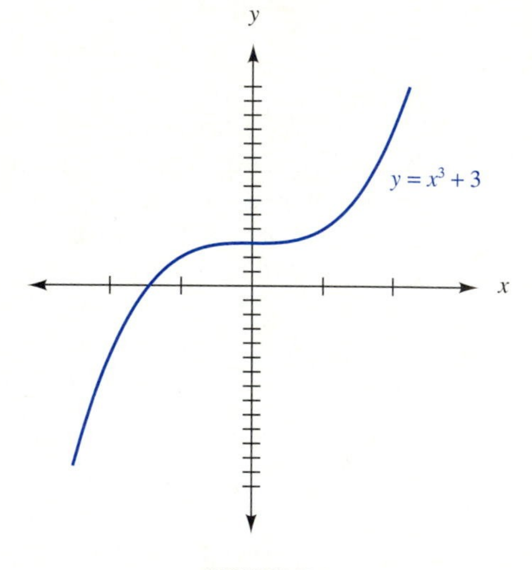

FIGURE 7.4

We plot these five points and connect them to form a smooth curve as shown in Figure 7.4. ■

At first, Example 2 may not appear to follow the pattern described for a cubic function. But if you imagine that the graph were made of wire, two bends would be required to shape it.

To sketch the graph of a polynomial function:

1. Choose several values for x. Five values will usually provide a good start toward the graph.
2. Calculate the corresponding function values.
3. Plot the points.
4. Connect them with a smooth curve going from left to right.
5. If the number of "bends" in the graph is one less than the degree of the polynomial, conclude that the graph is adequate. If the number of "bends" in the graph is not one less than the degree of the polynomial, examine more points until all of the bends are located or you are convinced that $|y|$ continues to increase as $|x|$ increases.

EXAMPLE 3 Sketch the graph of $y = x^4 - 3$.

Solution 1. We choose $-2, -1, 0, 1$, and 2 as x-values.

2. A table of values follows.

x	$x^4 - 3$	y
-2	$(-2)^4 - 3$	$16 - 3 = 13$
-1	$(-1)^4 - 3$	$1 - 3 = -2$
0	$0^4 - 3$	$0 - 3 = -3$
1	$1^4 - 3$	$1 - 3 = -2$
2	$2^4 - 3$	$16 - 3 = 13$

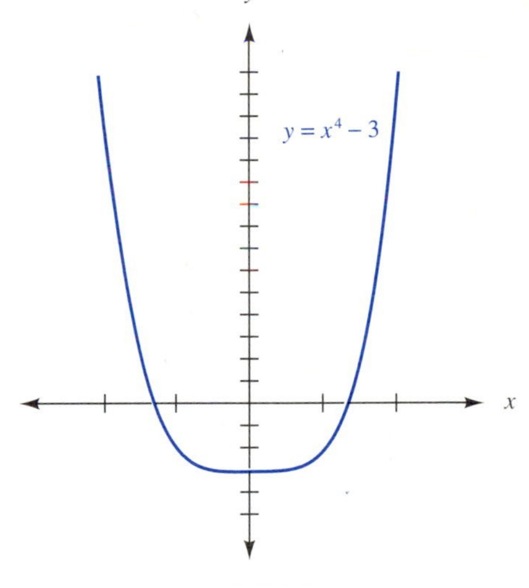

FIGURE 7.5

3–4. We plot these five points and connect them as shown in Figure 7.5.

5. Since only one "bend" appears in the graph, we consider the equation $y = x^4 - 3$. Notice that x^4 is nonnegative for all x. If x is allowed to assume values (positive or negative) that are larger and larger in absolute value, y will assume values that are larger and larger. Thus, we conclude that the graph has only one "bend." If there were another bend, y would begin to take on smaller values somewhere. ■

Putting together the techniques for finding rational and irrational roots, we are now able to summarize the procedure for finding the *real* roots of a polynomial equation.

> **To find the real roots of a polynomial equation $f(x) = 0$:**
>
> **1.** Determine the maximum number of positive and negative roots, using Descartes' rule of signs.
>
> **2.** Find rational roots, using the rational roots theorem and synthetic division, until
>
> **(a)** only a quadratic polynomial remains, or
>
> **(b)** the bottom row shows that it is not necessary to continue testing for rational roots, or
>
> **(c)** all numbers in the list of possible rational roots have been tested.
>
> **3.** If there is a possibility of one or more irrational roots, then graph the original equation to bracket the root or roots. Use the bisection method to find an approximate value for the root or roots.

EXAMPLE 4 Find the real roots of $2x^5 + x^4 - 4x^3 - 2x^2 - 4x - 2 = 0$.

Solution **1.** The number of variations in sign is one, so there is only one positive root. Replacing x with $-x$, we have

$$\begin{aligned} 2(-x)^5 + (-x)^4 - 4(-x)^3 - 2(-x)^2 - 4(-x) - 2 &= 0 \\ -2x^5 + x^4 + 4x^3 - 2x^2 + 4x - 2 &= 0. \end{aligned}$$

Since there are four variations in sign, the number of negative roots must be four, two, or zero.

2. The factors of the constant term are ± 1 and ± 2. The factors of the leading coefficient are 1 and 2. The rational roots must appear in the list $\pm 1/1$, $\pm 2/1$, $\pm 1/2$, $\pm 2/2$ or ± 1, ± 2, $\pm 1/2$. We seek the positive root first.

$$\begin{array}{r|rrrrrr} 1 & 2 & 1 & -4 & -2 & -4 & -2 \\ & & 2 & 3 & -1 & -3 & -7 \\ \hline & 2 & 3 & -1 & -3 & -7 & -9 \end{array} \qquad \begin{array}{r|rrrrrr} 2 & 2 & 1 & -4 & -2 & -4 & -2 \\ & & 4 & 10 & 12 & 20 & 32 \\ \hline & 2 & 5 & 6 & 10 & 16 & 30 \end{array}$$

$$\begin{array}{r|rrrrrr} 1/2 & 2 & 1 & -4 & -2 & -4 & -2 \\ & & 1 & 1 & -3/2 & -7/4 & -23/8 \\ \hline & 2 & 2 & -3 & -7/2 & -23/4 & -39/8 \end{array}$$

The positive root, then, is not rational, but from our division by $x - 2$, we know that it cannot exceed 2. We seek the negative roots.

$$\begin{array}{r|rrrrrr} -1 & 2 & 1 & -4 & -2 & -4 & -2 \\ & & -2 & 1 & 3 & -1 & 5 \\ \hline & 2 & -1 & -3 & 1 & -5 & 3 \end{array} \qquad \begin{array}{r|rrrrrr} -2 & 2 & 1 & -4 & -2 & -4 & -2 \\ & & -4 & 6 & -4 & 12 & -16 \\ \hline & 2 & -3 & 2 & -6 & 8 & -18 \end{array}$$

$$\begin{array}{r|rrrrrr} -1/2 & 2 & 1 & -4 & -2 & -4 & -2 \\ & & -1 & 0 & 2 & 0 & 2 \\ \hline & 2 & 0 & -4 & 0 & -4 & 0 \end{array}$$

Thus, we know that $-1/2$ is a root, and from our division by $x + 2$, we know that there are no roots smaller than -2.

3. We know that the positive root is irrational. Since there must be either two or four negative roots, but there was only one negative rational root, there must be at least one negative irrational root. At this stage, a graph helps (see Figure 7.6). The remainder theorem allows us to use the six synthetic divisions we have performed to identify six points on the graph. Additionally, we know that when x is 0, y is -2. A table of values for x between -2 and 2 follows.

x	-2	-1	$-1/2$	0	$1/2$	1	2
y	-18	3	0	-2	$-39/8$	-9	30

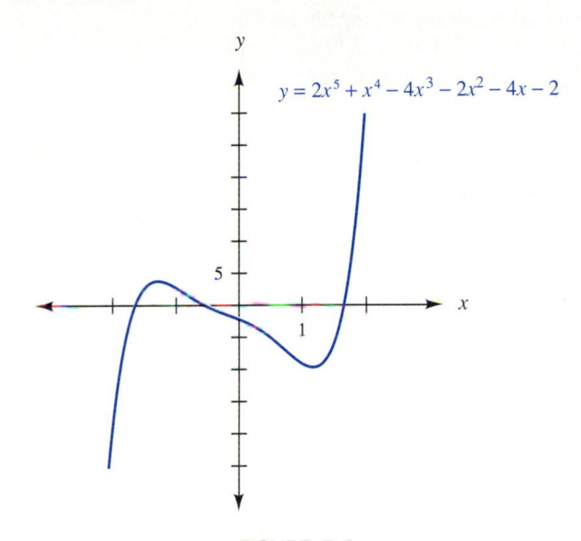

FIGURE 7.6

There is a negative root between -2 and -1 and a positive root between 1 and 2. The bisection method gives $x = -1.653$ and $x = 1.653$. ■

EXERCISES 7.4

In 1–12, find all real roots to three significant digits for each equation.

1. $2x^3 + 5x^2 - 6x - 4 = 0$

2. $x^4 + 2x^3 - 9x^2 - 8x + 20 = 0$

3. $2x^4 + 3x^3 - 3x^2 - 3x + 1 = 0$

4. $x^4 + x^3 - 2x^2 + x - 3 = 0$

5. $x^4 + x^3 + x^2 + 2x - 2 = 0$

6. $3x^4 - 2x^3 - 6x + 4 = 0$

7. $2x^4 + x^3 - 6x - 3 = 0$

8. $2x^4 + 3x^3 + 4x + 6 = 0$

9. $3x^4 + 2x^3 + 6x + 4 = 0$

10. $4x^4 - 2x^3 + 2x - 1 = 0$

11. $2x^4 - 3x^3 + 4x - 6 = 0$

12. $2x^4 - 2x^3 + 3x - 3 = 0$

---

13. A tank is to be constructed in the shape of a cylinder 10.0 m high topped by a hemisphere. If the volume is to be 19π m^3, find the radius. 1 2 4 7 8 3 5

14. Find the dimensions of a cube that, when topped by a pyramid 8 in high, forms a figure with volume 550 in^3. The volume of a pyramid is given by $V = (1/3)Bh$, where B is the area of the base and h is the height. 1 2 4 7 8 3 5

15. A series of squares, each having sides 1 cm longer than the previous square, are cut from a sheet of plastic, as shown in the accompanying figure. The areas, then, are $1^2, 2^2, 3^2, 4^2$, and so on. The sum of the areas of the first n squares is given by $A = \dfrac{2n^3 + 3n^2 + n}{6}$. How many such squares can be cut from a sheet of plastic that measures 25 cm $\times$ 26 cm? Assume that constructing a square by piecing together more than one piece of plastic is allowed, so that there is as little waste as possible. 1 2 4 7 8 3 5

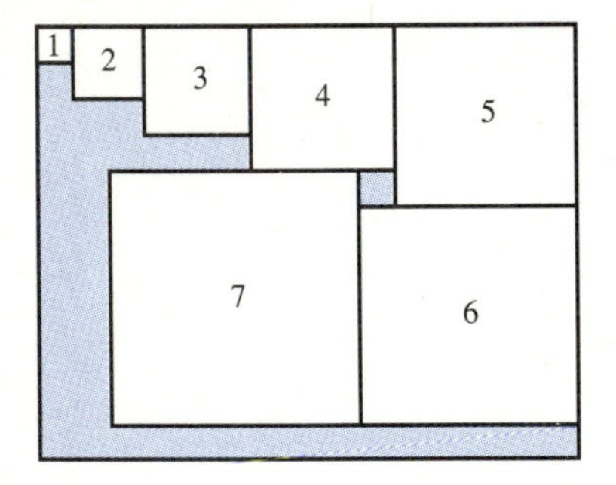

16. A box is to be constructed from a piece of metal 10.0 cm square by cutting a square from each corner and folding up the sides, as shown in the accompanying diagram. What size square should be removed if the volume of the box is to be 72.0 cm^3? 1 2 4 7 8 3 5

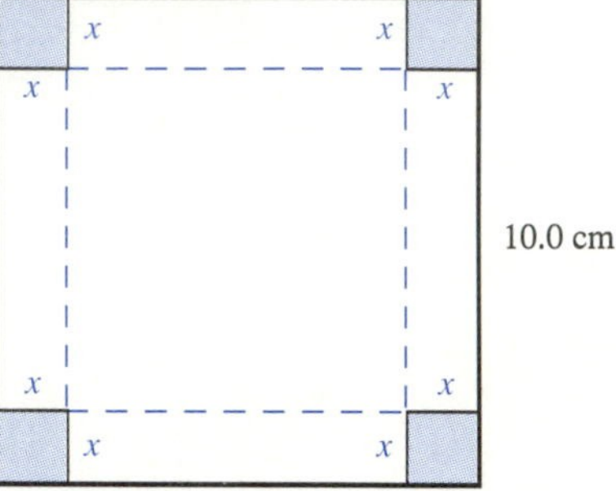

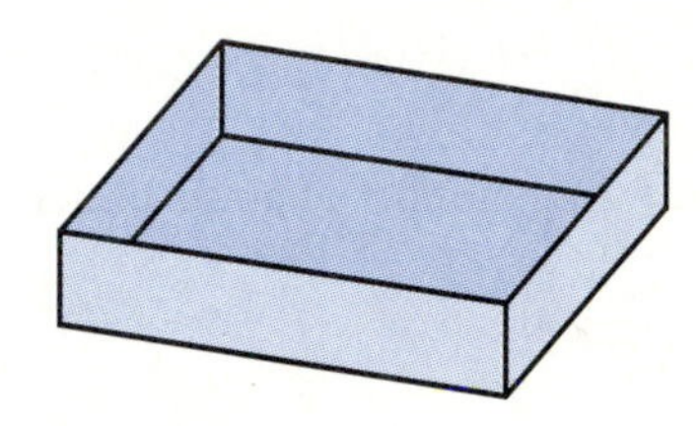

17. For a cantilever beam under a distributed load, the deflection at a distance x from the support is given by

$$y = \frac{-wx^4 + 4wbx^3 - 6wb^2x^2}{24EI} \quad \text{when } x \leq b.$$

Assume that $w = 0.02$ kg/mm (the load), $E = 4000$ kg/mm^2 (the modulus of elasticity), $I = 5$ mm^4 (the moment of inertia), and $b = 40$ mm (the length of the load). With the given values, the equation becomes

$$y = \frac{-x^4 + 160\,x^3 - 9600\,x^2}{24{,}000{,}000}.$$

Find the distance x at which the deflection is -0.002 mm. 1 2 4 8 7 9

---

1. Architectural Technology
2. Civil Engineering Technology
3. Computer Engineering Technology
4. Mechanical Drafting Technology
5. Electrical/Electronics Engineering Technology
6. Chemical Engineering Technology
7. Industrial Engineering Technology
8. Mechanical Engineering Technology
9. Biomedical Engineering Technology

7.5 CHAPTER REVIEW

KEY TERMS

7.2 root

multiple root

7.3 variation in sign

RULES AND FORMULAS

Remainder theorem:
If $f(x)$ and $q(x)$ are polynomials and R is a real number such that $f(x) = (x - b)[q(x)] + R$, then $f(b) = R$.
Factor theorem:
Let $f(x)$ be a polynomial. If $f(b) = 0$, then $x - b$ is a factor of $f(x)$. If $f(b) \neq 0$, then $x - b$ is not a factor of $f(x)$.
Descartes' rule of signs:
(a) If the terms of a polynomial equation $f(x) = 0$ are listed so that the exponents are in descending order, the number of positive roots of $f(x) = 0$ is either equal to the number of variations in sign occurring in the coefficients of $f(x)$ or is less than that number by an even number.
(b) If the terms of a polynomial equation $f(-x) = 0$ are listed so that the exponents are in descending order, the number of negative roots of $f(x) = 0$ is either equal to the number of variations in sign occurring in the coefficients of $f(-x)$ or is less than that number by an even number.
Rational roots theorem:
In the polynomial equation $a_nx^n + a_{n-1}x^{n-1} + \ldots + a_0 = 0$ where $a_n, a_{n-1}, \ldots, a_0$ are integers, and $a_n \neq 0$, each rational solution can be written as a fraction in lowest terms p/q where p is a factor of a_0 and q is a factor of a_n.

GEOMETRY CONCEPTS

$V = (1/3)\ Bh$ (volume of a pyramid or cone)
$V = s^3$ (volume of a cube)
$V = \pi r^2h$ (volume of a cylinder)
$V = lwh$ (volume of a rectangular solid)

SEQUENTIAL REVIEW

(Section 7.1) In 1–5, use synthetic division, if possible, to find the quotient and remainder for each division problem.

1. $(6x^3 + 3x^2 - x + 4) \div (x + 2)$

2. $(x^4 - x^2 + 1) \div (x^2 + 1)$

3. $(4x^4 + 2x^2 + x - 3) \div (2x - 3)$

4. $(x^4 + 2x^3 - 2x - 1) \div (x - 1)$

5. $(6x^3 - 16x^2 - 5x + 6) \div (x - 3)$

In 6–10, use synthetic division to determine whether the first polynomial is a factor of the second.

6. $x+3;\ x^4-2x^2+6$

7. $x-1;\ x^4+3x^3-2$

8. $x+2;\ x^4+x^3-2x^2+x+2$

9. $2x-3;\ 2x^4-3x^3+2x^2-5x+3$

10. $2x+3;\ 4x^4-2x^2+2x+1$

(Section 7.2) *In 11–14, find the indicated value for each polynomial using (a) the remainder theorem and (b) direct evaluation.*

11. $f(x)=4x^4+3x^2-2$; find $f(1)$

12. $f(x)=3x^4-2x^3+4$; find $f(2)$

13. $g(x)=2x^3+x-1$; find $g(-2)$

14. $g(x)=3x^3+x^2-x+2$; find $g(-1)$

In 15–18, use the remainder theorem to determine whether the first polynomial is a factor of the second.

15. $x-3;\ x^3-3x^2+7x-21$

16. $x+2;\ x^4+2x^2-3x-6$

17. $x+3;\ 2x^4-3x^2+2x-6$

18. $x-2;\ x^3+2x^2-4x-8$

In 19–22, find the remaining solutions of each equation, given the solution or solutions indicated.

19. $x=2;\ x^3-6x^2+12x-8=0$

20. $x=-1;\ x^3-4x^2+x+6=0$

21. $x=-1, x=2;\ x^4-x^3-11x^2+9x+18=0$

22. $x=1, x=-1;\ x^4+4x^3-7x^2-4x+6=0$

(Section 7.3) *In 23–25, determine the maximum number of positive roots of each equation.*

23. $3x^4+x^3-x^2-x+1=0$

24. $2x^3+3x^2+x-5=0$

25. $4x^4+2x^3+x-3=0$

In 26–28, determine the maximum number of negative roots of each equation.

26. $3x^4+x^3-x^2-x+1=0$

27. $2x^3+3x^2+x-5=0$

28. $4x^4+2x^3+x-3=0$

In 29–31, use the rational roots theorem to compile a list of all possible rational roots of each equation.

29. $x^3+2x^2-x+5=0$

30. $4x^4-3x+6=0$

31. $3x^4-2x^2+9=0$

In 32–34, find all rational roots of each equation.

32. $x^4-4x^3+2x^2+4x-3=0$

33. $2x^4-x^3+x^2-2x-6=0$

34. $2x^4-x^3-10x^2+2x+12=0$

(Section 7.4)

35. Consider $4x^4+2x^2+x-3=0$.
 (a) Is 1 a root?
 (b) Is it necessary to test values larger than 1 to find the roots?

36. Consider $x^4+3x^3-2=0$.
 (a) Is 2 a root?
 (b) Is it necessary to test values larger than 2 to find the roots?

37. Consider $x^3 - 6x^2 + 12x - 8 = 0$.
(a) Is 2 a root?
(b) Is it necessary to test values larger than 2 to find the roots?

38. Consider $4x^4 + 2x^2 + x - 3 = 0$.
(a) Is -1 a root?
(b) Is it necessary to test values smaller than -1 to find the roots?

39. Consider $x^4 + 3x^3 - 2 = 0$.
(a) Is -2 a root?
(b) Is it necessary to test values smaller than -2 to find the roots?

40. Consider $x^3 - 6x^2 + 12x - 8 = 0$.
(a) Is -2 a root?
(b) Is it necessary to test values smaller than -2 to find the roots?

In 41–45, find all real roots of each equation.

41. $x^3 + 5x^2 + 5x - 2 = 0$

42. $4x^3 - 7x - 3 = 0$

43. $2x^4 + 2x^3 + x + 1 = 0$

44. $2x^4 - 2x^3 - 3x + 3 = 0$

45. $2x^5 + 5x^4 - 3x^3 + 4x^2 + 10x - 6 = 0$

APPLIED PROBLEMS

1. The volume V of the portion of a cone between two parallel planes is given by

$$V = (1/3)\pi h(r_1^2 + r_2^2 + r_1r_2).$$

In the formula, h is the distance between planes, and r_1 and r_2 are the radii of the circular surfaces. Consider a conical shaped reservoir. Suppose it contains enough water that the radius of the water surface is R. Water is added so that the depth is increased by R, and the radius of the water surface is found to be 10.0 m.
(a) Write a formula for the volume V of water added in terms of R.
(b) Find V if the radius of the water surface was 5.0 m before water was added.

1 2 4 7 8 3 5

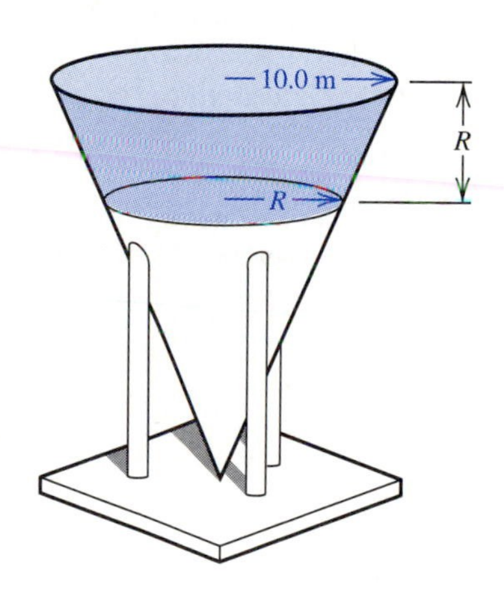

1. Architectural Technology
2. Civil Engineering Technology
3. Computer Engineering Technology
4. Mechanical Drafting Technology
5. Electrical/Electronics Engineering Technology
6. Chemical Engineering Technology
7. Industrial Engineering Technology
8. Mechanical Engineering Technology
9. Biomedical Engineering Technology

2. The amount of money B in an account that draws interest at the rate of R percent per year after 3 years is given by

$$B = P(R^3 + 3R^2 + 3R + 1),$$

where P is the original deposit, and R is written as a decimal. If $1000 is deposited initially into an account that draws 5% per year, how much money will be in the account after 3 years?
1 2 6 7 8 3 5 9

3. When a cubical box is filled with water to a depth of 2.0 cm from the top, the amount of water is 75 cm^3. Find the dimensions of the box.
1 2 4 7 8 3 5

4. A rectangular box that is 1.00″ × 2.00″ × 3.00″ is increased by the same amount on each edge. The new volume is 60.0 in^3. Determine the amount of increase. 1 2 4 7 8 3 5

5. A technician makes the mistake of adding fractions by adding numerators and adding denominators. Thus, he says

$$\frac{1}{x} + \frac{1}{x+2} + \frac{1}{x+3} = \frac{3}{3x+5}.$$

Multiplying both sides of this equation by the LCD leads to $8x^3 + 40x^2 + 62x + 30 = 0$. Are there any rational values of x for which this equation is true? If so, what are they?
1 2 6 7 8 3 5 9

6. A cubical container had its width increased by 2 ft, its depth decreased by 3 ft, and its height increased by 4 ft. The resulting container had a volume of 3876 ft^3. What was its original volume?
1 2 4 7 8 3 5

7. Three identical variable resistors are connected in parallel and set to a common value. The first is then adjusted upward by 1 ohm, the second by 2 ohms, and the third by 3 ohms. The resulting total resistance is found to be 39 ohms. In finding their original values, the following equation arises. Find R to the nearest tenth.

$$R^3 - 111R^2 - 457R - 423 = 0$$

(Hint: The root is irrational and is between 115 and 116.) 3 5 9 1 2 4 6 7 8

8. A 10.0 in × 19.0 in piece of aluminum is made into a parts chassis by cutting a square notch out of each corner and bending up the resulting rectangular projections to form a box. How much should be notched out to make a box containing 180 cubic inches? (Hint: One answer is 2.00. Find another.) 1 2 4 7 8 3 5

9. The owner of a welding shop wishes to purchase some new equipment for $30,000. She would like to borrow that amount at an annual interest rate that would allow her to make a payment each year for 3 years of not more than $12,000. The formula for the annual payment A_P required for such a transaction on a loan of P dollars for n years at a yearly interest rate i is, $A_P = \frac{Pi(1+i)^n}{(1+i)^n - 1}$. For the values given, the equation is $5i^3 + 13i^2 + 9i - 1 = 0$. Find i to the nearest thousandth. (Hint: The answer is between 0.09 and 0.10.)
1 2 6 7 8 3 5 9

10. In electronics the zeros of a filter function are those values for which the polynomial numerator of the function equals zero and the poles are those values for which the polynomial denominator is zero. Consider

$$F = \frac{w^3 - 16w^2 + 79w - 120}{w^4 - 14w^3 + 63w^2 - 106w + 56}.$$

Find the zeros. (Hint: All values are whole numbers less than 10. 3 5 9 4 7 8

11. Researchers who study gases under various conditions of pressure and temperature can use van der Waal's equation to compute the number of liters n, occupied by one standard mole of a gas under various temperatures T and pressures P. For oxygen this equation is given by $\left[P + \frac{1.36}{n^2}\right][n - 0.0318] = 0.082T$, where P is measured in atmospheres and T in degrees Kelvin with n in liters. How many liters would a mole of oxygen occupy at 5 atm pressure and 350° K? For the numbers given, this equation is $n^3 - 5.77n^2 + 0.272\,n - 0.00864 = 0$. Find n to the nearest tenth, given that n is between 5 and 6.
6 8 1 2 3 5 7 9

1. Architectural Technology
2. Civil Engineering Technology
3. Computer Engineering Technology
4. Mechanical Drafting Technology
5. Electrical/Electronics Engineering Technology
6. Chemical Engineering Technology
7. Industrial Engineering Technology
8. Mechanical Engineering Technology
9. Biomedical Engineering Technology

12. If water vapor is heated enough to break it down into oxygen and hydrogen, the fraction F of the total number of moles that does so can be found by solving the equation, $0.002087\left(\frac{1-F}{F}\right)^2 = \frac{2P}{F+2}$, where P is the pressure in atmospheres. If $P = 3$, the equation is $F^3 - 2874.94F^2 - 3F + 2 = 0$. Find F to the nearest hundredth, given that it is less than 1. 6 8 1 2 3 5 7 9

13. Over the first few half cycles a distorted sine wave of voltage v has a polynomial equation for its model given by the following equation: $v = 120t - 274t^2 + 225t^3 - 85t^4 + 15t^5 - t^6$. The accompanying figure shows the graph of this function for values of t between 0 and 2. Are there any positive roots greater than 2? 3 5 9 1 2 4 6 7 8

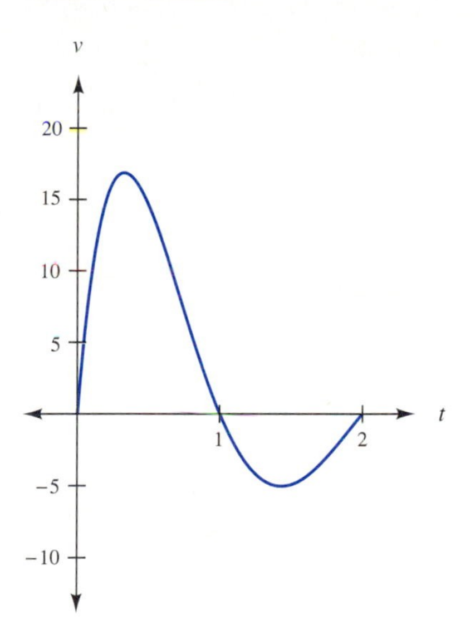

14. The volt ampere characteristic curve of a unijunction transistor is modeled by the equation $V = 0.002I^3 - 0.095I^2 + 1.15I$ when V is in volts and I in milliamps. For what values of I near 3 will V be 3 volts? Give your answer to the nearest tenth. 3 5 9 4 7 8

15. At a certain temperature a tunnel diode has a volt ampere characteristic curve modeled by the equation,

$$I = 100V^3 - 135V^2 + 45V + 10$$

with V in volts and I in ma. For what value of V near 0.05 will its current be equal to 12 ma? Give your answer to the nearest thousandth. 3 5 9 1 2 4 6 7 8

16. The moment of inertia of a rectangle of sides a and b about a diagonal is $I = \frac{a^3b^3}{6(a^2+b^2)}$. Assume $I = 2100$ when $a = 10$. For these values, the equation is $1{,}260{,}000 + 12{,}600b^2 - 1000b^3 = 0$. Find the value of b near 17. Give your answer to the nearest tenth. 1 2 4 8 7 9

17. The moment M induced in a beam of weight density D, depth h, and length l with parabolic sides b units thick when supported at its ends is given by $M = Dhb\left[\frac{5}{48}l^2 - \frac{x^2}{2} + \frac{x^4}{3l^2}\right]$ where x is the distance along the beam from its center. Assume $M = 1000$ ft·lb if $D = 500$ lb/ft^3, $h = 2$ ft, $b = 1$ ft, and $l = 20$ ft. For these values, the equation is $x^4 - 600x^2 + 48{,}800 = 0$. Find x to the nearest hundredth. 1 2 4 8 7 9

18. The equation of the curve assumed by the cable of a certain suspension bridge is given by $y = \frac{wx^2}{2T} + \frac{kx^4}{12T}$ where x is the horizontal distance from mid span, T the cable tension at mid span, w the weight per foot of bridge deck, and k a constant related to the shape of the deck. Assume $y = 50$ ft if $w = 5000$ lb/ft, $T = 10{,}000$ lb, and $k = 43{,}200$. For these values the equation is

$$50 = 0.250x^2 + 0.360x^4.$$

Find x to the nearest hundredth.
1 2 4 8 7 9

19. A cable of negligible weight if suspended from fixed points h vertical units and l horizontal units apart and supporting a distributed load which decreases uniformly from one support to the other will assume the shape of the curve $y = \frac{3h}{2}\left(\frac{x}{l}\right)^2\left(1 - \frac{x}{3l}\right)$. If $y = 10$ ft, $h = 20$ ft, and $l = 100$ ft, the equation is $x^3 - 300x^2 + 1{,}000{,}000 = 0$. Find x to the nearest tenth given that x is between 65 and 66.
1 2 4 8 7 9

RANDOM REVIEW

1. Given that $x = -4$ is a solution, find the remaining solutions of $2x^3 + 9x^2 + x - 12 = 0$.
2. Use synthetic division, if possible to find the quotient and remainder for the division $(x^3 + 2x^2 - x + 6) \div (x + 3)$.
3. Determine the maximum number of positive roots in the equation $5x^4 - x^3 + 2x^2 + x - 3 = 0$.
4. Find all rational roots of the equation $3x^3 + x^2 - 11x + 6 = 0$.
5. Determine whether $3x - 2$ is a factor of $3x^4 - 2x^3 + 3x^2 - 5x + 2$.
6. Determine the maximum number of negative roots in the equation $5x^4 - x^3 + 2x^2 + x - 3 = 0$.
7. If $f(x) = 2x^5 - 3x^4 + 5$, use the remainder theorem to find $f(-1)$.
8. Find all real roots of the equation $2x^3 + 7x^2 + x - 6 = 0$.
9. If $4x^4 - 3x^3 + x^2 - 4 = 0$, use the rational roots theorem to compile a list of all possible rational roots.
10. Given that $x = -2$ is a solution, find the remaining solutions of $x^3 - 3x^2 - 7x + 6 = 0$.

CHAPTER TEST

1. Use synthetic division, if possible, to find the quotient and remainder for the division $(x^6 - 6x^3 + 9) \div (x^3 - 3)$.
2. Determine whether $3x - 2$ is a factor of $3x^4 - 2x^3 + 3x^2 - 5x + 1$.
3. If $g(x) = x^4 - 3x^3 + 3x^2 - 5x + 1$, use the remainder theorem to find $g(2)$.
4. Given that $x = 1$ and $x = -1$ are solutions, find the remaining solutions of $x^4 + 3x^3 - 3x^2 - 3x + 2 = 0$.
5. Determine the maximum number of positive roots in the equation $7x^3 - 2x^2 + x - 1 = 0$.
6. Determine the maximum number of negative roots in the equation $7x^3 - 2x^2 + x - 1 = 0$.
7. Find all rational roots of the equation $6x^3 - x^2 - 11x + 6 = 0$.
8. Find all real roots of the equation $x^5 - 2x^4 + x^3 - x^2 + 2x - 1 = 0$.
9. An electronic circuit produces a voltage that increases as a function of time. If $v(t) = 2t^3 + 5t + 1$, find the time at which $v(t) = 12$ volts.
10. The moment of inertia I of a rectangular beam is given by $12I = bh^3$ where b and h are the dimensions of the cross section of the beam. If h is increased by 1, the equation is $12I = b(h^3 + 3h^2 + 3h + 1)$. Solve this equation for h, given $I = 16$ and $b = 3$.

INEQUALITIES CHAPTER 8

The problems that arise in science and technology do not always have a single correct solution. There may be a range of answers that are acceptable. For example, one might calculate the focusing range to be between 10 and 13 inches for a camera fitted with a certain type of close-up lens. The purpose of this chapter is to present the methods used to solve the types of inequalities that are common in technical fields.

8.1 LINEAR INEQUALITIES

A statement that two quantities are not equal is called an inequality.

INEQUALITY SYMBOLS

$\neq$ means "is not equal to"
$>$ means "is greater than"
$<$ means "is less than"
$\geq$ means "is greater than or equal to"
$\leq$ means "is less than or equal to"

NOTE ⇨⇨ ***Whenever two numbers are compared, the inequality symbol points toward the smaller number.*** Thus we can say $2 < 3$ or $3 > 2$. Both inequalities express the same relationship between 2 and 3. An inequality is said to be **linear** if the largest exponent of a variable is 1, and there are no fractional or negative exponents of variables. For the present, we will study only linear inequalities in one variable.

linear inequality

A number line may be used to illustrate an inequality. The statement $x > 3$ specifies all values that are to the right of 3. These values are illustrated in Figure 8.1. The open circle at 3 indicates that 3 does not satisfy the inequality.

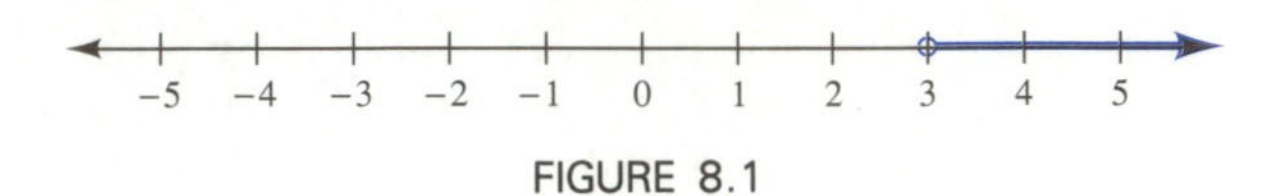

FIGURE 8.1

The inequality $x \geq 3$ is illustrated in Figure 8.2. The closed circle at 3 indicates that 3 does satisfy the inequality.

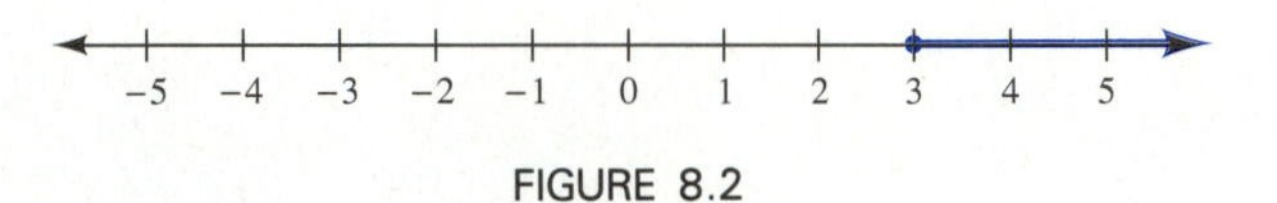

FIGURE 8.2

These remarks are generalized as follows:

To graph an inequality in x on the number line:

1. Write the inequality in the form

$$x > a,\ x < a,\ x \geq a, \text{ or } x \leq a.$$

2. Locate and mark the number a on the number line.
   (a) Use an open circle (○) at a for an inequality with $<$ or $>$.
   (b) Use a closed circle (•) at a for an inequality with $\leq$ or $\geq$.
3. Draw an arrow to indicate the solutions.
   (a) Use an arrow to the right for an inequality with $>$ or $\geq$.
   (b) Use an arrow to the left for an inequality with $<$ or $\leq$.

EXAMPLE 1 Graph $4 \geq x$ on a number line.

Solution

1. Write the inequality in the form $x \leq 4$. (The variable is on the left, but the inequality symbol still points toward x.)
2. The graph has a closed circle at 4.
3. The graph has an arrow to the left, as shown in Figure 8.3. ■

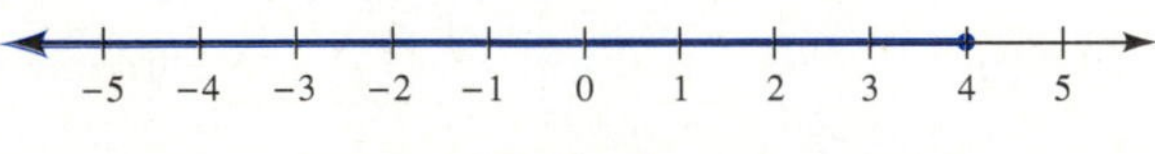

FIGURE 8.3

EXAMPLE 2 Graph $y > 7/5$ on a number line.

Solution The graph has an open circle at 7/5 or 1 2/5 and an arrow to the right, as shown in Figure 8.4. ■

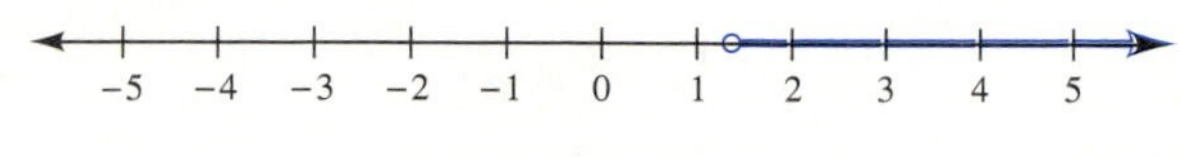

FIGURE 8.4

To solve an inequality means to find the values of the variable that make the inequality a true statement. Solving a linear inequality is almost like solving a linear equation. ***There is one additional rule: If you multiply or divide both sides of an inequality by a negative number, you must reverse the direction of the inequality symbol.*** For example, Figure 8.5 shows that $2 < 3$, but $-2 > -3$ or that $-3 < 2$, but $3 > -2$.

NOTE ⇨⇨

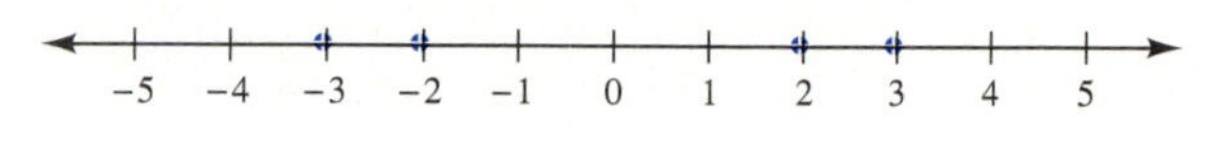

FIGURE 8.5

To solve a linear inequality in one variable:

1. Simplify each side.
   (a) Use the distributive property, if necessary.
   (b) Combine like terms.
2. Remove extra terms by using inverse operations. (Add or subtract the same number on both sides of the inequality.)
3. Remove extra factors or divisors by using inverse operations. (Multiply or divide both sides of the inequality by the same nonzero number.)
   (a) If you multiply or divide both sides by a *positive* number, the direction of the inequality symbol remains unchanged.
   (b) If you multiply or divide both sides by a *negative* number, the direction of the inequality symbol is reversed.

EXAMPLE 3 Solve $3m + 1 \geq 2$.

Solution

1. $3m + 1 \geq 2$ — Notice that each side is in simplest form.
2. $3m \geq 1$ — Subtract 1 from both sides.
3. $3m/3 \geq 1/3$ — Divide both sides by 3.
   $m \geq 1/3$.

We did not multiply or divide by a *negative* number, so the inequality continued to "point" in the same direction throughout the problem. ■

EXAMPLE 4 Solve $-3p + 1 < 2$.

Solution

$$-3p + 1 < 2 \quad \text{Notice that each side is in simplest form.}$$
$$-3p < 1 \quad \text{Subtract 1 from both sides.}$$
$$-3p/(-3) > 1/(-3) \quad \text{Divide both sides by } -3 \text{ and change } < \text{ to } >.$$
$$p > -1/3$$ ■

EXAMPLE 5 Solve $-4x + 3 > 2x + 1$.

Solution

$$-4x + 3 > 2x + 1 \quad \text{Notice that each side is in simplest form.}$$
$$-6x + 3 > 1 \quad \text{Subtract } 2x \text{ from both sides.}$$
$$-6x > -2 \quad \text{Subtract 3 from both sides.}$$
$$-6x/-6 < -2/-6 \quad \text{Divide both sides by } -6 \text{ and change } > \text{ to } <.$$
$$x < 1/3$$ ■

EXAMPLE 6 FHA regulations require the window area of a house to be at least 10 percent of the floor area. If a house has 10 windows, each measuring 0.838 m × 1.83 m, what can be said about the floor area?

Solution Let A = floor area. Each window has an area of (0.838)(1.83). Total window area is 10(0.838)(1.83). If the window area is at least 10 percent of the floor area, then it is 10 percent or more of the floor area.

$$10(0.838)(1.83) \geq 0.10\,A$$
$$\frac{10(0.838)(1.83)}{0.10} \geq A$$
$$153.354 \geq A \qquad \text{or} \qquad A \leq 153.354$$

The floor area, then, is 153 square meters or less. ■

It is sometimes helpful to illustrate the solution of an inequality on a number line.

EXAMPLE 7 Graph the solution of $1 - 3x < 2(x - 2)$.

Solution

$$1 - 3x < 2(x - 2)$$
$$1 - 3x < 2x - 4 \quad \text{Use the distributive property.}$$
$$1 - 5x < -4 \quad \text{Subtract } 2x \text{ from both sides.}$$
$$-5x < -5 \quad \text{Subtract 1 from both sides.}$$
$$-5x/-5 > -5/-5 \quad \text{Divide both sides by } -5.$$
$$x > 1$$

The graph has an open circle at 1 and an arrow to the right, as shown in Figure 8.6. ■

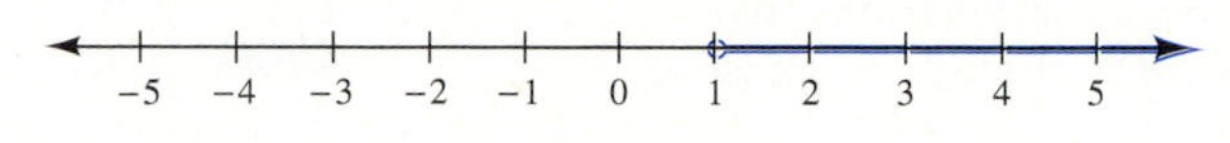

FIGURE 8.6

EXERCISES 8.1

In 1–14, graph each inequality on a number line.

1. $x > 1$
2. $x > -2$
3. $x \geq -3$
4. $x \geq 5$
5. $4 \geq x$
6. $-1 \geq x$
7. $2/3 > x$
8. $-3/4 > x$
9. $x < 2$
10. $x > -1/2$
11. $x \geq 1/2$
12. $x \leq -4$
13. $1/3 \leq x$
14. $-1/2 \leq x$

In 15–28, solve each inequality.

15. $3x - 5 > 7$
16. $2y - 4 \leq 3$
17. $z - 6 \geq 3$
18. $4x - 7 < 2$
19. $5x + 3 \leq 1$
20. $\frac{m}{2} - 5 < 2$
21. $\frac{2p}{3} - 3 > 1$
22. $\frac{3r}{4} + 2 \geq r$
23. $\frac{2s}{5} + 9 > 3$
24. $3x - 1 \geq x + 5$
25. $2y + 8 \leq 7y + 6$
26. $2(z + 3) \leq 3z - 1$
27. $m + 2 > 4(m + 1)$
28. $3(1 - x) < 5x + 2$

In 29–42, solve each inequality and graph the solution on a number line.

29. $3x + 6 \leq 8$
30. $y + 4 > 2$
31. $2z + 5 < 4$
32. $4x - 8 \geq 3$
33. $4x - 3 > 0$
34. $(2/3)m + 5 \geq 3$
35. $p/2 + 3 \leq 1$
36. $r/4 - 1 < 2r$
37. $(3/5)s + 7 \leq 4$
38. $2x - 2 < 3x - 4$
39. $3y - 7 > 6y + 7$
40. $3(z + 2) > 2z + 1$
41. $2m - 1 \leq 3(m - 2)$
42. $4(1 - 2x) \geq 3x - 1$

---

43. In 1970, the U.S. coal reserves were estimated at 5×10^{12} tons. If coal was consumed at the rate of 2.2×10^{9} tons per year between 1970 and 1990, what is the average amount that could be consumed each year if the reserves are to last at least 2000 additional years? 6 8 1 2 3 5 7 9

44. A child's dose of medicine is sometimes calculated by the formula $c = \left(\frac{A + 1}{24}\right) d$ where d is the adult dose, c is the child's dose, and A is the child's age in years. For what ages is the child's dose less than half of the adult dose? 6 9 2 4 5 7 8

---

1. Architectural Technology
2. Civil Engineering Technology
3. Computer Engineering Technology
4. Mechanical Drafting Technology
5. Electrical/Electronics Engineering Technology
6. Chemical Engineering Technology
7. Industrial Engineering Technology
8. Mechanical Engineering Technology
9. Biomedical Engineering Technology

45. The amount of time required for an average driver to react to danger and apply the brakes is 0.6 seconds. At what speeds (in mph) would a car travel less than 20 feet in 0.6 seconds?
4 7 8 | 1 2 3 5 6 9

46. For a simple beam under a concentrated load, the shearing load (in kg) is given by $s = P(l - a)/l$ where P is the load (in kg), l is the length of the beam (in mm), and a is the distance (in mm) of the load from a support. A load of 2 kg is to be placed 30 mm from a support. What length beams will have shearing loads of no more than 0.5 kg? 1 2 4 | 8 7 9

47. A 240 V appliance draws a current of 15 A. The energy (in kilowatt hours) is given by $W = 0.001IVt$ where I is the current (in amperes), V is the potential difference (in volts), and t is the time (in hours). How long can the appliance run without using more than 2.7 kwh of energy?
3 5 9 | 1 2 4 6 7 8

48. The area of a trapezoid is given by $A = (1/2)h(B + b)$ where B and b are the lengths of the bases. For a trapezoid in which the smaller base is 4 in long and the longer is 8 in long, what values of h would allow the area to be larger than the area of a square with sides of length 8 in?
1 2 4 7 8 | 3 5

49. If Problem 48 is generalized, so that $b = (1/2)B$, find a formula for h such that the area of the trapezoid is larger than the area of a square with sides of length B.

---

1. Architectural Technology
2. Civil Engineering Technology
3. Computer Engineering Technology
4. Mechanical Drafting Technology
5. Electrical/Electronics Engineering Technology
6. Chemical Engineering Technology
7. Industrial Engineering Technology
8. Mechanical Engineering Technology
9. Biomedical Engineering Technology

8.2 COMPOUND INEQUALITIES

compound inequality

When two or more inequalities are joined by the word **AND** or the word **OR**, the relation is called a **compound inequality.** The solution of a compound inequality with the word **AND** consists of those values that satisfy both inequalities. The solution of a compound inequality with the word **OR** consists of those values that satisfy one inequality or the other, or both.

> **To graph the solutions of a compound inequality:**
>
> **1.** Graph each separate inequality on the number line.
> **2. (a)** If the inequality is an **AND** inequality, locate and mark those values that are in *both* individual solutions.
> **(b)** If the inequality is an **OR** inequality, locate and mark those values that are in *at least one* individual solution.

EXAMPLE 1 Graph the solution of $x > -2$ and $x \leq 5$.

Solution **1.** The graph of $x > -2$ has an open circle at -2 and an arrow to the right. The graph of $x \leq 5$ has a closed circle at 5 and an arrow to the left. These graphs are shown just above the number line in Figure 8.7.

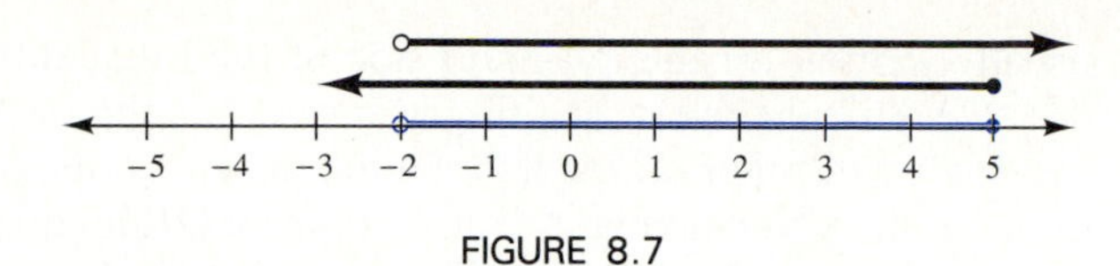

FIGURE 8.7

2. If the arrows were drawn on the number line itself, they would overlap between -2 and 5. The point -2 is only on one graph, but 5 is on both. Thus, -2 is not included in the solution of the compound inequality, but 5 is. The solution of the compound inequality is shown on the number line. ■

EXAMPLE 2 Graph the solution of $x \leq -1$ or $x > 3$.

Solution 1. The graph of $x \leq -1$ has a closed circle at -1 and an arrow to the left. The graph of $x > 3$ has an open circle at 3 and an arrow to the right. These graphs are shown just above the number line in Figure 8.8.

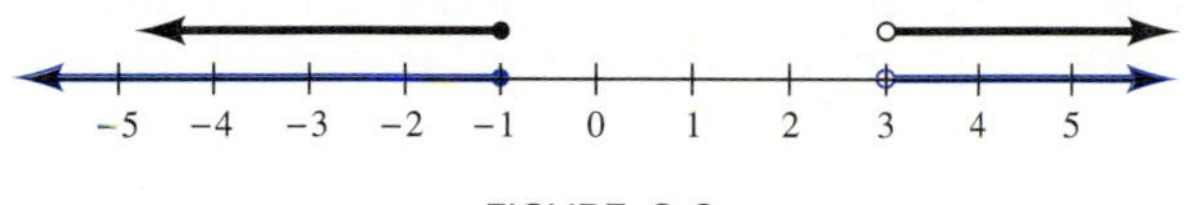

FIGURE 8.8

2. The solution of the compound inequality contains all of the points that are on either graph. The solution is shown on the number line. It must be drawn in two parts. ■

EXAMPLE 3 Graph the solution of $x > -3$ or $x < 2$.

Solution 1. The graph of $x > -3$ has an open circle at -3 and an arrow to the right. The graph of $x < 2$ has an open circle at 2 and an arrow to the left. These graphs are shown just above the number line in Figure 8.9.

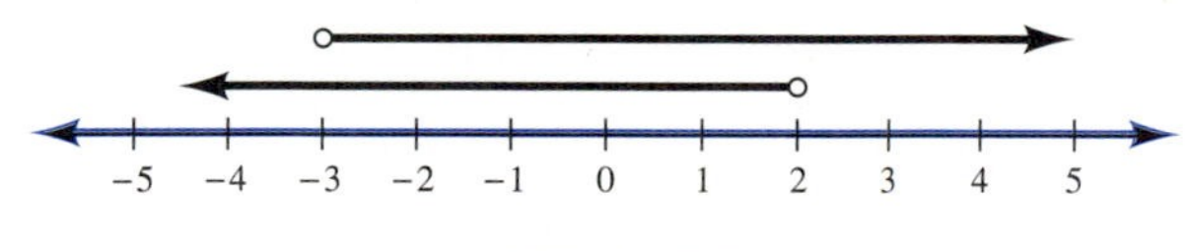

FIGURE 8.9

2. To be included in the solution of an **OR** inequality, a point must appear on at least one of the individual graphs. The solution of the compound inequality is the entire number line. Notice that -3 is included because it is on the graph of $x < 2$, and 2 is included because it is on the graph of $x > -3$. ■

For **AND** inequalities, it is possible to combine two inequalities into a single statement. The statement $a < x < b$ means $a < x$ and $x < b$. Or, since the variable is usually written on the left-hand side of the inequality $x > a$ and $x < b$. Although $b > x > a$ means the same thing as $a < x < b$, the inequality is usually written with the smaller number on the left. An inequality in one of the combined forms *always* indicates an **AND** inequality, rather than an **OR** inequality. ***A combined inequality is* never *written with the inequality symbols pointing in opposite directions, such as* $a < x > b$ or $a > x < b$.**

NOTE ⇨⇨

COMPOUND INEQUALITIES

$a < x < b$ means $x > a$ and $x < b$

$a \leq x < b$ means $x \geq a$ and $x < b$

$a < x \leq b$ means $x > a$ and $x \leq b$

$a \leq x \leq b$ means $x \geq a$ and $x \leq b$

EXAMPLE 4 Graph $-1 \leq x < 2$.

Solution The inequality $-1 \leq x < 2$ means $x \geq -1$ and $x < 2$. The individual graphs are shown above the number line in Figure 8.10. The solution of the compound inequality is shown on the number line. Such an inequality is sometimes read "x is between -1 and 2," but this phrase does not convey the fact that -1 is included in the solution while 2 is not. ***When the phrase "x is between" two values is used, further information about the end point should also be given.*** ■

NOTE ⇨⇨

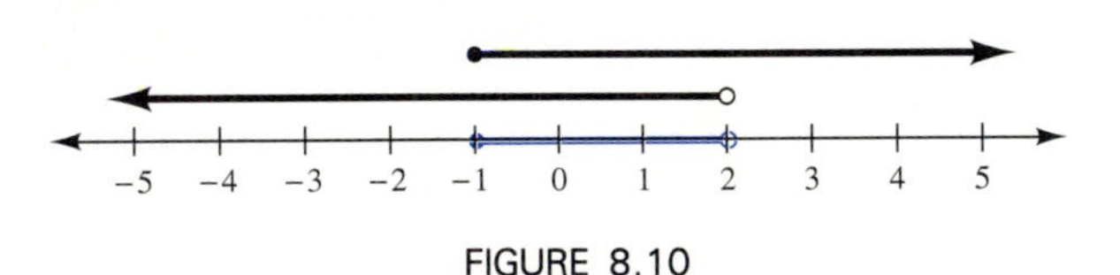

FIGURE 8.10

EXAMPLE 5 The equivalent resistance R of two resistors in parallel is given by $R = \dfrac{R_1 R_1}{R_1 + R_2}$, where R_1 and R_2 are the individual resistances. If $R_1 = 50\ \Omega$, what resistance is needed to reduce the equivalent resistance to between (but not equal to) $10\ \Omega$ and $20\ \Omega$?

Solution The problem is to find R_2 such that $10 < R < 20$.

$$R > 10 \qquad \text{and} \qquad R < 20$$

$$\frac{50\ R_2}{50 + R_2} > 10 \qquad\qquad \frac{50\ R_2}{50 + R_2} < 20$$

$$50\ R_2 > 10(50 + R_2) \qquad\qquad 50\ R_2 < 20(50 + R_2)$$

$$50\ R_2 > 500 + 10\ R_2 \qquad\qquad 50\ R_2 < 1000 + 20\ R_2$$

$$40\ R_2 > 500 \qquad\qquad 30\ R_2 < 1000$$

$$R_2 > 12.5 \qquad\qquad R_2 < 33.3$$

Figure 8.11 shows that a resistor that satisfies the inequality $12.5\ \Omega < R_2 < 33.3\ \Omega$ would lower the equivalent resistance to the desired range. ■

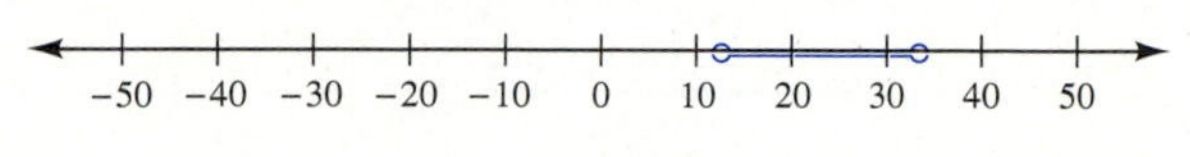

FIGURE 8.11

EXAMPLE 6 For each graph in Figure 8.12 and 8.13, write a compound inequality that specifies the values shown.

(a)

FIGURE 8.12

(b)

FIGURE 8.13

Solution **(a)** Since values between -2 and 0 are indicated, the solution may be written as $-2 < x \leq 0$. Notice that 0 is included, but -2 is not.

(b) Since the values are indicated on two separate sections of the number line, it takes two separate inequalities to specify the solution: $x < -2$ or $x \geq 0$. ■

EXERCISES 8.2

In 1–20, graph the solution of each compound inequality.

1. $x > -1$ and $x < 4$

2. $x > -2$ and $x \leq 3$

3. $x \leq -3$ or $x > 2$

4. $x \leq 0$ or $x \geq 1$

5. $x \leq 1$ or $x \geq 5$

6. $x \leq 2$ or $x > 4$

7. $x > -4$ and $x \leq 2$

8. $x > -3$ and $x < -1$

9. $x \geq -3$ and $x < 2$

10. $x \geq 0$ and $x \leq 1$

11. $x > -4$ or $x \leq -2$

12. $x > -1$ or $x < 4$

13. $x > -2$ or $x \leq 3$

14. $x \geq 1$ and $x \leq 5$

15. $2 < x < 5$

16. $1 < x \leq 4$

17. $-3 \leq x < 3$

18. $-2 \leq x \leq 2$

19. $-3 \leq x \leq -2$

20. $-4 \leq x < -1$

In 21–32, write a compound inequality that specifies the values shown for each graph.

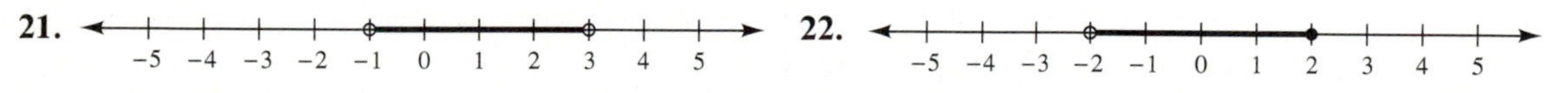

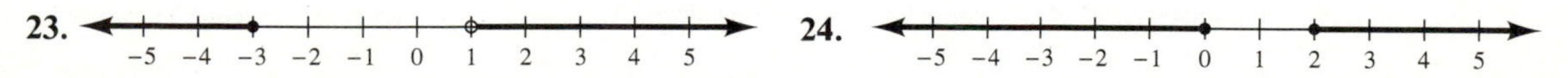

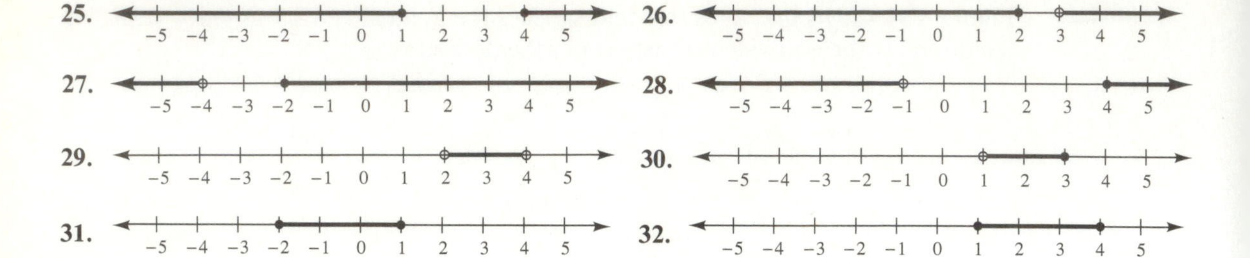

In 33–39, assume that "x is between a and b" means $a < x < b$.

33. The force F (in N) required to stretch a spring a distance of x (in cm) is given by $F = kx$ where k is a constant that depends on the spring. If $k = 20$, how far will the spring stretch if F is between 30 N and 40 N? **4 7 8** 1 2 3 5 6 9

34. A very young child's dose of medicine is sometimes calculated by the formula $c = (A/150)d$, where d is the adult dose, c is the child's dose, and A is the child's age in months. For what ages is the child's dose between 0.05 and 0.10 of an adult's dose? **6 9** 2 4 5 7 8

35. The formula $L = (l/k)(T - t)$ gives the length L (in m) of expansion of a highway at temperature T (in degrees Celsius). In the formula, l is the length (in m) before expansion, t is the temperature (in degrees Celsius) at which the highway was built, and k is a constant that depends on the construction of the highway. At what temperature will a 1200 m section of highway built at 20° C expand between 0.1 and 0.2 m in length? Use $k =$ 150,000. **1 2 4 6 8** 5 7 9

36. For a simple beam under a distributed load, the shearing load (in kg) is given by $s = w(l/2 - x)$. In the formula, w is the load (in kg/mm), l is the length (in mm) of the beam, and x is the distance (in mm) from a support at which the shearing load is measured. For a beam of length 40 mm, with a load of 0.05 kg/mm, at what distances x will the shearing load be between 0.5 and 0.8 kg? **1 2 4** 8 7 9

37. The formula $F = (9/5)C + 32$ gives the relationship between Fahrenheit temperature F and Celsius temperature C. For what Fahrenheit temperatures is the Celsius temperature different numerically by more than 10? (Hint: F can be either at least 10 more or at least 10 less than C.) **6 8** 1 2 3 5 7 9

38. The equivalent capacitance of two capacitors in series is given by $C = \dfrac{C_1C_2}{C_1 + C_2}$. If $C_1 =$ 3.00 μF, what capacitance is needed to reduce the equivalent capacitance to between 1.00 μF and 2.00 μF? **3 5 9** 4 7 8

39. For a telescope, the magnification is given by $m = f_1/f_2$ where f_1 and f_2 are the focal lengths of the objective and eyepiece, respectively. What focal lengths could be used (for the eyepiece) with an objective of focal length 1200 mm to produce a magnification of between 40 and 50? 3 5 6 9

40. To determine whether two values a and b are within an amount x of each other, if it does not matter which is larger, it is possible to have either $a - b < x$ or $b - a < x$. Express this relationship using an absolute value.

8.3 NONLINEAR INEQUALITIES

In previous chapters you have learned how to solve equations involving rational expressions, quadratic polynomials, and higher degree polynomials. To solve an inequality involving one of these types of nonlinear expressions, one method is to consider a graph. We illustrate the method in Example 1.

EXAMPLE 1 Solve $x^2 - 2x - 3 \geq 0$.

Solution Let $y = x^2 - 2x - 3$. Our goal is to find values for x that make y greater than or equal to zero. Figure 8.14 shows the graph of $y = x^2 - 2x - 3$. The y-values are greater than zero for those points that lie above the x-axis. The y-values are less than zero for those points that lie below the x-axis. The y-value is zero for those points that lie on the x-axis.

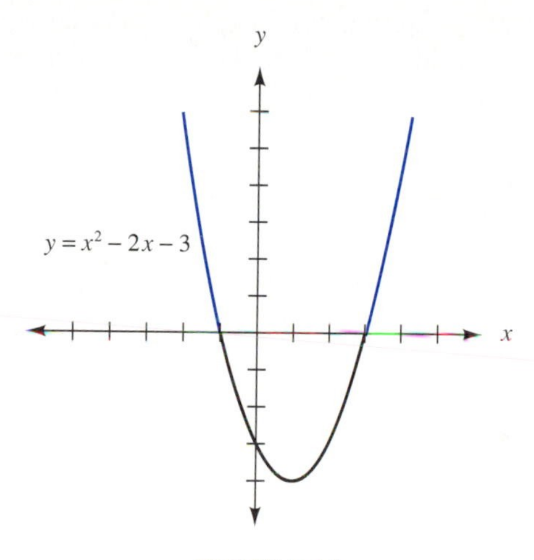

FIGURE 8.14

Notice that a section of the graph in which y is positive is separated from a section in which y is negative by a point for which y is zero. We need to find the points for which y is zero. The equation $0 = x^2 - 2x - 3$ can be solved as follows.

$$0 = x^2 - 2x - 3$$
$$0 = (x - 3)(x + 1)$$
$$0 = x - 3 \quad \text{or} \quad 0 = x + 1$$
$$x = 3 \quad \text{or} \quad x = -1.$$

From the graph, then, we know that $y \geq 0$ is true when $x \leq -1$ or that $y \geq 0$ is true when $x \geq 3$. ■

Some pocket calculators now have the capability to graph the types of functions we are considering. If a graph is readily available, an inequality may be solved by the preceding method. When a graph is not readily available, we need not take the time to graph the function. Instead, we examine only the x-axis, treating it as a number line. A section of the line in which the function takes on positive values is separated from a section in which the function takes on negative values by a point where the function takes on a value of zero.

To solve an inequality involving a rational expression, a quadratic polynomial, or a polynomial of higher degree:

1. Replace the inequality symbol with an equal sign and solve.
2. Locate and mark on a number line the solutions obtained in step 1.
   (a) Use a closed circle if the original inequality contains $\geq$ or $\leq$.
   (b) Use an open circle if the original inequality contains $>$ or $<$.
3. Locate and mark on the number line any value that would produce a denominator of zero in the inequality. Use an open circle. (If there is a closed circle already there, replace it with an open one.)
4. Notice that the n points located in steps 2 and 3 divide the number line into $n+1$ sections. For each section, choose any point within the section and substitute the value into the original inequality.
   (a) If the statement is true, all values in that section satisfy the inequality.
   (b) If the statement is false, no value in that section satisfies the inequality.

EXAMPLE 2 Solve $x^2 - x - 12 > 0$.

Solution

1. The equation $x^2 - x - 12 = 0$ can be solved by factoring.

$$x^2 - x - 12 = 0$$
$$(x-4)(x+3) = 0$$
$$x - 4 = 0 \quad \text{or} \quad x + 3 = 0$$
$$x = 4 \quad \text{or} \quad x = -3$$

2. The values 4 and -3 are marked on a number line. Since the original problem was a $>$ inequality, open circles are used, as shown in Figure 8.15.

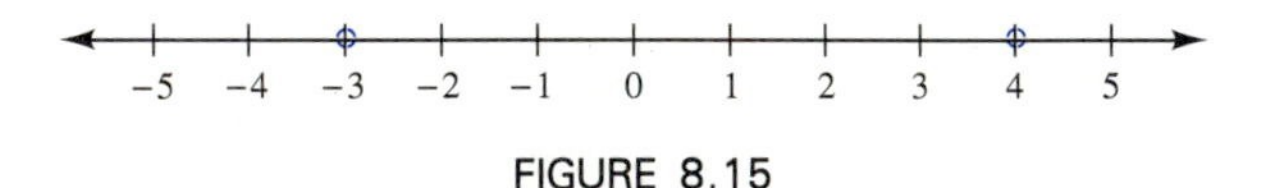

FIGURE 8.15

3. There are no values for which the expression $x^2 - x - 12$ has a denominator of zero.

4. The points at -3 and 4 divide the number line into three sections. From the first section, which represents $x < -3$, we choose -4 for x. Replacing x with -4 in

the original inequality gives

$$(-4)^2 - (-4) - 12 > 0$$
$$16 + 4 - 12 > 0.$$

Since the statement is true, $x < -3$ is part of the solution.

From the second section, which represents $-3 < x < 4$, we choose 0 for x. (Arithmetic is easy with 0.) The inequality gives

$$0^2 - 0 - 12 > 0$$
$$0 - 0 - 12 > 0.$$

Since the statement is false, $-3 < x < 4$ is *not* part of the solution.

From the third section, which represents $x > 4$, we choose 5 for x. The inequality gives

$$5^2 - 5 - 12 > 0$$
$$25 - 5 - 12 > 0.$$

Since the statement is true, $x > 4$ is part of the solution.

The solution of $x^2 - x - 12 > 0$ is shown in Figure 8.16.

The solution is written as $x < -3$ or $x > 4$. ■

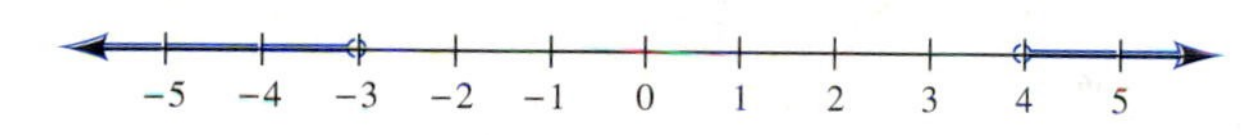

FIGURE 8.16

EXAMPLE 3 Solve $2x^2 + 3x - 2 \le 0$.

Solution 1. The equation $2x^2 + 3x - 2 = 0$ can be solved by factoring.

$$(2x - 1)(x + 2) = 0$$
$$2x - 1 = 0 \quad \text{or} \quad x + 2 = 0$$
$$x = 1/2 \quad \text{or} \quad x = -2$$

2. The values 1/2 and -2 are marked on a number line. Since the original problem was a $\le$ inequality, closed circles are used, as shown in Figure 8.17.

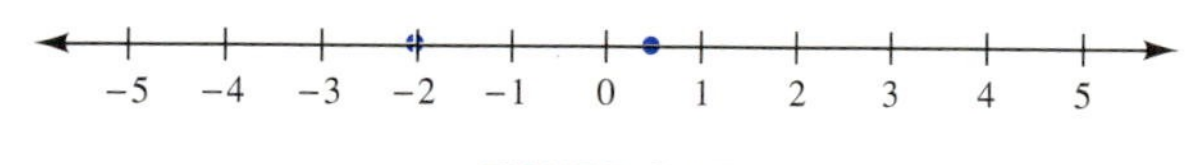

FIGURE 8.17

3. There are no values for which the expression $2x^2 + 3x - 2$ has a denominator of zero.

4. The points at 1/2 and -2 divide the number line into three sections. From the first section, $x \le -2$, we choose -3 for x. Replacing x with -3 in the original inequality gives

$$2(-3)^2 + 3(-3) - 2 \le 0$$
$$18 - 9 - 2 \le 0, \text{ which is false.}$$

From the second section, $-2 \leq x \leq 1/2$, we choose 0 for x. The inequality gives

$$2(0^2) + 3(0) - 2 \leq 0$$
$$0 + 0 - 2 \leq 0, \text{ which is true.}$$

Thus, $-2 \leq x \leq 1/2$ is part of the solution.
From the third section, $x \geq 1/2$, we choose 1 for x. The inequality gives

$$2(1^2) + 3(1) - 2 \leq 0$$
$$2 + 3 - 2 \leq 0\text{, which is false.}$$

The solution of $2x^2 + 3x - 2 \leq 0$ is shown in Figure 8.18.

The solution is written $-2 \leq x \leq 1/2$. ■

FIGURE 8.18

EXAMPLE 4 A bullet is fired with a horizontal velocity of 300 m/s. The force of gravity (9.8 m/s^2) acting on it will cause it to hit the target at a distance of $s = (1/2)gt^2$ below the point at which it is aimed, as shown in Figure 8.19. From what distance can the bullet be fired and still hit within 0.1 m of the target?

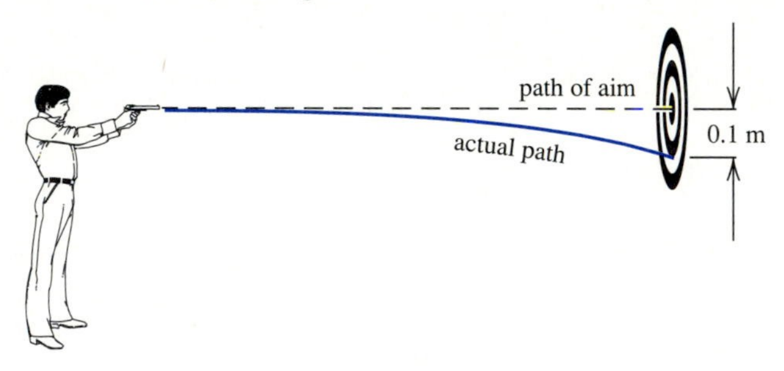

FIGURE 8.19

Solution The bullet strikes a point $s = (1/2)(9.8)t^2$ or $4.9t^2$ below the target. Thus,

$$0 \leq 4.9t^2 \leq 0.1\text{, or}$$

$4.9t^2 \geq 0$ and $4.9t^2 \leq 0.1$.

This statement is true for all values of t, since a square is always ≥ 0.

Solve $4.9t^2 = 0.1$

$$t = \pm\sqrt{0.1/4.9}$$
$$t = \pm 0.14 \text{ s}$$

We begin measuring time when the bullet is fired ($t = 0$), so negative values of t do not make sense in the context of the problem. We show 0, the smallest possible value, and 0.14 on the graph in Figure 8.20. We have already ruled out negative values of t. When t is 0.1, the inequality $4.9t^2 \leq 0.1$ is true. When t is 1, the inequality $4.9t^2 \leq 0.1$ is false. Thus, $0 \leq t \leq 0.14$.

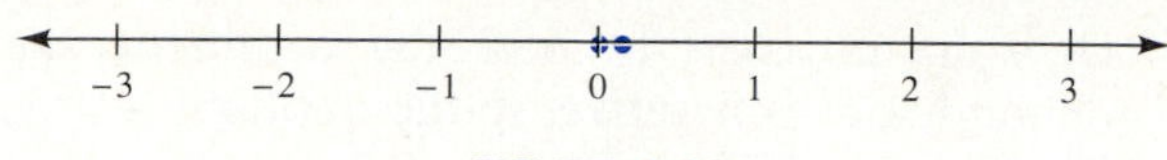

FIGURE 8.20

Since v is 300 m/s, in 0.14 s the bullet can travel a distance of (0.14)(300) m or 42 m. As long as the bullet is fired from a distance of 42 m or less, it will hit within 0.1 m of the target. ■

EXAMPLE 5 Solve $\dfrac{x^2-2x+1}{x-2} \leq 0$.

Solution 1. The equation $\dfrac{x^2-2x+1}{x-2} = 0$ is undefined when $x-2=0$. But if $x-2 \neq 0$, both sides of the equation can be multiplied by $x-2$.

$$x^2-2x+1=0$$
$$(x-1)^2=0$$
$$x-1=0$$
$$x=1$$

2. The value 1 is marked on the number line. Since the original inequality was a $\leq$ inequality, a closed circle is used.

3. Also, since the equation is undefined when $x-2=0$ (or x is 2), the point at 2 is marked on the number line in Figure 8.21 with an open circle.

FIGURE 8.21

4. The points at 1 and 2 divide the number line into three sections. From the first section, $x \leq 1$, we choose 0 for x. The inequality gives

$$\frac{0^2-2(0)+1}{0-2} \leq 0$$

$$1/-2 \leq 0, \text{ which is true.}$$

Thus, $x \leq 1$ is part of the solution.

From the second section, $1 \leq x < 2$, we choose 1 1/2 (or 3/2) for x. The inequality gives

$$\frac{(3/2)^2-2(3/2)+1}{(3/2)-2} \leq 0$$

$$\frac{9/4-3+1}{-1/2} \leq 0$$

$$\frac{9/4-12/4+4/4}{-1/2} \leq 0, \text{ which is true.}$$

(It is not necessary to finish the arithmetic; the numerator is positive and the denominator is negative, so the quotient is negative and the statement is true.) Thus, $1 \le x < 2$ is part of the solution.

From the third section, $x > 2$, we choose 3 for x. The inequality gives

$$\frac{3^2 - 2(3) + 1}{3 - 2} \le 0$$

$$9 - 6 + 1 \le 0, \text{ which is false.}$$

The solution of $\dfrac{x^2 - 2x + 1}{x - 2} \le 0$ is shown in Figure 8.22.

The solution is written $x < 2$. ■

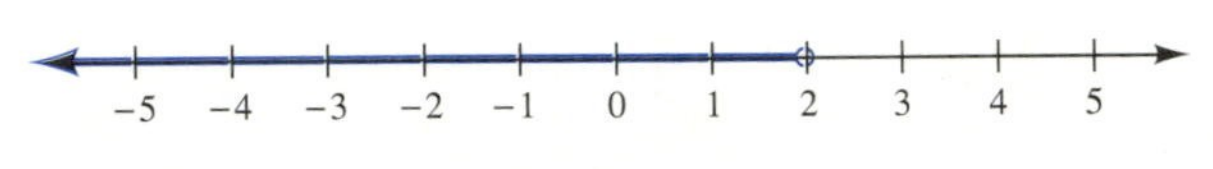

FIGURE 8.22

Because the solutions often consist of alternating sections as in Examples 2 and 3, students may mistakenly conclude that the solution of an inequality always consists of alternating sections. ***Example 6 illustrates the importance of testing a point in* each *section of the number line.***

NOTE ⇨⇨

EXAMPLE 6 Solve $x^4 - 5x^2 + 4 > 0$.

Solution

1. The equation $x^4 - 5x^2 + 4 = 0$ is in quadratic form and can be factored.

$$x^4 - 5x^2 + 4 = 0$$

$$(x^2 - 4)(x^2 - 1) = 0$$

$$(x - 2)(x + 2)(x - 1)(x + 1) = 0$$

$$x - 2 = 0,\ x + 2 = 0,\ x - 1 = 0, \text{ or } x + 1 = 0$$

$$x = 2,\ x = -2,\ x = 1, \text{ or } x = -1.$$

2. These values are marked on the number line in Figure 8.23.

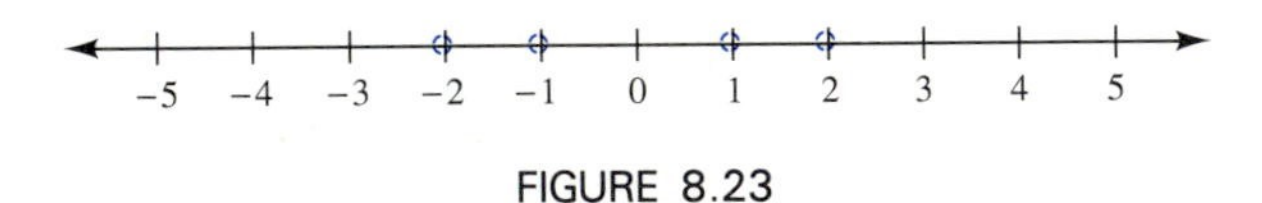

FIGURE 8.23

3. There are five sections.

Section 1: Choose $x = -3$: $\quad (-3)^4 - 5(-3)^2 + 4 > 0$

$81 - 45 + 4 > 0$ is true.

Thus, $x < -2$.

Section 2: Choose $x = -3/2$: $(-3/2)^4 - 5(-3/2)^2 + 4 > 0$

$81/16 - 45/4 + 4 > 0$

$81/16 - 180/16 + 64/16 > 0$ is false.

Section 3: Choose $x = 0$: $0^4 - 5(0)^2 + 4 > 0$

$0 - 0 + 4 > 0$ is true.

Thus, $-1 < x < 1$.

Section 4: Choose $x = 3/2$: $(3/2)^4 - 5(3/2)^2 + 4 > 0$

$81/16 - 45/4 + 4 > 0$

$81/16 - 180/16 + 64/16 > 0$ is false.

Section 5: Choose $x = 3$: $3^4 - 5(3)^2 + 4 > 0$

$81 - 45 + 4 > 0$ is true.

Thus, $x > 2$.

The solution of $x^4 - 5x^2 + 4 > 0$ is shown in Figure 8.24.

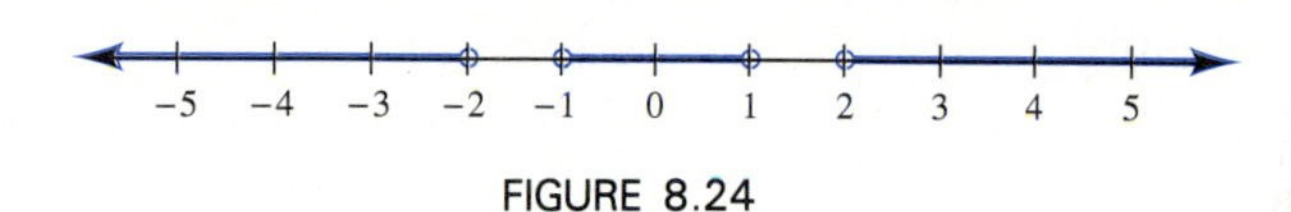

FIGURE 8.24

The solution is written $x < -2$ or $-1 < x < 1$ or $x > 2$. ■

EXERCISES 8.3

In 1–31, solve each inequality.

1. $x^2 - x - 6 > 0$
2. $x^2 + x - 12 \geq 0$
3. $x^2 - 2x - 24 \leq 0$
4. $6x^2 - 13x - 5 < 0$
5. $3x^2 - 5x + 2 < 0$
6. $x^2 + 5x - 24 > 0$
7. $x^2 + 2x - 15 \geq 0$
8. $2x^2 - 3x - 9 \leq 0$
9. $x^2 - 7x + 12 > 0$
10. $x^2 - 8x + 15 \geq 0$
11. $4x^2 + 4x + 1 \leq 0$
12. $9x^2 - 12x + 4 < 0$
13. $\dfrac{x+2}{x-1} < 0$
14. $\dfrac{x-3}{x+2} > 0$
15. $\dfrac{2x-1}{x-3} \geq 0$
16. $\dfrac{3x+7}{x+1} \leq 0$
17. $\dfrac{x+3}{x} \leq 0$
18. $\dfrac{x-2}{x} \geq 0$
19. $\dfrac{4x^2-1}{x-2} > 0$
20. $\dfrac{x^2-9}{3x-1} < 0$
21. $\dfrac{x^2-2x+1}{2x+1} < 0$
22. $\dfrac{x^2-6x+9}{3x+1} > 0$
23. $\dfrac{6x^2-7x-5}{3x+4} \geq 0$
24. $\dfrac{6x^2-11x+4}{2x-5} \leq 0$
25. $x^4 - 10x^2 + 9 \leq 0$
26. $x^4 - 13x^2 + 36 \geq 0$
27. $x^4 - 2x^2 + 1 > 0$
28. $2x^3 - 3x^2 + x < 0$
29. $x^3 - 6x^2 + 9x < 0$
30. $4x^5 - 9x^3 > 0$
31. $9x^5 - 4x^3 \geq 0$

Problems 32 and 33 use the following information. The formula $d + 0.05v^2 + v$ is used to estimate the distance d (in ft) required to bring to a complete stop a car traveling at a velocity of v (in mph).

32. At what speeds can a car be brought to a stop in 100 ft or less? 4 7 8 1 2 3 5 6 9

33. At what speeds will a car require more than 300 ft to stop? 4 7 8 1 2 3 5 6 9

Problems 34 and 35 use the following information. The distance s of an object from a fixed point is given by

$$s = s_0 + vt + (1/2)at^2.$$

In the formula, s_0 is the initial distance of the object from the fixed point, v is the initial velocity, a is a constant acceleration, and t is the time during which the object moves.

34. An object is thrown upward from a height of 24.0 ft with an initial velocity of 80.0 ft/s. The acceleration due to gravity is -32 ft/s². At what times will its height be greater than 120.0 ft? 4 7 8 1 2 3 5 6 9

35. An object is thrown upward from a height of 4.4 m with an initial velocity of 24.5 m/s. The acceleration due to gravity is -9.8 m/s². At what times will its height be no greater than 33.8 m? 4 7 8 1 2 3 5 6 9

36. Consider a cylinder of height 10.0 cm. For what values of the radius is the surface area of the cylinder (including the ends) less than the area of a sphere with the same radius? 1 2 4 7 8 3 5

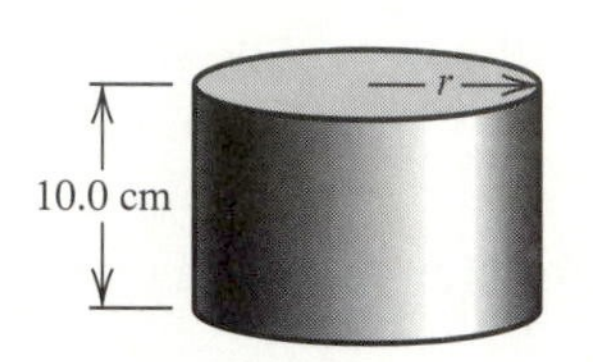

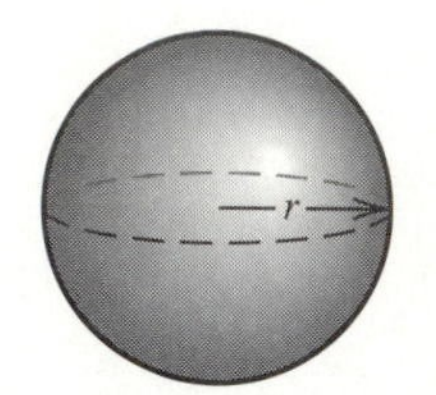

37. Consider a cylinder of height 10.0 cm. For what values of the radius is the surface area of the cylinder (including the ends) greater than the surface area of a cube with sides 10.0 cm in length? 1 2 4 7 8 3 5

38. Consider a cylinder of height h. For what values of the radius is the surface area of the cylinder (including the ends) greater than the surface area of a cube with sides h in length?

1. Architectural Technology
2. Civil Engineering Technology
3. Computer Engineering Technology
4. Mechanical Drafting Technology
5. Electrical/Electronics Engineering Technology
6. Chemical Engineering Technology
7. Industrial Engineering Technology
8. Mechanical Engineering Technology
9. Biomedical Engineering Technology

8.4 ABSOLUTE VALUE EQUATIONS AND INEQUALITIES

In Chapter 1, the absolute value of a nonzero number x was defined to be the positive number that is the same distance from zero on the number line as x. This definition has been adequate for working with numbers, but we need a more precise definition when variables are to be considered.

If x has a nonnegative value, it is permissible to say $|x| = x$. A variable, however, need not have a nonnegative value. If, for example, the value of x is -3, the statement $|x| = x$ says $|-3| = -3$, which is incorrect. That is, a variable may represent either a positive or negative number (or zero). We cannot assume that the representation of a negative number will always be prefaced by a negative sign. To work with the absolute value of a variable, we need to refine the concept of absolute value.

DEFINITION If x is a real number, the **absolute value** of x is given by

$$|x| = x \text{ if } x \geq 0$$
$$= -x \text{ if } x < 0.$$

EXAMPLE 1 For each value of x, verify that $|x|$ is nonnegative.

(a) $x = 2.5$ **(b)** $x = -3.7$ **(c)** $x = -2/3$

Solution **(a)**
$$x = 2.5$$
$$x > 0, \text{ so}$$
$$|x| = x$$
$$|2.5| = 2.5$$

(b)
$$x = -3.7$$
$$x < 0, \text{ so}$$
$$|x| = -x$$
$$|-3.7| = -(-3.7)$$
$$= 3.7$$

(c)
$$x = -2/3$$
$$x < 0, \text{ so}$$
$$|x| = -x$$
$$|-2/3| = -(-2/3)$$
$$= 2/3$$

In each case $|x| \geq 0$. ■

When an expression contains a variable, the expression might represent a number that is positive, negative, or zero, depending on the value of the variable. For example, $x - 3$ is positive if x is 5, but $x - 3$ is negative if x is 1. Generally, we do not know the value of the variable, so we do not know the sign of the number represented by the expression. To solve an equation containing the absolute value of a variable expression, it is necessary to account for both cases: $|x| = x$ and $|x| = -x$.

EXAMPLE 2 Solve $|x - 3| = 2$.

Solution There are two cases.

(a) If $x - 3 \geq 0$,
$$|x - 3| = x - 3$$
$$x - 3 = 2$$
$$x = 5$$

(b) If $x - 3 < 0$,
$$|x - 3| = -(x - 3) \text{ or } -x + 3$$
$$-x + 3 = 2$$
$$-x = -1, \text{ or } x = 1$$

Check: If x is 5, $|5 - 3| = |2| = 2$, and if x is 1, $|1 - 3| = |-2| = 2$. ■

Notice that in Example 2, the second equation $-(x - 3) = 2$ is equivalent to $x - 3 = -2$. The two equations, then, could be written as $x - 3 = \pm 2$. We make use of this fact to generalize the procedure.

To solve an absolute value equation:

1. Isolate the absolute value. That is, write an equivalent equation that has the absolute value on one side by itself and a nonnegative number on the other side.
2. Consider two cases.
   (a) Solve the equation obtained by omitting the absolute value symbol.
   (b) Solve the equation obtained by omitting the absolute value symbol and changing the sign on the other side of the equation.

EXAMPLE 3 Solve $|x^2 - 5x + 5| = 1$.

Solution 1. The equation is in the proper form.

2. (a)
$$\begin{aligned} x^2 - 5x + 5 &= 1 \\ x^2 - 5x + 4 &= 0 \\ (x - 4)(x - 1) &= 0 \\ x - 4 = 0 &\text{ or } x - 1 = 0 \\ x = 4 &\text{ or } \quad x = 1 \end{aligned}$$

(b)
$$\begin{aligned} x^2 - 5x + 5 &= -1 \\ x^2 - 5x + 6 &= 0 \\ (x - 2)(x - 3) &= 0 \\ x - 2 = 0 &\text{ or } x - 3 = 0 \\ x = 2 &\text{ or } \quad x = 3 \end{aligned}$$

The solutions then, are $x = 1$, $x = 2$, $x = 3$, or $x = 4$. ■

EXAMPLE 4 Solve $|3x - 2| = -7$.

Solution Regardless of what $3x - 2$ is equal to, the absolute value cannot be a negative number. There is no solution. ■

EXAMPLE 5 Solve $|2x + 1| = |x|$.

Solution 1. Consider $|2x + 1|$ to be the isolated absolute value and $|x|$ to be a nonnegative number.

2. Drop the absolute value bars from $|2x + 1|$ and solve the two equations that result from $2x + 1 = \pm|x|$. However, consider the meaning of $\pm|x|$.

$$\begin{aligned} |x| &= x \text{ if } x \geq 0 \\ &= -x \text{ if } x < 0 \end{aligned} \qquad \text{and} \qquad \begin{aligned} -|x| &= -x \text{ if } x \geq 0 \\ &= x \text{ if } x < 0. \end{aligned}$$

Solving $2x + 1 = -|x|$ leads to the same two equations as $2x + 1 = |x|$. Therefore, we solve $2x + 1 = \pm x$ as follows.

(a)
$$\begin{aligned} 2x + 1 &= x \\ x + 1 &= 0 \\ x &= -1 \end{aligned}$$

(b)
$$\begin{aligned} 2x + 1 &= -x \\ 3x + 1 &= 0 \\ 3x &= -1 \\ x &= -1/3 \end{aligned}$$
■

For many purposes, measurements are considered adequate if they are within a prescribed tolerance. That is, the actual measurement may be a little larger or a little smaller than its stated value. Since only the size of the difference (and not the sign) is important, absolute value inequalities are often used in error analysis.

EXAMPLE 6 Use an absolute value to state the condition that velocity V cannot differ from an average velocity of 55 mph by more than 5 mph.

Solution The difference between V and 55 is denoted $V - 55$. The quantity V, however, might be larger or smaller than 55 (or equal to 55), so $V - 55$ might be positive or negative (or zero). We have

$$|V - 55| \leq 5. \quad \blacksquare$$

Absolute value inequalities are solved using a technique similar to that used to solve nonlinear inequalities.

To solve an absolute value inequality:

1. Replace the inequality symbol with an equal sign and solve.
2. Locate and mark on a number line the solutions obtained in step 1.
   (a) Use closed circles if the original inequality contains $\geq$ or $\leq$.
   (b) Use open circles if the original inequality contains $>$ or $<$.
3. Locate and mark on the number line any value that would produce a denominator of zero in the inequality. Use an open circle. (If there is a closed circle already there, replace it with an open one.)
4. Notice that the n points located in steps 2 and 3 divide the number line into $n + 1$ sections. For each section, choose one point within that section and substitute the value into the original inequality.
   (a) If the statement is true, all values in that section satisfy the inequality.
   (b) If the statement is false, no value in that section satisfies the inequality.

EXAMPLE 7 Solve $|2x^2 + 8x + 7| \leq 1$.

Solution 1. Since the quantity on the right-hand side is nonnegative, the equation $|2x^2 + 8x + 7| = 1$ can be written as

$$\begin{array}{ccc} 2x^2 + 8x + 7 = 1, & \text{or} & 2x^2 + 8x + 7 = -1. \\ 2x^2 + 8x + 6 = 0 & & 2x^2 + 8x + 8 = 0 \\ 2(x^2 + 4x + 3) = 0 & & 2(x^2 + 4x + 4) = 0 \\ 2(x + 1)(x + 3) = 0 & & 2(x + 2)(x + 2) = 0 \\ x + 1 = 0 \quad \text{or } x + 3 = 0 & & x + 2 = 0 \\ x = -1, \quad x = -3 & \text{or} & x = -2 \end{array}$$

2. There are no values which produce a denominator of zero.
3. The numbers -3, -2, and -1 are located and marked with closed circles, as shown in Figure 8.25.

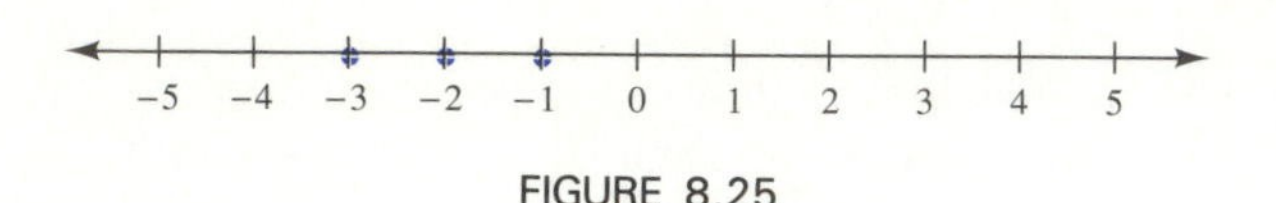

FIGURE 8.25

4. There are four sections of the number line.

Section 1: Choose $x=-4$:

$$\begin{aligned} |2(-4)^2+8(-4)+7| &\leq 1 \\ |32-32+7| &\leq 1 \\ |7| \leq 1 &\text{ is false.} \end{aligned}$$

Section 2: Since there are no integers in this section, we must choose a mixed number. A calculator, however, may be used to do the arithmetic.

Choose $x=-2.5$:

$$\begin{aligned} |2(-2.5)^2+8(-2.5)+7| &\leq 1 \\ |12.5-20+7| &\leq 1 \\ |-0.5| \leq 1 &\text{ is true, so } -3 \leq x \leq -2. \end{aligned}$$

Section 3: Choose $x=-1.5$:

$$\begin{aligned} |2(-1.5)^2+8(-1.5)+7| &\leq 1 \\ |4.5-12+7| &\leq 1 \\ |-0.5| \leq 1 &\text{ is true, so } -2 \leq x \leq -1. \end{aligned}$$

Section 4: Choose $x=0$:

$$\begin{aligned} |2(0)^2+8(0)+7| &\leq 1 \\ |0+0+7| &\leq 1 \\ |7| \leq 1 &\text{ is false.} \end{aligned}$$

The solution of $|2x^2+8x+7| \leq 1$ is shown on the graph in Figure 8.26. The solution is $-3 \leq x \leq -1$. ■

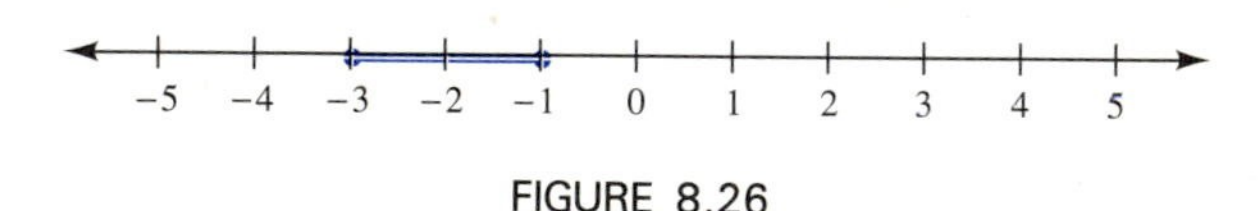

FIGURE 8.26

Once again, we see the importance of testing *each* section of the number line. As in Example 6 of the preceding section, the solution does *not* consist of alternating sections.

EXAMPLE 8 Solve $\left|\dfrac{x-1}{x+1}\right| \geq 2$.

Solution **1.** The equation $\left|\frac{x-1}{x+1}\right| = 2$ can be written as

$$\frac{x-1}{x+1} = 2 \qquad \text{or} \qquad \frac{x-1}{x+1} = -2.$$

If $x + 1 \neq 0$,

$$\begin{aligned} x - 1 &= 2(x+1) & \qquad x - 1 &= -2(x+1) \\ x - 1 &= 2x + 2 & x - 1 &= -2x - 2 \\ -x - 1 &= 2 & 3x - 1 &= -2 \\ -x &= 3 & 3x &= -1 \\ x &= -3 \quad \text{or} & x &= -1/3. \end{aligned}$$

2–3. The equation is undefined when $x + 1 = 0$ $(x = -1)$. The points at -3 and $-1/3$ are marked on the number line using closed circles, and the point at -1 is marked using an open circle, as shown in Figure 8.27.

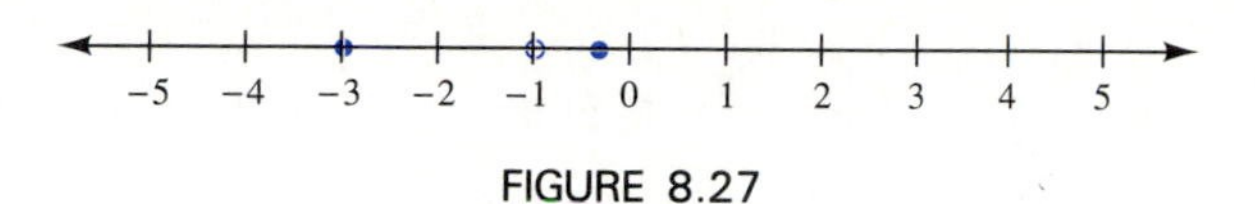

FIGURE 8.27

4. There are four sections of the number line.

Section 1: Choose $x = -4$: $\left|\frac{-4-1}{-4+1}\right| = \left|\frac{-5}{-3}\right| = \frac{5}{3} \geq 2$, which is false.

Section 2: Choose $x = -2$: $\left|\frac{-2-1}{-2+1}\right| = \left|\frac{-3}{-1}\right| = 3 \geq 2$.

The statement is true, so $-3 \leq x < -1$.

Section 3: Choose $x = -1/2$: $\left|\frac{-1/2-1}{-1/2+1}\right| = \left|\frac{-3/2}{1/2}\right| = |-3| \geq 2$.

The statement is true, so $-1 < x \leq -1/3$.

Section 4: Choose $x = 0$: $\left|\frac{0-1}{0+1}\right| = \left|\frac{-1}{1}\right| = |1| \geq 2$, which is false.

The graph of the solution of $\left|\frac{x-1}{x+1}\right| \geq 2$ is shown in Figure 8.28.

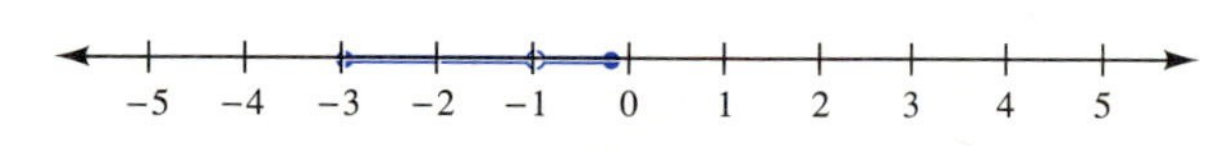

FIGURE 8.28

The solution is written as $-3 \leq x < -1$ or $-1 < x \leq -1/3$. ■

EXAMPLE 9 The diameter of a ball bearing must be within 0.01 mm of 2.00 mm as shown in Figure 8.29. What range of values is allowed?

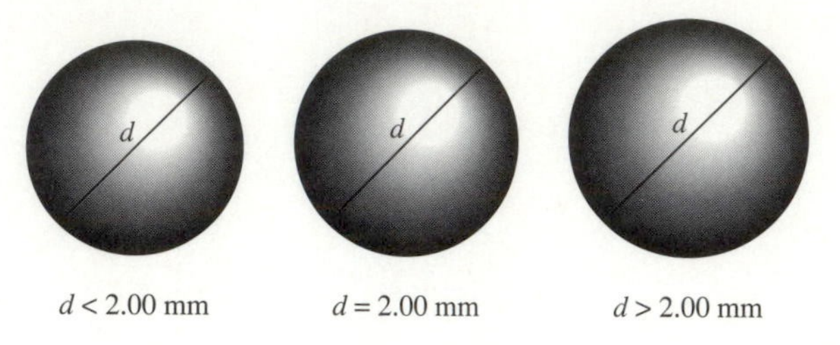

FIGURE 8.29

Solution Let $d =$ the diameter of the ball bearing. Then $|d - 2.00|$ is the difference between d and 2.00 regardless of which is larger. Thus, the condition is met if

$$|d - 2.00| \leq 0.01.$$

To solve $|d - 2.00| \leq 0.01$, first solve $|d - 2.00| = 0.01$.

$$d - 2.00 = 0.01 \quad \text{or} \quad d - 2.00 = -0.01$$
$$d = 2.01 \qquad\qquad d = 1.99$$

These two points divide the number line into three sections, as shown in Figure 8.30.

Section 1: Choose $d = 0$: $|0 - 2.00| \leq 0.01$ is false.

Section 2: Choose $d = 2$: $|2 - 2.00| \leq 0.01$ is true.

That is, $1.99 \leq d \leq 2.01$.

Section 3: Choose $d = 3$: $|3 - 2.00| \leq 0.01$ is false.

The range of values, then, is $1.99 \leq d \leq 2.01$. ■

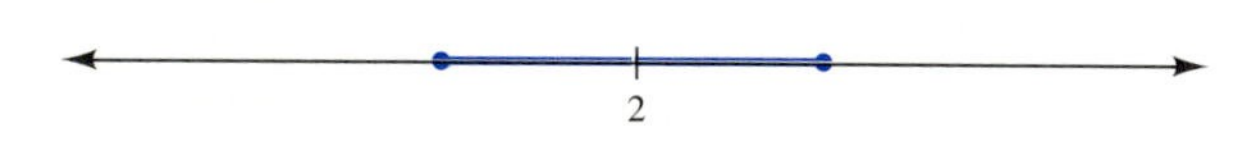

FIGURE 8.30

EXERCISES 8.4

In 1–21, solve each equation.

1. $|x + 3| = 7$

2. $|3x + 1| = 2$

3. $|4x + 3| = -2$

4. $|3x - 4| = -1$

5. $|2x - 3| = |x - 4|$

6. $|2x - 1| = |x + 5|$

7. $|x - 7| = |x + 7|$

8. $|x + 1| = |x - 1|$

9. $|3x + 2| = |2x - 3|$

10. $|x^2 + 3x - 4| = 6$

11. $|x^2 - 7x + 9| = 3$

12. $|2x^2 - x - 3| = 0$

13. $|2x^2 - 12x + 14| = 4$

14. $|3x^2 - 3x - 12| = 6$

15. $\left|\frac{x+2}{x-1}\right| = 2$

16. $\left|\frac{2x+1}{x+3}\right| = 1$

17. $\left|\frac{3x-1}{x+1}\right| = 1$

18. $\left|\frac{x-1}{x+1}\right| = 2$

19. $\left|\frac{2x+5}{x+3}\right| = 1$

20. $\left|\frac{2x+2}{x+5}\right| = 2$

21. $\left|\frac{x-5}{x+4}\right| = 2$

In 22–42, solve each inequality.

22. $|x+4| > 5$

23. $|3x+2| \geq 1$

24. $|2x-4| \leq |x-4|$

25. $|2x+1| < |x+5|$

26. $|x-7| > |x+7|$

27. $|4x+2| \geq -2$

28. $|3x-4| \leq -1$

29. $|x^2+x-9| \leq 3$

30. $|x^2+2x-9| < 6$

31. $|2x^2+4x-23| \leq 7$

32. $|2x^2-12x+3| \geq 17$

33. $|2x^2-8x-17| > 7$

34. $|x^2| > 9$

35. $|4x^2| \leq 1$

36. $\left|\frac{x+1}{x-1}\right| < 2$

37. $\left|\frac{2x+3}{x-3}\right| > 1$

38. $\left|\frac{2x+3}{x-1}\right| \geq 1$

39. $\left|\frac{6x}{x-4}\right| > 2$

40. $\left|\frac{2x+1}{x+4}\right| < 1$

41. $\left|\frac{3x-2}{2x}\right| \geq \frac{1}{3}$

42. $\left|\frac{2x-1}{3x}\right| > \frac{1}{3}$

For each problem, state the condition described as an absolute value inequality.

43. When aimed at a target 62 in off the ground, the bullet should strike vertically within 2 in of the target. **4 7 8** 1 2 3 5 6 9

44. The contents of a can labeled 15 oz must weigh within 0.5 oz of the labeled amount. **2 4 8** 1 5 6 7 9

45. A machine part must be manufactured to within 0.3 mm of 16.9 mm. **1 2 4** 8 7 9

46. The maximum deflection of a beam must be within 3 mm of the equilibrium position. **1 2 4** 8 7 9

47. The time required to perform a particular task must be within 6 seconds of 48 seconds. **4 7 8** 1 2 3 5 6 9

48. The time required for a plane flight is within 20 minutes of 2 hours and 45 minutes. **4 7 8** 1 2 3 5 6 9

49. In a computer program, it is sometimes desirable to stop execution of the program when two values are equal. For example, we might say that when a certain value of x makes a polynomial $f(x)$ equal to zero, we have solved the equation $f(x) = 0$. Because computers represent numbers using a finite number of decimal places, they cannot represent some real numbers exactly. For the equation $3x = 1$, a computer would give $x = 0.33333333$ as the solution, but when we ask if $3x$ is equal to 1 when x is 0.33333333, the answer would be, "no," since 0.99999999 and 1 are not exactly the same. Explain how you could use absolute values to avoid asking if two values are equal.

---

1. Architectural Technology
2. Civil Engineering Technology
3. Computer Engineering Technology
4. Mechanical Drafting Technology
5. Electrical/Electronics Engineering Technology
6. Chemical Engineering Technology
7. Industrial Engineering Technology
8. Mechanical Engineering Technology
9. Biomedical Engineering Technology

8.5 GRAPHING INEQUALITIES IN TWO VARIABLES

To represent a linear inequality in one variable, we have used a number line on which the solution of the corresponding equation could be found. Those points divided the number line into sections. Linear inequalities in two variables are represented on a plane by finding the solutions of the corresponding equation. Recall that the solutions of a linear *equation* in *two* variables fall along a straight line, and the line divides the plane into sections.

Since a line represents an equation of the form $y = mx + b$, we can say that the line represents ordered pairs (x, y) such that y and $mx + b$ are equal. For any given value of x, the points *above* the line have y-values greater than the y-value of the point *on* the line (that is, $y > mx + b$). Points *below* the line have y-values smaller than the y-value of the point *on* the line ($y < mx + b$). We use this observation to graph linear inequalities.

For instance, to graph $y \geq 2x - 3$, we first graph the equation $y = 2x - 3$. The solution of $y \geq 2x - 3$ includes the points *above* the line, as well as the points *on* the line, so we shade that area of the plane to represent the solution, as shown in Figure 8.31.

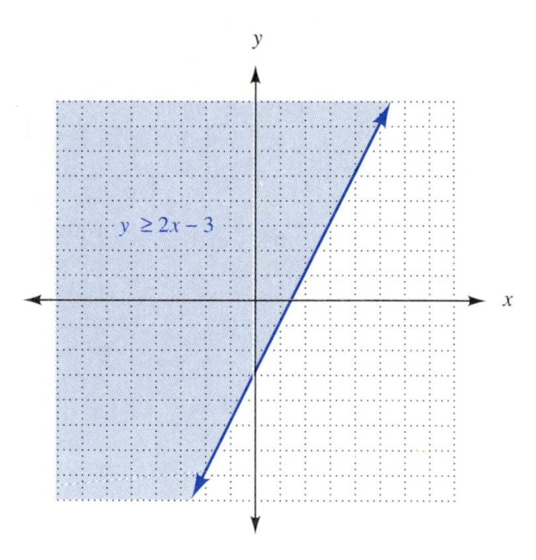

FIGURE 8.31

EXAMPLE 1 Sketch the graph of $y < (1/2)x - 2$.

Solution Sketch the graph of $y = (1/2)x - 2$ using a dashed line, since the values *on* the line are not included in the solution. Then shade the area *below* the line in order to show the values of y that are *less than* the y-value on the line for the corresponding value of x, as shown in Figure 8.32. ■

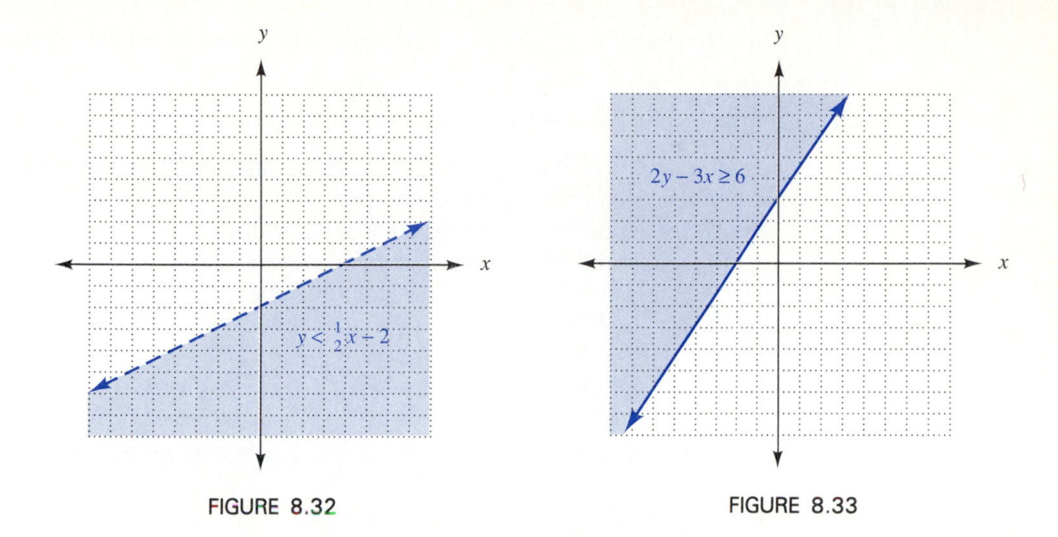

FIGURE 8.32

FIGURE 8.33

Linear inequalities are not always expressed with y on one side and $mx + b$ on the other. They are often written with both variables on the same side of the inequality.

EXAMPLE 2 Graph $2y - 3x \geq 6$.

Solution The boundary line is given by $2y - 3x = 6$. Rather than rewrite the equation as $y = (3/2)x + 3$ to use the slope and y-intercept, we keep the form $2y - 3x = 6$ to use both the x-intercept and y-intercept. Two points determine a line, and the two intercepts are particularly easy to work with. When x is 0, the line crosses the y-axis, giving the y-intercept. When y is 0, the line crosses the x-axis, giving the x-intercept.

To find the y-intercept for the line $2y - 3x = 6$, we let $x = 0$.

$$2y = 6, \text{ so } y = 3.$$

To find the x-intercept for the line $2y - 3x = 6$, we let $y = 0$.

$$-3x = 6, \text{ so } x = -2.$$

Thus, the graph has (0, 3) and (−2, 0) as two of its points. To decide which half of the graph to shade, we can write $2y - 3x \geq 6$ as $y \geq (3/2)x + 3$ to see that values above the boundary line should be shaded. An alternate procedure is to choose a point not on the boundary line and test it in the original inequality. Suppose we choose the origin (0, 0) as the test point. We have

$$2y - 3x \geq 6$$
$$2(0) - 3(0) \geq 6$$
$$0 \geq 6.$$

Since the ordered pair (0, 0) does not satisfy the inequality, the solutions must lie on the opposite side of the boundary line from (0, 0), as shown in Figure 8.33. ■

The steps used in graphing an inequality are summarized as follows:

To graph a linear inequality in two variables:

1. Replace the inequality symbol with an equal sign and find the two intercepts, or find the slope and y-intercept.
2. Graph the equation. It will be the boundary line.
   (a) Use a solid line for an inequality that contains $\geq$ or $\leq$.
   (b) Use a dashed line for an inequality that contains $>$ or $<$.
3. Write the original inequality so that it is in one of the following forms: $y > mx + b$, $y < mx + b$, $y \geq mx + b$, or $y \leq mx + b$.
   (a) Shade the area above the boundary line for an inequality that contains $\geq$ or $>$.
   (b) Shade the area below the boundary line for an inequality that contains $\leq$ or $<$.

or

3. Choose a test point not on the boundary line and substitute its coordinates into the original inequality.
   (a) If the statement is true, shade all points on the side of the boundary that contains the test point.
   (b) If the statement is false, shade all points on the opposite side of the boundary from the test point.

EXAMPLE 3 Sketch the graph of $4x - 3y < 12$.

Solution

1. Consider $4x - 3y = 12$. When x is 0, y is -4. When y is 0, x is 3.
2. Thus, $(0, -4)$ and $(3, 0)$ are two points on the boundary line. We use a dashed line.
3. Since the original inequality is equivalent to $y > (4/3)x - 4$, the area above the line is shaded as shown in Figure 8.34. ■

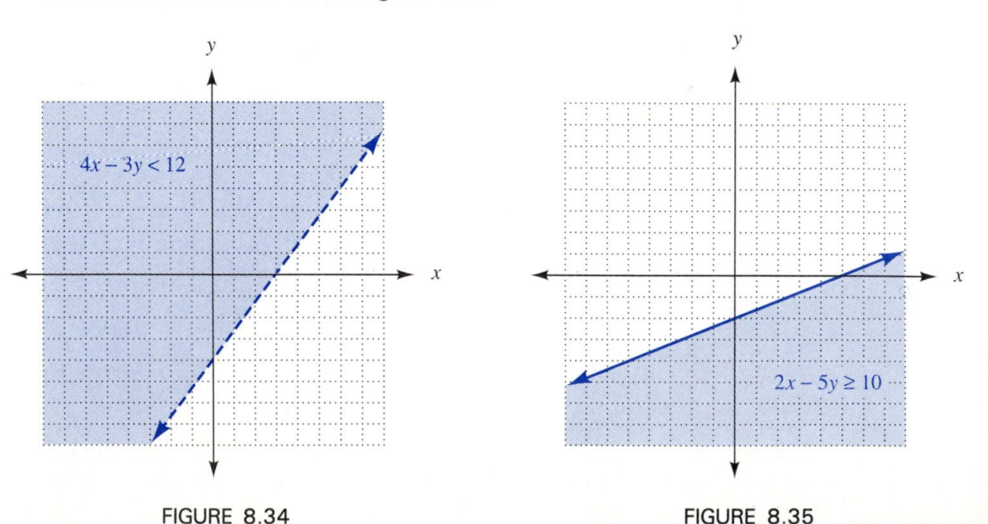

FIGURE 8.34

FIGURE 8.35

EXAMPLE 4 Sketch the graph of $2x - 5y \geq 10$.

Solution

1. Consider $2x - 5y = 10$. When x is 0, y is -2. When y is 0, x is 5.
2. The intercepts are $(0, -2)$ and $(5, 0)$. A solid line is used for the boundary line.
3. Since the original inequality is equivalent to $y \leq (2/5)x - 2$, the area below the line is shaded as shown in Figure 8.35. ■

Two or more inequalities may be graphed on the same plane.

To graph a compound inequality in two variables:

1. Graph each separate inequality on the plane.
2. (a) If the inequality is an AND inequality, locate and shade those points with coordinates that satisfy *both* of the inequalities.
   (b) If the inequality is an OR inequality, locate and shade those points with coordinates that satisfy *at least one* of the inequalities.

EXAMPLE 5 Graph $4x - 3y > 12$ and $2x - 5y \geq 10$.

Solution

1. Figure 8.36a shows the graphs of the individual inequalities.
2. Figure 8.36b shows the points with coordinates that satisfy $4x - 3y > 12$ and $2x - 5y \geq 10$. ■

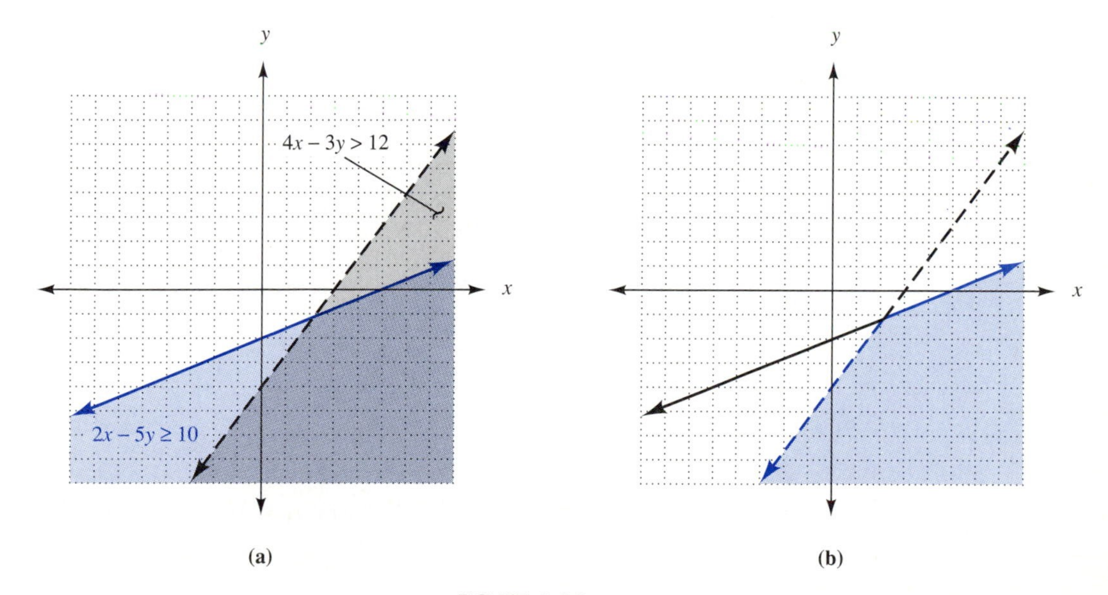

FIGURE 8.36

EXAMPLE 6 Graph $y \leq 2x - 1$ or $y > (1/3)x + 2$.

Solution 1. Figure 8.37a shows the graphs of the individual inequalities.

2. Figure 8.37b shows the points with coordinates that satisfy $y \leq 2x - 1$ or $y > (1/3)x + 2$. ■

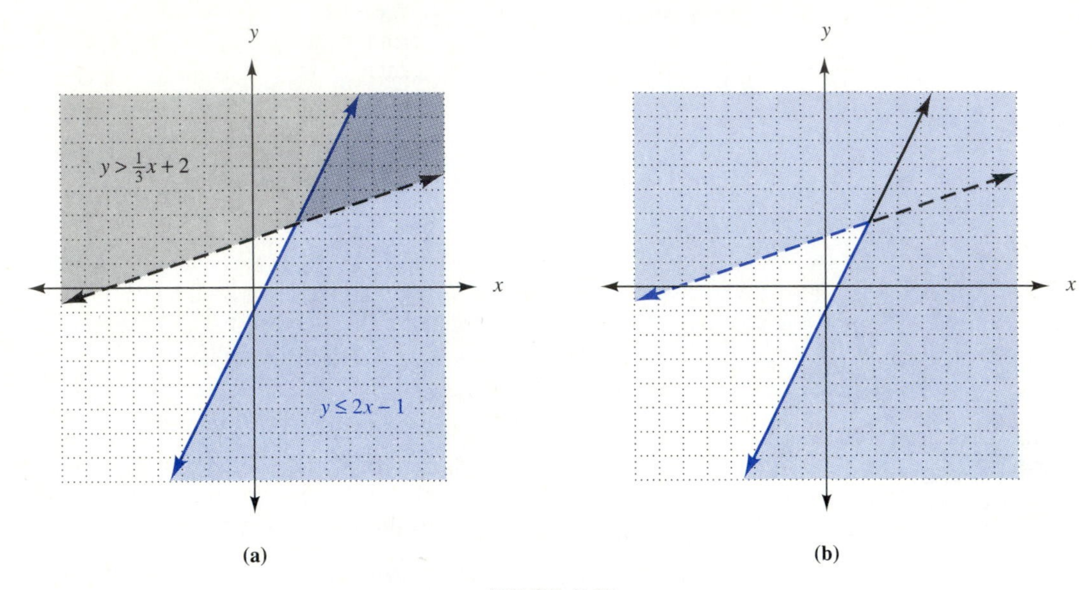

FIGURE 8.37

EXERCISES 8.5

In 1–24, sketch the graph of each inequality.

1. $y \geq 2x + 1$ **2.** $y < 3x - 2$ **3.** $y > (1/2)x - 1$ **4.** $y \leq (2/3)x + 3$

5. $y < x + 2$ **6.** $y \leq -x + 4$ **7.** $y \geq -2x - 3$ **8.** $x + y \geq 4$

9. $2x + 3y \leq 6$ **10.** $3x - 5y > 10$ **11.** $x - 2y < 12$ **12.** $2x + y \leq 4$

13. $x - y > 7$ **14.** $y \geq x$ **15.** $x \leq y$ **16.** $x > -3$

17. $y \leq 1$ **18.** $x < 2$ **19.** $y + 1 > x$ **20.** $x + 1 \leq y$

21. $3x - y \geq 2$ **22.** $3y - x < 2$ **23.** $y - 3 \leq x$ **24.** $x - 2 > y$

In 25–31, sketch the graph of each compound inequality.

25. $2x - y \leq 5$ and $x + 2y \leq 5$ **26.** $x - y \leq 1$ and $x + y \leq 5$

27. $3x - y \geq 4$ or $x + 3y \geq 8$ **28.** $2x + 5y \leq 9$ or $3x - 4y \leq 10$

29. $x + y \leq 6$ and $y - x \leq 2$ **30.** $x + 2y \leq 8$ and $2y \leq 3x$

31. $3y - x \leq 6$ or $3y + x \leq 18$ **32.** $3y + x \geq 9$ or $y + 2x \geq 8$

Problems 33–36 are known as linear programming *problems. In problems with two variables, the values that produce a maximum or minimum value for a particular quantity are always found at a vertex (corner) of a region graphed as a compound inequality. See the Extended Application at the end of this chapter for an example of a linear programming problem.*

33. A company makes granola by mixing cereal and fruit. They have 6000 pounds of cereal and 4000 pounds of fruit in stock. A regular box of granola contains 3/4 lb of cereal and 1/2 lb of fruit, while a premium box contains 1/2 lb of cereal and 1 lb of fruit. The profit on a regular box is \$0.20 and on a premium box it is \$0.50. Using x for the number of regular boxes and y for the number of premium boxes, these conditions can be expressed as follows:

$$x \geq 0,\ y \geq 0,\ (3/4)x + (1/2)y \leq 6000,$$
$$\text{and } (1/2)x + y \leq 4000.$$

Sketch these inequalities on a single graph with 1 block = 1000. Find the values of x and y that maximize profit by evaluating $P = 0.20x + 0.50y$ for the values at each vertex of the graph.
1 2 6 7 8 3 5 9

34. A baker wishes to enrich his breads by adding wheat germ and nonfat dry milk to the flour. One cup of wheat germ contains 1 1/2 mg of vitamin B_1 and 1/2 mg of vitamin B_2. One cup of nonfat dry milk contains 1/2 mg of vitamin B_1 and 1 1/2 mg of vitamin B_2. The baker wants his recipe to be enriched by at least 1 mg of each vitamin. Wheat germ costs \$0.41 per cup and nonfat dry milk costs \$0.34 per cup. Using x for the number of cups of wheat germ and y for the number of cups of dry milk, these conditions can be expressed as follows:

$$x \geq 0,\ y \geq 0,\ (3/2)x + (1/2)y \geq 1,$$
$$\text{and } (1/2)x + (3/2)y \geq 1.$$

Sketch these inequalities on a single graph with 6 blocks = 1. Find the values of x and y that minimize cost by evaluating $C = 0.41x + 0.34y$ for the values at each vertex of the graph.
1 2 6 7 8 3 5 9

35. A company that manufactures television sets finds that by testing its merchandise before shipping, it can increase profits. It takes 30 seconds to perform test A on a 13″ model and 40 seconds to perform test A on a 19″ model. It takes 15 seconds to perform test B on a 13″ model and 10 seconds to perform test B on a 19″ model. At most 36,000 seconds a day can be devoted to test A, and at most 15,000 seconds a day can be devoted to test B. The profit is increased on a 13″ model by \$1 and on a 19″ model by \$2. Using x for the number of 13″ models and y for the number of 19″ models, these conditions can be expressed as follows:

$$x \geq 0,\ y \geq 0,\ 30x + 40y \leq 36{,}000,$$
$$\text{and } 15x + 10y \leq 15{,}000.$$

Sketch these inequalities on a single graph with 1 block = 100. Find the values of x and y that maximize the increase in profit by evaluating $P = 1x + 2y$ for the values at each vertex of the graph.
1 2 6 7 8 3 5 9

36. A woman wants to lose weight by exercising. She has at most 6 hours a day to exercise and wants to golf at least as long as she swims. Golfing burns 250 calories per hour while swimming burns 350 calories per hour. Using x for the number of hours she golfs and y for the number of hours she swims, these conditions can be expressed as follows:

$$x \geq 0,\ x + y \leq 6, \text{ and } x \geq y.$$

Sketch these inequalities on a single graph with 1 block = 1. Find the values of x and y that maximize calories burned by evaluating $C = 750x + 350y$ for the values at each vertex of the graph.
1 2 6 7 8 3 5 9

8.6 CHAPTER REVIEW

KEY TERMS

8.1 linear inequality

8.2 compound inequality

8.4 absolute value

SYMBOLS

$\neq$ is not equal to

$>$ is greater than

$<$ is less than

$\geq$ is greater than or equal to

$\leq$ is less than or equal to

GEOMETRY CONCEPTS

$A = (h/2)(B + b)$ (area of a trapezoid)

SEQUENTIAL REVIEW

(Section 8.1) In 1–5, solve each inequality.

1. $2y + 4 > 5$

2. $(3/5)x - 1 < x$

3. $2(p - 3) \leq 3p + 2$

4. $3x - 2 < 7x + 6$

5. $2(y - 1) \geq 3(y + 1)$

In 6–9, solve each inequality and graph the solution on a number line.

6. $3x + 4 \geq 0$

7. $3m - 2 \leq 2(m + 1)$

8. $x/3 - 2 \geq -1$

9. $m + 3 < 2(m - 1)$

(Section 8.2) In 10–14, graph the solution of each compound inequality on a number line.

10. $x \geq -2$ and $x < 4$

11. $x > 2$ or $x \leq -4$

12. $1 \leq x \leq 3$

13. $-1 < x < 2$

14. $-3 < x \leq -2$

In 15–18, write a compound inequality that specifies the values shown for each graph.

15.

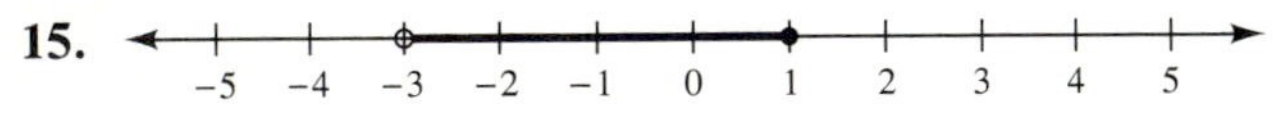

16.

17.

18.

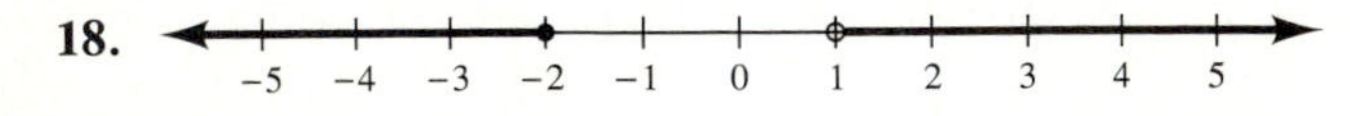

(Section 8.3) *In 19–27, solve each inequality.*

19. $2x^2 + 5x - 3 > 0$

20. $3x^2 - x - 2 \leq 0$

21. $2x^2 - 7x + 3 \geq 0$

22. $\dfrac{x^2 - 1}{x} \leq 0$

23. $\dfrac{x}{x - 3} < 0$

24. $\dfrac{x^2 + 2x - 24}{x} \geq 0$

25. $2x^3 + 5x^2 - 12x \geq 0$

26. $x^4 - 16x^2 < 0$

27. $x^4 - 8x^2 + 16 \leq 0$

(Section 8.4) *In 28–32, solve each equation.*

28. $|2x - 5| = 4$

29. $|x + 7| = -6$

30. $|x^2 + 2x - 1| = 2$

31. $\left|\dfrac{x}{2x - 1}\right| = 3$

32. $|2x + 1| = |x + 3|$

In 33–36, solve each inequality.

33. $|3x + 1| \geq 0$

34. $|2x - 5| > 3$

35. $\left|\dfrac{x + 5}{x - 3}\right| \leq 3$

36. $|2x^2 + x - 8| < 7$

(Section 8.5) *In 37–45, sketch the graph of each inequality.*

37. $y < 2x - 1$

38. $x - 2y \geq 4$

39. $3x + 2y \leq 12$

40. $2x + 5y > 10$

41. $y \geq -x$

42. $y < 3$

43. $y \leq 3$ and $x \geq 0$

44. $y \leq 2x - 1$ and $y \leq -3x + 2$

45. $y - x \geq 2$ and $y + x \geq 3$

APPLIED PROBLEMS

1. Consider a circular water tower. For what values of the radius will the circumference be more than 50.0 ft? 1 2 4 7 8 3 5

2. The formula $P = I^2R$ gives the power P (in watts) dissipated as heat when current flows through a resistor. In the formula, I is the current (in amperes), and R is the resistance (in ohms). What currents flowing through a 50 Ω resistor would cause more than 200 watts to be dissipated? 3 5 9 1 2 4 6 7 8

3. A civil engineer calculates the maximum safe load of a beam to be (48.3 ± 0.2) kg. Let x be the maximum safe load and state the condition as an absolute value inequality. 1 2 4 8 7 9

4. The volume V of a piece of cylindrical pipe is given by $V = 2\pi rLt + \pi Lt^2$. In the formula, r is the inside radius of the pipe, L is its length, and t is its thickness. Suppose a pipe of inside radius 4.170 inches and a length of 14.36 inches is to be made. Find the range of thicknesses that could be used if less than 40.00 in³ of metal is to be used to cast the pipe. 1 2 4 7 8 3 5

5. In working with electronic filters, the following equation occurs.

$$\frac{V_o}{V_i} = S^2 - S - 1.293.$$

In the equation, V_o and V_i are the output and

1. Architectural Technology
2. Civil Engineering Technology
3. Computer Engineering Technology
4. Mechanical Drafting Technology
5. Electrical/Electronics Engineering Technology
6. Chemical Engineering Technology
7. Industrial Engineering Technology
8. Mechanical Engineering Technology
9. Biomedical Engineering Technology

input voltages, respectively. For what nonnegative values of S will V_o/V_i be less than 0.800?
3 5 9 1 2 4 6 7 8

6. When a variable resistance is attached to a 24-volt supply with an internal resistance of 8 ohms, the current that flows is given by $I = \dfrac{24}{8 + R}$ amps. For what positive values of R will I be less than 2?
3 5 9 1 2 4 6 7 8

7. A 10-volt battery, a 2-ohm resistor, and a variable resistor of resistance R (in ohms) are connected in series. The voltage (in volts) developed across the resistor is given by the equation, $V = \dfrac{10R}{2 + R}$. For what positive values of R is it true that $V \leq 6$? 3 5 9 1 2 4 6 7 8

8. A particular transistor has a current equation given by $I = 24\left(1 + \dfrac{V}{3}\right)^2$, where I is in milliamps. If $I < 16$, find V, given $V > -3$. Use three significant digits. 3 5 9 4 7 8

9. A company makes 10.0-inch long solid cylindrical bars with a variety of diameters. These cylinders must be plated. One particular plating machine allows a fixed time for plating in which 245 sq inches or less of copper can be applied. What range of diameter cylinders could be plated by this machine? 1 2 4 7 8 3 5

10. A company can lease a PC for \$146 per month. The wages and overhead to operate the machine will cost \$18 per hour. How many hours per month could the PC be used if the total cost of operation is not to exceed \$5700 per month?
1 2 6 7 8 3 5 9

11. The time period T (in seconds) of a swinging pendulum 1 m long is given by $T = \dfrac{2\pi}{\sqrt{g}}$ when g (in m/s^2) is the acceleration due to gravity. On the earth with $g = 9.8$ m/s^2 the time is about 2 seconds. For what g values (for example, on other planets) is the time greater than 4 seconds?
4 7 8 1 2 3 5 6 9

12. The apparent frequency f_a (in hertz) of a sound heard by a stationary observer when the source is moving toward him is given by the equation $f_a = \dfrac{fv}{v - v_s}$ where v is the speed of sound in the medium around the source, v_s is the speed of the source, and f is the true frequency of the source. For what values of v will a 750 Hz sound moving at 100 ft/s appear to be a 900 Hz or greater sound?
1 3 4 5 6 8 9

13. A patrol car traveling at 55 mph receives a call from an air spotter that a car traveling at 75 mph is 600 ft ahead going in the same direction. The patrol car immediately accelerates at 10 mph/s to catch the speeder. The speeder, who has a police band radio, also hears the report and immediately accelerates but at only 3 mph/s. For how long will the speeder be ahead of the law? (Hint: For what t will the speeder's distance traveled be greater than the distance traveled by the law?) Use $s = s_0 + v_0 t + \dfrac{1}{2}at^2$. 4 7 8 1 2 3 5 6 9

14. The formula $\left|\omega - \dfrac{1}{\omega}\right| = 10$ appears in an electronics text chapter on resonant circuit theory. For what values of ω is it true? Give your answer to three significant digits. 3 5 9 1 2 4 6 7 8

15. The formula $\left|\omega - \dfrac{1}{\omega}\right| < 10$ appears in an electronics text chapter on resonant circuit theory. For what values of ω is it true? Give your answer to three significant digits. 3 5 9 1 2 4 6 7 8

16. The model circuit for a transistor amplifier consists of two series resistors of resistance R_1 and R_2 (in kilohms) carrying a constant current I (in amperes). The power lost to these resistors is given by $P = I^2R_1 + I^2R_2$. Assume that the current is a constant 3 amperes. Using x for the resistance of R_1 resistors and y for the resistance of R_2 resistors, these conditions can be expressed as follows:

$$x \geq 3,\ y \geq 10,\ x + y \leq 100.$$

Sketch these inequalities on a single graph with 1 block = 10. Find the maximum power by evaluating $P = 9x + 9y$ for the values of each vertex of the graph. 3 5 9 4 7 8

17. Factory A can produce 40,000 items each year and factory B can produce 28,000 of the same. Company policy dictates that between the two at least 5000 items must be made each month. The overall cost of operating factory A to make these items is \$2400 per hour, and factory B costs \$1700 per hour. Assume a normal work week of 40 hours for 50 weeks. Using x for the number of items produced by factory A and y for the number of items produced by factory B, these conditions can be expressed as follows:

$$x \geq 0,\ y \geq 0,\ x + y \geq 60{,}000,\ x \leq 40{,}000,$$
$$\text{and } y \leq 28{,}000.$$

Sketch these inequalities on a single graph with 1 block = 4000. Find the values of x and y that minimize the cost by evaluating $c = 120x + 121.43y$ for the values of each vertex of the graph. 1 2 6 7 8 3 5 9

18. An agri-technician has 1250 acres of land on which to plant crops A and B. Crop A yields \$300 per acre profit. Crop B yields only \$137. Planting contracts require that she plant at least 50% more acres of B than of A, and at least 200 acres of A. Using x for the number of acres planted with crop A and y for the number of acres planted with crop B, these conditions can be expressed as follows:

$$x \geq 0,\ y \geq 0,\ x + y \leq 1250,\ x \geq 200,$$
$$\text{and } y \geq 1.5x$$

Sketch these inequalities on a single graph with 1 block = 100. Find the values of x and y that maximize the profit by evaluating $P = 300x + 137y$ for the values of each vertex of the graph. 1 2 6 7 8 3 5 9

RANDOM REVIEW

1. Write a compound inequality that specifies the values shown on the graph.

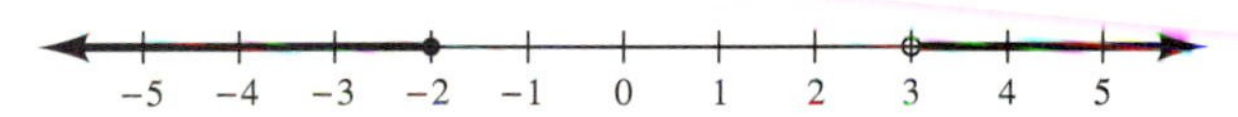

2. Solve $|x + 7| = 11$.

3. Sketch the graph of $y < x + 3$.

In 4–7, solve each inequality.

4. $3x - 1 \leq 5$

5. $3(m - 2) > 2m - 7$

6. $\dfrac{7x + 3}{x} > 0$

7. $|2x - 1| \leq -6$

8. Write a compound inequality that specifies the values shown on the graph.

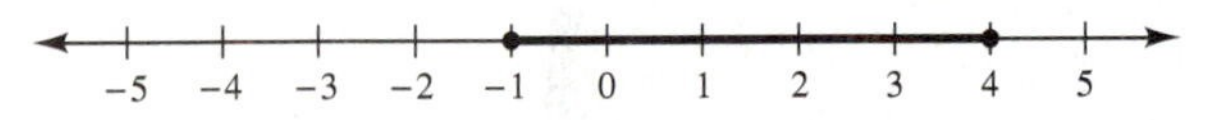

9. Solve $x^2 - 5x + 6 \leq 0$.

10. Sketch the graph of the compound inequality $y \geq 2x + 1$ and $y \geq x$.

CHAPTER TEST

1. Solve $(2/3)y + 1 > y - 3$.
2. Write a compound inequality that specifies the values shown on the graph.

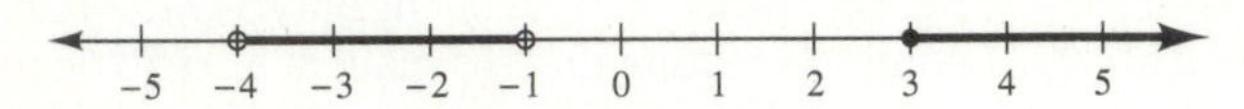

In 3–7, solve each equation or inequality.

3. $\dfrac{x+1}{x+2} > 0$
4. $x^3 - x \leq 0$
5. $|3x - 4| = 2$
6. $|3x - 1| = |x|$
7. $|x + 8| > 3$
8. Sketch the graph of $x - y \geq 4$.
9. An object dropped from rest falls a distance s (in meters) given by $s = 4.9t^2$, where t is the length of time it falls (in seconds). How long will it take for an object to fall at least 36 meters?
10. An industrial engineer claims that with an error of $\pm 1.5\%$, 97% of all parts manufactured by his company meet the given specifications. Let x be the actual percent, and state the condition as an absolute value inequality.

EXTENDED APPLICATION: CHEMICAL ENGINEERING TECHNOLOGY

WASTEWATER TREATMENT

In an *activated sludge system,* microorganisms are used to remove organic material from waste water. The microorganisms use part of the organic material as a carbon source to support growth. The rest is oxidized to provide energy for both growth and nongrowth functions. In order for growth to take place, there must be adequate supplies of certain elements. Generally, municipal wastewater contains the necessary elements, but industrial wastewaters sometimes require additional nitrogen and/or phosphorous. If the treatment process is to function properly, these elements must be added.

Consider a wastewater treatment plant that requires at least 600 pounds of nitrogen per day and at least 100 pounds of phosphorous per day. The wastewater itself supplies 208 pounds of nitrogen and 60 pounds of phosphorous, so at least 392 pounds of nitrogen and at least 40 pounds of phosphorous must be added.

Suppose two commercial mixtures of nitrogen and phosphorous are available. Brand A contains 28 pounds of nitrogen and 5 pounds of phosphorous. Brand B contains 56 pounds of nitrogen and 4 pounds of phosphorous. Because Brand B can be purchased in bulk, each bag costs the same amount as a bag of Brand A, and that amount is $6.00. The chemical engineer must decide how many bags of each mixture should be used daily to provide the required elements at minimum cost.

Problems of this type are called *linear programming* problems, because the conditions can be described by linear inequalities.

$$\text{Let } x = \text{number of bags of Brand A needed.}$$
$$\text{Let } y = \text{number of bags of Brand B needed.}$$

Both brands contain nitrogen, so the amount of nitrogen added is represented by $28x + 56y$, and this amount must be at least 392. Both brands contain phosphorous, so the amount of phosphorous added is represented by $5x + 4y$, and this amount must be at least 40. These conditions are expressed by the following inequalities.

$$28x + 56y \geq 392$$
$$5x + 4y \geq 40$$

Furthermore, since it makes no sense to purchase a negative number of bags, we know that x and y are both greater than or equal to zero.

$$x \geq 0 \text{ and } y \geq 0$$

Figure 8.38 shows the graph of these inequalities. The shaded region is called the *feasible* region, because the values represented within it satisfy the requirements, but only one such pair of values will do so at minimum cost. To find this pair, consider the cost C for x bags of Brand A and y bags of Brand B. We have $C = 6x + 6y$. Figure 8.39 shows the line $C = 6x + 6y$ plotted on the graph for various values of C.

EXTENDED APPLICATION, continued

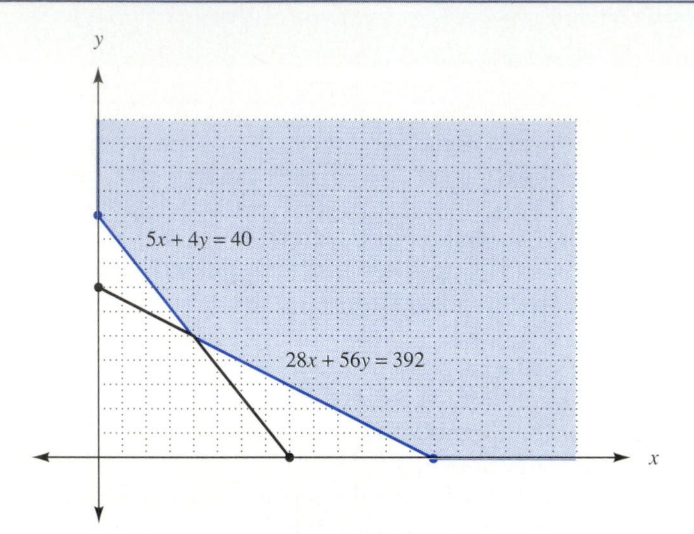

FIGURE 8.38

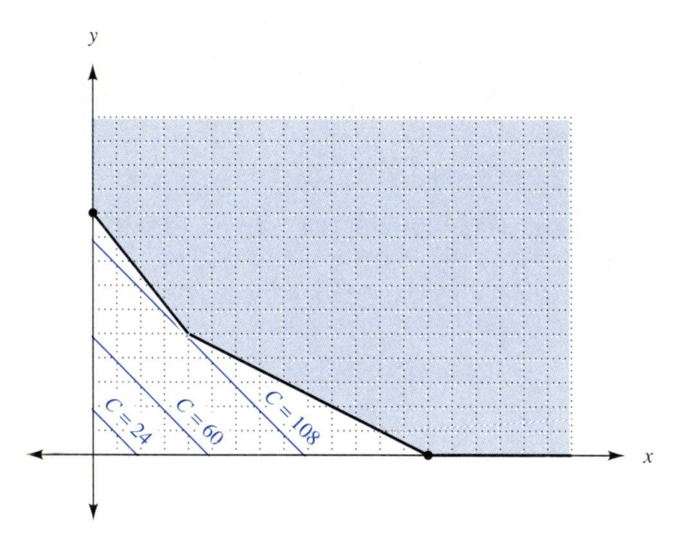

FIGURE 8.39

The smallest value of C for which the line touches the feasible region is the minimum cost, and it will always occur at a vertex, or corner point. The coordinates of the point at which it touches are the values of x and y sought. From the graph, we see that the solution is $x = 4$ and $y = 5$.

ANALYTIC GEOMETRY CHAPTER 9

When a cone is sliced by a plane, the shape of the cross section depends on the angle at which the plane is tilted in relation to the cone. Figure 9.1 illustrates the possibilities. If the cut is perpendicular to the axis of the cone, the cross section is a circle. If the cut is parallel to the edge of the cone, the cross section is a parabola. If the plane is tilted at

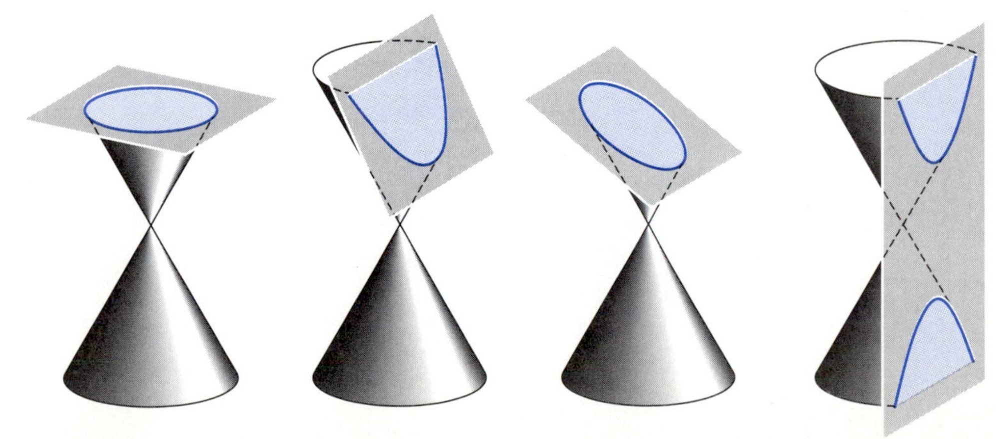

FIGURE 9.1

an angle between these positions, the cross section is an ellipse. And if the cut is parallel to the axis of the cone, the cross section is a shape called a hyperbola. In this chapter, we examine the mathematical properties of these figures.

9.1 THE STRAIGHT LINE AND THE DISTANCE FORMULA

Analytic geometry is the branch of mathematics that deals with the relationship between equations and their graphs. In Section 3.3, for example, by putting the equation in the form $y = mx + b$, we were able to obtain the slope and y-intercept of a straight line. Then the graph was sketched from this information. In preparation for the study of other types of figures, we examine the straight line in more detail.

The shortest path between two points lies along a straight line. We begin by considering the distance between two points (x_1, y_1) and (x_2, y_2). In Figure 9.2, line segment AB is the hypotenuse of triangle ABC. Let d represent its length. The length of side AC is simply the difference $x_2 - x_1$ of the x-values of A and C. The length of side BC is the difference $y_2 - y_1$ of the y-values of B and C.

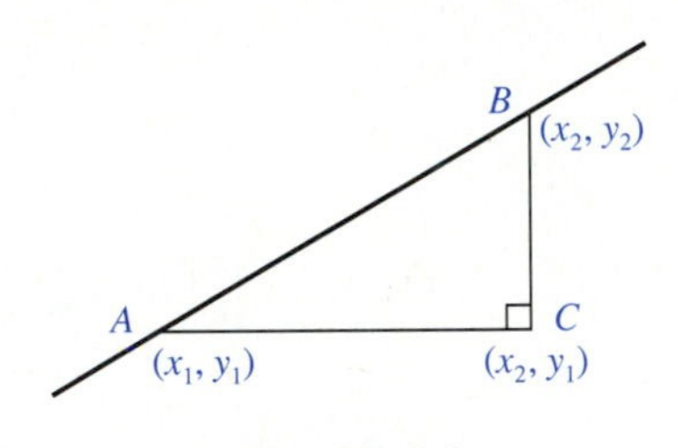

FIGURE 9.2

From the Pythagorean theorem, we have

$$d^2 = (x_2 - x_1)^2 + (y_2 - y_1)^2.$$

Although the equation has both a positive and a negative solution, only the positive solution makes sense as a length. Thus,

$$d = \sqrt{(x_2 - x_1)^2 + (y_2 - y_1)^2}.$$

We have derived the formula known as the *distance formula.*

THE DISTANCE FORMULA

If (x_1, y_1) and (x_2, y_2) are two points in a plane, then the distance d between them is given by

$$d = \sqrt{(x_2 - x_1)^2 + (y_2 - y_1)^2}.$$

EXAMPLE 1 In road construction, the term *grade* refers to the amount of change in elevation over a specified horizontal distance. A grade of 3% indicates that there is a vertical change of 3 ft over a horizontal distance of 100 ft. Consider a road with a grade of 3% over the first 320.0 ft (measured horizontally). The road then inclines with a grade of 2% over the next 530.0 ft. Find the length of the roadway with the 2% grade.

Solution Imagine that the roadway begins at the origin of a coordinate system, as shown in Figure 9.3. Over the 3% grade, the road rises a distance of (0.03)(320.0) or 9.600 ft. The 3% grade extends from (0, 0) to (320.0, 9.600). Over the 2% grade, the road rises a distance of (0.02)(530.0) or 10.60 ft. The 2% grade begins at (320.0, 9.600) and extends to (320.0 + 530.0, 9.600 + 10.60) or (850.0, 20.20).

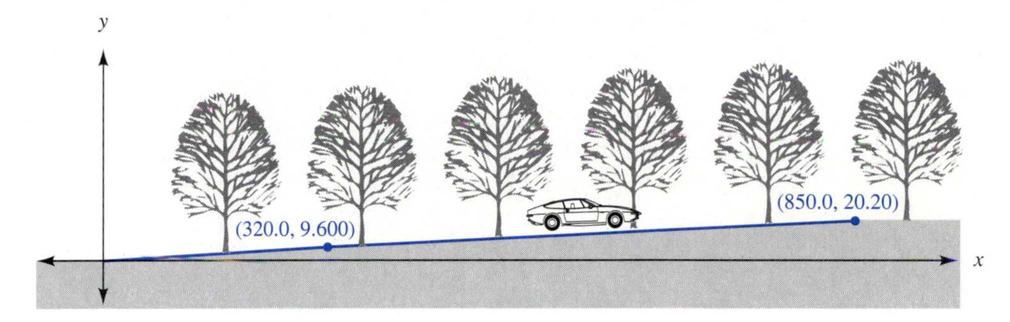

FIGURE 9.3

Let $(x_1, y_1) = (320.0, 9.600)$ and $(x_2, y_2) = (850.0, 20.20)$

$$d = \sqrt{(x_2 - x_1)^2 + (y_2 - y_1)^2}$$

$$d = \sqrt{(850.0 - 320.0)^2 + (20.20 - 9.600)^2}$$

$$= \sqrt{(530.0)^2 + (10.60)^2} \text{ or } 530.1 \text{ ft}$$ ■

Another important characteristic of a line is its slope. Recall from Section 3.3 that the slope of a line is a measure of its steepness. The larger the absolute value of the slope, the more nearly vertical the line is. The slope of a line can be calculated from any two points that lie on it.

DEFINITION If (x_1, y_1) and (x_2, y_2) are any two points on a nonvertical line, the slope is given by

$$m = \frac{y_2 - y_1}{x_2 - x_1}.$$

EXAMPLE 2 Find the slope of the line through (1, −2) and (2, 3).

Solution If $(x_1, y_1) = (1, -2)$ and $(x_2, y_2) = (2, 3)$, then

$$m = \frac{3 - (-2)}{2 - 1} = \frac{3 + 2}{1} = 5.$$ ■

Notice that in Example 2, it does not matter which point we call (x_1, y_1) and which we call $(x_2, y)_2$. If $(x_1, y_1) = (2, 3)$ and $(x_2, y_2) = (1, -2)$, then $m = \dfrac{-2-3}{1-2} = \dfrac{-5}{-1} = 5$. ***It is important, however, that the y-coordinate listed first in the numerator and the x-coordinate listed first in the denominator belong to the same point.***

NOTE ⇨⇨

The slope of a line can be measured between any two points on the line. Consider Figure 9.4. If horizontal and vertical lines are drawn as shown to form triangles ABC and $A'B'C'$, the two horizontal lines are parallel lines cut by a transversal. Thus, angle A and angle A' are equal. The two vertical lines are parallel lines cut by a transversal, so angle B and angle B' are equal. Angle C and angle C' are both right angles. Thus, triangle ABC and triangle $A'B'C'$ are similar.

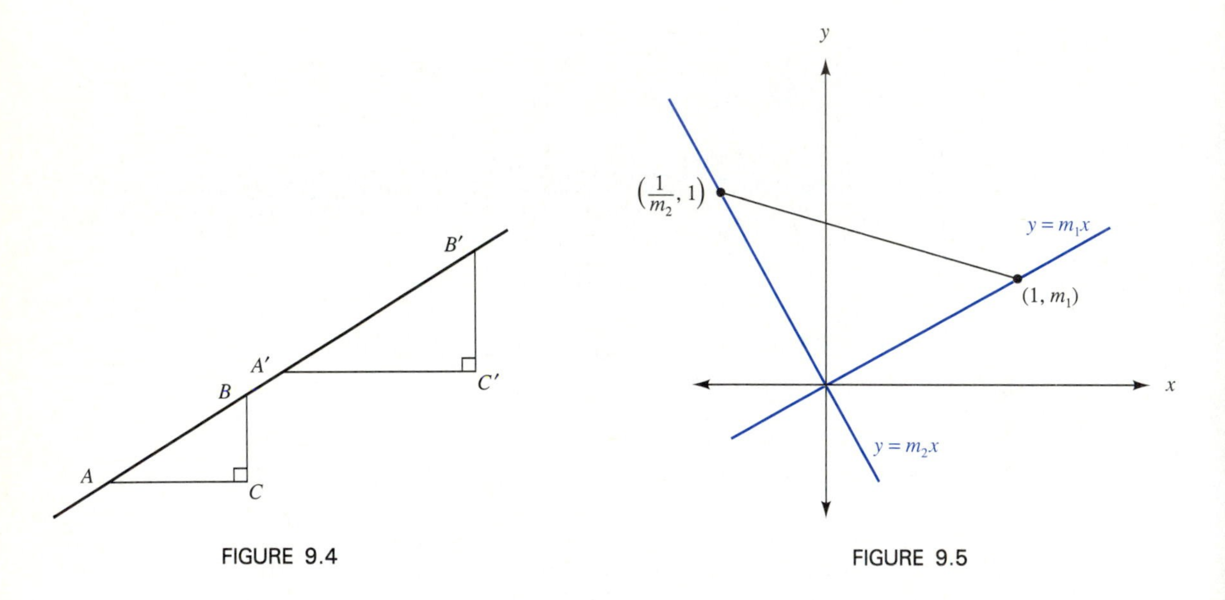

FIGURE 9.4

FIGURE 9.5

The ratio of vertical side to horizontal side is the same for both triangles and gives the slope of the line. That is, the slope is the same regardless of which points on the line are used to compute it.

Since slope is a measure of steepness, and two lines that are parallel have the same degree of steepness, their slopes must be equal. The slopes of perpendicular lines are also related. To discuss slopes of perpendicular lines, some new terminology is useful. The **reciprocal of a nonzero number** x is defined to be $1/x$. If the number x is written as a fraction, the reciprocal of x is simply the fraction obtained by interchanging numerator and denominator. Perpendicular lines have slopes that are negative reciprocals of each other. This relationship is verified as follows. Consider two lines that are perpendicular to each other. For convenience, we examine two lines that intersect at the origin, as in Figure 9.5, but the proof can be generalized.

reciprocal of a nonzero number

Let $y = m_1 x$ be the equation of one of these lines. If x is 1, then y is m_1, so the point $(1, m_1)$ is on this line. Let $y = m_2 x$ be the equation of the other line. If y is 1, then x is $1/m_2$, so the point $(1/m_2, 1)$ is on this line. Since the two lines are perpendicular, the points (0, 0), $(1, m_1)$, and $(1/m_2, 1)$ are the vertices of a right triangle. From the Pythagorean theorem, we have

$$(\sqrt{1^2 + (1/m_2)^2})^2 + (\sqrt{1^2 + m_1^2})^2 = (\sqrt{(1 - 1/m_2)^2 + (m_1 - 1)^2})^2$$
$$1^2 + (1/m_2)^2 + 1^2 + m_1^2 = (1 - 1/m_2)^2 + (m_1 - 1)^2$$
$$1 + 1/m_2^2 + 1 + m_1^2 = 1 - 2/m_2 + 1/m_2^2 + m_1^2 - 2m_1 + 1.$$

Subtract $1 + 1/m_2^2 + 1 + m_1^2$ from both sides, leaving

$$0 = -2/m_2 - 2m_1$$
$$2m_1 = -2/m_2$$
$$m_1 = -1/m_2 \text{ and } m_2 = -1/m_1.$$

In a similar manner, it follows that only perpendicular lines have slopes that are negative reciprocals of each other.

SLOPES OF PARALLEL OR PERPENDICULAR LINES

(a) Lines that are parallel have slopes that are equal.
(b) Lines that are perpendicular have slopes that are negative reciprocals of each other.

EXAMPLE 3 Verify that the line segments in Figure 9.6 form a parallelogram.

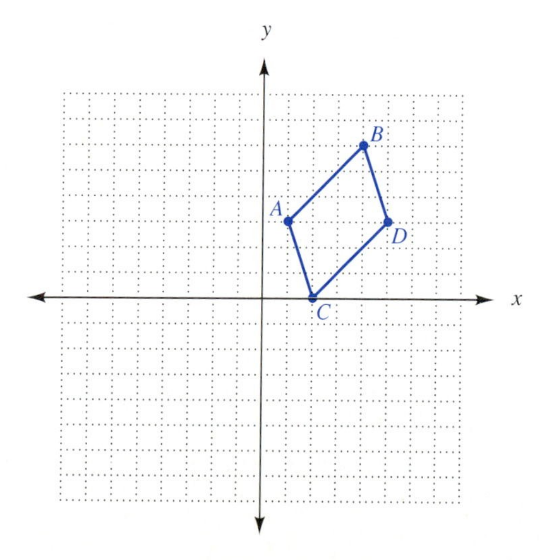

FIGURE 9.6

Solution If the opposite sides are parallel to each other, the figure is a parallelogram. The slope of line AB is $\frac{6-3}{4-1}$, which is $\frac{3}{3}$ or 1. The slope of line CD is $\frac{3-0}{5-2}$, which is $\frac{3}{3}$ or 1. Since the slopes are equal, the sides are parallel. Line AC has slope $\frac{3-0}{1-2}$, which is $\frac{3}{-1}$ or -3, and line BD has slope $\frac{6-3}{4-5}$, which is $\frac{3}{-1}$ or -3. Since the slopes are equal, the sides are parallel. ■

EXAMPLE 4 Find the equation of the line through (0, 4) that is

(a) parallel to the graph of $y = 2x - 3$.

(b) perpendicular to the graph of $y = 2x - 3$.

Solution **(a)** The slope of the line given by $y = 2x - 3$ is 2 or 2/1. The slope of a line parallel to this line is also 2. Since (0, 4) is on the y-axis, this point is the y-intercept. For the equation of the line sought, m is 2 and b is 4, so $y = 2x + 4$.

(b) The slope of the line given by $y = 2x - 3$ is 2. The slope of a line perpendicular to this line is $-1/2$. Since (0, 4) is on the y-axis, this point is the y-intercept. For the equation of the line sought, m is $-1/2$ and b is 4, so $y = (-1/2)x + 4$. ■

To use the formula $y = mx + b$ to write the equation of a line, the slope and y-intercept must be known. For this reason, the formula is called the *slope-intercept formula* for the equation of a line. A second formula, called the *point-slope formula,* permits us to write the equation of a line when the slope and any point (not necessarily the y-intercept) are known. The formula is obtained from the definition of the slope. Let (x_1, y_1) represent the point whose coordinates are known and let m represent the slope of the line. Let (x, y) represent a different, but unspecified, point. Since the coordinates of this point are variables, it can represent any point (other than (x_1, y_1)) on the line. Then

$$m = \frac{y - y_1}{x - x_1}.$$

Multiplying both sides of the equation by $x - x_1$, we have

$$m(x - x_1) = y - y_1.$$

THE POINT-SLOPE FORMULA FOR A STRAIGHT LINE

The equation of a line through the point (x_1, y_1) with slope m is

$$y - y_1 = m(x - x_1).$$

EXAMPLE 5 **(a)** Write the equation of the line through $(-1, 2)$ having slope $2/3$.

(b) Put the equation in slope-intercept form.

Solution **(a)** Since one point and the slope are known, the point-slope formula is used where m is $2/3$ and (x_1, y_1) is $(-1, 2)$. That is, x_1 is -1 and y_1 is 2.

$$y - 2 = (2/3)[x - (-1)]$$
$$y - 2 = (2/3)(x + 1)$$

(b) To put the equation in slope-intercept form, we solve for y.

$$y = (2/3)(x + 1) + 2 \quad \text{Add 2 to both sides.}$$
$$= (2/3)x + 2/3 + 2 \quad \text{Use the distributive property.}$$
$$= (2/3)x + 8/3 \quad \text{Replace } 2/3 + 2 \text{ or } 2/3 + 6/3 \text{ with } 8/3.$$
■

EXAMPLE 6 The relationship between the volume V (in cm^3) of a gas and the temperature T (in degrees Celsius) is linear. When T is $-3°$ C, V is 450 cm^3, and when T is $30°$ C, V is 505 cm^3. Use this information to write an equation that gives V in terms of T.

Solution The information given can be specified using ordered pairs (T, V). Since the relationship between T and V is linear, the points $(30, 505)$ and $(-3, 450)$ must lie on a line. To use either the slope-intercept formula or the point-slope formula, it is necessary to know the slope. Although it was not given, it can be calculated from the two given points.

$$m = \frac{505 - 450}{30 - (-3)} = \frac{55}{33} = \frac{5}{3}$$

Once the slope is known, the point-slope formula may be used. It does not matter which point is used as (T_1, V_1) in the formula. In fact, the problem is worked both ways to show you that the same equation is obtained.

If $(T_1, V_1) = (30, 505)$

$$V - 505 = (5/3)(T - 30)$$
$$V - 505 = (5/3)T - 50$$
$$V = (5/3)T - 50 + 505$$
$$V = (5/3)T + 455$$

If $(T_1, V_1) = (-3, 450)$

$$V - 450 = (5/3)[T - (-3)]$$
$$V - 450 = (5/3)(T + 3)$$
$$V = (5/3)T + 5 + 450$$
$$V = (5/3)T + 455$$
■

EXERCISES 9.1

In 1–8, find the distance between each pair of points.

1. $(2, -3)$ and $(7, -15)$

2. $(-3, 4)$ and $(-11, 19)$

3. $(-4, -3)$ and $(-10, 5)$

4. $(0, 1)$ and $(12, 10)$

5. $(-1, 0)$ and $(-7, 9)$

6. $(-2, 3)$ and $(5, 7)$

7. $(3, -4)$ and $(-6, -2)$

8. $(-4, -1)$ and $(5, 3)$

In 9–15, find the slope of the line through each pair of points.

9. (3, 5) and (7, 9)
10. (2, 6) and (4, 8)
11. (1, −7) and (10, 7)
12. (−4, 4) and (3, −2)
13. (−3, 5) and (−8, 1)
14. (2, −6) and (3, −4)
15. (1, −3) and (−3, 1)

In 16 and 17, verify that the figure formed by the line segments joining the given points is a parallelogram.

16. (−2, 1), (1, −1), (−3, −5), and (−6, −3)
17. (−1, 3), (2, 0), (1, −4), and (−2, −1)

In 18 and 19, verify that the figure formed by the line segments joining the given points is a right triangle.

18. (−3, 0), (0, −3), and (5, 2)
19. (−1, 2), (1, −2), and (7, 1)

In 20 and 21, verify that the figure formed by the line segments joining the given points is an isosceles triangle.

20. (−3, 3), (−1, −1), and (3, 1)
21. (1, 1), (0, −2), and (−3, −3)

In 22 and 23, verify that the figure formed by the line segments joining the given points is a square.

22. (0, 2), (2, 0), (0, −2), and (−2, 0)
23. (−2, 0), (1, −1), (2, 2), and (−1, 3)

In 24 and 25, verify that the figure formed by the line segments joining the given points is NOT a square, even though the sides are the same length.

24. (−2, 1), (1, −1), (4, 1), and (1, 3)
25. (−1, 0), (0, −2), (1, 0), and (0, 2)

In 26–41, find the equation of each line described.

26. through (2, 3) parallel to $y = (1/2)x - 3$
27. through (−1, 4) parallel to $3y - 2x = 6$
28. through (1, −2) perpendicular to $3x + 2y = 6$
29. through (−2, −3) perpendicular to $y = -3x + 2$
30. through (−1/2, 0) perpendicular to $y = (1/4)x + 3$
31. through (0, 2/3) parallel to $3x - 4y = 12$
32. through (1/4, 3/4) parallel to $y = 4x - 2$
33. through (1/3, −2/3) perpendicular to $y = -3x + 1$
34. through (4, 4) and (6, 10)
35. through (3, 5) and (7, 5)
36. through (2, −8) and (9, 8)
37. through (−3, 3) and (2, −1)
38. through (−2, 4) and (−9, 2)
39. through (3, −7) and (2, −3)
40. through (2, −4) and (−4, 2)
41. through (8, 8) and (2, 6)

---

42. The relationship between the temperature T (in degrees Celsius) at a particular height (in m) and the ground temperature t (in degrees Celsius) is linear. Given that at height 10 000 m the temperature is −80° C when the ground temperature is 20° C and $T = -90°$ C when $t = 10°$ C write an equation that gives T in terms of t.
6 8 1 2 3 5 7 9

43. In road construction, the grade is often given as a percent. A grade of 3% can also be expressed as a slope of 3/100. Determine the grade for a road that rises 5 ft over a horizontal distance of 250 ft.
1 2 4 6 8 5 7 9

44. The relationship between the depth (in m) of an object submerged in seawater and the pressure (in kg/cm²) is linear. An object submerged to a depth of 1 m is subjected to a pressure of 1.133 kg/cm^2, and an object submerged to a depth of 2 m is subjected to a pressure of 1.235 kg/cm^2. Write an equation that gives pressure p in terms of depth d. 4 8 1 2 6 7 9

45. The relationship between room temperature t (in degrees C) and the reduction in energy costs c below the cost at 22° C is linear. If lowering the temperature to 20° C saves 9% in energy costs and lowering it to 18° C saves 18%, write an equation that gives c in terms of t. 6 8 1 2 3 5 7 9

46. The relationship between angular velocity ω (in rad/s) and time t (in s) is linear when the angular acceleration is constant. A flywheel that has a velocity of $\omega = 150$ rad/s when $t = 5$ s has a velocity of 200 rad/s when $t = 10$ s. Write an equation that gives ω in terms of t. 4 7 8 1 2 3 5 6 9

47. The relationship between energy used (e) and time of use (t) is linear. A 75 W bulb used 0.90 kwh in 12 hours and 1.8 kwh in 24 hours. Use this information to write an equation that gives e in terms of t. 3 5 9 1 2 4 6 7 8

48. If an equation fits the pattern $y = mx + b$, the slope and y-intercept are immediately available. Show how a linear equation can be put in a form to show the x-intercept and y-intercept. 3 5 9 1 2 4 6 7 8

9.2 THE PARABOLA

In Section 6.3, we saw that a parabola given by the equation $y = ax^2 + bx + c$ can be graphed by plotting its vertex and just a few additional points. Our goal in this section is to emphasize the relationship between the coefficients of the equation and the vertex by putting the equation in a different form. We begin with a definition.

DEFINITION A **parabola** is the set of all points in a plane such that each is equidistant from a fixed line and a fixed point not on the line. The fixed line is called the **directrix** and the fixed point is called the **focus** of the parabola.

Consider the parabola shown in Figure 9.7. The distance d between two points is the length of the line segment that connects them. The distance from the focus F to a point (x, y) on the parabola could be computed using the distance formula. The distance between a point and a given line is understood to be the distance along the shortest path from the point to the line. This distance is the length of the line segment from the point perpendicular to the given line. In Figure 9.7, the distance between (x, y) and the directrix l is measured along the vertical line shown.

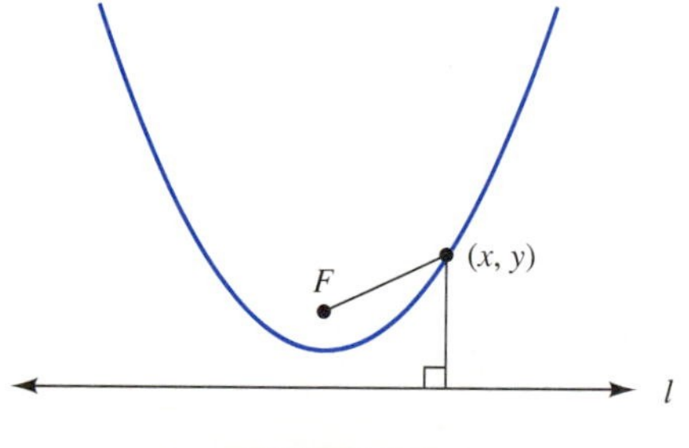

FIGURE 9.7

Now we are ready to write the equation for a parabola, based on the definition. For simplicity, we first consider a parabola that opens upward and whose vertex is at the origin, as in Figure 9.8. Suppose that the focus is p units above the vertex. Then its coordinates are $(0, p)$.

Since the vertex is a point on the parabola, it must be equidistant from focus and directrix. Thus, the directrix is located p units below the vertex. As a horizontal line, its equation is $y = -p$. Let (x, y) be any point on the parabola. The distance from the focus to (x, y) is equal to the distance from (x, y) to the directrix. Let us calculate these distances. From the focus $(0, p)$ to (x, y), we have

$$d_1 = \sqrt{(x-0)^2 + (y-p)^2}.$$

From the point (x, y) measured vertically to the point $(x, -p)$ on the directrix, the distance is simply the difference in y-values. That is,

$$d_2 = y - (-p) = y + p.$$

Since $d_2 = d_1$,

$$y + p = \sqrt{x^2 + (y-p)^2}.$$

Squaring both sides, we have

$$(y+p)^2 = x^2 + (y-p)^2$$
$$y^2 + 2yp + p^2 = x^2 + y^2 - 2yp + p^2.$$

Adding $2yp - y^2 - p^2$ to both sides and combining like terms leaves

$$4yp = x^2.$$

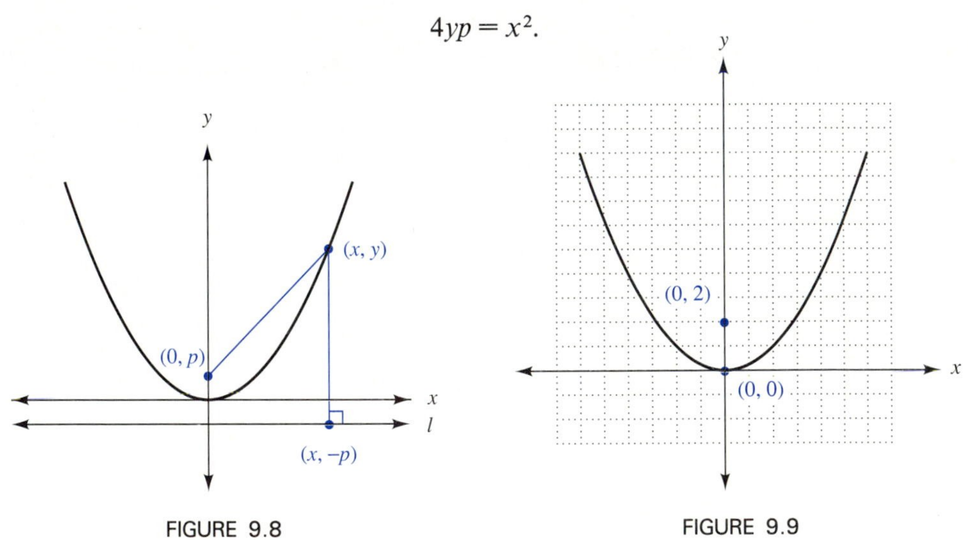

FIGURE 9.8

FIGURE 9.9

EXAMPLE 1 Find the equation of the parabola with its vertex at the origin and focus at $(0, 2)$.

Solution Since the vertex is at $(0, 0)$ and the focus is at $(0, 2)$, the distance between vertex and focus is 2, as shown in Figure 9.9. Thus, p is 2 and the equation is

$$4(2)y = x^2 \text{ or } 8y = x^2. \quad ■$$

EXAMPLE 2 Find the equation of the parabola with focus at $(0, -1)$ and directrix given by $y = 1$.

Solution Since the vertex must be halfway between focus and directrix, its y-coordinate will be halfway between -1 and 1, as shown in Figure 9.10.

The x-coordinate of the vertex will equal the x-coordinate of the focus, so $(0, 0)$ is the vertex. Previously, we let p represent the number of units *above* the vertex required to locate the focus. In this case, however, the focus is one unit *below* the vertex, so we say that p is -1. The equation, then, is

$$4(-1)y = x^2 \text{ or } -4y = x^2. \quad ■$$

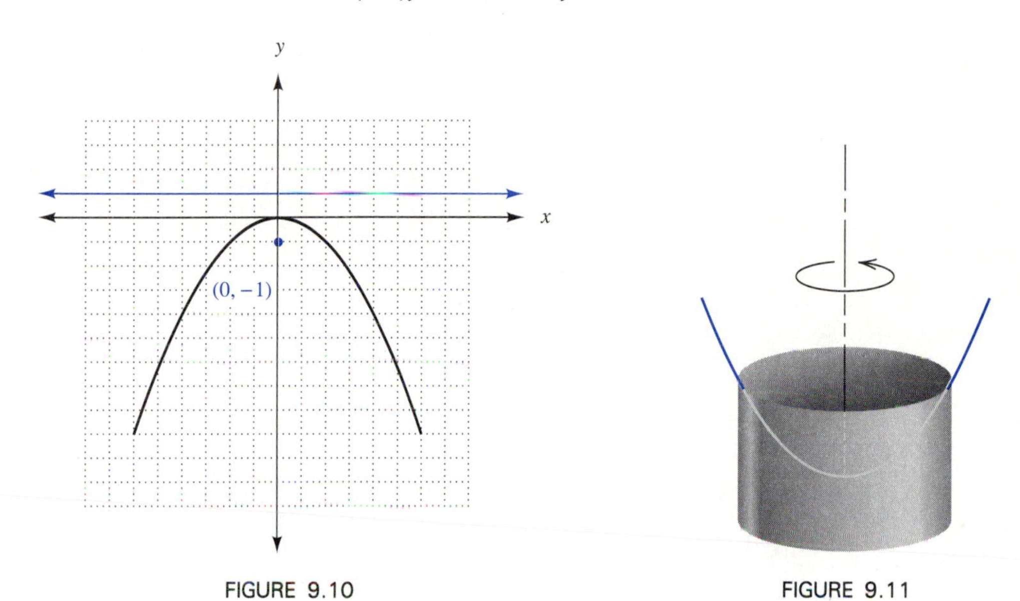

FIGURE 9.10

FIGURE 9.11

Because of its reflective properties, the parabola has applications in communications, electronic surveillance, radar, lighting, and astronomy. Imagine a parabola that is rotated through a disk of clay, as shown in Figure 9.11, so that the surface of the clay is shaped. A surface of this shape is said to be parabolic and has the property that light or radio waves striking its surface are all reflected to the focus.

EXAMPLE 3 A telescope has a focal length of 1200 mm. That is, the focus of the parabolic mirror is located at a distance of 1200 mm from the vertex. Assuming that the vertex is at the origin, write the equation of the parabola.

Solution For this parabola, $p = 1200$. Thus,

$$4py = x^2$$
$$4(1200)y = x^2$$
$$4800y = x^2. \quad ■$$

Most of the parabolas you have seen in the past did not have the vertex at the origin. Our next task is to modify the equation to account for the case when the vertex occurs at a point whose coordinates are (h, k).

translation of axes

Imagine a second set of axes whose origin is at (h, k). The locating of a second set of axes parallel to the first pair is sometimes referred to as a *translation of axes*. If we use x' to indicate the horizontal axis and y' to indicate the vertical axis, then there are really two coordinate systems, as shown in Figure 9.12.

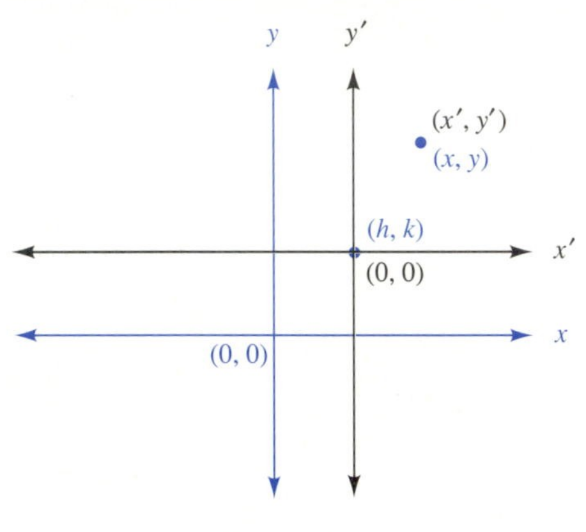

FIGURE 9.12

Consider the point (x, y) in the original system. Its coordinates in the new system are (x', y'). To convert coordinates in one system to coordinates in the other system, notice the relationship between (x, y) and (x', y'). Since x' is the point x' units over from the y'-axis, and the y'-axis is h units over from the y-axis, we have $x = x' + h$. Likewise, y' is the point y' units away from the x'-axis in the vertical direction. Since the x'-axis is k units above the x-axis, we have $y = y' + k$. That is,

$$x = x' + h \qquad \text{and} \qquad y = y' + k,$$

or

$$x - h = x' \qquad \text{and} \qquad y - k = y'.$$

A parabola whose vertex is at the origin of the $x'y'$ system has equation $4py' = x'^2$. In the xy system, the vertex is at (h, k), and the equation is $4p(y - k) = (x - h)^2$. ***Notice that the y-coordinate of the vertex is subtracted from y, and the x-coordinate of the vertex is subtracted from x.*** It is also important that $(y - k)$ is to the first power, but $(x - h)$ is squared.

NOTE ⇨⇨

THE VERTICAL PARABOLA

The equation of a parabola with vertex at (h, k) and focus p units away in the vertical direction is given by

$$4p(y - k) = (x - h)^2.$$

If $p > 0$, the parabola opens upward.
If $p < 0$, the parabola opens downward.

EXAMPLE 4 **(a)** Find the equation of the parabola with vertex at (2, 3) and focus at (2, 1).

(b) Write the equation in standard form ($y = ax^2 + bx + c$).

Solution Figure 9.13 shows the graph.

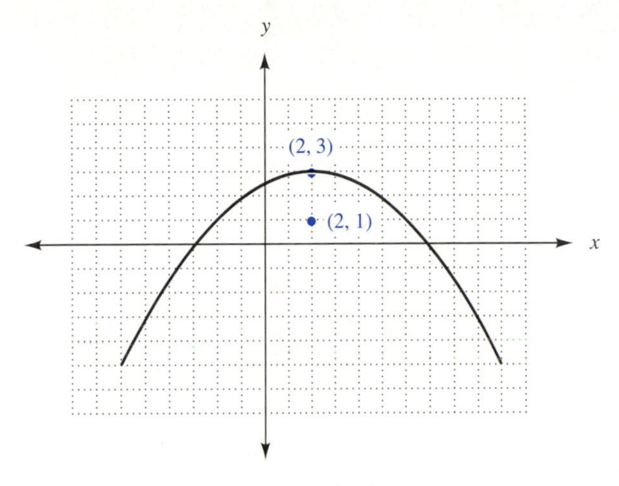

FIGURE 9.13

(a) Since the focus is 2 units *below* the vertex, p is -2. The equation, then, is

$$4(-2)(y-3) = (x-2)^2$$
$$-8(y-3) = (x-2)^2.$$

(b) To put the equation in standard form, we write

$$-8y + 24 = x^2 - 4x + 4 \quad \text{Simplify each side.}$$
$$-8y = x^2 - 4x - 20 \quad \text{Subtract 24 from both sides.}$$
$$y = (-1/8)x^2 + (1/2)x + 5/2 \quad \text{Divide both sides by } -8. \blacksquare$$

When a parabola is oriented in the horizontal direction, the roles of x and y are interchanged.

THE HORIZONTAL PARABOLA

The equation of a parabola with vertex at (h, k) and focus p units away in the horizontal direction is given by

$$4p(x-h) = (y-k)^2.$$

If $p > 0$, the parabola opens to the right.
If $p < 0$, the parabola opens to the left.

EXAMPLE 5 Find the equation of the parabola with focus at $(2, -2)$ and directrix given by $x = -4$.

Solution Since the vertex must be halfway between the focus and directrix $((-4 + 2)/2 = -1)$, its coordinates are $(-1, -2)$, as shown in Figure 9.14. Since the focus is three units to the right of the vertex, p is 3.

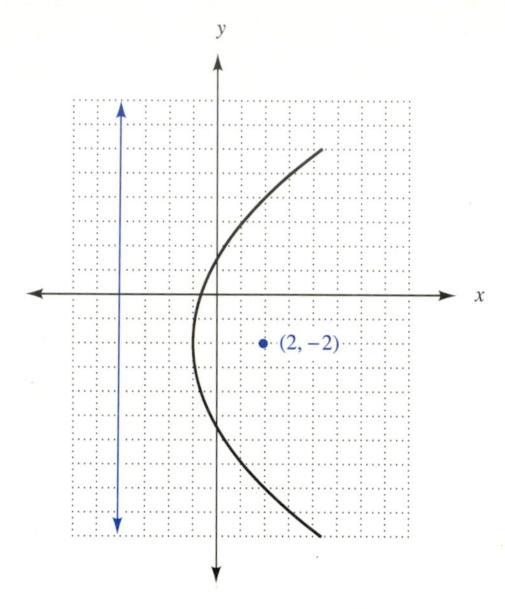

FIGURE 9.14

$$\begin{aligned}
4(3)[x - (-1)] &= [y - (-2)]^2 \\
12(x + 1) &= (y + 2)^2 \\
12x + 12 &= y^2 + 4y + 4 \\
12x &= y^2 + 4y - 8 \\
x &= (1/12)y^2 + (4/12)y - 8/12 \\
x &= (1/12)y^2 + (1/3)y - 2/3
\end{aligned}$$ ■

EXERCISES 9.2

In 1–9, write the equation of the parabola that satisfies the given conditions. Put the equation in standard form.

1. vertex at the origin, focus at $(0, 1)$

2. vertex at the origin, focus at $(0, 3)$

3. vertex at the origin, focus at $(0, -1/4)$

4. vertex at the origin, focus at $(0, -1/2)$

5. vertex at the origin, directrix $y = -1$

6. vertex at the origin, directrix $y = 2$

7. vertex at the origin, directrix $y = 3/4$

8. vertex at the origin, directrix $y = -1/8$

9. focus at $(0, 1/2)$, directrix $y = -1/2$

In 10–21, write the equation of the parabola that satisfies the given conditions. You need not put the equation in standard form.

10. focus at $(0, -3/8)$, directrix $y = 3/8$

11. focus at $(0, -3)$, directrix $y = 3$

12. focus at $(0, 2)$, directrix $y = -2$

13. vertex at the origin, focus at $(1, 0)$

14. vertex at the origin, focus at $(2, 0)$

15. vertex at the origin, focus at $(-1/4, 0)$

16. vertex at the origin, directrix $x = -1$

17. vertex at the origin, directrix $x = 2$

18. vertex at the origin, directrix $x = -1/2$

19. focus at $(-1, 0)$, directrix $x = 1$

20. focus at $(3/4, 0)$, directrix $x = -3/4$

21. focus at $(5/8, 0)$, directrix $x = -5/8$

In 22–31, write the equation of the parabola that satisfies the given conditions. Put the equation in standard form.

22. vertex at $(1, 2)$, focus at $(1, 3)$

23. vertex at $(2, 3)$, focus at $(2, 2)$

24. vertex at $(-1, 0)$, focus at $(-1, -1)$

25. vertex at $(0, -2)$, focus at $(0, -1)$

26. vertex at $(2, -1/2)$, directrix $y = 1/2$

27. vertex at $(-3, 1/4)$, directrix $y = 9/4$

28. vertex at $(0, 1)$, directrix $y = 0$

29. vertex at $(2, 0)$, directrix $y = -2$

30. focus at $(2, 4)$, directrix $y = 6$

31. focus at $(-1, 5)$, directrix $y = 1$

In 32–42, write the equation of the parabola that satisfies the given conditions. You need not put the equation in standard form.

32. focus at $(-3, -2)$, directrix $y = -1$

33. focus at $(5, -2)$, directrix $y = -4$

34. vertex at $(1, 2)$, focus at $(2, 2)$

35. vertex at $(2, 3)$, focus at $(0, 3)$

36. vertex at $(-1, 0)$, focus at $(0, 0)$

37. vertex at $(0, -2)$, directrix $x = 1/2$

38. vertex at $(-3, -1)$, directrix $x = -2$

39. vertex at $(1/2, -2)$, directrix $x = 1$

40. focus at $(1/2, 1/4)$, directrix $x = 3/2$

41. focus at $(3/4, -1/2)$, directrix $x = 1/4$

42. focus at $(-1/4, 3/4)$, directrix $x = 1/4$

---

43. In a fountain, water jets form parabolic arches. Consider the center of a certain fountain, which is elevated 5.0 ft off the ground, to be at the origin of a coordinate system. If the equation of the parabolic water arch is $y = -x^2 + 4x$, determine the radius of the basin necessary to catch the water at ground level. 4 8 1 2 6 7 9

44. A dam is constructed so that its horizontal cross section is a parabola. Write the equation of the dam shown in the accompanying figure. 2 1 9

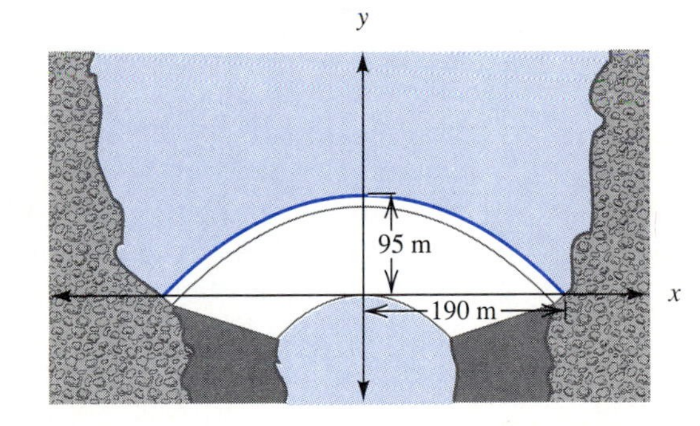

1. Architectural Technology
2. Civil Engineering Technology
3. Computer Engineering Technology
4. Mechanical Drafting Technology
5. Electrical/Electronics Engineering Technology
6. Chemical Engineering Technology
7. Industrial Engineering Technology
8. Mechanical Engineering Technology
9. Biomedical Engineering Technology

45. A fluorescent lamp has a reflector shaped so that its cross section is a parabola, as shown in the accompanying figure. The fluorescent tube is placed at the focus. If the equation of the parabola is $y = -3x^2$, how far from the vertex should the tube be placed? 9 1 2 3 5 6 7

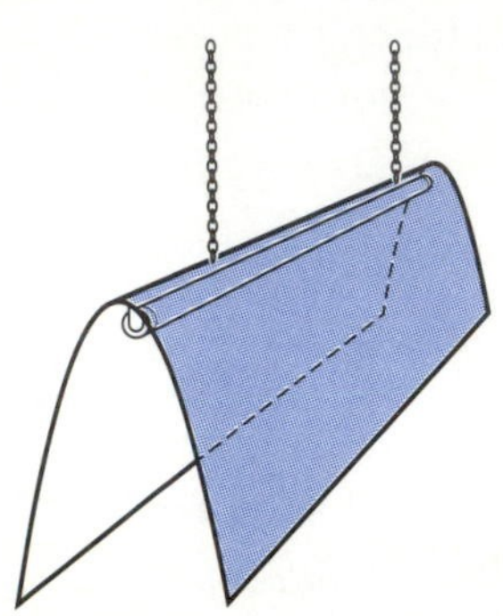

46. A parabolic satellite dish has a radius of 3.00 m and is 0.82 m deep. What is the focal length? 1 3 4 5 6 8 9

47. A parabolic arch is constructed as an architectural feature, and a light is to be hung at the focus. It is known that the height of the arch is 30.0 feet and the distance between the bases is 20.0 feet. How far below the vertex should the light be hung? 1 2 4 7 8 3 5

48. A flat road at an elevation of 1520.0 feet above sea level dips into a sag that approximates a parabola in shape. The lowest point in the sag, called the drain, is at the midpoint of the sag at an elevation of 1228.0 feet above sea level. The total horizontal span of the sag is 3245.0 feet. Assume that the drain is at the origin of a coordinate system, and write the equation of the parabola that approximates the road. 1 2 4 6 8 5 7 9

49. Generally, when a plane that is parallel to the edge of a cone slices the cone, the cross section is a parabola. Describe how the plane must be positioned to produce a cross section that is a straight line.

1. Architectural Technology
2. Civil Engineering Technology
3. Computer Engineering Technology
4. Mechanical Drafting Technology
5. Electrical/Electronics Engineering Technology
6. Chemical Engineering Technology
7. Industrial Engineering Technology
8. Mechanical Engineering Technology
9. Biomedical Engineering Technology

9.3 FINDING THE VERTEX BY COMPLETING THE SQUARE

In the previous section, we saw how to write the equation of a parabola, given any two of the following: focus, vertex, and directrix. Now it is necessary to see how we can use the equation of a parabola to find the vertex, focus, and directrix. That is, given an equation in the form $y = ax^2 + bx + c$, put it in the form $4p(y - k) = (x - h)^2$. The technique is completing the square and is similar to the procedure used in Section 6.1 to solve quadratic equations. In this section, the technique is applied to quadratic functions rather than to quadratic equations in one variable.

Recall that the key to completing the square is to add the square of one-half the coefficient of the linear term to both sides of the equation.

EXAMPLE 1 Use the procedure of completing the square to rewrite the given quadratic equation and quadratic function.

$$x^2 + 6x - 1 = 0 \qquad\qquad x^2 + 6x - 1 = y$$

Solution

Quadratic Equation	*Quadratic Function*	
$x^2 + 6x - 1 = 0$	$x^2 + 6x - 1 = y$	
$x^2 + 6x \quad = 1$	$x^2 + 6x \quad = y + 1$	Add 1 to both sides.
$x^2 + 6x + 9 = 1 + 9$	$x^2 + 6x + 9 = y + 1 + 9$	Add $[(1/2)6]^2$ to both sides.
$(x + 3)^2 = 10$	$(x + 3)^2 = y + 10$	Factor the left and combine $1 + 9$.

$$x + 3 = \pm\sqrt{10}$$
$$x = -3 \pm \sqrt{10}$$ ■

Special care must be taken when the coefficient of the square term is a number other than 1. When we solved quadratic equations, we divided both sides of the equation by the leading coefficient. When one side of the equation consists of y instead of 0, such a division would produce a fraction. To avoid the fraction, we factor rather than divide. Example 2 illustrates the procedure.

EXAMPLE 2 Write an equivalent equation that has a perfect square trinomial on the right-hand side: $y = 2x^2 + 4x$.

Solution Instead of dividing both sides by 2, we factor out 2. Thus, $y = 2(x^2 + 2x)$. To complete the square of $x^2 + 2x$, we add 1. If 1 is added inside parentheses, however, the distributive property applies to $2(x^2 + 2x + 1)$, and the expression is actually larger than $2(x^2 + 2x)$ by 2(1) or 2. Adding 1 inside the parentheses on the right is equivalent to adding 2 on the left. We have

$$y + 2 = 2(x^2 + 2x + 1), \text{ or}$$
$$y + 2 = 2(x + 1)^2$$ ■

The process is summarized as follows.

To complete the square of a quadratic function:

1. Write an equivalent equation that has only the square and linear terms of the same variable on one side of the equation.
2. Factor out the leading coefficient, leaving the square term with a coefficient of 1 in parentheses along with the linear term.
3. Add the square of one-half the coefficient of the linear term to the quantity within parentheses.
4. Multiply the term added within parentheses by the factor outside the parentheses and add the result to the other side of the equation.
5. Factor the trinomial as a perfect square and simplify the expression on the other side of the equation.

EXAMPLE 3 Complete the square of $y = 4x^2 + 16x + 19$.

Solution

1.	$y - 19 = 4x^2 + 16x$	Isolate the x-terms by subtracting 19 from both sides.
2.	$y - 19 = 4(x^2 + 4x)$	Factor out the leading coefficient.
3–4.	$y - 19 + 16 = 4(x^2 + 4x + 4)$	Add $[(1/2)(4)]^2$, or 4 inside parentheses; add 4(4) or 16 on the left.
5.	$y - 3 = 4(x + 2)^2$	Simplify both sides. ■

In Example 3, one more step will put the equation in the form

$$4p(y - k) = (x - h)^2.$$

That is, $y - 3 = 4(x + 2)^2$ is equivalent to $(1/4)(y - 3) = (x + 2)^2$. Thus, we have $4p = 1/4$ or $p = 1/16$ and the focus is 1/16 of a unit above the vertex. We also know that the vertex of the parabola is at $(-2, 3)$.

EXAMPLE 4 Determine the vertex of the parabola $y = -x^2 + 6x - 8$ by completing the square. Sketch the graph of the parabola.

Solution

$y + 8 = -x^2 + 6x$	Isolate the x-terms.
$y + 8 = -1(x^2 - 6x)$	Factor out -1.
$y + 8 - 9 = -(x^2 - 6x + 9)$	Add $[(1/2)(-6)]^2$ or 9 inside parentheses; add $-1(9)$ or -9 on the left.
$y - 1 = -(x - 3)^2$	Simplify both sides.

The vertex is at (3, 1). To sketch the graph, we find a few additional points as shown in the following table.

x	y
0	-8
1	-3
2	0
4	0
5	-3
6	-8

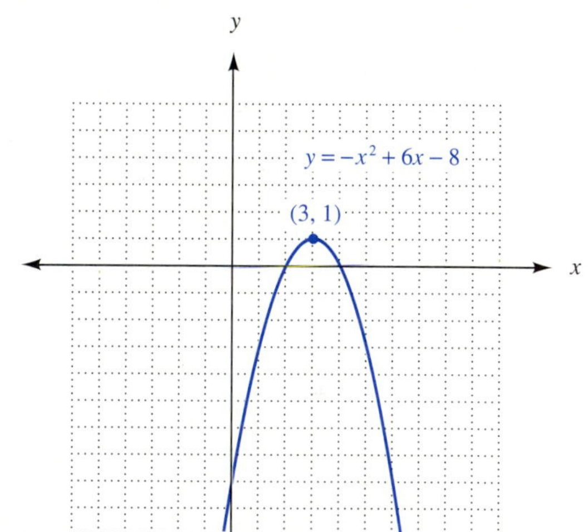

FIGURE 9.15

■

EXAMPLE 5 Determine the vertex of the parabola $x = (1/4)y^2 + 2y + 5$ by completing the square. Sketch the graph of the parabola.

Solution

$$x - 5 = (1/4)y^2 + 2y$$ Isolate the y-terms.

$$x - 5 = (1/4)(y^2 + 8y)$$ Factor out the leading coefficient.

$$x - 5 + 4 = (1/4)(y^2 + 8y + 16)$$ Add $[(1/2)(8)]^2$ or 16 inside parentheses; add $(1/4)(16)$ or 4 on the left.

$$x - 1 = (1/4)(y + 4)^2$$ Simplify both sides.

The vertex is at $(1, -4)$. Since the parabola is a horizontal parabola, we choose a few values for y and find the corresponding values of x.

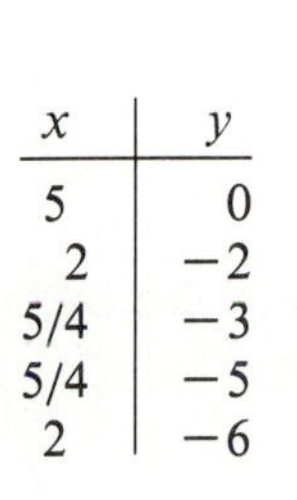

x	y
5	0
2	−2
5/4	−3
5/4	−5
2	−6

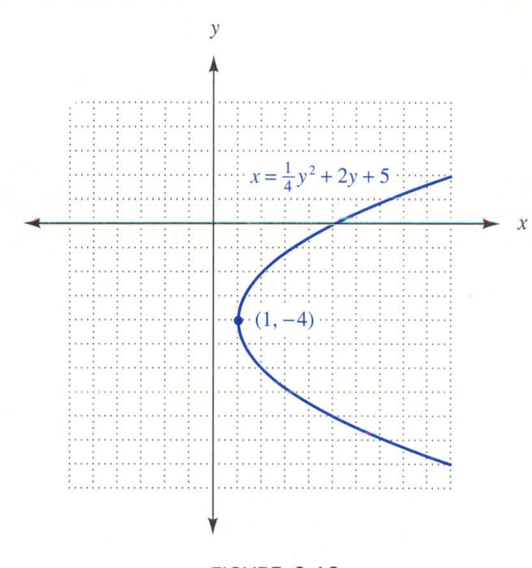

FIGURE 9.16

Sometimes it is easier to use the value of p (particularly when $p > 1$) to find additional points on a parabola. Recall that the value of p indicates the distance between focus and vertex. Consider the equation for a horizontal parabola.

$$4p(x - h) = (y - k)^2 \tag{9-1}$$

Let $x = p + h$. This is the value of x at the focus. To find the y-values on the parabola associated with this x-value, substitute $p + h$ into Equation 9-1.

$$4p[p + h - h] = (y - k)^2$$
$$4p^2 = (y - k)^2$$
$$(2p)^2 = (y - k)^2$$
$$\pm 2p = y - k$$
$$k \pm 2p = y$$

That is, two additional points on the parabola are found by counting from the vertex over p units then $2p$ units above and below the focus.

EXAMPLE 6 In Example 5, it was determined that the equation $x = (1/4)y^2 + 2y + 5$ can be written as $x - 1 = (1/4)(y + 4)^2$. Determine the vertex and use the value of p to sketch the graph.

Solution To determine the value of p, we use the form $4(x - 1) = (y + 4)^2$.

The vertex is at $(1, -4)$. Since $4p = 4$, $p = 1$. From the vertex, count over 1 to locate the focus, then count up 2 and down 2 from the focus to locate the points $(2, -2)$ and $(2, -6)$. A few additional points may be found by substitution. For example, to find the x-intercept, let y be 0. Then x is 5. Figure 9.17 shows the graph. ■

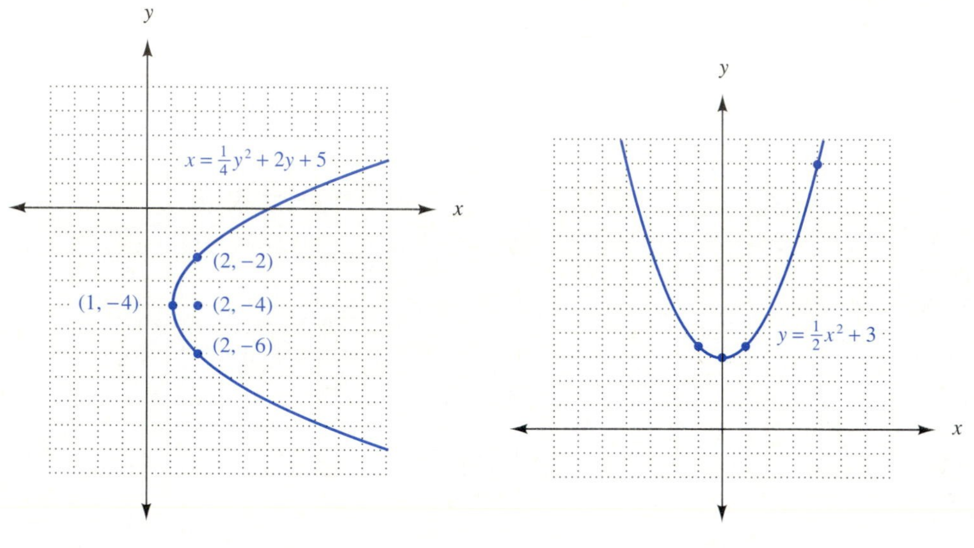

FIGURE 9.17

FIGURE 9.18

EXAMPLE 7 Sketch the graph of $y = (1/2)x^2 + 3$.

Solution

$$\begin{aligned} y &= (1/2)x^2 + 3 \\ y - 3 &= (1/2)x^2 \\ y - 3 &= (1/2)(x^2 + 0x) \\ y - 3 &= (1/2)(x^2 + 0x + 0) \\ y - 3 &= (1/2)(x + 0)^2 \\ 2(y - 3) &= (x + 0)^2 \end{aligned}$$

If you were able to see after the second line that completing the square would lead to $(x + 0)^2$ on the right, then you could have skipped down to the last line. The vertex is at $(0, 3)$, and we have $4p = 2$, or $p = 1/2$. Thus, additional points can be found by counting from the vertex up 1/2 to the focus and then from the focus over 1 to the left and over 1 to the right. That is, additional points are found at $(-1, 3\ 1/2)$ and $(1, 3\ 1/2)$. A few more points may be found by substitution. For example, when x is 4, $y = (1/2)(16) + 3$ or 11. Figure 9.18 shows the graph. ■

Example 8 illustrates a quadratic function in a technical situation.

EXAMPLE 8 When a load is dropped on the center of a simple beam, the deflection D (in mm) is given by $2HD_{st} = D^2 - 2D_{st}D$. In the formula, H is the height (in mm) from which the load is dropped, and D_{st} is the deflection (in mm) that would occur if the same load were placed, rather than dropped, on the beam. If D_{st} is 5 mm, the equation has two variables, H and D. A graph of H versus D is a parabola. Find the vertex.

Solution If D_{st} is 5, the equation is

$$2H(5) = D^2 - 2(5)D$$
$$10\,H = D^2 - 10\,D.$$

The coefficient of D is -10. One-half of this coefficient is -5. Adding 25 to both sides of the equation gives

$$10\,H + 25 = D^2 - 10\,D + 25$$
$$10\,H + 25 = (D - 5)^2$$
$$10(H + 2.5) = (D - 5)^2.$$

The vertex of the parabola, then, is at $(5, -2.5)$. ■

EXERCISES 9.3

In 1–24, determine the vertex of each parabola by completing the square. Sketch the graph of the parabola.

1. $y = x^2 + 4x + 6$
2. $y = -x^2 + 2x + 8$
3. $y = x^2 - 4x - 4$
4. $y = x^2 + 6x + 5$
5. $y = x^2 + 2x + 3$
6. $y = -x^2 + 2x + 3$
7. $y = x^2 - 1$
8. $y = 2x^2 - 9$
9. $y = -2x^2 + 4x$
10. $y = 2x^2 - 6x$
11. $y = 4x^2 - 4x + 1$
12. $y = 4x^2 - 4x - 3$
13. $x = y^2 - 4y + 3$
14. $x = y^2 - 2y + 1$
15. $x = -y^2 + 6y - 5$
16. $x = -y^2 + y + 2$
17. $x = 2y^2 + 4y + 3$
18. $x = 2y^2 - 2y + 1$
19. $x = -2y^2 + 2y - 5$
20. $x = -2y^2 - 4y + 3$
21. $x = (1/4)y^2 - 2y + 6$
22. $x = (-1/4)y^2 + y - 3$
23. $x = (-1/2)y^2 + y - 1$
24. $x = (1/2)y^2 + 2y + 3$

Problems 25–28 require the following information. The distance s of an object from a fixed point is given by $s = (1/2)gt^2 + vt + s_0$. In the formula, s_0 is the initial distance of the object from the fixed point, v is the initial velocity, a is a constant acceleration, and t is the time during which the object moves.

25. An object is thrown upward from a height of 4 ft with an initial velocity of 12 ft/s. The acceleration due to gravity is -32 ft/s^2. After what time does the object reach its maximum height?
4 7 8 1 2 3 5 6 9

26. A bullet is fired vertically from ground level with an initial velocity of 294 m/s. The acceleration due to gravity is -9.8 m/s^2. After what time does it reach its maximum height?
4 7 8 1 2 3 5 6 9

27. An object is thrown upward from a height of 4.4 m with an initial velocity of 24.5 m/s. The acceleration due to gravity is -9.8 m/s^2. After what time does it reach its maximum height?
4 7 8 1 2 3 5 6 9

28. An object is thrown upward from a height of 24.0 ft with an initial velocity of 80 ft/s. The acceleration due to gravity is -32 ft/s^2. After what time does it reach its maximum height?
4 7 8 1 2 3 5 6 9

29. A rectangle with dimensions 20.0 cm and $r + 10.0$ cm has a circle of radius r removed from it, as shown in the accompanying diagram at the right. For what value of r will the figure have maximum area? 1 2 4 7 8 3 5

30. A plant manager knows that 40 workers on a production line will produce an average of 20 items per worker per day. For each additional worker above 40, the average goes down by 0.4 item. How many workers should be hired to produce the maximum output? 1 2 6 7 8 3 5 9

31. A factory normally packages items in lots of 30. For a lot of exactly 30, the packaging cost is $0.60 per item. However, the cost per item is reduced by $0.01 for each item over 30. What size lot will produce the highest packaging costs?
1 2 6 7 8 3 5 9

32. Complete the square: $y = ax^2 + bx + c$.

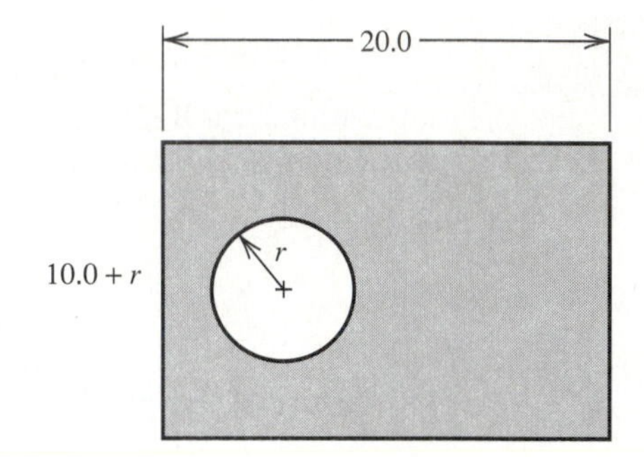

1. Architectural Technology
2. Civil Engineering Technology
3. Computer Engineering Technology
4. Mechanical Drafting Technology
5. Electrical/Electronics Engineering Technology
6. Chemical Engineering Technology
7. Industrial Engineering Technology
8. Mechanical Engineering Technology
9. Biomedical Engineering Technology

9.4 THE CIRCLE

In the previous two sections, you have seen how to write the equation of a parabola if any two of the vertex, focus, or directrix are known. You have also seen how to use the equation to determine the information needed to sketch the graph. The equation of a circle also can be used to display its special characteristics.

DEFINITION

A **circle** is the set of all points in a plane whose distance from a fixed point is a fixed constant. The fixed point is called the **center** of the circle. A line segment from the center to any point on the circle is a **radius** and the length of the radius is given by the fixed constant.

Consider a circle with its center at the origin and a radius of length r. Let (x, y) be a point on the circle. By definition, the distance between the center (0, 0) and the point (x, y) must be r. The distance formula gives

$$\sqrt{(x-0)^2+(y-0)^2}=r$$

$$\sqrt{x^2+y^2}=r, \text{ or squaring both sides,}$$

$$x^2+y^2=r^2.$$

Allowing for a translation of axes, just as we did for the parabola, it is possible to obtain the equation of a circle centered at (h, k) and having radius of length r.

THE CIRCLE

The equation of a circle with center (h, k) and radius of length r is given by

$$(x-h)^2+(y-k)^2=r^2.$$

EXAMPLE 1 Four pipes, each having a diameter of 6.0 inches, are positioned as shown in Figure 9.19. Find the equation for a single pipe having the same cross-sectional area as the four pipes.

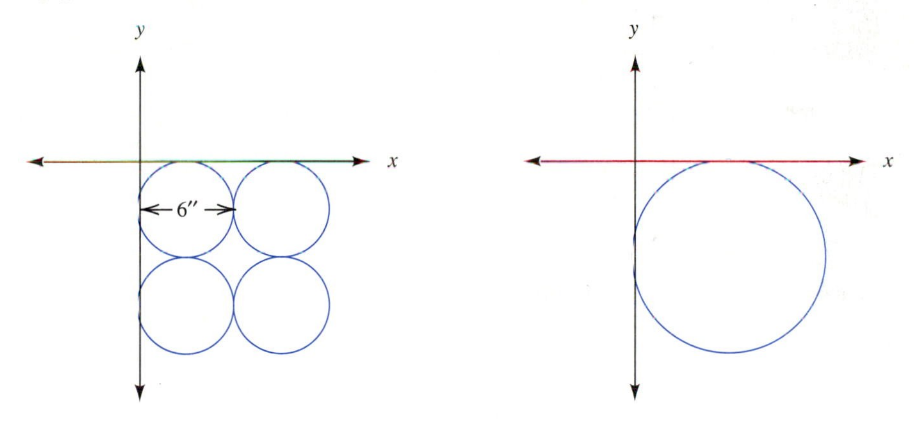

FIGURE 9.19

Solution The cross-sectional area of one 6.0-inch pipe is $\pi(3.0^2)$ or 9.0π. The area of four such pipes is 36.0π. A single pipe of radius r has area πr^2. If $\pi r^2 = 36.0\pi$, then $r = 6.0$. Since the pipe has a radius of 6.0, its center is at $(6.0, -6.0)$. Since $(h, k) = (6.0, -6.0)$ and $r = 6.0$, we have

$$(x-6.0)^2+[y-(-6.0)]^2=6.0^2$$

$$(x-6.0)^2+(y+6.0)^2=6.0^2$$

$$x^2-12.0x+36.0+y^2+12.0y+36.0=36.0$$

$$x^2-12.0x+y^2+12.0y+72.0=36.0$$

$$x^2-12.0x+y^2+12.0y+36.0=0 \quad \blacksquare$$

EXAMPLE 2 Find the equation of the circle that has center $(-2, 2)$ and passes through $(1, 6)$.

Solution Figure 9.20 shows the graph of the circle. Since the radius is measured from the center to any point on the circle, the distance from $(-2, 2)$ to $(1, 6)$ is the length of the radius. The distance formula gives

$$\begin{aligned} r &= \sqrt{[1-(-2)]^2 + (6-2)^2} \\ &= \sqrt{(1+2)^2 + (6-2)^2} \\ &= \sqrt{3^2 + 4^2} = \sqrt{9+16} \\ &= \sqrt{25} \text{ or } 5. \end{aligned}$$

The equation of the circle, then, is

$$\begin{aligned} [x-(-2)]^2 + (y-2)^2 &= 5^2 \\ (x+2)^2 + (y-2)^2 &= 5^2 \\ x^2 + 4x + 4 + y^2 - 4y + 4 &= 25 \\ x^2 + 4x + y^2 - 4y + 8 &= 25 \\ x^2 + 4x + y^2 - 4y - 17 &= 0. \end{aligned}$$ ■

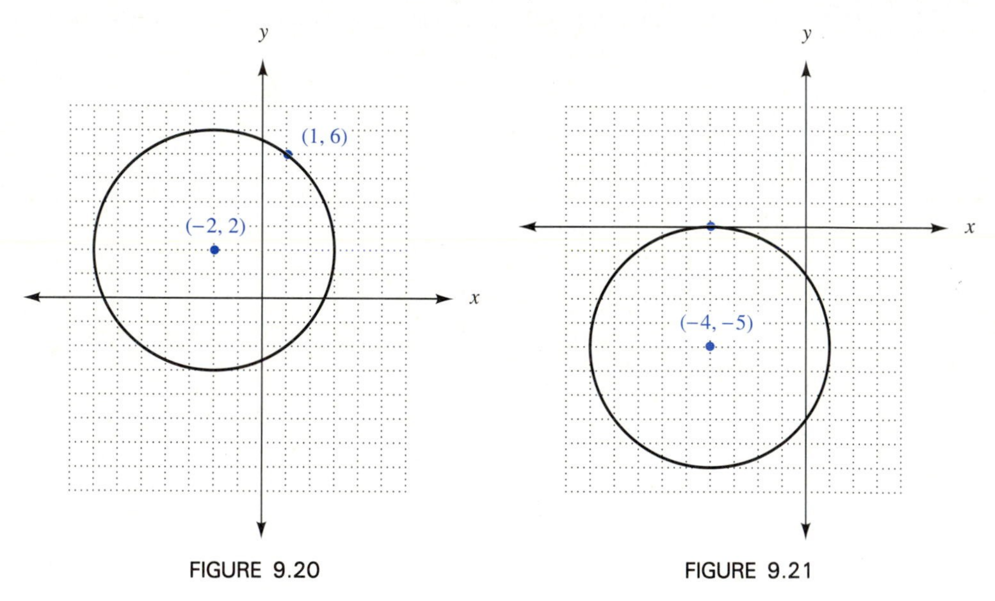

FIGURE 9.20

FIGURE 9.21

EXAMPLE 3 Find the equation of the circle that has center $(-4, -5)$ and is tangent to the x-axis.

Solution Figure 9.21 shows the graph of the circle. If the circle is tangent to the x-axis, the line segment from $(-4, -5)$ to $(-4, 0)$ is a radius, and its length is 5.

$$\begin{aligned} [x-(-4)]^2 + [y-(-5)]^2 &= 5^2 \\ (x+4)^2 + (y+5)^2 &= 5^2 \\ x^2 + 8x + 16 + y^2 + 10y + 25 &= 25 \\ x^2 + 8x + y^2 + 10y + 16 &= 0 \end{aligned}$$ ■

The technique of completing the square can be used to put the equation of a circle in the form

$$(x-h)^2+(y-k)^2=r^2$$

when it is not originally in that form. The first step, however, is to be able to identify the equation of a circle. In the equation of a parabola, only one variable has an exponent of two. In the equation of a circle, both variables have an exponent of two, and the coefficients of these terms are equal. When the coefficients are not equal, the graph is a type that will be covered in the next sections.

THE GENERAL FORM OF A SECOND-DEGREE EQUATION

When there are real solutions, the graph of an equation in the form

$$Ax^2+Bx+Cy^2+Dy+E=0 \text{ is}$$

(a) a vertical parabola, if $A \neq 0$, $C=0$, and $D \neq 0$.
(b) a horizontal parabola, if $C \neq 0$, $A=0$, and $B \neq 0$.
(c) a circle, if $A \neq 0$, $C \neq 0$, and $A=C$.

EXAMPLE 4 Determine which of the following equations represent circles.

(a) $2x^2+3x+4y^2+y-7=0$ **(b)** $2x^2+3x+y-7=0$

(c) $2x^2+2y^2+y-7=0$ **(d)** $2x^2+3x-2y^2+y-7=0$

Solution **(a)** The equation does not represent a circle, since the coefficient of x^2 is 2, but the coefficient of y^2 is 4.

(b) The equation does not represent a circle, since there is no y^2 term. The equation represents a vertical parabola.

(c) The equation does represent a circle, since the coefficient of x^2 is 2, and the coefficient of y^2 is also 2.

(d) The equation does not represent a circle, since the coefficient of x^2 is 2, but the coefficient of y^2 is -2. ■

To determine the center and radius of a circle:

1. Write an equivalent equation that has the variables on one side and the constant on the other.
2. Divide both sides of the equation by the coefficient of the square terms if that coefficient is not 1.
3. Complete the square using the x^2 and x terms.
4. Complete the square using the y^2 and y terms.
5. Factor so that the equation is in the form $(x-h)^2+(y-k)^2=r^2$.
6. Identify (h, k) as the center, and r as the length of the radius.

EXAMPLE 5 Sketch the graph of $2x^2 + 8x + 2y^2 + 4y - 8 = 0$.

Solution

1.	$2x^2 + 8x + 2y^2 + 4y = 8$	Add 8 to both sides.
2.	$x^2 + 4x + y^2 + 2y = 4$	Divide both sides by 2.
3–4.	$(x^2 + 4x) + (y^2 + 2y) = 4$	Prepare to complete the square for both variables.
	$(x^2 + 4x + \mathbf{4}) + (y^2 + 2y + \mathbf{1}) = 4 + \mathbf{4} + \mathbf{1}$	Add 4 and 1 to both sides.
5.	$(x + 2)^2 + (y + 1)^2 = 9$	Factor the expressions.

The center of the circle is at $(-2, -1)$ and the radius is 3, as shown in Figure 9.22. ■

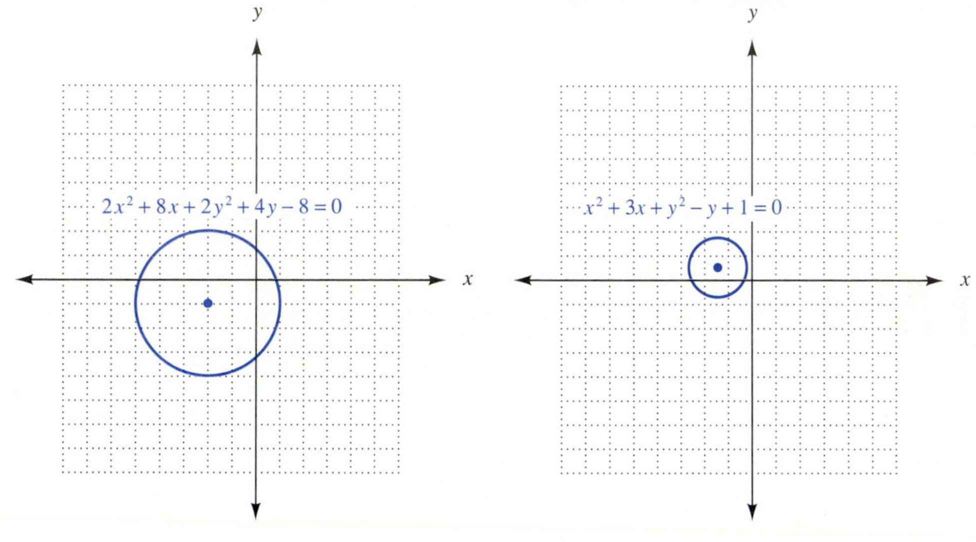

FIGURE 9.22

FIGURE 9.23

EXAMPLE 6 Sketch the graph of $x^2 + 3x + y^2 - y + 1 = 0$.

Solution

$$x^2 + 3x + y^2 - y + 1 = 0$$
$$x^2 + 3x + y^2 - y = -1$$
$$(x^2 + 3x) + (y^2 - y) = -1$$
$$(x^2 + 3x + \mathbf{9/4}) + (y^2 - y + \mathbf{1/4}) = -1 + \mathbf{9/4} + \mathbf{1/4}$$
$$(x + 3/2)^2 + (y - 1/2)^2 = 6/4$$

The center of the circle is at $(-3/2, 1/2)$, and the radius is $\sqrt{6/4} = \sqrt{6}/2$. The graph is shown in Figure 9.23. ■

EXAMPLE 7 Sketch the graph of $x^2 + y^2 = -3$.

Solution Since it is impossible to have $r^2 = -3$, there are no real solutions. ■

EXERCISES 9.4

In 1–10, write the equation of a circle with the given characteristics.

1. center at the origin, radius of length 7

2. center at the origin, radius of length 5

3. center at the origin, radius of length 3

4. center at (1, 2), radius of length 2

5. center at (−2, 3), passes through (1, −1)

6. center at (3, −1), passes through (0, 3)

7. center at (−4, 0), passes through (8, 5)

8. center at (3/2, 1/4), tangent to the y-axis

9. center at (−1/4, 1/2), tangent to the x-axis

10. center at (1/2, −3/4), tangent to the x-axis

In 11–24, determine which of the equations represent circles.

11. $x^2 + 2x + y^2 + 3y - 1 = 0$

12. $x^2 + 3x + y^2 - 4y - 2 = 0$

13. $x^2 - x + y^2 + 2y - 3 = 0$

14. $2x^2 + 3x + y^2 - 2y - 1 = 0$

15. $x^2 + 3y^2 + 3y - 2 = 0$

16. $x^2 - x - y^2 + y - 4 = 0$

17. $x^2 + 2x - y^2 - y - 5 = 0$

18. $x^2 + x + y - 3 = 0$

19. $x^2 + x + 2y - 1 = 0$

20. $x + y^2 + y - 4 = 0$

21. $x + y^2 - 3y - 2 = 0$

22. $2x^2 + 3x + 2y^2 - 2y - 1 = 0$

23. $3x^2 - x + 3y^2 + y - 2 = 0$

24. $4x^2 + 4y^2 - 16 = 0$

In 25–34, identify the center and radius of each circle.

25. $x^2 + y^2 - 36 = 0$

26. $x^2 - 2x + y^2 + 4y + 1 = 0$

27. $x^2 + 2x + y^2 - 6y + 9 = 0$

28. $x^2 - 4x + y^2 + 6y - 4 = 0$

29. $x^2 + 6x + y^2 + 4y - 12 = 0$

30. $5x^2 + 5y^2 - 20 = 0$

31. $4x^2 - 4x + 4y^2 + 4y + 1 = 0$

32. $2x^2 - 6x + 2y^2 - 2y - 5 = 0$

33. $3x^2 + 3y^2 - 2y - 1 = 0$

34. $2x^2 - 4x + 2y^2 - 4 = 0$

In 35–42, sketch the graph of each equation.

35. $(x - 1)^2 + (y + 2)^2 = 9$

36. $(x + 2)^2 + (y - 1)^2 = 16$

37. $(x - 3)^2 + (y - 2)^2 = 20$

38. $(x + 1)^2 + (y + 3)^2 = 7$

39. $(x - 1/2)^2 + y^2 = 4$

40. $x^2 + (y + 3/2)^2 = 9$

41. $x^2 + (y - 3/4)^2 = 1/4$

42. $(x - 2/3)^2 + y^2 = 4/9$

---

43. From the top of the Empire State Building, it is possible to see an area with a circumference of 320.0 km. Assuming that the building is at the origin of a coordinate system, write the equation for the circle that encloses the area that can be seen. 1 2 4 7 8 3 5

44. A building is to be constructed so that from its rooftop a distance of 30.0 km can be seen in all directions. Its location is 10.0 km east and 16.0 km south of the center of the city. Using the center of the city as the origin, write an equation for the circle that bounds the visible area. 1 2 4 7 8 3 5

---

1. Architectural Technology
2. Civil Engineering Technology
3. Computer Engineering Technology
4. Mechanical Drafting Technology
5. Electrical/Electronics Engineering Technology
6. Chemical Engineering Technology
7. Industrial Engineering Technology
8. Mechanical Engineering Technology
9. Biomedical Engineering Technology

45. A swinging pendulum bob sweeps out a sector of a circle. Write an equation for the circle if the pendulum is of length l and the rest position of the bob is the origin. 4 7 8 1 2 3 5 6 9

46. Two gears are meshed as shown in the accompanying diagram. The larger has radius 8.00 cm and the smaller has radius 4.00 cm. Consider the origin of a coordinate system to be at the center of the larger gear and write the equation of the smaller circle. 4 7 8 1 2 3 5 6 9

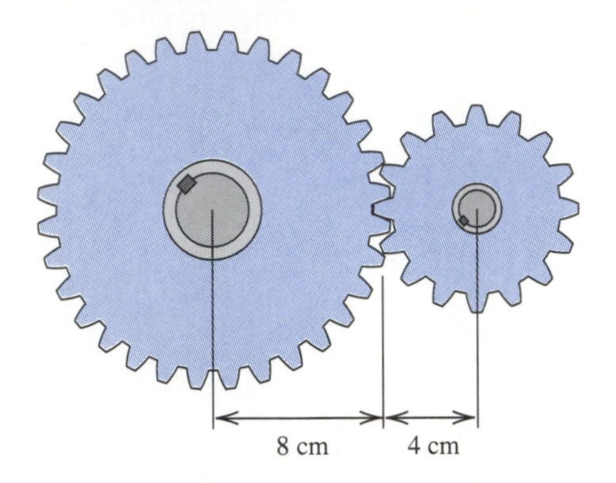

47. The equation $I^2 = I_R^2 + (I_C - I_L)^2$ gives the relationship between the currents in the 3 branches of the parallel circuit shown. If $I = 1.20$ A, and $I_L = 0.80$ A, sketch the graph of I_R versus I_C. 3 5 9 4 7 8

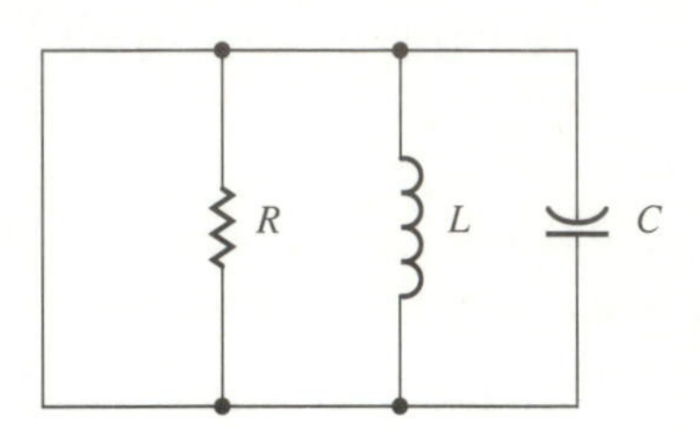

48. Under ideal conditions, a supersonic jet produces a conical shock wave. Assume that the earth is flat in the immediate vicinity of such a jet. Describe the path of a jet with a shock wave that would intersect the earth in a circle. Describe the path of a jet with a shock wave that would intersect the earth in a hyperbola.

1. Architectural Technology
2. Civil Engineering Technology
3. Computer Engineering Technology
4. Mechanical Drafting Technology
5. Electrical/Electronics Engineering Technology
6. Chemical Engineering Technology
7. Industrial Engineering Technology
8. Mechanical Engineering Technology
9. Biomedical Engineering Technology

9.5 THE ELLIPSE

You have learned to recognize the equation of a parabola or a circle from the coefficients of the square and linear terms. In the equation $Ax^2 + Bx + Cy^2 + Dy + E = 0$, if the coefficients of the square terms are not equal, the graph will be either an *ellipse* or a *hyperbola.* The graph is an ellipse when the coefficients have the same sign, but it is a hyperbola when the signs are different.

DEFINITION An **ellipse** is the set of all points in the plane such that the sum of the distances from a point on the ellipse to two fixed points is a fixed constant. The fixed points are called **foci** of the ellipse.

To see how this definition leads to an elliptical shape, drive two nails into a board and attach the ends of a piece of string to the nails. If the string is held taut by a pencil as it is moved, the string will restrict the movement of the pencil so that an ellipse is

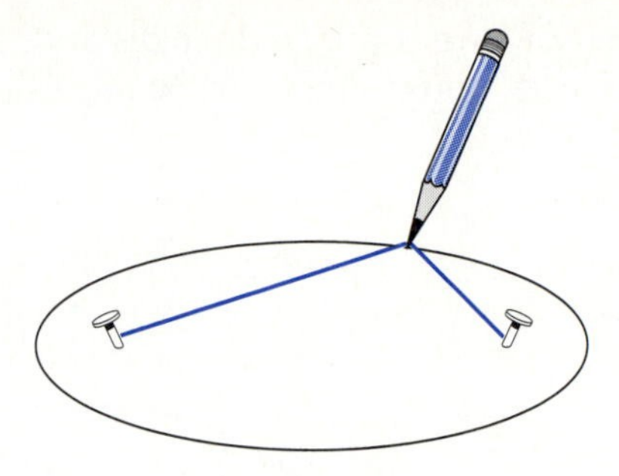

FIGURE 9.24

drawn with the foci at the nails, as shown in Figure 9.24. Although the distances from the foci to the pencil change as the pencil is moved, the sum of the distances always equals the length of the string.

We are now ready to quantify the discussion. For simplicity, we start with an ellipse with its center at the origin and foci at $(c, 0)$ and $(-c, 0)$. The ellipse will cross the x-axis at points we denote $(a, 0)$ and $(-a, 0)$. The line segment connecting these two points is called the **major axis** of the ellipse. The ellipse will cross the y-axis at points we denote $(0, b)$ and $(0, -b)$. The line segment connecting these two points is called the **minor axis** of the ellipse. Figure 9.25 shows these points.

major axis of an ellipse

minor axis of an ellipse

Furthermore, there is a relationship among a, b, and c. Since $(a, 0)$ is on the ellipse, the distance from F_1 to $(a, 0)$ plus the distance from F_2 to $(a, 0)$ must be the fixed constant referred to in the definition. From F_1 to $(a, 0)$, the distance is $a + c$. From F_2 to $(a, 0)$, the distance is $a - c$. The sum of these distances is $(a + c) + (a - c)$ or $2a$. Since the point $(0, b)$ is also on the ellipse, the sum of the distances from the foci to $(0, b)$ must be $2a$. In this case, however, the distance from F_1 to $(0, b)$ is the same as the distance from F_2 to $(0, b)$, so the distance from either focus to $(0, b)$ must be a. From Figure 9.26, it is evident that $b^2 + c^2 = a^2$.

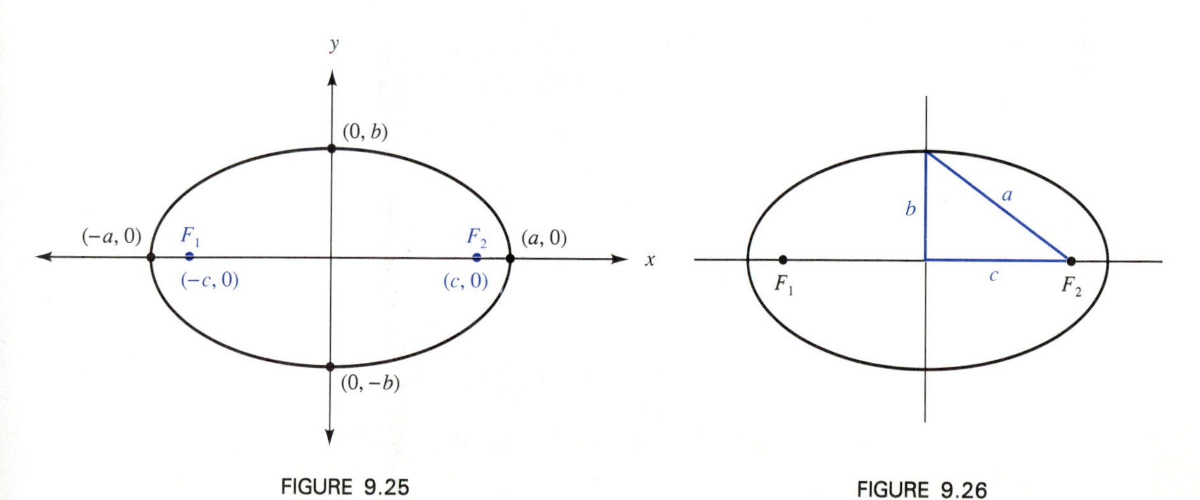

FIGURE 9.25

FIGURE 9.26

So far, we have considered the distance from the focus to a point on the ellipse only for special points. Let (x, y) be *any* point on the ellipse, as in Figure 9.27, for example.

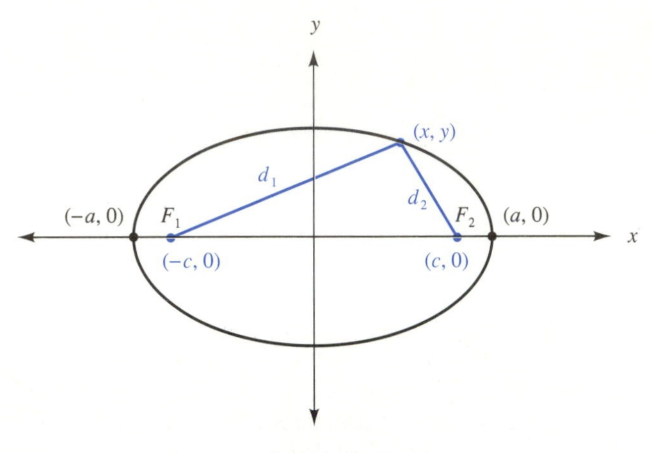

FIGURE 9.27

The distance formula can be used as follows.

$$d_1 + d_2 = 2a$$

$$\sqrt{(x+c)^2 + y^2} + \sqrt{(x-c)^2 + y^2} = 2a$$

$$\sqrt{(x-c)^2 + y^2} = 2a - \sqrt{(x+c)^2 + y^2}$$

Square both sides of the equation to obtain

$$(x-c)^2 + y^2 = 4a^2 - 4a\sqrt{(x+c)^2 + y^2} + (x+c)^2 + y^2$$

$$x^2 - 2xc + c^2 + y^2 = 4a^2 - 4a\sqrt{(x+c)^2 + y^2} + x^2 + 2xc + c^2 + y^2.$$

Now add the quantity $-4a^2 - x^2 - 2xc - c^2 - y^2$ to both sides.

$$-4a^2 - 4xc = -4a\sqrt{(x+c)^2 + y^2}$$

Divide both sides of the equation by -4, leaving

$$a^2 + xc = a\sqrt{(x+c)^2 + y^2}.$$

Since a radical remains, square both sides once more.

$$a^4 + 2a^2xc + (xc)^2 = a^2[(x+c)^2 + y^2]$$

$$a^4 + 2a^2xc + x^2c^2 = a^2[(x^2 + 2xc + c^2 + y^2]$$

$$a^4 + 2a^2xc + x^2c^2 = a^2x^2 + 2a^2xc + a^2c^2 + a^2y^2$$

Add the quantity $-2a^2xc - a^2c^2 - x^2c^2$ to both sides.

$$a^4 - a^2c^2 = a^2x^2 - x^2c^2 + a^2y^2$$
$$a^2(a^2 - c^2) = x^2(a^2 - c^2) + a^2y^2$$

Recall that $b^2 + c^2 = a^2$ or $b^2 = a^2 - c^2$. Making this substitution, we have

$$a^2(b^2) = x^2(b^2) + a^2y^2.$$

Then divide both sides by a^2b^2.

$$\frac{a^2b^2}{a^2b^2} = \frac{x^2b^2}{a^2b^2} + \frac{a^2y^2}{a^2b^2}$$
$$1 = \frac{x^2}{a^2} + \frac{y^2}{b^2}$$

If the ellipse has the major axis oriented in the vertical direction, a^2 is the denominator of the y^2 term.

$$1 = \frac{y^2}{a^2} + \frac{x^2}{b^2}$$

If we allow the center of the ellipse to be (h, k), we have the following information.

THE ELLIPSE

The graph of an equation of the form

$$\frac{(x-h)^2}{a^2} + \frac{(y-k)^2}{b^2} = 1 \quad \text{or} \quad \frac{(y-k)^2}{a^2} + \frac{(x-h)^2}{b^2} = 1$$

is an ellipse with the following characteristics.

(a) The center is at (h, k).
(b) If the larger denominator is a^2, a is the number of units from the center to the end points, called vertices, of the major axis.
(c) If the smaller denominator is b^2, b is the number of units from the center to the end points of the minor axis.
(d) The foci are located on the major axis at a distance of c units from the center where $c^2 = a^2 - b^2$.

EXAMPLE 1 Sketch the graph of $\dfrac{(x-2)^2}{9} + \dfrac{(y+3)^2}{4} = 1$.

Solution The graph is an ellipse with center at $(2, -3)$. The major axis is parallel to the x-axis, since the denominator of the x^2 term is larger. Since $a^2 = 9$ and $a = 3$, the vertices are 3 units from the center, so they are $(5, -3)$ and $(-1, -3)$. Because $b^2 = 4$ and $b = 2$, the end points of the minor axis are located two units above and below the center. Connecting these four points with a curve having the correct shape gives the graph, as shown in Figure 9.28. ■

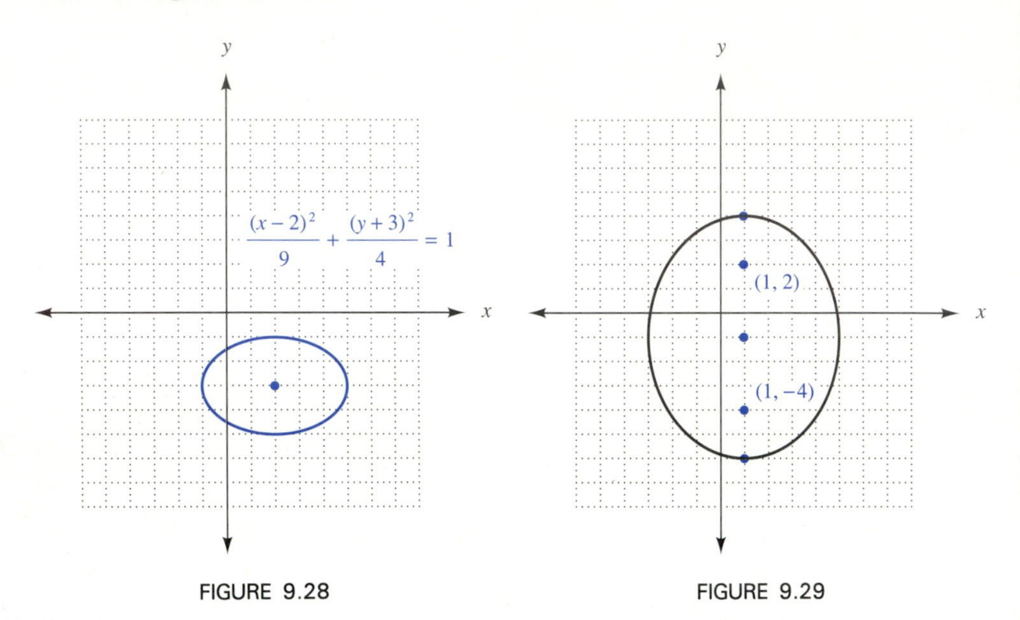

FIGURE 9.28

FIGURE 9.29

EXAMPLE 2 Write the equation of the ellipse with foci at (1, 2) and (1, −4) and major axis of length 10.

Solution Figure 9.29 shows the graph of the ellipse described.

To write the equation, it is necessary to know (h, k), a, and b. None of this information is given explicitly. The center, however, must be halfway between the foci, so $(h, k) = (1, -1)$. Since the total length of the major axis is 10, a is 5. Thus, a^2, the denominator of the y^2 term, is 25. To find b^2, we use the equation $c^2 = a^2 - b^2$. Since c is the distance from the center to a focus, c is 3. Thus $9 = 25 - b^2$, so $b^2 = 16$. The equation is

$$\frac{(y+1)^2}{25} + \frac{(x-1)^2}{16} = 1. \quad ■$$

EXAMPLE 3 Write the equation of an ellipse with center (2, 2) and vertex (6, 2) and that passes through (4, 5).

Solution Figure 9.30 shows the graph.

The center is given, so $(h, k) = (2, 2)$. The distance from the center (2, 2) to the vertex (6, 2) is 4, so a is 4. With the information obtained thus far, the equation is written as

$$\frac{(x-2)^2}{16} + \frac{(y-2)^2}{b^2} = 1.$$

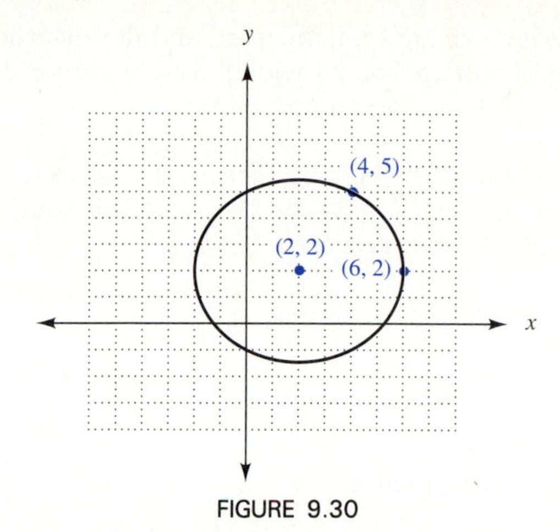

FIGURE 9.30

The equation $c^2 = a^2 - b^2$, which was used in Example 2, is of no help here. The point (4, 5), however, is a point (x, y) that satisfies the equation. That is

$$(4-2)^2/16 + (5-2)^2/b^2 = 1$$
$$4/16 + 9/b^2 = 1.$$

If both sides are multiplied by $16b^2$, we have

$$4b^2 + 144 = 16b^2$$
$$144 = 12b^2 \text{ and } 12 = b^2.$$

The equation, then, is $\dfrac{(x-2)^2}{16} + \dfrac{(y-2)^2}{12} = 1.$ ■

A ray emanating from one focus of an ellipse is reflected to the other focus. Because of this property, the ellipse has applications in the technologies.

EXAMPLE 4 A cup whose surface is the shape formed when an ellipse is rotated about its major axis is used to break up kidney stones. A spark is jumped through the gap of an electrode at one focus of the ellipse, as shown in Figure 9.31. A shock wave is reflected off the cup and through the body tissue to the kidney stone at the other focus.

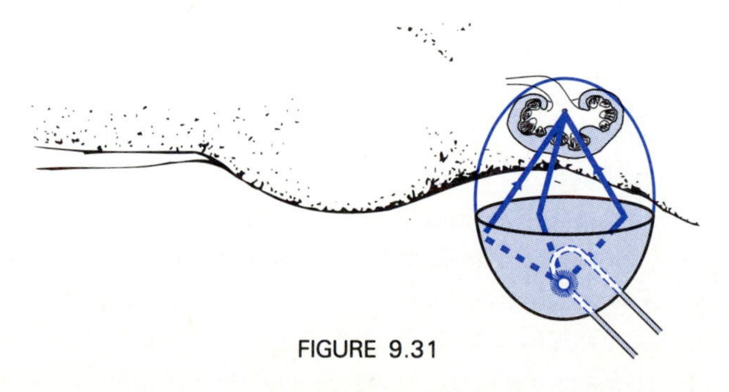

FIGURE 9.31

If the foci are 15.4 cm apart, and the electrode is to be placed 2.4 cm from the vertex of the cup, find the equation of the ellipse. Assume that the center of the ellipse is at the origin of the coordinate system.

Solution If the foci are 15.4 cm apart, then $c = 15.4/2$ or 7.7 cm. Since the vertex is 2.4 cm from the focus, $a = 7.7 + 2.4$ or 10.1 cm. Substituting these values into the equation $c^2 = a^2 - b^2$, we have the following.

$$7.7^2 = 10.1^2 - b^2$$
$$b^2 = 10.1^2 - 7.7^2$$
$$b = \sqrt{10.1^2 - 7.7^2}$$
$$b = 6.5 \text{ cm}$$

The equation, then, is

$$\frac{x^2}{6.5^2} + \frac{y^2}{10.1^2} = 1. \quad \blacksquare$$

EXERCISES 9.5

In 1–10, sketch the graph of each curve.

1. $\frac{(x+1)^2}{25} + \frac{(y-1)^2}{16} = 1$

2. $\frac{x^2}{9} + \frac{(y+2)^2}{4} = 1$

3. $\frac{(x-2)^2}{1} + \frac{y^2}{4} = 1$

4. $\frac{(x+3)^2}{9} + \frac{(y-3)^2}{16} = 1$

5. $\frac{(x-1)^2}{16} + \frac{(y+1)^2}{9} = 1$

6. $\frac{x^2}{25} + \frac{(y+3)^2}{9} = 1$

7. $\frac{(x-2)^2}{25} + \frac{(y-3)^2}{1} = 1$

8. $\frac{(x+2)^2}{1} + \frac{y^2}{36} = 1$

9. $\frac{x^2}{1} + \frac{y^2}{2} = 1$

10. $\frac{x^2}{3} + \frac{y^2}{1} = 1$

In 11–22, write the equation of each curve described.

11. an ellipse with foci at $(2, 4)$ and $(2, -2)$; major axis of length 10

12. an ellipse with foci at $(1, 3)$ and $(5, 3)$; major axis of length 8

13. an ellipse with foci at $(-2, 4)$ and $(2, 4)$; minor axis of length 4

14. an ellipse with foci at $(-3, 16)$ and $(-3, -8)$; minor axis of length 10

15. an ellipse with center at $(4, -1)$; focus at $(-8, -1)$; vertex at $(-9, -1)$

16. an ellipse with center at $(-2, -2)$; focus at $(2, -2)$; vertex at $(3, -2)$

17. an ellipse with center at $(0, 0)$; vertex at $(\sqrt{18}, 0)$; passes through $(4, 1)$

18. an ellipse with center at $(0, 0)$; vertex at $(0, \sqrt{12})$; passes through $(1, 3)$

19. an ellipse with major axis from $(-2, 2)$ to $(8, 2)$; minor axis from $(3, -1)$ to $(3, 5)$

20. an ellipse with major axis from $(-2, 2)$ to $(-2, -4)$; minor axis from $(-4, -1)$ to $(0, -1)$

21. an ellipse with major axis from $(-1, 3)$ to $(5, 3)$; minor axis from $(2, 2)$ to $(2, 4)$

22. an ellipse with major axis from $(5, -2)$ to $(5, 4)$; minor axis from $(3, 1)$ to $(7, 1)$

23. A spotlight is aimed from the balcony toward a stage at an angle as shown in the accompanying diagram. The area lit is elliptical in shape, is 1.3 m across the minor axis, and is 2.4 m across the major axis. Find the distance between the foci of the ellipse. 1 2 4 7 8 3 5

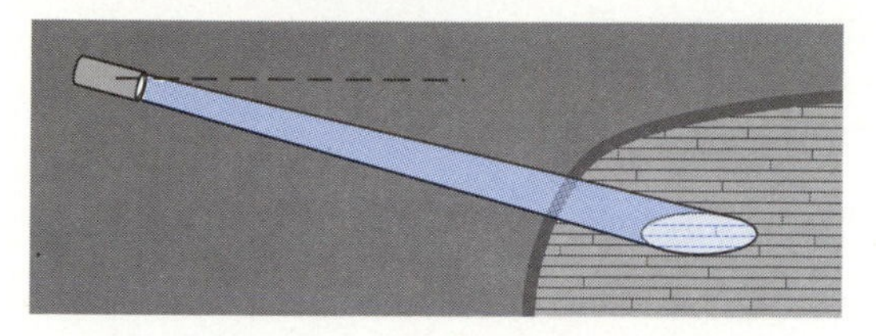

24. Two circular pipes that are each 6.0 inches in diameter are joined at a 45° angle as shown. Find the length of the major axis for the elliptical intersection. 1 2 4 7 8 3 5

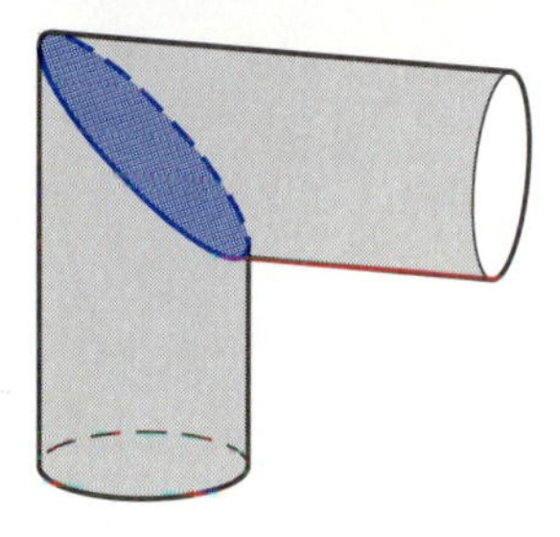

25. An oil tanker that has elliptical ends is to have hooks placed at the foci of the ellipse. The major axis of the ellipse is 1.5 m in length and the minor axis is 1.2 m in length. How far are the foci from the center? 1 2 4 7 8 3 5

26. A satellite is to make an elliptical orbit with the center of the earth at one focus. The distance between foci is 2046 km, and the radius of the earth is 6378 km. When it is at the vertex near the earth, the satellite is 34 865 km above the earth's surface. It is determined that the minor axis of the satellite's orbit is 82 486 km. The formula $P = \pi(a + b)$ is used to approximate the perimeter of a nearly circular ellipse. Find the distance traveled by the weather satellite in one revolution around the earth. 4 7 8 1 2 3 5 6 9

27. The Colosseum in Rome is elliptical in shape with a major axis of 186 m and a minor axis of 153 m. How far from the center are the foci? 1 2 4 7 8 3 5

28. The area of the cross section of an elliptical water pipe is smaller than the area of the cross section of a circular water pipe with the same circumference. If water freezes in an elliptical pipe, the pipe becomes more circular in shape, accommodating the expansion, and the pipe does not burst. Write the equation for the cross section of an elliptical pipe with major axis 5.5 inches long and minor axis 3.0 inches long. 1 2 4 6 8 5 7 9

29. To a person sitting at one focus of an elliptical room with mirrored walls, a light at the other focus seems to produce light coming from every direction. If the major axis is 8.0 m long and the minor axis is 6.0 m long, how far apart are the foci? 1 2 4 7 8 3 5

1. Architectural Technology
2. Civil Engineering Technology
3. Computer Engineering Technology
4. Mechanical Drafting Technology
5. Electrical/Electronics Engineering Technology
6. Chemical Engineering Technology
7. Industrial Engineering Technology
8. Mechanical Engineering Technology
9. Biomedical Engineering Technology

9.6 THE HYPERBOLA

The last geometric figure that we consider is the hyperbola.

DEFINITION A **hyperbola** is the set of all points in a plane such that the difference of the distances from a point on the hyperbola to two fixed points is a fixed constant. The fixed points are called **foci** of the hyperbola.

To derive the equation, once again, we begin with a figure whose center is the origin and foci are F_2 at $(c, 0)$ and F_1 at $(-c, 0)$, as in Figure 9.32. The hyperbola will cross the x-axis at points we denote $(a, 0)$ and $(-a, 0)$.

The distance from F_1 to $(a, 0)$ minus the distance from F_2 to $(a, 0)$ must be the fixed constant of the definition. From F_1 to $(a, 0)$, the distance is $a + c$. From F_2 to $(a, 0)$, the distance is $c - a$. Thus, the difference of the distances is $(a + c) - (c - a) = 2a$. Let (x, y) be any point on the hyperbola. Then the distance formula can be used.

$$d_1 - d_2 = 2a \text{ or } \sqrt{(x + c)^2 + y^2} - \sqrt{(x - c)^2 + y^2} = 2a$$

Algebraic manipulations similar to those used to derive the equation for an ellipse lead to

$$a^2(a^2 - c^2) = x^2(a^2 - c^2) + a^2y^2.$$

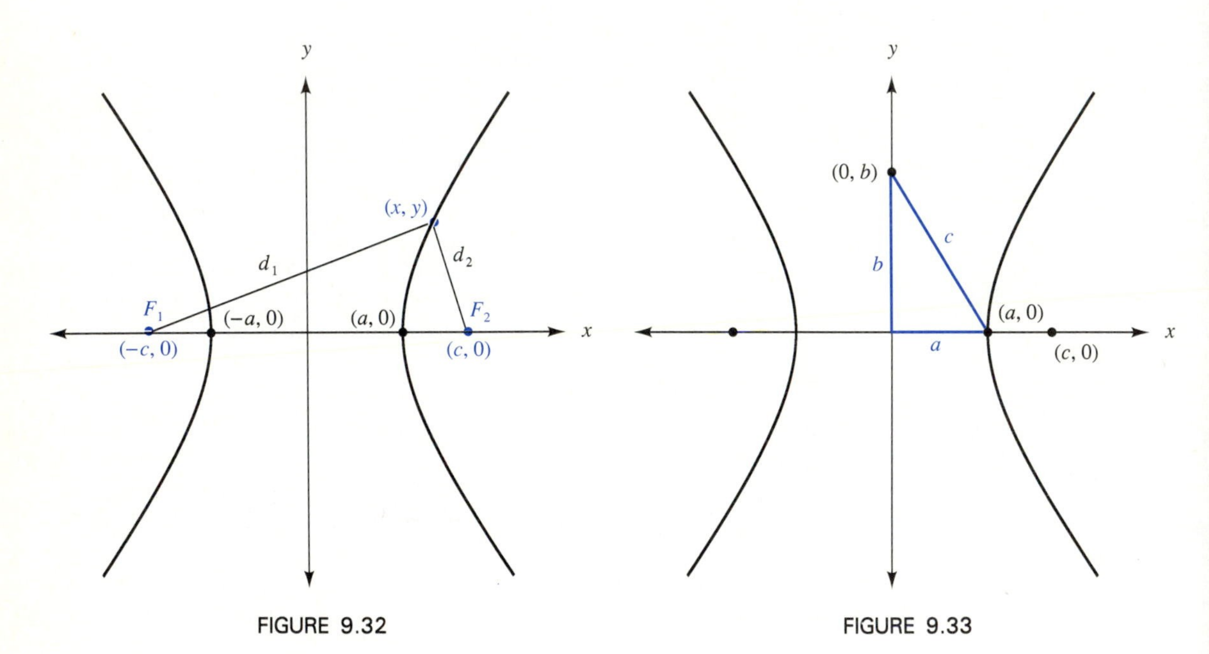

FIGURE 9.32

FIGURE 9.33

In Figure 9.33, the point $(0, b)$ is chosen so that $a^2 + b^2 = c^2$ or $a^2 - c^2 = -b^2$. Making this substitution,

$$a^2(-b^2) = x^2(-b^2) + a^2y^2. \tag{9-2}$$

Dividing both sides of Equation 9-2 by $-a^2b^2$, we have

$$\frac{-a^2b^2}{-a^2b^2} = \frac{-x^2b^2}{-a^2b^2} + \frac{a^2y^2}{-a^2b^2} \quad \text{or} \quad 1 = \frac{x^2}{a^2} - \frac{y^2}{b^2}.$$

transverse axis of hyperbola

conjugate axis of a hyperbola

asymptote

The line segment connecting the vertices $(a, 0)$ and $(-a, 0)$ is called the **transverse axis** of the hyperbola. The line segment connecting $(0, b)$ and $(0, -b)$ is called the **conjugate axis** of the hyperbola. If a rectangle is drawn as shown in Figure 9.34, using the transverse and conjugate axes as guides, the diagonals of the rectangle are called **asymptotes** for the hyperbola. The curve will never cross the asymptotes, but as x increases in absolute value, the hyperbola will get closer and closer to the asymptotes.

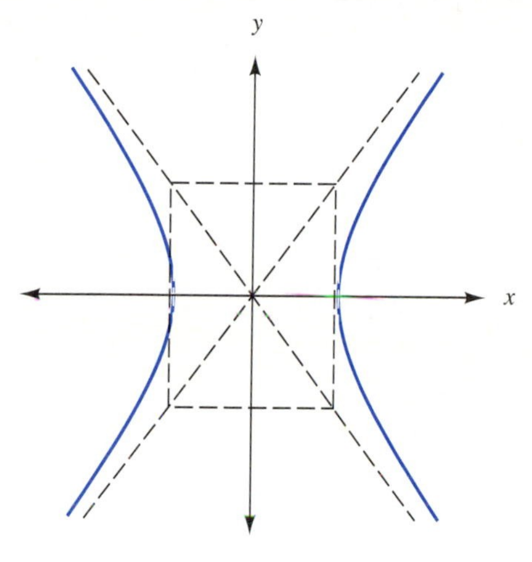

FIGURE 9.34

If the y^2 term has a positive coefficient, a^2 is the denominator of the y^2 term, and the hyperbola is oriented so that the vertices lie on a vertical line segment. If we allow the center of the hyperbola to be (h, k), we have the following information.

THE HYPERBOLA

The graph of an equation of the form

$$\frac{(x-h)^2}{a^2} - \frac{(y-k)^2}{b^2} = 1 \quad \text{or} \quad \frac{(y-k)^2}{a^2} - \frac{(x-h)^2}{b^2} = 1$$

is a hyperbola with the following characteristics.

(a) The center is at (h, k).

(b) If a^2 is the denominator of the square term with positive coefficient, the end points of the transverse axis, called vertices, are located a units from the center.

(c) If b^2 is the denominator of the square term with negative coefficient, the end points of the conjugate axis are located b units from the center.

(d) The foci are located on the transverse axis c units from the center, where $c^2 = a^2 + b^2$.

EXAMPLE 1 Sketch the graph of $\frac{(y-2)^2}{4} - \frac{(x+3)^2}{9} = 1$.

Solution The graph is a hyperbola with center at $(-3, 2)$. The transverse axis is parallel to the y-axis, since the y^2 term has the positive coefficient. Since a^2 is 4, the vertices are 2 units from the center. Thus, they are at $(-3, 4)$ and $(-3, 0)$. Because b^2 is 9, the end points of the conjugate axis are located 3 units to the left and right of the center. These points are $(-6, 2)$ and $(0, 2)$. Figure 9.35 shows the graph. The diagonals of the rectangle formed by the transverse and conjugate axes are the asymptotes of the hyperbola. ■

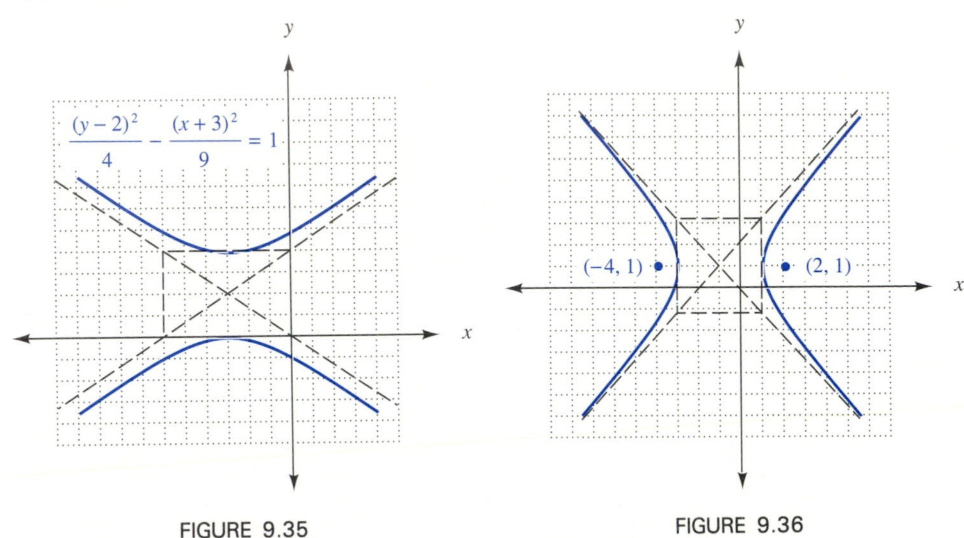

FIGURE 9.35

FIGURE 9.36

NOTE In Example 1, the transverse axis was shorter than the conjugate axis. ***In an ellipse, the value of a is used to locate the major axis, which not only contains the vertices, but is also the longer axis. For a hyperbola, the value of a is significant only in that it locates the vertices and thus indicates the orientation of the hyperbola.***

EXAMPLE 2 Write the equation of a hyperbola with foci at $(2, 1)$ and $(-4, 1)$ and transverse axis of length 4.

Solution In order to write the equation, it is necessary to know (h, k), a, and b. None of this information is given explicitly. The center, however, must be halfway between the foci. Figure 9.36 shows that $(h, k) = (-1, 1)$.

The transverse axis is parallel to the x-axis, so the x^2 term will have a positive coefficient. Since the total length of the transverse axis is 4, a is 2. Thus, a^2, the denominator of the x^2 term, is 4. To find b^2, we use the equation $c^2 = a^2 + b^2$. Since c is the distance from the center to a focus, c is 3. Thus $9 = 4 + b^2$, so $b^2 = 5$. The equation is

$$\frac{(x+1)^2}{4} - \frac{(y-1)^2}{5} = 1. \quad ■$$

The technique of completing the square is used with both the ellipse and the hyperbola to find the special characteristics of the curve.

THE GENERAL FORM OF A SECOND-DEGREE EQUATION

When there are real solutions, the graph of an equation in the form

$$Ax^2 + Bx + Cy^2 + Dy + E = 0 \text{ is}$$

(a) an ellipse if $A \neq 0$, $C \neq 0$, and A and C have the same sign, but $A \neq C$.
(b) a hyperbola if $A \neq 0$, $C \neq 0$, and A and C have opposite signs.

EXAMPLE 3 Determine what type of curve (if any) is represented by the equation $2x^2 - 4x - 4 + 3y^2 = 0$. Then write the equation to display its characteristics.

Solution Since the coefficients of the square terms are different, but have the same sign, the curve represents an ellipse, if it has real solutions. Complete the square.

$$2x^2 - 4x - 4 + 3y^2 = 0$$

$$2x^2 - 4x + 3y^2 = 4 \qquad \text{Add 4 to both sides.}$$

$$2(x^2 - 2x) + 3(y^2) = 4 \qquad \text{Factor the expressions.}$$

$$2(x^2 - 2x + 1) + 3(y^2) = 4 + 2 \qquad \text{Add 1 inside parentheses; add 2 on the right.}$$

$$2(x - 1)^2 + 3y^2 = 6$$

$$\frac{2(x-1)^2}{6} + \frac{3y^2}{6} = \frac{6}{6} \qquad \text{Divide both sides by 6.}$$

$$\frac{(x-1)^2}{3} + \frac{y^2}{2} = 1 \qquad \text{Reduce the fractions.} \blacksquare$$

EXAMPLE 4 Determine what type of curve (if any) is represented by the equation $3y^2 - 2x^2 - 4x - 4 = 0$. Then write the equation to display its characteristics.

Solution Since the coefficients of the square terms have opposite signs, the curve represents a hyperbola, if it has real solutions. Complete the square.

$$3y^2 - 2x^2 - 4x - 4 = 0$$

$$3y^2 - 2x^2 - 4x = 4 \qquad \text{Add 4 to both sides.}$$

$$3(y^2) - 2(x^2 + 2x) = 4 \qquad \text{Factor the expressions.}$$

$$3(y^2) - 2(x^2 + 2x + 1) = 4 - 2 \qquad \text{Add 1 inside parentheses; add } -2 \text{ on the right.}$$

$$3(y^2) - 2(x + 1)^2 = 2$$

$$\frac{3(y^2)}{2} - \frac{2(x+1)^2}{2} = \frac{2}{2} \qquad \text{Divide both sides by 2.}$$

$$\frac{(y^2)}{2/3} - \frac{(x+1)^2}{1} = 1 \qquad \text{Reduce the fractions } (1 \div 3/2 = 2/3). \blacksquare$$

EXAMPLE 5 Determine what type of curve (if any) is represented by the equation $4x^2 + 8x + y^2 - 4y + 12 = 0$. Then write the equation to display its characteristics.

Solution Since the coefficients of the square terms are different, but have the same sign, the curve represents an ellipse, if it has real solutions. Complete the square.

$$4x^2 + 8x + y^2 - 4y + 12 = 0$$

$$4x^2 + 8x + y^2 - 4y = -12 \qquad \text{Subtract 12 from both sides.}$$

$$4(x^2 + 2x) + y^2 - 4y = -12 \qquad \text{Factor.}$$

$$4(x^2 + 2x + 1) + y^2 - 4y + 4 = -12 + 4 + 4 \qquad \text{Add } 4 + 4 \text{ to both sides.}$$

$$4(x+1)^2 + (y-2)^2 = -4$$

$$-(x+1)^2 - \frac{(y-2)^2}{4} = 1 \qquad \text{Divide both sides by } -4.$$

Notice that for real values of x and y, both quantities on the left-hand side must be negative, but the quantity on the right is positive. Thus, there are no real solutions for this problem, so the equation does not represent a curve. ■

The general form of the second-degree equation as specified in this chapter does not contain an xy term. The presence of an xy term indicates a rotation of the axes. The simplest example is the equation $xy = 1$. The graph of an equation of the form $xy = c$, where c is a constant is a hyperbola with the axes rotated through a 45° angle.

EXAMPLE 6 Sketch the graph of $xy = 1$.

Solution Solve the equation for y, so that we have $y = 1/x$. Choosing several values for x and finding the corresponding y-value leads to the following table.

x	y
-2	$-1/2$
-1	-1
$-1/2$	-2
$-1/5$	-5
0	—
1/5	5
1/2	2
1	1
2	1/2

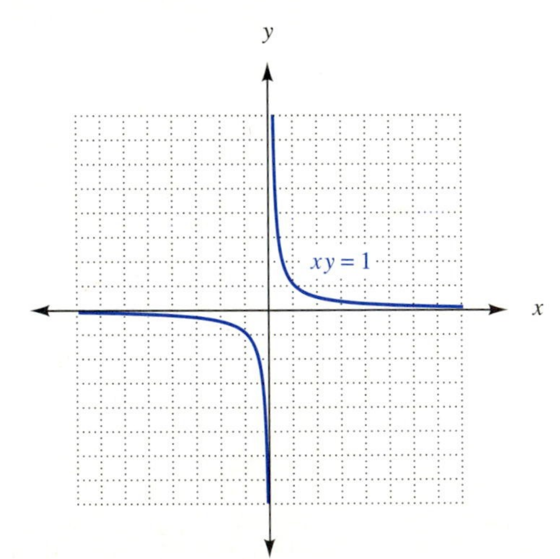

FIGURE 9.37

Notice that the equation is undefined when x is 0. Also as x gets larger and larger in absolute value, y gets smaller and smaller in absolute value. Hence, the x- and y-axes are asymptotes. The graph is shown in Figure 9.37. ■

EXERCISES 9.6

In 1–10, sketch the graph of each curve.

1. $\frac{(x-1)^2}{16} - \frac{(y+1)^2}{9} = 1$

2. $\frac{x^2}{25} - \frac{(y+3)^2}{9} = 1$

3. $\frac{(y-3)^2}{1} - \frac{(x-2)^2}{25} = 1$

4. $\frac{y^2}{36} - \frac{(x+2)^2}{1} = 1$

5. $\frac{(x-1)^2}{16} - \frac{(y-1)^2}{9} = 1$

6. $\frac{x^2}{4} - \frac{(y-2)^2}{9} = 1$

7. $\frac{y^2}{1} - \frac{(x+3)^2}{4} = 1$

8. $\frac{(y+3)^2}{9} - \frac{(x-3)^2}{16} = 1$

9. $\frac{y^2}{2} - \frac{x^2}{1} = 1$

10. $\frac{x^2}{3} - \frac{y^2}{1} = 1$

In 11–18, write the equation of each curve described.

11. a hyperbola with foci at $(17, 0)$ and $(-17, 0)$; transverse axis of length 16

12. a hyperbola with foci at $(12, 1)$ and $(-8, 1)$; conjugate axis of length 12

13. a hyperbola with center at $(-1, 2)$; focus at $(-1, 15)$; vertex at $(-1, 7)$

14. a hyperbola with center at $(-3, -2)$; focus at $(2, -2)$; vertex at $(0, -2)$

15. a hyperbola with center at $(0, 0)$; vertex $(1, 0)$; passes through $(2, \sqrt{3})$

16. a hyperbola with center at $(0, 0)$; vertex $(0, 1)$; passes through $(\sqrt{8}, 3)$

17. a hyperbola with transverse axis from $(-2, 3)$ to $(-2, 1)$; conjugate axis from $(-3, 2)$ to $(-1, 2)$

18. a hyperbola with transverse axis from $(2, -3)$ to $(6, -3)$; conjugate axis from $(4, -4)$ to $(4, -2)$

In 19–30, write each equation to display its special characteristics and determine what type of curve (if any) is represented.

19. $2x^2 - 8x + y^2 + 6y + 5 = 0$

20. $x^2 + 2x + 3y^2 + 6y - 8 = 0$

21. $2x^2 + 12x - 3y^2 + 6y + 9 = 0$

22. $3x^2 - 6x - 4y^2 - 16y - 37 = 0$

23. $2x^2 + 4y^2 - 24y + 28 = 0$

24. $3x^2 + 4y^2 - 8y + 10 = 0$

25. $x^2 + 3y^2 - 12y + 15 = 0$

26. $2x^2 - 8x + 3y^2 + 20 = 0$

27. $x^2 - 2x - y^2 + 2y - 1 = 0$

28. $x^2 - 4y^2 - 16y - 20 = 0$

29. $3x^2 - 6x + 2y^2 + 4y + 11 = 0$

30. $2x^2 + 4x - 3y^2 + 6y + 5 = 0$

31. The silhouette of the cooling tank of a nuclear reactor forms a hyperbola as shown in the accompanying figure. If the asymptotes are the lines given by $y = x$ and $y = -x$, find the equation of the hyperbola. 1 2 4 7 8 3 5

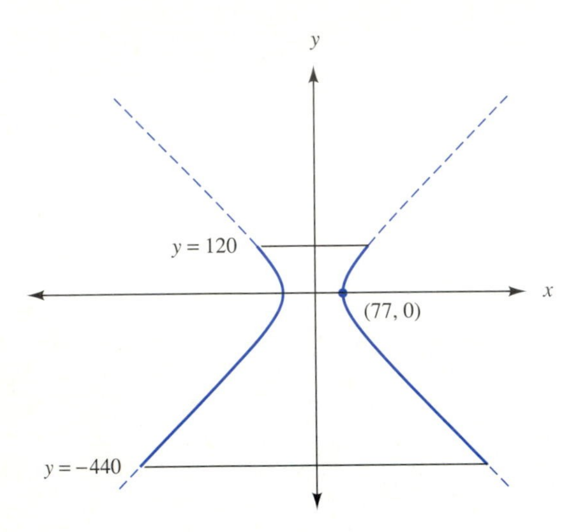

32. A Cassegrain reflecting telescope contains a large parabolic mirror and a smaller hyperbolic mirror, as shown in the accompanying figure. When a ray of light is directed toward one focus of the hyperbola, it is reflected toward the other focus. If the equation of the mirror's surface is $x^2/400 - y^2/441 = 1$, determine the location of the foci. 3 5 6 9

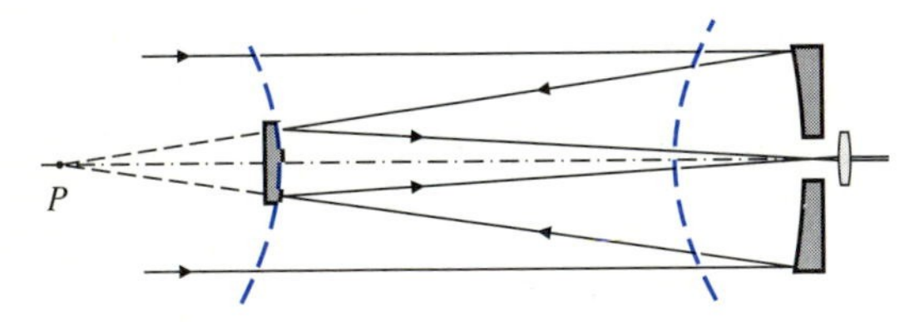

33. A charged particle that moves into the vicinity of another charged particle may be deflected so that its path is the branch of a hyperbola. Write the equation of the path shown in the accompanying diagram. 4 7 8 1 2 3 5 6 9

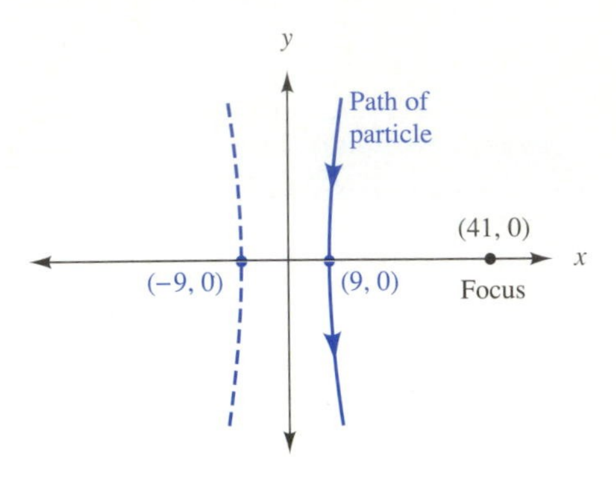

34. Consider two horizontal disks with holes spaced evenly around the circumference of each. When strings are threaded through the holes vertically to connect the disks, as shown in the accompanying figure, the strings form the outline of a cylinder. When one of the disks is rotated, the strings form the outline of a hyperbola. If the equation of one such hyperbola is $25x^2 - 4y^2 = 100$, find the vertical distance between disks, if each one has a radius of 4.00 inches. 1 2 4 7 8 3 5

35. A supersonic jet flying at a constant speed produces a conical shock wave. If the plane's path is parallel to flat terrain, the intersection of the cone with the ground is a hyperbola, as shown in the

accompanying figure. Find the equation of the hyperbola if the center is at the origin of a coordinate system, a vertex is at $(-35, 0)$, and the hyperbola passes through the point $(-70, 12\sqrt{3})$.
4 7 8 1 2 3 5 6 9

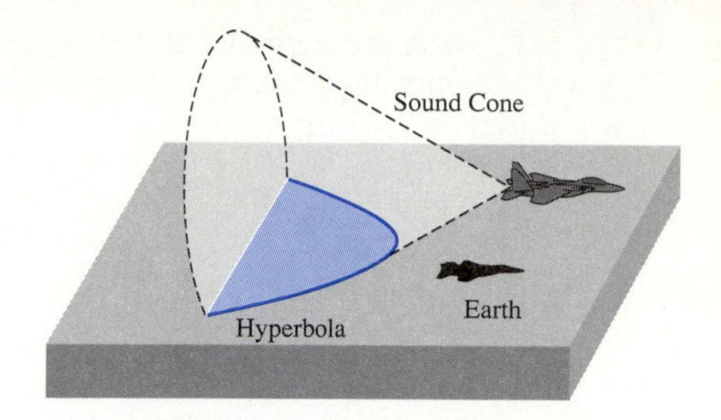

36. Ohm's law states that $V = IR$, where V is voltage, I is current, and R is resistance. If the voltage is a constant 12.0 volts, sketch the graph of I versus R for positive values of R. 3 5 9 1 2 4 6 7 8

9.7 POLAR COORDINATES

In Section 9.1, it was stated that analytic geometry is the branch of mathematics that deals with the relationship between equations and their graphs. A rectangular coordinate system is the only type of system that we have used thus far. The basic idea behind a coordinate system is that it is possible to represent any point in the plane by specifying its position from a fixed point. For a rectangular coordinate system, this position is specified by its horizontal and vertical distances from the origin. In a *polar coordinate system,* the position of a point is specified by a distance and an angle.

polar coordinate system

If a rectangular coordinate system is superimposed on a polar coordinate system, as in Figure 9.38, the positive x-axis coincides with what is called the **polar axis.** The origin coincides with what is called the **pole.** A point is specified by coordinates (r, θ). The angle θ is the angle between the polar axis and a line joining the pole and the point. The angle θ is often given in radians. The number r is the distance to the point from the pole.

polar axis

pole

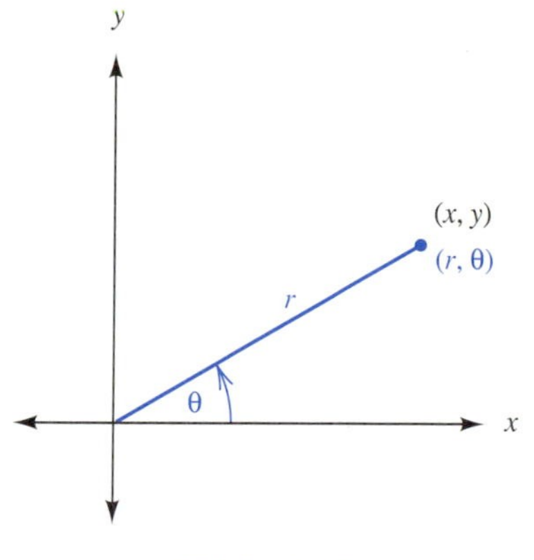

FIGURE 9.38

In a polar coordinate system, the plane is usually indicated by concentric circles whose radii represent different values of r. Lines are drawn through the pole to indicate various values of θ. Lines are often drawn at special angles such as 0, $\pi/6$, $\pi/3$,

$\pi/2$, and multiples of these angles, as shown in Figure 9.39a. Sometimes, however, only lines at 0, $\pi/4$, and multiples of $\pi/4$ are shown, as in Figure 9.39b.

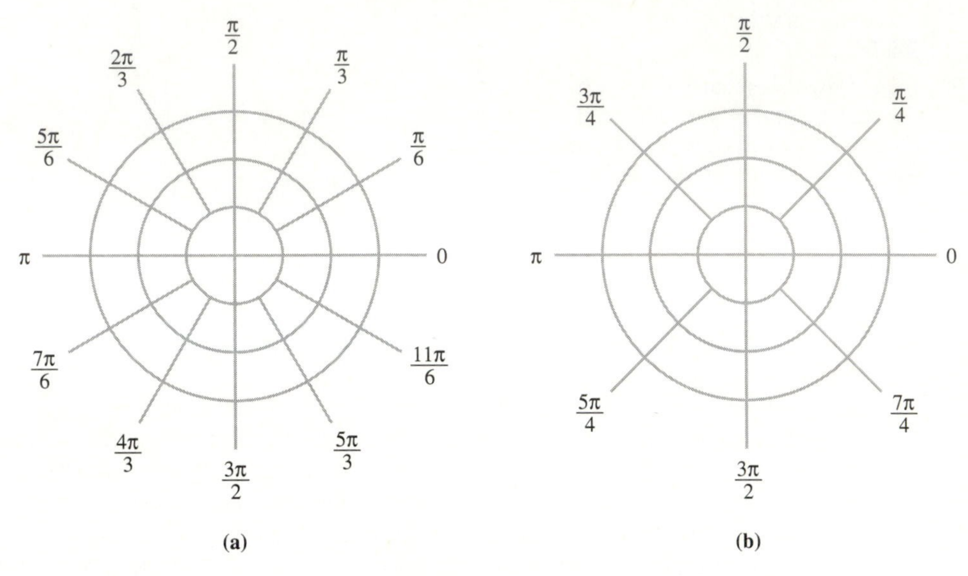

FIGURE 9.39

Although polar coordinates are listed with r first, it is usually easier to plot a point by finding θ first. Then r can be measured along the terminal side of θ. Example 1 illustrates how to plot a point on a polar coordinate system.

EXAMPLE 1 Plot the point $(3, \pi/4)$ on a polar coordinate system.

Solution Figure 9.40 shows the point. It is 3 units from the pole on the terminal side of the angle $\pi/4$ in standard position. ■

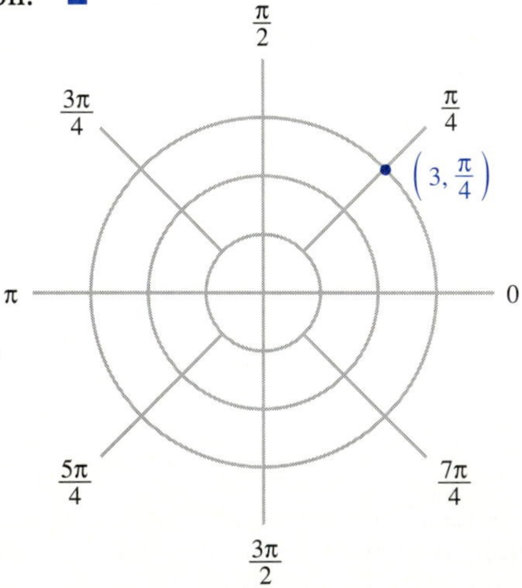

FIGURE 9.40

Negative angles are measured in the clockwise direction, just as they were in Chapter 4. When r is a negative number, the point is located in the opposite direction from $|r|$.

EXAMPLE 2 Plot the following points on a polar coordinate system.

(a) $(2, \pi/6)$ **(b)** $(2, -\pi/6)$ **(c)** $(-2, \pi/6)$ **(d)** $(-2, -\pi/6)$

Solution Figure 9.41 shows the points. ■

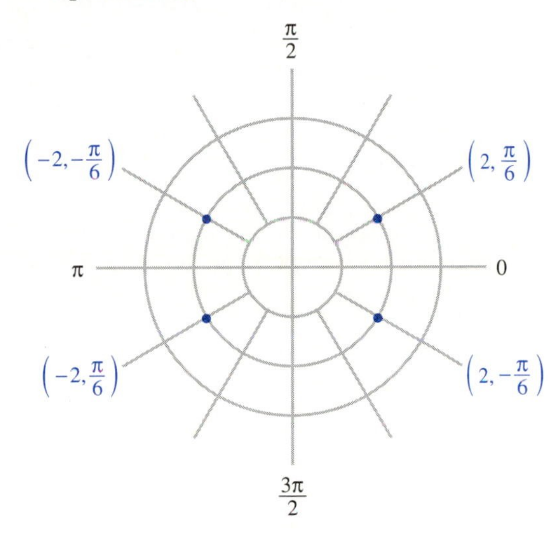

FIGURE 9.41

Although a point in a rectangular coordinate system can be specified by only one pair of coordinates, a point in a polar coordinate system can be specified in any number of ways. The point $(2, -\pi/6)$, for instance, could also be represented as $(-2, 5\pi/6)$. Likewise, $(-2, \pi/6)$ could be represented as $(2, 7\pi/6)$. Even the point $(2, \pi/6)$ could be represented as $(-2, 7\pi/6)$ or $(2, 13\pi/6)$.

It is possible to convert coordinates from one system to the other. Figure 9.38 should remind you of the trigonometric definitions in Section 4.4. We know that $\sin\theta = y/r$, $\cos\theta = x/r$, $\tan\theta = y/x$, and $x^2 + y^2 = r^2$. From these equations, we obtain the following formulas for converting from one coordinate system to the other.

RECTANGULAR-POLAR COORDINATE CONVERSION FORMULAS

(a) Given $(x\ y)$: $r = \sqrt{x^2 + y^2}$

$\tan\theta = y/x$

(b) Given (r, θ): $x = r\cos\theta$

$y = r\sin\theta$

EXAMPLE 3 Find the polar coordinates that correspond to the rectangular coordinates (3, 7).

Solution Figure 9.42 shows (x, y), r and θ. Since x is 3 and y is 7, we have

$$r = \sqrt{3^2 + 7^2} \text{ or } \sqrt{9 + 49}, \text{ which is } \sqrt{58} \text{ and } \tan\theta = 7/3.$$

A calculator may be used. In this example we use radian mode to find the value of θ for which $\tan\theta = 7/3$. The keystroke sequence is as follows.

58 [√x] **7.6157731**

7 [÷] 3 [=] [INV] [tan] **1.1659045**

We use the convention that approximate values are rounded to three significant digits when they are obtained from integer values that are understood to be exact. Thus, the coordinates are (7.62, 1.17). ■

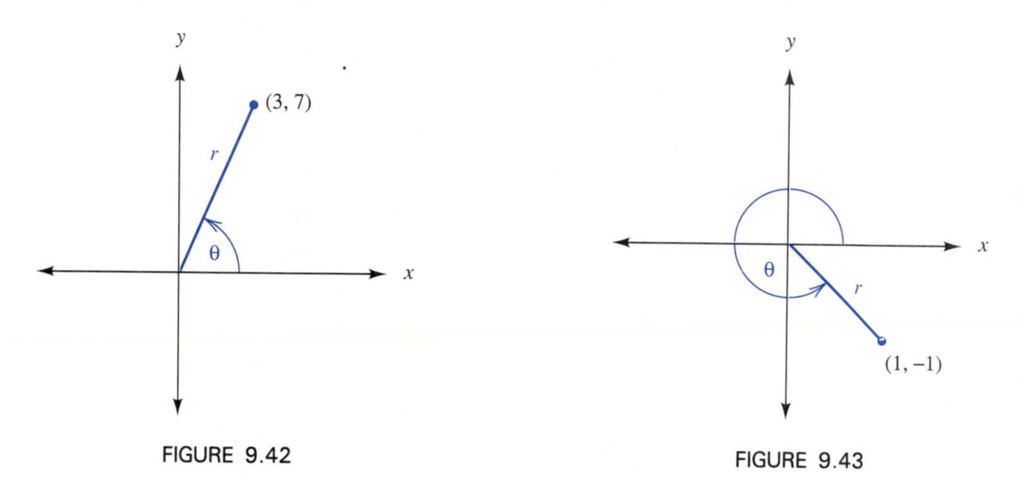

FIGURE 9.42

FIGURE 9.43

EXAMPLE 4 Find the polar coordinates that correspond to the rectangular coordinates (1, −1).

Solution Figure 9.43 shows (x, y), r and θ. Since x is 1 and y is −1, we have

$$r = \sqrt{1^2 + (-1)^2} \text{ or } \sqrt{2} \text{ and } \tan\theta = -1/1 \text{ or } -1.$$

Since θ is in the fourth quadrant, we use a calculator to find the reference angle θ_R associated with θ. Since $\theta_R = 45°$, $\theta = 315°$ or $7\pi/4$. ■

EXAMPLE 5 Find the rectangular coordinates that correspond to the polar coordinates $(3, \pi/3)$.

Solution Since $r = 3$ and $\theta = \pi/3$, we have

$$x = 3\cos\pi/3 \text{ and } y = 3\sin\pi/3.$$

Recall that $\pi/3 = 60°$ and $\cos 60° = 1/2$ while $\sin 60° = \sqrt{3}/2$. Thus,

$$x = 3(1/2) = 3/2 \quad \text{and} \quad y = 3(\sqrt{3}/2) = 3\sqrt{3}/2.$$

The coordinates are $(3/2, 3\sqrt{3}/2)$ or (1.50, 2.60). ■

EXERCISES 9.7

In 1–10, plot the given points on a polar coordinate system.

1. $(3, \pi/4)$ **2.** $(1, \pi/2)$ **3.** $(4, 3\pi/4)$ **4.** $(3, -\pi)$

5. $(2, -\pi/3)$ **6.** $(1, -2\pi/3)$ **7.** $(-2, 5\pi/6)$ **8.** $(-3, 3\pi/2)$

9. $(-1, -\pi/2)$ **10.** $(-4, -7\pi/6)$

In 11–20, find the polar coordinates that correspond to the given rectangular coordinates.

11. (3, 4) **12.** (5, −12) **13.** (−8, 15) **14.** (−1, 1)

15. (0, 1) **16.** (2, 2) **17.** $(\sqrt{3}, -1)$ **18.** (−3, 3)

19. (−1, −2) **20.** $(-1, -\sqrt{3})$

In 21–30, find the rectangular coordinates that correspond to the polar coordinates.

21. $(2, \pi/6)$ **22.** $(3, \pi/3)$ **23.** $(-1, \pi/4)$ **24.** $(-2, \pi/2)$

25. (4, 0) **26.** $(3, \pi)$ **27.** $(2, -3\pi/2)$ **28.** $(-1, -2\pi/3)$

29. $(-2, -4\pi/3)$ **30.** $(-3, -7\pi/6)$

Problems 31–34 refer to the accompanying diagram. The voltage drops in the circuit are as indicated. Plot these points.

31. $V_{R_1} = (4, 54°)$ volts 3 5 9 1 2 4 6 7 8

32. $V_L = (8, 143°)$ volts 3 5 9 1 2 4 6 7 8

33. $V_C = (24, -37°)$ volts 3 5 9 1 2 4 6 7 8

34. $V_{R_2} = (8, 53°)$ volts 3 5 9 1 2 4 6 7 8

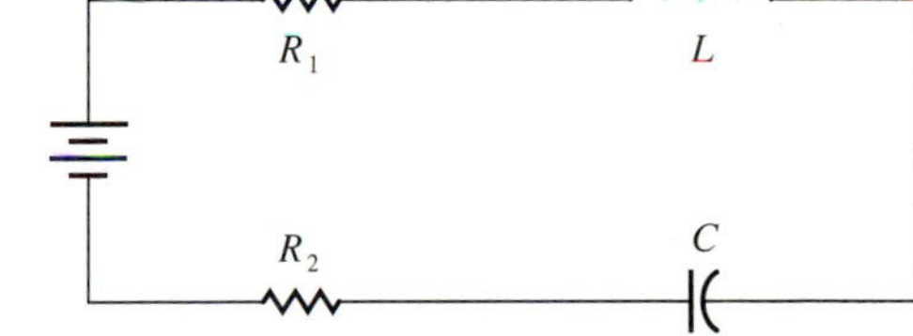

In 35–36, consider a force with the given magnitude and direction in polar coordinates. Convert to rectangular coordinates.

35. (8.0 N, −5.3°) 2 4 8 1 5 6 7 9 **36.** (6.2 lb, −30.6°) 2 4 8 1 5 6 7 9

In 37–38, consider a displacement indicated by rectangular coordinates. Convert to polar coordinates.

37. (4.1 ft, −5.3 ft) 4 7 8 1 2 3 5 6 9 **38.** (−2.4 cm, 3.9 cm) 4 7 8 1 2 3 5 6 9

---

1. Architectural Technology
2. Civil Engineering Technology
3. Computer Engineering Technology
4. Mechanical Drafting Technology
5. Electrical/Electronics Engineering Technology
6. Chemical Engineering Technology
7. Industrial Engineering Technology
8. Mechanical Engineering Technology
9. Biomedical Engineering Technology

9.8 CURVES IN POLAR COORDINATES

It is possible to convert not only single points, but also entire equations, from one coordinate system to the other. In many cases, an equation written in terms of r and θ has a simpler representation than the equivalent equation in x and y. The equation of a circle of radius 3, centered at the origin, for instance, is $x^2 + y^2 = 9$ in rectangular coordinates. To find the equation in polar coordinates, replace x with $r \cos \theta$ and replace y with $r \sin \theta$. That is,

$$(r \cos \theta)^2 + (r \sin \theta)^2 = 9.$$

Since $(\cos \theta)^2$ and $(\sin \theta)^2$ are usually written as $\cos^2 \theta$ and $\sin^2 \theta$, we have

$$r^2 \cos^2 \theta + r^2 \sin^2 \theta = 9$$
$$r^2(\cos^2 \theta + \sin^2 \theta) = 9.$$

It can be shown that $\cos^2 \theta + \sin^2 \theta = 1$ for any value of θ, so the equation becomes

$$r^2 = 9 \text{ or } r = \pm 3.$$

NOTE ⇨⇨ When we first began to graph equations in rectangular coordinates, we did it by plotting points. We do the same thing in polar coordinates. ***It is very important to use radian measure for θ when graphing equations in polar coordinates.***

EXAMPLE 1 Sketch the graph of $r = 2 \sin \theta$.

Solution We construct a table showing various values of θ and the corresponding value of r found by evaluating the expression $2 \sin \theta$. The letter n stands for the number of the entry in the table.

r	0	1	1.4	1.7	2	1.7	1.4	1	0	-1	-1.4	-1.7	-2
θ	0	$\pi/6$	$\pi/4$	$\pi/3$	$\pi/2$	$2\pi/3$	$3\pi/4$	$5\pi/6$	π	$7\pi/6$	$5\pi/4$	$4\pi/3$	$3\pi/2$
n	1	2	3	4	5	6	7	8	9	10	11	12	13

Figure 9.44 shows the graph obtained when these points are connected in order from small values of θ to larger values of θ. Each point plotted is numbered to correspond to its entry number n in the table. Thus, it is easy to see the order in which the points are connected to give the graph. Notice that the graph is a circle but it is not centered at the origin. ■

EXAMPLE 2 Sketch the graph of $\theta = \pi/3$.

Solution For any value of r, θ is $\pi/3$. Thus, the graph consists of those points along the terminal side of $\pi/3$, as shown in Figure 9.45. The points in the third quadrant are those for which r has a negative value. ■

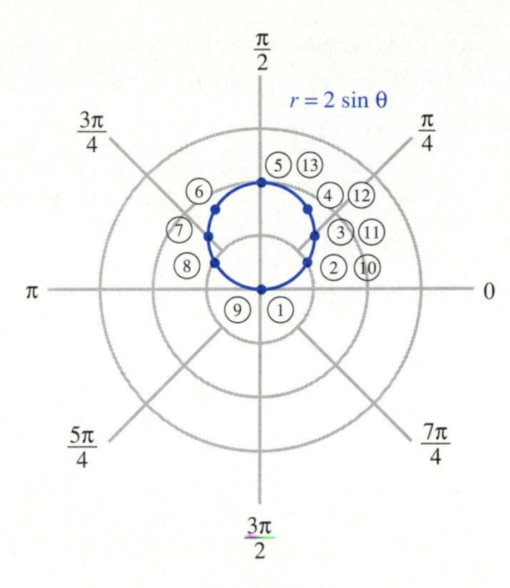

FIGURE 9.44

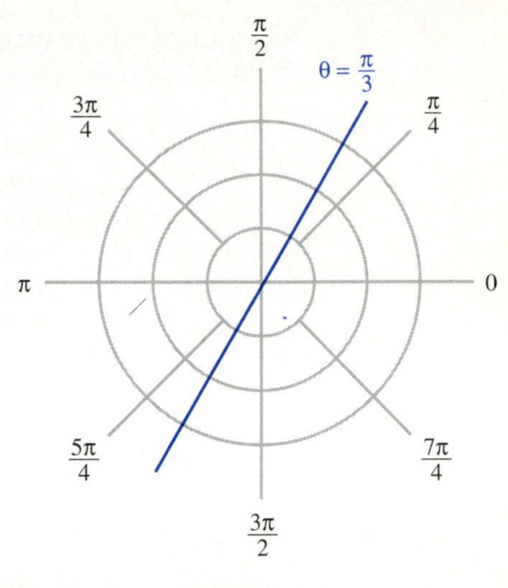

FIGURE 9.45

EXAMPLE 3 Graph $r = 1 - 2 \sin \theta$.

Solution We construct a table showing various values of θ and the corresponding value of r found by evaluating the expression $1 - 2 \sin \theta$.

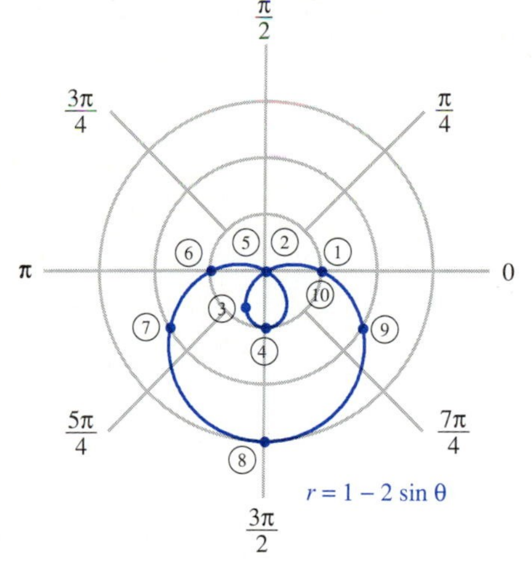

FIGURE 9.46

r	1	0	$-.414$	$-.732$	-1	$-.732$	$-.414$	0	1
θ	0	$\pi/6$	$\pi/4$	$\pi/3$	$\pi/2$	$2\pi/3$	$3\pi/4$	$5\pi/6$	π
n	1	2		3	4			5	6

r	2	2.414	2.732	3	2.732	2.414	2.00	1.00
θ	$7\pi/6$	$5\pi/4$	$4\pi/3$	$3\pi/2$	$5\pi/3$	$7\pi/4$	$11\pi/6$	2π
n	7			8			9	10

Figure 9.46 shows the graph obtained when these points are connected. ■

Graphs of equations in polar coordinates follow patterns just as those for rectangular coordinates do. It is not necessary for you to memorize these patterns. They are presented so that you are aware of the patterns that you will see as you plot the graphs of equations in polar coordinates.

GRAPHS OF EQUATIONS IN POLAR COORDINATES

In these equations, a represents a positive real number.

Straight Line

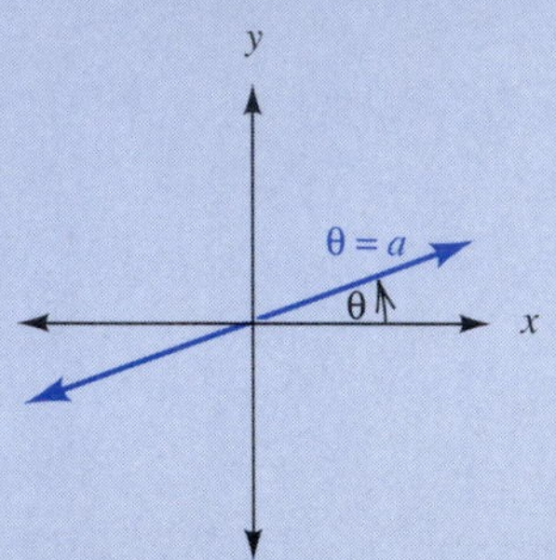

Circle of Radius *a*, Centered at the Pole

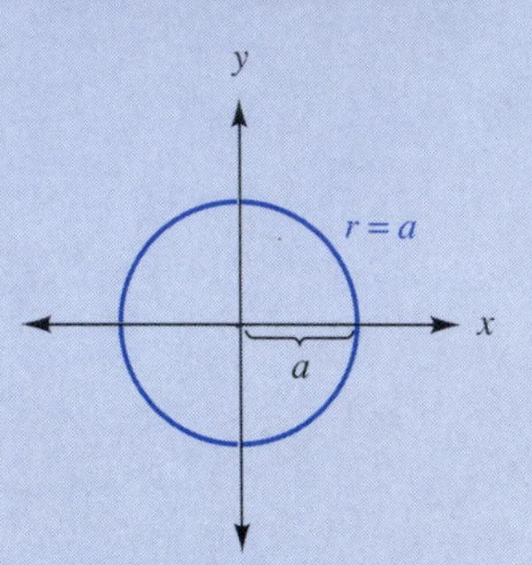

Spiral

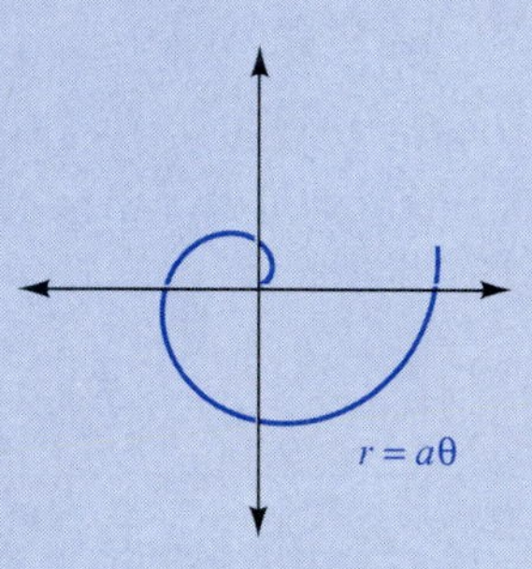

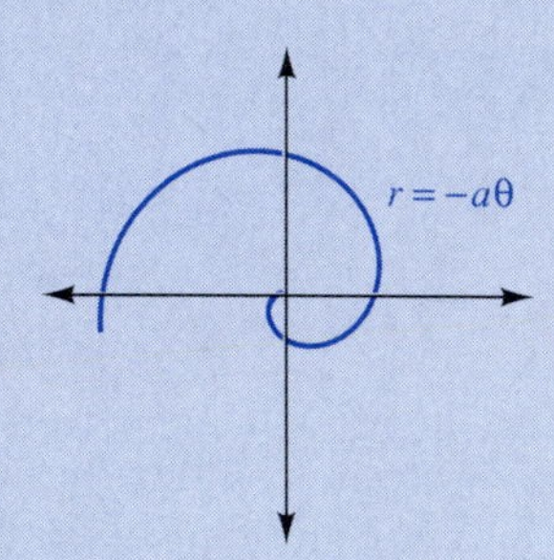

Three-Leaved Rose

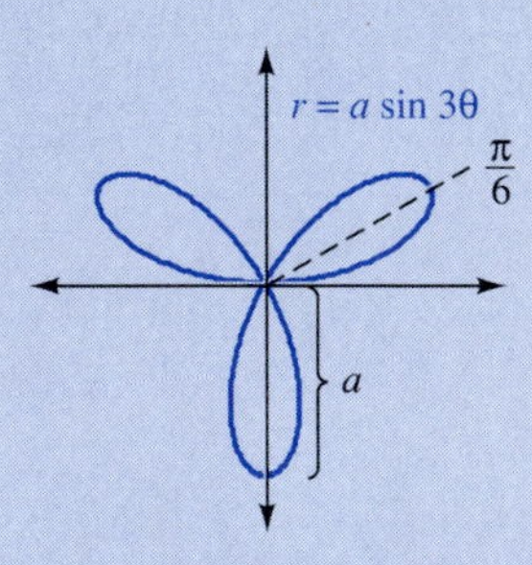

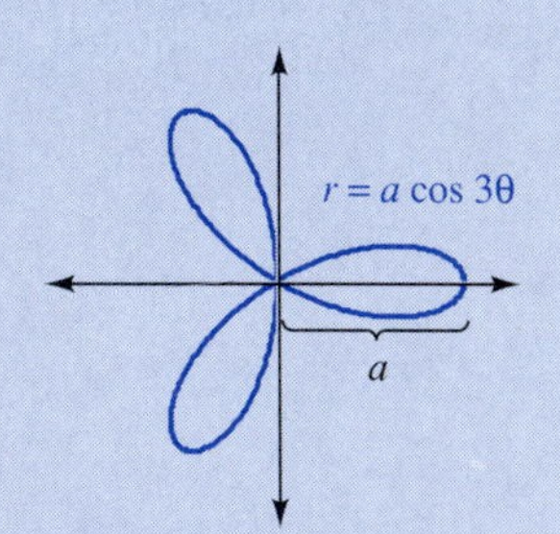

***n*-Petal Rose or 2*n*-Petal Rose**

$r = a \sin n\theta$
The first petal is along $\theta = \pi/(2n)$.

$r = a \cos n\theta$
The first petal is along $\theta = 0$.

The rose will have n petals when n is odd and $2n$ petals when n is even. The petals are equally spaced and have length a.

MORE GRAPHS OF EQUATIONS IN POLAR COORDINATES

In these equations, a represents a positive real number.

Cardioids

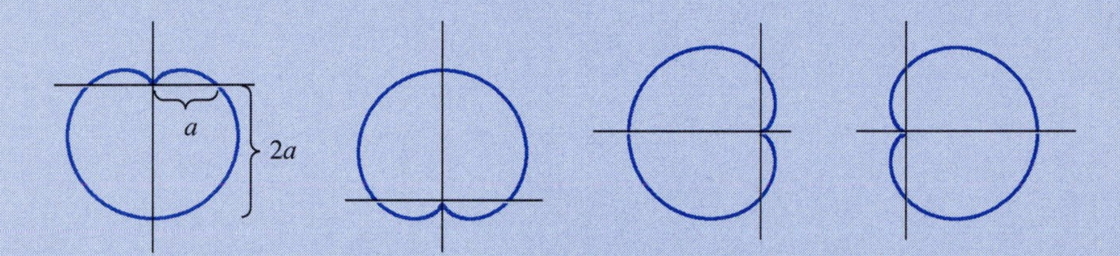

Limacons

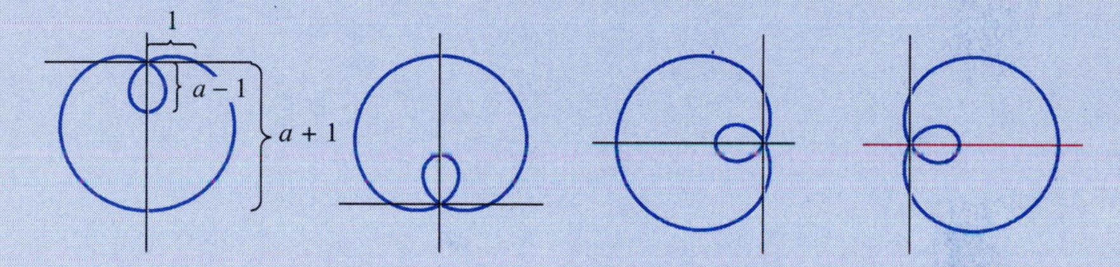

Lemniscates

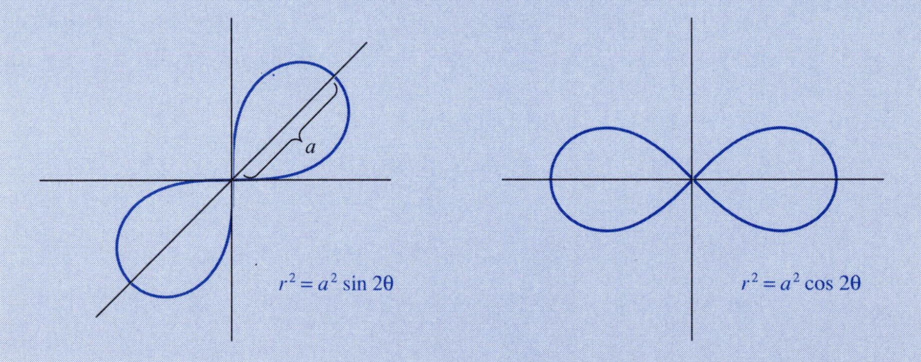

EXAMPLE 4 Sketch the graph of $r = 3\cos 2\theta$.

Solution The equation follows the pattern of a 4-petal rose. The first petal falls along $\theta = 0$. Since the petals are evenly spaced throughout the 2π radians of the circle, there are petals at $\pi/2$, π, and $3\pi/4$. Each petal has a length of 3. The following table shows various values of θ and the corresponding value of r found by evaluating the expression $3\cos 2\theta$.

r	3	1.5	0	−1.5	−3	−1.5	0	1.5	3	1.5	0	1.5	3
θ	0	$\pi/6$	$\pi/4$	$\pi/3$	$\pi/2$	$2\pi/3$	$3\pi/4$	$5\pi/6$	π	$7\pi/6$	$5\pi/4$	$4\pi/3$	$3\pi/2$
n	1	2	3	4	5	6		7	8	9		10	11

Figure 9.47 shows the graph. ■

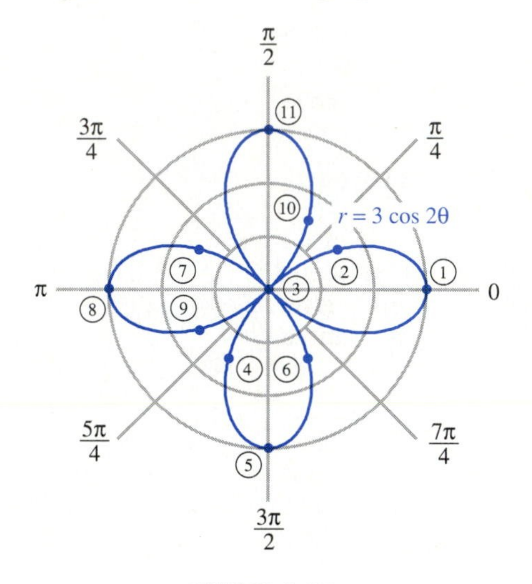

FIGURE 9.47

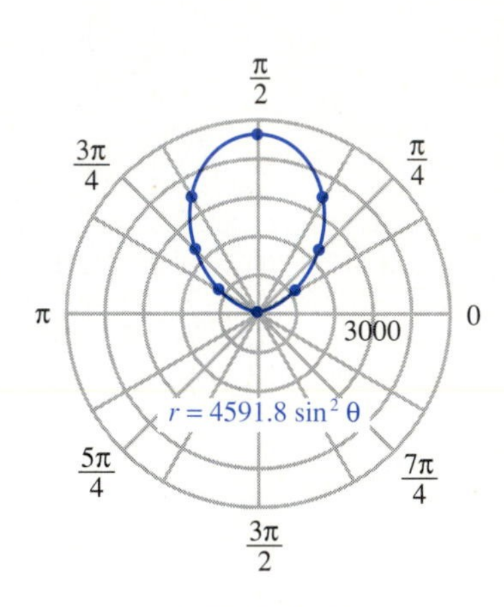

FIGURE 9.48

EXAMPLE 5 When a projectile is launched at an angle of θ, its maximum height is given by $r = [v_0^2/(2g)]\sin^2\theta$ where v_0 is the initial velocity and g is the acceleration due to gravity. If $g = 9.8$ m/s² and $v_0 = 300$ m/s, sketch a graph of r versus θ.

Solution The equation can be written as

$$r = \frac{(300)^2}{2(9.8)}\sin^2\theta \text{ or } r = 4591.8\sin^2\theta$$

A table of values lists values of r corresponding to various values of θ.

0	1148	2296	3444	4592	3444	2296	1148	0
0	$\pi/6$	$\pi/4$	$\pi/3$	$\pi/2$	$2\pi/3$	$3\pi/4$	$5\pi/6$	π

Figure 9.48 shows the graph. ■

EXERCISES 9.8

In 1–24, graph each equation.

1. $r = 2 \sin \theta$
2. $r = 3 \cos \theta$
3. $r = \cos 3\theta$
4. $r = \sin 3\theta$
5. $r = 3(1 - \sin \theta)$
6. $r = 2(1 + \cos \theta)$
7. $r = 1 - \cos \theta$
8. $r = 1 + \sin \theta$
9. $r = (1/2)\theta$
10. $r = 2\theta$
11. $r^2 = 4 \sin 2\theta$
12. $r^2 = 9 \cos 2\theta$
13. $r = 3$
14. $r = 2$
15. $\theta = \pi/3$
16. $\theta = \pi/4$
17. $r = 2(1 - \sin \theta)$
18. $r = 3(1 + \cos \theta)$
19. $r = 1 + \cos \theta$
20. $r = 1 - \sin \theta$
21. $r = (1/3)\theta$
22. $r = 3\theta$
23. $r^2 = 9 \cos 2\theta$
24. $r^2 = 4 \cos 2\theta$

25. When a pendulum bob is displaced from the rest position and released, its horizontal distance from the rest position is given by $r = l \sin \theta$ where l is the length of the pendulum and θ is the angle made by the pendulum with the vertical. Sketch the graph of r versus θ for a pendulum of length 2m and $0 < \theta < \pi/2$. **4 7 8** 1 2 3 5 6 9

26. When monochromatic light strikes a surface, the critical angle c is the angle of refraction for which all of the energy goes into the reflected beam. The critical angle is given by $\sin c = 1/n$ where n is the index of refraction of the medium that the light enters. Sketch the graph of n versus c for $0 < c < \pi/2$. 3 5 6 9

27. For a plane inclined at an angle of θ with the horizontal and an object resting on the plane, the coefficient of friction is given by $\mu = \tan \theta$. Sketch a graph of μ versus θ for $0 < \theta < \pi/2$. **4 7 8** 1 2 3 5 6 9

28. The power (in W) absorbed in an ac circuit is given by $P = IV \cos \theta$. If $I = 1.2$ A when $V = 120$ V sketch a graph of P versus θ for $0 < \theta \leq \pi/2$. **3 5 9** 1 2 4 6 7 8

29. In Great Britain, the length of a nautical mile is considered to vary with latitude. The length l (in m) is given by $l = 1823.1 - 9.3 \cos 2\phi$ where ϕ is the latitude. Sketch a graph of $9.3 \cos 2\phi$ versus ϕ for $0 < \phi < \pi/2$. **1 2 4 7 8** 3 5

30. When a projectile is launched at an angle of θ, its horizontal range is given by $r = (v_0^2/g) \sin 2\theta$ where v_0 is the initial velocity and g is the acceleration due to gravity. If $g = 9.8$ m/s^2 and $v_0 = 300$ m/s, sketch a graph of r versus θ. **4 7 8** 1 2 3 5 6 9

1. Architectural Technology
2. Civil Engineering Technology
3. Computer Engineering Technology
4. Mechanical Drafting Technology
5. Electrical/Electronics Engineering Technology
6. Chemical Engineering Technology
7. Industrial Engineering Technology
8. Mechanical Engineering Technology
9. Biomedical Engineering Technology

9.9 CHAPTER REVIEW

KEY TERMS

9.1 reciprocal of a nonzero number

9.2 parabola
directrix
focus
translation of axes

9.4 circle
center of a circle
radius of a circle

9.5 ellipse
foci of an ellipse
major axis of an ellipse
minor axis of an ellipse

9.6 hyperbola
foci of a hyperbola
transverse axis of a hyperbola
conjugate axis of a hyperbola
asymptote

9.7 polar coordinate system
polar axis
pole

RULES AND FORMULAS

Distance between two points: $d = \sqrt{(x_2 - x_1)^2 + (y_2 - y_1)^2}$

Slope of a line: $m = \dfrac{y_2 - y_1}{x_2 - x_1}$

Parabola: $4p(y - k) = (x - h)^2$ or $4p(x - h) = (y - k)^2$

Circle: $(x - h)^2 + (y - k)^2 = r^2$

Ellipse: $\dfrac{(x - h)^2}{a^2} + \dfrac{(y - k)^2}{b^2} = 1$ or $\dfrac{(y - k)^2}{a^2} + \dfrac{(x - h)^2}{b^2} = 1$ $(c^2 = a^2 - b^2)$

Hyperbola: $\dfrac{(x - h)^2}{a^2} - \dfrac{(y - k)^2}{b^2} = 1$ or $\dfrac{(y - k)^2}{a^2} - \dfrac{(x - h)^2}{b^2} = 1$ $(c^2 = a^2 + b^2)$

Given (x, y): $r = \sqrt{x^2 + y^2}$ and $\tan\theta = y/x$

Given (r, θ): $x = r\cos\theta$ and $y = r\sin\theta$

SEQUENTIAL REVIEW

(Section 9.1)

1. Find the distance between $(2, -1)$ and $(-3, 2)$.

In 2–4, find the equation of each line described.

2. through $(-2, 1)$; parallel to $y = (1/3)x + 2$
3. through $(1, 0)$; perpendicular to $y = -2x + 3$
4. through $(1, -2)$ and $(-2, -3)$
5. Find the slope of the line through $(2, -3)$ and $(4, -1)$.
6. Verify that the figure formed by the segments joining $(-3, 0)$, $(0, 4)$, $(3, 0)$, and $(0, -4)$ is not a square, even though the sides are the same length.

(Section 9.2) *In 7–12, find the equation of the parabola that satisfies the given conditions. Put the equation in standard form.*

7. vertex at the origin; focus at $(0, 4)$
8. focus at $(1, 0)$; directrix at $x = 3$
9. vertex at $(-1, 2)$; directrix at $x = -2$
10. focus at $(2, 2)$; directrix at $y = 6$
11. vertex at $(1, 1)$; focus at $(-1, 1)$
12. focus at the origin; directrix at $y = 2$

(Section 9.3) *In 13–16, determine the vertex of each parabola by completing the square.*

13. $y = x^2 + 8x + 5$

14. $y = 2x^2 + 5x$

15. $x = -y^2 + 6y + 2$

16. $x = (1/2)y^2 + 4y - 3$

In 17 and 18, sketch the graph of each curve.

17. $y = -x^2 + 6x - 5$

18. $x = 2y^2 - 9$

(Section 9.4)

19. Find the equation of a circle with center at $(0, -6)$ through $(8, 9)$.

In 20 and 21, sketch the graph of each curve.

20. $(x + 1)^2 + (y - 1)^2 = 5$

21. $(x + 2)^2 + y^2 = 16$

In 22–24, for each equation, determine whether it represents a circle. If the equation represents a circle, find the center and radius.

22. $x^2 + 2x + y^2 + 6y + 3 = 0$

23. $2x^2 + x + y^2 - 3y + 2 = 0$

24. $x^2 - 2x + y^2 + 4y - 4 = 0$

(Section 9.5) *In 25–29, write the equation of the ellipse described.*

25. foci at $(4, 2)$ and $(-2, 2)$; minor axis of length 4

26. center at the origin; vertical major axis of length 8; minor axis of length 6

27. center at $(2, 2)$; focus at $(5, 2)$; vertex at $(6, 2)$

28. foci at $(6, 1)$ and $(4, 1)$; minor axis of length 24

29. foci at $(2, 1)$ and $(2, 17)$; vertex at $(2, 24)$

(Section 9.6) *In 30–33, write the equation of the hyperbola described.*

30. center at the origin; horizontal transverse axis of length 8; conjugate axis of length 6

31. center at $(2, 2)$; vertex at $(5, 2)$; focus at $(7, 2)$

32. foci at $(-6, 1)$ and $(4, 1)$; conjugate axis of length 8

33. foci at $(2, 1)$ and $(2, 11)$; vertex at $(2, 3)$

In 34–37, write each equation to display its special characteristics and determine what type of curve (if any) is represented.

34. $2x^2 + 12x + 3y^2 - 6y - 27 = 0$

35. $x^2 + 3y^2 + 6 = 0$

36. $2x^2 + 3x + 2y^2 - 2y - 1 = 0$

37. $2x^2 - 8x - y^2 - 6y = 0$

(Section 9.7) *In 38 and 39, find the rectangular coordinates that correspond to the given polar coordinates*

38. $(-3, \pi/6)$

39. $(+4, 2\pi/3)$

In 40 and 41, find the polar coordinates that correspond to the given rectangular coordinates

40. $(2, -3)$

41. $(-1, 4)$

(Section 9.8) *In 42–45, graph each equation.*

42. $r = \sin\theta$

43. $r = 1 + 2\cos\theta$

44. $r = \sin 2\theta$

45. $r = 2$

APPLIED PROBLEMS

1. A city that has N-S streets numbered 1 through 50 and E-W streets lettered A through Z built a thoroughfare diagonally from the corner of 7th and B to the corner of 15th and H. How many blocks shorter is the new route than the old route?
1 2 4 7 8 3 5

2. The output voltage V_{out} (in volts) of a power supply is given by $V_{out} = V_{max} - R_T I_L$. In the formula, V_{max} is the maximum voltage (in volts) attained, I_L is the load (in amperes), and R_T is the Thevenin resistance (in ohms) of the supply. The accompanying diagram shows a graph of V_{out} versus I_L. Determine the value of R_T.
3 5 9 1 2 4 6 7 8

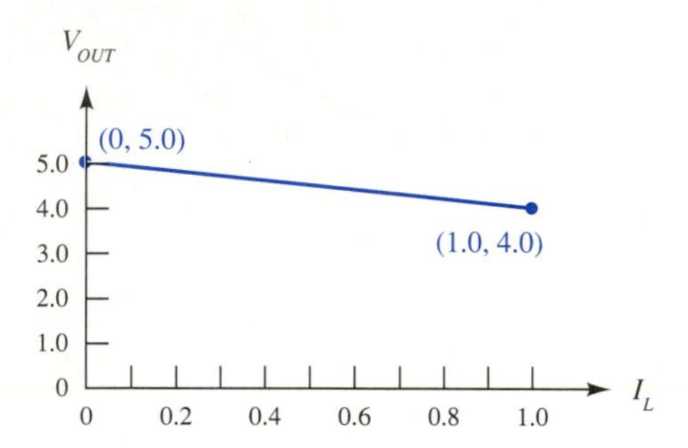

3. A telescope is constructed with a parabolic mirror. The equation of the parabola is $5600y = x^2$. The mirror has a radius of 7.60 cm, and it is 3.00 cm thick at the vertex, as shown in the accompanying figure. What is the thickness at the edge of the mirror? 3 5 6 9

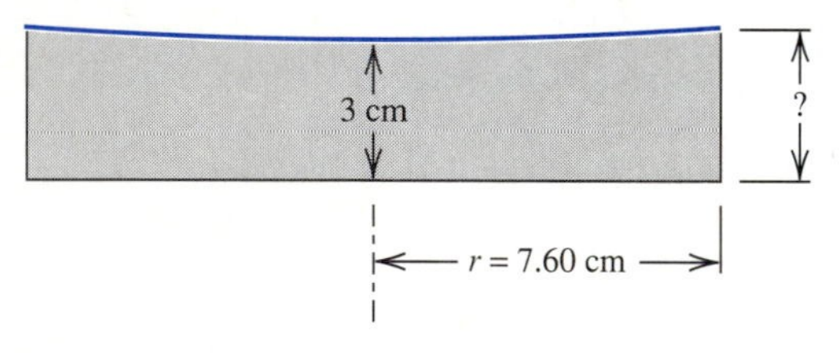

4. The formula for the electrical power P (in watts) dissipated in a circuit is $P = I^2R$, where I is the current (in amperes) and R is the resistance (in ohms). If R is a constant 250 Ω, plot P versus I for $0 < I < 10$. 3 5 9 1 2 4 6 7 8

5. A column that is 14.0 inches square is to be reinforced with 6 steel rods, located symmetrically within the column. If the rods are placed so that no rod is closer than 1.50 inches to an external face, write the equation of the circle along which the rods lie. Assume that the origin is at the lower left-hand corner of the square cross section of the column. 1 2 4 7 8 3 5

6. A printed circuit board has three terminal points that must be connected together by solder track. The coordinates of the points are $A = (3.175, 5.625)$, $B = (4.250, 3.115)$, and $C = (6.250, 3.115)$ all measured in inches from one corner of the board. How far is it from A to B? from B to C?
3 5 9

7. The ac-load line of a certain power amplifier intersects the vertical current axis at $I = 43$ ma and the horizontal voltage axis at $V = 17$ V on the graph of I versus V. Find the equation for this line.
3 5 9 1 2 4 6 7 8

8. A certain transistor has its current I (in milliamps) given by the equation for the parabola $I = 2.555V^2 + 15.333V + 23$ when V is measured in volts. Find the vertex and focus of this parabola.
3 5 9 4 7 8

9. The formula $I_t = -AI_0 + (I_0 - I_r)$ gives the light transmitted through a window. In the formula, I_0 is the original intensity of the light. I_r is the amount reflected back, and A is the absorption factor with a range of zero to one. All I's are measured in watts. For a given I_0 and I_r this becomes a linear equation.
   (a) What equation represents I_t if $I_0 = 150$ watts and $I_r = 12$ watts?
   (b) Use the equation in part a to find the value of A that corresponds to $I_t = 130$ watts.
9 1 2 3 5 6 7

10. The volume V of a certain metal part is thought to change in a linear manner with small temperature changes T. At 70° F the volume is measured to be 1.736 in³. At 90° F the volume expands to 1.743 in³. What linear equation describes this volume-temperature relationship? 1 2 4 7 8 3 5

1. Architectural Technology
2. Civil Engineering Technology
3. Computer Engineering Technology
4. Mechanical Drafting Technology
5. Electrical/Electronics Engineering Technology
6. Chemical Engineering Technology
7. Industrial Engineering Technology
8. Mechanical Engineering Technology
9. Biomedical Engineering Technology

11. An instructor remarked that the characteristic curve $I = 9V^3 + 9V^2 + 2V + 20$ for a certain device is the product of a straight line and a parabola. Find the curves he has in mind by factoring the expression on the right. 3 5 9 1 2 4 6 7 8

12. The floor plan of a concert stage shows a reflecting wall in the shape of a horizontal parabola opening toward the right as shown in the accompanying figure. The parabola's vertex is 8 ft from a flat wall which is parallel to the edge of the stage 50 ft away. A solo performer is to stand at the focus of the parabola. Find the coordinates of the focus and the equation for the directrix. 1 3 4 5 6 8 9

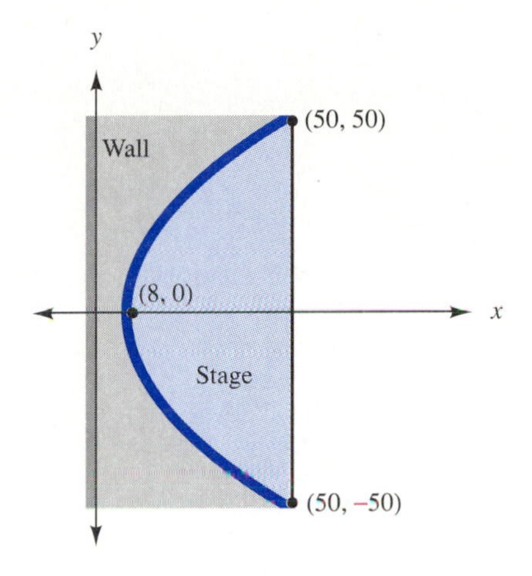

13. The fundamental frequency of vibration f (in Hz) of a stretched wire such as a guitar string is given by the formula $f^2 = \dfrac{8.048}{L^2}\left(\dfrac{T}{W}\right)$, where L is the string's length (in feet), T is the tension (in lbs), and W is its weight in (lbs/foot). Describe this equation's graph f versus T for fixed values of L and W. 1 3 4 5 6 8 9

14. The index of refraction of a certain glass used in fiber optic cables was given by the equation

$$n = 8.600 \times 10^{-7}\lambda^2 - 2.166 \times 10^{-3}\lambda + 2.500$$

where λ is the wavelength (in nanometers) of the light in the fiber. Find the vertex of the parabola it describes. 3 5 6 9

15. Consider a map having the center of the state as the origin of a coordinate system. Brownsville is centered at coordinates (26, 24). All measurements are in miles. Elm City has a radio paging system at (30, 18) that can reach out a maximum of 6 miles. Can someone in the center of Brownsville be paged from Elm City? 1 2 4 7 8 3 5

16. A specially designed hopper for emptying a large container of dry material with minimum clogging can be made by forming the sides to be hyperbolic in shape rather than simple flat surfaces. What equation would describe the sides of such a hopper on an xy-coordinate system with origin as shown if its height is 10 ft, its discharge hole width is 4 ft, and its entrance width is 7 ft? 1 2 4 7 8 3 5

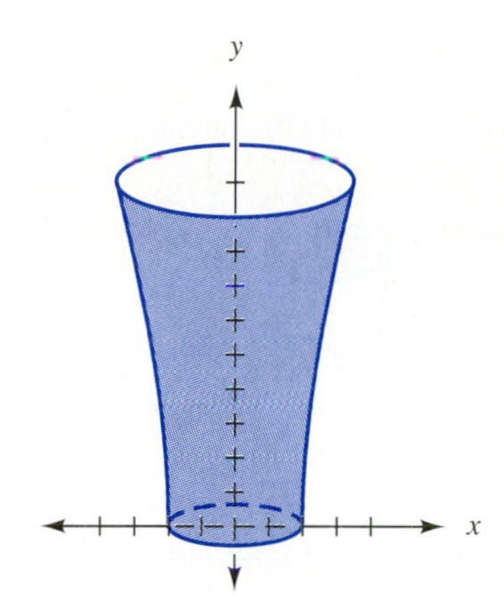

17. Sometimes very complicated equations are necessary to describe figures on ordinary xy-coordinate paper. Consider the cardioid or heart shaped figure describing an antenna radiation pattern from electronics given by the equation,

$$(x^2 + y^2 + x) - \sqrt{x^2 + y^2} = 0.$$

This equation has a very simple polar counterpart that is used extensively by engineers. Find the polar equation for this cardioid. 1 2 4 7 8 3 5

18. In fiber optics the radiation pattern of power emitting from a laser diode as a function of beam angle is sometimes plotted. Plot the radiation pattern given by $P = 100\ (\cos x)^{10}$ on polar coordinate paper for $0 < x < 2\pi$. 3 5 6 9

19. The distance between two points located on a polar coordinate map is given by the equation $D = \sqrt{r_1{}^2 + r_2{}^2 - 2r_1 r_2 \cos(\theta_2 - \theta_1)}$ where all distances are measured in miles from the pole (0, 0). Find the shortest distance between two cities A and B if A's coordinates are (45, 31°) and B's coordinates are (175, 75°). 1 2 4 7 8 3 5

RANDOM REVIEW

In 1–4, graph each equation.

1. $x^2 + 8x + y^2 + 10y + 16 = 0$
2. $y = x^2 - 4x - 3$
3. $9y^2 - 36y - 4x^2 - 24x - 36 = 0$
4. $r = 3 \cos \theta$
5. Write the equation of the line that goes through $(-3, -3)$ and is parallel to $y = 3x + 1$.
6. Convert $(-3, \pi/3)$ to rectangular coordinates.
7. Find the equation of the parabola with vertex at $(1, -2)$ and directrix at $y = 2$.
8. Write the equation of an ellipse with foci at $(2, 1)$ and $(-4, 1)$ and minor axis of length 8.
9. Convert $(8, -15)$ to polar coordinates.
10. Find the distance between $(-3, -3)$ and $(3, 3)$.

CHAPTER TEST

1. Write the equation of the line that goes through $(3, 3)$ and is perpendicular to the line $y = 3x + 1$.
2. Determine the focus of the parabola given by $4(y - 1) = (x + 3)^2$.
3. Find the vertex of the parabola given by $x = 2y^2 - 4y + 2$.
4. Write the equation of a circle that is tangent to the y-axis and has its center at $(2, -1/2)$.

In 5–7, graph each equation.

5. $\dfrac{x^2}{4} + \dfrac{(y - 1)^2}{9} = 1$
6. $(y + 2)^2 - (x - 3)^2 = 1$
7. $r = (1/3)\, \theta$
8. Convert the rectangular coordinates $(20, 21)$ to polar coordinates.
9. The cable of a suspension bridge hangs in the shape of a parabola when the load is uniformly distributed. A cable is suspended from two towers that rise 80.0 feet above the highway and are 2470 feet apart. Write the equation of the parabola, assuming that the vertex of the parabola is at the origin.
10. The Capitol in Washington, D.C., and the Mormon Tabernacle in Salt Lake City, Utah, are examples of buildings that contain whispering galleries. The ceiling of a whispering gallery is the shape formed when an ellipse is rotated about its major axis. A person standing at one focus can hear what is said at the other focus, although no one in between may hear. To construct a whispering gallery that is 40.0 ft long and has foci 30.0 ft apart, what equation should be used? Assume that the center of the ellipse is at the origin and the horizontal axis is the major axis.

VECTORS AND OBLIQUE TRIANGLES CHAPTER 10

Some quantities that occur in science and in technology are best described by giving both a size and a direction. The algebraic techniques used on such quantities are also applicable to triangles that are not right triangles. The concepts of this chapter will be used to solve problems in dynamics, such as those involving displacement, velocity, or acceleration. They will also be used to solve problems in statics, such as those involving equilibrium of forces or geometrical properties.

10.1 VECTORS AND VECTOR ADDITION

scalar

The question, "What is the temperature?" can be answered by specifying a single number. Such a quantity, called a *scalar,* is described by its size or magnitude. The question, "What is the wind's velocity?" cannot be answered by specifying a single number. The wind has both speed and direction.

DEFINITION A **vector** is a quantity that has both magnitude and direction.

A vector name is usually given in boldface type such as **v**. When handwritten, a vector name is usually specified by an arrow over the letter that names the vector, such as $\vec{v}$. A diagram is often useful when dealing with vectors. A vector may be represented as an arrow. The length of the arrow is an indication of magnitude, and the angle that it makes with a given reference line is an indication of its direction. A vector from point A to point B is said to have its *tail* at A and its *head* at B as shown in Figure 10.1.

tail
head

When two vectors are added, the sum is the single vector that produces the same result as the two vectors together. The vector sum is called the **resultant.** To find a vector sum, it is not sufficient to simply add the magnitudes, for the directions must also be considered.

resultant

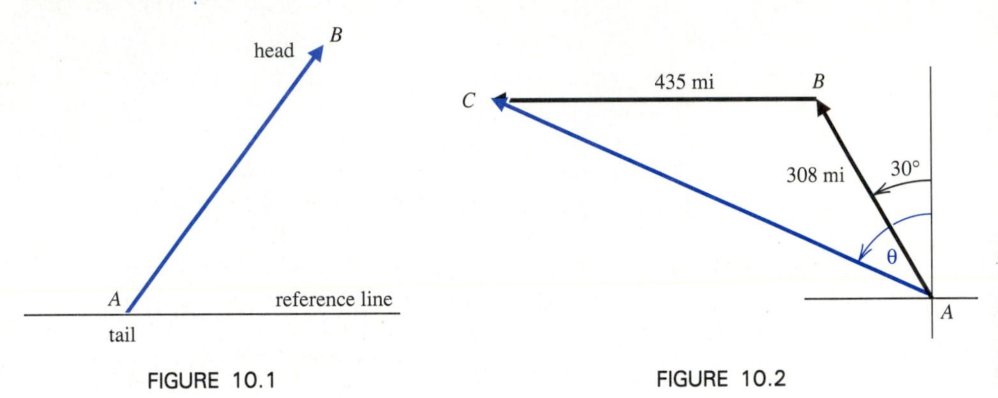

FIGURE 10.1

FIGURE 10.2

EXAMPLE 1 An airplane flies 308 miles N 30° W from A to B. After a brief stop, it flies 435 miles due west from point B to C. Draw a diagram to show the direct route the plane would take if the stop could be eliminated.

Solution Figure 10.2 shows vectors from A to B and from B to C. Since traveling from A to C produces the same result, the diagram shows a vector addition. The resultant is the vector from A to C. The direction could be specified as angle θ, measured from the north. ■

There are several common procedures for adding vectors diagrammatically. In any case, it is permissible to move a vector in the plane, provided that its magnitude (length) and direction (angle) are unchanged.

To add two vectors:

polygon method of adding vectors

(a) Polygon Method: If vectors **a** and **b** represent successive actions,
1. place **b** so that its tail is at the *head* of **a** as shown at left.
2. draw the resultant vector from the tail of **a** to the head of **b**.

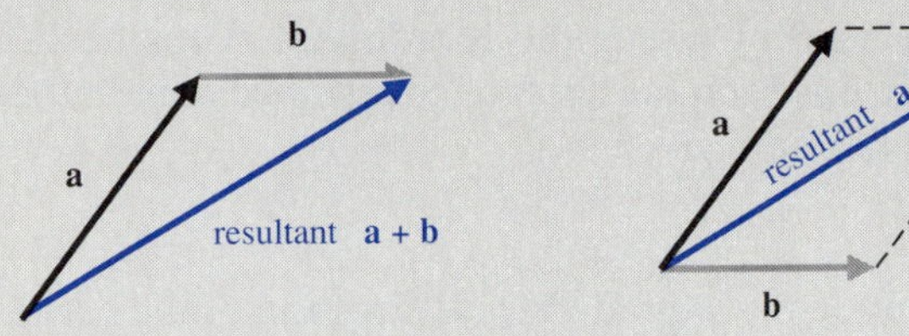

parallelogram method of adding vectors

(b) Parallelogram Method: If vectors **a** and **b** represent simultaneous actions,
1. place **b** so that its tail is at the *tail* of **a** as shown at right.
2. draw a parallelogram with **a** and **b** as two of its sides.
3. draw the resultant vector from the tails of **a** and **b** along the diagonal of the parallelogram.

displacement

Example 1 involved two *displacements.* A **displacement** is a change in position. That is, it is a distance with a specified direction. The distances were traveled successively, so the polygon method was used. Example 2 illustrates the parallelogram method with another important vector concept, force. A force may be considered as a push or a pull acting in a specified direction.

EXAMPLE 2 A ship is pulled by two tugboats. One pulls toward the east with a force of 105 newtons. The other pulls toward the north with a force of 208 newtons. What single force would have the same effect?

Solution The vectors represent forces that act simultaneously. We represent the vectors as in Figure 10.3. The resultant is the diagonal of parallelogram *ABDC*. The parallelogram is actually a rectangle for this example, so the length of the diagonal can be determined from the Pythagorean theorem. The vector has a magnitude of 233 newtons, and the direction is obtained from $\tan \theta = 208/105$, so $\theta = 63.2°$. ■

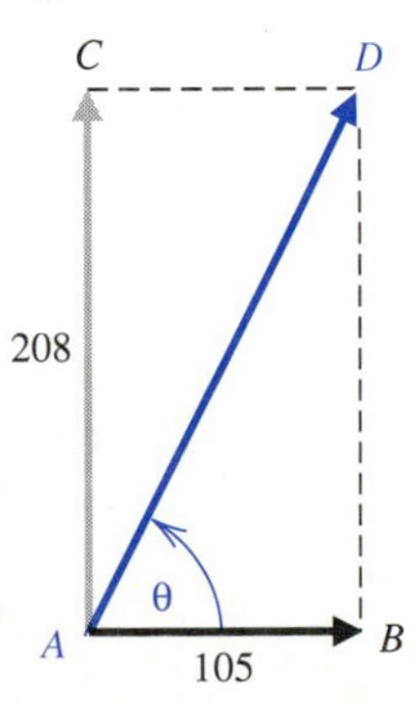

FIGURE 10.3

Example 2 involved a right triangle, so it was possible to find the magnitude and direction of the resultant vector. In the remainder of the chapter, we will examine some concepts that allow us to solve problems in which the triangles are *oblique* triangles. An **oblique triangle** does not have a right angle.

oblique triangle

So that you can practice drawing diagrams using both the polygon and parallelogram methods of addition, we consider vectors on the coordinate plane. A vector can be specified by giving the coordinates of the head and tail. Since a vector can be moved, however, the tail is often assumed to be at the origin and only the coordinates of the head are given.

EXAMPLE 3 Determine the magnitude and direction of the vector with its head at $(2, -3)$ and its tail at the origin.

Solution The vector is shown in Figure 10.4. Its length can be found from the Pythagorean theorem.

$$\sqrt{2^2 + (-3)^2} = \sqrt{4 + 9} = \sqrt{13} \text{ or about } 3.61$$

Coordinates that are integers are understood to be exact values, so the answer is given to three significant digits. The direction is specified as the angle the vector makes with a given reference line. If no reference line is specified, the positive x-axis is often used. The angle is found as follows.

$$\tan\theta = -3/2, \text{ so } \theta = 303.7°, \text{ to the nearest tenth of a degree}$$ ■

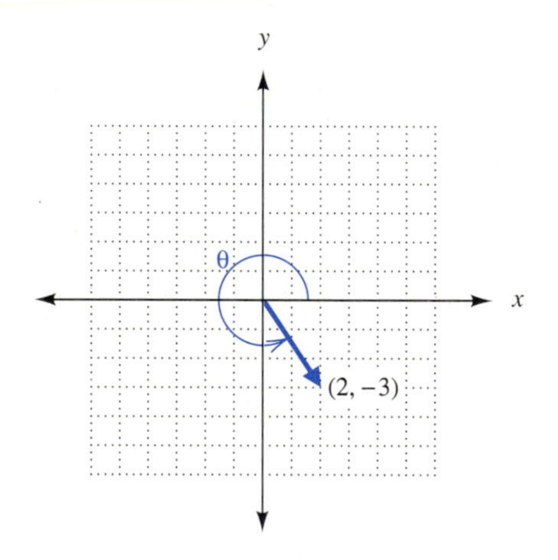

FIGURE 10.4

A vector with its head at (x, y) and its tail at $(0, 0)$ may be specified as $\langle x, y\rangle$ to distinguish it from the point (x, y). The steps used in Example 3 are generalized as follows.

To determine the magnitude and direction of a vector $\langle x, y\rangle$:

1. Put the tail of the vector at the origin of a coordinate plane so that the coordinates of the head are (x, y) as shown
2. Find the magnitude of **v**, denoted $|\mathbf{v}|$: $|\mathbf{v}| = \sqrt{x^2 + y^2}$.
3. Find the direction of **v**, denoted θ: $\tan\theta = y/x$.

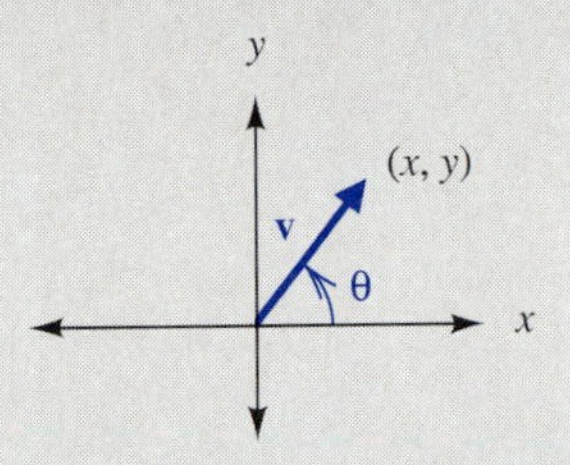

EXAMPLE 4 Use the parallelogram method to add the vectors $\mathbf{a} = \langle 2, 0\rangle$ and $\mathbf{b} = \langle 1, 4\rangle$.

Solution Figure 10.5 shows the resultant $\mathbf{r} = \langle 3, 4\rangle$.

$$|\mathbf{r}| = \sqrt{3^2 + 4^2} = 5 \qquad \tan\theta = 4/3, \text{ so } \theta = 53.1^\circ \quad ■$$

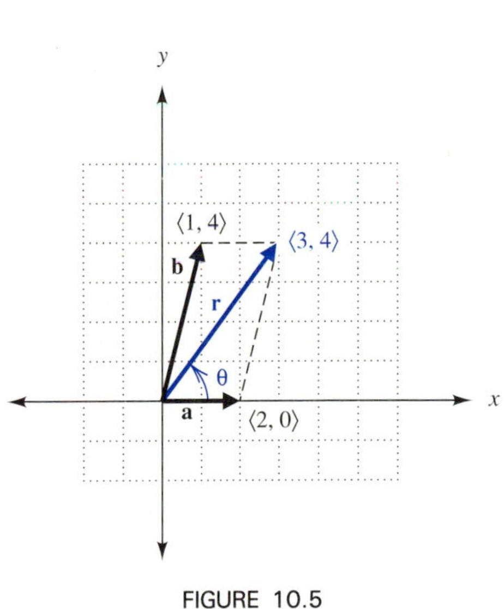

FIGURE 10.5

FIGURE 10.6

EXAMPLE 5 Use the polygon method to add the vectors $\mathbf{a} = \langle 3, 1\rangle$ and $\mathbf{b} = \langle 2, 11\rangle$.

Solution If we slide vector **b** over so that its tail is at the head of **a**, as in Figure 10.6, its head will be over two units and up 11 units from the tail. The resultant is the vector $\mathbf{r} = \langle 5, 12\rangle$.

$$|\mathbf{r}| = \sqrt{5^2 + 12^2} = 13 \qquad \tan\theta = 12/5 \text{ and } \theta = 67.4^\circ \quad ■$$

The polygon method may also be used to add more than two vectors.

EXAMPLE 6 Use the polygon method to add the vectors $\mathbf{a} = \langle 5, 2\rangle$, $\mathbf{b} = \langle 5, 5\rangle$, and $\mathbf{c} = \langle -2, 8\rangle$.

Solution Vector **b** is moved so that its tail is at the head of **a**. Then **c** is positioned so that its tail is at the head of **b**. The resultant is the vector $\mathbf{r} = \langle 8, 15\rangle$ shown in Figure 10.7.

$$|\mathbf{r}| = \sqrt{8^2 + 15^2} = 17 \qquad \tan\theta = 15/8 \text{ and } \theta = 61.9°$$ ■

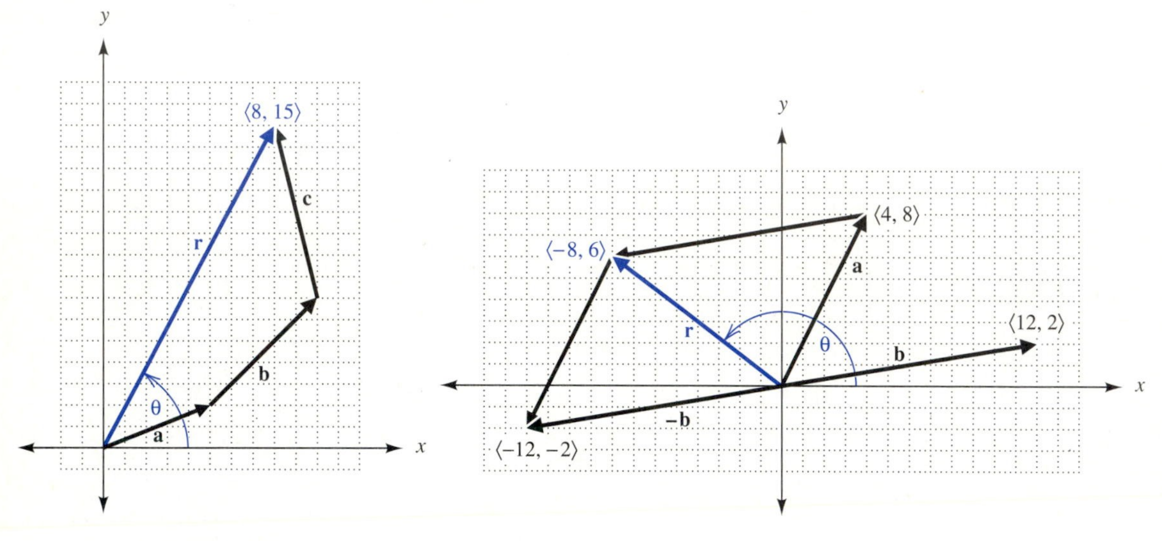

FIGURE 10.7

FIGURE 10.8

A calculator may be used to do the arithmetic. The keystroke sequence for Example 6 is as follows.

8 [x^2] [+] 15 [x^2] [=] **17**

15 [÷] 8 [=] [INV] [tan] **61.927513**

It is possible to subtract a vector **a** from a vector **b** by adding $-\mathbf{a}$ to **b**. Consider the vector $-\mathbf{a}$ to be the vector that has the same magnitude as **a**, but that has opposite direction. Thus, if **a** is specified by $\langle x, y\rangle$, $-\mathbf{a}$ is specified by $\langle -x, -y\rangle$.

EXAMPLE 7 Use the parallelogram method to find $\mathbf{a} - \mathbf{b}$ if $\mathbf{a} = \langle 4, 8\rangle$ and $\mathbf{b} = \langle 12, 2\rangle$.

Solution First, **a** and **b** are drawn as in Figure 10.8. The vector $-\mathbf{b} = \langle -12, -2\rangle$ is in the opposite direction from **b**. The figure shows $\mathbf{a} + (-\mathbf{b})$. The resultant is $\mathbf{r} = \langle -8, 6\rangle$.

$$|\mathbf{r}| = \sqrt{(-8)^2 + 6^2} = 10 \qquad \tan\theta = -(6/8) \text{ and } \theta = 143.1°$$ ■

EXERCISES 10.1

In 1–12, determine the magnitude and direction of each vector.

1. $\langle 5, 7\rangle$ **2.** $\langle -3, 6\rangle$ **3.** $\langle 4, -5\rangle$ **4.** $\langle 6, 6\rangle$

5. $\langle 2, 4\rangle$ **6.** $\langle -4, 6\rangle$ **7.** $\langle 7, -9\rangle$ **8.** $\langle 6, 8\rangle$

9. $\langle -7, -4\rangle$ **10.** $\langle -8, -6\rangle$ **11.** $\langle -6, -7\rangle$ **12.** $\langle -8, -4\rangle$

In 13–23, use the parallelogram method to add the vectors specified. Find the magnitude and direction of the resultant.

13. $\mathbf{a} = \langle 2, 1\rangle$ and $\mathbf{b} = \langle 1, 3\rangle$

14. $\mathbf{a} = \langle 3, 2\rangle$ and $\mathbf{b} = \langle 0, 2\rangle$

15. $\mathbf{a} = \langle -1, 2\rangle$ and $\mathbf{b} = \langle 4, 2\rangle$

16. $\mathbf{a} = \langle -1, 1\rangle$ and $\mathbf{b} = \langle 4, 3\rangle$

17. $\mathbf{a} = \langle 4, 2\rangle$ and $\mathbf{b} = \langle 1, 10\rangle$

18. $\mathbf{a} = \langle 5, 5\rangle$ and $\mathbf{b} = \langle 0, 7\rangle$

19. $\mathbf{a} = \langle -5, 5\rangle$ and $\mathbf{b} = \langle 10, 7\rangle$

20. $\mathbf{a} = \langle -2, 4\rangle$ and $\mathbf{b} = \langle 7, 8\rangle$

21. $\mathbf{a} = \langle 4, -2\rangle$ and $b = \langle 2, -6\rangle$

22. $\mathbf{a} = \langle -2, -5\rangle$ and $\mathbf{b} = \langle 8, -3\rangle$

23. $\mathbf{a} = \langle -3, -3\rangle$ and $\mathbf{b} = \langle 9, -5\rangle$

In 24–34, use the polygon method to add the vectors specified. Find the magnitude and direction of the resultant.

24. $\mathbf{a} = \langle -5, 2\rangle$ and $\mathbf{b} = \langle -1, 6\rangle$

25. $\mathbf{a} = \langle -5, 4\rangle$ and $\mathbf{b} = \langle -1, 4\rangle$

26. $\mathbf{a} = \langle -3, 2\rangle$ and $\mathbf{b} = \langle -3, 6\rangle$

27. $\mathbf{a} = \langle -3, 3\rangle$ and $\mathbf{b} = \langle -2, -15\rangle$

28. $\mathbf{a} = \langle -3, 2\rangle$ and $\mathbf{b} = \langle -2, -14\rangle$

29. $\mathbf{a} = \langle -4, -3\rangle$ and $\mathbf{b} = \langle -1, -9\rangle$

30. $\mathbf{a} = \langle -7, -8\rangle$ and $\mathbf{b} = \langle 2, -4\rangle$

31. $\mathbf{a} = \langle -1, 2\rangle$ and $\mathbf{b} = \langle 4, 2\rangle$

32. $\mathbf{a} = \langle 5, 5\rangle$ and $\mathbf{b} = \langle 0, 7\rangle$

33. $\mathbf{a} = \langle 4, -2\rangle$ and $\mathbf{b} = \langle 2, -6\rangle$

34. $\mathbf{a} = \langle 3, -6\rangle$ and $\mathbf{b} = \langle 3, -2\rangle$

In 35–42, use the polygon method to add the vectors specified. Find the magnitude and direction of the resultant.

35. $\mathbf{a} = \langle 5, 3\rangle$, $\mathbf{b} = \langle 2, 4\rangle$, and $\mathbf{c} = \langle 1, 8\rangle$

36. $\mathbf{a} = \langle 6, 2\rangle$, $\mathbf{b} = \langle 1, 3\rangle$, and $\mathbf{c} = \langle 1, 10\rangle$

37. $\mathbf{a} = \langle -2, 5\rangle$, $\mathbf{b} = \langle 5, 5\rangle$, and $\mathbf{c} = \langle 5, 5\rangle$

38. $\mathbf{a} = \langle -4, 4\rangle$, $\mathbf{b} = \langle 5, 5\rangle$, and $\mathbf{c} = \langle 7, 6\rangle$

39. $\mathbf{a} = \langle 2, 5\rangle$, $\mathbf{b} = \langle -5, -2\rangle$, and $\mathbf{c} = \langle -9, 2\rangle$

40. $\mathbf{a} = \langle -5, -5\rangle$, $\mathbf{b} = \langle -4, 7\rangle$, and $\mathbf{c} = \langle -3, 3\rangle$

41. $\mathbf{a} = \langle 3, 3\rangle$, $\mathbf{b} = \langle -15, 0\rangle$, and $\mathbf{c} = \langle 0, 2\rangle$

42. $\mathbf{a} = \langle -5, -2\rangle$, $\mathbf{b} = \langle -4, 3\rangle$, and $\mathbf{c} = \langle -3, 4\rangle$

---

43. A boat heads directly across a river at 24.0 km/hr. The 7.00 km/hr current in the river, however, acts to carry the boat downstream. What is the resultant velocity of the boat?
4 7 8 1 2 3 5 6 9

44. An airplane leaving Miami flies north for 252 miles then turns and flies east for 115 miles. What is its distance from the original point of departure? 1 2 4 7 8 3 5

45. An object is attached to a string and is swung in a circle. The object has both a tangential (along the tangent) acceleration **a** with $|\mathbf{a}| = 12.2$ m/s^2 and a centripetal (along the radius) acceleration **c** with $|\mathbf{c}| = 24.1$ m/s^2 as shown in the diagram. Find the

1. Architectural Technology
2. Civil Engineering Technology
3. Computer Engineering Technology
4. Mechanical Drafting Technology
5. Electrical/Electronics Engineering Technology
6. Chemical Engineering Technology
7. Industrial Engineering Technology
8. Mechanical Engineering Technology
9. Biomedical Engineering Technology

magnitude of the resultant acceleration and the angle between the resultant acceleration **r** and the tangential acceleration **a**. 4 7 8 1 2 3 5 6 9

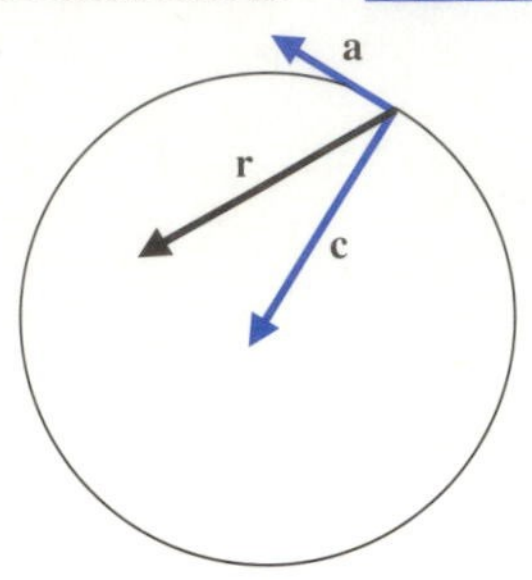

46. A child sitting in a swing is pushed horizontally with a force of 36.0 pounds. The weight of the child is 77.0 pounds, a gravitational force that acts downward as shown in the diagram. Find the resultant of these forces. 4 7 8 1 2 3 5 6 9

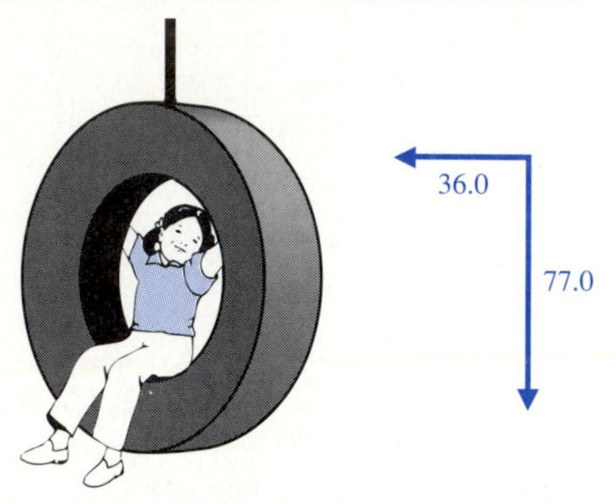

47. A picture hangs from a hook in the wall. The weight of the picture exerts a vertical force of 21.0 pounds on the hook and the wall exerts a horizontal force of 20.0 pounds on the hook. Find the magnitude of the resultant force on the hook. 2 4 8 1 5 6 7 9

48. The impedance of a circuit containing a resistor, a capacitor, and an inductor in series is the sum of the vector quantities **R** and $\mathbf{X_l} - \mathbf{X_c}$ shown in the diagram. Find the magnitude of the impedance and the angle θ. This angle is called the phase angle and is the angle by which the current leads or lags the applied voltage. 3 5 9 4 7 8

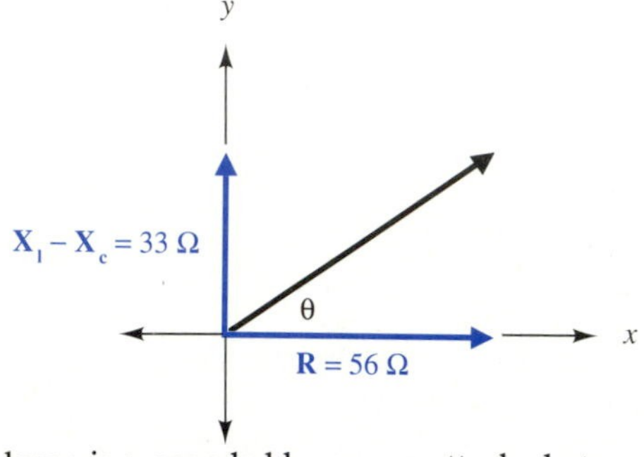

49. A lamp is suspended by a rope attached at equal heights to two walls. At each end, the rope makes an angle of 45° with the wall. In each part of the rope, the tension is 4.8 newtons, directed away from the lamp. Find the resultant tension. 2 4 8 1 5 6 7 9

50. Let $\mathbf{a} = \langle x_1, y_1 \rangle$ and $\mathbf{b} = \langle x_2, y_2 \rangle$.

(a) Find a formula for the magnitude of $\mathbf{c} = \mathbf{a} + \mathbf{b}$

(b) Find a formula for the direction θ of **c**.

1. Architectural Technology
2. Civil Engineering Technology
3. Computer Engineering Technology
4. Mechanical Drafting Technology
5. Electrical/Electronics Engineering Technology
6. Chemical Engineering Technology
7. Industrial Engineering Technology
8. Mechanical Engineering Technology
9. Biomedical Engineering Technology

10.2 VECTOR ADDITION BY COMPONENTS

In the previous section, it was possible to add vectors if they were at right angles to each other or if we knew the coordinates of the heads and tails. Most problems in the technologies do not meet these special conditions. Our goal in this section is to develop techniques that can be used in other situations to add vectors.

When two vectors are added, they produce a single resultant. The two vectors are called *components* of the resultant. For any given resultant vector, there are many possible vector additions that could have produced it. That is, there are many different pairs of components, as suggested by Figure 10.9.

components

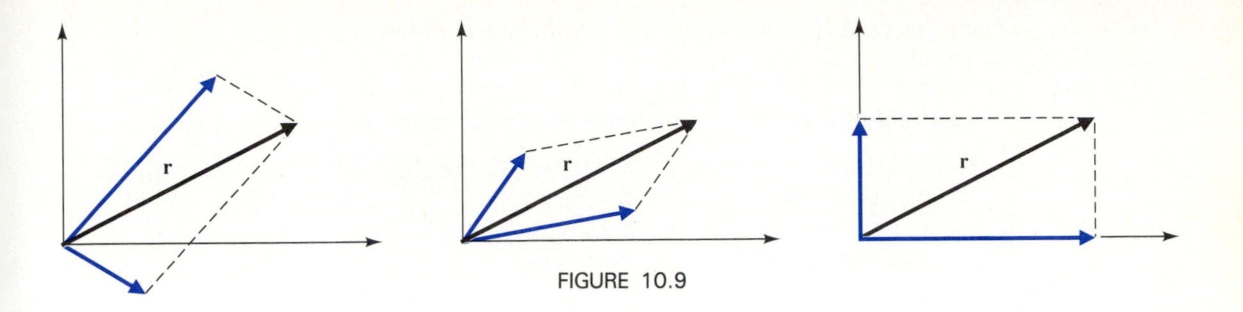

FIGURE 10.9

resolving a vector

The process of finding the components of a vector $\mathbf{v}$ is called **resolving** the vector. In particular, the pair that consists of a horizontal and a vertical vector will be important. We use $\mathbf{v_x}$ to represent the horizontal component and $\mathbf{v_y}$ to represent the vertical component. When a vector is resolved into horizontal and vertical components, the angle associated with $\mathbf{v_x}$ is understood to be 0°, and the angle associated with $\mathbf{v_y}$ is understood to be 90°. Thus, only a sign is needed to indicate direction for $\mathbf{v_x}$ or $\mathbf{v_y}$. If the sign of $\mathbf{v_x}$ is positive, then $\mathbf{v_x}$ lies along the positive x-axis, and if the sign of $\mathbf{v_x}$ is negative, then $\mathbf{v_x}$ lies along the negative x-axis. Vertical components are treated in an analogous manner.

EXAMPLE 1 Resolve the vector $\mathbf{v}$ with magnitude 5.00 and direction given by $\theta = 32.0°$ into its horizontal and vertical components.

Solution Consider the vector as shown in Figure 10.10. It is positioned with its tail at the origin of a coordinate system. The head of the horizontal component $\mathbf{v_x}$ is found by drawing a line from the head of $\mathbf{v}$ perpendicular to the x-axis. The head of the vertical component $\mathbf{v_y}$ is found by drawing a line from the head of $\mathbf{v}$ perpendicular to the y-axis.

Since angle θ is known to be 32.0°, $\mathbf{v_x}$ and $\mathbf{v_y}$ can be found using trigonometry.

$$\cos 32.0° = \mathbf{v_x}/5.00 \quad \text{and} \quad \sin 32.0° = \mathbf{v_y}/5.00$$
$$\mathbf{v_x} = 5.00 \cos 32.0° \quad \text{and} \quad \mathbf{v_y} = 5.00 \sin 32.0°$$
$$\mathbf{v_x} = 4.24 \quad \text{and} \quad \mathbf{v_y} = 2.65 \quad ■$$

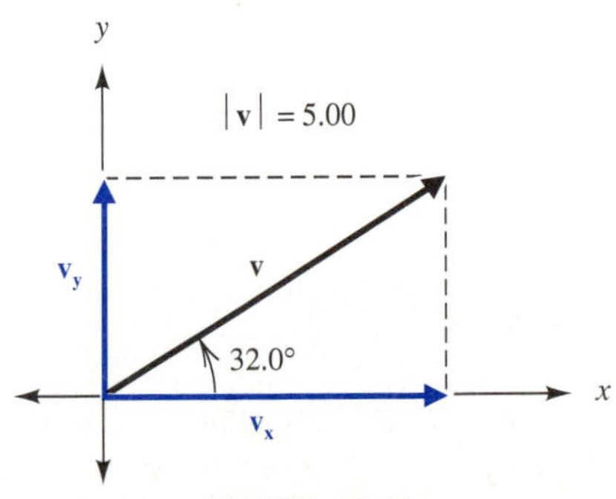

FIGURE 10.10

The steps used in Example 1 are generalized as follows.

To resolve a vector v into horizontal and vertical components:

1. Find the magnitude $|\mathbf{v}|$ of the vector and the angle θ that it makes with the *positive* x-axis.
2. Find the horizontal component: $\mathbf{v_x} = |\mathbf{v}| \cos \theta$.
3. Find the vertical component: $\mathbf{v_y} = |\mathbf{v}| \sin \theta$.

EXAMPLE 2 A weather balloon drifts at a constant speed of 12.3 ft/s at an angle of 30.0° with the vertical as shown in Figure 10.11.

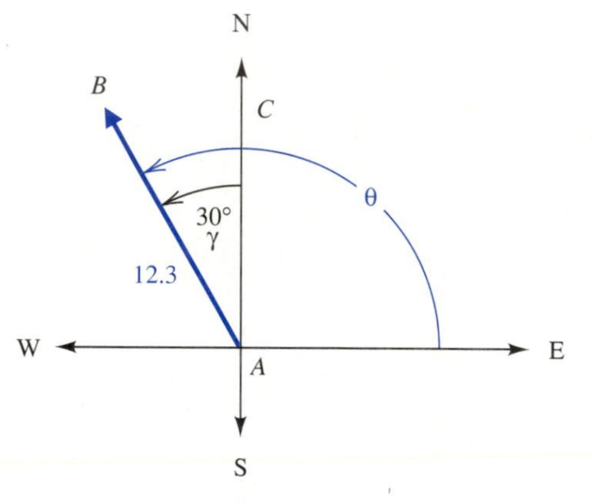

FIGURE 10.11

(a) Resolve the vector into its horizontal and vertical components.

(b) How long will it take to reach a height of 225 ft?

(c) What horizontal distance will it cover in this time?

Solution **(a)** If $\gamma = 30.0°$, then $\theta = 120.0°$.

$$\mathbf{v_x} = 12.3 \cos 120.0° = -6.15 \qquad \mathbf{v_y} = 12.3 \sin 120.0° = 10.7$$

A calculator can be used to obtain $\mathbf{v_x}$ and $\mathbf{v_y}$.
The negative value for $\mathbf{v_x}$ indicates a negative direction along the x-axis.

12.3 [×] 120 [cos] [=] **−6.15** 12.3 [×] 120 [sin] [=] **10.652113**

Answer: −6.15 Answer: 10.7

(b) Since the vertical component of velocity is 10.7 ft/s, the balloon will require 225 ÷ 10.7 or 21.0 seconds to reach a height of 225 feet.

(c) Since the horizontal component of velocity is −6.15 ft/s, the balloon will travel −6.15(21.0) or −129 feet, where the negative sign indicates motion toward the west.

Alternate Solution

(a) The angle γ can be used directly in triangle ABC. We know from the diagram that $\mathbf{v_x}$ is negative and $\mathbf{v_y}$ is positive. Since $\mathbf{v_x}$ is opposite γ, we have $|\mathbf{v_x}| = 12.3 \sin 30.0°$, and $\mathbf{v_x} = -6.15$. Since $\mathbf{v_y}$ is adjacent to γ, $|\mathbf{v_y}| = 12.3 \cos 30.0°$, and $\mathbf{v_y} = 10.7$. The direction can be specified from the positive x-axis by noticing that $\theta = 90° + \gamma = 120.0°$. ■

The components of vectors are useful in addition. When two vectors are both horizontal, placing the tail of one at the head of the other leads to a resultant that is also horizontal. Its magnitude is simply the arithmetic sum of the magnitudes of the two vectors, as shown in Figure 10.12. Two vertical vectors can be added in a similar manner. It is thus possible to add two vectors by adding their components.

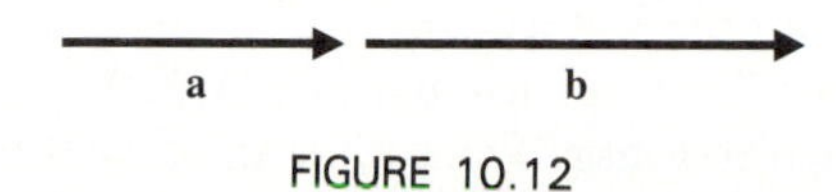

FIGURE 10.12

To add two vectors using components:

1. Resolve each vector into its horizontal and vertical components.
2. (a) Add the two horizontal components to obtain $\mathbf{r_x}$.
   (b) Add the two vertical components to obtain $\mathbf{r_y}$.
3. Find the magnitude $|\mathbf{r}|$ of the resultant using the Pythagorean theorem.

$$|\mathbf{r}| = \sqrt{|\mathbf{r_x}|^2 + |\mathbf{r_y}|^2}$$

4. Find the direction θ of the resultant from the positive x-axis: $\tan \theta = r_y/r_x$ where r_y is the number with the same sign and magnitude as $\mathbf{r_y}$, and $\mathbf{r_x}$ is the number with the same sign and magnitude as $\mathbf{r_x}$.

EXAMPLE 3 A beam is fixed at right angles to the wall as shown in Figure 10.13. The beam is supported by a tie making an angle of 40.2° with the beam. A load hangs from the end. The load exerts a force of 25.0 newtons and the tension in the tie is 38.7 newtons. Find the sum of these two forces.

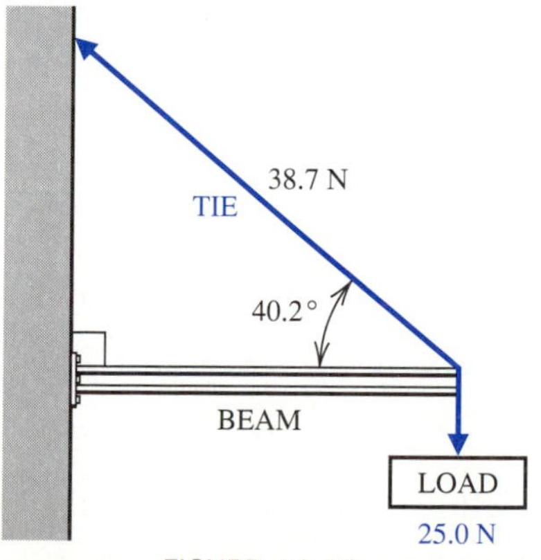

FIGURE 10.13

Solution The force of the load is its weight, due to the pull of gravity. The tension in the tie is a force pulling up at a 40.2° angle. Let **w** represent the weight of the load. Let **t** represent the tension in the tie. The vectors **w** and **t** are resolved into their horizontal and vertical components so that the resultant can be found.

1. $\mathbf{w_x} = 0.00$ $\mathbf{w_y} = -25.0$
   $\mathbf{t_x} = 38.7 \cos 139.8° = -29.6$ $\mathbf{t_y} = 38.7 \sin 139.8° = 25.0$
2. $\mathbf{r_x} = -29.6$ $\mathbf{r_y} = 0.0$
3. $|\mathbf{r}| = \sqrt{(29.6)^2 + (0.0)^2} = 29.6$
4. $\tan \theta = 0/-29.6 = 0.0$, thus, $\theta = 180.0°$ ■

In Example 3, the beam also exerts a force, called thrust, since it pushes outward from the wall. The thrust in the beam acts in the opposite direction from **r** and has the same magnitude as **r**. Thus, the sum of the three forces (weight, tension, and thrust) is zero. Since the forces balance each other, there is no motion. When the sum of the forces acting at a point is zero, the point is said to be in *equilibrium.* Also notice that because the weight of the load had no horizontal component, the thrust could be found by simply resolving the tension vector into its horizontal and vertical components. The horizontal component balances the thrust, and the vertical component balances the weight.

The procedure of adding vectors by components may be extended to include any number of vectors.

EXAMPLE 4 Add the following vectors, illustrated in Figure 10.14.

a with magnitude 10.0 and direction 53.1°
b with magnitude 8.00 and direction 0.0°
c with magnitude 6.00 and direction 157.4°

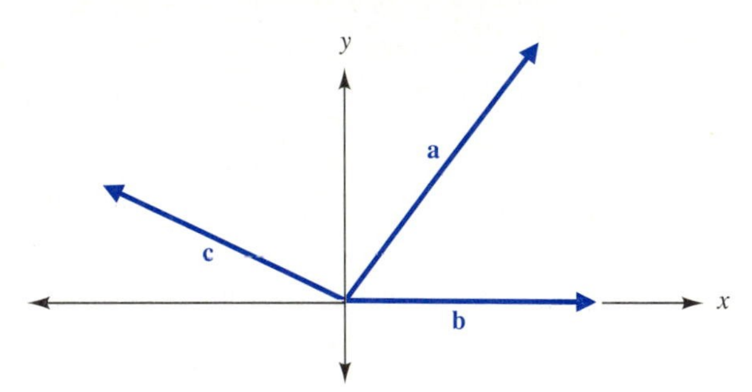

FIGURE 10.14

Solution

1. $\mathbf{a_x} = 10.0 \cos 53.1° = 6.00$ $\mathbf{a_y} = 10.0 \sin 53.1° = 8.00$
   $\mathbf{b_x} = 8.00$ $\mathbf{b_y} = 0$
   $\mathbf{c_x} = 6.00 \cos 157.4° = -5.54$ $\mathbf{c_y} = 6.00 \sin 157.4° = 2.31$
2. $\mathbf{r_x} = 8.46$ $\mathbf{r_y} = 10.31$
3. $|\mathbf{r}| = \sqrt{(8.46)^2 + (10.31)^2} = 13.3$
4. $\tan \theta = 10.31/8.46$, and θ is in the first quadrant, so $\theta = 50.6°$ ■

In Examples 3 and 4, intermediate results, as well as final results, were rounded so that you could see each step of the calculation. In practice, it is better not to round until the last step.

EXAMPLE 5 Solve the problem of Example 4 using a calculator and rounding only the final result.

Solution The strategy is as follows.

1. Compute $\mathbf{r_x}$ in one continuous calculation and store the result.
2. Compute $\mathbf{r_y}$ in one continuous calculation and record the result.
3. Square the value of $\mathbf{r_y}$ on the display. Recall $\mathbf{r_x}$ from calculator memory to complete the calculation for the magnitude of $\mathbf{r}$.
4. Compute θ using $\mathbf{r_x}$ and $\mathbf{r_y}$. Only $\mathbf{r_x}$ is available in memory, so the value of $\mathbf{r_y}$ recorded in step 2 is used.

Here, then, is the calculator keystroke sequence.

1. 10 [×] 53.1 [cos] [+] 8 [+] 6.0 [×] 157.4 [cos] [=] [STO] **8.464941**
2. 10.0 [×] 53.1 [sin] [+] 6.00 [×] 157.4 [sin] [=] **10.302619**

   Write down the number on the display for future use.
3. [x^2] [+] [RCL] [x^2] [=] **13.334136**
4. 10.302619 [÷] [RCL] [=] [INV] [tan] **50.592448**

The answer is rounded so that $|\mathbf{r}| = 13.3$ and $\theta = 50.6°$. ■

EXAMPLE 6 A ship sails 308 miles N 20.0° W from A to C. It then turns and sails 435 miles N 40.0° E from C to B. What is the distance and direction from A to B?

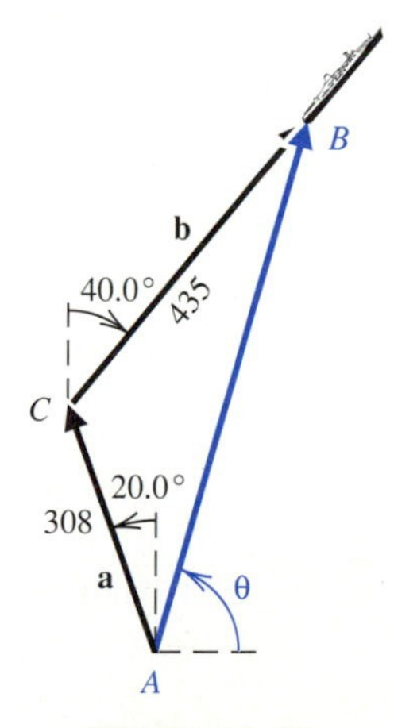

FIGURE 10.15

Solution Let **a** represent the trip from A to C. Let **b** represent the trip from C to B. Figure 10.15 illustrates the vectors. The two vectors can be resolved into components. The bearing N 20.0° W corresponds to an angle in standard position of 110.0°, and N 40.0° E corresponds to an angle in standard position of 50.0°.

$$\mathbf{a_x} = 308 \cos 110.0° \qquad \mathbf{a_y} = 308 \sin 110.0°$$
$$\mathbf{b_x} = 435 \cos 50.0° \qquad \mathbf{b_y} = 435 \sin 50.0°$$

Using a calculator to do the arithmetic, we find $|\mathbf{r}|$ and θ.

$$|\mathbf{r}| = 647 \quad \text{and} \quad \theta = 74.4° \quad ■$$

EXAMPLE 7 A catalog showroom has its warehouse on the upper level. There is to be a conveyor belt that carries merchandise from the upper level to the lower level. The largest item that will be placed on the belt weighs 300 pounds. At what angle would the force of friction be overcome by the force of gravity, so that the item will begin to slide down the incline, even if the conveyor belt is not running? It is known that the coefficient of friction μ is the ratio of the magnitudes of the force of friction to the component of weight that is perpendicular to the force of friction. It is also known that $\mu = 0.33$ for the materials used.

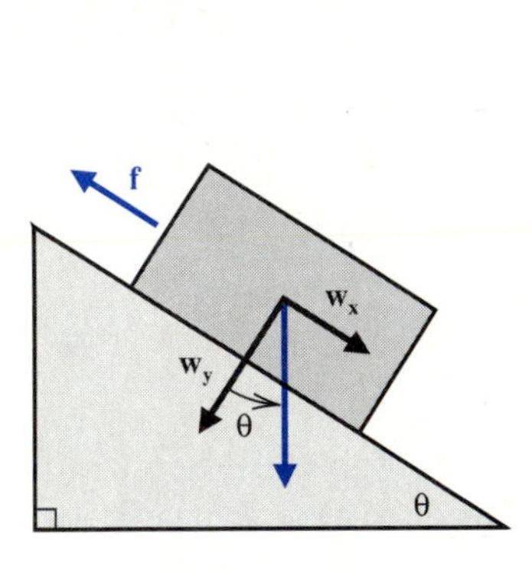

FIGURE 10.16

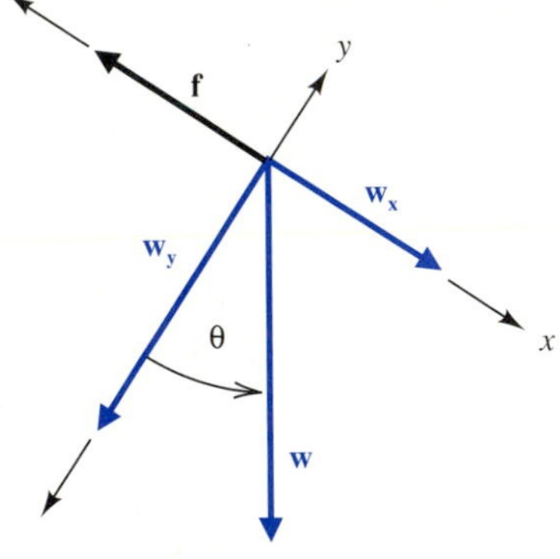

FIGURE 10.17

Solution Figure 10.16 shows the forces acting on the item. The weight **w** is due to a gravitational force pulling vertically downward. The force of friction **f** acts parallel to the plane, opposing the motion of the item sliding along the incline.

It is convenient to place the axes so that the x-axis is parallel to the surface, as shown in Figure 10.17. The vector **w** can be resolved into components $\mathbf{w_x}$ and $\mathbf{w_y}$. The force of friction **f** balances $\mathbf{w_x}$. Also, the angle between $\mathbf{w_y}$ and **w** is used to resolve **w**, because it must be equal to θ. (Why?)

Let $w_y = |\mathbf{w_y}|$, $w_x = |\mathbf{w_x}|$, and $f = |\mathbf{f}|$.

We were given the information that μ is the ratio of f to w_y. Thus, $\mu = f/w_y$. It follows that $\mu w_y = f$. Since **f** balances $\mathbf{w_x}$, we have $f = w_x$. Thus,

$$\mu w_y = w_x$$
$$\mu = w_x/w_y = \tan \theta$$
$$0.33 = \mu = \tan \theta, \text{ so } \theta = 18.26°. \quad ■$$

EXERCISES 10.2

In 1–8, resolve the given vector into horizontal and vertical components.

1. $|\mathbf{v}| = 32.0$, $\theta = 35.1°$

2. $|\mathbf{v}| = 12.2$, $\theta = 44.3°$

3. $|\mathbf{v}| = 78.4$, $\theta = 112.6°$

4. $|\mathbf{v}| = 5.98$, $\theta = 253.9°$

5. $|\mathbf{v}| = 153.6$, $\theta = 45.33°$

6. $|\mathbf{v}| = 244.5$, $\theta = 38.22°$

7. $|\mathbf{v}| = 335.4$, $\theta = -86.31°$

8. $|\mathbf{v}| = 26.54$, $\theta = -68.72°$

In 9–16, add each pair of vectors by adding horizontal and vertical components. Specify the magnitude and direction of the resultant.

9. $|\mathbf{a}| = 6.8$, $\theta_a = 18°$, $|\mathbf{b}| = 5.1$, $\theta_b = 98°$

10. $|\mathbf{a}| = 9.3$, $\theta_a = 23°$, $|\mathbf{b}| = 3.0$, $\theta_b = 74°$

11. $|\mathbf{a}| = 41.0$, $\theta_a = -36.7°$, $|\mathbf{b}| = 51.1$, $\theta_b = 25.4°$

12. $|\mathbf{a}| = 25.0$, $\theta_a = 62.3°$, $|\mathbf{b}| = 72.4$, $\theta_b = -50.9°$

13. $|\mathbf{a}| = 21.32$, $\theta_a = 44.47°$, $|\mathbf{b}| = 33.54$, $\theta_b = -59.31°$

14. $|\mathbf{a}| = 42.63$, $\theta_a = -68.40°$, $|\mathbf{b}| = 28.54$, $\theta_b = 77.72°$

15. $|\mathbf{a}| = 37.63$, $\theta_a = -86.81°$, $|\mathbf{b}| = 24.45$, $\theta_b = 74.22°$

16. $|\mathbf{a}| = 335.4$, $\theta_a = 83.31°$, $|\mathbf{b}| = 153.6$, $\theta_b = -45.33°$

In 17–22, add each group of vectors by adding horizontal and vertical components. Specify the magnitude and direction of the resultant.

17. $|\mathbf{a}| = 48.7$, $\theta_a = 22.5°$, $|\mathbf{b}| = 54.7$, $\theta_b = 51.9°$, $|\mathbf{c}| = 26.1$, $\theta_c = 27.2°$

18. $|\mathbf{a}| = 55.6$, $\theta_a = 25.8°$, $|\mathbf{b}| = 77.6$, $\theta_b = 110.2°$, $|\mathbf{c}| = 51.7$, $\theta_c = 38.6°$

19. $|\mathbf{a}| = 87.5$, $\theta_a = 101.1°$, $|\mathbf{b}| = 51.8$, $\theta_b = -112.6°$, $|\mathbf{c}| = 22.4$, $\theta_c = 45.5°$

20. $|\mathbf{a}| = 6.09$, $\theta_a = 121.7°$, $|\mathbf{b}| = 4.70$, $\theta_b = 292.4°$, $|\mathbf{c}| = 52.6$, $\theta_c = -77.4°$

21. $|\mathbf{a}| = 4.61$, $\theta_a = 201.5°$, $|\mathbf{b}| = 23.3$, $\theta_b = 44.1°$, $|\mathbf{c}| = 10.9$, $\theta_c = -15.6°$

22. $|\mathbf{a}| = 42.2$, $\theta_a = -53.2°$, $|\mathbf{b}| = 12.2$, $\theta_b = 35.3°$, $|c| = 12.5$, $\theta_c = 14.8°$

In 23–28, add the vectors shown in each diagram. Specify the magnitude and direction of the resultant.

23.

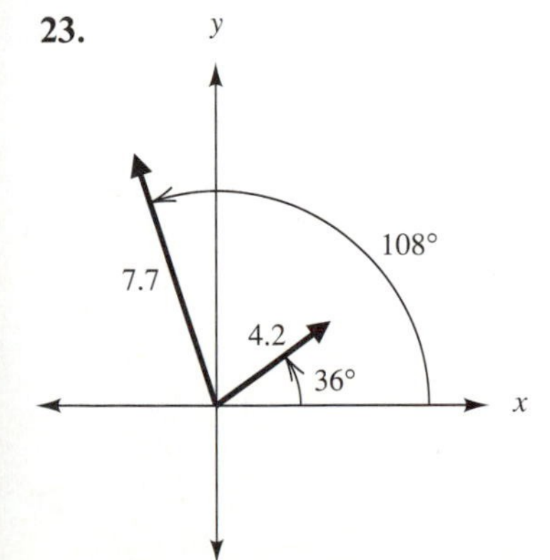

24.

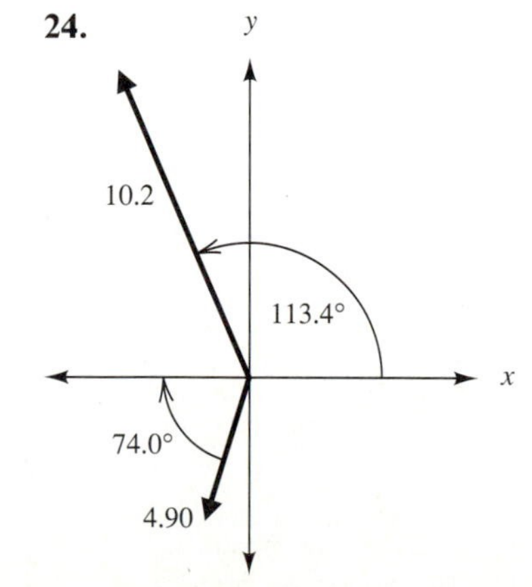

25.

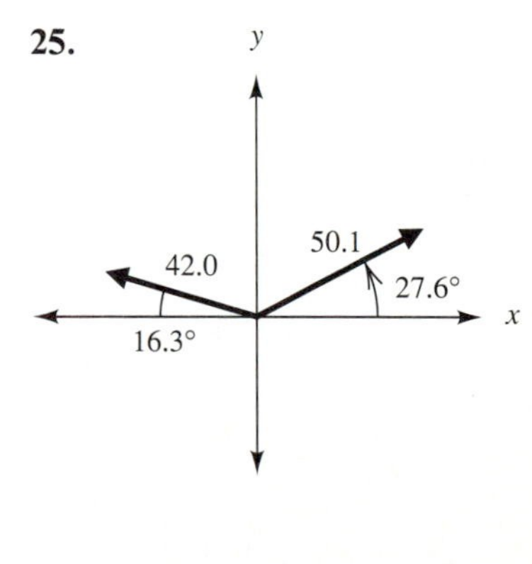

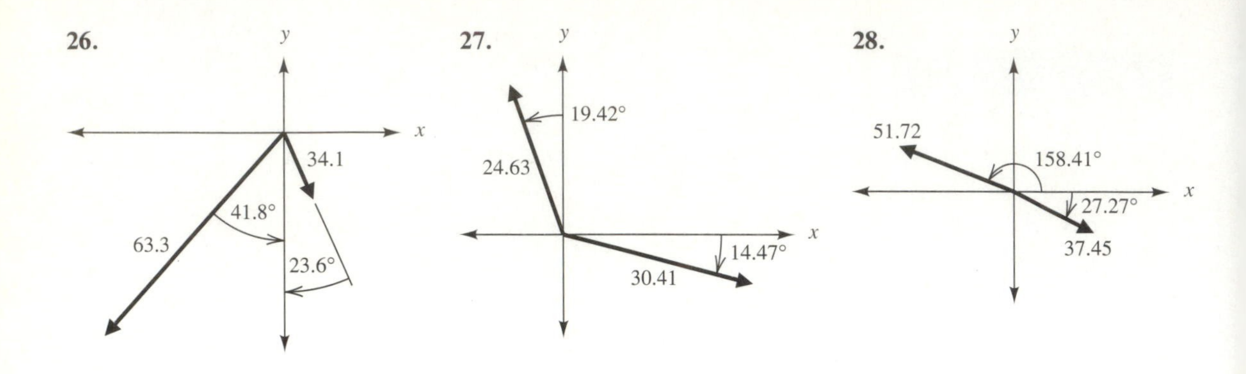

29. A wrecker tows a car with a force of 6000 pounds (to two significant digits) inclined at an angle of 30° with the horizontal. What is the horizontal component of the force pulling the car? 4 7 8 1 2 3 5 6 9

30. A ramp for the handicapped makes an angle of 20.0° with the horizontal. A man and his wheelchair weigh 203 pounds. What is the component of this weight perpendicular to the ramp? 2 4 8 1 5 6 7 9

31. An airplane that flies at 350 mph in still air heads N 45° E. If the wind is out of the south at 32 mph, find the speed and direction of the plane. 4 7 8 1 2 3 5 6 9

32. A ship that travels at 14 knots in still water heads S 25° W. There is a current of 3 knots due east. Find the speed and direction of the ship. 4 7 8 1 2 3 5 6 9

33. A large sign is supported by a guy wire that exerts a pull of 100 (to two significant digits) pounds on the top of the sign, as shown in the diagram. If the angle between the ground and the wire is 43°, find the horizontal and vertical components of the pull on the sign. 2 4 8 1 5 6 7 9

34. A pulley is supported by a wire as shown in the figure. Find the tension in the wire if the pulley supports a 175 pound man. (Hint: The sum of the vertical components of tension must balance the weight.) 2 4 8 1 5 6 7 9

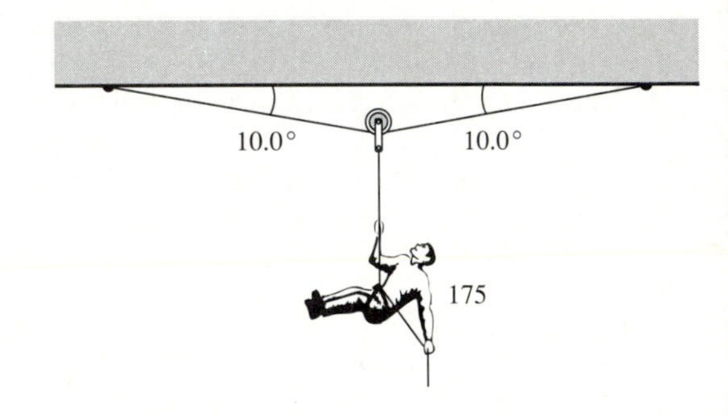

35. Two guy wires are fastened to an anchor in a foundation as shown (to three significant digits) in the figure. What pull does the bolt exert on the foundation? 2 4 8 1 5 6 7 9

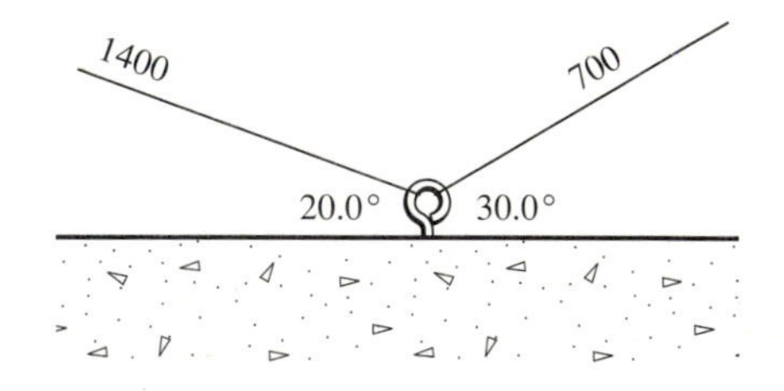

1. Architectural Technology
2. Civil Engineering Technology
3. Computer Engineering Technology
4. Mechanical Drafting Technology
5. Electrical/Electronics Engineering Technology
6. Chemical Engineering Technology
7. Industrial Engineering Technology
8. Mechanical Engineering Technology
9. Biomedical Engineering Technology

36. The impedance of a circuit consisting of a resistor, capacitor, and inductor connected in series to an ac power source is the vector sum $\mathbf{R} + \mathbf{X}_l - \mathbf{X}_c$. Find the magnitude and direction (measured from the horizontal) for the impedance shown in the figure if $|\mathbf{X}_c| = 250\ \Omega$, $|\mathbf{X}_l| = 35\ \Omega$, and $|\mathbf{R}| = 50\ \Omega$. 3 5 9 4 7 8

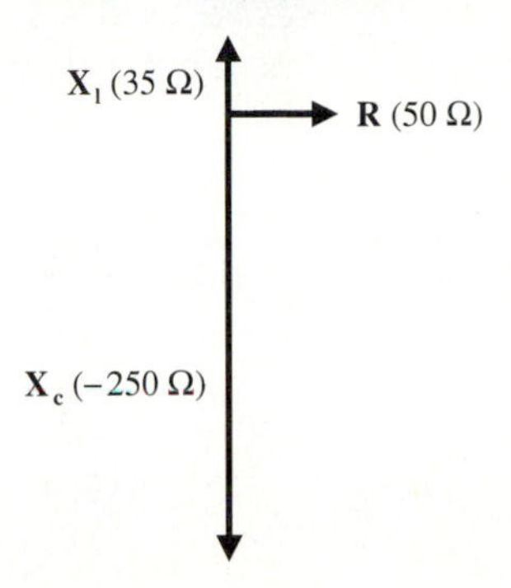

37. Rework problem 36, using $|\mathbf{X}_c| = 3.65\ \Omega$, $|\mathbf{X}_l| = 5.42\ \Omega$, and $|\mathbf{R}| = 4.78\ \Omega$. 3 5 9 4 7 8

38. A beam is fixed at right angles to the wall as shown in the figure. The beam is supported by a tie making an angle of 35.0° with the beam. A load hangs from the end. If the load exerts a force of 34.9 newtons and the thrust in the beam is 49.8 newtons, find the tension in the tie. (Hint: the vertical component of tension must balance the weight of the load.) 2 4 8 1 5 6 7 9

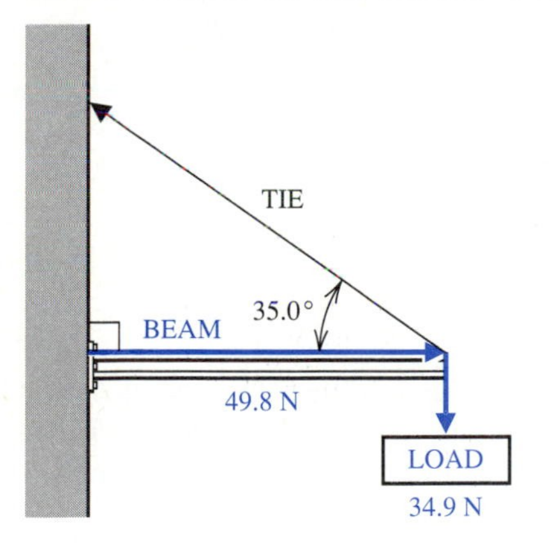

39. A beam is fixed at right angles to the wall as shown in the figure. The beam is supported by a brace making an angle of 35.0° with the beam. A load hangs from the end. If the load exerts a force of 35.0 newtons and the tension in the beam is 50.0 newtons, find the thrust in the brace. (Hint: the vertical component of thrust must balance the weight of the load.) 2 4 8 1 5 6 7 9

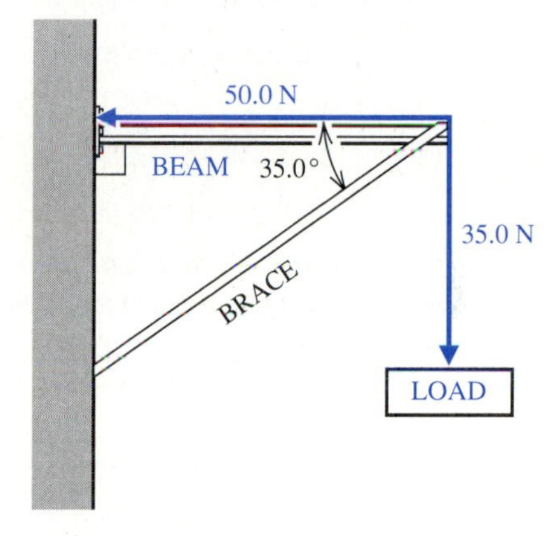

40. Two tugboats pull a barge. On one side the barge is pulled with a force of 108 newtons at an angle of 20.0° with the barge. On the other side, the force is 205 newtons at an angle of 30.0° with a line running the length of the barge. What single force would have the same effect? 4 7 8 1 2 3 5 6 9

41. A boat is pulled through a canal by a horizontal cable. The pull in the cable is 207 newtons, and the angle between the cable and the shore is 20.5°. Resolve the force vector into perpendicular components. 4 7 8 1 2 3 5 6 9

10.3 THE LAW OF SINES

Vector diagrams often involve triangles. The concepts of solving triangles can be extended to include oblique triangles as well as right triangles. Thus, it may not be necessary to work with the separate components of each vector.

It is possible to solve a given triangle if three parts are known, provided at least one of them is a side. The known parts will fall into one of three categories:

1. one side and two angles;
2. two sides and one angle;
   (a) The angle is opposite one of the two sides.
   (b) The angle is included between the two sides.
3. three sides.

Two important principles, the *law of sines* and the *law of cosines* are used to solve such triangles. The law of sines is presented first. After you gain some familiarity with the rule, its derivation will be given.

THE LAW OF SINES

For a triangle with angles A, B, C and sides a, b, c,

$$\frac{a}{\sin A} = \frac{b}{\sin B} = \frac{c}{\sin C}.$$

NOTE ⇨⇨ ***The law of sines is used to solve a triangle when the given parts consist of one side and two angles or two sides and the angle opposite one of them.***

Because only two ratios are needed for a proportion, the law of sines actually gives three proportions that can be used.

$$\frac{a}{\sin A} = \frac{b}{\sin B} \qquad \frac{a}{\sin A} = \frac{c}{\sin C} \qquad \frac{b}{\sin B} = \frac{c}{\sin C}$$

EXAMPLE 1 Solve the triangle for which $A = 32.8°$, $B = 44.6°$, and $a = 57.9$ m.

Solution Angle C can be determined from the sum of the angles.

$$32.8° + 44.6° + C = 180°$$
$$C = 180° - (32.8° + 44.6°) \text{ or } 102.6°$$

To apply the law of sines, two of the three ratios listed in the rule must be used. One side and the angle opposite it must be known for one of the ratios. Since both A and a are given, 57.9/sin 32.8° is one of the ratios that should be used. Since B is known, the ratio $b/\sin 44.6°$ should be used to find b.

$$\frac{57.9}{\sin 32.8°} = \frac{b}{\sin 44.6°} \text{ or } b = \frac{57.9 \sin 44.6°}{\sin 32.8°}, \text{ which is 75.0 m}$$

The law of sines may also be used to find c.

$$\frac{57.9}{\sin 32.8°} = \frac{c}{\sin 102.6°} \text{ or } c = \frac{57.9 \sin 102.6°}{\sin 32.8°}, \text{ which is 104 m}$$ ■

EXAMPLE 2 To measure the distance across a pond, as in Figure 10.18, a surveyor stands at point A, from which both sides of the pond are visible. Angle A is measured and found to be $54.3°$. The surveyor then walks 110.4 m to point B and measures angle B, which is $71.8°$. Find the distance across the pond and the distance from A to C.

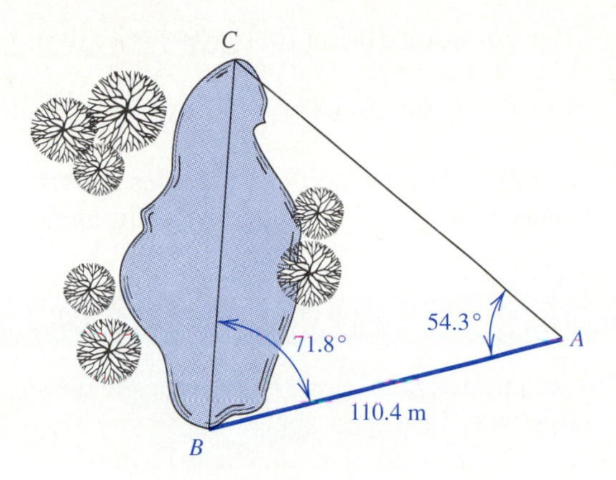

FIGURE 10.18

Solution Since two angles and a side are known, the law of sines is used. Side c is the only side given, so angle C must be known in order to use the law of sines. It is found from the sum of the angles.

$$54.3° + 71.8° + C = 180°$$

$$C = 180° - (54.3° + 71.8°) \text{ or } 53.9° \qquad \textbf{(10-1)}$$

Once both c and C are known, the law of sines may be used to find both a and b.

$$\frac{c}{\sin C} = \frac{a}{\sin A} \quad \text{and} \quad \frac{c}{\sin C} = \frac{b}{\sin B}$$

$$\frac{110.4}{\sin 53.9°} = \frac{a}{\sin 54.3°} \quad \text{and} \quad \frac{110.4}{\sin 53.9°} = \frac{b}{\sin 71.8°}$$

$$a = \frac{110.4 \sin 54.3°}{\sin 53.9°} \qquad \textbf{(10-2)}$$

$$a = 111 \text{ m}$$

$$b = \frac{110.4 \sin 71.8°}{\sin 53.9°} \qquad \textbf{(10-3)}$$

To three significant digits, $b = 130$ m

The distance across the pond is 111 m, and the distance from A to C is 130 m. ■

In Example 1, the ratio $a/\sin A$ was used to find two parts of the triangle. In Example 2, the ratio $c/\sin C$ was used to find two parts of the triangle. When using a calculator, you can store the ratio of the known side to the sine of its opposite angle in memory, since this ratio is used twice. It is then available for recall when needed.

Remember that you should solve an equation algebraically before you use a calculator to do the arithmetic. The calculator keystroke sequence for Example 2 follows.

For Equation 10-1: 180 [−] [(] 54.3 [+] 71.8 [)] [=] [STO] **53.9**

For Equation 10-2: 110.4 [÷] [RCL] [sin] [=] [STO] [×] 54.3 [sin] [=] **110.95934**

For Equation 10-3: [RCL] [×] 71.8 [sin] [=] **129.79979**

The following property of triangles is useful when it is necessary to know the relationship between the relative size of the angles and the relative lengths of the sides.

PROPERTY OF TRIANGLES

The shortest side of a triangle is opposite the smallest angle, and the longest side is opposite the largest angle.

EXAMPLE 3 Find the shortest side of the triangle for which $A = 32.4°$, $B = 11.6°$, and $c = 12.9$ cm.

Solution

$$32.4° + 11.6° + C = 180°$$

$$C = 180° - (32.4° + 11.6°) \text{ or } 136.0°$$

Since angle B is the smallest angle, side b is the shortest side. The law of sines is used.

$$\frac{b}{\sin B} = \frac{c}{\sin C}$$

$$\frac{b}{\sin 11.6°} = \frac{12.9}{\sin 136.0°} \text{ or } b = \frac{12.9 \sin 11.6°}{\sin 136.0°}, \text{ which is } 3.73 \text{ cm}$$ ■

EXAMPLE 4 A lamp is hung from a rope attached at two points 4.30 m apart. One end of the rope makes an angle of 30.0° with the horizontal and the other makes an angle of 20.0°. Find the length of the rope.

Solution Figure 10.19, shows that two angles and a side of a triangle are known.

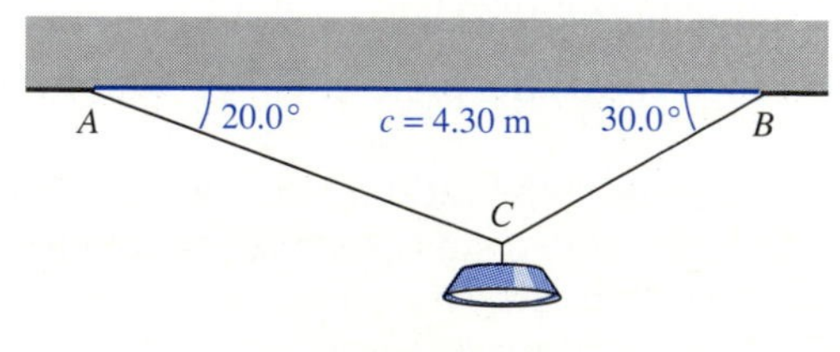

FIGURE 10.19

The law of sines can be applied to find the other two sides. Their sum is the length of the rope.

$$20.0° + 30.0° + C = 180°$$
$$C = 180° - (20.0° + 30.0°) \text{ or } 130.0°$$

$$\frac{4.30}{\sin 130.0°} = \frac{a}{\sin 20.0°} \quad \text{and} \quad \frac{4.30}{\sin 130.0°} = \frac{b}{\sin 30.0°}$$

$$a = \frac{4.30 \sin 20.0°}{\sin 130.0} \quad \text{and} \quad b = \frac{4.30 \sin 30.0°}{\sin 130.0°}$$

$$a = 1.92 \text{ m} \quad \text{and} \quad b = 2.81 \text{ m}$$

The length of the rope is $1.92 + 2.81$ or 4.73 m. ■

EXAMPLE 5 If the lamp in Example 4 exerts a force of 12.3 newtons, find the tension in each part of the rope.

Solution The force exerted by the lamp must be balanced by the tension in the rope. That is, the vector sum of the two tensions is equal to the force exerted by the lamp. The larger tension will be in the shorter rope, so the diagram is redrawn as in Figure 10.20. Consider triangle ABC, which shows the tension vectors and their resultant. The angles with the horizontal are the same as in Example 4, so the rope makes an angle of 130.0° at the lamp.

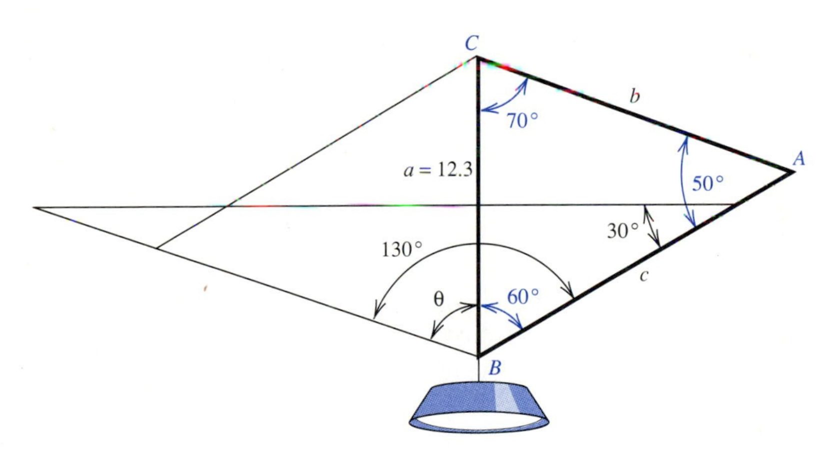

FIGURE 10.20

$B + 30.0° = 90°$, so $B = 60.0°$ $\qquad \theta = 130.0° - 60.0°$ or $70.0°$

Since C and θ are alternate interior angles, $C = 70.0°$. The sum of the angles is 180°, so $A = 50.0°$. The diagram shows that two angles and a side are known, so the law of sines is used.

$$\frac{12.3}{\sin 50.0^\circ} = \frac{b}{\sin 60.0^\circ} \quad \text{and} \quad \frac{12.3}{\sin 50.0^\circ} = \frac{c}{\sin 70.0^\circ}$$

$$b = \frac{12.3 \sin 60.0^\circ}{\sin 50.0^\circ} \quad \text{and} \quad c = \frac{12.3 \sin 70.0^\circ}{\sin 50.0^\circ}$$

$$b = 13.9 \quad \text{and} \quad c = 15.1$$

The tension, then, is 13.9 newtons in rope b and 15.1 newtons in rope c. ■

Now that you have seen how to use the law of sines, let us verify that it does, indeed, produce correct answers.

Consider triangle ABC as shown in Figure 10.21. A line segment h is drawn from C perpendicular to AB.

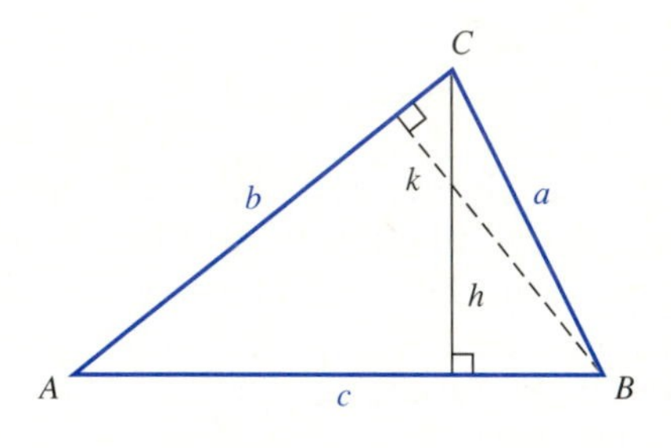

FIGURE 10.21

$$\sin A = h/b \text{ or } h = b \sin A \quad \text{and} \quad \sin B = h/a \text{ or } h = a \sin B$$

Since both $b \sin A$ and $a \sin B$ are equal to h, they are equal to each other.

$$a \sin B = b \sin A$$

Divide both sides of the equation by $\sin A$ ($\sin B$).

$$\frac{a}{\sin A} = \frac{b}{\sin B} \tag{10-4}$$

A line segment k is drawn from B perpendicular to AC.

$$\sin A = k/c \text{ or } k = c \sin A \quad \text{and} \quad \sin C = k/a \text{ or } k = a \sin C$$

$$a \sin C = c \sin A$$

$$\frac{a}{\sin A} = \frac{c}{\sin C} \tag{10-5}$$

Putting Equations 10-4 and 10-5 together, we have the law of sines. It can be shown that the law of sines also holds if one of the angles is obtuse.

The examples in this section involved triangles for which one side and two angles were known. If two sides and the angle opposite one of them are known, it is possible to obtain two solutions. This situation is called the ambiguous case and it will be covered in the next section.

EXERCISES 10.3

In 1–10, solve each triangle.

1. $A = 30.5°$, $B = 40.7°$, $a = 15.3$

2. $A = 35.7°$, $B = 42.4°$, $c = 24.5$

3. $A = 29.6°$, $B = 38.3°$, $b = 33.7$

4. $A = 26.7°$, $C = 36.8°$, $c = 4.51$

5. $A = 44.8°$, $C = 52.1°$, $b = 3.29$

6. 

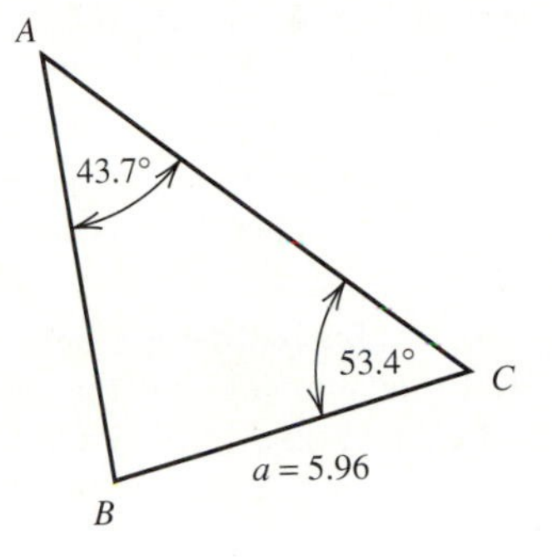

7. 

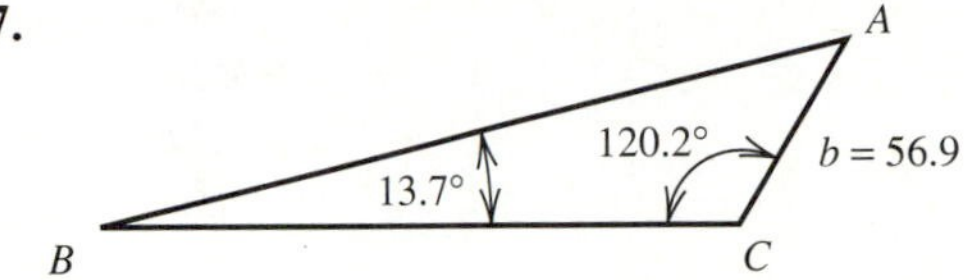

8.

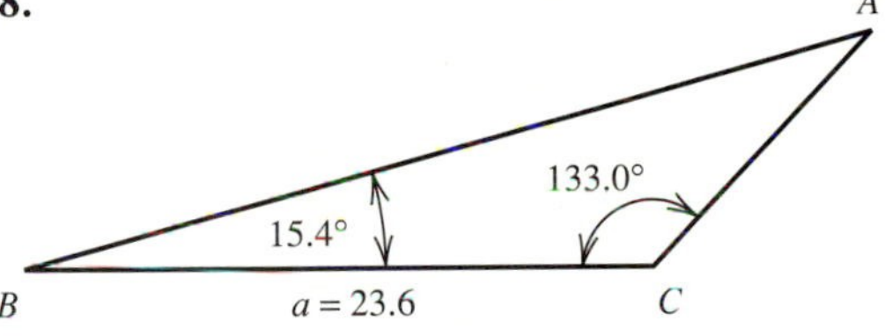

9.

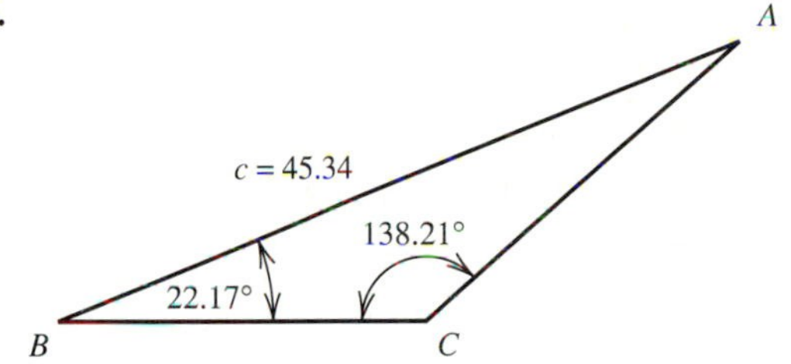

10.

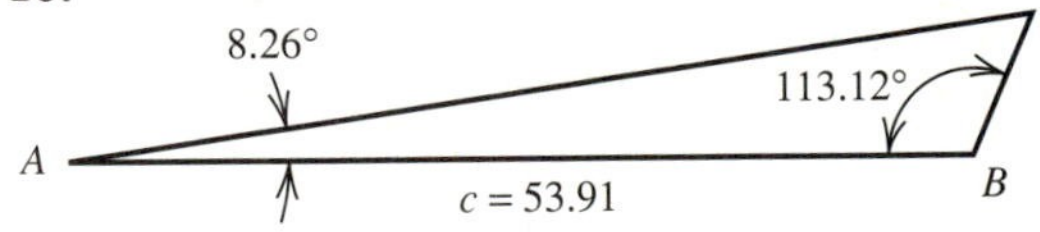

In 11–20, find the indicated part of each triangle.

11. b if $A = 21.6°$, $B = 59.6°$, and $a = 24.4$

12. b if $A = 26.8°$, $B = 51.3°$, and $c = 33.4$

13. c if $A = 10.7°$, $B = 47.2°$, and $b = 42.8$

14. a if $A = 17.8°$, $C = 47.7°$, and $c = 5.42$

15. c if $A = 35.9°$, $C = 61.0°$, and $b = 4.10$

16. b if $A = 34.8°$, $C = 62.3°$, and $a = 6.87$

17. a if $B = 4.8°$, $C = 139.1°$, and $b = 65.0$

18. b if $B = 6.5°$, $C = 142.9°$, and $a = 32.7$

19. c if $B = 13.26°$, $C = 127.32°$, and $a = 54.43$

20. b if $A = 7.17°$, $B = 122.23°$, and $a = 62.00$

21. A lighthouse is sighted in the direction of N 36.0° W from a ship. After the ship travels north for 826 m, the lighthouse is measured at N 54.0° W. How far is it from the ship to the lighthouse when the second sighting is made?
1 2 4 7 8 3 5

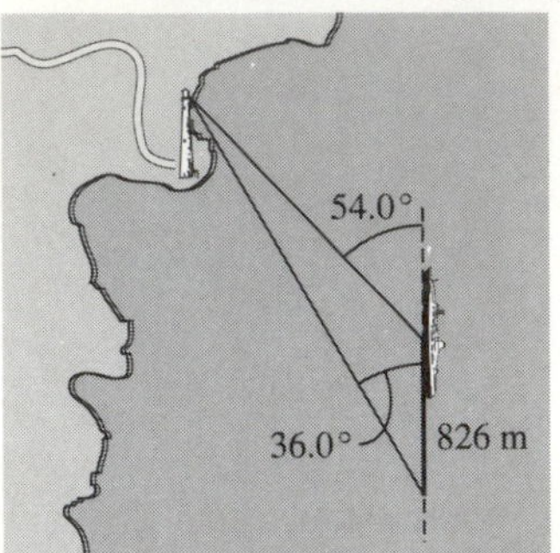

22. Find the lengths of the four sides of the plot of land shown in the accompanying figure.
1 2 4 7 8 3 5

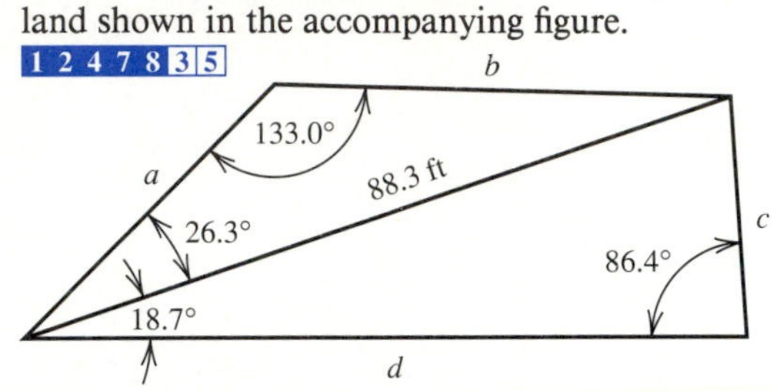

23. Two guy wires are attached to opposite sides of a radio tower. One of the wires makes an angle of 45.0° with the ground and the other makes an angle of 50.0° with the ground. Find the length of each wire if the distance between anchors is 48.0 m. 2 4 8 1 5 6 7 9

24. An airplane flies in the direction N 63° W. The 45 km/h wind blows at an angle of N 42° E. The effect is to change the plane's bearing to N 58° W. Find the speed of the plane that would be necessary to reach the same destination in the same length of time if there were no wind.
4 7 8 1 2 3 5 6 9

25. A lamp is hung from a rope attached at two points 5.34 m apart. One end of the rope makes an angle of 25.6° with the horizontal and the other makes an angle of 21.7°. Find the length of the rope.
2 4 8 1 5 6 7 9

26. If the lamp in problem 25 exerts a force of 10.7 newtons, find the tension in each part of the rope.
2 4 8 1 5 6 7 9

27. From a fire tower A, a fire is observed in the direction S 31.3° W. From a fire tower B, the same fire is observed in the direction S 11.5° E. Given that the distance from A to B is 34.6 km northwest of A, find the distance of the fire from tower A.
1 2 4 7 8 3 5

28. Show that the law of sines is also correct if one of the angles of the triangle is obtuse. See the accompanying figure.

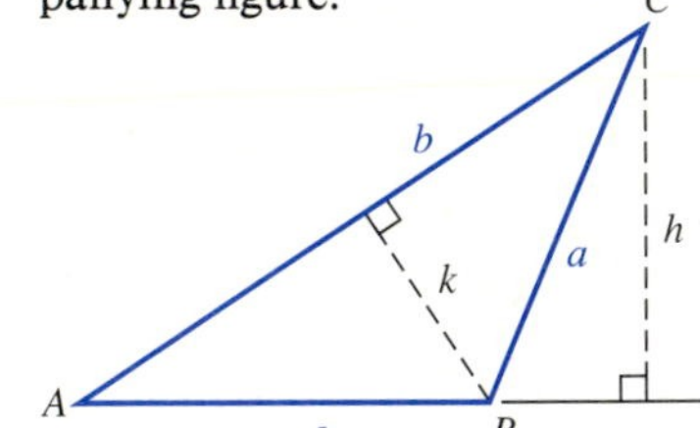

1. Architectural Technology
2. Civil Engineering Technology
3. Computer Engineering Technology
4. Mechanical Drafting Technology
5. Electrical/Electronics Engineering Technology
6. Chemical Engineering Technology
7. Industrial Engineering Technology
8. Mechanical Engineering Technology
9. Biomedical Engineering Technology

10.4 THE AMBIGUOUS CASE

The law of sines may be used to solve a triangle for which one side and two angles are known or for which two sides and the angle opposite one of them are known. The second case is called the *ambiguous case,* for there may be no solution, one solution, or two solutions. Imagine a triangle constructed with side a, side b, and angle A so that angle A is rigid, but sides a and b are joined by a hinge so that side a can be swung into different positions.

ambiguous case

If A is obtuse, there are three situations that can occur. If $a > b$, only one triangle is possible, as in Figure 10.22. But if $a < b$ or $a = b$, it is not possible to form a triangle from the given parts. Figures 10.23 and 10.24 illustrate these possibilities.

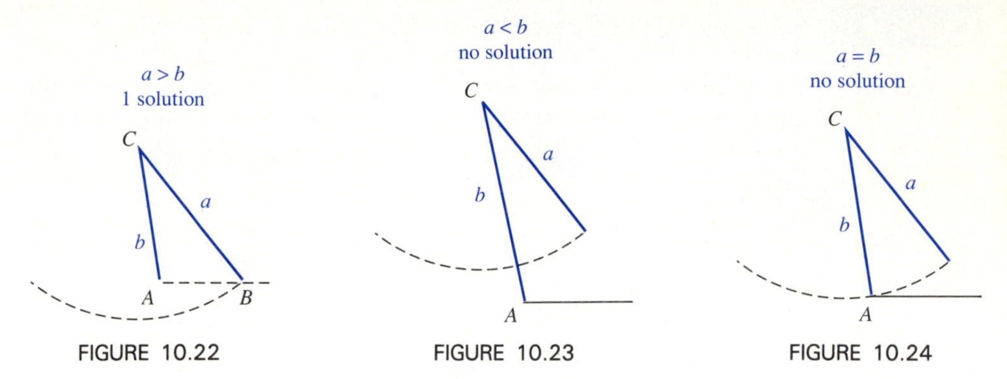

FIGURE 10.22 FIGURE 10.23 FIGURE 10.24

If angle A is an acute angle, there are five situations that can occur. If $a > b$ or $a = b$, only one solution is possible, as shown in Figures 10.25 and 10.26.

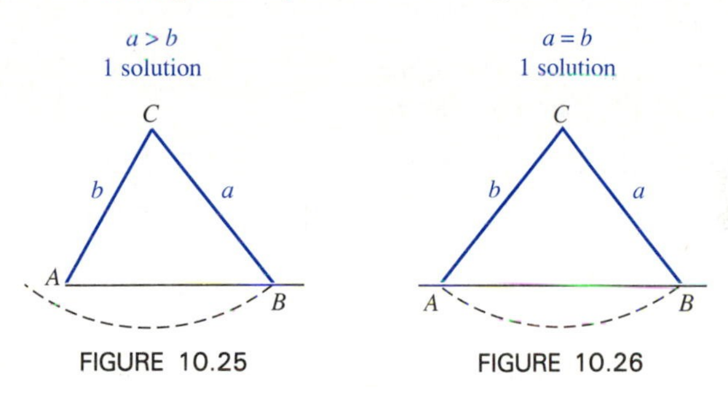

FIGURE 10.25 FIGURE 10.26

But if $a < b$, there may be zero, one, or two solutions. Figures 10.27–10.29 illustrate these situations.

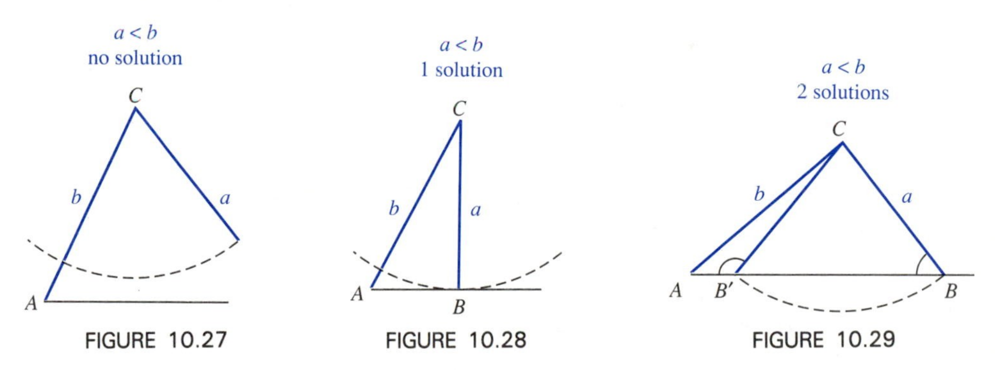

FIGURE 10.27 FIGURE 10.28 FIGURE 10.29

In Figure 10.29, side a touches side c in two different points, forming triangles ABC and $AB'C$. Angle B' is an obtuse angle such that $B' = 180° - B$.

It is not necessary to remember all of the different possibilities. Recall that the value of the sine function always falls between -1 and 1. When the law of sines leads to an equation in which $|\sin B| > 1$, no triangle satisfies the conditions. When the law of sines leads to an equation in which $|\sin B| = 1$, B is $90°$ and there is only one triangle that satisfies the conditions. ***The only case, then, that must be considered with unusual caution occurs when two sides and an acute angle opposite the shorter side are given.***

NOTE ⇨⇨

EXAMPLE 1 Solve the triangle for which $a = 15.3$, $c = 12.5$, and $C = 23.7°$.

Solution The information consists of two sides and the angle opposite one of them. Since the angle opposite c (the shorter given side) is acute, there may be two solutions.
First solution: The law of sines is used to find A, the angle opposite the longer given side.

$$\frac{15.3}{\sin A} = \frac{12.5}{\sin 23.7°} \text{ or } \sin A = \frac{15.3 \sin 23.7°}{12.5}, \text{ so } A = 29.5°$$

The sum of the angles is used to find B.

$$29.5° + B + 23.7° = 180°$$
$$B = 180° - (29.5° + 23.7°) \text{ or } 126.8°$$

The law of sines is used to find b.

$$\frac{12.5}{\sin 23.7°} = \frac{b}{\sin 126.8°} \text{ or } b = \frac{12.5 \sin 126.8°}{\sin 23.7°}, \text{ which is } 24.9$$

Second solution: Figure 10.30 shows ABC, the first solution, and $A'BC$, the second solution.

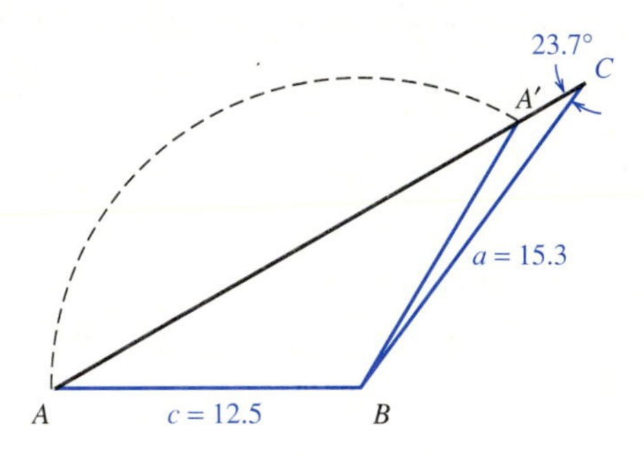

FIGURE 10.30

Since the first solution gave $A = 29.5°$, the second solution occurs when 29.5° is subtracted from 180°.

$$A' = 180° - 29.5° = 150.5°$$

The sum of the angles is used to find B.

$$150.5° + B + 23.7° = 180°$$
$$B = 180° - (150.5° + 23.7°) \text{ or } 5.8°$$

The law of sines is used to find b.

$$\frac{12.5}{\sin 23.7°} = \frac{b}{\sin 5.8°} \text{ or } b = \frac{12.5 \sin 5.8°}{\sin 23.7°}, \text{ which is } 3.14$$ ■

The procedure used in Example 1 is summarized as follows.

To solve a triangle when two sides and an acute angle opposite the shorter side are given:

(a) Find the first solution.
  1. Find the angle opposite the longer given side using the law of sines. Call it θ.
  2. Find the third angle using the sum of the angles.
  3. Find the third side using the law of sines.

(b) Find the second solution, if θ (from part (a), step 1) satisfies the condition $0° < \theta < 90°$.
  1. Find the angle opposite the longer given side by subtracting θ from 180°.
  2. Find the third angle using the sum of the angles.
  3. Find the third side using the law of sines.

EXAMPLE 2 Two ropes are used to anchor a hot air balloon. The ropes are fastened to the balloon and meet at a 30° angle. One rope is 30 ft long, and the stakes to anchor the ropes are placed 15 ft apart. How long is the second rope?

Solution Figure 10.31 shows that the information consists of two sides and the angle opposite one of them. Since the angle opposite a (the shorter side) is acute, there may be two solutions.

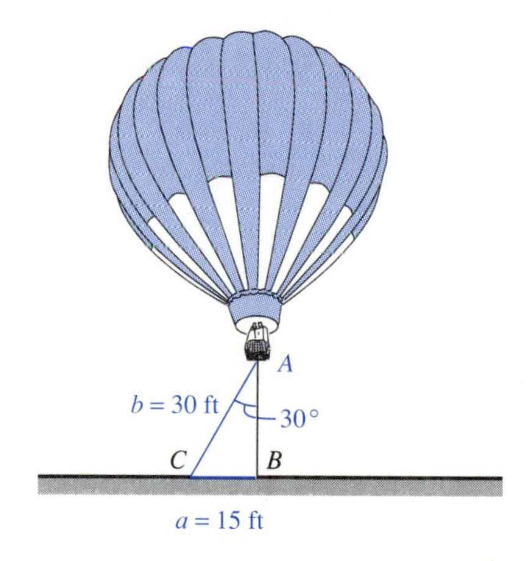

FIGURE 10.31

The law of sines is used to find B.

$$\frac{15}{\sin 30°} = \frac{30}{\sin B} \text{ or } \sin B = \frac{30 \sin 30°}{15}, \text{ so } B = 90°$$

Since $B = 90°$, there is only one solution. The triangle is a 30°–60°–90° triangle. The second rope has a length of $15\sqrt{3}$ or about 26 ft. ■

EXAMPLE 3 Solve the triangle for which $a = 4.6$, $b = 7.3$, and $A = 75°$.

Solution The given information consists of two sides and the angle opposite one of them. Since the angle opposite a (the shorter side) is acute, there may be two solutions. The law of sines is used to find B.

$$\frac{4.6}{\sin 75°} = \frac{7.3}{\sin B} \text{ or } \sin B = \frac{7.3 \sin 75°}{4.6}, \text{ which is } 1.53$$

Since $\sin B > 1$ is impossible, there is no solution. ■

EXAMPLE 4 Solve the triangle for which $b = 22.5$, $c = 38.6$, and $B = 122.8°$.

Solution The information consists of two sides and the angle opposite one of them. Since B is obtuse, there cannot be two solutions. The law of sines is used to find C.

$$\frac{22.5}{\sin 122.8°} = \frac{38.6}{\sin C} \text{ or } \sin C = \frac{38.6 \sin 122.8°}{22.5} = 1.44$$

There is no solution. ■

EXAMPLE 5 Solve the triangle for which $c = 18.4$, $b = 14.6$, and $C = 115.7°$.

Solution The information consists of two sides and the angle opposite one of them. Since the given angle is obtuse, there cannot be two solutions. The law of sines is used to find B.

$$\frac{18.4}{\sin 115.7°} = \frac{14.6}{\sin B} \text{ or } \sin B = \frac{14.6 \sin 115.7°}{18.4}, \text{ so } B = 45.6°$$

The sum of the angles is used to find A.

$$A + 45.6° + 115.7° = 180°$$

$$A = 180° - (45.6° + 115.7°) \text{ or } 18.7°$$

$$\frac{a}{\sin 18.7°} = \frac{18.4}{\sin 115.7°} \text{ or } a = \frac{18.4 \sin 18.7°}{\sin 115.7°}, \text{ which is } 6.55$$ ■

EXAMPLE 6 The distance from Richmond, Virginia, to St. Louis, Missouri, is 715.0 miles. The distance from St. Louis to Houston, Texas, is 667.0 miles. Determine the distance from Richmond to Houston as shown in Figure 10.32.

Solution The information consists of two sides and the acute angle opposite the shorter side. This situation suggests that there might be two solutions. For the first solution, H is acute and for the second solution H is obtuse. The map, however, clearly shows that the first solution is the correct solution. The law of sines is used to find H.

$$\frac{667.0}{\sin 31.88°} = \frac{715.0}{\sin H} \text{ or } \sin H = \frac{715.0 \sin 31.88°}{667.0}, \text{ so } H = 34.48°$$

The sum of the angles is used to find S.

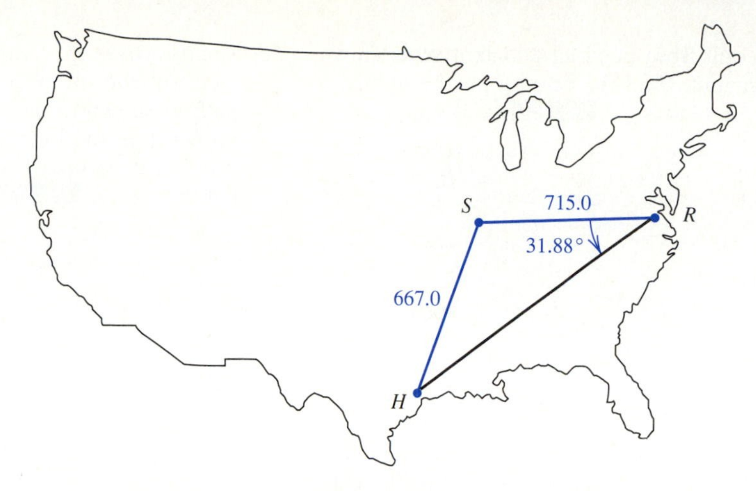

FIGURE 10.32

$$31.88° + 34.48° + S = 180°$$
$$S = 180° - (31.88° + 34.48°) \text{ or } 113.64°$$

The law of sines is used to find s.

$$\frac{667.0}{\sin 31.88°} = \frac{s}{\sin 113.64°}$$

$$s = \frac{667.0 \sin 113.64°}{\sin 31.88°}, \text{ which is 1157 miles}$$ ■

EXERCISES 10.4

In 1–14, identify each triangle that has two solutions.

1. $a = 7$, $b = 5$, and $A = 40°$

2. $b = 9$, $c = 7$, and $C = 30°$

3. $a = 3$, $b = 5$, and $C = 50°$

4. $a = 4$, $c = 6$, and $B = 35°$

5. $a = 3$, $b = 5$, and $c = 6$

6. $A = 20°$, $B = 40°$, and $c = 15$

7. $a = 5$, $c = 2$, and $C = 22°$

8. $a = 5$, $c = 2$, and $C = 112°$

9. $a = 9$, $b = 5$, and $A = 120°$

10. $b = 8$, $c = 10$, and $C = 32°$

11. $a = 15$, $b = 20$, and $A = 15°$

12. $a = 15$, $c = 20$, and $B = 42°$

13. $a = 12$, $c = 17$, and $A = 28°$

14. $a = 19$, $b = 14$, and $B = 33°$

In 15–24, solve each triangle.

15. $a = 13.1$, $b = 14.2$, and $A = 26.4°$

16. $a = 45.2$, $b = 32.7$, and $A = 32.9°$

17. $b = 22.6$, $c = 38.2$, and $C = 45.3°$

18. $b = 15.7$, $c = 13.6$, and $C = 38.2°$

19. $a = 34.6$, $c = 42.1$, and $A = 118.5°$

20. $a = 45.3$, $c = 24.2$, and $A = 126.9°$

21. $a = 32.7$, $b = 18.4$, and $A = 131.6°$

22. $a = 15.9$, $b = 23.6$, and $A = 127.4°$

23. $b = 14.3$, $c = 14.3$, and $C = 29.7°$

24. $b = 25.2$, $c = 25.2$, and $C = 129.7°$

25. A ship sails from one island to another as shown in the figure. Find the bearing for the ship as it sails from C to A. 1 2 4 7 8 3 5

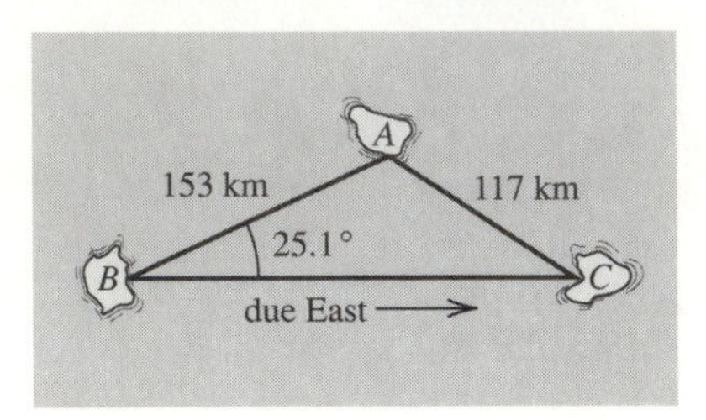

26. If it were vertical, the Leaning Tower of Pisa would be 55.0 m tall. At an observation point 65.0 m from the base, the angle of elevation to the top is measured to be 38.1°. How long a wire would be needed to reach the top of the tower from the observation point? See the accompanying figure. 2 4 8 1 5 6 7 9

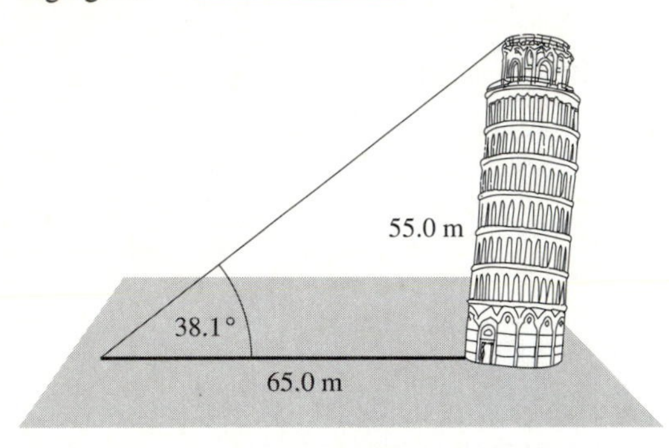

27. An observation tower at the top of a hill is 45.0 ft tall. The angle of depression to the base of the hill is 62.2° and the path up the hill is 98.0 ft long. How high is the hill? 1 2 4 7 8 3 5

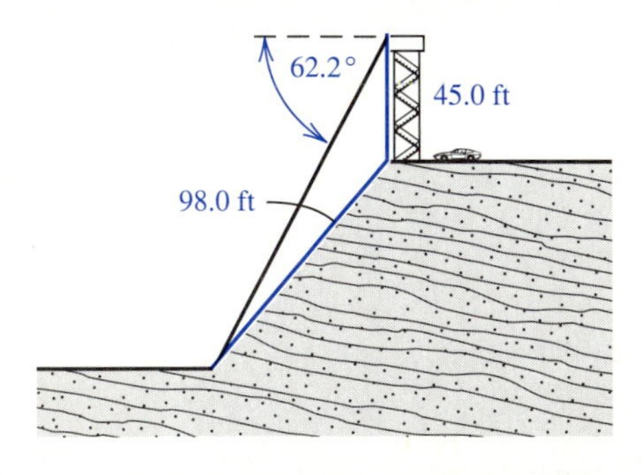

28. A ship leaves shore, sails 54.9 km and sights a lighthouse at an angle of 32.6° from the ship's path as shown in the accompanying figure. It is known that the lighthouse is 63.2 km from the point of departure. How far is the ship from the lighthouse? 1 2 4 7 8 3 5

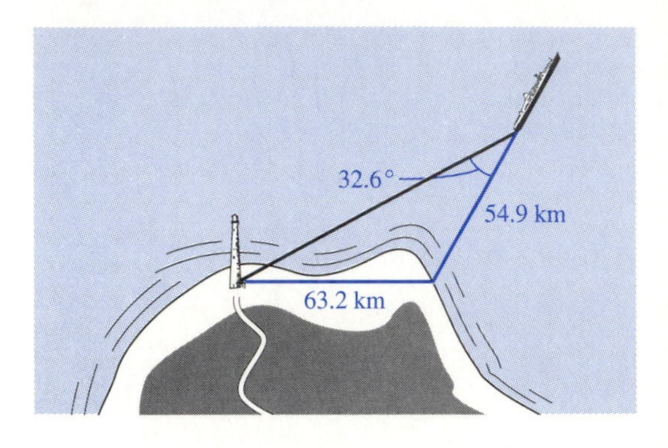

29. Find the dimensions of the parcel of land shown in the accompanying figure. 1 2 4 7 8 3 5

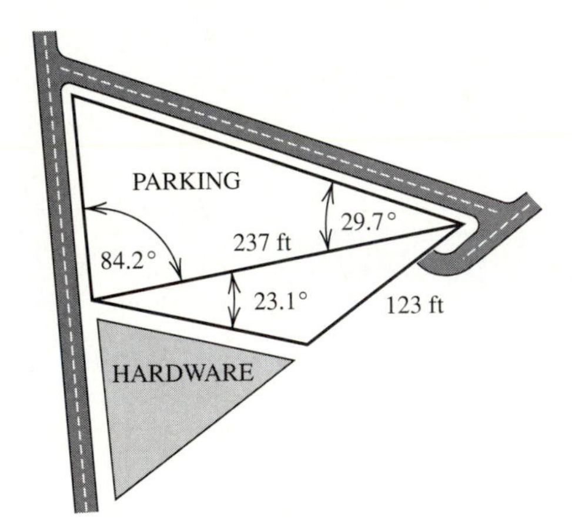

30. Using Figures 10.27–10.29, explain how to determine whether the number of solutions is zero, one, or two by comparing $b \sin A$ to a.

10.5 THE LAW OF COSINES

If we are given three sides or two sides and the included angle of a triangle, there is not enough information to use the law of sines. For example, consider a triangle with sides of lengths 4 m, 5 m, and 6 m, as shown in Figure 10.33.

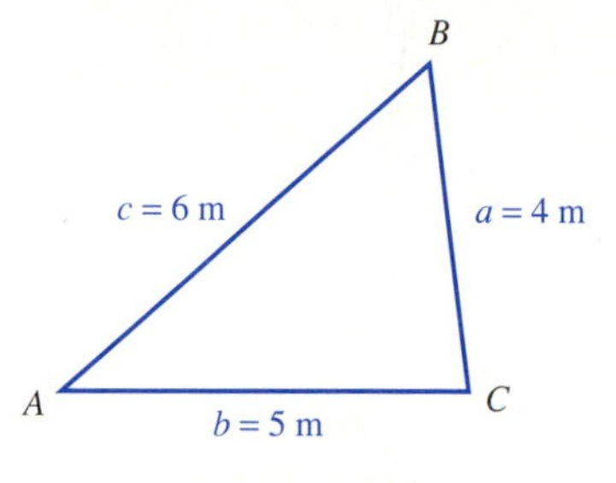

FIGURE 10.33

The law of sines gives

$$\frac{4}{\sin A} = \frac{5}{\sin B} = \frac{6}{\sin C}$$

But for any two of these ratios there are still two unknowns. It is impossible to use the law of sines to solve the triangle with the given information. The *law of cosines* is used to solve these types of problems. After you have gained some familiarity with the rule, its derivation will be given.

THE LAW OF COSINES

For a triangle with angles A, B, C and sides a, b, c,

$$c^2 = a^2 + b^2 - 2ab \cos C \quad \textbf{(10-6)}$$

$$a^2 = b^2 + c^2 - 2bc \cos A \quad \textbf{(10-7)}$$

$$b^2 = a^2 + c^2 - 2ac \cos B \quad \textbf{(10-8)}$$

NOTE ⇨⇨ ***The law of cosines is used to solve a triangle when the given parts consist of three sides or two sides and the included angle.***

It is not necessary to memorize all three equations, for they all follow the same pattern. The square of one side is equal to the sum of the squares of the other two sides minus twice the product of those two sides and the cosine of the angle between them. In each case, only one angle is involved. The formula has a term containing the cosine of that angle on one side of the equation and the square of the side opposite it on the other side of the equation.

When three sides are given, we do not have the ambiguous case. But once the first angle is found, you may want to use the law of sines with two sides and the acute angle opposite the shorter side. If there is an obtuse angle involved, it would be found only as a second solution. To avoid the problem, when using the law of cosines with three sides given, it is important to find the angles in a specified order. Since only one of the angles can be obtuse, the largest angle is found first. If there is an obtuse angle, we will be sure to find it using the law of cosines.

EXAMPLE 1 Solve the triangle that was shown in Figure 10.33.

Solution The given information consists of three sides. The law of cosines is used to find angle C, the largest angle. To find C, we use Equation 10-6, since it involves angle C.

$$6.00^2 = 4.00^2 + 5.00^2 - 2(4.00)(5.00)\cos C$$

$$6.00^2 - 4.00^2 - 5.00^2 = -2(4.00)(5.00)\cos C$$

$$\frac{6.00^2 - 4.00^2 - 5.00^2}{-2(4.00)(5.00)} = \cos C, \text{ so } C = 82.8^\circ$$

Once C is known, however, the law of sines, which is easier to apply, may be used to find A, the smaller unknown angle.

$$\frac{6.00}{\sin 82.8^\circ} = \frac{4.00}{\sin A} \quad \text{or} \quad \sin A = \frac{4.00 \sin 82.8^\circ}{6.00}, \text{ so } A = 41.4^\circ$$

The sum of the angles may be used to find the third angle.

$$41.4^\circ + B + 82.8^\circ = 180^\circ$$

$$B = 180^\circ - (82.8 + 41.4) \text{ or } 55.8^\circ \quad \blacksquare$$

The steps of Example 1 are generalized as follows.

To solve a triangle when three sides are given:

1. Find the largest angle, using the law of cosines.
2. Find the smaller unknown angle, using the law of sines.
3. Find the other angle, using the sum of the angles.

EXAMPLE 2 Solve the triangle shown in Figure 10.34.

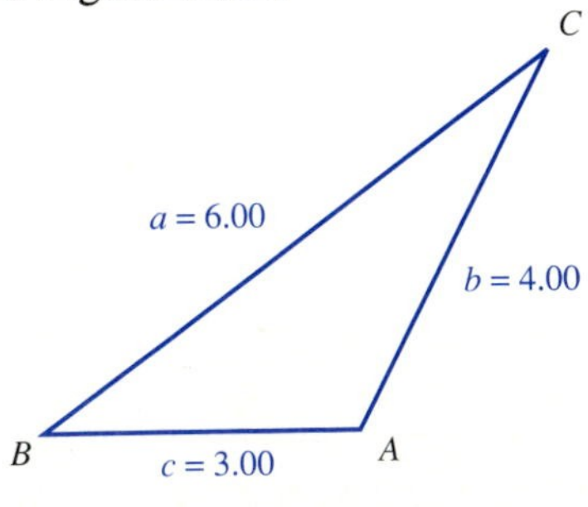

FIGURE 10.34

Solution The given information consists of three sides. The law of cosines (Equation 10-7) is used to find A, the largest angle.

$$6.00^2 = 3.00^2 + 4.00^2 - 2(3.00)(4.00)\cos A$$
$$6.00^2 - 3.00^2 - 4.00^2 = -2(3.00)(4.00)\cos A$$
$$\frac{6.00^2 - 3.00^2 - 4.00^2}{-2(3.00)(4.00)} = \cos A$$

Since $\cos A$ is negative, A is a second quadrant angle. The reference angle is 62.7°, so A is 117.3°. The law of sines is used to find C, the smaller unknown angle.

$$\frac{3.00}{\sin C} = \frac{6.00}{\sin 117.3°} \text{ or } \sin C = \frac{3.00 \sin 117.3°}{6.00}, \text{ so } C = 26.4°$$

The sum of the angles is used to find angle B.

$$117.3° + B + 26.4° = 180°$$
$$B = 180° - (117.3° + 26.4°) \text{ or } 36.3°$$ ■

NOTE ⇨⇨ ***Failure to find the angles in the specified order may lead to an incorrect solution.*** If we use the law of cosines to find C, then use the law of sines to find A in Example 2, we have the following result.

$$\frac{3.0}{\sin 26.4°} = \frac{6.0}{\sin A} \text{ or } \sin A = \frac{6.0 \sin 26.4°}{3.0}, \text{ so } A = 62.8°$$

Notice that the law of sines does not indicate that 62.8° is actually the reference angle for A and that A is 117.2°.

EXAMPLE 3 Specifications for a triangular metal plate call for sides of 13.6 cm, 27.3 cm, and 30.5 cm, as shown in Figure 10.35. Determine the angles that should be used.

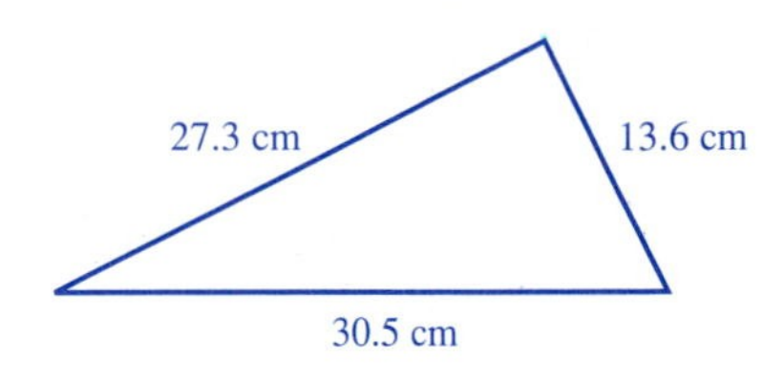

FIGURE 10.35

Solution Let $a = 13.6$, $b = 27.3$, and $c = 30.5$.
The law of cosines (Equation 10-6) is used to find C, the largest angle.

$$30.5^2 = 27.3^2 + 13.6^2 - 2(27.3)(13.6)\cos C$$
$$30.5^2 - 27.3^2 - 13.6^2 = -2(27.3)(13.6)\cos C$$
$$\frac{30.5^2 - 27.3^2 - 13.6^2}{-2(27.3)(13.6)} = \cos C, \text{ so } C = 90°$$

The law of sines is used to find A, the smaller unknown angle.

$$\frac{13.6}{\sin A} = \frac{30.5}{\sin 90^\circ} \text{ or } \sin A = \frac{13.6 \sin 90^\circ}{30.5}, \text{ so } A = 26.5^\circ$$

The sum of the angles is used to find B.

$$B = 180^\circ - (90^\circ + 26.5^\circ) \text{ or } 63.5^\circ$$ ■

Notice in Example 3, that when C is 90°, cos 90° is 0, and the law of cosines is simply the Pythagorean theorem.

When the given information consists of two sides and the included angle, we do not have the ambiguous case. But once the third side is found, you may want to use two sides and the acute angle opposite the shorter side. The data used at this point may lead to two solutions. Only one of them is correct, and we do not know which. The solution may involve an acute angle, or it may involve an obtuse angle. The easiest way to avoid the problem is to remember that it is impossible for two angles of a triangle to be obtuse. If there is an obtuse angle, it must be the largest angle. Once again, we use the law of sines to find the smaller unknown angle, knowing it is acute. The largest angle is found using the sum of the angles, and whether it is acute or obtuse, the answer will be correct.

EXAMPLE 4 Solve the triangle for which $a = 42.0$, $b = 66.0$, and $C = 38.0°$.

Solution The law of cosines (Equation 10-6) is used to find side c, since angle C is given.

$$c^2 = (42.0)^2 + (66.0)^2 - 2(42.0)(66.0)\cos 38.0^\circ, \text{ and } c = 41.8$$

The law of sines is used to find A, the smaller unknown angle.

$$\frac{42.0}{\sin A} = \frac{41.8}{\sin 38.0^\circ} \text{ or } \sin A = \frac{42.0 \sin 38.0^\circ}{41.8}, \text{ so } A = 38.2^\circ$$

The sum of the angles is used to find B.

$$38.2^\circ + B + 38.0^\circ = 180^\circ$$

$$B = 180^\circ - (38.2^\circ + 38.0^\circ) \text{ or } 103.8^\circ$$ ■

NOTE ⇨⇨ ***It is important to use the sum of the angles, rather than the law of sines, to find the third angle.*** The law of sines will yield the reference angle instead of the obtuse angle.

NOTE ⇨⇨ ***Failure to find the angles in the specified order may lead to an incorrect solution.*** If B is found before A in Example 4, we have the following result.

$$\frac{66.0}{\sin B} = \frac{41.8}{\sin 38.0^\circ} \text{ or } \sin B = \frac{66.0 \sin 38.0^\circ}{41.8}, \text{ so } B = 76.4^\circ$$

The correct answer is $B = 103.8°$.

The procedure used in Example 4 is summarized as follows.

To solve a triangle when two sides and the included angle are given:

1. Find the third side, using the law of cosines.
2. Find the smaller unknown angle, using the law of sines.
3. Find the third angle, using the sum of the angles.

Compare this procedure with the procedure used when three sides were given. In both cases, the law of cosines is used in the first step. Then the law of sines is applied to the *smaller* of the unknown angles. Finally, the third angle is found using the sum of the angles.

EXAMPLE 5 Solve the triangle shown in Figure 10.36.

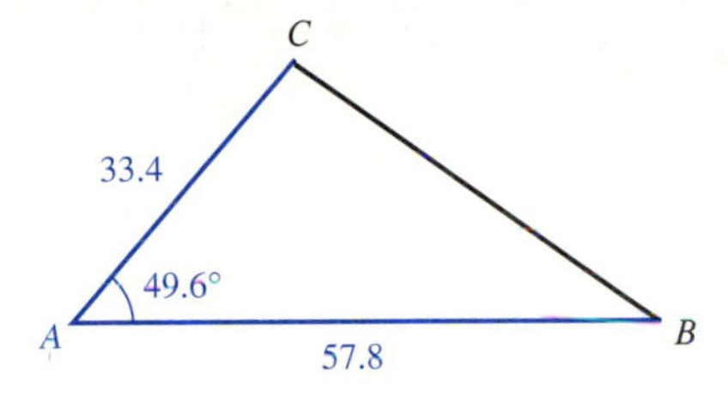

FIGURE 10.36

Solution The law of cosines (Equation 10-7) is used to find a.

$$a^2 = (33.4)^2 + (57.8)^2 - 2(33.4)(57.8)\cos 49.6°, \text{ and } a = 44.2$$

The law of sines is used to find B, the angle opposite the shorter side.

$$\frac{44.2}{\sin 49.6°} = \frac{33.4}{\sin B} \text{ or } \frac{33.4 \sin 49.6°}{44.2} = \sin B, \text{ so } B = 35.1°$$

The sum of the angles is used to find C.

$$49.6° + 35.1° + C = 180°$$

$$C = 180° - (49.6° + 35.1°) \text{ or } 95.3°$$ ■

The law of cosines may be used to solve applied problems.

EXAMPLE 6 The friction between a fixed pulley and the shaft is the resultant of the load **L** and applied force **P**. If the rope wrapped around the pulley subtends an angle of 120° at the center, and a force of 108 newtons is applied to lift a load of 75.3 newtons, find the magnitude of the frictional force **F**.

Solution Extending **L** and **P** beyond the pulley, as in Figure 10.37, shows that there is a 60° angle between them. Moving **L** so that its tail is at the head of **P**, a 120° angle is formed.

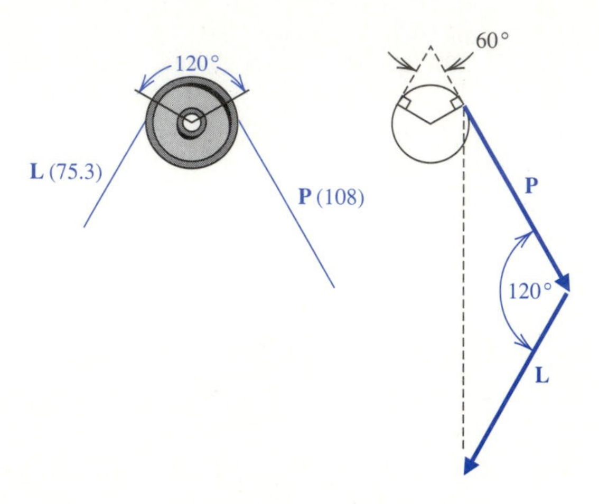

FIGURE 10.37

The law of cosines leads to the following equation.

$$|\mathbf{F}|^2 = 108^2 + (75.3)^2 - 2(108)(75.3)\cos 120°$$
$$|\mathbf{F}| = 160 \text{ newtons (to three significant digits)}$$ ■

To verify that the law of cosines will work for any triangle, it is necessary to show that $\sin^2 \theta + \cos^2 \theta = 1$ for any θ where $\sin^2 \theta$ means $(\sin \theta)^2$. If θ is an angle in standard position, $x^2 + y^2 = r^2$, where (x, y) is a point on the terminal side of θ. Divide both sides of the equation by r^2.

$$\frac{x^2}{r^2} + \frac{y^2}{r^2} = \frac{r^2}{r^2}$$

But this equation says that $\cos^2 \theta + \sin^2 \theta = 1$.

Consider triangle ABC as shown in Figure 10.38. A line segment h is drawn from B perpendicular to AC. Side b is split into two lengths: x and $b - x$.

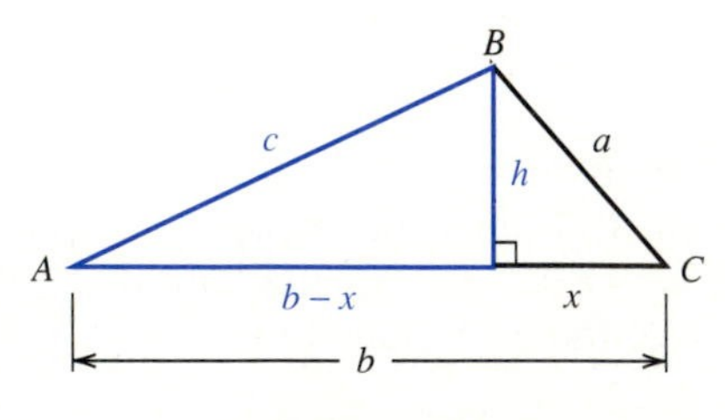

FIGURE 10.38

$$h = c \sin A \qquad \text{and} \qquad b - x = c \cos A$$

or

$$h = c \sin A \qquad \text{and} \qquad x = b - c \cos A.$$

By the Pythagorean theorem,

$$a^2 = h^2 + x^2.$$

Substituting for h and x, we have

$$a^2 = (c \sin A)^2 + (b - c \cos A)^2$$
$$a^2 = c^2 \sin^2 A + (b^2 - 2bc \cos A + c^2 \cos^2 A).$$

Rearranging terms, we have

$$a^2 = c^2 \sin^2 A + c^2 \cos^2 A + b^2 - 2bc \cos A$$
$$a^2 = c^2(\sin^2 A + \cos^2 A) + b^2 - 2bc \cos A$$
$$a^2 = c^2 + b^2 - 2bc \cos A.$$

Thus, we have Equation 10-7. Drawing h from C to side c, as shown in Figure 10.39, would lead to Equation 10-8. Likewise, drawing h from A to side a, as in Figure 10.40, would lead to Equation 10-6.

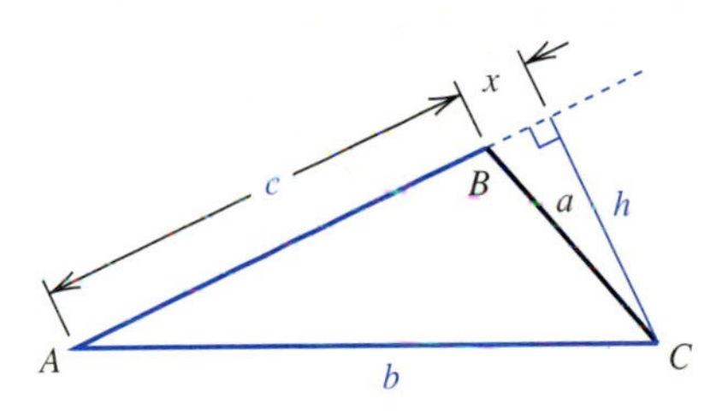

FIGURE 10.39

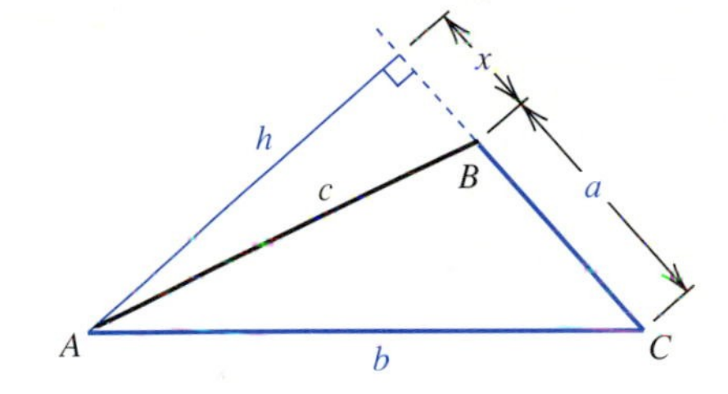

FIGURE 10.40

EXERCISES 10.5

In 1–14, decide whether the law of sines or the law of cosines should be used first in solving the triangle.

1. $a = 14$, $b = 19$, and $B = 24°$

2. $a = 17$, $c = 12$, and $A = 19°$

3. $a = 20$, $c = 15$, and $B = 33°$

4. $a = 20$, $b = 15$, and $A = 26°$

5. $b = 10$, $c = 8$, and $C = 23°$

6. $a = 5$, $b = 9$, and $A = 102°$

7. $a = 2$, $c = 5$, and $A = 123°$

8. $a = 2$, $c = 5$, and $C = 13°$

9. $A = 40°$, $B = 20°$, and $c = 26$

10. $a = 5$, $b = 3$, and $c = 7$

11. $a = 6$, $c = 4$, and $B = 26°$

12. $a = 5$, $b = 3$, and $C = 48°$

13. $b = 7$, $c = 9$, and $C = 29°$

14. $a = 5$, $b = 7$, and $A = 35°$

In 15–24, solve each triangle.

15. $a = 13.4$, $b = 14.6$, and $C = 20.2°$

16. $a = 23.6$, $b = 32.1$, and $c = 40.3$

17. $a = 12.6$, $b = 10.2$, and $c = 15.7$

18. $a = 4.7$, $c = 5.3$, and $B = 26°$

19. $b = 3.2$, $c = 4.6$, and $A = 41°$

20. $a = 36.0$, $b = 31.9$, and $c = 48.1$

21. $a = 30.8$, $b = 43.5$, and $c = 53.3$

22. $a = 5.31$, $b = 4.62$, and $C = 112.4°$

23. $a = 3.25$, $c = 8.76$, and $B = 126.3°$

24. $b = 12.17$, $c = 11.63$, and $A = 148.70°$

---

25. In the crankshaft shown in the figure, find length l of the connecting rod, given that the radius r of the crank is 10.6 cm, $x = 30.5$ cm, and $\theta = 120.0°$. 1 2 4 7 8 3 5

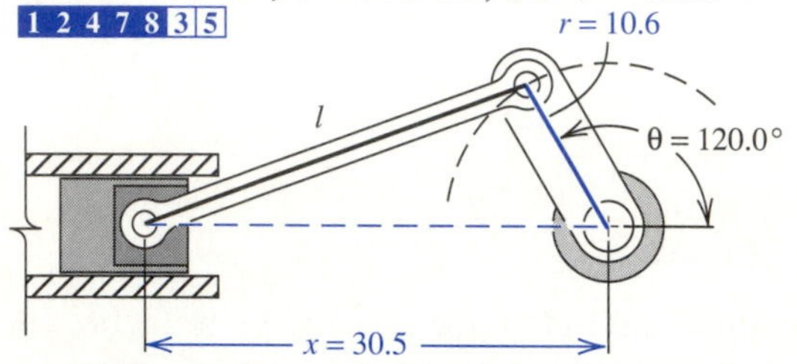

26. A lamp hangs on a rope. If the rope is anchored at two points a distance of 9.3 m apart and the sag in the rope is such that the lamp hangs as in the figure, find the angles it makes with the ceiling. 2 4 8 1 5 6 7 9

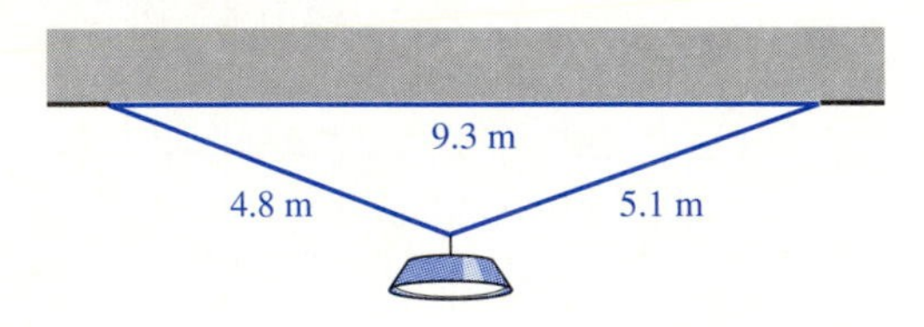

27. The distance from Mobile, Alabama, to Las Vegas, Nevada, is 1594 miles. The distance from Mobile to Albany, New York, is 1151 miles. The distance from Albany to Las Vegas is 2237 miles. Find the angle between routes to Albany and Las Vegas from Mobile. 1 2 4 7 8 3 5

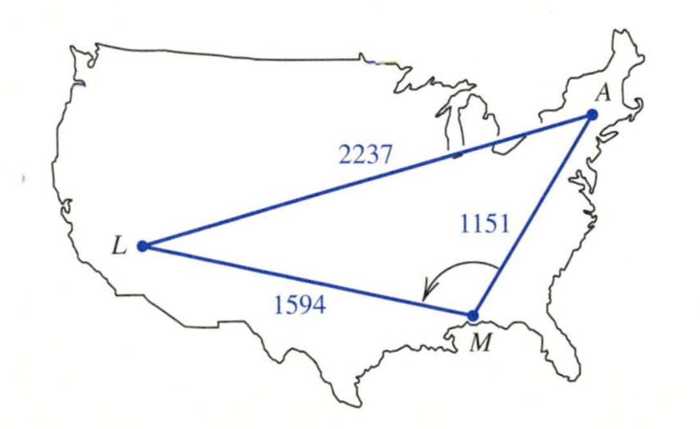

28. To measure the distance across a pond, the distance is measured from one point to each end as shown in the diagram. What is the distance across the pond? 1 2 4 7 8 3 5

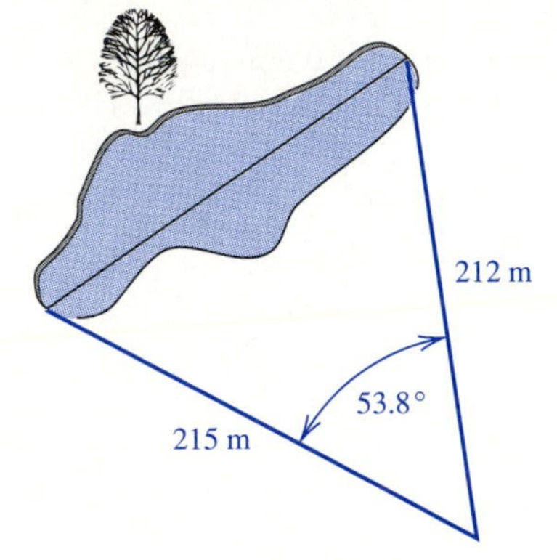

29. For the pulley shown, a force of 10.0 newtons is to be used to lift a load of 20.0 newtons and the friction is 24.0 newtons. Find angle θ. 4 7 8 1 2 3 5 6 9

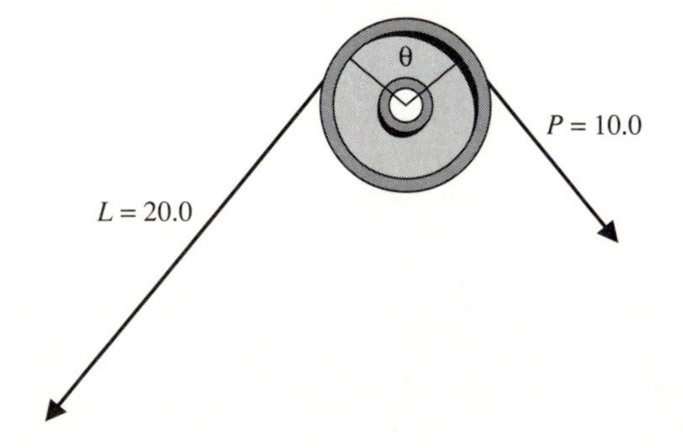

---

1. Architectural Technology
2. Civil Engineering Technology
3. Computer Engineering Technology
4. Mechanical Drafting Technology
5. Electrical/Electronics Engineering Technology
6. Chemical Engineering Technology
7. Industrial Engineering Technology
8. Mechanical Engineering Technology
9. Biomedical Engineering Technology

30. A ship leaves port and travels N 30° E for 10.7 km. A lighthouse known to be 20.1 km due north of the port is spotted. How far is the lighthouse from the ship? 1 2 4 7 8 3 5

31. In a three-phase ac circuit, voltages are generated 120° apart. If phase A is at 0°, phase B is at 120°, and phase C is at 240°, the potential difference between phase A and B is given by the vector $\mathbf{A} - \mathbf{B}$ as shown in the diagram. Find the magnitude of this vector if both A and B have a magnitude of 120 V. 3 5 9 1 2 4 6 7 8

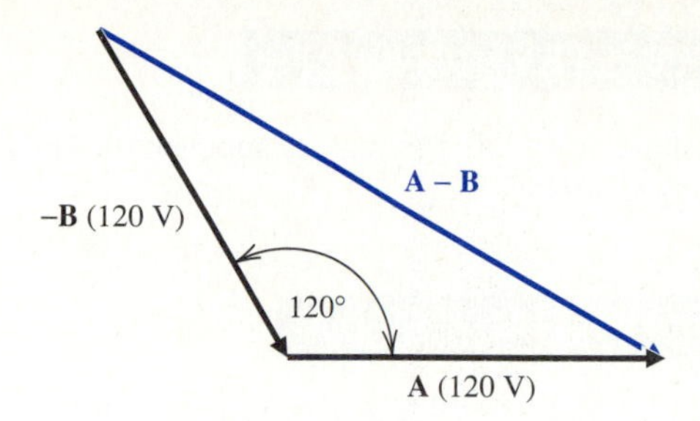

10.6 CHAPTER REVIEW

KEY TERMS

10.1 scalar
vector
tail
head
resultant
polygon method of adding vectors
parallelogram method of adding vectors
displacement
oblique triangle

10.2 components
resolving a vector

10.4 ambiguous case

SYMBOLS

$\mathbf{v}$ or $\vec{v}$	a vector
$\langle x, y\rangle$	a vector with its tail at the origin and head at (x, y)
$\lvert\mathbf{v}\rvert$	the magnitude of a vector
$\mathbf{v_x}, \mathbf{v_y}$	the horizontal and vertical components of a vector

RULES AND FORMULAS

If $\mathbf{v} = \langle x, y\rangle$, then $|\mathbf{v}| = \sqrt{x^2 + y^2}$ and $\tan\theta = y/x$.

$\mathbf{v_x} = |\mathbf{v}|\cos\theta$ and $\mathbf{v_y} = |\mathbf{v}|\sin\theta$

$|\mathbf{r}| = \sqrt{|\mathbf{r_x}|^2 + |\mathbf{r_y}|^2}$

For a triangle with angles A, B, C and sides a, b, c:

Law of Sines: $\dfrac{a}{\sin A} = \dfrac{b}{\sin B} = \dfrac{c}{\sin C}$

Law of Cosines: $a^2 = b^2 + c^2 - 2bc\cos A$
$b^2 = a^2 + c^2 - 2ac\cos B$
$c^2 = a^2 + b^2 - 2ab\cos C$

GEOMETRY CONCEPTS

The shortest side of a triangle is opposite the smallest angle, and the longest side is opposite the largest angle.

SEQUENTIAL REVIEW

(Section 10.1) *In 1–3, determine the magnitude and direction of each vector.*

1. $\langle -5, 9\rangle$ **2.** $\langle 6, -6\rangle$ **3.** $\langle -4, 7\rangle$

In 4–6, use the parallelogram method to add the given vectors. Specify the magnitude and direction of the resultant.

4. $\mathbf{a} = \langle 4, 2\rangle, \mathbf{b} = \langle 2, 6\rangle$ **5.** $\mathbf{a} = \langle 1, 5\rangle, \mathbf{b} = \langle -6, 7\rangle$ **6.** $\mathbf{a} = \langle 4, -1\rangle, \mathbf{b} = \langle 4, -5\rangle$

In 7–9, use the polygon method to add the vectors specified by the given coordinates. Specify the magnitude and direction of the resultant.

7. $\mathbf{a} = \langle 6, 3\rangle, \mathbf{b} = \langle 0, 5\rangle$ **8.** $\mathbf{a} = \langle -2, 8\rangle, \mathbf{b} = \langle -3, 4\rangle$ **9.** $\mathbf{a} = \langle 3, 4\rangle, \mathbf{b} = \langle 4, -5\rangle$

(Section 10.2) *In 10–12, resolve each vector into its horizontal and vertical components.*

10. $|\mathbf{v}| = 17.9, \theta = 22.1°$ **11.** $|\mathbf{v}| = 6.83, \theta = 108.4°$ **12.** $|\mathbf{v}| = 341.6, \theta = -72.65°$

In 13 and 14, add each pair of vectors by adding horizontal and vertical components. Specify the magnitude and direction of the resultant.

13. $|\mathbf{a}| = 37.3, \theta_a = 15.8°, |\mathbf{b}| = 83.8, \theta_b = 21.6°$ **14.** $|\mathbf{a}| = 16.0, \theta_a = 26.2°, |\mathbf{b}| = 3.70, \theta_b = 210.4°$

In 15 and 16, add each group of vectors by adding horizontal and vertical components. Specify the magnitude and direction of the resultant.

15. $|\mathbf{a}| = 59.6, \theta_a = 31.7°, |\mathbf{b}| = 45.8, \theta_b = 42.0°, |\mathbf{c}| = 35.2, \theta_c = 18.3°$

16. $|\mathbf{a}| = 78.4, \theta_a = 92.2°, |\mathbf{b}| = 42.9, \theta_b = -23.7°, |\mathbf{c}| = 31.5, \theta_c = 54.6°$

In 17 and 18, add each pair of vectors by adding horizontal and vertical components. Specify the magnitude and direction of the resultant.

17.

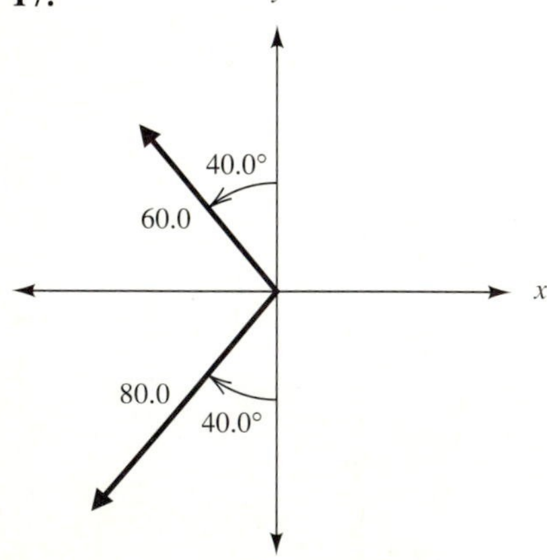

18.

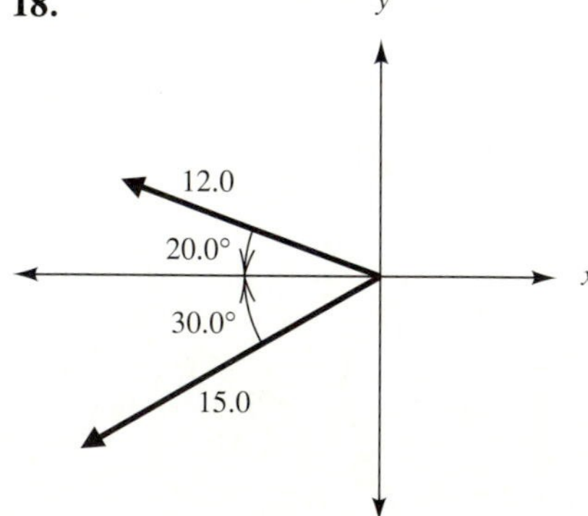

(Section 10.3) *In 19–25, find the indicated part of each triangle.*

19. Find b if $B = 54.2°$, $C = 31.6°$, and $a = 5.31$.

20. Find a if $A = 21.3°$, $B = 42.6°$, and $b = 12.7$.

21. Find a if $A = 8.8°$, $B = 88.8°$, and $c = 10.7$.

22. Find b if $A = 19.7°$, $C = 79.3°$, and $a = 22.1$.

23. Find c if $B = 26.3°$, $C = 14.2°$, and $a = 43.9$.

24. Find a if $A = 32.4°$, $C = 51.7°$, and $c = 21.6$.

25. Find a if $A = 42.5°$, $B = 17.3°$, and $b = 14.1$.

26. Solve the triangle shown in the accompanying diagram.

C
A 43.8° 31.5° B
$c = 24.2$

(Section 10.4) *In 27–30, identify each triangle that has two solutions.*

27. $a = 5$, $b = 3$, and $C = 26°$

28. $b = 6$, $c = 2$, and $A = 77°$

29. $a = 4$, $b = 1$, and $B = 43°$

30. $a = 4$, $c = 7$, and $A = 112°$

In 31–34, find the indicated part of each triangle.

31. Find A if $a = 7.36$, $c = 5.91$, and $C = 25.8°$.

32. Find A if $a = 9.52$, $b = 7.63$, and $B = 115.9°$.

33. Find A if $b = 15.9$, $c = 11.4$, and $B = 33.6°$.

34. Solve the following triangle: $b = 14.7$, $c = 20.3$, and $C = 42.5°$.

(Section 10.5) *In 35–41, decide whether the law of sines or the law of cosines should be used first in solving the given triangle.*

35. $a = 3.8$, $b = 6.2$, and $A = 135°$

36. $a = 3.7$, $b = 4.3$, and $c = 6.6$

37. $a = 3$, $b = 3$, and $C = 45°$

38. $a = 29.4$, $b = 12.8$, and $C = 13.1°$

39. $A = 38.9°$, $B = 32.7°$, and $c = 25.6$

40. $a = 14.3$, $b = 26.1$, and $c = 53.2$

41. $A = 4.6°$, $C = 110.4°$, and $c = 73.6$

In 42–44, find the missing part of each triangle.

42. Find A if $a = 3.1$, $b = 4.7$, and $c = 5.2$.

43. Find c if $a = 59.4$, $b = 52.6$, and $C = 42.8°$.

44. Find B if $a = 11.3$, $b = 18.7$, and $C = 7.2°$.

45. Solve the following triangle: $a = 10.5$, $b = 23.8$, and $C = 4.2°$.

APPLIED PROBLEMS

1. Two guy wires are attached to a pole as shown in the accompanying figure. If the tension in the longer wire is 87.4 newtons, find the tension in the shorter wire. 2 4 8 1 5 6 7 9

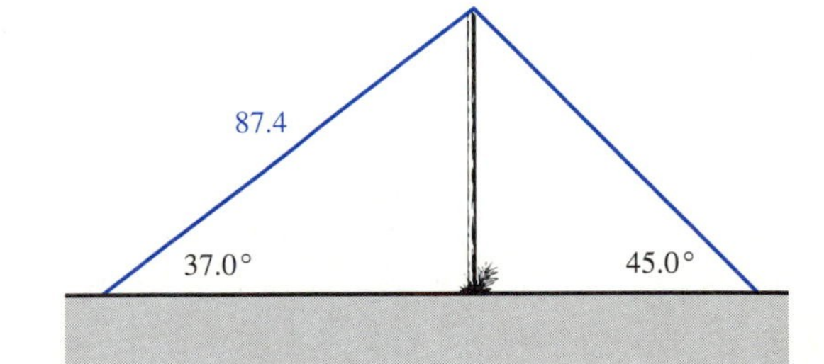

1. Architectural Technology
2. Civil Engineering Technology
3. Computer Engineering Technology
4. Mechanical Drafting Technology
5. Electrical/Electronics Engineering Technology
6. Chemical Engineering Technology
7. Industrial Engineering Technology
8. Mechanical Engineering Technology
9. Biomedical Engineering Technology

2. A column that is leaning is to be braced as shown in the diagram. If the column were vertical, it would be 10.0 ft tall. The brace is 8.00 ft long. If it is to be placed at an angle of 82.0° with the ground, how far from the column should the brace be placed? 2 4 8 1 5 6 7 9

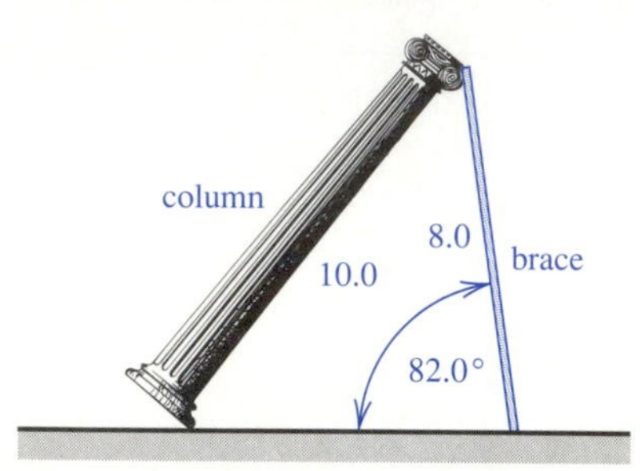

3. Point B is 253 km due south of point A. Point C is 117 km from A and 207 km from B. Find the angle (from the north) of a plane flying from B to C. 1 2 4 7 8 3 5

4. In surveying, latitudes and departures are used to calculate the accuracy of certain measurements. The latitude of a line is the product of its length and the cosine of its bearing. The departure of a line is the product of its length and the sine of its bearing. Find the latitude and departure for a track that is N 30.0°E and is 100 m long. 1 2 4 7 8 3 5

5. In a circuit that contains a resistor and an inductor in series, the sum of the voltages across the resistor and inductor is equal to the applied voltage when these voltages are assigned vector values $\mathbf{V_R}$, $\mathbf{V_L}$, and $\mathbf{V}$, respectively. If $\mathbf{V_R}$ has magnitude 104 and direction $-30.0°$ and $\mathbf{V_L}$ has magnitude 60.0 and direction 60.0°, find $\mathbf{V}$. 3 5 9 1 2 4 6 7 8

6. Consider a vector of magnitude $F = 975$ with its tail at $(-38.5, 55.0)$ as shown in the accompanying diagram.
   (a) Find the coordinates of point B.
   (b) Find the length of the vector that has its tail at the origin and its head at B.
   (c) Find the distance d (in mm).
   (d) Determine the moment M about the origin using the formula $M = Fd$.

   In the formula M is the moment (in N·m) of a coplanar force of magnitude F (in N), and d is in meters. 2 4 8 1 5 6 7 9

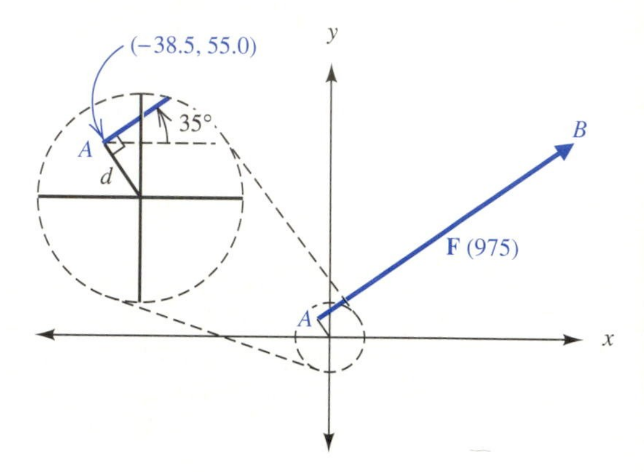

7. Two observers traveling down a highway 5000 ft apart spot a UFO in the morning sky directly ahead. The car farthest east estimates the craft to be 60° above the horizon. The west most car estimates the craft at 40°. Based on these observations determine how high the UFO is. Give your answer to two significant digits. 1 2 4 7 8 3 5

8. An observer in a traffic helicopter flying at 1500 ft observes an accident on the roadway ahead of him. His gaze must be cast down approximately 75° from the horizon to see it. Glancing up he spots another accident farther ahead. This time he must look down about 45°. How far apart are the accidents spaced? 1 2 4 7 8 3 5

9. An ac voltage is a vector quantity having both magnitude and angle associated with it. What ac voltage must be added to 35 volts at an angle of 90° to obtain a resultant voltage that has the same magnitude, but at an angle of zero degrees? 3 5 9 1 2 4 6 7 8

1. Architectural Technology
2. Civil Engineering Technology
3. Computer Engineering Technology
4. Mechanical Drafting Technology
5. Electrical/Electronics Engineering Technology
6. Chemical Engineering Technology
7. Industrial Engineering Technology
8. Mechanical Engineering Technology
9. Biomedical Engineering Technology

10. A hunter decides to walk around his camp. He walks 5 miles due north, then turns left 60° from north and walks 11 miles. Finally he turns 135° more to the left and walks 13 miles. In what direction should he now turn and how far should he walk to return to camp? 1 2 4 7 8 3 5

11. Determine the tension T (in lb) that would be placed on the 48.00-inch long string of a 50 lb pull hunting bow when the 28.00-inch arrow is pulled back 22.00 inches from the no tension position as shown in the accompanying figure. The string rests 6.00 inches from the bow with no pull.
1 2 4 8 7 9

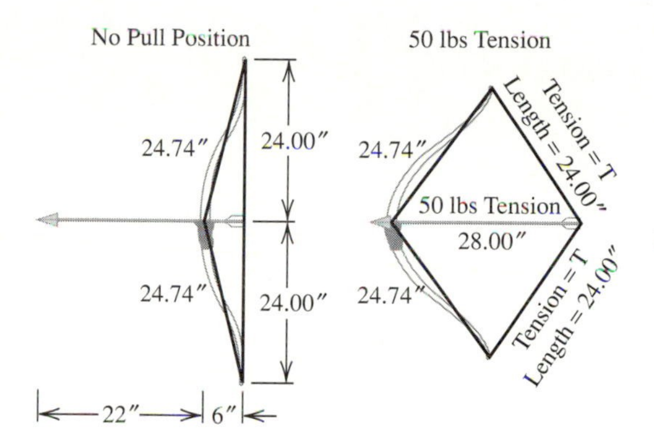

12. A car is attached to a tree with a rope 35 ft long. A man pulls the rope 2.0 inches to one side at its center with a force of 5.0 lb. How much force does this pull exert on the car? 4 7 8 1 2 3 5 6 9

13. The flight path of an airplane traveling at 390 mph crosses that of another traveling at 760 mph. The angle between their paths is 49°. How far apart will the planes be after one hour? Neglect the earth's curvature. 4 7 8 1 2 3 5 6 9

14. A cartographer notices city A on a map lies exactly 65 km east of the capital C while city B is positioned as shown in the accompanying figure. At what angle would a plane fly from the capital to pass over city B? 1 2 4 7 8 3 5

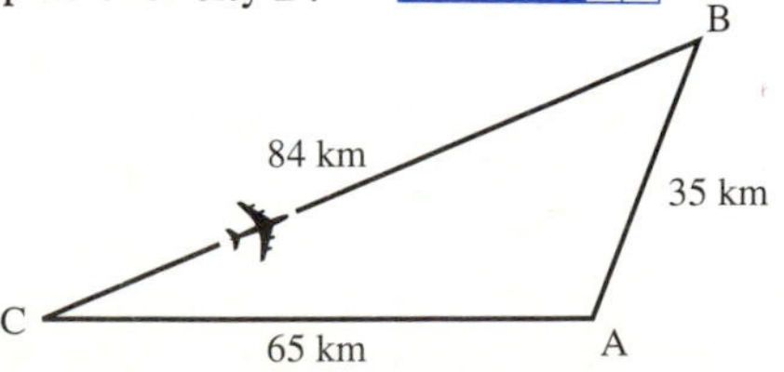

15. The Great Pyramid of Egypt has a present height of 458 ft from its base to the small flat surface at its top. This is not its original height, however, as the top is missing. The base of the pyramid is 761 ft long and the sides make a 51.83° angle with the ground. How tall was the missing piece?
1 2 4 7 8 3 5

16. An ac voltage is a vector with both magnitude and direction. When one positive ac voltage is added to another voltage twice as large, the resultant voltage is 100 volts making an angle of 35° with the first voltage. Find the size of the first voltage. Give your answer to three significant digits.
3 5 9 1 2 4 6 7 8

17. A parachutist jumps from a height of 10,000 ft. She falls vertically at 13 mph and drifts with a horizontal wind of 10 mph.

(a) How long will it take for her to hit the ground?

(b) What horizontal distance will she travel in this time? 4 7 8 1 2 3 5 6 9

18. The momentum P of a moving object of mass m and velocity v is a vector with magnitude mv and with an angle equal to its velocity's direction angle. If such an object should break apart, the vector sum of the momenta of the pieces will always equal the original momentum. A 15 kg mass moving east at 35 m/s breaks into two smaller pieces. A 5 kg piece moves away at 120 m/s in a direction 58° north of east. In what direction and how fast will the other piece move?
4 7 8 1 2 3 5 6 9

19. A 22,000 lb truck traveling north at 25 mph collides with a 3000 lb car traveling due east at 80 mph. If they lock together at the instant of collision, in what direction and at what speed is the wreckage traveling immediately after collision? Use the fact that the momentum (mv) of the car added to that of the truck must equal the vector representing the momentum of the wreckage.
4 7 8 1 2 3 5 6 9

RANDOM REVIEW

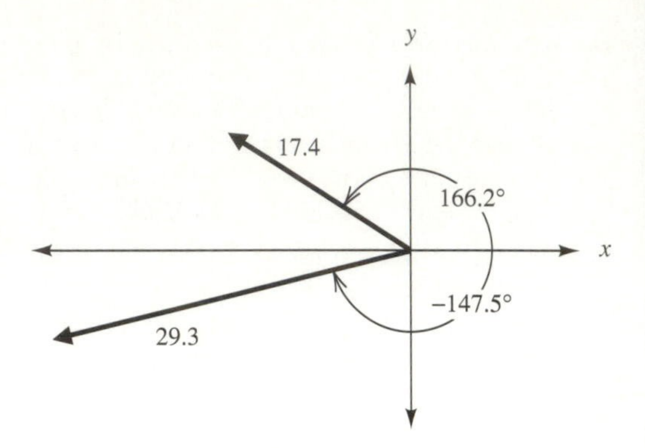

PROBLEM 3

1. Determine the magnitude and direction of $\langle 4, -7\rangle$.
2. Find C if $a = 4.6$, $b = 5.2$, and $c = 7.5$.
3. Add the vectors shown in the accompanying diagram. Specify the magnitude and direction of the resultant.
4. Solve the following triangle: $a = 4.9$, $b = 7.6$, and $B = 32.7°$.
5. Find b if $A = 42.4°$, $B = 29.3°$, and $a = 32.7$.

6. Find B if $a = 4.7$, $b = 7.1$, and $A = 126°$.
7. Find B if $a = 4$, $b = 4$, and $C = 45°$.
8. Resolve the following vector into its horizontal and vertical components: $|\mathbf{v}| = 84.1$, $\theta = -99.3°$.
9. Use the parallelogram method to add the following vectors. Specify the magnitude and direction of the resultant.

$$\mathbf{a} = \langle 7, 9\rangle,\ \mathbf{b} = \langle 8, -1\rangle$$

10. Find c if $A = 106.2°$, $B = 7.6°$, and $b = 18.3$.

CHAPTER TEST

1. Determine the magnitude and direction of $\langle 6, -8\rangle$.
2. Use the polygon method to add the following vectors. Specify the magnitude and direction of the resultant.

$$\mathbf{a} = \langle 8, 6\rangle,\ \mathbf{b} = \langle 0, 9\rangle$$

3. Add the following vectors by adding their horizontal and vertical components. Specify the magnitude and direction of the resultant. $|\mathbf{a}| = 4.11$, $\theta_a = 23.2°$, $|\mathbf{b}| = 5.23$, $\theta_b = 154.7°$
4. Find B if $a = 4.0$, $b = 5.6$, and $c = 6.1$.
5. Find c if $A = 51.5°$, $C = 30.4°$, and $a = 21.8$.
6. Find b if $a = 48.3$, $c = 63.7$, and $B = 33.7°$.
7. Find B if $a = 5.6$, $c = 6.2$, and $C = 133°$.
8. Solve the following triangle: $a = 77.9$, $b = 59.3$, and $B = 32.1°$.
9. A draftsman is to construct an isosceles triangle as shown in the accompanying diagram. Determine the angles of the triangle.
10. In a three-phase ac circuit, voltages are generated 120° apart. If phase A is at 30°, phase B is at 150°, and phase C is at 270°, the potential difference between A and B is given by the vector $\mathbf{A} - \mathbf{B}$ as shown in the accompanying diagram. Find the magnitude of this vector if $\mathbf{A}$ and $\mathbf{B}$ each have a magnitude of 120 volts.

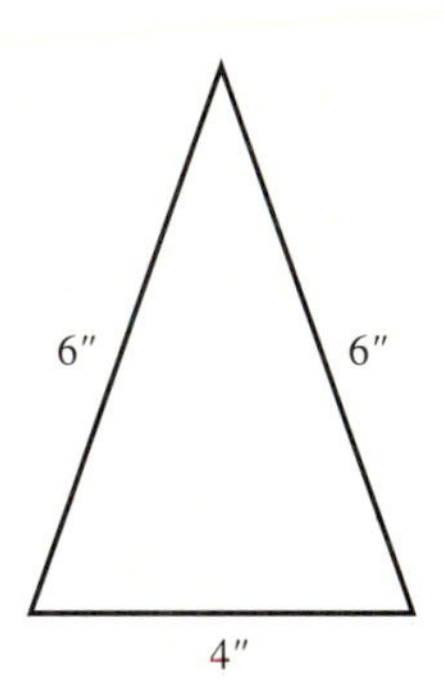

PROBLEM 9

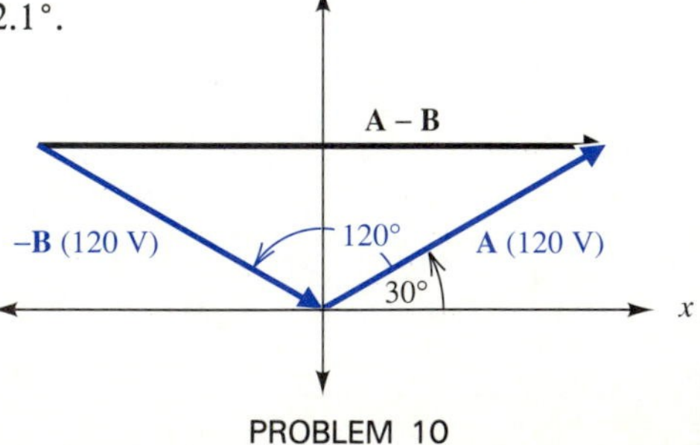

PROBLEM 10

EXTENDED APPLICATION: CIVIL ENGINEERING TECHNOLOGY

BANKING A CURVE OF A HIGHWAY

When an object moves along a circular path, the velocity continually changes in direction, even though its magnitude may remain constant. The inward force that must be applied to keep the object moving in a circle is called **centripetal force** and is given by

$$F = mv^2/r.$$

In the formula, m is the mass of the object, v is the magnitude of its velocity, and r is the radius of the circular path.

When a car rounds a curve on a level road, the force providing the centripetal acceleration is the force of friction between the wheels and the ground. Frictional force is given by

$$F = \mu mg,$$

where μ is the coefficient of friction, m is the mass of the object, and g is the acceleration due to gravity (32 ft/s^2).

Consider a car traveling at 60 mph (88 ft/s) on a wet concrete road. The coefficient of friction between the tires and the road would be about 0.4. The minimum turning radius is calculated as follows.

$$\begin{aligned} \text{Centripetal force} &= \text{frictional force} \\ mv^2/r &= \mu mg \\ v^2 &= \mu gr \\ \frac{v^2}{\mu g} &= r \\ \frac{88^2}{(0.4)(32)} &= r \\ 605 \text{ ft} &= r \end{aligned}$$

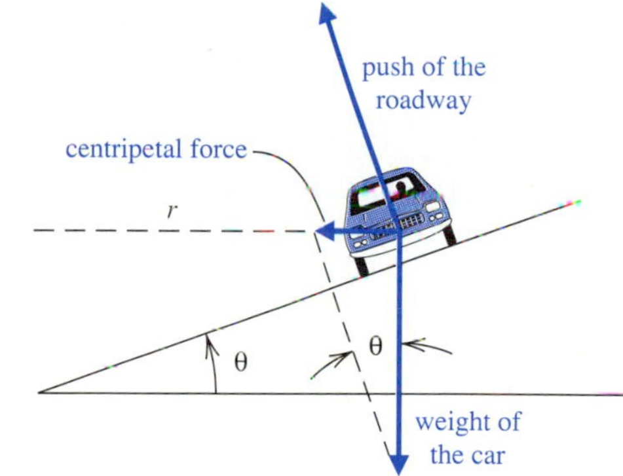

FIGURE 10.41

To eliminate the horizontal force of friction, highway curves are usually banked at an angle. For a car traveling at a specified velocity, the necessary centripetal force is supplied by the horizontal component of the force of the road acting against the wheels of the car. Figure 10.41 illustrates the forces involved.

To find the angle at which the road must be banked for a curve that is 605 feet in radius, notice that

$$\begin{aligned} \tan \theta &= \frac{mv^2/r}{mg} = \frac{v^2}{rg} \\ &= \frac{88^2}{605(32)} = 0.4 \end{aligned}$$

Thus, the road must be banked at an angle of 21.8°.

CHAPTER 11 GRAPHS OF THE TRIGONOMETRIC FUNCTIONS

Trigonometric functions are sometimes used to describe the behavior of scientific and technological occurrences that have a cyclic nature. Electronics, oceanography, and acoustics are a few of the fields in which cyclic phenomena are important. In this chapter, we examine the graphs of the trigonometric functions not only to see the applications, but also to gain an understanding of the properties of the functions.

11.1 AMPLITUDE AND PERIOD FOR SINE AND COSINE FUNCTIONS

In Chapter 4, we learned that when angles are specified in radian measure, the sine and cosine functions may be treated as having the real numbers as the domain. If the notation $y = \sin x$ is used, it is natural to graph the function on a rectangular coordinate system. It is necessary to plot enough points to get a clear picture of the graph, including all of its bends. Since the function values are known for many of the special angles (and hence the corresponding real numbers), it is convenient to use fractions of π for the x-values.

x	0	$\pi/6$	$\pi/4$	$\pi/3$	$\pi/2$	$2\pi/3$	$3\pi/4$	$5\pi/6$
y (exact)	0	1/2	$\sqrt{2}/2$	$\sqrt{3}/2$	1	$\sqrt{3}/2$	$\sqrt{2}/2$	1/2
y (decimal)	0	.500	.707	.866	1.00	.866	.707	.500

x	π	$7\pi/6$	$5\pi/4$	$4\pi/3$	$3\pi/2$	$5\pi/3$	$7\pi/4$	$11\pi/6$	2π
y (exact)	0	$-1/2$	$-\sqrt{2}/2$	$-\sqrt{3}/2$	-1	$-\sqrt{3}/2$	$-\sqrt{2}/2$	$-1/2$	0
y (decimal)	0	$-.500$	$-.707$	$-.866$	-1.00	$-.866$	$-.707$	$-.500$	0

When the points are plotted on a coordinate plane, it is easy to approximate fractions of π by observing that π is about 3, so $\pi/3$ is about 1, $\pi/6$ is about 1/2, and so on. The x-axis and y-axis are not required to have the same scale, but if it is convenient to make them the same, we will. Figure 11.1 shows the graph of $y = \sin x$.

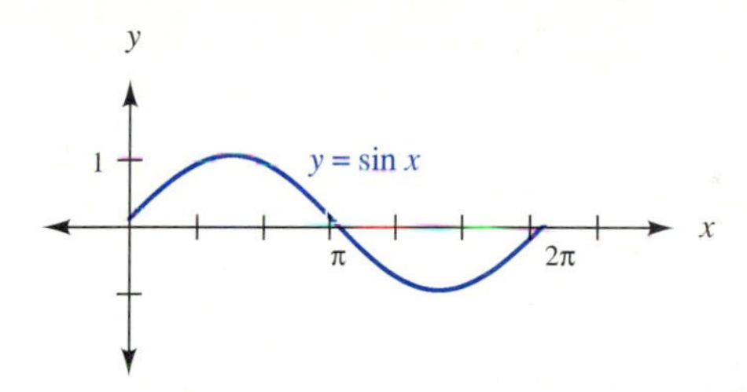

FIGURE 11.1

If values larger than 2π are used for x, the same y-values that occur for x-values between 0 and 2π will be repeated. Because of this repeating pattern of y-values, the function is said to be *periodic* or to have a **period** of 2π. In technical applications, periodic functions usually have time as the independent variable, so the period is a period of time. Mathematicians use the word periodic to describe *any* function in which the values of the dependent variable are repetitious. Since a continuation of the graph would simply be a repetition, periodic functions are often graphed through one period only. The graph through one period of a periodic function is called one *cycle*. Notice that for $y = \sin x$, the y-values are all between -1 and 1, inclusive.

periodic function
period
cycle

EXAMPLE 1 Sketch the graph of $y = 2 \sin x$.

Solution Using the same values for x that were used in $y = \sin x$, we have the following table of values.

x	0	$\pi/6$	$\pi/4$	$\pi/3$	$\pi/2$	$2\pi/3$	$3\pi/4$	$5\pi/6$
y (exact)	0	1	$\sqrt{2}$	$\sqrt{3}$	2	$\sqrt{3}$	$\sqrt{2}$	1
y (decimal)	0	1.0	1.4	1.7	2.0	1.7	1.4	1.0

x	π	$7\pi/6$	$5\pi/4$	$4\pi/3$	$3\pi/2$	$5\pi/3$	$7\pi/4$	$11\pi/6$	2π
y (exact)	0	-1	$-\sqrt{2}$	$-\sqrt{3}$	-2	$-\sqrt{3}$	$-\sqrt{2}$	-1	0
y (decimal)	0	-1.0	-1.4	-1.7	-2.0	-1.7	-1.4	-1.0	0

The graph in Figure 11.2 shows that the graph of the function looks very much like the graph of $y = \sin x$. ■

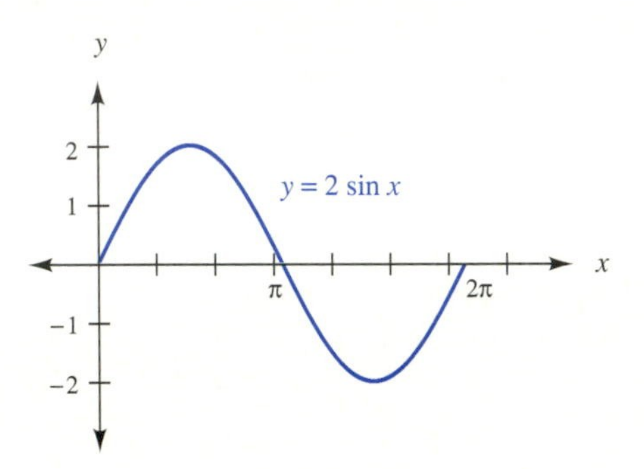

FIGURE 11.2

The function in Example 1 has period 2π. For this function, however, each y-value is twice the value of $\sin x$. Thus, the y-values are all between -2 and 2, inclusive. This observation can be generalized.

DEFINITION On the curve $y = a \sin x$, the highest point has a y-value of $|a|$ and the lowest point has a y-value of $-|a|$. The value $|a|$ is called the **amplitude** of the curve.

EXAMPLE 2 Sketch the graph of $y = (1/2) \sin x$.

Solution Since the period is 2π, and the shape is like that in Figures 11.1 and 11.2, the amplitude of 1/2 is used to locate the highest point on the curve $(\pi/2, 1/2)$ and the lowest point at $(3\pi/2, -1/2)$. The curve crosses the x-axis at $(0, 0)$, $(\pi, 0)$, and $(2\pi, 0)$. Thus, it is possible to sketch the curve without plotting lots of points. Figure 11.3 shows the graph. ■

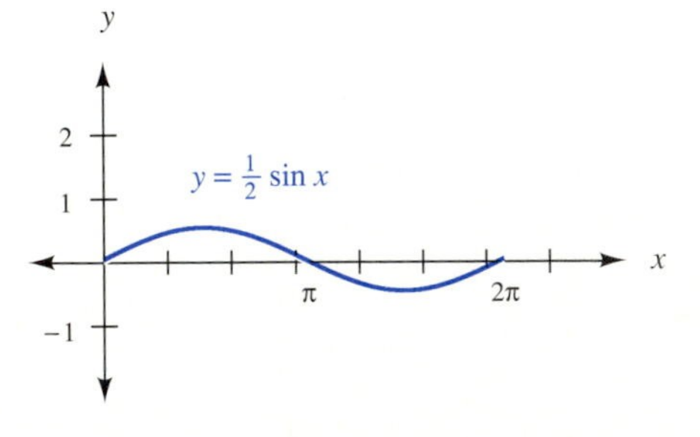

FIGURE 11.3

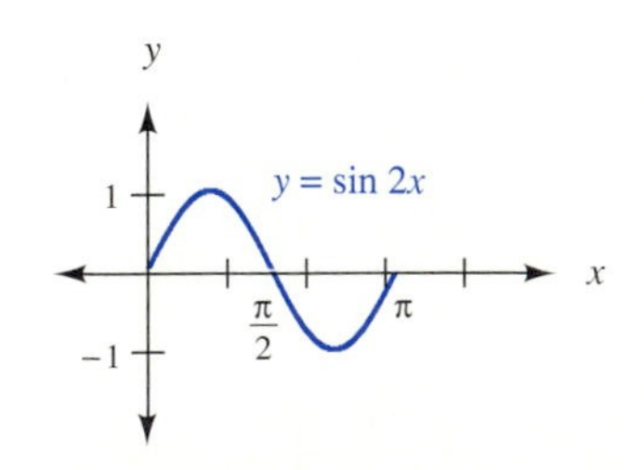

FIGURE 11.4

EXAMPLE 3 Sketch the graph of $y = \sin 2x$.

Solution The equation does not fit the pattern $y = a \sin x$, so we must plot points.

x	0	$\pi/6$	$\pi/4$	$\pi/3$	$\pi/2$	$2\pi/3$	$3\pi/4$	$5\pi/6$	π
y (exact)	0	$\sqrt{3}/2$	1	$\sqrt{3}/2$	0	$-\sqrt{3}/2$	-1	$-\sqrt{3}/2$	0
y (decimal)	0	.866	1.0	.866	0	$-.866$	-1.0	$-.866$	0

Figure 11.4 shows the graph. ■

Notice that as x increases from 0 to π, $2x$ increases from 0 to 2π. Therefore, as x increases from *0 to* π, $\sin 2x$ takes on the same y-values that $\sin x$ takes on between *0 and* 2π. That is, the graph of $y = \sin 2x$ completes one cycle over half the x-distance that $y = \sin x$ does. This observation can be generalized. If b is the coefficient of x, and b is positive, the period of the function $y = \sin bx$ is $2\pi/b$. The period and amplitude of a sine function are used to sketch the graph of the function.

To sketch the graph of $y = a \sin bx$ when $b > 0$:

1. Identify the period: $2\pi/b$.
2. Identify the amplitude: $|a|$.
3. Divide the period into four equal segments on the x-axis to locate the quarter points.

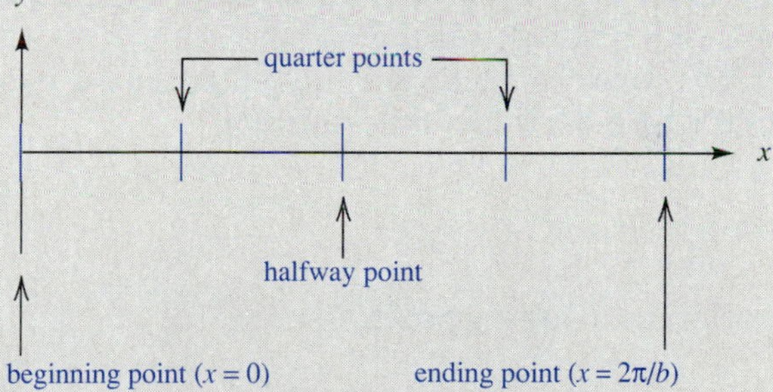

(a) At the beginning point, halfway point, and ending point, the curve crosses the x-axis.

(b) At the one-quarter point, the y-value is a.
At the three-quarter point, the y-value is $-a$.

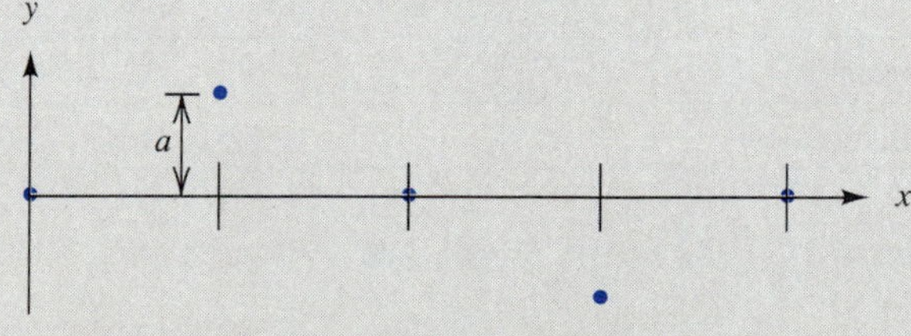

4. Connect the points with a curve that has the shape of $y = \sin x$.

EXAMPLE 4 Sketch the graph of $y = -2\sin[(1/2)x]$.

Solution

1. The period is $2\pi/(1/2)$ or 4π.
2. The amplitude is $|-2|$ or 2.
3. (a) The curve crosses the x-axis at 0, 2π, and 4π.

   (b) When x is π, the y-value is -2. When x is 3π, the y-value is $-(-2)$ or 2.
4. Figure 11.5 shows the graph of the function. ■

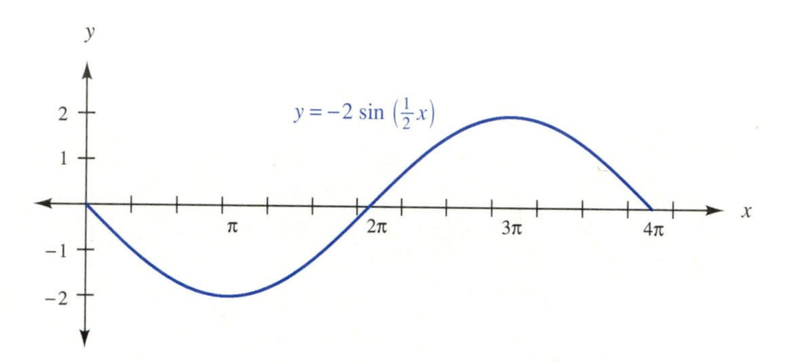

FIGURE 11.5

Notice that the effect of the negative sign in Example 4 was to "flip" the curve $y = 2\sin[(1/2)x]$ upside down.

The cosine curve is similar to the sine curve. It, too, has a period and amplitude. But the basic curve is a little different.

EXAMPLE 5 Sketch the graph of $y = \cos x$.

Solution We plot the following points.

x	0	$\pi/6$	$\pi/4$	$\pi/3$	$\pi/2$	$2\pi/3$	$3\pi/4$	$5\pi/6$
y (exact)	1	$\sqrt{3}/2$	$\sqrt{2}/2$	$1/2$	0	$-1/2$	$-\sqrt{2}/2$	$-\sqrt{3}/2$
y (decimal)	1	.866	.707	.500	0	$-.500$	$-.707$	$-.866$

x	π	$7\pi/6$	$5\pi/4$	$4\pi/3$	$3\pi/2$	$5\pi/3$	$7\pi/4$	$11\pi/6$	2π
y (exact)	-1	$-\sqrt{3}/2$	$-\sqrt{2}/2$	$-1/2$	0	$1/2$	$\sqrt{2}/2$	$\sqrt{3}/2$	1
y (decimal)	-1	$-.866$	$-.707$	$-.500$	0	.500	.707	.866	1

Figure 11.6 shows the graph. ■

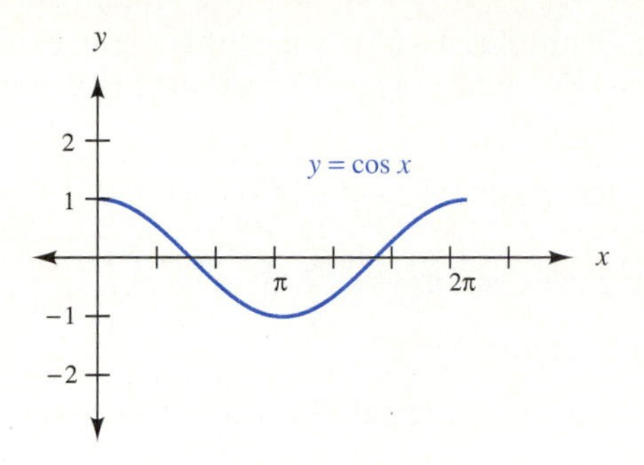

FIGURE 11.6

Generalizing, we have the following procedure.

To sketch the graph of $y = a \cos bx$ when $b > 0$:

1. Identify the period: $2\pi/b$.
2. Identify the amplitude: $|a|$.
3. Divide the period into four equal segments on the x-axis to locate the quarter points.

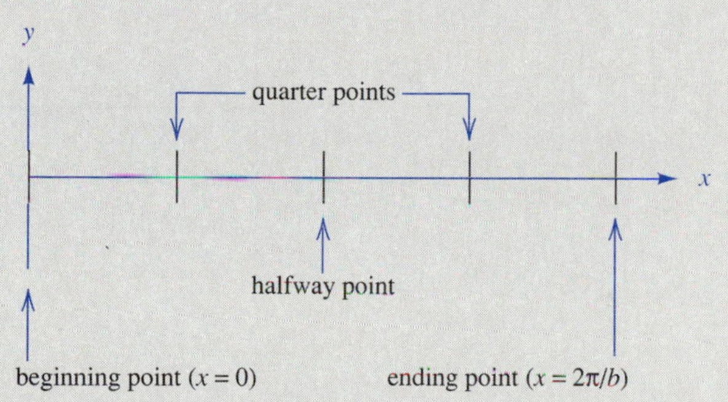

   (a) At the one-quarter point and three-quarter point, the curve crosses the x-axis.
   (b) At the beginning point, the y-value is a.
   At the halfway point, the y-value is $-a$.
   At the ending point, the y-value is a.

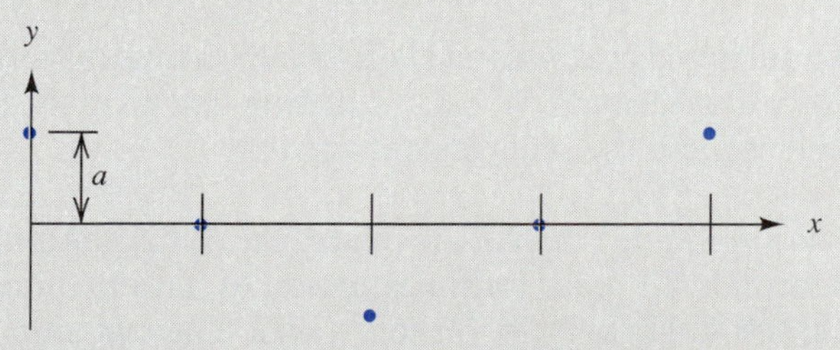

4. Connect the points with a curve that has the shape of $y = \cos x$.

In Examples 1–5, it was convenient to represent the period as a multiple or fraction of π. Example 6 shows that this is not always the case.

EXAMPLE 6 Sketch the graph of $y = (-1/3) \cos 4\pi x$.

Solution

1. The period is $2\pi/4\pi$ or $1/2$.
2. The amplitude is $1/3$.
3. (a) The curve crosses the x-axis at $(1/4)(1/2)$ or $1/8$ and at $(3/4)(1/2)$ or $3/8$.
   (b) Points on the graph include $(0, -1/3)$, $(1/4, 1/3)$, and $(1/2, -1/3)$.
4. Figure 11.7 shows the graph. ■

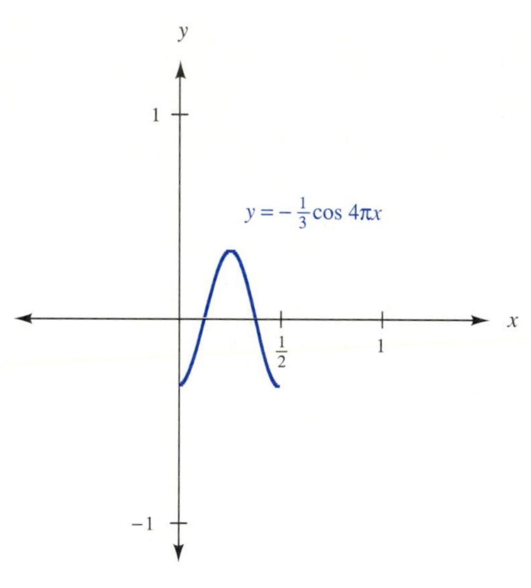

FIGURE 11.7

Periodic phenomena occur often in nature, science, and the technologies.

EXAMPLE 7 An anchored boat rises and falls through a total of 1.5 m, returning to the rest position every four seconds. Write an equation that gives the displacement as a sine function when time is the independent variable.

Solution The amplitude is half of the total range. Thus, the amplitude is 0.75 m. The period is 4 seconds. In order for the equation to have the form $y = a \sin(bx)$, we must have $2\pi/b = 4$. Thus, $2\pi = 4b$, so $b = \pi/2$. The equation is $d = 0.75 \sin(\pi x/2)$. ■

EXERCISES 11.1

In 1–21, identify the period and amplitude of each function.

1. $y = 3 \sin x$
2. $y = \frac{1}{2} \cos x$
3. $y = \frac{2}{3} \cos x$
4. $y = -3 \sin x$
5. $y = \frac{-2}{3} \sin 2x$
6. $y = \frac{-3}{2} \cos 2x$
7. $y = 2 \cos \frac{1}{3} x$
8. $y = 3 \sin \frac{1}{2} x$
9. $y = \frac{3}{2} \sin 4x$
10. $y = \frac{2}{3} \cos 4x$
11. $y = \cos 3x$
12. $y = \sin 3x$
13. $y = -\sin \frac{1}{2} x$
14. $y = -\cos \frac{1}{3} x$
15. $y = \frac{1}{2} \cos \frac{2}{3} x$
16. $y = \sin 2\pi x$
17. $y = -\cos 2\pi x$
18. $y = 2 \cos 4\pi x$
19. $y = 2 \sin \frac{1}{2} \pi x$
20. $y = \frac{1}{2} \sin \pi x$
21. $y = \frac{-1}{2} \cos \pi x$

In 22–42, sketch the graph of each function.

22. $y = \frac{1}{3} \cos x$
23. $y = 2 \sin x$
24. $y = \frac{3}{2} \sin x$
25. $y = \frac{-1}{3} \cos x$
26. $y = \frac{3}{2} \cos 2x$
27. $y = \frac{2}{3} \sin 2x$
28. $y = \frac{1}{2} \sin \frac{1}{3} x$
29. $y = \frac{1}{3} \cos \frac{1}{2} x$
30. $y = 2 \cos 4x$
31. $y = 3 \sin 4x$
32. $y = \sin 3x$
33. $y = \cos 3x$
34. $y = -\cos \frac{1}{2} x$
35. $y = -\sin \frac{1}{3} x$
36. $y = 2 \sin \frac{2}{3} x$
37. $y = \cos 2\pi x$
38. $y = -\sin 2\pi x$
39. $y = 2 \sin 4\pi x$
40. $y = 2 \cos \frac{1}{2} \pi x$
41. $y = \frac{1}{2} \cos \pi x$
42. $y = \frac{-1}{2} \sin \pi x$

---

43. In Great Britain, the nautical mile is defined as the length of a minute of arc of a meridian. Since the earth is flattened at the poles, a British nautical mile varies from the figure 6077 with latitude. If x is the latitude converted from degrees to radians, a good approximation for the variation y (in feet) is given by $y = -31 \cos(2x)$.

(a) Identify the period and amplitude of the function.

(b) Determine the maximum and minimum length of a nautical mile. 1 2 4 7 8 3 5

44. On March 21, the number of hours of daylight at a particular location is the same as the average num-

---

1. Architectural Technology
2. Civil Engineering Technology
3. Computer Engineering Technology
4. Mechanical Drafting Technology
5. Electrical/Electronics Engineering Technology
6. Chemical Engineering Technology
7. Industrial Engineering Technology
8. Mechanical Engineering Technology
9. Biomedical Engineering Technology

ber of hours of daylight per day. The seasonal variation from this average value may be approximated by a sine curve. In New Orleans, the average is about 11 2/3 hours and the variation is given by $y = 7/3 \sin(2\pi x/365)$ where x is the number of days after March 21.

(a) Identify the period and amplitude of this function.

(b) Determine the maximum and minimum number of hours of daylight. 9 1 2 3 5 6 7

45. The average yearly temperature at a particular location normally occurs on or about March 21. The seasonal variation from this average may be approximated by a sine curve. In Charleston, SC, the average temperature is about 66°F and the variation is given by $y = 26 \sin(2\pi x/365)$ where x is the number of days after March 21.

(a) Identify the period and amplitude of this function.

(b) Determine the maximum and minimum temperatures. 6 8 1 2 3 5 7 9

46. In charting biorhythms, a sine curve is used to indicate a physical cycle (strength and energy) of length 23 days, an emotional cycle (moodiness and creativity) of 28 days, and an intellectual cycle (alertness and memory) of 33 days. All cycles begin at birth. Write an equation for each curve using the number of days since birth as the independent variable. 6 9 2 4 5 7 8

47. Tides go up and down in a 12.4-hour period of time. The average depth of a certain river is 8 m and the depth ranges from 5 m to 11 m. The variation from the average may be approximated by a sine curve. Write an equation that gives the approximate variation y, if x is the number of hours after the average depth is recorded. 4 8 1 2 6 7 9

48. An airplane's electrical generator produces a time-varying output voltage (in volts) described by the equation $V = 120 \sin(2513t)$, where t is the time (in seconds).

(a) Identify the amplitude in volts.

(b) Identify the period in milliseconds.

(c) The frequency (in hertz) is the quotient of 1 and the period in seconds. Find the frequency. 3 5 9 1 2 4 6 7 8

49. The current (in amperes) in a particular 20 Hz ac circuit is given by $I = 6 \sin(40\pi t)$, where t is the time (in seconds). Identify the period and amplitude of this function. 3 5 9 1 2 4 6 7 8

50. Consider the equation $y = d + a \sin(bx)$.

(a) What effect does the value of d have on the graph?

(b) Use the data in problem 45 to write an equation that gives the approximate temperature for a date x days after March 21 (rather than the variation above or below average).

1. Architectural Technology
2. Civil Engineering Technology
3. Computer Engineering Technology
4. Mechanical Drafting Technology
5. Electrical/Electronics Engineering Technology
6. Chemical Engineering Technology
7. Industrial Engineering Technology
8. Mechanical Engineering Technology
9. Biomedical Engineering Technology

11.2 PHASE SHIFT FOR SINE AND COSINE FUNCTIONS

phase angle

In the equation $y = a \sin(bx + c)$, the number c is called the **phase angle,** and its effect is to shift the graph of $y = a \sin bx$ horizontally. The quantity $bx + c$ must range from 0 to 2π in order for $\sin(bx + c)$ to take on the same values that $\sin bx$ takes on between 0 and 2π. Since $bx + c = 0$ when x is $-c/b$, and $bx + c = 2\pi$ when x is $(2\pi - c)/b$ or $2\pi/b - c/b$, one cycle of the graph of $y = \sin(bx + c)$ begins c/b units from 0 and ends c/b units from $2\pi/b$. When b is positive, the quantity $-c/b$ is called the **phase shift** of the function. If the phase shift is positive, the graph of $y = a \sin bx$ is shifted to the right; if the phase shift is negative, the graph of $y = a \sin bx$ is shifted to the left. The following example is done by plotting points to verify the effect of c.

phase shift

EXAMPLE 1 Sketch the graph of $y = \sin(2x + \pi/3)$.

Solution The special angles between 0 and π are used for the graph.

x	0	$\pi/6$	$\pi/4$	$\pi/3$	$\pi/2$	$2\pi/3$	$3\pi/4$	$5\pi/6$	π
y (exact)	$\sqrt{3}/2$	$\sqrt{3}/2$	1/2	0	$-\sqrt{3}/2$	$-\sqrt{3}/2$	$-1/2$	0	$\sqrt{3}/2$
y (decimal)	.866	.866	.500	0	−.866	−.866	−.500	0	.866

Figure 11.8 shows the graph. ■

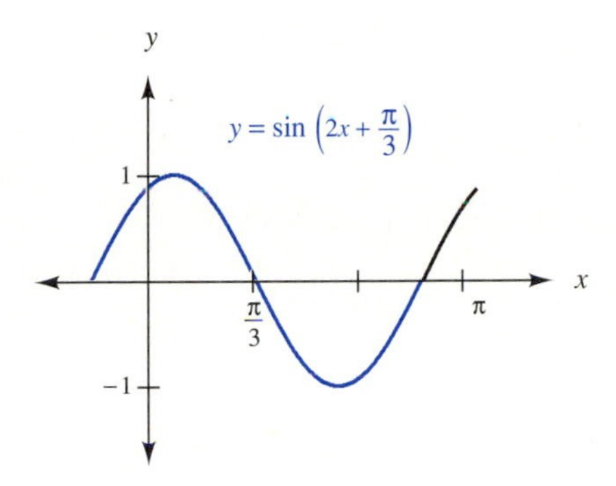

FIGURE 11.8

Since the curve can be extended to both the left and the right, the portion between x-values of $-\pi/6$ and $5\pi/6$ looks like the graph of $y = \sin 2x$ between 0 and 2π. That is, the graph of $y = \sin(2x + \pi/3)$ looks like the graph of $y = \sin 2x$ shifted by $\pi/6$ units to the left.

We make the following generalization.

To sketch the graph of $y = a \sin(bx + c)$ or $y = a \cos(bx + c)$ when $b > 0$:

1. Identify the period: $2\pi/b$.
2. Identify the amplitude: $|a|$.
3. Identify the phase shift: $-c/b$.
4. Locate the beginning point at $x = -c/b$.
   Locate the ending point at $x = 2\pi/b - c/b$.
5. Use the one-quarter, halfway, and three-quarter points to sketch the sine or cosine function with the appropriate amplitude.
6. Extend the curve to the left or right as necessary to show one full cycle starting with an x-value of 0.

EXAMPLE 2 Sketch the graph of $y = \frac{1}{2}\cos\left(2x - \frac{\pi}{2}\right)$.

Solution

1. The period is $\frac{2\pi}{2}$ or π.
2. The amplitude is $\frac{1}{2}$.
3. The phase shift is $-\left(\frac{-\pi}{2}\right) \div 2$ or $\frac{\pi}{4}$.
4. The beginning point is shifted $\frac{\pi}{4}$ units to the right of 0.

   The ending point is shifted $\frac{\pi}{4}$ units to the right of π.
5. Since the graph is a cosine curve, the y-value is $1/2$ ($a = 1/2$) at the beginning and ending points. Thus $(\pi/4, 1/2)$ and $(5\pi/4, 1/2)$ are on the graph. The y-value is $-1/2$ ($-a = -1/2$) at the halfway point. Thus $(3\pi/4, -1/2)$ is on the graph. The graph crosses the x-axis at the one-quarter point ($x = \pi/2$) and the three-quarter point ($x = \pi$). Figure 11.9 shows the graph. ■

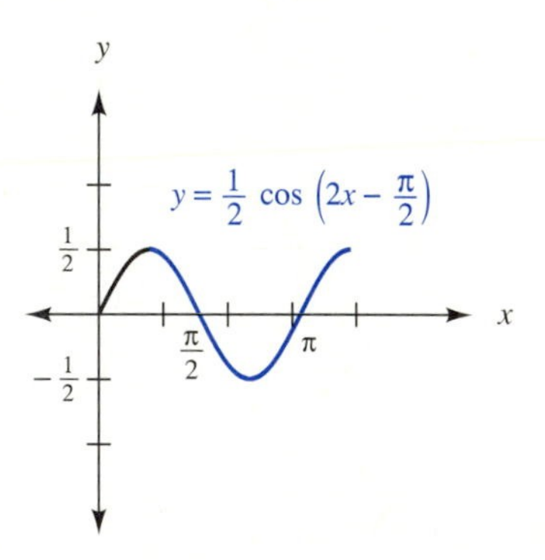

FIGURE 11.9

In applied problems, the term frequency is often used. If the period of a function is τ, its frequency is $1/\tau$. When the period is measured in seconds per cycle, the frequency is measured in cycles per second or hertz.

EXAMPLE 3 The current I (in amperes) in a certain 30 Hz ac circuit is given by $I = 12 \sin(60\pi t + \pi)$. Sketch the graph of this function.

Solution

1. The period is $\frac{2\pi}{60\pi}$ or $\frac{1}{30}$ second.
2. The amplitude is 12 amperes.

3. The phase shift is $-\left(\dfrac{\pi}{60\pi}\right)$ or $\dfrac{-1}{60}$ second.

4. The beginning point is at $t = \dfrac{-1}{60}$ second.

   The ending point is at $t = \dfrac{1}{30} - \dfrac{1}{60}$ or $\dfrac{1}{60}$ second.

5. Since the graph is a sine curve, the curve crosses the t-axis when t is $\dfrac{-1}{60}$, 0, or $\dfrac{1}{60}$.

   The I-value is 12 when t is $\dfrac{-1}{120}$, and the I-value is -12 when t is $\dfrac{1}{120}$.

6. Figure 11.10 shows the graph. ■

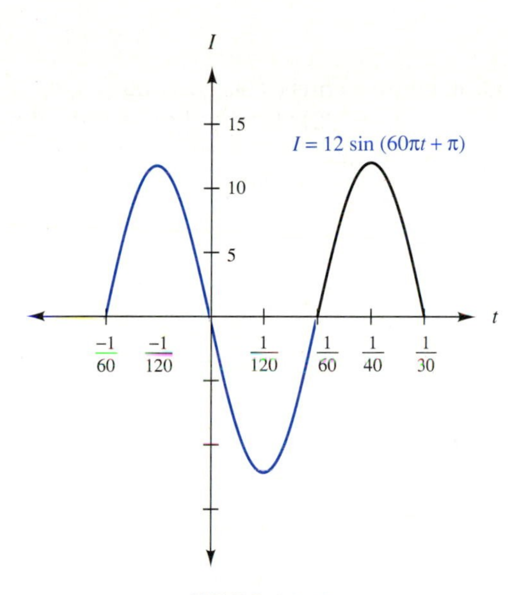

FIGURE 11.10

EXAMPLE 4 Sketch the graph of $y = -3 \cos\left(\dfrac{1x}{3} + \dfrac{\pi}{6}\right)$.

Solution

1. The period is $2\pi \div \dfrac{1}{3}$ or 6π.

2. The amplitude is 3.

3. The phase shift is $\dfrac{-\pi}{6} \div \dfrac{1}{3}$ or $\dfrac{-\pi}{6} \times 3$, which is $\dfrac{-\pi}{2}$.

4. The beginning point is at $x=\frac{-\pi}{2}$.

   The ending point is at $x=6\pi-\frac{\pi}{2}$ or $\frac{11\pi}{2}$.

5. Since the graph is a cosine curve, the y-value is -3, (because $a=-3$) at the beginning and ending points. Thus $(-\pi/2,-3)$ and $(11\pi/2,-3)$ are on the graph. The y-value is 3 (because $-a=-(-3)$) at the halfway point. Thus $(5\pi/2, 3)$ is on the graph. The graph crosses the x-axis at the one-quarter point when x is π and the three-quarter point when x is 4π.

6 Figure 11.11 shows the graph. ■

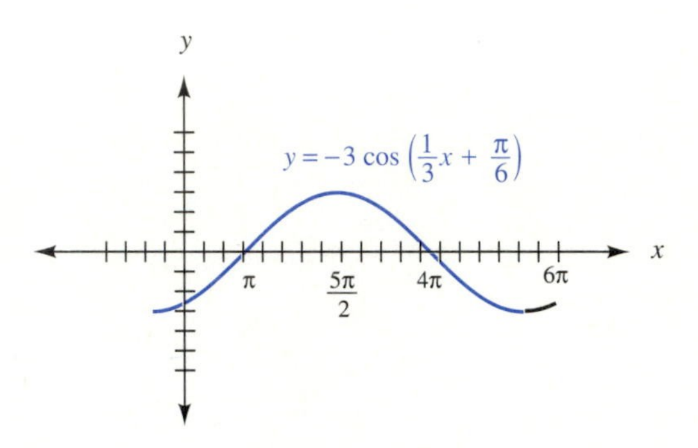

FIGURE 11.11

EXERCISES 11.2

In 1–14, identify the phase shift of each function.

1. $y=3\sin(x-\pi)$

2. $y=\frac{1}{2}\cos\left(x+\frac{\pi}{2}\right)$

3. $y=\frac{2}{3}\cos\left(x+\frac{\pi}{4}\right)$

4. $y=-3\sin(x-\pi)$

5. $y=\frac{-2}{3}\sin\left(2x-\frac{\pi}{2}\right)$

6. $y=\frac{-3}{2}\cos(2x+\pi)$

7. $y=2\cos\left(\frac{1}{3}x+\frac{\pi}{6}\right)$

8. $y=3\sin\left(\frac{1}{2}x-\frac{\pi}{2}\right)$

9. $y=\frac{3}{2}\sin(4x-2\pi)$

10. $y=\frac{2}{3}\cos(4x+2\pi)$

11. $y=\cos(3x+\pi)$

12. $y=\sin(3x-\pi)$

13. $y=-\sin\left(\frac{1}{2}x-\frac{\pi}{4}\right)$

14. $y=-\cos\left(\frac{1}{3}x+\frac{\pi}{6}\right)$

In 15–28, identify the period, amplitude, and phase shift of each function.

15. $y=\frac{1}{2}\cos\left(\frac{2}{3}x+\frac{\pi}{3}\right)$

16. $y=\frac{1}{2}\sin(2\pi x-\pi)$

17. $y=-\cos(2\pi x-\pi)$

18. $y = 2\cos(4\pi x + \pi)$

19. $y = 2\sin\left(\frac{1}{2}\pi x + \frac{\pi}{4}\right)$

20. $y = \frac{1}{2}\sin(\pi x - \pi)$

21. $y = \frac{-1}{2}\cos(\pi x - \pi)$

22. $y = \frac{1}{3}\cos\left(x + \frac{\pi}{2}\right)$

23. $y = 2\sin\left(x + \frac{\pi}{3}\right)$

24. $y = \frac{3}{2}\sin\left(x - \frac{\pi}{6}\right)$

25. $y = \frac{-1}{3}\cos\left(x - \frac{\pi}{4}\right)$

26. $y = \frac{3}{2}\cos\left(2x + \frac{\pi}{6}\right)$

27. $y = \frac{2}{3}\sin\left(2x + \frac{\pi}{6}\right)$

28. $y = \frac{1}{2}\sin\left(\frac{1}{3}x - \frac{\pi}{6}\right)$

In 29–42, sketch the graph of each function.

29. $y = \frac{1}{3}\cos\left(\frac{1}{2}x - \frac{\pi}{4}\right)$

30. $y = 2\cos(4x - \pi)$

31. $y = 3\sin(4x + \pi)$

32. $y = \sin(3x + \pi)$

33. $y = \cos(3x - \pi)$

34. $y = -\cos\left(\frac{1}{2}x - \frac{\pi}{4}\right)$

35. $y = -\sin\left(\frac{1}{3}x + \frac{\pi}{12}\right)$

36. $y = 2\sin\left(\frac{2}{3}x + \frac{\pi}{6}\right)$

37. $y = \cos(2\pi x - \pi)$

38. $y = -\sin(2\pi x - \pi)$

39. $y = 2\sin(4\pi x + \pi)$

40. $y = 2\cos\left(\frac{1}{2}\pi x + \frac{\pi}{4}\right)$

41. $y = \frac{1}{2}\cos\left(\pi x - \frac{\pi}{2}\right)$

42. $y = \frac{-1}{2}\sin\left(\pi x - \frac{\pi}{3}\right)$

---

43. The current (in amperes) in a certain 20 Hz ac circuit is given by $I = 6\sin\left(40\pi t - \frac{\pi}{3}\right)$ where t is the time (in seconds). Identify the period, amplitude, and phase shift of this function.
3 5 9 1 2 4 6 7 8

44. Modify the equation of problem 44 in Exercises 11.1 to begin on January 1, rather than March 21.
9 1 2 3 5 6 7

45. Modify the equation of problem 45 in Exercises 11.1 to begin on January 1, rather than March 21.
6 8 1 2 3 5 7 9

46. One way of producing ac voltage waves is to use a quadrature oscillator. Such an oscillator has two outputs: a sine wave and a cosine wave. What is the phase shift in degrees of the cosine function in relation to the sine function?
3 5 9 1 2 4 6 7 8

47. Rework problem 47 of Exercises 11.1 for the following additional conditions:
(a) x is the number of hours after midnight.
(b) high tide occurs at 8:00 A.M.
4 8 1 2 6 7 9

48. Rework problem 47 of Exercises 11.1 for the following additional conditions:
(a) x is the number of hours after midnight.
(b) high tide occurs at 8:00 A.M.
(c) the equation is a cosine function.
4 8 1 2 6 7 9

49. Explain how to calculate the phase shift for a biorhythm chart (see problem 46 of Exercises 11.1) if the starting point is to be the beginning of the current month, rather than the birth date.
6 9 2 4 5 7 8

50. The graph of the function $y = a\cos(bx)$ looks like the graph of $y = a\sin(bx)$, shifted. Write an equation that uses phase shift to relate the two functions.

---

1. Architectural Technology
2. Civil Engineering Technology
3. Computer Engineering Technology
4. Mechanical Drafting Technology
5. Electrical/Electronics Engineering Technology
6. Chemical Engineering Technology
7. Industrial Engineering Technology
8. Mechanical Engineering Technology
9. Biomedical Engineering Technology

11.3 GRAPHS OF THE OTHER TRIGONOMETRIC FUNCTIONS

If $f(x)$ represents one of the trigonometric functions $\tan x$, $\cot x$, $\sec x$, or $\csc x$, the function $y = af(bx + c)$ can be graphed by considering the effects of a, b, and c on the graph $y = f(x)$. The basic curves can be obtained by plotting the functions for the special angles and connecting points. They are shown in Figures 11.12–11.15.

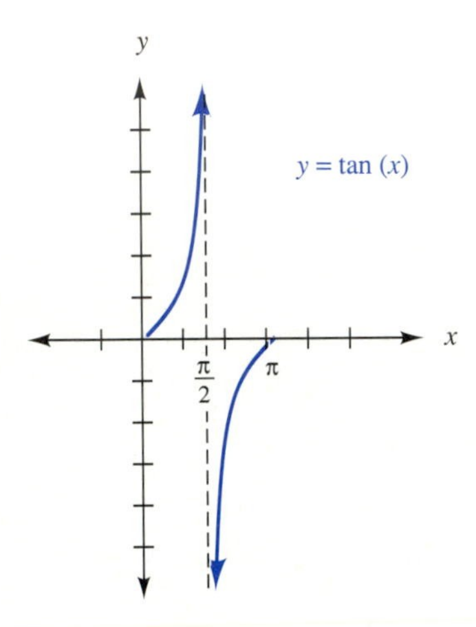

FIGURE 11.12

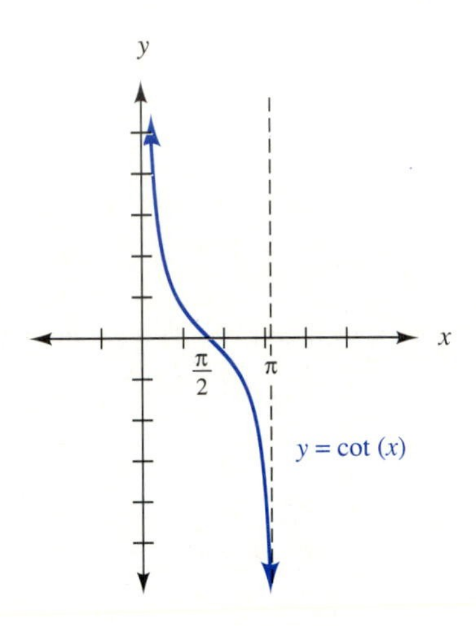

FIGURE 11.13

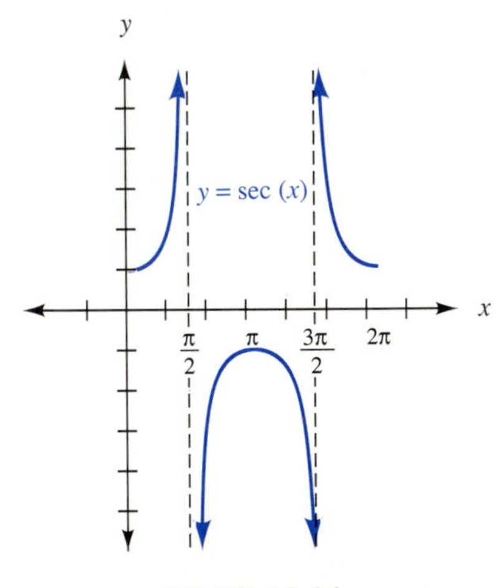

FIGURE 11.14

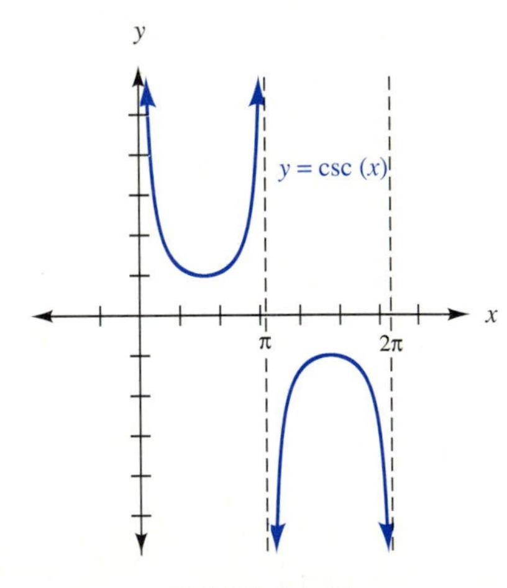

FIGURE 11.15

Analyzing Figure 11.12, we see that the period of the tangent function is π, rather than 2π. As x-values are chosen closer and closer to $\pi/2$, which is a little larger than 1.57, the y-values become larger and larger in absolute value, as shown in the following table.

x	1.30	1.40	1.50	1.51	1.52	1.53	1.54	1.55	1.56	1.57
y	3.60	5.80	14.1	16.4	19.7	24.5	32.5	48.1	92.6	1256

asymptote

The function is undefined when x is $\pi/2$. (Why?) On the graph, a dashed line, called an *asymptote,* is drawn vertically at $x = \pi/2$ to show that as x gets closer to $\pi/2$, and the y-values get larger and larger in absolute value, the curve approaches, but does not reach, the asymptote. Since there is no y-value that is higher or lower than all the others, the word amplitude is not used for the value $|a|$. There may be a phase shift for a tangent function.

A tangent function may be graphed by dividing the period into four parts, just as the periods of the sine and cosine functions are divided into four parts.

To sketch the graph of $y = a \tan(bx + c)$ when $b > 0$:

1. Identify the period: π/b.
2. Identify the value a.
3. Identify the phase shift: $-c/b$.
4. Plot the y-value 0 at the beginning point ($x = -c/b$).
   Plot the y-value 0 at the ending point ($x = \pi/b - c/b$).
5. Draw a vertical asymptote at the halfway point.
   Plot the y-value a at the one-quarter point.
   Plot the y-value $-a$ at the three-quarter point.
6. Connect the points with a smooth curve that has the shape of $y = \tan x$.

EXAMPLE 1 Sketch the graph of $y = 2 \tan [(1/2)x - \pi/6]$.

Solution

1. The period is $\pi \div (1/2)$ or 2π.
2. $a = 2$
3. The phase shift is $(\pi/6) \div (1/2)$ or $\pi/3$.
4. The curve crosses the x-axis when x is $\pi/3$ and when x is $2\pi + \pi/3$ or $7\pi/3$.
5. There is a vertical asymptote where x is $4\pi/3$, the halfway point. At the one-quarter point, the y-value is 2, and at the three-quarter point, the y-value is -2.
6. Figure 11.16 shows the graph. ■

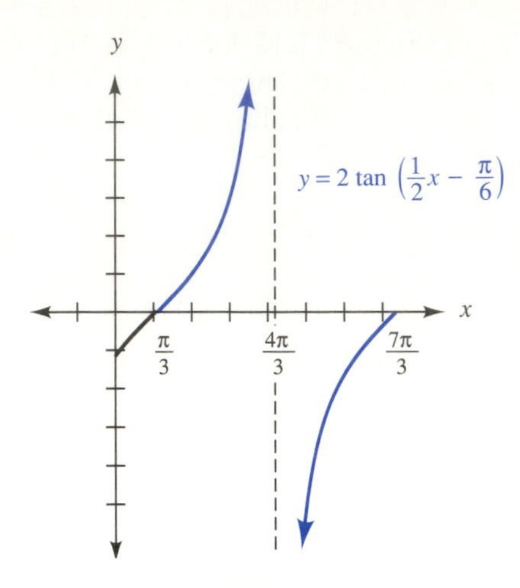

FIGURE 11.16

The cotangent function is the reciprocal of the tangent function. Therefore, when $\tan(bx + c)$ is 0, $\cot(bx + c)$ is undefined. Likewise, when $\tan(bx + c)$ is undefined, $\cot(bx + c)$ is 0. The graph of the cotangent function has asymptotes where a tangent function crosses the x-axis, and crosses the x-axis where a tangent function has asymptotes.

To sketch the graph of $y = a\cot(bx + c)$ when $b > 0$:

1. Consider $y = a\tan(bx + c)$.
2. Identify the period, the value a, and the phase shift.
3. Sketch the graph of $y = a\tan(bx + c)$ lightly, using the information from step 2.
4. **(a)** Plot points where the asymptotes of the tangent curve cross the x-axis.
   (b) Draw vertical asymptotes where the tangent curve crosses the x-axis.
5. Sketch a cotangent curve that goes through the quarter-points of the tangent curve and through the points plotted in step 4.

EXAMPLE 2 Sketch the graph of $y = (1/2)\cot(2x + \pi/3)$.

Solution

1. Consider $y = (1/2)\tan(2x + \pi/3)$.
2. The period is $\pi/2$, $a = 1/2$, and the phase shift is $-(\pi/3) \div 2$ or $-\pi/6$.
3. The tangent function is sketched lightly.
4. **(a)** The cotangent function crosses the x-axis at $\pi/12$.
   (b) There are vertical asymptotes at $x = -\pi/6$ and at $x = \pi/3$.
5. Figure 11.17 shows the graph. ■

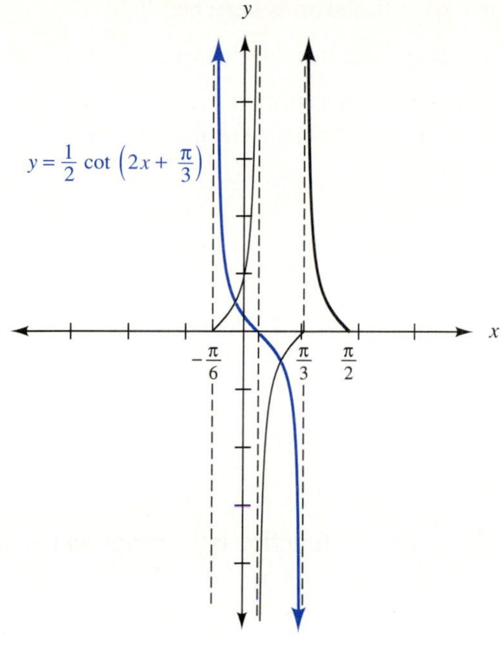

FIGURE 11.17

Analyzing Figure 11.14, we see that the period of the secant function is 2π. Since $\sec x = 1/\cos x$, $\sec x$ is undefined when $\cos x$ is 0. Thus, there are asymptotes where x is $\pi/2$ and where x is $3\pi/2$. The cosine curve may be used as an aid in sketching the secant curve.

To sketch the graph of $y = a \sec (bx + c)$ when $b > 0$:

1. Consider $y = a \cos (bx + c)$.
2. Identify the period, amplitude, and phase shift of the function in step 1.
3. Sketch the graph of $y = a \cos (bx + c)$ lightly, using the information from step 2.
4. Draw vertical asymptotes where the cosine curve crosses the x-axis.
5. Sketch a secant curve that just touches the cosine curve at its highest and lowest points.

EXAMPLE 3 Sketch the graph of $y = -2 \sec [(1/3)x + \pi/3]$.

Solution

1. Consider $y = -2 \cos[(1/3)x + \pi/3]$.
2. The period is $2\pi \div (1/3)$ or 6π, the amplitude is 2, and the phase shift is $-(\pi/3) \div (1/3)$ or $-\pi$.

3. The cosine function is sketched lightly.
4. Vertical asymptotes are drawn at x-values of $\pi/2$ and $7\pi/2$.
5. The secant curve is drawn so that it touches the cosine curve when x is $-\pi$, 2π, or 5π. Figure 11.18 shows the graph. ■

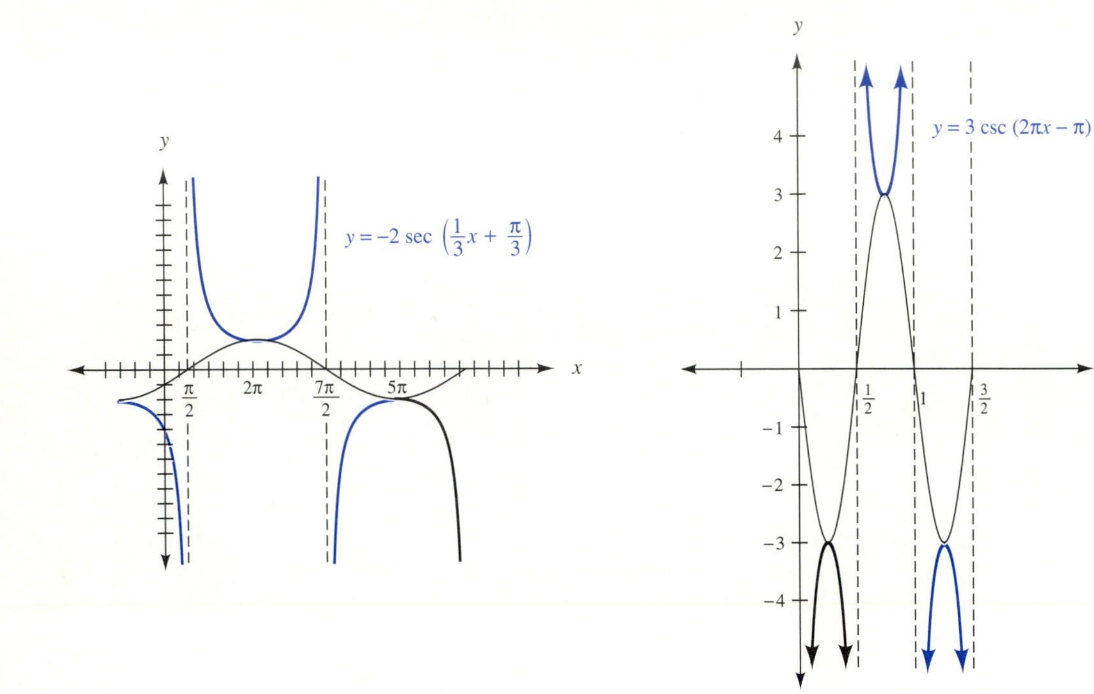

FIGURE 11.18

FIGURE 11.19

The cosecant and sine curves are related to each other in the same way as the secant and cosine curves.

EXAMPLE 4 Sketch the graph of $y = 3 \csc(2\pi x - \pi)$.

Solution
1. Consider $y = 3 \sin(2\pi x - \pi)$.
2. The period is $2\pi/(2\pi)$ or 1, the amplitude is 3, and the phase shift is $\pi/(2\pi)$ or $1/2$.
3. The sine curve is sketched lightly.
4. Vertical asymptotes are drawn at x-values of $1/2$, 1, and $3/2$.
5. The cosecant curve is drawn so that it just touches the sine curve when x is $1/4$, $3/4$, or $5/4$. Figure 11.19 shows the graph. ■

EXERCISES 11.3

In 1–28, sketch the graph of each function.

1. $y = 3 \csc (x - \pi)$

2. $y = \frac{1}{2} \sec \left(x + \frac{\pi}{2}\right)$

3. $y = \frac{2}{3} \sec \left(x + \frac{\pi}{4}\right)$

4. $y = -3 \csc (x - \pi)$

5. $y = \frac{-2}{3} \csc \left(2x - \frac{\pi}{2}\right)$

6. $y = \frac{-3}{2} \sec (2x + \pi)$

7. $y = 2 \sec \left(\frac{1}{3}x + \frac{\pi}{6}\right)$

8. $y = 3 \csc \left(\frac{1}{2}x - \frac{\pi}{2}\right)$

9. $y = \frac{3}{2} \csc (4x - 2\pi)$

10. $y = \frac{2}{3} \sec (4x + 2\pi)$

11. $y = \sec (3x + \pi)$

12. $y = \csc (3x - \pi)$

13. $y = -\csc \left(\frac{1}{2}x - \frac{\pi}{4}\right)$

14. $y = -\sec \left(\frac{1}{3}x + \frac{\pi}{6}\right)$

15. $y = \frac{1}{2} \tan \left(\frac{2}{3}x + \frac{\pi}{3}\right)$

16. $y = \frac{1}{2} \tan (2\pi x - \pi)$

17. $y = -\cot (2\pi x - \pi)$

18. $y = 2 \cot (4\pi x + \pi)$

19. $y = 2 \tan \left(\frac{1}{2}\pi x + \frac{\pi}{4}\right)$

20. $y = \frac{1}{2} \tan (\pi x - \pi)$

21. $y = \frac{-1}{2} \cot (\pi x - \pi)$

22. $y = \frac{1}{3} \cot \left(x + \frac{\pi}{2}\right)$

23. $y = 2 \tan \left(x + \frac{\pi}{3}\right)$

24. $y = \frac{3}{2} \tan \left(x - \frac{\pi}{6}\right)$

25. $y = \frac{-1}{3} \cot \left(x - \frac{\pi}{4}\right)$

26. $y = \frac{3}{2} \cot \left(2x + \frac{\pi}{6}\right)$

27. $y = \frac{2}{3} \tan \left(2x + \frac{\pi}{6}\right)$

28. $y = \frac{1}{2} \tan \left(\frac{1}{3}x - \frac{\pi}{6}\right)$

29. The height h (in m) of a cloud is given by $h = 1000 \tan \theta$. In the equation, θ is the angle of inclination to the cloud as seen by an observer 1000 m away. Sketch the graph of the equation for $0° < \theta < 60°$ (i.e., graph $h = 1000 \tan x$ for $0 < x < \pi/3$). 1 2 4 7 8 3 5

30. The index of refraction of the medium that light enters (in relation to the medium that it leaves) is given by $n = \csc \theta$ if θ is the critical angle. The critical angle is the angle of refraction for which the angle of incidence is 90°. Sketch the graph of the equation for $0° < \theta < 90°$ (i.e., graph $n = \csc x$ for $0 < x < \pi/2$). 3 5 6 9

31. When light passes from one medium into a vacuum at a 45° angle, the velocity (in m/s) in the original medium is given by $v = (2.12 \times 10^8) \csc \theta$ if θ is the angle of refraction in the vacuum. Choose an appropriate scale and sketch the graph for $0° < \theta < 90°$ (i.e., graph $v = (2.12 \times 10^8) \csc x$ for $0 < x < \pi/2$). 3 5 6 9

32. For a telescope, the distance f (in cm) of an object from the objective is given by $f = 2.19 \cot \theta$, if θ is the angle subtended by an image 25 cm away from the lens and the object subtends an angle of 5° when placed 25 cm from the eye. Sketch the graph of the curve for $0° < \theta < 90°$ (i.e., graph $f = 2.19 \cot x$ for $0 < x < \pi/2$). 3 5 6 9

1. Architectural Technology
2. Civil Engineering Technology
3. Computer Engineering Technology
4. Mechanical Drafting Technology
5. Electrical/Electronics Engineering Technology
6. Chemical Engineering Technology
7. Industrial Engineering Technology
8. Mechanical Engineering Technology
9. Biomedical Engineering Technology

33. For a certain object on an inclined plane, a force of 10 N will just start the body moving. The equation $N = 10 \cot \theta$ gives the force exerted by the object in a direction perpendicular to the plane, and θ is the angle of inclination of the plane. Sketch the graph of the curve for $0° < \theta < 90°$ (i.e., graph $N = 10 \cot x$ for $0 < x < \pi/2$).
4 7 8 1 2 3 5 6 9

34. For an object sliding down an inclined plane at constant velocity, the coefficient of friction is given by $\mu = \tan \theta$ if θ is the angle of inclination. Sketch the graph of the curve for $0° < \theta < 90°$ (i.e., graph $\mu = \tan x$ for $0 < x < \pi/2$). Why is it true that $\theta < 90°$? 4 7 8 1 2 3 5 6 9

1. Architectural Technology
2. Civil Engineering Technology
3. Computer Engineering Technology
4. Mechanical Drafting Technology
5. Electrical/Electronics Engineering Technology
6. Chemical Engineering Technology
7. Industrial Engineering Technology
8. Mechanical Engineering Technology
9. Biomedical Engineering Technology

11.4 ADDITION OF ORDINATES AND PARAMETRIC EQUATIONS

Sometimes a function may be considered as the sum of two functions. For example, $k(x) = x + \sin x$ may be considered as the sum of $f(x) = x$ and $g(x) = \sin x$. If the two functions have the same domain, $k(x) = f(x) + g(x)$. That is, the y-value associated with x for the sum is obtained by adding the y-values associated with x for each individual function. The addition may be done graphically. Since a y-value is sometimes called an *ordinate,* the method is called *addition of ordinates.* The method is particularly effective for complicated curves, but we begin with a few simple examples in order to explain the method.

ordinate

addition of ordinates

EXAMPLE 1 Sketch the graph of $y = x + 1$ by graphically adding $y = x$ and $y = 1$.

Solution Sketch the graphs of $y = x$ and $y = 1$ on the same coordinate plane. Figure 11.20 shows the graph.

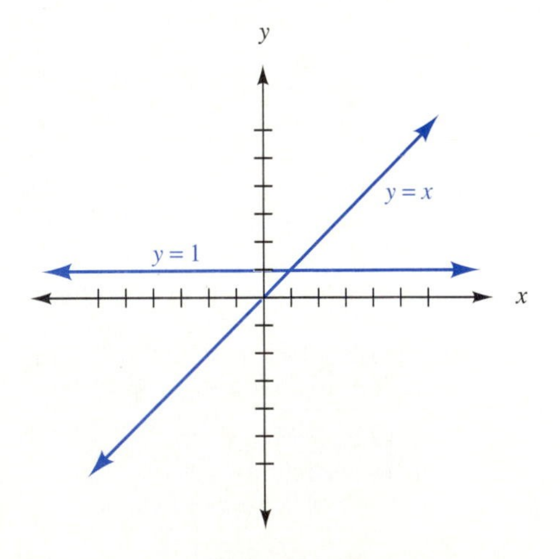

FIGURE 11.20

The graph of $y = 1$ is a simpler graph, so we choose it as the base curve. That is $y = x$ will be added to $y = 1$. The y-values on the graph of $y = x$ are measured from the x-axis. Positive values of y are above it, negative values of y, below it. When $y = x$ is added to $y = 1$, the y-values on the curve $y = x$ will be measured from the line $y = 1$.

Consider a few particular values. When x is 1, the distance from the x-axis to the line $y = x$ is 1. Measuring a distance of 1 above the line $y = 1$, we have a y-value of 2. When x is 2, the distance from the x-axis to the line $y = x$ is 2. Measuring a distance of 2 above the line $y = 1$, we have a y-value of 3, and so on. Figure 11.21 shows the graph. ■

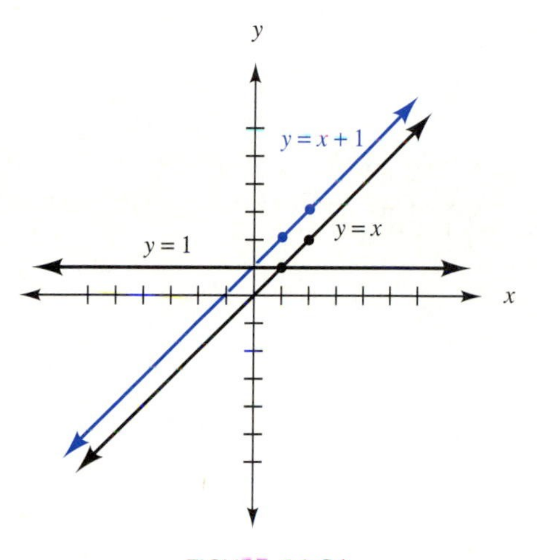

FIGURE 11.21

The method is summarized as follows.

To sketch the graph of $y = f(x) + g(x)$:

1. Sketch $y = f(x)$ and $y = g(x)$ on the same coordinate system.
2. Choose the less complicated curve as the first curve, to which the second curve will be added.
3. Choose several x-values, such as the quarter points for a trigonometric function.
4. Measure the distance from the x-axis to the second curve for each x-value chosen in step 3. Plot a point at the same distance above (or below) the first curve.
5. Connect the points obtained in step 3 to graph $y = f(x) + g(x)$.

EXAMPLE 2 Sketch the graph of $y = x + \sin x$ by the method of addition of ordinates.

Solution

1. Sketch the graph of $y = x$ and $y = \sin x$.
2. Choose $y = x$ as the first curve.

3. We choose 0, $\pi/2$, π, $3\pi/2$ and 2π as important x-values to examine, since they are the values used to sketch the sine curve.

4. For each of these x-values, the distance from the x-axis to the curve $y = \sin x$ is measured. The same distance is then measured from the line $y = x$. Notice that when x is $3\pi/2$, the distance to the curve $y = \sin x$ is measured *below* the x-axis. The same distance is then measured *below* the curve $y = x$, as in Figure 11.22(a).

5. Figure 11.22(b) shows the graph. ■

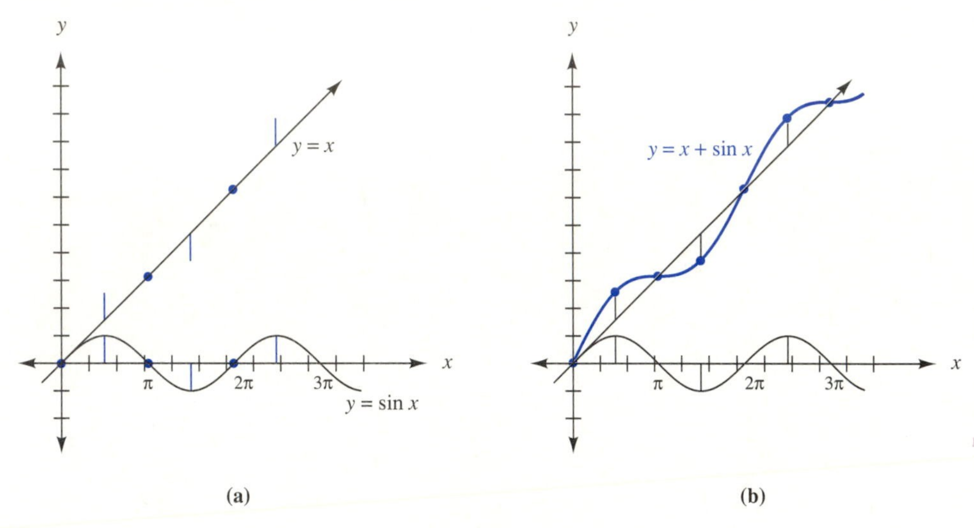

FIGURE 11.22

Two water, sound, or light waves sometimes combine, and the effect is to create a new wave with characteristics of the sum of the two waves. If the two waves peak simultaneously, for example, the sum has a higher peak than either individual wave. Whatever effect the peak has is thus exaggerated when the waves combine.

EXAMPLE 3 Two boats in a bay each create a wake. One creates waves that rise and fall through a vertical distance of 6 inches in a period of 2 seconds. The other creates waves that rise and fall through a vertical distance of 8 inches in 4 seconds. To say that the two waves are in phase means there is no phase shift. Sketch a graph to represent the sum of the two waves if they are in phase.

Solution The first wave has an amplitude of 3 and a period of 2. The second has an amplitude of 4 and a period of 4.

1. Sketch the curves.

2. Choose the first wave as the first curve.

3. Choose 0, 1/2, 1, 3/2, 2, 5/2, 3, 7/2, and 4, since they are important x-values for the two curves.

4. For each of the x-values chosen in step 3, the distance is measured from the x-axis to the second curve. The same distance is then measured from the first curve. For example, when x is 1/2, the second curve is about 2.8 units above the x-axis. Measure 2.8 units above the first curve, which has a y-value of 3, to find the point (1/2, 5.8). Likewise, when x is 1, the second curve is 4 units above the x-axis. Measure 4 units above the first curve, which has a y-value of 0, to find the point (1, 4).

5. Figure 11.23 shows the graph. ■

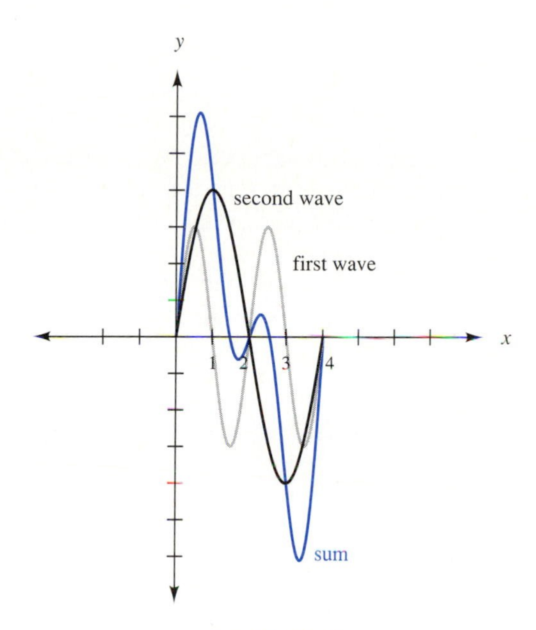

FIGURE 11.23

For a difference of functions $k(x) = f(x) - g(x)$, subtraction of ordinates may also be done, but it is usually easier to graph $y = f(x)$ and $y = -g(x)$ and treat the problem as one of addition.

EXAMPLE 4 Sketch the graph of $y = 2 \cos x - \sin 2x$.

Solution The graphs of $y = 2 \cos x$ and $y = -\sin 2x$ are sketched on the same coordinate system. The curve $y = 2 \cos x$ is used as the first curve. Figure 11.24 shows the graph. ■

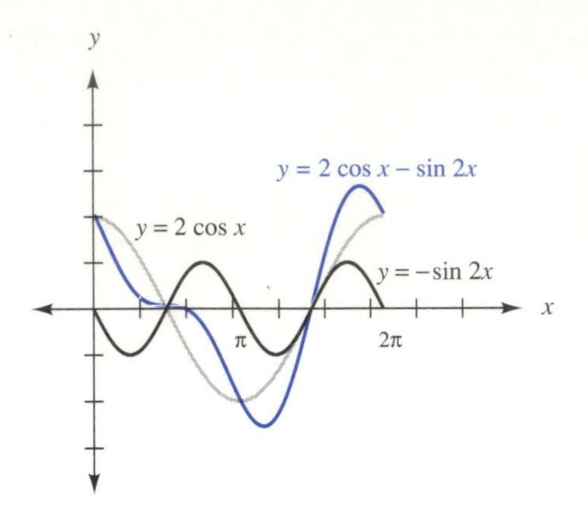

FIGURE 11.24

When both functions are periodic, be sure to extend the graph far enough to show a complete cycle of the sum. If greater accuracy is desired, a calculator may be used to add function values for additional values of x.

Sometimes it is easier to describe a curve by giving two equations instead of one. One equation describes how the x-values change and the other describes how the y-values change. Equations that specify distances x and y in terms of time as the third variable are common. Such equations are called *parametric equations.*

DEFINITION Equations that specify a curve by giving separate equations for its coordinates in terms of a third variable, called a **parameter,** are called **parametric equations.**

EXAMPLE 5 Sketch the graph of $x = (-1/2)t$ and $y = t^2$ for $t \geq 0$.

Solution A table shows the values associated with x and y for various values of t.

t	0	1	2	3	4	5
x	0	$-1/2$	-1	$-3/2$	-2	$-5/2$
y	0	1	4	9	16	25

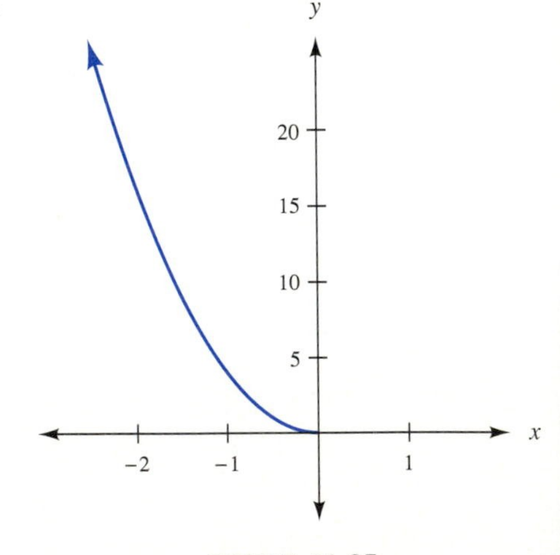

FIGURE 11.25

Plotting (x, y) for each t, we have the graph shown in Figure 11.25. ■

Lissajous figure

When the two functions x and y are sine or cosine functions of t, the resulting curve is called a *Lissajous figure.* Lissajous figures are used by engineers in the measurements of frequency and phase of wave motions.

EXAMPLE 6 Sketch the graph of $x = 2 \cos t$ and $y = \sin 2t$.

Solution A table of values shows the values of x and y associated with various values of t.

t	0	$\pi/6$	$\pi/3$	$\pi/2$	$2\pi/3$	$5\pi/6$	π	$7\pi/6$	$4\pi/3$	$3\pi/2$	$5\pi/3$	$11\pi/6$	2π
x	2	1.73	1	0	-1	-1.73	-2	-1.73	-1	0	1	1.73	2
y	0	.866	.866	0	$-.866$	$-.866$	0	.866	.866	0	$-.866$	$-.866$	0

Figure 11.26 shows the graph. The arrows on the figure indicate that the points are connected for successive values of t. Notice that the curve is not the graph of a function. ■

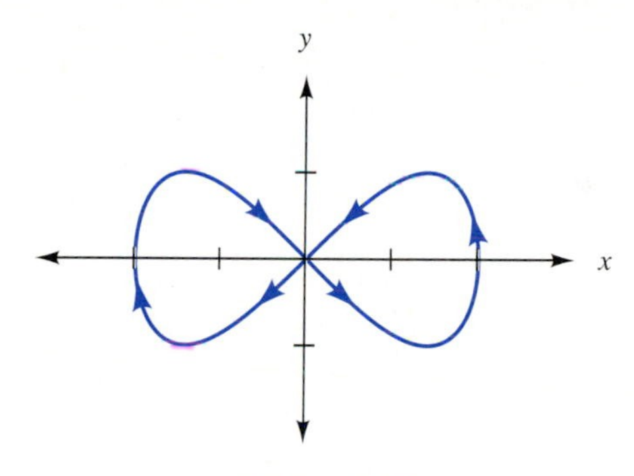

FIGURE 11.26

EXERCISES 11.4

In 1–21, sketch the graph of each function, using addition of ordinates.

1. $y = 1 + \sin 2x$
2. $y = 2 + 3 \cos x$
3. $y = 3 - 2 \cos x$
4. $y = -1 + \sin 3x$
5. $y = x + 2 \sin x$
6. $y = x + \cos x$
7. $y = x/2 - \cos x$
8. $y = x/2 - \sin 2x$
9. $y = x/3 + 3 \sin 2x$
10. $y = x/3 + 3 \cos 2x$
11. $y = \sin x + 2 \cos x$
12. $y = 2 \sin x + \cos x$
13. $y = \sin x - \cos 2x$
14. $y = \sin 2x - \cos x$
15. $y = 2 \sin 3x + 3 \cos 2x$
16. $y = 3 \sin 2x + 2 \cos 3x$
17. $y = 1/2 \sin 3\pi x + 1/3 \cos (2\pi x)$
18. $y = 3 \sin (\pi x/2) + 1/2 \cos (\pi x/3)$
19. $y = 2 \sin (2x - \pi/3) + 1/2 \cos (2x + \pi/3)$
20. $y = 1/2 \sin (2x + \pi/3) + 2 \cos (2x - \pi/3)$
21. $y = \sin (2x - \pi/2) + \cos (2x + \pi/2)$

In 22–37, plot the curve given by each pair of parametric equations for $t \geq 0$.

22. $x = 2t,\ y = t^2$ **23.** $x = t^2,\ y = 2t$ **24.** $x = t - 1,\ y = t + 2$ **25.** $x = t + 3,\ y = t - 2$

26. $x = t^2 - 1,\ y = t^2 + 1$ **27.** $x = 2t^2,\ y = -3t^2$ **28.** $x = 3t^2,\ y = -2t^2$

29. $x = t^2/2,\ y = t^3/2$ **30.** $x = \sin t,\ y = \cos t$ **31.** $x = \cos t,\ y = \sin t$

32. $x = \sin \pi t,\ y = \sin 3\pi t$ **33.** $x = \cos \pi t,\ y = \cos 3\pi t$ **34.** $x = 2 \sin t,\ y = 3 \cos t$

35. $x = \sin 2t,\ y = \cos 3t$ **36.** $x = 2 \sin \pi t,\ y = \sin 2\pi t$ **37.** $x = 2 \cos \pi t,\ y = \sin 2\pi t$

The figure shows a body of weight w being lifted by two equal and opposite forces acting through two wedges of angle β. If the friction angles at all surfaces in contact are θ, the equations in problems 38 and 39 hold. Use $\beta = 30°$ and $w = 100$ lb.

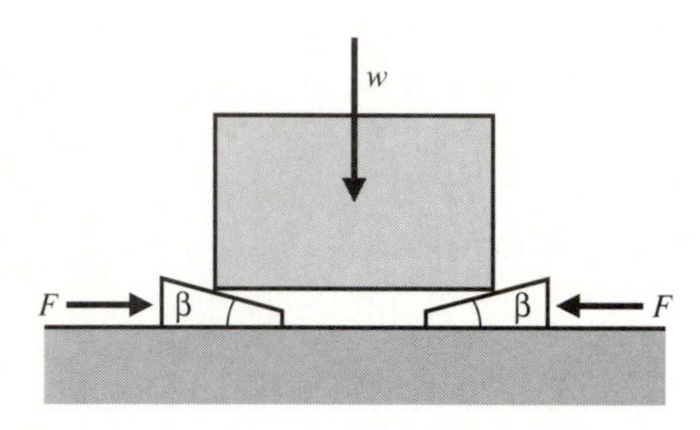

38. $F = (w/2) \tan(\theta + \beta) + (w/2) \tan \theta$ where F is the force (in ft·lb) necessary to lift the body. Let x represent the radian measure of θ and sketch the graph of the equation for $0 < x < \pi/6$. 4 7 8 1 2 3 5 6 9

39. $F = (w/2) \tan(\theta - \beta) + (w/2) \tan \theta$ where F is the force (in ft·lb) necessary to lower the body when $\theta > 15°$. Let x represent the radian measure of θ and sketch the graph of the equation for $\pi/12 < x < \pi/3$. 4 7 8 1 2 3 5 6 9

40. Under certain circumstances, when voltages from two sources are impressed upon a circuit, the resulting current is the sum of the currents that would be produced by each voltage source acting independently. Use Ohm's law (current = voltage/resistance) to find the current in a 5-ohm resistor produced by a dc source of 5 volts. The current (in amperes) in the 5-ohm resistor produced by an ac source of 2 volts amplitiude and frequency of 400 Hz is given by $i = 0.4 \sin(800\pi t)$. Sketch the graph of the sum of these currents versus time. 3 5 9 1 2 4 6 7 8

41. When an object suspended from a spring is attached to a support that oscillates in such a manner that the deflection (in cm) of the support is given by $y = a \sin \omega t$, the deflection of the object is given by

$$y = \frac{-a\omega\omega_n \sin \omega_n t}{\omega_n{}^2 - \omega^2} + \frac{a\omega_n{}^2 \sin \omega t}{\omega_n{}^2 - \omega^2}.$$

If $a = 2$ cm, $\omega = 5\pi$, and $\omega_n = 15.7\pi$, the equation is $y = -0.709 \sin(49.5t) + 2.23 \sin(15.7t)$. Sketch a graph of the equation for $0 < t < 1$. 4 7 8 1 2 3 5 6 9

42. One tuning fork vibrates, creating a wave with period 2 milliseconds and amplitude 2 mm. A second tuning fork vibrates, creating a wave with period 2.5 milliseconds and an amplitude of 2 mm. If the two waves are in phase and the waves combine, sketch the graph of the equation that represents the sum. 1 3 4 5 6 8 9

43. If an object moves so that its horizontal distance is given by the equation $x = 2t$, and its vertical distance is given by $y = 2t^2$, sketch a graph of its motion for $0 \leq t \leq 3$. 4 7 8 1 2 3 5 6 9

44. An object dropped from a plane moving at 490 km/hr moves a horizontal distance (in km) given by $x = 0.136t$ in t seconds. The vertical distance (in km) is given by $y = -0.0049t^2$. Sketch a graph of its motion for $50 \leq t \leq 100$. 4 7 8 1 2 3 5 6 9

45. If $x = B \cos 2\pi v_1 t$ and $y = A \sin 2\pi\, v_2 t$, then the resulting Lissajous figure may form a closed figure. Determine the relationship between v_1, v_2, and the number of loops.

---

1. Architectural Technology
2. Civil Engineering Technology
3. Computer Engineering Technology
4. Mechanical Drafting Technology
5. Electrical/Electronics Engineering Technology
6. Chemical Engineering Technology
7. Industrial Engineering Technology
8. Mechanical Engineering Technology
9. Biomedical Engineering Technology

11.5 APPLICATIONS

In the previous section, you saw that parametric equations can be used to specify x and y. Lissajous figures result when x and y represent *simple harmonic motion.*

To understand simple harmonic motion, imagine a wheel with a crank handle projecting from it near the edge as in Figure 11.27. If a light shines so that the edge of the wheel projects a shadow on the wall, the circular motion of the wheel will not be apparent from the shadow of the wheel itself. The shadow of the crank handle, however, will show a vertical motion and will oscillate between the top and bottom of the wheel's shadow. The motion of the handle's shadow is an example of simple harmonic motion.

FIGURE 11.27

DEFINITION

A point moves in **simple harmonic motion** if it moves along a diameter of a circle as though it were the projection of a point moving around the circle with constant angular velocity.

The position d of a particle moving under simple harmonic motion is given by $d = R \sin \omega t$, where R is the radius of the circle, ω is the angular velocity, and t is the time. To see that this is reasonable, consider a circle that has radius 1 such that a particle moves around the circle at a rate of 1 radian per second. It will take 2π seconds for it to go completely around the circle once. If we divide the circle into equal parts, as in Figure 11.28, we can examine the position of the particle every $\pi/6$ seconds after it starts its motion from the positive x-axis. The projection of the particle along a vertical, such as the y-axis, is simply the y-coordinate of the particle at that position. But for any angle θ, $y = \sin \theta$. Since the angle in radians is numerically equal to the time in this example, we have $y = \sin t$. We plot the vertical position of the particle ($y = \sin t$) for each time t by transferring the y-coordinate of the particle on the circle at time t to the graph beside the circle in Figure 11.28.

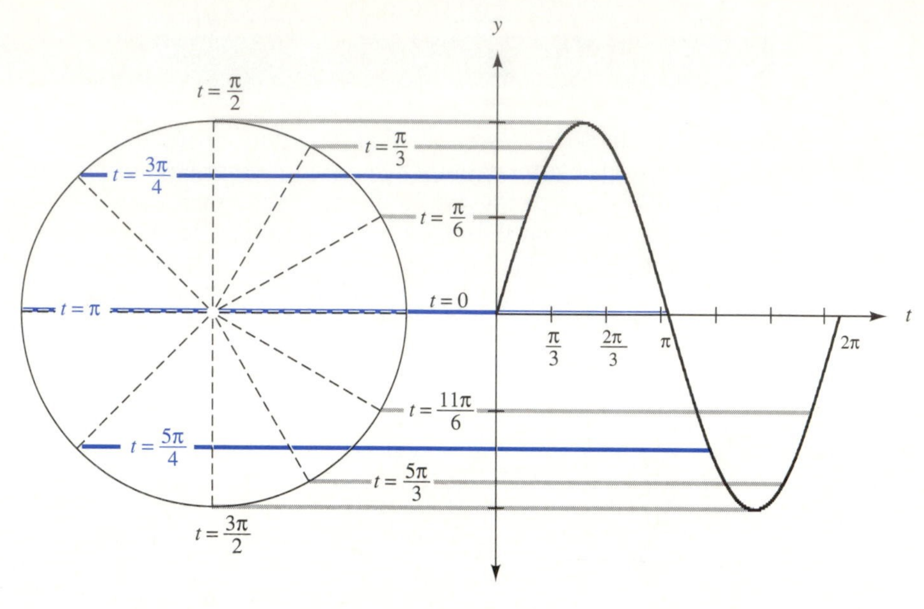

FIGURE 11.28

The effects of a, b, and c in the equation $y = a \sin(bt + c)$ on simple harmonic motion can be seen in Figures 11.29–11.31. For $y = a \sin t$, the circle about which the particle revolves must have radius a, as in Figure 11.29.

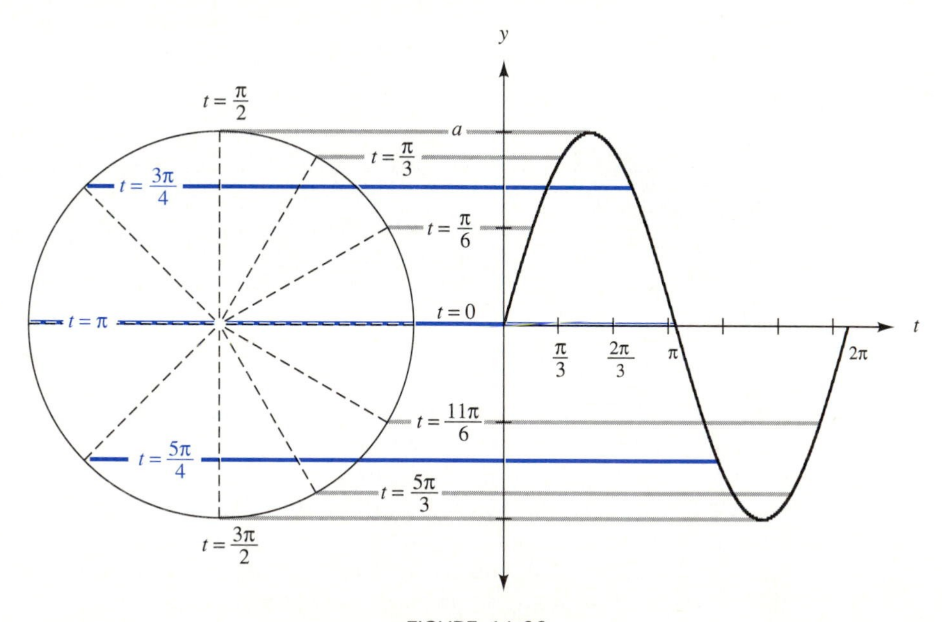

FIGURE 11.29

For $y = \sin bt$, the angular velocity of the particle is b radians per unit time. If the particle travels at the rate of 2 radians per second, it will make two revolutions around the circle in 2π seconds (i.e., as t ranges from 0 to 2π). That is, between 0 and 2π, the graph will show 2 cycles of the sine curve, so the period is π, as in Figure 11.30.

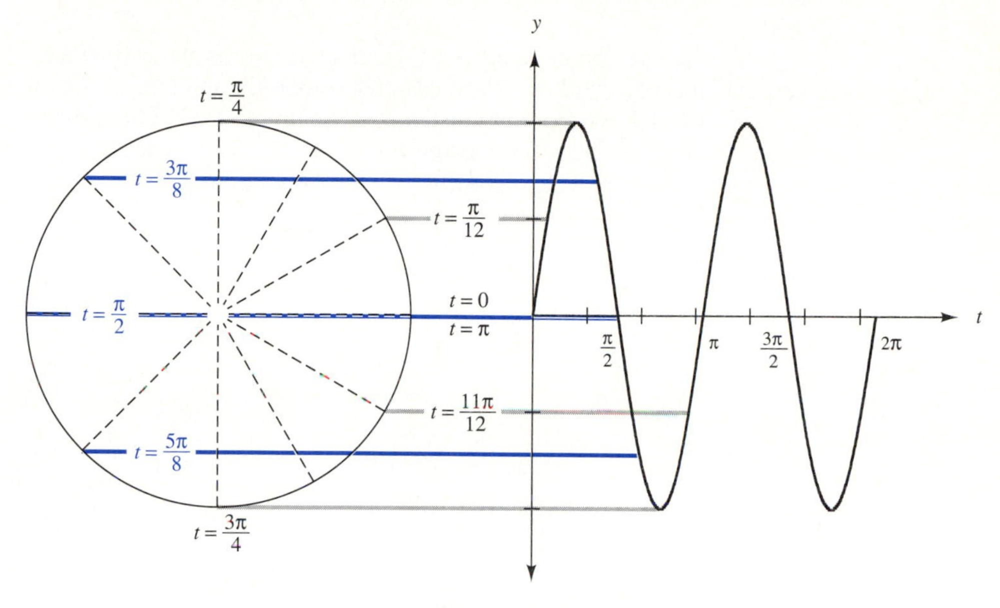

FIGURE 11.30

For $y = \sin(t + c)$, the value c indicates a starting position of $\theta = c$ for the particle, rather than $\theta = 0$. Thus, the graph shows a phase shift of $-c$, as in Figure 11.31.

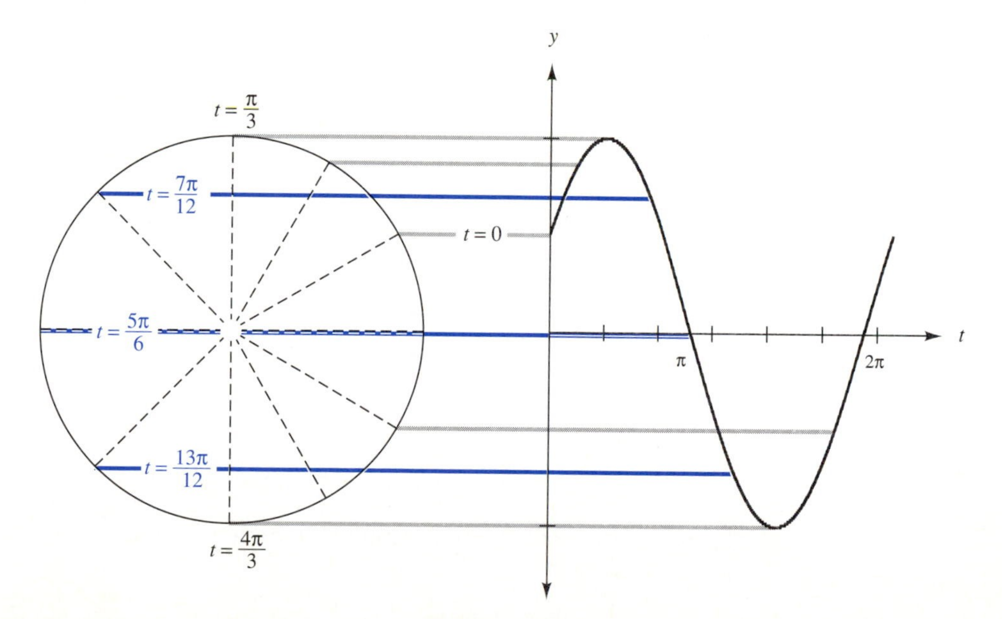

FIGURE 11.31

Simple harmonic motion occurs in the study of wave formation such as water waves, sound waves, and electromagnetic waves. In the study of alternating current, simple harmonic motion is important because of the way voltage and current change with time.

Alternating voltage is induced when a wire loop rotates through a magnetic field in a circular path with constant angular velocity. Each half of the loop alternatively passes up and down through the field, as in Figure 11.32. The resulting current in the loop, then, changes in both magnitude and direction twice as the loop makes one complete rotation. The current is given by the equation

$$I = I_{\max} \sin \omega t \tag{11-1}$$

where $I_{\max}$ is the maximum possible current, ω is the angular velocity of the rotating loop, and t is the time.

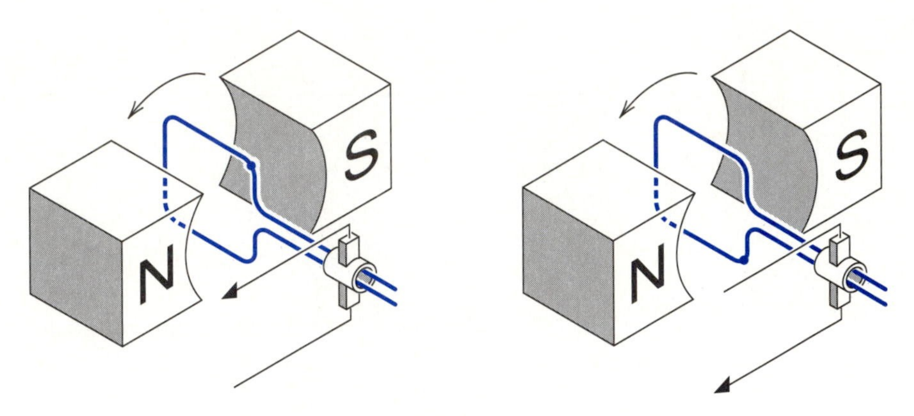

FIGURE 11.32

Likewise, the output voltage V of an ac generator is given by

$$V = V_{\max} \sin \omega t \tag{11-2}$$

where $V_{\max}$ is the maximum possible voltage, ω is the angular velocity, and t is the time.

The frequency of a wave is the number of cycles that pass a given point in one second. If the period is τ, then one cycle passes in τ seconds, and $1/\tau$ cycles pass in one second. Recall that $f = 1/\tau$, where f is the frequency and τ is the period. The unit of frequency is the *hertz.* One hertz = 1 cycle/second.

For Equation 11-1, the period is $2\pi/\omega$. Hence,

$$f = \frac{1}{2\pi/\omega} = \omega/(2\pi) \text{ or } 2\pi f = \omega.$$

The procedure for sketching the graph of a simple harmonic motion is as follows.

To sketch the graph of a simple harmonic motion given by $d = R \sin(\omega t + \theta)$:

1. Identify the period: $2\pi/\omega$ where $\omega = 2\pi f$.
2. Identify the amplitude: R.
3. Identify the phase shift: $-\theta/\omega$.
4. Choose appropriate scales and sketch the curve corresponding to the data in steps 1–3 with a horizontal t-axis and a vertical d-axis.

EXAMPLE 1 Sketch the graph of the voltage V (in volts) if

$$V = V_{max} \sin(\omega t + \theta)$$

when $V_{max} = 60$ V, $f = 30$ cycles/s, and $\theta = \pi/2$ s.

Solution

1. Since $\omega = 2\pi(30)$ or 60π, the period is $2\pi/(60\pi)$ or $1/30$ s.
2. The amplitude is 60 volts.
3. The phase shift is $(-\pi/2) \div 60\pi$ or $-1/120$ s.
4. Since the period is $1/30$, units will have to be large on the t-axis, but since the amplitude is 60, they will have to be small on the V-axis. Notice that the phase shift is $1/4$ of the period, so the t-axis is marked off in sections of $1/120$, which is $1/4$ of the period. Figure 11.33 shows the graph. ■

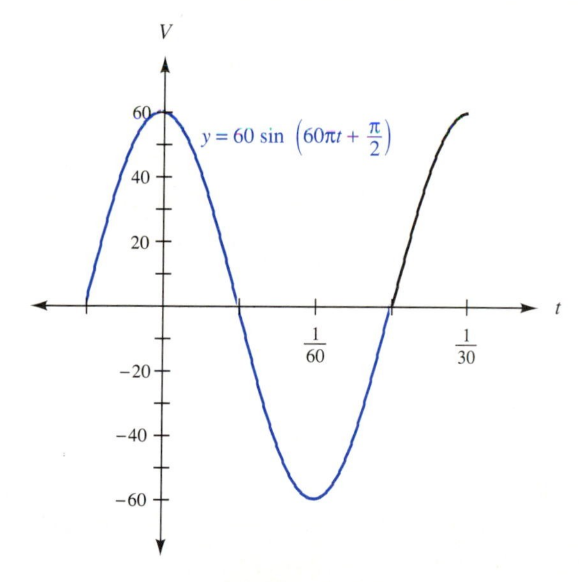

FIGURE 11.33

EXAMPLE 2 Sketch the graph of the current I (in amperes) if

$$I = I_{max} \sin(\omega t + \theta)$$

when $I_{max} = 4$ A, $f = 60$ Hz, and $\theta = \pi/3$ s.

Solution

1. Since $\omega = 2\pi(60)$ or 120π, the period is $2\pi/(120\pi)$ or $1/60$ s.
2. The amplitude is 4 amperes.
3. The phase shift is $(-\pi/3) \div 120\pi$ or $-1/360$ s.
4. Since the phase shift is 1/6 of the period, the t-axis is marked off in sections of 1/360, which is 1/6 of the period. Figure 11.34 shows the graph. ■

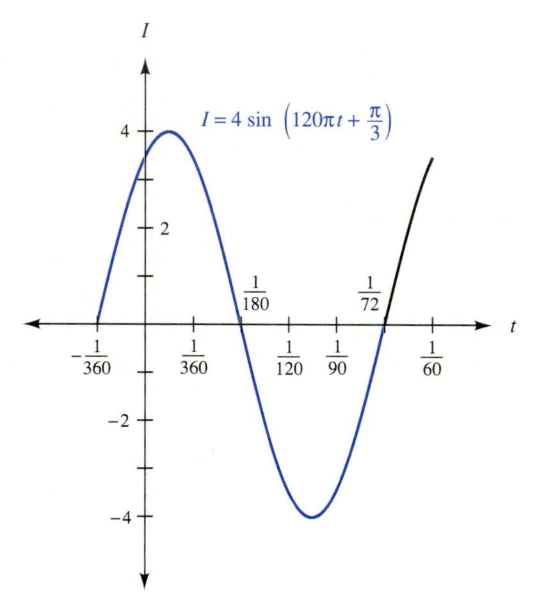

FIGURE 11.34

EXERCISES 11.5

In 1–7, sketch the graph of $V = V_{max} \sin(\omega t + \theta)$ for each set of values.

1. $V_{max} = 120$ V, $f = 30$ Hz, $\theta = \pi/2$ s
2. $V_{max} = 120$ V, $f = 30$ Hz, $\theta = -\pi/2$ s
3. $V_{max} = 60$ V, $f = 30$ Hz, $\theta = \pi/3$ s
4. $V_{max} = 30$ V, $f = 60$ Hz, $\theta = -\pi$ s
5. $V_{max} = 240$ V, $f = 20$ Hz, $\theta = \pi/4$ s
6. $V_{max} = 240$ V, $f = 20$ Hz, $\theta = -\pi/4$ s
7. $V_{max} = 12$ V, $f = 20$ Hz, $\theta = \pi/3$ s

In 8–14, sketch the graph of $I = I_{max} \sin(\omega t + \theta)$ for each set of values.

8. $I_{max} = 4$ A, $f = 60$ Hz, $\theta = \pi/2$ s

9. $I_{max} = 4$ A, $f = 30$ Hz, $\theta = \pi/3$ s

10. $I_{max} = 6$ A, $f = 60$ Hz, $\theta = \pi/4$ s

11. $I_{max} = 6$ A, $f = 30$ Hz, $\theta = -\pi/2$ s

12. $I_{max} = 8$ A, $f = 20$ Hz, $\theta = \pi/3$ s

13. $I_{max} = 10$ A, $f = 20$ Hz, $\theta = -\pi/3$ s

14. $I_{max} = 12$ A, $f = 60$ Hz, $\theta = -\pi/4$ s

When a spring with a mass of m hanging from it is stretched a distance of A and released, it oscillates with harmonic motion. The equation that gives the distance y (in m) from the rest position t seconds after release is $y = A \sin((\sqrt{k/m})t + \pi/2)$. In 15–19, identify the period, amplitude, and phase shift of the graph for each set of values.

15. $A = 0.32$ m, $k = 16$, $m = 0.80$ kg

16. $A = 0.24$ m, $k = 20$, $m = 0.25$ kg

17. $A = 0.18$ m, $k = 24$, $m = 0.12$ kg

18. $A = 0.12$ m, $k = 20$, $m = 0.05$ kg

19. $A = 0.08$ m, $k = 18$, $m = 0.09$ kg

When a pendulum of length l is displaced horizontally and released, it oscillates with harmonic motion. The equation that gives the distance y (in m) from the rest position t seconds after release is $y = A \sin((\sqrt{g/l})t + \pi/2)$ where $g = 9.8$ m/s². In 20–24, identify the period, amplitude, and phase shift of the graph for each set of values.

20. $A = 0.16$ m and $l = 0.98$ m

21. $A = 0.20$ m and $l = 0.49$ m

22. $A = 0.25$ m and $l = 1.4$ m

23. $A = 0.35$ m and $l = 0.98$ m

24. $A = 0.08$ m and $l = 1.4$ m

11.6 THE INVERSE TRIGONOMETRIC FUNCTIONS

Recall from Section 3.4 that the inverse of a function defined by $y = f(x)$ is obtained by interchanging x and y. Thus, if $y = \sin x$, the inverse relationship is given by $x = \sin y$. To rewrite the equation $x = \sin y$ so that it is solved for y, a new notation is needed. Two commonly used notations are $y = \arcsin x$ and $y = \sin^{-1} x$. It is important to recognize that the superscript $^{-1}$ is not used as an exponent, but rather as a symbol for the word inverse. The inverses of the other trigonometric functions can be specified in a similar manner. An expression such as $y = \sin^{-1} x$ is often read as, "y is the angle whose sine is x."

INVERSES OF THE TRIGONOMETRIC FUNCTIONS

$y = \sin^{-1} x$ or $y = \arcsin x$ means $x = \sin y$

$y = \cos^{-1} x$ or $y = \arccos x$ means $x = \cos y$

$y = \tan^{-1} x$ or $y = \arctan x$ means $x = \tan y$

$y = \cot^{-1} x$ or $y = \text{arccot } x$ means $x = \cot y$

$y = \sec^{-1} x$ or $y = \text{arcsec } x$ means $x = \sec y$

$y = \csc^{-1} x$ or $y = \text{arccsc } x$ means $x = \csc y$

The graphs of the inverses of the trigonometric functions could be obtained by plotting points. However, it is easier to use the fact that the graph of the inverse of a function is the mirror image about the line $y = x$ of the function. Figures 11.35–11.37 show the graphs of the sine, cosine, and tangent functions along with their inverses. The inverses of the cotangent, secant, and cosecant can be obtained in a similar manner.

It is clear from the graphs that the inverses are not functions. For example, on the graph of $y = \sin^{-1} x$, the y-axis is a vertical line that crosses the graph in more than one point. To treat the inverse of a trigonometric function as a function itself, it is necessary to restrict the range of the inverse. Equivalently, the domain of the original

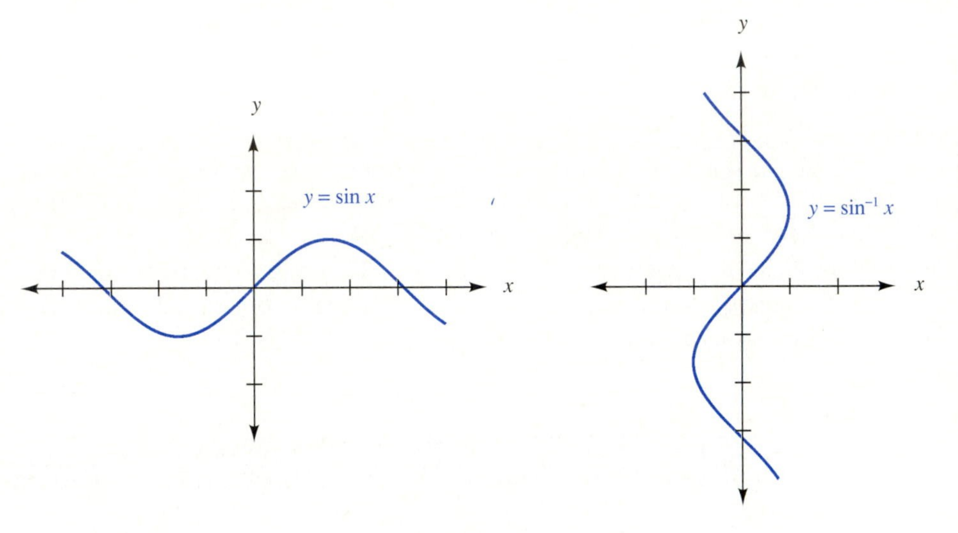

FIGURE 11.35

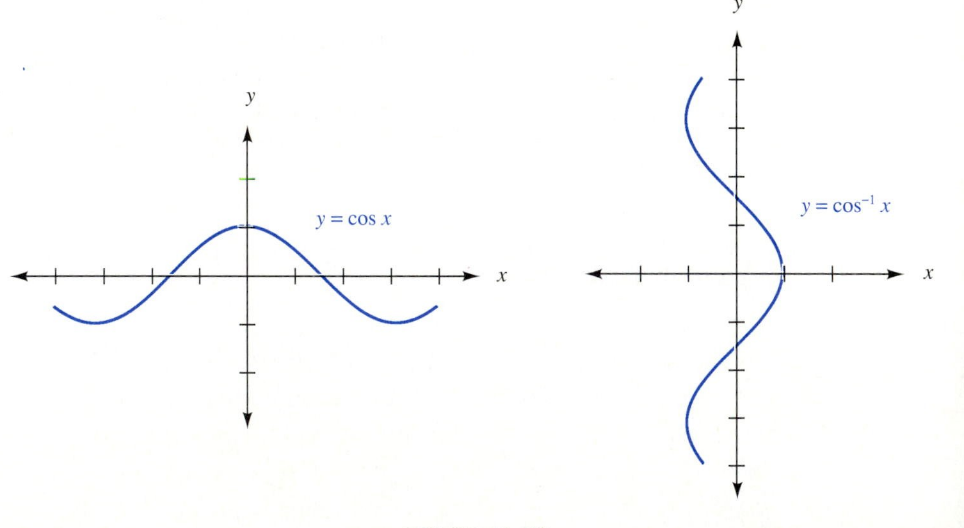

FIGURE 11.36

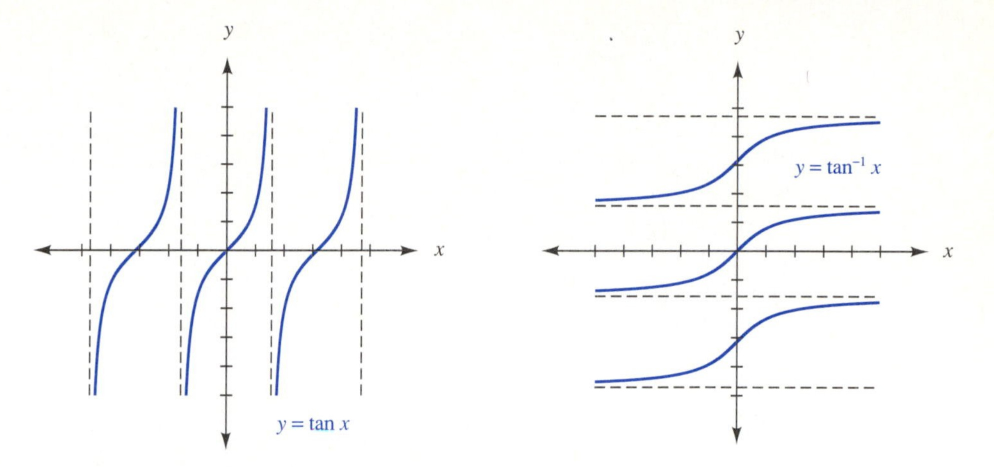

FIGURE 11.37

function may be restricted. The domain for each function is chosen as an interval of real numbers that includes 0 to $\pi/2$. It is necessary to exclude values for which the function is undefined, but, if possible, the interval is chosen so that within it, the graph has no gaps. When the inverses of the trigonometric functions are considered to be functions (that is, when the range is properly restricted), the function names are written beginning with upper case letters. The ranges of the common trigonometric functions are specified as follows.

THE COMMON INVERSE TRIGONOMETRIC FUNCTIONS

Function	Range
$y = \mathrm{Sin}^{-1}\, x$ or $y = \mathrm{Arcsin}\, x$	$-\pi/2 \le y \le \pi/2$

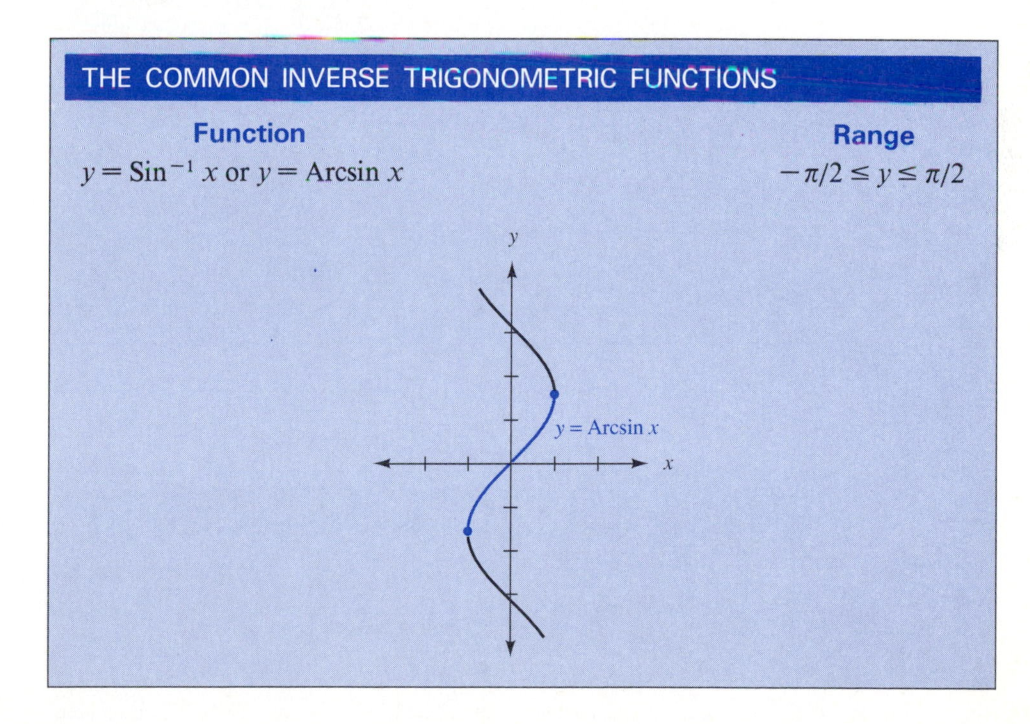

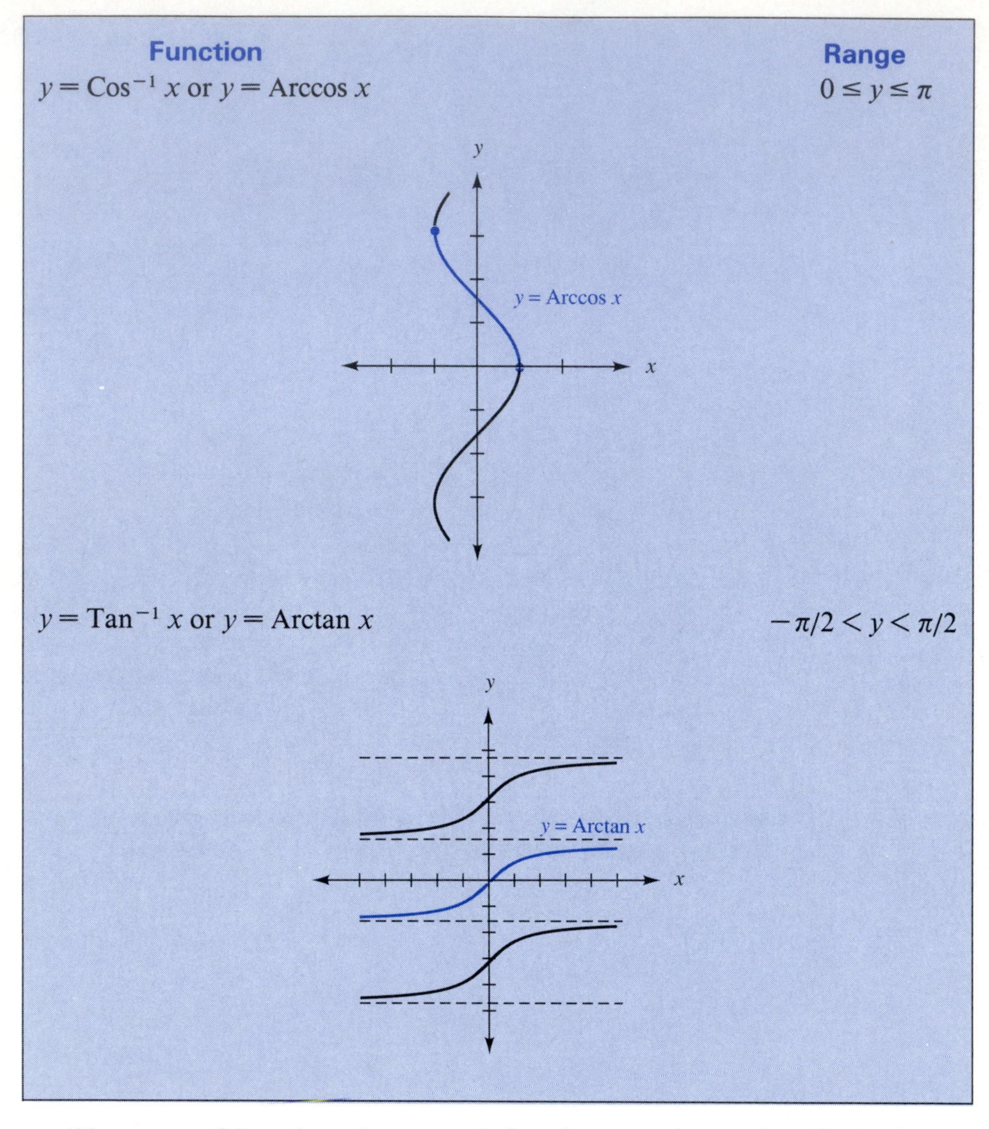

Function	Range
$y = \text{Cos}^{-1} x$ or $y = \text{Arccos } x$	$0 \leq y \leq \pi$
$y = \text{Tan}^{-1} x$ or $y = \text{Arctan } x$	$-\pi/2 < y < \pi/2$

The ranges of the other trigonometric functions may be restricted in the following manner.

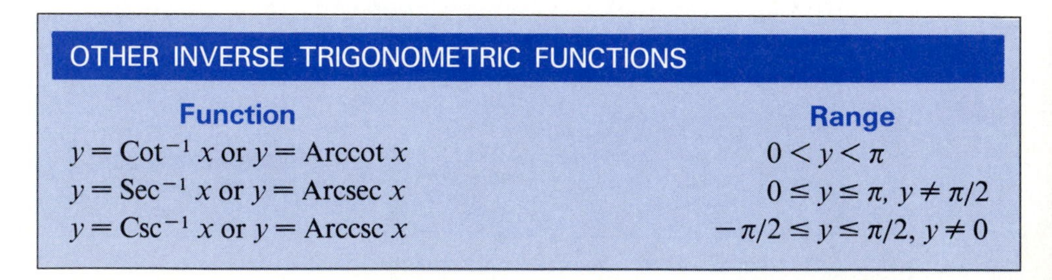

OTHER INVERSE TRIGONOMETRIC FUNCTIONS

Function	Range
$y = \text{Cot}^{-1} x$ or $y = \text{Arccot } x$	$0 < y < \pi$
$y = \text{Sec}^{-1} x$ or $y = \text{Arcsec } x$	$0 \leq y \leq \pi,\ y \neq \pi/2$
$y = \text{Csc}^{-1} x$ or $y = \text{Arccsc } x$	$-\pi/2 \leq y \leq \pi/2,\ y \neq 0$

EXAMPLE 1 Specify the correct value(s) for y.

(a) $y = \arcsin 0$ **(b)** $y = \text{Arcsin } 0$

Solution **(a)** Here, y represents an angle for which the sine is 0. Visualize the graph of $y = \sin x$. It crosses the x-axis at $0, \pi, 2\pi$, and at intervals of π before and thereafter. Hence, for $y = \arcsin 0$, $y = 0 \pm n\pi$ where $n = 1, 2, 3, 4. \ldots$

(b) Here, y represents the angle that lies in the interval from $-\pi/2$ to $\pi/2$, inclusive and such that the sine of the angle is 0. From Example 1(a), we know that the only angle that satisfies both conditions is 0. ■

EXAMPLE 2 Specify the correct value(s) for y.

(a) $y = \cos^{-1} 0$ **(b)** $y = \text{Cos}^{-1} 0$

Solution **(a)** Here, y represents an angle for which the cosine is 0. Visualize the graph of $y = \cos x$. It crosses the x-axis at $\pi/2$, $3\pi/2$, and at intervals of π before and thereafter. Hence, for $y = \cos^{-1} 0$, $y = \pi/2 \pm n\pi$ where $n = 1, 2, 3, 4. \ldots$

(b) Here, y represents the angle x that lies in the interval from 0 to π, inclusive and such that the cosine of the angle is 0. From Example 2(a), we know that the only angle that satisfies both conditions is $\pi/2$. ■

In Chapter 4, you used a calculator to evaluate inverse trigonometric functions. Recall that for a positive function value, an inverse trigonometric function key will return a first quadrant angle. For a negative function value, the absolute value of the function value is entered and the reference angle is returned. If the negative function value is entered directly, then the angle associated with the Arc function is returned. Thus, we make the following observations.

INVERSE TRIGONOMETRIC FUNCTIONS ON A CALCULATOR

(a) When the argument of an inverse trigonometric function is positive, the argument is entered, and a first quadrant angle is returned.
(b) When the argument of $\arcsin x$, $\arccos x$, or $\arctan x$ is negative, the absolute value of the argument is entered, and a reference angle is returned.
(c) When the argument of $\text{Arcsin } x$, $\text{Arccos } x$, or $\text{Arctan } x$ is negative, the argument may be entered directly, and the correct angle is returned.

EXAMPLE 3 Use a calculator to find the correct value for y in radians.

(a) $y = \arctan(-0.7631)$ and y is in the second quadrant

(b) $y = \text{Arctan}(-0.7631)$

Solution **(a)** The lower case *a* in arctan alerts us to enter 0.7631 as the argument. Put the calculator in radian mode. The keystroke sequence is as follows:

.7631 [INV] [tan] **0.6518325**

Since this is a reference angle and the angle sought is in the second quadrant, it is necessary to subtract the number on the display from π.

[STO] π [−] [RCL] [=] **2.4897601**

Answer: 2.490

(An alternate sequence would be [−] π [=] [+/−].)

(b) The upper case *A* in Arctan alerts us that a negative argument may be entered. The keystroke sequence is as follows.

.7631 [+/−] [INV] [tan] **−0.6518325**

Answer: −0.6518

Notice that the answers to Examples 3(a) and 3(b) are different. ■

EXAMPLE 4 The formula $\theta = \text{Tan}^{-1}[-Pa^2/(2EI)]$ gives the angle of deflection θ (in degrees) for a cantilever beam under a concentrated load. Find θ if $P = 2$ kg, $a = 30$ mm, $E = 4000$ kg/mm^2, and $I = 5$ mm^4. In the formula, P is the load (in kg), a is the distance (in mm) of the load from the support, E is the modulus of elasticity (in kg/mm^2), and I is the moment of inertia (in mm^4).

Solution The formula gives $\theta = \text{Tan}^{-1}\left(\dfrac{-2(30^2)}{2(4000)(5)}\right)$ or $\text{Tan}^{-1}(-0.045)$. The negative angle may be entered directly into a calculator in degree mode. The keystroke sequence is as follows.

.045 [+/−] [INV] [tan] **−2.5765718**

To the nearest whole degree, the angle of deflection is 3°, directed downward. ■

Recall from Section 3.4 that when the argument of a function is itself a function, we have a composition of functions. It is possible to find the composition of functions for a trigonometric function and an inverse trigonometric function.

EXAMPLE 5 Evaluate sin(Arccos x).

Solution In this example, Arccos x represents the function that is the argument of the sine function. Arccos x is the angle θ whose cosine is x. That is, there is a right triangle such that the side adjacent to θ has length x and the hypotenuse has length 1. By the Pythagorean theorem, the opposite side is $\sqrt{1 - x^2}$, as in Figure 11.38.

Since $\sin\theta$ is the ratio of the lengths of the side opposite θ and the hypotenuse, $\sin\theta$ is given by $\sin(\text{Arccos } x) = \sqrt{1 - x^2}$. ■

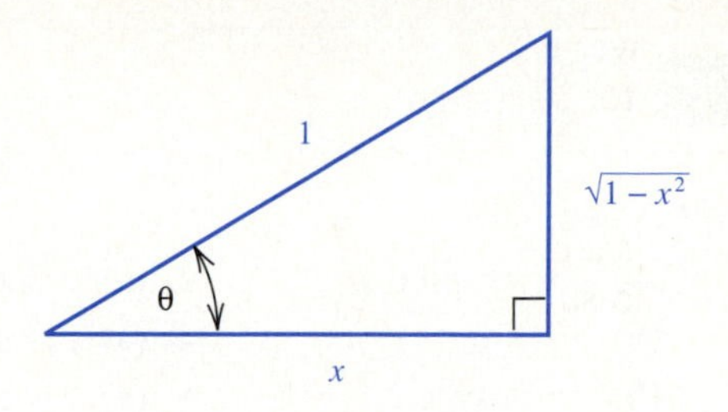

FIGURE 11.38

To evaluate $f[g(x)]$ where f is a trigonometric function and g is an inverse trigonometric function:

1. Draw a right triangle and label θ, the side opposite θ, the side adjacent to θ, and the hypotenuse as defined by $g(x)$.
2. Evaluate $f(\theta)$.

EXAMPLE 6 Evaluate sec(Arctan $2x$).

Solution 1. Arctan $2x$ is the angle θ whose tangent is $2x$. That is, there is a right triangle such that the side opposite θ has length $2x$ and the side adjacent to θ has length 1. The hypotenuse has length $\sqrt{1+(2x)^2}$ or $\sqrt{1+4x^2}$, as shown in Figure 11.39.

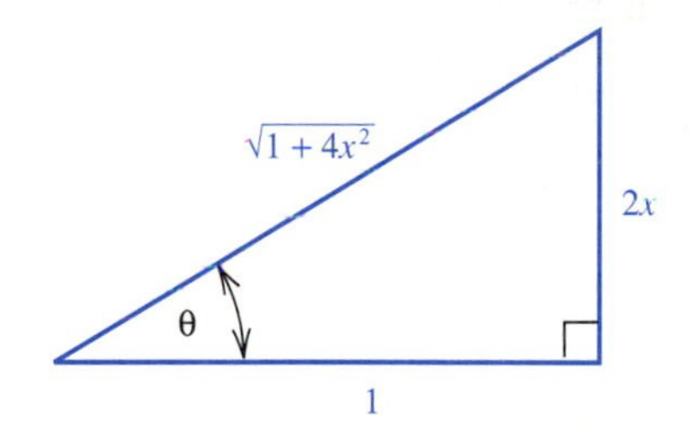

FIGURE 11.39

2. Since sec θ is the ratio of the lengths of the hypotenuse and the side adjacent to θ, sec θ is given by $\sec(\text{Arctan } 2x) = \sqrt{1+4x^2}$. ■

EXAMPLE 7 Evaluate $\sin(\text{Sin}^{-1} x)$.

Solution For inverse functions f and g, we know that $f \circ g(x) = x$. Since $y = \sin x$ and $y = \text{Sin}^{-1} x$ are inverse functions, $\sin(\text{Sin}^{-1} x) = x$ for all x for which $\text{Sin}^{-1} x$ is defined. ■

EXERCISES 11.6

In 1–14, specify the correct value(s) for y.

1. **(a)** $y = \arcsin 0.5$ **(b)** $y = \text{Arcsin } 0.5$

2. **(a)** $y = \arccos 0.5$ **(b)** $y = \text{Arccos } 0.5$

3. **(a)** $y = \tan^{-1} 1$ **(b)** $y = \text{Tan}^{-1} 1$

4. **(a)** $y = \cot^{-1} 1$ **(b)** $y = \text{Cot}^{-1} 1$

5. **(a)** $y = \text{arcsec } 2$ **(b)** $y = \text{Arcsec } 2$

6. **(a)** $y = \text{arccsc } 2$ **(b)** $y = \text{Arccsc } 2$

7. **(a)** $y = \sin^{-1}(-1)$ **(b)** $y = \text{Sin}^{-1}(-1)$

8. **(a)** $y = \cos^{-1}(-1)$ **(b)** $y = \text{Cos}^{-1}(-1)$

9. **(a)** $y = \arctan(-\sqrt{3})$ **(b)** $y = \text{Arctan}(-\sqrt{3})$

10. **(a)** $y = \text{arccot}(-\sqrt{3})$ **(b)** $y = \text{Arccot}(-\sqrt{3})$

11. **(a)** $y = \sec^{-1} 1$ **(b)** $y = \text{Sec}^{-1} 1$

12. **(a)** $y = \csc^{-1}(-1)$ **(b)** $y = \text{Csc}^{-1}(-1)$

13. **(a)** $y = \arcsin \sqrt{3}/2$ **(b)** $y = \text{Arcsin } \sqrt{3}/2$

14. **(a)** $y = \cos^{-1} \sqrt{3}/2$ **(b)** $y = \text{Cos}^{-1} \sqrt{3}/2$

In 15–28, use a calculator to find the correct value for y in radians.

15. **(a)** $y = \arcsin 0.5577$, quadrant II **(b)** $y = \text{Arcsin } 0.5577$

16. **(a)** $y = \arccos(-0.0407)$, quadrant II **(b)** $y = \text{Arccos}(-0.0407)$

17. **(a)** $y = \tan^{-1} 1.0899$, quadrant III **(b)** $y = \text{Tan}^{-1} 1.0899$

18. **(a)** $y = \cot^{-1} 0.8637$, quadrant IV **(b)** $y = \text{Cot}^{-1} 0.8637$

19. **(a)** $y = \text{arcsec } 3.5253$, quadrant IV **(b)** $y = \text{Arcsec } 3.5253$

20. **(a)** $y = \text{arccsc } 1.7931$, quadrant II **(b)** $y = \text{Arccsc } 1.7931$

21. **(a)** $y = \sin^{-1}(-0.8690)$, quadrant IV **(b)** $y = \text{Sin}^{-1}(-0.8690)$

22. **(a)** $y = \cos^{-1} 0.9706$, quadrant IV **(b)** $y = \text{Cos}^{-1} 0.9706$

23. **(a)** $y = \arctan(-0.9924)$, quadrant II **(b)** $y = \text{Arctan}(-0.9924)$

24. **(a)** $y = \text{arccot}(-0.2422)$, quadrant IV **(b)** $y = \text{Arccot}(-0.2422)$

25. **(a)** $y = \sec^{-1}(-1.3522)$, quadrant III **(b)** $y = \text{Sec}^{-1}(-1.3522)$

26. **(a)** $y = \csc^{-1}(-1.3321)$, quadrant III **(b)** $y = \text{Csc}^{-1}(-1.3321)$

27. **(a)** $y = \arcsin(-0.0783)$, quadrant III **(b)** $y = \text{Arcsin}(-0.0783)$

28. **(a)** $y = \cos^{-1}(-0.1814)$, quadrant III **(b)** $y = \text{Cos}^{-1}(-0.1814)$

In 29–42, evaluate each expression.

29. $\sin(\text{Arctan } x)$

30. $\cos(\text{Arcsin } x)$

31. $\tan(\text{Arccos } x)$

32. $\tan(\text{Arcsin } \sqrt{3}/2)$

33. $\cos(\text{Arctan } 1)$

34. $\sin(\text{Arccos } 1/2)$

35. $\cot(\text{Tan}^{-1} x)$

36. $\sec(\text{Cos}^{-1} x)$

37. $\csc(\text{Sin}^{-1} x)$

38. $\sin(\text{Sec}^{-1} \sqrt{2})$

39. $\cos(\text{Csc}^{-1} 2)$

40. $\tan(\text{Cot}^{-1} \sqrt{3})$

41. $\cos(\text{Cos}^{-1} x)$

42. $\tan(\text{Tan}^{-1} x)$

---

43. For a cantilever beam under a distributed load of w (in kg/mm), the angle of deflection is given by $\theta = \text{Tan}^{-1}[-wb^3/(6EI)]$. Find θ if $w = 0.02$ kg/mm, $b = 40$ mm, $E = 4000$ kg/mm^2, and $I = 5$ mm^4. In the formula, b is the length (in mm) of the load, E is the modulus of elasticity (in kg/mm^2), and I is the moment of inertia (in mm^4) of the beam. 1 2 4 8 7 9

44. The equation $\theta = \text{Cos}^{-1}(R/Z)$ gives the phase angle between **R** and **Z** in an ac circuit. In the formula, R is the magnitude of the resistance and Z is the magnitude of the impedance. Find θ if $R = 80\ \Omega$ and $Z = 100\ \Omega$. 3 5 9 1 2 4 6 7 8

45. The equation $c = \text{Sin}^{-1}(1/n)$ gives the critical angle for light entering a medium. In the formula,

n is the index of refraction of the medium that light enters in relation to the medium that it leaves. If $n = 4/3$, find c. 3 5 6 9

46. The equation $\theta_1 = \text{Tan}^{-1}(\omega L/R)$ gives the phase angle of impedance in the series portion of a distributed constant circuit. Find θ_1 if $\omega = 6280$ radians per second, $L = 0.49$ mH/km, and $R = 167\ \Omega$/km. In the formula, L is the inductance per unit length (in mH/km), and R is the resistance per unit length (in Ω/km). 3 5 9 1 2 4 6 7 8

47. The equation $\theta_2 = \text{Tan}^{-1}(\omega C/G)$ gives the phase angle of impedance in the parallel portion of a distributed constant circuit. Find θ_2 if $\omega = 380$ radians per second, $C = 0.05\ \mu$F/km, and $G = 1.66$ microsiemens/kilometer. In the formula, C is the capacitance per unit length (in μF/km), and G is the leakage conductance per unit length (in μS/km). 3 5 9 4 7 8

48. When a wide-angle lens is used on a camera to photograph an object that is not parallel to the film, to get the best "depth of field," the lens is turned at an angle α as shown in the diagram. The angle is given by the formula

$$\alpha = \text{Tan}^{-1}\left(\frac{x}{x+y}\tan\beta\right).$$

Find α when $x = 0.04$ m, $y = 12.00$ m, and $\beta = 82.0°$.* 3 5 6 9

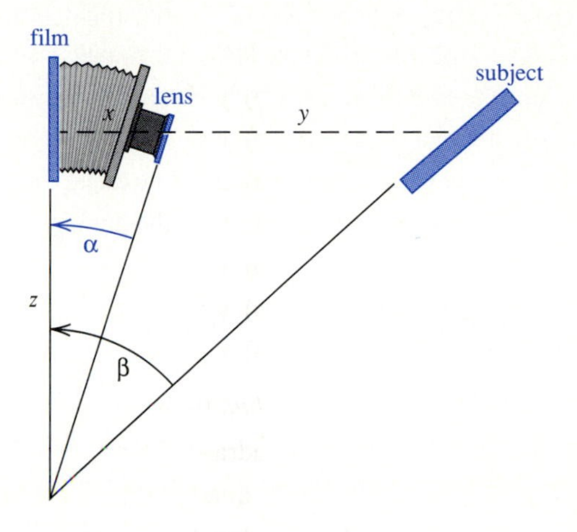

* From *A Sourcebook of Applications of School Mathematics*. Copyright © 1980 by The Mathematical Association of America. Reprinted by permission.

1. Architectural Technology
2. Civil Engineering Technology
3. Computer Engineering Technology
4. Mechanical Drafting Technology
5. Electrical/Electronics Engineering Technology
6. Chemical Engineering Technology
7. Industrial Engineering Technology
8. Mechanical Engineering Technology
9. Biomedical Engineering Technology

11.7 CHAPTER REVIEW

KEY TERMS

11.1 periodic functions
period
cycle
amplitude

11.2 phase angle
phase shift

11.3 asymptote

11.4 ordinate
addition of ordinates
parameter
parametric equations
Lissajous figures

11.5 simple harmonic motion

SYMBOLS

$y = \sin^{-1} x$ or $y = \arcsin x$	the angle y such that $x = \sin y$
$y = \cos^{-1} x$ or $y = \arccos x$	the angle y such that $x = \cos y$
$y = \tan^{-1} x$ or $y = \arctan x$	the angle y such that $x = \tan y$
$y = \text{Sin}^{-1} x$ or $y = \text{Arcsin } x$	the angle y such that $x = \sin y$ and $-\pi/2 \leq y \leq \pi/2$
$y = \text{Cos}^{-1} x$ or $y = \text{Arccos } x$	the angle y such that $x = \cos y$ and $0 \leq y \leq \pi$
$y = \text{Tan}^{-1} x$ or $y = \text{Arctan } x$	the angle y such that $x = \tan y$ and $-\pi/2 < y < \pi/2$

RULES AND FORMULAS

$y = a \sin (bx + c)$ has amplitude a, period $2\pi/b$, and phase shift $-c/b$

$y = a \cos (bx + c)$ has amplitude a, period $2\pi/b$, and phase shift $-c/b$

$y = a \tan (bx + c)$ has period π/b, and phase shift $-c/b$

$y = a \cot (bx + c)$ has period π/b, and phase shift $-c/b$

$y = a \sec (bx + c)$ has period $2\pi/b$, and phase shift $-c/b$

$y = a \csc (bx + c)$ has period $2\pi/b$, and phase shift $-c/b$

SEQUENTIAL REVIEW

(Section 11.1) *In 1–6, identify the amplitude and period of each function.*

1. $y = -2 \sin \frac{1}{2} x$

2. $y = \frac{1}{2} \sin 2x$

3. $y = \cos \frac{-2}{3} \cos 2x$

4. $y = \cos \frac{1}{3} x$

5. $y = 2 \sin 2\pi x$

6. $y = -\cos \frac{\pi x}{2}$

In 7 and 8, sketch the graph of each function.

7. $y = \frac{1}{2} \sin 4\pi x$

8. $y = -2 \cos 3x$

(Section 11.2) *In 9–14, identify the period, amplitude, and phase shift of each function.*

9. $y = \frac{1}{3} \sin(x - \pi)$

10. $y = 2 \cos \left(x + \frac{\pi}{2}\right)$

11. $y = -\cos \left(x - \frac{\pi}{3}\right)$

12. $y = 3 \sin \left(\frac{1}{2} x + \frac{\pi}{6}\right)$

13. $y = 2 \cos \left(\pi x - \frac{\pi}{2}\right)$

14. $y = \frac{1}{2} \sin \left(2\pi x + \frac{\pi}{2}\right)$

In 15 and 16, sketch the graph of each function.

15. $y = 3 \cos \left(\frac{1}{2} x - \frac{\pi}{3}\right)$

16. $y = \frac{-1}{2} \sin \left(2x + \frac{\pi}{6}\right)$

(Section 11.3) *In 17–19, identify the period and phase shift of each function.*

17. $y = -\tan\left(\frac{1}{2}x - \frac{\pi}{6}\right)$

18. $y = 2\cot\left(\pi x + \frac{\pi}{2}\right)$

19. $y = 1\sec\left(\frac{1}{2}x + \frac{\pi}{3}\right)$

In 20–23, sketch the graph of each function.

20. $y = \tan(3x - \pi)$

21. $y = 2\cot\left(\frac{1}{2}x + \frac{\pi}{6}\right)$

22. $y = -\csc\left(\pi x - \frac{\pi}{2}\right)$

23. $y = \frac{1}{3}\sec\left(x + \frac{\pi}{4}\right)$

(Section 11.4) *In 24–26, sketch the graph of each function, using addition of ordinates.*

24. $y = \frac{1}{2}x + \cos 2x$

25. $y = \sin x - 2\cos x$

26. $y = 2\cos\left(x - \frac{\pi}{2}\right) + \sin x$

In 27–29, plot the curve given by each pair of parametric equations for $t \geq 0$.

27. $x = \sqrt{t}$, $y = 2t$

28. $x = 2t$, $y = \cos t$

29. $x = \cos t$, $y = \cos 2t$

(Section 11.5) *In 30–32, identify the period, amplitude, and phase shift for the function given by $V = V_{max}\sin(\omega t + \theta)$ for each set of values.*

30. $V_{max} = 60$ V, $f = 30$ Hz, $\theta = -\pi/3$ s

31. $V_{max} = 30$ V, $f = 60$ Hz, $\theta = \pi$ s

32. $V_{max} = 240$ V, $f = 20$ Hz, $\theta = -\pi$ s

In 33 and 34, sketch the graph of $V = V_{max}\sin(\omega t + \theta)$ for each set of values.

33. $V_{max} = 12$ V, $f = 30$ Hz, $\theta = \pi/6$ s

34. $V_{max} = 9$ V, $f = 20$ Hz, $\theta = -\pi/3$ s

In 35 and 36, sketch the graph of $I = I_{max}\sin(\omega t + \theta)$ for each set of values.

35. $I_{max} = 8$A, $f = 30$ Hz, $\theta = \pi/2$ s

36. $I_{max} = 10$ A, $f = 30$ Hz, $\theta = -\pi/2$ s

(Section 11.6) *In 37 and 38, specify the correct value(s) for y.*

37. **(a)** $y = \text{arcsec}(-2)$ **(b)** $y = \text{Arcsec}(-2)$

38. **(a)** $y = \arcsin\sqrt{2}/2$ **(b)** $y = \text{Arcsin}\sqrt{2}/2$

In 39–41, use a calculator to find the correct value for y in radians.

39. **(a)** $y = \arccos 0.3721$, quadrant IV **(b)** $y = \text{Arccos } 0.3721$

40. **(a)** $y = \tan^{-1} 0.4632$, quadrant III **(b)** $y = \text{Tan}^{-1} 0.4632$

41. **(a)** $y = \sec^{-1}(-1.4731)$, quadrant II **(b)** $y = \text{Sec}^{-1}(-1.4731)$

In 42–45, evaluate each expression.

42. $\sin(\text{Arctan } 2x)$

43. $\cos(\text{Arcsin } 1/2)$

44. $\tan(\text{Sin}^{-1}\sqrt{3}/2)$

45. $\csc(\text{Tan}^{-1} x)$

APPLIED PROBLEMS

1. The average yearly temperature at a particular location normally occurs on or about March 21. The seasonal variation from this average may be approximated by a sine curve. In Baltimore, MD, if the highest temperature is 88°F and the lowest is about 26°F, write an equation to approximate the variation in temperature from average. Use x for the number of days after March 21.
6 8 1 2 3 5 7 9
2. Modify the equation of problem 1 to begin on January 1, rather than March 21.
6 8 1 2 3 5 7 9
3. In an ac circuit, a capacitor causes the voltage to peak after the current. In a certain circuit with a resistor and a capacitor, the current (in amperes) is given by $I = 5 \sin 377t$. Write the equation for voltage, assuming the voltage has an amplitude of 120 V and lags the current by 56°. (Hint: To say that the voltage lags the current by 56° means the phase angle is 56°.) 3 5 9 1 2 4 6 7 8
4. One tuning fork vibrates, creating a wave with period 2 ms and amplitude 2 mm. A second tuning fork vibrates, creating a wave with period 3 ms and amplitude 2 mm. If the two waves are in phase and combine, sketch the graph that represents the sum. 1 3 4 5 6 8 9
5. The function defined by $y = \frac{1}{2} + \frac{2}{\pi} \sin x + \frac{2}{3\pi} \sin 3x$ is used to approximate a square wave function. Sketch the graph of this function for $0 \leq x \leq 2\pi$. 5 3 4 7 8 9
6. When a voltage V (in volts) is applied to an RL circuit for time t (in seconds), a current I (in amps) results. The power (in watts) in this circuit is given by $P = VI$. What power results at $t = 5$ ms if $V = 170 \sin (377t)$ and $I = 18 \sin (377t - 58°)$? (Note: $377t$ is in radian measure.) Give your answer to three significant digits.
3 5 9 1 2 4 6 7 8
7. When a voltage V (in volts) is applied to an ordinary 100-watt light bulb, a current I (in amps) flows. The bulb's power P (in watts) is given by $P = IV$. If $I = 1.176 \sin (377t)$ and $V = 170 \sin 377t$, how much power will the bulb have **(a)** after one time period has passed? **(b)** after $\frac{1}{4}$ of a time period? 3 5 9 1 2 4 6 7 8
8. A voltage V (in volts) given by $V = 250 \sin 6000t$ is applied to a certain circuit. At what time will this voltage first reach 100 volts?
3 5 9 1 2 4 6 7 8
9. Assume the sun rises at 6:00 A.M. and sets at 6:00 P.M. and shines its brightest (100%) on a solar cell when directly overhead at noon. **(a)** What sinusoidal equation would describe this brightness as a function of time measured in hours from midnight? **(b)** At what time in the morning would the brightness be half its greatest value?
6 8 1 2 3 5 7 9
10. When a voltage V (in volts) is applied to a certain RL circuit the resulting current I (in amps) is given by $I = 18 \sin (377t - 58°)$. What power P (in watts) would be present at $t = 8$ ms if $V = 170 \sin 377t$ and $P = IV$. (Note: $377t$ is in radian measure.) 3 5 9 4 7 8
11. The voltage v (in volts) applied to the base of a certain transistor was given by the equation, $v = 5 \sin 2500t - 4.3$. Sketch this wave, labeling the t-axis in milliseconds. (The portion of the time period this voltage is greater than or equal to zero is called the duty cycle.) 3 5 9 4 7 8
12. In a capacitive ac circuit the instantaneous impedance Z can be found by taking the ratio of the voltage v to the current i at any instant. Suppose a voltage $v = 100 \sin 377t$ was applied to a capacitor resulting in a current $i = 20 \sin (377t + 90°)$. Find Z after $\frac{1}{8}$ of the time period. (Note: $377t$ is in radian measure.) 3 5 9 4 7 8

---

1. Architectural Technology
2. Civil Engineering Technology
3. Computer Engineering Technology
4. Mechanical Drafting Technology
5. Electrical/Electronics Engineering Technology
6. Chemical Engineering Technology
7. Industrial Engineering Technology
8. Mechanical Engineering Technology
9. Biomedical Engineering Technology

13. A plane flying level at 500 ft/s and 400 ft above the ground fires a missile that accelerates at 40 ft/s^2 away from the plane as it falls due to gravity. Its progress in the x-direction is given by $x = 500t + 20t^2$, and as it falls in the y-direction by $y = -16.1t^2$. When and where will it hit the ground? 4 7 8 1 2 3 5 6 9

14. In electronics, a certain type of radio wave is a sine wave of very high frequency with an amplitude that varies in a sinusoidal manner also, usually with a very low frequency compared to the carrier. The equation $v = (20 + 5 \sin 2000\pi t) \sin (1\,220\,000\pi t)$ represents such a wave.
(a) Find the range of its amplitude.
(b) Find the frequency f of the amplitude. Recall that frequency is the reciprocal of period.
1 3 4 5 6 8 9

15. When the frequency of a sine wave of voltage varies in a sinusoidal manner, the result is called frequency modulation (FM). $V = 25 \sin 2\pi (10^8 + 10^7 \sin 8 \times 10^4 \pi t)t$ represents such a wave. What is the range of its frequency? Recall that frequency is the reciprocal of period.
3 5 9 1 2 4 6 7 8

16. When a boat crosses a river, its velocity relative to the shore v_B is related to the speed of the current v_c by the equation $v = v_c \csc \theta$ where θ is the angle between a line that is perpendicular to the shore and a line that represents the boat's course. Find the speed with which a boat appears to travel with respect to an observer at the dock if the current flows at 7 mph and the angle of drift is 15°.
4 7 8 1 2 3 5 6 9

17. The formula $\theta = \text{Sec}^{-1}\left(\frac{F}{F_1}\right)$ gives the angle at which one of a given perpendicular pair of forces F_1 and F_2 must be applied to replace a fixed force F. What angle would be between F and F_1 for $F = 35$ lb and $F_1 = 21$ lb? 4 7 8 1 2 3 5 6 9

18. A company that builds swimming pools fills an unusual order by designing a pool whose shape is given by $y = |x| \pm \sqrt{1 - x^2}$. Use addition of ordinates to sketch the sum of $y = |x|$ and $y = \pm\sqrt{1 - x^2}$. What shape does the pool have?
1 2 4 7 8 3 5

RANDOM REVIEW

1. Identify the period and phase shift for $y = 2 \sec (3x - \pi)$.
2. Sketch the graph of $I = I_{max} \sin(\omega t + \theta)$ if $I_{max} = 8\text{A}$, $f = 60$ Hz, and $\theta = -\pi/4$.
3. Identify the period, amplitude, and phase shift for $y = 2 \sin (x + \pi)$.
4. Use a calculator to find the correct value of y in radians if $y = \arcsin (-0.3214)$ and y is in quadrant III.
5. Identify the period and amplitude of $y = 3 \cos 4x$.
6. Sketch the graph of $y = x + \cos \frac{1}{2} x$.
7. Use a calculator to find the correct value of y in radians if $y = \text{Arcsin} (-0.3214)$.
8. Sketch the graph of $y = \frac{1}{2} \tan \left(x + \frac{\pi}{4}\right)$.
9. Evaluate $\cot(\text{Tan}^{-1} x)$.
10. Sketch the graph of $y = -\cos (4\pi x + \pi)$.

CHAPTER TEST

1. Identify the period, amplitude, and phase shift for $y = \frac{1}{2}\cos\left(2x - \frac{\pi}{3}\right)$.
2. Identify the period and phase shift for $y = \cot(2\pi x + \pi/2)$.

In 3–5, sketch the graph of each equation.

3. $y = 3\cos(2x + \pi/2)$
4. $y = \csc\left(\frac{x}{2} - \frac{\pi}{6}\right)$
5. $x = t + 1$, $y = \sin t$

In 6 and 7, use a calculator to find the correct value of y in radians.

6. $y = \arctan(-1.5263)$ and y is in quadrant II
7. $y = \text{Arctan}(-1.5263)$
8. Evaluate $\sec(\text{Cos}^{-1} 0.5)$.
9. An object dropped from a helicopter moving at 88 ft/s moves a horizontal distance (in ft) given by $x = 88t$ in t seconds. The vertical distance fallen (in ft) is given by $y = -16t^2$. Sketch a graph of this motion for $0 \leq t \leq 0.25$.
10. A certain alternating current has a peak current of 10 A, a frequency of 60 Hz, and a phase angle of $\pi/4$. Write the equation that gives this current I in the form $I = I_{max} \sin(\omega t + \theta)$.

EXTENDED APPLICATION: BIOMEDICAL ENGINEERING TECHNOLOGY

CALIBRATING BLOOD FLOW

The measurement of blood flow poses special technical problems. Methods used in industry to measure flow of other liquids cannot be used, because they require cutting the blood vessel. In medical research with animals, a catheter may be inserted into the blood vessel. Another method is to lift the blood vessel up and attach it to a transducer, which is a device that converts energy to an electric signal. But these methods are not used routinely with humans.

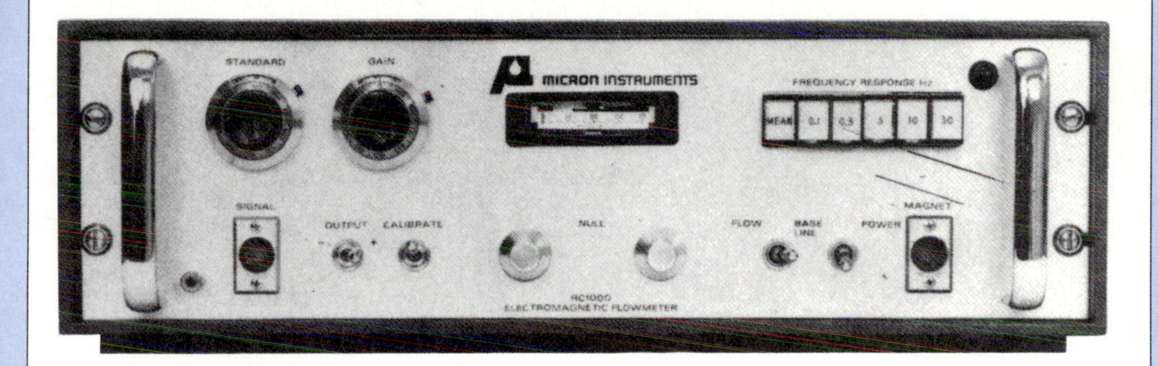

One method that is practical in clinical applications is to use a magnetic blood flow meter. The blood flow is measured in a given vessel, and that information is used to determine blood flow in an entire section of the body. A probe is placed around a blood vessel by slipping the vessel through a slot in the probe. That is, the vessel runs through the probe, like a thread through the eye of a needle. To operate properly, the probe must fit tightly around the blood vessel, so the probes used are made in various sizes. A probe can be chosen so that it will match the cross-sectional area of the vein at the point of measurement, and the transducer can be calibrated directly in liters/minute.

The probe contains an electromagnet, which generates a magnetic field perpendicular to the direction of blood flow. The electromagnet is used to induce a small voltage (such as a few microvolts) in the moving column of blood. A transducer, however, acts like a transformer because of the alternating current in the electromagnet. It induces a voltage that is sometimes quite large in comparison to the voltage to be measured. Certain types of amplifiers make it possible to recover the correct signal. The type of waveform used for the magnet current can

EXTENDED APPLICATION, continued

also help minimize the problem of induced voltage. When the magnet current is sinusoidal, the induced voltage is also sinusoidal but has a phase shift of 90°, as shown in Figure 11.40. Other waveforms are also useful. If a square waveform is used for the magnet current, the error signal of the voltage shows spikes when the current changes direction, but then levels off to zero, as shown in Figure 11.41. If a trapezoid wave form is used, the induced voltage appears as in Figure 11.42.

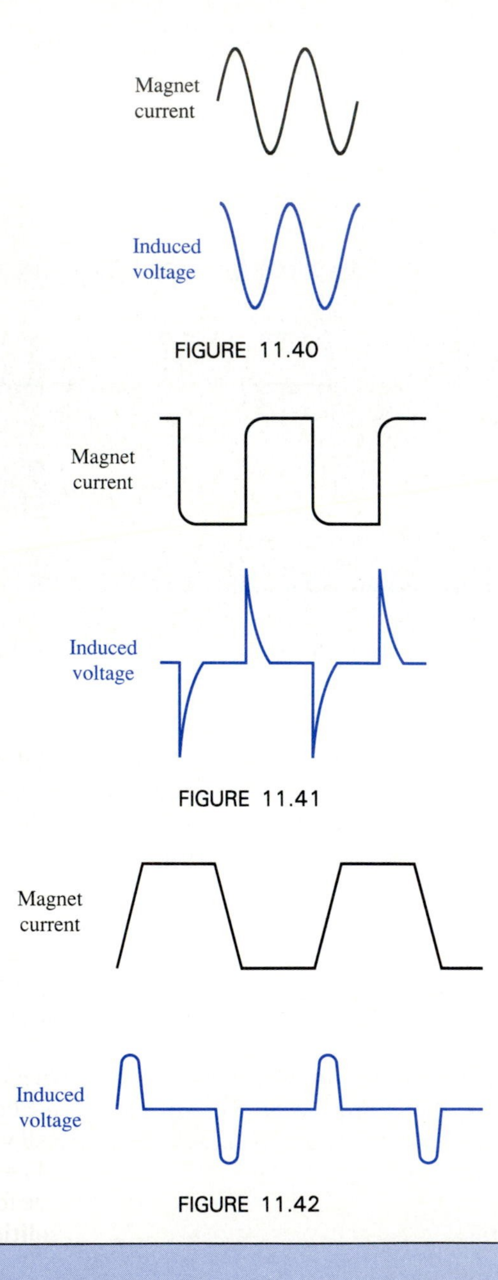

FIGURE 11.40

FIGURE 11.41

FIGURE 11.42

CHAPTER 12 TRIGONOMETRIC IDENTITIES AND EQUATIONS

In science and technology, equations involving trigonometric functions are common. Knowing certain relationships among the trigonometric functions often makes it possible to solve such equations or to simplify formulas. The purpose of this chapter is to introduce the techniques necessary to work effectively with trigonometric equations.

12.1 FUNDAMENTAL IDENTITIES

contradiction
identity
conditional equation

To solve an equation means to find the value (or values) of the variable for which the equation is true. Some equations, like $x = x + 1$, have no solution. Such an equation is called a **contradiction.** An equation that is true for all values for which the equation is defined is called an **identity.** The equation $x(x + 1) = x^2 + x$ is an example of an identity. An equation such as $2x + 1 = 3$, which is true for some value(s) (in this case, $x = 1$) but false for other values of x, is called a **conditional equation.**

In applied problems, it is sometimes necessary to solve equations involving trigonometric functions. Because it may be helpful to substitute a simpler expression for a complicated expression, trigonometric identities are particularly important. A trigonometric identity tells us that two expressions are equal for all valid values of the variable. Therefore, one expression may be replaced by the other.

Recall that the reciprocal of a nonzero real number x is $1/x$. Let θ be an angle in standard position determined by the point (x, y) in the plane. If the trigonometric functions are defined as they are in Section 4.4, we obtain the following identities by examining the reciprocal of each function.

$$\mathbf{1/\sin\theta} = 1 \div \sin\theta = 1 \div (y/r) = 1(r/y) = r/y = \mathbf{\csc\theta}$$
$$\mathbf{1/\cos\theta} = 1 \div \cos\theta = 1 \div (x/r) = 1(r/x) = r/x = \mathbf{\sec\theta}$$
$$\mathbf{1/\tan\theta} = 1 \div \tan\theta = 1 \div (y/x) = 1(x/y) = x/y = \mathbf{\cot\theta}$$
$$\mathbf{1/\cot\theta} = 1 \div \cot\theta = 1 \div (x/y) = 1(y/x) = y/x = \mathbf{\tan\theta}$$
$$\mathbf{1/\sec\theta} = 1 \div \sec\theta = 1 \div (r/x) = 1(x/r) = x/r = \mathbf{\cos\theta}$$
$$\mathbf{1/\csc\theta} = 1 \div \csc\theta = 1 \div (r/y) = 1(y/r) = y/r = \mathbf{\sin\theta}$$

In a similar fashion, it can be shown that

$$\tan\theta = \frac{\sin\theta}{\cos\theta} \text{ and } \cot\theta = \frac{\cos\theta}{\sin\theta}.$$

$$\frac{\sin\theta}{\cos\theta} = \sin\theta \div \cos\theta = \frac{y}{r} \div \frac{x}{r} = \frac{y}{r}\left(\frac{r}{x}\right) = \frac{y}{x} = \tan\theta$$

$$\frac{\cos\theta}{\sin\theta} = \cos\theta \div \sin\theta = \frac{x}{r} \div \frac{y}{r} = \frac{x}{r}\left(\frac{r}{y}\right) = \frac{x}{y} = \cot\theta$$

NOTE ⇨⇨ ***These identities are particularly important, because they, along with the reciprocal identities, make it possible to express each trigonometric function in terms of the sine and/or cosine functions.***

BASIC IDENTITIES

$\tan\theta = \sin\theta/\cos\theta$	$\sec\theta = 1/\cos\theta$
$\cot\theta = \cos\theta/\sin\theta$	$\csc\theta = 1/\sin\theta$

The Pythagorean theorem can be used to develop relationships among the trigonometric functions. If θ is an angle in standard position determined by the point (x, y), then

$$x^2 + y^2 = r^2 \qquad \textbf{(12-1)}$$

If both sides of the equation are divided by r^2,

$$\frac{x^2}{r^2} + \frac{y^2}{r^2} = \frac{r^2}{r^2} \text{ or } \left(\frac{x}{r}\right)^2 + \left(\frac{y}{r}\right)^2 = 1 \text{ or } \cos^2\theta + \sin^2\theta = 1.$$

Recall that $\sin^2\theta = (\sin\theta)^2$. If both sides of Equation 12-1 are divided by x^2 instead of r^2,

$$\frac{x^2}{x^2}+\frac{y^2}{x^2}=\frac{r^2}{x^2} \text{ or } 1+\left(\frac{y}{x}\right)^2=\left(\frac{r}{x}\right)^2 \text{ or } 1+\tan^2\theta=\sec^2\theta.$$

If both sides of Equation 12-1 are divided by y^2 instead of r^2,

$$\frac{x^2}{y^2}+\frac{y^2}{y^2}=\frac{r^2}{y^2} \text{ or } \left(\frac{x}{y}\right)^2+1=\left(\frac{r}{y}\right)^2 \text{ or } \cot^2\theta+1=\csc^2\theta.$$

These identities are known as Pythagorean identities.

PYTHAGOREAN IDENTITIES

$\sin^2\theta+\cos^2\theta=1$ $\qquad$ $1+\tan^2\theta=\sec^2\theta$ $\qquad$ $1+\cot^2\theta=\csc^2\theta$

The Pythagorean identities may be used in the following forms.

$\sin^2\theta = 1-\cos^2\theta$ $\qquad$ $\sec^2\theta-\tan^2\theta=1$ $\qquad$ $\csc^2\theta-\cot^2\theta=1$

$\cos^2\theta=1-\sin^2\theta$ $\qquad$ $\sec^2\theta-1=\tan^2\theta$ $\qquad$ $\csc^2\theta-1=\cot^2\theta$

The basic identities and Pythagorean identities are used to verify other identities. To work with identities, you must understand how to interpret a trigonometric expression that has no parentheses. For example, $\sin x \sin y$ means $(\sin x)(\sin y)$. Also, $\sin ax$ means $\sin(ax)$, but $\sin x + a$ is understood to be $(\sin x) + a$, instead of $\sin(x+a)$.

EXAMPLE 1 Prove that $\cot\theta\sec\theta=\csc\theta$.

Solution Since $\cot\theta=\cos\theta/\sin\theta$ and $\sec\theta=1/\cos\theta$, we make these substitutions.

$$\begin{aligned}\cot\theta\sec\theta &= \frac{\cos\theta}{\sin\theta}\left(\frac{1}{\cos\theta}\right)\\ &= \frac{1}{\sin\theta}\\ &= \csc\theta \quad \blacksquare\end{aligned}$$

Example 1 illustrates several important concepts. First, proving an identity is not the same thing as solving an equation. For an equation, it is known that the expressions on the two sides of the equation are equal, so it is permissible to add, subtract, multiply, and divide on both sides by the same nonzero number. To prove an identity, we may not assume that the expressions on the two sides of the equation are equal. The goal is to *show* that they are equal. Generally, a series of substitutions and algebraic manipulations are performed on only *one side of the equation.* It is usually easier to work with the more complicated expression, so the expression on the left-hand side was manipulated in Example 1.

A second important concept illustrated in Example 1 is that in working with trigonometric identities, a common technique is to express all functions in terms of the sine and cosine functions. The technique is not always the shortest or cleverest thing to use, but it is a useful tool if you don't see a better way to work the problem.

NOTE ⇨⇨ ***Finally, it is a common mistake to omit the argument of a function.*** It is incorrect to write, for example, $\sin^2 + \cos^2 = 1$. The correct notation is $\sin^2 \theta + \cos^2 \theta = 1$.

Some hints for proving identities follow.

HINTS FOR PROVING IDENTITIES

1. Choose the more complicated side to manipulate.
2. Look for obvious uses of the basic and Pythagorean identities.
3. If the functions on one side of the equation are related to the functions on the other side by a Pythagorean identity, consider making a substitution that will eliminate one of the functions.
4. Examine the pattern of terms and factors on each side of the equation. It is often possible to reduce two terms to one term by using a Pythagorean identity, to factor an expression with terms, or to add two fractions to obtain a single fraction.
5. Consider expressing all functions in terms of sine and cosine functions.

EXAMPLE 2 Prove the identity $\sin^2 x\,(1 + \cot^2 x) = 1$.

Solution The expression on the left is more complicated. The pattern $1 + \cot^2 x$ suggests the Pythagorean identity $1 + \cot^2 x = \csc^2 x$. Making the substitution, we have the following equation.

$$\begin{aligned} \sin^2 x\,(1 + \cot^2 x) &= \sin^2 x \csc^2 x \\ &= \sin^2 x\left(\frac{1}{\sin^2 x}\right) \\ &= 1 \quad \blacksquare \end{aligned}$$

Trigonometric identities can sometimes be used to simplify formulas in the technologies.

EXAMPLE 3 The brightness B of light at a point illuminated by a single source is given by $B = \dfrac{k \tan \theta}{r^2 \sec \theta}$. In the formula, k is a constant, and r is the distance from the source to the point at which B is measured. Show that $\dfrac{k \tan \theta}{r^2 \sec \theta} = \dfrac{k \sin \theta}{r^2}$.

Solution The expression on the left is more complicated. Rewriting the functions in terms of sine and cosine functions, we have the following equation.

$$\frac{k \tan \theta}{r^2 \sec \theta} = \frac{k\left(\dfrac{\sin \theta}{\cos \theta}\right)}{r^2\left(\dfrac{1}{\cos \theta}\right)}$$

$$= k\left(\frac{\sin \theta}{\cos \theta}\right) \div \left(\frac{r^2}{\cos \theta}\right)$$

$$= k\left(\frac{\sin \theta}{\cos \theta}\right)\left(\frac{\cos \theta}{r^2}\right)$$

$$= \frac{k \sin \theta}{r^2} \quad ■$$

The remaining examples illustrate the techniques of factoring and working with fractions.

EXAMPLE 4 Prove the identity $\sin y = \sin^3 y + \sin y \cos^2 y$.

Solution The expression on the right is more complicated, and it can be factored.

$$\sin^3 y + \sin y \cos^2 y = \sin y\,(\sin^2 y + \cos^2 y)$$
$$= \sin y\,(1)$$
$$= \sin y \quad ■$$

EXAMPLE 5 Prove the identity $\csc^2 \theta + 2 \cot \theta = (1 + \cot \theta)^2$.

Solution The expression on the left is more complicated. Since the cosecant function on the left and the cotangent on the right are related by a Pythagorean identity, we make the substitution $\csc^2 \theta = 1 + \cot^2 \theta$.

$$\csc^2 \theta + 2 \cot \theta = 1 + \cot^2 \theta + 2 \cot \theta$$

If the terms of the expression are rearranged, it is easy to see that the expression can be factored.

$$1 + \cot^2 \theta + 2 \cot \theta = 1 + 2 \cot \theta + \cot^2 \theta$$
$$= (1 + \cot \theta)^2 \quad ■$$

EXAMPLE 6 Prove the identity $\csc^2 x \sec x = \dfrac{1}{\cos x} + \dfrac{\cos x}{\sin^2 x}$.

Solution The expression on the right-hand side is more complicated. To add fractions, the lowest common denominator is needed. Both fractions are written with $\cos x \sin^2 x$ as the denominator.

$$\begin{aligned}\frac{1}{\cos x}+\frac{\cos x}{\sin^2 x} &= \frac{1}{\cos x}\left(\frac{\sin^2 x}{\sin^2 x}\right)+\frac{\cos x}{\sin^2 x}\left(\frac{\cos x}{\cos x}\right)\\ &= \frac{\sin^2 x+\cos^2 x}{\cos x\sin^2 x}\\ &= \frac{1}{\cos x\sin^2 x}\\ &= \frac{1}{\cos x}\left(\frac{1}{\sin^2 x}\right)\\ &= \sec x\csc^2 x \quad \blacksquare\end{aligned}$$

EXAMPLE 7 Prove the identity $\dfrac{\cos y}{1-\sin y}=\sec y+\tan y$.

Solution The expression on the left is more complicated. The pattern $1-\sin y$ suggests $1-\sin^2 y$. We multiply the fraction by $\dfrac{1+\sin y}{1+\sin y}$.

$$\begin{aligned}\frac{\cos y}{1-\sin y} &= \left(\frac{\cos y}{1-\sin y}\right)\left(\frac{1+\sin y}{1+\sin y}\right)\\ &= \frac{\cos y\,(1+\sin y)}{1-\sin^2 y}\\ &= \frac{\cos y\,(1+\sin y)}{\cos^2 y}\\ &= \frac{(1+\sin y)}{\cos y}\\ &= \frac{1}{\cos y}+\frac{\sin y}{\cos y}\\ &= \sec y+\tan y \quad \blacksquare\end{aligned}$$

EXERCISES 12.1

In 1–42, prove each identity.

1. $\sin\theta\csc\theta=1$

2. $\sec\theta\cos\theta=1$

3. $1=\tan x\cot x$

4. $\cot x=\cos x\csc x$

5. $\sin y\cot y=\cos y$

6. $\cos y\tan y=\sin y$

7. $\dfrac{\tan\theta}{\sec\theta}=\sin\theta$

8. $\dfrac{\cot\theta}{\cos\theta}=\csc\theta$

9. $\cos x = \dfrac{\cot x}{\csc x}$

10. $\sin x = \dfrac{\cos x}{\cot x}$

11. $\tan y = \dfrac{\sec y}{\csc y}$

12. $\cot y\,(1 + \tan^2 y) = \sec y \csc y$

13. $\tan\theta\,(1 + \cot^2\theta) = \sec\theta\csc\theta$

14. $\tan\theta = \tan\theta\,(\sin^2\theta + \cos^2\theta)$

15. $\cot^2 x = \csc^2 x\,(1 - \sin^2 x)$

16. $\tan^2 x = \sec^2 x\,(1 - \cos^2 x)$

17. $\dfrac{\sec^2 y - \tan^2 y}{\cos y} = \sec y$

18. $\dfrac{\csc^2 y - \cot^2 y}{\sin y} = \csc y$

19. $\csc^2\theta = \cot\theta\,(\tan\theta + \cot\theta)$

20. $\sec x = \cos x + \sin x\,(\tan x)$

21. $\tan x\,(\cos x + 1) = \sin x + \tan x$

22. $\cot y\,(\sin y + 1) = \cos y + \cot y$

23. $\tan y + \cot y = \sec y \csc y$

24. $\sin\theta\cos\theta\,(\tan\theta + \csc\theta) = \sin^2\theta + \cos\theta$

25. $\sin\theta + \cos^2\theta = \sin\theta\cos\theta\,(\sec\theta + \cot\theta)$

26. $\sin x = \csc x - \cot x \cos x$

27. $\sec x - \tan x \sin x = \cos x$

28. $\dfrac{\csc y}{\sin y} + \dfrac{\sec y}{\cos y} = \csc^2 y \sec^2 y$

29. $\cot^2 y \cos^2 y = \dfrac{\cot y}{\tan y} - \dfrac{\cos y}{\sec y}$

30. $\tan^2 x \sin^2 x = \dfrac{\tan x}{\cot x} - \dfrac{\sin x}{\csc x}$

31. $\dfrac{\sin x}{\csc x} + \dfrac{\cos x}{\sec x} = 1$

32. $\dfrac{\sin\theta}{1 + \cos\theta} = \dfrac{1 - \cos\theta}{\sin\theta}$

33. $\dfrac{1 - \sin\theta}{\cos\theta} = \dfrac{\cos\theta}{1 + \sin\theta}$

34. $\dfrac{\cos x}{\sec x - \tan x} = 1 + \sin x$

35. $1 + \cos x = \dfrac{\sin x}{\csc x - \cot x}$

36. $\cot y\,(\cos y + 1) = \dfrac{\sin y}{\sec y - 1}$

37. $\sin^4 y - \cos^4 y = \sin^2 y - \cos^2 y$

38. $\sec^4\theta - \tan^4\theta = \sec^2\theta + \tan^2\theta$

39. $\cos^2\theta = \cos^2\theta\sin^2\theta + \cos^4\theta$

40. $\cot^2 x = \cot^2 x \csc^2 x - \cot^4 x$

41. $\sec^4 x + \sec^2 x - 2 = (\sec^2 x + 2)\tan^2 x$

42. $\csc^4 y + 2\csc^2 y - 3 = (\csc^2 y + 3)\cot^2 y$

---

43. The maximum height h attained by a projectile launched with velocity v_0 at an angle of θ, is given by

$$h = v_0 \sin\theta\left(\frac{v_0 \sin\theta}{g}\right) - \frac{1}{2}g\left(\frac{v_0 \sin\theta}{g}\right)^2.$$

It is also given by $h = \dfrac{v_0^2 \sin^2\theta}{2g}$. In both formulas, g is the acceleration due to gravity. Show that the formulas are equivalent. 4 7 8 1 2 3 5 6 9

44. Due to atmospheric refraction, an astronomical object appears higher in the sky than it would if there were no air. If h is the apparent altitude in degrees, the refraction R in minutes of arc is given by the product of

$$\cos\left(\frac{h^2 + 4.4\,h + 7.31}{h + 4.4}\right)$$

and

$$\csc\left(\frac{h^2 + 4.4\,h + 7.31}{h + 4.4}\right).$$

Show that R can also be written as $R = \dfrac{1}{\tan\dfrac{h^2 + 4.4\,h + 7.31}{h + 4.4}}$. 3 5 6 9

45. The magnification of a telescope is given by $m = \left(\dfrac{\sin\alpha}{\sin\beta}\right)\left(\dfrac{\cos\beta}{\cos\alpha}\right)$ where β is the angular separa-

1. Architectural Technology
2. Civil Engineering Technology
3. Computer Engineering Technology
4. Mechanical Drafting Technology
5. Electrical/Electronics Engineering Technology
6. Chemical Engineering Technology
7. Industrial Engineering Technology
8. Mechanical Engineering Technology
9. Biomedical Engineering Technology

tion between two stars when α is the angular separation between their images. Show that $m = \tan \alpha / \tan \beta$. 3 5 6 9

46. Snell's law says that $n = \sin \theta_1 / \sin \theta_2$. In the formula, n is the index of refraction of the medium that light enters in relation to the medium that it leaves, θ_1 is the angle of incidence, and θ_2 is the angle of refraction. Show that $\dfrac{\sin \theta_1}{\sin \theta_2} = \dfrac{\tan \theta_1 \sec \theta_2}{\sec \theta_1 \tan \theta_2}$. 3 5 6 9

47. In an ac series circuit, the magnitude of the impedance Z (in ohms) is given by $Z = \sqrt{R^2 + X^2}$. If θ is the phase angle, then $R = Z \cos \theta$ and $X = Z \sin \theta$. Use these equations to derive the identity $\sin^2 \theta + \cos^2 \theta = 1$. 3 5 9 1 2 4 6 7 8

48. A cube rests on a wedge as shown in the diagram. The coefficient of friction between the cube and the wedge is μ, the angle of inclination of the wedge is θ, and g is the acceleration due to gravity. If a force acts to produce an acceleration on the wedge, the acceleration may be given by the two formulas that follow. Show that they are equivalent.

$$a = \frac{g(\mu - \tan \theta)}{1 + \mu \tan \theta} \quad \text{and} \quad a = \frac{g(\mu \cos \theta - \sin \theta)}{\cos \theta + \mu \sin \theta}$$

4 7 8 1 2 3 5 6 9

49. Write all six trigonometric functions in terms of $\tan \theta$ and $\cot \theta$.

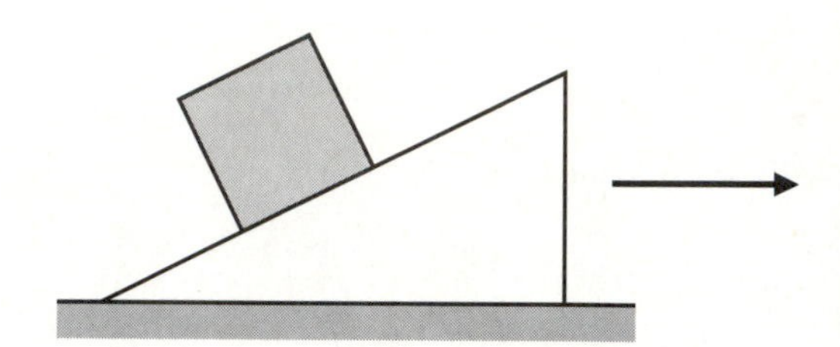

1. Architectural Technology
2. Civil Engineering Technology
3. Computer Engineering Technology
4. Mechanical Drafting Technology
5. Electrical/Electronics Engineering Technology
6. Chemical Engineering Technology
7. Industrial Engineering Technology
8. Mechanical Engineering Technology
9. Biomedical Engineering Technology

12.2 SINE AND COSINE OF THE SUM OR DIFFERENCE OF TWO ANGLES

The sine and cosine of the sum of two angles may be expressed using functions of the individual angles. These relationships are expressed in Equations 12-2 and 12-3 and are important trigonometric identities.

$$\sin (\alpha + \beta) = \sin \alpha \cos \beta + \cos \alpha \sin \beta \tag{12-2}$$

$$\cos (\alpha + \beta) = \cos \alpha \cos \beta - \sin \alpha \sin \beta \tag{12-3}$$

In this section, we will prove these identities, and also examine the sine and cosine of the difference of two angles.

To prove the identity of Equation 12-2, consider any two angles α and β as shown in Figure 12.1.

Let a, b, and c denote the lengths of the sides of right triangle ROS. Triangle POR, which is not a right triangle, has one side of length c. Let d and e denote the lengths of the other sides. Draw auxiliary lines as shown in Figure 12.2. Let f and g denote the lengths of the legs of triangle PRQ.

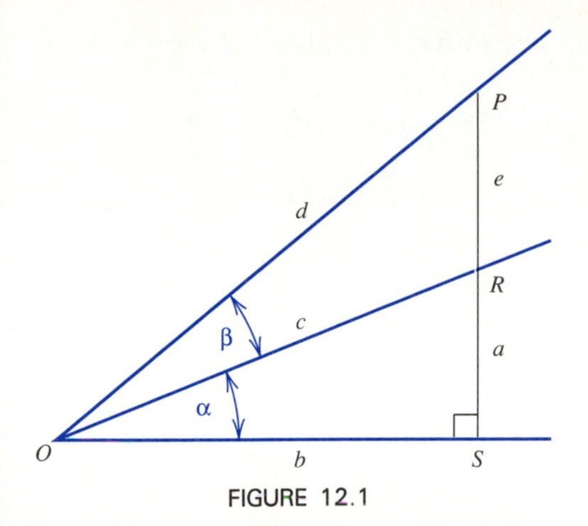

FIGURE 12.1

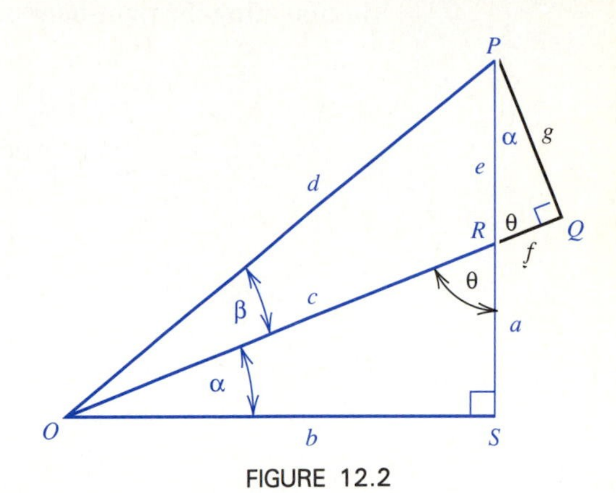

FIGURE 12.2

Consider triangle ROS and triangle PRQ. The angles θ at R are equal since they are vertical angles. Since $\theta + \alpha = 90°$ in triangle ROS, and triangle PRQ is a right triangle, angle P of this triangle must be α also. Figures 12.3a–12.3c highlight triangles ROS, PRQ, and POQ, respectively. We will refer to these triangles to rewrite the right-hand side of Equation 12-2.

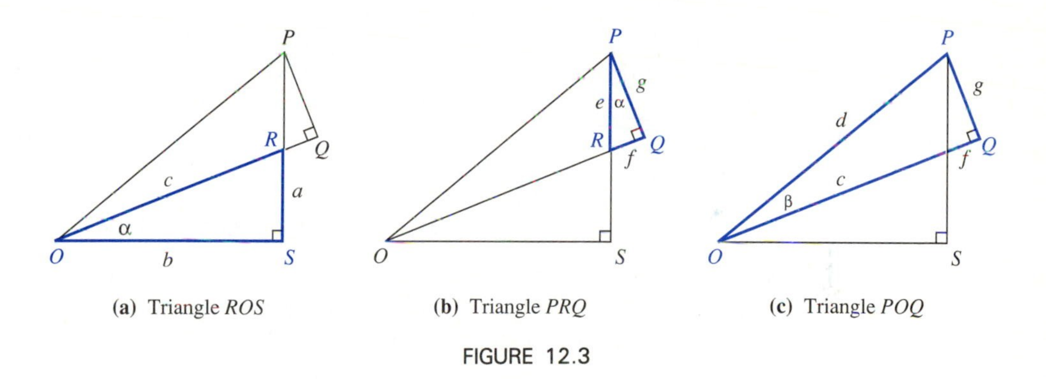

(a) Triangle *ROS* (b) Triangle *PRQ* (c) Triangle *POQ*

FIGURE 12.3

Consider the following function values.

$$\sin\alpha = \frac{a}{c} \qquad \cos\beta = \frac{c+f}{d} \qquad \cos\alpha = \frac{g}{e} \qquad \sin\beta = \frac{g}{d}$$

(from *ROS*) **(from *POQ*)** **(from *PRQ*)** **(from *POQ*)**

Substituting these values for the expressions in Equation 12-2, we have

$$\sin(\alpha + \beta) = \frac{a}{c}\left(\frac{c+f}{d}\right) + \frac{g}{e}\left(\frac{g}{d}\right).$$

Manipulating the right-hand side algebraically, we have the following series of steps.

$$\sin(\alpha+\beta)=\frac{a}{c}\left(\frac{c}{d}+\frac{f}{d}\right)+\frac{g^2}{ed}$$ Separate $\frac{c+f}{d}$ into two fractions.

$$=\frac{ac}{cd}+\frac{af}{cd}+\frac{g^2}{ed}$$ Apply the distributive property.

$$=\frac{ac}{cd}+\frac{f^2}{\mathrm{ed}}+\frac{g^2}{ed}$$ Replace $\frac{a}{c}$ with $\frac{f}{e}$ $\left(\sin\alpha=\frac{f}{e}\text{ in }PRQ\right)$.

$$=\frac{a}{d}+\frac{f^2}{ed}+\frac{g^2}{ed}$$ Simplify.

$$=\frac{a}{d}+\frac{f^2+g^2}{ed}$$ Add fractions with denominator *ed*.

$$=\frac{a}{d}+\frac{e^2}{ed}$$ $f^2+g^2=e^2$ in right triangle *PRQ*

$$=\frac{a}{d}+\frac{e}{d}$$ Reduce the second fraction.

$$=\frac{a+e}{d}$$ Add fractions.

Triangle *POS* in Figure 12.2 shows the expression on the right to be $\sin(\alpha+\beta)$. We have thus proved the identity. In a similar manner, Equation 12-3 can be verified. Start with the right-hand side and make the following substitutions. The remainder of the proof is left as an exercise.

$\cos\alpha=\frac{g}{e}$	$\cos\beta=\frac{c+f}{d}$	$\sin\alpha=\frac{a}{c}$	$\sin\beta=\frac{g}{d}$
(from *PRQ*)	**(from *POQ*)**	**(from *ROS*)**	**(from *POQ*)**

Since the proofs given are dependent on Figure 12-3, we have verified Equations 12-2 and 12-3 only for nonnegative angles α and β such that $\alpha+\beta<90°$. They are, however, valid for any angles α and β. To obtain formulas for $\sin(\alpha-\beta)$ and $\cos(\alpha-\beta)$, we simply treat $\alpha-\beta$ as the sum $\alpha+(-\beta)$. We have the following equations.

$$\sin(\alpha+(-\beta))=\sin\alpha\cos(-\beta)+\cos\alpha\sin(-\beta) \qquad \textbf{(12-4)}$$

$$\cos(\alpha+(-\beta))=\cos\alpha\cos(-\beta)-\sin\alpha\sin(-\beta). \qquad \textbf{(12-5)}$$

The expressions $\cos(-\beta)$ and $\sin(-\beta)$ can be replaced with simpler expressions. From Figure 12.4, we have

$$\cos(-\beta)=x/r=\cos\beta \quad\text{and}\quad \sin(-\beta)=-y/r=-\sin\beta.$$

Thus, Equations 12-4 and 12-5 can be written in the following manner.

$$\sin(\alpha-\beta)=\sin\alpha\cos\beta+\cos\alpha(-\sin\beta)=\sin\alpha\cos\beta-\cos\alpha\sin\beta \qquad \textbf{(12-6)}$$

$$\cos(\alpha-\beta)=\cos\alpha\cos\beta-\sin\alpha(-\sin\beta)=\cos\alpha\cos\beta+\sin\alpha\sin\beta \qquad \textbf{(12-7)}$$

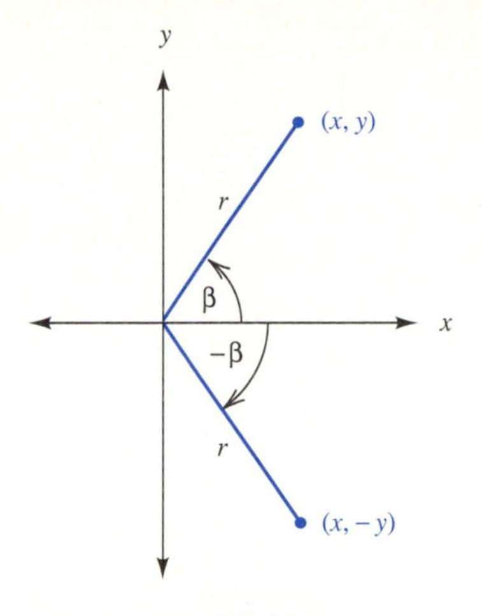

FIGURE 12.4

These formulas are summarized as follows.

SINE AND COSINE OF THE SUM OR DIFFERENCE OF ANGLES

$$\sin(\alpha+\beta) = \sin\alpha\cos\beta + \cos\alpha\sin\beta \quad (12\text{-}2)$$
$$\cos(\alpha+\beta) = \cos\alpha\cos\beta - \sin\alpha\sin\beta \quad (12\text{-}3)$$
$$\sin(\alpha-\beta) = \sin\alpha\cos\beta - \cos\alpha\sin\beta \quad (12\text{-}6)$$
$$\cos(\alpha-\beta) = \cos\alpha\cos\beta + \sin\alpha\sin\beta \quad (12\text{-}7)$$

With these formulas, it is possible to find the sine and cosine of many angles other than the special angles without the use of tables or a calculator. Recall from Sections 4.2 and 4.6 that we have the following values for special angles.

TRIGONOMETRIC FUNCTION VALUES FOR SPECIAL ANGLES

Radian	Degree	$\sin\theta$	$\cos\theta$	$\tan\theta$
0	0°	0	1	0
$\pi/6$	30°	$1/2$	$\sqrt{3}/2$	$\sqrt{3}/3$
$\pi/4$	45°	$\sqrt{2}/2$	$\sqrt{2}/2$	1
$\pi/3$	60°	$\sqrt{3}/2$	$1/2$	$\sqrt{3}$
$\pi/2$	90°	1	0	*

* indicates that the function is undefined

EXAMPLE 1 Find $\cos 75°$.

Solution Since $75° = 45° + 30°$, Equation 12-3 can be used.

$$\begin{aligned}\cos(45° + 30°) &= \cos 45° \cos 30° - \sin 45° \sin 30° \\ &= \frac{\sqrt{2}}{2}\left(\frac{\sqrt{3}}{2}\right) - \frac{\sqrt{2}}{2}\left(\frac{1}{2}\right) \\ &= \frac{\sqrt{6}}{4} - \frac{\sqrt{2}}{4} \\ &= \frac{\sqrt{6} - \sqrt{2}}{4}\end{aligned}$$ ■

A calculator confirms that the answer in Example 1 is correct, for we have the following keystroke sequence.

[(] 6 [√x] [−] 2 [√x] [)] [÷] 4 [=] **0.2588191**

The sequence 75 [cos] gives the same result.

EXAMPLE 2 Find $\sin 15°$.

Solution Since $15° = 45° - 30°$, Equation 12-6 can be used.

$$\begin{aligned}\sin(45° - 30°) &= \sin 45° \cos 30° - \cos 45° \sin 30° \\ &= \frac{\sqrt{2}}{2}\left(\frac{\sqrt{3}}{2}\right) - \frac{\sqrt{2}}{2}\left(\frac{1}{2}\right) \\ &= \frac{\sqrt{6}}{4} - \frac{\sqrt{2}}{4} \\ &= \frac{\sqrt{6} - \sqrt{2}}{4}\end{aligned}$$ ■

The sine and cosine of the sum and difference of two angles may be used in proving identities. Since proving identities generally involves working with the more complicated expression, it is important to be able to recognize the patterns in Equations 12-2, 12-3, 12-6, and 12-7. In the sine formulas, each term consists of two factors that are *different* functions. In the cosine formulas, each term consists of two factors that are the *same* function. Also, in the sine formulas, the sign between terms of the argument of the function on the left is the *same* as the sign between terms of the function value. In the cosine formulas, the sign between terms of the argument of the function on the left is *different* from the sign between terms of the function value.

EXAMPLE 3 Evaluate the following expression without the use of tables or a calculator:

$$\cos 33° \sin 12° + \sin 33° \cos 12°.$$

Solution The pattern fits one of the sine formulas, since each term involves both the sine and cosine functions. In fact, it is the sum formula, since there is a plus sign between terms. Thus, $\cos 33^\circ \sin 12^\circ + \sin 33^\circ \cos 12^\circ = \sin(33^\circ + 12^\circ) = \sin 45^\circ = \sqrt{2}/2$. ■

EXAMPLE 4 Evaluate the following expression without the use of tables or a calculator:

$$\cos 133^\circ \cos 43^\circ + \sin 133^\circ \sin 43^\circ.$$

Solution The pattern fits one of the formulas for a cosine function, since each term involves only one function. In fact, it is the difference formula, since there is a plus sign between terms. Thus, $\cos 133^\circ \cos 43^\circ + \sin 133^\circ \sin 43^\circ = \cos(133^\circ - 43^\circ) = \cos 90^\circ = 0$. ■

EXAMPLE 5 Prove the following identity.

$$\sin(x+y)\cos y - \cos(x+y)\sin y = \sin x$$

Solution The expression on the left is more complicated. It fits the pattern for a sine function. In fact, it is the difference formula.

$$\begin{aligned}\sin(x+y)\cos y - \cos(x+y)\sin y &= \sin[(x+y)-y] \\ &= \sin x \quad ■\end{aligned}$$

EXAMPLE 6 Prove the following identity.

$$\tan(x+y) = \frac{\tan x + \tan y}{1 - \tan x \tan y}$$

Solution Although the right-hand side is more complicated, the left-hand side suggests using the sine and cosine of a sum.

$$\begin{aligned}\tan(x+y) &= \frac{\sin(x+y)}{\cos(x+y)} \\ &= \frac{\sin x \cos y + \cos x \sin y}{\cos x \cos y - \sin x \sin y}\end{aligned}$$

Since we want the expression to show $\tan x + \tan y$ in the numerator, we need to divide it by $\cos x \cos y$. If both numerator and denominator are divided by the same nonzero expression, the value of the function is unchanged. Thus, we have the following equation.

$$\begin{aligned}\frac{\sin x \cos y + \cos x \sin y}{\cos x \cos y - \sin x \sin y} &= \frac{\dfrac{\sin x \cos y}{\cos x \cos y} + \dfrac{\cos x \sin y}{\cos x \cos y}}{\dfrac{\cos x \cos y}{\cos x \cos y} - \dfrac{\sin x \sin y}{\cos x \cos y}} \\ &= \frac{\tan x + \tan y}{1 - \tan x \tan y} \quad ■\end{aligned}$$

Identities arise in technological work as shown in Example 7.

EXAMPLE 7 At night, cloud height may be determined in the following manner. A light located at point B is directed at a cloud with a constant angle of elevation of 70°, as in Figure 12.5. An observer at point A, 1000 m from point B, measures θ, the angle of elevation to the spot where the light shines on the cloud. The height of the cloud is then given by

$$h = \frac{1000}{\cot 70^\circ + \cot \theta} \qquad \text{or} \qquad h = \frac{1000 \sin 70^\circ \sin \theta}{\sin (\theta + 70^\circ)}.$$

Show that the formulas are equivalent.

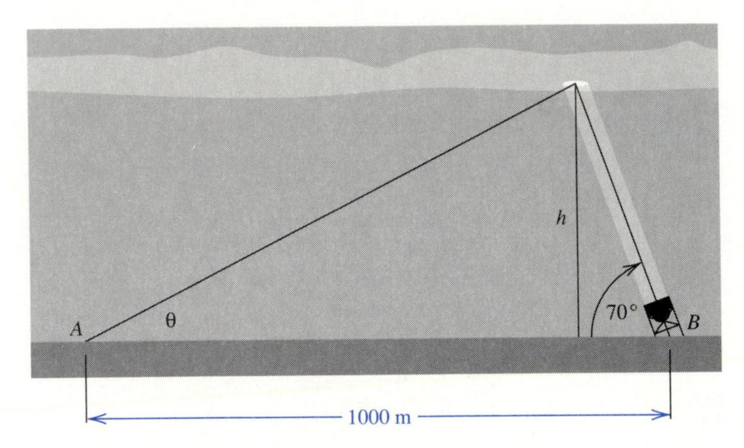

FIGURE 12.5

Solution

$$\frac{1000 \sin 70^\circ \sin \theta}{\sin (\theta + 70^\circ)} = \frac{1000 \sin 70^\circ \sin \theta}{\sin \theta \cos 70^\circ + \cos \theta \sin 70^\circ}$$

$$= \frac{\dfrac{1000 \sin 70^\circ \sin \theta}{\sin 70^\circ \sin \theta}}{\dfrac{\sin \theta \cos 70^\circ}{\sin 70^\circ \sin \theta} + \dfrac{\cos \theta \sin 70^\circ}{\sin 70^\circ \sin \theta}}$$

$$= \frac{1000}{\cot 70^\circ + \cot \theta} \qquad \blacksquare$$

EXERCISES 12.2

In 1–6, determine the value of each function.

1. $\sin 75^\circ$ using $75^\circ = 45^\circ + 30^\circ$

2. $\cos 15^\circ$ using $15^\circ = 45^\circ - 30^\circ$

3. $\cos 15^\circ$ using $15^\circ = 60^\circ - 45^\circ$

4. $\sin 15^\circ$ using $15^\circ = 60^\circ - 45^\circ$

5. $\sin 105^\circ$ using $105^\circ = 60^\circ + 45^\circ$

6. $\cos 105^\circ$ using $105^\circ = 60^\circ + 45^\circ$

In 7–14, evaluate each expression without the use of tables or a calculator.

7. $\sin 22° \cos 23° + \cos 22° \sin 23°$

8. $\sin 43° \cos 13° - \cos 43° \sin 13°$

9. $\cos 35° \cos 25° - \sin 35° \sin 25°$

10. $\cos 82° \cos 22° + \sin 82° \sin 22°$

11. $\cos 63° \cos 18° + \sin 63° \sin 18°$

12. $\cos 24° \cos 6° - \sin 24° \sin 6°$

13. $\sin 101° \cos 11° - \cos 101° \sin 11°$

14. $\sin 49° \cos 41° + \cos 49° \sin 41°$

In 15–29, prove each identity.

15. $\sin (90° - \theta) = \cos \theta$

16. $\cos (90° - \theta) = \sin \theta$

17. $\cos (x + \pi/2) = -\sin x$

18. $\sin (x + \pi/2) = \cos x$

19. $\sin (x + \pi) = -\sin x$

20. $\cos (\pi - x) = -\cos x$

21. $\sin (x - y) \cos y + \sin y \cos (x - y) = \sin x$

22. $\cos (x + y) \cos y + \sin (x + y) \sin y = \cos x$

23. $\cos (x - y) \cos y - \sin (x - y) \sin y = \cos x$

24. $\sin (A + B) + \sin (A - B) = 2 \sin A \cos B$

25. $\sin (A + B) - \sin (A - B) = 2 \cos A \sin B$

26. $\cos (A + B) + \cos (A - B) = 2 \cos A \cos B$

27. $\cos (A - B) - \cos (A + B) = 2 \sin A \sin B$

28. $\cos 2\theta = \cos^2 \theta - \sin^2 \theta$ (Hint: $2\theta = \theta + \theta$)

29. $\sin 2\theta = 2 \sin \theta \cos \theta$ (Hint: $2\theta = \theta + \theta$)

Problems 30–32 use the following variables. For an object resting on an inclined plane, w = the weight of the object, α = the angle of inclination of the plane, β = the angle between the plane and a force acting on the object, and θ = the angle between the resultant force on the object and the component of the object's weight perpendicular to the plane.

30. The force necessary to prevent slipping when $\alpha > \theta$ is given by

$$\frac{w \sin \alpha \cos \theta - w \cos \alpha \sin \theta}{\cos \beta \cos \theta - \sin \beta \sin \theta}$$

or

$$\frac{w \sin (\alpha - \theta)}{\cos (\beta + \theta)}.$$

Show that the two formulas are equivalent.

4 7 8 1 2 3 5 6 9

31. The force necessary to start the object moving up the plane is given by

$$\frac{w \sin \alpha \cos \theta + w \cos \alpha \sin \theta}{\cos \beta \cos \theta + \sin \beta \sin \theta}$$

or

$$\frac{w \sin (\alpha + \theta)}{\cos (\beta - \theta)}.$$

Show that the two formulas are equivalent.

4 7 8 1 2 3 5 6 9

32. The force necessary to start the object moving down the plane when $\alpha \leq \theta$ is given by

$$\frac{w \sin \theta \cos \alpha - w \cos \theta \sin \alpha}{\cos \theta \cos \beta - \sin \theta \sin \beta}$$

or

$$\frac{w \sin (\theta - \alpha)}{\cos (\theta + \beta)}.$$

Show that the two formulas are equivalent.

4 7 8 1 2 3 5 6 9

The accompanying figure is for problems 33 and 34. It shows a body of weight w being lifted by two equal and opposite forces acting through two wedges of angle β. If the friction angles at all surfaces in contact are θ, the equations in problems 33 and 34 hold.

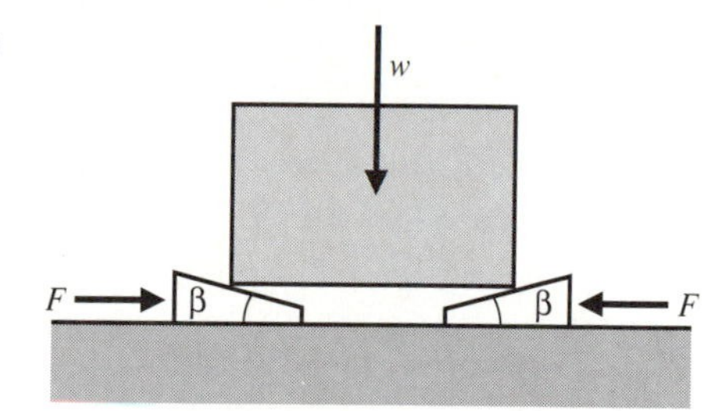

---

1. Architectural Technology
2. Civil Engineering Technology
3. Computer Engineering Technology
4. Mechanical Drafting Technology
5. Electrical/Electronics Engineering Technology
6. Chemical Engineering Technology
7. Industrial Engineering Technology
8. Mechanical Engineering Technology
9. Biomedical Engineering Technology

33. The force necessary to lift the body is given by

$$w/2\,[\tan(\theta+\beta)+\tan\theta]$$

or

$$\frac{w}{2}\left[\frac{\sin(2\theta+\beta)}{\cos\theta\cos(\theta+\beta)}\right].$$

Show that the two formulas are equivalent.
4 7 8 1 2 3 5 6 9

34. The force that must be applied to hold the body when $\beta > 2\theta$ is given by

$$w/2\,[\tan(\theta-\beta)-\tan\theta]$$

or

$$\frac{w}{2}\left[\frac{-\sin\beta}{\cos(\theta-\beta)\cos\theta}\right].$$

Show that the two formulas are equivalent.
4 7 8 1 2 3 5 6 9

35. If a jackscrew is used, the moment of the force necessary to lower a load of weight w can be computed. It is necessary to know the lead angle β, the friction angle θ, and the mean radius r of the thread. The moment is given by

$$wr\tan(\theta-\beta) \qquad \text{or} \qquad wr\left(\frac{\tan\theta-\tan\beta}{1+\tan\theta\tan\beta}\right).$$

Show that the formulas are equivalent.
4 7 8 1 2 3 5 6 9

36. To calculate the height h of the tide at a specified time t (using a 24-hour clock), the following formulas are used.

$$h=\frac{h_1+h_2}{2}-\frac{h_1-h_2}{2}\cos\left(\frac{\pi t}{6.2}-\frac{\pi t_2}{6.2}\right)$$

or

$$h=\frac{h_1+h_2}{2}-\frac{h_1-h_2}{2}\times$$

$$\left[\cos\left(\frac{\pi t}{6.2}\right)\cos\left(\frac{\pi t_2}{6.2}\right)+\sin\left(\frac{\pi t}{6.2}\right)\sin\left(\frac{\pi t_2}{6.2}\right)\right]$$

In the formula, h_1 = height at high tide, h_2 = height at low tide, and t_2 = time of low tide. Show that the two formulas are equivalent. 9 1 6

1. Architectural Technology
2. Civil Engineering Technology
3. Computer Engineering Technology
4. Mechanical Drafting Technology
5. Electrical/Electronics Engineering Technology
6. Chemical Engineering Technology
7. Industrial Engineering Technology
8. Mechanical Engineering Technology
9. Biomedical Engineering Technology

12.3 DOUBLE-ANGLE AND HALF-ANGLE IDENTITIES

The formulas for the sine and cosine of the sum of two angles may be used to find formulas for $\sin 2\theta$ and $\cos 2\theta$ in the following manner.

$$\sin 2\theta=\sin(\theta+\theta)=\sin\theta\cos\theta+\sin\theta\cos\theta=2\sin\theta\cos\theta$$

$$\cos 2\theta=\cos(\theta+\theta)=\cos\theta\cos\theta-\sin\theta\sin\theta=\cos^2\theta-\sin^2\theta$$

The formula for $\cos 2\theta$ may be written in terms of either the sine or cosine function. Since $\sin^2\theta+\cos^2\theta=1$, we have the following equations.

$$\cos^2\theta=1-\sin^2\theta \qquad \text{and} \qquad \sin^2\theta=1-\cos^2\theta$$

$$\begin{aligned}\cos 2\theta&=\cos^2\theta-\sin^2\theta\\&=(1-\sin^2\theta)-\sin^2\theta\\&=1-2\sin^2\theta\end{aligned}$$

$$\begin{aligned}\cos 2\theta&=\cos^2\theta-\sin^2\theta\\&=\cos^2\theta-(1-\cos^2\theta)\\&=\cos^2\theta-1+\cos^2\theta\\&=2\cos^2\theta-1\end{aligned}$$

The formulas developed thus far are summarized as follows.

DOUBLE-ANGLE IDENTITIES

$$\sin 2\theta = 2 \sin \theta \cos \theta \quad \text{(12-8)}$$

$$\cos 2\theta = \cos^2 \theta - \sin^2 \theta \quad \text{(12-9)}$$

$$\cos 2\theta = 1 - 2 \sin^2 \theta \quad \text{(12-10)}$$

$$\cos 2\theta = 2 \cos^2 \theta - 1 \quad \text{(12-11)}$$

EXAMPLE 1 Find $\sin 2\alpha$ if $\sin \alpha = -12/13$ for α in the third quadrant.

Solution If $\sin \alpha = -12/13$, then $y = -12$ and $r = 13$. Since α is in the third quadrant, $x < 0$, and $x = -\sqrt{13^2 - (-12)^2} = -5$. Thus,

$$\begin{aligned}\sin 2\alpha &= 2 \sin \alpha \cos \alpha \\ &= 2(-12/13)(-5/13) = 120/169.\end{aligned}$$ ■

EXAMPLE 2 Find $\cos 2\theta$ if $\cot \theta = -3/4$ and θ is in the second quadrant.

Solution If $\cot \theta = -3/4$, then $x = -3$ and $y = 4$. Notice that it is the fact that θ is in the second quadrant that tells us that $x < 0$ and $y > 0$. It follows that $r = \sqrt{(-3)^2 + 4^2} = 5$.

Thus, $$\begin{aligned}\cos 2\theta &= \cos^2 \theta - \sin^2 \theta \\ &= (-3/5)^2 - (4/5)^2 \\ &= 9/25 - 16/25 \\ &= -7/25.\end{aligned}$$ ■

The double-angle identities may be used to prove other identities.

EXAMPLE 3 When a projectile is launched at an angle of θ with a velocity of v_0, the horizontal distance x may be computed by the following formula, in which g represents the acceleration due to gravity.

$$x = v_0 \cos \theta \left(\frac{2\, v_0 \sin \theta}{g}\right)$$

Show that $x = \dfrac{{v_0}^2}{g} \sin 2\theta$.

Solution The problem is to prove the identity

$$v_0 \cos \theta \left(\frac{2\, v_0 \sin \theta}{g}\right) = \frac{{v_0}^2}{g} \sin 2\theta.$$

The factors of the expression on the left-hand side can be reordered and regrouped as follows

$$v_0 \cos\theta\left(\frac{2\,v_0 \sin\theta}{g}\right) = \frac{v_0 v_0}{g}(2\sin\theta\cos\theta)$$
$$= \frac{{v_0}^2}{g}(\sin 2\theta). \quad \blacksquare$$

EXAMPLE 4 Prove the identity $\sin 3x = 3 \sin x \cos^2 x - \sin^3 x$.

Solution Even though the expression on the right-hand side is more complicated, $\sin 3x$ suggests $\sin(x + 2x)$.

$$\begin{aligned} \sin 3x &= \sin(x + 2x) \\ &= \sin x \cos 2x + \cos x \sin 2x \\ &= \sin x(\cos^2 x - \sin^2 x) + \cos x(2 \sin x \cos x) \\ &= \sin x \cos^2 x - \sin^3 x + 2 \sin x \cos^2 x \\ &= 3 \sin x \cos^2 x - \sin^3 x \quad \blacksquare \end{aligned}$$

EXAMPLE 5 Prove the identity $\dfrac{\sin 3x}{\sin x} + \dfrac{\cos 3x}{\cos x} = 4 \cos 2x$.

Solution The expression on the left-hand side of the equation is more complicated. We begin by combining the two fractions by addition.

$$\frac{\sin 3x}{\sin x} + \frac{\cos 3x}{\cos x} = \frac{\sin 3x \cos x + \cos 3x \sin x}{\sin x \cos x}$$

Notice that the numerator fits the pattern for $\sin(3x + x) = \sin 4x$. Also, the double-angle formula may be used, since $\sin 4x = \sin 2 \cdot 2x$.

$$\begin{aligned} \frac{\sin 3x \cos x + \cos 3x \sin x}{\sin x \cos x} &= \frac{\sin 4x}{\sin x \cos x} \\ &= \frac{\sin 2 \cdot 2x}{\sin x \cos x} \\ &= \frac{2 \sin 2x \cos 2x}{\sin x \cos x} \\ &= \frac{2(2 \sin x \cos x) \cos 2x}{\sin x \cos x} \\ &= 4 \cos 2x \quad \blacksquare \end{aligned}$$

Just as the formulas for $\sin 2\alpha$ and $\cos 2\alpha$ are called double-angle identities, the formulas for $\sin \frac{\alpha}{2}$ and $\cos \frac{\alpha}{2}$ are called half-angle identities. To obtain these formulas, let $\theta = \frac{\alpha}{2}$ in Equations 12-10 and 12-11. Using Equation 12-10, we have the following steps.

$$\cos 2\theta = 1 - 2\sin^2\theta$$

$$\cos 2\left(\frac{\alpha}{2}\right) = 1 - 2\sin^2\frac{\alpha}{2}$$

$$\cos\alpha = 1 - 2\sin^2\frac{\alpha}{2}$$

$$\cos\alpha - 1 = -2\sin^2\frac{\alpha}{2}$$

$$\frac{\cos\alpha - 1}{-2} = \sin^2\frac{\alpha}{2}$$

Multiplying the expression on the left-hand side of the equation by $-1/-1$, we obtain the following equation.

$$\frac{1 - \cos\alpha}{2} = \sin^2\frac{\alpha}{2}$$

Recall from Section 2.5 that an equation of the form $x^2 = y$ has $x = \pm\sqrt{y}$ as its solutions. Thus,

$$\pm\sqrt{\frac{1 - \cos\alpha}{2}} = \sin\frac{\alpha}{2}.$$

The proper sign is determined by the quadrant in which $\frac{\alpha}{2}$ lies. Recall that the sine of an angle is positive in the first and second quadrants. In a similiar manner, $\cos\frac{\alpha}{2}$ is obtained from Equation 12-11.

The half-angle formulas are given by the following equations.

HALF-ANGLE IDENTITIES

$$\sin\frac{\alpha}{2} = \pm\sqrt{\frac{1 - \cos\alpha}{2}} \qquad \textbf{(12-12)}$$

$$\cos\frac{\alpha}{2} = \pm\sqrt{\frac{1 + \cos\alpha}{2}} \qquad \textbf{(12-13)}$$

EXAMPLE 6 Find $\sin\frac{\alpha}{2}$ if $\sin\alpha = \frac{-5}{13}$ and α is in the third quadrant.

Solution If $\sin\alpha = \frac{-5}{13}$, then $y = -5$ and $r = 13$. Since α is in the third quadrant, $x < 0$ and $x = -\sqrt{13^2 - 5^2} = -12$. We use Equation 12-12.

$$\sin\frac{\alpha}{2} = \pm\sqrt{\frac{1-\cos\alpha}{2}}$$

Because α is in the third quadrant, $180° < \alpha < 270°$ and, therefore, $90° < \frac{\alpha}{2} < 135°$. That is, $\frac{\alpha}{2}$ is in the second quadrant, so $\sin\frac{\alpha}{2} > 0$.

$$\begin{aligned}
\sin\frac{\alpha}{2} &= \sqrt{\frac{1-(-12/13)}{2}} \\
&= \sqrt{\frac{1+12/13}{2}} \\
&= \sqrt{\frac{25/13}{2}} \\
&= \sqrt{\frac{25}{26}} \\
&= \frac{5}{\sqrt{26}} \\
&= \frac{5\sqrt{26}}{26}
\end{aligned}$$

Multiply both numerator and denominator by $\sqrt{26}$ to rationalize the denominator. ■

Equations 12-12 and 12-13 may be required to prove identities.

EXAMPLE 7 Prove the identity $2\sin^2\left(\frac{x}{2}\right) + \cos x = 1$.

Solution Working with the expression on the left-hand side of the equation, we use Equation 12-12. Since $\frac{1-\cos x}{2} \geq 0$ for all x, it is true that $\sin^2\frac{x}{2} = \frac{1-\cos x}{2}$ for all x. Thus, we have the following steps.

$$\begin{aligned}
2\sin^2\left(\frac{x}{2}\right) + \cos x &= 2\left(\frac{1-\cos x}{2}\right) + \cos x \\
&= 1 - \cos x + \cos x \\
&= 1
\end{aligned}$$
■

EXERCISES 12.3

In 1–16, evaluate each function.

1. Find $\sin 2\theta$ if $\cos\theta = 12/13$ and θ is in the 4th quadrant.
2. Find $\sin 2\theta$ if $\sin\theta = -15/17$ and θ is in the 4th quadrant.
3. Find $\cos 2\theta$ if $\tan\theta = -7/24$ and θ is in the 2nd quadrant.
4. Find $\cos 2\theta$ if $\cot\theta = -20/21$ and θ is in the 2nd quadrant.
5. Find $\cos 2\theta$ if $\cot\theta = 12/35$ and θ is in the 3rd quadrant.
6. Find $\cos 2\theta$ if $\tan\theta = 9/40$ and θ is in the 3rd quadrant.
7. Find $\sin 2\theta$ if $\sin\theta = 45/53$ and θ is in the 1st quadrant.
8. Find $\sin 2\theta$ if $\cos\theta = 16/65$ and θ is in the 1st quadrant.
9. Find $\sin 2\theta$ if $\sin\theta = -11/61$ and θ is in the 4th quadrant.
10. Find $\cos 2\theta$ if $\cos\theta = -56/65$ and θ is in the 2nd quadrant.
11. Find $\sin\theta/2$ if $\sin\theta = 7/25$ and θ is in the 2nd quadrant.
12. Find $\sin\theta/2$ if $\cos\theta = -20/29$ and θ is in the 2nd quadrant.
13. Find $\sin\theta/2$ if $\cos\theta = -12/37$ and θ is in the 3rd quadrant.
14. Find $\sin\theta/2$ if $\sin\theta = -9/41$ and θ is in the 3rd quadrant.
15. Find $\cos\theta/2$ if $\cot\theta = 28/45$ and θ is in the 1st quadrant.
16. Find $\cos\theta/2$ if $\sin\theta = -11/61$ and θ is in the 4th quadrant.

In 17–30, prove each identity.

17. $\cos 3x = \cos^3 x - 3\sin^2 x\cos x$
18. $\sin 4x = 4\sin x\cos^3 x - 4\sin^3 x\cos x$
19. $(\sin y + \cos y)^2 = 1 + \sin 2y$
20. $(\cos y - \sin y)^2 = 1 - \sin 2y$
21. $\dfrac{\cos 2\theta}{\sin^2\theta} = \cot^2\theta - 1$
22. $\dfrac{\sin 2\theta}{\cos\theta - \sin\theta} = \tan 2\theta(\cos\theta + \sin\theta)$
23. $\dfrac{\sin 3x}{\sin x} - \dfrac{\cos 3x}{\cos x} = 2$
24. $\dfrac{\sin 3x}{\cos x} + \dfrac{\cos 3x}{\sin x} = \cot x - \tan x$
25. $2\cos^2\dfrac{y}{2} - \cos y = 1$
26. $\tan^2\dfrac{y}{2} = \left(\dfrac{1-\cos y}{\sin y}\right)^2$
27. $\cos^2\dfrac{y}{2} - \sin^2\dfrac{y}{2} = \cos y$
28. $\sin^2\dfrac{\theta}{2}\cos^2\dfrac{\theta}{2} = \dfrac{\sin^2\theta}{4}$
29. $\sec^2\dfrac{\theta}{2} = 2\csc^2\theta - 2\cot\theta\csc\theta$
30. $\tan^2\dfrac{x}{2} = \dfrac{2\sin x - \sin 2x}{2\sin x + \sin 2x}$

For problems 31 and 32, use the following information. In Great Britain, the nautical mile is defined as the length of a minute of arc of a meridian. Since the earth is flattened at the poles, a British nautical mile varies with latitude. If θ is the latitude in degrees, a good approximation for the length (in feet) is given by $y = 6077 - 31\cos 2\theta$.

31. Show that $y = 6046 + 62\sin^2\theta$. 1 2 4 7 8 3 5
32. Show that $y = 6108 - 62\cos^2\theta$. 1 2 4 7 8 3 5

1. Architectural Technology
2. Civil Engineering Technology
3. Computer Engineering Technology
4. Mechanical Drafting Technology
5. Electrical/Electronics Engineering Technology
6. Chemical Engineering Technology
7. Industrial Engineering Technology
8. Mechanical Engineering Technology
9. Biomedical Engineering Technology

For problems 33 and 34, use the following information. When a block of weight w_1 rests on a plane inclined at an angle of α, and a bar of weight w_2 rests horizontally with its end on the block as shown in the accompanying figure, the coefficient of friction between the block and the plane may be computed.

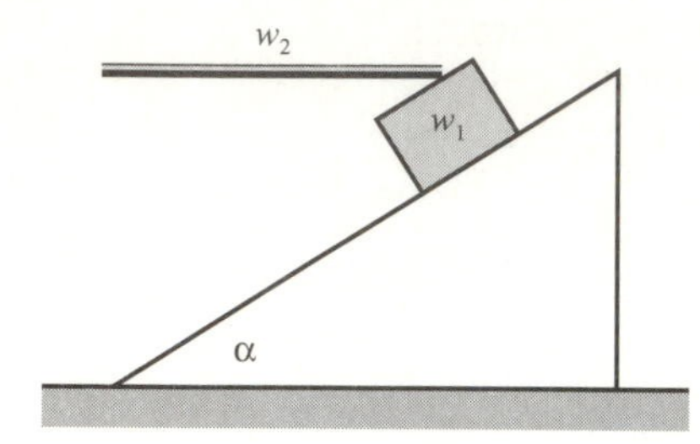

33. Show that $\dfrac{\sin 2\alpha}{1 + \dfrac{w_2}{w_1} + \cos 2\alpha} = \dfrac{2 \sin \alpha \cos \alpha}{\dfrac{w_2}{w_1} + 2 \cos^2 \alpha}$

2 4 8 1 5 6 7 9

34. Show that $\dfrac{\sin 2\alpha}{\dfrac{w_2}{w_1} + 2 \cos^2 \alpha} = \dfrac{2\, w_1 \tan \alpha}{w_2 \sec^2 \alpha + 2w_1}$

2 4 8 1 5 6 7 9

35. If a plane is passed through a beam at an angle α, the following equation holds.

$$\tau_1 + \sigma \sin \alpha \cos \alpha + \tau \sin^2 \alpha - \tau \cos^2 \alpha = 0$$

In the equation, τ_1 is the shear stress (in pascals) on the plane, and σ and τ are normal and shear stresses (in pascals) on the beam. Derive the following equation.

$$\tau_1 = \tau \cos 2\alpha - \frac{\sigma \sin 2\alpha}{2}$$

1 2 4 8 7 9

36. Show that $\tan 2\theta = \dfrac{2 \tan \theta}{1 - \tan^2\theta}$.

1. Architectural Technology
2. Civil Engineering Technology
3. Computer Engineering Technology
4. Mechanical Drafting Technology
5. Electrical/Electronics Engineering Technology
6. Chemical Engineering Technology
7. Industrial Engineering Technology
8. Mechanical Engineering Technology
9. Biomedical Engineering Technology

12.4 EQUATIONS WITH INVERSE TRIGONOMETRIC FUNCTIONS

Conditional trigonometric equations will be presented in the next section. Trigonometric identities are useful for changing the form of an equation. An understanding of the inverse trigonometric functions is also required to solve conditional trigonometric equations. In this section, then, we examine problems in which the concepts of trigonometric identities and inverse trigonometric functions come together. For the sake of review, the definitions of the important inverse trigonometric functions follow

INVERSE TRIGONOMETRIC FUNCTIONS

$y = \sin^{-1} x$ or $y = \arcsin x$ means $x = \sin y$
$y = \text{Sin}^{-1} x$ or $y = \text{Arcsin}\, x$ means $x = \sin y$ and $-\pi/2 \le y \le \pi/2$
$y = \cos^{-1} x$ or $y = \arccos x$ means $x = \cos y$
$y = \text{Cos}^{-1} x$ or $y = \text{Arccos}\, x$ means $x = \cos y$ and $0 \le y \le \pi$
$y = \tan^{-1} x$ or $y = \arctan x$ means $x = \tan y$
$y = \text{Tan}^{-1} x$ or $y = \text{Arctan}\, x$ means $x = \tan y$ and $-\pi/2 < y < \pi/2$

EXAMPLE 1 Find sec (2 Arcsin x).

Solution Since Arcsin x is the angle θ whose sine is x, $\sin\theta$ is $x/1$. That is, there is a right triangle such that the length of the side opposite θ is x and the length of the hypotenuse is 1. The length of the adjacent side is $\sqrt{1-x^2}$, as in Figure 12.6.

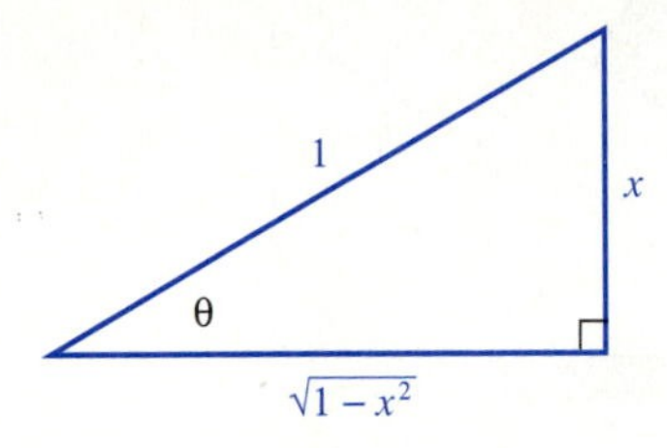

FIGURE 12.6

$$\begin{aligned}\sec(2 \text{ Arcsin } x) &= \sec 2\theta \\ &= \frac{1}{\cos 2\theta} && \text{Use the reciprocal identity.} \\ &= \frac{1}{2\cos^2\theta - 1} && \text{Use the double-angle identity.}\end{aligned}$$

From Figure 12.6, it can be seen that $\cos\theta$ is $\sqrt{1-x^2}$ and therefore $\cos^2\theta = 1 - x^2$.

$$\begin{aligned}\sec(2 \text{ Arcsin } x) &= \frac{1}{2(1-x^2)-1} \\ &= \frac{1}{2-2x^2-1} = \frac{1}{1-2x^2} \quad \blacksquare\end{aligned}$$

Compare Example 1 to the following examples, which do not require use of the double-angle identity.

EXAMPLE 2 Find sec (Arcsin $2x$).

Solution Since Arcsin $2x$ is the angle θ whose sine is $2x$, $\sin\theta$ is $2x/1$. That is, there is a right triangle such that the length of the side opposite θ is $2x$ and the length of the hypotenuse is 1, as in Figure 12.7.

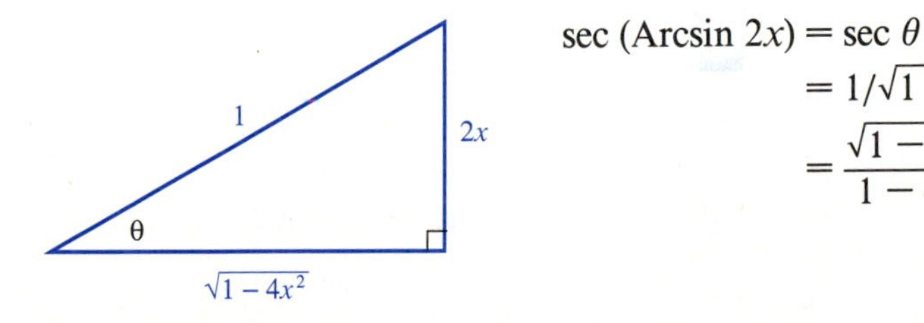

$$\begin{aligned}\sec(\text{Arcsin } 2x) &= \sec\theta \\ &= 1/\sqrt{1-4x^2} \\ &= \frac{\sqrt{1-4x^2}}{1-4x^2} \quad \blacksquare\end{aligned}$$

FIGURE 12.7

EXAMPLE 3 Find 2 sec (Arcsin x).

Solution Since Arcsin x is the angle θ whose sine is x, sin θ is x or $x/1$. That is, there is a right triangle such that the length of the side opposite θ is x and the length of the hypotenuse is 1, as in Figure 12.8.

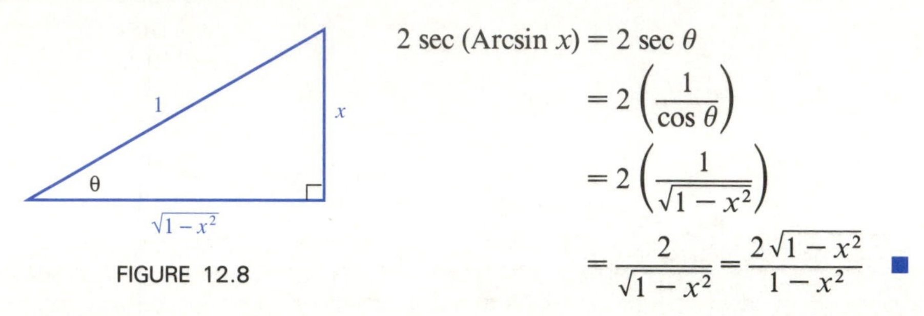

FIGURE 12.8

$$
\begin{aligned}
2 \sec(\text{Arcsin } x) &= 2 \sec \theta \\
&= 2\left(\frac{1}{\cos \theta}\right) \\
&= 2\left(\frac{1}{\sqrt{1-x^2}}\right) \\
&= \frac{2}{\sqrt{1-x^2}} = \frac{2\sqrt{1-x^2}}{1-x^2} \quad \blacksquare
\end{aligned}
$$

In a trigonometric equation, the variable is usually an angle. When a trigonometric equation is a conditional equation rather than an identity, our goal will be to first isolate the function. That is, we use algebraic techniques to write the equation with the trigonometric function on one side of the equation by itself. The equation can then be solved for the angle using an inverse function. It is important to remember, however, that there may be more than one angle that satisfies the equation.

EXAMPLE 4 Find all angles x in radians, $0 \le x < 2\pi$, for which $\sin x = 1/2$.

Solution If $\sin x = 1/2$, then $x = \arcsin 1/2$ gives all angles x for which the original equation is true. (Remember that the Arcsine function with an upper case A indicates a single value in the first or fourth quadrant.) Since we have the additional information that $0 \le x < 2\pi$, we know there are only two solutions for x. They lie in the first and second quadrants, since $\sin x$ is positive in those quadrants.

The equation $\sin x = 1/2$ suggests a triangle, such as the one in Figure 12.9, in which the length of the side opposite x is 1 and the length of the hypotenuse is 2.

Thinking of special angles, we see that x is a 30° angle (or $\pi/6$ in radians). This accounts for the angle in the first quadrant. To find the angle in the second quadrant, we use a 30° reference angle. Thus, x is 180° − 30°, which is 150° (or $\pi - \pi/6$, which is $5\pi/6$). The solutions are $\pi/6$ and $5\pi/6$. ■

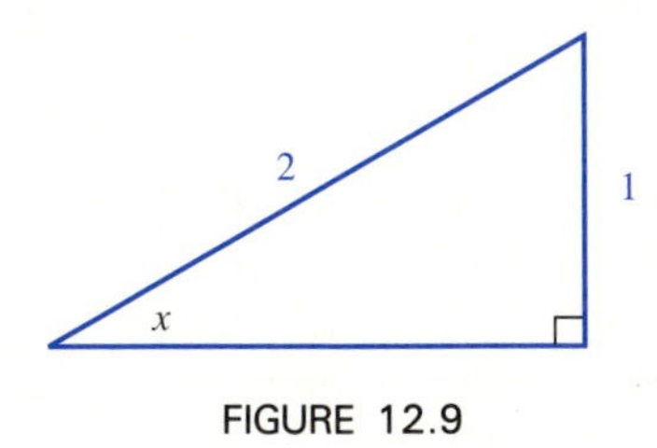

FIGURE 12.9

EXAMPLE 5 Find all angles x in radians, $0 \leq x < 2\pi$ for which $\cos 3x = -\sqrt{3}/2$.

Solution If $\cos 3x = -\sqrt{3}/2$, then $3x = \arccos(-\sqrt{3}/2)$ and $x = \dfrac{\arccos(-\sqrt{3}/2)}{3}$. Cos $3x$ is negative when $3x$ is in the second or third quadrant. The cosine of the reference angle is $|-\sqrt{3}/2|$ or $\sqrt{3}/2$. The equation $\cos 3x = \sqrt{3}/2$ suggests a triangle such as the one in Figure 12.10, in which the length of the side adjacent to $3x$ is $\sqrt{3}$, and the length of the hypotenuse is 2.

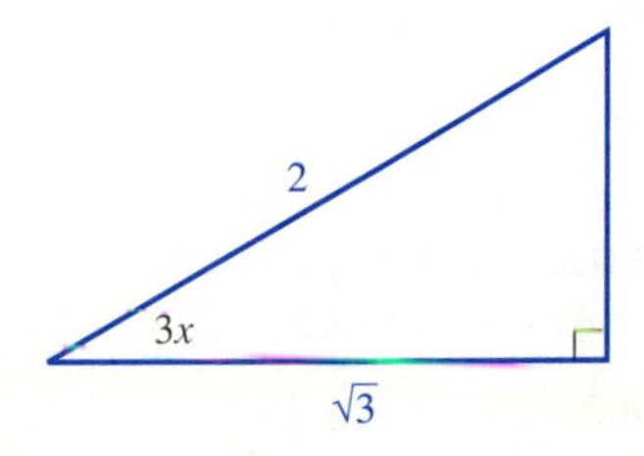

FIGURE 12.10

Thinking of special angles, we see that in the first quadrant $3x$ is $30°$ or $\pi/6$. This angle is the reference angle. In the second quadrant, the angle $3x$ is $\pi - \pi/6$ or $5\pi/6$. In the third quadrant, the angle $3x$ is $\pi + \pi/6$ or $7\pi/6$. However, these two angles are not the only ones that satisfy the equation. For x to range from 0 to 2π, $3x$ can range from 0 to 6π. Since adding a multiple of 2π to an angle produces a coterminal angle, $3x$ can also be $5\pi/6 + 2\pi$, $7\pi/6 + 2\pi$, $5\pi/6 + 4\pi$, or $7\pi/6 + 4\pi$. Dividing by 3 to find x, we have $x = 5\pi/18$, $7\pi/18$, $17\pi/18$, $19\pi/18$, $29\pi/18$, and $31\pi/18$. ■

The steps used in Example 5 can be summarized as follows.

To solve an equation of the form $f(nx) = a$ where $f(x)$ is a trigonometric function:

1. Rewrite the equation as $nx = f^{-1}(a)$ or $x = \dfrac{f^{-1}(a)}{n}$.
2. Determine the quadrants in which $f(nx)$ has the same sign as the number a.
3. Find the reference angle for which $f(nx) = a$.
4. Find the angles that satisfy the equation $f(nx) = a$ by using the reference angle from step 3 and the quadrants from step 2.
5. If $n > 1$, then find additional angles by adding multiples through $(n - 1)$ of 2π to the angles found in step 4.
6. Divide each angle from steps 4 and 5 by n.

The solution to a trigonometric equation will not always be a special angle. A calculator may be used to solve these problems.

EXAMPLE 6 Find all angles x in radians, $0 \le x < 2\pi$, for which $\tan 2x = 0.7107$.

Solution

1. If $\tan 2x = 0.7107$, then $2x = \tan^{-1}(0.7107)$ or $x = \dfrac{\tan^{-1}(0.7107)}{2}$.
2. Tan $2x$ is positive in the first and third quadrants.
3. The tangent of the reference angle is 0.7107. With the calculator in radian mode, the reference angle for $2x$ is obtained by the following keystroke sequence.

   0.7107 [INV] [tan] **0.6178711**

4. Since this value will be used several times, it should be stored in calculator memory. For future reference, we call this value R. The third quadrant angle, then, is $R + \pi$. That is, $2x = R$ or $2x = R + \pi$.
5. Additional solutions are obtained by adding 2π to the solutions found in step 4. That is, $2x = R + 2\pi$ or $2x = (R + \pi) + 2\pi$.
6. All solutions (R, $R + \pi$, $R + 2\pi$, and $R + 3\pi$) are divided by 2. It is a more efficient use of the calculator to combine steps 5 and 6. The reference angle for $2x$ is on the calculator display from step 2, so the following keystroke sequence is used to evaluate the expressions in steps 4 and 5.

   [STO] [÷] 2 [=] **0.3089356**

   [RCL] [+] π [=] [÷] 2 [=] **1.8797319**

   [RCL] [+] 2 [×] π [=] [÷] 2 [=] **3.4505282**

   [RCL] [+] 3 [×] π [=] [÷] 2 [=] **5.0213245**

That is, x can be 0.3089, 1.880, 3.451, or 5.021. ■

EXAMPLE 7 Find all angles x in radians, $0 \le x < 2\pi$, for which $\sin 3x = -0.8016$.

Solution Sin $3x$ is negative in the third and fourth quadrants. The sine of the reference angle R is 0.8016, so the keystroke sequence 0.8016 [INV] [sin] gives a reference angle R of 0.9310 in radians. We have $3x = \pi + R$ or $3x = 2\pi - R$.

Additional values are obtained by adding $1(2\pi)$ or $2(2\pi)$ to these solutions. That is, $3x$ can be $\pi + R$, $3\pi + R$, $5\pi + R$, $2\pi - R$, $4\pi - R$, or $6\pi - R$.

We have the following keystroke sequence.

STO + π = ÷ 3 = **1.3571864**

RCL + 3 × π = ÷ 3 = **3.4515815**

RCL + 5 × π = ÷ 3 = **5.5459766**

2 × π − RCL = ÷ 3 = **1.7844062**

4 × π − RCL = ÷ 3 = **3.8788013**

6 × π − RCL = ÷ 3 = **5.9731964**

That is, x can be 1.357, 3.452, 5.546, 1.784, 3.879, or 5.973. ■

Example 8 shows how these ideas may be applied.

EXAMPLE 8 A large sign that is exactly 10 feet high is to be supported on posts so that the base of the sign is 15 feet off the ground as shown in Figure 12.11. A floodlight is to be placed so that the angle of elevation to the top of the sign is 51.3°. Find the angle of elevation α to the bottom of the sign.

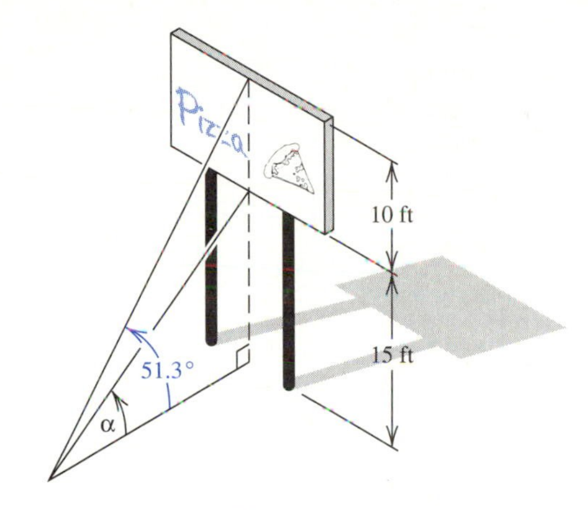

FIGURE 12.11

Solution Let z = the horizontal distance from the light to the sign.

$$\text{Tan } \alpha = 15/z$$

$$\alpha = \text{Arctan } 15/z \tag{12-14}$$

But $\tan 51.3° = 25/z$, so $z = 25/\tan 51.3°$.

Substituting this value of z into Equation 12-14, we have

$$\alpha = \text{Arctan}\left(\frac{15}{25/\tan 51.3°}\right)$$

$$= \text{Arctan}\left(\frac{15 \tan 51.3°}{25}\right) \text{ and, therefore, } \alpha = 36.8° \text{ or } 37°.$$

The angle of elevation to the bottom of the sign, then, is about 37°. ■

EXERCISES 12.4

In 1–16, evaluate each expression.

1. sin (Arctan $2x$)
2. sin (Arcsec $2x$)
3. cos (Arcsin $2y$)
4. cos (Arccot $2y$)
5. 2 tan (Arcsin z)
6. 2 tan (Arccos z)
7. 2 cot (Arcsec x)
8. 2 cot (Arctan x)
9. sec (2 Arccos y)
10. csc (2 Arccot y)
11. csc (2 Arctan w)
12. csc (2 Arcsin w)
13. sin (2 Arccos z)
14. sin (2 Arctan z)
15. cos (2 Arcsec x)
16. cos (2 Arccsc x)

In 17–32, solve each equation for x in radians, $0 \leq x < 2\pi$.

17. $\sin x = -\sqrt{3}/2$
18. $\cos x = \sqrt{2}/2$
19. $\tan x = \sqrt{3}$
20. $\sin x = \sqrt{2}/2$
21. $\cos 2x = -1/2$
22. $\tan 2x = -\sqrt{3}$
23. $\sin 2x = \sqrt{3}/2$
24. $\cos 2x = -\sqrt{2}/2$
25. $\tan 3x = 1$
26. $\sin 3x = -\sqrt{2}/2$
27. $\cos 3x = 0$
28. $\tan 3x = 0$
29. $\sec x = 1$
30. $\cot 2x = 1$
31. $\csc 3x = -1$
32. $\sec x = -1$

In 33–42, solve each equation for x in radians, $0 \leq x < 2\pi$.

33. $\sin x = 0.5247$
34. $\cos x = -0.6156$
35. $\tan x = -1.7065$
36. $\sin 2x = 0.8974$
37. $\cos 2x = 0.9883$
38. $\tan 2x = -2.0792$
39. $\sin 3x = -0.1601$
40. $\cos 3x = 0.2510$
41. $\tan 3x = 2.3429$
42. $\sin 4x = -0.4338$

---

43. A photographer has photographed a building from a distance of 70.0 m. The angle of elevation to the top of the building was 30.0°. A flagpole 15.0 m tall has been placed at the top of the building. If the photographer stands at the same place to take a photograph, what will the angle of elevation to the top of the pole be? See the accompanying figure. 1 2 4 7 8 3 5

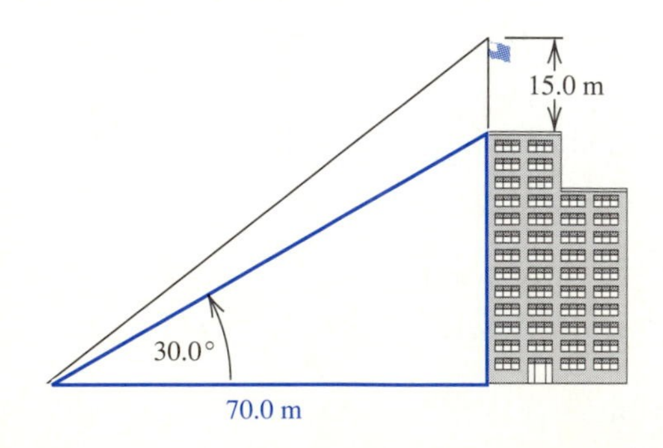

44. Find a general formula that the photographer in problem 43 could use to determine the angle of elevation α to the top of the pole. Use x for the distance to the building, y for the height of the pole, and β as the angle of elevation to the top of the building. 1 2 4 7 8 3 5

45. Find the angle α for the two pulleys shown in the accompanying diagram if $R_1 = 10.0$ cm, $R_2 = 5.00$ cm, and $d = 20.0$ cm. 4 7 8 1 2 3 5 6 9

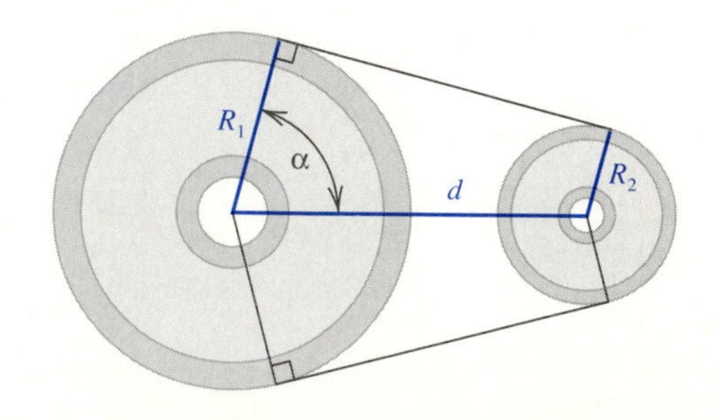

46. For two pulleys, as shown in problem 45, find a general formula for the angle α if the larger pulley has radius R_1, the smaller has radius R_2, and the distance between centers is d. 4 7 8 1 2 3 5 6 9

47. When a large-view camera is used to take a picture of an object that is not parallel to the film, the lens board should be turned so the planes containing the subject, the lens board, and the film intersect in a line, as shown in the accompanying figure. This gives the best "depth of field." Find the correct angle α of the lens board, given: $\beta = 82.0°$, $y = 12.00$ m, and $x = 0.04$ m.* 3 5 6 9

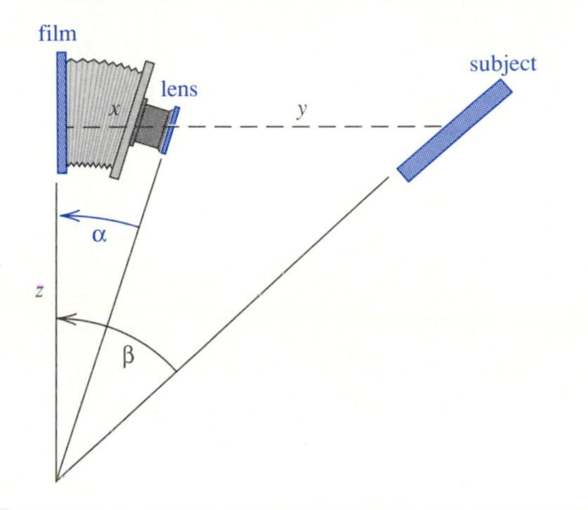

48. Find a general formula for the angle α of the lens board in problem 47 if β is the angle of the subject, y is the distance from the center of the subject to the lens, and x is the distance from the lens to the film.* 3 5 6 9

49. Some computer languages don't have a "built-in" arcsine or arccosine function. The arctangent function must be used. Show that

(a) $\arcsin x = \arctan (x/\sqrt{1 - x^2})$.

(b) $\arccos x = \arctan (\sqrt{1 - x^2})/x$.

* From *A Sourcebook of Applications of School Mathematics.* Copyright ©1980 by The Mathematical Association of America. Reprinted by permission.

1. Architectural Technology
2. Civil Engineering Technology
3. Computer Engineering Technology
4. Mechanical Drafting Technology
5. Electrical/Electronics Engineering Technology
6. Chemical Engineering Technology
7. Industrial Engineering Technology
8. Mechanical Engineering Technology
9. Biomedical Engineering Technology

12.5 CONDITIONAL TRIGONOMETRIC EQUATIONS

In the previous section, we solved some simple trigonometric equations. Since trigonometric equations may take different forms, it will be necessary to use techniques such as factoring, multiplying expressions on both sides of the equation by the same number, and squaring both sides. Examples that illustrate these techniques follow.

EXAMPLE 1 Solve the equation $3 \tan x - 3 = 0$ for $0 \le x < 2\pi$.

Solution

$$3 \tan x - 3 = 0$$

$$3 \tan x = 3 \quad \text{Add 3 to both sides.}$$

$$\tan x = 1 \quad \text{Divide both sides by 3.}$$

$$x = \tan^{-1} 1$$

Since $\tan x = 1$ in the first and third quadrants, and the reference angle is $\pi/4$, the solution is $x = \pi/4$ or $5\pi/4$. ■

Examples 2 and 3 illustrate the use of factoring to solve trigonometric equations.

EXAMPLE 2 Solve the equation $\sin^2 y - \sin y = 0$ for $0 \le y < 2\pi$.

Solution

$$\sin^2 y - \sin y = 0$$

$$\sin y(\sin y - 1) = 0 \qquad \text{Factor the expression on the left.}$$

$$\sin y = 0 \quad \text{or} \quad \sin y - 1 = 0 \qquad \text{Set each factor equal to zero.}$$

$$\sin y = 0 \quad \text{or} \quad \sin y = 1 \qquad \text{Add 1 to both sides of the second equation.}$$

$$y = \arcsin 0 \quad \text{or} \quad y = \arcsin 1$$

Sin $y = 0$ if y is 0 or π, and $\sin y = 1$ if y is $\pi/2$. The solution, then, is $y = 0, \pi/2$, or π. ■

EXAMPLE 3 Solve the equation $\sec^2 x - \sec x - 2 = 0$ for $0 \le x < 2\pi$.

Solution

$$\sec^2 x - \sec x - 2 = 0$$

$$(\sec x - 2)(\sec x + 1) = 0 \qquad \text{Factor the expression on the left.}$$

$$\sec x - 2 = 0 \quad \text{or} \quad \sec x + 1 = 0 \qquad \text{Set each factor equal to zero.}$$

$$\sec x = 2 \quad \text{or} \quad \sec x = -1 \qquad \text{Add 2 to both sides of the first equation and } -1 \text{ to the second.}$$

$$x = \text{arcsec } 2 \quad \text{or} \quad x = \text{arcsec}(-1)$$

The secant function is positive in quadrants I and IV. When $\sec x$ is 2, the reference angle is $\pi/3$. Thus, $\pi/3$ is the first quadrant solution and $5\pi/3$ is the fourth quadrant solution. Sec x is -1 if x is π. The solution, then, is $x = \pi/3, \pi$, or $5\pi/3$. ■

It may be necessary to use a trigonometric identity to change the form of an equation. Examples 4 and 5 illustrate this technique.

EXAMPLE 4 Solve the equation $\cos 2y - \cos^2 y = 0$ for $0 \le y < 2\pi$.

Solution

$$\cos 2y - \cos^2 y = 0$$

$$(2\cos^2 y - 1) - \cos^2 y = 0 \qquad \text{Use the double-angle identity } \cos 2y = 2\cos^2 y - 1 \text{ so that the same argument is used in both terms.}$$

$$\cos^2 y - 1 = 0 \qquad \text{Combine like terms.}$$

$$(\cos y - 1)(\cos y + 1) = 0 \qquad \text{Factor the expression on the left as the difference of squares.}$$

$$\cos y - 1 = 0 \quad \text{or} \quad \cos y + 1 = 0 \qquad \text{Set each factor equal to zero.}$$

$$\cos y = 1 \quad \text{or} \quad \cos y = -1 \qquad \text{Add 1 to both sides of the first equation and } -1 \text{ to the second.}$$

$$y = \cos^{-1} 1 \quad \text{or} \quad y = \cos^{-1}(-1)$$

Cos y is 1 if y is 0. Cos y is -1 if y is π. Hence, the solution is $y = 0$ or π. ■

EXAMPLE 5 Solve the equation $\cos \frac{x}{2} = \cos x$ for $0 \le x < 2\pi$.

Solution

$$\cos \frac{x}{2} = \cos x$$

$$\pm\sqrt{\frac{1+\cos x}{2}} = \cos x \qquad \text{Use the half-angle identity.}$$

$$\frac{1+\cos x}{2} = \cos^2 x \qquad \text{Square both sides of the equation.}$$

$$1 + \cos x = 2\cos^2 x \qquad \text{Multiply both sides of the equation by 2.}$$

$$0 = 2\cos^2 x - \cos x - 1 \qquad \text{Subtract } \cos x + 1 \text{ from both sides.}$$

$$0 = (2\cos x + 1)(\cos x - 1) \qquad \text{Factor the expression on the right.}$$

$$0 = 2\cos x + 1 \quad \text{or} \quad 0 = \cos x - 1 \qquad \text{Set each factor equal to zero.}$$

$$-1 = 2\cos x \quad \text{or} \quad 1 = \cos x \qquad \text{Add } -1 \text{ to both sides of the first equation and 1 to the second.}$$

$$\frac{-1}{2} = \cos x \quad \text{or} \quad 1 = \cos x \qquad \text{Divide both sides of the first equation by 2.}$$

$$x = \cos^{-1}\left(\frac{-1}{2}\right) \quad \text{or} \quad x = \cos^{-1} 1$$

Cos x is $-1/2$ in the second and third quadrants. The reference angle is $\pi/3$, so $x = \pi - \pi/3$ or $x = \pi + \pi/3$. Cos x is 1 for the quadrantal angle 0. Thus, it appears that $x = 0, 2\pi/3$, or $4\pi/3$. Whenever both sides of an equation are squared, there is the risk of introducing incorrect answers, called *extraneous roots*. ***It is imperative to check solutions in the original problem anytime both sides of an equation are squared.*** If x is 0, we have $\cos 0 = \cos 0$, which is true. If x is $2\pi/3$, we have $\cos \pi/3 = \cos 2\pi/3$, which is shown to be false when evaluated by a calculator. If x is $4\pi/3$, we have $\cos 2\pi/3 = \cos 4\pi/3$, which is shown to be true when evaluated by a calculator. The correct solution, then, is $x = 0$ or $4\pi/3$. ■

extraneous roots

NOTE ⇨⇨

It is interesting to see the graphs of $y = \cos x$ and $y = \cos x/2$ plotted on the same set of axes, as in Figure 12.12. Notice that the two curves cross when x is 0 or $4\pi/3$. A quick sketch is often useful to verify that a solution is reasonable.

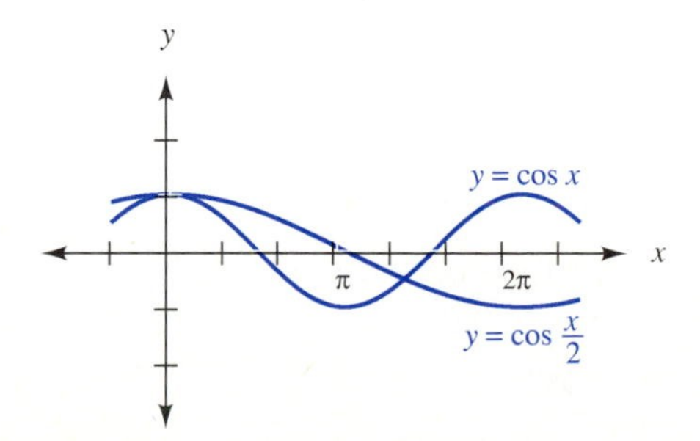

FIGURE 12.12

In Examples 1–5, the angles were special angles, and they were determined by inspection. It may be necessary to use tables or a calculator to solve an equation. Examples 6–8 illustrate the use of the calculator.

EXAMPLE 6 The horizontal distance d (in m) of a pendulum bob from the rest position is given by $d = l \sin \theta$, where l is the length of the pendulum and θ is the angle that the pendulum makes with the vertical. A pendulum of length 1.05 m is released from an angle of θ. What value of θ is necessary for the horizontal distance to be 0.750 m?

Solution Since $d = 0.750$ m and $l = 1.05$ m, the equation is

$$0.750 = 1.05 \sin \theta$$
$$0.750/1.05 = \sin \theta$$
$$\text{Arcsin } (0.750/1.05) = \theta.$$

The calculator keystroke sequence is as follows.

.75 [÷] 1.05 [=] [INV] [sin] **45.584691**

To the nearest tenth of a degree, the answer is 45.6°. ■

EXAMPLE 7 Solve the equation $\sin x \cos 2x + \cos x \sin 2x = 0.3000$ for x such that $0 \leq x < 2\pi$.

Solution $\sin x \cos 2x + \cos x \sin 2x = 3.000$

$$\sin 3x = 0.3000$$

Use the identity for the sine of the sum of two angles, since the expression on the left fits that pattern.

$$3x = \arcsin 0.3000$$
$$x = (\arcsin 0.3000)/3$$

Divide both sides of the equation by 3.

This problem has been reduced to the type covered in the last section. Sin $3x = 0.3000$ if $3x$ is in the first or second quadrants. To find the reference angle R, put the calculator into radian mode. Then use the following keystroke sequence.

.3 [INV] [sin] **0.3046926**

Thus, $3x$ can be $R, R + 2\pi, R + 4\pi, \pi - R, 3\pi - R$, or $5\pi - R$. The solutions, to four significant digits, are 0.1016, 2.196, 4.290, 0.9456, 3.040, or 5.134. These values are in radians, since $0 \leq x < 2\pi$. ■

EXAMPLE 8 Solve the equation $\cos y = \dfrac{1}{7 \cos y}$ for y, $0 \leq y < 2\pi$.

Solution To get rid of the denominator, multiply both sides of the equation by $7 \cos y$, assuming $\cos y \neq 0$. (If later steps lead to $y = \pi/2$ or $3\pi/2$, then $\cos y$ is 0 and that value will have to be eliminated from the solution.)

$$7 \cos^2 y = 1$$

$$\cos^2 y = \frac{1}{7} \qquad \text{Divide both sides by 7.}$$

$$\cos y = \pm\sqrt{\frac{1}{7}} \qquad \text{Recall that if } y^2 = x \text{, then } y = \pm\sqrt{x}.$$

$$y = \arccos \pm \sqrt{1/7}$$

Using a calculator in radian mode, the keystroke sequence is as follows.

1 [÷] 7 [=] [√x] [INV] [COS] [STO] **1.1831996**

The value was put into memory, because it is only the reference angle. Cos $y = +\sqrt{1/7}$ in the first and fourth quadrants. The displayed angle, then, is one solution. To obtain the fourth quadrant angle, the following keystroke sequence is used.

2 [×] π [−] [RCL] [=] **5.0999857**

Cos $y = -\sqrt{1/7}$ in the second and third quadrants. We have

π [−] [RCL] [=] **1.9583931**

π [+] [RCL] [=] **4.3247923**

The solutions to the equation, to four significant digits, are 1.183, 5.100, 1.958, or 4.325. ■

As you solve trigonometric equations, refer to the following hints.

HINTS FOR SOLVING CONDITIONAL TRIGONOMETRIC EQUATIONS

1. If there is only one term that is a trigonometric function, then isolate that term. Solve first for the function value. The angle can be found from the function value.
2. If there is more than one term that is a trigonometric function, look for the following patterns:
   - **(a)** it may be possible to factor out a common factor.
   - **(b)** it may be possible to factor the expression on one side of the equation if the same trigonometric function appears to both the first and second powers.
   - **(c)** it may be possible to use an identity to rewrite the equation so that it fits one of the patterns described in (a) and (b).
3. If there are fractions in the equation, consider multiplying both sides of the equation by the lowest common denominator of the fractions.
4. If the half-angle identities are used, eliminate square roots by squaring both sides of the equation. Check all solutions when both sides are squared. Also eliminate any value for which the original equation is undefined.

EXERCISES 12.5

In 1–30, solve each equation for x, $0 \le x < 2\pi$.

1. $2 \sin x - 1 = 0$
2. $3 \tan x - 2 = \tan x - 4$
3. $5 \cos x = 1 + 3 \cos x$
4. $4 \cos x = 2 - \cos x$
5. $2 \tan x - 5 = 0$
6. $\sin x - \sin x \cos x = 0$
7. $4 \tan^2 x = \tan x$
8. $3 \cos^2 x + \cos x = 0$
9. $\cos x = 2 \sin 2x$
10. $1 - \cos^2 x + \sin x = 0$
11. $\cos 2x - \cos x = 0$
12. $6 \sin^2 x + \sin x - 2 = 0$
13. $\tan^2 x - 1 = 0$
14. $4 \sin^2 x - 4 \sin x + 1 = 0$
15. $\sec^2 x - 3 \tan x = 5$
16. $\cos x \cos 2x - \sin x \sin 2x = 1$
17. $\sin 3x \cos x - \sin x \cos 3x = 0$
18. $\cos 2x + \sin^2 x = 0$
19. $\sin 2x - \sin x = 0$
20. $\cos x = 3 \sin 2x$
21. $\sin x = \dfrac{1}{2 \cos x}$
22. $\sin x = \dfrac{1}{3 \sin x}$
23. $\tan x - \dfrac{1}{\tan x} = 0$
24. $\dfrac{2 \sin x}{\cos x} = \dfrac{3}{\sin x}$
25. $\sin x = \cos x$
26. $\sin \dfrac{x}{2} = \cos x$
27. $\sin \dfrac{x}{2} = \cos \dfrac{x}{2}$
28. $\cos^4 2x - \sin^4 2x = 1$
29. $\sin \dfrac{x}{2} = \dfrac{\sin x}{2}$
30. $\cos x = 2 - 2 \cos^2 \dfrac{x}{2}$

---

31. In Great Britain, the nautical mile is defined as the length of a minute of arc of a meridian. Since the earth is flattened at the poles, a British nautical mile varies with latitude. A good approximation in feet is given by $6077 - 31 \cos 2\theta$ where θ is the latitude in degrees. In the United States, the nautical mile is defined everywhere as 6080.2 ft. At what latitude do the two definitions agree?* 1 2 4 7 8 3 5

* From *A Sourcebook of Applications of School Mathematics.*

Problems 32–34 use the following information. The maximum height h attained by a projectile is given by $h = \dfrac{v_0^2 \sin^2 \theta}{2g}$ *and its horizontal range is given by* $d = \dfrac{v_0^2 \sin 2\theta}{g}$. *In the formulas, g is the acceleration due to gravity, θ is the angle of launch, and v_0 is the launch velocity.*

32. Find the value of θ for which the maximum height and range are equal. 4 7 8 1 2 3 5 6 9
33. Find the value of θ for which the height is twice the range. 4 7 8 1 2 3 5 6 9

---

1. Architectural Technology
2. Civil Engineering Technology
3. Computer Engineering Technology
4. Mechanical Drafting Technology
5. Electrical/Electronics Engineering Technology
6. Chemical Engineering Technology
7. Industrial Engineering Technology
8. Mechanical Engineering Technology
9. Biomedical Engineering Technology

34. Find the value of θ for which the height is one-half the range. 4 7 8 1 2 3 5 6 9

35. A cube rests on a wedge. The coefficient of friction between the cube and the wedge is μ, the angle of inclination of the wedge is θ, and g is the acceleration due to gravity (9.8 m/s^2). A horizontal force that causes the cube to move relative to the surface gives the system an acceleration of $a = \dfrac{g(\mu - \tan\theta)}{1 + \mu\tan\theta}$. Find the value of θ that produces an acceleration of 1 m/s^2 when $\mu = 0.45$. 4 7 8 1 2 3 5 6 9

12.6 CHAPTER REVIEW

KEY TERMS

12.1 contradiction
identity
conditional equation

12.5 extraneous roots

RULES AND FORMULAS

Reciprocal Identities:

$\sin\theta = 1/\csc\theta$ $\qquad$ $\tan\theta = 1/\cot\theta$ $\qquad$ $\sec\theta = 1/\cos\theta$

$\cos\theta = 1/\sec\theta$ $\qquad$ $\cot\theta = 1/\tan\theta$ $\qquad$ $\csc\theta = 1/\sin\theta$

Basic Identities:

$\tan\theta = \sin\theta/\cos\theta$ $\qquad$ $\cot\theta = \cos\theta/\sin\theta$

Pythagorean Identities:

$\sin^2\theta + \cos^2\theta = 1$ $\qquad$ $1 + \tan^2\theta = \sec^2\theta$ $\qquad$ $1 + \cot^2\theta = \csc^2\theta$

Sum and Difference Identities:

$\sin(\alpha + \beta) = \sin\alpha\cos\beta + \cos\alpha\sin\beta$ $\qquad$ $\sin(\alpha - \beta) = \sin\alpha\cos\beta - \cos\alpha\sin\beta$

$\cos(\alpha + \beta) = \cos\alpha\cos\beta - \sin\alpha\sin\beta$ $\qquad$ $\cos(\alpha - \beta) = \cos\alpha\cos\beta + \sin\alpha\sin\beta$

Double-angle Identities:

$\sin 2\theta = 2\sin\theta\cos\theta$ $\qquad$ $\cos 2\theta = \cos^2\theta - \sin^2\theta$ $\qquad$ $\cos 2\theta = 1 - 2\sin^2\theta$ $\qquad$ $\cos 2\theta = 2\cos^2\theta - 1$

Half-angle Identities:

$\sin\dfrac{\alpha}{2} = \pm\sqrt{\dfrac{1 - \cos\alpha}{2}}$ $\qquad$ $\cos\dfrac{\alpha}{2} = \pm\sqrt{\dfrac{1 + \cos\alpha}{2}}$

GEOMETRY CONCEPTS

$A + B + C = 180°$ if A, B, and C are the angles of any triangle.
When two lines intersect, the vertical angles are equal. $\angle 1 = \angle 2$ and $\angle 3 = \angle 4$.

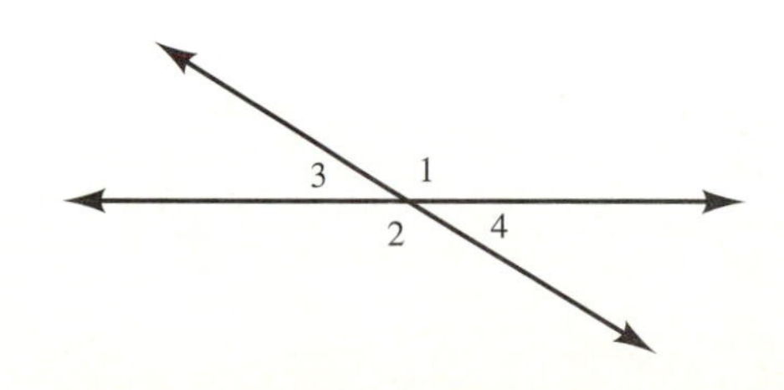

SEQUENTIAL REVIEW

(Section 12.1) *In 1–9, prove each identity.*

1. $\tan\theta \csc\theta = \sec\theta$

2. $\cos x(\sec x - \cos x) = \sin^2 x$

3. $\tan x = \sin x \sec x$

4. $\csc\theta = \sin\theta + \cos\theta \cot\theta$

5. $\dfrac{1 - \sin^2 x}{\cot^2 x} = \sin^2 x$

6. $1 + \sin x = \dfrac{\cos x}{\sec x - \tan x}$

7. $\dfrac{\sin\theta \sec\theta}{\tan^2\theta} = \cot\theta$

8. $\cos^2\theta\,(1 + \tan^2\theta) = 1$

9. $\tan^2 x = \tan^2 x \sec^2 x - \tan^4 x$

(Section 12.2) *In 10 and 11, determine the value of each function.*

10. $\cos(-15°)$, using $-15° = 30 - 45°$

11. $\sin(-15°)$, using $-15° = 45° - 60°$

In 12–14, evaluate each expression without the use of tables or a calculator.

12. $\sin 18° \cos 12° + \cos 18° \sin 12°$

13. $\cos 33° \cos 27° - \sin 33° \sin 27°$

14. $\cos 61° \cos 16° + \sin 61° \sin 16°$

In 15–18, prove each identity.

15. $\sin(\pi - x) = \sin x$

16. $\cos(x + \pi) = -\cos x$

17. $\sin(x - \pi/2) = -\cos x$

18. $\cos(\pi/2 - x) = \sin x$

(Section 12.3)

19. Find $\sin 2\theta$ if $\cos\theta = -12/13$ and θ is in the second quadrant.

20. Find $\cos 2\theta$ if $\sin\theta = 15/17$ and θ is in the second quadrant.

21. Find $\cos 2\theta$ if $\tan\theta = 45/28$ and θ is in the third quadrant.

22. Find $\sin\theta/2$ if $\cos\theta = -12/37$ and θ is in the second quadrant.

23. Find $\sin\theta/2$ if $\tan\theta = 7/24$ and θ is in the third quadrant.

24. Find $\cos\theta/2$ if $\cot\theta = -9/40$ and θ is in the fourth quadrant.

In 25–27, prove each identity.

25. $\dfrac{2}{1 + \cos 2x} = \sec^2 x$

26. $\sin 4\theta = 4 \sin\theta \cos^3\theta - 4 \sin^3\theta \cos\theta$

27. $\cos^2 \dfrac{x}{2} = \dfrac{\tan x + \sin x}{2 \tan x}$

(Section 12.4) *In 28–30, evaluate each expression.*

28. $\tan(\text{Arccos } 2x)$

29. $\sin(2 \text{ Arccos } x)$

30. $2 \cos(\text{Arctan } x)$

In 31–36, solve each equation for x in radians, $0 \le x < 2\pi$.

31. $\cos x = \dfrac{-\sqrt{2}}{2}$

32. $\tan 2x = \sqrt{3}$

33. $\sin x = -1/2$

34. $\sin x = 0.4356$

35. $\cos 2x = -0.5264$

36. $\tan 3x = 2.6157$

(Section 12.5) *In 37–45, solve each equation for x in radians, $0 \le x < 2\pi$.*

37. $5 \cos x - 2 = 0$

38. $3 \sec x - 2 = 2 \sec x$

39. $4(1 + \cos x) = \cos x + 2$

40. $4 \cot^2 x = \cot x$

41. $3 \sin^2 x + 2 \sin x = 0$

42. $1 + \tan^2 x + \sec x = 0$

43. $\sec x - \dfrac{1}{\sec x} = 0$

44. $\cos x = \dfrac{1}{2 \sin x}$

45. $\cos \dfrac{x}{2} = \sin x$

APPLIED PROBLEMS

1. The maximum height of a projectile launched with a velocity of v_0 at an angle of θ is given by the following two formulas, in which g is the acceleration due to gravity. Show that they are equivalent.

$$h = \frac{v_0{}^2 \sin^2 \alpha}{2g} \quad \text{and} \quad h = \frac{v_0{}^2 - v_0{}^2 \cos^2 \alpha}{2g}$$

4 7 8 1 2 3 5 6 9

2. The formula $Z = R \sec \theta$ gives the magnitude of the impedance in a series ac circuit. In the formula, R is the resistance and θ is the phase angle between R and Z. Show that $Z = R\sqrt{1 + \tan^2 \theta}$ for $-90° \leq \theta \leq 90°$.
3 5 9 1 2 4 6 7 8

3. The force P required to lift a load L is given by the equation $P = \dfrac{L(\sin \alpha + \mu \cos \alpha)}{\cos \alpha - \mu \sin \alpha}$ if the force acts at the mean diameter d of a jackscrew. The relationship between the angle α, the screwlead l, and the mean diameter d is shown in the accompanying diagram. Show that

$$P = \frac{L\dfrac{l}{\pi d} + \mu}{1 - \dfrac{\mu l}{\pi d}}.$$

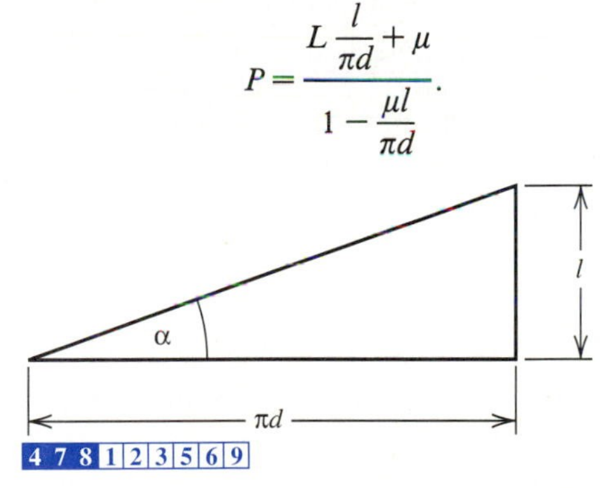

4 7 8 1 2 3 5 6 9

4. The angle that a conical pendulum makes with the vertical is given by $\text{Tan } \theta = \dfrac{V^2}{gr}$. In the formula, V is the linear velocity (in m/s) of the mass particle at the end of the pendulum, r is the radius (in m), and g is the acceleration due to gravity (9.8 m/s²).
   (a) Derive the given equation from
   $$T \cos \theta - W = 0 \text{ and } \frac{WV^2}{gr} = T \sin \theta.$$
   (b) Determine the length (in m) of the pendulum required for the mass particle to have a velocity of 4.1 m/s at a radius of 0.80 m.
4 7 8 1 2 3 5 6 9

5. The acceleration due to gravity is usually considered to be constant at the earth's surface. In fact, however, the acceleration varies slightly with latitude. If g represents the acceleration (in m/s²) and θ is the latitude (in degrees), g is the product of 9.78049 and
$$(1 + 0.005288 \sin^2 \theta - 0.000006 \sin^2 2\theta).$$
Write the formula for g so that θ is the only argument used. **4 7 8** 1 2 3 5 6 9

6. When an object is launched with an initial velocity of v_0 (in m/s) at an angle α (in degrees) from a height h (in m), its vertical distance y (in m) above ground at time t (in s) is given by the following formula.
$$y = v_0 t \sin \alpha - \frac{gt^2}{2} + h$$
Determine the angle α at which an object should be launched at a velocity of 130 m/s from ground level if it is to attain a height of 26 m after 0.50 s. Use $g = 9.8$ m/s². **4 7 8** 1 2 3 5 6 9

7. Students of electronics are often surprised to find that with ac voltages, it is possible for 2 voltages with amplitudes of 3 and 4 volts to have a sum that is equivalent to a voltage with an amplitude of 5. Show that
$$3 \sin 6500t + 4 \sin (6500t - 90°) = 5 \sin (6500t - 53.13°)$$
(Note: $6500t$ is in radian measure.)
3 5 9 1 2 4 6 7 8

1. Architectural Technology
2. Civil Engineering Technology
3. Computer Engineering Technology
4. Mechanical Drafting Technology
5. Electrical/Electronics Engineering Technology
6. Chemical Engineering Technology
7. Industrial Engineering Technology
8. Mechanical Engineering Technology
9. Biomedical Engineering Technology

8. When a sine wave of voltage v given by $v = \sin 2\pi ft$ is applied to a capacitor, the current i that flows is a cosine wave given by $i = \cos 2\pi ft$. The power p that results is given by $p = vi$ and appears to have a frequency that is twice as large as that of either the voltage or current. Show that this is to be expected. Recall that frequency is the reciprocal of period. 3 5 9 4 7 8

9. A paper on electronic wave form synthesis indicates that the voltage $v = 50 \sin^2 5000t$ can be created by combining a 25-volt battery and a 25-volt sine wave generator with a 90° phase angle and twice the frequency of the desired wave so that

$$v = 25 - 25 \cos 10\,000t.$$

Show that these two voltages are equivalent.

$\left(\text{Hint: } \sin^2 \theta = \frac{1}{2} - \frac{1}{2} \cos 2\theta.\right)$

3 5 9 1 2 4 6 7 8

10. A radio communications text refers to a sine wave created in a certain electronics circuit. The size of this wave is proportional to the product of the sum and difference of two other sine waves that are out of phase with one another by 90°. Show that this new wave has twice the frequency of the other two by proving the following identity.

$$k(\sin 2\pi ft + \cos 2\pi ft)(\sin 2\pi ft - \cos 2\pi ft) = -k \cos 4\pi ft.$$

Recall that frequency is the reciprocal of period. 1 3 4 5 6 8 9

11. The voltage v (in volts) of a certain device is given by $v = -5 \sin 250t$ where t is the time (in seconds) after 0. At what time will the voltage first equal -1.9 volts? 3 5 9 1 2 4 6 7 8

12. In a series ac circuit a relationship exists between the phase angle of the circuit θ and its quality factor Q. On one page of an ac text the equation $Q = \sqrt{\sec^2\theta - 1}$ appears. Later in the text the equation $\theta = \tan^{-1} Q$ appears. Show that if $Q = \sqrt{\sec^2\theta - 1}$, then it is true that $\theta = \tan^{-1} Q$. 3 5 9 1 2 4 6 7 8

13. A current, $i_1 = 24 \sin(157t + 35°)$, flows through one branch of a two-branch parallel circuit. A current, $i_2 = 48 \sin(157t - 43°)$, flows through the other. The total current i_T is given by

$$i_T = 54.76 \sin 157t - 18.97 \cos 157t.$$

Use this equation to find the value of θ in the following equation.

$$\begin{aligned} i_T &= k \sin(157t + \theta) \\ &= k \sin 157t \cos \theta + k \cos 157t \sin \theta \end{aligned}$$

3 5 9 1 2 4 6 7 8

14. The three large power lines that one sees crossing the countryside represent three-phase electricity. With a balanced load, sine waves of current of equal amplitude but with phase angles 120° apart are thus distributed. The equations

$$\begin{aligned} I_1 &= 350 \sin 377t \\ I_2 &= 350 \sin(377t + 120°) \\ I_3 &= 350 \sin(377t + 240°) \end{aligned}$$

would represent the currents in a typical line. Show that at any instant the sum of the currents in all three lines is zero. In the equations, $377t$ is radian measure. 3 5 9 1 2 4 6 7 8

15. Suppose two musical tones are sounded together, one with frequency $f = 25$ Hz, the other with $f = 30$ Hz. To determine how your brain, which adds these two sounds together, would perceive the resulting sound, use the following identity.

$$\sin A + \sin B = 2 \sin\left(\frac{A+B}{2}\right) \cos\left(\frac{A-B}{2}\right)$$

with $A = 2\pi(25t)$ and $B = 2\pi(30t)$

Identify the amplitude of the resulting sine wave. 1 3 4 5 6 8 9

16. The phase angle of an RLC circuit is given by $\theta = \text{Tan}^{-1} \frac{1}{R}\sqrt{\frac{L}{C}}$ in one text and by $\theta = \text{Sec}^{-1} \frac{1}{R}\sqrt{R^2 + \frac{L}{C}}$ in another. Show that these equations are both true. (Hint: Examine a right triangle with the leg adjacent to θ having length R.) 3 5 9 4 7 8

---

1. Architectural Technology
2. Civil Engineering Technology
3. Computer Engineering Technology
4. Mechanical Drafting Technology
5. Electrical/Electronics Engineering Technology
6. Chemical Engineering Technology
7. Industrial Engineering Technology
8. Mechanical Engineering Technology
9. Biomedical Engineering Technology

17. Put your calculator in radian mode and examine $\text{Sin}^{-1}x$ and $(\text{Sin } x)^{-1}$ for the following values of x:
$x = \frac{2\sqrt{2}}{3}$, $x = \frac{17}{18}$, and $x = \frac{3\pi}{10}$. Is it true that $\text{Sin}^{-1}x = (\text{Sin } x)^{-1}$? 3 5 9

18. The quality Q of a circuit is the tangent of its phase angle. Its reactive power factor F is the sine of the same angle. What formula relates the reactive power factor F to Q? (Hint: Draw a right triangle.) 3 5 9 1 2 4 6 7 8

19. The vibration of a machine part is given by the equation $x = 0.015 \sin 2\omega t$. In the formula x is the part's displacement, ω the radian frequency of a rotating part on the machine, and t time (in seconds). Find a formula for this vibration that contains ω rather than 2ω. 4 7 8 1 2 3 5 6 9

RANDOM REVIEW

In 1 and 2, prove each identity.

1. $\cos(\pi - x) = -\cos x$

2. $\sec^2\theta = \tan\theta(\tan\theta + \cot\theta)$

In 3 and 4, solve each equation for x in radians, $0 \le x < 2\pi$.

3. $3\sin^2 x + \sin x = 2$

4. $\cos 3x = 0.3333$

5. Prove: $\dfrac{2\tan\alpha}{1 + \tan^2\alpha} = \sin 2\alpha$.

6. Solve the following equation for x in radians, $0 \le x < 2\pi$.

$$\cos 2x - 1 = 3\cos 2x$$

7. If $\cos\theta = -15/17$ and θ is in the third quadrant, find $\sin 2\theta$.

8. Evaluate the following expression without the use of tables or a calculator.

$$\sin 121° \cos 76° - \sin 76° \cos 121°$$

9. Prove: $\dfrac{\csc y}{\tan y + \cot y} = \cos y$.

10. Evaluate $\cot(\text{Arcsin } 2x)$.

CHAPTER TEST

In 1 and 2, prove each identity.

1. $\csc x = \sin x + \cos x \cot x$

2. $\dfrac{\csc^2\theta}{\tan\theta} - \cot^3\theta = \cot\theta$

3. Determine the value of $\cos(-75°)$, using $-75° = -30° + (-45°)$.

In 4 and 5, prove each identity.

4. $\sin(\pi/2 - x) = \cos x$

5. $\sin^2\dfrac{x}{2} = \dfrac{\tan x - \sin x}{2\tan x}$

6. If $\cos\theta = -15/17$ and θ is in the second quadrant, find $\cos 2\theta$.

In 7 and 8, solve for x in radians, $0 \le x < 2\pi$.

7. $\tan 2x = 1.6648$

8. $2 \sin^2 x + 3 \sin x + 1 = 0$

9. In astronomical work, the azimuth A (in degrees) is an angle measured from the north toward the east along the horizon and, under certain conditions, is given by the following formula.

$$A = \mathrm{Cos}^{-1}\left(\frac{\sin \delta - \sin \psi \sin t}{\cos \psi \cos t}\right)$$

In the formula, ψ, δ, and t are the latitude, declination, and hour angle respectively and are measured in degrees. Calculate A if

$$\psi = 41°22', \delta = 12°15', \text{ and } t = 25°39'.$$

10. In a distributed constant circuit, the phase angle for the characteristic impedance is given by

$$\theta = \mathrm{Tan}^{-1}\, \omega L/R - \mathrm{Tan}^{-1}\, \omega C/G.$$

In the formula, R is the resistance per unit length (in Ω/km), L is the inductance per unit length (in mH/km), G is the leakage conductance per unit length (in μ℧/km), and C is the capacitance per unit length (in μF/km), where all of these quantities are measured between two conductors. Calculate θ if $\omega = 120\pi$/s, $L = 0.49$ mH/km, $R = 167\ \Omega$/km, $C = 0.05\ \mu$F/km, and $G = 1.66\ \mu$℧/km.

EXPONENTS AND RADICALS CHAPTER 13

You have already seen a few examples of applications of exponents and radicals to technical work. Only a basic knowledge of the topics was necessary to understand those problems. More advanced applications, however, require a broader and deeper familiarity with properties of exponents and operations on radicals.

13.1 FRACTIONAL EXPONENTS AND RADICALS

In Chapter 2, you saw how to simplify expressions with integral exponents. If we allow fractional exponents, most of the old rules should be valid under the extended range of exponents. Consider the problem $x^{1/2}(x^{1/2})$. Since $x^m(x^n) = x^{m+n}$,

$$x^{1/2}(x^{1/2}) = x^{1/2+1/2} = x^1 = x.$$

To ask "What does $x^{1/2}$ mean?" we could ask, "What number times itself is x?" The answer is, "the square root of x." We then say that $x^{1/2}$ must mean $\sqrt{x}$. By generalizing this example, we have for any positive integer n, $x^{1/n} = \sqrt[n]{x}$ when $\sqrt[n]{x}$ is a real number.

DEFINITION If x is a real number and n is a positive integer,

$$x^{1/n} = \sqrt[n]{x} \qquad \textbf{(13-1)}$$

when $\sqrt[n]{x}$ is a real number.

EXAMPLE 1 Simplify each expression.

(a) $25^{1/2}$ **(b)** $(-25)^{1/2}$ **(c)** $-(8^{1/3})$ **(d)** $(-8)^{1/3}$

Solution **(a)** $(25)^{1/2} = \sqrt{25}$. Although there are two square roots of 25 (5 and -5), the radical sign means the positive square root only. Hence $\sqrt{25} = 5$.

(b) $(-25)^{1/2} = \sqrt{-25}$, which is not a real number, because there is no real number that can be multiplied by itself to obtain -25.

(c) $-(8^{1/3})$ means to find $8^{1/3}$ and prefix the negative sign. Hence, we have $-\sqrt[3]{8} = -2$.

(d) $(-8)^{1/3} = \sqrt[3]{-8} = -2$, because $(-2)(-2)(-2) = -8$. ■

The rule $(x^m)^n$ may be used to simplify an expression having a fractional exponent with a numerator other than 1. Since $x^{mn} = (x^m)^n$, we have $x^{m/n} = (x^m)^{1/n} = \sqrt[n]{x^m}$.

This result leads to the following definition.

DEFINITION If x is a real number and m and n are integers such that $n > 0$,

$$x^{m/n} = (\sqrt[n]{x})^m = \sqrt[n]{(x^m)} \qquad \textbf{(13-2)}$$

when $\sqrt[n]{x}$ is a real number.

EXAMPLE 2 Simplify $8^{2/3}$.

Solution $8^{2/3} = (\sqrt[3]{8})^2 = 2^2 = 4$

We could also say $8^{2/3} = (8^2)^{1/3} = \sqrt[3]{(8^2)} = \sqrt[3]{64} = 4$. ■

The fractional exponent consists of two parts: the denominator, which tells us which root to take, and the numerator, which tells us which power to use. You should take the root first whenever possible. For example, consider $64^{4/3}$. If we take the root first, we have

$$64^{4/3} = (\sqrt[3]{64})^4 = 4^4 = 256.$$

But if we take the power first, we have

$$64^{4/3} = \sqrt[3]{(64^4)} = \sqrt[3]{16{,}777{,}216},$$

which is also 256, but you probably wouldn't recognize it as such.

EXAMPLE 3 Simplify $7^{3/5}$.

Solution This time we can't take the root first, because we don't know what $\sqrt[5]{7}$ is, but we do know that it is a real number. Therefore we take the power first, and say

$$7^{3/5} = \sqrt[5]{(7^3)} = \sqrt[5]{343}. \quad \blacksquare$$

A calculator may be used to evaluate an expression with a fractional exponent. The [y^x] key is used to signal the calculator that the next entry is an exponent. Since two numbers are required to produce an answer, the operation is binary and requires the use of the [=] key after entry of the second number.

EXAMPLE 4 Problems solved on a computer are often organized so that the data is arranged in rows and columns. Such an arrangement is called a matrix. The most efficient algorithms for inverting a matrix with N rows and N columns require approximately $N^{14/5}$ steps. Determine the number of steps required to invert a matrix with 10 rows and 10 columns.

Solution We must evaluate $10^{14/5}$. The keystroke sequence follows.

10 [y^x] [(] 14 [÷] 5 [)] [=] **630.957**

An alternate solution follows.

14 [÷] 5 [=] [STO] 10 [y^x] [RCL] [=] **630.957**

About 631 steps are required to invert a 10×10 matrix. ■

EXAMPLE 5 To calculate ship power (in horsepower), the formula $P = (D^2v^3/k)^{1/3}$ is used. In the formula, D is the displacement (in tons), v is the speed (in knots), and k is a constant based on similar ships. Calculate P if $D = 31{,}000$ tons, $v = 25$ knots, and $k = 370$.

Solution $$P = \left(\frac{(31{,}000)^2(25)^3}{370}\right)^{1/3}$$

The keystroke sequence is as follows.

31,000 [x^2] [×] 25 [y^x] 3 [÷] 370 [=] [y^x] [(] 1 [÷] 3 [)] [=] **3436.48**

To two significant digits, the answer is 3400 horsepower. ■

In Chapter 2, the exponent rules such as the product rule, quotient rule, and rules that pertain to expressions with parentheses, were presented for exponents that are integers. All of those rules are also valid for exponents that are rational numbers, provided the base is a positive number. ***The rules may not apply if the base is negative.*** For example, $[(-1)^2]^{1/2} \neq (-1)^1$. In this chapter you may assume that all variables represent nonnegative real numbers, and that no denominators are zero.

NOTE

EXAMPLE 6 Simplify $\left(\dfrac{x^2y^{-1/2}}{x^{-3/4}}\right)^{1/3}$.

Solution

$$\left(\frac{x^2y^{-1/2}}{x^{-3/4}}\right)^{1/3} = \frac{x^{2/3}y^{-1/6}}{x^{-1/4}}$$ Apply the power 1/3 to each factor in parentheses.

$$= \frac{x^{2/3}x^{1/4}}{y^{1/6}}$$ Move the factors with negative exponents across the fraction bar.

$$= \frac{x^{2/3+1/4}}{y^{1/6}}$$ Apply the rule $x^m x^n = x^{m+n}$.

$$= \frac{x^{11/12}}{y^{1/6}}$$ Add: $2/3 + 1/4 = 8/12 + 3/12$. ■

In Example 6, both numerator and denominator contain a single term. When an expression has several terms, the distributive property may be used to factor out the common factor. This procedure was used in Chapter 5 with exponents that were positive integers, but may also be used even when the exponents are not positive integers. It is an especially useful technique for certain types of calculus problems.

FACTORING PRINCIPLE

To apply the distributive property to an expression with terms, factor out the *lowest* power of each factor that is common to all terms.

EXAMPLE 7 Factor $a^2b^{-3} + a^3b^2$, and simplify the result.

Solution The lowest power of a is a^2. The lowest power of b is b^{-3}. The common factor, then, is a^2b^{-3}.

$$a^2b^{-3} + a^3b^2 = \boldsymbol{a^2b^{-3}}(1) + \boldsymbol{a^2b^{-3}}(ab^5)$$

Using the distributive property to factor a^2b^{-3} from both terms, we have

$$a^2b^{-3} + a^3b^2 = \boldsymbol{a^2b^{-3}}(1 + ab^5)$$
$$= \frac{a^2(1 + ab^5)}{b^3}. \quad ■$$

EXAMPLE 8 Factor $xy^{1/2} - x^{-3/4}y$, and simplify the result.

Solution The lowest power of x is $x^{-3/4}$ and the lowest power of y is $y^{1/2}$. The common factor, then, is $x^{-3/4}y^{1/2}$.

$$xy^{1/2} - x^{-3/4}y = \mathbf{x^{-3/4}y^{1/2}}(x^{7/4}) - \mathbf{x^{-3/4}y^{1/2}}(y^{1/2})$$

Using the distributive property, we have

$$\begin{aligned} xy^{1/2} - x^{-3/4}y &= \mathbf{x^{-3/4}y^{1/2}}(x^{7/4} - y^{1/2}) \\ &= \frac{y^{1/2}(x^{7/4} - y^{1/2})}{x^{3/4}}. \end{aligned}$$ ■

EXAMPLE 9 Factor $(8x + 1)^{-1}(4x + 3)^{-1/2} - 4(4x + 3)^{1/2}(8x + 1)^{-2}$, and simplify the result.

Solution The lowest power of $(8x + 1)$ is $(8x + 1)^{-2}$, and the lowest power of $(4x + 3)$ is $(4x + 3)^{-1/2}$. The common factor is $(8x + 1)^{-2}(4x + 3)^{-1/2}$.

$$\begin{aligned} &(8x + 1)^{-1}(4x + 3)^{-1/2} - 4(4x + 3)^{1/2}(8x + 1)^{-2} \\ &= \mathbf{(8x + 1)^{-2}(4x + 3)^{-1/2}}(8x + 1) - \mathbf{(8x + 1)^{-2}(4x + 3)^{-1/2}}4(4x + 3) \\ &= (8x + 1)^{-2}(4x + 3)^{-1/2}[8x + 1 - 16x - 12] \\ &= (8x + 1)^{-2}(4x + 3)^{-1/2}(-8x - 11) = \frac{-8x - 11}{(8x + 1)^2(4x + 3)^{1/2}} \end{aligned}$$ ■

EXAMPLE 10 Factor $2(3x + 2)^{-2}(2x - 1)^{-1/2} + 4(2x - 1)^{1/2}(3x + 2)^{-3}$, and simplify the result.

Solution The lowest power of $(3x + 2)$ is $(3x + 2)^{-3}$ and the lowest power of $(2x - 1)$ is $(2x - 1)^{-1/2}$. The common factor is $2(3x + 2)^{-3}(2x - 1)^{-1/2}$. Using the distributive property, we have the following.

$$\begin{aligned} &2(3x + 2)^{-2}(2x - 1)^{-1/2} + 4(2x - 1)^{1/2}(3x + 2)^{-3} \\ &= 2(3x + 2)^{-3}(2x - 1)^{-1/2}[(3x + 2) + 2(2x - 1)] \\ &= 2(3x + 2)^{-3}(2x - 1)^{-1/2}[3x + 2 + 4x - 2] \\ &= 2(3x + 2)^{-3}(2x - 1)^{-1/2}(7x) = \frac{14x}{(3x + 2)^3(2x - 1)^{1/2}} \end{aligned}$$ ■

EXERCISES 13.1

You may assume that all variables in these exercises represent nonnegative real numbers and no denominators are zero. In 1–15, simplify each expression.

1. $16^{1/2}$ **2.** $(-16)^{1/2}$ **3.** $-(16^{1/2})$ **4.** $27^{1/3}$ **5.** $(-27)^{1/3}$

6. $-(27^{1/3})$ **7.** $(64^{1/2})$ **8.** $64^{1/3}$ **9.** $(-64)^{1/3}$ **10.** $(-64)^{1/2}$

11. $64^{2/3}$ **12.** $27^{2/3}$ **13.** $4^{2/3}$ **14.** $5^{3/4}$ **15.** $81^{3/4}$

In 16–21, evaluate each expression, using a calculator. Give answers to three significant digits.

16. $4^{2/3}$ **17.** $5^{3/4}$ **18.** $(-9)^{3/7}$

19. $(-3)^{4/5}$ **20.** $13^{-2/3}$ **21.** $23^{-5/6}$

In 22–28, simplify each expression.

22. $\left(\frac{4ab^{-1/2}}{c^{-2}}\right)^{1/3}$ **23.** $\left(\frac{3x^{1/2}y^{-1}}{z^{2/3}}\right)^{-1/2}$ **24.** $\left(\frac{6x^{-1}y^{2/3}}{z^{1/3}}\right)^{-3}$ **25.** $\left(\frac{5a^{-2}}{b^{1/3}c^{-1/2}}\right)^{2}$

26. $\frac{5a^{-1/2}b^{2/3}}{a(b+1)^{-1/2}}$ **27.** $\frac{x^{-1/2}+2^{-2/3}}{(3x)^{-1}}$ **28.** $\frac{x^{2/3}+y^{3/4}}{xy^{-2}}$

In 29–34, factor each expression, and simplify the result.

29. $3a^{-1}b^{-2}+4a^{-2}b^{-1}$

30. $x^{1/2}y^{-1/2}+x^{-1/2}y^{1/2}$

31. $3x(2x+3)^{-1}(x-1)^{-2}+2x(2x+3)^{-2}(x-1)^{-1}$

32. $x(2x+1)^{-1/2}(x-2)^{-2}-3x(2x+1)^{1/2}(x-2)^{-1}$

33. $3x(x-4)^{-2/3}(x+1)^{-2}-x(x-4)^{1/3}(x+1)^{-1}$

34. $x(x-5)^{-1/3}(x+3)^{1/2}+2x(x+3)^{-1/2}(x-5)^{2/3}$

---

35. When a circuit contains an inductor and a resistor in series, and the battery is short-circuited, the current does not drop instantly. After t seconds, the current is given by $I = I_0e^{-t/\tau}$. In the formula, I_0 is the steady-state current, e is a constant that is approximately equal to 2.718, and τ is the time constant (i.e., the time that it will take for the current to fall to 37 percent of its original value). Find the decreased current if $I_0 = 2.0$ A, $t = 0.020$ s, and $\tau = 0.065$ s. **3 5 9** 1 2 4 6 7 8

36. Dosage for medicine is sometimes based on the surface area of the patient's body. If h and w are the patient's height (in cm) and weight (in kg), respectively, the following formula gives an approximation for body surface area (in m²).

DuBoise's formula:

$$A = 71.84(h^{29/40})(w^{17/40})(10^{-4})$$

Calculate the surface area for a person 162.6 cm tall and 46.4 kg in weight. **6 9** 2 4 5 7 8

37. If T is the temperature (in degrees Celsius) at which water boils at a particular location, an approximation for the elevation E (in m) above sea level for that location is given by

$$E = (10^3)(100 - T) + 5800(10^{-1})(100 - T)^2.$$

Factor the expression on the right-hand side of the equation. **6 8** 1 2 3 5 7 9

38. For a simple beam under a concentrated load, the deflection (in mm) is given by

$$y = P(l-a)x^3(6EIl)^{-1} + P(l-a)^3x(6EIl)^{-1} - P(l-a)xl^2(6EIl)^{-1}$$

if $x \le$ a. In the formula, P is the load (in kg), l is the length (in mm) of the beam, a is the distance (in mm) of the load from the support, x is the distance (in mm) from the support at which the deflection is measured, E is the modulus of elasticity (in kg/mm²), and I is the moment of inertia (in mm⁴). Factor the expression on the right-hand side of the equation. **1 2 4** 8 7 9

39. Consider the rectangle in the accompanying figure. The length and width can be determined from the Pythagorean theorem.

(a) Find the perimeter of the rectangle.

(b) Find the area of the rectangle. **1 2 4 7 8** 3 5

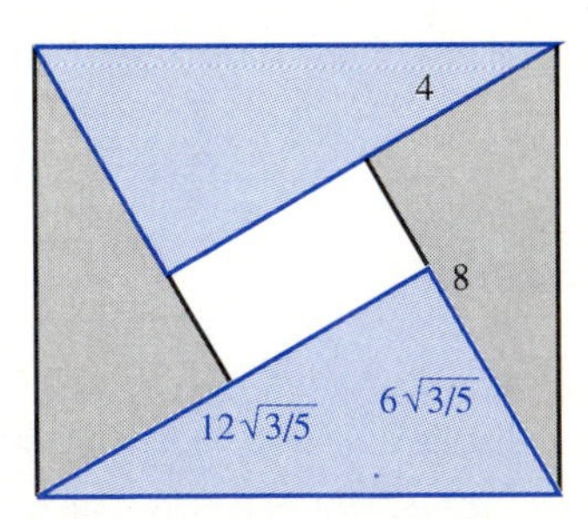

1. Architectural Technology
2. Civil Engineering Technology
3. Computer Engineering Technology
4. Mechanical Drafting Technology
5. Electrical/Electronics Engineering Technology
6. Chemical Engineering Technology
7. Industrial Engineering Technology
8. Mechanical Engineering Technology
9. Biomedical Engineering Technology

40. The formula $D_2 = (R_1 D_1^2/R_2)^{1/2}$ gives the diameter (D_2) of a wire having resistance R_2 if a wire of the same material with resistance R_1 has diameter D_1.
(a) Write the equation using a radical.
(b) Solve for D_2 if $R_2 = 1.5\ \Omega$, $R_1 = 2.5\ \Omega$, and $D_1 = 0.040$ in. 3 5 9 1 2 4 6 7 8

41. If it requires $N^{14/5}$ steps to invert an $N \times N$ matrix, how would the number of steps be affected if N were doubled? 3 5 9

42. For the problems in this chapter, the variables represent nonnegative real numbers so that $\sqrt{x^2} = x$.
(a) Explain why it is necessary to say $\sqrt{x^2} = |x|$ if x represents any real number.
(b) Simplify $\sqrt[3]{x^3}$ if x is any real number.

13.2 ADDITION, SUBTRACTION, AND MULTIPLICATION OF RADICALS

Recall that a number specified by a radical sign $\sqrt[n]{\ }$ is called a radical. Calculators are often used to evaluate radicals. In some circumstances, however, it may be preferable to work with radicals in exact form, rather than in approximate decimal form. Since an expression involving radicals can be written with fractional exponents, it is possible to simplify such expressions using rules that are similar to the exponent rules. Recall from Section 2.3 that we have the following rules. If x and y are real numbers and n is a positive integer, then

$$\sqrt[n]{xy} = \sqrt[n]{x}\sqrt[n]{y} \tag{13-3}$$

$$\sqrt[n]{\frac{x}{y}} = \frac{\sqrt[n]{x}}{\sqrt[n]{y}} \text{ if } y \neq 0 \tag{13-4}$$

when all of the radicals represent real numbers. These rules are reasonable if you notice that

$$\sqrt[n]{xy} = (xy)^{1/n} = x^{1/n}y^{1/n} = \sqrt[n]{x}\sqrt[n]{y} \text{ and}$$
$$\sqrt[n]{\frac{x}{y}} = \left(\frac{x}{y}\right)^{1/n} = \frac{x^{1/n}}{y^{1/n}} = \frac{\sqrt[n]{x}}{\sqrt[n]{y}}.$$

radicand
order

In order to discuss radicals, it is necessary to introduce some terminology. The number under the radical sign $\sqrt[n]{\ }$ is called the **radicand,** and the number n is called the **order** of the radical. If the exponent of the radicand and the order of the radical have a common factor, it is possible to write the radical with a smaller order. The procedure is called *reducing the order of the radical.*

To reduce the order of a radical

1. Write the radical expression as an equivalent expression using exponents.
2. Use the rule $(x^m)^n = x^{mn}$ to multiply exponents, assuming $x \geq 0$.
3. Reduce the fractional exponent to lowest terms.
4. Write the expression as a radical.

EXAMPLE 1 Reduce the order of each radical.

(a) $\sqrt[4]{9}$ **(b)** $\sqrt[6]{x^2y^4}$

Solution **(a)** Since 9 is 3^2, we have $\sqrt[4]{9} = \sqrt[4]{3^2}$.

The exponent of the radicand and the order of the radical are 2 and 4, respectively. They have a common factor of 2.

$$\sqrt[4]{9} = \sqrt[4]{3^2} = (3^2)^{1/4} = 3^{2/4} = 3^{1/2} = \sqrt{3}$$

(b) The radicand may be written with the exponent 2: $x^2y^4 = (xy^2)^2$.

$$\sqrt[6]{x^2y^4} = \sqrt[6]{(xy^2)^2} = [(xy^2)^2]^{1/6} = (xy^2)^{2/6} = (xy^2)^{1/3} = \sqrt[3]{xy^2}$$ ■

DEFINITION A radical is said to be in **simplest radical form** if it meets the following conditions:

1. No factor of the radicand has an exponent as large as the order of the radical.
2. There are no radicals in the denominator (or equivalently, the radicand contains no denominator.)
3. The order of the radical is reduced to lowest terms.

The rule $\sqrt[n]{xy} = \sqrt[n]{x}\sqrt[n]{y}$ is used to remove factors of the radicand when the exponent is greater than or equal to the order of the radical.

EXAMPLE 2 Write each expression in simplest form.

(a) $\sqrt{18}$ **(b)** $\sqrt[3]{32x^4y^7}$ **(c)** $\sqrt{27a^3b^5}$

Solution **(a)** To find $\sqrt{18}$, factor 18 as 9(2), since 9 is 3^2.

$\sqrt{18} = \sqrt{9(2)} = 3\sqrt{2}$

(b) $32x^4y^7$ is factored as $\mathbf{8} \cdot 4 \cdot \mathbf{x^3} \cdot x \cdot \mathbf{y^6} \cdot y$ so that as many factors as possible are cubes.

$\sqrt[3]{32x^4y^7} = \sqrt[3]{\mathbf{8} \cdot 4 \cdot \mathbf{x^3} \cdot x \cdot \mathbf{y^6} \cdot y} = \sqrt[3]{8x^3y^6}\sqrt[3]{4xy} = 2xy^2\sqrt[3]{4xy}$

(c) $\sqrt{27a^3b^5} = \sqrt{\mathbf{9} \cdot 3 \cdot \mathbf{a^2} \cdot a \cdot \mathbf{b^4} \cdot b} = \sqrt{9a^2b^4}\sqrt{3ab} = 3ab^2\sqrt{3ab}$ ■

To write an expression so that there are no radicals in the denominator, the denominator is rationalized. That is, both numerator and denominator are multiplied by the same number. This number is chosen so that the denominator will become the nth root of an nth power. For square roots, the process often leads to multiplying the denominator by itself.

EXAMPLE 3 For an ac circuit, the ratio of the maximum current to the effective current is $\sqrt{2}$. Find the effective current in a circuit for which the maximum current is 1.50 A.

Solution Let $I =$ effective current. Then $\frac{I_{max}}{I} = \sqrt{2}$. Solving for I, $I_{max} = \sqrt{2}I$, and $I = \frac{I_{max}}{\sqrt{2}} = \frac{I_{max}}{\sqrt{2}}\left(\frac{\sqrt{2}}{\sqrt{2}}\right) = \frac{I_{max}\sqrt{2}}{2} = \frac{1.50\sqrt{2}}{2} = 0.75\sqrt{2}$. ■

EXAMPLE 4 The current I (in A) in a resistor with resistance R (in Ω) and power rating P (in W) is given by $I = \sqrt{P/R}$. Find the current in a 300 Ω resistor rated at 1 W.

Solution An exact answer is obtained as follows.

$$I = \sqrt{1/300} = \frac{\sqrt{1}}{\sqrt{300}} = \frac{\sqrt{1}}{\sqrt{300}}\left(\frac{\sqrt{3}}{\sqrt{3}}\right) = \frac{\sqrt{3}}{30}\text{ A}$$

A decimal approximation may be obtained using a calculator. The keystroke sequence to evaluate $I = \sqrt{1/300}$ follows.

1 [÷] 300 [=] [√x] **0.057735**

A reasonable answer would be 0.06 A. ■

For a cube root or other nonsquare root, it is especially important to carefully choose the number by which to multiply numerator and denominator.

EXAMPLE 5 Rationalize each denominator.

(a) $\frac{1}{\sqrt[3]{2}}$ **(b)** $\frac{2}{\sqrt[4]{9}}$

Solution **(a)** We would like to have the cube root of a cube in the denominator. Since the factor 2 already appears in the radicand, it would be easy to make the denominator 2^3. We multiply by $\frac{\sqrt[3]{2^2}}{\sqrt[3]{2^2}}$.

$$\frac{1}{\sqrt[3]{2}}\left(\frac{\sqrt[3]{2^2}}{\sqrt[3]{2^2}}\right) = \frac{\sqrt[3]{2^2}}{\sqrt[3]{2^3}} = \frac{\sqrt[3]{2^2}}{2} = \frac{\sqrt[3]{4}}{2}$$

(b) Since the radicand in the denominator already contains 3^2, we make it become 3^4. That is, multiply by $\frac{\sqrt[4]{3^2}}{\sqrt[4]{3^2}}$.

$$\frac{2}{\sqrt[4]{9}}\left(\frac{\sqrt[4]{3^2}}{\sqrt[4]{3^2}}\right) = \frac{2\sqrt[4]{3^2}}{\sqrt[4]{81}} = \frac{2\sqrt[4]{9}}{3}$$

But $\sqrt[4]{9} = 9^{1/4} = (3^2)^{1/4} = 3^{2/4} = 3^{1/2} = \sqrt{3}$. The answer is $\frac{2\sqrt{3}}{3}$. ■

Now that you know how to simplify radicals, it is possible to add and subtract them. Adding and subtracting radical expressions is analogous to adding and subtracting polynomials. The rule for polynomials is add or subtract like terms. Just as like terms are different in at most the numerical factor, like radicals are different in at most the factor outside the radical. Just as x is understood to be $1x$, $\sqrt{2}$ is $1\sqrt{2}$.

DEFINITION **Like radicals** are radicals that have identical radicands and identical orders.

EXAMPLE 6 Add $\sqrt{3} + \sqrt{2} + \sqrt[3]{2} + 3\sqrt{2}$.

Solution $\sqrt{2}$ and $3\sqrt{2}$ are like radicals, since both are square roots and 2 is the radicand in both.

$$\sqrt{3} + \sqrt{2} + \sqrt[3]{2} + 3\sqrt{2} = \sqrt{3} + \sqrt[3]{2} + 1\sqrt{2} + 3\sqrt{2} = \sqrt{3} + \sqrt[3]{2} + 4\sqrt{2}$$ ■

NOTE ⇨⇨ ***Notice that in Example 6, $\sqrt{2}$ and $\sqrt[3]{2}$ are not like radicals, since they have different orders. Also $\sqrt{3}$ and $\sqrt{2}$ are not like radicals, since they have different radicands.*** Thus $\sqrt{3} + \sqrt{2}$ cannot be simplified.

EXAMPLE 7 Use a calculator to evaluate each expression.

(a) $\sqrt{3} + \sqrt{2}$ **(b)** $\sqrt{5}$

Solution **(a)** The keystroke sequence follows.

3 [√x] [+] 2 [√x] [=] **3.1462644**

(b) The keystroke sequence is 5 [√x] **2.236068** ■

NOTE ⇨⇨ Notice that the answers to Example 7 (a) and (b) are different. ***In general, $\sqrt{x} + \sqrt{y} \neq \sqrt{(x + y)}$.*** To add or subtract radical expressions without a calculator, first simplify each radical to determine if the radicals are like radicals.

EXAMPLE 8 Perform the indicated operations.

$$2\sqrt{32} - \sqrt{2} + \frac{6}{\sqrt{2}}$$

Solution

$$2\sqrt{32} - \sqrt{2} + \frac{6}{\sqrt{2}} = 2\sqrt{16(2)} - 1\sqrt{2} + \frac{6}{\sqrt{2}}\left(\frac{\sqrt{2}}{\sqrt{2}}\right)$$
$$= 2(4)\sqrt{2} - 1\sqrt{2} + \frac{6\sqrt{2}}{2}$$
$$= 8\sqrt{2} - 1\sqrt{2} + 3\sqrt{2}$$
$$= 10\sqrt{2} \quad ■$$

EXAMPLE 9 Perform the indicated operations.

$$\sqrt[3]{16} + 5\sqrt[3]{2} - \frac{4}{\sqrt[3]{4}}$$

Solution

$$\sqrt[3]{16} + 5\sqrt[3]{2} - \frac{4}{\sqrt[3]{4}} = \sqrt[3]{8(2)} + 5\sqrt[3]{2} - \frac{4}{\sqrt[3]{4}}\left(\frac{\sqrt[3]{2}}{\sqrt[3]{2}}\right)$$
$$= 2\sqrt[3]{2} + 5\sqrt[3]{2} - \frac{4\sqrt[3]{2}}{\sqrt[3]{8}}$$
$$= 2\sqrt[3]{2} + 5\sqrt[3]{2} - \frac{4\sqrt[3]{2}}{2}$$
$$= 2\sqrt[3]{2} + 5\sqrt[3]{2} - 2\sqrt[3]{2}$$
$$= 5\sqrt[3]{2} \quad ■$$

NOTE ⇨⇨ ***For an addition or subtraction problem, it is usually easier to simplify the radicals and then add or subtract. For a multiplication problem, it is usually easier to multiply first and then simplify.*** Multiplication of radicals with the same order is similar to multiplication of polynomials. When the two radical expressions, each consisting of a single term, are multiplied, the factors outside the radical are multiplied and the radicands are multiplied. Example 10 illustrates this principle.

EXAMPLE 10 Perform the indicated operations.

(a) $\sqrt{6}\sqrt{12}$ **(b)** $2\sqrt[3]{4}(5\sqrt[3]{4})$

Solution **(a)** $\sqrt{6}\sqrt{12} = \sqrt{72} = \sqrt{36(2)} = 6\sqrt{2}$

(b) $2\sqrt[3]{4}(5\sqrt[3]{4}) = 10\sqrt[3]{16} = 10\sqrt[3]{8(2)} = 10(2)\sqrt[3]{2} = 20\sqrt[3]{2}$ ■

When the expressions contain more than one term, the distributive property is used, as it is with polynomials.

EXAMPLE 11 Perform the indicated operations.

(a) $(2\sqrt{3} - \sqrt{5})(\sqrt{3} + 2\sqrt{5})$

(b) $(2\sqrt{3x} - 4)(2\sqrt{3x} + 4)$, assume that $x \geq 0$

Solution (a) $(2\sqrt{3}-\sqrt{5})(\sqrt{3}+2\sqrt{5}) = 2\sqrt{3}(\sqrt{3}) + 2\sqrt{3}(2\sqrt{5}) - \sqrt{5}(\sqrt{3}) - \sqrt{5}(2\sqrt{5})$
$= 2\sqrt{9} + 4\sqrt{15} - \sqrt{15} - 2\sqrt{25}$
$= 2(3) + 4\sqrt{15} - \sqrt{15} - 2(5)$
$= 6 + 3\sqrt{15} - 10$
$= -4 + 3\sqrt{15}$

(b) $(2\sqrt{3x}-4)(2\sqrt{3x}+4) = 2\sqrt{3x}(2\sqrt{3x}) + 2(4)\sqrt{3x} - 4(2\sqrt{3x}) - 4(4)$
$= 4\sqrt{9x^2} + 8\sqrt{3x} - 8\sqrt{3x} - 16$
$= 4(3x) - 16$
$= 12x - 16$ ■

EXERCISES 13.2

You may assume that all variables in these exercises represent nonnegative real numbers and that no denominators are zero. In 1–6, reduce the order of each radical and simplify, if possible.

1. $\sqrt[8]{16}$ **2.** $\sqrt[6]{8}$ **3.** $\sqrt[4]{4x^2y^6}$
4. $\sqrt[4]{9a^4b^8}$ **5.** $\sqrt[6]{27x^3y^6}$ **6.** $\sqrt[8]{16a^4b^8}$

In 7–20, write each expression in simplest radical form.

7. $\sqrt{20x^2y^5}$ **8.** $\sqrt{32a^3b^4}$ **9.** $\sqrt{50x^5y^7}$ **10.** $\sqrt[3]{24x^3y^5}$
11. $\sqrt[3]{81a^4b^6}$ **12.** $\sqrt[3]{128x^8y^7}$ **13.** $\sqrt[4]{32a^4b^9}$ **14.** $\sqrt[4]{48a^5b^{10}}$
15. $\dfrac{1}{\sqrt{2}}$ **16.** $\dfrac{3}{\sqrt{5}}$ **17.** $\dfrac{\sqrt[3]{2}}{\sqrt[3]{3}}$ **18.** $\dfrac{\sqrt[4]{5}}{3\sqrt[4]{2}}$
19. $\dfrac{7}{\sqrt[4]{49}}$ **20.** $\dfrac{-1}{\sqrt[3]{36}}$

In 21–42, perform the indicated operations. Do not use a calculator.

21. $3\sqrt{3} + 2\sqrt{2} - \sqrt{3}$ **22.** $5\sqrt{2} - 7\sqrt{3} + 4\sqrt{2}$ **23.** $4\sqrt[3]{3} - 2\sqrt[3]{3} + \sqrt{3}$
24. $2\sqrt[3]{5} + \sqrt[3]{5} - 3\sqrt{5}$ **25.** $2a\sqrt{b} + 2b\sqrt{a} - b\sqrt{a}$ **26.** $\sqrt[4]{4} + \sqrt{8} - \dfrac{4}{\sqrt{2}}$
27. $2\sqrt[4]{9} - 2\sqrt{12} + \dfrac{6}{\sqrt{3}}$ **28.** $\sqrt{12a^2b^3} + 2a\sqrt{27b^3} - \dfrac{ab^2}{\sqrt{3b}}$ **29.** $\sqrt[6]{9} - \sqrt[3]{24} + \dfrac{3}{\sqrt[3]{9}}$
30. $\sqrt[6]{4} - \sqrt[3]{54} + \dfrac{2}{\sqrt[3]{4}}$ **31.** $\sqrt[3]{40a^3b^4} + a\sqrt[3]{5b^4} - \dfrac{5ab^2}{\sqrt[3]{25b^2}}$ **32.** $\sqrt{3}\sqrt{12}$
33. $2\sqrt{20}(3\sqrt{5})$ **34.** $\sqrt[3]{3x}\sqrt[3]{9x^2}$ **35.** $\sqrt{2b}(\sqrt{3b} - 1)$
36. $2\sqrt{3}(\sqrt{2} + 3\sqrt{5})$ **37.** $\sqrt[3]{3a}(\sqrt[3]{9a^2} - 1)$ **38.** $(\sqrt{2x} - \sqrt{3y})(\sqrt{2x} + \sqrt{3y})$
39. $(3\sqrt{2} + \sqrt{5})(\sqrt{2} - 3\sqrt{5})$ **40.** $(3\sqrt{5} + 1)(2\sqrt{5} - 3)$
41. $(4\sqrt{7} - 3)(4\sqrt{7} + 3)$ **42.** $(6\sqrt{2} - \sqrt{3})(4\sqrt{2} + \sqrt{3})$

Problems 43 and 44 use the following information and accompanying diagram. For a well-tuned piano, the frequency (in cycles/second) of each note is $\sqrt[12]{2}$ *times the frequency of the note below it (black keys count, too). The note A above middle C is set at* 440.0 *cycles/sec.*

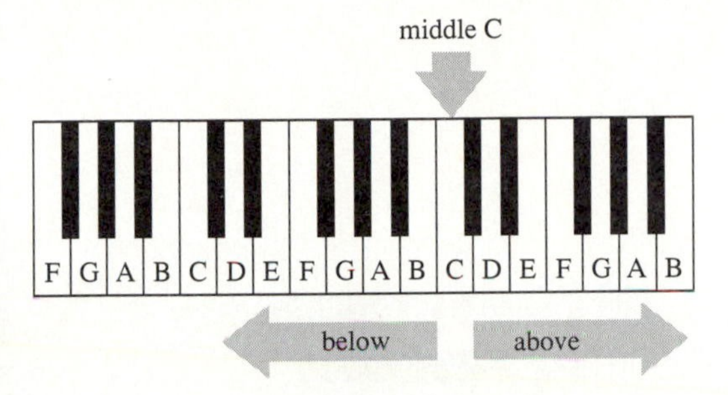

43. Find the frequency of middle C. 1 3 4 5 6 8 9

44. Find the frequency of F below middle C. 1 3 4 5 6 8 9

45. The formula $f_r = 1/(2\pi\sqrt{LC})$ gives resonant frequency f_r of a circuit containing an inductor and a capacitor in series. Find f_r if $L = 20 \times 10^{-3}$ Hz and $C = 1.30 \times 10^{-6}$ F. Write the answer in simplest radical form. 3 5 9 4 7 8

46. The radius of a sphere with volume V is given by $R = \sqrt[3]{3V/(4\pi)}$. Find the radius of a sphere that has the same volume as a cube 5 inches on each side.

(a) Give an exact answer in simplest radical form.

(b) Give a decimal approximation to three significant digits. 1 2 4 7 8 3 5

47. The formula $V = \sqrt{PR}$ gives the voltage for a circuit if P is the power and R is the resistance. Also, $P = I^2R$ where I is the current and R is the resistance. Show that $V = IR$ by substituting I^2R for P in the formula for V. 3 5 9 1 2 4 6 7 8

48. A small block of mass m is on a rotating turntable at a distance of r from the center. The maximum linear velocity the mass may have without slipping is given by $v = \sqrt{\mu g r}$ where μ is the coefficient of friction between the mass and the turntable, and g is the acceleration due to gravity. Use scientific notation to evaluate v if $\mu = 0.49$, $g = 980$ cm/s^2, and $r = 20$ cm. 4 7 8 1 2 3 5 6 9

49. The rule $\sqrt[n]{xy} = \sqrt[n]{x}(\sqrt[n]{y})$ was stated for real numbers x and y when all of the radicals represent real numbers.

(a) Give an example in which both of the values x and y are negative and the statement $\sqrt[n]{xy} = \sqrt[n]{x}(\sqrt[n]{y})$ is true.

(b) Give an example in which both of the values x and y are negative and the statement $\sqrt[n]{xy} = \sqrt[n]{x}(\sqrt[n]{y})$ is false.

1. Architectural Technology
2. Civil Engineering Technology
3. Computer Engineering Technology
4. Mechanical Drafting Technology
5. Electrical/Electronics Engineering Technology
6. Chemical Engineering Technology
7. Industrial Engineering Technology
8. Mechanical Engineering Technology
9. Biomedical Engineering Technology

13.3 DIVISION OF RADICALS

Division of radical expressions is often done by calculator.

EXAMPLE 1 Evaluate $\dfrac{4}{\sqrt{2} + \sqrt{3}}$ using a calculator.

Solution The problem may be done in several ways. Parentheses may be used to indicate the denominator. A second approach is to simplify the denominator and store it so that it can be recalled for the division. The keystroke sequence for each of these strategies follows.

(a) 4 [÷] [(] 2 [√x] [+] 3 [√x] [)] [=] *1.271349*

(b) 2 [√x] [+] 3 [√x] [=] [STO] 4 [÷] [RCL] [=] *1.271349*

Since the integers 2, 3, and 4 in this problem are not intended to be measurements, they may be assumed to be exact values. The answer is given as 1.27, to three significant digits. ■

EXAMPLE 2 Evaluate $\dfrac{\sqrt{3}+\sqrt{5}}{\sqrt{3}-\sqrt{5}}$ using a calculator.

Solution Two solutions are given.

(a) [(] 3 [√x] [+] 5 [√x] [)] [÷] [(] 3 [√x] [−] 5 [√x] [)] [=] *−7.8729833*

(b) 3 [√x] [−] 5 [√x] [=] [STO] 3 [√x] [+] 5 [√x] [=] [÷] [RCL] [=] *−7.8729833*

To three significant digits, the answer is −7.87. ■

EXAMPLE 3 Evaluate $\dfrac{2\sqrt{3}+\sqrt{5}}{4\sqrt{7}}$.

Solution Two solutions are given.

(a) [(] 2 [×] 3 [√x] [+] 5 [√x] [)] [÷] [(] 4 [×] 7 [√x] [)] [=] *0.5386154*

(b) 2 [×] 3 [√x] [+] 5 [√x] [=] [÷] 4 [÷] 7 [√x] [=] *0.5386154*

To three significant digits, the answer is 0.539. ■

When a radicand contains variables, it must be simplified algebraically rather than on a calculator. In such cases the denominator is rationalized. Recall that in this chapter, all variables represent nonnegative real numbers and no denominators are zero.

EXAMPLE 4 Perform the indicated operation.

$$\frac{2y}{3\sqrt{y}}$$

Solution $\dfrac{2y}{3\sqrt{y}}\left(\dfrac{\sqrt{y}}{\sqrt{y}}\right)=\dfrac{2y\sqrt{y}}{3\sqrt{y^2}}=\dfrac{2y\sqrt{y}}{3y}$

Because y is a factor in both the numerator and denominator, the fraction may be reduced.

$$\frac{2y\sqrt{y}}{3y}=\frac{2\sqrt{y}}{3}$$ ■

EXAMPLE 5 The current I (in A) in a circuit is given by $I = \sqrt{\frac{2W}{L}}$ if W is the potential energy (in joules) in an inductor, and L is the inductance (in H). Write the formula in simplest radical form.

Solution
$$I = \sqrt{\frac{2W}{L}} = \frac{\sqrt{2W}}{\sqrt{L}} = \frac{\sqrt{2W}}{\sqrt{L}}\left(\frac{\sqrt{L}}{\sqrt{L}}\right) = \frac{\sqrt{2WL}}{L}$$ ■

When a denominator contains two terms, it is necessary to use the *conjugate* of the denominator to rationalize the denominator. The conjugate of a radical expression contains terms with the same absolute values as the terms of the original expression, but one term is opposite in sign from the original.

DEFINITION If a, b, $\sqrt{x}$, and $\sqrt{y}$ are real numbers, radical expressions of the form $a\sqrt{x} + b\sqrt{y}$ and $a\sqrt{x} - b\sqrt{y}$ are **conjugates** of each other.

Recall that when we make a statement about real numbers, it is understood that the real numbers may be numerical values or variables that represent real numbers. Although numerical problems are usually done on a calculator, numerical problems are given in Examples 6 and 7 to illustrate the method before examples with variables are presented.

EXAMPLE 6 Perform the indicated operation.

$$\frac{1}{\sqrt{5} - \sqrt{3}}$$

Solution The conjugate of $\sqrt{5} - \sqrt{3}$ is $\sqrt{5} + \sqrt{3}$. Numerator and denominator are multiplied by $\sqrt{5} + \sqrt{3}$.

$$\left(\frac{1}{\sqrt{5} - \sqrt{3}}\right)\left(\frac{\sqrt{5} + \sqrt{3}}{\sqrt{5} + \sqrt{3}}\right) = \frac{\sqrt{5} + \sqrt{3}}{\sqrt{25} + \sqrt{15} - \sqrt{15} - \sqrt{9}} = \frac{\sqrt{5} + \sqrt{3}}{5 - 3} = \frac{\sqrt{5} + \sqrt{3}}{2}$$ ■

Notice that when the conjugates in Example 6 are multiplied, the product is the difference of squares.

$$(\sqrt{5} - \sqrt{3})(\sqrt{5} + \sqrt{3}) = (\sqrt{5})^2 - (\sqrt{3})^2 = 5 - 3 = 2$$

Thus, it may be possible to eliminate some of the steps.

EXAMPLE 7 Perform the indicated operation.

$$\frac{2\sqrt{7}-2}{3\sqrt{7}+2}$$

Solution $$\left(\frac{2\sqrt{7}-2}{3\sqrt{7}+2}\right)\left(\frac{3\sqrt{7}-2}{3\sqrt{7}-2}\right)=\frac{6(7)-4\sqrt{7}-6\sqrt{7}+4}{9(7)-4}=\frac{42-10\sqrt{7}+4}{63-4}$$
$$=\frac{46-10\sqrt{7}}{59} \quad \blacksquare$$

EXAMPLE 8 A cube rests on a wedge as shown in Figure 13.1. The coefficient of friction between the cube and the wedge is μ, the angle of inclination of the wedge is θ, and g is the acceleration due to gravity. If a horizontal force is applied to make the cube move relative to the surface, the system has an acceleration a given by

$$a=\frac{g(\mu-\tan\theta)}{1+\mu\tan\theta}.$$

Find a when $\theta=60°$, $\mu=1$, and $g=9.8$ m/sec^2.

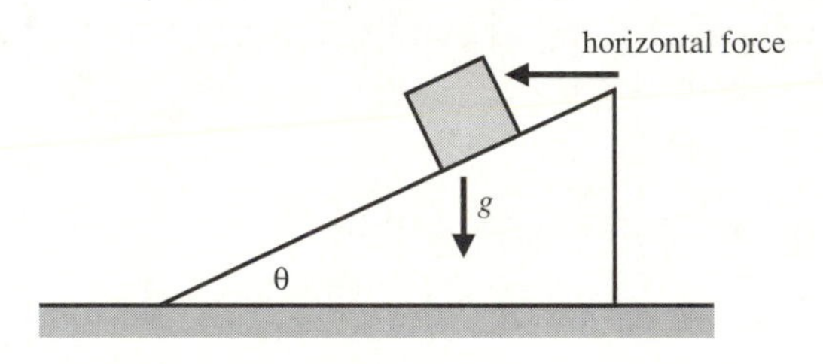

FIGURE 13.1

Solution When θ is 60°, $\tan\theta$ is $\sqrt{3}$. Thus,

$$a=\frac{9.8(1-\sqrt{3})}{1+\sqrt{3}}.$$

Rationalizing the denominator, we have the following.

$$a=\left(\frac{9.8(1-\sqrt{3})}{1+\sqrt{3}}\right)\left(\frac{1-\sqrt{3}}{1-\sqrt{3}}\right)=\frac{9.8(1-2\sqrt{3}+3)}{1-3}=\frac{9.8(4-2\sqrt{3})}{-2}$$
$$=-9.8(2-\sqrt{3}) \quad \blacksquare$$

Once the basic principles are understood, more complicated problems may be done.

EXAMPLE 9 Perform the indicated operation.

$$\frac{\sqrt{x}-3\sqrt{y}}{\sqrt{x}+3\sqrt{y}}$$

Solution $$\left(\frac{\sqrt{x}-3\sqrt{y}}{\sqrt{x}+3\sqrt{y}}\right)\left(\frac{\sqrt{x}-3\sqrt{y}}{\sqrt{x}-3\sqrt{y}}\right)=\frac{x-3\sqrt{xy}-3\sqrt{xy}+9y}{x-9y}=\frac{x-6\sqrt{xy}+9y}{x-9y}$$

NOTE ⇨⇨ ***It is a common mistake to try to reduce the fraction by crossing out 9y in both numerator and denominator. Fractions are reduced by dividing out* factors, *not* terms, *that are common to both numerator and denominator.*** ■

EXAMPLE 10 Perform the indicated operation.

$$\frac{\sqrt{a+b}}{2+\sqrt{a+b}}$$

Solution $$\left(\frac{\sqrt{a+b}}{2+\sqrt{a+b}}\right)\left(\frac{2-\sqrt{a+b}}{2-\sqrt{a+b}}\right)=\frac{2\sqrt{a+b}-(\sqrt{a+b})^2}{4-(\sqrt{a+b})^2}$$

$$=\frac{2\sqrt{a+b}-(a+b)}{4-(a+b)}$$

$$=\frac{2\sqrt{a+b}-a-b}{4-a-b}$$ ■

EXAMPLE 11 Perform the indicated operation.

$$\frac{2+4\sqrt{x}}{x-\dfrac{\sqrt{x}}{2}}$$

Solution $$\left(\frac{2+4\sqrt{x}}{2-\dfrac{\sqrt{x}}{2}}\right)\left(\frac{2+\dfrac{\sqrt{x}}{2}}{2+\dfrac{\sqrt{x}}{2}}\right)=\frac{4+\sqrt{x}+8\sqrt{x}+2x}{4-\dfrac{x}{4}}$$

$$=\frac{4+9\sqrt{x}+2x}{\dfrac{16-x}{4}}$$

Use a common denominator to add 4 and $-x/4$.

$$=(4+9\sqrt{x}+2x)\div\left(\frac{16-x}{4}\right)$$

Treat the fraction bar as a division symbol.

$$=(4+9\sqrt{x}+2x)\left(\frac{4}{16-x}\right)$$

Invert the divisor and multiply.

$$=\frac{16+36\sqrt{x}+8x}{16-x}$$

Use the distributive property. ■

EXERCISES 13.3

You may assume that all variables in these exercises represent nonnegative real numbers and that no denominators are zero. In 1–10, evaluate each expression using a calculator. Assume that integers represent exact values.

1. $\frac{\sqrt{3}}{\sqrt{5}}$ **2.** $\frac{\sqrt{2}}{\sqrt{5}}$ **3.** $\frac{-1}{\sqrt{2}}$ **4.** $\frac{3}{3+\sqrt{2}}$ **5.** $\frac{2}{4-\sqrt{3}}$

6. $\frac{-3}{\sqrt{5}-\sqrt{2}}$ **7.** $\frac{\sqrt{2}-\sqrt{3}}{\sqrt{2}+\sqrt{3}}$ **8.** $\frac{4+\sqrt{7}}{5-\sqrt{7}}$ **9.** $\frac{2\sqrt{3}+3\sqrt{2}}{2\sqrt{3}-3\sqrt{2}}$ **10.** $\frac{3\sqrt{2}-\sqrt{5}}{\sqrt{2}+2\sqrt{5}}$

In 11–42, perform each division by rationalizing the denominator.

11. $\frac{3}{4\sqrt{x}}$ **12.** $\frac{-2}{3\sqrt{y}}$ **13.** $\frac{-5}{2\sqrt{x}}$ **14.** $\frac{7}{\sqrt[3]{2z}}$

15. $\frac{\sqrt[3]{4}}{\sqrt[3]{9w^2}}$ **16.** $\frac{-\sqrt[3]{2}}{\sqrt[3]{4a^2}}$ **17.** $\frac{2}{\sqrt{3}-\sqrt{2}}$ **18.** $\frac{-1}{\sqrt{5}-\sqrt{2}}$

19. $\frac{-1}{2\sqrt{3}-5}$ **20.** $\frac{3}{5\sqrt{3}+2}$ **21.** $\frac{2}{5\sqrt{7}+\sqrt{2}}$ **22.** $\frac{3}{2\sqrt{7}-\sqrt{3}}$

23. $\frac{3\sqrt{2}-1}{2\sqrt{2}+3}$ **24.** $\frac{2\sqrt{3}+5}{5\sqrt{3}-1}$ **25.** $\frac{3\sqrt{5}-2}{2\sqrt{5}+3}$ **26.** $\frac{3\sqrt{5}-\sqrt{2}}{3\sqrt{5}+\sqrt{2}}$

27. $\frac{2\sqrt{7}+3\sqrt{2}}{2\sqrt{7}-3\sqrt{2}}$ **28.** $\frac{3\sqrt{5}+2\sqrt{3}}{2\sqrt{5}-\sqrt{3}}$ **29.** $\frac{1}{\sqrt{x}-2\sqrt{y}}$ **30.** $\frac{2}{3\sqrt{x}-5\sqrt{y}}$

31. $\frac{1-2\sqrt{x}}{2+3\sqrt{x}}$ **32.** $\frac{\sqrt{x}+\sqrt{y}}{\sqrt{x}-\sqrt{y}}$ **33.** $\frac{2\sqrt{a}-3\sqrt{b}}{3\sqrt{a}+2\sqrt{b}}$ **34.** $\frac{3\sqrt{a}+2\sqrt{b}}{2\sqrt{a}-\sqrt{b}}$

35. $\frac{\sqrt{x+2}}{1-\sqrt{x+2}}$ **36.** $\frac{\sqrt{y+3}}{2-\sqrt{y+3}}$ **37.** $\frac{x+\sqrt{y+1}}{x-\sqrt{y+1}}$ **38.** $\frac{2-\sqrt{a-3}}{2+\sqrt{a-3}}$

39. $\frac{1+\frac{\sqrt{2}}{3}}{1-\frac{\sqrt{2}}{3}}$ **40.** $\frac{2-\frac{\sqrt{3}}{5}}{2+\frac{\sqrt{3}}{5}}$ **41.** $\frac{3-\frac{\sqrt{3}}{2}}{3+\frac{\sqrt{3}}{2}}$ **42.** $\frac{1+\frac{\sqrt{5}}{3}}{1-\frac{\sqrt{5}}{3}}$

---

43. The diameter of a wire (in mils) is given by $d=\sqrt{\rho l/R}$. Express the formula in simplest radical form. In the formula, ρ is the resistivity (in $\Omega\cdot$cmils/ft), l is the length of the wire (in ft), and R is the resistance (in Ω). **3 5 9** 1 2 4 6 7 8

44. The velocity of sound (c) in a liquid is given by $c=\sqrt{\beta/\rho}$. Express the formula in simplest radical form. In the formula, β is the bulk modulus and ρ is the density of the liquid. **4 7 8** 1 2 3 5 6 9

45. The frequency f (in Hz) of an object attached to the end of a cantilever beam of length l is given by $f=\frac{1}{2\pi}\sqrt{\frac{3EI}{ml^3}}$. Express the formula in simplest radical form. In the formula, E is the tensile modulus of elasticity, I is the moment of inertia of the cross-sectional area, and m is the mass of the object. **6 9** 5 8

---

1. Architectural Technology
2. Civil Engineering Technology
3. Computer Engineering Technology
4. Mechanical Drafting Technology
5. Electrical/Electronics Engineering Technology
6. Chemical Engineering Technology
7. Industrial Engineering Technology
8. Mechanical Engineering Technology
9. Biomedical Engineering Technology

46. An object suspended from a spring is attached to a support that oscillates. When the deflection y of the support is given by $y = a \sin \omega t$, the deflection of the object is given by

$$x = \frac{a\omega\omega_n}{\omega_n^2 - \omega^2} \sin \omega_n t + \frac{a\omega_n^2}{\omega_n^2 - \omega^2} \sin \omega t.$$

The quantity ω_n is computed by $\omega_n = \sqrt{980k/w}$ where k is a constant that depends on the spring and w is the weight of the object in kg. Express the formula for ω_n in simplest radical form.
1 2 4 8 7 9

Problems 47 and 48 use the following formula.

$$l = \frac{2\pi yd}{\sqrt{2br - b^2}}$$

In the formula, l is the increase in length of a rope that is placed on a hemisphere of radius r and then moved downward, as in the accompanying figure. The rope is initially placed in a horizontal plane at a vertical distance b below the top and then moved downward. The vertical distance from the base to the final position of the rope is y, and d is the vertical distance that the rope is moved downward.

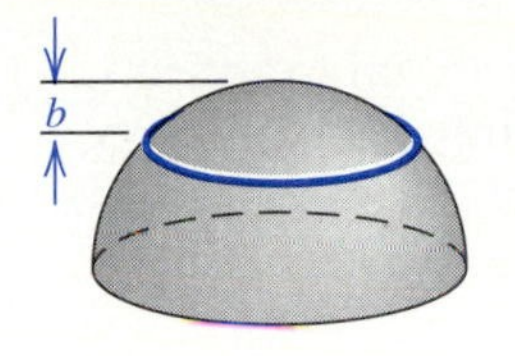

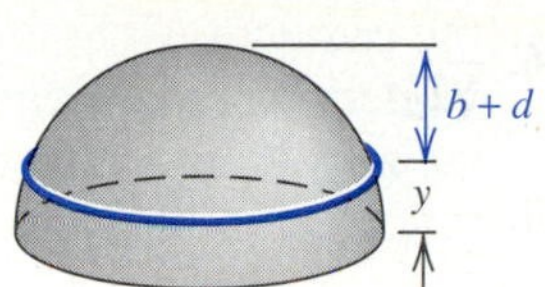

47. Let $r = 25.0$ cm, $b = 10.00$ cm, $d = 0.100$ cm, and $y = 14.9$ cm.
(**a**) Express l in simplest radical form when these values are used.
(**b**) Give a decimal approximation for l.
1 2 4 7 8 3 5

48. Let $r = 25.0$ cm, $b = 18.00$ cm, $d = 0.500$ cm, and $y = 6.50$ cm.
(**a**) Express l in simplest radical form when these values are used.
(**b**) Give a decimal approximation for l.
1 2 4 7 8 3 5

49. The maximum tension T_{max} (in ft·lb) in a horizontal cable under a uniform load w (in lb/ft) is given by

$$T_{max} = \sqrt{\frac{w^2L^4}{64s^2} + \frac{w^2L^2}{4}}.$$

Write this expression in simplest radical form. In the formula, L is the cable length, and s is the sag at the center. 2 4 8 1 5 6 7 9

50. Explain why the following calculator sequences for Example 3 will each give an incorrect answer.
(a) 2 [×] 3 [√x] [+] 5 [√x] [=] [÷] 4 [×] 7 [√x] [=]
(b) 2 [×] 3 [√x] [+] 5 [√x] [÷] [(] 4 [×] 7 [√x] [)] [=]

13.4 SIMPLIFYING RADICAL EXPRESSIONS

In all of the problems so far, you have used rules for simplifying radical expressions involving addition, subtraction, multiplication, or division only if the order was the same for both operands. You know, for example, how to multiply $\sqrt{2}$ and $\sqrt{3}$, but you have not yet seen how to multiply $\sqrt{2}$ and $\sqrt[3]{3}$. For addition and/or subtraction, there is the additional requirement that the radicand be the same for both operands. Thus, $5\sqrt{2} + 3\sqrt{2}$ can be simplified, but $\sqrt{2} + \sqrt{3}$ cannot.

A problem may involve radicals with two different orders. It may be possible to reduce the order of one of them. Recall that reducing the order is done by using the notation of fractional exponents.

EXAMPLE 1 Simplify $3\sqrt[4]{4} + 2\sqrt{2}$.

Solution The order may be reduced for $\sqrt[4]{4}$.

$$\sqrt[4]{4} = (4)^{1/4} = (2^2)^{1/4} = 2^{2/4} = 2^{1/2} = \sqrt{2}$$
$$3\sqrt[4]{4} + 2\sqrt{2} = 3\sqrt{2} + 2\sqrt{2} = 5\sqrt{2}$$ ■

In general, when a problem involves radicals with different orders, the simplification is done using exponent notation rather than radical notation. This principle applies whether the expression has one term or more than one term. It also applies to multiplication and division.

To simplify a radical expression when radicals of different orders are involved:

1. Rewrite the problem using fractional exponents.
2. Apply exponent rules to simplify the expression.
3. Express the answer as a radical in simplest radical form.

EXAMPLE 2 Simplify $\sqrt[3]{\sqrt{2}}$.

Solution

1. $\sqrt[3]{\sqrt{2}} = (2^{1/2})^{1/3}$ — Rewrite the problem using fractional exponents.
2. $= 2^{1/6}$ — Use the rule $(x^m)^n = x^{mn}$
3. $= \sqrt[6]{2}$ — Express the answer as a radical. ■

EXAMPLE 3 Simplify $\sqrt{2\sqrt[3]{4}}$.

Solution

1. $\sqrt{2\sqrt[3]{4}} = [2(4^{1/3})]^{1/2}$ — Rewrite the problem using fractional exponents.
2. $= 2^{1/2}4^{1/6}$ — Use the rules $(xy)^n = x^ny^n$ and $(x^m)^n = x^{mn}$.
   $= 2^{3/6}4^{1/6}$ — Find a common denominator for the exponents.
   $= (2^34)^{1/6}$ — Use the rule $(xy)^n = x^ny^n$.
   $= 32^{1/6}$ — Evaluate $2^34 = 8(4) = 32$.
3. $= \sqrt[6]{32}$ — Express the answer as a radical. ■

EXAMPLE 4 Figure 13.2 shows the frustum of a right pyramid. The formula $V = (1/3)(h)(A_1 + A_2 + \sqrt{A_1A_2})$ gives the volume V if A_1 and A_2 are the areas of the bases. Simplify the formula if $A_2 = \sqrt{A_1}$.

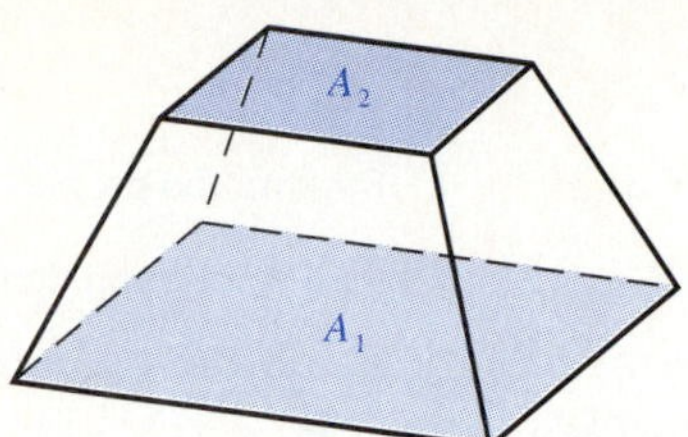

FIGURE 13.2

Solution

$V = (1/3)(h)(A_1 + A_2 + \sqrt{A_1 A_2})$

$= (1/3)(h)(A_1 + \sqrt{A_1} + \sqrt{A_1 \sqrt{A_1}}$

1. $= (1/3)(h)[A_1 + A_1^{1/2} + (A_1 A_1^{1/2})^{1/2}]$ — Rewrite using fractional exponents.
2. $= (1/3)(h)[A_1 + A_1^{1/2} + (A_1^{3/2})^{1/2}]$ — Apply exponent rules.

$= (1/3)(h)(A_1 + A_1^{1/2} + A_1^{3/4})$

$= (1/3)(h)(A_1 + \sqrt{A_1} + \sqrt[4]{A_1^3})$ — Express the answer as a radical. ■

EXAMPLE 5 Simplify $\sqrt{2}\sqrt[3]{3}$.

Solution

1. $\sqrt{2}\sqrt[3]{3} = 2^{1/2}3^{1/3}$ — Rewrite the problem using fractional exponents.
2. $= 2^{3/6}3^{2/6}$ — Find a common denominator for the exponents.

$= (2^3 3^2)^{1/6}$ — Use the rule $(xy)^n = x^n y^n$.

$= (72)^{1/6}$ — Evaluate $2^3(3^2) = 8(9) = 72$.

3. $= \sqrt[6]{72}$ — Express the answer as a radical. ■

EXAMPLE 6 Simplify $\dfrac{\sqrt[4]{2}}{\sqrt[3]{5^2}}$.

Solution Begin by rationalizing the denominator.

$$\frac{\sqrt[4]{2}}{\sqrt[3]{5^2}}\left(\frac{\sqrt[3]{5}}{\sqrt[3]{5}}\right) = \frac{\sqrt[4]{2}\sqrt[3]{5}}{5}$$

1. $= \dfrac{2^{1/4}5^{1/3}}{5}$ — Rewrite the problem using fractional exponents.
2. $= \dfrac{2^{3/12}5^{4/12}}{5}$ — Find a common denominator.

$= \dfrac{(2^3 5^4)^{1/12}}{5}$ — Use the rule $(xy)^n = x^n y^n$.

$= \dfrac{5000^{1/12}}{5}$ — Evaluate $2^3 5^4 = 8(625) = 5000$.

3. $= \dfrac{\sqrt[12]{5000}}{5}$ — Express the answer as a radical. ■

EXAMPLE 7 Simplify $\sqrt[5]{2} + \sqrt[3]{3}$.

Solution

1. $\sqrt[5]{2} + \sqrt[3]{3} = 2^{1/5} + 3^{1/3}$ Rewrite the problem using fractional exponents.
2. $= 2^{3/15} + 3^{5/15}$ Find a common denominator.

There is no exponent rule by which the two terms may be combined. The original expression is in simplest form. ■

Example 7 helps to illustrate the principle that radical expressions with different orders can be added or subtracted only when the order of one of them reduces so that they have the *same order* and the *same radicand.*

Now that you have learned to add, subtract, multiply, and divide radical expressions, you are ready to combine these operations. The answer should be in simplest radical form.

To simplify radical expressions:

1. Perform indicated multiplications, using the distributive property, if necessary.
2. Express each radical in simplest radical form.
3. Combine like radicals, using addition or subtraction.

EXAMPLE 8 Simplify $(\sqrt{2+\sqrt{3}})(\sqrt{2-\sqrt{3}})$.

Solution

$(\sqrt{2+\sqrt{3}})(\sqrt{2-\sqrt{3}})$

$= \sqrt{(2+\sqrt{3})(2-\sqrt{3})}$ Multiply, using the rule $\sqrt{x}\sqrt{y} = \sqrt{xy}$.

$= \sqrt{4-\sqrt{9}}$

$= \sqrt{4-3}$ Simplify the radical.

$= \sqrt{1} = 1$ ■

EXAMPLE 9 Find the area of rectangle *PQRS* shown in Figure 13.3.

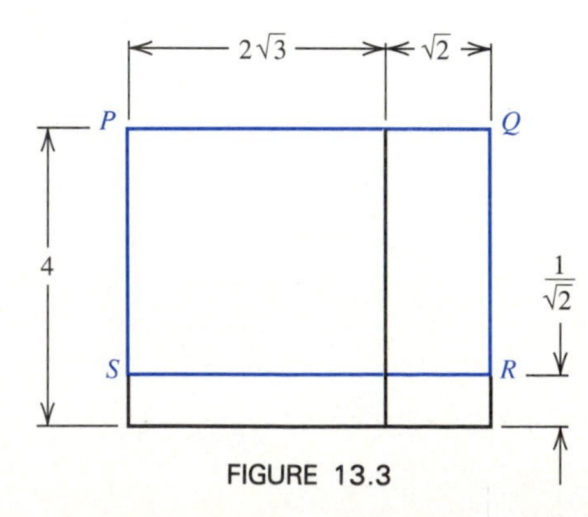

FIGURE 13.3

Solution

$$\begin{aligned} A &= (2\sqrt{3} + \sqrt{2})\left(4 - \frac{1}{\sqrt{2}}\right) \\ &= 8\sqrt{3} - \frac{2\sqrt{3}}{\sqrt{2}} + 4\sqrt{2} - 1 && \text{Multiply, using the distributive property.} \\ &= 8\sqrt{3} - \frac{2\sqrt{3}}{\sqrt{2}}\left(\frac{\sqrt{2}}{\sqrt{2}}\right) + 4\sqrt{2} - 1 && \text{Simplify radicals by rationalizing the denominator.} \\ &= 8\sqrt{3} - \frac{2\sqrt{6}}{2} + 4\sqrt{2} - 1 \\ &= 8\sqrt{3} - \sqrt{6} + 4\sqrt{2} - 1 \end{aligned}$$ ■

EXAMPLE 10 Simplify $(2\sqrt{3} - \sqrt{2})(\sqrt{6} + \sqrt{2})$.

Solution

$$\begin{aligned} &(2\sqrt{3} - \sqrt{2})(\sqrt{6} + \sqrt{2}) \\ &= 2\sqrt{18} + 2\sqrt{6} - \sqrt{12} - \sqrt{4} && \text{Multiply, using the distributive property.} \\ &= 2\sqrt{9 \cdot 2} + 2\sqrt{6} - \sqrt{4 \cdot 3} - \sqrt{4} && \text{Simplify radicals by factoring radicands.} \\ &= 2(3)\sqrt{2} + 2\sqrt{6} - 2\sqrt{3} - 2 \\ &= 6\sqrt{2} + 2\sqrt{6} - 2\sqrt{3} - 2 \end{aligned}$$ ■

EXAMPLE 11 Simplify $(\sqrt[4]{2} - \sqrt[4]{3})(\sqrt[4]{2} + \sqrt[4]{3})$.

Solution

$$\begin{aligned} &(\sqrt[4]{2} - \sqrt[4]{3})(\sqrt[4]{2} + \sqrt[4]{3}) \\ &= \sqrt[4]{4} + \sqrt[4]{6} - \sqrt[4]{6} - \sqrt[4]{9} && \text{Multiply, using the distributive property.} \\ &= 4^{1/4} - 9^{1/4} && \text{Simplify radicals by reducing the order.} \\ &= (2^2)^{1/4} - (3^2)^{1/4} \\ &= 2^{2/4} - 3^{2/4} \\ &= 2^{1/2} - 3^{1/2} \\ &= \sqrt{2} - \sqrt{3} \end{aligned}$$ ■

Although they may be done on a calculator, numerical examples have been used to illustrate the method. For numerical problems, it is possible to verify that the simplified form is equivalent to the original form by using a calculator.

EXAMPLE 12 Verify that $(\sqrt{2 + \sqrt{3}})(\sqrt{2 - \sqrt{3}} = 1$ in Example 8.

Solution Two solutions are given. The keystroke sequences follow.

(a) 2 [+] 3 [√x] [=] [√x] [STO] 2 [−] 3 [√x] [=] [√x] [×] [RCL] [=] **1**

(b) [(] [(] 2 [+] 3 [√x] [)] [√x] [)] [×] [(] [(] 2 [−] 3 [√x] [)] [√x] [)] [=] **1** ■

For problems containing variables, it is necessary to simplify the radicals using the procedures of this section. You may assume that variables represent nonnegative real numbers.

EXAMPLE 13 Simplify $\sqrt{27x^3y^4}(\sqrt{3xy^3} + 1) + \sqrt{3x}$.

Solution

$$\sqrt{27x^3y^4}(\sqrt{3xy^3} + 1) + \sqrt{3x}$$

$= \sqrt{81x^4y^7} + \sqrt{27x^3y^4} + \sqrt{3x}$ — Multiply.

$= \sqrt{81x^4y^6y} + \sqrt{9(3)x^2xy^4} + \sqrt{3x}$ — Simplify the radicals by factoring the radicands.

$= 9x^2y^3\sqrt{y} + 3xy^2\sqrt{3x} + \sqrt{3x}$

$= 9x^2y^3\sqrt{y} + (3xy^2 + 1)\sqrt{3x}$ — Combine like radicals. ■

EXAMPLE 14 Simplify $\sqrt[3]{4ab}(\sqrt[3]{16a^2b^2} - \sqrt[3]{2ab}) + 4\sqrt[3]{a^2b^2}$.

Solution

$$\sqrt[3]{4ab}(\sqrt[3]{16a^2b^2} - \sqrt[3]{2ab}) + 4\sqrt[3]{a^2b^2}$$

$= \sqrt[3]{64a^3b^3} - \sqrt[3]{8a^2b^2} + 4\sqrt[3]{a^2b^2}$ — Multiply.

$= 4ab - 2\sqrt[3]{a^2b^2} + 4\sqrt[3]{a^2b^2}$ — Simplify the radicals.

$= 4ab + 2\sqrt[3]{a^2b^2}$ — Combine like radicals. ■

EXERCISES 13.4

You may assume that all variables in these exercises represent nonnegative real numbers. In 1–42, simplify each expression, if possible.

1. $\sqrt{\sqrt[3]{5}}$ **2.** $\sqrt[3]{\sqrt[4]{x}}$ **3.** $\sqrt{\sqrt[5]{7y}}$ **4.** $\sqrt[3]{\sqrt[5]{2a}}$

5. $\sqrt[5]{\sqrt[4]{6b}}$ **6.** $\sqrt{3}\,\sqrt[3]{5}$ **7.** $\sqrt[3]{2\sqrt{7}}$ **8.** $\sqrt[4]{2\sqrt[3]{3}}$

9. $\sqrt{3\sqrt[3]{7}}$ **10.** $\sqrt{2\sqrt[4]{5}}$ **11.** $\sqrt[4]{3}\,\sqrt[3]{5}$ **12.** $\sqrt[3]{2}\,\sqrt{5}$

13. $\sqrt[4]{3}\,\sqrt{7}$ **14.** $\sqrt[3]{2}\,\sqrt[5]{3}$ **15.** $\sqrt{5}\,\sqrt[3]{7}$ **16.** $\sqrt{7}\,\sqrt[3]{5}$

17. $\sqrt[4]{7}\,\sqrt[3]{3}$ **18.** $\dfrac{\sqrt{2}}{\sqrt[3]{3}}$ **19.** $\dfrac{\sqrt[3]{2}}{\sqrt{3}}$ **20.** $\dfrac{\sqrt{5}}{\sqrt[3]{2}}$

21. $\dfrac{\sqrt[3]{5}}{\sqrt{2}}$ **22.** $\sqrt{2} + \sqrt[3]{3}$ **23.** $\sqrt[3]{2} + \sqrt{3}$ **24.** $\sqrt{5} + \sqrt[3]{2}$

25. $\sqrt[3]{5} + \sqrt{2}$ **26.** $(\sqrt{7} + 3)\left(2 - \dfrac{1}{\sqrt{7}}\right)$ **27.** $(\sqrt{5} + \sqrt{3})\left(2 + \dfrac{1}{\sqrt{3}}\right)$

28. $(\sqrt{11} + 2)\left(\sqrt{11} - \dfrac{2}{\sqrt{11}}\right)$ **29.** $(\sqrt{13} + 3)\left(\sqrt{13} - \dfrac{1}{\sqrt{13}}\right)$ **30.** $(\sqrt{5} - \sqrt{3})\left(\sqrt{5} + \dfrac{\sqrt{3}}{\sqrt{5}}\right)$

31. $(\sqrt{7} - 3)\left(2 + \dfrac{1}{\sqrt{7}}\right)$ **32.** $(\sqrt[6]{5} + \sqrt[6]{7})(\sqrt[6]{5} - \sqrt[6]{7})$ **33.** $(\sqrt[4]{7} + \sqrt[4]{5})(\sqrt[4]{7} - \sqrt[4]{5})$

34. $(\sqrt{1 + \sqrt{x}})(\sqrt{1 - \sqrt{x}})$ **35.** $(\sqrt{2 - \sqrt{y}})(\sqrt{2 + \sqrt{y}})$ **36.** $\sqrt{\sqrt{5} - \sqrt{3}}\,\sqrt{\sqrt{5} + \sqrt{3}}$

37. $\sqrt{\sqrt{5} + \sqrt{a}}\,\sqrt{\sqrt{5} - \sqrt{a}}$ **38.** $(\sqrt{11} - 3)(\sqrt{11} + 4) - 2\sqrt{11}$

39. $(\sqrt{13x} - \sqrt{2y})(\sqrt{13x} - \sqrt{2y}) + 2\sqrt{26xy}$ **40.** $\sqrt{6xy} + (\sqrt{3x} - \sqrt{2y})^2$

41. $(\sqrt{7a} - 5)^2 + (\sqrt{7a} + 5)^2$ **42.** $(2 - \sqrt{11b})(1 + 2\sqrt{11b}) + 2 + \sqrt{11b}$

43. The angle of elevation measured to the top of a building is found to be 30.0°. The distance to the base of the building is 50.0 m, as shown in the accompanying diagram.

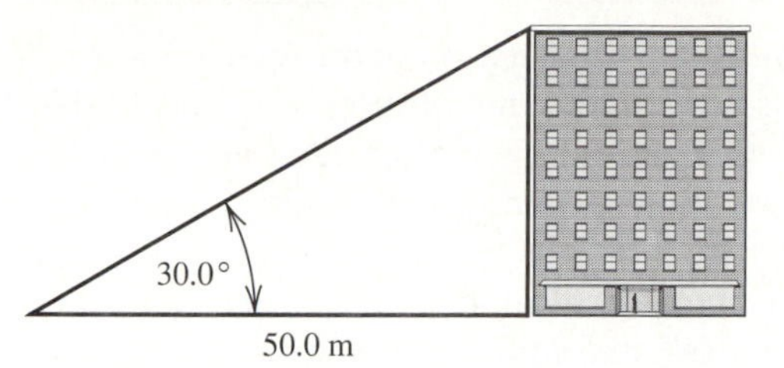

(a) What is the height of the building? Use special angle relationships and express the answer in simplest radical form.

(b) The maximum distance (in km) that can be seen from a tall building of height h (in km) is given by $(1117/10)\sqrt{h}$. Using your answer from (a), specify the maximum distance that can be seen. Express the answer in simplest radical form.

(c) Express the answer to (b) in decimal form. 1 2 4 7 8 3 5

44. (a) What is the radius of a sphere of volume 6π m^3? Express the answer in simplest radical form.

(b) The period T (in seconds) of a pendulum of length l(in m) is given by $T = (2\pi/3.13)\sqrt{l}$. Find the period of a pendulum that has a length equal to the radius in part (a). Express the answer in simplest radical form.

(c) Express the answer to (b) as a decimal approximation. 1 2 4 7 8 3 5

Problems 45–47 use the following formula for the area of a triangle of sides a, b, and c where $s = (a + b + c)/2$.

$$A = \sqrt{s(s-a)(s-b)(s-c)}$$

45. (a) Find the area of the triangle shown in the accompanying diagram. Express the answer in simplest radical form.

(b) Express the answer as a decimal approximation. 1 2 4 7 8 3 5

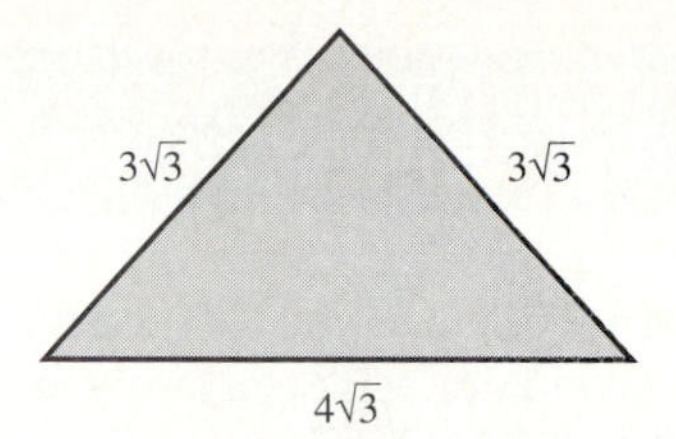

46. (a) Find the area of the triangle shown in the accompanying figure. Express the answer in simplest radical form.

(b) Express the answer as a decimal approximation. 1 2 4 7 8 3 5

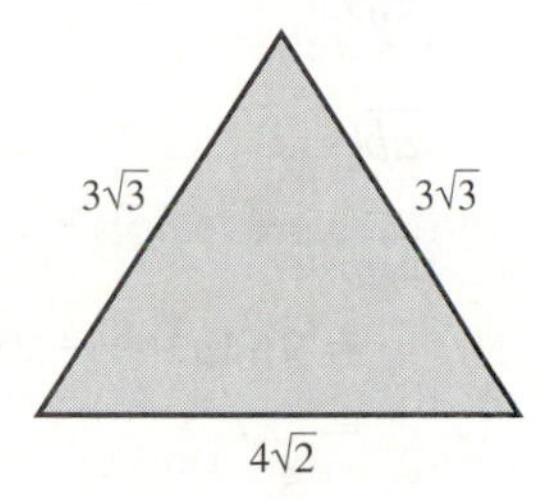

47. (a) Find the area of the triangle shown in the accompanying diagram. Express the answer in simplest radical form.

(b) Express the answer as a decimal approximation.

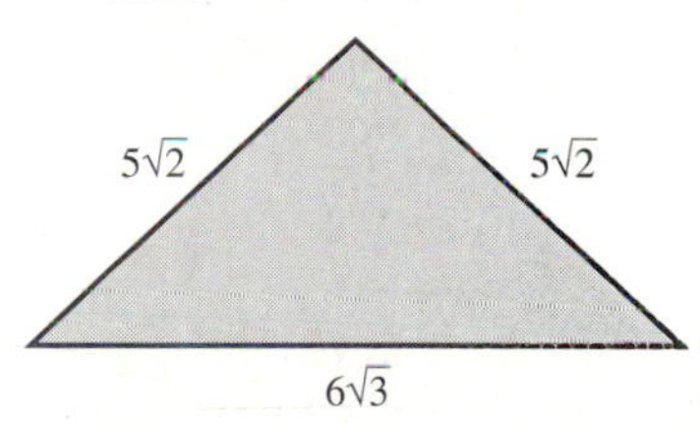

48. The formula $D = \sqrt[3]{2^4}\sqrt{T/(\pi S_s)}$ gives the required diameter D of a circular shaft to resist a given torque T, using a predetermined shearing stress S_s. Write the expression in simplest radical form.

49. Problems 45–47 involved lengths of $\sqrt{2}$, $\sqrt{3}$, etc. It is possible to construct lines of these lengths using the Pythagorean theorem. Explain how it is done.

13.5 EQUATIONS WITH RADICALS AND EQUATIONS IN QUADRATIC FORM

In Section 2.5, you learned that the solution to an equation of the form $\sqrt[n]{x} = b$ is b^n if n is odd or if n is even and $b \geq 0$. That is, we can find the solution by raising both sides of the equation to the nth power. If both sides of an equation are raised to a power, however, the resulting equation may not be equivalent to the original equation. That is, it may not have the same solutions. The solutions of the original equation will be included among the solutions to the altered equation, but the altered equation may have extra solutions that do not satisfy the original equation. Such solutions are called *extraneous solutions.* ***It is imperative to check each apparent solution to a radical equation.*** The steps are summarized as follows.

extraneous solutions

NOTE ⇨⇨

To solve an equation containing radicals:

1. Isolate the radical. That is, write the equation so that the term containing the radical is on one side of the equation by itself. If there is more than one radical, isolate one of them.
2. If the isolated radical is an nth root, raise both sides of the equation to the nth power.
3. Solve the resulting equation. If it contains one or more radicals, start again at step 1.
4. Check each and every possible solution from step 3 in the original equation.

EXAMPLE 1 Solve $3\sqrt{x} = 5$.

Solution

1. The term containing the radical is already isolated.
2. $(3\sqrt{x})^2 = 5^2$ — Square both sides.
   $9x = 25$ — Use the rule $(xy)^n = x^n y^n$.
3. $x = 25/9$ — Solve the equation for x.
4. Check: $3\sqrt{25/9} = 5$
   $3(5/3) = 5$, which is true. The solution is $x = 25/9$. ■

EXAMPLE 2 Solve $\sqrt[3]{3x+5} - 2 = 0$.

Solution

1. $\sqrt[3]{3x+5} = 2$ — Isolate the radical by adding 2 to both sides.
2. $(\sqrt[3]{3x+5})^3 = 2^3$ — Cube both sides.
   $3x + 5 = 8$ — Notice that the cube "undoes" the cube root.
3. $3x = 3$ — Solve the equation. Subtract 5 from both sides.
   $x = 1$ — Divide both sides by 3.

4. Check: $\sqrt[3]{3(1)+5} - 2 = 0$

$\sqrt[3]{8} - 2 = 0$, which is true. The solution is $x = 1$. ■

EXAMPLE 3 Solve $\sqrt{6-x} = x$.

Solution

1. The radical is already isolated.

2. $(\sqrt{6-x})^2 = x^2$ — Square both sides.

$6 - x = x^2$ — Notice that the square "undoes" the square root.

3. $0 = x^2 + x - 6$ — Solve the equation. Put the quadratic equation in standard form.

$0 = (x-2)(x+3)$ — Factor the expression on the right.

$x - 2 = 0$ or $x + 3 = 0$ — Set each factor equal to zero.

$x = 2$ or $x = -3$ — Solve each equation.

4. Check: If $x = 2$, $\sqrt{6-2} = 2$

$\sqrt{4} = 2$, which is true.

If $x = -3$, $\sqrt{6-(-3)} = -3$

$\sqrt{9} = -3$, which is not true.

Recall that the radical symbol indicates the positive square root. The solution is $x = 2$. ■

Example 3 illustrates how a discrepancy may arise if both sides of an equation are squared. There is a value of the variable for which the expressions on opposite sides of the equation have equal absolute values, but different signs. When both sides are squared, the equation is a true statement for that value of the variable, but the original equation is not. In this case, $3^2 = (-3)^2$, but $3 \neq -3$.

EXAMPLE 4 Solve $\sqrt{2x-3} = -4$.

Solution It is not necessary to solve the equation, because the radical on the left indicates a positive square root, while the number on the right is a negative number. There is no solution. ■

EXAMPLE 5 Solve $\sqrt{2x-1} + 1 = 2x$.

Solution

1. $\sqrt{2x-1} = 2x - 1$ — Isolate the radical.

2. $(\sqrt{2x-1})^2 = (2x-1)^2$ — Square both sides.

It is important that on the left-hand side of the equation, the square undoes the square root, but on the right-hand side, there is no square root. The expression must be squared as a binomial. The equation becomes

$$2x - 1 = 4x^2 - 4x + 1.$$

3. $0 = 4x^2 - 6x + 2$ — Solve the equation. Put the quadratic equation in standard form.

$0 = 2(2x^2 - 3x + 1)$ — Factor out the common factor.

$0 = 2(2x - 1)(x - 1)$ — Factor the remaining trinomial.

$2x - 1 = 0$ or $x - 1 = 0$ — Set each factor equal to zero.

$2x = 1$ or $x = 1$

$x = 1/2$ or $x = 1$ — Solve each equation.

4. Check: If $x = 1/2$, $\sqrt{2(1/2) - 1} + 1 = 2(1/2)$

$\sqrt{1 - 1} + 1 = 1$, which is true.

If $x = 1$, $\sqrt{2(1) - 1} + 1 = 2(1)$

$\sqrt{2 - 1} + 1 = 2$, which is true.

The solution is $x = 1/2$ or $x = 1$. ■

EXAMPLE 6 Solve $\sqrt{x + 2} + \sqrt{x - 3} = 5$.

Solution

1. $\sqrt{x + 2} = 5 - \sqrt{x - 3}$ — Isolate one of the radicals.

2. $(\sqrt{x + 2})^2 = (5 - \sqrt{x - 3})^2$ — Square both sides.

$x + 2 = (5 - \sqrt{x - 3})(5 - \sqrt{x - 3})$ — Notice that on the left, the square undoes the square root.

$x + 2 = 25 - 10\sqrt{x - 3} + (\sqrt{x - 3})^2$ — Perform the multiplication indicated on the right-hand side.

$x + 2 = 25 - 10\sqrt{x - 3} + x - 3$ — Notice that the square undoes the square root.

$x + 2 = x + 22 - 10\sqrt{x - 3}$ — Simplify the expression on the right.

1. $-20 = -10\sqrt{x - 3}$ — Isolate the radical on the right. Subtract $x + 22$ from both sides.

$2 = \sqrt{x - 3}$ — Divide both sides by -10.

2. $2^2 = (\sqrt{x - 3})^2$ — Square both sides.

$4 = x - 3$ — Notice that the square undoes the square root.

3. $7 = x$ — Solve the equation. Add 3 to both sides.

4. Check: $\sqrt{7 + 2} + \sqrt{7 - 3} = 5$

$\sqrt{9} + \sqrt{4} = 5$, which is true.

The solution is $x = 7$. ■

EXAMPLE 7 When an object suspended from a spring as in Figure 13.4 is pulled down and released, it oscillates with a frequency given by $f = \frac{1}{2\pi}\sqrt{\frac{k}{m}}$. In the formula, k is a constant that depends on the spring, and m is the mass of the object (in kg). If the value of k is 8.33 for a particular spring, find the mass that would be required to produce a frequency of 2.6 cycles per second.

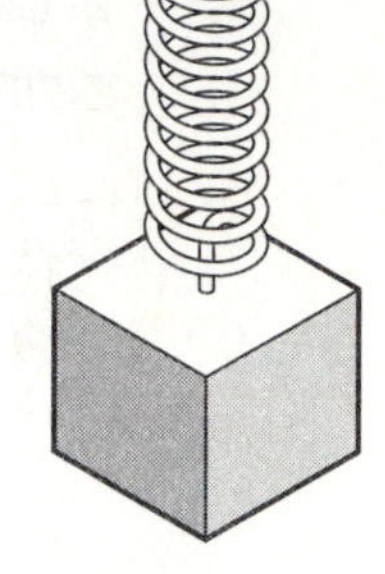

FIGURE 13.4

Solution $f = \frac{1}{2\pi}\sqrt{\frac{k}{m}}$, so $2.6 = \frac{1}{2\pi}\sqrt{\frac{8.33}{m}}$.

Squaring both sides, we have

$$6.76 = \frac{1}{4\pi^2}\left(\frac{8.33}{m}\right).$$

$$6.76(4\pi^2)m = 8.33$$

$$m = \frac{8.33}{6.76(4\pi^2)}$$

A calculator gives $m = 0.0312$ kg or 31.2 g. ■

When an equation contains radicals with two different orders, it may be an equation in quadratic form. Such equations do not actually fit the definition of a quadratic equation (see Section 5.1), but they are similar to quadratic equations in the pattern of exponents and coefficients.

DEFINITION An equation is said to be in **quadratic form** if it is in the form $ax^{2p} + bx^p + c = 0$ where a, b, c, and p are real numbers, $a \neq 0$, and x is a variable or variable expression.

The key to working with such equations is to recognize that the exponent of one term is twice the exponent of another. The term with the smaller exponent may be treated as the linear term of a quadratic equation.

EXAMPLE 8 Solve $2x + 5\sqrt{x} - 3 = 0$.

Solution The second term is $5\sqrt{x}$ or $5x^{1/2}$, and its exponent is 1/2. The first term is $2x$, and its exponent is 1, which is twice 1/2. The equation may be treated as a quadratic in the following manner.

Let $y = \sqrt{x}$. Then $y^2 = (\sqrt{x})^2 = x$. The equation $2x + 5\sqrt{x} - 3 = 0$ is thus equivalent to $2y^2 + 5y - 3 = 0$.

$$2y^2 + 5y - 3 = 0$$

$$(2y - 1)(y + 3) = 0$$

$$2y - 1 = 0 \quad \text{or} \quad y + 3 = 0$$

$$2y = 1 \quad \text{or} \quad y = -3$$

$$y = 1/2 \quad \text{or} \quad y = -3$$

The original equation requires a solution for x, not y. Recall that we let $y = \sqrt{x}$. Substitute $\sqrt{x}$ for y.

$$\sqrt{x} = 1/2 \quad \text{or} \quad \sqrt{x} = -3$$

The first of these equations may be solved for x, but the second has no solution. If $\sqrt{x} = 1/2$, then $(\sqrt{x})^2 = (1/2)^2$ or $x = 1/4$. Since both sides of the equation were squared, it is necessary to check the solution in the original equation.

Check: $2(1/4) + 5\sqrt{1/4} - 3 = 0$
$$1/2 + 5(1/2) - 3 = 0$$
$$1/2 + 5/2 - 3 = 0$$
$6/2 - 3 = 0$, which is true. The solution is $x = 1/4$. ■

The steps are generalized as follows.

To solve an equation in quadratic form $ax^{2p} + bx^p + c = 0$:

1. Let $y = x^p$, then $y^2 = x^{2p}$.
2. Substitute the expressions from step 1 for x^p and x^{2p} to obtain $ay^2 + by + c = 0$.
3. Solve the equation for y.
4. Substitute the expressions from step 1 for y and y^2 to obtain equations in x.
5. Solve the equations from step 4 for x.
6. Check the solutions in the original equation $ax^{2p} + bx^p + c = 0$.

EXAMPLE 9 Solve $x^{4/3} - 5x^{2/3} + 4 = 0$.

Solution

1. Let $y = x^{2/3}$ and $y^2 = x^{4/3}$.
2. $x^{4/3} - 5x^{2/3} + 4 = 0$ is equivalent to $y^2 - 5y + 4 = 0$.
3. $(y - 4)(y - 1) = 0$
$$y - 4 = 0 \quad \text{or} \quad y - 1 = 0$$
$$y = 4 \quad \text{or} \quad y = 1$$
4. $x^{2/3} = 4 \quad \text{or} \quad x^{2/3} = 1$
5. Since $x^{2/3} = (\sqrt[3]{x})^2$, the cube root may be undone with a cube.
$$(x^{2/3})^3 = 4^3 \quad \text{or} \quad (x^{2/3})^3 = 1^3$$
$$x^2 = 64 \quad \text{or} \quad x^2 = 1$$
$$x = \pm\sqrt{64} \quad \text{or} \quad x = \pm\sqrt{1}$$
$$x = \pm 8 \quad \text{or} \quad x = \pm 1$$

6. Check each solution:

If $x = 8$, $8^{4/3} - 5(8^{2/3}) + 4 = 0$ or $16 - 5(4) + 4 = 0$, which is true.
If $x = -8$, $(-8)^{4/3} - 5(-8)^{2/3} + 4 = 0$ or $16 - 5(4) + 4 = 0$, which is true.
If $x = 1$, $1^{4/3} - 5(1^{2/3}) + 4 = 0$ or $1 - 5(1) + 4 = 0$, which is true.
If $x = -1$, $(-1)^{4/3} - 5(-1)^{2/3} + 4 = 0$ or $1 - 5 + 4 = 0$, which is true.

The solutions, then are $x = 8, -8, 1$, or -1. ■

EXAMPLE 10 The numerical value of the effective area E of a chimney is given by $100E = 100A - 6\sqrt{A}$, where A is the numerical value of the actual area. Find the actual area necessary to produce an effective area of 80.46 square inches.

Solution If E is 80.46, the equation is

$$100(80.46) = 100\,A - 6\sqrt{A} \text{ or } 8046 = 100\,A - 6\sqrt{A}.$$

If it is written as $100\,A - 6\sqrt{A} - 8046 = 0$, it is in quadratic form.

1. Let $y = \sqrt{A}$ then $y^2 = A$

2. $100y^2 - 6y - 8046 = 0$

3. $2(50y^2 - 3y - 4023) = 0$

$2(y - 9)(50y + 447) = 0$

$y - 9 = 0$ or $50y + 447 = 0$

$y = 9$ or $50y = -447$

$y = -8.94$

4. Since $y = \sqrt{A}$, the negative answer is discarded. If y is 9, then we have $y^2 = A = 81$.

$$\text{Check: } 100(80.46) = 100(81) - 6\sqrt{81}$$
$$8046 = 8100 - 6(9)$$
$$8046 = 8100 - 54$$

The actual area necessary is 81 square inches. ■

EXERCISES 13.5

In 1–42, solve each equation.

1. $5\sqrt{x} = 3$
2. $7\sqrt{y} = 4$
3. $2\sqrt{y} = 9$
4. $\sqrt{x} + 1 = 4$
5. $\sqrt{x} + 3 = 0$
6. $2\sqrt{y} + 7 = 0$
7. $\sqrt{y} + 3 = 1$
8. $\sqrt{2x + 3} = x$
9. $\sqrt{3x + 7} = 2$
10. $\sqrt{2x + 11} + 2 = 5$
11. $\sqrt[3]{x - 2} = 2$
12. $\sqrt[3]{9x} = x$
13. $\sqrt[3]{-8y} - y = 0$
14. $2\sqrt[3]{y} + 7 = 9$
15. $\sqrt{x - 2} = \sqrt{2x - 7}$
16. $\sqrt[4]{x + 1} = \sqrt[4]{2x - 6}$
17. $\sqrt[3]{4x - 1} = \sqrt[3]{3x + 6}$
18. $7 + \sqrt{2x + 1} = x$
19. $\sqrt{6x + 1} = 2x - 9$
20. $\sqrt{5y - 4} = 8 - y$
21. $\sqrt{7y + 1} - 1 = y$
22. $\sqrt{x + 5} + \sqrt{5 - 4x} = 5$

23. $2\sqrt{2x+1} - \sqrt{6x+1} = 1$
24. $\sqrt{y+2} + 2 = \sqrt{3y+4}$
25. $2\sqrt{y-3} - 3 = \sqrt{4y-3}$
26. $\sqrt{10x+1} - 2\sqrt{2x-7} = 3$
27. $\sqrt{x+9} + \sqrt{x} = 9$
28. $\sqrt{7x} = \sqrt{3x+4} + 2$
29. $\sqrt{5y + \sqrt{4y}} = 0$
30. $\sqrt{7x + \sqrt{4x}} = 3$
31. $\sqrt{6x - \sqrt{x+7}} = 3$
32. $\sqrt{3x - \sqrt{x+3}} = 1$
33. $\sqrt{3y + \sqrt{y}} = 2$
34. $\sqrt{6 + \sqrt{3y}} = 3$
35. $\sqrt{3x - \sqrt{x+2}} = 2$
36. $x^{2/3} + 10x^{1/3} + 24 = 0$
37. $6x^{2/3} - 13x^{1/3} + 6 = 0$
38. $x^{4/3} - 13x^{2/3} + 36 = 0$
39. $x^{4/3} - 10x^{2/3} + 9 = 0$
40. $2x - 3\sqrt{x} - 9 = 0$
41. $x - 5\sqrt{x} + 4 = 0$
42. $\sqrt{x} - \sqrt[4]{x} - 2 = 0$

---

43. For a solenoid of length l (in m), the number of turns required in the coil is given by $N = \sqrt{\dfrac{Ll}{\mu A}}$. In the formula, L is the inductance (in H), μ is the magnetic permeability of the core material, and A is the cross-sectional area (in m^2). Find the length needed for a solenoid that is to have 400 turns, cross-sectional area 0.002 m^2, inductance 0.004 H, and $\mu = 1.26 \times 10^{-6}$ N/m. **3 5 7** 1 2 4 6 8 9

44. In a charged capacitor, the potential difference is given by $V = \sqrt{2W/C}$. In the formula, W is the potential energy (in joules) and C is the capacitance (in F). How much energy is stored in a 100 μF (μ means 10^{-6}) capacitor charged to a potential difference of 1000 V? **3 5 9** 4 7 8

45. In the two figures shown, the length of the edge of the square is the same length as the radius of the circle.

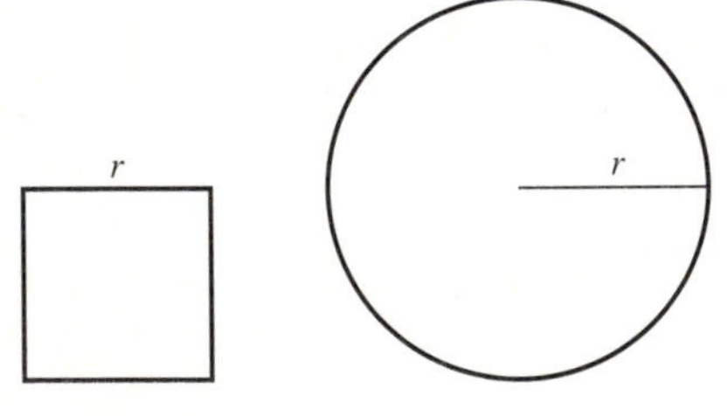

(a) Solve the formula $A = \pi r^2$ for r.
(b) Find the area of the circle if the area of the square is 7.00 cm^2. **1 2 4 7 8** 3 5

46. The diameter D_1 of a wire (in inches) of resistance R_1 is given by $D_1 = \sqrt{\dfrac{D_2{}^2 R_2}{R_1}}$ where D_2 and R_2 are the diameter and resistance, respectively, of another wire of the same material. If the resistance decreases by 0.1 Ω when the diameter increases by 0.01 in, find the original resistance of a wire of diameter 0.4 in. **3 5 9** 1 2 4 6 7 8

47. Find the dimensions of the box shown in the diagram. The diagonal from upper back left to lower front right is 13.0 cm. **1 2 4 7 8** 3 5

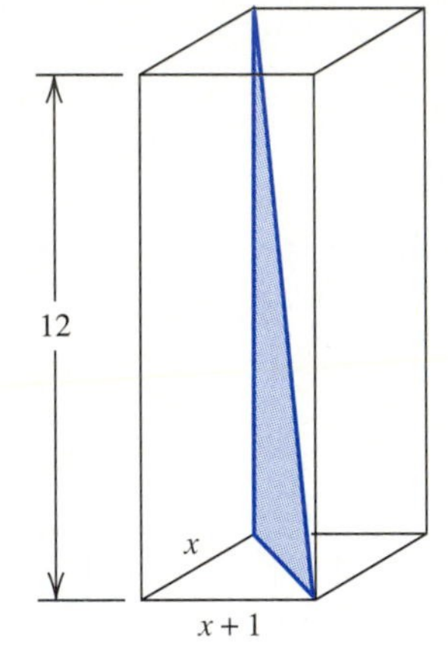

48. Find the dimensions of the box shown in the diagram. The diagonal from upper back left to lower front right is 17.0 cm. **1 2 4 7 8** 3 5

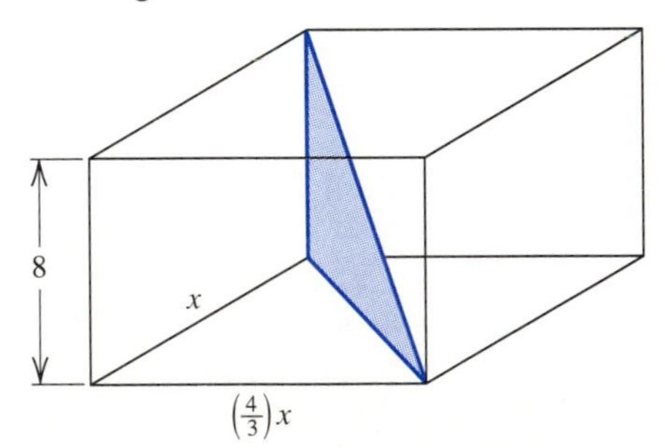

---

1. Architectural Technology
2. Civil Engineering Technology
3. Computer Engineering Technology
4. Mechanical Drafting Technology
5. Electrical/Electronics Engineering Technology
6. Chemical Engineering Technology
7. Industrial Engineering Technology
8. Mechanical Engineering Technology
9. Biomedical Engineering Technology

13.6 CHAPTER REVIEW

KEY TERMS

13.2 radicand
order
simplest radical form
like radicals

13.3 conjugates

13.5 extraneous solutions
quadratic form

SYMBOLS

$\sqrt{x}$ the square root of x

$\sqrt[n]{x}$ the nth root of x

RULES AND FORMULAS

If x is a real number and m and n are integers such that $n > 0$, then
$x^{1/n} = \sqrt[n]{x}$ when $\sqrt[n]{x}$ is a real number, and
$x^{m/n} = (\sqrt[n]{x})^m$ or $\sqrt[n]{(x^m)}$ when $\sqrt[n]{x}$ is a real number.
If x and y are real numbers and n is a positive integer, then $\sqrt[n]{xy} = \sqrt[n]{x}\sqrt[n]{y}$, and
$\sqrt[n]{\frac{x}{y}} = \frac{\sqrt[n]{x}}{\sqrt[n]{y}}$ if $y \neq 0$ and all of the radicals represent real numbers.

GEOMETRY CONCEPTS

$V = \frac{4}{3}\pi r^3$ (volume of a sphere)

$B = \frac{1}{3}Bh$ (volume of a pyramid)

$A = \pi r^2$ (area of a circle)

SEQUENTIAL REVIEW

You may assume that all variables in these exercises represent nonnegative real numbers and that no denominators are zero.

(Section 13.1) *In 1 and 2, simplify each expression.*

1. $(-8)^{2/3}$

2. $-8^{2/3}$

In 3 and 4, evaluate each expression, using a calculator. Give the answer to three significant digits.

3. $11^{3/7}$

4. $7^{2/9}$

In 5–7, simplify each expression.

5. $\left(\dfrac{2a^3b^{-1}}{c^{-1/2}}\right)^{2/3}$ **6.** $\left(\dfrac{4a^{-1}}{b^{1/2}c^{-2/3}}\right)^{-2}$ **7.** $\dfrac{4x^{1/2}y^{-2/3}}{(x+y)^{-1/3}}$

In 8 and 9, factor each expression, and simplify the result.

8. $2x^{-2}y^3 + 4x^2y^{-3}$ **9.** $x(x-2)^{-2}(4x+3)^{-1/2} + (x-2)^{-3}(4x+3)^{1/2}$

(Section 13.2) *In 10–12, reduce the order of each radical.*

10. $\sqrt[4]{49}$ **11.** $\sqrt[6]{9a^2b^4}$ **12.** $\sqrt[8]{81x^4y^{12}}$

In 13–15, write each expression in simplest radical form.

13. $\sqrt{63a^5b^6}$ **14.** $\sqrt[3]{40a^3b^6}$ **15.** $\dfrac{\sqrt{5}}{2\sqrt{7}}$

In 16–18, perform the indicated operation.

16. $4x\sqrt{5y} + 6y\sqrt{3} - 7x\sqrt{5y}$ **17.** $\sqrt[6]{8} + \sqrt{18} + \dfrac{3}{\sqrt{2}}$ **18.** $(2\sqrt{2} - 3)(\sqrt{2} + 1)$

(Section 13.3) *In 19–22, evaluate each expression using a calculator. Assume that integers represent exact values.*

19. $\dfrac{\sqrt{5}}{\sqrt{7}}$ **20.** $\dfrac{5}{1-\sqrt{5}}$ **21.** $\dfrac{7+\sqrt{5}}{3-\sqrt{5}}$ **22.** $\dfrac{2\sqrt{7}-3}{2\sqrt{7}+3}$

In 23–27, perform the division by rationalizing the denominator.

23. $\dfrac{-3}{x\sqrt{7}}$ **24.** $\dfrac{5}{\sqrt[3]{4a}}$ **25.** $\dfrac{7}{4\sqrt{3}+\sqrt{2}}$

26. $\dfrac{2\sqrt{5}-\sqrt{3}}{\sqrt{5}+2\sqrt{3}}$ **27.** $\dfrac{\sqrt{x+1}}{3+\sqrt{x+1}}$

(Section 13.4) *In 28–36, simplify each expression.*

28. $\sqrt[3]{3x}\sqrt{2}$ **29.** $\sqrt[3]{4}\sqrt{3}$

30. $\dfrac{\sqrt{3}}{\sqrt[3]{5}}$ **31.** $\sqrt{3} + \sqrt[3]{5}$

32. $(\sqrt{7} - 1)\left(3 + \dfrac{2}{\sqrt{7}}\right)$ **33.** $(\sqrt[3]{3} - \sqrt[3]{2})(\sqrt[3]{3} + \sqrt[3]{2})$

34. $\sqrt{3+\sqrt{7}}\,\sqrt{3-\sqrt{7}}$ **35.** $\sqrt{\sqrt{7}-a}\,\sqrt{\sqrt{7}+a}$

36. $\sqrt{5xy} + (\sqrt{5x} - \sqrt{y})^2$

(Section 13.5) *In 37–45, solve each equation.*

37. $3\sqrt{y} = 7$ **38.** $\sqrt[3]{3x-1} = 2$

39. $\sqrt{x+6} = 4 + x$ **40.** $\sqrt{3x+1} = -1$

41. $2\sqrt{x+1} - \sqrt{3x} = 1$ **42.** $\sqrt{3x - \sqrt{x+5}} = 3$

43. $\sqrt{4y - \sqrt{3y}} = 3$ **44.** $x^{2/3} - 5x^{1/3} + 6 = 0$

45. $\sqrt{x} - 3\sqrt[4]{x} - 4 = 0$

APPLIED PROBLEMS

1. The radioactivity A (in roentgens/hour) at time t (in days) for up to six months after a nuclear explosion may be approximated by $A = A_0 t^{-6/5}$. In the formula, A_0 is the radioactivity at unit time. Write the equation in simplest radical form. 6 9 5 8

2. The radius R of a sphere with volume V is given by $R = \sqrt[3]{\dfrac{3V}{4\pi}}$. Find the radius of a sphere with the same volume as the rectangular block in the accompanying figure. 1 2 4 7 8 3 5

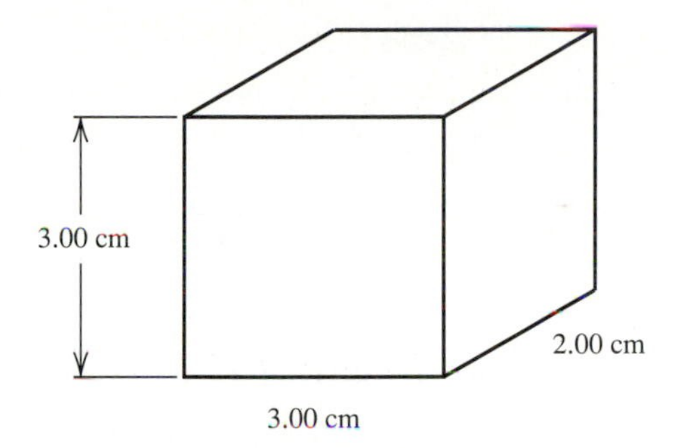

3. The formula $V = (PR)^{1/2}$ gives voltage V (in volts) if P is power (in watts) and R is resistance (in ohms). Compute the resistance if a 2000-watt heater is connected to a 115-volt circuit. The formula $P = I^2R$ gives the power P (in watts) if I is the current (in amperes) and R is the resistance (in ohms). Will the current exceed the normal maximum household current of approximately 15 amperes? 3 5 9 1 2 4 6 7 8

4. The formula $R = \left(\dfrac{3V}{4\pi}\right)^{1/3}$ gives the radius of a sphere if V is its volume.
   (a) Write the formula using a radical.
   (b) Solve for R if $V = 75.3\ \text{m}^3$.
   1 2 4 7 8 3 5

5. The formula $t = \dfrac{B}{U\sqrt{V_t}}$ gives the time (in min/ml) for the removal of urea from the body. In the formula, B is the urea concentration (in mg/ml) in the blood. U is the urea concentration (in mg/ml) in the urine, and V_t is the urine generation rate (in ml/min).
   (a) Calculate t if $B = 0.250$ mg/ml, $U = 91.2$ mg/ml, and $V_t = 1.80$ ml/min.
   (b) Write the formula in simplest radical form.
   6 9 2 4 5 7 8

6. In electronics, the formula for the resonant frequency of a circuit containing a resistor, inductor, and capacitor in series is given in all texts as $f = \dfrac{1}{2\pi\sqrt{LC}}$. Write this formula in simplest radical form. 3 5 9 1 2 4 6 7 8

7. The radius of curvature of a spherical lens is given by the equation $R = \dfrac{L^2}{6D} + \dfrac{D}{2}$ where L and D are given by a calibrating device. Find a simplified equation for R if $L = \sqrt{D}$. 3 5 6 9

8. If the angle between two vectors is C, and their magnitudes are a and b, their resultant has a magnitude given by

$$c = \sqrt{a^2 + b^2 - 2ab \cos C}.$$

   Find a simplified equation for c if $b = 10a$.
   2 4 8 1 5 6 7 9

9. The solution of a cubic equation of the form $x^3 + ax + b = 0$ involves finding an important constant that results from evaluating

$$-\sqrt[3]{\frac{b}{2} + \sqrt{\frac{b^2}{4} + \frac{a^3}{27}}}.$$

   Simplify this radical if $a = 3b$. 1 2 4 7 8 3 5

---

1. Architectural Technology
2. Civil Engineering Technology
3. Computer Engineering Technology
4. Mechanical Drafting Technology
5. Electrical/Electronics Engineering Technology
6. Chemical Engineering Technology
7. Industrial Engineering Technology
8. Mechanical Engineering Technology
9. Biomedical Engineering Technology

10. As time increased from zero, the current i (in amps) in a device increased from an initial value of 3 amps according to the equation $i = \sqrt{t+9}$, where t is measured in seconds. The voltage v (in volts) that caused this current change was given by $v = 3t$. What value will i have when $v = 6$? 3 5 9 1 2 4 6 7 8

11. In the study of parallel resonant circuits the frequency for the resonance phenomenon is given in terms of the circuit's components by the equation $f_p = \sqrt{\frac{L - CR^2}{4\pi^2 L^2 C}}$. Sometimes it is desirable to compare the parallel frequency f_p to the series frequency f_s of the same circuit. This can be done by factoring out the expression $\frac{1}{2\pi\sqrt{LC}}$ from the above radical. What formula results if $f_s = \frac{1}{2\pi\sqrt{LC}}$? 3 5 9 1 2 4 6 7 8

12. An approximation for the circumference of a certain type of elliptical conductor is given by $C = 2\pi\sqrt{\frac{a^2+b^2}{2}}$, where a and b are the semiaxes of the ellipse. Find a simplified equation for C if $b = 3a$. 3 5 9 1 2 4 6 7 8

13. A handbook gives an approximation for the length L of the arc of a parabola as $L = 2\sqrt{\frac{W^2}{4} + \frac{4}{3}H^2}$ where W is its width and H its height. Find L if $H = 3W$. 1 2 4 7 8 3 5

14. Two kinds of hazardous gases are stored in their own fenced areas that are 15.0 ft square. Since one of the gasses is not used much it is decided to reapportion the storage areas taking some square footage away from one area and giving it to the other. Redesign the storage areas in such a way as to save 20% of the fence needed to surround the old areas. Keep the new areas square. 1 2 4 7 8 3 5

15. In electronics, a formula found in communication theory uses the product $(e^x + e^{-x})(e^x - e^{-x})$, where e is an irrational number that is approximately 2.718. Simplify this expression. 3 5 9 1 2 4 6 7 8

16. The impedance of a circuit containing a resistor, an inductor, and a capacitor in series is given by $Z = \sqrt{R^2 + \left(\omega L - \frac{1}{\omega C}\right)^2}$. Solve this equation for R. 3 5 9 4 7 8

17. The impedance of a circuit containing a resistor, an inductor, and a capacitor in series is given by $Z = \sqrt{R^2 + (X_L - X_c)^2}$. Suppose $Z = 10R$ when $X_L = 10R$ and write an equivalent equation that does not contain a radical. 3 5 9 4 7 8

18. The volume of the early space capsules was approximated by the formula $V = \frac{h}{3}(A_1 + A_2 + \sqrt{A_1 A_2})$ where h was the height, A_1 the area of the heat shield on the top surface and A_2 the area of the heat shield on the bottom surface. Write an equivalent equation that does not contain a radical. 1 2 4 7 8 3 5

19. In the study of ac circuits the equation $\sin\alpha = \sqrt{\frac{1-\cos\beta}{2}}$ is encountered. Solve the equation for β. 3 5 9 1 2 4 6 7 8

---

1. Architectural Technology
2. Civil Engineering Technology
3. Computer Engineering Technology
4. Mechanical Drafting Technology
5. Electrical/Electronics Engineering Technology
6. Chemical Engineering Technology
7. Industrial Engineering Technology
8. Mechanical Engineering Technology
9. Biomedical Engineering Technology

RANDOM REVIEW

You may assume that all variables in these exercises represent nonnegative real numbers and that no denominators are zero.

1. Solve the equation $\sqrt{5+\sqrt{2x}} = 3$.
2. Perform the division $\dfrac{3+\sqrt{7}}{5-\sqrt{7}}$ by rationalizing the denominator.
3. Simplify the expression $\dfrac{(a+b)^{-1/2}}{8a^{1/3}b^{1/2}}$.
4. Simplify the expression $\sqrt[3]{7}\sqrt{2}$.
5. Write the following expression in simplest radical form.

$$\sqrt[4]{25a^4b^6}$$

6. Perform the division $\dfrac{4}{\sqrt{3}-2}$ using a calculator.
7. Solve the equation $x^{1/4} - 8x^{1/2} + 16 = 0$.
8. Perform the indicated operation.

$$\sqrt[4]{4} + \sqrt{50} - 1/\sqrt{2}$$

9. Simplify the expression $\sqrt{2x\sqrt{5}}$.
10. Factor $x(x+1)^{-1} + 3(x+1)^{-2}$, and simplify the result.

CHAPTER TEST

You may assume that all variables in these exercises represent nonnegative real numbers and that no denominators are zero.

1. Simplify the expression $\left(\dfrac{25m^2n^{-4}}{36p^6}\right)^{-1/2}$.
2. Factor $(x-3)^{-3}(x+2)^{-2} + (x-3)^{-2}(x+2)^{-3}$, and simplify the result.

In 3 and 4, write each expression in simplest radical form.

3. $\sqrt{44x^6y^{10}}$
4. $\dfrac{3\sqrt{7}}{5\sqrt{3}}$
5. Perform the indicated operation.

$$(3\sqrt{2}+1)(\sqrt{2}-3)$$

6. Perform the division $\dfrac{\sqrt{5}-1}{\sqrt{5}+3}$
   - **(a)** using a calculator.
   - **(b)** by rationalizing the denominator.

7. Perform the indicated operation.

$$(\sqrt{5} + 2)(1 + 3/\sqrt{5})$$

8. Solve the equation $\sqrt{x + 4} = x - 2$.

9. The radius of a circle of area A is given by $r = \sqrt{A/\pi}$. Find the radius of a circle with the same area as the figure shown.

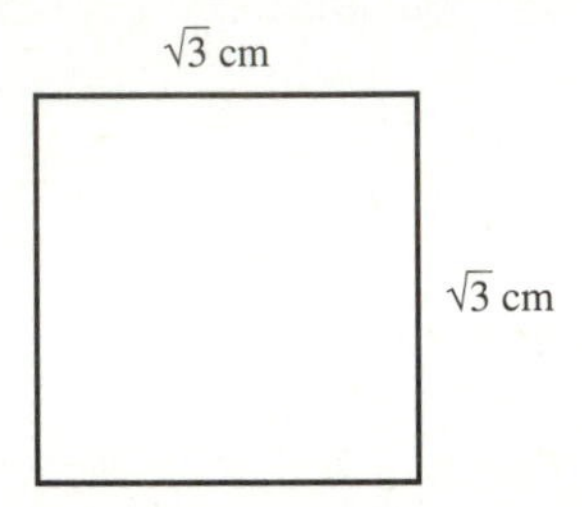

10. When a charged capacitor is discharged through a resistance, the charge Q (in coulombs) remaining after time t (in s) is given by $Q = Q_0 e^{-t/\tau}$. In the formula, Q_0 is the initial charge (in coulombs), e is a constant approximately equal to 2.718, and τ is the time constant (i.e., the number of seconds it will take for the charge to drop to 37% of its original value). Find the decreased charge when $Q_0 = 2.50 \times 10^{-5}$ coulombs, $t = 45.0$ s, and $\tau = 5.00 \times 10^3$ s.

EXTENDED APPLICATION: COMPUTER ENGINEERING TECHNOLOGY

DETECTING COMPUTER ERRORS

All information processed by a computer is translated, character by character, into a code that uses only two symbols: 0 and 1. The coding symbols, called bits, are stored on silicon memory chips. A memory chip is a square array of data storage cells. A 256 K chip, for example, has 256×2^{10} or 262,144 cells in an arrangement of 512 rows and 512 columns. Each cell is capable of storing one bit. When a 0 is supposed to be stored in a cell, the site is given a negative charge by increasing the number of electrons. When a 1 is supposed to be stored, the number of electrons is decreased.

In computer jargon, a word is the number of bits in the largest data value that can be transferred in one operation. Depending on the computer, a word may be from 8 to 60 bits. Occasionally, alpha particles enter the memory cells. If electrons are dislodged from one cell, they may be knocked into another, changing a 1 to a 0. To detect such errors, one or more extra bits are included in the coded representation of a character. If the extra bits are chosen so that the sum of the bits in any character is even, a single error is detectable. If two or more errors occur in the same word, the error is not so easily detected. It is possible, however to analyze the situation mathematically.

Let $n =$ the average number of errors required before two errors occur in the same word
$c =$ the capacity of the computer's memory in words
$b =$ the capacity of the computer's memory in bits
$t =$ the average time in days for an error that is undetected by the coding to occur

The formulas needed are as follows.

$$n \approx \sqrt{\pi c/2} \qquad \text{and} \qquad t \approx \frac{32(3.65 \times 10^8)}{39b}$$

Consider a 1 megabyte memory (2^{23} bits) that uses words that are 32 or 2^5 bits long.

$$b = 2^{23} \qquad \text{and} \qquad c = 2^{23}/2^5 = 2^{18} = 262{,}144$$

$$n \approx \sqrt{\pi(262{,}144)/2} \approx 642 \qquad \text{and} \qquad t \approx \frac{32(3.65 \times 10^8)}{39(2^{23})} \approx 35.7 \text{ days}$$

If about 642 errors occur before two of them occur in the same word, and errors happen at the rate of about 1 per 35.7 days, a computer with one megabyte of memory can be expected to work, on the average, for about 642×35.7 or 22,919 days (62.8 years) before an error that is undetected by the coding occurs.

■ This problem is based on the article "The Reliability of Computer Memories," by Robert J. McEliece, *Scientific American,* Vol. 252 No. 1 (January 1985), 88–95.

CHAPTER 14 COMPLEX NUMBERS

For many situations in the real world, the real numbers provide an adequate framework for problem solving. There are some situations, however, that are better described using complex numbers. In this chapter, we present the properties of complex numbers and apply them to problems in electronics.

14.1 ARITHMETIC WITH COMPLEX NUMBERS

pure imaginary number

In Section 1.1, a *complex number* was defined as any number that can be written in the form $a + b\sqrt{-1}$, where a and b are real numbers. All real numbers may be put in this form, for we may consider b to be zero for each real number a. A number for which a is zero and b is not zero is called a *pure imaginary number.* A complex number, then, may be a real number, a pure imaginary number, or a number composed of a real part and a pure imaginary part. Mathematicians traditionally use the symbol i to denote $\sqrt{-1}$. In technological work, i is often used to indicate current, so the letter j is used for $\sqrt{-1}$. Any complex number can be written in the form $a + bj$.

In general, it is possible to consider the complex number system to be an extension of the real number system. That is, the rules for doing arithmetic with real numbers hold for complex numbers. When the radicand is positive, squaring a number is the inverse operation of taking the square root. If this rule is valid for negative radicands, too, then it must be true that

$$(\sqrt{-1})^2 = -1. \quad \textbf{(14-1)}$$

Special care must be taken, however, when using the rule $\sqrt[n]{x}\sqrt[n]{y} = \sqrt[n]{xy}$. This rule was stated for real numbers x and y *when all of the radicals represent real numbers.* It does not apply to $(\sqrt{-1})^2 = \sqrt{-1}\sqrt{-1}$, in which the radical $\sqrt{-1}$ does not represent a real number. That is, $\sqrt{-1}\sqrt{-1} \neq \sqrt{1}$. Instead, $(\sqrt{-1})^2$ is defined to be -1, and the rule $\sqrt[n]{x}\sqrt[n]{y} = \sqrt[n]{xy}$ must have the restriction that x and y cannot both be negative when n is even.

Applying the restricted rule, we are able to write the square root of any negative real number as a pure imaginary number in standard form.

STANDARD FORM FOR A PURE IMAGINARY NUMBER

If $b \geq 0$, $\sqrt{-b} = \sqrt{-1(b)} = \sqrt{-1}\sqrt{b} = j\sqrt{b}$

EXAMPLE 1 Write each number in the form bj.

(a) $\sqrt{-4}$ **(b)** $\sqrt{-18}$ **(c)** $-\sqrt{-3}$

Solution **(a)** Since only one of the factors may be negative, we have $\sqrt{-4} = \sqrt{4(-1)} = \sqrt{4}\sqrt{-1} = 2j$

(b) $\sqrt{-18} = \sqrt{9(2)(-1)} = \sqrt{9}\sqrt{2}\sqrt{-1} = 3\sqrt{2}\,j = 3j\sqrt{2}$

Because of the possibility that $3\sqrt{2}\,j$ might be mistaken for $3\sqrt{2j}$, it is better to write the j to the left of the radical: $3j\sqrt{2}$.

(c) $-\sqrt{-3} = -\sqrt{-1}\sqrt{3} = -j\sqrt{3}$ ■

Mathematicians write complex numbers in the form $a + bj$ unless b is a radical, but engineers use the form $a + jb$, regardless of the value of b. If j is used to represent $\sqrt{-1}$, then j^2 represents $(\sqrt{-1})^2$ or -1 as in Equation 14-1. With this equation in mind, you can see how to multiply two pure imaginary numbers.

To multiply two pure imaginary numbers:

1. Express each number in the form bj.
2. Multiply.
3. Replace j^2 with -1.

NOTE ⇨⇨ ***Notice that while we interpret $(\sqrt{-1})^2$ as -1, we interpret $(\sqrt{(-1)^2}$ as $\sqrt{(-1)(-1)}$, which is 1.*** That is, the radicand must be evaluated before finding the square root.

EXAMPLE 2 Simplify each expression.

(a) $\sqrt{-3}\sqrt{-6}$ **(b)** $\sqrt{-3(-6)}$

Solution **(a)** $\sqrt{-3} = \sqrt{-1(3)} = j\sqrt{3}$ and $\sqrt{-6} = \sqrt{-1(6)} = j\sqrt{6}$

$\sqrt{-3}\sqrt{-6} = j\sqrt{3}(j\sqrt{6}) = j^2\sqrt{18} = -1\sqrt{18} = -1\sqrt{9(2)} = -1(3)\sqrt{2} = -3\sqrt{2}$

(b) $\sqrt{-3(-6)} = \sqrt{18} = \sqrt{9(2)} = \sqrt{9}\sqrt{2} = 3\sqrt{2}$ ■

You have seen that j^2 may be replaced with -1. It is interesting to note that any power of j may be replaced with j, $-j$, 1, or -1. Consider the following pattern.

$$j^1 = j \qquad j^5 = j^4(j) = 1(j) = j$$
$$j^2 = -1 \qquad j^6 = j^4(j^2) = 1(-1) = -1$$
$$j^3 = j(j^2) = j(-1) = -j \qquad j^7 = j^4(j^3) = 1(-j) = -j$$
$$j^4 = j^2(j^2) = -1(-1) = 1 \qquad j^8 = j^4(j^4) = 1(1) = 1$$

The pattern is a repeating one, so it is possible to replace any power of j with one of the first four powers listed.

EXAMPLE 3 Simplify j^{251}.

Solution Since $j^4 = 1$, we need to know how many factors of j^4 are in j^{251}. Since $251 \div 4 = 62$ with a remainder of 3, we have

$$j^{251} = (j^4)^{62}j^3 = 1^{62}j^3 = j^3. \quad ■$$

To simplify a power of *j*:

1. Divide the power by 4.
2. If the remainder is positive, use it as the new power of j. If the remainder is negative, add 4 and use the sum as the new power of j.
3. Evaluate the power of j according to the following pattern:

$$j^0 = 1, j^1 = j, j^2 = -1, \text{ and } j^3 = -j.$$

EXAMPLE 4 Simplify each power of j.

(a) j^{32} **(b)** j^{17} **(c)** j^{-6}

Solution **(a)** $32 \div 4 = 8$ with a remainder of 0. Thus, $j^{32} = j^0 = 1$.

(b) $17 \div 4 = 4$ with a remainder of 1. Thus, $j^{17} = j^1 = j$.

(c) $-6 \div 4 = -1$ with a remainder of -2. Since $-2 + 4 = 2$, $j^{-6} = j^2 = 1$. ■

We are now ready to perform arithmetic operations with complex numbers.

To add, subtract, or multiply complex numbers:

1. Express each number in the form $a + bj$ (or $a - bj$).
2. Add, subtract, or multiply as though the numbers were binomials with j as a variable.
3. Simplify any powers of j.

EXAMPLE 5 Perform the indicated operations.

(a) $(2 + 3j) + (-1 + 4j)$

(b) $(3 - 7j) - (-2 + 5j)$

Solution **(a)** $(2 + 3j) + (-1 + 4j) = 2 + 3j + (-1) + 4j = 1 + 7j$

Notice that $3j$ and $4j$ are treated as like terms.

(b) $(3 - 7j) - (-2 + 5j) = 3 - 7j + 2 - 5j = 5 - 12j$

Notice that the subtraction symbol applies to both terms of $-2 + 5j$. ■

EXAMPLE 6 Perform the indicated operations.

(a) $(3 + j^3)(2 - 7j^3)$

(b) $(1 - 2j)(1 + 2j)(2 - j)$

Solution **(a)**
$$
\begin{aligned}
(3 + j^3)(2 - 7j^3) &= (3 - j)(2 + 7j) && \text{Replace } j^3 \text{ with } -j. \\
&= 6 + 21j - 2j - 7j^2 && \text{Perform binomial multiplication.} \\
&= 6 + 19j - 7(-1) && \text{Replace } j^2 \text{ with } -1. \\
&= 6 + 19j + 7 \\
&= 13 + 19j
\end{aligned}
$$

(b)
$$
\begin{aligned}
\mathbf{(1 - 2j)(1 + 2j)}(2 - j) &= \mathbf{(1 - 4j^2)}(2 - j) \\
&= [1 - 4(-1)](2 - j) && \text{Replace } j^2 \text{ with } -1. \\
&= (1 + 4)(2 - j) \\
&= 5(2 - j) \\
&= 10 - 5j \quad ■
\end{aligned}
$$

NOTE ⇨⇨ ***When a radical contains a negative radicand, it is important to express the number in the form $a + bj$ before performing the operations.*** Furthermore, it is a common mistake to confuse addition and multiplication problems. Look carefully at the problems in Example 7.

EXAMPLE 7 Perform the indicated operations.

(a) $(3 + \sqrt{-2}) + (3 - \sqrt{-2})$

(b) $(3 + \sqrt{-2})(3 - \sqrt{-2})$

Solution **(a)** The problem is an addition problem.

$$\begin{aligned}(3 + \sqrt{-2}) + (3 - \sqrt{-2}) &= (3 + j\sqrt{2}) + (3 - j\sqrt{2}) \\ &= 3 + j\sqrt{2} + 3 - j\sqrt{2} \\ &= 6\end{aligned}$$

(b) The problem is a multiplication problem.

$$\begin{aligned}(3 + \sqrt{-2})(3 - \sqrt{-2}) &= (3 + j\sqrt{2})(3 - j\sqrt{2}) \\ &= 9 - 3j\sqrt{2} + 3j\sqrt{2} - j^2\sqrt{4} \\ &= 9 - (-1)(2) \qquad \text{Replace } j^2 \text{ with } -1. \\ &= 9 + 2 \\ &= 11 \end{aligned}$$

■

conjugate of a complex number

Two complex numbers of the form $a + bj$ and $a - bj$ are said to be **conjugates** of each other. Examine the pattern when a complex number is multiplied by its conjugate.

$$\begin{aligned}(a + bj)(a - bj) &= a^2 - abj + abj - b^2j^2 \\ &= a^2 - b^2(-1) \\ &= a^2 + b^2\end{aligned}$$

When two complex numbers are conjugates of each other, their product is the sum of squares. This observation allows us to perform division of complex numbers in a manner similar to division of radicals.

To divide complex numbers:

1. Express the quotient as a fraction with the numerator and denominator each in the form $a + bj$.
2. Multiply the numerator and denominator by the conjugate of the denominator.
3. Simplify both numerator and denominator, and reduce the fraction to lowest terms.

EXAMPLE 8 Perform the indicated operation.

$$\frac{2 + \sqrt{-5}}{4 - \sqrt{-5}}$$

Solution
$$\frac{2+\sqrt{-5}}{4-\sqrt{-5}}=\frac{2+j\sqrt{5}}{4-j\sqrt{5}}$$
Express $\sqrt{-5}$ as $\sqrt{-1(5)}=j\sqrt{5}$.

$$\frac{2+j\sqrt{5}}{4-j\sqrt{5}}=\left(\frac{2+j\sqrt{5}}{4-j\sqrt{5}}\right)\left(\frac{4+j\sqrt{5}}{4+j\sqrt{5}}\right)$$
Multiply both numerator and denominator by the conjugate of $4-j\sqrt{5}$.

$$=\frac{8+2j\sqrt{5}+4j\sqrt{5}+j^2(5)}{16+4j\sqrt{5}-4j\sqrt{5}-j^2(5)}$$
$$=\frac{8+6j\sqrt{5}+(-1)(5)}{16-(-1)(5)}$$
Replace j^2 with -1.

$$=\frac{8+6j\sqrt{5}-5}{16+5}$$
$$=\frac{3+6j\sqrt{5}}{21}$$

Remember that fractions are reduced by dividing *factors,* not *terms,* that are common to both numerator and denominator. The numerator of this expression may be factored using the distributive property so that the fraction can be reduced.

$$\frac{3+6j\sqrt{5}}{21}=\frac{3(1+2j\sqrt{5})}{21}=\frac{1+2j\sqrt{5}}{7}$$

The answer could be expressed in the form $a+bj$ as $\frac{1}{7}+\frac{2\sqrt{5}}{7}j$, but it is generally left as a single fraction. ■

In electric/electronic circuits in which currents and voltages are sinusoidal and of the same frequency, complex numbers are used so that analogies may be made between dc circuits and ac circuits. In an ac circuit with two or more sections in series, the total impedance **Z** (in ohms) is the sum of the impedances of each individual section. The current **I** (in amperes) is the same through each section of the circuit, and Ohm's law ($\mathbf{V}=\mathbf{IZ}$) gives the voltage **V** (in volts).

EXAMPLE 9 Calculate the current that would flow if 240 V were applied across a circuit with an impedance of $8.7+j9.5\ \Omega$.

Solution Since $\mathbf{V}=\mathbf{IZ}$, $240=\mathbf{I}(8.7+j9.5)$.

$$\frac{240}{8.7+j9.5}=\mathbf{I}$$
$$\left(\frac{240}{8.7+j9.5}\right)\left(\frac{8.7-j9.5}{8.7-j9.5}\right)=\frac{2088-j2280}{75.69-j^2 90.25}$$
$$=\frac{2088-j2280}{165.94}=13-j14\text{ A}$$ ■

EXERCISES 14.1

In 1–6, express each number in the form bj.

1. $\sqrt{-25}$
2. $\sqrt{-32}$
3. $\sqrt{-27}$
4. $-\sqrt{-7}$
5. $-\sqrt{-11}$
6. $-\sqrt{-17}$

In 7–18, simplify each expression.

7. **(a)** $\sqrt{-5}\sqrt{-10}$ **(b)** $\sqrt{-5(-10)}$
8. **(a)** $\sqrt{-6}\sqrt{-8}$ **(b)** $\sqrt{-6(-8)}$
9. **(a)** $\sqrt{-2}\sqrt{-6}$ **(b)** $\sqrt{-2(-6)}$
10. **(a)** $\sqrt{-2}\sqrt{-10}$ **(b)** $\sqrt{-2(-10)}$
11. **(a)** $(\sqrt{-10})^2$ **(b)** $\sqrt{(-10)^2}$
12. **(a)** $(\sqrt{-8})^2$ **(b)** $\sqrt{(-8)^2}$
13. j^{18}
14. j^{40}
15. j^{-25}
16. j^{-35}
17. j^{208}
18. j^{157}

In 19–42, perform the indicated operations.

19. $(4+5j)+(2-3j)$
20. $(\sqrt{2}+j)+(4\sqrt{2}-j)$
21. $(3+\sqrt{-12})+(-3+\sqrt{-48})$
22. $(-7+\sqrt{-27})+(-6+\sqrt{-75})$
23. $(3.2+4.1j)+(5.7-2.6j)$
24. $(-1.7+8.2j)+(2.3-6.4j)$
25. $(3+7j^3)-(j^8-4j)$
26. $(\sqrt{3}+2j)-(2\sqrt{3}-3j^5)$
27. $(4+\sqrt{-20})-(-3+\sqrt{-45})$
28. $(-6+\sqrt{-64})-(-5-\sqrt{-81})$
29. $(5.6+9.1j)-(3.2+7.3j)$
30. $(5+4j)(5-4j)$
31. $(7-j)(7+j)$
32. $(\sqrt{3}+3\sqrt{-5})(2\sqrt{3}+\sqrt{-5})$
33. $(\sqrt{7}-\sqrt{-4})(\sqrt{7}+3\sqrt{-4})$
34. $(\sqrt{2}+3\sqrt{-16})(-\sqrt{2}+2\sqrt{-9})$
35. $(2-j)(2+j)(3-j)$
36. $(3-2j)^3$
37. $\dfrac{1-2j}{1+2j}$
38. $\dfrac{7-3j}{5+2j}$
39. $\dfrac{3+4j}{2-5j}$
40. $\dfrac{6+j}{6-j}$
41. $\dfrac{1-\sqrt{-1}}{1+\sqrt{-1}}$
42. $\dfrac{\sqrt{3}-\sqrt{-9}}{\sqrt{3}+\sqrt{-9}}$

Problems 43–47 use the following information. For two sections of a circuit connected in series, current is the same in each section, voltage (in V) is given by $\mathbf{V}=\mathbf{V}_1+\mathbf{V}_2$, and impedance (in Ω) is given by $\mathbf{Z}=\mathbf{Z}_1+\mathbf{Z}_2$.

43. Calculate the voltage that would occur in a circuit containing two sections in series if the individual voltages are $8.96+j2.53$ V and $1.73+j1.89$ V, respectively. 3 5 9 1 2 4 6 7 8

44. Calculate the total impedance that would occur in a circuit containing two sections in series if the individual impedances are $25.0-j\,36.0$ Ω and $13.0-j22.0$ Ω, respectively. 3 5 9 1 2 4 6 7 8

45. Calculate the voltage produced if $0.191 - 0.0590j$ A flows through a circuit with an impedance of $2.58 + 15.4j\ \Omega$. 3 5 9 1 2 4 6 7 8

46. Calculate the current that would flow if 120 V were applied across a circuit with an impedance of $7.8 + j5.9\ \Omega$. 3 5 9 1 2 4 6 7 8

47. Calculate the voltage produced if $0.191 - j0.595$ A flows through a circuit containing two sections in series if the individual impedances are $2.58 + j15.4\ \Omega$ and $4.30 + j14.5\ \Omega$, respectively. 3 5 9 1 2 4 6 7 8

Problems 48–49 use the following information. For two sections of a circuit connected in parallel, current (in A) is given by $\mathbf{I} = \mathbf{I}_1 + \mathbf{I}_2$, *and impedance (in Ω) is given by* $1/\mathbf{Z} = 1/\mathbf{Z}_1 + 1/\mathbf{Z}_2$.

48. Calculate the current that would flow through a circuit containing two sections in parallel if the individual currents are $0.678 + j0.563$ A and $0.850 + j0.0745$ A, respectively. 3 5 9 1 2 4 6 7 8

49. Calculate the impedance that would occur in a circuit containing two sections in parallel if the individual impedances are $25.0 - j31.0\ \Omega$ and $13.0 - j27.0\ \Omega$, respectively. 3 5 9 1 2 4 6 7 8

50. Consider the complex number $a + bj$ and its conjugate $a - bj$.
(a) What type of number is the sum?
(b) What type of number is the difference?
(c) What type of number is the product?
(d) What type of number is $a + bj$ if it is equal to its conjugate?

1. Architectural Technology
2. Civil Engineering Technology
3. Computer Engineering Technology
4. Mechanical Drafting Technology
5. Electrical/Electronics Engineering Technology
6. Chemical Engineering Technology
7. Industrial Engineering Technology
8. Mechanical Engineering Technology
9. Biomedical Engineering Technology

14.2 GRAPHICAL REPRESENTATION OF COMPLEX NUMBERS

Recall from Section 1.1 that real numbers may be represented graphically as points on a line. Since a complex number is composed of both a real part and a pure imaginary part, two axes are used to give a graphical interpretation. If a rectangular coordinate system is used, a complex number is represented by using the horizontal axis (x-axis) to locate the real part and the vertical axis (y-axis) to locate the pure imaginary part. When used to represent complex numbers, the coordinate plane is called the **complex plane.** The horizontal axis is called the **real axis** and the vertical axis is called the **imaginary axis.**

complex plane
real axis
imaginary axis

GRAPHICAL REPRESENTATION OF A COMPLEX NUMBER

The graphical representation of the complex number $a + bj$ is the point (a, b) of a rectangular coordinate system.

EXAMPLE 1 Locate each complex number in the complex plane.

(a) $2 + 3j$ **(b)** $-1 + 2j$

Solution **(a)** See Figure 14.1 **(b)** See Figure 14.2. ■

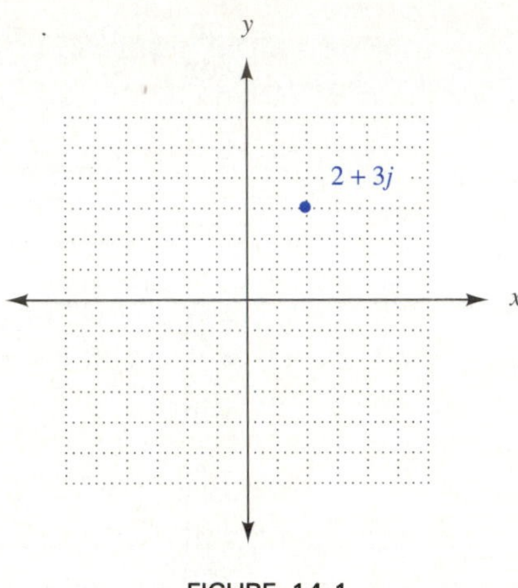

FIGURE 14.1

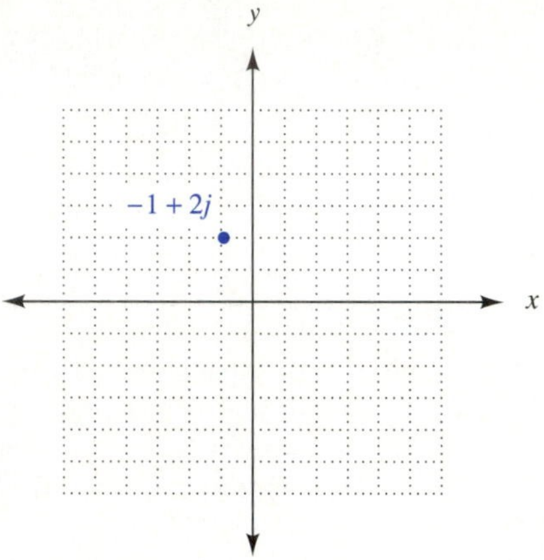

FIGURE 14.2

EXAMPLE 2 Locate each complex number in the complex plane.

(a) 2 **(b)** $-j$

Solution **(a)** See Figure 14.3 **(b)** See Figure 14.4. ■

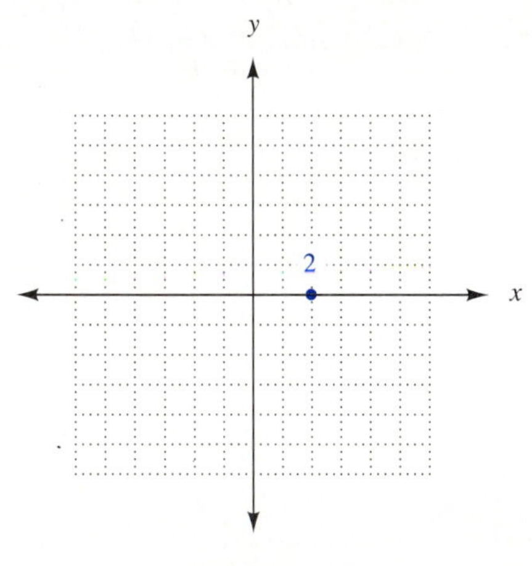

FIGURE 14.3

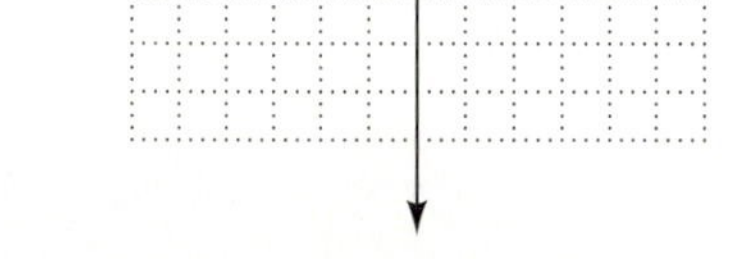

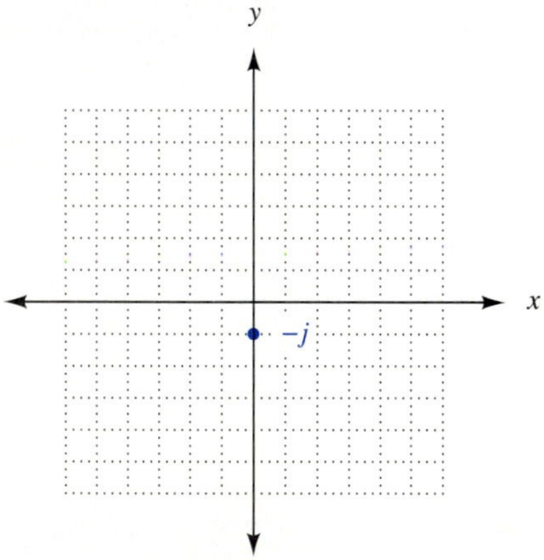

FIGURE 14.4

In some applications in the technologies, it is convenient to think of a complex number as a *vector*.

DEFINITION A **vector** is a quantity that has both magnitude and direction.

As a vector, a complex number $a + bj$ is represented as an arrow from the origin to the point (a, b). The length of the arrow is an indication of magnitude and the angle that it makes with the positive real axis is an indication of its direction. When two complex numbers are added, the real parts are added to obtain the real part of the sum, and the pure imaginary parts are added to obtain the pure imaginary part of the sum. But the vector sum may also be found graphically. The interpretation of vector addition may be stated specifically for complex numbers as follows.

To add two complex numbers graphically:

1. Represent each complex number $a + bj$ by drawing an arrow from $(0, 0)$ to (a, b).
2. Draw a parallelogram with the two arrows from step 1 as adjacent sides of the parallelogram. Draw the other two sides with dashed lines.
3. Draw the diagonal from $(0, 0)$ to the opposite vertex to represent the sum.

EXAMPLE 3 Add each pair of complex numbers graphically.

(a) $(4 - 3j) + (-2 - j)$ **(b)** $(-3 + 2j) + (1 - 3j)$

Solution **(a)** See Figure 14.5. The sum is $2 - 4j$, which agrees with the sum obtained by adding the numbers like binomials.

(b) See Figure 14.6. The sum is $-2 - j$. Once again, the sum is the expected result. ■

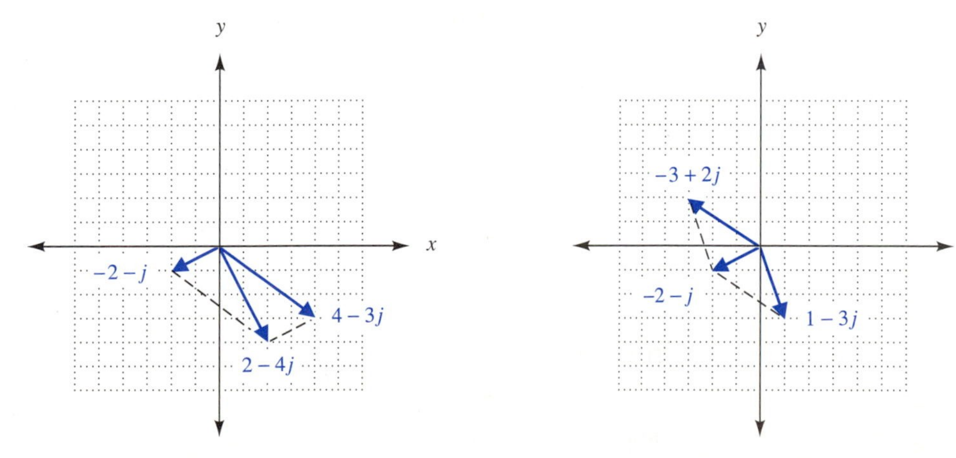

FIGURE 14.5 FIGURE 14.6

EXAMPLE 4 Add each pair of complex numbers graphically.

(a) $(2 + 3j) + (2 - 3j)$ **(b)** $(-3 + j) + (-3 - j)$

Solution **(a)** See Figure 14.7. The sum is 4.

(b) See Figure 14.8. The sum is -6. ■

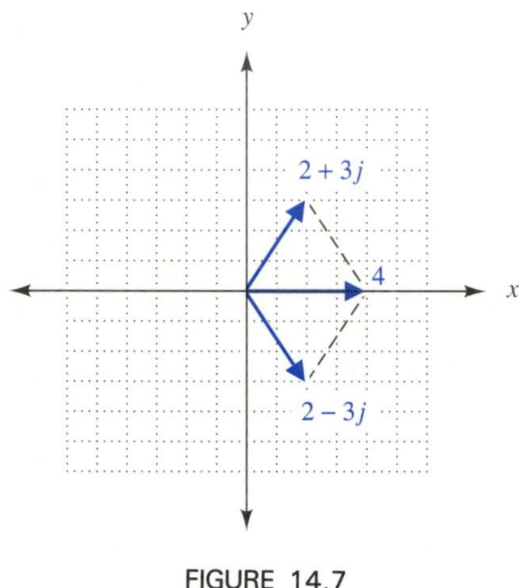

FIGURE 14.7

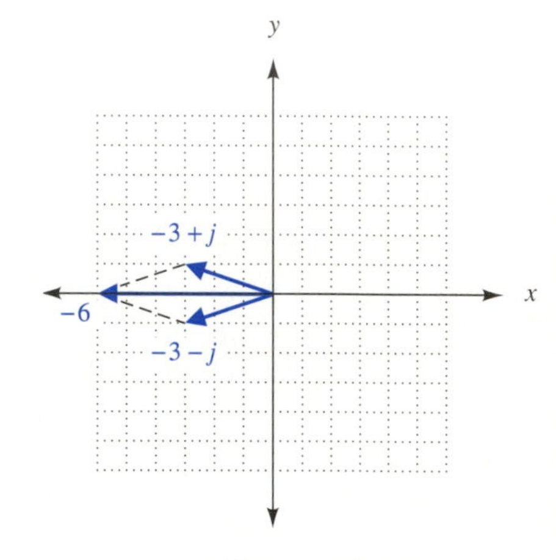

FIGURE 14.8

Example 4 illustrates the fact that the sum of a complex number and its conjugate is a real number. This result is to be expected, since $(a + bj) + (a - bj) = 2a$.

Subtraction of complex numbers is accomplished graphically by using the principle that subtracting a number is equivalent to adding the number that is opposite in sign. If $c + dj$ is a complex number, subtract $(c + dj)$ by adding its opposite, $-(c + dj)$ or $-c - dj$.

To subtract two complex numbers graphically:

1. Write the subtraction as an equivalent addition:

$$(a + bj) - (c + dj) = (a + bj) + (-c - dj).$$

2. Represent $(a + bj)$ as an arrow from $(0, 0)$ to (a, b).
3. Represent $(-c - dj)$ as an arrow from $(0, 0)$ to $(-c, -d)$.
4. Add as complex numbers.

EXAMPLE 5 Perform the indicated operations.

(a) $(3 + 2j) - (3 - j)$ **(b)** $(-3 + j) - (2 + 3j)$

Solution **(a)** Add $(3 + 2j) + (-3 + j)$ as in Figure 14.9. The sum is $3j$.

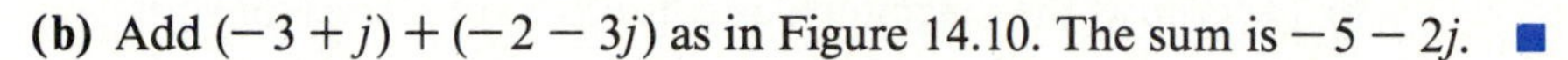
(b) Add $(-3 + j) + (-2 - 3j)$ as in Figure 14.10. The sum is $-5 - 2j$. ■

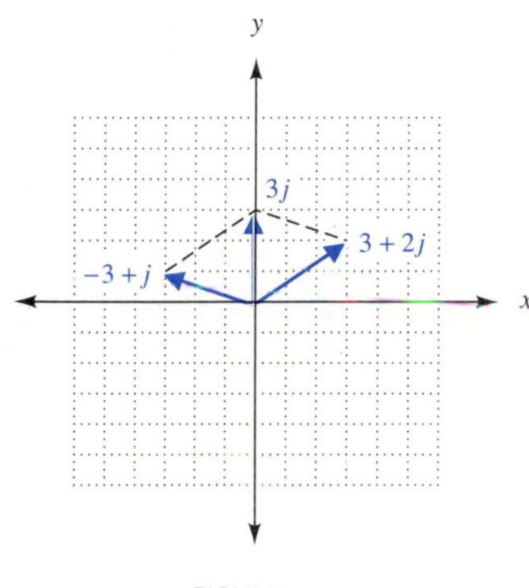

FIGURE 14.9

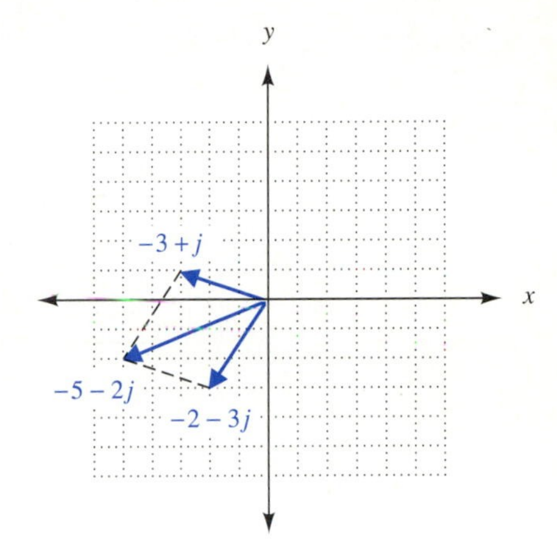

FIGURE 14.10

EXAMPLE 6 Perform the indicated operations graphically. $(2 - j) + (-1 + 2j) - (3 + j)$

Solution $(2 - j) + (-1 + 2j) = 1 + j$, as in Figure 14.11.
$(1 + j) - (3 + j) = (1 + j) + (-3 - j) = -2$, as in Figure 14.12. ■

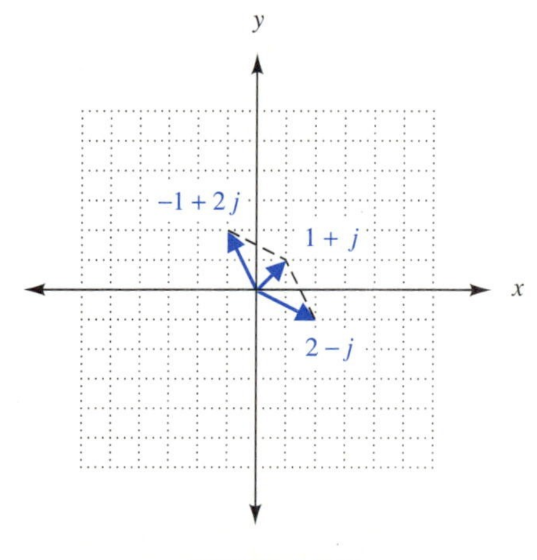

FIGURE 14.11

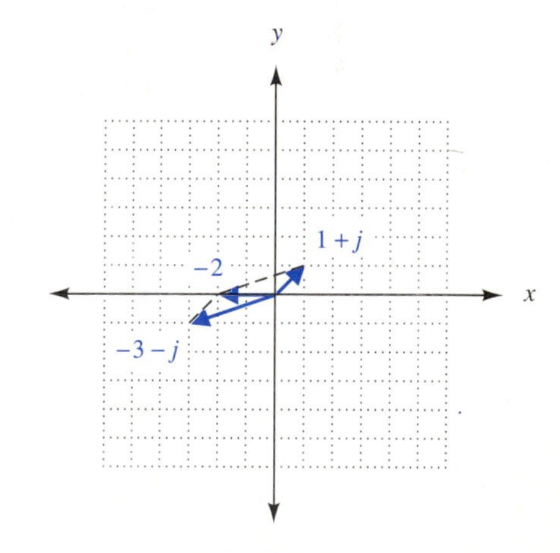

FIGURE 14.12

Of particular interest in the technologies are problems in which a real number and a pure imaginary number are added.

EXAMPLE 7 **(a)** Add a resistance of 3 Ω and a reactance of $j4$ Ω graphically.

(b) Determine the magnitude and direction of the vector sum.

Solution **(a)** See Figure 14.13. The sum is $3 + j4$.

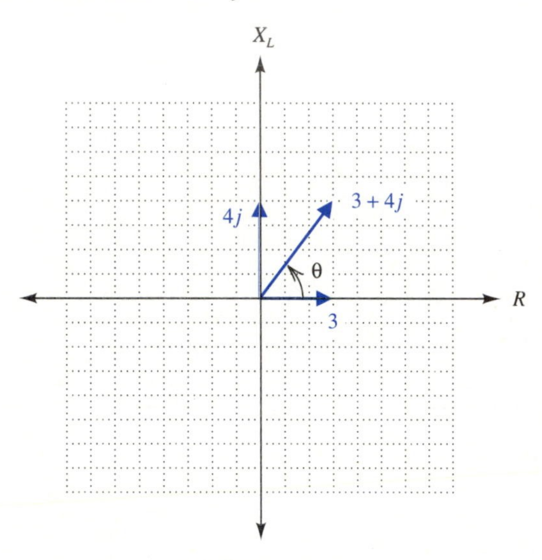

FIGURE 14.13

(b) The magnitude can be found by using the Pythagorean theorem.

$$a^2 + b^2 = c^2 \quad \text{or} \quad 3^2 + 4^2 = c^2$$
$$9 + 16 = c^2 \quad \text{so} \quad 25 = c^2 \text{ and } 5 = c$$

The direction is found by finding angle θ. Tan $\theta = 4/3$, so $\theta = 53.1°$. ■

Even though integers are often intended to be exact values rather than approximations, it may still be necessary to round either r or θ (or both). For such problems, a reasonable procedure is to round lengths to three significant digits and angles to the nearest tenth of a degree, as in Example 7.

It is also possible to illustrate graphically the multiplication of a complex number and a real number.

EXAMPLE 8 Illustrate $3(1 - 2j)$ graphically.

Solution $3(1 - 2j)$ is a vector in the same direction as $1 - 2j$, but three times as long and is indicated by laying three arrows end-to-end as in Figure 14.14. The product is $3 - 6j$. ■

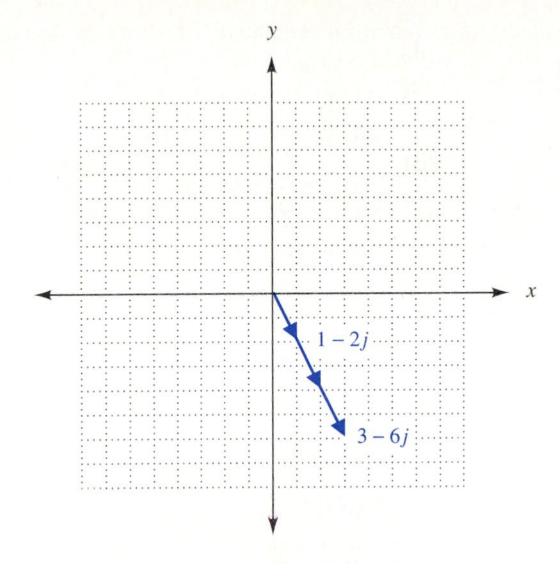

FIGURE 14.14

The number 3 in Example 8 is not considered as a complex number, but rather as a number by which $1 - 2j$ is magnified. A **scalar** is a quantity that has magnitude, but not direction. The type of multiplication illustrated in Example 8 is called scalar multiplication.

EXERCISES 14.2

In 1–7, plot each complex number on the complex plane.

1. $3 + 4j$
2. $2 - j$
3. $-2 + j$
4. $-3 - 4j$
5. $-1 - 2j$
6. $-1 + 2j$
7. $4 - 3j$

In 8–35, perform the indicated operations graphically.

8. $(2 - 3j) + (-3 + 2j)$
9. $(3 - j) + (3 + j)$
10. $(2 - 3j) + (2 + 3j)$
11. $(4 - j) + (-1 + 4j)$
12. $(0 + 2j) + (0 + 2j)$
13. $(3 + 0j) + (3 + 0j)$
14. $(1 + j) + (-1 - j)$
15. $(3 + 4j) - (4 - 3j)$
16. $(2 - 3j) - (-3 + 2j)$
17. $(2 - j) - (2 + j)$
18. $(0 + j) - (0 - j)$
19. $(3 + 0j) - (0 + 3j)$
20. $(1 + j) - (-2 + 3j)$
21. $(-2 + 3j) - (3 - j)$
22. $(1 + j) + (2 - j) + (3 + 2j)$
23. $(-2 + 3j) + (1 - 2j) + (1 + j)$
24. $(1 + 2j) + (-2 + j) - (1 + 0j)$
25. $(3 + j) + (-3 - j) - (0 + j)$
26. $(-2 + 3j) - (2 - 3j) + (1 - j)$
27. $(1 + 2j) - (2 - j) + (3 + j)$
28. $(-3 + 2j) - (-3 - 2j) - (2 + 3j)$
29. $3(1 - j)$
30. $2(-1 + 2j)$
31. $3(2 + j)$
32. $-2(3 - j)$
33. $-1(2 - 3j)$
34. $2(-2 + 3j)$
35. $3(0 - j)$

In 36–42, add graphically, then determine the magnitude and direction of the vector sum. Specify angles to the nearest tenth of a degree.

36. $(2 + j) + (1 + 3j)$
37. $(3 + 2j) + (0 + 2j)$
38. $(-1 + 2j) + (4 + 2j)$
39. $(-1 + j) + (4 + 3j)$
40. $(4 + 2j) + (1 + 10j)$
41. $(5 + 5j) + (0 + 7j)$
42. $(-5 + 5j) + (10 + 7j)$

In 43–46, illustrate the given quantities and the sum graphically.

43. a resistance of 3.00 Ω and a capacitive reactance of $-j4.00$ Ω 3 5 9 1 2 4 6 7 8

44. a resistance of 8.00 Ω and a capacitive reactance of $-j6.00$ Ω 3 5 9 1 2 4 6 7 8

45. a resistance of 12.00 Ω and an inductive reactance of $j5.00$ Ω 3 5 9 1 2 4 6 7 8

46. a resistance of 7.5 Ω and an inductive reactance of $j4.00$ Ω 3 5 9 1 2 4 6 7 8

In 47–49, illustrate the given quantities and the sum graphically.

47. an inductive reactance of $j6.00$ Ω and a capacitive reactance of $-j8.00$ Ω 3 5 9 4 7 8

48. an inductive reactance of $j5.00$ Ω and a capacitive reactance of $-j7.50$ Ω 3 5 9 4 7 8

49. an inductive reactance of $j5.00$ Ω and a capacitive reactance of $-j5.00$ Ω 3 5 9 4 7 8

50. Consider the complex number $a + bj$. If its opposite and its conjugate are graphed,
(a) about what axis is the conjugate a mirror image of the number?
(b) about what axis is the opposite a mirror image of the conjugate?

---

1. Architectural Technology
2. Civil Engineering Technology
3. Computer Engineering Technology
4. Mechanical Drafting Technology
5. Electrical/Electronics Engineering Technology
6. Chemical Engineering Technology
7. Industrial Engineering Technology
8. Mechanical Engineering Technology
9. Biomedical Engineering Technology

14.3 POLAR AND EXPONENTIAL FORM

rectangular form

A complex number can be written using several different notations. A complex number in the form $x + yj$ is said to be in **rectangular form.** Since we will examine equality of forms, it is important to know that two complex numbers are considered to be equal if their real parts are equal and their pure imaginary parts are equal. When a complex number is considered as a vector and represented as in Figure 14.15, the vector has magnitude determined by $r = \sqrt{x^2 + y^2}$ and direction θ determined by $\tan \theta = y/x$. Since $\sin \theta = y/r$ and $\cos \theta = x/r$, it is also true that $y = r \sin \theta$ and $x = r \cos \theta$. Thus,

$$x + yj = r \cos \theta + j(r \sin \theta) = r(\cos \theta + j \sin \theta).$$

DEFINITION

The **polar form** of a complex number $x + yj$ is

$$r(\cos \theta + j \sin \theta) \text{ where}$$

$r = \sqrt{x^2 + y^2}$ and $\tan \theta = y/x$. The value r is called the **modulus** and the angle θ is called the **argument** of the complex number.

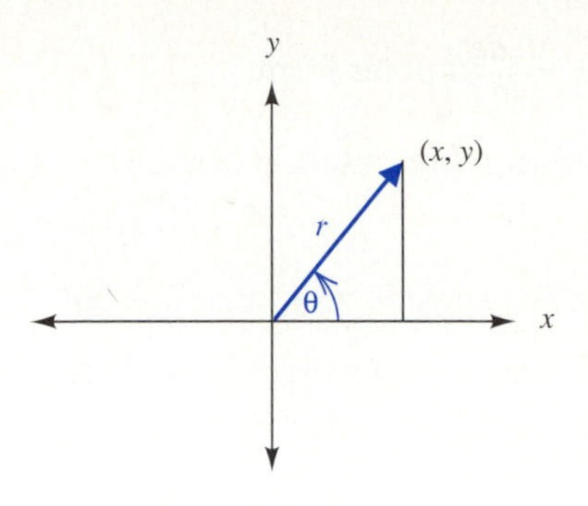

FIGURE 14.15

EXAMPLE 1 Express $-5.0 + 12j$ in polar form.

Solution The graphical representation is shown in Figure 14.16.

$$r = \sqrt{(-5.0)^2 + 12^2} = 13 \text{ and } \tan\theta = 12/-5$$

The reference angle is 67°, and θ is 113°. Thus,

$$-5 + 12j = 13(\cos 113° + j \sin 113°).$$ ■

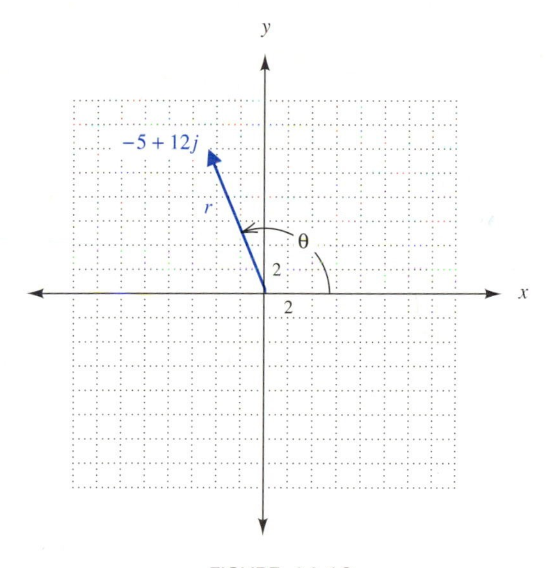

FIGURE 14.16

In Example 1, lengths were specified to two significant digits, so the angle was rounded to the nearest degree. Recall that when integers are intended to be exact values, lengths may be rounded to three significant digits and angles to the nearest tenth of a degree.

EXAMPLE 2 Express $3 - 4j$ in polar form.

Solution The graphical representation is shown in Figure 14.17.

$$r = \sqrt{3^2 + (-4)^2} = 5 \text{ and } \tan\theta = -4/3$$

The reference angle is 53.1° and θ is 306.9°. Thus,

$$3 - 4j = 5(\cos 306.9° + j \sin 306.9°).$$ ■

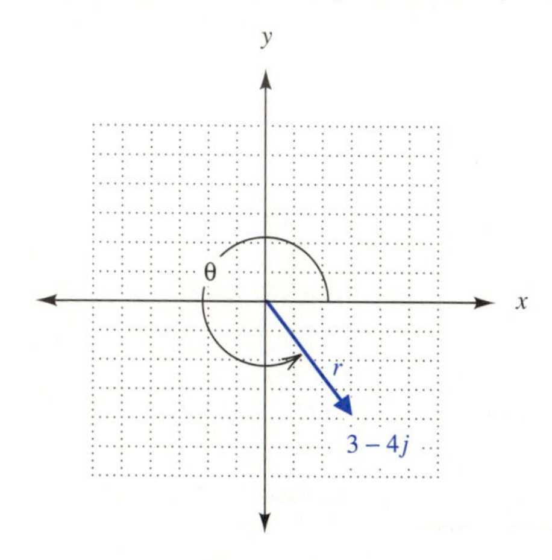

FIGURE 14.17

EXAMPLE 3 Express each number in polar form.

(a) 3 **(b)** $-4j$

Solution **(a)** From Figure 14.18, we have $r = x = 3$ and $\theta = 0°$, so

$$3 = 3(\cos 0° + j \sin 0°).$$

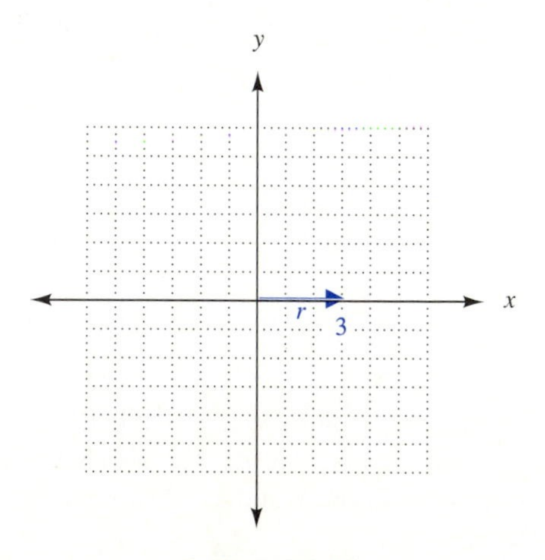

FIGURE 14.18

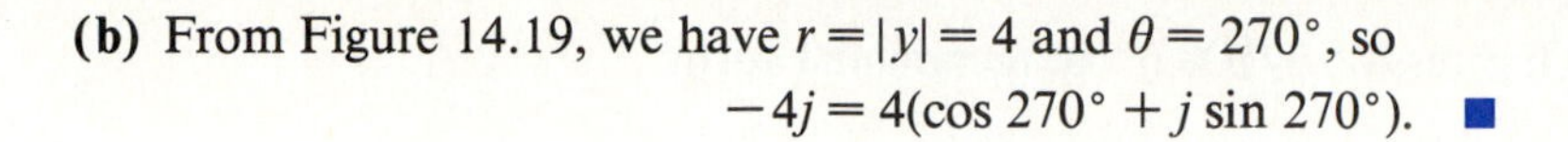

(b) From Figure 14.19, we have $r = |y| = 4$ and $\theta = 270°$, so

$$-4j = 4(\cos 270° + j \sin 270°). \quad ■$$

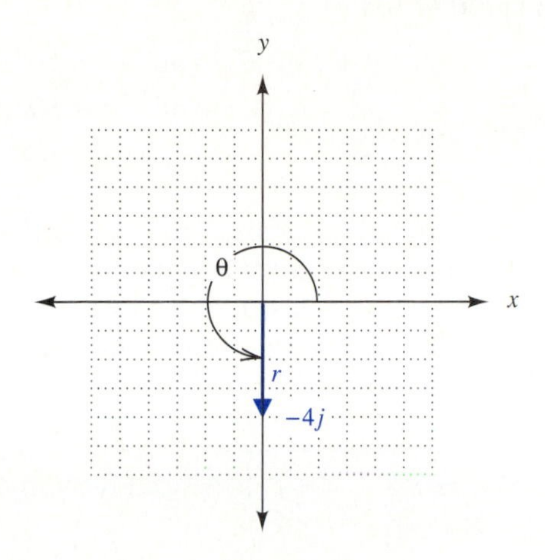

FIGURE 14.19

It is sometimes necessary to convert a complex number from polar form to rectangular form.

EXAMPLE 4 Express $33.22(\cos 6.44° + j \sin 6.44°)$ in rectangular form.

Solution

$$\begin{aligned} x + yj &= 33.22(\cos 6.44° + j \sin 6.44°) \\ &= 33.22 \cos 6.44° + 33.22j \sin 6.44° \\ &= 33.01 + 3.73j \quad ■ \end{aligned}$$

The notations $r \text{ cis } \theta$ or $r\underline{/\theta}$ are sometimes used as abbreviations for the polar form of a complex number.

EXAMPLE 5 Express $3.12\underline{/12.7°}$ in rectangular form.

Solution $r = 3.12$ and $\theta = 12.7°$

$$\begin{aligned} r(\cos \theta + j \sin \theta) &= 3.12(\cos 12.7° + j \sin 12.7°) \\ &= 3.12 \cos 12.7° + 3.12j \sin 12.7° \\ &= 3.04 + 0.686j \quad ■ \end{aligned}$$

EXAMPLE 6 Express $46.3\underline{/108.6^\circ}$ in rectangular form.

Solution $r = 46.3$ and $\theta = 108.6^\circ$

$$\begin{aligned} r(\cos\theta + j\sin\theta) &= 46.3(\cos 108.6^\circ + j\sin 108.6^\circ) \\ &= 46.3\cos 108.6^\circ + 46.3j\sin 108.6^\circ \\ &= -14.8 + 43.9j \end{aligned}$$ ■

If θ is given in radians, r and θ may be used to express a complex number in what is known as exponential form. Exponential form involves the number e, which is an irrational number like π. Just as you may have used π without knowing how the value 3.141 is obtained, for the present, you will need to accept the value of e to be approximately 2.718.

> **DEFINITION** The **exponential form** of a complex number $x + yj$ is $re^{j\theta}$ where
>
> $$r = \sqrt{x^2 + y^2} \text{ and } \tan\theta = y/x \text{ with } \theta \text{ in radians.}$$

EXAMPLE 7 Express $2.00\underline{/120^\circ}$ in exponential form.

Solution $r = 2.00$ and $\theta = 120^\circ$
To change 120° to radians, multiply by $\pi/180^\circ$.

$$120^\circ(\pi/180^\circ) = 2\pi/3$$

Thus, $2.00\underline{/120^\circ} = 2.00\ e^{j2\pi/3}$. ■

EXAMPLE 8 Express $-8 - 15j$ in exponential form.

Solution $x = -8$ and $y = -15$
Assuming that -8 and -15 are exact values, $r = \sqrt{(-8)^2 + (-15)^2}$ or 17. Tan θ is 15/8, so using a calculator *in radian mode,* the reference angle is 1.08. The angle is a third quadrant angle, so θ is $\pi + 1.08$ or 4.22. We have $-8 - 15j = 17e^{4.22j}$. ■

EXAMPLE 9 Express $4j$ in exponential form.

Solution $x = 0$ and $y = 4$, so $r = 4$ and $\theta = \pi/2$.

$$4 = 4e^{j\pi/2}$$ ■

Conversions from one form to another are often required in technical work.

EXAMPLE 10 Express a current of $2.71\ e^{j4.66}$ A in polar form.

Solution $r = 2.71$ and $\theta = 4.66$
To change 4.66 radians to degrees, multiply by $180°/\pi$.

$$4.66(180°/\pi) = 267.0°$$
$$2.71\ e^{j4.66}\ \text{A} = 2.71(\cos 267.0° + j \sin 267.0°)\ \text{A}$$ ■

EXAMPLE 11 Express $3e^{j3.6}$ V in rectangular form.

Solution $r = 3$ and $\theta = 3.6$

$$3e^{j3.6}\ \text{V} = 3\ (\cos 3.6 + j \sin 3.6)\ \text{V}$$
$$= 3 \cos 3.6 + j3 \sin 3.6\ \text{V}$$

Using a calculator *in radian mode* to evaluate the expression,

$$3e^{j3.6}\ \text{V} = -2.7 - j1.3\ \text{V}.$$ ■

EXERCISES 14.3

For these problems, assume that integers are exact values. In 1–10, express each complex number in polar form.

1. $-6 + 8j$
2. $12 - 5j$
3. $-1.62 + 2.53j$
4. $3.44 - 0.71j$
5. $4.35 - 9.82j$
6. -2
7. $3j$
8. $-5j$
9. 4
10. 0

In 11–20, express each complex number in rectangular form.

11. $3\angle 45°$
12. $5\angle 120°$
13. $14.3\angle 17.2°$
14. $2.55\angle 96.3°$
15. $22.3\angle 105.8°$
16. $7(\cos 30° + j \sin 30°)$
17. $4(\cos 150° + j \sin 150°)$
18. $6.42(\cos 23.1° + j \sin 23.1°)$
19. $5.37(\cos 143.2° + j \sin 143.2°)$
20. $4.86(\cos 312.7° + j \sin 312.7°)$

In 21–30, express each number in exponential form.

21. $7\angle 30°$
22. $9\angle 60°$
23. $12.6\angle 54.6°$
24. $2.73\angle 76.8°$
25. $145\angle 119.3°$
26. $17.4(\cos 37.8° + j \sin 37.8°)$
27. $8.32(\cos 110.6° + j \sin 110.6°)$
28. $125(\cos 214.6° + j \sin 214.6°)$
29. $6(\cos 30° + j \sin 30°)$
30. $12(\cos 135° + j \sin 135°)$

In 31–36, express each number in polar form.

31. $4e^{j\pi}$
32. $5e^{j\pi/3}$
33. $15e^{j2\pi}$
34. $4.11e^{5.23j}$
35. $9.27e^{4.68j}$
36. $86.3e^{2.85j}$

In 37–42, express each number in rectangular form.

37. $3e^{j\pi/2}$
38. $7e^{j3\pi/4}$
39. $22e^{j\pi/6}$
40. $3.07e^{0.322j}$
41. $8.73e^{0.576j}$
42. $5.76e^{1.29j}$

In 43–49, convert each quantity to a complex number in the indicated form.

43. $9.87 + j3.44$ V to polar form 3 5 9 1 2 4 6 7 8

44. $4.2 - j5.7$ kΩ to polar form 3 5 9 1 2 4 6 7 8

45. $0.22\underline{/-20.3°}$ A to rectangular form 3 5 9 1 2 4 6 7 8

46. $0.753\underline{/5.45°}$ A to rectangular form 3 5 9 1 2 4 6 7

47. $10.8\underline{/23.7°}$ Ω to rectangular form 3 5 9 1 2 4 6 7 8

48. $2.37 + j14.6$ Ω to exponential form 3 5 9 1 2 4 6 7 8

49. $0.57 - j0.87$ mV to exponential form 3 5 9 1 2 4 6 7 8

50. Which of the arithmetic operations (addition, subtraction, multiplication, and division) are easier to perform
(a) with complex numbers in rectangular form?
(b) with complex numbers in exponential form?

1. Architectural Technology
2. Civil Engineering Technology
3. Computer Engineering Technology
4. Mechanical Drafting Technology
5. Electrical/Electronics Engineering Technology
6. Chemical Engineering Technology
7. Industrial Engineering Technology
8. Mechanical Engineering Technology
9. Biomedical Engineering Technology

14.4 DEMOIVRE'S THEOREM

Although complex numbers may be written in rectangular, polar, or exponential form, each has advantages and disadvantages. Numbers written in rectangular form are easy to add and subtract, but multiplication and division may be cumbersome.

Numbers written in exponential form have the advantage of being easy to manipulate, because the exponent rules may be used. Exponential form is often used to obtain generalized or theoretical results. For example, exponential form may be used to show that multiplication, division, powers, and roots are easily done when the numbers are written in polar form. We begin by examining multiplication and division.

Consider complex numbers in exponential form: $r_1 e^{j\theta_1}$ and $r_2 e^{j\theta_2}$. Because the exponent rules apply,

$$(r_1 e^{j\theta_1})(r_2 e^{j\theta_2}) = r_1 r_2 e^{j\theta_1 + j\theta_2} = r_1 r_2 e^{j(\theta_1 + \theta_2)}.$$

In polar form, this result is expressed as follows.

$$r_1(\cos\theta_1 + j\sin\theta_1)r_2(\cos\theta_2 + j\sin\theta_2)$$
$$= r_1 r_2[\cos(\theta_1 + \theta_2) + j\sin(\theta_1 + \theta_2)] \text{ or}$$
$$(r_1\underline{/\theta_1})(r_2\underline{/\theta_2}) = r_1 r_2\underline{/(\theta_1 + \theta_2)}$$

Likewise

$$\frac{r_1 e^{j\theta_1}}{r_2 e^{j\theta_2}} = \left(\frac{r_1}{r_2}\right) e^{j\theta_1 - j\theta_2} = \left(\frac{r_1}{r_2}\right) e^{j(\theta_1 - \theta_2)}, \text{ and}$$

$$\frac{r_1(\cos\theta_1 + j\sin\theta_1)}{r_2(\cos\theta_2 + j\sin\theta_2)} = \frac{r_1}{r_2}[\cos(\theta_1 - \theta_2) + j\sin(\theta_1 - \theta_2)], \text{ or}$$

$$\frac{r_1 \underline{/\theta_1}}{r_2 \underline{/\theta_2}} = \frac{r_1}{r_2} \underline{/(\theta_1 - \theta_2)}.$$

These results are summarized as follows.

ARITHMETIC WITH COMPLEX NUMBERS IN POLAR FORM

If $r_1(\cos\theta_1 + j\sin\theta_1)$ and $r_2(\cos\theta_2 + j\sin\theta_2)$ are two complex numbers such that r_2 is not 0, then

$$r_1(\cos\theta_1 + j\sin\theta_1) r_2(\cos\theta_2 + j\sin\theta_2) = r_1 r_2[\cos(\theta_1 + \theta_2) + j\sin(\theta_1 + \theta_2)] \quad \textbf{(14-2)}$$

and

$$\frac{r_1(\cos\theta_1 + j\sin\theta_1)}{r_2(\cos\theta_2 + j\sin\theta_2)} = \left(\frac{r_1}{r_2}\right)[\cos(\theta_1 - \theta_2) + j\sin(\theta_1 - \theta_2)]. \quad \textbf{(14-3)}$$

EXAMPLE 1 Perform the indicated operation.

$$3.24(\cos 65.1^\circ + j\sin 65.1^\circ)\, 4.35(\cos 46.2^\circ + j\sin 46.2^\circ)$$

Solution Using Equation 14-2, we have

$$3.24(\cos 65.1^\circ + j\sin 65.1^\circ)\, 4.35(\cos 46.2^\circ + j\sin 46.2^\circ) = 3.24(4.35)[\cos(65.1^\circ + 46.2^\circ) + j\sin(65.1^\circ + 46.2^\circ)].$$

A calculator is used to simplify the expression. The answer is given in the same form as the numbers in the problem.

$$14.1(\cos 111.3^\circ + j\sin 111.3^\circ) \quad \blacksquare$$

EXAMPLE 2 Perform the indicated operation.

$$14.6\underline{/22.1^\circ}(23.9\underline{/85.7^\circ})$$

Solution

$$14.6\underline{/22.1^\circ}(23.9\underline{/85.7^\circ}) = 14.6(23.9)\underline{/(22.1^\circ + 85.7^\circ)} = 349\underline{/107.8^\circ} \quad \blacksquare$$

EXAMPLE 3 Perform the indicated operation $\dfrac{3.24(\cos 65.1^\circ + j \sin 65.1^\circ)}{4.35(\cos 46.2^\circ + j \sin 46.2^\circ)}$.

Solution Using Equation 14-3, we have

$$\begin{aligned}\frac{3.24(\cos 65.1^\circ + j \sin 65.1^\circ)}{4.35(\cos 46.2^\circ + j \sin 46.2^\circ)} &= \frac{3.24}{4.35}[\cos(65.1^\circ - 46.2^\circ) + j \sin(65.1^\circ - 46.2^\circ)] \\ &= 0.745(\cos 18.9^\circ + j \sin 18.9^\circ). \quad \blacksquare\end{aligned}$$

EXAMPLE 4 Perform the indicated operation $\dfrac{14.6\underline{/22.1^\circ}}{23.9\underline{/85.7^\circ}}$.

Solution

$$\frac{14.6\underline{/22.1^\circ}}{23.9\underline{/85.7^\circ}} = \left(\frac{14.6}{23.9}\right)\underline{/(22.1^\circ - 85.7^\circ)} = 0.611\underline{/-63.6^\circ} \quad \blacksquare$$

EXAMPLE 5 Use Ohm's law to calculate the current that would flow if 240 V were applied across a circuit with an impedance of $8.6 + j6.8\ \Omega$.

Solution Ohm's law says $\mathbf{V} = \mathbf{IZ}$. That is, voltage is the product of current and impedance. Solving for $\mathbf{I}$, we have $\mathbf{I} = \mathbf{V}/\mathbf{Z}$. When $\mathbf{V}$ is 240 and $\mathbf{Z}$ is $8.6 + j6.8$, we have

$$\mathbf{I} = \frac{240}{8.6 + j6.8}$$

We will divide, using the polar form of the numbers. When x is 8.6 and y is 6.8, we have

$$\sqrt{8.6^2 + 6.8^2} = 11 \text{ and } \tan\theta = 6.8/8.6.$$

That is, r is 11 and θ is 38.3°. Thus, $(8.6 + 6.8j) = 11(\cos 38.3^\circ + j \sin 38.3^\circ)$. Also, $240 = 240(\cos 0^\circ + j \sin 0^\circ)$.

$$\begin{aligned}\mathbf{I} = \frac{240(\cos 0^\circ + j \sin 0^\circ)}{11(\cos 38.3^\circ + j \sin 38.3^\circ)} &= \frac{240}{11}(\cos - 38.3^\circ + j \sin - 38.3^\circ) \\ &= 17 - j14 \text{ A} \quad \blacksquare\end{aligned}$$

To develop a formula for the nth power of a complex number, consider

$$(re^{j\theta})^n = r^n e^{jn\theta}.$$

This equation is stated in polar form as

$$r(\cos\theta + j \sin\theta)^n = r^n(\cos n\theta + j \sin n\theta), \text{ or}$$
$$(r\underline{/\theta})^n = r^n\underline{/n\theta}.$$

This result is known as DeMoivre's theorem.

DEMOIVRE'S THEOREM

If $r(\cos\theta + j\sin\theta)$ is a complex number, and n is a positive integer, then

$$[r(\cos\theta + j\sin\theta)]^n = r^n(\cos n\theta + j\sin n\theta). \quad \textbf{(14-4)}$$

EXAMPLE 6 Perform the indicated operation.

$$[4(\cos 30° + j\sin 30°)]^2$$

Solution Using DeMoivre's theorem, we have

$$\begin{aligned}[4(\cos 30° + j\sin 30°)]^2 &= 4^2\{\cos[2(30°)] + j\sin[2(30°)]\}\\ &= 16(\cos 60° + j\sin 60°). \quad ■\end{aligned}$$

EXAMPLE 7 Perform the indicated operation.

$$(3-4j)^6$$

Solution Although six factors of $(3-4j)$ could be multiplied in rectangular form, the multiplication would be cumbersome. Using polar form, we have,

$$r = \sqrt{3^2 + (-4)^2} = 5 \text{ and } \tan\theta = -4/3.$$

The reference angle is 53.1°, and since the angle is a fourth quadrant angle, θ is 306.9°.

$$\begin{aligned}(3-4j) &= [5(\cos 306.9° + j\sin 306.9°)]^6\\ &= 5^6\{\cos[6(306.9°)] + j\sin[6(306.9°)]\}\\ &= 15{,}625(\cos 1841.4° + j\sin 1841.4°)\end{aligned}$$

Since the problem was in rectangular form, we express the answer in rectangular form also. Using a calculator and three significant digits, we have

$$(3-4j)^6 = 11{,}700 + 10{,}300j. \quad ■$$

In Example 7, angles were rounded to the nearest tenth of a degree. Since 306.9° is actually an intermediate value and must be multiplied by 6, it would be better to use the following calculator keystroke sequence, in which the value of θ is used before it is rounded.

4 [÷] 3 [=] [INV] [tan] [STO] *53.130102*

360 [−] [RCL] [=] [×] 6 [=] [STO] *1841.2194*

[cos] [×] 5 [y^x] 6 [=] *11753.001*

[RCL] [sin] [×] 5 [y^x] 6 [=] *10295.999*

Thus, to three significant digits, $(3-4j)^6 = 11{,}800 + 10{,}300j$.

Although it is easy to multiply, divide, and raise to a power using complex numbers in polar form, it is generally easier to add or subtract complex numbers if they are in rectangular form.

EXAMPLE 8 Perform the indicated operation.

$$6(\cos 45^\circ + j \sin 45^\circ) + 3(\cos 30^\circ + j \sin 30^\circ)$$

Solution Express the complex numbers in rectangular form.

$$\begin{aligned} &6(\cos 45^\circ + j \sin 45^\circ) + 3(\cos 30^\circ + j \sin 30^\circ) \\ &\quad = 6 \cos 45^\circ + 6j \sin 45^\circ + 3 \cos 30^\circ + 3j \sin 30^\circ \\ &\quad = 4.24 + 4.24j + 2.60 + 1.50j \\ &\quad = 6.84 + 5.74j \end{aligned}$$

Intermediate results were rounded for the purpose of illustration. In practice, however, $6 \cos 45^\circ + 3 \sin 30^\circ$ would be added on a calculator before rounding the real part to 6.84. Likewise, $6 \sin 45^\circ + 3 \sin 30^\circ$ would be added before rounding the imaginary part to 5.74. Since the problem was in polar form, we express the answer in polar form.

$$\sqrt{(6.84)^2 + (5.74)^2} = 8.93 \text{ and } \tan \theta = 5.74/6.84$$

That is, r is 8.93 and θ is 40.0°. Thus,

$$\begin{aligned} &6(\cos 45^\circ + j \sin 45^\circ) + 3(\cos 30^\circ + j \sin 30^\circ) \\ &\quad = 8.93(\cos 40.0^\circ + j \sin 40.0^\circ). \end{aligned}$$ ■

The last operation we will examine is finding the nth root of a complex number. DeMoivre's theorem, which was stated for exponents that are positive integers, can be extended so that roots of complex numbers may be computed.

EXTENSION OF DEMOIVRE'S THEOREM

If $r(\cos \theta + j \sin \theta)$ is a complex number and n is a positive integer, then

$$[r(\cos \theta + j \sin \theta)]^{1/n} = \sqrt[n]{r}\left[\cos\left(\frac{\theta + k360^\circ}{n}\right) + j \sin\left(\frac{\theta + k360^\circ}{n}\right)\right]$$

where $k = 0, 1, 2, \ldots, n - 1$.

Notice that $\sqrt[n]{r}$ will always be a positive real number, since r is positive.

EXAMPLE 9 Find the fourth roots of -1.

Solution The number -1 can be written as $-1 + 0j$. The graph in Figure 14.20 shows that r is 1 and θ is 180°. Thus, -1 can be written in polar form as $1(\cos 180° + j \sin 180°)$.

To apply the extension of DeMoivre's theorem, notice that n is 4 and $\sqrt[4]{1}$ is 1. Let r_0, r_1, r_2, and r_3 denote the four fourth roots of -1.

$$\text{when } k \text{ is } 0,\ r_0 = 1\left[\cos\left(\frac{180° + 0(360°)}{4}\right) + j\sin\left(\frac{180° + 0(360°)}{4}\right)\right]$$

$$\text{when } k \text{ is } 1,\ r_1 = 1\left[\cos\left(\frac{180° + 1(360°)}{4}\right) + j\sin\left(\frac{180° + 1(360°)}{4}\right)\right]$$

$$\text{when } k \text{ is } 2,\ r_2 = 1\left[\cos\left(\frac{180° + 2(360°)}{4}\right) + j\sin\left(\frac{180° + 2(360°)}{4}\right)\right]$$

$$\text{when } k \text{ is } 3,\ r_3 = 1\left[\cos\left(\frac{180° + 3(360°)}{4}\right) + j\sin\left(\frac{180° + 3(360°)}{4}\right)\right]$$

Using a calculator to evaluate each expression, we have the following.

$$r_0 = 0.707 + 0.707j$$
$$r_1 = -0.707 + 0.707j$$
$$r_2 = -0.707 - 0.707j$$
$$r_3 = 0.707 - 0.707j \quad \blacksquare$$

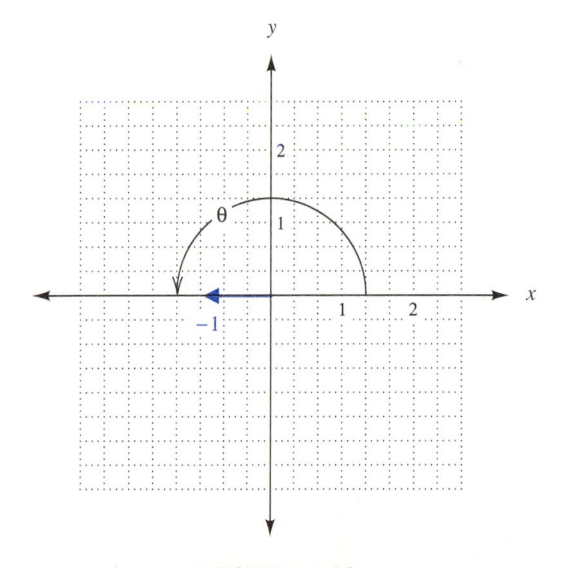

FIGURE 14.20

It is interesting to observe that when all of the nth roots of the complex number $r(\cos\theta + j\sin\theta)$ are graphed, the roots are equally spaced around a circle of radius $r^{1/n}$. Figure 14.21 shows that the four fourth roots of -1 are equally spaced around a circle with a radius of 1.

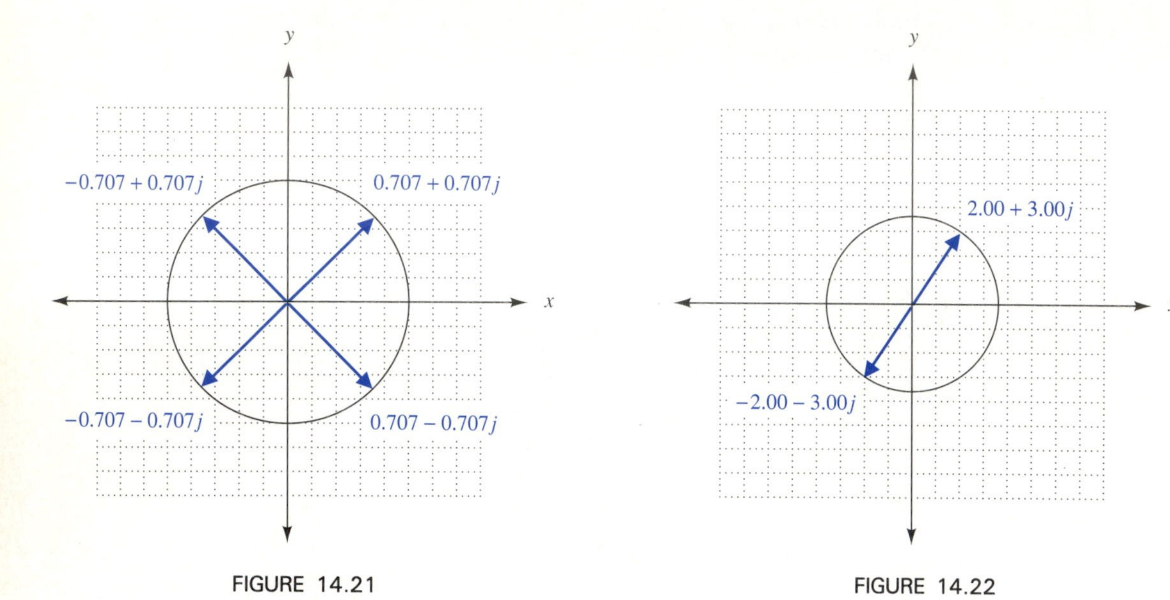

FIGURE 14.21

FIGURE 14.22

EXAMPLE 10 Find the two square roots of $-5 + 12j$.

Solution $\sqrt{(-5)^2 + 12^2} = 13$ and $\tan\theta = -12/5$

That is, r is 13 and θ is 112.6°, since θ is in the second quadrant. Let r_0 and r_1 denote the two square roots of $-5 + 12j$.

$$\text{when } k \text{ is } 0,\ r_0 = \sqrt{13}\left[\cos\left(\frac{112.6^\circ + 0(360^\circ)}{2}\right) + j\sin\left(\frac{112.6^\circ + 0(360^\circ)}{2}\right)\right]$$

$$\text{when } k = 1,\ r_1 = \sqrt{13}\left[\cos\left(\frac{112.6^\circ + 1(360^\circ)}{2}\right) + j\sin\left(\frac{112.6^\circ + 1(360^\circ)}{2}\right)\right]$$

Using a calculator, we have the following.

$$r_0 = 2.00 + 3.00j$$
$$r_1 = -2.00 - 3.00j$$

The graph in Figure 14.22 shows that the roots are equally spaced around a circle with a radius of $\sqrt{13}$. ■

EXERCISES 14.4

In 1–35, perform the indicated operations. In the final answer, angles should be less than 360° or 2π.

1. $4.31(\cos 42.7^\circ + j\sin 42.7^\circ)3.38(\cos 53.4^\circ + j\sin 53.4^\circ)$
2. $14.9(\cos 22.6^\circ + j\sin 22.6^\circ)13.6(\cos 22.6^\circ + j\sin 22.6^\circ)$
3. $22.4(\cos 123.2^\circ + j\sin 123.2^\circ)4.50(\cos 6.4^\circ + j\sin 6.4^\circ)$
4. $1.26(\cos 212.7^\circ + j\sin 212.7^\circ)2.55(\cos 156.5^\circ + j\sin 156.5^\circ)$
5. $0.58(\cos 18^\circ + j\sin 18^\circ)4.3(\cos 96^\circ + j\sin 96^\circ)$
6. $74\angle 22^\circ\ (1.6\angle 41^\circ)$
7. $26\angle 39^\circ\ (5.9\angle 54^\circ)$
8. $2.58\angle 33.7^\circ\ (1.96\angle 64.5^\circ)$
9. $3.49\angle 106.8^\circ\ (1.27\angle 88.3^\circ)$
10. $5.81\angle 214.6^\circ\ (2.13\angle 6.4^\circ)$
11. $\dfrac{4.31(\cos 42.7^\circ + j\sin 42.7^\circ)}{3.38(\cos 53.4^\circ + j\sin 53.4^\circ)}$
12. $\dfrac{14.9(\cos 22.6^\circ + j\sin 22.6^\circ)}{13.6(\cos 22.6^\circ + j\sin 22.6^\circ)}$
13. $\dfrac{22.4(\cos 123.2^\circ + j\sin 123.2^\circ)}{4.50(\cos 6.4^\circ + j\sin 6.4^\circ)}$
14. $\dfrac{1.26(\cos 212.7^\circ + j\sin 212.7^\circ)}{2.55(\cos 156.5^\circ + j\sin 156.5^\circ)}$
15. $\dfrac{0.58(\cos 18^\circ + j\sin 18^\circ)}{4.3(\cos 96^\circ + j\sin 96^\circ)}$
16. $\dfrac{74\angle 22^\circ}{1.6\angle 41^\circ}$
17. $\dfrac{26\angle 39^\circ}{5.9\angle 54^\circ}$
18. $\dfrac{2.58\angle 33.7^\circ}{1.96\angle 64.5^\circ}$
19. $\dfrac{3.49\angle 106.8^\circ}{1.27\angle 88.3^\circ}$
20. $\dfrac{5.81\angle 214.6^\circ}{2.13\angle 6.4^\circ}$
21. $[5.23(\cos 51.8^\circ + j\sin 51.8^\circ)]^3$
22. $[23.8(\cos 13.5^\circ + j\sin 13.5^\circ)]^3$
23. $[31.5(\cos 214.1^\circ + j\sin 214.1^\circ)]^4$
24. $[2.17(\cos 301.6^\circ + j\sin 301.6^\circ)]^4$
25. $[4.9(\cos 27^\circ + j\sin 27^\circ)]^5$
26. $(2 + 5j)^5$
27. $(-1 + 4j)^5$
28. $(3 - 2j)^4$
29. $(2 - 3j)^4$
30. $(-3 + j)^6$
31. $14.9(\cos 22.6^\circ + j\sin 22.6^\circ) + 13.6(\cos 22.6^\circ + j\sin 22.6^\circ)$
32. $1.26(\cos 212.7^\circ + j\sin 212.7^\circ) + 2.55(\cos 156.5^\circ + j\sin 156.5^\circ)$
33. $74\angle 22^\circ - 1.6\angle 41^\circ$
34. $2.58\angle 33.7^\circ - 1.96\angle 64.5^\circ$
35. $5.81\angle 214.6^\circ + 2.13\angle 6.4^\circ$

In 36–42, find the indicated roots.

36. square roots of $-j$
37. cube roots of -1
38. cube roots of 1
39. cube roots of $-3 + 4j$
40. square roots of $3 - 4j$
41. fourth roots of $-4 + 3j$
42. fourth roots of $4 - 3j$

Problems 43–48 use the following formulas, in which **V** *represents voltage,* **I** *represents current, and* **Z** *represents impedance.* $\mathbf{V} = \mathbf{IZ}$

For sections in series: $\mathbf{V} = \mathbf{V}_1 + \mathbf{V}_2$, $\mathbf{I} = \mathbf{I}_1 = \mathbf{I}_2$, *and* $\mathbf{Z} = \mathbf{Z}_1 + \mathbf{Z}_2$

For sections in parallel: $\mathbf{V} = \mathbf{V}_1 = \mathbf{V}_2$, $\mathbf{I} = \mathbf{I}_1 + \mathbf{I}_2$, *and* $1/\mathbf{Z} = 1/\mathbf{Z}_1 + 1/\mathbf{Z}_2$

43. Calculate the voltage that would occur in a circuit containing two sections in series if the individual voltages are $118\angle 10.0°$ V and $97.0\angle 15.0°$ V. 3 5 9 1 2 4 6 7 8

44. Calculate the total impedance that would occur in a circuit containing two sections in series if the individual impedances are $0.82\angle 39.6°$ Ω and $1.3\angle 24.3°$ Ω, respectively. 3 5 9 1 2 4 6 7 8

45. Calculate the voltage produced if $0.23 - j0.074$ A flow through a circuit with an impedance of $1.69 + j14.6$ Ω. 3 5 9 1 2 4 6 7 8

46. Calculate the voltage produced if $0.237 - j0.0641$ A flow through a circuit containing two sections in series if the individual impedances are $2.69 + j14.6$ Ω and $4.51 + j19.6$ Ω, respectively. 3 5 9 1 2 4 6 7 8

47. Calculate the current that would flow through a circuit containing two sections in parallel if the individual currents are $0.742 + j0.132$ A and $0.921 + j0.106$ A, respectively. 3 5 9 1 2 4 6 7 8

48. Calculate the impedance that would occur in a circuit containing two sections in parallel if the individual impedances are $19.3 - j22.6$ Ω and $12.8 - j24.1$ Ω, respectively. 3 5 9 1 2 4 6 7 8

1. Architectural Technology
2. Civil Engineering Technology
3. Computer Engineering Technology
4. Mechanical Drafting Technology
5. Electrical/Electronics Engineering Technology
6. Chemical Engineering Technology
7. Industrial Engineering Technology
8. Mechanical Engineering Technology
9. Biomedical Engineering Technology

14.5 APPLICATION: THE *J*-OPERATOR

An electric current occurs when electrons flow through a conductor, such as a wire. The force that moves the electrons is called a potential difference. Current is measured in amperes and the potential difference in volts.

An electric circuit may contain devices to impede the flow of electrons. Resistors, capacitors, and inductors are such devices. A resistor is used to intentionally create an obstruction to the flow of current. Capacitors and inductors serve other purposes, but when current flows, they build up a force that tends to counter or reduce the voltage force that moves the electrons and the current. A capacitor basically consists of two parallel plates. Current does not actually flow between the plates, but rather each plate stores electrons, allowing a charge to build up momentarily creating an electric field. A charge on one plate induces a charge on the other, so that current is not stopped. An inductor is a coil of wire wrapped around a core of insulating material. The voltage produced by an inductor depends on the number of loops or turns in the wire, the cross-sectional area of the coil, and the type of material inside the coil. Information about these devices is summarized as follows.

ELEMENTS OF AN ELECTRIC CIRCUIT

Device	Symbol	Diagram Symbol	Unit of Measurement
Resistor	R	—ᴧᴧᴧ—	Ohms (Ω)
Capacitor	C	—|(—	Farads (F)
Inductor	L	—ᴖᴖᴖ—	Henries (H)

resistance
reactance

The **resistance** is the opposition to the flow of electrons created by a resistor. The term **reactance** refers to the opposition to the flow of electrons created by a capacitor or inductor. A reactance is sometimes called an effective resistance since the effect is the same for both resistance and reactance, and both are measured in ohms.

The current and voltage in an electric circuit are affected not only by the resistance and reactance, but also by the *phase angle.* To develop an intuitive understanding of the phase angle, consider the equations for current and voltage. The magnitude of the current at any given time t is given by the equation

$$I = I_{max} \sin \omega t \qquad \textbf{(14-5)}$$

where I_{max} is the maximum value that the current attains, and ω is the constant angular velocity of a wire loop rotating through a magnetic field to induce the current. Likewise, the magnitude of the voltage at any given time t is given by the equation

$$V = V_{max} \sin (\omega t + \theta) \qquad \textbf{(14-6)}$$

phase angle

where V_{max} is the maximum value that the voltage attains, and θ is the **phase angle.** Since current and voltage both vary with time, they can be plotted on a rectangular coordinate system and compared for a given time as in Figures 14.23 and 14.24.

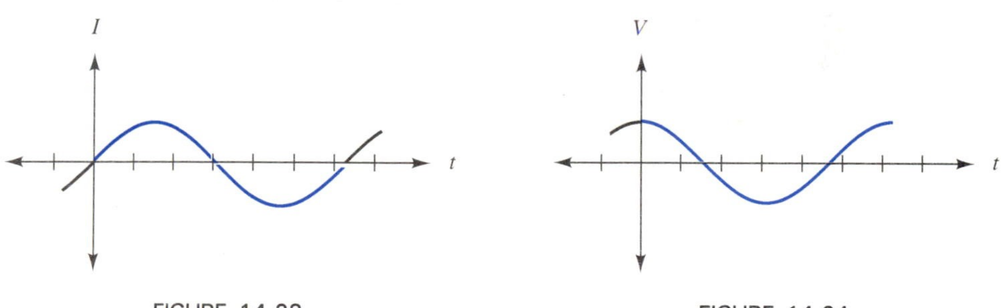

FIGURE 14.23

FIGURE 14.24

in phase

If the two functions reach their peak values for the same value of t, they are said to be **in phase.** If the voltage reaches its peak before the current does, voltage is said to *lead* the current (or current *lags* the voltage). If the voltage reaches its peak after the current does, voltage is said to *lag* the current (or current *leads* the voltage). If ωt in Equations 14.5 and 14.6 represents an angle, then the lead or lag may be measured in degrees.

Placing a resistor in an ac circuit does not affect the phase relation between current and voltage. Placing a capacitor in an ac circuit causes the voltage to lag the current by 90°. Placing an inductor in an ac circuit causes the voltage to lead the current by 90°. Figures 14.25–14.27 illustrate these relationships.

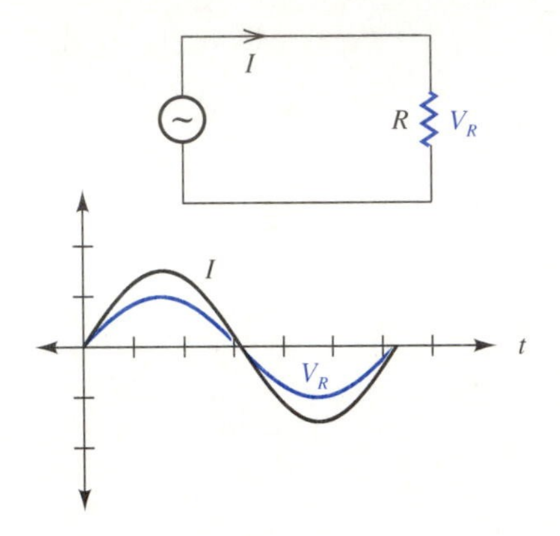

FIGURE 14.25

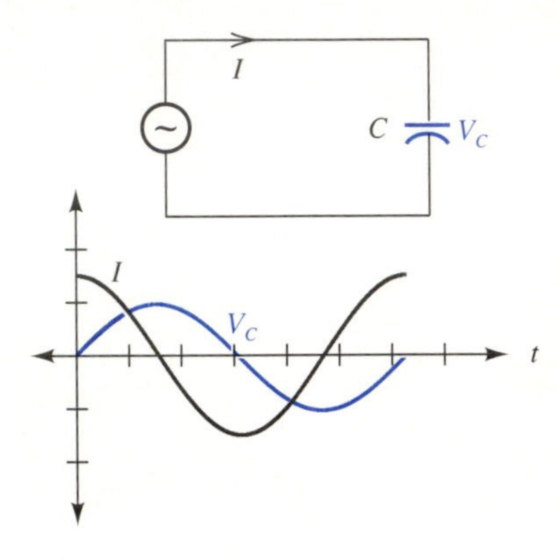

FIGURE 14.26

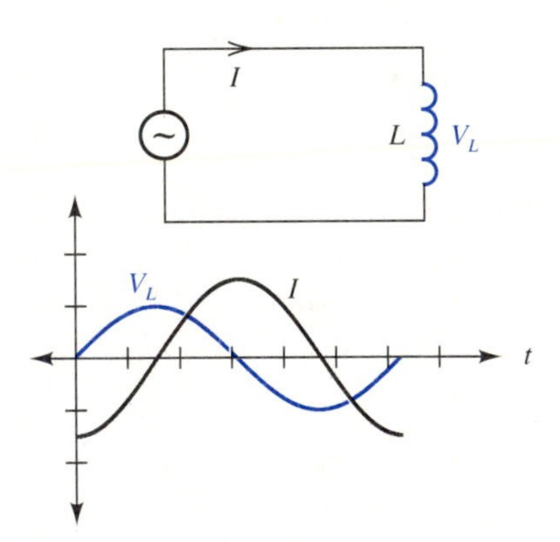

FIGURE 14.27

We are now ready to consider the mathematical relationships among these quantities. A resistance is denoted by **R**. A reactance created by a capacitor is denoted by $\mathbf{X_C}$, while a reactance created by an inductor is denoted by $\mathbf{X_L}$. Using $\mathbf{V_R}$, $\mathbf{V_C}$, and $\mathbf{V_L}$ to denote voltages (in volts) of a resistor, capacitor, and inductor, respectively, and **I** to denote the current (in amperes), we have the following equations.

$$\mathbf{V_R = IR} \qquad \mathbf{V_C = IX_C} \qquad \mathbf{V_L = IX_L}$$

Complex numbers provide a useful way to represent resistance and reactance. Although the scheme described here is not the only one possible, it is a common one. An angle of 0° is used to indicate no difference in phase for the current through and the voltage across a resistor. That is, resistance is represented by R on the real axis. An angle of $-90°$ is used to indicate that the voltage across a capacitor lags the current by 90°. That is, capacitive reactance is represented by $-jX_C$ on the negative imaginary axis. Likewise, an angle of 90° is used to indicate that the voltage across an inductor leads the current by 90°, and inductive reactance is thus represented by jX_L on the positive imaginary axis.

phasor

Reactance is an example of a *phasor.* A **phasor** is a quantity that has magnitude and direction with respect to time. Phasors are used to show phase relationships. Multiplying a phasor by j has the effect of rotating it through 90°, just as multiplying a real number by j has the effect of moving it through 90° on the plane. The term ***j*-operator** is sometimes used in reference to rotating a phasor.

j-operator

impedance

The sum of all the resistances and reactances in series in a circuit is called the **impedance** and is denoted **Z**. Impedance, then, is the total opposition created by all elements in an electric circuit to the flow of current, and the total voltage is given by **V** = **IZ**. Impedance is computed by adding the complex numbers R, $-jX_C$, and jX_L as in Figure 14.28.

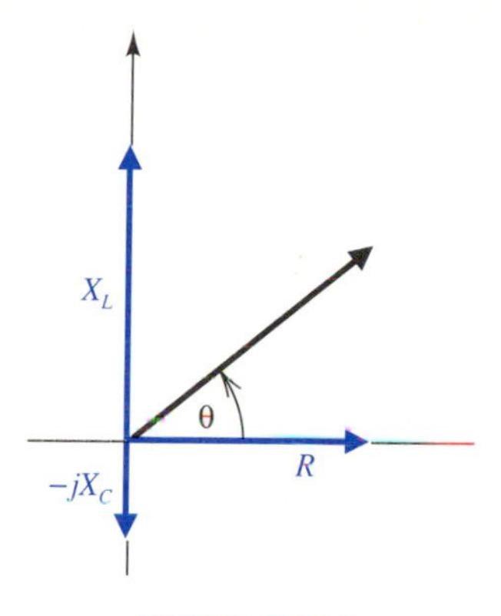

FIGURE 14.28

IMPEDANCE IN AN AC CIRCUIT

The impedance **Z** in an ac circuit containing a resistor, inductor, and capacitor in series is given by

$$\mathbf{Z} = R + (-jX_C) + jX_L = R + j(X_L - X_C)$$

where X_C and X_L are the magnitudes of the capacitive and inductive reactance, respectively, and R is the resistance. The magnitude of **Z**, denoted $|\mathbf{Z}|$, is given by $|\mathbf{Z}| = \sqrt{R^2 + (X_L - X_C)^2}$. The phase angle θ between the current and voltage is given by

$$\tan\theta = \frac{X_L - X_C}{R}.$$

EXAMPLE 1 For a series circuit in which $R = 15.0\ \Omega$, $X_L = 12.0\ \Omega$, and $X_C = 4.00\ \Omega$, find the magnitude of the impedance and the phase angle between the current and the voltage.

Solution Figure 14.29 shows the complex numbers R and $j(X_L - X_C)$.

$$|\mathbf{Z}| = \sqrt{R^2 + (X_L - X_C)^2}$$
$$= \sqrt{(15.0)^2 + (12.0 - 4.00)^2}$$

The calculator keystroke sequence follows.

15 [x^2] [+] [(] 12 [−] 4 [)] [x^2] [=] [$\sqrt{x}$] **17**

$$|\mathbf{Z}| = 17.0\ \Omega$$
$$\tan\theta = (X_L - X_C)/R$$
$$= (12.0 - 4.00)/15.0 \text{ and } \theta = 28.1^\circ$$ ■

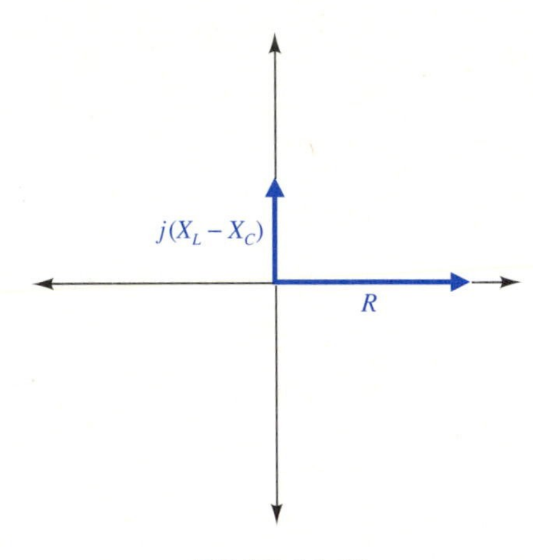

FIGURE 14.29

EXAMPLE 2 For a series circuit in which $R = 4.00\ \Omega$, $X_L = 8.00\ \Omega$, and $X_C = 5.00\ \Omega$, find the magnitude of the impedance and the phase angle between the current and the voltage.

Solution

$$|\mathbf{Z}| = \sqrt{R^2 + (X_L - X_C)^2}$$
$$= \sqrt{(4.00)^2 + (8.00 - 5.00)^2}$$
$$= 5.00\ \Omega$$
$$\tan\theta = (X_L - X_C)/R$$
$$= (8.00 - 5.00)/4.00 \text{ and } \theta = 36.9^\circ$$ ■

EXAMPLE 3 If $R = 6.00\ \Omega$, $X_L = 8.00\ \Omega$, and $I = 2.00$ A in a series circuit, find the magnitude of the voltage.

Solution $\mathbf{V} = \mathbf{IZ}$ for the circuit, so it is necessary to find $\mathbf{Z}$.

$$\begin{aligned}\mathbf{Z} &= R + j(X_L - X_C)\\ &= 6.00 + j(8.00 - 0)\\ &= (6.00 + j8.00)\ \Omega\end{aligned}$$

Since $\mathbf{V} = \mathbf{IZ}$, $\mathbf{V} = 2.00(6.00 + j8.00)$
$= 12.0 + j16.0$

The magnitude of $\mathbf{V}$ is $\sqrt{12.0^2 + 16.0^2}$ or 20.0 volts. ■

EXAMPLE 4 For a circuit in which the current is $3.00 - j4.00$ amperes and the impedance is $4.00 + j1.00$ ohms, find the magnitude of the voltage.

Solution $\mathbf{V} = \mathbf{IZ}$. Multiplication is usually easier if the numbers are in polar form.

$\mathbf{I} = 5.00(\cos 306.9° + j \sin 306.9°)$

$\mathbf{Z} = \sqrt{17.0}\,(\cos 14.0° + j \sin 14.0°)$

$\mathbf{V} = 5.00(\cos 306.9° + j \sin 306.9°)\sqrt{17.0}\,(\cos 14.0° + j \sin 14.0°)$

$\mathbf{V} = 5.00\sqrt{17.0}(\cos 320.9° + j \sin 320.9°)$

The magnitude of $\mathbf{V}$ is $5.00\sqrt{17.0}$ or 20.6 volts. ■

Alternating current is produced when a wire loop rotates through a magnetic field in a circular path with constant angular velocity. The angular velocity is denoted ω and is measured in radians per second. The rate of rotation may also be measured in rotations (or cycles) per second. This rate, denoted f, is called the frequency and is measured in hertz (Hz). The relationship between the angular velocity and the frequency is given by $\omega = 2\pi f$. The reactance of a capacitor or an inductor can be determined from the following equations.

EQUATIONS FOR REACTANCE

Capacitive reactance: $X_C = 1/(\omega C)$ **(14-7)**

Inductive reactance: $X_L = \omega L$ **(14-8)**

where $\omega = 2\pi f$

EXAMPLE 5 In a series circuit for which $R = 10.0\ \Omega$, $L = 0.200$ H, $C = 300\ \mu$F, and $\omega = 60.0$ rad/s, find the magnitude of the impedance and the phase difference between the current and voltage.

Solution $\mathbf{Z} = R + j(X_L - X_C)$
To determine X_L, we use Equation 14-8.

$$X_L = 60.0(0.200) = 12.0\ \Omega$$

To determine X_C, we use Equation 14-7 and change 300 μF to 300×10^{-6} F.

$$X_C = \frac{1}{60.0(300 \times 10^{-6})} = 55.6\ \Omega$$

$\mathbf{Z} = 10.0 + j(12.0 - 55.6)\ \Omega$ or $10.0 - j43.6\ \Omega$
$|\mathbf{Z}| = \sqrt{10.0^2 + (-43.6)^2}$ or $44.7\ \Omega$
Finally, $\tan\theta$ is $-43.6/10.0$ so the reference angle is 77.1°, and θ is $-77.1°$. Thus, the voltage lags the current by 77.1°. ■

EXERCISES 14.5

In 1–10, determine the magnitude of the impedance and the phase angle between current and voltage for a series circuit with the given values of R, X_C, and X_L.

1. $R = 6.00\ \Omega$, $X_C = 4.00\ \Omega$

2. $R = 10.0\ \Omega$, $X_C = 5.00\ \Omega$

3. $R = 8.00\ \Omega$, $X_L = 4.00\ \Omega$

4. $R = 6.75\ \Omega$, $X_L = 5.00\ \Omega$

5. $R = 8.25\ \Omega$, $X_C = 4.15\ \Omega$

6. $R = 10.5\ \Omega$, $X_C = 7.35\ \Omega$, $X_L = 6.21\ \Omega$

7. $R = 10.2\ \Omega$, $X_C = 6.42\ \Omega$, $X_L = 5.74\ \Omega$

8. $R = 9.55\ \Omega$, $X_C = 8.91\ \Omega$, $X_L = 6.45\ \Omega$

9. $R = 7.82\ \Omega$, $X_C = 9.33\ \Omega$, $X_L = 4.83\ \Omega$

10. $R = 9.37\ \Omega$, $X_C = 5.24\ \Omega$, $X_L = 5.27\ \Omega$

In 11–38, determine the magnitude of the voltage for a series circuit with the given values of R, X_L, X_C, and I.

11. $R = 6.00\ \Omega$, $X_L = 6.14\ \Omega$, $I = 2.00$ A

12. $R = 8.00\ \Omega$, $X_L = 5.24\ \Omega$, $I = 3.00$

13. $R = 10.0\ \Omega$, $X_L = 4.92\ \Omega$, $I = 3.00$ A

14. $R = 6.12\ \Omega$, $X_C = 6.52\ \Omega$, $I = 2.00$ A

15. $R = 7.41\ \Omega$, $X_C = 4.31\ \Omega$, $I = 3.00$ A

16. $R = 10.0\ \Omega$, $X_L = 6.14\ \Omega$, $X_C = 4.31\ \Omega$, $I = 2.00$ A

17. $R = 6.12\ \Omega$, $X_L = 5.24\ \Omega$, $X_C = 6.52\ \Omega$, $I = 2.00$ A

18. $R = 7.41\ \Omega$, $X_L = 4.92\ \Omega$, $X_C = 7.31\ \Omega$, $I = 3.00$ A

In 19–25, determine the magnitude of the voltage for a series circuit with the given values of **I** *and* **Z**.

19. $\mathbf{I} = 3.00 - j2.00$ A, $\mathbf{Z} = 5.00 + j3.00\Omega$

20. $\mathbf{I} = 3.00 - j2.00$ A, $\mathbf{Z} = 7.00 - j2.00\ \Omega$

21. $\mathbf{I} = 2.15 + j3.42$ A, $\mathbf{Z} = 4.23 - j3.67\ \Omega$

22. $\mathbf{I} = 0.57 + j0.63$ A, $\mathbf{Z} = 3.9 + j8.7\ \Omega$

23. $\mathbf{I} = 0.83 - j1.6$ A, $\mathbf{Z} = 5.3 - j6.9\ \Omega$

24. $\mathbf{I} = 3.11 - j1.69$ A, $\mathbf{Z} = 4.6 - j7.32\ \Omega$

25. $\mathbf{I} = 2.87 + j1.22$ A, $\mathbf{Z} = 3.63 + j8.09\ \Omega$

In 26–30, determine the magnitude of the impedance to three significant digits and the phase angle between current and voltage for a series circuit with the given values of R, L, C, and ω.

26. $R = 4.00\ \Omega$, $L = 0.300$ H, $C = 350\ \mu$F, and $\omega = 60.0$ rad/sec

27. $R = 6.00\ \Omega$, $L = 0.350$ H, $C = 250\ \mu$F, and $\omega = 60.0$ rad/sec

28. $R = 8.00\ \Omega$, $L = 0.400$ H, $C = 200\ \mu$F, and $\omega = 80.0$ rad/sec

29. $R = 5.52\ \Omega$, $L = 250$ mH, $C = 300\ \mu$F, and $\omega = 80.0$ rad/sec

30. $R = 7.38\ \Omega$, $L = 450$ mH, $C = 400\ \mu$F, and $\omega = 50.0$ rad/sec

31. In a parallel ac circuit, total impedance is calculated by determining the impedance in each branch. Use $1/\mathbf{Z} = 1/\mathbf{Z}_1 + 1/\mathbf{Z}_2$ to determine the total impedance and the phase angle for each circuit shown when $R_1 = 8.00\ \Omega$, $R_2 = 7.00\ \Omega$, $X_C = 15.0\ \Omega$, and $X_L = 13.0\ \Omega$.

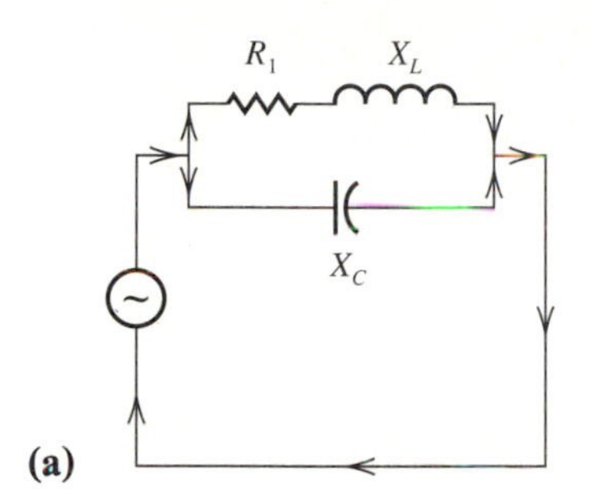

(a)

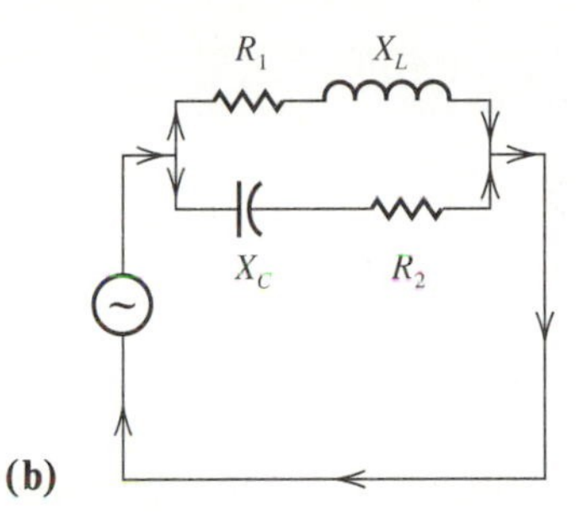

(b)

14.6 CHAPTER REVIEW

KEY TERMS

14.1 pure imaginary number
complex number
conjugate of a complex number

14.2 complex plane
real axis
imaginary axis
vector

14.3 rectangular form
polar form
modulus
argument
exponential form

14.5 resistance
reactance
phase angle
in phase
phasor
j-operator
impedance

SYMBOLS

j	$\sqrt{-1}$
j^2	-1
—\/\/\/—	resistor
—|(—	capacitor
—⌒⌒⌒—	inductor
$r\angle\theta$	$(r \cos \theta + j \sin \theta)$
R	resistance
$\mathbf{X_C}$, $\mathbf{X_L}$	capacitive reactance and inductive reactance
$\mathbf{V_C}$, $\mathbf{V_R}$, $\mathbf{V_L}$	voltage across a capacitor, resistor, or inductor
$\mathbf{Z}$	impedance
$\lvert\mathbf{Z}\rvert$	magnitude of impedance

RULES AND FORMULAS

$(r_1\angle\theta_1)(r_2\angle\theta_2) = r_1 r_2 \angle\theta_1 + \theta_2$

$$\frac{r_1\angle\theta_1}{r_2\angle\theta_2} = \frac{r_1}{r_2}\angle\theta_1 - \theta_2$$

DeMoivre's theorem: $(r\angle\theta)^n = r^n\angle n\theta$

Extension of DeMoivre's theorem:

$$(r\angle\theta)^{1/n} = r^{1/n}\angle\frac{\theta + k\,360°}{n} \text{ for } k = 1, 2, \ldots, n - 1$$

The impedance $\mathbf{Z}$ in an ac circuit containing a resistor, inductor, and capacitor in series is given by $\mathbf{Z} = R + j(X_L - X_C)$.

$$|\mathbf{Z}| = \sqrt{R^2 + (X_L - X_C)^2}$$

$$\mathbf{V} = \mathbf{IZ} \text{ (Ohm's law)}$$

$$X_C = \frac{1}{\omega C} \text{ and } X_L = \omega L \text{ where } \omega = 2\pi f$$

SEQUENTIAL REVIEW

(Section 14.1) *In 1–9, simplify each expression.*

1. $-\sqrt{-45}$

2. (a) $\sqrt{-3(-27)}$ **(b)** $\sqrt{-3}\sqrt{-27}$

3. (a) $(\sqrt{-5})^2$ **(b)** $\sqrt{(-5)^2}$

4. j^{97}

5. $(4 + j) + (2 + 3j)$

6. $(6 + 2j) - (1 - 7j^3)$

7. $(3 - \sqrt{-18})(-3 - \sqrt{-32})$

8. $(2 + 3j)(4 - 5j)$

9. $\dfrac{\sqrt{2} - \sqrt{-4}}{\sqrt{2} + \sqrt{-4}}$

(Section 14.2) *In 10–12, plot each complex number in the complex plane.*

10. $2 + 4j$ **11.** $-3 - j$ **12.** $3j$

In 13–16, perform the indicated operations graphically and state the answer.

13. $(3 + 5j) + (2 - 3j)$

14. $(4 - 3j) - (1 - 2j)$

15. $3(2 - j)$

16. $-4(1 + j)$

In 17 and 18, add graphically, then determine the magnitude and direction of the vector sum.

17. $(4 + 16j) + (4 - j)$

18. $(5 + 3j) + (1 + 5j)$

(Section 14.3)

19. Express $-2.73 + 3.45j$ in polar form.

20. Express $-3j$ in polar form.

21. Express $14.9\angle 36.8^\circ$ in rectangular form.

22. Express $5.3(\cos 117^\circ + j \sin 117^\circ)$ in rectangular form.

23. Express $8\angle 45^\circ$ in exponential form.

24. Express $7.25(\cos 44.6^\circ + j \sin 44.6^\circ)$ in exponential form.

25. Express $7.23\, e^{3.41j}$ in polar form.

26. Express $8ej^{\pi/4}$ in polar form.

27. Express $4e^{j3\pi/2}$ in rectangular form.

(Section 14.4) *In 28–36, perform the indicated operations. In the final answer, angles should be less than* 360° *or* 2π.

28. $3.23(\cos 33.5^\circ + j \sin 33.5^\circ)4.17(\cos 27.4^\circ + j \sin 27.4^\circ)$

29. $5.3\angle 19^\circ\ (7.2\angle 21^\circ)$

30. $\dfrac{63\angle 32^\circ}{6.1\angle 14^\circ}$

31. $\dfrac{7(\cos 12^\circ + j \sin 12^\circ)}{5(\cos 7^\circ + j \sin 7^\circ)}$

32. $(1 - 4j)^4$

33. $[12.6\,(\cos 12.6^\circ + j \sin 12.6^\circ)]^3$

34. $4.73\angle 115.3^\circ + 3.21\angle 4.6^\circ$

35. Find the three cube roots of j.

36. Find the two square roots of $5 - 12j$.

(Section 14.5)
In 37–40, determine the magnitude of the impedance and the phase angle between current and voltage for a series circuit with the given values.

37. $R = 8.00\ \Omega$, $X_C = 5.00\ \Omega$

38. $R = 7.25\ \Omega$, $X_C = 4.50\ \Omega$

39. $R = 6.28\ \Omega$, $X_C = 5.31\ \Omega$, $X_L = 6.14\ \Omega$

40. $R = 7.44\ \Omega$, $X_C = 6.02\ \Omega$, $X_L = 7.13\ \Omega$

In 41–45, determine the magnitude of the voltage for a series circuit with the given values.

41. $R = 6.31\ \Omega$, $X_L = 4.78\ \Omega$, $I = 2.00$ A

42. $R = 5.49\ \Omega$, $X_L = 3.97\ \Omega$, $I = 2.00$ A

43. $R = 7.53\ \Omega$, $X_C = 4.26\ \Omega$, $I = 3.00$ A

44. $\mathbf{I} = (2.00 - j3.00)$ A, $\mathbf{Z} = (6.00 - j2.00)\ \Omega$

45. $\mathbf{I} = (3.15 - j2.15)$ A, $\mathbf{Z} = (5.15 - j3.12)\ \Omega$

APPLIED PROBLEMS

1. Determine the magnitude of the impedance and the phase angle between the current and voltage for a series circuit if $R = 8.14\ \Omega$, $L = 0.400$ H, $C = 250\ \mu$F (to three significant digits), and $\omega = 60.0$ rad/s. 3 5 9 1 2 4 6 7 8
2. Determine the magnitude of the impedance and the phase angle between the current and voltage for a series circuit if $R = 6.00\ \Omega$, $L = 0.375$ H, $C = 250\ \mu$F (to three significant digits), and $\omega = 80.0$ rad/s. 3 5 9 1 2 4 6 7 8
3. Find the impedance from a resistor, a capacitor, and an inductor connected in series if $R = 5.00\ \Omega$, $X_L = 17.0\ \Omega$, and $X_C = 27.0\ \Omega$. Give the answer in polar form. 3 5 9 1 2 4 6 7 8
4. Find the current in a parallel circuit containing two branches: a branch with a resistor and an inductor with impedance $6800 + j4700\ \Omega$; and a branch with a resistor and a capacitor with an impedance of $5600 + j3900\ \Omega$. The applied voltage is 120 V. Give the answer in rectangular form. (Hint: For parallel circuits, $\mathbf{I} = \mathbf{I}_1 + \mathbf{I}_2$.) 3 5 9 1 2 4 6 7 8
5. Find the impedance in a series circuit containing a resistance, capacitance, and inductance such that $R = 56\ \Omega$, $X_C = 98\ \Omega$, and $X_L = 102\ \Omega$. 3 5 9 1 2 4 6 7 8
6. When two or more concurrent forces act on a body, the resultant force can be found by considering each force in polar form and adding. Suppose two people pull on a tree stump. One pulls with 75 lb force due north, the other with 96 lb at an angle of 35° north of east. What one force could replace these two? 2 4 8 1 5 6 7 9
7. The sum of two or more displacements can be found by considering each displacement in polar form and adding. A plane leaving Chicago flies 35 miles at a heading of 27° north of east, followed by 47 miles at 35° south of east, followed by 187 miles due west. How far is the plane from its starting point? Give your answer to the nearest mile. 4 7 8 1 2 3 5 6 9
8. The resultant of two velocity vectors can be found by considering each vector in polar form and adding. A boat that can travel at 6.0 mph in still water heads directly north across a stream which flows at 3.0 mph to the west. What speed will the boat have relative to the bank? 4 7 8 1 2 3 5 6 9
9. The sum of two or more displacements can be found by considering each displacement in polar form and adding. A hunter walks 4.0 miles from camp at an angle of 43° east of north then 7 miles at an angle of 121° west of north followed by 3.0 miles at an angle of 50.0° south of east. How far and at what angle measured from north would he walk to return to camp? 4 7 8 1 2 3 5 6 9
10. The impedance Z_T (in ohms) of a two-branch parallel ac circuit with one branch containing a 24-ohm resistor in series with a 36-ohm inductor and the other branch containing a 15-ohm resistor and a 27-ohm capacitor in series is given by
$$Z_T = \frac{(24 + j36)(15 - j27)}{(24 + j36) + (15 - j27)}.$$
Simplify this expression. 3 5 9 1 2 4 6 7 8
11. To find the resistance that would have to be placed in series with a 75-ohm capacitor to make the total impedance have an angle of −65°, one uses $R - j75$. Write this number in polar form. 3 5 9 4 7 8
12. Complex number math can be used to reduce an electronic circuit to its simplest form containing only two series components $R + jX$ and $R - jX$ ohms. The technique involves reducing the lengthy equation for total circuit impedance down to a single complex number. Find the two series components that could be used to replace a circuit with total impedance given by
$$Z_T = (4 - j11) + \frac{(21 - j3)(13 + j11)}{(21 - j3) + (13 + j11)} + (8 + j12).$$
Use three significant digits in your answer. 3 5 9 1 2 4 6 7 8

1. Architectural Technology
2. Civil Engineering Technology
3. Computer Engineering Technology
4. Mechanical Drafting Technology
5. Electrical/Electronics Engineering Technology
6. Chemical Engineering Technology
7. Industrial Engineering Technology
8. Mechanical Engineering Technology
9. Biomedical Engineering Technology

13. What impedance $R \pm jX$ would be placed in parallel with an impedance of $26 - j55$ ohms to yield a total impedance of $15 + j18$ ohms? Use $\frac{1}{Z_T} = \frac{1}{Z_1} + \frac{1}{Z_2}$.
3 5 9 1 2 4 6 7 8

14. What impedance $R \pm jX$ would be placed in series with $75 - j26$ ohms to yield an impedance of $143 + j65$ ohms? Use $Z_T = Z_1 + Z_2$.
3 5 9 1 2 4 6 7 8

15. An ac circuit will resonate at a frequency ω when its total impedance is purely resistive. For what value of ω will a series circuit with $Z = R + j\omega L - j\frac{1}{\omega c}$ resonate? (Hint: The inductive and capacitive reactances must have a sum of zero.)
3 5 9 1 2 4 6 7 8

16. To resonate, a circuit must have the pure imaginary part of its total impedance be zero. What reactance would you add in series with a circuit with impedance given by $Z_T = (2.6 + j5.9) + \left(\frac{8 - j11}{2 + j4.6}\right)$ to make it resonate? (Hint: Simplify Z_T.)
3 5 9 1 2 4 6 7 8

17. Near resonance, the voltage across a capacitor increases tremendously. This is desirable in small signal circuits such as radio receivers but can be dangerous in circuits with large voltages present. To compute the voltage present across the capacitor in a series circuit consisting of a 3-ohm resistor, a 500-ohm coil, and a 504-ohm capacitor when fed by a source voltage $V = 120\angle 0°$ volts, simplify $I = \frac{120 + j0}{3 + j500 - j504}$ and find $IX_C = I(0 - j504)$. Give your answer to three significant digits.
3 5 9 4 7 8

18. Find the total impedance Z_T (in ohms) of a parallel circuit consisting of an 8.00-ohm resistance and series 275-ohm coil in parallel with a 275.5-ohm capacitor, by simplifying

$$\frac{1}{Z_T} = \frac{1}{8 + j275} + \frac{1}{0 - j275.5}.$$

3 5 9 1 2 4 6 7 8

19. When supplied with $V\angle 0°$ volts the true power P in an RL circuit is the real part of the power represented as the complex number $P = \frac{V^2}{Z}$. Find the true power in a circuit with $Z = 36.0 + j17.3$ ohms if its voltage is $120\angle 0°$ volts.
3 5 9 1 2 4 6 7 8

RANDOM REVIEW

1. Add $(4 - 3j)$ and $(3 + 4j)$ graphically, and determine the magnitude and direction of the vector sum.
2. Find the product of $3.82\angle 104.5°$ and $2.35\angle 5.4°$.
3. Express $3.64 - 4.39j$ in polar form.
4. Simplify $\sqrt{-32(-50)}$.
5. Express $23.8\angle 42.7°$ in exponential form.
6. Simplify $j^{43} + j^{59}$.
7. Find $(2 + 3j)^9$.
8. Express $3e^{j\pi/3}$ in rectangular form.
9. Simplify $\frac{\sqrt{3} - \sqrt{-9}}{\sqrt{3} + \sqrt{-9}}$.
10. Find the three cube roots of j.

CHAPTER TEST

In 1 and 2, simplify each expression.

1. $\sqrt{-32}\sqrt{-50}$

2. $(3 + 2j^3)(j - 9j^{12})$

3. Express $4.53 + 3.47j$ in exponential form.

4. Express $3.19\underline{/31.8^\circ}$ in rectangular form.

5. Express $e^{j\pi/12}$ in polar form.

In 6–8, perform the indicated operations.

6. $4.2\,(\cos 88^\circ + j \sin 88^\circ)\;3.9\,(\cos 41^\circ + j \sin 41^\circ)$

7. $(4 + 7j)^6$

8. $\dfrac{13\,\underline{/15^\circ}}{7\underline{/12^\circ}}$

9. Determine the magnitude of the impedance and the phase angle between current and voltage for a series circuit with the given values.

$$R = 6.58\ \Omega,\ X_C = 5.93\ \Omega,\ \text{and}\ X_L = 6.89\ \Omega$$

10. Determine the magnitude of the voltage for a series circuit with the given values.

$$\mathbf{I} = (3.00 - j2.00)\ \text{A},\ \mathbf{Z} = (5.00 - j1.00)\ \Omega$$

EXPONENTIAL AND LOGARITHMIC FUNCTIONS

CHAPTER 15

In this chapter, we introduce exponential functions, which are used to describe relationships of growth (increase) and decay (decrease). We also examine the related logarithmic functions. Although logarithms were once used for simplifying computations, widespread use of computers and calculators has lessened the importance of this application. Significant applications of these functions, however, still occur. We will see applied problems involving exponential and logarithmic functions in such diverse fields as electronics, money management, civil engineering, and medicine.

15.1 EXPONENTIAL FUNCTIONS

For a fixed base b, $b > 0$, the equation $y = b^x$ defines a function. The independent variable (x) is an exponent, and the function is called an *exponential function.* All functions of the form $y = b^x$ share certain characteristics. Before making some general observations about the function, we examine a few specific examples.

EXAMPLE 1 Sketch the graph of $y = 2^x$.

Solution Consider a few values of x. We choose -2, -1, 0, 1, and 2.

x	$y = 2^x$
-2	$2^{-2} = 1/(2^2) = 1/4$
-1	$2^{-1} = 1/(2^1) = 1/2$
0	$2^0 = 1$
1	$2^1 = 2$
2	$2^2 = 4$

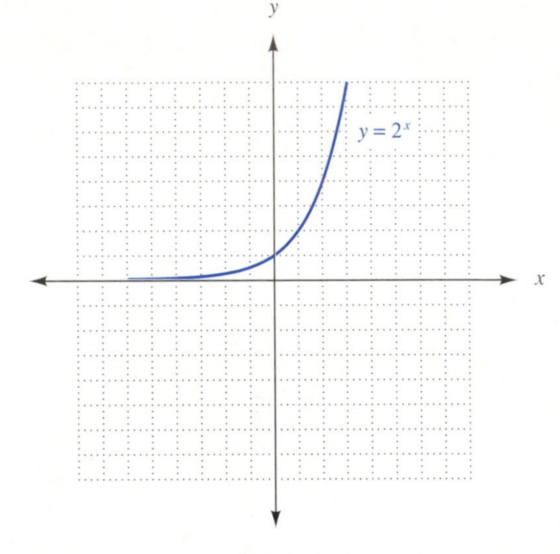

FIGURE 15.1

These points are plotted and the graph is completed by connecting them with a smooth curve from left to right, as shown in Figure 15.1 ■

Notice that as x increases, y increases. When $x < 0$, the values of y are fractions less than one. As x increases in absolute value in the negative direction, the denominators are increasing. The larger the denominator, however, the smaller the fraction. It is impossible for y to equal zero in the equation $y = 2^x$, and therefore the graph does not cross the x-axis. As points are plotted farther and farther to the left, though, the graph gets closer and closer to the x-axis. When a curve approaches a line in this manner, the line is called an *asymptote* of the curve.

asymptote

In Example 1, we used integers for x, but the $\boxed{y^x}$ (or $\boxed{x^y}$) key on a calculator may be used to evaluate the function when x is not an integer. ***It is essential to consider the order of operations, since exponentiation is performed before division.*** If $x = 1/3$, $(2)^{1/3}$ would be entered with the following keystroke sequence.

NOTE ⇨⇨

2 $\boxed{y^x}$ $\boxed{(}$ 1 $\boxed{\div}$ 3 $\boxed{)}$ $\boxed{=}$ **1.25992**

If the parentheses are inadvertently omitted, the expression $2^1 \div 3$ will be evaluated. The result (0.6666667) is incorrect. Because the $\boxed{y^x}$ key carries out a binary operation (that is, it requires both an x and a y to be entered), it is necessary to use the $\boxed{=}$ key to get the final answer. Example 2 illustrates the use of the $\boxed{y^x}$ key on a calculator.

EXAMPLE 2 Sketch the graph of $y = (4.14)^x$.

Solution The following table gives the calculator keystroke sequence used to obtain y for each value of x. Since 4.14 will be used repeatedly, it is entered into memory before the first computation is performed.

x	*Keystroke sequence for y*	*Display*
−2	4.14 [STO] [y^x] 2 [+/−] [=]	0.0583444
−1.5	[RCL] [y^x] 1.5 [+/−] [=]	0.118713
−1	[RCL] [1/x]	0.2415459
0	none (Don't forget that $x^0 = 1$ for any $x \neq 0$.)	
2	[RCL] [x^2]	17.1396
2.5	[RCL] [y^x] 2.5 [=]	34.8739

Figure 15.2 shows the graph. ■

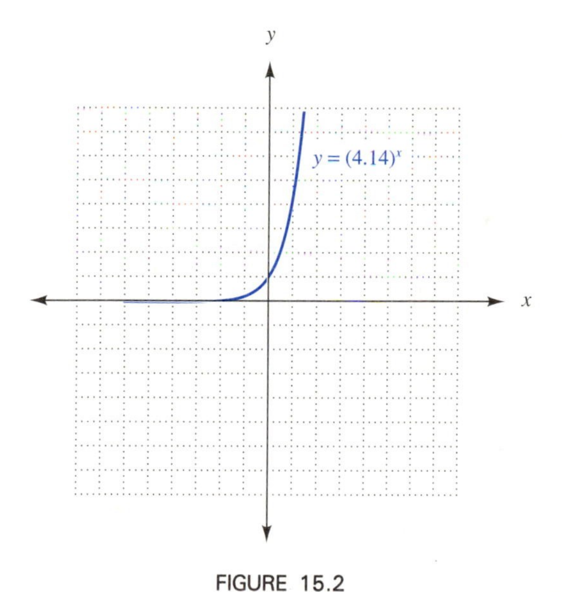

FIGURE 15.2

The functions in Examples 1 and 2 are both of the form $y = b^x$. Notice the characteristics that both graphs have. Each passes through the points (0, 1) and (1, b), and has the x-axis as an asymptote. An exponential function may contain constants other than the base. That is, an exponential function may take the form $y = a(b^{cx})$. In Examples 3 and 4, we examine the effect of the numbers a and c on the graph.

EXAMPLE 3 Sketch the graph of $y = 2^{2x}$.

Solution Consider a few values of x. We choose $-2, -1, 0, 1$, and 2.

x	$y = 2^{2x}$
-2	$2^{-4} = 1/(2^4) = 1/16$
-1	$2^{-2} = 1/(2^2) = 1/4$
0	$2^0 = 1$
1	$2^2 = 4$
2	$2^4 = 16$

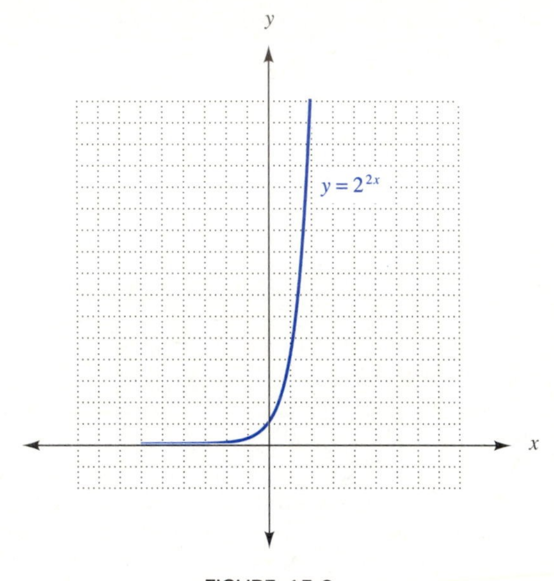

FIGURE 15.3

These points are plotted and the graph is completed by connecting them with a smooth curve from left to right, as shown in Figure 15.3. ■

The graph in Example 3 is similar in shape to the one in Example 1. Instead of passing through the point (1, 2), however, it passes through $(1, 2^2)$.

EXAMPLE 4 Sketch the graph of $y = -2^x$.

Solution Consider a few values of x. We choose $-2, -1, 0, 1$, and 2. Recall that -2^x means $-(2^x)$ rather than $(-2)^x$.

x	$y = -2^x$
-2	$-2^{-2} = -1/(2^2) = -1/4$
-1	$-2^{-1} = -1/(2^1) = -1/2$
0	$-2^0 = -1$
1	$-2^1 = -2$
2	$-2^2 = -4$

These points are plotted and the graph is completed by connecting them with a smooth curve from left to right, as shown in Figure 15.4. ■

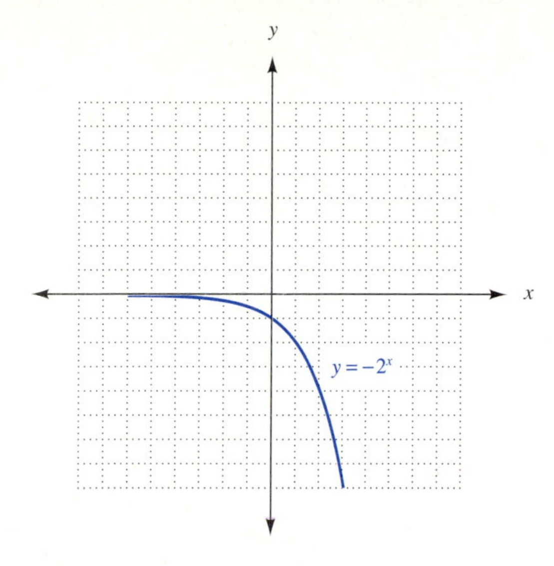

FIGURE 15.4

The graph in Example 4 is similar in shape to the one in Example 1 also. Instead of passing through the points (0, 1) and (1, 2), however, it passes through (0, −1) and (1, −2). Generalizing from these four examples, we make the following observations.

NOTE ⇨⇨ ***A graph of the form $y = a(b^{cx})$ has the x-axis as an asymptote and passes through the points $(0, a)$ and $(1, ab^c)$.*** The procedure for sketching the graph of an exponential function is as follows.

To sketch the graph of an exponential function:

1. Find the y-value associated with an x-value of 0. (This point is the y-intercept.)
2. Find the y-value associated with an x-value of 1.
3. Plot the points obtained in steps 1 and 2 and connect them with a smooth curve from left to right, with the x-axis as an asymptote.
4. If a more accurate graph is desired, find additional points by substituting values of x to find the corresponding values of y.

EXAMPLE 5 Sketch the graph of $y = 3(1/2)^x$.

Solution

1. When x is 0, $y = 3(1/2)^0$ or 3(1), which is 3.
2. When x is 1, $y = 3(1/2)$ or 3/2.
3. Figure 15.5 shows the graph. ■

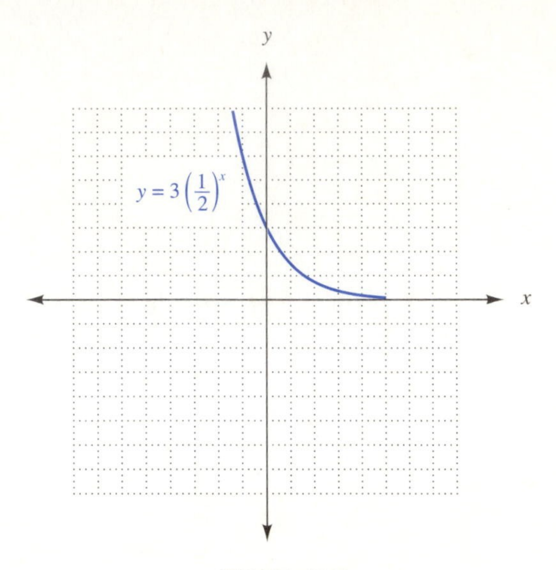

FIGURE 15.5

Example 6 illustrates an exponential function that is important in electronics.

EXAMPLE 6 In a circuit containing a resistor and an inductor in series, if the battery is short-circuited, the current does not fall instantly to zero. The current drops gradually, so that the current I (in A) is given by $I = I_0 e^{-t/\tau}$ where I_0 is the steady-state current (in A), $\tau = L/R$ (inductance/resistance), t is the time elapsed (in seconds), and $e = 2.718$. Sketch the graph of I versus t for $R = 3.0\ \Omega$, $I_0 = 2.0$ A, and $L = 0.30$ H.

Solution Since $\tau = L/R$, we have $\tau = 0.30/3.0$ or 0.10 s.

1. When t is 0, $I = 2.0(2.718)^{-0}$ or 2.0.
2. When t is 1, $I = 2.0(2.718)^{-10}$ or 0.000091.

Before sketching the graph, we find a few additional values, listed in the following table.

t	I
0.1	$2.0(2.718)^{-1} = 0.74$
0.2	$2.0(2.718)^{-2} = 0.27$
0.3	$2.0(2.718)^{-3} = 0.10$
0.4	$2.0(2.718)^{-4} = 0.037$

3. Both t and I have small values, but the range of I-values is greater. It would also be appropriate to use different scales on the axes of the graph, unlike Figure 15.6. ■

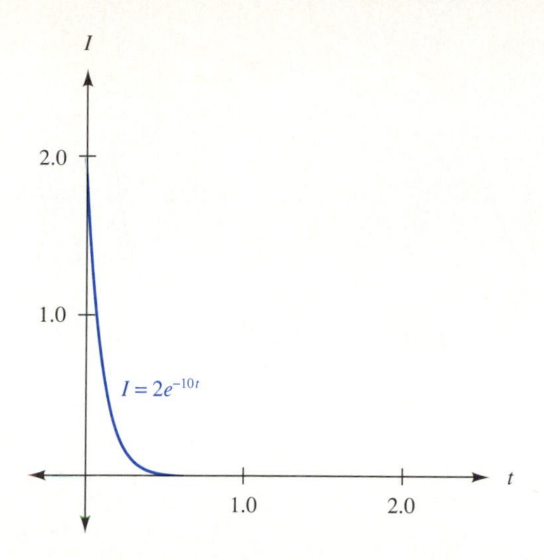

FIGURE 15.6

EXERCISES 15.1

In 1–14, sketch the graph of each function.

1. $y = 4^x$
2. $y = (1/2)3^x$
3. $y = (1/3)2^x$
4. $y = 2^{-x}$
5. $y = 3(5^{-x})$
6. $y = -1(2^x)$
7. $y = (2/3)^x$
8. $y = 3(1/4)^x$
9. $y = 5(1/2)^x$
10. $y = 4(1/3)^{-x}$
11. $y = -2(1/5)^x$
12. $y = 2.78^x$
13. $y = 3.14^{-x}$
14. $y = -1(1.73^x)$

In 15–24, find the letter of the graph that corresponds to the indicated function.

15. $y = 3^x$
16. $y = (1/2)3^x$
17. $y = 3^{2x}$
18. $y = 3^{x/2}$
19. $y = -2(3^x)$
20. $y = (1/2)^x$
21. $y = 3(1/2)^x$
22. $y = -(1/2)^x$
23. $y = (1/2)^{3x}$
24. $y = 3(1/2)^{3x}$

a.

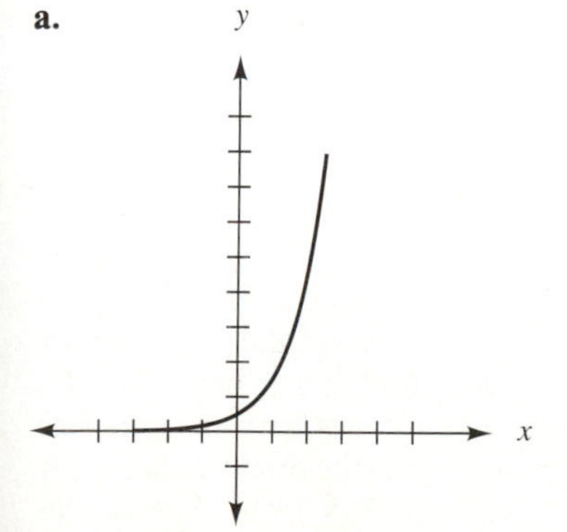

b.

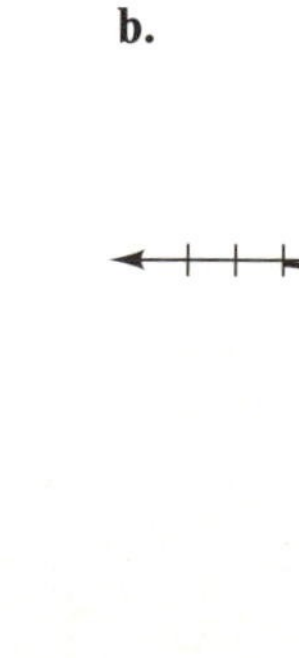

c.

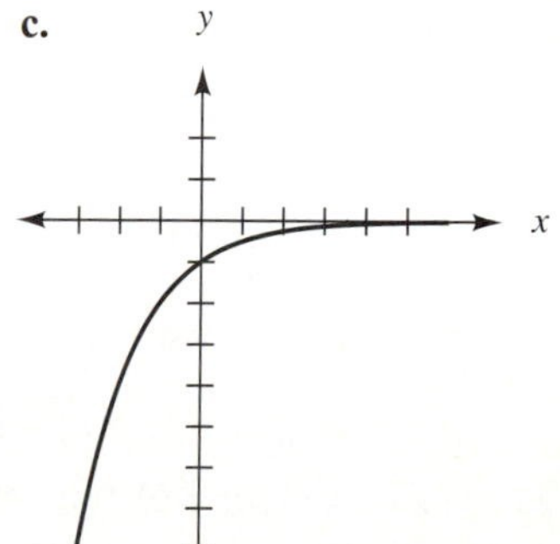

d.

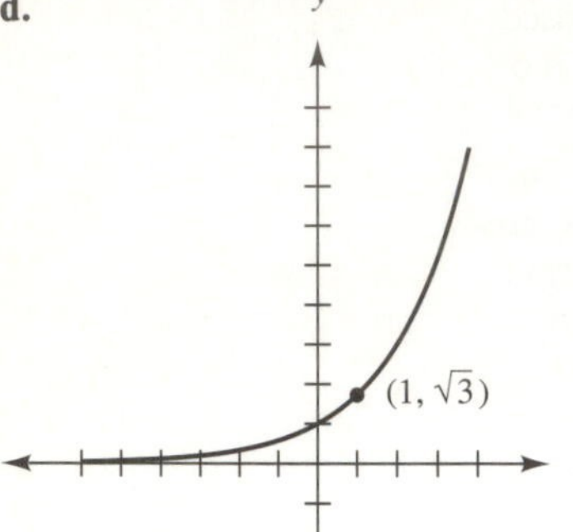

e.

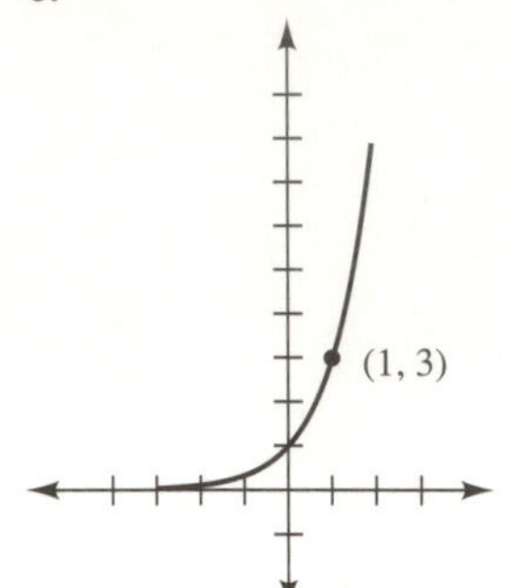

f.

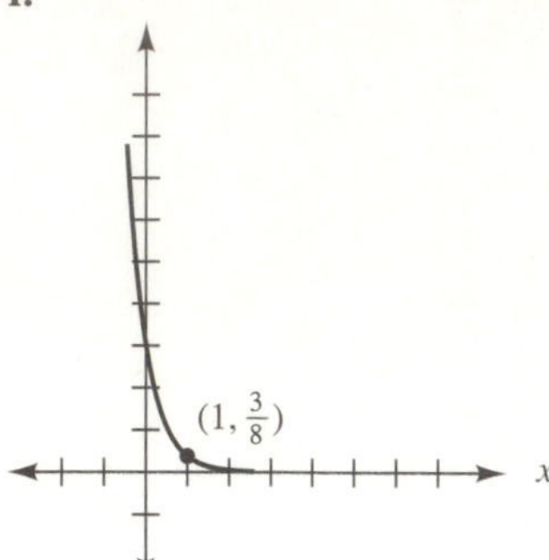

g.

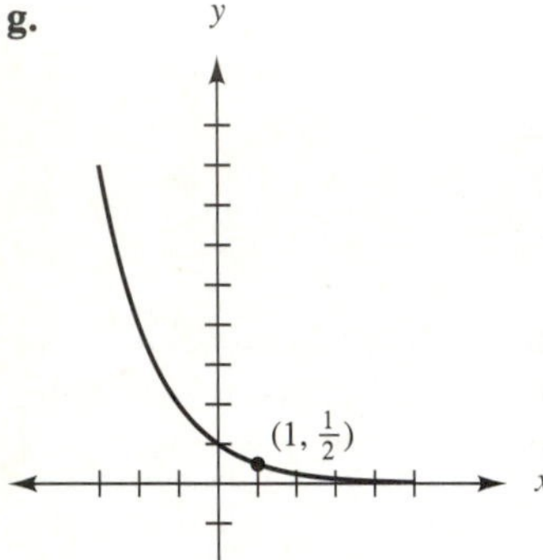

h.

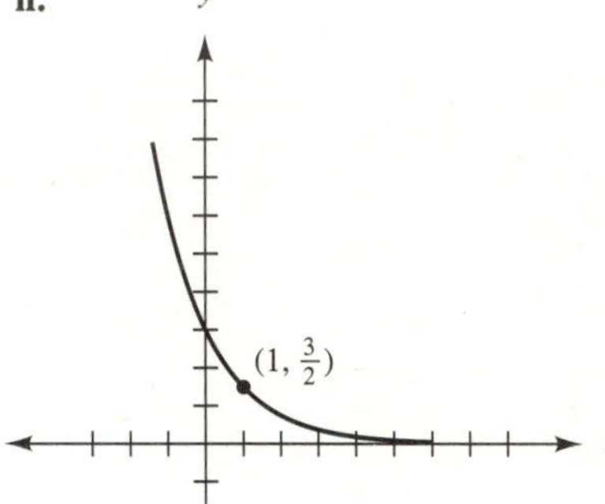

i.

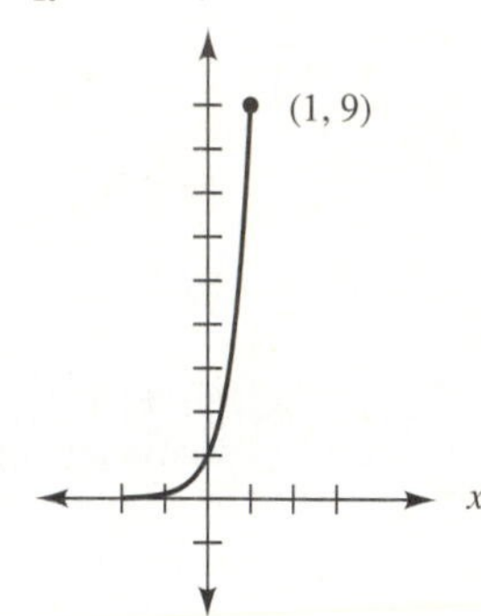

j.

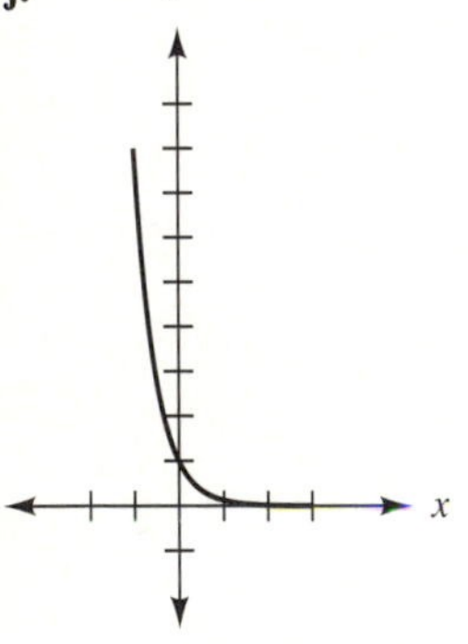

---

25. The intensity I of an earthquake is given by $I = I_0(10^R)$ where I_0 is a standard minimum intensity, and R is the magnitude of the earthquake on the Richter scale. Sketch the graph of I/I_0 versus R. 1 3 4 5 6 8 9

26. The formula $B = P(1 + r)^n$ gives the balance B in an account after n compounding periods if the principal is P and r is the rate of interest (as a decimal) per compounding period. If \$1 is invested at a 5 percent annual percentage rate, sketch the graph of B versus n. 1 2 6 7 8 3 5 9

27. The formula $Q = Q_0 e^{-t/\tau}$ gives the charge Q (in coulombs) on a capacitor after it has been discharging through a resistance for time t (in seconds). In the formula, Q_0 is the original charge (in coulombs), $\tau = RC$, and $e = 2.718$. Sketch the graph of Q versus t for $0 \le t \le 0.01$ when $Q_0 = 3.6 \times 10^{-5}$ coulombs and $\tau = 0.002$ seconds. 3 5 9 4 7 8

28. In a circuit containing a resistor and an inductor in series, if the battery is short-circuited, the current does not fall instantly to zero. The current drops gradually, so that the current I (in A) is given by $I = I_0 e^{-t/\tau}$ where I_0 is the steady-state current (in A), $\tau = L/R$ (inductance/resistance), t is the time elapsed (in seconds), and $e = 2.718$. Sketch the graph of I versus t for $R = 2.0\ \Omega$, $I_0 = 2.0$ A, and $L = 0.40$ H. 3 5 9 1 2 4 6 7 8

29. The half-life of carbon 14 is 5570 years. That is, after 5570 years, half of any size sample of carbon 14 will have decayed. The formula $P = (1/2)^{t/5570}$ gives the percent (as a decimal) of a sample remaining after time t (in years). Sketch the graph of P versus t for $0 \le t \le 20{,}000$. 6 9 5 8

1. Architectural Technology
2. Civil Engineering Technology
3. Computer Engineering Technology
4. Mechanical Drafting Technology
5. Electrical/Electronics Engineering Technology
6. Chemical Engineering Technology
7. Industrial Engineering Technology
8. Mechanical Engineering Technology
9. Biomedical Engineering Technology

15.2 LOGARITHMIC FUNCTIONS

Recall from Section 3.4 that when the values of x and y are interchanged in an equation that defines a function, the resulting equation defines the inverse of the function. In Example 1 of the previous section, we examined the graph of $y = 2^x$. We now consider the graph of $x = 2^y$.

EXAMPLE 1 Sketch the graph of $x = 2^y$.

Solution We substitute values for y and evaluate 2^y to obtain x. Figure 15.7 shows the graph.

y	$x = 2^y$
-2	$2^{-2} = 1/(2^2) = 1/4$
-1	$2^{-1} = 1/(2^1) = 1/2$
0	$2^0 = 1$
1	$2^1 = 2$
2	$2^2 = 4$

■

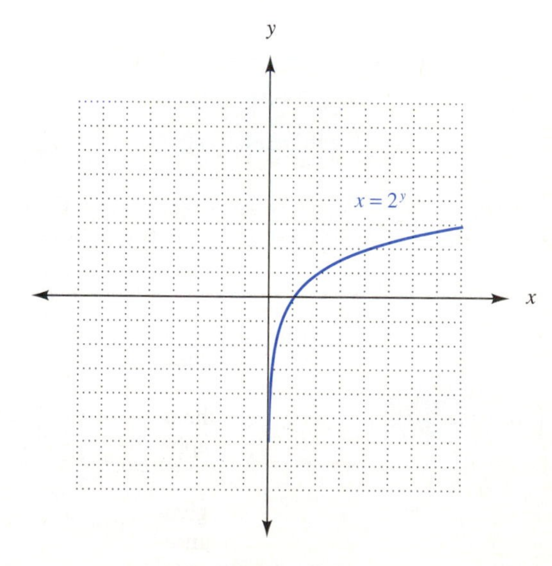

FIGURE 15.7

Recall that the graph of the inverse of a function is a mirror image (about the line $y = x$) of the graph of the function. The graphs of the equations $y = 2^x$ and $x = 2^y$ confirm the inverse relationship of $y = b^x$ and $x = b^y$ for the special case when b is 2. Figure 15.8 shows both graphs on a single coordinate plane.

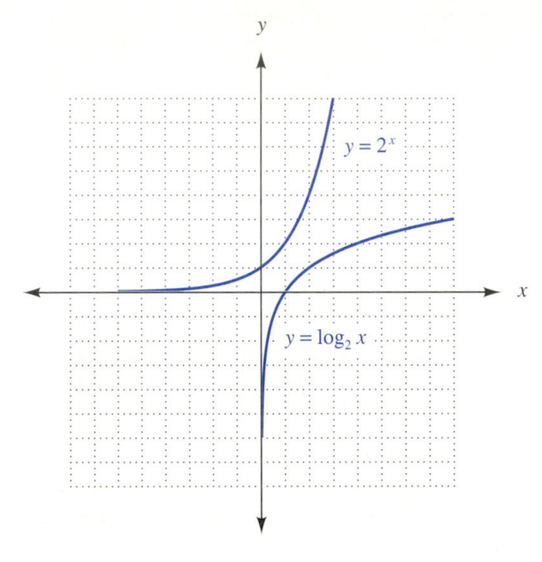

FIGURE 15.8

To solve the equation $x = b^y$ for y, we introduce a new notation. If $x = b^y$, we say that y is the logarithm of x to the base b. This relationship is abbreviated as $y = \log_b x$. Think of the logarithm as an exponent. The equation $y = \log_b x$ says that y is the exponent applied to base b.

DEFINITION

If $\mathbf{y = \log_b x}$ where $b > 0$, **(15-1)**
then $x = b^y$. **(15-2)**

NOTE ⇨⇨ logarithmic form exponential form

It is important that Equations 15-1 and 15-2 express the same relationship among x, b, and y. Equation 15-1 is said to be in *logarithmic form* and Equation 15-2 is said to be in *exponential form.* The relationship holds if $b > 0$. If $b < 0$ and y is a fraction with an even denominator, then b^y is not a real number. For instance if b is -2 and y is 1/2, then $(-2)^{1/2}$ is not a real number.

EXAMPLE 2 Express each equation in logarithmic form.

(a) $3^2 = 9$ **(b)** $2^3 = 8$

Solution **(a)** Since 2 is the exponent, or logarithm, of 9 to the base 3, we have $\log_3 9 = 2$.

(b) Since 3 is the exponent, or logarithm, of 8 to the base 2, we have $\log_2 8 = 3$. ■

EXAMPLE 3 Express each equation in exponential form.

(a) $\log_{10} 100 = 2$ **(b)** $\log_5 1 = 0$

Solution **(a)** Since 2 is the logarithm, or exponent, of 100 to the base 10, we have $10^2 = 100$.

(b) Since 0 is the logarithm, or exponent, of 1 to the base 5, we have $5^0 = 1$. ■

Logarithmic functions in the technologies are sometimes converted into exponential equations.

EXAMPLE 4 The formula $\log_v (P/k) = 3$ expresses the relationship between the power P (in watts) generated by a windmill if v is the velocity (in mph) of the wind, and k is a constant that depends on characteristics of the windmill. Write the equation in exponential form.

Solution Since the base is v, and the logarithm is 3, we have $P/k = v^3$ or $P = kv^3$. ■

Just as the equation $y = b^x$ defines a function for a fixed b, the equation $y = \log_b x$ also defines a function. In the equation $y = \log_b x$, x is the independent variable, and the function is called a *logarithmic function*. ***Since $y = \log_b x$ is equivalent to $x = b^y$, the function defined by $y = \log_b x$ is the inverse of the function defined by $y = b^x$.***

logarithmic function

NOTE ⇨⇨

Logarithmic functions, like exponential functions, may contain constants other than the base. A logarithmic function may take the form $y = a \log_b cx$.

To graph a logarithmic function, first write it in exponential form. The techniques used are summarized as follows.

To sketch the graph of a logarithmic function:

1. Write the equation in exponential form.
2. Find the value of x associated with $y = 0$. (This point is the x-intercept.)
3. Find the value of x associated with $y = 1$.
4. Plot the points obtained in steps 2 and 3 and connect them with a smooth curve from left to right with the y-axis as an asymptote.

EXAMPLE 5 Sketch the graph of $y = \log_5 \left(\frac{1}{2}x\right)$.

Solution 1. Write the equation in exponential form.

$$\frac{1}{2}x = 5^y$$

$$x = 2(5^y)$$

2. When y is 0, x is 2(1) or 2.

3. When y is 1, x is 2(5) or 10.

4. Figure 15.9 shows the graph. ■

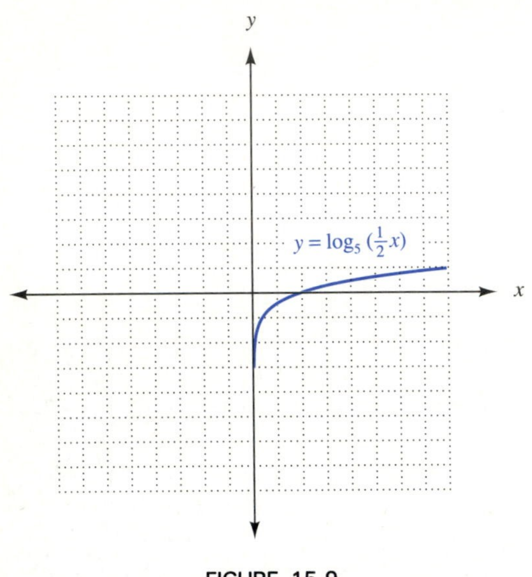

FIGURE 15.9

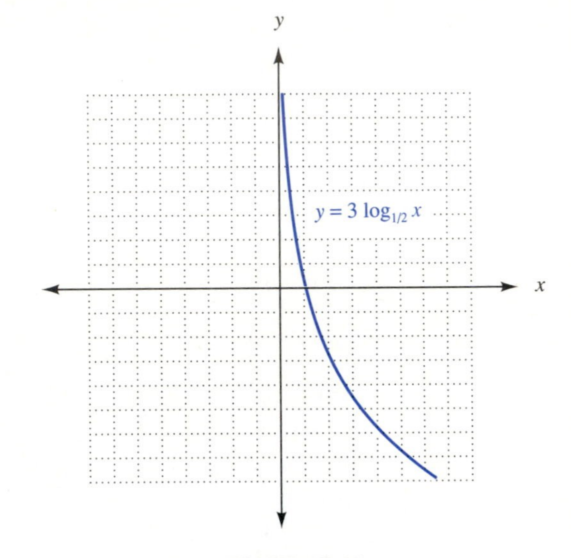

FIGURE 15.10

EXAMPLE 6 Sketch the graph of $y = 3 \log_{1/2} x$.

Solution

1. To put the equation in exponential form, we divide both sides by 3; and $y/3 = \log_{1/2} x$ is equivalent to $(1/2)^{y/3} = x$.

2. When y is 0, $x = (1/2)^0$ or 1.

3. When y is 1, $x = (1/2)^{1/3}$, which is about 0.79.

4. Figure 15.10 shows the graph. ■

EXAMPLE 7 Sketch the graph of $y = (-1/2) \log_3 2x$.

Solution

1. To put the equation in exponential form, multiply both sides by -2.

$$-2y = \log_3 2x$$
$$3^{-2y} = 2x$$
$$\frac{3^{-2y}}{2} = x$$

2. When y is 0, $x = 1/2$.

3. When y is 1, $x = \frac{3^{-2}}{2}$ or $\frac{1}{2(3^2)}$, which is $\frac{1}{18}$.

4. Figure 15.11 shows the graph. ■

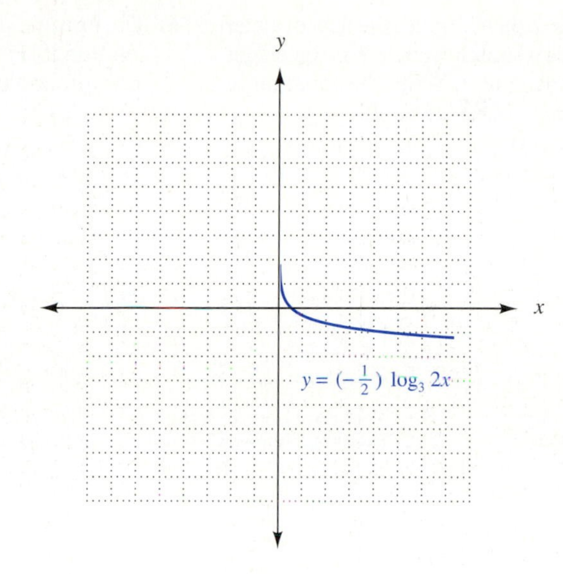

FIGURE 15.11

EXERCISES 15.2

In 1–10, write each equation in logarithmic form.

1. $3^4 = 81$
2. $2^5 = 32$
3. $7^2 = 49$
4. $4^3 = 64$
5. $2^{-3} = 1/8$
6. $3^{-2} = 1/9$
7. $4^{-1} = 1/4$
8. $64^{1/3} = 4$
9. $27^{2/3} = 9$
10. $16^{1/2} = 4$

In 11–20, write each equation in exponential form.

11. $\log_{10} 1000 = 3$
12. $\log_2 8 = 3$
13. $\log_3 9 = 2$
14. $\log_{10} 1 = 0$
15. $\log_2 1/2 = -1$
16. $\log_2 1/8 = -3$
17. $\log_3 1/9 = -2$
18. $\log_4 2 = 1/2$
19. $\log_8 4 = 2/3$
20. $\log_{16} 2 = 1/4$

In 21–32, sketch the graph of each function.

21. $y = \log_4 x$
22. $y = 3 \log_4 x$
23. $y = 2 \log_3 x$
24. $y = -2 \log_3 x$
25. $y = -1 \log_2 x$
26. $y = (1/2) \log_5 x$
27. $y = (2/3) \log_4 x$
28. $y = \log_{1.51} x$
29. $y = \log_{2.73} x$
30. $y = 0.1 \log_{1.51} x$
31. $y = 0.2 \log_{1.1} x$
32. $y = 1.5 \log_{0.5} x$

33. The formula $d = 111.7\sqrt{h}$ gives the distance (in km) that can be seen from a building of height h (in km). Write the equation in logarithmic form with h as the base. 1 2 4 7 8 3 5

34. The formula $Q = 2.48\,h^{2.54}$ gives the flow of water Q (in ft³/s) over a V-notch weir, if h is the height (in ft) of flow above the V. Write the equation in logarithmic form. 4 8 1 2 6 7 9

35. The void ratio of a soil is the ratio of the volume of voids in the soil mass to the volume of the solids in the soil mass. When a soil is loaded with foundation loads, the soil will exhibit a certain amount of settlement associated with a decrease in the void ratio. The change Δe in the void ratio of the soil is given by $\Delta e = C_c \log_{10}(P_2/P_1)$. Write the formula in exponential form. In the formula, C_c is the compression index, P_1 is the initial soil pressure (in tons/ft²), and P_2 is the final soil pressure (in tons/ft²). 1 2 4 6 8 5 7 9

36. The formula $\text{pH} = -\log_{10}[\text{H}^+]$ gives the pH of a solution if $[\text{H}^+]$ is the concentration of hydrogen ions (in moles/liter). Sketch the graph of pH versus $[\text{H}^+]$. 6 9 2 4 5 7 8

37. The formula $G = -4.58\,T\log_{10} k$ gives the standard free energy change G (in Calories) of a reaction where k is the equilibrium constant and T is the temperature (in degrees Kelvin). If $T = 300°$ K, sketch the graph of G versus k. 6 9 2 4 5 7 8

1. Architectural Technology
2. Civil Engineering Technology
3. Computer Engineering Technology
4. Mechanical Drafting Technology
5. Electrical/Electronics Engineering Technology
6. Chemical Engineering Technology
7. Industrial Engineering Technology
8. Mechanical Engineering Technology
9. Biomedical Engineering Technology

15.3 PROPERTIES OF LOGARITHMS

In Section 2.4, rules were introduced for the purpose of simplifying algebraic expressions with exponents that were integers. The rules are valid, however, for exponents that are real numbers. Three of those rules are repeated here for review.
If x, m, and n are real numbers, then

$$x^m(x^n) = x^{m+n}. \tag{15-3}$$

If x is a nonzero real number and m and n are real numbers, then

$$x^m/x^n = x^{m-n}. \tag{15-4}$$

If x, m, and n are real numbers, then

$$(x^m)^n = x^{mn}. \tag{15-5}$$

These rules can also be stated as follows.
Assuming that the base is the same for each expression:

The *exponent* of a product is the sum of the *exponents.*

The *exponent* of a quotient is the difference of the *exponents.*

The *exponent* of a power is the product of the *exponents.*

Since a logarithm is an exponent, it is reasonable that the following statements are true.

Assuming that the base is the same for each expression:

The *logarithm* of a product is the sum of the *logarithms.*

The *logarithm* of a quotient is the difference of the *logarithms.*

The *logarithm* of a power is the product of the *logarithms.*

Using the standard notation, we have the following properties of logarithms.

PROPERTIES OF LOGARITHMS

$$\log_b (xy) = \log_b x + \log_b y \text{ if } x > 0 \text{ and } y > 0 \quad \textbf{(15-6)}$$

$$\log_b (x/y) = \log_b x - \log_b y \text{ if } x > 0 \text{ and } y > 0 \quad \textbf{(15-7)}$$

$$\log_b x^y = y \log_b x \text{ if } x > 0 \quad \textbf{(15-8)}$$

To verify that these rules are valid, we rely on the rules for exponents. To examine $\log_b (xy)$, we want to work with a base of b. Let

$$u = \log_b x \quad \textbf{(15-9)}$$

and

$$v = \log_b y. \quad \textbf{(15-10)}$$

Equations 15-9 and 15-10 are equivalent to $x = b^u$ and $y = b^v$. Thus,

$$xy = (b^u)(b^v)$$
$$= b^{u+v}.$$

Putting this equation in logarithmic form, we have

$$\log_b (xy) = u + v.$$

Equations 15-9 and 15-10 allow the following substitutions for u and v.

$$\log_b (xy) = \log_b x + \log_b y$$

Thus, we arrive at Equation 15-6.

Similarly, to examine $\log_b (x/y)$, consider the following.

$$x/y = b^u/b^v$$
$$= b^{u-v}$$

In logarithmic form, the equation is $\log_b (x/y) = u - v$. Equations 15-9 and 15-10 allow the following substitutions for u and v.

$$\log_b (x/y) = \log_b x - \log_b y$$

Finally, to examine $\log_b x^y$, consider the following.

$$x^y = (b^u)^y$$
$$= b^{uy}$$
$$\log_b x^y = uy$$
$$\log_b x^y = y \log_b x$$

The properties of logarithms will be especially important when we solve logarithmic equations in the last section of this chapter.

EXAMPLE 1 Write each expression as the logarithm of a single quantity.

(a) $\log_3 4 + \log_3 5$

(b) $\log_5 6 - \log_5 2$

(c) $2 \log_5 3$

Solution **(a)** $\log_3 4 + \log_3 5 = \log_3 (4 \cdot 5) = \log_3 20$

(b) $\log_5 6 - \log_5 2 = \log_5 (6/2) = \log_5 3$

(c) $2 \log_5 3 = \log_5 3^2 = \log_5 9$ ■

EXAMPLE 2 Write each expression as the logarithm of a single quantity.

(a) $\log_{10} x - \log_{10} y$

(b) $2 \log_x 4$

(c) $\log_2 y^3 + \log_2 y^2$

Solution **(a)** $\log_{10} x - \log_{10} y = \log_{10} (x/y)$

(b) $2 \log_x 4 = \log_x 4^2 = \log_x 16$

(c) $\log_2 y^3 + \log_2 y^2 = \log_2 y^3(y^2) = \log_2 y^5$
or $\log_2 y^3 + \log_2 y^2 = 3 \log_2 y + 2 \log_2 y = 5 \log_2 y = \log_2 y^5$ ■

EXAMPLE 3 The formula $A = 17(\log_{10} P_0 - \log_{10} P_1)$ gives the altitude gain (in km) of an unpressurized vehicle if P_0 is the atmospheric pressure (in millibars) at the lower altitude and P_1 is the atmospheric pressure (in millibars) at the higher altitude. Write the expression on the right-hand side of the equation as the logarithm of a single quantity.

Solution

$$\begin{aligned} A &= 17(\log_{10} P_0 - \log_{10} P_1) \\ &= 17 \log_{10} (P_0/P_1) \\ &= \log_{10} (P_0/P_1)^{17} \end{aligned}$$ ■

EXAMPLE 4 A population of bacteria grows in such a way as to double in number at regular intervals. The number n of intervals elapsed after time t is given by $n = 3.32(\log_{10} b_n - \log_{10} b_0)$. In the formula, b_0 is the initial population and b_n is the population after time t. Write the expression on the right-hand side of the equation as the logarithm of a single quantity.

Solution

$$\begin{aligned} 3.32(\log_{10} b_n - \log_{10} b_0) &= 3.32 \log_{10} (b_n/b_0) \\ &= \log_{10}(b_n/b_0)^{3.32} \end{aligned}$$ ■

NOTE ***Since $\log_b b^n$ means the logarithm of b^n when b is the base, we have $\log_b b^n = n$.*** This observation is useful for simplifying logarithmic expressions.

EXAMPLE 5 Rewrite each expression and simplify, if possible.

(a) $\log_3 18$

(b) $\log_2 (5/8)$

Solution **(a)** We can factor 18 as $2(3^2)$. It is desirable to have 3 as a factor, since 3 is the base of the logarithms.

$$\begin{aligned} \log_3 18 &= \log_3 (9 \cdot 2) \\ &= \log_3 9 + \log_3 2 \\ &= 2 + \log_3 2 \end{aligned}$$

(b) $$\begin{aligned} \log_2 5/8 &= \log_2 5 - \log_2 8 \\ &= \log_2 5 - 3 \end{aligned}$$

NOTE ***Note that $\log_2 5 - 3$ should not be mistaken for $\log_2 (5 - 3)$.*** ■

Recall that $x^{1/n} = \sqrt[n]{x}$ when $\sqrt[n]{x}$ is a real number. Properties of logarithms may therefore be used with radical expressions.

EXAMPLE 6 Rewrite each expression and simplify, if possible.

(a) $\log_{10} \sqrt[3]{5}$

(b) $\log_8 \sqrt{8y}$, $y > 0$

Solution **(a)** $$\begin{aligned} \log_{10} \sqrt[3]{5} &= \log_{10} 5^{1/3} \\ &= (1/3)\log_{10} 5 \end{aligned}$$

(b) $$\begin{aligned} \log_8 \sqrt{8y} &= \log_8 (8y)^{1/2} \\ &= (1/2)\log_8 (8y) \\ &= (1/2)[\log_8 8 + \log_8 y] \\ &= (1/2)[1 + \log_8 y] \\ &= 1/2 + (1/2)\log_8 y \end{aligned}$$ ■

EXERCISES 15.3

In 1–20, write each expression as the logarithm of a single quantity.

1. $\log_2 5 + \log_2 7$
2. $\log_7 9 + \log_7 8$
3. $\log_6 3 + \log_6 y$
4. $\log_{10} 15 - \log_{10} x$
5. $\log_4 24 - \log_4 6$
6. $\log_9 35 - \log_9 7$
7. $2 \log_5 4$
8. $3 \log_4 x$
9. $-1 \log_b 2$
10. $3 \log_x 4 + \log_x 2$
11. $2 \log_y 5 + \log_y 3$
12. $\log_b 8 - 2 \log_b 2$

13. $2 \log_a x - \log_a 4$

14. $\log_2 7 + 3 \log_2 3 - \log_2 5$

15. $2 \log_3 5 + \log_3 7 - \log_3 2$

16. $2 \log_5 3 - \log_5 7 + 3 \log_5 2$

17. $3 \log_7 2 - \log_7 5 + 2 \log_7 3$

18. $3 \log_{10} 3 + 2 \log_{10} 5 - \log_{10} 2$

19. $\log_a 2 + 3 \log_a 5 - 2 \log_a 7$

20. $2 \log_b 5 - \log_b 3 + 3 \log_b 2$

In 21–40, rewrite each expression and simplify, if possible.

21. $\log_2 20$

22. $\log_2 24$

23. $\log_3 36$

24. $\log_3 45$

25. $\log_5 75$

26. $\log_5 15$

27. $\log_7 (5/7)$

28. $\log_7 (2/49)$

29. $\log_{10} (3/10)$

30. $\log_{10} (2/100)$

31. $\log_x (3/x^2)$

32. $\log_x (x^3/2)$

33. $\log_y 7y$

34. $\log_y 6y^2$

35. $\log_2 \sqrt{2y}, y > 0$

36. $\log_2 \sqrt{2x}, x > 0$

37. $\log_a \sqrt[3]{5a}, a > 0$

38. $\log_b \sqrt[3]{3b}, b > 0$

39. $\log_{10} \sqrt[3]{10x}, x > 0$

40. $\log_{10} \sqrt{10y}, y > 0$

---

41. Open parallel transmission lines have a characteristic impedance given by the formula $Z_0 = 276(\log_{10} a - \log_{10} b)$. Write the expression on the right-hand side as the logarithm of a single quantity. **3 5 9** 1 2 4 6 7 8

42. The formula $A = -10(\log_{10} P_I - \log_{10} P_O)$ gives the power gain (in dB) if P_I is the power input and P_O is the power output. Write the expression on the right-hand side as the logarithm of a single quantity. **3 5 9** 1 2 4 6 7 8

43. The magnitudes m and n of two stars are compared using the formula $\log_{10} b_m - \log_{10} b_n = (n - m)\log_{10} r$. In the formula, b_m and b_n are the actual brightnesses, and r is the ratio of the brightness of two stars whose magnitudes differ by 1. Write the expression on each side of the equation as the logarithm of a single quantity. **9** 1 2 3 5 6 7

44. For a flat belt drive at equilibrium when slippage is impending, the formula $132(\log_{10} T_L - \log_{10} T_S) = f\theta$ gives the relationship between tension T_L (in lbs) on the tight side, tension T_S (in lbs) on the loose side, the coefficient of friction f, and the angle θ (in degrees) of contact. Write the expression on the left-hand side of the equation as the logarithm of a single quantity. **2 4 8** 1 5 6 7 9

45. The formula $R = \log_{10} I - \log_{10} I_0$ gives the magnitude of an earthquake where I is the intensity of the quake and I_0 is a standard minimum intensity. Write the expression on the right as the logarithm of a single quantity. 1 3 4 5 6 8 9

46. The formula $S = C_v \log_{10} T + R \log_{10} V + \log_{10} a$ gives the entropy of a thermodynamic system. In the formula, C_v is the molecular heat at constant volume, T is the temperature, and a is a constant called the entropy constant. Write the expression on the right-hand side of the equation as the logarithm of a single quantity. **6 8** 1 2 3 5 7 9

---

1. Architectural Technology
2. Civil Engineering Technology
3. Computer Engineering Technology
4. Mechanical Drafting Technology
5. Electrical/Electronics Engineering Technology
6. Chemical Engineering Technology
7. Industrial Engineering Technology
8. Mechanical Engineering Technology
9. Biomedical Engineering Technology

15.4 COMMON LOGARITHMS

You have probably noticed a key on your calculator labeled LOG. When no base is indicated, the base is understood to be 10. That is, $\log x$ means $\log_{10} x$. When a number is entered and followed by the use of the [log] key, the number displayed is the logarithm to the base 10 of the number entered. Since only one number is entered, it is not necessary to use the [=] key. The logarithm is displayed automatically.

EXAMPLE 1 Use a calculator to evaluate each expression.

(a) $\log 42.1$

(b) $\log 0.321$

Solution The keystroke sequence is as follows.

(a) 42.1 [log] ***1.6242821***

(b) 0.321 [log] ***−0.493495*** ■

In Example 1, we were given a number and we found its logarithm. If we are given a logarithm, it is possible to find the number whose logarithm it is. In the equation $y = \log x$, x is called the *antilogarithm* of y. To use a calculator to solve $y = \log x$ for x, we enter a value for y, followed by the [INV] [log] keystroke sequence.

antilogarithm

EXAMPLE 2 **(a)** If $\log x = 2.547$, find x.

(b) If $\log x = -1.532$, find x.

Solution The keystroke sequence follows.

(a) 2.547 [INV] [log] ***352.37087***

(b) 1.532 [+/−] [INV] [log] ***0.0293765*** ■

Base 10 logarithms are called **common logarithms** to distinguish them from logarithms to other bases.

DEFINITION The **common logarithm** of x is denoted $\log x$ and

$$\log x = \log_{10} x.$$

Logarithms were first used in the seventeenth century as an aid to computation. Logarithms to base 10 proved to be especially convenient and were used extensively for long and involved calculations until calculators became readily available. Although logarithms are rarely used for their original purpose, the logarithmic function still occurs in many technical applications and therefore an understanding of it is essential. Example 3 illustrates a technical application of common logarithms.

EXAMPLE 3 The void ratio of a soil is the ratio of the volume of voids in the soil mass to the volume of the solids in the soil mass. When a soil is loaded with foundation loads, the soil will exhibit a certain amount of settlement associated with a decrease in the void ratio. The change Δe in the void ratio of the soil can be calculated as

$$\Delta e = C_c \log (P_2/P_1).$$

Find Δe if $C_c = 0.22$, $P_1 = 0.50$ tons/ft^2, and $P_2 = 2.4$ tons/ft^2. In the formula, C_c is the compression index, P_1 is the initial soil pressure (in tons/ft^2), and P_2 is the final soil pressure (in tons/ft^2).

Solution

$$\begin{aligned}\Delta e &= C_c \log (P_2/P_1)\\ &= 0.22 \log (2.4/0.50)\end{aligned}$$

To evaluate the expression, the following keystroke sequence is used.

2.4 [÷] 0.5 [=] [log] [×] 0.22 [=] **0.1498731**

The change in the void ratio is 0.15, to two significant digits. ■

characteristic
mantissa

NOTE

The integer portion of a logarithm is called the **characteristic,** and the decimal portion is called the **mantissa.** ***When a logarithm is the final answer to a problem, it should be expressed so that the mantissa has the same number of significant digits as the antilogarithm.*** As always, you should carry as many digits as possible during calculations.

Sometimes it is necessary to solve a formula for one of the variables used in a logarithmic expression. Example 4 illustrates the procedure.

EXAMPLE 4 Solve the formula $\Delta e = C_c \log (P_2/P_1)$ for P_2. (This is the formula that was used in Example 3.)

Solution First, divide both sides of the equation by C_c to isolate the logarithm.

$$\frac{\Delta e}{C_c} = \log \left(\frac{P_2}{P_1}\right)$$

Then write the equation in exponential form.

$$10^{\Delta e/C_c} = \frac{P_2}{P_1}$$

Finally, multiply both sides by P_1 to solve the equation for P_2.

$$P_1 \cdot 10^{\Delta e/C_c} = P_2 \quad ■$$

NOTE ⇨⇨ Common logarithms are often used to solve equations that contain exponential expressions. ***If two quantities are equal, their common logarithms are equal.*** Therefore, it is possible to set the logarithm of the expression on one side of the equation equal to the logarithm of the expression on the other side. The properties of logarithms are then used to simplify the result.

EXAMPLE 5 Solve $3^x = 7$.

Solution

$$3^x = 7$$

$$\log 3^x = \log 7 \qquad \text{Set logarithms of equal quantities equal to each other.}$$

$$x \log 3 = \log 7 \qquad \text{Write the logarithm of a power as a product of logarithms.}$$

At this point log 3 and log 7 are treated just like any other real numbers. It is neither necessary nor desirable to evaluate these expressions yet. The equation is solved as a linear equation.

$$x = \frac{\log 7}{\log 3} \qquad \text{Divide both sides by log 3.}$$

Now that the algebraic steps have been completed, the arithmetic may be done with a calculator.

7 [log] [÷] 3 [log] [=] *1.7712438*

Assuming that 3 and 7 were exact (rather than approximate) numbers, x is 1.77, to three significant digits. ■

Although the steps will differ somewhat from one problem to the next, a few general principles can be stated about solving exponential equations.

To solve an exponential equation:

1. Try to isolate the factor containing the exponent. That is, write the equation so that only a base and exponent appear on one side of the equation.
2. Take the logarithm of the expressions on both sides of the equation and set these quantities equal to each other.
3. Use the properties of logarithms to simplify the equation. In particular, use the property that the logarithm of a power is the product of logarithms, so that the variable is no longer an exponent.
4. Solve the equation for the variable, treating logarithms as real numbers (without evaluating them).
5. Evaluate the expression found in step 4, using a calculator.

EXAMPLE 6 Solve $3(2^{x-3}) = 5$.

Solution

1. $2^{x-3} = 5/3$ — Divide both sides by 3 to isolate 2^{x-3}.

2. $\log 2^{x-3} = \log (5/3)$ — Set the logarithms of equal quantities equal to each other.

3. $(x - 3)\log 2 = \log (5/3)$ — Write the logarithm of a power as a product of logarithms.

4. $x \log 2 - 3 \log 2 = \log (5/3)$ — Use the distributive property.

$x \log 2 = \log (5/3) + 3 \log 2$ — Add 3 log 2 to both sides.

$x = \dfrac{\log (5/3) + 3 \log 2}{\log 2}$ — Divide both sides by log 2.

5. A calculator is used to do the arithmetic.

5 [÷] 3 [=] [log] [+] 3 [×] 2 [log] [=] [÷] 2 [log] [=] **3.7369656**

To three significant digits, x is 3.74. ■

EXERCISES 15.4

In 1–28, use a calculator to evaluate each expression. The number and the mantissa of its logarithm should have the same number of significant digits.

1. log 321
2. log 794
3. log 503
4. log 60.3
5. log 78.5
6. log 51.2
7. log 9.67
8. log 4.21
9. log 2.49
10. log 0.158
11. log 0.330
12. log 0.876
13. log 0.0694
14. log 4.0412
15. x if $\log x = 0.3214$
16. x if $\log x = 0.3043$
17. x if $\log x = 1.1342$
18. x if $\log x = 1.4321$
19. x if $\log x = 2.4150$
20. y if $\log y = -2.2459$
21. y if $\log y = -2.5438$
22. y if $\log y = -1.5267$
23. y if $\log y = -1.3566$
24. y if $\log y = -0.6545$
25. x if $\log x = 0.6374 - 1$
26. x if $\log x = 1.4673 - 2$
27. x if $\log x = 1.7652 - 2$
28. x if $\log x = 2.7481 - 3$

In 29–37, solve each equation.

29. $2^x = 3$
30. $5^x = 9$
31. $3^x = 12$
32. $2^{x+1} = 7$
33. $3^{x+2} = 15$
34. $5^{2x-1} = 50$
35. $3(7^x) = 27$
36. $2(5^{x+1}) = 13$
37. $5(3^{x+2}) = 31$

38. The formula $\text{pH} = -\log [\text{H}^+]$ gives the acidity of a solution when $[\text{H}^+]$ is the concentration of hydrogen ions (in moles/liter). Find the acidity if $[\text{H}^+] = 0.00030$. 6 9 2 4 5 7 8

39. If a population of bacteria grows in such a way as to double in number at regular intervals, the number n of intervals elapsed after time t is given by $n = 3.32 \log (b_n/b_0)$ where b_0 is the initial popula-

tion and b_n is the population after time t. Find n if the population grows from 1000 to 100,000.
6 9 2 4 5 7 8

40. The formula $A = 20 \log (V_O/V_I)$ gives the voltage gain A (in dB) of an amplifier, if V_I is the input voltage (in volts) and V_O is the output voltage (in volts). Find the voltage gain of an amplifier whose input voltage is 2.5 millivolts and whose output voltage is 1 volt. **3 5 9** 1 2 4 6 7 8

41. The formula $A = 10 \log P_O/P_I$ gives the power gain (in dB) if P_O is the power output (in W) and P_I is the power input (in W). Find the ratio P_O/P_I for an amplifier that has a power gain of 30 dB.
3 5 9 1 2 4 6 7 8

42. The formula $R = \log (I/I_O)$ gives the magnitude of an earthquake where I is the intensity of the quake and I_0 is a standard minimum intensity. Find the ratio I/I_0 for an earthquake that measures 8.25 on the Richter scale. (The San Francisco earthquake of 1906 measured 8.25.)
1 3 4 5 6 8 9

43. The formula $A = 17 \log (P_0/P_1)$ gives the altitude gain A (in km) of an unpressurized vehicle if P_0 is the atmospheric pressure (in millibars) at the lower altitude and P_1 is the atmospheric pressure (in millibars) at the higher altitude. Find P_0 if P_1 is 550 millibars, and A is 4.5 km.
2 4 8 1 5 6 7 9

44. The formula $B = 10 \log (I/I_0)$ gives the acoustic intensity level (in dB) of a sound of physical intensity I, if I_0 is the physical intensity of a barely audible sound. Given that $I_0 = 10^{-12}\ w/m^2$, find the physical intensity of the noise from a street, which registers 73.2 dB. 1 3 4 5 6 8 9

1. Architectural Technology
2. Civil Engineering Technology
3. Computer Engineering Technology
4. Mechanical Drafting Technology
5. Electrical/Electronics Engineering Technology
6. Chemical Engineering Technology
7. Industrial Engineering Technology
8. Mechanical Engineering Technology
9. Biomedical Engineering Technology

15.5 NATURAL LOGARITHMS

Another commonly used base for logarithms is the number e, which we encountered in Section 14.3. Recall that the value of e is approximately 2.718. Although it may seem to be a strange choice for a base of logarithms, the choice was not made arbitrarily. Logarithms were introduced in 1614 by John Napier, but his definition had nothing to do with exponents. Recall that in contemporary notation, the equation $y = \log_b x$ is equivalent to $x = b^y$. When y is increased by adding 1, x is multiplied by b. Napier developed a relationship between lengths of line segments—one increasing by successive additions, the other decreasing by successive multiplications by a fraction less than 1. The tables he constructed were used to perform a multiplication by performing a related addition. The idea of defining logarithms as exponents began to develop at the end of the seventeenth century. Napier's logarithms were found to be closely related to logarithms with a base of $1/e$. The base e arose in the development of calculus.

Logarithms to the base e are called *natural logarithms.* Natural logarithms, like common logarithms, are rarely used now to perform numerical calculations. They are important because the logarithmic and exponential functions occur in the technologies.

Just as common logarithms were important enough to have their own notation, the natural logarithm of x is indicated by $\ln x$.

DEFINITION The **natural logarithm** of x is denoted $\ln x$ and

$$\ln x = \log_e x,$$

where e is approximately 2.718.

A scientific calculator has an [ln] key. It is used just like the [log] key. Antilogarithms are found by the keystroke sequence [INV] [ln] or [e^x].

EXAMPLE 1 Use a calculator to evaluate each expression.

(a) $\ln 42.1$

(b) $\ln 0.321$

Solution The keystroke sequence is as follows.

(a) 42.1 [ln] **3.7400477**

(b) .321 [ln] **−1.1363142** ■

EXAMPLE 2 **(a)** If $\ln x = 2.547$, find x.

(b) If $\ln x = -1.532$, find x.

Solution **(a)** 2.547 [INV] [ln] **12.76874**

(b) 1.532 [+/−] [INV] [ln] **0.216103** ■

Natural logarithms, like common logarithms, have technical applications.

EXAMPLE 3 The formula $f = (f_0 - f_c)e^{-kt} + f_c$ gives the rate of infiltration f (in in/hr) of rainwater into the soil at time t (in hrs) after a rainstorm begins. Write the equation in logarithmic form. In the formula, f_0 is the infiltration rate (in in/hr) at the beginning of the storm, f_c is a constant infiltration rate (in in/hr) that is attained after a period of time, and k is a constant, depending on the field conditions.

Solution To write the formula in logarithmic form, we must first solve the equation for e^{-kt}.

$$f - f_c = (f_0 - f_c)e^{-kt} \quad \text{Subtract } f_c \text{ from both sides.}$$

$$\frac{f - f_c}{f_0 - f_c} = e^{-kt} \quad \text{Divide both sides by } f_0 - f_c.$$

Now we can say that $-kt$ is the exponent or logarithm of the quantity on the left-hand side of the equation when the base is e. That is,

$$\ln\left(\frac{f - f_c}{f_0 - f_c}\right) = -kt. \quad ■$$

It is natural to ask whether there is a relationship between natural logarithms and common logarithms. There is, and in fact, the relationship is more general. There is a relationship between logarithms of *any* two bases. To develop that relationship, consider the logarithm of a number x in the two bases, a and b.

$$\text{Let } u = \log_b x. \qquad \textbf{(15-11)}$$
$$\text{Then } b^u = x.$$

Taking the logarithm to the base a of the expressions on both sides of the equation, we have the following.

$$\log_a b^u = \log_a x$$

$$u \log_a b = \log_a x \qquad \text{Write the log of a power as the product of logs.}$$

$$u = \frac{\log_a x}{\log_a b} \qquad \text{Divide both sides of the equation by } \log_a b.$$

$$\log_b x = \frac{\log_a x}{\log_a b} \qquad \text{Substitute } \log_b x \text{ for } u.$$

The formula is especially useful for converting logarithms to either base 10 or base e. A scientific calculator does not have a $\boxed{\log_b x}$ key, so such a function could be evaluated by computing $\frac{\log x}{\log b}$. Likewise, computer languages have certain functions that are built-in. That is, there are certain functions that can be used automatically. The functions $\log x$ and $\ln x$ are common computer functions. However, the more general $\log_b x$ is not. In Pascal, only $\ln x$ is available. A computer program requiring that $y = \log_b x$ be evaluated could use $\frac{\ln x}{\ln b}$.

CONVERSIONS OF LOGARITHMS TO BASE 10 AND BASE *e*

$$\log_b x = \frac{\log x}{\log b} \qquad \textbf{(15-12)}$$

$$\log_b x = \frac{\ln x}{\ln b} \qquad \textbf{(15-13)}$$

NOTE ⇨⇨ ***Notice that in each formula, both numerator and denominator are logarithms in the same base, either* 10 *or e. The numerator is the logarithm of the* number *in the original expression and the denominator is the logarithm of the* base *in the original expression.***

EXAMPLE 4 A certain camera filter transmits 40% of the light striking it. Find a formula for the number of filters needed to transmit p percent.

Solution Let n = number of filters used. Since 40% is transmitted for *each* filter, we have

$$p = (0.40)^n.$$

The equation can be written in logarithmic form, rather than exponential form.

$$\log_{0.40} p = n$$

A formula based on logarithms to base 0.40 is of little use. Therefore, we can use
$n = \dfrac{\log p}{\log 0.40}$ or $n = \dfrac{\ln p}{\ln 0.40}$. ■

EXAMPLE 5 **(a)** Write $\log_3 x$ in terms of common logarithms and evaluate the expression when x is 212.

(b) Write $\log_2 x$ in terms of natural logarithms and evaluate the expression when x is 0.411.

Solution **(a)** $\log_3 x = \dfrac{\log x}{\log 3}$

When x is 212, we have $\log_3 212 = \dfrac{\log 212}{\log 3}$. The keystroke sequence is as follows.

212 [log] [÷] 3 [log] [=] **4.8757751**

Since the number has three significant digits, the mantissa is given with three significant digits: $\log_3 212 = 4.876$.

(b) $\log_2 x = \dfrac{\ln x}{\ln 2}$

When x is 0.411, $\log_2 0.411 = \dfrac{\ln 0.411}{\ln 2}$. The keystroke sequence is as follows.

.411 [ln] [÷] 2 [ln] [=] **−1.2827897**

Since the number has three significant digits, the mantissa is given with three significant digits: $\log_2 0.411 = -1.283$. ■

Although a scientific calculator and some computer languages will handle the function defined by $y = a^x$, some computer languages (such as Pascal) do not. To evaluate a^x in such a language, the function is written in terms of e^x. Since the logarithm of a power is the product of logarithms, we have

$$\ln a^x = x \ln a.$$

Writing this equation in exponential form, we have

$$a^x = e^{x \ln a}.$$

CONVERSION OF AN EXPONENTIAL FUNCTION TO BASE e

$$a^x = e^{x \ln a} \tag{15-14}$$

EXAMPLE 6 **(a)** Write 52.1^x in terms of e and evaluate the expression when x is 2.41.

(b) Write 3.88^x in terms of e and evaluate the expression when x is 17.3.

Solution **(a)** $52.1^x = e^{x \ln 52.1}$. When x is 2.41, $52.1^{2.41} = e^{2.41 \ln 52.1}$. The keystroke sequence is as follows.

2.41 [×] 52.1 [ln] [=] [e^x] **13727.102**

Some calculators do not have an [e^x] key, but the sequence [INV] [ln] is equivalent to [e^x]. Thus, $52.1^{2.41} = 13{,}700$, to three significant digits.

(b) $3.88^x = e^{x \ln 3.88}$. When x is 17.3, $3.88^{17.3} = e^{17.3 \ln 3.88}$

The keystroke sequence is as follows.

17.3 [×] 3.88 [ln] [=] [e^x] **1.5374 10**

Thus, $3.88^{17.3} = 1.54 \times 10^{10}$ to three significant digits. ■

EXERCISES 15.5

In 1–24, use a calculator to evaluate each expression.

1. ln 321 **2.** ln 794 **3.** ln 60.3 **4.** ln 78.5

5. ln 51.2 **6.** ln 9.67 **7.** ln 4.21 **8.** ln 0.158

9. ln 0.330 **10.** ln 0.876 **11.** ln 0.0694 **12.** ln 4.0412

13. x if $\ln x = 0.321$ **14.** x if $\ln x = 1.1342$

15. x if $\ln x = 1.4321$ **16.** x if $\ln x = 2.4150$

17. y if $\ln y = -2.2459$ **18.** y if $\ln y = -2.543$

19. y if $\ln y = -1.3566$ **20.** y if $\ln y = -0.6545$

21. x if $\ln x = 0.6374 - 1$ **22.** x if $\ln x = 1.4673 - 2$

23. x if $\ln x = 1.7652 - 2$ **24.** x if $\ln x = 2.7481 - 3$

In 25–30, write each expression in terms of the common logarithm function and evaluate it for the given value of x.

25. $\log_2 x$, $x = 51$ **26.** $\log_2 x$, $x = 95$

27. $\log_3 x$, $x = 6.2$ **28.** $\log_3 x$, $x = 8.7$

29. $\log_5 x$, $x = 0.77$ **30.** $\log_5 x$, $x = 0.84$

In 31–37, write each expression in terms of the natural logarithm function and evaluate it for the given value of x.

31. $\log_7 x$, $x = 953$

32. $\log_7 x$, $x = 16.2$

33. $\log_7 x$, $x = 59.5$

34. $\log_9 x$, $x = 4.01$

35. $\log_9 x$, $x = 0.31$

36. $\log_{11} x$, $x = 0.317$

37. $\log_{11} x$, $x = 0.380$

In 38–43, write each expression in terms of the exponential function $y = e^x$ and evaluate it for the given value of x.

38. 18.4^x, $x = 5.39$

39. 54.7^x, $x = 4.48$

40. 0.366^x, $x = 12.7$

41. 0.456^x, $x = 14.6$

42. 2.75^x, $x = 7.54$

43. 3.64^x, $x = 6.32$

Equations of the form $y = ae^{bx}$ are often solved for x by taking the natural logarithm of both sides rather than by taking the common logarithm of both sides. Apply this principle to the equations in 44–47.

44. The formula $Q = Q_0 e^{-t/\tau}$ gives the charge Q (in coulombs) on a capacitor after it has been discharging through a resistance for time t (in s). In the formula, Q_0 is the original charge (in coulombs), and $\tau = RC$ (resistance × capacitance). Find a formula for t. 3 5 9 4 7 8

45. In a circuit containing a resistor and an inductor in series, if the battery is short-circuited, the current does not fall immediately to zero. The current drops gradually, so that the current I (in A) is given by $I = I_0 e^{-t/\tau}$ where I_0 is the steady-state current (in A), $\tau = L/R$ (inductance/resistance), and t is the time elapsed (in s). Find a formula for t. 3 5 9 1 2 4 6 7 8

46. When a belt passes over a rough pulley, the tensions in the belt will be different on the two sides. When the system is in equilibrium, but slippage is impending, the formula $T_1 = T_2 e^{\mu\alpha}$ gives the larger tension T_1. T_2 is the smaller tension, μ is the coefficient of friction, and α is the angle of wrap in radians. Find a formula for μ. 2 4 8 1 5 6 7 9

47. The formula $C = C_0 e^{-t}$ gives the flow C (in liters/min) of blood through a person's arteries. In the formula, C_0 is the initial velocity, and t is the time elapsed (in minutes). Find a formula for t. 6 9 2 4 5 7 8

48. For a well-tuned piano, the frequency of each note is $\sqrt[12]{2}$ times that of the note below it. The note A above middle C is set at 440 cycles/second. Find a formula that gives the number of notes n above A of a note with frequency f. Express the final answer using natural logarithms. 1 3 4 5 6 8 9

49. A particular vacuum pump is designed so that after each stroke, 97% of the gas remains in the chamber. Find a formula for the number of strokes n necessary to leave only p% of the gas. (Hint: To express p% as a decimal, use $0.01\,p$.) Express the final answer using natural logarithms. 4 8 1 2 6 7 9

50. The half-life of a radioactive element is the time needed for a sample of any size to retain exactly half of the radioactivity it had at the beginning of the period. Let A_0 be the initial amount present, let t be the time elapsed, let A be the amount present after time t, and let n be the half-life of the element (expressed in the same unit as t). Find a formula for n, and express the final answer using natural logarithms. 6 9 5 8

51. Some scientific calculators do not have an e^x key. Explain how to find the value of e with such a calculator.

---

1. Architectural Technology
2. Civil Engineering Technology
3. Computer Engineering Technology
4. Mechanical Drafting Technology
5. Electrical/Electronics Engineering Technology
6. Chemical Engineering Technology
7. Industrial Engineering Technology
8. Mechanical Engineering Technology
9. Biomedical Engineering Technology

15.6 LOGARITHMIC AND EXPONENTIAL EQUATIONS

When an equation contains one or more logarithms, there are several techniques that can be used. If there is a single logarithmic term, it may be possible to write the equation in exponential form and solve it by inspection. That is, one simply examines the problem and looks for an obvious solution.

EXAMPLE 1 Solve each equation by inspection.

(a) $\log_x 2 = 1/2$

(b) $\log_3 1/9 = x$

Solution **(a)** If $\log_x 2 = 1/2$, then $x^{1/2} = 2$. Recalling that $x^{1/2}$ means $\sqrt{x}$, we quickly recognize that 4 is the solution of $\sqrt{x} = 2$.

(b) If $\log_3 1/9 = x$, then $3^x = 1/9$. A little thought shows that -2 is the solution. ■

EXAMPLE 2 Solve each equation by inspection.

(a) $x = \log_{1/2} 1/4$

(b) $\log_4 x = -1/2$

(c) $\log_x 4 = 2/3$

Solution **(a)** If $x = \log_{1/2} 1/4$, then $(1/2)^x = 1/4$.
Since $(1/2)(1/2) = 1/4$, we must have $x = 2$.

(b) If $\log_4 x = -1/2$, then $4^{-1/2} = x$.

$$\begin{aligned} x &= 1/4^{1/2} \\ &= 1/\sqrt{4} \\ &= 1/2 \end{aligned}$$

(c) If $\log_x 4 = 2/3$, then $x^{2/3} = 4$.

We are looking for a value whose cube root is squared to produce 4. Since 2^2 is 4, 2 is the cube root, and 8 is the number. That is, x is 8. Another way the problem could be solved would be to cube both sides of the equation $x^{2/3} = 4$.

$$\begin{aligned} (x^{2/3})^3 &= 4^3 \\ x^2 &= 64 \\ x &= \pm\sqrt{64} \text{ or } \pm 8 \end{aligned}$$

Since the base must be positive in a logarithmic expression, we have only $x = 8$. ■

It is not always possible to solve a logarithmic equation by inspection. But if the equation can be written with a single logarithmic term, it may be possible to write it in exponential form and use a calculator to evaluate exponential expressions. If there are several logarithmic terms in the equation, the properties of logarithms are used to simplify the equation.

EXAMPLE 3 The formula $A = 20 \log (V_O/V_I)$ gives the voltage gain A (in dB) of an amplifier if V_O is the output voltage and V_I is the input voltage. How much input voltage is required to produce a 10-volt output from an amplifier whose voltage gain is 38 dB?

Solution

$$A = 20 \log (V_O/V_I)$$

$$38 = 20 \log (10/V_I)$$

$$38/20 = \log (10/V_I)$$ Divide both sides by 20.

$$1.9 = \log (10/V_I)$$ Reduce 38/20 to 19/10, which is 1.9.

$$10^{1.9} = 10/V_I$$ Write in exponential form.

$$V_I(10^{1.9}) = 10$$ Multiply both sides by V_I.

$$V_I = 10/10^{1.9} \text{ or } 10^{-0.9}$$ Divide both sides by $10^{1.9}$.

A calculator shows that $10^{-0.9}$ is about 0.13. An input voltage of 0.13 volts would produce a 10-volt output. ■

If the equation contains logarithms with a base other than 10 or e, Equations 15–12 or 15–13 ($\log_b x = \log x/\log b$ and $\log_b x = \ln x/\ln b$) may be needed to change the base so that the [log] or [ln] keys on a calculator may be used.

EXAMPLE 4 Solve $\log_3 8 = x - \log_3 4$.

Solution

$$\log_3 8 = x - \log_3 4$$

$$\log_3 8 + \log_3 4 = x$$ Add $\log_3 4$ to both sides.

$$\log_3 (32) = x$$ Write the sum of logarithms as the logarithm of a product.

$$x = \log 32/\log 3$$ Replace $\log_3 32$ with $\log 32/\log 3$.

$x = 3.15$, to three significant digits. ■

It may not be necessary to write the equation with a single logarithm. If two logarithms are equal, their antilogarithms must be equal. If the equation can be written so that there is a single term on each side, and each term is a logarithm, it may help to apply this principle.

EXAMPLE 5 Solve $2 \ln x = \ln 2x$.

Solution

$$2 \ln x = \ln 2x$$

$$\ln x^2 = \ln 2x$$ Write the product of logarithms as the logarithm of a power.

$x^2 = 2x$	Set the antilogarithms equal, since the logarithms are equal.
$x^2 - 2x = 0$	Subtract $2x$ from both sides.
$x(x-2) = 0$	Factor the left-hand side.
$x = 0$ or $x - 2 = 0$	Set both factors equal to 0.
$x = 0$ or $x = 2$	Solve the two linear equations.

Recall that logarithms are defined only for positive numbers. The only valid solution, then, is given by $x = 2$. ■

EXAMPLE 6 Solve $\log 6 = \log (x^2 + 4x) - 1$.

Solution

$\log 6 = \log (x^2 + 4x) - 1$	
$0 = \log (x^2 + 4x) - \log 6 - 1$	Subtract log 6 from both sides.
$0 = \log \dfrac{(x^2 + 4x)}{6} - 1$	Write the difference of logarithms as the logarithm of a quotient.
$1 = \log \dfrac{x^2 + 4x}{6}$	Add 1 to both sides.
$10 = \dfrac{x^2 + 4x}{6}$	Set the antilogs equal, since $1 = \log 10$.
$60 = x^2 + 4x$	Multiply both sides by 6.
$0 = x^2 + 4x - 60$	Subtract 60 from both sides.
$0 = (x + 10)(x - 6)$	Factor the expression on the right.
$x + 10 = 0$ or $x - 6 = 0$	Set each factor equal to zero.
$x = -10$ or $x = 6$	Solve each linear equation.

A check shows that both solutions are correct:
When x is -10, $\log 6 = \log ((-10)^2 + 4(-10)) - 1 = \log (60) - 1$.
When x is 6, $\log 6 = \log (6^2 + 4(6)) - 1 = \log (60) - 1$. ■

In many cases, an equation with two or more variables may be solved for one variable in terms of the others.

EXAMPLE 7 Solve $3 \log x = 2 - \log y$ for y.

Solution

$3 \log x = 2 - \log y$	
$\log x^3 = 2 - \log y$	Write the product of logarithms as the logarithm of a power.
$\log x^3 + \log y = 2$	Add $\log y$ to both sides.
$\log x^3 y = 2$	Write the sum of logarithms as the logarithm of a product.
$x^3 y = 10^2$	Write the equation in exponential form.
$x^3 y = 100$	Replace 10^2 with 100.
$y = 100/x^3$	Divide both sides by x^3. ■

EXAMPLE 8 The formula $R = \log (I/I_0)$ gives the magnitude of an earthquake where I is the intensity of the quake and I_0 is a standard minimum intensity. How many times greater is the intensity of an earthquake that measures 8.2 than one that measures 4.1?

Solution Let $I_1 =$ the intensity of the 8.2 earthquake.
Let $I_2 =$ the intensity of the 4.1 earthquake.
To determine how many times greater I_1 is than I_2, we must find the ratio I_1/I_2. Consider $R = \log (I/I_0)$ for $I = I_1$ and for $I = I_2$. Then

$$8.2 = \log (I_1/I_0) \qquad \text{and} \qquad 4.1 = \log (I_2/I_0).$$

These equations can be solved for I_1 and I_2, respectively.

$$\begin{aligned} 8.2 &= \log I_1 - \log I_0 \\ 8.2 + \log I_0 &= \log I_1 \\ \log 10^{8.2} + \log I_0 &= \log I_1 \\ \log (10^{8.2} I_0) &= \log I_1 \end{aligned} \qquad \begin{aligned} 4.1 &= \log I_2 - \log I_0 \\ 4.1 + \log I_0 &= \log I_2 \\ \log 10^{4.1} + \log I_0 &= \log I_2 \\ \log (10^{4.1} I_0) &= \log I_2 \end{aligned}$$

When the logarithms are equal, the antilogarithms are equal. Thus,

$$10^{8.2} I_0 = I_1 \qquad \text{and} \qquad 10^{4.1} I_0 = I_2.$$

To determine how many times greater I_1 is than I_2, consider the ratio I_1/I_2.

$$\frac{I_1}{I_2} = \frac{10^{8.2} I_0}{10^{4.1} I_0} = \frac{10^{8.2}}{10^{4.1}} = 10^{4.1} = 13{,}000, \text{ to two significant digits.}$$

Thus, an earthquake that measures 8.2 is about 13,000 times greater in intensity than one that measures 4.1. ■

In Section 15.4, we saw that exponential equations can be solved by taking the common logarithm of both sides. Sometimes the properties of logarithms may be applied to simplify exponential equations.

EXAMPLE 9 Solve $2^{x-1} = 3^x$.

Solution

$$\log 2^{x-1} = \log 3^x$$ Set logarithms of equal quantities equal to each other.

$$(x - 1)\log 2 = x \log 3$$ Write the logarithm of a power as the product of logarithms.

$$x \log 2 - \log 2 = x \log 3$$ Use the distributive property.

$$x \log 2 = x \log 3 + \log 2$$ Add log 2 to both sides.

$$x \log 2 - x \log 3 = \log 2$$ Subtract x log 3 from both sides.

$$x(\log 2 - \log 3) = \log 2$$ Factor out x.

$$x = \frac{\log 2}{\log 2 - \log 3}$$ Divide both sides by log 2 − log 3.

Although we could evaluate this expression for x, the computation will be shortened slightly if $\log 2 - \log 3$ is replaced by $\log (2/3)$.

$$x = \frac{\log 2}{\log (2/3)}$$

Assuming that the integer values in the equation are exact values, the answer is given to three significant digits, so x is -1.71. ■

When e appears in an exponential equation, we take the *natural logarithm* of both sides.

EXAMPLE 10 Solve $2e^{3x} = 5(3^{x+1})$.

Solution

$$2e^{3x} = 5(3^{x+1})$$

$$e^{3x} = \frac{5(3^{x+1})}{2}$$ Divide both sides by 2.

$$\ln e^{3x} = \ln[2.5(3^{x+1})]$$ Set logarithms of equal quantities equal to each other.

$$3x = \ln[2.5(3^{x+1})]$$ Replace $\ln e^{3x}$ with $3x$.

$$3x = \ln 2.5 + \ln 3^{x+1}$$ Write the logarithm of a product as the sum of logarithms.

$$3x = \ln 2.5 + (x + 1)\ln 3$$ Write the logarithm of a power as the product of logarithms.

$$3x = \ln 2.5 + x \ln 3 + \ln 3$$ Use the distributive property.

$$3x - x \ln 3 = \ln 2.5 + \ln 3$$ Subtract $x \ln 3$ from both sides.

$$x(3 - \ln 3) = \ln 2.5 + \ln 3$$ Use the distributive property.

$$x = \frac{\ln 2.5 + \ln 3}{3 - \ln 3}$$ Divide both sides by $3 - \ln 3$.

$$x = \frac{\ln [3(2.5)]}{3 - \ln 3}$$ Write the sum of logarithms as the logarithm of a product.

To three significant digits, x is 1.06. ■

EXERCISES 15.6

In 1–8, solve each equation by inspection.

1. $x = \log_3 9$ **2.** $x = \log_4 1/4$ **3.** $y = \log_{3/4} 9/16$ **4.** $y = \log_{2/3} 8/27$

5. $-1 = \log_a 1/3$ **6.** $1/2 = \log_b 5$ **7.** $1/2 = \log_x 6$ **8.** $-2 = \log_y 1/4$

In 9–34, solve each equation.

9. $\log_5 14 - x = \log_5 2$ **10.** $\log_3 6 + 2x = \log_3 36$ **11.** $\ln 7 = \ln 2x - \ln 3$

12. $2 \ln 9 = 3x - \ln 2$ **13.** $3 \log (2x + 1) = 6$ **14.** $2 \log (x + 2) = 10$

15. $2 \ln 5 = x - \ln 15$

16. $5 \ln (2x - 1) = 15$

17. $\log 5 = \log (x^2 - 5x) - 1$

18. $\log 4 + 1 = \log(x^2 - 3x)$

19. $1 + \log 3 = \log(x^2 + x)$

20. $\log y = \log x - \log 3$ for y

21. $\log y + \log x = \log 5$ for y

22. $4 \log y + 2 \log x = 4$ for y

23. $3 \log y - 1 = \log (x + 1)$ for y

24. $2 \log y - 3 = \log (2x - 1)$ for y

25. $5^{x-1} = 2^x$

26. $3^{x-1} = 5^x$

27. $3^{x+2} = 4^x$

28. $2^{x-1} = 7^{2x}$

29. $5^{x+2} = 2^{3x}$

30. $2e^{x+1} = 5$

31. $5e^{x-1} = 10$

32. $3e^{3x} = 2^{x+1}$

33. $(1/2)e^{2x} = 3^{x-1}$

34. $(1/3)e^{x-1} = 5(2^x)$

---

35. The formula $B = P(1 + r)^n$ gives the balance B in an account after n compounding periods if the principal was P and r is the rate of interest (as a decimal) per compounding period.

(a) If \$1000 is invested at a 5 percent annual percentage rate, how long will it take for the balance to reach \$1500?

(b) How many years will it take for an investment of \$1000 to earn \$200 at a 4% annual percentage rate?

(c) How many years will it take for an investment of any size to double at a 6% annual percentage rate? 1 2 6 7 8 3 5 9

36. A copy machine can be set to reduce or enlarge a document. The copy can then be reduced or enlarged, and so on.

(a) If the machine is set to reduce to 75%, how many copies must be made to ensure that the final copy is no more than 25% of its original size?

(b) If the machine is set to reduce to 64%, how many copies must be made to ensure that the final copy is no more than 25% of its original size?

(c) If the machine is set to enlarge a document by 25%, how many copies must be made to ensure that the final copy is at least twice as large as the original? 1 2 4 7 8 3 5

37. The reliability R of a roller bearing is given by the following formula.

$$R = e^{-[L/(6.84\,L_{10})]^{1.17}}$$

Solve the equation for L_{10}. In the formula, L is the required life (in hours) and L_{10} is the rated life of the bearing (in hours). 4 7 8 1 2 3 5 6 9

38. When a capacitor is charged through a circuit with a resistance, it does not reach the full charge immediately. The charge increases gradually so that the charge Q (in coulombs) at time t (in seconds) is given by $Q = Q_0(1 - e^{-t/\tau})$ where Q_0 is the final charge and $\tau = RC$ (resistance × capacitance).

(a) If $Q_0 = 9 \times 10^{-4}$ coulombs, and $\tau = 0.04$ s, find the time t required for the charge to reach 4×10^{-4} coulombs.

(b) If $Q_0 = 9 \times 10^{-4}$ coulombs, and $\tau = 0.04$ s, find the time t required for the charge to reach 63% of Q_0. 3 5 9 4 7 8

39. The formula $B = 10 \log (I/I_0)$ gives the acoustic intensity level (in dB) of a sound of physical intensity I, if I_0 is the physical intensity of a sound that is barely audible to an average person.

(a) A noisy office has a decibel level of about 70. A machine shop has a decibel level of about 100. How many times greater is the intensity of sound in the machine shop than in the office?

(b) If the physical intensity of one noise is twice that of another, find the difference in their decibel levels. 1 3 4 5 6 8 9

40. The equation $\Delta e = C_c \log(P_2/P_1)$ gives the change Δe in the void ratio of a soil that is loaded with a foundation load. If the change in void ratio

is 0.20 and $C_c = 0.30$, how many times greater is P_2 than P_1? In the formula, C_c is a constant that depends on the soil, P_1 is the initial soil pressure in tons/ft^2, and P_2 is the final soil pressure in tons/ft^2.
1 2 4 6 8 5 7 9

41. Assume that equipment depreciates in value at a rate of 10% a year.

(a) How many years will it take for equipment purchased for \$5000 to depreciate to a value of \$3000?

(b) How many years will it take for equipment to depreciate to half of its original value?

(c) Find a formula for the number of years n for equipment to depreciate from V_0 to V_d if r is the annual rate of depreciation (as a decimal).
1 2 6 7 8 3 5 9

15.7 GRAPHS ON SEMILOGARITHMIC AND LOGARITHMIC PAPER

In Sections 15.1 and 15.2, the graphs of logarithmic and exponential functions were sketched. Very often, the y-values were spread over a larger range than the x-values. In such cases, a different scale may be used on the two axes.

EXAMPLE 1 Sketch the graph of $y = 4e^{2x}$.

Solution Several x-values are chosen and the corresponding y-values computed. The graph is shown in Figure 15.12.

x	y
-2	0.0733
-1	0.541
0	4
1	29.6
2	218

■

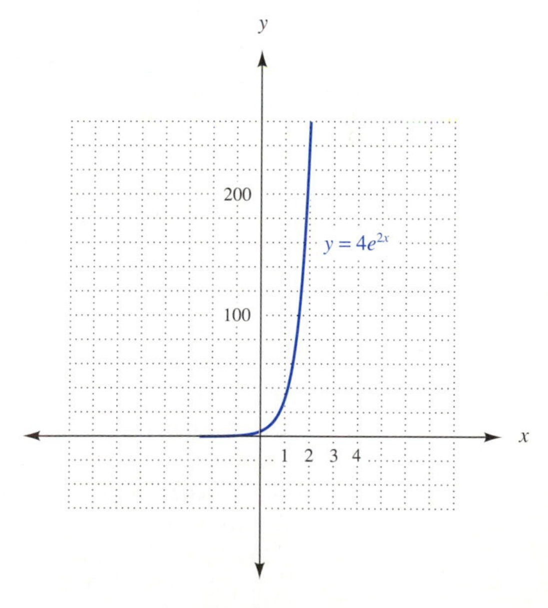

FIGURE 15.12

Since the units are so small on the y-axis, it is difficult to see the difference in y-values for x-values of -2, -1, and 0. The y-values are so small in comparison to the scale used that all three values appear at approximately the same height in the vertical direction.

To show such subtle differences, a different type of graph paper is useful. Up to this point, we have used graphs for which equal increments are represented by equal distances anywhere on the same axis. For example, the distance between 4.5 and 5.0 is the same as the distance between 7.0 and 7.5. The axis for such a graph is said to be uniformly scaled. On *semilogarithmic graph paper* (sometimes called semilog paper) the y-axis is scaled so that between 1 and 10, the units are labeled where their logarithms would be if the axis were uniformly scaled from 0 to 1. Figure 15.13 shows a y-axis that is uniformly scaled and logarithmically scaled.

semilogarithmic graph paper

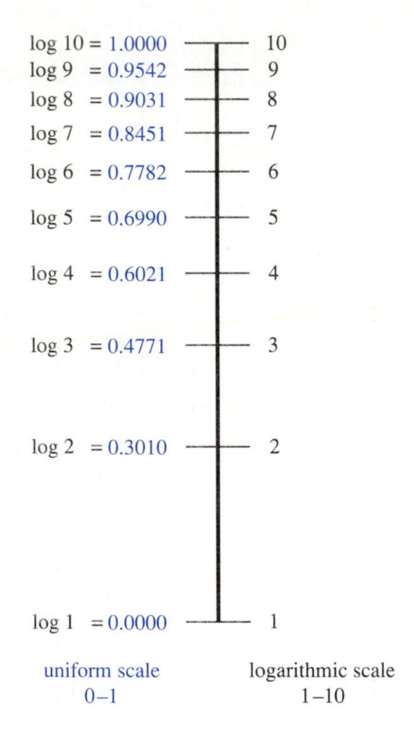

FIGURE 15.13

cycle

Semilog paper characteristically has either two or three *cycles*. Each cycle is used to represent values that are larger than those of the previous cycle by a factor of 10. The range of the first cycle is determined from the data to be graphed. If the first cycle has y-values that run from 1 to 10, the second cycle contains values that run from 10 to 100, and the third cycle contains values that run from 100 to 1000. If, on the other hand, it is decided that the first cycle should be used to represent values from 0.1 to 1, the second cycle contains values that run from 1 to 10, and the third cycle contains values that run from 10 to 100.

EXAMPLE 2 Sketch the graph of $y = 4e^{2x}$ on semilog paper.

Solution Since this is the same function graphed in Example 1, the table of values from that example is used. Before we begin to plot points, we must label the axes. The x-values are to range from -2 to 2. The x-axis is uniformly scaled on semilog paper, so the vertical axis may be conveniently placed in the middle of the graph. The y-values range from 0.0733 to 218. That is, they range from hundredths to hundreds, so five cycles (hundredths, tenths, units, tens, and hundreds) would be necessary to graph all of the values in the table. Using 3-cycle paper, we can graph the first cycle from 0.01 to 0.1, the second cycle from 0.1 to 1, and the third cycle from 1 to 10. Or, we can graph the first cycle from 0.1 to 1, the second cycle from 1 to 10, and the third cycle from 10 to 100. Both possibilities exclude two values listed on the table. We choose the second option, since it includes a larger range of values. Figure 15.14 shows the graph. ■

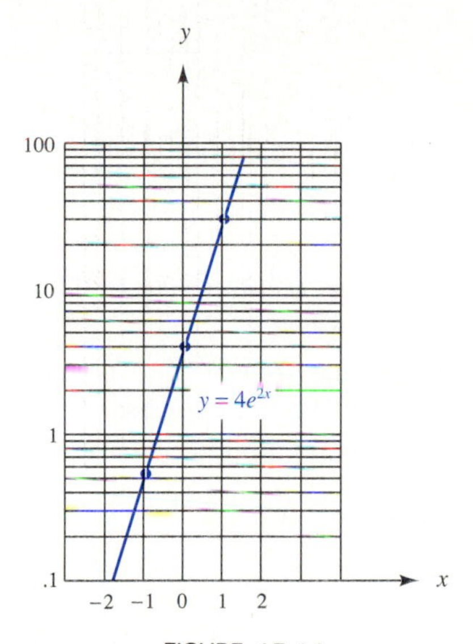

FIGURE 15.14

The graph in Example 2 appears to be a straight line, and, in fact, it is. This fact is not a coincidence. ***Graphs of functions of the form $y = ab^x$ are straight lines when plotted on semilog paper.*** To see why this is so, consider the equation $y = ab^x$. On semilog paper, the y-values of the function are plotted where their logarithms would be if a uniform scale were superimposed on the graph. Thus, on the uniform scale, we have plotted $y = \log ab^x$. But this equation can be written

NOTE ⇨⇨

$$y = \log a + \log b^x$$
$$y = \log a + x \log b.$$

Since $\log a$ and $\log b$ are constants, this equation is a linear equation.

The steps used in Example 2 are summarized as follows.

To sketch a graph on semilogarithmic paper:

1. Choose a few values for x and find the corresponding y-values.
2. Place the vertical axis in a position that will allow all or most of the x-values in the table to be shown. Place the horizontal axis at the bottom of the graph.
3. Choose the range for the first cycle of y-values. Each cycle must contain y-values that are larger than those in the previous cycle by a factor of 10.
4. Plot the points and connect them.

EXAMPLE 3 Sketch the graph of $y = 3(2^{-x})$ on semilog paper.

Solution 1. We choose a few values for x and find the corresponding y-values.

x	$y = 3(2^{-x})$
-2	$3(2^2) = 12$
-1	$3(2^1) = 6$
0	$3(2^0) = 3$
1	$3(2^{-1}) = 3/2 = 1.5$
2	$3(2^{-2}) = 3/4 = 0.75$

2. The x-values range from -2 to 2, so the vertical axis is conveniently located in the middle of the graph.
3. The y-values range from 0.75 to 12, that is, from tenths to tens, so three cycles (tenths, units, and tens) are necessary. The first cycle runs from 0.1 to 1, the second cycle runs from 1 to 10, and the third cycle runs from 10 to 100.
4. Figure 15.15 shows the graph. ■

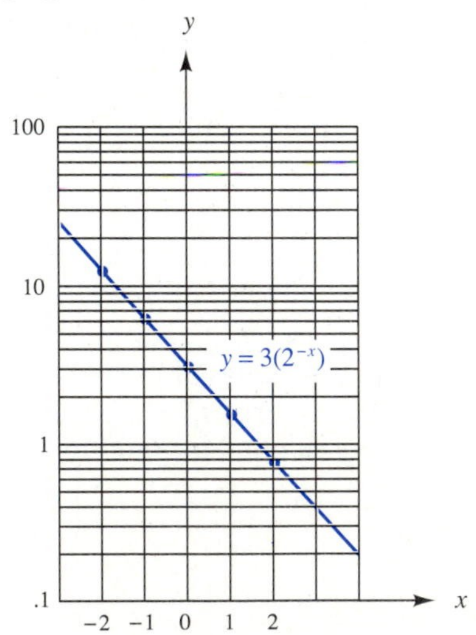

FIGURE 15.15

NOTE

logarithmic graph paper

Semilogarithmic paper has one axis (usually the vertical) marked with a logarithmic scale and one axis marked with a uniform scale and is used when there is a large range of values for one of the variables. Logarithmic graph paper (sometimes called log-log paper) is marked with a logarithmic scale on both axes and is used when there is a large range of values for both of the variables.

To sketch a graph on logarithmic paper:

1. Choose a few values for x and find the corresponding y-values.
2. Place the vertical axis at the left-hand side of the graph. Place the horizontal axis at the bottom of the graph.
3. Choose the domain for the first cycle of x-values. Choose the range for the first cycle of y-values. Each cycle must contain values that are larger than those in the previous cycle by a factor of 10.
4. Plot the points and connect them.

EXAMPLE 4 Sketch the graph of $y = 3x^3$ on log-log paper.

Solution Since only positive x-values can be shown on log-log paper, a few positive values are chosen for x and the corresponding y-values are found. Figure 15.16 shows the graph.

x	$y = 3x^3$
0.5	$3(0.5)^3 = 3(0.125) = 0.375$
0.6	$3(0.6)^3 = 3(0.216) = 0.648$
1	$3(1^3) = 3(1) = 3$
1.2	$3(1.2)^3 = 3(1.728) = 5.184$ ■

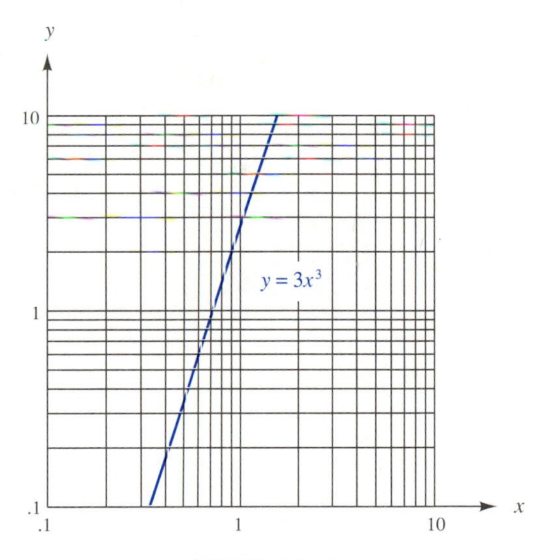

FIGURE 15.16

Just as the graph of a function of the form $y = ab^x$ is a straight line on semilog paper, the graph of a function of the form $y = ax^b$ is a straight line on log-log paper. ***You should not, however, get the idea that graphs of all functions are straight lines when plotted on logarithmic scales.***

NOTE

EXAMPLE 5 Sketch the graph of $y = x^3 + 5x$ on log-log paper for values of x between 1 and 10.

Solution The following table shows a few x-values and the corresponding y-values. Figure 15.17 shows the graph.

x	$y = x^3 + 5x$
1.5	$1.5^3 + 5(1.5) = 10.9$
2	$2^3 + 5(2) = 18$
3	$3^3 + 5(3) = 42$
4	$4^3 + 5(4) = 84$
5	$5^3 + 5(5) = 150$
6	$6^3 + 5(6) = 246$
8	$8^3 + 5(8) = 552$
9	$9^3 + 5(9) = 774$

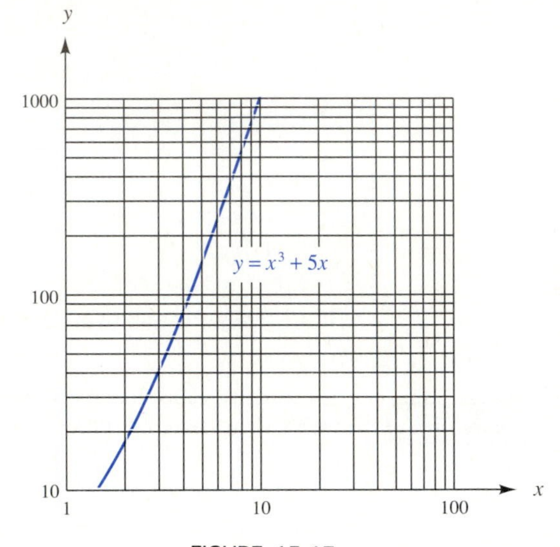

FIGURE 15.17

Only one cycle is necessary for the x-values, but two cycles are necessary for the y-values. ■

EXERCISES 15.7

In 1–12, sketch the graph of each function on semilog paper.

1. $y = 3^x$ **2.** $y = 4^x$ **3.** $y = (1/4)5^x$

4. $y = 4(2^{-x})$ **5.** $y = 5^{-x}$ **6.** $y = (5/4)^{-x}$

7. $y = 4(1/3)^x$ **8.** $y = 2(1/5)^x$ **9.** $y = 5(1/2)^x$

10. $y = 3e^{2x}$ **11.** $y = 2e^{3x}$ **12.** $y = 3^{-x}$

In 13–24, sketch the graph of each function on log-log paper.

13. $y = 2x^3$ **14.** $y = 3x^2$ **15.** $y = 2x^5$

16. $y = x^{-1}$ **17.** $y = 3x^4$ **18.** $y = 4x^{-3}$

19. $y = 5x^{-2}$ **20.** $y = x^3 + 3x$ **21.** $y = 3x^3 + x^2$

22. $y = 3x^4 - x$ **23.** $y = x^3 - 2x$ **24.** $y = 4x^3 - 3x^2$

25. The formula $Q = Q_0 e^{-t/\tau}$ gives the charge Q (in coulombs) on a capacitor after it has been discharging through a resistance for time t (in seconds). In the formula, Q_0 is the original charge (in coulombs), and $\tau = RC$ (resistance × capacitance). Use semilog paper and sketch the graph of Q versus t for $0 \le t \le 0.01$ when $Q_0 = 3.6 \times 10^{-5}$ coulombs and $\tau = 0.002$ s. 3 5 9 4 7 8

26. In a circuit containing a resistor and an inductor in series, if the battery is short-circuited, the current does not fall instantly to zero. The current drops gradually, so that the current I (in A) is given by $I = I_0 e^{-t/\tau}$ where I_0 is the steady-state current (in A), $\tau = L/R$ (inductance/resistance), and t is the time elapsed (in s). Use semilog paper and sketch the graph of I versus t for $R = 2.0\ \Omega$, $I_0 = 2.0$ A, and $L = 0.40$ H. 3 5 9 1 2 4 6 7 8

27. The formula $\text{pH} = -\log_{10}[\text{H}^+]$ gives the pH of a solution if $[\text{H}^+]$ is the concentration of hydrogen ions (in moles/liter). Use semilog paper and sketch the graph of pH versus $[\text{H}^+]$ for $0.001 < [\text{H}^+] < 0.005$. 6 9 2 4 5 7 8

28. The formula $B = 10 \log_{10}(I/I_0)$ gives the acoustic intensity level (in dB) of a sound of physical intensity I, if I_0 is the physical intensity of a sound that is barely audible to the average person. Use semilog paper and sketch the graph of B versus I/I_0 for $2 < I/I_0 < 6$. 1 3 4 5 6 8 9

29. The formula $G = -4.58\ T \log k$ gives the standard free energy change G (in Calories) of a reaction where k is the equilibrium reaction constant and T is the temperature (in degrees Kelvin). If $T = 300°$ K, use log-log paper and sketch the graph of G versus k for $0.01 < k < 1$. 6 8 1 2 3 5 7 9

30. The half-life of carbon 14 is 5570 years. That is, after 5570 years, half of any size sample of carbon 14 will have decayed. The formula $P = (1/2)^{t/5570}$ gives the fraction of a sample remaining after the time t (in years). Use log-log paper and sketch the graph of P versus t for $0 \le t \le 20{,}000$. 6 9 5 8

31. The flow of water Q (in ft³/s) through a certain flume is given by $Q = \log 0.992 + 1.547 \log H$, where H is the height of flow (in ft). Use semilog paper and sketch the graph of Q versus H for $0 < H \le 4.0$. 4 8 1 2 6 7 9

1. Architectural Technology
2. Civil Engineering Technology
3. Computer Engineering Technology
4. Mechanical Drafting Technology
5. Electrical/Electronics Engineering Technology
6. Chemical Engineering Technology
7. Industrial Engineering Technology
8. Mechanical Engineering Technology
9. Biomedical Engineering Technology

15.8 CHAPTER REVIEW

KEY TERMS

15.1 exponential function
asymptote
15.2 logarithmic form
exponential form
logarithmic function
15.4 antilogarithm
common logarithm
characteristic
mantissa
15.5 natural logarithm
15.7 semilogarithmic (semilog) graph paper
cycle
logarithmic (log-log) graph paper

SYMBOLS

$\log_b x = y$	$x = b^y$ where $b > 0$
$\log x$	$\log_{10} x$
$\ln x$	$\log_e x$

RULES AND FORMULAS

$\log_b xy = \log_b x + \log_b y$ if $x > 0$ and $y > 0$

$\log_b x/y = \log_b x - \log_b y$ if $x > 0$ and $y > 0$

$\log_b x^y = y \log_b x$ if $x > 0$

$\log_b x = \dfrac{\log_a x}{\log_a b}$ if $x > 0$

$a^x = e^{x \ln a}$ if $a > 0$

SEQUENTIAL REVIEW

(Section 15.1) *In 1–6, sketch the graph of each function.*

1. $y = 2(1/3)^{-x}$ **2.** $y = -(1/2)^x$ **3.** $y = -(3^x)$

4. $y = (1/3)2^{-x}$ **5.** $y = -2^{3x}$ **6.** $y = 0.32^{2x}$

(Section 15.2) *In 7 and 8, write each equation in logarithmic form.*

7. $125^{2/3} = 25$ **8.** $36^{1/2} = 6$

In 9 and 10, write each equation in exponential form.

9. $\log_3 81 = 4$ **10.** $\log_5 25 = 2$

In 11–14, sketch the graph of each function.

11. $y = 3 \log_2 x$ **12.** $y = (-1/2) \log_3 x$ **13.** $y = \log_{1/2} x$ **14.** $y = \log_3 2x$

(Section 15.3) *In 15–18, write each expression as the logarithm of a single quantity.*

15. $3 \log_b 2 - \log_b 6$ **16.** $\log_7 4 + \log_7 3$

17. $4 \log_2 5$ **18.** $2 \log_6 4 + 3 \log_6 2 - \log_6 3$

In 19–22, rewrite each expression and simplify by inspection, if possible.

19. $\log_3 45$ **20.** $\log_x 5/x^3$ **21.** $\log_b \sqrt{2b}, b > 0$ **22.** $\log_4 48$

(Section 15.4) *In 23–26, use a calculator to evaluate each expression.*

23. $\log 0.0421$ **24.** $\log 14.37$

25. x if $\log x = -1.2817$ **26.** y if $\log y = 2.8911$

In 27–30, solve each equation.

27. $7^x = 35$ **28.** $5^{x+1} = 12$ **29.** $5(3^x) = 35$ **30.** $4(3^{x+1}) = 24$

(Section 15.5) *In 31 and 32, use a calculator to evaluate each expression.*

31. ln(0.321)

32. x if $\ln x = -1.3368$

In 33 and 34, write each expression in terms of the common logarithm function and evaluate it for the given value of x.

33. $\log_3 x$, $x = 19$

34. $\log_7 x$, $x = 25.1$

In 35 and 36, write each expression in terms of the natural logarithm function and evaluate it for the given value of x.

35. $\log_2 x$, $x = 33.0$

36. $\log_3 x$, $x = 0.119$

In 37 and 38, write each expression in terms of the exponential function and evaluate it for the given value of x.

37. 9.31^x, $x = 0.241$

38. 0.321^x, $x = 6.35$

(Section 15.6) *In 39–42, solve each equation.*

39. $1/3 = \log_b 8$

40. $3e^{2x} = 15$

41. $\log_4 6 + 3x = \log_4 36$

42. $3 \ln (2x + 1) = 6$

(Section 15.7) *In 43–45, sketch the graph of each function on the type of paper indicated.*

43. $y = (1/2)(3^x)$ on semilog paper

44. $y = 2(1/3)^x$ on semilog paper

45. $y = 4x^3$ on log-log paper

APPLIED PROBLEMS

1. The formula $R = \log (I/I_0)$ gives the magnitude of an earthquake where I is the intensity of the quake and I_0 is a standard minimum intensity. How many times greater is the intensity of an earthquake that measures 6.3 than one that measures 2.1? 1 3 4 5 6 8 9

2. The formula $v_c = v_0(1 - e^{-t/\tau})$ gives the voltage (in V) across a capacitor that has been charging through a resistance for time t (in s). In the formula, v_0 is the voltage (in V) of the battery, and $\tau = RC$ (resistance $\times$ capacitance). Find the time required to charge a 10 μF capacitor to 9.9 volts from a 10-volt battery, if the resistance is 1000 Ω. 3 5 9 4 7 8

3. The formula $N = n^{2.81}$ is sometimes used to estimate the number N of multiplications required to evaluate an $n \times n$ determinant, if n is large. If a computer can perform one million multiplications per second, what size determinant could be evaluated in one second? 3 5 9

4. Some tumors grow in such a way as to double in diameter at a regular interval of time known as the "doubling time." If the doubling time of a tumor is six weeks, how many weeks will it take for a tumor that is 1 cm in diameter to become 5 cm in diameter? 6 9 2 4 5 7 8

5. World population growth has followed (approximately) the equation $P = P_0e^{kt}$. In the formula, P is the population at the time of interest, P_0 is the population at some initial time, t is the time (in years) elapsed, and k is a constant. In 1987 when the world population was 5 billion, it was projected that the population would double in 40 years.

1. Architectural Technology
2. Civil Engineering Technology
3. Computer Engineering Technology
4. Mechanical Drafting Technology
5. Electrical/Electronics Engineering Technology
6. Chemical Engineering Technology
7. Industrial Engineering Technology
8. Mechanical Engineering Technology
9. Biomedical Engineering Technology

(a) Use this information to solve for k in the equation.
(b) Predict the population in the year 2007.
9 1 6

6. A businessman purchases a $50,000 certificate of deposit that renews monthly with an interest rate of 9% per year. The equation which gives the value A of this certificate after t years is $A = 50{,}000\left[1 + \frac{.09}{12}\right]^{12t}$. How much will his investment be worth after one month? 1 2 6 7 8 3 5 9

7. When infrared light passes through a piece of fiber optic cable, its power decreases as it travels because of losses in the glass. This loss is indicated by the equation $P = P_0e^{-kx}$ where P is the power (in watts) at any distance x (in m) from the transmitter, P_0 is the initial power (in watts) at the transmitter, and k is a constant depending on the particular glass employed. Suppose an initial burst of power $P_0 = 25\ \mu\text{W}$ is injected into a 1.0-m long fiber and 18 μW exits its end. How much power would exit from a 2.0-m long piece of this same fiber? 9 1 2 3 5 6 7

8. The power P (in watts) at any point x down a fiber optic cable is given by $P = P_0e^{-kx}$. At a break in a long fiber the incident power is reflected back to the source. If 25 μW are sent down a fiber with $k = 0.3285$ and 1.3 nanowatts return, how far is it to the break? (Hint: Let P_1 represent the power at the break when P_0 is 25 μW, then use P_1 as the power sent back to the source, which receives 1.3 nW.) 3 5 9 1 2 4 6 7 8

9. The voltage v (in volts) on a capacitor in a deactivated RLC circuit dies out as time passes with instantaneous value given by,

$$v = v_0e^{-Rt/(2L)}\cos\left[\sqrt{\frac{1}{LC} - \left(\frac{R}{2L}\right)^2}\right]t.$$

In the formula C is the capacitance (in farads), L is the inductance (in henries), R is the resistance (in ohms), v_0 is the initial voltage (in volts), and t is the time (in seconds). What voltage would reside on a 500-μF capacitor initially charged to 250 volts if it discharges through a 140-ohm resistor and a 7-henry inductor, 50 ms after deactivation. (Hint: Don't forget to put your calculator in radian mode!) 3 5 9 4 7 8

10. When a load is parachuted from a plane, its velocity first increases from zero as it falls, then when the parachute opens its speed levels off to some constant value. The velocity at any time can be approximated by the following equation: $V = V_F(1 - e^{-kt})$. In the formula, V_F is the terminal velocity, k is a constant that depends on the type of parachute, and t is time. Experimental data for a certain type of parachute indicated that its terminal velocity was 10 ft/second and that it had reached 4 ft/second after 2 seconds. What is k for this parachute? Give your answer to two significant digits. 4 7 8 1 2 3 5 6 9

11. The number of homes N sold in a new suburb per month can be approximated by the following equation.

$$N = \frac{M}{1 + \left(\frac{M}{N_0} - 1\right)e^{-KMt}}$$

In the formula M is the maximum number of lots available, N_0 is the number of homes occupied when heavy advertising begins, K is the growth rate (in homes/month), and t is time (in months). Suppose a builder wants to predict when a development will be filled and knows there are presently 100 homes sold of a maximum 1000 lots. When will there be 500 homes sold? Use $K = 0.000131$ per month. 1 2 6 7 8 3 5 9

12. The amount of corrosion R (in mg) on a surface undergoing chemical cleaning was given by the following equation.

$$R = 26.3e^{-0.3t} + 6e^{-0.05t}$$

What value will R have after 4 seconds? 6 9 2 4 5 7 8

13. When a resort city with existing population N takes in a large group of temporary new residents of number M, the population gradually returns to normal according to an equation of the form $P = Me^{-kt} + N$ where k is a constant and t represents

time (in days). Suppose it is known that a town with population 750 takes in 1000 new residents. After 2 days, the population is found to be 1200. On what day will the last visitor leave town? (Hint: There are 751 residents on the day the last visitor leaves.) 1 2 6 7 8 3 5 9

14. When light of initial intensity I_0 (in watts) passes through a transparent material of thickness x (in cm) it loses power as described by the equation $I = I_0 e^{-kx}$ where k is a constant. What value of k does a window material 5 cm thick have if 35 microwatts of light is reduced to 27 microwatts after passing through? Give your answer to three significant digits. 9 1 2 3 5 6 7

15. The decibel gain A_v' of a certain type of circuit is given by $A_v' = 20\log_{10}\left(\frac{R_f + R_i}{R_i}\right)$. What value of R_f will yield a gain of 30 dB if $R_i = 600\ \Omega$? Give your answer to two significant digits. 3 5 9 1 2 4 6 7 8

16. The time period T (in seconds) of a certain type of oscillator is given by $T = -RC \ln\left(1 - \frac{V_{BO}}{V}\right)$ where V_{BO} is the breakover voltage (in volts), V the power supply voltage (in volts), and RC is the circuit's time constant (in seconds). What value of V_{BO} would be needed for a time period of 2.0 seconds from a 36-volt circuit if the time constant is 5.0 seconds? 3 5 9 4 7 8

17. An object cools at a rate proportional to its temperature.

$$T = T_0 e^{-kt}$$

In the equation, T_0 is the initial temperature, k is a constant, and t is the time (in hours). Find k for a material that cools from 600° F to 100° F in 3 hours. Give your answer to three significant digits. 6 8 1 2 3 5 7 9

18. A chemical process takes place such that the concentration of free ions decreases with time according to the equation

$$C = C_0 e^{-0.05t}$$

where C_0 is the initial concentration and t is time in seconds. Another process with twice the initial concentration takes 3 times as long as the first. When will both processes have equal concentrations? Give your answer to three significant digits. 6 9 2 4 5 7 8

19. Consider a rope that is wrapped around a drum n times. The force T_2 on one end of the rope needed to hold a force T_1 on the other end is given by $T_2 = T_1 e^{-2\pi\mu n}$ where n is the number of wraps and μ the coefficient of friction between the rope and the drum. How many times would a cowboy have to wrap his horse's reins around a hitching post to hold him if he can pull with a 300 lb (T_1) force? Assume μ for leather on wood is 0.4 and the weight of the free end of the rein is 1 ounce. 1 2 4 6 8 5 7 9

RANDOM REVIEW

1. Rewrite the following expression and simplify it by inspection, if possible. $\log_3 \sqrt[3]{3y}$, $y > 0$
2. Sketch the graph of $y = 3(1/2)^x$ on semilog paper.
3. Sketch the graph of $y = 1/2(3^{-x})$.
4. Write 26.3^x in terms of the exponential function and evaluate it when x is 4.28.
5. Write $\log_8 2 = 1/3$ in exponential form.
6. Use a calculator to evaluate log 15.73.
7. Write $8^2 = 64$ in logarithmic form.
8. Solve the following equation for x. $\log y = \log x + \log 3$
9. Use a calculator to find y if $\ln y = -2.414$.
10. Rewrite the following expression as the logarithm of a single quantity. $2 \log_7 5 - \log_7 3$

CHAPTER TEST

1. Write $6^{-2} = 1/36$ in logarithmic form.
2. Sketch the graph of $y = -3 \log_2 x$.

In 3 and 4, rewrite each expression and simplify by inspection, if possible.

3. $\log_5 50$
4. $\log_y (45/y)$
5. Use a calculator to find x if $\log x = 1.325$.
6. Write $\log_5 x$ in terms of the natural logarithm function and evaluate it for $x = 0.33$.
7. Solve the following equation. $5(3^{x-1}) = 2^x$
8. Sketch the graph of $y = 3x^2$ on log-log paper.
9. If a signal passes through two particular amplifiers in series, the total voltage gain (in dB) is given by the following expression.

$$20 \log 150 + 20 \log 200$$

   (a) Write the expression using a single logarithm.

   (b) Evaluate the expression.

10. In a manually adjusted camera, the f-stop indicates how wide the opening in the diaphragm is. Each step on the f-stop setting allows twice as much light to enter the camera as the preceding f-stop.

   (a) If a photographer wants the light to be only 28% as bright as his previous shot, determine how many f-stops the camera should be opened by solving the equation $2^x = 0.28$.

   (b) By how many f-stops should the camera be opened if the light is to be only 14% as bright as the first shot?

CHAPTER 16 SEQUENCES AND SERIES

In engineering, it is sometimes necessary to examine how one quantity changes with respect to another. As a car is braked to a stop, for example, the distance that it travels during each second can be tabulated. Such measurements often exhibit patterns with mathematical properties that can be used to analyze the data and to make predictions. In this chapter, we introduce some of the common patterns and illustrate their use in technological settings.

16.1 ARITHMETIC SEQUENCES AND SERIES

Carl Friedrich Gauss, born in 1777 in Germany, was one of the most famous mathematicians of all time. One of the stories often told about his childhood is that when he was about ten, his school class was given the assignment of adding the numbers from 1 to 100. The task was intended to keep the boys busy, but the teacher was barely through giving instructions when Gauss wrote the single number 5050 on his slate and turned it in. He reasoned this way: $100 + 1 = 101, 99 + 2 = 101, 98 + 3 = 101$, etc. Since there are 50 such pairs, the sum is 50(101) or 5050.

In this section, we will generalize Gauss' reasoning to solve a type of problem that has applications in many areas. First we consider the terminology.

DEFINITION An **arithmetic sequence or progression** (abbreviated AP) is a list of numbers in which each term after the first is obtained by adding a given constant, called the **common difference,** to the preceding term. An AP may have a finite or an infinite number of terms.

EXAMPLE 1 Write the first four terms of the arithmetic progression that has 3 as the first term and a common difference of 4.

Solution The first term is 3. The second term is $3 + 4$ or 7. The third term is $7 + 4$ or 11, and the fourth term is $11 + 4$ or 15. The progression, then, is 3, 7, 11, 15. ■

EXAMPLE 2 For each sequence that is arithmetic, determine the common difference.

(a) 1, 8, 15, 22, 29, 36, . . .

(b) 2, 4, 7, 11, 16, 22

(c) 1, 2, 4, 8, 16, 32, . . . , 1024

Solution **(a)** arithmetic: each term is obtained by adding 7 to the previous term. The common difference may be found by subtracting any term from the following term. That is,

$$8 - 1 = 7;\ 15 - 8 = 7;\ 22 - 15 = 7;\ 29 - 22 = 7;\ 36 - 29 = 7.$$

(b) not arithmetic: $4 - 2 = 2$; $7 - 4 = 3$; $11 - 7 = 4$; etc. That is, the difference between terms is not constant.

(c) not arithmetic: $2 - 1 = 1$; $4 - 2 = 2$; $8 - 4 = 4$; etc. That is, the difference between terms is not constant. ■

The letters a and d are often used to represent the first term and the common difference, respectively. Since each term after the first is obtained by adding the common difference to the preceding term, the second term would be $a + d$. The third term would be $(a + d) + d$ or $a + 2d$. The fourth term would be $(a + 2d) + d$ or $a + 3d$, and so on. The coefficient of d in each case is one less than the number of the term. We have the following pattern.

number of term	1	2	3	4	5	. . .	n
term	a	$a + d$	$a + 2d$	$a + 3d$	$a + 4d$	. . .	$a + (n - 1)d$

FORMULA FOR THE nTH TERM a_n OF AN ARITHMETIC SEQUENCE

$$a_n = a + (n - 1)d$$

where a is the first term of the sequence and d is the common difference.

EXAMPLE 3 Find the twentieth term of the arithmetic progression that has a first term of 10 and a common difference of 2.

Solution For this progression, a is 10, d is 2, and n is 20.

$$a_{20} = 10 + (20 - 1)2$$
$$a_{20} = 10 + 19(2)$$
$$a_{20} = 10 + 38 \text{ or } 48 \quad ■$$

EXAMPLE 4 Find the tenth term of the arithmetic sequence 3, 7, 11, 15, 19, . . .

Solution The common difference is 7 − 3 or 4. For this progression, a is 3, d is 4, and n is 10.

$$a_{10} = 3 + (10 - 1)4$$
$$= 3 + 9(4)$$
$$= 3 + 36 \text{ or } 39 \quad ■$$

EXAMPLE 5 A hotel stands on a beach that is 80.0 ft wide. An average of 6.0 ft is lost each year to erosion. How wide is the beach after 6 years?

Solution If the width of the beach is listed for each year, the numbers form an arithmetic sequence with 6 terms. The first term is the width of the beach *after* the first year, and the other terms are the widths after each of the succeeding 5 years. The first term is 74.0, and the common difference d is −6.0, since the beach width is decreasing. Thus,

$$a_6 = 74.0 + (6 - 1)(-6.0)$$
$$= 74.0 + 5(-6.0)$$
$$= 74.0 - 30.0 \text{ or } 44.0.$$

At the end of 6 years, the beach will be 44.0 ft wide. ■

If a specific term of an arithmetic sequence is known, the formula for the nth term of the sequence can be used to determine the position of the term in the sequence.

EXAMPLE 6 In the arithmetic progression 7, 10, 13, 16, . . . , 49 what is the number of the term 49?

Solution In this progression, a is 7, d is 3, and a_n is 49. We must find n. The formula for a_n gives

$$49 = 7 + (n - 1)3$$
$$49 = 7 + 3n - 3$$
$$49 = 4 + 3n$$
$$45 = 3n$$
$$15 = n$$ ■

arithmetic series

NOTE ⇨⇨

When the terms of an arithmetic sequence or progression are added, the expression that represents the sum is called an **arithmetic series.** ***The terms of an arithmetic sequence are separated by commas, but the terms of an arithmetic series are separated by plus signs.***

We are now ready to develop a formula that can be used to evaluate an arithmetic series that has n terms. We generalize the reasoning that Gauss used to evaluate $1 + 2 + 3 + \cdots + 100$. The nth term, a_n, is represented by the letter l, since it is the *last* term to be included in the sum. The variable S_n represents the sum.

$$S_n = a + (a + d) + (a + 2d) + (a + 3d) + \cdots + (l - d) + l \qquad \textbf{(16-1)}$$

Since the sum is the same when the terms are listed in reverse order,

$$S_n = l + (l - d) + \cdots + (a + 3d) + (a + 2d) + (a + d) + a. \qquad \textbf{(16-2)}$$

Adding Equations 16-1 and 16-2, we obtain

$$2S_n = (a + l) + (a + l) + (a + l) + \cdots + (a + l).$$

Since S_n contains n terms, $2S_n$ contains n terms, each equal to $(a + l)$.

$$2S_n = n(a + l) \text{ and } S_n = (n/2)(a + l).$$

FORMULA FOR THE SUM S_n OF THE FIRST n TERMS OF AN ARITHMETIC SERIES

$$S_n = (n/2)(a + l)$$

where a is the first term of the series, and l is the last term of the series.

EXAMPLE 7 Evaluate the series $1 + 2 + 3 + \cdots + 100$.

Solution This problem is the one that Gauss solved. For the series, a is 1, l is 100, and n is 100. Thus,

$$S_{100} = (100/2)(1 + 100)$$
$$= 50(101) \text{ or } 5050.$$ ■

EXAMPLE 8 Find the sum of the first 12 positive multiples of 3.

Solution The positive multiples of 3 are 3(1), 3(2), 3(3), 3(4), 3(5), and so on. The sequence 3, 6, 9, 12, 15, . . . is an arithmetic sequence with common difference 3. The twelfth term is given by

$$\begin{aligned} l &= 3 + (12 - 1)3 \\ &= 3 + 33 \text{ or } 36. \end{aligned}$$

The problem, then, is to find $3 + 6 + 9 + \cdots + 36$. This expression is an arithmetic series.

$$\begin{aligned} S_{12} &= (12/2)(3 + 36) \\ &= 6(39) \text{ or } 234 \end{aligned}$$ ■

EXAMPLE 9 A fence is to be constructed along a hill so that the top is horizontal as shown in Figure 16.1. Vertical posts are to be placed so that each one is 3 inches taller than the previous one. The first one is to be 3′ 4″ and the total length of lumber used is to be 76′ 3″. How many posts are needed?

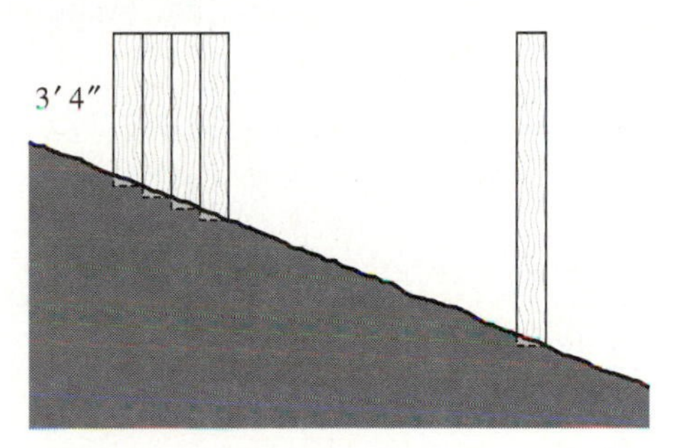

FIGURE 16.1

Solution The lengths form an AP with $a = 3' 4''$ or $40''$ and common difference $3''$. The sum is 76′ 3″ or 915″. The length (in inches) of the nth post is given by

$$40 + (n - 1)3 = 40 + 3n - 3 = 37 + 3n.$$

The formula for S_n gives

$915 = (n/2)[40 + (37 + 3n)].$	
$1830 = n[77 + 3n]$	Multiply both sides by 2.
$1830 = 77n + 3n^2$	Use the distributive property.
$0 = 3n^2 + 77n - 1830$	Subtract 1830 from both sides.
$0 = (3n + 122)(n - 15)$	Factor the expression on the right.
$3n + 122 = 0 \text{ or } n - 15 = 0$	Set both factors equal to zero.
$n = -122/3 \text{ or } n = 15.$	

Since the number of posts cannot be negative, n must be 15. ■

EXERCISES 16.1

In 1–7, find the first four terms of the arithmetic progression with given first term and common difference.

1. $a = 4, d = 5$
2. $a = 30, d = 3$
3. $a = 2, d = -4$
4. $a = 11, d = -3$
5. $a = -20, d = 7$
6. $a = -13, d = -6$
7. $a = -40, d = -1$

In 8–14, determine which of the following sequences are arithmetic.

8. 3, 6, 9, 12, 15, 18
9. 100, 95, 90, 85, 80, 75, . . .
10. 2, 3, 5, 8, 13, 21
11. 3, 4, 7, 11, 18, 29
12. 3, 4, 6, 9, 13, 18, . . .
13. 12, 8, 4, 0, −4, −8
14. 1, 3, 9, 27, 81, 243, . . .

In 15–21, find the indicated term of each arithmetic progression.

15. $a = 3, d = 5, n = 20$
16. $a = 20, d = -4, n = 12$
17. $a = 5, d = -2, n = 8$
18. tenth term of 7, 15, 23, 31, . . .
19. fifth term of 40, 37, 34, 31, . . .
20. twelfth term of 13, 5, −3, −11, . . .
21. twentieth term of 50, 45, 40, 35, . . .

In 22–28, a_n represents the nth term of an arithmetic sequence. Find n.

22. $a_n = 58$: 2, 4, 6, 8, 10, . . .
23. $a_n = 49$: 5, 9, 13, 17, 21, . . .
24. $a_n = 85$: 7, 13, 19, 25, 31, . . .
25. $a_n = 111$: 13, 15, 17, 19, 21, . . .
26. $a_n = -44$: 19, 16, 13, 10, 7, . . .
27. $a_n = -103$: 17, 5, −7, −19, −31, . . .
28. $a_n = 135$: 200, 195, 190, 185, 180, . . .

In 29–35, evaluate each arithmetic series.

29. $n = 50: 2 + 4 + 6 + 8 + 10 + \cdots + a_n$
30. $n = 15: 15 + 10 + 5 + 0 + -5 + \cdots + a_n$
31. $n = 20: 200 + 195 + 190 + 185 + 180 + \cdots + a_n$
32. $n = 10: 100 + 103 + 106 + 109 + 112 + \cdots + a_n$
33. $n = 25: 3 + 8 + 13 + 18 + 23 + \cdots + a_n$
34. $n = 15: 20 + 17 + 14 + 11 + 8 + \cdots + a_n$
35. $n = 100: 50 + 48 + 46 + 44 + 42 + \cdots + a_n$

In 36–42, find the number of terms of each series that were added to obtain the sum.

36. $S_n = 190: 1 + 5 + 9 + 13 + 17 + \cdots + a_n$
37. $S_n = 301: 2 + 5 + 8 + 11 + 14 + \cdots + a_n$
38. $S_n = -165: 10 + 5 + 0 + (-5) + (-10) + \cdots + a_n$
39. $S_n = -615: 8 + 1 + (-6) + (-13) + (-20) + \cdots + a_n$
40. $S_n = 1332: 45 + 57 + 69 + 81 + 93 + \cdots + a_n$
41. $S_n = -712: 38 + 27 + 16 + 5 + (-6) + \cdots + a_n$
42. $S_n = -130: 26 + 20 + 14 + 8 + 2 + \cdots + a_n$

43. As part of a therapy program, a patient is instructed to do a certain exercise for 10 minutes the first day. Thereafter the duration is to be increased by 3 minutes each day until 31 minutes is reached. On what day will the patient reach 31 minutes? **6 9** 2 4 5 7 8

44. A retaining wall is being pushed over at the rate of 0.3° per year. If it is currently 4.5° off the vertical, in how many years will it be 6.9° off the vertical? **4 7 8** 1 2 3 5 6 9

45. An auditorium is built so that each row of seats contains 2 more chairs than the previous row. The first row contains 20 seats and there are 20 rows. How many chairs are needed? **1 2 4 7 8** 3 5

46. A computer program to sort a list of numbers from largest to smallest uses the following strategy. The first number is assumed to be largest. It is then compared to the next number in the list. If that number is larger, the two numbers switch places in the list. After the entire list has been through the comparison, the largest number will be first. The process then starts again with the second number in the list and so on. How many comparisons must be made to sort the list if there are 50 numbers in the list? (Hint: There will be only 49 comparisons the first time through the list, because the first number is not compared to itself.) **3 5** 9

47. A plot of land in the shape of a trapezoid has the dimensions shown in the diagram. It is to be roped off into 4 segments of equal width. How much rope is needed for the five lengths indicated if there is a common difference between lengths? **1 2 4 7 8** 3 5

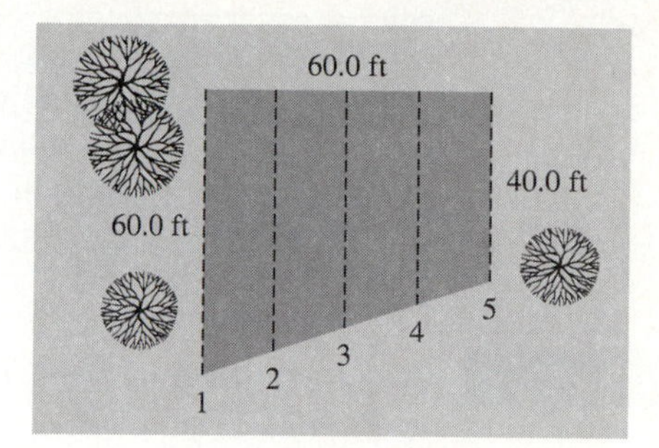

48. The figure shown is to be incorporated into an architectural design. Each segment of the drawing is 1/2 inch longer than the previous segment. If the figure can be carved at the rate of 1″ in two hours, how long will it take to finish the design? **1 2 4 7 8** 3 5

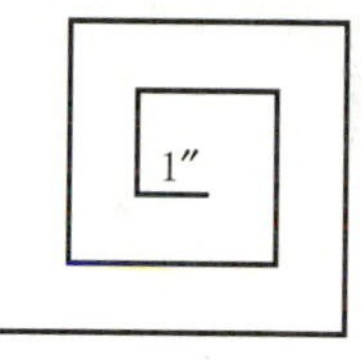

49. A businessman finds that the cost of one of his materials increases at the rate of \$2.00 per dozen each month. If he buys a dozen at \$22.00 in January, and one dozen each month thereafter, what is the total paid by the end of December? **1 2 6 7 8** 3 5 9

50. The sequence 1, 1, 2, 3, 5, 8, 13, . . . is not an AP, but it does follow a pattern. Determine the next four numbers in the sequence.

---

1. Architectural Technology
2. Civil Engineering Technology
3. Computer Engineering Technology
4. Mechanical Drafting Technology
5. Electrical/Electronics Engineering Technology
6. Chemical Engineering Technology
7. Industrial Engineering Technology
8. Mechanical Engineering Technology
9. Biomedical Engineering Technology

16.2 GEOMETRIC SEQUENCES AND SERIES

In Example 2 of Section 16.1, you saw some sequences that were not arithmetic sequences. A second type of sequence, called a *geometric sequence or progression* also has applications in the technologies.

DEFINITION A **geometric sequence or progression** (abbreviated GP) is a list of numbers in which each term after the first is obtained by multiplying the preceding term by a fixed constant, called the **common ratio.** A geometric progression may have a finite or an infinite number of terms.

EXAMPLE 1 Write the first four terms of a geometric progression with first term 5 and common ratio of 2.

Solution The first term is 5. The second term is 5(2) or 10. The third term is 10(2) or 20, and the fourth term is 20(2) or 40. The progression, then, is 5, 10, 20, 40. ■

EXAMPLE 2 For each sequence that is geometric, determine the common ratio.

(a) 1, 2, 3, 5, 8, 13, . . .

(b) 3, 9, 27, 81, 243, 729, . . .

(c) $1, -1/2, 1/4, -1/8, 1/16, -1/32, \ldots$

Solution **(a)** not geometric: The ratio of the second term to the first term is 2/1, but the ratio of the third term to the second term is 3/2, the ratio of the fourth term to the third term is 5/3, and so on. That is, the ratio between terms is not constant.

(b) geometric: Each term is obtained by multiplying the previous term by 3. The common ratio may be found by dividing any term by the previous term. That is, $9/3 = 3$; $27/9 = 3$; $81/27 = 3$; etc.

(c) geometric: Each term is obtained by multiplying the previous term by $-1/2$. That is, $-1/2 \div 1 = -1/2$; $1/4 \div -1/2 = -1/2$; $-1/8 \div 1/4 = -1/2$; etc. ■

The letters a and r are often used to represent the first term and the common ratio, respectively. Since each term after the first is obtained by multiplying the preceding term by r, the second term would be ar. The third term would be $(ar)r$ or ar^2. The fourth term would be $(ar^2)r$ or ar^3, and so on. The exponent of r in each case is one less than the number of the term. We have the following pattern.

number of term	1	2	3	4	5	. . .	n
term	a	ar	ar^2	ar^3	ar^4	. . .	ar^{n-1}

FORMULA FOR THE nTH TERM a_n OF A GEOMETRIC SEQUENCE

$$a_n = ar^{n-1}$$

where a is the first term of the sequence and r is the common ratio.

EXAMPLE 3 Find the tenth term of the geometric progression that has a first term of 7 and a common ratio of 2/3.

Solution For this progression, a is 7, r is 2/3, and n is 10.

$$a_{10} = 7(2/3)^{10-1}$$
$$a_{10} = 7(2/3)^9$$
$$a_{10} = 7(2^9)/3^9 \text{ or } 3584/19683$$ ■

EXAMPLE 4 Find the twentieth term of the geometric sequence 1, 2, 4, 8, 16, . . .

Solution The common ratio is 2. For this progression, a is 1, r is 2, and n is 20.

$$a_{20} = 1(2^{20-1})$$
$$= 1(2^{19})$$

A calculator shows that $a_{20} = 524{,}288$. ■

EXAMPLE 5 A bank pays interest of 6% per year, compounded quarterly. This means that interest is calculated at 1.5% and deposited into the account four times each year. Because interest is deposited into the account, the principal on which interest is calculated grows each quarter. If $1200 is deposited in this bank, determine the balance after 5 1/2 years.

Solution Since interest is paid quarterly, there are 5 1/2 × 4 or 22 interest periods in 5 1/2 years. After each quarter, 1.5% interest is paid on the previous balance. The new balance is 100% of the previous balance plus 1.5% of the previous balance, or 101.5% of the previous balance. If decimal notation is used, the balance for each quarter is obtained by multiplying the previous balance by 1.015. We have a geometric progression for which r is 1.015, and n is 22. Because the first term is the amount *after* the first interest period, a is 1200(1.015) or 1218. The balance after 5 1/2 years is

$$a_{22} = 1218(1.015)^{21}.$$

Using a calculator, we have the following keystroke sequence.

1218 [×] 1.015 [y^x] 21 [=] **1665.0791**

Rounding the answer to the nearest cent, the balance is $1665.08. ■

EXAMPLE 6 Find the twelfth term of the geometric sequence

$$1 + j, 2, 2 - 2j, -4j, \ldots$$

Solution The common ratio can be obtained by taking the ratio of any two consecutive terms. We could find $\frac{a_2}{a_1}$, which is $\frac{2}{1+j}$, but it is easier to use $\frac{a_3}{a_2}$, which is $\frac{2-2j}{2}$ or $1 - j$.

Thus, the twelfth term is $a_{12} = (1 + j)(1 - j)^{11}$.

Recall that powers of complex numbers are easier to use in polar form.

$$1 + j = \sqrt{2}(\cos 45° + j \sin 45°) \text{ and } 1 - j = \sqrt{2}(\cos 315° + j \sin 315°)$$

$$\begin{aligned}(1 - j)^{11} &= (\sqrt{2})^{11}[\cos [11(315°)] + j \sin [11(315°)]] \\ &= (\sqrt{2})^{11}[\cos 3465° + j \sin 3465°] \\ (1 + j)(1 - j)^{11} &= \sqrt{2}(\cos 45° + j \sin 45°)(\sqrt{2})^{11}[\cos 3465° + j \sin 3465°] \\ &= (\sqrt{2})^{12}[\cos (45° + 3465°) + j \sin (45° + 3465°)] \\ &= (2^{1/2})^{12}[\cos 3510° + j \sin 3510°] \\ &= 2^6(0 - j) \\ &= -64j \quad ■\end{aligned}$$

geometric series

NOTE ⇨⇨

When the terms of a geometric sequence or progression are added, the expression that represents the sum is called a **geometric series.** ***The terms of a geometric sequence are separated by commas, but the terms of a geometric series are separated by plus signs.***

We are now ready to develop a formula to evaluate a geometric series with n terms. The argument is similar to that used to evaluate an arithmetic series. Using S_n for the sum, we have

$$S_n = a + ar + ar^2 + ar^3 + ar^4 + \cdots + ar^{n-2} + ar^{n-1}. \qquad \textbf{(16-3)}$$

Multiplying both sides of Equation 16-3 by r, we have

$$S_n r = ar + ar^2 + ar^3 + ar^4 + \cdots + ar^{n-1} + ar^n. \qquad \textbf{(16-4)}$$

Notice that the series in Equations 16-3 and 16-4 are alike except for the first term in Equation 16-3 and the last term in Equation 16-4. Subtracting Equation 16-4 from Equation 16-3, we have the following.

$$S_n - S_n r = a - ar^n$$

$$S_n(1 - r) = a - ar^n \qquad \text{Factor the expression on the left-hand side.}$$

$$S_n = \frac{a - ar^n}{1 - r} \qquad \text{Divide both sides by } 1 - r.$$

FORMULA FOR THE SUM S_n OF THE FIRST n TERMS OF A GEOMETRIC SERIES

$$S_n = \frac{a - ar^n}{1 - r}$$

where a is the first term of the series, and r is the common ratio.

EXAMPLE 7 Evaluate the series $1 + 2 + 4 + 8 + \cdots + a_{10}$.

Solution For the series, a is 1, r is 2, and n is 10.

$$S_{10} = \frac{1 - 1(2^{10})}{1 - 2} = \frac{1 - 1024}{-1} = \frac{-1023}{-1} \text{ or } 1023$$

EXAMPLE 8 Evaluate the series

$$6 - 18 + 54 - 162 + \cdots + a_{12}.$$

Solution For this series, a is 6, r is -3, and n is 12.

$$S_{12} = \frac{6 - 6(-3)^{12}}{1 - (-3)}$$

Some calculators will not accept a negative base with the y^x key. An even power of a negative base, however, will be positive, so the following calculator keystroke sequence is used

6 [−] 6 [×] 3 [y^x] 12 [=] [÷] 4 [=] **−797160**

EXAMPLE 9 A ball is dropped from a height of 6.00 m. On each upward bounce, it reaches 2/3 of the height from which it last fell. Find the total distance traveled when the ball touches the ground for the fifth time. Give the answer to two decimal places.

Solution Figure 16.2 shows that two series are involved. The distances traveled in falling down form a geometric progression for which a is 6.00 m, r is 2/3, and n is 5.

$$S_5 = \frac{6 - 6(2/3)^5}{1 - 2/3}$$

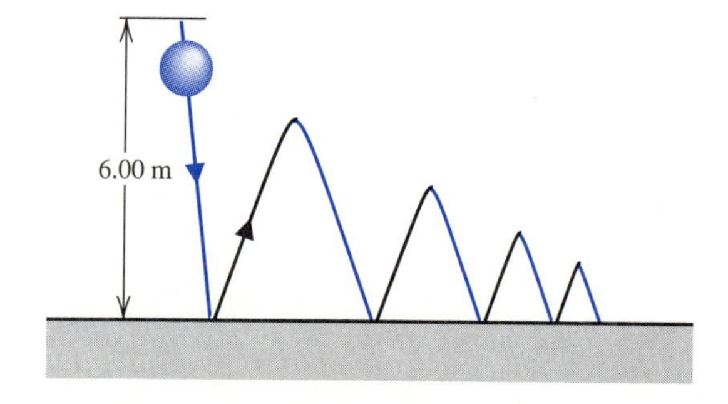

FIGURE 16.2

The following calculator keystroke sequence is used.

6 [−] 6 [×] [(] 2 [÷] 3 [)] [y^x] 5 [=] [÷] [(] 1 [÷] 3 [)] [=] [STO] **15.629634**

The distances traveled in bouncing up form a geometric progression with $a = 4.00$ m, $r = 2/3$, and $n = 4$.

$$S_4 = \frac{4 - 4(2/3)^4}{1 - 2/3}$$

The calculator keystroke sequence is as follows.

4 [−] 4 [×] [(] 2 [÷] 3 [)] [y^x] 4 [=] [÷] [(] 1 [÷] 3 [)] [=] **9.629628**

The total distance is obtained by adding the two sums. Since the first sum was put into memory, we have

[+] [RCL] [=] **25.259262**

The distance to two decimal places, then, is 25.26 m. ■

EXERCISES 16.2

In 1–10, find the first four terms of the geometric progression with the given first term and common ratio.

1. $a = 3, r = 2$
2. $a = 4, r = 3$
3. $a = 5, r = -10$
4. $a = 12, r = -3$
5. $a = 8, r = 1/4$
6. $a = 12, r = 1/3$
7. $a = 20, r = -1/4$
8. $a = 100, r = -2/5$
9. $a = 40, r = -0.2$
10. $a = 80, r = 0.1$

In 11–20, for each sequence that is geometric, determine the common ratio.

11. 2, 4, 8, 16, 32, . . .
12. 3, 9, 27, 81, 243
13. 100, −500, 2500, −12500, 62500
14. 1, −2, 6, −24, 120
15. 3, 9, 36, 180, 1080, 7560, . . .
16. 2, 4, 12, 48, 240, 1440
17. 3, 6, 18, 72, 360
18. 12, −3, 3/4, −3/16, 3/64
19. 8, −4, 2, −1, 1/2, . . .
20. 1, 2/3, 4/9, 8/27, 16/81, . . .

In 21–32, find the indicated term of each geometric progression.

21. $a = 20, r = 3, n = 10$
22. $a = 3, r = 5, n = 20$
23. $a = 20, r = -4, n = 12$
24. $a = 5, r = 1/2, n = 8$
25. tenth term of 8, 2, 1/2, 1/8, . . .
26. fifth term of 40, −120, 360, −1080, . . .
27. twelfth term of 14, −7, 7/2, −7/4, . . .
28. twentieth term of −50, 10, −2, 2/5, . . .
29. $a = 1, r = j, n = 20$
30. $a = 1 + j, r = -j, n = 50$
31. tenth term of $1 - j, 1 + j, -1 + j, -1 - j$, . . .
32. twentieth term of $1 - j, 2, 2 + 2j, 4j$, . . .

In 33–42, evaluate each geometric series. If the series has fractional terms, round the sum to five decimal places.

33. $n = 10$: $2 + 4 + 8 + 16 + \cdots + a_n$

34. $n = 6$: $15 + 5 + 5/3 + 5/9 + \cdots + a_n$

35. $n = 8$: $12 + 8 + 16/3 + 32/9 + \cdots + a_n$

36. $n = 10$: $200 - 100 + 50 - 25 + \cdots + a_n$

37. $n = 10$: $125 - 25 + 5 - 1 + \cdots + a_n$

38. $n = 12$: $3 + 2 + 4/3 + 8/9 + \cdots + a_n$

39. $n = 8$: $243 + 162 + 108 + 72 + \cdots + a_n$

40. $n = 10$: $50 - 5 + 0.5 - 0.05 + \cdots + a_n$

41. $n = 6$: $1 + j - 1 - j + \cdots + a_n$

42. $n = 8$: $(1 + j) + (-1 + j) + (-1 - j) + (1 - j) + \cdots + a_n$

---

43. A special type of filter allows only 2/5 of the light reaching it to be transmitted. How much light is transmitted when four of these filters are used together? 9 1 2 3 5 6 7

44. For a well-tuned piano, the frequency of each note is $\sqrt[12]{2}$ times the frequency of the note below it. There are 12 keys in an octave. Find the ratio of the largest frequency to the smallest frequency in an octave. 1 3 4 5 6 8 9

45. Carbon 14 is said to have a half-life of 5570 years, because when living matter dies, the carbon 14 atoms decay at such a rate that there will be only half as many of them every 5570 years. A piece of wood is alleged to be 27,850 years old. If a similar piece of a living tree has 5.63×10^4 carbon 14 atoms, how many atoms would be found in the sample, assuming the age estimate is correct? 6 9 5 8

46. A vacuum pump is designed to remove 4% of the air in the chamber on each stroke. If there are originally 3.00 liters in the chamber, how much air remains after 7 strokes? 4 8 1 2 6 7 9

47. A bank account pays 5% annual interest. If $1000 is deposited at the beginning of each year for 12 years, how much money is in the account at the end of the twelve years? 1 2 6 7 8 3 5 9

48. A piece of paper 0.001″ thick is folded in two. The folded paper is then folded again, and so on. After 8 folds, how thick is the folded paper? 1 2 4 6 8 5 7 9

49. The following problem can be solved by evaluating a series, but there is a much simpler way to solve it. Try to solve it the easy way. Two cars 100 miles apart start at the same time and travel toward each other at 40 mph and 60 mph respectively. At the moment they start, a fly takes off from the front of the slower car and travels toward the faster car at 30 mph. After reaching the faster car, the fly instantly turns around and flies toward the slower car, continuing in this manner until the cars meet. How far does the fly travel?

---

1. Architectural Technology
2. Civil Engineering Technology
3. Computer Engineering Technology
4. Mechanical Drafting Technology
5. Electrical/Electronics Engineering Technology
6. Chemical Engineering Technology
7. Industrial Engineering Technology
8. Mechanical Engineering Technology
9. Biomedical Engineering Technology

16.3 INFINITE GEOMETRIC SERIES

In Example 9 of Section 16.2, we examined the total distance that the bouncing ball would travel on the first few bounces. But if the ball bounces up 2/3 of the height from which it last fell, the series never terminates. It is always possible to calculate the distance traveled, however small, on the next bounce. The sum of the distances traveled downward is given by

$$6 + 4 + 8/3 + 16/9 + 32/27 + 64/81 + 128/243 + \cdots.$$

In theory, then, the ball continues to bounce forever. In practice, however, at some point in time the distance traveled on a given bounce becomes so small that all motion afterwards may be ignored. Thus, we are able to find a good approximation for the total distance traveled.

To see how we arrive at such an approximate sum, consider the two series in Example 1.

EXAMPLE 1 Compute the next four terms of each geometric series.

(a) $2 + 4 + 8 + 16 + \cdots$

(b) $2 + 1 + 1/2 + 1/4 + \cdots$

Solution **(a)** The common ratio is 2.

$$2 + 4 + 8 + 16 + \mathbf{32 + 64 + 128 + 256}$$

(b) The common ratio is 1/2.

$$2 + 1 + 1/2 + 1/4 + \mathbf{1/8 + 1/16 + 1/32 + 1/64} \quad \blacksquare$$

In the first series, the terms are getting successively larger. Whenever $|r| > 1$, the terms will increase in absolute value. In the second series, however, the terms are getting successively smaller. Whenever $|r| < 1$, the terms will decrease in absolute value. Consider $S_n = \dfrac{a - ar^n}{1 - r}$ where $|r| < 1$. Since $a - ar^n = a(1 - r^n)$, when n is large enough, r^n is small enough that its value may be ignored. Thus, $a - ar^n$ is very close to a. As an approximation for $S_n = \dfrac{a - ar^n}{1 - r}$, we may therefore use $S = \dfrac{a}{1 - r}$. The resulting error should be so small that it is no worse than the error that occurs when rounding answers. ***It is important to realize that this approximation holds only if $|r| < 1$. If $|r| \geq 1$, r^n is too large to be ignored in the formula.***

NOTE ⇨⇨

EXAMPLE 2 Evaluate the following series

$$1 + 1/2 + 1/4 + 1/8 + 1/16 + \cdots + a_{23}$$

(a) using the formula $S_n = \dfrac{a - ar^n}{1 - r}$.

(b) using the formula $S = \dfrac{a}{1 - r}$ as an approximation.

Solution (a) $S_{23} = \dfrac{1 - 1(1/2)^{23}}{1 - 1/2} = \dfrac{1 - 0.0000001}{0.5} = 1.9999998$

(b) $S = \dfrac{1}{1 - 1/2} = 2$ ■

If we are to evaluate a geometric series containing an infinite number of terms, the argument is essentially the same as in the preceding argument. Consider the case in which $0 < r < 1$. At some point in the series, the terms will become so small that all terms beyond that point may be ignored without producing a significant error. The more terms we consider, the closer the sum is to the value $\dfrac{a}{1-r}$. The number n is allowed to take on larger and larger values without bound. We say *n approaches infinity,* and write $n \to \infty$. As n increases, the sum S_n also increases, but never quite reaches the value $\dfrac{a}{1-r}$. We say that this value is the *limit of the sequence* of sums S_n, call it S, and write $S = \lim_{n \to \infty} S_n = \dfrac{a}{1-r}$. The argument can be generalized for any value of r with absolute value less than 1.

limit of a sequence

FORMULA FOR THE SUM S OF THE TERMS OF AN INFINITE GEOMETRIC SERIES

$$S = \lim_{n \to \infty} S_n = \frac{a}{1-r} \text{ if } |r| < 1$$

where a is the first term of the series, r is the common ratio, and S_n is the sum of the first n terms.

EXAMPLE 3 Evaluate the geometric series.

$$2 + 1 + 1/2 + 1/4 + 1/8 + \cdots$$

Solution In the series, a is 2 and r is 1/2. Since $|1/2| < 1$,

$$S = \frac{2}{1 - 1/2} = \frac{2}{1/2} = 4. \quad ■$$

EXAMPLE 4 A ball is dropped from a height of 6.00 m. On each upward bounce, it reaches 2/3 of the height from which it last fell, as shown in Figure 16.3. Find the total distance the ball travels.

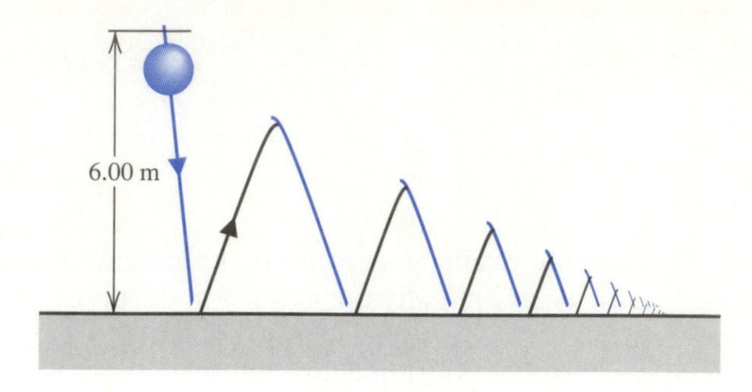

FIGURE 16.3

Solution The distances traveled on the downward bounces form a geometric sequence with $a = 6.00$ and $r = 2/3$. Let S_D represent the sum of these distances.

$$S_D = \frac{6.00}{1 - 2/3} = \frac{6.00}{1/3} = 6.00(3) = 18.00 \text{ m}$$

The distances traveled on the upward bounces form a geometric sequence with $a = 4.00$ and $r = 2/3$. Let S_U represent the sum of these distances.

$$S_U = \frac{4.00}{1 - 2/3} = \frac{4.00}{1/3} = 4.00(3) = 12.00 \text{ m}$$

The total distance traveled is 18.00 m + 12.00 m or 30.00 m. ■

Recall that in Section 1.1, a rational number was defined as any number that can be written in the form a/b, where a and b are integers, and b is not 0. It was stated that infinitely repeating decimals are rational because they *can* be written in this form. If an infinitely repeating decimal is written as a geometric series, it is possible to determine the equivalent fraction.

EXAMPLE 5 Write $0.\overline{42}$ as a fraction.

Solution Recall that the bar above digits indicates repeating digits.

$$\begin{aligned} 0.\overline{42} &= 0.42 + 0.0042 + 0.000042 + \cdots \\ &= 42/100 + 42/10{,}000 + 42/1{,}000{,}000 + \cdots \end{aligned}$$

Since the expression is a geometric series for which a is 42/100 and r is 1/100, we have

$$S = \frac{42/100}{1 - 1/100} = \frac{42/100}{99/100} = \left(\frac{42}{100}\right)\left(\frac{100}{99}\right) = \frac{14}{33}. \quad ■$$

EXAMPLE 6 Write $0.52\overline{631}$ as a fraction.

Solution
$$0.52\overline{631} = 0.52 + 0.00631 + 0.00000631 + \cdots$$
$$= 52/100 + 631/100{,}000 + 631/100{,}000{,}000 + \cdots$$

Starting with the second term, we have a geometric series for which a is 631/100,000 and r is 1/1000. For this series,

$$S = \frac{631/100{,}000}{1 - 1/1000} = \frac{631/100{,}000}{999/1000} = \left(\frac{631}{100{,}000}\right)\left(\frac{1000}{999}\right) = \frac{631}{99{,}900},$$

and $0.52\overline{631} = \dfrac{52}{100} + \dfrac{631}{99{,}900} = \dfrac{52(999) + 631}{99{,}900} = \dfrac{52{,}579}{99{,}900}$. ■

EXAMPLE 7 A calculator display shows 2.1851852. Assuming that this number is actually $2.\overline{185}$ rounded, what fraction does it represent?

Solution
$$2.\overline{185} = 2 + 0.185 + 0.000185 + 0.000000185 + \cdots$$
$$= 2 + 185/1000 + 185/1{,}000{,}000 + \cdots$$

Starting with the second term, we have a geometric series for which a is 185/1000 and r is 1/1000. For this series,

$$S = \frac{185/1000}{1 - 1/1000} = \frac{185/1000}{999/1000} = \left(\frac{185}{1000}\right)\left(\frac{1000}{999}\right) = \frac{185}{999}$$

Thus, $2.\overline{185} = 2\ 185/999$ or $2\ 5/27$, which is 59/27. ■

EXERCISES 16.3

In 1–10, consider each geometric series.
(a) Find S_{10}. Round the sum to five decimal places.
(b) Find S_{20}. Round the sum to five decimal places.
(c) Find S. Specify the sum as a fraction.

1. $1 + 1/3 + 1/9 + 1/27 + \cdots$

2. $1 + 2/5 + 4/25 + 8/125 + \cdots$

3. $2 - 4/3 + 8/9 - 16/27 + \cdots$

4. $-2 + 2/5 - 2/25 + 2/125 - \cdots$

5. $3/4 + 9/16 + 27/64 + \cdots$

6. $a = 12$, $r = 1/2$

7. $a = 12$, $r = 1/3$

8. $a = 16$, $r = -3/4$

9. $a = 20$, $r = -3/5$

10. $a = 24$, $r = -1/6$

In 11–20, write each repeating decimal as an equivalent fraction.

11. $0.\overline{12}$

12. $0.\overline{471}$

13. $0.\overline{8}$

14. $0.1\overline{3}$

15. $0.0\overline{5}$

16. $0.23\overline{9}$

17. $0.4\overline{76}$

18. $5.\overline{2316}$

19. $6.45\overline{83}$

20. $9.61\overline{032}$

21. A ball is dropped from a height of 10.0 m. On each upward bounce, it reaches 3/5 of the height from which it last fell. Find the total vertical distance the ball travels before coming to rest.
4 7 8 | 1 | 2 | 3 | 5 | 6 | 9

22. An object suspended from a spring is pulled down and released. It springs up a distance of 40.0 mm and oscillates in such a way that the distance traveled during each oscillation is 3/4 of the distance traveled during the last one (that is, each time the direction changes, the next distance is 3/4 of the last distance). Find the total distance that the object travels before coming to rest after its release.
4 7 8 | 1 | 2 | 3 | 5 | 6 | 9

23. A pendulum bob swings in such a way that it travels a distance of 40.0 cm on the first oscillation and the distance traveled during each subsequent oscillation is 7/10 of the distance traveled during the last one (that is, each time the direction changes, the next distance is 7/10 of the last distance). Find the total distance that the pendulum bob travels before coming to rest.
4 7 8 | 1 | 2 | 3 | 5 | 6 | 9

24. A string is stretched between two supports. It is pulled at a point in the middle and released. On the first oscillation, the point travels a distance of 4.00 cm. On subsequent oscillations, it travels 3.20 cm, 2.56 cm, and so on. How far will it travel before coming to rest after its release?
4 7 8 | 1 | 2 | 3 | 5 | 6 | 9

25. A child is photographed in one mirror while holding another, so that the child's image is reflected over and over ad infinitum. If the first image of the child occupies 144.0 cm^2 of the photograph, and each of the reflections occupies 1/4 of the area of the one before, what is the total area of the photograph occupied by images of the child?
1 2 4 7 8 | 3 | 5

26. A calculator display shows 0.3888889 as an answer. Assuming that this number is actually $0.3\overline{8}$ rounded, what fraction does it represent?
3 5 | 9

27. It is possible to convert an infinitely repeating decimal to a fraction using the fact that the number of nines in the denominator is the same as the number of barred digits, and the number of zeros in the denominator is the same as the number of unbarred digits after the decimal point. How can the barred and unbarred digits be used to determine the numerator?

1. Architectural Technology
2. Civil Engineering Technology
3. Computer Engineering Technology
4. Mechanical Drafting Technology
5. Electrical/Electronics Engineering Technology
6. Chemical Engineering Technology
7. Industrial Engineering Technology
8. Mechanical Engineering Technology
9. Biomedical Engineering Technology

16.4 THE BINOMIAL EXPANSION WITH POSITIVE INTEGRAL POWERS

In Section 2.2, we considered multiplication of algebraic expressions. An expression of the form $(a + b)^n$, where n is a positive integer, indicates a product of n factors of $(a + b)$. When the product is written as a sum of terms, we say that the expression is in expanded form. Our goal in this section is to examine the expanded form of such expressions to find a pattern. Consider the following expressions.

Expression	**Expanded form**
$(a+b)^0 =$	1
$(a+b)^1 =$	$a+b$
$(a+b)^2 =$	$a^2+2ab+b^2$
$(a+b)^3 =$	$a^3+3a^2b+3ab^2+b^3$
$(a+b)^4 =$	$a^4+4a^3b+6a^2b^2+4ab^3+b^4$
$(a+b)^5 =$	$a^5+5a^4b+10a^3b^2+10a^2b^3+5ab^4+b^5$

Pascal's triangle

Figure 16.4 shows the coefficients in an arrangement called *Pascal's triangle.* Even though it was used as early as 1303 in China and by several Europeans before Pascal, the triangle is named in honor of Blaise Pascal, who developed and applied its properties in the mid-seventeenth century. The coefficients follow a pattern in which each row after the second begins and ends with 1, and the other entries are obtained by adding the two entries to the left and right of it on the previous row. Figure 16.5 illustrates how the entries of the fifth row are obtained from those of the fourth row.

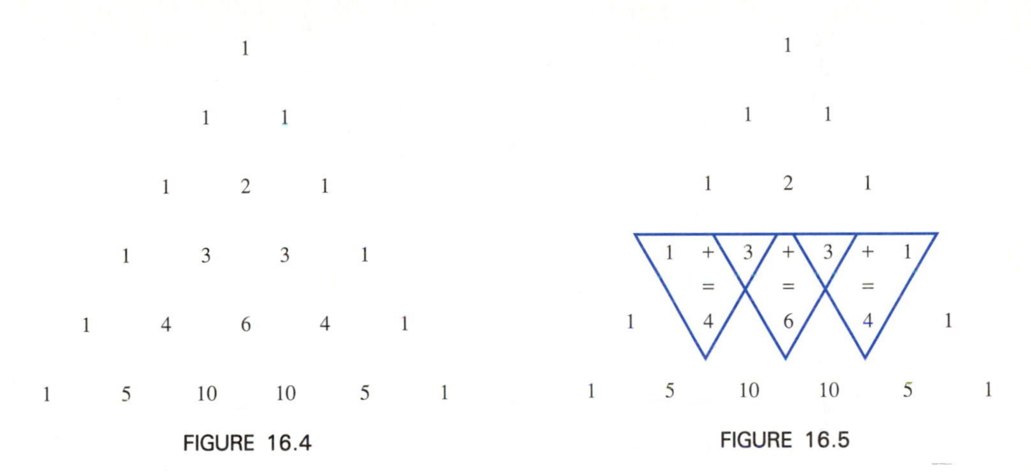

FIGURE 16.4

FIGURE 16.5

An examination of the products $(a+b)^n$ leads to the following observations.

BINOMIAL EXPANSION OF $(a+b)^n$

(a) There are $n+1$ terms.

(b) The coefficients are the numbers in the row of Pascal's triangle that has n as the second entry.

(c) In the first term, the exponent of a is n. In each term thereafter, the exponent of a *decreases* by 1.

(d) In the first term, the exponent of b is 0. ($b^0 = 1$, so $a^nb^0 = a^n$.) In each term thereafter, the exponent of b *increases* by 1.

EXAMPLE 1 Write the expanded form of $(x+y)^6$.

Solution There are 7 terms. Pascal's triangle from Figure 16.4 is extended to find the sixth row.

$$\begin{array}{ccccccccccccc} & 1 & & 5 & & 10 & & 10 & & 5 & & 1 & \\ 1 & & 6 & & 15 & & 20 & & 15 & & 6 & & 1 \end{array}$$

Combining the coefficients and the powers of x, we have

$$1x^6 \quad 6x^5 \quad 15x^4 \quad 20x^3 \quad 15x^2 \quad 6x \quad 1$$

Inserting the powers of y and the plus signs between terms, we have

$$(x+y)^6 = 1x^6 + 6x^5y + 15x^4y^2 + 20x^3y^3 + 15x^2y^4 + 6xy^5 + 1y^6. \quad \blacksquare$$

EXAMPLE 2 Write the expanded form of $(2x+y)^4$.

Solution There are 5 terms. From Pascal's triangle, we see that the coefficients are

$$1 \quad 4 \quad 6 \quad 4 \quad 1.$$

Inserting the powers of $2x$, we have

$$1(2x)^4 \quad 4(2x)^3 \quad 6(2x)^2 \quad 4(2x)^1 \quad 1.$$

Inserting powers of y and the plus signs, we have

$$\begin{aligned}(2x+y)^4 &= 1(2x)^4 + 4(2x)^3y + 6(2x)^2y^2 + 4(2x)^1y^3 + 1y^4 \\ &= 16x^4 + 32x^3y + 24x^2y^2 + 8xy^3 + y^4. \quad \blacksquare\end{aligned}$$

EXAMPLE 3 Write the expanded form of $(2x-3y)^3$.

Solution $(2x-3y)^3 = [2x+(-3y)]^3$
There are 4 terms.
The coefficients are

$$1 \quad 3 \quad 3 \quad 1.$$

Inserting powers of $(2x)$ gives

$$1(2x)^3 \quad 3(2x)^2 \quad 3(2x) \quad 1.$$

Inserting powers of $(-3y)$ and plus signs gives

$$\begin{aligned}(2x-3y)^3 &= 1(2x)^3 + 3(2x)^2(-3y) + 3(2x)(-3y)^2 + 1(-3y)^3 \\ &= 8x^3 - 36x^2y + 54xy^2 - 27y^3. \quad \blacksquare\end{aligned}$$

Pascal's triangle is convenient for small powers. But to find $(x+2y)^{10}$, for example, it would be cumbersome to have to write down 11 rows of the triangle. The following rule is generalized from the expansions of $(a+b)^n$ for the first few values of n.

To find the coefficient of the $(k + 1)$st term of $(a + b)^n$:

(a) If $k = 0$, the coefficient of the first term is 1.

(b) If $k > 0$,

1. Multiply the coefficient of the kth term by the exponent of a in the kth term.
2. Divide by k to obtain the coefficient of the next term.

EXAMPLE 4 Find the first four terms of $(a + b)^{25}$.

Solution The first term is $1a^{25}$.
The coefficient of the second term is obtained as follows.

$1\ a^{25}$ → These two numbers are multiplied. → $\dfrac{25(1)}{1} = 25$

Term (1) → This number is divided into the product.

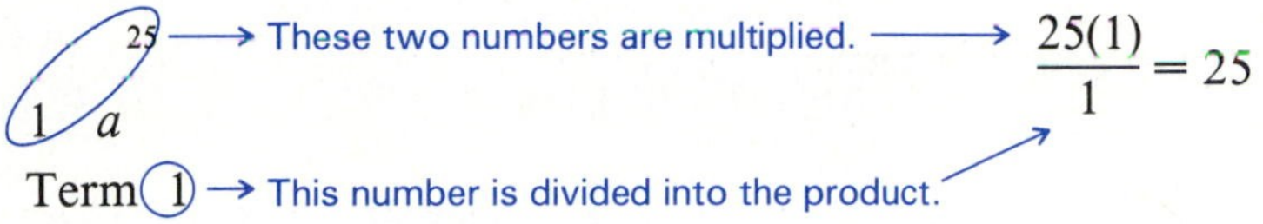

The second term is $25a^{24}b$.
The coefficient of the third is obtained as follows.

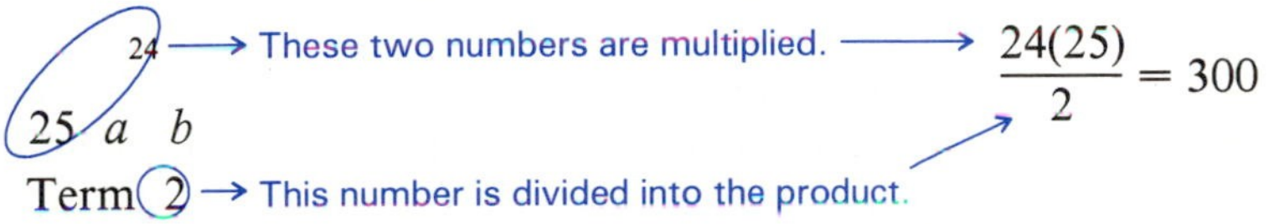

$25\ a^{24}\ b$ → These two numbers are multiplied. → $\dfrac{24(25)}{2} = 300$

Term (2) → This number is divided into the product.

The third term is $300a^{23}b^2$.
The coefficient of the fourth term is obtained as follows.

$300\ a^{23}\ b^2$ → These two numbers are multiplied. → $\dfrac{23(300)}{3} = 2300$

Term (3) → This number is divided into the product.

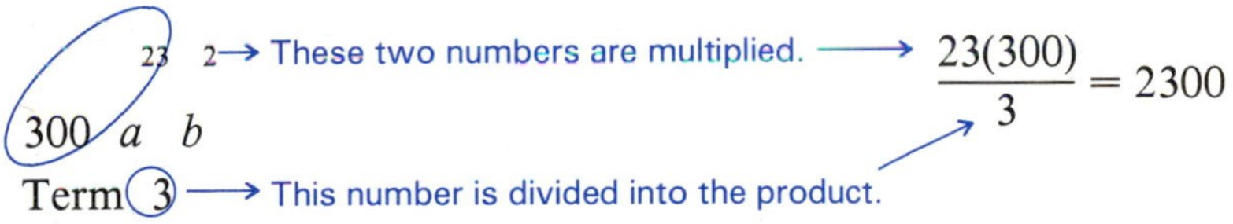

The fourth term is $2300a^{22}b^3$. ■

In practice, the arithmetic expressions may be written down and then a calculator can be used to evaluate the coefficients successively, as shown in Example 5.

EXAMPLE 5 Find the first four terms of $(a + b)^{25}$.

Solution Term number: 1 2 3 4

$$(a + b)^{25} = a^{25} + \frac{25a^{24}b}{1} + \frac{24(25)a^{23}b^2}{2(1)} + \frac{23(24)(25)a^{22}b^3}{3(2)(1)}$$

$$= a^{25} + 25a^{24}b + 300a^{23}b^2 + 2300a^{22}b^3$$ ■

factorial

The denominators in Example 5 follow a pattern. The notation $n!$ (read *n factorial*) is often used to indicate the product of all positive integers less than or equal to n. For example, 3! means 3(2)(1) or 6 and 5! means 5(4)(3)(2)(1) or 120. Many calculators have a [!] key that may be used to make data entry more efficient. The fifth term of the expansion in Example 5 would be $\frac{22(23)(24)(25)}{4!}a^{21}b^4$. The following calculator keystroke sequence is used.

22 [×] 23 [×] 24 [×] 25 [÷] 4 [!] [=] **12650**

EXAMPLE 6 Find the first four terms of $(3x - 5)^{10}$.

Solution The terms are found as follows.

$$(3x)^{10} + \frac{10}{1}(3x)^9(-5) + \frac{9(10)}{2(1)}(3x)^8(-5)^2 + \frac{8(9)(10)}{3(2)(1)}(3x)^7(-5)^3$$

$$59{,}049x^{10} - 984{,}150x^9 + 7{,}381{,}125x^8 - 32{,}805{,}000x^7 \quad \blacksquare$$

When a variable changes by a small amount, the Greek letter Δ (delta) is often used with the variable to indicate the change. Thus, Δx would represent the amount by which x changes. The quantity $x + \Delta x$ would represent the new value. Such quantities are sometimes used in technological problems.

EXAMPLE 7 When a car travels at speed s, the distance d required to stop the car as shown in Figure 16.6 may be approximated by the formula $d = ks^2$. Assume that s increases by Δs and write a formula for d when the speed is $s + \Delta s$.

FIGURE 16.6

Solution When the speed is $s + \Delta s$,

$$\begin{aligned} d &= k(s + \Delta s)^2 \\ &= k[s^2 + 2s\Delta s + (\Delta s)^2] \\ &= ks^2 + 2ks\Delta s + k(\Delta s)^2. \quad \blacksquare \end{aligned}$$

EXAMPLE 8 For a simple beam supported at both ends, a concentrated load produces a maximum deflection given by $d = \frac{Pl^3}{48EI}$. Assume that the load P, modulus of elasticity E, and the moment of inertia I all remain constant. Assume that the length l of the beam is increased by Δl and write a formula for d when the length is $l + \Delta l$.

Solution When the length is $l + \Delta l$,

$$d = \frac{P(l + \Delta l)^3}{48EI} = \frac{P}{48EI}[l^3 + 3l^2(\Delta l) + 3l(\Delta l)^2 + (\Delta l)^3] \quad \blacksquare$$

EXERCISES 16.4

In 1–10, write the expanded form of each expression.

1. $(x + 3)^3$ **2.** $(x + 2)^4$ **3.** $(3x - 1)^4$ **4.** $(1 - 5y)^3$ **5.** $(3a + b)^5$
6. $(a + 5b)^6$ **7.** $(3a + 2b)^6$ **8.** $(2a - 5b)^5$ **9.** $(5a - 3b)^6$ **10.** $(2a - 3b)^6$

In 11–20, find the first four terms of each expression.

11. $(y - 2)^{10}$ **12.** $(a + 3)^{10}$ **13.** $(1 + 3b)^{10}$ **14.** $(2y - 1)^{12}$ **15.** $(x + 3y)^{12}$
16. $(x - 5y)^{12}$ **17.** $(2x - 5y)^{14}$ **18.** $(3x - 2y)^{11}$ **19.** $(a - b)^{20}$ **20.** $(x - y)^{25}$

In 21–26, assume that the given quantity changes as indicated, and use a binomial expansion to write a formula for the new value of the variable on the left.

21. The volume V of a sphere is given by $V = (4/3)\pi r^3$. Assume that the radius r increases by Δr. 1 2 4 7 8 3 5

22. The maximum load L for a circular column is given by $L = (k/h^2)d^4$. Assume that the height h remains constant, but the diameter d increases by Δd. 1 2 4 8 7 9

23. For a cantilever beam under a distributed load, the maximum deflection of the beam is given by $d = \frac{w}{8EI} l^4$. Assume that the weight per unit length of the load w, modulus of elasticity E, and the moment of inertia I all remain constant, but that the length l of the beam decreases by Δl. 1 2 4 8 7 9

24. The power P generated by a windmill is given by $P = kv^3$. Assume that the velocity v of the wind increases by Δv. 4 7 8 1 2 3 5 6 9

25. The power P to propel a ship is given by $P = \frac{D^{2/3}}{k} v^3$. Assume that the displacement D remains constant, but that the velocity of the ship v decreases by Δv. 4 7 8 1 2 3 5 6 9

26. The elevation above sea level for a location can be approximated by the formula $E = 580T^2 - 117{,}000T + 5{,}900{,}000$. Assume that the temperature T at which water boils increases by ΔT. 6 8 1 2 3 5 7 9

1. Architectural Technology
2. Civil Engineering Technology
3. Computer Engineering Technology
4. Mechanical Drafting Technology
5. Electrical/Electronics Engineering Technology
6. Chemical Engineering Technology
7. Industrial Engineering Technology
8. Mechanical Engineering Technology
9. Biomedical Engineering Technology

16.5 THE BINOMIAL EXPANSION WITH REAL POWERS

As early as 1676, Isaac Newton suspected that the pattern of exponents and coefficients that holds for the expansion of $(a + b)^n$ when n is a positive integer will also hold for rational and negative powers. Although he did not prove the result, he was correct in his hypothesis. When n is a fraction or a negative number, the series will be infinite. For an expression of the form $(1 + x)^n$ where $|x| < 1$, the expansion produces an infinite series, called the *binomial series,* such that the sum approaches a limit as the number of terms summed approaches infinity.

binomial series

THE BINOMIAL SERIES

$$(1 + x)^n = 1 + \frac{nx}{1!} + \frac{n(n-1)x^2}{2!} + \frac{n(n-1)(n-2)x^3}{3!} + \cdots$$

where n is a real number and $|x| < 1$.

EXAMPLE 1 Find the first four terms of the expansion of $(1 + x)^{-3}$.

Solution

$$\begin{aligned}(1 + x)^{-3} &= 1 + \frac{(-3)x}{1!} + \frac{(-3)(-4)x^2}{2!} + \frac{(-3)(-4)(-5)x^3}{3!} + \cdots \\ &= 1 - 3x + 6x^2 - 10x^3 + \cdots\end{aligned}$$

Calculators and computers do not store tables of square roots and cube roots. When instructed to return such a value, the machine calls upon a built-in program to compute an approximate value using a number of terms of an infinite series. Although other types of series are actually used, the binomial series illustrates the principle involved.

EXAMPLE 2 **(a)** Find an approximate value for $\sqrt{1.004}$ using four terms of a binomial series.

(b) Compare the answer in part (a) to the square root obtained using the square root key on a calculator.

Solution **(a)** $\sqrt{1.004}$ must be written in the form $(1 + x)^n$ with $|x| < 1$.

$$\begin{aligned}\sqrt{1.004} &= (1 + 0.004)^{0.5} \\ &= 1 + \frac{0.5}{1!}(0.004) + \frac{0.5(-0.5)}{2!}(0.004)^2 + \frac{0.5(-0.5)(-1.5)}{3!}(0.004)^3 \\ &= 1 + 0.002 - 0.000002 + 0.000000004 \\ &= 1.001998004\end{aligned}$$

(b) The calculator keystroke sequence 1.004 $\boxed{\sqrt{x}}$ yields 1.001998.

EXAMPLE 3 Find an approximate value for $\sqrt[3]{17.882}$ using four terms of a binomial series.

Solution $\sqrt[3]{17.882} = (17.882)^{1/3}$. It is necessary to use an expression of the form $(1 + x)^n$ with $|x| < 1$. If we write $(1 + 16.882)^{1/3}$, the expression fails to meet the condition $|x| < 1$. We can circumvent the problem by writing the number so that a cube may be factored out. That is,

$$17.882 = 8 + 9.882 \text{ or } 8(1 + 1.23525).$$

But this expression, too, fails to meet the condition that $|x| < 1$. If we use 27 as the cube, however,

$$\begin{aligned} 17.882 &= 27 - 9.118 \text{ or } 27(1 - 0.3377037), \text{ and} \\ (17.882)^{1/3} &= [27(1 - 0.3377037)]^{1/3} \\ &= 27^{1/3}\left[1 + \frac{(1/3)}{1!}(-0.3377037) + \frac{(1/3)(-2/3)}{2!}(-0.3377037)^2 \right. \\ &\qquad \left. + \frac{(1/3)(-2/3)(-5/3)}{3!}(-0.3377037)^3\right] \\ &= 3[1 - 0.1125679 - 0.0126715 - 0.0023773] \\ &= 2.6171499 \quad \blacksquare \end{aligned}$$

The steps used in Example 3 are summarized as follows.

To calculate the *n*th root of a positive number using a binomial series:

1. Write the number in the form $a^n + b$ or $a^n - b$ where a and b are positive and $b < a^n$.
2. Factor out a^n, so that the number is in the form $a^n(1 + x)$.
3. Express $[a^n(1 + x)]^{1/n}$ as $a(1 + x)^{1/n}$.
4. Expand $(1 + x)^{1/n}$ using the binomial series and multiply by a.

EXAMPLE 4 Find an approximate value for $\sqrt{234}$ using four terms of a binomial series.

Solution We need a perfect square fairly close to 234 so that 234 can be written in the form $a^2 + b$ or $a^2 - b$. The squares closest to 234 are 15^2 or 225 and 16^2 or 256, so we have the following.

$$\begin{aligned} 234 &= 225 + 9 & \text{or} \qquad 234 &= 256 - 22 \\ &= 225(1 + 9/225) & &= 256(1 - 22/256) \\ &= 225(1 + 0.04) & &= 256(1 - 0.0859375) \end{aligned}$$

Both of these expressions meet the condition $|x| < 1$, so we choose the first because the value of x has fewer digits.

$$\begin{aligned}
\sqrt{234} &= [225(1+0.04)]^{1/2} \\
&= 225^{1/2}[1+0.04]^{0.5} \\
&= 15\left[1+\frac{(0.5)}{1!}(0.04)+\frac{(0.5)(-0.5)}{2!}(0.04)^2+\frac{(0.5)(-0.5)(-1.5)}{3!}(0.04)^3\right] \\
&= 15[1 + 0.02 - 0.0002 + 0.000004] \\
&= 15.29706 \quad \blacksquare
\end{aligned}$$

EXAMPLE 5 Use four terms of a binomial series to find the approximate dimensions of a cubical box that has a volume of 57 cm^3.

Solution Let s be the length of one edge of the box, as in Figure 16.7.

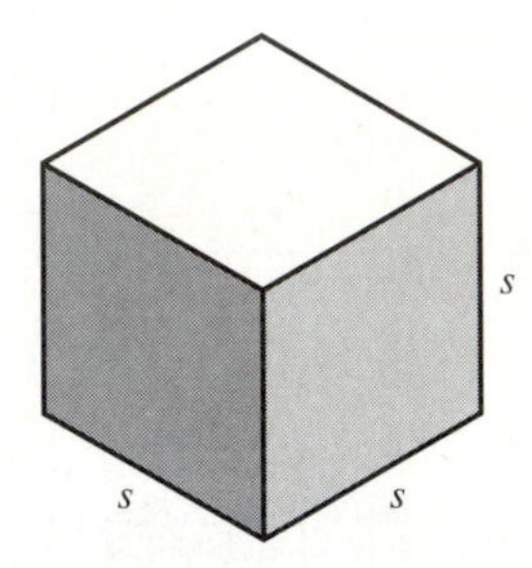

FIGURE 16.7

Then $s^3 = 57$ and $s = \sqrt[3]{57}$.
The cubes closest to 57 are 3^3 or 27 and 4^3 or 64. But $57 = 27 + 30$ fails to meet the condition $b < a^3$, so we use $57 = 64 - 7 = 64(1 - 7/64)$.

$$\begin{aligned}
\sqrt[3]{57} &= [64(1-0.109375)]^{1/3} \\
&= 64^{1/3}(1-0.109375)^{1/3} \\
&= 4(1-0.109375)^{1/3} \\
&= 4\left[1+\frac{(1/3)}{1!}(-0.109375)+\frac{(1/3)(-2/3)}{2!}(-0.109375)^2\right. \\
&\qquad \left. +\frac{(1/3)(-2/3)(-5/3)}{3!}(-0.109375)^3\right] \\
&= 4[1-0.0364583-0.0013292-0.0000808] \\
&= 3.8485268 \quad \blacksquare
\end{aligned}$$

In both this section and the previous section, the binomial expansion was considered by examining terms from the first on. It is possible, however, to find a single term of a binomial expansion. Following the pattern developed for particular values of n, the general expression $(a + b)^n$ can be written as follows.

$$(a+b)^n = a^n + \frac{n}{1!}a^{n-1}b + \frac{n(n-1)}{2!}a^{n-2}b^2 + \frac{n(n-1)(n-2)}{3!}a^{n-3}b^3 + \cdots$$

A table of values will help to isolate the pattern of a single term.

Term number	*Coefficient*	*Exponent of* a	*Exponent of* b
1	1	n	0
2	$n/1!$	$n-1$	1
3	$n(n-1)/2!$	$n-2$	2
4	$n(n-1)(n-2)/3!$	$n-3$	3

In term number 2, the number 1 appears in several places. In term number 3, the number 2 appears in several places. In term number 4, the number 3 appears in several places, etc. In term number $k+1$, we would expect k to appear in several places. For consistency, mathematicians say that $0! = 1$, and then term number 1 has a coefficient of $1/0!$. Following this pattern, the coefficient of term $k+1$ is a product starting with n and multiplying successively smaller numbers until there are k factors, and the denominator is $k!$. The coefficient can be written as $n(n-1)(n-2)\ .\ .\ .\ (n-k+1)/k!$ The exponent of a is $n-k$, and the exponent of b is k.

THE $(k+1)$st TERM OF THE EXPANSION $(a+b)^n$

(a) When $k = 0$, the first term is given by a^n.

(b) When $k > 0$, the $(k+1)$st term is given by

$$\frac{n(n-1)(n-2)\cdots(n-k+1)}{k!}\,a^{n-k}\,b^k$$

where the numerator contains k factors.

EXAMPLE 6 Find the fifth term in the expansion of $(2x+y)^9$.

Solution If term number 5 is term number $k+1$, then k is 4. Since n is 9, the term is

$$\frac{9(8)(7)(6)}{4!}(2x)^5y^4 = 4032x^5y^4. \quad ■$$

EXAMPLE 7 Find the fourth term in the expansion of $(1+0.5)^{-3}$.

Solution If $k+1$ is 4, then k is 3. Since n is -3, the term is

$$\frac{(-3)(-4)(-5)}{3!}(1)^{-6}(0.5)^3 = -1.25. \quad ■$$

experiment
probability

In the study of probability, a situation under study is called an **experiment.** If all outcomes of an experiment are considered, the **probability** of an event is the ratio of the number of outcomes in which the event can occur to the total number of outcomes. When a single experiment is performed successively n times such that each trial has no effect on the others, and there are only two possible outcomes for each trial, the probability that the event will occur exactly k times is given by term number $k + 1$ of the binomial expansion $(a + b)^n$. In the formula, a is the probability that the event does not occur and b is the probability that it does.

EXAMPLE 8 When machine parts are manufactured, their dimensions must fall within acceptable ranges. The probability that the size of a particular part is within an acceptable error tolerance is 0.80 and the probability that it is outside the acceptable tolerance is 0.20. If six such parts are picked at random, what is the probability that exactly four of them are outside the acceptable error tolerance?

Solution If the event (being outside the error tolerance) does not occur, then the size is within the error tolerance. The probability that a part is inside the error tolerance is 0.80, so a is 0.80. The probability that a part is outside the error tolerance is 0.20, so b is 0.20. We also know that n is 6, and k is 4. The probability is given by the fifth term of the expansion

$$(0.8 + 0.2)^6.$$

The probability is

$$\frac{6(5)(4)(3)}{4!}(0.80)^2(0.20)^4 \text{ or } 0.01536,$$

which is 0.015, to two significant digits. ■

EXERCISES 16.5

In 1–8, find the first four terms in the expansion of each expression.

1. $(1 + x)^{-4}$ **2.** $(1 + x)^{-5}$ **3.** $(1 - x)^{-5}$

4. $(1 - x)^{-6}$ **5.** $(1 + x)^{1/2}$ **6.** $(1 - x)^{1/2}$

7. $(1 - x)^{-1/3}$ **8.** $(1 + x)^{-1/3}$

In 9–16, find an approximate value for each number using the first four terms of a binomial expansion. Round answers to five decimal places.

9. $\sqrt{1.07}$ **10.** $\sqrt{0.85}$ **11.** $\sqrt{4.6}$

12. $\sqrt{8.3}$ **13.** $\sqrt[3]{1.07}$ **14.** $\sqrt[3]{8.2}$

15. $\sqrt[3]{25.75}$ **16.** $\sqrt[3]{127}$

In 17–26, find the indicated term in the expansion of each expression.

17. third term of $(3a + 1)^5$

18. third term of $(a + 5b)^6$

19. fourth term of $(3a + 2b)^6$

20. fourth term of $(y - 2)^{10}$

21. fifth term of $(1 + 3b)^{10}$

22. fifth term of $(5x - 1)^{10}$

23. fifth term of $(1 + 0.3)^{-2}$

24. fifth term of $(1 + 0.3)^{-4}$

25. fourth term of $(1 + 0.5)^{1/2}$

26. fourth term of $(1 + 0.5)^{-1/2}$

---

27. When a baby is born, the probability that it is a girl is 0.49 and the probability that it is a boy is 0.51. If a family has three children, what is the probability that exactly two of them are girls? 6 9 2 4 5 7 8

28. When a baby is born, the probability that it is a girl is 0.49 and the probability that it is a boy is 0.51. If a family has four children, what is the probability that none of them are boys? 6 9 2 4 5 7 8

29. If the probability that an item coming off the assembly line of a factory is defective is 0.01 and the probability that it is good is 0.99, what is the probability that of 5 items picked at random, exactly one of them is defective? 1 2 6 7 8 3 5 9

30. If the probability that an item coming off the assembly line of a factory is defective is 0.01 and the probability that it is good is 0.99, what is the probability that of 3 items picked at random, none of them is defective? 1 2 6 7 8 3 5 9

31. If a computer printer has 7 DIP switches, and they are set on or off at random, the probability that each is set correctly is 1/2. What is the probability that all of them are set correctly? 3 5 9

32. When a particular type of missile is fired, the probability that it hits within an acceptable radius of the intended target is 0.9 and the probability that it is outside is 0.1. If four such missiles are fired, find the probability that exactly one of them hits within the accepted range. 4 7 8 1 2 3 5 6 9

33. Adapt the expression for term number $k + 1$ of the expansion $(a + b)^n$ using a factorial in the numerator as well as in the denominator. (Hint: The denominator will have to be written differently.)

1. Architectural Technology
2. Civil Engineering Technology
3. Computer Engineering Technology
4. Mechanical Drafting Technology
5. Electrical/Electronics Engineering Technology
6. Chemical Engineering Technology
7. Industrial Engineering Technology
8. Mechanical Engineering Technology
9. Biomedical Engineering Technology

16.6 CHAPTER REVIEW

KEY TERMS

16.1 arithmetic sequence or progression (AP)
common difference
arithmetic series

16.2 geometric sequence or progression (GP)
common ratio
geometric series

16.3 limit of a sequence

16.4 Pascal's triangle
factorial

16.5 binomial series
experiment
probability

SYMBOLS

a_n the nth term of a sequence

S_n the sum of the first n terms of a series

S the sum of the terms of an infinite geometric series

$n \to \infty$ n approaches infinity

$n!$ n factorial

Δx the change in x

RULES AND FORMULAS

$a_n = a + (n-1)d$ the nth term of an AP

$S_n = (n/2)(a + l)$ the sum of the first n terms of an arithmetic series

$a_n = ar^{n-1}$ the nth term of a GP

$S_n = \dfrac{a - ar^n}{1 - r}$ the sum of the first n terms of a geometric series

$S = \dfrac{a}{1 - r}$ if $|r| < 1$ the sum of the terms of an infinite geometric series

$(1 + x)^n = 1 + \dfrac{nx}{1!} + \dfrac{n(n-1)x^2}{2!} + \dfrac{n(n-1)(n-2)}{3!}x^3 + \cdots$ where n is real number and $|x| < 1$

$(k + 1)$st term of $(a + b)^n = \dfrac{n(n-1)(n-2)\ .\ .\ .\ (n-k+1)}{k!}a^{n-k}b^k$

GEOMETRY CONCEPTS

$V = (4/3)\ \pi r^3$ (volume of a sphere)

SEQUENTIAL REVIEW

(Section 16.1)

1. Write the first five terms of the arithmetic sequence that has 200 as the first term and 4 as the common difference.
2. Determine which of the following sequences are arithmetic.
   (a) 2, 4, 6, 8, 10, 12, . . .
   (b) 2, 3, 5, 8, 12, 17
   (c) 8, 4, 2, 1, 1/2, 14, . . .
3. Find the tenth term of the sequence 20, 17, 14, 11, . . .
4. If 45 is the nth term of the AP 3, 6, 9, 12, . . . , find n.
5. If -21 is the nth term of the AP 7, 3, -1, -5, -9, . . . , find n.
6. Evaluate the series $17 + 21 + 25 + 29 + 33 + \cdots + a_n$ for $n = 12$.

7. Evaluate the series $48 + 45 + 42 + 39 + \cdots + a_n$ for $n = 40$.

8. Find the number of terms of the series $1 + 10 + 19 + 28 + \cdots$ that will give a sum of 606.

9. Find the number of terms of the series $3 + 9 + 15 + 21 + 27 + \ldots$ that will give a sum of 1200.

(Section 16.2) *In 10 and 11, find the first four terms of the geometric progression with given first term and common ratio.*

10. $a = 6$, $r = -1/2$.

11. $a = 4$, $r = 4$

In 12 and 13, find the common ratio for each geometric progression.

12. 12, 6, 3, 3/2, 3/4, . . .

13. 7, −21, 63, −189, 567, . . .

In 14–16, find the indicated term of each geometric progression.

14. $a = 7$, $r = 2$, $n = 8$

15. fifteenth term of 20, −40, 80, −160, . . .

16. tenth term of $2j, -4, -8j, 16, \ldots$

In 17 and 18, evaluate each geometric series. If the series has fractional terms, round the sum to five decimal places.

17. $n = 12$: $4 + 12 + 36 + 108 + \cdots + a_n$

18. $n = 7$: $5 + 1 + 1/5 + 1/25 + 1/125 + \cdots + a_n$

(Section 16.3) *In 19–21, evaluate each infinite geometric series.*

19. $3 + 3/2 + 3/4 + 3/8 + \cdots$

20. $16 + 4 + 1 + 1/4 + 1/16 + \cdots$

21. $1 + 2/7 + 4/49 + 8/343 + 16/2401 + \cdots$

In 22–24, evaluate the infinite geometric series with given first term and common ratio.

22. $a = 12$, $r = -1/3$

23. $a = 36$, $r = 1/9$

24. $a = 1$, $r = 2/3$

In 25–27, write each repeating decimal as an equivalent fraction.

25. $0.\overline{25}$

26. $0.\overline{4}$

27. $4.3\overline{72}$

(Section 16.4) *In 28–31, write the expanded form of each expression.*

28. $(y - 5)^4$

29. $(x + 1)^5$

30. $(a - 2b)^4$

31. $(2a + 3b)^5$

In 32–36, find the first four terms of each expression.

32. $(y - 3)^{12}$

33. $(y + 2)^8$

34. $(1 + 2x)^{14}$

35. $(5a - 3b)^5$

36. $(2a + 5b)^6$

(Section 16.5) *In 37–39, find the first four terms of each expression.*

37. $(1 - x)^{-4}$

38. $(1 + x)^{1/3}$

39. $(1 + x)^{-1/2}$

In 40–42, find an approximate value for each number using the first four terms of a binomial expansion. Round the answer to five decimal places.

40. $\sqrt{1.12}$

41. $\sqrt{6.9}$

42. $\sqrt[3]{10}$

In 43–45, find the indicated term in the expansion of each series.

43. third term of $(2a + 3)^7$

44. fifth term of $(3x - 1)^{10}$

45. fourth term of $(1 - 2b)^8$

APPLIED PROBLEMS

1. In a particular locality, buildings appreciate at the rate of 8% per year. If a building initially cost \$100,000.00, find its value after 7 years. 1 2 6 7 8 3 5 9

2. A hotel is to be constructed in such a way that each story has two fewer rooms than the previous story. There are to be 15 stories and the first should have 40 rooms. How many rooms will the hotel have? 1 2 4 7 8 3 5

3. When an object is dropped from rest, the distance s (in m) that it travels during time t (in s) is given by $s = 4.9t^2$.
   (a) Compute the distance traveled during the twelfth second. (Hint: Consider the difference between the distance traveled over a 12-second interval and an 11-second interval.)
   (b) Compute the distance traveled during the first second, the second second, and the third second. Examine the numbers for a pattern and use the pattern to predict how far the object will fall during the twelfth second. 4 7 8 1 2 3 5 6 9

4. When the temperature of a flat object is increased by ΔT (in degrees Celsius), the object expands in such a manner that the area A (in m^2) is given by $A = A_0[1 + \alpha(\Delta T)]^2$. In the formula, A_0 is the area (in m^2) before expansion and α is the coefficient of linear expansion (for steel $\alpha = 1.2 \times 10^{-5}$ per °C). Explain why the formula $A = A_0[1 + 2\alpha(\Delta T)]$ gives a very good approximation for A. 6 8 1 2 3 5 7 9

5. A calculator display shows 7.6767677 as an answer. Assuming that this number is actually $7.\overline{67}$ rounded, what fraction does it represent? 3 5 9

Problems 6 and 7 use the following information. The decibel (dB) is a unit of amplifier gain used in electronics. Each amplifier in a multi-stage amplifier circuit adds its dB gain to the previous amplifier's efforts.

6. Find the total decibel gain that would result for a three-stage amplifier with gains of 10 dB, 15 dB, and 20 dB. 3 5 9 1 2 4 6 7 8

7. What dB gain would each of 20 amplifiers need to sequentially boost a 10 dB signal to 100 dB? 3 5 9 1 2 4 6 7 8

8. A string of connectors in a long run of fiber optic cables each introduce 2 dB loss to the signal. It is known that after the sixth connector, the remaining signal is 5 dB. What was the original signal? 3 5 6 9

9. A high frequency television signal travels through a string of tandem attenuators each with attenuation of 1.7 dB. The total attenuation is additive. What total loss would be experienced by a signal of 10 dB initial strength passing through 5 such attenuators? 5 3 4 7 8 9

Problems 10–12 use the following information. The signal gains of electronic amplifiers defined by the ratio $\frac{V_{out}}{V_{in}}$ multiply. Thus, an amplifier with a gain of 20 *feeding another amplifier with a gain of* 30 *would have a total gain of* 20×30 *or* 600.

10. What geometric series describes a 10 mV signal as it progresses through amplifier states each with $\frac{V_{out}}{V_{in}}$ gains of 25? 3 5 9 1 2 4 6 7 8

11. What gain would each of 7 tandem amplifiers need to boost a 3.0-μv signal to 4.3 volts? 3 5 9 1 2 4 6 7 8

12. How many amplifiers, each with a gain of 36, would be needed to make a 45-μv signal look approximately like a 2-volt signal? 3 5 9 1 2 4 6 7 8

13. A surface refinishing process starts with a coating 1 mm thick from which $\frac{1}{2}$ mm is removed and replaced with $\frac{1}{4}$ mm of a second coat. When this

1. Architectural Technology
2. Civil Engineering Technology
3. Computer Engineering Technology
4. Mechanical Drafting Technology
5. Electrical/Electronics Engineering Technology
6. Chemical Engineering Technology
7. Industrial Engineering Technology
8. Mechanical Engineering Technology
9. Biomedical Engineering Technology

cures, $\frac{1}{8}$ mm is removed and $\frac{1}{16}$ mm of new material is applied. If this process were continued indefinitely, how thick would the final coating be?
1 2 4 7 8 3 5

14. The anode current in a cathode ray tube is given by

$$I = 50\left(1 + \frac{V}{300}\right)^{3/2}.$$

Find the first three terms of the binomial expansion of the expression on the right. Find I using both equations when $V = 10$ volts.
5 3 4 7 8 9

15. A transistor's leakage current increases 1% per degree Celsius. If it is 1.5 nA at 20° C how much will it be at 30° C? (Hint: An increase of 1% means the amount is 1.01 times as much as it was originally.) 3 5 9 4 7 8

16. A fiber optic cable is known to absorb 53% of the infrared radiation injected into it. What percent of the original radiation would remain if the light had to travel through 6 of these identical cables? Neglect any losses at the connectors. 3 5 6 9

17. Corrosion builds faster on metal as the temperature rises. Suppose a certain rust is known to build 5% faster for each 10° C rise in temperature. By what percent will the rust formation rate increase if the temperature increases 75° C?
1 2 4 6 8 5 7 9

18. On a very cold winter's day a steel billet is heated to 2000° F and allowed to cool. How hot will it be after 3 hours if it loses 25% of its temperature every 10 minutes? 1 2 4 6 8 5 7 9

19. Damascus steel blades are made by folding the metal over and over itself, pounding it to its original thickness each time. Suppose a sheet of steel 0.015 in thick is folded and pounded 20 times. (a) How many layers thick is the resulting blade? (b) How thick is each layer? (Hint: Each pounding makes the layers half as thick as the previous pounding.) 1 2 4 6 8 5 7 9

20. A technician wishes to have $500,000 accumulated in her retirement fund over a period of 40 years. How much should she deposit in a CD each January 1 for 40 years if the CD earns 8.5% interest annually? 1 2 6 7 8 3 5 9

RANDOM REVIEW

1. Write the expanded form of $(2x - 0.1)^4$.
2. Evaluate the infinite geometric series $6 + 3 + 3/2 + 3/4 + \cdots$
3. Find the twelfth term of 14, 12, 10, 8, 6, . . .
4. Evaluate the following series if $n = 18$.
   $3 + 7 + 11 + 15 + \cdots + a_n$
5. Find an approximate value for $\sqrt{1.49}$ using the first four terms of a binomial expansion. Give your answer to five decimal places.
6. Find the first four terms of the expansion of $(1 - x)^{-3}$.
7. Find the number of terms of the series $2 + 9 + 16 + 23 + \cdots + a_n$ that will give a sum of 1370.
8. Evaluate the following series for $n = 10$.
   $6 - 12 + 24 - 48 + 96 + \cdots + a_n$
9. Find the eighth term of $j, -2, -4j, 8, 16j, \ldots$
10. Find the fourth term in the expansion of $(2x + 1)^{12}$.

CHAPTER TEST

1. Find the fifth term of the AP that has first term $1 + j$ and common difference $-j$.
2. Evaluate the following series for $n = 12$.
   $12 + 10 + 8 + 6 + 4 + \cdots + a_n$
3. Find the fourth term of the geometric sequence for which $a = 243$, and $r = -1/3$.
4. Evaluate the following series for $n = 10$.
   $5 - 5/2 + 5/4 - 5/8 + 5/16 + \cdots + a_n$
5. Evaluate the infinite geometric series $6 + 0.6 + 0.06 + 0.006 + \cdots$
6. Find the first four terms of the expansion of $(1 + x)^{-1/4}$.
7. Find an approximate value for $\sqrt[3]{2.33}$ using the first four terms of the binomial expansion. Give your answer to five decimal places.
8. Find the sixth term in the expansion of $(x + 0.2)^{10}$.
9. The Leaning Tower of Pisa was 15.5 feet off the perpendicular in 1830. In 1910, it was 16.5 feet off the perpendicular. Assuming that it fell the same amount in each decade, how far off the perpendicular was it in 1930?
10. An inductor and a resistor are connected in series. After a period of time, the battery is short-circuited. The following table of values shows the current I (in amperes) in the inductor after time t (in seconds). Does this sequence for I represent an arithmetic progression, a geometric progression, or neither?

t	0.1	0.2	0.3	0.4
I	0.3	0.06	0.012	0.0024

CHAPTER 17 SYSTEMS OF EQUATIONS

You have seen many examples in which a technical problem is represented by an equation that can be solved algebraically. Often, however, a single equation cannot convey all of the relevant information about a problem. Two or more equations may be required to describe the situation. In this chapter, we examine techniques for solving such problems when there are two equations with two variables or three equations with three variables.

17.1 SYSTEMS OF TWO LINEAR EQUATIONS: GRAPHING AND ADDITION

system of equations

In Section 2.5, we solved equations. Remember that to solve an equation means to find the values of the variables that make the equation true. In this chapter we consider *systems* of equations. A **system of equations** is a set of two or more equations. To solve the system means to find the values that make *all* of the equations true at the same time.

linear equation

We begin with systems of two *linear equations.* An equation is **linear** if each variable is in a separate term and has an exponent of 1. Consider $x + y = 5$ and $x - y = 1$. For each equation individually there are infinitely many ordered pairs (x, y) that will make the equation true. The equation $x + y = 5$, for example, is true for $(1, 4)$, $(6, -1)$, and $(0, 5)$, to name just three possibilities. The equation $x - y = 1$ is true for $(5, 4)$, $(-1, -2)$, and $(0, -1)$, among other values. The ordered pair $(3, 2)$ is the solution of this system, because it is a solution of *both* equations.

A geometric interpretation for solving a system of equations is possible if we sketch both equations on the same coordinate plane. Figure 17.1 shows the graph for the system we have examined. Each point on the graph of $x + y = 5$ represents a solution of that equation. Likewise, each point on the graph of $x - y = 1$ represents a solution of that equation. The point $(3, 2)$ is the only point that lies on both graphs.

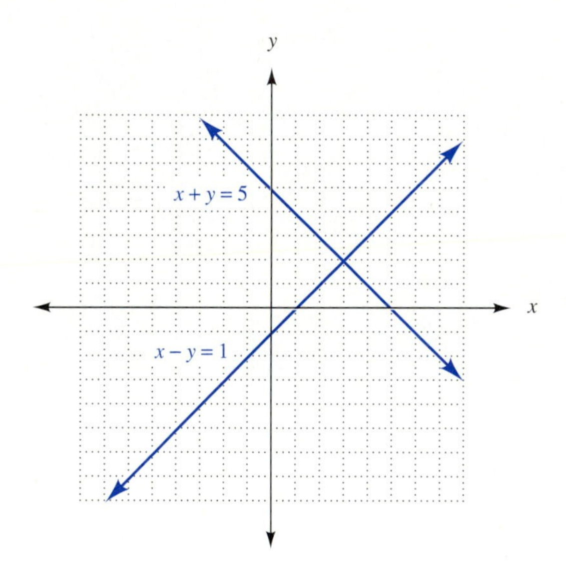

FIGURE 17.1

To solve a system of two simultaneous linear equations in two variables by graphing:

1. Graph the first equation on a rectangular coordinate system.
2. Graph the second equation on the same coordinate system.
3. Identify the point of intersection of the two lines.

EXAMPLE 1 Solve $2x - \quad y = 5$

$x + 2y = 5.$

Solution

1. Graph $2x - y = 5$. When x is 0, y is -5. When y is 0, x is 5/2.
2. Graph $x + 2y = 5$. When x is 0, y is 5/2. When y is 0, x is 5. Figure 17.2 shows the graph.
3. The solution is (3, 1). ■

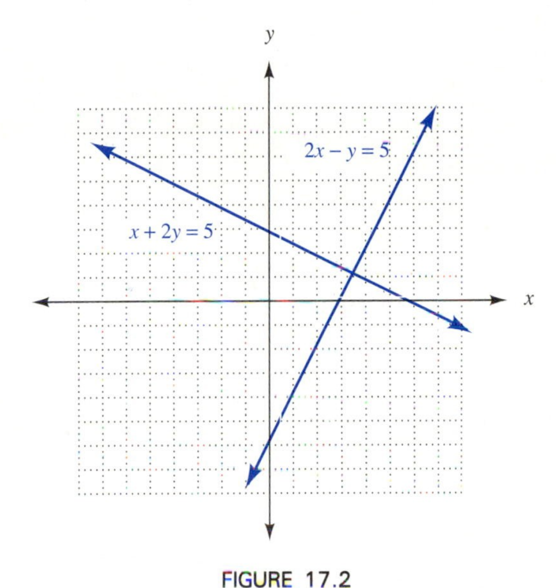

FIGURE 17.2

Graphing as a method of finding the solution of a system of equations has two serious disadvantages.

1. It is difficult to read the graph when the coordinates of the solution are fractions.
2. It is difficult to generalize to three or more variables.

Example 2 illustrates the difficulty of reading fractions on the graph.

EXAMPLE 2 Solve $x + 2y = 1$

$7x + \quad y = 2.$

Solution Figure 17.3 shows the graph.
It is impossible to read the solution (3/13, 5/13) from the graph. ■

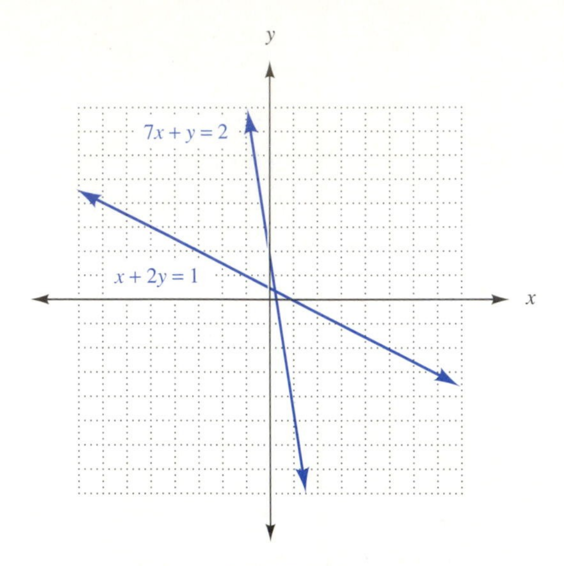

FIGURE 17.3

An algebraic solution allows us to *calculate* the solution. The method presented here is often called the **addition method.**

addition method

Suppose we reconsider the system $x + y = 5$
$x - y = 1.$

If we add the second equation to the first one, we have added equal numbers to both sides of the first equation. The resulting equation is $2x = 6$. It follows that x is 3. Since the solution of the system satisfies *both* equations, we can calculate the y-value by substituting 3 for x in either one. Using the first equation, $3 + y = 5$, so y is 2.

The fact that we had $+y$ and $-y$ in this system led to the elimination of the variable y from the system. We could then solve the resulting equation as a linear equation in one variable.

EXAMPLE 3 Solve $x + 2y = 1$ **(17-1)**

$7x + \quad y = 2.$ **(17-2)**

Solution We would like to eliminate one of the variables. However, we must first replace one of the equations with an equivalent one. It is important to see that we can eliminate *either* variable. If we choose to eliminate x, we must force Equation 17-1 to contain $-7x$. We can make this modification if we multiply both sides of the equation by -7.

$$\begin{array}{r} -7x - 14y = \ -7 \\ 7x + \quad y = \quad 2 \\ \hline -13y = \ -5 \\ y = 5/13 \end{array}$$

Add the two equations.

Solve for y.

If we had chosen to eliminate y instead of x, we would have forced Equation 17-2 to contain $-2y$ by multiplying both sides of that equation by -2.

$$\begin{aligned} x + 2y &= 1 \\ -14x - 2y &= -4 \\ \hline -13x &= -3 \quad \text{Add the two equations.} \\ x &= 3/13 \quad \text{Solve for } x. \end{aligned}$$

Eliminating x produces the y-value while eliminating y produces the x-value. It does not matter which is done first. To check the solution, we must check both equations.

$$(3/13) + 2(5/13) = 3/13 + 10/13 = 13/13 = 1$$
$$7(3/13) + 5/13 = 21/13 + 5/13 = 26/13 = 2 \quad ■$$

Sometimes it is necessary to replace both equations in order to eliminate one of the variables.

EXAMPLE 4 Solve $2x + 5y = 1$ **(17-3)**

$3x - 4y = -10.$ **(17-4)**

Solution To eliminate x, force Equation 17-3 to contain $+6x$ and force Equation 17-4 to contain $-6x$. That is, multiply Equation 17-3 by 3 and Equation 17-4 by -2.

$$\begin{aligned} 6x + 15y &= 3 \\ -6x + 8y &= 20 \\ \hline 23y &= 23 \quad \text{Add the two equations.} \\ y &= 1 \quad \text{Solve for } y. \end{aligned}$$

To eliminate y, multiply Equation 17-3 by 4 and Equation 17-4 by 5 to get $20y$ and $-20y$. Notice that the signs are already opposites.

$$\begin{aligned} 8x + 20y &= 4 \\ 15x - 20y &= -50 \\ \hline 23x &= -46 \quad \text{Add the two equations.} \\ x &= -2 \quad \text{Solve for } x. \end{aligned}$$

The solution, then, is $(-2, 1)$. ■

EXAMPLE 5 Solve $5x + 3y = 7$ **(17-5)**

$4x - 2y = 5.$ **(17-6)**

Solution To eliminate x, multiply Equation 17-5 by -4 and Equation 17-6 by 5.

$$\begin{aligned} -20x - 12y &= -28 \\ 20x - 10y &= 25 \\ \hline -22y &= -3 \quad \text{Add the two equations.} \\ y &= 3/22 \quad \text{Solve for } y. \end{aligned}$$

To eliminate y, multiply Equation 17-5 by 2 and Equation 17-6 by 3.

$$\begin{aligned} 10x + 6y &= 14 \\ 12x - 6y &= 15 \\ \hline 22x &= 29 \quad \text{Add the two equations.} \\ x &= 29/22 \quad \text{Solve for } x. \end{aligned}$$

The solution is (29/22, 3/22).

Check: $5(29/22) + 3(3/22) = 145/22 + 9/22 = 154/22 = 7$

$4(29/22) - 2(3/22) = 116/22 - 6/22 = 110/22 = 5$ ■

When the first value is a fraction (as in Examples 3 and 5), rather than an integer, we solve the system by going through the elimination process for each variable. When the first value is an integer, as it was in Example 4, it is probably easier to find the second value by substituting the first into *either* equation. The two equations will produce the same answer, so choose the one that looks like it will be easier to solve. The procedure can be generalized.

To solve a system of two simultaneous linear equations in two variables by the addition method:

1. If the equations are not in the form $ax + by = c$, write equivalent equations that are in that form.
2. Choose one of the variables as the one to eliminate. If necessary, multiply one or both equations in order to make the coefficients of this variable have the same absolute value, but opposite signs.
3. Add the two equations.
4. Solve for the variable.
5. Return to the original problem and solve for the other variable by
   (a) substituting the value from step 4 into one of the equations, or
   (b) repeating steps 2–4 to eliminate the variable solved for in step 4.

EXAMPLE 6 Solve $3x + 4y = 10$ **(17-7)**

$5x - 7 = 3y.$ **(17-8)**

Solution

1. Replace Equation 17-8 with the equivalent equation $5x - 3y = 7$. **(17-9)**
2. Choose y as the variable to eliminate. (We could also work the problem by choosing x.)

$$\begin{aligned} 9x + 12y &= 30 \quad \text{Multiply Equation 17-7 by 3.} \\ 20x - 12y &= 28 \quad \text{Multiply Equation 17-9 by 4.} \end{aligned}$$

3. $29x = 58$ Add the two equations.
4. $x = 2$ Solve for x.

5. Since 2 is an integer, we substitute 2 for x in the first equation.

$$\begin{aligned} 3(2) + 4y &= 10 \\ 6 + 4y &= 10 \\ 4y &= 4 \\ y &= 1 \end{aligned}$$

The solution is (2, 1). ■

Example 7 illustrates how systems of equations arise in technological settings.

EXAMPLE 7 Gold is measured by troy weight. One troy pound of 10 kt gold contains 5 troy ounces of pure gold and 7 troy ounces of another metal such as copper, silver, or zinc. Each troy pound of 18 kt gold contains 9 troy ounces of pure gold and 3 troy ounces of another metal. How many troy pounds of 10 kt gold and how many troy pounds of 18 kt gold can be made from 40 troy ounces of pure gold and 24 troy ounces of other metals?

Solution Let

$$\begin{aligned} x &= \text{no. of troy pounds of 10 kt gold} \\ y &= \text{no. of troy pounds of 18 kt gold} \end{aligned}$$

Each pound of 10 kt gold requires 5 ounces of pure gold, so $5x$ is the number of ounces of pure gold needed to make x pounds. Each pound of 18 kt gold requires 9 ounces of pure gold, so $9y$ is the number of ounces of pure gold needed to make y pounds. Thus, $5x + 9y$ is the number of ounces of pure gold needed to make x pounds of 10 kt gold and y pounds of 18 kt gold. Since the amount of gold to be used is 40 ounces, we have $5x + 9y = 40$. Likewise, each pound of 10 kt gold requires 7 ounces of other metals, so $7x$ is the number of ounces of other metals needed to make x pounds. Each pound of 18 kt gold requires 3 ounces of other metals, so $3y$ is the number of ounces of other metals needed to make y pounds. Thus, $7x + 3y$ is the number of ounces of other metals needed to make x pounds of 10 kt gold and y pounds of 18 kt gold. Since the amount of other metals to be used is 24, we have $7x + 3y = 24$. Thus, we have a system of equations.

$$\begin{aligned} 5x + 9y &= 40 \\ 7x + 3y &= 24 \end{aligned}$$

To solve it, we choose to eliminate y. Multiply the second equation by -3.

$$\begin{array}{rcrcr} 5x & + & 9y & = & 40 \\ -21x & - & 9y & = & -72 \\ \hline -16x & & & = & -32 \\ x & & & = & 2 \end{array}$$

To find y, substitute the value 2 for x in the equation $5x + 9y = 40$.

$$10 + 9y = 40$$
$$9y = 30$$
$$y = 30/9 \text{ or } 10/3$$

That is, 2 pounds of 10 kt gold and 3 1/3 pounds of 18 kt gold can be made. ■

EXAMPLE 8 Solve

$$3x + 2y - 5 = 0 \quad \textbf{(17-10)}$$
$$6x + 4y = 10. \quad \textbf{(17-11)}$$

Solution Replace Equation 17-10 with the equivalent equation

$$3x + 2y = 5. \quad \textbf{(17-12)}$$

Choose x as the variable to eliminate.

$$-6x - 4y = -10 \quad \text{Multiply Equation 17-12 by } -2.$$
$$6x + 4y = 10 \quad \text{Copy Equation 17-11.}$$
$$0 = 0 \quad \text{Add the two equations.}$$

dependent system

Eliminating x also eliminated y. Every solution of one equation is a solution of the other. Thus, there are an infinite number of solutions of the system. We say that the system is *dependent*. Graphically, one equation would produce the same line as the other, as shown in Figure 17.4. This happens because one equation is a multiple of the other (i.e., they are equivalent equations). ■

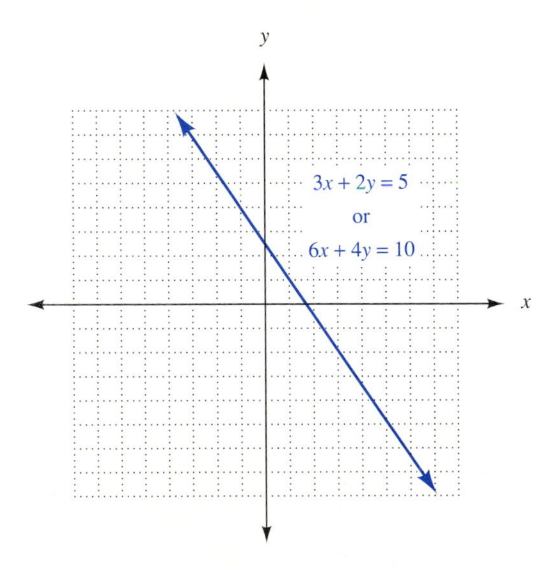

FIGURE 17.4

EXAMPLE 9 Solve

$$-2x + 4y = 7 \quad \textbf{(17-13)}$$
$$x - 2y = 3. \quad \textbf{(17-14)}$$

Solution The equations are in the correct form. Choose to eliminate x.

$$\begin{aligned} -2x + 4y &= 7 && \text{Copy Equation 17-13.} \\ 2x - 4y &= 6 && \text{Multiply Equation 17-14 by 2.} \\ \hline 0 &= 13 && \text{Add the two equations.} \end{aligned}$$

Eliminating x simultaneously eliminated y again, but this time, the resulting statement $(0 = 13)$ is false. There are no values of x and y for which the equation will be true. We say that the system is *inconsistent*. Graphically, the two equations will produce lines that never intersect, because they are parallel to each other, as shown in Figure 17.5. ■

inconsistent system

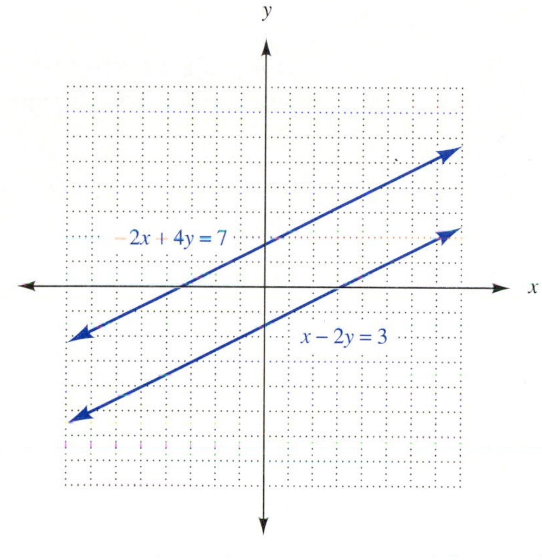

FIGURE 17.5

Example 10 shows how a calculator may be used to simplify the computation required to solve a system of equations.

EXAMPLE 10 Figure 17.6a shows a 500.0 N weight supported by a structure consisting of a rigid body making a 50.00° angle with the horizontal and a cable making a 20.00° angle with the horizontal. By analyzing the free-body diagram in Figure 17.6b, an engineer obtains the following system of linear equations. Solve for A and B.

$$0 = B \cos 50° - A \cos 20°$$
$$0 = -500 + B \sin 50° - A \sin 20°$$

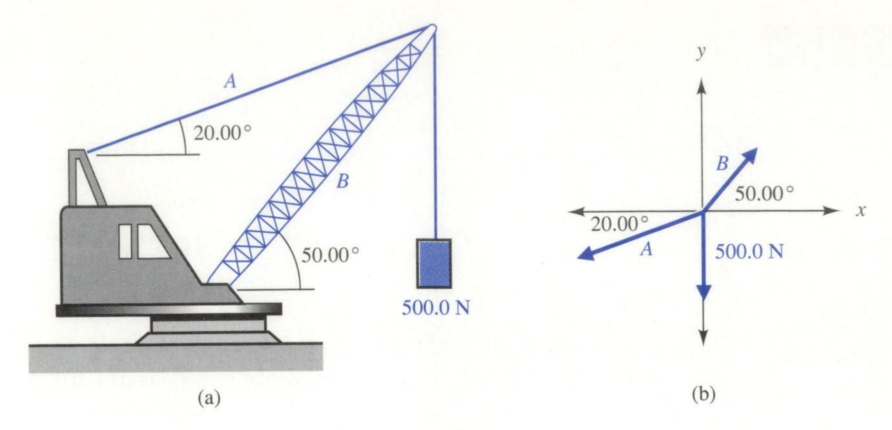

FIGURE 17.6

Solution We choose to eliminate A. Multiply the first equation by sin 20° and the second equation by $-\cos 20°$.

$$0 = B \sin 20° \cos 50° - A \sin 20° \cos 20°$$
$$-500 \cos 20° = -B \sin 50° \cos 20° + A \sin 20° \cos 20°$$

To use the calculator efficiently, the strategy is to compute both B-coefficients, add, and store the sum. The sum of the A-terms should be zero, so the next step is to compute both constant terms and add. Then A can be found. The keystroke sequence is as follows.

20 [sin] [×] 50 [cos] [−] 50 [sin] [×] 20 [cos] [=] [STO] **−0.5**

500 [×] 20 [cos] [+/−] [=] [÷] [RCL] [=] **939.69261**

To eliminate B, return to the original system of equations and multiply the first equation by sin 50° and the second equation by $-\cos 50°$.

$$0 = B \sin 50° \cos 50° - A \sin 50° \cos 20°$$
$$-500 \cos 50° = -B \sin 50° \cos 50° + A \cos 50° \sin 20°$$

The keystroke sequence follows.

50 [sin] [×] 20 [cos] [+/−] [+] 50 [cos] [×] 20 [sin] [=] [STO] **−0.5**

500 [×] 50 [cos] [+/−] [=] [÷] [RCL] [=] **642.7876**

The solution, then, to four significant digits, is $A = 642.8$ N and $B = 939.7$ N. ■

EXAMPLE 11 An airplane traveled a distance of 1054 miles in 3 hours 6 minutes with a tailwind. The return trip required 3 hours 24 minutes. Assuming the speed of the wind was the same on both trips, find the speed of the plane in still air and the speed of the wind.

Solution Let x = speed of the plane in still air
y = speed of the wind
$x + y$ = speed of the plane on original trip
$x - y$ = speed of the plane on return trip
1054 = distance for each trip
3.1 hours = time of original trip
3.4 hours = time of return trip
$D = RT$ (Distance = rate × time)

For the original trip: $1054 = 3.1(x + y)$
For the return trip: $1054 = 3.4(x - y)$

Thus, we have a system of two equations with two unknowns. The distributive property is used to put the equations in the form to eliminate one of the variables.

$$1054 = 3.1x + 3.1y$$
$$1054 = 3.4x - 3.4y$$

To eliminate y, multiply the first equation by 3.4 and the second by 3.1.

$$(3.4)(1054) = (3.4)(3.1)x + (3.4)(3.1)y$$
$$(3.1)(1054) = (3.1)(3.4)x - (3.1)(3.4)y$$

A calculator may be used. The keystroke sequence follows.

3.4 [×] 3.1 [+] 3.1 [×] 3.4 [=] [STO] **21.08**
3.4 [×] 1054 [+] 3.1 [×] 1054 [=] [÷] [RCL] [=] **325**

To eliminate x, multiply the first equation by 3.4 and the second by -3.1.

$$(3.4)(1054) = (3.4)(3.1)x + (3.4)(3.1)y$$
$$(-3.1)(1054) = (-3.1)(3.4)x - (-3.1)(3.4)y$$

3.4 [×] 3.1 [+] 3.1 [×] 3.4 [=] [STO] **21.08**
3.4 [×] 1054 [+] 3.1 [+/-] [×] 1054 [=] [÷] [RCL] [=] **15**

The speed of the plane in still air is 325 mph, and the speed of the wind is 15 mph. ■

EXERCISES 17.1

In 1–10, graph each pair of equations and determine the solution of the system from the graph.

1. $x + y = 8$, $2x - y = 1$

2. $2x + y = 8$, $x + 2y = 10$

3. $3x - 2y = 1$, $4x - y = 1$

4. $2x + 3y = 1$, $3x - 2y = -5$

5. $x + 2y = 1$, $2x - y = 0$

6. $6x + 6y = 7$, $6x - 6y = 1$

7. $x + 3y = 4$, $2x + 6y = 7$

8. $2x - 5y = 10$, $4x - 10y = 20$

9. $3x - 4y = 12$, $9x - 12y = 36$

10. $x + 7y = 7$, $2x + 14y = 15$

In 11–30, solve each system of equations by the addition method. Round answers to the nearest tenth, if necessary.

11. $x + y = 8$
$2x - y = 1$

12. $2x + y = 8$
$x + 2y = 10$

13. $3x - 2y = 1$
$4x - y = 1$

14. $2x + 3y = 1$
$3x - 2y = -5$

15. $x + 2y = 1$
$2x = y$

16. $6x + 6y = 7$
$6x = 6y + 1$

17. $x + 3y = 4$
$2x + 6y = 7$

18. $2x = 5y + 10$
$4x = 10y + 20$

19. $3x - 4y = 12$
$9x - 12y = 36$

20. $x + 7y - 7 = 0$
$2x + 14y - 15 = 0$

21. $x + 3y = 7$
$2x + y + 1 = 0$

22. $2x = 3y + 8$
$5x + 2y = 1$

23. $3x + 2y = 0$
$3x = 2y$

24. $3x - 4y = 1$
$4x = 3y - 1$

25. $1.9x + 5y = 2.6$
$4x - 2.7y = 3.8$

26. $4.5x + 9.1y = 2.8$
$3.7x - 7.8y = 6.0$

27. $3.6x + 7y = 4.4$
$2x - 4.5y = 5.6$

28. $2.7x + 6y = 3.5$
$3x - 3.6y = 4.7$

29. $3.6x + 8.2y = 1.9$
$2.8x - 6.9y = 5.1$

30. $4.5x + 8y = 5.3$
$x - 5.4y = 6.5$

---

31. A draftsman purchased some pens and some pencils. Each pen cost \$0.35 and each pencil cost \$0.10. He put them in with other supplies before he wrote down how many of each he bought. He couldn't remember how many of each item there were, but the invoice said he bought 20 items for \$5.00. He figured out how many of each item he purchased. Show how he did it. **1 2 6 7 8** 3 5 9

32. An engineering firm charged \$108 for a job that required two hours of an engineer's time and 3 hours of her assistant's time. Assuming that the engineer's hourly wage is three times the assistant's, find the hourly wage of each. **1 2 6 7 8** 3 5 9

33. A factory has room to install 8 machines to stamp out parts. Two models are available. One model produces 100 parts per hour while the other model produces 120 parts per hour. In order to match production with other divisions of the factory, 900 parts per hour must be produced. How many machines should be installed? **1 2 6 7 8** 3 5 9

34. An airplane traveled a distance of 1224 miles in 3 hours 36 minutes into the wind. The return trip required 3 hours 24 minutes. Assuming the speed of the wind was the same on both trips, find the speed of the plane in still air and the speed of the wind. **4 7 8** 1 2 3 5 6 9

35. Two alloys are made by a factory. For the first alloy, each ton produced requires 300 lbs of metal A and 500 pounds of metal B. For the second alloy, each ton produced requires 500 pounds of metal A and 500 pounds of metal B. The factory can obtain 1900 pounds of A and 2500 pounds of B per day. How many tons of each alloy can be produced per day? **1 2 4 6 8** 5 7 9

36. For a laboratory experiment, students need a 15-Ω resistor. They must make up this resistor from eight resistors connected in parallel. When resistors are connected in parallel, the reciprocal of the equivalent resistance is the sum of the reciprocals of the individual resistances. Two values of resistors are available, 100 Ω and 180 Ω. How many resistors of each value are needed? **3 5 9** 1 2 4 6 7 8

37. When a see-saw is balanced by two objects, the product of the weight of one object and its distance from the fulcrum must equal the product of the weight of the other object and its distance from the fulcrum. Suppose a see-saw is balanced by a weight of 100 lbs at one end and 300 lbs at the other. If 100 lbs is added to the existing 100 lbs, the other weight must be moved a distance of 1 ft farther away from the fulcrum to keep the see-saw in balance. How far were the 100-lb and 300-lb weights from the fulcrum originally? **2 4 8** 1 5 6 7 9

38. Solve $a_1x + b_1y = c_1$
$a_2x + b_2y = c_2$.

---

1. Architectural Technology
2. Civil Engineering Technology
3. Computer Engineering Technology
4. Mechanical Drafting Technology
5. Electrical/Electronics Engineering Technology
6. Chemical Engineering Technology
7. Industrial Engineering Technology
8. Mechanical Engineering Technology
9. Biomedical Engineering Technology

17.2 SYSTEMS OF TWO LINEAR EQUATIONS: SUBSTITUTION

If we use the addition method to solve a system of equations, the variables must be listed in the same order for each equation. We saw that it was possible to replace a system that does not follow this pattern with an equivalent system that does follow it. Often, however, it is easier to use the *method of substitution*. This method is particularly effective when one variable is expressed in terms of the other in one of the equations.

substitution method

EXAMPLE 1 Solve by substitution.

$$x + 2y = 3$$
$$y = 2x - 1$$

Solution The *second* equation gives an expression for y in terms of x. Since this equation says that y and $2x - 1$ represent the same number, we merely substitute $2x - 1$ for y in the *first* equation. That is,

$$x + 2\mathbf{y} = 3$$

becomes

$$x + 2(\mathbf{2x - 1}) = 3.$$

The resulting equation is a linear equation in one variable and can be solved accordingly.

$$x + 4x - 2 = 3 \quad \text{Use the distributive property.}$$
$$5x - 2 = 3 \quad \text{Combine like terms.}$$
$$5x = 5 \quad \text{Add 2 to both sides.}$$
$$x = 1 \quad \text{Divide both sides by 5.}$$

Since we know that $y = 2x - 1$, it is true that when x is 1,

$$y = 2(1) - 1$$
$$= 2 - 1 \text{ or } 1.$$

The solution, then, is (1, 1). ■

EXAMPLE 2 A piece of wire 18.0 cm long is used to construct a rectangle with a length twice the width. Find the dimensions of the rectangle.

Solution Since we are to find the *dimensions*, we are looking for both length and width. Let

$$l = \text{length}$$
$$w = \text{width}$$

Since the wire is 18.0 cm long, the perimeter of the rectangle is 18.0 cm. One equation is

$$2l + 2w = 18.0$$

Because the length is twice the width, we have $l = 2w$. The system of equations, then, can be written as

$$2l + 2w = 18.0$$
$$l = 2w.$$

Since the second equation is solved for l in terms of w, substitution is a good method to use. In the first equation, we replace l with $2w$.

$$2(2w) + 2w = 18.0$$
$$4w + 2w = 18.0$$
$$6w = 18.0 \quad \text{Combine like terms.}$$
$$w = 3.0 \quad \text{Divide both sides by 6.}$$

Substitute 3.0 for w in the equation $l = 2w$, to obtain $l = 2(3.0) = 6.0$. The dimensions, then, are 3.0 cm and 6.0 cm. ■

It is not necessary to have one variable expressed in terms of the other in order to use the substitution method.

EXAMPLE 3 Solve by substitution. $x + 3y = 7$
$3x - 2y = 1$

Solution Neither equation gives one variable in terms of the other. Notice, however, that *either* equation could be solved for *either* variable.

Solve the first equation for x

$$x = 7 - 3y$$

Solve the first equation for y

$$3y = 7 - x$$
$$y = \frac{7 - x}{3}$$

Solve the second equation for x

$$3x = 1 + 2y$$
$$x = \frac{1 + 2y}{3}$$

Solve the second equation for y

$$-2y = 1 - 3x$$
$$y = \frac{1 - 3x}{-2}$$

NOTE By choosing the first case, we can avoid fractions. ***Since $x = 7 - 3y$ is the result of solving the first equation, this expression for x must be substituted into the second equation.***

$$3(7 - 3y) - 2y = 1 \quad \text{Replace } x \text{ with } 7 - 3y.$$
$$21 - 9y - 2y = 1 \quad \text{Use the distributive property.}$$
$$21 - 11y = 1 \quad \text{Combine like terms.}$$
$$-11y = -20 \quad \text{Subtract 21 from both sides.}$$
$$y = 20/11 \quad \text{Divide both sides by } -11.$$

Since $x = 7 - 3y$, we have $x = 7 - 3(20/11)$

$$= 7 - 60/11 = 77/11 - 60/11 \text{ or } 17/11.$$

The solution, then, is (17/11, 20/11). ■

The procedure is generalized as follows.

To solve a system of linear equations by the substitution method:

1. Solve either equation for either variable. Say the variable solved for is on the left-hand side of the equation.
2. Use the expression on the right-hand side of the equation in step 1 to replace its variable in the other equation.
3. Solve the resulting equation as an equation in one unknown.
4. Return to the equation obtained in step 1 and substitute the value found in step 3 for the variable.

EXAMPLE 4 Solve by the substitution method.

$$3x - 2y = 7$$
$$2x + 3y = 5$$

Solution

1. Either equation can be solved for either variable. There is no clear best choice. We choose to solve the second equation for x, because the coefficient of that variable is both positive and small.

$$2x + 3y = 5$$
$$2x = 5 - 3y$$
$$x = \frac{5 - 3y}{2} \qquad \textbf{(17-15)}$$

2. Since the equation resulted from solving the *second* equation for x, this value must be substituted into the *first* equation.

$$3\left(\frac{5 - 3y}{2}\right) - 2y = 7 \qquad \text{Replace } x \text{ with } \frac{5 - 3y}{2}.$$

3. Solve this equation for x.

$$2\left[3\left(\frac{5 - 3y}{2}\right) - 2y\right] = 2(7) \qquad \text{Multiply both sides by 2.}$$

$$(2)(3)\left(\frac{5 - 3y}{2}\right) - 2(2y) = 2(7) \qquad \text{Use the distributive property.}$$

$$3(5 - 3y) - 4y = 14 \qquad \text{Divide out the 2's on the left.}$$

$$15 - 9y - 4y = 14 \qquad \text{Use the distributive property.}$$

$$15 - 13y = 14 \qquad \text{Combine like terms.}$$

$$-13y = -1 \qquad \text{Subtract 15 from both sides.}$$

$$y = 1/13 \qquad \text{Divide both sides by } -13.$$

4. Once the value of y is known, the value of x can be found by returning to Equation 17-15, which gives x in terms of y.

$$\begin{aligned} x &= \frac{5 - 3y}{2} \\ &= \frac{5 - 3(1/13)}{2} \\ &= \frac{5 - 3/13}{2} \\ &= \frac{65/13 - 3/13}{2} \\ &= 62/13 \div 2 \\ &= 62/13 \times 1/2 \text{ or } 31/13 \end{aligned}$$ ■

The system of equations in Example 4 would have been easier to solve by the addition method. The advantage of knowing more than one method is in being able to choose the easiest method for any particular problem.

EXAMPLE 5 A company buys computer diskettes priced so that the first 100 diskettes are sold at one price, but the price of each additional diskette is \$0.54 less than a diskette in the first 100. The company is quoted a price of \$534 for 500 diskettes. Find the two prices for diskettes.

Solution Let $x =$ price per diskette for the first 100
$y =$ price per diskette for additional diskettes
Since the additional diskettes cost \$0.54 less per diskette, we have $y = x - 0.54$. The price of the first 100 diskettes is $100x$. The price of the 400 additional diskettes is $400y$. Since the total price is \$534, we have $100x + 400y = 534$. Thus, we have a system of equations.

$$\begin{aligned} y &= x - 0.54 \\ 100x + 400y &= 534 \end{aligned}$$

We solve this system by the substitution method. Replace y in the second equation with $x - 0.54$.

$$\begin{aligned} 100x + 400(x - 0.54) &= 534 \\ 100x + 400x - 216 &= 534 && \text{Use the distributive property.} \\ 500x - 216 &= 534 && \text{Combine like terms.} \\ 500x &= 750 && \text{Add 216 to both sides.} \\ x &= 1.5 && \text{Divide both sides by 500.} \end{aligned}$$

Once the value of x is known, the value of y can be found.

$$y = x - 0.54 \text{ or } y = 1.5 - 0.54, \text{ which is } 0.96.$$

That is, the first 100 diskettes cost \$1.50 each, and additional diskettes cost \$0.96 each. ■

EXAMPLE 6 A chemist has two bottles of hydrochloric acid. One bottle is marked 40% HCL and the other is marked 50% HCL. How much of each solution should be mixed to produce 24 ml of a solution that is 42% HCL?

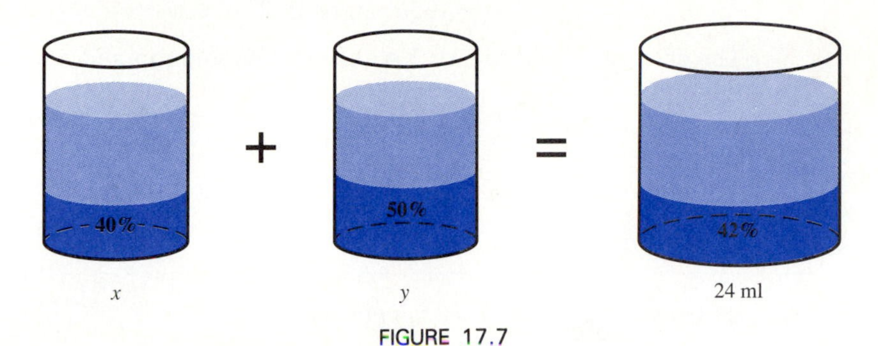

FIGURE 17.7

Solution Since we are to find the number of milliliters of each solution, let $x =$ number of ml of 40% solution, and $y =$ number of ml of 50% solution. A diagram, such as Figure 17.7, might help to organize the data. The mixture is to contain 24 ml, so we have $x + y = 24$. The second equation comes from the percentages. The amount of HCL will be the same whether it is in two separate bottles or mixed in one. That is, 40% of x mixed with 50% of y must equal 42% of 24.

$$0.40x + 0.50y = 0.42(24)$$

The system of equations is

$$x + y = 24$$
$$0.40x + 0.50y = 10.08.$$

Multiplying both sides of the second equation by 100, we have

$$40x + 50y = 1008.$$

Now we are ready to solve the system.

$$x + y = 24 \qquad \textbf{(17-16)}$$
$$40x + 50y = 1008 \qquad \textbf{(17-17)}$$

We choose to eliminate x.

$$-40x - 40y = -960 \qquad \text{Multiply Equation 17-16 by } -40.$$
$$40x + 50y = 1008 \qquad \text{Copy Equation 17-17.}$$
$$10y = 48 \qquad \text{Add the equations.}$$
$$y = 48/10 \text{ or } 4.8 \qquad \text{Solve for } y.$$

We eliminate y.

$-50x - 50y = -1200$	Multiply Equation 17-16 by -50.
$40x + 50y = 1008$	Copy Equation 17-17.
$-10x \quad = -192$	Add the equations.
$x \quad = -192/-10$ or 19.2	Solve for x.

The chemist should use 19.2 ml of the 40% solution and 4.8 ml of the 50% solution. ■

EXERCISES 17.2

In 1–14, solve each system of equations by the substitution method.

1. $x + y = 8$; $2x - y = 1$
2. $2x + y = 8$; $x + 2y = 10$
3. $3x - 2y = 1$; $x - y = 1$
4. $2x + 3y = 1$; $3x - 2y = -5$
5. $x + 2y = 1$; $2x = y$
6. $6x + 6y = 7$; $6x = 6y + 1$
7. $x + 3y = 4$; $2x + 6y = 7$
8. $2x = 5y + 10$; $4x = 10y + 20$
9. $3x - 4y = 12$; $9x - 12y = 36$
10. $x + 7y - 7 = 0$; $2x + 14y - 15 = 0$
11. $x + 3y = 7$; $2x + y + 1 = 0$
12. $2x = 3y + 8$; $5x + 2y = 1$
13. $3x + 2y = 0$; $3x = 2y$
14. $3x - 4y = 1$; $4x = 3y - 1$

In 15–24, solve each system of equations by the most appropriate method.

15. $2x - y = 7$; $x + y = 2$
16. $3x - y = 7$; $-x + 3y = 9$
17. $2x + 2y = 0$; $y = 3x - 2$
18. $x - 3y = 0$; $x = 2y - 3$
19. $-x + y = 2$; $x = 2y$
20. $5x - 6y = 5$; $5x + 6y = 1$
21. $-x + 3y = 5$; $x - 6y = 6$
22. $x = 4y - 9$; $3x = 9y - 5$
23. $y = 8 - 3x$; $7x + 9y = 15$
24. $2x + 5y = 0$; $y = 3x + 1$

25. A piece of wire 24.0 cm long is to be used to construct an isosceles triangle with a base 1/2 the length of any other side. Find the dimensions of the triangle. 1 2 4 7 8 3 5

26. A wire 30.0 cm long is in the shape of a circle. If it is reshaped into a rectangle with a width equal to the radius of the circle, what is the length of the rectangle? 1 2 4 7 8 3 5

27. Two cables support a 125-lb weight, as shown in the accompanying diagram. The tensions x and y (in lb) in the cables can be found from the following system of equations.

$$x \sin 60° + y \sin 30° = 125$$
$$x \cos 60° = y \cos 30°$$

Find x and y. 2 4 8 1 5 6 7 9

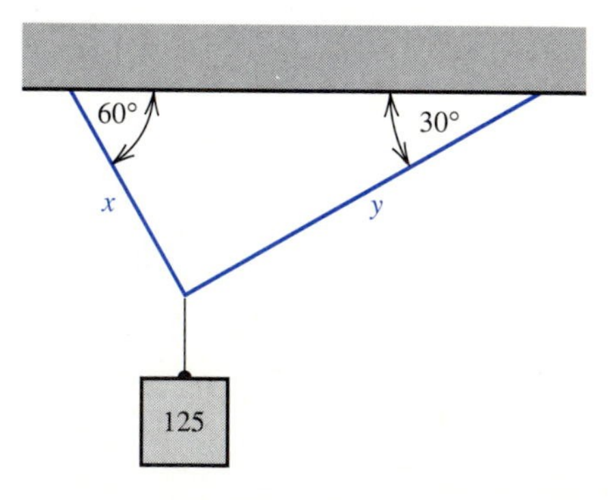

28. A chemist needs 50 liters of a solution that is 25% HCL. He has in stock a 20% HCL solution and a 40% HCL solution. How many liters of each should be mixed to obtain the desired solution?
6 9 2 4 5 7 8

29. A 10% saline solution and a 5% saline solution are to be mixed to produce 80 ml of an 8% solution. How many ml of each solution should be used?
6 9 2 4 5 7 8

30. An electrical circuit is designed with 3 resistors in series. Resistances in series are added. One resistor is to have twice the resistance of each of the others, and the total equivalent resistance is to be 8 Ω. What resistances should be used?
3 5 9 1 2 4 6 7 8

31. A boat travels a distance of 80.25 km with the current in 2 hours 30 minutes. On the return trip, the rate of the current is doubled and the trip takes 3 hours 36 minutes. Find the rate of the boat in still water and the rate of the current on the original trip. Give your answer to the nearest tenth.
4 7 8 1 2 3 5 6 9

32. A lab test is done in two parts. The first part takes 3 minutes longer than the second part. Five tests can be run in one hour. How long does it take to run each part? 1 2 6 7 8 3 5 9

33. Two resistors connected as shown in the accompanying diagram are called a voltage divider.

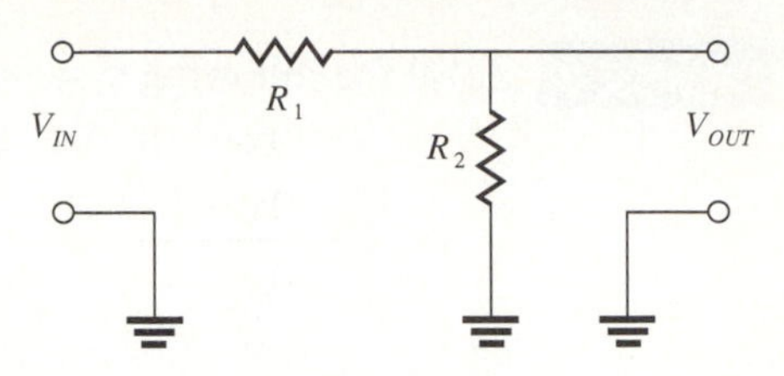

The output voltage V_{out} of such a circuit is given by

$$V_{out} = V_{in}\frac{R_2}{R_1 + R_2}.$$

Find values of R_1 and R_2 so that the resistance at the input terminals is 33 kΩ, and $V_{out} = 0.17\ V_{in}$. Since the resistors are in series, the resistance at the input terminals is the sum of the individual resistances. Ignore the effects of any circuit that may be connected to the output terminals.
3 5 9 1 2 4 6 7 8

34. The following equations represent the forces (in newtons) analyzed in a free-body diagram. Solve the system of equations.

$0 = -N_{AB} + N_A$

$0 = -5 + F_A + F_{AB}$

$F_{AB} = 0.18\ N_{AB}$

$F_A = 0.14N_A$

$0 = -R_{bx} + N_{AB}$

$0 = -3 - F_{AB} + R_{by}$

$0 = 18\ R_{bx} - 8\ F_s$

$0 = 10\ F_s - 18\ R_{ax}$

$0 = R_{ay} - R_{by}$

1. Architectural Technology
2. Civil Engineering Technology
3. Computer Engineering Technology
4. Mechanical Drafting Technology
5. Electrical/Electronics Engineering Technology
6. Chemical Engineering Technology
7. Industrial Engineering Technology
8. Mechanical Engineering Technology
9. Biomedical Engineering Technology

17.3 SYSTEMS OF THREE LINEAR EQUATIONS

Our strategy in solving a system of three equations in three variables is to consider the equations two at a time. By eliminating one variable from a pair, we reduce the pair of equations to one equation with two variables. Once we have eliminated the same variable from two pairs, we can find the solution to the system formed by the resulting equations.

EXAMPLE 1 Solve the following system of three equations with three variables.

$$x - 3y + 2z = 5 \qquad (17\text{-}18)$$
$$x + 2y + 3z = 7 \qquad (17\text{-}19)$$
$$2x - y + z = 4 \qquad (17\text{-}20)$$

Solution We want to eliminate one of the variables from all three equations. Since x is listed first in each equation, y is listed second in each, and z is listed third in each, this system is ready for the elimination step. We choose x as the variable to eliminate. Working with the equations two at a time, we pair the first two to get the system

$$x - 3y + 2z = 5 \qquad (17\text{-}18)$$
$$x + 2y + 3z = 7. \qquad (17\text{-}19)$$

$$-x + 3y - 2z = -5 \qquad \text{Multiply Equation 17-18 by } -1.$$
$$x + 2y + 3z = 7 \qquad \text{Copy Equation 17-19 and add.}$$
$$5y + z = 2 \qquad (17\text{-}21)$$

We have not yet used Equation 17-20 of the original system. We must eliminate x from it also. We can pair it with either of the other two equations. We choose the first, so that we have the system

$$x - 3y + 2z = 5 \qquad (17\text{-}18)$$
$$2x - y + z = 4. \qquad (17\text{-}20)$$

$$-2x + 6y - 4z = -10 \qquad \text{Multiply Equation 17-18 by } -2.$$
$$2x - y + z = 4 \qquad \text{Copy Equation 17-20 and add.}$$
$$5y - 3z = -6 \qquad (17\text{-}22)$$

The results of these two elimination procedures give us a system of two equations in two variables. That is, we have

$$5y + z = 2 \qquad (17\text{-}21)$$
$$5y - 3z = -6. \qquad (17\text{-}22)$$

If we solve this system for y and z, we will have two of the three values in the original system.
Eliminate y.

$$5y + z = 2 \qquad \text{Copy Equation 17-21.}$$
$$-5y + 3z = 6 \qquad \text{Multiply Equation 17-22 by } -1.$$
$$4z = 8 \qquad \text{Add the equations.}$$
$$z = 2 \qquad \text{Solve for } z.$$

We solve for y by substituting the value 2 for z in Equation 17-21. Thus, $5y + 2 = 2$, so $5y = 0$ and $y = 0$.

Now that we have the values of y and z, we return to the original system of three equations to solve for x. We can use any one of the three equations, because all three will produce the same solution. We choose the first.

$$x - 3y + 2z = 5 \text{ becomes } x - 3(0) + 2(2) = 5, \text{ or}$$
$$x + 4 = 5$$
$$x = 1.$$

We now have the complete solution, which can be written as (1, 0, 2). To check the solution, we substitute the values into the original system.

$$1 - 3(0) + 2(2) = 5 \quad \text{or} \quad 1 + 4 = 5$$
$$1 + 2(0) + 3(2) = 7 \quad \text{or} \quad 1 + 6 = 7$$
$$2(1) - 0 + 2 = 4 \quad \text{or} \quad 2 + 2 = 4$$

■

This procedure is summarized as follows.

To solve a system of three linear equations in three variables:

1. Arrange the terms of each equation so that the variables appear in the same order for each equation.
2. Choose a variable to eliminate. Use two of the three equations to eliminate this variable. (Multiply one or both equations to make the coefficients have the same absolute value but different signs, and add.)
3. Pair the remaining equation of the system with one of the other two equations in the original system and eliminate the same variable that was eliminated in step 2.
4. Pair the resulting equations from steps 2 and 3. Solve as a system of two equations in two variables.
5. Return to the original system of three equations. Choose one of the equations and substitute the values found in step 4. Solve the equation for the remaining variable.

EXAMPLE 2 Solve the system.

$$2x + y - 3z = 2 \tag{17-23}$$
$$3x - y + 4z = 9 \tag{17-24}$$
$$4x + 2y - 3z = 7 \tag{17-25}$$

Solution

1. The variables appear in the same order for each equation, so the system is ready for the elimination step.
2. We choose to eliminate y. Pair the first and second equations.

$$\begin{array}{r} 2x + y - 3z = 2 \\ \underline{3x - y + 4z = 9} \\ 5x \qquad + z = 11 \end{array} \tag{17-26}$$

3. Pair Equation 17-25 with Equation 17-23 and eliminate y.

$$-4x - 2y + 6z = -4 \quad \text{Multiply Equation 17-23 by } -2.$$
$$4x + 2y - 3z = 7 \quad \text{Copy Equation 17-25.}$$
$$3z = 3 \quad \text{Add the equations.}$$
$$z = 1 \qquad (17\text{-}27)$$

4. We pair the results of steps 2 and 3.

$$5x + z = 11 \qquad (17\text{-}26)$$
$$z = 1 \qquad (17\text{-}27)$$

Solving this system, we have $z = 1$ and $5x + 1 = 11$, or $5x = 10$, so x is 2.

5. We return to the original system and choose the first equation. Substituting 2 for x and 1 for z (from step 4), we have

$$2(2) + y - 3(1) = 2$$
$$4 + y - 3 = 2$$
$$1 + y = 2$$
$$y = 1.$$

The complete solution is (2, 1, 1). ■

EXAMPLE 3 A drug preparation calls for 2 mg of solute, 5 mg of preservative, and 4 mg of filler. One gram of product A contains 1, 3, and 4 mg of solute, preservative, and filler, respectively. One gram of product B contains 2 mg of each, and one gram of product C contains only 2 mg of solute and 4 mg of preservative. How many grams of each product should be combined to get the required dosage?

Solution

Let a = number of grams of product A
let b = number of grams of product B
let c = number of grams of product C

When products A, B, and C are combined, some of the solute comes from each product. Each gram of A contributes 1 mg, each gram of B contributes 2 mg, and each gram of C contributes 2 mg. The total amount of solute in the mixture is $1a + 2b + 2c$, which we know should be 2. Consider the preservative. Each gram of A contributes 3 mg, each gram of B contributes 2 mg, and each gram of C contributes 4 mg. The total amount of the preservative in the mixture is $3a + 2b + 4c$, which we know should be 5. For the filler, each gram of A contributes 4 mg, each gram of B contributes 2 mg, and each gram of C contributes 0 mg. The total amount of filler in the mixture is $4a + 2b + 0c$ or $4a + 2b$, which we know should be 4. Thus, we have the following system of equations.

$$1a + 2b + 2c = 2 \qquad (17\text{-}28)$$
$$3a + 2b + 4c = 5 \qquad (17\text{-}29)$$
$$4a + 2b = 4 \qquad (17\text{-}30)$$

We choose to eliminate c.
From Equations 17-28 and 17-29 we have the following system.

$$\begin{aligned} -2a - 4b - 4c &= -4 \quad \text{Multiply Equation 17-28 by } -2. \\ 3a + 2b + 4c &= 5 \quad \text{Copy Equation 17-29.} \\ \hline a - 2b \quad &= 1 \end{aligned} \qquad (17\text{-}31)$$

From Equation 17-30: It is not necessary to pair Equation 17-30 with another, since c does not appear. We have

$$4a + 2b = 4. \qquad (17\text{-}32)$$

Pairing Equations 17-31 and 17-32, we have

$$a - 2b = 1 \qquad (17\text{-}31)$$
$$4a + 2b = 4. \qquad (17\text{-}32)$$

We choose to eliminate a.

$$\begin{aligned} -4a + 8b &= -4 \\ 4a + 2b &= 4 \\ \hline 10b &= 0 \\ b &= 0 \end{aligned}$$

From Equation 17-31, we have $a - 2(0) = 1$, so a is 1.
Finally, Equation 17-28 gives us

$$\begin{aligned} 1(1) + 2(0) + 2c &= 2 \text{ or} \\ 1 + 2c &= 2 \\ 2c &= 1 \\ c &= 1/2. \end{aligned}$$

To get the required dosage, mix 1 gram of product A and 1/2 gram of product C. None of product B should be used. ■

Systems of linear equations are used in circuit analysis. They arise as a result of applying Kirchhoff's laws. Kirchhoff's laws are stated in terms of *nodes* and *loops*. A **node** is a point in a network where two or more paths branch off. A **loop** is a complete circuit path. See Figure 17.8.

node
loop

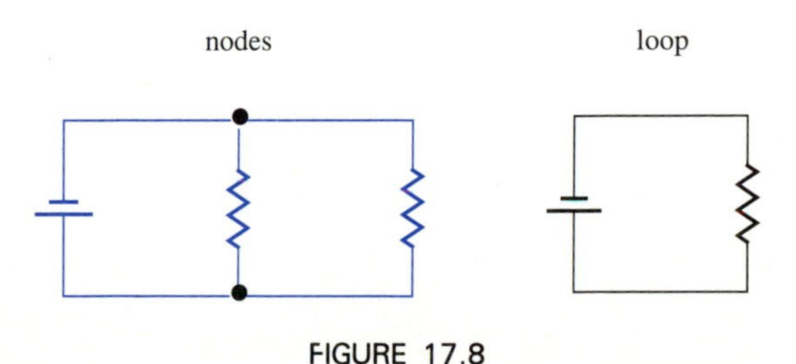

FIGURE 17.8

KIRCHHOFF'S LAWS

(a) Current law: The algebraic sum of all currents at a node must be zero.
(b) Voltage law: The algebraic sum of all voltage sources and voltage drops around a closed loop must be zero.

To apply Kirchhoff's laws, it is necessary to adopt a few conventions. The ones that follow are common, but they are not the only ones in use.

To apply the current law:

1. The direction of current flow through each branch containing a resistor (^\/\/\/\) is indicated by an arrow. It is not necessary to trace the flow of current through a circuit. A common procedure is to mark any current in a horizontal branch as left to right and any current in a vertical branch as top to bottom. It is inevitable that some branches will be marked incorrectly, but in the final solution, a positive current will be considered to flow in the direction indicated on the diagram. A negative current will be considered to flow in the opposite direction from that indicated in the diagram. The arrows are then labeled with variable names for current, such as I_1, I_2, I_3, etc.
2. At any node that joins three or more branches, an equation is written to show the algebraic sum of the currents. A current flowing toward a node is added while current flowing away from a node is subtracted in the equation.
3. If there are n such nodes, the current law is used to obtain $n - 1$ equations. Each current should appear in at least one equation.

To apply the voltage law:

1. Place a + where current enters a resistor and a − where it leaves a resistor. (We are using what is called *conventional* current flow.)
2. Ohm's law ($V = IR$) is used to obtain the algebraic sum of the voltages. In tracing a clockwise path through a loop of the circuit, the voltage across a resistor is considered to have the same sign as the end approached.
3. If there are m loops in the circuit, the voltage law is used $m - 1$ times to obtain $m - 1$ equations. Each current should be included in at least one equation.

EXAMPLE 4 Find the current in each branch of the circuit shown in Figure 17.9.

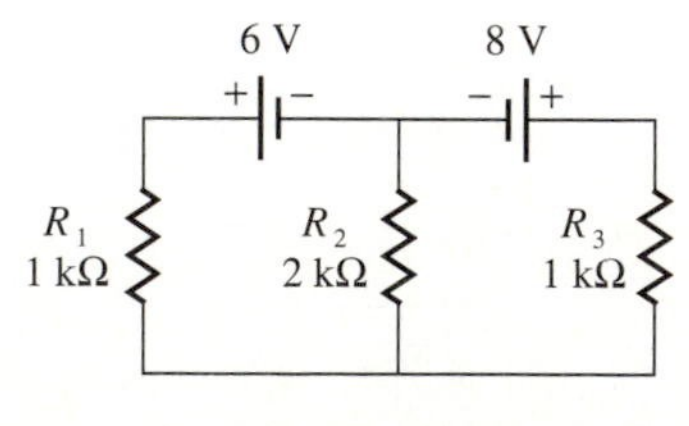

FIGURE 17.9

Solution To apply the current law:

1. In Figure 17.10, the current is marked from left to right or top to bottom in each branch containing a resistor. I_1 flows through R_1, I_2 flows through R_2, and I_3 flows through R_3.

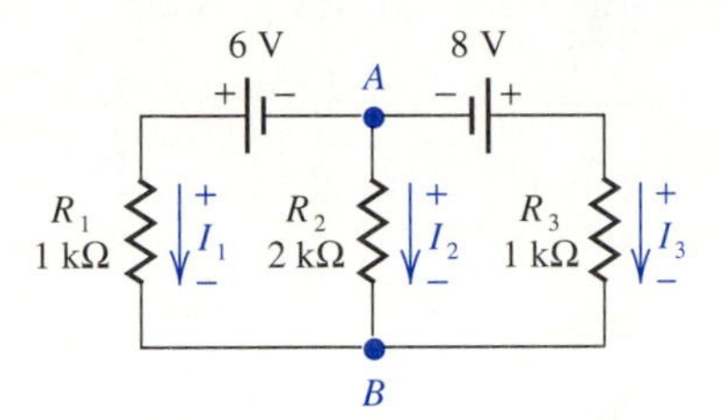

FIGURE 17.10

2. The nodes are labeled A and B. Consider A. Since I_1, I_2, and I_3 all flow away from A, the equation is

$$I_1 + I_2 + I_3 = 0. \quad \textbf{(17-33)}$$

3. Since there are only two nodes, Equation 17-33 is the only equation obtained using the current law.

To apply the voltage law:

1. The resistors are marked with a + on the left or top and a − on the right or bottom to indicate direction of current flow.

2. There are actually three loops in the circuit (two small square ones and one large outer rectangular one). Kirchhoff's laws need only be applied to two of them. We choose the two smaller loops. In the loop on the left, as a clockwise path is traced from the upper left-hand corner, the voltage at the battery (—|⊢—) is 6 V. Ohm's law, $V = IR$, holds if R is in ohms and I is in amperes. But when R is in kilohms, I is in milliamperes. (Why?) The voltage across R_2 is given by $V = I_2R_2$ where R_2 is 2 kΩ. The voltage is positive, because the positive end of R_2 is encountered first. Finally across R_1, $V = I_1R_1$ where R_1 is 1 kΩ. The voltage is negative, because the negative end of R_1 is encountered first. Thus, $6 + 2I_2 - I_1 = 0$. Similarly, the right-hand loop yields $-8 + I_3 - 2I_2 = 0$.

The equations obtained from Kirchhoff's laws form a system of equations.

$$I_1 + I_2 + I_3 = 0 \quad \textbf{(17-34)}$$
$$-I_1 + 2I_2 = -6 \quad \textbf{(17-35)}$$
$$-2I_2 + I_3 = 8 \quad \textbf{(17-36)}$$

The system can be solved as follows.
Eliminate I_1.

$$I_1 + I_2 + I_3 = 0 \quad \text{Copy Equation 17-34.}$$
$$\underline{-I_1 + 2I_2 \quad = -6} \quad \text{Copy Equation 17-35 and add.}$$
$$3I_2 + I_3 = -6 \quad \textbf{(17-37)}$$

Pair the resulting equation with Equation 17-36.

$$-2I_2 + I_3 = 8 \qquad (17\text{-}36)$$
$$3I_2 + I_3 = -6 \qquad (17\text{-}37)$$

Eliminate I_3 from this system.

$$\begin{array}{rcl} -2I_2 + I_3 & = & 8 \\ -3I_2 - I_3 & = & 6 \\ \hline -5I_2 & = & 14 \\ I_2 & = & -14/5 \\ & = & -2.8 \end{array}$$

Eliminate I_2 from this system.

$$\begin{array}{rcl} -6I_2 + 3I_3 & = & 24 \\ 6I_2 + 2I_3 & = & -12 \\ \hline 5I_3 & = & 12 \\ I_3 & = & 12/5 \\ & = & 2.4 \end{array}$$

From Equation 17-34 or 17-35, it follows that I_1 is 0.4. The solution, then, is given by $I_1 = 0.4$ mA, $I_2 = -2.8$ mA, and $I_3 = 2.4$ mA. The negative in I_2 indicates that the current actually flows in the opposite direction from that marked on the diagram. ■

EXERCISES 17.3

In 1–12, solve each system of equations.

1. $\begin{aligned} x - y + 2z &= -1 \\ 2x + y - z &= 9 \\ x - 2y - z &= 0 \end{aligned}$

2. $\begin{aligned} 2x + y + 3z &= 0 \\ x - 3y + 2z &= 7 \\ 3x - 2y - z &= 7 \end{aligned}$

3. $\begin{aligned} 2x - 3y + z &= 0 \\ 3x + 2y - z &= 4 \\ 2x - 2y + z &= 1 \end{aligned}$

4. $\begin{aligned} 3x + y + 2z &= 3 \\ 2x - y - 3z &= 15 \\ 2x + 2y + 3z &= -3 \end{aligned}$

5. $\begin{aligned} 3x + 2z &= 3 \\ x - y - 3z &= 15 \\ x + y &= 0 \end{aligned}$

6. $\begin{aligned} x + y - z &= 2 \\ x - y + z &= 4 \\ -x + y + z &= 6 \end{aligned}$

7. $\begin{aligned} 2x + 2y + z &= 0 \\ x - y - 1 &= 0 \\ 3x - 3y + 2z &= 3 \end{aligned}$

8. $\begin{aligned} 2x + 3y &= z + 2 \\ 4x - 3y &= 1 - z \\ 2x - 3y &= z - 2 \end{aligned}$

9. $\begin{aligned} x + z &= y + 3 \\ 2x + z &= 3 - y \\ 2x - z &= 2y \end{aligned}$

10. $\begin{aligned} 3x + 4y + z - 6 &= 0 \\ 6x - 4y - 2z - 7 &= 0 \\ 3x - 4y - 3z - 4 &= 0 \end{aligned}$

11. $\begin{aligned} 3x - z + 2y &= 3 \\ 2x - y + 3z &= 8 \\ 3y + 2z + x &= 1 \end{aligned}$

12. $\begin{aligned} x + 2y + 2z &= 2 \\ 2x + y - 7 &= 2z \\ 2x + 2y &= z + 8 \end{aligned}$

---

13. The forces acting on bodies A and B are shown in the accompanying diagram. By analyzing the free-body diagram shown, an engineer obtains the following system of linear equations involving forces (in newtons) acting on A. Assume that μ, the coefficient of friction between the surfaces of A and B, is 0.35, and solve for F_{AB}, T, and N_{AB}. Give your answer to three significant digits.

$$0 = F_{AB} - T\cos 18°$$
$$0 = N_{AB} - 40 - T\sin 18°$$
$$F_{AB} = \mu N_{AB}$$

4 7 8 1 2 3 5 6 9

Problem 13

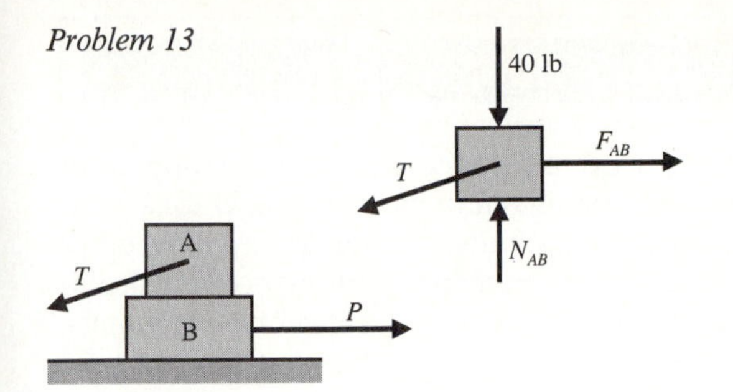

In 14 and 15 use Kirchhoff's laws to find the current (in mA) in each resistor.

14.

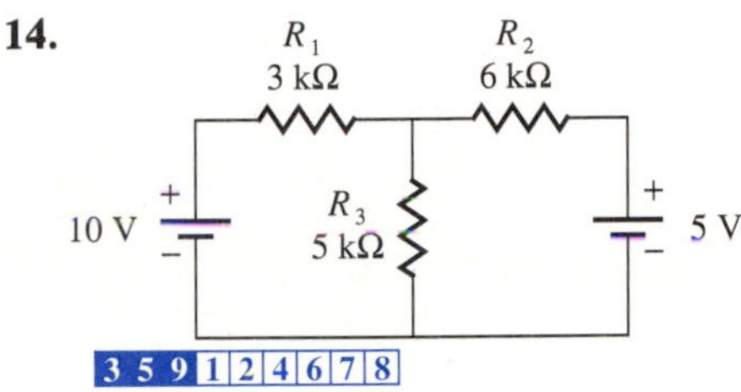

3 5 9 1 2 4 6 7 8

15.

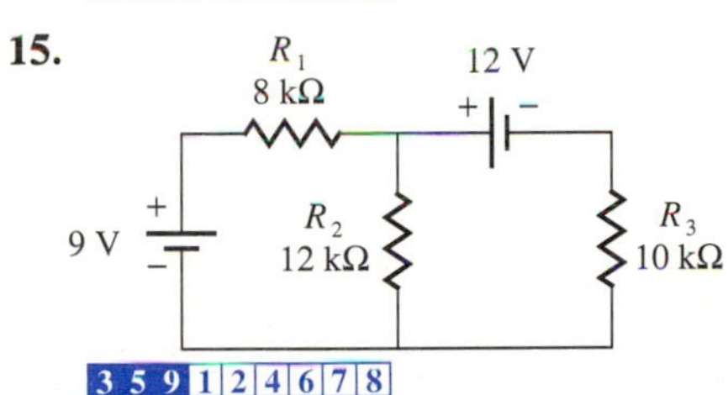

3 5 9 1 2 4 6 7 8

16. The relationship between the speed v of a car (in mph) and the distance d (in ft) required to bring it to a complete stop is known to be quadratic. That is, it follows the pattern $d = av^2 + bv + c$. Use the following data to determine the values of a, b, and c. When v is 10.0, d is 15.0. When v is 20.0, d is 40.0, and when v is 30.0, d is 75.0.
4 7 8 1 2 3 5 6 9

17. A piece of wire 24.0 cm long is to be bent into the shape of a triangle. The sum of the lengths of the two shortest sides is to be 6.0 cm longer than the third side, and the middle side is to be half as long as the sum of the lengths of the other two sides. Find the dimensions that satisfy the conditions. 1 2 4 7 8 3 5

18. A drug preparation calls for 7 mg of solute, 6 mg of preservative, and 14 mg of filler. One gram of product A contains 2, 2, and 3 mg of solute, preservative, and filler respectively. One gram of product B contains only 1 mg of solute and 5 mg of filler, and one gram of product C contains 1 mg of each. How many grams of each product should be combined to get the required dosage?
6 9 2 4 5 7 8

19. An industrial plant needs to store machine parts. They have small, medium, and large boxes available. A small box holds 1 motor and two shafts. A medium box holds 1 motor, 1 casing, and 1 shaft. A large box holds 1 motor, 2 casings, and 1 shaft. How many boxes of each size are necessary to hold exactly 18 motors, 15 casings, and 24 shafts?
1 2 6 7 8 3 5 9

20. We represent three variables using ordered triples in space with three axes. In a system of three equations with three variables, each equation represents a plane. You might visualize two walls and the floor as an example of three planes that intersect in a point, as shown in the accompanying figure.

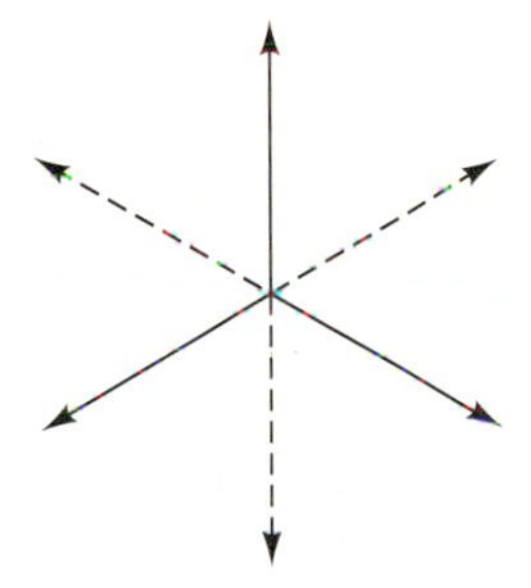

(a) What type of figure is formed by the intersection of two planes?
(b) In a system of three equations, how many pairs of equations are there?
(c) Why do we consider only two pairs in the solution of a system of three equations?

1 Architectural Technology
2 Civil Engineering Technology
3 Computer Engineering Technology
4 Mechanical Drafting Technology
5 Electrical/Electronics Engineering Technology
6 Chemical Engineering Technology
7 Industrial Engineering Technology
8 Mechanical Engineering Technology
9 Biomedical Engineering Technology

17.4 SYSTEMS OF NONLINEAR EQUATIONS

nonlinear equation

The methods that we have used to solve systems of linear equations may be applied in some instances to systems of equations that contain one or more *nonlinear equations.* An equation is **nonlinear** if it contains at least one variable that has an exponent other than 1, or if it contains a product of variables. Computer programs or pocket calculators with graphics capability can be used to speed up the process of drawing graphs, but an understanding of the graphing method is needed to use these tools. If you have studied Chapter 9, you will be able to sketch the graphs of some of the equations using the methods of that chapter. But the method used here for nonlinear equations is that described in Section 3.4. Choose several values to replace x (or y), in the equation and find the corresponding values of y (or x). Plot the points and connect them to sketch the graph.

EXAMPLE 1 Use the graphing method to solve the system of equations.

$$x^2 + y^2 = 25$$
$$3x - y = 5$$

Solution The graph of the equation $3x - y = 5$ is a straight line. Rewriting the equation as $y = 3x - 5$, we see that the slope is 3 and the y-intercept is -5. To sketch the graph of the equation $x^2 + y^2 = 25$, we solve the equation for y. (If you were going to use a computer or calculator to plot the graph, you would probably need to put the equation in this form, too.)

$$y^2 = 25 - x^2$$
$$y = \pm\sqrt{25 - x^2}$$

We now choose several values for x and make a table to show the corresponding y-values. Notice that if we choose values of x that are larger than 5 in absolute value, x^2 is larger than 25, and the quantity under the radical sign is negative. Since the coordinate plane represents points with coordinates that are real numbers, these values of x are not allowed.

x	-5	-4	-3	-2	-1	0	1	2	3	4	5
y	0	± 3	± 4	$\pm\sqrt{21}$	$\pm\sqrt{24}$	± 5	$\pm\sqrt{24}$	$\pm\sqrt{21}$	± 4	± 3	0

The graph of the two equations is shown in Figure 17.11. The curves appear to cross at (3, 4) and at (0, -5). Checking these values in the original system vertifies that they are correct.

$$3^2 + 4^2 = 25 \qquad 0^2 + (-5)^2 = 25$$
$$3(3) - 4 = 5 \qquad 3(0) - (-5) = 5$$

■

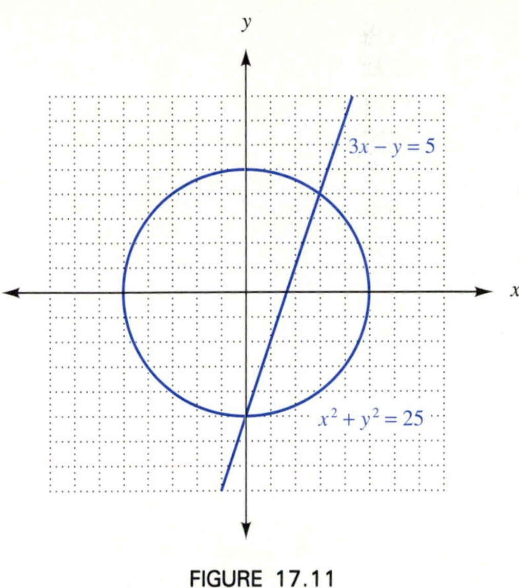

FIGURE 17.11

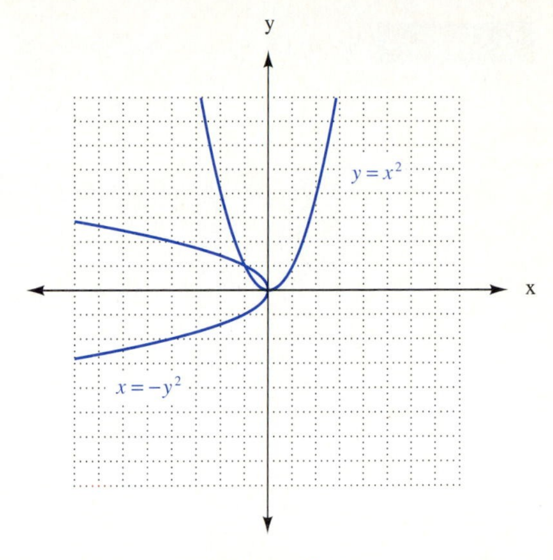

FIGURE 17.12

EXAMPLE 2 Solve the system of equations.

$$y = x^2$$
$$x = -y^2$$

Solution The following table gives values of y for various values of x if $y = x^2$.

x	-3	-2	-1	0	1	2	3
y	9	4	1	0	1	4	9

The following table gives values of x for various values of y if $x = -y^2$.

x	-9	-4	-1	0	-1	-4	-9
y	-3	-2	-1	0	1	2	3

The graph of the two equations is shown in Figure 17.12. The curves appear to cross at (0, 0) and at $(-1, 1)$. Checking these values in the original system verifies that they are correct.

$$0 = 0^2 \qquad 1 = (-1)^2$$
$$0 = -0^2 \qquad -1 = -(1^2) \quad \blacksquare$$

With the graphing method, we will often have to be content with an approximate solution.

EXAMPLE 3 Solve the system of equations.

$$y = 2^x$$
$$xy = 1$$

Solution The following table gives values of y for various values of x if $y = 2^x$.

x	-3	-2	-1	0	1	2	3
y	1/8	1/4	1/2	1	2	4	8

The equation $xy = 1$ may be solved for y. That is, $y = 1/x$. Notice that $1/x$ is undefined when x is 0. The following table gives values of y for various values of x if $y = 1/x$.

x	-3	-2	-1	0	1	2	3
y	$-1/3$	$-1/2$	-1	–	1	1/2	1/3

The graph of the two equations is shown in Figure 17.13. The curves appear to cross at about (0.6, 1.6). ■

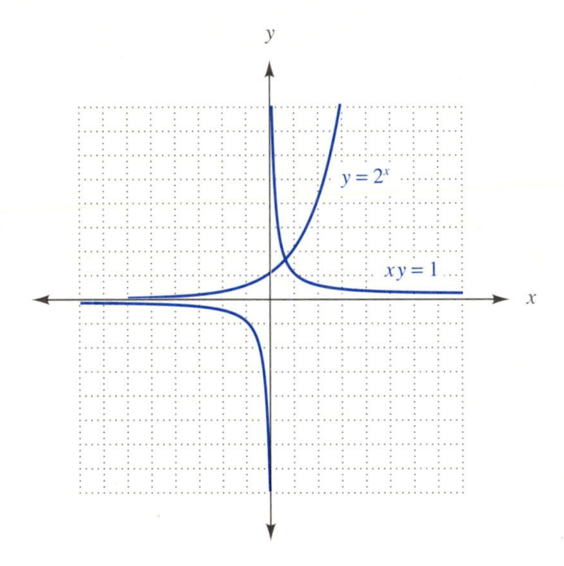

FIGURE 17.13

EXAMPLE 4 A manufacturing plant buys parts from a company that uses the following price structure. If exactly 40 parts are purchased, the price is 50¢ per part. If more than 40 parts are purchased, the price of every part is reduced by 1¢ for each part over 40. A competitor's policy is that if exactly 50 parts are purchased, the price is 60¢ per part, but the price of every part is reduced by 2¢ for each part over 50. Write an equation to give the total cost for x parts from each company. Then graph the two equations to find the approximate number of parts for which the total cost is the same under the two pricing structures.

Solution Let x = number of parts ordered.
Let y = total cost for x parts.
For the first company:
The number of additional parts over 40 is $x - 40$.
The reduction in price for *each* part is $1(x - 40)$.
The cost of one part is $50 - (x - 40)$ or $50 - x + 40$, which is $90 - x$.
Total cost is the product of number of parts and cost per part.
The total cost, then is given by $y = x(90 - x)$ or $y = -x^2 + 90x$.
For the competitor:
The number of additional parts over 50 is $x - 50$.
The reduction in price for each part is $2(x - 50)$.
The cost of one part is $60 - 2(x - 50)$ or $60 - 2x + 100$, which is $160 - 2x$.
The total cost, then is given by $y = x(160 - 2x)$ or $y = -2x^2 + 160x$.
The system of equations is given by

$$y = -x^2 + 90x$$
$$y = -2x^2 + 160x.$$

Figure 17.14 shows the graph.

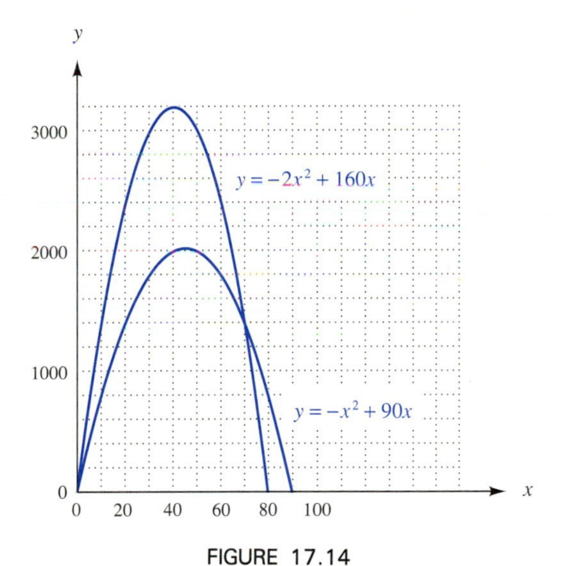

FIGURE 17.14

Since the two graphs intersect when x is 70, we conclude that if 70 parts are ordered, the cost y is the same. ■

Algebraic methods are sometimes applicable to systems of nonlinear equations. Examples 5 and 6 illustrate how substitution may be used.

EXAMPLE 5 Solve the system of equations in Example 1 algebraically.

$$x^2 + y^2 = 25$$
$$3x - y = 5$$

Solution It is important to recognize that there are no like terms that can be eliminated by addition or subtraction in this system. The second equation, however, can be solved for y quickly.

$$y = 3x - 5.$$

Substituting $3x - 5$ for y in the first equation, we have the following.

$$x^2 + (3x - 5)^2 = 25$$
$$x^2 + 9x^2 - 30x + 25 = 25$$
$$10x^2 - 30x = 0$$
$$10x(x - 3) = 0$$
$$10x = 0 \quad \text{or} \quad x - 3 = 0$$
$$x = 0 \quad \text{or} \quad x = 3$$

When x is 0, the equation $y = 3x - 5$ gives $y = 3(0) - 5$ or -5, and when x is 3, $y = 3(3) - 5$ or 4. The solutions, then, are $(0, -5)$ and $(3, 4)$. ■

EXAMPLE 6 A projectile launched with an initial velocity of v_0 at an angle of θ with the horizontal travels a horizontal distance given by $x = v_0(\cos\theta)t$ and a vertical distance given by $y = v_0(\sin\theta)t - 4.9t^2$. In the formulas, t is the time (in seconds) of travel. Figure 17.15 shows the position of the projectile at time t. For a bullet launched at a 55° angle with an initial velocity of 300 m/s, find the time of travel when the horizontal and vertical distances are equal.

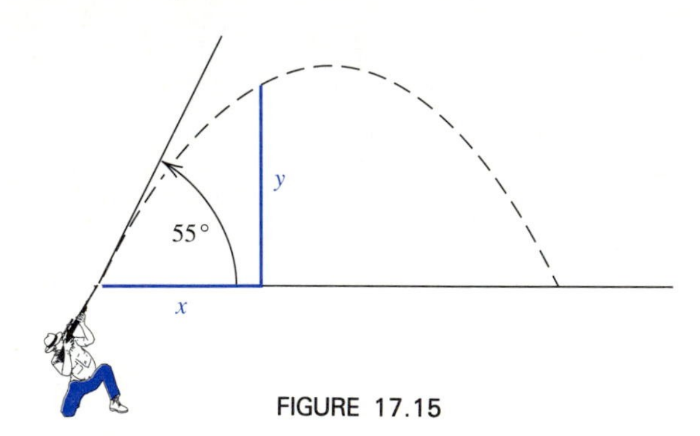

FIGURE 17.15

Solution Let t = time of travel
let d = distance
Since θ is 55°, $\cos\theta$ is 0.57358 and $\sin\theta$ is 0.81915. Also, we know that v_0 is 300, so we have the following system of equations.

$$d = 300(0.57358)t \quad \text{or} \quad d = 172.1t$$
$$d = 300(0.81915)t - 4.9t^2 \quad \quad d = 245.7t - 4.9t^2$$

Substituting $172.1t$ for d in the second equation, we have

$$172.1t = 245.7t - 4.9t^2$$
$$0 = 73.6t - 4.9t^2$$
$$0 = t(73.6 - 4.9t)$$
$$t = 0 \text{ or } 73.6 - 4.9t = 0$$
$$73.6 = 4.9t$$
$$15.0 = t$$

The distances are equal after 15 seconds of travel. ■

Some nonlinear systems are readily solved by the addition method.

EXAMPLE 7 Solve the system of equations.

$$x^2 + y^2 = 32$$
$$x^2 - y^2 = 18$$

Solution If the equations are added, the y^2 terms can be eliminated. Thus,

$$2x^2 = 50$$
$$x^2 = 25$$
$$x = \pm 5.$$

Since $y^2 = 32 - x^2$, it follows that $y = \pm\sqrt{32 - x^2}$. When x is 5, $y = \pm\sqrt{32 - 25}$ or $\pm\sqrt{7}$. Thus, $(5, \sqrt{7})$ and $(5, -\sqrt{7})$ are solutions. When x is -5, $y = \pm\sqrt{32 - 25}$ or $\pm\sqrt{7}$. Thus, $(-5, \sqrt{7})$ and $(-5, -\sqrt{7})$ are also solutions.
Figure 17.16 shows that those solutions are reasonable. ■

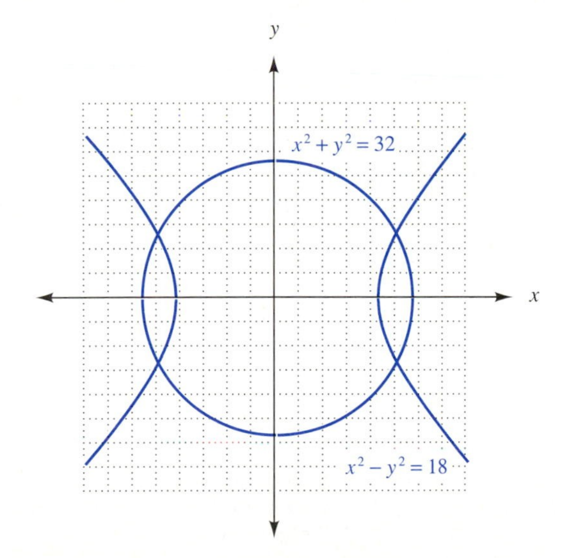

FIGURE 17.16

EXERCISES 17.4

In 1–12, solve each system of equations by graphing. It may be necessary to give an approximate solution.

1. $x^2 + y^2 = 25$, $3x + 4y = 0$

2. $x^2 + y^2 = 20$, $y = 2x$

3. $y = x^2 + 1$, $y = 2x$

4. $y = x^2 + 2x$, $y = x + 2$

5. $xy = 2$, $y = x + 1$

6. $xy = 6$, $y = 2x - 1$

7. $y = 3^x$, $xy = 1$

8. $y = 2^x$, $xy = 2$

9. $4x^2 + y^2 = 4$, $y = x + 1$

10. $x^2 + 4y^2 = 4$, $x + y = 0$

11. $x^2 + y^2 = 4$, $y = x + 3$

12. $y = x^2$, $y = x - 2$

In 13–23, solve each system of equations algebraically.

13. $x^2 + y^2 = 25$, $3x + 4y = 0$

14. $x^2 + y^2 = 20$, $y = 2x$

15. $y = x^2 + 1$, $y = 2x$

16. $y = x^2 + 2x$, $y = x + 2$

17. $xy = 2$, $y = x + 1$

18. $xy = 6$, $y = 2x - 1$

19. $2x^2 + y^2 = 9$, $x^2 - y^2 = 3$

20. $x^2 + y^2 = 3$, $x^2 - 2y^2 = 6$

21. $2x^2 + y^2 = 2$, $x = y^2 + 1$

22. $3x^2 + y^2 = 1$, $x = -y^2 - 1$

23. $x^2 + 3y^2 = 1$, $y = -x^2 - 1$

---

24. Fourteen 56-Ω resistors are to be connected so that there are several parallel groups of resistors, and each group of resistors consists of several resistors connected in series. The equivalent resistance of the system is 16 Ω. Let x be the number of parallel groups, and let y be the number of resistors in each group. Find x and y. For resistors in series, the equivalent resistance is the sum of the individual resistances. For resistors in parallel, the reciprocal of the equivalent resistance is the sum of the reciprocals of the individual resistances. **3 5 9** 1 2 4 6 7 8

25. A manufacturing plant buys parts from a company that uses the following price structure. If exactly 30 parts are purchased, the price is 60¢ per part. If more than 30 parts are purchased, the price of every part is reduced by 1¢ for each part over 30. A competitor's policy is that if exactly 40 parts are purchased, the price is 60¢ per part, but the price of every part is reduced by 2¢ for each part over 40. Write an equation to give the total cost y for x parts from each company. Then graph the two equations to determine the approximate number of parts for which the total cost is the same under the two pricing structures. **1 2 6 7 8** 3 5 9

26. An architect specifies that the dimensions of a rectangular room are such that the length is twice the width and the area is 288 ft^2. Find the dimensions. **1 2 4 7 8** 3 5

27. A helicopter travels in a straight-line path given by the equation $y = -25x + 101280$. A missile follows the path given by $y = -0.00007x^2 + 0.6x$, where it is assumed that the missile is fired from the point (0, 0). Graph the two equations and determine the approximate coordinates of intersection of the paths. **4 7 8** 1 2 3 5 6 9

---

1. Architectural Technology
2. Civil Engineering Technology
3. Computer Engineering Technology
4. Mechanical Drafting Technology
5. Electrical/Electronics Engineering Technology
6. Chemical Engineering Technology
7. Industrial Engineering Technology
8. Mechanical Engineering Technology
9. Biomedical Engineering Technology

28. If a bullet is launched at a 65° angle with an initial velocity of 320 m/s, find the time of travel when the horizontal and vertical distances are equal. Refer to Example 6 for additional information about this problem. 4 7 8 1 2 3 5 6 9

29. If a builder has 100 m of fencing on hand, find the dimensions of a rectangle that can be surrounded with the fencing to provice 600m^2 of enclosed space. 1 2 4 7 8 3 5

17.5 CHAPTER REVIEW

KEY TERMS

17.1 system of equations
linear equation
addition method
dependent system
inconsistent system

17.2 substitution method

17.3 node
loop

17.4 nonlinear equation

SYMBOLS

resistor
battery

RULES AND FORMULAS

Kirchhoff's laws

Current law: The algebraic sum of all currents at a node must be zero.

Voltage law: The algebraic sum of all voltage sources and voltage drops around a closed loop must be zero.

GEOMETRY CONCEPTS

$C = \pi d$ (circumference of a circle)

$P = 2l + 2w$ (perimeter of a rectangle)

$P = 2a + b$ (perimeter of an isosceles triangle)

SEQUENTIAL REVIEW

(Section 17.1) In 1–5, graph each pair of equations and determine the solution of the system from the graph.

1. $x - 2y = 0$
$2x + y = 10$

2. $3x - 2y = 1$
$2x - 3y = 9$

3. $2x + 3y = 5$
$2x - y = 1$

4. $x + y = 10$
$2x + y = 16$

5. $2x - 3y = 2$
$3x + 2y = 16$

In 6–11, solve each system of equations by the addition method.

6. $3x - 2y = 1$, $2x - 3y = 5$

7. $y = x + 1$, $x = y + 1$

8. $x + y = 7$, $2x - y = 4$

9. $3x + y = 5$, $x - 3y = 1$

10. $3.1x + 1.7y = 8.9$, $2.5x - 4.6y = 5.2$

11. $2.2x + 0.8y = 16.5$, $3.4x + 2.8y = 34.1$

(Section 17.2) *In 12–15, solve each system of equations by the substitution method.*

12. $y = 4$, $x + y = 6$

13. $x + y = 5$, $x - 3y = 0$

14. $x = 4 - 2y$, $3x + 6y = 12$

15. $x - 3y = 2$, $x + 2y = 7$

In 16–23, solve each system of equations by the most appropriate method.

16. $2x + 2y = 7$, $x + 3y = 8$

17. $x = y + 1$, $x - 3y = 14$

18. $2x + 2y = 7$, $y = 8 - 3x$

19. $5x - y = 0$, $3x - 2y = 7$

20. $2x + 3y = 4$, $x - y = -3$

21. $3x + 5y = 2$, $5x - 3y = 1$

22. $7x - y = 5$, $x = 3y$

23. $1.1x + 2.2y = 9.9$, $3.2x - 3.2y = -9.6$

(Section 17.3) *In 24–34, solve each system of equations.*

24. $x - 2y - 3z = 2$, $2x + 3y - z = 6$, $3x - y - 2z = 4$

25. $2x + y - 2z = 9$, $x + 2y + 2z = 0$, $2x - y + 2z = -1$

26. $x + y - z = 5$, $x - y + z = 1$, $x - y - z = 5$

27. $2x + y - 3z = -13$, $x - 3y + z = 8$, $3x + 2y - z = -10$

28. $2x + y - 2z = 9$, $x + 2y + 2z = -6$, $2x - 2y - z = 9$

29. $x + 3y + 3z = -12$, $3x - y - 3z = 10$, $3x - 3y + z = 0$

30. $3x + y - 5z = 9$, $x - 5y + 2z = 12$, $2x + 3y - z = 6$

31. $x + y + 2z = 5$, $2x + y + 3z = 7$, $x - 3y + z = -4$

32. $x + y = 3$, $y + z = 1$, $x + z = 0$

33. $x - 2z = 6$, $2x + y = 5$, $2y + z = 0$

34. $2x - 3y + z = -3$, $x + 3z = 10$, $2y + z = 6$

(Section 17.4) *In 35–39, solve each system of equations by graphing.*

35. $x^2 + y^2 = 25$, $4x - 3y = 0$

36. $xy = 4$, $y = 2x - 7$

37. $2y = x^2$, $y = 2x$

38. $x = y^2 + 3$, $y = x - 5$

39. $x^2 + y^2 = 5$, $y = 2x$

In 40–45, solve each system of equations algebraically.

40. $x^2 + y^2 = 25$, $4x - 3y = 0$

41. $xy = 4$, $y = 2x - 7$

42. $x^2 + y^2 = 25$, $x^2 - y^2 = 7$

43. $y = x^2 + 4$, $y = 2x^2$

44. $y = x^2 - x - 6$, $y = x + 2$

45. $x^2 + y^2 = 9$, $y = x^2 - 9$

APPLIED PROBLEMS

1. Use Kirchhoff's laws to find the current (in mA) in each resistor. 3 5 9 1 2 4 6 7 8

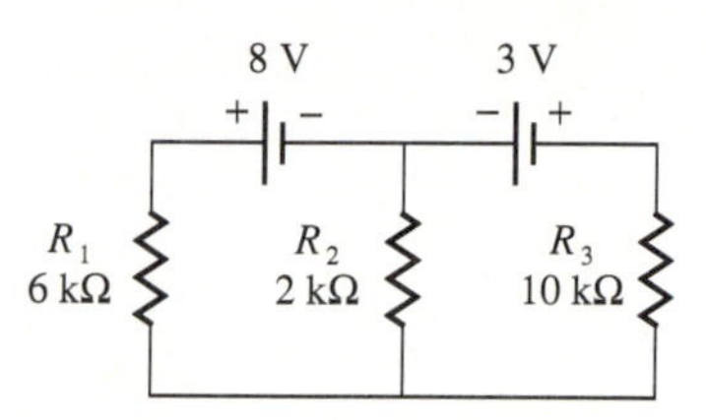

2. All alloy that is 80% silver is mixed with an alloy that is 98% silver to produce 1100 grams of an alloy that is 85% silver. Find the quantity of each alloy that is used. 1 2 4 6 8 5 7 9

3. The following three equations result from studying the forces (in lbs) on a body. Solve for R_{ax}, R_{ay}, and R_b.

$$0 = R_b \cos 40° - R_{ax} - 100$$
$$0 = R_{ay} - R_b \sin 40° - 200$$
$$0 = 6\ R_b \sin 40° - 1800 \cos 30° - 1200$$

2 4 8 1 5 6 7 9

4. A piece of wire that is 24.0 cm long is bent into a rectangular shape. It can be reshaped into an equilateral triangle with sides equal to the length of the rectangle. Determine the dimensions of the rectangle. 1 2 4 7 8 3 5

5. To determine the moment about the origin of the 975 newton force shown in the accompanying figure, it is necessary to solve the following system of equations. Find x and y.

$$y - 110 = (\tan 35°)(x - 40)$$
$$y = (-\tan 55°)x$$

1 2 4 8 7 9

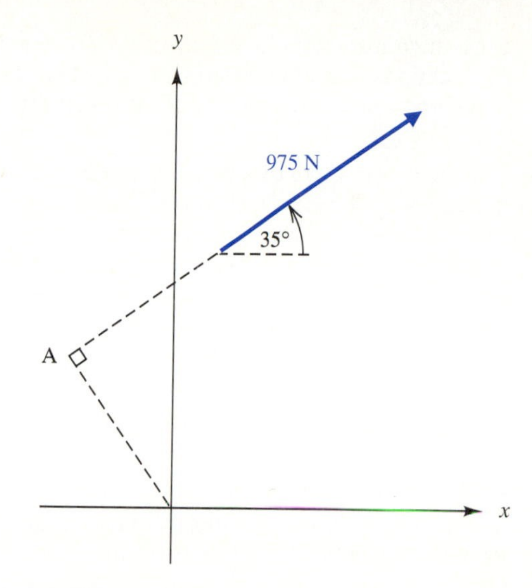

6. Company A charges a flat \$17.00 handling fee to purchasers of a model XYZ motor but discounts this charge \$1.00 for each motor bought. Company B charges only \$12.00 but only discounts \$0.055 for each purchase. How many motors should be ordered to make the total price the same regardless of which company supplies them? 1 2 6 7 8 3 5 9

7. An alloy of gold and copper was made in a 25-gram sample size. A second 25-gram sample was made in which the quantity of gold was reduced 10% and the copper increased by 30%. How much gold and how much copper were in the first sample? 1 2 4 6 8 5 7 9

8. The transfer characteristic I (in milliamps) of a certain transistor is given by the equation $I = 9V + 36$. This transistor is in a circuit that requires that $I = -6V$. What values of I and V will satisfy both equations? 3 5 9 4 7 8

9. A radio transmitter with a range of 25 miles is located on a map at coordinates (15, 35) miles. A plane flies from coordinates (0, 0) miles toward a

town at coordinates (45, 60) miles. What system of equations gives the coordinates at which the plane first will be within range of the transmitter? 1 3 4 5 6 8 9

10. The flight path of an experimental rocket follows the equation $h = 25r - r^2$ where h is its altitude (in miles) and r is the distance (in miles) down range. An interceptor missle is fired at an angle such that its altitude (in miles) is given by $h = 9r - 0.25r^2$. How far down range will their trajectories intersect? Give your answer to three significant digits. 4 7 8 1 2 3 5 6 9

11. A calculator and its case cost $105. The calculator costs $100 more than the case. How much does each cost? 1 2 6 7 8 3 5 9

12. If R_0 is the resistance of a metal conductor at 0°, its resistance at any other temperature T can be determined from the equation $R = R_0\left(1 - \dfrac{T}{T_A}\right)^2$ where T_A is the temperature at which $R = 0$. For what value of T (in terms of T_A) would a second material with a resistance of $\dfrac{1}{4} R_0$ at 0° and a zero temperature of $\dfrac{3}{2} T_A$ have the same resistance as the first? (Hint: You can solve the equation by dividing by R_0 and taking the square root of both sides.) 3 5 9 1 2 4 6 7 8

13. In finding the characteristic curve of a tunnel diode, the coefficients of its $I = f(V)$ equation had to be found by solving the following equations.

$$5000 = 27a + 90b + 300c$$
$$4000 = 125a + 250b + 500c$$
$$3000 = 343a + 490b + 700c$$

Find a, b, and c to three significant digits. 3 5 9 4 7 8

14. Because of external circuit elements, the current I (in milliamps) in a certain diode must obey the equation $I = -0.416V + 4$. The physics of the diode require it to obey the equation $I = 61.9V^3 - 92.86V^2 + 38.95V$. What one equation relates values for V and I that satisfy both of these conditions? 3 5 9 4 7 8

15. A radio transceiver with a range of 5 miles is located on a map at coordinates (0, 0) miles. Another with the same range is located at (4, 7) miles. To determine where a third would be located to be exactly 5 miles from each of the others, it is necessary to solve the following system of equations.

$$x^2 + y^2 = 25$$
$$(x - 4)^2 + (y - 7)^2 = 25$$

Find an approximate solution by graphing. 1 3 4 5 6 8 9

16. A service technician borrowed money from two different banks to purchase equipment. Each loan was paid back in one year. Loan A had an interest rate of 11%. Loan B's interest rate was 12.6%. The total amount borrowed was $22,500. The total interest paid was $2619. How large was each loan? 1 2 6 7 8 3 5 9

17. When connected in parallel, two resistors R_1 and R_2 have a total resistance $R_P = \dfrac{R_1 R_2}{R_1 + R_2}$. In series, it is $R_S = R_1 + R_2$. A technician has two unmarked resistors. In series they measure 900 ohms, in parallel 200 ohms. What are their individual values? 3 5 9 1 2 4 6 7 8

18. It takes exactly 850 ft of chain link fence to surround a rectangular storage yard of 37,500 sq ft area. What are the dimensions of the lot? 1 2 4 7 8 3 5

1. Architectural Technology
2. Civil Engineering Technology
3. Computer Engineering Technology
4. Mechanical Drafting Technology
5. Electrical/Electronics Engineering Technology
6. Chemical Engineering Technology
7. Industrial Engineering Technology
8. Mechanical Engineering Technology
9. Biomedical Engineering Technology

RANDOM REVIEW

Solve each system of equations by the indicated method. If no method is indicated, use the most appropriate method.

1. $2x + y = 7$ (by graphing)
$y = 3x - 18$

2. $x + 2y = 6$
$2x - y = 4$

3. $xy = 4$ (algebraically)
$y^2 - x^2 = 15$

4. $y = x^2 - 5x + 6$ (by graphing)
$y = 2x - 6$

5. $2.3x + 3.2y - 2.5 = 0$ (by the addition method)
$4.7x - y + 6.2 = 0$

6. $x + 7y - 2z = 8$
$3x - 5y + z = 3$
$2x + y - 9z = 31$

7. $3x - 3y = 11$ (by the substitution method)
$3x + y = 5$

8. $y = 2x - 1$
$x + y = 3$

9. $4x - 3y = 1$ (by the addition method)
$3x + 5y = 9$

10. $2x - 3z = 1$
$3y + z = 0$
$x - 2y = 7$

CHAPTER TEST

1. Solve the following system of equations by graphing.

$$2x + y = 4$$
$$x - y = -1$$

2. Solve the following system of equations by the addition method.

$$2.4x + 4.1y = 1$$
$$3.1x + 5.7y = 2.1$$

3. Solve the following system of equations by the substitution method.

$$3x - y = 7$$
$$y = 2x - 2$$

In 4–7, solve each system of equations by the most appropriate method.

4. $2x - 7 = -3y$
$6y = 9 - 4x$

5. $x + y - 2z = 3$
$x + 3y + z = 0$
$-2x + y - 3z = 1$

6. $2x + y - z = 0$
$x + 2y + 3z = 11$
$2x + 4y + 6z = 22$

7. $3x^2 + 2y^2 = 14$
$x^2 - y^2 = 3$

8. Solve the following system of equations by graphing.

$$y = -x^2 + 4x$$
$$y = 2x^2 - 8x + 9$$

9. Use Kirchhoff's laws to find the current (in mA) in each resistor of the circuit shown.

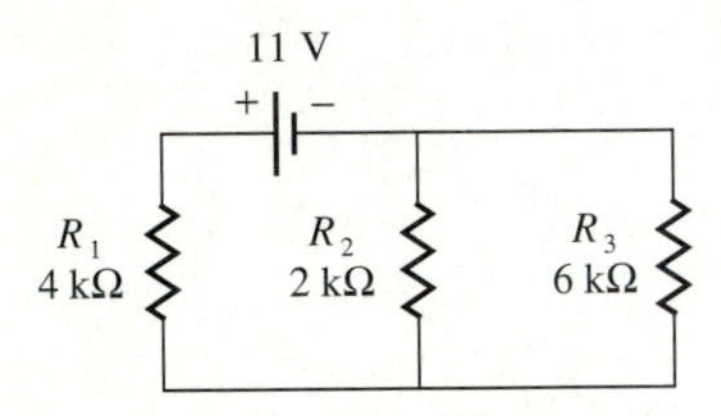

10. A quantity of paint that is 98% pure solvent is to be mixed with thinner that is 80% pure solvent to produce 100 gallons of a paint that is 85% pure solvent. Find the amount of paint and the amount of thinner that should be used.

DETERMINANTS CHAPTER 18

In Chapter 17, we solved systems of two and three linear equations. The methods that we used there can be generalized for solving larger systems of equations. However, larger systems are usually solved by methods that do not require writing down the variables repetitiously. In this chapter, we introduce the concept of determinants and use determinants to solve systems of linear equations.

18.1 DETERMINANTS AND CRAMER'S RULE

If the variables occur in the same order in each equation for a system of linear equations, we can obtain a solution by manipulating the coefficients. In the late seventeenth century, Gottfried Wilhelm von Leibniz in Germany and Seki Kowa in Japan independently discovered one such method. Their idea was not widely used, however, until 1750 when Gabriel Cramer from Switzerland rediscovered the method and popularized it. The method is now known as *Cramer's rule.*

Cramer's rule is based on the concept of a *determinant.*

DEFINITION An $n \times n$ **determinant** is a real number represented by a square array (or arrangement) of n^2 numbers.

An $n \times n$ determinant is said to have dimensions n by n. A pair of bars is used with the array to indicate that the numbers are to be considered as an array and not as individual numbers.

For example, $\begin{vmatrix} 2 & 3 \\ 1 & 2 \end{vmatrix}$ is a 2×2 determinant.

$\begin{vmatrix} 3 & 2 & 0 \\ 0 & 1 & 3 \\ 2 & 0 & 1 \end{vmatrix}$ is a 3×3 determinant.

main diagonal

We say that the entries along a slanted line from upper left to lower right are on the *main diagonal.*

M a i n d i a g o n a l

Evaluating a 2×2 determinant is easier than evaluating a 3×3 determinant, so we consider that case first.

To evaluate a 2×2 determinant:

1. Form the product of the entries along the main diagonal.
2. Form the product of the entries along the other diagonal (from upper right to lower left).
3. Subtract the product found in step 2 from the product found in step 1.

EXAMPLE 1 Evaluate $\begin{vmatrix} 5 & -1 \\ -5 & 2 \end{vmatrix}$.

Solution

$$\begin{vmatrix} 5 & -1 \\ -5 & 2 \end{vmatrix}$$

$$(5)(2) - (-1)(-5)$$

1. We have 5(2) or 10 from the main diagonal.
2. We have $(-1)(-5)$ or 5 from the other diagonal.
3. Subtracting, we have $10 - 5$ or 5. The value of the determinant is 5. ■

EXAMPLE 2 Evaluate $\begin{vmatrix} 2 & -1 \\ -5 & 2 \end{vmatrix}$.

Solution $2(2) - (-1)(-5) = 4 - 5 = -1$. The value of the determinant is -1. ■

To evaluate a 3×3 determinant, we copy the first two columns to the right of the original determinant. There are now two diagonals, each containing three entries, parallel to the main diagonal. We find the product of the entries along *each* of these diagonals, as well as the product of the entries along *each* of the three-entry diagonals running from upper right to lower left. The procedure is known as Sarrus' rule and is summarized as follows.

To evaluate a 3×3 determinant:

1. Copy the first two columns to the right of the original determinant.
2. Find the product of the entries along the main diagonal and along each of the two 3-entry diagonals parallel to it. Add these products.
3. Find the product of the entries along each of the three 3-entry diagonals from upper right to lower left. Add these products.
4. Subtract the sum found in step 3 from the sum found in step 2.

EXAMPLE 3 Evaluate $\begin{vmatrix} 1 & -3 & 2 \\ 1 & 2 & 3 \\ 2 & -1 & 1 \end{vmatrix}$.

Solution 1. We write: $\begin{vmatrix} 1 & -3 & 2 \\ 1 & 2 & 3 \\ 2 & -1 & 1 \end{vmatrix} \begin{matrix} 1 & -3 \\ 1 & 2 \\ 2 & -1 \end{matrix}$

2. Using the main diagonal and diagonals parallel to it, we have:

$$\begin{vmatrix} 1 & -3 & 2 \\ 1 & 2 & 3 \\ 2 & -1 & 1 \end{vmatrix} \begin{matrix} 1 & -3 \\ 1 & 2 \\ 2 & -1 \end{matrix}$$

$$\begin{aligned} 2(1)(-1) &= -2 \\ -3(3)(2) &= -18 \\ 1(2)(1) &= \underline{2} \\ & -18 \end{aligned}$$

3. Using the other diagonals, we have:

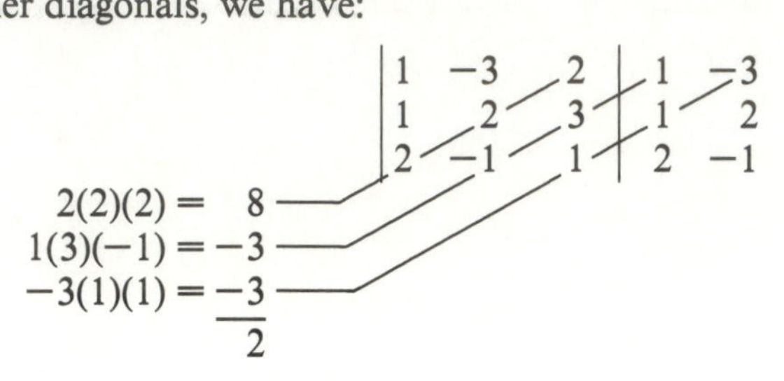

4. The value of the determinant is $-18 - 2 = -20$. ■

EXAMPLE 4 Evaluate $\begin{vmatrix} 5 & -3 & 2 \\ 7 & 2 & 3 \\ 4 & -1 & 1 \end{vmatrix}$.

Solution

$$\begin{vmatrix} 5 & -3 & 2 \\ 7 & 2 & 3 \\ 4 & -1 & 1 \end{vmatrix} \begin{matrix} 5 & -3 \\ 7 & 2 \\ 4 & -1 \end{matrix}$$

$$10 \quad -36 \quad -14 = -40$$

$$16 \quad -15 \quad -21 = -20$$

The value of the determinant is $-40 - (-20) = -40 + 20 = -20$. ■

Cramer's rule says that the value of each variable of a system of linear equations can be written as a ratio of determinants. Each ratio has the same denominator, but the numerator is obtained differently for each one. The equations must be written so that the variables occur in the same order for each equation, and the constant term is on the other side of the equation. The procedure is explained in Example 5.

EXAMPLE 5 Use Cramer's rule to solve the system of equations.

$$\begin{aligned} x - 3y + 2z &= 5 \\ x + 2y + 3z &= 7 \\ 2x - y + z &= 4 \end{aligned}$$

Solution All three variables will have the same denominator. It is the determinant formed by listing the coefficients of x in the first column, the coefficients of y in the second column, and the coefficients of z in the third column. The denominator, then, is the

determinant $\begin{vmatrix} 1 & -3 & 2 \\ 1 & 2 & 3 \\ 2 & -1 & 1 \end{vmatrix}$.

To solve for x, we replace the first column (the x column) with the constants from the right-hand side of the equation to form the determinant that is the numerator.

That is, we have $\begin{vmatrix} 5 & -3 & 2 \\ 7 & 2 & 3 \\ 4 & -1 & 1 \end{vmatrix}$.

To solve for y, we replace the second column (the y column) with the constants; and to solve for z, we replace the third column (the z column) with the constants.

Constants replace coefficients of x
Constants replace coefficients of y
Constants replace coefficients of z

$$x = \frac{\begin{vmatrix} 5 & -3 & 2 \\ 7 & 2 & 3 \\ 4 & -1 & 1 \end{vmatrix}}{\begin{vmatrix} 1 & -3 & 2 \\ 1 & 2 & 3 \\ 2 & -1 & 1 \end{vmatrix}} \qquad y = \frac{\begin{vmatrix} 1 & 5 & 2 \\ 1 & 7 & 3 \\ 2 & 4 & 1 \end{vmatrix}}{\begin{vmatrix} 1 & -3 & 2 \\ 1 & 2 & 3 \\ 2 & -1 & 1 \end{vmatrix}} \qquad z = \frac{\begin{vmatrix} 1 & -3 & 5 \\ 1 & 2 & 7 \\ 2 & -1 & 4 \end{vmatrix}}{\begin{vmatrix} 1 & -3 & 2 \\ 1 & 2 & 3 \\ 2 & -1 & 1 \end{vmatrix}}$$

The denominator was evaluated in Example 3. Its value is -20. The numerator of x was evaluated in Example 4. Its value is -20. Thus, x is $-20/-20$ or 1. We evaluate the numerator of y.

$$\left|\begin{matrix} 1 & 5 & 2 \\ 1 & 7 & 3 \\ 2 & 4 & 1 \end{matrix}\right| \begin{matrix} 1 & 5 \\ 1 & 7 \\ 2 & 4 \end{matrix} \qquad (7 + 30 + 8) - (28 + 12 + 5) = 45 - 45 \text{ or } 0$$

The value of y, then, is $0/-20$ or 0.
We evaluate the numerator of z.

$$\left|\begin{matrix} 1 & -3 & 5 \\ 1 & 2 & 7 \\ 2 & -1 & 4 \end{matrix}\right| \begin{matrix} 1 & -3 \\ 1 & 2 \\ 2 & -1 \end{matrix} \qquad (8 - 42 - 5) - (20 - 7 - 12) = -39 - 1 \text{ or } -40$$

The value of z, then, is $-40/-20$ or 2. The solution of the system is $(1, 0, 2)$. ■

At this point, we have considered evaluation of 2×2 and 3×3 determinants only. In the next section, higher order determinants will be covered. Cramer's rule can also be used with these higher order determinants to solve systems of n equations in n variables. The procedure is generalized as follows.

To solve a system of linear equations by Cramer's rule:

1. If the variables do not appear in the same order in each equation, write an equivalent system in which they do. If a variable is missing from one of the equations, use zero as the coefficient.
2. Write the value of each variable as the ratio of two determinants.
   (a) The determinant in the denominator contains the coefficients of the first variable of each equation in the first column, the coefficients of the second variable of each equation in the second column, and so on.
   (b) The determinant in the numerator is like the one in the denominator, except the constants replace the entries in the column that represents the variable sought.
3. Evaluate the determinants. If a denominator is zero and any numerator has a nonzero value, the system is inconsistent. If the system is dependent, all of the numerators, as well as the denominator, are equal to zero.

Cramer's rule may be used to solve applied problems.

EXAMPLE 6 An engineering firm charged \$180 for a job that required 4 hours of the engineer's time and 2 hours of the assistant's time. Assuming that the engineer's hourly wage is 2.5 times the assistant's, find the hourly wage of each.

Solution Let $x =$ engineer's hourly wage
$y =$ assistant's hourly wage

One equation is based on the charge. The engineer's wage is 4 hours at x dollars per hour or $4x$. The assistant's wage is 2 hours at y dollars per hour or $2y$. Thus, $4x + 2y = 180$. The second equation is based on the relationship between the wages. We have $x = 2.5\,y$.

We must solve the system $4x + 2y = 180$.
$x - 2.5y = 0$.

$$x = \frac{\begin{vmatrix} 180 & 2 \\ 0 & -2.5 \end{vmatrix}}{\begin{vmatrix} 4 & 2 \\ 1 & -2.5 \end{vmatrix}} \qquad y = \frac{\begin{vmatrix} 4 & 180 \\ 1 & 0 \end{vmatrix}}{\begin{vmatrix} 4 & 2 \\ 1 & -2.5 \end{vmatrix}}$$

$$x = \frac{-450 - 0}{-10 - 2} = \frac{-450}{-12} = 37.50$$

$$y = \frac{0 - 180}{-10 - 2} = \frac{-180}{-12} = 15$$

The engineer's hourly wage is \$37.50 and the assistant's is \$15.00. ■

EXAMPLE 7 Use Kirchhoff's laws and Cramer's rule to find the current in each resistor of the circuit shown in Figure 18.1.

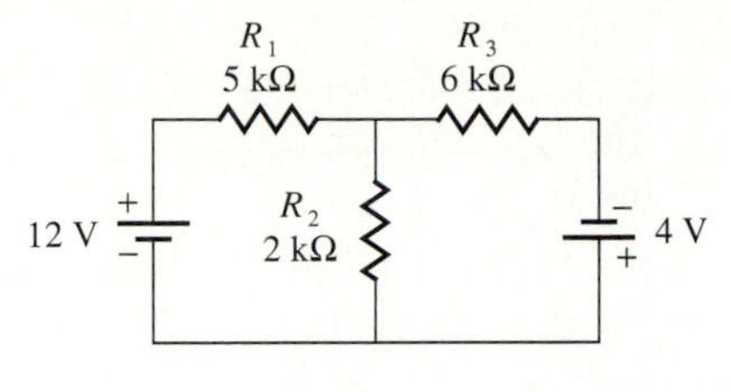

FIGURE 18.1

Solution We have the following system of equations.

$$\begin{aligned} I_1 - I_2 - I_3 &= 0 \\ 5I_1 + 2I_2 \phantom{{}+6I_3} &= 12 \\ -2I_2 + 6I_3 &= 4 \end{aligned}$$

$$I_1 = \frac{\begin{vmatrix} 0 & -1 & -1 \\ 12 & 2 & 0 \\ 4 & -2 & 6 \end{vmatrix}}{\begin{vmatrix} 1 & -1 & -1 \\ 5 & 2 & 0 \\ 0 & -2 & 6 \end{vmatrix}} \qquad I_2 = \frac{\begin{vmatrix} 1 & 0 & -1 \\ 5 & 12 & 0 \\ 0 & 4 & 6 \end{vmatrix}}{\begin{vmatrix} 1 & -1 & -1 \\ 5 & 2 & 0 \\ 0 & -2 & 6 \end{vmatrix}} \qquad I_3 = \frac{\begin{vmatrix} 1 & -1 & 0 \\ 5 & 2 & 12 \\ 0 & -2 & 4 \end{vmatrix}}{\begin{vmatrix} 1 & -1 & -1 \\ 5 & 2 & 0 \\ 0 & -2 & 6 \end{vmatrix}}$$

Evaluate the determinant in the denominator.

$$\left|\begin{matrix} 1 & -1 & -1 \\ 5 & 2 & 0 \\ 0 & -2 & 6 \end{matrix}\right|\begin{matrix} 1 & -1 \\ 5 & 2 \\ 0 & -2 \end{matrix} \qquad (12 + 0 + 10) - (0 + 0 - 30) = 22 - (-30) = 52$$

Evaluate the numerator of I_1.

$$\left|\begin{matrix} 0 & -1 & -1 \\ 12 & 2 & 0 \\ 4 & -2 & 6 \end{matrix}\right|\begin{matrix} 0 & -1 \\ 12 & 2 \\ 4 & -2 \end{matrix} \qquad (0 + 0 + 24) - (-8 + 0 - 72) = 24 - (-80) = 104$$

Evaluate the numerator of I_2.

$$\left|\begin{matrix} 1 & 0 & -1 \\ 5 & 12 & 0 \\ 0 & 4 & 6 \end{matrix}\right|\begin{matrix} 1 & 0 \\ 5 & 12 \\ 0 & 4 \end{matrix} \qquad (72 + 0 - 20) - (0 + 0 - 0) = 52 - 0 = 52$$

Evaluate the numerator of I_3.

$$\left|\begin{matrix} 1 & -1 & 0 \\ 5 & 2 & 12 \\ 0 & -2 & 4 \end{matrix}\right|\begin{matrix} 1 & -1 \\ 5 & 2 \\ 0 & -2 \end{matrix} \qquad (8 + 0 + 0) - (0 - 24 - 20) = 8 - (-44) = 52$$

Therefore we have $I_1 = 104/52$ or 2, $I_2 = 52/52$ or 1, and $I_3 = 52/52$ or 1. Since the unit of resistance was kΩ, the unit of current is mA. Thus, the currents are 2 mA, 1 mA, and 1 mA.

$$\begin{aligned} \text{Check: } 2 - 1 - 1 &= 0 &\quad \text{or} \quad 2 - 2 &= 0 \\ 5(2) + 2(1) &= 12 &\quad \text{or} \quad 10 + 2 &= 12 \\ -2(1) + 6(1) &= 4 &\quad \text{or} \quad -2 + 6 &= 4 \end{aligned}$$ ■

The method for evaluating determinants introduced in this section cannot be generalized for determinants with dimensions larger than 3×3. There are more general methods that can be used on such determinants. Cramer's rule, however, is generally not an efficient method to use on larger systems. Consider a system of 12 equations with 12 variables (not uncommon in practice). Cramer's rule would require evaluating 13 determinants. There would be over 6×10^9 terms, each obtained by 11 multiplications. That is, there would be almost 7×10^{10} multiplications. A computer capable of performing 100,000 multiplications per second, operating 24 hours a day, would take over a week to solve the system. Gaussian elimination, which is similar to the addition method used in Chapter 17, would require 794 multiplications and would take about 0.008 seconds.

EXERCISES 18.1

In 1–10, evaluate each determinant.

1. $\begin{vmatrix} 2 & 3 \\ 1 & -1 \end{vmatrix}$

2. $\begin{vmatrix} 3 & -1 \\ 2 & 3 \end{vmatrix}$

3. $\begin{vmatrix} 1 & -2 \\ -3 & 4 \end{vmatrix}$

4. $\begin{vmatrix} 1 & 0 \\ 1 & -1 \end{vmatrix}$

5. $\begin{vmatrix} -1 & 2 \\ 3 & -4 \end{vmatrix}$

6. $\begin{vmatrix} -7 & -5 \\ 6 & 8 \end{vmatrix}$

7. $\begin{vmatrix} 2 & 1 & -1 \\ -1 & 2 & 1 \\ 1 & -1 & 2 \end{vmatrix}$

8. $\begin{vmatrix} 2 & 3 & 0 \\ 1 & -2 & 3 \\ 3 & 0 & -1 \end{vmatrix}$

9. $\begin{vmatrix} 1 & 0 & 0 \\ 0 & 1 & 0 \\ 0 & 0 & 1 \end{vmatrix}$

10. $\begin{vmatrix} 0 & -1 & 3 \\ 1 & -3 & 0 \\ 0 & 3 & -1 \end{vmatrix}$

In 11–24, use Cramer's rule to solve each system of equations.

11. $x + y = 8$
$2x - y = 1$

12. $3x - 2y = 1$
$4x - y = 1$

13. $x + 2y = 1$
$2x - y = 0$

14. $x + 3y = 4$
$2x + 6y = 7$

15. $2x - 5y = 10$
$4x - 10y = 20$

16. $2x + 3y = 7$
$3x - y = 5$

17. $3x + 2y = 0$
$x - 5y = 9$

18. $2x - 3y + z = 0$
$3x + 2y - z = 4$
$2x - 2y + z = 1$

19. $2x + 2y + z = 0$
$3x + 2y - z = 4$
$2x - 2y + z = 1$

20. $x + y - z = 2$
$x - y + z = 4$
$-x + y + z = 6$

21. $x + 3y - 2z = 6$
$2x - 2y + z = -1$
$2x + 6y - 4z = 3$

22. $x - y + z = 3$
$2x + y + z = 3$
$2x - 2y + z = 0$

23. $2x - y + z = 5$
$x + 3y - z = -1$
$x + y + 4z = 1$

24. $2x + y - z = 4$
$x - 3y + z = -3$
$4x + 2y - 2z = 8$

25. Two cars leave points A and B, which are 222 miles apart, at the same time. They travel toward each other, and one car travels five miles per hour faster than the other. After two hours, the cars have passed each other and are separated by 10 miles. Find the rates of the two cars. 4 7 8 | 1 2 3 5 6 9

26. A chemist has a 10% acid solution and a 50% acid solution available. How many milliliters of each should be mixed to obtain 40 ml of a 30% solution? 6 9 | 2 4 5 7 8

27. Two cables support a 350-lb weight as shown in the accompanying diagram. Find the tension in the cables by solving the following system of equations for T_1 and T_2.

$$T_1 \sin 70° + T_2 \sin 20° = 350$$
$$T_1 \cos 70° - T_2 \cos 20° = 0$$

2 4 8 | 1 5 6 7 9

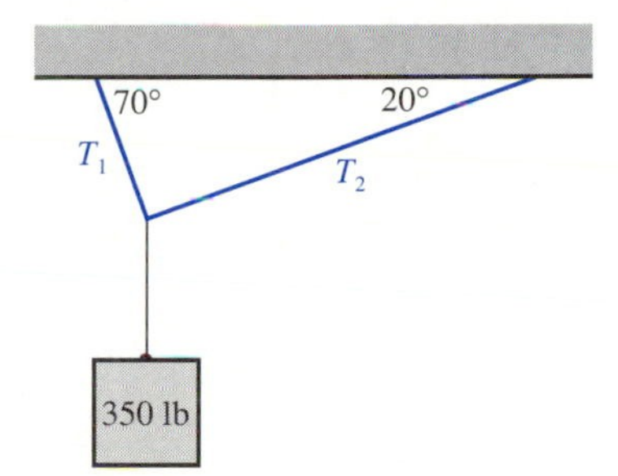

28. A consulting firm decides to base its price structure as follows. The charge per hour for an on-site consultation is 1 1/2 times the charge per hour for office time. They want to charge $720 for an average job, which requires 12 hours in the office and 4 hours on site. How much should they charge for each type of work? 1 2 6 7 8 | 3 5 9

29. A draftsman is instructed to draw a triangle in which the second angle is twice as large as the first, and the third is twice as large as the second. How large should the angles be? 1 2 4 7 8 | 3 5

30. When a triangle is drawn on a Cartesian plane so that its vertices are (x_1, y_1), (x_2, y_2), and (x_3, y_3), its area is given by

$$A = \left(\frac{1}{2}\right)\begin{vmatrix} x_1 & y_1 & 1 \\ x_2 & y_2 & 1 \\ x_3 & y_3 & 1 \end{vmatrix}.$$

Find the area of a triangle whose vertices are $(2, -5)$, $(6, 2)$, and $(4, 1)$. 1 2 4 7 8 | 3 5

31. Use Kirchhoff's laws and Cramer's rule to find the current (in mA) in the resistors shown in the accompanying diagram. 3 5 9 | 1 2 4 6 7 8

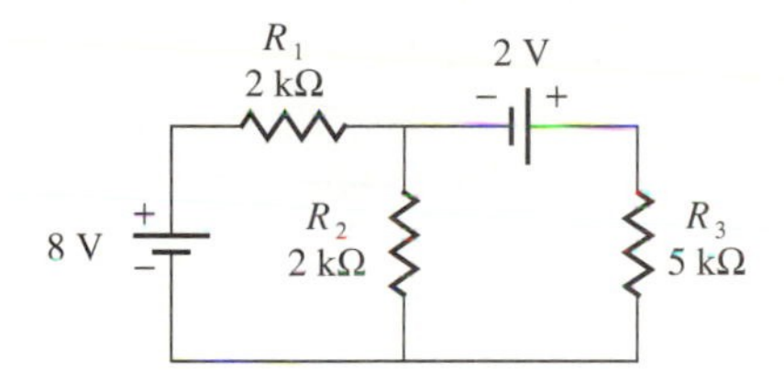

1. Architectural Technology
2. Civil Engineering Technology
3. Computer Engineering Technology
4. Mechanical Drafting Technology
5. Electrical/Electronics Engineering Technology
6. Chemical Engineering Technology
7. Industrial Engineering Technology
8. Mechanical Engineering Technology
9. Biomedical Engineering Technology

18.2 HIGHER ORDER DETERMINANTS: EXPANSION BY COFACTORS

The method of expansion by cofactors is to replace an $n \times n$ determinant with a sum of n determinants having dimensions $(n - 1) \times (n - 1)$. There are shorter methods (see problem 25), but the notation used in expansion by cofactors is often preferred by mathematicians.

Before we examine the method, a comment about notation is in order. A determinant has both horizontal rows and vertical columns. A double subscript indicates the row and column of a particular entry. For example, a_{11} indicates the entry in the first row and first column, while a_{23} indicates the entry in the second row and third column. When two determinants are compared, entries that have the same row number and the same column number are called corresponding entries.

EXAMPLE 1 Consider $A = \begin{vmatrix} 1 & -2 & 3 \\ 4 & 5 & -6 \\ -7 & 8 & 9 \end{vmatrix}$.

(a) Determine the entry a_{21}.

(b) Determine the entry a_{13}.

Solution **(a)** The entry a_{21} is in the second row and first column.

First column ↓

$$\begin{vmatrix} 1 & -2 & 3 \\ \textcircled{4} & 5 & -6 \\ -7 & 8 & 9 \end{vmatrix} \leftarrow \text{Second row}$$

Thus, a_{21} is 4.

(b) The entry a_{13} is in the first row and third column.

Third column ↓

$$\begin{vmatrix} 1 & -2 & \textcircled{3} \\ 4 & 5 & -6 \\ -7 & 8 & 9 \end{vmatrix} \leftarrow \text{First row}$$

Thus, a_{13} is 3. ■

Now that the notation is familiar, we introduce the necessary terminology to discuss expansion by cofactors.

DEFINITION The **minor** of an entry a_{ij} of a determinant A is the determinant obtained by deleting row i and column j of A.

EXAMPLE 2 Given $A = \begin{vmatrix} 2 & 3 & 1 \\ 6 & 4 & 2 \\ 0 & 1 & 7 \end{vmatrix}$,

(a) find the minor of 4.

(b) find the minor of 0.

Solution **(a)** Since 4 is in row 2 and column 2, we delete that row and column.

$$\begin{vmatrix} 2 & 3 & 1 \\ 6 & 4 & 2 \\ 0 & 1 & 7 \end{vmatrix}$$

The remaining entries form the determinant $\begin{vmatrix} 2 & 1 \\ 0 & 7 \end{vmatrix}$. It is the minor of 4.

(b) Since 0 is in row 3 and column 1, we delete that row and column.

$$\begin{vmatrix} 2 & 3 & 1 \\ 6 & 4 & 2 \\ 0 & 1 & 7 \end{vmatrix}$$

The remaining entries form the determinant $\begin{vmatrix} 3 & 1 \\ 4 & 2 \end{vmatrix}$. It is the minor of 0. ■

cofactor

The *cofactor* of an entry is simply the minor of that entry with an appropriate sign attached. Imagine a checkerboard pattern of +'s and −'s with + in the upper left-hand corner. If this pattern is superimposed on a determinant, the sign associated with each position in the determinant is the sign associated with the cofactor of the entry in that position.

$$\begin{vmatrix} + & - & + \\ - & + & - \\ + & - & + \end{vmatrix}$$

EXAMPLE 3 Given $A = \begin{vmatrix} 2 & 3 & 1 \\ 6 & 4 & 2 \\ 0 & 1 & 7 \end{vmatrix}$,

(a) find the cofactor of 6.

(b) find the cofactor of 7.

Solution **(a)** We find the minor of 6. $\begin{vmatrix} 2 & 3 & 1 \\ 6 & 4 & 2 \\ 0 & 1 & 7 \end{vmatrix}$ It is $\begin{vmatrix} 3 & 1 \\ 1 & 7 \end{vmatrix}$.

The sign associated with 6 is negative, since the checkerboard pattern of signs would place a negative in row 2, column 1. Thus, the cofactor of 6 is $-\begin{vmatrix} 3 & 1 \\ 1 & 7 \end{vmatrix}$, which is $-[3(7) - 1(1)]$ or -20.

(b) The minor of 7 is $\begin{vmatrix} 2 & 3 \\ 6 & 4 \end{vmatrix}$.

The sign associated with 7 is positive. Thus, the cofactor of 7 is $\begin{vmatrix} 2 & 3 \\ 6 & 4 \end{vmatrix}$, which is $2(4) - 3(6)$ or -10. ■

The alternating pattern of signs can be extended to a determinant of any size. Since the sum of two even integers or two odd integers is even, and the sum of an even integer and an odd integer is odd, it is possible to make the following observation. The sign of the cofactor of an entry is positive if the sum of the row number and column number is even, but the sign is negative if the sum of the row number and column number is odd. The formal definition of a cofactor is based on this principle.

DEFINITION The cofactor of an entry a_{ij} of a determinant A is
(a) the minor of that entry if the sum of the row number and column number is even.
(b) -1 times the minor of that entry if the sum of the row number and column number is odd.

To method of expansion by cofactors is summarized as follows.

To evaluate a determinant using expansion by cofactors:

1. Choose *any* row or column.
2. Multiply each entry in that row or column by its cofactor.
3. Add the products found in step 2 to obtain the value of the determinant.

EXAMPLE 4 Evaluate using expansion by cofactors. $\begin{vmatrix} 5 & -3 & 2 \\ 7 & 2 & 3 \\ 4 & -1 & 1 \end{vmatrix}$

Solution

1. We arbitrarily choose the first row. It is important, however, that any row or column can be chosen for the expansion.
2. The cofactor of 5 is $+\begin{vmatrix} 2 & 3 \\ -1 & 1 \end{vmatrix}$. The product is $5\begin{vmatrix} 2 & 3 \\ -1 & 1 \end{vmatrix}$. The cofactor of -3 is $-\begin{vmatrix} 7 & 3 \\ 4 & 1 \end{vmatrix}$. The product is $-3\left(-\begin{vmatrix} 7 & 3 \\ 4 & 1 \end{vmatrix}\right)$. The cofactor of 2 is $+\begin{vmatrix} 7 & 2 \\ 4 & -1 \end{vmatrix}$. The product is $2\begin{vmatrix} 7 & 2 \\ 4 & -1 \end{vmatrix}$.
3. The determinant, then, is given by

$$\begin{vmatrix} 5 & -3 & 2 \\ 7 & 2 & 3 \\ 4 & -1 & 1 \end{vmatrix} = 5\begin{vmatrix} 2 & 3 \\ -1 & 1 \end{vmatrix} + 3\begin{vmatrix} 7 & 3 \\ 4 & 1 \end{vmatrix} + 2\begin{vmatrix} 7 & 2 \\ 4 & -1 \end{vmatrix}$$

$$\begin{aligned} &= 5(2+3) + 3(7-12) + 2(-7-8) \\ &= 5(5) + 3(-5) + 2(-15) \\ &= 25 - 15 - 30 \\ &= -20. \quad \blacksquare \end{aligned}$$

Expansion by cofactors is sometimes used as a mnemonic device for formulas.

EXAMPLE 5 The velocity of a point P on a body rotating about an axis is calculated as the cross product of the vector representing the radius from the axis to P and the vector representing the angular velocity. If $\langle x_1, y_1, z_1 \rangle$ and $\langle x_2, y_2, z_2 \rangle$ are two vectors, the coordinates of the cross product are given by the coefficients of i, j, and k when the determinant $\begin{vmatrix} i & j & k \\ x_1 & y_1 & z_1 \\ x_2 & y_2 & z_2 \end{vmatrix}$ is expanded about the first row. Give a formula for the coordinates of the cross product.

Solution

$$\begin{vmatrix} i & j & k \\ x_1 & y_1 & z_1 \\ x_2 & y_2 & z_2 \end{vmatrix} = i \begin{vmatrix} y_1 & z_1 \\ y_2 & z_2 \end{vmatrix} - j \begin{vmatrix} x_1 & z_1 \\ x_2 & z_2 \end{vmatrix} + k \begin{vmatrix} x_1 & y_1 \\ x_2 & y_2 \end{vmatrix}$$

$$= i(y_1 z_2 - y_2 z_1) - j(x_1 z_2 - x_2 z_1) + k(x_1 y_2 - x_2 y_1)$$

$$= i(y_1 z_2 - y_2 z_1) + j(x_2 z_1 - x_1 z_2) + k(x_1 y_2 - x_2 y_1)$$

The coordinates are $\langle y_1 z_2 - y_2 z_1, x_2 z_1 - x_1 z_2, x_1 y_2 - x_2 y_1 \rangle$ ■

EXAMPLE 6 Evaluate using expansion by cofactors. $\begin{vmatrix} 3 & 0 & 4 \\ 2 & 1 & -3 \\ -1 & 0 & 5 \end{vmatrix}$

Solution Since arithmetic with 0 is easy, we choose the second column for the expansion by cofactors.

$$\begin{vmatrix} 3 & \mathbf{0} & 4 \\ 2 & \mathbf{1} & -3 \\ -1 & \mathbf{0} & 5 \end{vmatrix} = \mathbf{-0} \begin{vmatrix} 2 & -3 \\ -1 & 5 \end{vmatrix} \mathbf{+1} \begin{vmatrix} 3 & 4 \\ -1 & 5 \end{vmatrix} \mathbf{-0} \begin{vmatrix} 3 & 4 \\ 2 & -3 \end{vmatrix}$$

It is not necessary to evaluate the determinants that are multiplied by 0, since the product is 0, regardless of the value of the determinant. Thus,

$$\begin{vmatrix} 3 & 0 & 4 \\ 2 & 1 & -3 \\ -1 & 0 & 5 \end{vmatrix} = -0 + 1(15 + 4) - 0 = 19. \quad ■$$

EXAMPLE 7 Evaluate using expansion by cofactors. $\begin{vmatrix} 4 & 3 & 2 & 1 \\ 2 & 4 & 1 & 3 \\ 1 & 2 & 3 & 4 \\ 3 & 1 & 4 & 2 \end{vmatrix}$

Solution It makes little difference which row or column is chosen for the expansion, so we choose the first row.

$$\begin{vmatrix} \mathbf{4} & \mathbf{3} & \mathbf{2} & \mathbf{1} \\ 2 & 4 & 1 & 3 \\ 1 & 2 & 3 & 4 \\ 3 & 1 & 4 & 2 \end{vmatrix} = \mathbf{4}\begin{vmatrix} 4 & 1 & 3 \\ 2 & 3 & 4 \\ 1 & 4 & 2 \end{vmatrix} \mathbf{-3}\begin{vmatrix} 2 & 1 & 3 \\ 1 & 3 & 4 \\ 3 & 4 & 2 \end{vmatrix} \mathbf{+2}\begin{vmatrix} 2 & 4 & 3 \\ 1 & 2 & 4 \\ 3 & 1 & 2 \end{vmatrix} \mathbf{-1}\begin{vmatrix} 2 & 4 & 1 \\ 1 & 2 & 3 \\ 3 & 1 & 4 \end{vmatrix}$$

Each 3×3 determinant must be evaluated. Expanding each about the first row, we have the following.

$$\begin{aligned}
\begin{vmatrix} 4 & 3 & 2 & 1 \\ 2 & 4 & 1 & 3 \\ 1 & 2 & 3 & 4 \\ 3 & 1 & 4 & 2 \end{vmatrix} &= 4\left(4\begin{vmatrix} 3 & 4 \\ 4 & 2 \end{vmatrix} - 1\begin{vmatrix} 2 & 4 \\ 1 & 2 \end{vmatrix} + 3\begin{vmatrix} 2 & 3 \\ 1 & 4 \end{vmatrix}\right) \\
&\quad -3\left(2\begin{vmatrix} 3 & 4 \\ 4 & 2 \end{vmatrix} - 1\begin{vmatrix} 1 & 4 \\ 3 & 2 \end{vmatrix} + 3\begin{vmatrix} 1 & 3 \\ 3 & 4 \end{vmatrix}\right) \\
&\quad +2\left(2\begin{vmatrix} 2 & 4 \\ 1 & 2 \end{vmatrix} - 4\begin{vmatrix} 1 & 4 \\ 3 & 2 \end{vmatrix} + 3\begin{vmatrix} 1 & 2 \\ 3 & 1 \end{vmatrix}\right) \\
&\quad -1\left(2\begin{vmatrix} 2 & 3 \\ 1 & 4 \end{vmatrix} - 4\begin{vmatrix} 1 & 3 \\ 3 & 4 \end{vmatrix} + 1\begin{vmatrix} 1 & 2 \\ 3 & 1 \end{vmatrix}\right) \\
&= 4[4(6-16) - 1(4-4) + 3(8-3)] \\
&\quad -3[2(6-16) - 1(2-12) + 3(4-9)] \\
&\quad +2[2(4-4) - 4(2-12) + 3(1-6)] \\
&\quad -1[2(8-3) - 4(4-9) + 1(1-6)] \\
&= 4[4(-10) - 1(0) + 3(5)] \\
&\quad -3[2(-10) - 1(-10) + 3(-5)] \\
&\quad +2[2(0) - 4(-10) + 3(-5)] \\
&\quad -1[2(5) - 4(-5) + 1(-5)] \\
&= 4(-40 - 0 + 15) \\
&\quad -3(-20 + 10 - 15) \\
&\quad +2(0 + 40 - 15) \\
&\quad -1(10 + 20 - 5) \\
&= 4(-25) - 3(-25) + 2(25) - 1(25) \\
&= -100 + 75 + 50 - 25 \\
&= 0 \quad \blacksquare
\end{aligned}$$

EXERCISES 18.2

In 1–11, evaluate each determinant using expansion by cofactors.

1. $\begin{vmatrix} -1 & 1 & 2 \\ 1 & 2 & -1 \\ 2 & -1 & 1 \end{vmatrix}$ **2.** $\begin{vmatrix} 2 & 1 & 3 \\ 3 & -2 & 0 \\ 0 & 3 & -1 \end{vmatrix}$ **3.** $\begin{vmatrix} 1 & 0 & 0 \\ 0 & 1 & 0 \\ 0 & 0 & 1 \end{vmatrix}$ **4.** $\begin{vmatrix} 3 & 0 & -1 \\ -1 & -3 & 3 \\ 0 & 1 & 0 \end{vmatrix}$

5. $\begin{vmatrix} 4 & 2 & -1 \\ -1 & 3 & 5 \\ 6 & -1 & 0 \end{vmatrix}$

6. $\begin{vmatrix} -1 & 2 & 0 & -2 \\ 2 & -3 & 1 & 3 \\ 4 & -2 & 3 & -1 \\ 3 & -1 & 2 & 0 \end{vmatrix}$

7. $\begin{vmatrix} 3 & 2 & -1 & 0 \\ 1 & -1 & 2 & 3 \\ 0 & 1 & -3 & 2 \\ -1 & 3 & 0 & 1 \end{vmatrix}$

8. $\begin{vmatrix} 2 & 1 & 0 & 4 \\ 4 & -1 & 2 & 2 \\ 3 & 1 & 0 & 3 \\ 1 & 3 & -2 & 5 \end{vmatrix}$

9. $\begin{vmatrix} 4 & -1 & 3 & 2 \\ -2 & 2 & -1 & -3 \\ 3 & 0 & 2 & 1 \\ -1 & -2 & 0 & 3 \end{vmatrix}$

10. $\begin{vmatrix} 4 & -2 & 0 & -1 & 3 \\ 2 & 0 & -2 & 1 & 0 \\ 0 & 2 & -3 & 0 & -1 \\ -2 & 4 & 0 & 5 & -3 \\ -4 & 0 & -2 & 1 & 1 \end{vmatrix}$

11. $\begin{vmatrix} 3 & -1 & 1 & 0 & 4 \\ 3 & 1 & -1 & 2 & -1 \\ -1 & 3 & -5 & 3 & 0 \\ 1 & 2 & -3 & 4 & -4 \\ -5 & 1 & 3 & -2 & 2 \end{vmatrix}$

In 12–18, use Cramer's rule and expansion by cofactors to solve each system of equations.

12. $\begin{aligned} x+2y+z&=6 \\ 2x-y+3z&=3 \\ x+3y-z&=6 \end{aligned}$

13. $\begin{aligned} -x-y+2z&=3 \\ -2x+y-z&=5 \\ -3x-2y+z&=2 \end{aligned}$

14. $\begin{aligned} 2x+3y-2z&=8 \\ 3x-2y+3z&=6 \\ 2x+2y-3z&=9 \end{aligned}$

15. $\begin{aligned} x+2y-3z&=3 \\ 2x-3y+z&=6 \\ 3x-y+2z&=1 \end{aligned}$

16. $\begin{aligned} x+y+2z-w&=8 \\ 2x-y+z+2w&=3 \\ x-2y-z+w&=-3 \\ x+y-2z-2w&=5 \end{aligned}$

17. $\begin{aligned} 2x-y+3z-w&=9 \\ 3x+y-2z+w&=2 \\ x+2y+z-3w&=4 \\ x+3y+z-2w&=2 \end{aligned}$

18. $\begin{aligned} 2x+3y-z+w&=5 \\ 3x-y+z+2w&=14 \\ -x+y+2z+3w&=2 \\ x+2y+3z-w&=-1 \end{aligned}$

---

19. A piece of wire 36.0 cm long is bent into the shape of a triangle. The sum of the lengths of the two shortest sides is 4.0 cm longer than the third side. The middle side is half as long as the sum of the lengths of the other two sides. Find the length of each side. 1 2 4 7 8 3 5

20. Gold is measured by troy weight. One pound of 24 kt gold contains 12 troy ounces of pure gold and no other metals. Each troy pound of 10 kt gold contains 5 troy ounces of pure gold and 7 troy ounces of another metal. Each troy pound of 14 kt gold contains 7 troy ounces of pure gold and 5 troy ounces of another metal. Each troy pound of 18 kt gold contains 9 troy ounces of pure gold and 3 troy ounces of another metal. It is possible to mix 10 kt, 14 kt, and 24 kt gold to form 18 kt gold. Show that there is more than one way to mix x pounds of 10 kt gold, y pounds of 14 kt gold, and z pounds of 24 kt gold to form 1 pound of 18 kt gold. (Hint: Write a system of equations and show that it is dependent.) 1 2 4 6 8 5 7 9

21. An engineer has analyzed the free-body diagram in the accompanying figure on the next page. The following equations describe the forces (in lbs). Solve the system of equations.

$$F\cos 0^\circ + N\cos 90^\circ + T\cos 165^\circ + 100\cos 0^\circ + 400\cos 270^\circ = 0$$

$$F\sin 0^\circ + N\sin 90^\circ + T\sin 165^\circ + 100\sin 0^\circ + 400\sin 270^\circ = 0$$

$$F = 0.35N$$

(Hint: $100\cos 0^\circ$ and $400\cos 270^\circ$ are both constant terms.) 2 4 8 1 5 6 7 9

---

1. Architectural Technology
2. Civil Engineering Technology
3. Computer Engineering Technology
4. Mechanical Drafting Technology
5. Electrical/Electronics Engineering Technology
6. Chemical Engineering Technology
7. Industrial Engineering Technology
8. Mechanical Engineering Technology
9. Biomedical Engineering Technology

Problem 21

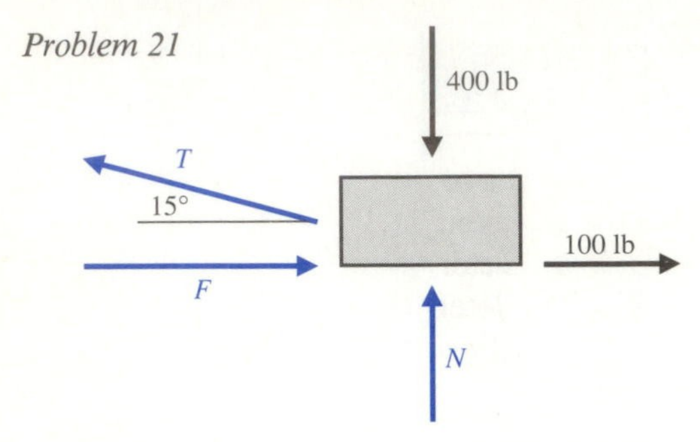

Problem 24

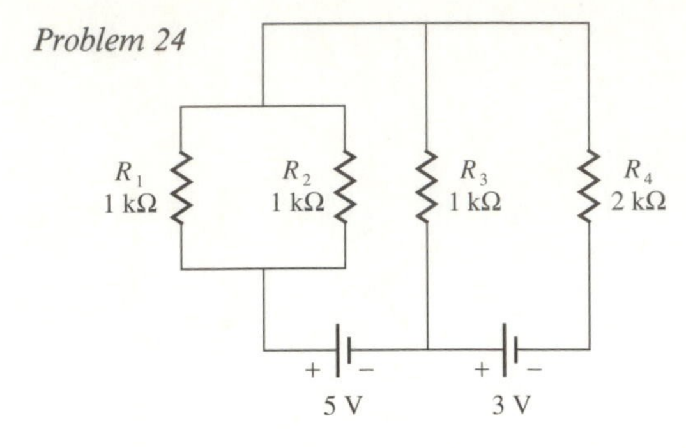

Problems 22 and 23 use the following information. A tetrahedron is a solid figure with four triangular faces, as shown in the accompanying figure. When a tetrahedron is formed in space so that its vertices are at (x_1, y_1, z_1), (x_2, y_2, z_2), (x_3, y_3, z_3), and (x_4, y_4, z_4), its volume is given by

$$V = \left(\frac{1}{6}\right)\begin{vmatrix} x_1 & y_1 & z_1 & 1 \\ x_2 & y_2 & z_2 & 1 \\ x_3 & y_3 & z_3 & 1 \\ x_4 & y_4 & z_4 & 1 \end{vmatrix}.$$

22. Find the volume of a tetrahedron whose vertices are (2, 0, 1), (1, 2, 0), (0, 1, 2), and (0, 0, 0).
1 2 4 7 8 3 5

23. Find the volume of a tetrahedron whose vertices are (2, 0, 0), (0, 0, 2), (0, 0, 0), and (0, 2, 0).
1 2 4 7 8 3 5

24. Applying Kirchhoff's laws to the circuit shown leads to the following system of equations. Solve it using Cramer's rule. The current is in mA.

$$\begin{aligned} I_1 + I_2 + I_3 + I_4 &= 0 \\ -I_1 + I_2 &= 0 \\ -I_2 + I_3 &= 5 \\ -I_3 + 2I_4 &= 3 \end{aligned}$$

3 5 9 1 2 4 6 7 8

25. One general method for evaluating determinants larger than 2×2 is known as Chio's rule. If A is an $n \times n$ determinant with $a_{11} \neq 0$, then $A = 1/(a_{11})^{n-2} B$, where B is an $(n-1) \times (n-1)$ determinant with entries

$$b_{ij} = \begin{vmatrix} a_{11} & a_{1(j+1)} \\ a_{(i+1)1} & a_{(i+1)(j+1)} \end{vmatrix}.$$

With a little practice, you will see the pattern used to locate the entries of each b_{ij}, and since 2×2 determinants are easy to evaluate, you can quickly write down B using Chio's rule. Then Chio's rule can be applied to B to obtain an $(n-2) \times (n-2)$ determinant, etc. Consider the example.

$$\begin{vmatrix} 2 & -1 & 0 & 3 \\ 1 & 0 & -3 & 2 \\ 0 & 3 & -2 & 1 \\ -3 & 2 & 1 & 0 \end{vmatrix} = \frac{1}{4}\begin{vmatrix} 1 & -6 & 1 \\ 6 & -4 & 2 \\ 1 & 2 & 9 \end{vmatrix}$$

$$= \frac{1}{4}\begin{vmatrix} 32 & -4 \\ 8 & 8 \end{vmatrix}$$

$$= 1/4\,(256 + 32) = 1/4(288) = 72$$

Use Chio's rule to evaluate each of the following determinants.

(a) $\begin{vmatrix} -1 & 2 & 0 & -2 \\ 2 & -3 & 1 & 3 \\ 4 & -2 & 3 & -1 \\ 3 & -1 & 2 & 0 \end{vmatrix}$

(b) $\begin{vmatrix} 3 & 2 & -1 & 0 \\ 1 & -1 & 2 & 3 \\ 0 & 1 & -3 & 2 \\ -1 & 3 & 0 & 1 \end{vmatrix}$

18.3 PROPERTIES OF DETERMINANTS

equivalent determinant

Two determinants are said to be **equivalent** if they have the same value. Sometimes the arithmetic is easier if a determinant that is equivalent to the original determinant is evaluated instead of the original determinant. Determinants have a number of properties that make it possible to find equivalent determinants.

In Examples 1–4, we develop some of the most useful properties.

EXAMPLE 1 Evaluate $\begin{vmatrix} -1 & 2 & -5 \\ 3 & -1 & 4 \\ 3 & -1 & 4 \end{vmatrix}$.

Solution Use Sarrus' rule to evaluate the determinant.

$$\left|\begin{matrix} -1 & 2 & -5 \\ 3 & -1 & 4 \\ 3 & -1 & 4 \end{matrix}\right| \begin{matrix} -1 & 2 \\ 3 & -1 \\ 3 & -1 \end{matrix} \qquad (4 + 24 + 15) - (15 + 4 + 24) = 0$$ ■

Notice that row 2 and row 3 of the determinant in Example 1 are the same. If two rows (or columns) of a determinant are identical for each corresponding entry, the value of the determinant is 0.

EXAMPLE 2 Evaluate $\begin{vmatrix} 2 & -6 & 2 \\ -3 & 1 & -3 \\ 7 & -4 & 7 \end{vmatrix}$.

Solution Column 1 and column 3 are identical, so the value of the determinant is 0. ■

To develop additional properties of determinants, we examine what happens if we make certain changes to a determinant. Consider

$$A = \begin{vmatrix} 0 & 1 & -3 \\ 2 & -2 & 4 \\ 5 & 0 & 7 \end{vmatrix}.$$

Using expansion by cofactors about row 1, we have

$$\begin{aligned} A = \begin{vmatrix} 0 & 1 & -3 \\ 2 & -2 & 4 \\ 5 & 0 & 7 \end{vmatrix} &= 0\begin{vmatrix} -2 & 4 \\ 0 & 7 \end{vmatrix} - 1\begin{vmatrix} 2 & 4 \\ 5 & 7 \end{vmatrix} - 3\begin{vmatrix} 2 & -2 \\ 5 & 0 \end{vmatrix} \\ &= 0 - 1(14 - 20) - 3(0 + 10) \\ &= 0 - 1(-6) - 3(10) \\ &= 0 + 6 - 30 \\ &= -24. \end{aligned}$$

EXAMPLE 3 Evaluate determinant B, which is obtained from A by interchanging the first two rows.

$$B = \begin{vmatrix} 2 & -2 & 4 \\ 0 & 1 & -3 \\ 5 & 0 & 7 \end{vmatrix}$$

Solution Evaluate the determinant, using Sarrus' rule.

$$\left|\begin{matrix} 2 & -2 & 4 \\ 0 & 1 & -3 \\ 5 & 0 & 7 \end{matrix}\right| \begin{matrix} 2 & -2 \\ 0 & 1 \\ 5 & 0 \end{matrix} \qquad (14 + 30 + 0) - (20 + 0 + 0) = 44 - 20 \text{ or } 24$$ ■

The value of A is -24, and the value of B is $+24$. It is possible to show that if any two rows (or columns) of a determinant are interchanged, the value of the resulting determinant is -1 times the value of the original determinant.

EXAMPLE 4 Evaluate determinant C, which is obtained from A by a multiplication of the second row by 3.

$$C = \begin{vmatrix} 0 & 1 & -3 \\ 6 & -6 & 12 \\ 5 & 0 & 7 \end{vmatrix}$$

Solution Evaluate the determinant using Sarrus' rule.

$$\left|\begin{matrix} 0 & 1 & -3 \\ 6 & -6 & 12 \\ 5 & 0 & 7 \end{matrix}\right| \begin{matrix} 0 & 1 \\ 6 & -6 \\ 5 & 0 \end{matrix} \qquad (0 + 60 + 0) - (90 + 0 + 42) = 60 - 132 \text{ or } -72$$ ■

The value of A is -24, and the value of C is -72. The value of C is 3 times the value of A. It is possible to show that if each entry of a row (or column) of a determinant is multiplied by a constant k, the value of the determinant is multiplied by k.

The properties we have examined are summarized as follows.

PROPERTIES OF DETERMINANTS

1. If two rows (or columns) of a determinant are identical for each corresponding entry, the value of the determinant is zero.
2. If two rows (or columns) of a determinant are interchanged, the value of the resulting determinant is -1 times the value of the original determinant.
3. If each entry of a row (or column) is multiplied by a constant k, the value of the determinant is multiplied by k.

In this section, we examine one more property of determinants.

PROPERTY OF DETERMINANTS

4. If each entry of any row (or column) is multiplied by the same number k and added to the corresponding entry of another row (or column), the value of the resulting determinant is equal to the value of the original determinant.

This property makes the arithmetic less cumbersome when evaluating a determinant using expansion by cofactors. The entries of the determinant are manipulated until one row (or column) consists mostly of zeros. Then, the expansion about that row (or column) is particularly easy.

EXAMPLE 5 Evaluate $\begin{vmatrix} 0 & 1 & -3 \\ 2 & -2 & 4 \\ 5 & 0 & 7 \end{vmatrix}$.

Solution The entry 1 in the first row, second column is especially easy to change with a multiplication by the number k referred to in property 4. Therefore we choose to replace the second row with a row that has a zero in the second column below the entry 1. To find this replacement row, multiply the first row by 2 and add the result to the second row.

0	2	−6	2 × row 1
2	−2	4	row 2
2	0	−2	sum

Expand the new determinant about the second column.

$$\begin{vmatrix} 0 & 1 & -3 \\ 2 & 0 & -2 \\ 5 & 0 & 7 \end{vmatrix} = -1\begin{vmatrix} 2 & -2 \\ 5 & 7 \end{vmatrix} + 0\begin{vmatrix} 0 & -3 \\ 5 & 7 \end{vmatrix} - 0\begin{vmatrix} 0 & -3 \\ 2 & -2 \end{vmatrix}$$

This part does not have to be written down because multiplication by zero produces zero.

$$= -1[14 - (-10)]$$
$$= -1(24) \text{ or } -24 \quad ■$$

This property is especially helpful for larger determinants.

EXAMPLE 6 Evaluate $\begin{vmatrix} 0 & 3 & -4 & 2 \\ -3 & 4 & 2 & -3 \\ 2 & 0 & 0 & 5 \\ 4 & -2 & 3 & 4 \end{vmatrix}$.

Solution There is no 1 in this determinant, but the third row has two zeros, so we choose to expand about that row. It would be nice if we could obtain another 0 in that row. To obtain 0 in the fourth column, use property 4. That is, multiply the first column by a constant and add the result to the fourth column. The constant we need is $-5/2$, since the product of $-5/2$ and 2 is -5.

$$\begin{aligned}(-5/2)(0) \quad &+2 &&= 0+2 &&= 2\\ (-5/2)(-3)&+(-3) &&= 15/2 - 6/2 &&= 9/2\\ (-5/2)(2) \quad &+5 &&= -5+5 &&= 0\\ (-5/2)(4) \quad &+4 &&= -10+4 &&= -6\end{aligned}$$

$$\begin{vmatrix} 0 & 3 & -4 & 2\\ -3 & 4 & 2 & -3\\ 2 & 0 & 0 & 5\\ 4 & -2 & 3 & 4\end{vmatrix} = \begin{vmatrix} 0 & 3 & -4 & 2\\ -3 & 4 & 2 & \frac{9}{2}\\ 2 & 0 & 0 & 0\\ 4 & -2 & 3 & -6\end{vmatrix}$$

$$= 2\begin{vmatrix} 3 & -4 & 2\\ 4 & 2 & \frac{9}{2}\\ -2 & 3 & -6\end{vmatrix} \quad \text{when expanded about row 3}$$

Now the 3×3 determinant must be evaluated. Once again, there is no entry of 1, so we arbitrarily choose to replace the first column with a column that has mostly zeros. If we multiply the third row by 1 and add to the first row, the entry in the first row will be 1.

$$\begin{array}{rrrl} -2 & 3 & -6 & 1 \times \text{row } 3\\ 3 & -4 & 2 & \text{row } 1\\ \hline 1 & -1 & -4 & \text{sum}\end{array}$$

$$2\begin{vmatrix} 3 & -4 & 2\\ 4 & 2 & \frac{9}{2}\\ -2 & 3 & -6\end{vmatrix} = 2\begin{vmatrix} 1 & -1 & -4\\ 4 & 2 & \frac{9}{2}\\ -2 & 3 & -6\end{vmatrix}$$

It is now possible to get zeros below the 1 in the first column with minimal difficulty. Multiply the first row by -4 and add to the second row.

$$\begin{array}{rrrl} -4 & 4 & 16 & -4 \times \text{row } 1\\ 4 & 2 & \frac{9}{2} & \text{row } 2\\ \hline 0 & 6 & \frac{41}{2} & \text{sum}\end{array}$$

Multiply the first row by 2 and add to the third row.

$$\begin{array}{rrrl} 2 & -2 & -8 & 2 \times \text{row } 1\\ -2 & 3 & -6 & \text{row } 3\\ \hline 0 & 1 & -14 & \text{sum}\end{array}$$

$$2\begin{vmatrix} 1 & -1 & -4 \\ 4 & 2 & \frac{9}{2} \\ -2 & 3 & -6 \end{vmatrix} = 2\begin{vmatrix} 1 & -1 & -4 \\ 0 & 6 & \frac{41}{2} \\ 0 & 1 & -14 \end{vmatrix} = 2\begin{vmatrix} 6 & \frac{41}{2} \\ 1 & -14 \end{vmatrix}$$

$$= 2[6(-14) - 1(41/2)]$$
$$= 2(-84 - 41/2)$$
$$= 2(-168/2 - 41/2)$$
$$= 2(-209/2)$$
$$= -209 \quad ■$$

When you solve applied problems using determinants, you may want to employ the properties of this section to make the computations simpler.

EXAMPLE 7 When a triangle is drawn on a Cartesian plane so that its vertices are (x_1, y_1), (x_2, y_2), and (x_3, y_3), its area is given by

$$A = \left(\frac{1}{2}\right)\begin{vmatrix} x_1 & y_1 & 1 \\ x_2 & y_2 & 1 \\ x_3 & y_3 & 1 \end{vmatrix}.$$

Consider a design that shows a triangular metal plate with vertices at (1, 1), (5, 6), and (3, 8). Find the area of the plate.

Solution Let $(x_1, y_1) = (1, 1)$

$(x_2, y_2) = (5, 6)$

$(x_2, y_3) = (3, 8)$

Then $A = \left(\frac{1}{2}\right)\begin{vmatrix} 1 & 1 & 1 \\ 5 & 6 & 1 \\ 3 & 8 & 1 \end{vmatrix}$.

We choose to replace the second and third rows with rows that each have a zero in the third column. Multiply the first row by -1 and add the result to the second row.

-1	-1	-1	$-1 \times$ row 1
5	6	1	row 2
4	5	0	sum

Multiply the first row by -1 and add the result to the third row.

-1	-1	-1	$-1 \times$ row 1
3	8	1	row 3
2	7	0	sum

Expand the new determinant about the third column.

$$\begin{vmatrix} 1 & 1 & 1 \\ 4 & 5 & 0 \\ 2 & 7 & 0 \end{vmatrix} = 1\begin{vmatrix} 4 & 5 \\ 2 & 7 \end{vmatrix} = 28 - 10 \text{ or } 18 \quad ■$$

EXERCISES 18.3

In 1–24, evaluate each determinant.

1. $\begin{vmatrix} 2 & -5 & 7 \\ 1 & 0 & -3 \\ 2 & -5 & 7 \end{vmatrix}$

2. $\begin{vmatrix} 3 & -9 & 12 \\ 0 & 7 & -8 \\ 0 & 0 & 4 \end{vmatrix}$

3. $\begin{vmatrix} 1 & 2 & 1 \\ 3 & -1 & 3 \\ -8 & 6 & -8 \end{vmatrix}$

4. $\begin{vmatrix} 10 & 2 & -5 \\ 0 & -4 & -6 \\ 0 & 0 & -1 \end{vmatrix}$

5. $\begin{vmatrix} 1 & 1 & 2 \\ 0 & 5 & 8 \\ 0 & 0 & 3 \end{vmatrix}$

6. $\begin{vmatrix} 0 & 1 & -2 & 1 \\ 3 & 4 & 5 & 4 \\ 6 & -7 & 8 & -7 \\ 7 & 3 & 9 & 3 \end{vmatrix}$

7. $\begin{vmatrix} 2 & 0 & -3 & 2 \\ -1 & 6 & 7 & 2 \\ 2 & -3 & 0 & -5 \\ 2 & 0 & -3 & 2 \end{vmatrix}$

8. $\begin{vmatrix} -1 & 0 & 0 & 0 \\ 4 & 3 & 0 & 0 \\ 5 & -6 & 7 & 0 \\ 8 & 2 & -10 & 9 \end{vmatrix}$

9. $\begin{vmatrix} 0 & 2 & 4 \\ 6 & -3 & 1 \\ -2 & 5 & -7 \end{vmatrix}$

10. $\begin{vmatrix} 0 & 4 & -1 \\ -7 & 5 & 2 \\ 3 & -1 & 6 \end{vmatrix}$

11. $\begin{vmatrix} 0 & 1 & -8 & 7 \\ 9 & -2 & 3 & 0 \\ -1 & 2 & 4 & -5 \\ 6 & 3 & -4 & 6 \end{vmatrix}$

12. $\begin{vmatrix} 0 & -2 & 7 & 8 \\ -1 & 1 & 4 & -9 \\ 0 & 3 & -3 & 6 \\ 4 & -2 & 5 & -5 \end{vmatrix}$

13. $\begin{vmatrix} 2 & -1 & 4 \\ 3 & 0 & -5 \\ -6 & 0 & 8 \end{vmatrix}$

14. $\begin{vmatrix} 1 & -2 & 3 \\ 0 & 0 & 0 \\ -4 & 5 & -6 \end{vmatrix}$

15. $\begin{vmatrix} 2 & 2 & 3 & 6 \\ 0 & 1 & 0 & 0 \\ 8 & 5 & 7 & 1 \\ 0 & 9 & 4 & 9 \end{vmatrix}$

16. $\begin{vmatrix} -4 & 5 & -3 \\ 0 & 1 & 0 \\ 7 & -6 & 2 \end{vmatrix}$

17. $\begin{vmatrix} 3 & 7 & 0 \\ 0 & 1 & 0 \\ 7 & -6 & 2 \end{vmatrix}$

18. $\begin{vmatrix} 6 & 8 & 0 & 9 \\ 3 & 4 & 0 & 2 \\ 5 & 7 & 1 & 6 \\ 4 & 2 & 0 & 1 \end{vmatrix}$

19. $\begin{vmatrix} 1 & 4 & -3 \\ 6 & -5 & 2 \\ 4 & 3 & 1 \end{vmatrix}$

20. $\begin{vmatrix} -3 & 1 & 7 \\ 6 & 4 & -2 \\ 2 & -3 & 5 \end{vmatrix}$

21. $\begin{vmatrix} 7 & -2 & 5 \\ -4 & 6 & 0 \\ 8 & 3 & -1 \end{vmatrix}$

22. $\begin{vmatrix} 2 & 3 & -7 & 4 \\ -4 & 5 & 3 & 6 \\ -6 & 2 & 2 & -3 \\ 8 & -4 & 9 & 5 \end{vmatrix}$

23. $\begin{vmatrix} 5 & 2 & 3 & 0 \\ 4 & -6 & 7 & 4 \\ -3 & 0 & -2 & 8 \\ 2 & 4 & 5 & -2 \end{vmatrix}$

24. $\begin{vmatrix} 4 & -5 & 2 & -7 \\ -3 & 9 & 5 & 2 \\ 6 & 3 & 9 & 15 \\ 2 & 7 & -3 & 9 \end{vmatrix}$

---

25. The cross product of two vectors **B** and **C** (See Example 5 of Section 18.2) is denoted $\mathbf{B} \times \mathbf{C}$. The dot product of two vectors **B** and **C** (defined as the product of the magnitudes and the cosine of the angle between the vectors) is denoted $\mathbf{B} \cdot \mathbf{C}$. If $\mathbf{A} = \langle x_a, y_a, z_a \rangle$, $\mathbf{B} = \langle x_b, y_b, z_b \rangle$, and $\mathbf{C} = \langle x_c, y_c, z_c \rangle$, then

$$\mathbf{A} \cdot (\mathbf{B} \times \mathbf{C}) = \begin{vmatrix} x_a & y_a & z_a \\ x_b & y_b & z_b \\ x_c & y_c & z_c \end{vmatrix}.$$

Write $\mathbf{B} \cdot (\mathbf{C} \times \mathbf{A})$ in determinant form. Is it true that $\mathbf{A} \cdot (\mathbf{B} \times \mathbf{C}) = \mathbf{B} \cdot (\mathbf{C} \times \mathbf{A})$? 2 4 8 1 5 6 7 9

26. The double subscript notation is used to analyze stress relationships. The first subscript indicates the surface on which the stress acts. The second denotes the direction in which the stress acts.

The formula $\theta_3 = \begin{vmatrix} \tau_{11} & \tau_{21} & \tau_{31} \\ \tau_{12} & \tau_{22} & \tau_{32} \\ \tau_{13} & \tau_{23} & \tau_{33} \end{vmatrix}$ gives an in-

variant of the stress tensor. Interchange row 1 with column 1, row 2 with column 2, and row 3 with column 3. How is the new determinant related to the original determinant?
1 2 4 8 7 9

27. The formula $\phi_2 = \begin{vmatrix} e_{11} & e_{12} & e_{31} \\ e_{12} & e_{22} & e_{23} \\ e_{31} & e_{23} & e_{33} \end{vmatrix}$ is used to find an invariant of a strain tensor. Interchange row 1 with column 1, row 2 with column 2, and row 3 with column 3. How is the new determinant related to the original determinant?
1 2 4 8 7 9

Problems 28 and 29 use the following information. The determinant

$$\begin{vmatrix} (I_{xx} - I') & -I_{xy} & -I_{xz} \\ -I_{xy} & (I_{yy} - I') & -I_{yz} \\ -I_{xx} & -I_{yz} & (I_{zz} - I') \end{vmatrix}$$

is used to find principal moments of inertia.

28. Consider determinants A and B as follows. Compare row 1 of A with row 3 of B, and row 3 of A with row 1 of B. Are the determinants equal?

$$A = \begin{vmatrix} (I_{xx} - I') & -I_{xy} & -I_{xz} \\ -I_{xy} & (I_{yy} - I') & -I_{yz} \\ -I_{xx} & -I_{yz} & (I_{zz} - I') \end{vmatrix}$$

$$B = \begin{vmatrix} -I_{xx} & -I_{yz} & (I_{zz} - I') \\ -I_{xy} & (I_{yy} - I') & -I_{yz} \\ (I_{xx} - I') & -I_{xy} & -I_{xz} \end{vmatrix}$$

1 2 4 8 7 9

29. Consider determinants A and C as follows. Compare corresponding rows of the two determinants. Are the determinants equal?

$$A = \begin{vmatrix} (I_{xx} - I') & -I_{xy} & -I_{xz} \\ -I_{xy} & (I_{yy} - I') & -I_{yz} \\ -I_{xx} & -I_{yz} & (I_{zz} - I') \end{vmatrix}$$

$$C = \begin{vmatrix} (I' - I_{xx}) & I_{xy} & I_{xz} \\ I_{xy} & (I' - I_{yy}) & I_{yz} \\ I_{xx} & I_{yz} & (I' - I_{zz}) \end{vmatrix}$$

1 2 4 8 7 9

---

1. Architectural Technology
2. Civil Engineering Technology
3. Computer Engineering Technology
4. Mechanical Drafting Technology
5. Electrical/Electronics Engineering Technology
6. Chemical Engineering Technology
7. Industrial Engineering Technology
8. Mechanical Engineering Technology
9. Biomedical Engineering Technology

18.4 SIMPLIFYING DETERMINANTS

Consider $A = \begin{vmatrix} 2 & 3 & 4 \\ 0 & 5 & 6 \\ 0 & 0 & 7 \end{vmatrix}$. If we expand about column 1, we have

$$\begin{vmatrix} 2 & 3 & 4 \\ 0 & 5 & 6 \\ 0 & 0 & 7 \end{vmatrix} = 2 \begin{vmatrix} 5 & 6 \\ 0 & 7 \end{vmatrix}$$
$$= 2[5(7) - 6(0)]$$
$$= 2(5)(7) \text{ or } 70.$$

Notice that the determinant has zeros for each entry below the main diagonal. Also notice that the value of the determinant is 2(5)(7), which is the product of the entries along the main diagonal. This observation is sometimes generalized as a property of determinants.

THE DIAGONAL PROPERTY OF DETERMINANTS

If each entry below (or above) the main diagonal of a determinant is 0, then the value of the determinant is the product of the entries along the main diagonal.

Computer programs to evaluate determinants generally use properties 2 and 4 of the previous section to form a determinant that has zeros below the main diagonal, so that the diagonal property may be used.

EXAMPLE 1 Use the diagonal property of determinants to evaluate

$$\begin{vmatrix} 1 & 2 & -5 \\ 3 & -1 & 4 \\ 2 & 2 & -2 \end{vmatrix}.$$

Solution If there are to be zeros below the main diagonal, the first column should contain zeros in the second and third rows. We leave the first row as it is: 1 2 −5. To obtain 0 as the first entry in the second row, multiply the first row by −3 and add the result to the second row. (We are using property 4 of the previous section.)

$$\begin{array}{rrrl} -3 & -6 & 15 & -3 \times \text{row 1} \\ 3 & -1 & 4 & \text{row 2} \\ \hline 0 & -7 & 19 & \text{sum} \end{array}$$

To obtain 0 as the first entry in the third row, multiply the first row by −2 and add the result to the third row.

$$\begin{array}{rrrl} -2 & -4 & 10 & -2 \times \text{row 1} \\ 2 & 2 & -2 & \text{row 3} \\ \hline 0 & -2 & 8 & \text{sum} \end{array}$$

$$\begin{vmatrix} 1 & 2 & -5 \\ 3 & -1 & 4 \\ 2 & 2 & -2 \end{vmatrix} = \begin{vmatrix} 1 & 2 & -5 \\ 0 & -7 & 19 \\ 0 & -2 & 8 \end{vmatrix}$$

The first column now has the desired form, but the second column does not. At first, you might think that we should add row 1 to row 3, because the sum of 2 and −2 is 0. This move, however, would make the first entry of the third row become 1. Since we do not want to undo what we have done, we multiply the second row by a constant and add the result to the third row. The constant we need is −2/7, since the product of −7 and −2/7 is 2.

$$\begin{array}{rrll} 0 & 2 & -38/7 & (-2/7) \times \text{row 2} \\ 0 & -2 & 8 \text{ (or } 56/7) & \text{row 3} \\ \hline 0 & 0 & 18/7 & \text{sum} \end{array}$$

$$\begin{vmatrix} 1 & 2 & -5 \\ 0 & -7 & 19 \\ 0 & -2 & 8 \end{vmatrix} = \begin{vmatrix} 1 & 2 & -5 \\ 0 & -7 & 19 \\ 0 & 0 & \frac{18}{7} \end{vmatrix}$$

The determinant now has zeros below the main diagonal. Its value is the product of 1, −7, and 18/7.

$$1(-7)(18/7) = -18. \quad ■$$

We are now ready to generalize the procedure used in Example 1.

To use the diagonal property of determinants:

If the following procedure calls for division by 0 at any stage, interchange two rows. This move changes the value of the determinant by a factor of −1.

1. (a) For each row i after row 1, multiply row 1 by the quotient of the first entry of row i and the first entry of row 1.
   (b) Change the sign of each entry.
   (c) Add the result to row i.
2. (a) For each row i after row 2, multiply row 2 by the quotient of the second entry of row i and the second entry of row 2.
   (b) Change the sign of each entry.
   (c) Add the result to row i.
3. (a) In general, for each row i after row k, multiply row k by the quotient of the kth entry of row i and the kth entry of row k.
   (b) Change the sign of each entry.
   (c) Add the result to row i.

EXAMPLE 2 Evaluate $\begin{vmatrix} 0 & 8 & 2 & 6 \\ 2 & -5 & 7 & -1 \\ -4 & 10 & -18 & 10 \\ 0 & 4 & 1 & 9 \end{vmatrix}$.

Solution Since we cannot divide row 1 by a_{11}, which is 0, we interchange rows 1 and 2.

$$\begin{vmatrix} 0 & 8 & 2 & 6 \\ 2 & -5 & 7 & -1 \\ -4 & 10 & -18 & 10 \\ 0 & 4 & 1 & 9 \end{vmatrix} = -1 \begin{vmatrix} 2 & -5 & 7 & -1 \\ 0 & 8 & 2 & 6 \\ -4 & 10 & -18 & 10 \\ 0 & 4 & 1 & 9 \end{vmatrix}$$

Consider column 1. Row 2 and row 4 each have zero as the entry. To obtain a zero in row 3, we multiply the first row by a constant and add the result to the third row. The constant we need is the fraction whose numerator is the entry in row 3 with the sign changed, and whose denominator is the entry in row 1. That is, we multiply the first row by 4/2 or 2 and add the result to row 3.

$$\begin{array}{rrrrl} 4 & -10 & 14 & -2 & 2 \times \text{row } 1 \\ -4 & 10 & -18 & 10 & \text{row } 3 \\ \hline 0 & 0 & -4 & 8 & \text{sum} \end{array}$$

$$-1\begin{vmatrix} 2 & -5 & 7 & -1 \\ 0 & 8 & 2 & 6 \\ -4 & 10 & -18 & 10 \\ 0 & 4 & 1 & 9 \end{vmatrix} = -1\begin{vmatrix} 2 & -5 & 7 & -1 \\ 0 & 8 & 2 & 6 \\ 0 & 0 & -4 & 8 \\ 0 & 4 & 1 & 9 \end{vmatrix}$$

Consider column 2. Row 3 has zero as the entry. To obtain a zero in row 4, we must find the fraction whose numerator is the entry in row 4 with the sign changed, and whose denominator is the entry in row 2. That is, we multiply the second row by $-4/8$ or $-1/2$ and add the result to row 4.

$$\begin{array}{rrrrl} -0 & -4 & -1 & -3 & (-1/2) \times \text{row } 2 \\ 0 & 4 & 1 & 9 & \text{row } 4 \\ \hline 0 & 0 & 0 & 6 & \text{sum} \end{array}$$

$$-1\begin{vmatrix} 2 & -5 & 7 & -1 \\ 0 & 8 & 2 & 6 \\ 0 & 0 & -4 & 8 \\ 0 & 4 & 1 & 9 \end{vmatrix} = -1\begin{vmatrix} 2 & -5 & 7 & -1 \\ 0 & 8 & 2 & 6 \\ 0 & 0 & -4 & 8 \\ 0 & 0 & 0 & 6 \end{vmatrix}$$

The determinant now has the desired form. Its value is the product of the entries along the main diagonal: $2(8)(-4)(6)$ or -384. The original determinant, then, has a value of $-1(-384)$ or 384. ■

EXAMPLE 3 Alloys containing 15% zinc, 30% zinc, and 45% zinc are available. The three alloys weigh a total of 90 kg. If the first two are mixed, the alloy is 24% zinc. If the first and third are mixed, the alloy is 35% zinc. Find the amount of the 30% zinc alloy.

Solution Let $x =$ weight of 15% zinc alloy

$y =$ weight of 30% zinc alloy

$z =$ weight of 45% zinc alloy

We know the total weight of all three alloys is $x + y + z$ or 90. If the first two are mixed, the total weight is $x + y$. Consider the amount of zinc in the mixture.

$$0.15x + 0.30y = 0.24(x + y)$$

If the first and third are mixed, the total weight is $x + z$. Consider the amount of zinc in the mixture.

$$0.15x + 0.45z = 0.35(x + z)$$

The system of equations is as follows.

$$\begin{array}{lcrrrl} x + y + z = 90 & & x + & y + & z = & 90 \\ 0.15x + 0.30y = 0.24(x + y) & \text{or} & -0.09x + & 0.06y & = & 0 \\ 0.15x + 0.45z = 0.35(x + z) & & -0.20x & & +0.10z = & 0 \end{array}$$

Multiplying the second equation by 100 and the third by 10, we have the following system.

$$\begin{aligned} x + y + z &= 90 \\ -9x + 6y &= 0 \\ -2x + z &= 0 \end{aligned}$$

To find the amount of 30% zinc alloy, solve for y.

$$y = \frac{\begin{vmatrix} 1 & 90 & 1 \\ -9 & 0 & 0 \\ -2 & 0 & 1 \end{vmatrix}}{\begin{vmatrix} 1 & 1 & 1 \\ -9 & 6 & 0 \\ -2 & 0 & 1 \end{vmatrix}} = \frac{\begin{vmatrix} 1 & 90 & 1 \\ 0 & 810 & 9 \\ 0 & 180 & 3 \end{vmatrix}}{\begin{vmatrix} 1 & 1 & 1 \\ 0 & 15 & 9 \\ 0 & 2 & 3 \end{vmatrix}} \quad \begin{array}{l} 9 \times \text{row } 1 + \text{row } 2 \\ 2 \times \text{row } 1 + \text{row } 3 \\ \\ 9 \times \text{row } 1 + \text{row } 2 \\ 2 \times \text{row } 1 + \text{row } 3 \end{array}$$

$$= \frac{\begin{vmatrix} 1 & 90 & 1 \\ 0 & 810 & 9 \\ 0 & 0 & 1 \end{vmatrix}}{\begin{vmatrix} 1 & 1 & 1 \\ 0 & 15 & 9 \\ 0 & 0 & \frac{9}{5} \end{vmatrix}} \quad \begin{array}{l} (-2/9) \times \text{row } 2 + \text{row } 3 \\ \\ (-2/15) \times \text{row } 2 + \text{row } 3 \end{array}$$

$$= \frac{1(810)(1)}{1(15)(9/5)} \text{ or } \frac{810}{27}, \text{ which is } 30$$ ■

EXERCISES 18.4

In 1–15, evaluate each determinant.

1. $\begin{vmatrix} 4 & -3 & 1 \\ 0 & 2 & 7 \\ 0 & 0 & -5 \end{vmatrix}$

2. $\begin{vmatrix} 4 & -3 & 8 \\ 0 & 7 & -2 \\ 0 & 0 & 3 \end{vmatrix}$

3. $\begin{vmatrix} 2 & 0 & 0 \\ 3 & -5 & 0 \\ -6 & 8 & 7 \end{vmatrix}$

4. $\begin{vmatrix} 4 & 0 & 0 \\ 8 & -8 & 0 \\ -5 & 1 & 8 \end{vmatrix}$

5. $\begin{vmatrix} 0 & 0 & 6 \\ 2 & -7 & 5 \\ 0 & 3 & -9 \end{vmatrix}$

6. $\begin{vmatrix} 4 & -7 & 0 \\ 8 & 0 & 0 \\ -3 & 2 & 9 \end{vmatrix}$

7. $\begin{vmatrix} 8 & 0 & -3 & 6 \\ 0 & 3 & -8 & 2 \\ 0 & 0 & 3 & -5 \\ 0 & 0 & 0 & 2 \end{vmatrix}$

8. $\begin{vmatrix} 9 & -6 & 9 & 1 \\ 0 & 2 & -1 & 7 \\ 0 & 0 & 2 & 1 \\ 0 & 0 & 0 & 9 \end{vmatrix}$

9. $\begin{vmatrix} 2 & -2 & 1 & 0 \\ 9 & 0 & 0 & 0 \\ -9 & 2 & -5 & 2 \\ 1 & 2 & 0 & 0 \end{vmatrix}$

10. $\begin{vmatrix} -7 & 1 & 0 & 0 \\ 5 & 2 & -5 & 8 \\ 1 & -5 & 2 & 0 \\ 3 & 0 & 0 & 0 \end{vmatrix}$

11. $\begin{vmatrix} 9 & 1 & 9 \\ 6 & 2 & 0 \\ 4 & 1 & 9 \end{vmatrix}$

12. $\begin{vmatrix} 2 & 18 & 16 \\ 1 & 10 & 15 \\ 2 & 19 & 28 \end{vmatrix}$

13. $\begin{vmatrix} 8 & 2 & 0 & 0 \\ 14 & 4 & 0 & 0 \\ 3 & 10 & 3 & 5 \\ 4 & 10 & 4 & 10 \end{vmatrix}$

14. $\begin{vmatrix} 4 & 2 & 16 & 8 \\ 14 & 3 & 3 & 0 \\ 15 & 6 & 9 & 0 \\ 1 & 1 & 8 & 4 \end{vmatrix}$

15. $\begin{vmatrix} 3 & 16 & 9 & 2 \\ 10 & 15 & 2 & 0 \\ 2 & 14 & 4 & 0 \\ 1 & 9 & 7 & 2 \end{vmatrix}$

16. Alloys containing 20% copper, 30% copper, and 50% copper are available. The three alloys weigh a total of 120 kg. If the first two are mixed, the alloy is 25% copper. If the last two are mixed, the alloy is 40% copper. Find the amount of the third alloy. 1 2 4 6 8 5 7 9

17. A petroleum engineer made the following observation while testing different gasoline mixtures in a car. Using the second fuel, the car traveled 30 miles farther than when it used the first fuel. Using the third fuel, it traveled 30 miles farther than when it used the second. The total distance traveled was 960 miles. Find the distance traveled on the first fuel. 6 9 2 4 5 7 8

18. The relationship between the number T of degrees Fahrenheit below 212 at which water boils and the elevation s (in ft) is known to be quadratic. That is, it follows, the pattern $s = aT^2 + bT + c$. Use the following data to determine the values of a, b, and c. When T is 0, s is 0. When T is 1, s is 518. When T is 2, s is 1038. 6 8 1 2 3 5 7 9

Problems 19 and 20 use the following information. When a triangle is drawn on a Cartesian plane so that its vertices are (x_1, y_1), (x_2, y_2), and (x_3, y_3), its area A is given by

$$A = \frac{1}{2}\begin{vmatrix} x_1 & y_1 & 1 \\ x_2 & y_2 & 1 \\ x_3 & y_3 & 1 \end{vmatrix}.$$

19. Find the area of a triangle with vertices at (2, 3), (−2, 3), and (0, 1). 1 2 4 7 8 3 5

20. Find the area of a triangle with vertices at (0, 0), (−2, 8), and (−3, −4). 1 2 4 7 8 3 5

21. For the circuit shown in the accompanying figure, the following system of equations applies. Find the current I_4 (in A) in R_4, given $I_4 = I_1 - I_2$.

$$\begin{aligned} -1500 I_1 + 5000 I_2 - 1500 I_3 &= 0 \\ 2500 I_1 - 1500 I_2 \qquad\qquad &= 10 \\ -1500 I_2 + 2500 I_3 &= -18 \end{aligned}$$

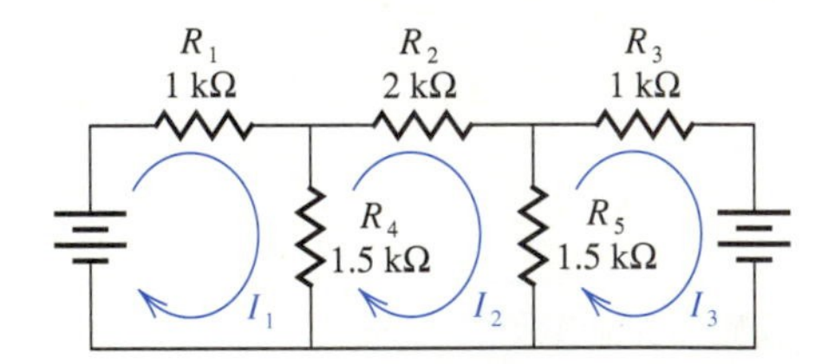

3 5 9 1 2 4 6 7 8

1. Architectural Technology
2. Civil Engineering Technology
3. Computer Engineering Technology
4. Mechanical Drafting Technology
5. Electrical/Electronics Engineering Technology
6. Chemical Engineering Technology
7. Industrial Engineering Technology
8. Mechanical Engineering Technology
9. Biomedical Engineering Technology

18.5 CHAPTER REVIEW

KEY TERMS

18.1 determinant
main diagonal

18.2 minor
cofactor

18.3 equivalent determinant

SYMBOLS

$|\ |$ a determinant, if a square array of numbers is included between bars

a_{ij} the entry in row i, column j of a determinant

RULES AND FORMULAS

If two rows (or columns) of a determinant are identical for each corresponding entry, the value of the determinant is zero.

If two rows (or columns) of a determinant are interchanged, the value of the resulting determinant is -1 times the value of the original determinant.

If each entry of a row (or column) is multiplied by a constant k, the value of the determinant is multiplied by k.

If each entry of any row (or column) is multiplied by the same number k and added to the corresponding entry of another row (or column), the value of the resulting determinant is equal to the value of the original determinant.

The Diagonal Property of Determinants: If each entry below (or above) the main diagonal of a determinant is 0, then the value of the determinant is the product of the entries along the main diagonal.

GEOMETRY CONCEPTS

$A + B + C = 180°$, if A, B, and C are the angles of any triangle.

SEQUENTIAL REVIEW

(Section 18.1) *In 1–4, evaluate each determinant.*

1. $\begin{vmatrix} 5 & -1 \\ 7 & 4 \end{vmatrix}$ **2.** $\begin{vmatrix} 4 & -1 \\ -5 & 7 \end{vmatrix}$ **3.** $\begin{vmatrix} -3 & 9 \\ 5 & -5 \end{vmatrix}$ **4.** $\begin{vmatrix} 6 & 2 \\ 8 & -3 \end{vmatrix}$

In 5–8, evaluate each determinant using Sarrus' rule.

5. $\begin{vmatrix} 2 & -8 & 2 \\ 6 & 4 & 0 \\ -4 & 8 & -6 \end{vmatrix}$ **6.** $\begin{vmatrix} 3 & 2 & -8 \\ -1 & 6 & 5 \\ 4 & -7 & 9 \end{vmatrix}$ **7.** $\begin{vmatrix} 5 & 7 & -5 \\ -9 & 9 & 1 \\ 3 & 7 & -3 \end{vmatrix}$ **8.** $\begin{vmatrix} -2 & 4 & 9 \\ 2 & 5 & -6 \\ 7 & 4 & -8 \end{vmatrix}$

In 9–12, use Cramer's rule to solve each system of equations.

9. $2x + 3y = 3$
$3x - y = 10$

10. $3x + 5y = 22$
$5x - 3y = 14$

11. $2x + y - 3z = 0$
$3x - 2y + z = 9$
$x + 3y + 2z = 1$

12. $3x + 2y - 4z = 12$
$4x - 3y + z = 4$
$2x + 4y + 3z = 5$

(Section 18.2) *In 13–18, evaluate each determinant using expansion by cofactors.*

13. $\begin{vmatrix} 2 & -8 & 2 \\ 6 & 4 & 0 \\ -4 & 8 & -6 \end{vmatrix}$ **14.** $\begin{vmatrix} 3 & 2 & -8 \\ -1 & 6 & 5 \\ 4 & -7 & 9 \end{vmatrix}$ **15.** $\begin{vmatrix} 5 & 7 & -5 \\ -9 & 9 & 1 \\ 3 & 7 & -3 \end{vmatrix}$

16. $\begin{vmatrix} -2 & 4 & 9 \\ 2 & 5 & -6 \\ 7 & 4 & -8 \end{vmatrix}$ **17.** $\begin{vmatrix} -2 & 4 & -3 & 9 \\ 2 & 5 & 7 & -6 \\ 1 & -5 & 8 & 0 \\ 7 & 4 & -8 & 1 \end{vmatrix}$ **18.** $\begin{vmatrix} -2 & 0 & 3 & 1 \\ 2 & 1 & 4 & -3 \\ 5 & 0 & -2 & 3 \\ 1 & 2 & 1 & 2 \end{vmatrix}$

In 19–23, use Cramer's rule to solve each system of equations.

19. $\begin{aligned} x + 2y &= 1 \\ y + 2z &= 1 \\ 2x + z &= -2 \end{aligned}$

20. $\begin{aligned} 2x + 3z &= 7 \\ -4y - 5z &= 7 \\ -x + 6z &= 4 \end{aligned}$

21. $\begin{aligned} x + y - z &= -5 \\ x - y + z &= 11 \\ x - y - z &= 1 \end{aligned}$

22. $\begin{aligned} 2x - y + 3z &= 13 \\ x + 2y - 3z &= -5 \\ x - 3y + 2z &= 10 \end{aligned}$

23. $\begin{aligned} 3x - y + z &= 1 \\ x + 2y - z &= 3 \\ 2x + y - 3z &= 8 \end{aligned}$

(Section 18.3) *In 24–29, find the value of each determinant, using properties of determinants and the fact that* $\begin{vmatrix} 2 & 1 & -3 \\ 1 & 3 & 2 \\ 3 & -2 & 1 \end{vmatrix} = 52$.

24. $\begin{vmatrix} 3 & -2 & 1 \\ 1 & 3 & 2 \\ 2 & 1 & -3 \end{vmatrix}$ **25.** $\begin{vmatrix} 2 & 1 & -3 \\ 3 & -2 & 1 \\ 1 & 3 & 2 \end{vmatrix}$ **26.** $\begin{vmatrix} 4 & 2 & -6 \\ 1 & 3 & 2 \\ 3 & -2 & 1 \end{vmatrix}$

27. $\begin{vmatrix} -2 & 1 & -3 \\ -1 & 3 & 2 \\ -3 & -2 & 1 \end{vmatrix}$ **28.** $\begin{vmatrix} 2 & 2 & -3 \\ 1 & 6 & 2 \\ 3 & -4 & 1 \end{vmatrix}$ **29.** $\begin{vmatrix} 2 & -3 & 1 \\ 1 & 2 & 3 \\ 3 & 1 & -2 \end{vmatrix}$

In 30–34, evaluate each determinant, using properties of determinants to simplify the computation.

30. $\begin{vmatrix} 1 & -7 & 9 \\ 3 & 5 & -1 \\ 1 & -7 & 9 \end{vmatrix}$ **31.** $\begin{vmatrix} 2 & -8 & 2 \\ -6 & 4 & -6 \\ 10 & 0 & 10 \end{vmatrix}$ **32.** $\begin{vmatrix} 2 & -3 & 5 \\ 7 & 9 & 11 \\ 2 & -3 & 5 \end{vmatrix}$

33. $\begin{vmatrix} 1 & 0 & 7 & -5 \\ 2 & -3 & 4 & 8 \\ 5 & 9 & -3 & 6 \\ -1 & 0 & -7 & 5 \end{vmatrix}$ **34.** $\begin{vmatrix} 4 & -3 & 8 & 2 \\ 5 & 7 & -4 & 5 \\ 0 & 0 & 1 & 0 \\ 2 & -6 & 9 & -3 \end{vmatrix}$

(Section 18.4) *In 35–45, use the diagonal property of determinants to evaluate each determinant.*

35. $\begin{vmatrix} 7 & 3 & 1 \\ 0 & -4 & 5 \\ 0 & 0 & 2 \end{vmatrix}$ **36.** $\begin{vmatrix} 3 & 4 & -7 \\ 0 & 5 & 8 \\ 0 & 0 & -2 \end{vmatrix}$ **37.** $\begin{vmatrix} 2 & 0 & 0 \\ -9 & -4 & 0 \\ 6 & 3 & 8 \end{vmatrix}$

38. $\begin{vmatrix} 1 & 5 & -2 & 8 \\ 0 & -1 & 2 & 7 \\ 0 & 0 & 1 & 4 \\ 0 & 0 & 0 & -1 \end{vmatrix}$ **39.** $\begin{vmatrix} -2 & 0 & 0 & 0 \\ 1 & 5 & 0 & 0 \\ 4 & 7 & -3 & 0 \\ 9 & 8 & 6 & 7 \end{vmatrix}$ **40.** $\begin{vmatrix} 2 & -8 & 2 \\ 6 & 4 & 0 \\ -4 & 2 & -6 \end{vmatrix}$

41. $\begin{vmatrix} 3 & 2 & -8 \\ -6 & 6 & 4 \\ 9 & 1 & 9 \end{vmatrix}$ **42.** $\begin{vmatrix} 3 & 7 & -5 \\ -9 & 7 & 1 \\ 6 & 7 & -3 \end{vmatrix}$ **43.** $\begin{vmatrix} -2 & 4 & 9 \\ 2 & 6 & -6 \\ 8 & 4 & -8 \end{vmatrix}$

44. $\begin{vmatrix} -2 & 4 & 3 & 9 \\ 2 & 5 & 9 & -6 \\ 4 & -5 & 8 & 0 \\ 8 & 2 & -8 & 1 \end{vmatrix}$

45. $\begin{vmatrix} -2 & 0 & 3 & 1 \\ 2 & 1 & 4 & -3 \\ 6 & 0 & -2 & 3 \\ 4 & 2 & 1 & 2 \end{vmatrix}$

APPLIED PROBLEMS

1. The forces acting on bodies A and B are shown in the accompanying figure. By analyzing the free-body diagram shown, an engineer obtains the following system of equations involving forces (in newtons) acting on B. Assume that μ, the coefficient of friction between the surfaces of A and B is 0.350, and solve for P, F_B, and N_B.

$$0 = P - F_B - 15.8$$
$$0 = N_B - 115$$
$$F_B = \mu N_B$$

6 9 2 4 5 7 8

2. Ten carat gold is 41.7% gold, fourteen carat gold is 58.3% gold, and eighteen carat gold is 75.0% gold. How many ounces of 10 kt gold and 18 kt gold must be mixed to produce 4.00 ounces of 14 kt gold? 2 4 8 1 5 6 7 9

3. Use Kirchhoff's laws and Cramer's rule to find the current (in mA) in each resistor of the circuit shown. 3 5 9 1 2 4 6 7 8

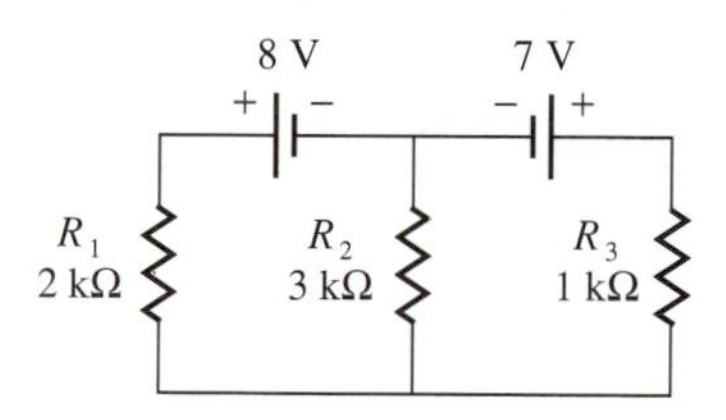

4. Two resistors in series have an equivalent resistance of 600 Ω. (For resistors in series, resistances are added.) If the resistance in the first resistor is doubled while the resistance in the second is halved, the equivalent resistance is 450 Ω. Find the resistance in each resistor. 3 5 9 1 2 4 6 7 8

5. Airline regulations specify that for a checked bag, the sum of the length, width, and height cannot exceed 55 inches. An engineer plans to ship some equipment in a box whose dimensions total 50.0 inches. He finds that if the length and width are both decreased by 2.0 inches, the height can be doubled and the dimensions still total 50.0 inches. If the width is halved, the length can be increased by 8.0 inches and the height by 2.0 inches, and the dimensions will again total 50.0 inches. Find the original dimensions. 1 2 4 7 8 3 5

6. A solder manufacturer accumulated a total of 2950 lbs of tin in three stockpiles. The first is 94% pure, the second 97% pure, and the third is 92% pure. If the three grades are mixed, the result is 95% pure. He mixes half his middle-grade material with sixty percent of his high-grade material and three-quarters of his low grade to obtain 1930 lbs of tin. How much tin of each purity was in his original stockpiles? 1 2 4 6 8 5 7 9

7. A determinant appearing in a paper on phased linear radar systems was $\begin{vmatrix} 15+\lambda & 12 \\ 11+3\lambda & 15 \end{vmatrix} = 0$. Find λ to three significant digits. 1 3 4 5 6 8 9

8. A manufacturing company invested a \$75,000 surplus in three CDs paying yearly interest of $9\frac{1}{2}$%, $8\frac{1}{4}$%, and 7%. After one year the total interest

1. Architectural Technology
2. Civil Engineering Technology
3. Computer Engineering Technology
4. Mechanical Drafting Technology
5. Electrical/Electronics Engineering Technology
6. Chemical Engineering Technology
7. Industrial Engineering Technology
8. Mechanical Engineering Technology
9. Biomedical Engineering Technology

earned was \$6687.50. Company policy requires that the amount invested in the highest interest CD be twice the total invested in the two low paying CDs. How much was invested in each CD? 1 2 6 7 8 3 5 9

9. For a chemical balance scale to stabilize, the products of all the weights on the left and their respective distances from the center added together must equal the products of all the weights on the right and their respective distances added together. Three weights A, B, and C totaling 100 grams are placed on a balance scale. A and B differ by 12.0 grams. A is placed 30.0 cm from the center, and B is placed 20.0 cm from the center. Both A and B are on the left. C is placed on the right 35.0 cm from the center. Find A, B, and C if the scale is balanced. 6 9 2 4 5 7 8

10. Tickets for a technical society trade show cost \$25 for members and \$5 for each guest. Student tickets cost \$2 each. A total of 7800 tickets were sold for \$141,200. How many people in each category attended if members outnumbered all others by two to one? 1 2 6 7 8 3 5 9

11. The value of an important resistance R in the design of an electronic thermometer is given by

$$\begin{vmatrix} 2 & 7 & R \\ 3 & -4 & 8 \\ -1 & R & 5 \end{vmatrix} = -9.$$

Find R. 3 5 9 1 2 4 6 7 8

12. For a certain type of transistor the following equations hold. C_{in}, C_{out}, and C_{GD} represent certain values of capacitance.

$$C_{in} + C_{out} = 12.00 \text{ PF}$$
$$C_{GD} + C_{out} = 2.60 \text{ PF}$$
$$C_{GD} + C_{in} = 10.70 \text{ PF}$$

Find C_{in}, C_{out}, and C_{GD}. 3 5 9 4 7 8

13. A company manufactures three systems A, B, and C. The hours needed for assembly, testing, and packing are shown in the accompanying table.

System	*Assembly*	*Test*	*Packing*
A	7	3	1
B	5	2.5	0.5
C	11	4	1.5

How many of each system should be manufactured if 119 hrs are available for assembly, 51 hrs for testing, and 15 hrs for packing? 1 2 6 7 8 3 5 9

14. A company makes three different size equipment cabinets A, B, and C. It ships these cabinets in three different size trucks T_1, T_2, and T_3. The capacities of the various trucks for items A, B, and C are shown in the accompanying table.

Item	T_1	T_2	T_3
A	8	11	24
B	15	9	19
C	10	5	13

How many of each size truck would be needed to ship 189 of type A, 195 of type B, and 124 of type C? 1 2 6 7 8 3 5 9

15. A research lab has four types of animal feed with protein, fat, water, and filler. The percent of each is shown in the accompanying table.

Type	*% Protein*	*% Fat*	*% Water*	*% Filler*
A	56	5	24	15
B	44	8	28	20
C	57	7	19	17
D	66	3	16	15

How much of each mixture would be needed to make a 10,000-lb sample containing 58% protein, 5.3% fat, 21% water, and 15.7% filler? Give your answer to the nearest pound. 1 2 6 7 8 3 5 9

16. A company sells a 750-lb mixture of grinding sand for \$4000. The mixture contains three types of sand. Type one sells individually for \$6.20 per

pound. Type two sells for \$4.00 per pound. And type three sells for \$1.75 per pound. The sale mixture contained three times as much of the expensive sand as the total weight of the other two. How much of each grade of sand was sold?
1 2 6 7 8 3 5 9

17. A metalurgical company purchased a supply of copper at \$1.33/lb, lead at \$0.51/lb, zinc at \$0.87/lb, and tin at \$3.67/lb, for a total of \$14,424. The total amount purchased was 13,200 lb. The company purchased one-half as much zinc as the combined weight of lead and copper and two and one-half times as much copper as tin. How much of each was purchased? 1 2 6 7 8 3 5 9

18. A municipal water system has three storage tanks that are emptied and filled on a daily basis. On Monday, 18.3 million gallons were consumed, emptying tank one three times, tank two fifteen times, and tank three nine times. On Tuesday, tank one was emptied twice, tank two thirteen times, and tank three ten times for a total of 16.9 million gallons. On Wednesday, tank one was emptied three times, tank two seventeen times, and tank three twelve times for a total of 21.7 million gallons. Find the capacity of each tank.
1 2 4 7 8 3 5

RANDOM REVIEW

1. If $\begin{vmatrix} 2 & -1 & 0 \\ 3 & 0 & -4 \\ -5 & 1 & 2 \end{vmatrix} = -6$, find the value of $\begin{vmatrix} 2 & -1 & 0 \\ -5 & 1 & 2 \\ 3 & 0 & -4 \end{vmatrix}$.

2. Use Sarrus' rule to evaluate $\begin{vmatrix} 3 & -5 & 0 \\ -4 & 0 & 2 \\ 0 & 1 & 7 \end{vmatrix}$.

3. Use the diagonal property of determinants to evaluate $\begin{vmatrix} 5 & 0 & 0 \\ 1 & -3 & 0 \\ 4 & 2 & 7 \end{vmatrix}$.

4. If $\begin{vmatrix} 2 & -1 & 0 \\ 3 & 0 & -4 \\ -5 & 1 & 2 \end{vmatrix} = -6$, find the value of $\begin{vmatrix} 2 & -1 & 0 \\ 3 & 0 & 8 \\ -5 & 1 & -4 \end{vmatrix}$.

5. Use Cramer's rule to solve the following system of equations.

$$\begin{aligned} x - y &= 6 \\ 2x - 3y &= 13 \end{aligned}$$

6. Evaluate $\begin{vmatrix} 1 & 0 \\ 2 & -7 \end{vmatrix}$.

7. Use Cramer's rule to solve the following system of equations.

$$\begin{aligned} x + 2y + z &= 3 \\ x - y + z &= 4 \\ 2x + 4y + 2z &= 2 \end{aligned}$$

8. Evaluate the following determinant, using expansion by cofactors. $\begin{vmatrix} 4 & -6 & 1 \\ -3 & 0 & 5 \\ 1 & 0 & 8 \end{vmatrix}$

9. Use the diagonal property of determinants to evaluate $\begin{vmatrix} 2 & -4 & 8 \\ -6 & 8 & 3 \\ 6 & 0 & -8 \end{vmatrix}$.

10. Evaluate the following determinant, using properties of determinants to simplify the computation.

$$\begin{vmatrix} 1 & -4 & 3 \\ -2 & 7 & 5 \\ 1 & -4 & 3 \end{vmatrix}$$

CHAPTER TEST

In 1–4, evaluate each determinant.

1. $\begin{vmatrix} 4 & 1 \\ 5 & -2 \end{vmatrix}$

2. $\begin{vmatrix} 3 & 6 \\ -6 & 7 \end{vmatrix}$

3. $\begin{vmatrix} 3 & 7 & -2 \\ 1 & 0 & 5 \\ 4 & -1 & 6 \end{vmatrix}$

4. $\begin{vmatrix} 2 & 8 & -3 \\ 2 & -1 & 6 \\ 3 & 0 & 5 \end{vmatrix}$

5. Evaluate the following determinant, using expansion by cofactors.

$$\begin{vmatrix} 1 & 9 & -4 \\ 3 & -2 & 7 \\ 2 & 1 & 4 \end{vmatrix}$$

6. Use the diagonal property of determinants to evaluate the following determinant.

$$\begin{vmatrix} 1 & 0 & 2 & -6 \\ 4 & 1 & -5 & 0 \\ 0 & -4 & 5 & 3 \\ 5 & 8 & 0 & 7 \end{vmatrix}$$

In 7 and 8, use Cramer's rule to solve each system of equations.

7.
$$\begin{aligned} 2x + y &= 13 \\ x - y &= 2 \end{aligned}$$

8.
$$\begin{aligned} x + 2y - z &= 2 \\ 3x - y + 2z &= 1 \\ 4x + y - 5z &= 9 \end{aligned}$$

9. The distance from Los Angeles to San Francisco is 480 miles. A plane makes the flight one way in 1 hour, 30 minutes and the return trip in 1 hour, 36 minutes. Assuming that the velocity of the wind is the same for both flights, find the speed of the plane in still air and the speed of the wind.

10. An industrial engineer has some disinfectant that is 70% pure and she also has some disinfectant that is 90% pure. How much of each disinfectant should be mixed to obtain 100 gallons of a mixture that is 75% pure?

DESIGNING A CONTAINER FOR CREAM

Cream is to be packaged in individual containers for restaurant use. An economical way to make such a container is to start with a rectangular piece of paper, glue the edges to form a cylinder, and flatten the ends of the cylinder by making a vertical seal on one end and a horizontal seal on the other, as shown in Figure 18.2.

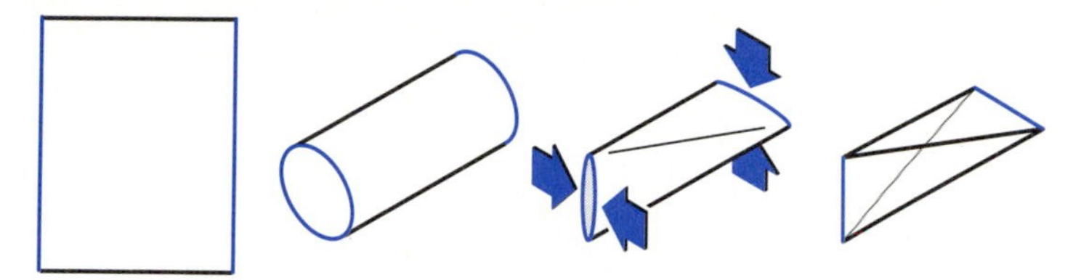

FIGURE 18.2

When the edges are creased, the solid formed has four triangular faces and is called a *tetrahedron.* It is possible to find the dimensions of a rectangle to produce a tetrahedron with a given volume.

Consider a three-dimensional coordinate system, as illustrated in Figure 18.3. If the tetrahedron is placed in the plane with its vertices at (x_1, y_1, z_1), (x_2, y_2, z_2), (x_3, y_3, z_3), and (x_4, y_4, z_4), its volume is given by the formula

$$V = \frac{1}{6}\begin{vmatrix} x_1 & y_1 & z_1 & 1 \\ x_2 & y_2 & z_2 & 1 \\ x_3 & y_3 & z_3 & 1 \\ x_4 & y_4 & z_4 & 1 \end{vmatrix}.$$

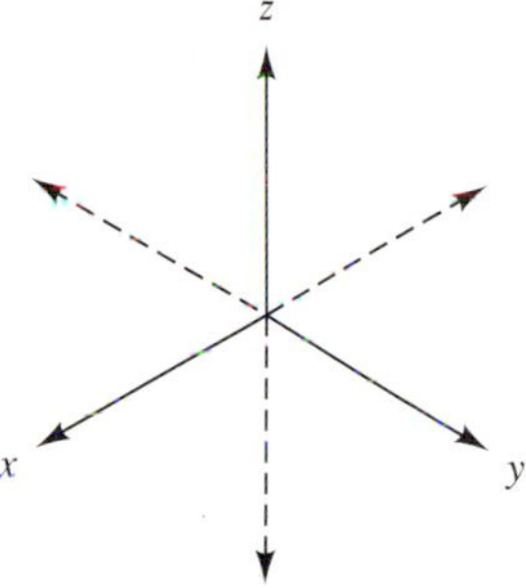

FIGURE 18.3

Consider the special case in which the tetrahedron is a *regular* tetrahedron. That is, the triangular faces are equilateral triangles. Let $2s$ represent the length of an edge of an equilateral triangle. Then the width of the rectangle needed is $s\sqrt{3}$, and the length is $4s$, as shown in Figure 18.4.

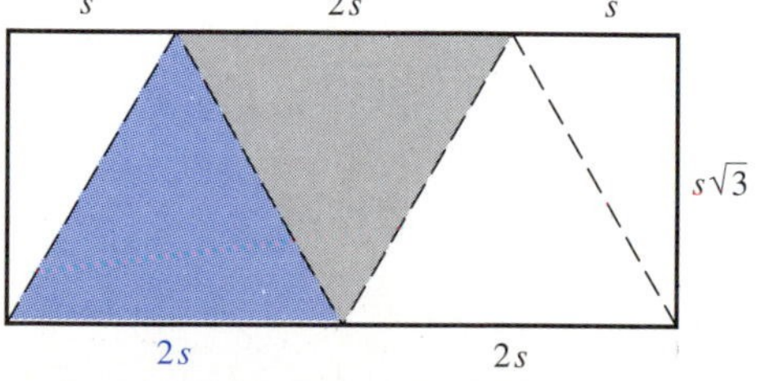

FIGURE 18.4

EXTENDED APPLICATION, continued

Place the tetrahedron so that one vertex is at the origin and one sealed edge lies along the x-axis, as shown in Figure 18.5. The coordinates of the two vertices on this "bottom edge" are $C = (0, 0, 0)$ and $D = (2s, 0, 0)$.

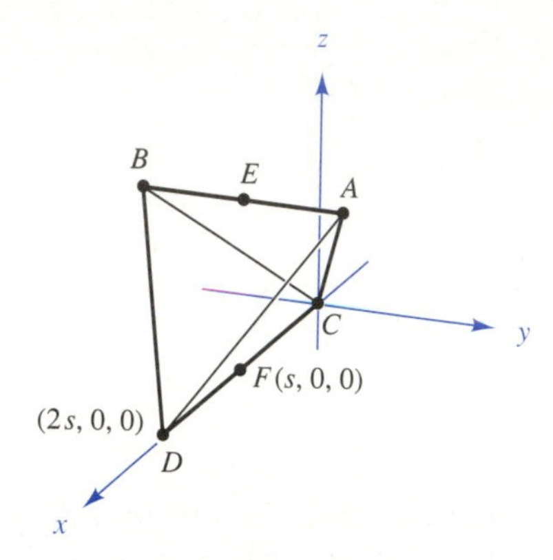

FIGURE 18.5

The other sealed edge is parallel to the y-axis and is centered over the midpoint of the bottom edge. The two vertices on this "top edge" each have an x-coordinate of s and a y-coordinate of either s or $-s$. To find the z-coordinate, let E and F represent the midpoints of the top and bottom edges, respectively. The length, h, of line segment EF can be found using the Pythagorean theorem. Consider triangle EBF. Line segment EB has length s, and the length of the hypotenuse (BF) is the height of a triangular face, so its length is $s\sqrt{3}$. Thus,

$$(s\sqrt{3})^2 - s^2 = h^2, \text{ and } h = s\sqrt{2}.$$

The upper vertices, then, are $(s, s, s\sqrt{2})$, and $(s, -s, s\sqrt{2})$. We have

$$V = \frac{1}{6}\begin{vmatrix} s & s & s\sqrt{2} & 1 \\ s & -s & s\sqrt{2} & 1 \\ 2s & 0 & 0 & 1 \\ 0 & 0 & 0 & 1 \end{vmatrix} = \frac{2}{3}s^3\sqrt{2}.$$

For a given volume, then, it is possible to find s, so that a rectangle of length $4s$ and width $s\sqrt{3}$ can be used to construct a tetrahedron with that volume.

In a similar manner, the formula can be generalized for the volume of a tetrahedron formed by a rectangle of length l and width w to obtain $V = \left(\frac{1}{96}\right) l^2\sqrt{16w^2 - l^2}$.

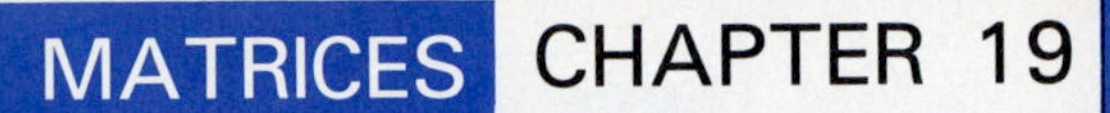

MATRICES CHAPTER 19

In Chapters 17 and 18, we solved systems of linear equations by various methods. Technical problems often involve many equations with many variables and are generally solved on a computer by matrix methods. We are now ready to examine two methods that use matrix concepts to solve systems of equations. Because of the importance of matrix operations to applications in economics, atomic physics, and statistics, these methods are presented as a means of introducing the branch of mathematics known as linear algebra.

19.1 MATRIX OPERATIONS: ADDITION, SUBTRACTION, AND MULTIPLICATION

The word *matrix* was introduced in 1850 by the mathematician J. J. Sylvester. He wanted to refer to an array of numbers as a single object, but he could not use the word *determinant,* because the array did not have a value. Matrices (the plural of matrix) eventually became more important than determinants in the study of mathematics. To distinguish a matrix from a determinant, brackets are used instead of bars.

DEFINITION An **m × n matrix** is a rectangular array, or arrangement, of numbers with m horizontal rows and n vertical columns.

For example, $\begin{bmatrix} 1 & -2 & 3 \\ -4 & 5 & 6 \end{bmatrix}$ is a 2×3 matrix and

$\begin{bmatrix} 1 & 0 \\ 2 & -1 \\ -3 & 6 \end{bmatrix}$ is a 3×2 matrix.

Matrices are sometimes used to organize and display information.

EXAMPLE 1 A lab tested three brands of apple juice for traces of daminozide, measured in parts per million (ppm). One year later they repeated the test. The first year, Brand A showed 0.04 ppm, Brand B showed 0.05 ppm, and Brand C showed 0.07 ppm. The second year, Brand A showed 0.00 ppm, Brand B showed 0.02 ppm, and Brand C showed 0.11 ppm, Organize this information in matrix form.

Solution Let each column represent the results from one year's test. Let each row represent the results from one brand.

	TEST I	TEST II
BRAND A	0.04	0.00
BRAND B	0.05	0.02
BRAND C	0.07	0.11

■

The entries of a matrix are often written using double subscripts, just as they are for determinants. The notation a_{ij} means the entry in row i and column j. In the matrix $\begin{bmatrix} 1 & -2 & 3 \\ -4 & 5 & 6 \end{bmatrix}$, a_{12} is -2 and a_{23} is 6.

If we are to add two matrices, the two matrices must have the same dimensions. That is, the number of rows is the same in each and the number of columns is the same in each. We say that *corresponding entries* are added. **Corresponding entries** have the same row number and the same column number. Two matrices are equal if corresponding entries are equal.

corresponding entries

To add two matrices A and B with the same dimensions:

Add each entry a_{ij} of matrix A to the corresponding entry b_{ij} of matrix B to get the entry of c_{ij} of the sum.

Matrices can be subtracted by subtracting corresponding entries.

EXAMPLE 2 Add, if possible $\begin{bmatrix} 1 & 2 \\ 0 & -3 \end{bmatrix} + \begin{bmatrix} -4 & 5 \\ -1 & 0 \end{bmatrix}$.

Solution $$\begin{bmatrix} 1 & 2 \\ 0 & -3 \end{bmatrix} + \begin{bmatrix} -4 & 5 \\ -1 & 0 \end{bmatrix} = \begin{bmatrix} 1+(-4) & 2+5 \\ 0+(-1) & -3+0 \end{bmatrix} = \begin{bmatrix} -3 & 7 \\ -1 & -3 \end{bmatrix}$$ ■

EXAMPLE 3 Subtract, if possible $\begin{bmatrix} 4 & -2 & 3 \\ -1 & 0 & 1 \end{bmatrix} - \begin{bmatrix} 2 & 0 & 5 \\ 3 & -1 & 4 \end{bmatrix}$.

Solution $$\begin{bmatrix} 4 & -2 & 3 \\ -1 & 0 & 1 \end{bmatrix} - \begin{bmatrix} 2 & 0 & 5 \\ 3 & -1 & 4 \end{bmatrix} = \begin{bmatrix} 2 & -2 & -2 \\ -4 & 1 & -3 \end{bmatrix}$$ ■

EXAMPLE 4 Some computer languages, such as BASIC, allow the use of matrices. It is necessary to specify the dimensions of matrices used in computer programs. DIM M(3, 4) creates a 3×4 matrix named M. A statement such as MAT $A = B + C$ directs the computer to add matrices named B and C and to store the result in a matrix named A. Likewise, MAT $A = B - C$ directs the computer to subtract matrix C from B and to store the result in A. Which of the following matrix operations are valid based on the given dimension statements?

DIM $A(3, 4)$ DIM $B(4, 3)$ DIM $C(3, 4)$ DIM $D(4, 3)$

(a) MAT $E = A + B$

(b) MAT $F = A + C$

(c) MAT $G = B - D$

(d) MAT $H = C - D$

Solution

(a) Invalid; Matrix A is a 3×4 matrix, but B is a 4×3 matrix.

(b) Valid; both matrices are 3×4 matrices.

(c) Valid; both matrices are 4×3 matrices.

(d) Invalid; matrix C is a 3×4 matrix, but D is a 4×3 matrix. ■

In order to multiply two matrices, the first matrix must have exactly as many columns as the second has rows. The product of an $m \times n$ matrix and an $n \times p$ matrix is an $m \times p$ matrix. That is, when

an $m \times n$ matrix and an $n \times p$ matrix are multiplied,

these numbers must agree and

these numbers give the dimensions of the product.

Thus, we can multiply a 2×3 matrix by a 3×2 matrix, but we cannot multiply a 2×3 matrix by another 2×3 matrix. Example 5 illustrates the process of multiplication.

EXAMPLE 5 Find the product of $\begin{bmatrix} 1 & 0 \\ 3 & -2 \\ 0 & 4 \end{bmatrix}$ and $\begin{bmatrix} 1 & 0 & 2 \\ 0 & -3 & 4 \end{bmatrix}$.

Solution Begin by partitioning the first matrix into rows and the second matrix into columns.

$$\begin{bmatrix} 1 & 0 \\ \hline 3 & -2 \\ \hline 0 & 4 \end{bmatrix} \left[\begin{array}{c|c|c} 1 & 0 & 2 \\ 0 & -3 & 4 \end{array}\right]$$

To find the entry that goes in row **1**, column **1** of the product, multiply successive entries in *row 1* of the *first* matrix and *column 1* of the *second* matrix, then add the results.

Multiply first entries.

$$\begin{bmatrix} 1 & 0 \\ 3 & -2 \\ 0 & 4 \end{bmatrix} \begin{bmatrix} 1 & 0 & 2 \\ 0 & -3 & 4 \end{bmatrix} = \begin{bmatrix} 1(1) + 0(0) & & \\ & & \\ & & \end{bmatrix}$$

Multiply second entries.

To find the entry that goes in row **1**, column **2** of the product, multiply successive entries in *row 1* of the *first* matrix and *column 2* of the *second* matrix, then add the results.

$$\begin{bmatrix} 1 & 0 \\ 3 & -2 \\ 0 & 4 \end{bmatrix} \begin{bmatrix} 1 & 0 & 2 \\ 0 & -3 & 4 \end{bmatrix} = \begin{bmatrix} 1(1) + 0(0) & 1(0) + 0(-3) & \\ & & \\ & & \end{bmatrix}$$

To find the entry that goes in row **1**, column **3** of the product, multiply successive entries in *row 1* of the *first* matrix and *column 3* of the *second* matrix, then add the results.

$$\begin{bmatrix} 1 & 0 \\ 3 & -2 \\ 0 & 4 \end{bmatrix} \begin{bmatrix} 1 & 0 & 2 \\ 0 & -3 & 4 \end{bmatrix} = \begin{bmatrix} 1(1) + 0(0) & 1(0) + 0(-3) & 1(2) + 0(4) \\ & & \\ & & \end{bmatrix}$$

In general, to find the entry that goes in row i, column j of the product, multiply successive entries in ***row i*** of the *first* matrix and ***column j*** of the *second* matrix, then add the results. The final result in this example is the matrix that follows.

$$\begin{bmatrix} 1(1) + 0(0) & 1(0) + 0(-3) & 1(2) + 0(4) \\ 3(1) + (-2)(0) & 3(0) + (-2)(-3) & 3(2) + (-2)(4) \\ 0(1) + 4(0) & 0(0) + 4(-3) & 0(2) + 4(4) \end{bmatrix}$$

or

$$\begin{bmatrix} 1 & 0 & 2 \\ 3 & 6 & -2 \\ 0 & -12 & 16 \end{bmatrix}$$ ■

To multiply two matrices A and B:

1. Be sure that the number of columns in the first matrix is equal to the number of rows in the second matrix.
2. Partition the first matrix into rows and the second matrix into columns.
3. Multiply successive entries in *row* ***i*** of the *first* matrix and *column* ***j*** of the *second* matrix, then add the results to find the entry that goes in row i, column j of the product.
4. Repeat step 3 for each row i and column j of the product matrix until it has as many rows as the first matrix and as many columns as the second matrix.

EXAMPLE 6 Multiply $\begin{bmatrix} 1 & -2 \\ 0 & 3 \end{bmatrix}\begin{bmatrix} 2 & 3 \\ 0 & 1 \end{bmatrix}$.

Solution

1. The first matrix is a 2×2 matrix; the second is a 2×2 matrix. The multiplication can be done, and the product will be a 2×2 matrix.
2. Partition the first matrix into rows and the second into columns.

$$\left[\begin{array}{cc} 1 & -2 \\ \hline 0 & 3 \end{array}\right]\left[\begin{array}{c|c} 2 & 3 \\ 0 & 1 \end{array}\right]$$

3–4. $$\left[\begin{array}{cc} 1 & -2 \\ \hline 0 & 3 \end{array}\right]\left[\begin{array}{c|c} 2 & 3 \\ 0 & 1 \end{array}\right] = \begin{bmatrix} 1(2) + (-2)(0) & 1(3) + (-2)(1) \\ 0(2) + 3(0) & 0(3) + 3(1) \end{bmatrix}$$

The product is $\begin{bmatrix} 2 & 1 \\ 0 & 3 \end{bmatrix}$. ■

EXAMPLE 7 When the terminal side of an angle in standard position is rotated through an angle θ as shown in Figure 19.1, the point (x, y) is rotated to a position (x', y'), given by

$$[x', y'] = [x \quad y]\begin{bmatrix} \cos\theta & \sin\theta \\ -\sin\theta & \cos\theta \end{bmatrix}.$$

If $(x, y) = (3, 4)$, and $\theta = 30°$, find (x', y').

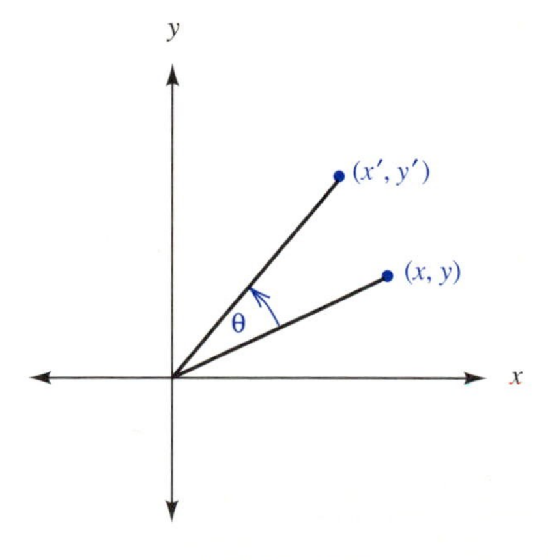

FIGURE 19.1

Solution The first matrix is a 1×2 matrix; the second is a 2×2 matrix. The multiplication can be done, and the product will be a 1×2 matrix.

$$[x', y'] = [3 \quad 4]\begin{bmatrix} \cos 30° & \sin 30° \\ -\sin 30° & \cos 30° \end{bmatrix}$$
$$= [3 \cos 30° - 4 \sin 30° \quad 3 \sin 30° + 4 \cos 30°]$$
$$= [0.598 \quad 4.96]$$

The new position is (0.598, 4.96). ■

It is interesting to see that matrix multiplication is not commutative. For example, $\begin{bmatrix} 1 & 0 \\ 3 & -2 \\ 0 & 4 \end{bmatrix}\begin{bmatrix} 1 & 0 & 2 \\ 0 & -3 & 4 \end{bmatrix} = \begin{bmatrix} 1 & 0 & 2 \\ 3 & 6 & -2 \\ 0 & -12 & 16 \end{bmatrix}$, but when we reverse the order, $\begin{bmatrix} 1 & 0 & 2 \\ 0 & -3 & 4 \end{bmatrix}\begin{bmatrix} 1 & 0 \\ 3 & -2 \\ 0 & 4 \end{bmatrix} = \begin{bmatrix} 1 & 8 \\ -9 & 22 \end{bmatrix}$.

Even when the two matrices have the same dimensions, order affects the product. For instance,

$$\begin{bmatrix} 1 & -2 \\ 0 & 3 \end{bmatrix}\begin{bmatrix} 2 & 3 \\ 0 & 1 \end{bmatrix} = \begin{bmatrix} 2 & 1 \\ 0 & 3 \end{bmatrix},$$

but

$$\begin{bmatrix} 2 & 3 \\ 0 & 1 \end{bmatrix}\begin{bmatrix} 1 & -2 \\ 0 & 3 \end{bmatrix} = \begin{bmatrix} 2 & 5 \\ 0 & 3 \end{bmatrix}.$$

A square matrix is one that has the same number of rows as columns. An $n \times n$ matrix with ones along the main diagonal and zeros for all of the other entries is called the **identity matrix** for $n \times n$ matrices. When the product of two square matrices is the identity, each matrix is the **inverse** of the other.

identity matrix

inverse of a matrix

EXAMPLE 8 Multiply $\begin{bmatrix} 7 & -1 \\ -6 & 1 \end{bmatrix}\begin{bmatrix} 1 & 1 \\ 6 & 7 \end{bmatrix}$.

Solution $\begin{bmatrix} 7 & -1 \\ -6 & 1 \end{bmatrix}\begin{bmatrix} 1 & 1 \\ 6 & 7 \end{bmatrix} = \begin{bmatrix} 1 & 0 \\ 0 & 1 \end{bmatrix}$

The matrices $\begin{bmatrix} 7 & -1 \\ -6 & 1 \end{bmatrix}$ and $\begin{bmatrix} 1 & 1 \\ 6 & 7 \end{bmatrix}$ are inverses of each other. ■

EXAMPLE 9 Multiply $\begin{bmatrix} 1 & 0 & 3 \\ 1 & 1 & 3 \\ 0 & 2 & 2 \end{bmatrix}\begin{bmatrix} -2 & 3 & -\frac{3}{2} \\ -1 & 1 & 0 \\ 1 & -1 & \frac{1}{2} \end{bmatrix}$.

Solution $\begin{bmatrix} 1 & 0 & 3 \\ 1 & 1 & 3 \\ 0 & 2 & 2 \end{bmatrix}\begin{bmatrix} -2 & 3 & -\frac{3}{2} \\ -1 & 1 & 0 \\ 1 & -1 & \frac{1}{2} \end{bmatrix} = \begin{bmatrix} 1 & 0 & 0 \\ 0 & 1 & 0 \\ 0 & 0 & 1 \end{bmatrix}$

The two matrices that were multiplied are inverses of each other. ■

NOTE ⇨⇨ ***When the two factors of a matrix multiplication are inverses, or when one of the factors is the identity matrix, the multiplication is commutative, but remember that matrix multiplication, in general, is not.***

EXERCISES 19.1

In 1–11, perform the indicated matrix additions and subtractions, if possible.

1. $\begin{bmatrix} 2 & -1 \\ 3 & 0 \end{bmatrix} + \begin{bmatrix} 0 & 3 \\ 7 & -5 \end{bmatrix}$

2. $\begin{bmatrix} 12 & 5 \\ -3 & 4 \end{bmatrix} - \begin{bmatrix} 6 & -2 \\ 1 & 7 \end{bmatrix}$

3. $\begin{bmatrix} 10 & 6 \\ -4 & 6 \end{bmatrix} - \begin{bmatrix} 7 & -4 \\ 2 & 5 \end{bmatrix}$

4. $\begin{bmatrix} 8 & 7 \\ -5 & 8 \end{bmatrix} + \begin{bmatrix} 8 & -6 \\ 3 & 3 \end{bmatrix}$

5. $\begin{bmatrix} 3 & 0 & -5 \\ 2 & -4 & 7 \\ 6 & 3 & -8 \end{bmatrix} + \begin{bmatrix} 7 & -8 & 5 \\ 4 & 2 & -9 \\ -6 & 9 & 1 \end{bmatrix}$

6. $\begin{bmatrix} -2 & 3 & 5 \\ -6 & 8 & 0 \\ 3 & -4 & 4 \end{bmatrix} - \begin{bmatrix} 7 & 2 & -4 \\ 7 & -7 & 8 \\ -5 & 5 & 9 \end{bmatrix}$

7. $\begin{bmatrix} 3 & -8 \\ -1 & 7 \\ 2 & -3 \end{bmatrix} - \begin{bmatrix} 6 & 9 \\ 1 & -4 \\ 0 & 6 \end{bmatrix}$

8. $\begin{bmatrix} 2 & -9 \\ 0 & 6 \\ 3 & -4 \end{bmatrix} + \begin{bmatrix} 7 & 8 \\ 0 & -3 \\ 1 & 5 \end{bmatrix}$

9. $\begin{bmatrix} 9 & -7 & 5 \\ 0 & 3 & -1 \end{bmatrix} - \begin{bmatrix} -2 & 0 & 1 \\ 1 & -6 & 3 \end{bmatrix}$

10. $\begin{bmatrix} 1 & 3 & -2 \\ 0 & -5 & 6 \end{bmatrix} + \begin{bmatrix} 1 & 3 \\ -2 & 0 \\ -5 & 6 \end{bmatrix}$

11. $\begin{bmatrix} 1 & 4 \\ 9 & 0 \end{bmatrix} + \begin{bmatrix} 2 & 3 \\ 5 & 6 \\ 8 & 7 \end{bmatrix}$

In 12–29, perform the matrix multiplications, if possible.

12. $\begin{bmatrix} 9 & -1 \\ 1 & -9 \end{bmatrix}\begin{bmatrix} 4 & 6 \\ -2 & 1 \end{bmatrix}$

13. $\begin{bmatrix} 4 & 6 \\ -2 & 1 \end{bmatrix}\begin{bmatrix} 9 & -7 \\ 1 & -9 \end{bmatrix}$

14. $\begin{bmatrix} 3 & 5 \\ 5 & 8 \end{bmatrix}\begin{bmatrix} -8 & 5 \\ 5 & -3 \end{bmatrix}$

15. $\begin{bmatrix} 2 & -1 & 5 \\ -3 & 0 & 4 \\ 1 & 1 & 7 \end{bmatrix}\begin{bmatrix} -4 \\ 3 \\ 6 \end{bmatrix}$

16. $\begin{bmatrix} 1 & 0 & 2 \\ -1 & 1 & 3 \\ 3 & 2 & 4 \end{bmatrix}\begin{bmatrix} 2 \\ 1 \\ 5 \end{bmatrix}$

17. $\begin{bmatrix} 5 & 9 \\ 4 & 7 \end{bmatrix}\begin{bmatrix} -7 & 9 \\ 4 & -5 \end{bmatrix}$

18. $\begin{bmatrix} 1 & 0 \\ 2 & -3 \end{bmatrix}\begin{bmatrix} 1 & 2 \\ 0 & -3 \end{bmatrix}$

19. $\begin{bmatrix} 2 & 3 \\ 4 & 0 \\ 1 & -1 \end{bmatrix}\begin{bmatrix} 2 & 4 & 1 \\ 3 & 0 & -1 \end{bmatrix}$

20. $\begin{bmatrix} 2 & 4 & 1 \\ 3 & 0 & -1 \end{bmatrix}\begin{bmatrix} 2 & 3 \\ 4 & 0 \\ 1 & -1 \end{bmatrix}$

21. $\begin{bmatrix} 1 & 5 & 0 \\ 4 & -1 & 0 \end{bmatrix}\begin{bmatrix} 3 & 2 \\ 3 & 1 \\ 0 & -1 \end{bmatrix}$

22. $\begin{bmatrix} 1 & 0 & 2 \\ 3 & 1 & -1 \end{bmatrix}\begin{bmatrix} 4 \\ -7 \\ 5 \end{bmatrix}$

23. $\begin{bmatrix} 0 & 1 & 1 \\ 4 & 0 & 0 \end{bmatrix}\begin{bmatrix} 5 \\ -8 \\ 6 \end{bmatrix}$

24. $\begin{bmatrix} 1 & 0 & 2 \\ 3 & 2 & -1 \\ -2 & 0 & 3 \end{bmatrix}\begin{bmatrix} 0 & 1 & -1 \\ -1 & 0 & 1 \\ 1 & -1 & 0 \end{bmatrix}$

25. $\begin{bmatrix} 1 & 0 & 1 \\ 0 & -1 & 1 \\ -1 & 1 & 0 \end{bmatrix}\begin{bmatrix} 1 & 0 & -1 \\ 0 & -1 & 1 \\ 1 & 1 & 0 \end{bmatrix}$

26. $\begin{bmatrix} 0 & 1 & -1 \\ -1 & 0 & 1 \\ 1 & -1 & 0 \end{bmatrix}\begin{bmatrix} 1 & 0 & 2 \\ 3 & 2 & -1 \\ -2 & 0 & 3 \end{bmatrix}$

27. $\begin{bmatrix} 1 & 0 & 0 \\ 0 & 1 & 0 \\ 0 & 0 & 1 \end{bmatrix}\begin{bmatrix} 1 & 2 & 3 \\ 4 & 5 & 6 \\ 7 & 8 & 9 \end{bmatrix}$

28. $\begin{bmatrix} 2 & 2 & 0 \\ 3 & 0 & 1 \\ 3 & 1 & 1 \end{bmatrix}\begin{bmatrix} \frac{1}{2} & 1 & -1 \\ 0 & -1 & 1 \\ -\frac{3}{2} & -2 & 3 \end{bmatrix}$

29. $\begin{bmatrix} \frac{1}{2} & 1 & -1 \\ 0 & -1 & 1 \\ -\frac{3}{2} & -2 & 3 \end{bmatrix}\begin{bmatrix} 2 & 2 & 0 \\ 3 & 0 & 1 \\ 3 & 1 & 1 \end{bmatrix}$

---

30. A 64K memory chip is a square array of data storage cells with 256 rows and 256 columns. Each cell can store 1 bit.
(a) How many cells are there in a 64K chip?
(b) K is short for "kilo." For computer applications, kilo means 1024 (or 2^{10}) rather than 1000. Use powers of 2 to verify that $256 \times 256 = 64(1024)$.
(c) How many rows and columns would you expect to find on a 256K chip? 3 5 9

31. Samples are analyzed in a laboratory where each sample is tested for lead, zinc, and iron. Three samples are tested and the levels are 0.01, 0.20, and 0.03, respectively for the first sample; 0.02, 0.10, and 0.02 for the second sample; and 0.03, 0.30, and 0.01 for the third sample. Write this information in matrix form.
1 2 4 6 8 5 7 9

32. Suppose a company operates two plants, each of which produces both gold and silver. Each plant incurs costs in purchasing the metals, processing, and transportation. For each plant, form a matrix in which row 1 represents costs involved for gold and row 2 represents costs involved for silver. At one plant, the costs for purchasing, processing, and transporting gold are \$150, \$150, and \$8. For silver, these costs are \$20, \$20, and \$6. At the other plant, the costs for purchasing, processing, and transporting gold are \$150, \$125, and \$12. For silver, these costs are \$18, \$15, and \$8. Show, in matrix form, the costs for the two plants combined. 1 2 6 7 8 3 5 9

33. Which of the following matrix operations are valid based on the given dimension statements? DIM $A(2, 5)$ DIM $B(5, 2)$ DIM $C(2, 5)$ DIM $D(5, 2)$
(a) MAT $E = A + B$
(b) MAT $F = A + C$
(c) MAT $G = B - D$
(d) MAT $H = C - D$ 3 5 9

34. In BASIC, a statement such as MAT $A = B * C$ directs the computer to multiply matrices named B and C and to store the result in a matrix named A. Which of the following matrix operations are valid based on the given dimension statements? DIM $A(3, 4)$ DIM $B(4, 3)$ DIM $C(3, 4)$ DIM $D(4, 3)$
(a) MAT $E = A * B$
(b) MAT $F = A * C$
(c) MAT $G = B * D$
(d) MAT $H = C * D$ 3 5 9

Problems 35 and 36 use the following information. In a cylindrical coordinate system, the location of a point P in space is specified by the three quantities (r, θ, z) shown in the illustration.

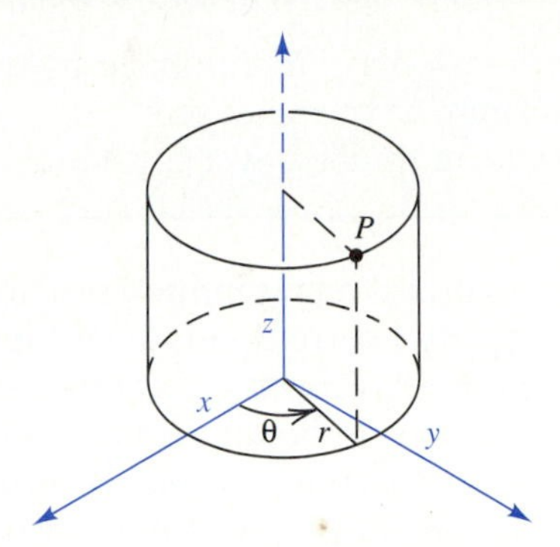

Cylindrical coordinates are useful for such things as calculating moments of inertia of a cylinder or in problems of heat conduction in cylindrical bodies. To convert P, which is specified by (x, y, z) in rectangular coordinates, to cylindrical coordinates, find

$$[x \quad y \quad z]\begin{bmatrix} \cos\theta & \sin\theta & 0 \\ -\sin\theta & \cos\theta & 0 \\ 0 & 0 & 1 \end{bmatrix}$$

35. If $P = (3, 4, 8)$ in a rectangular coordinate system, find the coordinates of P in a cylindrical coordinate system. (Hint: Notice that $\tan\theta = y/x$.)
1 2 4 7 8 3 5

36. If $P = (5, 12, -1)$ in a rectangular coordinate system, find the coordinates of P in a cylindrical coordinate system. (Hint: Notice that $\tan\theta = y/x$.)
1 2 4 7 8 3 5

19.2 SOLVING SYSTEMS OF LINEAR EQUATIONS

If a system of equations is written so that the variables occur in the same order in each equation, then it is not necessary to write down the variables. The coefficients may be specified in matrix form. For instance, a system of three equations in three variables written in the form

$$\begin{aligned} a_1x + b_1y + c_1z &= d_1 \\ a_2x + b_2y + c_2z &= d_2 \\ a_3x + b_3y + c_3z &= d_3 \end{aligned}$$

can be represented by the 3×4 matrix $\begin{bmatrix} a_1 & b_1 & c_1 & d_1 \\ a_2 & b_2 & c_2 & d_2 \\ a_3 & b_3 & c_3 & d_3 \end{bmatrix}$. The concept is easily generalized to any number of equations. The first column simply contains the coefficients of the first variable, the second column contains the coefficients of the second variable, and so on. The last column contains the constants from the right-hand side of the equation.

Matrices, like determinants, have properties that allow us to find equivalent matrices. The manipulations allowed are called *elementary row operations.*

ELEMENTARY ROW OPERATIONS FOR MATRICES

1. Each entry of a row may be multiplied by the same nonzero constant.
2. Any two rows may be interchanged.
3. Any row may be multiplied by a constant and added to another row.

If you think of a matrix as representing a system of equations, these properties are quite reasonable. Property 1 corresponds to multiplying both sides of an equation by the same nonzero number. Property 2 corresponds to changing the order of the equations, and property 3 corresponds to the technique used in the method of Section 17.3. ***These properties are not to be confused with the properties of determinants.*** Property 3 is the only one of the three properties that matrices and determinants share.

NOTE

We can solve a system of linear equations by taking a matrix that contains the coefficients of the system of linear equations and using elementary row operations to rewrite it so that the first n columns look like the columns of an identity matrix. The $(n+1)$st column represents the values of the variables. For example, a 3×3 matrix in the form $\begin{bmatrix} 1 & 0 & 0 & e_1 \\ 0 & 1 & 0 & e_2 \\ 0 & 0 & 1 & e_3 \end{bmatrix}$ represents the equations $\begin{matrix} x = e_1 \\ y = e_2 \\ z = e_3 \end{matrix}$. That is, the values e_1, e_2, and e_3 are the values of the variables x, y, and z, respectively. The method of solving a system of equations by transforming a matrix of the form

$$\begin{bmatrix} a_1 & b_1 & c_1 & d_1 \\ a_2 & b_2 & c_2 & d_2 \\ a_3 & b_3 & c_3 & d_3 \end{bmatrix} \text{to the form} \begin{bmatrix} 1 & 0 & 0 & e_1 \\ 0 & 1 & 0 & e_2 \\ 0 & 0 & 1 & e_3 \end{bmatrix}$$

is called the *Gauss-Jordan method.*

Gauss-Jordan method

It is explained in Example 1. ***As we transform each column to look like a column of the identity matrix, it is important that we first change the entry that is to become a 1.*** The row that contains the entry 1 is called the **pivot row** for the transformation of the other rows of the matrix.

NOTE

pivot row

EXAMPLE 1 Solve the following system of linear equations.

$$\begin{aligned} x - 3y + 2z &= 5 \\ x + 2y + 3z &= 7 \\ 2x - y + z &= 4 \end{aligned}$$

Solution The system is represented in matrix form as $\begin{bmatrix} 1 & -3 & 2 & 5 \\ 1 & 2 & 3 & 7 \\ 2 & -1 & 1 & 4 \end{bmatrix}$.

We begin by changing the first column to $\begin{matrix} 1 \\ 0 \\ 0 \end{matrix}$. Since the first row has 1 as the first entry, the first row of the new matrix is the same as the first row of the original matrix. To get 0 as the first entry of the second row, multiply the first row, which is the pivot row, by -1 and add the result to the second row. The arithmetic is shown to the right of the new matrix, which follows. To get 0 as the first entry of the third row, multiply the first row by -2 and add the result to the third row. The new matrix, then, is written as follows.

$$\begin{bmatrix} 1 & -3 & 2 & 5 \\ 0 & 5 & 1 & 2 \\ 0 & 5 & -3 & -6 \end{bmatrix}$$

pivot row (row 1)

$-1 \times$ pivot row + row 2

-1	3	-2	-5
1	2	3	7
0	5	1	2

$-2 \times$ pivot row + row 3

-2	6	-4	-10
2	-1	1	4
0	5	-3	-6

Column 1 now looks like the first column of the identity matrix. Next we must change the second column of this matrix to $\begin{matrix}0\\1\\0\end{matrix}$. The entry in the second row must be replaced with 1. We therefore divide the second row by 5 to get the pivot row: 0 1 1/5 2/5. To get a 0 as the second entry of the first row, we multiply the pivot row by 3 and add the result to the first row. To get a 0 as the second entry of the third row, we multiply the pivot row by -5 and add the result to the third row. The new matrix is written as follows.

$$\begin{bmatrix} 1 & 0 & \frac{13}{5} & \frac{31}{5} \\ 0 & 1 & \frac{1}{5} & \frac{2}{5} \\ 0 & 0 & -4 & -8 \end{bmatrix}$$

$3 \times$ pivot row + row 1

0	3	3/5	6/5
1	-3	2	5
1	0	13/5	31/5

pivot row (divide row 2 by 5)

$-5 \times$ pivot row + row 3

0	-5	-1	-2
0	5	-3	-6
0	0	-4	-8

Column 2 now looks like the second column of the identity matrix. Next we must change the third column of this matrix to $\begin{matrix}0\\0\\1\end{matrix}$. The entry in the third row must be replaced with 1. We therefore divide the third row by -4 to get the pivot row: 0 0 1 2. To get a 0 as the third entry of the first row, we multiply the pivot row by $-13/5$ and add the result to the first row. To get a 0 as the third entry of the second row, we multiply the pivot row by $-1/5$ and add the result to the second row. The new matrix is written as follows.

$$\begin{bmatrix} 1 & 0 & 0 & 1 \\ 0 & 1 & 0 & 0 \\ 0 & 0 & 1 & 2 \end{bmatrix}$$

(−13/5) × pivot row + row 1

0	0	−13/5	−26/5
1	0	13/5	31/5
1	0	0	1

(−1/5) × pivot row + row 2

0	0	−1/5	−2/5
0	1	1/5	2/5
0	1	0	0

pivot row (divide row 3 by −4)

Since this matrix represents the equations $x = 1$, $y = 0$, and $z = 2$, we have the solution to the system of equations.

Check: $1 - 3(0) + 2(2) = 1 - 0 + 4 = 5$

$1 + 2(0) + 3(2) = 1 + 0 + 6 = 7$

$2(1) - 0 + 2 = 2 - 0 + 2 = 4$ ■

The steps used in Example 1 are summarized as follows.

To solve a system of simultaneous linear equations in *n* variables using the Gauss-Jordan method:

1. If the variables do not appear in the same order for each equation, write an equivalent system in which they do. If a variable is missing from one of the equations, use zero for the coefficient.
2. Form a matrix that contains the coefficients of the first variable of each equation in the first column, the coefficients of the second variable in the second column, and so on.
3. Repeat steps (a) and (b) for each row k starting with row 1, until the first n columns look like the n columns of an $n \times n$ identity matrix.
   - **(a)** For row k, divide each entry by the kth entry of that row. (If this step requires division by zero, first interchange the row with one of those below it.) Use the resulting row as row k of the new matrix and as the pivot row.
   - **(b)** For each row i above or below row k, multiply the pivot row by the product of -1 and the kth entry in row i and add to row i. Use the resulting row as row i of the new matrix.

EXAMPLE 2 When a seesaw is balanced by two objects, the product of the weight of one object and its distance from the fulcrum must equal the product of the weight of the other object and its distance from the fulcrum. When a 50 lb boy and his 25 lb sister sit together on one side, the seesaw is balanced by an 80 lb child on the other side, as shown in Figure 19.2. If the 50 lb boy sits without his sister, the 80 lb child must move 1 foot 3 inches closer to the fulcrum to balance it as in Figure 19.3. Find the original distances of the children from the fulcrum.

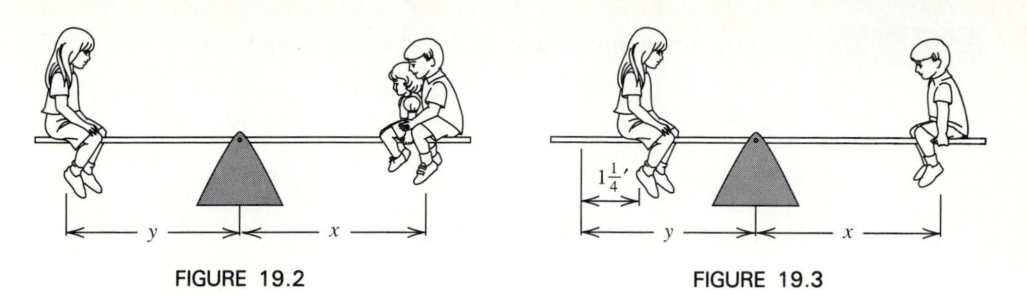

FIGURE 19.2

FIGURE 19.3

Solution Let x = distance of 50 lb and 25 lb children
let y = distance of 80 lb child
$75x = 80y$

One foot, 3 inches is 1 3/12 ft or 1 1/4 ft. The 80 lb child's new distance is $y - 5/4$, but the 50 lb child's distance is still x.

$$50x = 80(y - 5/4) \qquad \text{or} \qquad 50x = 80y - 100$$

The system of equations, then, is

$$\begin{aligned} 75x &= 80y \\ 50x &= 80y - 100. \end{aligned}$$

To apply the Gauss-Jordan method, we follow the steps listed in the summary box.

1. Rewrite the equations in the correct form. $\begin{aligned} 75x - 80y &= 0 \\ 50x - 80y &= -100 \end{aligned}$

2. Form a matrix containing the coefficients. $\begin{bmatrix} 75 & -80 & 0 \\ 50 & -80 & -100 \end{bmatrix}$

3. Apply elementary row operations in the order listed by the number to the right of each row.

$$\begin{bmatrix} 1 & -\frac{16}{15} & 0 \\ 0 & -\frac{80}{3} & -100 \end{bmatrix}$$

pivot row (divide row 1 by 75) **1**

$-50 \times$ pivot row + row 2 **2**

$$\begin{array}{ccc} -50 & 160/3 & 0 \\ 50 & -80 & -100 \\ \hline 0 & -80/3 & -100 \end{array}$$

$$\begin{bmatrix} 1 & 0 & 4 \\ 0 & 1 & \frac{15}{4} \end{bmatrix}$$

(16/15) × pivot row + row 1 **4**

$$\begin{array}{ccc} 0 & 16/15 & 4 \\ 1 & -16/15 & 0 \\ \hline 1 & 0 & 4 \end{array}$$

pivot row (divide row 2 by $-80/3$) **3**

The solution to the system is $x = 4$ ft and $y = 15/4$ or 3 3/4 ft.

Check: $75(4) = 80(15/4)$ or $300 = 300$

$50(4) = 80(15/4 - 5/4)$ or $200 = 80(5/2)$ ■

EXAMPLE 3 Solve the following system by the Gauss-Jordan method.

$$\begin{aligned} 2x + y - 3z &= 2 \\ 3x - y + 4z &= 9 \\ 4x + 2y - 3z &= 7 \end{aligned}$$

Solution The system is represented by the following matrix.

$$\begin{bmatrix} 2 & 1 & -3 & 2 \\ 3 & -1 & 4 & 9 \\ 4 & 2 & -3 & 7 \end{bmatrix}$$

Apply elementary row operations in the order indicated by the numbers to the right of each row in the following matrix.

$$\begin{bmatrix} 1 & \frac{1}{2} & -\frac{3}{2} & 1 \\ 0 & -\frac{5}{2} & \frac{17}{2} & 6 \\ 0 & 0 & 3 & 3 \end{bmatrix}$$

pivot row (divide row 1 by 2) **1**

$-3 \times$ pivot row + row 2

-3	$-3/2$	$9/2$	-3
3	-1	4	9
0	$-5/2$	$17/2$	6

2

$-4 \times$ pivot row + row 3 **3**

-4	-2	6	-4
4	2	-3	7
0	0	3	3

$$\begin{bmatrix} 1 & 0 & \frac{1}{5} & \frac{11}{5} \\ 0 & 1 & -\frac{17}{5} & -\frac{12}{5} \\ 0 & 0 & 3 & 3 \end{bmatrix}$$

$(-1/2) \times$ pivot row + row 1 **5**

0	$-1/2$	$17/10$	$6/5$
1	$1/2$	$-3/2$	1
1	0	$1/5$	$11/5$

pivot row (divide row 2 by $-5/2$) **4**

row 3 (no change) **6**

$$\begin{bmatrix} 1 & 0 & 0 & 2 \\ 0 & 1 & 0 & 1 \\ 0 & 0 & 1 & 1 \end{bmatrix}$$

$(-1/5) \times$ pivot row + row 1 **8**

0	0	$-1/5$	$-1/5$
1	0	$1/5$	$11/5$
1	0	0	2

$17/5 \times$ pivot row + row 2 **9**

0	0	$17/5$	$17/5$
0	1	$-17/5$	$-12/5$
0	1	0	1

pivot row (divide row 3 by 3) **7**

Thus, the solution is $x = 2$, $y = 1$, and $z = 1$. ■

EXERCISES 19.2

In 1–22, solve each equation using the Gauss-Jordan method.

1. $\begin{aligned} x + y &= 8 \\ 2x - y &= 1 \end{aligned}$

2. $\begin{aligned} 2x + y &= 8 \\ x + 2y &= 10 \end{aligned}$

3. $\begin{aligned} 3x - 2y &= 1 \\ 4x - y &= 1 \end{aligned}$

4. $\begin{aligned} 2x + 3y &= 1 \\ 3x - 2y &= -5 \end{aligned}$

5. $\begin{aligned} x + 2y &= 1 \\ 2x - y &= 0 \end{aligned}$

6. $\begin{aligned} 6x + 6y &= 7 \\ 6x - 6y &= 1 \end{aligned}$

7. $\begin{aligned} x + 3y &= 4 \\ 2x + 6y &= 7 \end{aligned}$

8. $\begin{aligned} 2x - 5y &= 10 \\ 4x - 10y &= 20 \end{aligned}$

9. $\begin{aligned} 3x - 4y &= 12 \\ 9x - 12y &= 36 \end{aligned}$

10. $\begin{aligned} x + 7y &= 7 \\ 2x + 14y &= 15 \end{aligned}$

11. $\begin{aligned} x - y + 2z &= -1 \\ 2x + y - z &= 9 \\ x - 2y - z &= 0 \end{aligned}$

12. $\begin{aligned} 2x + y + 3z &= 0 \\ x - 3y + 2z &= 7 \\ 3x - 2y - z &= 7 \end{aligned}$

13. $\begin{aligned} 2x - 3y + z &= 0 \\ 3x + 2y - z &= 4 \\ 2x - 2y + z &= 1 \end{aligned}$

14. $\begin{aligned} 3x + y + 2z &= 3 \\ 2x - y - 3z &= 15 \\ 2x + 2y + 3z &= -3 \end{aligned}$

15. $\begin{aligned} 3x \qquad + 2z &= 3 \\ x - y - 3z &= 15 \\ x + y \qquad &= 0 \end{aligned}$

16. $\begin{aligned} x + y - z &= 2 \\ x - y + z &= 4 \\ -x + y + z &= 6 \end{aligned}$

17. $\begin{aligned} 2x + 2y + z &= 0 \\ x - y - 1 &= 0 \\ 3x - 3y + 2z &= 3 \end{aligned}$

18. $\begin{aligned} 2x + 3y &= z + 2 \\ 4x - 3y &= 1 - z \\ 2x - 3y &= z - 2 \end{aligned}$

19. $\begin{aligned} x + z &= y + 3 \\ 2x + z &= 3 - y \\ 2x - z &= 2y \end{aligned}$

20. $\begin{aligned} 3x + 4y + z - 6 &= 0 \\ 6x - 4y - 2z - 7 &= 0 \\ 3x - 4y - 3z - 4 &= 0 \end{aligned}$

21. $\begin{aligned} 3x - z + 2y &= 3 \\ 2x - y + 3z &= 8 \\ 3y + 2z + x &= 1 \end{aligned}$

22. $\begin{aligned} x + 2y + 2z &= 2 \\ 2x + y - 7 &= 2z \\ 2x + 2y &= z + 8 \end{aligned}$

---

23. Use Kirchhoff's laws to find the current (in mA) in each resistor of the circuit shown in the accompanying diagram. 3 5 9 1 2 4 6 7 8

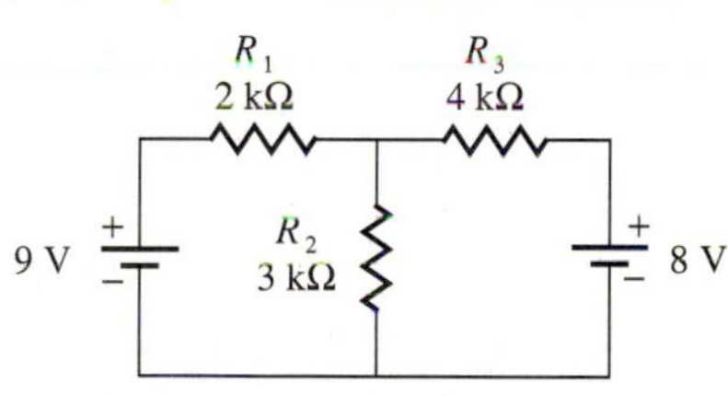

24. An engineer has to travel 808 miles on business. The trip will be made partly by plane, partly by train, and partly by automobile. He knows that he will have to travel twice as for by plane as by train and three times as far by train as by automobile. How far will he have to travel by each mode of transportation? 1 2 6 7 8 3 5 9

25. It is known that when an object is thrown vertically upward, the relationship between height of the object and time of travel is quadratic. That is, the equation relating distance and time is of the form $d = at^2 + bt + c$. Find a, b, and c, given that when $t = 1.00$ s, $d = 30.0$ ft, when $t = 2.00$ s, $d = 88.0$ ft, and when $t = 3.00$ s, $d = 178$ ft. 4 7 8 1 2 3 5 6 9

26. An alloy is to be made from three alloys: a nickel-copper alloy that is 30% nickel and 40% copper; a nickel-zinc alloy that is 60% nickel; and a copper-zinc alloy that is 30% copper. How much of each alloy should be mixed to make 300 pounds of an alloy that is 31% nickel and 30% copper? 1 2 4 6 8 5 7 9

27. An industrial plant is designed to use robots of three different types. A model 1 robot costs \$8000, a model 2 robot costs \$10,000, and a model 3 robot costs \$12,000. The plant can accommodate 24 robots and there must be just as many model 1 robots as the total of model 2 and model 3 robots. How many of each model can be purchased for a total of \$220,000? 1 2 6 7 8 3 5 9

28. Find the current (in mA) in each resistor of the circuit shown in the accompanying diagram. 3 5 9 1 2 4 6 7 8

---

1. Architectural Technology
2. Civil Engineering Technology
3. Computer Engineering Technology
4. Mechanical Drafting Technology
5. Electrical/Electronics Engineering Technology
6. Chemical Engineering Technology
7. Industrial Engineering Technology
8. Mechanical Engineering Technology
9. Biomedical Engineering Technology

Problem 28

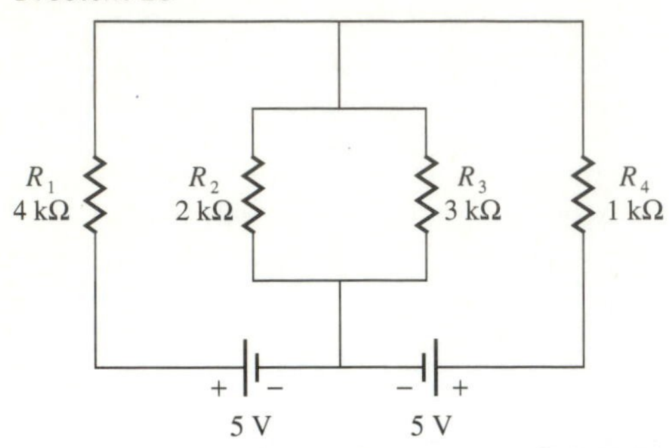

29. How many multiplications and divisions are generally required to solve a system of n equations in n variables by the Gauss-Jordan method?

19.3 THE INVERSE OF A MATRIX

In Section 19.1, we saw several examples of matrices that are inverses of each other. In this section, we will see how to find the inverse of a square matrix.

To find the inverse of a matrix, we write the matrix and the identity matrix side by side. To make the computation easier, we can include both matrices within a single pair of brackets. For example, to find the inverse of $\begin{bmatrix} 2 & 5 \\ 1 & 3 \end{bmatrix}$ we write $\left[\begin{array}{cc|cc} 2 & 5 & 1 & 0 \\ 1 & 3 & 0 & 1 \end{array}\right]$.

As the matrix on the left is transformed into the identity by using elementary row operations, the identity matrix will be transformed into the inverse of the matrix on the left. Zeros and ones are obtained just as they were in Section 19.2. We begin by

NOTE ⇨⇨ changing the first column to $\begin{matrix} 1 \\ 0 \end{matrix}$ and then we change the second column to $\begin{matrix} 0 \\ 1 \end{matrix}$. ***As we work with each column, it is important to first change the entry that is to become 1, so that it can be used to replace the other entries of the column with zeros.***

EXAMPLE 1 Find the inverse of $\begin{bmatrix} 2 & 5 \\ 1 & 3 \end{bmatrix}$.

Solution The identity is written besides the matrix. $\left[\begin{array}{cc|cc} 2 & 5 & 1 & 0 \\ 1 & 3 & 0 & 1 \end{array}\right]$

We begin by changing column 1. Divide each entry of row 1 by 2 to get the pivot row: 1 $\frac{5}{2}$ $\frac{1}{2}$ 0. Multiply the pivot row by -1 and add the result to row 2 to get the second row of the new matrix.

$$\left[\begin{array}{cc|cc} 1 & \frac{5}{2} & \frac{1}{2} & 0 \\ 0 & \frac{1}{2} & -\frac{1}{2} & 1 \end{array}\right]$$

pivot row (divide row 1 by 2)
$-1 \times$ pivot row + row 2

−1	−5/2	−1/2	−0
1	3	0	1
0	1/2	−1/2	1

The first column now has the desired form, so we use this matrix as we change the second column. We begin by finding a new pivot row. We must change the second entry of row 2 to 1. Divide each entry of the row by 1/2 (or multiply by 2) to obtain 0 1 −1 2. Multiply the pivot row by −5/2 and add the result to row 2.

$$\left[\begin{array}{cc|cc} 1 & 0 & 3 & -5 \\ & & & \\ 0 & 1 & -1 & 2 \end{array}\right]$$

−5/2 × pivot row + row 1

−0	−5/2	5/2	−5
1	5/2	1/2	0
1	0	3	−5

pivot row (multiply row 2 by 2)

The second column is now in the desired form, and the solution is complete, for we have the identity matrix on the left. That means the matrix on the right, $\begin{bmatrix} 3 & -5 \\ -1 & 2 \end{bmatrix}$, is the inverse of $\begin{bmatrix} 2 & 5 \\ 1 & 3 \end{bmatrix}$. We can verify that these two matrices are inverses of each other by checking to see that their product is the identity matrix.

$$\begin{bmatrix} 2 & 5 \\ 1 & 3 \end{bmatrix}\begin{bmatrix} 3 & -5 \\ -1 & 2 \end{bmatrix} = \begin{bmatrix} 6-5 & -10+10 \\ 3-3 & -5+6 \end{bmatrix} = \begin{bmatrix} 1 & 0 \\ 0 & 1 \end{bmatrix}. \quad ■$$

EXAMPLE 2 Find the inverse of $\begin{bmatrix} 2 & 1 & -3 \\ 3 & -1 & 4 \\ 4 & 2 & -3 \end{bmatrix}$.

Solution

$$\left[\begin{array}{ccc|ccc} 2 & 1 & -3 & 1 & 0 & 0 \\ 3 & -1 & 4 & 0 & 1 & 0 \\ 4 & 2 & -3 & 0 & 0 & 1 \end{array}\right]$$

Change the first column of this matrix. Apply row operations in the order indicated by the numbers to the right of each row in the following matrix.

$$\left[\begin{array}{ccc|ccc} 1 & \frac{1}{2} & -\frac{3}{2} & \frac{1}{2} & 0 & 0 \\ 0 & -\frac{5}{2} & \frac{17}{2} & -\frac{3}{2} & 1 & 0 \\ & & & & & \\ 0 & 0 & 3 & -2 & 0 & 1 \end{array}\right]$$

pivot row (divide row 1 by 2) **1**

(−3) × pivot row + row 2 **2**

−3	−3/2	9/2	−3/2	0	0
3	−1	4	0	1	0
0	−5/2	17/2	−3/2	1	0

(−4) × pivot row + row 3 **3**

−4	−2	6	−2	0	0
4	2	−3	0	0	1
0	0	3	−2	0	1

Change the second column of this matrix, as follows.

$$\left[\begin{array}{ccc|ccc} 1 & 0 & \frac{1}{5} & \frac{1}{5} & \frac{1}{5} & 0 \\ & & & & & \\ 0 & 1 & -\frac{17}{5} & \frac{3}{5} & -\frac{2}{5} & 0 \\ 0 & 0 & 3 & -2 & 0 & 1 \end{array}\right]$$

(−1/2) × pivot row + row 1 **5**

−0	−1/2	17/10	−3/10	1/5	0
1	1/2	−3/2	1/2	0	0
1	0	1/5	1/5	1/5	0

pivot row (multiply row 2 by −2/5) **4**

(no change) **6**

Change the third column of this matrix, as follows.

$$\left[\begin{array}{ccc|ccc} 1 & 0 & 0 & \frac{1}{3} & \frac{1}{5} & -\frac{1}{15} \\ 0 & 1 & 0 & -\frac{5}{3} & -\frac{2}{5} & \frac{17}{15} \\ 0 & 0 & 1 & -\frac{2}{3} & 0 & \frac{1}{3} \end{array}\right]$$

(− 1/5) × pivot row + row 1 **8**

0	0	−1/5	2/15	0	−1/15
1	0	1/5	1/5	1/5	0
1	0	0	1/3	1/5	−1/15

17/5 × pivot row + row 2 **9**

0	0	17/5	−34/15	0	17/15
0	1	−17/5	3/5	−2/5	0
0	1	0	−5/3	−2/5	17/15

pivot row (divide row 3 by 3) **7**

The solution is complete.

The inverse of $\begin{bmatrix} 2 & 1 & -3 \\ 3 & -1 & 4 \\ 4 & 2 & -3 \end{bmatrix}$ is $\begin{bmatrix} \frac{1}{3} & \frac{1}{5} & -\frac{1}{15} \\ -\frac{5}{3} & -\frac{2}{5} & \frac{17}{15} \\ -\frac{2}{3} & 0 & \frac{1}{3} \end{bmatrix}$.

That these two matrices are inverses of each other can be verified by checking their product.

Check: $\begin{bmatrix} 2 & 1 & -3 \\ 3 & -1 & 4 \\ 4 & 2 & -3 \end{bmatrix} \begin{bmatrix} \frac{1}{3} & \frac{1}{5} & -\frac{1}{15} \\ -\frac{5}{3} & -\frac{2}{5} & \frac{17}{15} \\ -\frac{2}{3} & 0 & \frac{1}{3} \end{bmatrix} = \begin{bmatrix} 1 & 0 & 0 \\ 0 & 1 & 0 \\ 0 & 0 & 1 \end{bmatrix}$ ■

NOTE ⇨⇨ If there is a 0 in the position that a 1 should be in, it is impossible to divide by 0 to change it to 1. It is permissible, however, to interchange the row with another. ***So that you do not undo what you have already done, you should interchange the offending row with one below it.***

EXAMPLE 3 Find the inverse of $\begin{bmatrix} 0 & 2 \\ 3 & 1 \end{bmatrix}$.

Solution $\left[\begin{array}{cc|cc} 0 & 2 & 1 & 0 \\ 3 & 1 & 0 & 1 \end{array}\right]$

To get 1 in the first position, it is impossible to divide by 0, so we interchange the rows.

$$\left[\begin{array}{cc|cc} 3 & 1 & 0 & 1 \\ 0 & 2 & 1 & 0 \end{array}\right]$$

Change column 1. $\left[\begin{array}{cc|cc} 1 & \frac{1}{3} & 0 & \frac{1}{3} \\ 0 & 2 & 1 & 0 \end{array}\right]$ pivot row (1/3 × row 1) **1**; no change **2**

Change column 2. $\left[\begin{array}{cc|cc} 1 & 0 & -\frac{1}{6} & \frac{1}{3} \\ 0 & 1 & \frac{1}{2} & 0 \end{array}\right]$ (− 1/3) × pivot row + row 1 **4**; pivot row (1/2 × row 2) **3**

So $\begin{bmatrix} -\frac{1}{6} & \frac{1}{3} \\ \frac{1}{2} & 0 \end{bmatrix}$ is the inverse of $\begin{bmatrix} 0 & 2 \\ 3 & 1 \end{bmatrix}$.

Check: $\begin{bmatrix} -\frac{1}{6} & \frac{1}{3} \\ \frac{1}{2} & 0 \end{bmatrix} \begin{bmatrix} 0 & 2 \\ 3 & 1 \end{bmatrix} = \begin{bmatrix} 1 & 0 \\ 0 & 1 \end{bmatrix}$ ■

Some matrices do not have inverses. Even interchanging rows in such a matrix will not produce an inverse.

EXAMPLE 4 Find the inverse, if it exists, of $\begin{bmatrix} 1 & -1 & 0 \\ 0 & 1 & 1 \\ -1 & 1 & 0 \end{bmatrix}$.

Solution $\left[\begin{array}{ccc|ccc} 1 & -1 & 0 & 1 & 0 & 0 \\ 0 & 1 & 1 & 0 & 1 & 0 \\ -1 & 1 & 0 & 0 & 0 & 1 \end{array}\right]$

Change column 1. $\left[\begin{array}{ccc|ccc} 1 & -1 & 0 & 1 & 0 & 0 \\ 0 & 1 & 1 & 0 & 1 & 0 \\ 0 & 0 & 0 & 1 & 0 & 1 \end{array}\right]$ pivot row (row 1) **1**; no change **2**; pivot row + row 3 **3**

$\left[\begin{array}{ccc|ccc} 1 & 0 & 1 & 1 & 1 & 0 \\ 0 & 1 & 1 & 0 & 1 & 0 \\ 0 & 0 & 0 & 1 & 0 & 1 \end{array}\right]$ pivot row + row 2 **5**; pivot row (no change) **4**; no change **6**

Now we are ready to change column 3. Zero appears in the third position, and there is no row below row 3 to interchange with it. You may have forseen this problem after changing the first column. If so, it was not necessary to continue. ■

If we are interested only in whether an inverse exists, rather than in finding it if it does, we may use the following criterion.

To determine whether a square matrix has an inverse:

(a) If the determinant of a square matrix is not zero, then the matrix has an inverse.

(b) If the determinant is zero, then the matrix does not have an inverse.

EXAMPLE 5 Maxwell's equations, used in electromagnetic theory, require the use of a matrix of the form $\begin{bmatrix} -1 & 0 & 0 & 0 \\ 0 & -1 & 0 & 0 \\ 0 & 0 & -1 & 0 \\ 0 & 0 & 0 & 1 \end{bmatrix}$. Does this matrix have an inverse?

Solution The determinant is $\begin{vmatrix} -1 & 0 & 0 & 0 \\ 0 & -1 & 0 & 0 \\ 0 & 0 & -1 & 0 \\ 0 & 0 & 0 & 1 \end{vmatrix}$.

Since all the entries below the main diagonal are zero, the value of the determinant is the product of the entries along the main diagonal. That is, the determinant is -1, and the matrix has an inverse. ■

EXAMPLE 6 Verify that $\begin{bmatrix} 1 & -1 & 0 \\ 0 & 1 & 1 \\ -1 & 1 & 0 \end{bmatrix}$ does not have an inverse.

Solution The determinant is $\begin{vmatrix} 1 & -1 & 0 \\ 0 & 1 & 1 \\ -1 & 1 & 0 \end{vmatrix}$.

Using Sarrus' rule to evaluate it, we have

$$\left|\begin{array}{rrr|rr} 1 & -1 & 0 & 1 & -1 \\ 0 & 1 & 1 & 0 & 1 \\ -1 & 1 & 0 & -1 & 1 \end{array}\right. \qquad (0+1+0)-(0+1+0)$$

$$= 1 - 1$$
$$= 0.$$

Since the determinant is zero, the matrix does not have an inverse. ■

EXERCISES 19.3

In 1–20, find the inverse of each matrix.

1. $\begin{bmatrix} 5 & 4 \\ 6 & 5 \end{bmatrix}$ **2.** $\begin{bmatrix} 3 & -2 \\ -2 & 1 \end{bmatrix}$ **3.** $\begin{bmatrix} 2 & 1 \\ -5 & -3 \end{bmatrix}$ **4.** $\begin{bmatrix} 2 & 3 \\ 3 & 4 \end{bmatrix}$

5. $\begin{bmatrix} 1 & 3 \\ 2 & 3 \end{bmatrix}$ **6.** $\begin{bmatrix} 1 & 2 \\ -3 & 0 \end{bmatrix}$ **7.** $\begin{bmatrix} 2 & 4 \\ 0 & -1 \end{bmatrix}$ **8.** $\begin{bmatrix} \frac{1}{2} & 1 \\ -1 & \frac{1}{2} \end{bmatrix}$

9. $\begin{bmatrix} \frac{1}{3} & -2 \\ 2 & \frac{1}{3} \end{bmatrix}$ **10.** $\begin{bmatrix} -\frac{2}{3} & 1 \\ 1 & \frac{2}{3} \end{bmatrix}$ **11.** $\begin{bmatrix} 1 & 1 & 1 \\ 1 & 3 & 3 \\ 2 & 4 & 3 \end{bmatrix}$ **12.** $\begin{bmatrix} 1 & 0 & 1 \\ 0 & 2 & 1 \\ 3 & 2 & 3 \end{bmatrix}$

13. $\begin{bmatrix} 1 & 2 & 1 \\ 2 & 2 & 2 \\ 2 & 3 & 3 \end{bmatrix}$ **14.** $\begin{bmatrix} 5 & 0 & -2 \\ 0 & -2 & 1 \\ -7 & 0 & 3 \end{bmatrix}$ **15.** $\begin{bmatrix} 2 & 3 & 2 \\ 3 & 5 & 2 \\ 4 & 6 & 5 \end{bmatrix}$ **16.** $\begin{bmatrix} 6 & -10 & -7 \\ -1 & 2 & 1 \\ -5 & 11 & 5 \end{bmatrix}$

17. $\begin{bmatrix} 1 & 4 & 2 \\ 2 & 2 & 1 \\ 8 & 1 & 1 \end{bmatrix}$ **18.** $\begin{bmatrix} 3 & 1 & 2 \\ 1 & 2 & 3 \\ 2 & 3 & 1 \end{bmatrix}$ **19.** $\begin{bmatrix} 1 & 3 & 2 \\ 1 & 4 & 3 \\ 1 & 3 & 3 \end{bmatrix}$ **20.** $\begin{bmatrix} 4 & 1 & 4 \\ 3 & 3 & 3 \\ 2 & 5 & 2 \end{bmatrix}$

In 21–30 determine whether the inverse of each matrix exists.

21. $\begin{bmatrix} 1 & 0 \\ 0 & -1 \end{bmatrix}$ **22.** $\begin{bmatrix} \frac{1}{2} & \frac{1}{3} \\ 3 & 2 \end{bmatrix}$ **23.** $\begin{bmatrix} \frac{1}{2} & \frac{1}{3} \\ 2 & 3 \end{bmatrix}$ **24.** $\begin{bmatrix} -\frac{1}{4} & 2 \\ 1 & -8 \end{bmatrix}$

25. $\begin{bmatrix} \frac{1}{2} & -\frac{1}{3} & \frac{1}{6} \\ 0 & 6 & 0 \\ -3 & 2 & -1 \end{bmatrix}$

26. $\begin{bmatrix} 1 & -1 & 0 \\ 2 & 3 & 0 \\ -1 & 1 & 0 \end{bmatrix}$

27. $\begin{bmatrix} 1 & -2 & 3 \\ -3 & 1 & -2 \\ 2 & -3 & 1 \end{bmatrix}$

28. $\begin{bmatrix} 1 & 0 & 0 \\ 0 & 1 & 2 \\ -1 & \frac{1}{2} & 0 \end{bmatrix}$

29. $\begin{bmatrix} \frac{1}{2} & \frac{1}{3} & -\frac{1}{4} \\ 1 & 0 & 0 \\ -6 & -4 & 3 \end{bmatrix}$

30. $\begin{bmatrix} 1 & 0 & 1 \\ 0 & 2 & 1 \\ 3 & 2 & 3 \end{bmatrix}$

Problems 31–33 use the following information. In Section 19.1, Example 7 used the matrix $A = \begin{bmatrix} \cos\theta & \sin\theta \\ -\sin\theta & \cos\theta \end{bmatrix}$ *as the rotation factor for a point (x, y) on the terminal side of an angle in standard position. If the terminal side of the angle is rotated through* $-\theta$*, the rotation factor is* $B = \begin{bmatrix} \cos\theta & -\sin\theta \\ \sin\theta & \cos\theta \end{bmatrix}$.

31. When $\theta = 45°$, find A. Then find its inverse using the method of this chapter. Does the formula for B give the same result? 1 2 4 7 8 3 5

32. When $\theta = 30°$, find A. Then find its inverse using the method of this chapter. Does the formula for B give the same result? 1 2 4 7 8 3 5

33. Show that the two matrices are inverses of each other. (Hint: $\sin^2\theta + \cos^2\theta = 1$.) 1 2 4 7 8 3 5

Problems 34–36 use the following information. In Section 19.1, problems 35 and 36 used

$$A = \begin{bmatrix} \cos\theta & \sin\theta & 0 \\ -\sin\theta & \cos\theta & 0 \\ 0 & 0 & 1 \end{bmatrix}$$

as the conversion factor from rectangular to cylindrical coordinates. The conversion factor from cylindrical to rectangular coordinates is

$$B = \begin{bmatrix} \cos\theta & -\sin\theta & 0 \\ \sin\theta & \cos\theta & 0 \\ 0 & 0 & 1 \end{bmatrix}.$$

34. When $\theta = 30°$, find A. Then find its inverse using the method of this chapter. Does the formula for B give the same result? 1 2 4 7 8 3 5

35. When $\theta = 60°$, find A. Then find its inverse using the method of this chapter. Does the formula for B give the same result? 1 2 4 7 8 3 5

36. Show that these two matrices are inverses of each other. (Hint: $\sin^2\theta + \cos^2\theta = 1$.) 1 2 4 7 8 3 5

37. The following method can be used to find the inverse of a matrix A. Work problem 15 using it.
1. Find the determinant of the matrix A.
2. Form a new matrix B in which each entry is the cofactor of the corresponding entry in the determinant of A.
3. Form a new matrix C in which rows and columns of B are interchanged. That is, row 1 becomes column 1, row 2 becomes column 2, and so on.
4. Divide each entry of C by the determinant found in step 1. The resulting matrix is A^{-1}.

---

1. Architectural Technology
2. Civil Engineering Technology
3. Computer Engineering Technology
4. Mechanical Drafting Technology
5. Electrical/Electronics Engineering Technology
6. Chemical Engineering Technology
7. Industrial Engineering Technology
8. Mechanical Engineering Technology
9. Biomedical Engineering Technology

19.4 MATRIX EQUATIONS

Matrix inverses can be used to solve a system of equations, although the method is rarely used in practice. In general, the process of finding the inverse, even using a computer, requires significantly more computation than other methods. An understanding of the method is required, however, for the following reasons. The matrix inverse is sometimes used in the technologies in the statement of a problem, even though another method, such as Gaussian elimination, might be used to find the solution. Also, some calculators are programmed to solve systems of linear equations, and the equations are entered in matrix form.

A system of n equations in n variables can be written in matrix form $\boldsymbol{AX}=\boldsymbol{B}$. Matrix $\boldsymbol{A}$ is a square matrix whose entries are the coefficients of the variables in the system, matrix $\boldsymbol{X}$ is a single column whose entries are the variables of the system, and matrix $\boldsymbol{B}$ is a single column whose entries are the constants of the system.

EXAMPLE 1 Consider the forces acting on bodies A and B, as shown in Figure 19.4. Analysis of the free-body diagram leads to the following system of linear equations involving forces (in newtons) acting on A. Assume that μ, the coefficient of friction between the surfaces of A and B, is 0.35. Write a matrix equation to represent the system of equations.

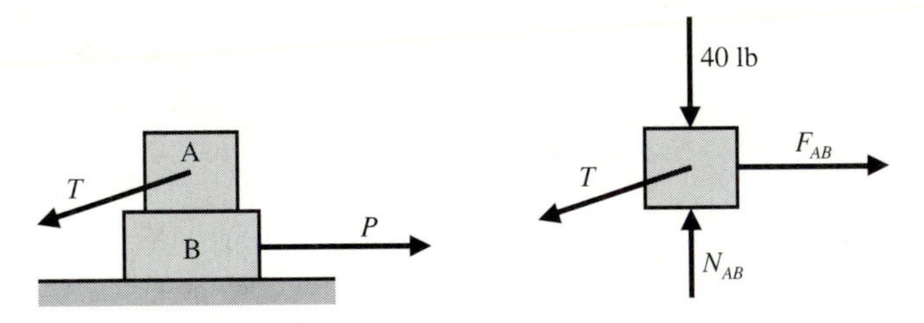

FIGURE 19.4

$$\begin{aligned} F_{AB} \qquad\quad - T\cos 18^\circ &= 0 \\ N_{AB} - T\sin 18^\circ &= 40 \\ F_{AB} - \mu N_{AB} \qquad\qquad &= 0 \end{aligned}$$

Solution The variables are F_{AB}, N_{AB}, and T. Thus, we have

$$\begin{bmatrix} 1 & 0 & -\cos 18^\circ \\ 0 & 1 & -\sin 18^\circ \\ 1 & -0.35 & 0 \end{bmatrix} \begin{bmatrix} F_{AB} \\ N_{AB} \\ T \end{bmatrix} = \begin{bmatrix} 0 \\ 40 \\ 0 \end{bmatrix}. \quad \blacksquare$$

Consider the equation $\boldsymbol{AX}=\boldsymbol{B}$. If we multiply both sides by the inverse of $\boldsymbol{A}$ (denoted $\boldsymbol{A}^{-1}$), we have

$$\boldsymbol{A}^{-1}\boldsymbol{AX}=\boldsymbol{A}^{-1}\boldsymbol{B}.$$

NOTE Remember that multiplication of matrices is not, in general, commutative. ***Thus, if we use the factor A^{-1} as the first factor on the left-hand side of the equation, we use A^{-1} as the first factor on the right-hand side also.*** Since $A^{-1}A = I$, the identity matrix, we have

$$IX = A^{-1}B.$$

But $IX = X$, so

$$X = A^{-1}B.$$

EXAMPLE 2 Solve the system $\begin{matrix} x + y = 0 \\ 6x + 7y = 1 \end{matrix}$ using a matrix equation.

Solution The system can be written as $\begin{bmatrix} 1 & 1 \\ 6 & 7 \end{bmatrix}\begin{bmatrix} x \\ y \end{bmatrix} = \begin{bmatrix} 0 \\ 1 \end{bmatrix}$. The inverse of $\begin{bmatrix} 1 & 1 \\ 6 & 7 \end{bmatrix}$ is $\begin{bmatrix} 7 & -1 \\ -6 & 1 \end{bmatrix}$. We multiply both sides of the equation by this matrix. We have

$$\begin{bmatrix} 7 & -1 \\ -6 & 1 \end{bmatrix}\begin{bmatrix} 1 & 1 \\ 6 & 7 \end{bmatrix}\begin{bmatrix} x \\ y \end{bmatrix} = \begin{bmatrix} 7 & -1 \\ -6 & 1 \end{bmatrix}\begin{bmatrix} 0 \\ 1 \end{bmatrix}.$$

$$\begin{bmatrix} 1 & 0 \\ 0 & 1 \end{bmatrix}\begin{bmatrix} x \\ y \end{bmatrix} = \begin{bmatrix} -1 \\ 1 \end{bmatrix}$$

$$\begin{bmatrix} x \\ y \end{bmatrix} = \begin{bmatrix} -1 \\ 1 \end{bmatrix}$$

Since two matrices are equal if their corresponding entries are equal, x is -1 and y is 1. The solution of the system, then, is $(-1, 1)$.

Check: $-1 + 1 = 0$

$6(-1) + 7(1) = -6 + 7 = 1$ ■

To solve a system of *n* equations in *n* variables using a matrix equation:

1. If the variables are not in the same order for each equation, write an equivalent system in which they are.
2. Write the matrix equation $AX = B$ which represents the system.
   (a) A is the matrix of coefficients of the variables, in order.
   (b) X is a column matrix containing the variables, in order.
   (c) B is a column matrix containing the constants of the equations, in order.
3. Find the inverse (A^{-1}) of matrix A.
4. Multiply both sides of the matrix equation by A^{-1}, using A^{-1} as the first factor.
5. Set each variable equal to the corresponding entry in the product matrix $A^{-1}B$.

EXAMPLE 3 Solve the system using a matrix equation. $4x + 7y = 1$
$3x + 5y = 1$

Solution

1. The equations are in the correct form.
2. Write the equations as a matrix equation. $\begin{bmatrix} 4 & 7 \\ 3 & 5 \end{bmatrix}\begin{bmatrix} x \\ y \end{bmatrix} = \begin{bmatrix} 1 \\ 1 \end{bmatrix}$
3. Find the inverse of $\begin{bmatrix} 4 & 7 \\ 3 & 5 \end{bmatrix}$. $\left[\begin{array}{cc|cc} 4 & 7 & 1 & 0 \\ 3 & 5 & 0 & 1 \end{array}\right]$

   Change column 1. $\left[\begin{array}{cc|cc} 1 & \frac{7}{4} & \frac{1}{4} & 0 \\ 0 & -\frac{1}{4} & -\frac{3}{4} & 1 \end{array}\right]$

   Change column 2. $\left[\begin{array}{cc|cc} 1 & 0 & -5 & 7 \\ 0 & 1 & 3 & -4 \end{array}\right]$

   Thus, the inverse of $\begin{bmatrix} 4 & 7 \\ 3 & 5 \end{bmatrix}$ is $\begin{bmatrix} -5 & 7 \\ 3 & -4 \end{bmatrix}$.
4. Multiply both sides of the matrix equation by this inverse.

$$\begin{bmatrix} -5 & 7 \\ 3 & -4 \end{bmatrix}\begin{bmatrix} 4 & 7 \\ 3 & 5 \end{bmatrix}\begin{bmatrix} x \\ y \end{bmatrix} = \begin{bmatrix} -5 & 7 \\ 3 & -4 \end{bmatrix}\begin{bmatrix} 1 \\ 1 \end{bmatrix}$$

   We have $\begin{bmatrix} x \\ y \end{bmatrix} = \begin{bmatrix} 2 \\ -1 \end{bmatrix}$.
5. The solution of the system is $(2, -1)$, which is easily verified.

$$4(2) + 7(-1) = 1 \text{ or } 8 - 7 = 1$$
$$3(2) + 5(-1) = 1 \text{ or } 6 - 5 = 1 \quad ■$$

EXAMPLE 4 Solve the system
$$\begin{aligned} x + y \quad &= 3 \\ y + z &= 1. \\ x \quad + z &= 0 \end{aligned}$$

Solution We have $\begin{bmatrix} 1 & 1 & 0 \\ 0 & 1 & 1 \\ 1 & 0 & 1 \end{bmatrix}\begin{bmatrix} x \\ y \\ z \end{bmatrix} = \begin{bmatrix} 3 \\ 1 \\ 0 \end{bmatrix}$.

We find the inverse of $\begin{bmatrix} 1 & 1 & 0 \\ 0 & 1 & 1 \\ 1 & 0 & 1 \end{bmatrix}$.

$$\left[\begin{array}{ccc|ccc} 1 & 1 & 0 & 1 & 0 & 0 \\ 0 & 1 & 1 & 0 & 1 & 0 \\ 1 & 0 & 1 & 0 & 0 & 1 \end{array}\right]$$

Change column 1. $\left[\begin{array}{ccc|ccc} 1 & 1 & 0 & 1 & 0 & 0 \\ 0 & 1 & 1 & 0 & 1 & 0 \\ 0 & -1 & 1 & -1 & 0 & 1 \end{array}\right]$

Change column 2. $\left[\begin{array}{ccc|ccc} 1 & 0 & -1 & 1 & -1 & 0 \\ 0 & 1 & 1 & 0 & 1 & 0 \\ 0 & 0 & 2 & -1 & 1 & 1 \end{array}\right]$

Change column 3. $\left[\begin{array}{ccc|ccc} 1 & 0 & 0 & \frac{1}{2} & -\frac{1}{2} & \frac{1}{2} \\ 0 & 1 & 0 & \frac{1}{2} & \frac{1}{2} & -\frac{1}{2} \\ 0 & 0 & 1 & -\frac{1}{2} & \frac{1}{2} & \frac{1}{2} \end{array}\right]$

Thus, $\begin{bmatrix} \frac{1}{2} & -\frac{1}{2} & \frac{1}{2} \\ \frac{1}{2} & \frac{1}{2} & -\frac{1}{2} \\ -\frac{1}{2} & \frac{1}{2} & \frac{1}{2} \end{bmatrix}$ is the inverse of $\begin{bmatrix} 1 & 1 & 0 \\ 0 & 1 & 1 \\ 1 & 0 & 1 \end{bmatrix}$.

Both sides of the matrix equation are multiplied by this inverse.

$$\begin{bmatrix} \frac{1}{2} & -\frac{1}{2} & \frac{1}{2} \\ \frac{1}{2} & \frac{1}{2} & -\frac{1}{2} \\ -\frac{1}{2} & \frac{1}{2} & \frac{1}{2} \end{bmatrix}\begin{bmatrix} 1 & 1 & 0 \\ 0 & 1 & 1 \\ 1 & 0 & 1 \end{bmatrix}\begin{bmatrix} x \\ y \\ z \end{bmatrix} = \begin{bmatrix} \frac{1}{2} & -\frac{1}{2} & \frac{1}{2} \\ \frac{1}{2} & \frac{1}{2} & -\frac{1}{2} \\ -\frac{1}{2} & \frac{1}{2} & \frac{1}{2} \end{bmatrix}\begin{bmatrix} 3 \\ 1 \\ 0 \end{bmatrix}$$

$$\begin{bmatrix} x \\ y \\ z \end{bmatrix} = \begin{bmatrix} 1 \\ 2 \\ -1 \end{bmatrix}$$

The solution of the system is $(1, 2, -1)$.

Check: $1 + 2 = 3$

$2 + (-1) = 1$

$1 + (-1) = 0$ ■

EXERCISES 19.4

In 1–15, use a matrix equation to solve each system of equations.

1. $2x + y = 8$
$x + 2y = 10$

2. $2x + 3y = 1$
$3x - 2y = -5$

3. $6x + 6y = 7$
$6x - 6y = 1$

4. $2x - 3y = 1$
$5x - 7y = 3$

5. $x - 5y = 6$
$2x - 9y = 11$

6. $x + y = 1$
$2x - y = 1$

7. $2x + 3y = 2$
$2x - 3y = 0$

8. $2x + 2y = 1$
$4x + y = 1$

9. $x + 3y + 3z = 2$
$2x + 4y + 3z = 5$
$x + 2y + 3z = 1$

10. $3x + 3y + 3z = 6$
$4x + 3y + 3z = 7$
$3x + 3y + 2z = 7$

11. $x + 2y + z = 3$
$3x + 4y + 2z = 8$
$x + y + z = 2$

12. $x + 3y + 2z = 1$
$x + 4y + 3z = 2$
$x + 3y + 3z = 4$

13. $2x + 3y + z = 15$
$4x + 3y + 2z = 21$
$3x + 3y + z = 19$

14. $x + z = 5$
$2y + z = -1$
$3x + 2y + 3z = 5$

15. $5x - 2z = -7$
$-2y + z = 3$
$-7x + 3z = 10$

16. The currents in the unbalanced Wheatstone bridge shown in the accompanying diagram are given by the following system of equations.

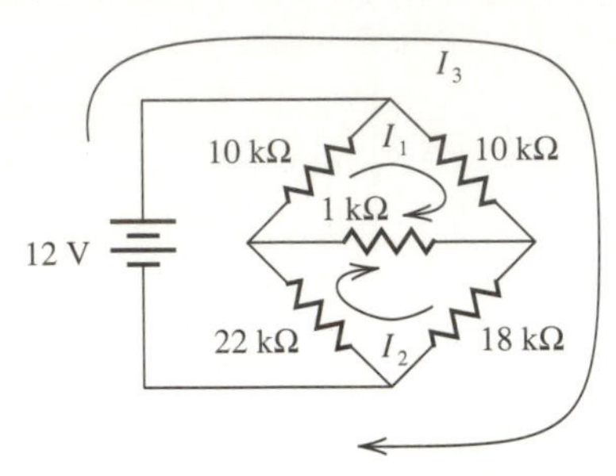

$$21{,}000\ I_1 - 1000\ I_2 + 10{,}000\ I_3 = 0$$
$$-1000\ I_1 + 41{,}000\ I_2 + 18{,}000\ I_3 = 0$$
$$10{,}000\ I_1 + 18{,}000\ I_2 + 28{,}000\ I_3 = 12$$

Write this system as a matrix equation.

3 5 9 1 2 4 6 7 8

17. The following three equations result from studying the forces (in lbs) on a body. Write the system of equations as a matrix equation.

$$R_b \cos 40° - R_{ax} - 100 = 0$$
$$R_{ay} - R_b \sin 40° - 200 = 0$$
$$6\ R_b \sin 40° - 1800 \cos 30° - 1200 = 0$$

2 4 8 1 5 6 7 9

18. In Section 19.1, Example 7 used the equation $[x', y'] = [x \quad y]\begin{bmatrix} \cos\theta & \sin\theta \\ -\sin\theta & \cos\theta \end{bmatrix}$ to find the coordinates of the point (x, y) on the terminal side of an angle that has been rotated through angle θ. Solve the equation for $[x \quad y]$, given the inverse of

$$\begin{bmatrix} \cos\theta & \sin\theta \\ -\sin\theta & \cos\theta \end{bmatrix} \quad \text{is} \quad \begin{bmatrix} \cos\theta & -\sin\theta \\ \sin\theta & \cos\theta \end{bmatrix}$$

1 2 4 7 8 3 5

19. In Section 19.1, problems 35 and 36 used the equation

$$[x' \ y' \ z'] = [x \ y \ z]\begin{bmatrix} \cos\theta & \sin\theta & 0 \\ -\sin\theta & \cos\theta & 0 \\ 0 & 0 & 1 \end{bmatrix}$$

to find the cylindrical coordinates of a point whose rectangular coordinates are (x, y, z). Solve the equation for $[x \ y \ z]$, given the inverse of

$$\begin{bmatrix} \cos\theta & \sin\theta & 0 \\ -\sin\theta & \cos\theta & 0 \\ 0 & 0 & 1 \end{bmatrix} \text{ is } \begin{bmatrix} \cos\theta & -\sin\theta & 0 \\ \sin\theta & \cos\theta & 0 \\ 0 & 0 & 1 \end{bmatrix}.$$

1 2 4 7 8 3 5

20. The general stress-strain relationship for anisotropic media is

$$\begin{bmatrix} \tau_{11} \\ \tau_{22} \\ \tau_{33} \\ \tau_{12} \\ \tau_{23} \\ \tau_{13} \end{bmatrix} = \begin{bmatrix} C_{11} & C_{12} & \cdots & C_{16} \\ C_{21} & C_{22} & \cdots & C_{26} \\ C_{31} & C_{32} & \cdots & C_{36} \\ C_{41} & C_{42} & \cdots & C_{46} \\ C_{51} & C_{52} & \cdots & C_{56} \\ C_{61} & C_{62} & \cdots & C_{66} \end{bmatrix}\begin{bmatrix} e_{11} \\ e_{22} \\ e_{33} \\ e_{12} \\ e_{23} \\ e_{13} \end{bmatrix}.$$

Write the system of equations represented by this matrix equation. 1 2 4 8 7 9

21. The following matrix equation is used to compute angular momentum. Write the system of equations that it represents.

$$\begin{bmatrix} h_1 \\ h_2 \\ h_3 \end{bmatrix} = \begin{bmatrix} I_{xx} & -I_{xy} & -I_{xz} \\ -I_{xy} & I_{yy} & -I_{yz} \\ -I_{xz} & -I_{yz} & -I_{zz} \end{bmatrix}\begin{bmatrix} \omega_1 \\ \omega_2 \\ \omega_3 \end{bmatrix}$$

4 7 8 1 2 3 5 6 9

1. Architectural Technology
2. Civil Engineering Technology
3. Computer Engineering Technology
4. Mechanical Drafting Technology
5. Electrical/Electronics Engineering Technology
6. Chemical Engineering Technology
7. Industrial Engineering Technology
8. Mechanical Engineering Technology
9. Biomedical Engineering Technology

19.5 CHAPTER REVIEW

KEY TERMS

19.1 $m \times n$ matrix
corresponding entries
identity matrix
inverse of a matrix

19.2 Gauss-Jordan method
pivot row

SYMBOLS

[ ] a matrix, if a rectangular array of numbers is included between the brackets

RULES AND FORMULAS

Elementary Row Operations for Matrices

1. Each entry of a row may be multiplied by the same nonzero constant.
2. Any two rows may be interchanged.
3. Any row may be multiplied by a constant and added to another row.

SEQUENTIAL REVIEW

(Section 19.1) *In 1–6, perform the matrix addition and subtraction, if possible.*

1. $\begin{bmatrix} 1 & 0 \\ 2 & -2 \end{bmatrix} + \begin{bmatrix} 2 & -3 \\ 0 & 3 \end{bmatrix}$

2. $\begin{bmatrix} 2 & -5 & 0 \\ 1 & 7 & 3 \end{bmatrix} + \begin{bmatrix} 3 & -6 & 1 \\ 0 & 8 & -2 \end{bmatrix}$

3. $\begin{bmatrix} 1 & 1 \\ 0 & -2 \\ 4 & 7 \end{bmatrix} - \begin{bmatrix} 1 & 0 \\ 2 & 5 \\ 0 & -3 \end{bmatrix}$

4. $\begin{bmatrix} 9 & 5 \\ 3 & -6 \end{bmatrix} - \begin{bmatrix} -4 & 2 \\ 0 & 8 \end{bmatrix}$

5. $\begin{bmatrix} 1 & 2 & 0 \\ 3 & -1 & 4 \end{bmatrix} + \begin{bmatrix} 1 & 2 \\ 2 & 5 \\ -1 & 4 \end{bmatrix}$

6. $\begin{bmatrix} 5 & 9 \\ 1 & -3 \\ 7 & 2 \end{bmatrix} + \begin{bmatrix} -4 & 0 & 6 \\ 8 & 2 & -3 \end{bmatrix}$

In 7–12, perform the matrix multiplication, if possible.

7. $\begin{bmatrix} 1 & 0 \\ 3 & 2 \end{bmatrix}\begin{bmatrix} 2 & -5 \\ 6 & 0 \end{bmatrix}$

8. $\begin{bmatrix} 2 & 9 \\ 4 & -1 \end{bmatrix}\begin{bmatrix} 3 & 4 \\ -7 & 1 \end{bmatrix}$

9. $\begin{bmatrix} 3 & 8 \\ 5 & -2 \\ -6 & 9 \end{bmatrix}\begin{bmatrix} 4 & 7 & -1 \\ 6 & -3 & 2 \end{bmatrix}$

10. $\begin{bmatrix} 4 & 7 & -1 \\ 6 & -3 & 2 \end{bmatrix}\begin{bmatrix} 3 & 8 \\ 5 & -2 \\ -6 & 9 \end{bmatrix}$

11. $\begin{bmatrix} 2 & 3 & 0 \\ 1 & -2 & 4 \\ 3 & 0 & -1 \end{bmatrix}\begin{bmatrix} 3 & -2 & 0 \\ 1 & 0 & -1 \\ 2 & 3 & 4 \end{bmatrix}$

12. $\begin{bmatrix} 1 & -4 & 1 \\ 0 & 3 & 5 \\ 2 & 1 & 0 \end{bmatrix}\begin{bmatrix} 2 & 3 & 9 \\ 0 & -1 & 2 \\ 1 & 4 & -3 \end{bmatrix}$

(Section 19.2) *In 13–23, solve each system of equations using the Gauss-Jordan method.*

13. $\begin{aligned} 2x + 3y &= 10 \\ x - y &= -5 \end{aligned}$

14. $\begin{aligned} x + 3y &= 14 \\ 2x - y &= 7 \end{aligned}$

15. $\begin{aligned} 2x - 5y &= 9 \\ 5x + 3y &= 7 \end{aligned}$

16. $\begin{aligned} 5x - 7y &= 29 \\ 6x + 4y &= 10 \end{aligned}$

17. $\begin{aligned} 4x + 3y &= 13 \\ 3x - 5y &= 46 \end{aligned}$

18. $\begin{aligned} x - 7y &= 20 \\ 7x + y &= 40 \end{aligned}$

19. $\begin{aligned} x - 2y - 3z &= 2 \\ 2x + 3y - z &= 6 \\ 3x - y - 2z &= 4 \end{aligned}$

20. $\begin{aligned} x - y - z &= 4 \\ -x + y - z &= 4 \\ -x - y + z &= 0 \end{aligned}$

21. $\begin{aligned} x + y &= 3 \\ y + z &= 5 \\ x + z &= 4 \end{aligned}$

22. $\begin{aligned} x &= y + 5 \\ y &= z - 1 \\ x &= z + 4 \end{aligned}$

23. $\begin{aligned} 2x + y - 2z &= 9 \\ x + 2y + 2z &= 0 \\ 2x - y + 2z &= -1 \end{aligned}$

(Section 19.3) *In 24–30, find the inverse of each matrix.*

24. $\begin{bmatrix} 2 & 5 \\ 1 & 3 \end{bmatrix}$ **25.** $\begin{bmatrix} 1 & 4 \\ 2 & 9 \end{bmatrix}$ **26.** $\begin{bmatrix} 3 & 4 \\ 0 & 2 \end{bmatrix}$ **27.** $\begin{bmatrix} -4 & 3 \\ 9 & 1 \end{bmatrix}$

28. $\begin{bmatrix} 1 & -2 & 3 \\ 2 & -5 & 10 \\ -1 & 2 & -2 \end{bmatrix}$ **29.** $\begin{bmatrix} 2 & 4 & 1 \\ 1 & 1 & 0 \\ 3 & 4 & 1 \end{bmatrix}$ **30.** $\begin{bmatrix} 1 & 5 & 2 \\ -1 & 3 & 1 \\ -3 & 4 & 1 \end{bmatrix}$

In 31–34, determine whether the inverse of each matrix exists.

31. $\begin{bmatrix} -3 & 2 \\ 6 & -4 \end{bmatrix}$ **32.** $\begin{bmatrix} 4 & -1 \\ 7 & 6 \end{bmatrix}$ **33.** $\begin{bmatrix} 1 & 2 & 3 \\ 4 & 5 & 6 \\ 7 & 8 & 9 \end{bmatrix}$ **34.** $\begin{bmatrix} 2 & 0 & 1 \\ 3 & -6 & 5 \\ 8 & 9 & -7 \end{bmatrix}$

(Section 19.4) *In 35–45, use a matrix equation to solve each system of equations.*

35. $\begin{aligned} x + 2y &= 3 \\ 2x + 3y &= 4 \end{aligned}$

36. $\begin{aligned} 2x + y &= 15 \\ 3x - 2y &= -2 \end{aligned}$

37. $\begin{aligned} x + y &= 2 \\ 2x + 3y &= 4 \end{aligned}$

38. $\begin{aligned} 2x - y &= 4 \\ -x + y &= 5 \end{aligned}$

39. $\begin{aligned} 3x - 4y &= 30 \\ 5x + 3y &= 21 \end{aligned}$

40. $\begin{aligned} x + 2y &= 14 \\ 3x - y &= 7 \end{aligned}$

41. $\begin{aligned} x + 4y + 2z &= 7 \\ 2x + 2y + z &= 5 \\ 8x + y + z &= 10 \end{aligned}$

42. $\begin{aligned} x + y + z &= 2 \\ x + 3y + 3z &= 4 \\ 2x + 4y + 3z &= 4 \end{aligned}$

43. $\begin{aligned} 2x - 2y - 3z &= -7 \\ 5y + 2z &= 2 \\ -7x - 6y + 10z &= 24 \end{aligned}$

44. $\begin{aligned} x + z &= 4 \\ 2y + z &= 9 \\ 3x + 2y + 3z &= 20 \end{aligned}$

45. $\begin{aligned} 2x + 3y + 2z &= 11 \\ 3x + 5y + 2z &= 19 \\ 4x + 6y + 5z &= 21 \end{aligned}$

APPLIED PROBLEMS

1. A vitamin tablet provides Vitamins B-1, B-2, and B-12. The amount of each vitamin is measured in mg and in percent of U.S. Recommended Daily Allowance (USRDA %). The tablet provides 3 mg of B-1, which is 200 USRDA %, 3.4 mg of Vitamin B-2, which is 200 USRDA %, and 9 mg of B-12, which is 150 USRDA %. Organize and display this information in matrix form with information about B-1 in column 1, B-2 in column 2, and B-12 in column 3. **6 9** 2 4 5 7 8

2. Use Kirchhoff's laws and a matrix equation to find the current (in mA) in each branch of the circuit shown. **3 5 9** 1 2 4 6 7 8

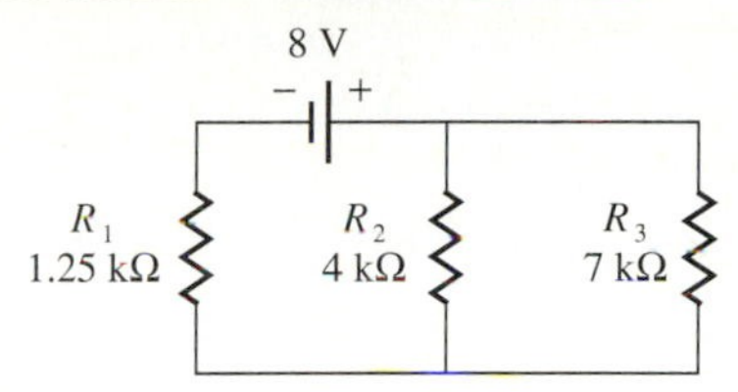

3. Two grades of gasoline are combined. One grade of gasoline is 98% pure gasoline, and the other is 85% pure gasoline. How many gallons of each grade are needed to produce 100 gallons of a mixture that is 90% pure gasoline? Use a matrix equation to solve the problem. **6 9** 2 4 5 7 8

4. An engineer driving home noticed that the temperature on a bank sign was shown in both Fahrenheit and Celsius degress. The digits were the same, but reversed. Later, he couldn't remember the temperatures, but he remembered that the sum of the temperatures was his mother's age, and the difference of the temperatures was his age. If he was 45, and his mother was 77, find the two temperatures. [Hint: Let $10x + y$ be one temperature and let $10y + x$ be the other.] **6 8** 1 2 3 5 7 9

5. A particular computer is available in several models. Model I has 240 kilobytes of memory, an 8 megaHertz processor, and a 20 megabyte disk drive. Model II has a 1 megabyte memory, a 10 megaHertz processor, and a 30 megabyte disk drive. Model III has a 2 megabyte memory, a 16 megaHertz processor, and a 60 megabyte disk drive. Organize this data in matrix form. Let each row represent a different model. Column 1 should show memory, column 2 should show processor frequency, and column 3 disk drive capacity. **3 5** 9

6. For finding a straight line to represent some laboratory data, the following equations had to be solved.

$$12a + 240b = 384$$
$$240a + 5708b = 7098$$

Use a matrix equation to solve for a and b. Give your answer to two significant digits. **1 2 6 7 8** 3 5 9

7. For finding a curve to represent the data of an experiment, the following equations had to be solved.

$$7a + 28c = 808$$
$$28b = 465$$
$$28a + 196c = 3415$$

Use a matrix equation to solve for a, b, and c. Give your answer to three significant digits. **1 2 6 7 8** 3 5 9

8. A 16,000 lb truck hangs from two cables A and B as shown. The cable tensions T_A and T_B (in lbs) are related to the truck's weight by the matrix equation,

$$\begin{bmatrix} -0.42 & 0.69 \\ 0.91 & 0.72 \end{bmatrix} \begin{bmatrix} T_A \\ T_B \end{bmatrix} = \begin{bmatrix} 0 \\ 16{,}000 \end{bmatrix}.$$

Find T_A and T_B. **2 4 8** 1 5 6 7 9

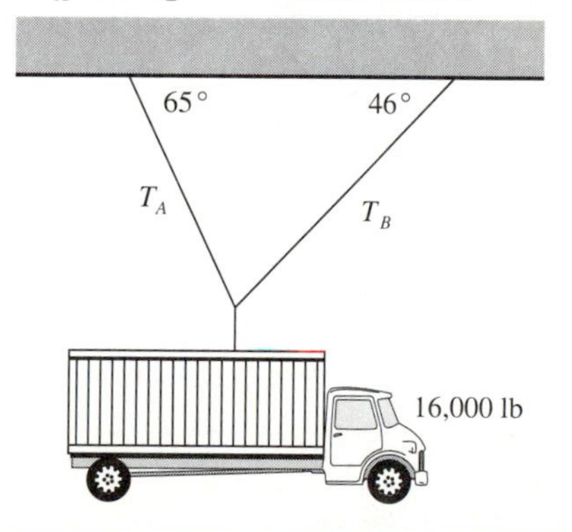

1. Architectural Technology
2. Civil Engineering Technology
3. Computer Engineering Technology
4. Mechanical Drafting Technology
5. Electrical/Electronics Engineering Technology
6. Chemical Engineering Technology
7. Industrial Engineering Technology
8. Mechanical Engineering Technology
9. Biomedical Engineering Technology

9. When a matrix A, consisting of a single row, is multiplied by matrix B, having a single column, the result C is a matrix having one row and one column. This single number is sometimes called the scalar or dot product of A and B. If matrix B is multiplied by A, however, the result is another matrix called the tensor product. Tensors are used in advanced studies. Find (a) the dot product and (b) tensor product of the matrices.

$$A = |9 \quad 3 \quad 8 \quad 7|$$

$$B = \begin{vmatrix} 2 \\ 6 \\ 1 \\ 5 \end{vmatrix}$$

3 5 9 | 4 | 7 | 8

10. Three tanker trucks of differing capacity worked for three days hauling gasoline as detailed in the accompanying table of loads per day.

	MON	TUE	WED
T_1	5	7	4
T_2	3	4	2
T_3	4	6	7

On Monday 146,500 gal were delivered, on Tuesday 209,500 gal, and on Wednesday a total of 171,000 gal were delivered. Use a matrix equation to find the capacity of each truck.

1 2 6 7 8 | 3 | 5 | 9

11. One of a pair of equations needed to represent a moving mechanical system was given by

$$\begin{bmatrix} y_1 \\ y_2 \end{bmatrix} = \begin{bmatrix} 1 & 0 & 0 & 0 \\ 0 & 0 & 1 & 0 \end{bmatrix} \begin{vmatrix} x_1 \\ x_2 \\ x_3 \\ x_4 \end{vmatrix}.$$

Evaluate $\begin{bmatrix} y_1 \\ y_2 \end{bmatrix}$ when $\begin{matrix} x_1 = 2 \\ x_2 = 5 \\ x_3 = -3 \\ x_4 = 0 \end{matrix}$.

4 7 8 | 1 | 2 | 3 | 5 | 6 | 9

12. In electronic control theory, a representation of a dc motor system is given by $y = [1 \quad 0] \begin{bmatrix} x_1 \\ x_2 \end{bmatrix}$.

Solve this equation for $\begin{matrix} x_1 = \frac{\pi}{2} \\ x_2 = 0.5 \end{matrix}$.

3 5 9 | 1 | 2 | 4 | 6 | 7 | 8

13. A mechanical system with coordinates (x'', y'', z'') when studied from a system with coordinates (x', y', z') located a distance a units away and tilted by an angle θ' will have new coordinates given in the matrix equation

$$\begin{bmatrix} x' \\ y' \\ z' \\ 1 \end{bmatrix} = \begin{bmatrix} \cos\theta' & -\sin\theta' & 0 & a \\ \sin\theta' & \cos\theta' & 0 & 0 \\ 0 & 0 & 1 & 0 \\ 0 & 0 & 0 & 1 \end{bmatrix} \begin{bmatrix} x'' \\ y'' \\ z'' \\ 1 \end{bmatrix}.$$

Find x', y', and z' given the following values.

$$x'' = 6 \quad a = 20$$
$$y'' = 9 \quad \theta' = 50^\circ$$
$$z'' = -3$$

4 7 8 | 1 | 2 | 3 | 5 | 6 | 9

14. A text on control engineering gave the matrix equation

$$x_N = \begin{bmatrix} 1 \\ 5 \end{bmatrix} \left\{ [1 \quad 5] \begin{bmatrix} 1 \\ 5 \end{bmatrix} \right\}^{-1}.$$

Find x_N to three significant digits.

1 2 6 7 8 | 3 | 5 | 9

15. Evaluate the following matrix equation from a technical paper on robotics. Give your answer to three significant digits.

$$M = \left\{ \begin{bmatrix} 1 & 1 & 1 \\ 1 & 2 & 4 \end{bmatrix} \begin{bmatrix} 1 & 1 \\ 1 & 2 \\ 1 & 4 \end{bmatrix} \right\}^{-1} \begin{bmatrix} 1 & 1 & 1 \\ 1 & 2 & 4 \end{bmatrix} \begin{bmatrix} 1 \\ 2 \\ 2 \end{bmatrix}$$

3 5 9 | 1 | 2 | 4 | 6 | 7 | 8

16. Show that the matrix equation

$$y = [1 \quad 0 \quad 0] \begin{bmatrix} 1 & 1 & 1 \\ -1 & -2 & -3 \\ 1 & 4 & 9 \end{bmatrix} \begin{bmatrix} x_1 \\ x_2 \\ x_3 \end{bmatrix}$$

can be written $y = [1 \quad 1 \quad 1] \begin{bmatrix} x_1 \\ x_2 \\ x_3 \end{bmatrix}$.

1 2 6 7 8 | 3 | 5 | 9

1. Architectural Technology
2. Civil Engineering Technology
3. Computer Engineering Technology
4. Mechanical Drafting Technology
5. Electrical/Electronics Engineering Technology
6. Chemical Engineering Technology
7. Industrial Engineering Technology
8. Mechanical Engineering Technology
9. Biomedical Engineering Technology

17. It is known that automobile A takes 210 bolts, 35 grommets, and 350 washers to assemble. Automobile B takes 175 bolts, 18 grommets, and 200 washers, while automobile C takes 112 bolts, 45 grommets, and 126 washers. At the end of a day's production 28,070 bolts were used, 7240 grommets, and 37,360 washers. Find the number of each type of car produced. 1 2 6 7 8 3 5 9

18. The matrix equation for the currents in an electronic network is

$$\begin{bmatrix} 15 & -7 \\ -7 & 22 \end{bmatrix}\begin{bmatrix} I_1 \\ I_2 \end{bmatrix} = \begin{bmatrix} 24 \\ 36 \end{bmatrix},$$

where 24 and 36 represent the voltages of two batteries. How would I_1 and I_2 be affected if the 24 V battery and 36 V battery were interchanged? 3 5 9 4 7 8

19. For the circuit in the accompanying figure, Kirchhoff's law yields the following equations.

$$36 - V_1 = V_1 + V_1 - V_2$$
$$V_1 - V_2 = V_2 + V_2 - V_3$$
$$V_2 - V_3 = V_3 + V_3 - 24$$

Use a matrix equation to find V_1, V_2, and V_3 to three significant digits. 3 5 9 1 2 4 6 7 8

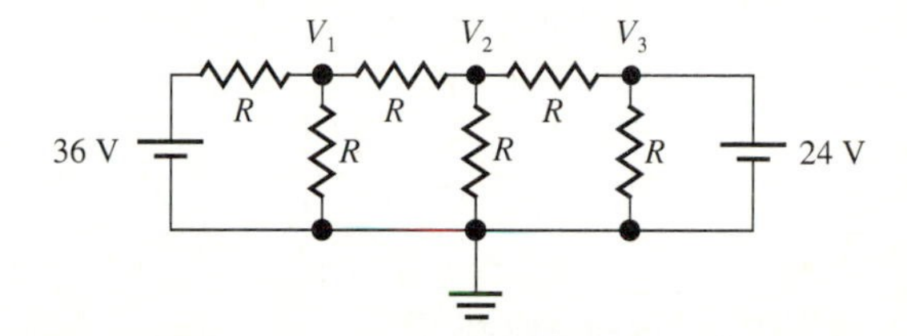

RANDOM REVIEW

1. Determine whether the inverse of the following matrix exists.

$$\begin{bmatrix} 10 & 0 & 10 \\ 5 & 10 & 15 \\ -5 & 10 & 5 \end{bmatrix}$$

2. Perform the indicated operation, if possible.

$$\begin{bmatrix} 1 & 4 & 3 \\ 5 & 7 & -1 \end{bmatrix}\begin{bmatrix} 2 & 8 \\ -1 & 0 \\ 6 & 9 \end{bmatrix}$$

3. Solve the following system of equations by the Gauss-Jordan method.

$$3x + y = 1$$
$$4x - 6y = 38$$

4. Find the inverse of the following matrix. $\begin{bmatrix} -2 & 2 & 3 \\ 0 & -5 & -2 \\ -7 & 6 & 10 \end{bmatrix}$

5. Solve the following system of equations by the Gauss-Jordan method.

$$3x + y - 2z = -1$$
$$4x - 5y + z = -14$$
$$2x + 4y - z = 6$$

6. Find the inverse of the following matrix. $\begin{bmatrix} 5 & 10 \\ 7 & -3 \end{bmatrix}$

7. Perform the indicated operation, if possible.

$$\begin{bmatrix} 3 & -5 & 2 \\ 1 & 0 & -1 \end{bmatrix} + \begin{bmatrix} 4 & 0 \\ 7 & -6 \\ 1 & 5 \end{bmatrix}$$

8. Use a matrix equation to solve the following system of equations.

$$\begin{aligned} 2x + 5y &= 46 \\ 7x - 2y &= 5 \end{aligned}$$

9. Use a matrix equation to solve the following system of equations.

$$\begin{aligned} x - 2y + 3z &= 11 \\ 2x - 5y + 10z &= 31 \\ -x + 2y - 2z &= -9 \end{aligned}$$

10. Perform the indicated operation, if possible

$$\begin{bmatrix} 1 & 0 & 0 \\ 0 & 1 & 0 \\ 0 & 0 & 1 \end{bmatrix}\begin{bmatrix} 4 & 8 & 8 \\ 5 & 1 & 8 \\ 6 & 9 & 1 \end{bmatrix}.$$

CHAPTER TEST

In 1–4, perform the indicated operation, if possible.

1. $\begin{bmatrix} 2 & 3 \\ 4 & -1 \end{bmatrix} + \begin{bmatrix} 3 & 1 \\ -2 & 5 \end{bmatrix}$

2. $\begin{bmatrix} 3 & -2 & 5 \\ 1 & 4 & -7 \\ -2 & 5 & 1 \end{bmatrix} - \begin{bmatrix} 3 & 2 \\ 0 & 7 \end{bmatrix}$

3. $\begin{bmatrix} 2 & 5 \\ 1 & -4 \end{bmatrix}\begin{bmatrix} 3 & 4 \\ 6 & -9 \end{bmatrix}$

4. $\begin{bmatrix} 2 & 8 \\ -1 & 0 \\ 6 & 9 \end{bmatrix}\begin{bmatrix} 1 & 4 & 3 \\ 5 & 7 & -1 \end{bmatrix}$

5. Solve the following system of equations using the Gauss-Jordan method.

$$\begin{aligned} 2x + 3y - 5z &= 7 \\ 3x - 5y - 2z &= 5 \\ -5x + 2y - 3z &= -2 \end{aligned}$$

6. Find the inverse of the following matrix. $\begin{bmatrix} -2 & 9 & 4 \\ 1 & -5 & 2 \\ -2 & 1 & 5 \end{bmatrix}$

7. Write the following system of equations as a matrix equation. Do not solve it.

$$\begin{aligned} 3x + 2y + z &= 6 \\ x - 3y - 2z &= -4 \\ -2x + y + 3z &= 2 \end{aligned}$$

8. Use a matrix equation to solve the following system of equations.

$$\begin{aligned} x + y - z &= 0 \\ x - 2y + 2z &= 9 \\ -2x + 2y + z &= -9 \end{aligned}$$

9. Use Kirchhoff's laws to find the current (in mA) in each branch of the circuit shown.

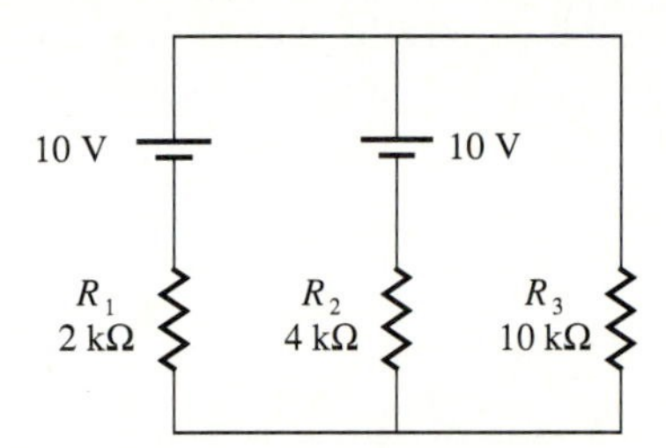

10. When a crane supports a load as in the accompanying figure, there is tension in the supporting cable and compression in the boom at the point P.

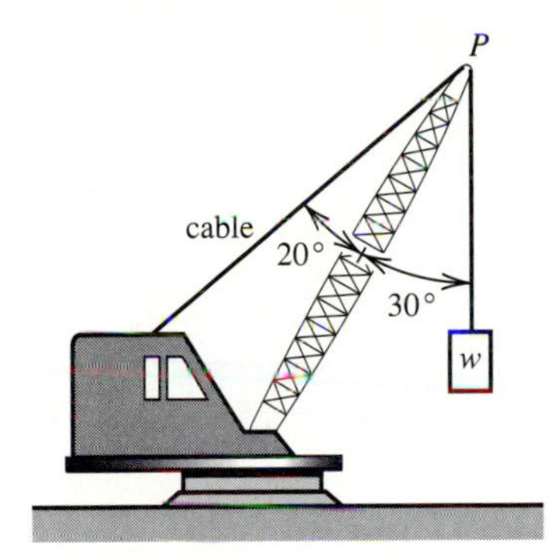

If w is the weight of the load, and R is the magnitude of the resultant force due to load w and tension of magnitude T, the values of R and T can be found from the following system of equations.

$$\begin{aligned} R \cos 60° &= T \cos 40° \\ R \sin 60° - T \sin 40° &= w \end{aligned}$$

Write this system of equations as a matrix equation.

LADDER NETWORKS

In an electrical circuit, two resistors are said to be in **series** if they are connected so that the same current flows through each one in turn. The circuit diagram in Figure 19.5 shows two resistors in series. Two resistors are said to be in **parallel** if they are connected so that the current must divide, some going through one resistor, some through the other. Figure 19.6 shows two resistors that are in parallel.

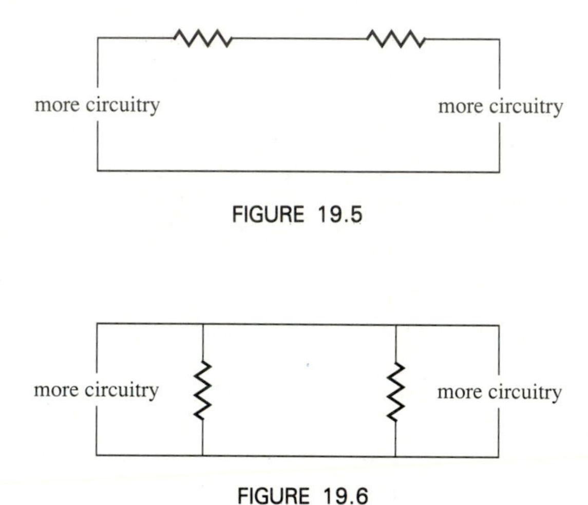

FIGURE 19.5

FIGURE 19.6

A **ladder network** is a circuit diagram that can be drawn as a string of boxes, each containing one resistor, connected so that the two output wires for one box are the two input wires for the next box, as in Figure 19.7. Matrix multiplication may be used to find the voltage drop (v_o) and the current output (i_o) in a ladder network.

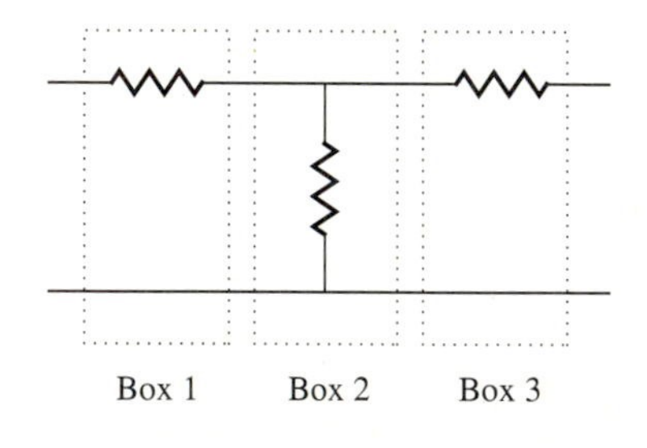

FIGURE 19.7

Each resistor in series is represented by a matrix of the form $\begin{bmatrix} 1 & -R \\ 0 & 1 \end{bmatrix}$, where R is the resistance. Each resistor in parallel is represented by a matrix of the form $\begin{bmatrix} 1 & 0 \\ -\frac{1}{R} & 1 \end{bmatrix}$ where R is the resistance. The voltage drop and output current are obtained by multiplying the matrices representing the boxes together with the matrix $\begin{bmatrix} v_i \\ i_i \end{bmatrix}$, containing the voltage and current entering the circuit. Since matrix multiplication is not commutative, it is important to multiply the matrices in the correct order. The factors should be taken in reverse order from the order in which the boxes are encountered.

Consider the network in Figure 19.8. It consists of four boxes. The 10-Ω and 20-Ω resistors are in parallel, and the 5-Ω and 6-Ω resistors are in series.

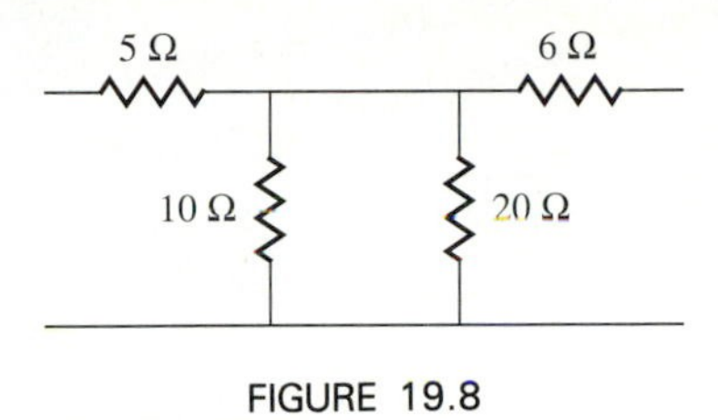

FIGURE 19.8

If $i_i = 2$ amperes and $v_i = 12$ volts, then

$$\begin{bmatrix} v_o \\ i_o \end{bmatrix} = \begin{bmatrix} 1 & -6 \\ 0 & 1 \end{bmatrix} \begin{bmatrix} 1 & 0 \\ -\frac{1}{20} & 1 \end{bmatrix} \begin{bmatrix} 1 & 0 \\ -\frac{1}{10} & 1 \end{bmatrix} \begin{bmatrix} 1 & -5 \\ 0 & 1 \end{bmatrix} \begin{bmatrix} 12 \\ 2 \end{bmatrix}$$

$$= \begin{bmatrix} \frac{13}{10} & -6 \\ -\frac{1}{20} & 1 \end{bmatrix} \begin{bmatrix} 1 & 0 \\ -\frac{1}{10} & 1 \end{bmatrix} \begin{bmatrix} 1 & -5 \\ 0 & 1 \end{bmatrix} \begin{bmatrix} 12 \\ 2 \end{bmatrix}$$

$$= \begin{bmatrix} \frac{19}{10} & -6 \\ -\frac{3}{20} & 1 \end{bmatrix} \begin{bmatrix} 1 & -5 \\ 0 & 1 \end{bmatrix} \begin{bmatrix} 12 \\ 2 \end{bmatrix}$$

$$= \begin{bmatrix} \frac{19}{10} & -\frac{31}{2} \\ -\frac{3}{20} & \frac{7}{4} \end{bmatrix} \begin{bmatrix} 12 \\ 2 \end{bmatrix}$$

$$= \begin{bmatrix} -\frac{41}{5} \\ \frac{17}{10} \end{bmatrix}$$

Thus, the voltage drop is given by $v_o = -8.2$ volts, and the output current is given by $i_o = 1.7$ amperes.

■ From "Matrix Multiplication and DC Ladder Circuits" by Philip M. Tuchinsky in *The UMAP Journal,* Volume 7, Number 3, 1986, pp. 235–247.

STATISTICS CHAPTER 20

As engineering technology evolves, it is not always possible to go to textbooks for answers to the questions that arise. For example, to make automobiles more fuel-efficient, some metal parts have been replaced by lighter plastic parts. Engineers must decide whether the plastic part is as durable as its metal counterpart. Many of the plastic parts are tested. They do not all fail under the same circumstances, but there should be enough similarity that it is possible to determine what percent of them will withstand a given condition. Statistics is the branch of mathematics devoted to summarizing, analyzing, and interpreting data. In this chapter, we consider both numerical and graphical ways to make sense from masses of data.

20.1 MEASURES OF CENTRAL TENDENCY

If your grades in math are 85, 93, 87, 89, and 86 for the term, your teacher would probably say your grade for the term is 88. This number is obtained by adding the scores and dividing by 5. Some of the grades are higher than 88 and some are lower than 88; 88 is a typical grade that represents the entire list, since the grades cluster around 88. Such a typical value is often called an average or a measure of **central tendency.** We will consider three different averages or measures of central tendency. The one used in the example above is the *arithmetic mean* and is the one most often used in statistical studies.

central tendency

population

variate

statistical population

The set of all objects having a certain characteristic is called a **population.** If the characteristic under study can be measured, the measurement for each object in the population is called a **variate.** The set of all variates for members of a population is called a **statistical population.**

DEFINITION

The **arithmetic mean,** denoted μ, for a statistical population is the sum of the variates divided by the number of variates.

EXAMPLE 1 Find the arithmetic mean of the statistical population: 34, 23, 42, 27, 41, and 13.

Solution The sum of the numbers is $34 + 23 + 42 + 27 + 41 + 13$ or 180. Since there are six variates in the population, we have $\mu = 180/6$ or 30. ■

EXAMPLE 2 Find the arithmetic mean of the statistical population: 101, 203, 491, 321, and 114.

Solution The sum of the numbers is 1230. Since there are five variates in the population, we have $\mu = 1230/5$ or 246. ■

The data for technological problems are sometimes obtained by measuring quantities. A technique to minimize errors associated with measurement is to take a number of measurements of the same quantity. The mean of the measurements is used instead of a single measurement.

EXAMPLE 3 The diameter of a steel rod is measured with a micrometer at intervals along the rod. Find the mean diameter. The five measurements are 9.047 mm, 9.065 mm, 9.060 mm, 9.068 mm, and 9.055 mm.

Solution $(9.047 + 9.065 + 9.060 + 9.068 + 9.055)/5 = 45.295/5 = 9.059$ mm ■

Suppose that five people have annual incomes of $20,000, $23,000, $19,000, $18,000, and $500,000 respectively. The arithmetic mean is $116,000 for this population. The arithmetic mean, however, does not seem very representative of the numbers in this population. Since the arithmetic mean gives equal weight to all of the variates, one or two extreme values (such as $500,000) may result in misleading conclusions about the distribution. In this example, a different measure of central tendency called the *median* gives a more accurate picture. Half of the numbers in a statistical population are greater than or equal to the median and half are less than or equal to the median.

> **DEFINITION** For a statistical population containing an odd number of variates, the **median** is the number in the middle of the list when the variates are arranged in ascending (or descending) order. For a statistical population containing an even number of variates, the median is the arithmetic mean of the two middle numbers.

The median income for the five people in the example would be the middle number in the distribution when the variates are ordered: {\$18,000, \$19,000, **\$20,000**, \$23,000, \$500,000}. The median is \$20,000.

EXAMPLE 4 Find the median for the statistical population: 53, 22, 47, 61, 65, 35, 47.

Solution First we list the variates in order: 22, 35, 47, **47**, 53, 61, 65. The number in the middle is 47, so 47 is the median. ■

EXAMPLE 5 Find the median for the statistical population: 68, 51, 73, 62, 70, 81.

Solution First, we list the variates in order: 51, 62, 68, 70, 73, 81. Since there are six numbers, we find the arithmetic mean of the two variates in the middle. The median is $(68 + 70)/2$ or $138/2$, which is 69. ■

If the manager of a hardware store decides to package air filters, he/she might watch sales of air filters for a few days to determine how many a typical customer buys at once. Suppose sales of air filters for one day are as follows: 1, 1, 2, 2, 2, 2, 2, 3, 4, 4, 4, 4, 5, 7, 9. The arithmetic mean is $52/15 = 3\ 7/15$. The manager certainly would not package air filters with 3 7/15 air filters in a package. The median is 3; yet only one person bought 3 air filters. The manager might very well decide to package air filters in groups of 2, since more people bought two than any other number. This measure of central tendency is called the *mode.* A statistical population is said to be *bimodal* if there are two variates that are listed the same number of times, but more than the other variates. If each number in the population occurs only once, we say there is no mode.

bimodal

> **DEFINITION** The **mode,** if there is one, of a statistical population, is the variate that appears most often in the population.

EXAMPLE 6 Find the mode of the statistical population: 53, 22, 47, 61, 65, 35, 47.

Solution Listing the variates in order, we have: 22, 35, 47, 47, 53, 61, 65. Since 47 occurs twice, and each other variate occurs only once, 47 is the mode. ■

EXAMPLE 7 Find the mode of the statistical population: 3, 4, 4, 5, 6, 6, 7.

Solution There are two modes: 4 and 6. ■

We have seen three different types of averages or measures of central tendency. Sometimes one type of measurement is more appropriate than the others. The mean is the best choice when every variate is to be given equal weight. The median is the best choice when there are one or two extreme values. The mode is the best choice for something that can only be measured in certain increments, such as clothing sizes.

EXAMPLE 8 Determine which type of average is most appropriate for each statistical population.

(a) \$76,000, \$45,000, \$72,000, \$74,000, and \$73,000 where the variates are the prices of homes in one neighborhood.

(b) 4, 5, 5 1/2, 6, 6, 6 1/2, 6 1/2, 6 1/2, 7, 7, 7 1/2, and 8 where the variates are sizes of shoes sold by a salesperson.

Solution **(a)** The median is the most appropriate average, because it is not affected by the one relatively small value.

(b) The mode is the most appropriate average, because shoe sizes occur only in whole and half sizes. ■

EXAMPLE 9 Determine which type of average is most appropriate for each statistical population.

(a) 92, 93, 94, 95, and 100 where the variates are scores of students on a test.

(b) \$22,000, \$20,000, \$28,000, \$19,000, and \$4500 where the variates are the annual incomes of five people.

Solution **(a)** The arithmetic mean is the most appropriate average, because it is affected by every score.

(b) The median is the most appropriate average, because it is not affected by the one relatively small value. ■

Averages, by themselves, often give an incomplete picture from which erroneous conclusions are drawn. For example, it was widely reported in 1984, that for the first time, the average price of a new home exceeded \$100,000. Many people concluded that it would be impossible for them to buy an average home. At the time, however, interest rates were high. Builders were constructing larger, more expensive homes that would appeal to affluent buyers for whom interest rate was not a primary consideration. The median price for resale homes was \$73,300 for the same period of time.

EXERCISES 20.1

In 1–11, find the arithmetic mean of each statistical population.

1. 7, 15, 12, 19, 13, 8, 10
2. 32, 19, 42, 30, 19, 40, 28
3. 41, 32, 50, 40, 55, 43, 32, 18, 58
4. 14, 5, 17, 12, 10, 5, 17, 8
5. 23, 10, 8, 10, 2, 25
6. 15, 2, 30, 11, 2, 30, 10, 16
7. 1, 1, 2, 3, 5, 8, 13, 21
8. 2, 3, 5, 7, 11, 13, 17, 19
9. 0.02, 0.04, 0.08, 0.16, 0.32
10. 1.1, 3.0, 2.2, 7.0, 1.0, 1.6, 3.0
11. 2.78, 3.14, 1.73, 1.41, 2.53

In 12–22, find the median of each statistical population.

12. 7, 15, 12, 19, 13, 8, 10
13. 32, 19, 42, 30, 19, 40, 28
14. 41, 32, 50, 40, 55, 43, 32, 18, 58
15. 14, 5, 17, 12, 10, 5, 17, 8
16. 23, 10, 8, 10, 2, 25
17. 15, 2, 30, 11, 2, 30, 10, 16
18. 1, 1, 2, 3, 5, 8, 13, 21
19. 2, 3, 5, 7, 11, 13, 17, 19
20. 0.02, 0.04, 0.08, 0.16, 0.32
21. 1.1, 3.0, 2.2, 7.0, 1.0, 1.6, 3.0
22. 2.78, 3.14, 1.73, 1.41, 2.53

In 23–33, find the mode of each statistical population.

23. 7, 15, 12, 19, 13, 8, 10
24. 32, 19, 42, 30, 19, 40, 28
25. 41, 32, 50, 40, 55, 43, 32, 18, 58
26. 14, 5, 17, 12, 10, 5, 17, 8
27. 23, 10, 8, 10, 2, 25
28. 15, 2, 30, 11, 2, 30, 10, 16
29. 1, 1, 2, 3, 5, 8, 13, 21
30. 2, 3, 5, 7, 11, 13, 17, 19
31. 0.02, 0.04, 0.08, 0.16, 0.32
32. 1.1, 3.0, 2.2, 7.0, 1.0, 1.6, 3.0
33. 2.78, 3.14, 1.73, 1.41, 2.53

In 34–43, determine the most appropriate average to use in each case.

34. Each amount is the price of an ounce of a certain chemical: $3, $4, $3.50, $4.25, $20.

35. Each number is the time (in seconds) for an assembly line robot to solder a seam: 10.3, 9.6, 10.4, 13.6, 10.6, 10.7, 10.8, 10.1.

36. Each number is the run time in seconds for a computer program used with different sets of data: 2.3, 1.9, 2.1, 1.8, 1.8, 2.0.

37. Each number is a hat size: 6 5/8, 6 7/8, 7 1/4, 7 1/2, 6 7/8.

38. Each number is the age of a patient admitted to the hospital: 3, 7, 19, 22, 25, 45, 67, 72, 75.

39. Each number is the length of a hospital stay in days: 1, 2, 2, 3, 4, 4, 107.

40. Each number is the systolic blood pressure of a patient in the hospital: 110, 120, 120, 130, 130, 135, 140, 160.

41. Each number is the number of workers sharing an office: 1, 2, 2, 2, 2, 3, 3, 4, 4.

42. Each number is the number of electrical outlets in offices of a building: 2, 2, 2, 3, 3, 4, 4, 4, 4, 4, 4.

43. Each number is the number of miles driven in one year by sales representatives for a certain company: 12,000, 15,000, 22,000, 24,000, 110,000.

Find the mean, median, and mode of each statistical population.

44. Each number is the time (in minutes: seconds) required to evacuate a building during a fire drill: 5:18, 4:36, 6:22, 5:14. 1 2 4 7 8 3 5

45. Each number is the breaking strain (in lbs) of an iron chain made of 1 3/8″ round iron: 88,301, 88,305, 88,296, 88,307. 1 2 4 8 7 9

46. Each number is current (in A) measured with an ammeter: 1.3, 1.2, 1.2, 1.4, 1.1. 3 5 9 1 2 4 6 7 8

47. Each number is potential difference (in V) measured with a voltmeter: 40, 42, 38, 42, 41, 44. 3 5 9 1 2 4 6 7 8

48. Each number is the number of white blood cells present in a 1 mm^3 sample of blood: 7000, 7500, 8000, 7500, 7500. 6 9 2 4 5 7 8

49. Each number is the number of hydrogen ions present in one mole of a solution: 2.23×10^{18}, 2.22×10^{18}, 2.21×10^{18}, 2.21×10^{18}, 2.23×10^{18}, 2.21×10^{18}. 6 9 2 4 5 7 8

50. Each number is the soil pressure (in tons/ft^2) measured at a building site: 0.51, 0.52, 0.48, 0.56, 0.57. 1 2 4 6 8 5 7 9

51. What happens to the mean, median, and mode, if each variate in the statistical population is
(a) increased by 4?
(b) multiplied by 4?

1. Architectural Technology
2. Civil Engineering Technology
3. Computer Engineering Technology
4. Mechanical Drafting Technology
5. Electrical/Electronics Engineering Technology
6. Chemical Engineering Technology
7. Industrial Engineering Technology
8. Mechanical Engineering Technology
9. Biomedical Engineering Technology

20.2 MEASURES OF DISPERSION

A measure of central tendency is useful for summarizing the data in a statistical population, but it does not give a complete picture. Consider the two populations.

39, 40, 41, 42, 43 and 1, 21, 41, 61, 81.

Each has a mean of 41, but the values in the first population are clustered more closely around that mean. The values in the second population are widely scattered on both sides of the mean. There are ways to measure this scatter or *dispersion*.

dispersion

DEFINITION The **range** is the difference between the largest and smallest numbers in a statistical population.

EXAMPLE 1 Find the mean and the range of each statistical population.

(a) 39, 40, 41, 42, 43

(b) 1, 21, 41, 61, 81

Solution **(a)** $\mu = 205/5$ or 41, and the range is $43 - 39$ or 4.

(b) $\mu = 205/5$ or 41, and the range is $81 - 1$ or 80. ■

EXAMPLE 2 Find the mean and the range of each statistical population.

(a) 47, 48, 49, 50, 51, 52, 53 **(b)** 1, 48, 49, 50, 51, 52, 99

Solution **(a)** $\mu = 350/7$ or 50, and the range is $53 - 47$ or 6.

(b) $\mu = 350/7$ or 50, and the range is $99 - 1$ or 98. ■

In Example 2, the two distributions have the same mean, and most of the values in each population are clustered around that mean. The two extreme values of 1 and 99 in the second population, however, produce a large range. A mean of 50 and a range of 98 could also describe the distribution

1, 2, 3, 50, 97, 98, 99.

In this distribution the values tend to be clustered near the largest and smallest values in the population.

A measure of dispersion that takes into account how much each score deviates from the mean would be more useful than the range, which is affected only by the largest and smallest values in the distribution. The *variance* is one such measure of dispersion.

DEFINITION The **deviation** is the difference between a single number in a statistical population and the mean of that population. The **variance** is the arithmetic mean of the squared deviations for all numbers in a distribution.

EXAMPLE 3 Find the variance of the following statistical population. Each number is the percent of instructions that access memory in one of the computer programs used by a particular company.

39, 40, 41, 42, 43

Solution The mean is 205/5 or 41.

The deviations are $39 - 41$, $40 - 41$, $41 - 41$, $42 - 41$, $43 - 41$ or -2, -1, 0, 1, and 2.

The squares of the deviations are 4, 1, 0, 1, 4.

The variance is given by $(4 + 1 + 0 + 1 + 4)/5 = 10/5$, or 2. ■

EXAMPLE 4 Find the variance of each distribution.

(a) 47, 48, 49, 50, 51, 52, 53 **(b)** 1, 48, 49, 50, 51, 52, 99

(c) 1, 2, 3, 50, 97, 98, 99

Solution **(a)** The mean is 50. The deviations are -3, -2, -1, 0, 1, 2, 3.

The squared deviations are 9, 4, 1, 0, 1, 4, 9.

The variance is given by $(9 + 4 + 1 + 0 + 1 + 4 + 9)/7 = 28/7$, or 4.

(b) The mean is 50. The deviations are $-49, -2, -1, 0, 1, 2, 49$.
The squared deviations are 2401, 4, 1, 0, 1, 4, 2401.
The variance is given by $(2401 + 4 + 1 + 0 + 1 + 4 + 2401)/7 = 4812/7$, or 687, to the nearest whole number.

(c) The mean is 50. The deviations are $-49, -48, -47, 0, 47, 48, 49$. The squared deviations are 2401, 2304, 2209, 0, 2209, 2304, 2401. The variance is given by $(2401 + 2304 + 2209 + 0 + 2209 + 2304 + 2401)/7 = 13{,}828/7$, or 1975, to the nearest whole number. ■

In Examples 4(b) and 4(c), the mean is 50 and the range is 98 for both populations. The variance indicates that the values in the population of 4(c) are more widely scattered.

One difficulty in using the variance as a measure of dispersion is that squaring the deviations also squares any units that are being used. For example, if our population is a group of college students, and the statistical population is their heights, the deviations will be expressed in inches, but the variance will be expressed in square inches, making comparison with the mean (expressed in inches) difficult. The most widely used measure of dispersion, the *standard deviation,* avoids this problem.

DEFINITION The **standard deviation,** denoted σ, for a statistical population is the square root of the variance.

EXAMPLE 5 Find the standard deviation of each statistical population.

(a) 39, 40, 41, 42, 43

(b) 1, 21, 41, 61, 81

Solution **(a)** The variance was calculated in Example 3. Since the variance is 2, the standard deviation is $\sqrt{2}$. Assuming that the integers denote exact values, $\sqrt{2}$ is 1.41, to three significant digits.

(b) The mean is 41. The variance is given by

$$(40^2 + 20^2 + 0^2 + 20^2 + 40^2)/5 = 4000/5 \text{ or } 800.$$

Since the variance is 800, the standard deviation is $\sqrt{800}$ or 28.3, to three significant digits. ■

EXAMPLE 6 Find the standard deviation of the statistical population: 1, 6, 11, 16, 21, 26, 31.

Solution The mean is 112/7 or 16. The variance is given by

$$(225 + 100 + 25 + 0 + 25 + 100 + 225)/7 = 700/7, \text{ or } 100.$$

Therefore σ is $\sqrt{100}$ or 10. ■

EXAMPLE 7 To study the effectiveness of measures taken to reduce sound levels, the decibel level in a typical office building was measured over a period of days. The readings were 68, 74, 67, 72, and 69. Find the mean and the standard deviation.

Solution The mean is 350/5 or 70.
The variance is given by $(4 + 16 + 9 + 4 + 1)/5 = 34/5$, or 6.8. Therefore σ is 2.6, to two significant digits. ■

Standard deviation is sometimes computed by using the formula $\sigma = \sqrt{\text{mean of (variates)}^2 - (\text{mean of variates})^2}$. A computer program that uses this formula would require only one "pass" through the data. Round-off error, however, is more serious with this formula than with the definition. Some calculators have special keys for finding the mean and standard deviation. You may want to consult your owner's manual to learn how to use the statistical keys on your calculator.

There is an interesting relationship between the range and the standard deviation. The standard deviation is always less than or equal to one-half of the range. You may wish to use this relationship to help detect arithmetic errors made in calculating σ. It will not show every error, but will show some of the worst.

EXERCISES 20.2

In 1–12, find the mean and the range of each statistical population.

1. 9, 10, 12, 13, 16
2. 11, 24, 25, 30, 35
3. 0, 1, 12, 23, 24
4. 4, 23, 25, 35, 38
5. 26, 41, 42, 44, 46, 59
6. 32, 36, 40, 44, 48, 52
7. 29, 30, 31, 55, 56, 57
8. 45, 46, 51, 53, 57, 61, 72
9. 25, 47, 50, 53, 56, 62, 92
10. 1.1, 1.2, 1.3, 8.5, 8.7, 8.8, 8.9
11. 2.0, 2.1, 2.2, 9.4, 9.6, 9.7, 9.8
12. 1.6, 3.8, 4.9, 4.4, 4.6, 4.7, 5.3, 8.3

In 13–24, find the variance for each statistical population.

13. 9, 10, 12, 13, 16
14. 11, 24, 25, 30, 35
15. 0, 1, 12, 23, 24
16. 4, 23, 25, 35, 38
17. 26, 41, 42, 44, 46, 59
18. 32, 36, 40, 44, 48, 52
19. 29, 30, 31, 55, 56, 57
20. 45, 46, 51, 53, 57, 61, 72
21. 25, 47, 50, 53, 56, 62, 92
22. 1.1, 1.2, 1.3, 8.5, 8.7, 8.8, 8.9
23. 2.0, 2.1, 2.2, 9.4, 9.6, 9.7, 9.8
24. 2.6, 3.8, 4.9, 4.4, 4.6, 4.7, 5.3, 9.3

In 25–36, find the standard deviation for each statistical population.

25. 4, 6, 8, 10
26. 7, 10, 16, 19
27. 1, 2, 12, 13
28. 10, 13, 14, 15
29. 19, 21, 22, 23, 25
30. 1, 4, 22, 40, 43
31. 3, 5, 10, 12, 17, 19
32. 1, 5, 8, 12, 15, 19
33. 2, 3, 5, 7, 11, 14
34. 1.1, 1.2, 1.3, 8.5, 8.7, 8.8, 8.9
35. 2.0, 2.1, 2.2, 9.4, 9.6, 9.7, 9.8
36. 2.6, 3.8, 4.9, 4.4, 4.6, 4.7, 5.3, 9.3

Find the mean and standard deviation for each population.

37. Each number is the deflection (in inches) for a stick of white pine 1″ × 1/2″, supported at points 24″ apart and loaded in the middle with 38 pounds: 0.77, 0.75, 0.79, 0.78, 0.76. 1 2 4 8 7 9

38. Each number is the effective pressure (in psi) on a piston of a steam engine: 59.75, 60.05, 59.50, 59.60, 60.15. 2 4 8 1 5 6 7 9

39. Each number is the volume (in yd^3) occupied by crushed rock that had a volume of 1 yd^3 before it was crushed: 2.22, 1.43, 1.83, 1.83, 1.96. 1 2 4 7 8 3 5

40. Each number is the amount of sun-induced expansion (in mm) of a steel I-beam 7.82 m long before expansion: 7.80, 7.75, 7.70, 7.82, 7.74. 1 2 4 6 8 5 7 9

41. Each number is the diameter (in cm) of a small steel shaft as measured with a micrometer: 1.442, 1.441, 1.443, 1.442, 1.441. 1 2 4 7 8 3 5

42. Each number is an estimated reduction (in percent) in heating costs when the thermostat is lowered from 22° C to 20° C: 9, 10, 9, 9, 8. 1 2 6 7 8 3 5 9

43. Each number is the pH of a patient's blood: 7.38, 7.40, 7.39, 7.41, 7.42. 6 9 2 4 5 7 8

44. What happens to the range and standard deviation if
(a) each variate is increased by 4?
(b) each variate is multiplied by 4?

1. Architectural Technology
2. Civil Engineering Technology
3. Computer Engineering Technology
4. Mechanical Drafting Technology
5. Electrical/Electronics Engineering Technology
6. Chemical Engineering Technology
7. Industrial Engineering Technology
8. Mechanical Engineering Technology
9. Biomedical Engineering Technology

20.3 HISTOGRAMS AND THE NORMAL CURVE

The data used in statistics may be classified as either discrete or continuous. For example, the number of television sets per household is a discrete variable. If the number of television sets increases, it must increase by 1 or 2 or so on, and not by 1/2 or 3/4. That is, measurements can be made only in certain increments. Height, on the other hand, is a continuous variable, because height does not have to be measured in intervals of 1″, but can assume any value between whole inches, such as 5′ 4 1/2″.

frequency distribution

A graphical interpretation of the data in a statistical population is often useful. One type of graph is called a *frequency distribution.* The horizontal axis shows the possible values for the variable, and the vertical axis indicates the number of times a value occurs in the population. Figure 20.1 shows a frequency distribution for the number of heads obtained when five coins were tossed 64 times.

class intervals

The horizontal axis is divided into intervals called *class intervals.* For a continuous variable, the center for an interval may be chosen arbitrarily. Each class interval, however must have the same width. For a discrete variable, each class interval is usually centered on one of the values the variable can assume. For example, an interval would be centered at 2, but would run from 1.5 to 2.5. A rectangle is drawn with the class interval as the base. The height of the rectangle is the number on the vertical axis that indicates how many variates fall within that class interval. Such a graph is called a *histogram.*

histogram

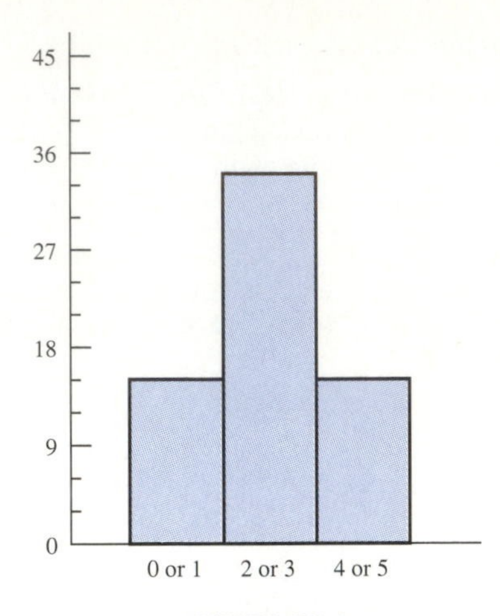

FIGURE 20.1

The histogram in Figure 20.1 illustrates the frequency with which different numbers of heads occur when five coins are tossed 64 times. The class intervals are each two units wide. If we show the same data on a histogram with class intervals of only 1 unit, as in Figure 20.2, the histogram begins to approach a shape called the *normal curve.*

normal curve

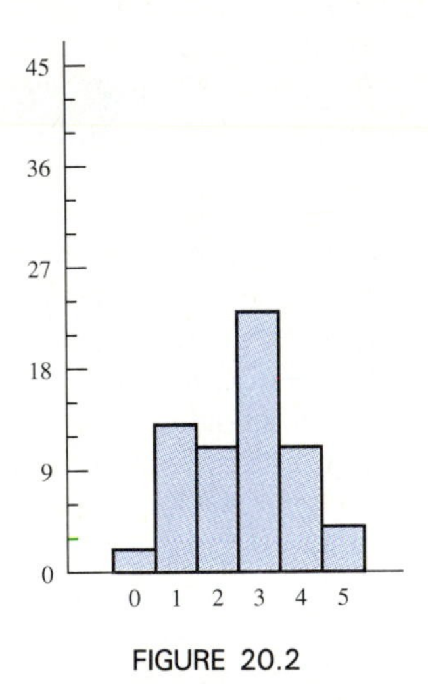

FIGURE 20.2

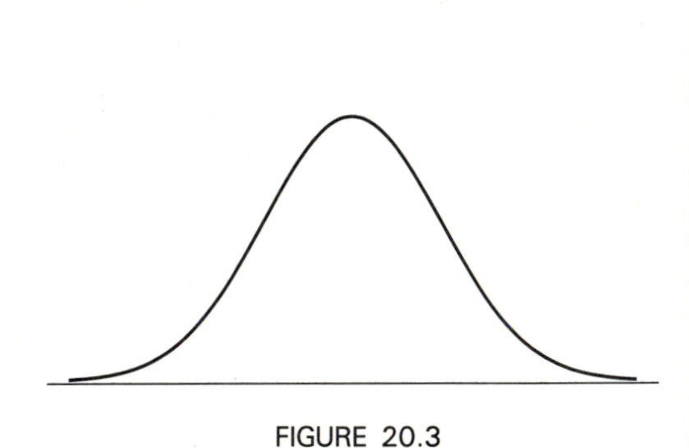

FIGURE 20.3

bell curve

The normal curve, shown in Figure 20.3, is sometimes called a *bell curve.* It is often familiar to students, as the scores on many standardized tests are distributed according to this curve. At one time it was thought that any data taken from nature, such as height or weight, would be normally distributed. Although many exceptions have been found, the normal curve still has important applications in statistical theory.

The normal curve is like a histogram for continuous variables, where each point on the horizontal axis represents a class interval. The vertical axis of a histogram does not have to show a scale. The graph can be drawn so that the area (rather than the height) of each rectangle represents the number of variates within that class interval. When drawn in this manner, the area under the entire histogram totals 100 percent.

There are several important characteristics of the normal curve.

(a) It is symmetric about the center line. That is, the left-hand side is a mirror image of the right-hand side.

(b) The area under the curve represents 100% of the population.

(c) The curve is completely specified by the mean and standard deviation of the population.

We say that the curve is completely specified by the mean and standard deviation of the population, because the curve is centered at the mean and approximately 68% of the area under the curve lies within 1 standard deviation on either side of the mean. Approximately 95% of the area under the curve lies within two standard deviations on either side of the mean; and approximately 99% of the area under the curve lies within three standard deviations of the mean. By placing the center of the normal curve at the mean, and locating the points along the horizontal axis that are ± 1, ± 2, and ± 3 standard deviations from the mean, we can draw the normal curve.

raw score

In order to use the normal curve for statistical analysis, it is necessary to convert the units of the horizontal axis (called *raw scores*) to units called *z-scores*, or *standard scores*. A z-score is a measure of how many standard deviations a number is from the mean. Thus, the mean, μ, has a z-score of 0. A z-score of 2 represents two standard deviations to the right of the mean. Negative z-scores indicate measurements to the left of the mean.

DEFINITION

For a variate x in a statistical population, the **standard score,** or ***z*-score,** denoted z, is given by

$$z = \frac{x - \mu}{\sigma}$$

where μ is the mean of the population and σ is the standard deviation of the population.

EXAMPLE 1

If a normally distributed population has a mean of 50 and a standard deviation of 5, find the z-score for each measurement.

(a) $x = 53$ **(b)** $x = 42$

Solution

(a) $z = (53 - 50)/5 = 3/5 = 0.6$

(b) $z = (42 - 50)/5 = -8/5 = -1.6$ ■

EXAMPLE 2 A chemical company packages sulfur in boxes. If the weights of the boxes are normally distributed with a mean of 12.0 oz and a standard deviation of 0.300 oz, by how many standard deviations is a box weighing 11.5 oz underweight?

Solution $z = (11.5 - 12)/0.3 = -0.5/0.3 = -5/3 = -1.67$, to three significant digits. ■

Table 20.1 shows what percent (written as a decimal) of the area under the normal curve lies between the mean and $|z|$ for various values of $|z|$. It is possible to use this table to answer many questions about a particular normal distribution.

TABLE 20.1 Area Under the Normal Curve
To find the percent of area under the normal curve between the mean and a particular *z*-score, locate the number in the left-hand column that has the same units and tenths digits as the *z*-score. Locate the entry in this row that is in the column headed by the hundredths digit of the *z*-score. Change this decimal entry to a percent.

	0	1	2	3	4	5	6	7	8	9
0.0	0.000	0.004	0.008	0.012	0.016	0.020	0.024	0.028	0.032	0.036
0.1	0.040	0.044	0.048	0.052	0.056	0.060	0.064	0.068	0.071	0.075
0.2	0.079	0.083	0.087	0.091	0.095	0.099	0.103	0.106	0.110	0.114
0.3	0.118	0.122	0.126	0.129	0.133	0.137	0.141	0.144	0.148	0.152
0.4	0.155	0.159	0.163	0.166	0.170	0.174	0.177	0.181	0.184	0.188
0.5	0.192	0.195	0.199	0.202	0.205	0.209	0.212	0.216	0.219	0.222
0.6	0.226	0.229	0.232	0.236	0.239	0.242	0.245	0.249	0.252	0.255
0.7	0.258	0.261	0.264	0.267	0.270	0.273	0.276	0.279	0.282	0.285
0.8	0.288	0.291	0.294	0.297	0.300	0.302	0.305	0.308	0.311	0.313
0.9	0.316	0.319	0.321	0.324	0.326	0.329	0.332	0.334	0.337	0.339
1.0	0.341	0.344	0.346	0.349	0.351	0.353	0.355	0.358	0.360	0.362
1.1	0.364	0.367	0.369	0.371	0.373	0.375	0.377	0.379	0.381	0.383
1.2	0.385	0.387	0.389	0.391	0.393	0.394	0.396	0.398	0.400	0.402
1.3	0.403	0.405	0.407	0.408	0.410	0.412	0.413	0.415	0.416	0.418
1.4	0.419	0.421	0.422	0.424	0.425	0.427	0.428	0.429	0.431	0.432
1.5	0.433	0.435	0.436	0.437	0.438	0.439	0.441	0.442	0.443	0.444
1.6	0.445	0.446	0.447	0.448	0.450	0.451	0.452	0.453	0.454	0.455
1.7	0.455	0.456	0.457	0.458	0.459	0.460	0.461	0.462	0.463	0.463
1.8	0.464	0.465	0.466	0.466	0.467	0.468	0.469	0.469	0.470	0.471
1.9	0.471	0.472	0.473	0.473	0.474	0.474	0.475	0.476	0.476	0.477
2.0	0.477	0.478	0.478	0.479	0.479	0.480	0.480	0.481	0.481	0.482
2.1	0.482	0.483	0.483	0.483	0.484	0.484	0.485	0.485	0.485	0.486
2.2	0.486	0.486	0.487	0.487	0.488	0.488	0.488	0.488	0.489	0.489
2.3	0.489	0.490	0.490	0.490	0.490	0.491	0.491	0.491	0.491	0.492
2.4	0.492	0.492	0.492	0.493	0.493	0.493	0.493	0.493	0.493	0.494
2.5	0.494	0.494	0.494	0.494	0.495	0.495	0.495	0.495	0.495	0.495
2.6	0.495	0.496	0.496	0.496	0.496	0.496	0.496	0.496	0.496	0.496
2.7	0.497	0.497	0.497	0.497	0.497	0.497	0.497	0.497	0.497	0.497
2.8	0.497	0.498	0.498	0.498	0.498	0.498	0.498	0.498	0.498	0.498
2.9	0.498	0.498	0.498	0.498	0.498	0.498	0.499	0.499	0.499	0.499
3.0	0.499	0.499	0.499	0.499	0.499	0.499	0.499	0.499	0.499	0.499
3.1	0.499	0.499	0.499	0.499	0.499	0.499	0.499	0.499	0.499	0.499
3.2	0.499	0.499	0.499	0.499	0.499	0.499	0.499	0.500	0.500	0.500

We might ask what percent of the boxes of sulfur (in Example 2) weigh more than 12.7 oz? We convert 12.7 oz to a z-score.

$$z = (12.7 - 12)/0.3 = 0.7/0.3 = 7/3 \text{ or about } 2.33$$

We are asking what percent of the area lies to the right of this z-score. In Figure 20.4, this area is shaded.

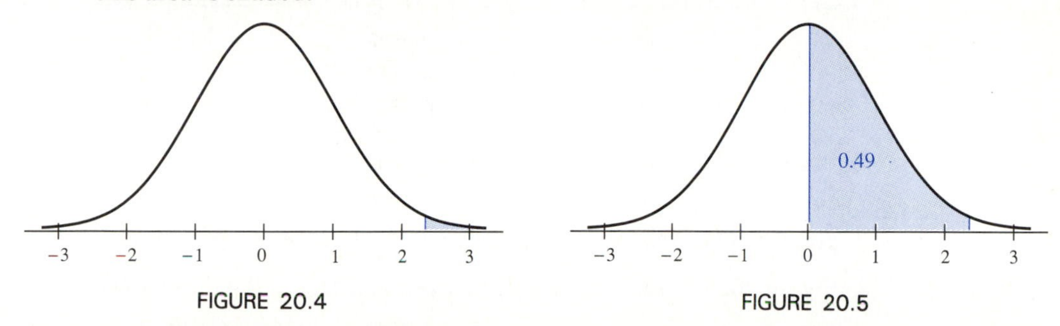

FIGURE 20.4

FIGURE 20.5

Table 20.1 tells us that 0.49 or 49 percent of the area under the curve lies between z-scores of 0 and 2.33 (that is, between 12 and 12.7), as shown in Figure 20.5. Since 50 percent of the area lies to the right of the mean, the area we want is $0.50 - 0.49$ or 0.01. Therefore 1 percent of the boxes weigh more than 12.7 oz.

To find the percent of variates in a population that satisfy a condition:

1. Convert raw scores to standard scores.
2. Sketch the normal curve and shade in the area that satisfies the condition. The mean is at the center of the normal curve.
3. Use Table 20.1 to determine the area under the curve between the mean and the boundary of the shaded area, and sketch the normal curve with this area shaded.
4. Compare the two sketches from steps 2 and 3. Remember that the area from the mean to the far right (or left) of the normal curve is 50% of the area under the curve.
5. Add or subtract as necessary to determine the area shaded in step 2.

EXAMPLE 3 What percent of the boxes of sulfur (see Example 2) weigh between 11.8 and 12.3 oz?

Solution

1. We want the area between 11.8 and 12.3.
   The z-score associated with 11.8 is given by

$$z = (11.8 - 12)/0.3 = -0.2/0.3 = -0.67$$
$$|z| = 0.67$$

The z-score associated with 12.3 is given by

$$z = (12.3 - 12)/0.3 = 0.3/0.3 = 1.00$$
$$|z| = 1.00$$

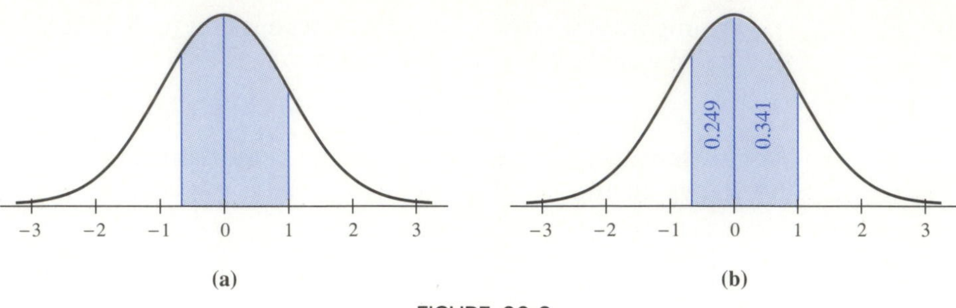

FIGURE 20.6

2. Figure 20.6(a) illustrates the area we want.
3. Table 20.1 gives the area between the mean and a z-score of 0.67 as 0.249 or 24.9 percent. The area between the mean and a z-score of 1.00 is 0.341 or 34.1 percent.
4. Figure 20.6(b) shows the area between z-scores of 0.67 and 1.00.

We conclude that the area between −0.67 and 1.00 is 24.9% + 34.1% or 59.0% of the boxes weigh between 11.8 and 12.3 oz. ■

EXAMPLE 4 What percent of the boxes of sulfur weigh less than 11.5 oz?

Solution The z-score associated with 11.5 is given by

$$z = (11.5 - 12)/0.3 = -5/3 = -1.67.$$
$$|z| = 1.67$$

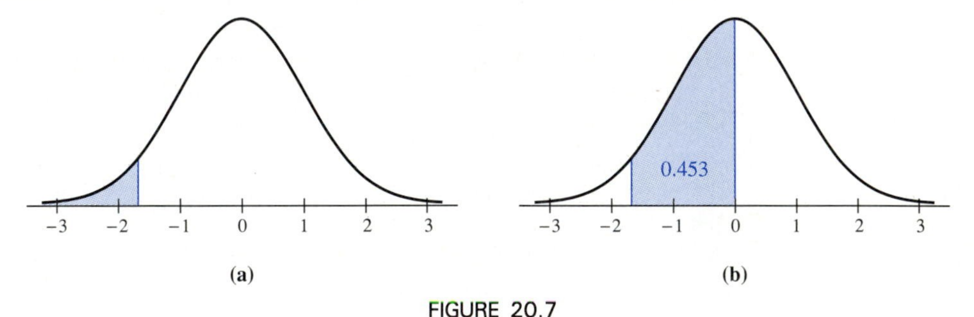

FIGURE 20.7

Table 20.1 gives the area between the mean and a z-score of 1.67 as 0.453 or 45.3 percent. Figure 20.7 illustrates the area we want and the area given by Table 20.1. The percent of boxes that weigh less than 11.5 oz is 50% − 45.3% or 4.7%. ■

EXERCISES 20.3

In 1–5, find the z-score for each measurement (to the nearest hundredth). The normally distributed population has a mean of 45 *and a standard deviation of* 4.

1. $x = 51$ **2.** $x = 39$ **3.** $x = 55$ **4.** $x = 35$ **5.** $x = 34$

In 6–10, find the z-score for each measurement (to the nearest hundredth). The normally distributed population has a mean of 32 and a standard deviation of 7.

6. $x = 42$ **7.** $x = 22$ **8.** $x = 15$ **9.** $x = 49$ **10.** $x = 52$

In 11–13, assume that a company packages printer paper so that the lengths of the rolls are normally distributed with an average (mean) length of 100 feet and a standard deviation of 3 feet. 1 2 4 7 8 3 5

11. What percent of the rolls are over 105 feet long?

12. What percent of the rolls are less than 98 feet long?

13. What percent of the rolls are between 96 and 104 feet long?

In 14–16, assume that a company manufactures ball bearings with diameters that are normally distributed with an average diameter of 0.1200 cm and a standard deviation of 0.0050 cm. 1 2 4 7 8 3 5

14. What percent of the ball bearings have a diameter greater than 0.1300 cm?

15. What percent of the ball bearings have a diameter less than 0.1125 cm?

16. What percent of the ball bearings are between 0.1175 cm and 0.1225 cm in diameter?

In 17–19, assume that the times it takes lab technicians to perform a certain test are normally distributed with a mean of 1 hour 15 minutes and a standard deviation of 7 minutes. 1 2 6 7 8 3 5 9

17. What percent of the technicians take longer than 1 hour 25 minutes?

18. What percent of the technicians take less than 1 hour?

19. What percent of the technicians take between 1 hour 12 minutes and 1 hour 18 minutes?

In 20–22, assume that the typing speeds of the secretaries in a typing pool are normally distributed with a mean of 80 words per minute and a standard deviation of 10. 1 2 6 7 8 3 5 9

20. What percent of them type between 75 and 90 words per minute?

21. What percent of them type between 72 and 82 words per minute?

22. What percent of them type between 68 and 88 words per minute?

In 23–25, assume that a hardware company finds that the weights of boxes of nails are normally distributed with an average weight of 16 oz and a standard deviation of 2 oz. 1 2 6 7 8 3 5 9

23. What percent of the boxes weigh between 11 oz and 19 oz?

24. What percent of the boxes weigh between 15 oz and 18 oz?

25. What percent of the boxes weigh between 13 oz and 17 oz?

In 26–28, assume that the scores for a test are normally distributed with a mean of 100 and a standard deviation of 10. 1 2 6 7 8 3 5 9

26. What percent of the scores are between 90 and 105?

27. What percent of the scores are between 98 and 107?

28. What percent of the scores are between 83 and 113?

In 29–31, assume that a computer engineer finds that when he tallies the percent of instructions that access memory in the programs used by a particular company, the percents are normally distributed with a mean of 40 and a standard deviation of 3.7. 3 5 9

29. What percent of the programs access memory for between 42% and 46% of the instructions?

30. What percent of the programs access memory for between 38% and 44% of the instructions?

31. What percent of the programs access memory for between 32% and 34% of the instructions?

1. Architectural Technology
2. Civil Engineering Technology
3. Computer Engineering Technology
4. Mechanical Drafting Technology
5. Electrical/Electronics Engineering Technology
6. Chemical Engineering Technology
7. Industrial Engineering Technology
8. Mechanical Engineering Technology
9. Biomedical Engineering Technology

20.4 REGRESSION LINES

descriptive statistics

inferential statistics

The study of gathering, analyzing, and interpreting data through measures of central tendency and measures of dispersion is called *descriptive statistics.* The study of using collected data to make predictions is called *inferential statistics.*

empirical data

scatter diagram

In the technologies, inferential statistics is used to study the relationship between two variables. When the value of one variable is thought to affect the value of the other variable, a series of observations is made. Data obtained through observation or experimentation is said to be *empirical.* The values for the variables are plotted on a rectangular coordinate system, and the graph is called a *scatter diagram.* The horizontal axis is used for the variable that is thought to influence a change in the other variable. It may be possible to write an equation that gives the relationship between variables by finding the equation of the line connecting the points. The points of a scatter diagram, however, rarely fall in a perfectly straight line. Due to inaccuracies involved in measurement, the points will probably appear like those in Figure 20.8.

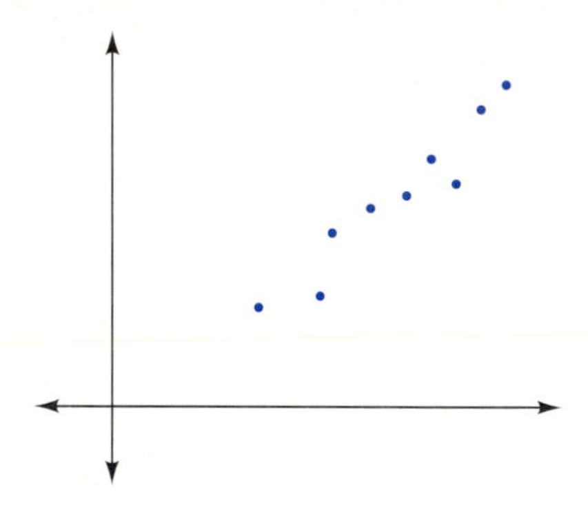

FIGURE 20.8

regression line

Our goal is to find the equation of a line that minimizes the discrepancy for most of the points. The most obvious approach would be to seek a line such that the sum of the vertical distances between each point and the line would be as small as possible. If we consider points above the line to have a positive deviation and points below the line to have a negative deviation, then the positive deviations and negative deviations could very well produce a sum of zero. In fact, many lines can be found to produce a sum of zero, so this approach is unacceptable. The solution is to square the deviations, just as we did in finding a standard deviation. We then seek the line that minimizes the sum of the squared deviations. This line is called the *regression line.* The regression line on a scatter diagram is analogous to the average in a list of numbers. It tells us that although a different value may be obtained for the dependent variable each time a particular value of the independent variable is used, the average value should occur on the regression line.

To write an equation that best describes the relationship, a technique often employed in the social sciences is used. The *coefficient of correlation* is a measure of how closely the points of a scatter diagram cluster around a line. The closer the coefficient of correlation is to 1 in absolute value, the more closely the points tend to

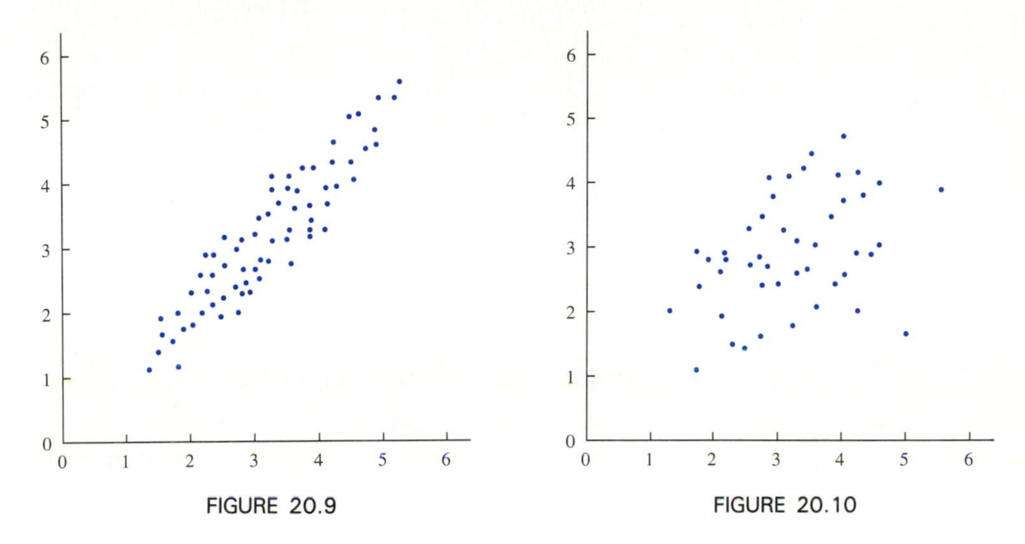

FIGURE 20.9

FIGURE 20.10

fall along a line. Figure 20.9 shows data with a coefficient of correlation of about 0.90 while Figure 20.10 shows data with a coefficient of correlation of about 0.10.

coefficient of correlation

The **coefficient of correlation** is the average (mean) of the products xy when the coordinates of each ordered pair (x, y) are converted to z-scores. The coefficient of correlation is also the slope of the regression line when the values of both x and y are converted to z-scores.

Recall that when the slope m and y-intercept b of a line are known, it is possible to write the equation of the line. The y-intercept is found by using the fact that the regression line must go through $(\bar{x}, \bar{y})$ where $\bar{x}$ denotes the mean of the x-values and $\bar{y}$ denotes the mean of the y-values. We have $y = mx + b$ for all values of x and y. Thus,

$$\bar{y} = m\bar{x} + b \text{ for } (\bar{x}, \bar{y},).$$

From this equation, it follows that

$$b = \bar{y} - m\bar{x}.$$

The slope of the regression line $(m = \overline{xy})$ and the y-intercept $(b = \bar{y} - m\bar{x})$ are easy to remember in this form, but it requires a great deal of computation to convert all values to z-scores in order to calculate m and b. In practice, the following formulas, which are equivalent to those using z-scores, are used with the raw scores. The formulas are more complicated in this form, but the computation is less complicated. The symbol Σ in Equation 20-1 is read "the sum of all." The regression line obtained by using these formulas is sometimes called the line of least-squares-best-fit.

SLOPE AND INTERCEPT OF THE REGRESSION LINE

$$m = \frac{n(\Sigma\, xy) - (\Sigma\, x)(\Sigma\, y)}{n(\Sigma\, x^2) - (\Sigma\, x)^2} \tag{20-1}$$

$$b = \bar{y} - m\bar{x} \tag{20-2}$$

where $\bar{x}$ and $\bar{y}$ are the averages of the x-values and y-values, respectively, and Σ means "the sum of all."

It is important to pay attention to the parentheses in the formula. For example, $(\Sigma\, x^2)$ means to find x^2 for each x and then sum the results, but $(\Sigma\, x)^2$ means to sum the values of x and then square the result. To use the formulas, the following steps are generally used.

To find the equation of the regression line:

1. Set up a table with column headings x, y, xy, and x^2.
2. Enter the values of x, y, xy, and x^2 in the appropriate columns of the table for each x and y.
3. Calculate the sum of the entries in each column.
4. Calculate the average x-value $\bar{x}$ and the average y-value $\bar{y}$.
5. Calculate m and b using Equations 20-1 and 20-2.
6. Interpret $y = mx + b$ as the equation for the regression line.

EXAMPLE 1 The following measurements are given for two variables.

x	2	4	7	8	9	10
y	6	10	13	16	19	21

(a) Construct a scatter diagram. **(b)** Write the equation for the regression line.
(c) Plot the regression line on the scatter diagram.

Solution **(a)** The scatter diagram in Figure 20.11 shows the relationship to be approximately linear.

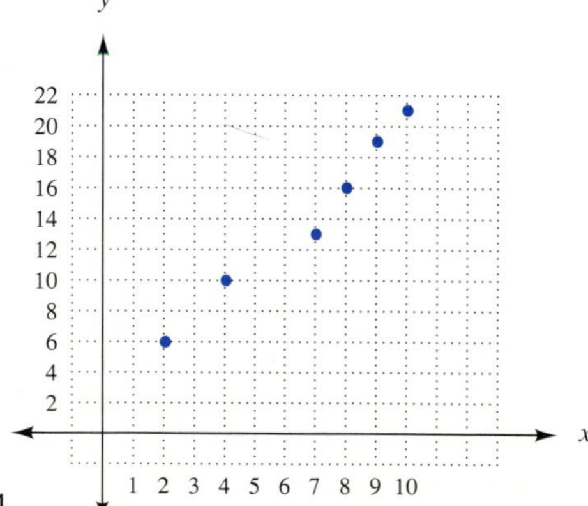

FIGURE 20.11

(b) The steps are numbered to correspond with the steps in the summary box.

	x	y	xy	x^2	
1.					Write down column headings.
2.	2	6	12	4	Record the values of x, y, xy, and x^2.
	4	10	40	16	
	7	13	91	49	
	8	16	128	64	
	9	19	171	81	
	10	21	210	100	
3.	40	85	652	314	Calculate the column sums.

4. $\bar{x} = \dfrac{\Sigma x}{6}$ or $\dfrac{40}{6}$ and $\bar{y} = \dfrac{\Sigma y}{6}$ or $\dfrac{85}{6}$ Calculate $\bar{x}$ and $\bar{y}$.

5. Before calculating m and b, we list the values needed: $n = 6$, $\Sigma\, xy = 652$, $\Sigma\, x = 40$, $\Sigma\, y = 85$, $\Sigma\, x^2 = 314$, $\bar{x} = 40/6$, and $\bar{y} = 85/6$. Since x and y were given as integers that are assumed to be exact in this problem, calculations will be carried to three significant digits. Thus,

$$m = \frac{6(652) - 40(85)}{6(314) - (40)^2}.$$

The denominator is calculated first and put into memory. Then the numerator is calculated. Finally, the division is performed. The calculator keystroke sequence follows.

6 [×] 314 [−] 40 [x^2] [=] [STO]

6 [×] 652 [−] 40 [×] 85 [=] [÷] [RCL] [=] **1.8028169**

Using this value of m on the display, b is calculated. $b = (85/6) - m(40/6) = 4.23$

The keystroke sequence is

display [×] 40 [÷] 6 [=] [+/−] [+] 85 [÷] 6 [=] **2.1478874**

6. The equation, then, is $y = 1.80x + 2.15$.

(c) Figure 20.12 shows the line. ■

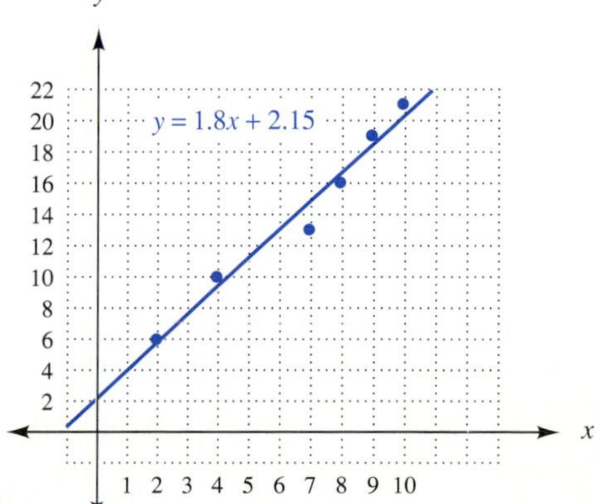

FIGURE 20.12

EXAMPLE 2 As the voltage across a 5-Ω resistor changes, the measured current changes. The following data were collected.

Voltage (V)	0.0	10.0	20.0	30.0	40.0	50.0
Current (A)	0.0	2.1	3.9	6.0	7.9	10.0

(a) Construct a scatter diagram.

(b) Write the equation for the regression line.

(c) Plot the regression line on the scatter diagram.

Solution **(a)** The scatter diagram in Figure 20.13 shows the relationship to be approximately linear.

(b)

x	y	xy	x^2
0.0	0.0	0	0
10.0	2.1	21	100
20.0	3.9	78	400
30.0	6.0	180	900
40.0	7.9	316	1600
50.0	10.0	500	2500
150.0	29.9	1095	5500

From the table, we find the values needed in the formulas.

$n = 6$
$\Sigma xy = 1095$
$\Sigma x = 150$
$\Sigma y = 29.9$
$\Sigma x^2 = 5500$
$\bar{x} = 150/6$
$\bar{y} = 29.9/6$

$$m = \frac{6(1095) - 150(29.9)}{6(5500) - (150)^2} \text{ or } 0.20, \text{ to two significant digits}$$

$$b = (29.9/6) - m(150/6) = -0.02$$

Since the equation will be used to estimate y, and y-values are given to two significant digits, we have $y = 0.20x - 0.02$.

(c) Figure 20.14 shows the line. ■

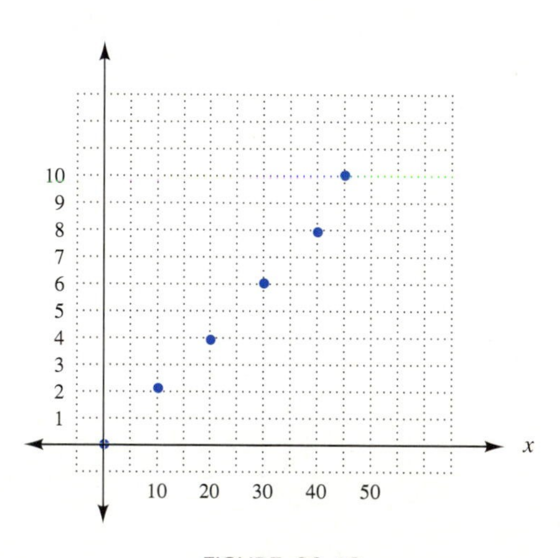

FIGURE 20.13

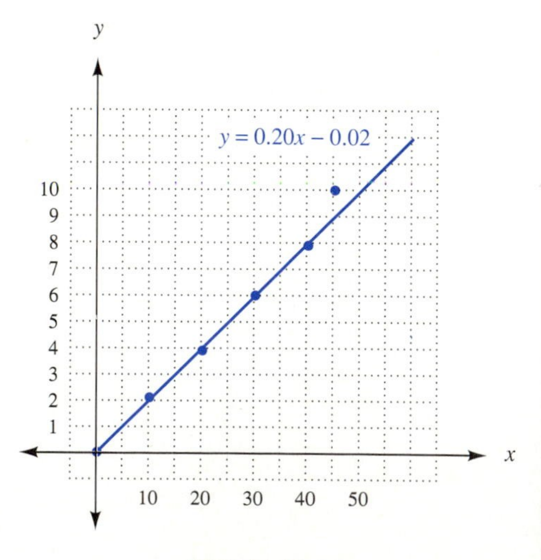

FIGURE 20.14

Before finding the equation of the regression line, it is important to determine whether the association between variables appears to be linear. A regression line can be found for any scatter diagram, but would be quite misleading when the points do not cluster around a line. The points in Figure 20.15 show a nonlinear association. One further cautionary note is in order. Even if the association appears to be linear, it is unwise to continue the line beyond the end points of the interval that contains the empirical values, for the association may not be linear outside that interval.

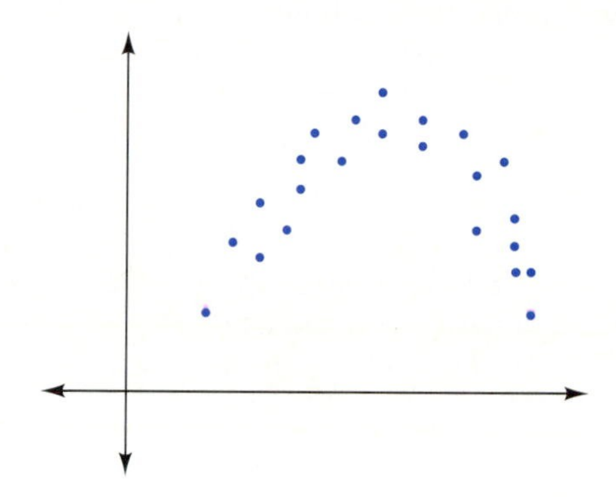

FIGURE 20.15

EXERCISES 20.4

In 1–10, construct a scatter diagram, write the equation for the regression line, and plot the line on the diagram for each set of values.

1.

x	3	6	9	12	15
y	10	15	25	35	30

2.

x	2	5	8	10	16
y	9	16	19	26	29

3.

x	4	7	10	13	16
y	11	14	20	24	31

4.

x	1	4	7	10	13
y	8	15	27	35	28

5.

x	2	6	7	10	11
y	4	10	11	16	18

6.

x	4	5	7	9	10	11
y	6	9	10	15	17	20

7.

x	1.8	2.5	2.8	3.3	3.6	4.2
y	5.4	6.3	5.4	6.1	6.8	7.0

8.

x	3.7	4.4	4.7	5.2	5.5	6.1
y	4.7	5.6	4.7	5.4	5.1	6.3

9.

x	3.1	4.2	6.0	7.3	8.4	9.6
y	2.0	3.9	8.0	10.4	12.2	14.4

10.

x	1.8	4.1	7.8	10.6	12.4	14.2
y	5.0	6.1	8.0	9.2	10.3	11.5

11. The table shows a correspondence between the temperature T (in degrees C) of a copper wire and its resistance R (in Ω). Write the equation for the regression line.

T	0.0	20.0	40.0	60.0	80.0
R	9.0	10.0	10.6	11.6	12.6

3 5 9 1 2 4 6 7 8

12. The table shows a correspondence between the distance d (in cm) that a spring is stretched and the stretching forces F (in newtons). Write the equation for the regression line.

d	1.25	2.50	3.75	5.00
F	24.0	51.0	76.0	98.0

3 5 9 1 2 4 6 7 8

13. The table shows a correspondence between the depth d (in m) below the surface of a liquid and the pressure p (in kg/cm^2) exerted by the liquid on a point at that depth. Write the equation for the regression line.

d	0.0	5.0	10.0	15.0	20.0
p	0.0	18.0	35.0	53.0	70.0

2 4 8 1 5 6 7 9

14. The table shows a correspondence between the elevation h (in m) and the temperature T (in degrees C) at that height if the ground temperature is 0° C. Write the equation for the regression line.

h	100	500	1000	1500	2000
T	0	−5.0	−9.0	−16.0	−20.0

6 8 1 2 3 5 7 9

15. The table shows a correspondence between the time of use t (in hours) and the energy used e (in kwh) for a 75-watt light bulb. Write the equation for the regression line.

t	0	8	12	20	24
e	0	0.6	0.8	1.6	1.8

3 5 9 1 2 4 6 7 8

16. The table shows a correspondence between the temperature T (in degrees C) and the volume V (in cm^3) of a gas. Write the equation for the regression line.

T	15	30	45	60
V	480	500	530	560

6 8 1 2 3 5 7 9

17. The table shows a correspondence between the time t (in microseconds) for the failure of a computer motherboard and the temperature T (in degrees C). Write the equation for the regression line.

t	104	331	546	769	986
T	102	96	89	85	80

3 5 9

1. Architectural Technology
2. Civil Engineering Technology
3. Computer Engineering Technology
4. Mechanical Drafting Technology
5. Electrical/Electronics Engineering Technology
6. Chemical Engineering Technology
7. Industrial Engineering Technology
8. Mechanical Engineering Technology
9. Biomedical Engineering Technology

20.5 CHAPTER REVIEW

KEY TERMS

20.1 central tendency
arithmetic mean
population
variate
statistical population
median
mode
bimodal

20.2 dispersion
range
deviation
variance
standard deviation

20.3 frequency distribution
class intervals
histogram
normal or bell curve
raw score
standard score or z-score

20.4 descriptive statistics
inferential statistics
empirical data
scatter diagram
regression line
coefficient of correlation

SYMBOLS

μ the mean of a population

σ standard deviation of a population

z z-score

Σ the sum of all

$\bar{x}$ the mean of a sample of a population

$\bar{y}$ the mean of a sample of a population

RULES AND FORMULAS

$$z = \frac{x - \mu}{\sigma}$$

For the regression line $m = \dfrac{n(\Sigma xy) - (\Sigma x)(\Sigma y)}{n(\Sigma x^2) - (\Sigma x)^2}$ and $b = \bar{y} - m\bar{x}$

SEQUENTIAL REVIEW

(Section 20.1) *In 1–3, find the arithmetic mean for each population.*

1. 4, 9, 11, 4, 7

2. 3, 10, 10, 5, 6, 8

3. 2, 11, 9, 6, 5

In 4–6, find the median for each population.

4. 4, 9, 11, 4, 7

5. 3, 10, 10, 5, 6, 8

6. 2, 11, 9, 6, 5

In 7–9, find the mode for each population.

7. 4, 9, 11, 4, 7

8. 3, 10, 10, 5, 6, 8

9. 2, 11, 9, 6, 5

In 10 and 11, determine the most appropriate average to use.

10. Each amount is the cost of performing a test at a local laboratory: \$25, \$28, \$30, \$32, \$75.

11. Each number is the number of hours spent by students on a homework assignment: 4, 3, 2, 4, 2.

(Section 20.2) *In 12–15, find the mean and the range for each population.*

12. 27, 40, 43, 43, 47, 58

13. 6, 13, 20, 21, 21, 23, 29

14. 54, 80, 86, 86, 94

15. 3, 6, 10, 10, 11, 12, 15

In 16–19, find the variance for each population.

16. 5, 22, 26, 34, 38

17. 26, 41, 42, 44, 45, 60

18. 13, 21, 22, 23, 23, 30

19. 11, 45, 53, 65, 76

In 20–23, find the standard deviation for each population.

20. 2, 5, 23, 41, 44

21. 0, 1, 3, 6, 9, 9, 10, 10

22. 3, 9, 45, 81, 87

23. 3, 5, 11, 16, 17, 19, 20

(Section 20.3) *In 24–26, find the z-score for each raw score x. The normally distributed population has a mean of* 38 *and a standard deviation of* 6.

24. $x = 29$

25. $x = 52$

26. $x = 41$

In 27–29, find the z-score for each raw score x. The normally distributed population has a mean of 19 *and a standard deviation of* 4.

27. $x = 15$ **28.** $x = 25$ **29.** $x = 17$

In 30–32, assume a normally distributed population with a mean of 66 *and a standard deviation of* 8.

30. What percent of the variates are larger than 78?

31. What percent of the variates are smaller than 58?

32. What percent of the variates are between 50 and 82?

In 33–35, assume a normally distributed population with a mean of 35.0 *and a standard deviation of* 0.6.

33. What percent of the variates are larger than 34.4?

34. What percent of the variates are smaller than 35.9?

35. What percent of the variates are between 34.4 and 35.9?

(Section 20.4) *In 36 and 37, construct a scatter diagram, find the equation for the regression line, and plot it on the diagram for each set of values.*

36.

x	2	4	6	8
y	3	4	5	9

37.

x	1	3	5	9
y	−1	2	5	10

In 38–41, find the slope of the regression line.

38.

x	2	3	4	5	6
y	7	9	11	15	17

39.

x	10	20	30	40	50
y	15	45	60	75	100

40.

x	0.00	0.50	1.0	1.5	2.0	2.5
y	0.00	0.70	1.7	2.3	3.0	3.7

41.

x	0.00	0.20	0.40	0.60	0.80	1.00
y	0.30	0.40	0.40	0.60	0.70	0.70

In 42–45, find the equation of the regression line.

42.

x	1.1	2.2	3.3	4.4
y	5.6	6.4	7.7	8.9

43.

x	2.0	3.0	5.0	6.0
y	2.1	3.4	5.5	6.5

44.

x	12	18	24	30
y	0.01	0.04	0.03	0.04

45.

x	100	200	300	400
y	50	95	155	200

APPLIED PROBLEMS

1. Find the mean, median, and mode. Each number is the amount of radioactivity (in roentgens/hour) present in the vicinity of a nuclear explosion after a three-hour period of time: 15.8, 15.7, 15.6, 15.8, 15.9, 16.0. 6 9 5 8

2. Find the mean and standard deviation. Each number is the amount (in mg) of mercury compounds found in a kiloliter of water from a reservoir: 6.8, 7.1, 8.2, 7.1. 6 9 2 4 5 7 8

1. Architectural Technology
2. Civil Engineering Technology
3. Computer Engineering Technology
4. Mechanical Drafting Technology
5. Electrical/Electronics Engineering Technology
6. Chemical Engineering Technology
7. Industrial Engineering Technology
8. Mechanical Engineering Technology
9. Biomedical Engineering Technology

In 3–5, assume that a chemical is bottled so that the weights of the bottles are normally distributed with an average weight of 12 *oz and a standard deviation of* 0.5 *oz.*

3. What percent of the bottles weigh more than 13.5 oz? **6 9** 2 4 5 7 8

4. What percent of the bottles weigh less than 11.0 oz? **6 9** 2 4 5 7 8

5. What percent of the bottles weigh between 11.75 and 12.75 oz? **6 9** 2 4 5 7 8

6. A prototype coffee machine was given a trial run and dispensed the following amounts of coffee in ounces.

6.01, 6.50, 4.98, 7.00, 6.00, 7.85, 8.00

Compute the mean and standard deviation of this sample. **1 2 6 7 8** 3 5 9

7. The following table gives the current and voltage for a certain resistor. The technician taking the data wishes to know the mean and standard deviation for both I and V. Find all four quantities.

I	8.7	17.4	26.1	30.4	43.5	52.2	65.2
V	2	4	6	7	10	12	15

3 5 9 1 2 4 6 7 8

8. In a certain assembly plant, automobile transmissions are built by three teams of various experience. Team A can build a unit in 23 minutes. Team B in 16 minutes, and team C in 12 minutes.

(a) What is the average time per unit for all three teams? (Hint: Determine how many units each team can make in 1 minute, then determine the total number of units made in 1 minute.)

(b) How many transmissions could this group build in one 7-hour day? **1 2 6 7 8** 3 5 9

9. An exam was given to 25 people. The mean score was 68 with a standard deviation of 13. The instructor gives B's to any student whose grade lies between 1 and 2 standard deviations above the mean. How many B's would he expect to give if the population is normally distributed? **1 2 6 7 8** 3 5 9

10. A statics instructor gave a test on which the average grade was 53% with a standard deviation of 17. Anyone with a grade lower than two standard deviations below the mean failed. How many people in a class of 20 could be expected to fail if the population is normally distributed? **1 2 6 7 8** 3 5 9

11. One brand of heat pump has a life expectancy of 12 years with a standard deviation of 3 years. If this heat pump is guaranteed for 7 years, what percent of them will require replacement? What if the warranty were extended to 10 years? Give your answer to the nearest tenth of a percent. **1 2 6 7 8** 3 5 9

12. A manufacturer has 100,000 washers having average diameter 0.950 inches with a standard deviation of 0.003 inches. If the population is normally distributed, how many could they sell legally if they advertise them as having diameter $0.950 \pm .001$ inch? Give your answer to two significant digits. **1 2 6 7 8** 3 5 9

13. A random sample of 500 resistors yielded an average value of 820 ohms with standard deviation 62 ohms.

(a) How many resistors will have resistance greater than 900 ohms?

(b) Find the percent of "good" resistors if good is defined as average $\pm 10\%$. **1 2 6 7 8** 3 5 9

In 14–16, assume that the speeds of 25 *automobiles clocked by a highway survey crew were normally distributed with an average of* 58 *mph and a standard deviation of* 6 *mph. Give your answers to the nearest tenth of a percent.*

14. What percent of automobiles were traveling over 60 mph? **1 2 6 7 8** 3 5 9

15. What percent of automobiles were traveling under 48 mph? **1 2 6 7 8** 3 5 9

16. What percent were within ± 1 mph of the 55 mph speed limit? **1 2 6 7 8** 3 5 9

17. A random sample of 600 circuit boards yielded an average of 14 solder defects per board with a standard deviation of 3 defects. Assume a normal distribution. Give your answer to the nearest tenth of a percent.

(a) What percent of the boards would have no more than 6 defects?

(b) What percent of the boards would have more than 20 defects? **1 2 6 7 8** 3 5 9

18. A technician measured the voltage V (in volts) and corresponding current I (in milliamps) for a certain type of diode and recorded the following data:

V	0.500	0.600	0.700	0.800	0.900	1.00
I	1.10	5.00	4.20	11.5	25.0	29.0

Find the equation that gives the regression line for this data. 3 5 9 4 7 8

19. The yearly coal production of a certain region was monitored every ten years from 1940 to 1980. The results (in thousands of tons) are given in the following table.

Year	1940	1950	1960	1970	1980
Amount	1000	850	750	450	400

Find the equation of the regression line for this data. 1 2 6 7 8 3 5 9

20. The following data, taken in the lab, show how a particular transistor's current decreases with decreasing voltage.

V	0	−1	−2	−3	−4	−5
I	0.065	0.040	0.010	0.0025	0.005	0.001

Find the equation of the regression line for this data using three significant digits for m and b. 3 5 9 4 7 8

1. Architectural Technology
2. Civil Engineering Technology
3. Computer Engineering Technology
4. Mechanical Drafting Technology
5. Electrical/Electronics Engineering Technology
6. Chemical Engineering Technology
7. Industrial Engineering Technology
8. Mechanical Engineering Technology
9. Biomedical Engineering Technology

RANDOM REVIEW

1. Assume a normally distributed population with a mean of 305 and a standard deviation of 20. What percent of the variates are smaller than 290?
2. Find the range for the following population.
   14.2, 16.1, 21.4, 21.5, 23.2
3. Find the standard deviation for the following population.
   3.1, 7.2, 10.3, 12.4, 14.3, 17.2, 21.1
4. Find the arithmetic mean for the following population.
   2, 16, 17, 17, 19, 22, 26
5. Find the median for the following population.
   2, 16, 17, 17, 19, 22, 26
6. Find the mode for the following population.
   2, 16, 17, 17, 19, 22, 26
7. Assume a normally distributed population with a mean of 305 and a standard deviation of 20. What percent of the variates are larger than 335?
8. Find the equation of the regression line, given the following data.

x	4.51	5.03	5.52	6.04
y	3.14	5.27	7.38	9.55

9. Find the slope of the regression line, given the following data.

x	15	16	17	18	19
y	0.2	0.4	1.0	1.6	1.6

10. Find the z-score for a raw score of 88.7, if the mean of the population is 82.4 and the standard deviation is 3.1.

CHAPTER TEST

1. Find the arithmetic mean for the following population.
   1, 3, 3, 8, 10, 12, 18, 23
2. Find the mode for the following population.
   1, 3, 3, 8, 10, 12, 18, 23
3. Find the median for the following population.
   1, 3, 3, 8, 10, 12, 18, 23
4. Find the range of the following population.
   3.1, 7.2, 10.3, 12.4, 14.3, 17.2, 21.1
5. Find the standard deviation of the following population.
   14.2, 16.1, 21.4, 21.5, 23.2
6. Find the z-score for a raw score of 79.3, if the mean of the population is 82.4 and the standard deviation is 3.1.
7. Assume a normally distributed population with a mean of 35.1 and a standard deviation of 1.8. What percent of the variates fall between 33.3 and 38.7?
8. Construct a scatter diagram and find the equation of the regression line, given the following data.

x	0	1	2	3	4	5
y	1	2	5	8	8	11

9. The manager of an electronics store finds that the stereo receivers in stock are rated (in watts) as follows: 47, 51, 76, 68, 50, 46, 58, 59. Find the mean and standard deviation of this population.
10. Most programming languages for microcomputers provide a random number generator. In testing one such random number generator, a computer engineer repeatedly generates groups of 50 random numbers and computes the mean of each group. These means fall into a normal distribution with a mean of 0.500 and a standard deviation of 0.041.
    (a) What percent of the groups have a mean smaller than 0.400?
    (b) What percent of the groups have a mean smaller than 0.600?

EXTENDED APPLICATION: INDUSTRIAL ENGINEERING TECHNOLOGY

QUALITY CONTROL FOR BATTERY PRODUCTION

Quality control is an area in which industrial engineering technicians are often involved. Consider, for example, a company that manufactures batteries. The company has determined that the batteries should have a mean lifetime of 1000 hours of continuous use. To ensure that this standard is maintained, a random sample of batteries is tested periodically. The type of test used is called a one-sided test, because the company will be concerned only if the lifetime of the batteries is *below* the expected 1000 hours. The hypothesis that the mean of the population is 1000 hours is tested against the hypothesis that the mean is less than 1000 hours. The hypothesis $\mu = 1000$ is called the *null hypothesis,* and the hypothesis $\mu < 1000$ is called the *alternative hypothesis.*

Suppose a sample of 49 batteries chosen at random shows a mean of 916 hours with a standard deviation of 280 hours. The quality control personnel must determine the probability that a sample of size 49 would have an average of 916 or less if the mean is, in fact, 1000. The company will choose to "reject" the null hypothesis that μ is 1000 if they believe that the low sample mean is indicative of a low population mean. They will choose not to reject the null hypothesis if they believe that the low sample mean is due to chance error in choosing the sample. The **level of significance** is the probability of rejecting a true hypothesis, and it is set in advance. Often the 0.05 level or the 0.01 level is used.

The analysis is done as follows. It is a well-known principle of statistics that for any population, if all samples of size n are considered, the mean of the sample means will equal the population mean. Furthermore, the standard deviation of these sample means may be approximated by finding $\sigma/\sqrt{n}$. For the sample of batteries, this figure is $280/\sqrt{49}$ or 40. The z-score for 916 is given by

$$z = (916 - 1000)/40 = -84/40 = -2.1.$$

Figure 20.16 shows that if z is -2.1, then only 1.8% of the sample means could be smaller than 916. That is, the probability of obtaining a sample mean of 916 or less when the population mean is 1000, is 0.018 or 1.8%.

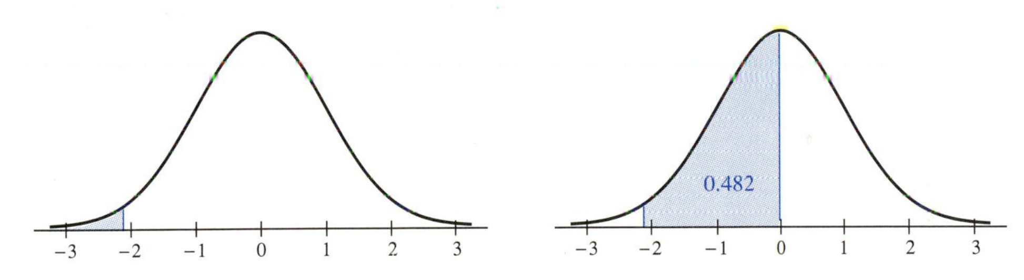

FIGURE 20.16

If the 0.05 level has been specified, the company would reject the null hypothesis, since the probability (0.018) of rejecting a true hypothesis would be less than 0.05. They would then investigate the production process to try to determine why the batteries have a shorter lifetime than expected. If the 0.01 level has been specified, the company would not reject the null hypothesis. That is, if they did reject it, the probability of rejecting a true hypothesis would be greater than 0.01.

DERIVATIVES CHAPTER 21

In Chapter 3, we explored the concept of variation, the mathematical relationship between two quantities that behave in such a way that a change in one of them is accompanied by a change in the other. To study the rates at which such changes take place, we use a branch of mathematics called differential calculus. Developed in the seventeenth century to study motion, differential calculus can be applied to problems involving such diverse fields as gravitation, heat, light, sound, electricity, magnetism, hydrology, and aeronautics. The methods of calculus can be used on many problems that would be difficult or impossible to solve by algebra and trigonometry alone. In this chapter, we introduce the basic concepts of this powerful branch of mathematics.

21.1 LIMITS AND CONTINUITY

Suppose that an object is dropped from the top of a tower. The formula $s = (1/2)gt^2$ gives the number of feet through which the object has fallen after t seconds. In the formula, g is the acceleration due to gravity and is equal to 32 ft/s². If the tower is 144 feet tall, the object will hit the ground three seconds after it is released, as the following equation shows.

$$144 = (1/2)(32)t^2 \text{ or } 144 = 16t^2, \text{ so } 9 = t^2$$

A negative time does not make sense, so 3 is the only solution. In the past, we have computed velocity using the formula $V = D/T$ or $v = s/t$. It is important to realize, however, that this formula delivers an *average* velocity for the period of time under consideration. Sometimes it is more important to know the velocity at a given instant of time, rather than the average over a period of time. In this example, the velocity at the time of impact is a factor in determining the amount of damage that will be done to the object when it strikes the ground. The velocity at a given time is called the *instantaneous velocity.*

A graph of distance fallen versus time is shown in Figure 21.1. It is important to realize that the graph does not show the path of the object. A graph of its path would show movement from a height of 144 ft to 0 ft. Although the object ceases to move when it strikes the ground after three seconds, the mathematical characteristics of the graph may be considered beyond that value, so an extended graph is shown. When the average velocity is calculated over a period of time, that period of time is represented by an interval on the t-axis. Since the average velocity is given by the formula $v = s/t$, for the three second interval as t ranges from 0 to 3, v is 144/3 or 48 ft/s.

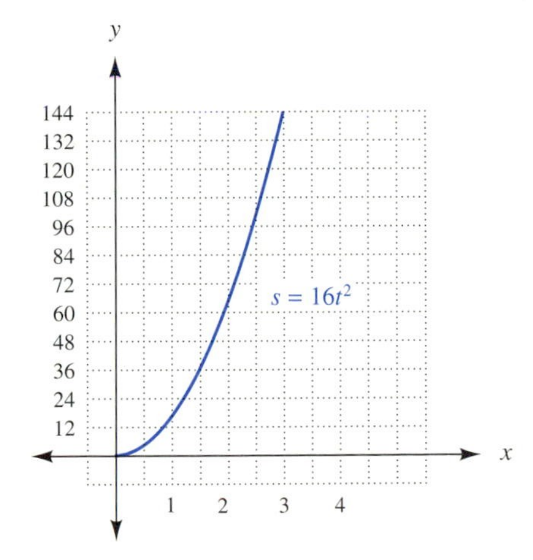

FIGURE 21.1

If we try to compute the instantaneous velocity at the three-second mark, however, the period of time is represented by a point on the t-axis. That is, the length of the interval is 0. We know that $s/0$ is undefined. We could estimate the velocity when t is 3 by calculating the average velocity over the interval as t ranges from 2 to 3. Since the distance traveled during this time is $144 - 64$ or 80 ft, and the time elapsed is 1 second, v is 80 ft/s.

By considering an even smaller interval of time, such as from 2 1/2 to 3, we could find a better estimate. When t is 5/2, s is 16(25/4) or 100 ft. Thus, v is $(144 - 100)/0.5$ or 88 ft/s. Table 21.1 shows values of v obtained by examining smaller and smaller intervals of time near a t-value of 3.

TABLE 21.1

t (s)	$s = 16t^2$ (ft)	Distance Fallen between t and 3 (ft)	Length of Time from t to 3 (s)	$v = s/t$ (ft/s)
2	64	80	1	80
2.5	100	44	0.5	88
2.8	125.44	18.56	0.2	92.8
2.9	134.56	9.44	0.1	94.4
2.95	139.24	4.76	0.05	95.2
2.98	142.0864	1.9136	0.02	95.68
2.99	143.0416	0.9584	0.01	95.84
2.995	143.5204	0.4796	0.005	95.92
2.998	143.80806	0.191936	0.002	95.97
2.999	143.90402	0.095984	0.001	95.98

As t gets closer and closer to 3, v appears to get closer and closer to 96. Assuming that v does, in fact, get closer and closer to 96, mathematicians, physicists, and engineers would say that as t approaches 3, v approaches 96, or the limit of v as t approaches 3 is 96.

A formal definition of a limit is not needed at this time, but the foregoing discussion should give you an intuitive idea of what limits are and why it is necessary to consider them. Informally, we say that if $f(x)$ becomes arbitrarily close to a single number L as x approaches the number c from either direction, then L is the limit of $f(x)$ as x approaches c. This relationship is expressed as $\lim_{x \to c} f(x) = L$. For the velocity example, we have $\lim_{t \to 3} v = 96$. The procedure we have used, however, is not entirely satisfactory for computing limits, because we have no assurance that 96 is, in fact, the correct answer; it merely appears to be the correct answer.

Eventually we will develop some more reliable mathematical methods for computing limits, but first we consider the graphical interpretation of a limit.

EXAMPLE 1 Sketch the graph of $f(x) = x^2 - 1$ and determine $\lim_{x \to 1} f(x)$.

Solution The graph of $f(x)=x^2-1$ (or $y=x^2-1$) is a parabola with vertex at $(0,-1)$. The graph is shown in Figure 21.2.

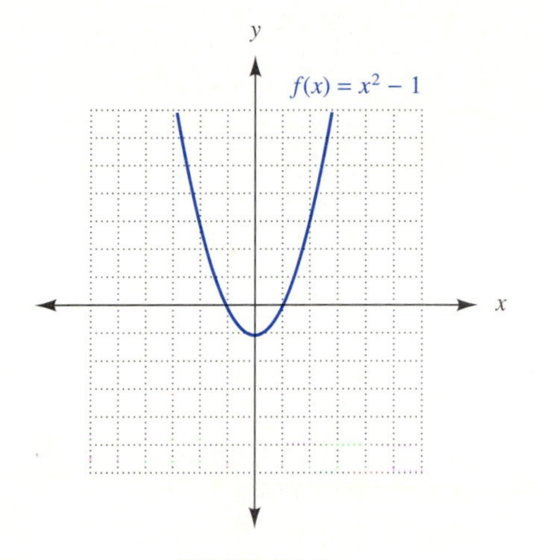

FIGURE 21.2

Looking at the graph, one can see that as x approaches 1 from either the left or the right, the value $f(x)$ approaches 0. Thus, $\lim_{x \to 1} f(x) = 0$. ■

continuous over an interval

continuous

The function in Example 1 is an example of a *continuous function.* Mathematicians need a much more precise definition of a continuous function, but for now it is sufficient to say that a function is **continuous over an interval** if its graph can be drawn without lifting the pencil from the paper over that interval. When a function is described as **continuous,** it is continuous over all real numbers. Example 1 illustrates that as x approaches any real number c, the limit of $f(x)$ is simply the function value when x is c, if the function is continuous. This statement is written mathematically as follows.

THE LIMIT OF A FUNCTION THAT IS CONTINUOUS OVER AN INTERVAL

If f is a function that is continuous over an interval, and c is a real number in that interval, then $\lim_{x \to c} f(x) = f(c)$.

point of discontinuity

More difficult to handle are functions that have one or more *points of discontinuity.* When there is a value for which the function is undefined or for which the function value suddenly jumps from one number to another without taking on the values in between, we say there is a **point of discontinuity** at that value.

EXAMPLE 2 Identify the point of discontinuity for each function.

(a)

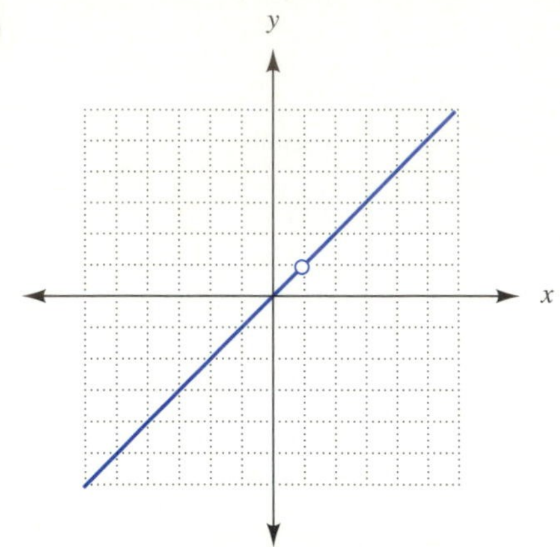

FIGURE 21.3

(b)

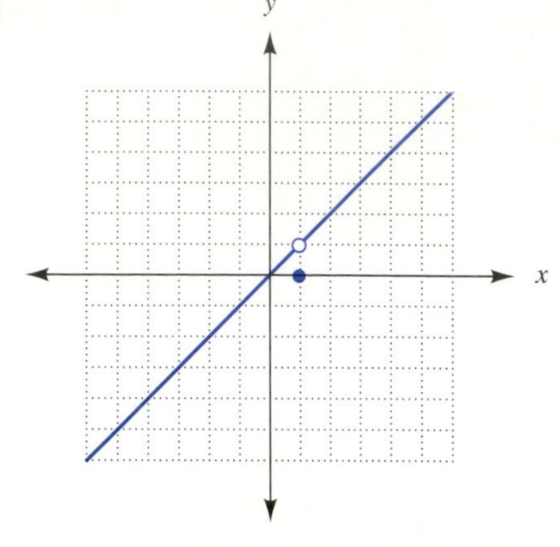

FIGURE 21.4

(c)

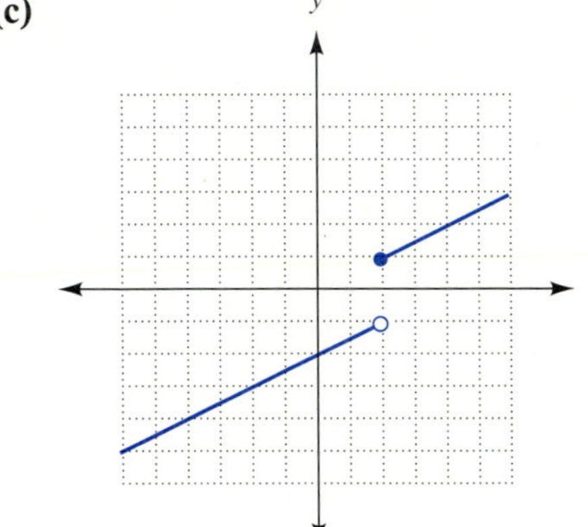

FIGURE 21.5

(d)

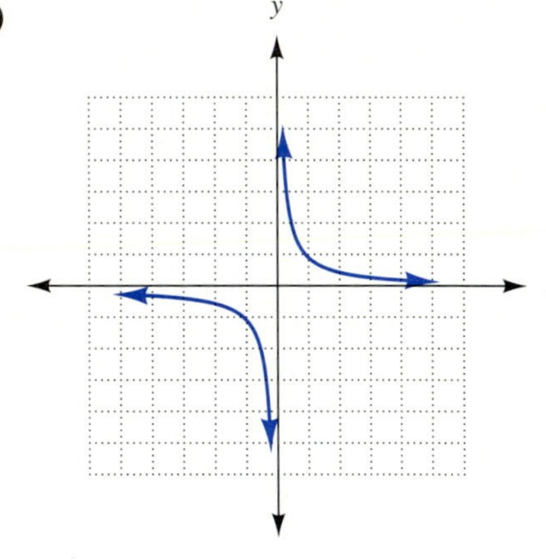

FIGURE 21.6

Solution **(a)** There is a "hole" in the graph when x is 1. The open circle indicates that the function is undefined when x is 1. Thus, there is a point of discontinuity when x is 1.

(b) The graph looks very much like the graph in Example 2(a), but this function is defined when x is 1. When x is 1, the value of the function suddenly jumps from about 1 to 0. Thus, there is a point of discontinuity when x is 1.

(c) When x is 2, the function value suddenly jumps from about -1 to 1. Thus, there is a point of discontinuity when x is 2.

(d) The graph of the function does not cross the y-axis. That is, the function is undefined when x is 0. Thus, there is a point of discontinuity when x is 0. ■

The concept of a limit is especially important as x approaches a point of discontinuity. It is important to distinguish between the function value when x is c and the limit of $f(x)$ as x approaches c.

EXAMPLE 3 For each function, find $f(0)$ and $\lim_{x\to 0} f(x)$, if possible.

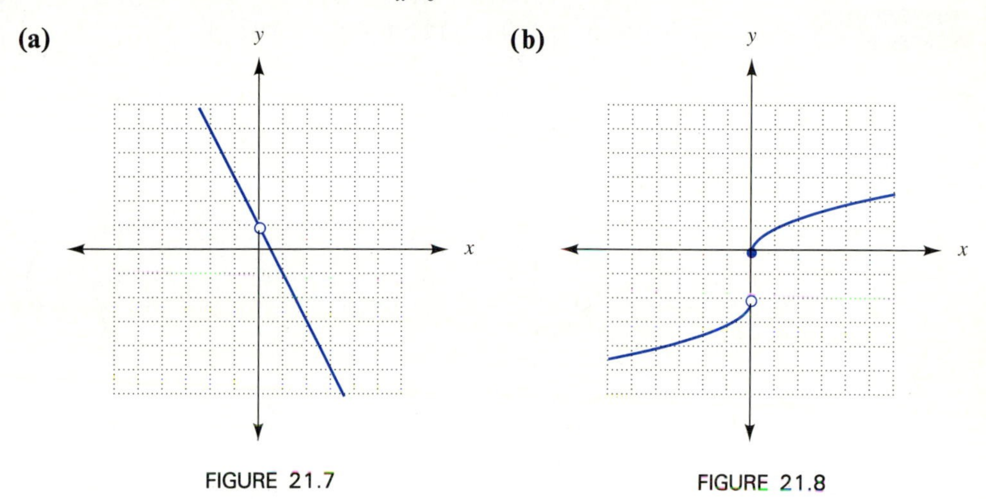

FIGURE 21.7

FIGURE 21.8

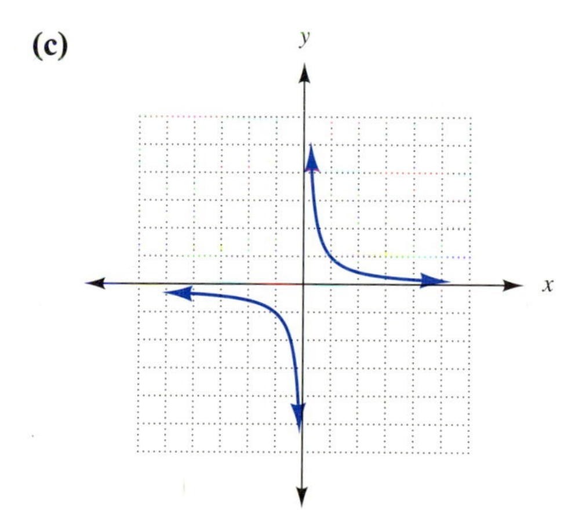

FIGURE 21.9

Solution **(a)** The function is not defined when x is 0, so $f(0)$ does not exist. As x gets closer and closer to 0 from both the left and the right, however, the function values get closer and closer to 1. Thus $\lim_{x\to 0} f(x) = 1$.

(b) When x is 0, the function value is 0. Thus $f(0) = 0$. The limit, however, does not exist, because as x approaches 0 from the right, $f(x)$ approaches 0, but as x

approaches 0 from the left, $f(x)$ approaches -2 instead of 0. In order to find $\lim_{x \to c} f(x)$, $f(x)$ must approach the same value from *both* sides as x approaches c.

(c) Neither $f(0)$ nor $\lim_{x \to 0} f(x)$ exists. ■

EXAMPLE 4 For each function, find $f(1)$ and $\lim_{x \to 1} f(x)$, if possible.

(a)

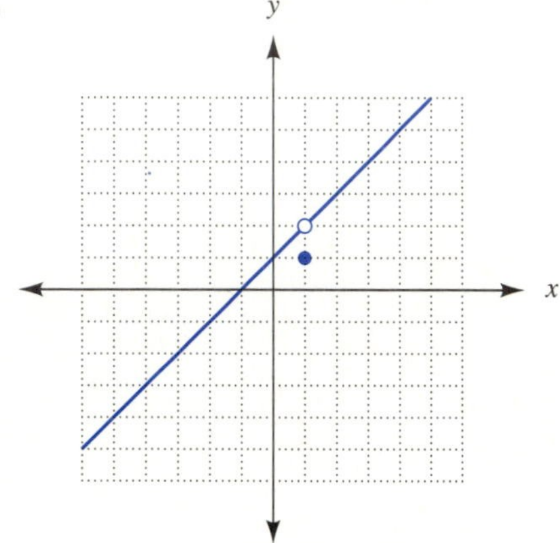

FIGURE 21.10

(b)

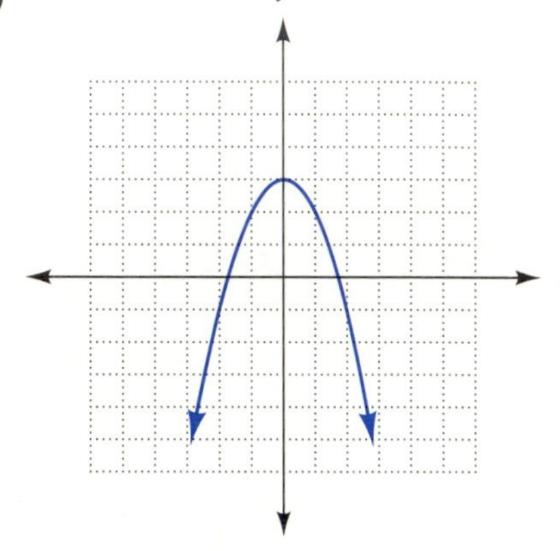

FIGURE 21.11

Solution **(a)** The point at (1, 1) indicates that $f(1)$ is 1. As x gets closer and closer to 1 from both the left and the right, however, the function values get closer and closer to 2. Thus, $\lim_{x \to 1} f(x) = 2$. Both the function value and the limit exist, but they are not equal.

(b) Since the point (1, 2) is on the graph, we know that $f(1)$ is 2. This function is a continuous function. Thus, $\lim_{x \to 1} f(x) = 2$. ■

Examples 3 and 4 illustrate that when the function value at c is compared with the limit as x approaches c, it is possible to have the following situations:

1. The limit exists, but the function value does not. (Example 3a)
2. The function value exists, but the limit does not. (Example 3b)
3. Neither the function value nor the limit exists. (Example 3c)
4. Both the function and the limit exist, but are not equal. (Example 4a)
5. Both exist and are equal. (Example 4b)

Sometimes it is necessary to see what happens to the dependent variable as the independent variable takes on larger and larger values, without bound. We say that x *approaches infinity* and write $\lim_{x \to \infty} f(x)$. ***It is important to realize that*** ∞ ***is not a number. It is merely an abstraction to indicate that there is no largest value for x.*** The

NOTE ⇨⇨

variable x can always be assigned a larger value. When we consider $\lim_{x \to \infty} f(x)$, we examine the function values as x-values are chosen farther and farther to the right on a graph. When we consider $\lim_{x \to -\infty} f(x)$, we examine the function values as x-values are chosen farther and farther to the left.

EXAMPLE 5 When the current is switched on in a circuit containing a resistor and an inductor in series, the current I rises gradually as shown in Figure 21.12. Find $\lim_{t \to \infty} I$.

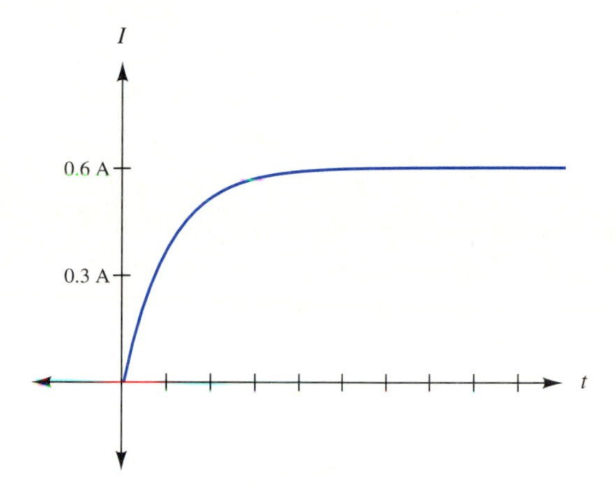

FIGURE 21.12

Solution As the t-values increase, I appears to get closer and closer to 0.6 A. Thus $\lim_{t \to \infty} I =$ 0.6 A. ■

We are now ready to compute limits mathematically. There are several techniques that are useful. Sometimes factoring an expression will make it easier to find a limit. Example 6 illustrates the procedure.

EXAMPLE 6 Find $\lim_{x \to 1} \dfrac{x^2 - 1}{x - 1}$.

Solution The notation $\lim_{x \to 1} \dfrac{x^2 - 1}{x - 1}$ means that x gets very close to 1, but is not equal to 1. If we treat the fraction algebraically,

$$\frac{x^2 - 1}{x - 1} = \frac{(x - 1)(x + 1)}{x - 1} \text{ or } x + 1,$$

for all values of x except 1. Since the fraction $\dfrac{x^2 - 1}{x - 1}$ has a denominator of 0 when x is

1, the division by $x - 1$ is not valid when x is 1, but it is valid for all other values of x. If we let

$$f(x) = \frac{x^2 - 1}{x - 1} \text{ and } g(x) = x + 1,$$

then $f(x) = g(x)$ for all values of x except 1. The graph of $f(x) = \dfrac{x^2 - 1}{x - 1}$ is just like the graph of $g(x) = x + 1$, except that it has a point of discontinuity when x is 1. The graphs are shown in Figure 21.13. For both graphs, as x approaches 1 from either the left or the right, $f(x)$ and $g(x)$ approach 2. That is,

$$\lim_{x \to 1} \frac{x^2 - 1}{x - 1} = \lim_{x \to 1} (x + 1) = 2. \quad ■$$

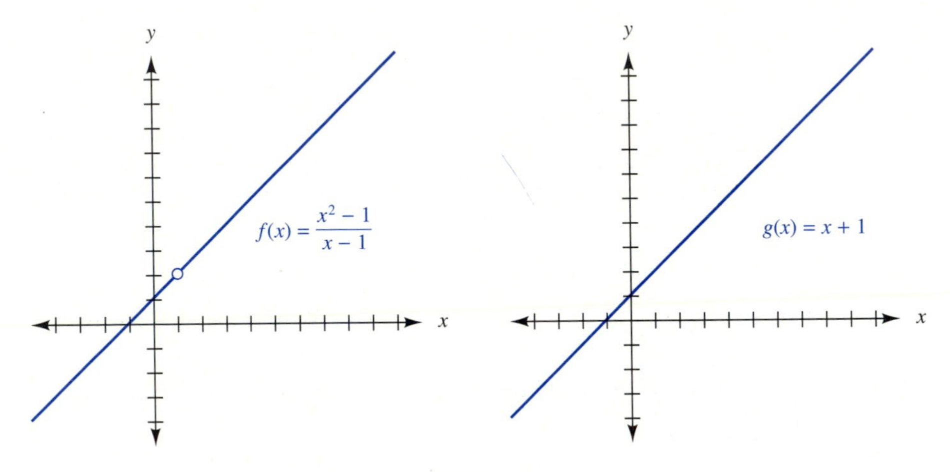

FIGURE 21.13

A table of values such as Table 21.1 may be helpful.

EXAMPLE 7 Find $\lim_{x \to \infty} \dfrac{1}{x}$.

Solution Consider the following table of values.

x	1	2	3	4	5	6	7	8
$\frac{1}{x}$	$\frac{1}{1}$	$\frac{1}{2}$	$\frac{1}{3}$	$\frac{1}{4}$	$\frac{1}{5}$	$\frac{1}{6}$	$\frac{1}{7}$	$\frac{1}{8}$

Intuitively, one can see that as the denominator gets larger, the fraction must get smaller. We also know that if x is a positive number, $1/x$ must be positive. Thus,

$\lim_{x\to\infty} \frac{1}{x} = 0$, as shown in Figure 21.14, but it should be noted that there is no value of x for which $\frac{1}{x}$ is 0. ■

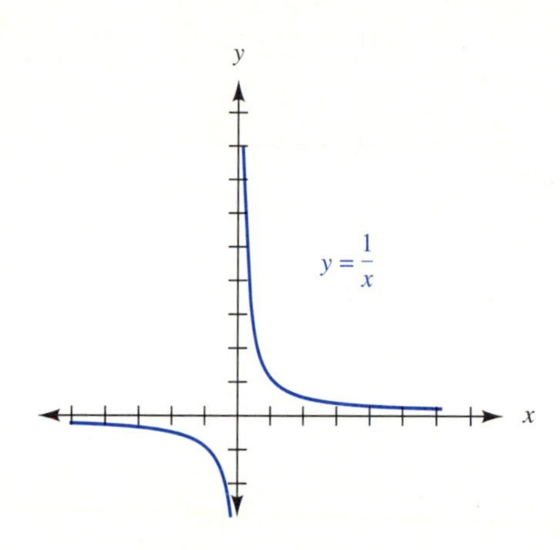

FIGURE 21.14

The observation that the value of a fraction with a constant numerator will approach 0 as the denominator becomes large in absolute value is useful in evaluating other limits.

EXAMPLE 8 Find $\lim_{x\to\infty} \frac{2x^2+1}{3x^2-2}$.

Solution Since both numerator and denominator become large as x becomes large, we cannot yet tell much about the behavior of the function. A large numerator tends to make the fraction large, but a large denominator tends to make the fraction small. The limit depends on how fast the numerator and denominator increase, relative to each other. To find the limit, we divide both numerator and denominator by x^2, the highest power of the variable that appears in the fraction.

$$\lim_{x\to\infty} \frac{2x^2+1}{3x^2-2} = \lim_{x\to\infty} \frac{\frac{2x^2+1}{x^2}}{\frac{3x^2-2}{x^2}}$$

$$= \lim_{x\to\infty} \frac{\frac{2x^2}{x^2}+\frac{1}{x^2}}{\frac{3x^2}{x^2}-\frac{2}{x^2}}$$

$$= \lim_{x\to\infty} \frac{2+\frac{1}{x^2}}{3-\frac{2}{x^2}}$$

But as x increases, both $\frac{1}{x^2}$ and $\frac{2}{x^2}$ approach 0, since these fractions have constant numerators. Thus $\lim_{x\to\infty} \frac{2x^2+1}{3x^2-2} = \frac{2}{3}$. ■

EXERCISES 21.1

In 1–6, determine whether each function is continuous. Find the given function value and limit, if possible.

1. $f(-2)$ and $\lim_{x\to -2} f(x)$

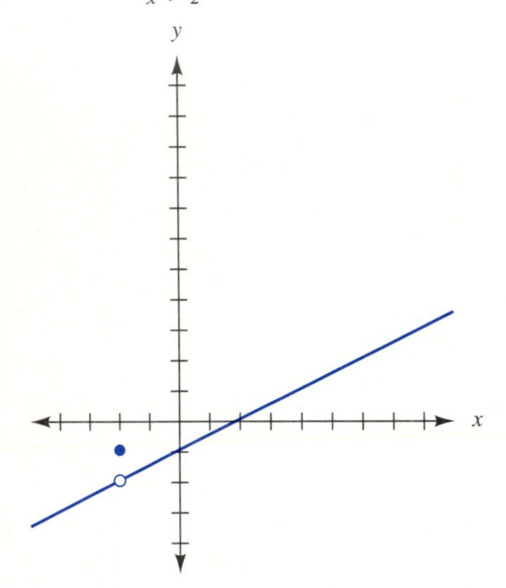

2. $f(3)$ and $\lim_{x\to 3} f(x)$

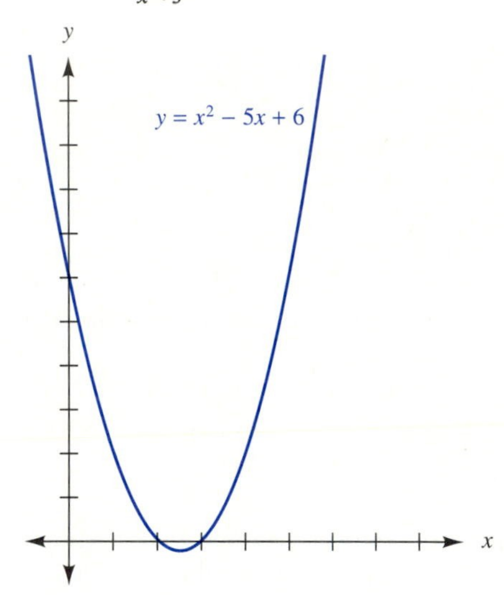

3. $f(\pi)$ and $\lim_{x\to \pi} f(x)$

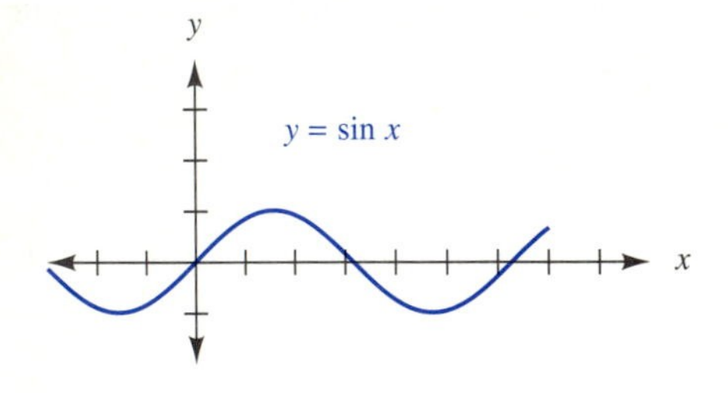

4. $f(\pi)$ and $\lim_{x\to \pi} f(x)$

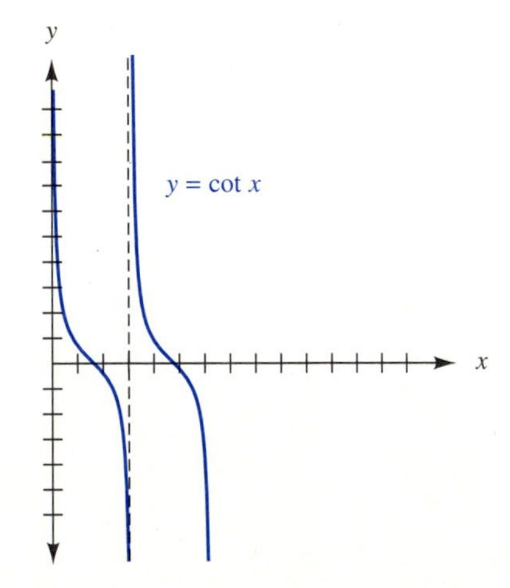

5. $f(0)$ and $\lim_{x \to 0} f(x)$

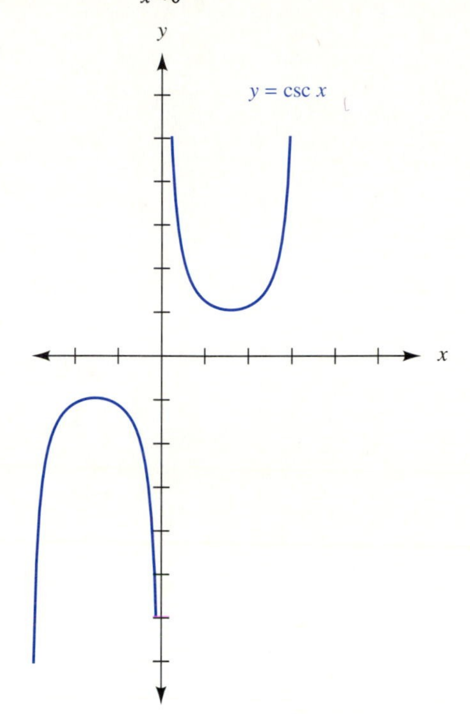

6. $f(1)$ and $\lim_{x \to 1} f(x)$

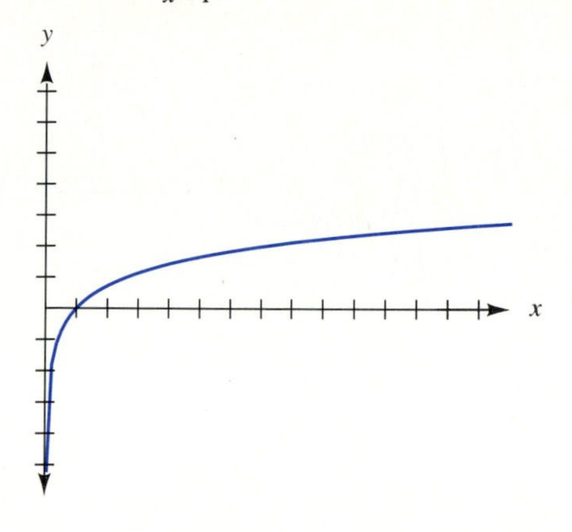

In 7–10, determine whether each function is continuous.

7. $y = \dfrac{1}{x-2}$

8. $y = 3x^2 + 2$

9. $y = 4x + 3$

10. $y = \dfrac{2x+3}{3x+1}$

In 11–20, identify the points of discontinuity for each function.

11.

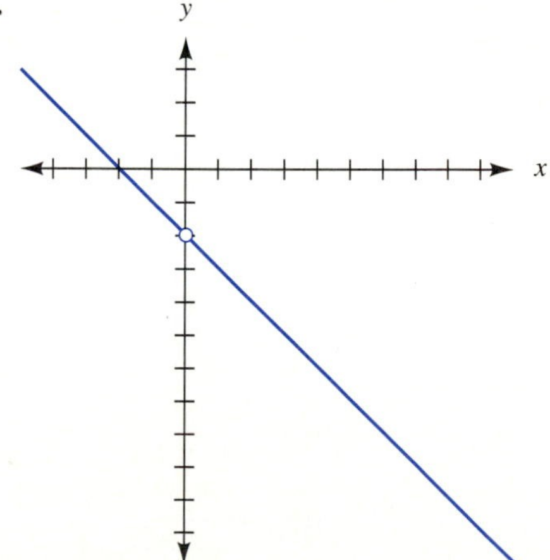

12.

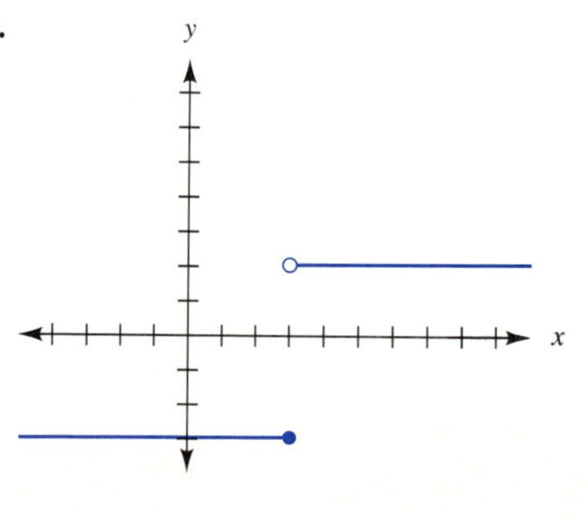

13.

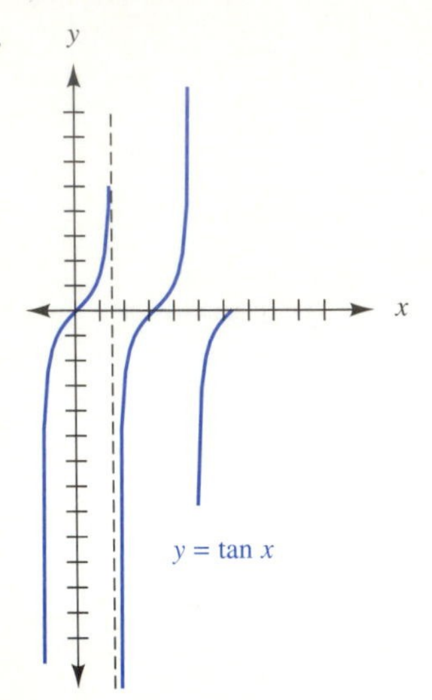

14.

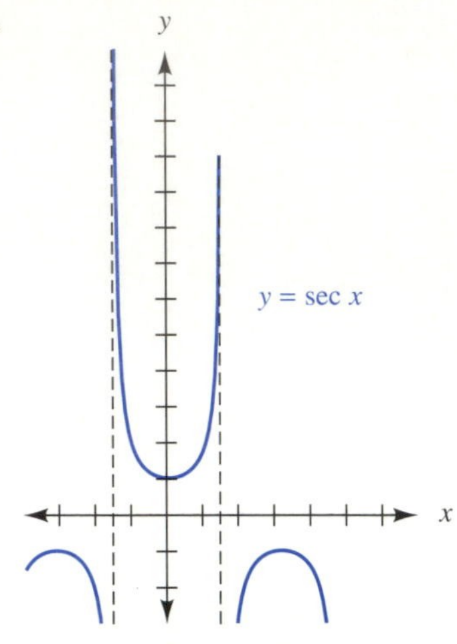

15.

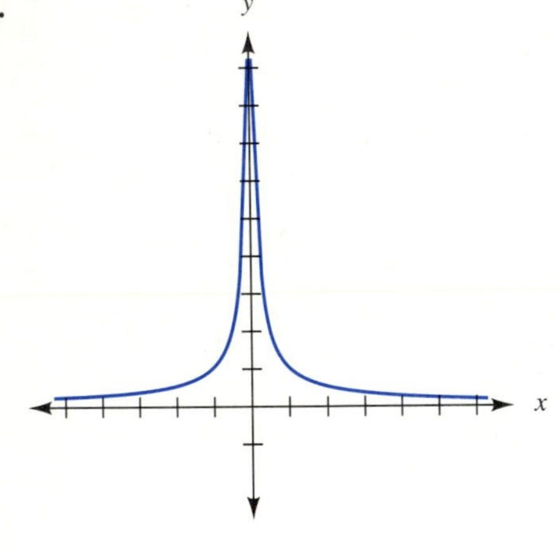

16.

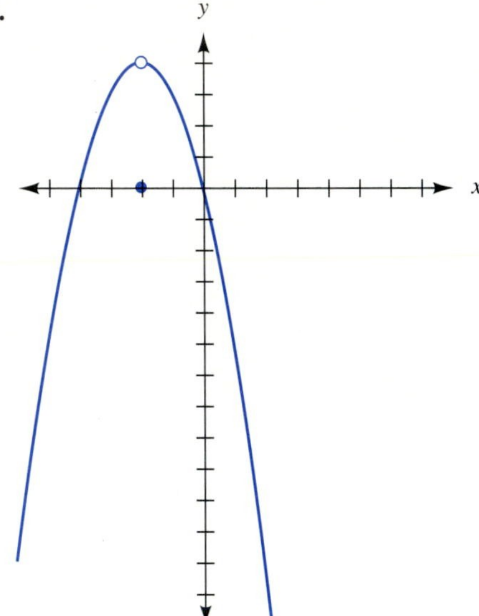

17. $y = \dfrac{1}{x}$ **18.** $y = \dfrac{2x+1}{x-3}$ **19.** $y = \dfrac{x+1}{2x-3}$ **20.** $y = \dfrac{2x+3}{3x+1}$

In 21–40, evaluate each limit.

21. $\lim\limits_{x\to 2} \dfrac{x^2+x-6}{x-2}$ **22.** $\lim\limits_{x\to 3} \dfrac{x^2-2x-3}{x-3}$ **23.** $\lim\limits_{x\to 0} \dfrac{x^2-2x}{x}$

24. $\lim\limits_{x\to 0} \dfrac{x^3+x^2}{x^2}$ **25.** $\lim\limits_{x\to -1} \dfrac{x^2+2x+1}{x+1}$ **26.** $\lim\limits_{x\to 1/2} \dfrac{4x^2-1}{2x-1}$

27. $\lim\limits_{x\to 1/3} \dfrac{3x^2-x}{3x-1}$ **28.** $\lim\limits_{x\to -1/3} \dfrac{9x^2+6x+1}{3x+1}$ **29.** $\lim\limits_{x\to -2} \dfrac{x^2+4x+4}{x+2}$

30. $\lim_{x \to -3} \frac{x^2 + 4x + 3}{x + 3}$

31. $\lim_{x \to \infty} \frac{2x^2 + x}{2x^2 + 1}$

32. $\lim_{x \to \infty} \frac{3x - 1}{2x + 1}$

33. $\lim_{x \to \infty} \frac{2x^3 - 1}{x^3 + x}$

34. $\lim_{x \to \infty} \frac{3x^2 + x}{2x^3 - 3}$

35. $\lim_{x \to \infty} \frac{2x^2 + 4}{x^3 - x}$

36. $\lim_{x \to \infty} \frac{x^2 - 3x}{2x^3 + 1}$

37. $\lim_{x \to \infty} \frac{2x^2 - x + 3}{3x^2 + x - 1}$

38. $\lim_{x \to \infty} \frac{4x^2 + 3x - 1}{2x^2 - x + 3}$

39. $\lim_{x \to \infty} \frac{x^2 - 2x + 1}{2x^2 - x + 3}$

40. $\lim_{x \to \infty} \frac{2x - 1}{3x^2 + 2x - 1}$

---

41. The graph shows the percent of initial charge C at time t on a capacitor that is disconnected from the charging battery at time $t = 0$. Find $\lim_{t \to \infty} C$.

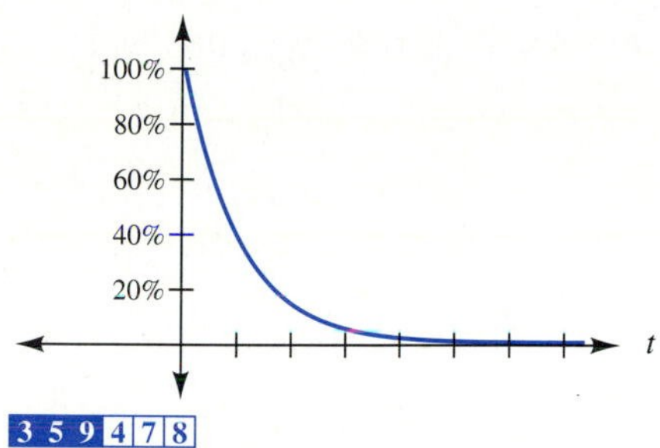

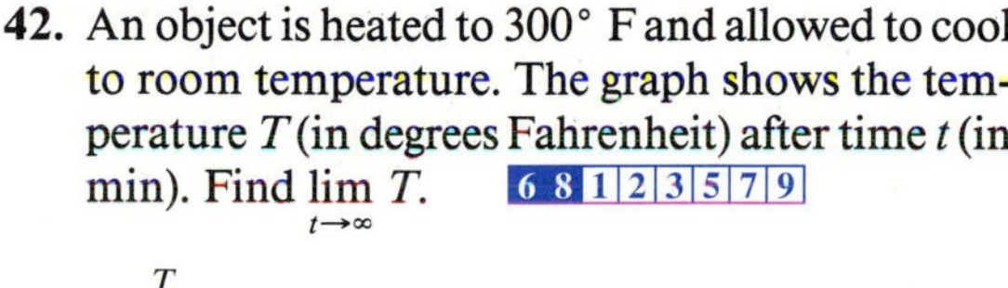
3 5 9 | 4 7 8

42. An object is heated to 300° F and allowed to cool to room temperature. The graph shows the temperature T (in degrees Fahrenheit) after time t (in min). Find $\lim_{t \to \infty} T$. 6 8 | 1 2 3 5 7 9

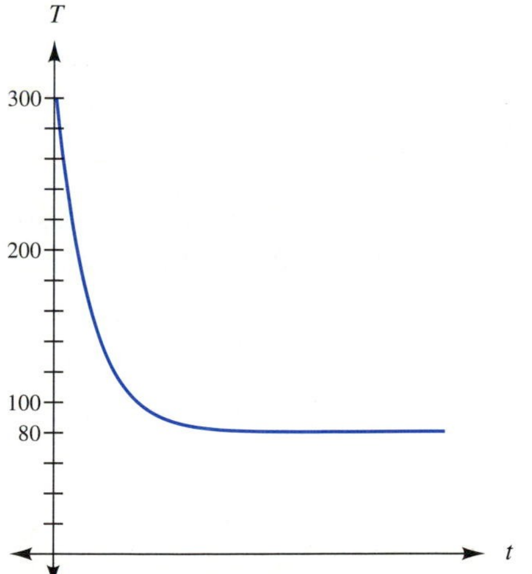

43. The graph shows the vertical displacement S of an oscillating spring. Find $\lim_{t \to \infty} S$.

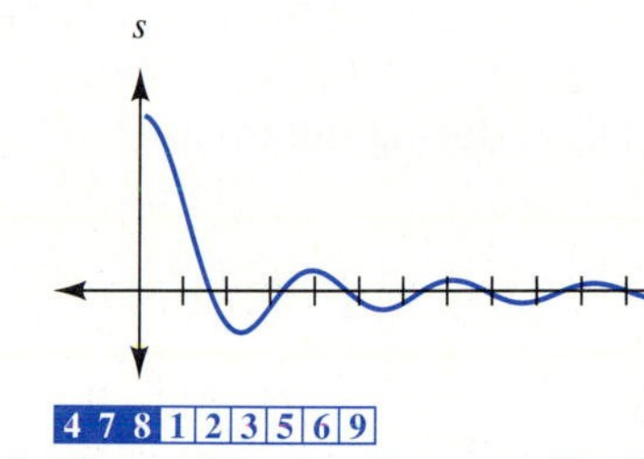

4 7 8 | 1 2 3 5 6 9

44. The total load resistance r (in Ω) seen by a transistor amplifier is given by the formula

$$r = \frac{R_l R_d}{R_l + R_d}.$$

Find the total load resistance if R_l "burns open" (that is, R_l approaches infinity).
3 5 9 | 1 2 4 6 7 8

45. The voltage gain A (in dB) of a certain amplifier is given by the equation

$$A = \frac{R^3 - 8}{R - 2}$$

if R is a resistance (in kΩ). Find $\lim_{R \to 2} \frac{R^3 - 8}{R - 2}$.
3 5 9 | 1 2 4 6 7 8

---

1. Architectural Technology
2. Civil Engineering Technology
3. Computer Engineering Technology
4. Mechanical Drafting Technology
5. Electrical/Electronics Engineering Technology
6. Chemical Engineering Technology
7. Industrial Engineering Technology
8. Mechanical Engineering Technology
9. Biomedical Engineering Technology

21.2 RATE OF CHANGE AND THE SLOPE OF A TANGENT TO A CURVE

To introduce the idea of a "rate of change," we return to the problem of computing instantaneous velocity at any time for an object falling from a tower. Velocity is sometimes said to be the "rate of change of distance with respect to time." This phrase means that as time changes, distance changes. Recall from the previous section that the formula $s = 16t^2$ gives the distance that an object has fallen from a tower after t seconds. In that section, we considered an object falling from a height of 144 ft. We looked at the limit of the average velocities for various intervals of time as t approached 3 to find the instantaneous velocity when t is 3.

Our goal at this time is to develop a method for finding the instantaneous velocity at any time t. Let Δt represent a small interval of time and let Δs represent the distance fallen during the interval of time from t to $t + \Delta t$, as shown in Figure 21.15.

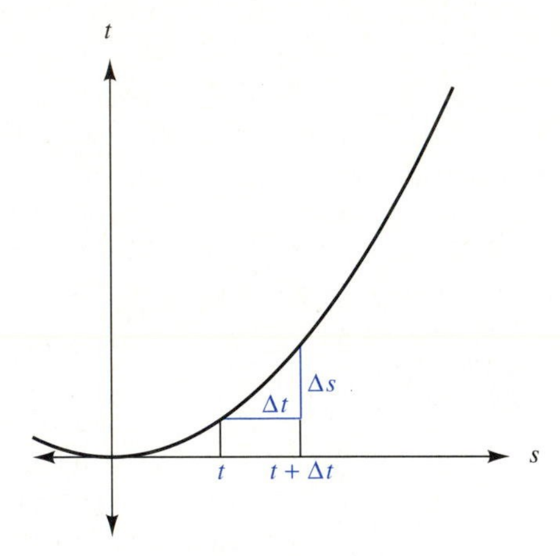

FIGURE 21.15

The distance traveled from the time 0 seconds to t seconds, is $16t^2$. Also, the distance fallen from 0 seconds to $t + \Delta t$ seconds is $16(t + \Delta t)^2$. Thus, the distance traveled between t seconds and $t + \Delta t$ seconds is given by

$$\Delta s = 16(t + \Delta t)^2 - 16t^2.$$

The average velocity over this period of time is $\Delta s/\Delta t$. As Δt gets smaller and smaller, the average velocity approaches the instantaneous velocity at time t.

$$\lim_{\Delta t \to 0} \frac{\Delta s}{\Delta t} = \lim_{\Delta t \to 0} \frac{16(t + \Delta t)^2 - 16t^2}{\Delta t}$$

$$= \lim_{\Delta t \to 0} \frac{16[t^2 + 2t(\Delta t) + (\Delta t)^2] - 16t^2}{\Delta t} \qquad \text{Expand } (t + \Delta t)^2.$$

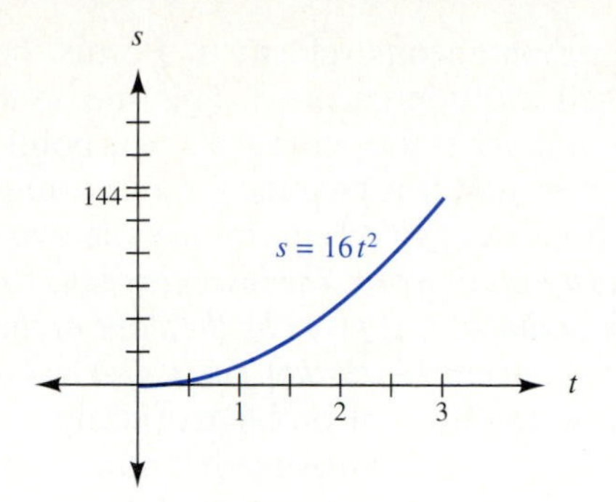

FIGURE 21.16

We have chosen the scale so that the graphs will be easy to interpret in the following discussion. If we want to compute the instantaneous velocity at time t, we consider the average velocity between t seconds and $t + \Delta t$ seconds. If we let P be (t, s) and let Q be $(t + \Delta t, s + \Delta s)$, Figure 21.17 shows what happens as Δt approaches 0. As Δt gets smaller, point Q gets closer to point P. The average velocity is given by $\Delta s/\Delta t$, and this quotient can be interpreted as the slope of the line segment connecting the points P and Q.

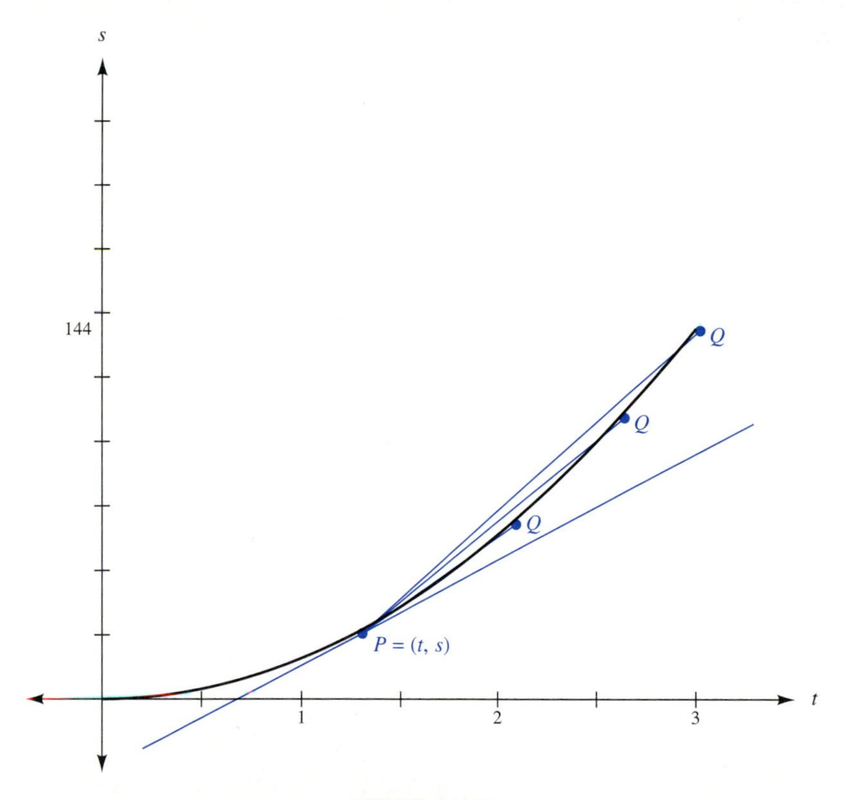

FIGURE 21.17

The instantaneous velocity at P must be the slope of the limiting line that is approached as Q approaches P. This line is said to be **tangent** to P. For this example, the tangent touches the curve at a single point and does not cross the curve. You will see, however, that it is possible for a line tangent to a curve to cross the curve.

tangent

It is important to recognize that the two ideas of this section are tied together. ***When y is a function of x, the instantaneous rate of change of y with respect to x, for a particular value of x, is given by the slope of the tangent to the graph of the function at the point associated with that particular value of x.***

NOTE

We now examine the problem of computing the slope of a line tangent to a curve. We will use (x, y) notation, since it is more familiar to you. ***You should realize that the choice of variable name is immaterial.***

NOTE

EXAMPLE 3 Find a formula for the slope of the line tangent to the curve defined by $y = 2x^2 + 3$ at any point on the curve. Then find the slope of the tangent line when x is 1.

Solution Figure 21.18 shows the graph. Let $P = (x, y)$ be a point on the curve and let Q be a point located some small distance from P. We say its coordinates are $(x + \Delta x, y + \Delta y)$, where Δx and Δy indicate small changes in the x- and y-values, respectively, from the position (x, y).

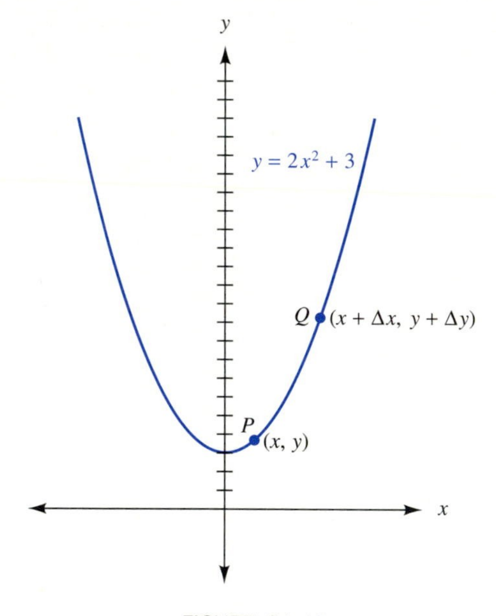

FIGURE 21.18

The slope m of the line connecting (x, y) and $(x + \Delta x, y + \Delta y)$ is given by

$$m = \frac{(y + \Delta y) - y}{(x + \Delta x) - x} = \frac{\Delta y}{\Delta x}.$$

We find $y + \Delta y$ by noting that the equation of the curve is

$$y = 2x^2 + 3, \qquad \textbf{(21-1)}$$

and $(x + \Delta x, y + \Delta y)$ is a point on this curve. Thus,

$$y + \Delta y = 2(x + \Delta x)^2 + 3. \qquad \textbf{(21-2)}$$

Subtracting Equation 21-1 from Equation 21-2, we have

$$(y + \Delta y) - y = [2(x + \Delta x)^2 + 3] - (2x^2 + 3) \text{ or}$$
$$\Delta y = [2(x + \Delta x)^2 + 3] - (2x^2 + 3).$$

As point Q is chosen closer and closer to point P, Δx gets smaller and smaller. Thus, the slope of the line tangent at P is given by the following.

$$\lim_{\Delta x \to 0} \frac{\Delta y}{\Delta x} = \lim_{\Delta x \to 0} \frac{[2(x + \Delta x)^2 + 3] - (2x^2 + 3)}{\Delta x}$$

$$= \lim_{\Delta x \to 0} \frac{2(x + \Delta x)^2 + 3 - 2x^2 - 3}{\Delta x}$$ Rewrite $-(2x^2 + 3)$.

$$= \lim_{\Delta x \to 0} \frac{2[x^2 + 2x(\Delta x) + (\Delta x)^2] - 2x^2}{\Delta x}$$ Expand $(x + \Delta x)^2$.

$$= \lim_{\Delta x \to 0} \frac{2x^2 + 4x(\Delta x) + 2(\Delta x)^2 - 2x^2}{\Delta x}$$ Distribute 2.

$$= \lim_{\Delta x \to 0} \frac{4x(\Delta x) + 2(\Delta x)^2}{\Delta x}$$ Combine like terms.

But just as we were able to eliminate the common factor from both numerator and denominator in Examples 1 and 2, we can eliminate Δx in this example.

$$\lim_{\Delta x \to 0} \frac{\Delta y}{\Delta x} = \lim_{\Delta x \to 0} \frac{(\Delta x)[4x + 2(\Delta x)]}{\Delta x}$$
$$= \lim_{\Delta x \to 0} 4x + 2(\Delta x)$$

As Δx approaches 0, $2(\Delta x)$ approaches 0. Thus,

$$\lim_{\Delta x \to 0} 4x + 2(\Delta x) = 4x.$$

When x is 1, the slope of the tangent is 4(1) or 4. ■

The procedure we have used can be generalized as follows.

To find the slope of the line tangent to a curve at (x, y):

1. Let $P = (x, y)$ and let $Q = (x + \Delta x, y + \Delta y)$.
2. Compute $y + \Delta y$ by substituting $x + \Delta x$ into the equation for the curve.
3. Find $\Delta y = (y + \Delta y) - y$.
4. Evaluate $\lim_{\Delta x \to 0} \Delta y/\Delta x$ where $\Delta y/\Delta x$ is the slope of the line connecting P and Q.

EXAMPLE 4 Find the slope of the line tangent to the curve defined by $y = x^3 - x$

(a) when x is 0. **(b)** when x is -1.

Solution

1. Let $P = (x, y)$ and let $Q = (x + \Delta x, y + \Delta y)$.
2. $y + \Delta y = (x + \Delta x)^3 - (x + \Delta x)$
3. Subtract $y = x^3 - x$ from $y + \Delta y = (x + \Delta x)^3 - (x + \Delta x)$.
   $(y + \Delta y) - y = [(x + \Delta x)^3 - (x + \Delta x)] - (x^3 - x)$
   Use the binomial theorem from Section 16.4 to expand $(x + \Delta x)^3$.

$$\begin{aligned}\Delta y &= x^3 + 3x^2(\Delta x) + 3x(\Delta x)^2 + (\Delta x)^3 - x - \Delta x - x^3 + x \\ &= 3x^2(\Delta x) + 3x(\Delta x)^2 + (\Delta x)^3 - \Delta x \qquad \text{Combine like terms.}\end{aligned}$$

4.
$$\begin{aligned}\lim_{\Delta x \to 0} \frac{\Delta y}{\Delta x} &= \lim_{\Delta x \to 0} \frac{3x^2(\Delta x) + 3x(\Delta x)^2 + (\Delta x)^3 - \Delta x}{\Delta x} \\ &= \lim_{\Delta x \to 0} \frac{(\Delta x)[3x^2 + 3x(\Delta x) + (\Delta x)^2 - 1]}{\Delta x} \qquad \text{Factor out } \Delta x. \\ &= \lim_{\Delta x \to 0} 3x^2 + 3x(\Delta x) + (\Delta x)^2 - 1 \qquad \text{Reduce the fraction.} \\ &= 3x^2 - 1\end{aligned}$$

(a) When x is 0, the slope of the tangent is $3(0^2) - 1$ or -1.

(b) When x is -1, the slope of the tangent is $3(-1)^2 - 1$ or 2.

Figure 21.19 shows that these answers are reasonable. ■

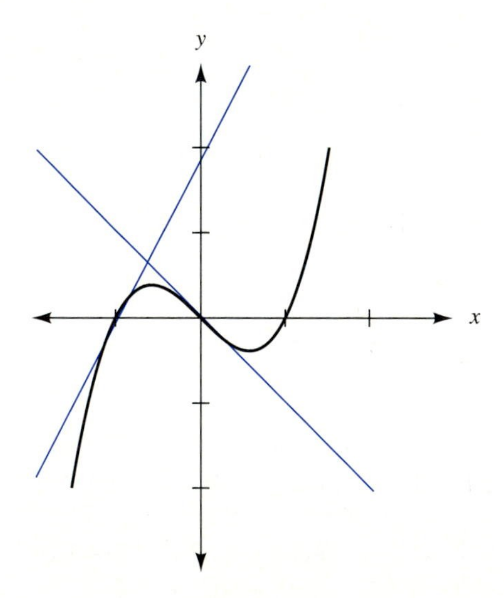

FIGURE 21.19

EXERCISES 21.2

In 1–14, find an expression for the slope of the line tangent to each curve at x, and determine the slope of the tangent at the given point.

1. $y = 2x^2$ at $x = 2$
2. $y = -3x^2$ at $x = 2$
3. $y = 3 - x^2$ at $x = 1.5$
4. $y = x^2 - 4$ at $x = 1.5$
5. $y = 2x^2 + x$ at $x = -1$
6. $y = x^2 - 3x$ at $x = -1$
7. $y = x^2 + 2x + 1$ at $x = 0$
8. $y = 2x^2 + x - 3$ at $x = 0$
9. $y = 3x^2 - 2x$ at $x = 1$
10. $y = x^3 - x$ at $x = 2$
11. $y = x^3 + 2x$ at $x = 1$
12. $y = 2x^3 + 3$ at $x = -2$
13. $y = 1 - x^3$ at $x = 1$
14. $y = x - 2x^3$ at $x = -1$

---

15. A circular colony of bacteria is expanding in area. Find a formula for the rate of change of area with respect to the radius of the circle. 9 1 6

16. The potential energy W (in joules) stored in an inductor is given by $W = (1/2)LI^2$, where I is the current (in amperes) and L is the inductance (in henries). Find a formula for the rate of change of energy with respect to current.
3 5 9 1 2 4 6 7 8

17. The potential energy W (in joules) stored in a capacitor is given by $W = (1/2)CV^2$, where C is the capacitance (in farads) and V is voltage (in volts). Find a formula for the rate of change of energy with respect to voltage. 3 5 9 4 7 8

18. The kinetic energy K (in joules) of a moving body is given by $K = (1/2)mv^2$, where m is the mass (in kg) and v is the velocity (in m/s). Find a formula for the rate of change of K with respect to v.
4 7 8 1 2 3 5 6 9

19. A metal sphere shrinks as it cools. Find a formula for the rate of change of volume with respect to radius. 1 2 4 6 8 5 7 9

20. The power P (in watts) generated by a windmill is given by $P = kV^3$, where V is the velocity (in mph) of the wind and k is a constant that depends on characteristics of the windmill. Find a formula for the rate of change of P with respect to V.
4 7 8 1 2 3 5 6 9

21. The resistance R (in ohms) of a resistor is given by $R = (1/P)(V^2)$ where the power dissipated is P (in watts) and the potential difference is V (in volts). Find a formula for the rate of change of R with respect to V. 3 5 9 1 2 4 6 7 8

22. Sketch the graph of $y = (1/3)x^3 - 4x$. Draw the tangent to the curve at $x = -2$. At how many points does the tangent intersect the curve?

---

1. Architectural Technology
2. Civil Engineering Technology
3. Computer Engineering Technology
4. Mechanical Drafting Technology
5. Electrical/Electronics Engineering Technology
6. Chemical Engineering Technology
7. Industrial Engineering Technology
8. Mechanical Engineering Technology
9. Biomedical Engineering Technology

21.3 THE DERIVATIVE

In the previous section, we used the same procedure to examine instantaneous rates of change and to find the slope of a tangent to a curve. For both types of problems, we began with a function. We now use function notation to describe the procedure that was used, and we make the following observations.

1. $y = f(x)$ **(21-3)**
2. When we compute $y + \Delta y$ by substituting $x + \Delta x$ into the equation that defines the function, we have

$$y + \Delta y = f(x + \Delta x). \qquad \textbf{(21-4)}$$

3. To compute Δy, we used $(y + \Delta y) - y$. Subtracting Equation 21-3 from Equation 21-4, we have

$$(y + \Delta y) - y = f(x + \Delta x) - f(x).$$

4. $\lim_{\Delta x \to 0} \frac{\Delta y}{\Delta x} = \lim_{\Delta x \to 0} \frac{f(x + \Delta x) - f(x)}{\Delta x}$ **(21-5)**

The entire procedure is summed up in Equation 21-5. Since there is only one limit associated with each value of x, this limit defines a new function called the *derivative* of the function f.

DEFINITION

Let f be a function. Then the **derivative** of f is the function f' defined by

$$f'(x) = \lim_{\Delta x \to 0} \frac{f(x + \Delta x) - f(x)}{\Delta x}$$

for all x for which this limit exists.

A word about notation is in order. We have used $f'(x)$ in the definition. This notation is informative, because it shows that f is the original function, and that x is the independent variable. If $y = f(x)$, the notation $\frac{dy}{dx}$ is often used to denote the derivative. The notation is suggestive of the quotient $\frac{\Delta y}{\Delta x}$, but it is important to remember that ***the derivative itself is not the quotient*** $\frac{\Delta y}{\Delta x}$, ***but rather is the* limit *obtained from this quotient.*** Sometimes y' is used. This notation is more concise, but it has the disadvantage of not showing what the independent variable is. All three notations are commonly used and will be employed in this chapter so that you will gain some familiarity with all of them. Other symbols such as $D_x f$ may also be used to represent the derivative.

NOTE ⇨⇨

NOTATION FOR THE DERIVATIVE

If $y = f(x)$ defines a function, the derivative may be denoted as

$$f'(x), \frac{dy}{dx}, \frac{df}{dx}, \text{ or } y'.$$

differentiation

The process of finding a derivative is called **differentiation,** and we are now ready to see some examples of it. Pay close attention to the instructions in the following examples, so that you can see the different ways that questions about the derivative may be phrased.

EXAMPLE 1 Find the derivative of the function defined by $f(x) = 2x^2 + x$.

Solution $f(x + \Delta x) = 2(x + \Delta x)^2 + (x + \Delta x)$ and $f(x) = 2x^2 + x$

$$f'(x) = \lim_{\Delta x \to 0} \frac{f(x + \Delta x) - f(x)}{\Delta x}$$

$$= \lim_{\Delta x \to 0} \frac{[2(x + \Delta x)^2 + (x + \Delta x)] - (2x^2 + x)}{\Delta x}$$

$$= \lim_{\Delta x \to 0} \frac{2[x^2 + 2x(\Delta x) + (\Delta x)^2] + (x + \Delta x) - (2x^2 + x)}{\Delta x}$$

Expand $(x + \Delta x)^2$.

$$= \lim_{\Delta x \to 0} \frac{2x^2 + 4x(\Delta x) + 2(\Delta x)^2 + x + \Delta x - 2x^2 - x}{\Delta x}$$

Use the distributive property.

$$= \lim_{\Delta x \to 0} \frac{4x(\Delta x) + 2(\Delta x)^2 + \Delta x}{\Delta x}$$

Combine like terms.

$$= \lim_{\Delta x \to 0} \frac{(\Delta x)[4x + 2(\Delta x) + 1]}{\Delta x}$$

Factor out Δx.

$$= \lim_{\Delta x \to 0} 4x + 2(\Delta x) + 1$$

Reduce the fraction.

$$= 4x + 1$$

Take the limit. ■

EXAMPLE 2 If $f(x) = \frac{1}{x}$, find $f'(2)$.

Solution Before we find $f'(2)$, we find $f'(x)$.

$$f(x + \Delta x) = \frac{1}{x + \Delta x} \text{ and } f(x) = \frac{1}{x}$$

$$f'(x) = \lim_{\Delta x \to 0} \frac{f(x + \Delta x) - f(x)}{\Delta x}$$

To avoid a complex fraction, we consider multiplication by $1/\Delta x$ instead of division by Δx. Thus,

$$f'(x) = \lim_{\Delta x \to 0} \left[\frac{1}{x + \Delta x} - \frac{1}{x}\right] \frac{1}{\Delta x}.$$

Using $(x + \Delta x)(x)$ as a common denominator for the two expressions within brackets, we have the following.

$$f'(x) = \lim_{\Delta x \to 0} \left[\frac{x - (x + \Delta x)}{(x + \Delta x)(x)}\right] \frac{1}{\Delta x}$$

$$= \lim_{\Delta x \to 0} \left[\frac{x - x - \Delta x}{(x + \Delta x)(x)}\right] \frac{1}{\Delta x}$$ Rewrite $-(x + \Delta x)$.

$$= \lim_{\Delta x \to 0} \left[\frac{-\Delta x}{(x + \Delta x)(x)}\right] \frac{1}{\Delta x}$$ Combine like terms.

$$= \lim_{\Delta x \to 0} \frac{-1}{(x + \Delta x)(x)}$$ Reduce the fraction

$$= \frac{-1}{x^2}$$ Take the limit.

Since $f'(x) = \frac{-1}{x^2}$, $f'(2) = \frac{-1}{2^2} = \frac{-1}{4}$. ■

NOTE ⇨⇨ Because the process of finding a derivative is exactly the same process we used on the problems of the previous section, we make the following observations. ***The instantaneous rate of change of a function of x with respect to x is given by the derivative of the function with respect to x. Also, the slope of the tangent to the graph of a function at x is given by the value of the derivative of the function at x.***

EXAMPLE 3 Find the slope of the tangent to the curve $f(x) = 2x^3 + 1$ when x is $-1/2$.

Solution Since the slope of the tangent to the curve is given by the derivative, we find $f'(-1/2)$.

$$f(x + \Delta x) = 2(x + \Delta x)^3 + 1 \text{ and } f(x) = 2x^3 + 1$$

$$f'(x) = \lim_{\Delta x \to 0} \frac{f(x + \Delta x) - f(x)}{\Delta x}$$

$$= \lim_{\Delta x \to 0} \frac{2(x + \Delta x)^3 + 1 - (2x^3 + 1)}{\Delta x}$$

$$= \lim_{\Delta x \to 0} \frac{2[x^3 + 3x^2(\Delta x) + 3x(\Delta x)^2 + (\Delta x)^3] + 1 - (2x^3 + 1)}{\Delta x}$$ Expand $(x + \Delta x)^3$.

$$= \lim_{\Delta x \to 0} \frac{2x^3 + 6x^2(\Delta x) + 6x(\Delta x)^2 + 2(\Delta x)^3 + 1 - 2x^3 - 1}{\Delta x}$$ Use the distributive property.

$$= \lim_{\Delta x \to 0} \frac{6x^2(\Delta x) + 6x(\Delta x)^2 + 2(\Delta x)^3}{\Delta x}$$ Combine like terms.

$$= \lim_{\Delta x \to 0} \frac{(\Delta x)[6x^2 + 6x(\Delta x) + 2(\Delta x)^2]}{\Delta x}$$ Factor out Δx.

$$= \lim_{\Delta x \to 0} [6x^2 + 6x(\Delta x) + 2(\Delta x)^2]$$ Reduce the fraction.

$$= 6x^2$$ Take the limit.

Since $f'(x) = 6x^2, f'(-1/2) = 6(-1/2)^2 = 6(1/4) = 3/2$. Thus, for the curve defined by $f(x) = 2x^3 + 1$, the slope of the tangent is $3/2$ when x is $-1/2$. The graph is shown in Figure 21.20. ■

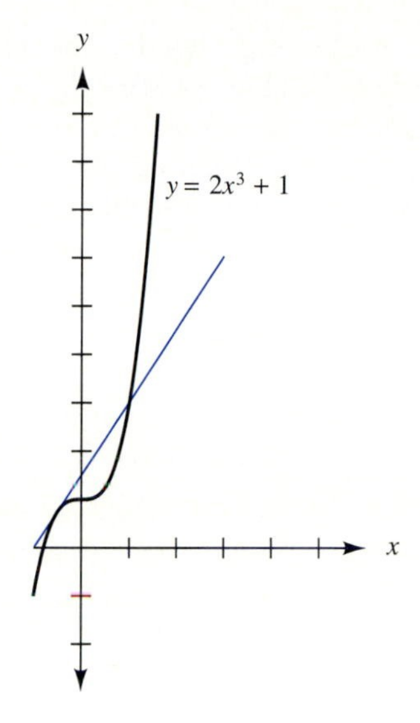

FIGURE 21.20

EXAMPLE 4 In a laboratory, a certain growing culture appears as an expanding circle. Find the instantaneous rate of change of area with respect to the radius when the radius is 0.01 mm.

Solution The area, given by $A = \pi r^2$, may be considered to be a function of the radius. Since an instantaneous rate of change is a derivative, we want to find $\frac{dA}{dr}$.

$$\frac{dA}{dr} = \lim_{\Delta r \to 0} \frac{\pi(r + \Delta r)^2 - \pi r^2}{\Delta r}$$

$$= \lim_{\Delta r \to 0} \frac{\pi[r^2 + 2r(\Delta r) + (\Delta r)^2] - \pi r^2}{\Delta r} \qquad \text{Expand } (r + \Delta r)^2.$$

$$= \lim_{\Delta r \to 0} \frac{\pi r^2 + 2\pi r(\Delta r) + \pi(\Delta r)^2 - \pi r^2}{\Delta r} \qquad \text{Distribute } \pi.$$

$$= \lim_{\Delta r \to 0} \frac{2\pi r(\Delta r) + \pi(\Delta r)^2}{\Delta r} \qquad \text{Combine like terms.}$$

$$= \lim_{\Delta r \to 0} \frac{(\Delta r)\{2\pi r + \pi(\Delta r)]}{\Delta r} \qquad \text{Factor out } \Delta r.$$

$$= \lim_{\Delta r \to 0} [2\pi r + \pi(\Delta r)] \qquad \text{Reduce the fraction.}$$

$$= 2\pi r \qquad \text{Take the limit.}$$

Since $\frac{dA}{dr} = 2\pi r$, the rate of change when $r = 0.01$ mm is given by $\frac{dA}{dr} = 2\pi(0.01)$ or 0.06 mm/mm, to one significant digit. ■

The process of computing derivatives using the definition can be rather long. In the remainder of the chapter, we will see that derivatives of many functions follow patterns that make the computation much easier. By working a few problems the long way, however, you should gain some familiarity with the notation, an understanding that the derivative is a limit, and an appreciation for the shortcuts you will soon learn.

EXERCISES 21.3

In 1–4, use the definition to find the derivative of each function.

1. $f(x) = x^2 + 1$

2. $f(x) = 2 - x^2$

3. $f(x) = 2 - 3x^3$

4. $f(x) = x^3 + 2$

In 5–7, use the definition of derivative to find y' for each function.

5. $y = \frac{2}{x}$

6. $y = \frac{1}{x+2}$

7. $y = \frac{-1}{2x}$

8. If $f(x) = 2x^2 - 1$, find $f'(0)$.

9. If $g(x) = x^2 + 1$, find $g'(-1)$.

10. If $g(x) = 3x - x^3$, find $g'(-2)$.

11. If $g(x) = -1/x^2$, find $g'(2)$.

12. If $g(x) = \frac{-2}{3x}$, find $g'(-1)$.

In problems 13–19, find the slope of the tangent to each curve at the given point. Sketch the graph of the function and the tangent to see if your answer seems reasonable.

13. $f(x) = -x^2 + 4x$ at $x = 1$

14. $f(x) = x^2 - 6x$ at $x = 3$

15. $f(x) = x^2 + 2x + 3$ at $x = -1$

16. $f(x) = -x^2 + 2x + 3$ at $x = 0$

17. $y = x^3 - 4x$ at $x = 1$

18. $y = x^3 - 3$ at $x = -1$

19. $y = x^3 + 2x^2$ at $x = -2$

---

20. For an automobile traveling at speed v (in mph) when the brakes are applied, the distance d (in ft) required to stop the car may be approximated by the formula $d = 0.05\ v^2 + v$. Find the instantaneous rate of change of distance with respect to velocity when the speed is 50 mph.
4 7 8 1 2 3 5 6 9

21. The angle θ (in radians) through which a rotating object moves is given by $\theta = (1/2)\alpha t^2 + \omega_0 t$, where α is the angular acceleration (in rad/s^2), t is the time (in seconds) and ω_0 is the initial angular velocity (in rad/s). Find the rate of change of angle with respect to time if $\alpha = 2.00$ rad/s^2, $t = 3.00$ s, and $\omega_0 = 9.00$ rad/s. 4 7 8 1 2 3 5 6 9

---

1. Architectural Technology
2. Civil Engineering Technology
3. Computer Engineering Technology
4. Mechanical Drafting Technology
5. Electrical/Electronics Engineering Technology
6. Chemical Engineering Technology
7. Industrial Engineering Technology
8. Mechanical Engineering Technology
9. Biomedical Engineering Technology

22. As a spherical balloon is inflated, the surface area expands. Find the rate of change of surface area with respect to the radius. 1 2 4 7 8 3 5

23. The power P (in W) generated by a particular windmill is given by $P = 0.015\ V^3$ where V is the velocity of the wind (in mph). Find the instantaneous rate of change of power with respect to velocity when the velocity is 10.0 mph. 4 7 8 1 2 3 5 6 9

24. A certain tumor that is approximately spherical grows in such a way as to maintain its shape. Find the instantaneous rate of change of volume of the sphere with respect to the radius when the radius is 1 cm. 6 9 2 4 5 7 8

25. A circular pipe is tapered from a large diameter to a smaller diameter over a short length of pipe. Find a formula for the rate of change of the pipe area with respect to its diameter. 1 2 4 7 8 3 5

21.4 BASIC DIFFERENTIATION RULES

Our goal in this section is to develop some rules or short cuts for differentiating functions so that we do not have to depend on the definition to work out each derivative. The first case we consider is the constant function defined by $f(x) = c$. A constant function always has the same value. Thus,

$$f(x + \Delta x) = c \text{ and } f(x) = c.$$

$$f'(x) = \lim_{\Delta x \to 0} \frac{f(x + \Delta x) - f(x)}{\Delta x} = \lim_{\Delta x \to 0} \frac{c - c}{\Delta x} = \lim_{\Delta x \to 0} \frac{0}{\Delta x}$$

Regardless of how small Δx becomes, as long as Δx is not equal to zero, $0/\Delta x$ is 0. Thus, we have the following rule.

THE DERIVATIVE OF A CONSTANT

If f is the constant function defined by $f(x) = c$, then $\dfrac{df}{dx} = 0$.

This result is reasonable, since the graph of $f(x) = c$ is a horizontal line, as in Figure 21.21. At any point, the slope of the tangent at that point is 0.

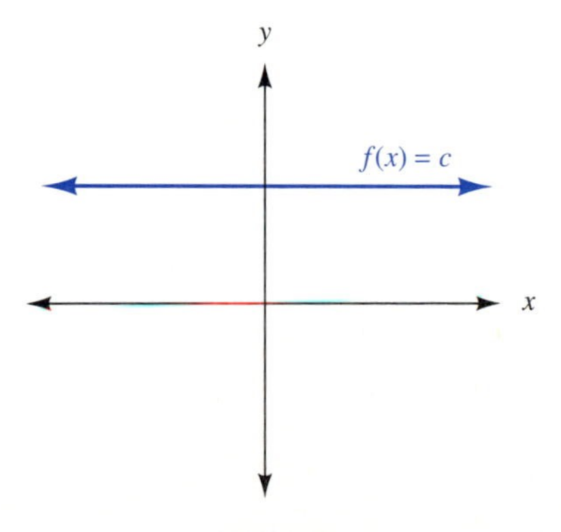

FIGURE 21.21

EXAMPLE 1 If $f(x) = 3$, find $f'(2)$.

Solution The function defined by $f(x) = 3$ is a constant function. Therefore, $f'(x) = 0$ for all x, and $f'(2) = 0$. ■

We now consider the derivative of the function defined by $f(x) = ax^n$, where n is a positive integer and a is a real number. $f(x + \Delta x) = a(x + \Delta x)^n$ and $f(x) = ax^n$.

$$f'(x) = \lim_{\Delta x \to 0} \frac{f(x + \Delta x) - f(x)}{\Delta x} = \lim_{\Delta x \to 0} \frac{a(x + \Delta x)^n - ax^n}{\Delta x}$$

The binomial expansion is used to write $(x + \Delta x)^n$.

$$\begin{aligned} &a(x + \Delta x)^n - ax^n \\ &\quad = a\left[x^n + nx^{n-1}(\Delta x) + \frac{(n-1)(n)x^{n-2}}{2}(\Delta x)^2 + \cdots + (\Delta x)^n\right] - ax^n \end{aligned}$$

Distributing a gives the following equation.

$$\begin{aligned} &a(x + \Delta x)^n - ax^n \\ &\quad = ax^n + anx^{n-1}(\Delta x) + \frac{a(n-1)(n)x^{n-2}}{2}(\Delta x)^2 + \cdots + a(\Delta x)^n - ax^n \end{aligned}$$

Combining like terms shortens the right-hand side.

$$a(x + \Delta x)^n - ax^n = anx^{n-1}(\Delta x) + \frac{an(n-1)x^{n-2}}{2}(\Delta x)^2 + \cdots + a(\Delta x)^n$$

Multiply both sides of this equation by $1/\Delta x$, and take the limit as Δx approaches 0.

$$\begin{aligned} &\lim_{\Delta x \to 0} \frac{a(x + \Delta x)^n - ax^n}{\Delta x} \\ &\quad = \lim_{\Delta x \to 0} \frac{1}{\Delta x}\left[anx^{n-1}(\Delta x) + \frac{an(n-1)x^{n-2}}{2}(\Delta x)^2 + \cdots + a(\Delta x)^n\right] \\ &\quad = \lim_{\Delta x \to 0} \frac{\Delta x}{\Delta x}\left[anx^{n-1} + \frac{an(n-1)x^{n-2}}{2}(\Delta x) + \cdots + a(\Delta x)^{n-1}\right] \end{aligned}$$

Inside the brackets, each term after the first has a factor of Δx. As Δx approaches 0, each term after the first will therefore approach 0. Thus, if

$$f(x) = ax^n, f'(x) = anx^{n-1}. \qquad \textbf{(21-6)}$$

NOTE ⇨⇨ ***Notice the pattern: The exponent in the original function becomes a factor in the derivative, and the exponent in the derivative is obtained by subtracting 1 from the exponent in the original function.*** Although we have justified the rule only for powers that are positive integers, it can be shown to be valid for rational powers of x. If the exponent in the derivative is negative, however, the derivative is undefined when x is 0. The rule is stated as follows.

THE DERIVATIVE OF A POWER

If $f(x) = ax^n$, where a is a real number and n is a rational number, then

$$\frac{df}{dx} = anx^{n-1}$$

for all x for which x^{n-1} is defined.

EXAMPLE 2 If $f(x) = 3x^8$, find $f'(0)$.

Solution

Exponent of original function

$$f'(x) = \mathbf{8}(3x^{\mathbf{8-1}}) = 24x^7$$

Exponent decreased by 1

$$f'(0) = 24(0^7) = 0 \quad ■$$

EXAMPLE 3 If $f(x) = -3x^{-4/3}$, find $f'(1)$.

Solution

Exponent of original function

$$f'(x) = \mathbf{(-4/3)}(-3x^{\mathbf{-4/3-1}}) = 4x^{-7/3}$$

Exponent decreased by 1

$$f'(1) = 4(1)^{-7/3} = 4 \quad ■$$

EXAMPLE 4 The radioactivity A (in roentgens/hr) at time t (in hours) after an explosion may be approximated by $A = A_0 t^{-1.2}$ where A_0 is the radioactivity immediately after the explosion, and t is less than 6 months. Find a formula for the instantaneous rate of change of radioactivity with respect to time.

Solution If $A = A_0 t^{-1.2}$, then $A' = -1.2 A_0 t^{-2.2}$. Notice that the exponent of the original function is a factor of the derivative, and the exponent of the derivative is obtained by subtracting 1 from -1.2, the exponent of the original function. ■

Since a derivative is a limit, properties of derivatives depend to some extent, on properties of limits. Up to this point, we have assumed certain properties of limits, because they seem quite natural in the context of working out specific problems. It is now necessary to state them formally (but without proof), for we will need to use them to justify differentiation rules in this section and the next.

If $\lim_{x\to a} f(x)$ and $\lim_{x\to a} g(x)$ exist, then

$$\lim_{x\to a} [f(x) + g(x)] = \left[\lim_{x\to a} f(x)\right] + \left[\lim_{x\to a} g(x)\right], \text{ and} \tag{21-7}$$

$$\lim_{x\to a} [f(x)\, g(x)] = \left[\lim_{x\to a} f(x)\right]\left[\lim_{x\to a} g(x)\right]. \tag{21-8}$$

Equations 21-7 and 21-8 may be stated informally in words as follows. The limit of a sum is the sum of the limits. The limit of a product is the product of the limits.

Now we are ready to consider a function that is obtained by adding two functions. The notation $(f+g)(x)$, means $f(x) + g(x)$. Therefore,

$$(f+g)(x+\Delta x) = f(x+\Delta x) + g(x+\Delta x).$$

The derivative of the function $(f+g)$ is obtained as follows.

$$(f+g)'(x) = \lim_{\Delta x\to 0} \frac{(f+g)(x+\Delta x) - (f+g)(x)}{\Delta x}$$

$$= \lim_{\Delta x\to 0} \frac{[f(x+\Delta x) + g(x+\Delta x)] - [f(x) + g(x)]}{\Delta x}$$

$$= \lim_{\Delta x\to 0} \frac{f(x+\Delta x) + g(x+\Delta x) - f(x) - g(x)}{\Delta x}$$

$$= \lim_{\Delta x\to 0} \frac{f(x+\Delta x) - f(x) + g(x+\Delta x) - g(x)}{\Delta x}$$

Change the order of the terms in the numerator.

$$= \lim_{\Delta x\to 0} \left[\frac{f(x+\Delta x) - f(x)}{\Delta x} + \frac{g(x+\Delta x) - g(x)}{\Delta x}\right]$$

Write the fraction as two fractions.

$$= \lim_{\Delta x\to 0} \frac{f(x+\Delta x) - f(x)}{\Delta x} + \lim_{\Delta x\to 0} \frac{g(x+\Delta x) - g(x)}{\Delta x}$$

The limit of the sum is the sum of the limits.

But these limits define the derivatives of the functions f and g. Thus, we have the following rule.

THE DERIVATIVE OF A SUM OF FUNCTIONS

If f and g are functions that have derivatives, then

$$\frac{d(f+g)}{dx} = \frac{df}{dx} + \frac{dg}{dx}. \tag{21-9}$$

Informally, we say that the derivative of a sum is the sum of the derivatives. The principle generalizes to the sum of any finite number of functions. Because a polynomial in x is a finite sum of terms of the form ax^n, where a is a real number and n is a nonnegative integer, a polynomial may be differentiated using Equation 21-9.

EXAMPLE 5 If $y = 2x^2 + 3x - 1$, find dy/dx.

Solution Since y is a sum, it can be differentiated term by term. The derivative of $2x^2$ is $4x$, the derivative of $3x$ is 3, and the derivative of -1 is 0, since -1 is a constant. Thus,

$$dy/dx = 4x + 3. \quad \blacksquare$$

EXAMPLE 6 If $f(x) = 3x^3 - 2x + 4$, find $f'(x)$.

Solution Since $f(x)$ is a sum, it can be differentiated term by term. The derivative of $3x^3$ is $9x^2$, the derivative of $-2x$ is -2, and the derivative of 4 is 0, since 4 is a constant. Thus,

$$dy/dx = 9x^2 - 2. \quad \blacksquare$$

EXAMPLE 7 If the deflection y (in inches) of a cantilever beam under a concentrated load is measured at a distance of x (in inches) from the support,

$$y = \frac{Pa^3}{6EI} - \frac{Pxa^2}{2EI}.$$

As the load is moved, the amount of deflection at a given point changes. Find dy/da when a is 51.6 inches if $P = 950$ lb, $E = 1.1 \times 10^6$ psi, $I = 536$ in^4, and $x = 64.8$ inches. In the formula, a is the distance (in inches) of the load from the support, P is the weight of the load (in lb), E is the modulus of elasticity (in psi), and I is the moment of inertia (in in^4).

Solution Substituting the given values, we have

$$y = \frac{950a^3}{6(1.1)(10^6)(536)} - \frac{950(64.8)a^2}{2(1.1)(10^6)(536)}.$$

Differentiating with respect to a gives

$$\frac{dy}{da} = \frac{2{,}850a^2}{3{,}537{,}600{,}000} - \frac{123{,}120\,a}{1{,}179{,}200{,}000}.$$

When a is 51.6 inches, the equation becomes

$$\frac{dy}{da} = \frac{(2{,}850)(51.6)^2}{3{,}537{,}600{,}000} - \frac{(123{,}120)(51.6)}{1{,}179{,}200{,}000}$$

$$= -0.0032 \text{ inches per inch, to two significant digits.} \quad \blacksquare$$

EXAMPLE 8 A certain highway curve is described by $y = 0.28x^2 - 0.05x$. In the formula, y is the vertical distance from the beginning of the curve in feet, and x is the horizontal distance from the beginning of the curve measured in 100-foot increments, called stations. Find the slope of the line tangent to the curve when x is 1 station.

Solution Since the slope of the line tangent to the curve at any point is given by the derivative at that point, we have

$$dy/dx = 0.56x - 0.05.$$

When x is 1, $dy/dx = 0.56(1) - 0.05$ or 0.51 ft/station. Figure 21.22 shows the graph. ■

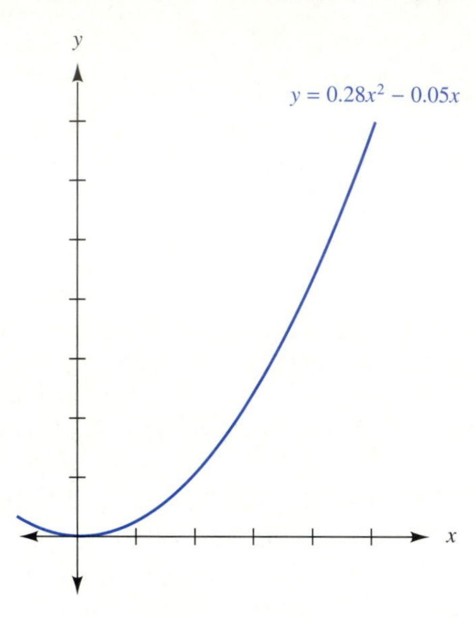

FIGURE 21.22

EXERCISES 21.4

In 1–4, find the derivative of the given function.

1. $y = 4x^3 + 3x^2 + 1$

2. $y = 2x^3 - x + 5$

3. $y = 3x^4 - 2x^3 + x$

4. $y = x^4 - 3x^2 + 2$

5. If $y = x^5 - 4x^3 + 1$, find y'.

6. If $y = 2x^5 - x^2 + x$, find y'.

7. If $y = 2x^7 - 1$, find y'.

8. If $y = 3x^6 + 5$, find y'.

9. If $f(x) = x^4 - 3x^2 + 2x - 1$, find $f'(0)$.

10. If $f(x) = 2x^4 + x^3 - 3x^2 + 4$, find $f'(1)$.

11. If $g(x) = 3x^5 - 4x^3 + x - 2$, find $g'(-1)$.

12. If $g(x) = 2x^5 + x^4 - 3x^2 + 1$, find $g'(-2)$.

13. If $f(x) = 2x^{-3}$ find $f'(1)$.

14. If $f(x) = 2x^{-5}$, find $f'(-2)$.

15. If $g(x) = 3x^{2/3}$, find $g'(8)$.

16. If $g(x) = 2x^{1/2}$, find $g'(4)$.

In 17–24, find the slope of the line tangent to the curve at the given point.

17. $y = x^2 - 4x + 3$ at $(0, 3)$

18. $y = -x^2 + 2$ at $(1, 1)$

19. $y = 2x^2 - 3x + 1$ at $(2, 3)$

20. $y = 3x^3 - x - 2$ at $(-1, -4)$

21. $y = x^3 - 3x^2 - 2$ at $(-1, -2)$

22. $y = 2x^3 + x^2 - x$ at $(0, 0)$

23. $y = x^4 - 3x^3 + 2x - 1$ at $(1, -1)$

24. $y = 2x^4 + x^3 - 3x^2 + 4$ at $(-1, 2)$

25. The formula $W = Fx$ gives the work done (in joules) by a constant force (in newtons) applied through a distance of x (in meters). Find the rate of change of work done with respect to distance when an object is in the position of having been moved through 12.0 m by a force of 2.0 N. 4 7 8 1 2 3 5 6 9

26. The reactance X_L (in ohms) of an inductor is given by $X_L = 2\pi fL$. Find the rate of change of reactance with respect to frequency when $f = 1.1 \times 10^6$ and $L = 4.0$ mH. In the formula, f is the frequency (in cycles/s) of a coil of wire rotating through a magnetic field, and L is the inductance (in henries) of the inductor. 3 5 9 1 2 4 6 7 8

27. The velocity of water (in ft/s) at the point of discharge is given by $v = 12.16\sqrt{P}$, where P is the pressure (in lb/in^2) of the water at the point of discharge. Find the rate of change of the velocity with respect to pressure if the pressure is 50.00 lb/in^2. 4 8 1 2 6 7 9

28. A balloon used in a surgical procedure is approximately cylindrical in shape. As it expands outward, assume that the length remains a constant 30.0 mm and find the rate of change of surface area with respect to radius when the radius is 0.080 mm. 1 2 4 7 8 3 5

29. The gravitational force F (in newtons) of attraction between two bodies of mass M_1 and M_2 (in kg) is given by $F = GM_1M_2/r^2$ where r is the distance (in m) between the bodies. G is a constant and is equal to 6.67×10^{-11}. As the bodies move closer together, r decreases, and F increases. Find the rate of change of F with respect to r for two bodies, each of mass 1.50 kg when the distance is 1.50 m. 4 7 8 1 2 3 5 6 9

30. The formula $Q = 5.22\sqrt{h}$ gives the flow rate Q (in ft/s) of water flowing through a circular-shaped hole in the side of a tank when h is the height (in ft) of the water surface above the hole. Find the rate of change of the flow rate with respect to h when h is 2.60 ft. 4 8 1 2 6 7 9

31. The specific energy E (in ft) of the flow of water in a rectangular channel is given by $E = y + q^2/(64.4y^2)$. Find dE/dy when $q = 4.00$ ft^3/s per foot and $y = 0.500$ ft. In the formula, q is the flow in a one foot wide section of the channel, and y is the flow depth. 4 8 1 2 6 7 9

1. Architectural Technology
2. Civil Engineering Technology
3. Computer Engineering Technology
4. Mechanical Drafting Technology
5. Electrical/Electronics Engineering Technology
6. Chemical Engineering Technology
7. Industrial Engineering Technology
8. Mechanical Engineering Technology
9. Biomedical Engineering Technology

21.5 DIFFERENTIATION RULES FOR PRODUCTS AND QUOTIENTS

Sometimes it is necessary to differentiate a function that is the product of two functions. The function defined by $y = x^2(2x - 1)$ is such a function. It can be written as $y = 2x^3 - x^2$, and the derivative found by the methods of the previous section. Later, however, we will encounter functions such as $y = e^x \sin x$, which must be treated as products. ***It is therefore important for you to learn to work with products, even though you may see a way to avoid them in this section.***

NOTE ⇨⇨

Because the limit of a product is the product of the limits, it is a common mistake to assume that the derivative of a product is the product of the derivatives. The function defined by $y = x^2(2x - 1)$ may be used to show that this assumption is incorrect. The product of $x^2(2x - 1)$ is $2x^3 - x^2$, and the derivative of the product is $6x^2 - 2x$. The derivative of x^2 is $2x$, and the derivative of $2x - 1$ is 2. The product of these derivatives is $4x$. Clearly, $6x^2 - 2x$ is not the same as $4x$.

We have used function notation for most of the derivatives up to this point, so that we could name one function f and the other g. But just as y is often used in place of $f(x)$, the letters u and v are often used to represent functions when there are two functions. Thus, we might have $u = f(x)$, and $v = g(x)$. This notation has the advantage of being more compact. We use it to derive the rule for the derivative of a product.

$$\frac{d(uv)}{dx} = \lim_{\Delta x \to 0} \frac{(u + \Delta u)(v + \Delta v) - uv}{\Delta x}$$

$$= \lim_{\Delta x \to 0} \frac{uv + u(\Delta v) + (\Delta u)v + (\Delta u)(\Delta v) - uv}{\Delta x}$$ Expand $(u + \Delta u)(v + \Delta v)$.

$$= \lim_{\Delta x \to 0} \frac{u(\Delta v) + (\Delta u)v + (\Delta u)(\Delta v)}{\Delta x}$$ Combine like terms.

$$= \lim_{\Delta x \to 0} \left[\frac{u(\Delta v)}{\Delta x} + \frac{(\Delta u)v}{\Delta x} + \frac{(\Delta u)(\Delta v)}{\Delta x}\right]$$ Write as 3 fractions.

$$= \lim_{\Delta x \to 0} \frac{u(\Delta v)}{\Delta x} + \lim_{\Delta x \to 0} \frac{(\Delta u)v}{\Delta x} + \lim_{\Delta x \to 0} \frac{(\Delta u)(\Delta v)}{\Delta x}$$ Write the limit of a sum as a sum of the limits.

This equation is equivalent to

$$\frac{d(uv)}{dx} = \lim_{\Delta x \to 0} u\left(\frac{\Delta v}{\Delta x}\right) + \lim_{\Delta x \to 0} v\left(\frac{\Delta u}{\Delta x}\right) + \lim_{\Delta x \to 0} \Delta u\left(\frac{\Delta v}{\Delta x}\right).$$

Each limit, however, is the limit of a product. As Δx approaches 0, $\lim_{\Delta x \to 0} u = u$ and $\lim_{\Delta x \to 0} v = v$. Furthermore, Δu and Δv both approach 0. Thus,

$$\frac{d(uv)}{dx} = u \lim_{\Delta x \to 0} \frac{\Delta v}{\Delta x} + v \lim_{\Delta x \to 0} \frac{\Delta u}{\Delta x} + 0 \lim_{\Delta x \to 0} \frac{\Delta v}{\Delta x}.$$

The remaining limits define derivatives of u or v. That is,

$$\frac{d(uv)}{dx} = u\frac{dv}{dx} + v\frac{du}{dx} + 0\frac{dv}{dx}$$

$$= u\frac{dv}{dx} + v\frac{du}{dx}.$$

This statement is known as the product rule for derivatives.

THE PRODUCT RULE FOR DERIVATIVES

If u and v are functions that have derivatives, then

$$\frac{d(uv)}{dx} = u\frac{dv}{dx} + v\frac{du}{dx}. \quad \textbf{(21-10)}$$

The product rule for derivatives, may be stated informally. The derivative of a product of two functions is the first times the derivative of the second plus the second times the derivative of the first.

EXAMPLE 1 If $y = x^2(2x - 1)$, find dy/dx using the product rule.

Solution The first factor in the product is x^2. Its derivative is $2x$. The second factor in the product is $2x - 1$. Its derivative is 2. Thus, the derivative of the product is given by

First × derivative of second

$$dy/dx = x^2(2) + (2x - 1)(2x)$$

Second × derivative of first

$$\begin{aligned} dy/dx &= 2x^2 + 4x^2 - 2x \\ &= 6x^2 - 2x. \end{aligned}$$ ■

Notice that the result in Example 1 agrees with the result obtained in the second paragraph of this section. While the derivative of this function could be found without using the product rule, Example 2, which follows, illustrates the importance of learning the product rule now, so that you will be able to use it when it becomes necessary to do so.

EXAMPLE 2 If $y = x^2 \sin x$, find $\frac{dy}{dx}$.

Solution $$\frac{dy}{dx} = x^2 \frac{d(\sin x)}{dx} + 2x \sin x$$

If we accept, temporarily, that $\frac{d(\sin x)}{dx} = \cos x$, a fact that we will prove later, then we have the following result.

$$\frac{dy}{dx} = x^2 \cos x + 2x \sin x$$ ■

The product rule may be used to show that the derivative of a constant times a function is the constant times the derivative of the function. Consider $y = cf(x)$ or $y = cu$.

The first factor is c. Its derivative is 0.

The second factor is u. Its derivative is du/dx.

Thus, $\frac{d(cu)}{dx} = c\frac{du}{dx} + u(0) = c\frac{du}{dx}$. This statement is known as the constant rule.

THE CONSTANT RULE FOR DERIVATIVES

If u is a function that has a derivative, then

$$\frac{d(cu)}{dx} = c\left(\frac{du}{dx}\right). \qquad \textbf{(21-11)}$$

EXAMPLE 3 If $y = 2(3x^2 + 4x - 2)$, find y'.

Solution The constant rule says that the derivative of y is given by the product of 2 and the derivative of $(3x^2 + 4x - 2)$. That is, $y' = 2(6x + 4)$ or $12x + 8$. ■

The quotient of two functions $u = f(x)$ and $v = g(x)$ is written as $\frac{u}{v}$.

$$\begin{aligned}
\frac{d\left(\frac{u}{v}\right)}{dx} &= \lim_{\Delta x \to 0} \frac{1}{\Delta x}\left[\frac{(u + \Delta u)}{(v + \Delta v)} - \frac{u}{v}\right] \\
&= \lim_{\Delta x \to 0} \frac{1}{\Delta x}\left[\frac{v(u + \Delta u) - u(v + \Delta v)}{v(v + \Delta v)}\right] && \text{Use a common denominator.} \\
&= \lim_{\Delta x \to 0} \frac{1}{\Delta x}\left[\frac{vu + v(\Delta u) - uv - u(\Delta v)}{v(v + \Delta v)}\right] && \text{Use the distributive property.} \\
&= \lim_{\Delta x \to 0} \frac{1}{\Delta x}\left[\frac{v(\Delta u) - u(\Delta v)}{v(v + \Delta v)}\right] && \text{Combine like terms.}
\end{aligned}$$

This equation is equivalent to the following equation.

$$\frac{d\left(\frac{u}{v}\right)}{dx} = \lim_{\Delta x \to 0}\left[\frac{v(\Delta u) - u(\Delta v)}{\Delta x} \cdot \frac{1}{v(v + \Delta v)}\right]$$

Finding the individual limits, we have

$$\begin{aligned}
\frac{d\left(\frac{u}{v}\right)}{dx} &= \left[\lim_{\Delta x \to 0} v\left(\frac{\Delta u}{\Delta x}\right) - \lim_{\Delta x \to 0} u\left(\frac{\Delta v}{\Delta x}\right)\right] \lim_{\Delta x \to 0}\left[\frac{1}{v(v + \Delta v)}\right] \\
&= \left[v \lim_{\Delta x \to 0} \frac{\Delta u}{\Delta x} - u \lim_{\Delta x \to 0} \frac{\Delta v}{\Delta x}\right] \lim_{\Delta x \to 0} \frac{1}{v(v + \Delta v)} \\
&= \left[v\left(\frac{du}{dx}\right) - u\left(\frac{dv}{dx}\right)\right]\frac{1}{v^2}.
\end{aligned}$$

This statement is known as the quotient rule for derivatives and is more commonly written as follows.

THE QUOTIENT RULE FOR DERIVATIVES

If u and v are functions that have derivatives and $v \neq 0$, then

$$\frac{d\left(\frac{u}{v}\right)}{dx} = \frac{v\left(\frac{du}{dx}\right) - u\left(\frac{dv}{dx}\right)}{v^2}. \qquad \textbf{(21-12)}$$

Informally, we may say that the derivative of a quotient of functions is the denominator times the derivative of the numerator minus the numerator times the derivative of the denominator, all divided by the square of the denominator.

EXAMPLE 4 If $y = \frac{1}{x+2}$, find $\frac{dy}{dx}$.

Solution The numerator is 1. Its derivative is 0.
The denominator is $x + 2$. Its derivative is 1.
Thus, the derivative of the quotient is given by the following equation.

The denominator
Derivative of numerator
Numerator
Derivative of denominator

$$\frac{dy}{dx} = \frac{(x+2)(0) - 1(1)}{(x+2)^2} \qquad \leftarrow \text{Square of the denominator}$$

$$= \frac{-1}{(x+2)^2} \quad ■$$

EXAMPLE 5 If $f(x) = \frac{x^3 + 3x - 1}{3}$, find $f'(2)$.

Solution Although the quotient rule could be used, it is neither necessary nor desirable to use it.

Since $f(x) = \frac{x^3 + 3x - 1}{3} = \frac{1}{3}(x^3 + 3x - 1)$, we apply the constant rule.

$$f'(x) = \frac{1}{3}(3x^2 + 3) = \frac{3}{3}(x^2 + 1) = x^2 + 1$$

$$f'(2) = 2^2 + 1 = 5 \quad ■$$

EXAMPLE 6 Find the slope of the tangent to the curve $y = \dfrac{x^3 + x + 1}{x^2 - 1}$ when x is -2.

Solution The slope of the tangent when x is -2 is given by the value of the derivative of the function when x is -2. The numerator of the function is $x^3 + x + 1$. Its derivative is $3x^2 + 1$. The denominator is $x^2 - 1$. Its derivative is $2x$. Thus, the derivative of the quotient is given by the following equation.

$$y' = \frac{\overbrace{(x^2 - 1)}^{\text{The denominator}}\overbrace{(3x^2 + 1)}^{\text{Derivative of numerator}} - \overbrace{(x^3 + x + 1)}^{\text{Numerator}}\overbrace{(2x)}^{\text{Derivative of denominator}}}{(x^2 - 1)^2} \quad \leftarrow \text{Square of the denominator}$$

$$= \frac{3x^4 + x^2 - 3x^2 - 1 - 2x^4 - 2x^2 - 2x}{(x^2 - 1)^2} \quad \text{Expand the numerator.}$$

$$= \frac{x^4 - 4x^2 - 2x - 1}{(x^2 - 1)^2} \quad \text{Combine like terms.}$$

When x is -2, $y' = \dfrac{16 - 4(4) - 2(-2) - 1}{(4 - 1)^2} = \dfrac{3}{9} = \dfrac{1}{3}$.

The graph is shown in Figure 21.23. ■

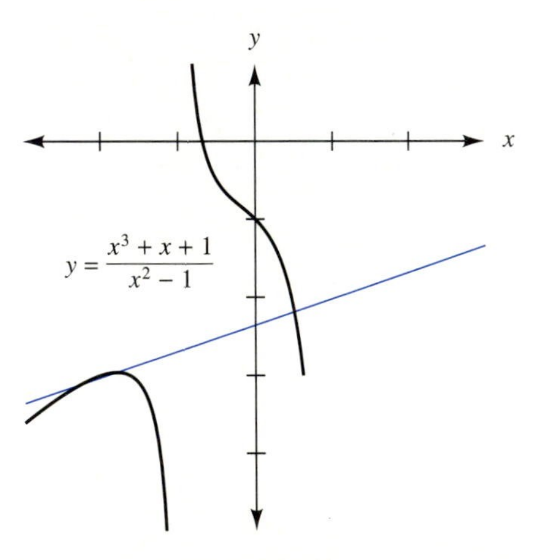

FIGURE 21.23

The quotient rule can be used to justify the formula for the derivative of a power when the power is negative. Example 7 illustrates the procedure for a specific exponent.

EXAMPLE 7 Let $y = x^{-2} = \frac{1}{x^2}$. Find y' as the derivative of a power and as the derivative of a quotient.

Solution As a power: $y' = -2(x^{-3}) = \frac{-2}{x^3}$

As a quotient: $y' = \frac{x^2(0) - 1(2x)}{(x^2)^2}$

$$= \frac{-2x}{x^4}$$

$$= \frac{-2}{x^3} \quad \blacksquare$$

When a function can be treated as a power or as a quotient, it is usually more efficient to treat it as a power, as in Example 7.

EXERCISES 21.5

In 1–8, use the product rule to find the derivative of each function. You may check your work by finding the product first and then differentiating.

1. $y = x^3(2x + 3)$

2. $y = 4x^2(2x - 1)$

3. $y = (3x^2 + 1)(x + 3)$

4. $y = (2x^2 - 3)(3x + 4)$

5. $f(x) = (x^3 - x)(x^2 + 1)$

6. $f(x) = (2x^3 + 1)(x^2 - 3x)$

7. $f(x) = (x^2 + 1)(x^2 - 1)$

8. $f(x) = (2x^2 - x)(x + 4)$

In 9–12, find the derivative of each function. You need to know that if $y = \sin x$, then $y' = \cos x$, and if $y = \cos x$, then $y' = -\sin x$.

9. $f(x) = 4x^2 \sin x$

10. $f(x) = 2x^3 \cos x$

11. $f(x) = (3x + 1) \cos x$

12. $f(x) = \sin x \cos x$

In 13–22, use the quotient rule to evaluate the derivative of each function for the given value.

13. $y = \frac{2x + 1}{3x^2}$ at $x = 1$

14. $y = \frac{3x - 2}{x^2}$ at $x = 1$

15. $y = \frac{x^2}{2x - 1}$ at $x = -1$

16. $y = \frac{3x^2}{2x + 3}$ at $x = -1$

17. $y = \frac{x - 4}{2x + 3}$ at $x = 0$

18. $y = \frac{3x - 1}{2x - 1}$ at $x = 0$

19. $f(x) = \frac{2x - 3}{x^2 + 1}$ at $x = 2$

20. $f(x) = \frac{3x + 2}{2x^2 - 3}$ at $x = 2$

21. $f(x) = \frac{2x^3 + x + 1}{x^2 - 3}$ at $x = -2$

22. $f(x) = \frac{x^3 - x^2 + 3}{3x^2 + 1}$ at $x = -2$

In 23–28, find the slope of the tangent to the curve at the given point.

23. $y = (x^2 + 2)(x - 1)$ at $x = 0$

24. $y = (2x - 3)(x^2 + 1)$ at $x = 2$

25. $y = \dfrac{x^2 - 1}{2}$ at $x = 1$

26. $y = \dfrac{3x + 2}{4}$ at $x = -1$

27. $y = \dfrac{x^2 + 3}{2x}$ at $x = -1$

28. $y = \dfrac{2x - 1}{3x^2}$ at $x = 2$

---

29. The potential difference in a circuit is given by $V = IR$. If I and R both change with time, find a formula for dV/dt. 3 5 9 1 2 4 6 7 8

30. Newton's second law of motion is sometimes stated as follows: "The rate of change of the momentum of a body is directly proportional to the force acting on the body." Momentum is given by mv, the product of the mass and velocity of the body. For most purposes, it is sufficient to consider m to be a constant, but in relativity theory, both m and v are functions of time.

(a) Use the product rule to find $F = \dfrac{d}{dt}(mv)$.

(b) Show that if m is constant, the result in (a) agrees with the result when the constant rule is used. 4 7 8 1 2 3 5 6 9

31. A rectangular steel plate expands as it is heated. Find the rate of change of area with respect to temperature T when the width is 1.3 cm and the length is 2.6 cm if $\dfrac{dl}{dT} = 1.7 \times 10^{-5}\ \dfrac{\text{cm}}{°\text{C}}$ and $\dfrac{dw}{dT} = 8.5 \times 10^{-6}\ \dfrac{\text{cm}}{°\text{C}}$. 1 2 4 6 8 5 7 9

32. For two resistors connected in parallel, the equivalent resistance R is given by $\dfrac{1}{R} = \dfrac{1}{R_1} + \dfrac{1}{R_2}$ where R_1 and R_2 are the resistances of the two resistors. Assume that R_1 is a variable resistance and R_2 is a constant resistance of 20.0 Ω. Solve the equation for R, and find the rate of change of R with respect to R_1 when R_1 is 40.0 Ω. 3 5 9 1 2 4 6 7 8

33. For a thin lens of focal length f, $\dfrac{1}{f} = \dfrac{1}{x} + \dfrac{1}{y}$, where x and y are the distances from the lens to the object and image, respectively. As the position of the object changes, the position of the image changes, but f is constant for a given lens. Solve the equation for y, and find dy/dx for a lens of focal length 11.0 mm when $x = 3.71$ mm. 3 5 6 9

34. A spike of voltage (in volts) given by the equation $v = 30.0t^{1.6}$ is applied to a circuit element whose resistance varies according to the equation $r = 5.0 + 0.010\,t$, where t is the time in seconds from the start of the experiment. Use Ohm's law ($i = v/r$) to find an expression for current, then find the rate of change of current with respect to time when $t = 1.3$ seconds. 3 5 9 1 2 4 6 7 8

35. Show that $(fgh)' = f'gh + fg'h + fgh'$.

---

1. Architectural Technology
2. Civil Engineering Technology
3. Computer Engineering Technology
4. Mechanical Drafting Technology
5. Electrical/Electronics Engineering Technology
6. Chemical Engineering Technology
7. Industrial Engineering Technology
8. Mechanical Engineering Technology
9. Biomedical Engineering Technology

21.6 THE CHAIN RULE

So far we have been able to differentiate only functions of the form ax^n, and combinations of such functions formed by addition, subtraction, multiplication, and division. Technological problems often lead to more complicated functions. In particular, *compositions of functions* are often encountered. In Chapter 3, we found the composition, given two functions. In this chapter, we will find the two functions, given the

composition of functions

composition. A **composition of functions** occurs when a function y can be expressed in terms of u, where u is a function that can be expressed in terms of x. Of course, the letters used are immaterial. For example, when a balloon is blown up, the volume v changes as the radius r changes, and the radius changes as the time t changes. Thus, v is a function of r, and r is a function of t.

EXAMPLE 1 Show that $y = \sqrt{x^2 + 1}$ is a composition of functions.

Solution Let $u = x^2 + 1$. Then $y = \sqrt{u}$. Thus, y is a function of u, and u is a function of x. ■

EXAMPLE 2 When a projectile is launched with an initial velocity of 300 m/s at an angle of θ degrees, the range x is given by

$$x = 9184 \sin 2\theta.$$

Show that x is a composition of functions.

Solution Let $u = 2\theta$. Then $x = 9184 \sin u$. Thus, x is a function of u, and u is a function of θ. ■

Let us now consider the instantaneous rate of change dy/dx of a function y that is a composition of functions. For example, if dy/du is 2, then as u increases, y increases twice as fast. If we also know that du/dx is 3, then as x increases, u increases three times as fast. But if y increases twice as fast as u, and u increases three times as fast as x, then y must increase six times as fast as x. This result can be generalized and is known as the chain rule for derivatives. It is stated symbolically as follows.

THE CHAIN RULE FOR DERIVATIVES

If y is a function of u, and u is a function of x, and if y and u have derivatives, then

$$\frac{dy}{dx} = \frac{dy}{du}\left(\frac{du}{dx}\right).$$

The steps for deriving the chain rule are similar to those you have seen for the product and quotient rules, but the proof is longer and will not be presented. Although we have not simply cancelled the quantity du, the apparent cancellation makes this formulation of the chain rule easy to remember.

EXAMPLE 3 If $y = (x^2 + 2x + 1)^9$, find dy/dx.

Solution Let $u = x^2 + 2x + 1$. Then $y = u^9$.

$$\frac{dy}{du} = 9u^8 \text{ and } \frac{du}{dx} = 2x + 2$$

Since, $\frac{dy}{dx} = \frac{dy}{du}\left(\frac{du}{dx}\right)$, $\frac{dy}{dx} = 9u^8(2x + 2)$.

Rewriting u in terms of x, $\frac{dy}{dx} = 9(x^2 + 2x + 1)^8(2x + 2)$. ■

Although Example 3 could have been done by expanding $(x^2 + 2x + 1)^9$, the chain rule is used to simplify the differentiation process. The chain rule is also used to differentiate functions that cannot be differentiated using methods covered up to this point.

EXAMPLE 4 Let $y = \sqrt{x^2 + 1}$. Find dy/dx.

Solution This is the function of Example 1.

Since $u = x^2 + 1$ and $y = \sqrt{u}$ or $u^{1/2}$, $\frac{du}{dx} = 2x$ and $\frac{dy}{du} = (1/2)u^{-1/2}$.

$$\frac{dy}{dx} = \frac{dy}{du}\left(\frac{du}{dx}\right) = (1/2)u^{-1/2}(2x) = (1/2)(x^2 + 1)^{-1/2}\,(2x) = \frac{x}{\sqrt{x^2 + 1}} \quad ■$$

A composition of functions f and g is sometimes written as $f \circ g(x)$, which means $f(g(x))$. If f and g are functions having derivatives, such that f is defined for all numbers that are values of $g(x)$, the chain rule can be expressed as

$$(f \circ g)'(x) = f'[g(x)]g'(x).$$

If we call f the "outer" function and g the "inner" function, the chain rule may be stated informally as "the derivative of a composition of functions is the derivative of the outer function times the derivative of the inner function." This formulation of the chain rule will help you to find derivatives without having to go through the intermediate step of replacing one of the functions with the letter u.

EXAMPLE 5 Find the derivative of $y = \frac{2}{(3x^2 + 1)}$.

Solution Although the quotient rule could be used, it is also possible to write $y = 2(3x^2 + 1)^{-1}$ and to use the constant rule. Ignore the quantity $3x^2 + 1$ inside parentheses for a moment and notice the pattern $y = 2(\quad)^{-1}$. The derivative is given by $y' = -2(\quad)^{-2}$. That is, the derivative of the outer function is $-2(3x^2 + 1)^{-2}$. The quantity in parentheses is the inner function, and its derivative is $6x$.

$$\text{Thus, } y' = -2(3x^2 + 1)^{-2}(6x) = \frac{-12x}{(3x^2 + 1)^2}. \quad ■$$

EXAMPLE 6 The formula $l = \frac{2\pi yd}{\sqrt{r^2 - y^2}}$ gives the approximate increase l in the length of a rope that is placed on a hemisphere of radius r and then moved downward a distance d, as shown in Figure 21.24.

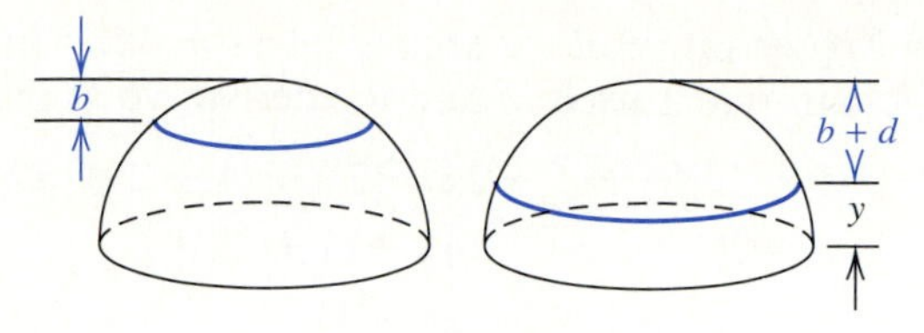

FIGURE 21.24

Assume that l and y are variables, but that all other quantities are constants. Find the instantaneous rate of change of l with respect to y.

Solution $l = \dfrac{2\pi yd}{\sqrt{r^2 - y^2}} = \dfrac{2\pi yd}{(r^2 - y^2)^{1/2}}$

The quotient rule gives the following result.

Denominator; Derivative of numerator; Numerator; Derivative of denominator (for outer function) (for inner function)

$$l' = \frac{\underbrace{(r^2 - y^2)^{1/2}}_{\text{Denominator}}\underbrace{(2\pi d)}_{\text{Derivative of numerator}} - \underbrace{(2\pi yd)}_{\text{Numerator}}\underbrace{(1/2)(r^2 - y^2)^{-1/2}}_{\text{for outer function}}\underbrace{(-2y)}_{\text{for inner function}}}{r^2 - y^2} \quad \longleftarrow \text{Square of the denominator}$$

$$= \frac{(r^2 - y^2)^{1/2}(2\pi d) + 2\pi y^2 d(r^2 - y^2)^{-1/2}}{r^2 - y^2}$$

$$= \frac{(r^2 - y^2)^{-1/2}(2\pi d)\,[(r^2 - y^2) + y^2]}{r^2 - y^2} \qquad \text{Factor out } (r^2 - y^2)^{-1/2}(2\pi d).$$

$$= \frac{(2\pi d)(r^2)}{(r^2 - y^2)^{3/2}} \qquad \text{Combine like terms in brackets and move } (r^2 - y^2)^{1/2} \text{ to the denominator.}$$

■

EXAMPLE 7 Find the slope of the line tangent to the curve $y = \dfrac{2x}{(1 + 2x)^2}$ when x is -1.

Solution The slope of the tangent when x is -1 is the derivative when x is -1. The function may be rewritten as a product, so that the product rule for derivatives applies.

$$y = \frac{2x}{(1 + 2x)^2} = 2x(1 + 2x)^{-2}$$

$y' = 2x$[derivative of the second] $+ (1 + 2x)^{-2}$ [derivative of the first]
The chain rule is used to find the derivative of the second function.

$$
\begin{aligned}
y' &= 2x[(-2)(1 + 2x)^{-3}(2)] + (1 + 2x)^{-2}(2) \\
&= 2(1 + 2x)^{-3}[-4x + (1 + 2x)] && \text{Factor out } 2(1+2x)^{-3}. \\
&= 2(1 + 2x)^{-3}(-2x + 1) && \text{Combine like terms.} \\
&= \frac{2(-2x+1)}{(1+2x)^3} && (1+2x)^{-3} = \frac{1}{(1+2x)^3}
\end{aligned}
$$

When $x = -1$, $y' = \dfrac{2(2+1)}{(1-2)^3} = \dfrac{2(3)}{(-1)^3} = \dfrac{6}{-1} = -6$. ■

If we accept without proof, for the present, that if $f(x) = \sin x$, then $f'(x) = \cos x$, the chain rule may be applied to problems involving the sine function.

EXAMPLE 8 If $y = \sin 2x$, find y'.

Solution Let $u = 2x$ and $y = \sin u$. Then $\dfrac{dy}{du} = \cos u$, and $\dfrac{du}{dx} = 2$. Thus, $\dfrac{dy}{dx} = \dfrac{dy}{du}\left(\dfrac{du}{dx}\right) = (\cos u)(2) = 2\cos 2x$. ■

EXERCISES 21.6

In 1–10, show that each function can be written as the composition of functions by writing y as a function of u, and identify u.

1. $y = (x^3 - 1)^4$
2. $y = (3x^2 + 2)^3$
3. $y = \sqrt{2x - 3}$
4. $y = \dfrac{1}{4x + 5}$
5. $y = \dfrac{1}{\sqrt{3x + 2}}$
6. $y = \sin x^2$
7. $y = \cos(2x + 3)$
8. $y = \ln(x^2 + 4)$
9. $y = \ln(\sin x)$
10. $y = e^{x^2/2}$

In 11–16, find the derivative of each function.

11. $y = (x^2 + 1)^5$
12. $y = (2x^2 - 3)^6$
13. $y = (2x - 4)^2$
14. $y = (3x^2 + 2x - 4)^5$
15. $y = (2x^2 - 3x + 5)^4$
16. $y = (x^2 + x - 1)^3$

In 17–20, find the derivative of each function. You need to know that if $f(x) = \sin x$, then $f'(x) = \cos x$, and if $f(x) = \cos x$, then $f'(x) = -\sin x$.

17. $y = \cos 3x$
18. $y = \sin 4x$
19. $y = \sin x^3$
20. $y = \cos x^2$

In 21–24, find the derivative of each function for the given value of x.

21. $y = (1 - 4x)^{-3}$ at $x = 1$

22. $y = (3x - 2x^2)^{-2}$ at $x = 2$

23. $y = (2 - x)^{-4}$ at $x = 0$

24. $y = (x^4 + 3x^2)^{-1}$ at $x = -1$

In 25–34, find the derivative of each function.

25. $y = (x^3 - 1)^{2/3}$

26. $y = (2x^3 + 2)^{3/2}$

27. $y = \sqrt{x^2 + 4}$

28. $y = \sqrt{3x - 1}$

29. $y = \sqrt[3]{2x + x^2}$

30. $y = (x^4 - 3)^{-1/2}$

31. $y = (x - 2)^{-1/2}$

32. $y = \dfrac{1}{\sqrt{2x^2 + 1}}$

33. $y = \dfrac{-1}{\sqrt{4x^2 - 3}}$

34. $y = \dfrac{2}{\sqrt[3]{2x + 5}}$

In 35–42, find the derivative of each function using the quotient rule. You may check the problem using the product rule.

35. $y = \dfrac{x^2}{(1 - 3x)^3}$

36. $y = \dfrac{2}{(1 - x)^2}$

37. $y = \dfrac{2x}{(x^2 - 4)^3}$

38. $y = \dfrac{3x + 1}{(x^3 - 5)^2}$

39. $y = \dfrac{x}{\sqrt{2x - 3}}$

40. $y = \dfrac{4x + 3}{\sqrt{x^2 - 1}}$

41. $y = \dfrac{2x + 1}{\sqrt{x - 4}}$

42. $y = \dfrac{\sqrt{x + 1}}{\sqrt{x - 1}}$

---

43. The impedance in an electrical circuit containing a resistor, capacitor, and inductor is given by $Z = \sqrt{R^2 + (X_L - X_C)^2}$. Find a formula for dZ/dR. 3 5 9 1 2 4 6 7 8

44. The formula $n = \sqrt{1 - k/f^2}$ gives the index of refraction for radio waves entering the ionosphere. In the formula, f is the frequency of the wave (in hertz) and k is a constant. Find a formula for dn/df. 5 3 4 7 8 9

45. The formula $E = 1000(100 - T) + 580(100 - T)^2$ is used to approximate the elevation (in meters) above sea level at which water boils at a temperature of T (in degrees Celsius). Find the rate of change of E with respect to T for a temperature of 96° C. 6 8 1 2 3 5 7 9

46. A circular oil slick spreads so that as its radius changes, its area changes. Both the radius r and the area A change with respect to time. If dr/dt is found to be 3.20 m/hr, find dA/dt when $r = 26.1$ m. 1 2 4 7 8 3 5

47. The charge Q (in coulombs) on a capacitor is given by $Q = CV$, where C is the capacitance (in farads) and V is the voltage (in volts). As V changes, Q changes. Find dQ/dt for a 5.0 μF capacitor, if $dV/dt = 0.0006$ volts/s. 3 5 9 4 7 8

48. When an amount of heat Q (in kcal) is added to a unit mass (in kg) of a substance, the temperature rises by an amount T (in degrees Celsius). The quantity dQ/dT, called the specific heat, is 0.42 for wood. If $dQ/dt = 10.0$ kcal/min for a 1 kg sample of wood at 20.0° C, find dT/dt for this same sample. (Hint: $dQ/dt = (dQ/dT)(dT/dt)$.) 6 8 1 2 3 5 7 9

---

1. Architectural Technology
2. Civil Engineering Technology
3. Computer Engineering Technology
4. Mechanical Drafting Technology
5. Electrical/Electronics Engineering Technology
6. Chemical Engineering Technology
7. Industrial Engineering Technology
8. Mechanical Engineering Technology
9. Biomedical Engineering Technology

21.7 THE CHAIN RULE AND IMPLICIT DIFFERENTIATION

Thus far, all of the functions we have considered involved only two variables. If, for instance, the variables are x and y, we write dy/dx to indicate that y is differentiated "with respect to x." It may be necessary to use the chain rule. Examples 1 and 2 illustrate the importance of the phrase "with respect to" in a differentiation problem.

EXAMPLE 1 Consider y to be a function of x and differentiate z with respect to x.

(a) $z = 3x^2$ **(b)** $z = 3y^2$

Solution **(a)** This is an example of straightforward differentiation. $dz/dx = 6x$

(b) To differentiate the function with respect to x, y must be considered to be a function of x. Use the chain rule. That is, differentiate $3y^2$ with respect to y then differentiate y with respect to x.

$$dz/dx = 6y(dy/dx)$$ ■

EXAMPLE 2 Consider y to be a function of x and differentiate z with respect to x.

(a) $z = xy$ **(b)** $z = x^2y$

Solution **(a)** Use the product rule: $dz/dx = x(dy/dx) + y(1) = x(dy/dx) + y$

(b) Use the product rule: $dz/dx = x^2(dy/dx) + y(2x)$ ■

Up to this point, functions have been stated *explicitly*. A function is said to be an **explicit function** if the equation that defines it has the dependent variable on one side by itself. If the dependent variable is not on one side by itself, the function is said to be an **implicit function.** When written as $y - x^2 = 2x - 1$, the equation defines y implicitly, but when written as $y = x^2 + 2x - 1$, the equation defines y explicitly. This example shows that it is sometimes possible to rewrite an implicit function as an explicit function.

explicit function

implicit function

Although an equation such as $x^2 + y^2 = 16$ is not the equation of a function, it can be written as $y = \pm\sqrt{16 - x^2}$. The equation can be considered to define two functions: $y = \sqrt{16 - x^2}$ and $y = -\sqrt{16 - x^2}$. We therefore say that the equation $x^2 + y^2 = 16$ defines y as an implicit function of x.

It is not always possible to solve an equation for the dependent variable. The equation $x^2 + xy + y^3 = 7$ is an example of an implicit function that cannot be solved explicitly. Nevertheless, it is possible to find dy/dx. When dy/dx is found without solving for y first, the process is called **implicit differentiation.** Even when it is possible to solve the equation explicitly, it may be easier to use implicit differentiation. It is important to treat y as a function of x in such problems. The next example illustrates implicit differentiation.

implicit differentiation

EXAMPLE 3 Find $\frac{dy}{dx}$ if $x^3 + x + y^3 = 2$.

Solution Differentiate both sides of the equation with respect to x. Be sure to treat y as a function of x.

Differentiate x^3.
Differentiate x.
Differentiate y^3, using the chain rule.
Differentiate the constant 2.

$$3x^2 + 1 + 3y^2\left(\frac{dy}{dx}\right) = 0$$

$$3y^2\left(\frac{dy}{dx}\right) = -3x^2 - 1 \qquad \text{Subtract } 3x^2 + 1 \text{ from both sides.}$$

$$\frac{dy}{dx} = \frac{-3x^2 - 1}{3y^2} \qquad \text{Solve for } \frac{dy}{dx}.$$

The procedure of Example 3 is summarized as follows.

To find $\frac{dy}{dx}$ when y is defined as an implicit function of x:

1. Differentiate both sides of the defining equation with respect to x. Use the chain rule for terms containing a y. That is, differentiate with respect to y and multiply by $\frac{dy}{dx}$.
2. Solve the resulting equation for $\frac{dy}{dx}$.
3. Evaluate the derivative at the given value, if necessary.

EXAMPLE 4 Find dy/dx at (2, 1) if $x^2 + xy + y^3 = 7$.

Solution 1. Differentiate both sides of the equation with respect to x. Be sure to treat y as a function of x.

Differentiate x^2.
Differentiate xy, using the product rule.
Differentiate y^3, using the chain rule.
Differentiate the constant 7.

$$2x + x\left(\frac{dy}{dx}\right) + y(1) + 3y^2\left(\frac{dy}{dx}\right) = 0$$

2. Solve the equation for $\frac{dy}{dx}$.

$$\frac{dy}{dx}(x + 3y^2) + 2x + y = 0 \qquad \text{Factor out } \frac{dy}{dx} \text{ from two terms.}$$

$$\frac{dy}{dx}(x + 3y^2) = -2x - y \qquad \text{Subtract } 2x + y \text{ from both sides.}$$

$$\frac{dy}{dx} = \frac{-2x - y}{x + 3y^2} \qquad \text{Solve for } \frac{dy}{dx}.$$

3. At (2, 1) $\frac{dy}{dx} = \frac{-4 - 1}{2 + 3} = \frac{-5}{5} = -1$. ■

EXAMPLE 5 Find the slope of the line tangent to the curve $x^2 + y^2 = 25$ at (4, 3).

Solution Use implicit differentiation to find the slope of the tangent at (4, 3).

$$2x + 2y\left(\frac{dy}{dx}\right) = 0$$

$$2y\left(\frac{dy}{dx}\right) = -2x$$

$$\frac{dy}{dx} = \frac{-2x}{2y} \text{ or } \frac{-x}{y}$$

Thus, the slope of the tangent line at (4, 3) is $\frac{-4}{3}$. The graph is shown in Figure 21.25. ■

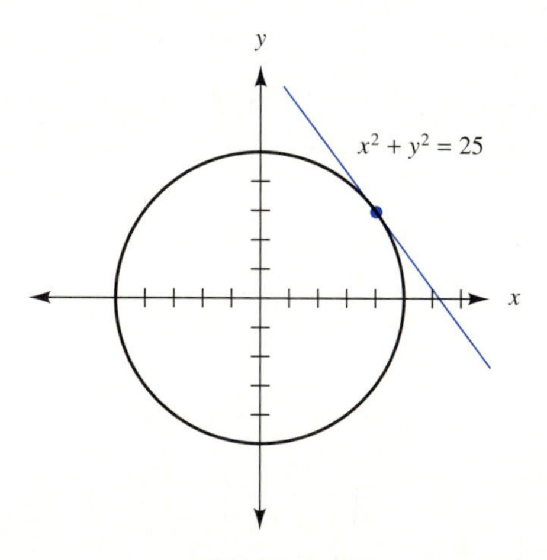

FIGURE 21.25

EXAMPLE 6 If $(2x + 1)^4 = xy^4$, find $\frac{dy}{dx}$.

Solution

Differentiate $(2x + 1)^4$, using the chain rule.

Differentiate xy^4, using the product rule.

x (derivative of y^4) + y^4 (derivative of x).

$$4(2x + 1)^3(2) = x(4y^3)\left(\frac{dy}{dx}\right) + y^4(1)$$

$$4(2x + 1)^3(2) - y^4 = x(4y^3)\left(\frac{dy}{dx}\right)$$ Subtract y^4 from both sides.

$$\frac{4(2x + 1)^3(2) - y^4}{4xy^3} = \frac{dy}{dx}$$ Divide both sides by $4xy^3$.

$$\frac{8(2x + 1)^3 - y^4}{4xy^3} = \frac{dy}{dx}$$ Simplify. ■

EXERCISES 21.7

In 1–10, consider y to be a function of x, and differentiate z with respect to x.

1. **(a)** $z = 2x^3$ **(b)** $z = 2y^3$

2. **(a)** $z = x^4$ **(b)** $z = y^4$

3. **(a)** $z = 3x^2 + 2x$ **(b)** $z = 3y^2 + 2y$

4. **(a)** $z = x^3 + 2x^2 - 1$ **(b)** $z = y^3 + 2y^2 - 1$

5. $z = xy^2$

6. $z = 2x^3y$

7. $z = -2x^2y$

8. $z = -3x^2y^2$

9. $z = -x^2y^3$

10. $z = x^3y^3$

In 11–22, use implicit differentiation to find dy/dx.

11. $xy + y^2 = 4$

12. $3x^2 - xy + y^2 = 2$

13. $2x^2 + 3xy - y^2 = 1$

14. $(x + 2)^3 = x^3y^2$

15. $(2x - 3)^2 = 2x^2y^2$

16. $(3x + 1)^2 - 3xy^3 = 0$

17. $x^3 + y^3 = 6$

18. $x^3 + 2x + y^2 = 0$

19. $x^4 + 3x + y^4 = 7$

20. $2x^3 + 4 = 3y^2 - y$

21. $2x^4 = y^3 - 2y^2$

22. $x^5 - x^3 = y^2 - y$

In 23–26, find the slope of the line tangent to the indicated curve at the given point.

23. $x^2 + y^2 = 100$ at $(6, 8)$

24. $x^2 + y^2 = 169$ at $(-5, 12)$

25. $2x^2 + 3y^2 = 30$ at $(3, -2)$

26. $y^2 - x^2 = 9$ at $(4, 5)$

21.8 CHAPTER REVIEW

KEY TERMS

21.1 continuous over an interval
continuous
point of discontinuity

21.2 instantaneous rate of change
tangent

21.3 derivative
differentiation

21.6 composition of functions

21.7 explicit function
implicit function
implicit differentiation

SYMBOLS

$\lim_{x \to c} f(x)$	the limit of $f(x)$ as x approaches c
$\lim_{x \to \infty} f(x)$	the limit of $f(x)$ as x approaches infinity
$f'(x)$	the derivative of f at x
dy/dx	the derivative of y with respect to x
df/dx	the derivative of f with respect to x
y'	the derivative of y

RULES AND FORMULAS

$\lim_{x \to c} f(x) = f(c)$ for a continuous function.

If $f(x) = c$, then $df/dx = 0$.

If $f(x) = ax^n$, where a is a real number and n is a rational number, then $df/dx = anx^{n-1}$ for all x for which x^{n-1} is defined.

If f and g are functions that have derivatives at x, then

$$\frac{d(f+g)}{dx} = \frac{df}{dx} + \frac{dg}{dx}.$$

If u and v are functions of x for which the derivatives are defined, then

$$\frac{d}{dx}(cu) = c\frac{du}{dx}$$ the constant rule

$$\frac{d}{dx}(uv) = u\frac{dv}{dx} + v\frac{du}{dx}$$ the product rule

$$\frac{d}{dx}\left(\frac{u}{v}\right) = \frac{v\left(\frac{du}{dx}\right) - u\left(\frac{dv}{dx}\right)}{v^2}$$ the quotient rule

If y is a function of u, and u is a function of x, and if y and u have derivatives, then

$$\frac{dy}{dx} = \frac{dy}{du}\left(\frac{du}{dx}\right)$$ the chain rule

GEOMETRY CONCEPTS

$V = s^3$ (volume of a cube)
$A = \pi r^2$ (area of a circle)
$V = (4/3)\pi r^3$ (volume of a sphere)
$A = 4\pi r^2$ (surface area of a sphere)
$A = lw$ (area of a rectangle)

SEQUENTIAL REVIEW

(Section 21.1) In 1–3, determine whether each function is continuous. If it is not continuous, identify the points of discontinuity.

1.

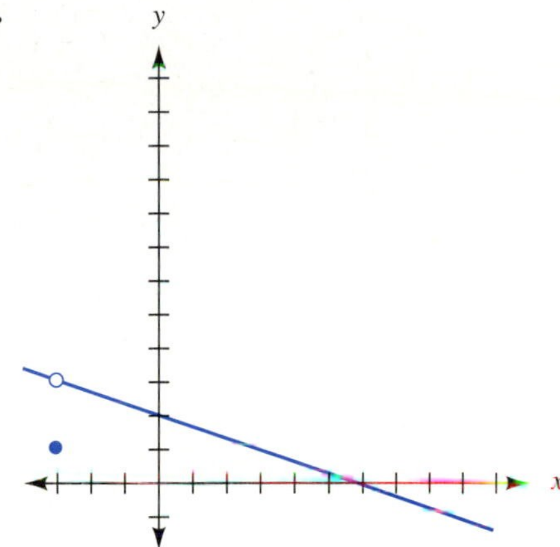

2.

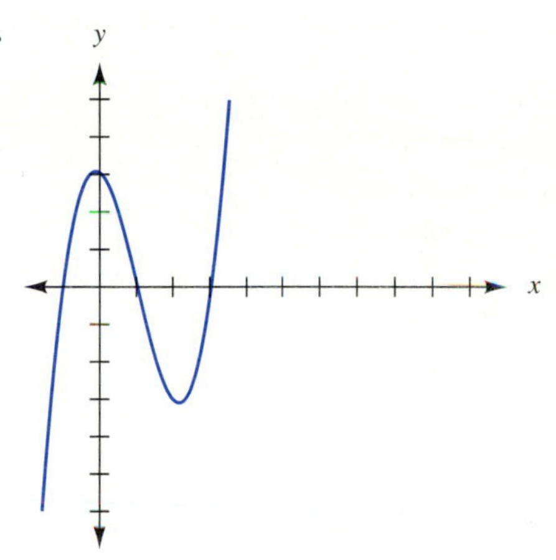

3. $y = \dfrac{x+1}{x-1}$

4. Find $\lim_{x \to -3} f(x)$ for the function in problem 1.

In 5–7, evaluate each limit.

5. $\lim_{x \to -1} \dfrac{x^2 - 2x - 3}{x+1}$

6. $\lim_{x \to 0} \dfrac{3x^2 + x}{x}$

7. $\lim_{x \to \infty} \dfrac{3x+1}{2x-1}$

(Section 21.2)

8. Find the slope of the tangent to the curve $y = 1 - 3x^2$ when x is 1.

(Section 21.3) In 9–11, use the definition to find the derivative of each function.

9. $g(x) = 1 - 2x^2$

10. $f(x) = 2x^3 - 1$

11. $f(x) = -1/x$

In 12–14, use the definition of derivative to find y′ for each function.

12. $y = 3x^2 + 2$

13. $y = 3/x$

14. $y = x^3 + x$

15. If $f(x) = 1 - x^2$, find $f'(2)$.

16. If $g(x) = \dfrac{1}{x-1}$, find $g'(0)$.

17. If $f(x) = x^3$, find $f'(1)$.

(Section 21.4) *In 18–20, find the slope of the tangent to each curve at the given point.*

18. $f(x) = 2x^2 + 3$ at $x = 2$ **19.** $y = -x^2 + 3x$ at $x = 1$ **20.** $y = 1 - 2x^3$ at $x = 0$

In 21–23, find the derivative of each function.

21. $y = 2x^{-3}$ **22.** $y = 4x^3 - 2x^2 + 1$ **23.** $f(x) = 3x^7 + 2$

24. If $f(x) = 3x^4 - 2x^2 + 1$, find $f'(0)$. **25.** If $g(x) = 4x^{3/4}$, find $g'(1)$.

In 26 and 27, find the slope of the tangent to the curve at the given point.

26. $y = -x^2 + 3x$ at $x = 1$ **27.** $y = x^3 - 2x + 3$ at $x = 0$

(Section 21.5) *In 28 and 29, use the product rule to find the derivative of each function.*

28. $y = 2x^3(3x - 1)$ **29.** $y = (x^3 + 3x)(3x^2 - 2)$

30. If $f(x) = (x^2 - 2)(x^2 + 2)$, find $f'(0)$.

31. Use the quotient rule to find the derivative of y if $y = \dfrac{3x^2}{2x + 1}$.

32. If $f(x) = \dfrac{x^2 + 1}{2x - 3}$, find $f'(1)$.

33. If $y = \dfrac{1}{3x + 1}$, find y' when $x = 1$.

(Section 21.6) *In 34 and 35, show that y can be written as a function of u and identify u.*

34. $y = (2x^2 - 1)^6$ **35.** $y = \sin(2x^2 - 3)$

In 36 and 37, find the derivative of each function.

36. $y = (3x + 2)^5$ **37.** $y = \dfrac{3}{\sqrt{3x^2 - 2}}$

In 38 and 39, find the derivative of each function for the given value.

38. $y = (3x + 2)^{-3}$ when $x = 0$ **39.** $y = \sqrt{x^3 + x}$ when $x = 1$

(Section 21.7) *In 40 and 41, consider y to be a function of x, and differentiate z with respect to x.*

40. (a) $z = x^3 + 2x$ **(b)** $z = y^3 + 2y$ **41.** $z = 2x^2y^2$

In 42 and 43, use implicit differentiation to find dy/dx.

42. $y^2 - xy = 3$ **43.** $(x - 3)^3 = 2xy^3$

In 44 and 45, find the slope of the line tangent to the indicated curve at the given point.

44. $x^2 + y^2 = 289$ at $(8, 15)$ **45.** $y^2 - x^2 = 16$ at $(3, 5)$

APPLIED PROBLEMS

1. For a lens of focal length f, $f = \dfrac{xy}{x + y}$, where x and y are the distances from the lens to the object and image, respectively. As the position of the object changes, the position of the image changes. For astronomical work, the object distance is considered to approach infinity. Find $\lim_{x \to \infty} f$.
3 5 6 9

2. The distance x (in km) that can be seen from height h (in km) is given by $x = 111.7\sqrt{h}$. As h increases, x increases. Find dx/dh.
1 2 4 7 8 3 5

3. The formula $I = \sqrt{\frac{P}{2R}}$ gives the current (in amps) for a series circuit containing 2 resistors, each of resistance R (in ohms), if P is the power (in watts). Find dI/dR for $P = 15.0$ W and $R = 150\ \Omega$. 3 5 9 1 2 4 6 7 8

4. The deflection y (in inches) of a cantilever beam bearing a uniform load is given by $y = c(l^2 - x^2)^2$. Treat l and c as constants and find a formula for the rate of change of y with respect to x. In the formula, x is the distance (in inches) from the free end of the beam to the point of interest, l is the length (in inches) of the beam, and c is a constant that depends on the beam and its load.
1 2 4 8 7 9

5. The velocity v of water in a circular pipe at a distance r from the center of a pipe is given by $v = v_{max}(1 - r^2/R^2)$. Treat v_{max} and R as constants, and find dv/dr. In the formula, v_{max} is the velocity along the center of the pipe, and R is the radius of the pipe. 4 8 1 2 6 7 9

6. The velocity v (in ft/s) of sound in air is given by $v = 49.02\sqrt{T + 460}$ where T is the temperature in degrees Fahrenheit. Find the rate (in ft/s per °F) at which v changes with respect to T when T is 70.0° F. 1 3 4 5 6 8 9

7. The energy loss E (in joules/kilogram) due to friction when water flows through a pipe is given by $E = 0.020(L/D)v^2$. In the formula, L is the pipe length (in m), D is the pipe diameter (in m), and v is the water velocity (in m/s). Find dE/dv for a 30 meter long pipe that is 25 cm in diameter if the water flows at 10 m/s. 4 8 1 2 6 7 9

8. A rectangular weir can be used to measure water flow rates. The discharge rate R (in ft^3/s) is given by the formula $R = 5.35\ Lh^{3/2}$, where L is the length of weir (in ft) and h is the height of fluid (in ft). Find dR/dh when h is 4.60 inches for a 2.00 ft long weir. 4 8 1 2 6 7 9

9. The formula $f = \frac{1}{2\pi}\sqrt{\frac{1}{LC} - \frac{R^2}{4L}}$ gives the resonant frequency f (in Hz) of a parallel tank circuit. In the formula, R is the resistance (in ohms), L is the inductance (in henries), and C is the capacitance (in farads). Find df/dR if $R = 2480$ ohms, $L = 0.40$ H, and $C = 0.65\ \mu$F.
3 5 9 1 2 4 6 7 8

10. The buckling load F (in lb) for a long column is given by $F = 4\pi^2 MA/(L/r)^2$. In the formula, M is the modulus of elasticity (in lb/in^2), A is the cross-sectional area (in in^2), L is the column length (in inches), and r is the radius of gyration (in inches). Find dF/dr if $M = 27 \times 10^6$ lb/in^2, $A = 1.5$ in^2, $L = 4.0$ ft, and $r = 0.0465$ in. 1 2 4 8 7 9

11. A chemical process proceeds in such a way that the temperature C (in degrees Celsius) is given by $C = 10t^{2.32} + 20$ when t is measured in hours. How fast will the temperature in Fahrenheit degrees change when t is 2.00 hours? [Hint: $F = (9/5)C + 32$.] Give your answer to the nearest degree per hour. 6 9 2 4 5 7 8

12. A washer is made by stamping out a circle of radius r from the center of a metal disk of radius 4 inches. This washer has an area A (in in^2) given by $A = \pi(4 + r)(4 - r)$. How fast is A changing with respect to r when r is 1 inch? 1 2 4 7 8 3 5

13. A voltage source of 12 volts can supply a voltage V (in volts) through a 72-ohm line to a variable resistance R (in ohms) given by the following equation. $V = \frac{12\ R}{72 + R}$. How fast does V change with respect to R when R is 72 ohms?
3 5 9 1 2 4 6 7 8

14. The formula $\alpha = \frac{\beta}{1 + \beta}$ gives the fraction of the emitter current in a transistor that flows in the collector if β is the current gain. Find the rate of change of α with respect to β if β is 200.0.
3 5 9 4 7 8

15. The formula $V_{rms} = (V_T^2 - V_{dc}^2)^{1/2}$ gives the RMS value V_{rms} (in volts) of the ac-voltage component of a signal having a dc level. In the formula, V_T is the RMS value of the total voltage and V_{dc} is the dc level. Find dV_{rms}/dV_{dc} when $V_T = 170$ volts and $V_{dc} = 150$ volts. 3 5 9 1 2 4 6 7 8

16. The formula $V_r = \dfrac{r}{1 + r\sqrt{3}} V_p$ relates the peak voltage V_P, the ripple voltage V_r, and the ripple factor r of a power supply used in an electronics lab. Find dV_r/dr if $V_p = 36$ and $r = 0.080$. 3 5 9 1 2 4 6 7 8

17. For a certain transistor, the current I (in amperes) is related to the voltage V (in volts) by the equation $I = 0.043[1 + V/3]^2$. Find dI/dV when V is -1.5 volts. 3 5 9 4 7 8

18. The formula $C = \dfrac{35 \times 10^{-12}}{\sqrt[3]{V + 0.7}}$ gives the capacitance C (in farads) of a particular diode if V is its voltage (in volts). Find dC/dV if V is -2 volts. 3 5 9 4 7 8

19. The constant e that appears in many scientific formulas can be found by taking the following limit

$$\lim_{x \to 0} \frac{1}{(1 - x)^{1/x}}$$

Find the value of this limit to four significant digits by using smaller and smaller values of x, such as 0.1, 0.01, 0.001, and so on. 1 2 4 7 8 3 5

1. Architectural Technology
2. Civil Engineering Technology
3. Computer Engineering Technology
4. Mechanical Drafting Technology
5. Electrical/Electronics Engineering Technology
6. Chemical Engineering Technology
7. Industrial Engineering Technology
8. Mechanical Engineering Technology
9. Biomedical Engineering Technology

RANDOM REVIEW

1. Evaluate $\lim\limits_{x \to \infty} \dfrac{2x^2 - 3}{3x^2 + 2}$.

3. Find the slope of the tangent to the curve given by $y = 3x^2 - 2x + 1$ when x is 0.

3. Use the definition of derivative to find y' if $y = 2x^2 - x$.

4. Find $f'(0)$ if $f(x) = (3x^2 + 2)^5$.

5. Show that y can be written as a function of u, and identify u if $y = (3x^3 - 2x)^7$.

6. Use implicit differentiation to find dy/dx if $x^2 + y^2 = 2xy$.

7. Find y' if $y = (x^3 - 2)(x^3 + 2x)$.

8. Consider y to be a function of x and differentiate z with respect to x if $z = 2xy^3$.

9. Find $f'(x)$ if $f(x) = 3x^3 + 2/x^2$.

10. Find the derivative of y if $y = \dfrac{x + 1}{x - 1}$.

CHAPTER TEST

1. Consider the function shown on the next page.
(a) Is it continuous?
(b) If not, identify the point of discontinuity.
(c) Find $\lim\limits_{x \to 0} f(x)$.
(d) Find $f(0)$.

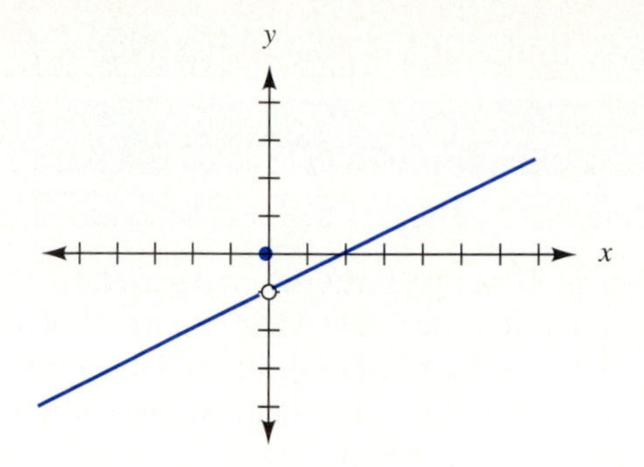

2. Evaluate $\lim_{x \to -2} \dfrac{2x^2 - x - 10}{x + 2}$.

3. Use the definition of derivative to find y' if $y = 3x^2 + 2x$.

4. Find the derivative of y if $y = 3x^{-4}$.

5. Find the slope of the tangent to the curve given by $y = -2x^3 + 3x$ when x is 1.

6. Show that y can be written as a function of u, and identify u, if $y = \ln(3x - 1)$.

7. Find $f'(x)$ if $f(x) = \dfrac{x - 1}{x^2 + 3}$.

8. Use implicit differentiation to find dy/dx if $(3x + 2)^2 = y^2$.

9. A silicon diode has the characteristic curve shown.

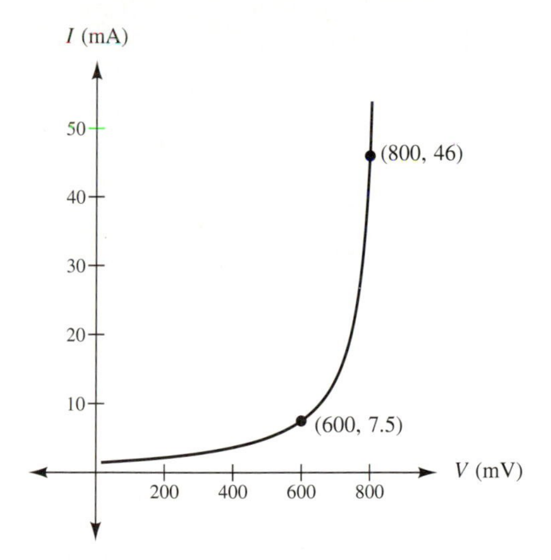

The diode's ac resistance is defined as the average rate of change of V with respect to I. That is, $R = \Delta V/\Delta I$. Find the resistance over the interval from 600 mV to 800 mV.

10. A wire expands when heated. Find the rate of change of volume ($V = \pi r^2 l$) with respect to time if $dr/dt = 6.69 \times 10^{-5}$ in/hr and $dl/dt = 6.69 \times 10^{-5}$ in/hr when $r = 0.0320$ inches and $l = 17.3$ inches.

EXTENDED APPLICATION: ELECTRONICS ENGINEERING TECHNOLOGY

POWER TRANSFER

All circuits require an applied voltage as a source of electrical energy. Batteries and generators, for example, are sources of electrical energy. The potential difference across the terminals of a battery or other source when no current flows is called the **electromotive force,** abbreviated emf. When a current I flows, the potential difference V is actually less than the electromotive force V_{emf}. The difference is due to the internal resistance r of the source. Ohm's law tells us that the potential difference for this internal resistance is given by $V_i = Ir$. The terminal voltage, then, is given by $V = V_{emf} - V_i$.

The part of the circuit connected to the voltage source is called the **load resistance.** The power P transferred from a source emf V_{emf} with internal resistance R_i to a load resistance r_L is given by

$$P = \frac{r_L\, V_{emf}^2}{(r_L + R_i)^2}$$

For a battery of emf 12 volts and an internal resistance of 0.2 ohms, a sketch of P versus r_L will show that there is a value of r_L for which P is at its maximum. We consider

$$P = \frac{144\, r_L}{(r_L + 0.2)^2}.$$

Since r_L must be nonnegative, it is not necessary to consider values of r_L less than zero. A table shows the correspondence of values.

r_L	0.0	0.1	0.2	0.3	0.4	0.6	0.8	1	2	3	4
P	0	160	180	173	160	135	115	100	60	42	33

The table shows only a limited number of values. To investigate what happens if r_L is allowed to increase without bound, consider

$$\lim_{r_L \to \infty} \frac{144\, r_L}{(r_L + 0.2)^2} = \lim_{r_L \to \infty} \frac{144\, r_L}{r_L{}^2 + 0.4r_L + 0.04}$$

$$= \lim_{r_L \to \infty} \frac{\dfrac{144\, r_L}{r_L{}^2}}{\dfrac{r_L{}^2}{r_L{}^2} + \dfrac{0.4r_L}{r_L{}^2} + \dfrac{0.04}{r_L{}^2}}$$

$$= 0$$

That is, as r_L gets larger and larger, P gets closer and closer to 0. Thus, the horizontal axis is an asymptote. In fact, the horizontal axis is an asymptote whenever the function is a ratio of two polynomials, with the polynomial of larger degree in the denominator. The graph is shown in Figure 21.26. From the graph, it appears that the maximum power occurs when r_L is 0.2 ohms. In general, it is true that power is at its maximum when r_L is the same as R_i.

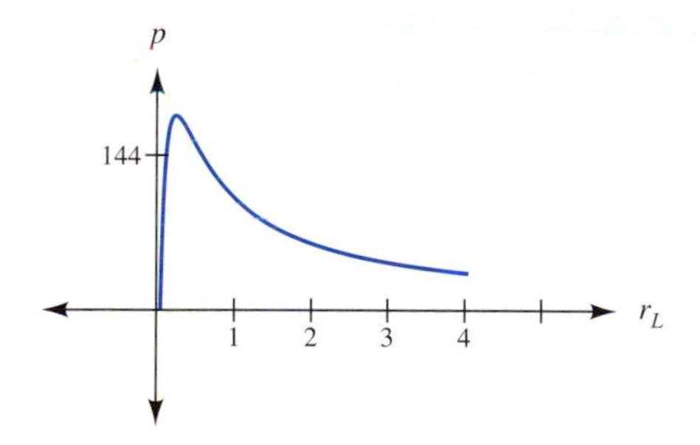

FIGURE 21.26

APPLICATIONS OF THE DERIVATIVE CHAPTER 22

In Chapter 21, we presented a number of applications of the derivative. Our main goal there, however, was to explain what it means to find a derivative and to present the rules for differentiating various types of functions. In this chapter, the emphasis will be on using derivatives to solve applied problems in the technologies.

22.1 EQUATIONS OF TANGENTS AND NORMALS

The problem of finding the tangent to a curve at a particular point was one of the motivating factors behind the development of the calculus in the seventeenth century. The tangent is important as a purely geometric concept, and writing the equation for the tangent is therefore necessary in the study of geometry. The tangent is also useful in technology, and after examining a few geometric problems, we will consider some of the applications.

EXAMPLE 1 Find the equation of the line tangent to the curve $y = x^2 - 5x + 6$ when x is 2.

Solution To write the equation of the line, we need to know the slope of the line and one point on the line. The point-slope form of the equation of a line is

$$y - y_1 = m(x - x_1) \qquad \textbf{(22-1)}$$

where m is the slope of the line and (x_1, y_1) is a point on the line. The slope of the tangent sought is given by the derivative of y when x is 2.

$$y' = 2x - 5$$

When x is 2, we have $y' = 4 - 5$ or -1. The point at which the tangent touches the curve is a point on both the tangent and the curve. The y-coordinate of this point may therefore be obtained from the equation for the curve. When x is 2, we have $y = 2^2 - 5(2) + 6$ or 0. Thus, the tangent is a line through (2, 0) with slope -1. Substituting (2, 0) for (x_1, y_1) and -1 for m, the point-slope formula (Equation 22-1) gives $y - 0 = -1(x - 2)$ or $y = -x + 2$.
Figure 22.1 shows both the curve and the tangent. ■

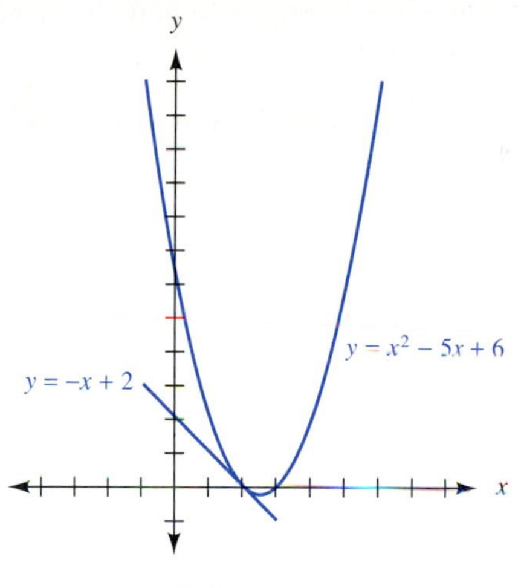

FIGURE 22.1

EXAMPLE 2 Find the equation of the line tangent to the curve $y = 1/x$ when x is 3.

Solution Recall that the slope of the tangent is given by the derivative. Since $y = 1/x$ or x^{-1}, we have $y' = -x^{-2}$ or $-1/x^2$. When x is 3, y' is $-1/9$, so the slope of the tangent at that point is $-1/9$. When x is 3, y is 1/3, so the point (3, 1/3) is a point on the line tangent to the curve.

The point-slope formula gives

$$y - 1/3 = (-1/9)(x - 3)$$
$$y - 1/3 = (-1/9)x + 1/3$$
$$y = (-1/9)x + 2/3.$$

Figure 22.2 shows the curve and the tangent. ■

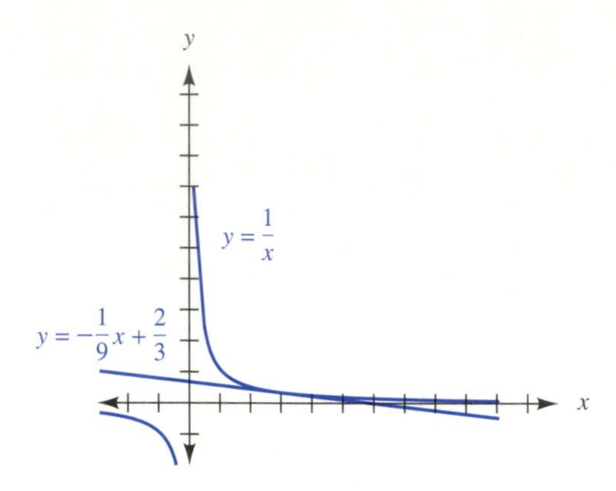

FIGURE 22.2

Sometimes it is necessary to find the equation of the *normal* to a curve. A line is said to be the **normal** to a curve at a point if it is perpendicular to the tangent at that point. Since perpendicular lines have slopes that are negative reciprocals of each other, the slope of the normal to a curve at a point is given by the negative of the reciprocal of the derivative at that point.

normal

To find the equation of a tangent or a normal to a curve $y = f(x)$ when x has a value of x_1:

1. Find $f'(x)$.
2. Find the slope m of the line.
   (a) For a tangent, $m = f'(x_1)$.
   (b) For a normal, $m = -1/f'(x_1)$, if $f'(x_1) \neq 0$.
3. Find $y_1 = f(x_1)$
4. Use the point-slope formula $y - y_1 = m(x - x_1)$ to find the equation of the line.

EXAMPLE 3 Find the equation of the normal to the curve $x^2 + y^2 = 25$ at (3, 4).

Solution The graph of the equation is a circle. To find the slope of the normal, we first need to find the slope of the tangent. Although it is possible to solve the equation for y and find dy/dx, this type of problem lends itself to implicit differentiation. Differentiating with respect to x on both sides of the equation $x^2 + y^2 = 25$, we have the following.

$$2x + 2y\left(\frac{dy}{dx}\right) = 0$$

$$2y\left(\frac{dy}{dx}\right) = -2x$$

$$\frac{dy}{dx} = \frac{-2x}{2y} \text{ or } \frac{-x}{y}$$

When x is 3 and y is 4, $\dfrac{dy}{dx}$ is $\dfrac{-3}{4}$.

Since the slope of the tangent is $-3/4$ when x is 3, the slope of the normal at that point is 4/3.

$$y - 4 = (4/3)(x - 3)$$

$$y - 4 = (4/3)x - 4$$

$$y = (4/3)x$$

Figure 22.3 shows the graph. If you have studied geometry, you might remember that a line tangent to a circle at a given point is perpendicular to the radius of the circle through that point. The example agrees with this result from geometry. ■

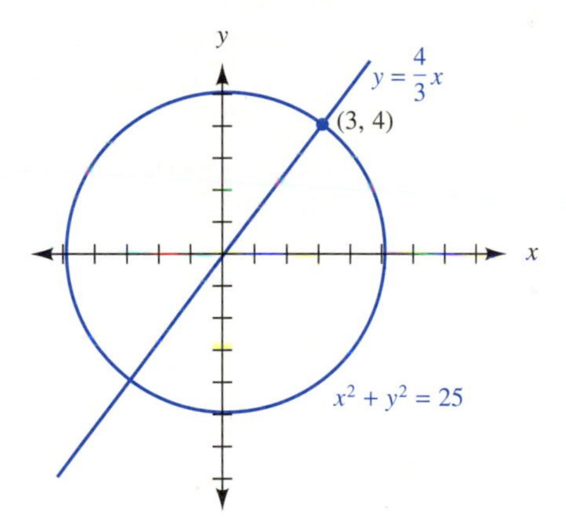

FIGURE 22.3

The tangent and normal are used in technological problems. If an object moves along a curved path, the tangent at a point represents the direction in which the object is headed at that point.

EXAMPLE 4 A bridge is built so that its shape is approximated by the equation $y = -0.01x^2 + 36$ with the center of the bridge at $x = 0$ and its ends at $x = -40$ and $x = 40$. Figure 22.4 shows the graph.

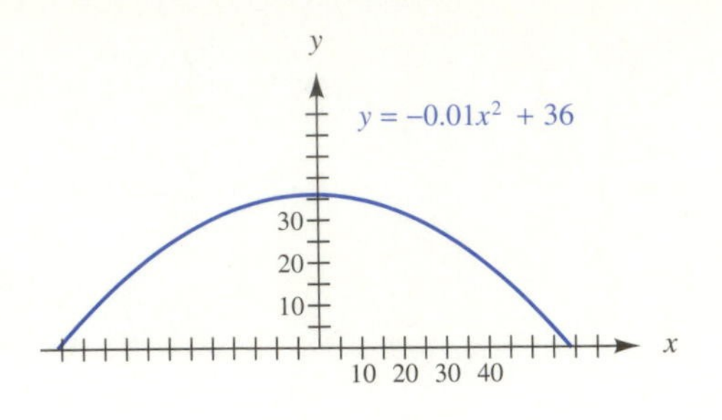

FIGURE 22.4

(a) Find the slope of the road at $x = -30$.

(b) Find the slope of the road at $x = 40$.

Solution The slope of the road at any point is given by the slope of the line that is tangent at that point. Thus, we find the derivative.

$$y' = -0.02x$$

(a) If x is -30, then $y' = -0.02(-30)$, which is 0.6 or 3/5. That is, when x is -30, the bridge is rising at a rate of 3 feet for every 5 feet traveled horizontally.

(b) If x is 40, $y' = -0.02(40)$, which is -0.8 or $-4/5$. When x is 40, the bridge is descending at a rate of 4 feet for every 5 feet traveled horizontally. ■

Normals occur in the study of optics. When light waves strike a surface, they are reflected in such a way that the angle of incidence is equal to the angle of reflection, where both angles are measured from the normal as shown in Figure 22.5 for a flat surface. If a parabola is rotated about its axis, which is a line through the focus and vertex, it sweeps out what is called a parabolic surface. This type of surface serves to concentrate the rays and is widely used in reflectors for light, sound, and radio waves. If a light source is placed at the focus of the parabola, all rays from it striking the parabolic surface are reflected out parallel to the axis of revolution. Likewise, rays that are parallel to the axis of revolution and that strike a parabolic surface are reflected to the focus.

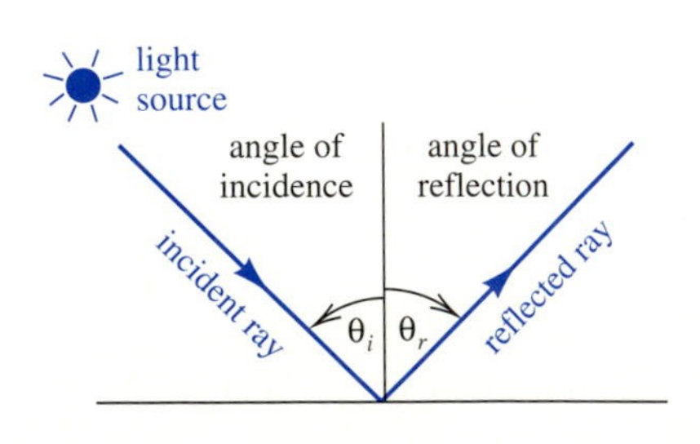

FIGURE 22.5

EXAMPLE 5 The cross section of a parabolic reflector is shown in Figure 22.6. The equation of the parabola is $y = (1/4)x^2$. The angle between the incident ray, which is parallel to the y-axis, and the reflected ray is 40.1°. Find the equation of the normal through the point at which the ray strikes the surface.

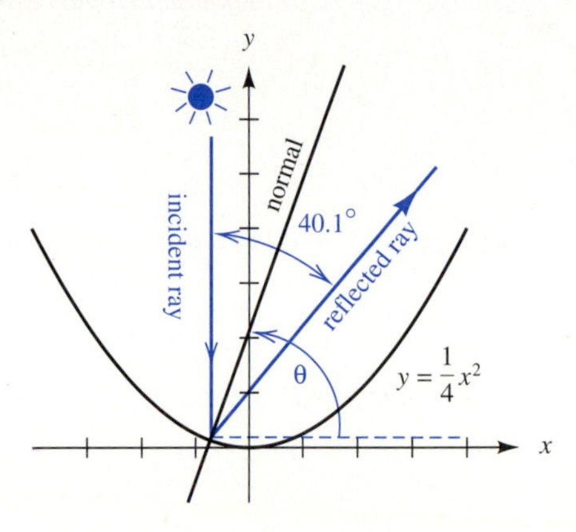

FIGURE 22.6

Solution The slope of the normal may be computed in two ways. The first way is to use the derivative.

$$y = (1/4)x^2, \text{ so } y' = (1/2)x = x/2.$$

Thus, for the normal, $m = -2/x$. But we do not know the value of x. The second way to compute the slope is to use the fact that $m = \tan \theta$, where θ is the angle between the horizontal and the normal. To find θ, notice that the angle of reflection is half of 40.1° or 20.05°. Angle θ is obtained by subtracting 20.05° from 90°, so θ is 69.95°. The slope of the normal is tan 69.95°, which is 2.7400. Now that we know m, we must find (x_1, y_1), the coordinates of the point at which the ray strikes the surface. Since the slope was obtained by two different methods, we can equate the two expressions obtained earlier. That is,

$$-2/x = 2.7400$$
$$-2 = 2.7400\,x$$
$$-0.7299 = x.$$

The y-coordinate is obtained from the equation $y = (1/4)x^2$.
When x is -0.7299, y is 0.1332.
Summarizing, the slope of the normal is 2.7400, and the point $(-0.7299, 0.1332)$ is on the normal. The equation of the normal, then, is

$$y - 0.1332 = 2.7400(x + 0.7299).$$
$$y = 2.7400(x + 0.7299) + 0.1332$$
$$y = 2.7400x + 2.1331$$

Using three significant digits, the equation is $y = 2.74x + 2.13$. ■

EXERCISES 22.1

In 1–5, find the equation of the line tangent to the indicated curve at the given point. Sketch the curve and the tangent.

1. $y = x^2 - x$ at $x = 1$

2. $y = x^2 + 2x - 3$ at $x = 0$

3. $y = 2x^2 + 1$ at $x = -1$

4. $y = x^3 + 4x^2 + 4x$ at $x = -2$

5. $y = x^3 - 2x^2 + x$ at $x = 0$

In 6–9, find the equation of the line tangent to the indicated curve at the given point.

6. $x^2 + y^2 = 5$ at $(1, -2)$

7. $x^2 + y^2 = 8$ at $(-2, 2)$

8. $xy = 2$ at $(2, 1)$

9. $xy = -1$ at $(3, -1/3)$

In 10–14, find the equation of the line normal to the indicated curve at the given point. Sketch the curve and the normal.

10. $y = x - x^2$ at $x = 0$

11. $y = 1 - 4x^2$ at $x = -1$

12. $y = x^3 - x$ at $x = -2$

13. $x^2 + y^2 = 10$ at $(1, 3)$

14. $x^2 - y^2 = 3$ at $(2, -1)$

In 15–18, find the equation of the line normal to the indicated curve at the given point.

15. $x^2 - y^2 = 5$ at $(3, -2)$

16. $x^2 + y^2 = 8$ at $(-2, 2)$

17. $xy = 4$ at $(1, 4)$

18. $xy = -3$ at $(3, -1)$

---

19. The cross section of a parabolic reflector is given by the equation $y = (1/4)x^2$. A ray leaving a light source at the focus of the parabola strikes the point (2, 1). Find the equation of the normal at that point. 3 5 6 9

20. The cross section of a parabolic reflector is given by the equation $y = (1/8)x^2$. The angle between the incident ray, which is parallel to the y-axis, and the reflected ray is 62.4°. Find the equation of the normal through the point at which the ray strikes the surface. 3 5 6 9

21. The cables of a suspension bridge take the shape of a parabola. At each point there is a tension in the cable directed along the tangent to the curve. If the equation for the shape of the cable is $y = x^2/240$, find the direction of the tension (specify an angle with the horizontal) when x is 60. 1 2 4 8 7 9

22. A sign is to be placed at the top of a hill so that the headlights of a car coming over the hill illuminate the sign, as shown in the diagram. The equation for the shape of the hill is $y = \sqrt{5329 - x^2}$. Find the height at which the sign should be placed so that it is illuminated when the vertical distance between the car and top of the hill is 25 feet. (Hint: For the car, $y = 73 - 25$. (Why?) Solve for x and find the equation of the tangent at this point. Use the y-intercept to determine the height of the hill.) 1 2 4 7 8 3 5

---

1. Architectural Technology
2. Civil Engineering Technology
3. Computer Engineering Technology
4. Mechanical Drafting Technology
5. Electrical/Electronics Engineering Technology
6. Chemical Engineering Technology
7. Industrial Engineering Technology
8. Mechanical Engineering Technology
9. Biomedical Engineering Technology

23. A helicopter travels in a straight line path as shown in the accompanying diagram. A missile fired from point A follows the path given by $y = -0.00007x^2 + 0.6x$, where point A is considered to be the origin of the coordinate system. The missile scores a direct hit (i.e., the path of the helicopter is normal to the path of the missile) when the x-coordinate is 4000. Find the equation of the path of the helicopter.
4 7 8 1 2 3 5 6 9

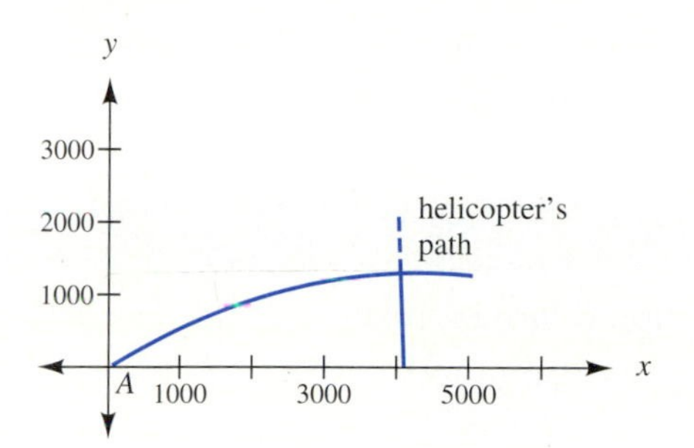

24. A light is placed in a tunnel that has a semi-elliptical archway, as shown in the diagram. The light will shine along a normal to the curve as shown in the diagram. The equation of the ellipse is $2x^2 + y^2 = 96$, and the light is at (4, 8). Find the equation of the line along which the light shines.
3 5 6 9

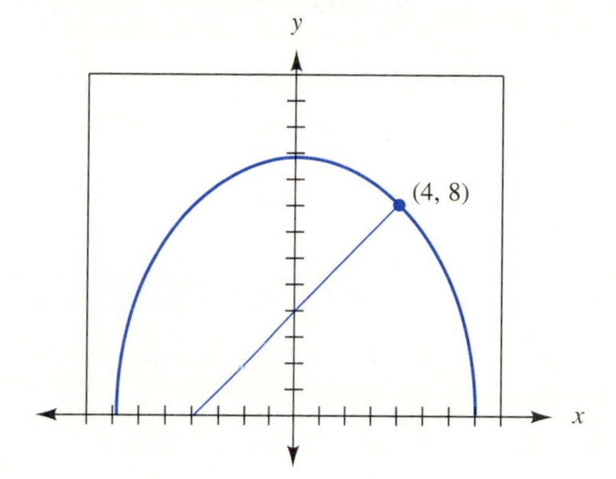

22.2 VELOCITY AND ACCELERATION

We have considered velocity to be the rate of change of distance with respect to time. Average velocity is the rate of change over an interval of time and instantaneous velocity is the rate of change at a particular instant. If the movement is one-dimensional, that is, if it takes place in a straight line, it is said to be **rectilinear.** Mathematicians, physicists, and engineers use signs to indicate direction for this type of motion. Along a horizontal line, distance traveled to the right is usually considered to be positive, and distance traveled to the left is considered to be negative. Likewise, along a vertical line, distance traveled upward is considered to be positive, and distance traveled downward is considered to be negative.

rectilinear

Signs are used with velocity also. A positive velocity indicates motion to the right (or up) while a negative velocity indicates motion to the left (or down). Speed is the absolute value of velocity. Speed merely indicates rate of motion; velocity indicates rate of motion in a specific direction.

If a car starts moving to the right from rest (a velocity of 0) and reaches a velocity of 88 ft/s in 20 seconds, its velocity is constantly changing. The rate of change of velocity with respect to time is called **acceleration.** In this example, the average acceleration over the 20-second interval is $\frac{88 \text{ ft/s}}{20 \text{ s}}$, which is written as 4.4 ft/s^2. Because acceleration is the rate of change of velocity, you might guess that we could find the instantaneous acceleration at a given time by finding the value of the derivative of the velocity function at that time. That is, the function that gives distance in terms of time is differentiated to obtain velocity, and the resulting function is differentiated to obtain acceleration. Acceleration is therefore an example of a *second derivative.*

DEFINITION

The **second derivative** of a function f defined by $y = f(x)$ is the derivative of $f'(x)$, and is denoted by

$$y'', f''(x), \text{ or } d^2y/dx^2.$$

NOTE ⇨⇨ ***It is important to realize that d^2y/dx^2 does not mean $(dy/dx)^2$.*** For example, if $y = x^2$, $dy/dx = 2x$ and $d^2y/dx^2 = 2$, but $(dy/dx)^2 = (2x)^2$ or $4x^2$.

VELOCITY AND ACCELERATION

If $s = f(t)$ gives the distance s as a function of time t, then
velocity at time t is given by $v = f'(t)$ or ds/dt, and
acceleration at time t is given by $a = f''(t)$ or d^2s/dt^2.

velocity

acceleration

EXAMPLE 1 A car travels in such a way that the distance s (in ft) traveled in time t (in seconds) is given by $s = 1.1t^2 + 5t$. Find the velocity and acceleration when t is 10 s.

Solution $v = s'(t) = 2.2t + 5$, and $a = s''(t) = 2.2$ ft/s^2.
When t is 10, $v = 2.2(10) + 5 = 22 + 5 = 27$ ft/s, and $a = 2.2$ ft/s^2. ■

EXAMPLE 2 A car travels in such a way that the distance s (in ft) traveled in time t (in seconds) is given by $s = -1.1t^2 - 5t$. Find the velocity and acceleration when t is 10 s.

Solution $v = s'(t) = -2.2t - 5$, and $a = s''(t) = -2.2$ ft/s^2.
When t is 10, $v = -2.2(10) - 5 = -22 - 5 = -27$ ft/s, and $a = -2.2$ ft/s^2. ■

curvilinear

When motion is two-dimensional, that is, when it takes place in a plane, the motion is said to be **curvilinear.** Velocity, acceleration, and displacement are vector quantities, which we encountered in Chapter 10. Both magnitude and direction are needed to specify vectors. Because of the increased complexity of such problems, the horizontal and vertical components of motion are often specified separately by using parametric equations, such as those in Section 11.4. Since the direction of the horizontal or vertical component of a vector $\mathbf{r}$ may be specified by a sign, these components may be represented by signed numbers r_x and r_y, where r_x is the number with the same sign and magnitude as the horizontal component $\mathbf{r}_x$, and r_y is the number with the same sign and magnitude as the vertical component $\mathbf{r}_y$. The magnitude $|\mathbf{r}|$ of the resultant vector is given by

$$|\mathbf{r}| = \sqrt{r_x^2 + r_y^2}.$$

EXAMPLE 3 If a particle follows a path specified by $x = 2t^2$ and $y = (1/3)t^3$, find the velocity and acceleration when t is 4.

Solution Begin by finding the first and second derivatives of x and y.
Since $x = 2t^2$, $x' = 4t$, and $x'' = 4$.
Since $y = (1/3)t^3$, $y' = t^2$, and $y'' = 2t$.
The velocity has a horizontal component given by the derivative of x when t is 4 and a vertical component given by the derivative of y when t is 4.

$$v_x = x' = 4(4) = 16 \qquad v_y = y' = 4^2 = 16$$

Thus, the magnitude of the velocity is

$$|\mathbf{v}| = \sqrt{(v_x)^2 + (v_y)^2} = \sqrt{256 + 256} = \sqrt{512} = 16\sqrt{2} \text{ or about } 22.6.$$

Recall that the direction of a vector $\mathbf{r}$ is specified by the angle θ from the horizontal such that $\tan\theta = r_y/r_x$. The direction θ_v of the velocity vector, then, is given by

$$\tan\theta_v = v_y/v_x.$$

Since v_x and v_y are both positive in this example, θ_v is a first quadrant angle.

$$\tan\theta_v = 16/16 = 1, \text{ so } \theta_v = 45°$$

The vectors $\mathbf{v_x}$, $\mathbf{v_y}$, and $\mathbf{v}$ are shown in Figure 22.7.

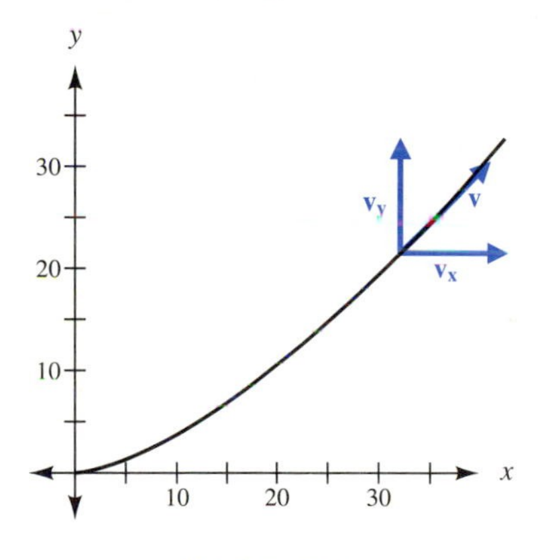

FIGURE 22.7

The acceleration has a horizontal component given by the second derivative of x and a vertical component given by the second derivative of y when t is 4.

$$a_x = x'' = 4 \qquad a_y = y'' = 2(4) \text{ or } 8$$

Thus, the magnitude of the acceleration is given by

$$|\mathbf{a}| = \sqrt{(a_x)^2 + (a_y)^2} = \sqrt{16 + 64} = \sqrt{80} = 4\sqrt{5} \text{ or about } 8.94.$$

The direction θ_a of the acceleration vector is given by $\tan\theta_a = a_y/a_x$. Since a_x and a_y are both positive, θ_a is a first quadrant angle.

$$\tan\theta_a = 8/4 = 2, \text{ so } \theta_a = 63.4° \quad ■$$

The procedure used in Example 3 is summarized as follows.

To find the velocity and acceleration at time t of a particle whose x and y components are specified by parametric equations:

1. Express the horizontal and vertical components of velocity as
$$v_x = dx/dt \qquad \text{and} \qquad v_y = dy/dt.$$
2. Express the horizontal and vertical components of acceleration as
$$a_x = d^2x/dt^2 \qquad \text{and} \qquad a_y = d^2y/dt^2.$$
3. Evaluate the expressions found in steps 1 and 2 for the given time t.
4. Find the magnitude of both the velocity and acceleration vectors.
$$|\mathbf{v}| = \sqrt{v_x^2 + v_y^2} \qquad \text{and} \qquad |\mathbf{a}| = \sqrt{a_x^2 + a_y^2}$$
5. Find the direction of both the velocity and acceleration vectors.
$$\tan \theta_v = v_y/v_x \qquad \text{and} \qquad \tan \theta_a = a_y/a_x$$

NOTE ⇨⇨ ***Be sure to differentiate velocity to find acceleration* before *you substitute a specific value of t into the velocity equations.***

A missile launched with a velocity of 700 ft/s at an angle of 45° with the horizontal follows a path given by
$$x = 495t \text{ and } y = 495t - 16t^2.$$
Find the magnitude and direction of the acceleration when t is 5 seconds.

Solution
1. $v_x = dx/dt = 495$ and $v_y = dy/dt = 495 - 32t$
2. $a_x = d^2x/dt^2 = 0$ and $a_y = d^2y/dt^2 = -32$
3. When $t = 5$, $a_x = 0$ and $a_y = -32$.

4–5. Since the horizontal component is 0, acceleration is directed vertically downward. We can see immediately that its magnitude is 32 and the direction is $-90°$. ■

EXERCISES 22.2

In 1–8, s is the distance (in ft) traveled in time t (in s) by a particle. Find the velocity and acceleration at the given time.

1. $s = 2t^2 - 5t, t = 3$
2. $s = t^2 - 4t, t = 1$
3. $s = -t^2 + 3t, t = 1$
4. $s = 2t^2 - 5t, t = 1$

5. $s = t^2 - 4t, t = 2$

6. $s = \dfrac{1}{t+3}, t = 2$

7. $s = \sqrt{t^2 - 5}, t = 3$

8. $s = \sqrt{t^2 + 7}, t = 3$

In 9–16, the parametric equations specify the path of a moving particle. Find the magnitude and direction of the velocity and acceleration at the given time.

9. $x = (1/2)t^2, y = 3t^3, t = 2$

10. $x = -2t^2, y = t^3, t = 2$

11. $x = (-1/3)t^3, y = 2t^2, t = 1$

12. $x = t^2 + 2t, y = 1 - t^2, t = 1$

13. $x = 1 - t^3, y = t^3 + t, t = 4$

14. $x = 2 - t^2, y = 2t^3 - 4, t = 4$

15. $x = 1/t^2, y = \sqrt{t}, t = 1$

16. $x = \sqrt{t}, y = \dfrac{1}{t-1}, t = 4$

Problems 17–19 use the following information. A bullet launched with a velocity of 300 m/s at an angle of 10° with the horizontal follows a path given by

$$x = 295t \text{ and } y = 52t - 4.9t^2,$$

where t is the time of travel (in s).

17. Find the magnitude and direction of the acceleration when t is 4.0. 4 7 8 1 2 3 5 6 9

18. If the bullet strikes an object 3 m off the ground, find the magnitude of the velocity with which it strikes. [Hint: First find the value of t when y is 3. (Why?)] 4 7 8 1 2 3 5 6 9

19. If the bullet misses the target, find the angle at which it strikes the ground. [Hint: First find the value of t when y is 0. (Why?)] 4 7 8 1 2 3 5 6 9

Problems 20–22 use the following information. An object thrown upward with a velocity of 12 ft/s at an angle of 20° with the horizontal follows a path given by

$$x = 11.3t \text{ and } y = 4.1 - 16t^2,$$

where t is the time traveled.

20. Find the magnitude of the velocity when the object strikes the ground. 4 7 8 1 2 3 5 6 9

21. At what time is the direction of the velocity given by an angle of 4.6° with the horizontal? 4 7 8 1 2 3 5 6 9

22. At what time is the magnitude of the horizontal component of velocity equal to the magnitude of the vertical component of velocity? 4 7 8 1 2 3 5 6 9

23. Speed is the absolute value of velocity. For rectilinear motion, if velocity and acceleration have the same sign at a given time, speed is increasing. If velocity and acceleration have different signs at a given time, speed is decreasing. Determine whether speed is increasing or decreasing in Examples 1 and 2.

---

1. Architectural Technology
2. Civil Engineering Technology
3. Computer Engineering Technology
4. Mechanical Drafting Technology
5. Electrical/Electronics Engineering Technology
6. Chemical Engineering Technology
7. Industrial Engineering Technology
8. Mechanical Engineering Technology
9. Biomedical Engineering Technology

22.3 RELATED RATES

We have seen many examples in which two quantities are related in such a way that a change in one of them is accompanied by a change in the other. For instance, Ohm's law ($V = IR$) tells us that if the voltage changes across a fixed resistance, then the current changes. If we consider the rates at which the changes take place, the rates of change are also related. If it is possible to specify an equation that relates the quantities, it may be possible to specify an equation that relates their rates of change. The procedure is similar to implicit differentiation. With implicit differentiation, however, we differentiated with respect to one of the variables in the equation. Here, we will differentiate with respect to t, a variable that does not appear in the equation. It will be necessary to use the chain rule.

EXAMPLE 1 The equation $P = I^2R$ relates the power P (in watts) to the current I (in amperes) for a fixed resistance R (in ohms). Write the equation that relates the rate of change of power to the rate of change of current for a 50 Ω resistor.

Solution The equation is $P = 50I^2$
Differentiate both sides with respect to t, treating both P and I as functions of t.

$$\frac{dP}{dt} = 50(2I)\left(\frac{dI}{dt}\right) \text{ or } \frac{dP}{dt} = 100I\left(\frac{dI}{dt}\right) \quad ■$$

If one rate of change is known, it may be possible to find the other.

EXAMPLE 2 Find the rate of change of current in Example 1 when the current is 2 A if power is changing at the rate of 20 W/s.

Solution Since $\frac{dP}{dt} = 100I\left(\frac{dI}{dt}\right)$, we can solve for $\frac{dI}{dt}$.

$$\frac{dI}{dt} = \frac{1}{100I}\left(\frac{dP}{dt}\right)$$

Substituting the known values, we have

$$\frac{dI}{dt} = \frac{1}{100(2)}(20) \text{ or } 0.1 \text{ A/s.} \quad ■$$

The procedure for solving problems involving related rates is summarized as follows.

To find the rate of change of a quantity that is related to another quantity:

1. Write an equation that relates the quantities.
2. Differentiate both sides of the equation with respect to time t. *Be sure to use the chain rule.*
3. Solve this equation for the unknown rate of change.
4. Substitute the values that are known for the given instant in time and evaluate.

EXAMPLE 3 A colony of bacteria that is approximately circular in shape grows in such a way that the area increases at a rate of 1.00 mm^2/hour. Find the rate at which the radius is changing when the radius is 2.00 mm.

Solution

1. For a circle, $A = \pi r^2$.
2. To find dr/dt, differentiate both sides of the equation with respect to t.

$$\frac{dA}{dt} = 2\pi r\left(\frac{dr}{dt}\right)$$

3. Solving this equation for dr/dt, we have

$$\frac{1}{2\pi r}\left(\frac{dA}{dt}\right) = \frac{dr}{dt}.$$

4. We know that dA/dt is 1.00 mm^2/hour, and that r is 2.00 mm. Substituting these values, we have

$$\frac{1.00}{4.00\pi} = \frac{dr}{dt} \text{ or } \frac{dr}{dt} = 0.0796 \text{ mm/hour.} \quad \blacksquare$$

EXAMPLE 4 A radar observation post at point B tracks a severe storm located 40 miles due north of the post and headed due north at 35 mph. A mobile unit leaves point B traveling west at 50 mph. Find the rate of change of distance between the storm and the mobile unit after one hour.

Solution Figure 22.8 shows the configuration. The storm is located at C. The mobile unit is at A.

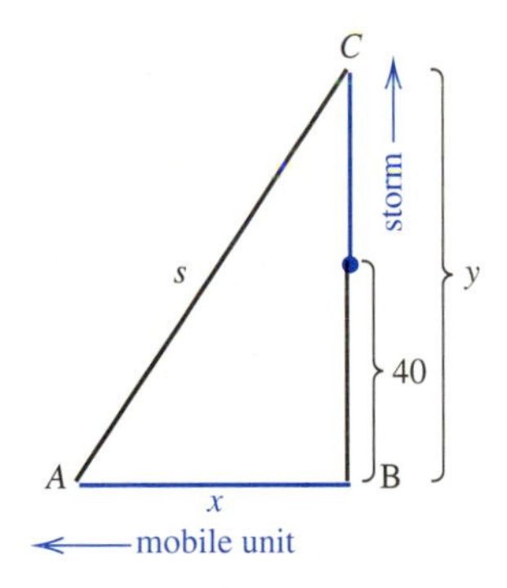

FIGURE 22.8

From the Pythagorean theorem, we have

$$s^2 = x^2 + y^2.$$

We must find ds/dt, so we differentiate both sides of the equation with respect to t.

$$2s\left(\frac{ds}{dt}\right) = 2x\left(\frac{dx}{dt}\right) + 2y\left(\frac{dy}{dt}\right) \text{ or } s\left(\frac{ds}{dt}\right) = x\left(\frac{dx}{dt}\right) + y\left(\frac{dy}{dt}\right)$$

Solving for ds/dt, we have

$$\frac{ds}{dt} = \frac{1}{s}\left[x\left(\frac{dx}{dt}\right) + y\left(\frac{dy}{dt}\right)\right].$$

We know that dx/dt is 50 mph, and dy/dt is 35 mph. Also, after one hour, the distance x is 50 miles, and y is 40 + 35 or 75 miles, so s is given by $\sqrt{50^2 + 75^2}$, which is about 90.1 miles. Making these substitutions, we have

$$\frac{ds}{dt} = \frac{1}{90.1}[50(50) + 75(35)] \text{ or 60 mph, to one significant digit.}$$ ■

EXAMPLE 5 A tank in the shape of an inverted right circular cone has a radius of 8.00 m at the top and a height of 15.0 m. At the instant when the water in the tank is 5.00 m deep, the surface level is rising at the rate of 0.50 m/min. Find the rate (in m^3/min) at which water is being added. That is, find the rate at which volume changes with respect to time.

Solution Figure 22.9 shows the tank with h as the surface level of the water at time t and r as the associated radius of the surface.

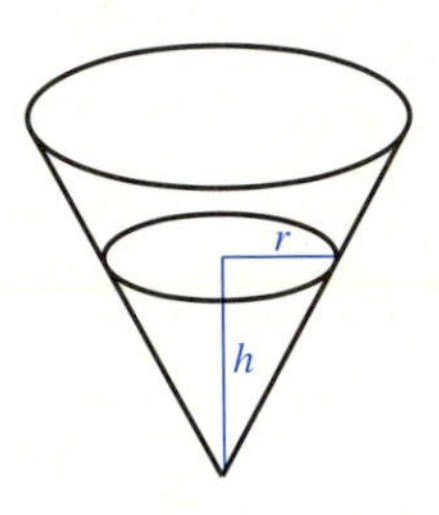

FIGURE 22.9

The volume of the water at time t is given by

$$V = (1/3)\pi r^2 h.$$

We must find dV/dt when $h = 5.00$ and $dh/dt = 0.50$. Since $(1/3)\pi$ is a constant, differentiate r^2h using the product rule. Both r and h change with t, so

$$\frac{dV}{dt} = (1/3)\pi\left[r^2\left(\frac{dh}{dt}\right) + 2r\left(\frac{dr}{dt}\right)h\right]. \tag{22-2}$$

Even though r and h change, they remain in the proportion $\frac{r}{h} = \frac{8.00}{15.00}$, because the tank has radius 8.00 m and height 15.00 m. Thus,

$$r = \left(\frac{8.00}{15.00}\right)h \text{ and } \frac{dr}{dt} = \frac{8.00}{15.00}\left(\frac{dh}{dt}\right).$$

Recall that we also know that h is 5.00 and dh/dt is 0.50. Thus,

$$r = \frac{8.00}{15.00}(5.00) \text{ or } \frac{8.00}{3.00} \text{ and } \frac{dr}{dt} = \frac{8.00}{15.00}(0.50) \text{ or } \frac{4.00}{15.00}.$$

These values may be used in Equation 22-2.

$$\frac{dV}{dt} = (1/3)\pi\left[\left(\frac{8.00}{3.00}\right)^2(0.50) + 2\left(\frac{8.00}{3.00}\right)\left(\frac{4.00}{15.00}\right)(5.00)\right]$$
$$= 11.2 \text{ in}^3/\text{min} \quad \blacksquare$$

In the previous section, we considered velocity and acceleration for an object whose path is specified by parametric equations. Sometimes, however, the path is specified by an equation that gives y in terms of x, instead of t. We may treat these problems as related rates problems.

EXAMPLE 6 A particle follows a path given by $y = x^2 - 2x$. If the vertical component of velocity is a constant 4 m/s, find the magnitude of the velocity when the particle is at (2, 0).

Solution Figure 22.10 shows the graph of the function, as well as the velocity vector **v** and its components.

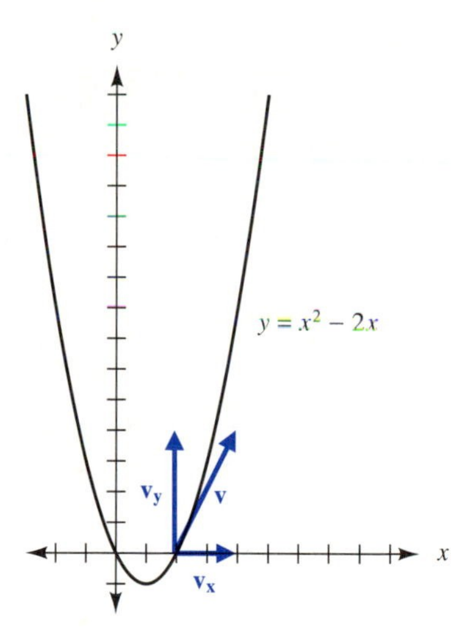

FIGURE 22.10

Recall that $|\mathbf{v}|$ is given by

$$|\mathbf{v}| = \sqrt{(v_x)^2 + (v_y)^2}$$

where $v_x = dx/dt$ and $v_y = dy/dt$.

We know that dy/dt is 4 m/s. To find dx/dt, differentiate both sides of the equation $y = x^2 - 2x$ with respect to t.

$$\frac{dy}{dt} = 2x\left(\frac{dx}{dt}\right) - 2\left(\frac{dx}{dt}\right)$$

$$\frac{dy}{dt} = \frac{dx}{dt}(2x - 2) \qquad \text{Factor out } \frac{dx}{dt}.$$

$$4 = \frac{dx}{dt}(4 - 2) \qquad \text{Substitute 4 for } \frac{dy}{dt} \text{ and 2 for } x.$$

$$2 = \frac{dx}{dt} \qquad \text{Solve for } \frac{dx}{dt}.$$

Since the magnitude of the velocity is given by $|\mathbf{v}| = \sqrt{(v_x)^2 + (v_y)^2}$,

$$|\mathbf{v}| = \sqrt{2^2 + 4^2} = \sqrt{4 + 16} = \sqrt{20} = 2\sqrt{5} \text{ m/s or about 4.47 m/s.}$$ ■

EXAMPLE 7 A particle follows a path given by $y = x^3 + 3x$. If the horizontal component of velocity is a constant 2 m/s, find the magnitude of the velocity when the particle is at (1, 4).

Solution We know that dx/dt is 2. To find dy/dt, differentiate both sides of the equation $y = x^3 + 3x$ with respect to t.

$$\frac{dy}{dt} = 3x^2\left(\frac{dx}{dt}\right) + 3\left(\frac{dx}{dt}\right)$$

$$\frac{dy}{dt} = \frac{dx}{dt}(3x^2 + 3) \qquad \text{Factor out } \frac{dx}{dt}.$$

$$\frac{dy}{dt} = 2(3 + 3) \qquad \text{Substitute 2 for } \frac{dx}{dt} \text{ and 1 for } x.$$

$$\frac{dy}{dt} = 12$$

Since the magnitude of the velocity is given by $|\mathbf{v}| = \sqrt{(v_x)^2 + (v_y)^2}$,

$$|\mathbf{v}| = \sqrt{2^2 + 12^2} = \sqrt{4 + 144} = \sqrt{148} = 2\sqrt{37} \text{ m/s or about 12.2 m/s.}$$ ■

EXAMPLE 8 A car follows a path specified by $y = 2x^2 + 3x + 1$. The horizontal component of velocity is a constant 2 ft/s. Find the magnitude and direction of the acceleration at the point (1, 6).

Solution We know that $\frac{dx}{dt} = 2$, and therefore $\frac{d^2x}{dt^2} = 0$. We need $\frac{d^2y}{dt^2}$. Differentiate both sides of the equation $y = 2x^2 + 3x + 1$ with respect to t.

$$\frac{dy}{dt} = 4x\left(\frac{dx}{dt}\right) + 3\left(\frac{dx}{dt}\right) + 0$$

Differentiate y. Differentiate $2x^2$. Differentiate $3x$. Differentiate 1.

$$\frac{dy}{dt} = \frac{dx}{dt}(4x + 3)$$

Factor out $\frac{dx}{dt}$.

The second derivative of y is found by the product rule.

$$\frac{d^2y}{dt^2} = \left(\frac{dx}{dt}\right)(\text{derivative of } 4x + 3) + (4x + 3)\left(\text{derivative of } \frac{dx}{dt}\right)$$

$$\frac{d^2y}{dt^2} = \frac{dx}{dt}\left(4\frac{dx}{dt}\right) + (4x + 3)\frac{d^2x}{dt^2}$$

$$\frac{d^2y}{dt^2} = 4\left(\frac{dx}{dt}\right)^2 + (4x + 3)\frac{d^2x}{dt^2}$$

NOTE ⇨ ***Remember that*** $\left(\frac{dx}{dt}\right)^2$ ***is the square of the first derivative, but*** $\frac{d^2x}{dt^2}$ ***is the second derivative.*** Since $\frac{dx}{dt} = 2$ and $\frac{d^2x}{dt^2} = 0$, the equation can be written

$$\frac{d^2y}{dt^2} = 4(2)^2 + (4x + 3)(0) = 4(4) = 16 \text{ ft/s}^2.$$

The horizontal component of the acceleration is 0, so the acceleration of 16 ft/s² acts parallel to the y-axis. ■

EXERCISES 22.3

Solve each problem.

1. The water velocity v (in ft/s) at the point of discharge is given by $v = 12.16\,p^{1/2}$, where p is the pressure (in psi). Find the acceleration (dv/dt) of the water if pressure is changing at the rate of 0.2500 psi/second when $p = 50.00$ psi.
 4 8 1 2 6 7 9

2. The potential energy W (in joules) of an inductor of inductance L (in henries) carrying a current i (in amperes) is given by $W = (1/2)Li^2$. Find the rate at which current is changing if potential energy is changing at the rate of 0.075 J/s when the current is 150 mA and L is 250 mH.
 3 5 9 1 2 4 6 7 8

3. A rectangle that changes size is incorporated into the design of a flashing electric sign. The original size of the rectangle is 2.00 m by 3.00 m. Find the rate at which the area begins to decrease (dA/dt) if

1. Architectural Technology
2. Civil Engineering Technology
3. Computer Engineering Technology
4. Mechanical Drafting Technology
5. Electrical/Electronics Engineering Technology
6. Chemical Engineering Technology
7. Industrial Engineering Technology
8. Mechanical Engineering Technology
9. Biomedical Engineering Technology

the length changes at the rate of 0.20 m/s and the width changes at the rate of 0.10 m/s.
1 2 4 7 8 3 5

4. Two perpendicular walls of a stage set are moved in such a way as to increase the size of the rectangular set. If the width changes at the rate of 1.5 ft/s, find the rate at which the length must change so that when the set measures 16.0 ft by 12.0 ft, the area will change at the rate of 30.0 ft^2/s.
1 2 4 7 8 3 5

5. Two rescue ships approach a craft in distress. One ship approaches from the north at a rate of 15.0 knots. The other ship approaches from the east at 12.0 knots. Find the rate at which the distance between the two ships is changing when the southbound ship is 25.0 nautical miles from the craft and the westbound ship is 32.0 nautical miles from the craft. 4 7 8 1 2 3 5 6 9

6. A television cameraman films a reporter walking during an interview. The reporter and cameraman move at right angles to each other as shown in the accompanying diagram. The reporter walks at the rate of 4.00 ft/s. Find the rate at which the cameraman must move so that the distance between them changes at the rate of 5.00 ft/s when they are in the positions shown.

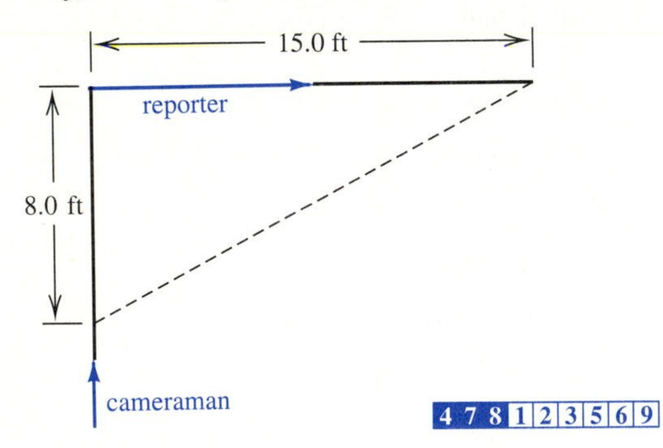

4 7 8 1 2 3 5 6 9

7. A container in the shape of an inverted right circular cone has a radius of 4.00 inches at the top and a height of 5.00 inches. At the instant when the water in the container is 2.00 inches deep, the surface level is falling at the rate of 0.500 in/s. Find the rate (in in^3/s) at which water is being drained. 1 2 4 7 8 3 5

8. A clay cylinder is rolled so that as its radius decreases, its height increases while the volume remains constant. Find the rate of change of height if the radius decreases by 0.250 mm/s when r is 30.0 mm and h is 80.0 mm. 1 2 4 7 8 3 5

9. If a resistor and inductor are connected in series, the impedance Z (in ohms) is given by $Z^2 = R^2 + X_L^2$, where R is the resistance (in ohms) and X_L is the inductive reactance (in ohms). Find the rate by which X_L must vary with respect to R if R changes in such a way as to keep the impedance constant at 60.0 ohms.
3 5 9 1 2 4 6 7 8

10. The equivalent resistance R (in ohms) of two resistors in parallel is given by $R = \dfrac{R_1 R_2}{R_1 + R_2}$, where R_1 and R_2 are the individual resistances (in ohms). If R_2 is a constant 20 ohms, find dR_1/dt when $R_1 = 50\ \Omega$ and $dR/dt = -0.50\ \Omega$/minute.
3 5 9 1 2 4 6 7 8

11. A particle follows a path given by $y = x^3 + 1$. If the horizontal component of velocity is a constant 3 m/s, find the magnitude of the velocity when the particle is at (2, 9). 4 7 8 1 2 3 5 6 9

12. A particle follows a path given by $y = 3x^2 - 2x$. If the vertical component of velocity is a constant 4 m/s, find the magnitude of the velocity when the particle is at (1, 1). 4 7 8 1 2 3 5 6 9

13. A particle follows a path given by $y = x^2 - 2x$. If the horizontal component of velocity is given by $v_x = 2x$, find the vertical component of velocity when the particle is at (3, 3). 4 7 8 1 2 3 5 6 9

14. A particle follows a path given by $y = x^3 - x$. If the vertical component of velocity is given by $v_y = 2x$, find the horizontal component of velocity when the particle is at (2, 6).
4 7 8 1 2 3 5 6 9

15. Find the magnitude and direction of the acceleration for the particle in problem 11 when it is at (1, 2). 4 7 8 1 2 3 5 6 9

16. Find the magnitude and direction of the acceleration for the particle in problem 12 when it is at (2, 8). 4 7 8 1 2 3 5 6 9

22.4 CURVE SKETCHING

In Section 6.3, you learned to sketch a parabola by finding its vertex. The vertex is an important point, because it is the highest or lowest point on the curve. For graphing polynomial equations of degree higher than 2, we have seen no short cuts. We sketched those curves by plotting points one-by-one until the shape of the graph was apparent. Computers and even some calculators are capable of producing graphs, but a rough sketch is often all that is necessary. Even when a computer-generated graph is used, a rough sketch may help determine the interval over which the function should be graphed. Calculus can be used to simplify the procedure by identifying certain characteristics of a curve.

Figure 22.11 shows the graph of a cubic equation.

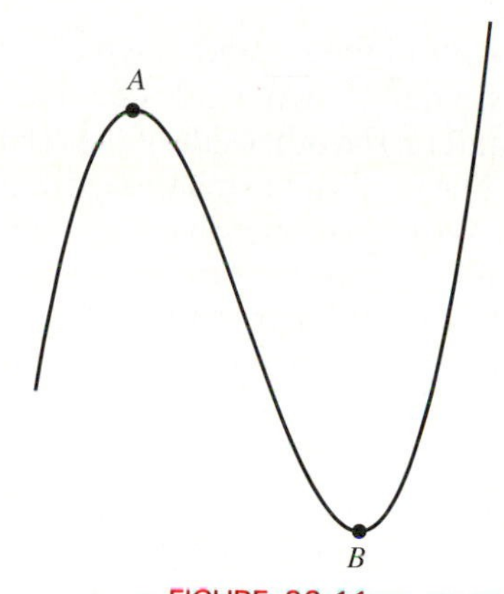

FIGURE 22.11

relative maximum

Point A is called a *relative maximum.* Notice that the point is not the highest point on the graph. There is no highest point, because as one follows the curve farther and farther to the right, the curve goes higher and higher. Point A, however, is higher than points on either side of it, in the immediate vicinity of point A. That is, the function value at point A is at its maximum, relative to nearby values. Likewise, point B is called a *relative minimum.* The term *relative extrema* refers to either relative maxima or relative minima.

relative minimum

relative extrema

increasing/decreasing

We refer to a curve as *increasing* or *decreasing* over an interval on the x-axis. To understand these terms from an intuitive point of view, trace the curve in Figure 22.11 from left to right. As you move upward, the curve increases until point A is reached. At point A, the curve begins decreasing as you move downward, and it continues decreasing until point B is reached. From point B on, the curve increases.

The slope of the curve at different points provides a convenient way to quantify the terms increasing and decreasing. Figure 22.12 shows tangents to the curve at several points. As one traces the curve from the left toward point A, the slope of the tangent at each point is positive. Since the tangents become less steep as point A is approached, the slopes of the tangents decrease until at point A, the slope of the tangent is 0. Immediately to the right of point A, the slopes of the tangents are negative.

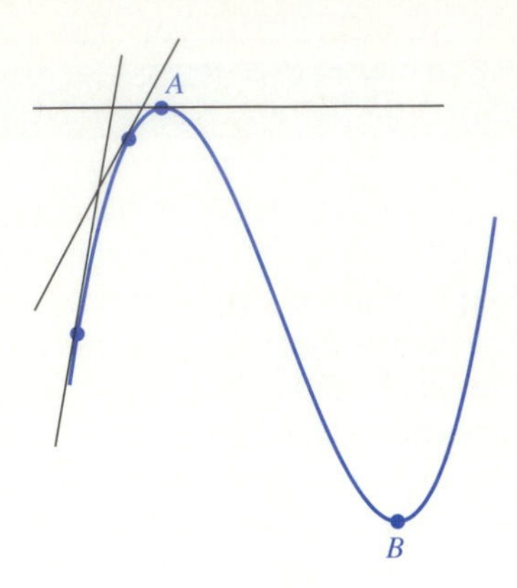

FIGURE 22.12

The behavior of these slopes is typical of slopes near relative extrema of continuous functions. That is, on one side of a relative extremum, the slopes of the tangents are positive, and on the other side of the relative extremum, the slopes of the tangents are negative. The slope at the relative extremum is 0, as it is in this example, or it does not exist. The x-values for which the derivative is 0 or does not exist are called *critical numbers*. The points on the graph associated with these values are called *critical points*.

critical numbers

critical points

Since the slope of a tangent is given by the derivative, we can say that at any relative maximum or minimum point on the graph of a function, the derivative of the function either has a value of 0 or does not exist. It is important to realize, however, that if the derivative of the function is 0 or does not exist, the point may or may not be a relative extremum. That is, if we find the critical points, we have a list that contains all of the relative extrema, and possibly some points that are not relative extrema.

In Figure 22.13, the derivative is 0 at points A and B; yet only point B is a relative extremum. Notice that to the left of point A, slopes of tangents to the curve are negative. Immediately to the right of point A, slopes of tangents are also negative. Immediately to the left of point B, slopes of tangents are negative, but to the right of point B, slopes of tangents are positive. That is, at point B, the curve stops decreasing and starts increasing.

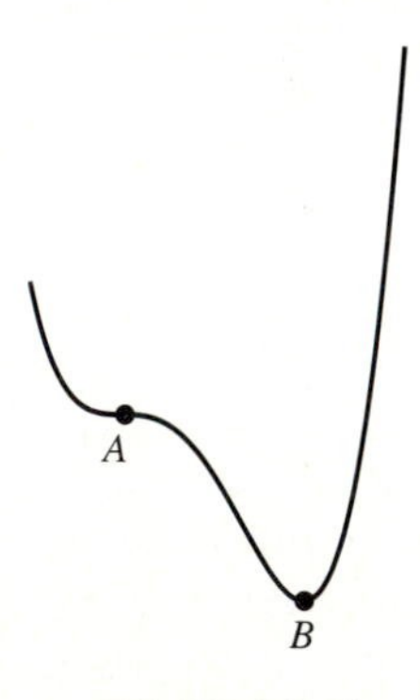

FIGURE 22.13

first derivative test

These observations can be generalized to provide a procedure, called the *first derivative test,* for determining whether a critical point is a relative maximum, minimum, or neither. In the remainder of the chapter, we will assume only continuous functions, although the first derivative test may be applied with minor adjustments to functions that have a few discontinuities or to equations that do not represent functions.

To apply the first derivative test to a continuous function f:

1. Find the critical numbers by finding the values of x for which $f'(x) = 0$ or does not exist.
2. Let the critical numbers divide the x-axis into sections. Evaluate $f'(x)$ for one number in each section.
   (a) If the derivative changes from positive to negative (so that f increases and then decreases), the critical point is a relative maximum.
   (b) If the derivative changes from negative to positive (so that f decreases and then increases), the critical point is a relative minimum.
   (c) If the derivative has the same sign on both sides of the critical point, the point is neither a relative maximum nor a relative minimum.

Example 1 shows how to use the first derivative test to sketch the graph of a polynomial function.

EXAMPLE 1 Sketch the graph of $f(x) = x^6 - 6x^4$.

Solution The critical numbers are the solutions of $f'(x) = 0$ or

$$6x^5 - 24x^3 = 0$$

$$6x^3(x^2 - 4) = 0 \qquad \text{Factor out } 6x^3.$$

$$6x^3(x-2)(x+2) = 0 \qquad \text{Factor } x^2 - 4.$$

$$6x^3 = 0,\ x - 2 = 0, \text{ or } x + 2 = 0 \qquad \text{Set each factor equal to zero.}$$

$$x = 0,\ x = 2, \text{ or } x = -2 \qquad \text{Solve each equation.}$$

To apply the first derivative test, we choose -3 as a value less than -2, -1 as a value between -2 and 0, 1 as a value that is between 0 and 2, and 3 as a value that is larger than 2. Recall that

$$f'(x) = 6x^5 - 24x^3$$

$$f'(-3) = 6(-3)^5 - 24(-3)^3 = -1458 + 648 = -810$$

If $x < -2$, the curve decreases.

$$f'(-1) = 6(-1)^5 - 24(-1)^3 = -6 + 24 = 18$$

If $-2 < x < 0$, the curve increases.

$$f'(1) = 6(1)^5 - 24(1)^3 = 6 - 24 = -18$$

If $0 < x < 2$, the curve decreases.

$$f'(3) = 6(3)^5 - 24(3)^3 = 1458 - 648 = 810$$

If $x > 2$, the curve increases.

Next, we find the y-value associated with each critical number.

$$f(x) = x^6 - 6x^4$$
$$f(-2) = (-2)^6 - 6(-2)^4 = 64 - 96 = -32$$
$$f(0) = (0)^6 - 6(0)^4 = 0$$
$$f(2) = (2)^6 - 6(2)^4 = 64 - 96 = -32$$

Figure 22.14 shows the graph. ■

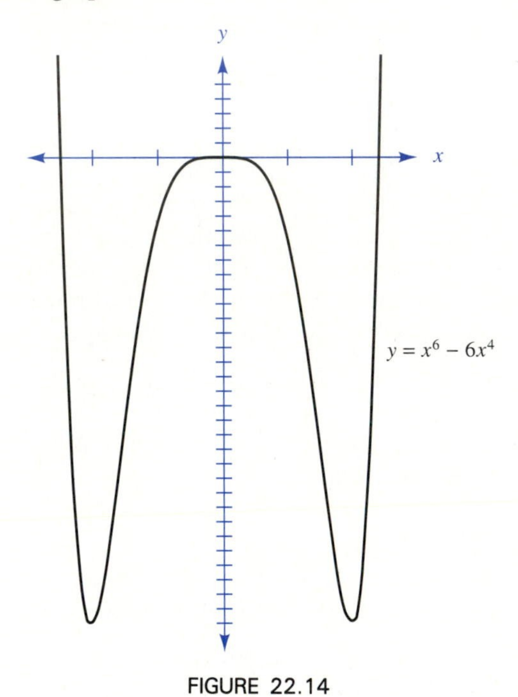

FIGURE 22.14

Just as the first derivative of a function gives information about how the curve is increasing or decreasing, the second derivative of the function gives information about how the slopes are increasing or decreasing. In Figure 22.15, the slopes are decreasing. That is, at the far left, slopes are fairly large, but as one moves to the right, the slopes gradually decrease, reaching 0 at about the middle of the graph. On the right-hand side of the graph, the slopes are negative, and continue to decrease. Figure 22.16 shows increasing slopes.

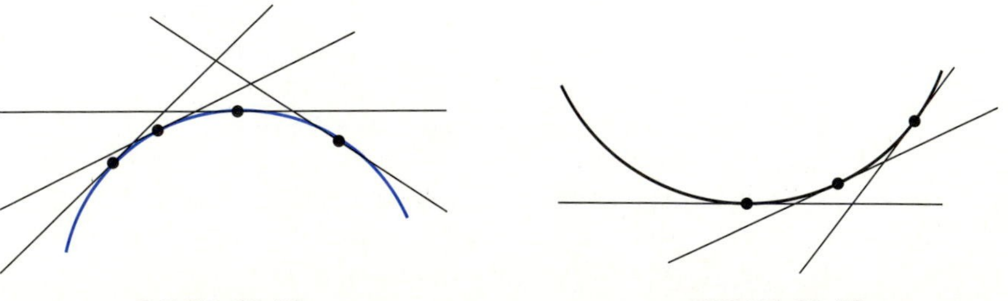

FIGURE 22.15

FIGURE 22.16

We conclude that the sign of the second derivative gives an indication of the direction a curve bends. When a curve bends up, the slopes are increasing, and the second derivative is positive. When a curve bends down, the slopes are decreasing, and the second derivative is negative. When the curve changes from bending up to bending down (or vice versa), the second derivative is 0 or does not exist. A curve that is bending up is said to be **concave upward,** and a curve that is bending down is said to be **concave downward.** It is a common mistake to associate the words "upward" with "maximum" and "downward" with "minimum." ***Notice, however, that when a curve is concave* upward *as in Figure 22.16, the critical point is a relative* minimum, *and when the curve is concave* downward *as in Figure 22.15, the critical point is a relative* maximum.**

concave upward

concave downward

NOTE

The second derivative provides an alternate method for determining which critical points are relative extrema. Called the *second derivative test,* this procedure is shorter than the first derivative test, but sometimes it is inconclusive, and one must revert to the first derivative test.

second derivative test

To apply the second derivative test to a continuous function f:

Evaluate $f''(x)$ for each critical number.

(a) If $f''(x) > 0$, the curve is concave upward, and the critical point is a relative minimum.

(b) If $f''(x) < 0$, the curve is concave doward, and the critical point is a relative maximum.

(c) If $f''(x) = 0$ or does not exist, the point may be a relative maximum, a relative minimum, or neither.

NOTE

When the second derivative test is inconclusive (condition (c)), the first derivative test may be used to determine how the function behaves at the point in question.

EXAMPLE 2 Sketch the graph of $y = 3x^4 - 4x^3$.

Solution Find the critical numbers: $y' = 12x^3 - 12x^2 = 0$

$$12x^2(x - 1) = 0$$

$$12x^2 = 0 \text{ or } x - 1 = 0$$

$$x = 0 \text{ or } x = 1$$

Apply the second derivative test: $y'' = 36x^2 - 24x$

When x is 1, $y'' = 36 - 24 = 12$.

The curve is concave upward, and there is a relative minimum at $x = 1$.

When x is 0, $y'' = 0$, and the test is inconclusive.

We know that the curve is decreasing to the right of 0, since it has a relative minimum at $x = 1$. We examine y' for a value to the left of 0.

When x is -1, $y' = -12 - 12 = -24$, so the curve is decreasing to the left of 0 also. The critical points are $(1, -1)$ and $(0, 0)$ (but only $(1, -1)$ is a relative extremum). Figure 22.17 shows the graph. ■

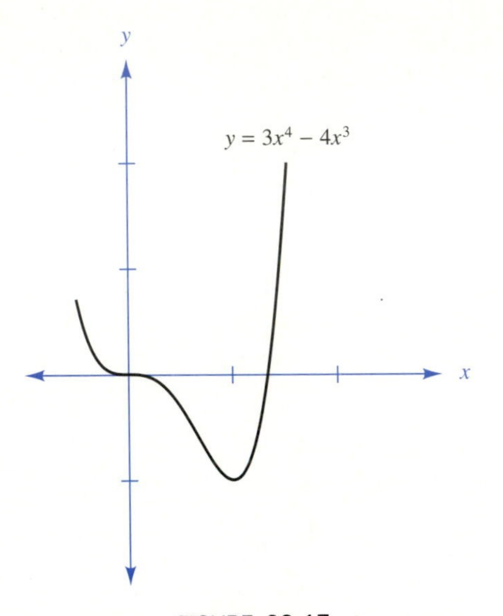

FIGURE 22.17

inflection point

The second derivative may also be used to find possible *inflection points.* An **inflection point** is a point at which a curve changes concavity, as in Figure 22.18. Since concavity changes from positive to negative (or vice versa) at an inflection point, the second derivative must be 0 or not exist at an inflection point. A list of values where the second derivative is 0 or does not exist is a list of *possible* inflection points. The situation is thus analogous to finding the *possible* relative extrema using the first derivative.

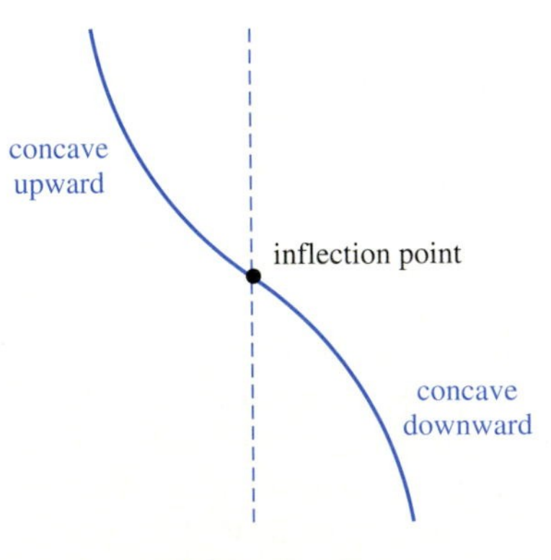

FIGURE 22.18

EXAMPLE 3 Find the inflection points for the curve in Example 2.

Solution The second derivative of $y = 3x^4 - 4x^3$ is given by $y'' = 36x^2 - 24x$. Possible inflection points occur when x satisfies the following equation.

$$0 = 36x^2 - 24x$$
$$0 = 12x(3x - 2)$$
$$x = 0 \text{ or } x = 2/3$$

To determine whether either or both of these values produce an inflection point, choose -1 to the left of 0, 1/3 between 0 and 2/3, and 1 to the right of 2/3.
When x is -1, $y'' = 36(-1)^2 - 24(-1) = 36 + 24 = 60$.
To the left of 0, the curve is concave upward.
When x is 1/3, $y'' = 36(1/3)^2 - 24(1/3) = 4 - 8 = -4$.
Between 0 and 2/3, the curve is concave downward.
When x is 1, $y'' = 36(1)^2 - 24(1) = 36 - 24 = 12$.
To the right of 2/3, the curve is concave upward.
The curve changes concavity at both $x = 0$ and $x = 2/3$.
Next we find the y-values associated with these values. When x is 0, y is 0, and when x is 2/3, y is $-16/27$. Thus, (0, 0) and (2/3, $-16/27$) are inflection points. Refer to Figure 22.17 to see the change in concavity at these points. ■

EXERCISES 22.4

In 1–15, sketch the graph of each function.

1. $y = 3x^3 - 16x + 1$
2. $y = x^3 - 3x + 2$
3. $y = x^3 - 12x + 7$
4. $y = 4x^3 - 27x - 7$
5. $y = 2x^3 + 3x^2 - 12x + 4$
6. $y = 2x^3 - 3x^2 + 5$
7. $y = x^4 - 4x^2 + 5$
8. $y = x^4 - 18x^2 + 10$
9. $y = 9x^4 - 8x^3 + 1$
10. $y = 3x^4 + 2x^3 - 1$
11. $y = 3x^4 - 5x^3 + 2$
12. $y = 3x^5 - 5x^3$
13. $y = x^5 - 5x^4 + 5x^3$
14. $y = x^3 - 3x^2 + 3x + 2$
15. $y = x^3 - 6x^2 + 12x - 4$

---

16. A tank is to be constructed in the shape of a cylinder 10.0 m high topped by a hemisphere. If the radius is r, the volume V (in m^3) is given by $V = (2/3)\pi r^3 + 10\pi r^2$. Sketch the graph of V versus r. 1 2 4 7 8 3 5

17. When a cube with side of length s (in inches) is topped by a pyramid 8″ high, the volume V (in in^3) is given by $V = s^3 + (8/3)s^2$. Sketch the graph of V versus s. 1 2 4 7 8 3 5

18. A series of squares, each having sides 1 cm longer than the previous square, are cut from a sheet of plastic. The areas, then, are 1^2, 2^2, 3^2, 4^2, and so on. The sum S (in cm^2) of the areas of the first n squares is given by $S = (1/6)(2n^3 + 3n^2 + n)$. Sketch the graph of S versus n. 1 2 4 7 8 3 5

1. Architectural Technology
2. Civil Engineering Technology
3. Computer Engineering Technology
4. Mechanical Drafting Technology
5. Electrical/Electronics Engineering Technology
6. Chemical Engineering Technology
7. Industrial Engineering Technology
8. Mechanical Engineering Technology
9. Biomedical Engineering Technology

19. A box is to be constructed from a piece of metal 10.0 cm square by cutting a square with side of length s (in cm) from each corner and folding up the sides. The volume V (in cm^3) of the box is given by $V = 4s^3 - 40s^2 + 100s$. Sketch the graph of V versus s. 1 2 4 7 8 3 5

20. For a cantilever beam under a distributed load, the deflection at a distance x from the support is given by

$$y = \frac{-wx^4 + 4wbx^3 - 6wb^2x^2}{24EI} \text{ when } x \leq b.$$

Assume that $w = 0.02$ kg/mm (the load), $E = 4000$ kg/mm^2 (the modulus of elasticity), $I = 5$ mm^4 (the moment of inertia), and $b = 40$ mm (the length of the load). With the given values, the equation becomes

$$y = \frac{-x^4 + 160\,x^3 - 9600\,x^2}{24\,000\,000}$$

Sketch the graph of y versus x for $0 \leq x \leq 40$. 1 2 4 8 7 9

1. Architectural Technology
2. Civil Engineering Technology
3. Computer Engineering Technology
4. Mechanical Drafting Technology
5. Electrical/Electronics Engineering Technology
6. Chemical Engineering Technology
7. Industrial Engineering Technology
8. Mechanical Engineering Technology
9. Biomedical Engineering Technology

22.5 ADDITIONAL TECHNIQUES FOR CURVE SKETCHING

When information obtained by algebraic means is considered, the sketch can be made with more accuracy. The intercepts, for instance, can sometimes be found algebraically. Remember that the y-intercept of a function is located where x is 0 and the x-intercept is located where y is 0.

symmetric with respect to the y-axis

symmetric with respect to the x-axis

symmetric with respect to the origin

Symmetry conditions can also be checked algebraically. We say that a curve is *symmetric with respect to the y-axis* if replacing x with $-x$ in the equation yields an equivalent equation. When a curve is symmetric with respect to the y-axis, the part of the curve to the left of the y-axis is a reflection or "mirror image" of the part of the curve to the right of the y-axis. A curve is *symmetric with respect to the x-axis* if replacing y with $-y$ in the equation yields an equivalent equation. When a curve is symmetric with respect to the x-axis, the part of the curve below the x-axis is a reflection of the part of the curve about the x-axis. A curve is *symmetric with respect to the origin* if replacing x with $-x$ and y with $-y$ in the equation yields an equivalent equation. When a curve is symmetric with repect to the origin, the part of the curve in the third quadrant is a reflection of the part of the curve in the first quadrant, and the part of the curve in the fourth quadrant is a reflection of the part in the second quadrant.

EXAMPLE 1 Find the x and y intercepts, and determine the symmetry conditions for the graph of $f(x) = x^6 - 6x^4$.

Solution First, we write the equation as $y = x^6 - 6x^4$.

x-intercept: let $y = 0$	y-intercept: let $x = 0$
$0 = x^6 - 6x^4$	$y = 0^6 - 6(0^4)$
$0 = x^4(x^2 - 6)$	$y = 0$
$x^4 = 0$ or $x^2 - 6 = 0$	
$x = 0$ or $x = \pm\sqrt{6}$	

The intercepts, then, occur at $(-\sqrt{6}, 0)$, $(0, 0)$, and $(\sqrt{6}, 0)$.

y-axis symmetry:	x-axis symmetry:	origin symmetry:
replace x with $-x$	replace y with $-y$	replace x with $-x$ and y with $-y$
$y = (-x)^6 - 6(-x)^4$	$-y = x^6 - 6x^4$	$-y = (-x)^6 - 6(-x)^4$
$y = x^6 - 6x^4$	$y = -x^6 + 6x^4$	$-y = x^6 - 6x^4$
		$y = -x^6 + 6x^4$

The curve is symmetric with respect to the y-axis, but not with respect to the x-axis or the origin. This is the same function that we graphed in Example 1 of the previous section. Examine Figure 22.19 to see how this information is incorporated into the graph. ■

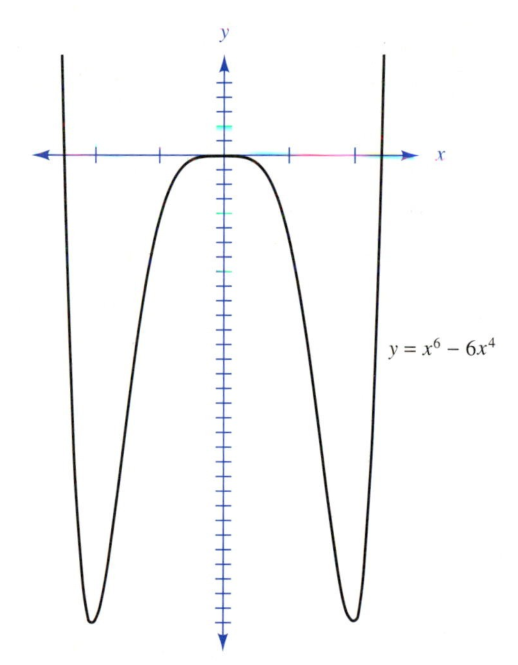

FIGURE 22.19

It is impossible for a *function* to be symmetric with respect to the x-axis (why?), so in the remaining examples, we will not test for x-axis symmetry. A function that is represented by a rational expression may have a horizontal asymptote. If the degree of the denominator is greater than or equal to the degree of the numerator, the horizontal asymptote can be found by examining the limit of y as x approaches infinity. (See the Extended Application at the end of Chapter 21 for an explanation.) The steps for sketching a curve are summarized as follows.

To sketch the graph of a continuous function defined by $y = f(x)$:

1. Find and plot the x- and y-intercepts.
2. Determine the symmetry conditions, if any.
3. Identify the horizontal asymptote, if there is one.
4. Find and plot the relative extrema and other critical points.
5. Connect the plotted points with a smooth curve.

EXAMPLE 2 Sketch the graph of $f(x) = \dfrac{x}{x^2 + 1}$.

Solution

1. Find and plot the intercepts.

x-intercept: let $y = 0$

$$0 = \frac{x}{x^2 + 1}$$

$$0 = x$$

y-intercept: let $x = 0$

$$y = \frac{0}{0^2 + 1}$$

$$y = 0$$

The only intercept, then, occurs at (0, 0).

2. Check for symmetry conditions.

y-axis symmetry:
replace x with $-x$

$$y = \frac{-x}{(-x)^2 + 1}$$

$$y = \frac{-x}{x^2 + 1}$$

origin symmetry:
replace x with $-x$ and y with $-y$

$$-y = \frac{-x}{(-x)^2 + 1}$$

$$-y = \frac{-x}{x^2 + 1}$$

The curve is symmetric about the origin.

3. To find the horizontal asymptote, we examine $\lim\limits_{x \to \infty} \dfrac{x}{x^2 + 1}$.

$$\lim_{x \to \infty} \frac{x}{x^2 + 1} = \lim_{x \to \infty} \frac{\dfrac{x}{x^2}}{\dfrac{x^2}{x^2} + \dfrac{1}{x^2}} = \frac{0}{1 + 0} = 0$$

The horizontal line given by $y = 0$ (the x-axis) is an asymptote.

4. The critical numbers are the solutions of $f'(x) = 0$. Since $f(x) = x(x^2 + 1)^{-1}$, we apply the product rule to find $f'(x)$.

$$
\begin{aligned}
f'(x) &= x(-1)(x^2 + 1)^{-2}(2x) + (x^2 + 1)^{-1}(1) \\
&= -2x^2(x^2 + 1)^{-2} + (x^2 + 1)^{-1} && \text{Simplify.} \\
&= (x^2 + 1)^{-2}[-2x^2 + (x^2 + 1)] && \text{Factor out } (x^2 + 1)^{-2}. \\
&= \frac{-x^2 + 1}{(x^2 + 1)^2} && \text{Combine like terms and rewrite.}
\end{aligned}
$$

Now we solve $f'(x) = 0$.

$$
\begin{aligned}
\frac{-x^2 + 1}{(x^2 + 1)^2} &= 0 && \text{Notice that the denominator cannot be zero.} \\
-x^2 + 1 &= 0 && \text{Multiply both sides by } (x^2 + 1)^2. \\
1 &= x^2 && \text{Add } x^2 \text{ to both sides.} \\
\pm 1 &= x && \text{Solve for } x.
\end{aligned}
$$

If the curve can be sketched for positive values of x, we can use the symmetry conditions to draw the portion for negative values of x. To apply the second derivative test, consider $f''(1)$ and $f''(-1)$.

$$
\begin{aligned}
f'(x) &= (x^2 + 1)^{-2}(-x^2 + 1) \\
f''(x) &= (x^2 + 1)^{-2}(-2x) + (-x^2 + 1)(-2)(x^2 + 1)^{-3}(2x) \\
&= (-2x)(x^2 + 1)^{-3}[(x^2 + 1) + 2(-x^2 + 1)] \\
&= (-2x)(x^2 + 1)^{-3}[x^2 + 1 - 2x^2 + 2] \\
&= (-2x)(x^2 + 1)^{-3}(-x^2 + 3) \\
&= \frac{(-2x)(-x^2 + 3)}{(x^2 + 1)^3}
\end{aligned}
$$

Thus, $f''(1) = \frac{-2(2)}{2^3}$ or $\frac{-1}{2}$.

The curve is concave downward, so there is a maximum when x is 1. Now we find the y-value associated with $x = 1$.

$$f(x) = \frac{x}{x^2 + 1}$$

$$f(1) = \frac{1}{1^2 + 1} = \frac{1}{1 + 1} = \frac{1}{2}$$

Because of the symmetry condition, we know that $f(-1) = \frac{-1}{2}$.

5. Plot these points. Figure 22.20 on the following page shows the graph. ■

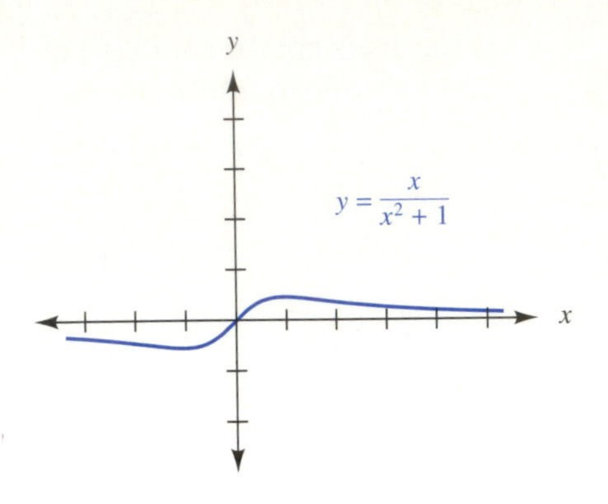

FIGURE 22.20

EXAMPLE 3 Sketch the graph of $f(x) = x^3 - 2x^2 + x + 2$.

Solution

1. x-intercept: let $y = 0$

   $0 = x^3 - 2x^2 + x + 2$

   The equation cannot be solved by factoring, but $f(-1) = -2$, and $f(0) = 2$, so $f(x) = 0$ for some value of x between -1 and 0.

   y-intercept: let $x = 0$

   $y = 0^3 - 2(0)^2 + 0 + 2$

   $y = 2$

2. Check for symmetry conditions.

   y-axis symmetry:

   replace x with $-x$

   $y = (-x)^3 - 2(-x)^2 + (-x) + 2$

   $y = -x^3 - 2x^2 - x + 2$

   origin symmetry:

   replace x with $-x$ and y with $-y$

   $-y = (-x)^3 - 2(-x)^2 + (-x) + 2$

   $-y = -x^3 - 2x^2 - x + 2$

   $y = x^3 + 2x^2 + x + 2$

   The curve does not show either type of symmetry.

3. Since y is not a rational expression, there are no horizontal asymptotes.

4. The critical numbers are the solutions of $f'(x) = 0$.

$$f'(x) = 3x^2 - 4x + 1 = 0$$
$$(3x - 1)(x - 1) = 0$$
$$3x - 1 = 0 \text{ or } x - 1 = 0$$
$$x = 1/3 \text{ or } x = 1$$

Consider $f''(x) = 6x - 4$. Then $f''(1/3) = 6(1/3) - 4$ or -2, and $f''(1) = 6(1) - 4$ or 2. Thus, there is a relative maximum when x is 1/3 and a relative minimum when x is 1. Next, we find the y-value associated with each critical number.

$$
\begin{aligned}
f(x) &= x^3 - 2x^2 + x + 2 \\
f(1/3) &= (1/3)^3 - 2(1/3)^2 + (1/3) + 2 \\
&= 1/27 - 2/9 + 1/3 + 2 \\
&= 1/27 - 6/27 + 9/27 + 54/27 \\
&= 58/27 \text{ or } 2\ 4/27 \\
f(1) &= 1^3 - 2(1)^2 + 1 + 2 \\
&= 1 - 2 + 1 + 2 \\
&= 2
\end{aligned}
$$

5. Figure 22.21 shows the graph. ■

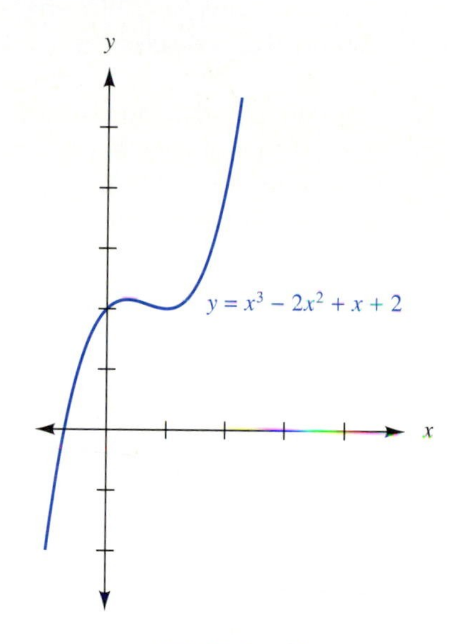

FIGURE 22.21

EXERCISES 22.5

In 1–6, find the x- and y-intercepts, if possible. Also check for symmetry conditions, and find any asymptotes of the graph.

1. $y = 3x^3 - 16x + 1$ (compare with problem 1 in Exercises 22.4)

2. $y = x^3 - 3x + 2$ (compare with problem 2 in Exercises 22.4)

3. $y = x^4 - 4x^2 + 5$ (compare with problem 7 in Exercises 22.4)

4. $y = x^4 - 18x^2 + 10$ (compare with problem 8 in Exercises 22.4)

5. $y = 3x^5 - 5x^3$ (compare with problem 12 in Exercises 22.4)

6. $y = x^5 - 5x^4 + 5x^3$ (compare with problem 13 in Exercises 22.4)

In 7–17, sketch the graph of each function.

7. $y = 3x^3 - 4x$

8. $y = x^3 + 2x^2 - 6x$

9. $y = x^4 - 8x^2 + 16$

10. $y = x^4 + x^3$

11. $y = x^5 - x$

12. $y = \dfrac{3x}{x^2 + 1}$

13. $y = \dfrac{-2x}{x^2 + 1}$

14. $y = \dfrac{3x^2}{x^2 + 1}$

15. $y = \dfrac{-2x^2}{x^2 + 1}$

16. $y = \dfrac{x}{x^2 + 2}$

17. $y = \dfrac{x^2}{x^2 + 2}$

22.6 FINDING MAXIMUM AND MINIMUM VALUES

Sometimes it is enough to find the maximum or minimum value of a function, or the value for which the function reaches its maximum or minimum value, without sketching the curve. **NOTE** ***Pay particular attention to whether such a problem calls for the extremum or the value that* produces *the extremum.***

EXAMPLE 1 What are the dimensions of the rectangle with the largest area that can be enclosed with a perimeter of 16.0 m?

Solution The perimeter P is given by $P = 2L + 2W$. Thus,

$$16.0 = 2L + 2W$$
$$8.0 = L + W$$
$$8.0 - L = W.$$

The area A is given by $A = LW$. Thus,

$$A = L(8.0 - L) = 8.0L - L^2.$$

To find the value of L for which A is at its maximum, we determine when the derivative of A with respect to L is 0. That is

$$dA/dL = 8.0 - 2L$$
$$0 = 8.0 - 2L$$
$$2L = 8.0, \text{ and } L = 4.0 \text{ m}.$$

Since $W = 8.0 - L$, W is 4.0 m also, and we see that the largest area is enclosed by a square having sides of length 4.0 m. ■

EXAMPLE 2 High speed computer systems cost more than lower speed systems. The rate at which the cost increases with respect to increased processor speed limits actual designs. The cost C of a particular system as a function of processor speed is estimated as $C = 10S^2 - 4S + 1600$, where S is the processor speed in MHz. Find the processor speed for which cost is at a minimum.

Solution To find the value of S for which C is at its minimum, we determine when the derivative of C with respect to S is 0. That is

$$dC/dS = 20S - 4$$
$$0 = 20S - 4$$
$$4 = 20S, \text{ and } S = 0.2 \text{ MHz.}$$

(The second derivative test confirms that this value produces a minimum.) ■

EXAMPLE 3 The formula $s = -16t^2 + v_0 t$ gives the distance s (in ft) that an object thrown vertically upward with an initial velocity of v_0 (in ft/s) will travel in time t (in seconds). If a bullet is fired straight up with an initial velocity of 300 ft/s, find the maximum height that the bullet will reach.

Solution The question is to find the maximum height rather than the time for which height is at its maximum. But if we know the time at which the bullet reaches its maximum height, we can determine that height. To find the value of t for which s is at its maximum, we determine when the derivative of s with respect to t is 0.

$$ds/dt = -32t + v_0$$
$$0 = -32t + 300$$
$$32t = 300 \text{ or } t = 9.375 \text{ seconds}$$

To find s, we use the equation $s = -16t^2 + v_0 t$.

$$s = -16(9.375)^2 + 300(9.375) \text{ or } 1400 \text{ ft, to two significant digits.}$$ ■

EXAMPLE 4 For a simply supported beam with a uniformly distributed load, the bending moment M (in ft·lb) at a distance of x (in ft) from the left end of the beam is given by $M = (1/2)(wlx - wx^2)$. Determine the location of the maximum bending moment. In the formula, w is the weight of the load (in lb/ft), and l is the length (in ft) of the beam.

Solution To find the value of x for which M is at its maximum, we determine when the derivative of M with respect to x is 0. That is

$$dM/dx = (1/2)(wl - 2wx)$$
$$0 = (1/2)(wl - 2wx)$$
$$2wx = wl$$
$$x = (wl)/(2w) = l/2$$

The second derivative test confirms that this value produces a maximum, rather than a minimum.

$$d^2M/dx^2 = (1/2)(-2w) = -w.$$ ■

EXAMPLE 5 Equal squares are cut off at each corner of a rectangular piece of cardboard 9 inches wide by 14 inches long. The sides are turned up to form an open-topped box. Find the length x of the sides of the squares that must be cut off to form a box of maximum volume.

Solution Figure 22.22 shows that the length of the box is $14 - 2x$, and the width of the box is $9 - 2x$. The height of the box is x, so

$$V = (14 - 2x)(9 - 2x)(x)$$
$$V = 126x - 46x^2 + 4x^3.$$

The maximum volume occurs when the derivative of the volume function is 0.

$$dV/dx = 126 - 92x + 12x^2$$
$$0 = 63 - 46x + 6x^2$$

The quadratic formula gives

$$x = \frac{46 \pm \sqrt{46^2 - 4(6)(63)}}{12}, \text{ so } x = 5.88 \text{ or } 1.79.$$

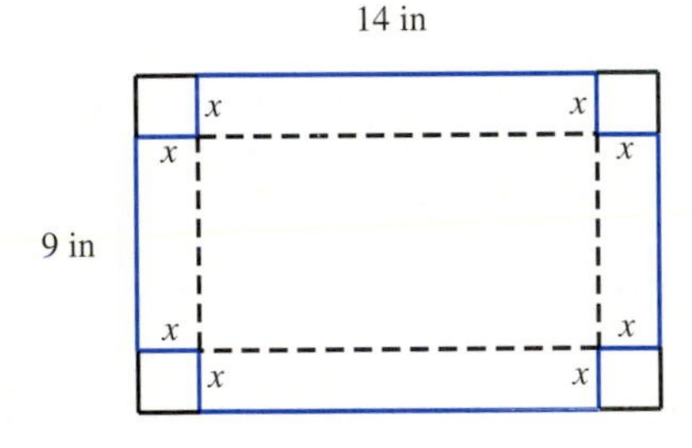

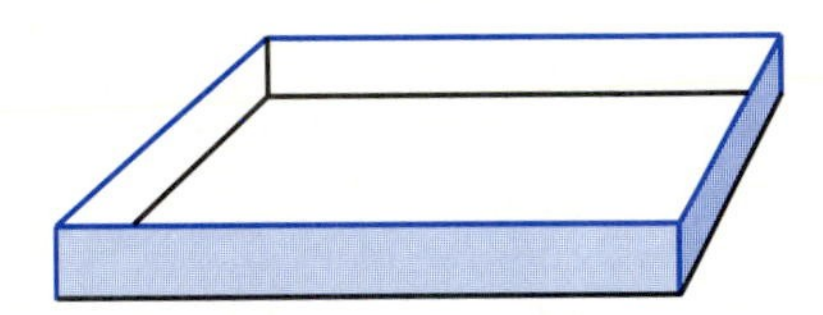

FIGURE 22.22

The rectangle is not wide enough to cut off two 5.88 inch squares, so 1.79 inches must be the answer. The second derivative test confirms that the value produces a maximum, rather than a minimum. ■

EXERCISES 22.6

1. The velocity of a particle (in ft/s) is given by $v = t^2 - 3t + 4$, where t is the time (in seconds) for which it has traveled. Find the time at which the velocity is at a minimum. **4 7 8** 1 2 3 5 6 9
2. The formula $s = (-1/2)gt^2 + v_0t$ gives the distance that an object thrown vertically upward with an initial velocity of v_0 will travel in time t. If the initial velocity is 26.2 m/s and $g = 9.8$ m/s^2, find the time t (in s) at which the object reaches its maximum height. **4 7 8** 1 2 3 5 6 9
3. Find the dimensions that produce the maximum floor area for a one-story house that is rectangular in shape and has a perimeter of 160 ft. **1 2 4 7 8** 3 5

4. A cylindrical tin cup with open top is to hold a volume of 10.0 cm^3. Find the dimensions of the cup for which the surface area is a minimum. (Hint: Write a formula for volume and use $V=10.0$ to solve for h in terms of r. Then write a formula for surface area in terms of r.) 1 2 4 7 8 3 5

5. In an electrical circuit that contains a resistor, an inductor, and a capacitor in series, the magnitude of the impedance Z (in ohms) is given by

$$Z=\sqrt{R^2+\left(\omega L-\frac{1}{\omega C}\right)^2}$$

where R is the resistance (in ohms), L is the inductance (in henries), and C is the capacitance (in farads). As ω changes, Z changes. Find a formula for the minimum value of Z and the value of ω for which it occurs. 3 5 9 1 2 4 6 7 8

6. For a cantilever beam under a concentrated load, the deflection y at a point located a distance of x from the support depends on the distance a of the load from the support and is given by $y=\frac{Pa^2}{6EI}(3x-a)$ when $a<x$. One would expect the minimum deflection to occur when a is 0. Show that this is the case. In the formula, P is the load, E is the modulus of elasticity, and I is the moment of inertia. 1 2 4 8 7 9

7. Ship A is 42 km due west of ship B and is sailing north at the rate of 32 km/hr, while ship B is sailing west at the rate of 22 km/hr, as shown in the accompanying diagram. If the ships continue on their respective courses, find the minimum distance between them and the time at which it occurs. (Hint: If $t=$ time, then $x=42-22t$.) 4 7 8 1 2 3 5 6 9

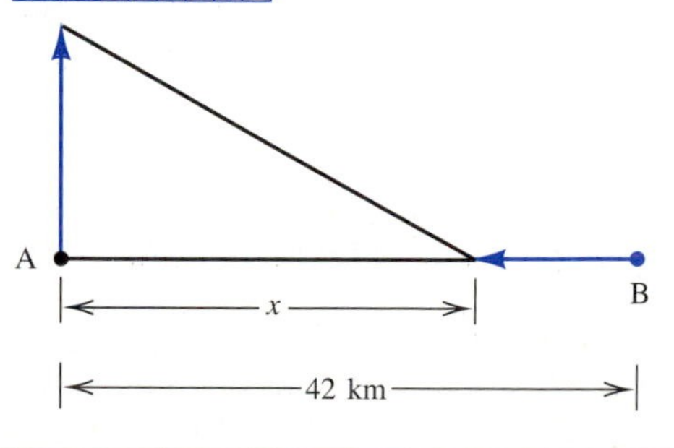

8. For a certain tunnel diode, $I=0.714V^3-1.286V^2+0.564\ V+0.150$ where V is the voltage (in volts) and I is the current (in amperes). Determine the value of V for which I is at its maximum. 3 5 9 1 2 4 6 7 8

9. For a simply supported beam with a load that increases uniformly from left to right, the bending moment M (in ft·lb) at a distance of x (in ft) from the left end is given by $M=(1/6)(wl^2x-wx^3)$. Determine the location of the maximum bending moment. In the formula, w is the rate of load increase (in lb/ft) and l is the length (in ft) of the beam. 1 2 4 8 7 9

10. The price P of a certain computer system decreases immediately after its introduction and then increases. If the price P is estimated by the formula $P=150t^2-2000t+6500$, where t is the time in months from its introduction, find the time until the minimum price is reached. 3 5 9 1 2 4 6 7 8

11. A trough has a cross section that is in the shape of an isosceles triangle with equal sides of length 14.0 inches. Find the length of the base of the triangle if the cross-sectional area is to be at its maximum. 1 2 4 7 8 3 5

12. An architect wants to construct an enclosure with two walls at a right angle, using an existing wall as the third side, as shown in the accompanying diagram. Find the dimensions x and y that will maximize the area enclosed if $x+y=12.0$ m. 1 2 4 7 8 3 5

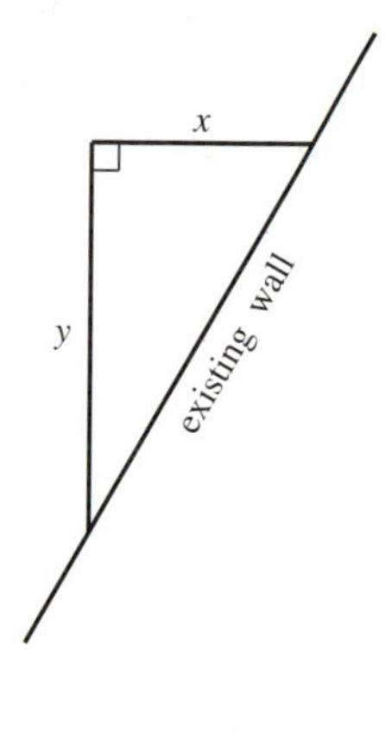

1. Architectural Technology
2. Civil Engineering Technology
3. Computer Engineering Technology
4. Mechanical Drafting Technology
5. Electrical/Electronics Engineering Technology
6. Chemical Engineering Technology
7. Industrial Engineering Technology
8. Mechanical Engineering Technology
9. Biomedical Engineering Technology

13. An engineer notices that certain numbers exceed their squares. Find the number that exceeds its square by the greatest amount. 1 2 6 7 8 3 5 9

14. Equal squares are cut off at each corner of a square piece of cardboard 16 inches on each side. The sides are turned up to form an open-topped box. Find the length of the sides of the square that must be cut off to form a box of maximum volume. 1 2 4 7 8 3 5

15. A box is to be made with a square bottom and vertical sides so that it has a volume of 4.0 ft^3. Find the height of the sides that will satisfy these conditions with the minimum amount of material (i.e., minimum surface area). 1 2 4 7 8 3 5

16. A piece of molding 164 cm long is to be cut to form a rectangular picture frame. What dimensions will enclose the largest area? 1 2 4 7 8 3 5

17. An architect needs to design a rectangular room with an area of 88 ft^2. What dimensions should he use in order to minimize the perimeter? 1 2 4 7 8 3 5

18. A draftsman is working with a rectangle that has two of its vertices on the x-axis and the other two above the x-axis and on the graph of the parabola given by $y = 16 - x^2$, as shown in the accompanying diagram. Find the dimensions of the rectangle that has maximum area. 1 2 4 7 8 3 5

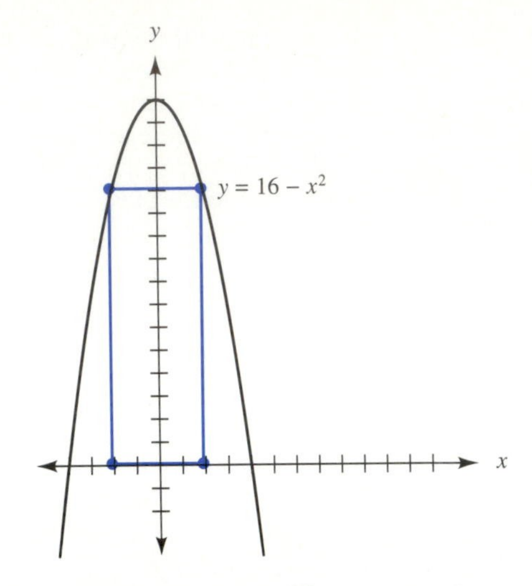

19. A manufacturing plant buys parts from a company that uses the following price structure. If exactly 40 parts are purchased, the price is 50¢ per part. If more than 40 parts are purchased, the price of every part is reduced by 1¢ for each part over 40. Find the maximum price that can be paid under this system. 1 2 6 7 8 3 5 9

20. The owner of a restaurant determines that if the seating capacity is 40 to 80 people, the weekly profit will be \$8 per seat. If the seating capacity is more than 80, however, the weekly profit on every seat is decreased by \$0.04 for each seat over 80. What seating capacity should be used to obtain the greatest weekly profit? 1 2 6 7 8 3 5 9

1. Architectural Technology
2. Civil Engineering Technology
3. Computer Engineering Technology
4. Mechanical Drafting Technology
5. Electrical/Electronics Engineering Technology
6. Chemical Engineering Technology
7. Industrial Engineering Technology
8. Mechanical Engineering Technology
9. Biomedical Engineering Technology

22.7 NEWTON'S METHOD FOR SOLVING EQUATIONS

You have learned algebraic methods for solving linear and quadratic equations. Third and fourth degree equations can also be solved by algebraic techniques, but the general procedures are rather complicated. Often, it is not necessary to obtain an exact solution to these or other types of equations. Approximate solutions are often obtained by an iterative process, such as the bisection method of Section 6.4. Calculus

Newton's method

can be used to develop an iterative method known as *Newton's method.* Newton's method, like the bisection method, requires an initial estimate, which is used to obtain a better approximation, and so on.

Let $f(x) = 0$ be an equation that has s as a solution. Then the graph of $y = f(x)$ crosses the x-axis at s, as shown in Figure 22.23.

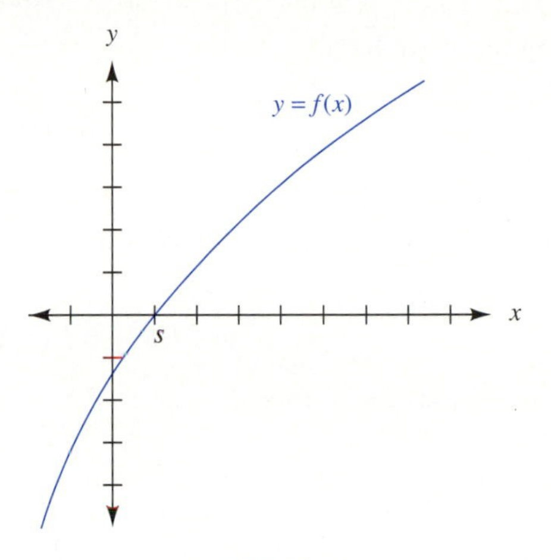

FIGURE 22.23

We choose as an initial estimate, a value of x that is reasonably close to s. Call it x_0. The y-coordinate of this point can be denoted $f(x_0)$. A particularly effective way to choose this initial value is to examine a graph. The method may fail if there is a relative extremum or an inflection point near the root, so a graph will be helpful not only in choosing x_0, but also in determining if the method is feasible for the problem at hand. Newton's method will, however, work for most simple functions. Draw the line tangent to the curve at x_0, as shown in Figure 22.24.

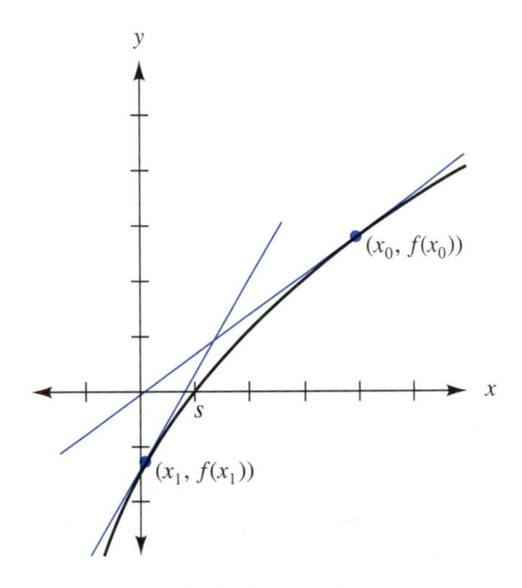

FIGURE 22.24

Since $(x_0, f(x_0))$ is a point on the line, and its slope is $f'(x_0)$, the equation of the line is

$$y - f(x_0) = f'(x_0)(x - x_0).$$

This line crosses the x-axis when y is 0, so we write

$$0 - f(x_0) = f'(x_0)(x - x_0).$$

Solving for x, we have

$$\frac{-f(x_0)}{f'(x_0)} = x - x_0 \text{ or } x = x_0 - \frac{f(x_0)}{f'(x_0)}.$$

We now use this new value of x as the starting point. Call it x_1. The line tangent to the curve at x_1 goes through $(x_1, f(x_1))$ with a slope of $f'(x_1)$. The equation of this tangent line is

$$y - f(x_1) = f'(x_1)(x - x_1).$$

The line crosses the x-axis at

$$x = x_1 - \frac{f(x_1)}{f'(x_1)}.$$

Generalizing from these first two iterations, we have Newton's method.

To solve an equation $f(x) = 0$ by Newton's method:

1. Sketch a graph to locate the vicinity of the solution s.
2. Choose an initial value x_0 (x_n with $n = 0$) reasonably close to s and verify that $f'(x) \neq 0$ and $f''(x) \neq 0$ near s.
3. Calculate the next value of x using

$$x_{n+1} = x_n - \frac{f(x_n)}{f'(x_n)}.$$

4. Repeat step 3 until x_n and x_{n+1} are equivalent when rounded to the desired number of decimal places.

EXAMPLE 1 Solve for positive x and round to hundredths.

$$x^3 + 5x^2 - 3x - 15 = 0.$$

Solution 1. Sketch the graph as in Figure 22.25 in order to find the vicinity of the solution. We have $f(x) = x^3 + 5x^2 - 3x - 15$ and $f'(x) = 3x^2 + 10x - 3$.

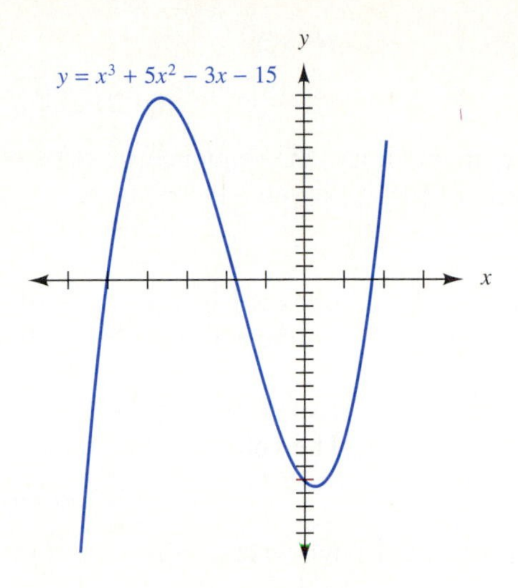

FIGURE 22.25

2. The graph crosses the positive x-axis near 2, so the function takes on a value of 0 near 2. Therefore, we choose $x_0 = 2$ to begin the iteration.

3–4. It is helpful to organize the data in table form.

n	x_n	$f'(x_n)$	$f(x_n)$	$x_n - f(x_n)/f'(x_n)$
0	2	29	7	1.7586207
1	1.7586207	23.864447	0.6268399	1.732354
2	1.732354	23.326691	0.0070717	1.7320509

Thus, $x = 1.73$ when rounded to hundredths. ■

As with the bisection method, it is important to recognize that the answer is approximate. That is, if the intermediate results are rounded differently, the final result may be 1.72 or 1.74, instead of 1.73.

A calculator may be used to do the calculations. It is necessary to record some of the intermediate results. An efficient sequence is to store x_n in calculator memory, compute $f'(x_n)$ and record the value, compute $f(x_n)$, divide by $f'(x_n)$, change the sign, and add to x_n. The keystroke sequence for the first iteration of Example 1 follows.

2 [STO]

Calculate $f'(x_n)$:

[x^2] [×] 3 [+] 10 [×] [RCL] [−] 3 [=] **29**

Calculate $f(x_n)$:

[RCL] [x^2] [×] [RCL] [+] [RCL] [x^2] [×] 5 [−] 3 [×] [RCL] [−] 15 [=] **7**

Calculate $x_n - f(x_n)/f'(x_n)$:

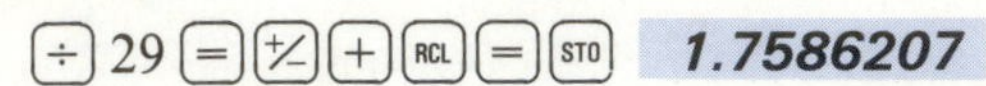

With x_1 in memory, the sequence of steps is repeated from the point where the calculation of $f'(x_n)$ began.

EXAMPLE 2 A tank is to be constructed in the shape of a cylinder 10.0 m high topped by a hemisphere. If the volume is to be 19π m^3, find the radius.

Solution The volume of a hemisphere is given by $V = (2/3)\pi r^3$, and the volume of a cylinder is given by $V = \pi r^2 h$. The volume of the tank, then, is given by

$$V = (2/3)\pi r^3 + \pi r^2 h.$$

Substituting the known values of $h = 10$, and $V = 19\pi$, we have the following.

$$19\pi = (2/3)\pi r^3 + 10\pi r^2$$

$$0 = (2/3)\pi r^3 + 10\pi r^2 - 19\pi \quad \text{Add } -19\pi \text{ to both sides.}$$

$$0 = 2r^3 + 30r^2 - 57 \quad \text{Multiply both sides by } 3/\pi.$$

Let $f(r) = 2r^3 + 30r^2 - 57$. Then $f'(r) = 6r^2 + 60r$.
From Figure 22.26, which shows the graph for $r \geq 0$, we see that the solution is near $r = 1$.

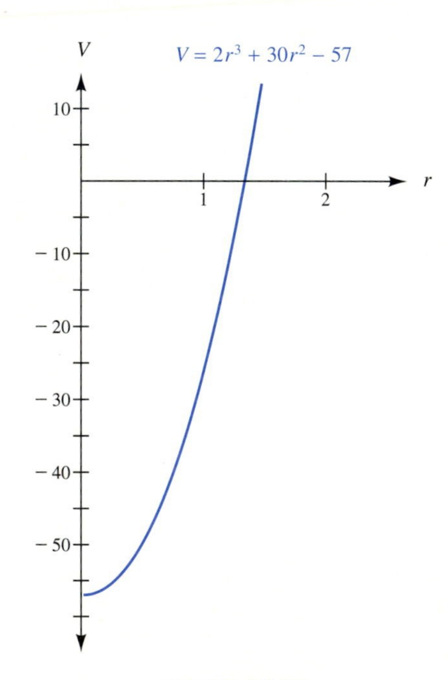

FIGURE 22.26

The solution by Newton's method is shown in the following table.

n	x_n	$f'(x_n)$	$f(x_n)$	$x_n - f(x_n)/f'(x_n)$
0	1	66	−25	1.3787879
1	1.3787879	94.133608	5.2739855	1.3227613
2	1.3227613	89.863861	0.1197859	1.3214283

To three significant digits, the answer is 1.32 m. ■

EXERCISES 22.7

In 1–12, solve each equation for x and round to thousandths.

1. $x^3 - 3x + 1 = 0$ near $x = 2$

2. $4x^3 - 27x + 3 = 0$ near $x = 3$

3. $2x^3 - 3x^2 + 4 = 0$ near $x = -1$

4. $3x^3 - 16x + 2 = 0$ near $x = 2$

5. $x^3 - 12x + 5 = 0$ near $x = 3$

6. $2x^3 - 3x^2 + 3 = 0$ near $x = -1$

7. $x^4 - 18x^2 + 8 = 0$ near $x = -1$

8. $x^4 - 4x^2 + 3 = 0$ near $x = 2$

9. $x^3 + 4x^2 - 11x + 8 = 0$ for $x < 0$

10. $x^3 - x^2 - 5x + 3 = 0$ for $x < 0$

11. $4x^3 + x^2 - 4x + 2 = 0$ for $x < 0$

12. $4x^3 + 7x^2 - 6x - 2 = 0$ for $x > 0$

13. Find the dimensions of a cube that, when topped by a pyramid 8 inches high, forms a figure with volume 550 in³. The volume of a pyramid is given by $V = (1/3)Bh$, where B is the area of the base and h is the height. Use 8 for the initial estimate. 1 2 4 7 8 3 5

14. A series of squares, each having sides 1 cm longer than the previous square, are cut from a sheet of plastic. The areas, then, are 1^2, 2^2, 3^2, 4^2, and so on. The sum of the areas of the first n squares is given by $A = (1/6)(2n^3 + 3n^2 + n)$. Assume that squares may be pieced together so that you are concerned only with total area of plastic used. How many squares can be cut from a sheet of plastic that measures 25 cm by 26 cm? Use 10 for the initial estimate. 1 2 4 7 8 3 5

15. A box is to be constructed from a piece of metal 10.0 cm square by cutting a square from each corner and folding up the sides. What size square should be removed if the volume of the box is to be 72.0 cm³? Use 3 for the initial estimate. 1 2 4 7 8 3 5

16. To solve a system of n equations with n variables, the number of multiplications and divisions required by the method of Gaussian elimination is given by $M = (1/6)(2n^3 + 9n^2 + n)$. If a computer is capable of performing 1,000,000 multiplications and divisions per second, what is the largest number of equations that can be solved in three ten-thousandths of a second? Use 20 for the initial estimate. 3 5 9

17. For a cantilever beam under a concentrated load, the deflection y (in mm) is given by $y = [1/(6EI)](Px^3 - 3Pax^2)$. For this problem, $P = 2.00$ kg (the weight of the load), $a = 30.00$ mm (the distance of the load from the support), $E = 4000$ kg/mm² (the modulus of elasticity), $I = 5.00$ mm⁴ (the moment of inertia), and x is the distance from the support at which the deflection is measured. With the values given, the equation is $42000 = x^3 - 90x^2$. Solve it for x. Use 95 for the initial estimate. 1 2 4 8 7 9

1. Architectural Technology
2. Civil Engineering Technology
3. Computer Engineering Technology
4. Mechanical Drafting Technology
5. Electrical/Electronics Engineering Technology
6. Chemical Engineering Technology
7. Industrial Engineering Technology
8. Mechanical Engineering Technology
9. Biomedical Engineering Technology

22.8 CHAPTER REVIEW

KEY TERMS

22.1 normal

22.2 rectilinear
second derivative
velocity
acceleration
curvilinear

22.4 relative maximum
relative minimum
relative extrema
increasing
decreasing
critical numbers
critical points
first derivative test
concave upward
concave downward
second derivative test
inflection point

22.5 symmetric with respect to the y-axis
symmetric with respect to the x-axis
symmetric with respect to the origin

22.7 Newton's method

SYMBOLS

y''	the second derivative of y
$f''(x)$	the second derivative of $f(x)$ with respect to x
d^2y/dx^2	the second derivative of y with respect to x
v_x, v_y	magnitude and sign of the horizontal and vertical components of velocity
a_x, a_y	magnitude and sign of the horizontal and vertical components of acceleration

RULES AND FORMULAS

$|\mathbf{v}| = \sqrt{v_x^2 + v_y^2}$

$\tan \theta_v = v_y/v_x$

$|\mathbf{a}| = \sqrt{a_x^2 + a_y^2}$

$\tan \theta_a = a_y/a_x$

GEOMETRY CONCEPTS

$V = (1/3)\pi r^2 h$ (volume of a cone)

$V = \pi r^2 h$ (volume of a cylinder)

$V = (4/3)\pi r^3$ (volume of a sphere)

$A = s^2$ (area of a square)

$V = lwh$ (volume of a rectangular solid)

$A = lw$ (area of a rectangle)

$P = 2L + 2W$ (perimeter of a rectangle)

$a^2 + b^2 = c^2$ Pythagorean theorem

$V = (1/3)\ Bh$ (volume of a pyramid)

SEQUENTIAL REVIEW

(Section 22.1) In 1–4, find the equation of the line tangent to the indicated curve at the given point.

1. $y = 3x^2 - 2$ at $(1, 1)$

2. $y = 2x^2 - x + 1$ at $(0, 1)$

3. $x^2 + 3y^2 = 4$ at $(-1, 1)$

4. $xy = 3$ at $(1, 3)$

In 5–7, find the equation of the line normal to the indicated curve at the given point.

5. $y = 2x^3 + x^2 - 3$ at $(-1, -4)$

6. $xy = 6$ at $(2, 3)$

7. $x^2 + y^2 = 13$ at $(3, -2)$

(Section 22.2)

8. Find the velocity at time $t = 2$ for a particle if $s = -2t^2 + 5$. In the formula, s is the distance traveled (in ft), and t is the time (in seconds).

9. Find the velocity at time $t = 3$ for a particle if $s = 1/(t - 2)$. In the formula, s is the distance traveled (in ft), and t is the time (in seconds).

10. Find the acceleration at time $t = 1$ for a particle if $s = 3t^2 + 2t$. In the formula, s is the distance traveled (in ft), and t is the time (in seconds).

11. Find the acceleration at time $t = 4$ for a particle if $s = \sqrt{t^2 - 7}$. In the formula, s is the distance traveled (in ft), and t is the time (in seconds).

12. If the parametric equations $x = 3t^3$ and $y = 1/(t + 1)$ specify the path of a moving particle, find the magnitude and direction of the velocity when t is 1.

In Problems 13 and 14, the parametric equations $x = (1/8)t^3$ and $y = 1 - \sqrt{t}$ specify the path of a moving particle.

13. Find the magnitude and direction of the velocity when t is 4.

14. Find the magnitude and direction of the acceleration when t is 4.

(Section 22.3)

15. If $z = 7x^4$, find the rate at which z changes when x is 2, if x increases at the rate of 3 m/s.

16. If $z = x^2y$, find the rate at which z changes when x is 2 and y is 3, if x and y change at a constant rate of 6 cm/s.

17. A particle follows a path given by $y = 2x^3 + x$. If the horizontal component of velocity is a constant 4 m/s, find the magnitude of the velocity when the particle is at $(1, 3)$.

18. A particle follows a path given by $y = 1/x$. If the vertical component of velocity is a constant 2 m/s, find the magnitude of the velocity when the particle is at $(1, 1)$.

19. A particle follows a path given by $y = 2x^3 - x$. If the horizontal component of velocity is given by $v_x = 3x$, find the magnitude of the vertical component of velocity when the particle is at $(1, 1)$.

20. Find the magnitude and direction of the acceleration of the particle in problem 17 when it is at $(1, 3)$.

(Section 22.4)

21. Determine whether the function defined by $y = 3x^3 + 2x^2 - 1$ is increasing or decreasing at $(-1, -2)$.

22. Determine whether the function defined by $y = 2x^4 + 3x$ is increasing or decreasing at $(1, 5)$.

In 23 and 24, find the critical points for the graph of each function.

23. $y = 4x^3 + 15x^2 - 18x$

24. $y = x^5 - 80x$

25. Sketch the graph of $y = x^3 - 6x^2 + 3$.

26. Sketch the graph of $y = x^4 - 1$.

27. Sketch the graph of $y = x^2/(x^2 + 1)$.

(Section 22.5)

In 28 and 29, find the inflection points for the graph of each function.

28. $y = x^3 - 9x^2 + 24x - 6$

29. $y = x^3 - 2x^2 - 7$

30. Determine whether the graph of the function defined by $y = 3x^3 - 2x^2 - 7$ is concave upward or concave downward at $(-1, -2)$.

31. Determine whether the graph of the function defined by $y = x^2/(x^2 + 1)$ is concave upward or concave downward at $(1, 1/2)$.

32. Refine the graph that you drew in problem 25 to show the inflection points.

33. Refine the graph that you drew in problem 27 to show the inflection points.

(Section 22.6) *In 34 and 35, determine the maximum value of y (if there is one) for each function.*

34. $y = -x^2 + 5x - 6$

35. $y = x^4 - 6x + 1$

In 36 and 37, determine the minimum value of y (if there is one) for each function.

36. $y = 2x^2 - 8x + 7$

37. $y = x^4 - 3x^2 + 2$

In 38 and 39, determine the value of x that produces the maximum value for y.

38. $y = -x^2 + 7x - 3$

39. $y = 3x^4 - x^3 + 5$

(Section 22.7)

40. Newton's method for solving an equation gives $x_n =$ ____________________________.

In 41–45, solve each equation by Newton's method and express the answer to tenths.

41. $x^3 - 4x^2 + 3 = 0$ near $x = 3$

42. $x^3 - 3x + 4 = 0$ $(-3 < x < -2)$

43. $x^3 - 29 = 0$

44. $x^4 - 68 = 0$ (x is positive)

45. $x^4 + 2x^3 - 1$ $(0 < x < 1)$

APPLIED PROBLEMS

1. The cables of a suspension bridge take the shape of a parabola. At each point, there is a tension in the cable directed along the tangent to the curve. If the equation for the shape of the cable is $y = x^2/240$, find the direction of the tension (specify an angle with the horizontal) at $x = 40$. 1 2 4 8 7 9

2. For a cantilever beam under a distributed load, the deflection y depends on the length of the load b and is given by $y = [1/(24EI)](wb^4 - 4wb^3x)$ when $x > b$. One would expect the minimum deflection to occur when $b = 0$. Show that this is the case. In the formula, P is the load, E is the modulus of elasticity, I is the moment of inertia, and x is the distance of the point of interest from the support. 1 2 4 8 7 9

3. When electric power is removed from a certain large motor, its rotor's angular velocity ω (in radians/second) is given by $\omega = (175 \times 10^{-6})(t - 600)^2$. Find the time (in seconds) at which the velocity is at its minimum. 4 7 8 1 2 3 5 6 9

4. In highway traffic studies, it has been observed that the volume of traffic q (vehicles per hour) is related to the density of traffic k (vehicles per mile) by the formula $q = k(50 - k/3.5)$. Determine the maximum value of q. 1 2 6 7 8 3 5 9

5. A particle follows a path given by $y = x^3 + 2x - 1$. Find the x-coordinate (to the nearest tenth) when the y-coordinate is 3. 4 7 8 1 2 3 5 6 9

6. In electronics, sometimes it is necessary to find the line tangent to the graph of current versus voltage for a device. Find the equation of the line tangent to the diode current curve given by $I = 0.00016\ V^{3/2}$ at the point where V is 20.0 volts. 3 5 9 1 2 4 6 7 8

7. Find the equation of the line tangent to the following transistor curve when V is 0 volts. $I = 25(1 + V/3.0)^2$ 3 5 9 4 7 8

8. A ladder leans against a storage tank as shown. How far is the end of the ladder from the center line of the tank? (Hint: Find the equation of a straight line tangent to the circle at a point.) 1 2 4 7 8 3 5

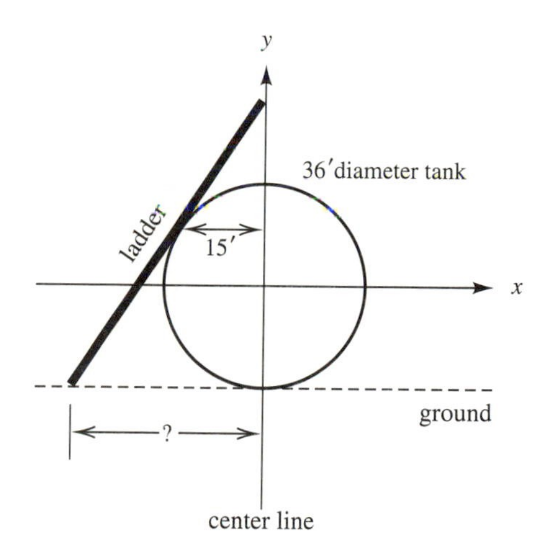

9. The distance s (in inches) between the drill and a piece of work in a drill press is given by the equation $s = 1/(3t^2 - 2t + 1)$ where t is time (in seconds). How fast is this drill moving when t is 1/4 second? Give your answer to three significant digits. 4 7 8 1 2 3 5 6 9

10. The formula $v = 331.5\sqrt{1 + T/273}$ gives the velocity v (in m/s) of sound at a temperature T (in degrees Celsius). How will a sound wave's velocity change with time t (in seconds) if the temperature changes according to the equation $T = 273 + 5t$, and t is 6.0 seconds? Give your answer to three significant digits. 4 7 8 1 2 3 5 6 9

11. Motion picture analysis of the movement of two masses shows the distance s (in cm) between them to be given by the equation $s = 0.05400t^3 - 0.9343t^2 + 3.920t + 1.060$, where t is time (in seconds). For $t = 13.00$ seconds, find
    (a) the distance between the masses.
    (b) the relative velocity.
    (c) the relative acceleration.
    4 7 8 1 2 3 5 6 9

12. A plane dives with an instantaneous velocity of 200 mph along a hyperbolic path given by $y = 500{,}000/x$. Both x and y are given in feet. What is the plane's velocity in the x-direction when it is 500 ft above the ground? Give your answer to the nearest mile per hour. (Hint: Find the angle θ between the velocity vector and the horizontal using dy/dx.) 4 7 8 1 2 3 5 6 9

13. The end of a rotating machine part travels clockwise around the circle $x^2 + y^2 = 65$ at a rate of 35 in/s. Find its velocity in the x-direction as it passes the point (4, 7). Give your answer to three significant digits. (Hint: Find the angle between the velocity vector and the horizontal using dy/dx.) 4 7 8 1 2 3 5 6 9

14. A hurricane's distance s (in miles) from a tracking station is observed to follow the equation $s = 2.40t^2 - 66.9t + 679$ when t is time (in hours). How fast is the hurricane traveling toward the east if it proceeds away from the station at an angle of 63.0° north of east 20.0 hours after the observation begins? 4 7 8 1 2 3 5 6 9

15. The clamping force F (in lb) of a certain vice is given by $F = \dfrac{100(10 - x/2)}{\sqrt{40x - x^2}}$ if x is the width of the

1. Architectural Technology
2. Civil Engineering Technology
3. Computer Engineering Technology
4. Mechanical Drafting Technology
5. Electrical/Electronics Engineering Technology
6. Chemical Engineering Technology
7. Industrial Engineering Technology
8. Mechanical Engineering Technology
9. Biomedical Engineering Technology

jaw opening (in inches). How fast (in lb/s) does F change if x changes at the rate of -2 in/s when the distance between the jaws is 3 inches? Give your answer to three significant digits.
4 7 8 1 2 3 5 6 9

16. A 20-foot ladder leans against a building with its base 12 feet from the wall. A worker pulls on the ladder's base so that its distance x (in ft) from the wall is given by $x = 3t^2 + 12$ if t is in seconds. How fast will the top of the ladder be falling after 1.00 second? Give your answer to the nearest tenth of a foot per second. (Hint: $y = \sqrt{400 - x^2}$.)
4 7 8 1 2 3 5 6 9

17. The moment of inertia J (in in^4) of a solid circular shaft is given by $J = \pi d^4/32$, where d is the shaft diameter (in inches). Suppose heat changes the diameter of a 1-inch test shaft by an amount d (in inches) given by $d = (3.6 \times 10^{-6})T^2 + 1$, where T is the temperature (in degrees Celsius). Find the rate of change of J with respect to T when T is 115° C. 1 2 4 8 7 9

18. The formula $y = -3x^5 + 10L^2x^3 - 7L^4x$ gives the deflection y (in inches) of a certain loaded elastic beam of length L at a distance of x (in inches) from the end. For what value of x (in terms of L) is the deflection at its maximum? Give your answer to three significant digits.
1 2 4 8 7 9

19. One half of a hollow concrete cylinder is used to make an arched bridge across a stream. The distance D (in ft) from the center of gravity of this arch to the water is given by $D = \frac{4}{3\pi}\left[\frac{R^3 - r^3}{R^2 - r^2}\right]$.

In the formula, R is the outside radius (in ft) of the arch and r is the inside radius (in ft). What value of r would this arch have if $R = 12$ ft and $D = 7$ ft? (Hint: Use Newton's method to solve for r to three significant digits, knowing that the answer is near 10.) 1 2 4 7 8 3 5

1. Architectural Technology
2. Civil Engineering Technology
3. Computer Engineering Technology
4. Mechanical Drafting Technology
5. Electrical/Electronics Engineering Technology
6. Chemical Engineering Technology
7. Industrial Engineering Technology
8. Mechanical Engineering Technology
9. Biomedical Engineering Technology

RANDOM REVIEW

1. Consider $y = x^3 - 3x + 4$.
   (a) Find the critical points. **(b)** Sketch the graph.
2. Solve $x^5 - 37 = 0$ by Newton's method if $x > 0$. Give the answer to tenths.
3. The parametric equations $x = \sqrt{t}$ and $y = t^3$ specify the path of a moving particle. Find the magnitude and direction of the acceleration when t is 1.
4. Find the equation of the line normal to the curve $xy = 6$ at (2, 3).
5. A particle follows a path given by $y = x^3 - 2x$. If the horizontal component of velocity is a constant 2 m/s, find the magnitude of the velocity when the particle is at (2, 4).
6. Find the inflection points for the graph of the curve given by $y = x^3 + 3x + 4$.
7. Find the velocity at time $t = 2$ for a particle if $s = -3t^2 + 4$. In the formula, s is the distance traveled (in ft) and t is the time (in seconds).
8. Determine the value of x that produces a minimum for y if $y = x^4 - 6x + 1$.
9. Find the magnitude and direction of the acceleration of the particle in problem 5.
10. Find the equation of the line tangent to the curve given by $2x^2 + 3y^2 = 14$ at the point (1, 2).

CHAPTER TEST

1. Find the equation of the line tangent to the curve given by $4x^2 + 9y^2 = 25$ at the point (2, 1).
2. Find the equation of the line normal to the curve given by $y = 2x^3 - 3x$ at the point (2, 10).
3. For a certain particle, the distance traveled (in ft) is given by $s = \sqrt{t^2 - 7}$, if t is the time (in seconds) of travel.
   (a) Find the magnitude of the velocity when t is 4.
   (b) Find the magnitude of the acceleration when t is 4.
4. The parametric equations $x = 1/t$ and $y = 1 - t^2$ specify the path of a moving particle.
   (a) Find the magnitude and direction of the velocity when t is 2.
   (b) Find the magnitude and direction of the acceleration when t is 2.
5. A certain particle follows a path given by $y = x^3 - 2x^2$. If the vertical component of velocity is a constant 3 m/s, find the magnitude of the velocity when the particle is at (2, 0).
6. Consider $f(x) = x/(x^2 + 1)$.
   (a) Find the critical points. **(b)** Sketch the graph.
7. Consider $f(x) = 2x^4 - x + 3$.
   (a) Find the critical points. **(b)** Find the inflection points.
8. Solve the equation $x^3 - 2x + 3 = 0$ by Newton's method and express the answer to tenths, if x is near -2.
9. The formula $P = I^2 R$ gives the power P (in watts) dissipated through a resistance R (in ohms) when the current is I (in amperes). If the resistance is a constant 480 Ω, find the rate of change of power with respect to time if the current is changing at a rate of 0.1 A/s when $I = 0.25$ A.
10. A woman 5 feet tall walks directly away from the base of a lamppost that is 25 feet tall at the rate of 4 ft/s. At what rate is the length of her shadow increasing? (Hint: Let y be her distance from the lamp at time t as shown in the accompanying diagram and set up a proportion using similar triangles.)

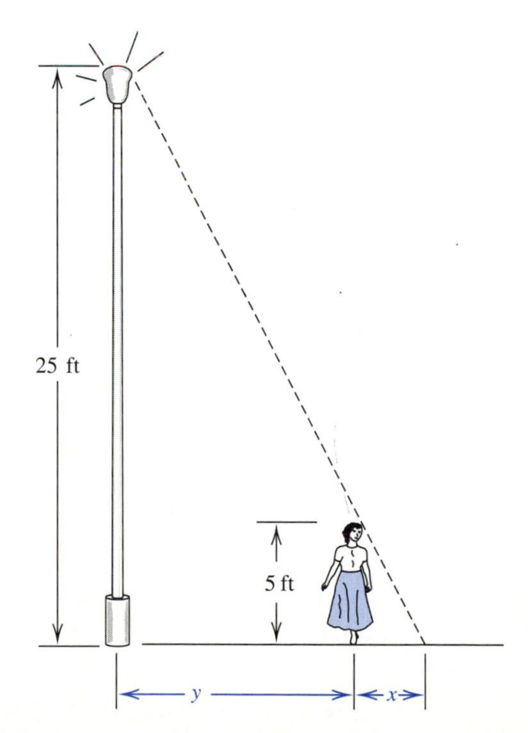

EXTENDED APPLICATION: ARCHITECTURAL ENGINEERING TECHNOLOGY

THE MAXIMUM DEFLECTION OF A BEAM

In engineering, assumptions about physical conditions are often introduced to simplify calculations. Each simplification, however, introduces a source of error. The errors may balance out, or they may accumulate and become significant. A safety factor is often used to allow for such errors. In beam analysis, the following assumptions are often made.

(a) All forces lie in the same plane along the length of the beam, through the centroids of the cross section.
(b) The beam has uniform cross sections throughout its length.
(c) A concentrated load acts at a single point, and a distributed load acts along a line.
(d) The forces are applied without impact.

Consider a beam that is clamped at one end and supported at the same height at the other end, as shown in Figure 22.27. Even if the beam bears no load, other than its own weight, it sags between the supports. If the supports were of the same type, the maximum deflection would occur at the midpoint. Because the supports are of different types, however, the shape of the beam is not symmetrical about its midpoint.

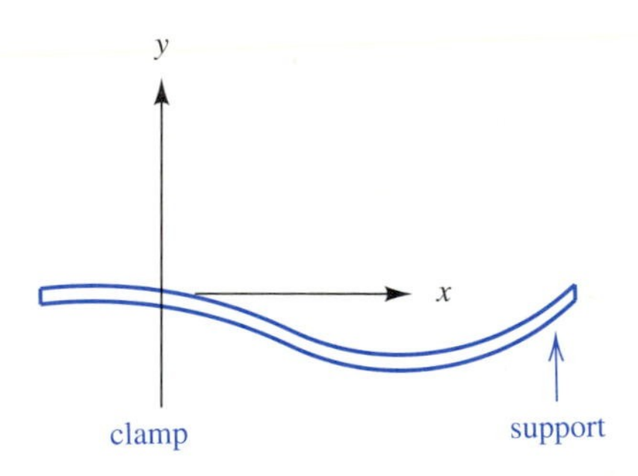

FIGURE 22.27

The amount of deflection y at a distance x from the clamped end is given by

$$y = \frac{wx^2}{48EI}(2x^2 - 5xl + 3l^2) = \frac{1}{48EI}(2wx^4 - 5wx^3l + 3wx^2l^2)$$

where w is the weight of the beam, E is the modulus of elasticity, I is the moment of inertia, and l is the length of the beam.

To determine where the maximum deflection occurs, we find the value of x for which the derivative is zero.

$$y' = \frac{1}{48EI}(8wx^3 - 15wx^2l + 6wxl^2)$$

$$0 = \frac{1}{48EI}(8wx^3 - 15wx^2l + 6wxl^2)$$

$$0 = 8wx^3 - 15wx^2l + 6wxl^2$$

$$0 = x(8wx^2 - 15wxl + 6wl^2)$$

$$x = 0 \text{ or } 8wx^2 - 15wxl + 6wl^2 = 0$$

When x is 0, there is no deflection, so this value is not where the maximum deflection occurs. Using the quadratic formula with $a = 8w$, $b = -15wl$, and $c = 6wl^2$, we have the following.

$$\begin{aligned} x &= \frac{15wl \pm \sqrt{225w^2l^2 - 192w^2l^2}}{16w} \\ &= \frac{15wl \pm \sqrt{33w^2l^2}}{16w} \\ &= \frac{15wl \pm wl\sqrt{33}}{16w} \\ &= \frac{15l \pm l\sqrt{33}}{16} \\ &= 1.29l \text{ or } 0.57l \end{aligned}$$

If l is the length of the beam, it makes no sense to measure deflection at $1.29l$, so we suspect that the maximum deflection occurs at $0.57l$. The second derivative test shows that this interpretation is correct.

$$y'' = \frac{1}{48EI}(24wx^2 - 30wxl + 6wl^2)$$

When x is $0.57l$,

$$\begin{aligned} y'' &= \frac{1}{48EI}[24w(0.57\,l)^2 - 30w(0.57\,l)l + 6wl^2] \\ &= \frac{-3.3024\,wl^2}{48EI}. \end{aligned}$$

The variables are all positive, so the second derivative is negative, and the value does produce a maximum.

CHAPTER 23 INTEGRALS

In the seventeenth century, two of the motivating factors behind the development of calculus were the need to understand motion, particularly the motion of the planets, and the need to work with tangents to curves. As you have seen, differentiation may be used to solve problems of both types. The problems of finding areas and volumes were also important topics in the seventeenth century. Surprisingly these problems are related to the other two. As you will see, areas, volumes, and many other quantities may be found by a process that is the reverse of differentiation.

23.1 AREA UNDER A CURVE

You have learned to compute the area of a straight-sided figure, such as a triangle or a rectangle. Other straight-sided figures can be partitioned into rectangles or triangles so that the total area can be computed by adding the areas of the individual parts, as shown in Figure 23.1.

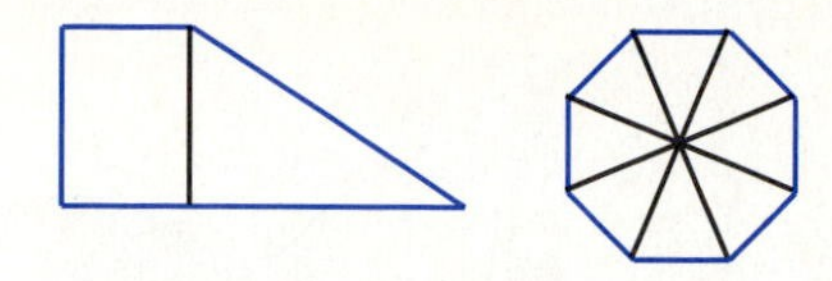

FIGURE 23.1

To find the area of a figure with a curved boundary, we also partition the figure into smaller parts. We begin in the coordinate plane by considering the area of a figure bounded by the x-axis, the lines $x = a$ and $x = b$, and the curve $y = f(x)$, as shown in Figure 23.2. Sometimes the phrase *area under the curve* is used to indicate the area of a region that is between a curve and the x-axis.

area under a curve

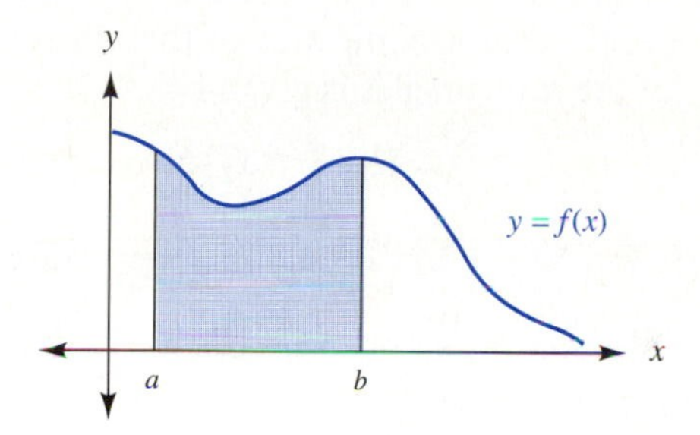

FIGURE 23.2

We can obtain an approximation for such an area in the following manner. Divide the interval from a to b into n equal lengths, marking the division points x_1, $x_2, x_3, \ldots, x_{n-1}$. We also use $a = x_0$ and $b = x_n$, for consistency in the notation. Then draw a vertical line segment up to the curve at each x_i, as shown in Figure 23.3.

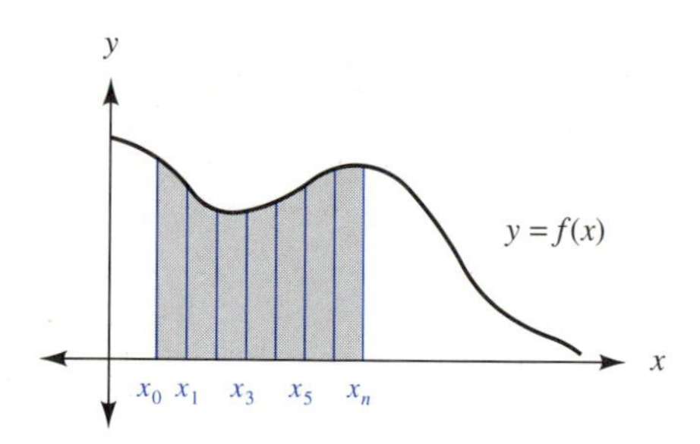

FIGURE 23.3

Each little strip is bounded by straight lines on three sides. The area of each strip is approximately equal to the area of the rectangle that has three of its sides along those lines and that touches the curve at its upper right-hand corner as shown in Figure 23.4.

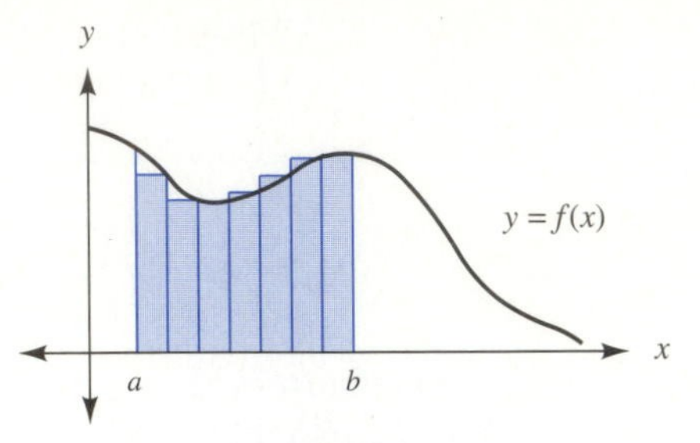

FIGURE 23.4

The length of each rectangle is its height, which in Figure 23.4, is given by the y-coordinate of the curve on the right-hand side of the rectangle. For the first rectangle, the length is $f(x_1)$, for the second, it is $f(x_2)$, and so on. If we let Δx denote the width of each rectangle, the area of the nth rectangle is given by $f(x_n)\Delta x$. The total area A_n of the n rectangles is given by

$$A_n = f(x_1)\Delta x + f(x_2)\Delta x + f(x_3)\Delta x + \cdots + f(x_n)\Delta x.$$

The notation $\sum_{i=1}^{n} f(x_i)\Delta x$ is used to abbreviate this sum. The symbol $\sum$ is the capital Greek letter sigma, and the notation means to find the sum of the quantities that are obtained by replacing i in the expression for $f(x_i)$ with each integer from 1 to n, inclusive.

Since we divided the interval from a to b into n parts, each of width Δx, we have $\Delta x = (b - a)/n$. The more rectangles we use, the narrower they are, and the more closely the total area of the rectangles A_n approximates the actual area A. Figure 23.5 shows an example with five rectangles and with ten.

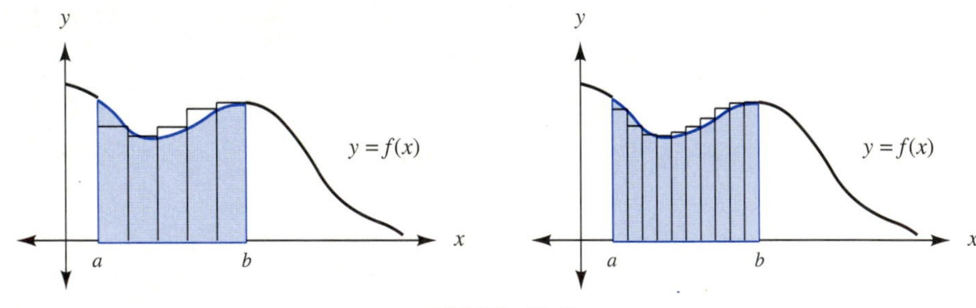

FIGURE 23.5

The condition $n \to \infty$ implies the condition $\Delta x \to 0$. Thus, it is reasonable to say that the area A is given by

$$A = \lim_{n\to\infty} A_n = \lim_{\Delta x\to 0} \sum_{i=1}^{n} f(x_i)\Delta x.$$

This limit is called an *integral* and is denoted $\int_a^b f(x)dx$. The symbol $\int$, called an *integral sign,* is an elongated S for "sum." The numbers a and b are called the *limits of integration,* but the word limit in this context merely means the end-points of the

limits of integration

interval. In our discussion, we have used the function value at the right-hand side of each strip. This point was chosen arbitrarily. Although it is not obvious, our conclusion would have been the same had we used the function value at any point in the interval.

> **DEFINITION** For a continuous function defined by $y = f(x)$ in the interval $a \le x \le b$, the **integral** of $f(x)$ from a to b is given by
>
> $$\int_a^b f(x)\,dx = \lim_{\Delta x \to 0} \sum_{i=1}^{n} f(x_i)\Delta x.$$

If $f(x)$ is nonnegative for all values between a and b, as it is in Figures 23.2–23.5, the integral can be interpreted as the area of the region bounded by the x-axis, the lines $x = a$ and $x = b$, and the graph of $y = f(x)$. If $f(x)$ is negative for all values between a and b, the integral is the negative of the area.

Eventually, you will learn several techniques for evaluating integrals, but many functions do not lend themselves to easy calculations. For these problems, integrals are found using procedures that lead to approximate solutions. We begin our study of integration, the process of finding integrals, by considering approximations.

EXAMPLE 1 Find an approximate value for $\int_1^4 (x + 3)\,dx$

(a) using three rectangles. **(b)** using six rectangles.

(c) using the formula for the area of a trapezoid: $A = (1/2)h(b_1 + b_2)$.

Solution The graph of $y = x + 3$ is a straight line. We interpret the integral as the area of the figure bounded by the x-axis, the lines $x = 1$ and $x = 4$, and the graph of $y = x + 3$.

(a) Divide the interval from 1 to 4 into three equal lengths as shown in Figure 23.6.

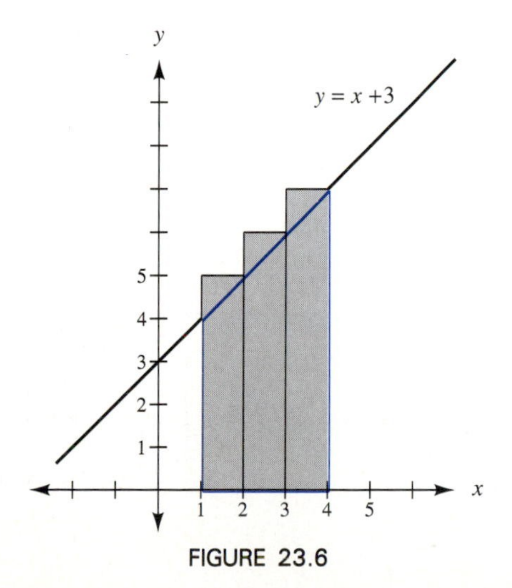

FIGURE 23.6

The points (2, 5), (3, 6), and (4, 7) are on the curve, so the heights of the rectangles are 5, 6, and 7. The width of each rectangle is 1, and the sum of the areas of the three rectangles is given by $A_3 = 5(1) + 6(1) + 7(1) = 18$.

(b) Divide the interval from 1 to 4 into six equal lengths as shown in Figure 23.7.

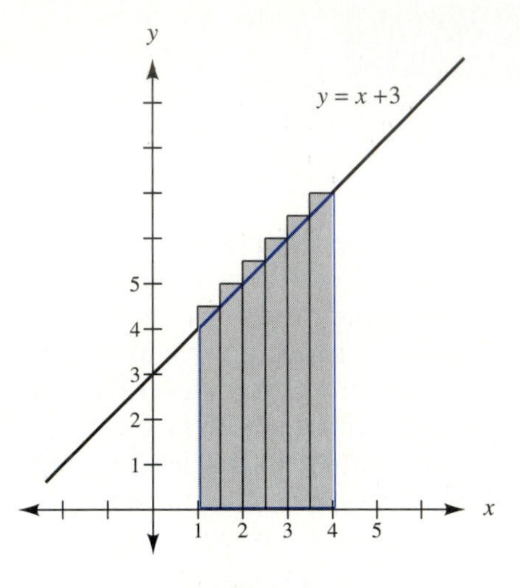

FIGURE 23.7

The points (3/2, 9/2), (2, 5), (5/2, 11/2), (3, 6), (7/2, 13/2), (4, 7) are on the curve, so the heights of the rectangles are 9/2, 5, 11/2, 6, 13/2, and 7. The width of each rectangle is 1/2, and the sum of the areas of the six rectangles is given by

$$\begin{aligned} A_6 &= (9/2)(1/2) + 5(1/2) + (11/2)(1/2) + 6(1/2) + (13/2)(1/2) + 7(1/2) \\ &= 9/4 + 5/2 + 11/4 + 6/2 + 13/4 + 7/2 \\ &= 9/4 + 10/4 + 11/4 + 12/4 + 13/4 + 14/4 \\ &= 69/4 \text{ or } 17\ 1/4. \end{aligned}$$

(c) The height, h, is the distance between parallel sides and is 3 in this case. The two bases have lengths 4 and 7. Thus,

$$\begin{aligned} A &= (1/2)(3)(4 + 7) \\ &= (3/2)(11) \\ &= 33/2 \text{ or } 16\ 1/2. \end{aligned}$$

Notice that this answer is an exact answer, rather than an approximation. ■

EXAMPLE 2 Find an approximate value for $\int_0^3 x^2\, dx$, using four rectangles.

Solution The graph is a parabola, as shown in Figure 23.8. Divide the interval from 0 to 3 into four equal lengths.

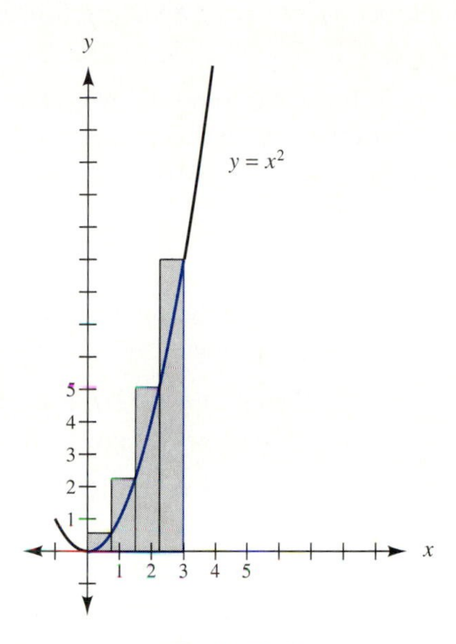

FIGURE 23.8

The points (3/4, 9/16), (3/2, 9/4), (9/4, 81/16), and (3, 9) are on the curve, so the heights of the rectangles are 9/16, 9/4, 81/16, and 9. The width of each rectangle is 3/4, and the sum of the areas of the four rectangles is given by

$$A_4 = (9/16)(3/4) + (9/4)(3/4) + (81/16)(3/4) + (9)(3/4).$$

Using a calculator to do the arithmetic, we obtain 12.65625. Since the figure is only an approximation, we might round it off and say the area is about 13. ■

The interpretation of the integral $\int_a^b f(x)\,dx$ as an area is reasonable for a function whose graph is above the x-axis for all values between a and b. If the graph is below the x-axis for all values between a and b, then $f(x) < 0$, and $\lim_{\Delta x \to 0} \sum_{i=1}^{n} f(x_i)\Delta x < 0$, so $\int_a^b f(x)\,dx < 0$. The area is given by $-\int_a^b f(x)\,dx$. That is, the area must be a positive number, but the integral can be negative. When the graph is above the x-axis for some values and below it for others, the area is considered in parts. Example 3 illustrates a problem of this type.

EXAMPLE 3 **(a)** Use four rectangles to find the approximate area bounded by the x-axis, the line $x = 2$, and the graph of $y = x^3 - 2x^2 - x + 2$ in the first and fourth quadrants.

(b) Use four rectangles to find an approximate value for $\int_0^2 (x^3 - 2x^2 - x + 2)\,dx$.

Solution **(a)** Figure 23.9 shows the graph. Between 0 and 1, the curve is above the axis, so the integral gives the area. Between 1 and 2, however, the curve is below the axis, so the integral gives the negative of the area. That is, the area is given by

$$A = \int_0^1 (x^3 - 2x^2 - x + 2)\,dx - \int_1^2 (x^3 - 2x^2 - x + 2)\,dx.$$

There are two rectangles between 0 and 1, and two rectangles between 1 and 2. The width of each rectangle is 1/2, so we have

$$\begin{aligned} A &\approx [(1/2)f(1/2) + (1/2)f(1)] - [(1/2)f(3/2) + (1/2)f(2)] \\ &= [(1/2)(9/8) + (1/2)(0)] - [(1/2)(-5/8) + (1/2)(0)] \\ &= 9/16 + 5/16 \text{ or } 14/16, \text{ which is } 0.875. \end{aligned}$$

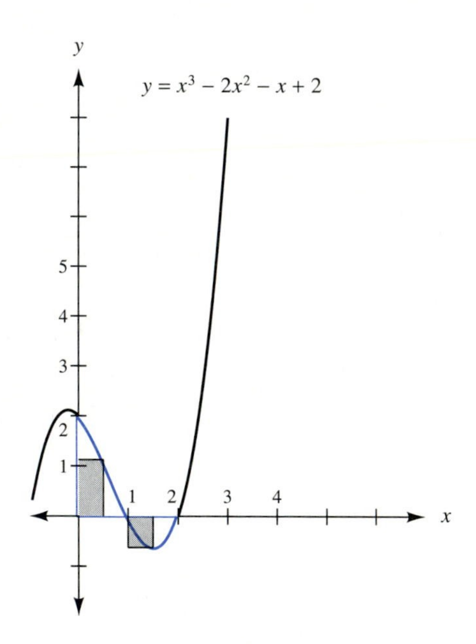

FIGURE 23.9

(b) The integral is given by

$$\begin{aligned} \int_0^2 (x^3 - 2x^2 - x + 2)\,dx &\approx (9/8)(1/2) + (-5/8)(1/2) \\ &= 9/16 - 5/16 \text{ or } 4/16, \text{ which is } 0.25. \end{aligned}$$ ■

EXERCISES 23.1

In 1–6, find an approximate value for each integral, using the number of rectangles specified.

1. $\int_1^5 (2x - 1)\,dx$ Use four rectangles.

2. $\int_0^2 (3x + 2)\,dx$ Use six rectangles.

3. $\int_{-1}^1 (x^2 + 1)\,dx$ Use four rectangles.

4. $\int_{-1}^2 (x^2 - 2x + 4)\,dx$ Use four rectangles.

5. $\int_0^3 (x^3 - 3x^2 - 4x + 12)dx$ Use six rectangles.

6. $\int_0^3 (x^3 - x^2 - 9x + 9)dx$ Use ten rectangles.

In 7–12, use the given number (n) of rectangles to find the approximate area bounded by the x-axis, the lines $x = a$ and $x = b$, and the graph of $y = f(x)$.

7. $y = 2x - 4$ $a = 0, b = 4, n = 4$

8. $y = x$ $a = -2, b = 2, n = 6$

9. $y = 2x^2 - x$ $a = -1, b = 1, n = 4$

10. $y = x^2 - 1$ $a = -2, b = 2, n = 4$

11. $y = x^3 - 3x^2 - 4x + 12$ $a = 0, b = 3, n = 6$ (compare with problem 5)

12. $y = x^3 - x^2 - 9x + 9$ $a = 0, b = 3, n = 6$ (compare with problem 6)

23.2 ANTIDERIVATIVES

In Chapters 21 and 22, when we solved problems, we were given a function and we found the function that was its derivative. For a large class of functions we will be able to reverse the process. That is, if we are given a function, we may be able to find functions for which the given function is the derivative. We call this process "antidifferentiation."

DEFINITION An **antiderivative** of a function f is a function F whose derivative is f.

EXAMPLE 1 Which of the following functions are antiderivatives of the function defined by $f(x) = 2x + 1$?

(a) $F(x) = x^2 + x$

(b) $F(x) = x^2 + 1$

(c) $F(x) = x^2 + x + 5$

Solution **(a)** The derivative of $x^2 + x$ is $2x + 1$, so $x^2 + x$ is an antiderivative of $2x + 1$.

(b) The derivative of $x^2 + 1$ is $2x$, so $x^2 + 1$ is *not* an antiderivative of $2x + 1$.

(c) The derivative of $x^2 + x + 5$ is $2x + 1$, so $x^2 + x + 5$ is an antiderivative of $2x + 1$. ■

The function f in Example 1 has more than one antiderivative. Two functions, such as $x^2 + x$ and $x^2 + x + 5$, that differ only by a constant, have the same derivative. We verify this statement as follows. Let f and F be two functions such that $F'(x) = f(x)$. If $u = F(x)$ and $v = F(x) + C$, where C is a constant, then $u' = f(x)$, and $v' = f(x) + 0$ or $f(x)$. Thus, both $F(x)$ and $F(x) + C$ are antiderivatives of $f(x)$. In fact, over a particular interval of real numbers, every antiderivative of $f(x)$ can be written in the form $F(x) + C$ where C is a constant. For this reason, antiderivatives are often specified with an arbitrary constant added.

EXAMPLE 2 Which of the following functions are antiderivatives of the function defined by $f(x) = 3x^2$?

(a) $F(x) = 3x^3$

(b) $F(x) = x^3 + C$

(c) $F(x) = x^3 + x$

Solution **(a)** $F'(x) = 9x^2$, so $3x^3$ is *not* an antiderivative of $3x^2$.

(b) $F'(x) = 3x^2$, so $x^3 + C$ represents antiderivatives of $3x^2$.

(c) $F'(x) = 3x^2 + 1$, so $x^3 + x$ is not an antiderivative of $3x^2$. ■

We now turn our attention to the problem of finding antiderivatives. We began our study of derivatives with a function of the form $y = ax^n$. The derivative is obtained from this function by making two changes.

1. Multiply the coefficient by the exponent.
2. Decrease the exponent by 1.

To find an antiderivative of a function, reverse the steps to undo the differentiation. That is,

1. Increase the exponent by 1.
2. Divide the coefficient by the new exponent.

EXAMPLE 3 If $f(x) = x^2$, find the antiderivatives.

Solution We increase the exponent by 1 and divide the function by 3. Thus, the antiderivatives are given by $F(x) = x^3/3 + C$. ■

EXAMPLE 4 If $f(x) = 3\sqrt{x}$, find the antiderivatives.

Solution $f(x) = 3\sqrt{x} = 3x^{1/2}$
We increase the exponent by 1 $(1/2 + 1 = 3/2)$ and divide the coefficient by 3/2 $[3 \div (3/2) = 2]$. Thus, the antiderivatives are given by $F(x) = 2x^{3/2} + C$. ■

Just as the derivative of a sum is the sum of the derivatives, an antiderivative of a sum is the sum of the antiderivatives.

EXAMPLE 5 If $f(x) = 3x^2 + 1/x^2$, find the antiderivatives.

Solution $f(x) = 3x^2 + 1/x^2 = 3x^2 + x^{-2}$
The antiderivatives are given by

$$F(x) = \frac{3x^3}{3} + \frac{x^{-1}}{-1} + C = x^3 - x^{-1} + C = x^3 - \frac{1}{x} + C. \quad ■$$

In our study of differentiation, we saw that the chain rule is used to differentiate a *function* raised to a power. It is often possible to reverse the process to find antiderivatives that are functions raised to powers.

POWER RULE FOR ANTIDERIVATIVES

If $g(x) = [f(x)]^n f'(x)$, then the antiderivatives of g are given by

$$G(x) = \frac{[f(x)]^{n+1}}{n+1} + C \text{ if } n \neq -1.$$

Notice that although both $f(x)$ and $f'(x)$ appear in the function $g(x)$, only $f(x)$ appears in the antiderivative $G(x)$. Before we use this rule to work problems, we verify its correctness by finding the derivative of $y = \dfrac{[f(x)]^{n+1}}{n+1} + C$.

$$y' = \underbrace{\frac{(n+1)[f(x)]^n}{n+1}}\underbrace{\left[f'(x)\right]} + 0$$

Use the exponent as a factor and reduce the exponent by 1.

Apply the chain rule.

Find the derivative of the constant C.

Thus, $y' = [f(x)]^n f'(x)$.

NOTE ⇨⇨ ***To use the power rule, one must be able to identify in the problem both a function raised to a power and the derivative of that function.***

EXAMPLE 6 If $g(x) = (3x + 5)^2(3)$, find the antiderivatives.

Solution The quantity $3x + 5$ is raised to the second power, so we must also be able to see the derivative of the function defined by $f(x) = 3x + 5$. Since $f'(x) = 3$, we are able to find the antiderivatives.

$$g(x) = \underbrace{(3x + 5)^2}_{[f(x)]^2}\underbrace{3}_{f'(x)}$$

Thus, $G(x) = \dfrac{[f(x)]^3}{3} + C$ or $G(x) = \dfrac{(3x + 5)^3}{3} + C$. ■

EXAMPLE 7 If $g(x) = \dfrac{3x^2 + 2}{\sqrt{x^3 + 2x}}$, find the antiderivatives.

Solution The first step is to write the denominator using an exponent.

$$g(x) = \frac{3x^2 + 2}{\sqrt{x^3 + 2x}} = (x^3 + 2x)^{-1/2}\,(3x^2 + 2)$$

The quantity $x^3 + 2x$ is raised to the power $-1/2$, so we must also be able to see the derivative of the function defined by $f(x) = x^3 + 2x$. Since $f'(x) = 3x^2 + 2$, we are able to perform the antidifferentiation. We have

$$g(x) = \underbrace{(x^3 + 2x)^{-1/2}}_{[f(x)]^{-1/2}}\underbrace{(3x^2 + 2)}_{f'(x)}$$

$$G(x) = \frac{[f(x)]^{1/2}}{1/2} + C \text{ or } G(x) = \frac{(x^3 + 2x)^{1/2}}{1/2} + C \text{ or } 2(x^3 + 2x)^{1/2} + C. \quad ■$$

EXERCISES 23.2

In 1–10, determine which of the functions listed are antiderivatives of $f(x)$. Recall that if $y = \sin x$, $y' = \cos x$, and if $y = \cos x$, then $y' = -\sin x$.

1. $f(x) = 4x$
(a) $F(x) = 4x^2$ **(b)** $F(x) = 2x^2$ **(c)** $F(x) = 2x^2 + 3$

2. $f(x) = 4x^3$
(a) $F(x) = x^4$ **(b)** $F(x) = x^4 - 1$ **(c)** $F(x) = 4x^4$

3. $f(x) = 3x^2$
(a) $F(x) = x^3 + 2$ **(b)** $F(x) = x^3$ **(c)** $F(x) = 3x^3$

4. $f(x) = x^4$
(a) $F(x) = x^5$ **(b)** $F(x) = 5x^5$ **(c)** $F(x) = (1/5)x^2 + 7$

5. $f(x) = 3x^2 - 8x$
(a) $F(x) = x^3 - 4x^2$ **(b)** $F(x) = x^3 - 4x^2 + 3$ **(c)** $F(x) = 3x^3 - 3x^2$

6. $f(x) = 4x^3 - 2$
(a) $F(x) = 4x^4 - 4x$ **(b)** $F(x) = x^4 - 2$ **(c)** $F(x) = x^4 - 2x - 1$

7. $f(x) = 4 \cos 4x$
(a) $F(x) = 4 \sin x$ **(b)** $F(x) = 4 \sin 4x$ **(c)** $F(x) = \sin 4x$

8. $f(x) = 2x \cos x^2$
(a) $F(x) = \sin x^2$ **(b)** $F(x) = 2x \sin x^2$ **(c)** $F(x) = 1 + \sin x^2$

9. $f(x) = 3 \cos 3x$
(a) $F(x) = 3 \cos^2 3x$ **(b)** $F(x) = 3 \cos 3x^2$ **(c)** $F(x) = \sin 3x$

10. $f(x) = -2 \cos x \sin x$
(a) $F(x) = \cos^2 x$ **(b)** $F(x) = \cos^2 x + 1$ **(c)** $2 \cos x \sin x$

In 11–28, find an expression to represent all antiderivatives F of the function f.

11. $f(x) = 6x^2$
12. $f(x) = 8x^3$
13. $f(x) = 5x^4$
14. $f(x) = 4x^3 - 3x^2$
15. $f(x) = 5x^4 + 8x^3$
16. $f(x) = 2x - 1$
17. $f(x) = 3x^{-2} + 4$
18. $f(x) = 2x^{-3} + x^{-2}$
19. $f(x) = 2x^{1/2} + 3x^{1/3}$
20. $f(x) = x^{2/3} + x^{-2/3}$
21. $f(x) = (x^4 - 1)^2(4x^3)$
22. $f(x) = (x^2 - 2)^{1/2}(2x)$
23. $f(x) = (2x^3 - 1)^{-4}(6x^2)$
24. $f(x) = (x^2 + 3)^{-2}(2x)$
25. $f(x) = \dfrac{6x}{(3x^2 - 1)^2}$
26. $f(x) = \dfrac{3}{(3x + 4)^3}$
27. $f(x) = \dfrac{1}{\sqrt{x - 1}}$
28. $f(x) = \dfrac{2}{\sqrt{2x + 3}}$

23.3 THE DEFINITE INTEGRAL AND THE FUNDAMENTAL THEOREM OF CALCULUS

In Section 23.1, we spent a great deal of time on the study of area. We will now use area to make a connection between differentiation and integration. Let f be a continuous function defined by $y = f(x)$. In the following discussion, we assume that $f(x) \geq 0$ for all x in the interval $a \leq x \leq b$ and that $a \geq 0$. The discussion can, however, be extended to include functions whose graphs drop below the x-axis and for negative values of a.

Consider the function whose graph is shown in Figure 23.10.

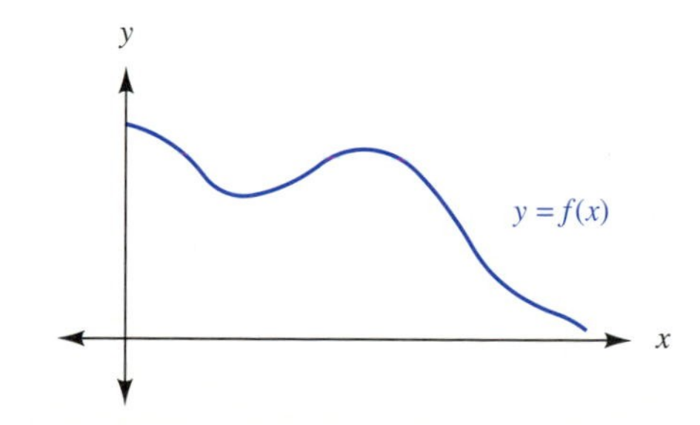

FIGURE 23.10

With each value of x, such as x_1 or x_2, we associate the area under the curve between 0 and x as shown in Figure 23.11. The phrase "area under the curve" should be understood to mean the area between the x-axis and the curve.

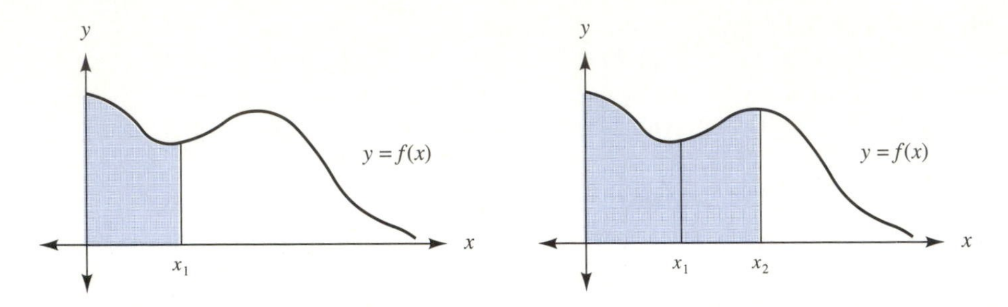

FIGURE 23.11

This relationship is a function, since each value of x has associated with it, a single number that is the area. We denote this function by F. There are, then, two functions under consideration. The function defined by $y = f(x)$ associates with x the height of the graph at x, and the function defined by $A = F(x)$ associates with x the area under the curve from 0 up to x. Consider an arbitrary value of x. If x increases by some small amount Δx, the value of A increases by some amount, which we denote ΔA. The vertical strip with boundaries at x and $x + \Delta x$, shown shaded in Figure 23.12, has area ΔA. Once again we are using the word strip to denote a region that is bounded by the function at the top or bottom and by straight lines elsewhere.

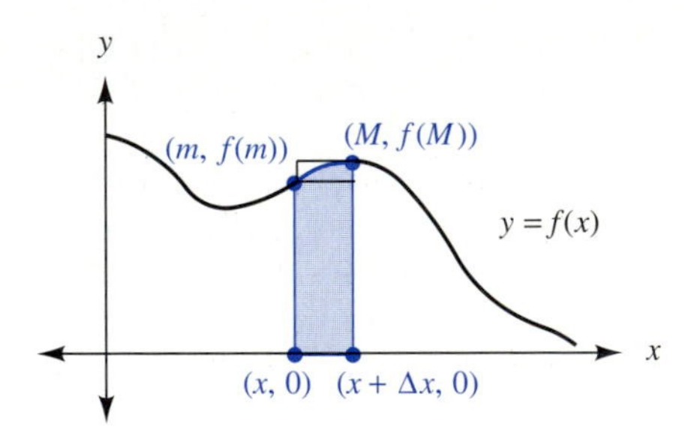

FIGURE 23.12

Find the lowest point on the curve within the strip, and denote the x-coordinate of that point m. The height of the strip at that point is $f(m)$. Draw a rectangle with height $f(m)$ and width Δx. Find the highest point on the curve within the strip, and denote the x-coordinate of that point M. The height of the strip at that point is $f(M)$. Draw a rectangle with height $f(M)$ and width Δx. The actual area of the strip, ΔA, is greater than the area of the small rectangle, but less than the area of the large rectangle. Thus,

$$[f(m)](\Delta x) \leq \Delta A \leq [f(M)](\Delta x).$$

Divide by Δx.

$$f(m) \leq \Delta A/\Delta x \leq f(M)$$

The notation $\Delta A/\Delta x$ should remind you of a derivative. Suppose we find $\lim\limits_{\Delta x \to 0} \Delta A/\Delta x$. We have

$$\lim_{\Delta x \to 0} f(m) \leq \lim_{\Delta x \to 0} \Delta A/\Delta x \leq \lim_{\Delta x \to 0} f(M). \quad \textbf{(23-1)}$$

Notice that as Δx approaches 0, the rectangles shrink so that the heights of both the larger and smaller rectangle approach the same number. That is

$$\lim_{\Delta x \to 0} f(m) = f(x) \text{ and } \lim_{\Delta x \to 0} f(M) = f(x).$$

The inequalities in Equation 23-1 can be written as

$$f(x) \leq dA/dx \leq f(x). \quad \textbf{(23-2)}$$

But for $f(x) \leq dA/dx$ and $dA/dx \leq f(x)$ to be true, we must have

$$f(x) = dA/dx. \quad \textbf{(23-3)}$$

Since the area function was denoted $A = F(x)$, $dA/dx = F'(x)$, then Equation 23-3 can be written

$$f(x) = F'(x). \quad \textbf{(23-4)}$$

Consider Figure 23.13. The area function defined by $A = F(x)$ gives the area under the curve between 0 and x. Thus, the area between 0 and b is given by $F(b)$, and the area between 0 and a is given by $F(a)$. The area between a and b is given by $F(b) - F(a)$.

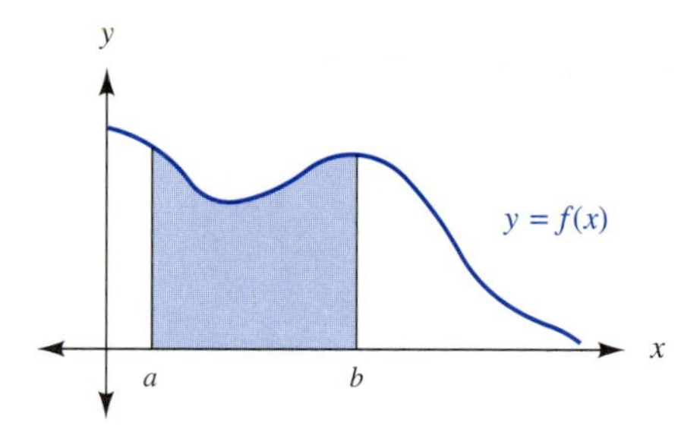

FIGURE 23.13

Since we assumed that the graph of $y = f(x)$ is above the x-axis, the area under the curve between a and b is also given by $\int_a^b f(x)\,dx$. Because the area can be written in these two different ways, we have

$$\int_a^b f(x)\,dx = F(b) - F(a) \text{ where } F'(x) = f(x). \quad \textbf{(23-5)}$$

Equation 23-5 is sometimes written

$$\int_a^b f(x)\,dx = F(x)\Big|_a^b.$$

The notation $\Big|_a^b$ indicates that the function value when x is a is to be subtracted from the function value when x is b.

It is the fact that $f(x) = F'(x)$ that makes Equation 23-5 significant. Equation 23-5 is known as the *fundamental theorem of calculus.*

THE FUNDAMENTAL THEOREM OF CALCULUS

If f and F are two functions such that f is continuous for $a \le x \le b$ and $F'(x) = f(x)$, then

$$\int_a^b f(x)\,dx = F(b) - F(a).$$

NOTE ⇨⇨ ***The importance of the fundamental theorem is that it gives a connection between differentiation and integration, two processes that at first appear to be totally unrelated.*** The theorem says that we can integrate a function f if we can find a function F that has f as its derivative. That is, when we are *doing* an integration we are *undoing* a differentiation.

NOTE ⇨⇨ ***The fundamental theorem of calculus enables us to evaluate an integral $\int_a^b f(x)\,dx$ without having to consider area under the curve whenever we can find an antiderivative of the function f.***

EXAMPLE 1 Evaluate $\int_0^3 x^2\,dx.$

Solution Recall from Section 23.2 that $x^3/3$ is an antiderivative of x^2.

$$\int_0^3 x^2\,dx = x^3/3\Big|_0^3 = 27/3 - 0/3 = 9$$ ■

Our observations are stated as follows.

INTEGRATION RULE

$$\int_a^b x^n\,dx = \frac{x^{n+1}}{n+1}\Big|_a^b \text{ if } n \ne -1$$

The antiderivative $x^3/3$ was used in Example 1, but any function that differs from this one only by a constant is also an antiderivative and could have been used. For instance,

$$\int_0^3 x^2\,dx = x^3/3 + 2\Big|_0^3 = (27/3 + 2) - (0/3 + 2) = 11 - 2 = 9.$$

In the problems that follow, we will follow the convention of choosing to evaluate the antiderivative for which C is 0.

Integrals have the following properties, which are stated here without proof.

PROPERTIES OF INTEGRALS

$$\int_a^b cf(x)\,dx = c\int_a^b f(x)\,dx \qquad \textbf{(23-6)}$$

$$\int_a^b [f(x) + g(x)]\,dx = \int_a^b f(x)\,dx + \int_a^b g(x)\,dx \qquad \textbf{(23-7)}$$

Informally, we say that the integral of a constant times a function is the constant times the integral of the function, and that the integral of a sum is the sum of the integrals.

EXAMPLE 2 Evaluate $\int_0^1 (x^3 - 2x^2 - x + 2)\,dx$.

Solution Equation 23-7 tells us to integrate the polynomial term by term. For each term, the numerical coefficient is treated like c in Equation 23-6.

$$\int_0^1 (x^3 - 2x^2 - x + 2)\,dx = \int_0^1 x^3\,dx - 2\int_0^1 x^2\,dx - \int_0^1 x\,dx + \int_0^1 2\,dx$$

Generally, however, it is not necessary to write four separate integrals. The process of separating the problem into four parts is done mentally, and the problem is written as follows.

$$\begin{aligned}\int_0^1 (x^3 - 2x^2 - x + 2)\,dx &= \frac{x^4}{4} - \frac{2x^3}{3} - \frac{x^2}{2} + 2x\Big|_0^1\\ &= (1/4 - 2/3 - 1/2 + 2) - (0/4 - 0/3 - 0/2 + 0)\\ &= 13/12 \text{ or } 1\ 1/12\text{, which is about } 1.08\end{aligned}$$ ■

EXAMPLE 3 Evaluate $\int_1^2 8x^{-5}\,dx$.

Solution

$$\int_1^2 8x^{-5}\,dx = \frac{8x^{-4}}{-4}\Big|_1^2 = \frac{-2}{x^4}\Big|_1^2 = \frac{-2}{16} - (-2) = \frac{-1}{8} + 2 = \frac{15}{8}$$ ■

The integral has many applications as an antiderivative.

EXAMPLE 4 A certain object moves in such a way that its velocity (in m/s) after time t (in s) is given by $v = 2t^2 + 3t$. Find the distance (in m) traveled during the first four seconds.

Solution Since velocity is the derivative of distance with respect to time, distance is the integral of velocity with respect to time. Between $t = 0$ and $t = 4$ we have the following.

$$\int_0^4 (2t^2 + 3t)\,dt = \frac{2t^3}{3} + \frac{3t^2}{2}\Bigg|_0^4 = \left[\frac{2(4^3)}{3} + \frac{3(4^2)}{2}\right] - \left[\frac{2(0^3)}{3} + \frac{3(0^2)}{2}\right] = 66.7 \text{ m}$$ ■

EXAMPLE 5 Find the area bounded by the x-axis, the lines $x = 1$ and $x = 3$, and the graph of $y = x^3 - 6x^2 + 11x - 6$.

Solution Since we are to find an area, we must know where the graph crosses the x-axis. The graph is shown in Figure 23-14. It appears to cross the x-axis at $x = 1$, $x = 2$, and $x = 3$. Evaluating y for each of these x-values, we see that y is zero in each case. The area, then, can be computed by evaluating integrals.

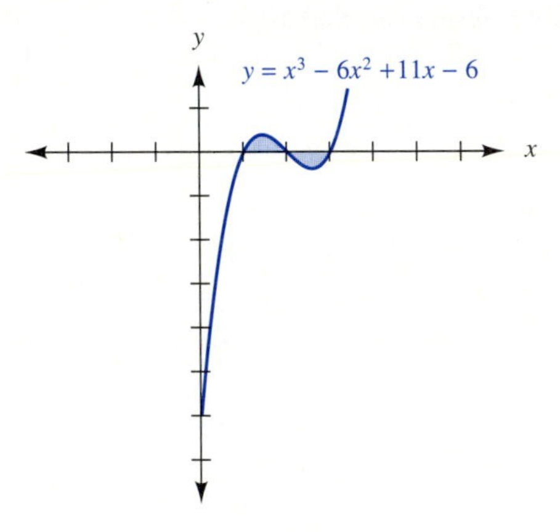

FIGURE 23.14

$$\begin{aligned}
\text{Area} &= \int_1^2 (x^3 - 6x^2 + 11x - 6)\,dx - \int_2^3 (x^3 - 6x^2 + 11x - 6)\,dx \\
&= \left[\frac{x^4}{4} - \frac{6x^3}{3} + \frac{11x^2}{2} - 6x\right]\Bigg|_1^2 - \left[\frac{x^4}{4} - \frac{6x^3}{3} + \frac{11x^2}{2} - 6x\right]\Bigg|_2^3 \\
&= \left[\left(\frac{16}{4} - \frac{48}{3} + \frac{44}{2} - 12\right) - \left(\frac{1}{4} - \frac{6}{3} + \frac{11}{2} - 6\right)\right] \\
&\qquad - \left[\left(\frac{81}{4} - \frac{162}{3} + \frac{99}{2} - 18\right) - \left(\frac{16}{4} - \frac{48}{3} + \frac{44}{2} - 12\right)\right] \\
&= [-2 - (-2.25)] - [-2.25 - (-2)] = 0.50
\end{aligned}$$ ■

EXERCISES 23.3

In 1–10, evaluate each integral.

1. $\int_1^5 (2x-1)\,dx$

2. $\int_0^2 (3x+2)\,dx$

3. $\int_{-1}^1 (x^2+1)\,dx$

4. $\int_{-1}^2 (x^2-2x+4)\,dx$

5. $\int_0^3 (x^3-3x^2-4x+12)\,dx$

6. $\int_0^3 (x^3-x^2-9x+9)\,dx$

7. $\int_1^4 2x^{-3}\,dx$

8. $\int_{-3}^{-2} 3x^{-2}\,dx$

9. $\int_{-3}^{-1} 6x^{-2}\,dx$

10. $\int_{-2}^{-1} 3x^{-4}\,dx$

In 11–14, find the area bounded by the x-axis, the lines $x=a$ and $x=b$, and the graph of $y=f(x)$.

11. $y=x^2-3x+2$, $a=0$, $b=2$

12. $y=x^2-5x+6$, $a=0$, $b=3$

13. $y=x^2-x$, $a=-1$, $b=3$

14. $y=x^2-1$, $a=-1$, $b=2$

---

15. A certain object moves in such a way that its velocity (in m/s) after time t (in s) is given by $v=-t^2+4t$. Find the distance traveled during the first four seconds by evaluating $\int_0^4 (-t^2+4t)\,dt$.

4 7 8 1 2 3 5 6 9

16. A certain object moves in such a way that its velocity (in m/s) after time t (in s) is given by $v=t^2-4t+5$. Find the distance traveled during the first four seconds by evaluating $\int_0^4 (t^2-4t+5)\,dt$.

4 7 8 1 2 3 5 6 9

Problems 17 and 18 use the following information. Since current is given by the derivative of charge with respect to time, charge is given by the integral of current with respect to time.

17. For a particular circuit, the current (in amperes) after time t (in seconds) at a certain point P is given by $i=0.005t^{0.25}$. Find the charge (in coulombs) that passes point P during the first second by evaluating $\int_0^1 (0.005t^{0.25})\,dt$.

3 5 9 1 2 4 6 7 8

18. For a particular circuit, the current (in amperes) after time t (in seconds) at a certain point P is given by $i=0.0005(t^{-1/2}+t^{1/2})$. Find the charge (in coulombs) that passes point P during the first second by evaluating $\int_0^1 0.0005(t^{-1/2}+t^{1/2})\,dt$.

3 5 9 1 2 4 6 7 8

Problems 19 and 20 use the following information. Since force is the derivative of work with respect to distance, work is the integral of force with respect to distance.

19. A force acts on a certain object in such a way that when the object has moved a distance of r (in m), the force f (in newtons) is given by $f=3r^{-2}$. Find the work (in joules) done through the second four meters by evaluating $\int_4^8 3r^{-2}\,dr$.

4 7 8 1 2 3 5 6 9

---

1. Architectural Technology
2. Civil Engineering Technology
3. Computer Engineering Technology
4. Mechanical Drafting Technology
5. Electrical/Electronics Engineering Technology
6. Chemical Engineering Technology
7. Industrial Engineering Technology
8. Mechanical Engineering Technology
9. Biomedical Engineering Technology

20. A force acts on a certain object in such a way that when the object has moved a distance of r (in m), the force f (in newtons) is given by $f = 3r^2 + 2r$. Find the work (in joules) done through the first four meters by evaluating $\int_0^4 (3r^2 + 2r)\, dr$.

4 7 8 1 2 3 5 6 9

21. In Section 23.6, we will use the formula for the area under a parabolic arc. $A = (\Delta x/3)[f(x_0) + 4f(x_1) + f(x_2)]$. Derive this formula using the following steps.

1. Let $f(x) = ax^2 + bx + c$ represent the parabola.
2. Evaluate $\int_{-h}^{h} (ax^2 + bx + c)\, dx$ to find the area under the curve between $-h$ and h as shown in the accompanying diagram.
3. Find $(\Delta x/3)[f(x_0) + 4f(x_1) + f(x_2)]$ where $x_0 = -h$, $x_1 = 0$, $x_2 = h$, and $\Delta x = h$.
4. Show that the quantities obtained in steps 2 and 3 are equal.

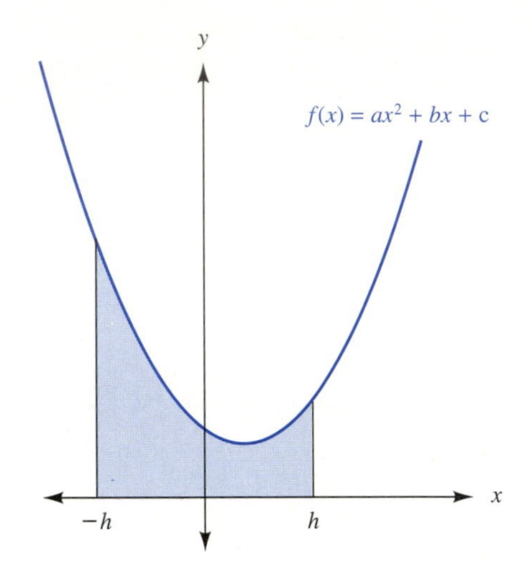

1. Architectural Technology
2. Civil Engineering Technology
3. Computer Engineering Technology
4. Mechanical Drafting Technology
5. Electrical/Electronics Engineering Technology
6. Chemical Engineering Technology
7. Industrial Engineering Technology
8. Mechanical Engineering Technology
9. Biomedical Engineering Technology

23.4 THE INDEFINITE INTEGRAL

In the previous section, you learned that the first step in evaluating an integral $\int_a^b f(x)\, dx$ is to find an antiderivative F of the function f. The second step was to evaluate $F(x)$ at the limits of integration. The symbol $\int f(x)\, dx$, written without limits of integration, is used to denote the antiderivatives of a function f. We write $\int f(x)\, dx = F(x) + C$, where F is an antiderivative of f, and C is an arbitrary constant. We write $F(x) + C$, instead of $F(x)$, because two functions that differ only by a constant have the same derivative. Whereas $\int_a^b f(x)\, dx$ is called a *definite integral,* $\int f(x)\, dx$ is called an *indefinite integral.* It is important to see that these two notations represent related, but different, concepts.

definite integral

indefinite integral

DEFINITE AND INDEFINITE INTEGRALS

If f and F are two functions such that $F'(x) = f(x)$ for $a \le x \le b$, then

$\int_a^b f(x)\,dx$ is called a **definite integral.**

$\int_a^b f(x)\,dx = F(b) - F(a)$

$\int_a^b f(x)\,dx$ is a number.

$\int_a^b f(x)\,dx$ is interpreted as

$\lim_{\Delta x \to 0} \sum_{i=1}^{n} [f(x_i)](\Delta x)$.

$\int f(x)\,dx$ is called an **indefinite integral.**

$\int f(x)\,dx = F(x) + C$

$\int f(x)\,dx$ is a function.

$\int f(x)\,dx$ is interpreted as the antiderivatives of f.

EXAMPLE 1 Find $\int x^2\,dx$.

Solution From Section 23.2, we know that an antiderivative of x^2 is obtained by adding 1 to the exponent and dividing the expression by the new exponent. Thus,

$$\int x^2\,dx = x^3/3 + C. \quad ■$$

Recall that in Section 23.2 we were able, in some cases, to find antiderivatives of functions that were raised to powers. Hence, we make the following observation.

POWER RULE FOR INTEGRATION

$$\int [f(x)]^n f'(x)\,dx = \frac{[f(x)]^{n+1}}{n+1} + C \text{ if } n \ne -1$$

EXAMPLE 2 Find $\int (2x + 1)^3 2\,dx$.

Solution The quantity $2x + 1$ is raised to the third power, so we must be able to see the derivative of the function defined by $f(x) = 2x + 1$. Since $f'(x) = 2$, we are able to perform the integration.

$$(2x + 1)^3 2\,dx = \frac{(2x + 1)^4}{4} + C \quad ■$$

EXAMPLE 3 Find $\int \frac{2x+1}{\sqrt{x^2+x}}\,dx$.

Solution The first step is to write the denominator using an exponent.

$$\int \frac{2x+1}{\sqrt{x^2+x}}\,dx = \int (x^2+x)^{-1/2}(2x+1)\,dx$$

The quantity x^2+x is raised to the power $-1/2$, so we must be able to see the derivative of the function defined by $f(x)=x^2+x$. Since $f'(x)=2x+1$, we are able to perform the integration.

$$\int (x^2+x)^{-1/2}(2x+1)\,dx = \frac{(x^2+x)^{1/2}}{1/2} + C = 2(x^2+x)^{1/2} + C \quad \blacksquare$$

Examples 2 and 3 were exceptionally "nice." The derivative that we need does not always appear in the problem. If, however, we lack only a constant factor of having the necessary derivative, the problem can be salvaged as shown in Example 4.

EXAMPLE 4 Find $\int (x^3+1)^4 x^2\,dx$.

Solution The quantity x^3+1 is raised to the fourth power, so we must be able to see the derivative of the function defined by $f(x)=x^3+1$. The derivative f' is given by $f'(x)=3x^2$, but we have only x^2 in the problem. If we could multiply x^2 by 3, we would have the necessary derivative. Such a multiplication would change the problem. But multiplication by 3(1/3) would not. (Why?) Thus,

$$\int (x^3+1)^4 x^2\,dx = \int (x^3+1)^4(1/3)(3x^2)\,dx.$$

Recall that $\int_a^b cf(x)\,dx = c\int_a^b f(x)\,dx$. This property of integrals also holds for indefinite integrals. Thus,

$$\int (x^3+1)^4(1/3)(3x^2)\,dx = (1/3)\int (x^3+1)^4(3x^2)\,dx.$$

Now it is possible to proceed with the integration.

$$1/3\int (x^3+1)^4(3x^2)\,dx = (1/3)\left[\frac{(x^3+1)^5}{5} + C_1\right] = \frac{(x^3+1)^5}{15} + (1/3)C_1 \quad \blacksquare$$

NOTE ⇨⇨ We have used C_1 to represent the constant of integration. ***It makes no difference what name is given to a constant, but generally C is used for the constant that appears in the final answer.*** Since C_1 is a constant, $(1/3)C_1$ is also a constant, which can be represented as C. The answer would be written as

$$\frac{(x^3+1)^5}{15} + C.$$

As a short cut, the adjustments to the problem can be made by inserting one factor to the right of the integral sign and the reciprocal of that factor to the left of it. In symbols, this statement is written as follows.

PROPERTY OF INTEGRALS

$$\int f(x)\,dx = (1/c)\int cf(x)\,dx \text{ if } c \neq 0$$

EXAMPLE 5 Find $\int 3x^3\sqrt{x^4+1}\,dx$.

Solution $\int 3x^3\sqrt{x^4+1}\,dx = \int (x^4+1)^{1/2}(3x^3)\,dx$

The quantity $x^4 + 1$ is raised to the power $1/2$, so we must be able to see the derivative of the function defined by $f(x) = x^4 + 1$. The derivative f' is given by $f'(x) = 4x^3$, but we see $3x^3$ in the problem. The necessary compensating factor is $4/3$. Therefore, we place a factor of $4/3$ to the right of the integral sign, and a factor of $3/4$ to the left of it.

$$\begin{aligned}\int (x^4+1)^{1/2}(3x^3)\,dx &= (3/4)\int (x^4+1)^{1/2}(4/3)(3x^3)\,dx \\ &= (3/4)\int (x^4+1)^{1/2}(4x^3)\,dx \\ &= (3/4)\left[\frac{(x^4+1)^{3/2}}{3/2} + C_1\right] \\ &= (1/2)(x^4+1)^{3/2} + C \quad ■\end{aligned}$$

A different notation is sometimes used to simplify the statement of integration rules. Up to this point, we have not used dx or dy as separate symbols. The notation dy/dx has been treated as a single symbol, as is $\int_a^b f(x)\,dx$ or $\int f(x)\,dx$. In each case, dx tells us which variable the differentiation or integration is "with respect to." Physicists and engineers often use dx instead of Δx to indicate a very small change in the variable. This substitution causes no problem in applied situations, for which the data consists of measurements, which are approximate anyway. From a mathematical point of view, however, we must be careful not to introduce inconsistencies. If we allow dx and dy to be used separately by defining dx to be equal to Δx, then we can multiply both sides of the equation

$$dy/dx = f'(x)$$

by dx to obtain the equivalent equation

$$dy = f'(x)\,dx.$$

Mathematicians call dy and dx *differentials.*

DEFINITION

The **differential of x,** denoted dx, is defined by $dx = \Delta x$, where Δx represents a small change in the value of the variable x. The **differential of y,** denoted dy, is defined by $dy = f'(x)\,dx$, where $y = f(x)$.

Technically, dy and Δy are not equal to each other. Consider Figure 23.15, in which $dx = \Delta x$. The tangent at x is considered to have slope dy/dx, so dy is the change associated with the y-values on the *tangent* between x and $x + \Delta x$ (or $x + dx$). The quantity Δy is the change associated with the y-values on the *graph of* $y = f(x)$ between x and $x + \Delta x$. When dx is small enough, however, dy and Δy are approximately equal, and hence using dy as an approximation for Δy is acceptable in many cases.

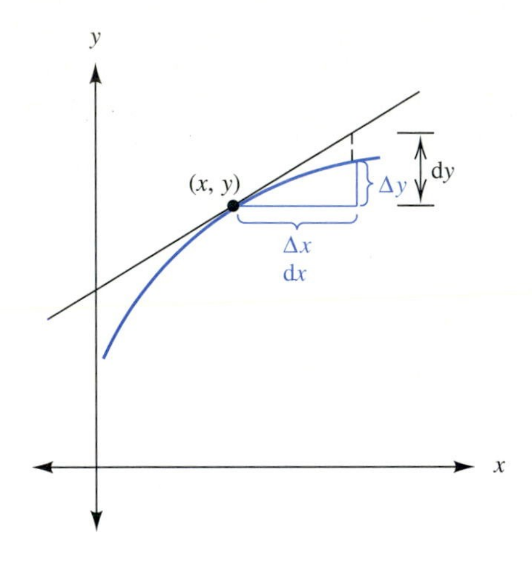

FIGURE 23.15

The concept of differentials is used to state the power rule and other integration rules.

THE POWER RULE FOR INTEGRALS

If u is a function, then

$$\int u^n\,du = \frac{u^{n+1}}{n+1} + C \qquad \text{if } n \neq -1 \tag{23-8}$$

When the power rule is stated as in Equation 23-8, it is important to identify du properly.

EXAMPLE 6 Find $\int 2(2x+1)^3\,dx$.

Solution The quantity $2x+1$ is raised to the third power, so $u = 2x+1$, and $n = 3$. To find du, find the derivative of u with respect to x, then multiply by dx. That is, $du = 2\,dx$.

$$\int 2(2x+1)^3 dx = \int \underbrace{(2x+1)^3}_{u^n}\underbrace{2dx}_{du} = \frac{(2x+1)^4}{4} + C$$

Notice that the integral in Example 6 is the same one that we found in Example 2.

EXAMPLE 7 Find $\int x^2(x^3+1)^4\,dx$.

Solution The quantity x^3+1 is raised to the fourth power. Thus, $u = x^3+1$, and $n = 4$. The du in this example is $3x^2\,dx$. Thus, the problem must be adjusted.

$$\begin{aligned}\int x^2(x^3+1)^4 dx &= (1/3)\int 3x^2(x^3+1)^4 dx \\ &= (1/3)\int \underbrace{(x^3+1)^4}_{u^n}\underbrace{3x^2dx}_{du} \\ &= (1/3)\left[\frac{(x^3+1)^5}{5} + C_1\right] \\ &= \frac{(x^3+1)^5}{15} + C\end{aligned}$$

Indefinite integrals, like definite integrals, are used in the technologies.

EXAMPLE 8 The voltage V (in volts) across a capacitor of capacitance C (in farads) is given by $V = \frac{1}{C}\int i\,dt$ where i is the current (in amperes) and t is the time (in seconds). Find a formula for V, if $i = 1 + 10t$.

Solution $V = \frac{1}{C}\int (1+10t)\,dt = \frac{1}{C}\left[t + \frac{10t^2}{2} + C_1\right] = \frac{1}{C}[t + 5t^2] + C_2$

We use C_1 and C_2 as constants of integration, since C already appears in the formula as a constant.

EXERCISES 23.4

In 1–32, find each integral.

1. $\int (x^3 + 3x)\, dx$
2. $\int (3x^3 - 4x)\, dx$
3. $\int (2x^2 - 5x)\, dx$
4. $\int (4x^2 + 1)\, dx$
5. $\int 2x^{-3}\, dx$
6. $\int 4x^{-2}\, dx$
7. $\int (5x^{-3} + 2x^{-2})\, dx$
8. $\int (3x^2 - x^{-2})\, dx$
9. $\int 2x(x^2 + 1)^3\, dx$
10. $\int 3x^2(x^3 + 2)^{-2}\, dx$
11. $\int 3(3x - 5)^{-3}\, dx$
12. $\int 2\sqrt{2x - 3}\, dx$
13. $\int \sqrt{x - 1}\, dx$
14. $\int 2x\sqrt{x^2 + 4}\, dx$
15. $\int \frac{2x}{\sqrt{x^2 - 1}}\, dx$
16. $\int \frac{4x}{\sqrt{2x^2 - 3}}\, dx$
17. $\int x^3(x^4 - 1)^{-2}\, dx$
18. $\int x^2(x^3 + 2)^{-4}\, dx$
19. $\int 2x^2(x^3 + 1)^5\, dx$
20. $\int 3x(x^2 - 4)^7\, dx$
21. $\int x\sqrt{x^2 - 1}\, dx$
22. $\int x^2\sqrt{x^3 + 2}\, dx$
23. $\int \frac{3x}{(2x^2 + 3)^3}\, dx$
24. $\int \frac{3x^2}{\sqrt{2x^3 - 3}}\, dx$
25. $\int (2x + 1)(x^2 + x)^8\, dx$
26. $\int (x^2 - 1)(x^3 - 3x)^6\, dx$
27. $\int (x^2 + 1)(x^3 + 3x + 2)^7\, dx$
28. $\int (4x - 3)\sqrt{2x^2 - 3x}\, dx$
29. $\int (3x + 1)\sqrt{3x^2 + 2x}\, dx$
30. $\int \frac{3x^2 + x}{\sqrt{2x^3 + x^2}}\, dx$
31. $\int \frac{3x^2 - 4x}{\sqrt{x^3 - 2x^2}}\, dx$
32. $\int \frac{3x^2 - 2x}{\sqrt{x^3 - x^2 + 3}}\, dx$

---

33. The hard disk in a personal computer must be brought from rest to a high rotational speed when the system is turned on. Find a formula for the speed s (in rpm) of the disk at time t (in minutes) if $s = 5000 \int (0.6 - 0.3t)\, dt$. 3 5 9

34. The current (in amperes) in an inductor of inductance L (in henries) is given by $i = \frac{1}{L}\int V\, dt$, where V is the voltage (in volts) and t is the time (in seconds). Find a formula for i, if $V = 9t(t^2 - 8)$. 3 5 9

35. The angular displacement θ (in radians) of an object moving with an angular velocity ω (in radians/second) is given by $\theta = \int \omega\, dt$, where t is the time (in seconds). Find a formula for θ, if $\omega = t^2(t^3 - 1)^{-2}$. 4 7 8 1 2 3 5 6 9

36. The velocity v (in m/s) of an object moving with acceleration a (in m/s²) is given by $v = \int a\, dt$, where t is the time (in seconds). Find a formula for v, if $a = t(t^2 + 1)^3$. 4 7 8 1 2 3 5 6 9

37. The work W (in joules) done by a force F (in newtons) moving an object through a distance x (in meters) is given by $W = \int F\,dx$. Find a formula for W, if $F = kx$ and k is a constant. 4 7 8 1 2 3 5 6 9

38. The buoyant force F (in newtons) on a body of volume V (in m^3) immersed in a fluid of density ρ is given by $F = \int \rho g\,dV$. Find a formula for F if ρ and g are constants. 2 4 8 1 5 6 7 9

39. The resistance R (in ohms) of a certain thermistor is given by $R = \int (0.009T^2 + 0.02T - 0.7)\,dT$, where T is the temperature (in degrees Celsius). Find a formula for R. 3 5 9 1 2 4 6 7 8

40. Let W be the work done in charging a capacitor with capacitance c. If v is the voltage drop across the capacitor, and i is the current flowing into the capacitor, then the following relationships hold. $W = \int p\,dt$, $p = vi$, and $i = c(dv/dt)$. Show that $W = (1/2)cv^2$ if the constant of integration is 0.

1. Architectural Technology
2. Civil Engineering Technology
3. Computer Engineering Technology
4. Mechanical Drafting Technology
5. Electrical/Electronics Engineering Technology
6. Chemical Engineering Technology
7. Industrial Engineering Technology
8. Mechanical Engineering Technology
9. Biomedical Engineering Technology

23.5 NUMERICAL INTEGRATION: THE TRAPEZOIDAL RULE

There are many functions for which it is impossible to find explicit antiderivatives. The function defined by $f(x) = (\cos x)/x$ is an example. Although we cannot find an indefinite integral of such a function, we may approximate a definite integral of such a function using the method of Section 23.1. Some degree of error is introduced because of the overlap and underlap created by replacing each little strip (having a curved boundary) with a rectangle (having a straight horizontal boundary). One way to reduce the error is to use a large number of very thin rectangles. Another way is to replace each little strip with a thin trapezoid instead of a rectangle. A trapezoid fits the curve better than a rectangle, as shown in Figure 23.16. Since there is less overlap and underlap with trapezoids than with rectangles, fewer trapezoids are required to find an approximation with an acceptable degree of error.

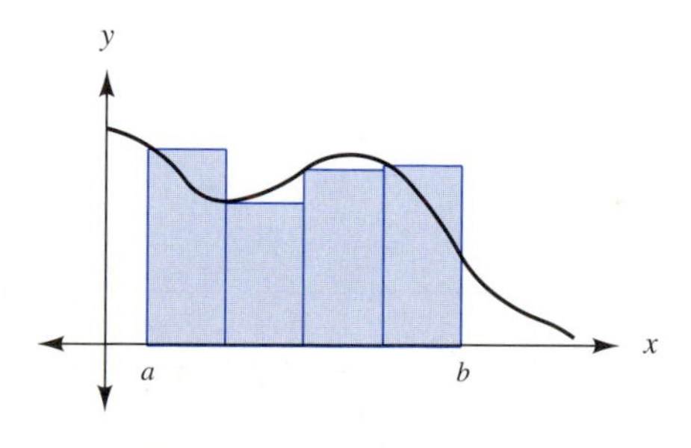

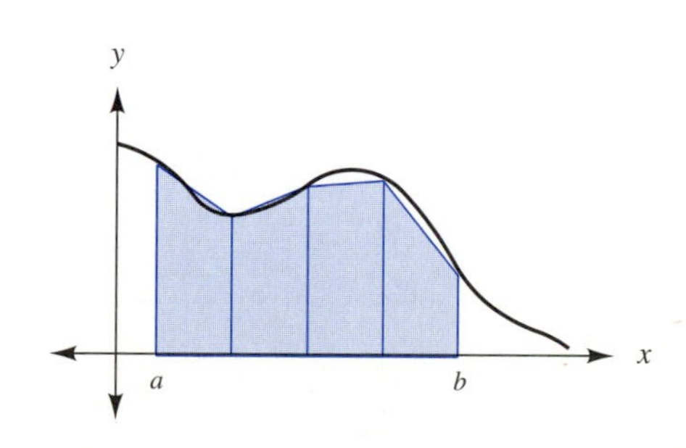

FIGURE 23.16

The area of a trapezoid is given by $A = (h/2)(b_1 + b_2)$, where b_1 and b_2 are the lengths of the two parallel sides of the trapezoid, and h is the distance between the two parallel sides. The area of a figure bounded by the x-axis, the lines $x = a$ and $x = b$, and the curve $y = f(x)$ can be found in the following manner. Divide the interval from a to b into n equal segments, marking the division points $x_1, x_2, x_3, \ldots, x_{n-1}$ ($x_0 = a$ and $x_n = b$). Draw a vertical line segment up to the curve at each x_i. The width of each strip, which we denote Δx, is $\dfrac{b-a}{n}$. the area of the ith little strip is approximately equal to the area of a trapezoid with height Δx and parallel sides of length $f(x_{i-1})$ and $f(x_i)$. That is,

$$\text{area of one strip} = (\Delta x/2)[f(x_{i-1}) + f(x_i)].$$

An approximation for the area A under the curve is given by

$$A \approx (\Delta x/2)[f(x_0) + f(x_1)] + (\Delta x/2)[f(x_1) + f(x_2)] + \cdots + (\Delta x/2)[f(x_{n-2}) + f(x_{n-1})] + (\Delta x/2)[f(x_{n-1}) + f(x_n)].$$

Factoring out $\Delta x/2$, we have

$$A \approx (\Delta x/2)[f(x_0) + f(x_1) + f(x_1) + f(x_2) + \cdots + f(x_{n-2}) + f(x_{n-1}) + f(x_{n-1}) + f(x_n)].$$

Notice that the values $f(x_0)$ and $f(x_n)$ each occur only once inside the brackets, but the values $f(x_1)$ through $f(x_{n-1})$ each occur twice. Thus,

$$A = (\Delta x/2)[f(x_0) + 2f(x_1) + 2f(x_2) + \cdots + 2f(x_{n-1}) + f(x_n)]. \qquad \textbf{(23-9)}$$

Generalizing this result, we can see that Equation 23-9 can be used to compute integrals.

$$\int_a^b f(x)\,dx \approx (\Delta x/2)[f(x_0) + 2f(x_1) + 2f(x_2) + \cdots + 2f(x_{n-1}) + f(x_n)] \qquad \textbf{(23-10)}$$

Equation 23-10 is called the *trapezoidal rule.*

THE TRAPEZOIDAL RULE

If f is a function that is continuous for $a \le x \le b$, then

$$\int_a^b f(x)\,dx \approx (\Delta x/2)[f(x_0) + 2f(x_1) + 2f(x_2) + \cdots + 2f(x_{n-1}) + f(x_n)]$$

where $\Delta x = (b - a)/n$, $x_0 = a$, $x_{i+1} = x_i + \Delta x$, and $x_n = b$.

EXAMPLE 1 Use the trapezoidal rule with $n = 4$ to find an approximate value for $\int_0^3 x^2\,dx$.

Solution This is the integral from Example 2 of Section 23.1. The distance from 0 to 3 is divided into 4 equal lengths, so Δx is 3/4, and $\Delta x/2$ is 3/8. The function must be evaluated for $x_0 = 0$, $x_1 = 3/4$, $x_2 = 3/2$, $x_3 = 9/4$, and $x_4 = 3$.

$$\int_0^3 x^2\,dx \approx (3/8)[(0)^2 + 2(3/4)^2 + 2(3/2)^2 + 2(9/4)^2 + 3^2]$$

Using a calculator to do the arithmetic, we see that $\int_0^3 x^2\,dx$ is 9.28, to three significant digits. ■

EXAMPLE 2 Use the trapezoidal rule to find the approximate area bounded by the x-axis, the lines $x = -2$ and $x = 2$, and the graph of $y = x\sqrt{x^2 + 2}$.

Solution The graph is shown in Figure 23.17. The curve is symmetric with respect to the origin (why?), so if we find the area in the first quadrant and double it, we will have the entire area.

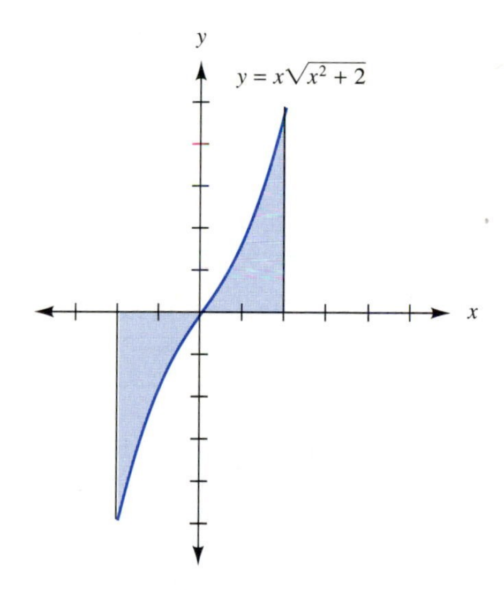

FIGURE 23.17

$$A = 2\int_0^2 (x\sqrt{x^2 + 2})\,dx$$

We arbitrarily choose $n = 5$. Since the distance from 0 to 2 is 2, Δx is 2/5 or 0.4, and $\Delta x/2$ is 0.2. The computation can be done with a calculator. For each value of x_i, $2f(x_i)$ is computed using the following keystroke sequence.

x_i [x²] [+] 2 [=] [√x] [×] x_i [×] 2 [=]

Most calculators allow you to keep a running total by adding a number to the value that is stored in memory. Whereas a key labeled [STO] or [M] erases memory before storing the displayed value, a key labeled [SUM] or [M+] adds the displayed number to the number that is already in memory. If your calculator does not have a special key to perform this task, you can use the following keystroke sequence to add the displayed number to the number in memory.

[+] [RCL] [=] [STO]

The table that follows shows the value of x_i, $f(x_i)$, and the cumulative sum of the $2f(x_i)$ values (with $f(x_0)$ and $f(x_5)$). It is not necessary to record all of this information when you do the calculations. It is shown so that you can perform the calculations yourself and verify that you understand the procedure.

i	x_i	$f(x_i)$	Sum
0	0	0.0	0.0
1	0.4	0.5878775	1.1757551
2	0.8	1.2998461	3.7754474
3	1.2	2.2256684	8.2267842
4	1.6	3.416665	15.060114
5	2.0	4.8989795	19.959094

Since $\Delta x/2 = 0.2$, the value of the integral is 0.2(19.959094) or 3.9918188. Thus,

$2\int_0^2 (x\sqrt{x^2+2})\,dx \approx 2(3.9918188)$ or 7.98, to three significant digits. ■

We have arbitrarily set the number of intervals for each problem, but in practice, it is necessary to have an idea of how accurate the final answer is. Because the procedure involves performing the same series of operations repetitively, it is well suited for use on a computer. A common procedure for computer applications is to use more decimal places than desired to find an approximation with $n = 2$. The procedure is then repeated again and again, doubling n each time, until two successive approximations agree when rounded to the desired number of decimal places.

It is sometimes necessary to find an integral when the equation of the function is unknown. Experiments or observations are often used to collect the data for such problems. Example 3 illustrates this type of problem.

EXAMPLE 3 One pound of grass seed is required for 500 ft^2 of lawn. Determine the approximate amount of seed necessary for the plot of land shown in Figure 23.18.

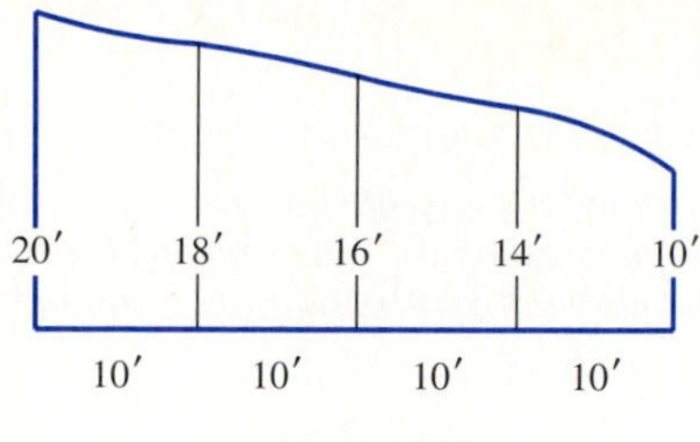

FIGURE 23.18

Solution Use the trapezoidal rule to find the approximate area.

$$A \approx 5[20 + 2(18) + 2(16) + 2(14) + 10] \approx 5(126) \text{ or } 630 \text{ ft}^2$$

Since 1 pound of seed is required for every 500 ft^2 of lawn, the amount of seed needed is $630 \div 500 = 1.26$ pounds. ■

EXAMPLE 4 The following table gives the approximate percent y per mm of the population having the given systolic blood pressure x.

x	121	127	133
y	3.3	3.0	2.0

Find the percent of people who have systolic blood pressure between 121 and 133. Blood pressure is normally distributed.

Solution The percent having systolic blood pressures between 121 and 133 is given by the area under the normal curve between 121 and 133, as shown in Figure 23.19.

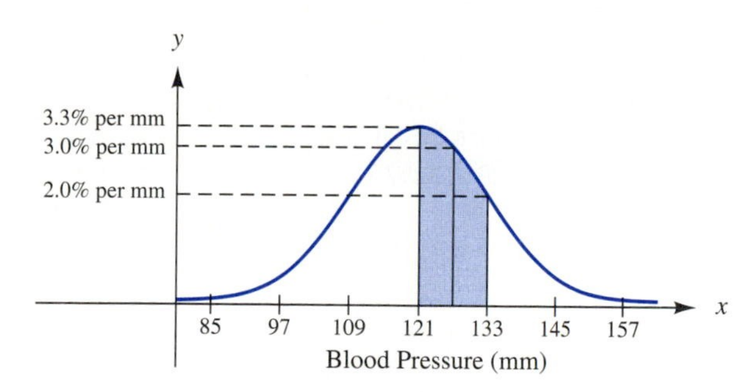

FIGURE 23.19

Since we do not have the equation that gives the curve, we use the trapezoidal rule. With two trapezoids, $n = 2$, and $\Delta x = 127 - 121$.

$$\Delta x = 6, \text{ and } \Delta x/2 = 3$$

$$A \approx 3(3.3 + 2(3.0) + 2.0) \approx 3(11.3) \text{ or } 33.9\% \quad ■$$

EXERCISES 23.5

In 1–6, find an approximate value for each integral, using the trapezoidal rule with n intervals. Round each answer to the nearest tenth. These integrals are the same as those in problems 1–6 of Section 23.1. Compare the approximations obtained by the trapezoidal rule with those obtained by using rectangles.

1. $\int_1^5 (2x - 1)\,dx \quad n = 4$

2. $\int_0^2 (3x + 2)\,dx \quad n = 6$

3. $\int_{-1}^1 (x^2 + 1)\,dx \quad n = 4$

4. $\int_{-1}^2 (x^2 - 2x + 4)\,dx \quad n = 4$

5. $\int_0^3 (x^3 - 3x^2 - 4x + 12)\,dx \quad n = 6$

6. $\int_0^3 (x^3 - x^2 - 9x + 9)\,dx \quad n = 6$

In 7–12, use the trapezoidal rule with n intervals to find an approximate value for each integral. Round the answer to the number of decimal places specified.

7. $\int_{-1}^1 (x^3 - x)\,dx$, $n = 4$, round to the nearest whole number

8. $\int_0^2 (2x^3 + x^2)\,dx$, $n = 4$, round to the nearest whole number

9. $\int_0^3 \frac{1}{2x + 1}\,dx$, $n = 10$, round to the nearest tenth

10. $\int_1^3 \frac{1}{x^2}\,dx$, $n = 6$, round to the nearest tenth

11. $\int_1^4 \sqrt{x^2 + 1}\,dx$, $n = 8$, round to the nearest tenth

12. $\int_1^4 x\sqrt{x^3 + 1}\,dx$, $n = 10$, round to the nearest whole number

---

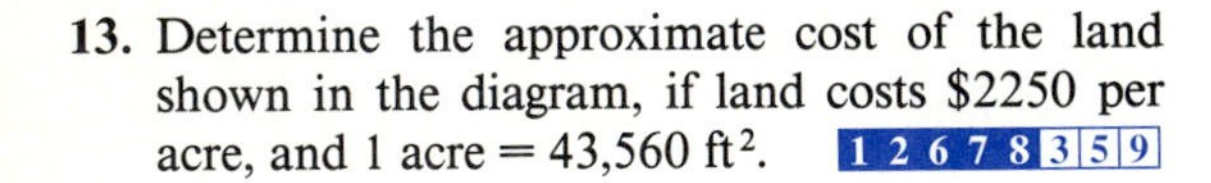

13. Determine the approximate cost of the land shown in the diagram, if land costs $2250 per acre, and 1 acre = 43,560 ft². 1 2 6 7 8 3 5 9

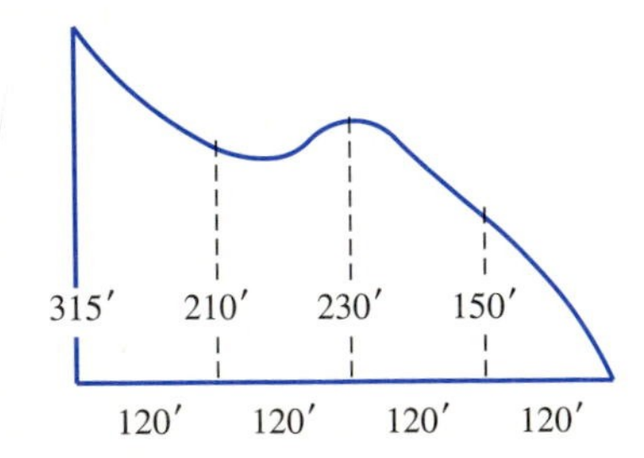

14. A 2.2 μF capacitor is charged by a variable current. The following table gives the current measured at 10 ms intervals.

t (in ms)	0	10	20	30
i (in mA)	0	1.0	1.3	1.5

Find the approximate voltage if $V = \frac{1}{C}\int_0^{0.03} i\,dt$.

3 5 9 4 7 8

15. Determine the approximate area of the dam that has the cross section shown in the accompanying diagram. 1 2 4 8 7 9

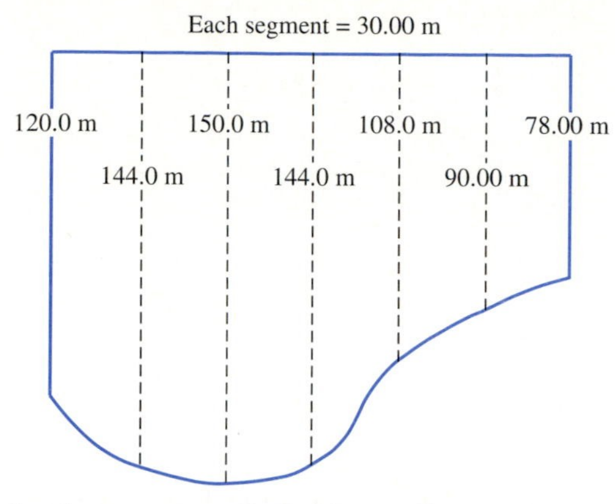

16. It takes two seconds for the cooling system of a computer to reach its maximum level. The graph shows air flow rate versus time. Determine the total volume of air delivered in the first four seconds by finding the area under the curve between $t = 0$ and $t = 4$. 3 5 9

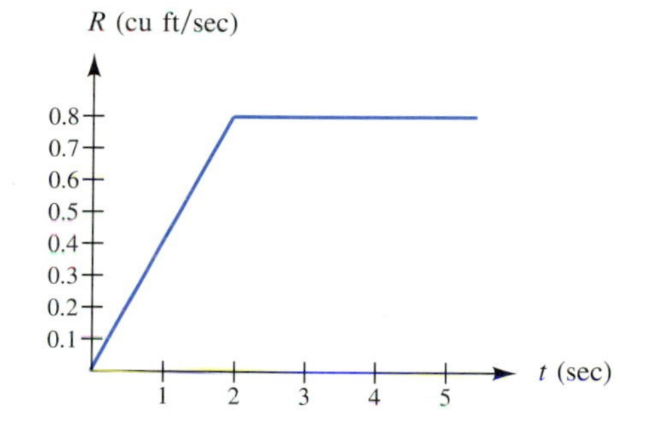

17. A certain parabolic reflector has a diameter of 4.00 inches. The depth y of the cross section is measured at the edge and at distances x, 1.00 inch apart across the diameter. The measurements are given in the following table. Find the approximate area of the cross section.

x	0.00	1.00	2.00	3.00	4.00
y	0.750	0.375	0.250	0.375	0.750

3 5 6 9

18. Determine the approximate volume of an elliptical oil tank that is 5.0 m long, if the cross section is given by $x^2 + 4y^2 = 4$. [Hint: To find the cross-sectional area, find the area in quadrant I and multiply by 4.] 1 2 4 7 8 3 5

19. The hydrograph shown is a graph of storm water runoff rate versus time at a given point on a stream for a particular storm. Use five intervals to determine the total water runoff by finding the area under the curve shown. Because the vertical axis shows in³/hour, water runoff is in cubic inches. 2 1 9

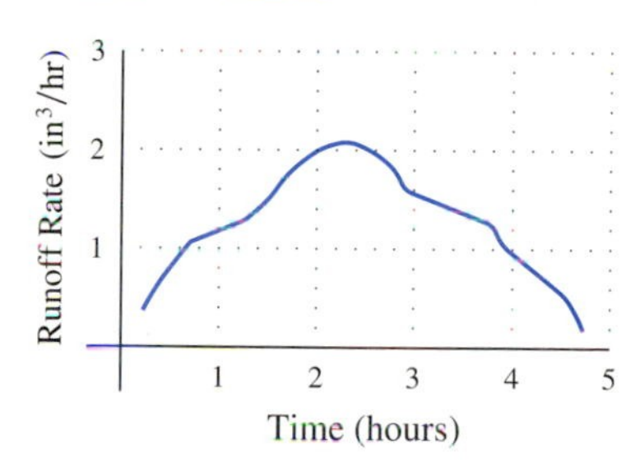

1. Architectural Technology
2. Civil Engineering Technology
3. Computer Engineering Technology
4. Mechanical Drafting Technology
5. Electrical/Electronics Engineering Technology
6. Chemical Engineering Technology
7. Industrial Engineering Technology
8. Mechanical Engineering Technology
9. Biomedical Engineering Technology

23.6 NUMERICAL INTEGRATION: SIMPSON'S RULE

We have seen that when a figure is divided into strips, a trapezoid more closely approximates the shape of a strip with a curved boundary than a rectangle does. There is another shape that usually provides an even better approximation to the shape of the strip. The curved boundary is approximated by the arc of a parabola. It is known that for any three points, there is a parabola that passes through them. If we consider two adjacent strips, there are three end points, and thus a parabola that can be used as a fourth boundary for each of the two strips, as shown in Figure 23.20.

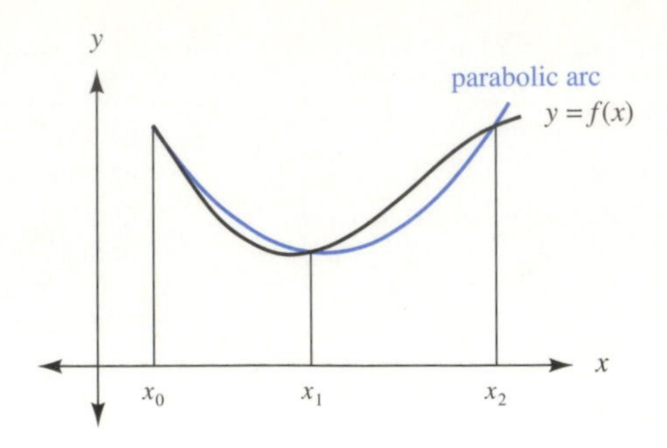

FIGURE 23.20

The method we are about to develop is called *Simpson's rule.* It is based on the following formula for the area of two adjacent strips when the fourth boundary is a parabolic arc. In problem 21 of Section 23.3, you were asked to derive this formula.

$$\text{area} = (\Delta x/3)[f(x_0) + 4f(x_1) + f(x_2)]$$

To find the area bounded by the x-axis, the lines $x = a$, $x = b$, and the graph of $y = f(x)$, divide the interval from a to b into n equal segments of width Δx $[\Delta x = (b - a)/n]$, marking the division points $x_1, x_2, x_3, \ldots, x_{n-1}$ ($x_0 = a$ and $x_n = b$). It is important that n be an even number, since we need two intervals for each parabolic arc. Draw a vertical line segment up to the curve at each x_i as in Figure 23.21.

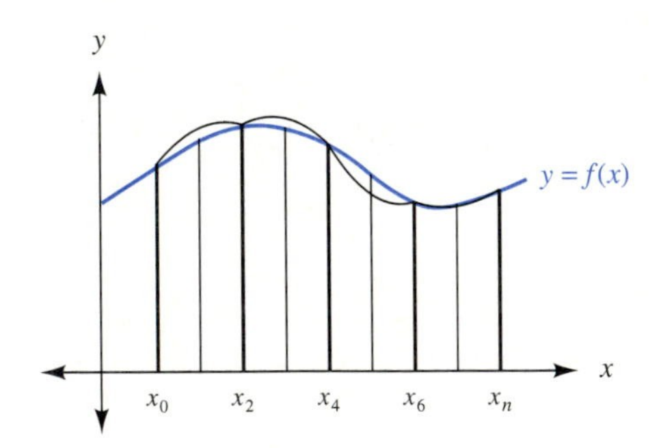

FIGURE 23.21

Group the strips into pairs, so that the top of each pair can be approximated by a parabolic arc. An approximation for the total area A under the curve is given by

$$A = (\Delta x/3)[f(x_0) + 4f(x_1) + f(x_2)] + (\Delta x/3)[f(x_2) + 4f(x_3) + f(x_4)] + \cdots + (\Delta x/3)[f(x_{n-2}) + 4f(x_{n-1}) + f(x_n)].$$

Factoring out $(\Delta x/3)$, we have

$$A = (\Delta x/3)[f(x_0) + 4f(x_1) + f(x_2) + f(x_2) + 4f(x_3) + f(x_4) + \cdots + f(x_{n-2}) + 4f(x_{n-1}) + f(x_n)].$$

When like terms are combined, the formula is written as

$$A = (\Delta x/3)[f(x_0) + 4f(x_1) + 2f(x_2) + 4f(x_3) + 2f(x_4) + \cdots + 2f(x_{n-2}) + 4f(x_{n-1}) + f(x_n)].$$

NOTE ⇨⇨ Notice the pattern of coefficients. ***Inside brackets, the first and last function values each have a coefficient of* 1. *The other coefficients alternate between* 4 *and* 2, *beginning and ending with* 4.** We generalize this result, just as we have previously generalized from areas to integrals.

SIMPSON'S RULE

If f is a function that is continuous for $a \le x \le b$, then

$$\int_a^b f(x)\,dx \approx (\Delta x/3)[f(x_0) + 4f(x_1) + 2f(x_2) + 4f(x_3) + 2f(x_4) + \cdots + 2f(x_{n-2}) + 4f(x_{n-1}) + f(x_n)]$$

where n is an even number, $\Delta x = (b - a)/n$, $x_0 = a$, $x_{i+1} = x_i + \Delta x$, and $x_n = b$.

EXAMPLE 1 Use Simpson's rule with $n = 4$ to find an approximate value for $\int_0^3 x^3\,dx$.

Solution Since the distance from 0 to 3 is 3, and there are 4 intervals, $\Delta x = 3/4$ and $\Delta x/3 = 1/4$ or 0.25. The function must be evaluated for $x_0 = 0$, $x_1 = 3/4$, $x_2 = 3/2$, $x_3 = 9/4$, and $x_4 = 3$.

$$\int_0^3 x^3\,dx = (0.25)[f(0) + 4f(3/4) + 2f(3/2) + 4f(9/4) + f(3)]$$
$$= (0.25)[0^3 + 4(3/4)^3 + 2(3/2)^3 + 4(9/4)^3 + 3^3]$$

Using a calculator to do the arithmetic, we see that the sum is 20.25. ■

In Example 1, the value of the integral is, in fact, exactly 20.25. It is surprising that Simpson's rule yields an exact answer whenever the function integrated is a cubic polynomial. It is not surprising that Simpson's rule yields an exact value for a quadratic function also, since the formula is based on a shape with a parabolic boundary.

Simpson's rule is generally conceded to be the best numerical method in elementary calculus. Some calculators have a key that calls up a built-in program to perform integration using Simpson's rule. As with other numerical methods, it is important to have some knowledge of how accurate the result is. There are, in fact, two sources of error in the trapezoidal rule and in Simpson's rule. One source of error is the formula itself. Each formula is based on a geometric figure that, in general, only approximates the figure whose area is sought. The second source of error is round-off error that results from the calculations. The first type of error may be reduced by increasing the number of intervals. As the number of intervals increases, however, the number of calculations also increases, so that the round-off error can accumulate. In some cases, the round-off error is so severe that a more accurate result is obtained when fewer intervals are used. In more advanced courses, these errors are studied in detail. For our purposes, the procedure described for the trapezoidal rule is a good one to use with Simpson's rule also. That is, repeat the procedure again and again, doubling n each time, until two successive approximations agree to the desired number of decimal places.

EXAMPLE 2 Use a calculator and Simpson's rule with $n = 4$ and also with $n = 8$ to find $\int_0^1 \frac{1}{x^2+1}\,dx$, and state the result to the appropriate degree of precision.

Solution If the calculations are done with a calculator, it is not necessary to write down the intermediate results, as is done here. The solution is shown in table form so that you can perform the calculations yourself and verify that you understand the procedure. The keystroke sequence for each $f(x_i)$ is

$$x_i \; \boxed{x^2} \; \boxed{+} \; 1 \; \boxed{=} \; \boxed{1/x}.$$

When n is 4, Δx is 1/4 or 0.25.

i	x_i	$f(x_i)$	Sum
0	0	1	1
1	0.25	0.9411765	4.7647059
2	0.50	0.8	6.3647059
3	0.75	0.64	8.9247059
4	1.00	0.5	9.4247059

Multiplying by $\Delta x/3$, we see that the value of the integral is about 0.7853922. When n is 8, Δx is 1/8 or 0.125.

i	x_i	$f(x_i)$	Sum
0	0	1	1
1	0.125	0.9846154	4.9384615
2	0.25	0.9411765	6.8208145
3	0.375	0.8767123	10.327664
4	0.5	0.8	11.927664
5	0.625	0.7191011	14.804068
6	0.75	0.64	16.084068
7	0.875	0.5663717	18.349555
8	1.0	0.5	18.849555

Multiplying by $\Delta x/3$, we see that the value of the integral is about 0.7853981.

Since the two calculations agree when rounded to four decimal places, we say that

$$\int_0^1 \frac{1}{x^2+1}\,dx = 0.7854, \text{ to four decimal places.} \quad ■$$

EXERCISES 23.6

In 1–6, find an approximate value for each integral, using Simpson's rule with n intervals. Round each answer to the number of decimal places specified. These integrals were given in problems 7–12 of Section 23.5. Compare answers.

1. $\int_{-1}^{1} (x^3 - x)\,dx$, $n = 4$, give an exact answer
2. $\int_{0}^{2} (2x^3 + x^2)\,dx$, $n = 4$, give an exact answer
3. $\int_{0}^{3} \frac{1}{2x+1}\,dx$, $n = 10$, round to the nearest hundredth
4. $\int_{1}^{3} \frac{1}{x^2}\,dx$, $n = 6$, round to the nearest hundredth
5. $\int_{1}^{4} \sqrt{x^2+1}\,dx$, $n = 8$, round to the nearest thousandth
6. $\int_{1}^{4} x\sqrt{x^3+1}\,dx$, $n = 10$, round to the nearest thousandth

In 7–12, find an approximate value for each integral, using Simpson's rule with 4 intervals and also with 8 intervals. Round each answer to the appropriate number of decimal places.

7. $\int_{-3}^{1} x^3 + 2x\,dx$
8. $\int_{-1}^{3} \frac{2x}{\sqrt{x+3}}\,dx$
9. $\int_{-1}^{2} \sqrt{x+1}\,dx$
10. $\int_{-1}^{1} \sqrt[3]{x+3}\,dx$
11. $\int_{0}^{\pi} 2\sin x\,dx$
12. $\int_{-1}^{1} 4^x\,dx$

---

13. A Gothic arch is formed by 2 arcs, each one-sixth of a circle. Find the approximate area of the Gothic arch window shown in the accompanying diagram. **1 2 4 7 8** 3 5

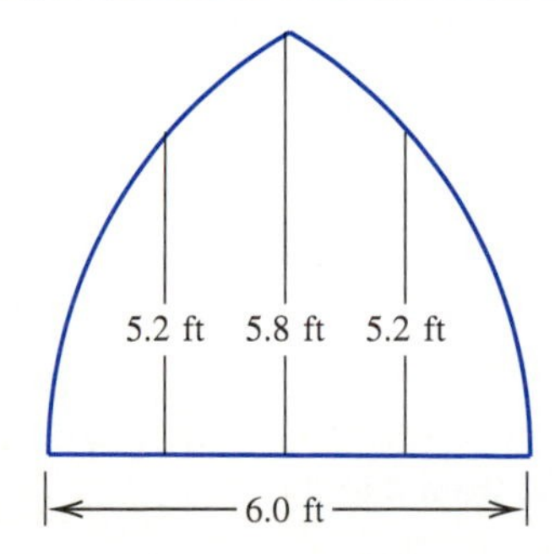

---

1. Architectural Technology
2. Civil Engineering Technology
3. Computer Engineering Technology
4. Mechanical Drafting Technology
5. Electrical/Electronics Engineering Technology
6. Chemical Engineering Technology
7. Industrial Engineering Technology
8. Mechanical Engineering Technology
9. Biomedical Engineering Technology

14. Find the approximate area of the lake shown in the accompanying diagram. 1 2 4 7 8 3 5

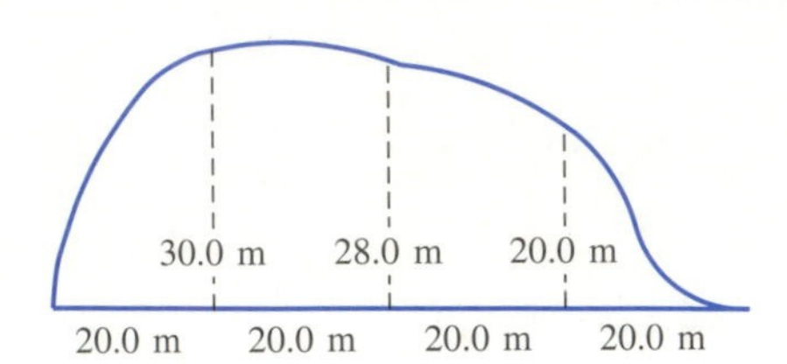

15. Find the approximate number of cubic meters of crushed rock required to make a road 0.5 km in length, if the cross section is as shown in the accompanying diagram. (Hint: Volume = cross sectional area × length of road.
1 2 4 6 8 5 7 9

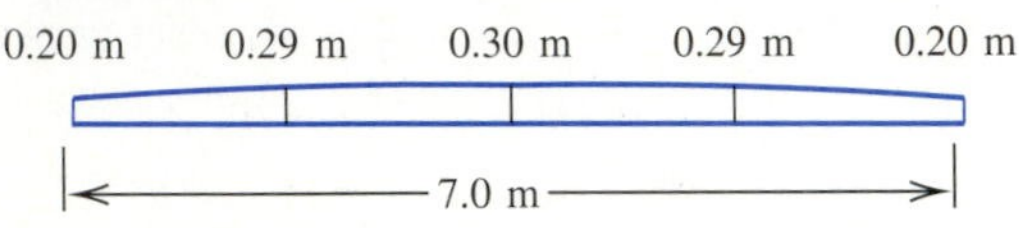

16. Find the approximate volume of gasoline that is stored in a cylindrical tank of radius 1.0 m and length 3.0 m when the depth of the gasoline is 0.50 m at the deepest point. The accompanying diagram shows the cross section. (Hint: Volume = cross-sectional area × length of tank.)
1 2 4 7 8 3 5

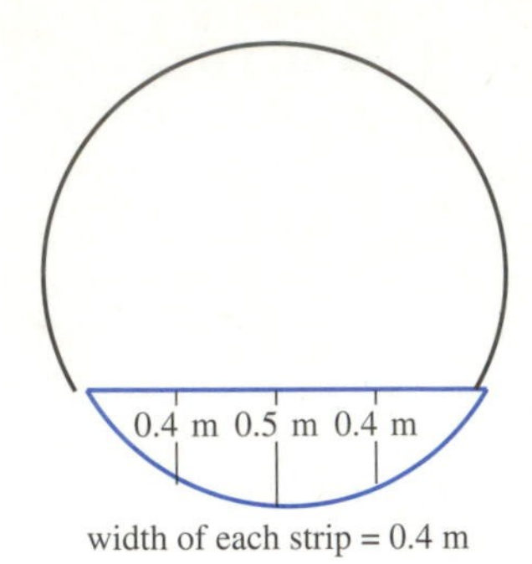

width of each strip = 0.4 m

17. The cross section of a parabolic reflector with a diameter of three inches is given by the equation $y = (1/4)x^2 + 1/2$. Find the approximate cross-sectional area if the x-axis represents the base of the parabola. 3 5 6 9

18. Energy pours into an integrated circuit memory chip at a rate given by the following table.

t (seconds)	0	0.01	0.02	0.03	0.04
P (joules/second)	0	12	20	26	26

Graph the data, and determine an approximate value for the energy delivered in the first 40 milliseconds by finding the area under the curve between 0 and 0.04. 3 5 9

19. For what type of graph does the trapezoidal rule yield a better approximation than Simpson's rule?

1. Architectural Technology
2. Civil Engineering Technology
3. Computer Engineering Technology
4. Mechanical Drafting Technology
5. Electrical/Electronics Engineering Technology
6. Chemical Engineering Technology
7. Industrial Engineering Technology
8. Mechanical Engineering Technology
9. Biomedical Engineering Technology

23.7 CHAPTER REVIEW

KEY TERMS

23.1 area under a curve
limits of integration
integral of $f(x)$ from a to b

23.2 antiderivative

23.4 definite integral
indefinite integral
differential of x
differential of y

SYMBOLS

$\int_a^b f(x)\,dx$ the definite integral

$\int f(x)\,dx$ the indefinite integral

dx the differential of x

Δx the change in x

dy the differential of y

Δy the change in y

RULES AND FORMULAS

Power Rule for Antiderivatives

If $g(x) = [f(x)]^n f'(x)$, then the antiderivatives of g are given by

$$G(x) = \frac{[f(x)]^{n+1}}{n+1} + C \text{ if } n \neq -1.$$

The Fundamental Theorem of Calculus

If f and F are two functions such that f is continuous for $a \leq x \leq b$, and $F'(x) = f(x)$, then

$$\int_a^b f(x)\,dx = F(b) - F(a).$$

Integration Rule

$$\int x^n\,dx = \frac{x^{n+1}}{n+1} \text{ if } n \neq -1$$

Properties of Integrals

$$\int_a^b cf(x)\,dx = c\int_a^b f(x)\,dx$$

$$\int_a^b [f(x) + g(x)]\,dx = \int_a^b f(x)\,dx + \int_a^b g(x)\,dx$$

The Power Rule for Integration

$$\int [f(x)]^n f'(x)\,dx = \frac{[f(x)]^{n+1}}{n+1} + C \text{ if } n \neq -1$$

The Trapezoidal Rule

If f is a function that is continuous for $a \leq x \leq b$, then

$$\int_a^b f(x)\,dx \approx (\Delta x/2)[f(x_0) + 2f(x_1) + 2f(x_2) + \cdots + 2f(x_{n-1}) + f(x_n)]$$

where $\Delta x = (b - a)/n$, $x_0 = a$, $x_{i+1} = x_i + \Delta x$, and $x_n = b$.

Simpson's Rule

If f is a function that is continuous for $a \leq x \leq b$, then

$$\int_a^b f(x)\,dx \approx (\Delta x/3)[f(x_0) + 4f(x_1) + 2f(x_2) + 4f(x_3) + 2f(x_4) + \cdots + 2f(x_{n-2}) + 4f(x_{n-1}) + f(x_n)]$$

where n is an even number, $\Delta x = (b - a)/n$, $x_0 = a$, $x_{i+1} = x_i + \Delta x$, and $x_n = b$.

GEOMETRY CONCEPTS

$a = lw$ (area of a rectangle)

$A = (h/2)(b_1 + b_2)$ (area of a trapezoid)

SEQUENTIAL REVIEW

(Section 23.1) *In 1–4, find an approximate value for each integral, using the number of rectangles specified.*

1. $\int_0^3 (3x - 2)\,dx$ Use three rectangles.

2. $\int_1^6 (x^2 - x)\,dx$ Use five rectangles.

3. $\int_0^4 (x^2 - 3x)\,dx$ Use four rectangles.

4. $\int_{-1}^1 (x^3 + x^2)\,dx$ Use six rectangles.

In 5–7, use the given number (n) of rectangles to find the approximate area bounded by the x-axis, the lines $x = a$ and $x = b$, and the graph of $y = f(x)$.

5. $y = 3x + 1$, $a = 0$, $b = 8$, $n = 4$

6. $y = 2x^2 + 1$, $a = -3$, $b = 2$, $n = 5$

7. $y = x^3 - x$, $a = -2$, $b = 1$, $n = 6$

(Section 23.2) *In 8–11, determine which of the functions listed are antiderivatives of $f(x)$.*

8. $f(x) = 2x^2 + 5x$
(a) $2x^3 + 5x^2$ **(b)** $2x^3/3 + 5x^2/2 + 3$ **(c)** $4x + 5$

9. $f(x) = 3x^2 - 5$
(a) $x^3 - 5x$ **(b)** $6x$ **(c)** $3x^3 + 5x + 4$

10. $f(x) = 4x^3 - 1$
(a) $12x^2$ **(b)** $4x^4 - x$ **(c)** $x^4 - x + 6$

11. $f(x) = 8x^3 + 2x$
(a) $2x^4 + x^2 - 1$ **(b)** $24x^2 + 2$ **(c)** $8x^4 + 2x^2 + 3$

In 12–15, find an expression to represent all antiderivatives F of the function f.

12. $f(x) = 3x + 2$

13. $f(x) = x^3 + 6x^2$

14. $f(x) = 2x^{-3} + x^{-1/3}$

15. $f(x) = 3/\sqrt{3x - 2}$

(Section 23.3) *In 16–20, evaluate each integral.*

16. $\int_0^4 (x^2 - 3x)\,dx$

17. $\int_{-1}^1 (x^3 + x^2)\,dx$

18. $\int_{-2}^{-1} 2x^{-3}\,dx$

19. $\int_4^9 x^{1/2}\,dx$

20. $\int_0^3 2x(x^2 + 2)^2\,dx$

In 21–23, find the exact area bounded by the x-axis, the lines $x = a$ and $x = b$, and the graph of $y = f(x)$.

21. $y = x^2 - 4$, $a = 0$, $b = 4$

22. $y = x^2 - 2x$, $a = 0$, $b = 2$

23. $y = x^3 + x$, $a = -1$, $b = 2$

(Section 23.4) In 24–31, find each integral.

24. $\int (4x^3 + x)\,dx$

25. $\int (x^{1/2} + 3x^{-3})\,dx$

26. $\int 2x(x^2 + 5)^4\,dx$

27. $\int (2x - 3)^{-2}\,dx$

28. $\int 3x\sqrt[3]{2x^2 - 7}\,dx$

29. $\int (3x + 4)^{-3}\,dx$

30. $\int 2x^2\sqrt{x^3 + 5}\,dx$

31. $\int \frac{(2x + 1)}{\sqrt{x^2 + x}}\,dx$

(Section 23.5) In 32–36, find an approximate value for each integral using the trapezoidal rule with n intervals. Round each answer to the nearest tenth.

32. $\int_0^4 (x^2 - 3x)\,dx \quad n = 4$

33. $\int_{-1}^1 (x^3 + x^2)\,dx \quad n = 6$

34. $\int_1^3 1/x\,dx \quad n = 4$

35. $\int_1^4 \sqrt{x^2 - 1}\,dx \quad n = 6$

36. $\int_0^3 3^x\,dx \quad n = 6$

In 37 and 38, assume that the tables represent experimental data. Use the trapezoidal rule with $n = 5$ to find an approximate value for the given integral.

37.

x	0	1	2	3	4	5
y	1	2	1	3	2	3

Find $\int_0^5 y\,dx$.

38.

x	0	2	4	6	8	10
y	0	3	7	13	15	20

Find $\int_0^{10} y\,dx$.

(Section 23.6) In 39–41, find an approximate value for each integral using Simpson's rule with n intervals. Round each answer to the nearest tenth.

39. $\int_{-1}^1 (x^3 + x^2)\,dx \quad n = 6$

40. $\int_1^3 1/x\,dx \quad n = 4$

41. $\int_1^4 \sqrt{x^2 - 1}\,dx \quad n = 6$

In 42 and 43, find an approximate value for each integral, using Simpson's rule with 4 intervals and also with 8 intervals. Round each answer to the appropriate number of decimal places.

42. $\int_1^4 \ln x\,dx$

43. $\int_0^{2\pi} \cos(x/4)\,dx$

In 44 and 45, assume that the tables represent experimental data. Use Simpson's rule with $n = 4$ to find an approximate value for the given integral.

44.

x	0	1	2	3	4
y	2	4	1	5	9

Find $\int_0^4 y\,dx$.

45.

x	2	4	6	8	10
y	0	1	2	4	4

Find $\int_2^{10} y\,dx$.

APPLIED PROBLEMS

1. The moment of inertia of a circular cylinder with respect to an axis through the center and perpendicular to the base is given by $I = \frac{m}{2}\int_0^2 x^3\,dx$ for a cylinder of radius 2, height 4, and mass m. Evaluate the integral. 1 2 4 8 7 9

2. The distance s (in ft) of an object moving with a velocity v (in ft/s) is given by $s = \int v\,dt$, where t is the time (in seconds). Find a formula for s, if $v = 4t + 12$. 4 7 8 1 2 3 5 6 9

3. A cantilever beam supports a load of 300.0 lb/ft of its length. Determine the total load on the first 6.50 feet of the beam by evaluating $W = \int_0^{6.50} 300.0\,dx$. 1 2 4 8 7 9

4. During an experiment, the power output P (in mW) of a certain fiber optic cable was given by $P = 1/(t^2 - 20t + 101)$ if t is in seconds. Sketch a graph of P versus t for $6 \le t \le 13$. Use the trapezoidal rule with $n = 6$ to find the approximate total energy W (in mJ) contained in the pulse between $t = 7$ and $t = 13$, given that $W = \int_7^{13} P\,dt$. 3 5 9 1 2 4 6 7 8

5. Use Simpson's rule to calculate the area in acres (1 acre = 43,560 square feet) of the lot shown in the diagram. 1 2 4 7 8 3 5

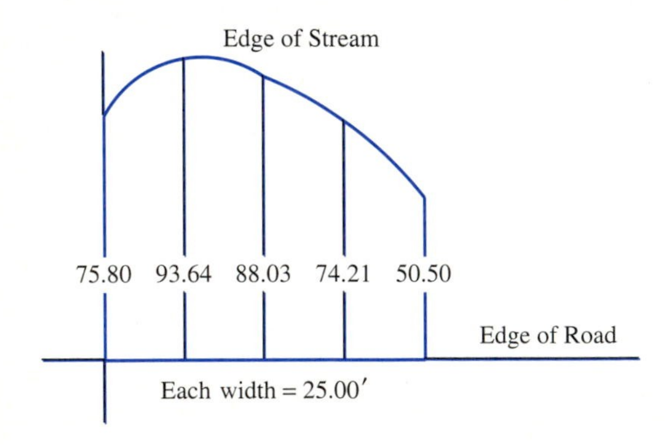

6. A businessman uses a new computer program to fit a curve to data giving him sales S (in dollars/hour) versus the number n of hours his store is open. The equation that results is

$$S = 0.1300n^5 - 4.170n^4 + 44.30n^3 - 183.3n^2 + 270.8n.$$

The area under the graph of S versus n gives the total sales for the first n hours the store is open. Find the total sales to the nearest dollar for the first 10 hours the store is open. 1 2 6 7 8 3 5 9

7. The formula $v = 3t^{0.65}$ gives the rate (in gal/min) at which water is pumped into an empty tank after time t (in minutes). The full tank alarm will ring when a total of 2500 gal has been pumped. The area under the graph of rate versus time gives the volume filled in time t. When will the alarm ring? Give your answer to the nearest minute. 4 8 1 2 6 7 9

8. The formula $F = 500\,s^{1.1}$ gives the force F (in lbs) needed to compress an automobile spring a distance s (in inches) from the rest position. Assume a spring is compressed two inches past the rest position. The work W (in inch·lbs) done in compression is given by $W = \int F\,ds$. Determine how many additional inches x of compression would require 6000 inch·lbs of work. (Hint: The limits of integration are 2 and x.) 4 7 8 1 2 3 5 6 9

9. The RMS value of a voltage pulse is defined as $V_{RMS} = \left[\frac{1}{b-a}\int_a^b [f(t)]^2\,dt\right]^{1/2}$, where $f(t)$ is the voltage as a function of time, and a and b are the voltage pulse widths. Find the RMS value of the voltage pulse V (in volts) given by $V = 150 - 150(t-6)^2$ when a is 5 and b is 7. Give your answer to three significant digits. 3 5 9 1 2 4 6 7 8

10. The length L along a curve given by $y = f(x)$ from $x = a$ to $x = b$ is given by $L = \int_a^b \sqrt{1 + (dy/dx)^2}\,dx$. Consider a cable that reaches from the base of one building across an

alley 30 feet wide to a point on the wall of another building. Find the length of the cable if it hangs in such a way that when x is the horizontal distance across the alley, its height y is given by $y = x^{3/2}$. Give your answer to three significant digits.

1 2 4 7 8 3 5

11. The force F (in lbs) needed to keep a large truck rolling at a constant speed over certain terrain is listed in the following table for various distances s (in ft). Use the trapezoidal rule to estimate the work W (in ft·lb) done in moving the truck 1000 ft, given that $W = \int F\,ds$. Give your answer to two significant digits.

s	0	200	400	600	800	1000
F	0	25	150	150	250	250

4 7 8 1 2 3 5 6 9

12. A plane flies in a straight line away from an airport. Its speed v (in mph) is listed in the following table for various times t (in hours). Use the trapezoidal rule to estimate the distance s (in miles) the plane travels during the time interval shown, given that $s = \int v\,dt$. Give your answer to the nearest mile.

t	0	1	2	3	4	5	6	7
v	400	300	350	200	300	250	375	500

4 7 8 1 2 3 5 6 9

13. The current i (in milliamps) in a resistance is listed in the following table for various times t (in seconds). Use the trapezoidal rule to estimate the charge q (in millicoulombs) passing through the resistor during the interval shown, given that $q = \int i\,dt$. Give your answer to three significant digits.

t	1	2	3	4	5	6
i	2236	786	347	179	103	65

3 5 9 1 2 4 6 7 8

14. Depth measurements D (in ft) made at various distances W (in ft) from shore across a 320 ft wide river that flows at 2.7 mph are listed in the following table. Estimate the number N of cubic feet of water passing by in one 24-hour period. (Hint: Use Simpson's rule to estimate the cross-sectional area of the river.)

W	0	80	160	240	320
D	7	25	8	25	5

1 2 4 7 8 3 5

15. Measurements of output power P (in milliwatts) for various wavelengths λ (in nanometers) of the light from an LED (light emitting diode) are listed in the following table. Estimate the average output power P_{ave} (in milliwatts) due to all wavelengths, given that $P_{ave} = \frac{1}{\lambda_2 - \lambda_1} \int_{\lambda_1}^{\lambda_2} P\,d\lambda$. Give your answer to two significant digits.

λ	600	620	640	660	680
P	40	100	50	25	8

9 1 2 3 5 6 7

16. The width W (in ft) of a twenty mile long finger lake is measured every four miles along the length of the lake. The results are shown in the following table. Estimate the surface area (in square miles) of the lake. Give your answer to two significant digits.

L	0	4	8	12	16	20
W	1520	2500	2275	2475	2510	2200

1 2 4 7 8 3 5

1. Architectural Technology
2. Civil Engineering Technology
3. Computer Engineering Technology
4. Mechanical Drafting Technology
5. Electrical/Electronics Engineering Technology
6. Chemical Engineering Technology
7. Industrial Engineering Technology
8. Mechanical Engineering Technology
9. Biomedical Engineering Technology

17. Measurements of the pressure P (in lb/in^2) for various volumes V (in in^3) for one cylinder of an engine are listed in the following table. Estimate the work W (in ft·lb) done by this cylinder as the gas expands, given that $W = P\int dV$. Give your answer to two significant digits.

V	5	10	15	20	25	30
P	347	150	97	66	55	40

3 5 7 1 2 4 6 8 9

18. A loud speaker cone has the cross-sectional areas A (in in^2) shown in the following table for various distances x (in inches) from the magnet. Estimate the volume V (in in^3) of the cone, given $V = \int A\,dx$. Give your answer to three significant digits.

d	0	0.5	1.0	1.5	2.0	2.5	3.0
A	7	8.70	14.7	28.3	55.0	100	175

1 2 4 7 8 3 5

1. Architectural Technology
2. Civil Engineering Technology
3. Computer Engineering Technology
4. Mechanical Drafting Technology
5. Electrical/Electronics Engineering Technology
6. Chemical Engineering Technology
7. Industrial Engineering Technology
8. Mechanical Engineering Technology
9. Biomedical Engineering Technology

RANDOM REVIEW

1. Use Simpson's rule with 4 intervals to evaluate $\int_6^8 1/(x-5)\,dx$. Round your answer to the nearest tenth.

2. Find $\int (x+2)\sqrt{x^2+4x}\,dx$.

3. Find an expression to represent all antiderivatives F of the function defined by $f(x) = (2x^2+3)^{-4}(4x)$.

4. Use the trapezoidal rule with 4 intervals to evaluate $\int_6^8 1/(x-5)\,dx$. Round your answer to the nearest tenth.

5. Determine which of the functions listed are antiderivatives of $f(x) = 6x^5 + 4x^3 - 3$.
(a) $F(x) = 6x^6 + 4x^4 - 3x$ **(b)** $F(x) = x^6 + x^4 - 3x + 2$ **(c)** $F(x) = 30x^4 + 12x^2$

6. Evaluate $\int_{-2}^3 (3x^5 + x^3 + 1)\,dx$.

7. Evaluate $\int_4^9 (-1/4)x^{-1/2}\,dx$

8. Use four rectangles to find the approximate area bounded by the x-axis, the lines $x = -3$ and $x = 1$, and the graph of $y = \sqrt{x+3}$. Round your answer to the nearest tenth.

9. Find $\int (4x^3 + x)\,dx$.

10. Assume that the following table represents experimental data. Use the trapezoidal rule with $n = 5$ to find an approximate value for the given integral.

x	0	1	2	3	4	5
y	1	2	1	3	2	3

Find $\int_0^5 y\,dx$.

CHAPTER TEST

In 1 and 2, evaluate each integral.

1. $\int_{1}^{2} (2x^3 + 3x)\, dx$

2. $\int_{-3}^{-2} 4x^{-3}\, dx$

3. Find an expression to represent all antiderivatives F of the function f defined by $f(x) = x(x^2 + 1)^3$.

4. State the fundamental theorem of calculus.

5. Find $\int \frac{3x^2 - 2}{\sqrt{x^3 - 2x}}\, dx$

In 6–8, find an approximate value for each integral, as directed.

6. $\int_{-2}^{3} (x^3 - x^2 - 6x)\, dx$ Use five rectangles.

7. $\int_{0}^{2} (x^3 - 2x^2 + 4x - 8)\, dx$ Use four trapezoids.

8. $\int_{-1}^{2} (x^4 - 5x^2 + 4)\, dx$ Use Simpson's rule with $n = 6$.

9. Determine the load W (in pounds) on the first 8 feet of a cantilever beam supporting a distributed load of 250 lb/ft by evaluating $W = \int_{0}^{8} 250\, x^{2.5}\, dx$.

10. During a rainstorm, the rate of rainfall intensity is measured (in inches/hr) at 15-minute intervals as 0, 0.4, 0.8, 1.1, 1.2, 1.3, and 1.5. Plot the rainfall intensity versus time, and use Simpson's rule to calculate the area under the curve.

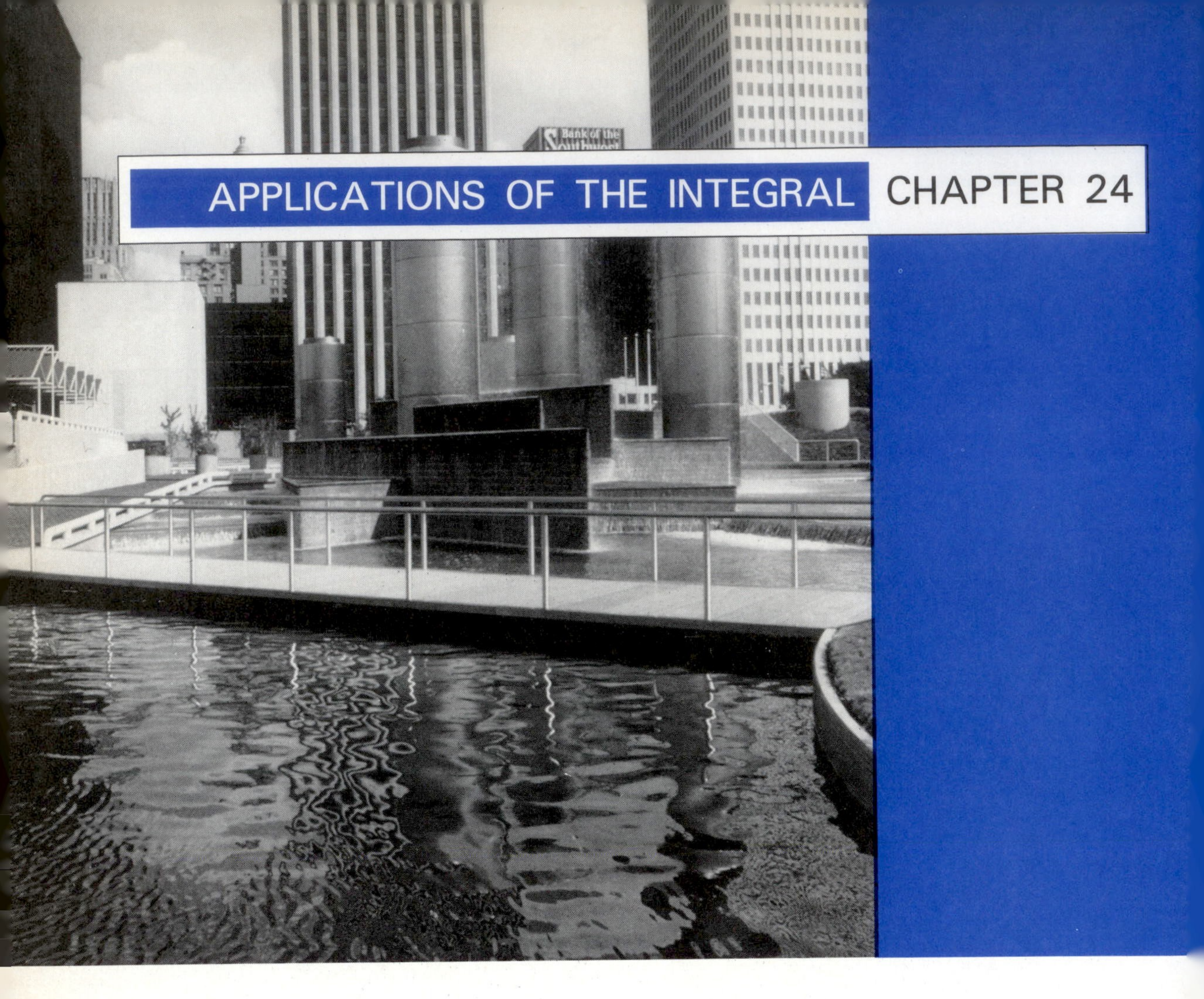

APPLICATIONS OF THE INTEGRAL CHAPTER 24

In Chapter 23, we used an integral to find the area in a variety of situations. There are many quantities in addition to area that can be computed using integration. The versatility of the integral is due to its interpretation as a sum of infinitely many parts of the form $f(x)\,dx$. We are now ready to solve problems of area, volume, velocity, acceleration, averages, work, and fluid pressure.

24.1 AREA BETWEEN CURVES

In Chapter 23, we used integration to find the area of the region bounded by the x-axis, the lines $x = a$ and $x = b$, and the graph of $y = f(x)$. Integration can also be used to find the area of a region that has more than one nonlinear boundary. A region is said to be bounded by two or more curves when it is completely enclosed by the curves. The region can be partitioned into n vertical strips, each of which has an area approximately equal to the area of a rectangle, as shown in Figure 24.1.

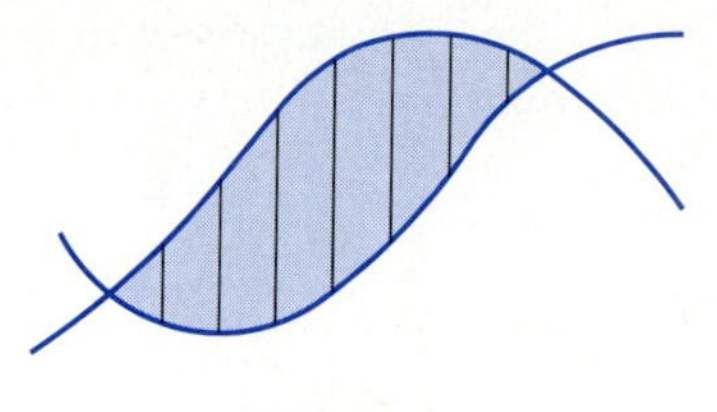

FIGURE 24.1

The sum of the areas of the rectangles approximates the area sought. The exact area is taken to be the limit of the sum of the rectangular areas as the number of rectangles becomes arbitrarily large. Whenever we can view a problem as finding the limit of a sum of parts as the number of parts approaches infinity, integration can be used.

Notice from Figure 24.2 that subtracting the y-coordinate of the lower curve from the y-coordinate of the upper curve yields a positive difference, regardless of whether both curves are above the x-axis, below the x-axis, or one above and one below.

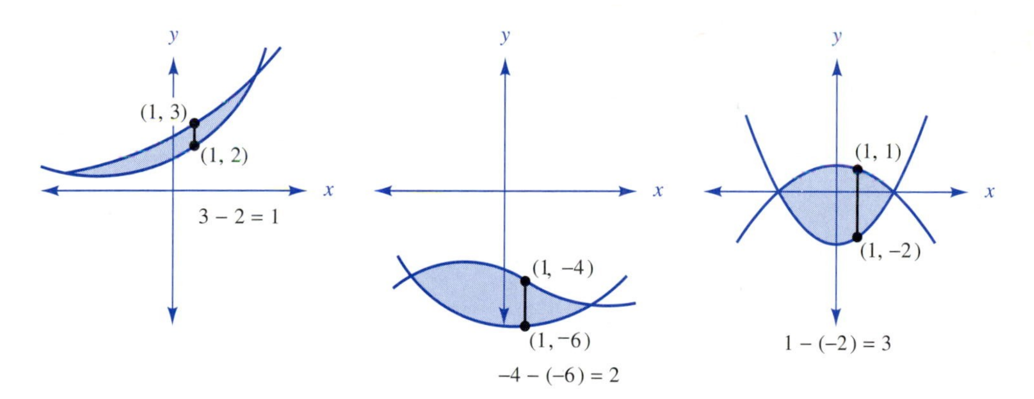

FIGURE 24.2

We now begin to take the short cut used in applied mathematics. Think of the integral sign as denoting a sum, and think of each rectangle as having width dx and height at x obtained by subtracting the y-coordinates at x. If we let $y = u(x)$ denote the upper curve and $y = l(x)$ denote the lower curve, then $u(x) - l(x)$ denotes the height of the rectangle at x. Thus, we go directly to the integral without using Δx and the limit process. We make the following generalization.

To find the area of a region between curves:

1. Identify the x-coordinate at the left-hand side of the region, and call it a. Identify the x-coordinate at the right-hand side of the region, and call it b.
2. Let $y = u(x)$ be the equation of the upper curve and $y = l(x)$ be the equation of the lower curve.
3. Evaluate $\int_a^b [u(x) - l(x)]\,dx$ to find the area.

EXAMPLE 1 Find the area of the region in the third quadrant bounded by the graphs of $y = x^3$ and $y = x$, as shown in Figure 24.3.

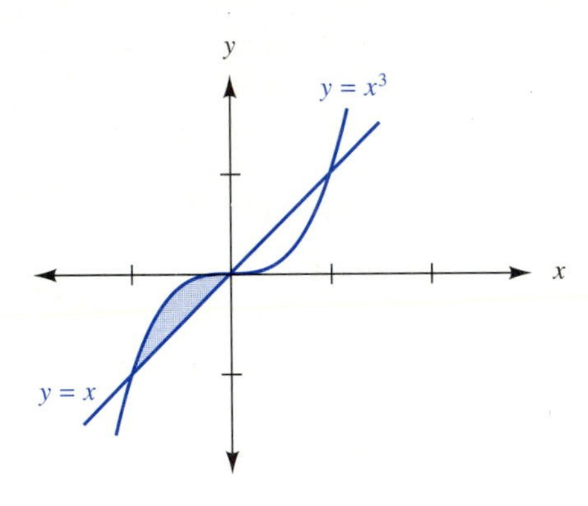

FIGURE 24.3

Solution 1. The curves intersect when they have the same y-coordinate. That is,

$$x^3 = x.$$

$$x^3 - x = 0 \quad \text{Subtract } x \text{ from both sides.}$$

$$x(x^2 - 1) = 0 \quad \text{Factor out } x.$$

$$x(x-1)(x+1) = 0 \quad \text{Factor } x^2 - 1.$$

$$x = 0,\ x - 1 = 0,\ \text{or } x + 1 = 0 \quad \text{Set each factor equal to zero.}$$

$$x = 0,\ x = 1,\ \text{or } x = -1 \quad \text{Solve for } x.$$

2. The curve $y = x^3$ is above the line $y = x$ throughout the third quadrant.
3. Using $y = x^3$ as the upper curve and $y = x$ as the lower curve, we have

$$\int_{-1}^{0} (x^3 - x)\,dx = \frac{x^4}{4} - \frac{x^2}{2}\Bigg|_{-1}^{0} = 0 - \left[\frac{1}{4} - \frac{1}{2}\right] = \frac{1}{4}. \quad \blacksquare$$

EXAMPLE 2 Find the area of the region bounded by $y = x^2 - 4$ and $y = x + 2$.

Solution The graphs of $y = x^2 - 4$ and $y = x + 2$ are shown in Figure 24.4.

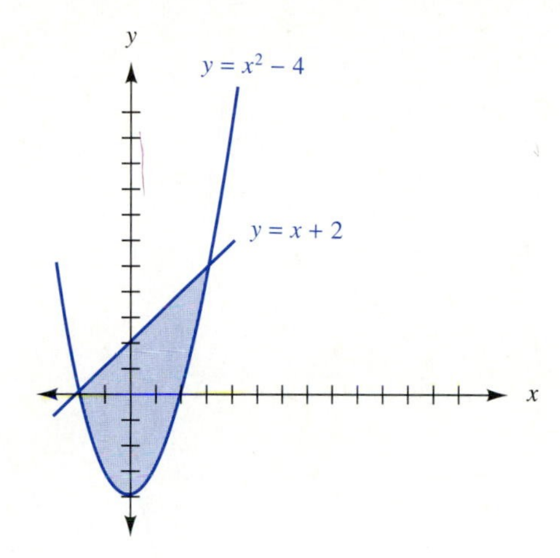

FIGURE 24.4

1. Since the area bounded by the curves is the area that is enclosed by the curves, it is necessary to find the x-values of the points of intersection.

$$x^2 - 4 = x + 2$$
$$x^2 - x - 6 = 0 \quad \text{Subtract } x + 2 \text{ from both sides.}$$
$$(x - 3)(x + 2) = 0 \quad \text{Factor.}$$
$$x - 3 = 0 \text{ or } x + 2 = 0 \quad \text{Set each factor equal to zero.}$$
$$x = 3 \text{ or } x = -2 \quad \text{Solve.}$$

2. The line is above the parabola throughout the region, so the line is the upper curve, and the parabola is the lower curve.
3. The area is given by

$$\int_{-2}^{3} [(x + 2) - (x^2 - 4)]\, dx = \int_{-2}^{3} (-x^2 + x + 6)\, dx$$
$$= \frac{-x^3}{3} + \frac{x^2}{2} + 6x \Big|_{-2}^{3}$$
$$= [-9 + 9/2 + 18] - [8/3 + 2 - 12]$$
$$= 27/2 - (-22/3)$$
$$= 81/6 + 44/6 \text{ or } 125/6, \text{ which is } 20\ 5/6. \quad ■$$

Sometimes the curves cross in the region under consideration. In such cases, the area should be computed in two or more parts. Example 3 illustrates this type of problem.

EXAMPLE 3 Find the area of the region bounded by $y = x^2 - 1$, $y = 1 - x^2$, and the lines $x = 0$ and $x = 2$.

Solution The graphs are shown in Figure 24.5.

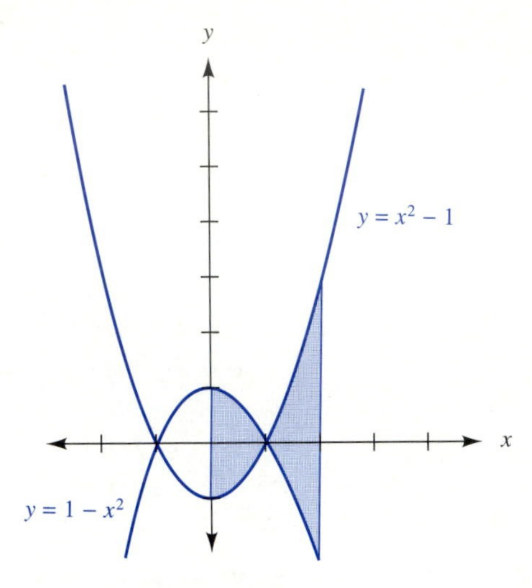

FIGURE 24.5

1. Find the x-values of the points of intersection.

$$x^2 - 1 = 1 - x^2$$
$$2x^2 - 2 = 0 \qquad \text{Add } -1 + x^2 \text{ to both sides.}$$
$$2(x^2 - 1) = 0 \qquad \text{Factor out 2.}$$
$$2(x - 1)(x + 1) = 0 \qquad \text{Factor } x^2 - 1.$$
$$x - 1 = 0 \text{ or } x + 1 = 0 \qquad \text{Set each factor equal to zero.}$$
$$x = 1 \text{ or } x = -1 \qquad \text{Solve for } x.$$

2. Between $x = 0$ and $x = 1$, the graph of $y = 1 - x^2$ is the upper curve, but between $x = 1$ and $x = 2$, the graph of $y = x^2 - 1$ is the upper curve.

3. The area is given by

$$\int_0^1 [(1 - x^2) - (x^2 - 1)]\, dx + \int_1^2 [(x^2 - 1) - (1 - x^2)]\, dx$$
$$= \int_0^1 (2 - 2x^2)\, dx + \int_1^2 (2x^2 - 2)\, dx$$
$$= \left[2x - \frac{2x^3}{3}\right]\Bigg|_0^1 + \left[\frac{2x^3}{3} - 2x\right]\Bigg|_1^2$$
$$= [(2 - 2/3) - 0] + [(16/3 - 4) - (2/3 - 2)]$$
$$= 4/3 + [4/3 - (-4/3)]$$
$$= 4/3 + 8/3 \text{ or } 12/3, \text{ which is } 4. \quad ■$$

EXAMPLE 4 A lens that is 6.00 cm in diameter is made with a parabolic surface on each side. The equations of the two parabolas are $y = 0.020x^2 + 0.500$ and $y = -0.025x^2 - 0.500$. Find the cross-sectional area as shown in Figure 24.6.

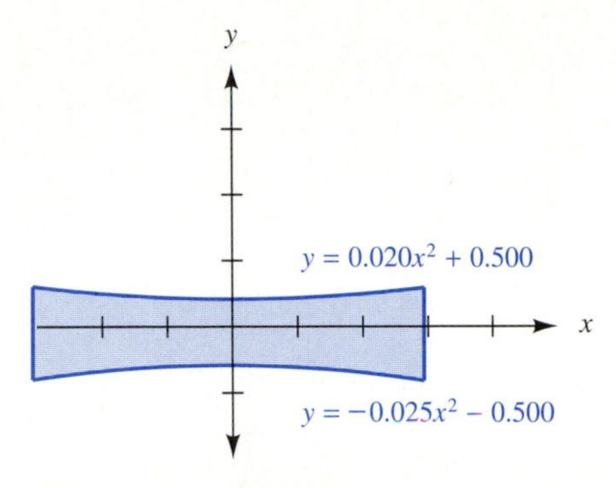

FIGURE 24.6

Solution Since the diameter of the lens is 6.00 cm, the edges are given by the lines $x = -3.00$ and $x = 3.00$. The area sought is given by

$$\int_{-3.00}^{3.00} [(0.020x^2 + 0.500) - (-0.025x^2 - 0.500)]\, dx$$

$$= \int_{-3.00}^{3.00} (0.045x^2 + 1.00)\, dx$$

$$= \frac{0.045x^3}{3} + x \Bigg|_{-3.00}^{3.00}$$

$$= 0.015x^3 + x \Bigg|_{-3.00}^{3.00}$$

$$= (0.405 + 3.00) - (-0.405 - 3.00)$$

$$= 6.81 \text{ cm}^2. \quad \blacksquare$$

EXERCISES 24.1

In 1–15, find the areas bounded by the given curves.

1. $y = x^2$ and $y = 4$
2. $y = x^2$ and $y = 9$
3. $y = x^2 - 3$ and $y = -2x$
4. $y = -x^2 + 4$ and $y = 3x$
5. $y = x^2$ and $y = x^3$
6. $y = x^2$ and $y = x^4$
7. $y = x^3 - x$ and $y = 3x$
8. $y = x^3 - 2x$ and $y = -x$
9. $y = x^2$ and $y = \sqrt{x}$
10. $y = \sqrt{x}$, $y = -\sqrt{x}$, and $x = 4$
11. $y = x$, $y = 2 - x$, and $y = 0$
12. $y = x + 1$, $y = -x - 1$, and $x = 1$
13. $y = x^3$, $y = 8$, and $x = -2$
14. $y = x^4$ and $y = 2x^2 - 1$
15. $y = x^4$ and $y = 8x^2 - 16$

16. Find the area of the metal plate shown in the accompanying diagram. Measurements are in cm.
1 2 4 7 8 3 5

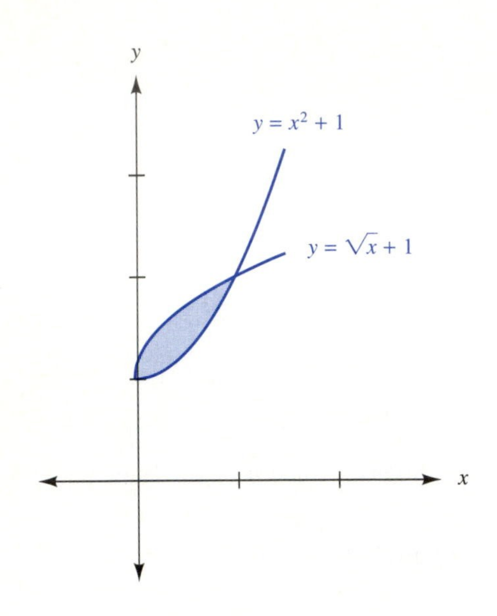

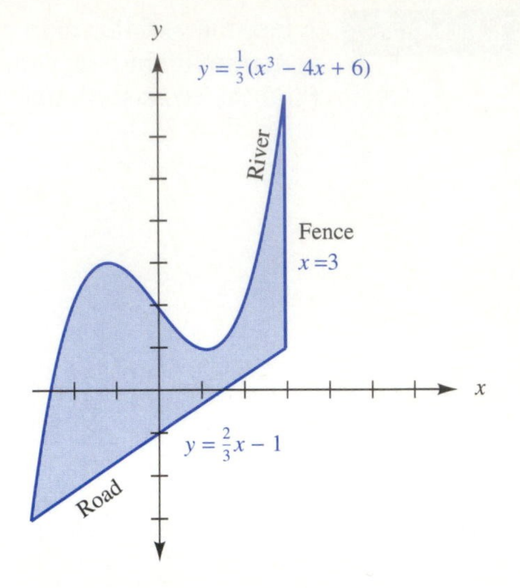

17. Find the cross-sectional area of the dam shown in the accompanying diagram. Measurements are in meters. 1 2 4 7 8 3 5

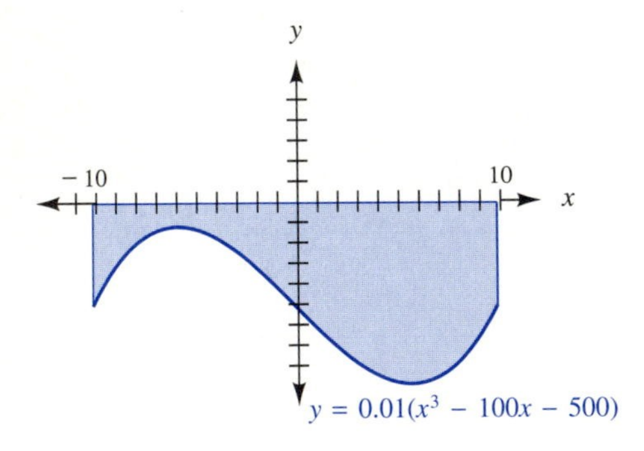

18. A farm is bounded by a river, a road, and a fence, as shown in the accompanying diagram. Find the area if measurements are in rods.
1 2 4 7 8 3 5

19. Consider a building on a hill with a slope of 1/2. Determine the area of a parabolic ($y = -x^2 + 10$) window as shown in the accompanying diagram, if measurements are in feet. Use three significant digits. 1 2 4 7 8 3 5

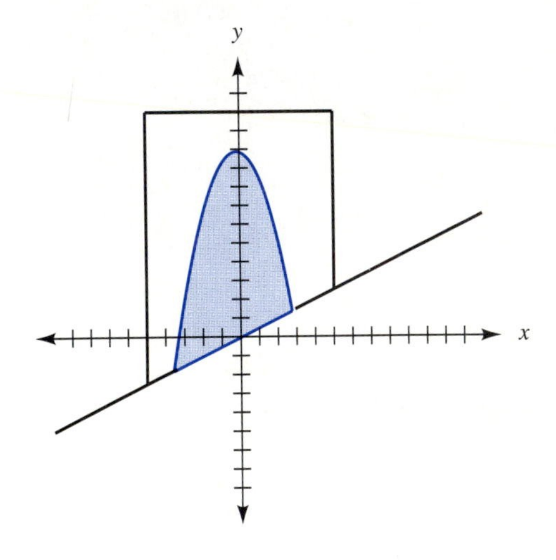

20. A lens of diameter 6.00 cm is made with a parabolic surface on each side. The equations of the two parabolas are $y = 0.250x^2 + 0.300$ and $y = -0.250x^2 - 0.400$. Find the cross-sectional area.
3 5 6 9

21. A parking lot has boundaries given by $y = 0.25x^2 + 1$, $y = 3x^2 + 2$, $x = 0$, and $y = 5$. Draw a sketch and find the area, if measurements are in rods. Use three significant digits. 1 2 4 7 8 3 5

22. The cross section of a stream bed can be described by the equation $7y = x^2$. Draw a sketch and find the cross-sectional area of the stream if it is 7 ft deep at the center. 2 1 9

23. In physics, energy W (in joules) is defined as the integral of power with respect to time. How much energy would be contained in a pulse of laser light whose power (in watts) is given by the equation $p = 1400 - 1400(t - 6.0)^2$? (Hint: Find the area under the curve for nonnegative p.) 5 3 4 7 8 9

1. Architectural Technology
2. Civil Engineering Technology
3. Computer Engineering Technology
4. Mechanical Drafting Technology
5. Electrical/Electronics Engineering Technology
6. Chemical Engineering Technology
7. Industrial Engineering Technology
8. Mechanical Engineering Technology
9. Biomedical Engineering Technology

24.2 VOLUME OF A SOLID OF REVOLUTION: DISK METHOD

NOTE

You have seen a number of problems in which area was found by evaluating an integral. We are now ready to find other quantities that can be computed using integration. ***The key concept is that the integral can be thought of as the sum of infinitely many parts of the form f(x) dx.*** In this section, we examine how the volume of many geometric solids may be obtained in this manner.

solid of revolution

A **solid of revolution** is a figure that can be generated by rotating a plane figure about a line in a coordinate plane. Consider the region formed by the line $x = 3$, the x-axis, and the parabola $y = (1/2)x^2$ as shown in Figure 24.7a. Now imagine that the shaded region is hinged along the x-axis and visualize the region rotating around the x-axis. The region sweeps out a shape like that in Figure 24.7b.

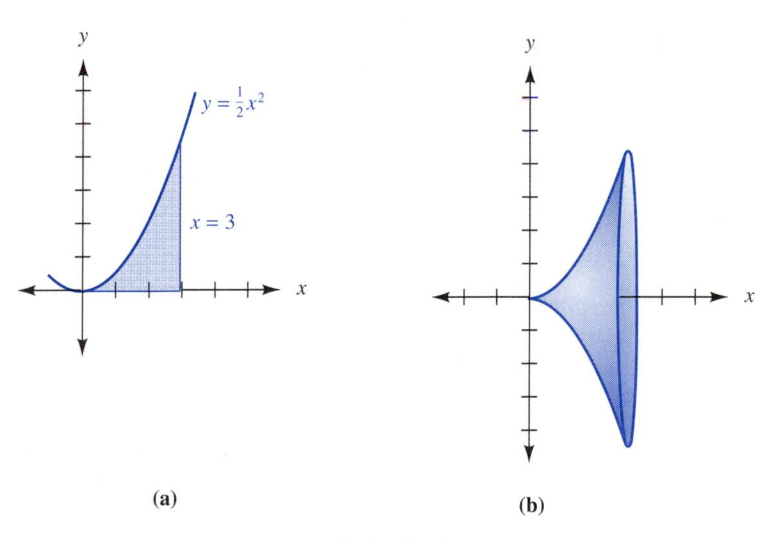

FIGURE 24.7

Now slice the area of Figure 24.7a into many vertical slices, as shown in Figure 24.8a. Each slice can be approximated by a rectangle of width dx. As the rectangles are rotated about the x-axis, each one sweeps out a disk or cylinder whose height is dx and radius is equal to the y-coordinate at the top of the generating rectangle. Figure 24.8b shows that the figure formed is a stack of disks like weights on the end of a barbell.

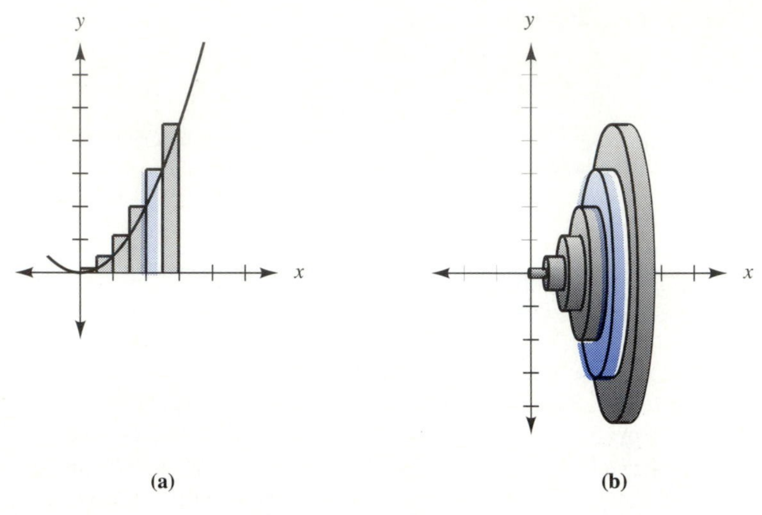

FIGURE 24.8

Each disk has a volume given by the formula $V = \pi r^2 h$. The volume of the solid in Figure 24.7b is approximately equal to the sum of the volumes of all the disks in Figure 24.8b. As the number of disks becomes arbitrarily large, the sum is an integral.

EXAMPLE 1 Find the volume of the solid generated by rotating the area bounded by $y = (1/2)x^2$, the x-axis, and $x = 3$ about the x-axis.

Solution This is the problem we have been examining. Each disk has a width of dx and a radius given by a y-coordinate of the parabola. The volume dV of each disk is given by

$$dV = \pi r^2 h \text{ with } r = y \text{ and } h = dx.$$

$$dV = \pi[(1/2)x^2]^2\, dx$$

The volume of the solid is given by

$$V = \int_0^3 \pi[(1/2)x^2]^2\, dx.$$

$$= \int_0^3 (\pi/4)x^4\, dx \qquad \text{Replace } [(1/2)x^2]^2 \text{ with } (1/4)x^4.$$

$$= \frac{\pi}{4}\left[\frac{x^5}{5}\right]\Bigg|_0^3 \qquad \text{Treat } \pi/4 \text{ as a constant and integrate } x^4.$$

$$= \frac{\pi x^5}{20}\Big|_0^3 \qquad \text{Multiply } \frac{\pi}{4}\left(\frac{x^5}{5}\right).$$

$$= \frac{243\pi}{20} \qquad \text{Evaluate the integral.}$$

The same principle can be used when an area is rotated about the y-axis. Example 2 illustrates the procedure.

EXAMPLE 2 Compute the volume of the cone generated by rotating the triangle bounded by the x-axis, the y-axis, and the line $y = -2x + 6$ about the y-axis.

Solution Since rotation is about the y-axis, slice the area horizontally, as shown in Figure 24.9a. The solid generated is shown in Figure 24.9b.

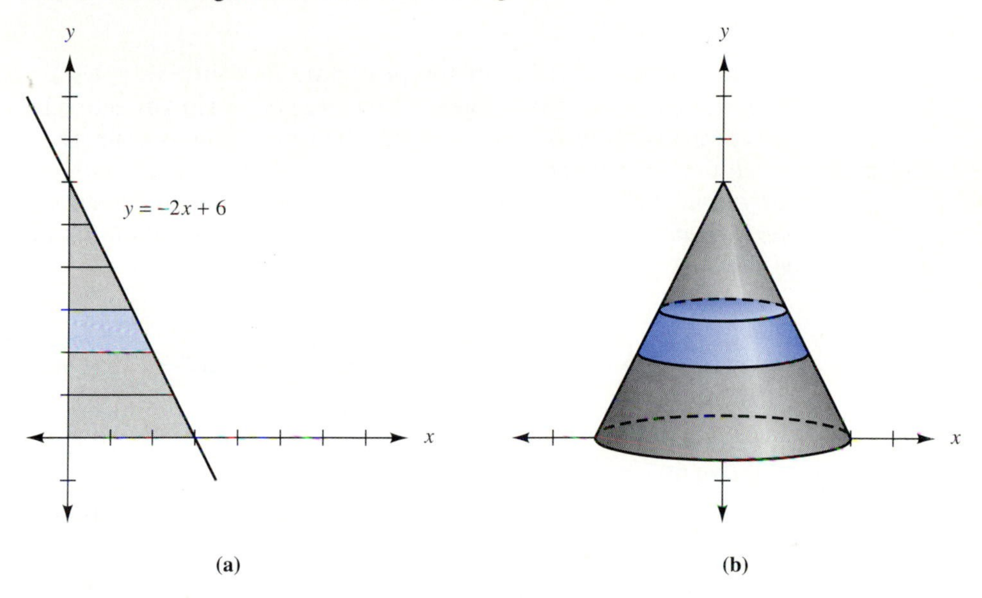

FIGURE 24.9

Each disk has radius given by an x-coordinate on the line $y = -2x + 6$, so we solve the equation for x.

$$-2x = y - 6$$
$$x = \frac{y-6}{-2}$$
$$= (-1/2)(y) + 3$$

Each disk, then, has a volume dV given by

$$dV = \pi r^2 h \text{ with } r = x \text{ and } h = dy.$$
$$dV = \pi[(-1/2)y + 3]^2\, dy$$

NOTE ⇨⇨ ***Since the disks are stacked up vertically, the limits of integration must be on the vertical axis.*** That is, the sum of the volumes is found between $y = 0$ and $y = 6$. Hence, the volume of the cone is given by

$$V = \int_0^6 \pi[(-1/2)y + 3]^2\,dy \text{ or } \pi \int_0^6 [(-1/2)y + 3]^2\,dy.$$

Multiplying by -2 to the left of the integral sign and by $-1/2$ to the right of it, we are able to use the power rule.

$$\begin{aligned} V &= -2\pi \int_0^6 [(-1/2)y + 3]^2(-1/2)\,dy \\ &= -2\pi \left.\frac{[(-1/2)y + 3]^3}{3}\right|_0^6 \\ &= -2\pi[0 - 9] \text{ or } 18\pi \quad \blacksquare \end{aligned}$$

To see the generality of the procedure used in Examples 1 and 2, it is very important to realize that the generating area is sliced in a direction that is *perpendicular* to the axis of revolution, and that the summation is done *parallel to* the axis of revolution. ***Thus, when a figure is sliced vertically, the summation is done horizontally, and the limits of integration are x-values. Likewise, when a figure is sliced horizontally, the summation is done vertically, and the limits of integration are y-values.***

NOTE ⇨⇨

To find the volume of a solid of revolution by the disk method:

If a plane figure bounded by the graph of $y = f(x)$ and one of the coordinate axes is rotated about that axis to form a solid that can be sliced into disks, follow the steps listed for the appropriate axis.

	Axis of Revolution	
	x-axis	**y-axis**
1. Slice the plane figure:	vertically	horizontally
2. Identify the radius r of 1 disk:	y	x
3. Identify the height h of 1 disk:	dx	dy
4. Express the volume of 1 disk: $dV = \pi r^2 h$	$dV = \pi y^2\,dx$	$dV = \pi x^2\,dy$
5. Identify the limits of integration:	$x = a, x = b$	$y = c, y = d$
6. Evaluate the integral:	$V = \pi \int_a^b y^2\,dx$ with y expressed in terms of x	$V = \pi \int_c^d x^2\,dy$ with x expressed in terms of y

EXAMPLE 3 Find the volume of the solid generated by rotating the area bounded by $y = \sqrt{4 - x}$, the y-axis, and the x-axis about the x-axis.

Solution The generating region is shown in Figure 24.10a, and the solid generated is shown in Figure 24.10b.

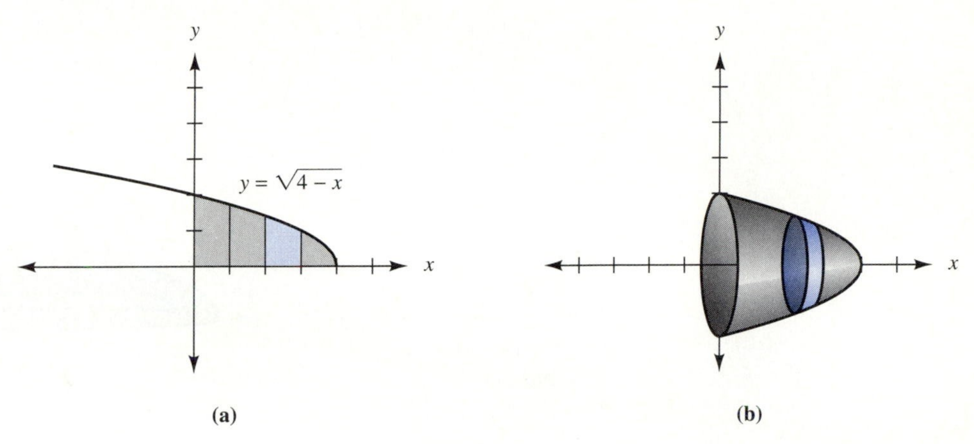

FIGURE 24.10

1. Slice the figure vertically, as shown in Figure 24.10a.

2–3. The radius of each disk is y, and the height is dx.

4. Each disk has a volume dV given by

$$dV = \pi y^2 \, dx$$
$$dV = \pi(\sqrt{4 - x})^2 \, dx$$
$$= \pi(4 - x) \, dx.$$

5. Since the disks were sliced vertically, the sum must be taken horizontally. The left-hand limit is 0, since the figure is bounded by the y-axis. The right-hand limit occurs where the curve intersects the x-axis. That is,

$$0 = \sqrt{4 - x}, \text{ so } x = 4.$$

The limits of integration are 0 and 4.

6. The volume of the solid is given by

$$V = \int_0^4 \pi(4 - x) \, dx.$$
$$= \pi \int_0^4 (4 - x) \, dx$$
$$= \pi[4x - (1/2)x^2] \Big|_0^4$$
$$= \pi[(16 - 8) - (0 - 0)] \text{ or } 8\pi. \quad ■$$

EXAMPLE 4 Find the volume of the solid generated by rotating the area bounded by $y = \sqrt{4 - x}$, the y-axis, and the x-axis about the y-axis.

Solution Slice the figure horizontally, as shown in Figure 24.11a. The solid generated is shown in Figure 24.11b.

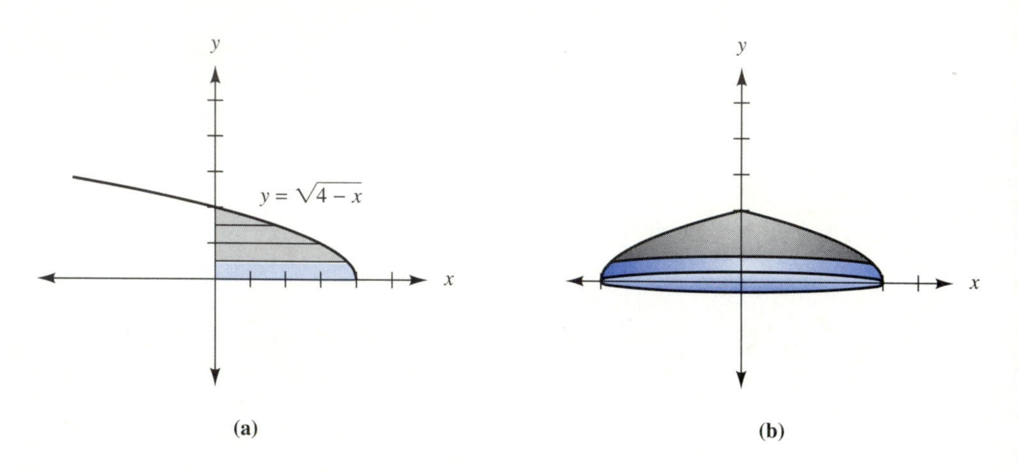

FIGURE 24.11

Since the y-axis is the axis of revolution, the radius of each disk is an x-value.

$$y = \sqrt{4 - x}$$

$$y^2 = 4 - x \quad \text{Square both sides.}$$

$$x = 4 - y^2 \quad \text{Solve for } x.$$

Each disk has a volume dV given by

$$dV = \pi x^2 \, dy$$

$$dV = \pi(4 - y^2)^2 \, dy.$$

Since the disks were sliced horizontally, the sum must be taken vertically. The lower limit is given by $y = 0$. The upper limit occurs where the curve intersects the y-axis. That is, $y = \sqrt{4 - 0}$ or 2.

The volume of the solid is given by

$$V = \int_0^2 \pi(4 - y^2)^2 \, dy.$$

The power rule does not apply to this integral (why not?), so the expression is expanded.

$$V = \pi \int_0^2 (16 - 8y^2 + y^4)\, dy$$

$= \pi[16y - (8/3)y^3 + (1/5)y^5] \Big|_0^2$	Integrate each term.
$= \pi[(32 - 64/3 + 32/5) - (0 - 0 + 0)]$	Evaluate the integral.
$= \pi(32/3 + 32/5)$	Simplify the result.
$= 32\pi(8/15)$	Factor out 32 and add (1/3 + 1/5).
$= 256\pi/15$ ■	

Volumes of revolution are useful in technical applications.

EXAMPLE 5 A pond has a surface that is approximately circular, and the bottom of the pond has the approximate shape of the parabola $y = 0.004x^2$. Find the volume of water in the pond if it is 10 ft deep in the center.

Solution Revolve the region in Figure 24.12 about the y-axis. Slice the generating area horizontally and find the sum vertically from 0 to 10.

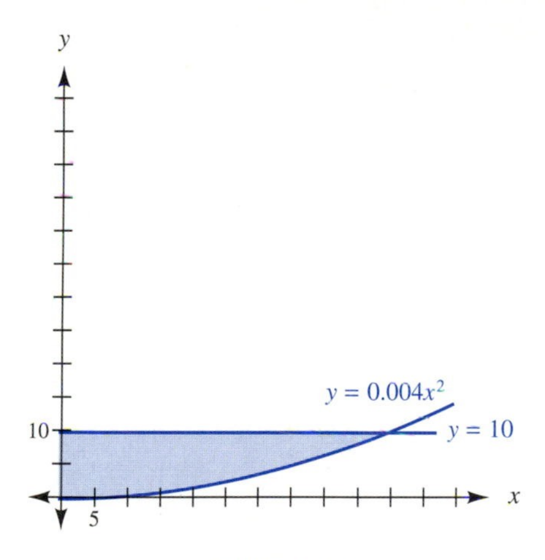

FIGURE 24.12

$$dV = \pi x^2\, dy$$

Since $y = 0.004x^2$, we have $x^2 = y/0.004$.

$$V = \int_0^{10} \frac{\pi y}{0.004}\, dy$$

$$= \frac{\pi y^2}{0.008} \Big|_0^{10}$$

$= 4 \times 10^4$ ft³ or 40,000 ft³, to one significant digit ■

EXERCISES 24.2

In 1–8, find the volume of the solid generated by rotating the area bounded by the given curves about the x-axis.

1. $y = (1/2)x$, $x = 6$, and the x-axis.
2. $y = (-1/3)x$, $x = -6$, and the x-axis
3. $y = -x^2 + 4$, and the x-axis
4. $y = 1 - x^2$ and the x-axis
5. $y = x^3 + 3$, $x = 0$, $x = 2$, and the x-axis
6. $y = x^4 + 2$, $x = 0$, $x = 1$, and the x-axis
7. $y = \sqrt{x - 1}$, $x = 5$, and the x-axis
8. $y = \sqrt{2x}$, $x = 8$, and the x-axis

In 9–15, find the volume of the solid generated by rotating the area bounded by the given curves about the y-axis.

9. $y = -3x$, $y = 2$, and the y-axis
10. $y = (1/2)x$, $y = 3$, and the y-axis
11. $y = -x^2 + 2$ in the first quadrant
12. $y = -2x^2$, $y = -4$ in the fourth quadrant
13. $y = x^2 + 1$, $y = 3$ in the first quadrant
14. $y = \sqrt{9 - x}$, the x-axis, and the y-axis
15. $y = 3\sqrt{x}$, $y = 6$, and the y-axis

---

16. A basin for a fountain is made by rotating the region shown in the accompanying diagram about a vertical axis through wet cement. Find the volume of the basin, if the measurements are in feet. 1 2 4 7 8 3 5

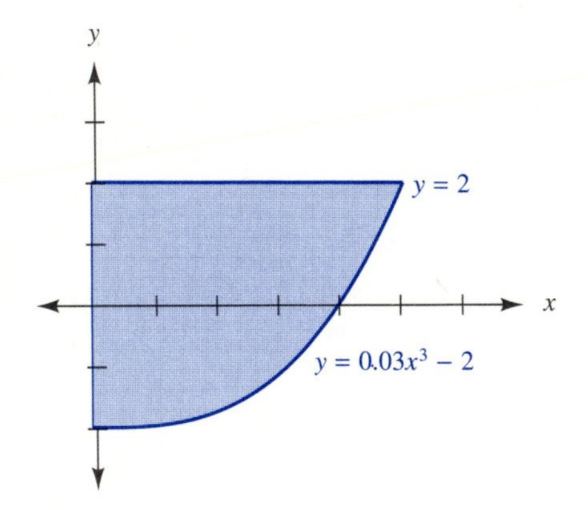

17. Find the volume of a funnel that occupies the space swept out by rotating the region bounded by the x-axis and $y = 1/x^2$ between $x = 1/3$ and $x = 4$ about the x-axis. Measurements are in centimeters. Use three significant digits. 1 2 4 7 8 3 5

18. Find the volume of a container that occupies the space swept out by rotating the region bounded by the y-axis and $x = y^2 + 6$ between $y = 0$ and $y = 2$ about the y-axis. Measurements are in centimeters. Use three significant digits. 1 2 4 7 8 3 5

19. The cooling tank for a nuclear reactor occupies the space swept out by rotating the region bounded by the y-axis and one branch of the hyperbola given by $x^2 - y^2 = 5929$ between $y = -440$ to $y = 120$ about the y-axis. Find the volume of the tank, if measurements are in feet. 6 9 5 8

20. Find the weight of water that would fill a barrel 1.00 m tall if the equation that gives the side of the barrel is $x = -0.10\ y^2 + 0.50$. Assume that the center of the barrel is at the origin. Water weighs 9800 N/m³. 2 1 9

21. Find the volume of a barrel that occupies the space swept out by rotating the region bounded by $y = -2$, $y = 2$, the y-axis, and the ellipse $9x^2 + 4y^2 = 36$ about the y-axis. Measurements are in feet. Use three significant digits. 1 2 4 7 8 3 5

---

1. Architectural Technology
2. Civil Engineering Technology
3. Computer Engineering Technology
4. Mechanical Drafting Technology
5. Electrical/Electronics Engineering Technology
6. Chemical Engineering Technology
7. Industrial Engineering Technology
8. Mechanical Engineering Technology
9. Biomedical Engineering Technology

24.3 VOLUME OF A SOLID OF REVOLUTION: SHELL METHOD

Consider the area bounded by the x-axis, the line $x = 4$, and the graph of $y = \sqrt{x}$. If this area, shown in Figure 24.13a, is rotated about the y-axis, it sweeps out a solid that has a depression in the center as shown in Figure 24.13b.

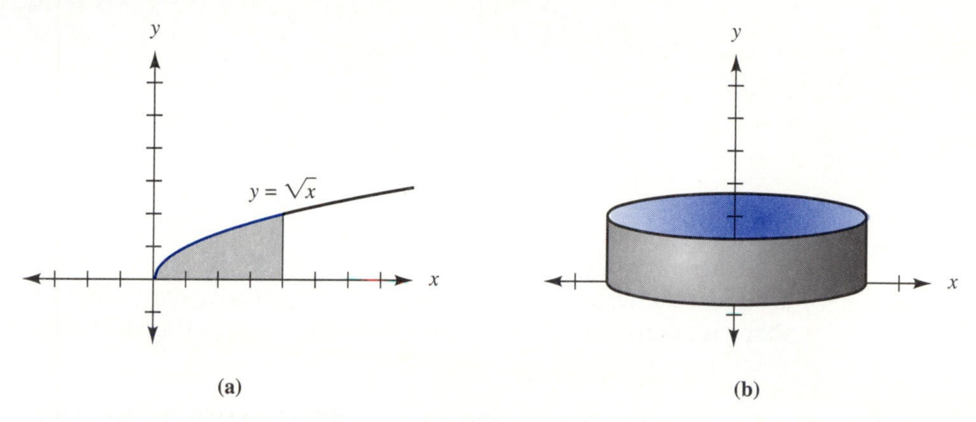

FIGURE 24.13

If the area is sliced horizontally, as in Figure 24.14a, each strip sweeps out a ring-shaped figure rather than a disk, as shown in Figure 24.14b. The disk method could be modified for such problems. We could consider the ring-shaped figure to be formed by removing a small disk from a larger disk. The volume could be found by subtracting the volume of the small disk from the volume of the larger disk.

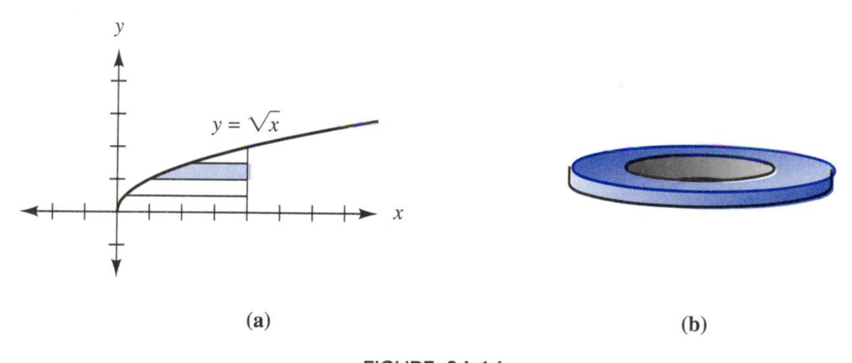

FIGURE 24.14

On problems of this type, however, it is often easier to use another approach. Instead of slicing the generating area horizontally, we slice it vertically, as in Figure 24.15a. The figure is generated by rotating each vertical strip about the y-axis. If a rectangle replaces each strip, each rectangle generates a hollow cylinder, called a shell, as shown in Figure 24.15b.

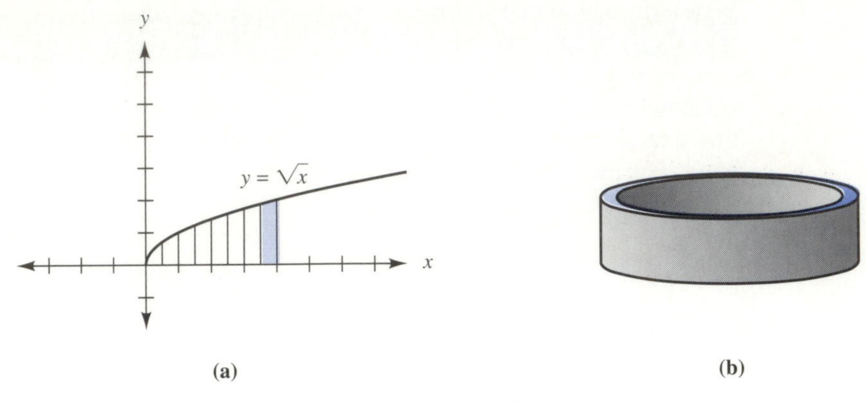

FIGURE 24.15

The hollow cylinders are nested, one inside the other, like layers of insulation around a hot water heater. If one of the shells (or layers) is split and opened up, its shape is approximately that of a rectangular box, as shown in Figure 24.16.

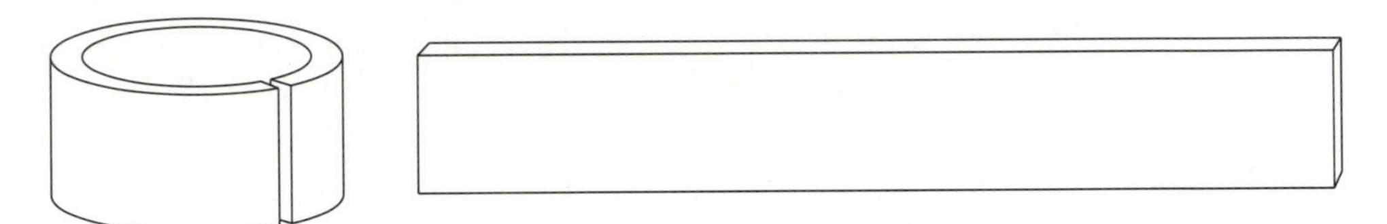

FIGURE 24.16

The volume of a rectangular box is given by $V = lwh$. The length l of this box is equal to the circumference of the cylindrical shell from which it came. Circumference is given by $C = 2\pi r$, and the radius of one of these shells is equal to the x-coordinate on the left-hand side of the generating rectangle. Thus, $l = 2\pi x$. The height h of the box is equal to the y-coordinate at the top of the generating rectangle, so $h = y$. The width w of the box is equal to the width of the generating rectangle. That is, $w = dx$. Thus, the volume of one of the cylindrical shells is given by

$$dV = 2\pi xy\, dx.$$

The volume of the original figure, then, is given by

$$V = \int_a^b 2\pi xy\, dx = 2\pi \int_a^b xy\, dx.$$

Because the integration is with respect to x, y is expressed in terms of x, and a and b are the x-values of the left and right boundaries of the generating area.

The method of cylindrical shells can also be used when the axis of revolution is the x-axis. ***It is important to realize that the plane figure is sliced in a direction that is parallel to the axis of revolution, and the summation is done in a direction that is perpendicular to the axis of revolution.***

NOTE ⇨⇨

To find the volume of a solid of revolution by the shell method:

If a plane figure bounded by the graph of $y=f(x)$ and one of the coordinate axes is rotated about the other axis to form a solid that can be partitioned into shells, follow the steps listed for the appropriate axis.

	Axis of Revolution	
	x-axis	**y-axis**
1. Slice the plane figure:	horizontally	vertically
2. Identify the radius r of 1 shell:	y	x
3. Identify the height h of 1 shell:	x	y
4. Identify the thickness t of 1 shell:	dy	dx
5. Express the volume of 1 shell as $dV=2\pi rht$	$dV=2\pi yx\,dy$	$dV=2\pi xy\,dx$
6. Identify the limits of integration:	$y=c,\ y=d$	$x=a,\ x=b$
7. Evaluate the integral:	$V=2\pi\int_c^d xy\,dy$ with x expressed in terms of y	$V=2\pi\int_a^b xy\,dx$ with y expressed in terms of x

EXAMPLE 1 Find the volume of the solid generated if the figure bounded by the graph of $y=\sqrt{x}$, $x=4$, and the x-axis is rotated about the y-axis.

Solution This is the problem we have been considering. Refer back to Figure 24.15a. The shell method, as described, applies.

1. The area is sliced vertically.

2–4. The radius is given by an x-coordinate, the height is given by a y-coordinate, and the thickness is dx.

5. Each shell has a volume given by

$$dV=2\pi xy\,dx=2\pi x\sqrt{x}\,dx.$$

6. The limits of integration are given by the x-values 0 and 4.

7. The volume of the solid is given by

$$V=2\pi\int_0^4 x\sqrt{x}\,dx.$$

$$=2\pi\int_0^4 x^{3/2}\,dx \qquad \text{Replace } x\sqrt{x} \text{ with } x(x^{1/2}) \text{ or } x^{3/2}.$$

$$=2\pi\left(\frac{x^{5/2}}{5/2}\right)\Bigg|_0^4 \qquad \text{Use the power rule to integrate.}$$

$$= 2\pi(2/5)x^{5/2}\Big|_0^4$$ Replace 1/(5/2) with 2/5.

$$= (4\pi/5)[4^{5/2} - 0^{5/2}]$$ Evaluate the integral.

$$= (4\pi/5)(32) \text{ or } 128\pi/5$$ Simplify the result. ■

EXAMPLE 2 A lens with a parabolic surface is made from glass. When the base of the lens is on the x-axis, the equation for the parabola is $y = 0.04x^2 + 0.50$, and the lens has a radius of 1.50 in. Find the weight of the lens if glass weighs 1.72 oz/in³.

Solution Figure 24.17a shows the graph of the generating area. It is bounded by the line $x = 1.50$, the graph of $y = 0.04x^2 + 0.50$, and the x-axis. The area is rotated about the y-axis, as shown in Figure 24.17b.

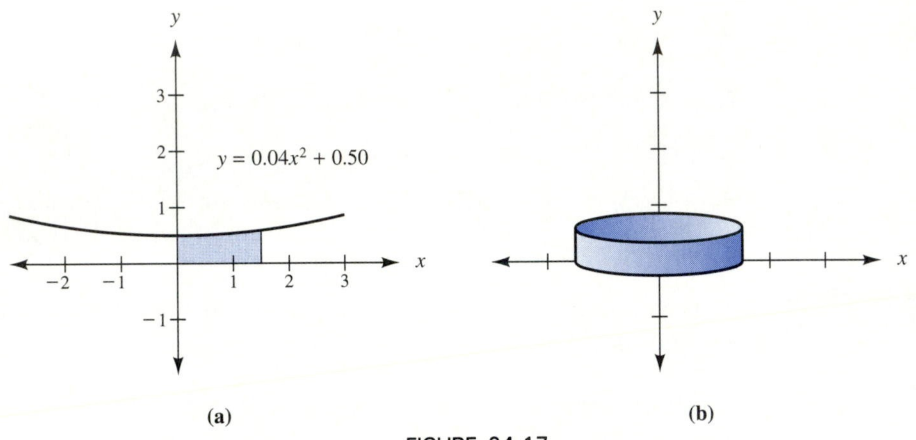

FIGURE 24.17

Slice the area vertically. The volume of each shell is given by

$$dV = 2\pi xy\,dx = 2\pi x(0.04x^2 + 0.50)\,dx.$$

The volume of the lens is given by

$$V = 2\pi \int_0^{1.50} x(0.04x^2 + 0.50)\,dx.$$

$$= 2\pi \int_0^{1.50} (0.04x^3 + 0.50x)\,dx$$ Distribute x.

$$= 2\pi \left(\frac{0.04x^4}{4} + \frac{0.50x^2}{2}\right)\Bigg|_0^{1.50}$$ Integrate each term.

$$= 2\pi[0.01x^4 + 0.25x^2]\Big|_0^{1.50}$$ Simplify the result.

$$= 2\pi([0.01(1.50)^4 + 0.25(1.50)^2] - [0 + 0])$$ Evaluate the integral.

$$= 3.8524 \text{ in}^3$$

The weight, then, is $3.8524(1.72) = 6.63$ oz. ■

NOTE The method of finding volumes of revolution by shells can be adapted for use when the generating area lies between two curves. ***When the y-axis is the axis of revolution, the height of each shell is the difference between two y-values, and when the x-axis is the axis of revolution, the height of each shell is the difference between two x-values.***

EXAMPLE 3 If the region bounded by the line $y = x$, and the graph of $y = x^2$ is rotated about the x-axis, find the volume of the solid generated.

Solution Slice the area horizontally as shown in Figure 24.18a. Figure 24.18b shows the solid of revolution.

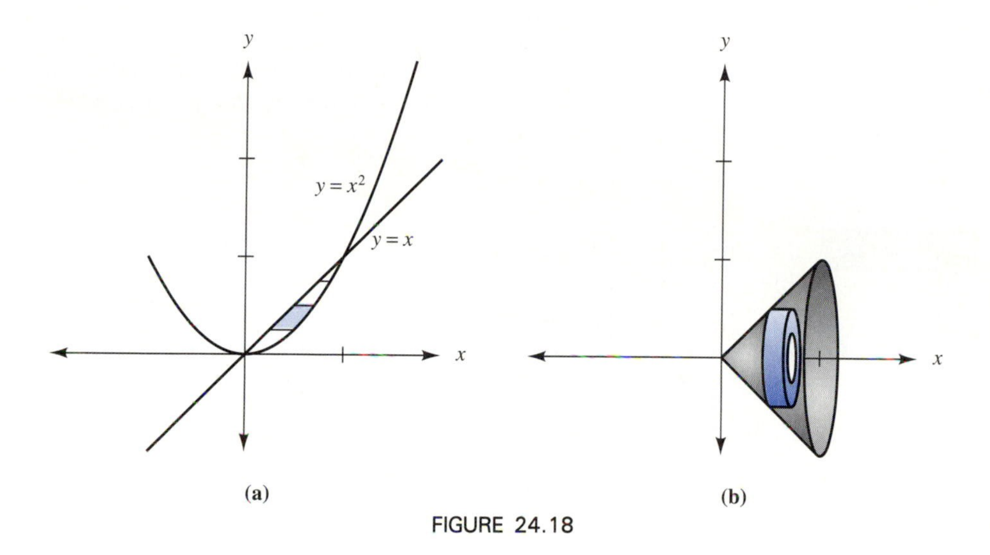

FIGURE 24.18

Each shell has radius y, thickness dy, and a height that is the difference of x-values. But we must express that height in terms of y. For the parabola $y = x^2$, we have $x = \pm\sqrt{y}$. Since the generating area is in the first quadrant, x cannot be negative, so we use $x = \sqrt{y}$. Using $x = \sqrt{y}$ as the right-hand curve and $x = y$ as the left-hand curve, the height is $\sqrt{y} - y$. Hence, the volume dV of each shell is given by

$$dV = 2\pi rht$$
$$dV = 2\pi y(\sqrt{y} - y)\, dy.$$

Since the region bounded by the two curves is the region enclosed by them, we must find the points at which the two curves intersect. The equation $x^2 = x$ has 0 and 1 as its solutions. (You should verify the solution.) The y-coordinates associated with these points are 0 and 1, respectively. Thus, $y = 0$ is at the bottom of the region and $y = 1$ is at the top. The volume of the figure is given by

$$V = 2\pi \int_0^1 y(\sqrt{y} - y)\, dy.$$

$$
\begin{aligned}
V &= 2\pi \int_0^1 y(y^{1/2} - y)\, dy && \text{Replace with } \sqrt{y} \text{ with } y^{1/2}. \\
&= 2\pi \int_0^1 (y^{3/2} - y^2)\, dy && \text{Distribute } y. \\
&= 2\pi \left(\frac{y^{5/2}}{5/2} - \frac{y^3}{3} \right) \Big|_0^1 && \text{Integrate each term.} \\
&= 2\pi \left(\frac{2y^{5/2}}{5} - \frac{y^3}{3} \right) \Big|_0^1 && \text{Replace } 1 \div (5/2) \text{ with } 2/5. \\
&= 2\pi[(2/5 - 1/3) - (0 - 0)] && \text{Evaluate the integral.} \\
&= 2\pi/15 \quad \blacksquare
\end{aligned}
$$

The method of finding volumes by shells may be modified to handle problems in which the axis of revolution is not one of the coordinate axes, but is parallel to either the x-axis or the y-axis.

EXAMPLE 4 If the region bounded by the line $y = x$, and the graph of $y = x^2$ is rotated about the line $y = 1$, find the volume of the solid generated.

Solution Slice the area horizontally as shown in Figure 24.19a. The volume of revolution is shown in Figure 24.19b.

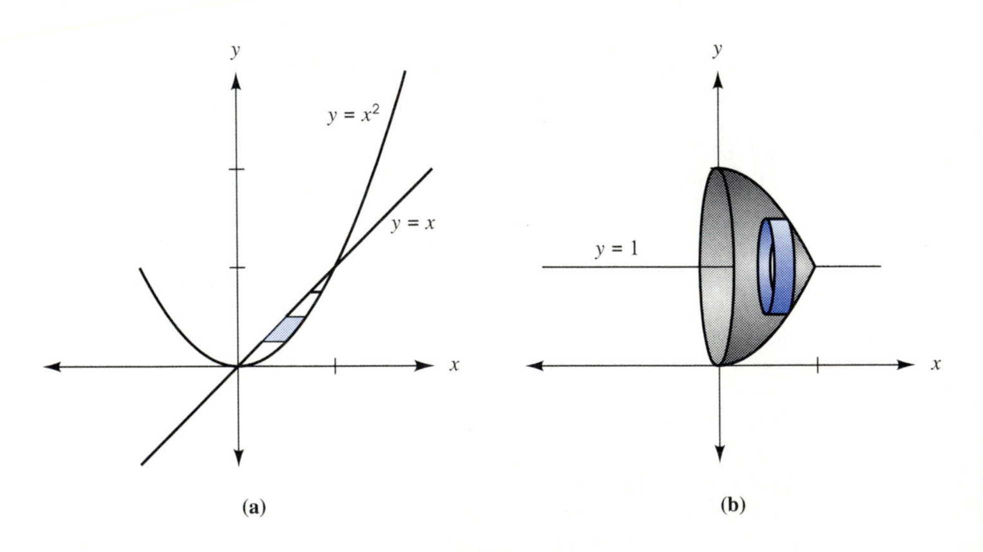

FIGURE 24.19

Since the axis of revolution is the line $y = 1$, the radius is measured from this line instead of from the x-axis. The radius, then, is $1 - y$. The height of each shell is given by a difference of x-coordinates. As in Example 3, we have the height of each shell given by $\sqrt{y} - y$. Each shell has a thickness of dy. The volume dV of each shell is given by

$$dV = 2\pi rht$$
$$dV = 2\pi(1 - y)(\sqrt{y} - y)\,dy$$
$$= 2\pi(y^{1/2} - y - y^{3/2} + y^2)\,dy.$$

The limits of integration are the y-coordinates of the points of intersection. The curves cross when x is 0 and when x is 1, as shown in Example 3. The y-coordinates associated with these values are $y = 0$ and $y = 1$. The volume of the figure is given by

$$V = 2\pi \int_0^1 (y^{1/2} - y - y^{3/2} + y^2)\,dy.$$
$$= 2\pi\left(\frac{y^{3/2}}{3/2} - \frac{y^2}{2} - \frac{y^{5/2}}{5/2} + \frac{y^3}{3}\right)\Bigg|_0^1$$
$$= 2\pi\left(\frac{2y^{3/2}}{3} - \frac{y^2}{2} - \frac{2y^{5/2}}{5} + \frac{y^3}{3}\right)\Bigg|_0^1$$
$$= 2\pi[(2/3 - 1/2 - 2/5 + 1/3) - (0 - 0 - 0 + 0)]$$
$$= \pi/5 \quad \blacksquare$$

EXERCISES 24.3

In 1–8, use the method of shells to find the volume of the solid generated by rotating the region bounded by the given curves about the x-axis.

1. $y = (1/2)x + 1$, $y = 3$, and the y-axis

2. $y = 3x + 2$, $y = 4$, and the y-axis

3. $y = x^2$ and $y = 4$

4. $y = 2x^2$ and $y = 4$

5. $y = (1/2)x^2$ and $y = 1$

6. $y = \sqrt{x}$, $y = 3$, and the y-axis

7. $y = \sqrt{x}$ and $y = x^2$

8. $y = x^3$ and $y = x^2$

In 9–16, use the method of shells to find the volume of the solid generated by rotating the region bounded by the given curves about the given axis of revolution.

9. $y = x - 3$, $x = 3$, $x = 6$, and the x-axis about the y-axis

10. $y = -2x + 4$, $x = 2$, and $y = 4$ about $y = 4$

11. $y = -x^2 + 4x - 3$ in the first quadrant about the y-axis

12. $y = (x - 1)^2$ and $y = 1$ about the y-axis

13. $y = \sqrt{x - 2}$, $x = 6$, and the x-axis about $x = 2$

14. $y = \sqrt{x}$ and $y = x^2$ about the y-axis

15. $y = x^3$ and $y = x^2$ about the y-axis

16. $y = x^2 - 2x$ and $y = 3$ about $x = -1$

17. A basin for a fountain is made by rotating the region shown in the accompanying diagram about a vertical axis through wet cement. Find the volume of the basin if measurements are in feet. 1 2 4 7 8 3 5

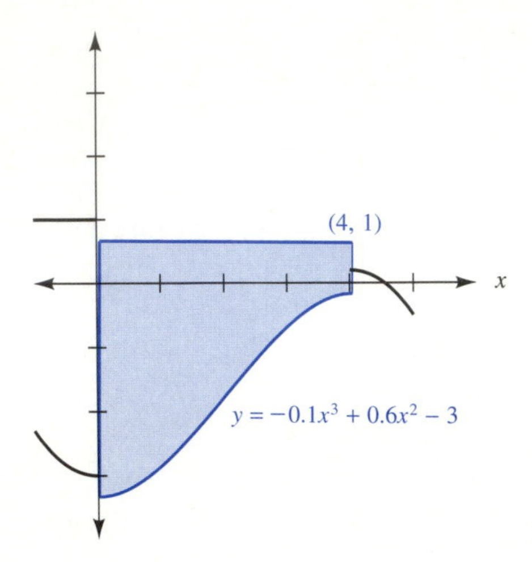

18. An aluminum machine part occupies the space swept out by rotating the region bounded by $y = 1/x$, the x-axis, $x = 1$, and $x = 3$ about the y-axis. Assume that measurements are in inches and find the weight of the part if aluminum weighs 1.5 oz per in^3. 1 2 4 7 8 3 5

19. A lawn sprinkler is designed to be circular with a cross section whose top is the parabola $y = -2x^2 + 16x - 30$, as shown in the accompanying diagram. Assume that measurements are in inches, and find the volume of water that will fill it. Use two significant digits. 1 2 4 7 8 3 5

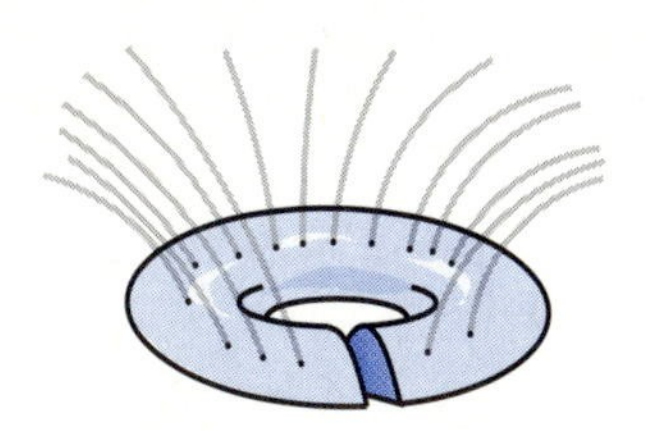

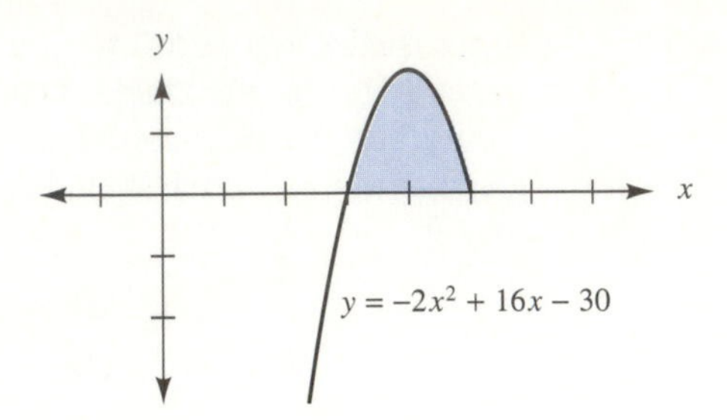

20. A lens with a parabolic surface is made from glass. When the base of the lens is on the x-axis, the equation for the parabola is $y = 0.02x^2 + 0.30$, and the lens has a diameter of 10.0 cm. Use the fact that glass has a density of 0.002 kg/cm^3 to find the mass of the lens. 3 5 6 9

21. Bottles are often made with the surface raised on the inside at the bottom. Find the volume of a bottle that occupies the space swept out by the region shown in the accompanying figure about the y-axis. Measurements are in inches. Use three significant digits. 1 2 4 7 8 3 5

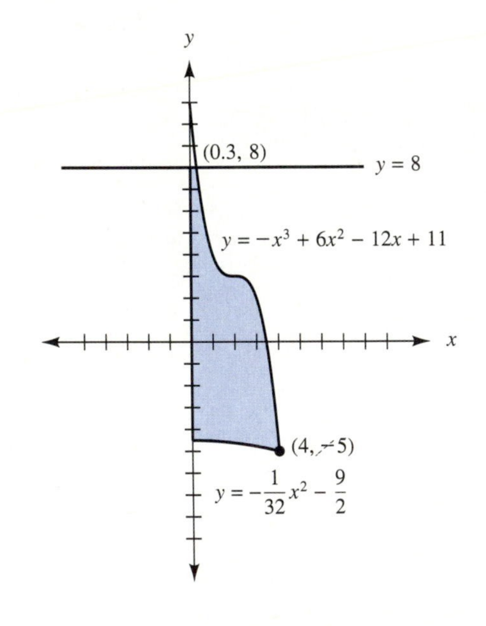

1. Architectural Technology
2. Civil Engineering Technology
3. Computer Engineering Technology
4. Mechanical Drafting Technology
5. Electrical/Electronics Engineering Technology
6. Chemical Engineering Technology
7. Industrial Engineering Technology
8. Mechanical Engineering Technology
9. Biomedical Engineering Technology

22. A cast iron bell occupies the space swept out by rotating the region shown between 0 and 2.5 about the y-axis. Assume that measurements are given in inches, and find the weight of the bell. Cast iron weighs 0.26 pounds/in^3.
1 2 4 7 8 3 5

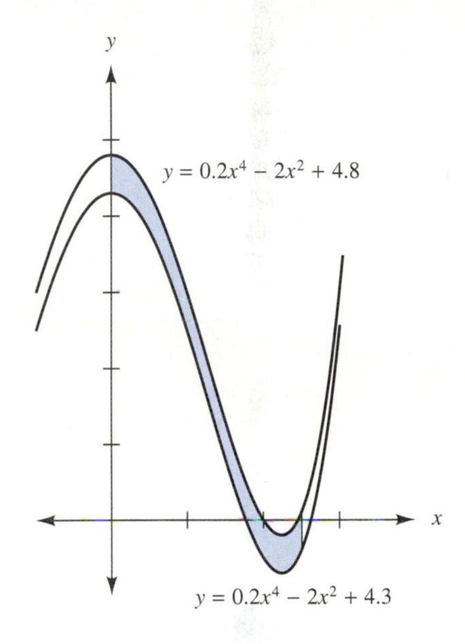

23. **(a)** Draw an example of a region for which the disk method is more appropriate than the shell method when the axis of revolution is the x-axis.
(b) Draw an example of a region for which the disk method is more appropriate than the shell method when the axis of revolution is the y-axis.
(c) Draw an example of a region for which the shell method is more appropriate than the disk method when the axis of revolution is the x-axis.
(d) Draw an example of a region for which the shell method is more appropriate than the disk method when the axis of revolution is the y-axis.

24.4 AVERAGE VALUE AND WORK

If a student has grades of 85, 93, 87, 89, and 86, he/she computes the average grade by finding the sum of the grades and dividing by the number of grades. This type of average is called an arithmetic mean. It is possible to consider the test grade to be a function of the number of tests taken. The function would associate the first test grade of 85 with 1, the second test grade of 93 with 2, and so on. In finding the average grade, the student finds the average value of the function.

Most of the functions we have dealt with, however, do not have just a finite number of values. Consider a car traveling a distance of 110 miles in 2 hours. Unless the car travels at a constant speed of 55 mph, it actually has an infinite number of speeds. If it accelerates from 0 mph to 60 mph, it must travel (at least momentarily) at each speed between 0 and 60. The average value of a continuous function is related to the idea of an arithmetic mean by using an integral.

Let f be a function that is continuous for $a \leq x \leq b$. Divide the interval into n equal segments, each of width $dx = (b - a)/n$, as shown in Figure 24.20. Let x_i be a value in the ith interval. Then $f(x_i)$ is the function value at x_i.

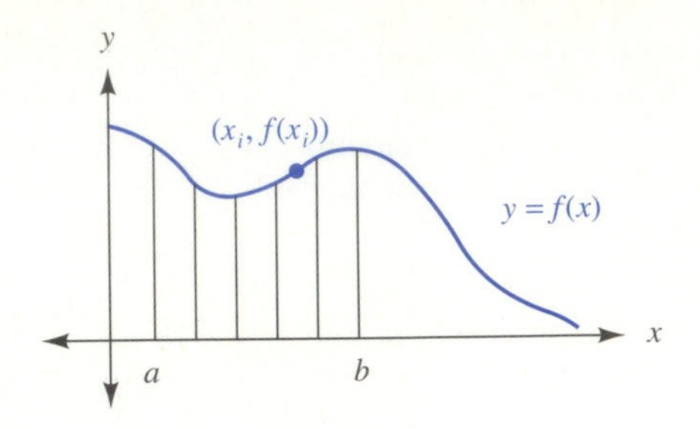

FIGURE 24.20

The arithmetic mean of these function values is given by

$$\frac{1}{n}\sum_{i=1}^{n} f(x_i).$$

Since $dx = \dfrac{b-a}{n}$, $n = \dfrac{b-a}{dx}$, and $\dfrac{1}{n} = \dfrac{dx}{b-a}$.

Thus, the average function value is

$$\frac{dx}{b-a}\sum_{i=1}^{n} f(x_i) = \frac{1}{b-a}\sum_{i=1}^{n} f(x_i)\,dx.$$

Recall that a sum of terms of the form $f(x)\,dx$ can be thought of as an integral if the number of parts approaches infinity, so the average value of f is given by

$$f_{ave} = \frac{1}{b-a}\int_a^b f(x)\,dx.$$

AVERAGE VALUE OF A FUNCTION

If f is a function that is continuous for $a \le x \le b$, the average value of $f(x)$, denoted f_{ave}, is given by

$$f_{ave} = \frac{1}{b-a}\int_a^b f(x)\,dx.$$

EXAMPLE 1 The radioactivity A at time t after a nuclear explosion is given by $A = A_0 t^{-6/5}$. If an explosion occurs at 3:00 P.M., and A_0, the radioactivity after 1 hour, is 15.8 roentgens/hr, find the average radioactivity between 4:00 P.M. and 7:00 P.M.

Solution For the time from 4:00 P.M. until 7:00 P.M., t ranges from 1 to 4. Thus,

$$A_{ave} = \frac{1}{4-1}\int_1^4 15.8t^{-6/5}\,dt.$$

$$= \frac{1}{3}\left(\frac{15.8t^{-1/5}}{-1/5}\right)\Bigg|_1^4$$

$$= \frac{-5(15.8)\,t^{-1/5}}{3}\Bigg|_1^4$$

$$= \frac{-5(15.8)}{3}(4^{-1/5} - 1^{-1/5})$$

Using a calculator to do the computation, we find that

$$A_{ave} \approx 6.38 \text{ roentgens/hr.} \quad \blacksquare$$

EXAMPLE 2 The potential energy W (in joules) of an inductor with inductance L (in henries) carrying current I (in amperes) is given by $W = (1/2)LI^2$. Find the average potential energy stored by a 0.20 H inductor if the current ranges from 0.27 A to 0.95 A.

Solution

$$W_{ave} = \frac{1}{0.95 - 0.27}\int_{0.27}^{0.95} (1/2)(0.20)I^2\,dI$$

$$= \frac{1}{0.95 - 0.27}\left(\frac{0.10\,I^3}{3}\right)\Bigg|_{0.27}^{0.95}$$

$$= \frac{1}{0.95 - 0.27}\left(\frac{0.10}{3}\right)[(0.95)^3 - (0.27)^3]$$

Using a calculator to do the computation, we see that

$$W_{ave} \approx 0.041 \text{ J} \quad \blacksquare$$

The concept of work can be used to determine the amount of energy needed to perform certain tasks. If a constant force F is applied to an object, the work done in moving the object a distance D in the direction of the force is defined to be the product of F and D. If a variable force is applied to an object, the work is calculated using an integral. Work is expressed in units such as foot-pounds. When metric units are used, a newton-meter is often called a joule.

Suppose an object is being pushed along a straight line from a to b by a variable force. We consider the force F to depend on the distance x from some initial point. The force, then, is a function of x. If the distance from a to b is divided into n equal segments, each of length dx, the force may vary somewhat as the object moves across each segment. The smaller dx is, however, the smaller the variation is. Thus, for small values of dx, F may be treated as though it is constant across each segment. If x_i is in the ith segment, then $F(x_i)$ represents the force for that segment. The work done in moving the object through the ith segment is given by

$$dW = F(x_i)\,dx.$$

The work done in moving the object from a to b is given by the sum of the individual values of dW. In the limit, as dx approaches zero, this sum is an integral.

WORK REQUIRED TO MOVE AN OBJECT

If an object is moved along a straight line by a force that varies with distance x from some initial point, then the work done by the force in moving the object from a to b is given by

$$W = \int_a^b F(x)\, dx.$$

EXAMPLE 3 The force F (in newtons) required to stretch a spring a distance of x (in cm) from the rest position is given by $F = kx$, where k is a constant that depends on the characteristics of the spring. If $k = 0.0020$, find the work (in joules) done in stretching a 5.2 cm spring from a length of 5.7 cm to a length of 5.9 cm, as shown in Figure 24.21.

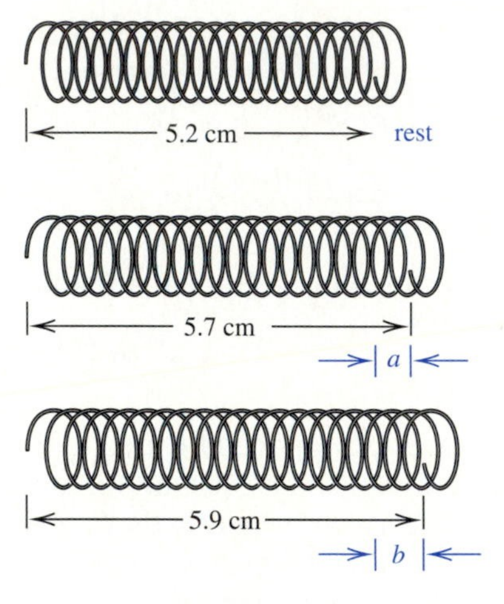

FIGURE 24.21

Solution Although the unit of work is the joule or Newton-meter, the constant k in this example has been adjusted to accommodate the limits of integration in centimeters. The force is given by $F = 0.0020x$. Let dx be a small distance over which the force acts. The work done is given by

$$W = \int_a^b 0.0020x\, dx.$$

It is important that the distance is measured from the rest position. The spring is 5.2 cm long when in that position. The spring is stretched from the rest position, and we are interested in what happens as it moves from 5.7 cm to 5.9 cm. The limits of integration are

$$a = 5.7 - 5.2 \text{ or } 0.5 \qquad \text{and} \qquad b = 5.9 - 5.2 \text{ or } 0.7.$$

Hence,

$$\begin{aligned} W &= \int_{0.5}^{0.7} 0.0020x\,dx \\ &= \frac{0.0020x^2}{2}\Big|_{0.5}^{0.7} \\ &= 0.0010[0.7^2 - 0.5^2] \\ &= 0.00024 \text{ joules, to two significant digits.} \end{aligned}$$ ■

EXAMPLE 4 A 200-lb man is hoisted up 20 ft to a helicopter. Find the work done in lifting the man and the rope if the rope weighs 10 lb. Treat the man's weight as though it is concentrated at the end of the rope. Since he can lift himself up when he reaches the helicopter, consider him to be lifted when his hands reach the helicopter.

Solution The force necessary to lift an object is equal to the weight of the object. When y feet of rope have been lifted up to the helicopter, $20 - y$ feet remain to be lifted, as shown in Figure 24.22.

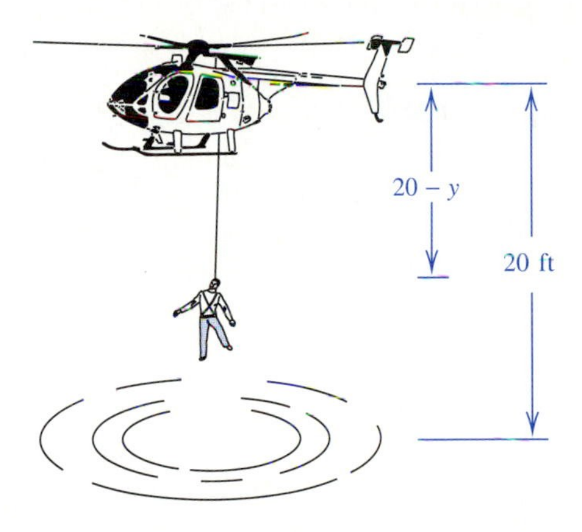

FIGURE 24.22

The weight of the remaining rope depends on its length. Assuming that the weight of the rope is evenly distributed, it weighs 10 lb/20 ft or 0.5 lb/ft. Thus, at any time, the weight of the rope yet to be lifted is $0.5(20 - y)$, and the weight of the man is a constant 200 lb. When y feet of rope have been lifted, the weight yet to be lifted is $0.5(20 - y) + 200$.

Imagine raising the rope in very small increments dy, over which the weight (force) is treated as though it is constant. As this weight is moved a distance of dy, the work done is

$$dW = [0.5(20 - y) + 200]\,dy.$$

The work done in lifting the rope and man all the way is given by

$$W = \int_0^{20} [0.5(20 - y) + 200]\, dy.$$

$$= \int_0^{20} [10 - 0.5y + 200]\, dy \qquad \text{Distribute 0.5.}$$

$$= \int_0^{20} [210 - 0.5y]\, dy \qquad \text{Simplify the result.}$$

$$= 210y - \frac{0.5y^2}{2}\Big|_0^{20} \qquad \text{Integrate each term.}$$

$$= (4200 - 100) - (0 - 0) \qquad \text{Evaluate the integral.}$$

$$= 4100 \text{ ft}\cdot\text{lb or } 4000 \text{ ft}\cdot\text{lb, to one significant digit.}$$ ■

EXERCISES 24.4

1. The distance d (in ft) required to bring an automobile to a complete stop when it is traveling at a speed s (in mph) is given by $d = 0.0500s^2 + s$. Find the average stopping distance for speeds between 50.0 and 60.0 mph. **4 7 8** 1 2 3 5 6 9

2. It is determined that for a particular section of highway 100.0 m long, the amount of expansion I (in m) is given by $I = 0.007128(T - 20.00)$ where T is the temperature (in degrees Celsius). Find the average amount of expansion for temperatures between 30.00° C and 40.00° C. **1 2 4 6 8** 5 7 9

3. The power (in W) dissipated by a 50.0-Ω resistor is given by $P = 50.0I^2$ where I is the current (in A). Find the average power dissipated for currents between 1.2 and 2.1 A. **3 5 9** 1 2 4 6 7 8

4. The velocity V (in ft/s) of water discharged is given by $V = 12.16\sqrt{P}$, where P is the water pressure (in lb/in^2) at the point of discharge. Find the average velocity of water discharged between a pressure of 40.00 lb/in^2 and 60.00 lb/in^2. **4 8** 1 2 6 7 9

5. A certain tumor grows in such a way as to maintain an approximately spherical shape. Find the average volume of the tumor as it grows from a diameter of 2.00 cm to 6.00 cm. **6 9** 2 4 5 7 8

6. An electronic wave form of voltage is given the equation $v = 3t^2 + 6$ over a 3-second time period. Find the average value (in volts) of this waveform. **5** 3 4 7 8 9

7. The force (in lb) required to stretch a particular 8.00-inch long steel rod is given by $F = 2{,}900{,}000x$. Find the work done in stretching the rod from a length of 8.002 inches to a length of 8.008 inches. The unit of work is ft·lb. **4 7 8** 1 2 3 5 6 9

8. The force (in lb) required to compress a certain 12-inch spring a distance of x inches is given by $F = 600x$. Find the work done in compressing it from 10″ to 8″. The unit of work is ft·lb. **4 7 8** 1 2 3 5 6 9

9. The force F (in lb) required to accelerate a particular car at a constant rate of 4.40 ft/s^2 is given by $F = 9680\, d^2$, where d is the distance (in ft)

1. Architectural Technology
2. Civil Engineering Technology
3. Computer Engineering Technology
4. Mechanical Drafting Technology
5. Electrical/Electronics Engineering Technology
6. Chemical Engineering Technology
7. Industrial Engineering Technology
8. Mechanical Engineering Technology
9. Biomedical Engineering Technology

through which the car has moved. Find the work required to accelerate the car through the first 20.0 feet. 4 7 8 1 2 3 5 6 9

10. The force F (in lb) required to stop a particular car is given by $F = 10{,}000x$ where x is the distance (in ft) traveled after the brake is applied. Find the work required to stop the car in 180 ft. 4 7 8 1 2 3 5 6 9

11. A chain that is 12.0 ft long weighs 3.00 lb/ft and hangs over the edge of a roof. Find the work required to pull the chain up onto the roof. 4 7 8 1 2 3 5 6 9

12. A 20-ft chain that weighs 80 lbs is lying in a coil on the ground. How much work is required to lift one end of the chain so that the chain is fully extended? 4 7 8 1 2 3 5 6 9

13. How much work is required to lift an elevator a distance of 50.0 feet when it is loaded to capacity (total weight = 4400 pounds)? Assume that the cable weighs 10.0 ft/lb and is wound up as the elevator is lifted. 4 7 8 1 2 3 5 6 9

14. A cylindrical tank that is 10.0 m high and has a radius of 4.0 m is filled to a depth of 6.0 m. How much work is done if water is pumped out of the top of the tank until the depth is only 3.0 m? Water weighs 9800 N/m^3. (Hint: Imagine lifting the water out in disks as shown in the accompanying diagram.) 4 7 8 1 2 3 5 6 9

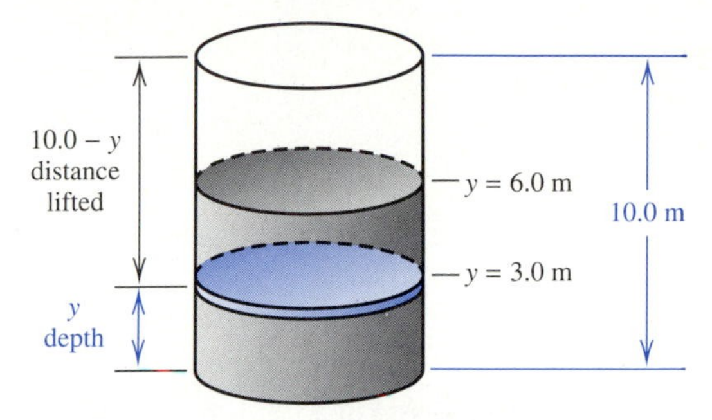

15. Would it require more work to lower the water level of the tank in problem 14 from 6.0 m to 5.0 m or from 4.0 m to 3.0 m? Explain your answer.

24.5 VELOCITY AND FLUID PRESSURE

When an object moves at a constant velocity, the distance s traveled is given by the product of velocity v and time t. When the velocity varies, it is possible to consider the time to be divided into small increments of length dt. The velocity can be treated as though it is constant during that length of time and

$$ds = v\, dt$$

The total distance traveled is given by the sum of the individual distances for each interval of time dt. In the limit, as dt approaches 0, the sum is an integral.

DISTANCE TRAVELED

If v is the velocity of an object, the distance s traveled is given by

$$s = \int v\, dt.$$

Notice that this example confirms our previous ideas about the nature of differentiation and integration. You learned in Section 22.2 that velocity is the derivative of distance. Hence, it is reasonable that distance should be the integral of velocity.

EXAMPLE 1 Find an expression for distance s traveled by an object in t seconds if $v = 3t^2 + 2$, and $s = 16$ when $t = 2$.

Solution The instruction asks for an *expression* for distance, not a specific distance. Thus, we need an antiderivative, or indefinite integral.

$$\begin{aligned} s &= \int (3t^2 + 2)\, dt \\ &= \frac{3t^3}{3} + 2t + C \\ &= t^3 + 2t + C \end{aligned}$$

Since we know that $s = 16$ when $t = 2$, it is possible to determine the value of C.

$$\begin{aligned} 16 &= 2^3 + 2(2) + C \\ 16 &= 8 + 4 + C \\ 4 &= C \end{aligned}$$

Thus, $s = t^3 + 2t + 4$. ■

Recall that acceleration is given by the derivative of velocity. Therefore, it is possible to treat velocity as the antiderivative of acceleration.

VELOCITY OF AN OBJECT

If a is the acceleration of a moving object, its velocity is given by

$$v = \int a\, dt.$$

EXAMPLE 2 The acceleration due to gravity is -9.8 m/s^2. Determine an expression for distance fallen in t seconds if an object is dropped from rest.

Solution First, determine the velocity.

$$\begin{aligned} v &= \int a\, dt \\ &= \int -9.8\, dt \\ &= -9.8\, t + C_1 \end{aligned}$$

Since the object is dropped from rest, we know that its initial velocity is 0. That is, when t is 0, v is 0.

$$\begin{aligned} 0 &= -9.8(0) + C_1 \\ 0 &= C_1 \end{aligned}$$

The velocity is given by $v = -9.8t$.

The distance is given by

$$s = \int v\,dt.$$
$$= \int -9.8t\,dt$$
$$= \frac{-9.8t^2}{2} + C_2$$
$$= -4.9\,t^2 + C_2$$

Since the distance s fallen at time $t = 0$ is 0, we have

$$0 = -4.9\,(0^2) + C_2$$
$$0 = C_2.$$

The distance is given by $s = -4.9t^2$. ■

A flat object submerged in a fluid may be in either a horizontal or vertical position like the circular plate in Figure 24.23. The weight of the fluid exerts a force on the object.

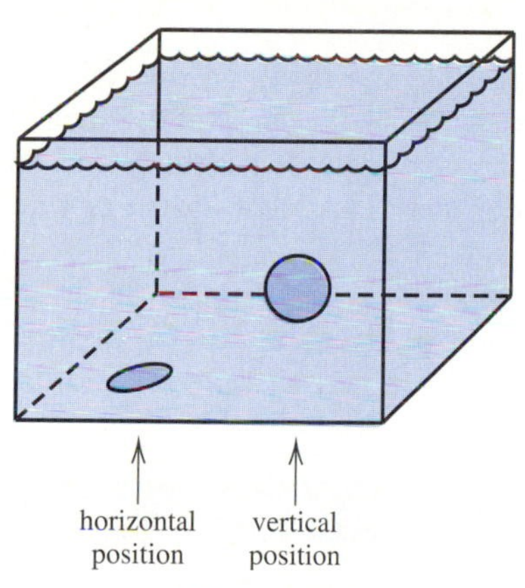

FIGURE 24.23

When the object is submerged horizontally, the force F of the fluid is given by the product of weight per unit volume w, and the volume v of a column of the liquid above the object. The volume of the column of liquid is given by the product of area A and depth y of the object. Thus, $F = wv = wyA$. Force per unit area is called **pressure** and is given by the formula $P = wy$.

pressure

When an object is placed into a fluid, the pressure on that object increases as the object is submerged to greater and greater depths. It is surprising that the pressure exerted by a fluid at a particular depth is equal in all directions. Thus, the horizontal pressure on a vertical object at a given depth is equal to the vertical pressure on a horizontal object at the same depth. This principle is known as **Pascal's principle.**

Pascal's principle

When a flat object, such as a circular plate, is submerged vertically, the depth varies from point to point, and therefore the horizontal pressure varies from point to point. If the interval from c to d is divided into n segments of width dy, as shown in Figure 24.24, then the area may be considered as the sum of the areas of horizontal strips of width dy. The force may be treated as though it is constant throughout each strip.

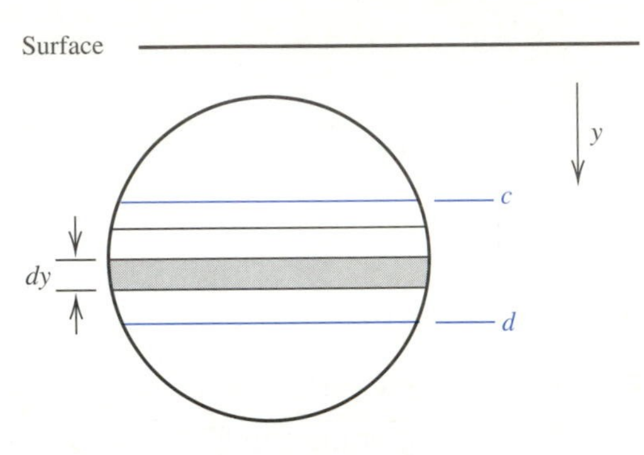

FIGURE 24.24

To calculate the force on one of these strips, we need to know its depth and surface area. The surface area of the strip that lies at depth y can be approximated by the area of a rectangle of width dy and length l, the horizontal distance across the object at depth y. Thus, $A = l\,dy$. For one strip, force may be denoted dF, and

$$dF = wyA = wyl\,dy.$$

The total force on the surface of the object is given by the sum of the individual values of dF. In the limit, as dy approaches 0, this sum is an integral.

FORCE DUE TO PRESSURE OF A FLUID

For a flat object submerged vertically in a liquid whose weight per unit volume is w, between horizontal lines at depths c and d, the force F on the surface due to the pressure of the liquid is

$$F = \int_c^d wyl\,dy = w\int_c^d yl\,dy$$

where l is the horizontal distance across the object at depth y.

EXAMPLE 3 A cylindrical tank is half full of gasoline, as shown in Figure 24.25. The radius of the tank is 1.00 m and gasoline weighs 7430 N/m^3. Find the force exerted on one end of the tank.

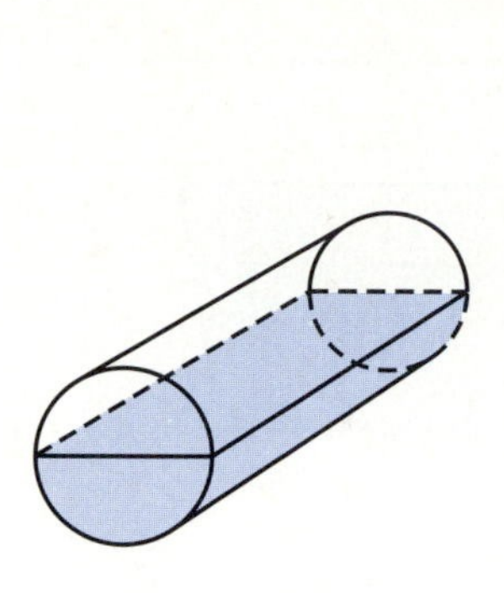

FIGURE 24.25

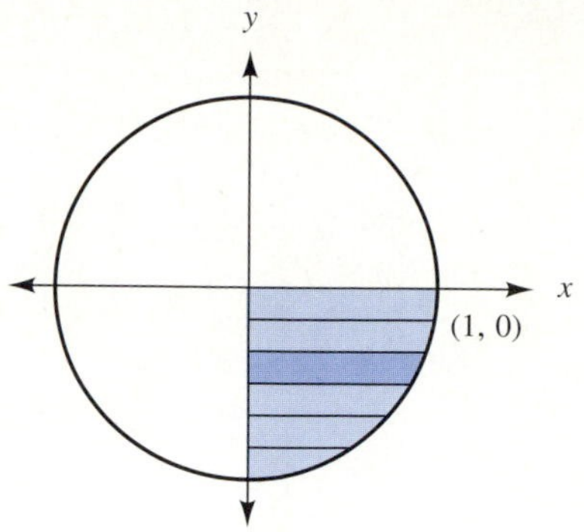

FIGURE 24.26

Solution Figure 24.26 shows the end of the tank placed on a coordinate plane. For simplicity, we find the force on the area in quadrant IV and double it. Computation is easier if the positive y-axis is directed downward. There is nothing wrong with using this orientation, as long as we use it consistently throughout the problem. The equation of the curve is $x^2 + y^2 = 1$. The length of one horizontal strip is given by $x = \sqrt{1 - y^2}$. The force on one strip at depth y is given by

$$dF = 7430y\sqrt{1 - y^2}\, dy,$$

and the force on the semicircle in quadrants III and IV is given by

$$F = 2\int_0^1 7430y\sqrt{1 - y^2}\, dy.$$

$$= 2(7430)\int_0^1 y\sqrt{1 - y^2}\, dy$$

$$= (-1/2)(2)(7430)\int_0^1 -2y(1 - y^2)^{1/2}\, dy$$ Multiply by $-1/2$ and by -2.

$$= -7430\, \frac{(1 - y^2)^{3/2}}{3/2}\Bigg|_0^1$$ Use the power rule to integrate.

$$= \frac{-14\,860}{3}(1 - y^2)^{3/2}\Bigg|_0^1$$ Replace $\frac{-7430}{3/2}$ with $\frac{-14\,860}{3}$.

$$= \frac{-14\,860}{3}(0 - 1)$$ Evaluate the integral.

$$\approx 4950 \text{ newtons}$$ ■

EXAMPLE 4 A vertical floodgate in a dam has the shape of an isosceles trapezoid 10.0 ft across the top and 8.00 ft across the bottom. The height is 6.00 ft. Find the total force against the gate if the top of the gate is 3.00 ft below the surface of the water. Water weighs 62.4 lb/ft^3.

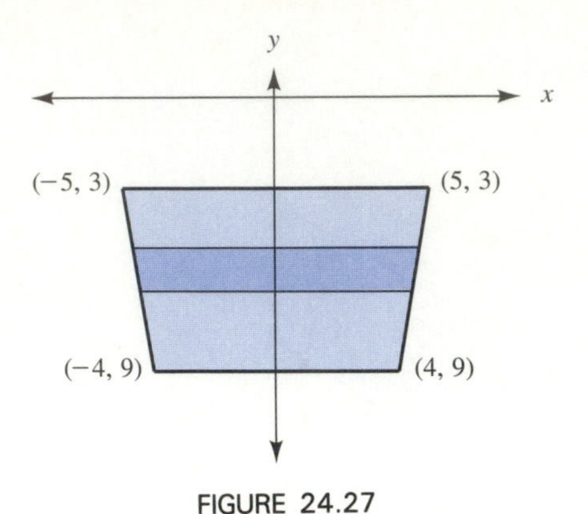

FIGURE 24.27

Solution An isosceles trapezoid has base angles that are equal. Figure 24.27 shows the trapezoid on a coordinate plane with the x-axis and y-axis located to make calculations convenient. To simplify the explanation, we will use integer values, but remember that the final answer should be specified to three significant digits. Once again, the positive y-axis is directed downward.

To find the length of a rectangular strip, we must write the equation for the line through (5, 3) and (4, 9). The slope is given by

$$m = \frac{9-3}{4-5} = \frac{6}{-1} = -6.$$

Thus,

$$\begin{aligned} y - 3 &= -6(x-5) \\ y - 3 &= -6x + 30 \\ y - 33 &= -6x \\ (-1/6)(y-33) &= x. \end{aligned}$$

The area of a single strip at depth y is the product of the width dy and twice the length x or $2(-1/6)(y-33)dy$, and the force on it is given by

$$dF = 62.4y(-1/3)(y-33)\,dy.$$

The total force is given by

$$\begin{aligned} F &= \int_3^9 62.4y(-1/3)(y-33)\,dy \\ &= -20.8\int_3^9 (y^2 - 33y)\,dy \\ &= -20.8\left(\frac{y^3}{3} - \frac{33y^2}{2}\right)\Bigg|_3^9 \\ &= -20.8\left[\left(\frac{9^3}{3} - \frac{33(9^2)}{2}\right) - \left(\frac{3^3}{3} - \frac{33(3^2)}{2}\right)\right] \\ &= -20.8[-1093.5 - (-139.5)] \text{ or } 19{,}800 \text{ lbs, to three significant digits.} \end{aligned}$$

■

EXERCISES 24.5

1. Find an expression for distance s traveled by an object in t seconds if $v = -32t$, and $s = 16$ when $t = 1$. 4 7 8 1 2 3 5 6 9
2. Find an expression for distance s traveled by an object in t seconds if $v = 10 + 9.8t$, and $s = 1500$ when $t = 10$. 4 7 8 1 2 3 5 6 9
3. Find an expression for distance s traveled by an object in t seconds if $v = 9.8t$, and $s = 0$ when $t = 0$. 4 7 8 1 2 3 5 6 9
4. Find an expression for the velocity of an object after t seconds if $a = 3\sqrt{t}$, and $v = 16$ when $t = 4$. 4 7 8 1 2 3 5 6 9
5. Find an expression for the velocity of an object after t seconds if $a = -1.67t$, and $v = 0$ when $t = 0$. 4 7 8 1 2 3 5 6 9
6. Find an expression for the velocity of an object after t seconds if $a = 4t + 3$, and $v = 8$ when $t = 1$. 4 7 8 1 2 3 5 6 9

In 7–12, use the fact that water weighs 9800 *newtons/m*^{3} *or* 62.4 *lb/ft*^{3}, *and oil weighs* 9400 *newtons/m*^{3} *or* 59.9 *lb/ft*^{3}.

7. A vertical dam has the shape of a rectangle 25 m long and 5.0 m deep. Find the force on the dam if the water level is at the top of the dam. 2 4 8 1 5 6 7 9
8. A swimming pool is built so that the bottom is an inclined plane, and the sides are vertical. The pool is 2.00 feet deep at one end, and 12.0 feet deep at the other. If the pool is 30.0 feet long, find the total force against one side that runs the length of the pool. 2 4 8 1 5 6 7 9
9. A river bed that is approximately parabolic in shape is dammed. Find the force on the dam if the equation of the parabola is $y = 0.050x^2$, and the water is 30.0 ft deep at the center of the dam. 2 4 8 1 5 6 7 9
10. The end of a water trough has the shape of an isosceles triangle with the apex pointed downward. If the trough is 2.00 ft deep at the center and 3.00 ft wide at the top, find the force on one end when it is full of water. 2 4 8 1 5 6 7 9
11. A circular porthole with diameter 0.30 m is half submerged in the ocean. The weight of sea water is 1.025 times that of fresh water. Find the force exerted by the water on the porthole. 2 4 8 1 5 6 7 9
12. An oil tanker has ends in the shape of an ellipse with horizontal axis 8.00 m long and vertical axis 6.00 m long. Find the force on the end if the tanker is half full of oil. 2 4 8 1 5 6 7 9

1. Architectural Technology
2. Civil Engineering Technology
3. Computer Engineering Technology
4. Mechanical Drafting Technology
5. Electrical/Electronics Engineering Technology
6. Chemical Engineering Technology
7. Industrial Engineering Technology
8. Mechanical Engineering Technology
9. Biomedical Engineering Technology

24.6 CHAPTER REVIEW

KEY TERMS

24.2 solid of revolution

24.5 pressure

Pascal's principle

RULES AND FORMULAS

Volume of a solid of revolution by the method of disks:

$V = \pi \int_a^b y^2\, dx$ if the area is rotated about the x-axis

$V = \pi \int_c^d x^2\, dy$ if the area is rotated about the y-axis

Volume of a solid of revolution by the method of shells:

$V = 2\pi \int_c^d xy\, dy$ if the area is rotated about the x-axis

(x is expressed in terms of y)

$V = 2\pi \int_a^b xy\, dx$ if the area is rotated about the y-axis

(y is expressed in terms of x)

The average value of a function: $f_{ave} = \frac{1}{b-a} \int_a^b f(x)\, dx$

Work done by a force: $W = \int_a^b F(x)\, dx$

Distance: $s = \int v\, dt$

Velocity: $v = \int a\, dt$

Force due to pressure of a fluid: $F = w \int_c^d yl\, dy$

GEOMETRY CONCEPTS

$V = \pi r^2 h$ (volume of a cylinder)

$V = lwh$ (volume of a rectangular solid)

$V = (4/3)\pi r^3$ (volume of a sphere)

SEQUENTIAL REVIEW

(Section 24.1) *In 1–9, find the area bounded by the given curves.*

1. $y = x^2 - 1$ and $y = 3$

2. $y = 2x^2 + 1$ and $y = 3$

3. $y = \sqrt{x}$ and $y = (1/2)x$

4. $y = x^2$ and $y = x$

5. $y = x^3 - x$ and $y = (-3/4)x$

6. $y = x^3$ and $y = x^4$

7. $y = x^2 - 1$ and $y = -x^2 + 1$

8. $y = 2x^2 - 1$ and $y = 2x + 3$

9. $y = x + 2$, $y = -x + 2$, and $y = 0$

(Section 24.2) In 10–14, use the method of disks to find the volume of the solid generated by rotating the area bounded by the given curves about the x-axis.

10. $y = 2x$, $x = 3$, and the x-axis

11. $y = (x - 3)^2$, the y-axis, and the x-axis

12. $y = x^2$, $y = (x - 2)^2$, and the x-axis

13. $y = \sqrt{x + 2}$, $x = 2$, and the x-axis

14. $y = \sqrt{x}$, $y = -x + 6$ in the first quadrant

In 15–18, use the method of disks to find the volume of the solid generated by rotating the area bounded by the given curves about the y-axis.

15. $y = -2x$, $y = 4$, and the y-axis

16. $y = -x^2 + 3$, $y = 2$, $y = 0$, and the y-axis

17. $y = x^3$, $y = 8$, and the y-axis

18. $y = \sqrt{x}$, $y = 2$, and the y-axis

(Section 24.3) In 19–23, use the method of shells to find the volume of the solid generated by rotating the area bounded by the given curves about the y-axis.

19. $y = x + 2$, $x = 4$, the x-axis, and the y-axis

20. $y = x^2 + 2$, $x = 4$, the x-axis, and the y-axis

21. $y = \sqrt{x}$, $x = 9$, and the x-axis

22. $y = -x^2 + 2x$ and the x-axis

23. $y = \sqrt{x}$ and $y = x^3$

In 24–27, use the method of shells to find the volume of the solid generated by rotating the area bounded by the given curves about the x-axis.

24. $y = 2x + 1$, $y = 2$, and the y-axis

25. $y = \sqrt{x}$, $y = 4$, and the y-axis

26. $y = x + 1$, $y = x + 2$, $y = 4$, and the x-axis

27. $y = \sqrt{x}$ and $y = x^3$

(Section 24.4) In 28–33, find the average value of the given function.

28. $y = -x + 4$ between $x = 0$ and $x = 4$

29. $y = \sqrt{x}$ between $x = 1$ and $x = 4$

30. $y = 2x^2$ between $x = 1$ and $x = 3$

31. $y = 4x^{1.5}$ between $x = 0$ and $x = 4$

32. $y = 2x^2 + 3x$ between $x = 0$ and $x = 1$

33. $y = 3\sqrt{x}$ between $x = 4$ and $x = 9$

In 34–36, find the work done by the given force F in moving an object along a straight line as indicated.

34. $F(x) = x + 3$ from $x = 0$ to $x = 3$

35. $F(x) = 2x$ from $x = 1$ to $x = 4$

36. $F(x) = 3\sqrt{x}$ from $x = 0$ to $x = 4$

(Section 24.5) In 37–39, find an expression for distance traveled by an object with the given velocity.

37. $v = -9.8t$ and $s = 25.4$ m when $t = 2.0$ s

38. $v = t^2 + 2t$ and $s = 21$ ft when $t = 3$ s

39. $v = 2t/\sqrt{t^2 + 1}$ and $s = 2$ cm when $t = 0$ s

In 40–42, find an expression for the velocity of an object with the given acceleration.

40. $a = 3t + 4$ and $v = 6$m/s when $t = 0$ s

41. $a = 2t^{1.3}$ and $v = 0$ m when $t = 0$ s

42. $a = 3t^2 + 2t + 1$ and $v = 4$ ft/s when $t = 1$ s

In 43–45, find the force on the surface of the flat object described, if it is submerged vertically in water (62.4 lb/ft³) as indicated.

43. A rectangle 5.00 ft wide and 3.00 ft high with the top 2.00 ft below the surface of the water.

44. An isosceles triangle with a 6.00-ft base and 4.00-ft height pointed downward with the base at the surface of the water.

45. A semicircle with diameter 10.0 ft with the curved part downward and the diameter at the surface of the water.

APPLIED PROBLEMS

1. A chain that is 40.0 m long and weighs 4.00 N/m lies on the ground. Find the work done in winding up 10.0 m of it. 4 7 8 1 2 3 5 6 9

2. A circular garden area is to be constructed with an outer concrete wall and a central concrete island, as shown in the diagram. The equation for the cross section of the island is given by $y = -0.25x^2 + 4.0$. Find the volume of soil needed to fill the area between the wall and the island, if the top surface of the soil is to be at an elevation of 3.5 ft. 1 2 4 7 8 3 5

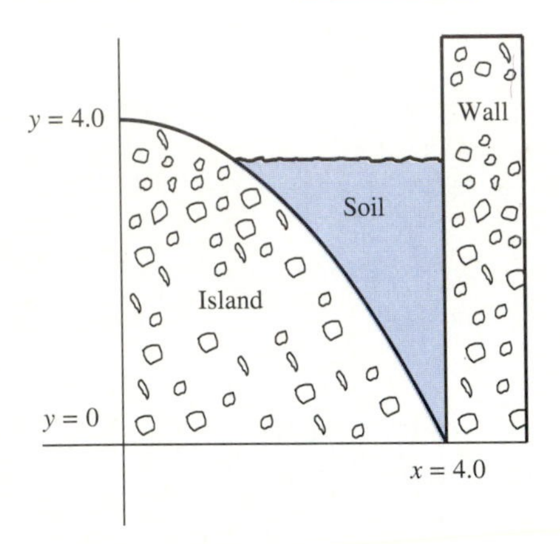

3. A rectangular floodgate is 3.0 m wide and 2.0 m high. Find the total force against the gate if the top is 1.0 m below the surface. Water weighs 9800 N/m³. 2 4 8 1 5 6 7 9

4. Find the volume of the basin swept out when the area shown is rotated about the y-axis.

1 2 4 7 8 3 5

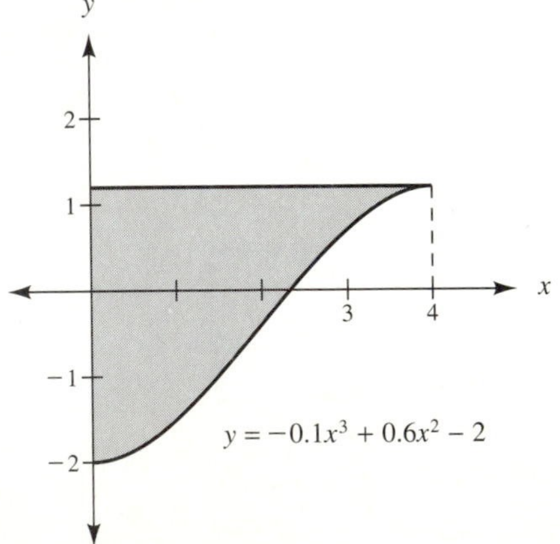

5. The formula $V = 3t^2 + 6$ gives voltage (in volts) after time t (in seconds). Find the average value of the voltage over the time interval $0 \leq t \leq 3$. Give your answer to the nearest volt. 3 5 9 1 2 4 6 7 8

6. The current i (in amperes) in an inductor after time t (in seconds) is given by the equation $i = (1/25)\int 500\,000t\,dt$. Experiment shows $i =$ 2.00 amperes when t is 0 seconds. Find i when t is 5.00 milliseconds. 3 5 9 4 7 8

7. An experimental antenna consists of a frame having the shape of the parabola given by $y = -0.04x^2 + 4x$. Both x and y are measured in feet. How many square feet of copper mesh would it take to cover this frame from the ground ($y = 0$) to the top? (Hint: Sketch the curve.) 1 3 4 5 6 8 9

8. A suspension bridge cable hangs with the shape of the parabola given by $y = 0.00125x^2 + 25$, where y is the distance (in ft) from the bridge deck up to the cable, and x is the distance (in ft) toward the end of the deck. Plastic sheeting is used to cover the side of the bridge to shield workers from the wind. How many rolls of plastic will it take if the bridge is 400 ft long and plastic comes in $12' \times 100'$ rolls? Assume there is no waste. (Hint: Sketch the curve.) 1 2 4 8 7 9

9. The formula $F = 4.049(s - 5.52)^2$ gives a force F (in lbs) acting over a distance s (in ft) when applied to a large cylinder of compressed gas. The force compresses the gas a distance of 4 additional feet. How much work W (in ft·lbs) is required? Give your answer to three significant digits. 6 8 1 2 3 5 7 9

10. The cylinder of a compressor is filled with gas at atmospheric pressure. The piston compresses the gas from a height of 7.000 inches to a height of 1.000 inch. How much work (in in·lbs) was done if the force F (in lb) is given by $F = 43.98\,s^{1.4}$, where s is the distance (in inches) through which the piston moves. 6 8 1 2 3 5 7 9

11. A 500-pound bucket containing 3000 pounds of sand is lifted up from the bottom of a quarry that is 100 ft deep. The sand leaks out at the rate of 25 lb/ft as it rises. How much work will be done in lifting the bucket from the quarry? Assume the force of gravity is constant. (Hint: When the bucket has risen a distance of s feet, the amount of

sand that has leaked out is $25s$.) Give your answer to three significant digits. 4 7 8 | 1 2 3 5 6 9

12. A force F (in lb) given by $F = 12.5s^2 - 50.0s + 50.0$ is applied to a bullet in a 2.00 ft long gun barrel. How much work is done on the bullet while it travels the length of the barrel? 4 7 8 | 1 2 3 5 6 9

13. The force F (in N) of attraction between two particular objects is given by $F = 2\,000\,000/(s + 200)^2$ where s is the distance (in m) between the objects. Find the work W (in joules) needed to completely separate the objects by evaluating $\lim_{x\to\infty} \int_0^x \frac{2\,000\,000}{(s+200)^2}\, ds.$ 4 7 8 | 1 2 3 5 6 9

Problems 14 and 15 use the following information. The formula $I = 126.5\sqrt{2t - 0.001}$ gives a pulse of current I (in amperes) if t is time (in seconds).

14. The current is applied to a resistance of 8 ohms. Find the average current between 0.001 and 0.002 second. Give your answer to three significant digits. 3 5 9 | 1 2 4 6 7 8

15. Power P (in watts) in a resistance of R (in ohms) is given by $P = I^2R$. Find the average power in an 8-ohm resistor for the given current between 0.001 and 0.002 second. Give your answer to three significant digits. 3 5 9 | 1 2 4 6 7 8

16. The velocity v (in cm/s) of gas molecules depends on the density d (in g/cm³) of the gas. For gas at 0° C and one atmosphere pressure, the equation relating velocity to density is $v = 1743.27/\sqrt{d}$. Find the average velocity of gas molecules for densities from 0.001 to 0.002 g/cm³. Give your answer to three significant digits. 6 8 | 1 2 3 5 7 9

17. Above 20 km, the density d (in mg/cm³) of the earth's atmosphere varies with altitude A (in km) according to the equation $d = 2.456 \times 10^{11}/A^{8.637}$. Find the average density for $80.00 \text{ km} \le A \le 140.0 \text{ km}$. 1 2 4 6 8 | 5 7 9

18. For a certain thermistor, the rate of change of the resistance R (in ohms) with respect to temperature T (in degrees Celsius) is given by $dR/dT = 0.009\,T^2 + 0.02\,T - 0.7$. What value will R have when T is 30° C if R is known to be 2 ohms at 10° C. Give your answer to two significant digits. 3 5 9 | 1 2 4 6 7 8

19. A lump of coal takes 5 seconds to fall to the bottom of a mine shaft. Determine the depth of the mine. $\left(\text{Hint: Use } a = 32 \text{ ft/s}^2,\ v = \int a\, dt, \text{ and } s = \int v\, dt.\right)$ 4 7 8 | 1 2 3 5 6 9

1. Architectural Technology
2. Civil Engineering Technology
3. Computer Engineering Technology
4. Mechanical Drafting Technology
5. Electrical/Electronics Engineering Technology
6. Chemical Engineering Technology
7. Industrial Engineering Technology
8. Mechanical Engineering Technology
9. Biomedical Engineering Technology

RANDOM REVIEW

1. Find the work done by a force $F(x) = x^2 + 2$ in moving an object along a straight line from $x = 2$ to $x = 6$.
2. Find the area bounded by $y = x^2 - 2$ and $y = x$.
3. Find the volume of the solid generated by rotating the region bounded by $y = x^2$, $x = 3$, and the x-axis about the y-axis.
4. Find the average value of the function defined by $y = \sqrt{x - 2}$ between $x = 2$ and $x = 6$.
5. Find the volume of the solid generated by rotating the region bounded by $y = \sqrt[3]{x}$, $x = 8$, and the x-axis about the x-axis.
6. Find an expression for the velocity of an object that has an acceleration given by $a = \sqrt{t + 1}$, if $v = 6$ when $t = 3$.

7. Find the volume of the solid generated by rotating the region bounded by $y = x$, $y = x + 2$, $y = 4$, and the x-axis about the x-axis.
8. Find the force on the surface of a square 2.0 m on each side if it is submerged vertically in oil (9400 N/m^3) with the top 1.0 m below the surface.
9. Find the volume of the solid generated by rotating the region bounded by $y = x^3 + 1$, $y = 9$, and the y-axis about the y-axis.
10. Find an expression for distance traveled by an object if the velocity is given by $v = 4t^{-3}$ and $v = 0$ when $t = 2$.

CHAPTER TEST

1. Find the area bounded by $y = \sqrt{x + 2}$ and $y = x^2 + (1/2)x - 3$, given that the curves intersect when x is -2 and when x is 2.
2. Use the method of disks to find the volume of the solid generated by rotating the area bounded by $y = (1/9)x^3$, $x = 3$, and the x-axis about the x-axis.
3. Use the method of disks to find the volume of the solid generated by rotating the area bounded by $y = (1/4)x^2$, $y = 2$, and the y-axis about the y-axis.
4. Use the method of shells to find the volume of the solid generated by rotating the area bounded by $y = (1/8)x^3$, $y = 8$, and the y-axis about the x-axis.
5. Use the method of shells to find the volume of the solid generated by rotating the area bounded by $y = (1/4)x^4$, $x = 2$, and the x-axis about the y-axis.
6. Find the average value of the function given by $y = 3x^3$ between $x = 2$ and $x = 5$.
7. Find an expression for distance traveled by an object if the velocity is given by $v = 12 + 32t$, and $s = 0$ when $t = 0$.
8. Find an expression for the velocity of an object if the acceleration is given by $a = 3t - 2$ and $v = 6$ when $t = 2$.
9. The force (in lb) required to stop a particular car is given by $F(x) = 12{,}000x$ where x is the distance (in ft) traveled after the brake is applied. Find the work done to stop the car in 120 ft.
10. An elliptical-shaped culvert has a cross section given by $x^2/9 + y^2/3 = 1$. Find the force on one end if the water surface is level with the center of the culvert. (For water, $w = 62.4$ lb/ft^3.)

EXTENDED APPLICATION: CIVIL ENGINEERING TECHNOLOGY

MOMENT OF INERTIA FOR A BEAM

To analyze the ability of a beam to resist bending, it is necessary to consider its cross-sectional area. Two important properties of a plane figure, such as a beam cross section, are its *centroid* and its *moment of inertia.*

The center of gravity* of an object is the point at which a support could be placed to balance the object. The center of gravity of a plane area is called a ***centroid.*** For a symmetrical figure, such as a rectangle or a circle, the centroid coincides with the geometrical center of the figure. The **moment of inertia** I of a plane figure about the y-axis is defined to be

$$I = \int x^2 \, dA$$

where dA represents the area of a vertical strip of the figure, and x is the distance from the y-axis to the centroid of the strip. The moment of inertia is found in the following manner.

Suppose the plane figure is bounded by the upper curve $y = u(x)$ and the lower curve $y = l(x)$ between $x = a$ and $x = b$, as shown in Figure 24.28. Then the region may be divided into vertical strips, each of width dx. Consider one such strip. Let x be the x-coordinate of its centroid. The area of the strip (denoted dA) is approximately equal to $[u(x) - l(x)]dx$, which is the area of a rectangle.

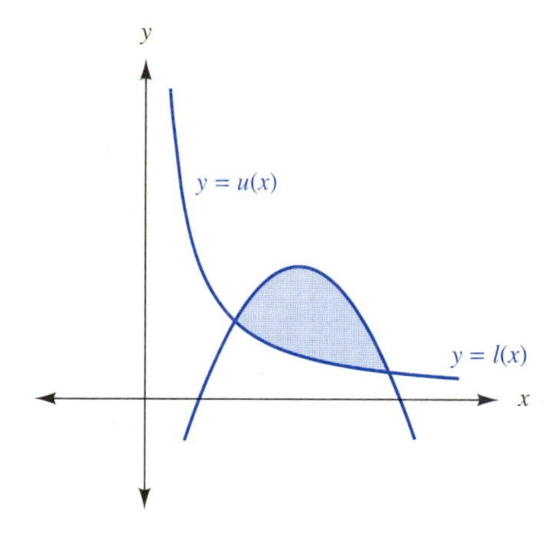

FIGURE 24.28

* The center of gravity is sometimes called the center of mass. If the acceleration due to gravity is the same in magnitude and direction at all points on the body, the center of mass and the center of gravity coincide.

EXTENDED APPLICATION, continued

The moment of inertia about the y-axis is

$$\int_a^b x^2\,dA = \int_a^b x^2[u(x) - l(x)]\,dx.$$

Likewise the moment of inertia about the x-axis is defined to be

$$I = \int y^2\,dA$$

where dA represents the area of a horizontal strip of the figure, and y is the y-coordinate of the centroid of the strip. If the plane figure is bounded by the right-hand curve $y = r(x)$ and the left-hand curve $y = l(x)$ between $y = c$ and $y = d$, the moment of inertia is given by

$$I = \int_c^d y^2[r(x) - l(x)]\,dy.$$

The moments of inertia, as defined here, are sometimes called the *second moments of area* to distinguish them from the integrals $I = \int x\,dA$ and $I = \int y\,dA$, called the *first moments of area.* In technical applications, the moment of inertia is usually obtained from reference tables. A reference table would show that a standard 2×4 has a moment of inertia of 5.36×10^4 in^4.

CHAPTER 25

DIFFERENTIATION OF TRANSCENDENTAL FUNCTIONS

algebraic functions

transcendental functions

All of our work with integrals and most of our work with derivatives has been restricted to functions that can be formed by a finite number of additions, subtractions, multiplications, divisions, and extractions of roots using only a single variable and constants. Such functions are called *algebraic functions.* For instance, $f(x) = 2x^2 + 3x - 1$ and $g(x) = \sqrt{x+2}/(x-3)$ are algebraic functions. Many functions common in technological work cannot be formed in this way. A function that is not algebraic is said to be *transcendental.* In this chapter, we develop formulas for the derivatives of the trigonometric, exponential, and logarithmic functions. We will also examine the use of these derivatives in problem solving.

25.1 DERIVATIVES OF THE SINE AND COSINE FUNCTIONS

The theory we have developed for algebraic functions applies to other functions also. That is, the sum, product, and quotient rules for derivatives apply, as does the chain rule. We have not yet established the derivatives of the transcendental functions that are commonly used in technical settings. We begin with an intuitive look at the sine function. Examine the graph of $y = \sin x$, shown in Figure 25.1.

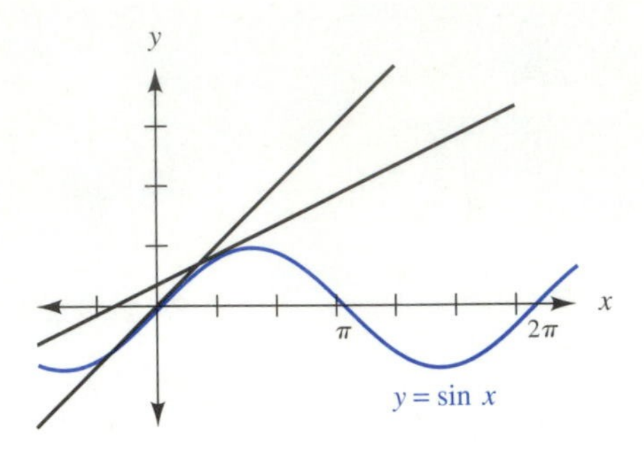

FIGURE 25.1

Recall that the derivative of a function defined by $y = f(x)$ can be interpreted as a function that gives the slope of the line tangent to the graph of $y = f(x)$ at x. When x is 0, the slope of the line tangent to the graph of $y = \sin x$ appears to be about 1. As x increases, the slope of the tangent at x decreases until x is $\pi/2$. When x is $\pi/2$, the tangent line is horizontal, so its slope is 0. Continuing in this manner, we are able to make a number of observations about the derivative of the sine function. Table 25.1 summarizes the information.

TABLE 25.1

When x is	$\frac{d(\sin x)}{dx}$ *appears to be*
0	1
between 0 and $\pi/2$	positive and decreasing
$\pi/2$	0
between $\pi/2$ and π	negative and decreasing
π	-1
between π and $3\pi/2$	negative and increasing
$3\pi/2$	0
between $3\pi/2$ and 2π	positive and increasing
2π	1

Observation shows that the cosine function, shown in Figure 25.2, satisfies these conditions. An educated guess that

$$\frac{d(\sin x)}{dx} = \cos x$$

proves to be correct.

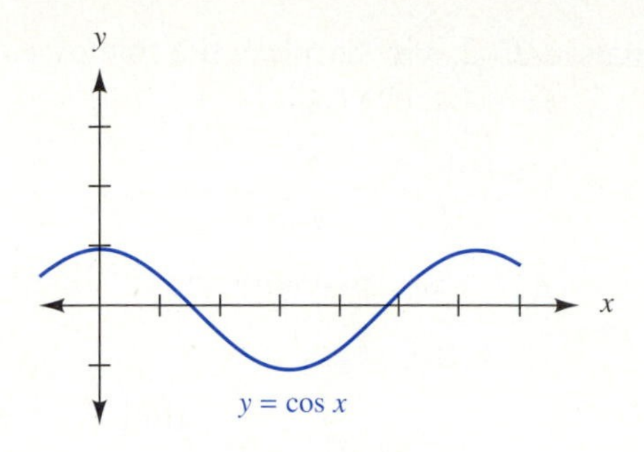

FIGURE 25.2

The conventional mathematical argument, which we will present shortly, requires the following theorem.

$$\text{When } x \text{ is measured in radians, } \lim_{x\to 0} \frac{\sin x}{x} = 1.$$

Although we will use this theorem without proof, notice that the result seems reasonable, if one examines Figure 25.3. For a circle of radius 1, the arc length x is equal to the measure of the angle subtended. The vertical line has length $\sin x$. As x gets smaller and smaller, there is less and less difference between the lengths of the vertical line and the arc length. Thus,

$$\lim_{x\to 0} \frac{\sin x}{x} = 1. \tag{25-1}$$

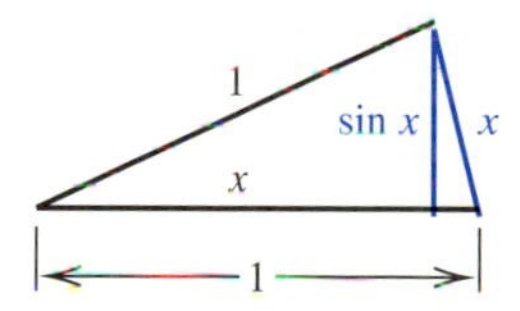

FIGURE 25.3

With this background, we proceed to show that $\dfrac{d \sin x}{dx} = \cos x$.

By definition,

$$\frac{d\,(\sin x)}{dx} = \lim_{\Delta x\to 0} \frac{\sin (x+\Delta x) - \sin x}{\Delta x}. \tag{25-2}$$

In order to rewrite $\sin (x+\Delta x) - \sin x$, it is necessary to use the following trigonometric identity.

$$\sin A - \sin B = 2 \sin \left(\frac{A-B}{2}\right) \cos \left(\frac{A+B}{2}\right) \tag{25-3}$$

In Equation 25-2, $x + \Delta x$ plays the role of A and x plays the role of B.

$$\frac{d(\sin x)}{dx} = \lim_{\Delta x \to 0} \frac{2 \sin\left(\frac{x + \Delta x - x}{2}\right) \cos\left(\frac{x + \Delta x + x}{2}\right)}{\Delta x}$$

Since $2 = 1/(1/2)$, we can rewrite the "2" in the numerator as "1/2" in the denominator. Thus,

$$\frac{d(\sin x)}{dx} = \lim_{\Delta x \to 0} \frac{\sin\left(\frac{\Delta x}{2}\right) \cos\left(x + \frac{\Delta x}{2}\right)}{\Delta x/2}$$

$$= \lim_{\Delta x \to 0} \frac{\sin\left(\frac{\Delta x}{2}\right)}{\Delta x/2} \lim_{\Delta x \to 0} \cos\left(x + \frac{\Delta x}{2}\right). \qquad \text{(Why?)}$$

As Δx approaches 0, $\Delta x/2$ approaches 0, so the first limit is equivalent to

$$\lim_{\Delta x/2 \to 0} \frac{\sin \Delta x/2}{\Delta x/2}.$$

From Equation 25-1, we know that this limit is 1. As Δx approaches 0, $x + \Delta x/2$ approaches x, and the second limit is $\cos x$. Thus,

$$\frac{d(\sin x)}{dx} = \cos x.$$

This proof that $\frac{d(\sin x)}{dx} = \cos x$ assumes that the angles are measured in radians, because the development of Equation 25-1 was based on radian measure. If θ measures degrees, then $\theta^\circ = \theta\left(\frac{\pi}{180}\right)$ radians.

$$\sin \theta^\circ = \sin(\theta\pi/180)$$

$$\frac{d(\sin \theta^\circ)}{d\theta} = \left(\frac{\pi}{180}\right) \cos\left(\frac{\theta\pi}{180}\right) \qquad \text{Apply the chain rule.}$$

$$\frac{d(\sin \theta^\circ)}{d\theta} = \left(\frac{\pi}{180}\right) \cos \theta^\circ \qquad \text{Replace } \frac{\theta\pi}{180} \text{ with } \theta^\circ$$

It is reasonable that the derivative is different if degree measure is used. The derivative is a rate of change, and a change in x of 1 radian could produce a bigger change in $\sin x$ (and hence a bigger derivative) than a change in x of 1 degree.

Before we consider some examples, a few comments are in order. The chain rule was stated informally in Section 21.6 as, "the derivative of a composition of functions is the derivative of the outer function times the derivative of the inner function." ***This rule is valid for any composition of functions, not just for powers of functions.*** We therefore make the following generalization.

NOTE ⇨⇨

THE DERIVATIVE OF THE SINE FUNCTION

If u is a function of x, then $\dfrac{d(\sin u)}{dx} = \cos u \dfrac{du}{dx}$.

EXAMPLE 1 If $y = \sin(3x)$, find y'.

Solution The outer function is defined by $y = \sin u$. The derivative is $\cos u$. The inner function is defined by $u = 3x$. The derivative du/dx is 3.

$$\begin{aligned} y' &= [\cos(u)](du/dx) \\ &= [\cos(3x)](3) \text{ or } 3\cos(3x) \quad ■ \end{aligned}$$

EXAMPLE 2 If $y = \sin^2 x$, find y'.

Solution Recall that $y = \sin^2 x$ means $y = (\sin x)^2$. The outer function is defined by $y = u^2$. The derivative is $2u$. The inner function is defined by $u = \sin x$. The derivative du/dx is $\cos x$.

$$\begin{aligned} y' &= 2u(du/dx) \\ &= 2\sin x \cos x \quad ■ \end{aligned}$$

EXAMPLE 3 If $y = \sin^3 2x$, find y'.

Solution The outer function is defined by $y = u^3$. The derivative is $3u^2$. The inner function is defined by $u = \sin 2x$. This function is itself a composition of functions, so its derivative du/dx is $(\cos 2x)2$ or $2\cos(2x)$.

$$\begin{aligned} y' &= 3u^2(du/dx) \\ &= (3\sin^2 2x)(2\cos 2x) \\ &= 6(\sin^2 2x)(\cos 2x) \quad ■ \end{aligned}$$

To find the derivative of the cosine function, notice that the graph of $y = \cos x$, shown in Figure 25.4 is simply the graph of $y = \sin x$ with a phase shift of $-\pi/2$.

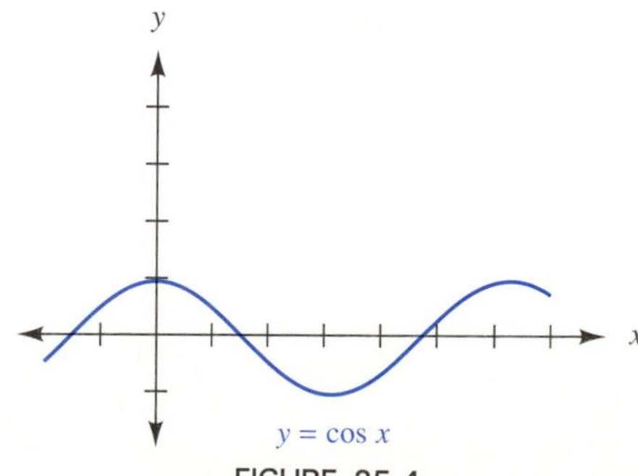

FIGURE 25.4

That is,

$$\cos x = \sin (x + \pi/2).$$

Thus,

$$\frac{d(\cos x)}{dx} = \frac{d[\sin (x + \pi/2)]}{dx}$$
$$= \cos (x + \pi/2).$$

But the graph of $y = \cos (x + \pi/2)$, shown in Figure 25.5, is also the graph of $y = -\sin x$.

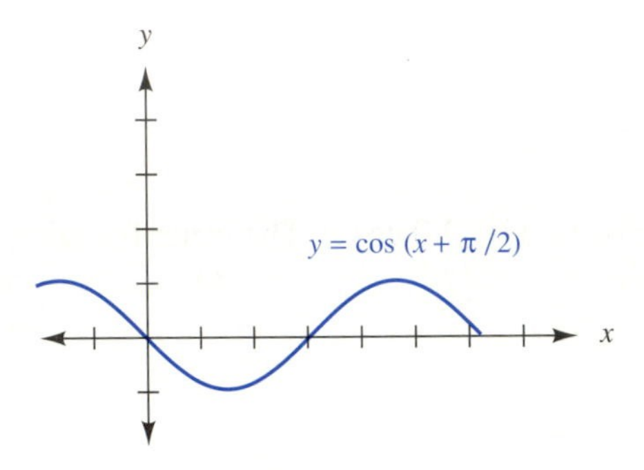

FIGURE 25.5

Thus, we have the following generalization.

DERIVATIVE OF THE COSINE FUNCTION

If u is a function of x, $\dfrac{d(\cos u)}{dx} = -\sin u \dfrac{du}{dx}$.

EXAMPLE 4 If $y = 3 \cos^4(2x^3)$, find y'.

Solution The outer function is defined by $y = 3u^4$. The derivative is $12u^3$. The inner function is defined by $u = \cos (2x^3)$. The derivative is $[-\sin (2x^3)](6x^2)$.

$$y' = 12u^3(du/dx)$$
$$= [12 \cos^3(2x^3)][-\sin (2x^3)](6x^2)$$
$$= -72x^2 \cos^3(2x^3) \sin (2x^3) \quad ■$$

EXAMPLE 5 If $y = \sin^2 x \cos x$, find y'.

Solution The product rule yields

$$\begin{aligned} y' &= (\sin^2 x)(-\sin x) + (\cos x)(2 \sin x \cos x) \\ &= -\sin^3 x + 2 \sin x \cos^2 x \end{aligned}$$ ■

Example 6 illustrates how derivatives of transcendental functions occur in technological problems.

EXAMPLE 6 The equation for a certain ac circuit is given by an electronics engineer as $I = 12.0 \sin(60.0\pi t + 60.0°)$. This equation is equivalent to the mathematical notation $I = 12.0 \sin(60.0\pi t + \pi/3)$. Find the voltage drop v across a 3.00-H inductor, given that $v = L\, dI/dt$. In the formula, I is the current (in amperes), L is the inductance (in henries), and t is the time (in seconds).

Solution If $I = 12.0 \sin(60.0\pi t + \pi/3)$, the outer function is defined by $I = 12.0 \sin u$. The derivative is $12.0 \cos u$. The inner function is defined by $u = 60.0\pi t + \pi/3$. The derivative is 60.0π. Thus, $dI/dt = [12.0 \cos(60.0\pi t + \pi/3)](60.0\pi)$. Since $v = L\, dI/dt$, and $L = 3.00$ H, we have $v = (3.00)[12.0 \cos(60.0\pi t + \pi/3)](60.0\pi)$ or $2160\pi[\cos(60.0\pi t + \pi/3)]$. ■

EXERCISES 25.1

In 1–27, find the derivative of each function.

1. $y = \sin 4x$
2. $y = \sin x^2$
3. $y = \cos x^5$
4. $y = \cos(3x - 1)$
5. $y = \cos(5x^2 + 2)$
6. $y = 3 \sin 2x^6$
7. $y = 2 \sin x^4$
8. $y = \sin^2(3x + 1)$
9. $y = \sin^3(2x - 3)$
10. $y = 4 \cos^3(2x - 1)$
11. $y = 5 \cos^2(3x + 1)$
12. $y = -\cos^4 x^2$
13. $y = -2 \cos^5 2x^3$
14. $y = -3 \sin^3 \sqrt{x}$
15. $y = 3x \sin 2x$
16. $y = 2x^2 \sin x$
17. $y = x^3 \cos 2x^2$
18. $y = 2 \sin x \cos 3x$
19. $y = 3 \sin 2x \cos x$
20. $y = \sin 3x^2 \cos(2x - 1)$
21. $y = 2 \sin^2 x \cos 2x$
22. $y = \sin(2x^2 + x) \cos 4x$
23. $y = \sin^3 3x^2 \cos^2 2x^3$
24. $y = \dfrac{\sin x}{x}$
25. $y = \dfrac{\cos x}{x^2}$
26. $y = \dfrac{1 - 2x}{\sin 3x}$
27. $y = \dfrac{2 + 3x^2}{\cos^2 x}$

28. The voltage in an ac circuit is given by $V = V_m \sin \omega t$. Find the expression for the charging current I of a capacitor, given that $I = C(dv/dt)$.
3 5 9 1 2 4 6 7 8

29. A certain object moves in such a way that its distance is given by $s = 2 \sin^2 t$. Find an expression for its velocity. **4 7 8** 1 2 3 5 6 9

30. A British nautical mile varies with latitude. The equation $l = 6077 - 31 \cos 2\theta$, where θ is measured in radians, gives the length of a nautical mile (in ft) at latitude θ. Find the rate at which l changes with respect to θ when $\theta = \pi/6$.
1 2 4 7 8 3 5

31. When a projectile is fired at an angle of θ from the horizontal, the horizontal distance it travels is given by $x = (v_0^2/g)\sin 2\theta$. In the equation, v_0 is the initial velocity, and g is the acceleration due to gravity (9.8 m/s^2). For a projectile launched at 300.0 m/s, find the rate at which x changes with respect to θ when $\theta = \pi/15$. **4 7 8** 1 2 3 5 6 9

32. A particular hill has a cross section given by $y = \sin^2(x/10)$ between $x = 0$ and $x = 31.5$ m as shown in the accompanying diagram. Find the slope of the tangent when x is 7.9 m.
1 2 4 7 8 3 5

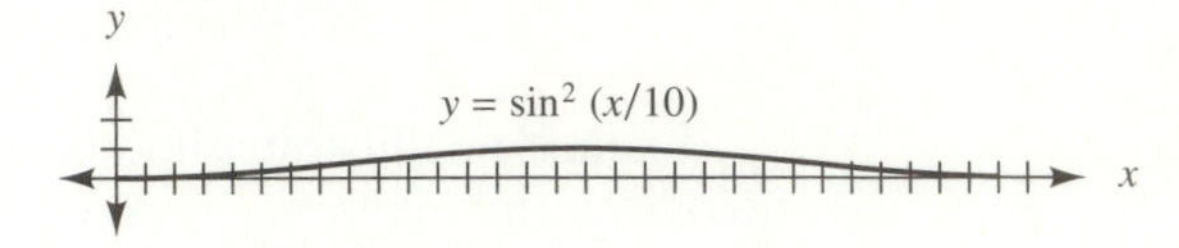

33. The voltage v (in volts) available from the wall outlets of a residential dwelling is given by $v = 170.0 \sin 377t$, where t is the time (in seconds). Find the rate of change of voltage with respect to time when t is 4.00 milliseconds.
3 5 9 1 2 4 6 7 8

34. To prove the identity in Equation 25-3, find $\sin(\alpha + \beta) - \sin(\alpha - \beta)$. Let $A = \alpha + \beta$ and let $B = \alpha - \beta$. Solve this system of equations for α and β and substitute into the equation with which you started.

1. Architectural Technology
2. Civil Engineering Technology
3. Computer Engineering Technology
4. Mechanical Drafting Technology
5. Electrical/Electronics Engineering Technology
6. Chemical Engineering Technology
7. Industrial Engineering Technology
8. Mechanical Engineering Technology
9. Biomedical Engineering Technology

25.2 DERIVATIVES OF THE OTHER TRIGONOMETRIC FUNCTIONS

The derivatives of trigonometric functions other than sine and cosine are found by expressing the functions in terms of the sine and cosine functions and applying the product or quotient rule for derivatives. Since $\tan x = \dfrac{\sin x}{\cos x}$, we apply the quotient rule.

$$\frac{d(\tan x)}{dx} = \frac{(\cos x)(\cos x) - (\sin x)(-\sin x)}{\cos^2 x}$$

$$= \frac{\cos^2 x + \sin^2 x}{\cos^2 x}$$

$$= \frac{1}{\cos^2 x}$$

$$= \sec^2 x$$

The derivation for the derivative of the cotangent is similar.

$$\cot x = \frac{\cos x}{\sin x}$$

$$\begin{aligned}\frac{d(\cot x)}{dx} &= \frac{(\sin x)(-\sin x) - (\cos x)(\cos x)}{\sin^2 x}\\ &= \frac{-\sin^2 x - \cos^2 x}{\sin^2 x}\\ &= \frac{-1(\sin^2 x + \cos^2 x)}{\sin^2 x}\\ &= \frac{-1}{\sin^2 x}\\ &= -\csc^2 x\end{aligned}$$

The derivative of the secant function is found as follows.

$$\sec x = \frac{1}{\cos x}$$

$$\begin{aligned}\frac{d(\sec x)}{dx} &= \frac{(\cos x)(0) - 1(-\sin x)}{\cos^2 x}\\ &= \frac{\sin x}{\cos^2 x}\\ &= \left(\frac{\sin x}{\cos x}\right)\left(\frac{1}{\cos x}\right)\\ &= \tan x \sec x\end{aligned}$$

Similarly for the cosecant, we have

$$\csc x = \frac{1}{\sin x}$$

$$\begin{aligned}\frac{d(\csc x)}{dx} &= \frac{(\sin x)(0) - 1(\cos x)}{\sin^2 x}\\ &= \frac{-\cos x}{\sin^2 x}\\ &= \left(\frac{-\cos x}{\sin x}\right)\left(\frac{1}{\sin x}\right)\\ &= -\cot x \csc x.\end{aligned}$$

Generalizing these results, we note that the argument of each function may itself be a function, in which case the chain rule for differentiation applies.

DERIVATIVES OF THE TRIGONOMETRIC FUNCTIONS

If u is a function of x,

$$\frac{d(\sin u)}{dx} = \cos u \frac{du}{dx} \qquad \frac{d(\cos u)}{dx} = -\sin u \frac{du}{dx}$$

$$\frac{d(\tan u)}{dx} = \sec^2 u \frac{du}{dx} \qquad \frac{d(\cot u)}{dx} = -\csc^2 u \frac{du}{dx}$$

$$\frac{d(\sec u)}{dx} = \sec u \tan u \frac{du}{dx} \qquad \frac{d(\csc u)}{dx} = -\csc u \cot u \frac{du}{dx}$$

EXAMPLE 1 If $y = 3 \tan 5x$, find y'.

Solution The outer function is defined by $y = 3 \tan u$. The derivative is $3 \sec^2 u$. The inner function is defined by $u = 5x$. The derivative is 5.

$$y' = (3 \sec^2 u)(du/dx) = [3 \sec^2 (5x)]5 \text{ or } 15 \sec^2 (5x). \quad ■$$

EXAMPLE 2 If $y = 3x \cot x^2$, find y'.

Solution The function is a product of two functions. The first is $3x$. Its derivative is 3. The second is $\cot x^2$. Its derivative is $(-\csc^2 x^2)(2x)$.

$$y' = 3x(-\csc^2 x^2)(2x) + (\cot x^2)(3) = -6x^2(\csc^2 x^2) + 3(\cot x^2) \quad ■$$

EXAMPLE 3 Find the slope of the line tangent to the curve $y = -\csc^3 2x$ at $x = \pi/4$.

Solution The slope of the tangent line is given by the derivative. The outer function is defined by $y = -u^3$. The derivative is $-3u^2$. The inner function is defined by $u = \csc 2x$. Its derivative is given by $[-\cot 2x \csc 2x](2)$. Thus,

$$y' = [-3 \csc^2 2x][-\cot 2x \csc 2x](2) = 6 \csc^3 2x \cot 2x.$$

When x is $\pi/4$, $y' = 6 \csc^3 (\pi/2) \cot (\pi/2)$ or $6(1)^3(0)$, which is 0. Figure 25.6 shows the graph of the function and the tangent line. ■

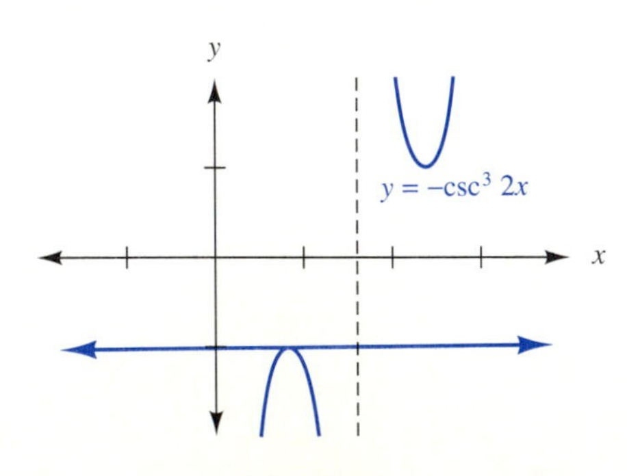

FIGURE 25.6

EXAMPLE 4 If $y = \dfrac{\cos 3x}{1 + \cot(2x)}$, find y'.

Solution The function is the quotient of two functions.

$$y' = \frac{[1 + \cot 2x][-3 \sin 3x] - \cos 3x[-\csc^2 2x]2}{(1 + \cot 2x)^2}$$

$$= \frac{-3 \sin 3x - 3 \cot 2x \sin 3x + 2 \cos 3x \csc^2 2x}{(1 + \cot 2x)^2} \quad ■$$

EXAMPLE 5 The phase angle θ (in radians) of a circuit containing a resistor and an inductor in series is given by $\tan \theta = 2\pi fL/R$. By differentiating both sides of the equation, find the rate of change of θ with respect to f when $f = 200$ Hz, $L = 350$ mH (or 0.350 H), and $R = 400\ \Omega$. In the formula, f is the applied frequency (in hertz), L is the inductance (in henries), and R the resistance (in ohms).

Solution

$$\tan \theta = 2\pi fL/R$$

$$(\sec^2 \theta)(d\theta/df) = 2\pi L/R \qquad \text{Differentiate with respect to } f.$$

$$d\theta/df = (2\pi L/R) \div \sec^2 \theta \qquad \text{Divide both sides by } \sec^2 \theta$$

$$= (2\pi L/R)(\cos^2 \theta) \qquad \text{Replace } (1/\sec^2 \theta) \text{ with } \cos^2 \theta.$$

To find θ, use $\tan \theta = 2\pi(200)(0.350)/400$ or 1.0996, so θ is 0.833 radians. Substituting this value and the given values of f, L, and R, $d\theta/df = [2\pi(0.350)/400(\cos^2 0.833)$ or about 2×10^{-3} radians/Hz. ■

EXERCISES 25.2

In 1–28, find the derivative of each function.

1. $y = \tan 3x$
2. $y = 2 \tan(2x - 1)$
3. $y = 2 \cot(4x - 3)$
4. $y = \cot 5x$
5. $y = 3 \sec 2x$
6. $y = -\sec(1 - x)^2$
7. $y = \csc \sqrt{x - 1}$
8. $y = \csc \sqrt{2x + 3}$
9. $y = \tan^3 2x$
10. $y = \sqrt{\tan x}$
11. $y = 2 \cot^4 3x$
12. $y = -\sec^3 4x$
13. $y = \csc^2 x^3$
14. $y = \sqrt{\csc 2x}$
15. $y = 1 + \tan^2 x + x$
16. $y = 1 - \csc^2 x + x^2$
17. $y = \sec^2 x + \tan x$
18. $y = \csc^2 x - \cot x$
19. $y = 2x \tan 2x$
20. $y = x^2 \tan x^2$
21. $y = \dfrac{\cot x}{2x}$
22. $y = \dfrac{4x}{\cot 2x}$
23. $y = \sec x \tan x$
24. $y = 3 \sec x \sin 2x$
25. $y = 3 \csc x \cos 2x$
26. $y = -\csc x \cot x$
27. $y = \dfrac{\tan^2 x}{2 + \sin x}$
28. $y = \dfrac{\cot^3 x}{1 - \cos x}$

29. The index of refraction of the medium that light enters (in relation to the medium that it leaves) is given by $n = \csc\theta$ if θ is the critical angle. Find the rate of change of n with respect to θ when θ is $\pi/6$. 3 5 6 9

30. In an ac circuit with resistance 3.00 Ω, the impedance is given by $Z = 3.00 \sec\theta$ for $0° \le \theta < 90.0°$. Find the rate of change of Z with respect to θ when θ is $\pi/3$. 3 5 9 1 2 4 6 7 8

31. For a cantilever beam under a concentrated load, the angle of deflection θ at a distance of x from the support is given by $\tan\theta = Px(x - 2a)/(2EI)$ when $x \le a$. Differentiate both sides of the equation to find the rate of change of θ with respect to x when $x = 30.0$ in, $a = 51.6$ in, $P = 950$ lb, $E = 1.1 \times 10^6$ psi, and $I = 536$ in^4. In the formula, P is the load (in lb), E is the modulus of elasticity (in psi), and I is the moment of inertia (in in^4). 1 2 4 8 7 9

32. For a cantilever beam under a concentrated load, the angle of deflection θ at a distance of x from the support is given by $\tan\theta = -Pa^2/(2EI)$ when $x > a$. Differentiate both sides of the equation to find the rate of change of θ with respect to a when $a = 51.6$ in, $P = 950$ lb, $E = 1.1 \times 10^6$ lb/in^2, and $I = 536$ in^4. In the formula, a is the distance of the load from the support (in inches), P is the load (in lb), E is the modulus of elasticity (in lb/in^2), and I is the moment of inertia (in in^4). 1 2 4 8 7 9

1. Architectural Technology
2. Civil Engineering Technology
3. Computer Engineering Technology
4. Mechanical Drafting Technology
5. Electrical/Electronics Engineering Technology
6. Chemical Engineering Technology
7. Industrial Engineering Technology
8. Mechanical Engineering Technology
9. Biomedical Engineering Technology

25.3 DERIVATIVES OF THE INVERSE TRIGONOMETRIC FUNCTIONS

Recall that the function denoted $y = \text{Arcsin } x$ is called the inverse sine function, or the arcsine function. To find the derivative of this function, we rely on the relationship between a function and its inverse.

The equation $y = \text{Arcsin } x$ or $y = \text{Sin}^{-1} x$ means

$$x = \sin y \text{ for } -\pi/2 \le y \le \pi/2. \tag{25-4}$$

Assuming that dy/dx exists, we differentiate both sides of this equation with respect to x.

$$1 = (\cos y)\, dy/dx$$

Divide both sides of this equation by $\cos y$, if $\cos y \ne 0$.

$$dy/dx = 1/\cos y \tag{25-5}$$

However, because we are looking for the derivative of y with respect to x, it is necessary to rewrite Equation 25-5 with the expression on the right-hand side in terms of x. Since

$$\sin^2 y + \cos^2 y = 1,$$
$$\cos^2 y = 1 - \sin^2 y \text{ and}$$
$$\cos y = \pm\sqrt{1 - \sin^2 y}.$$

From Equation 25-4 we know that $\sin y = x$, so we replace $\sin y$ with x.

$$\cos y = \pm\sqrt{1 - x^2}$$

From Equation 25-4, we also know that $-\pi/2 \leq y \leq \pi/2$. In quadrants I and IV, $\cos y$ is nonnegative, so we use the positive square root.

$$\cos y = \sqrt{1 - x^2}$$

Thus, Equation 25-5 can be written

$$\frac{d(\text{Arcsin } x)}{dx} = \frac{1}{\cos y} = \frac{1}{\sqrt{1 - x^2}}.$$

The derivative of the inverse cosine is found in a similar manner. The equation $y = \text{Arccos } x$ or $\text{Cos}^{-1} x$ means

$$x = \cos y \text{ for } 0 \leq y \leq \pi.$$

Assuming dy/dx exists, we differentiate both sides of this equation with respect to x.

$$1 = (-\sin y)\, dy/dx \text{ and } dy/dx = \frac{-1}{\sin y}$$

From the identity $\sin^2 y + \cos^2 y = 1$ and the equation $x = \cos y$, we have

$$\sin y = \pm\sqrt{1 - \cos^2 y} = \pm\sqrt{1 - x^2}.$$

For $0 \leq y \leq \pi$ (quadrants I and II), $\sin y$ is nonnegative. Thus,

$$\frac{d(\text{Arccos } x)}{dx} = \frac{-1}{\sin y} = \frac{-1}{\sqrt{1 - x^2}}.$$

To find the derivative of $y = \text{Arctan } x$, recall that $y = \text{Arctan } x$ or $y = \text{Tan}^{-1} x$ means

$$x = \tan y \text{ for } -\pi/2 < y < \pi/2.$$

$$1 = (\sec^2 y)\, dy/dx \qquad \text{Differentiate both sides of the equation.}$$

$$dy/dx = 1/\sec^2 y \qquad \text{Divide both sides by } \sec^2 y.$$

$$= \frac{1}{1 + \tan^2 y} \qquad \text{Replace } \sec^2 y \text{ with } 1 + \tan^2 y.$$

$$= \frac{1}{1 + x^2}. \qquad \text{Replace } \tan y \text{ with } x.$$

NOTE ⇨⇨ ***If the argument of one of the inverse trigonometric functions is itself a function, the chain rule applies.*** The derivatives of the Arccotangent, Arcsecant, and Arccosecant can be obtained in a similar manner. They are included in the table which follows, although they are far less common in technological work than the derivatives of the Arcsine, Arccosine, and Arctangent.

DERIVATIVES OF THE INVERSE TRIGONOMETRIC FUNCTIONS

if u is a function of x,

$$\frac{d(\text{Arcsin } u)}{dx}=\frac{1}{\sqrt{1-u^2}}\frac{du}{dx} \qquad \frac{d(\text{Arccsc } u)}{dx}=\frac{-1}{\sqrt{u^2(u^2-1)}}\frac{du}{dx}$$

$$\frac{d(\text{Arccos } u)}{dx}=\frac{-1}{\sqrt{1-u^2}}\frac{du}{dx} \qquad \frac{d(\text{Arcsec } u)}{dx}=\frac{1}{\sqrt{u^2(u^2-1)}}\frac{du}{dx}$$

$$\frac{d(\text{Arctan } u)}{dx}=\frac{1}{1+u^2}\frac{du}{dx} \qquad \frac{d(\text{Arccot } u)}{dx}=\frac{-1}{1+u^2}\frac{du}{dx}$$

EXAMPLE 1 If $y=\text{Arccos}^2\,(2x-1)$, find y'.

Solution The outer function is defined by $y=u^2$. The inner function is defined by $u=\text{Arccos}$ $(2x-1)$. Thus,

$$y'=\underbrace{2[\text{Arccos}\,(2x-1)]}_{\text{Derivative of the outer function}}\underbrace{\frac{-1}{\sqrt{1-(2x-1)^2}}(2)}_{\text{Derivative of the inner function}}.$$

$$y'=\frac{-4\text{ Arccos}\,(2x-1)}{\sqrt{-4x^2+4x}}$$

$$=\frac{-4\text{ Arccos}\,(2x-1)}{2\sqrt{-x^2+x}}$$

$$=\frac{-2\text{ Arccos}\,(2x-1)}{\sqrt{-x^2+x}}$$ ■

EXAMPLE 2 For a cantilever beam under a concentrated load P (in kg), the angle of deflection θ is given by $\theta=\text{Tan}^{-1}[-Pa^2/(2EI)]$. Find the rate at which θ changes with respect to a when $a=6.0$ m, $P=650$ kg, $E=2.11\times10^{10}$ kg/m², and $I=2.81\times10^{-5}$ m⁴. In the formula, a is the distance of the load from the support, E is the modulus of elasticity, and I is the moment of inertia.

Solution $\theta=\text{Tan}^{-1}[-Pa^2/(2EI)]$
Let $u=-Pa^2/(2EI)$, then $du/da=-Pa/(EI)$.

$$\frac{d\theta}{da}=\frac{1}{1+[(-Pa^2/(2EI)]^2}\left(\frac{-Pa}{EI}\right)$$

For the values given,

$$\frac{d\theta}{da} = \frac{1}{1 + [-650(6.0^2)/(2(2.11)(10^{10})(2.81)(10^{-5}))]^2}\left(\frac{-650(6.0)}{(2.11)(10^{10})(2.81)(10^{-5})}\right)$$
$$= -6.6 \times 10^{-3} \text{ rad/m.} \quad \blacksquare$$

EXAMPLE 3 If $y = x^2$ Arctan $3x$, find y'.

Solution The function is the product of two functions. The first factor is x^2. The second factor is Arctan $3x$.

$$y' = x^2 \frac{1}{1 + (3x)^2}(3) + (\text{Arctan } 3x)(2x)$$

First × Derivative of second + Second × Derivative of first

$$y' = \frac{3x^2}{1 + 9x^2} + 2x \text{ Arctan } 3x \quad \blacksquare$$

EXERCISES 25.3

In 1–28, find the derivative of each function.

1. $y = \text{Arcsin } 2x^3$

2. $y = 3 \text{ Arcsin } 4x$

3. $y = -\text{Arccos } (3x + 2)$

4. $y = \text{Arccos } (x^2 - 1)$

5. $y = 4 \text{ Arctan } (2x^3 + x)$

6. $y = 2 \text{ Arctan } (x^2 - x)$

7. $y = \text{Arcsin } (x - 1)$

8. $y = \text{Arcsin } \sqrt{1 - x}$

9. $y = \text{Arccos } \sqrt{2x}$

10. $y = 3 \text{ Arccos } (1/x)$

11. $y = 2 \text{ Arctan } (2x + 1)$

12. $y = \text{Arctan } (1/\sqrt{x})$

13. $y = -\text{Arcsin}^2\, 3x$

14. $y = \sqrt{\text{Arcsin } x}$

15. $y = \sqrt{\text{Arccos } x}$

16. $y = \text{Arccos}^3\, 2x$

17. $y = \text{Arctan}^{-1}\, x^2$

18. $y = \text{Arctan}^{-2}\, (3x + 1)$

19. $y = x^2 \text{ Arcsin } x$

20. $y = 2x \text{ Arcsin } x^2$

21. $y = 3x \text{ Arccos } (1 - x)$

22. $y = (1 - x)\text{Arccos } 3x$

23. $y = x \text{ Arctan } (2x - 3)$

24. $y = (x^2 + 1)/\text{Arctan } x$

25. $y = \dfrac{\text{Arcsin } x}{\text{Arccos } x}$

26. $y = \dfrac{\text{Arccos } x}{3x}$

27. $y = \dfrac{\text{Arctan } x}{x^2 + 1}$

28. $y = \dfrac{3x}{\text{Arcsin } 2x}$

29. The equation $\theta_1 = \text{Tan}^{-1}(\omega L/R)$ gives the phase angle of the impedance in the series portion of a distributed constant circuit. Find a formula for the rate of change of θ_1 with respect to L. In the formula, L is the inductance per unit length (in mH/km), and R is the resistance per unit length (in Ω/km). 3 5 9 1 2 4 6 7 8

30. The equation $\theta_2 = \text{Tan}^{-1}(\omega C/G)$ gives the phase angle of the impedance in the parallel portion of a distributed constant circuit. Find a formula for the rate of change of θ_2 with respect to C. In the formula, C is the capacitance per unit length (in μF/km), and G is the leakage conductance per unit length (in μS/km). 3 5 9 1 2 4 6 7 8

31. The equation $\theta = \text{Cos}^{-1}(R/Z)$ gives the phase angle between R and **Z** in an ac circuit. Find the rate at which θ is changing with respect to Z when R is 80 Ω and Z is 100 Ω. In the formula, R is the magnitude of the resistance and Z is the magnitude of the impedance. 3 5 9 1 2 4 6 7 8

32. For a cantilever beam under a distributed load of W (in kg/mm), the angle of deflection is given by $\theta = \text{Tan}^{-1}[Wb^3/(6EI)]$. Find the rate at which θ is changing with respect to b when $b = 10.0$ in, $W = 37.5$ lb/in, $E = 3.0 \times 10^6$ lb/in², and $I = 797$ in⁴. In the formula, b is the length (in inches) of the load, E is Young's modulus, and I is the moment of inertia (in in⁴) of the beam. 1 2 4 8 7 9

33. Derive the following formulas.

(a) $$\frac{d(\text{Arccot } u)}{dx} = \frac{-1}{1+u^2}\frac{du}{dx}$$

(b) $$\frac{d(\text{Arcsec } u)}{dx} = \frac{1}{\sqrt{u^2(u^2-1)}}\frac{du}{dx}$$

(c) $$\frac{d(\text{Arccsc } u)}{dx} = \frac{-1}{\sqrt{u^2(u^2-1)}}\frac{du}{dx}$$

1. Architectural Technology
2. Civil Engineering Technology
3. Computer Engineering Technology
4. Mechanical Drafting Technology
5. Electrical/Electronics Engineering Technology
6. Chemical Engineering Technology
7. Industrial Engineering Technology
8. Mechanical Engineering Technology
9. Biomedical Engineering Technology

25.4 DERIVATIVES OF THE EXPONENTIAL AND LOGARITHMIC FUNCTIONS

Recall that the functions defined by $y = e^x$ and $y = \ln x$ are inverses of each other. If we know the derivative of one of these functions, we can find the derivative of the other using a process similar to that used with the inverse trigonometric functions.

To find the derivative of the exponential function, we must find

$$y' = \lim_{\Delta x \to 0} \frac{e^{x+\Delta x} - e^x}{\Delta x}.$$

The number e has not yet been formally defined. We have accepted its value to be approximately 2.718. In fact, if we investigate the more general function defined by $y = b^x$ where b is any positive real number, we will find the derivative of the exponential function $y = e^x$ and also solve the problem of defining e.

$$\frac{d(b^x)}{dx} = \lim_{\Delta x \to 0} \frac{b^{x+\Delta x} - b^x}{\Delta x}$$

If we factor out b^x, we have.

$$\frac{d(b^x)}{dx} = \lim_{\Delta x \to 0} b^x\left(\frac{b^{\Delta x} - 1}{\Delta x}\right)$$

The limit of a product is the product of the limits. Thus,

$$\frac{d(b^x)}{dx} = (\lim_{\Delta x \to 0} b^x)\left(\lim_{\Delta x \to 0} \frac{b^{\Delta x} - 1}{\Delta x}\right). \tag{25-6}$$

The first limit is b^x, since Δx does not appear in the expression. To evaluate the second limit, notice that $\frac{b^{\Delta x} - 1}{\Delta x}$ gives the slope of the line between the points $(\Delta x, b^{\Delta x})$ and $(0, 1)$, as shown in Figure 25.7. Thus, $\lim_{\Delta x \to 0} \frac{b^{\Delta x} - 1}{\Delta x}$ is the slope of the line tangent to the curve $y = b^x$ at $(0, 1)$. The slope of the line tangent at this point depends on the value of b.

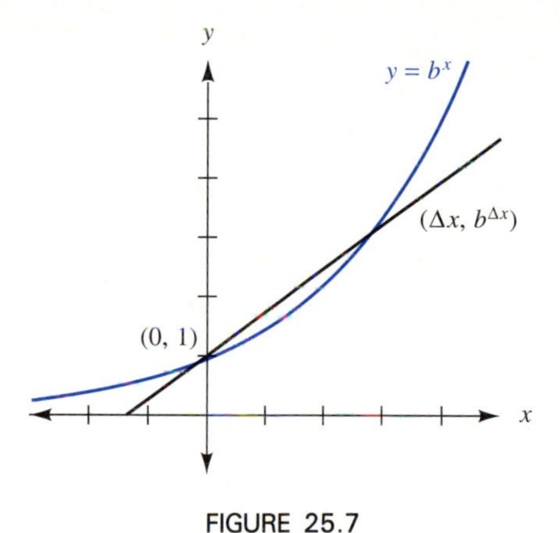

FIGURE 25.7

When b is chosen to make the slope equal 1, the value of b is defined to be the number e. For this value of b, Equation 25-6 becomes

$$\frac{d(e^x)}{dx} = (\lim_{\Delta x \to 0} e^x)\left(\lim_{\Delta x \to 0} \frac{e^{\Delta x} - 1}{\Delta x}\right) \tag{25-7}$$

$$= e^x(1) \text{ or } e^x.$$

Since the chain rules applies, the more general form is

$$\frac{d(e^u)}{dx} = e^u \frac{du}{dx}.$$

Once the derivative of the exponential function has been established, we have the inverse function $y = \ln x$, which means

$$x = e^y.$$

Assuming that dy/dx exists, we differentiate both sides of the equation with respect to x. We have

$$1 = e^y(dy/dx) \text{ and } dy/dx = 1/e^y.$$

Since $x = e^y$, the equation can be written as

$$\frac{d(\ln x)}{dx} = \frac{1}{x}.$$

This formula, too, can be generalized for use with the chain rule.

DERIVATIVES OF THE EXPONENTIAL AND LOGARITHMIC FUNCTIONS

If u is a function of x, then

$$\frac{d(e^u)}{dx} = e^u \frac{du}{dx} \text{ and } \frac{d(\ln u)}{dx} = \frac{1}{u}\left(\frac{du}{dx}\right).$$

EXAMPLE 1 If $y = e^{2x}$, find dy/dx.

Solution The outer function is defined by $y = e^u$. The inner function is defined by $u = 2x$.

$$dy/dx = \underbrace{e^{2x}}_{\text{Derivative of the outer function}}\underbrace{(2)}_{\text{Derivative of the inner function}}$$

Generally, we would write $dy/dx = 2e^{2x}$. ■

Both exponential and logarithmic functions occur in technical applications.

EXAMPLE 2 For a resistor and capacitor in series, the voltage v across the capacitor is given by $v = v_0(1 - e^{-t/\tau})$. Find a formula for the rate of change of v with respect to t. In the formula, v_0 is the initial voltage (in volts), t is the time elapsed (in seconds), and τ is the product of resistance (in ohms) and capacitance (in farads).

Solution The outer function is defined by $v = v_0(1 - e^u)$. The inner function is defined by $u = -t/\tau$.

$$v' = \underbrace{v_0(-e^{-t/\tau})}_{\text{Derivative of the outer function}}\underbrace{(-1/\tau)}_{\text{Derivative of the inner function}}$$

Thus, $v' = (v_0/\tau)(e^{-t/\tau})$. ■

EXAMPLE 3 If a signal passes through an amplifier, the total voltage gain A (in dB) is given by $A = 20 \log (v_O/v_I)$, where v_O is the output voltage and v_I is the input voltage. Find the rate of change of A with respect to v_I when $v_I = 2.5$ mV (or 2.5×10^{-3} V).

Solution Recall that $\log x = \ln x/\ln 10$ or $0.4343 \ln x$. Thus,

$$\begin{aligned} A &= 20 \log (v_O/v_I) \\ &= 20[0.4343 \ln (v_O/v_I)] \\ &= 8.686 \ln (v_O/v_I). \end{aligned}$$

The outer function is defined by $A = 8.686 \ln u$. The inner function is defined by $u = v_O v_I^{-1}$.

$$\begin{aligned} \frac{dA}{dv_I} &= 8.686 \left(\frac{1}{v_O v_I^{-1}}\right)(-1 v_O v_I^{-2}) \\ &= -8.686\, v_I^{-1} \end{aligned}$$

When v_I is 2.5 mV, $\dfrac{dA}{dv_I} = \dfrac{-8.686}{2.5 \times 10^{-3}}$ or -3.47×10^3 dB/V. ■

It may be necessary to use the chain rule more than once.

EXAMPLE 4 If $y = \ln (\sin x^2)$, find y'.

Solution The outer function is defined by $y = \ln u$. The inner function is defined by $u = \sin x^2$, which is itself a composition of functions.

$$y' = \underbrace{\frac{1}{\sin x^2}}_{\text{Derivative of the outer function}} \underbrace{(\cos x^2)(2x)}_{\text{Derivative of the inner function}}$$

Thus, $y' = 2x \cot x^2$. ■

EXAMPLE 5 If $y = \text{Arctan } e^{x^2}$, find y'.

Solution The outer function is defined by $y = \text{Arctan } u$. The inner function is defined by $u = e^{x^2}$, which is itself a composition of functions.

$$y' = \underbrace{\left(\frac{1}{1 + (e^{x^2})^2}\right)}_{\text{Derivative of the outer function}} \underbrace{e^{x^2}(2x)}_{\text{Derivative of the inner function}}$$

$$y' = \frac{2xe^{x^2}}{1 + e^{2x^2}} \quad ■$$

It is possible to classify the functions we have considered up to this point into two categories, based on the use of exponents. Either the base has been a variable expression and the exponent a constant, or the base has been the constant e with a variable expression as the exponent. The functions defined by $y = 3x^3 + 2x$, $y = (x + 1)^2$, and $y = \sin x$ are examples of the first type. The functions defined by $y = e^{2\tan x}$, $y = 3e^{2x}$ and $y = \ln \sin e^x$ are examples of the second type.

It is also possible to have a constant other than e as the base with a variable expression as the exponent. The key to differentiating a function of the type $y = a^x$, where a is a constant, is to recall from Chapter 15 that

$$a^x = e^{x \ln a}.$$

Thus, if $y = a^x$, the function is of a type we have considered. The outer function is defined by $y = e^u$. The inner function is defined by $u = x \ln a$. Since $\ln a$ is a constant, we have

$$y' = e^{x \ln a}(\ln a) = a^x(\ln a).$$

This rule may be generalized for use with the chain rule.

THE DERIVATIVE OF THE GENERAL EXPONENTIAL FUNCTION

If u is a function of x, then

$$\frac{d\,(a^u)}{dx} = a^u\,(\ln a)\,\frac{du}{dx}.$$

EXAMPLE 6 If $y = 2^{3x^2}$, find y'.

Solution The outer function is defined by $y = 2^u$. The inner function is defined by $u = 3x^2$.

$$y' = \underbrace{(\ln 2)(2^{3x^2})}_{\text{Derivative of the outer function}}\underbrace{(6x)}_{\text{Derivative of the inner function}}$$

$$y' = (6x \ln 2)(2^{3x^2}) \quad ■$$

Another type of problem occurs when both base and exponent are variable expressions. A common practice is to take the logarithm of the expression on each side of the equation and then differentiate both sides.

EXAMPLE 7 If $y = x^{\sin x}$, find y'.

Solution

$$y = x^{\sin x}, \text{ so}$$
$$\ln y = \ln(x^{\sin x})$$
$$\ln y = (\sin x)(\ln x)$$

Differentiating with respect to x, we use the product rule on the right-hand side of the equation.

$$\frac{1}{y}\left(\frac{dy}{dx}\right) = (\sin x)\left(\frac{1}{x}\right) + (\ln x)(\cos x)$$

$$\frac{dy}{dx} = y\left[(\sin x)\left(\frac{1}{x}\right) + (\ln x)(\cos x)\right]$$ Multiply both sides by y.

$$\frac{dy}{dx} = x^{\sin x}\left[(\sin x)\left(\frac{1}{x}\right) + (\ln x)(\cos x)\right]$$ Replace y with $x^{\sin x}$. ■

Of course, a logarithmic or exponential function may appear as part of a sum, difference, product, or quotient of functions.

EXAMPLE 8 If $y = x^2(2^{\sin x})$, find y'.

Solution The function is a product of functions.

$$y' = x^2(\ln 2)(2^{\sin x})(\cos x) + (2^{\sin x})(2x)$$

First × Derivative of the second + Second × Derivative of the first

Factoring out the common factor of $x(2^{\sin x})$, we have

$$y' = x(2^{\sin x})[(x \ln 2)(\cos x) + 2]. \quad ■$$

Sometimes it is easier to find the derivative of a function if the properties of logarithms are applied before the differentiation rules are applied. Example 9 illustrates this concept.

EXAMPLE 9 If $y = \dfrac{\ln (x+1)^2}{x+1}$, find y'.

Solution

NOTE ⇨⇨ ***Do not confuse*** $\ln(x+1)^2$ ***with*** $[\ln (x+1)]^2$. ***The first expression says to square*** $(x+1)$ ***and find the natural logarithm of the square. The second expression says to find the natural logarithm of*** $(x+1)$ ***and square the logarithm.***

$$\frac{\ln (x+1)^2}{x+1} = \frac{2 \ln (x+1)}{(x+1)}$$

The quotient rule for derivatives gives the following.

Denominator × Derivative of numerator − Numerator × Derivative of denominator

$$y' = \frac{(x+1)\dfrac{2}{x+1} - [2 \ln(x+1)](1)}{(x+1)^2}$$

Square of the denominator

$$y' = \frac{2 - 2 \ln (x+1)}{(x+1)^2} \quad ■$$

EXERCISES 25.4

In 1–30, find the derivative of each function.

1. $y = e^{3x^2}$
2. $y = 2\, e^{\cos x}$
3. $y = e^{2x^2} + x$
4. $y = e^{1/x}$
5. $y = \cos e^{2x}$
6. $y = \ln 4x^3$
7. $y = 2 \ln (\tan x)$
8. $y = \ln (2x^3 - x^2)$
9. $y = (\ln x)^3$
10. $y = \ln(\cos e^{x^2})$
11. $y = 3^{2x}$
12. $y = 2^{x-1}$
13. $y = 4^{\sin x}$
14. $y = 10^{x^2}$
15. $y = x^{2x}$
16. $y = (1 + 2^x)^4$
17. $y = (\sin x)^x$
18. $y = (\ln x)^{x+1}$
19. $y = x^{\ln x}$
20. $y = (x + 1)^{1/x}$
21. $y = e^x \text{ Arcsin } x$
22. $y = x^2 \ln x$
23. $y = 2x \ln (x + 1)$
24. $y = (e^{\sin x})(\sin x)$
25. $y = 2^x \cos x$
26. $y = 3e^x/(2e^x + 1)$
27. $y = \ln (x + 1)^3$
28. $y = \ln (e^x \cos x)$
29. $y = \ln (x^2 \sin x)$
30. $y = \ln (e^x/\cos x)$

---

31. The formula $Q = Q_0\, e^{-t/\tau}$ gives the charge on a capacitor after it has been discharging through a resistance for time t (in s). In the formula, Q_0 is the original charge (in coulombs) and $\tau = RC$ (resistance × capacitance). Find a formula for rate of change of charge with respect to time. **3 5 9** 4 7 8

32. When a circuit contains an inductor and a resistor in series, and the current is switched on, the current does not rise immediately to its steady-state value. It rises gradually, so that the current I (in amperes) is given by $I = I_0(1 - e^{-t/\tau})$, where I_0 is the steady-state current (in amperes), $\tau = L/R$ (inductance/resistance), and t is the time elapsed (in seconds). Find a formula for the rate of change of I with respect to t. **3 5 9** 1 2 4 6 7 8

33. When a belt passes over a rough pulley, the tensions in the belt will be different on the two sides. When the system is in equilibrium, but slippage is impending, the formula $T_1 = T_2\, e^{\mu\alpha}$ gives the larger tension T_1 if T_2 is the smaller tension, μ is the coefficient of friction, and α is the angle of wrap in radians. Find a formula for the rate of change of T_1 with respect to α. **2 4 8** 1 5 6 7 9

34. The formula $P = 15\, e^{-0.0004h}$ gives the atmospheric pressure P (in lb/in^2) at an altitude of h (in ft) above sea level. Find the rate of change of P with respect to h for an airplane that is at 20,000 feet. **2 4 8** 1 5 6 7 9

35. The formula $V = kx^2 \ln (1/x)$ gives the speed V of the signal in a submarine telegraph cable. Find a formula for the rate of change of V with respect to x. In the formula, k is a constant, and x is the ratio of the radius of the core to the thickness of the winding. **4 7 8** 1 2 3 5 6 9

---

1. Architectural Technology
2. Civil Engineering Technology
3. Computer Engineering Technology
4. Mechanical Drafting Technology
5. Electrical/Electronics Engineering Technology
6. Chemical Engineering Technology
7. Industrial Engineering Technology
8. Mechanical Engineering Technology
9. Biomedical Engineering Technology

25.5 APPLICATIONS

The derivative has been used with polynomial functions to find maximum and minimum values, to solve equations by Newton's method, to determine velocity and acceleration, and to write equations of tangents and normals to curves. The techniques presented to solve these problems may also be used when the functions are not polynomial functions.

Recall that values of x for which the derivative of a function is zero may produce relative maxima or minima on a graph of the function, and values of x for which the second derivative is zero may produce inflection points.

EXAMPLE 1 Sketch the graph of $y = e^x \sin x$ for $0 \leq x \leq 2\pi$.

Solution **1.** Find the intercepts.

x-intercept: let $y = 0$

$$0 = e^x \sin x$$
$$0 = \sin x \; (e^x \neq 0)$$
$$x = 0, \pi, 2\pi$$

y-intercept: let $x = 0$

$$y = e^0(\sin 0) = 1(0) = 0$$

The points $(0, 0)$, $(\pi, 0)$, and $(2\pi, 0)$ are on the graph.

2. The curve is not symmetric with respect to the y-axis or the origin.

3. Find the critical points.
The derivative is given by

$$y' = (e^x)(\cos x) + (\sin x)(e^x)$$
$$= e^x(\cos x + \sin x).$$

When $y' = 0$, we have

$$0 = e^x(\cos x + \sin x)$$
$$e^x = 0 \text{ or } \cos x + \sin x = 0.$$

The equation $e^x = 0$ has no solution, but $\cos x + \sin x = 0$ when

$$\sin x = -\cos x \text{ or}$$
$$\tan x = -1.$$

Thus, $x = 3\pi/4$ or $7\pi/4$.
Next, we use the second derivative test to determine whether these values give relative extrema.

$$y'' = e^x(-\sin x + \cos x) + (\cos x + \sin x)e^x \qquad \text{Apply the product rule to } y'.$$
$$= -e^x \sin x + e^x \cos x + e^x \cos x + e^x \sin x \qquad \text{Distribute } e^x.$$
$$= 2e^x \cos x \qquad \text{Combine like terms.}$$

If x is $3\pi/4$, $y'' = 2e^{3\pi/4} \cos(3\pi/4)$, which is negative. The curve, then, is concave downward when x is $3\pi/4$.
If x is $7\pi/4$, $y'' = 2e^{7\pi/4} \cos(7\pi/4)$, which is positive. The curve, then, is concave upward when x is $7\pi/4$.
Recall that $y = e^x(\sin x)$.
If x is $3\pi/4$ or 2.36, $y = e^{3\pi/4}(\sin 3\pi/4)$ or 7.46.
If x is $7\pi/4$ or 5.48, $y = e^{7\pi/4}(\sin 7\pi/4)$ or -173.
Thus, there is a relative maximum at $(2.36, 7.46)$ and a relative minimum at $(5.48, -173)$.

4. Find possible inflection points.

$$y'' = 2e^x \cos x$$
$$0 = 2e^x \cos x \text{ when } \cos x = 0, \text{ so}$$
$$x = \pi/2 \text{ or } 3\pi/2$$

We know that y'' is negative when x is $3\pi/4$ and positive when x is $7\pi/4$. Thus, the curve must change concavity when x is $3\pi/2$, which is between these two values. To determine whether there is an inflection point when x is $\pi/2$, consider y'' at 0.

$$y'' = 2e^0 \cos 0 \text{ or } 2(1)(1), \text{ which is } 2$$

Since y'' is positive when x is 0 and negative when x is $3\pi/4$, the curve must change concavity when x is $\pi/2$, which is between these two values. We conclude that both values produce inflection points.

$$\text{If } x \text{ is } \pi/2,\ y = e^{\pi/2} \sin \pi/2$$
$$= e^{\pi/2} \text{ or } 4.81.$$
$$\text{If } x \text{ is } 3\pi/2,\ y = e^{3\pi/2} \sin 3\pi/2$$
$$= -e^{3\pi/2} \text{ or } -111.$$

Figure 25.8 shows the graph. ■

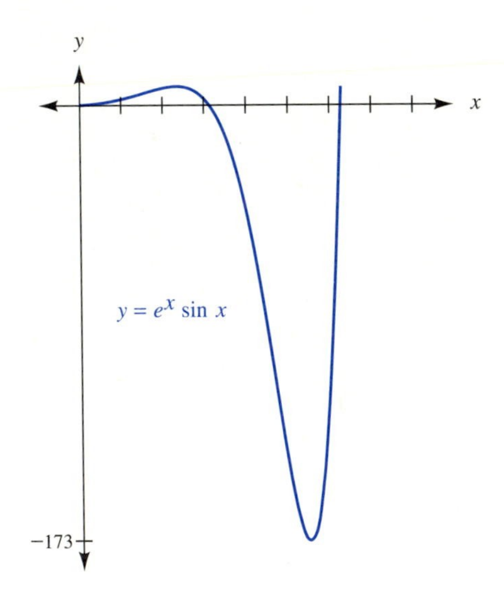

FIGURE 25.8

Sometimes it is necessary to determine when a function takes on a maximum or minimum value, but it is not necessary to sketch a graph of the entire function. The derivative may be used to solve such problems.

EXAMPLE 2 When a projectile is fired on an angle $\theta\,(0 \le \theta \le 90°)$, with velocity v_0, the horizontal distance it travels is given by

$$y = \frac{v_0^2}{g} \sin 2\theta \text{ where } g \text{ is the acceleration due to gravity.}$$

Determine the angle at which it should be launched to travel the maximum horizontal distance.

Solution Let x be the radian measure of θ. The maximum occurs when $dy/dx = 0$.

$$\frac{dy}{dx} = \frac{v_0^2}{g}(\cos 2x)(2)$$

$$0 = \frac{2v_0^2}{g}\cos 2x$$

$$\cos 2x = 0 \text{ when } 2x = \pi/2 \text{ or } 3\pi/2 \text{ and therefore } x = \pi/4 \text{ or } 3\pi/4$$

But the angle of launch should be between 0° and 90°. We suspect that $\pi/4$ will produce the maximum value. We apply the second derivative test.

$$\frac{d^2y}{dx^2} = \frac{2v_0^2}{g}(-\sin 2x)(2) \text{ or } \frac{-4v_0^2}{g}\sin 2x$$

When x is $\pi/4$, $\dfrac{d^2y}{dx^2}$ is $\dfrac{-4v_0^2}{g}\sin\dfrac{\pi}{2}$, or $\dfrac{-4v_0^2}{g}$, which is negative. We conclude that the maximum does occur at $\pi/4$, and the projectile should be launched at an angle of 45° to travel the maximum horizontal distance. ■

In the following example, we use Newton's method to solve an equation.

EXAMPLE 3 Solve the equation $3x - 1 = 2 \sin x$ to thousandths.

Solution Recall that Newton's method gives an approximation for successive values of x using

$$x_{n+1} = x_n - f(x_n)/f'(x_n).$$

The equation $3x - 1 = 2 \sin x$ is equivalent to

$$0 = 2 \sin x - 3x + 1$$

so let $f(x) = 2 \sin x - 3x + 1$.
Then $f'(x) = 2 \cos x - 3$.
A graph of $y = 3x - 1$ and $y = 2 \sin x$, as shown in Figure 25.9, shows that the solution is near $x = 1$. Thus, we choose $x_1 = 1$.

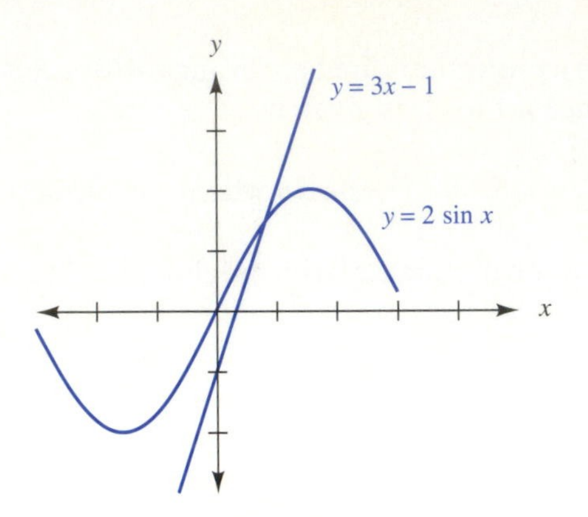

FIGURE 25.9

The following table shows the successive values of x_n.

n	x_n	$f'(x_n)$	$f(x_n)$	$x_n - f(x_n)/f'(x_n)$
1	1	−1.9193954	−0.317058	0.8348136
2	0.8348136	−1.6573684	−0.022098	0.8214804
3	0.8214804	−1.6377239	−0.0001313	0.8214003

Since two consecutive values agree when rounded to thousandths, we know that $x = 0.821$. ■

EXAMPLE 4 Find the magnitude and direction of the velocity of an object when t is 0.25 seconds if the position at time t (in seconds) is given by

$$x = \sin \pi t \text{ and } y = \cos \pi t.$$

Solution $dx/dt = \pi \cos \pi t$ and $dy/dt = -\pi \sin \pi t$

$$\begin{aligned} |\mathbf{v}| &= \sqrt{\pi^2 \cos^2 \pi t + \pi^2 \sin^2 \pi t} \\ &= \sqrt{\pi^2(\cos^2 \pi t + \sin^2 \pi t)} \\ &= \sqrt{\pi^2(1)} \\ &= \pi \end{aligned}$$

The direction θ is given by

$$\begin{aligned} \tan \theta &= \frac{-\pi \sin \pi t}{\pi \cos \pi t} \\ &= -\tan \pi t. \end{aligned}$$

When $t = 0.25$ or $1/4$, $\tan \theta = -\tan \pi/4$ or -1.
Tan $\theta = -1$ for $\theta = -\pi/4$ or $-45°$. ■

EXAMPLE 5 Find the equation of a line that is normal to the curve $y = \text{Arctan } x$ when x is 2.

Solution Recall that the derivative of $y = f(x)$ gives the slope of the line tangent to the curve $y = f(x)$ at x, and that the normal is perpendicular to the tangent.

$$\text{Since } y' = \frac{1}{1 + x^2},$$

the slope of the tangent at $x = 2$ is $\dfrac{1}{1 + 4} = \dfrac{1}{5}$.

The slope of the normal, then, is -5.
Arctan $2 = 1.11$, so $(2, 1.11)$ is on the line. The equation of the line is

$$y - 1.11 = -5(x - 2)$$
$$y - 1.11 = -5x + 10 \text{ or } y = -5x + 11.1. \quad \blacksquare$$

EXERCISES 25.5

In 1–10, sketch the graph of each function.

1. $y = \sin 3x^2$ for $0 \le x \le \pi/4$

2. $y = \cos(1/x)$ for $\pi/12 < x \le \pi$

3. $y = \cos e^x$ for $-\pi/2 \le x \le \pi/2$

4. $y = \sin e^{-x}$ for $-\pi/2 \le x \le \pi/2$

5. $y = e^{-x} \cos x$ for $0 \le x \le \pi$

6. $y = e^{\sin x}$ for $0 \le x \le \pi$

7. $y = \ln(\sin x)$ for $0 < x < \pi$

8. $y = \ln(\cos x)$ for $0 \le x < \pi/2$

9. $y = e^{-\cos x}$ for $0 \le x \le \pi$

10. $y = e^{-x \sin x}$ for $-\pi/2 \le x \le \pi/2$

In 11–15, use Newton's method to solve each equation.

11. $x - 3 \sin x = 0$ near $x = 2$

12. $\tan x = 2x$ near $x = 1$

13. $e^x = \ln x + 3$ near $x = 1$

14. $e^x = \sin x$ near $x = -3$

15. $\sin(x - 1) = \tan x$ near $x = -1$

In 16–18, find the magnitude and direction of the velocity of an object whose position at time t is given by the specified equations.

16. $x = \cos 2\pi t$, $y = \sin \pi t$, $t = 1.00$

17. $x = \cos \pi t$, $y = -\sin \pi t$, $t = 2.00$

18. $x = t^2$, $y = \cos t$, $t = 1.00$

In 19 and 20, find the magnitude of the acceleration of an object whose position at time t is given by the specified equations.

19. $x = \sin t$, $y = 2t$, $t = 1.00$

20. $x = 2t$, $y = \cos 2t$, $t = 2.00$

In 21–23, find the equation of the line tangent to the given curve at the point specified.

21. $y = e^x$ at $(0, 1)$

22. $y = \ln x$ at $(1, 0)$

23. $y = \text{Arctan } x$ at $(1, \pi/4)$

In 24 and 25, find the equation of the line normal to the given curve at the point specified.

24. $y = e^x \sin x$ at $(0, 0)$

25. $y = \dfrac{\ln x}{x}$ at $(1,0)$

26. The equation for the normal curve used in statistics is $y = (1/\sqrt{2\pi})e^{-x^2/2}$. Sketch the graph.
1 2 4 7 8 3 5

27. An object on the end of a spring moves so that its displacement y (in cm) is given by $y = e^{-0.5t}(0.4 \cos 6t - 0.2 \sin 6t)$. Find the maximum displacement that occurs in the interval $0 \leq t \leq \pi/3$.
4 7 8 1 2 3 5 6 9

28. A beam of rectangular cross section is to be cut from a log of radius 20.0 cm. Find the cross-sectional area for the largest beam that can be cut. (Hint: Consider a circle with its center at the origin of a coordinate plane, as shown in the accompanying diagram. $A = 4xy$ with $x = 20.0 \cos \theta$ and $y = 20.0 \sin \theta$.) 1 2 4 8 7 9

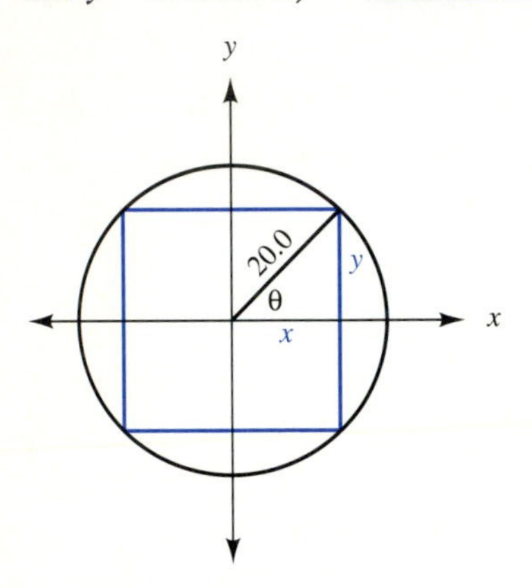

29. The maximum height of a projectile is given by $y = (v_0^2 \sin^2 \theta)/(2g)$, and the horizontal range is given by $x = (v_0^2 \sin 2\theta)/g$. Find the value of θ for which x and y are equal. In the formula, v_0 is the velocity of launch, θ is the angle of launch, and g is the acceleration due to gravity.
4 7 8 1 2 3 5 6 9

30. The formula for an amplitude modulated voltage wave form is $v = A(1 + m \sin \omega t/n)(\sin \omega t)$. Find the rate of change of v with respect to t (in volts/second) if $A = 35$, $m = 0.5$, $\omega = 30\pi \times 10^6$, $n = 10$, and $t = 200$ nanoseconds (or 200×10^{-9} seconds). 3 5 9 1 2 4 6 7 8

1. Architectural Technology
2. Civil Engineering Technology
3. Computer Engineering Technology
4. Mechanical Drafting Technology
5. Electrical/Electronics Engineering Technology
6. Chemical Engineering Technology
7. Industrial Engineering Technology
8. Mechanical Engineering Technology
9. Biomedical Engineering Technology

25.6 CHAPTER REVIEW

KEY TERMS

algebraic functions
transcendental functions

RULES AND FORMULAS

If u is a function of x,

$$\frac{d(\sin u)}{dx} = \cos u \frac{du}{dx} \qquad \frac{d(\cos u)}{dx} = -\sin u \frac{du}{dx}$$

$$\frac{d(\tan u)}{dx} = \sec^2 u \frac{du}{dx} \qquad \frac{d(\cot u)}{dx} = -\csc^2 u \frac{du}{dx}$$

$$\frac{d(\sec u)}{dx} = \sec u \tan u \frac{du}{dx} \qquad \frac{d(\csc u)}{dx} = -\csc u \cot u \frac{du}{dx}$$

$$\frac{d(\text{Arcsin } u)}{dx} = \frac{1}{\sqrt{1-u^2}} \frac{du}{dx} \qquad \frac{d(\text{Arccos } u)}{dx} = \frac{-1}{\sqrt{1-u^2}} \frac{du}{dx}$$

$$\frac{d(\text{Arctan } u)}{dx} = \frac{1}{1+u^2} \frac{du}{dx} \qquad \frac{d(\text{Arccot } u)}{dx} = \frac{-1}{1+u^2} \frac{du}{dx}$$

$$\frac{d(\text{Arcsec } u)}{dx} = \frac{1}{\sqrt{u^2(u^2-1)}} \frac{du}{dx} \qquad \frac{d(\text{Arccsc } u)}{dx} = \frac{-1}{\sqrt{u^2(u^2-1)}} \frac{du}{dx}$$

$$\frac{d(e^u)}{dx} = e^u \frac{du}{dx} \qquad \frac{d(a^u)}{dx} = a^u (\ln a) \frac{du}{dx}$$

$$\frac{d(\ln u)}{dx} = \frac{1}{u} \frac{du}{dx}$$

SEQUENTIAL REVIEW

(Section 25.1) *In 1–9, find the derivative of each function.*

1. $y = \cos 2x$ **2.** $y = \sin x^3$ **3.** $y = 3 \cos (4x^2 + 1)$

4. $y = \sin^2 2x^3$ **5.** $y = 4 \sin^3 \sqrt{x}$ **6.** $y = 2 \sin x \cos^2 3x$

7. $y = 3x^2 \cos^3 2x$ **8.** $y = \dfrac{1+x}{\cos 2x}$ **9.** $y = \dfrac{\cos (1-x)}{2x}$

(Section 25.2) *In 10–18, find the derivative of each function.*

10. $y = \cot 4x$ **11.** $y = \sec x^2$ **12.** $y = \tan^2 (2x - 1)$

13. $y = 2 \cot (3x + 1)^2$ **14.** $y = x^2 + \sec^2 2x$ **15.** $y = 3x \csc x$

16. $y = 3x^2 \sec 2x$ **17.** $y = \dfrac{3x}{\tan 3x}$ **18.** $y = \dfrac{\csc^2 x}{1 - \cos x}$

(Section 25.3) *In 19–27, find the derivative of each function.*

19. $y = 4 \text{ Arcsin } 3x$ **20.** $y = -2 \text{ Arctan } 4x$ **21.** $y = \text{Arcsin } (x^2 + 3x)$

22. $y = \text{Arccos } \sqrt{x+1}$ **23.** $y = 3 \text{ Arctan } \sqrt{x}$ **24.** $y = x \text{ Arctan}^2\, 2x$

25. $y = x^2 \text{ Arccos } 2x$ **26.** $y = \dfrac{\text{Arcsin } 2x}{2x}$ **27.** $y = \dfrac{2x}{\text{Arctan } x}$

(Section 25.4) *In 28–36, find the derivative of each function.*

28. $y = 2e^{\sin 3x}$ **29.** $y = \sin e^{2x}$ **30.** $y = 2 \ln \sqrt{x}$

31. $y = [\ln(3x + 1)]^2$ **32.** $y = 4^x$ **33.** $y = 2^{\cos x}$

34. $y = x^{\cos x}$ **35.** $y = (1 - x)^x$ **36.** $y = e^x \sin x$

(Section 25.5)

37. Find the equation of the line tangent to the curve $y = x \cos x$ at $(\pi, -\pi)$.

38. Find the equation of the line normal to the curve $y = \text{Arcsin } x$ at $(0, 0)$.

39. Find the magnitude and direction when t is 1.00 of the velocity of an object whose position at time t is given by the equations $x = \sin t$ and $y = 3t$.

40. Find the magnitude and direction when t is 1.00 of the acceleration of the object in problem 39.

41. If $y = e^x$, find dy/dt when x is 2, given $dx/dt = 3$.

42. Sketch the graph of $y = x^{-1} \sin x$ for $0 < x < \pi$.

43. Find the maximum value, if there is one, of $y = \sin e^x$ for $0 \le x \le 2$.

44. Find the minimum value, if there is one, of $y = \sin e^x$ for $0 \le x \le 2$.

45. Use Newton's method to solve $e^x = \cos x$ near $x = -1$.

APPLIED PROBLEMS

1. When a projectile is given an initial velocity of v_0 from a height h at an angle of θ with the horizontal, the vertical distance y after time t is given by $y = v_0 t \sin \theta - 4.9t^2 + h$. Find an expression for the maximum height attained. 4 7 8 | 1 2 3 5 6 9

2. The voltage v developed across an inductor is related to its current i by the formula $v = -L\, di/dt$ where L is the inductance in henries. Find a formula for voltage if a current of $36e^{-0.2t}$ flows through a 15-H coil. 3 5 9 | 1 2 4 6 7 8

3. Consider a circuit that contains a resistor and an inductor in series. When the battery is disconnected, the direct current I gradually dies out and is given by $I = 1.5e^{-15t}$ at time t (in seconds). At what time close to 0.01 is this current equal to an alternating current given by $I = 2 \cos 20\pi t$? Give the answer to the nearest thousandth of a second. 3 5 9 | 1 2 4 6 7 8

4. A formula for the dB (decibel) attenuation of a two-resistor voltage divider is given by $A = 20 \log (R_2/(R_1 + R_2))$. Find the approximate rate of change in A with respect to R_2 when $R_1 = 100\ \Omega$ and $R_2 = 50\ \Omega$. (Hint: Recall that $\log x = \ln x/\ln 10$.) 3 5 9 | 1 2 4 6 7 8

5. The total light power P (in watts) collected by a lens is given in a paper on fiber optics as $P = 2\pi BA(1 - \cos 2\theta)$. In the formula, B is the radiance (in watts/cm^2) of the source, A is the surface area (in cm^2) of the source, and θ is the collection half angle of the lens. Find $dP/d\theta$ if $B = 1.7$, $A = 0.60$, and $\theta = 33°$. 3 5 6 9

6. The formula $\lambda = \dfrac{3.46 \times 10^{-7}}{1 - e^{1-W}}$ gives the spectral line width λ (in m) of a certain solid state laser. In the formula, W is the energy level (in electron volts) of the electrons. Find $d\lambda/dW$ if $W = 1.7$ electron volts. 3 5 6 9

7. The formula $T = T_0 e^{-x^2/(r_1^2 + r_2^2)}$ gives the power transmission coefficient T for a fiber optic cable splice with a lateral offset x (in meters). In the formula, T_0 is the transmission with no offset, and r_1 and r_2 are the radii (in meters) of the fibers. Find dT/dx if $T_0 = 0.93$, $x = 23\ \mu$m, $r_1 = 50\ \mu$m, and $r_2 = 70\ \mu$m. Give your answer to two significant digits. 3 5 6 9

8. The formula $v_F = -v_E \ln\left[1 - \frac{m}{m+M}\right]$ gives the final speed v_F (in m/s) after burnout for a rocket. In the formula, M is the empty mass (in kg), m is the mass (in kg) of the fuel, and v_E is the velocity (in m/s) of the exhaust gas. Find dv_F/dm if $v_E = 1500$ m/s, $m = 2600$ kg, and $M = 7500$ kg.
4 7 8 1 2 3 5 6 9

9. When a radioactive substance decays, the number N of grams remaining from an initial mass N_0 (in grams) is given by $N = N_0(1/2)^n$, where n is the number of half-lives for which the substance has decayed. Given that the half-life for tritium is 12.5 years, find the rate in (grams/half-life) at which a 100-gram initial mass of radioactive tritium decays after 40 years. Give your answer to three significant digits. **6 9** 5 8

10. Rayleigh's formula for the angular separation θ (in radians) needed to resolve two objects on photographic film is $\theta = \text{Sin}^{-1}(1.22\lambda/D)$. In the formula, λ is the wavelength of the light (in meters), and D is the lens diameter (in m). Find $d\theta/dD$ if $\lambda = 600$ nm and $D = 40$ mm. Give your answer to two significant digits. 3 5 6 9

11. The index of refraction n of quartz is given in a handbook as $n = 1.51e^{-5.84\times10^{-6}\lambda}$ where λ is the light's wavelength (in angstroms). How does n change with respect to λ if $\lambda = 4500$ angstroms?
3 5 6 9

12. The energy W (in joules) stored in the magnetic field around a coaxial conductor is given by $W = 7.08 \times 10^{-13}i^2l)\ln(b/a)$. In the formula, a is the inner radius (in meters), b is the outer radius (in meters), l is the conductor length (in meters), and i is the current (in amperes). How does W change with respect to b for a 1000 m long conductor carrying a current of 2.5 milliamperes if $a = 0.45$ cm and $b = 1.3$ cm? **3 5 9** 4 7 8

13. When a gas expands with constant temperature, its new volume V is given by the equation $V = V_0e^{W/(8.314T)}$. In the formula, V_0 is the initial volume (in liters), W is the work (in joules) done by the gas, and T is the temperature (in degrees Kelvin). Find the rate of change of V with respect to W for a gas at 298° K that has done 1000 joules of work. Its initial volume was 5 liters.
6 8 1 2 3 5 7 9

14. Between 0° C and 10° C, the density D (in kg/m³) of water is given almost exactly as $D = 1000 \cos(0.00418T - 0.0167)$ where T is in degrees Celsius. Find the temperature within this interval at which the density is at its maximum.
6 8 1 2 3 5 7 9

15. A vertical cable of a suspension bridge vibrates according to the equation $x = 2.5 \sin 0.12y \cos 640t$ where x is the horizontal displacement (in cm) and y is the vertical distance (in cm) from the bridge deck. What is the velocity of a point on the cable 150 cm above the deck when t is 4.0 ms?
1 2 4 8 7 9

16. When a particular circuit containing a resistor, an inductor, and a capacitor in series is connected to a battery, the current i (in amperes) is given by $i = 24e^{-3t}(e^{2.6t} - e^{-2.6t})$ where t is the time (in seconds). Find the time at which the maximum current occurs. **3 5 9** 4 7 8

17. A voltage V (in volts) given by $v = 1.5 - e^{\cos 3t}$ is applied to an electronic circuit for 2.0 seconds. Find the maximum voltage during this time.
3 5 9 1 2 4 6 7 8

18. The field pattern of a certain antenna is given by $r = 10/(1 + 4\cos^2\theta)$.
   (a) For what values of θ $(0 \le \theta < 2\pi)$ is r a maximum?
   (b) a minimum?
1 3 4 5 6 8 9

RANDOM REVIEW

In 1–4, find the derivative of each function.

1. $y = \dfrac{\text{Arcsin } x}{x^2}$

2. $y = x \sin 3x^2$

3. $y = \ln(\sin x)$

4. $y = \sec^2 2x$

5. Find the equation of the line tangent to the curve $y = e^x \ln x$ at $(1, 0)$.

6. Sketch the graph of $y = \dfrac{e^x}{1 + e^x}$

In 7–10, find the derivative of each function.

7. $y = \tan \sqrt{x + 1}$

8. $y = 5^{\sin x}$

9. $y = \cos^2 (3x + 2)$

10. $y = x^2 \text{ Arctan } x^2$

CHAPTER TEST

In 1–8, find the derivative of each function.

1. $y = \dfrac{\sin^2 4x}{4x^2}$

2. $y = x^{1/2} \cos 2x$

3. $y = x \tan x$

4. $y = 2 \text{ Arcsin}^{-3} \sqrt{x}$

5. $y = \dfrac{\text{Arctan } x}{2x}$

6. $y = 3\, e^{\sin 2x}$

7. $y = x^2 \ln (\sin x)$

8. $y = 5 \cos x^2$

9. Sketch the graph of $y = e^{\cos 2x}$ for $0 < x < \pi$.

10. Find the magnitude and direction at time $t = 1.00$ of the acceleration of an object whose position at time t is given by $x = \cos t$ and $y = t^2$.

TECHNIQUES OF INTEGRATION CHAPTER 26

Now that we have encountered the derivatives of the common transcendental functions, we can find many integrals using the process of antidifferentiation. Although there are some functions (such as $y = e^{-x^2}$, $1/(\ln x)$, and $y = \sin x^2$) for which we cannot find antiderivatives, there are techniques that can be applied to integrate a variety of functions. In this chapter, we introduce a few of those techniques along with integrals of the common transcendental functions.

26.1 THE POWER FORMULA

In Section 23.4, we encountered the power rule for integrals. That rule is

$$\int [f(x)]^n f'(x)\,dx = \frac{[f(x)]^{n+1}}{n+1} + C \text{ if } n \neq -1.$$

NOTE ⇨⇨ ***Formulas for integrals are usually stated in differential form.*** It is important to remember that if $u = f(x)$, then $du = f'(x)\,dx$.

THE POWER RULE FOR INTEGRALS

If u is a function, then

$$\int u^n\,du = \frac{u^{n+1}}{n+1} + C \text{ if } n \neq -1. \qquad \textbf{(26-1)}$$

integrand

The function that is integrated is called the **integrand.** Up to this point, we have applied the power rule only to integrands that are polynomial functions. It is valid, however, for all other types of functions as well. Now that you are familiar with derivatives of the trigonometric, inverse trigonometric, exponential, and logarithmic functions, you should be able to recognize many more integrals that fit the pattern of Equation 26-1. Recognizing the pattern is an important skill. ***The rule applies when the integrand is the product of any function that is raised to a numerical power (other than −1) and the derivative of that function.***

NOTE

EXAMPLE 1 Find $\int -\cos^5 x \sin x\,dx$.

Solution The function defined by $f(x) = \cos x$ is raised to the fifth power. The derivative of this function is given by $f'(x) = -\sin x$. Thus,

$$\int -\cos^5 x \sin x\,dx = \int \underbrace{(\cos x)^5}_{u^5}\,\underbrace{(-\sin x)\,dx}_{du} = \frac{\cos^6 x}{6} + C. \quad ■$$

EXAMPLE 2 Find $\int \frac{[\text{Arctan } x]^3}{1+x^2}\,dx$.

Solution The function defined by $f(x) = \text{Arctan } x$ is raised to the third power.

Since $f'(x) = \frac{1}{1+x^2}$, we have

$$\int \frac{[\text{Arctan } x]^3}{1+x^2}\,dx = \int \underbrace{[\text{Arctan } x]^3}_{u^3}\,\underbrace{\left(\frac{1}{1+x^2}\right)dx}_{du} = \frac{[\text{Arctan } x]^4}{4} + C. \quad ■$$

EXAMPLE 3 Find $\int e^{3x}\sqrt{1+2e^{3x}}\,dx$.

Solution The function defined by $f(x) = 1 + 2e^{3x}$ is raised to the power 1/2. Since $f'(x) = 2e^{3x}(3)$ or $6e^{3x}$, the factor 6 is missing. The missing factor, however, is a constant, so the integrand can be modified to include it.

$$\int e^{3x}\sqrt{1+2e^{3x}}\,dx = 1/6\int \underbrace{(1+2e^{3x})^{1/2}}_{u^{1/2}}\,\underbrace{6e^{3x}\,dx}_{du} = (1/6)\,\frac{(1+2e^{3x})^{3/2}}{3/2} + C$$

$$\int e^{3x}\sqrt{1+2e^{3x}}\,dx = \frac{(1+2e^{3x})^{3/2}}{9} + C \quad ■$$

Sometimes a trigonometric identity can be used to transform an integrand that does not appear to follow the power formula pattern.

EXAMPLE 4 Find $\int [\ln(\sin x)](\cot x)\,dx$.

Solution The function defined by $f(x) = \ln(\sin x)$ is raised to the first power. Since $f'(x) = \dfrac{1}{\sin x}(\cos x)$, we have

$$\begin{aligned}\int [\ln(\sin x)](\cot x)\,dx &= \int [\ln(\sin x)]\left(\frac{\cos x}{\sin x}\right)dx\\ &= \int [\ln(\sin x)]\left(\frac{1}{\sin x}\right)(\cos x)\,dx\\ &= \frac{[\ln(\sin x)]^2}{2} + C. \quad ■\end{aligned}$$

It may be necessary to evaluate a definite integral or to solve applied problems using the power formula.

EXAMPLE 5 Evaluate $\int_{\pi/6}^{\pi/4} \tan x \sec^2 x\,dx$.

Solution This problem may be approached in either of two ways. One way is to let $f(x) = \tan x$. Then $f'(x) = \sec^2 x$, and

$$\begin{aligned}\int_{\pi/6}^{\pi/4} \tan x \sec^2 x\,dx &= \left.\frac{\tan^2 x}{2}\right|_{\pi/6}^{\pi/4}\\ &= (1/2)[\tan^2(\pi/4) - \tan^2(\pi/6)]\\ &= (1/2)[1^2 - (1/\sqrt{3})^2]\\ &= (1/2)(1 - 1/3)\\ &= (1/2)(2/3)\\ &= 1/3.\end{aligned}$$

The other way is to let $f(x) = \sec x$. Then $f'(x) = \sec x \tan x$, and

$$\begin{aligned}
\int_{\pi/6}^{\pi/4} \tan x \sec^2 x \, dx &= \int_{\pi/6}^{\pi/4} (\sec x)[\sec x \tan x] \, dx \\
&= \frac{\sec^2 x}{2} \Bigg|_{\pi/6}^{\pi/4} \\
&= (1/2)[\sec^2 (\pi/4) - \sec^2 (\pi/6)] \\
&= (1/2)[(\sqrt{2})^2 - (2/\sqrt{3})^2] \\
&= (1/2)(2 - 4/3) \\
&= (1/2)(2/3) \\
&= 1/3.
\end{aligned}$$

EXAMPLE 6 Find the area bounded by the curve $y = \dfrac{\ln x}{x}$, the x-axis, and the lines $x = 1$ and $x = e$.

Solution The graph is shown in Figure 26.1.

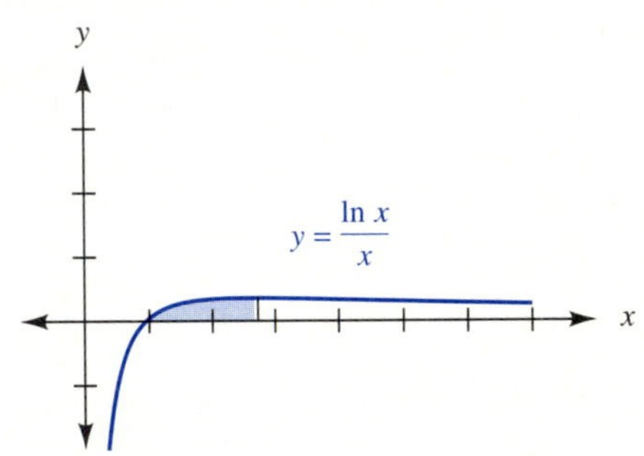

FIGURE 26.1

The area is given by

$$\int_1^e \frac{\ln x}{x} \, dx.$$

The function defined by $y = \ln x$ is raised to the first power. Since $f'(x) = 1/x$, we have

$$\begin{aligned}
\int_1^e \frac{\ln x}{x} \, dx &= \int_1^e (\ln x)\left(\frac{1}{x}\right) dx \\
&= \frac{[\ln x]^2}{2} \Bigg|_1^e \\
&= \frac{[\ln e]^2}{2} - \frac{[\ln 1]^2}{2} \\
&= 1/2 - 0/2 \\
&= 1/2.
\end{aligned}$$

EXERCISES 26.1

In 1–28, find each integral.

1. $\int \sin^3 x \cos x\, dx$

2. $\int \cos^2 x \sin x\, dx$

3. $\int \frac{\sin x}{\cos^2 x}\, dx$

4. $\int (1 + \tan x)^3 \sec^2 x\, dx$

5. $\int \csc^2 x \sqrt{\cot x - 1}\, dx$

6. $\int -2 \cot^{-3} x \csc^2 x\, dx$

7. $\int \sec^3 2x \tan 2x\, dx$

8. $\int x \sec^2 x^2 \tan x^2\, dx$

9. $\int \csc^4(x-1) \cot(x-1)\, dx$

10. $\int \frac{(\text{Arcsin } x)^4}{\sqrt{1-x^2}}\, dx$

11. $\int \frac{(\text{Arcsin } x)^{-3}}{\sqrt{1-x^2}}\, dx$

12. $\int \frac{\text{Arccos } 3x}{\sqrt{1-9x^2}}\, dx$

13. $\int \frac{(\text{Arcsin } 2x)^2}{\sqrt{1-4x^2}}\, dx$

14. $\int \frac{\text{Arctan } 4x}{1+16x^2}\, dx$

15. $\int \frac{\sqrt{\text{Arctan } 3x}}{1+9x^2}\, dx$

16. $\int \frac{(\ln x)^4}{x}\, dx$

17. $\int \frac{[\ln(2x+1)]^{-3}}{2x+1}\, dx$

18. $\int \frac{(1-\ln 3x)^2}{x}\, dx$

19. $\int \frac{\ln(3x-2)}{3x-2}\, dx$

20. $\int (1+2e^x)^3 2e^x\, dx$

21. $\int e^x \sqrt{1-e^x}\, dx$

22. $\int (3e^{2x}-1)^{1/2} e^{2x}\, dx$

23. $\int (e^x + e^{-x})^2 (e^x - e^{-x})\, dx$

24. $\int_0^{\pi/2} \sin^2 x \cos x\, dx$

25. $\int_{\pi/6}^{\pi/3} (\tan x)^{-2} \sec^2 x\, dx$

26. $\int_0^1 \frac{(\text{Arctan } x)^2}{1+x^2}\, dx$

27. $\int_1^e \frac{(1+3\ln x)}{x}\, dx$

28. $\int_0^1 \frac{e^x}{\sqrt{1+e^x}}\, dx$

---

29. In aerodynamics, the angle (in radians) of attack under certain conditions is given by the integral $\frac{2}{\pi}\int_0^1 \frac{x\xi}{\sqrt{1-x^2}}\, dx$. Treat ξ as a constant and evaluate the integral. In the formula, x is the chord station and ξ is the mean camber. 4 7 8 1 2 3 5 6 9

30. The formula $T = \frac{1}{2} wa \sqrt{1 + \frac{a^2}{16d^2}}$ gives the tension T (in lbs) in the cable at the end of a suspension bridge. If $w = 800.0$ and $d = 40.0$, find the area under the graph of T versus a, between $a = 0$ and $a = 600.0$. In the formula, w is the load (in lb/ft), a is the length of the span (in ft), and d is the sag (in ft). 2 4 8 1 5 6 7 9

31. A rope is placed on a hemisphere of radius r so that it rests horizontally, as shown in the accompanying diagram. When it is moved down so that it is a vertical distance of y above the base of the hemisphere, the increase in the length of the rope is given by $l = \int_a^b \frac{-2\pi y}{\sqrt{r^2-y^2}}\, dy$. Find l (in cm) for

1. Architectural Technology
2. Civil Engineering Technology
3. Computer Engineering Technology
4. Mechanical Drafting Technology
5. Electrical/Electronics Engineering Technology
6. Chemical Engineering Technology
7. Industrial Engineering Technology
8. Mechanical Engineering Technology
9. Biomedical Engineering Technology

a hemisphere of radius 8.00 cm when the rope is moved from $a = 5.0$ cm to $b = 4.5$ cm.
1 2 4 8 7 9

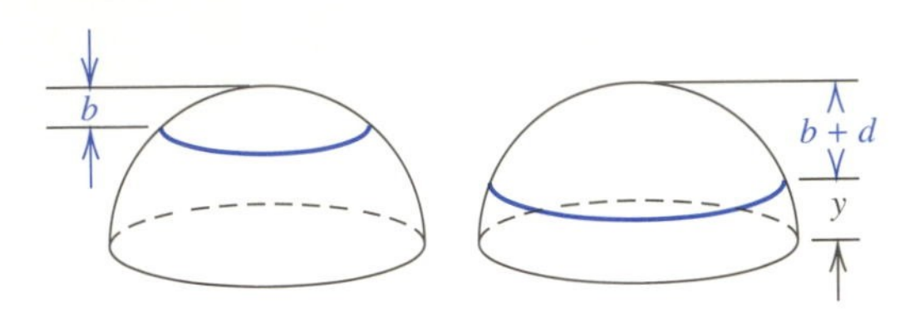

32. The formula $d = \dfrac{2v_0^2 \cos\theta \sin\theta}{g}$ gives the range d of a projectile launched with a velocity of v_0 at an angle of θ where g is the acceleration due to gravity (9.8 m/s²). Find the average range (in meters) of projectiles launched with an initial velocity of 300.0 m/s at angles between 0° and 30° ($0 \le x \le \pi/6$). 4 7 8 1 2 3 5 6 9

33. When R_1 is calculated to match impedance Z_0 with loss minimized, the formula is $R_1 = Z_0\sqrt{1 - Z_1/Z_0}$. Find the average value of R_1 (in ohms) for $175\ \Omega \le Z_1 \le 225\ \Omega$ when $Z_0 = 500.0\ \Omega$. 3 5 9 1 2 4 6 7 8

1. Architectural Technology
2. Civil Engineering Technology
3. Computer Engineering Technology
4. Mechanical Drafting Technology
5. Electrical/Electronics Engineering Technology
6. Chemical Engineering Technology
7. Industrial Engineering Technology
8. Mechanical Engineering Technology
9. Biomedical Engineering Technology

26.2 EXPONENTIAL FORMS

From Section 25.4, we know that if $y = e^{f(x)}$, then $y' = e^{f(x)}f'(x)$. Taking the antiderivatives of this function with respect to x, we have

$$\int e^{f(x)} f'(x)\, dx = e^{f(x)} + C.$$

If $u = f(x)$, then the differential is given by $du = f'(x)\, dx$, and the formula is written as follows:

THE EXPONENTIAL FORMULA FOR INTEGRALS

If u is a function, then

$$\int e^u\, du = e^u + C. \qquad \textbf{(26-2)}$$

NOTE ⇨⇨ Once again, recognizing the pattern is important. ***The formula applies when the integrand is the product of e raised to a power that is a function and the derivative of that function.***

EXAMPLE 1 Find $\int 3e^{3x}\, dx$.

Solution The exponent of e is the function defined by $f(x) = 3x$. The derivative of this function is given by $f'(x) = 3$. The integrand fits the pattern of Equation 26-2. Thus,

$$\int 3e^{3x}\,dx = \int \underbrace{e^{3x}}_{e^u} \underbrace{3\,dx}_{du} = e^{3x} + C. \quad ■$$

EXAMPLE 2 Find $\int e^{x^3} x^2\,dx$.

Solution The exponent of e is the function defined by $f(x) = x^3$. The derivative of this function is given by $f'(x) = 3x^2$. The factor 3 is missing in the integrand, but since the missing factor is a constant, the integrand can be adjusted. Thus,

$$\int e^{x^3} x^2\,dx = 1/3 \int \underbrace{e^{x^3}}_{e^u} \underbrace{3x^2\,dx}_{du} = (1/3)\,e^{x^3} + C. \quad ■$$

EXAMPLE 3 Find $\int \frac{\cos x}{e^{\sin x}}\,dx$.

Solution The integral can be written as $\int \frac{\cos x}{e^{\sin x}}\,dx = \int e^{-\sin x} \cos x\,dx$.

The exponent of e is the function defined by $f(x) = -\sin x$. The derivative of this function is given by $f'(x) = -\cos x\,dx$. Thus,

$$\int e^{-\sin x} \cos x\,dx = -\int \underbrace{e^{-\sin x}}_{e^u} \underbrace{(-\cos x)\,dx}_{du} = -e^{-\sin x} + C. \quad ■$$

EXAMPLE 4 Find $\int \frac{4e^{x+1} - 3e^{-x}}{e^x}\,dx$.

Solution

$$\begin{aligned}
\int \frac{4e^{x+1} - 3e^{-x}}{e^x}\,dx &= \int (4e - 3e^{-2x})\,dx \\
&= \int 4e\,dx - 3\int e^{-2x}\,dx \\
&= \int 4e\,dx + \frac{3}{2}\int -2\,e^{-2x}\,dx \\
&= \int 4e\,dx + \frac{3}{2}\int \underbrace{e^{-2x}}_{e^u} \underbrace{(-2)\,dx}_{du}
\end{aligned}$$

$$\int \frac{4e^{x+1} - 3e^{-x}}{e^x}\,dx = 4ex + \frac{3}{2}\,e^{-2x} + C \quad ■$$

Exponential functions may also be used with definite integrals and applied problems.

EXAMPLE 5 Find $\int_0^1 e^{x^2} 2x\, dx$.

Solution $\int_0^1 e^{x^2} 2x\, dx = e^{x^2}\Big|_0^1 = e^1 - e^0$ or about $2.718 - 1$, which is 1.718. ■

EXAMPLE 6 The voltage v (in volts) across a capacitor being charged by a current i (in amperes) is given by $v = \frac{1}{C}\int i\, dt$, where C is the capacitance (in farads) and t is the time (in seconds). Find a formula for the voltage across a 100 μF capacitor if $i = 8(e^{-1.5t} - e^{-5.5t})$, and $v = 0$ when $t = 0$.

Solution $100\ \mu\text{F} = 100 \times 10^{-6}$ F or 10^{-4} F

$$\begin{aligned}
v &= \frac{1}{10^{-4}}\int 8(e^{-1.5t} - e^{-5.5t})\, dt \\
&= 10^4(8)\int (e^{-1.5t} - e^{-5.5t})\, dt \\
&= 10^4(8)\int e^{-1.5t}\, dt - 10^4(8)\int e^{-5.5t}\, dt \\
&= \frac{10^4(8)}{-1.5}\int e^{-1.5t}(-1.5)\, dt - \frac{10^4(8)}{-5.5}\int e^{-5.5t}(-5.5)\, dt \\
&= \left(\frac{10^4(8)}{-1.5}\right)(e^{-1.5t}) - \left(\frac{10^4(8)}{-5.5}\right)(e^{-5.5t}) + C_1 \\
&= -10^4\left(\frac{16}{3}\right)e^{-1.5t} + 10^4\left(\frac{16}{11}\right)e^{-5.5t} + C_1
\end{aligned}$$

We use C_1 as the constant of integration, since C stands for capacitance in this problem. Using the values $v = 0$ and $t = 0$, we have

$$\begin{aligned}
0 &= -10^4\left(\frac{16}{3}\right)e^0 + 10^4\left(\frac{16}{11}\right)e^0 + C_1 \\
0 &= -10^4\left(\frac{16}{3}\right) + 10^4\left(\frac{16}{11}\right) + C_1 \\
0 &= 10^4\left(\frac{-176}{33} + \frac{48}{33}\right) + C_1 \\
0 &= 10^4\left(\frac{-128}{33}\right) + C_1
\end{aligned}$$

$C_1 = 10^4(4)$, to one significant digit.

$$v = -10^4\left(\frac{16}{3}\right)e^{-1.5t} + 10^4\left(\frac{16}{11}\right)e^{-5.5t} + 4(10^4)$$

$$= 10^4\left[-\left(\frac{16}{3}\right)e^{-1.5t} + \left(\frac{16}{11}\right)e^{-5.5t} + 4\right] \quad ■$$

EXERCISES 26.2

In 1–26, find each integral.

1. $\int 5e^{5x}\,dx$

2. $\int 7e^{7x-3}\,dx$

3. $\int e^{2x+1}\,dx$

4. $\int 3e^{4x}\,dx$

5. $\int x^3e^{x^4}\,dx$

6. $\int x^4e^{x^5-1}\,dx$

7. $\int xe^{-x^2}\,dx$

8. $\int x^2e^{-2x^3}\,dx$

9. $\int e^{x^2+x}(2x+1)\,dx$

10. $\int e^{x^3-x}(3x^2-1)\,dx$

11. $\int_0^1 e^{4x}\,dx$

12. $\int_1^2 6x^2e^{2x^3}\,dx$

13. $\int_1^2 xe^{x^2}\,dx$

14. $\int_0^1 e^{-x^3+x}(-3x^2+1)\,dx$

15. $\int e^{\tan x}\sec^2 x\,dx$

16. $\int \frac{\csc^2 x}{e^{\cot x}}\,dx$

17. $\int e^{\cos x}\sin x\,dx$

18. $\int \frac{\sin x}{e^{\cos x}}\,dx$

19. $\int \frac{e^x - e^{-x}}{e^x}\,dx$

20. $\int \frac{e^{2x}+e^x}{e^{-x}}\,dx$

21. $\int \frac{e^x + e^{-x}}{e^{x+1}}\,dx$

22. $\int \frac{e^{2x}-e^x}{e^{2x-1}}\,dx$

23. $\int 2e^x(e^x-1)\,dx$

24. $\int 4e^x(e^x+1)\,dx$

25. $\int_0^{\pi} e^{\sec x}(\sec x\tan x)\,dx$

26. $\int_0^1 \frac{e^{\operatorname{Arctan} x}}{x^2+1}\,dx$

27. The current i (in amperes) in an inductor is given by $i = \frac{1}{L}\int v\,dt$ where v is the voltage (in volts), L is the inductance (in henries), and t is the time (in seconds). Find a formula for the current in a 0.4-mH inductor, if $v = 10^{-4}e^{-0.3t}$ and $i = 0$ when $t = 0$. 3 5 9 4 7 8

28. When a charged capacitor is discharged through a resistance, the charge (in coulombs) remaining after time t (in seconds) is given by $Q = Q_0e^{-t/\tau}$. Find the average value of Q between $t = 0.00$ and $t = 0.01$ if $Q_0 = 3.6 \times 10^{-5}$ and $\tau = 0.002$. In the formula, Q_0 is the original charge (in coulombs), and $\tau = RC$ (resistance × capacitance). 3 5 9 4 7 8

29. If the acceleration of a moving particle at time t is given by $a = e^{\sin t}\cos t$, find an expression for its velocity at time t, given that $v = 0$ when $t = 0$. 4 7 8 1 2 3 5 6 9

30. The formula $q = \int i\,dt$ gives the total charge (in coulombs) delivered to a capacitor if i is the current (in amperes) and t is the time (in seconds) of charging. A certain capacitor is charged with a current i given by $i = 10e^{-4t}$. Find a formula for the total charge, if $q = 0$ when $i = 0$. 3 5 9 4 7 8

1. Architectural Technology
2. Civil Engineering Technology
3. Computer Engineering Technology
4. Mechanical Drafting Technology
5. Electrical/Electronics Engineering Technology
6. Chemical Engineering Technology
7. Industrial Engineering Technology
8. Mechanical Engineering Technology
9. Biomedical Engineering Technology

26.3 LOGARITHMIC FORMS

The exponential formula and the power formula are not the only integral forms that are necessary. The power rule does not apply to $\int x^{-1}\,dx$. If we tried to use it, we would have $\int x^{-1}\,dx = \frac{x^0}{0} + C$, which is undefined. It is possible, however, to integrate $\int x^{-1}\,dx$. Recall that if $y = \ln(x)$, $y' = 1/x$. But we cannot simply say that $\int \frac{1}{x}\,dx = \ln x + C$, because $\ln x$ is not defined for negative values of x. In Section 8.4, we defined the absolute value of a real number x as follows.

$$\begin{aligned} |x| &= x \text{ if } x \geq 0 \\ &= -x \text{ if } x < 0 \end{aligned}$$

Consider the derivative of $y = \ln|x|$.

When $x > 0$, $y = \ln|x| = \ln x$ and $y' = \frac{1}{x} = x^{-1}$

When $x < 0$, $y = \ln|x| = \ln(-x)$ and $y' = \frac{1}{-x}(-1) = \frac{1}{x}$.

Thus, for all nonzero x, we have $\frac{d(\ln|x|)}{dx} = \frac{1}{x}$, so $\int \frac{1}{x}\,dx = \ln|x| + C$.

Generalizing, we have $\frac{d[\ln|f(x)|]}{dx} = \frac{1}{f(x)} f'(x)$, so $\int \frac{1}{f(x)} f'(x)\,dx = \ln|f(x)| + C$.

In differential notation, this statement is as follows.

THE LOGARITHMIC FORMULA FOR INTEGRALS

If u is a function,

$$\int \frac{1}{u}\,du = \ln|u| + C$$

or

$$\int u^{-1}\,du = \ln|u| + C. \qquad \textbf{(26-3)}$$

EXAMPLE 1 Find $\int \frac{1}{2x+3}\,dx$.

Solution Before we apply Equation 26-3, the integrand is rewritten as the product $1/f(x)$, where $f(x) = (2x+3)$, and the derivative of f, which is given by $f'(x) = 2$. The constant factor 2 is missing.

$$\int \frac{1}{2x+3}\,dx = \frac{1}{2}\int \underbrace{\frac{1}{2x+3}}_{\frac{1}{u}}\,\underbrace{2\,dx}_{du} = \frac{1}{2}\ln|2x+3| + C \quad \blacksquare$$

EXAMPLE 2 Find $\int \frac{e^x}{1+3e^x}\,dx$.

Solution

$$\int \frac{e^x}{1+3e^x}\,dx = \int (1+3e^x)^{-1}\,e^x\,dx$$

$$= \frac{1}{3}\int \underbrace{(1+3e^x)^{-1}}_{u^{-1}}\,\underbrace{3e^x\,dx}_{du} = \frac{1}{3}\ln|1+3e^x| + C \quad \blacksquare$$

Example 3 illustrates how a definite integral might require the use of the logarithmic form.

EXAMPLE 3 Evaluate $\int_2^3 \frac{x}{x^2-1}\,dx$.

Solution

$$\int_2^3 \frac{x}{x^2-1}\,dx = \int_2^3 x(x^2-1)^{-1}\,dx.$$

$$= \frac{1}{2}\int_2^3 2x(x^2-1)^{-1}\,dx$$

$$= \frac{1}{2}\ln|x^2-1|\Big|_2^3$$

$$= \frac{1}{2}(\ln|9-1| - \ln|4-1|)$$

$$= \frac{1}{2}(\ln|8| - \ln|3|)$$

$= 0.490$, to three significant digits $\blacksquare$

EXERCISES 26.3

In 1–26, find each integral.

1. $\int \frac{3}{1+3x}\,dx$

2. $\int \frac{2}{3-2x}\,dx$

3. $\int \frac{1}{2-x}\,dx$

4. $\int \frac{1}{1+4x}\,dx$

5. $\int \frac{x^2}{x^3+1}\,dx$

6. $\int \frac{2x^3}{x^4-3}\,dx$

7. $\int \frac{x}{2x^2+3}\,dx$

8. $\int \frac{5x^4}{2x^5+3}\,dx$

9. $\int \frac{3x^2+2x}{x^3+x^2}\,dx$

10. $\int \frac{2x^2-1}{2x^3-3x}\,dx$

11. $\int_1^4 \frac{1}{1+2x}\,dx$

12. $\int_2^3 \frac{3x^2-1}{x^3-x}\,dx$

13. $\int_2^3 (\ln x)^{-1}\frac{1}{x}\,dx$

14. $\int_0^1 \frac{2e^x}{2e^x-1}\,dx$

15. $\int (2+\sin x)^{-1}\cos x\,dx$

16. $\int (3-\cot x)^{-1}\csc^2 x\,dx$

17. $\int (1+\tan x)^{-1}\sec^2 x\,dx$

18. $\int (3\cos x+2)^{-1}\sin x\,dx$

19. $\int \frac{\sin x}{1-\cos x}\,dx$

20. $\int \frac{\sec x\tan x}{2\sec x}\,dx$

21. $\int \frac{e^x}{1+e^x}\,dx$

22. $\int \frac{e^x}{3+2e^x}\,dx$

23. $\int \frac{dx}{x\ln 2x}$

24. $\int \frac{dx}{(3+\ln x)x}$

25. $\int_2^3 \frac{2x+3}{x^2+3x}\,dx$

26. $\int_{\pi/4}^{\pi/3} \frac{\cos x}{1-\sin x}\,dx$

---

27. If the force (in lb) acting on an object is given by $F=\frac{1}{2x+3}$, where x is the distance (in ft) from the initial position, find the work (in ft·lb) done in moving the object from $x = 5.00$ to $x = 7.00$.
4 7 8 1 2 3 5 6 9

28. The work (in joules) done by a gas during an isothermal expansion from an initial volume of V_1 (in m³) to a final volume of V_2 (in m³) is given by $W=\frac{mRT}{M}\int_{V_1}^{V_2}\frac{dV}{V}$. Perform the integration. In the formula, V is the volume of the gas, m is the mass of the gas (in grams), M is the molecular weight of the gas, T is the absolute temperature (in degrees Kelvin), and R is a constant.
6 8 1 2 3 5 7 9

29. For an object moving through a resisting medium, the time t is given by $t=\int \frac{dv}{1-kv}$, where v is the velocity and k is a constant. Perform the integration.
4 7 8 1 2 3 5 6 9

30. The formula $X_C = 1/(2\pi fC)$ gives capacitive reactance X_C (in ohms) of a capacitor if f is the frequency (in hertz) and C is the capacitance (in farads). A 0.5 μF capacitor is attached to a sweep frequency generator and the frequency f allowed to sweep from 200 Hz to 700 Hz. Find the average reactance over this interval.
3 5 9 4 7 8

---

1. Architectural Technology
2. Civil Engineering Technology
3. Computer Engineering Technology
4. Mechanical Drafting Technology
5. Electrical/Electronics Engineering Technology
6. Chemical Engineering Technology
7. Industrial Engineering Technology
8. Mechanical Engineering Technology
9. Biomedical Engineering Technology

26.4 TRIGONOMETRIC FORMS

Recall from Sections 25.1 and 25.2 that the derivatives of the trigonometric functions are given by the following formulas.

$$\frac{d(\sin) x}{dx} = \cos x \qquad \frac{d(\cos x)}{dx} = -\sin x$$

$$\frac{d(\tan x)}{dx} = \sec^2 x \qquad \frac{d(\cot x)}{dx} = -\csc^2 x$$

$$\frac{d(\sec x)}{dx} = \sec x \tan x \qquad \frac{d(\csc x)}{dx} = -\csc x \cot x$$

Each of these formulas can be reversed using the antidifferentiation process.

$$\int \cos x\,dx = \sin x + C \qquad \int -\sin x\,dx = \cos x + C$$

$$\int \sec^2 x\,dx = \tan x + C \qquad \int -\csc^2 x\,dx = \cot x + C$$

$$\int \sec x \tan x\,dx = \sec x + C \qquad \int -\csc x \cot x\,dx = \csc x + C$$

Each formula is valid when the argument of the trigonometric function is itself a function, so the formulas are usually written in differential form, so that if $u = f(x)$, then $du = f'(x)\,dx$. NOTE ***That is, for one of these formulas to apply, the integrand must contain the derivative of the argument of the trigonometric function as a factor.*** For the three equations on the right, both sides are usually multiplied by -1. Thus, we have the following trigonometric forms.

BASIC TRIGONOMETRIC FORM FOR INTEGRALS

If u is a function, then

$$\int \cos u\,du = \sin u + C \qquad \int \sin u\,du = -\cos u + C$$

$$\int \sec^2 u\,du = \tan u + C \qquad \int \csc^2 u\,du = -\cot u + C$$

$$\int \sec u \tan u\,du = \sec u + C \qquad \int \csc u \cot u\,du = -\csc u + C$$

EXAMPLE 1 Find $\int \cos 2x\,dx$.

Solution Since the argument of the function is $2x$, the integrand must contain the derivative of $2x$, which is 2, as a factor.

$$\int \cos 2x\, dx = \frac{1}{2}\int 2\cos 2x\, dx = \frac{1}{2}\int \underbrace{\cos 2x}_{\cos u}\, \underbrace{2\, dx}_{du} = \frac{1}{2}\sin 2x + C \quad ■$$

EXAMPLE 2 Find $\int 2xe^{x^2} \sec e^{x^2} \tan e^{x^2}\, dx$.

Solution The argument of the function is e^{x^2}, and the integrand contains $e^{x^2}(2x)$, which is the derivative of e^{x^2}, as a factor.

$$\int 2xe^{x^2} \sec e^{x^2} \tan e^{x^2}\, dx = \int \underbrace{\sec e^{x^2}}_{\sec u}\, \underbrace{\tan e^{x^2}}_{\tan u}\, \underbrace{2xe^{x^2}\, dx}_{du} = \sec e^{x^2} + C \quad ■$$

EXAMPLE 3 When a projectile is launched at an angle of θ with the horizontal, the range of the projectile, as shown in Figure 26.2, is given by $x = \frac{v_0^2}{g}\sin 2\theta$, where v_0 is the initial velocity, and g is the acceleration due to gravity (9.8 m/s^2). If projectiles are launched with an initial velocity of 300 m/s at each angle between 30° and 45°, find the average range.

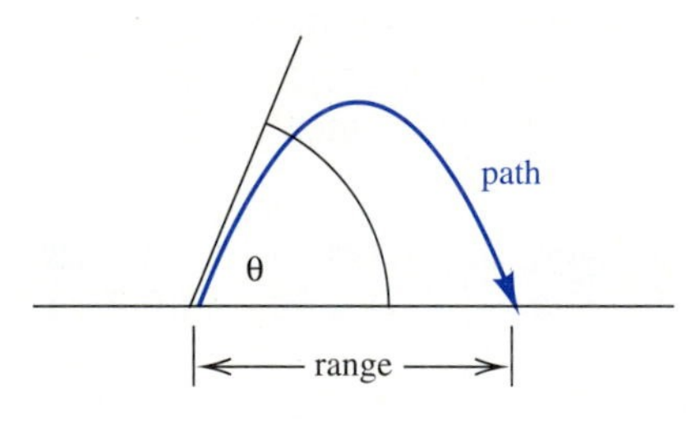

FIGURE 26.2

Solution Derivatives and integrals assume radian measure, so the angles range from $\pi/6$ to $\pi/4$. The average value of the function is given by

$$\frac{\int_{\pi/6}^{\pi/4} \frac{300^2}{9.8}\sin 2\theta\, d\theta}{\pi/4 - \pi/6} = \frac{\frac{300^2}{9.8}\int_{\pi/6}^{\pi/4} \sin 2\theta\, d\theta}{\pi/4 - \pi/6}$$

$$= \frac{\left(\frac{1}{2}\right)\left(\frac{300^2}{9.8}\right)\int_{\pi/6}^{\pi/4} 2\sin 2\theta\, d\theta}{\pi/4 - \pi/6}$$

$$= \frac{\left(\frac{1}{2}\right)\left(\frac{300^2}{9.8}\right)(-\cos 2\theta)\Big|_{\pi/6}^{\pi/4}}{\pi/4 - \pi/6}$$

$$= \frac{\left(\frac{1}{2}\right)\left(\frac{300^2}{9.8}\right)(-\cos \pi/2 + \cos \pi/3)}{\pi/4 - \pi/6}$$

$$= \frac{\left(\frac{1}{2}\right)\left(\frac{300^2}{9.8}\right)\left(0 + \frac{1}{2}\right)}{\pi/4 - \pi/6}$$

$= 9000$ m, to one significant digit ■

The integrals for other trigonometric functions can be written in terms of familiar integrals, as illustrated in Example 4.

EXAMPLE 4 Find $\int \tan x \, dx$.

Solution

$$\int \tan x \, dx = \int \frac{\sin x}{\cos x} \, dx$$

$$= \int \sin x (\cos x)^{-1} \, dx$$

Since $\cos x$ is raised to the power -1, the integral is a logarithmic form if the derivative of $\cos x$ is also a factor of the integral. The derivative of $\cos x$ is $-\sin x$, so we have

$$\int \sin x (\cos x)^{-1} \, dx = -1 \int -\sin x (\cos x)^{-1} \, dx$$

$$= -1 \int \underbrace{(\cos x)^{-1}}_{u^{-1}} \underbrace{(-\sin x) \, dx}_{du}.$$

$$\int \sin x (\cos x)^{-1} \, dx = -\ln |\cos x| + C \quad ■$$

EXAMPLE 5 Find $\int \frac{\sec^2 x + \sec x \tan x}{\sec x + \tan x} \, dx$.

Solution

$$\int \frac{\sec^2 x + \sec x \tan x}{\sec x + \tan x} \, dx = \int (\sec x + \tan x)^{-1}(\sec^2 x + \sec x \tan x) \, dx$$

If $f(x) = \sec x + \tan x$, then $f'(x) = \sec x \tan x + \sec^2 x$. Thus, the integrand is the product of a function to the power -1 and the derivative of that function. Therefore, it fits the pattern of a logarithmic function.

$$\int \underbrace{(\sec x + \tan x)^{-1}}_{u^{-1}} \underbrace{(\sec^2 x + \sec x \tan x) \, dx}_{du} = \ln |\sec x + \tan x| + C \quad ■$$

The integrand in Example 5 is actually equal to sec x. That is,

$$\frac{\sec^2 x + \sec x \tan x}{\sec x + \tan x} = \frac{(\sec x)(\sec x + \tan x)}{\sec x + \tan x} = \sec x.$$

Thus, we have

$$\int \sec x\, dx = \ln |\sec x + \tan x| + C.$$

In a manner similar to that of Examples 4 and 5, the integrals of the cotangent and cosecant functions may be found. These integrals are not as important in technical applications as the other trigonometric integrals that we have considered, but they are listed here for completeness.

INTEGRALS OF THE TRIGONOMETRIC FUNCTIONS

If u is a function, then

$$\int \sin u\, du = -\cos u + C \qquad \int \cos u\, du = \sin u + C$$

$$\int \tan u\, du = -\ln |\cos u| + C \qquad \int \cot u\, du = \ln |\sin u| + C$$

$$\int \sec u\, du = \ln |\sec u + \tan u| + C \qquad \int \csc u\, du = -\ln |\csc u + \cot u| + C$$

EXERCISES 26.4

In 1–28, find each integral.

1. $\int \sin 3x\, dx$ **2.** $\int \cos (3x + 1)\, dx$ **3.** $\int x \cos x^2\, dx$ **4.** $\int x^2 \sin x^3\, dx$

5. $\int e^x \sin e^x\, dx$ **6.** $\int \frac{\cos (\ln x)}{x}\, dx$ **7.** $\int \sec^2 2x\, dx$ **8.** $\int x \sec x^2 \tan x^2\, dx$

9. $\int e^x \sec e^x \tan e^x\, dx$ **10.** $\int (3x + 1) \sec^2 (3x^2 + 2x)\, dx$

11. $\int \cos x \sec^2 (\sin x)\, dx$ **12.** $\int \frac{\sec e^{-x} \tan e^{-x}}{e^x}\, dx$

13. $\int \csc^2 4x\, dx$ **14.** $\int (3x^2 + 4x) \csc^2 (x^3 + 2x^2)\, dx$

15. $\int x^2 \csc x^3 \cot x^3\, dx$

16. $\int \frac{\csc \sqrt{x} \cot \sqrt{x}}{\sqrt{x}}\, dx$

17. $\int e^{2x} \csc e^{2x} \cot e^{2x}\, dx$

18. $\int \frac{\csc^2 x^{-1}}{x^2}\, dx$

19. $\int \cot x\, dx$

20. $\int \csc x\, dx$

21. $\int x \tan x^2\, dx$

22. $\int e^x \tan e^x\, dx$

23. $\int x^2 \sec x^3\, dx$

24. $\int \frac{1}{x^2} \sec x^{-1}\, dx$

25. $\int \frac{1}{\sqrt{x}} \cot \sqrt{x}\, dx$

26. $\int \cos x \cot (\sin x)\, dx$

27. $\int \frac{\csc (\ln x)}{x}\, dx$

28. $\int \frac{\csc e^{-x}}{e^x}\, dx$

---

29. The formula $Z = R \sec \theta$ gives the impedance Z (in ohms) in an ac circuit with resistance R (in ohms) and phase angle θ (in degrees). Find the average value of Z for an 80.0-Ω resistor as θ ranges from 15.0° to 45.0° ($Z = R \sec x$ for $\pi/12 \le x \le \pi/4$). 3 5 9 1 2 4 6 7 8

30. The length of a British nautical mile varies with latitude according to the formula $l = 6077 - 31 \cos \theta$, where θ is the latitude in degrees and l is in feet. Find the average length of a nautical mile for latitudes between 30° and 45° ($l = 6077 - 31 \cos x$ for $\pi/6 \le x \le \pi/4$). 1 2 4 7 8 3 5

31. An oscilloscope with a circular screen 10.0 inches in diameter shows the equations $y = \sin x$ and $y = \cos x$ with a scale of 1 inch = 1 unit. What percent of the screen's area is bounded by these two curves between the first two points of intersection with positive x-coordinates? 1 2 4 7 8 3 5

32. If the velocity (in ft/s) of a particle is given by $v = \tan 2t$, find the distance traveled as time t changes from 2.0 s to 3.0 s. 4 7 8 1 2 3 5 6 9

33. A half-wave rectified voltage wave form consists of a voltage v given by $v = v_m \sin \theta$ for one half of the cycle and $v = 0$ for the other half cycle. Find a formula for the average voltage over the complete cycle ($0 \le \theta \le 2\pi$). 3 5 9 1 2 4 6 7 8

34. The force F (in newtons) acting on an object is given by $F = \csc (x/3)$ where x is the distance (in meters) from the rest position. Find the work done in moving the object from $x = 0.75$ to $x = 1.5$. 4 7 8 1 2 3 5 6 9

35. Derive the following formulas.

(a) $\int \cot x\, dx = \ln |\sin x| + C$

(b) $\int \csc x\, dx = -\ln |\csc x + \cot x| + C$

1. Architectural Technology
2. Civil Engineering Technology
3. Computer Engineering Technology
4. Mechanical Drafting Technology
5. Electrical/Electronics Engineering Technology
6. Chemical Engineering Technology
7. Industrial Engineering Technology
8. Mechanical Engineering Technology
9. Biomedical Engineering Technology

26.5 TRIGONOMETRIC SUBSTITUTION AND THE INVERSE TRIGONOMETRIC FUNCTIONS

Trigonometric identities are sometimes used to replace complicated integrands with simpler ones.

EXAMPLE 1 Find $\int \frac{1}{\sqrt{1-\sin^2\theta}}\,d\theta$.

Solution Although the integrand may be written as $(1-\sin^2\theta)^{-1/2}$, the power rule does not apply. The derivative of $1-\sin^2\theta$ is not a factor of the integrand. Consider the following identity.

$$\begin{aligned}\cos^2\theta+\sin^2\theta&=1\\ \cos^2\theta&=1-\sin^2\theta\\ \cos\theta&=\pm\sqrt{1-\sin^2\theta}\end{aligned}$$

The sign is determined by the quadrant in which θ lies. To simplify notation, we consider only the case for which $0<\theta<\pi/2$, although the result may be generalized for all θ for which the integral is defined. For θ between 0 and $\pi/2$, $\cos\theta>0$, and we have the following equation.

$$\begin{aligned}\int \frac{1}{\sqrt{1-\sin^2\theta}}\,d\theta&=\int\frac{1}{\cos\theta}\,d\theta\\ &=\int \sec\theta\,d\theta\\ &=\ln|\sec\theta+\tan\theta|+C\end{aligned}$$ ■

Trigonometric identities may also be used to simplify algebraic expressions. The technique is often appropriate for integrals involving radicals of the form $\sqrt{a^2-x^2}$, $\sqrt{a^2+x^2}$, or $\sqrt{x^2-a^2}$.

The algebraic expression a^2-x^2 is similar to the left-hand side of the trigonometric identity

$$1-\sin^2\theta=\cos^2\theta. \qquad \textbf{(26-4)}$$

The algebraic expression a^2+x^2 is similar to the left-hand side of the trigonometric identity

$$1+\tan^2\theta=\sec^2\theta. \qquad \textbf{(26-5)}$$

The algebraic expression x^2-a^2 is similar to the left-hand side of the trigonometric identity

$$\sec^2\theta-1=\tan^2\theta. \qquad \textbf{(26-6)}$$

To use the method of trigonometric substitution, we replace a variable with a trigonometric function, then use an identity to simplify the integral. Example 2 illustrates the technique.

EXAMPLE 2 Find $\int \frac{1}{x\sqrt{x^2+1}}\,dx$.

Solution The quantity $x^2 + 1$ in the denominator is reminiscent of Equation 26-5. We let $x = \tan\theta$, once again assuming that $0 < \theta < \pi/2$. Then $dx = \sec^2\theta\, d\theta$. The integral is

$$\begin{aligned}
\int \frac{1}{x\sqrt{x^2+1}}\,dx &= \int \frac{\sec^2\theta}{\tan\theta\sqrt{\tan^2\theta+1}}\,d\theta \\
&= \int \frac{\sec^2\theta}{\tan\theta\,(\sec\theta)}\,d\theta \\
&= \int \frac{\sec\theta}{\tan\theta}\,d\theta \\
&= \int \frac{1/\cos\theta}{\sin\theta/\cos\theta}\,d\theta \\
&= \int 1/\sin\theta\,d\theta \\
&= \int \csc\theta\,d\theta \\
&= -\ln|\csc\theta + \cot\theta| + C.
\end{aligned}$$

In the original problem, the integration was to be done with respect to x. To restore x to the answer, examine Figure 26.3, which shows a triangle for which $\tan\theta = x$. Since the side opposite θ must be x, $\csc\theta = \sqrt{x^2+1}/x$, and $\cot\theta = 1/x$.

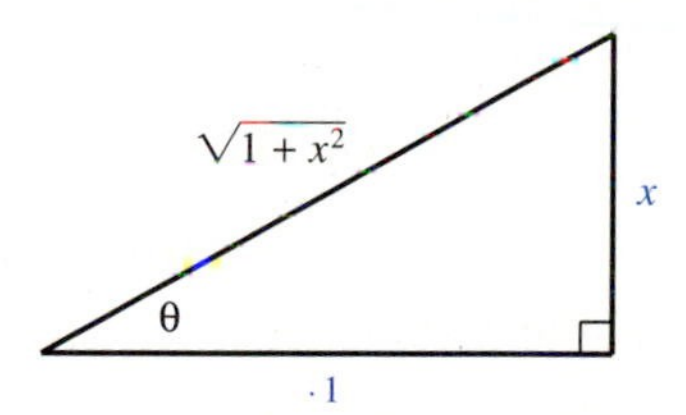

FIGURE 26.3

$$\begin{aligned}
\int \frac{1}{x\sqrt{x^2+1}}\,dx &= -\ln\left|\frac{\sqrt{x^2+1}}{x} + \frac{1}{x}\right| + C \\
&= -\ln\left|\frac{\sqrt{x^2+1}+1}{x}\right| + C \quad ■
\end{aligned}$$

Equations 26-4 through 26-6 each contain the constant 1, but the technique we have used may also be used when a^2 appears in the integrand. We make the following generalizations.

To integrate functions containing radicals:

In each case below, it is assumed that $a > 0$ and $0 < \theta < \pi/2$.

(a) If the radical is of the form $\sqrt{a^2 - x^2}$

1. Check to see if the power rule applies.
2. If not, let $x = a \sin \theta$, and substitute for x and dx.
3. Integrate with respect to θ.
4. Draw the appropriate triangle and restore x to the problem.

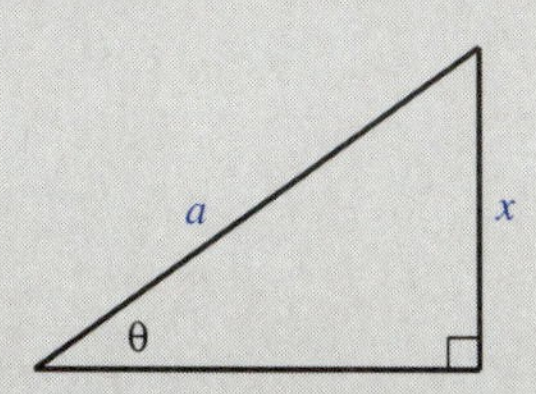

(b) If the radical is of the form $\sqrt{a^2 + x^2}$

1. Check to see if the power rule applies.
2. If not, let $x = a \tan \theta$, and substitute for x and dx.
3. Integrate with respect to θ.
4. Draw the appropriate right triangle and restore x to the problem.

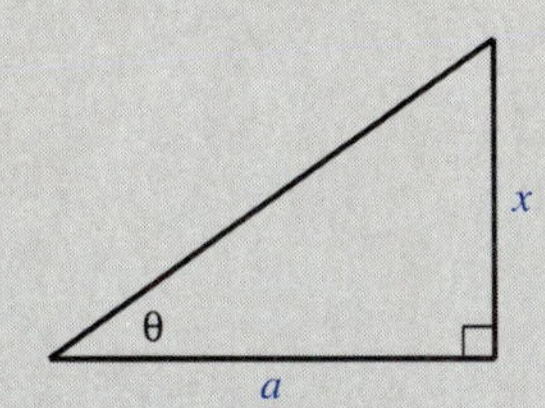

(c) If the radical is of the form $\sqrt{x^2 - a^2}$

1. Check to see if the power rule applies.
2. If not, let $x = a \sec \theta$, and substitute for x and dx.
3. Integrate with respect to θ.
4. Draw the appropriate right triangle and restore x to the problem.

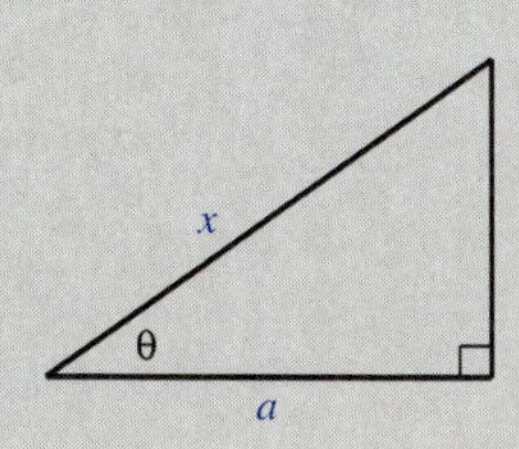

EXAMPLE 3 Find $\int \frac{\sqrt{x^2 - 16}}{2x}\,dx$.

Solution Let $x = 4 \sec \theta$. Then $dx = 4 \sec \theta \tan \theta\, d\theta$.

$$\begin{aligned}\int \frac{\sqrt{x^2 - 16}}{2x}\,dx &= \int \frac{\sqrt{16 \sec^2 \theta - 16}\,(4 \sec \theta \tan \theta)}{2(4 \sec \theta)}\,d\theta \\ &= \frac{\sqrt{16 \tan^2 \theta}\,(\tan \theta)}{2}\,d\theta \\ &= \int \frac{4 \tan \theta\,(\tan \theta)}{2}\,d\theta \\ &= \int 2 \tan^2 \theta\, d\theta\end{aligned}$$

To integrate $\tan^2 \theta$, the identity $\tan^2 \theta = \sec^2 \theta - 1$ is helpful, because the integral of $\sec^2 \theta$ may be found by antidifferentiation.

$$\begin{aligned}\int 2 \tan^2 \theta\, d\theta &= \int 2(\sec^2 \theta - 1)\, d\theta \\ &= \int 2 \sec^2 \theta\, d\theta - \int 2\, d\theta \\ &= 2 \tan \theta - 2\theta\end{aligned}$$

To restore x, examine Figure 26.4, which shows a triangle having $x = 4 \sec \theta$ or $\sec \theta = x/4$.

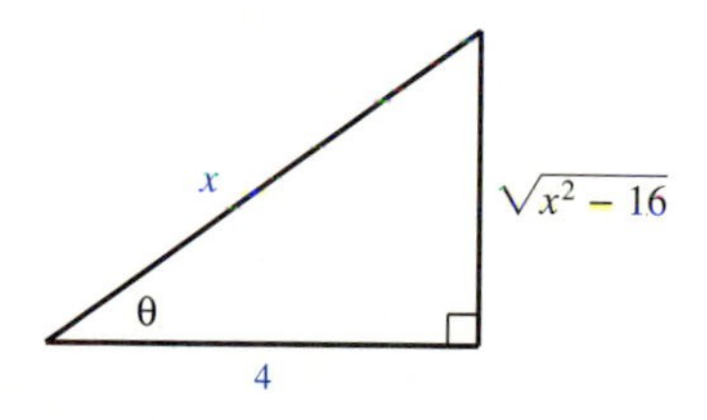

FIGURE 26.4

From the Pythagorean theorem, we see that the side opposite θ must be $\sqrt{x^2 - 16}$. Thus, $\tan \theta = \sqrt{x^2 - 16}/4$ and $\theta = \text{Arcsec}\, x/4$.

$$\begin{aligned}\int \frac{\sqrt{x^2 - 16}}{2x}\,dx &= 2\,\frac{\sqrt{x^2 - 16}}{4} - 2\,\text{Arcsec}\,\frac{x}{4} + C \\ &= \frac{\sqrt{x^2 - 16}}{2} - 2\,\text{Arcsec}\,\frac{x}{4} + C \quad \blacksquare\end{aligned}$$

When trigonometric substitution is used with a definite integral, the limits of integration may be omitted in the intermediate steps. The limits in the original problem are x-values. The trigonometric substitution eliminates x and introduces θ as the variable. When x is restored at the end, the limits of integration are restored also.

EXAMPLE 4 Evaluate $\displaystyle\int_1^2 \frac{1}{x^2\sqrt{4-x^2}}\,dx$.

Solution Let $x = 2\sin\theta$. Then $dx = 2\cos\theta\,d\theta$.

$$\begin{aligned}
\int \frac{1}{x^2\sqrt{4-x^2}}\,dx &= \int \frac{2\cos\theta\,d\theta}{(4\sin^2\theta)\sqrt{4-4\sin^2\theta}} \\
&= \int \frac{2\cos\theta\,d\theta}{(4\sin^2\theta)\sqrt{4\cos^2\theta}} \\
&= \int \frac{2\cos\theta\,d\theta}{(4\sin^2\theta)(2\cos\theta)} \\
&= \int \frac{d\theta}{4\sin^2\theta} \\
&= \int \frac{\csc^2\theta\,d\theta}{4} \\
&= \frac{-\cot\theta}{4} + C
\end{aligned}$$

Figure 26.5 shows a triangle for which $x = 2\sin\theta$ or $\sin\theta = x/2$. The side adjacent to θ must be $\sqrt{4-x^2}$. Thus, $-\cot\theta = -\sqrt{4-x^2}/x$.

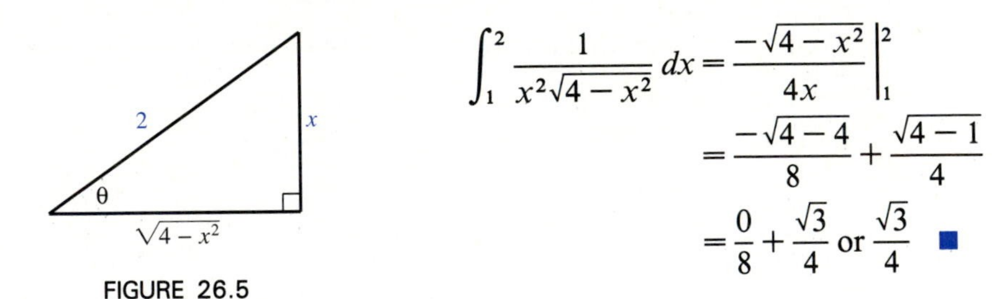

FIGURE 26.5

$$\begin{aligned}
\int_1^2 \frac{1}{x^2\sqrt{4-x^2}}\,dx &= \left.\frac{-\sqrt{4-x^2}}{4x}\right|_1^2 \\
&= \frac{-\sqrt{4-4}}{8} + \frac{\sqrt{4-1}}{4} \\
&= \frac{0}{8} + \frac{\sqrt{3}}{4} \text{ or } \frac{\sqrt{3}}{4}
\end{aligned}$$ ■

Trigonometric substitution may be used to verify the formulas for the integrals that yield the inverse trigonometric functions.

EXAMPLE 5 Find $\displaystyle\int \frac{dx}{\sqrt{a^2-x^2}}$.

Solution Let $x = a\sin\theta$. Then $dx = a\cos\theta\,d\theta$.

$$\begin{aligned}
\int \frac{dx}{\sqrt{a^2-x^2}} &= \int \frac{a\cos\theta\,d\theta}{\sqrt{a^2-a^2\sin^2\theta}} \\
&= \int \frac{a\cos\theta\,d\theta}{\sqrt{a^2\cos^2\theta}}
\end{aligned}$$

$$= \int \frac{a \cos \theta \, d\theta}{a \cos \theta}$$

$$= \int d\theta$$

$$= \theta + C$$

But we let $x = a \sin \theta$, so $\sin \theta = x/a$ and $\theta = \text{Arcsin } x/a$. Thus,

$$\int \frac{dx}{\sqrt{a^2 - x^2}} = \text{Arcsin } \frac{x}{a} + C. \quad \blacksquare$$

Similarly, we can find the integral that yields Arctan x/a. Generalizing for the chain rule, we have the following formulas.

INTEGRALS THAT YIELD THE INVERSE TRIGONOMETRIC FUNCTIONS

If u is a function, then

$$\int \frac{du}{\sqrt{a^2 - u^2}} = \text{Arcsin } \frac{u}{a} + C \qquad \textbf{(26-7)}$$

$$\int \frac{du}{a^2 + u^2} = \frac{1}{a} \text{Arctan } \frac{u}{a} + C \qquad \textbf{(26-8)}$$

A trigonometric substitution may lead to an integral of the form

$$\int \sin^2 \theta \, d\theta \qquad \text{or} \qquad \int \cos^2 \theta \, d\theta.$$

The double angle identities from Section 12.3 are used to rewrite the integrand.

$$\cos 2\theta = 1 - 2 \sin^2 \theta \qquad \text{and} \qquad \cos 2\theta = 2 \cos^2 \theta - 1$$

$$2 \sin^2 \theta = 1 - \cos 2\theta \qquad \text{and} \qquad 2 \cos^2 \theta = \cos 2\theta + 1$$

$$\sin^2 \theta = \frac{1}{2}(1 - \cos 2\theta) \qquad \text{and} \qquad \cos^2 \theta = \frac{1}{2}(\cos 2\theta + 1)$$

Then we find $\int \sin^2 \theta \, d\theta$ and $\int \cos^2 \theta \, d\theta$ as follows.

$$\int \sin^2 \theta \, d\theta = \frac{1}{2} \int (1 - \cos 2\theta) \, d\theta$$

$$\int \sin^2 \theta \, d\theta = \frac{1}{2} \int d\theta - \frac{1}{2} \int \cos 2\theta \, d\theta$$

$$\int \sin^2 \theta \, d\theta = \frac{\theta}{2} - \frac{1}{2}\left(\frac{1}{2}\right) \int (\cos 2\theta) 2 \, d\theta$$

$$\int \sin^2 \theta \, d\theta = \frac{\theta}{2} - \frac{1}{4} \sin 2\theta + C$$

$$\int \cos^2 \theta \, d\theta = \frac{1}{2} (\cos 2\theta + 1) \, d\theta$$

$$\int \cos^2 \theta \, d\theta = \frac{1}{2} \int \cos 2\theta \, d\theta + \frac{1}{2} \int d\theta$$

$$\int \cos^2 \theta \, d\theta = \frac{1}{2}\left(\frac{1}{2}\right) \int (\cos 2\theta) 2 \, d\theta + \frac{\theta}{2}$$

$$\int \cos^2 \theta \, d\theta = \frac{1}{4} \sin 2\theta + \frac{\theta}{2} + C$$

EXERCISES 26.5

In 1–16, find each integral, using a trigonometric substitution.

1. $\int \frac{\sqrt{9-x^2}}{x^2}\,dx$
2. $\int \sqrt{1-x^2}\,dx$
3. $\int \frac{dx}{x^2\sqrt{x^2+25}}\,dx$
4. $\int \frac{x}{\sqrt{16+x^2}}\,dx$
5. $\int \frac{\sqrt{x^2-4}}{x}\,dx$
6. $\int \frac{1}{\sqrt{x^2-9}}\,dx$
7. $\int \frac{dx}{x\sqrt{1-x^2}}$
8. $\int \frac{\sqrt{4-x^2}}{x^2}\,dx$
9. $\int \frac{x^3}{\sqrt{16+x^2}}\,dx$
10. $\int \frac{dx}{(x^2+9)^{3/2}}$
11. $\int_0^5 \frac{x^2\,dx}{\sqrt{25-x^2}}$
12. $\int_1^2 \frac{dx}{x^2\sqrt{9+x^2}}$
13. $\int_0^1 \frac{x}{\sqrt{4+x^2}}\,dx$
14. $\int_1^4 \frac{\sqrt{1+x^2}}{x^4}\,dx$
15. $\int_5^6 \frac{dx}{x\sqrt{x^2-25}}$
16. $\int_4^5 \frac{dx}{x^2\sqrt{x^2-16}}$

17. The cross section of an elliptical oil tank 8.0 ft long is given by $x^2+4y^2=16$, where measurements are in feet. Find the volume of oil in the tank when the depth of the oil is 1.5 ft in the center. 1 2 4 7 8 3 5

18. A machine part occupies the space swept out by rotating the area bounded by $y=1/\sqrt{4+x^2}$ between $x=-3$ and $x=3$ about the x-axis. If the measurements are in inches, find the weight of the part if it is made of steel, which weighs 0.3 lb/in^3. 1 2 4 7 8 3 5

19. The formula $P=v^2/R$ gives the power P (in watts) in a resistance of R (in ohms) due to an applied voltage v (in volts). The electrical energy (in joules) converted to heat is given by $W=\int P\,dt$.

How much electrical energy would be converted to heat in 10 seconds, if the voltage applied is given by $v=170\sin 377t$ and $R=8$ ohms? 3 5 9 1 2 4 6 7 8

20. The force (in newtons) acting on an object is given by $F=1/\sqrt{x^2+16}$ where x is the distance (in meters) from the initial position. Find the work done (in joules) in moving the object from $x=0$ to $x=2$. 4 7 8 1 2 3 5 6 9

21. Show that $\int \frac{dx}{a^2+x^2}=\frac{1}{a}\text{Arctan}\,\frac{x}{a}+C.$

1. Architectural Technology
2. Civil Engineering Technology
3. Computer Engineering Technology
4. Mechanical Drafting Technology
5. Electrical/Electronics Engineering Technology
6. Chemical Engineering Technology
7. Industrial Engineering Technology
8. Mechanical Engineering Technology
9. Biomedical Engineering Technology

26.6 INTEGRATION BY PARTS

Many of the forms for integrals have been obtained by reversing derivative formulas. Starting with the product rule for derivatives, we are led to a procedure called "integration by parts."

Recall that the derivative of a product is given by

$$\frac{d[f(x)g(x)]}{dx}=f(x)g'(x)+g(x)f'(x).$$

Thus,

$$\int [f(x)g'(x) + g(x)f'(x)]\,dx = f(x)g(x) + C_1 \text{ or}$$
$$\int f(x)g'(x)\,dx + \int g(x)f'(x)\,dx = f(x)g(x) + C_1.$$

Subtracting $\int g(x)f'(x)\,dx$ from both sides of the equation, we have

$$\int f(x)g'(x)\,dx = f(x)g(x) + C_1 - \int g(x)f'(x)\,dx.$$

At this point, we drop the constant C_1, because another constant of integration will arise from the integration of $g(x)f'(x)$. The two constants may be combined to produce a single constant, which may be denoted C in the final answer.

$$\int f(x)g'(x)\,dx = f(x)g(x) - \int g(x)f'(x)\,dx$$

Since integral formulas are usually stated in differential form, let $u = f(x)$ and $v = g(x)$. Then $du = f'(x)\,dx$ and $dv = g'(x)\,dx$. Using this notation, the formula is written in the following manner.

INTEGRATION BY PARTS

If u and v are functions, then

$$\int u\,dv = uv - \int v\,du. \qquad \textbf{(26-9)}$$

Integration by parts is the procedure of applying Equation 26-9. This formula is useful when $\int u\,dv$ is not one of the basic forms we have considered, but $\int v\,du$ is.

EXAMPLE 1 Find $\int x \sin x\,dx$.

Solution To apply Equation 26-9, we must identify u and dv in the integrand. Once these functions are determined, we must integrate dv to obtain v and differentiate u to obtain du. The derivative and integral of $\sin x$ are of equal complexity. The derivative of x, however, is simpler than the integral of x. Thus, we choose $u = x$ and $dv = \sin x\,dx$.

$$\text{Let } u = x \qquad dv = \sin x\,dx$$
$$du = 1\,dx \qquad v = -\cos x$$

(It is not necessary to include a constant when v is found by integration. If a constant is included, the integration by parts procedure is such that this constant eventually is subtracted out.) Then

$$\int \underbrace{x}_{u} \underbrace{\sin x\,dx}_{dv} = \underbrace{x}_{u}\underbrace{(-\cos x)}_{v} - \int \underbrace{-\cos x}_{v}\,\underbrace{dx}_{du} = -x\cos x + \sin x + C. \quad \blacksquare$$

To use integration by parts:

1. Identify u and dv.
   (a) If possible, choose u so that the derivative of u is a simpler expression than u.
   (b) If possible, choose dv so that the integral of dv is easy to find.
2. Find $du = u'\,dx$ and $v = \int dv$.
3. Substitute u, dv, v, and du into the equation
$$\int u\,dv = uv - \int v\,du.$$

EXAMPLE 2 Find $\int xe^x\,dx$.

Solution Since the derivative and integral of e^x are equally simple in form, but the derivative of x is simpler than the integral of x, we identify u and dv as follows.

$$\text{Let } u = x \qquad dv = e^x\,dx$$
$$du = 1\,dx \qquad v = e^x$$

Then

$$\int xe^x\,dx = xe^x - \int e^x\,dx$$
$$= xe^x - e^x + C. \quad \blacksquare$$

EXAMPLE 3 Evaluate $\int_1^3 x^2 \ln x\,dx$.

Solution The derivative of x^2 is a simpler form than the integral of x^2. If $u = x^2$, then $dv = \ln x\,dx$. But $\int \ln x\,dx$ is not easily obtainable. Thus, we identify u and dv as follows.

$$\text{Let } u = \ln x \qquad dv = x^2$$
$$du = \frac{1}{x}\,dx \qquad v = \frac{x^3}{3}$$

Then

$$\begin{aligned}
\int_1^3 x^2 \ln x\,dx &= \left(\frac{x^3}{3}\right)\ln x\,\Big|_1^3 - \int_1^3 \frac{x^3}{3}\left(\frac{1}{x}\right)dx \\
&= \frac{x^3 \ln x}{3}\Big|_1^3 - \int_1^3 \frac{x^2}{3}\,dx \\
&= \frac{x^3 \ln x}{3}\Big|_1^3 - \frac{x^3}{9}\Big|_1^3 \\
&= \left(\frac{27 \ln 3}{3} - \frac{\ln 1}{3}\right) - \left(\frac{27}{9} - \frac{1}{9}\right) \text{ or about 7.} \quad ■
\end{aligned}$$

Sometimes integration by parts can be used when there appears to be only one function in the integrand.

EXAMPLE 4 Find $\int \text{Arccos}\, x\,dx$.

Solution Let $u = \text{Arccos}\, x \qquad dv = dx$

$$du = \frac{-1}{\sqrt{1-x^2}}\,du \qquad v = x$$

$$\begin{aligned}
\int \text{Arccos}\, x\,dx &= x\,\text{Arccos}\, x - \int \frac{-x}{\sqrt{1-x^2}}\,dx \\
&= x\,\text{Arccos}\, x - \int -x(1-x^2)^{-1/2}\,dx && \sqrt{1-x^2} = (1-x^2)^{1/2} \\
&= x\,\text{Arccos}\, x - \frac{1}{2}\int -2x(1-x^2)^{-1/2}\,dx && \text{Multiply by 1/2 and 2.} \\
&= x\,\text{Arccos}\, x - \frac{1}{2}\,[2(1-x^2)^{1/2} + C_1] && \text{Use the power rule.} \\
&= x\,\text{Arccos}\, x - \sqrt{1-x^2} + C && (1-x^2)^{1/2} = \sqrt{1-x^2} \quad ■
\end{aligned}$$

Integration by parts is a powerful technique that has been used to obtain many of the integrals in tables of integrals, which we will encounter in the next section. For some problems, it is necessary to use the technique twice. Example 5 illustrates this type of problem in an applied situation.

EXAMPLE 5 A particle moves so that its acceleration a is given by $a = e^t \cos t\, dt$. Find a formula for its velocity.

Solution To find the velocity, we must find $\int e^t \cos t\, dt$.

$$\text{Let } u = e^t \qquad dv = \cos t\, dt$$
$$du = e^t\, dt \qquad v = \sin t$$

Then

$$\int e^t \cos t\, dt = e^t \sin t - \int e^t \sin t\, dt. \tag{26-10}$$

Now we use integration by parts to find $\int e^t \sin t\, dt$.

$$\text{Let } u = e^t \qquad dv = \sin t\, dt$$
$$du = e^t\, dt \qquad v = -\cos t$$

$$\int e^t \sin t\, dt = -e^t \cos t - \int -e^t \cos t\, dt$$

$$= -e^t \cos t + \int e^t \cos t\, dt \tag{26-11}$$

Notice that now we are faced with finding $\int e^t \cos t\, dt$. But the original problem was to find this integral. We have not, however, returned to the starting point. Recall that the expression on the right-hand side of Equation 26-11 was to be substituted into Equation 26-10. We have

$$\int e^t \cos t\, dt = e^t \sin t - [-e^t \cos t + \int e^t \cos t\, dt]$$
$$\int e^t \cos t\, dt = e^t \sin t + e^t \cos t - \int e^t \cos t\, dt.$$

The term $\int e^t \cos t\, dt$ appears on both sides of the equation. Adding this expression to both sides gives

$$2 \int e^t \cos t\, dt = e^t \sin t + e^t \cos t$$
$$\int e^t \cos t\, dt = (1/2)[e^t \sin t + e^t \cos t].$$

Now we factor out e^t and restore the constant of integration.

$$\int e^t \cos t\, dt = (1/2)e^t[\sin t + \cos t] + C \quad ■$$

EXERCISES 26.6

In 1–16, find each integral.

1. $\int x \cos 2x \, dx$

2. $\int x \sec^2 x \, dx$

3. $\int x \sec x \tan x \, dx$

4. $\int xe^{-x} \, dx$

5. $\int \ln x \, dx$

6. $\int_1^3 x \ln x \, dx$

7. $\int_4^9 x \ln \sqrt{x} \, dx$

8. $\int_0^1 x\sqrt{1-x} \, dx$

9. $\int_0^3 x\sqrt{x+1} \, dx$

10. $\int_0^3 \frac{x}{\sqrt{x+1}} \, dx$

11. $\int \text{Arcsin } x \, dx$

12. $\int \text{Arctan } x \, dx$

13. $\int e^x \sin x \, dx$

14. $\int e^{2x} x^2 \, dx$

15. $\int x^2 e^{-x} \, dx$

16. $\int \sin(\ln x) \, dx$

---

17. The formula $i = \frac{1}{L}\int v \, dt$ gives the current i (in amperes) in an inductor when v is the voltage (in volts) and L is the inductance (in henries). Find a formula for the current in a 0.4-mH inductor if $v = 10^{-4}e^{0.3t}t$, and $i = 0$ when $t = 0$. 3 5 9 1 2 4 6 7 8

18. A particle moves so that its velocity (in m/s) is given by $v = 2te^{-t}$, where t is the time (in seconds). Find the distance traveled between $t = 0$ and $t = 4$. 4 7 8 1 2 3 5 6 9

19. Find the area of a metal plate that occupies the region bounded by $y = 2x \sin x$ and the x-axis between $x = 0$ and $x = \pi$, if measurements are in centimeters. 1 2 4 7 8 3 5

20. The voltage v (in volts) induced in a tape head is given by $v = t^2e^{3t}$, where t is the time (in seconds). Find the average value of v over the three-second interval from $t = 0$ to $t = 3$. 3 5 9 1 2 4 6 7 8

21. For a cantilever beam under a concentrated load P (in kg), the angle of deflection θ is given by $\theta = \text{Tan}^{-1}[-Pa^2/(2EI)]$. Find the average angle of deflection as P increases from 600 kg to 650 kg, if $a = 4.0$ m, $E = 2.11 \times 10^{10}$ kg/m², and $I = 2.81 \times 10^{-5}$ m⁴. In the formula, a is the distance of the load from the support, E is the modulus of elasticity, and I is the moment of inertia. 1 2 4 8 7 9

22. The charge q (in coulombs) delivered by a current i (in amperes) is given by $q = \int i \, dt$, where t is the time (in seconds). A damped-out periodic wave form has current given by $i = e^{-3t} \cos 5t$. Find a formula for the charge delivered over time t. 3 5 9 1 2 4 6 7 8

23. Rework Example 1 to show that if a constant of integration is included when v is found, the constant will eventually be subtracted out.

---

1. Architectural Technology
2. Civil Engineering Technology
3. Computer Engineering Technology
4. Mechanical Drafting Technology
5. Electrical/Electronics Engineering Technology
6. Chemical Engineering Technology
7. Industrial Engineering Technology
8. Mechanical Engineering Technology
9. Biomedical Engineering Technology

26.7 USING TABLES OF INTEGRALS

As one acquires knowledge of more and more formulas for integrals, the ability to recognize patterns becomes increasingly important. There are many forms for integrals that we have not yet encountered. Many of these forms, however, do follow patterns. These patterns have been tabulated and are often listed in tables of integrals. There is a short table of integrals in the Appendix of this book.

In preparation for using tables, we consider the inverse trigonometric forms, which we encountered in Section 26.5. In some cases an inverse trig form is very similar to the power rule or the logarithmic form. Practice in distinguishing these forms makes it easier to distinguish other forms that we will eventually encounter. We now compare and contrast some similar formulas.

SIMILAR INTEGRAL FORMS

$$\int \frac{u\,du}{a^2+u^2} = \frac{1}{2}\ln|a^2+u^2| + C \qquad \text{logarithmic form} \qquad (26\text{-}12)$$

$$\int \frac{du}{a^2+u^2} = \frac{1}{a}\operatorname{Arctan}\frac{u}{a} + C \qquad (26\text{-}13)$$

$$\int \frac{u\,du}{\sqrt{a^2-u^2}} = -\sqrt{a^2-u^2} + C \qquad \text{power formula} \qquad (26\text{-}14)$$

$$\int \frac{du}{\sqrt{a^2-u^2}} = \operatorname{Arcsin}\frac{u}{a} + C \qquad (26\text{-}15)$$

EXAMPLE 1 Find each integral.

(a) $\displaystyle\int \frac{dx}{\sqrt{1-x^2}}$ **(b)** $\displaystyle\int \frac{dx}{1+x^2}$

Solution The integrand in part (a) contains a square root in the denominator. Examine Equations 26-14 and 26-15. It appears that u is x and a is 1. Since $dy = dx$, the integral in part (a) is like Equation 26-15. We therefore replace a with 1 and u with x. Thus,

$$\int \frac{dx}{\sqrt{1-x^2}} = \operatorname{Arcsin} x + C.$$

The integrand in part (b) contains a sum of squares in the denominator. Examine Equations 26-12 and 26-13. It appears that u is x and a is 1. Since $du = dx$, the integral in part (b) is like Equation 26-13. We therefore replace a with 1 and u with x. Thus,

$$\int \frac{dx}{1+x^2} = \operatorname{Arctan} x + C. \quad ■$$

EXAMPLE 2 Find each integral.

(a) $\int \frac{2\,dx}{1 + 4x^2}$ **(b)** $\int \frac{4x\,dx}{1 + 4x^2}$

Solution Notice the similarity of the forms. The denominator of each integrand is $1 + 4x^2$. It appears that u is $2x$ and a is 1. Since $du = 2\,dx$, the integral in (a) is the Arctangent form (Equation 26-13). We therefore replace u with $2x$, and a with 1.

$$\int \frac{2\,dx}{1 + 4x^2} = \text{Arctan } 2x + C$$

The numerator in (b) is $2x(2\,dx)$, so it is the logarithmic form (Equation 26-12). We replace u with $2x$ and a with 1.

$$\int \frac{4x\,dx}{1 + 4x^2} = \int \frac{2x(2\,dx)}{1 + 4x^2} = \frac{1}{2} \ln |1 + 4x^2| + C \quad ■$$

EXAMPLE 3 Find each integral.

(a) $\int \frac{x\,dx}{\sqrt{4 - 9x^2}}$ **(b)** $\int \frac{1\,dx}{\sqrt{4 - 9x^2}}$

Solution Once again, we have two very similar problems. The denominator of each integral is $\sqrt{4 - 9x^2}$. It appears that a is 2 and u is $3x$. Since $du = 3\,dx$, the numerator of the integrand in (a) needs a factor of 3 to contain du. It also needs a factor of 3 to contain u. To make it fit the pattern of Equation 26-14, we multiply by 9 and by 1/9.

$$\int \frac{x\,dx}{\sqrt{4 - 9x^2}} = \frac{1}{9} \int \frac{3x(3\,dx)}{\sqrt{4 - 9x^2}} = -\frac{1}{9} \sqrt{4 - 9x^2} + C$$

The numerator of the integrand in (b) must be adjusted by a constant factor of 3 to make it fit the pattern of Equation 26-15.

$$\int \frac{1\,dx}{\sqrt{4 - 9x^2}} = \frac{1}{3} \int \frac{3\,dx}{\sqrt{4 - 9x^2}} = \frac{1}{3} \text{Arcsin } \frac{3x}{2} + C \quad ■$$

We are now ready to begin comparing the forms of integrals against those listed in the Appendix.

EXAMPLE 4 Find $\int \frac{x}{(3 + 2x)^2}\,dx$.

Solution Since the denominator involves an expression of the form $a + bu$, we scan the first group of integrals in the table. The integral has the form of number 3 in the tables. Here $u = x$, $du = dx$, $a = 3$, and $b = 2$. Hence

$$\int \frac{x}{(3 + 2x)^2}\,dx = \frac{1}{4}\left[\frac{3}{3 + 2x} + \ln |3 + 2x|\right] + C. \quad ■$$

EXAMPLE 5 Find $\int 3x^2 \sin^2 x^3\, dx$.

Solution The integral is a trigonometric form, so we scan the sixth part of the tables. The integral has the form of number 30 with $u = x^3$, $du = 3x^2\, dx$, and $n = 2$. Hence

$$\begin{aligned}\int 3x^2 \sin^2 x^3\, dx &= \frac{-\sin x^3 \cos x^3}{2} + \frac{1}{2}\int (\sin^0 x^3)(3x^2)\, dx \\ &= \frac{-\sin x^3 \cos x^3}{2} + \frac{1}{2}\left(\frac{3x^3}{3}\right) + C \\ &= \frac{-\sin x^3 \cos x^3}{2} + \frac{x^3}{2} + C.\end{aligned}$$

EXAMPLE 6 Find $\int \frac{e^{2x}\, dx}{\sqrt{9 + e^{4x}}}$.

Solution The denominator is an expression of the form $\sqrt{a^2 + u^2}$, so scanning the third part of the tables, we see that the integral has the form of number 14. Here, $u = e^{2x}$, $du = 2e^{2x}\, dx$, and $a = 3$.

$$\int \frac{e^{2x}\, dx}{\sqrt{9 + e^{4x}}} = \frac{1}{2}\int \frac{2e^{2x}\, dx}{\sqrt{9 + e^{4x}}} = \frac{1}{2}\ln|e^{2x} + \sqrt{9 + e^{4x}}| + C$$

EXERCISES 26.7

In 1–4, find each integral using the inverse trigonometric forms (Equations 26-13 and 26-15).

1. $\int \frac{3\, dx}{\sqrt{1 - 9x^2}}$ **2.** $\int \frac{2\, dx}{\sqrt{25 - 4x^2}}$ **3.** $\int \frac{dx}{16 + x^2}$ **4.** $\int \frac{dx}{9x^2 + 25}$

In 5–16, identify the form of each integral as (a) Arcsine, (b) Arctangent, (c) logarithmic, (d) power rule, or (e) none of these.

5. $\int \frac{2\, dx}{1 + 4x^2}$ **6.** $\int \frac{8x\, dx}{1 + 4x^2}$ **7.** $\int \frac{4\, dx}{1 - 4x^2}$

8. $\int \frac{x\, dx}{1 + 9x^2}$ **9.** $\int \frac{dx}{\sqrt{1 - 9x^2}}$ **10.** $\int \frac{4\, dx}{\sqrt{1 - 16x^2}}$

11. $\int \frac{2x\, dx}{\sqrt{9 - 4x^2}}$ **12.** $\int \frac{dx}{\sqrt{9x^2 - 1}}$ **13.** $\int \frac{5x\, dx}{9 + 25x^2}$

14. $\int \frac{x\, dx}{1 - x^2}$ **15.** $\int \frac{dx}{1 + 25x^2}$ **16.** $\int \frac{dx}{x\sqrt{1 - x^2}}$

In 17–36, find each integral, using the tables in the Appendix.

17. $\int \frac{2x\,dx}{x^2(1+3x^2)}$

18. $\int \frac{x\,dx}{(2+3x)^2}$

19. $\int \frac{x\,dx}{\sqrt{1+x}}$

20. $\int \frac{dx}{x\sqrt{2+3x}}$

21. $\int 2x^3\sqrt{5+x^2}\,dx$

22. $\int \frac{\sin x \cos x\,dx}{\sqrt{1+\sin x}}$

23. $\int \frac{\cos x\,dx}{\sqrt{1+\sin^2 x}}$

24. $\int \frac{-\sin x\,dx}{9-\cos^2 x}$

25. $\int \sqrt{16+e^{2x}}\,e^x\,dx$

26. $\int \sqrt{4-\cos^2 x}\,(\sin x)\,dx$

27. $\int \frac{\sqrt{25-x^4}}{x^2}\,2x\,dx$

28. $\int \frac{e^x\,dx}{\sqrt{e^{2x}-9}}$

29. $\int \cos^3 2x\,dx$

30. $\int e^x \tan^2 e^x\,dx$

31. $\int \frac{x^2\,dx}{x^3(1+2x^3)}$

32. $\int \frac{\cos 2x\,dx}{(\sin 2x)\sqrt{4+\sin^2 2x}}$

33. $\int \sin^3 x\,dx$

34. $\int \cot^3 x\,dx$

35. $\int 2x \operatorname{Arcsin} x^2\,dx$

36. $\int \operatorname{Arctan} 3x\,dx$

---

37. Consider a cylindrical tank that is 4.0 ft in diameter and 6.0 ft long, lying horizontally as shown in the diagram. Determine the volume of liquid in the tank when it is 1.0 ft deep in the center.
1 2 4 7 8 | 3 5

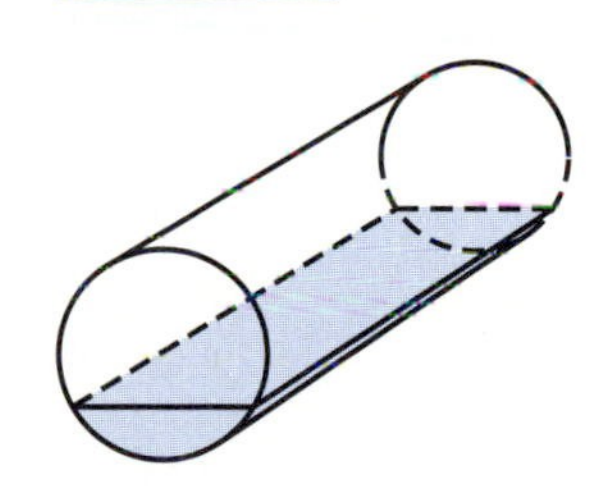

38. The cross section of an elliptical oil tank 8.00 m long is given by $4x^2 + 9y^2 = 36$, where measurements are in meters. Find the volume of oil in the tank when the depth of the oil is 1.00 m in the center. 1 2 4 7 8 | 3 5

39. The acceleration of an object is given by $a = \frac{t}{1+3t}$. Find a formula for velocity if $v = 0$ when $t = 0$. 4 7 8 | 1 2 3 5 6 9

40. The force (in newtons) acting on an object is given by $F = \frac{1}{9-4x^2}$ where x is the distance (in meters) from the initial position. Find the work done in moving the object from $x = 0$ to $x = 1.00$.
4 7 8 | 1 2 3 5 6 9

41. The formula $i = \frac{1}{L}\int v\,dt$ gives the current (in amperes) in an inductor when v is the voltage (in volts) and L is the inductance (in henries). Find a formula for the current in a 0.4-mH inductor if $v = 10^{-4}\,e^{0.3t}t$, and $i = 0$ when $t = 0$.
3 5 9 | 1 2 4 6 7 8

26.8 CHAPTER REVIEW

KEY TERMS

26.1 integrand

RULES AND FORMULAS

If u is a function, then

$$\int u^n\,du = \frac{u^{n+1}}{n+1} + C$$

$$\int e^u\,du = e^u + C$$

$$\int \frac{1}{u}\,du = \ln|u| + C \qquad \text{or} \qquad \int u^{-1}\,du = \ln|u| + C$$

$$\int \cos u\,du = \sin u + C \qquad \int \sin u\,du = -\cos u + C$$

$$\int \sec^2 u\,du = \tan u + C \qquad \int \csc^2 u\,du = -\cot u + C$$

$$\int \sec u \tan u\,du = \sec u + C \qquad \int \csc u \cot u\,du = -\csc u + C$$

$$\int \tan u\,du = -\ln|\cos u| + C \qquad \int \cot u\,du = \ln|\sin u| + C$$

$$\int \sec u\,du = \ln|\sec u + \tan u| + C \qquad \int \csc u\,du = -\ln|\csc u + \cot u| + C$$

$$\int \frac{du}{\sqrt{a^2 - u^2}} = \text{Arcsin}\,\frac{u}{a} + C \qquad \int \frac{du}{a^2 + u^2} = \frac{1}{a}\,\text{Arctan}\,\frac{u}{a} + C$$

$$\int u\,dv = uv - \int v\,du$$

GEOMETRY CONCEPTS

$A = \pi r^2$ (area of a circle)

SEQUENTIAL REVIEW

(Section 26.1) *In 1–7, find each integral.*

1. $\int (1 + \cos 2x)^2 \sin 2x\,dx$

2. $\int \csc^3 3x \cot 3x\,dx$

3. $\int \frac{\sqrt{\ln x}}{x}\,dx$

4. $\int_0^1 \frac{e^{2x}}{(1+e^{2x})^3}\,dx$ 5. $\int xe^{3x^2-1}\,dx$ 6. $\int \frac{\cos x}{\sin^2 x}\,dx$

7. $\int \sqrt{\frac{\text{Arcsin } 2x}{1-4x^2}}\,dx$

(Section 26.2) *In 8–13, find each integral.*

8. $\int_0^1 6x\,e^{3x^2}\,dx$ 9. $\int x^2\,e^{-x^3}\,dx$ 10. $\int_0^{\pi} e^{\sin x}\cos x\,dx$

11. $\int_0^1 \frac{e^x - e^{-x}}{e^{-x}}\,dx$ 12. $\int 3e^x(e^x+2)\,dx$ 13. $\int \frac{e^{\ln x}}{x}\,dx$

(Section 26.3) *In 14–18, find each integral.*

14. $\int \frac{5}{2+5x}\,dx$ 15. $\int_2^3 \frac{4x^3-3x^2}{x^4-x^3}\,dx$ 16. $\int \frac{\cos x}{1+\sin x}\,dx$

17. $\int_0^{\pi/4} \frac{\sec^2 x}{1+\tan x}\,dx$ 18. $\int \frac{e^x}{2-e^x}\,dx$

(Section 26.4) *In 19–25, find each integral.*

19. $\int_0^{\pi} \sin 4x\,dx$ 20. $\int e^x \cos e^x\,dx$ 21. $\int \frac{\tan x^{-1}\sec x^{-1}}{x^2}\,dx$

22. $\int (3x^2+2)\sec^2(x^3+2x)\,dx$ 23. $\int e^{2x}\sec e^{2x}\,dx$ 24. $\int \frac{\tan(\ln x)}{x}\,dx$

25. $\int_0^1 x^2\tan x^3\,dx$

(Section 26.5) *In 26–31, find each integral.*

26. $\int \frac{\sqrt{1-x^2}}{x^2}\,dx$ 27. $\int \frac{\sqrt{x^2-9}}{x}\,dx$ 28. $\int \frac{\sqrt{x^2+16}}{x}\,dx$

29. $\int \frac{1}{x^2\sqrt{1-x^2}}\,dx$ 30. $\int_0^4 \frac{x^2}{\sqrt{16-x^2}}\,dx$ 31. $\int_0^1 \frac{1}{(x^2+1)^{3/2}}\,dx$

(Section 26.6) *In 32–37, find each integral.*

32. $\int x\sin 3x\,dx$ 33. $\int x\csc^2 x\,dx$ 34. $\int x^2\,e^{3x}\,dx$

35. $\int x^3\ln x\,dx$ 36. $\int \frac{x}{\sqrt{1-x}}\,dx$ 37. $\int \text{Arccos } x\,dx$

(Section 26.7) *In 38–41, identify the form of each integral as (a) Arcsine, (b) Arctangent, (c) logarithmic, (d) power rule, or (e) none of these.*

38. $\int \frac{2x}{1-4x^2}\,dx$ 39. $\int \frac{4x^2}{1+4x^2}\,dx$ 40. $\int \frac{2x}{\sqrt{1-x^2}}\,dx$ 41. $\int \frac{2x}{\sqrt{1+x^2}}\,dx$

In 42–45, find each integral, using the tables in the Appendix.

42. $\int x\,\text{Arcsin } x^2\,dx$ 43. $\int \frac{dx}{x(2+3x)}$ 44. $\int \frac{dx}{x\sqrt{x^2-4}}$ 45. $\int 2^x\,dx$

APPLIED PROBLEMS

1. The equation $f = v/\lambda$ gives the frequency (in hertz) of a wave if v is the velocity (in m/s) and λ is the wavelength (in m). Find the average frequency for lightwaves with wavelengths between 4×10^{-7} m and 7×10^{-7} m (visible light). The speed of light is 3×10^8 m/s. 5 3 4 7 8 9

2. The current i (in amperes) in an inductor is given by $i = \frac{1}{L} \int v\, dt$, where L is the inductance (in henries), v is the voltage (in volts), and t is the time (in seconds). Find a formula for the current if $v = \sqrt{9 + 2t}$ and $L = 5$ H. Assume that $i = 0$ when $t = 0$. 3 5 9 1 2 4 6 7 8

3. A machine part occupies the space swept out by rotating the area bounded by $y = \frac{\sqrt{x}}{x + 9}$ between $x = 0$ and $x = 3$ about the x-axis. If the measurements are in cm, find the volume of the part. 1 2 4 7 8 3 5

4. When a charged capacitor is discharged through a resistance, the charge (in coulombs) remaining after time t (in seconds) is given by $Q = Q_0 e^{-t/\tau}$. Find the average value of Q between $t = 0.00$ and $t = 0.02$, if $Q_0 = 2.5 \times 10^{-5}$ and $\tau = 0.002$. In the formula, Q_0 is the original charge (in coulombs) and $\tau = RC$ (resistance $\times$ capacitance). 3 5 9 4 7 8

5. The 500-volt source charging a capacitor is removed in such a way that the current i (in amperes) is given by $i = \sqrt{1 - 2t}$ where t is the discharge time (in seconds). Use $v = 1/C \int i\, dt$ to determine the equation that gives the capacitor's voltage if $C = 2.5 \times 10^{-3}$. (Hint: When t is 0, v is 500.) 3 5 9 4 7 8

6. Use $i = 1/L \int v\, dt$ to determine the current that would cause a voltage given by $v = \sin 377t \cos 377t$ across a coil with an inductance of 265 henries and an initial current of 10^{-4} amperes. 3 5 9 4 7 8

7. Find a formula for the voltage applied to a capacitor with $C = 0.47\ \mu$F that would cause its current i (in amperes) to follow the equation $i = 5t\, e^{3+t^2}$. Use $v = 1/C \int i\, dt$. Assume v is 0 when t is 0. 3 5 9 4 7 8

8. The intensity of the light in a fiber optic cable decreases with distance according to Lambert's law $P = P_0 e^{-\alpha x}$, where P is the power (in watts), x is the distance (in m) down the cable, and α is a constant that depends on the cable. Find the average power in a cable if the initial power P_0 is 500 μW over a run of 3000 meters and α is 0.0015. Give your answer to two significant digits. 3 5 6 9

9. An isothermal gas expansion can do an amount of work W (in joules) given by $W = \int_{V_1}^{V_2} P\, dV$, where V_1 is the initial volume (in liters), V_2 the final volume (in liters), and P is the pressure (in N/m^2). How much work could be done by a 3-liter gas expanding to 3.5 liters if its pressure is given by $P = 4/V$? 6 8 1 2 3 5 7 9

10. The amplitude A (in cm) of vibration of a stretched cable dies out according to the equation $A = 15.0\ e^{-t/6}$, where t is the time (in seconds). Find the average amplitude over the interval $2.00 \le t \le 5.00$. 1 3 4 5 6 8 9

11. A certain object undergoing simple harmonic motion has an acceleration a (in cm/s^2) given by $a = 35\,000 \cos(2450t)$. Find its average velocity over the interval $0.50 \text{ ms} \le t \le 0.80$ ms. Assume v is 0 when t is 0. (Hint: Since acceleration is in cm/s^2, change milliseconds to seconds.) 4 7 8 1 2 3 5 6 9

1. Architectural Technology
2. Civil Engineering Technology
3. Computer Engineering Technology
4. Mechanical Drafting Technology
5. Electrical/Electronics Engineering Technology
6. Chemical Engineering Technology
7. Industrial Engineering Technology
8. Mechanical Engineering Technology
9. Biomedical Engineering Technology

12. The velocity of an object falling through air will reach a maximum value v_T because of air drag. The velocity v (in ft/s) of an object falling from rest at any time t is given by $v = v_T(1 - e^{-kt})$. In the formula, v_T is the terminal velocity (in ft/s), k is a constant that depends on the object's shape, and t is the time (in seconds) traveled. How far would an object fall from rest during the first 20 seconds if k is 0.16 and v_T is 147 ft/s?
4 7 8 1 2 3 5 6 9

13. The formula $a = R_F V_G/(M - R_F t)$ gives the acceleration a (in m/s²) of a rocket in space. In the formula, R_F is the rate (in kg/s) at which fuel is burned, V_G is the velocity (in m/s) of exhaust gas, M is the total mass (in kg) of rocket and fuel, and t is the time (in seconds) of burn. Find the velocity at the end of 20.0 seconds of a rocket starting from rest if $R_F = 5.00$ kg/s, $V_G = 30.0$ m/s, and $M = 750.0$ kg. 4 7 8 1 2 3 5 6 9

14. The formula $V = 12.0\ R/(4.00 + R)$ gives the voltage on a resistor when a variable load R (in ohms) is attached to a 12.0-volt supply with internal resistance 4.00 ohms. Find the average value of V as R varies from 2.00 to 6.00 ohms.
3 5 9 1 2 4 6 7 8

15. The formula $K = \dfrac{2 \times 10^{-25}\,\Delta\lambda}{\lambda\,(\Delta\lambda + \lambda)}$ gives the kinetic energy K (in joules) imparted to the carbon electrons when x-rays are scattered from a carbon block. In the formula λ is the wavelength of the x-rays (in meters) and $\Delta\lambda$ is the Compton shift in meters. Find the average value of K for x-rays with $\lambda = 0.1$ nanometer and $\Delta\lambda$ between 0.002 and 0.003 nanometers. 4 7 8 1 2 3 5 6 9

16. The intensity pattern for a certain diffraction grating is given as $I = 100(1 + 4\cos\phi + 4\cos^2\phi)$. Find the average value of I over the interval $0.26 \le \phi \le 0.35$. Give your answer to two significant digits. 3 5 6 9

17. The formula $\theta = \text{Sin}^{-1}\ 1.22\lambda/d$ gives an important angle for a circular diffraction pattern under certain conditions. In the formula, λ is the wavelength (in m) of the light and d is the diameter (in m) of the lens aperture. Find the average value of θ for a lens of diameter 2.5 cm if λ varies from 650 nm to 675 nm. 3 5 6 9

18. A current i (in amperes) flows through a resistance r (in ohms). The current is given by $i = (1 - e^{-2t})$, and the resistance is given by $r = 5/(t + 2)$, if t is the time (in seconds) of flow. Use $p = i^2 r$ to find the average power between $t = 1$ and $t = 2$. Give your answer to three significant digits.
3 5 9 1 2 4 6 7 8

19. The formula $\theta = \text{Cos}^{-1}\ R/\sqrt{R^2 + X^2}$ gives the phase angle θ (in radians) of a circuit containing a resistor and an inductor in series. In the formula, R is the circuit's resistance (in ohms) and X is the circuit's reactance (in ohms). Find the average value of θ (in degrees) if R is 75 ohms and X varies from 30 ohms to 60 ohms. 3 5 9 4 7 8

RANDOM REVIEW

In 1–10, find each integral.

1. $\displaystyle\int e^x \cot e^x\, dx$

2. $\displaystyle\int (1 + 3e^x)^4 3e^x\, dx$

3. $\displaystyle\int_0^2 \frac{dx}{\sqrt{4 + x^2}}$

4. $\displaystyle\int e^x \ln e^x\, dx$ (Use the tables of integrals in the Appendix.)

5. $\displaystyle\int (4 - \cos x)^{-1} \sin x\, dx$

6. $\displaystyle\int x \sin 3x\, dx$

7. $\int_0^1 e^{2x}\,dx$

8. $\int_{-\pi/2}^{\pi/2} \frac{\cos x}{\sin^3 x}\,dx$

9. $\int \frac{e^{\text{Arcsin}\,x}}{\sqrt{1-x^2}}\,dx$

10. $\int 3e^x(e^x - 3)\,dx$

CHAPTER TEST

In 1–8, find each integral.

1. $\int \frac{\sec^2 x}{\sqrt{\tan x}}\,dx$

2. $\int x^2 e^{x^3+1}\,dx$

3. $\int_0^{\pi/4} \frac{\sec x \tan x}{\sec x + 1}\,dx$

4. $\int_0^{\pi/4} x \tan x^2\,dx$

5. $\int \frac{\sqrt{x^2 - 25}}{x}\,dx$

6. $\int \frac{1}{\sqrt{9 - x^2}}\,dx$

7. $\int_1^3 x \ln x^2\,dx$

8. $\int \frac{e^x}{\sqrt{4 + e^{2x}}}\,dx$

9. A capacitor is charged by a current I (in mA) given by $I = (0.15 + 3t)(0.1t + t^2)^{0.56}$ at time t (in seconds). Find the average value of this function for values of t between 0 and 0.1 s.

10. If the velocity of a moving particle at time t is given by $v = \cos t\, e^{\sin t}$, find an expression for s, the distance traveled, given that $s = 0$ when $t = 0$.

INFINITE SERIES CHAPTER 27

You have seen that the transcendental functions have many applications. One of the early applications of calculus was to evaluate such functions for values other than the special angles (0°, 30°, 45°, 60°, 90°, etc.). Today such values are usually obtained by calculator. When a function is called upon, the calculator computes the value using procedures similar to those of this chapter. These procedures may also be used in certain integration problems. There are some integrals arising in engineering that involve integrands that do not have antiderivatives. If a definite integral is needed, we may be able to find an approximate value using numerical integration. If an indefinite integral is needed, it may be possible to approximate the integrand with the first few terms of an infinite series that can be integrated term by term.

27.1 MACLAURIN SERIES

Recall from Section 2.2 that a polynomial in x is an expression of the form $a_nx^n + a_{n-1}x^{n-1} + \cdots + a_0$ where $a_n, a_{n-1}, a_{n-2}, \ldots, a_0$ are real numbers and each exponent is a nonnegative integer. If we put the constant term first, and allow an infinite number of terms, we have a *power series.*

DEFINITION A **power series** is an expression of the form

$$a_0 + a_1x + a_2x^2 + a_3x^3 + \cdots$$

where $a_0, a_1, a_2, a_3, \ldots$ are real numbers.

Many functions that are not originally in the form of a power series may be represented in that form. For example, long division shows that $\frac{1}{1-x} = 1 + x + x^2 + x^3 + \cdots$. The division is shown below.

$$\begin{array}{r|l} & 1 + x + x^2 + \cdots \\ \hline 1 - x & 1 \\ & \underline{1 - x} \\ & \quad x \\ & \quad \underline{x - x^2} \\ & \qquad x^2 \end{array}$$

There is no apparent advantage to writing this function as a power series, but it was chosen to illustrate the concept because it is familiar to you. The series is a geometric series in which the first term is 1 and the common ratio is x. From Section 16.3, we know that if $S_n = 1 + x + x^2 + x^3 + \cdots + x^n$, then $\lim_{n\to\infty} S_n = \frac{1}{1-x}$, provided $-1 < x < 1$. A finite number of terms of the series are used to provide an approximate value for the function. The more terms used, the better the approximation. Consider the following table of values when x is 1/2.

1/(1 − x) Number of Terms	*Value of Series when x = 1/2*
1	1
2	1 + 1/2 = 1.5
3	1 + 1/2 + 1/4 = 1.75
4	1 + 1/2 + 1/4 + 1/8 = 1.875
5	1 + 1/2 + 1/4 + 1/8 + 1/16 = 1.9375
$n \to \infty$	2

Calculus is used to write a power series expansion for a function such as a trigonometric, logarithmic, or exponential function. A finite number of terms of the power series can then be used to approximate the function.

converge
diverge

The series is said to **converge** if the sum of the first n terms approaches a limit as n approaches infinity. The series is said to **diverge** if the limit does not exist. A power series may converge for some values of x and diverge for other values of x. Some power series converge only when x is 0. Others converge for all real values of x. An **interval of convergence** is the set of values for which the series converges. For our purposes, it will not be necessary to compute the interval of convergence. In this chapter, you may assume that the series converges for the indicated values unless specified otherwise.

interval of convergence

We now obtain a power series expansion for an arbitrary function f that is defined at 0 and whose derivatives are defined at 0. We want to find an expression for $f(x)$ that has the form required by the definition. Let

$$f(x) = a_0 + a_1x + a_2x^2 + a_3x^3 + a_4x^4 + \cdots + a_nx^n + \cdots \quad \textbf{(27-1)}$$

To find the values of the coefficients, we must first find $f'(x)$. Then we find $f''(x)$, and so on, using $f^{(n)}(x)$ to represent the nth derivative of f at x. We begin by repeating Equation 27-1.

$$f(x) = a_0 + a_1x + a_2x^2 + a_3x^3 + a_4x^4 + \cdots + a_nx^n + \cdots \quad \textbf{(27-1)}$$

$$f'(x) = a_1 + 2a_2x + 3a_3x^2 + 4a_4x^3 + \cdots + na_nx^{n-1} + \cdots \quad \textbf{(27-2)}$$

$$f''(x) = 2a_2 + 2\cdot 3a_3x + 3\cdot 4a_4x^2 + \cdots + (n-1)na_nx^{n-2} + \cdots$$

$$f^{(3)}(x) = 2\cdot 3a_3 + 2\cdot 3\cdot 4a_4x + \cdots + (n-2)(n-1)na_nx^{n-3} + \cdots$$

$$f^{(4)}(x) = 2\cdot 3\cdot 4a_4 + \cdots + (n-3)(n-2)(n-1)na_nx^{n-4} + \cdots$$

Next, we evaluate the function and each derivative when x is 0. We can then determine the coefficients from the equations that result. From Equation 27-1, we have $f(0) = a_0$. From Equation 27-2, we have $f'(0) = a_1$. Continuing in this manner, we have

$$f''(0) = 2a_2, \text{ and therefore } \frac{f''(0)}{2} = a_2.$$

$$f^{(3)}(0) = 2\cdot 3a_3, \text{ and therefore } \frac{f^{(3)}(0)}{3\cdot 2} = a_3.$$

$$f^{(4)}(0) = 2\cdot 3\cdot 4a_4, \text{ and therefore } \frac{f^{(4)}(0)}{4\cdot 3\cdot 2} = a_4.$$

Generalizing, $a_n = \dfrac{f^{(n)}(0)}{n(n-1)(n-2)\cdots 1}$, which can be written more compactly using the factorial notation that was introduced in Section 16.4. Recall that $n!$ is the product of the positive integers less than or equal to n, and that $0!$ is defined to be 1. Thus,

$$a_n = \frac{f^{(n)}(0)}{n!}. \quad \textbf{(27-3)}$$

MacLaurin series

The function f, then, may be represented as a power series in which the coefficient of x^n is a_n, as determined by Equation 27-3. Such a power series is called a *MacLaurin series.* In general, a MacLaurin series converges on an interval centered about 0.

THE MACLAURIN SERIES FOR A FUNCTION

$$f(x) = \frac{f(0)}{0!} + \frac{f'(0)x}{1!} + \frac{f''(0)x^2}{2!} + \frac{f^{(3)}(0)x^3}{3!} + \cdots + \frac{f^{(n)}(0)x^n}{n!} + \cdots$$

provided $f(x)$ and all of its derivatives exist when x is 0 and x is in the interval of convergence.

EXAMPLE 1 Find the MacLaurin series for the sine function. The series converges for all real values of x.

Solution Let $f(x) = \sin x$. We begin by finding the value of the function and its derivatives at $x = 0$.

$$\begin{aligned} f(x) &= \sin x & f(0) &= 0 \\ f'(x) &= \cos x & f'(0) &= 1 \\ f''(x) &= -\sin x & f''(0) &= 0 \\ f^{(3)}(x) &= -\cos x & f^{(3)}(0) &= -1 \end{aligned}$$

At this point, we start to cycle through the same pattern of functions.

$$\begin{aligned} f^{(4)}(x) &= \sin x & f^{(4)}(0) &= 0 \\ f^{(5)}(x) &= \cos x & f^{(5)}(0) &= 1 \\ f^{(6)}(x) &= -\sin x & f^{(6)}(0) &= 0 \\ f^{(7)}(x) &= -\cos x & f^{(7)}(0) &= -1 \end{aligned}$$

Thus,

$$\begin{aligned} \sin x &= \frac{0}{0!} + \frac{1x}{1!} + \frac{0x^2}{2!} - \frac{1x^3}{3!} + \frac{0x^4}{4!} + \frac{1x^5}{5!} + \frac{0x^6}{6!} - \frac{1x^7}{7!} + \cdots \\ &= x - \frac{x^3}{3!} + \frac{x^5}{5!} - \frac{x^7}{7!} + \cdots . \quad \blacksquare \end{aligned}$$

EXAMPLE 2 Find the first four nonzero terms of the MacLaurin series for e^x. The series converges for all real values of x.

Solution Let $f(x) = e^x$.

$$\begin{aligned} f(x) &= e^x & f(0) &= 1 \\ f'(x) &= e^x & f'(0) &= 1 \\ f''(x) &= e^x & f''(0) &= 1 \\ f^{(3)}(x) &= e^x & f^{(3)}(0) &= 1, \text{ and so on.} \end{aligned}$$

Thus,

$$e^x \approx 1 + \frac{x}{1!} + \frac{x^2}{2!} + \frac{x^3}{3!}. \quad \blacksquare$$

EXAMPLE 3 Find the first four nonzero terms of the MacLaurin series for the function defined by $y = (1 + x)^{1/2}$.

Solution

$$f(x) = (1 + x)^{1/2} \qquad f(0) = 1$$
$$f'(x) = (1/2)(1 + x)^{-1/2} \qquad f'(0) = 1/2$$
$$f''(x) = (-1/4)(1 + x)^{-3/2} \qquad f''(0) = -1/4$$
$$f^{(3)}(x) = (3/8)(1 + x)^{-5/2} \qquad f^{(3)}(0) = 3/8$$

Thus,

$$(1 + x)^{1/2} \approx 1 + \frac{(1/2)x}{1!} - \frac{(1/4)x^2}{2!} + \frac{(3/8)x^3}{3!}.$$

Although the series may be written as

$$(1 + x)^{1/2} = 1 + \frac{x}{2} - \frac{x^2}{8} + \frac{x^3}{16} + \cdots,$$

it is a general practice to leave the factorial notation so that the pattern of the coefficients is not obscured. The interval of convergence can be shown to be $-1 \leq x \leq 1$. ■

Notice that the series in Example 3 is the same as that obtained using the binomial expansion.

$$(1 + x)^{1/2} = 1^{1/2} + (1/2)(1^{-1/2})x - (1/8)(1^{-3/2})x^2 + (1/16)(1^{-5/2})x^3 + \cdots.$$

EXERCISES 27.1

In 1–18, find the first three nonzero terms of the MacLaurin series for the given function.

1. $f(x) = \cos x$
2. $f(x) = \cos 2x$
3. $f(x) = \sin 2x$
4. $f(x) = x \sin x$
5. $f(x) = x \cos x$
6. $f(x) = e^{2x}$
7. $f(x) = e^{-x}$
8. $f(x) = xe^{x}$
9. $f(x) = xe^{-2x}$
10. $f(x) = e^{x} \sin x$
11. $f(x) = e^{-x} \cos x$
12. $f(x) = (1 + x)^{1/3}$
13. $f(x) = \dfrac{1}{(1 - x)^2}$
14. $f(x) = \dfrac{1}{\sqrt{1 - x}}$
15. $f(x) = \dfrac{2}{3 - x}$
16. $f(x) = \dfrac{1}{x + 3}$
17. $f(x) = \ln (1 + 4x)$
18. $f(x) = \ln (1 - 2x)$

In 19 and 20, find the first two nonzero terms of the MacLaurin series for each function.

19. $f(x) = \tan x$
20. $f(x) = \sec x$

In 21–24, find the first four (zero and nonzero) terms of the MacLaurin series for each function.

21. $f(x) = \text{Arctan } x$
22. $f(x) = \sin x^2$
23. $f(x) = \cos e^{x}$
24. $f(x) = x^2 \cos x$

27.2 TAYLOR SERIES

When the value of a function is computed using a MacLaurin series, the more terms used, the greater the accuracy of the results. Consider, for example, the first few terms of the MacLaurin series for $y = \sin x$.

TABLE 27.1

n	*n terms of MacLaurin series*
1	$y = x$
2	$y = x - x^3/3!$
3	$y = x - x^3/3! + x^5/5!$

Figure 27.1 shows a graph of $y = \sin x$ and also the graphs of the three functions from Table 27.1. Each additional term produces a graph that more closely matches the curve $y = \sin x$. For values of x near zero, the MacLaurin series produces a fairly close match with very few terms. For values of x much larger than 0 in absolute value, however, a large number of terms of the expansion may be necessary to produce a close match.

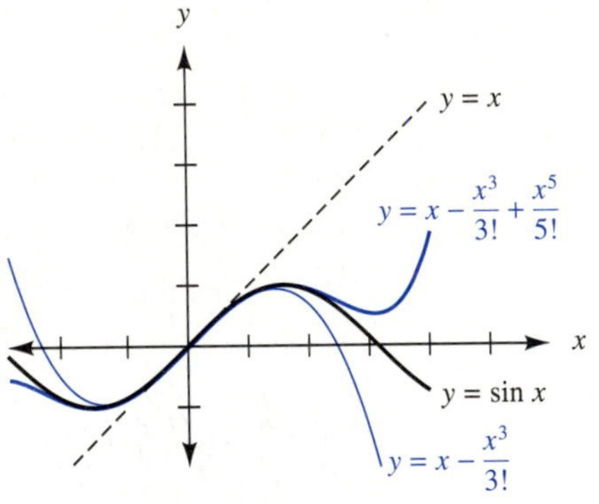

FIGURE 27.1

Taylor series

In general, to evaluate a function for a value of x that is not close to 0, a *Taylor series* is used. The Taylor series expansion is derived in a manner similar to that in which we derived the MacLaurin series expansion. Our goal is to find an expansion of the form

$$f(x) = b_0 + b_1(x - a) + b_2(x - a)^2 + b_3(x - a)^3 + \cdots + b_n(x - a)^n + \cdots \quad \textbf{(27-4)}$$

To find the values of the coefficients, we must first find $f(x)$. Then we find $f'(x)$, $f''(x)$, and so on. We begin by repeating Equation 27-4.

$$f(x) = b_0 + b_1(x - a) + b_2(x - a)^2 + b_3(x - a)^3 + \cdots + b_n(x - a)^n + \cdots \quad \textbf{(27-4)}$$

$$f'(x) = b_1 + 2b_2(x - a) + 3b_3(x - a)^2 + \cdots + nb_n(x - a)^{n-1} + \cdots \quad \textbf{(27-5)}$$

$$f''(x) = 2b_2 + 2 \cdot 3b_3(x - a) + \cdots + (n - 1)nb_n(x - a)^{n-2} + \cdots$$

Next, we evaluate the function and each derivative when x is a. We can then determine the coefficients from the equations that result. From Equation 27-4, we have $f(a) = b_0$. From Equation 27-5, we have $f'(a) = b_1$. Continuing in this manner, we have

$$f''(a) = 2b_2, \text{ so } \frac{f''(a)}{2} = b_2.$$

We can generalize to find the nth coefficient.

THE TAYLOR SERIES FOR A FUNCTION

$$f(x) = f(a) + f'(a)(x-a) + \frac{f''(a)(x-a)^2}{2!} + \cdots + \frac{f^{(n)}(a)(x-a)^n}{n!} + \cdots$$

provided $f(x)$ and all of its derivatives exist when x is a and x is in the interval of convergence.

Notice that the MacLaurin series is the special case of the Taylor series in which a is 0.

EXAMPLE 1 Find the Taylor series for $\sin x$ with $a = \pi$. The series converges for all real values of x.

Solution

$$\begin{aligned} f(x) &= \sin x & f(\pi) &= 0 \\ f'(x) &= \cos x & f'(\pi) &= -1 \\ f''(x) &= -\sin x & f''(\pi) &= 0 \\ f^{(3)}(x) &= -\cos x & f^{(3)}(\pi) &= 1 \end{aligned}$$

Thus,

$$\begin{aligned} \sin x &= 0 - (x-\pi) + \frac{0(x-\pi)^2}{2!} + \frac{1(x-\pi)^3}{3!} + \cdots \\ &= -(x-\pi) + \frac{(x-\pi)^3}{3!} - \frac{(x-\pi)^5}{5!} + \cdots . \end{aligned}$$

■

EXAMPLE 2 Find the first four nonzero terms of the Taylor series that would be used to evaluate e^x near $x = 4$.

Solution To evaluate e^x near $x = 4$, we should write the Taylor series for e^x in powers of $(x-4)$. That is, $a = 4$.

$$\begin{aligned} f(x) &= e^x & f(4) &= e^4 \\ f'(x) &= e^x & f'(4) &= e^4 \\ f''(x) &= e^x & f''(4) &= e^4 \\ f^{(3)}(x) &= e^x & f^{(3)}(4) &= e^4 \end{aligned}$$

$$e^x \approx e^4 + e^4(x-4) + \frac{e^4(x-4)^2}{2!} + \frac{e^4(x-4)^3}{3!}$$

■

EXAMPLE 3 Find a series expansion for $f(x) = \ln x$ for values of x that are small in absolute value.

Solution If we let $a = 0$, it would appear that we need a MacLaurin series. But neither the function nor its derivatives are defined at $x = 0$. Thus, we choose to use a Taylor series in which a is 0.5.

$$\begin{aligned} f(x) &= \ln x & f(0.5) &= -0.693 \\ f'(x) &= x^{-1} & f'(0.5) &= 2 \\ f''(x) &= -x^{-2} & f''(0.5) &= -4 \\ f^{(3)} &= 2x^{-3} & f^{(3)}(0.5) &= 16 \\ f^{(4)} &= -6x^{-4} & f^{(4)}(0.5) &= -96 \end{aligned}$$

$$\ln x = -0.693 + 2(x - 0.5) - \frac{4(x-0.5)^2}{2!} + \frac{16(x-0.5)^3}{3!} - \frac{96(x-0.5)^4}{4!} + \cdots$$

■

NOTE Both Taylor series and MacLaurin series are based on derivatives, and our formulas for differentiation are based on radian measure. ***When a Taylor (or MacLaurin) series is used to evaluate a trigonometric function for a particular value, it is therefore important to use radian measure.*** When a function value can be computed using either a MacLaurin series or a Taylor series, one series will usually give a more accurate result than the other using the same number of terms.

EXAMPLE 4 Calculate $\sin(3\pi/4)$

(a) using a calculator.

(b) using the first three nonzero terms of the MacLaurin series.

(c) using the first three nonzero terms of the Taylor series with $a = \pi$.

Solution **(a)** $\sin(3\pi/4) = 0.7071068$

(b) In Example 1 of the previous section, we found the MacLaurin series for $\sin x$.

$$\sin x = x - \frac{x^3}{3!} + \frac{x^5}{5!}$$

$$\begin{aligned} \sin(3\pi/4) &= (3\pi/4) - \frac{(3\pi/4)^3}{3!} + \frac{(3\pi/4)^5}{5!} \\ &= (3\pi/4) - \frac{27\pi^3/64}{6} + \frac{243\pi^5/1024}{120} \\ &= (3\pi/4) - \frac{27\pi^3}{6(64)} + \frac{243\pi^5}{120(1024)} \\ &= 0.7812305 \end{aligned}$$

(c) In Example 1 of this section, we found the Taylor series for $\sin x$ with $a = \pi$.

$$\sin x = -(x-\pi) + \frac{(x-\pi)^3}{3!} - \frac{(x-\pi)^5}{5!}$$

$$\begin{aligned}\sin(3\pi/4) &= -(3\pi/4-\pi) + \frac{(3\pi/4-\pi)^3}{3!} - \frac{(3\pi/4-\pi)^5}{5!}\\ &= -(-\pi/4) + \frac{(-\pi/4)^3}{3!} - \frac{(-\pi/4)^5}{5!}\\ &= \pi/4 - \frac{\pi^3/64}{3!} + \frac{\pi^5/1024}{5!}\\ &= \pi/4 - \frac{\pi^3}{64(6)} + \frac{\pi^5}{1024(120)}\\ &= 0.7071431 \quad \blacksquare\end{aligned}$$

Notice that the Taylor series ($a = \pi$) gives a value that is closer to the value given by the calculator than the value given by the MacLaurin series ($a = 0$). The Taylor series approximation is closer because π is closer to $3\pi/4$ than 0 is.

EXAMPLE 5 A silicon diode has a characteristic curve of current I (in amperes) as a function of voltage v (in volts) given by the equation $I = 10^{-13}(e^{39.6v} - 1)$. Find a linear equation for this function that can be used to approximate the function for values of v near 0.6.

Solution Since I is a function of v, we use the notation $I = I(v)$.

$$\begin{aligned}I(v) &= 10^{-13}(e^{39.6v}) - 10^{-13} & I(0.6) &= 10^{-13}(e^{23.76}) - 10^{-13} = 0.00208\\ I'(v) &= 10^{-13}(39.6)(e^{39.6v}) & I'(0.6) &= 10^{-13}(39.6)(e^{23.76}) = 0.0825\\ I''(v) &= 10^{-13}(39.6)^2(e^{39.6v}) & I''(0.6) &= 10^{-13}(39.6)^2(e^{23.76}) = 3.27\\ I^{(3)}(v) &= 10^{-13}(39.6)^3(e^{39.6v}) & I^{(3)}(0.6) &= 10^{-13}(39.6)^3(e^{23.76}) = 129\end{aligned}$$

The Taylor series is given by

$$\begin{aligned}I &= I(0.6) + I'(0.6)(v-0.6) + \frac{I''(0.6)(v-0.6)^2}{2!} + \frac{I^{(3)}(0.6)(v-0.6)^3}{3!} + \cdots\\ &= 0.00208 + 0.0825(v-0.6) + \frac{3.27(v-0.6)^2}{2!} + \frac{129(v-0.6)^3}{3!} + \cdots .\end{aligned}$$

The nonlinear terms are small when v is close to 0.6, so the first two terms give a fairly good linear approximation.

$$\begin{aligned}I &= 0.00208 + 0.0825(v-0.6)\\ &= 0.00208 + 0.0825v - 0.0495\\ &= 0.0825v - 0.0474 \quad \blacksquare\end{aligned}$$

EXERCISES 27.2

In 1–12, find the first three nonzero terms of the Taylor series for the given function and value of a.

1. $f(x) = \cos x\ (a = \pi)$
2. $f(x) = \cos 3x\ (a = \pi/2)$
3. $f(x) = \sin 2x\ (a = \pi/4)$
4. $f(x) = e^{-x}\ (a = 1)$
5. $f(x) = e^{2x}\ (a = 3)$
6. $f(x) = e^{x+1}\ (a = 2)$
7. $f(x) = \sqrt{x}\ (a = 4)$
8. $f(x) = \sqrt[3]{x}\ (a = 1)$
9. $f(x) = 1/x\ (a = 3)$
10. $f(x) = \ln(1 + x)\ (a = 3)$
11. $f(x) = \ln(1 - 2x)\ (a = -2)$
12. $f(x) = \ln x\ (a = 4)$

In 13–22, calculate each function value and show the result to seven significant digits
(a) using a calculator.
(b) using the first two nonzero terms of a MacLaurin series, if possible.
(c) using the first two nonzero terms of a Taylor series with the given value of a.

13. $\tan \pi/12$, $a = \pi/6$ (See problem 19 of Exercises 27.1.)

14. $\tan \pi/8$, $a = \pi/4$ (See problem 19 of Exercises 27.1.)

15. $e^{2.8}$, $a = 3$ (See Example 2 of Section 27.1.)

16. $\sqrt{17}$, $a = 16$
17. $\sqrt{27}$, $a = 25$
18. $\sqrt{8}$, $a = 9$
19. $\sqrt[3]{9}$, $a = 8$
20. $\sqrt[3]{2}$, $a = 1$
21. $\sqrt[3]{25}$, $a = 27$
22. $\ln 3$, $a = e$

---

23. A cable that is attached at both ends but bears no weight (other than its own) falls naturally into the shape called a catenary. The equation for a catenary is $y = (1/2)c(e^{x/c} + e^{-x/c})$ where c is a constant. Find the MacLaurin series for y.
2 4 8 | 1 5 6 7 9

24. The equation of the normal curve used in statistics is $y = (1/\sqrt{2\pi})e^{-x^2/2}$. Find the first two nonzero terms of the MacLaurin series for y.
1 2 4 7 8 | 3 5

25. An important factor found in the equation for parallel resonance is $\alpha = \sqrt{1 - R^2C/L}$. Show that when R is very small, this equation may be approximated as $\alpha = 1 - (1/2)R^2C/L$.
3 5 9 | 4 7 8

26. The equation $i = 0.005(1 + v/5000)^{3/2}$ gives the current i (in amperes) flowing from the cathode to the anode of a certain cathode ray tube as a function of voltage (in volts). Find a linear equation for this function that can be used to approximate the function for values of v near 5000.
3 5 9 | 1 2 4 6 7 8

27. The velocity of an object is given by $v = \sqrt{1 + t^2}$, where t is the time of travel. Use the first two nonzero terms of the MacLaurin series for v to find a formula that can be used to approximate velocity.
4 7 8 | 1 2 3 5 6 9

28. Explain why it is necessary to use a Taylor series (instead of a MacLaurin series) with the following functions.
(a) $y = \ln x$
(b) $y = \csc x$
(c) $y = x^{-1}$
(d) $y = \sqrt{x}$

1. Architectural Technology
2. Civil Engineering Technology
3. Computer Engineering Technology
4. Mechanical Drafting Technology
5. Electrical/Electronics Engineering Technology
6. Chemical Engineering Technology
7. Industrial Engineering Technology
8. Mechanical Engineering Technology
9. Biomedical Engineering Technology

27.3 OPERATIONS WITH SERIES

In Section 27.1, we found the MacLaurin series directly for each function. For example, to find sin $3x$, we evaluated sin $3x$ and its derivatives when x is 0.

$$\begin{aligned} f(x) &= \sin 3x & f(0) &= 0 \\ f'(x) &= 3 \cos 3x & f'(0) &= 3 \\ f''(x) &= -9 \sin 3x & f''(0) &= 0 \\ f^{(3)}(x) &= -27 \cos 3x & f^{(3)}(0) &= -27 \\ f^{(4)}(x) &= 81 \sin 3x & f^{(4)}(0) &= 0 \\ f^{(5)}(x) &= 243 \cos 3x & f^{(5)}(0) &= 243 \end{aligned}$$

Thus,

$$\sin 3x = 0 + 3x + \frac{0x^2}{2!} - \frac{27x^3}{3!} + \frac{0x^4}{4!} + \frac{243x^5}{5!} + \cdots . \qquad \textbf{(27-6)}$$

Notice, however, that we may treat $y = \sin 3x$ as a composition of functions. That is, if $f(x) = \sin x$ and $g(x) = 3x$, then

$$y = f \circ g(x) = f[g(x)] = \sin (3x).$$

Example 1, which follows, shows how to use the composition of functions to determine the MacLaurin series.

EXAMPLE 1 Find the MacLaurin series for $y = \sin 3x$ by treating y as a composition of functions.

Solution We know from Example 1 of Section 27.1 that

$$\sin x = x - \frac{x^3}{3!} + \frac{x^5}{5!} - \cdots .$$

We may replace x with $3x$ in this equation.

$$\sin 3x = 3x - \frac{(3x)^3}{3!} + \frac{(3x)^5}{5!} - \cdots$$

But this result agrees precisely with Equation 27-6. ■

The MacLaurin series for the sine, cosine, exponential, and logarithmic functions are particularly important. Each of these four functions may be treated as a composition of functions. The MacLaurin series for each of these series was developed in Section 27.1, either in the examples or in the exercises. They are repeated here for reference.

MACLAURIN SERIES FOR COMMON FUNCTIONS

$$e^x = 1 + x + \frac{x^2}{2!} + \frac{x^3}{3!} + \cdots \quad \text{for all real } x \qquad \textbf{(27-7)}$$

$$\sin x = x - \frac{x^3}{3!} + \frac{x^5}{5!} - \cdots \quad \text{for all real } x \qquad \textbf{(27-8)}$$

$$\cos x = 1 - \frac{x^2}{2!} + \frac{x^4}{4!} - \cdots \quad \text{for all real } x \qquad \textbf{(27-9)}$$

$$\ln (1 + x) = x - \frac{x^2}{2} + \frac{x^3}{3} - \frac{x^4}{4} + \cdots \quad \text{for } |x| < 1 \qquad \textbf{(27-10)}$$

EXAMPLE 2 Find the MacLaurin series for $y = e^{x^2}$.

Solution We replace x with x^2 in the MacLaurin series for e^x (Equation 27-7).

$$e^x = 1 + x + \frac{x^2}{2!} + \frac{x^3}{3!} + \cdots \qquad \textbf{(27-7)}$$

$$e^{x^2} = 1 + x^2 + \frac{(x^2)^2}{2!} + \frac{(x^2)^3}{3!} + \frac{(x^2)^4}{4!} \cdots$$

$$e^{x^2} = 1 + x^2 + \frac{x^4}{2!} + \frac{x^6}{3!} + \frac{x^8}{4!} \cdots \quad ■$$

When a Taylor series is desired, it is generally better to find the series directly, as illustrated in Example 3.

EXAMPLE 3 Find the Taylor series for $\cos x$ near π.

Solution

$$\begin{aligned} f(x) &= \cos x & f(\pi) &= -1 \\ f'(x) &= -\sin x & f'(\pi) &= 0 \\ f''(x) &= -\cos x & f''(\pi) &= 1 \end{aligned}$$

Thus,

$$\cos x = -1 + 0(x - \pi) + \frac{(x-\pi)^2}{2!} + \cdots . \quad ■$$

NOTE ⇨⇨ ***It is important to recognize that the series obtained in Example 3 is not the same series that results from replacing x with $(x - \pi)$ in the Maclaurin series for cos x (Equation 27-9).*** Such a substitution does produce a series in powers of $(x - \pi)$, but it is a series for $\cos (x - \pi)$, not a series for $\cos x$.

$$\cos (x - \pi) = 1 - \frac{(x-\pi)^2}{2!} + \frac{(x-\pi)^4}{4!} - \cdots$$

We have observed that a polynomial shares some characteristics with a power series. The operations of addition, subtraction, multiplication, and division are valid for power series, just as they are for polynomials. Power series are added or subtracted by combining like terms. Multiplication is performed by repeated application of the distributive property, using a finite number of terms of each series. Division, too, is performed like polynomial division.

EXAMPLE 4 Find a MacLaurin series for $\cos\theta + j\sin\theta$.

Solution

$$\cos\theta = 1 - \frac{\theta^2}{2!} + \frac{\theta^4}{4!} - \frac{\theta^6}{6!} \cdots \text{ and}$$

$$\sin\theta = \theta - \frac{\theta^3}{3!} + \frac{\theta^5}{5!} - \frac{\theta^7}{7!} \cdots$$

Multiplying the series for $\sin\theta$ by j, we have

$$j\sin\theta = j\theta - \frac{j\theta^3}{3!} + \frac{j\theta^5}{5!} - \frac{j\theta^7}{7!} \cdots .$$

$$\cos\theta + j\sin\theta = 1 + j\theta - \frac{\theta^2}{2!} - \frac{j\theta^3}{3!} + \frac{\theta^4}{4!} + \frac{j\theta^5}{5!} - \frac{\theta^6}{6!} \cdots \tag{27-11}$$

■

Recall from Section 14.3 that we accepted the equation

$$e^{j\theta} = \cos\theta + j\sin\theta$$

without justification. But now it is possible to see why this equation holds. Consider the MacLaurin series for $e^{j\theta}$. From Equation 27-7, we obtain

$$e^{j\theta} = 1 + j\theta + \frac{(j\theta)^2}{2!} + \frac{(j\theta)^3}{3!} + \frac{(j\theta)^4}{4!} + \cdots .$$

Replacing each power of j with j, $-j$, 1, or -1, we have

$$e^{j\theta} = 1 + j\theta - \frac{\theta^2}{2!} - \frac{j\theta^3}{3!} + \frac{\theta^4}{4!} + \cdots . \tag{27-12}$$

It is clear from Equations 27-11 and 27-12 that $e^{j\theta} = \cos\theta + j\sin\theta$.

Our next example illustrates the multiplication of a finite number of terms of two series to obtain the initial terms of the series that is the product.

EXAMPLE 5 Find a Taylor series for $\dfrac{e^x}{1+x}$ near 2, using three terms of the series for e^x and three terms of the series for $\dfrac{1}{1+x}$.

Solution First, we find the series for e^x.

$$f(x) = e^x \qquad f(2) = e^2$$
$$f'(x) = e^x \qquad f'(2) = e^2$$
$$f''(x) = e^x \qquad f''(2) = e^2$$

Thus,

$$e^x = e^2 + e^2(x-2) + \frac{e^2(x-2)^2}{2!} + \cdots .$$

Next, we find the series for $\dfrac{1}{1+x}$.

$$f(x) = (1+x)^{-1} \qquad f(2) = 1/3$$
$$f'(x) = -1(1+x)^{-2} \qquad f'(2) = -1/9$$
$$f''(x) = 2(1+x)^{-3} \qquad f''(2) = 2/27$$

Thus,

$$\frac{1}{1+x} = \frac{1}{3} - \frac{1(x-2)}{9} + \frac{2(x-2)^2}{27(2!)} + \cdots .$$

The product is given by

$$\begin{aligned}
\frac{e^x}{1+x} &= \left[e^2 + e^2(x-2) + \frac{e^2(x-2)^2}{2!}\right]\left[\frac{1}{3} - \frac{1(x-2)}{9} + \frac{2(x-2)^2}{27(2!)}\right] \\
&= \frac{e^2}{3} - \frac{e^2(x-2)}{9} + \frac{2e^2(x-2)^2}{27(2!)} \\
&\quad + \frac{e^2(x-2)}{3} - \frac{e^2(x-2)^2}{9} + \frac{2e^2(x-2)^3}{27(2!)} \\
&\quad + \frac{e^2(x-2)^2}{3\cdot 2!} - \frac{e^2(x-2)^3}{9\cdot 2!} + \frac{2e^2(x-2)^4}{27\cdot 2!\cdot 2!} \\
&= \frac{e^2}{3} + \frac{2e^2(x-2)}{9} + \frac{5e^2(x-2)^2}{54} - \frac{e^2(x-2)^3}{54} + \frac{e^2(x-2)^4}{54}. \quad ■
\end{aligned}$$

When a function cannot be integrated directly, it may be possible to write the function as a series and integrate term by term. Although numerical techniques, such as the trapezoidal rule or Simpson's rule can be used to approximate definite integrals, integration of a series allows us to approximate an indefinite integral.

EXAMPLE 6 Find $\displaystyle\int \frac{\ln(1+x)}{x}\,dx$.

Solution The MacLaurin series for $\ln(1+x)$ is one of the basic series and is given in Equation 27-10.

$$\ln(1+x) = x - \frac{x^2}{2} + \frac{x^3}{3} - \frac{x^4}{4} + \cdots$$

If each term of this series is divided by x, the series for $\dfrac{\ln(1+x)}{x}$ is obtained. Thus

$$\int \frac{\ln(1+x)}{x}\,dx = \int \left(1 - \frac{x}{2} + \frac{x^2}{3} - \frac{x^3}{4} + \cdots\right) dx$$
$$= x - \frac{x^2}{4} + \frac{x^3}{9} - \frac{x^4}{16} + \cdots\,. \quad ■$$

EXAMPLE 7 The equation $p = e^{-400t^2}$ gives the power p (in watts) of a burst of light emitted by a certain laser, if t is the duration of the emission (in seconds). The energy W (in joules) is given by $W = \int p\,dt$. Find the approximate amount of energy delivered during the 30 milliseconds immediately following the laser's peak power output. The peak output occurs when t is 0.

Solution Since $p = e^{-400t^2}$, we have $W = \int_0^{0.030} e^{-400t^2}\,dt$. The integral, however, is not one of the forms we have studied. Using four terms of a power series, we have

$$e^{-400t^2} = 1 + (-400t^2) + \frac{(-400t^2)^2}{2!} + \frac{(-400t^2)^3}{3!}$$
$$= 1 - 400t^2 + 80\,000\,t^4 - 10\,666\,667\,t^6.$$

Now we may perform the integration.

$$W = \int_0^{0.030} e^{-400t^2}\,dt$$
$$W \approx \int_0^{0.030} (1 - 400t^2 + 80\,000\,t^4 - 10\,666\,667\,t^6)\,dt$$
$$\approx t - \frac{400t^3}{3} + \frac{80\,000\,t^5}{5} - \frac{10\,666\,667t^7}{7}\Bigg|_0^{0.030}$$
$$\approx 0.268 \text{ joules} \quad ■$$

EXERCISES 27.3

In 1–8, find the MacLaurin series for each function using Equations 27-7 through 27-10.

1. $f(x) = e^{x^2}$
2. $f(x) = e^{-2x}$
3. $f(x) = \sin\sqrt{x}$
4. $f(x) = \sin x^2$
5. $f(x) = \cos 2x$
6. $f(x) = \cos(x/2)$
7. $f(x) = \ln(1-x)$
8. $f(x) = \ln(1+2x)$

9. Find the first four nonzero terms of the MacLaurin series for $f(x) = (1/2)(e^x - e^{-x})$. (This function is called the hyperbolic sine function and has applications in engineering.)

10. Find the first four nonzero terms of the MacLaurin series for $f(x) = (1/2)(e^x + e^{-x})$. (This function is called the hyperbolic cosine function and has applications in engineering.)

11. Find the first four nonzero terms of the MacLaurin series for $f(x) = e^x \sin x^2$.

12. Find the first three nonzero terms of the Taylor series for $f(x) = \dfrac{\sin x}{1+x}$ near $\pi/2$.

13. Find the first three nonzero terms of the Taylor series for $f(x) = \dfrac{\cos x}{1-x}$ near $\pi/4$.

14. Find the first three nonzero terms of the Taylor series for $f(x) = \sin x \cos x$ near $\pi/2$.

15. Find the first three nonzero terms of the Taylor series for $f(x) = e^x \ln(1+x)$ near 2.

16. Find the first three nonzero terms of the Taylor series for $f(x) = \dfrac{\ln(1+x)}{1+x}$ near 3.

In 17–24, express each integral as a series.

17. $\displaystyle\int \frac{\sin x}{x}\,dx$ **18.** $\displaystyle\int x \cos x\,dx$ **19.** $\displaystyle\int \cos x^2\,dx$

20. $\displaystyle\int \sin\sqrt{x}\,dx$ **21.** $\displaystyle\int \frac{e^x}{x^2}\,dx$ **22.** $\displaystyle\int e^{x^2}\,dx$

23. $\displaystyle\int \ln(1+x^2)\,dx$ **24.** $\displaystyle\int \ln(1+\sqrt{x})\,dx$

---

25. The formula $x = \dfrac{v_0^2}{g}\sin 2\theta$ gives the range x (in m) of a projectile launched at an angle θ with an initial velocity of v_0 (in m/s) when g is the acceleration due to gravity. Use the first three nonzero terms of a MacLaurin series to find x if $v_0 = 300.0$ m/s, $g = 9.8$ m/s^2, and $\theta = 10.0°$. (Hint: Change θ to radian measure. Why?) **4 7 8** 1 2 3 5 6 9

26. The formula $t = 1.443T\ln(1 + 8.33\,R)$ gives the age t (in years) of a sample of rock. In the formula, T is the half-life of potassium 40 (1.26×10^9 years), and R is the ratio of the amount of Argon 40 present to the amount of potassium 40 present. Use the first three nonzero terms of a MacLaurin series to find an approximate value for t if $R = 0.100$. **1 2 4 6 8** 5 7 9

27. The formula $i = \dfrac{1}{L}\displaystyle\int v\,dt$ gives the current i (in amperes) in an ac circuit when v is the voltage (in volts). Use the first three nonzero terms of the MacLaurin series to find a formula for the current in a 0.4-mH inductor if $v = 10^{-4}e^{0.3t}t$ and $i = 0$ when $t = 0$. **3 5 9** 1 2 4 6 7 8

28. The formula $a = -A\omega^2\cos(\omega t - \theta)$ gives the vertical acceleration a of a mass hanging from a spring subjected to a certain periodic force. In the formula, $\omega = 2\pi f$ where f is the frequency of the applied force, θ is the angle by which the displacement lags behind the force, and A is the amplitude of the oscillations. Find the first four nonzero terms of the MacLaurin series in powers of $(\omega t - \theta)$. **4 7 8** 1 2 3 5 6 9

---

1. Architectural Technology
2. Civil Engineering Technology
3. Computer Engineering Technology
4. Mechanical Drafting Technology
5. Electrical/Electronics Engineering Technology
6. Chemical Engineering Technology
7. Industrial Engineering Technology
8. Mechanical Engineering Technology
9. Biomedical Engineering Technology

29. The equation for the normal curve used in statistics is $y = (1/\sqrt{2\pi})e^{-x^2/2}$. Use the first four nonzero terms of a MacLaurin series to find the average value of this function between 0 and 1.
1 2 4 7 8 3 5

30. The formula $Q = Q_0 e^{-t/\tau}$ gives the charge Q (in coulombs) on a capacitor after it has been discharging through a resistance for time t (in seconds). In the formula, Q_0 is the original charge (in coulombs), and $\tau = RC$ (reistance × capacitance). Use the first three nonzero terms of a MacLaurin series to find the average value of Q for values of t between 0 and 0.010 if $Q_0 = 3.6 \times 10^{-5}$ coulombs and $\tau = 0.0020$ seconds.
3 5 9 4 7 8

31. **(a)** If the MacLaurin series for $\cos x$ is integrated term by term, is the result equal to the series for $\sin x$?
(b) If the MacLaurin series for $\sin x$ is integrated term by term, is the result equal to the series for $-\cos x$?
(c) If the answer is "no" in (a) or (b), explain why.

27.4 COMPUTATIONS WITH SERIES

One of the early applications of calculus was the calculation of values for functions such as the trigonometric, exponential, and logarithmic functions. Even today computers and calculators do not store tables of values, but rather compute values when the function is called upon. Although the procedures are not identical to the ones used here, the underlying principle is the same. You know that a Taylor series in powers of $(x - a)$ is used to evaluate functions when x is close to a. (When a is 0, the series is a MacLaurin series.)

When a finite number of terms are used, the function value is approximate, but by using sufficiently many terms of the series, it is possible to obtain any degree of accuracy desired. For practical applications, it is necessary to answer the following two questions.

1. If a predetermined number of terms of a series are used, how accurate is the approximation?
2. How many terms of a series should be used to attain a given degree of accuracy?

alternating series

We will consider only the first question. The simplest case to consider is a series in which the terms alternate in sign. Such a series is called an **alternating series.** A formula for the error estimate is given as follows.

ERROR IN AN ALTERNATING TAYLOR SERIES

If a function value $f(x)$ is approximated by a finite number of terms of an alternating Taylor series, then the absolute value of the error R is less than the absolute value of the first term omitted.

EXAMPLE 1 Find the approximate value of sin 5° using two nonzero terms of a MacLaurin series and estimate the error.

Solution The first step is to change 5° to $\pi/36$ radians. Then

$$\sin x = x - \frac{x^3}{3!}$$

$$\sin\left(\frac{\pi}{36}\right) = \left(\frac{\pi}{36}\right) - \frac{\left(\frac{\pi}{36}\right)^3}{6}$$

$$= 0.871557, \text{ when evaluated with a calculator.}$$

To determine the accuracy of the result, notice that the power series for $\sin x$ is an alternating series. The first term omitted in the series was the fifth power term. The error R satisfies the following condition.

$$|R| \leq \left|\frac{1}{5!}\left(\frac{\pi}{36}\right)^5\right| = 0.000000042 \quad ■$$

To translate the error into the language of decimal places, we use the following definition.

DEFINITION An approximation is said to be accurate to at least k decimal places if $|R| < 0.5 \times 10^{-k}$, where R is the error.

Since $0.000000042 \leq 0.00000005 = 0.5 \times 10^{-7}$, the value in Example 1 would be accurate to at least seven decimal places, if we had carried it that far. Since we gave the answer to only six places, it is accurate to the number of places shown. You should compare this value with the value of sin 5° on your calculator.

Example 1 actually involves two sources of error in the approximation. The first type of error, called truncation error, is the error that results from using a finite number of terms of the series to approximate the infinite series. The second source of error is the round-off error introduced when the decimal approximation for $\pi/36$ is used in the calculation.

Error approximation is a complex subject, which is dealt with extensively by the branch of mathematics called numerical analysis.

EXAMPLE 2 Suppose that a computer program is written to find an approximate value of cos 92° using two nonzero terms of a Taylor series. Determine the accuracy of the result.

Solution Since 92° is close to 90°, the Taylor series is given in powers of $(x - \pi/2)$, and $x = 92°(\pi/180°)$ or $23\pi/45$.

$$\begin{aligned} f(x) &= \cos x & f(\pi/2) &= 0 \\ f'(x) &= -\sin x & f'(\pi/2) &= -1 \\ f''(x) &= -\cos x & f''(\pi/2) &= 0 \\ f^{(3)}(x) &= \sin x & f^{(3)}(\pi/2) &= 1 \end{aligned}$$

$$\cos x = 0 - (x - \pi/2) + \frac{0(x - \pi/2)^2}{2!} + \frac{1(x - \pi/2)^3}{3!} + \cdots$$

$$= -(x - \pi/2) + \frac{(x - \pi/2)^3}{3!}$$

Thus,

$$\cos(23\pi/45) = -(23\pi/45 - \pi/2) + \frac{(23\pi/45 - \pi/2)^3}{6}$$

$$= -0.0348995.$$

Once again, we have an alternating series.

$$|R| \le \left| \frac{1}{5!}\left(\frac{23\pi}{45} - \frac{\pi}{2}\right)^5 \right| = 4.3187 \times 10^{-10} \le 5 \times 10^{-10} = 0.5 \times 10^{-9}$$

The result would be accurate to at least nine decimal places, if we had carried it that far. ■

EXAMPLE 3 Find the approximate value of ln 1.3 using four nonzero terms of a MacLaurin series and determine the accuracy of the result.

Solution

$$\ln(1 + x) = x - \frac{x^2}{2} + \frac{x^3}{3} - \frac{x^4}{4} + \cdots$$

$$\ln(1 + 0.3) = 0.3 - \frac{(0.3)^2}{2} + \frac{(0.3)^3}{3} - \frac{(0.3)^4}{4} + \cdots$$

$$= 0.261975$$

The series is an alternating series. The next term would be $(0.3)^5/5$.

$$|R| \le |(0.3)^5/5| = 0.000486 \le 0.0005 = 0.5 \times 10^{-3}$$

The result is accurate to at least three decimal places. To three decimal places, the result is 0.262. A calculator gives 0.2623643, which is also 0.262 to three decimal places. ■

Up to this point, you have accepted the values of e and π. It is possible to compute these constants to any desired degree of accuracy using a series. The value for e is found here, and you will find the value of π in the exercises.

EXAMPLE 4 Compute the value of e using five nonzero terms of a MacLaurin series.

Solution From Section 27.3, we know that

$$e^x = 1 + x + \frac{x^2}{2!} + \frac{x^3}{3!} + \frac{x^4}{4!} + \cdots .$$

Thus,

$$e^1 = 1 + 1 + \frac{1^2}{2} + \frac{1^3}{6} + \frac{1^4}{24} + \cdots$$
$$= 2.7083333.$$

The series is not an alternating series, so we cannot evaluate the error by considering the first term omitted. Although there are ways to estimate the error in this type of series, we will not consider them here. ■

EXERCISES 27.4

In 1–10, find an approximate value of each function for the given value of x using the specified number of nonzero terms of a MacLaurin series. Determine the accuracy of the result for each alternating series.

1. $\sin x$ ($x = 1°$, 2 terms)
2. $\sin x$ ($x = 2°$, 2 terms)
3. $\cos x$ ($x = 3°$, 2 terms)
4. $\cos x$ ($x = 4°$, 2 terms)
5. e^x ($x = 0.5$, 3 terms)
6. e^x ($x = 1.5$, 3 terms)
7. e^x ($x = -1$, 3 terms)
8. $\ln(1 + x)$ ($x = 0.1$, 5 terms)
9. $\ln(1 + x)$ ($x = 0.9$, 5 terms)
10. $\ln(1 + x)$ ($x = 0.2$, 4 terms)

In 11–16, find an approximate value of each function for the given x using the specified number of nonzero terms of a Taylor series in powers of $(x - a)$. Compare your answer to that obtained on a calculator. Notice that when x is close to the given value of a, the results agree closely.

11. $\sin x$ ($x = 32°$, 2 terms with $a = \pi/6$)
12. $\cos x$ ($x = 4°$, 2 terms with $a = \pi/4$)
13. e^x ($x = 3.1$, 3 terms with $a = 3$)
14. e^x ($x = 3.3$, 3 terms with $a = 3$)
15. $\ln(x + 1)$ ($x = 7.3$, 5 terms with $a = 7$)
16. $\ln(x + 1)$ ($x = 7.5$, 5 terms with $a = 7$)

---

17. A computer must be programmed to compute the value of π. Find the approximate value of π using three nonzero terms of a MacLaurin series for Arctan x, and the fact that Arctan $1 = \pi/4$. (Three terms of this series will not give a very good approximation.) 3 5 9

---

1. Architectural Technology
2. Civil Engineering Technology
3. Computer Engineering Technology
4. Mechanical Drafting Technology
5. Electrical/Electronics Engineering Technology
6. Chemical Engineering Technology
7. Industrial Engineering Technology
8. Mechanical Engineering Technology
9. Biomedical Engineering Technology

18. It is possible to program a computer to obtain a better approximation for π with fewer terms than are required using the series of problem 17.
   (a) Given that $\tan(x+y) = \dfrac{\tan x + \tan y}{1 - \tan x \tan y}$, show that $\pi/4 = \text{Arctan}(1/2) + \text{Arctan}(1/3)$. (Hint: Let $x = \text{Arctan } 1/2$ and let $y = \text{Arctan } 1/3$. Then $\tan x = 1/2$ and $\tan y = 1/3$.)
   (b) Compute the value of π using three terms of a MacLaurin series for y. 3 5 9

19. The equation $y = 0.5(e^{0.05x} + e^{-0.05x}) + 24$ gives the height y (in meters) above the ground of a certain long wire shortwave antenna hanging between two equally high supports, if x is the distance (in meters) along the ground from the center of the sag to the point of interest.
   (a) Find two nonzero terms of a series expansion for the antenna's height near the bottom of the sag.
   (b) Use the series to compute the height when x is 0.5 meter. 2 4 8 1 5 6 7 9

20. When a capacitor and resistor are connected in series, the voltage V_c (in volts) across the capacitor is given by $V_c = V(1 - e^{-t/\tau})$. In the formula, V is the voltage of the power source, t is the time in seconds, and τ is the product of resistance (in ohms) and capacitance (in farads). Find three terms of a series expansion for V_c when $R = 1$ megohm, $C = 1$ microfarad, and $V = 24$ volts and use it to compute V_c when t is 100 milliseconds. 3 5 9 4 7 8

21. If the value of $f(x)$ is to be accurate to n decimal places, then it is necessary to make the error R less than 0.5×10^{-n} in absolute value. The error R in a Taylor series in powers of $(x - a)$ is given by $|R| \le \dfrac{M|x-a|^{m+1}}{(m+1)!}$, where M is the maximum value of the function. and $m + 1$ is the number of terms used. For a given problem, if M and a are known, the value of n can be determined by trial and error. A computer must be programmed to compute $\cos x$ for values of x near 0 using a series. How many terms of a MacLaurin series are necessary to produce $\cos 10°$ to six decimal places?

27.5 FOURIER SERIES

We have seen that when a function is approximated by a finite number of terms of a MacLaurin series or a Taylor series, the approximation is a polynomial function. The graphs of such functions are continuous. In physics and engineering, there are many phenomena that are represented by graphs that are made up of disjointed curves or straight line segments, as illustrated by the graph in Figure 27.2.

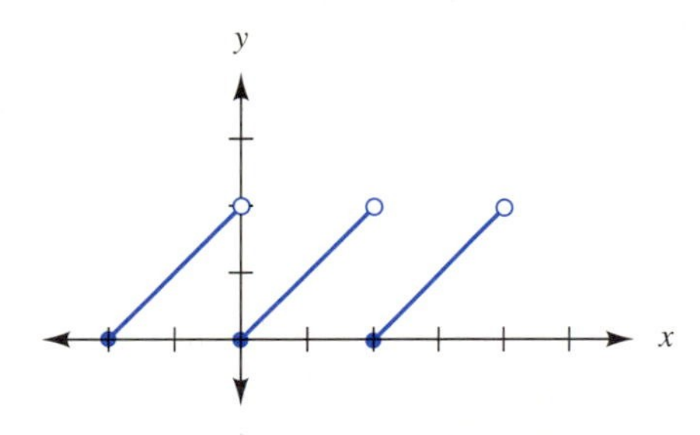

FIGURE 27.2

Because such functions are often periodic, a series based on periodic functions usually provides a more effective approximation than a power series does. The sine and cosine functions, of course, are examples of periodic functions. Most other

Fourier series

functions can be written as a series of terms involving sines and cosines. We will study the Fourier series expansion. A *Fourier series* is a representation of a function in the form

$$f(x) = a_0 + a_1 \cos x + a_2 \cos 2x + a_3 \cos 3x + \cdots + a_n \cos nx + \cdots$$
$$+ b_1 \sin x + b_2 \sin 2x + b_3 \sin 3x + \cdots + b_n \sin nx + \cdots .$$

It should be noted that the Fourier series for a function may not converge, and that even if it does converge for some x, it may not converge to $f(x)$, but we will not consider these difficult cases. Our goal is to find the coefficients $a_0, a_1, a_2, \ldots, a_n, \ldots, b_1, b_2, \ldots, b_n, \ldots$ for the class of functions that have a Fourier series representation and to apply the series to the functions of the type typically encountered in engineering applications.

Eventually, we will show how to derive the formulas for the coefficients of the Fourier series, but first, the formulas are merely stated and applied in several examples. Postponing the derivations allows you to become somewhat familiar with the concept of a Fourier series before tackling the details of the theory.

THE COEFFICIENTS OF A FOURIER SERIES

If $f(x)$ is defined over the interval $-\pi \leq x \leq \pi$, then

$$f(x) = a_0 + a_1 \cos x + a_2 \cos 2x + a_3 \cos 3x + \cdots + a_n \cos nx + \cdots$$
$$+ b_1 \sin x + b_2 \sin 2x + b_3 \sin 3x + \cdots + b_n \sin nx + \cdots$$

where

$$a_0 = \frac{1}{2\pi} \int_{-\pi}^{\pi} f(x)\, dx \qquad \textbf{(27-13)}$$

$$a_n = \frac{1}{\pi} \int_{-\pi}^{\pi} f(x) \cos nx\, dx \qquad \textbf{(27-14)}$$

$$b_n = \frac{1}{\pi} \int_{-\pi}^{\pi} f(x) \sin nx\, dx \qquad \textbf{(27-15)}$$

EXAMPLE 1 Find the Fourier series representation of the function defined by

$$f(x) = \begin{cases} 1 \text{ if } -\pi \leq x < 0 \\ -1 \text{ if } 0 \leq x < \pi. \end{cases}$$

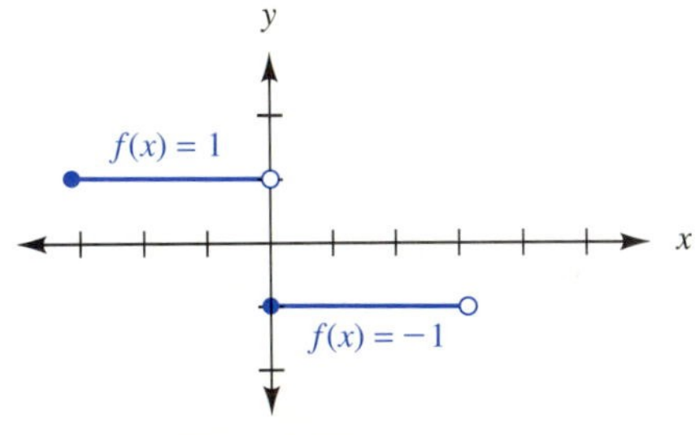

FIGURE 27.3

Solution Figure 27.3 shows the graph of the function. To find a_0, use Equation 27-13. Because the function is defined differently in the interval $-\pi \leq x < 0$ from the way it is defined in the interval $0 \leq x < \pi$, the integration is done in two parts.

$$\begin{aligned} a_0 &= \frac{1}{2\pi}\int_{-\pi}^{0} 1\,dx + \frac{1}{2\pi}\int_{0}^{\pi} -1\,dx \\ &= \left[\frac{1}{2\pi}(x)\right]\Bigg|_{-\pi}^{0} + \left[\frac{1}{2\pi}(-x)\right]\Bigg|_{0}^{\pi} \\ &= \frac{1}{2\pi}(0+\pi) + \frac{1}{2\pi}(-\pi-0) \\ &= \frac{1}{2} - \frac{1}{2} \text{ or } 0 \end{aligned}$$

To find a_1, use Equation 27-14.

$$\begin{aligned} a_1 &= \frac{1}{\pi}\int_{-\pi}^{\pi} f(x)\cos x\,dx \\ &= \frac{1}{\pi}\int_{-\pi}^{0} 1\cos x\,dx + \frac{1}{\pi}\int_{0}^{\pi} -1\cos x\,dx \\ &= \left[\frac{1}{\pi}(\sin x)\right]\Bigg|_{-\pi}^{0} - \left[\frac{1}{\pi}(\sin x)\right]\Bigg|_{0}^{\pi} \\ &= \frac{1}{\pi}(0-0) - \frac{1}{\pi}(0-0) = 0 \end{aligned}$$

In fact, $a_n = 0$ for all n, as the following discussion shows.

$$\begin{aligned} a_n &= \frac{1}{\pi}\int_{-\pi}^{\pi} f(x)\cos nx\,dx \\ &= \frac{1}{\pi}\int_{-\pi}^{0} 1\cos nx\,dx + \frac{1}{\pi}\int_{0}^{\pi} -1\cos nx\,dx \\ &= \frac{1}{\pi}\left(\frac{1}{n}\right)\int_{-\pi}^{0} n\cos nx\,dx + \frac{1}{\pi}\left(\frac{1}{n}\right)\int_{0}^{\pi} -n\cos nx\,dx \\ &= \left[\frac{1}{\pi n}(\sin nx)\right]\Bigg|_{-\pi}^{0} - \left[\frac{1}{\pi n}(\sin nx)\right]\Bigg|_{0}^{\pi} \\ &= \frac{1}{\pi n}(0-0) - \frac{1}{\pi n}(0-0) = 0 \end{aligned}$$

Let us then find the b_n's, using Equation 27-15.

$$\begin{aligned} b_1 &= \frac{1}{\pi}\int_{-\pi}^{\pi} f(x)\sin x\,dx \\ &= \frac{1}{\pi}\int_{-\pi}^{0} 1\sin x\,dx + \frac{1}{\pi}\int_{0}^{\pi} -1\sin x\,dx \end{aligned}$$

$$= \left[\frac{1}{\pi}(-\cos x)\right]\Bigg|_{-\pi}^{0} + \left[\frac{1}{\pi}(\cos x)\right]\Bigg|_{0}^{\pi}$$
$$= \frac{1}{\pi}(-1-1) + \frac{1}{\pi}(-1-1)$$
$$= -\frac{2}{\pi} - \frac{2}{\pi} \text{ or } \frac{-4}{\pi}$$
$$b_2 = \frac{1}{\pi}\int_{-\pi}^{\pi} f(x)\sin 2x\,dx$$
$$= \frac{1}{\pi}\int_{-\pi}^{0} 1 \sin 2x\,dx + \frac{1}{\pi}\int_{0}^{\pi} -1 \sin 2x\,dx$$
$$= \frac{1}{\pi}\left(\frac{1}{2}\right)\int_{-\pi}^{0} 2\sin 2x\,dx + \frac{1}{\pi}\left(\frac{1}{2}\right)\int_{0}^{\pi} -2\sin 2x\,dx$$
$$= \left[\frac{1}{2\pi}(-\cos 2x)\right]\Bigg|_{-\pi}^{0} + \left[\frac{1}{2\pi}(\cos 2x)\right]\Bigg|_{0}^{\pi}$$
$$= \frac{1}{2\pi}(-1+1) + \frac{1}{2\pi}(1-1) = 0$$
$$b_3 = \frac{1}{\pi}\int_{-\pi}^{\pi} f(x)\sin 3x\,dx$$
$$= \frac{1}{\pi}\int_{-\pi}^{0} 1 \sin 3x\,dx + \frac{1}{\pi}\int_{0}^{\pi} -1 \sin 3x\,dx$$
$$= \frac{1}{\pi}\left(\frac{1}{3}\right)\int_{-\pi}^{0} 3\sin 3x\,dx + \frac{1}{\pi}\left(\frac{1}{3}\right)\int_{0}^{\pi} -3\sin 3x\,dx$$
$$= \left[\frac{1}{3\pi}(-\cos 3x)\right]\Bigg|_{-\pi}^{0} + \left[\frac{1}{3\pi}(\cos 3x)\right]\Bigg|_{0}^{\pi}$$
$$= \frac{1}{3\pi}(-1-1) + \frac{1}{3\pi}(-1-1)$$
$$= \frac{-2}{3\pi} + \frac{-2}{3\pi} \text{ or } \frac{-4}{3\pi}$$

Thus, $f(x)$ can be approximated by

$$f(x) = \frac{-4}{\pi}\sin x - \frac{4}{3\pi}\sin 3x.$$

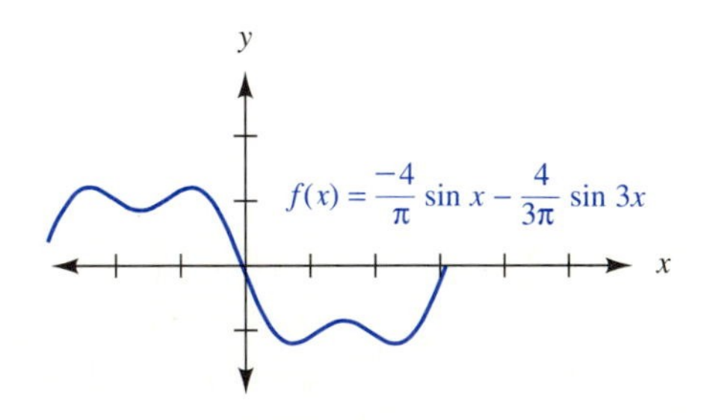

FIGURE 27.4

Figure 27.4 shows the graph of this function. Compare it to the graph of $f(x)$ shown in Figure 27.3. ■

In applied problems, curves such as the one in Figure 27.3 are common. Voltage, for instance, might have a graph that follows this pattern.

In Example 1, we evaluated each coefficient of the Fourier series individually. Often, it is possible to find formulas for a_n and b_n. By evaluating each formula for various values of n, we obtain the Fourier coefficients. Example 2 illustrates the method.

EXAMPLE 2 Find the Fourier series representation of the function defined by

$$f(x) = \begin{cases} x+1 \text{ for } -\pi \leq x < 0 \\ x \text{ for } 0 \leq x < \pi \end{cases}.$$

Solution

$$a_0 = \frac{1}{2\pi} \int_{-\pi}^{\pi} f(x)\, dx$$

$$a_0 = \frac{1}{2\pi} \int_{-\pi}^{0} (x+1)\, dx + \frac{1}{2\pi} \int_{0}^{\pi} x\, dx$$

$$= \left[\frac{1}{2\pi}\left(\frac{x^2}{2} + x\right)\right]\Bigg|_{-\pi}^{0} + \left[\frac{1}{2\pi}\left(\frac{x^2}{2}\right)\right]\Bigg|_{0}^{\pi}$$

$$= \frac{1}{2\pi}\left[0 - \left(\frac{\pi^2}{2} - \pi\right)\right] + \frac{1}{2\pi}\left[\frac{\pi^2}{2} - 0\right]$$

$$= \frac{-\pi}{4} + \frac{1}{2} + \frac{\pi}{4} \text{ or } \frac{1}{2}$$

$$a_n = \frac{1}{\pi} \int_{-\pi}^{\pi} f(x) \cos nx\, dx$$

$$= \frac{1}{\pi} \int_{-\pi}^{0} (x+1) \cos nx\, dx + \frac{1}{\pi} \int_{0}^{\pi} x \cos nx\, dx$$

$$= \frac{1}{\pi} \int_{-\pi}^{0} (x \cos nx + \cos nx)\, dx + \frac{1}{\pi} \int_{0}^{\pi} x \cos nx\, dx$$

$$= \frac{1}{\pi} \int_{-\pi}^{0} x \cos nx\, dx + \frac{1}{\pi} \int_{-\pi}^{0} \cos nx\, dx + \frac{1}{\pi} \int_{0}^{\pi} x \cos nx\, dx$$

$$= \frac{1}{\pi}\left(\frac{1}{n^2}\right) \int_{-\pi}^{0} nx(\cos nx)n\, dx + \frac{1}{\pi}\left(\frac{1}{n}\right) \int_{-\pi}^{0} (\cos nx)n\, dx +$$
$$\frac{1}{\pi}\left(\frac{1}{n^2}\right) \int_{0}^{\pi} nx(\cos nx)n\, dx$$

$$= \frac{1}{n^2\pi} [\cos nx + nx \sin nx]\Bigg|_{-\pi}^{0} + \frac{1}{n\pi} [\sin nx]\Bigg|_{-\pi}^{0} + \frac{1}{n^2\pi} [\cos nx + nx \sin nx]\Bigg|_{0}^{\pi}$$

$$= \frac{1}{n^2\pi} [1 - \cos(-n\pi)] + \frac{1}{n^2\pi} (\cos n\pi - 1)$$

$$= \frac{1}{n^2\pi} - \frac{\cos(-n\pi)}{n^2\pi} + \frac{\cos n\pi}{n^2\pi} - \frac{1}{n^2\pi}$$
$$= \frac{-\cos(-n\pi)}{n^2\pi} + \frac{\cos n\pi}{n^2\pi}$$
$$= \frac{-\cos(n\pi)}{n^2\pi} + \frac{\cos n\pi}{n^2\pi}$$
$$= 0$$

That is, $a_n = 0$ for all n. Let us then find a formula for b_n, using Equation 27-15.

$$b_n = \frac{1}{\pi}\int_{-\pi}^{\pi} f(x)\sin nx\,dx$$
$$= \frac{1}{\pi}\int_{-\pi}^{0}(x+1)\sin nx\,dx + \frac{1}{\pi}\int_{0}^{\pi} x\sin nx\,dx$$
$$= \frac{1}{\pi}\int_{-\pi}^{0} x\sin nx\,dx + \frac{1}{\pi}\int_{-\pi}^{0}\sin nx\,dx + \frac{1}{\pi}\int_{0}^{\pi} x\sin nx\,dx$$
$$= \frac{1}{\pi}\left(\frac{1}{n^2}\right)\int_{-\pi}^{0} nx(\sin nx)n\,dx + \frac{1}{\pi}\left(\frac{1}{n}\right)\int_{-\pi}^{0}(\sin nx)n\,dx$$
$$+ \frac{1}{\pi}\left(\frac{1}{n^2}\right)\int_{0}^{\pi} nx(\sin nx)n\,dx$$
$$= \frac{1}{n^2\pi}[(\sin nx - nx\cos nx)]\Big|_{-\pi}^{0} + \frac{1}{n\pi}[-\cos nx]\Big|_{-\pi}^{0}$$
$$+ \frac{1}{n^2\pi}[(\sin nx - nx\cos nx)]\Big|_{0}^{\pi}$$
$$= \frac{1}{n^2\pi}[-\pi n\cos(-\pi n)] + \frac{1}{n\pi}[-1+\cos(-n\pi)] + \frac{1}{n^2\pi}(-\pi n\cos n\pi)$$

However, $\cos(-n\pi) = \cos n\pi$, which is 1 for even n, but -1 for odd n. Thus, we can say that b_n is $\frac{-2}{n}$ if n is even, and $\frac{2}{n} - \frac{2}{n\pi}$ if n is odd.

Now that we have a general formula for b_n, we can find b_1, b_2, b_3, and so on.

$$b_1 = 2 - \frac{2}{\pi}$$
$$b_2 = -1$$

Thus, $f(x) = \frac{1}{2} + \left(2 - \frac{2}{\pi}\right)\sin x - \sin 2x + \cdots$. ■

To derive the formulas for the coefficients of the Fourier series, it is necessary to use the trigonometric identities that follow.

$$(1/2)[\cos(\alpha+\beta)+\cos(\alpha-\beta)]=\cos\alpha\cos\beta \tag{27-16}$$
$$(-1/2)[\cos(\alpha+\beta)-\cos(\alpha-\beta)]=\sin\alpha\sin\beta \tag{27-17}$$
$$(1/2)[\sin(\alpha+\beta)+\sin(\alpha-\beta)]=\sin\alpha\cos\beta \tag{27-18}$$
$$(1/2)[\sin(\alpha+\beta)-\sin(\alpha-\beta)]=\cos\alpha\sin\beta \tag{27-19}$$

To derive Equations 27-13 through 27-15, we must find $a_0, a_1, a_2, \ldots, a_n,$ $b_1, b_2, \ldots, b_n, \ldots$ such that

$$f(x)=a_0+a_1\cos x+a_2\cos 2x+\cdots+a_n\cos nx+\cdots$$
$$+b_1\sin x+b_2\sin 2x+\cdots+b_n\sin nx+\cdots. \tag{27-20}$$

Because we are interested in functions that might have sharp turns or discontinuities, where the derivative does not exist, the differentiation process used to find the coefficients of the MacLaurin and Taylor series is not helpful. Integration over one period of the function, however, can be used. Each term of the series (other than a_0) represents a function that completes one or more periods in 2π units. Thus, the sum will also repeat in 2π units. We therefore choose to integrate from $-\pi$ to π. In the following discussion, we indicate the steps to be carried out but do not show how the integrations are done. The coefficient a_0 is obtained by direct integration of the function in Equation 27-20.

$$\int_{-\pi}^{\pi} f(x)\,dx=\int_{-\pi}^{\pi}(a_0+a_1\cos x+a_2\cos 2x+\cdots+a_n\cos nx+\cdots$$
$$+b_1\sin x+b_2\sin 2x+\cdots+b_n\sin nx+\cdots)\,dx$$

When the integration is done, each term after the first is equal to 0. Thus,

$$\int_{-\pi}^{\pi} f(x)\,dx=\int_{-\pi}^{\pi} a_0\,dx=a_0x\Big|_{-\pi}^{\pi}=a_0\pi-a_0(-\pi)\text{ or }2a_0\pi.$$

Solving for a_0, we find

$$a_0=\frac{1}{2\pi}\int_{-\pi}^{\pi} f(x)\,dx.$$

To find a general formula for a_n, multiply both sides of Equation 27-20 by $\cos mx$ and integrate the function.

$$\int_{-\pi}^{\pi} f(x)\cos mx=\int_{-\pi}^{\pi}(a_0+a_1\cos x+a_2\cos 2x+\cdots+a_n\cos nx+\cdots$$
$$+b_1\sin x+b_2\sin 2x+\cdots+b_n\sin nx+\cdots)(\cos mx)\,dx \tag{27-21}$$

Integrating term by term, each term after the first takes the form

$$\int_{-\pi}^{\pi} a_n\cos nx\cos mx\,dx \qquad\text{or}\qquad \int_{-\pi}^{\pi} b_n\sin nx\cos mx\,dx.$$

To evaluate $\int_{-\pi}^{\pi} a_n \cos nx \cos mx\, dx$, we invoke Equation 27-16 with $\alpha = nx$ and $\beta = mx$, and first consider the case for which $m \neq n$.

$$\int_{-\pi}^{\pi} a_n \cos nx \cos mx\, dx = \int_{-\pi}^{\pi} a_n(1/2)[\cos(nx + mx) + \cos(nx - mx)]\, dx$$

It can be shown that this integral has a value of 0. Thus,

$$\int_{-\pi}^{\pi} a_n \cos nx \cos mx\, dx = 0. \tag{27-22}$$

Next, we consider the case for which $m = n$.

$$\int_{-\pi}^{\pi} a_n \cos nx \cos nx\, dx = a_n \int_{-\pi}^{\pi} \cos^2 nx\, dx$$

But we found this integral in Section 26.5. We have

$$a_n \int_{-\pi}^{\pi} \cos^2 nx\, dx = a_n \left(\frac{1}{n}\right) \int_{-\pi}^{\pi} (\cos^2 nx) n\, dx = a_n \pi. \tag{27-23}$$

To evaluate $\int_{-\pi}^{\pi} b_n \sin nx \cos mx\, dx$, we invoke Equation 27-18 with $\alpha = nx$ and $\beta = mx$, and first consider the case for which $m \neq n$.

$$\int_{-\pi}^{\pi} b_n \sin nx \cos mx\, dx = \int_{-\pi}^{\pi} b_n(1/2)[\sin(nx + mx) + \sin(nx - mx)]\, dx$$

It can be shown that this integral has a value of 0. Thus,

$$\int_{-\pi}^{\pi} b_n \sin nx \cos mx\, dx = 0. \tag{27-24}$$

Next, we consider the case for which $m = n$. It can be shown that

$$\int_{-\pi}^{\pi} b_n \sin nx \cos nx\, dx = b_n \left(\frac{1}{n}\right) \int_{-\pi}^{\pi} n \sin nx \cos nx\, dx = 0. \tag{27-25}$$

Notice that all of the integrals are equal to zero, except the one in Equation 27-23. When m and n are equal, we have

$$\int_{-\pi}^{\pi} a_n \cos nx \cos nx\, dx = \pi a_n.$$

We can replace m with n in Equation 27-21 and write it as

$$\int_{-\pi}^{\pi} f(x) \cos nx = \pi a_n.$$

Solving for a_n, we have

$$a_n = \frac{1}{\pi} \int_{-\pi}^{\pi} f(x) \cos nx\, dx.$$

The b_n's are found by returning to Equation 27-20. This time, multiply both sides of the equation by sin mx and integrate term by term. Each integral will be of the form

$$\int_{-\pi}^{\pi} (a_n \cos nx \sin mx)\, dx \qquad \text{or} \qquad \int_{-\pi}^{\pi} (b_n \sin nx \sin mx)\, dx.$$

Equations 27-17 and 27-19 can then be used.

We have assumed that $f(x)$ is defined on the interval from $-\pi$ to π. It is possible to adjust the formula for the coefficients of the Fourier series for a function defined on a different interval, although we will not consider that problem here.

EXERCISES 27.5

In 1–12, find the first few terms (at least through $n = 2$) of the Fourier series representation for the given functions, and sketch two cycles of the function.

1. $f(x) = \begin{cases} 0 & \text{if } -\pi \le x < 0 \\ 1 & \text{if } 0 \le x < \pi \end{cases}$

2. $f(x) = \begin{cases} 1 & \text{if } -\pi \le x < 0 \\ 0 & \text{if } 0 \le x < \pi \end{cases}$

3. $f(x) = \begin{cases} -1 & \text{if } -\pi \le x < 0 \\ 1 & \text{if } 0 \le x < \pi \end{cases}$

4. $f(x) = \begin{cases} x & \text{if } -\pi \le x < 0 \\ 0 & \text{if } 0 \le x < \pi \end{cases}$

5. $f(x) = \begin{cases} 0 & \text{if } -\pi \le x < 0 \\ x & \text{if } 0 \le x < \pi \end{cases}$

6. $f(x) = \begin{cases} x & \text{if } -\pi \le x < 0 \\ x - 1 & \text{if } 0 \le x < \pi \end{cases}$

7. $f(x) = \begin{cases} -x & \text{if } -\pi \le x < 0 \\ x & \text{if } 0 \le x < \pi \end{cases}$

8. $f(x) = \begin{cases} 0 & \text{if } -\pi \le x < 0 \\ x^2 & \text{if } 0 \le x < \pi \end{cases}$

9. $f(x) = x + 1$ for all x

10. $f(x) = x^2$ for all x

11. $f(x) = 1$ for all x

12. $f(x) = x - 1$ for all x

---

13. A half-wave rectifier is an electronic device that allows current to flow in only one direction. When an alternating voltage is applied, the resulting current is 0 for half of a cycle and a sine wave for the remaining half cycle. The equation for current is given by

$$f(t) = \begin{cases} 0 & \text{if } -\pi \le t < 0 \\ \sin t & \text{if } 0 \le t < \pi. \end{cases}$$

Graph one cycle of the function and find the Fourier series for $f(t)$. **3 5 9** 1 2 4 6 7 8

14. A full-wave rectifier is an electronic device that allows sinusoidal current to flow through a load for one half the time period of the ac supply and then, using diodes, routes the source's negative half cycle current through the load in the same direction as the first positive half cycle. The current equation is given by

$$f(t) = \begin{cases} -\sin t \text{ if } -\pi \le t < 0 \\ \sin t \text{ if } 0 \le t < \pi. \end{cases}$$

Graph one cycle of the function and find the Fourier series for $f(t)$. **3 5 9** 1 2 4 6 7 8

---

1. Architectural Technology
2. Civil Engineering Technology
3. Computer Engineering Technology
4. Mechanical Drafting Technology
5. Electrical/Electronics Engineering Technology
6. Chemical Engineering Technology
7. Industrial Engineering Technology
8. Mechanical Engineering Technology
9. Biomedical Engineering Technology

Problems 15 and 16 use the following information. In many applications, especially those relating to the periodic wave forms seen on an oscilloscope in the field of electronics, the functions of time t are specified over an interval $0 \le t < 2P$, where 2P is the period of the wave. The Fourier coefficients are given by the following equations.

$$a_0 = \frac{1}{2P}\int_0^{2P} f(t)\, dt$$

$$a_n = \frac{1}{P}\int_0^{2P} f(t) \cos \frac{n\pi t}{P}\, dt$$

$$b_n = \frac{1}{P}\int_0^{2P} f(t) \sin \frac{n\pi t}{P}\, dt$$

15. Find the first term of the Fourier series for the half-wave rectifier whose output voltage is given by

$$f(x) = \begin{cases} 170 \sin 120\pi t & \text{if } 0 \le t < 1/120 \\ 0 & \text{if } 1/120 \le t < 1/60. \end{cases}$$

3 5 9 1 2 4 6 7 8

16. Find the first term of the Fourier series for the full-wave rectifier in problem 14 for

$$f(t) = \begin{cases} 170 \sin 120\pi t \text{ if } 0 \le t < 1/120 \\ -170 \sin 120\pi t \text{ if } 1/120 \le t < 1/60. \end{cases}$$

3 5 9 1 2 4 6 7 8

17. Evaluate the integrals in Equations 27-22 through 27-24 to show that

$$\int_{-\pi}^{\pi} a_n(1/2)[\cos(nx + mx) + \cos(nx - mx)]\, dx = 0 \text{ if } m \ne n.$$

$$\int_{-\pi}^{\pi} a_n \cos^2 nx\, dx = a_n\pi.$$

$$\int_{-\pi}^{\pi} b_n(1/2)[\sin(nx + mx) + \sin(nx - mx)]dx = 0, \text{ if } m \ne n.$$

$$\int_{-\pi}^{\pi} b_n(\sin nx \cos nx)\, dx = 0.$$

1. Architectural Technology
2. Civil Engineering Technology
3. Computer Engineering Technology
4. Mechanical Drafting Technology
5. Electrical/Electronics Engineering Technology
6. Chemical Engineering Technology
7. Industrial Engineering Technology
8. Mechanical Engineering Technology
9. Biomedical Engineering Technology

27.6 CHAPTER REVIEW

KEY TERMS

27.1 power series
converge
diverge
interval of convergence
MacLaurin series

27.2 Taylor series

27.4 alternating series

27.5 Fourier series

SYMBOLS

$f^{(n)}(x)$ the nth derivative of f with respect to x

RULES AND FORMULAS

The MacLaurin Series:

$$f(x) = \frac{f(0)}{0!} + \frac{f'(0)x}{1!} + \frac{f''(0)(x^2)}{2!} + \frac{f^{(3)}(0)x^3}{3!} + \cdots + \frac{f^{(n)}(0)x^n}{n!} + \cdots$$

provided $f(x)$ and all of its derivatives exist when x is 0 and x is in the interval of convergence.

The Taylor Series:

$$f(x) = f(a) + f'(a)(x-a) + \frac{f''(a)(x-a)^2}{2!} + \cdots + \frac{f^{(n)}(a)(x-a)^n}{n!} + \cdots$$

provided $f(x)$ and all of its derivatives exist when x is a and x is in the interval of convergence.

MacLaurin Series for Common Functions:

$$e^x = 1 + x + \frac{x^2}{2!} + \frac{x^3}{3!} + \cdots \text{ for all real } x.$$

$$\sin x = x - \frac{x^3}{3!} + \frac{x^5}{5!} - \cdots \text{ for all real } x.$$

$$\cos x = 1 - \frac{x^2}{2!} + \frac{x^4}{4!} - \cdots \text{ for all real } x.$$

$$\ln(1+x) = x - \frac{x^2}{2} + \frac{x^3}{3} - \frac{x^4}{4} + \cdots \text{ for } |x| < 1.$$

The Coefficients of a Fourier Series:
If $f(x)$ is defined over the interval $-\pi \le x \le \pi$, then

$$\begin{aligned} f(x) = a_0 &+ a_1 \cos x + a_2 \cos 2x + a_3 \cos 3x + \cdots + a_n \cos nx + \cdots \\ &+ b_1 \sin x + b_2 \sin 2x + b_3 \sin 3x + \cdots + b_n \sin nx + \cdots \end{aligned}$$

where

$$a_0 = \frac{1}{2\pi}\int_{-\pi}^{\pi} f(x)\,dx$$

$$a_n = \frac{1}{\pi}\int_{-\pi}^{\pi} f(x)\cos nx\,dx$$

$$b_n = \frac{1}{\pi}\int_{-\pi}^{\pi} f(x)\sin nx\,dx$$

SEQUENTIAL REVIEW

(Section 27.1) *In 1–6, use the definition to find the first three nonzero terms of the MacLaurin series for the given function.*

1. $f(x) = \sin(1/2)x$
2. $f(x) = \cos(3x)$
3. $f(x) = xe^{-x}$
4. $f(x) = x \sin 2x$
5. $f(x) = e^{-2x}$
6. $f(x) = 1/(x-1)^2$

In 7–9, use the definition to find the first two nonzero terms of the MacLaurin series for the given function.

7. $f(x) = (3+x)^{1/3}$
8. $f(x) = \cos x^2$
9. $f(x) = \ln(1-3x)$

(Section 27.2) In 10–15, find the first three nonzero terms of the Taylor series for the given function and value of a.

10. $f(x) = \sin 3x \quad (a = \pi)$
11. $f(x) = \cos x \quad (a = \pi/2)$
12. $f(x) = e^x \quad (a = 2)$
13. $f(x) = e^{x-1} \quad (a = 2)$
14. $f(x) = \sqrt{x} \quad (a = 1)$
15. $f(x) = \ln(1 - x) \quad (a = -2)$

In 16–18, calculate each function value using the first two nonzero terms of a Taylor series with the given value of a. Show the result to five significant digits.

16. $\sqrt[3]{28}\ (a = 27)$
17. $\cos (3\pi/8)\ (a = \pi/2)$
18. $\sqrt{2}\ (a = 1)$

(Section 27.3) In 19–22, find the MacLaurin series for each function.

19. $f(x) = \sin(x/2)$
20. $f(x) = \cos x^2$
21. $f(x) = e^{\sqrt{x}}$
22. $f(x) = \ln (1 + x^2)$

In 23 and 24, find two nonzero terms of the Taylor series for the given function and value of a.

23. $f(x) = e^x \sin x\ (a = \pi)$
24. $f(x) = \dfrac{\cos x}{1 + x}\ (a = \pi)$

In 25–27, express each integral as a series.

25. $\displaystyle\int \frac{\cos x}{x}\, dx$
26. $\displaystyle\int x^2 e^x\, dx$
27. $\displaystyle\int \ln(1 - \sqrt{x})\, dx$

(Section 27.4) In 28–30, find an approximate value of each function for the given x using the specified number of nonzero terms of a MacLaurin series.

28. $\sin x\ (x = 3°, 2 \text{ terms})$
29. $e^x\ (x = 2.5, 3 \text{ terms})$
30. $\ln (1 + x)\ (x = 0.3, 3 \text{ terms})$
31. Determine the accuracy of the result in problem 28, if possible.
32. Determine the accuracy of the result in problem 29, if possible.
33. Determine the accuracy of the result in problem 30, if possible.

In 34–36, find an approximate value of each function for the given x using the specified number of nonzero terms of a Taylor series in powers of (x − a).

34. $\sin 28°$ (3 terms, $a = \pi/6$)
35. $\cos 50°$ (3 terms, $a = \pi/4$)
36. $\ln (2.5 + 1)$ (2 terms, $a = 2$)

(Section 27.5) In 37–41, consider $f(x) = \begin{cases} 1 & \text{if} -\pi \le x < 0 \\ x & \text{if}\ 0 \le x < \pi \end{cases}$.

37. Find the value of a_0 in the Fourier series representation of $f(x)$.
38. Find a formula for a_n in the Fourier series representation of $f(x)$.
39. Find a formula for b_n in the Fourier series representation of $f(x)$.
40. Find the first few terms (at least through $n = 2$) of the Fourier representation of $f(x)$.
41. Sketch two cycles of the function.

In 42–45, consider $f(x) = \begin{cases} x & \text{if } -\pi \le x < 0 \\ -x & \text{if } 0 \le x < \pi \end{cases}$.

42. Find the value of a_0 in the Fourier series representation of $f(x)$.

43. Find a formula for a_n in the Fourier series representation of $f(x)$.

44. Find a formula for b_n in the Fourier series representation of $f(x)$.

45. Find the first few terms (at least through $n = 2$) of the Fourier representation of $f(x)$.

APPLIED PROBLEMS

1. Suppose that a computer program is written to find an approximate value of cos 88° using two nonzero terms of a Taylor series in powers of $(x - \pi/2)$. Determine the accuracy of the result. 3 5 9

2. The formula $dW/dh = (-k/R^2)(1 + h/R)^{-2}$ is used to find the approximate work done by gravity in pulling an object from a height of h above the earth's surface to the surface. Find the first two nonzero terms of the MacLaurin series for dW/dh, and integrate to find W. In the formula, k is a constant, and R is the radius of the earth. 4 7 8 1 2 3 5 6 9

3. The voltage V supplied by U.S. power companies is given by $V = 169 \sin 120\pi t$. Find the first three nonzero terms of the MacLaurin series for this function. 3 5 9 1 2 4 6 7 8

4. Determine the accuracy of the result in problem 3. 3 5 9 1 2 4 6 7 8

5. A silicon diode of a certain type has its current i (in amperes) given by $i = 10^{-8}(e^{23.03v} - 1)$, where v is its voltage (in volts). Find the first three nonzero terms of a Taylor series for this current near $v = 0.7$ volt. Use four significant digits. 3 5 9 1 2 4 6 7 8

6. The force F (in lb) of the wind on the front of a truck of a certain design is given by $F = 0.001503\, v^{3.15}$ where v is its speed (in mph). Find the first three nonzero terms of a Taylor series to express this force near $v = 55$ mph. Use four significant digits. 4 7 8 1 2 3 5 6 9

7. In electronics, it is often necessary to evaluate a function for very small values of time t. If t is in picoseconds, microseconds, or even milliseconds, the numbers are close enough to zero to allow use of the MacLaurin series to represent some very complicated voltage and current equations. Find the first three nonzero terms of the MacLaurin series to represent $i = 3\, e^{3t^2 - 2t + 0.6}$. Use three significant digits. 3 5 9 4 7 8

8. The formula $\alpha = \beta/(1 + \beta)$ gives the α of a transistor in terms of its current gain β. Find the first three nonzero terms of a Taylor series for α near $\beta = 100$. Use two significant digits. 3 5 9 4 7 8

9. The formula $i = e^{2v-1}$, given in a technical paper on a new electronic device gives the current i (in amperes) of the device as a function of its voltage v (in volts) if v is near the turn on voltage of 1.3 volts. Another paper, discussing the same device, gives $i = 9.906\, v^2 - 15.85\, v + 8.816$. Show that the second function gives the first three terms of the Taylor series for the first function. 3 5 9 4 7 8

10. For very small values of time t, the formula $i = 30(1 + v/5)^{1.63}$ gives the beam current i (in milliamps) in a cathode ray tube when v is in volts. Suppose $v = \cos t$, and find the first three nonzero terms of the Taylor series for i. Use four significant digits. (Hint: If t is very small, then v is near 1.) 3 5 9 4 7 8

11. The formula $n = T^{3/2}\, e^{-4348/T}$ gives the density n (in coulombs/m^3) of charge carriers in a semiconductor when T is temperature (in degrees Kelvin). Find the first two nonzero terms of the Taylor series for n if T is near 273° K. Use three significant digits. 3 5 9 4 7 8

1. Architectural Technology
2. Civil Engineering Technology
3. Computer Engineering Technology
4. Mechanical Drafting Technology
5. Electrical/Electronics Engineering Technology
6. Chemical Engineering Technology
7. Industrial Engineering Technology
8. Mechanical Engineering Technology
9. Biomedical Engineering Technology

12. The power P (in watts) in a certain load resistance is given by $P = (1 - e^{-0.03t})/t$ where t is time (in seconds). Use $W = \int P\,dt$ and three nonzero terms of a MacLaurin series for $1 - e^{-0.03t}$ to find the energy dissipated between $t = 0$ and $t = 0.30$ s. Give your answer to one significant digit. 3 5 9 1 2 4 6 7 8

13. Consider a voltage given by $v = 0.0192 \sin 377t$. The approximation based on two nonzero terms of the MacLaurin series for $\sin 377t$ is accurate to four significant digits if $t \leq 0.0009$. Find these two terms and evaluate v if $t = 0.0001$. 3 5 9 1 2 4 6 7 8

14. For values of t such that $3.736 \leq t \leq 4.415$, the first two terms of the Taylor series approximation for $p = e^t$ about $t = 3$ is within 10% of the true value. Find two terms of this series. Use three significant digits. 3 5 9 1 2 4 6 7 8

15. For values of t such that $-1.05346 \leq t \leq 1.05346$, the first two terms of the MacLaurin series approximation for $p = \cos t$ is at least 90% of the true value. Find two terms of this series. 3 5 9 1 2 4 6 7 8

16. When a taut cable of length L is deflected a distance M and allowed to vibrate freely, it will do so with harmonic vibrations. The amplitude y of the vibrations at a point a distance of x from one end is given by

$$y = 8M/\pi^2 \sum_{k=1}^{\infty} = \frac{(-1)^{(k-1)/2}}{k^2} \sin (k\pi x/L).$$

If a 5-meter cable is deflected 0.02 m at a distance $x = (1/3)L$ from one end, find the terms of the series associated with $k = 1$, 3, and 5. 4 7 8 1 2 3 5 6 9

17. Consider the periodic function shown in the accompanying figure. Write the equation for the function. 3 5 9 1 2 4 6 7 8

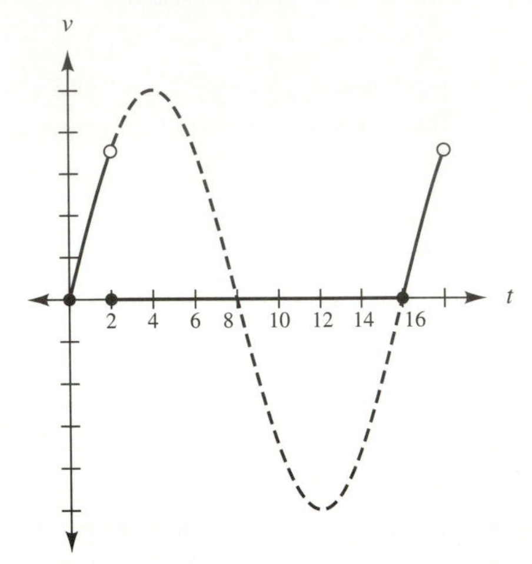

18. Consider the periodic function shown in the accompanying figure. Write the equation for the function. 3 5 9 1 2 4 6 7 8

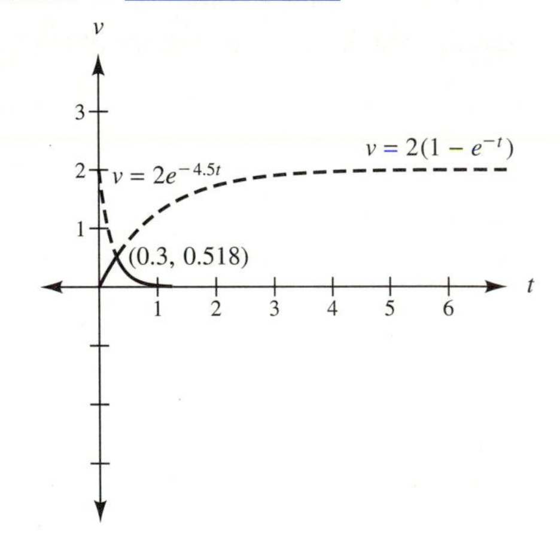

1. Architectural Technology
2. Civil Engineering Technology
3. Computer Engineering Technology
4. Mechanical Drafting Technology
5. Electrical/Electronics Engineering Technology
6. Chemical Engineering Technology
7. Industrial Engineering Technology
8. Mechanical Engineering Technology
9. Biomedical Engineering Technology

RANDOM REVIEW

1. Find the first three nonzero terms of the MacLaurin series for $f(x) = \cos \sqrt{x}$.
2. Find the first three nonzero terms of the Taylor series for $f(x) = e^{x-1}$ near 1.
3. Find an approximate value for sin 25° using two nonzero terms of a Taylor series with $a = \pi/6$.
4. Use the definition to find the first three nonzero terms of the MacLaurin series for $f(x) = \dfrac{2}{3+x}$.
5. Consider $f(x) = \begin{cases} x \text{ if } -\pi \le x < 0 \\ 1 \text{ if } 0 \le x < \pi \end{cases}$.

   Find a formula for b_n in the Fourier series representation of the function.
6. Find an approximate value for cos 1° using two nonzero terms of the MacLaurin series.
7. Consider $f(x) = \begin{cases} x^2 \text{ if } -\pi \le x < 0 \\ 0 \text{ if } 0 \le x < \pi \end{cases}$.

   Find the value of a_0 in the Fourier series representation of the function.
8. Express as a series: $\int x \sin x$.
9. Calculate $\sqrt[3]{29}$ using two nonzero terms of a Taylor series with $a = 27$.
10. Find the first three nonzero terms of the MacLaurin series for $f(x) = e^{-2x}$.

CHAPTER TEST

In 1 and 2, find the first three nonzero terms of the MacLaurin series for the given function.

1. $f(x) = \cos \sqrt{x}$
2. $f(x) = e^x(\cos x)$

In 3 and 4, find the first three nonzero terms of the Taylor series for the given function and value of a.

3. $f(x) = \sqrt{x} \quad (a = 9)$
4. $f(x) = \cos 2x \quad (a = \pi/3)$

In 5 and 6, express each integral as a series.

5. $\int \sin x^2 \, dx$
6. $\int e^{\sqrt{x}} \, dx$
7. Find an approximate value for cos 2° using the first two nonzero terms of the MacLaurin series, and determine the accuracy of the result.
8. Consider $f(x) = \begin{cases} x \text{ if } -\pi \le x < 0 \\ 1 \text{ if } 0 \le x < \pi \end{cases}$.

   Find the first few terms (at least through $n = 2$) of the Fourier series representation for $f(x)$.
9. The velocity of discharge V (in ft/s) of water is given by $V = 12.16\sqrt{P}$, where P is the pressure of the water at the point of discharge. Use two nonzero terms of the Taylor series with $a = 4$ to find a linear equation that can be used to approximate V for small values of P.
10. Suppose that a computer program is written to find an approximate value of sin 88° using two nonzero terms of a Taylor series. Determine the accuracy of the result.

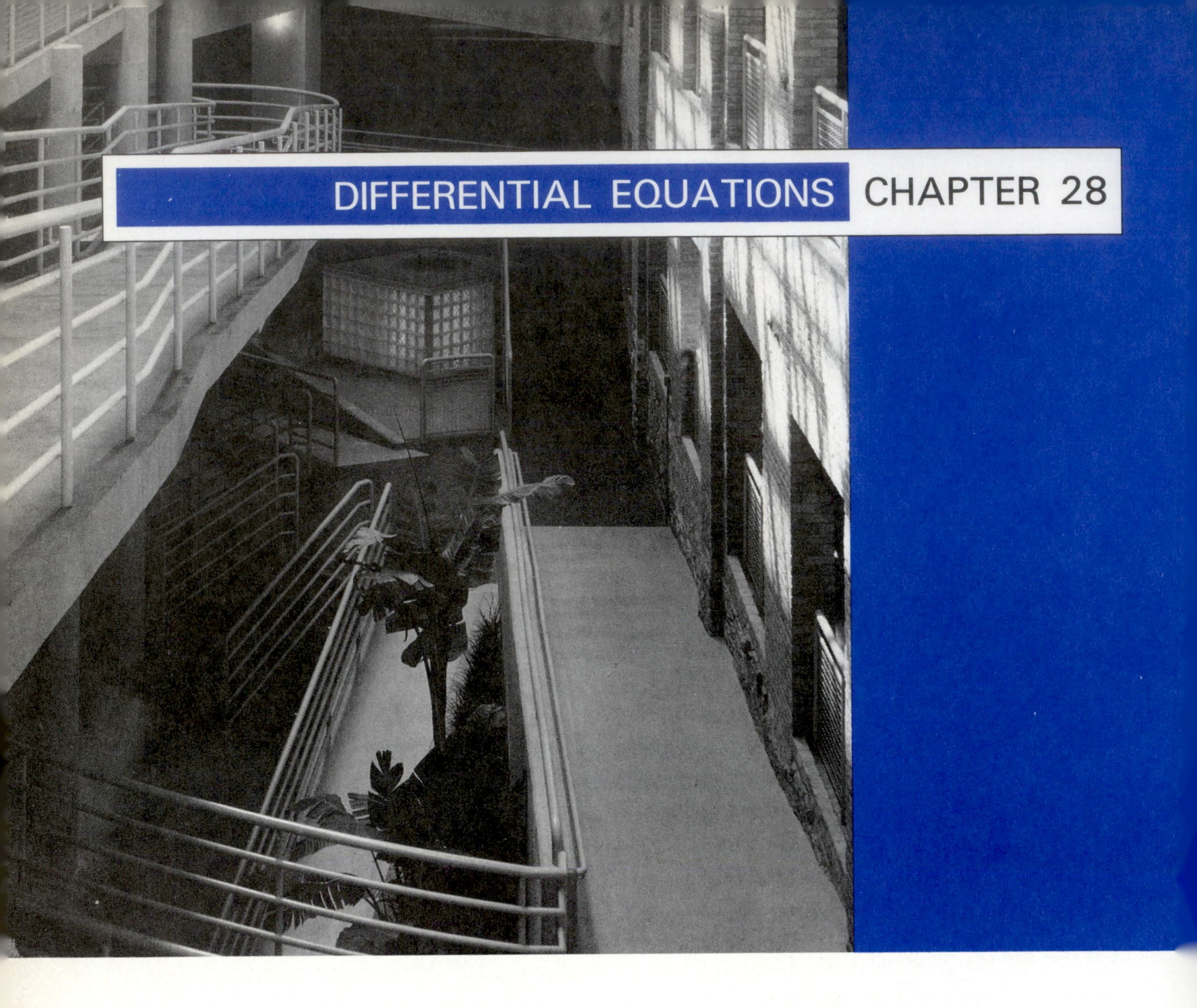

CHAPTER 28 DIFFERENTIAL EQUATIONS

In engineering and technology, the relationship between two quantities often involves the rate of change of one quantity with respect to the other. Such relationships are expressed using equations called differential equations. There is no one method that will allow us to solve all differential equations. There are, however, certain standard types of differential equations for which routine methods can be used. In this chapter, we consider common types of differential equations and their applications in technical fields.

28.1 TYPES OF DIFFERENTIAL EQUATIONS AND THEIR SOLUTIONS

The concept of solving an algebraic equation is familiar to you. In an algebraic equation, the unknown is a number, typically denoted x. To solve the equation means to find the values of the variable that make the equation a true statement. A differential equation

differential equation is an equation that contains one or more derivatives or differentials. In a differential equation, the unknown is a function, typically denoted y. To solve a differential equation means to find the functions that make the equation an identity.

We begin by examining how differential equations may arise in practical situations. Consider a 20-gallon tank initially filled with gasoline. The liquid is pumped out at the rate of 2 gal/min, but the tank is simultaneously refilled at the same rate with alcohol. When the process begins, there is only gasoline in the tank, but eventually, there is only alcohol. The problem is to find a function, expressed as $y = f(t)$, that gives the amount of alcohol in the tank at time t, assuming that the two liquids are mixed uniformly at all times.

Consider the alcohol in the tank. The rate at which alcohol enters the tank is a constant 2 gal/min. As liquid is pumped out, however, the proportion of alcohol in the tank changes continuously. Since we let $f(t)$ represent the amount of alcohol in the tank at any given time t, the fraction of alcohol in the tank at any given time is $f(t)/20$. At any given time, the fraction of alcohol in the liquid leaving the tank is the same fraction as the fraction of alcohol in the tank. The rate at which alchol leaves the tank is the product of the fraction of alcohol leaving the tank and the rate at which liquid leaves the tank. That is, the rate at which alcohol leaves the tank is $[f(t)/20] \times 2$ or $f(t)/10$.

The amount $f(t)$ of alcohol in the tank at any time is the difference between the amount of alcohol that has come in and the amount of alcohol that has gone out. The rate at which $f(t)$ changes is the difference between the rate at which alcohol flows in and the rate at which it flows out. The derivative of $f(t)$ also gives the rate at which $f(t)$ changes. Thus,

$$f'(t) = 2 - f(t)/10. \tag{28-1}$$

Equation 28-1 is an example of a differential equation. A differential equation may contain only first derivatives, as does Equation 28-1, or it may contain higher order derivatives instead of, or in addition to, first derivatives. An equation containing only first derivatives is called a **first order** differential equation. An equation containing only second derivatives or first and second derivatives is called a **second order** differential equation. Higher order differential equations contain higher derivatives. The *order* of a differential equation is not to be confused with the *degree.* Just as each variable in an algebraic equation may be raised to a power, each derivative in a differential equation may be raised to a power.

DEFINITION The **order of a differential equation** is the order of the highest order derivative in the equation. The **degree of a differential equation** is the highest power of the highest order derivative.

EXAMPLE 1 Determine the order and the degree of each differential equation.

(a) $y' + y/10 = 2$ **(b)** $y' + y'' = 2$ **(c)** $y' + (y'')^2 = 1$

Solution (a) Since the first derivative is the only derivative, the equation is a first order differential equation. Since y', the highest derivative, is to the first power, the equation is a first degree differential equation.

(b) Since the equation contains both first and second derivatives, the equation is a second order differential equation. Since y'', the highest derivative, is to the first power, the equation is a first degree differential equation.

(c) Since the equation contains both first and second derivatives, the equation is a second order differential equation. Since y'', the highest derivative, is to the second power, the equation is a second degree differential equation. ■

We will restrict our attention in this chapter to first degree differential equations. A solvable differential equation has what is called a *general* solution and *particular* solutions. The general solution of an nth order differential equation is specified in a form containing arbitrary constants, so that it represents all functions that are solutions. For instance, the general solution of the equation $y''=0$ is given by $y=ax+b$, where a and b are constants. (For any function of this form, $y'=a$, and $y''=0$.) The equations $y=2x+3$ and $y=3x-1$, with values of a and b specified, are particular solutions. A solution is said to contain n constants if it cannot be reduced to a form containing fewer constants.

DEFINITION The **general solution** of an nth order differential equation is a function, generally containing n arbitrary constants, which yields an identity when substituted into the original equation. When at least one of the constants has a specific value, the solution is said to be a **particular solution.***

EXAMPLE 2 Verify that $y=18-Be^{-x/6}$ is a general solution of the differential equation $3-(1/6)y=y'$ and that $y=18-18e^{-x/6}$ is a particular solution.

Solution In order to substitute the solution into the original equation, $3-(1/6)y=y'$, we must know both y and y'.

If $y=18-Be^{-x/6}$, then $y'=(-1/6)(-Be^{-x/6})$ or $(1/6)Be^{-x/6}$.

Making the substitution, we have

$$3-(1/6)\underbrace{(18-Be^{-x/6}}_{y}=\underbrace{(1/6)Be^{-x/6}}_{y'}$$

$$3-3+(1/6)Be^{-x/6}=(1/6)Be^{-x/6}$$

$$(1/6)Be^{-x/6}=(1/6)Be^{-x/6}\text{, so the equation is an identity.}$$

* In some cases, there are further solutions, called singular solutions, that cannot be obtained by assigning values to the constants.

If $y = 18 - 18e^{-x/6}$, then $y' = (-1/6)(-18e^{-x/6})$ or $3e^{-x/6}$. Making the substitution, we have

$$3 - (1/6)(18 - 18e^{-x/6}) = 3e^{-x/6}$$
$$3 - 3 + 3e^{-x/6} = 3e^{-x/6}$$
$$3e^{-x/6} = 3e^{-x/6}, \text{ so the equation is an identity.} \blacksquare$$

NOTE ⇨⇨ ***It is not always possible to express the solution of a differential equation as an equation that is solved for y.*** Recall that a function so expressed is said to be defined explicitly or to be an explicit function.

EXAMPLE 3 Verify that $x^2 = 2y^2 \ln y$ $(y > 0)$ is a solution of $y' = \dfrac{xy}{x^2 + y^2}$.

Solution To find y', differentiate both sides of the equation $x^2 = 2y^2 \ln y$ with respect to x.

$$2x = \frac{2y^2y'}{y} + (\ln y)(4y)(y')$$ Apply the product rule on the right.

$$x = y'(y + 2y \ln y)$$ Factor out $2y'$ on the right and divide both sides by 2.

$$\frac{x}{y + 2y \ln y} = y'$$ Divide both sides by $y + 2y \ln y$.

To remove the factor $\ln y$ from the second term of the denominator, notice that from the proposed solution $x^2 = 2y^2 \ln y$, we can conclude that $\ln y = \dfrac{x^2}{2y^2}$. Thus,

$$\frac{x}{y + 2y\left(\dfrac{x^2}{2y^2}\right)} = y'$$ Replace $\ln y$ with $\dfrac{x^2}{2y^2}$.

$$\frac{x}{y + \dfrac{x^2}{y}} = y'$$ Simplify the denominator.

$$\frac{xy}{y^2 + x^2} = y'.$$ Replace $y + \dfrac{x^2}{y}$ with $\dfrac{y^2 + x^2}{y}$ and divide.

Substituting this expression for y' in the original equation shows that the equation is an identity.

$$\frac{xy}{y^2 + x^2} = \frac{xy}{x^2 + y^2} \quad \blacksquare$$

Sometimes properties of logarithms may be used to simplify the solution of a differential equation. Example 4 illustrates the technique.

EXAMPLE 4 Show that $(1/3)\ln(x^3-1)+\ln y=C_1$ can be shortened to $(x^3-1)y^3=C^3$.

Solution For each value of C_1 there is a constant C such that $\ln C=C_1$. Thus, we have the following simplification.

$$(1/3)\ln(x^3-1)+\ln y=C_1$$

$$(1/3)\ln(x^3-1)+\ln y=\ln C \quad \text{Replace } C_1 \text{ with } \ln C.$$

$$\ln(x^3-1)+3\ln y=3\ln C \quad \text{Multiply by 3.}$$

$$\ln(x^3-1)+\ln y^3=\ln C^3 \quad \text{Use } \ln m^n=n\ln m.$$

$$\ln[(x^3-1)y^3]=\ln C^3 \quad \text{Use } \ln(mn)=\ln m+\ln n.$$

$$(x^3-1)y^3=C^3 \quad \text{If the logarithms are equal, the numbers are equal.}$$

EXERCISES 28.1

In 1–10, determine the order and degree of each differential equation.

1. $y'+y/5=x$
2. $xy'+y^2=x$
3. $yy'+x^2=y$
4. $yy'+y^2-xy=x$
5. $y+y''=x$
6. $y+xy''=x^2$
7. $y^2-yy''=y$
8. $(y')^2+y'=x$
9. $(y')^2+y'=x^2$
10. $y(y')^2+y^2=x$

In 11–20, verify that the first function given, implicitly or explicity, is a solution of the given differential equation.

11. $y=12-Be^{-x/3}$; $4-(1/3)y=y'$
12. $y=4-Ce^{-x/2}$; $y'+(1/2)y=2$
13. $y=Ae^{2x}+Be^{-2x}$; $y''-4y=0$
14. $y=A\sin x-B\cos x$; $y''+y=0$
15. $y=A\sin 2x+B\cos 2x$; $y''+4y=0$
16. $y=Ce^{y/x}$; $y'=\dfrac{y^2}{xy-x^2}$
17. $xy=\ln y+C$; $y'=\dfrac{y^2}{1-xy}$
18. $x/y=\ln(1/y)+C$; $\dfrac{y}{x-y}=y'$
19. $y+\sin y=x$; $(y\cos y-\sin y+x)y'=y$
20. $y+\cos y=x$; $(x-y\sin y-\cos y)y'=y$

In 21–26, each expression represents the solution of a differential equation. Use the properties of logarithms to simplify each solution.

21. $\ln(2x-1)+\ln y=C_1$
22. $\ln(x-3)-3\ln y=C_1$
23. $(1/2)\ln(x^2+1)+\ln y=C_1$
24. $2\ln(x^3+1)+\ln y=C_1$
25. $3\ln(x-1)-2\ln y=C_1$
26. $(1/3)\ln(x^2-1)-\ln y=C_1$

28.2 SEPARATION OF VARIABLES

The simplest type of differential equation is one that has the form $y' = f(x)$. Any solution must be an antiderivative of the function f. That is, $y = \int f(x)\, dx$. Recall that the indefinite integral denotes an entire class of antiderivatives, each of which differs from the others by a constant.

EXAMPLE 1 Find a general solution of $y' = e^x$.

Solution

$$\begin{aligned} y &= \int e^x\, dx \\ &= e^x + C \quad \blacksquare \end{aligned}$$

In Example 1, particular solutions would be obtained by replacing the arbitrary constant C with particular values. Thus, $e^x + 2$ and $e^x - 10$ are examples of particular solutions.

In a first order, first degree differential equation, each term may contain factors that are functions of x (e.g., x^2, e^x, $\sin x$) and factors that are functions of y (e.g., y^2, y^{-1}, $\ln y$). Many such equations cannot be put in the form $y' = f(x)$. If it is possible, however, to write the equation so that all of the x's appear on one side of the equation and all of the y's on the other, it may be possible to solve the equation by the method of *separation of variables*, and such an equation is said to be separable. Example 2 illustrates the technique.

separation of variables

EXAMPLE 2 Solve $y' = x^2/y^3$.

Solution Generally, the procedure is shortened by using differential notation. Thus, $y' = \dfrac{x^2}{y^3}$ is written as $\dfrac{dy}{dx} = \dfrac{x^2}{y^3}$. Then dx and dy are treated as differentials, which means they may be manipulated individually. If both sides of the equation are multiplied by $y^3\, dx$, we obtain

$$y^3\, dy = x^2\, dx.$$

Then both sides may be integrated.

$$\int y^3\, dy = \int x^2\, dx$$

Since the left-hand side calls for integration with respect to y, and the right-hand side calls for integration with respect to x, we have the following.

$$\frac{y^4}{4} + C_1 = \frac{x^3}{3} + C_2 \qquad \text{Apply the power rule.}$$

$$\frac{y^4}{4} = \frac{x^3}{3} + C_2 - C_1 \qquad \text{Subtract } C_1 \text{ from both sides.}$$

$$y^4 = 4\left(\frac{x^3}{3}\right) + 4(C_2 - C_1) \qquad \text{Multiply both sides by 4.}$$

$$y^4 = 4\left(\frac{x^3}{3}\right) + C \qquad \text{Let } C = 4(C_2 - C_1).$$

$$y = \pm\sqrt[4]{\frac{4x^3}{3} + C} \qquad \text{Take the 4th root.}$$

■

NOTE

Notice that in Example 2, the constant of integration was added during the integration step. It must be considered in the algebraic manipulations that follow, so it is not sufficient to simply tack on "$+C$" at the end of the problem. It is often possible, however, to combine constants and replace them with a single constant, as in this example. In the future, we will only add the constant of integration on one side of the equation, since a constant on the other side could be subtracted and combined with the first constant. Also, it is possible to express y explicitly in this example, but it may be that y must be given implicitly.

In practical applications, variables are usually positive quantities. It is traditional in problems involving the logarithmic form to omit the absolute value. That is, $\int u^{-1}\, du$ is given as $\ln u + C$. If there is any doubt as to the validity of the resulting solution, it may be substituted into the original equation and checked. We will also ignore the possibility that some solutions may be lost when we multiply by a denominator that could be equal to zero, as well as some of the finer points regarding differentiability and continuity.

To solve a first order, first degree differential equation by the method of separation of variables:

1. Write the equation using the dy/dx notation.
2. Write the equation so that it has only x's (including dx) on one side and only y's (including dy) on the other.
3. Integrate both sides of the equation, if possible, introducing a constant on only one side of the equation.
4. Solve for y, if possible; otherwise leave the solution in implicit form.

EXAMPLE 3 Solve the differential equation $y' = 2 - \frac{y}{10}$. This is the equation that we obtained in Section 28.1 to describe the 20-gallon gasoline tank being refilled with alcohol.

Solution $\dfrac{dy}{dx} = 2 - \dfrac{y}{10}$ Write the equation using dy/dx notation.

To separate the variables, we need to remove dx on the left and $2 - \dfrac{y}{10}$ on the right.

$$\frac{dy}{2 - \frac{y}{10}} = dx$$ Multiply both sides by $\dfrac{dx}{2 - \frac{y}{10}}$.

$$\frac{10\,dy}{20 - y} = dx$$ Simplify the expression on the left.

$$\int \frac{10\,dy}{20 - y} = \int dx$$ Integrate both sides.

$$-10 \int \frac{-dy}{20 - y} = \int dx$$ Use $\int cf(x)\,dx = c\int f(x)\,dx$.

$$-10 \ln (20 - y) = x + C_1$$ Integrate, using the logarithmic form.

$$\ln (20 - y) = \frac{x + C_1}{-10}$$ Divide both sides by -10.

$$20 - y = e^{(x+c_1)/-10}$$ Use the definition of natural logarithm.

$$20 - y = e^{-x/10} e^{-c_1/10}$$ Use properties of exponents.

$$20 - y = Ce^{-x/10}$$ Replace $e^{-C_1/10}$ with the constant C.

$$20 - Ce^{-x/10} = y$$ Solve for y. ■

Notice that had the constant been added to the left-hand side during the integration step, we would have had

$$-10 \ln (20 - y) + C_1 = x$$

$$-10 \ln (20 - y) + -10 \ln C_2 = x$$ Let $C_1 = -10 \ln C_2$.

$$-10 [\ln (20 - y) + \ln C_2] = x$$ Factor out -10.

$$-10 \ln C_2(20 - y) = x$$ Use the property $\ln m + \ln n = \ln mn$.

$$\ln C_2(20 - y) = \frac{x}{-10}$$ Divide both sides by -10.

$$C_2(20 - y) = e^{-x/10}$$

$$20 - y = \frac{1}{C_2} e^{-x/10}$$

$$20 - y = C\,e^{-x/10}$$ Let $C = 1/C_2$.

$$y = 20 - C\,e^{-x/10}$$ Solve for y.

This is the same answer that we obtained before. It is possible, however, for two different procedures to lead to two answers that appear to be different.

Sometimes one encounters equations that are already written using the notation of differentials.

EXAMPLE 4 Solve the differential equation $dy = x\,dx + x^2\,dy$.

Solution We begin by combining the terms containing dy.

$$dy - x^2\,dy = x\,dx \qquad \text{Subtract } x^2\,dy \text{ from both sides.}$$

$$(1 - x^2)\,dy = x\,dx \qquad \text{Factor out } dy.$$

$$dy = \frac{x\,dy}{1 - x^2} \qquad \text{Multiply both sides by } \frac{1}{1 - x^2}.$$

$$\int dy = \int \frac{x\,dx}{1 - x^2} \qquad \text{Integrate both sides.}$$

$$\int dy = \frac{-1}{2}\int \frac{-2x\,dx}{1 - x^2} \qquad \text{Multiply on the right by } \frac{-1}{2} \text{ and } -2.$$

$$y = \frac{-1}{2}\ln(1 - x^2) + C_1 \qquad \text{Use the logarithmic form to integrate.}$$

Or, using the properties of logarithms, we could write the following.

$$y = \frac{-1}{2}\ln(1 - x^2) + \frac{-1}{2}\ln C \qquad \text{Let } C_1 = (-1/2)\ln C.$$

$$= \frac{-1}{2}[\ln(1 - x^2) + \ln C] \qquad \text{Factor out } -1/2.$$

$$= \left(\frac{-1}{2}\right)\ln C(1 - x^2) \qquad \text{Use } \ln mn = \ln m + \ln n. \quad ■$$

Trigonometric identities and properties of exponents may also be used to rewrite an equation so that the variables may be separated.

EXAMPLE 5 Solve the differential equation $e^{\cos x} = y' \csc x$.

Solution

$$e^{\cos x} = \frac{dy}{dx}\csc x \qquad \text{Write in differential form.}$$

$$e^{\cos x}\,dx = (\csc x)\,dy \qquad \text{Multiply both sides by } dx.$$

$$\frac{e^{\cos x}}{\csc x}\,dx = dy \qquad \text{Divide both sides by } \csc x.$$

$$\int \frac{e^{\cos x}}{\csc x}\,dx = \int dy \qquad \text{Integrate both sides.}$$

$$\int (e^{\cos x})(\sin x)\,dx = \int dy \qquad \text{Replace } 1/\csc x \text{ with } \sin x.$$

$$-e^{\cos x} + C = y \quad ■$$

Particular solutions of differential equations arise when there are additional conditions, called initial conditions or boundary conditions, specified. Applied problems often involve this type of additional data.

EXAMPLE 6 The instantaneous charging current i (in amperes) for a capacitor connected to a source of varying voltage v (in volts) is given by $i = C\,dv/dt$. Find the voltage across a 100 μF capacitor after it has charged at a constant current of 1 mA for 50 ms if the voltage is 0 when t is 0. In the formula, C is the capacitance (in farads) and t is the time (in seconds).

Solution Substituting the known values i and C into the equation $i = C\,dv/dt$, we have

$$10^{-3} = 100 \times 10^{-6}\frac{dv}{dt}$$ Recall that 1 mA $= 10^{-3}$ A and 100 μF $= 100 \times 10^{-6}$ F.

$$10^{-3} = 10^{-4}\frac{dv}{dt}$$ Simplify $100 \times 10^{-6} = 10^2 \times 10^{-6}$.

$$10 = \frac{dv}{dt}$$ Divide both sides by 10^{-4}.

$$\int 10\,dt = \int dv$$ Separate the variables.

$$10t = v + C_1$$ Perform the integration.

Knowing that v is 0 when t is 0 allows us to find C_1.

$$0 = 0 + C_1$$

Hence C_1 is 0, and the equation is

$$10t = v.$$

Thus, when t is 50 ms (or 50×10^{-3} s), we have

$$10(50)(10^{-3}) = v$$
$$0.5 \text{ volts} = v. \quad ■$$

EXAMPLE 7 Newton's law of cooling states that the rate at which the temperature of an object changes is proportional to the difference between its temperature and that of the surrounding medium. (Both theory and experiments support a more complicated law, but Newton's law is a good approximation.) An object with an initial temperature of 105° C is allowed to cool in a room at a temperature of 20° C. After 15 minutes, the temperature of the object has dropped to 65° C. What is the temperature after 30 minutes?

Solution Let y be the temperature after t minutes. Newton's law of cooling states that

$$\frac{dy}{dt} = k(y - 20).$$

Separating the variables, we have

$$\int \frac{dy}{(y - 20)} = \int k\,dt.$$

Thus, $\ln(y-20) = kt + C_1$. Use the logarithmic form to integrate.

$y - 20 = e^{kt+C_1}$ Use the definition of natural logarithm.

$y = 20 + e^{kt+C_1}$ Add 20 to both sides.

$= 20 + e^{kt}e^{C_1}$ Use properties of exponents.

$= 20 + Ce^{kt}$ Let $C = e^{C_1}$.

The equation $y = 20 + Ce^{kt}$ represents the general solution of the differential equation $\frac{dy}{dt} = k(y-20)$. Substituting the initial conditions $y = 105$ when $t = 0$, we obtain a particular solution.

$$105 = 20 + Ce^0 \text{ or } 85 = C$$

Thus, $y = 20 + 85e^{kt}$ is a particular solution. It is also possible to determine the value of k from the given information. We know that $y = 65$ when $t = 15$, so

$65 = 20 + 85e^{15k}$.

$\frac{45}{85} = e^{15k}$ Subtract 20 and divide by 85 on both sides.

$\ln(45/85) = 15k$ Use the definition of natural logarithm.

$\frac{\ln(45/85)}{15} = k$ Divide both sides by 15.

$-0.042 = k$

Thus, $y = 20 + 85e^{-0.042t}$.
When $t = 30$, $y = 20 + 85e^{-0.042(30)}$ or about 44° C. ■

EXERCISES 28.2

In 1–16, find a general solution for each differential equation.

1. $3x = yy'$
2. $x^2 = yy'$
3. $y^2 = y'$
4. $y = xy'$
5. $1 + x^3yy' = 0$
6. $x^2 + y^3y' = 0$
7. $y' = y + 2$
8. $y' - 2x = 3$
9. $y^2e^x\,dx + (e^x + 1)\,dy = 0$
10. $e^{2x}\,dy + e^x\,dx = 0$
11. $e^{x-y}\,dx - dy = 0$
12. $x^2\,dy = e^y\,dx$
13. $\cos x \csc y\,dx = \cos^2 x\,dy$
14. $(e^{\tan x} - 1)\,dx = \cos^2 x\,dy$
15. $dy = y^4 \sin x\,dx$
16. $\csc y\,dx = (1 - y)\,dy$

In 17–24, find the particular solution of each differential equation for the indicated conditions.

17. $y^2y' + x^3 = 0$; $y = 3$ when $x = 0$
18. $y'(2x + 3) = y$; $y = 1$ when $x = 3$
19. $y' = xy$; $y = e$ when $x = 0$
20. $y' + 4x = 3$; $y = 1$ when $x = 0$
21. $y'(x^2 + 4) = y$; $y = 1$ when $x = 0$
22. $y' - 3x^2 = 1$; $y = 2$ when $x = 0$
23. $x^2y' + y^2 = 0$; $y = 1$ when $x = 1$
24. $xy' = 2y$; $y = 16$ when $x = 2$

25. A 10-liter container is filled with acid. Liquid is pumped out at the rate of 1 liter per minute, and water is pumped in at the same rate.
 (a) Find a formula for the amount of water in the tank at time t (in minutes), assuming that the two liquids are uniformly mixed at all times.
 (b) Find the amount of water in the tank after 5 minutes. **4 8** 1 2 6 7 9

26. In radioactive decay, the rate of decay is directly proportional to the mass of material present. Suppose that 10.0 grams of a substance decays at a rate equal to 1/3 of the amount of the substance present.
 (a) Find a formula for the amount of the substance present after t hours. (Hint: Since the problem involves a decreasing amount, use a negative constant in the proportion.)
 (b) Find the amount present when $t = 5.00$. **6 9** 5 8

27. In a particular culture, the number of bacteria increases at a rate that is directly proportional to the number of bacteria present. Suppose that the number of bacteria rose from 200 to 2000 in 8 hours.
 (a) Find a formula for the number of bacteria present after t hours.
 (b) Find the number of bacteria present after 4 hours. **9** 1 6

28. The velocity v of an object falling through a resisting medium satisfies the equation $m\,dv/dt = mg - cv$. In the formula, m is the mass of the object, g is the acceleration due to gravity, and $-c$ is the deceleration due to air resistance. Suppose that a 98-kg paratrooper falls from rest with $g = 9.8\ \text{m/s}^2$ and $c = 27.3\ \text{m/s}^2$.
 (a) Find a formula for the paratrooper's velocity v at time t.
 (b) Find the velocity of the paratrooper after 1.0 second. **4 7 8** 1 2 3 5 6 9

29. A bowl of soup is heated to a temperature of 185° F and then allowed to cool in a room at a temperature of 68° F. After one-half hour, the temperature of the soup has dropped to 110° F.
 (a) Find a formula for the temperature after time t (in hours).
 (b) Find the temperature after 1 hour. **6 8** 1 2 3 5 7 9

30. Find an equation for the curve that has a slope of x/y and passes through the point (0, 4). (Hint: Solve $dy/dx = x/y$, and use the fact that (0, 4) is on the graph to evaluate the constant of integration.) **1 2 4 7 8** 3 5

1. Architectural Technology
2. Civil Engineering Technology
3. Computer Engineering Technology
4. Mechanical Drafting Technology
5. Electrical/Electronics Engineering Technology
6. Chemical Engineering Technology
7. Industrial Engineering Technology
8. Mechanical Engineering Technology
9. Biomedical Engineering Technology

28.3 FIRST ORDER LINEAR DIFFERENTIAL EQUATIONS

Not all first order, first degree differential equations are separable. Many different methods are necessary to solve such equations. The technique for solving a specific equation depends on the form of the equation. Particularly important in applied problems is the first order *linear* differential equation. A first order differential equation is said to be **linear** if it can be written in the form

first order linear differential equation

$$y' + f(x)\,y = g(x).$$

Both y and y' will always have exponents of 1. Since f and g are functions of x, however, they may contain any power of x, or they may be exponential, logarithmic, or trigonometric functions. In differential notation, a first order linear differential equation is written as

$$dy + f(x)\,y\,dx = g(x)\,dx. \qquad \textbf{(28-2)}$$

To solve an equation of this form, multiply both sides by $e^{\int f(x)\,dx}$.

$$e^{\int f(x)dx}\,dy + e^{\int f(x)dx} f(x)\,y\,dx = e^{\int f(x)dx} g(x)\,dx \tag{28-3}$$

We simplify this equation by observing that

$$\text{if } u = y\,e^{\int f(x)dx}, \text{ then } du = y\,d(e^{\int f(x)dx}) + (e^{\int f(x)dx})\,dy \text{ or}$$

$$du = y\,e^{\int f(x)dx} f(x)\,dx + e^{\int f(x)dx}\,dy. \tag{28-4}$$

The right-hand side of Equation 28-4 is equivalent to the left-hand side of Equation 28-3. Therefore, putting together these equations, we obtain

$$du = e^{\int f(x)dx} g(x)\,dx. \text{ Thus,}$$

$$\int du = \int e^{\int f(x)dx} g(x)\,dx \qquad \text{Integrate both sides.}$$

$$u = \int e^{\int f(x)dx} g(x)\,dx. \qquad \text{Replace } \int du \text{ with } u.$$

Finally, we replace u with $y\,e^{\int f(x)dx}$.

$$y\,e^{\int f(x)dx} = \int e^{\int f(x)dx} g(x)\,dx \tag{28-5}$$

Equation 28-5 is the solution of Equation 28-2. The right-hand side of this equation involves only x, so it may be possible to integrate it directly. Many differential equations, however, have no direct solution. When such equations arise in real world problems, they must be solved by numerical methods, which we will not consider.

To solve a first order linear differential equation:

1. Write the equation in the form

$$dy + f(x)\,y\,dx = g(x)\,dx. \tag{28-2}$$

2. Identify $f(x)$ and $g(x)$.
3. Find $\int f(x)\,dx$.
4. Find the solution by substituting $g(x)$ and $\int f(x)\,dx$ into the equation

$$y\,e^{\int f(x)dx} = \int e^{\int f(x)dx} g(x)\,dx. \tag{28-5}$$

EXAMPLE 1 Solve the differential equation $x\,dy + x^3 y\,dx = 2x^3\,dx$.

Solution 1. Divide both sides by x to obtain $dy + x^2 y\,dx = 2x^2\,dx$.

2. This equation fits the pattern of Equation 28-2 with $f(x) = x^2$ and $g(x) = 2x^2$.

3. $\int f(x)\,dx = \int x^2\,dx = x^3/3 + C_1$

$e^{\int f(x)dx} = e^{x^3/3 + C_1} = e^{x^3/3}e^{C_1} = Ce^{x^3/3}$

Including the constant of integration C_1 means that a constant factor C will show up on both sides of Equation 28-5 when $Ce^{x^3/3}$ replaces $e^{\int f(x)dx}$. For that reason, the constant of integration is generally omitted. Thus,

$$\int f(x)\,dx = x^3/3 \text{ and } e^{\int f(x)dx} = e^{x^3/3}.$$

4. The solution is $y\,e^{x^3/3} = \int e^{x^3/3}(2x^2)\,dx$, or

$$y\,e^{x^3/3} = 2\int e^{x^3/3}(x^2)\,dx.$$

The integral on the right is of the form $\int e^u\,du$, with $u = x^3/3$. We have

$$y\,e^{x^3/3} = 2e^{x^3/3} + C.$$

Both sides of the equation are divided by $e^{x^3/3}$ to produce

$$y = 2 + Ce^{-x^3/3}. \quad ■$$

In first order linear differential equations, the function $f(x)$ is often a function of the form $1/u$. Thus, $\int f(x)\,dx = \ln u$, and $e^{\int f(x)dx}$ is of the form $e^{\ln u}$. From the definition of a natural logarithm, we know that $e^{\ln u} = u$.

EXAMPLE 2 Solve the differential equation $y' + y/x = 2x$.

Solution In differential form, the equation is $dy + y/x\,dx = 2x\,dx$. This equation follows the pattern of Equation 28-2 with $f(x) = 1/x$ and $g(x) = 2x$. $\int f(x)\,dx = \ln x$, so $e^{\int f(x)dx} = e^{\ln x} = x$.

$$yx = \int x(2x)\,dx \text{ or } yx = \int 2x^2\,dx$$

$$yx = (2/3)x^3 + C \text{ or } y = (2/3)x^2 + Cx^{-1} \quad ■$$

The functions f and g may be trigonometric functions.

EXAMPLE 3 Solve the differential equation $dy - y \tan x\, dx = dx$ for $0 \leq x \leq \pi/2$.

Solution Let $f(x) = -\tan x$, and $g(x) = 1$.

$$\int f(x)\, dx = \ln|\cos x| \text{ and } e^{\int f(x)\, dx} = e^{\ln|\cos x|} = |\cos x|$$

But for $0 \leq x \leq \pi/2$, $|\cos x| = \cos x$. Thus,

$$y \cos x = \int \cos x\, dx$$
$$= \sin x + C.$$

Dividing both sides by $\cos x$, we obtain

$$y = \frac{\sin x}{\cos x} + \frac{C}{\cos x}$$
$$= \tan x + C \sec x. \quad ■$$

EXAMPLE 4 Consider the circuit shown in Figure 28.1. The inductance is L (in henries), the resistance is R (in ohms), the current is I (in amperes), and the voltage is V (in volts). The quantities are related by the differential equation

$$dI/dt + RI/L = V/L.$$

Solve the equation for I.

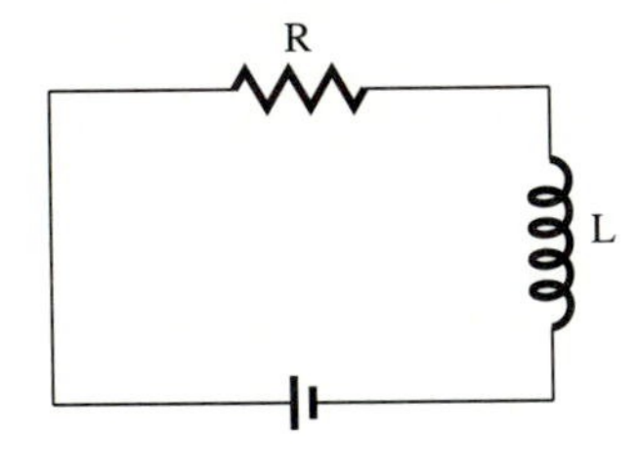

FIGURE 28.1

Solution To put the equation in the form of Equation 28-2, note that I is the dependent variable and t the independent variable. Multiply by dt.

$$dI + (RI/L)\, dt = (V/L)\, dt$$

Thus,

$$f(t) = R/L, \text{ and } g(t) = V/L.$$

Since $\int f(t)\, dt = (R/L)t$, we have $e^{\int f(t)\, dt} = e^{Rt/L}$.

Substituting into Equation 28-5, we find the solution.

$$I\,e^{Rt/L} = \int e^{Rt/L}\,(V/L)\,dt$$

$$I\,e^{Rt/L} = (V/L)\int e^{Rt/L}\,dt \qquad \text{Use } \int cu\,du = c\int u\,du.$$

$$I\,e^{Rt/L} = (V/R)\int e^{Rt/L}\,(R/L)\,dt \qquad \text{Multiply by } L/R \text{ and } R/L.$$

$$I\,e^{Rt/L} = (V/R)e^{Rt/L} + C \qquad \text{Let } u = Rt/L \text{ and integrate } \int e^{u}\,du.$$

Dividing both sides of the equation by $e^{Rt/L}$, we obtain

$$I = (V/R) + C\,e^{-Rt/L}. \quad ■$$

constant coefficients

The first order linear differential equation in which $f(x)$ and $g(x)$ are constant functions is said to have *constant coefficients.* For many differential equations arising in applied problems, $g(x)$ is 0. That is, the equation follows the pattern $y' + ky = 0$.

EXAMPLE 5 Solve the equation $y' + ky = 0$.

Solution In differential form, the equation is

$$dy + ky\,dx = 0\,dx.$$

Since the equation follows the form of Equation 28-2 with $f(x) = k$ and $g(x) = 0$, we have $\int f(x)\,dx = \int k\,dx$, which is kx. Thus,

$$ye^{kx} = \int e^{kx}\,(0)\,dx.$$

$$ye^{kx} = \int 0\,dx$$

$$ye^{kx} = C \text{ where } C \text{ is the constant of integration}$$

$$y = Ce^{-kx} \quad ■$$

The equation in Example 5 could have been solved using separation of variables. This equation is typical of problems involving growth or decay. In applied problems, initial values are often known and allow us to find particular solutions of differential equations. Example 6 illustrates the procedure.

EXAMPLE 6 When interest is compounded continuously, the balance grows at a rate that is directly proportional to the balance at any given time. If a \$12,000 deposit grows to \$13,000 in 2.2 years (803 days), find the balance after 60 days.

Solution Let y represent the balance and t the time in days. Then dy/dt is the rate of growth of balance and $dy/dt = ky$. Rewrite the equation as

$$dy - ky\,dt = 0.$$

The equation fits the form of Equation 28-2 with $f(t) = -k$ and $g(t) = 0$. Thus,

$$ye^{-kt} = \int e^{-kt}(0)\,dt.$$

$ye^{-kt} = C$ where C is the constant of integration

$$y = Ce^{kt}$$

When t is 0, y is 12,000.

$$12{,}000 = Ce^{(0)}$$

$$12{,}000 = C, \text{ and } y = 12{,}000e^{kt}$$

When t is 803, y is 13,000.

$$13{,}000 = 12{,}000e^{k(803)}$$

$$13/12 = e^{k(803)}$$

$$\ln(13/12) = k(803)$$

$$\ln(13/12) \div 803 = k \text{ and } y = 12{,}000e^{0.0000997t}$$

When t is 60, $y = 12{,}000e^{0.0000997(60)}$ or \$12,071.98. ■

EXERCISES 28.3

In 1–16, find a solution of each first order linear differential equation.

1. $x\,dy + x^2y\,dx = 3x^2\,dx$
2. $dy - x^3y\,dx = 4x^3\,dx$
3. $dy + y\,dx = 2x\,dx$
4. $x\,dy + 2xy\,dx = x^2\,dx$
5. $dy + y\,dx = e^x\,dx$
6. $dy + 2y\,dx = e^{-x}\,dx$
7. $x\,dy + 3xy\,dx = 2x\,dx$
8. $x^2\,dy - x^2y\,dx = x^2\,dx$
9. $x\,dy + 3y\,dx = x^3\,dx$
10. $dy + \frac{2}{x}y\,dx = 3x\,dx$
11. $y' + \frac{2}{x}y = \cos x$
12. $xy' + 2xy = x\sin 2x$
13. $y'\cos x - y\sin x = 1, 0 \le x \le \pi/2$
14. $y' + y\sec^2 x = \sec^2 x$
15. $y' - y\csc^2 x = \csc^2 x$
16. $y' + y\cot x = 1, 0 < x \le \pi/2$

In 17–24, find the particular solution of each first order linear differential equation for the given initial conditions.

17. $dy + y\,dx = x\,dx;\ y = 1$ when $x = 0$
18. $dy + y\,dx = x\,dx;\ y = 3$ when $x = 0$
19. $dy + 2y\,dx = x\,dx;\ y = 0$ when $x = 0$
20. $dy + y\sin x\,dx = \sin x\,dx;\ y = 2$ when $x = 0$
21. $dy - 2y\,dx = e^{2x}\,dx;\ y = 2$ when $x = 0$
22. $3\,dy + y\,dx = x\,dx;\ y = 1$ when $x = 0$
23. $y' + y\sec x\tan x = \sec x\tan x;\ y = 1$ when $x = 0$
24. $x^2y' + 2xy = x\cos x;\ y = 1$ when $x = 0$

25. In a chemical reaction, a particular compound changes into another compound at a rate that is directly proportional to the unchanged amount. Initially there are 20.0 g of a substance present, and after one hour, only 16 g are unchanged.
(a) Find a formula for the amount unchanged after time t (in hours).
(b) How much remains unchanged after 2.00 hours?
(c) Find the time t at which 1/2 of the substance remains. 6 9 2 4 5 7 8

26. As light strikes a translucent material, the equation $dA/dh = -kA$ gives the relationship between the amount A of light transmitted and the depth h of the material.
(a) Find a formula for the amount of transmitted light at a depth of h if 1/2 of the light is transmitted at a depth of 1 cm. [Hint: Use A_0 for the amount of light when $h = 0$. Then $(1/2)A_0$ is transmitted when h is 1.]
(b) What fraction of light is transmitted at a depth of 2 cm? 9 1 2 3 5 6 7

27. When interest is compounded continuously, the balance increases at a rate that is directly proportional to the balance at any given time.
(a) If a deposit of exactly \$10,000 grows to exactly \$12,000 in 2.5 years (912.5 days), find a formula for the balance after t days.
(b) Find the balance after 60 days.
(c) Find the time t required to double the initial balance. 1 2 6 7 8 3 5 9

28. A room containing 1800 cubic feet of air has a smoke content of 5 parts per million. An air conditioner removes air from the room and replaces it with fresh air at a rate of 800 cubic feet per minute.
(a) Find a formula for the volume v of smoke in the room at time t (in minutes).
(b) Find a formula for the number n of parts per million of smoke in the room at time t (in minutes). 4 8 1 2 6 7 9

29. For the circuit shown in the accompanying diagram, R is the resistance (in ohms), q is the charge (in coulombs) on the capacitor, C is the capacitance (in farads), and V is the potential difference (in volts). The quantities are related by the differential equation $R\frac{dq}{dt} + \frac{g}{C} = V$. 3 5 9 1 2 4 6 7 8

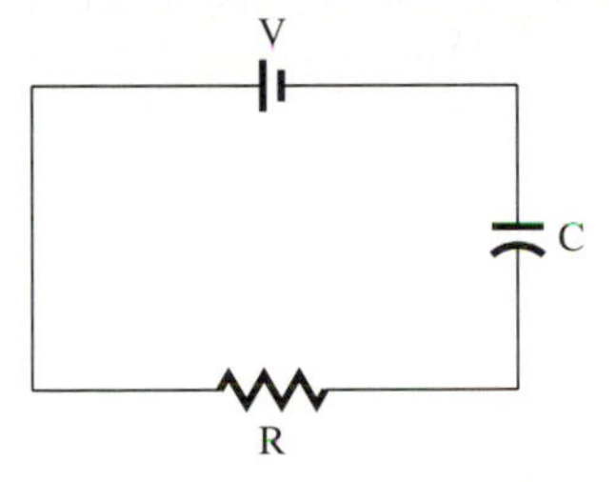

(a) Solve the equation for q, given that $q = 0$ when $t = 0$.
(b) If $V = 50$ volts, $C = 0.006$ farads, and $R = 25\ \Omega$, find q when $t = 0.3$ seconds.

Problems 30 and 31 use the following information. When two families (or collections) of curves are such that each curve of one family intersects each member of the other family at right angles, the two families of curves are said to be orthogonal trajectories of each other. In electrostatics, orthogonal trajectories are called equipotential curves and lines of force. In thermodynamics, they are called isothermal lines and heat flow lines. In hydrostatics, they are called velocity potential curves and flow lines or stream lines. The orthogonal trajectories for a given family of curves are found as follows.

1. *Find an expression for the slope of the curves in the given family. If the expression for slope contains a constant, eliminate it by substituting an expression for the constant obtained from the original equation.*
2. *Find an expression for the slope of the curves in the other family. Curves that are perpendicular have slopes that are negative reciprocals of each other.*
3. *Solve the equation that equates dy/dx with the expression from step* 2.

30. Find the orthogonal trajectories for the family of ellipses $4x^2 + y^2 = K$, $K \geq 0$. 6 8 1 2 3 5 7 9

31. Find the orthogonal trajectories for the family of hyperbolas $y = K/x$, $K \geq 0$. 6 8 1 2 3 5 7 9

1. Architectural Technology
2. Civil Engineering Technology
3. Computer Engineering Technology
4. Mechanical Drafting Technology
5. Electrical/Electronics Engineering Technology
6. Chemical Engineering Technology
7. Industrial Engineering Technology
8. Mechanical Engineering Technology
9. Biomedical Engineering Technology

28.4 SECOND ORDER HOMOGENEOUS LINEAR DIFFERENTIAL EQUATIONS WITH CONSTANT COEFFICIENTS

second order linear differential equation

A second order differential equation is said to be **linear** if it can be written in the form

$$y'' + f(x)y' + g(x)y = r(x). \qquad \textbf{(28-6)}$$

The derivatives of y and y itself will always have 1 as the exponent. Since f, g, and r are functions of x, they may therefore contain any power of x or they may be exponential, logarithmic, or trigonometric functions. The functions f and g are called the coefficients of the equation. We will restrict our discussion of second order linear differential equations to those in which f and g are constant functions.

homogeneous nonhomogeneous

Linear differential equations are generally divided into two categories, *homogeneous* and *nonhomogeneous.* If $r(x)$ is 0 in Equation 28-6, the equation is said to be **homogeneous.** If $r(x)$ is not 0, the equation is said to be **nonhomogeneous.** Homogeneous linear equations have somewhat simpler properties, so we consider those first. Thus, we will consider only equations that can be put in the form

$$y'' + by' + cy = 0. \qquad \textbf{(28-7)}$$

In Example 5 of the previous section, we saw that the solution of the first order homogeneous linear differential equation with constant coefficients $y' + ky = 0$ is given by $y = Ce^{-kx}$. We might make an educated guess that the solution of Equation 28-7 also involves the exponential function. We try $y = e^{mx}$ to see if it is a solution.

$$\text{If } y = e^{mx}, \text{ then } y' = me^{mx}, \text{ and } y'' = m^2e^{mx}.$$

Replacing y, y', and y'' in Equation 28-7 by these expressions, we have

$$m^2e^{mx} + bme^{mx} + ce^{mx} = 0$$

$$e^{mx}(m^2 + bm + c) = 0.$$

Since it is impossible for e^{mx} to be zero for any value of m, the other factor must be 0. That is,

$$m^2 + bm + c = 0.$$

NOTE ⇨⇨

This equation is called the characteristic equation, or auxiliary equation, of Equation 28-7. ***Notice that the characteristic equation can be determined from its differential equation simply by replacing y'' with m^2, y' with m, and y with 1.*** We now show that this observation is valid even if the equation has the form $ay'' + by' + cy = 0$. Such an equation can be written

$$y'' + (b/a)y' + (c/a)y = 0$$

with characteristic equation

$$m^2 + (b/a)m + (c/a),$$

which can be multiplied by a to obtain

$$am^2 + bm + c = 0.$$

Returning to the problem at hand, then, $m^2 + bm + c = 0$ can be solved by the quadratic formula. Thus, the two solutions are given by

$$m_1 = \frac{-b + \sqrt{b^2 - 4c}}{2} \quad \text{and} \quad m_2 = \frac{-b - \sqrt{b^2 - 4c}}{2}.$$

Thus, $y = e^{m_1 x}$ and $y = e^{m_2 x}$ are both solutions of Equation 28-7. It can be shown (although we will not do so here) that the general solution of the equation is given by

$$y = C_1 e^{m_1 x} + C_2 e^{m_2 x}.$$

To solve a second order homogeneous linear equation with constant coefficients:

1. Write the equation in the form $ay'' + by' + cy = 0$.
2. Solve the characteristic equation $am^2 + bm + c = 0$ to find m_1 and m_2.
3. Find the solution by substituting the values of m_1 and m_2 from step 2 into the equation $y = C_1 e^{m_1 x} + C_2 e^{m_2 x}$, if $m_1 \neq m_2$.

EXAMPLE 1 Solve the differential equation $2y'' + y' - 3y = 0$.

Solution The characteristic equation is

$$2m^2 + m - 3 = 0$$
$$(2m + 3)(m - 1) = 0$$
$$m = -3/2 \text{ or } m = 1.$$

The general solution is $y = C_1 e^{(-3/2)x} + C_2 e^x$. ■

EXAMPLE 2 Solve the differential equation $\frac{d^2y}{dx^2} - 4\frac{dy}{dx} = 0$, and find the particular solution that satisfies the initial conditions that y is 5 and y' is 8 when x is 0.

Solution First, we find the general solution. The equation can be written in the form

$$y'' - 4y' = 0.$$

Its characteristic equation is solved as follows.

$$m^2 - 4m = 0$$
$$m(m - 4) = 0 \quad \text{Factor.}$$
$$m = 0 \text{ or } m = 4 \quad \text{Set each factor equal to zero and solve.}$$

Thus, the general solution is

$$y = C_1 e^{0x} + C_2 e^{4x}$$
$$= C_1 + C_2 e^{4x}.$$

The initial conditions are used to evaluate y and y'.

$$\text{If } y = C_1 + C_2 e^{4x}, \text{ then } y' = 4C_2 e^{4x}.$$

When x is 0, y is 5, so $5 = C_1 + C_2$.
Also, when x is 0, y' is 8, so $8 = 4C_2$.
These two equations form a system of equations, which we can solve.

$$\begin{aligned} 5 &= C_1 + C_2 \\ 8 &= \quad\;\; 4C_2 \end{aligned}$$

From the second equation, we know that C_2 is 2. Substituting this value into the first equation, we learn that C_1 is 3. The particular solution sought is

$$y = 3 + 2e^{4x}. \quad ■$$

EXAMPLE 3 The equation $d^2i/dt^2 + 21\ di/dt + 20\ i = 0$ describes the current for a particular circuit. Solve the equation for i, if i is 1.1, and i' is -31.5 when t is 0.

Solution The equation can be written $i'' + 21\ i' + 20\ i = 0$. Its characteristic equation is

$$\begin{aligned} m^2 + 21m + 20 &= 0 \\ (m + 20)(m + 1) &= 0 \\ m = -20 \text{ or } m &= -1. \end{aligned}$$

Thus, the general solution is $i = C_1 e^{-20t} + C_2 e^{-t}$.
The initial conditions are used to evaluate i and i'.

$$i' = -20\ C_1 e^{-20t} - C_2 e^{-t}$$

When t is 0, i is 1.1, and

$$1.1 = C_1 e^0 + C_2 e^0 \text{ or } 1.1 = C_1 + C_2.$$

When t is 0, i' is -31.5, and

$$-31.5 = -20\ C_1 e^0 - C_2 e^0 \text{ or } -31.5 = -20\ C_1 - C_2.$$

We now have a system of equations, which can be solved as follows.

$$\begin{aligned} 1.1 &= \quad\;\; C_1 + C_2 \\ -31.5 &= -20\ C_1 - C_2 \\ \hline -30.4 &= -19\ C_1 \text{ or } C_1 = 1.6 \end{aligned}$$

Since $1.1 = C_1 + C_2$, we conclude that $C_2 = -0.5$. The particular solution, then, is $i = 1.6\ e^{-20t} - 0.5\ e^{-t}$. ■

A quadratic equation may have

(a) two distinct real roots
(b) a double real root, or
(c) two complex conjugate roots.

So far, in each example of this section, the characteristic equation has had two distinct real roots. If the characteristic equation falls into category (b) or (c), the solution must be modified. Consider, for instance,

$$y'' - 6y' + 9 = 0.$$

Its characteristic equation is

$$m^2 - 6m + 9 = 0$$
$$(m - 3)(m - 3) = 0.$$

Thus, $m = 3$ is a double root. If we try to use $y = C_1 e^{3x} + C_2 e^{3x}$ as the solution, it can be simplified to $y = (C_1 + C_2)e^{3x}$. The two constants can be combined, so that $y = Ce^{3x}$. Since the equation can be written with a single arbitrary constant, this solution is not the general solution, which should have two such constants. Second order homogeneous linear equations whose characteristic equations fall into category (b) or (c) may be treated by methods similar to those used in Examples 1 – 3. The formulas are stated for completeness, but are not justified here.

GENERAL SOLUTION OF $ay'' + by' + cy = 0$

(a) If m_1 and m_2 are distinct real roots of the characteristic equation $am^2 + bm + c = 0$, then the general solution is $y = C_1 e^{m_1 x} + C_2 e^{m_2 x}$.
(b) If m_1 and m_2 are equal real roots of the characteristic equation $am^2 + bm + c = 0$, then the general solution is $y = C_1 e^{m_1 x} + C_2 x e^{m_2 x}$.
(c) If m_1 and m_2 are complex roots of the characteristic equation $am^2 + bm + c = 0$ with $m_1 = \alpha + \beta j$ and $m_2 = \alpha - \beta j$, then the general solution is $y = C_1 e^{\alpha x}(\cos \beta x) + C_2 e^{\alpha x}(\sin \beta x)$.

EXAMPLE 4 Solve $y'' - 6y' + 9 = 0$.

Solution Since the characteristic equation $m^2 - 6m + 9 = 0$ has a double root ($m = 3$), the general solution is obtained from (b) in the box with $m_1 = m_2 = 3$. We have

$$y = C_1 e^{3x} + C_2 x e^{3x}. \quad ■$$

EXAMPLE 5 Solve $y'' - 2y' + 2 = 0$.

Solution The characteristic equation is $m^2 - 2m + 2 = 0$. Since the polynomial cannot be factored, we use the quadratic formula.

$$m = \frac{2 \pm \sqrt{4 - 8}}{2} = \frac{2 \pm \sqrt{-4}}{2} = \frac{2 \pm 2j}{2} = 1 \pm j$$

That is, $m_1 = 1 + j$ and $m_2 = 1 - j$. The solution is given by

$$y = C_1 e^{\alpha x}(\cos \beta x) + C_2 e^{\alpha x}(\sin \beta x) \text{ with } \alpha = 1 \text{ and } \beta = 1.$$
$$y = C_1 e^{x}(\cos x) + C_2 e^{x}(\sin x) \quad ■$$

EXERCISES 28.4

In 1–10, find the general solution of each differential equation.

1. $y'' - y' - 12y = 0$
2. $y'' + 5y' + 6y = 0$
3. $2y'' - 3y' - 2y = 0$
4. $\frac{2d^2y}{dx^2} + \frac{dy}{dx} - 3y = 0$
5. $\frac{2d^2y}{dx^2} + \frac{dy}{dx} = 0$
6. $\frac{3d^2y}{dx^2} - \frac{2dy}{dx} = 0$
7. $y'' + 4y' + 4y = 0$
8. $y'' - 2y' + y = 0$
9. $\frac{d^2y}{dx^2} + \frac{dy}{dx} + y = 0$
10. $\frac{2d^2y}{dx^2} + \frac{dy}{dx} + 2y = 0$

In 11–20, find the particular solution of each differential equation for the given initial conditions.

11. $y'' - y' = 0$; $y = 5$ and $y' = -3$ when $x = 0$
12. $y'' - y' - 12y = 0$; $y = 2$ and $y' = 1$ when $x = 0$
13. $3y'' - 4y' - 4y = 0$; $y = 4$ and $y' = 0$ when $x = 0$
14. $\frac{3d^2y}{dx^2} - \frac{dy}{dx} = 0$; $y = 5$ and $y' = 3$ when $x = 0$
15. $\frac{2d^2y}{dx^2} + \frac{4dy}{dx} = 0$; $y = 0$ and $y' = 2$ when $x = 0$
16. $\frac{2d^2y}{dx^2} - \frac{5dy}{dx} - 3y = 0$; $y = 1$ and $y' = 3$ when $x = 0$
17. $y'' + 2y' + y = 0$; $y = 1$ and $y' = 0$ when $x = 0$
18. $y'' - 4y' + 4y = 0$; $y = 0$ and $y' = -3$ when $x = 0$
19. $y'' + y = 0$; $y = 1$ and $y' = 1$ when $x = 0$
20. $y'' - 2y' + 10y = 0$; $y = 4$ and $y' = 1$ when $x = 0$

Problems 21–24 use the following information. Consider an object with mass m fastened to a spring whose constant is k. Assume that the mass of the spring is negligible in comparison to the mass of the object and that there is no resistance to the motion when the object is pulled down from the rest position and released. That is, there is no damping force. The equation $m\, d^2x/dt^2 + kx = 0$ describes the motion of the object. In reality, there is a damping force opposing the motion of the object, and it is directly proportional to the velocity of the object. The equation $m\, d^2x/dt^2 + \beta\, dx/dt + kx = 0$ describes the motion in such cases.

21. A particular spring has spring constant $k = 27$ if the distances are measured in centimeters and the mass in kilograms. An object with mass of 3 kg is attached. Find a general formula for the position of the object at time t, assuming no damping force. 4 7 8 1 2 3 5 6 9
22. Rework problem 21, assuming a damping force with $\beta = 18$. 4 7 8 1 2 3 5 6 9
23. Rework problem 21, assuming a damping force with $\beta = 30$. 4 7 8 1 2 3 5 6 9
24. Rework problem 21, assuming a damping force with $\beta = 14.4$. 4 7 8 1 2 3 5 6 9

25. In the study of planetary motion, the equation $\frac{d^2z}{d\theta^2} + z = k$ is used to determine $z = 1/r$, where r is the distance of the planet from the sun, k is a constant determined by observation, and θ is the angle through which the planet has traveled. Find the general solution for this equation, if k is 0.
4 7 8 1 2 3 5 6 9

26. For small angles, the angle θ (in radians) that a pendulum makes with the vertical at time t (in seconds) may be approximated by the equation $L\,d^2\theta/dt^2 + g\theta/L = 0$.

In the formula, L is the length of the pendulum, and g is the acceleration due to gravity (-9.8 m/s^2). Find the general solution.
4 7 8 1 2 3 5 6 9

1. Architectural Technology
2. Civil Engineering Technology
3. Computer Engineering Technology
4. Mechanical Drafting Technology
5. Electrical/Electronics Engineering Technology
6. Chemical Engineering Technology
7. Industrial Engineering Technology
8. Mechanical Engineering Technology
9. Biomedical Engineering Technology

28.5 SECOND ORDER NONHOMOGENEOUS LINEAR DIFFERENTIAL EQUATIONS WITH CONSTANT COEFFICIENTS

We now turn our attention to second order nonhomogeneous linear equations. Consider the differential equation

$$y'' + by' + cy = r(x). \tag{28-8}$$

It can be shown that if y_p is a particular solution of this equation and y_h is the general solution of the corresponding homogeneous equation (in which $r(x)$ is 0), then $y = y_p + y_h$ is the general solution of Equation 28-8. We will not show that any solution must take this form, but we do verify that a solution of this form is correct. If $y = y_p + y_h$, then y'', y', and y can be replaced in Equation 28-8.

$$(y_p + y_h)'' + b(y_p + y_h)' + c(y_p + y_h) = r(x)$$
$$y_p'' + y_h'' + by_p' + by_h' + cy_p + cy_h = r(x)$$

Reordering and regrouping terms, we have

$$(y_p'' + by_p' + cy_p) + (y_h'' + by_h' + cy_h) = r(x). \tag{28-9}$$

But y_p is a particular solution of $y'' + by' + cy = r(x)$, so

$$y_p'' + by_p' + cy_p = r(x),$$

and y_h is the general solution of $y'' + by' + cy = 0$, so

$$y_h'' + by_h' + cy_h = 0.$$

Equation 28-9 is therefore $r(x) + 0 = r(x)$, and we have verified that $y = y_p + y_h$ is a solution of Equation 28-8.

Finding a particular solution of a nonhomogeneous equation can be difficult. For the special cases in which $r(x)$ contains only sums or products of functions of the form x^n, e^{mx}, $\cos\beta x$, or $\sin\beta x$, the method of *undetermined coefficients* can be used. This method is illustrated in Example 1.

undetermined coefficients

EXAMPLE 1 Find the general solution of

$$y'' - 2y' - 3y = 2\cos x. \qquad \textbf{(28-10)}$$

Solution The corresponding homogeneous equation is $y'' - 2y' - 3y = 0$. Its characteristic equation is

$$\begin{aligned} m^2 - 2m - 3 &= 0 \\ (m-3)(m+1) &= 0 \\ m = 3 \text{ or } m &= -1. \end{aligned}$$

Thus, $y = C_1 e^{-x} + C_2 e^{3x}$ is a solution. Next, we try to find a particular solution of the nonhomogeneous equation (Equation 28-10). We choose $y_p = A\cos x + B\sin x$, so that the solution will contain trigonometric functions, making it similar in form to $r(x) = 2\cos x$.

$$\begin{aligned} \text{If } y_p &= A\cos x + B\sin x, \text{ then} \\ y_p' &= -A\sin x + B\cos x, \text{ and} \\ y_p'' &= -A\cos x - B\sin x. \end{aligned}$$

Substituting y_p'', y_p', and y_p into Equation 28-10, we have $(-A\cos x - B\sin x) - 2(-A\sin x + B\cos x) - 3(A\cos x + B\sin x) = 2\cos x$. Reordering and regrouping terms, we have

$$\begin{aligned} (-A\cos x - 2B\cos x - 3A\cos x) + (-B\sin x + 2A\sin x - 3B\sin x) &= 2\cos x \\ (-4A\cos x - 2B\cos x) + (2A\sin x - 4B\sin x) &= 2\cos x \\ (-4A - 2B)\cos x + (2A - 4B)\sin x &= 2\cos x. \end{aligned}$$

Equating coefficients of $\cos x$ on both sides and coefficients of $\sin x$ on both sides, we obtain the following system of equations.

$$\begin{array}{rcl} -4A - 2B = 2 & \text{or} & 8A + 4B = \ \ -4 \\ 2A - 4B = 0 & & \underline{2A - 4B = \quad 0} \\ & & 10A \qquad = \ \ -4 \\ & & A \qquad = -2/5 \end{array}$$

Since

$$\begin{aligned} 2A - 4B &= 0, \\ -4/5 - 4B &= 0 \\ -4B &= 4/5 \\ B &= -1/5. \end{aligned}$$

The particular solution sought, then is $y_p = (-2/5)\cos x - (1/5)\sin x$. The general solution is

$$y = y_h + y_p = C_1 e^{-x} + C_2 e^{3x} - (1/5)\sin x - (2/5)\cos x. \quad ■$$

The method of undetermined coefficients is summarized as follows.

To solve a second order nonhomogeneous linear equation by the method of undetermined coefficients:

1. Write the equation in the form $ay'' + by' + cy = r(x)$.
2. Find the general solution of the corresponding homogeneous equation $ay'' + by' + cy = 0$. Call it y_h.
3. Find a particular solution of the nonhomogeneous equation $ay'' + by' + cy = r(x)$. Call it y_p.
   (a) If $f(x) = Cx^k$, then $y_p = A_0 + A_1x + A_2x^2 + \cdots + A_kx^k$.
   (b) If $r(x) = Ce^{mx}$, then $y_p = Ae^{mx}$.
   (c) If $r(x) = C \sin \beta x$ or $r(x) = C \cos \beta x$, then $y_p = A \cos \beta x + B \sin \beta x$.
4. If y_p and y_h contain a like term, multiply the terms in y_p by the smallest power of x that removes the duplication.
5. Substitute y_p and its derivatives into the nonhomogeneous equation and solve for the coefficients.
6. Specify the general solution $y = y_h + y_p$.

The procedure may be modified if $r(x)$ is a sum of terms.

EXAMPLE 2 Solve $y'' + 2y' = x - e^x$.

Solution Begin by solving $y'' + 2y' = 0$. Its characteristic equation is

$$m^2 + 2m = 0$$
$$m(m + 2) = 0$$
$$m = 0 \text{ or } m = -2.$$
$$y_h = C_1e^{0x} + C_2e^{-2x} \text{ or } y_h = C_1 + C_2e^{-2x}$$

Next, we find a particular solution of $y'' + 2y' = x - e^x$. Think of $r(x)$ as a sum $r(x) = r_1(x) + r_2(x)$, with $r_1(x) = x$ and $r_2(x) = -e^x$. Then y_p is the sum $y_p = y_{p_1} + y_{p_2}$, with y_{p_1} based on the form $r_1(x)$ and y_{p_2} based on the form of $r_2(x)$. We have

$$y_p = \underbrace{(A_0 + A_1x)}_{y_{p1}} + \underbrace{Ae^x}_{y_{p2}}.$$

But both y_h and y_p contain constant terms (C_1 in y_h and A_0 in y_p). The duplication is removed if, instead of multiplying all of y_p by x, we multiply y_{p_1} by x, leaving y_{p_2} as is. That is, we multiply the polynomial $A_0 + A_1x$ by x, leaving Ae^x as is. Thus,

$$y_p = (A_0x + A_1x^2) + Ae^x.$$

It follows that

$$y_p' = (A_0 + 2A_1x) + Ae^x,$$

and

$$y_p'' = 2A_1 + Ae^x.$$

Replacing y'', y', and y in the nonhomogeneous equation, we have

$$2A_1 + Ae^x + 2(A_0 + 2A_1x + Ae^x) = x - e^x$$
$$(2A_1 + 2A_0) + 4A_1x + (Ae^x + 2Ae^x) = x - e^x$$
$$(2A_1 + 2A_0) + 4A_1x + 3Ae^x = x - e^x.$$

Equating like terms on the left and right, we obtain the following system of equations.

$$\begin{aligned} 2A_1 + 2A_0 &= 0 \\ 4A_1 \quad\quad &= 1 \\ 3A &= -1 \end{aligned}$$

From the second and third equations, we conclude that $A = -1/3$ and $A_1 = 1/4$. The first equation tells us that

$$2(1/4) + 2A_0 = 0$$
$$2A_0 = -1/2 \text{ or } A_0 = -1/4.$$

Thus, $$y_p = (-1/4)x + (1/4)x^2 - (1/3)e^x.$$

The general solution is $y = C_1 + C_2e^{-2x} - (1/4)x + (1/4)x^2 - (1/3)e^x$. ■

The methods you have seen may be extended for use with nth order linear differential equations. An nth order linear differential equation can be written in the form

$$f_n(x)\frac{d^n y}{dx^n} + f_{n-1}(x)\frac{d^{n-1} y}{dx^{n-1}} + \cdots + f_1(x)\frac{dy}{dx} + f_0(x)\, y = r(x).$$

If each function $f_i(x)$ is a constant function, we say that the equation has constant coefficients. If $r(x)$ is 0, the equation is said to be homogeneous, and if $r(x)$ is not 0, it is nonhomogeneous.

EXERCISES 28.5

In 1–12, find the general solution of each differential equation.

1. $y'' - 4y = 2x$
2. $y'' - 2y' - 3y = x^2 + x$
3. $2y'' + 3y' + y = x^2 - x$
4. $2y'' + y' = 3x + 2$
5. $y'' - 4y' = e^{2x}$
6. $y'' + 3y' = e^{-x}$
7. $2y'' + 5y' + 3y = 3e^{2x} + x$
8. $2y'' + y' - 6y = 2e^x - x$
9. $y'' + y' - 6y = 3\cos x$
10. $2y'' - 3y' = 2\sin 2x$
11. $3y'' + y' = -\sin 2x$
12. $6y'' + y' - 2y = \cos 3x$

In 13–16, find the particular solution of each differential equation for the given initial conditions.

13. $y'' + 25y = 5x$; $y = 5$ and $y' = -4.8$ when $x = 0$
14. $y'' - 4y' + 3y = 4e^{3x}$; $y = -1$ and $y' = 3$ when $x = 0$
15. $y'' - y' - 2y = 10\sin x$; $y = 1$ and $y' = -3$ when $x = 0$
16. $y'' + y' - 2y = 3e^x$; $y = 2$ and $y' = 3$ when $x = 0$

17. In problems 21–24 of the previous section, we considered an object with mass m fastened to a spring acted upon by a constant force. If a variable force F acts on the system, the equation of motion is given by

$$m\, d^2x/dt^2 + \beta\, dx/dt + kx = F$$

Find the general solution if $m = 1$, $\beta = 4$, $k = 3$, and $F = 32.5 \cos 2t$. 4 7 8 1 2 3 5 6 9

Problems 18–21 use the following information. In the circuit shown, the charge q (in coulombs) on the capacitor at time t is given by the equation

$$L\, d^2q/dt^2 + R\, dq/dt + q/C = V.$$

In the equation L is the inductance (in henries), R is the resistance (in ohms), C is the capacitance (in farads), and V is the voltage (in volts).

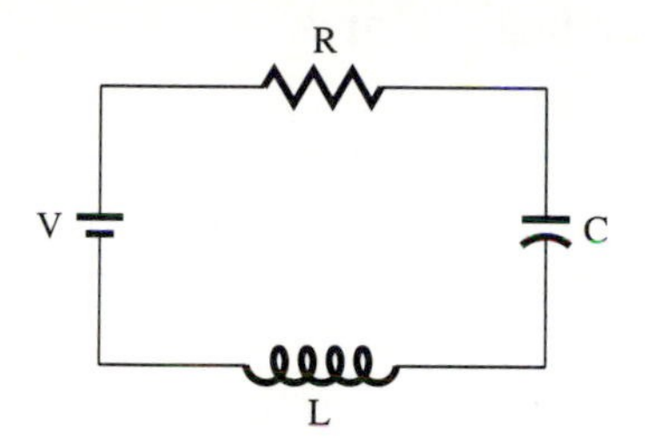

18. Find the general solution for q, assuming that the characteristic equation has two distinct real roots and that V is a constant. 3 5 9 1 2 4 6 7 8

19. If V is of the form $V_0 \cos \omega t$ and the roots of the characteristic equation are complex, the solution can be written as $q = e^{-\alpha t}\,(C_1 \cos \beta t + C_2 \sin \beta t) + A \cos \omega t + B \sin \omega t$ with $\alpha = R/(2L)$. Part of this solution tends toward zero as time increases. This part of the solution, called the "transient term" is often dropped. The remaining part, called the "steady-state term" remains. Identify the transient term and the steady-state term. 3 5 9 1 2 4 6 7 8

20. Find the general solution if $L = 10$ H, $R = 20\ \Omega$, $C = 0.05$ F, and $V = 50 \cos t$ volts. 3 5 9 1 2 4 6 7 8

21. Find the general solution if $L = 5$ H, $R = 20\ \Omega$, $C = 10^{-2}$ F, and $V = 340 \cos 4t$ volts. 3 5 9 1 2 4 6 7 8

1. Architectural Technology
2. Civil Engineering Technology
3. Computer Engineering Technology
4. Mechanical Drafting Technology
5. Electrical/Electronics Engineering Technology
6. Chemical Engineering Technology
7. Industrial Engineering Technology
8. Mechanical Engineering Technology
9. Biomedical Engineering Technology

28.6 LAPLACE TRANSFORMS

Differential equations arising in engineering and electronics are often solved by means of *Laplace transforms.* The technique essentially consists of three steps. First, the differential equation is transformed into an algebraic equation. The resulting equation is then solved algebraically. Finally, the solution is rewritten in a form that is appropriate for the solution of the original differential equation. One of the main advantages of using Laplace transforms is that it reduces the problem of solving a differential equation to that of solving an algebraic equation. Another advantage is that when initial values are given, it is not necessary to obtain a general solution before obtaining the particular solution that satisfies the initial conditions. Before we actually apply the technique, a few preliminaries are necessary. We begin with the notion of transformations in general.

We have previously encountered the idea of transforming one function into another. For instance, the operation of differentiation transforms a function f into its derivative f'. If the letter D is used to denote differentiation, the notation

$$D[f(x)] = f'(x)$$

denotes the transformation. Likewise,

$$I[f(x)] = \int f(x)\,dx$$

denotes the transformation of a function into its antiderivatives. The Laplace transformation is the transformation that produces

$$\int_0^\infty e^{-sx}f(x)\,dx$$

from $f(x)$. Since we have not previously defined integrals with infinite limits of integration, it is necessary to state that such integrals are found by evaluating the limit of the integral as the upper limit approaches infinity. That is,

$$\int_0^\infty e^{-sx}f(x)\,dx = \lim_{c\to\infty} \int_0^c e^{-sx}f(x)\,dx.$$

Thus, we have the following definition.

DEFINITION

The **Laplace transformation** L is the operation that produces a function F, defined in terms of s, from a function f, defined in terms of x as follows.

$$L[f(x)] = \int_0^\infty e^{-sx}f(x)\,dx = F(s)$$

provided the integral exists. The function F is called the **Laplace transform** of f.

It is customary to denote a function of x by a lower case letter (such as f) and its transform by the same letter in upper case (such as F). Before working problems, we note that there are certain limits that arise in working with Laplace transforms. Common limits are listed in Equations 28-11 through 28-13.

$$\lim_{c\to\infty} (e^{-sc}) = 0 \text{ for } s > 0 \qquad \textbf{(28-11)}$$

$$\lim_{c\to\infty} (e^{-sc})(c^n) = 0 \text{ for } s > 0 \qquad \textbf{(28-12)}$$

$$\lim_{c\to\infty} (e^{-sc})(sc) = 0 \text{ for } s > 0 \qquad \textbf{(28-13)}$$

We will not justify these limits, but using a calculator to evaluate the expressions for a few values of c, such as 10, 50, and 100 should make the limits seem reasonable.

EXAMPLE 1 Find the Laplace transform of the function defined by $f(x) = x$ for $x \geq 0$.

Solution

$$L[f(x)] = \int_0^\infty xe^{-sx}\,dx$$
$$= \lim_{c\to\infty} \int_0^c xe^{-sx}\,dx$$

This integral can be determined from the table of integrals in the Appendix. Although s is a variable, it can be treated as a constant, since the integration is with respect to x. Formula 46 with $u = -sx$ and $du = -s\,dx$ gives

$$L[f(x)] = \lim_{c\to\infty} \frac{1}{s^2} \int_0^c s^2xe^{-sx}\,dx$$
$$= \lim_{c\to\infty} \frac{1}{s^2} [e^{-sx}(-sx-1)]\Big|_0^c$$
$$= \lim_{c\to\infty} \frac{1}{s^2} [e^{-sc}(-sc-1) - e^0(-1)]$$
$$= \lim_{c\to\infty} \frac{1}{s^2} [e^{-sc}(-sc-1) + 1].$$

From Equations 28-13 and 28-11, we know that $\lim_{c\to\infty} e^{-sc}(sc) = 0$ and $\lim_{c\to\infty} e^{-sc} = 0$ if $s > 0$. Thus,

$$\lim_{c\to\infty} \frac{1}{s^2} [e^{-sc}(-sc-1) + 1] = \frac{1}{s^2}.$$

Thus, $L[f(x)] = F(s) = \dfrac{1}{s^2}$ if $s > 0$. ■

EXAMPLE 2 Find the Laplace transform of the function defined by $f(x) = \cos x$.

Solution

$$L[f(x)] = \int_0^\infty e^{-sx}(\cos x)\,dx$$
$$= \lim_{c\to\infty} \int_0^c e^{-sx}(\cos x)\,dx$$

Formula 48 with $u = x$, $a = -s$, and $b = 1$, gives

$$L[f(x)] = \lim_{c\to\infty} \left[\frac{e^{-sx}(-s\cos x + \sin x)}{s^2+1}\Big|_0^c\right]$$
$$= \lim_{c\to\infty} \left[\frac{e^{-sc}(-s\cos c + \sin c)}{s^2+1} - \frac{e^0(-s\cos 0 + \sin 0)}{s^2+1}\right]$$
$$= \lim_{c\to\infty} \left[\frac{e^{-sc}(-s\cos c + \sin c)}{s^2+1} - \frac{1(-s+0)}{s^2+1}\right].$$

From Equation 28-11, we know that e^{-sc} approaches 0 as c increases if $s > 0$. The values of cos c and sin c fluctuate between 1 and -1. Hence the quantity $-s\cos c + \sin c$ is bounded. Since $s^2 + 1$ is also bounded, the limit of the first fraction is zero if $s > 0$. We have

$$L[f(x)] = 0 - \left(\frac{-s}{s^2+1}\right) = \frac{s}{s^2+1}.$$

$$L[f(x)] = F(s) = \frac{s}{s^2+1} \text{ for } s > 0. \quad ■$$

Once a few basic Laplace transforms have been developed using the techniques of Examples 1 and 2, others may be found by applying properties of Laplace transforms. One important property is the *linearity property.* A transformation T is said to be linear if

$$T[f(x) + g(x)] = T[f(x)] + T[g(x)] \text{ and}$$
$$T[af(x)] = aT[f(x)]$$

for all functions f and g for which the transform is defined and all real numbers a. Informally, this statement says that the transform of a sum is the sum of the transforms, and the transform of a product of a constant and a function is the product of the constant and the transform of the function. These two statements are often combined into one.

LINEARITY PROPERTY

$$L[af(x) + bg(x)] = aL[f(x)] + bL[g(x)]$$

where L is a Laplace transformation, f and g are functions, and a and b are constants.

Tables of Laplace transforms have been constructed. The following short table will be adequate for our purposes. More extensive tables can be found in reference books. Many of these formulas can be obtained by the linearity property. Others are obtained by the shifting property, which says

$$L[e^{ax}f(x)] = F(s - a) \text{ where } F(s) = L[f(x)].$$

TABLE 28.1 Table of Laplace Transforms

Function $f(x)$	Laplace transform $F(s)$, $s > 0$	Function $(f(x)$	Laplace transform $F(s)$, $s > 0$
1. 1	$\frac{1}{s}$	9. $e^{ax} - e^{bx}$	$\frac{a-b}{(s-a)(s-b)}$ $s > a$, $s > b$
2. x	$\frac{1}{s^2}$	10. $-ae^{ax} + be^{bx}$	$\frac{s(-a+b)}{(s-a)(s-b)}$ $s > a$, $s > b$
3. x^n	$\frac{n!}{s^{n+1}}$	11. xe^{ax}	$\frac{1}{(s-a)^2}$ $s > a$
4. e^{ax}	$\frac{1}{s-a}$ $s > a$	12. $e^{ax}(1 + ax)$	$\frac{s}{(s-a)^2}$ $s > a$
5. $\cos ax$	$\frac{s}{s^2 + a^2}$	13. $x \sin ax$	$\frac{2as}{(s^2 + a^2)^2}$
6. $\sin ax$	$\frac{a}{s^2 + a^2}$	14. $x \cos ax$	$\frac{s^2 - a^2}{(s^2 + a^2)^2}$
7. $1 - \cos ax$	$\frac{a^2}{s(s^2 + a^2)}$	15. $e^{ax} \sin bx$	$\frac{b}{(s-a)^2 + b^2}$ $s > a$
8. $ax - \sin ax$	$\frac{a^3}{s^2(s^2 + a^2)}$	16. $e^{ax} \cos bx$	$\frac{s-a}{(s-a)^2 + b^2}$ $s > a$

EXAMPLE 3 Find the Laplace transform of $2x + 3 \sin x$.

Solution

$$\text{Let } f(x) = x \text{ and let } g(x) = \sin x.$$

$$L[2x + 3 \sin x] = 2L[x] + 3L[\sin x]$$

But $L[x]$ and $L[\sin x]$ can be determined from Table 28.1.

$$\begin{aligned} L[2x + 3 \sin x] &= 2\left(\frac{1}{s^2}\right) + 3\left(\frac{1}{s^2 + 1}\right) \\ &= \frac{2}{s^2} + \frac{3}{s^2 + 1} \\ &= \frac{2(s^2 + 1)}{s^2(s^2 + 1)} + \frac{3s^2}{s^2(s^2 + 1)} \\ &= \frac{2s^2 + 2 + 3s^2}{s^2(s^2 + 1)} \\ &= \frac{5s^2 + 2}{s^2(s^2 + 1)} \text{ if } s > 0 \quad \blacksquare \end{aligned}$$

Before we can solve differential equations using Laplace transforms, we must examine one more concept. The *inverse Laplace transform* allows us to find a function if its Laplace transform is known.

DEFINITION If $L[f(x)] = F(s)$, then the **inverse Laplace transformation,** is given by

$$L^{-1}[F(s)] = f(x),$$

and the function f is said to be the **inverse Laplace transform** of the function F.

Since some transformations, such as integration, do not produce a transform that is a single function, it is reasonable to ask if it is possible for a function to have more than one inverse Laplace transform. The answer is no.

EXAMPLE 4 Find $L^{-1}\left[\dfrac{3}{s^2+9}\right]$.

Solution From Table 28.1, it is clear that $\dfrac{a}{s^2+a^2}$ is the Laplace transform of sin ax. Thus,

$$L^{-1}\left[\frac{3}{s^2+9}\right] = \sin 3x \text{ if } s > 0. \quad ■$$

NOTE ***In Example 4, it is important to remember that s is a variable and a is a constant.*** Compare Example 4 and Example 5.

EXAMPLE 5 Find $L^{-1}\left[\dfrac{s}{s^2+9}\right]$.

Solution From Table 28.1, it is clear that $\dfrac{s}{s^2+a^2}$ is the Laplace transform of cos ax. Thus,

$$L^{-1}\left[\frac{s}{s^2+9}\right] = \cos 3x \text{ if } s > 0. \quad ■$$

EXERCISES 28.6

In 1–7, use the definition to find the Laplace transform for each function.

1. $f(x) = 1$
2. $f(x) = ax$
3. $f(x) = e^{-ax}$
4. $f(x) = \cos ax$
5. $f(x) = \sin ax$
6. $f(x) = xe^{ax}$
7. $f(x) = e^{ax} \sin bx$

In 8–14, use the linearity property to find the Laplace transform of each function.

8. $f(x) = 1 + e^{ax}$ **9.** $f(x) = 1 - \cos ax$ **10.** $f(x) = 1 + \sin ax$

11. $f(x) = ax - \sin ax$ **12.** $f(x) = e^{ax} - e^{bx}$ **13.** $f(x) = \sin ax - \cos ax$

14. $f(x) = \sin ax + \cos ax$

In 15–24, use Table 28.1 to find the inverse Laplace transform of each function.

15. $F(s) = \dfrac{1}{s-2}$ **16.** $F(s) = \dfrac{s}{s^2+4}$

17. $F(s) = \dfrac{1}{s^2+1}$ **18.** $F(s) = \dfrac{s}{(s+3)^2}$

19. $F(s) = \dfrac{1}{(s+3)^2}$ **20.** $F(s) = \dfrac{1}{(s+2)(s+3)}$

21. $F(s) = \dfrac{-s}{(s+2)(s+3)}$ **22.** $F(s) = \dfrac{s-1}{(s-1)^2+4}$

23. $F(s) = \dfrac{s^2-4}{(s^2+4)^2}$ **24.** $F(s) = \dfrac{3}{(s+1)^2+9}$

28.7 SOLVING DIFFERENTIAL EQUATIONS BY LAPLACE TRANSFORMS

A Laplace transform is used to find a particular solution of a differential equation by applying the Laplace transformation to both sides of the equation. The transforms $L[y']$ and $L[y'']$ arise in the process, so before we consider a specific example, let us examine these two transforms.

$$L[y'] = \int_0^\infty e^{-sx} y' \, dx$$
$$= \lim_{c\to\infty} \int_0^c e^{-sx} y' \, dx$$

Using integration by parts, we have the following equations.

$$u = e^{-sx} \qquad dv = y' \, dx$$
$$du = -se^{-sx} \, dx \qquad v = y$$

Then $L[y'] = \lim\limits_{c\to\infty} \left[e^{-sx} y \Big|_0^c - \int_0^c y(-se^{-sx}) \, dx \right]$

$$= \lim_{c\to\infty} e^{-sx} y \Big|_0^c - \lim_{c\to\infty} \int_0^c y(-se^{-sx}) \, dx.$$

If we let y_c denote the value of y when x is c and let y_0 denote the value of y when x is 0, we have

$$L[y'] = \lim_{c\to\infty} (e^{-sc} y_c - e^0 y_0) - \lim_{c\to\infty} \int_0^c y(-se^{-sx}) \, dx.$$

But as c approaches infinity, $e^{-sc}y_c$ approaches zero for most y_c (e.g., polynomial, trigonometric, or logarithmic functions). We also know that e^0 is 1, so the first limit is $0 - y_0$. Furthermore,

$$-\lim_{c\to\infty}\int_0^c y(-se^{-sx})\,dx = \int_0^\infty y(se^{-sx})\,dx,$$

and is, by definition, $L[sy]$, which can be written as $sL[y]$. Thus,

$$L[y'] = -y_0 + sL[y] \text{ or}$$
$$L[y'] = sL[y] - y_0. \qquad \textbf{(28-14)}$$

Substituting y'' for y' and y' for y in Equation 28-14, we have

$$L[y''] = sL[y'] - y'_0,$$

where y'_0 is the value of y' when x is 0. Now we can use Equation 28-14 to substitute for $L[y']$ in this equation.

$$L[y''] = s(sL[y] - y_0) - y'_0$$
$$L[y''] = s^2L[y] - sy_0 - y'_0 \qquad \textbf{(28-15)}$$

Some differential equations may be solved with Laplace transforms when initial values are known.

To solve a differential equation using Laplace transforms:

1. Apply the Laplace transformation to both sides of the equation.
2. Use the linearity property of transformations to rewrite the equation.
3. Make the following substitutions:

$$L[y''] = s^2L[y] - sy_0 - y'_0. \qquad \textbf{(28-15)}$$
$$L[y'] = sL[y] - y_0. \qquad \textbf{(28-14)}$$

4. Combine $L[y]$ terms.
5. Use Table 28.1 to substitute algebraic formulas for any Laplace transforms other than $L[y]$ and substitute given values of y_0 and y'_0.
6. Solve the equation for $L[y]$.
7. Use Table 28.1 to find the inverse Laplace transformation of $L[y]$.

EXAMPLE 1 Solve $y'' + y = x$ if $y = 0$ and $y' = 1$ when $x = 0$.

Solution 1. Apply the Laplace transformation to both sides of the equation.

$$L[y'' + y] = L[x]$$

2. Use the linearity property.

$$L[y''] + L[y] = L[x]$$

3. Use Equation 28-15 to rewrite the equation.

$$s^2L[y] - sy_0 - y'_0 + L[y] = L[x]$$

4. Combine the two terms containing $L[y]$ to obtain

$$(s^2 + 1)L[y] - sy_0 - y'_0 = L[x].$$

5. But $y_0 = 0$ and $y'_0 = 1$ were given as initial conditions. Also, formula 2 of Table 28-1 of the previous section shows that $L[x] = 1/s^2$. Thus, we have the following algebraic equation

$$(s^2 + 1)L[y] - 1 = \frac{1}{s^2}.$$

6. Solve the equation for $L[y]$.

$$(s^2 + 1)L[y] = \frac{1}{s^2} + 1$$

$$(s^2 + 1)L[y] = \frac{1 + s^2}{s^2}$$

$$L[y] = \frac{1 + s^2}{s^2} \frac{1}{(1 + s^2)}$$

$$= \frac{1}{s^2}$$

7. Since $L[y] = 1/s^2$, we can determine y by finding $L^{-1}[1/s^2]$. Formula 2 of Table 28.1 shows that if $s > 0$, then $y = x$, and this is the solution of the original differential equation. ■

Frequently, the algebraic equation produces a quantity that bears little resemblance to any of the Laplace transforms in Table 28.1. It is often possible to rewrite the expression in one of the familiar forms. Examples 2 and 3 illustrate some of the techniques that are used.

EXAMPLE 2 Solve $y'' + y = 0$ if $y = 1$ and $y' = 2$ when x is 0.

Solution

$$L[y'' + y] = L[0] \quad \text{Take the Laplace transform.}$$

$$L[y''] + L[y] = L[0] \quad \text{Apply the linearity property.}$$

$$s^2L[y] - sy_0 - y'_0 + L[y] = L[0] \quad \text{Replace } L[y''] \text{ with } s^2L[y] - sy_0 - y'_0.$$

$$(s^2 + 1)L[y] - sy_0 - y'_0 = L[0] \quad \text{Combine } L[y] \text{ terms.}$$

$$(s^2 + 1)L[y] - s - 2 = 0 \quad \text{Substitute known values.}$$

$$(s^2 + 1)L[y] = s + 2 \quad \text{Add } s + 2 \text{ to both sides.}$$

$$L[y] = \frac{s + 2}{(s^2 + 1)} \quad \text{Divide both sides by } s^2 + 1.$$

The algebraic expression on the right does not fit any of the patterns in Table 28.1. But the fraction can be split into two fractions. Since the inverse Laplance transform, like the Laplace transform, has the linearity property, we may apply the transformation to each fraction separately.

$$L[y] = \frac{s}{(s^2+1)} + \frac{2}{(s^2+1)}$$

$$y = L^{-1}\left[\frac{s}{(s^2+1)}\right] + L^{-1}\left[\frac{2}{(s^2+1)}\right]$$

$$y = L^{-1}\left[\frac{s}{(s^2+1)}\right] + 2L^{-1}\left[\frac{1}{(s^2+1)}\right]$$

Formulas 5 and 6 of Table 28.1 show that if $s > 0$,

$$y = \cos x + 2 \sin x. \quad ■$$

Example 3 illustrates how completing the square is used to obtain a denominator in the form of formula 15 of Table 28.1.

EXAMPLE 3 Solve $y'' + 2y' + 5y = 0$ if $y = 0$ and $y' = 2$ when $x = 0$.

Solution

$$L[y'' + 2y' + 5y] = L[0]$$ Take the Laplace transform.

$$L[y''] + 2L[y'] + 5L[y] = L[0]$$ Apply the linearity property.

$$s^2L[y] - sy_0 - y'_0 + 2(sL[y] - y_0) + 5L[y] = L[0]$$ Replace $L[y'']$ with $s^2L[y] - sy_0 - y'_0$ and $L[y']$ with $sL[y] - y_0$.

$$(s^2 + 2s + 5)L[y] - (s + 2)y_0 - y'_0 = L[0]$$ Combine like terms.

$$(s^2 + 2s + 5)L[y] - 2 = 0$$ Substitute known values.

$$L[y] = \frac{2}{(s^2 + 2s + 5)}$$ Solve for $L[y]$.

Since the denominator does not factor as in formulas 9–12 of Table 28.1, write it as

$$s^2 + 2s + 5 = (s^2 + 2s + 1) + 4$$
$$= (s + 1)^2 + 2^2.$$

Thus,

$$L[y] = \frac{2}{(s + 1)^2 + 2^2}.$$

Using formula 15 of Table 28.1 with $a = -1$ and $b = 2$, we have

$$y = L^{-1}\left[\frac{2}{(s + 1)^2 + 2^2}\right] = e^{-x} \sin 2x \text{ if } s > -1. \quad ■$$

Laplace transforms can be used to solve applied problems.

EXAMPLE 4 In radioactive decay, the rate of decay is directly proportional to the mass of material present. Suppose that 8 grams of a substance decays at a rate equal to 1/4 of the substance present. Find a formula for the amount of the substance present after t hours.

Solution Let y represent the amount of material present at time t.

Then $y' = -0.25\,y$.

$$sL[y] - y_0 = L[-0.25\,y] \quad \text{Take the Laplace transform.}$$

$$sL[y] + 0.25L[y] = y_0 \quad \text{Use } L[-0.25y] = -0.25\,L[y].$$

$$(s + 0.25)L[y] = y_0 \quad \text{Factor out } L[y].$$

$$L[y] = \frac{y_0}{s + 0.25} \quad \text{Solve for } L[y].$$

$$L[y] = \frac{8}{s + 0.25} \quad \text{Replace } y_0 \text{ with 8.}$$

$$y = L^{-1}\left[\frac{8}{s + 0.25}\right] = 8e^{-0.25t} \quad \text{Use formula 4 of Table 28.1 with } a = -0.25, \text{ if } s > 0.$$ ■

EXERCISES 28.7

In 1–16, use Laplace transforms to solve each differential equation.

1. $y' - y = 0$; $y = 1$ when $x = 0$
2. $2y' - y = 0$; $y = 1$ when $x = 0$
3. $y' + 2y = 0$; $y = 2$ when $x = 0$
4. $y' + y = 0$; $y = 1/2$ when $x = 0$
5. $y'' + 4y = 0$; $y = 0$ and $y' = 2$ when $x = 0$
6. $y'' + y = 0$; $y = 2$ and $y' = 0$ when $x = 0$
7. $y'' + y = 0$; $y = 2$ and $y' = 1$ when $x = 0$
8. $y'' + y = 0$; $y = -1$ and $y' = 1$ when $x = 0$
9. $y'' - 2y' - 3y = -3e^{2x}$; $y = 2$ and $y' = 5$ when $x = 0$
10. $y'' - 2y' = e^x$; $y = 0$ and $y' = 1$ when $x = 0$
11. $y'' + 4y = 4\cos 2x$; $y = 0$ and $y' = 0$ when $x = 0$
12. $y'' + y + 2\sin x = 0$; $y = 0$ and $y' = 1$ when $x = 0$
13. $y'' - 2y' + 2y = 0$; $y = 0$ and $y' = 1$ when $x = 0$
14. $2y'' - y' - y = 0$; $y = -1$ and $y' = -1$ when $x = 0$
15. $y'' + 2y' + y = 0$; $y = 0$ and $y' = 1$ when $x = 0$
16. $y'' - 2y' + 5y = 0$; $y = 1$ and $y' = 1$ when $x = 0$

17. A 5.3-liter radiator is filled initially with a solution that is 50.0% water and 50.0% antifreeze. The solution is pumped out at the rate of 1.00 liter per minute, and water is pumped in at the same rate. Assuming that the temperature is uniform throughout, what percent of the mixture in the radiator is antifreeze after 3.00 minutes? 4 8 1 2 6 7 9

18. A roast is cooked to a temperature of 82° C and then allowed to cool at a room temperature of 43° C. After 30 minutes, the temperature of the roast is 50° C. Assume that the temperature is uniform throughout the roast and determine the temperature after 60 minutes. Show your answer to two significant digits. (Hint: Use Newton's law of cooling.) 6 8 1 2 3 5 7 9

19. The velocity v of an object falling through a resisting medium satisfies the equation $m\ dv/dt = mg - cv$. In the equation, m is the mass of the object, g is the acceleration due to gravity, and $-c$ is the deceleration due to air resistance. Suppose that a 90.0 kg paratrooper jumps with a velocity of 4 m/s from a plane. Find the velocity of the paratrooper after 2.0 seconds. Assume $g = 9.8$ m/s^2 and $c = 27.3$ m/s^2. 4 7 8 1 2 3 5 6 9

20. Consider an object with mass m fastened to a spring whose constant is k and whose mass is negligible in comparison to the mass of the object. If it is assumed that there is no resistance to the motion when the object is pulled down from the rest position and released, the equation

$$m\frac{d^2x}{dt^2} + kx = 0$$

describes the position x of the object. An object with a mass of 2 kg is attached to a spring for which $k = 30$ if distances are measured in centimeters and the mass in kilograms. Find the position of the object aftrer 0.1 second if it is pulled down 5 cm from the rest position and released with an initial velocity of 0. 4 7 8 1 2 3 5 6 9

1. Architectural Technology
2. Civil Engineering Technology
3. Computer Engineering Technology
4. Mechanical Drafting Technology
5. Electrical/Electronics Engineering Technology
6. Chemical Engineering Technology
7. Industrial Engineering Technology
8. Mechanical Engineering Technology
9. Biomedical Engineering Technology

28.8 CHAPTER REVIEW

KEY TERMS

28.1 differential equation
order of a differential equation
degree of a differential equation
general solution
particular solution

28.2 separation of variables

28.3 first order linear differential equation
constant coefficients

28.4 second order linear differential equation
homogeneous
nonhomogeneous

28.5 undetermined coefficients

28.6 Laplace transformation
Laplace transform
inverse Laplace transformation
inverse Laplace transform

SYMBOLS

$L[f(x)]$ the Laplace transform of the function f

$L^{-1}[F(s)]$ the inverse Laplace transform of the function F

RULES AND FORMULAS

The Linearity Property: $L[a\,f(x) + b\,g(x)] = a\,L[f(x)] + b\,L[g(x)]$ where L is a Laplace transformation, f and g are functions, and a and b are constants.
Laplace Transforms Used to Solve Differential Equations:

$$L[y''] = s^2\,L[y] - sy_0 - y'_0$$

$$L[y'] = s\,L[y] - y_0$$

SEQUENTIAL REVIEW

(Section 28.1) *In 1–3, determine the order and degree of each differential equation.*

1. $xy'' + x = y^2$

2. $(y')^2 + y = y''$

3. $(y'')^2 - 2y' = xy$

In 4 and 5, verify that the first function given, implicitly or explicitly, is a particular solution of the given differential equation.

4. $y = 3 + 2e^{4x}$; $\quad y'' - 4y' = 0$

5. $y = 1 + 3\sin 2x$; $\quad y'' + 4y = 4$

In 6–8, each expression represents the solution of a differential equation. Use the properties of logarithms to simplify each solution.

6. $\ln(3x + 1) + \ln y = C_1$

7. $\ln(x^2 - 1) - 3\ln y = C_1$

8. $2\ln(x^2 + 1) - 3\ln y = C_1$

(Section 28.2) *In 9–12, find a general solution for each differential equation.*

9. $2x + 1 = yy'$

10. $1 + x^2yy' = 0$

11. $e^x\,dy + e^{-x}\,dx = 0$

12. $\cos^2 y\,dx - 3\,dy = 0$

In 13–16, find the particular solution of each differential equation for the indicated conditions.

13. $y' - 3x = 1$; $y = 1$ when $x = 0$

14. $y' = y + 2$; $y = -1$ when $x = 3$

15. $x^2\,dy = e^y\,dx$; $y = 0$ when $x = 1$

16. $2y' - x^2 = 3$; $y = 12$ when $x = 3$

(Section 28.3) *In 17–20, find a general solution of each first order linear differential equation.*

17. $y' - y\sin x = \cos x \sin x$

18. $y' + 2xy = 2x$

19. $dy + xy\,dx = 2x\,dx$

20. $dy - y\cos x\,dx = \cos x\,dx$

In 21–23, find the particular solution of each first order linear differential equation for the given initial conditions.

21. $dy - y\,dx = dx$; $y = 2$ when $x = 0$

22. $2\,dy + y\,dx = 2x\,dx$; $y = 1$ when $x = 0$

23. $y' + y/x = 3x$; $y = 4$ when $x = 1$

(Section 28.4) *In 24–27, find the general solution of each differential equation.*

24. $y'' + 4y' + 3y = 0$

25. $2y'' - 3y' = 0$

26. $\dfrac{d^2y}{dx^2} + 12\dfrac{dy}{dx} + 9y = 0$

27. $9\dfrac{d^2y}{dx^2} + 6\dfrac{dy}{dx} + y = 0$

In 28 and 29, find the particular solution of each differential equation for the given initial conditions.

28. $2y'' - 3y' + y = 0$; $y = 3$ and $y' = 2$ when $x = 0$

29. $y'' - 3y' = 0$; $y = 5$ and $y' = 6$ when $x = 0$

(Section 28.5) *In 30 and 31, find the general solution of each differential equation.*

30. $y'' - 3y = x^2$

31. $\dfrac{d^2y}{dx^2} - 3\dfrac{dy}{dx} + 2y = \sin x$

In 32 and 33, find the particular solution of each differential equation for the given initial conditions.

32. $y'' + 2y' = e^x$; $y = 1$ and $y' = -1/3$ when $x = 0$

33. $y'' - 4y' + 4y = x$; $y = 1$ and $y' = 2$ when $x = 0$

(Section 28.6) *In 34 and 35, use the definition to find the Laplace transform for each function.*

34. $y = \sin x$

35. $y = x$

In 36–38, use the linearity property and Table 28.1 to find the Laplace transform of each function.

36. $f(x) = ax + \cos ax$

37. $f(x) = 1 - e^{-ax}$

38. $f(x) = e^{ax} + x \cos ax$

In 39 and 40, use Table 28.1 to find the inverse Laplace transform of each function.

36. $F(s) = \dfrac{4}{s^2 + 16}$

40. $F(s) = \dfrac{4}{s(s^2 + 4)}$

(Section 28.7) *In 41–45, use Laplace transforms to solve each differential equation.*

41. $y' + y = 0$; $y = 1$ when $x = 0$

42. $y'' + 4y = 0$; $y = 4$ and $y' = 4$ when $x = 0$

43. $y'' - 3y' = 2e^{2x}$; $y = 0$ and $y' = 1$ when $x = 0$

44. $2y'' - y' - 6y = 0$ if $y = 1$ and $y' = -3/2$ when $x = 0$

45. $y'' + 2y' + 5y = 0$ if $y = 0$ and $y' = 2$ when $x = 0$

APPLIED PROBLEMS

1. In the circuit shown, the charge q (in coulombs) on the capacitor at time t is given by the equation

$$L\frac{d^2q}{dt^2} + R\frac{dq}{dt} + \frac{q}{C} = V.$$

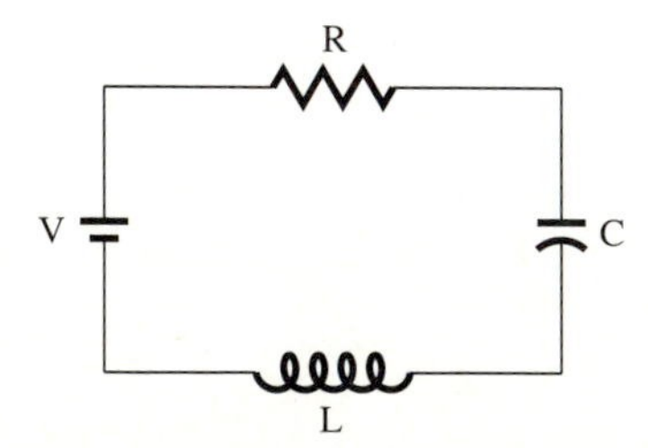

In the formula, L is the inductance (in henries), R is the resistance (in ohms), C is the capacitance (in farads), and V is the voltage (in volts). Find the charge on a 0.2 F capacitor after 2 seconds if $R = 5\ \Omega$, $V = 12$ volts and $L = 1.3$ H. Assume $q = 0$ and $q' = -60/13$ when $t = 0$. 3 5 9 1 2 4 6 7 8

2. When interest is compounded continuously, the balance grows at a rate that is directly proportional to the balance at any given time. If a \$6000 deposit grows to \$12,000 in 7.2 years (2628 days), find the balance after 1 year (365 days). 1 2 6 7 8 3 5 9

3. In a conventional automotive ignition system, as shown in the accompanying diagram, the current $i(t)$ in the inductor after the switch (distributor points) opens is given by $i(t) = 6e^{-t} \cos(707t) + 5.66 \times 10^{-3} e^{-t} \sin(707\,t)$. Find the Laplace transform for this function. 3 5 9 1 2 4 6 7 8

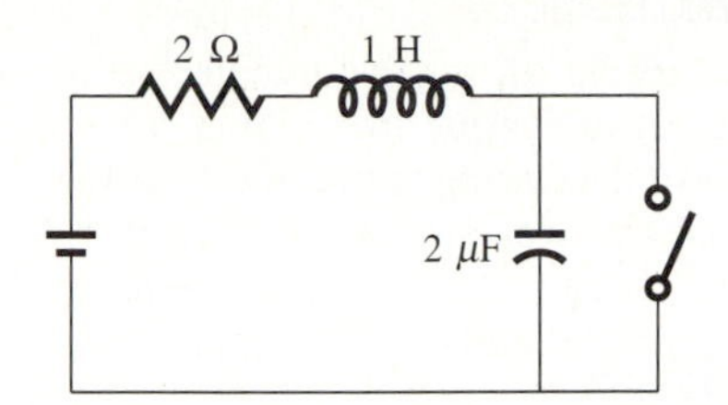

4. The equation for the voltage across the load resistor in the low-pass filter shown in the accompanying diagram is given by $V = iR$.

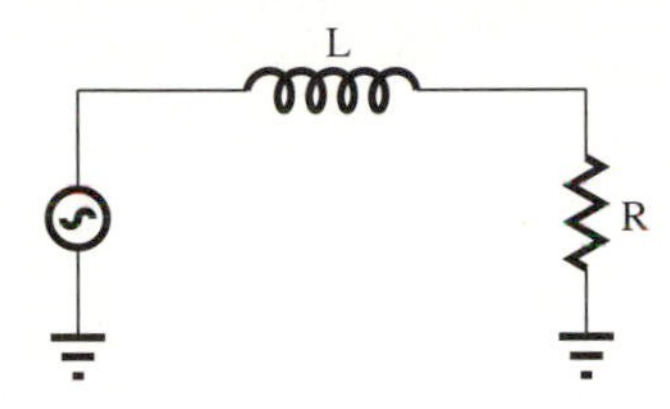

The total voltage is given by $v = L\,di/dt + iR$. Solve this equation for i, if $v = A \sin \omega t$, where A is the amplitude of the applied voltage.
3 5 9 1 2 4 6 7 8

5. The current i (in amps) in a certain circuit is given by $d^2i/dt^2 + 5\,di/dt + 4i = 0$ when t is time (in seconds). Verify that $i = 7e^{-4t} + 3e^{-t}$ is a solution of this differential equation.
3 5 9 1 2 4 6 7 8

6. A relationship in fluid mechanics is given by $ph^2 = C_1 p + C_2$. Verify that this equation is a solution of the differential equation $2\,ph + h^2p' = C_1 p'$.
4 8 1 2 6 7 9

7. Use the impedance triangle shown in the accompanying figure to find R. Then show that $dR/d\theta + R \tan \theta = 0$. 3 5 9 1 2 4 6 7 8

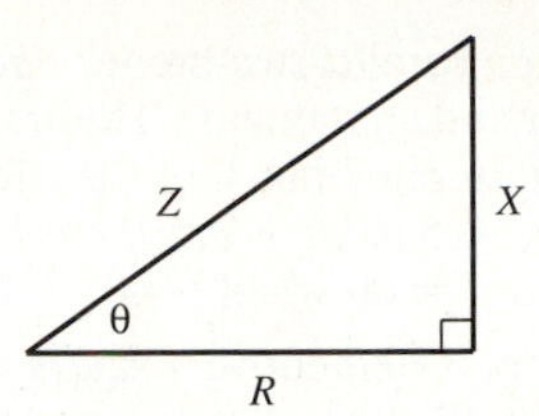

8. The relationship between a transistor's current i (in milliamps) and its voltage V (in volts) is given by $dI/dV = I^2/(20V^3)$. When V is 1.5 volts, I is 16.36 milliamps. Solve for I. 3 5 9 4 7 8

9. Consider a sliding machine part that moves on a rotating rod. It travels in such a way that its distance r (in inches) from the starting point is given by $dr/d\theta = r/\theta^2$ where θ (in radians) is the angle of the rotating part. It is known that $r = 26.0$ inches when θ is 45.0°. Solve for r.
4 7 8 1 2 3 5 6 9

10. Light energy traveling down a fiber optic cable is lost to imperfections in the glass. This loss (in joules per meter) is expressed as dW/dl. In an experiment this loss was investigated and thought to be proportional to both the distance l (in meters) and the amount of energy W (in joules). That is, $dW/dl = kWl$, where k is a constant. If this assumption is valid, find a formula for W if W is 3 millijoules when l is 0, and W is 1.2 millijoules when l is 250 m. 9 1 2 3 5 6 7

11. Consider a 1 henry coil placed in series with a 3 ohm resistance. The resulting circuit is then attached to a supply voltage v (in volts). If t is time (in seconds), and $v = \cos t$, then the current i (in amps) that flows is given by $di/dt + 3i = \cos t$. Solve for i, given that i is 0 when t is 0.
3 5 9 4 7 8

12. An inductor and a resistor are connected in series and attached to a supply voltage. The current i (in amperes) that results is given by $L\,di/dt + Ri = V$. In this equation, L is the inductance (in henries), R is the resistance (in ohms), i is the current (in amperes), t is the time (in seconds), and V is the voltage (in volts). Suppose $L = 2t$, $R = 2 + 4t$, and $V = 2te^{-2t}$. Solve for i if i is 1 ampere when t is 1 second. 3 5 9 4 7 8

13. Consider a parallel two-branch circuit containing a resistor and an inductor. The branch current (in amperes) in one branch of the circuit is given by $d^2i/dt^2 + 125\ di/dt + 2150i = 175$. Solve for i if $i = 0$ and $i' = 40$ when $t = 0$. 3 5 9 4 7 8

14. A resistance element in a faulty thermostat has resistance R (in ohms) given by $d^2R/dt^2 + 0.33\ dR/dt + 0.125R = 0$, if t is time (in seconds). Solve for R if R is 400 ohms when t is 0 and R is 300 ohms when t is 5 seconds. 3 5 9 1 2 4 6 7 8

15. A voltage v (in volts) given by $v = e^{8t}$ is applied to a circuit containing an inductor and a resistor in series. The resulting current i (in amps) is given by $di/dt - 3i = e^{8t}$. Find the Laplace transform $L[i]$ if $i = 4$ when $t = 0$. 3 5 9 4 7 8

16. The charge q (in coulombs) in a certain circuit is given by $q'' - 2q' - 8q = 0$. Find the Laplace transform $L[q]$ if $q = 3$ and $q' = 6$ when $t = 0$. 3 5 9 1 2 4 6 7 8

17. The needle on a galvanometer settles to some value by oscillating around that value with vibrations of decreasing amplitude. The angle θ (in radians) of swing is given by $d^2\theta/dt^2 + 2\ d\theta/dt + 15(\theta - \pi/2) = 0$. Find the Laplace transform $L[\theta]$ if θ is 0 and θ' is 10 when t is 0. 3 5 9 4 7 8

1. Architectural Technology
2. Civil Engineering Technology
3. Computer Engineering Technology
4. Mechanical Drafting Technology
5. Electrical/Electronics Engineering Technology
6. Chemical Engineering Technology
7. Industrial Engineering Technology
8. Mechanical Engineering Technology
9. Biomedical Engineering Technology

RANDOM REVIEW

1. Find a general solution of $y'' - 4y = e^x$.
2. Determine the order and degree of $(y')^2 - 2y'' + 3y = 0$.
3. Use Laplace transforms to solve $y'' + y = 0$, if $y = 4$ and $y' = 2$ when $x = 0$.
4. Use Table 28.1 to find $L^{-1}\left[\dfrac{1}{(s-1)^2}\right]$.
5. Find the particular solution of $dy = y^3 e^x\,dx$ if $y = 1$ when $x = 0$.
6. Verify that $y = 8 - 2e^{-x}$ is a particular solution of $y' + y = 8$.
7. Find a general solution of $y^2 y' - 2x^4 = 0$.
8. Find a general solution of $2y'' + 3y' - 5y = 0$.
9. Find the particular solution of $dy - 3y\,dx = e^x\,dx$ if $y = 3$ when $x = 0$.
10. Use Table 28.1 to find $L\,[x + e^{ax}]$.

CHAPTER TEST

1. Determine the order and degree of $xy'' + x^2 = y$.
2. Find a general solution of $dy = y \cos x\,dx$.
3. Find the particular solution of $e^x\,dy = e^{2x}\,dx$ if $y = 2$ when $x = 0$.
4. Find the particular solution of $x\,dy - x^2 y\,dx = 2x^2\,dx$ if $y = 2$ when $x = 0$.
5. Find the general solution of $3\dfrac{d^2y}{dx^2} + 5\dfrac{dy}{dx} - 2y = 0$.

6. Find the general solution of $y'' - 2y' + y = 2e^x$.
7. Use Table 28.1 to find the Laplace transform of the function defined by $f(x) = e^{ax} + \sin ax$.
8. Use Laplace transforms to solve $y'' + y' = 1$ if $y = 1$ and $y' = 1$ when $x = 0$.
9. A 16 gallon (64 quart) tank is filled with gasoline. Liquid is pumped out at the rate of 1 quart per minute, and alcohol is pumped in at the same rate. Find the amount of alcohol in the tank after 3 minutes.
10. The equation for the voltage v (in volts) across an inductor fed by a varying current i (in amperes) is $v = L\, di/dt$, if L is the inductance (in henries). Find the equation for the current in a 10 mH inductor if the equation for the voltage is $v = 5 \sin (377t)$.

MAKING A TELESCOPE MIRROR

A small telescope mirror is made from a glass disk by grinding and polishing it until the surface is parabolic in shape. The surface is then covered with a reflective coating. A mirror for a research observatory, however, must be so large that a solid disk of glass is too heavy to be practical. Solid glass also adjusts so slowly to temperature changes that air currents are created by the differing temperatures of mirror and atmosphere, causing distorted images.

Modern technology allows the creation of large mirrors by *spin-casting.* For this process, a mold is made of ceramic fiberboard, so that it looks like a giant honeycomb cut in the shape of a disk, as in Figure 28.2.

Each cell of the honeycomb is individually machined so that the surface of the honeycomb, as a whole, is parabolic. Chunks of borosilicate are placed into, and on top of, the cells of the honeycomb. The mold, which is placed in a furnace, is then heated. The whole furnace rotates,

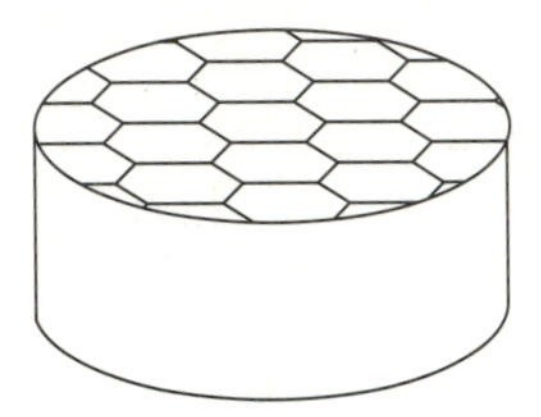

FIGURE 28.2

and as the glass melts, inertial forces push the liquid up the sides of the mold, forming a parabolic surface. The furnace continues spinning as the glass cools, so that the glass maintains its shape.

Differential equations may be used to show that the surface of a spinning liquid is indeed parabolic. Consider a liquid in a cylindrical mold rotating with a constant angular velocity of ω. If the vessel is situated as shown in Figure 28.3, the y-axis is the axis of rotation, and the x-axis runs through the lowest point on the surface.

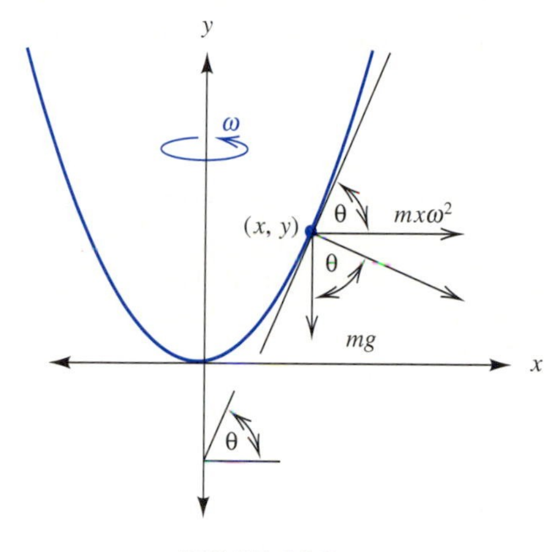

FIGURE 28.3

A particle at (x, y) is pulled downward by the force due to gravity (mg) and outward by a centrifugal force ($mx\omega^2$). When the particle is in equilibrium, the tangent to the surface at (x, y) makes an angle of θ with the horizontal.

Since $\tan\theta = \dfrac{mx\omega^2}{mg}$ or $\dfrac{x\omega^2}{g}$, and the slope of the line tangent at (x, y) is given by $\tan\theta$ as well as by the derivative, it is true that $\dfrac{dy}{dx} = \dfrac{x\omega^2}{g}$. Separating the variables, we have

$$\int g\,dy = \int x\omega^2\,dx$$

$$gy = \frac{x^2\omega^2}{2} + C_1$$

$$y = \frac{x^2\omega^2}{2g} + \frac{C_1}{g} \quad \text{or} \quad \frac{x^2\omega^2}{2g} + C.$$

Since the equation describes the vertical position y on the surface for any value of x, we see that the surface is necessarily parabolic. Since y is 0 when x is 0, we can also determine that the constant of integration is 0. A typical value of ω is 8.5 revolutions per minute, or about 0.9 radians per second.

APPENDIX: INTEGRALS

It is assumed that no denominators are equal to zero and that all radicals represent real numbers.

I. Forms Involving $a + bu$

1. $\displaystyle\int \frac{u\,du}{a+bu} = \frac{1}{b^2}\left[(a+bu) - a\ln|a+bu|\right] + C$

2. $\displaystyle\int \frac{u^2\,du}{a+bu} = \frac{1}{b^3}\left[\frac{1}{2}(a+bu)^2 - 2a(a+bu) + a^2\ln|a+bu|\right] + C$

3. $\displaystyle\int \frac{u\,du}{(a+bu)^2} = \frac{1}{b^2}\left[\frac{a}{a+bu} + \ln|a+bu|\right] + C$

4. $\displaystyle\int \frac{du}{u(a+bu)} = \frac{-1}{a}\ln\left|\frac{a+bu}{u}\right| + C$

5. $\displaystyle\int \frac{du}{u^2(a+bu)} = \frac{-1}{au} + \frac{b}{a^2}\ln\left|\frac{a+bu}{u}\right| + C$

6. $\displaystyle\int \frac{du}{u(a+bu)^2} = \frac{1}{a(a+bu)} - \frac{1}{a^2}\ln\left|\frac{a+bu}{u}\right| + C$

II. Forms Involving $\sqrt{a + bu}$

7. $\displaystyle\int u\sqrt{a+bu}\,du = \frac{2(3bu-2a)(a+bu)^{3/2}}{15b^2} + C$

8. $\displaystyle\int \frac{u\,du}{\sqrt{a+bu}} = \frac{2(bu-2a)\sqrt{a+bu}}{3b^2} + C$

9. $\displaystyle\int \frac{du}{u\sqrt{a+bu}} = \frac{1}{\sqrt{a}}\ln\left|\frac{\sqrt{a+bu}-\sqrt{a}}{\sqrt{a+bu}+\sqrt{a}}\right| + C \text{ (if } a > 0)$

10. $\displaystyle\int \frac{\sqrt{a+bu}\,du}{u} = 2\sqrt{a+bu} + a\int \frac{du}{u\sqrt{a+bu}}$

11. $\displaystyle\int u^2\sqrt{a+bu}\,du = \frac{2(8a^2 - 12abu + 15b^2u^2)(a+bu)^{3/2}}{105b^3} + C$

III. Forms Involving $a^2 + u^2$ or $\sqrt{a^2 + u^2}$

12. $\displaystyle\int \frac{du}{a^2+u^2} = \frac{1}{a}\,\mathrm{Tan}^{-1}\frac{u}{a} + C$

13. $\displaystyle\int \sqrt{a^2+u^2}\,du = \frac{u}{2}\sqrt{a^2+u^2} + \frac{a^2}{2}\ln|u + \sqrt{a^2+u^2}| + C$

14. $\displaystyle\int \frac{du}{\sqrt{a^2+u^2}} = \ln|u+\sqrt{a^2+u^2}| + C$

15. $\displaystyle\int \frac{du}{u\sqrt{a^2+u^2}} = \frac{-1}{a}\ln\left|\frac{\sqrt{a^2+u^2}+a}{u}\right| + C$

16. $\displaystyle\int \frac{\sqrt{a^2+u^2}\,du}{u} = \sqrt{a^2+u^2} - a\ln\left|\frac{a+\sqrt{a^2+u^2}}{u}\right| + C$

17. $\displaystyle\int u^2\sqrt{a^2+u^2}\,du = \frac{u}{8}(a^2+2u^2)\sqrt{a^2+u^2} - \frac{a^4}{8}\ln|u+\sqrt{a^2+u^2}| + C$

IV. Forms Involving $a^2 - u^2$ or $\sqrt{a^2-u^2}$

18. $\displaystyle\int \frac{du}{a^2-u^2} = \frac{1}{2a}\ln\left|\frac{a+u}{a-u}\right| + C$

19. $\displaystyle\int \sqrt{a^2-u^2}\,du = \frac{u}{2}\sqrt{a^2-u^2} + \frac{a^2}{2}\,\text{Sin}^{-1}\frac{u}{a} + C$

20. $\displaystyle\int \frac{du}{\sqrt{a^2-u^2}} = \text{Sin}^{-1}\frac{u}{a} + C$

21. $\displaystyle\int \frac{du}{u\sqrt{a^2-u^2}} = \frac{-1}{a}\ln\left|\frac{a+\sqrt{a^2-u^2}}{u}\right| + C$

22. $\displaystyle\int \frac{\sqrt{a^2-u^2}\,du}{u} = \sqrt{a^2-u^2} - a\ln\left|\frac{a+\sqrt{a^2-u^2}}{u}\right| + C$

23. $\displaystyle\int u^2\sqrt{a^2-u^2}\,du = \frac{u(2u^2-a^2)\sqrt{a^2-u^2}}{8} + \frac{a^4}{8}\,\text{Sin}^{-1}\frac{u}{a} + C$

V. Forms Involving $u^2 - a^2$ or $\sqrt{u^2-a^2}$

24. $\displaystyle\int \frac{du}{u^2-a^2} = \frac{1}{2a}\ln\left|\frac{u-a}{u+a}\right| + C$

25. $\displaystyle\int \sqrt{u^2-a^2}\,du = \frac{u}{2}\sqrt{u^2-a^2} - \frac{a^2}{2}\ln|u+\sqrt{u^2-a^2}| + C$

26. $\displaystyle\int \frac{du}{\sqrt{u^2-a^2}} = \ln|u+\sqrt{u^2-a^2}| + C$

27. $\displaystyle\int \frac{du}{u\sqrt{u^2-a^2}} = \frac{1}{a}\,\text{Sec}^{-1}\frac{u}{a} + C$

28. $\displaystyle\int \frac{\sqrt{u^2-a^2}\,du}{u} = \sqrt{u^2-a^2} - a\,\text{Sec}^{-1}\frac{u}{a} + C$

29. $\displaystyle\int u^2\sqrt{u^2-a^2}\,du = \frac{u}{8}(2u^2-a^2)\sqrt{u^2-a^2} - \frac{a^4}{8}\ln|u+\sqrt{u^2-a^2}| + C$

VI. Trigonometric and Inverse Trigonometric Forms (n is an integer and $n \geq 2$)

30. $\int \sin^n u \, du = \frac{-\sin^{n-1} u \cos u}{n} + \frac{n-1}{n} \int \sin^{n-2} u \, du$

31. $\int \cos^n u \, du = \frac{\cos^{n-1} u \sin u}{n} + \frac{n-1}{n} \int \cos^{n-2} u \, du$

32. $\int \tan^n u \, du = \frac{1}{n-1} \tan^{n-1} u - \int \tan^{n-2} u \, du$

33. $\int \cot^n u \, du = \frac{-1}{n-1} \cot^{n-1} u - \int \cot^{n-2} u \, du$

34. $\int \sec^n u \, du = \frac{1}{n-1} \sec^{n-2} u \tan u + \frac{n-2}{n-1} \int \sec^{n-2} u \, du$

35. $\int \csc^n u \, du = \frac{-1}{n-1} \csc^{n-2} u \cot u + \frac{n-2}{n-1} \int \csc^{n-2} u \, du$

36. $\int u \sin u \, du = \sin u - u \cos u + C$

37. $\int u \cos u \, du = \cos u + u \sin u + C$

38. $\int \sin au \sin bu \, du = \frac{\sin (a-b)u}{2(a-b)} - \frac{\sin (a+b)u}{2(a+b)} + C$

39. $\int \sin au \cos bu \, du = -\frac{\cos (a-b)u}{2(a-b)} - \frac{\cos (a+b)u}{2(a+b)} + C$

40. $\int \cos au \cos bu \, du = \frac{\sin (a-b)u}{2(a-b)} + \frac{\sin (a+b)u}{2(a+b)} + C$

41. $\int \sin^m u \cos^n u \, du = \frac{\sin^{m+1} u \cos^{n-1} u}{m+n} + \frac{m-1}{m+n} \int \sin^m u \cos^{n-2} u \, du \quad (\text{if } m \geq 0)$

42. $\int \text{Sin}^{-1} u \, du = u \, \text{Sin}^{-1} u + \sqrt{1-u^2} + C$

43. $\int \text{Cos}^{-1} u \, du = u \, \text{Cos}^{-1} u - \sqrt{1-u^2} + C$

44. $\int \text{Tan}^{-1} u \, du = u \, \text{Tan}^{-1} u - \frac{1}{2} \ln |1+u^2| + C$

VII. Exponential and Logarithmic Forms

45. $\int a^u \, du = \frac{a^u}{\ln a} + C \quad (\text{if } a > 0)$

46. $\int u e^u \, du = e^u(u-1) + C$

47. $\int e^{au} \sin bu \, du = \frac{e^{au}(a \sin bu - b \cos bu)}{a^2+b^2} + C$

48. $\displaystyle\int e^{au} \cos bu \, du = \frac{e^{au}(a \cos bu + b \sin bu)}{a^2 + b^2} + C$

49. $\displaystyle\int \ln u \, du = u \ln u - u + C$

50. $\displaystyle\int u^n \ln u \, du = u^{n+1}\left[\frac{\ln u}{n+1} - \frac{1}{(n+1)^2}\right] + C$ (if n is an integer and $n \geq 1$)

GLOSSARY OF APPLIED TERMS

absolute temperature The temperature of an object on the *Kelvin* scale.

ac See *alternating current.*

acceleration The rate at which *velocity* changes with respect to time.

acceleration, angular The rate at which *angular velocity* changes with respect to time.

acceleration, centripetal *Acceleration* directed toward the center of a circle for an object traveling in a circular path.

acceleration due to gravity The *acceleration* of a freely falling object near the earth's surface. Denoted by g, its value is 32 feet/second^2 or 9.8 meters/second^2.

acceleration, initial The *acceleration* of an object at the beginning of the time interval under consideration.

acceleration, linear The rate at which *linear velocity* changes with respect to time.

acceleration, tangential *Acceleration* directed along the tangent to a curve for an object traveling in a curved path.

acoustic intensity *Energy* per unit area of sound waves traveling in a fluid.

acre A measure of area. One acre = 43,560 feet^2.

aerodynamics The study of the motion of air and other gaseous fluids and of the motion of bodies relative to such fluids.

algorithm A sequence of unambiguous steps that will produce a solution to a given problem in a finite number of steps.

alloy A metal that is a mixture of two or more metals, or of a metal and something else.

alternating current (ac) An electric *current* that has one direction during one part of a generating cycle and the opposite direction during the remainder of the cycle.

alveolar membrane A membrane surrounding an air pocket in the lung.

ammeter An instrument for measuring the strength of an electric *current.*

ampere A unit of electric *current:* 1 ampere = 1 *coulomb*/second.

amplifier A device for increasing the strength of a signal. An amplifier consists of one or more *transistors* (or vacuum tubes) and associated *circuits.*

angle, critical See *critical angle.*

angle, friction See *friction angle.*

angle, lead See *lead angle.*

angle of attack The angle that the bottom of an airplane wing (or other body in motion) makes with the approaching airstream.

angle of incidence See *incidence, angle of.*

angle of reflection See *reflection, angle of.*

angle of refraction See *refraction, angle of.*

angle, phase See *phase angle.*

angular velocity See *velocity, angular.*

anisotropic Exhibiting properties with different values when measured along axes in different directions.

annealed copper wire Wire made of copper that has been heated then cooled to make it less brittle.

anode The positive *terminal* of an electroplating apparatus or vacuum tube.

atmospheric pressure The *pressure* exerted by the earth's atmosphere.

atmospheric refraction See *refraction, atmospheric.*

attenuator An adjustable device for reducing the strength of an audio or radio signal with minimal distortion of the signal.

ball bearings See *bearings, ball.*

BASIC Beginners' All Purpose Symbolic Instruction Code. A computer language.

beam, cantilever A beam that is anchored at one end to an object, such as a wall.

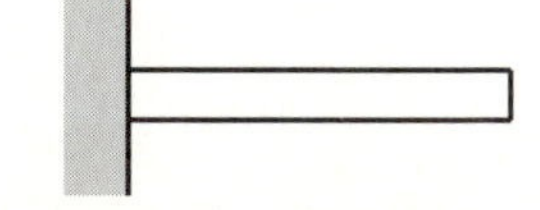

beam, simple A beam that is anchored or supported at each end.

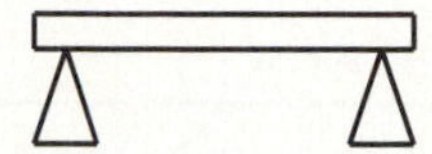

bearing An acute angle measured from north or south toward the east or west that is used to specify direction for navigational purposes.

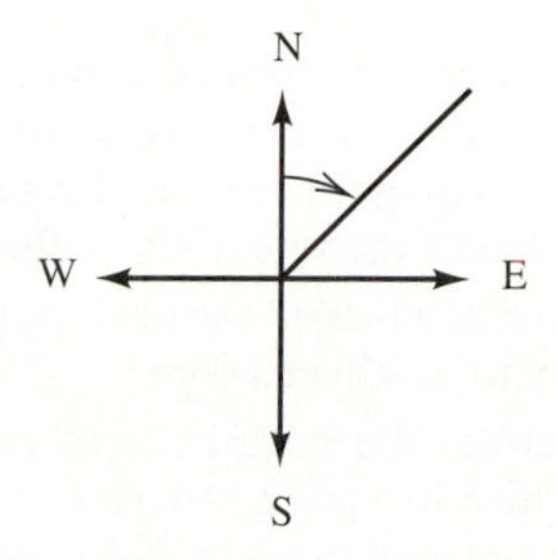

bearings, ball Metal balls upon which part of a machine turns or slides with minimal *friction.*

bearings, roller Rollers upon which part of a machine turns or slides with minimal *friction.*

biorhythm Three approximately monthly cycles used to explain high and low levels of an individual's emotional, intellectual, and physical achievement.

blood pressure, systolic See *systolic blood pressure.*

borosilicate A pyrex-like glass.

Btu British Thermal Unit. A unit of heat. One Btu is the amount of heat necessary to raise the temperature of one pound of water by one degree *Fahrenheit.*

bulk modulus The ratio of the *pressure* applied to a sample of a material to the corresponding fractional decrease in its volume.

buoyant force See *force, buoyant.*

calorie A unit of heat. One calorie is the amount of heat necessary to raise the temperature of one *gram* of water by one degree *Celsius.*

Calorie A unit of heat. One Calorie is the amount of heat necessary to raise the temperature of one *kilogram* of water by one degree *Celsius.*

camber The arch in the surface of an airplane wing or other airborne body.

cantilever beam See *beam, cantilever.*

capacitance The property of a material that allows it to store electric *charge.*

capacitance, equivalent The capacitance of a single *capacitor* that can be substituted for a set of interconnected *capacitors* without changing the properties of the *circuit.*

capacitive reactance See *reactance, capacitive.*

capacitor A device for storing electric *energy* in the form of an electric field. A capacitor consists of conducting plates separated by layers of an insulator. The symbol —)(— is used in *circuit* diagrams to represent a capacitor.

capillaries Minute blood vessels that form a network in close contact with all living cells in a body.

carat A measure of the fraction of pure gold in an *alloy.* One carat = 1/24.

carbon-14 A radioactive form of carbon used to determine age of substances in archaeological and geological applications.

cathode The negative *terminal* of an electroplating apparatus or vacuum tube.

cathode ray tube A vacuum tube in which an electron beam controlled by electric or *magnetic fields* traces out an image on a fluorescent screen.

cell, dry See *dry cell.*

Celsius The temperature scale that assigns 0 to the freezing point of water and 100 to the boiling point of water, with 100 equal degrees between.

centrifugal force See *force, centrifugal.*

centroid The point on a plane area at which a support could be placed to balance the object.

characteristic impedance See *impedance, characteristic.*

charge An accumulation of electricity of only one kind, such as positive.

chip, memory A small piece of silicon which contains data storage cells for a computer.

chord The straight line joining the leading and trailing edges of a body in motion relative to the surrounding air.

circuit The complete path of an electric *current,* usually including the source of electric *energy.*

circuit, integrated See *integrated circuit.*

cmil Circular mil. A unit of area. One cmil is the area of a circle whose diameter is 0.001 inch. The cmil is sometimes used in electrical applications to measure a round conductor.

coefficient of friction The ratio of the *force* needed to overcome *friction* between two surfaces to the force pressing the surfaces together.

coefficient of linear expansion The ratio of the change in length of a solid to the product of the original length and the change in temperature that caused the change in length.

computer monitor See *monitor, computer.*

concentrated load See *load, concentrated.*

condition, initial See *initial condition.*

conductance The property of a material that allows it to conduct *current.*

conductor A material through which an electric *charge* is transferred.

cone A solid generated by moving a straight line passing through a fixed point so that another point on the line traces a circle or ellipse on a flat surface. The word cone often refers to a right circular cone, in which the figure traced is a circle and the fixed point is on a line through the center of the circle.

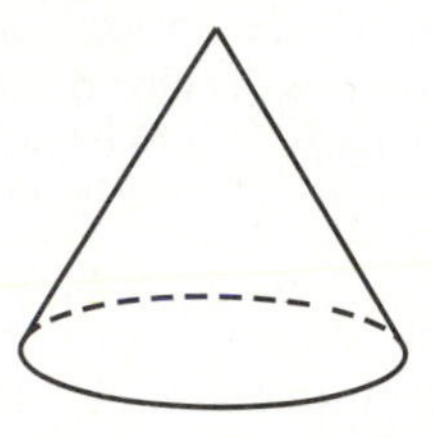

constant, distributed See *distributed constant.*

constant, equilibrium See *equilibrium constant.*

constant, Lame's See *Lame's constant.*

coulomb A unit of electric *charge.* One coulomb is the *charge* on 6.25×10^{18} electrons.

crankshaft A shaft driven by or driving a crank.

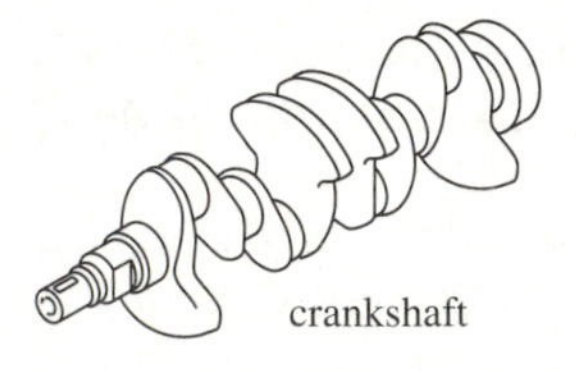

critical angle The limiting *angle of incidence* θ in the optically denser medium that results in a 90° *angle of refraction.*

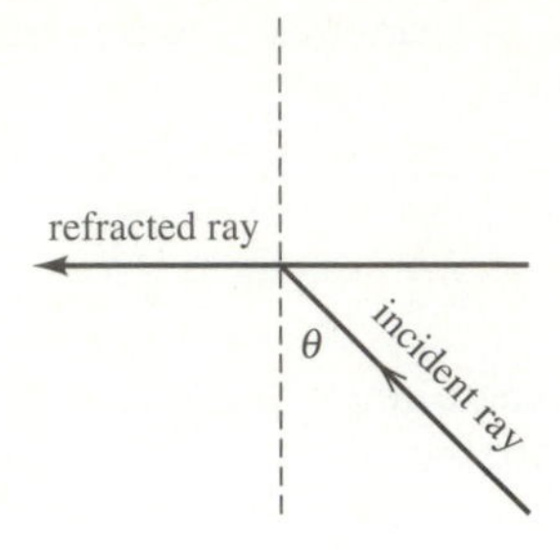

cross-product For two vectors, a particular vector that is perpendicular to the plane of the two vectors and has a magnitude that is the product of the two vectors and the sine of the included angle.

current A flow of charged particles, for example, a flow of electrons along a wire.

current, effective The magnitude of an *alternating current,* which in a given *resistance* produces heat at the same average rate as a *direct current* with the same magnitude.

current, steady-state See *steady-state current.*

curvature, radius of See *radius of curvature.*

cylindrical coordinates Coordinates that specify the position of a point in space. Denoted (r, θ, z), the cylindrical coordinates coincide with the polar coordinates (r, θ) in the xy plane, and the rectangular coordinate z.

daminozide A chemical compound used as a plant growth regulator and to keep harvested fruit firm during storage.

damp To diminish progressively in vibration or *oscillation.*

dc See *direct current.*

decibel A unit of *acoustic intensity.* One decibel is the *acoustic intensity* of a sound that has a *physical intensity* of 1.259×10^{-12} *watts*/meter^2. The decibel is also used to measure *voltage* gain (or loss) in an electronic device.

deflection The movement of an object from a specified position.

density *Mass* per unit volume of a substance. The word density sometimes refers to weight per unit volume.

depth of field The range of distances in which an object may be photographed to produce a focused image.

determinant A real number represented by a square array of numbers.

diagram, free-body See *free-body diagram.*

difference, potential See *potential difference.*

digital multimeter See *multimeter, digital.*

DIM Dimension. A statement used in computer programming to specify the dimensions of a *matrix.*

diode A vacuum tube containing only two *electrodes* —a filament or heater and a plate.

DIP switch Dual in-line package switch. A switch that is set to provide instructions from a computer to a peripheral device, such as a printer.

direct current (dc) A *current* in which the movement of electrons is in one direction only.

displacement A change of position, specified by a distance and direction.

distributed constant A *circuit* parameter that exists along the length of a hollow metal tube that guides *microwave energy* from one point to another.

distributed load See *load, distributed.*

divider, voltage See *voltage divider.*

dot product For two vectors, the scalar quantity that is the product of the magnitudes of the two vectors and the cosine of the included angle.

drain The lowest point of a *sag* or dip in a roadway.

dry cell A portable device for changing chemical *energy* into electric *energy* by the action of two dissimilar metals immersed in a *conductor* that is of a nonspillable consistency.

dyne A unit of *force.* One dyne is the force necessary to give a mass of one *gram* an *acceleration* of one centimeter per second in one second.

effective current See *current, effective.*

effective voltage See *voltage, effective.*

efficiency The ratio, given as a percent, of *work* done by a machine to the *energy* supplied to the machine.

elasticity, modulus of See *modulus of elasticity.*

electrode Either of the two *terminals* of an electric source.

electromotive force The *potential difference* across the *terminals* of a battery, generator, or other source of electrical *energy* when no current flows.

electrostatics The study of phenomena due to attractions or repulsions of electric *charges.*

elimination, Gaussian See *Gaussian elimination.*

emf Electromotive *force.*

energy The ability to do *work.* The units of energy are the same as the units of *work.*

energy, kinetic See *kinetic energy.*

energy, potential See *potential energy.*

entropy A measure of the unavailability of thermal *energy* for conversion into mechanical *energy.*

equilibrium A state of balance between two or more opposing *forces* or actions.

equilibrium constant For a chemical reaction, a fixed constant that depends only on the temperature of the reaction mixture, once it has reached equilibrium.

equipotential curves A family of curves such that each member is perpendicular to all members of a family of curves representing electric *force.*

equivalent capacitance See *capacitance, equivalent.*

equivalent resistance See *resistance, equivalent.*

eyepiece The *lens* or system of lenses nearest the observer's eye in a telescope or microscope.

Fahrenheit The temperature scale that assigns 32 to the freezing point of water and 212 to the boiling point of water, with 180 equal degrees between.

farad A unit of *capacitance.* 1 farad = 1 *coulomb/volt.*

filter (electrical) An electrical *circuit* used to prevent the flow of some frequencies of *alternating current* but to permit the passage of others.

filter (optical) A piece of colored glass placed in front of a beam of light to block the passage of light of certain *wavelengths.*

floodgate A gate for shutting out, admitting, or releasing a body of water.

flume An inclined chute or trough for conveying water from one place to another.

flywheel A heavy wheel for opposing any fluctuation of speed in the machinery to which the wheel is attached.

focal length The distance of the principal focus from the center of a *lens* or curved mirror.

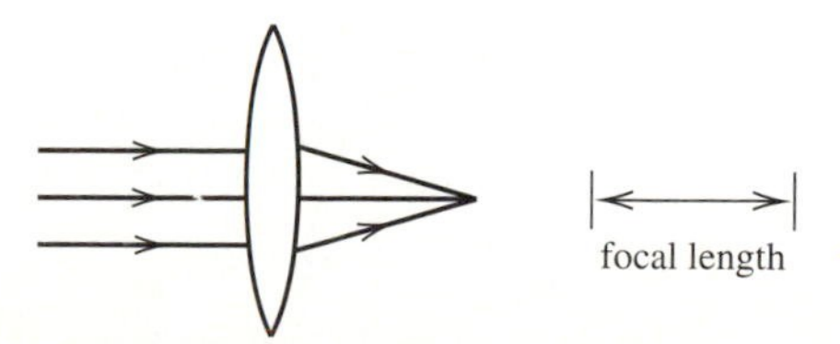

force An influence, such as a push or a pull, that either produces or prevents motion of an object.

force, buoyant The upward *force* exerted on an object in a fluid. The buoyant force is equal to the weight of the fluid displaced by the object.

force, centrifugal The *force* equal and opposite to the *force* exerted on an object at right angles to its motion to make it move in a curved path.

force, electromotive See *electromotive force.*

foundation load The weight per unit area that the foundation of a structure such as a building or a bridge is designed to support.

free-body diagram A sketch showing an object and a representation of all active and reactive *forces* that affect the object.

free energy The *thermodynamic* quantity that relates the heat content of a substance and the randomness (or disorder) of the substance.

frequency Number of vibrations per unit time.

frequency, resonant See *resonant frequency.*

friction A *force* that resists the motion of objects that are in contact with one another.

friction angle The angle that a resultant *force* makes with a line perpendicular to the surface over which a body is sliding when *friction* is present.

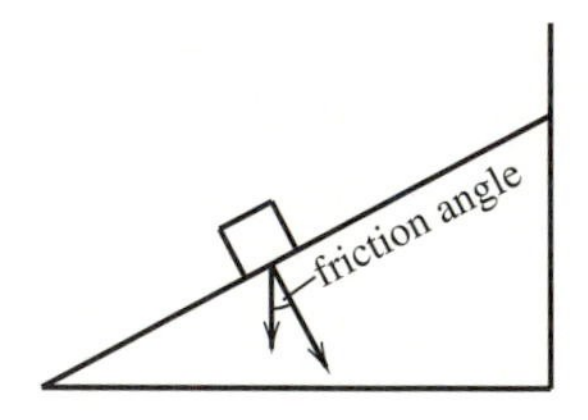

friction, coefficient of See *coefficient of friction.*

frustum The part of a solid between the base and a plane that is parallel to the base and intersects the solid.

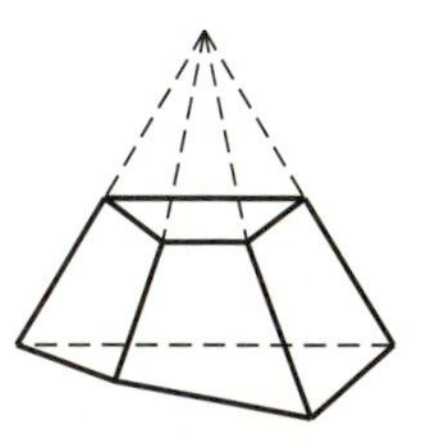

fulcrum A support about which a rigid bar resting on it is allowed to pivot.

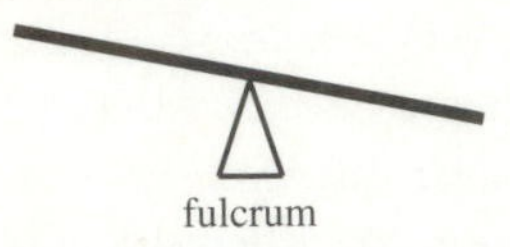

full-wave rectifier An electronic device for allowing sinusoidal *current* to flow through a *load* for one half the time period of the *ac* supply and then, using *diodes,* routes the source's negative half cycle *current* through the *load* in the same direction as the first positive half cycle.

Gaussian elimination A method for solving a system of linear equations. If the variables appear in the same order in each equation, the first variable is eliminated from equations after the first, the second variable is eliminated from equations after the second, and so on.

generator A device for converting mechanical *energy* into electrical *energy.*

grade The ratio, given as a percent, of change in elevation of a roadway over a specified horizontal distance.

gram A unit of mass. One gram is one-thousandth of the mass of the international prototype of the kilogram.

gravity, acceleration due to See *acceleration due to gravity.*

guy wire A wire attached to an object as a brace or support.

H_3O A chemical compound composed of hydrogen and oxygen.

half-life The time required for half of the radioactive atoms in a given sample of a substance to decay.

half-wave rectifier An electronic device for allowing *current* to flow in only one direction. When an *alternating current* is applied, the resulting *current* is 0 for half of a cycle and a sine wave for the remaining half cycle.

heat, molecular See *molecular heat.*

heat, specific See *specific heat.*

hemisphere Half of a sphere.

hemoglobin The red coloring matter of blood responsible for carrying oxygen from the lungs to the tissues of the body and carbon monoxide from the tissues to the lungs.

henry A unit of *inductance.* 1 henry = 1 *volt*-second/*ampere.*

Hertz A unit of *frequency.* 1 Hertz = 1 cycle/second.

hopper A receptacle, usually funnel-shaped, for feeding material such as grain into a container.

hydraulic Operated by means of water.

hydrograph A graph of storm water runoff rate versus time.

hydrostatics The study of the characteristics of liquids at rest.

ideal gas A theoretical gas consisting of infinitely small molecules that exert no *forces* on each other.

illumination The *density* of *energy* per unit time, when the *energy* is from a source that is capable of producing light.

impedance The effect that is produced jointly by a *reactance* and a *resistance* to oppose *current* in an *ac circuit.*

impedance, characteristic The ratio of *voltage* to *current* of a wave traveling in either direction on a transmission line.

incandescent lamp A lamp in which the light is produced by an electric *current* passing through a filament contained in a vacuum.

incidence, angle of The angle θ between the ray striking a surface and a line perpendicular to the surface at the point of impact.

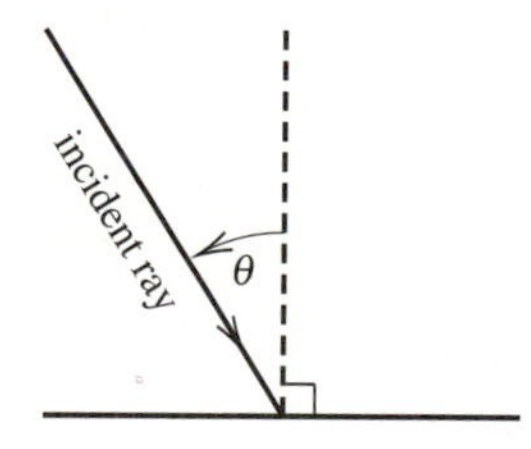

index of refraction See *refraction, index of.*

inductance The property of an electric *circuit* that allows a changing current to produce an *electromotive force* in that *circuit* or a neighboring *circuit.*

inductive reactance See *reactance, inductive.*

inductor A device for producing an *electromotive force* in a *circuit.* The symbol ⁀⁀⁀ is used in *circuit* diagrams to represent an inductor.

inertia, moment of See *moment of inertia.*

infiltration rate The rate at which rainwater penetrates a depth of soil with respect to time.

initial condition The values of certain variables in a formula at the beginning of the time period under consideration.

initial velocity See *velocity, initial.*

integrated circuit A *circuit* that contains components such as *transistors, resistors,* and *capacitors* in one miniaturized package.

intensity, physical *Power* per unit area.

invariant A quantity that remains unchanged while a given process occurs.

ion An electrically charged atom or group of atoms.

ionosphere The part of the earth's atmosphere (from about 25 miles to 250 miles above the earth's surface) that contains free electrically charged particles that are able to reflect radio waves.

isothermal Having constant or equal temperatures.

jackscrew A screw-operated device for exerting *pressure* or lifting a heavy body a short distance.

joist Any of the beams running parallel from wall to wall in a building to support the floor or ceiling.

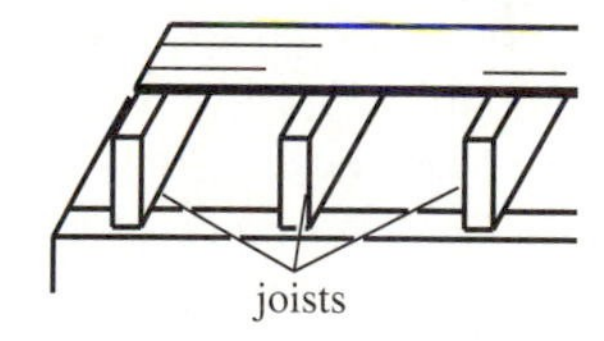

joule A unit of *work.* One joule is the amount of *work* done by one *newton* acting through a distance of one meter. *Energy* is also measured in joules.

Kelvin The temperature scale that assigns 0 to $-273.16°C$ and uses scale divisions of the same magnitude as *Celsius* degrees.

kilogram A unit of *mass.* 1 kilogram = 1000 *grams.* Sometimes used as a measure of weight. 1 kilogram = 2.2 pounds.

kinetic energy The *energy* that an object has due to its motion.

Kirchhoff's laws *Current* law: The algebraic sum of all *currents* at a *node* must be zero. *Voltage* law: The algebraic sum of all *voltage* sources and *voltage* drops around a closed *loop* must be zero.

knot A unit of *velocity.* One knot = 1 *nautical mile/* hour.

ladder network A *circuit* that can be drawn as a string of boxes connected so that the two output wires for one box are the two input wires for the next box.

Lame's constant A constant used in the analysis of *stress-strain* relationships. Denoted by λ, its value is given by $\lambda = (3k - 2\mu)/3$, where k is the *bulk modulus* and μ is the *shear modulus.*

laser A device for producing a narrow, monochromatic beam of light for which there is a fixed phase relationship between waves emitted.

latitude Angular distance north or south of the earth's equator.

lead angle An angle used in measuring the placement of teeth on a gear.

length, focal See *focal length.*

lens A transparent object, regular in shape, that can produce an image of an object placed before it.

lens board In a camera, a board that supports the *lens.*

lever A bar used for prying or dislodging an object.

linear expansion, coefficient of See *coefficient of linear expansion.*

linear velocity See *velocity, linear.*

liter A unit of capacity. One liter is the amount of a substance that fills a cubical box that measures 10 centimeters on each edge.

load (electrical) The part of an electrical *circuit* that draws *current* from the source.

load, concentrated A load that is positioned so that the effect of its weight is localized in a small area. Load may be specified in newtons, pounds, or kilograms.

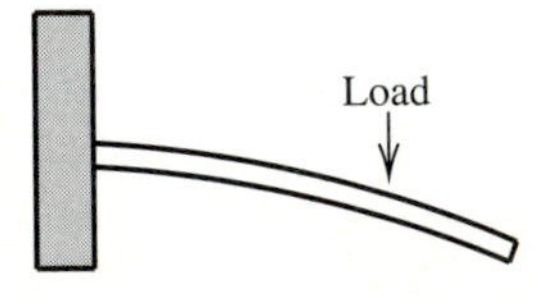

load, distributed A load that is positioned so that the effect of its weight is spread out over an area. Load may be specified in newtons, pounds, or kilograms.

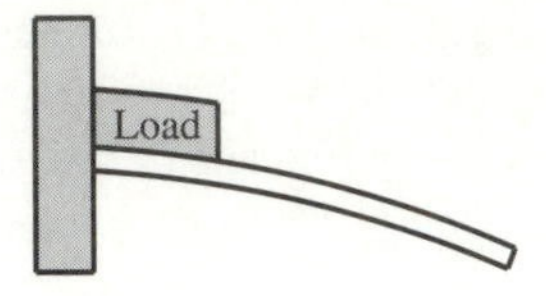

longitudinal Directed along the length of an object such as a bar or rod.

loop (computer) A set of instructions that may be executed more than once (usually with changes in the values of the variables) before further action takes place.

loop (electrical) A complete path in an electric *circuit.*

lumen A unit of luminous flux. One lumen is the amount of light emitted in a unit solid angle by a uniform point source of one candle.

magnetic field A region of space in which a *force* can be exerted at every point on a magnetic, but not on a nonmagnetic, object.

magnification For a lens or system of lenses, the ratio of the angle subtended by the image to the angle subtended by the object when it is placed at a distance of 25 cm from the unaided eye.

mass The property of an object that tends to resist changes in its linear motion.

matrix A rectangular array of numbers.

memory chip See *chip, memory.*

meridian A circle on the surface of the earth that is the intersection of the earth with a plane passing through its center. Meridians are used to measure angular distances east and west of the zero position located near London, England.

micrometer An instrument for measuring small distances.

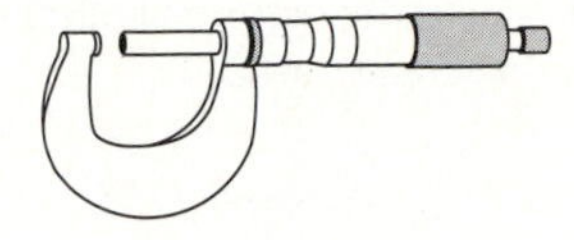

micron A unit of length. 1 micron = 1/1,000,000 meter.

microorganism Any microscopic animal or vegetable, most often bacteria.

microwave A very short electromagnetic wave. Waves between 0.3 cm and 30 cm long are classified as microwaves.

mil A unit of length. One mil = 0.001 inch.

millibar A unit of pressure. 1 millibar = 100 *newtons*/meter^2. The millibar is used in meteorology.

modulate To vary the amplitude, *frequency*, or *phase angle* of a wave or signal.

modulus, bulk See *bulk modulus*.

modulus of elasticity The ratio of *stress* to *strain* for a material subjected to a *stress* that does not cause a permanent deformation.

modulus, shear See *shear modulus*.

modulus, Young's See *Young's modulus*.

mole An amount of a substance that contains 6.023×10^{23} elementary particles, such as atoms, *molecules, ions,* etc.

molecular heat The ratio of heat absorbed by one *mole* of a substance to the increase in temperature caused by the absorption.

molecular weight The sum of the atomic weights of the atoms in the formula of a chemical compound. The atomic weights may be obtained from a table of values.

molecule A group of atoms that stick together strongly enough to act as a single particle.

moment The product of a force's magnitude and the perpendicular distance from the line of action of the *force* to a pivot point.

moment of inertia The property of a body that tends to resist change in its rotational motion about a given axis. The magnitude of the moment of inertia depends on the way in which the *mass* is distributed about the axis.

momentum The vector that is the product of *mass* and *velocity* of a moving body.

monitor, computer A high-resolution screen (similar to a television set) for displaying information that is fed into or retrieved from a computer.

monochromatic light Light consisting of only one color or *wavelength*.

motherboard A board that contains data paths for conveying information from one part of a microcomputer to another and slots for attaching such things as printers, disk drives, *memory chips,* and *integrated circuits*.

multimeter, digital A meter that measures several different quantities, such as *voltage, current,* and *resistance* and displays the measurements on a digital display rather than on a gauge.

nautical mile, British A unit of length. One nautical mile is the length of a minute of arc of a *meridian*. Since the earth is not exactly spherical, the length of a British nautical mile varies with latitude. The British nautical mile is used by English ships and aircraft.

nautical mile, U.S. A unit of length. One nautical mile = 6080.2 ft. The U.S. nautical mile is used by American ships and aircraft.

network A combination of components, such as *resistors* and *capacitors,* interconnected in any way.

network, ladder See *ladder network*.

newton A unit of *force*. A newton is the *force* necessary to give a *mass* of one *kilogram* an *acceleration* of one meter per second in one second.

Newton's law of cooling The principle that the rate at which the temperature of an object changes is proportional to the difference between its temperature and that of the surrounding medium.

node A point in an electrical *network* where two or more paths branch.

objective The principle image-forming component of a telescope or other optical instrument.

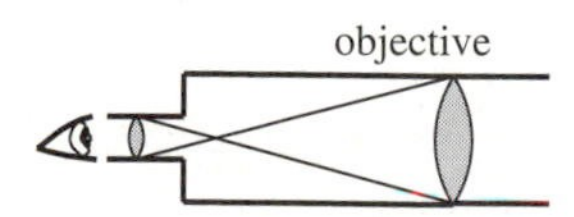

ohm A unit of electrical *resistance*. One ohm = 1 *volt/ampere*.

Ohm's law The law that states that the *current* I that flows in a *circuit* is directly proportional to the applied *voltage* V and inversely proportional to the *resistance* R. $I = V/R$.

orthogonal trajectory A curve that intersects each member of a family of curves at right angles.

oscillation A fluctuation between a maximum and a minimum value.

oscillator, quadrature A *circuit* that generates *alternating current* at a *frequency* determined by the values of its components and shifts the phase of a signal 90°.

oscilloscope An instrument with a fluorescent screen on which variations in a fluctuating electrical signal appear as a visible *waveform.*

parallel A type of electrical *circuit* in which two or more electrical devices are connected so that the *current* must divide, part going through one device and part through another. For two *resistors* in parallel, $1/R = 1/R_1 + 1/R_2$. For two *capacitors* in parallel, $C = C_1 + C_2$.

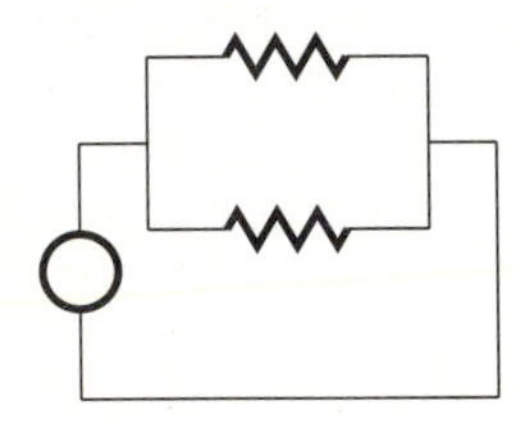

pascal A unit of *pressure.* 1 pascal = 1 *newton/*meter^2.

pendulum A body suspended so that it can swing back and forth freely.

pendulum, period of See *period of a pendulum.*

period of a pendulum The time for a *pendulum* to make one complete swing back and forth.

permeability The property of a material that tends to change the concentration of imaginary lines, called lines of *force,* that describe a magnetic field.

pH A scale for expressing hydrogen *ion* concentration of a solution.

phase angle The angle θ between *voltage* and *current* vectors.

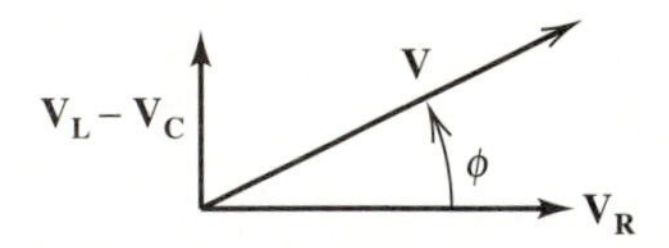

physical intensity See *intensity, physical.*

piston A cylinder attached to a rod and closely fitted into another cylinder within which the inner cylinder moves back and forth.

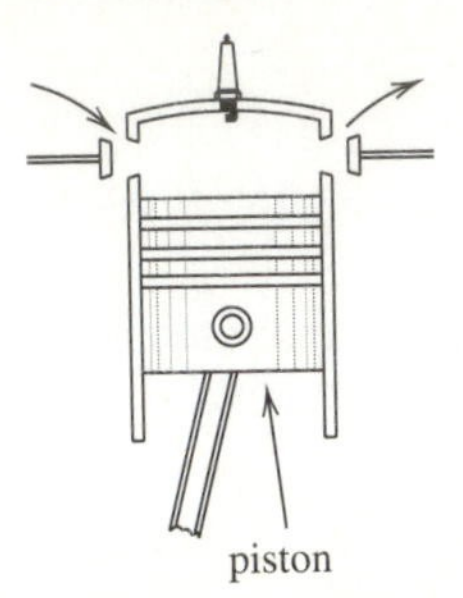

Poisson ratio A constant used in the analysis of *stress-strain* relationships. Denoted by σ, its value is given by $\sigma = (E - 2\mu)/(2\mu)$, where E is *Young's modulus* and μ is the *shear modulus.*

potential difference The *work* done per unit *charge* as a *charge* is moved between two points in an electric field.

potential energy The stored *energy* that an object has due to its position. For example, an object at a height has gravitational potential energy.

potentiometer A variable *resistor* connected as a *series voltage divider.*

pound A unit of weight. One pound is the *force* necessary to give a *mass* of one *slug* an *acceleration* of one foot per second in one second.

power The rate at which *work* is done with respect to time.

preamplifier A *voltage amplifier* for receiving the output from a low level device, such as a microphone, and amplifying it before sending the signal to additional *amplifier circuits.*

pressure *Force* per unit area. Sometimes the word pressure refers to *mass* per unit area.

prism A solid figure with two parallel bases of the same size and shape whose edges form the bases of faces that are parallelograms.

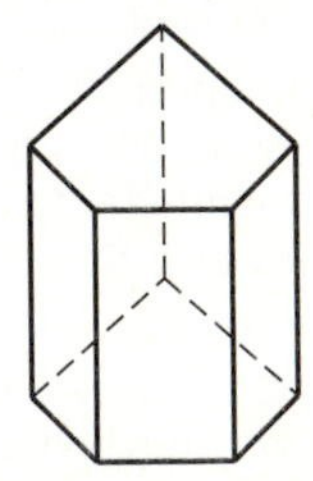

processor The program or equipment used in a computer for assembling, compiling, translating, generating, and computing operations.

product, cross See *cross-product.*

product, dot See *dot product.*

projectile An object that is thrown or shot forward.

pulley A wheel with a grooved rim in which a rope runs making it possible to lift a weight on one end of the rope by pulling down on the other end.

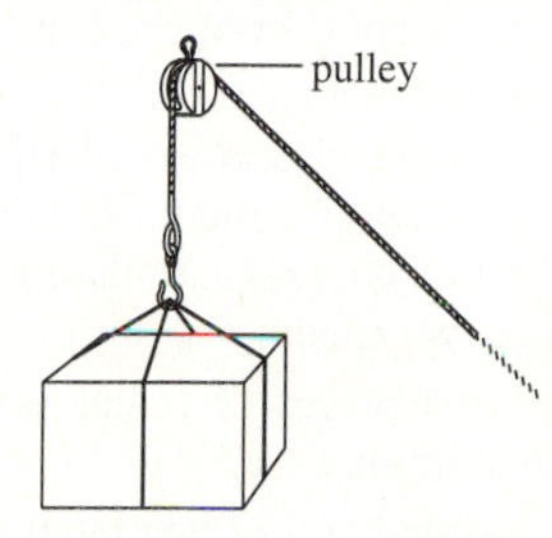

pyramid A solid figure with a base whose edges form the bases of triangular faces meeting at a common point.

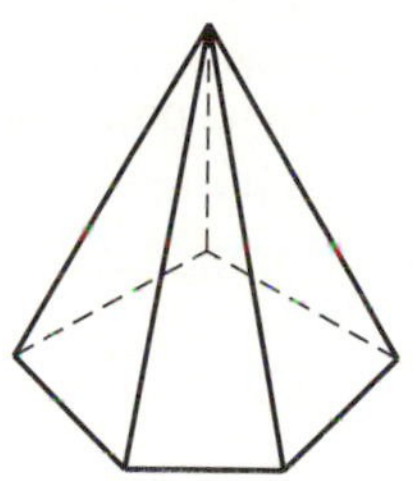

quadrature oscillator See *oscillator, quadrature.*

radar An electronic device for detecting distant objects by sending out very short radio waves and timing the return of the echo.

radioactivity The spontaneous decay of an atomic nucleus with the emission of alpha particles, beta particles, and gamma rays as a new kind of atom is formed.

radius of curvature The radius R of a circle that would coincide with the surface of a circular lens or mirror.

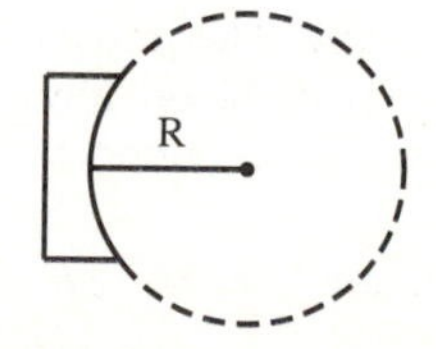

rate of infiltration See *infiltration rate.*

ratio, void See *void ratio.*

reactance The property of a *circuit* that tends to oppose the flow of *alternating current* and is produced by the combined effects of *capacitance* and *inductance.*

reactance, capacitive The property of a *circuit* that tends to oppose the flow of *alternating current* produced by *capacitance.*

reactance, inductive The property of a *circuit* that tends to oppose the flow of *alternating current* produced by *inductance.*

rectifier, full-wave See *full-wave rectifier.*

rectifier, half-wave See *half-wave rectifier.*

reflection, angle of The angle θ between a ray bouncing off a surface and the line that is perpendicular to the surface at the point of impact.

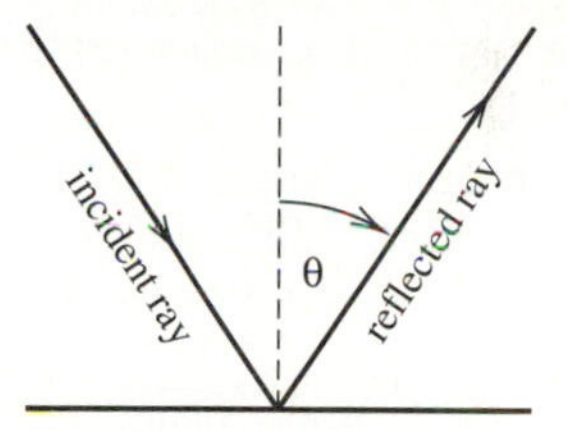

refraction, angle of The angle θ between a ray that is bent as it passes from one medium to another and the line that is perpendicular to the surface of the second medium at the point at which the ray enters the medium.

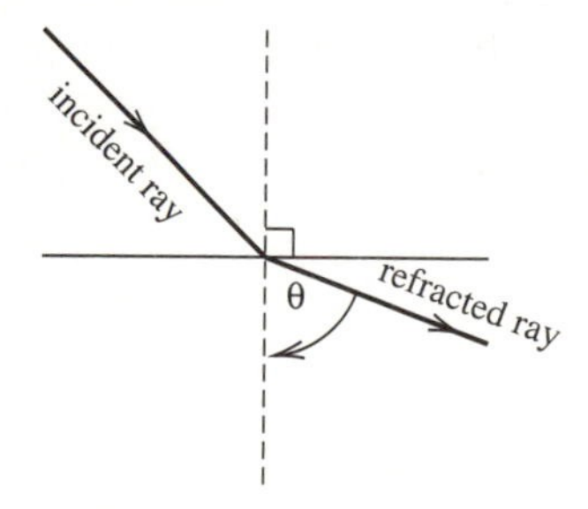

refraction, atmospheric The bending, caused by the earth's atmosphere, of light rays from celestial objects.

refraction, index of The ratio of the speed of light in a vacuum to its speed in a given substance.

relativity An interdependence of matter, time, and space that avoids the contradictions of nineteenth-century classical physics.

resistance The property of a substance that tends to oppose the flow of electricity.

resistance, equivalent The *resistance* of a single *resistor* that can be substituted for a set of interconnected *resistors* without changing the properties of any *circuit.*

resistance, specific See *specific resistance.*

resistivity The *resistance* of a block of material 1 cm long and 1 cm^2 in cross section, usually at 20°C. Resistivity is sometimes called *specific resistance.*

resistor A device for opposing the flow of *current.* The symbol /\/\/ is used in *circuit* diagrams to represent a resistor.

resonance The state at which the natural *frequency* of *oscillation* of two electrical *circuits* is the same, and the maximum transfer of *energy* can occur.

resonant frequency The *frequency* at which the *impedance* is at a minimum in an *ac circuit.*

Richter scale A scale that measures the intensity of earthquakes.

roentgen A unit of radiation. One roentgen is the amount of gamma radiation that, when absorbed in one gram of air, will produce 1.61×10^{12} pairs of ions.

roller bearing See *bearing, roller.*

rotor The rotating component of a generator or motor.

sag A dip in a roadway.

saline A mixture of salt and water.

series A type of electrical *circuit* in which two or more electrical devices are connected so that the *current* must pass through one device before it passes through the other. For two *resistors* in series, $R = R_1 + R_2$. For two *capacitors* in series, $1/C = 1/C_1 + 1/C_2$.

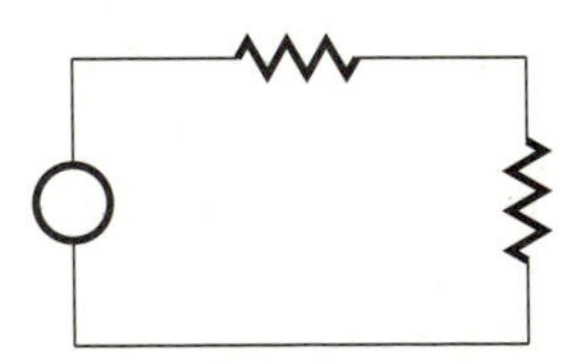

shaft A long narrow opening sunk into the earth.

shear The state of a system in which equal and opposite *forces* that do not act along the same straight line are applied to a body.

shear modulus The ratio of *shear stress* to *shear strain.*

shortwave A radio wave of *wavelength* 60 meters or less.

siemen A unit of *conductance.* 1 siemen = 1 *ampere/volt.*

simple beam See *beam, simple.*

slug A unit of *mass.* One slug is the *mass* necessary for a *force* of one pound to produce an *acceleration* of one foot per second in one second.

Snell's law The law that states that the ratio between the sine of the *angle of incidence* of a light ray upon a boundary between two media and the sine of the *angle of refraction* is equal to the ratio of the speeds of light in the two media.

solder To join two pieces of metal with a melted *alloy,* which solidifies.

solenoid A cylindrical coil of insulated wire.

solute A substance that is dissolved in another substance.

specific heat The amount of heat necessary to raise a unit mass of a material through one degree of temperature.

specific resistance See *resistivity.*

speed The rate, without regard to direction, at which *displacement* changes with time.

steady-state current A pure *direct current* or periodic *alternating current.*

stepping motor A mechanical device that rotates by a fixed amount each time that it is pulsed. Stepping motors are often used in disk drives and digital plotters.

superelevation The ratio of vertical rise of a roadway to horizontal distance covered.

strain The ratio of change in size to original size when that change is caused by *stress.*

strength, tensile See *tensile strength.*

stress *Force* per unit area, when the *force* is a deforming *force.*

substrate The bottom layer of a multilayer *circuit* structure.

systolic blood pressure The *pressure* exerted by the heart's contraction on blood, forcing it to flow into an artery that has been temporarily blocked.

temperature, absolute See *absolute temperature.*

tensile strength The *force* necessary to break a rod or wire of unit cross-sectional area.

tension A stretching *force.*

terminal A connecting device placed at the end of a wire, appliance, machine, etc. so that a connection can be made to it.

thermistor A type of *resistor* whose *resistance* increases as temperature increases.

thermodynamic Relating to *thermodynamics.*

thermodynamics The study of quantitative relationships between heat and other forms of *energy.*

thermodynamic system A collection of masses among which *thermodynamic* reactions can be observed.

thrust The forward *force* exerted upon an airplane by its propulsion system.

torque A turning or twisting *force.*

trajectory, orthogonal See *orthogonal trajectory.*

transformer An electrical device for changing the *voltage* of an *alternating current.*

transistor A semiconductor device for amplifying electrical signals.

troy weight A system of weights for gold, silver, and gems. One troy pound = 12 troy ounces.

tuning fork A two-pronged metal implement that gives a fixed tone when struck and is used for tuning musical instruments.

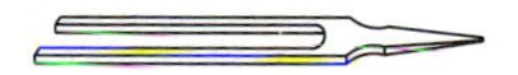

velocity The rate at which *displacement* changes with respect to time.

velocity, angular The rate at which angular position changes with respect to time.

velocity, initial The *velocity* of an object at the beginning of the time period under consideration.

velocity, linear The rate at which distance changes with respect to time.

velocity potential curves A family of curves such that each member is perpendicular to all members of a family of curves representing motion of a fluid.

vertex The highest or lowest point on a surface.

VHF Very High Frequency. Radio waves with *frequencies* between 54 and 216 megaHertz are classified as VHF.

v-notch weir A device for backing up or diverting water in a channel.

void ratio The ratio of the volume of voids in a soil mass to the volume of the solids in the soil mass.

volt A unit of electrical *potential difference.* One volt is the *potential difference* necessary to produce one *ampere* of *current* through a *resistance* of one *ohm.*

voltage See *potential difference.*

voltage divider A *series circuit* for providing a *voltage* less than the source *voltage.*

voltage, effective The *equivalent dc voltage* that when multiplied by the *effective current,* gives the average *power* in an *ac circuit.*

voltmeter An instrument for measuring *electromotive force* or a difference in *potential energy* in *volts.*

watt A unit of *power.* 1 watt = 1 *joule*/second.

waveform The shape of a wave.

wavelength The distance from any point on a wave to the corresponding point on the next wave.

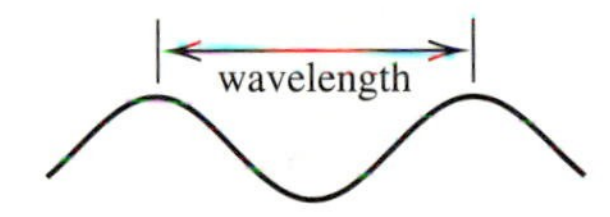

weir, v-notch See *v-notch weir.*

Wheatstone bridge An instrument for measuring *resistance.*

winding One or more turns of wire forming a continuous coil for a *transformer* or other electrical device.

work The product of the magnitudes of a *displacement* and the component of *force* in the direction of the *displacement.*

Young's modulus The ratio of the *longitudinal stress* to the associated *longitudinal strain* of a rod or wire.

CREDITS

Unless otherwise acknowledged, all photographs are the property of Scott, Foresman.

p. 1 Acrylic Wall Sculpture (Detail) by Zelda W. Werner/Photo: Ron Testa

p. 41 Chicago Photographic Company

p. 95 Brent Jones

p. 137 Brent Jones

p. 191 David Belle

p. 236 Milt & Joan Mann/Cameramann International, Ltd.

p. 269 Milt & Joan Mann/Cameramann International, Ltd.

p. 299 Milt & Joan Mann/Cameramann International, Ltd.

p. 337 Chicago Photographic Company

p. 395 Brent Jones

p. 440 Milt & Joan Mann/Cameramann International, Ltd.

p. 487 Courtesy Micron Instruments, Los Angeles, CA

p. 489 Milt & Joan Mann/Cameramann International, Ltd.

p. 529 Courtesy Hoechst-Celanese Chemical Group, Pasadena, TX/Milt & Joan Mann/ Cameramann International, Ltd.

p. 568 Milt & Joan Mann/Cameramann International, Ltd.

p. 609 Chicago Photographic Company

p. 655 Milt & Joan Mann/Cameramann International, Ltd.

p. 689 Chicago Photographic Company

p. 729 David Belle

p. 765 Brent Jones

p. 800 Milt & Joan Mann/Cameramann International, Ltd.

p. 828 Courtesy Eveready Battery Company, Inc.

p. 830 Brent Jones

p. 888 Milt & Joan Mann/Cameramann International, Ltd.

p. 938 Brent Jones

p. 982 Milt & Joan Mann/Cameramann International, Ltd.

p. 1025 Milt & Joan Mann/Cameramann International, Ltd.

p. 1057 Chicago Photographic Company

p. 1095 Brent Jones

p. 1130 David Belle

p. 1174 Photo Courtesy The University of Arizona, Tucson

ANSWERS TO SELECTED EXERCISES

This section contains the answers to all of the *odd-numbered* problems in the section exercises and in the sequential review and applied portions of the chapter reviews, as well as the answers to *all* of the random review and chapter test problems.

If you need further help with this course, you may want to obtain a copy of the *Student's Solutions Manual* that accompanies this textbook. This manual provides detailed step-by-step solutions to the odd-numbered exercises in the textbook and can help you study and understand the course material. Your college bookstore either has this manual or can order it for you.

CHAPTER 1

Section 1.1 (page 7)

1 **(a)** 1/4 **(b)** 1/4 **3.** **(a)** -7 **(b)** -7 **5.** **(a)** -3.14 **(b)** -3.14 **7.** **(a)** $\sqrt{5}$ **(b)** $-\sqrt{5}$ **9.** **(a)** -2.18 **(b)** 2.18 **11.** $-9, \sqrt{5}, 3, \pi, |-5|, 19$ **13.** $-7, \sqrt{7}, |-3|, |\pi|, 5, 21$ **15.** $-1/2, -1/3, -1/4, 1/4, 1/3, 1/2$ **17.** $-|-1/3|, -|0.33|, -0.3, |0.03|, 0.3, 1/3, \sqrt{3}$ **19.** **(a)** True **(b)** True **21.** **(a)** False **(b)** True **23.** **(a)** True **(b)** True **25.** **(a)** False **(b)** False **27.** Integer, Rational, Real, Complex **29.** Integer, Rational, Real, Complex **31.** Real, Complex **33.** Rational, Real, Complex **35.** Complex **37.** Complex **39.** Real, Complex **41.** Rational, Real, Complex **43.** Integer **45.** Rational **47.** Real **49.** **(a)** area is multiplied by 4 **(b)** area is the same **(c)** area is 1/4 as large

Section 1.2 (page 15)

1. 20 **3.** -5 **5.** -4 **7.** -20 **9.** -2 **11.** 78 **13.** 120 **15.** -4 **17.** -3 **19.** 14 **21.** 2 **23.** 98 **25.** -1 **27.** -4 **29.** 16 **31.** 486 **33.** -44 **35.** 0 **37.** 21 **39.** -12 **41.** -4 **43.** 1.62 lb **45.** **(a)** 35.6 m/s **(b)** 51.2 m/s **47.** 38 in^3 **49.** **(a)** 0.12 Ω **(b)** 30.0°

Section 1.3 (page 22)

1. 21.27 **3.** -4.68 **5.** 128 **7.** 3529 **9.** 3 **11.** 4 **13.** 1 **15.** 4 **17.** 2 **19.** 2 **21.** 0.001 **23.** 14 **25.** 600 **27.** 4.5×10^4 **29.** 2.31×10^8 **31.** 6×10^7 **33.** 2.7×10^{-5} **35.** 6700 **37.** 0.0835 **39.** 0.0091 **41.** 62.04 **43.** **(a)** 0.0000000000027 F **(b)** 3800 V **45.** **(a)** 1.459×10^4 grams **(b)** 4.45×10^5 dynes **47.** 305.8732 **49.** **(a)** 3.6 newtons **(b)** 11 pounds

Section 1.4 (page 27)

1. 78 **3.** 67 **5.** 290 **7.** 75 **9.** 11 **11.** 29 **13.** 6868 **15.** 1470 **17.** Associative **19.** Commutative **21.** Commutative **23.** Distributive **25.** Commutative **27.** Commutative and Associative **29.** Commutative **31.** Commutative **33.** 3.1 ohms **35.** 8.39 newtons **37.** 2.2 cm **39.** 1.2 μF

Section 1.5 (page 33)

1. -4 **3.** 65 **5.** 432 **7.** 16.5 **9.** 0.0667 **11.** 41 **13.** 8.21 **15.** 22.7 **17.** 13 **19.** 11 **21.** 103 or 100 to two significant digits **23.** 507 **25.** 88 **27.** 34.7 **29.** 11.30 to two decimal places, or 11.3 to three significant digits **31.** -9.62 **33.** -70.5 **35.** 1.4 **37.** 3.5 **39.** 4.5 **41.** 5.6×10^{18} **43.** 0.16 Ω **45.** 40.6 kN/m^2 or 41 kN/m^2 **47.** 1.3 atmospheres **49.** 0.6 nF

Section 1.6 (page 36)

Sequential Review: 1. 5.91 **3.** $-11, -\pi, \sqrt{7}, |-3|, 5, 17$ **5.** False **7.** True **9.** Real, Complex **11.** Complex **13.** -26 **15.** 3 **17.** -1 **19.** 4 **21.** 0.12 **23.** 2 **25.** 5 **27.** 7.53×10^{-3} **29.** 62,000 **31.** 82 **33.** 7979 **35.** Associative **37.** Distributive **39.** -6.14 **41.** 976 to three significant digits **43.** 81 **45.** 60.6
Applied Problems: 1. Real **3.** 2 200 000 Ω **5.** 8° **7.** 11 cm^2 **9.** 9.4 cm/s **11.** 0.675 μm **13.** 4 $kg \cdot m^2$ **15.** 110 cm^3 **17.** 35° C **19.** 0.477
Random Review: 1. (a) Associative **(b)** Commutative and associative **2. (a)** False **(b)** True **3. (a)** 3 **(b)** 2 **4.** $-8, |-2|, 15/7, \sqrt{5}, 3, \pi$ **5.** 113 **6. (a)** -29 **(b)** -7 **7.** -96.0 **8. (a)** -6.33 **(b)** 12.55 **9. (a)** 11 **(b)** 1/7 **10. (a)** 11 **(b)** 7 **11. (a)** 8.4362×10^6 **(b)** 7.31×10^{-4} **12. (a)** Rational, Real, Complex **(b)** Real, Complex **13. (a)** 15.0 **(b)** 2.87 **14. (a)** 0.0006415 **(b)** 732,100 **15.** 30, to two significant digits
Chapter Test: 1. (a) False **(b)** True **2. (a)** Complex **(b)** Rational, Real, Complex **3. (a)** -2.00 **(b)** -0.250 **4. (a)** 2; 7.2×10^{-4} **(b)** 1; 6×10^2 **5. (a)** 3000 c **(b)** 0.0072 1 **6. (a)** 0.7 **(b)** 29 **7. (a)** Distributive **(b)** Commutative and Associative **8. (a)** 0.333 **(b)** 10,000, to two significant digits **9.** Real **10.** 180 cm

CHAPTER 2

Section 2.1 (page 46)

1. 2 factors **3.** 2 terms **5.** 3 terms **7.** 2 factors **9.** 2 terms **11.** $3x + 2y$ **13.** $3x + y - 3$ **15.** $4x^2 - x + 1$ **17.** $5x + 3y - z + w$ **19.** $x^2 + 4x - 5$ **21.** $4y - 1$ **23.** $11m + 1$ **25.** $2x + 3$ **27.** -4 **29.** $2z^2 + 3$ **31.** $2y + 2$ **33.** 1 **35.** $3ax + b - 1$ **37.** $3 - y$ **39.** $-2z + 3$ **41.** -3 **43.** $-4x + 2$ **45.** $7y + 7$

Section 2.2 (page 52)

1. $6x^5$ **3.** $14y^7$ **5.** $48x^9$ **7.** $45x^3y^3$ **9.** $2x^2 - 2x$ **11.** $3x^3 - 3x^2 + 12x$ **13.** $2x^2 - x - 3$ **15.** $15y^2 - 23y + 4$ **17.** $16a^2 + 8ab + b^2$ **19.** $z^3 - 1$ **21.** $8y^3 + 12y^2 + 6y + 1$ **23.** $y^4 + y^3 - 7y^2 + 7y - 2$ **25.** $x^2 + x - 1$ **27.** $x^2 - x - 7$ **29.** $x^2 - 4$ **31.** $6x^2 - 6$ **33.** 49 **35.** $2x/y^2$ **37.** $3/(2mn^2)$ **39.** $3a/(2b)$ **41.** $2x/y - 3 + 1/(2y)$ **43.** $2a^2 - 3a + 1$ **45.** $4mn + 3 - 1/(mn)$ **47.** $3x - 2$ **49.** $2x - 1$ **51.** $x^2 + 2x + 1$ **53.** $x^2 + 1$ **55.** $T = t - h/100$ **57.** $(a + b)^2$ or $a^2 + 2ab + b^2$ **59.** 10 MPa

Section 2.3 (page 58)

1. x^3y^3 **3.** $9y^2$ **5.** $16a^2b^2$ **7.** $x^3 + 3x^2y + 3xy^2 + y^3$ **9.** $y^2 + 6y + 9$ **11.** $16a^2 + 8ab + b^2$ **13.** 9 **15.** 3 **17.** -4 **19.** not a real number **21.** -2 **23.** $5\sqrt{2} = 7.07$ **25.** $2\sqrt{10} = 6.32$ **27.** $2\sqrt[3]{5}$ **29.** $5\sqrt{10} = 15.8$ **31.** $2\sqrt{3}/3 = 1.15$ **33.** $\sqrt{3}/9 = 0.192$ **35.** 10 **37.** 5 **39.** 8

41. **(a)** $0.55\sqrt{681}$ **(b)** 14 **43.** $A = \pi r^2 + 4\pi r + 4\pi$ **45.** $P = 8kV^3$
47. **(a)** $(x+y)^3 = x^3 + 3x^2y + 3xy^2 + y^3$ **(b)** 4 **(c)** decrease from 3 to 0 **(d)** increase from 0 to 3
(e) 1 3 3 1 **(f)** $(2a+3b)^3 = 8a^3 + 36a^2b + 54ab^2 + 27b^3$

Section 2.4 (page 65)

1. 1 **3.** 1/1000 **5.** $56/x$ **7.** $18b^6c/a^5$ **9.** $4y^2/z$ **11.** $1/(2a^4)$ **13.** $y^3/10^4$ **15.** $72ax^2/(by)$
17. 3^{20} **19.** x^3y^3 **21.** $125x^3y^3$ **23.** 9/16 **25.** 8/27 **27.** $3x^6/y^3$ **29.** $1/(5xy)$ **31.** $9b^2/16$
33. y^3/x **35.** $3a/(5b^2)$ **37.** $b^3c^6/(8a^3)$ **39.** 1 **41.** $4(b+3)/(a^3b^3)$ **43.** $y^3/x^3 + y^2/x$
45. **(a)** 5×10^4 joules **(b)** 2×10^5 joules **(c)** 4.2×10^5 joules **47.** $P = I^2R$ **49.** $P = \pi^2EI/l^2$ **51.** no

Section 2.5 (page 72)

1. $x = 4/5$ **3.** $z = -2/3$ **5.** $y = 8/3$ **7.** $p = -16$ **9.** $x = 0$ **11.** $x = 0$ **13.** $p = 7/8$
15. $n = 4/3$ **17.** $y = -8/7$ **19.** $x = 5.5$ **21.** $x = -0.4925$ **23.** $x = -14.5$ **25.** **(a)** $x = 2$
(b) $x = 10^5$ **27.** **(a)** $x = \pm 2$ **(b)** $x = 16$ **29.** $F = w/d$ **31.** $I = V/R$ **33.** $l = \mu N^2A/L$
35. $I_2 = I_1d_1^2/d_2^2$ **37.** $N = PV/(kT)$ **39.** $V_{emf} = I(r+R)$ **41.** $h = \sqrt[3]{12I/b}$ **43.** $v = \sqrt{30fd}$
45. $P = V^2/R$ **47.** $(1/2)bh + (1/2)Bh = (1/2)h(b+B)$

Section 2.6 (page 79)

1. 2/3 **3.** 8/3 **5.** 3/8 **7.** 12 knots **9.** 6 joules **11.** 4.35 **13.** 9.10 **15.** $x = 0.11$
17. $z = 470$ **19.** $y = 90$ **21.** 45 kg **23.** 41 mi **25.** 50.0 l **27.** 11 km/l **29.** 5×10^{-10} F
31. 240.0 Ω **33.** 240 V **35.** 27 1/2 mi **37.** 37.5 cm **39.** $d = 28.0$ m **41.** 1.4 mi
43. 40,000,000 gallons **45.** 5×10^{12} **47.** numerator = denominator

Section 2.7 (page 86)

1. 99 **3.** 7.2 grams **5.** 1 liter **7.** 1.5 hours **9.** 261 miles, 84 miles **11.** 4 **13.** 1.24 m
15. 160 per column, 280 per row **17.** −40° **19.** 1.80 m **21.** 0.42A **23.** 8420 W **25.** 75 kg
27. 21.4 lb **29.** 5.6% **31.** 349

Section 2.8 (page 89)

Sequential Review: **1.** 2 terms **3.** 3 terms **5.** $-x-2$ **7.** $-2z+6$ **9.** $2z^4 + 3z^3 + z^2 + 3z - 1$
11. $-6x+6$ **13.** $x^2 + x + 2$ R -6 **15.** $9x^2y^3$ **17.** $8a^3 + 36a^2 + 54a + 27$ **19.** $4\sqrt{3} = 6.93$
21. $2\sqrt{7}/7$ **23.** $12/(x^2y)$ **25.** $3x^2/y^4$ **27.** $9/(25x^6y^4)$ **29.** $m = 5/2$ **31.** $x = -2$
33. $w = V/(lh)$ **35.** $y = 243$ **37.** 3/4 **39.** $y = 742$ **41.** 8.9 km **43.** 2.2 lb
45. 50 km/hr, 55 km/hr
Applied Problems: **1.** 2^{18} **3.** 30 cm **5.** 40 rpm **7.** 68/5 **9.** $P = NA_vS_v$ **11.** 10^6
13. 28.3 ft **15.** 21.9° C **17.** 57.5° C **19.** 62 500 m
Random Review: **1.** **(a)** $x = 4.86$ **(b)** $y = 1.1$ **2.** **(a)** $9x^8y^2$ **(b)** $9x^8 + 6x^4y + y^2$ **3.** 86,551 lb
4. **(a)** $2x^2 + 2x - 1$ **(b)** $-4x$ **5.** **(a)** 9 **(b)** $7x/(2y)$ **6.** **(a)** $2x^2 - 5x$ **(b)** $x - 4$ **7.** **(a)** $x = 81$
(b) x is not a real number **8.** **(a)** 1 **(b)** $x^3y + x^2y^2$ **9.** **(a)** $8\sqrt{2}$ **(b)** $5\sqrt{3}/3$ **10.** **(a)** $x = 5/4$
(b) $y = (z-x)/2$
Chapter Test: **1.** **(a)** $x^2 - 3x - 11$ **(b)** $-2x^2 + 2x + 4$ **2.** **(a)** $24x^4y^3$
(b) $m^4 - m^3 + 4m^2 - 3m + 8$ **3.** **(a)** $x^2 - x + 3$ **(b)** $2x^3 - 3x^2y$ **4.** **(a)** $(y^6z^3)/(64x^3)$
(b) $x^3y^3 + 2x^2y^4 + xy^5$ **5.** $x = -4/3$ **6.** $x = 3.2$ **7.** $z = 17.1$ **8.** **(a)** 19 cm **(b)** 75 kg
9. 130 cm **10.** 68.2 min

CHAPTER 3

Section 3.1 (page 99)

1. yes **3.** yes **5.** yes **7.** yes **9.** no **11.** 0 **13.** 0 **15.** 5 **17.** −1 **19.** 19 **21.** 1 **23.** 6 **25.** 5 **27.** no **29.** yes; real numbers, $x \neq 1$ **31.** yes; real numbers, $x \geq 4$ **33.** yes; real numbers **35.** yes; real numbers **37.** 30 to two significant digits **39.** 1.6 **41.** 0.479 **43.** yes **45.** yes **47.** real numbers, $d \geq 0$ **49.** real numbers, $R_1 \geq 0$

Section 3.2 (page 106)

1. **3.**

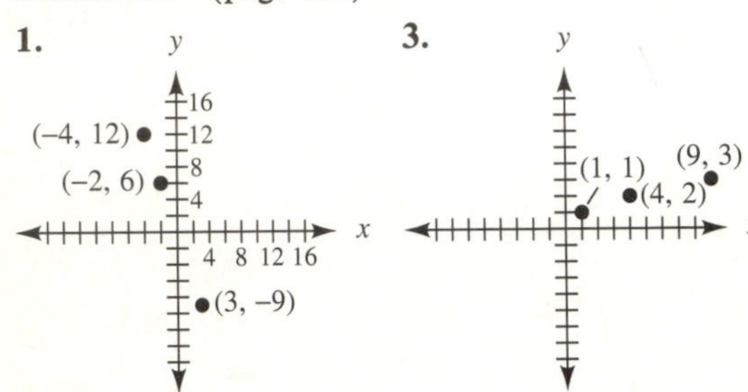

5.

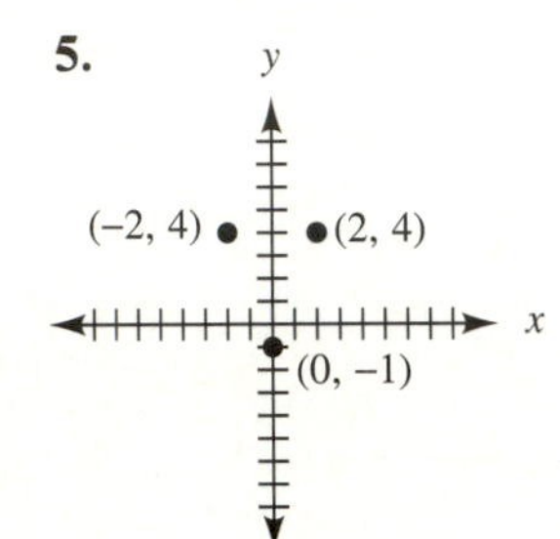

7.

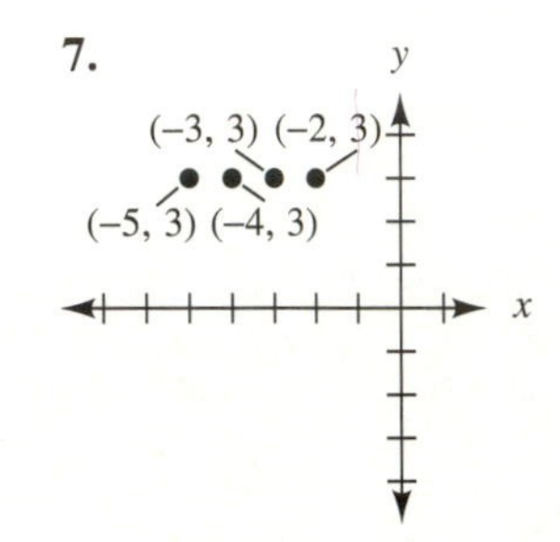

9.

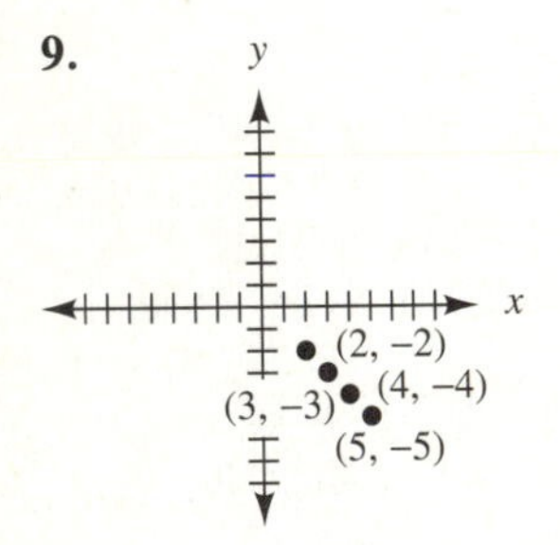

11.

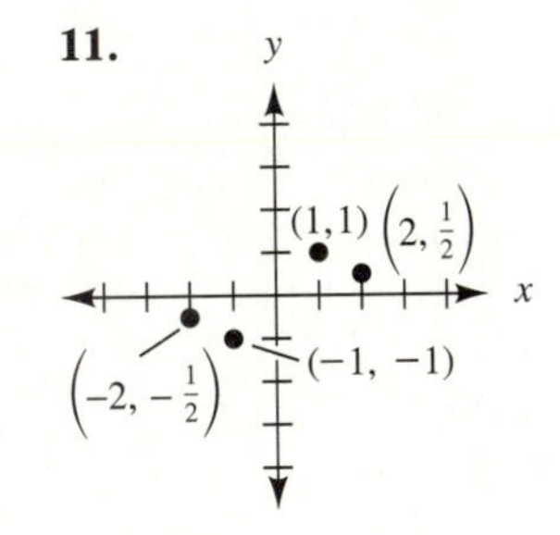

13.

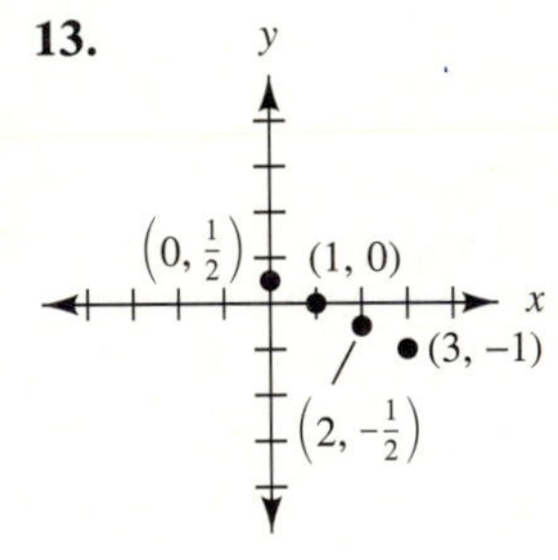

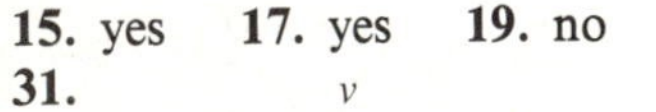

15. yes **17.** yes **19.** no **21.** yes **23.** yes **25.** yes **27.** yes **29.** no

31.

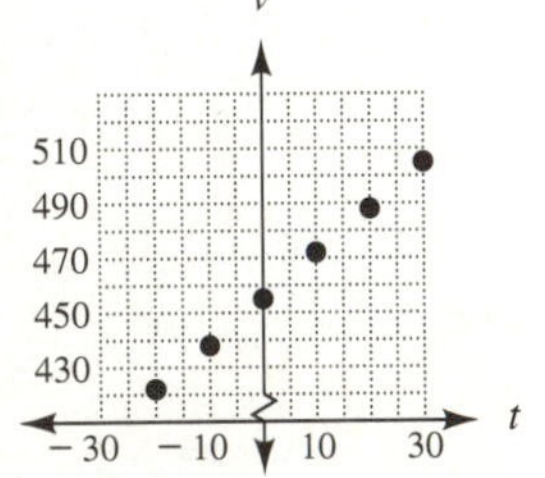

33.

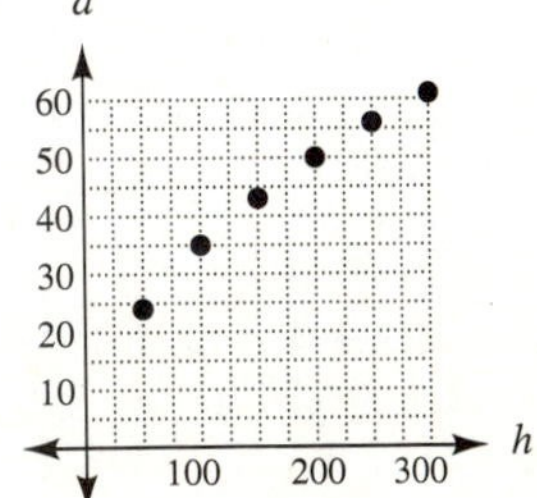

35.

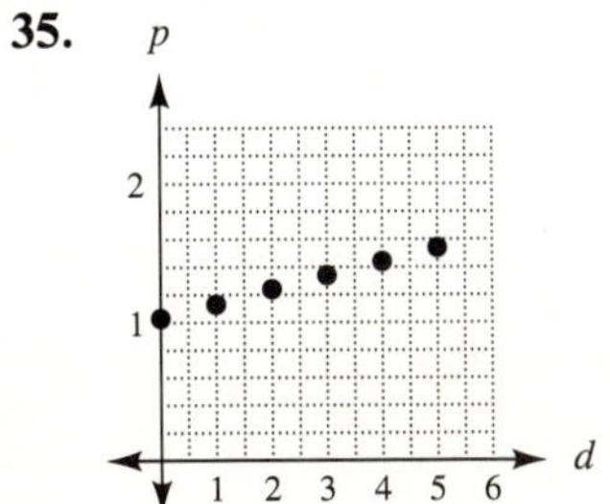

Section 3.3 (page 115)

1.

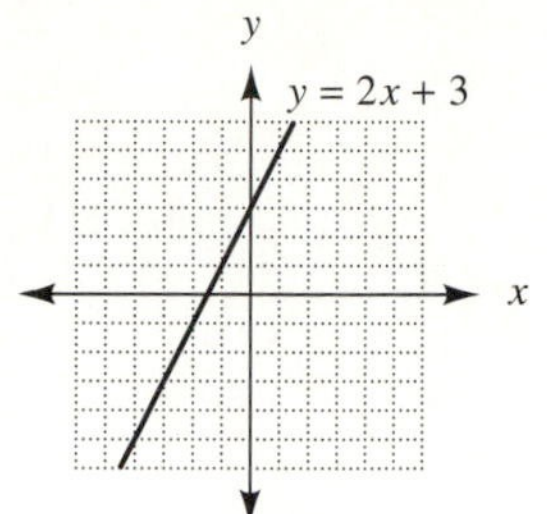

3.

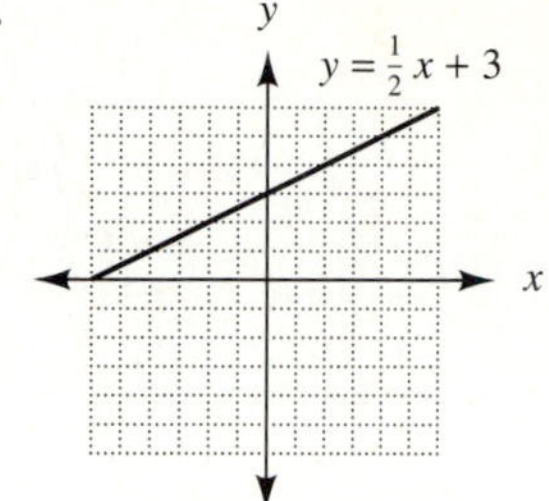

5.

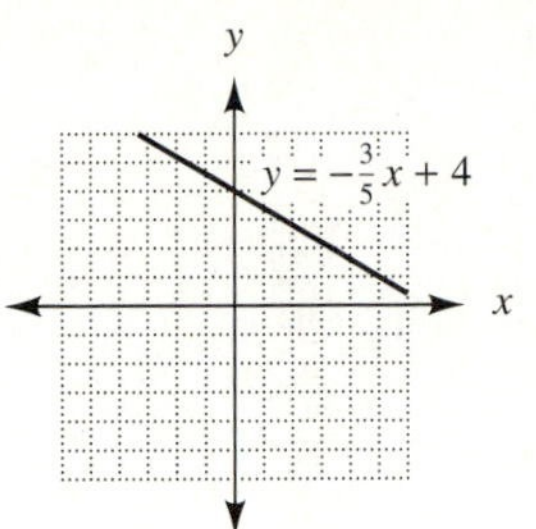

7.

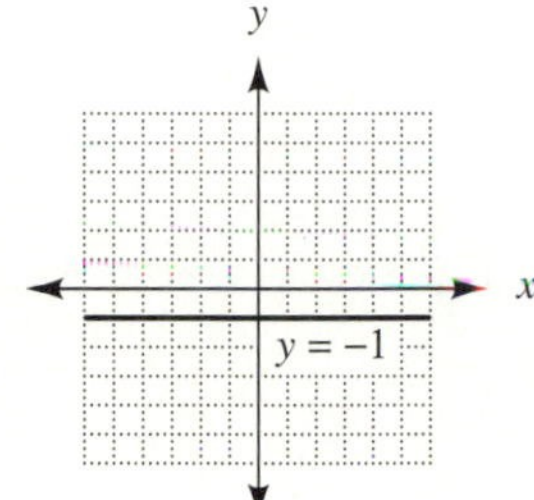

9.

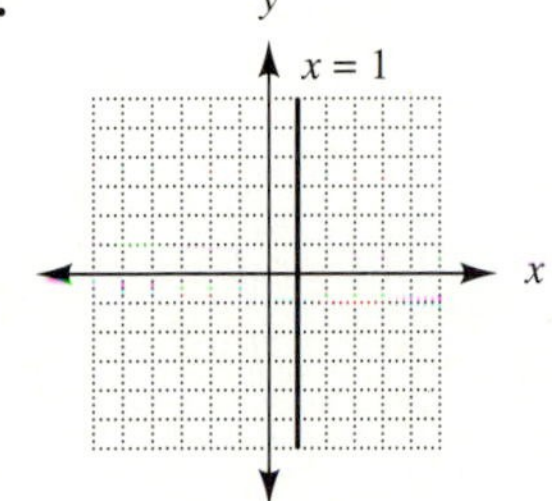

11.

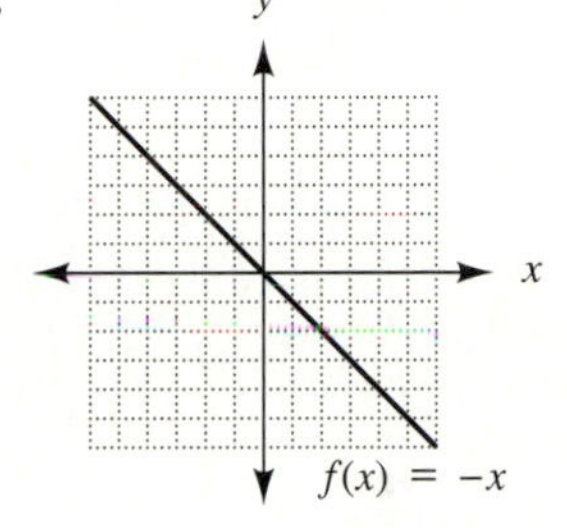

13.

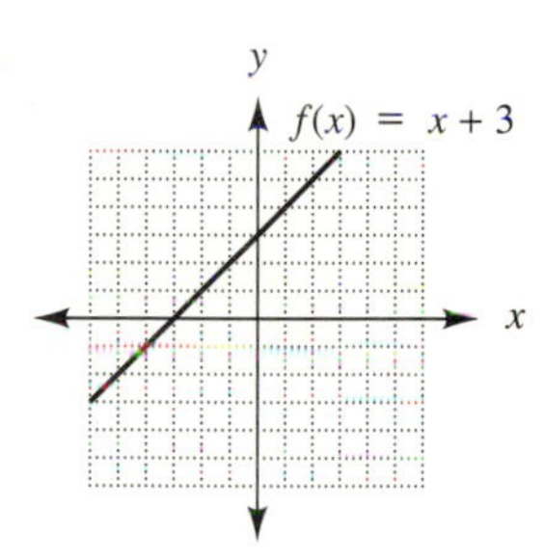

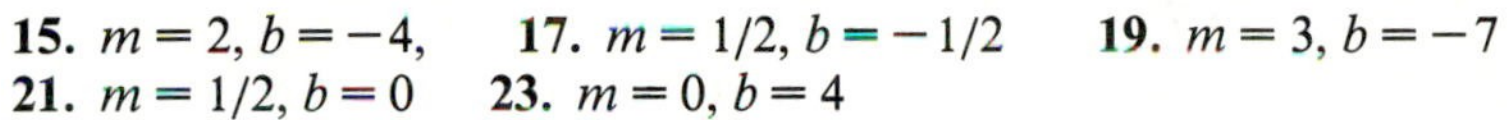

15. $m = 2, b = -4$, **17.** $m = 1/2, b = -1/2$ **19.** $m = 3, b = -7$
21. $m = 1/2, b = 0$ **23.** $m = 0, b = 4$

25.

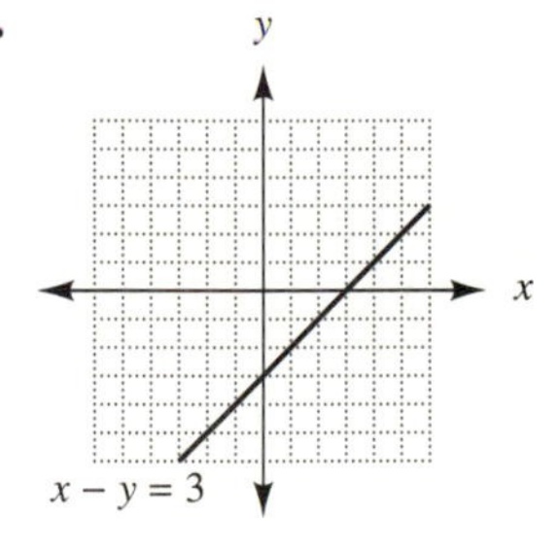

27.

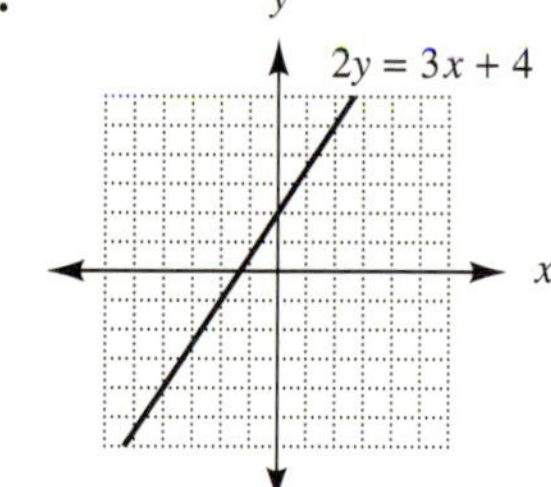

29.

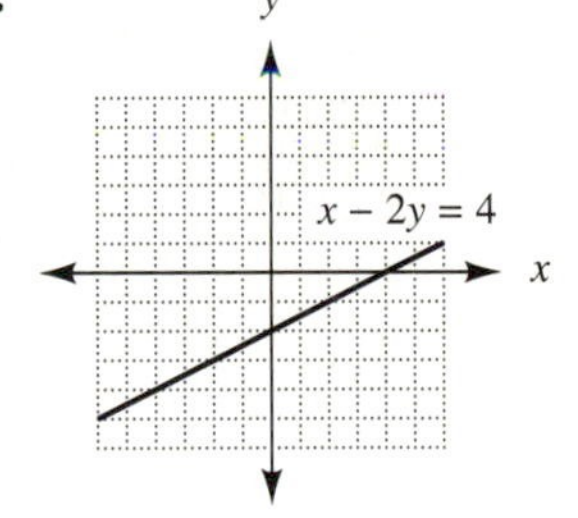

31.

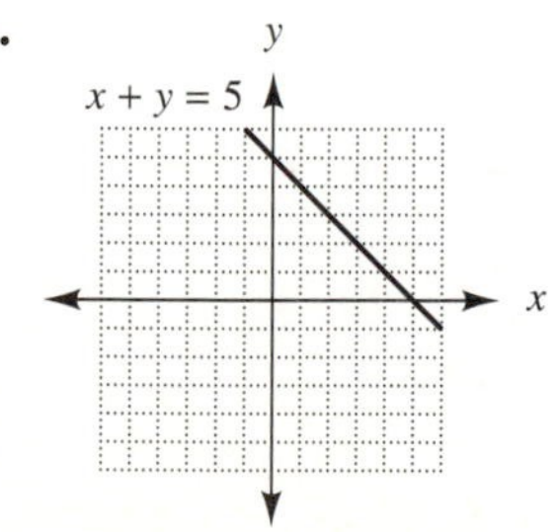

33.

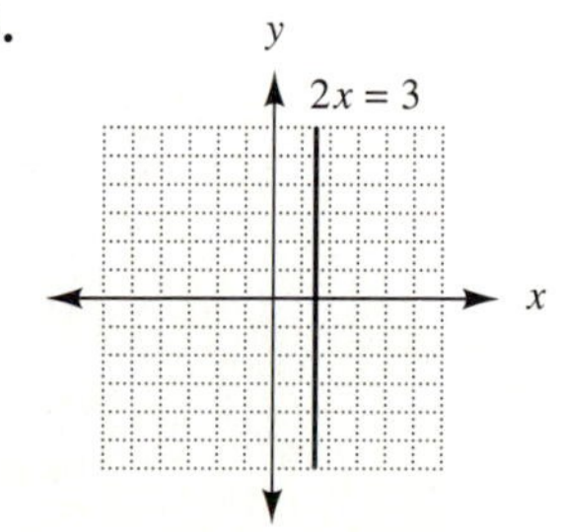

35. 3°

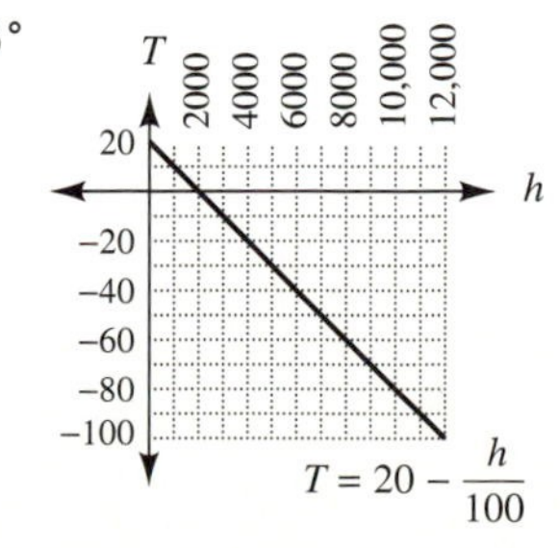

37.

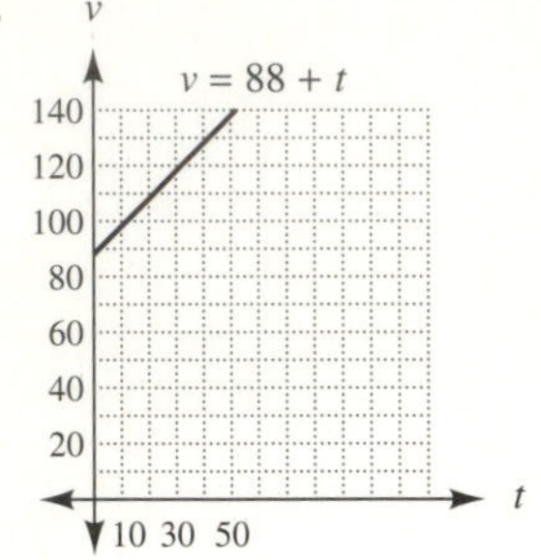

39. 3

Section 3.4 (page 124)

1. $y = x^2 + 7x + 12$

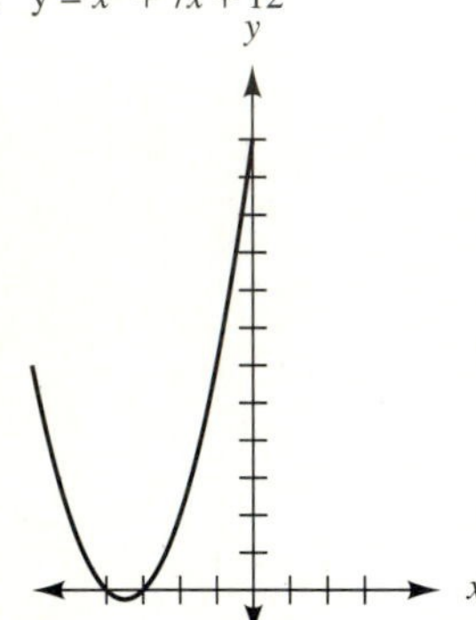

3.

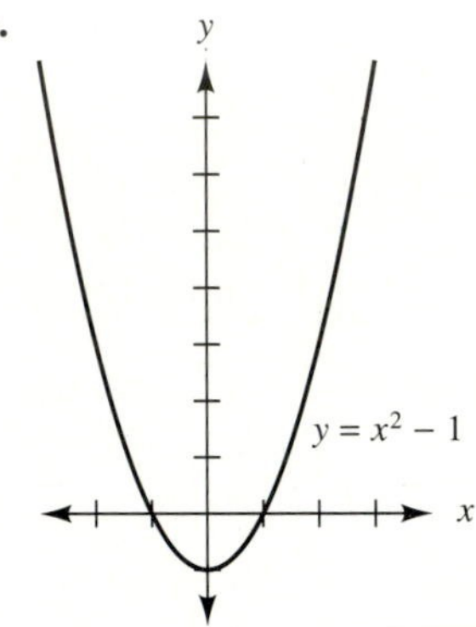

5.

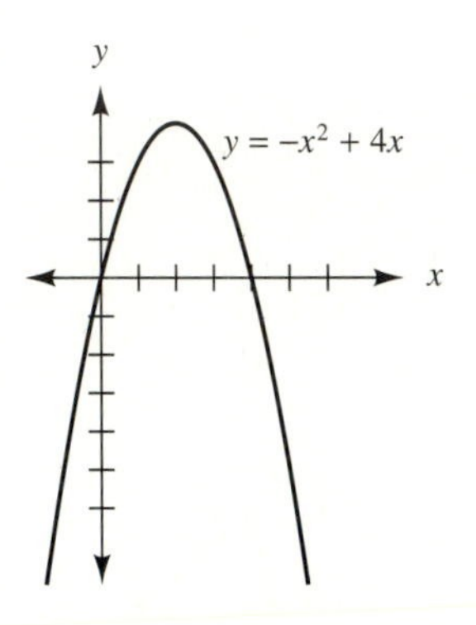

7.

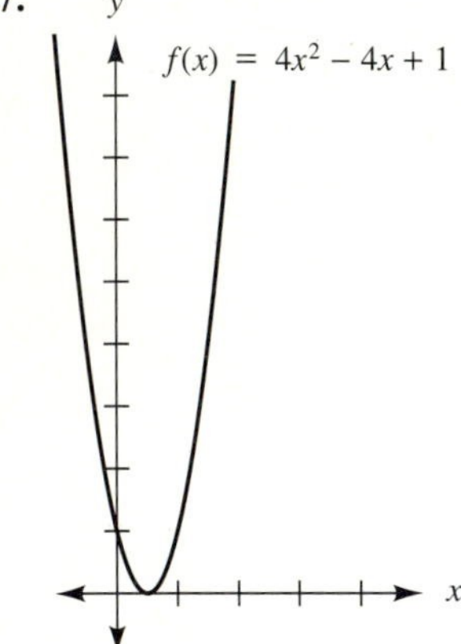

9.

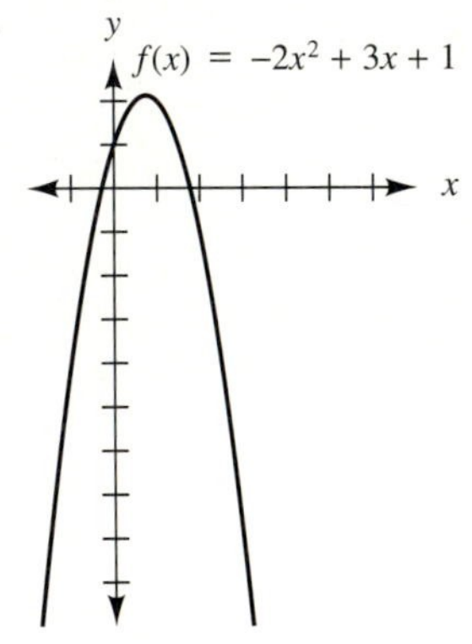

11. $y = x^3 + x^2 - 9x - 9$

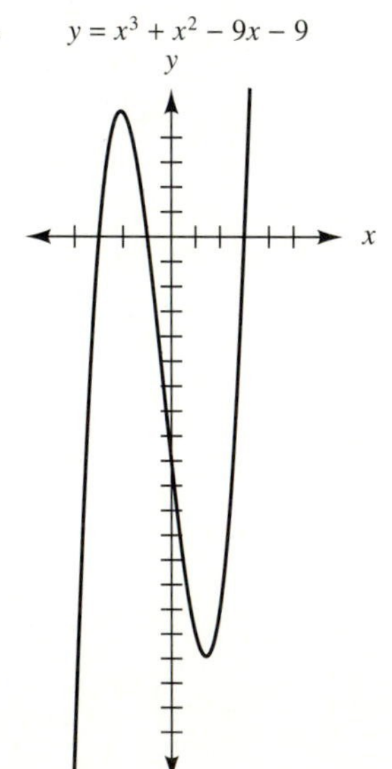

13.

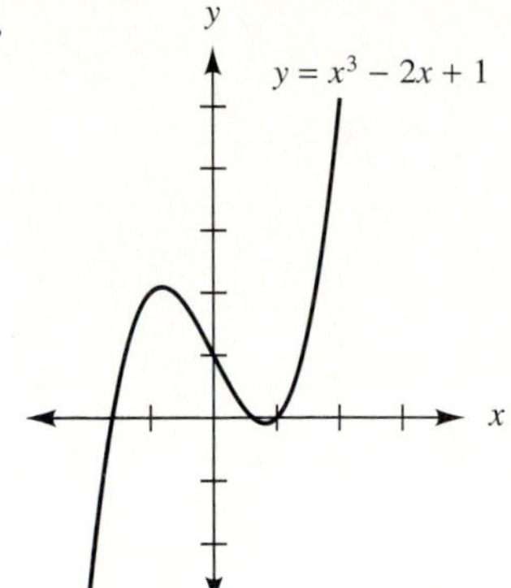

15.

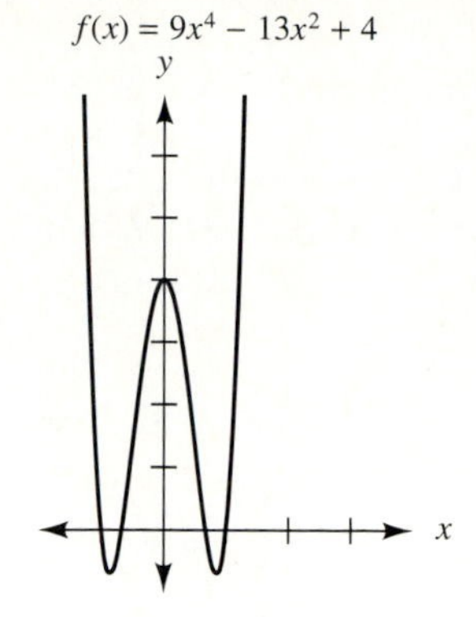

17.

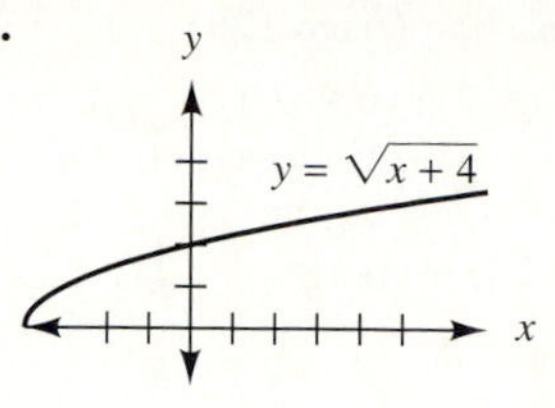

19.

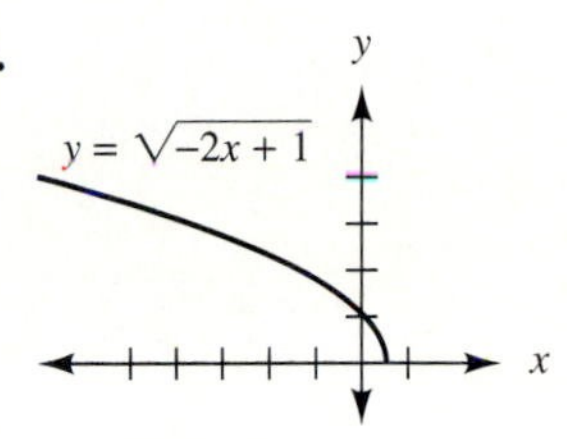

21.

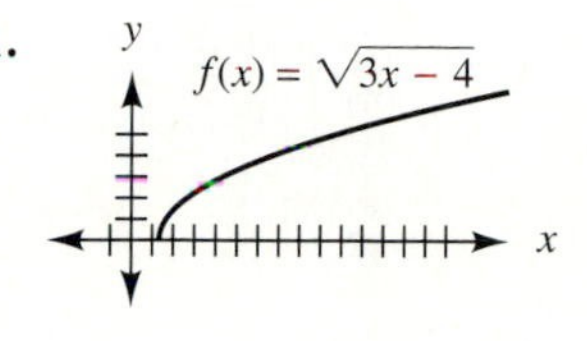

23. yes **25.** no **27.** yes **29.** yes

31. yes

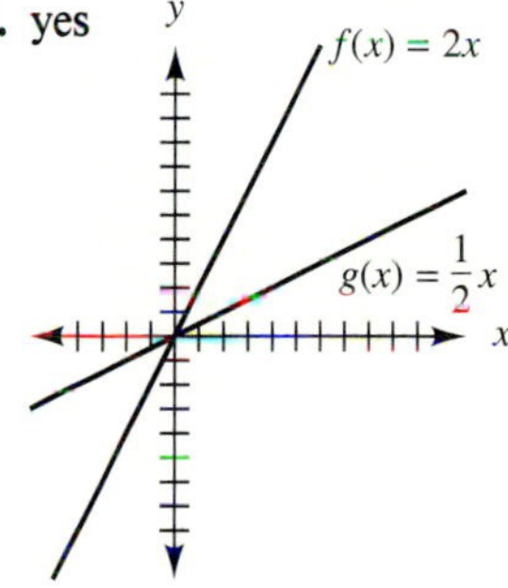

33. no

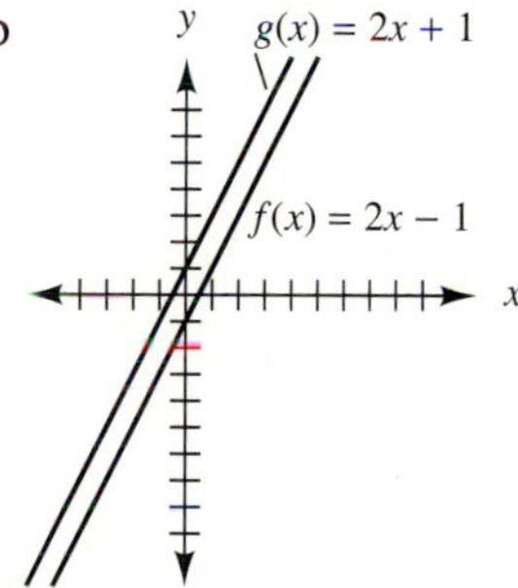

35. yes

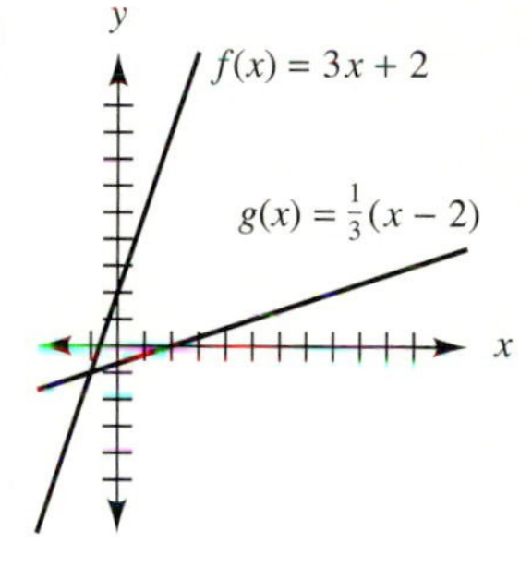

37. yes

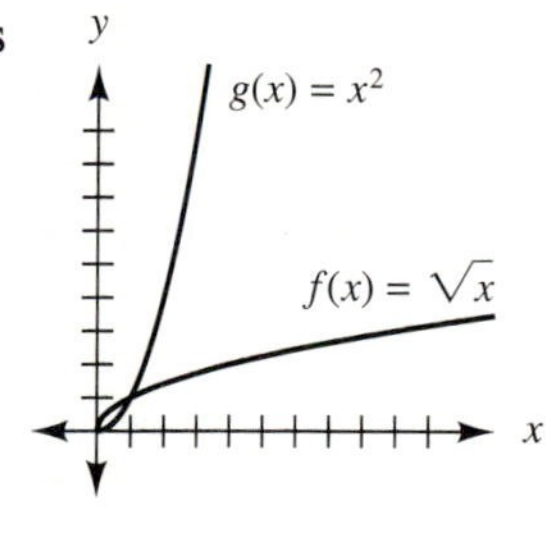

39.

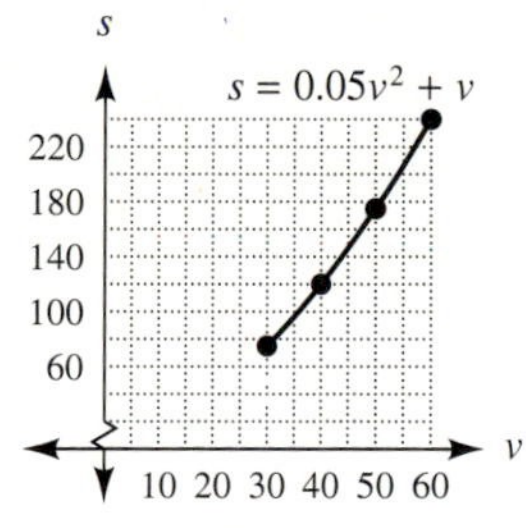

41.

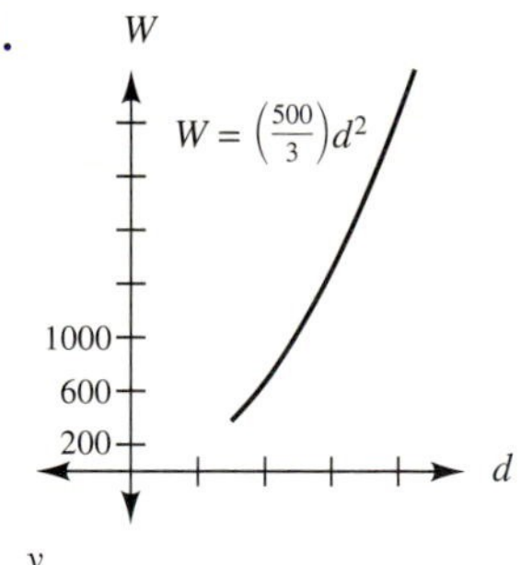

43.

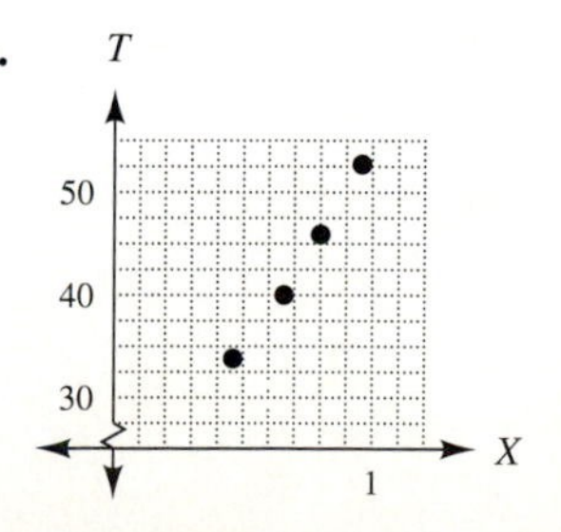

45. x is continuous, y is not

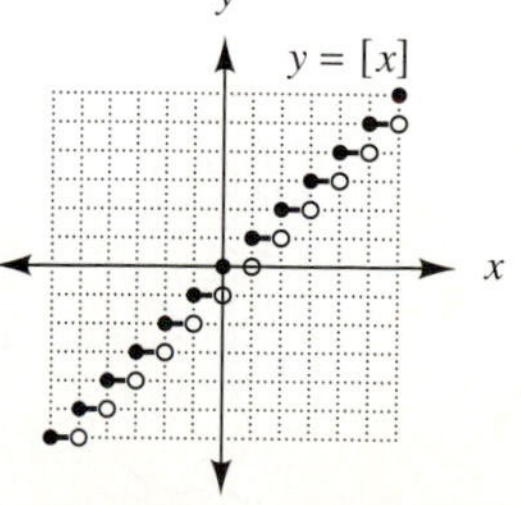

Section 3.5 (page 128)

1. $y = (3/2)x$, $y = 7\ 1/2$ **3.** $z = (-5/6)w$, $w = -8\ 2/5$ **5.** $v = -3\sqrt{u}$, $v = -9$ **7.** $s = 12/t$, $t = 2$
9. $w = -243/r^2$, $w = -60\ 3/4$ **11.** $z = 3xy$, $z = 12$ **13.** $w = (-1/60)uv$, $w = -6\ 2/3$
15. $t = (3/2)f\sqrt{g}$, $f = 8/9$ **17.** $q = 1.74$ **19.** $z = 12$ **21.** 32.4 Ω **23.** 256 ft
25. 70.0 lumens/m² **27.** 325 Hz **29.** 3220 lb **31.** 9.78 in

Section 3.6 (page 131)

Sequential Review: 1. yes **3.** yes **5.** −8 **7.** no **9.** 2.96

11.

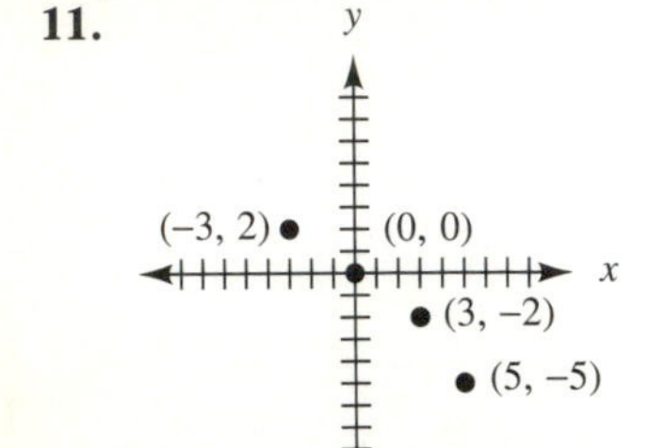

13.

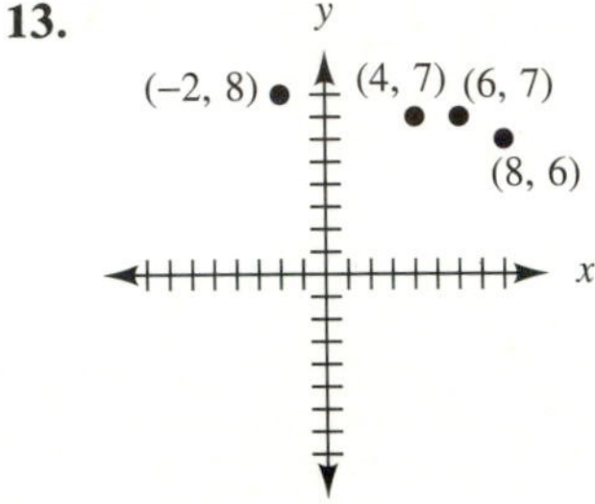

15. no **17.** yes

19.

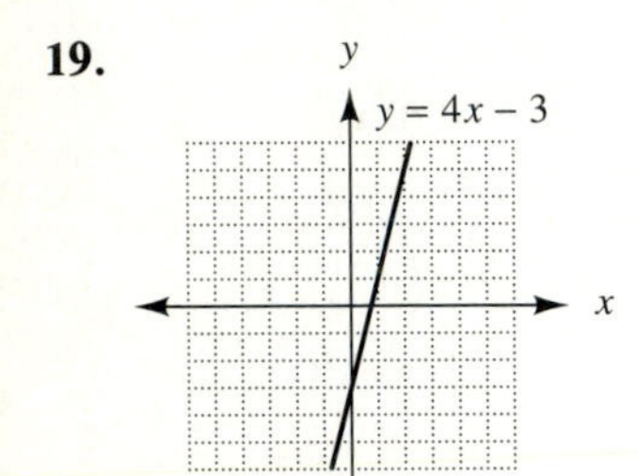

21.

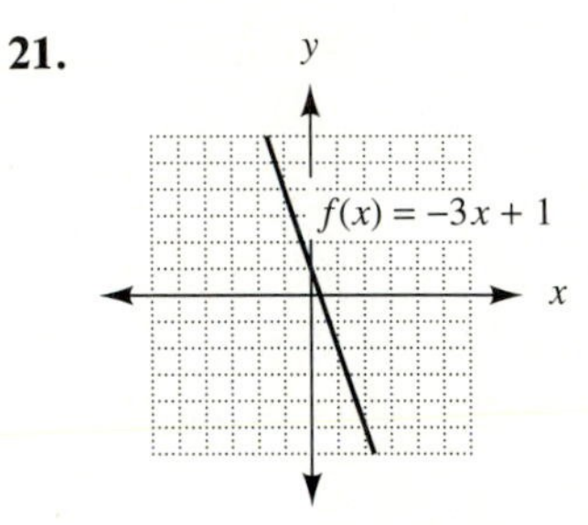

23.

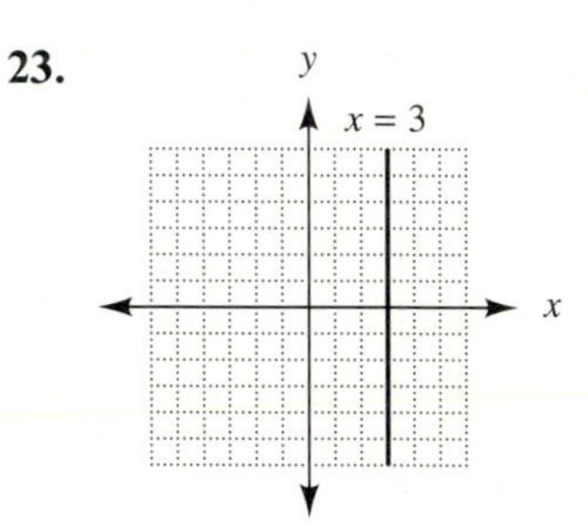

25. $m = 1/3$, $b = 0$
27. $m = 3/2$, $b = -7/2$

29.

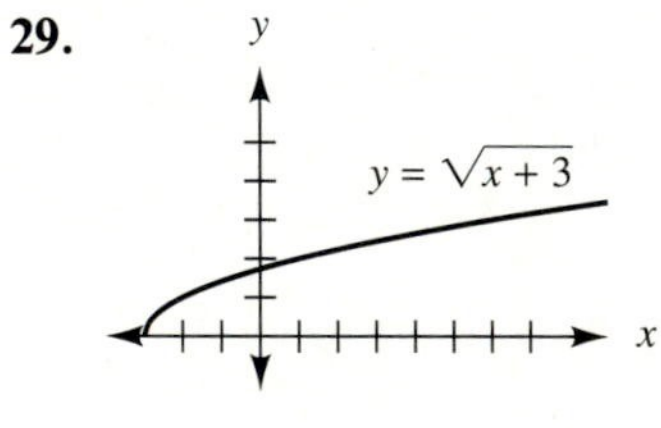

31. yes
33. yes
35. no

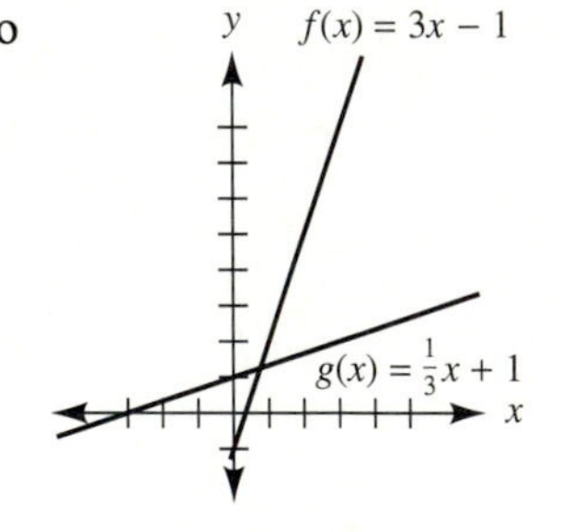

37. $y = 8x$, 2.4 **39.** $y = (1/3)x$, 2 **41.** $w = 32/z^2$, 3 5/9 **43.** $z = x^2y$, 54 **45.** $x = 0.346\ tv^2$, 14

Applied Problems: 1.

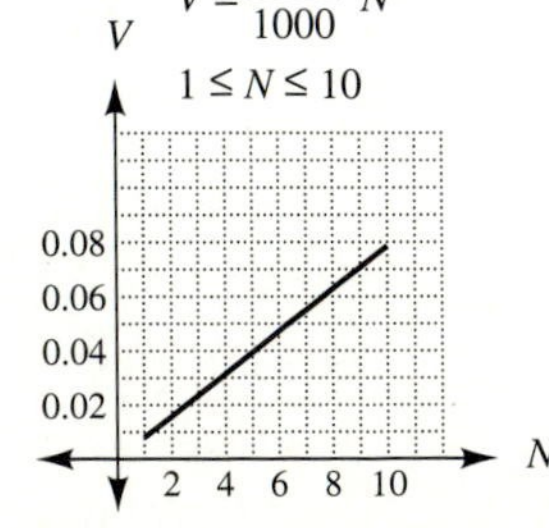

3. 8.4 gigahertz **5.** $v(t) = (-3/4)t + 27/4$

7.

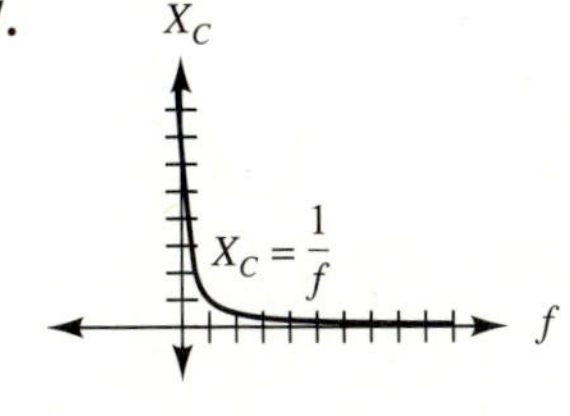

9. 30,400 ft **11.** 7700 calories **13.** 1200 lb **15.** 106 mph **17.** 44% less **19.** 24 ft

Random Review: 1. (a) 5 **(b)** 2 **2.** 1 **3. (a)** yes **(b)** yes **4.**

5. 48 **6.**

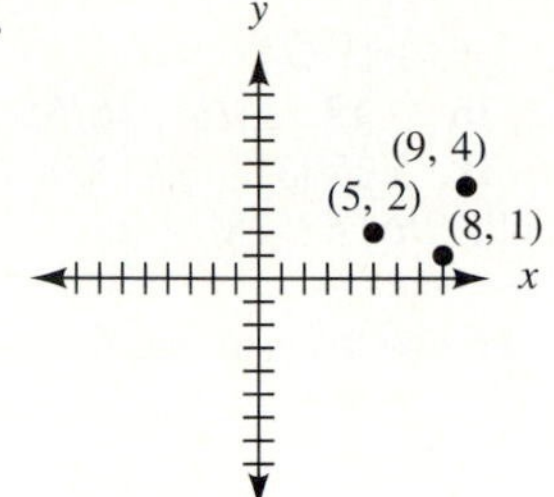

7. (a) $m = 3/2$, $b = 3/2$

(b)

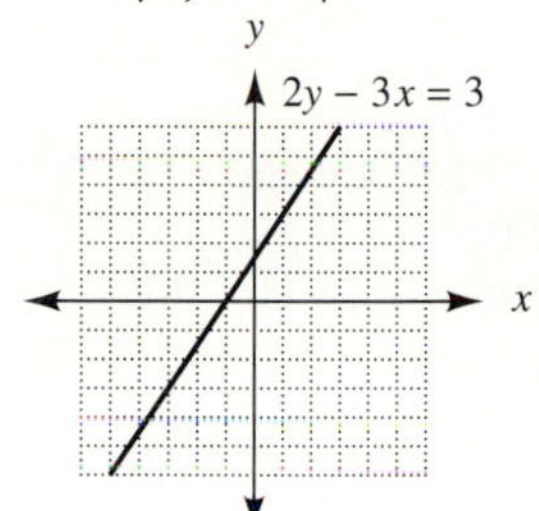

8. yes **9. (a)** yes **(b)** no **10.**

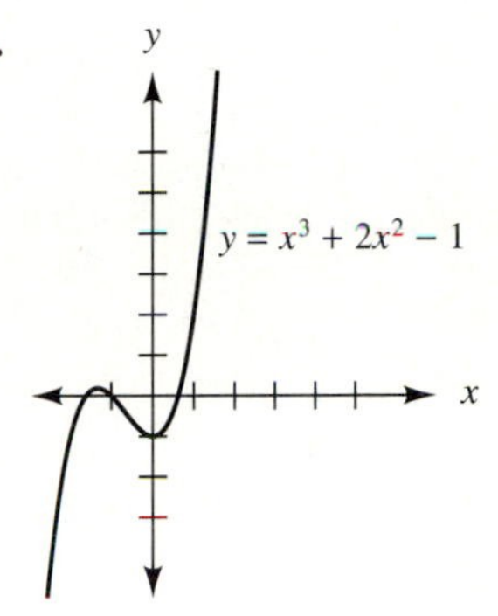

Chapter Test: 1. (a) yes **(b)** no **2. (a)** yes **(b)** no **3.**

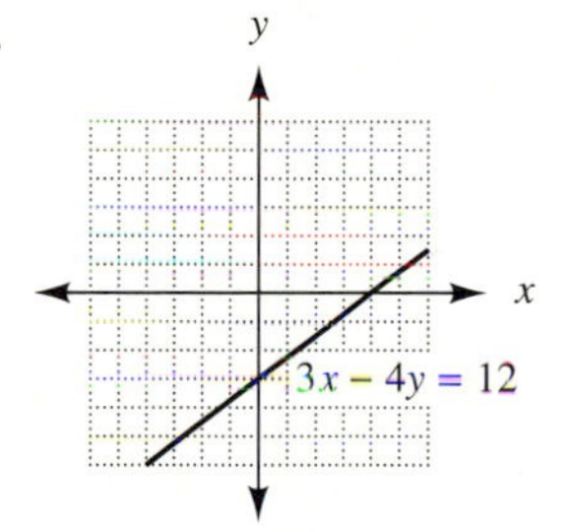

4.

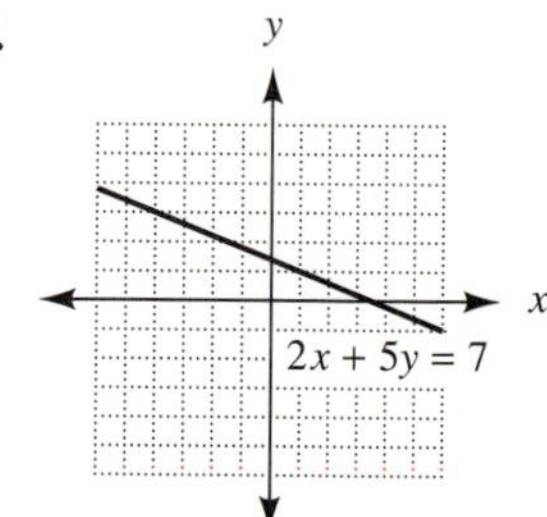

5.

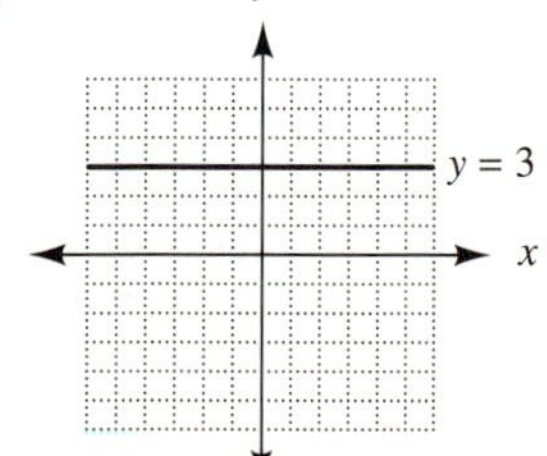

6.

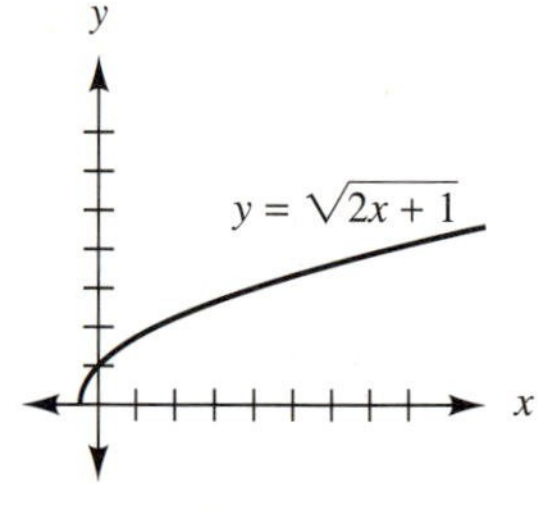

7. (a) no **(b)** yes **8. (a)** 5 1/3 **(b)** 3 **9.** 16.0 inches **10.** 2.7 lb

CHAPTER 4

Section 4.1 (page 143)

1. 5.3 cm **3.** 6.80 in **5.** 5.8 mm **7.** 7.7 km **9.** 12.3 m **11.** 3/5, 3/5, 3/4
13. 24/7, 25/7, 25/7 **15.** 60/61, 11/61, 60/11 **17.** 77/36, 85/77, 85/36 **19.** 33/65, 56/65, 56/33
21. 0.7157 **23.** 0.8890 **25.** 2.8878 **27.** 6.6252 **29.** 2.2412 **31.** 5.3205 **33.** 1.0° or 1°0′
35. 17.6° or 17°36′ **37.** 43.7° or 43°42′ **39.** 9.5° or 9°30′ **41.** 42.3° or 42°18′
43. 40° to the nearest degree **45.** 37.7 ft **47.** 0.6° **49.** 1′5″
51. $\sin\theta = \sin\theta$, $\cos\theta = \cos\theta$, $\tan\theta = \sin\theta/\cos\theta$, $\cot\theta = \cos\theta/\sin\theta$, $\sec\theta = 1/\cos\theta$, $\csc\theta = 1/\sin\theta$

Section 4.2 (page 150)

1. $A = 57.2°$, $B = 32.8°$, $C = 90°$ **3.** $A = 26.5°$, $B = 63.5°$, $C = 90°$
5. $A = 60.5°$, $B = 29.5°$, $c = 19.3$ in **7.** $A = 31.8°$, $B = 58.2°$, $c = 37.9$ m
9. $B = 45°$, $b = 4.7$ cm, $c = 6.6$ cm **11.** $A = 63.2°$, $B = 26.8°$, $c = 2.33$ m
13. $A = 31.6°$, $b = 46.8$ in, $c = 55.0$ in **15.** $A = 35.1°$, $a = 2.28$ m, $b = 3.25$ m
17. $A = 64°$, $B = 26°$, $c = 4.8$ in **19.** $A = 50°$, $a = 6.2$ mm, $c = 8.1$ mm **21.** 4.6 cm **23.** 4.7 in
25. 35.0° **27.** 32.9° **29.** 27.05 m **31.** $B = 60°$, $b = 6\sqrt{3}$ ft, $c = 12$ ft
33. $B = 30°$, $a = 10\sqrt{3}$ m, $c = 20$ m **35.** $A = 60°$, $a = 2\sqrt{3}$ mm, $b = 2$ mm
37. $B = 45°$, $b = 6$ ft, $c = 6\sqrt{2}$ ft **39.** $A = 45°$, $a = 10$ mm, $b = 10$ mm
41. **(a)** $B = 90° - A$, $a = c\sin A$, $b = c\cos A$ **(b)** $\sin B = b/c$, $\cos A = b/c$, $a = \sqrt{c^2 - b^2}$

Section 4.3 (page 157)

1. 18.4 m **3.** 5° **5.** 7.4 ft **7.** 75.5° **9.** just under 79° **11.** 75.8 mi
13. 1160 m to three significant digits **15.** 1280 m **17.** 764 m **19.** 142 m **21.** 6′8″
23. 34.17′ from B, 76.49′ from A

Section 4.4 (page 164)

	$\sin\theta$	$\cos\theta$	$\tan\theta$	$\cot\theta$	$\sec\theta$	$\csc\theta$
1.	−5/13	12/13	−5/12	−12/5	13/12	−13/5
3.	7/25	−24/25	−7/24	−24/7	−25/24	25/7
5.	−35/37	−12/37	35/12	12/35	−37/12	−37/35
7.	45/53	28/53	45/28	28/45	53/28	53/45
9.	−11/61	60/61	−11/60	−60/11	61/60	−61/11
11.	−55/73	−48/73	55/48	48/55	−73/48	−73/55
13.	99/101	−20/101	−99/20	−20/99	−101/20	101/99

15. + **17.** + **19.** + **21.** + **23.** + **25.** + **27.** + **29.** + **31.** + **33.** + **35.** +
37. − **39.** − **41.** + **43.** II **45.** III **47.** IV **49.** II **51.** II **53.** II **55.** IV **57.** 50°
59. 290° **61.** 30°, 2 **63.** 315° **65.** no significance for x, indicates downward aim for y
67. 0.97 **69.** Think positive.

Section 4.5 (page 170)

1. 0.766 **3.** −1.19 **5.** −2.00 **7.** −5.76 **9.** −5.67 **11.** 0.985 **13.** −0.174 **15.** 150.45°
17. 256.70° **19.** 245.52° **21.** 216.15° **23.** 171.81° **25.** 314.03° **27.** 241.38° **29.** 90°
31. 0° or 180° **33.** 0° **35.** 270° **37.** 0° or 180° **39.** 90° or 270° **41.** 180° **43.** 60°
45. 50° **47.** 15.4°

Section 4.6 (page 175)

1. 90° **3.** 330° **5.** 30° **7.** −90° **9.** −150° **11.** $5\pi/6$ **13.** $11\pi/6$ **15.** π **17.** $23\pi/9$ **19.** $-3\pi/2$ **21.** 1.12 **23.** 0.778 **25.** 1.95 **27.** 7.02 **29.** −1.49 **31.** 0.9998 **33.** 0.2293 **35.** 1.111 **37.** −1.140 **39.** −0.6971 **41.** −0.1910 **43.** 0.8643 **45.** 0.3972 **47.** 2.647 **49.** 4.184 **51.** 2.086
53. The numerators of $\sin\theta$ are the square roots of 0, 1, 2, 3, and 4, and each denominator is 2. The numerators of $\cos\theta$ are the square roots of 4, 3, 2, 1, and 0.

Section 4.7 (page 180)

1. 3.67 cm **3.** 11.8 ft **5.** 14.1 m **7.** 8.38 ft **9.** 10.1 m **11.** 3.2 radians **13.** 12.8 cm^2 **15.** 29.5 ft^2 **17.** 21.2 m^2 **19.** 101 ft^2 **21.** 3.67 m **23.** 3.57 radians **25.** 24.9 m/s **27.** 4.0 ft/min **29.** 322 m/min **31.** 190 ft/min **33.** 0.6 radians/s **35.** 3.8 m **37.** approximately 740 ft **39.** about 530 **41.** 2.36 in **43.** 229′, 233′

Section 4.8 (page 183)

Sequential Review: 1. 14 **3.** 28/53 **5.** 53/45 **7.** 3.8118 **9.** 77.4° **11.** $A = 61°$, $B = 29°$, $c = 8.4$ **13.** 38° **15.** $A = 60°$ **17.** $\sqrt{2}/2$ **19.** 9.9 m
21. $\sin\theta = -35/37$, $\cos\theta = 12/37$, $\tan\theta = -35/12$, $\cot\theta = -12/35$, $\sec\theta = 37/12$, $\csc\theta = -37/35$ **23.** + **25.** + **27.** II **29.** 0.3640 **31.** −1.7434 **33.** 244.68° **35.** 0°, 180° **37.** 660° **39.** −0.7764 **41.** 3.67 cm **43.** 4.71 m^2 **45.** 3.72 cm/s
Applied Problems: 1. 8.57 cm **3.** 51′ **5.** 1100 ft **7.** 3470 rpm **9.** 13° **11.** 4.9 ft/s **13.** 0.5° **15.** 27,000 **17.** 500,000 **19.** 0.50
Random Review: 1. 9.2 radians **2.** 14.7026 **3.** 0.8192 **4.** $A = 50°$ **5.** − **6.** II **7.** 1/2 **8.** 90°, 270° **9.** 15.3° **10.** 75 in/min
Chapter Test: 1. (a) 91/109 **(b)** 60/91 **2.** $A = 65.3°$, $b = 2.9$ cm, $c = 6.9$ cm **3.** $A = 29.9°$, $B = 60.1°$, $c = 18.2$ in **4. (a)** III **(b)** IV **5. (a)** 320.5° **(b)** 5.594 radians **6. (a)** 348.69° **(b)** 6.086 **7. (a)** 90°, 270° **(b)** 0°, 180° **8. (a)** π or 3.14 in **(b)** 9.42 in^2 **9.** 31.72° **10.** $X_C = 55.8$ kΩ

CHAPTER 5

Section 5.1 (page 197)

1. $2x(x+1)$ **3.** $(x+2)(x+3)$ **5.** $(p+4)(p+6)$ **7.** $(2z-3)(3z-2)$ **9.** $(3p+4)(4p-3)$ **11.** cannot be factored **13.** cannot be factored **15.** $(p+4)(p+5)$ **17.** $2(a^2-12a+2)$ **19.** $(2m+3)(m-3)$ **21.** $(3x-2)(3x+2)$ **23.** −3, −4 **25.** −6, 4 **27.** 0, 2 **29.** 1/2, −3/2 **31.** 3/2, −1/3 **33.** 1, −1 **35.** 6/5, −6/5 **37.** 1/2, −3 **39.** 5, −2 **41.** −2/3, −5 **43.** $[wx/(24EI)](l^3 - x^3 - 2x^2 l)$ **45.** $c(a+2b)$ **47.** $(p/l^3)(l-a)^2(l+2a)$ **49.** 210° F

Section 5.2 (page 202)

1. $(2x-1)^2$ **3.** $(3z-2)(3z+2)$ **5.** $4(p^2+25)$ **7.** $(2z+5)^2$ **9.** $(4x+1)(x+1)$ **11.** $(3m-2)(9m^2+6m+4)$ **13.** $2(z+4)(z^2-4z+16)$ **15.** cannot be factored **17.** $(7p-5)(7p+5)$ **19.** $(5x+3)^2$ **21.** $(y-9)(y-1)$ **23.** $9(m-8)(m-1)$ **25.** cannot be factored **27.** $4(4x-3)(4x+3)$ **29.** $(z-4)(z^2+4z+16)$ **31.** $4(16p^2+25)$ **33.** $(3b-5)(9b^2+15b+25)$ **35.** $(2y-5)(4y^2+10y+25)$ **37.** $(a^2+4)(a^4-4a^2+16)$ **39.** $(m-3)(m^2+3m+9)(m+3)(m^2-3m+9)$ **41.** $(4x^2+9)(16x^4-36x^2+81)$ **43.** $a^2+2ab+b^2$ **45.** $k(v_1-v_2)(v_1^2+v_1v_2+v_2^2)$ **47.** $[P/(3EI)](l_1-l_2)(l_1^2+l_1l_2+l_2^2)$ **49.** $(1/P)V^2 = (1/P)(V_1^2+2V_1V_2+V_2^2) = (1/P)(V_1+V_2)^2$

Section 5.3 (page 207)

1. $(x+1)(x+2)$ **3.** $(4z-2)(z+1)$ **5.** $(y+4)(2y-1)$ **7.** $(2p+1)(2p-1)$ **9.** $(b-1)(2b+5)$
11. $(2p-1)(-p+5)$ **13.** $(3y+4)(y-2)(y+2)$ **15.** $(y+2)(y-1)(y+1)$ **17.** $(p-1)(p^2+4)$
19. $(2x+3)(x-2)(x+2)$ **21.** $(b+2)(2b-1)(2b+1)$ **23.** $(x+3-y)(x+3+y)$
25. $(p-q-4)(p-q+4)$ **27.** $(x+3-y)(x+3+y)$
29. $(x+6-2y)(x+6+2y)$ **31.** $(y+3)(y-2)$ **33.** $(y-6)(y-2)$ **35.** $(m-3)(3m+2)$
37. $(3m+2)(2m+3)$ **39.** $(x+3)(2x-3)$ **41.** $3(y+2)^2$ **43.** $s(s-a)(s-b)(s-c)$
45. $(I-I_C+I_L)(I+I_C-I_L)$ **47.** $R(2I+1)$ **49.** $R_2+R_1=t^2-1$

Section 5.4 (page 215)

1. 2/3 **3.** 4/15 **5.** 1/3 **7.** 3/5 **9.** $\frac{y+3}{y-3}$ **11.** $\frac{x+2}{x+3}$ **13.** $\frac{z-4}{z-5}$ **15.** $\frac{y+2}{y+1}$ **17.** 1/6

19. 18/49 **21.** 1/6 **23.** 1 **25.** $\frac{y-2}{y+3}$ **27.** $\frac{m+5}{m-3}$ **29.** $\frac{(a-3)(a-1)}{(a+3)(a+1)}$ **31.** 1 **33.** $\frac{x-3}{x+4}$

35. 3/4 **37.** m^2 **39.** $\frac{x+1}{x-2}$ **41.** $I=V/R;\ P=(V^2/R^2)(R)=V^2/R$ **43.** $2/V$ **45.** $r/3$

47. $\frac{2(2l+w)}{lw}$ **49.** $\frac{3\sigma}{1+\sigma}$

Section 5.5 (page 222)

1. $\frac{a^2+b^2}{ab}$ **3.** $\frac{p^3-q^2}{p^2q}$ **5.** $\frac{a^2b+a}{b^2}$ **7.** $\frac{2p-1}{p^2q}$ **9.** $\frac{3w+z^2-wz}{w^2z^2}$ **11.** $\frac{y^2+1}{(y-1)^2(y+1)}$

13. $\frac{6x+9-x^2}{x(2x-3)(2x+3)}$ **15.** $\frac{2(y^2+1)}{(y+1)(y-1)}$ **17.** $\frac{p^2+1}{2p(p-1)(p+1)}$ **19.** $\frac{2(4m+3)}{(m-2)(m+2)}$

21. $\frac{-x-1}{(3x+1)^2}$ **23.** $\frac{-z+5}{z(z+1)}$ **25.** $\frac{a^2+4a+2}{(a+1)(a-1)}$ **27.** $\frac{-m^2+9m+13}{(m-4)(m+2)}$ **29.** $\frac{y}{2z}$ **31.** $\frac{y(1-x)}{1+x}$

33. $\frac{x^2+1}{x^3}$ **35.** $\frac{1}{x+1}$ **37.** $\frac{y+x}{y-x}$ **39.** $\frac{1}{x^2+x-1}$ **41.** $\frac{f(s+v)}{s}$

43. $C=\frac{C_1C_2C_3}{C_2C_3+C_1C_3+C_1C_2}$ **45.** $\frac{RTV^2-aV+ab}{V^2(V-b)}$

Section 5.6 (page 229)

1. $-9/10$ **3.** 2 **5.** 8/5 **7.** 1/4 **9.** no solution **11.** 1/6 **13.** no solution **15.** -5.2
17. -4 **19.** 33/13 **21.** $6, -25/4$ **23.** 1 **25.** $-1/3$ **27.** 1 **29.** no solution

31. no solution **33.** 1/4 **35.** 1 5/7 hours **37.** **(a)** 3 μF **(b)** $C_1=\frac{CC_2}{C_2-C}$

39. 3 hours and 2 hours or 7 6/7 hours and 6 6/7 hours **41.** 50 mph

Section 5.7 (page 231)

Sequential Review: 1. $(a+4)(a+7)$ **3.** $5x(x+2)$ **5.** $4(m^2+5m-3)$ **7.** $y=\pm 3$ **9.** $(3x+1)^2$ **11.** $4(x^2+25)$ **13.** $(z-9)(z+9)$ **15.** $(3z+1)(9z^2-3z+1)$ **17.** $(3x-2)(x+5)$

19. $(m+2)(2m^2+1)$ **21.** $(x-2)(8x^2-1)$ **23.** $(m+n-3)(m+n+3)$ **25.** $\dfrac{2x+1}{x+2}$ **27.** $4/15$

29. $(y-3)(y-1)$ **31.** $\dfrac{x^3+y^2}{x^2y}$ **33.** $\dfrac{z^3+x^3+1}{x^2z^2}$ **35.** $\dfrac{z^2-z-1}{(2z+3)(z-1)}$ **37.** $\dfrac{1}{2yz}$ **39.** $5/3$

41. $5/2$ **43.** no solution **45.** no solution

Applied Problems: 1. $V(1-e^{-t/\tau})$ **3.** $(i+o)/(io)$ **5.** $w=\dfrac{\theta M_x}{3lx-4x^2}$ **7.** $C=(5/9)(F-32)$

9. $25(4-\pi/m^2)$ **11.** $(p-5)(p-12)$ **13.** $R_T=R(1+N)/3$

15. $L=L_1L_2L_3/(L_2L_3+L_1L_3+L_1L_2)$ **17.** $F=(9/5)(C+40)-40$

Random Review: 1. $\dfrac{x+3}{x-2}$ **2.** $a=3/2, -5$ **3.** $(2x+3)(5x-1)$ **4.** $\dfrac{19}{6(x+3)}$

5. $(3x-2)(9x^2+6x+4)$ **6.** $x=-27/5$ **7.** $(2x-9)(2x+9)$ **8.** $(m+9)(m-6)$ **9.** $\dfrac{1}{1-x}$

10. $\dfrac{x^2+x+1}{x(x+1)}$

Chapter Test: 1. $(m+9)(m-7)$ **2.** $(x^2+4)(x^4-4x^2+16)$ **3.** $(x+y-9z)(x+y+9z)$

4. $x=1/2, -1/3$ **5.** $x=-6$ **6.** $\dfrac{(y-2)(y+2)}{2y}$ **7.** $\dfrac{x^2-y^3}{xy^2}$ **8.** $\dfrac{(x-1)^2}{(x+1)^2}$

9. $M_x=\dfrac{-wl^4+6wxl^4-6wl^2x^2}{12l^2}$ **10.** $R_2=\dfrac{RR_1}{R_1-R}$

CHAPTER 6

Section 6.1 (page 240)

1. $-4, -6$ **3.** $2\pm\sqrt{2}$ **5.** $3, -2$ **7.** $-1, -4$ **9.** $-1/2, -3/2$ **11.** $\dfrac{3\pm\sqrt{15}}{3}$ **13.** $\dfrac{-4\pm\sqrt{22}}{2}$

15. $\dfrac{9\pm 2\sqrt{21}}{3}$ **17.** $\dfrac{5+\sqrt{19}}{2}$ **19.** $\dfrac{-1\pm\sqrt{2}}{2}$ **21.** $\dfrac{1\pm\sqrt{21}}{10}$ **23.** $5/2, 2$ **25.** $2/5, -1$

Section 6.2 (page 246)

1. $2\pm\sqrt{2}$ **3.** $\dfrac{5\pm\sqrt{13}}{2}$ **5.** $\dfrac{-5\pm\sqrt{33}}{4}$ **7.** $\pm\sqrt{21}$ **9.** $0, 7/2$ **11.** $\dfrac{1\pm\sqrt{13}}{6}$ **13.** $7, -1$

15. $-4\pm\sqrt{15}$ **17.** $\dfrac{-2\pm\sqrt{6}}{2}$ **19.** $\dfrac{-5\pm\sqrt{17}}{2}$ **21.** $\dfrac{-1\pm\sqrt{13}}{6}$ **23.** $3, 0$ **25.** 6.50 cm

27. 343 lb/ft² **29.** 99.7° C **31.** $(2\pm\sqrt{2})L$

Section 6.3 (page 252)

1. $(-2, 4)$

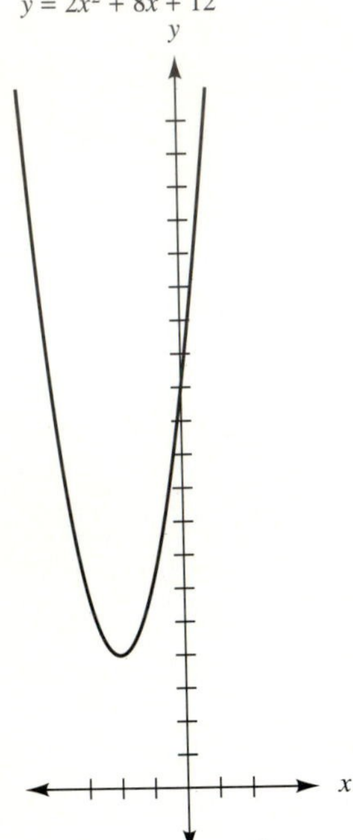

3. $(2, -8)$

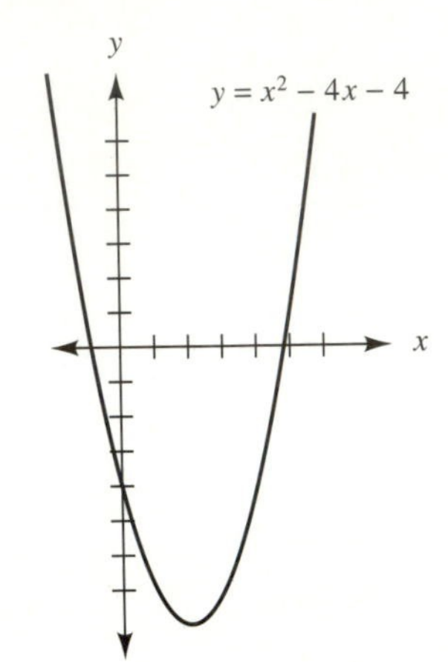

5. $(0, -1)$

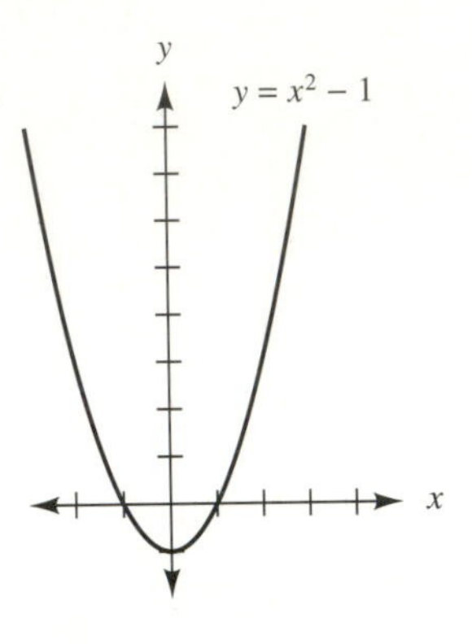

7. $(1, 2)$

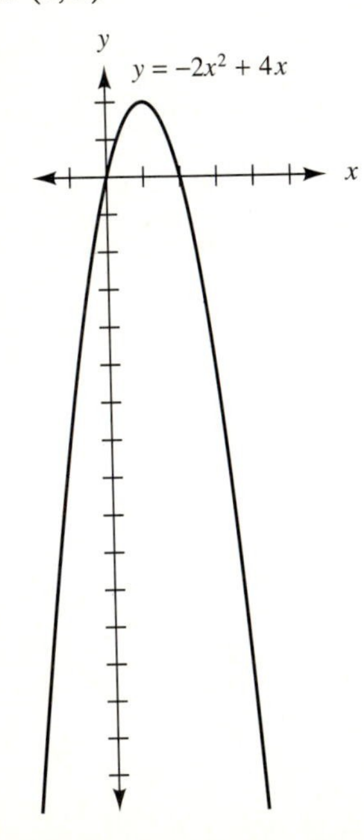

9. $(1/2, 0)$

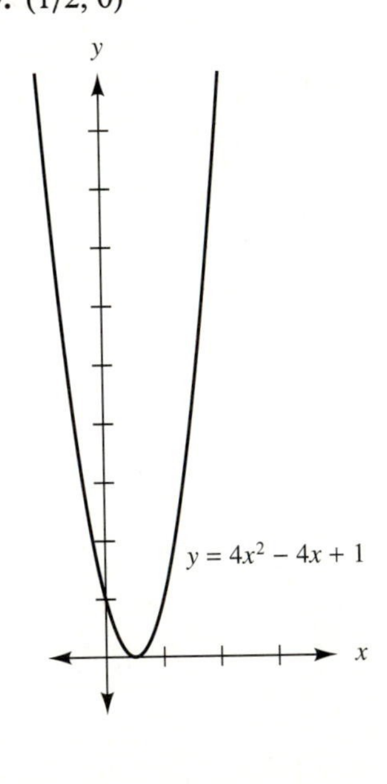

11. $(-1, 2)$

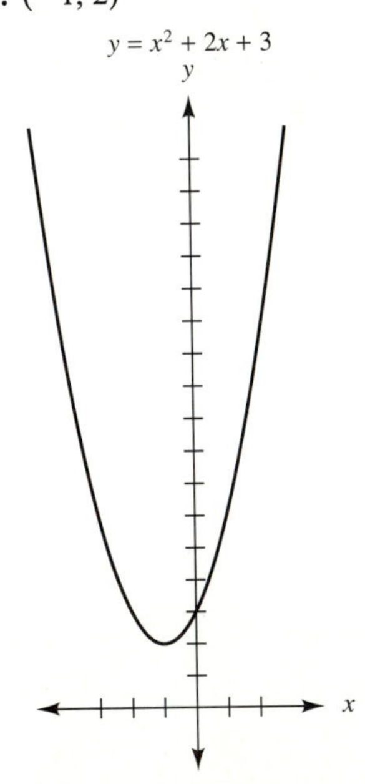

13. (1, 2)

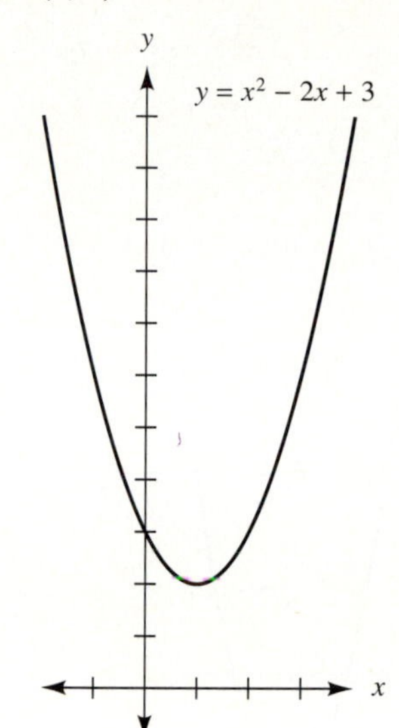

15. (−3/4, −1/8)

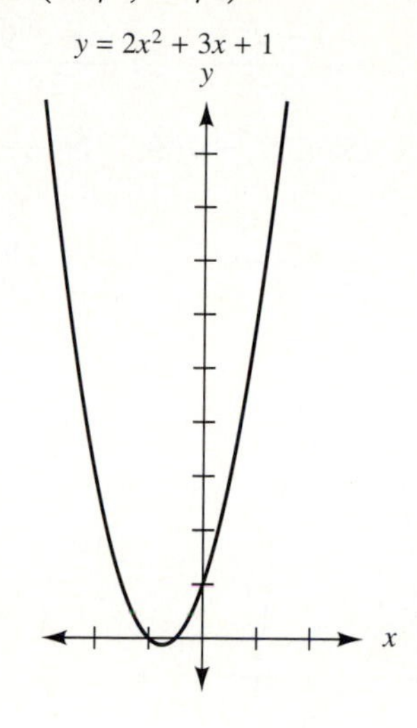

17. (3/4, 17/8)

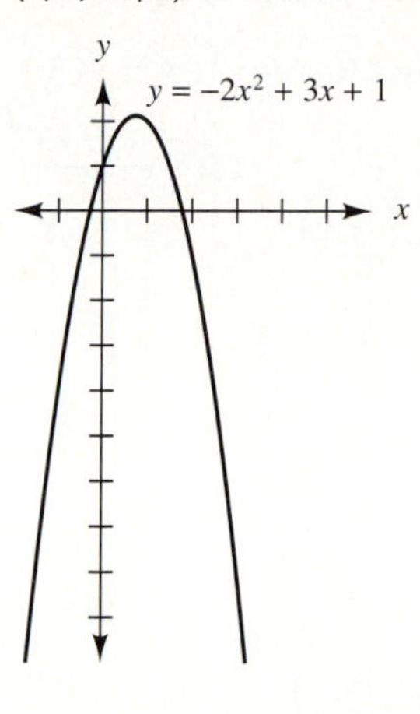

19. (1/2, −3/4)

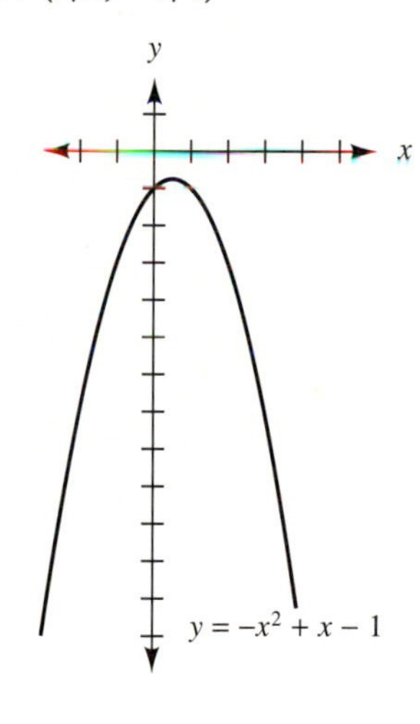

21. (a)

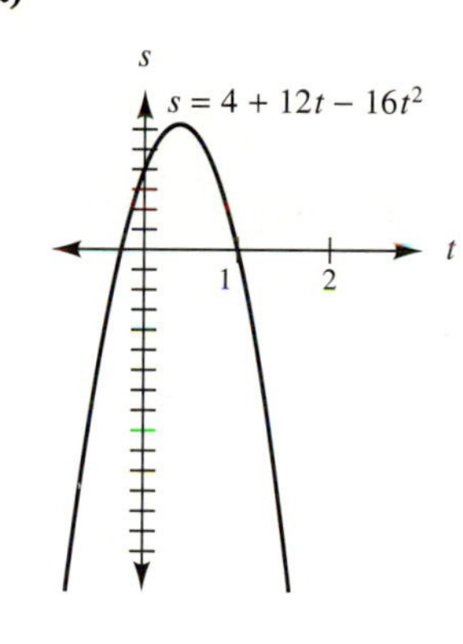

(b) 1 second

23. 57 sec, 2.9 sec.

25.

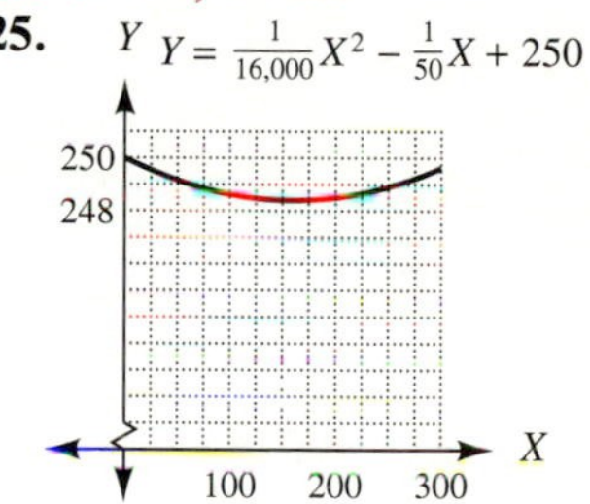

Section 6.4 (page 256)

1. 3, −3, 1, −1 **3.** 1/2, −1/2, 1, −1 **5.** 1, −1 **7.** 1/3, 1/4 **9.** 1/3, 1 **11.** 2, 1/2 **13.** −1
15. −4, 6 **17.** 1, 3 **19.** 1/3, 1/4 **21.** 0, 2, 3 **23.** 0, 1/2, 1 **25.** 0, 3/2, 2/3
27. 0, $(1 \pm \sqrt{13})/6$ **29.** 0, 3, −3 **31. (a)** $V = (4/3)\pi r^3 + 10\pi r^2$ **(b)** 0.0317 m **33.** 99.8° C
35. $0 = v_0^2 \sin^2\theta - 4v_0^2 \sin\theta\cos\theta + 4000g$; $a = v_0^2$, $b = -4v_0^2\cos\theta$, $c = 4000g$ **37.** 11.52 ft or 0.78 ft

Section 6.5 (page 262)

1. 0.6 **3.** 0.3 **5.** 2.8 **7.** 2.23 **9.** 1.46 **11.** 2.3 **13.** 4.6 s **15.** 1.58 ft

Section 6.6 (page 263)

Sequential Review: 1. $4 \pm \sqrt{13}$ **3.** $1, -8$ **5.** $\frac{-7 \pm \sqrt{51}}{2}$ **7.** $\frac{9 \pm \sqrt{79}}{2}$ **9.** $\frac{5 \pm \sqrt{13}}{6}$ **11.** 7, 0
13. $\frac{-3 \pm \sqrt{17}}{4}$ **15.** $\frac{-1 \pm \sqrt{37}}{6}$ **17.** $\pm\sqrt{7}$ **19.** $(3/2, -9/2)$ **21.** $(-3/2, -31/4)$
23. $(3/4, 31/8)$
25.

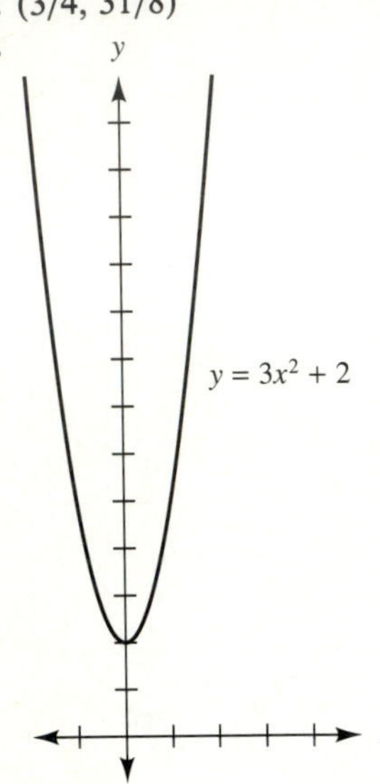

27.

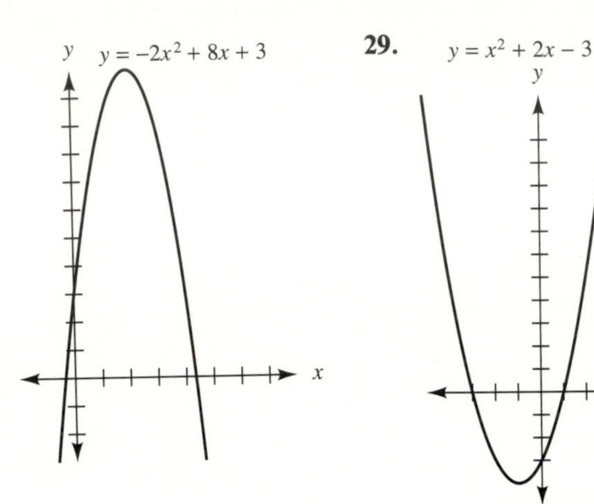

29.

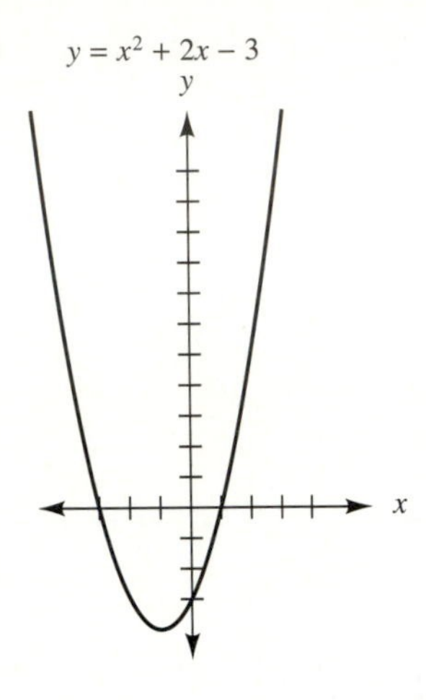

31. $3, -3$ **33.** $2, -1/2$ **35.** $2, -4$ **37.** $1/2, -1/2, 3, -3$ **39.** $0, -1/3, 3/2$ **41.** 2.3 or 2.4
43. 1.3 **45.** 2.65
Applied Problems: 1. 12 years **3.** 1.41 s **5.** 395 m **7.** 0.173 A, 4.63 A **9.** 660 nm, 620 nm
11. 3.43 in or 14.6 in **13.** 14.3 ft **15.** 1.9 ft **17.** 6.73 ft × 6.73 ft **19.** 79.2 Ω, 20.8 Ω

Random Review: 1.

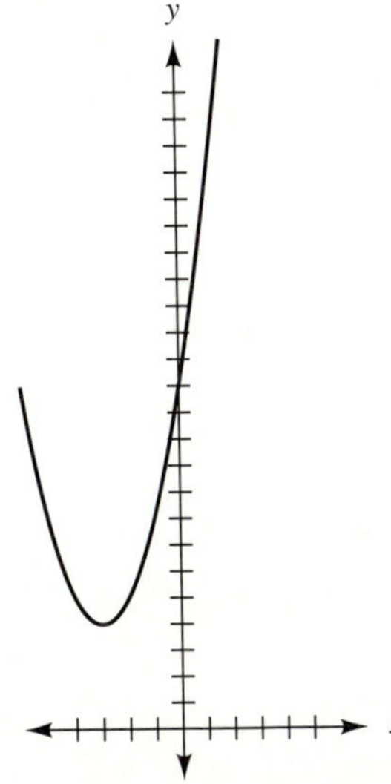

2. 3.87 **3.** $\frac{-7 \pm \sqrt{65}}{2}$
4. $(3/4, -9/8)$ **5.** 1 **6.** $\frac{3 \pm \sqrt{6}}{3}$
7.

y
y = 3x² − 2
x

8. 0, $\dfrac{-1 \pm \sqrt{37}}{6}$ **9.** 0.5 or 0.6 **10.** 2, 1/2

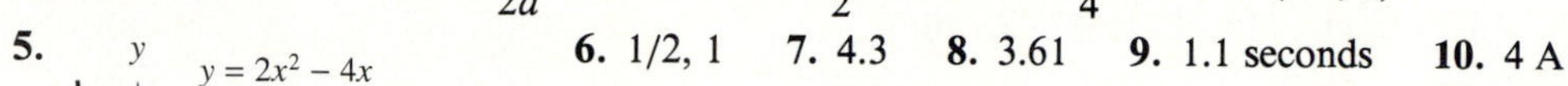

Chapter Test: 1. $x = \dfrac{-b \pm \sqrt{b^2 - 4ac}}{2a}$ **2.** $\dfrac{7 \pm \sqrt{33}}{2}$ **3.** $\dfrac{5 \pm \sqrt{17}}{4}$ **4.** $(-2, 0)$

5.

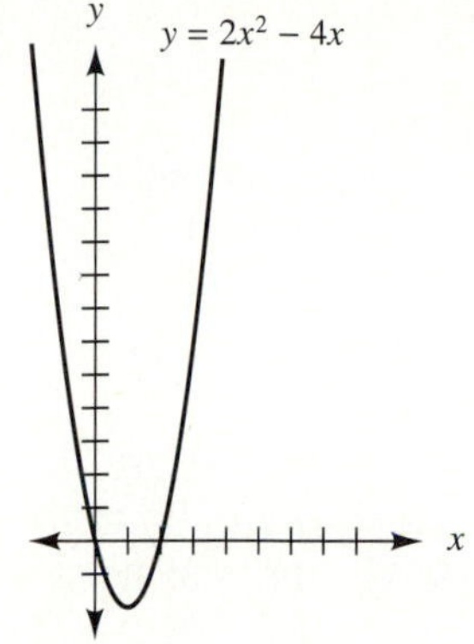

6. 1/2, 1 **7.** 4.3 **8.** 3.61 **9.** 1.1 seconds **10.** 4 A

CHAPTER 7

Section 7.1 (page 273)

1. not possible **3.** $x^3 - 5x^2 + 8x - 10$ R. 11 **5.** $x^2 + 4x + 3$ R. 0 **7.** $6x^2 + x - 2$ R. 0 **9.** not possible **11.** $2x^4 + 6x^3 + 12x^2 + 37x + 111$ R. 330 **13.** $x^2 - 2x + 4$ R. 0 **15.** $x^2 + x + 1$ R. 0 **17.** $x - 1$ R. 0 **19.** $3x^4 + 6x^3 + x + 2$ R. 0 **21.** not possible **23.** yes **25.** no **27.** no **29.** no **31.** no **33.** yes **35.** yes **37.** no **39.** no **41.** yes
43. **(a)** no **(b)** yes; If a and b have no common factor, a is a factor of the quotient, but if a and b have a common factor, a need not be a factor of the quotient.

Section 7.2 (page 279)

1. **(a)** 103 **(b)** 103 **3.** **(a)** -76 **(b)** -76 **5.** **(a)** -55 **(b)** -55 **7.** **(a)** 21 **(b)** 21 **9.** **(a)** 0 **(b)** 0 **11.** **(a)** -117 **(b)** -117 **13.** **(a)** 2 **(b)** 2 **15.** yes **17.** no **19.** no **21.** yes **23.** yes **25.** yes **27.** no **29.** $\dfrac{-1 \pm \sqrt{5}}{2}$ **31.** -3, 1 **33.** $\dfrac{-3 \pm \sqrt{13}}{2}$ **35.** $\dfrac{\pm\sqrt{2}}{2}$ **37.** $\dfrac{-3 \pm \sqrt{17}}{4}$ **39.** 1/3, -1 **41.** $\pm\sqrt{6}/2$ **43.** **(a)** 17,202,000 **(b)** 17,202,000 **45.** 94% **47.** 0.0833 mm

Section 7.3 (page 285)

1. 3, 1 **3.** 2, 2, **5.** 1, 2 **7.** 2, 2 **9.** 5, 0 **11.** 0, 0 **13.** 1, 2 **15.** $\pm 1, \pm 2$ **17.** $\pm 1, \pm 2, \pm 4$ **19.** $\pm 1, \pm 3, \pm 1/2, \pm 3/2$ **21.** $\pm 1, \pm 2, \pm 3, \pm 6, \pm 1/2, \pm 3/2$ **23.** $\pm 1, \pm 3, \pm 1/3$ **25.** $\pm 1, \pm 2, \pm 5, \pm 10, \pm 1/2, \pm 5/2, \pm 1/4, \pm 5/4$ **27.** $\pm 1, \pm 3, \pm 5, \pm 15, \pm 1/2, \pm 3/2, \pm 5/2, \pm 15/2, \pm 1/3, \pm 5/3, \pm 1/6, \pm 5/6$ **29.** 2, 1/2 **31.** 1, 3/2, -1 **33.** 1, -1, 2, 3 **35.** 1, 3, $-1/2$ **37.** no rational roots **39.** 2, -2 **41.** -2, 1/2

Section 7.4 (page 291)

1. $-1/2, -1 \pm \sqrt{5}$ **3.** $1, -1, \dfrac{-3 \pm \sqrt{17}}{4}$ **5.** $-1.62, 0.618$ **7.** $-1/2, 1.44$ **9.** $-2/3, -1.26$

11. $3/2, -1.26$ **13.** 1.32 m **15.** 12 **17.** 2.3 mm

Section 7.5 (page 293)

Sequential Review: 1. $6x^2 - 9x + 17$ R. -30 **3.** not possible **5.** $6x^2 + 2x + 1$ R. 9 **7.** no **9.** yes **11. (a)** 5 **(b)** 5 **13. (a)** -19 **(b)** -19 **15.** yes **17.** no **19.** 2 (triple root) **21.** $3, -3$ **23.** 2 **25.** 1 **27.** 2 **29.** $\pm 5, \pm 1$ **31.** $\pm 1, \pm 1/3, \pm 3, \pm 9$ **33.** $-1, 3/2$

35. (a) no **(b)** no **37. (a)** yes **(b)** no **39. (a)** no **(b)** yes **41.** $-2, \dfrac{-3 \pm \sqrt{13}}{2}$

43. $-1, -0.794$ **45.** $1/2, -3, -1.26$

Applied Problems: 1. (a) $V = (\pi/3)(R^3 + 10R^2 + 100R)$ **(b)** 916 m^3 **3.** 5 cm on each edge **5.** $-1, -3/2, -5/2$ **7.** 115.0 Ω **9.** 0.097 **11.** 5.7 **13.** yes **15.** 0.052 V **17.** 9.85 ft **19** 65.3 ft

Random Review: 1. $-3/2, 1$ **2.** $x^2 - x + 2$ R.0 **3.** 3 **4.** 2/3 **5.** yes **6.** 1 **7.** 0

8. $-3, \dfrac{-1 \pm \sqrt{17}}{4}$ **9.** $\pm 1, \pm 2, \pm 4, \pm 1/2, \pm 1/4$ **10.** $\dfrac{5 \pm \sqrt{13}}{2}$

Chapter Test: 1. not possible **2.** no **3.** -5 **4.** $\dfrac{-3 \pm \sqrt{17}}{2}$ **5.** 3 **6.** none **7.** $1, 2/3, -3/2$

8. 1 is a triple root **9.** 1.31 seconds **10.** 3

CHAPTER 8

Section 8.1 (page 303)

1.

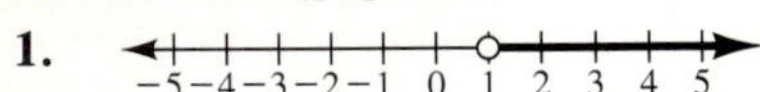

3.

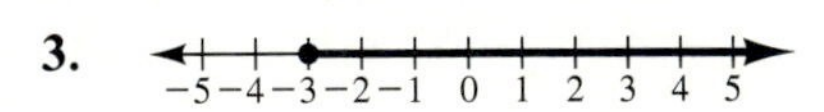

5.

7.

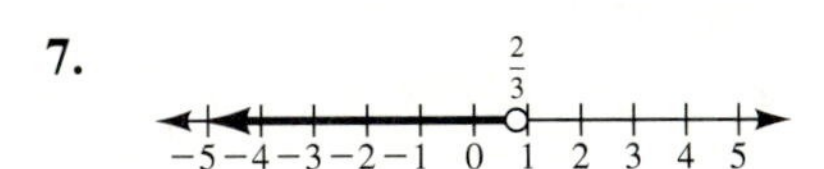

9.

11.

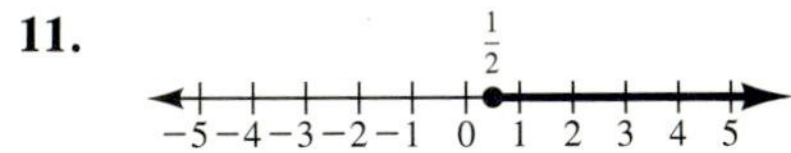

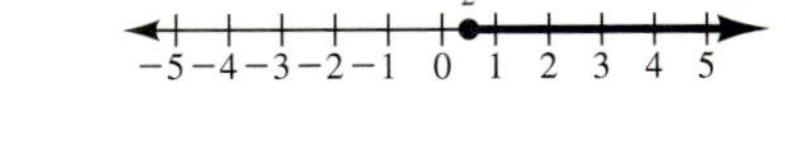

13.

15 $x > 4$ **17.** $z \geq 9$ **19.** $x \leq -2/5$ **21.** $p > 6$ **23.** $s > -15$ **25.** $y \geq 2/5$ **27.** $m < -2/3$

29.

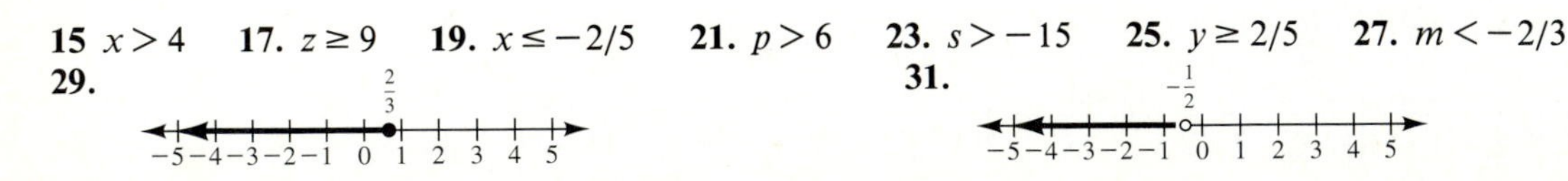

31.

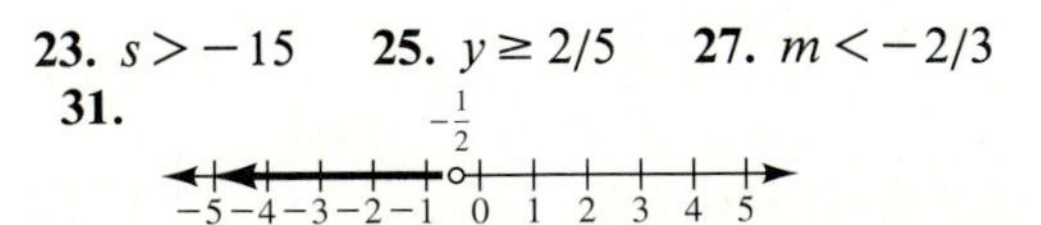

33. 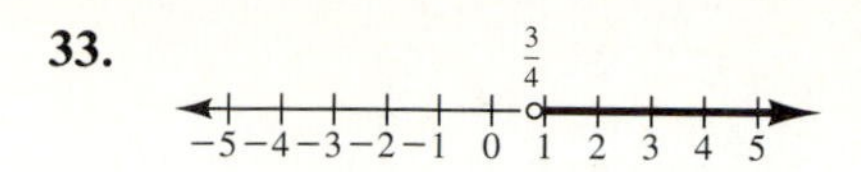

35. 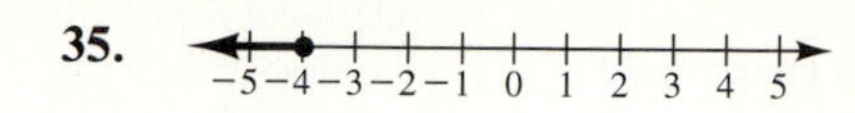

37. 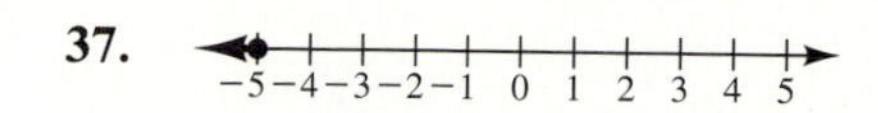

39. 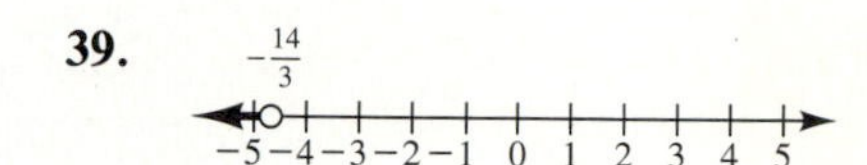

41. 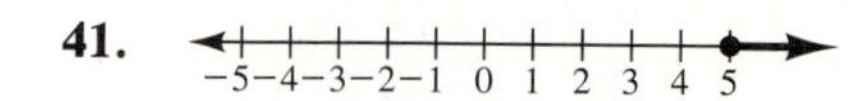

43. $x \leq 2.5 \times 10^9$ tons **45.** $v < 20$ mph, to one significant digit **47.** $t < 45$ min **49.** $h > (4/3)B$

Section 8.2 (page 307)

1. 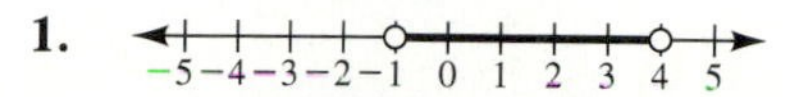

3.

5.

7.

9.

11. 

13.

15.

17.

19.

21. $-1 < x < 3$ **23.** $x \leq -3$ or $x > 1$ **25.** $x \leq 1$ or $x \geq 4$ **27.** $x < -4$ or $x \geq -2$
29. $2 < x < 4$ **31.** $-2 \leq x \leq 1$ **33.** $1.5 \text{ cm} < x < 2 \text{ cm}$ **35.** $32.5° \text{ C} < T < 45° \text{ C}$
37. $F < -62.5°$ or $F > -17.5°$ **39.** $24 \text{ mm} < f_2 < 30 \text{ mm}$

Section 8.3 (page 315)

1. $x < -2$ or $x > 3$ **3.** $-4 \leq x \leq 6$ **5.** $2/3 < x < 1$ **7.** $x \leq -5$ or $x \geq 3$ **9.** $x < 3$ or $x > 4$
11. $x = -1/2$ **13.** $-2 < x < 1$ **15.** $x \leq 1/2$ or $x > 3$ **17.** $-3 \leq x < 0$
19. $-1/2 < x < 1/2$ or $x > 2$ **21.** $x < -1/2$ **23.** $-4/3 < x \leq -1/2$ or $x \geq 5/3$
25. $-3 \leq x \leq -1$ or $1 \leq x \leq 3$ **27.** $x < -1$ or $-1 < x < 1$ or $x > 1$ **29.** $x < 0$
31. $-2/3 \leq x \leq 0$ or $x \geq 2/3$ **33.** $v > 70$ mph, to one significant digit **35.** $t \leq 2$ s or $t \geq 3$ s
37. $r > 5.98$ cm

Section 8.4 (page 322)

1. $4, -10$ **3.** no solution **5.** $-1, 7/3$ **7.** 0 **9.** $-5, 1/5$ **11.** $6, 1, 3, 4$ **13.** $5, 1, 3$
15. $4, 0$ **17.** $1, 0$ **19.** $-2, -8/3$ **21.** $-13, -1$ **23.** $x \leq -1$ or $x \geq -1/3$ **25.** $-2 < x < 4$
27. all real numbers **29.** $-4 \leq x \leq -3$ or $2 \leq x \leq 3$ **31.** $-5 \leq x \leq -4$ or $2 \leq x \leq 3$
33. $-1 < x < 5$ or $x < -2$ or $x > 6$ **35.** $-1/2 \leq x \leq 1/2$ **37.** $x < -6$ or $0 < x < 3$ or $x > 3$
39. $x < -2$ or $1 < x < 4$ or $x > 4$ **41.** $x < 0$ or $0 < x < 6/11$ or $x > 6/7$ **43.** $|h - 62''| \leq 2$
45. $|x - 16.9| \leq 0.3$ **47.** $|t - 48| \leq 6$
49. Decide what error ϵ is acceptable; Let a = number sought; let b = number compared to; Ask $|a - b| < \epsilon$?

Section 8.5 (page 328)

1.

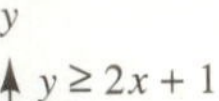
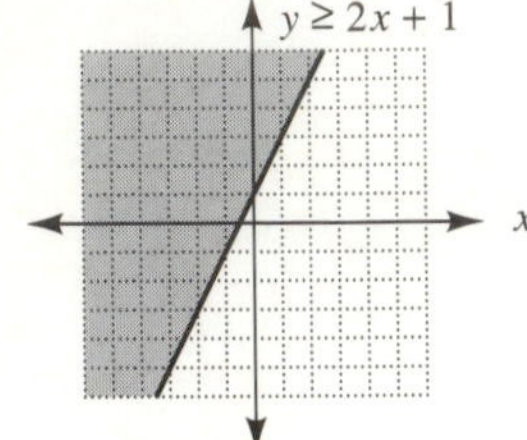

3.

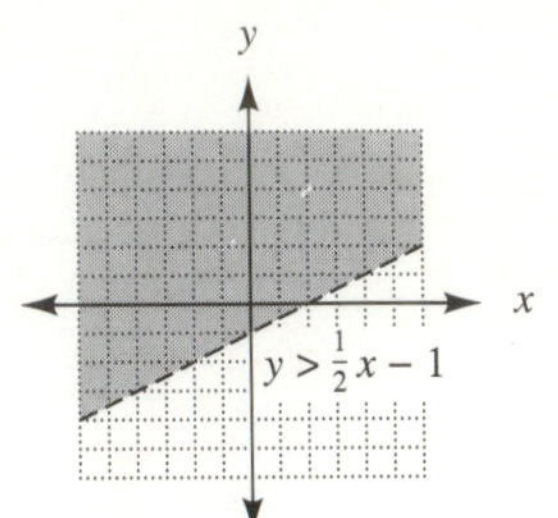

5.

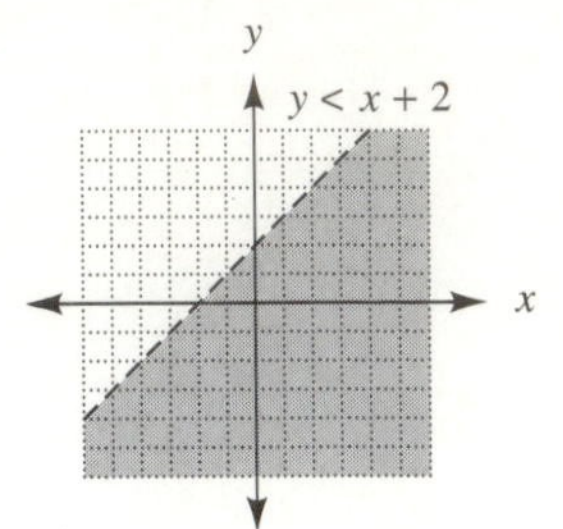

7.

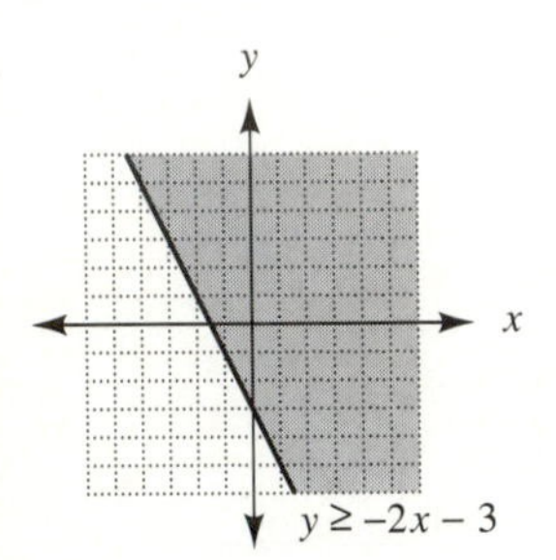

9.

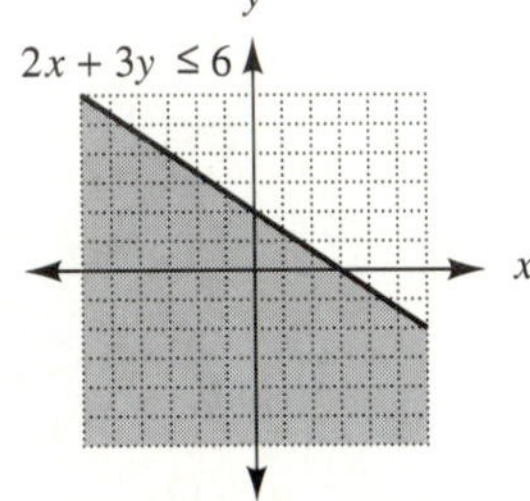

11.

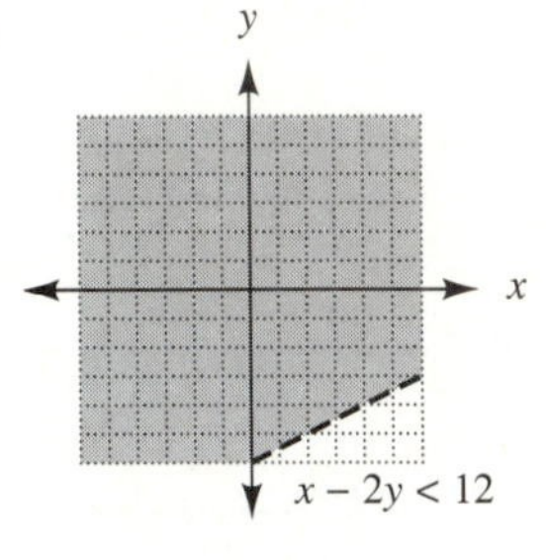

13.

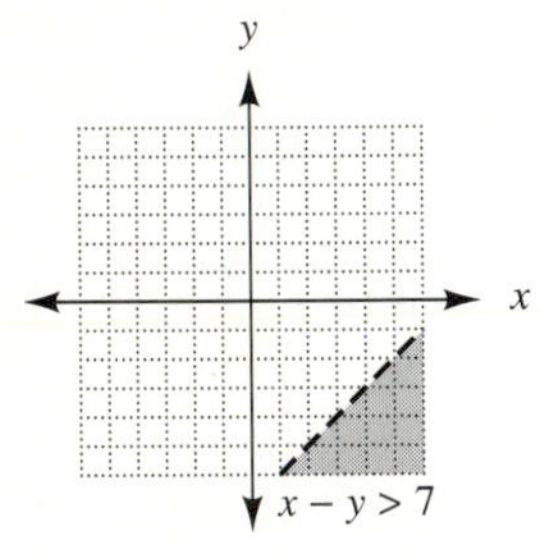

15.

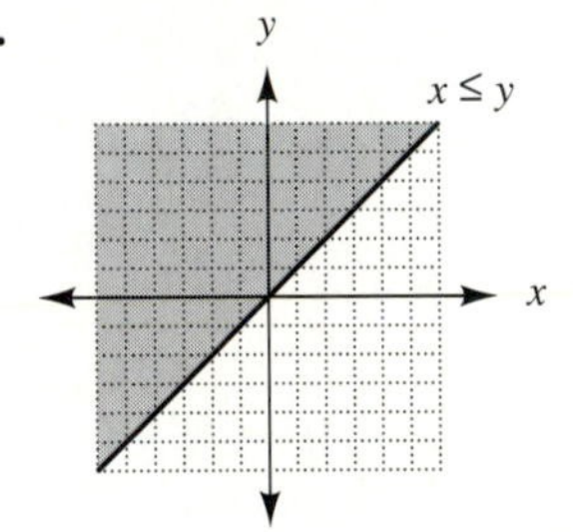

17.

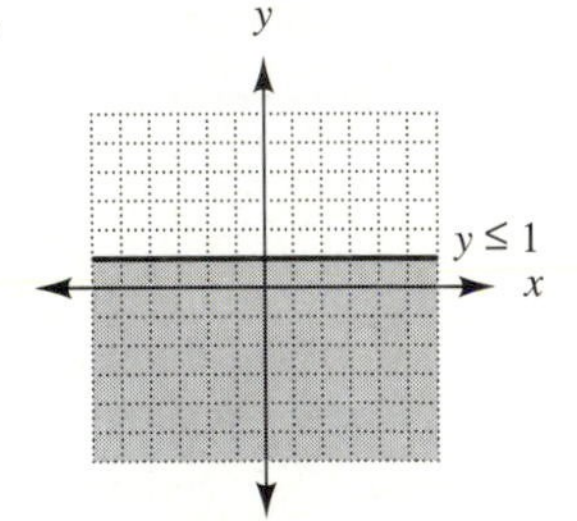

19.

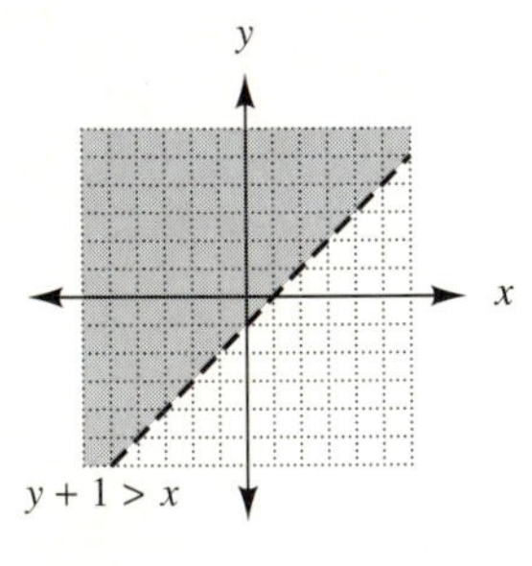

21.

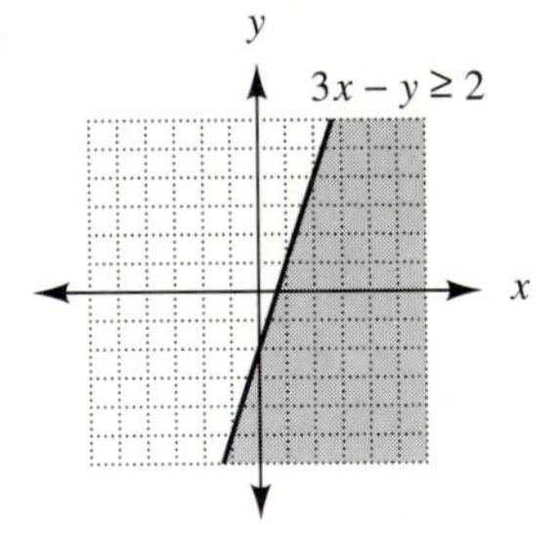

23.

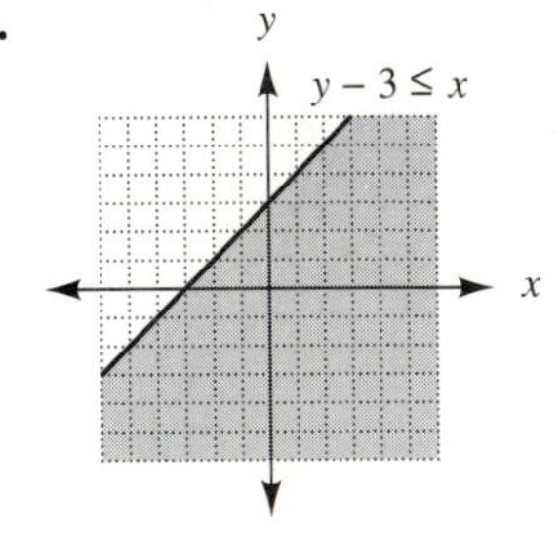

25.

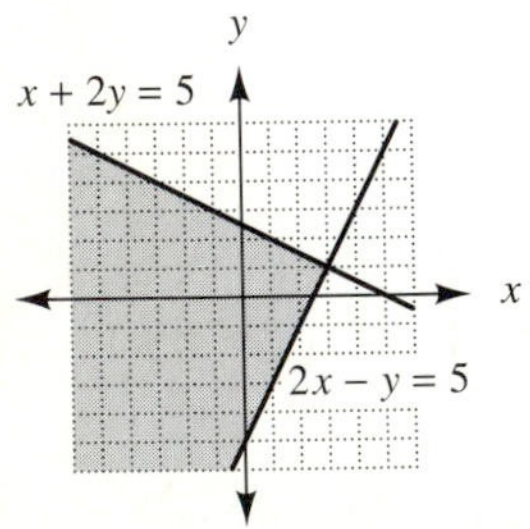

27.

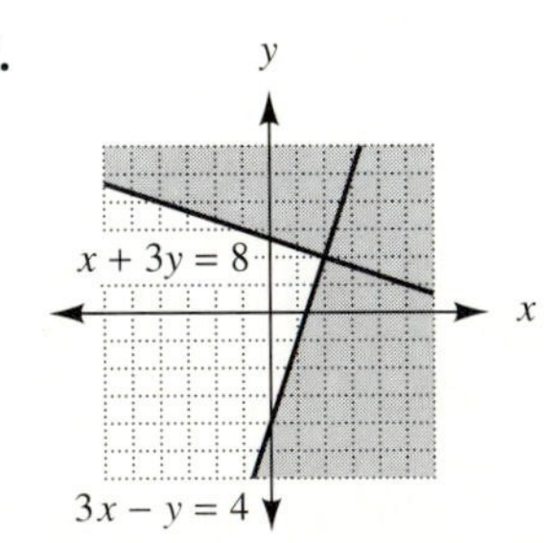

29.

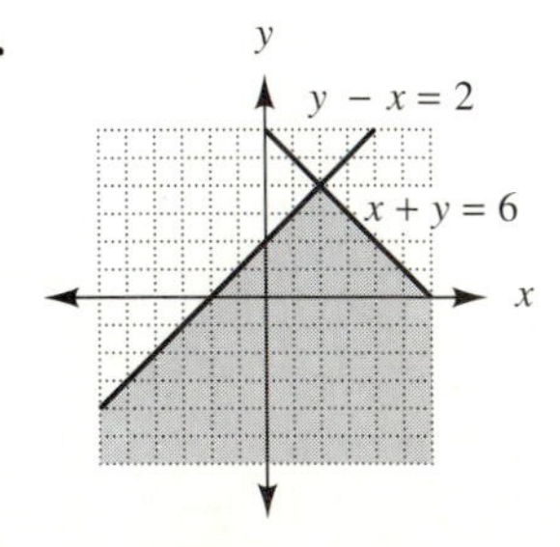

31.

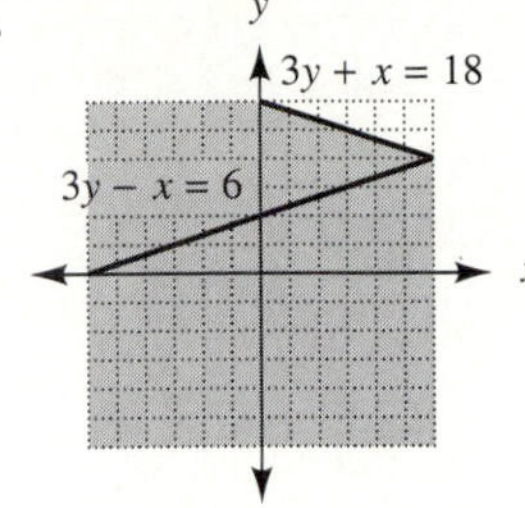

33. 0 regular, 4000 premium

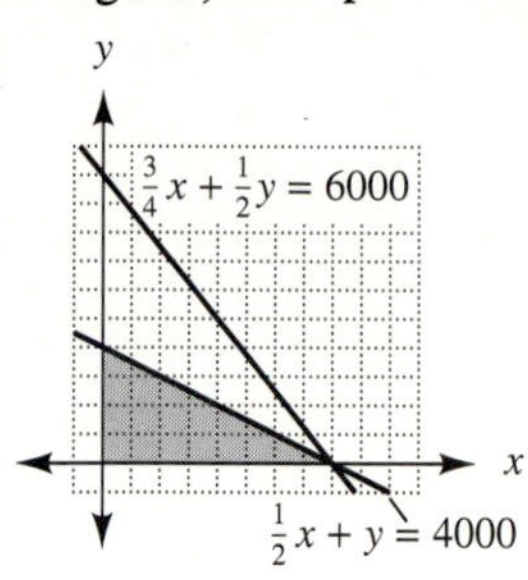

35. 0 13″-models, 900 19″-models

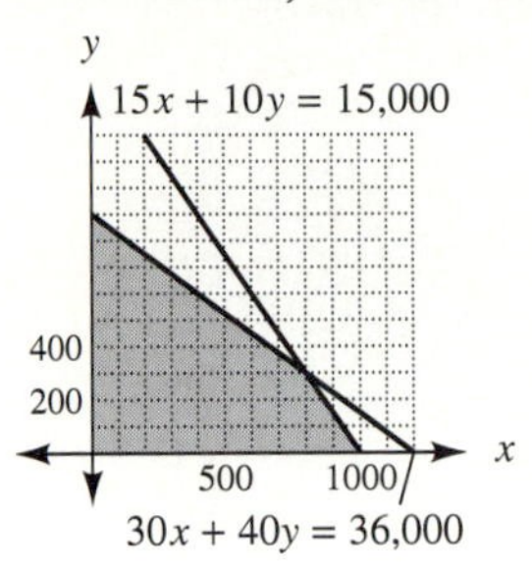

Section 8.6 (page 330)

Sequential Review: 1. $y > 1/2$ **3.** $p \geq -8$ **5.** $y \leq -5$

7. $m \leq 4$

9. $m > 5$

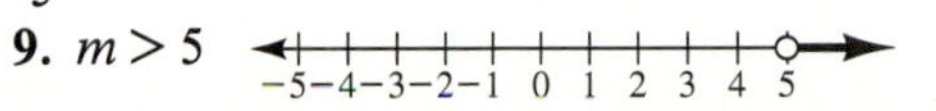

11.

13.

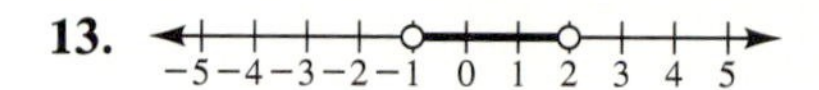

15. $-3 < x \leq 1$ **17.** $-1 < x \leq 2$ **19.** $x < -3$ or $x > 1/2$ **21.** $x \leq 1/2$ or $x \geq 3$ **23.** $0 < x < 3$
25. $-4 \leq x \leq 0$ or $x \geq 3/2$ **27.** $x = \pm 2$ **29.** no solution **31.** 3/5, 3/7 **33.** all real numbers
35. $x \leq 1$ or $x \geq 7$

37.

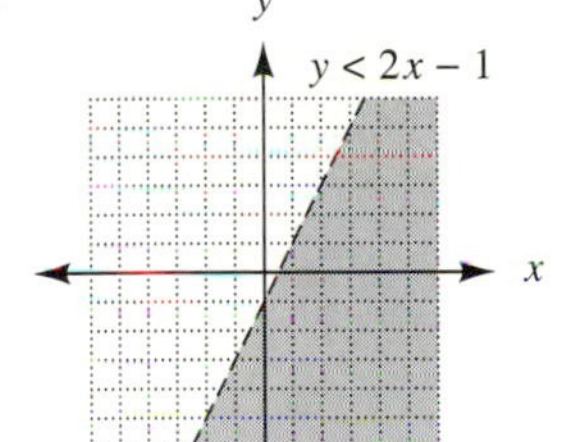

39.

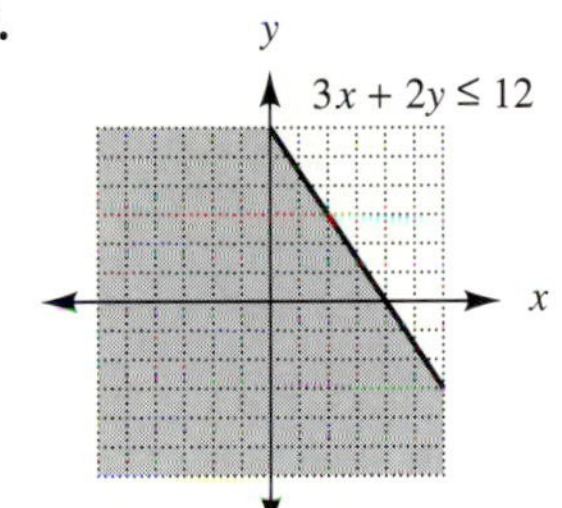

41.

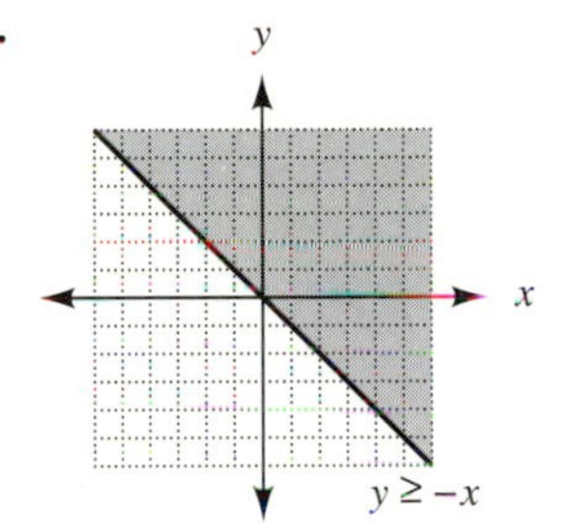

43.

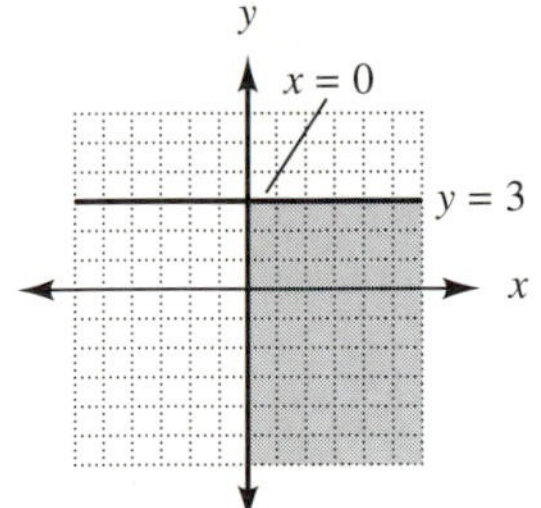

45.

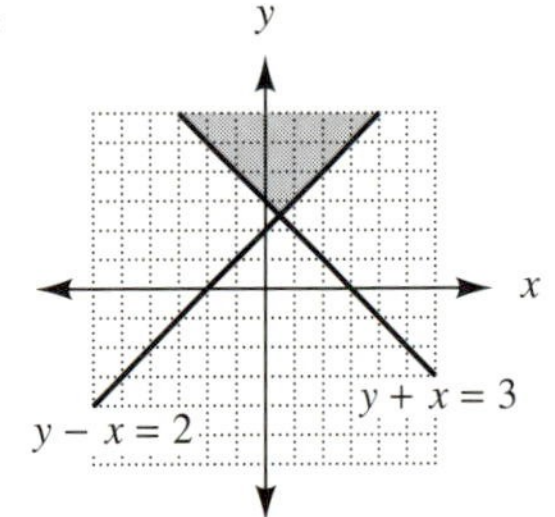

Applied Problems: 1. $r > 7.96$ **3.** $|48.3 - x| \leq 0.2$ **5.** $0 \leq s < 2.03$ **7.** $0 < R \leq 3$
9. $0 < d \leq 6.00$ **11.** $0 < g < 2.5$ **13.** less than 0.004 hours or about 14 s
15. $-10.1 < \omega < -0.990$ or $0.990 < \omega < 10.1$ **17.** $x = 40{,}000$ and $y = 20{,}000$

Random Review: 1. $x \leq -2$ or $x > 3$ **2.** 4, -18

3.

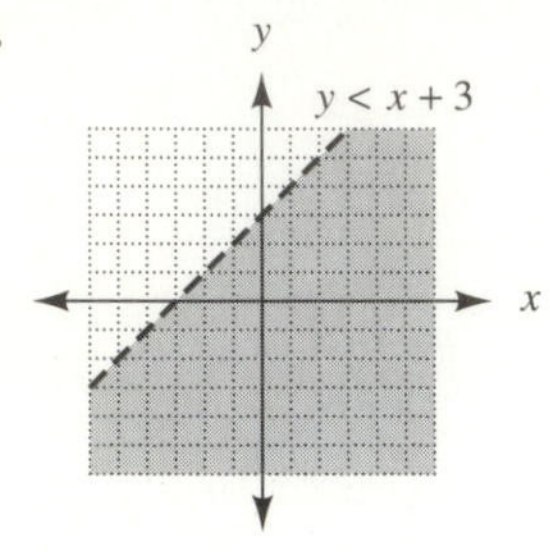

4. $x \leq 2$ **5.** $m > -1$ **6.** $x < -3/7$ or $x > 0$
7. no solution **8.** $-1 \leq x \leq 4$ **9.** $2 \leq x \leq 3$
10.

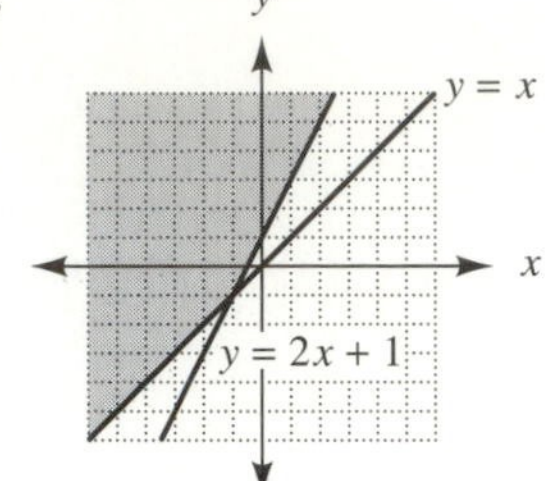

Chapter Test: 1. $y < 12$ **2.** $-4 < x < -1$ or $x \geq 3$ **3.** $x < -2$ or $x > -1$ **4.** $x \leq -1$ or $0 \leq x \leq 1$
5. 2, 2/3 **6.** 1/2, 1/4 **7.** $x < -11$ or $x > -5$
8.

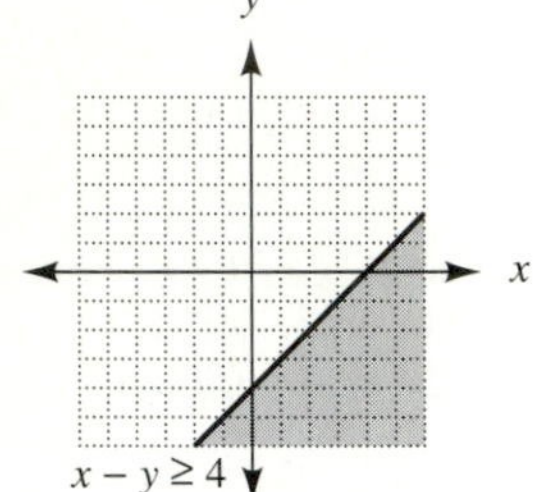

9. $t > 2.71$ s **10.** $|x - 97| < 1.5$

CHAPTER 9

Section 9.1 (page 343)

1. 13 **3.** 10 **5.** $\sqrt{117}$ **7.** $\sqrt{85}$ **9.** 1 **11.** 14/9 **13.** 4/5 **15.** -1
17. One pair of opposite sides have slopes of -1 and the other pair have slopes of 4.
19. The slopes of adjacent sides are -2 and 1/2. **21.** Two sides have length $\sqrt{10}$.
23. The sides are all $\sqrt{10}$ in length, and adjacent sides have slopes of $-1/3$ and 3.
25. The slopes of adjacent sides are -2 and 2. **27.** $y = (2/3)x + 14/3$ **29.** $y = (1/3)x - 7/3$
31. $y = (3/4)x + 2/3$ **33.** $y = (1/3)x - 7/9$ **35.** $y = 5$ **37.** $y = (-4/5)x + 3/5$
39. $y = -4x + 5$ **41.** $y = (1/3)x + 16/3$ **43.** 2% **45.** $c = (-9/2)t + 99$ **47.** $e = (3/40)t$

Section 9.2 (page 350)

1. $y = (1/4)x^2$ **3.** $y = -x^2$ **5.** $y = (1/4)x^2$ **7.** $y = (-1/3)x^2$ **9.** $y = (1/2)x^2$ **11.** $-12y = x^2$
13. $4x = y^2$ **15.** $-x = y^2$ **17.** $-8x = y^2$ **19.** $-4x = y^2$ **21.** $(5/2)x = y^2$
23. $y = (-1/4)x^2 + x + 2$ **25.** $y = (1/4)x^2 - 2$ **27.** $y = (-1/8)x^2 - (3/4)x - 7/8$
29. $y = (1/8)x^2 - (1/2)x + 1/2$ **31.** $y = (1/8)x^2 + (1/4)x + 25/8$ **33.** $4(y + 3) = (x - 5)^2$
35. $-8(x - 2) = (y - 3)^2$ **37.** $-2x = (y + 2)^2$ **39.** $-2(x - 1/2) = (y + 2)^2$
41. $(x - 1/2) = (y + 1/2)^2$ **43.** 5.0 ft **45.** 1/12 of a unit **47.** 0.833 ft below the vertex
49. tangent to surface

Section 9.3 (page 357)

1. $(-2, 2)$

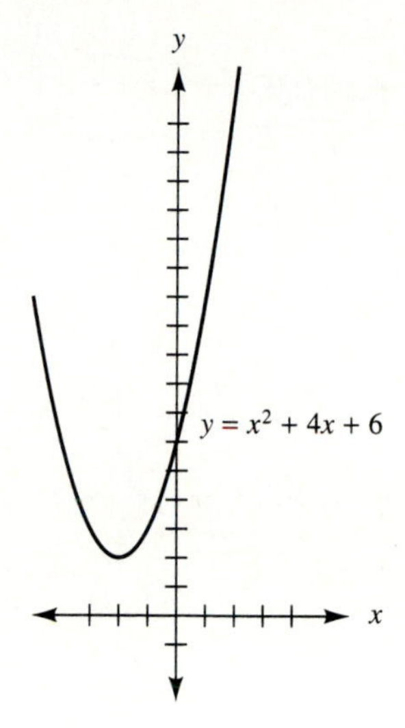

3. $(2, -8)$

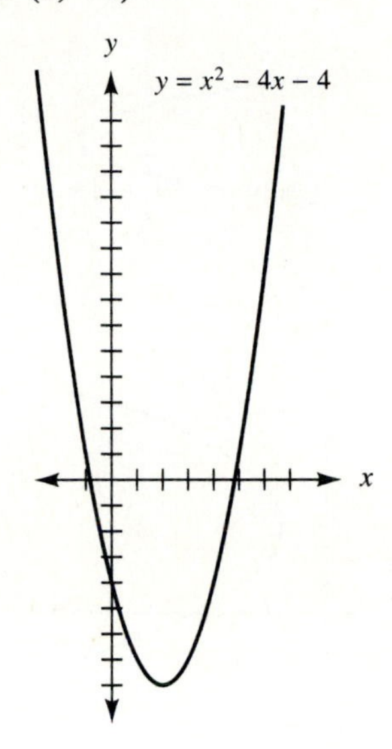

5. $(-1, 2)$

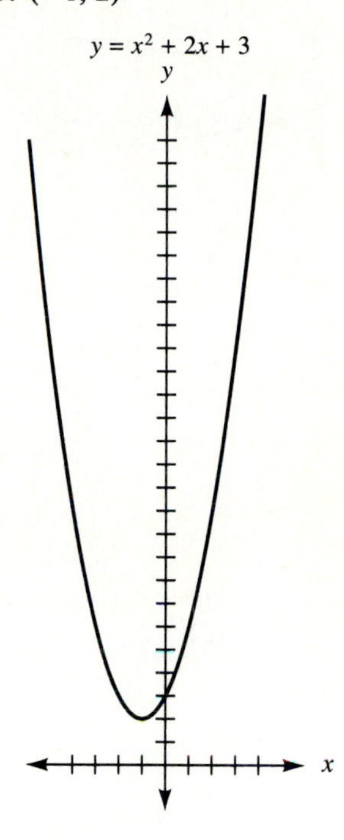

7. $(0, -1)$

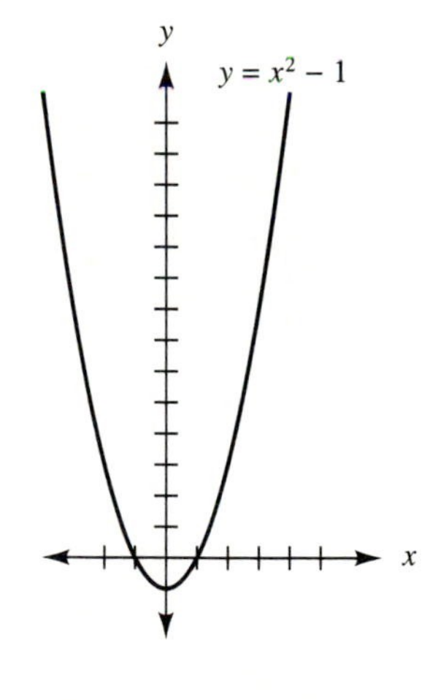

9. $(1, 2)$

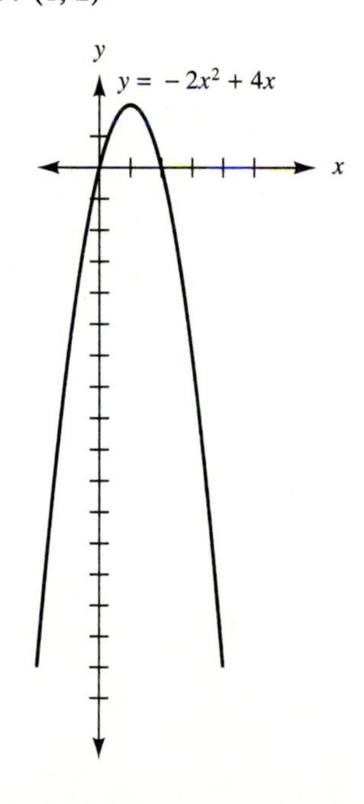

11. $(1/2, 0)$

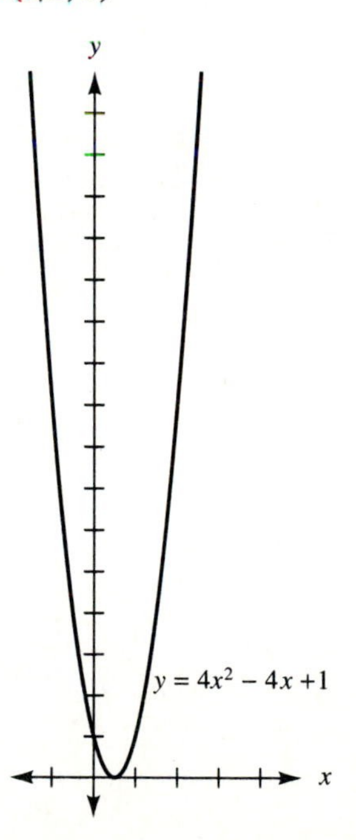

13. $(-1, 2)$

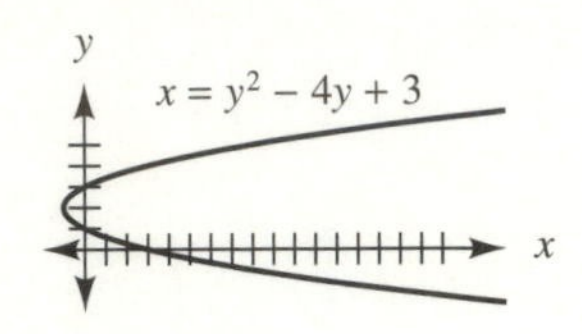

15. $(4, 3)$

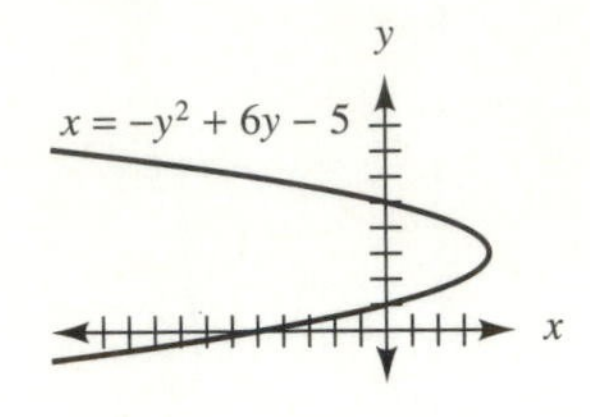

17. $(1, -1)$

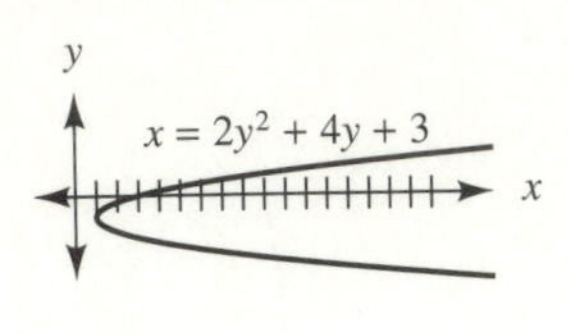

19. $(-9/2, 1/2)$

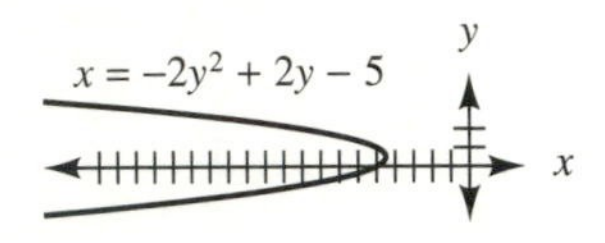

21. $(2, 4)$

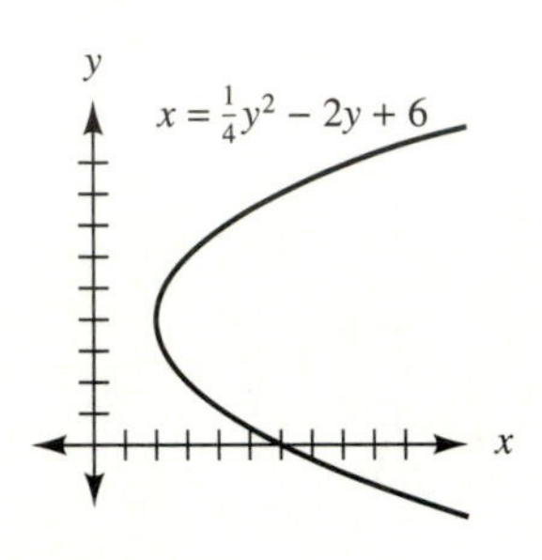

23. $(-1/2, 1)$

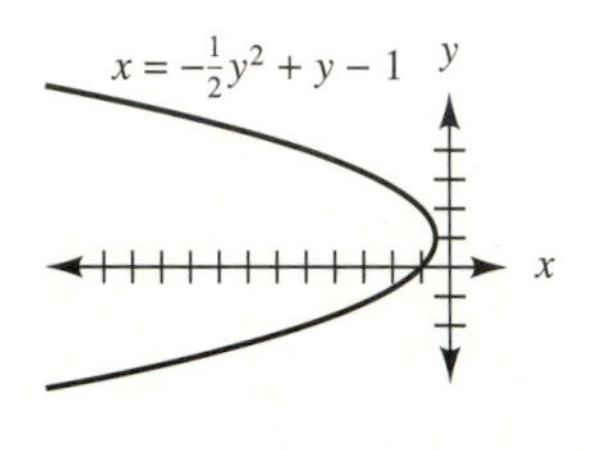

25. 3/8 s **27.** 2.5 s **29.** 3.18 cm **31.** 45

Section 9.4 (page 363)

1. $x^2 + y^2 = 49$ **3.** $x^2 + y^2 = 9$ **5.** $x^2 + 4x + y^2 - 6y - 12 = 0$ **7.** $x^2 + 8x + y^2 - 153 = 0$
9. $16x^2 + 8x + 16y^2 - 16y + 1 = 0$ **11.** yes **13.** yes **15.** no **17.** no **19.** no **21.** no
23. yes **25.** $(0, 0)$ $r = 6$ **27.** $(-1, 3)$ $r = 1$ **29.** $(-3, -2)$ $r = 5$ **31.** $(1/2, -1/2)$ $r = 1/2$
33. $(0, 1/3)$ $r = 2/3$

35. $(x - 1)^2 + (y + 2)^2 = 9$

37. $(x - 3)^2 + (y - 2)^2 = 20$

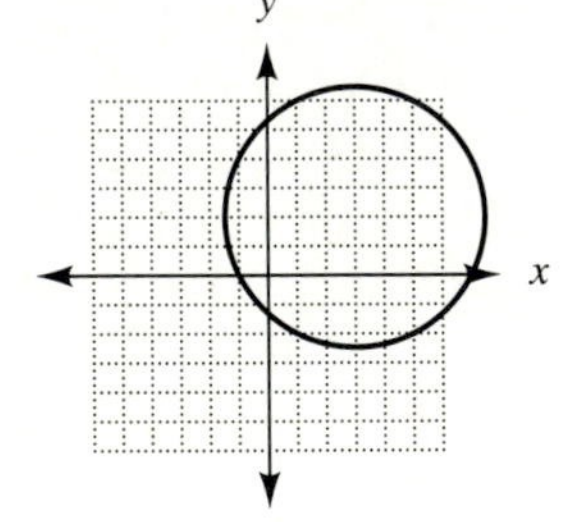

39. $\left(x - \frac{1}{2}\right)^2 + y^2 = 4$

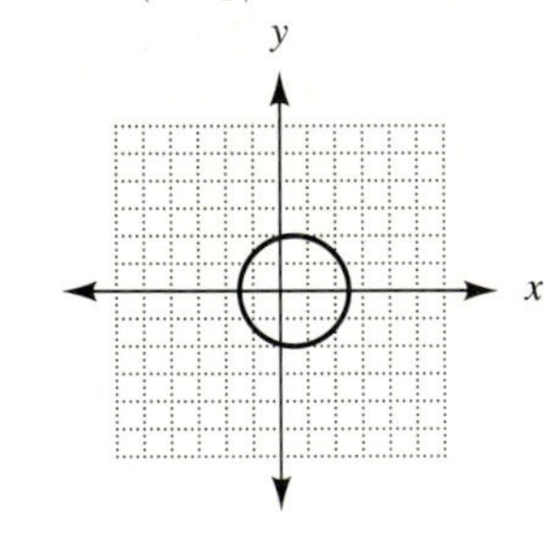

41. $x^2 + \left(y - \frac{3}{4}\right)^2 = \frac{1}{4}$

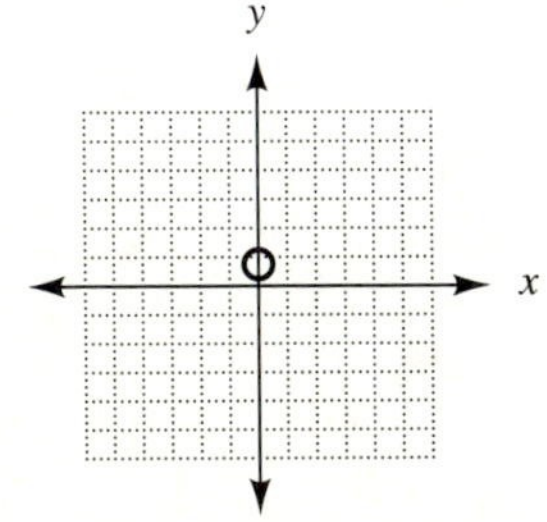

43. $x^2 + y^2 = 2594$
45. $x^2 + y^2 - 2yl = 0$

47. $I_R^2 + (I_C - 0.80)^2 = (1.20)^2$

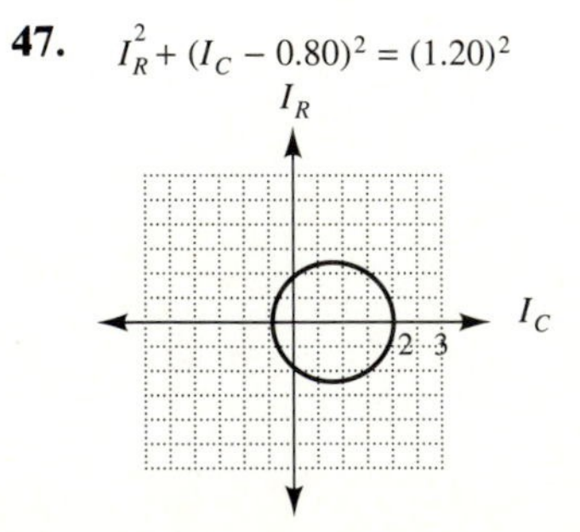

Section 9.5 (page 370)

1. $\frac{(x+1)^2}{25} + \frac{(y-1)^2}{16} = 1$

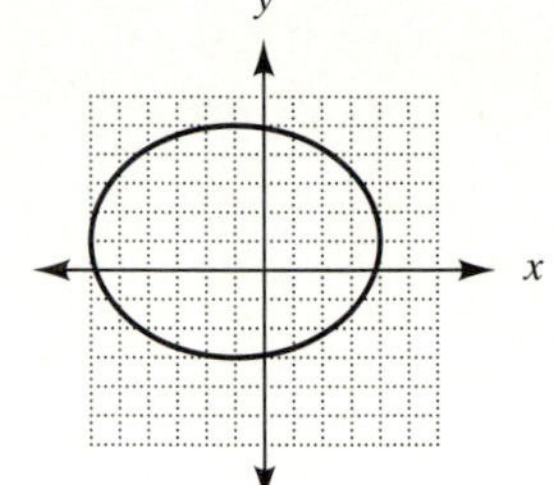

3. $\frac{(x-2)^2}{1} + \frac{y^2}{4} = 1$

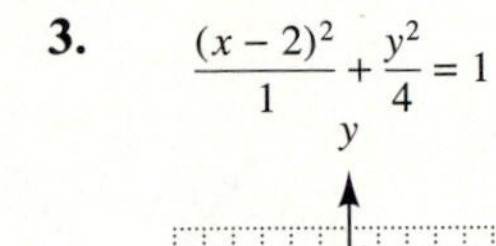

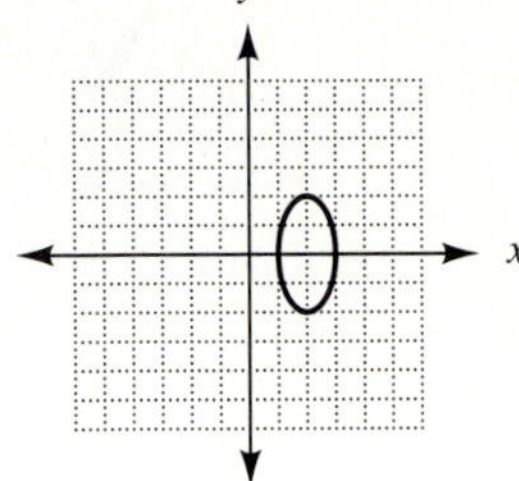

5. $\frac{(x-1)^2}{16} + \frac{(y+1)^2}{9} = 1$

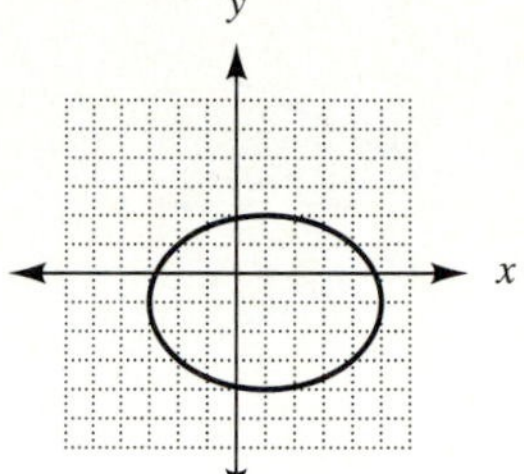

7. $\frac{(x-2)^2}{25} + \frac{(y-3)^2}{1} = 1$

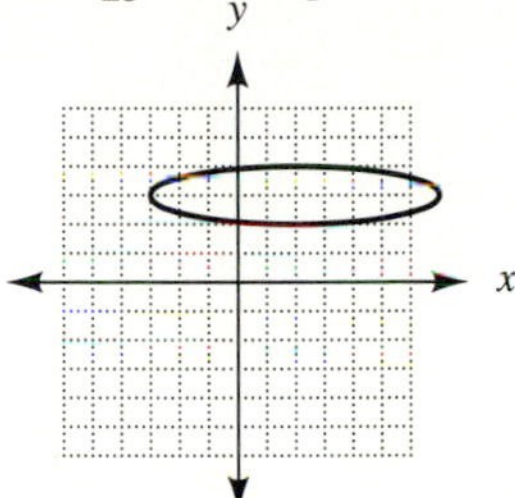

9. $\frac{x^2}{1} + \frac{y^2}{2} = 1$

11. $\frac{(x-2)^2}{4^2} + \frac{(y-1)^2}{5^2} = 1$ **13.** $\frac{x^2}{8} + \frac{(y-4)^2}{4} = 1$ **15.** $\frac{(x-4)^2}{13^2} + \frac{(y+1)^2}{5^2} = 1$ **17.** $\frac{x^2}{18} + \frac{y^2}{9} = 1$

19. $\frac{(x-3)^2}{5^2} + \frac{(y-2)^2}{3^2} = 1$ **21.** $\frac{(x-2)^2}{9} + \frac{(y-3)^2}{1} = 1$ **23.** 2.0 m **25.** 0.45 m **27.** 52.9 m

29. 5.3 m

Section 9.6 (page 377)

1. $\frac{(x-1)^2}{16} - \frac{(y+1)^2}{9} = 1$

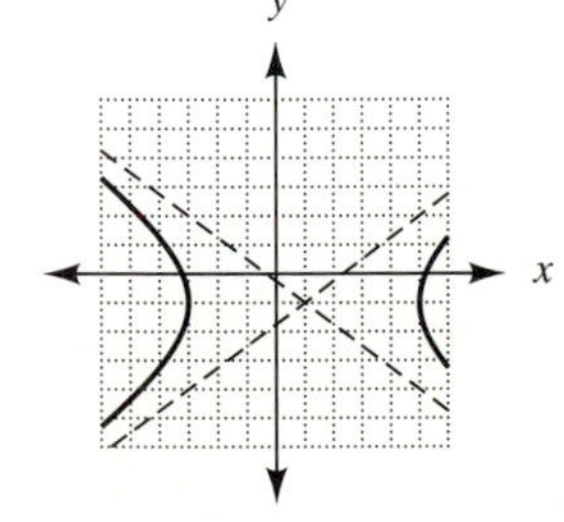

3. $\frac{(y-3)^2}{1} - \frac{(x-2)^2}{25} = 1$

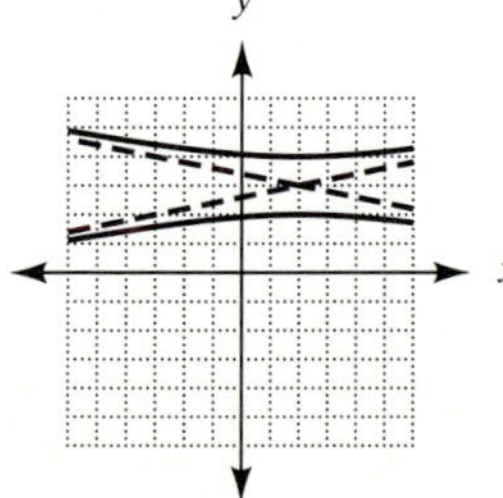

5. $\frac{(x-1)^2}{16} - \frac{(y-1)^2}{9} = 1$

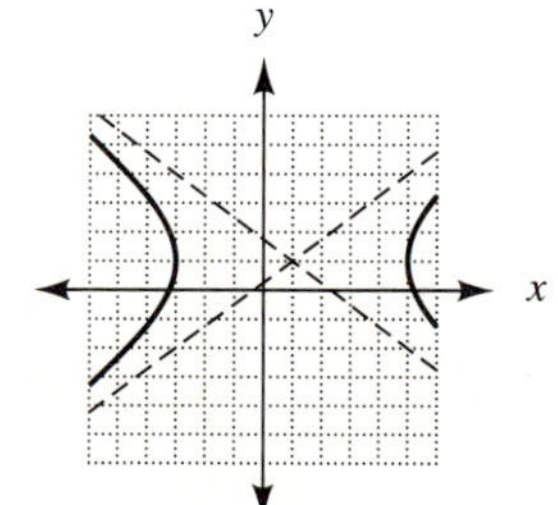

7. $\frac{y^2}{1} - \frac{(x+3)^2}{4} = 1$

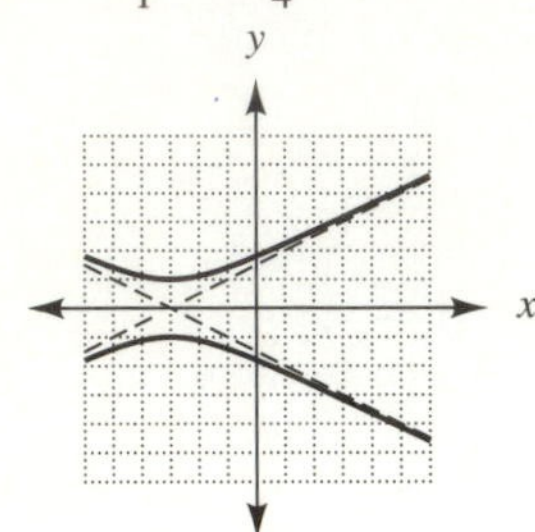

9. $\frac{y^2}{2} - \frac{x^2}{1} = 1$

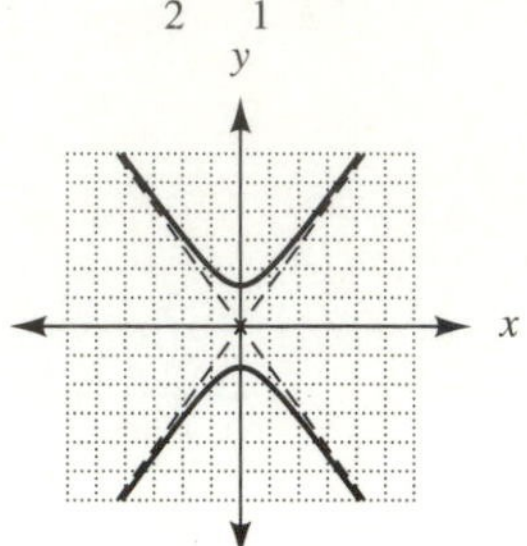

11. $\frac{x^2}{8^2} - \frac{y^2}{15^2} = 1$ **13.** $\frac{(y-2)^2}{5^2} - \frac{(x+1)^2}{12^2} = 1$ **15.** $x^2 - y^2 = 1$ **17.** $(y-2)^2 - (x+2)^2 = 1$

19. $\frac{(x-2)^2}{6} + \frac{(y+3)^2}{12} = 1$, ellipse **21.** $\frac{(x+3)^2}{3} - \frac{(y-1)^2}{2} = 1$, hyperbola

23. $\frac{x^2}{4} + \frac{(y-3)^2}{2} = 1$, ellipse **25.** $x^2 + 3(y-2)^2 = -3$, no curve

27. $(x-1)^2 - (y-1)^2 = 1$, hyperbola **29.** $3(x-1)^2 + 2(y+1)^2 = -6$, no curve

31. $x^2 - y^2 = 5929$ **33.** $\frac{x^2}{81} - \frac{y^2}{1600} = 1$ **35.** $\frac{x^2}{1225} - \frac{y^2}{144} = 1$

Section 9.7 (page 383)

1. A

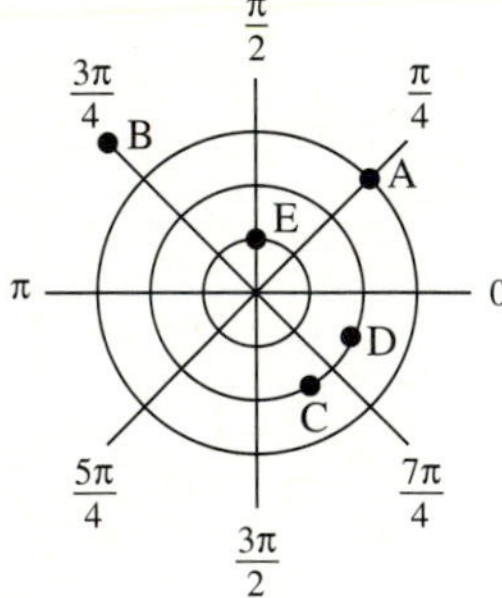

3. B **5.** C **7.** D **9.** E

11. (5, 0.927) **13.** (17, 2.06) **15.** $(1, \pi/2)$ **17.** $(2, 11\pi/6)$ **19.** (2.24, 4.25) **21.** $(\sqrt{3}, 1)$
23. $(-\sqrt{2}/2, -\sqrt{2}/2)$ **25.** (4, 0) **27.** (0, 2) **29.** $(1, -\sqrt{3})$

31.

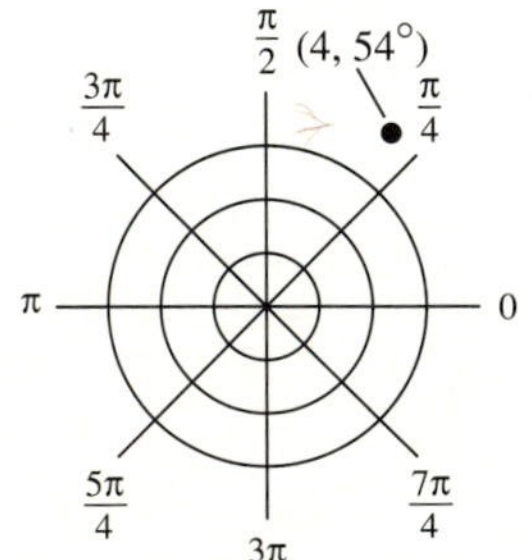

33.

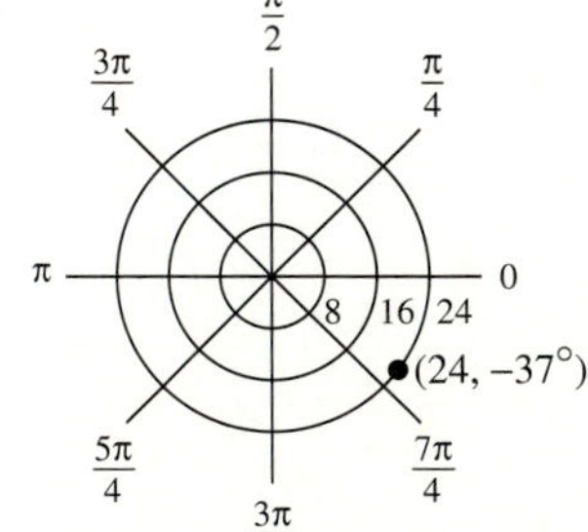

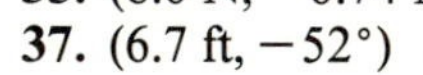
35. (8.0 N, −0.74 N)
37. (6.7 ft, −52°)

Section 9.8 (page 389)

1. $r = 2 \sin \theta$

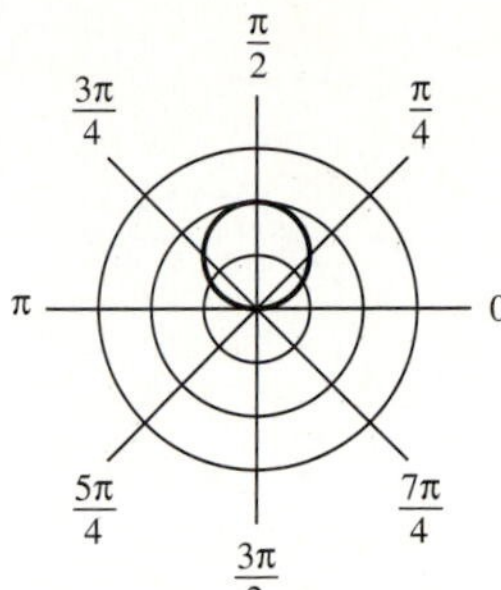

3. $r = \cos 3\theta$

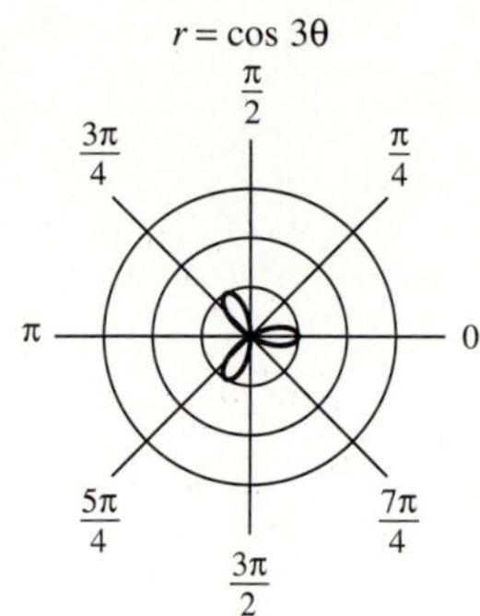

5. $r = 3(1 - \sin \theta)$

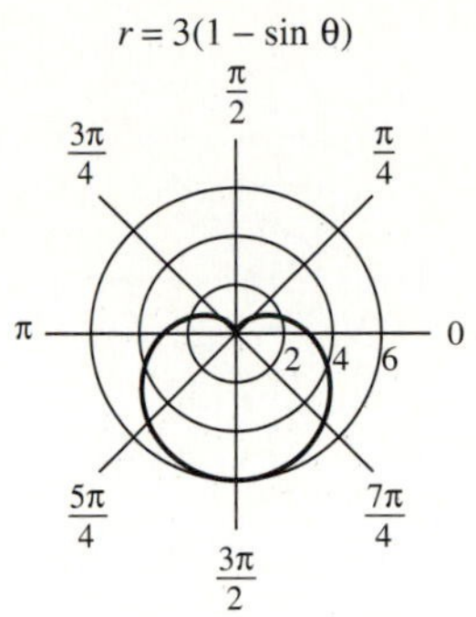

7. $r = 1 - \cos \theta$

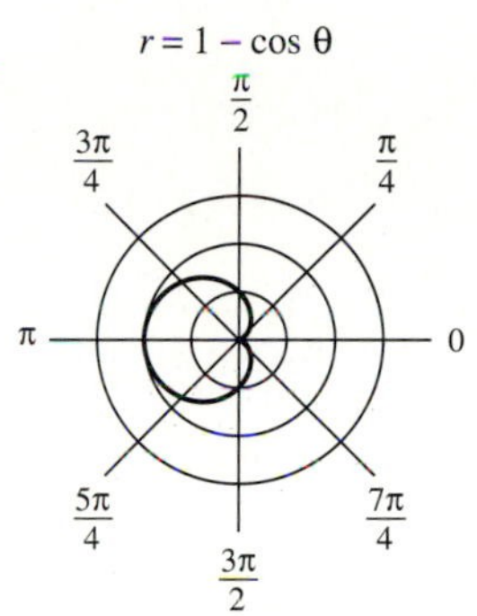

9. $r = \frac{1}{2}\theta$

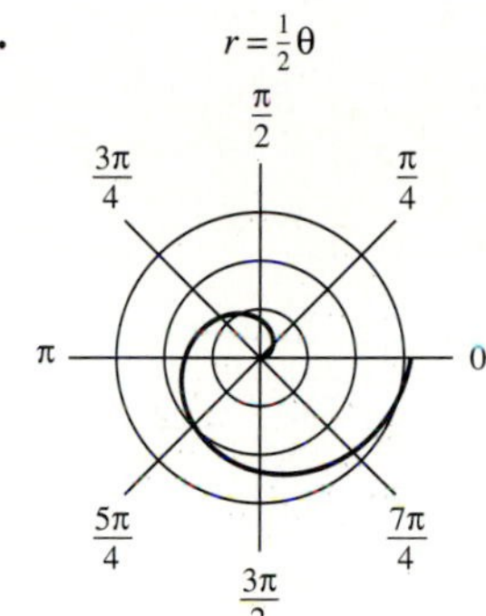

11. $r^2 = 4 \sin 2\theta$

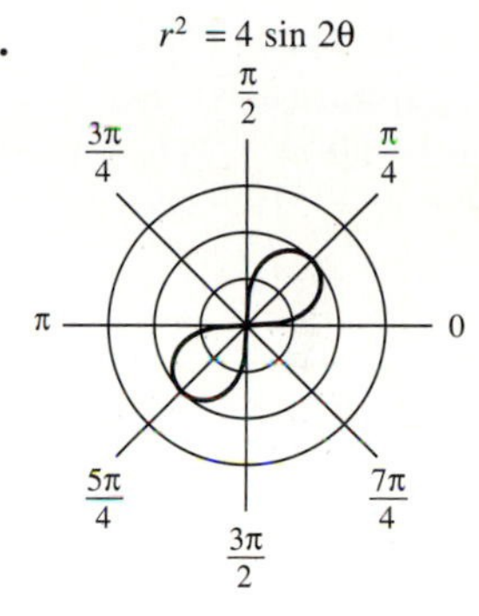

13. $r = 3$

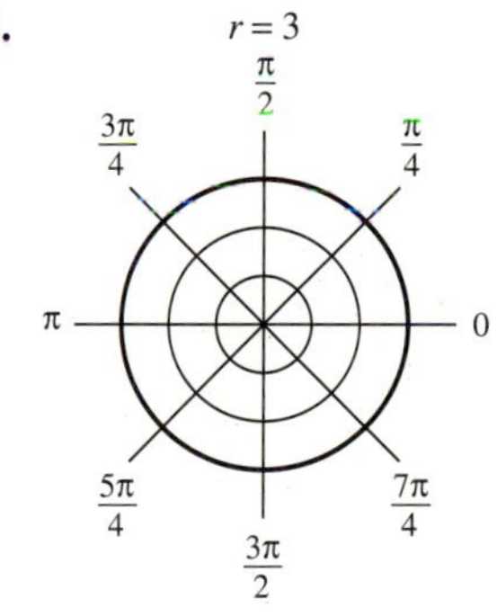

15. $\theta = \frac{\pi}{3}$

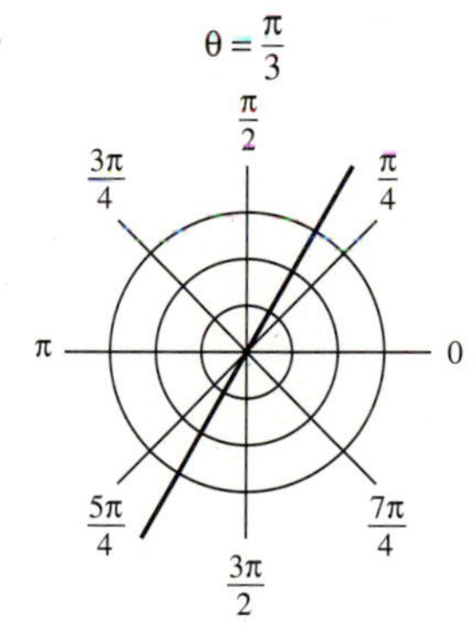

17. $r = 2(1 - \sin \theta)$

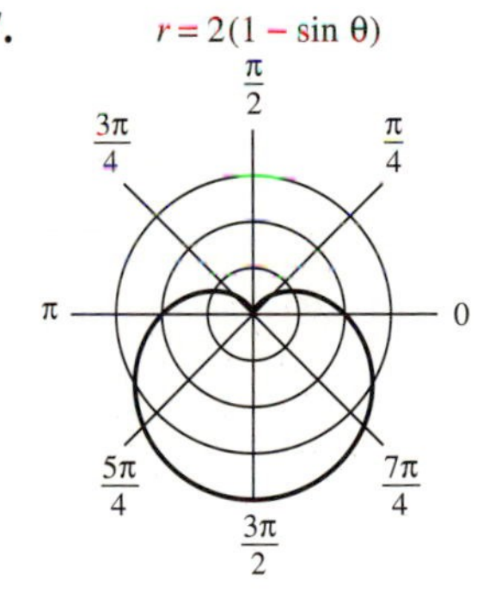

19. $r = 1 + \cos \theta$

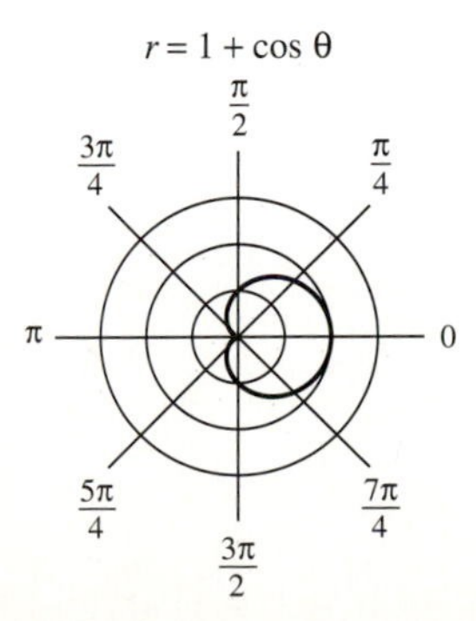

21. $r = \frac{1}{3}\theta$

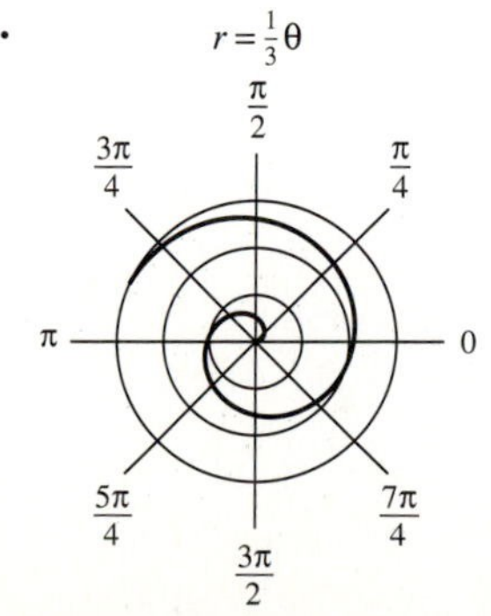

23. $r^2 = 9 \cos 2\theta$

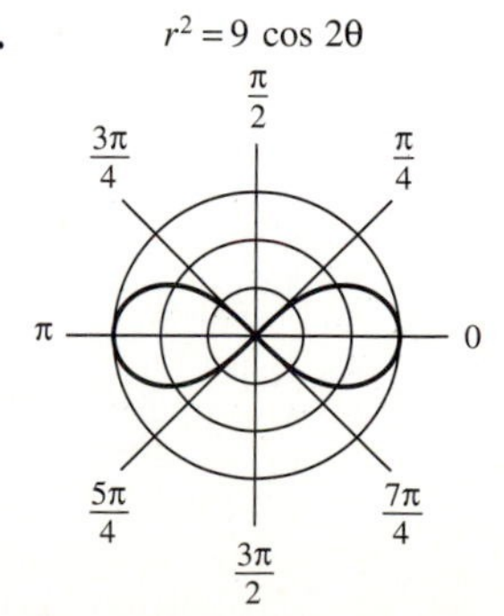

25. $r = 2\sin\theta, 0 < \theta < \frac{\pi}{2}$

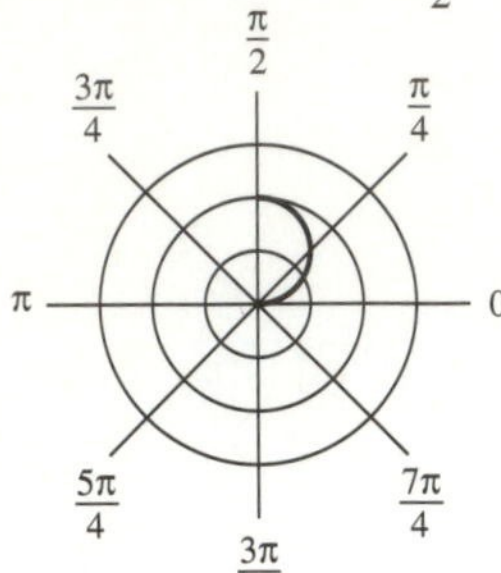

27. $\mu = \tan\theta, 0 < \theta < \frac{\pi}{2}$

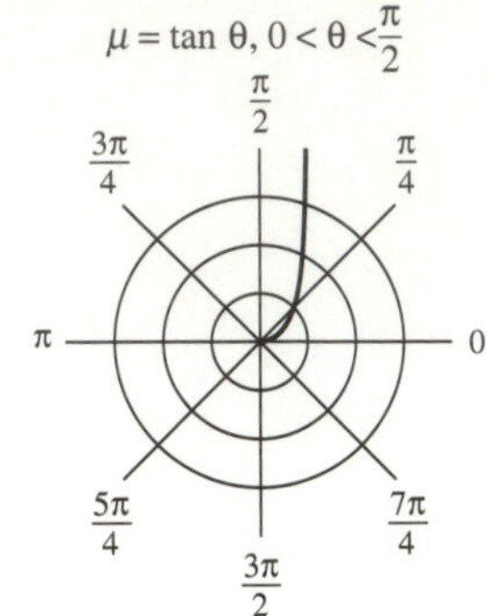

29. $r = 9.3\cos 2\phi, 0 < \phi < \frac{\pi}{2}$

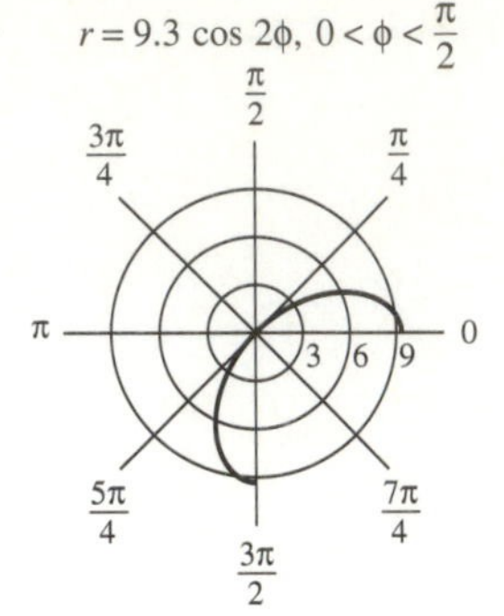

Section 9.9 (page 390)

Sequential Review: 1. $\sqrt{34}$ **3.** $y = (1/2)x - 1/2$ **5.** 1 **7.** $y = (1/16)x^2$ **9.** $x = (1/4)y^2 - y$ **11.** $x = (-1/8)y^2 + (1/4)y + 7/8$ **13.** $(-4, -11)$ **15.** $(11, 3)$ **17.**

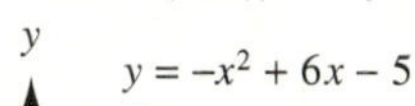

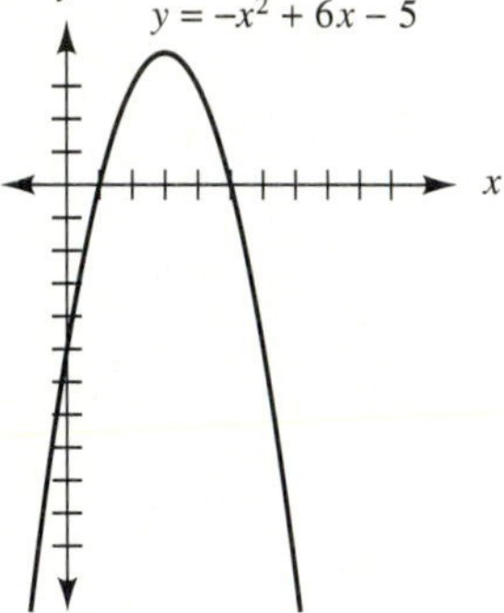

19. $x^2 + y^2 + 12y - 253 = 0$ **21.** $(x + 2)^2 + y^2 = 16$

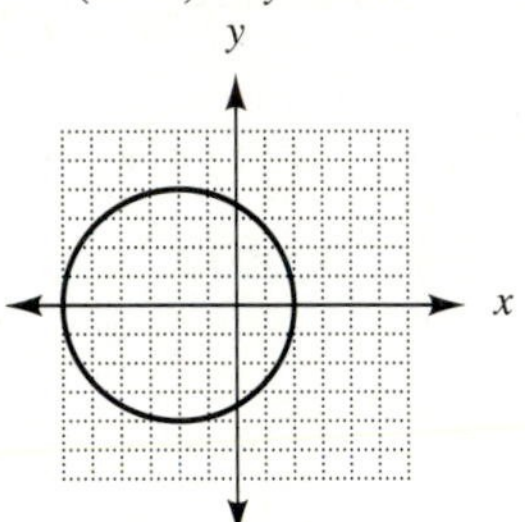

23. no **25.** $\frac{(x-1)^2}{13} + \frac{(y-2)^2}{4} = 1$ **27.** $\frac{(x-2)^2}{16} + \frac{(y-2)^2}{7} = 1$ **29.** $\frac{(x-2)^2}{161} + \frac{(y-9)^2}{225} = 1$

31. $\frac{(x-2)^2}{9} - \frac{(y-2)^2}{16} = 1$ **33.** $\frac{(y-6)^2}{9} - \frac{(x-2)^2}{16} = 1$ **35.** $x^2 + 3y^2 = -6$, no curve

37. $(y+3)^2 - \frac{(x-2)^2}{0.5} = 1$, hyperbola **39.** $(-2.00, 3.46)$ **41.** $(\sqrt{17}, 1.82)$

43. $r = 1 + 2\cos\theta$

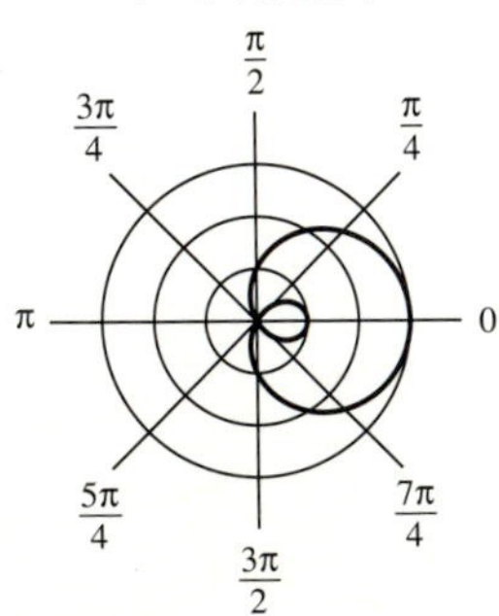

45. $r = 2$

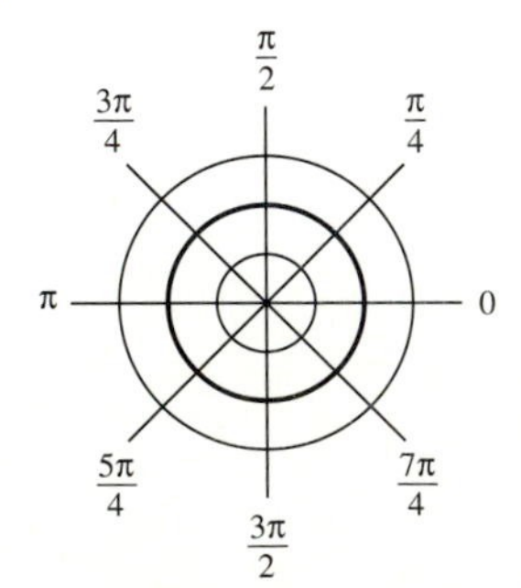

Applied Problems: 1. 4 blocks **3.** 3.01 cm **5.** $(x-7)^2+(y-7)^2=5.5^2$ **7.** $I=-2.53V+43$
9. (a) $It=-150A+138$ **(b)** 0.053 **11.** $y=3V+5$ and $y=4-2V+3V^2$
13. horizontal parabola with vertex at (0, 0), opens to the right, and has focus at $(2.012/(WL^2), 0)$
15. no **17.** $r=1-\cos\theta$ **19.** 150 miles, to two significant digits

Random Review:

1. $x^2+8x+y^2+10y+16=0$

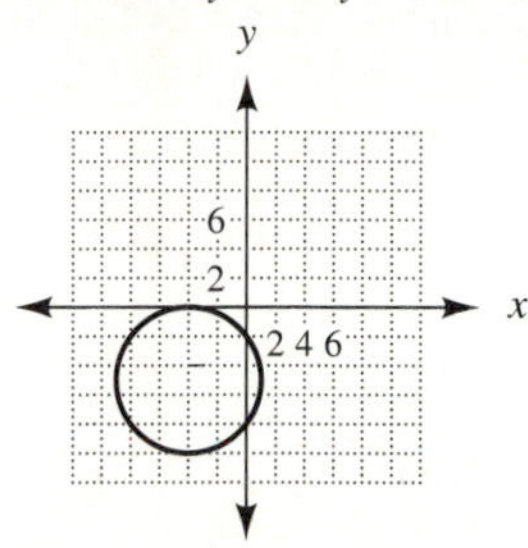

2. $y=x^2-4x-3$

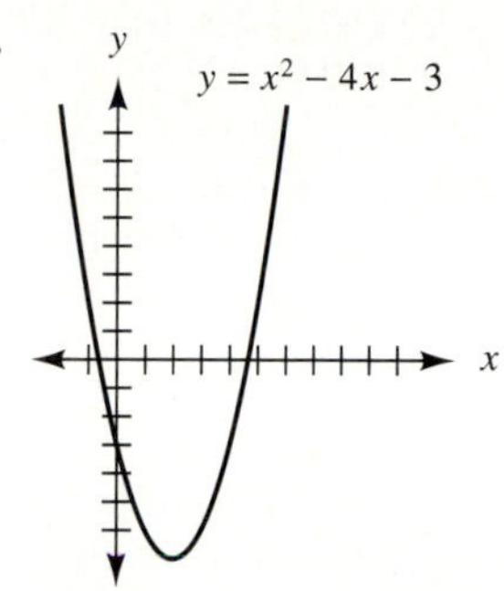

3. $9y^2-36y-4x^2-24x-36=0$

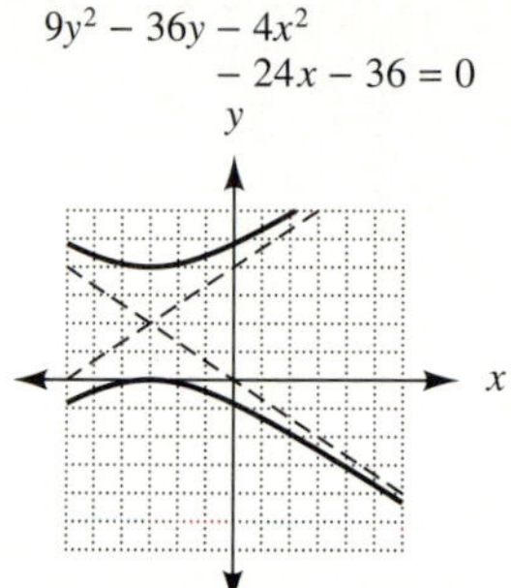

4. $r=3\cos\theta$

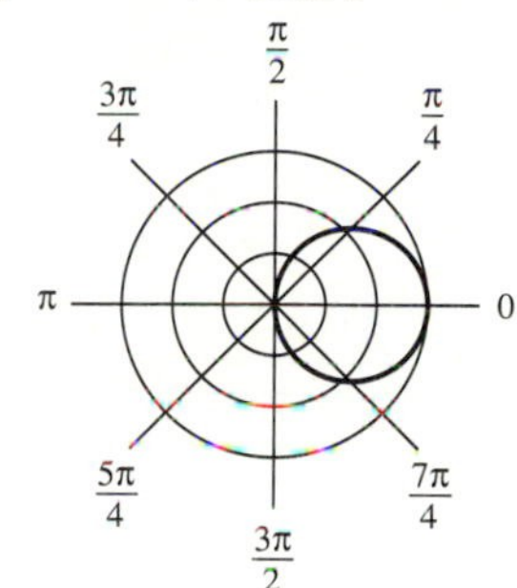

5. $y=3x+6$ **6.** $(-1.50, -2.60)$ **7.** $y=(1/16)x^2-(1/8)x-31/16$
8. $\dfrac{(x+1)^2}{25}+\dfrac{(y-1)^2}{16}=1$ **9.** (17, 5.20) **10.** $6\sqrt{2}$

Chapter Test: 1. $y=(-1/3)(x+4)$ **2.** $(-3, 2)$ **3.** (0, 1)
4. $(x-2)^2+(y+1/2)^2=4$ or $4x^2-16x+4y^2+4y+1=0$

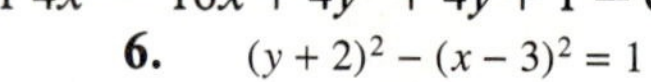

5. $\dfrac{x^2}{4}+\dfrac{(y-1)^2}{9}=1$

6. $(y+2)^2-(x-3)^2=1$

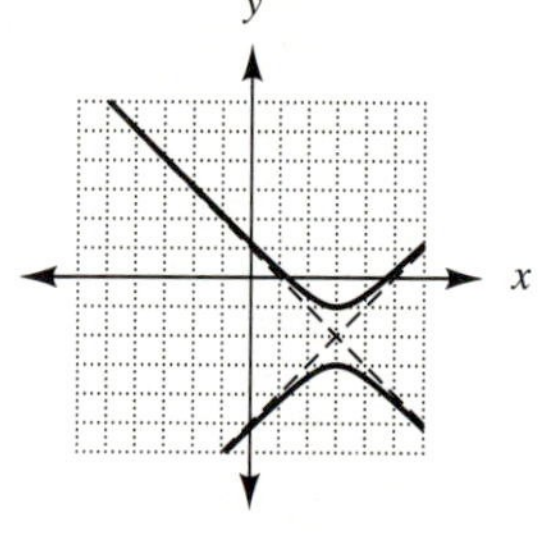

7. $r=\frac{1}{3}\theta$

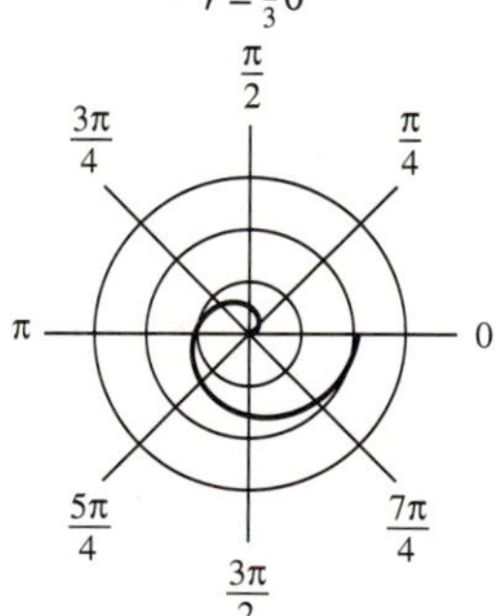

8. (29.0, 0.810) **9.** $19065y=x^2$ **10.** $\dfrac{x^2}{400}+\dfrac{y^2}{175}=1$

CHAPTER 10

Section 10.1 (page 401)

1. $|\mathbf{r}| = 8.60$, $\theta = 54.5°$ **3.** $|\mathbf{r}| = 6.40$, $\theta = 308.7°$ **5.** $|\mathbf{r}| = 4.47$, $\theta = 63.4°$ **7.** $|\mathbf{r}| = 11.4$, $\theta = 307.9°$
9. $|\mathbf{r}| = 8.06$, $\theta = 209.7°$ **11.** $|\mathbf{r}| = 9.22$, $\theta = 229.4°$ **13.** $|\mathbf{r}| = 5$, $\theta = 53.1°$ **15.** $|\mathbf{r}| = 5$, $\theta = 53.1°$
17. $|\mathbf{r}| = 13$, $\theta = 67.4°$ **19.** $|\mathbf{r}| = 13$, $\theta = 67.4°$ **21.** $|\mathbf{r}| = 10$, $\theta = 306.9°$ **23.** $|\mathbf{r}| = 10$, $\theta = 306.9°$
25. $|\mathbf{r}| = 10$, $\theta = 126.9°$ **27.** $|\mathbf{r}| = 13$, $\theta = 247.4°$ **29.** $|\mathbf{r}| = 13$, $\theta = 247.4°$ **31.** $|\mathbf{r}| = 5$, $\theta = 53.1°$
33. $|\mathbf{r}| = 10$, $\theta = 306.9°$ **35.** $|\mathbf{r}| = 17$, $\theta = 61.9°$ **37.** $|\mathbf{r}| = 17$, $\theta = 61.9°$ **39.** $|\mathbf{r}| = 13$, $\theta = 157.4°$
41. $|\mathbf{r}| = 13$, $\theta = 157.4°$ **43.** 25.0 km/h at an angle of 73.7° with the bank
45. $|\mathbf{r}| = 27$ m/s^2, $\theta = 63.2°$ **47.** 29 lbs **49.** 6.8 N directed vertically upward

Section 10.2 (page 409)

1. $\mathbf{v}_x = 26.2$, $\mathbf{v}_y = 18.4$ **3.** $\mathbf{v}_x = -30.1$, $\mathbf{v}_y = 72.4$ **5.** $\mathbf{v}_x = 107.98$, $\mathbf{v}_y = 109.24$
7. $\mathbf{v}_x = 21.59$, $\mathbf{v}_y = -334.7$ **9.** $|\mathbf{r}| = 9.2$, $\theta = 51°$ **11.** $|\mathbf{r}| = 79.1$, $\theta = 358.1°$
13. $|\mathbf{r}| = 35.20$, $\theta = 336.73°$ **15.** $|\mathbf{r}| = 16.54$, $\theta = 301.91°$ **17.** $|\mathbf{r}| = 126$, $\theta = 35.8°$
19. $|\mathbf{r}| = 58.0$, $\theta = 111.3°$ **21.** $|\mathbf{r}| = 25.7$, $\theta = 26.8°$ **23.** $|\mathbf{r}| = 9.8$, $\theta = 84°$
25. $|\mathbf{r}| = 35.2$, $\theta = 83.3°$ **27.** $|\mathbf{r}| = 26.38$, $\theta = 36.33°$ **29.** 5200 lbs **31.** 370 mph N 42° E
33. 73 lbs horizontally, 68 lbs vertically **35.** 1090 lbs at an angle of 130.6° from the horizontal
37. 5.1 at 20° **39.** 61.0 N **41.** 194 N and 72.5 N

Section 10.3 (page 417)

1. $C = 108.8°$, $b = 19.7$, $c = 28.5$ **3.** $C = 112.1°$, $a = 26.9$, $c = 50.4$ **5.** $B = 83.1°$, $a = 2.34$, $c = 2.62$ **7.** $A = 46.1°$, $a = 173$, $c = 208$ **9.** $A = 19.62°$, $a = 22.85$, $b = 25.67$ **11.** 57.2
13. 49.4 **15.** 3.61 **17.** 458 **19.** 68.17 **21.** 1570 m **23.** 34.1 m, 36.9 m **25.** 5.83 m
27. 28.1 km

Section 10.4 (page 423)

1. no **3.** no **5.** no **7.** yes **9.** no **11.** yes **13.** yes
15. First solution: $B = 28.8°$, $C = 124.8°$, $c = 24.2$; Second solution: $B = 151.2°$, $C = 2.4°$, $c = 1.24$
17. $A = 109.8°$, $B = 24.9°$, $a = 50.6$ **19.** no solution **21.** $B = 24.9°$, $C = 23.5°$, $c = 17.4$
23. $A = 120.6°$, $B = 29.7°$, $a = 24.8$ **25.** N 56.3° W **27.** 74.9 ft **29.** 128 ft, 258 ft, 123 ft, and 137 ft

Section 10.5 (page 431)

1. sines **3.** cosines **5.** sines **7.** sines **9.** sines **11.** cosines **13.** sines
15. $A = 66.4°$, $B = 93.4°$, $c = 5.05$ **17.** $A = 53.2°$, $B = 40.4°$, $C = 86.4°$
19. $B = 44°$, $C = 95°$, $a = 3.0$ **21.** $A = 35.3°$, $B = 54.7°$, $C = 90°$
23. $A = 13.8°$, $C = 39.9°$, $b = 11.0$ **25.** 26.8 cm **27.** 108.08° **29.** 101.0° **31.** 208 V

Section 10.6 (page 434)

Sequential Review: 1. $|\mathbf{r}| = 10.3$, $\theta = 119.1°$ **3.** $|\mathbf{r}| = \sqrt{65}$, $\theta = 119.7°$ **5.** $|\mathbf{r}| = 13$, $\theta = 112.6°$
7. $|\mathbf{r}| = 10$, $\theta = 53.1°$ **9.** $|\mathbf{r}| = 5\sqrt{2}$, $\theta = 351.9°$ **11.** $\mathbf{v}_x = -2.16$, $\mathbf{v}_y = 6.48$
13. $|\mathbf{r}| = 121$, $\theta = 19.8°$ **15.** $|\mathbf{r}| = 139$, $\theta = 31.7°$ **17.** $|\mathbf{r}| = 91.3$, $\theta = 189.7°$ **19.** 4.32 **21.** 1.65
23. 16.6 **25.** 32.0 **27.** no **29.** no **31.** 32.8° or 147.2° **33.** 123.0° **35.** law of sines
37. law of cosines **39.** law of sines **41.** law of sines **43.** 41.4 **45.** $A = 3.3°$, $B = 172.5°$, $c = 13.4$

Applied Problems: 1. 98.7 N **3.** 27.1° **5.** 120 V at 0.0° **7.** 8100 ft **9.** 49.5 at an angle of −45° **11.** about 45 lb **13.** about 580 miles **15.** 26 ft **17. (a)** 8.7 min **(b)** 1.5 miles **19.** $|\mathbf{v}| = 24$ mph, $\theta = 24°$

Random Review: 1. $|\mathbf{r}| = 8.06$, $\theta = 299.7°$ **2.** 100° **3.** $|\mathbf{r}| = 43.2$, $\theta = 195.6°$ **4.** $A = 20°$, $C = 127°$, $c = 11$ **5.** 23.7 **6.** no solution **7.** 67° **8.** $\mathbf{v_x} = -13.6$, $\mathbf{v_y} = -83.0$ **9.** $|\mathbf{r}| = 17$, $\theta = 28.1°$ **10.** 127

Chapter Test: 1. $|\mathbf{r}| = 10$, $\theta = 306.9°$ **2.** $|\mathbf{r}| = 17$, $\theta = 61.9°$ **3.** $|\mathbf{r}| = 3.97$, $\theta = 103.9°$ **4.** 63° **5.** 14.1 **6.** 35.7 **7.** 6° **8.** First solution: $A = 44.3°$, $C = 103.6°$, $c = 108.5$; Second solution: $A = 135.7°$, $C = 12.2°$, $c = 23.6$ **9.** $A = B = 71°$, $C = 38°$ **10.** 208 volts

CHAPTER 11

Section 11.1 (page 447)

1. 2π, 3 **3.** 2π, 2/3 **5.** π, 2/3 **7.** 6π, 2 **9.** $\pi/2$, 3/2 **11.** $2\pi/3$, 1 **13.** 4π, 1 **15.** 3π, 1/2 **17.** 1, 1 **19.** 4, 2 **21.** 2, 1/2

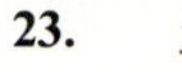

23.

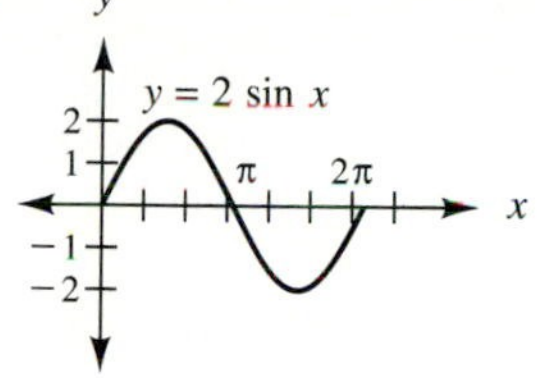

25.

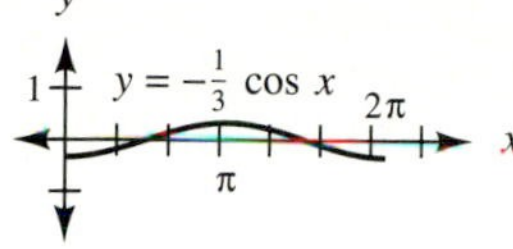

27.

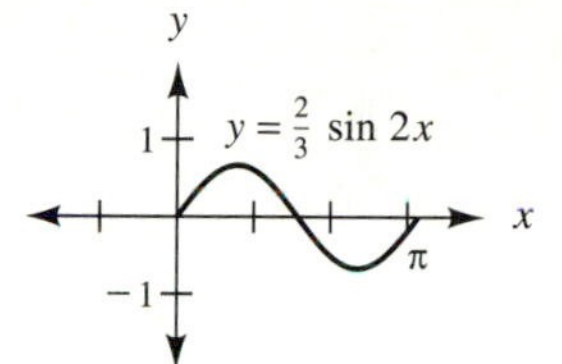

29.

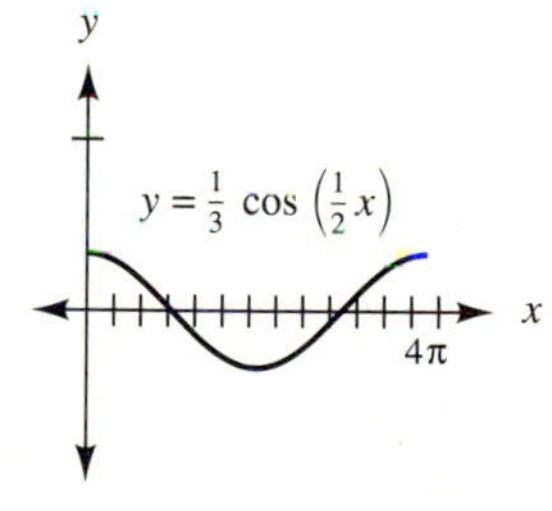

31.

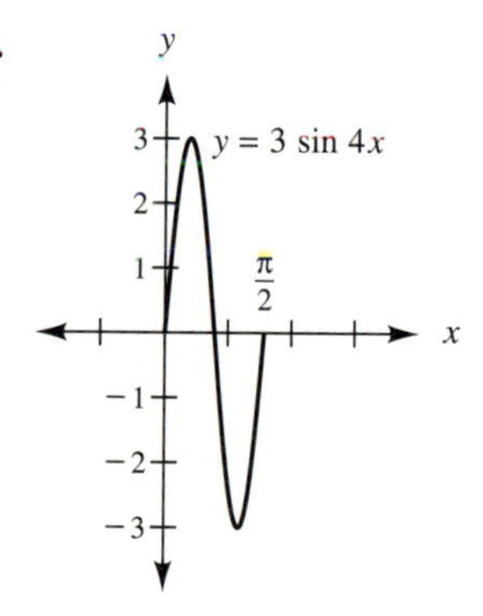

33.

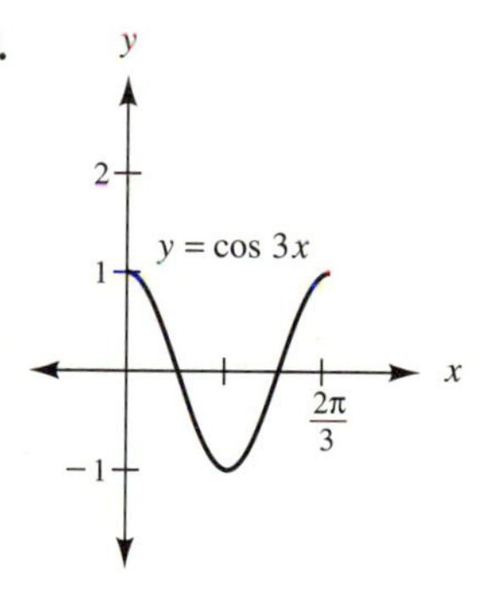

35.

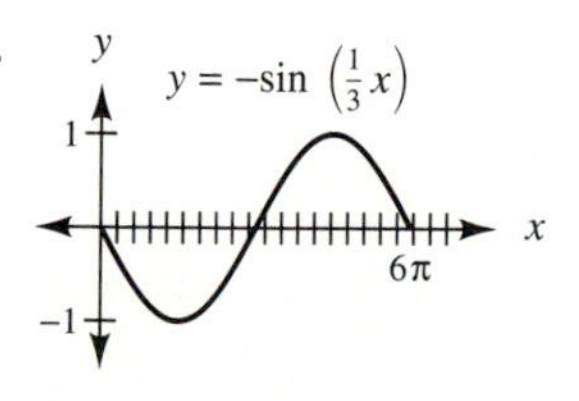

37.

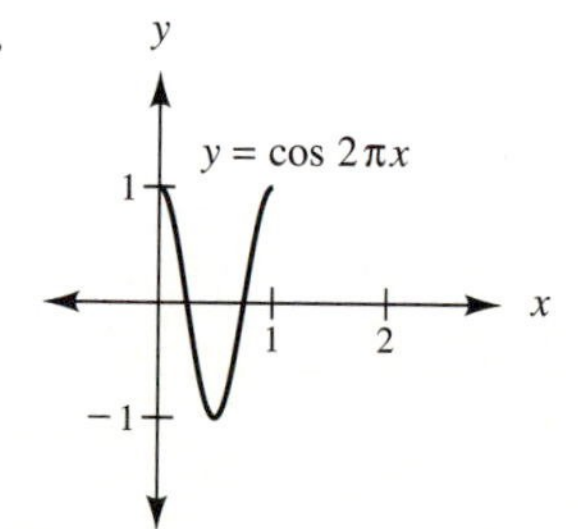

39.

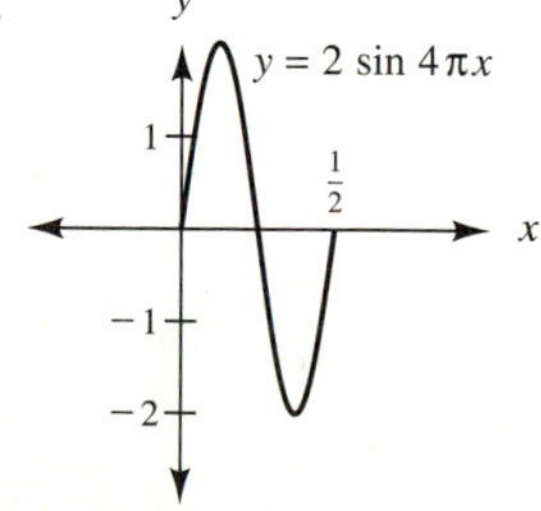

41.

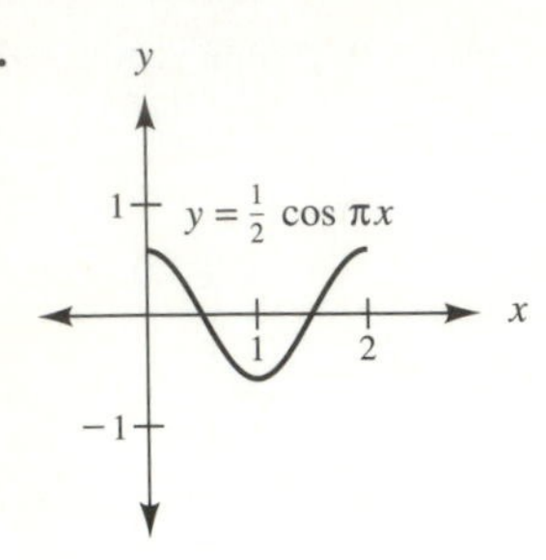

43. **(a)** π, 31 **(b)** maximum = 6108 ft, minimum = 6046 ft
45. **(a)** 365 days, 26 **(b)** maximum = 92°, minimum = 40°
47. $y = 3\sin(\pi x/6.2)$ **49.** 1/20, 6

Section 11.2 (page 452)

1. π **3.** $-\pi/4$ **5.** $\pi/4$ **7.** $-\pi/2$ **9.** $\pi/2$ **11.** $-\pi/3$ **13.** $\pi/2$ **15.** $3\pi, 1/2, -\pi/2$
17. 1, 1, 1/2 **19.** 4, 2, −1/2 **21.** 2, 1/2, 1 **23.** $2\pi, 2, -\pi/3$ **25.** $2\pi, 1/3, \pi/4$
27. $\pi, 2/3, -\pi/12$

29.

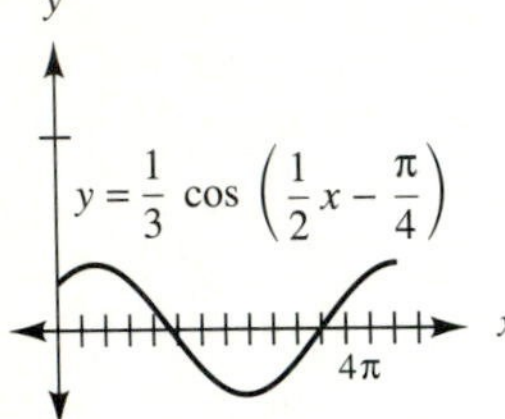

31.

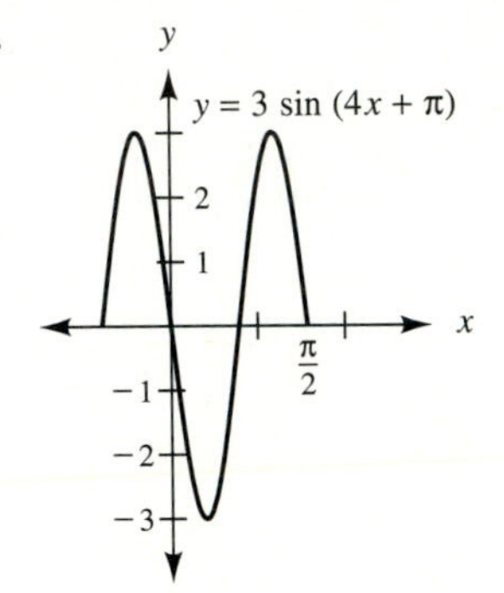

33.

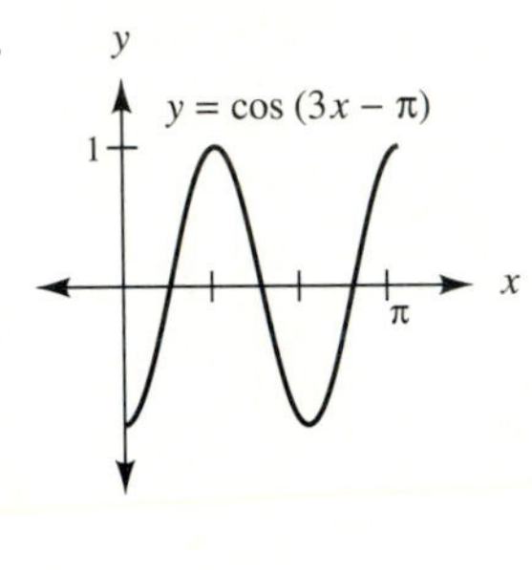

35.

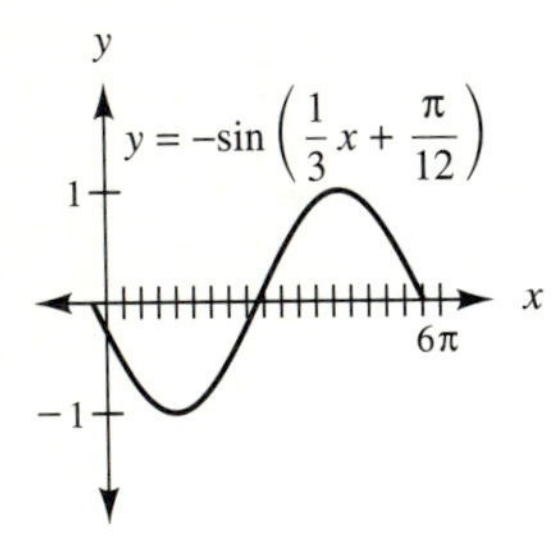

37.

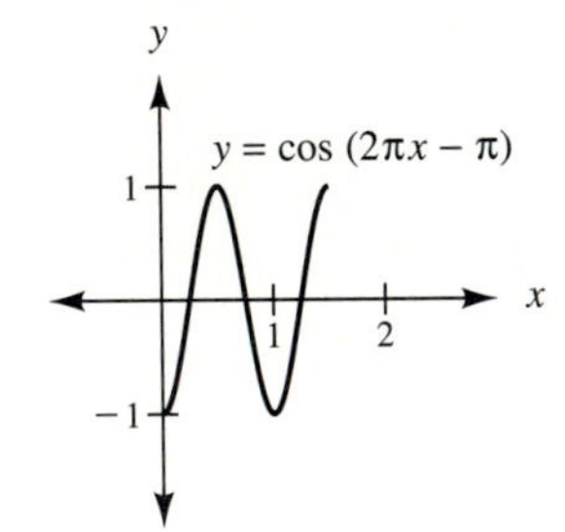

39. $y = 2\sin(4\pi x + \pi)$

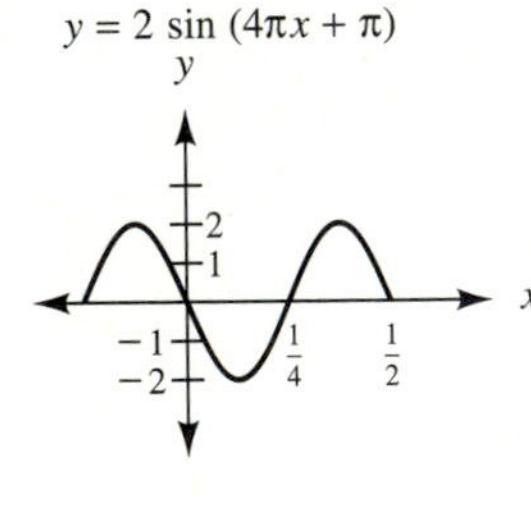

41.

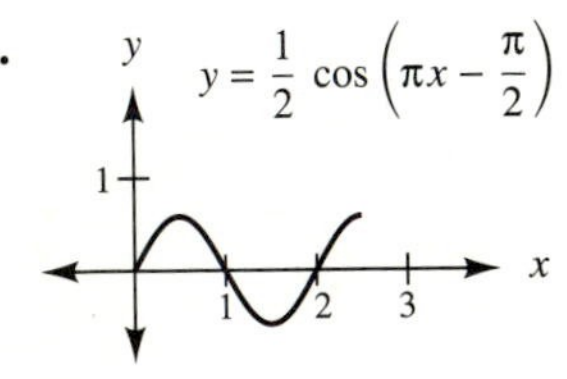

43. 1/20, 6, 1/120 **45.** $y = 26\sin(2\pi x/365 + 160\pi/365)$
47. $y = 3\sin(\pi x/6.2 - 4.9\pi/6.2)$ **49.** Compute age in days, divide by the period, and use the remainder as the phase shift in the negative direction

Section 11.3 (page 459)

1.

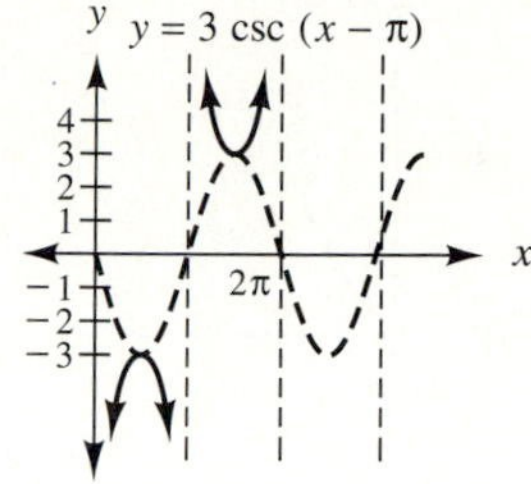

3.

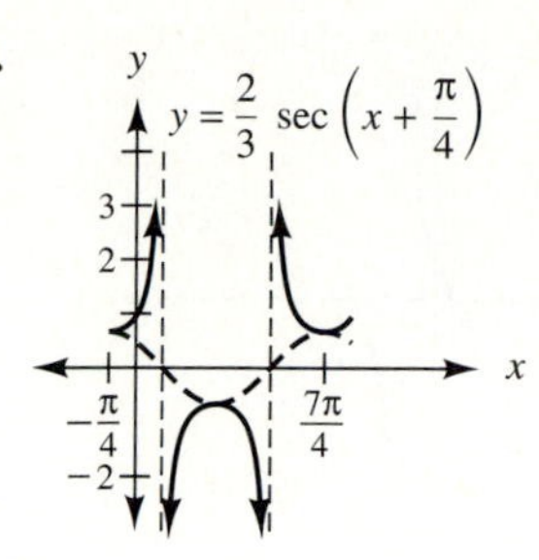

5.

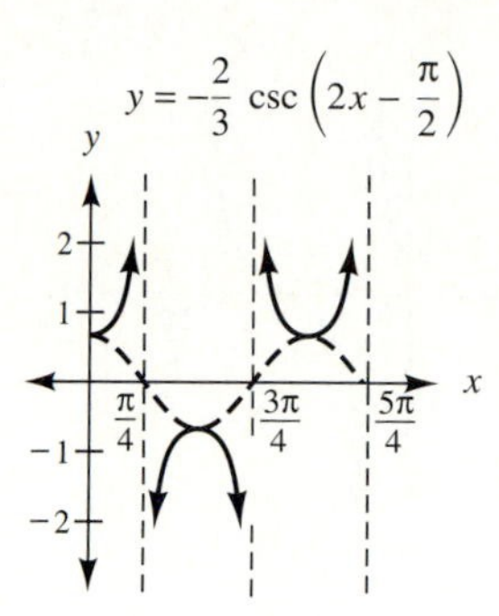

7.

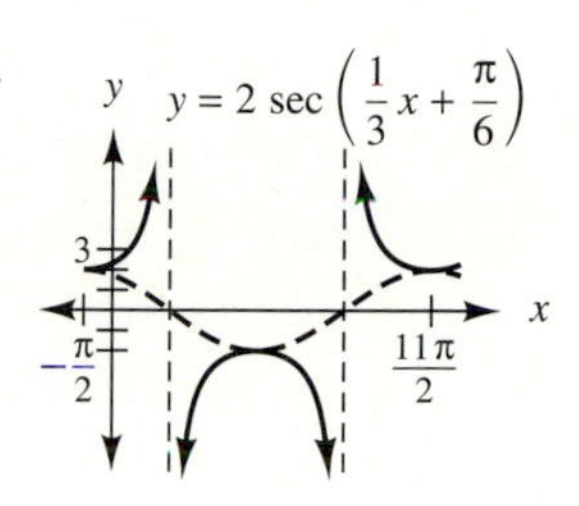

9.

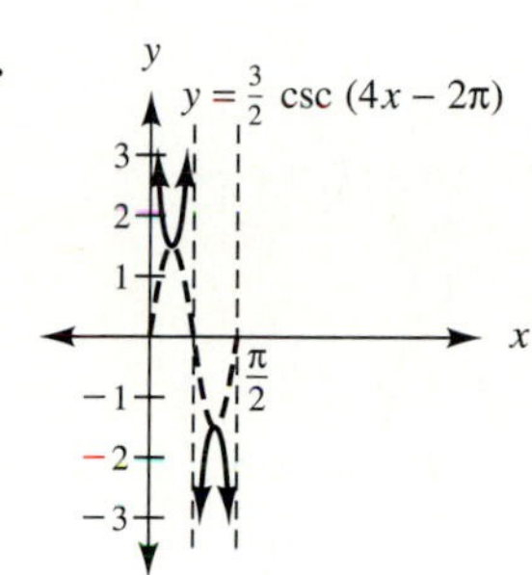

11.

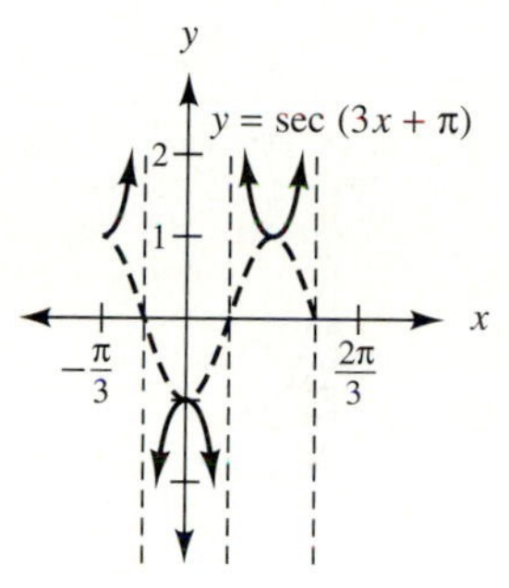

13.

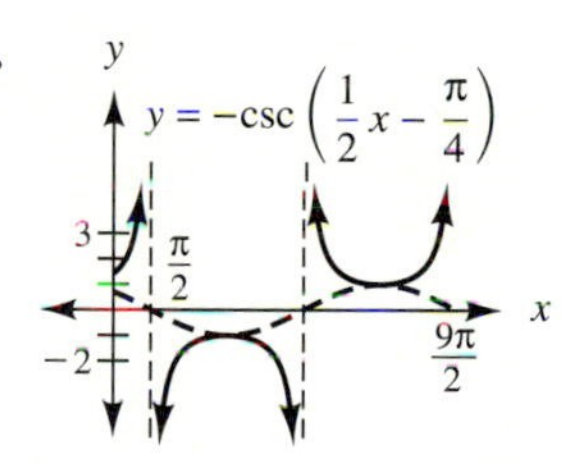

15.

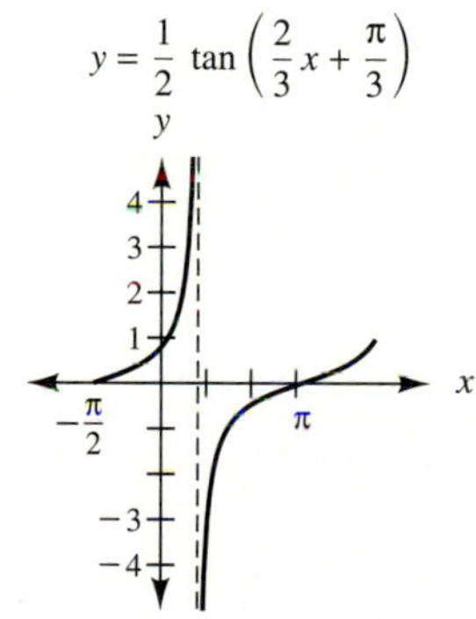

17.

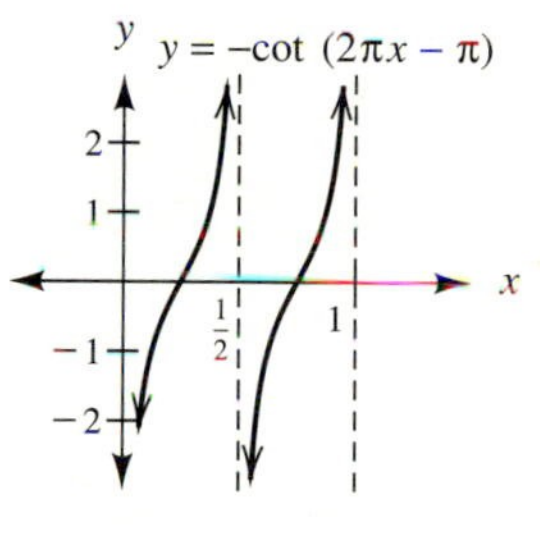

19.

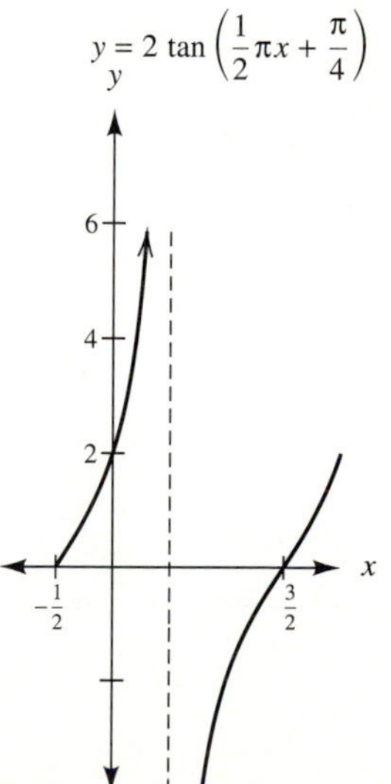

21.

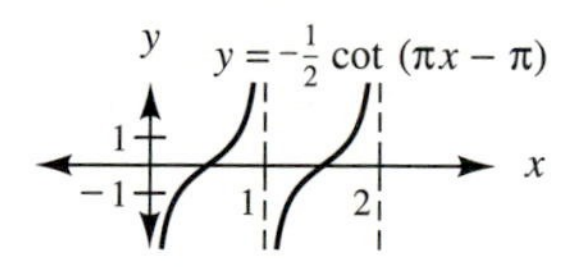

23.

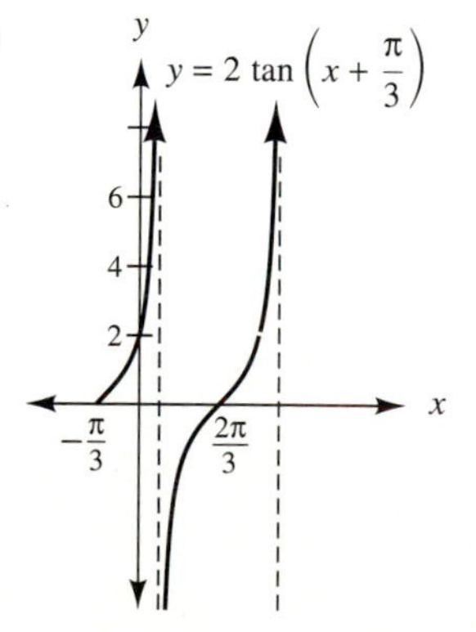

25. $y = -\frac{1}{3}\cot\left(x - \frac{\pi}{4}\right)$

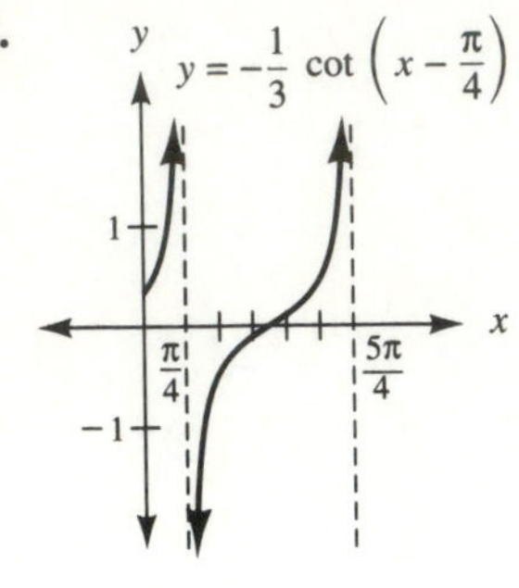

27. $y = \frac{2}{3}\tan\left(2x + \frac{\pi}{6}\right)$

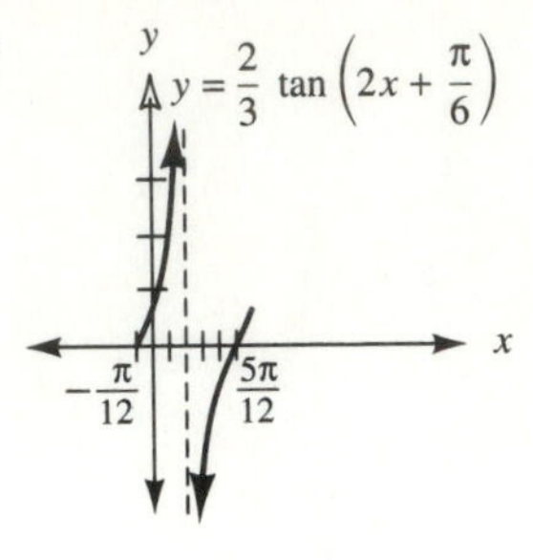

29. $h = 1000 \tan x,\ 0 < x < \frac{\pi}{3}$

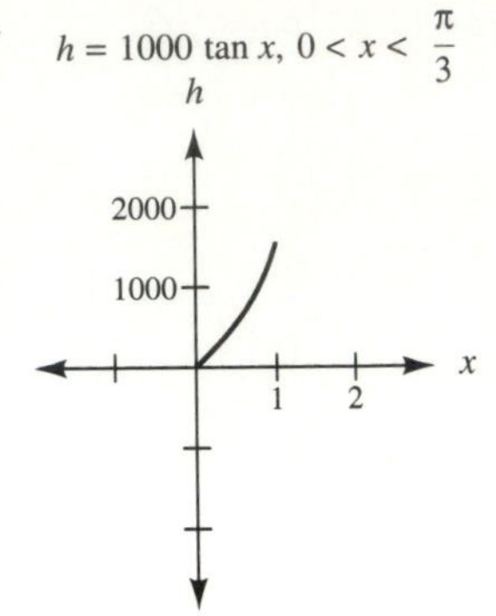

31. $v = (2.12 \times 10^8) \csc x$
$0 < x < \frac{\pi}{2}$

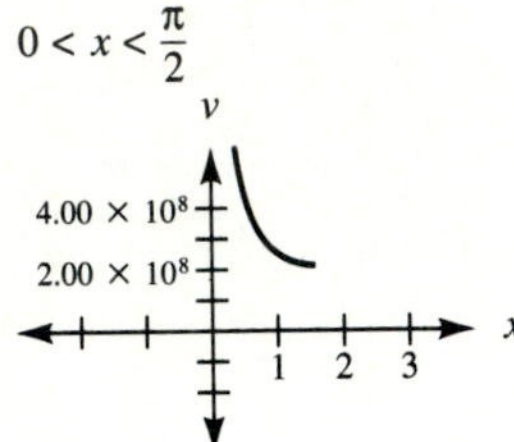

33. $N = 10 \cot x,\ 0 < x < \frac{\pi}{2}$

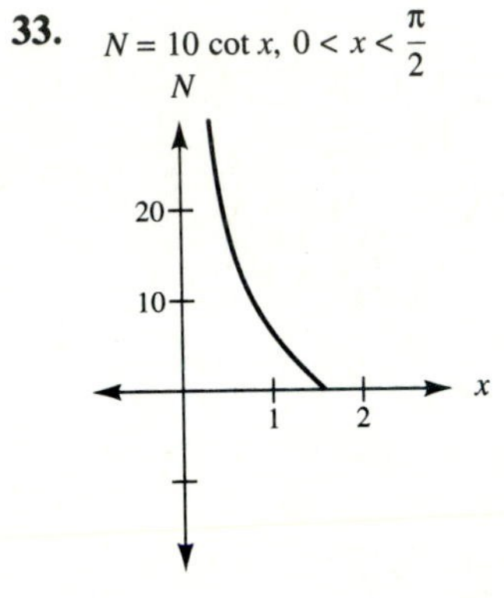

Section 11.4 (page 465)

1. $y = 1 + \sin 2x$

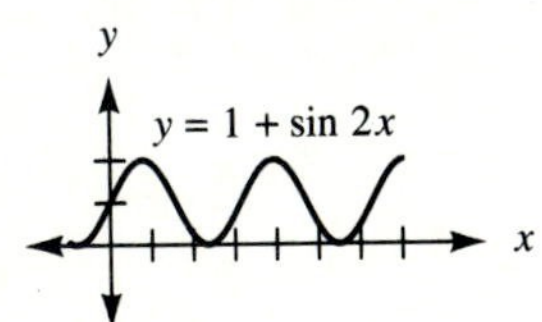

3. $y = 3 - 2\cos x$

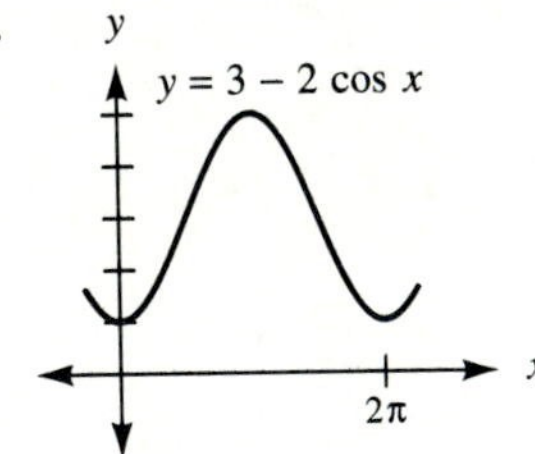

5. $y = x + 2\sin x$

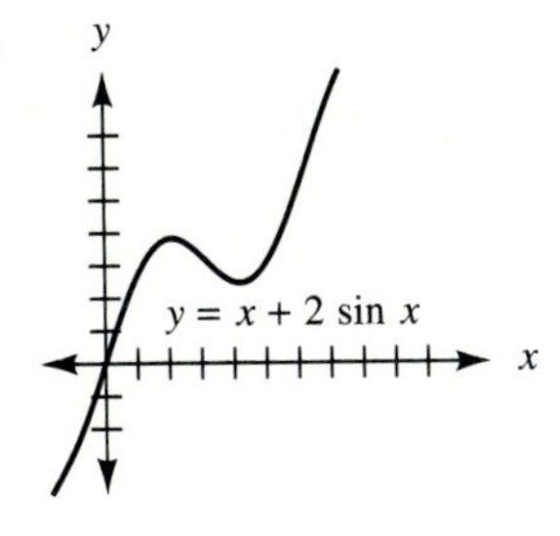

7. $y = \frac{1}{2}x - \cos x$

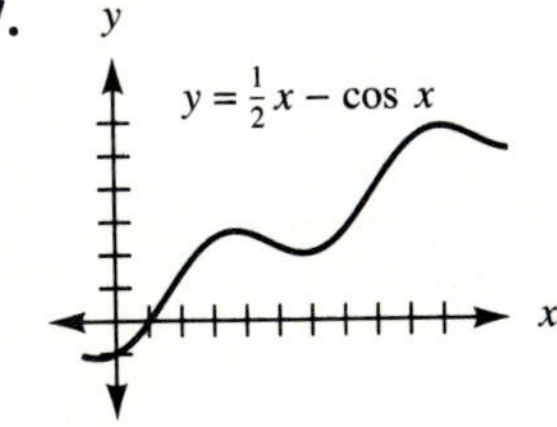

9. $y = \frac{1}{3}x + 3\sin 2x$

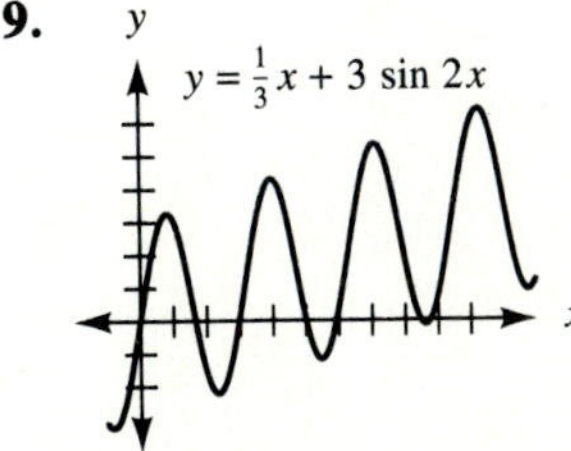

11. $y = \sin x + 2\cos x$

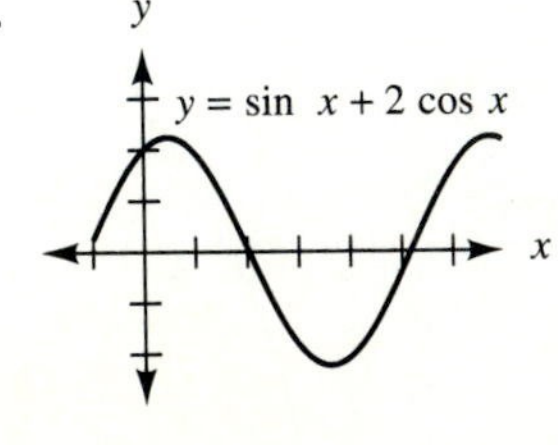

13.

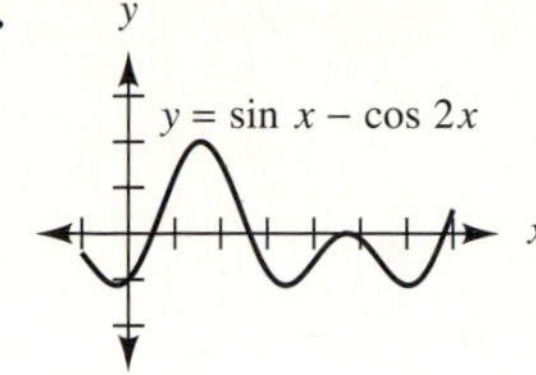

15.

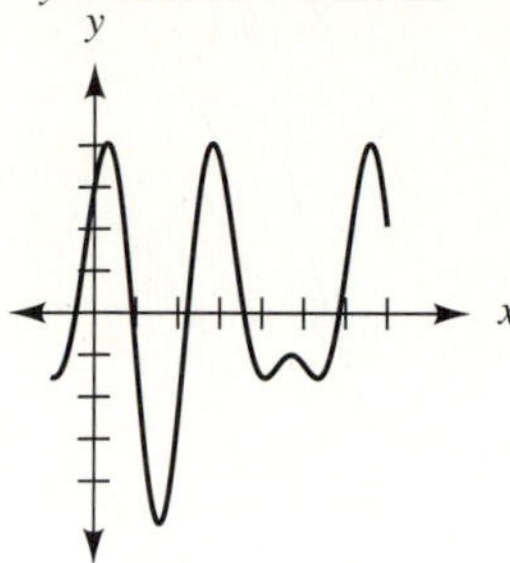

17.

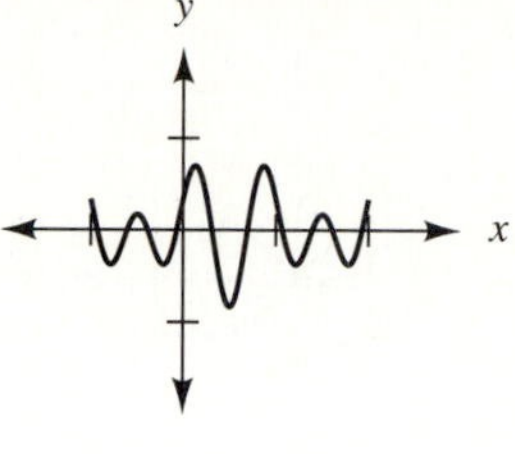

19.

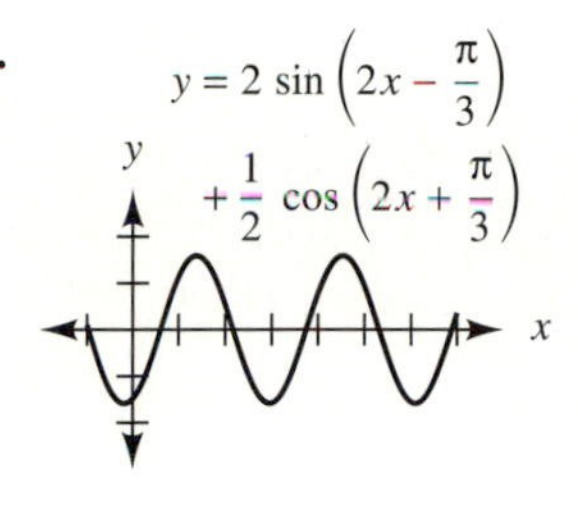

21.

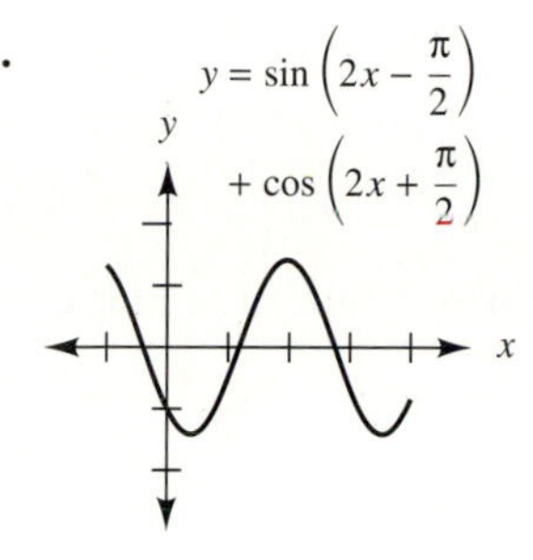

23.

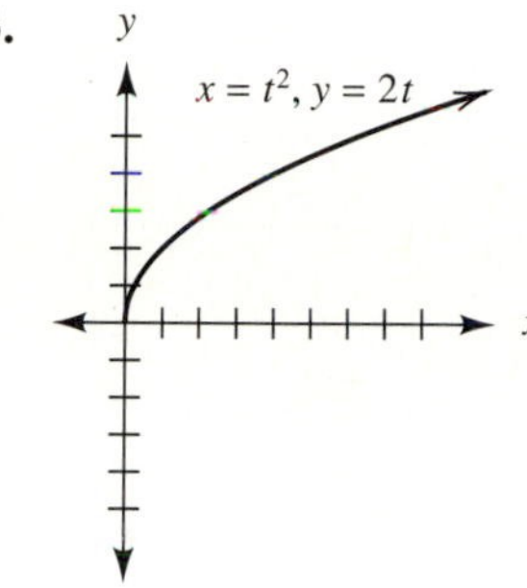

25.

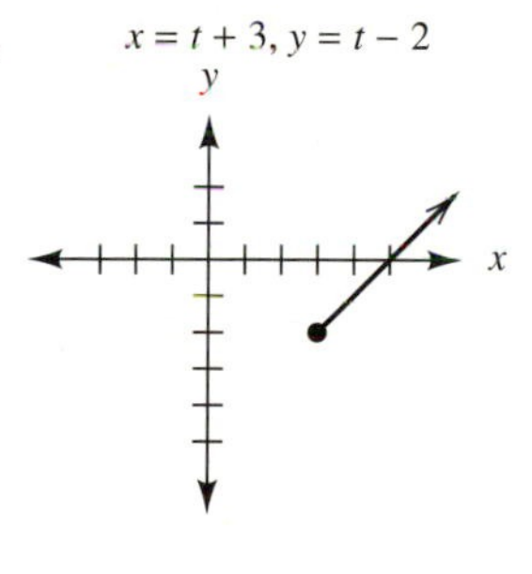

27.

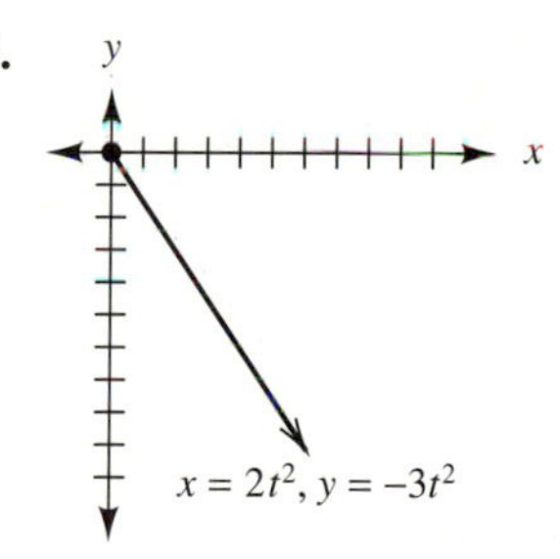

29.

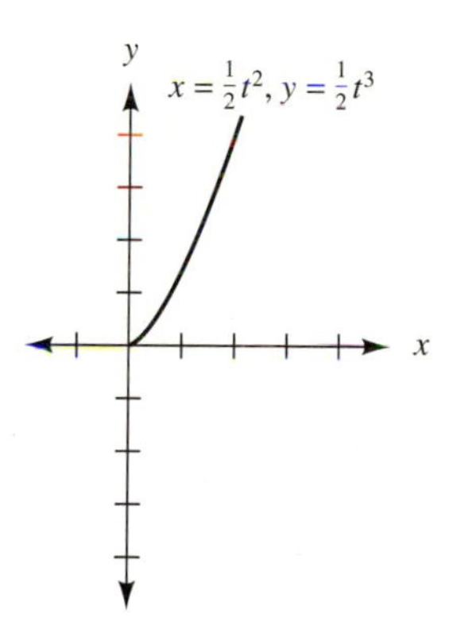

31.

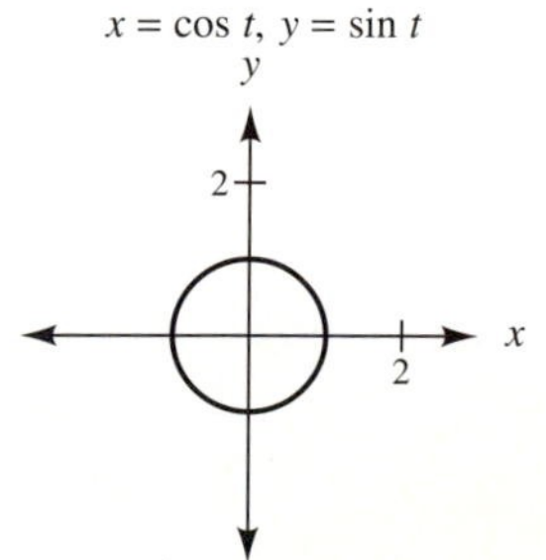

33.

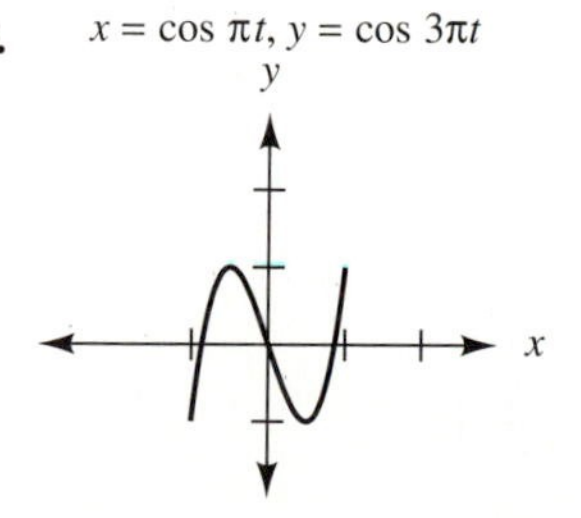

35.

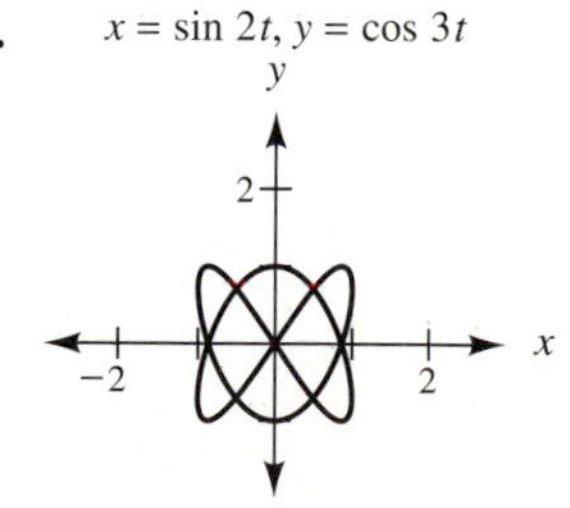

37. $x = 2\cos \pi t,\ y = \sin 2\pi t$

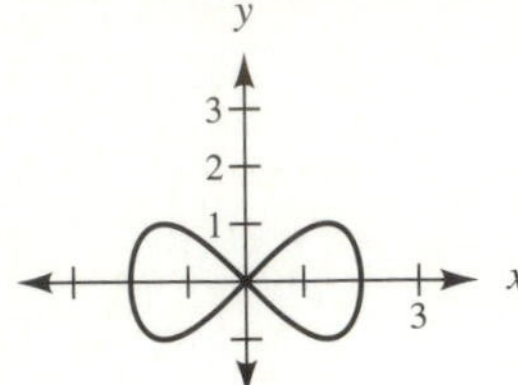

39. $F = 50 \tan\left(x - \frac{\pi}{6}\right) + 50 \tan x$

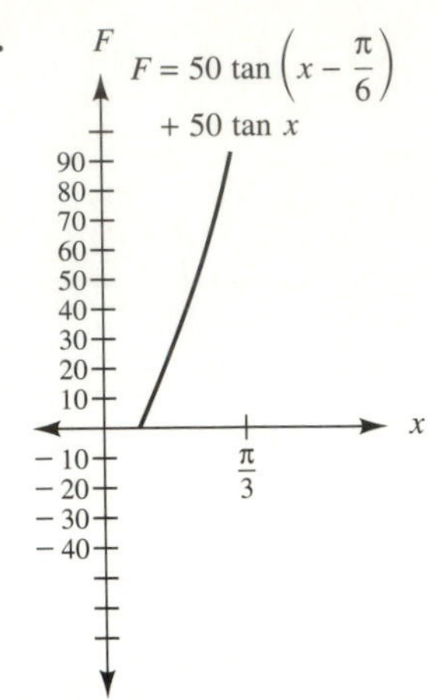

41. $y = -0.709 \sin(49.5t) + 2.23 \sin(15.7t)$
$0 < t < 1$

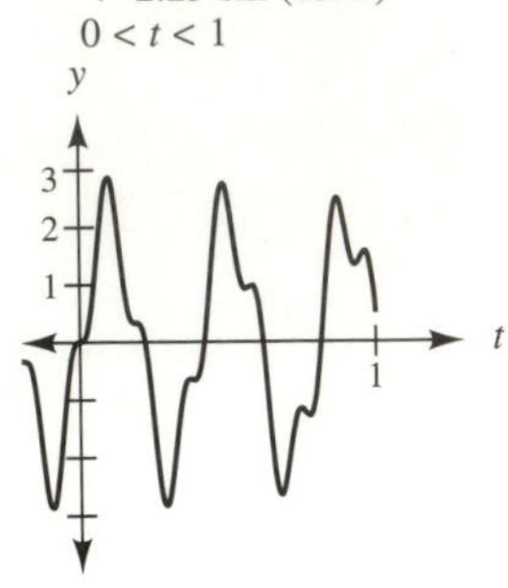

43. $x = 2t,\ y = 2t^2,\ 0 \le t \le 3$

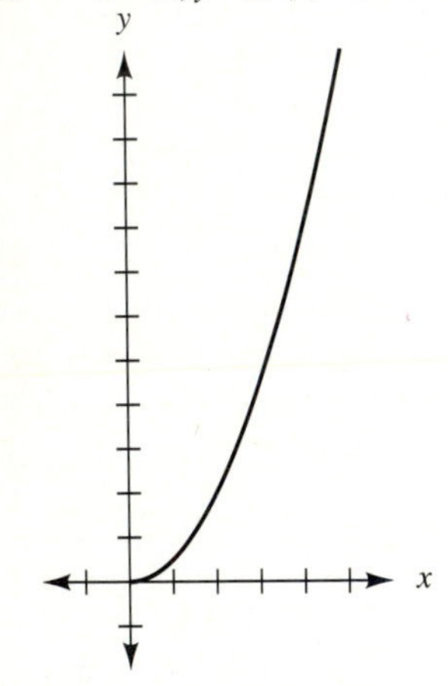

45. Let n = number of loops, then $v_2 = nv_1$

Section 11.5 (page 472)

1. $V = 120 \sin\left(60\pi t + \frac{\pi}{2}\right)$

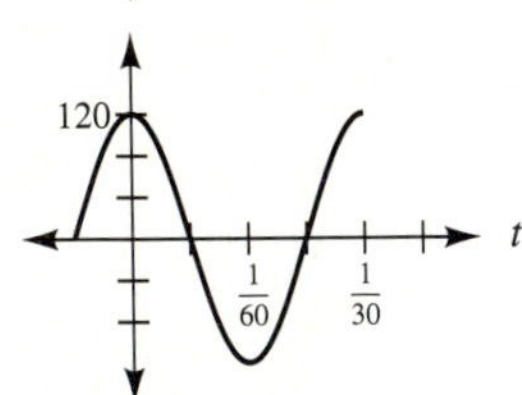

3. $V = 60 \sin\left(60\pi t + \frac{\pi}{3}\right)$

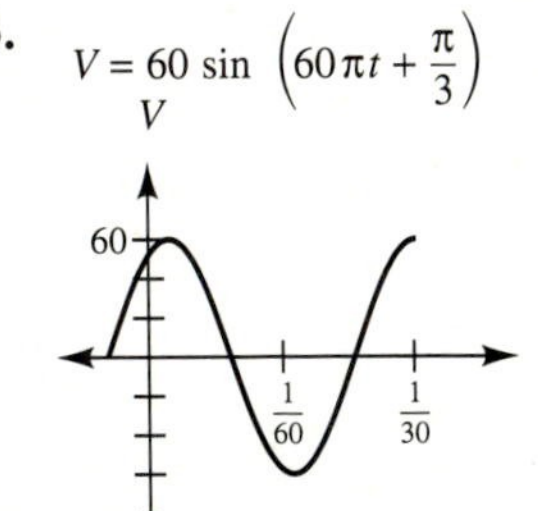

5. $V = 240 \sin\left(40\pi t + \frac{\pi}{4}\right)$

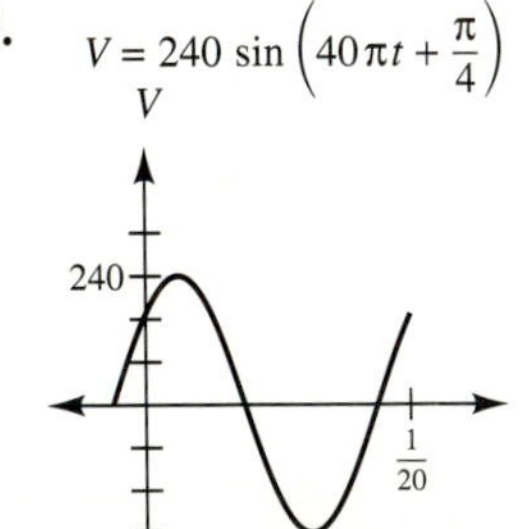

7.

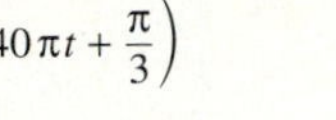

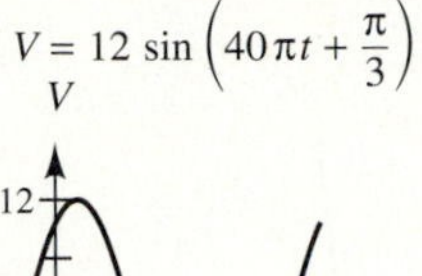

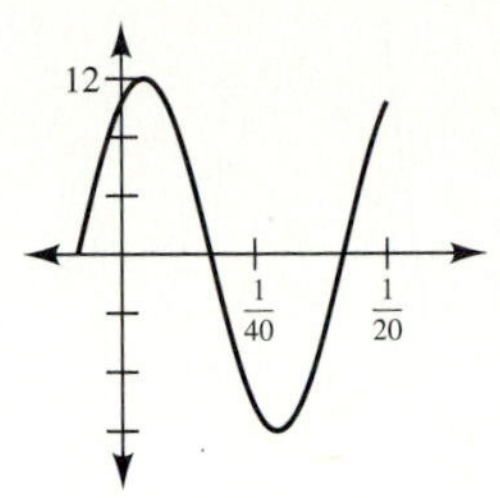

9.

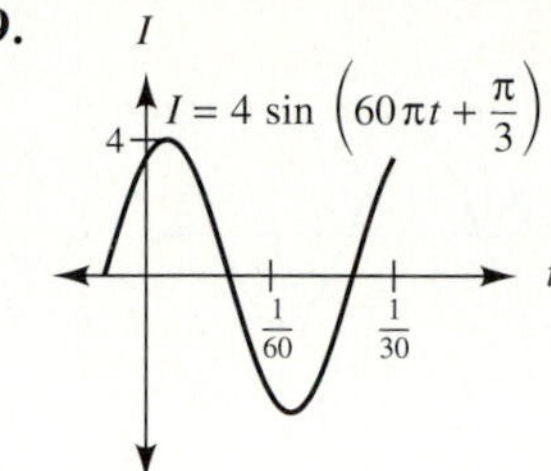

11.

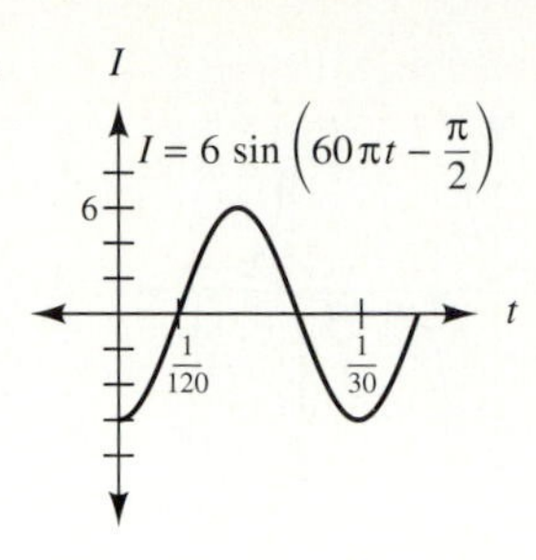

13.

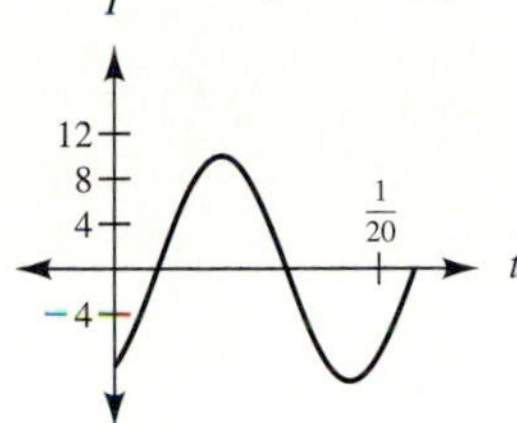

15. 1.40, 0.32, −0.35 **17.** 0.44, 0.18, −0.11 **19.** 0.44, 0.08, −0.11 **21.** 1.4, 0.20, −0.35 **23.** 2.0, 0.35, −0.50

Section 11.6 (page 480)

1. (a) $\pi/6 \pm 2\pi n$, $5\pi/6 \pm 2\pi n$ **(b)** $\pi/6$ **3. (a)** $\pi/4 \pm \pi n$ **(b)** $\pi/4$ **5. (a)** $\pi/3 \pm 2\pi n$, $5\pi/3 \pm 2\pi n$ **(b)** $\pi/3$ **7. (a)** $3\pi/2 \pm 2\pi n$ **(b)** $-\pi/2$ **9. (a)** $2\pi/3 \pm \pi n$ **(b)** $-\pi/3$ **11. (a)** $0 \pm 2\pi n$ **(b)** 0 **13. (a)** $\pi/3 \pm 2\pi n$, $2\pi/3 \pm 2\pi n$ **(b)** $\pi/3$ **15. (a)** 2.550 **(b)** 0.5916 **17. (a)** 3.9700 **(b)** 0.82839 **19. (a)** 5.0000 **(b)** 1.2832 **21. (a)** 5.230 **(b)** −1.053 **23. (a)** 2.360 **(b)** −0.7816 **25. (a)** 3.8800 **(b)** 2.4032 **27. (a)** 3.220 **(b)** −0.07838 **29.** $x\sqrt{x^2+1}/(x^2+1)$ **31.** $\sqrt{1-x^2}/x$ **33.** $\sqrt{2}/2$ **35.** $1/x$ **37.** $1/x$ **39.** $\sqrt{3}/2$ **41.** x **43.** −1°, to the nearest degree **45.** 48.6° **47.** 85°

Section 11.7 (page 482)

Sequential Review: 1. 2, 4π **3.** 2/3, π **5.** 2, 1

7.

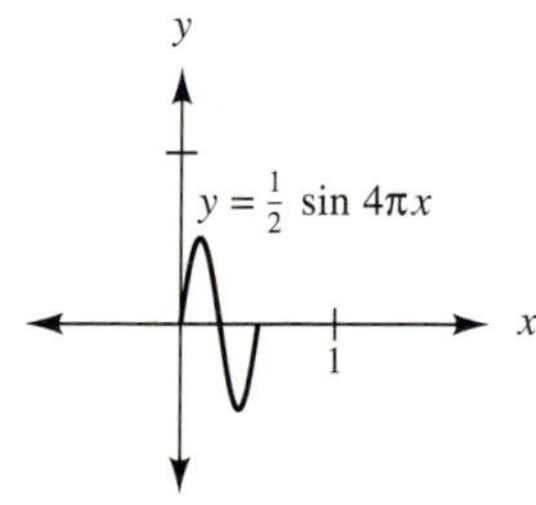

9. 2π, 1/3, π **11.** 2π, 1, $\pi/3$ **13.** 2, 2, 1/2

15. $y = 3\cos\left(\frac{1}{2}x - \frac{\pi}{3}\right)$ ($\frac{2\pi}{3}$, 4π)

17. 2π, $\pi/3$ **19.** 4π, $-2\pi/3$

21. $y = 2\cot\left(\frac{1}{2}x + \frac{\pi}{6}\right)$ ($-\frac{\pi}{3}$, $\frac{5\pi}{3}$, $\frac{11\pi}{3}$)

23.

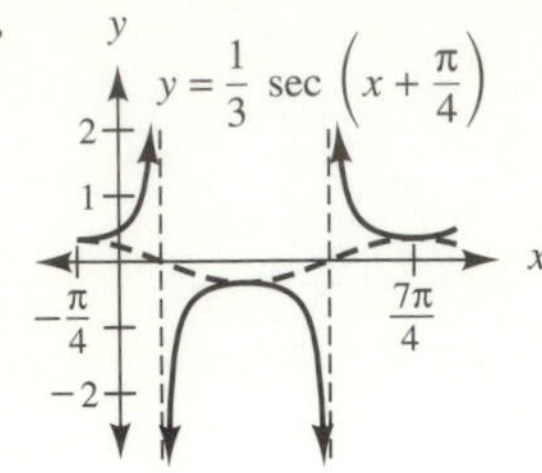

25.

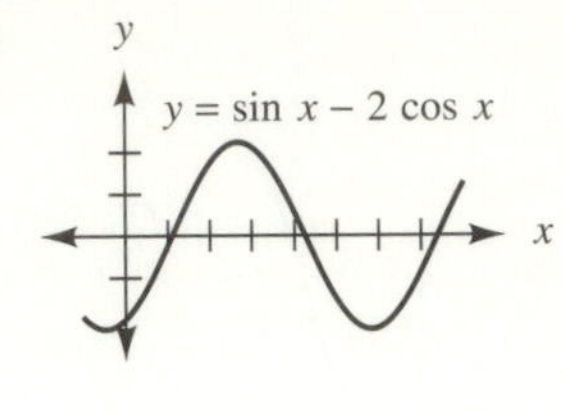

27.

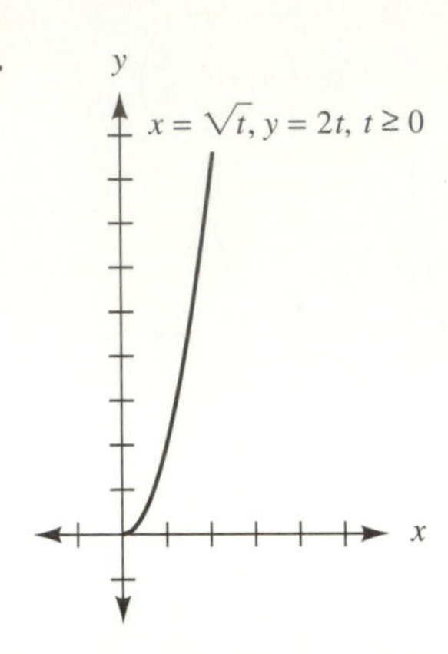

29.

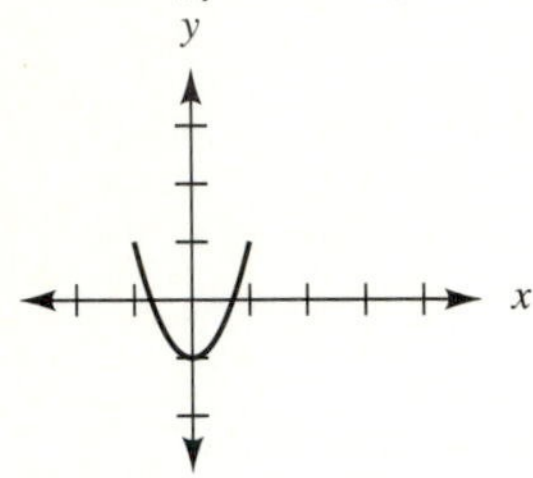

31. 1/60, 30, −1/120

33.

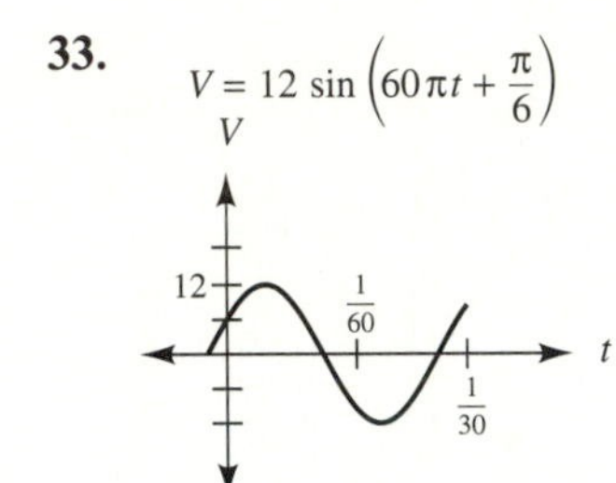

35.

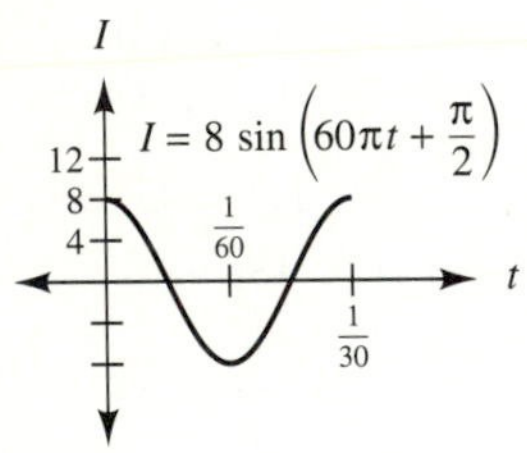

37. (a) $2\pi/3 \pm 2\pi n$, $4\pi/3 \pm 2\pi n$ **(b)** $2\pi/3$ **39. (a)** 5.094 **(b)** 1.190 **41. (a)** 2.3170 **(b)** 2.3170 **43.** $\sqrt{3}/2$ **45.** $\sqrt{1+x^2}/x$

Applied Problems: 1. $y = 31 \sin (2\pi x/365)$ **3.** $v = 120 \sin (377t + 0.977)$ **5.**

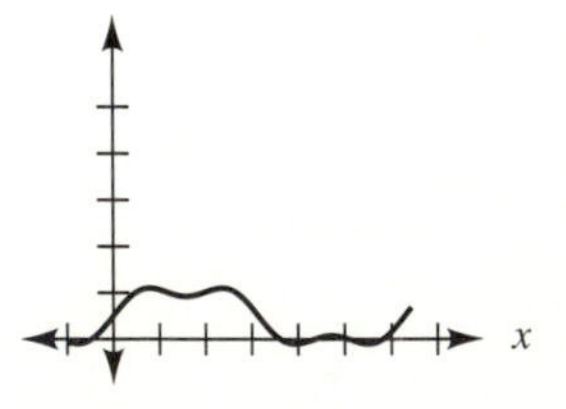

7. (a) 0 **(b)** 200 W

9. (a) $B = 100 \sin (\pi/12\, t - \pi/2)$ or $B = 100 \sin (\pi/12\, t + \pi/2)$ **(b)** 8:00 a.m.

11.

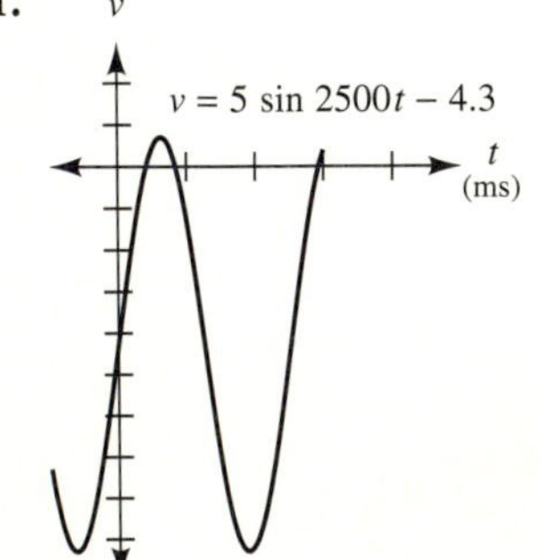

13. after approximately 5 seconds, it will hit the ground about 3000 ft away
15. 9×10^7 Hz to 11×10^7 Hz **17.** 53°

Random Review: 1. $2\pi/3$, $\pi/3$ **2.** $I = 8 \sin\left(120\pi t - \frac{\pi}{4}\right)$ **3.** 2π, 2, $-\pi$ **4.** 3.469 **5.** $\pi/2$, 3

6.

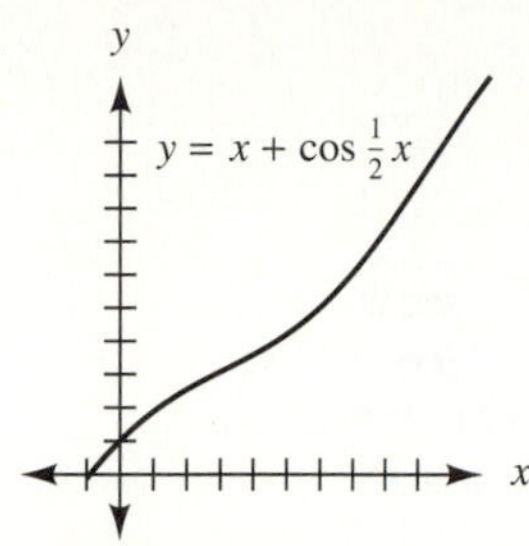

7. -0.3272

8.

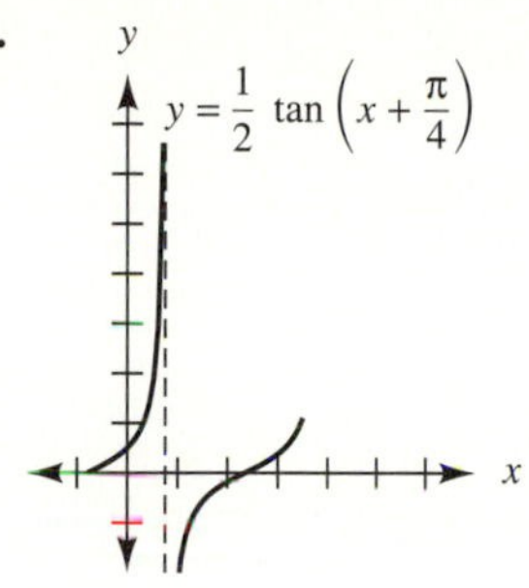

9. $1/x$

10.

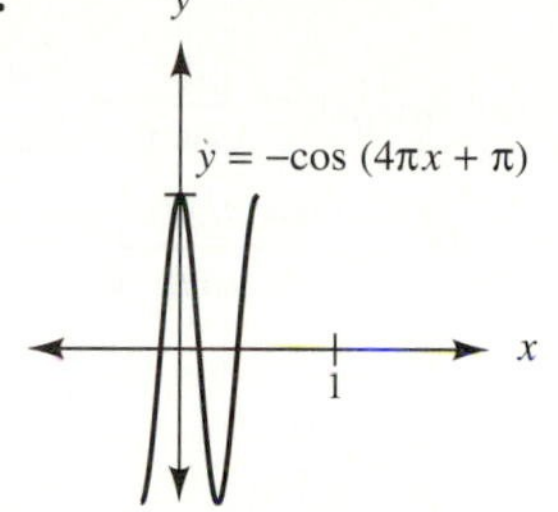

Chapter Test: 1. π, $1/2$, $\pi/6$ **2.** $1/2$, $-1/4$

3.

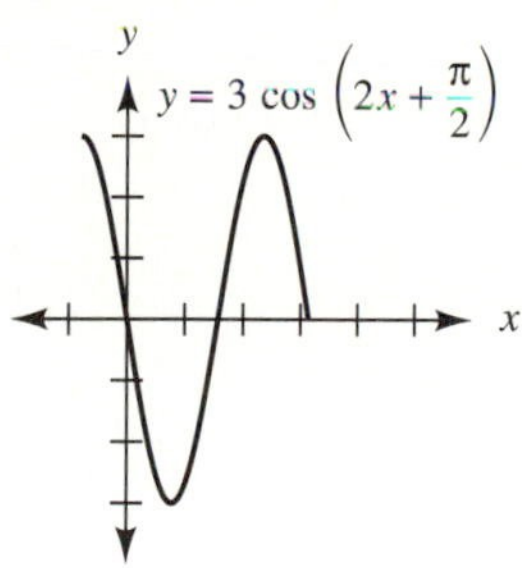

4.

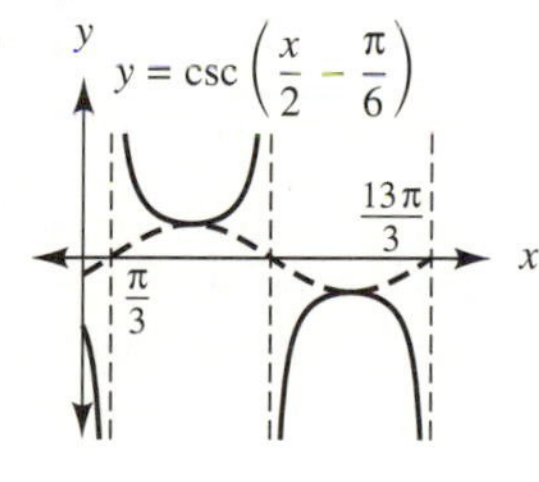

5.

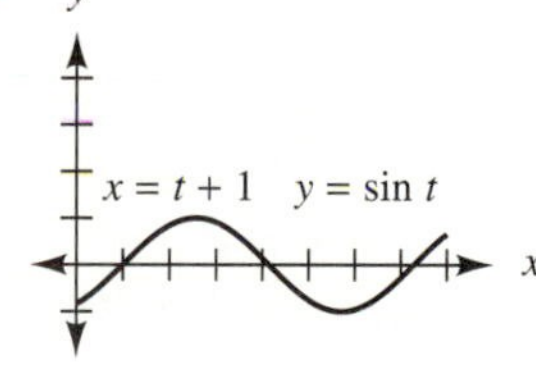

6. 2.151 **7.** -0.9908 **8.** 2 **9.**

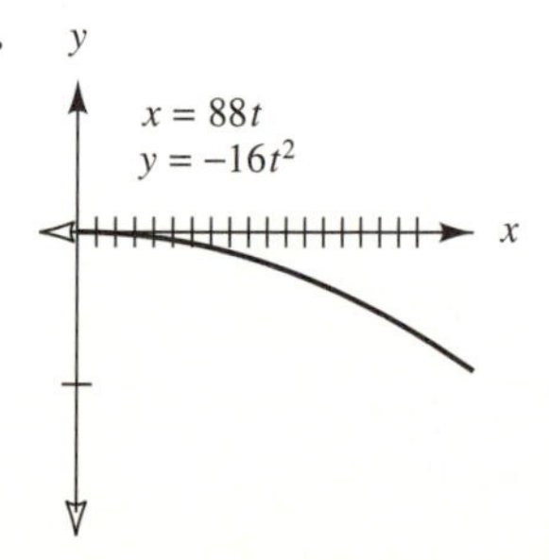

10. $I = 10 \sin(120\pi t + \pi/4)$

CHAPTER 12

Section 12.1 (page 494)

1. $\sin\theta\csc\theta = \sin\theta\,(1/\sin\theta) = 1$ **3.** $\tan x\cot x = \tan x\,(1/\tan x) = 1$
5. $\sin y\cot y = \sin y\,(\cos y/\sin y) = \cos y$
7. $\tan\theta/\sec\theta = (\sin\theta/\cos\theta) \div (1/\cos\theta) = (\sin\theta/\cos\theta)(\cos\theta) = \sin\theta$
9. $\cot x/\csc x = (\cos x/\sin x) \div (1/\sin x) = (\cos x/\sin x)(\sin x) = \cos x$
11. $\sec y/\csc y = (1/\cos y) \div (1/\sin y) = (1/\cos y)(\sin y) = \tan y$
13. $\tan\theta\,(1+\cot^2\theta) = \tan\theta\csc^2\theta = (\sin\theta/\cos\theta)(1/\sin^2\theta) = (1/\cos\theta)(1/\sin\theta) = \sec\theta\csc\theta$
15. $\csc^2 x\,(1-\sin^2 x) = (1/\sin^2 x)(\cos^2 x) = \cot^2 x$ **17.** $(\sec^2 y - \tan^2 y)/\cos y = 1/\cos y = \sec y$
19. $\cot\theta\,(\tan\theta+\cot\theta) = \cot\theta\tan\theta + \cot^2\theta = (1/\tan\theta)(\tan\theta) + \cot^2\theta = 1+\cot^2\theta = \csc^2\theta$
21. $\tan x\,(\cos x + 1) = \tan x\cos x + \tan x = (\sin x/\cos x)(\cos x) + \tan x = \sin x + \tan x$

23. $\tan y + \cot y = \dfrac{\sin y}{\cos y} + \dfrac{\cos y}{\sin y} = \dfrac{\sin^2 y + \cos^2 y}{\cos y\sin y} = \dfrac{1}{\cos y\sin y} = \sec y\csc y$

25. $\sin\theta\cos\theta\,(\sec\theta+\cot\theta) = \sin\theta\cos\theta\left(\dfrac{1}{\cos\theta} + \dfrac{\cos\theta}{\sin\theta}\right) = \sin\theta + \cos^2\theta$

27. $\sec x - \tan x\sin x = \dfrac{1}{\cos x} - \left(\dfrac{\sin x}{\cos x}\right)\sin x = \dfrac{1-\sin^2 x}{\cos x} = \dfrac{\cos^2 x}{\cos x} = \cos x$

29. $\dfrac{\cot y}{\tan y} - \dfrac{\cos y}{\sec y} = \dfrac{(\cos y/\sin y)}{(\sin y/\cos y)} - \dfrac{\cos y}{(1/\cos y)} = \dfrac{\cos^2 y}{\sin^2 y} - \cos^2 y = \dfrac{\cos^2 y - \cos^2 y\sin^2 y}{\sin^2 y} =$
$\dfrac{\cos^2 y\,(1-\sin^2 y)}{\sin^2 y} = \dfrac{\cos^2 y}{\sin^2 y}(1-\sin^2 y) = \cot^2 y\,(\cos^2 y)$

31. $\dfrac{\sin x}{\csc x} + \dfrac{\cos x}{\sec x} = \dfrac{\sin x}{(1/\sin x)} + \dfrac{\cos x}{(1/\cos x)} = \sin^2 x + \cos^2 x = 1$

33. $\dfrac{\cos\theta}{1+\sin\theta}\left(\dfrac{1-\sin\theta}{1-\sin\theta}\right) = \dfrac{\cos\theta\,(1-\sin\theta)}{1-\sin^2\theta} = \dfrac{\cos\theta\,(1-\sin\theta)}{\cos^2\theta} = \dfrac{1-\sin\theta}{\cos\theta}$

35. $\dfrac{\sin x}{\csc x - \cot x}\left(\dfrac{\csc x + \cot x}{\csc x + \cot x}\right) = \dfrac{\sin x\,(\csc x + \cot x)}{\csc^2 x - \cot^2 x} = \sin x\,(1/\sin x + \cos x/\sin x) = 1 + \cos x$

37. $\sin^4 y - \cos^4 y = (\sin^2 y - \cos^2 y)(\sin^2 y + \cos^2 y) = \sin^2 y - \cos^2 y$
39. $\cos^2\theta\sin^2\theta + \cos^4\theta = \cos^2\theta\,(\sin^2\theta + \cos^2\theta) = \cos^2\theta$
41. $\sec^4 x + \sec^2 x - 2 = (\sec^2 x + 2)(\sec^2 x - 1) = (\sec^2 x + 2)(\tan^2 x)$

43. $(v_0\sin\theta)\left(\dfrac{v_0\sin\theta}{g}\right) - \dfrac{1}{2}g\left(\dfrac{v_0\sin\theta}{g}\right)^2 = \dfrac{2v_0^2\sin^2\theta}{2g} - \dfrac{v_0^2\sin^2\theta}{2g} = \dfrac{v_0^2\sin^2\theta}{2g}$ **45.** $\dfrac{\sin\alpha}{\sin\beta}\left(\dfrac{\cos\beta}{\cos\alpha}\right) =$
$\dfrac{\sin\alpha}{\cos\alpha}\left(\dfrac{\cos\beta}{\sin\beta}\right) = \tan\alpha\cot\beta = \tan\alpha\left(\dfrac{1}{\tan\beta}\right) = \dfrac{\tan\alpha}{\tan\beta}$ **47.** $Z = \sqrt{Z^2\cos^2\theta + Z^2\sin^2\theta} =$
$\sqrt{Z^2(\cos^2\theta + \sin^2\theta)} = Z\sqrt{(\cos^2\theta + \sin^2\theta)}$, so $1 = \sqrt{(\cos^2\theta + \sin^2\theta)}$, and $1 = \cos^2\theta + \sin^2\theta$
49. $\sin\theta = \pm 1/\sqrt{1+\cot^2\theta}$, $\cos\theta = \pm 1/\sqrt{1+\tan^2\theta}$, $\tan\theta = \tan\theta$, $\cot\theta = \cot\theta$, $\sec\theta = \pm\sqrt{1+\tan^2\theta}$, $\csc\theta = \pm\sqrt{1+\cot^2\theta}$

Section 12.2 (page 502)

1. $(\sqrt{6}+\sqrt{2})/4$ **3.** $(\sqrt{2}+\sqrt{6})/4$ **5.** $(\sqrt{6}+\sqrt{2})/4$ **7.** $\sqrt{2}/2$ **9.** $1/2$ **11.** $\sqrt{2}/2$ **13.** 1
15. $\sin(90^\circ - \theta) = \sin 90^\circ\cos\theta - \cos 90^\circ\sin\theta = \cos\theta$

17. $\cos(x + \pi/2) = \cos x \cos \pi/2 - \sin x \sin \pi/2 = -\sin x$
19. $\sin (x + \pi) = \sin x \cos \pi + \cos x \sin \pi = -\sin x$
21. $\sin (x - y) \cos y + \sin y \cos (x - y) = \sin [(x - y) + y] = \sin x$
23. $\cos (x - y) \cos y - \sin(x - y) \cos y = \cos [(x - y) + y] = \cos x$
25. $\sin (A + B) - \sin (A - B) = \sin A \cos B + \cos A \sin B - (\sin A \cos B - \cos A \sin B) = \sin A \cos B + \cos A \sin B - \sin A \cos B + \cos A \sin B = 2 \cos A \sin B$
27. $\cos (A - B) - \cos (A + B) = \cos A \cos B + \sin A \sin B - (\cos A \cos B - \sin A \sin B) = \cos A \cos B + \sin A \sin B - \cos A \cos B + \sin A \sin B = 2 \sin A \sin B$
29. $\sin 2\theta = \sin (\theta + \theta) = \sin \theta \cos \theta + \sin \theta \cos \theta = 2 \sin \theta \cos \theta$

31. $\dfrac{w[\sin (\alpha + \theta)]}{\cos (\beta - \theta)} = \dfrac{w(\sin \alpha \cos \theta + \cos \alpha \sin \theta)}{\cos \beta \cos \theta + \sin \beta \sin \theta} = \dfrac{w \sin \alpha \cos \theta + w \cos \alpha \sin \theta}{\cos \beta \cos \theta + \sin \beta \sin \theta}$

33. $\dfrac{w}{2}\left[\dfrac{\sin (2\theta + \beta)}{\cos \theta \cos (\theta + \beta)}\right] = \dfrac{w}{2}\left[\dfrac{\sin [\theta + (\theta + \beta)]}{\cos \theta \cos (\theta + \beta)}\right] = \dfrac{w}{2}\left[\dfrac{\sin \theta \cos (\theta + \beta) + \cos \theta \sin (\theta + \beta)}{\cos \theta \cos (\theta + \beta)}\right] =$
$\dfrac{w}{2}[\tan \theta + \tan (\theta + \beta)]$

35. $wr \tan(\theta - \beta) = wr \tan [\theta + (-\beta)] = wr\left[\dfrac{\tan \theta + \tan (-\beta)}{1 - \tan \theta \tan (-\beta)}\right] = wr\left[\dfrac{\tan \theta - \tan \beta}{1 + \tan \theta \tan \beta}\right]$

Section 12.3 (page 509)

1. $-120/169$ **3.** $527/625$ **5.** $-1081/1369$ **7.** $2520/2809$ **9.** $-1320/3721$ **11.** $7\sqrt{50}/50$
13. $7\sqrt{74}/74$ **15.** $9\sqrt{106}/106$
17. $\cos (3x) = \cos(2x + x) = \cos 2x \cos x - \sin 2x \sin x = (\cos^2 x - \sin^2 x)\cos x - 2 \sin x \cos x \sin x = \cos^3 x - \sin^2 x \cos x - 2 \sin^2 x \cos x = \cos^3 x - 3 \sin^2 x \cos x$
19. $(\sin y + \cos y)^2 = \sin^2 y + 2 \sin y \cos y + \cos^2 y = 1 + \sin 2y$
21. $\cos 2\theta/\sin^2 \theta = (\cos^2 \theta - \sin^2 \theta)/\sin^2 \theta = \cos^2 \theta/\sin^2 \theta - 1 = \cot^2 \theta - 1$

23. $\dfrac{\sin 3x}{\sin x} - \dfrac{\cos 3x}{\cos x} = \dfrac{\sin 3x \cos x - \cos 3x \sin x}{\sin x \cos x} = \dfrac{\sin (3x - x)}{\sin x \cos x} = \dfrac{\sin 2x}{\sin x \cos x} = \dfrac{2 \sin x \cos x}{\sin x \cos x} = 2$

25. $2 \cos^2 (y/2) - \cos y = 2\left(\dfrac{1 + \cos y}{2}\right) - \cos y = 1 + \cos y - \cos y = 1$

27. $\cos^2 (y/2) - \sin^2 (y/2) = \cos [2(y/2)] = \cos y$

29. $\sec^2 (\theta/2) = 1/\cos^2 (\theta/2) = 1 \div \left(\dfrac{1 + \cos \theta}{2}\right) = \dfrac{2}{1 + \cos \theta}\dfrac{(1 - \cos \theta)}{(1 - \cos \theta)} = \dfrac{2(1 - \cos \theta)}{1 - \cos^2 \theta} = \dfrac{2 - 2 \cos \theta}{\sin^2 \theta} =$
$\dfrac{2}{\sin^2 \theta} - \dfrac{2 \cos \theta}{\sin^2 \theta} = 2 \csc^2 \theta - 2 \cot \theta \csc \theta$ **31.** $y = 6077 - 31 \cos 2\theta = 6077 - 31(1 - 2 \sin^2 \theta) = 6077 - 31 + 62 \sin^2 \theta = 6046 + 62 \sin^2 \theta$ **33.** $\dfrac{\sin 2\alpha}{1 + \dfrac{w_2}{w_1} + \cos 2\alpha} = \dfrac{2 \sin \alpha \cos \alpha}{1 + \dfrac{w_2}{w_1} + 2 \cos^2 \alpha - 1} = \dfrac{2 \sin \alpha \cos \alpha}{\dfrac{w_2}{w_1} + 2 \cos^2 \alpha}$

35. $\tau_1 + \sigma \sin \alpha \cos \alpha + \tau \sin^2 \alpha - \tau \cos^2 \alpha = 0$
$\tau_1 = \tau \cos^2 \alpha - \tau \sin^2 \alpha - \sigma \sin \alpha \cos \alpha$
$\tau_1 = \tau (\cos^2 \alpha - \sin^2 \alpha) - \sigma \sin \alpha \cos \alpha$
$\tau_1 = \tau \cos 2\alpha - \dfrac{\sigma \sin 2\alpha}{2}$

Section 12.4 (page 516)

1. $\dfrac{2x\sqrt{1+4x^2}}{1+4x^2}$ **3.** $\sqrt{1-4y^2}$ **5.** $\dfrac{2z\sqrt{1-z^2}}{1-z^2}$ **7.** $\dfrac{2\sqrt{x^2-1}}{x^2-1}$ **9.** $\dfrac{1}{2y^2-1}$ **11.** $\dfrac{1+w^2}{2w}$

13. $2z\sqrt{1-z^2}$ **15.** $\dfrac{2}{x^2}-1$ **17.** $4\pi/3, 5\pi/3$ **19.** $\pi/3, 4\pi/3$
21. $\pi/3, 2\pi/3, 4\pi/3, 5\pi/3$ **23.** $\pi/6, \pi/3, 7\pi/6, 4\pi/3$ **25.** $\pi/12, 5\pi/12, 3\pi/4, 13\pi/12, 17\pi/12, 7\pi/4$
27. $\pi/6, \pi/2, 5\pi/6, 7\pi/6, 3\pi/2, 11\pi/6$ **29.** 0 **31.** $\pi/2, 7\pi/6, 11\pi/6$ **33.** 0.5524, 2.589
35. 2.101, 5.242 **37.** 0.0766, 3.065, 3.218, 6.207 **39.** 1.101, 2.041, 3.195, 4.135, 5.290, 6.230
41. 0.3891, 1.436, 2.484, 3.531, 4.578, 5.625 **43.** 38.4° **45.** 75.5° **47.** 1.4°
49. If $\sin\theta = x$, $\tan\theta = x/\sqrt{1-x^2}$ and Arctan $(x/\sqrt{1-x^2})$ = Arcsin x; If $\cos\theta = x$, $\tan\theta = \sqrt{1-x^2}/x$ and Arctan $(\sqrt{1-x^2}/x)$ = Arccos x

Section 12.5 (page 522)

1. $\pi/6, 5\pi/6$ **3.** $\pi/3, 5\pi/3$ **5.** 1.1903, 4.3319 **7.** $0, \pi$, 0.2450, 3.3866
9. $\pi/2, 3\pi/2$, 0.2527, 2.8889 **11.** $0, 2\pi/3, 4\pi/3$ **13.** $\pi/4, 3\pi/4, 5\pi/4, 7\pi/4$
15. 1.3258, 4.4674, $3\pi/4, 7\pi/4$ **17.** $0, \pi/2, \pi, 3\pi/2$ **19.** $0, \pi, \pi/3, 5\pi/3$ **21.** $\pi/4, 5\pi/4$
23. $\pi/4, 5\pi/4, 3\pi/4, 7\pi/4$ **25.** $\pi/4, 5\pi/4$ **27.** $\pi/2$ **29.** 0 **31.** 48° **33.** 82.9° **35.** 18°

Section 12.6 (page 524)

Sequential Review: 1. $\tan\theta\csc\theta = (\sin\theta/\cos\theta)(1/\sin\theta) = 1/\cos\theta = \sec\theta$
3. $\sin x\sec x = \sin x\,(1/\cos x) = \sin x/\cos x = \tan x$
5. $(1-\sin^2 x)/\cot^2 x = \cos^2 x/(\cos^2 x/\sin^2 x) = \sin^2 x$
7. $[\sin\theta\sec\theta]/\tan^2\theta = [(\sin\theta)(1/\cos\theta)]/\tan^2\theta = \tan\theta/\tan^2\theta = 1/\tan\theta = \cot\theta$
9. $\tan^2 x\sec^2 x - \tan^4 x = \tan^2 x(\sec^2 x - \tan^2 x) = (\tan^2 x)(1) = \tan^2 x$ **11.** $(\sqrt{2}-\sqrt{6})/4$ **13.** 1/2
15. $\sin(\pi - x) = \sin\pi\cos x - \cos\pi\sin x = 0 - (-\sin x) = \sin x$
17. $\sin(x-\pi/2) = \sin x\cos\pi/2 - \cos x\sin\pi/2 = 0 - (\cos x)(1) = -\cos x$
19. $-120/169$ **21.** $-1241/2809$ **23.** $7\sqrt{50}/50$
25. $2/(1+\cos 2x) = 2/(1+\cos^2 x - \sin^2 x) = 2/(2\cos^2 x) = 1/\cos^2 x = \sec^2 x$

27. $\cos^2\dfrac{x}{2} = \dfrac{1+\cos x}{2} = \left(\dfrac{1+\cos x}{2}\right)\left(\dfrac{\tan x}{\tan x}\right) = \dfrac{\tan x + \cos x\tan x}{2\tan x} = \dfrac{\tan x + \sin x}{2\tan x}$

29. $2x\sqrt{1-x^2}$ **31.** $3\pi/4, 5\pi/4$ **33.** $7\pi/6, 11\pi/6$ **35.** 1.063, 4.204, 2.079, 5.221
37. 1.159, 5.124 **39.** 2.30, 3.98 **41.** $0, \pi$, 3.87, 5.55 **43.** $0, \pi$ **45.** $\pi/3, 5\pi/3, \pi$
Applied Problems: 1. $(v_0{}^2 - v_0{}^2\cos^2\alpha)/(2g) = v_0{}^2(1-\cos^2\alpha)/(2g) = v_0{}^2\sin^2\alpha/(2g)$

3. $P = \dfrac{L(\sin\alpha + \mu\cos\alpha)}{\cos\alpha - \mu\sin\alpha} = L\dfrac{\dfrac{\sin\alpha}{\cos\alpha} + \dfrac{\mu\cos\alpha}{\cos\alpha}}{\dfrac{\cos\alpha}{\cos\alpha} - \dfrac{\mu\sin\alpha}{\cos\alpha}} = \dfrac{L(\tan\alpha + \mu)}{1 - \mu\tan\alpha} = \dfrac{L\left(\dfrac{l}{\pi d} + \mu\right)}{1 - \dfrac{\mu l}{\pi d}}$

5. $g = 9.78049(1 + 0.005288\sin^2\theta - 0.000006\sin^2 2\theta) = 9.78049(1 + 0.005288\sin^2\theta - 0.000024\sin^2\theta\cos^2\theta)$ **7.** $5\sin(6500t - 53.13°) = 5\sin 6500t\cos(53.13°) - 5\cos 6500\,t\sin(53.13°) = 3.000\sin 6500t - 4.000\cos 6500t = 3\sin 6500t + 4\sin(6500t - 90°)$ **9.** $\sin^2\theta = 1/2 - 1/2\cos 2\theta$, so $50\sin^2 5000t = 50\,(1/2 - 1/2\cos 10\,000t) = 25 - 25\cos 10\,000t$ **11.** 0.0016 s or 1.6 ms **13.** $-19.1°$
15. varying amplitude of $2\cos(-5\pi t)$ **17.** no **19.** $0.03\sin\omega t\cos\omega t$
Random Review: 1. $\cos(\pi - x) = \cos\pi\cos x + \sin\pi\sin x = -1\cos x + 0 = -\cos x$ **2.** $\tan\theta(\tan\theta + \cot\theta) = \tan^2\theta + 1 = \sec^2\theta$ **3.** $3\pi/2$, 0.730, 2.412

4. 0.410, 2.505, 4.599, 1.684, 3.778, 5.873 **5.** $2\tan\alpha/(1+\tan^2\alpha) = 2\tan\alpha/\sec^2\alpha = (2\sin\alpha/\cos\alpha) \div (1/\cos^2\alpha) = (2\sin\alpha/\cos\alpha)(\cos^2\alpha) = 2\sin\alpha\cos\alpha = \sin 2\alpha$ **6.** $\pi/3, 2\pi/3, 4\pi/3, 5\pi/3$ **7.** 240/289 **8.** $\sqrt{2}/2$ **9.** $\csc y/(\tan y + \cot y) = (1/\sin y) \div (\sin y/\cos y + \cos y/\sin y) = (1/\sin y) \div [(\sin^2 y + \cos^2 y)/(\cos y \sin y)] = (1/\sin y) \div [1/(\cos y \sin y)] = (1/\sin y)(\cos y \sin y) = \cos y$
10. $\sqrt{1-4x^2}/2x$
Chapter Test: 1. $\sin x + \cos x \cot x = \sin x + \cos x(\cos x/\sin x) = (\sin^2 x + \cos^2 x)/\sin x = 1/\sin x = \csc x$ **2.** $\csc^2\theta/\tan\theta - \cot^3\theta = \csc^2\theta\cot\theta - \cot^3\theta = \cot\theta(\csc^2\theta - \cot^2\theta) = \cot\theta\,(1) = \cot\theta$
3. $(\sqrt{6}-\sqrt{2})/4$ **4.** $\sin(\pi/2 - x) = \sin\pi/2\cos x - \sin x\cos\pi/2 = 1(\cos x) - 0 = \cos x$
5. $\sin^2 x/2 = (1-\cos x)/2 = [(1-\cos x)/2](\tan x/\tan x) = (\tan x - \cos x\tan x)/(2\tan x) = (\tan x - \sin x)/(2\tan x)$ **6.** 161/289 **7.** 0.51494, 2.0857, 3.6565, 5.2273 **8.** $3\pi/2, 7\pi/6, 11\pi/6$ **9.** 96° 16′
10. −55°

CHAPTER 13

Section 13.1 (page 533)

1. 4 **3.** −4 **5.** −3 **7.** 8 **9.** −4 **11.** 16 **13.** $2\sqrt[3]{2}$ **15.** 27 **17.** 3.34 **19.** 2.41
21. 0.0733 **23.** $\dfrac{y^{1/2}z^{1/3}}{(3^{1/2}x^{1/4})}$ **25.** $\dfrac{25c}{a^4b^{2/3}}$ **27.** $3x^{1/2} + \dfrac{3x}{2^{2/3}}$ **29.** $\dfrac{3a+4b}{a^2b^2}$ **31.** $\dfrac{x(8x+7)}{(2x+3)^2(x-1)^2}$
33. $\dfrac{x(-x^2+3x+7)}{(x-4)^{2/3}(x+1)^2}$ **35.** 1.5 A **37.** $(100-T)(10^{-1})(590\,000 - 5800\,T)$ **39.** **(a)** $8\sqrt{5} + 12\sqrt{3}$
(b) $24\sqrt{15}$ **41.** about 7 times as many

Section 13.2 (page 540)

1. $\sqrt{2}$ **3.** $y\sqrt{2xy}$ **5.** $y\sqrt{3x}$ **7.** $2xy^2\sqrt{5y}$ **9.** $5x^2y^3\sqrt{2xy}$ **11.** $3ab^2\sqrt[3]{3a}$ **13.** $2ab^2\sqrt[4]{2b}$
15. $\sqrt{2}/2$ **17.** $\sqrt[3]{18}/3$ **19.** $\sqrt{7}$ **21.** $2\sqrt{3}+2\sqrt{2}$ **23.** $2\sqrt[3]{3}+\sqrt{3}$ **25.** $2a\sqrt{b}+b\sqrt{a}$ **27.** 0
29. 0 **31.** $2ab\sqrt[3]{5b}$ **33.** 60 **35.** $b\sqrt{6}-\sqrt{2b}$ **37.** $3a-\sqrt[3]{3a}$ **39.** $-9-8\sqrt{10}$ **41.** 103
43. 261.6 cycles/s **45.** $10^5\sqrt{65}/(260\pi)$ **47.** $V = \sqrt{I^2R^2} = IR$ **49.** **(a)** $\sqrt[3]{-8}\,\sqrt[3]{-1} = \sqrt[3]{8}$
(b) $\sqrt{-4}\sqrt{-9} \neq \sqrt{36}$

Section 13.3 (page 546)

1. 0.775 **3.** −0.707 **5.** 0.882 **7.** −0.101 **9.** −9.90 **11.** $3\sqrt{x}/(4x)$ **13.** $-5\sqrt{x}/(2x)$
15. $\sqrt[3]{12w}/(3w)$ **17.** $2\sqrt{3}+2\sqrt{2}$ **19.** $(2\sqrt{3}+5)/13$ **21.** $(10\sqrt{7}-2\sqrt{2})/173$ **23.** $-15+11\sqrt{2}$
25. $(36-13\sqrt{5})/11$ **27.** $(23+6\sqrt{14})/5$ **29.** $(\sqrt{x}+2\sqrt{y})/(x-4y)$ **31.** $(2-7\sqrt{x}+6x)/(4-9x)$
33. $(6a+6b-13\sqrt{ab})/(9a-4b)$ **35.** $(\sqrt{x+2}+x+2)/(-1-x)$
37. $(x^2+y+1+2x\sqrt{y+1})/(x^2-y-1)$ **39.** $(11+6\sqrt{2})/7$ **41.** $(13-4\sqrt{3})/11$ **43.** $\sqrt{\rho IR}/R$
45. $\sqrt{3EIlm}/(2\pi l^2 m)$ **47.** **(a)** 0.149π **(b)** 0.468 cm **49.** $[wL/(8s)]\sqrt{L^2+16s^2}$

Section 13.4 (page 552)

1. $\sqrt[6]{5}$ **3.** $\sqrt[10]{7y}$ **5.** $\sqrt[20]{6b}$ **7.** $\sqrt[6]{28}$ **9.** $\sqrt[6]{189}$ **11.** $\sqrt[12]{135}$ **13.** $\sqrt[4]{147}$ **15.** $\sqrt[6]{6125}$
17. $\sqrt[12]{27{,}783}$ **19.** $\sqrt[6]{108}/3$ **21.** $\sqrt[6]{200}/2$ **23.** $\sqrt[3]{2}+\sqrt{3}$ **25.** $\sqrt[3]{5}+\sqrt{2}$
27. $2\sqrt{5}+\sqrt{15}/3+2\sqrt{3}+1$ **29.** $12+36\sqrt{13}/13$ **31.** $11\sqrt{7}/7-5$ **33.** $\sqrt{7}-\sqrt{5}$ **35.** $\sqrt{4-y}$
37. $\sqrt{5-a}$ **39.** $13x+2y$ **41.** $14a+50$ **43.** **(a)** $50.0\sqrt{3}/3$ m **(b)** $1117\,\sqrt[4]{108}/6$
(c) 600 km, to three significant digits **45.** **(a)** $6\sqrt{5}$ **(b)** 13.4 **47.** **(a)** $3\sqrt{69}$ **(b)** 24.9
49. Construct a right triangle with legs of length 1 so that the hypotenuse is of length $\sqrt{2}$; Construct a right triangle with legs of length $\sqrt{2}$ and 1 so that the hypotenuse is of length $\sqrt{3}$, and so on.

Section 13.5 (page 559)

Note: Extraneous solutions are shown in parentheses

1. 9/25 **3.** 81/4 **5.** no solution, (9) **7.** no solution, (4) **9.** -1 **11.** 10 **13.** 0 **15.** 5 **17.** 7 **19.** 8, (5/2) **21.** 0, 5 **23.** 0, 4 **25.** no solution, (3) **27.** 16 **29.** 0, (4/25) **31.** 2, (37/36) **33.** 1, (16/9) **35.** 2, (7/9) **37.** 27/8, 8/27 **39.** $\pm 1, \pm 27$ **41.** 16, 1 **43.** 0.10 m **45. (a)** $\sqrt{A\pi}/\pi$ **(b)** 7π cm^2 **47.** 3 cm $\times$ 4 cm $\times$ 12 cm

Section 13.6 (page 561)

Sequential Review: 1. 4 **3.** 2.79 **5.** $2^{2/3}a^2c^{1/3}/b^{2/3}$ **7.** $4x^{1/2}(x+y)^{1/3}/y^{2/3}$ **9.** $\dfrac{(x^2+2x+3)}{(x-2)^3(4x+3)^{1/2}}$ **11.** $\sqrt[3]{3ab^2}$ **13.** $3a^2b^3\sqrt{7a}$ **15.** $\sqrt{35}/14$ **17.** $11\sqrt{2}/2$ **19.** 0.845 **21.** 12.1 **23.** $-3\sqrt{7}/(7x)$ **25.** $\dfrac{28\sqrt{3}-7\sqrt{2}}{46}$ **27.** $\dfrac{3\sqrt{x+1}-x-1}{8-x}$ **29.** $\sqrt[6]{432}$ **31.** $\sqrt{3}+\sqrt[3]{5}$ **33.** $\sqrt[4]{3}-\sqrt[4]{2}$ **35.** $\sqrt{7-a^2}$ **37.** 49/9 **39.** -2, (-5 is an extraneous solution) **41.** 3 **43.** 3, (27/16 is an extraneous solution) **45.** 256

Applied Problems: 1. $(A_0/t^2)\sqrt[5]{t^4}$ **3.** $R = 7\ \Omega$, $I > 15$ A **5. (a)** 0.0020 min/ml **(b)** $t = B\sqrt{v_t}/(uv_t)$ **7.** $R = (1+3D)/6$ **9.** $-\sqrt[3]{(b/2)(1+\sqrt{1+4b}}$ **11.** $f_p = f_s\sqrt{(L-CR^2)/L}$ **13.** $L = 7W$ **15.** $e^{2x} - e^{-2x}$ **17.** $0 = R^2 - 20RX_C + X_C^2$ **19.** $\beta = \text{Cos}^{-1}(1 - 2\sin^2\alpha)$

Random Review: 1. 8 **2.** $(11+4\sqrt{7})/9$ **3.** $1/[8a^{1/3}b^{1/2}(a+b)^{1/2}]$ **4.** $\sqrt[6]{392}$ **5.** $ab\sqrt{5b}$ **6.** -14.9 **7.** no solution **8.** $11\sqrt{2}/2$ **9.** $\sqrt[4]{20x^2}$ **10.** $(x^2+x+3)/(x+1)^2$

Chapter Test: 1. $6n^2p^3/(5m)$ **2.** $(2x-1)/[(x-3)^3(x+2)^3]$ **3.** $2x^3y^5\sqrt{11}$ **4.** $\sqrt{21}/5$ **5.** $3-8\sqrt{2}$ **6. (a)** 0.236 **(b)** $-2+\sqrt{5}$ **7.** $\dfrac{11\sqrt{5}}{5}+5$ **8.** 5, (0 is an extraneous solution) **9.** 0.551 cm **10.** 2.48×10^{-5} coulombs

CHAPTER 14

Section 14.1 (page 574)

1. $5j$ **3.** $3j\sqrt{3}$ **5.** $-j\sqrt{11}$ **7. (a)** $-5\sqrt{2}$ **(b)** $5\sqrt{2}$ **9. (a)** $-2\sqrt{3}$ **(b)** $2\sqrt{3}$ **11. (a)** -10 **(b)** 10 **13.** -1 **15.** $-j$ **17.** 1 **19.** $6+2j$ **21.** $6j\sqrt{3}$ **23.** $8.9+1.5j$ **25.** $2-3j$ **27.** $7-j\sqrt{5}$ **29.** $2.4+1.8j$ **31.** 50 **33.** $19+4j\sqrt{7}$ **35.** $15-5j$ **37.** $(-3-4j)/5$ **39.** $(-14+23j)/29$ **41.** $-j$ **43.** $(10.69+j4.42)$ V **45.** $(1.40+j2.79)$ V **47.** $(19.1+j1.62)$ V **49.** $(8.96-j14.7)\ \Omega$

Section 14.2 (page 581)

1. **3.** **5.**

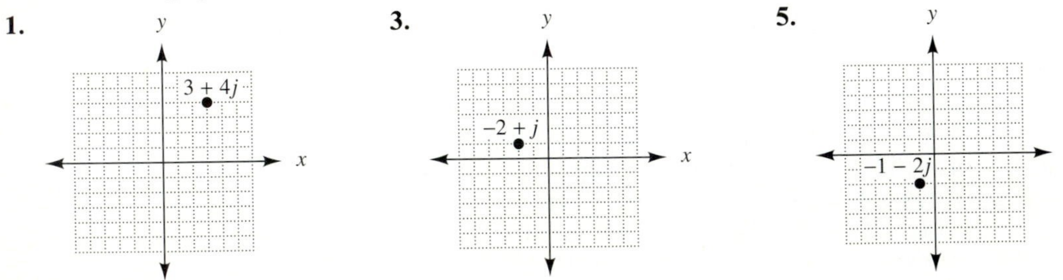

7.

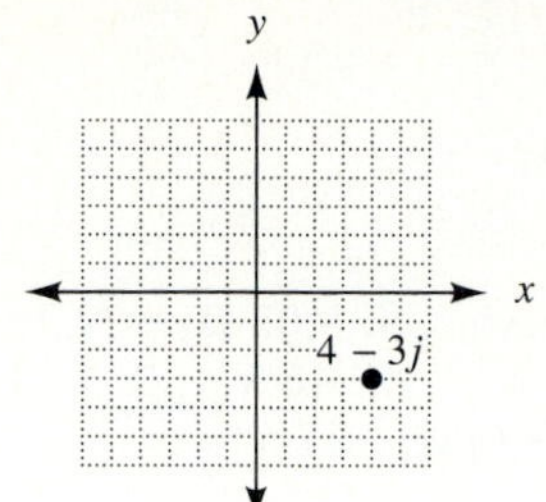

9. 6 **11.** $3 + 3j$ **13.** 6 **15.** $-1 + 7j$ **17.** $-2j$ **19.** $3 - 3j$
21. $-5 + 4j$ **23.** $2j$ **25.** $-j$ **27.** $2 + 4j$ **29.** $3 - 3j$
31. $6 + 3j$ **33.** $-2 + 3j$ **35.** $-3j$ **37.** $3 + 4j$, 5, 53.1°
39. $3 + 4j$, 5, 53.1° **41.** $5 + 12j$, 13, 67.4°

43.

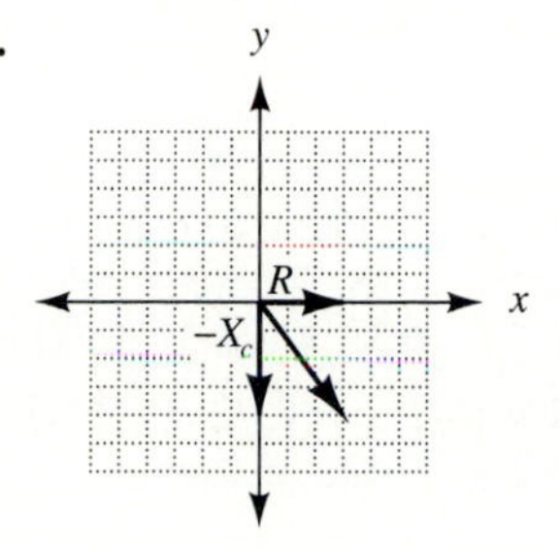

45.

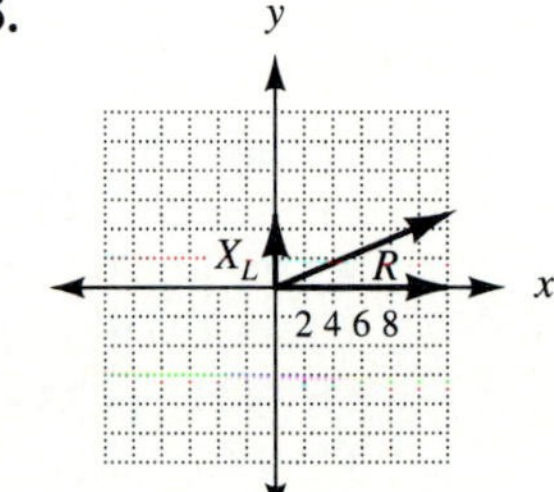

47.

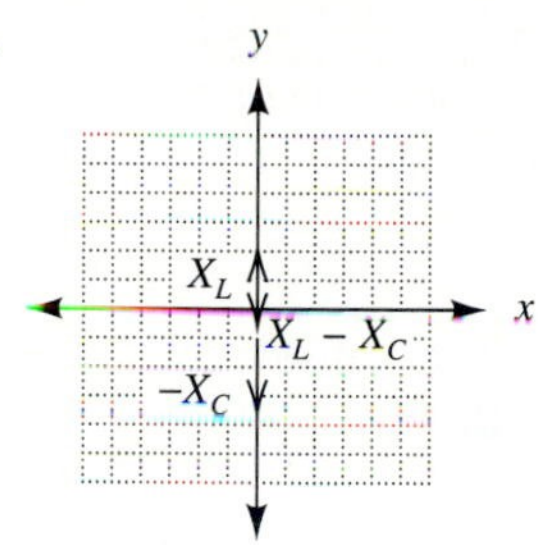

49.

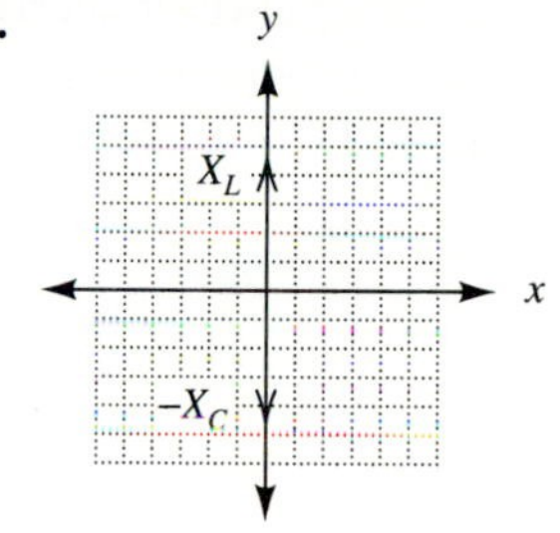

Section 14.3 (page 587)

1. $10(\cos 126.9^\circ + j \sin 126.9^\circ)$ **3.** $3.00(\cos 122.6^\circ + j \sin 122.6^\circ)$
5. $10.7(\cos 293.9^\circ + j \sin 293.9^\circ)$ **7.** $3(\cos 90^\circ + j \sin 90^\circ)$ **9.** $4(\cos 0^\circ + j \sin 0^\circ)$
11. $3\sqrt{2}/2 + (3\sqrt{2}/2)j$ **13.** $13.7 + 4.23j$ **15.** $-6.07 + 21.5j$ **17.** $-2\sqrt{3} + 2j$ **19.** $-4.30 + 3.22j$
21. $7e^{j\pi/6}$ **23.** $12.6e^{0.953j}$ **25.** $145e^{2.08j}$ **27.** $8.32e^{1.93j}$ **29.** $6e^{j\pi/6}$
31. $4(\cos 180^\circ + j \sin 180^\circ)$ **33.** $15(\cos 360^\circ + j \sin 360^\circ)$ **35.** $9.27(\cos 268.1^\circ + j \sin 268.1^\circ)$
37. $3j$ **39.** $11\sqrt{3} + 11j$ **41.** $7.32 + 4.76j$ **43.** $10.5(\cos 19.2^\circ + j \sin 19.2^\circ)$ V
45. $(0.21 - j0.076)$ A **47.** $(9.89 + j4.34)\ \Omega$ **49.** $1.0e^{j5.3}$ mV

Section 14.4 (page 595)

1. $14.6(\cos 96.1^\circ + j \sin 96.1^\circ)$ **3.** $101(\cos 129.6^\circ + j \sin 129.6^\circ)$ **5.** $2.5(\cos 114^\circ + j \sin 114^\circ)$
7. $150(\cos 93^\circ + j \sin 93^\circ)$ **9.** $4.43(\cos 195.1^\circ + j \sin 195.1^\circ)$
11. $1.28[\cos(-10.7^\circ) + j \sin(-10.7^\circ)]$ **13.** $4.98(\cos 116.8^\circ + j \sin 116.8^\circ)$
15. $0.13[\cos(-78^\circ) + j \sin(-78^\circ)]$ **17.** $4.4\angle -15^\circ$ **19.** $2.75\angle 18.5^\circ$
21. $143(\cos 155.4^\circ + j \sin 155.4^\circ)$ **23.** $985{,}000(\cos 136.4^\circ + j \sin 136.4^\circ)$
25. $2800(\cos 135^\circ + j \sin 135^\circ)$ **27.** $-1120 + 404j$ **29.** $-119 + 120j$
31. $28.5(\cos 22.7^\circ + j \sin 22.7^\circ)$ **33.** $72.5\angle 21.6^\circ$ **35.** $4.06\angle 229.0^\circ$
37. $1/2 + j\sqrt{3}/2$, -1, $1/2 - j\sqrt{3}/2$ **39.** $1.26 + 1.15j$, $-1.63 + 0.520j$, $0.364 - 1.67j$

41. $1.21 + 0.874j, -0.874 + 1.21j, -1.21 - 0.874j, 0.874 - 1.21j$ **43.** $215\angle 12°$ V
45. $(1.5 - j3.2)$ V **47.** $(1.66 + j0.238)$ A

Section 14.5 (page 602)

1. 7.21 Ω, −33.7° **3.** 8.94 Ω, 26.6° **5.** 9.23 Ω, −26.7° **7.** 10.2 Ω, −3.8° **9.** 9.02 Ω, −29.9°
11. 17.2 V **13.** 33.4 V **15.** 25.7 **17.** 12.5 V **19.** 21.0 V **21.** 22.6 V **23.** 16 V
25. 27.7 V **27.** 46.1 Ω, −82.5° **29.** 22.4 Ω, −75.7° **31.** **(a)** $(26.5 - j8.4)$ Ω, −17.6°
(b) $(16.7 + j0.293)$ Ω, 1.0°

Section 14.6 (page 604)

Sequential Review: 1. $-3j\sqrt{5}$ **3.** **(a)** −5 **(b)** 5 **5.** $6 + 4j$ **7.** $-33 - 3j\sqrt{2}$ **9.** $-1/3 - (2\sqrt{2}/3)j$
11.

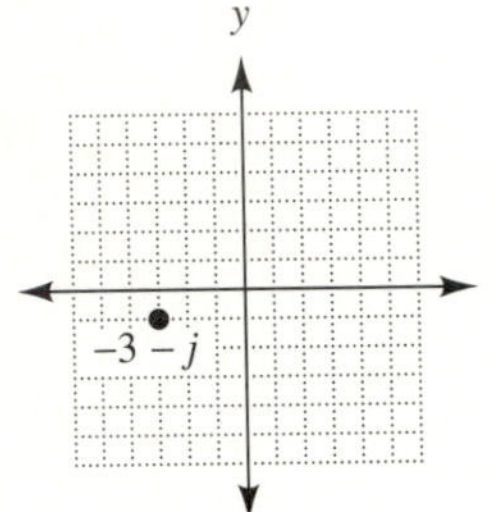

13. $5 + 2j$ **15.** $6 - 3j$ **17.** $8 + 15j$, 17, 61.9° **19.** 4.40(cos 128.4° + j sin 128.4°) **21.** $11.9 + 8.93j$ **23.** $8e^{j\pi/4}$ **25.** 7.23(cos 195.4° + j sin 195.4°) **27.** $0 - 4j$ **29.** $38\angle 40°$ **31.** 1.4(cos 5° + j sin 5°)
33. 2000(cos 37.8° + j sin 37.8°) **35.** $3\sqrt{2} + (1/2)j, -\sqrt{3}/2 + (1/2)j, -j$
37. 9.43, −32.0° **39.** 6.33, 7.5° **41.** 15.8 V **43.** 26.0 volts
45. 23.0 volts
Applied Problems: 1. 43.4 Ω, −79.2° **3.** $(11.2\angle -63.4°)$ Ω
5. $z = 56 + j4$ **7.** 118 miles **9.** 3.4 miles at an angle of 20° E of N
11. $R - j75 = \sqrt{R^2 + 75^2}\angle -65°$ **13.** $(7.6 + j18)$ Ω
15. $\sqrt{1/(LC)}$ or $1/\sqrt{LC}$ **17.** $I = (14.4 + j19.2)$ A, $IX_C = (9680 - j7260)$ V **19.** 325 W
Random Review: 1. $7 + j$ **2.** $8.977\angle 109.9°$ **3.** 5.70(cos 309.7° + j sin 309.7°) **4.** 40
5. $23.8\, e^{0.745j}$ **6.** $-2j$ **7.** $-86{,}200 + 56{,}400\, j$, to three significant digits **8.** $1.5 + 3j\sqrt{3}/2$
9. $(-1 - j\sqrt{3})/2$ **10.** $0.866 + 0.5\, j, -0.866 + 0.5\, j, -j$
Chapter Test: 1. −40 **2.** $-25 + 21j$ **3.** $5.71\, e^{0.654j}$ **4.** $2.71 + 1.68j$ **5.** 1(cos 15° + j sin 15°)
6. 16.38(cos 129° + j sin 129°) **7.** $275\,000 + 7340j$, to three significant digits **8.** $1.86\angle 3°$
9. 6.65, 8.3° **10.** 18.4 volts

CHAPTER 15

Section 15.1 (page 615)

1.

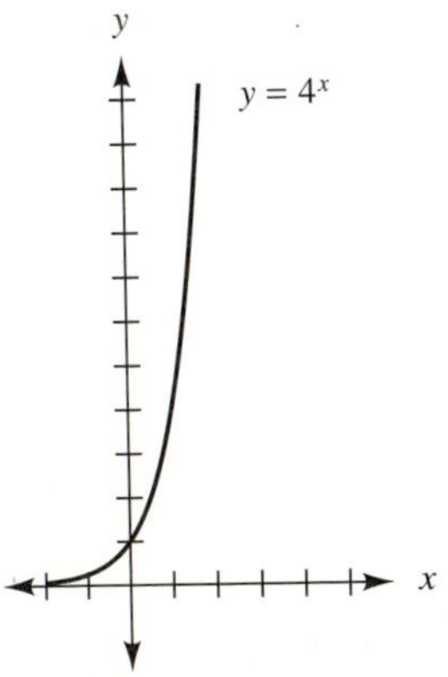

3.

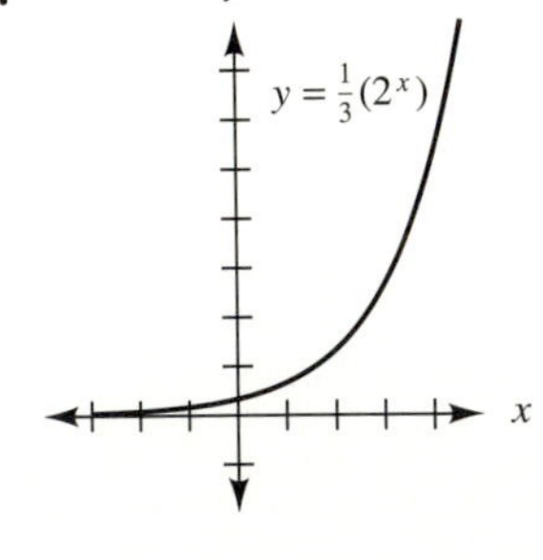

5.

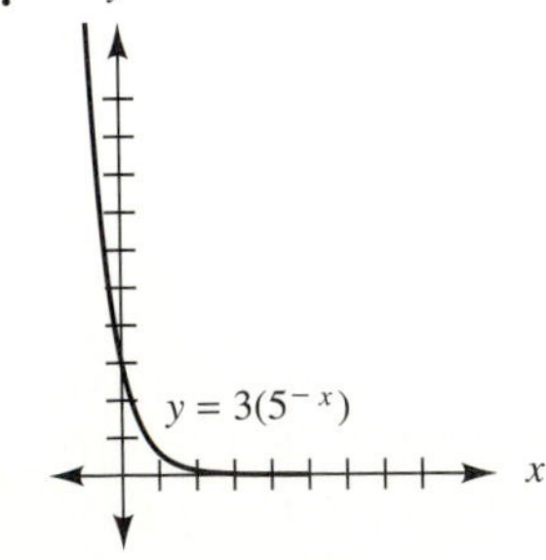

7.

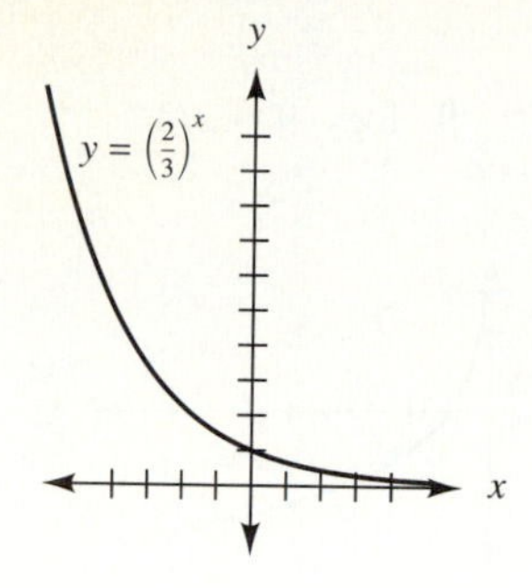

9.

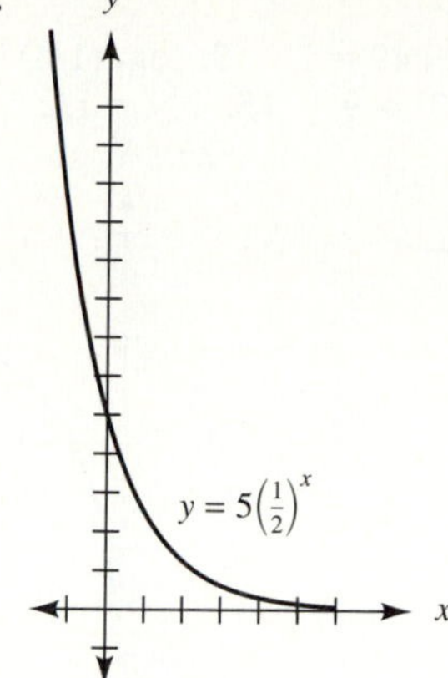

11.

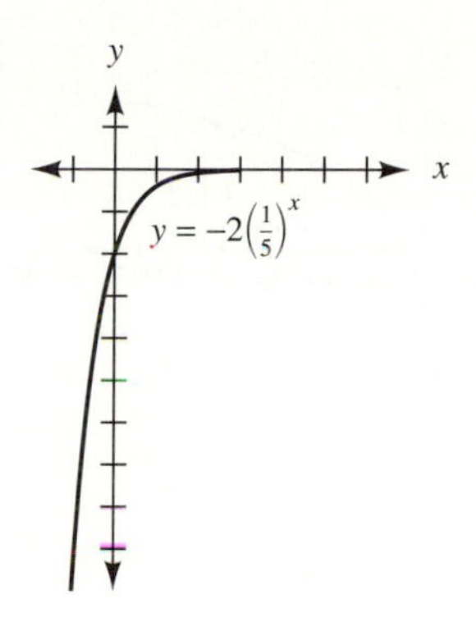

13.

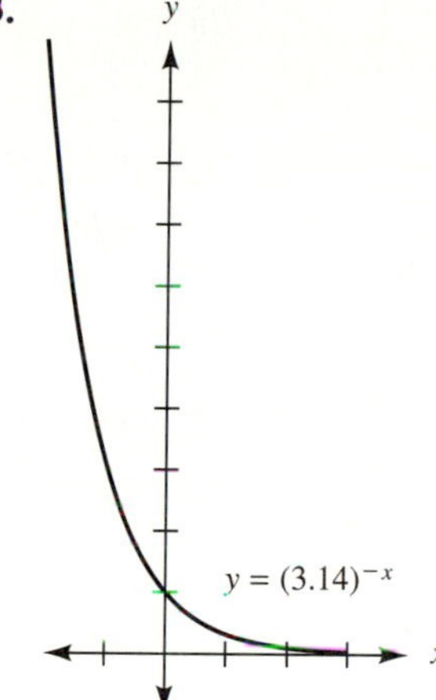

15. e **17.** i
19. b **21.** h **23.** j

25.

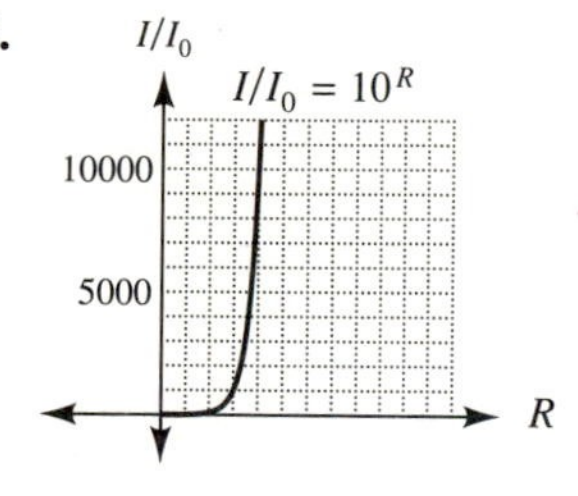

27. $Q = 3.6 \times 10^{-5}e^{-t/.002}$
$0 \le t \le 0.01$

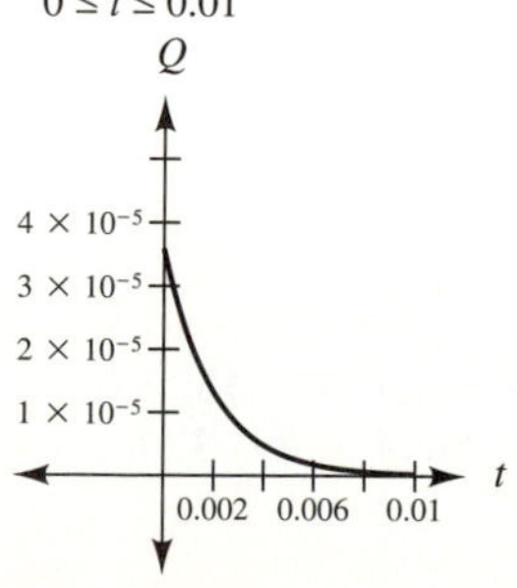

29.

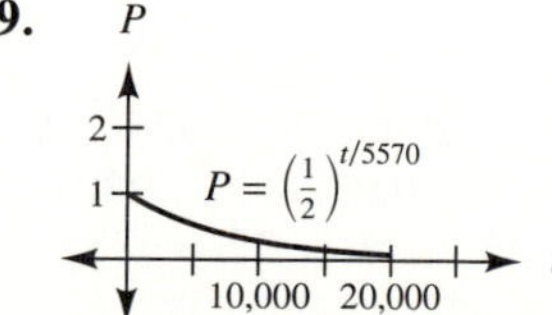

Section 15.2 (page 621)

1. $\log_3 81 = 4$ **3.** $\log_7 49 = 2$ **5.** $\log_2 (1/8) = -3$ **7.** $\log_4 (1/4) = -1$ **9.** $\log_{27} 9 = 2/3$
11. $10^3 = 1000$ **13.** $3^2 = 9$ **15.** $2^{-1} = 1/2$ **17.** $3^{-2} = 1/9$ **19.** $8^{2/3} = 4$

21. **23.** **25.**

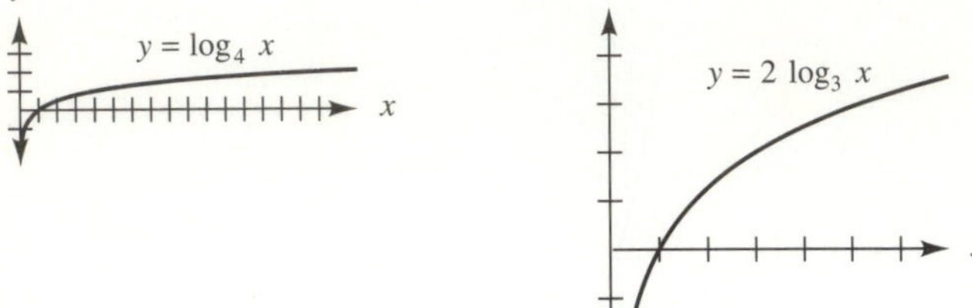

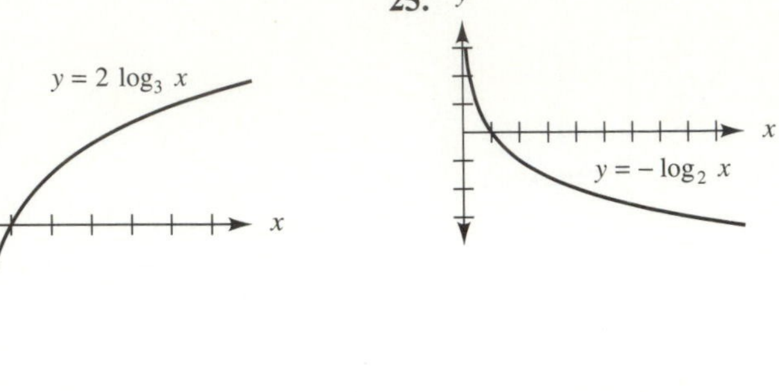

27.

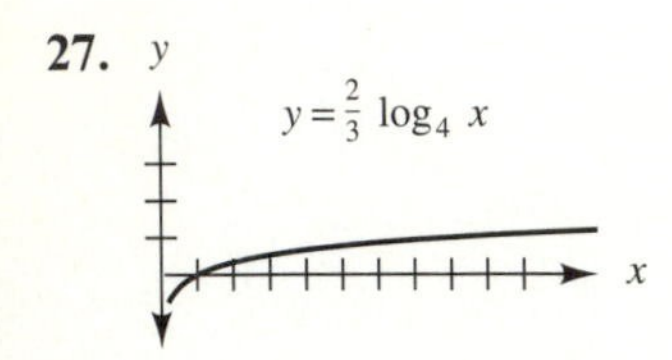

29.

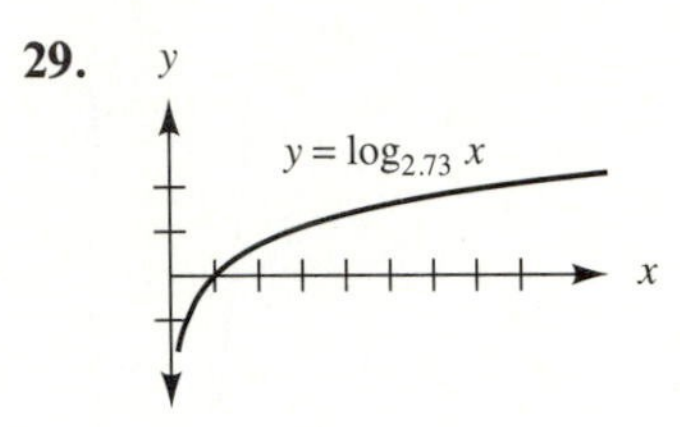

31.

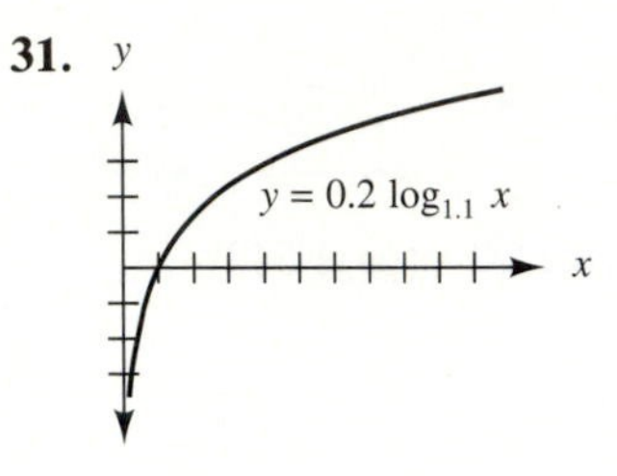

33. $\log_h (d/111.7) = 1/2$ **35.** $P_2 = P_1(10^{\Delta e/C_c})$ **37.**

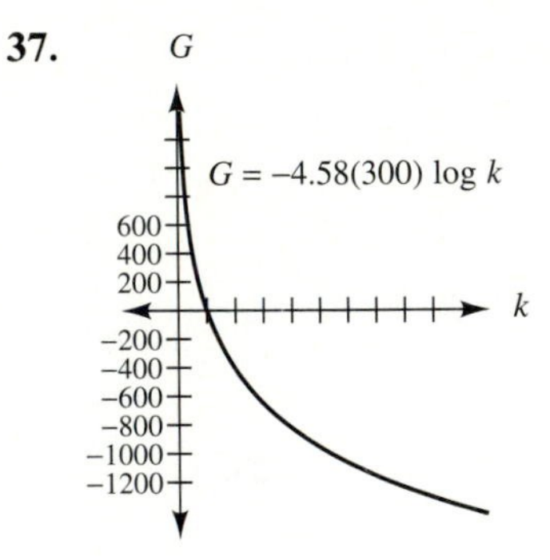

Section 15.3 (page 625)

1. $\log_2 35$ **3.** $\log_6 3y$ **5.** $\log_4 4$ **7.** $\log_5 16$ **9.** $\log_b 1/2$ **11.** $\log_y 75$ **13.** $\log_a x^2/4$
15. $\log_3 87.5$ **17.** $\log_7 14.4$ **19.** $\log_a (250/49)$ **21.** $2 + \log_2 5$ **23.** $2 + \log_3 4$ **25.** $2 + \log_5 3$
27. $\log_7 5 - 1$ **29.** $\log_{10} 3 - 1$ **31.** $\log_x 3 - 2$ **33.** $\log_y 7 + 1$ **35.** $1/2 + (1/2) \log_2 y$
37. $(1/3) \log_a 5 + 1/3$ **39.** $1/3 + (1/3) \log_{10} x$ **41.** $\log_{10} (a/b)^{276}$ **43.** $\log_{10} (b_m/b_n) = \log_{10} r^{n-m}$
45. $\log_{10} (I/I_0)$

Section 15.4 (page 630)

1. 2.507 **3.** 2.702 **5.** 1.895 **7.** 0.985 **9.** 0.396 **11.** −0.481 **13.** −1.159 **15.** 2.096
17. 13.62 **19.** 260.0 **21.** 0.002859 **23.** 0.04399 **25.** 0.4339 **27.** 0.5824 **29.** 1.58
31. 2.26 **33.** 0.465 **35.** 1.13 **37.** −0.339 **39.** 6.64 **41.** 1000
43. 1000 millibars, to two significant digits

Section 15.5 (page 635)

1. 5.771 **3.** 4.0993 **5.** 3.936 **7.** 1.437 **9.** -1.109 **11.** -2.668 **13.** 1.38 **15** 4.187 **17.** 0.1058 **19.** 0.2575 **21.** 0.6959 **23.** 0.7907 **25.** $\frac{\log x}{\log 2}$, 5.67 **27.** $\frac{\log x}{\log 3}$, 1.66 **29.** $\frac{\log x}{\log 5}$, -0.16 **31.** $\frac{\ln x}{\ln 7}$, 3.525 **33.** $\frac{\ln x}{\ln 7}$, 2.100 **35.** $\frac{\ln x}{\ln 9}$, -0.53 **37.** $\frac{\ln x}{\ln 11}$, -0.404 **39.** $y = e^{x\ln 54.7}$, 6.11×10^7 **41.** $y = e^{x\ln 0.456}$, 1.05×10^{-5} **43.** $y = e^{x\ln 3.64}$, 3520 **45.** $t = -\tau \ln (I/I_0)$ **47.** $t = -\ln (C/C_0)$ **49.** $n = \frac{\ln (0.01p)}{\ln (0.97)}$ **51.** Find the value of INV LN 1

Section 15.6 (page 641)

1. 2 **3.** 2 **5.** 3 **7.** 36 **9.** 1.21 **11.** 10.5 **13.** 49.5 **15.** 5.93 **17.** 10, -5 **19.** -6, 5 **21.** $5/x$ **23.** $\sqrt[3]{10x + 10}$ **25.** 1.76 **27.** 7.64 **29.** 6.85 **31.** 1.69 **33.** -0.450 **35.** **(a)** 8.3 years **(b)** 4.6 years **(c)** 11.9 years **37.** $L_{10} = 0.146L(-\ln R)^{-0.855}$ **39.** **(a)** 1000 times greater **(b)** 3.01 dB **41.** **(a)** 4.8 years **(b)** 6.6 years **(c)** $n = \frac{\log (V_d/V)}{\log(1 - r)}$

Section 15.7 (page 648)

1.

3.

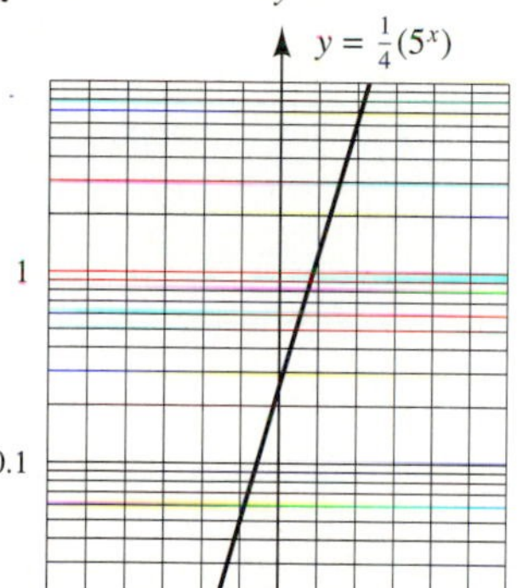

5.

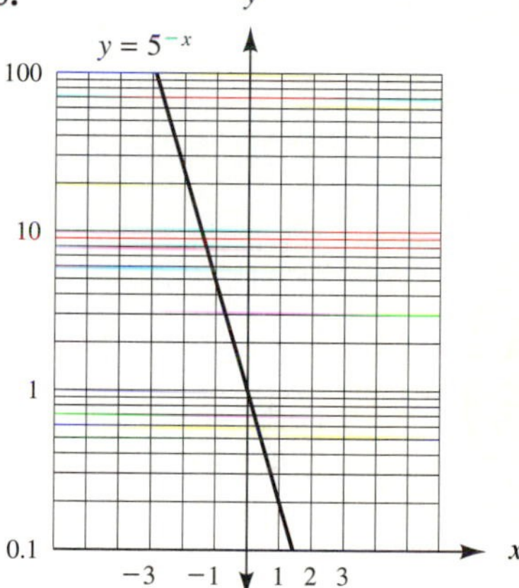

7.

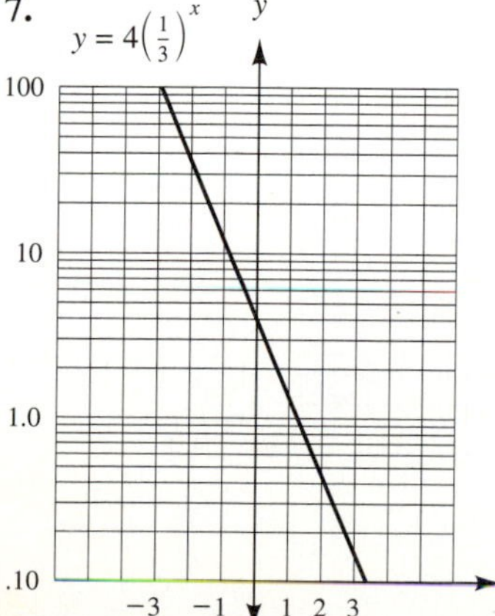

9.

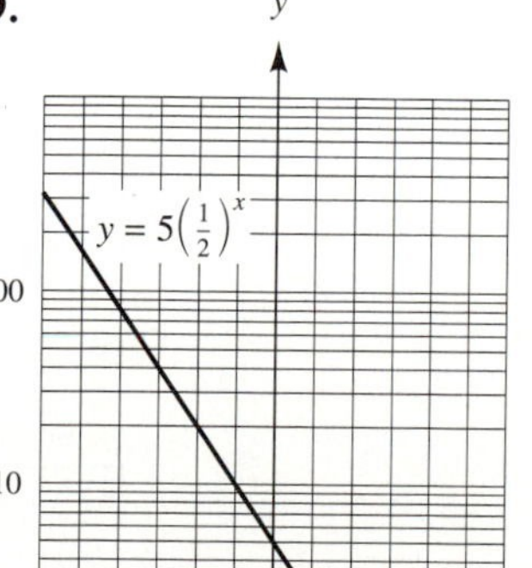
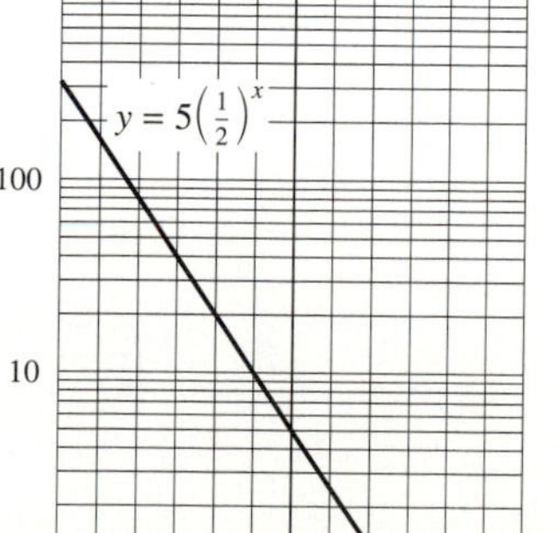

11.

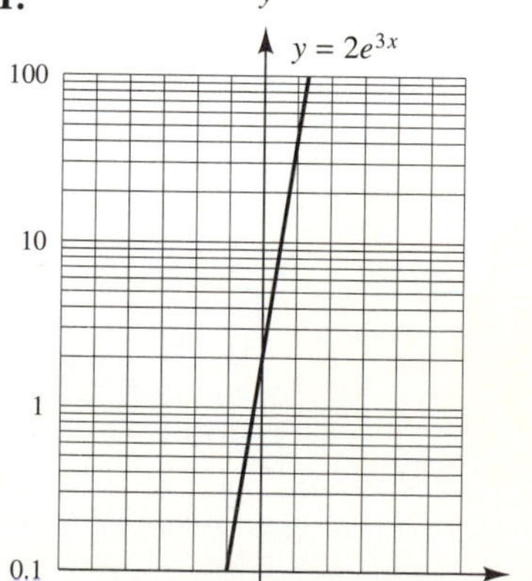

13.

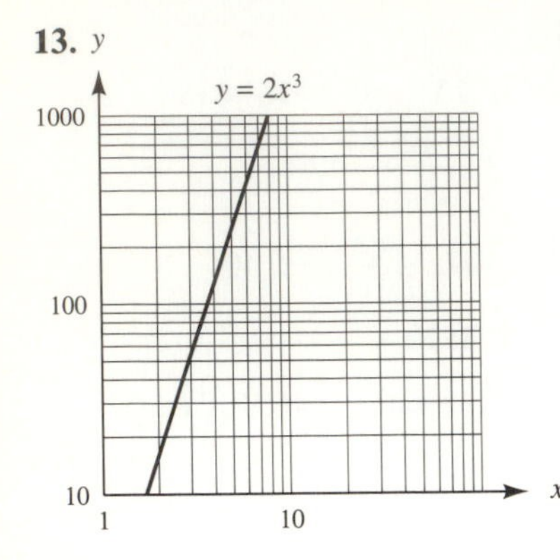

15.

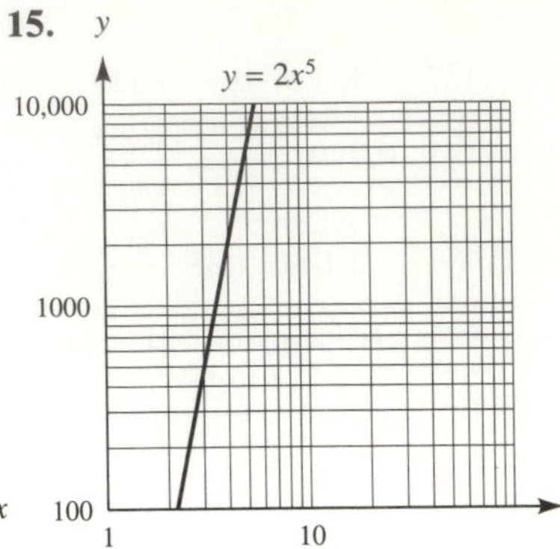

17.

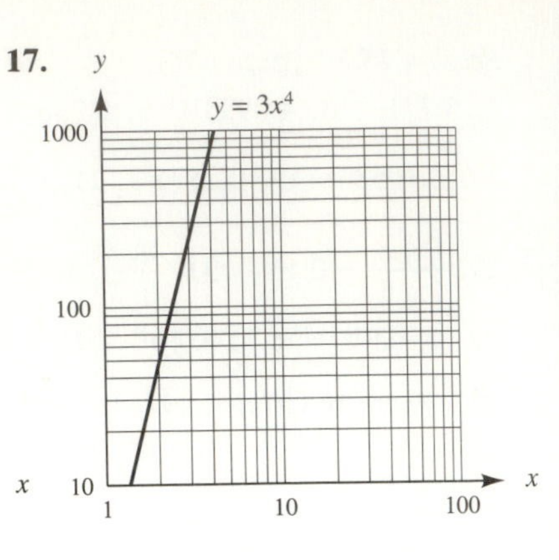

19.

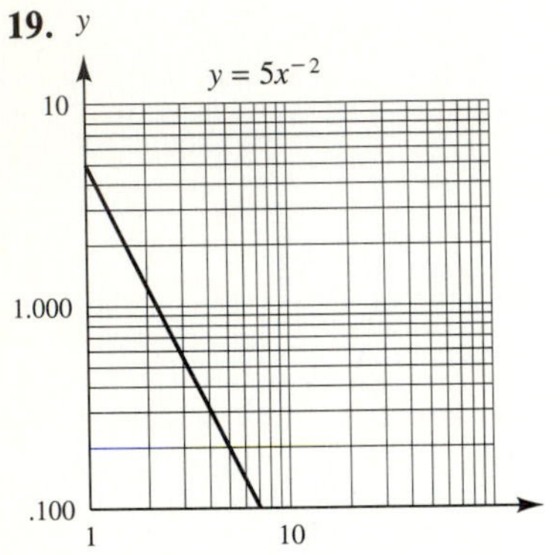

21.

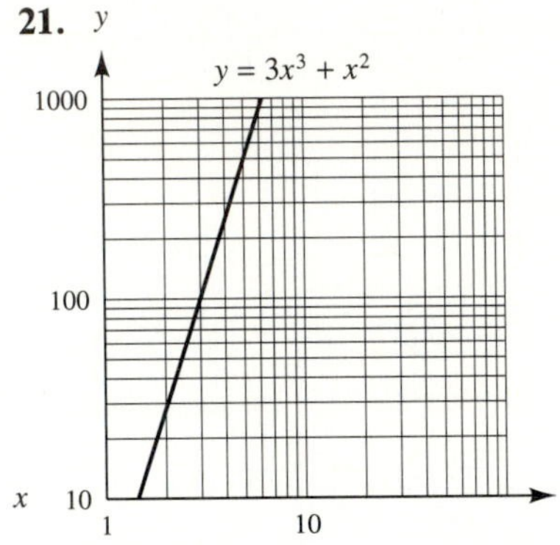

23.

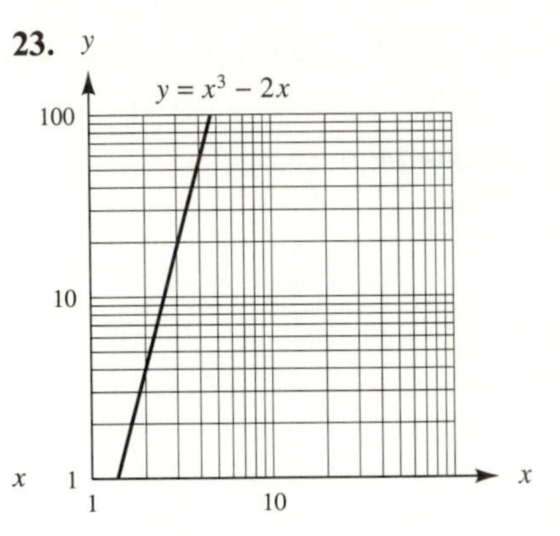

25. $Q = 3.6 \times 10^{-5}e^{-t/.002},\ 0 \le t \le 0.01$

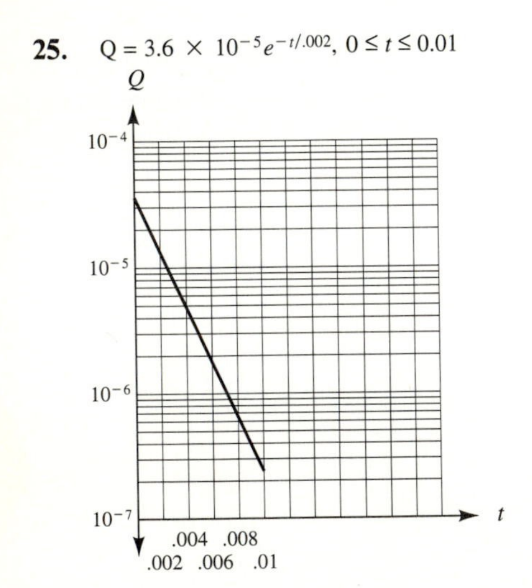

27. $\mathrm{pH} = -\log_{10}[\mathrm{H}^+],\ 0.001 < [\mathrm{H}^+] < 0.005$

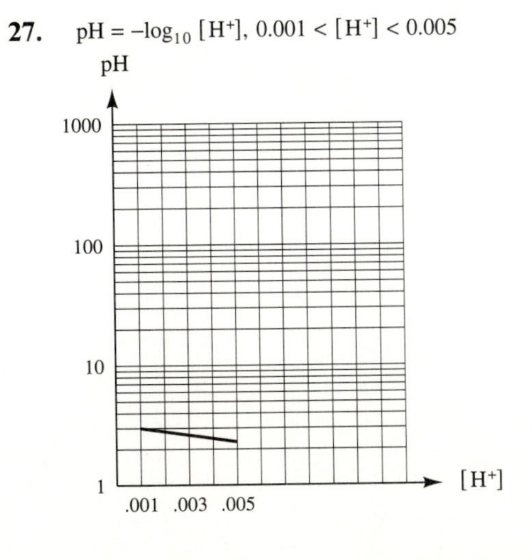

29. $G = -4.58(300) \log k$, $0.01 < k < 1$

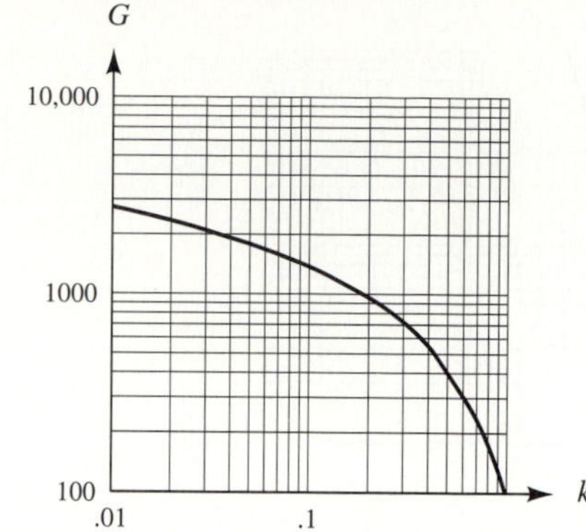

31. $Q = \log 0.992 + 1.547 \log H$, $1 < H \leq 4$

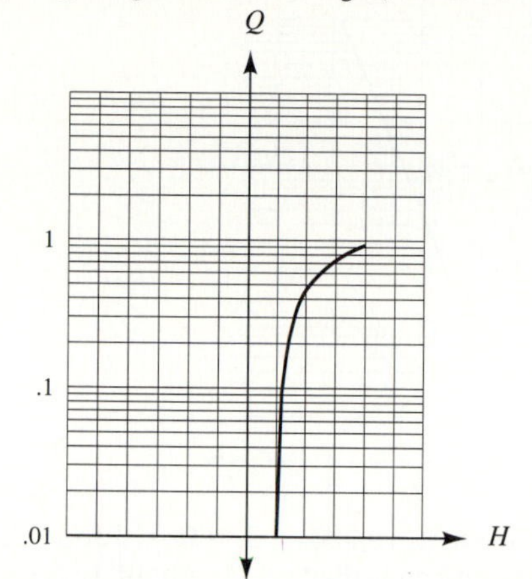

Section 15.8 (page 650)

Sequential Review:

1.

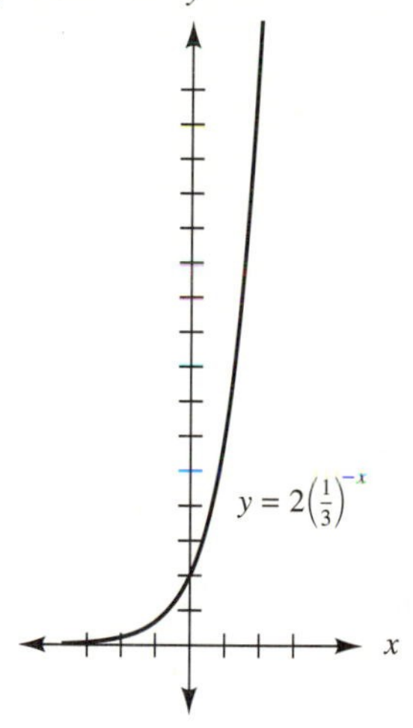

3.

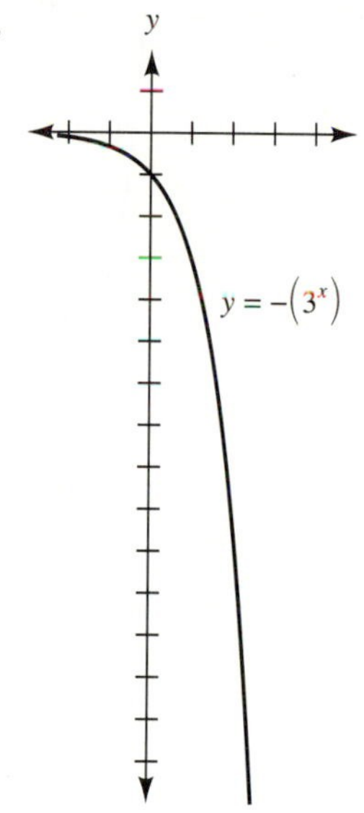

5.

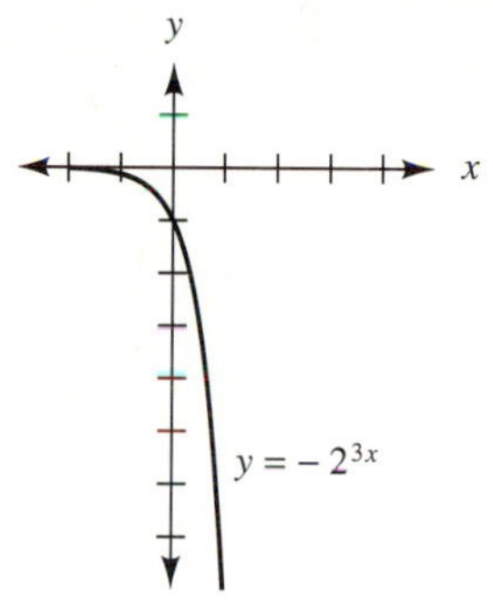

7. $\log_{125} 25 = 2/3$ **9.** $3^4 = 81$

11.

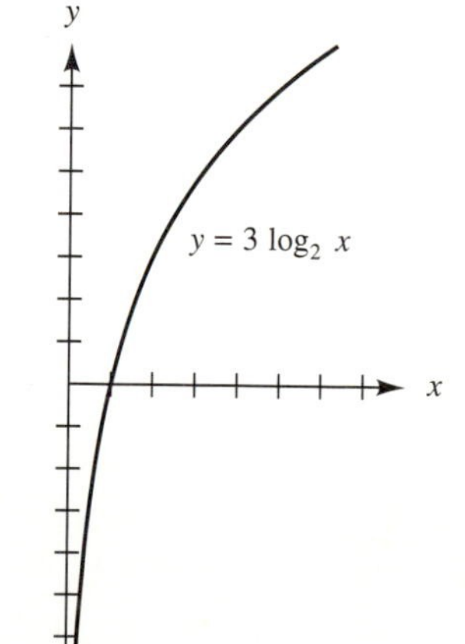

13.

y

x

$y = \log_{1/2} x$

15. $\log_b (4/3)$ **17.** $\log_2 625$
19. $2 + \log_3 5$ **21.** $(1/2) \log_b 2 + 1/2$
23. -1.38 **25.** 0.05228
27. 1.83 **29.** 1.77 **31.** -1.136
33. $\log x/\log 3$, 2.68
35. $\ln x/\ln 2$, 5.04 **37.** $e^{x \ln 9.31}$, 1.71
39. 512 **41.** 0.431

43.

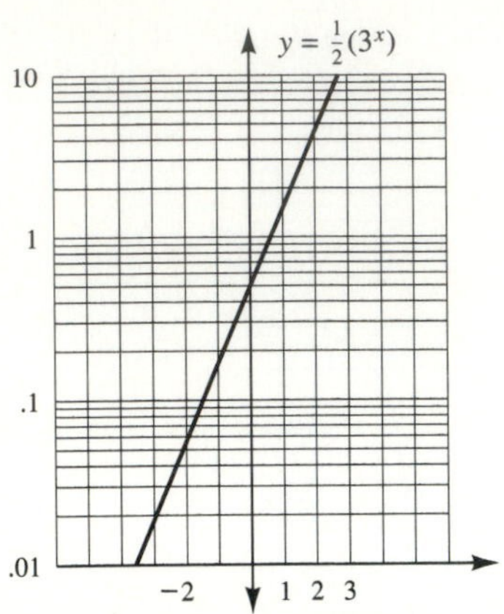

45.

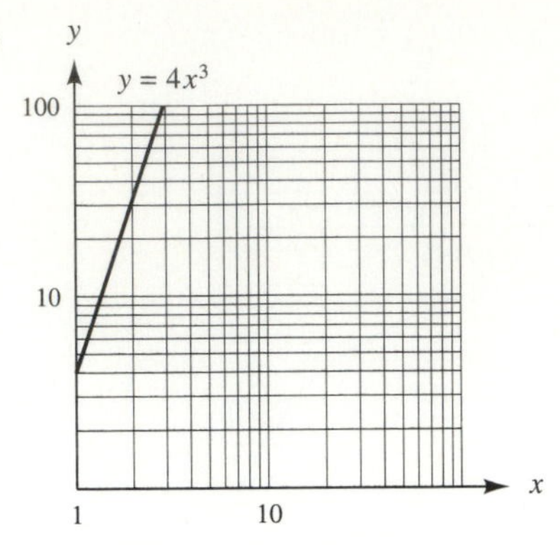

Applied Problems: 1. 16,000 **3.** 136 × 136 **5. (a)** 0.0173 **(b)** 7 billion **7.** 13 μW **9.** 118 V or 100 V, to one significant digit **11.** about 16.8 months **13.** on the 18th day **15.** 18 000 Ω or 18 kΩ **17.** 0.597 **19.** about 3

Random Review: 1. $1/3 + (1/3)(\log_3 y)$

2.

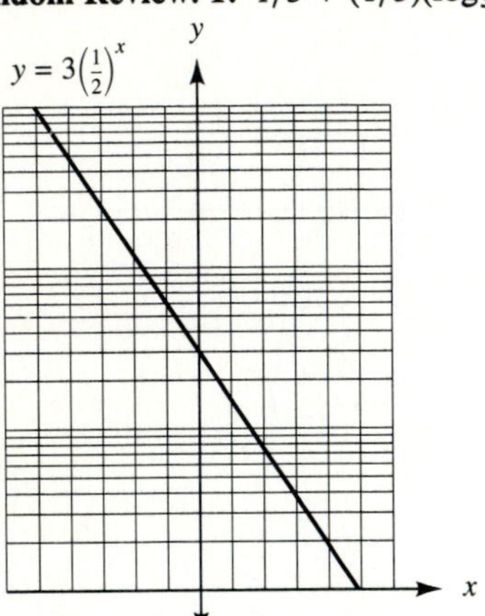

3.

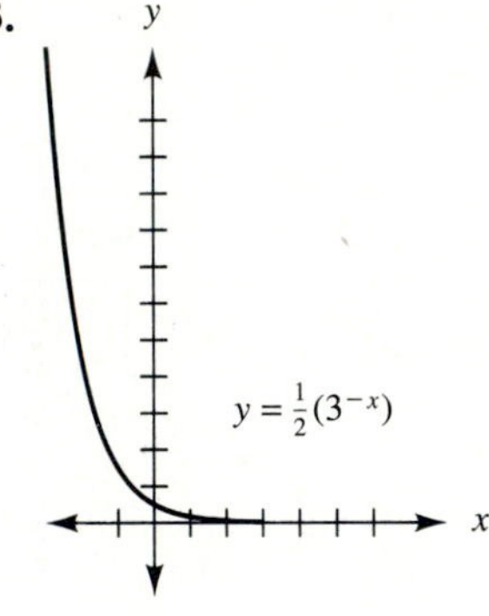

4. $e^{x \ln 26.3}$, 1,200,000 to three significant digits **5.** $8^{1/3} = 2$ **6.** 1.197 **7.** $\log_8 64 = 2$ **8.** $x = y/3$ **9.** 0.08946 **10.** $\log_7 25/3$

Chapter Test: 1. $\log_6 1/36 = -2$

2.

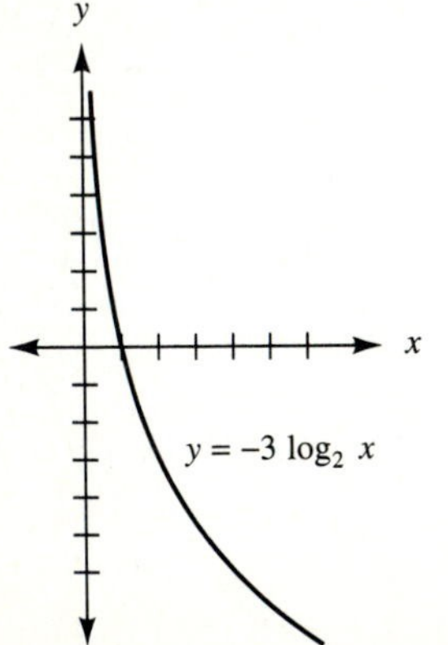

3. $2 + \log_5 2$ **4.** $\log_y 45 - 1$ **5.** 21.13 **6.** $\ln x/\ln 5$, −0.689 **7.** −1.26 **8.**

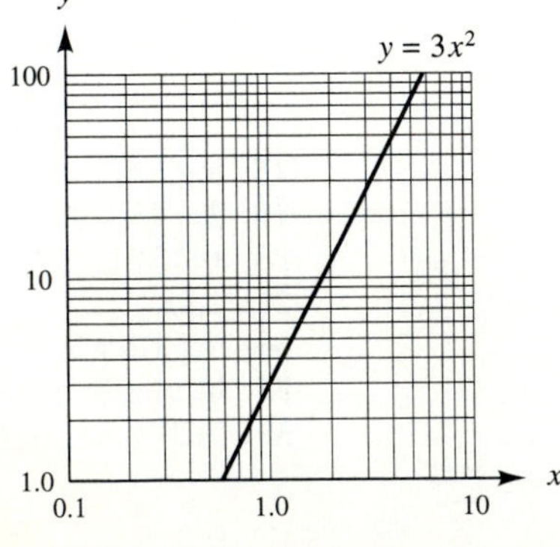

9. (a) log $30{,}000^{20}$ **(b)** 89.5 **10. (a)** -1.84 **(b)** -2.84

CHAPTER 16

Section 16.1 (page 660)

1. 4, 9, 14, 19 **3.** 2, −2, −6, −10 **5.** −20, −13, −6, 1 **7.** −40, −41, −42, −43 **9.** −5 **11.** not arithmetic **13.** −4 **15.** 98 **17.** −9 **19.** 28 **21.** −45 **23.** 12 **25.** 50 **27.** 11 **29.** 2550 **31.** 3050 **33.** 1575 **35.** −4900 **37.** 14 **39.** 15 **41.** 16 **43.** eighth day **45.** 780 **47.** 250.0 ft **49.** $396.00

Section 16.2 (page 666)

1. 3, 6, 12, 24 **3.** 5, −50, 500, −5000 **5.** 8, 2, 1/2, 1/8 **7.** 20, −5, 5/4, −5/16 **9.** 40, −8, 1.6, −0.32 **11.** 2 **13.** −5 **15.** not geometric **17.** not geometric **19.** −1/2 **21.** 393,660 **23.** −83,886,080 **25.** 1/32,768 **27.** −7/1024 **29.** $-j$ **31.** $1+j$ **33.** 2046 **35.** 34.60 **37.** 104.17 **39.** 700.56 **41.** $1+j$ **43.** 16/625 **45.** 1.76×10^3 **47.** $16,712.98 **49.** 30 miles

Section 16.3 (page 671)

1. (a) 1.49997 **(b)** 1.50000 **(c)** 3/2 **3. (a)** 1.17919 **(b)** 1.19964 **(c)** 6/5 **5. (a)** 2.83106 **(b)** 2.99049 **(c)** 3 **7. (a)** 17.99970 **(b)** 18.00000 **(c)** 18 **9. (a)** 12.42442 **(b)** 12.49954 **(c)** 25/2 **11.** 4/33 **13.** 8/9 **15.** 1/18 **17.** 236/495 **19.** 31,969/4950 **21.** 40.0 m **23.** 133.3 cm **25.** 192.0 cm^2

27. The number formed by the barred and unbarred digits minus the number formed by the unbarred digits

Section 16.4 (page 677)

1. $x^3 + 9x^2 + 27x + 27$ **3.** $81x^4 - 108x^3 + 54x^2 - 12x + 1$

5. $243a^5 + 405a^4b + 270a^3b^2 + 90a^2b^3 + 15ab^4 + b^5$

7. $729a^6 + 2916a^5b + 4860a^4b^2 + 4320a^3b^3 + 2160a^2b^4 + 576ab^5 + 64b^6$

9. $15{,}625a^6 - 56{,}250a^5b + 84{,}375a^4b^2 - 67{,}500a^3b^3 + 30{,}375a^2b^4 - 7290ab^5 + 729b^6$

11. $y^{10} - 20y^9 + 180y^8 - 960y^7$ **13.** $1 + 30b + 405b^2 + 3240b^3$

15. $x^{12} + 36x^{11}y + 549x^{10}y^2 + 5940x^9y^3$

17. $16{,}384x^{14} - 573{,}440x^{13}y + 9{,}318{,}400x^{12}y^2 - 93{,}184{,}000x^{11}y^3$

19. $a^{20} - 20a^{19}b + 190a^{18}b^2 - 1140a^{17}b^3$ **21.** $V = (4\pi/3)(r^3 + 3r^2(\Delta r) + 3r(\Delta r)^2 + (\Delta r)^3)$

23. $d = (w/8EI)(l^4 - 4l^3(\Delta l) + 6l^2(\Delta l)^2 - 4l(\Delta l)^3 + (\Delta l)^4)$

25. $P = (D^{2/3}/k)[v^3 - 3v^2(\Delta v) + 3v(\Delta v)^2 - (\Delta v)^3]$

Section 16.5 (page 682)

1. $1 - 4x + 10x^2 - 20x^3$ **3.** $1 + 5x + 15x^2 + 35x^3$ **5.** $1 + (1/2)x - (1/8)x^2 + (1/16)x^3$ **7.** $1 + (1/3)x + (2/9)x^2 + (14/81)x^3$ **9.** 1.03441 **11.** 2.14480 **13.** 1.02281 **15.** 2.95297 **17.** $270a^3$ **19.** $4320a^3b^3$ **21.** $17{,}010b^4$ **23.** 0.0405 **25.** 0.0078125 **27.** 0.36735

29. 0.04803 **31.** 0.00781 **33.** $\dfrac{n!}{k!(n-k)!}\, a^{n-k}b^k$

Section 16.6 (page 684)

Sequential Review: 1. 200, 204, 208, 212, 216 **3.** -7 **5.** 8 **7.** -420 **9.** 20 **11.** 4, 16, 64, 256 **13.** -3 **15.** 327,680 **17.** 1,062,880 **19.** 6 **21.** 1.4 **23.** 40.5 **25.** 25/99 **27.** 481/110 **29.** $x^5 + 5x^4 + 10x^3 + 10x^2 + 5x + 1$ **31.** $32a^5 + 240a^4b + 720a^3b^2 + 1080a^2b^3 + 810ab^4 + 243b^5$ **33.** $y^8 + 16y^7 + 112y^6 + 448y^5$ **35.** $3125a^5 - 9375a^4b + 11{,}250a^3b^2 - 6750a^2b^3$ **37.** $1 + 4x + 10x^2 + 20x^3$ **39.** $1 - (1/2)x + (3/8)x^2 - (5/16)x^3$ **41.** 2.62720 **43.** $6048a^5$ **45.** $-448b^3$

Applied Problems: 1. \$171,382.00 **3. (a)** $(705.6 - 592.9)$ ft $= 112.7$ ft **(b)** $(4.9 + 11(9.8))$ ft $= 112.7$ ft **5.** 760/99 **7.** 4.5 dB **9.** 8.5 dB **11.** 7.6 dB **13.** 2/3 mm **15.** 1.7 nA **17.** 44% **19. (a)** 1,048,576 **(b)** 1.4×10^{-8} inches

Random Review: 1. $16x^4 + 3.2x^3 + 0.24x^2 + 0.008x + 0.0001$ **2.** 12 **3.** -8 **4.** 666 **5.** 1.22234 **6.** $1 - 3x + 6x^2 - 5x^3$ **7.** 20 **8.** -2046 **9.** 128 **10.** $112{,}640x^9$

Chapter Test: 1. $1 - 3j$ **2.** 12 **3.** -9 **4.** 1705/512 **5.** 20/3 **6.** $1 - (1/4)x + (5/32)x^2 - (15/128)x^3$ **7.** 1.37192 **8.** $0.08064x^5$ **9.** 16.75 ft **10.** geometric progression

CHAPTER 17

Section 17.1 (page 699)

1. (3, 5)

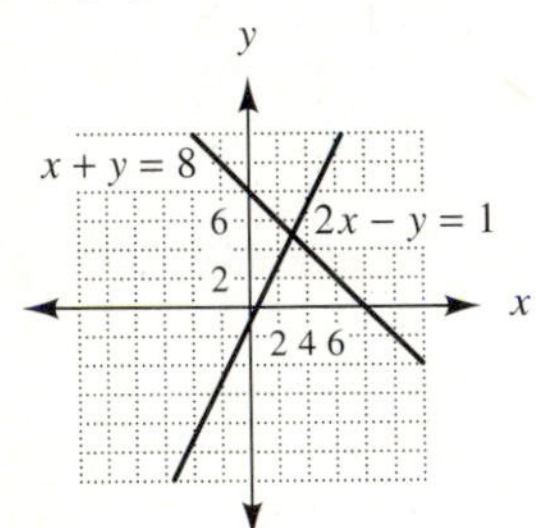

3. (1/5, −1/5)

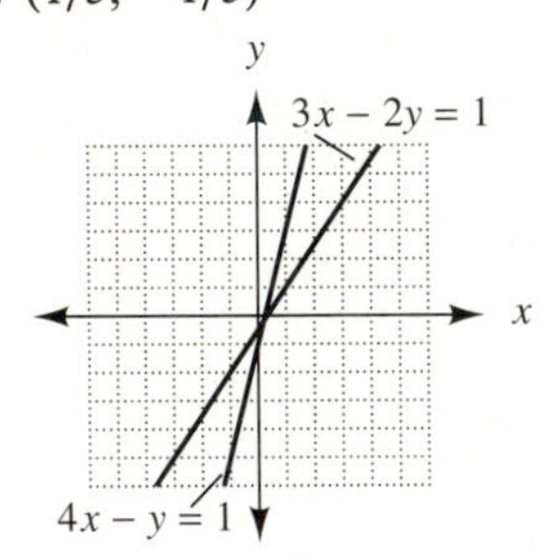

5. (1/5, 2/5)

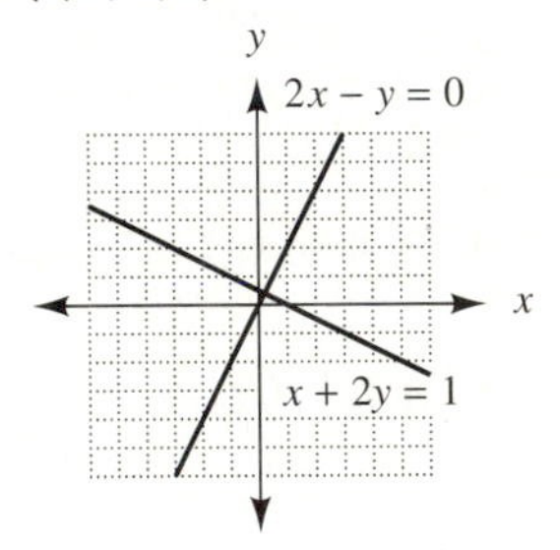

7. inconsistent

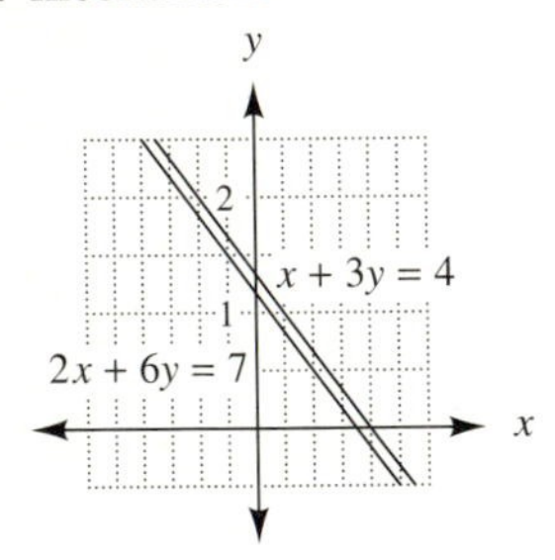

9. dependent

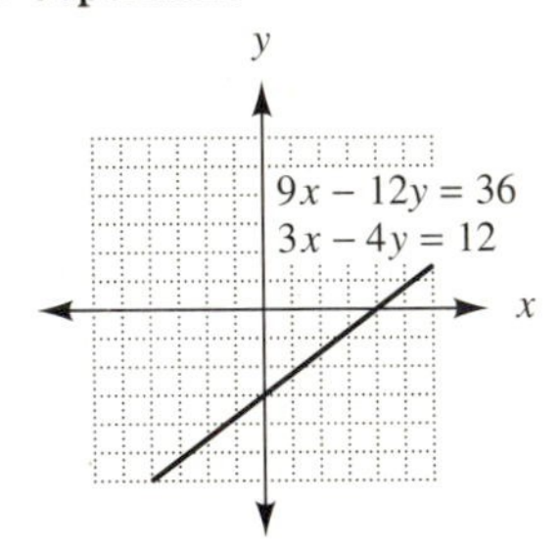

11. (3, 5) **13.** (1/5, −1/5) **15.** (1/5, 2/5) **17.** inconsistent **19.** dependent **21.** (−2, 3) **23.** (0, 0) **25.** (1.0, 0.1) **27.** (2.0, −0.4) **29.** (1.1, −0.3) **31.** $x + y = 20$; $0.35x + 0.10y = 5.00$; $x = 12$, $y = 8$ **33.** 3 at 100/h, 5 at 120/h **35.** 3 tons of the first alloy, 2 tons of the second **37.** 3 ft for 100 lb weight, 1 ft for 300 lb weight

Section 17.2 (page 706)

1. (3, 5) **3.** (1/5, −1/5) **5.** (1/5, 2/5) **7.** inconsistent **9.** dependent **11.** (−2, 3) **13.** (0, 0)
15. (3, −1) **17.** (1/2, −1/2) **19.** (−4, −2) **21.** (−16, −11/3) **23.** (57/20, −11/20)
25. 9.6 cm, 9.6 cm, and 4.8 cm **27.** $x = 108$ lb, $y = 62.5$ lb
29. 48 ml of 10% solution, 32 ml of 8% solution **31.** 28.8 km/h for the boat, 3.3 km/h for the current
33. $R_1 = 27$ kΩ, $R_2 = 5.6$ kΩ

Section 17.3 (page 714)

1. (3, 2, −1) **3.** (1, 1, 1) **5.** (3, −3, −3) **7.** (1/2, −1/2, 0) **9.** (2/3, −1/3, 2) **11.** (2, −1, 1)
13. $F_{AB} = 15.8$ N, $N_{AB} = 45.1$ N, $T = 16.6$ N **15.** $I_1 = 0.182$ mA, $I_2 = 0.628$ mA, $I_3 = -0.446$ mA
17. 7.0 cm, 8.0 cm, and 9.0 cm **19.** 6 small, 9 medium, 3 large

Section 17.4 (page 722)

1. (−4, 3), (4, −3)

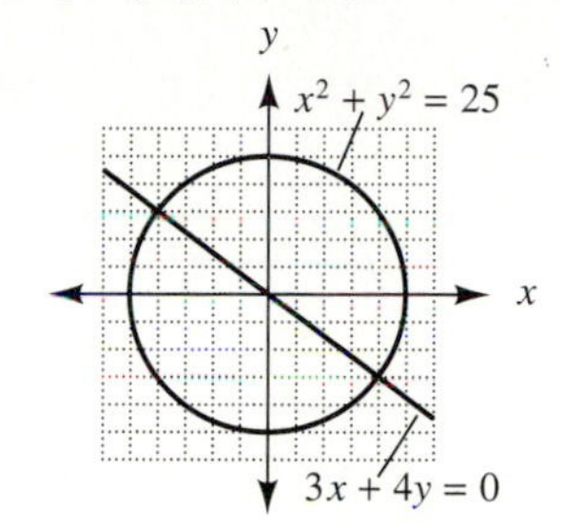

3. (1, 2)

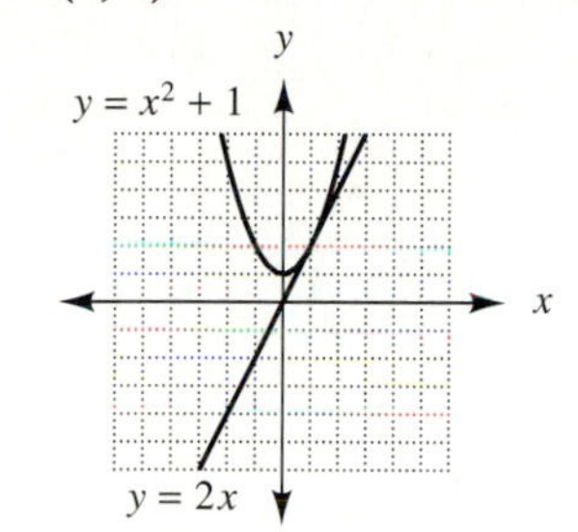

5. (−2, −1), (1, 2)

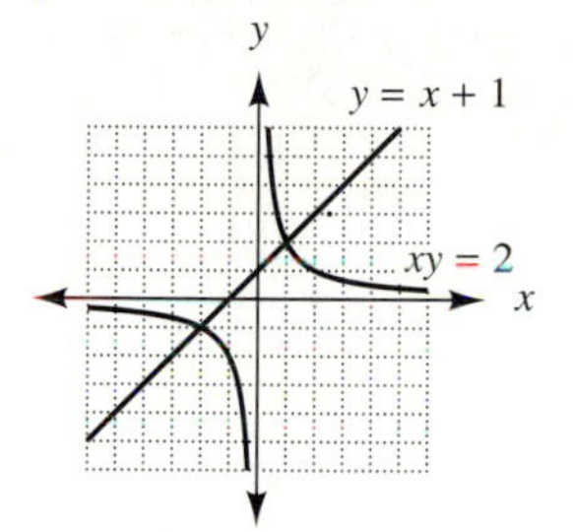

7. (0.5, 1.7)

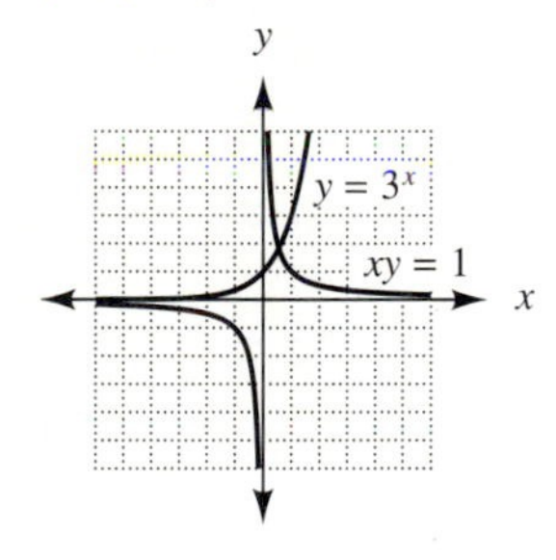

9. (−1, 0), (0.6, 1.6)

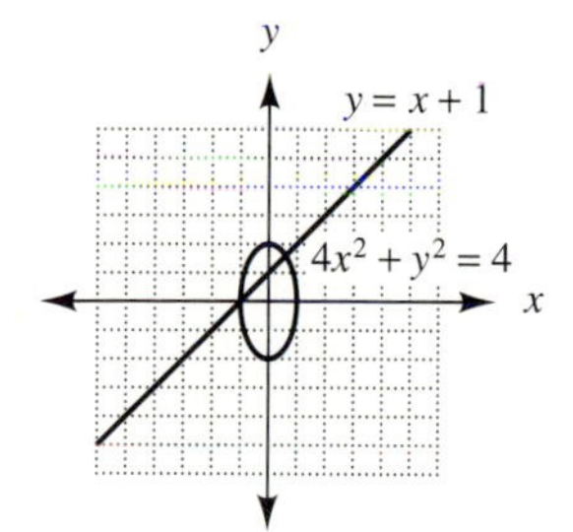

11. no real solution

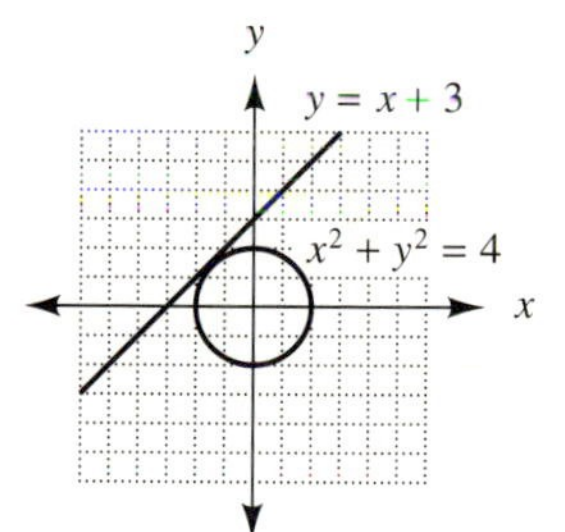

13. (4, −3), (−4, 3) **15.** (1, 2) **17.** (1, 2), (−2, −1) **19.** (2, 1), (2, −1), (−2, 1), (−2, −1)
21. (1, 0) **23.** no real solution
25. 50

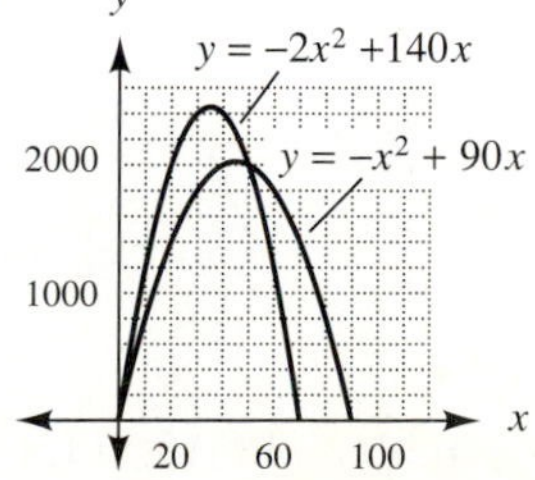

27. (4000, 1280)

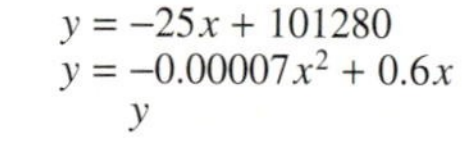

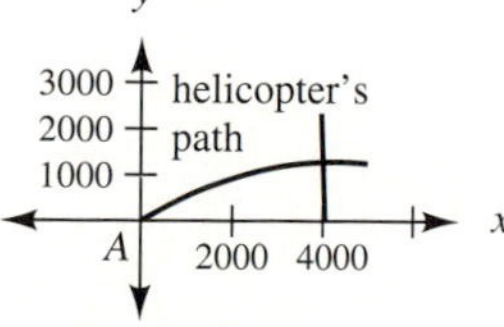

29. 20 m × 30 m

Section 17.5 (page 723)

Sequential Review:

1. (4, 2) **3.** (1, 1) **5.** (4, 2)

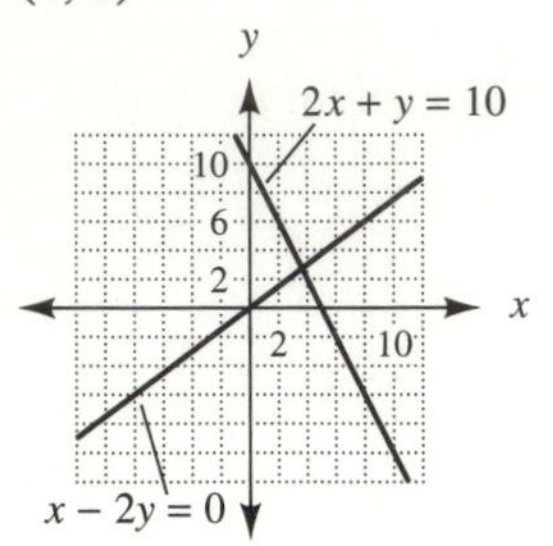

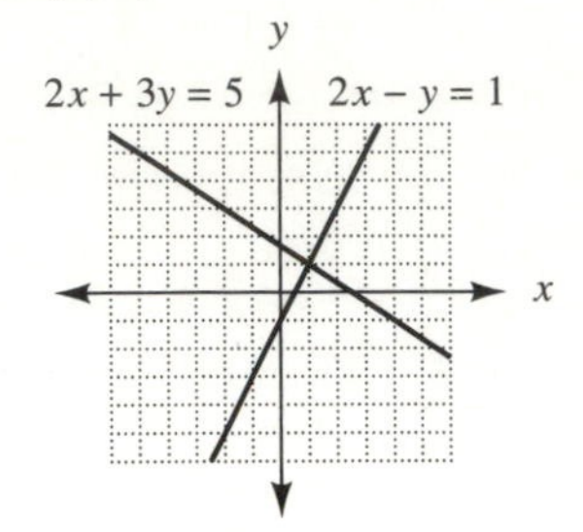

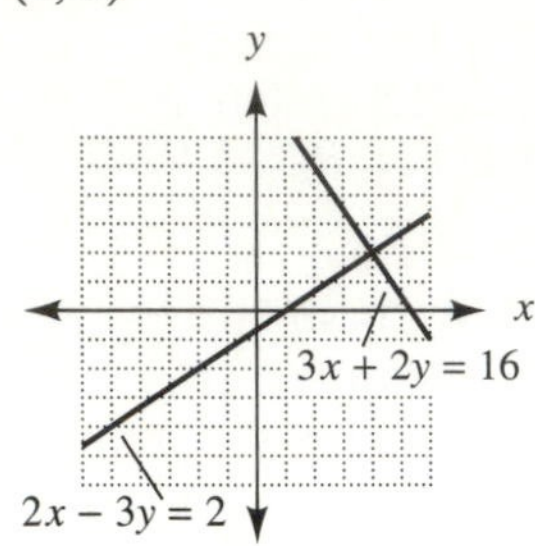

7. inconsistent **9.** (8/5, 1/5) **11.** (5.5, 5.5) **13.** (15/4, 5/4) **15.** (5, 1) **17.** (−11/2, −13/2) **19.** (−1, −5) **21.** (11/34, 7/34) **23.** (1, 4) **25.** (2, 1, −2) **27.** (−1, −2, 3) **29.** (0, −1, −3) **31.** (1, 2, 1) **33.** (2, 1, −2)

35. (3, 4), (−3, −4) **37.** (0, 0), (4, 8) **39.** (1, 2), (−1, −2)

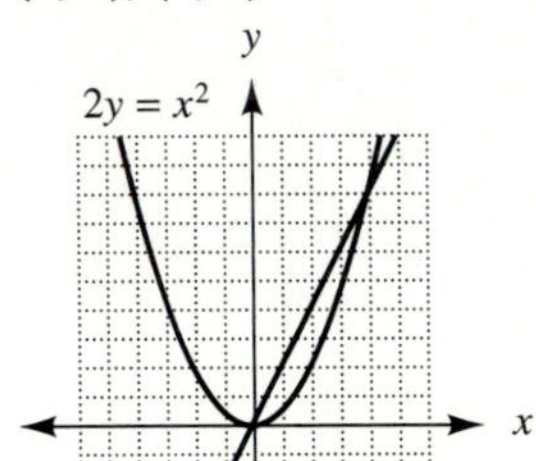

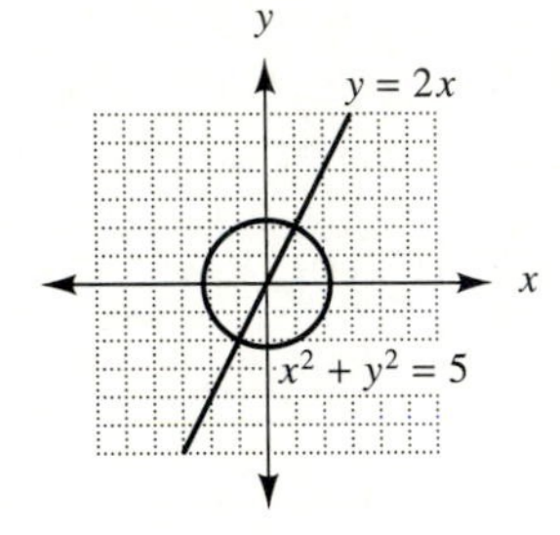

41. (4, 1), (−1/2, −8) **43.** (2, 8), (−2, 8) **45.** (−3, 0), (3, 0), ($\sqrt{8}$, −1), ($-\sqrt{8}$, −1)

Applied Problems: 1. $I_1 = 0.98$ mA, $I_2 = -1.07$ mA, $I_3 = 0.09$ mA
3. $R_b = 700$ lb, $R_{ax} = 400$ lb, $R_{ay} = 700$ lb **5.** $x = -38.5$, $y = 55.0$ **7.** $C = 6.3$ g, $G = 18.8$ g
9. $(x - 15)^2 + (y - 35)^2 = 25^2$, $y = (4/3)x$ **11.** \$2.50 for the case, \$102.50 for the calculator
13. $a = 61.9$, $b = -92.9$, $c = 39.0$ **15.** about (4 1/2, 2) or (−1/2, 5) **17.** 300 Ω and 600 Ω

Random Review:

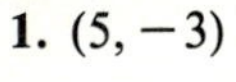

1. (5, −3)

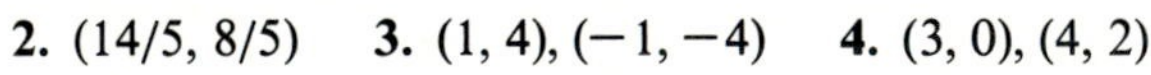

2. (14/5, 8/5) **3.** (1, 4), (−1, −4) **4.** (3, 0), (4, 2)

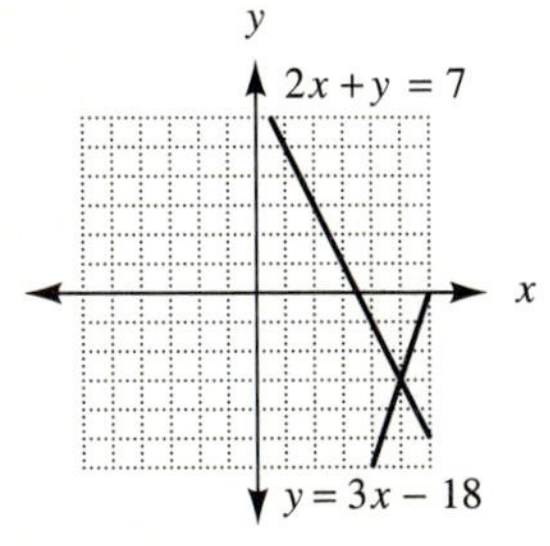

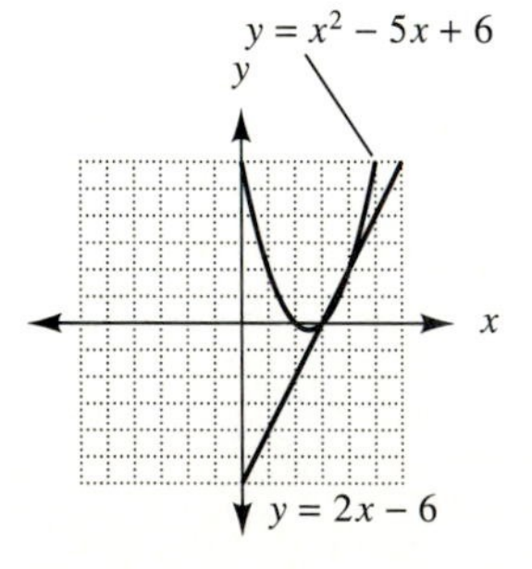

5. (−1, 1.5) **6.** (2, 0, −3) **7.** (13/6, −3/2) **8.** (4/3, 5/3) **9.** (32/29, 33/29) **10.** (5, −1, 3)

Chapter Test: 1. (1, 2)

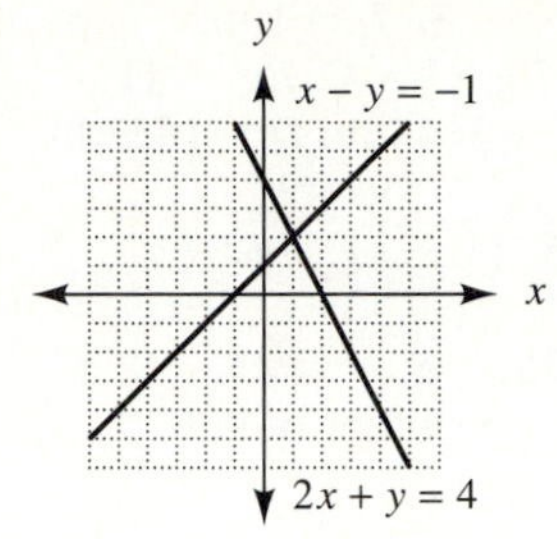

2. (−3, 2) **3.** (5, 8) **4.** inconsistent system
5. (1, 0, −1) **6.** dependent
7. (2, −1), (2, 1), (−2, 1), (−2, −1)
8. (1, 3), (3, 3)

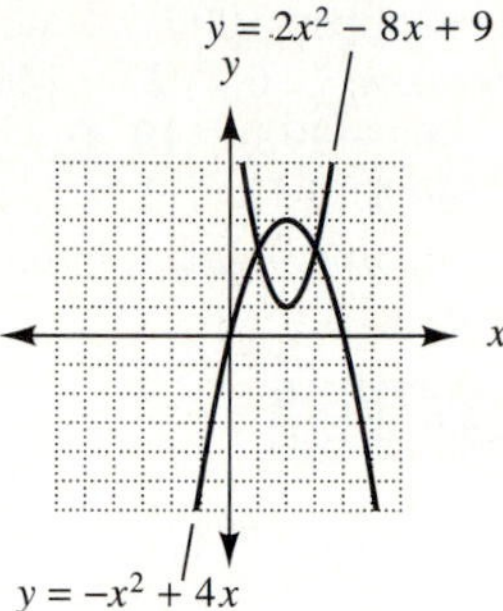

9. $I_1 = 2$ mA, $I_2 = -1.5$ mA, $I_3 = -0.5$ mA
10. 27.8 gallons of paint, 72.2 gallons of thinner

CHAPTER 18

Section 18.1 (page 736)

1. −5 **3.** −2 **5.** −2 **7.** 14 **9.** 1 **11.** (3, 5) **13.** (1/5, 2/5) **15.** dependent
17. (18/17, −27/17) **19.** (1, −1/4, −3/2) **21.** inconsistent **23.** (2, −1, 0)
25. 60.5 mph and 55.5 mph **27.** $T_1 = 329$ lb, $T_2 = 120$ lb **29.** 25.7°, 51.4°, 102.9°
31. $I_1 = 2.5$ mA, $I_2 = 1.5$ mA, $I_3 = 1.0$ mA

Section 18.2 (page 742)

1. −14 **3.** 1 **5.** 97 **7.** 152 **9.** −6 **11.** −386 **13.** (−2, 3, 2) **15.** (1, −2, −2)
17. (2, −1, 1, −1) **19.** 8.0 cm, 12.0 cm, 16.0 cm
21. $F = 119$ newtons, $N = 341$ newtons, $T = 227$ newtons **23.** 4/3 **25. (a)** 6 **(b)** 152

Section 18.3 (page 750)

1. 0 **3.** 0 **5.** 15 **7.** 0 **9.** 176 **11.** −437 **13.** −6 **15.** 94 **17.** 6 **19.** −117
21. −334 **23.** 1888 **25.** $\begin{vmatrix} x_b & y_b & z_b \\ x_c & y_c & z_c \\ x_a & y_a & z_a \end{vmatrix}$, yes **27.** They are equal.
29. No: the determinants have opposite signs.

Section 18.4 (page 755)

1. −40 **3.** −70 **5.** 36 **7.** 144 **9.** −36 **11.** 90 **13.** 40 **15.** 64 **17.** 290 miles
19. 4 **21.** 0.0046 A

Section 18.5 (page 757)

Sequential Review: 1. 27 **3** −30 **5.** −208 **7.** 112 **9.** (3, −1) **11.** (2, −1, 1) **13.** −208
15. 112 **17.** −7854 **19.** (−1, 1, 0) **21.** (3, −3, 5) **23.** (1, 0, −2) **25.** −52 **27.** −52
29. −52 **31.** 0 **33.** 0 **35.** −56 **37.** −64 **39.** 210 **41.** 810 **43** −440 **45.** −196

Applied Problems: 1. $P = 56.1$ N, $F_B = 40.3$ N, $N_B = 115$ N **3.** $I_1 = 1$ mA, $I_2 = -2$ mA, $I_3 = 1$ mA **5.** $w = 20$ in, $h = 4$ in **7.** 4.43 **9.** $A = 34.7$ g, $B = 22.7$ g, $C = 42.7$ g **11.** 12 Ω **13.** $A = 5, B = 8, C = 9$ **15.** $A = 4857$ lb, $B = 143$ lb, $C = 3143$ lb, $D = 1857$ lb **17.** 3000 lb copper, 5000 lb lead, 4000 lb zinc, 1200 lb tin

Random Review: 1. 6 **2.** -146 **3.** -105 **4.** 12 **5.** $(5, -1)$ **6.** -7 **7.** inconsistent system **8.** -174 **9.** -392 **10.** 0

Chapter Test: 1. -13 **2.** 57 **3.** 115 **4.** 45 **5.** -25 **6.** -2021 **7.** $(5, 3)$ **8.** $(1, 0, -1)$ **9.** plane: 310 mph, wind: 10 mph **10.** 75 gallons of 70%, 25 gallons of 90%

CHAPTER 19

Section 19.1 (page 771)

1. $\begin{bmatrix} 2 & 2 \\ 10 & -5 \end{bmatrix}$ **3.** $\begin{bmatrix} 3 & 10 \\ -6 & 1 \end{bmatrix}$ **5.** $\begin{bmatrix} 10 & -8 & 0 \\ 6 & -2 & -2 \\ 0 & 12 & -7 \end{bmatrix}$ **7.** $\begin{bmatrix} -3 & -17 \\ -2 & 11 \\ 2 & -9 \end{bmatrix}$ **9.** $\begin{bmatrix} 11 & -7 & 4 \\ -1 & 9 & -4 \end{bmatrix}$

11. not defined **13.** $\begin{bmatrix} 42 & -82 \\ -17 & 5 \end{bmatrix}$ **15.** $\begin{bmatrix} 19 \\ 36 \\ 41 \end{bmatrix}$ **17.** $\begin{bmatrix} 1 & 0 \\ 0 & 1 \end{bmatrix}$ **19.** $\begin{bmatrix} 13 & 8 & -1 \\ 8 & 16 & 4 \\ -1 & 4 & 2 \end{bmatrix}$

21. $\begin{bmatrix} 18 & 7 \\ 9 & 7 \end{bmatrix}$ **23.** $\begin{bmatrix} -2 \\ 20 \end{bmatrix}$ **25.** $\begin{bmatrix} 2 & 1 & -1 \\ 1 & 2 & -1 \\ -1 & -1 & 2 \end{bmatrix}$ **27.** $\begin{bmatrix} 1 & 2 & 3 \\ 4 & 5 & 6 \\ 7 & 8 & 9 \end{bmatrix}$ **29.** $\begin{bmatrix} 1 & 0 & 0 \\ 0 & 1 & 0 \\ 0 & 0 & 1 \end{bmatrix}$

31. lead zinc iron

$$\begin{bmatrix} 0.01 & 0.20 & 0.03 \\ 0.02 & 0.10 & 0.02 \\ 0.03 & 0.30 & 0.01 \end{bmatrix}$$

33. (a) invalid **(b)** valid **(c)** valid **(d)** invalid

35. $(-7/5, 24/5, 8)$

Section 19.2 (page 779)

1. $x = 3, y = 5$ **3.** $x = 1/5, y = -1/5$ **5.** $x = 1/5, y = 2/5$ **7.** inconsistent **9.** dependent **11.** $x = 3, y = 2, z = -1$ **13.** $x = 1, y = 1, z = 1$ **15.** $x = 3, y = -3, z = -3$ **17.** $x = 1/2, y = -1/2, z = 0$ **19.** $x = 2/3, y = -1/3, z = 2$ **21.** $x = 2, y = -1, z = 1$ **23.** $I_1 = 1.5$ mA, $I_2 = 2.0$ mA, $I_3 = -0.5$ mA **25.** $a = 16, b = 10, c = 4$ **27.** 12 type 1, 10 type 2, 2 type 3 **29.** $(n^3 + 3n^2)/2$

Section 19.3 (page 784)

1. $\begin{bmatrix} 5 & -4 \\ -6 & 5 \end{bmatrix}$ **3.** $\begin{bmatrix} 3 & 1 \\ -5 & -2 \end{bmatrix}$ **5.** $\begin{bmatrix} -1 & 1 \\ \frac{2}{3} & -\frac{1}{3} \end{bmatrix}$ **7.** $\begin{bmatrix} \frac{1}{2} & 2 \\ 0 & -1 \end{bmatrix}$ **9.** $\begin{bmatrix} \frac{3}{37} & \frac{18}{37} \\ -\frac{18}{37} & \frac{3}{37} \end{bmatrix}$

11. $\begin{bmatrix} \frac{3}{2} & -\frac{1}{2} & 0 \\ -\frac{3}{2} & -\frac{1}{2} & 1 \\ 1 & 1 & -1 \end{bmatrix}$ **13.** $\begin{bmatrix} 0 & \frac{3}{2} & -1 \\ 1 & -\frac{1}{2} & 0 \\ -1 & -\frac{1}{2} & 1 \end{bmatrix}$ **15.** $\begin{bmatrix} 13 & -3 & -4 \\ -7 & 2 & 2 \\ -2 & 0 & 1 \end{bmatrix}$ **17.** $\begin{bmatrix} -\frac{1}{3} & \frac{2}{3} & 0 \\ -2 & 5 & -1 \\ \frac{14}{3} & -\frac{31}{3} & 2 \end{bmatrix}$

19. $\begin{bmatrix} 3 & -3 & 1 \\ 0 & 1 & -1 \\ -1 & 0 & 1 \end{bmatrix}$ **21.** yes **23.** yes **25.** no **27.** yes **29.** no

31. $A = \begin{bmatrix} \frac{\sqrt{2}}{2} & \frac{\sqrt{2}}{2} \\ -\frac{\sqrt{2}}{2} & \frac{\sqrt{2}}{2} \end{bmatrix}$ $A^{-1} = \begin{bmatrix} \frac{\sqrt{2}}{2} & -\frac{\sqrt{2}}{2} \\ \frac{\sqrt{2}}{2} & \frac{\sqrt{2}}{2} \end{bmatrix}$, yes **33.** The product is $\begin{bmatrix} 1 & 0 \\ 0 & 1 \end{bmatrix}$

35. $A = \begin{bmatrix} \frac{1}{2} & \frac{\sqrt{3}}{2} & 0 \\ -\frac{\sqrt{3}}{2} & \frac{1}{2} & 0 \\ 0 & 0 & 1 \end{bmatrix}$ $A^{-1} = \begin{bmatrix} \frac{1}{2} & -\frac{\sqrt{3}}{2} & 0 \\ \frac{\sqrt{3}}{2} & \frac{1}{2} & 0 \\ 0 & 0 & 1 \end{bmatrix}$, yes **37.** $\begin{bmatrix} 13 & -3 & -4 \\ -7 & 2 & 2 \\ -2 & 0 & 1 \end{bmatrix}$

Section 19.4 (page 789)

1. $x = 2, y = 4$ **3.** $x = 2/3, y = 1/2$ **5.** $x = 1, y = -1$ **7.** $x = 1/2, y = 1/3$
9. $x = 2, y = 1, z = -1$ **11.** $x = 2, y = 1, z = -1$ **13.** $x = 4, y = 3, z = -2$

15. $x = -1, y = -1, z = 1$ **17.** $\begin{bmatrix} 0 & \cos 40° & -1 \\ 1 & -\sin 40° & 0 \\ 0 & 6 \sin 40° & 0 \end{bmatrix} \begin{bmatrix} R_{ay} \\ R_b \\ R_{ax} \end{bmatrix} = \begin{bmatrix} 100 \\ 200 \\ 1800 \cos 30° + 1200 \end{bmatrix}$

19. $[x, y, z] = [x', y', z'] \begin{bmatrix} \cos\theta & -\sin\theta & 0 \\ \sin\theta & \cos\theta & 0 \\ 0 & 0 & 1 \end{bmatrix}$ **21.** $\begin{aligned} I_{xx}\,\omega_1 - I_{xy}\,\omega_2 - I_{xz}\,\omega_3 &= h_1 \\ -I_{xy}\,\omega_1 + I_{yy}\,\omega_2 - I_{yz}\,\omega_3 &= h_2 \\ -I_{xz}\,\omega_1 - I_{yz}\,\omega_2 - I_{zz}\,\omega_3 &= h_3 \end{aligned}$

Section 19.5 (page 791)

Sequential Review: 1. $\begin{bmatrix} 3 & -3 \\ 2 & 1 \end{bmatrix}$ **3.** $\begin{bmatrix} 0 & 1 \\ -2 & -7 \\ 4 & 10 \end{bmatrix}$ **5.** not defined **7.** $\begin{bmatrix} 2 & -5 \\ 18 & -15 \end{bmatrix}$

9. $\begin{bmatrix} 60 & -3 & 13 \\ 8 & 41 & -9 \\ 30 & -69 & 24 \end{bmatrix}$ **11.** $\begin{bmatrix} 9 & -4 & -3 \\ 9 & 10 & 18 \\ 7 & -9 & -4 \end{bmatrix}$ **13.** $(-1, 4)$ **15.** $(2, -1)$ **17.** $(7, -5)$

19. $(1, 1, -1)$ **21.** $(1, 2, 3)$ **23.** $(2, 1, -2)$ **25.** $\begin{bmatrix} 9 & -4 \\ -2 & 1 \end{bmatrix}$ **27.** $\begin{bmatrix} -\frac{1}{31} & \frac{3}{31} \\ \frac{9}{31} & \frac{4}{31} \end{bmatrix}$

29. $\begin{bmatrix} -1 & 0 & 1 \\ 1 & 1 & -1 \\ -1 & -4 & 2 \end{bmatrix}$ **31.** no **33.** no **35.** $(-1, 2)$ **37.** $(2, 0)$ **39.** $(6, -3)$ **41.** $(1, 1, 1)$

43. $(-2, 0, 1)$ **45.** $(2, 3, -1)$

Applied Problems: 1. $\begin{bmatrix} 3 & 3.4 & 9 \\ 700 & 200 & 150 \end{bmatrix}$ **3.** 38.5 gal of 98% pure, 61.5 gal of 85% pure

5. $\begin{bmatrix} 0.240 & 8 & 20 \\ 1 & 10 & 30 \\ 2 & 16 & 60 \end{bmatrix}$ **7.** $a = 107, b = 16.6, c = 2.18$ **9.** $\begin{bmatrix} 18 & 6 & 16 & 14 \\ 54 & 18 & 48 & 42 \\ 9 & 3 & 8 & 7 \\ 45 & 15 & 40 & 35 \end{bmatrix}$ **11.** $y_1 = 2, y_2 = -3$

13. $x' = 16.96, y' = 10.38, z' = -3$ **15.** $\begin{bmatrix} 1.00 \\ 0.286 \end{bmatrix}$ **17.** 50 bolts, 30 grommets, 110 washers

19. $V_1 = 14.9, V_2 = 8.57, V_3 = 10.9$

Random Review: 1. no **2.** $\begin{bmatrix} 16 & 35 \\ -3 & 31 \end{bmatrix}$ **3.** $(2, -5)$ **4.** $\begin{bmatrix} 38 & 2 & -11 \\ -14 & -1 & 4 \\ 35 & 2 & -10 \end{bmatrix}$ **5.** $(-1, 2, 0)$

6. $\begin{bmatrix} \frac{3}{85} & \frac{2}{17} \\ \frac{7}{85} & -\frac{1}{17} \end{bmatrix}$ **7.** not defined **8.** $(3, 8)$ **9.** $(3, -1, 2)$ **10.** $\begin{bmatrix} 4 & 8 & 8 \\ 5 & 1 & 8 \\ 6 & 9 & 1 \end{bmatrix}$

Chapter Test: 1. $\begin{bmatrix} 5 & 4 \\ 2 & 4 \end{bmatrix}$ **2.** not defined **3.** $\begin{bmatrix} 36 & -37 \\ -21 & 40 \end{bmatrix}$ **4.** $\begin{bmatrix} 42 & 64 & -2 \\ -1 & -4 & -3 \\ 51 & 87 & 9 \end{bmatrix}$

5. $(1, 0, -1)$ **6.** $\begin{bmatrix} \frac{3}{7} & \frac{41}{63} & -\frac{38}{63} \\ \frac{1}{7} & \frac{2}{63} & -\frac{8}{63} \\ \frac{1}{7} & \frac{16}{63} & -\frac{1}{63} \end{bmatrix}$ **7.** $\begin{bmatrix} 3 & 2 & 1 \\ 1 & -3 & -2 \\ -2 & 1 & 3 \end{bmatrix}\begin{bmatrix} x \\ y \\ z \end{bmatrix} = \begin{bmatrix} 6 \\ -4 \\ 2 \end{bmatrix}$ **8.** $(3, -2, 1)$

9. $I_1 = -0.588$ mA, $I_2 = -0.294$ mA, $I_3 = 0.882$ mA **10.** $\begin{bmatrix} \cos 60° & -\cos 40° \\ \sin 60 & -\sin 40° \end{bmatrix}\begin{bmatrix} R \\ T \end{bmatrix} = \begin{bmatrix} 0 \\ w \end{bmatrix}$

CHAPTER 20

Section 20.1 (page 804)

1. 12 **3.** 41 **5.** 13 **7.** 6.75 **9.** 0.12 **11.** 2.32 **13.** 30 **15.** 11 **17.** 13 **19.** 9 **21.** 2.2 **23.** none **25.** 32 **27.** 10 **29.** 1 **31.** none **33.** none **35.** mean **37.** mode **39.** median **41.** mode **43.** median **45.** 88,302 lb, 88,303 lb, none **47.** 41 V, 42 V, 42 V **49.** 2.22×10^{18}, 2.22×10^{18}, 2.21×10^{18} **51. (a)** each increased by 4 **(b)** each multiplied by 4

Section 20.2 (page 808)

1. 12, 7 **3.** 12, 24 **5.** 43, 33 **7.** 43, 28 **9.** 55, 67 **11.** 6.4, 7.8 **13.** 6 **15.** 106 **17.** 93.3 **19.** 170 **21.** 345 **23.** 14, to two significant digits **25.** 2.24 **27.** 5.52 **29.** 2 **31.** 5.80 **33.** 4.28 **35.** 3.7 **37.** 0.77 in, 0.01 in **39.** 1.85 yd³, 0.255 yd³ **41.** 1.442 cm, 0.00075 cm **43.** 7.40, 0.0141

Section 20.3 (page 814)

1. 1.50 **3.** 2.50 **5.** -2.75 **7.** -1.43 **9.** 2.43 **11.** 4.7% **13.** 81.6% **15.** 6.7% **17.** 7.6% **19.** 33.2% **21.** 36.7% **23.** 92.7% **25.** 62.5% **27.** 33.7% **29.** 24.2% **31.** 3.8%

Section 20.4 (page 821)

1.

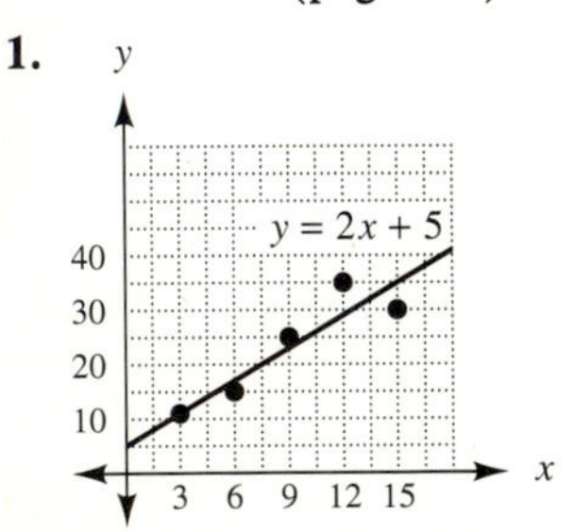

3.

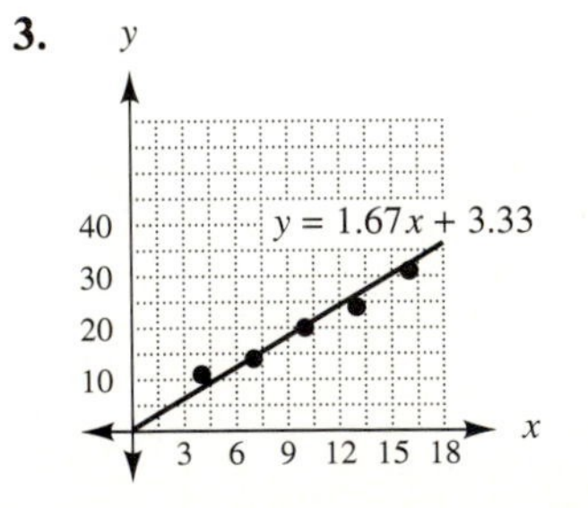

5.

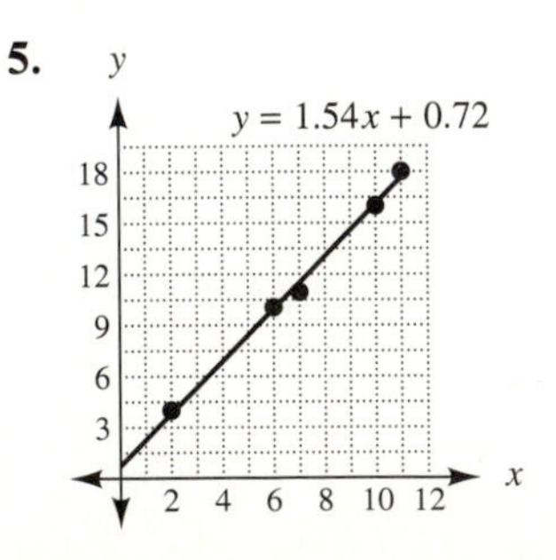

7.

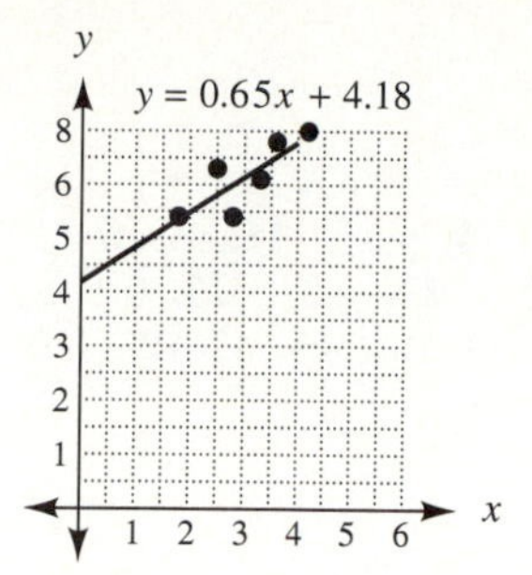

9.

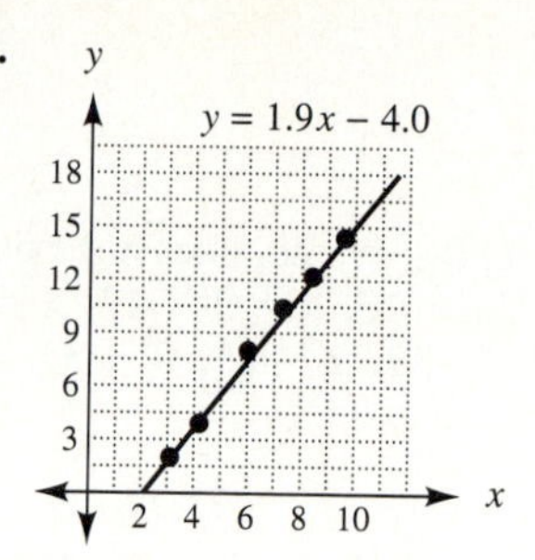

11. $R = 0.04T + 9.00$ **13.** $p = 3.50d + 0.20$ **15.** $e = 0.08t - 0.03$ **17.** $T = -0.025t + 104$

Section 20.5 (page 823)

Sequential Review: 1. 7 **3.** 6.6 **5.** 7 **7.** 4 **9.** no mode **11.** mean **13.** 19, 23 **15.** 9 4/7, 12 **17.** 98 **19.** 491.2 **21.** 3.87 **23.** 6.30 **25.** 2.33 **27.** −1 **29.** −0.5 **31.** 15.9% **33.** 84.1% **35.** 77.4% **37.** $y = 1.37x - 2.17$ **39.** 2.0 **41.** 0.4

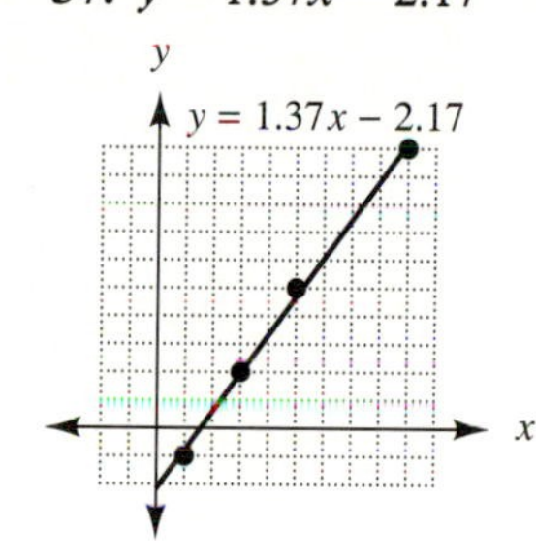

43. $y = 1.1x + 0.02$ **45.** $y = 0.51x - 2.5$

Applied Problems: 1. 15.8 roentgens/hour, each answer **3.** 0.1% **5.** 62.5% **7.** $\mu_I = 34.8$, $\sigma_I = 18.4$, $\mu_v = 8$, $\sigma_v = 4.2$ **9.** 3 people **11.** 4.7%, 25.1% **13. (a)** 49 **(b)** 81.4% **15.** 4.7% **17. (a)** 0.4% **(b)** 2.3% **19.** $y = -16x + 32{,}000$

Random Review: 1. 22.7% **2.** 9 **3.** 5.6 **4.** 17 **5.** 17 **6.** 17 **7.** 6.7% **8.** $y = 4.20x - 15.8$ **9.** 0.4 **10.** 2.03

Chapter Test: 1. 9.75 **2.** 3 **3.** 9 **4.** 18 **5.** 3.48 **6.** −1 **7.** 81.8% **8.** $y = 2.03x + 0.762$ **9.** 56.9, 9.96 **10. (a)** 0.7% **(b)** 99.3%

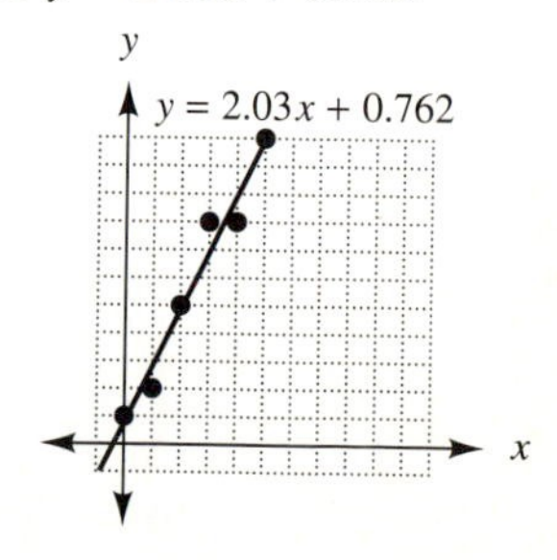

CHAPTER 21

Section 21.1 (page 840)

1. no, $-1, -2$ **3.** yes, 0, 0 **5.** no, does not exist, does not exist **7.** no **9.** yes **11.** $x=0$
13. $x=\pm\pi/2, \pm 3\pi/2, \ldots$ **15.** $x=0$ **17.** $x=0$ **19.** $x=3/2$ **21.** 5 **23.** -2 **25.** 0
27. 1/3 **29.** 0 **31.** 1 **33.** 2 **35.** 0 **37.** 2/3 **39.** 1/2 **41.** 0 **43.** 0 **45.** 12 dB

Section 21.2 (page 851)

1. $4x$, 8 **3.** $-2x, -3$ **5.** $4x+1, -3$ **7.** $2x+2$, 2 **9.** $6x-2$, 4 **11.** $3x^2+2$, 5
13. $-3x^2, -3$ **15.** $dA/dr = 2\pi r$ **17.** $dW/dV = CV$ **19.** $dV/dr = 4\pi r^2$ **21.** $dR/dV = 2V/P$

Section 21.3 (page 856)

1. $2x$ **3.** $-9x^2$ **5.** $-2/x^2$ **7.** $1/(2x^2)$ **9.** -2 **11.** 1/4

13. 2

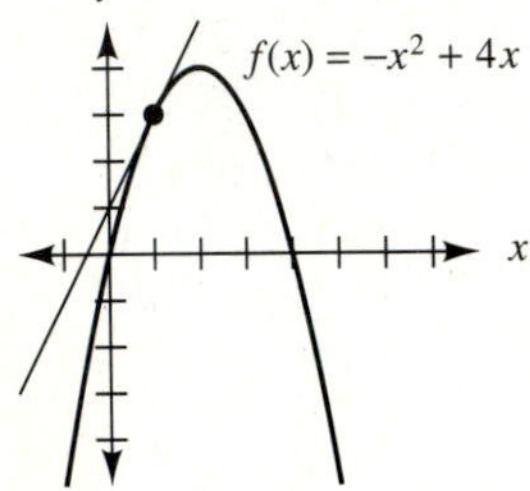

15. 0

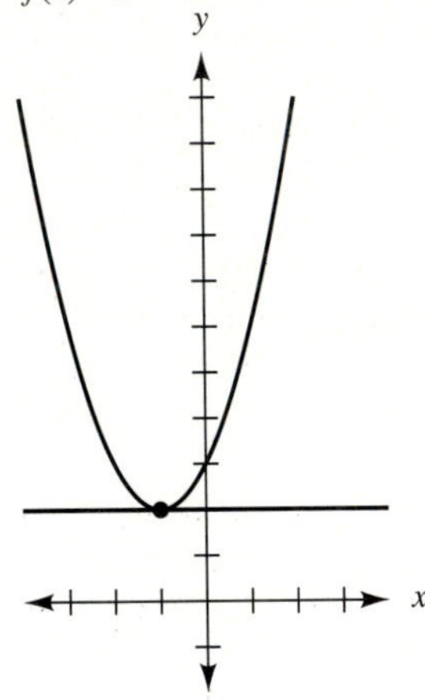

17. -1

y
$y = x^3 - 4x$
x

19. 4

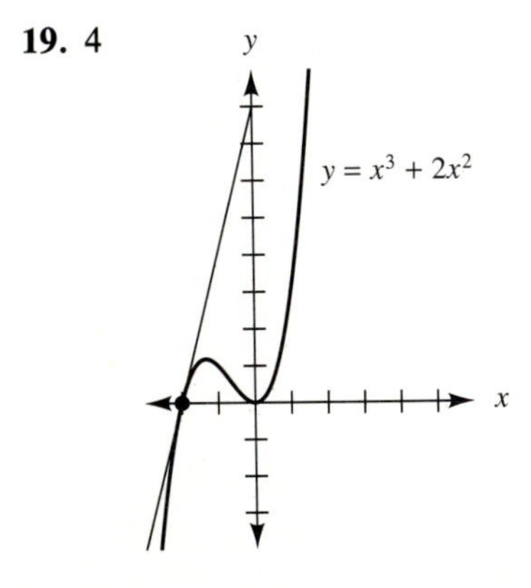

21. 15.0 radians/s **23.** 4.5 W/mph **25.** $dA/dd = 0.5\pi d$

Section 21.4 (page 862)

1. $12x^2 + 6x$ **3.** $12x^3 - 6x^2 + 1$ **5.** $5x^4 - 12x^2$ **7.** $14x^6$ **9.** 2 **11.** 4 **13.** -6 **15.** 1 **17.** -4 **19.** 5 **21.** 9 **23.** -3 **25.** 2.0 J/m **27.** 0.8598 ft/s per lb/in^2 **29.** -8.89×10^{-11} N/m **31.** -2.98 ft/ft

Section 21.5 (page 869)

1. $8x^3 + 9x^2$ **3.** $9x^2 + 18x + 1$ **5.** $5x^4 - 1$ **7.** $4x^3$ **9.** $4x^2 \cos x + 8x \sin x$ **11.** $-(3x + 1)\sin x + 3 \cos x$ **13.** $-4/3$ **15.** 4/9 **17.** 11/9 **19.** 6/25 **21.** -43 **23.** 2 **25.** 1 **27.** -1 **29.** $dV/dt = I\, dR/dt + R\, dI/dt$ **31.** 4.4×10^{-5} cm^2/degree **33.** -2.28 mm/mm **35.** $(fgh)' = (fg)h' + (fg)'h = fgh' + [fg' + f'g]h = fgh' + fg'h + f'gh$

Section 21.6 (page 874)

1. $y = u^4, u = x^3 - 1$ **3.** $y = \sqrt{u}, u = 2x - 3$ **5.** $y = 1/u, u = \sqrt{3x + 2}$ or $y = 1/\sqrt{u}, u = 3x + 2$ **7.** $y = \cos u, u = 2x + 3$ **9.** $y = \ln u, u = \sin x$ **11.** $10x(x^2 + 1)^4$ **13.** $8x - 16$ **15.** $(16x - 12)(2x^2 - 3x + 5)^3$ **17.** $-3 \sin 3x$ **19.** $3x^2 \cos x^3$ **21.** 4/27 **23.** 1/8 **25.** $2x^2(x^3 - 1)^{-1/3}$ **27.** $x(x^2 + 4)^{-1/2}$ **29.** $(1/3)(2 + 2x)(2x + x^2)^{-2/3}$ **31.** $y = (-1/2)(x - 2)^{-3/2}$ **33.** $4x(4x^2 - 3)^{-3/2}$ **35.** $(3x^2 + 2x)/(1 - 3x)^4$ **37.** $2(-5x^2 - 4)/(x^2 - 4)^4$ **39.** $(x - 3)/(2x - 3)^{3/2}$ **41.** $(2x - 17)/[2(x - 4)^{3/2}]$ **43.** $dZ/dR = R/\sqrt{R^2 + (x_L - x_C)^2}$ **45.** -5600 m/°C, to two significant digits **47.** 3×10^{-9} C/s

Section 21.7 (page 879)

1. **(a)** $6x^2$ **(b)** $6y^2\, dy/dx$ **3.** **(a)** $6x + 2$ **(b)** $(6y + 2)dy/dx$ **5.** $2xy\, dy/dx + y^2$ **7.** $-2x^2\, dy/dx - 4xy$ **9.** $-3x^2y^2\, dy/dx - 2xy^3$ **11.** $-y/(x + 2y)$ **13.** $(-4x - 3y)/(3x - 2y)$ **15.** $(2x - 3 - xy^2)/(x^2y)$ **17.** $-x^2/y^2$ **19.** $(-4x^3 - 3)/(4y^3)$ **21.** $8x^3/(3y^2 - 4y)$ **23.** $-3/4$ **25.** 1

Section 21.8 (page 881)

Sequential Review: 1. discontinuous at $x = -3$ **3.** discontinuous at $x = 1$ **5.** -4 **7.** 3/2 **9.** $-4x$ **11.** $1/x^2$ **13.** $-3/x^2$ **15.** -4 **17.** 3 **19.** 1 **21.** $-6/x^4$ **23.** $21x^6$ **25.** 3 **27.** -2 **29.** $15x^4 + 21x^2 - 6$ **31.** $(6x^2 + 6x)/(2x + 1)^2$ **33.** $-3/16$ **35.** $y = \sin u, u = 2x^2 - 3$ **37.** $-9x/(\sqrt{3x^2 - 2})^3$ **39.** $\sqrt{2}$ **41.** $4x^2y\, dy/dx + 4xy^2$ **43.** $[3(x - 3)^2 - 2y^3]/(6xy^2)$ **45.** 3/5
Applied Problems: 1. y **3.** -0.00075 A/Ω **5.** $dv/dr = -2rv_{max}/R^2$ **7.** 50 joules/kg per m/s **9.** -5.3 Hz/Ω **11.** 104°F/hour **13.** 0.042 V/Ω **15.** -1.9 volts/volt **17.** 0.014 A/V **19.** 2.718
Random Review: 1. 2/3 **2.** -2 **3.** $4x - 1$ **4.** 0 **5.** $y = u^7, u = 3x^3 - 2x$ **6.** 1 **7.** $6x^5 + 8x^3 - 6x^2 - 4$ **8.** $6xy^2\, dy/dx + 2y^3$ **9.** $9x^2 - 4/x^3$ **10.** $-2/(x - 1)^2$
Chapter Test: 1. **(a)** no **(b)** (0, 0) **(c)** -1 **(d)** 0 **2.** -9 **3.** $6x + 2$ **4.** $-12x^{-5}$ **5.** -3 **6.** $y = \ln u, u = 3x - 1$ **7.** $(-x^2 + 2x + 3)/(x^2 + 3)^2$ **8.** $3(3x + 2)/y$ **9.** 5.2 Ω **10.** 2.33×10^{-4} in^3/hr

CHAPTER 22

Section 22.1 (page 894)

1. $y = x - 1$

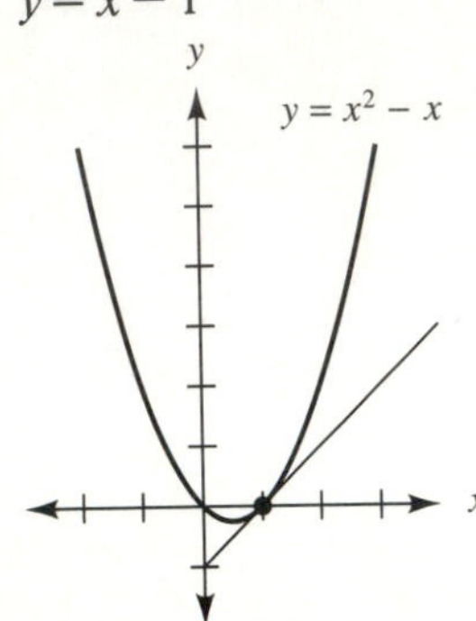

3. $y = -4x - 1$

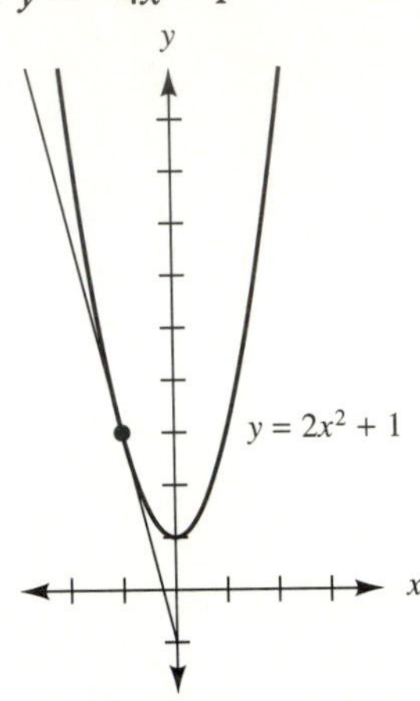

5. $y = x$

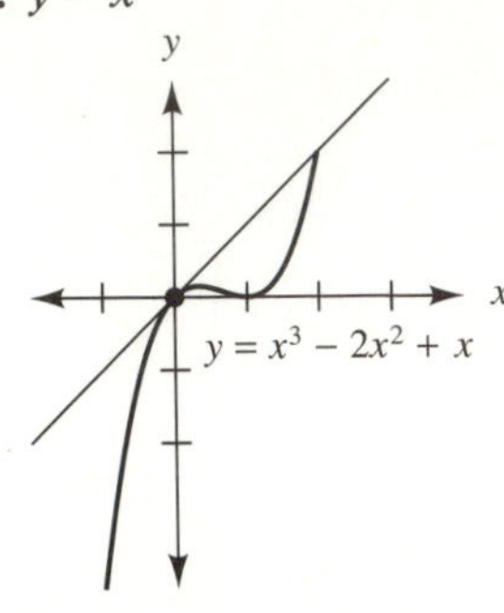

7. $y = x + 4$ **9.** $y = (1/9)x - 2/3$
11. $y = (-1/8)x - 25/8$

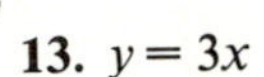
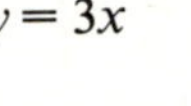

13. $y = 3x$

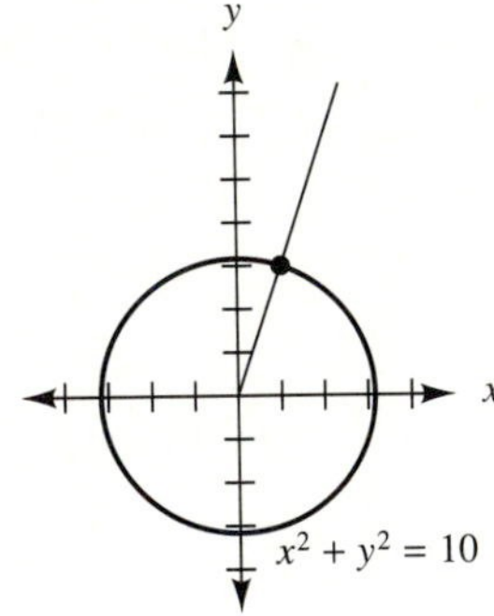

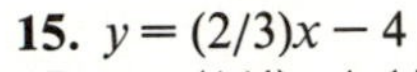

15. $y = (2/3)x - 4$
17. $y = (1/4)x + 15/4$
19. $y = -x + 3$ **21.** about 27°
23. $y = -25x + 101{,}280$

Section 22.2 (page 898)

1. $v = 7$ ft/s, $a = 4$ ft/s^2 **3.** $v = 1$ ft/s, $a = -2$ ft/s^2 **5.** $v = 0$ ft/s, $a = 2$ ft/s^2
7. $v = 1.5$ ft/s, $a = -5/8$ ft/s^2 **9.** $|\mathbf{v}| = 36.1$, $\theta_v = 86.8°$, $|\mathbf{a}| = 36.0$, $\theta_a = 88.4°$
11. $|\mathbf{v}| = 4.12$, $\theta_v = 104.0°$, $|\mathbf{a}| = 4.47$, $\theta_a = 116.6°$ **13.** $|\mathbf{v}| = 68.6$, $\theta_v = 134.4°$ $|\mathbf{a}| = 33.9$, $\theta_a = 135.0°$
15. $|\mathbf{v}| = 2.06$, $\theta_v = 166.0°$, $|\mathbf{a}| = 6.00$, $\theta_a = 357.6°$ **17.** $|\mathbf{a}| = 9.8$, $\theta_a = 270°$
19. $\theta_v = 10°$ with the horizontal **21.** $t = 0.028$ second **23.** Increasing in both examples

Section 22.3 (page 905)

1. 0.2150 ft/s^2 **3.** 0.70 m^2/s **5.** decreasing at 18.7 knots **7.** 4.02 in^3/s **9.** $-R/X_L$
11. 36.1 m/s **13.** $v_y = 24$ **15.** 54, 90°

Section 22.4 (page 913)

1.

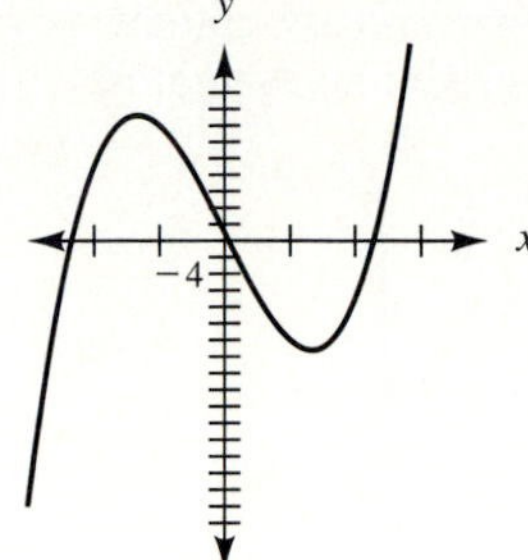

3.

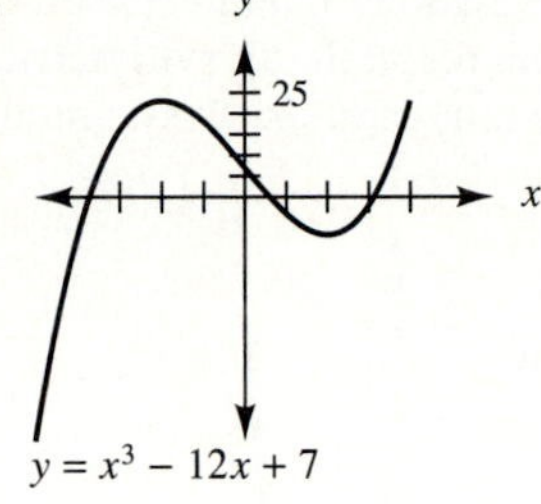

5.

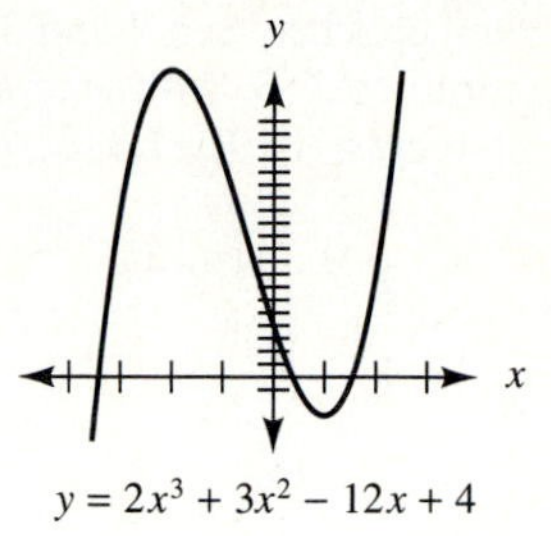

7.

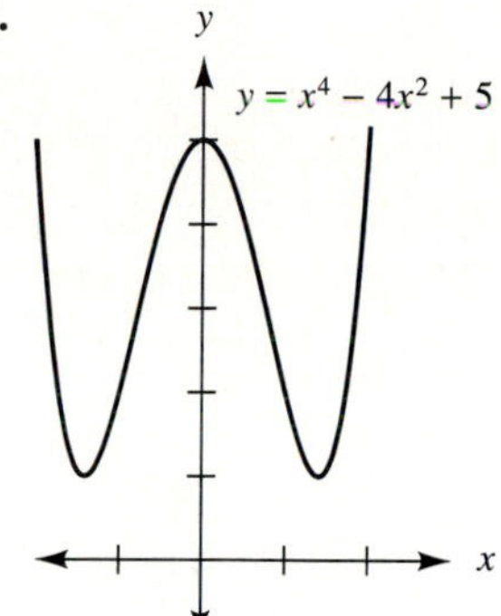

9.

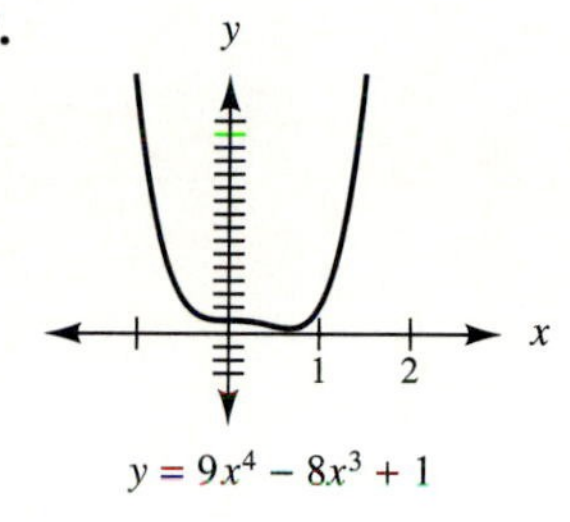

11.

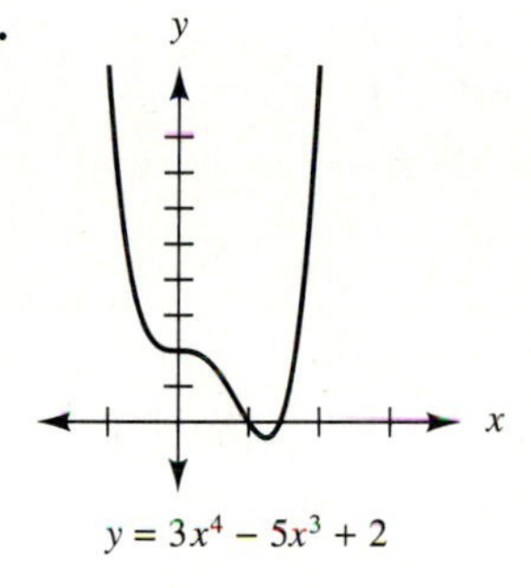

13.

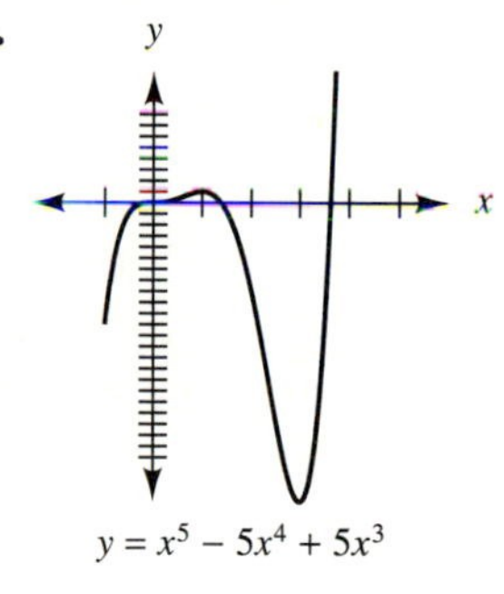

15.

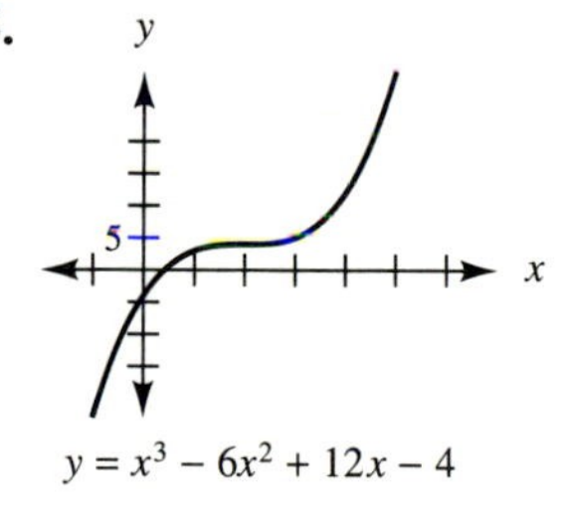

17.

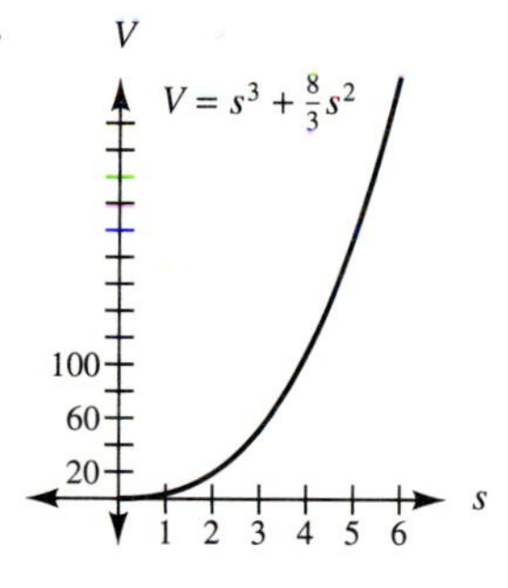

19.

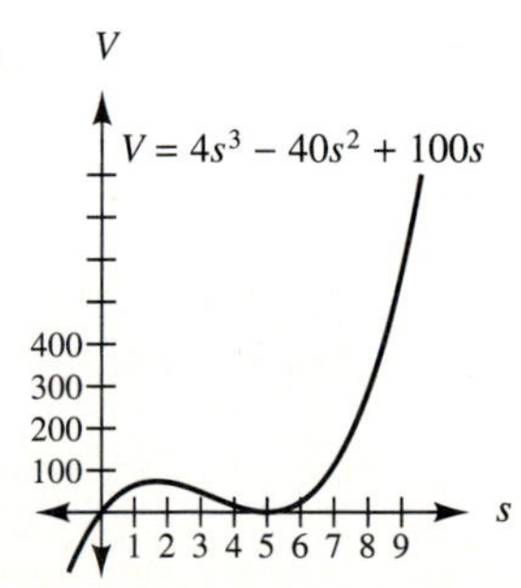

Section 22.5 (page 919)

1. x-intercepts between 2 and 3, between -2 and -3, and between 0 and 1; y-intercept: (0, 1); no symmetry; no asymptotes **3.** x-intercept: none; y-intercept: (0, 5); symmetric about the y-axis; no asymptotes **5.** x-intercepts: (0, 0), (1.3, 0), $(-1.3, 0)$; y-intercept: (0, 0); symmetric about the origin; no asymptotes

7.

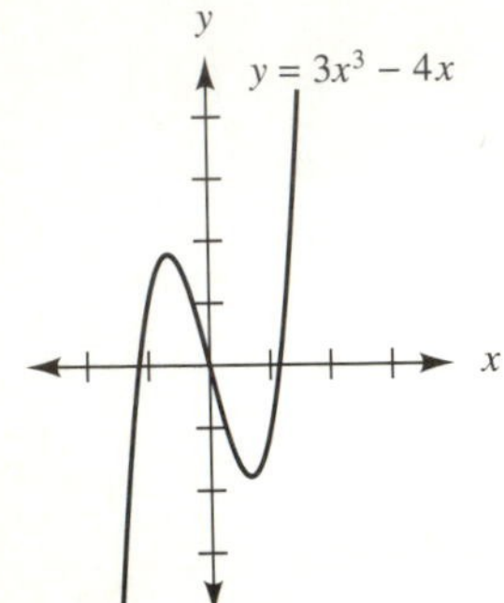

9.

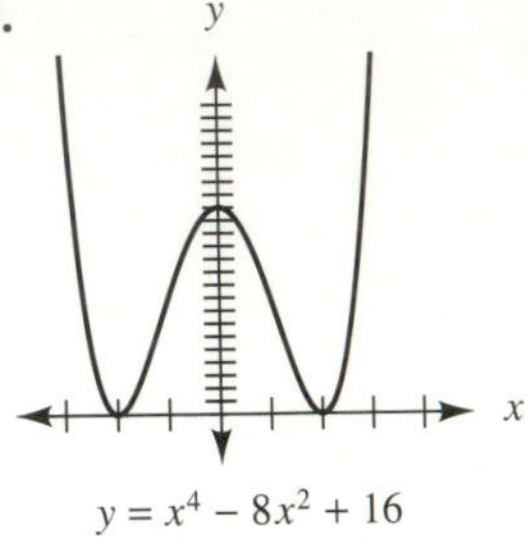

11.

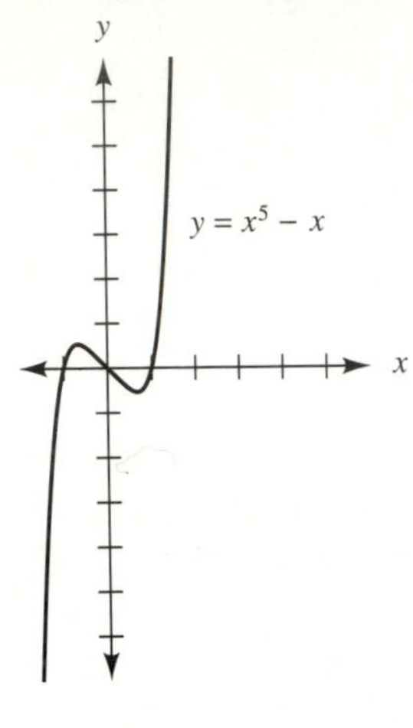

13.

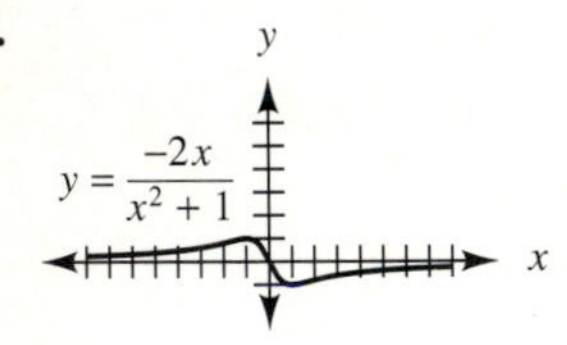

15.

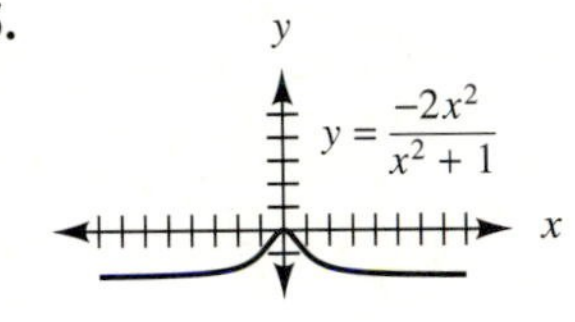

17.

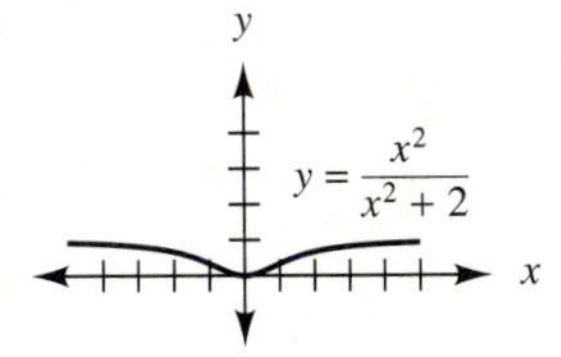

Section 22.6 (page 922)

1. 1.5 s **3.** 40 ft × 40 ft **5.** $Z = R$, $\omega = \sqrt{1/(LC)}$ **7.** $s = 35$ km, $t = 0.61$ hr **9.** $x = l\sqrt{3}/3$ **11.** 19.8″ **13.** 1/2 **15.** 1.0 ft **17.** 9.38 ft × 9.38 ft **19.** $20.25

Section 22.7 (page 929)

1. 1.532 **3.** -0.911 **5.** 3.233 **7.** -0.675 **9.** -6.040 **11.** -1.307 **13.** 7.4 inches **15.** 2.0 cm × 2.0 cm **17.** 94.7 mm, to three significant digits

Section 22.8 (page 931)

Sequential Review: 1. $y = 6x - 5$ **3.** $y = (1/3)x + 4/3$ **5.** $y = (-1/4)x - 17/4$ **7.** $y = (-2/3)x$ **9.** -1 ft/s **11.** $-7/27$ ft/s^2 **13.** $|\mathbf{v}| = 6.01$, $\theta_v = 177.6°$ **15.** 672 m/s **17.** 28.3 m/s **19.** 15 **21.** increasing **23.** $(1/2, -4\ 3/4)$, $(-3, 81)$

25.

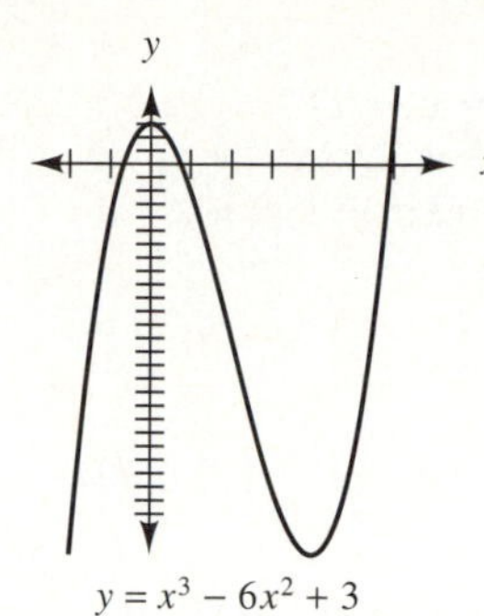

27.

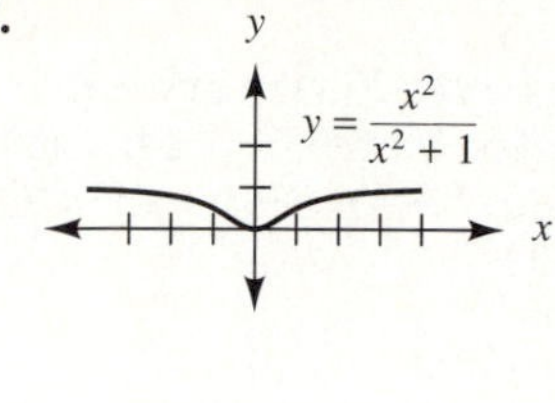

29. (2/3, −205/27)
31. concave downward
33.

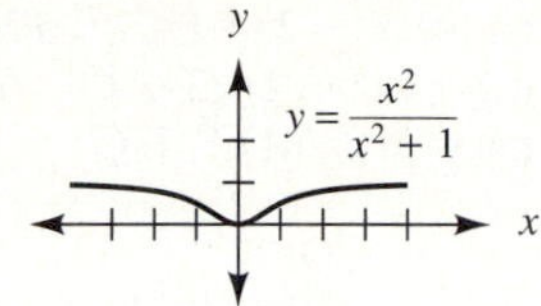

35. not a maximum **37.** $y = -1/4$ **39.** no maximum **41.** 3.8 **43.** 3.1 **45.** 0.7
Applied Problems: 1. 18.4° **3.** $t = 600$ s **5.** 1.2 **7.** $I = 17V + 2.5$ **9.** 1.06 in/s
11. **(a)** $s = 12.76$ cm **(b)** 7.006 cm/s **(c)** 2.343 cm/s² **13.** 30.4 in/s **15.** 34.2 lb/s
17. 0.00037 in⁴/°C **19.** 9.93
Random Review: 1. **(a)** (1, 2), (−1, 6) **(b)**

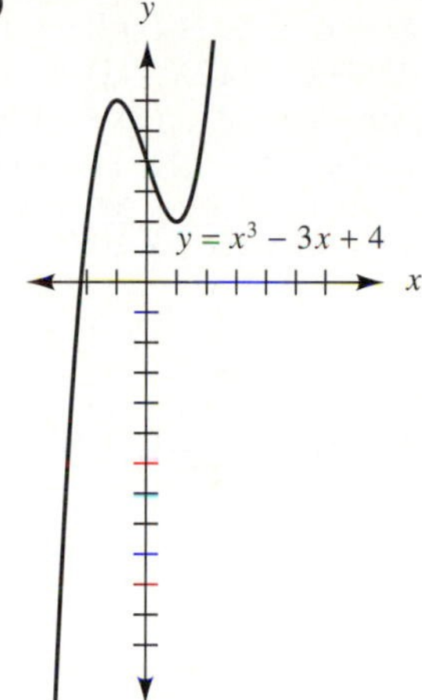

2. 2.1 **3.** $|\mathbf{a}| = 6.01$ $\theta_a = 92.4°$ **4.** $y = (2/3)x + 5/3$ **5.** 20.1 m/s **6.** (0, 4) **7.** −12 ft/s
8. $\sqrt[3]{3/2}$ **9.** $|\mathbf{a}| = 24$, $\theta_a = 90°$. **10.** $y = (-1/3)x + 7/3$
Chapter Test: 1. $y = (-8/9)x + 25/9$ **2.** $y = (-1/21)x + 212/21$ **3.** **(a)** 4/3 ft/s **(b)** −7/27 ft/s²
4. **(a)** $|\mathbf{v}| = 4.00$, $\theta_v = 266.4°$ **(b)** $|\mathbf{a}| = 2.02$, $\theta_a = -82.9°$ **5.** 3.09 m/s **6.** **(a)** (1, 1/2), (−1, −1/2)
(b)

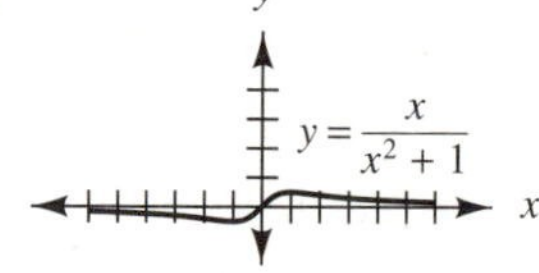

7. **(a)** (1/2, 21/8) **(b)** none **8.** −1.9 **9.** 24 W/s **10.** 1 ft/s

CHAPTER 23

Section 23.1 (page 945)

1. 24 **3.** 2.75 **5.** 8.4375 **7.** 8 **9.** 1 **11.** 9.5625

Section 23.2 (page 948)

1. b, c **3.** a, b **5.** a, b **7.** c **9.** c **11.** $F(x) = 2x^3 + C$ **13.** $F(x) = x^5 + C$
15. $F(x) = x^5 + 2x^4 + C$ **17.** $F(x) = -3/x + 4x + C$ **19.** $F(x) = (4/3)x^{3/2} + (9/4)x^{4/3} + C$
21. $F(x) = (x^4 - 1)^3/3 + C$ **23.** $F(x) = (2x^3 - 1)^{-3}/-3 + C$ **25.** $F(x) = -(3x^2 - 1)^{-1} + C$
27. $F(x) = 2(x - 1)^{1/2} + C$

Section 23.3 (page 955)

1. 20 **3.** 2 2/3 **5.** 11 1/4 **7.** 15/16 **9.** 4 **11.** 1 **13.** 5 2/3 **15.** 10.7 m **17.** 0.004 C
19. 0.375 J
21. $\int_{-h}^{h} f(x)\,dx = ax^3/3 + bx^2/2 + cx\Big|_{-h}^{h} = 2ah^3/3 + 2ch$ and $(\Delta x/3)[f(-h) + 4f(0) + f(h)] =$ $2ah^3/3 + 2ch$

Section 23.4 (page 962)

1. $x^4/4 + 3x^2/2 + C$ **3.** $2x^3/3 - 5x^2/2 + C$ **5.** $-1/x^2 + C$ **7.** $-5/(2x^2) - 2/x + C$
9. $(x^2 + 1)^4/4 + C$ **11.** $-1/[2(3x - 5)^2] + C$ **13.** $(2/3)(x - 1)^{3/2} + C$ **15.** $2(x^2 - 1)^{1/2} + C$
17. $-1/[4(x^4 - 1)] + C$ **19.** $(x^3 + 1)^6/9 + C$ **21.** $(x^2 - 1)^{3/2}/3 + C$ **23.** $-3/[8(2x^2 + 3)^2] + C$
25. $(x^2 + x)^9/9 + C$ **27.** $(x^3 + 3x + 2)^8/24 + C$ **29.** $(1/3)(3x^2 + 2x)^{3/2} + C$
31. $2(x^3 - 2x^2)^{1/2} + C$ **33.** $s = 3000t - 750t^2 + C$ **35.** $\theta = (-1/3)(t^3 - 1)^{-1} + C$
37. $W = kx^2/2 + C$ **39.** $R = 0.003\,T^3 + 0.01\,T^2 - 0.7\,T + C$

Section 23.5 (page 968)

1. 20.0 **3.** 2.8 **5.** 11.4 **7.** 0 **9.** 1.0 **11.** 8.1 **13.** \$4630 **15.** 22 050 m^2 **17.** 1.75 in^2
19. 5.8 in^3

Section 23.6 (page 973)

1. 0 **3.** 0.97 **5.** 8.146 **7.** −28 **9.** 3.4 **11.** 4.0 **13.** 27 ft^2 **15.** 970 m^3 **17.** 2 in^2
19. a straight line

Section 23.7 (page 976)

Sequential Review: 1. 12 **3.** 0 **5.** 128 **7.** 21/16 **9.** a **11.** a **13.** $x^4/4 + 2x^3 + C$
15. $2(3x - 2)^{1/2} + C$ **17.** 2/3 **19.** 12 2/3 **21.** 16 **23.** 6 3/4 **25.** $(2/3)x^{3/2} - (3/2)x^{-2} + C$
27. $(-1/2)(2x - 3)^{-1} + C$ **29.** $(-1/6)(3x + 4)^{-2} + C$ **31.** $2(x^2 + x)^{1/2} + C$ **33.** 19/27 **35.** 6.6
37. 10 **39.** 2/3 **41.** 6.7 **43.** 4.00 **45.** 18 2/3
Applied Problems: 1. $I = 2m$ **3.** 1950 lb **5.** 0.1863 acres **7.** 80 min **9.** 110 volts
11. 140,000 ft·lb **13.** 2570 millicoulombs **15.** 54 mV (Simpson's rule) **17.** 234 ft·lb
Random Review: 1. 1.1 **2.** $(1/3)(x^2 + 4x)^{3/2} + C$ **3.** $(-1/3)(2x^2 + 3)^{-3} + C$ **4.** 1.1 **5.** b
6. 353 3/4 **7.** −1/2 **8.** 6.1 **9.** $x^4 + x^2/2 + C$ **10.** 10
Chapter Test: 1. 12 **2.** −5/18 **3.** $(1/8)(x^2 + 1)^4 + C$ **4.** If f and F are two functions such that f is continuous for $a \le x \le b$, and $F'(x) = f(x)$, then $\int_a^b f(x)\,dx = F(b) - F(a)$. **5.** $2(x^3 - 2x)^{1/2} + C$
6. −10 **7.** −9.25 **8.** 3.625 **9.** 100,000 lb, to two significant digits **10.** 1.4 inches

CHAPTER 24

Section 24.1 (page 987)

1. 10 2/3 **3.** 10 2/3 **5.** 1/12 **7.** 8 **9.** 1/3 **11.** 1 **13.** 32 **15.** 34 2/15 **17.** 100 m^2
19. 42.6 ft^2 **21.** 8.67 sq. rods

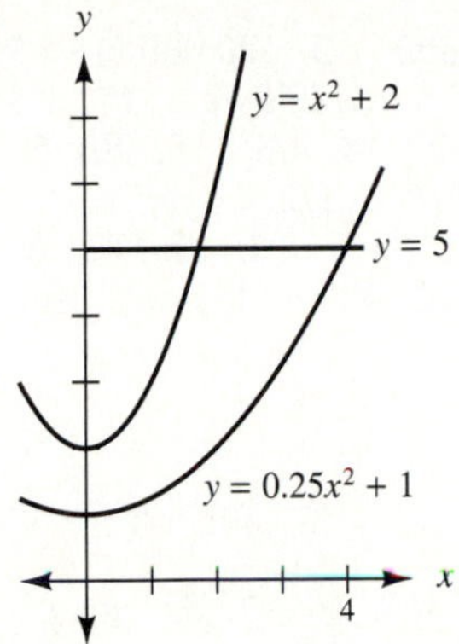

23. 1866 2/3 joules or 1900 joules, to two significant digits

Section 24.2 (page 996)

1. 18π **3.** $512\pi/15$ **5.** $422\pi/7$ **7.** 8π **9.** $8\pi/27$ **11.** 2π **13.** 2π **15.** $96\pi/5$
17. 28.3 cm^3 **19.** 1.0×10^8 ft^3 **21.** 42.8 ft^3

Section 24.3 (page 1003)

1. $56\pi/3$ **3.** $256\pi/5$ **5.** $8\pi\sqrt{2}/5$ **7.** $3\pi/10$ **9.** 45π **11.** $16\pi/3$ **13.** $128\pi/5$ **15.** $\pi/10$
17. 90 ft^3, to one significant digit **19.** 67 in^3 **21.** 309 in^3
23.

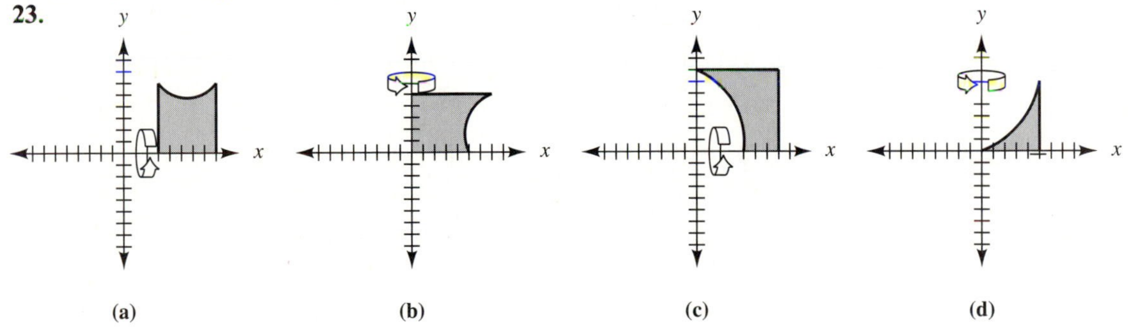

Section 24.4 (page 1010)

1. 207 ft **3.** 140 W **5.** 41.9 cm^3 **7.** 7.3 ft·lb **9.** 25,800,000 ft·lb **11.** 216 ft·lb
13. 233,000 ft·lb

Section 24.5 (page 1017)

1. $s = -16t^2 + 32$ **3.** $s = 4.9t^2$ **5.** $v = -0.835\ t^2$ **7.** 3 100 000 N **9.** 730,000 lb **11.** 23 N

Section 24.6 (page 1018)

Sequential Review: 1. 10 2/3 **3.** 4/3 **5.** 1/32 **7.** 8/3 **9.** 4 **11.** $243\pi/5$ **13.** 8π **15.** $16\pi/3$ **17.** $96\pi/5$ **19.** $224\pi/3$ **21.** $972\pi/5$ **23.** $2\pi/5$ **25.** 128π **27.** $5\pi/14$ **29.** 14/9 **31.** 12 4/5 **33.** 7 3/5 **35.** 15 **37.** $s = -4.9t^2 + 45$ **39.** $s = 2\sqrt{t^2+1}$ **41.** $v = 2t^{2.3}/2.3$ **43.** 3276 lb **45.** 5200 lb
Applied Problems: 1. 1400 J (three significant digits) **3.** 120 000 N **5.** 15 V **7.** 7000 ft^2 **9.** 222 ft·lb **11.** 225,000 ft·lb **13.** 10 000 J **15.** 256 W **17.** 1.546×10^{-6} mg/cm^3 **19.** 400 ft
Random Review: 1. 77 1/3 **2.** 4 1/2 **3.** $81\pi/2$ **4.** 4/3 **5.** $96\pi/5$ **6.** $v = (2/3)(t+1)^{3/2} + 2/3$ **7.** 32π **8.** 75 200 N **9.** $96\pi/5$ **10.** $s = -2t^{-2} + 1/2$
Chapter Test: 1. 12 **2.** $27\pi/7$ **3.** 8π **4.** $1536\pi/7$ **5.** $16\pi/3$ **6.** 152 1/4 **7.** $s = 12t + 16t^2$ **8.** $v = (3/2)t^2 - 2t + 4$ **9.** 86,400,000 ft·lb **10.** 400 lb, to one significant digit

CHAPTER 25

Section 25.1 (page 1031)

1. $4 \cos 4x$ **3.** $-5x^4 \sin x^5$ **5.** $(-10x) \sin (5x^2+2)$ **7.** $8x^3 \cos x^4$
9. $6 \sin^2 (2x-3) \cos (2x-3)$ **11.** $-30 \cos(3x+1) \sin (3x+1)$ **13.** $60x^2 \cos^4 (2x^3) \sin (2x^3)$
15. $6x \cos 2x + 3 \sin 2x$ **17.** $-4x^4 \sin (2x^2) + 3x^2 \cos (2x^2)$ **19.** $-3 \sin 2x \sin x + 6 \cos x \cos 2x$
21. $-4 \sin^2 x \sin 2x + 4 \sin x \cos x \cos 2x$
23. $6x[\sin^2 (3x^2) \cos (2x^3)][-2x \sin 3x^2 \sin 2x^3 + 3 \cos 2x^3 \cos 3x^2]$
25. $(-2 \cos x - x \sin x)/x^3$ **27.** $2[3x \cos x + (2+3x^2)\sin x]/\cos^3 x$ **29.** $4 \sin t \cos t$
31. 17 000 m/radian **33.** 4020 V/s

Section 25.2 (page 1035)

1. $3 \sec^2 3x$ **3.** $-8 \csc^2 (4x-3)$ **5.** $6 \sec 2x \tan 2x$ **7.** $(-\csc\sqrt{x-1}) \cot \sqrt{x-1})/(2\sqrt{x-1})$
9. $6 \tan^2 2x \sec^2 2x$ **11.** $-24 \cot^3 3x \csc^2 3x$ **13.** $-6x^2 \csc^2 x^3 \cot x^3$ **15.** $2 \tan x \sec^2 x + 1$
17. $(\sec^2 x)(2 \tan x + 1)$ **19.** $4x \sec^2 2x + 2 \tan 2x$ **21.** $(-2x \csc^2 x - 2 \cot x)/4x^2$
23. $\sec^3 x + \sec x \tan^2 x$ **25.** $(-3 \csc x)(2 \sin 2x + \cos 2x \cot x)$
27. $[\tan x][(4 + 2 \sin x)\sec^2 x - \sin x]/(2 + \sin x)^2$ **29.** -3.46/radian **31.** -3.5×10^{-5} radian/in

Section 25.3 (page 1039)

1. $6x^2/\sqrt{1-4x^6}$ **3.** $3/\sqrt{1-(3x+2)^2}$ **5.** $(24x^2+4)/(1+4x^6+4x^4+x^2)$ **7.** $1/\sqrt{-x^2+2x}$
9. $-1/\sqrt{2x-4x^2}$ **11.** $2/(2x^2+2x+1)$ **13.** $(-6 \text{ Arcsin } 3x)/\sqrt{1-9x^2}$
15. $-1/(2\sqrt{(1-x^2)\text{Arccos } x})$ **17.** $-2x/[(\text{Arctan}^2 x^2)(1+x^4)]$ **19.** $x^2/\sqrt{1-x^2} + 2x \text{ Arcsin } x$
21. $3x/\sqrt{2x-x^2} + 3 \text{ Arccos } (1-x)$ **23.** $x/(2x^2-6x+5) + \text{Arctan } (2x-3)$ **25.** $(\text{Arccos } x + \text{Arcsin } x)/[\sqrt{1-x^2} \text{ Arccos}^2 x]$ **27.** $1 - 2x \text{ Arctan } x/(1+x^2)^2$ **29.** $d\theta_1/dL = \omega R/(R^2 + \omega^2 L^2)$
31. 0.01 radian/Ω **33.** outlines: **(a)** $x = \cot y$, $dy/dx = -1/\csc^2 y = -1/(1+x^2)$
(b) $x = \sec y$, $dy/dx = 1/(\sec y \tan y) = 1/[x(\pm\sqrt{x^2-1})]$, for $0 \le y \le \pi$, use +
(c) $x = \csc y$, $dy/dx = -1/(\csc y \cot y)$, $dy/dx = -1/[x(\pm\sqrt{x^2-1})] = -1/[\pm\sqrt{x^2(x^2-1)}]$

Section 25.4 (page 1046)

1. $6xe^{3x^2}$ **3.** $4xe^{2x^2} + 1$ **5.** $-2e^{2x} \sin e^{2x}$ **7.** $2 \sec^2 x/\tan x$ or $4/\sin 2x$ **9.** $(3/x)(\ln x)^2$
11. $(2 \ln 3)(3^{2x})$ **13.** $4^{\sin x}(\ln 4)(\cos x)$ **15.** $2x^{2x} + 2x^{2x}\ln x$ **17.** $(\sin x)^x[x \cot x + \ln(\sin x)]$
19. $x^{-1+\ln x}(2 \ln x)$ **21.** $e^x(1/\sqrt{1-x^2}) + e^x \text{ Arcsin } x$ **23.** $2x/(x+1) + 2 \ln (x+1)$
25. $2^x[\ln 2 \cos x - \sin x]$ **27.** $3/(x+1)$ **29.** $\cot x + 2/x$ **31.** $dQ/dt = (-Q_0/\tau)e^{-t/\tau}$
33. $dT_1/d\alpha = \mu T_2 e^{\mu\alpha}$ **35.** $dV/dx = -kx[1 - 2 \ln(1/x)]$

Section 25.5 (page 1051)

1.

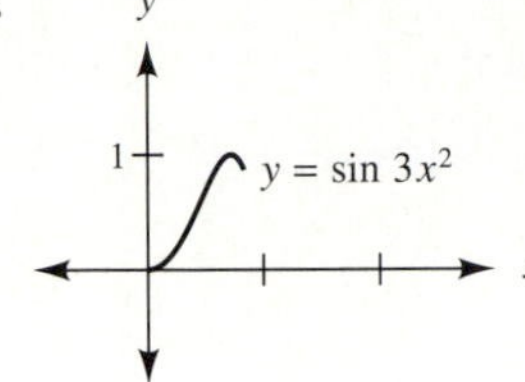

3.

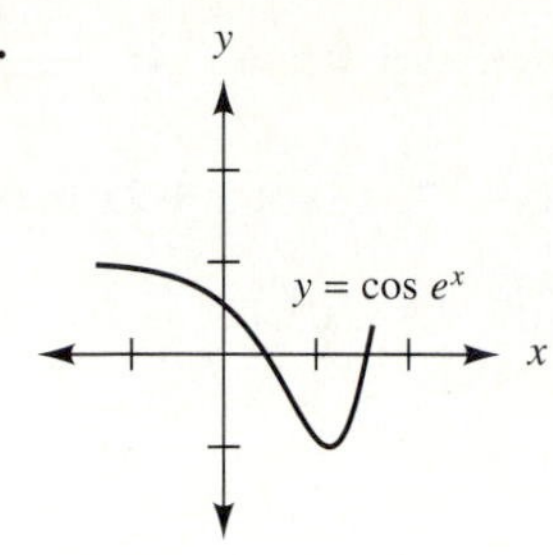

5.

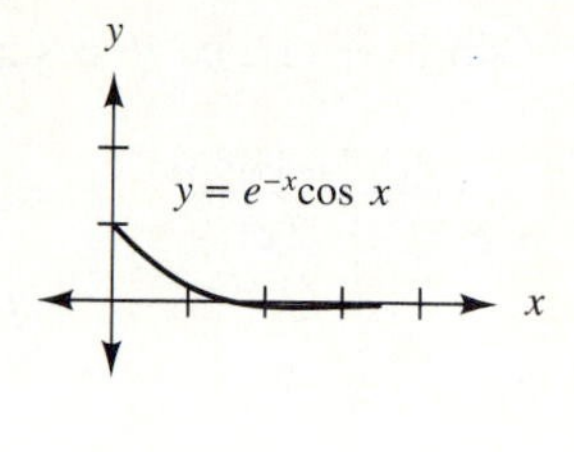

7.

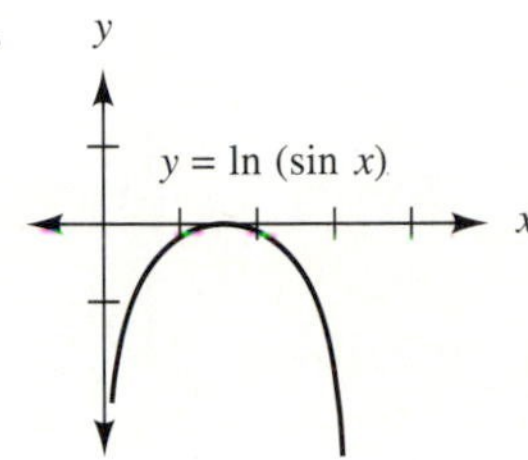

9.

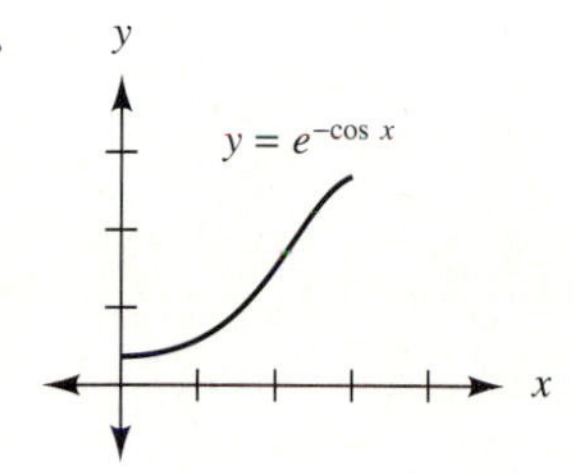

11. 2.28 **13.** 1.14 **15.** -0.77 **17.** π, 270° **19.** 0.841 **21.** $y = x + 1$
23. $y = (1/2)x + (\pi - 2)/4$ **25.** $y = -x + 1$ **27.** 0.3 cm **29.** 1.33

Section 25.6 (page 1053)

Sequential Review: 1. $-2 \sin 2x$ **3.** $-24x \sin(4x^2 + 1)$ **5.** $(6/\sqrt{x}) \sin^2 \sqrt{x} \cos \sqrt{x}$
7. $-18x^2 \cos^2 2x \sin 2x + 6x \cos^3 2x$ **9.** $[x \sin(1 - x) - \cos(1 - x)]/(2x^2)$
11. $2x \sec x^2 \tan x^2$ **13.** $-12(3x + 1) \csc^2 (3x + 1)^2$ **15.** $-3x \csc x \cot x + 3 \csc x$
17. $(3 \tan 3x - 9x \sec^2 3x)/\tan^2 3x$ **19.** $12/\sqrt{1 - 9x^2}$ **21.** $(2x + 3)/\sqrt{1 - (x^2 + 3x)^2}$
23. $3\sqrt{x}/[2x(x + 1)]$ **25.** $-2x^2\sqrt{1 - 4x^2} + 2x \text{ Arccos } 2x$
27. $2/\text{Arctan } x - 2x/[(1 + x^2)\text{Arctan}^2 x]$ **29.** $2e^{2x} \cos e^{2x}$ **31.** $[6 \ln(3x + 1)]/(3x + 1)$
33. $-2^{\cos x}(\sin x)(\ln 2)$ **35.** $(1 - x)^{x-1}[-x + (1 - x)\ln(1 - x)]$ **37.** $y = -x$
39. $|\mathbf{v}| = 3.05$, $\theta_v = 79.8°$ **41.** 22.2 **43.** $y = 1$ **45.** -1.29
Applied Problems: 1. $(v_0^2 \sin^2 \theta)/19.6 + h$ **3.** $t = 0.015$ **5.** 12 W/degree **7.** -5400/m
9. -7.54 g/half-life **11.** -8.6×10^{-6} per Å **13.** 0.003 liters/J **15.** 660 cm/s **17.** 1.1 V

Random Review: 1. $\dfrac{1}{x^2\sqrt{1 - x^2}} - \dfrac{2 \text{ Arcsin } x}{x^3}$ **2.** $6x^2 \cos 3x^2 + \sin 3x^2$ **3.** $\cot x$

4. $4 \sec^2 2x \tan 2x$ **5.** $y = ex - e$ **6.**

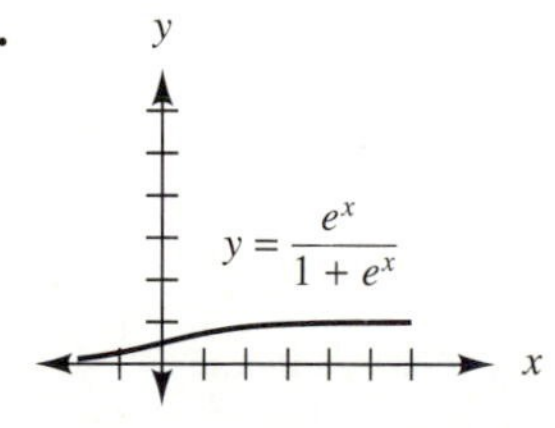

7. $\dfrac{\sec^2 \sqrt{x + 1}}{2\sqrt{x + 1}}$ **8.** $(\ln 5)(\cos x)(5^{\sin x})$ **9.** $-6 \cos(3x + 2) \sin(3x + 2)$ **10.** $\dfrac{2x^3}{1 + x^4} + 2x \text{ Arctan } x^2$

Chapter Test: **1.** $(4x \sin 4x \cos 4x - \sin^2 4x)/(2x^3)$
2. $-2x^{1/2} \sin 2x + (1/2)x^{-1/2} \cos 2x$ **3.** $x \sec^2 x + \tan x$ **4.** $\dfrac{-3}{\sqrt{x(1-x)}\ \text{Arcsin}^4 \sqrt{x}}$
5. $\dfrac{1}{2x(1+x^2)} - \dfrac{\text{Arctan}\ x}{2x^2}$ **6.** $6e^{\sin 2x} \cos 2x$ **7.** $x^2 \cot x + 2x \ln(\sin x)$ **8.** $-10x \sin x^2$
9.

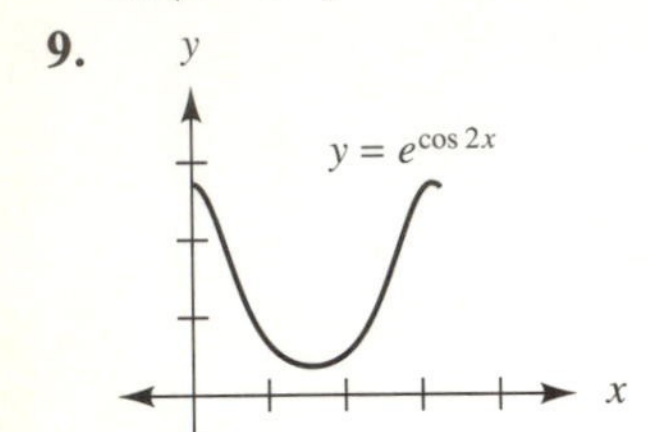

10. $|\mathbf{a}| = 2.07$, $\theta_a = 105.1°$

CHAPTER 26

Section 26.1 (page 1061)
1. $(\sin^4 x)/4 + C$ **3.** $\sec x + C$ **5.** $(-2/3)(\cot x - 1)^{3/2} + C$ **7.** $(1/6)\sec^3 2x + C$
9. $[-\csc^4(x-1)]/4 + C$ **11.** $-1/(2\ \text{Arcsin}^2 x) + C$ **13.** $(\text{Arcsin}\ 2x)^3/6 + C$
15. $(2/9)(\text{Arctan}\ 3x)^{3/2} + C$ **17.** $(-1/4)[\ln(2x+1)]^{-2} + C$
19. $(1/6)[\ln(3x-2)]^2 + C$ **21.** $(-2/3)(1 - e^x)^{3/2} + C$ **23.** $(e^x + e^{-x})^3/3 + C$ **25.** $2\sqrt{3}/3$
27. 5/2 **29.** $2\xi/\pi + 5/6$ **31.** 2.3 cm **33.** 387 Ω

Section 26.2 (page 1065)
1. $e^{5x} + C$ **3.** $(1/2)e^{2x+1} + C$ **5.** $(1/4)e^{x^4} + C$ **7.** $(-1/2)e^{-x^2} + C$ **9.** $e^{x^2+x} + C$ **11.** 13.4
13. 25.9 **15.** $e^{\tan x} + C$ **17.** $-e^{\cos x} + C$ **19.** $x + (1/2)e^{-2x} + C$ **21.** $x/e - (1/2)e^{-2x-1} + C$
23. $e^{2x} - 2e^x + C$ **25.** -2.35 **27.** $i = (-5/6)e^{-0.3t} + 5/6$ **29.** $v = e^{\sin t} - 1$

Section 26.3 (page 1068)
1. $\ln|1 + 3x| + C$ **3.** $-\ln|2 - x| + C$ **5.** $(1/3)\ln|x^3 + 1| + C$ **7.** $(1/4)\ln|2x^2 + 3| + C$
9. $\ln|x^3 + x^2| + C$ **11.** 0.549 **13.** 0.461 **15.** $\ln|2 + \sin x| + C$ **17.** $\ln|1 + \tan x| + C$
19. $\ln|1 - \cos x| + C$ **21.** $\ln|1 + e^x| + C$ **23.** $\ln|\ln 2x| + C$ **25.** 0.588 **27.** 0.134 ft·lb
29. $(-1/k)\ln|1 - kV| + C$

Section 26.4 (page 1072)
1. $(-1/3)\cos 3x + C$ **3.** $(1/2)\sin x^2 + C$ **5.** $-\cos e^x + C$ **7.** $(1/2)\tan 2x + C$ **9.** $\sec e^x + C$
11. $\tan(\sin x) + C$ **13.** $(-1/4)\cot 4x + C$ **15.** $(-1/3)\csc x^3 + C$ **17.** $(-1/2)\csc e^{2x} + C$
19. $\ln|\sin x| + C$ **21.** $-\ln|\cos x^2|/2 + C$ **23.** $(1/3)\ln|\sec x^3 + \tan x^3| + C$
25. $2\ln|\sin x^{1/2}| + C$ **27.** $-\ln|\csc(\ln x) + \cot(\ln x)| + C$ **29.** 94.2 Ω **31.** 3.6%
33. $V_{ave} = V_m/\pi$ **35.** **(a)** $\int \cot x\, dx = \int \cos x/\sin x\, dx = \int (\sin x)^{-1} \cos x\, dx = \ln|\sin x| + C$
(b) $\csc x(\csc x + \cot x)/(\csc x + \cot x)\, dx = -1\int -1(\csc^2 x + \cot x \csc x)(\csc x + \cot x)^{-1}\, dx =$
$-\ln|\csc x + \cot x| + C$

Section 26.5 (page 1080)

1. $-\sqrt{9-x^2}/x - \text{Arcsin}\,(x/3) + C$ **3.** $(-1/25)\sqrt{x^2+25}/x + C$ **5.** $\sqrt{x^2-4} - 2\ \text{Arccos}\,|2/x| + C$
7. $-\ln|(1+\sqrt{1-x^2})/x| + C$ **9.** $(x^2-32)\sqrt{16+x^2}/3 + C$ **11.** $25\pi/4$ **13.** 0.236 **15.** 0.117
17. 69 ft^3 **19.** about 18 000 joules **21.** Let $x = a\tan\theta$, $\int dx/(a^2+x^2)\,dx =$

$\int a\sec^2\theta/(a^2+a^2\tan^2\theta)\,d\theta = \int (1/a)\,d\theta = (1/a)\theta + C = (1/a)\ \text{Arctan}\,(x/a) + C$

Section 26.6 (page 1085)

1. $(x/2)\sin 2x + (1/4)\cos 2x + C$ **3.** $x\sec x - \ln|\sec x + \tan x| + C$ **5.** $x\ln x - x + C$ **7.** 30.8
9. 7.73 **11.** $x\ \text{Arcsin}\,x + (1-x^2)^{1/2} + C$ **13.** $(1/2)[e^x\sin x - e^x\cos x] + C$
15. $-e^{-x}(x^2+2x+2) + C$ **17.** $i = (5/6)e^{0.3t}(t - 10/3) + 25/9$ **19.** 2π or 6.28 cm^2
21. downward about 0.008 radian or 0.5°

23. $x(-\cos x + C_1) - \int(-\cos x + C_1)dx = -x\cos x + xC_1 - (-\sin x + xC_1 + C_2) =$

$-x\cos x + xC_1 + \sin x - xC_1 - C_2 = -x\cos x + \sin x + C$

Section 26.7 (page 1088)

1. $\text{Arcsin}\,3x + C$ **3.** $(1/4)\ \text{Arctan}\,(x/4) + C$ **5.** Arctangent **7.** none **9.** Arcsine **11.** power

13. logarithm **15.** Arctangent **17.** $-\ln|(1+3x^2)/x^2| + C$ **19.** $\dfrac{2(x-2)\sqrt{1+x}}{3} + C$

21. $2(3x^2-10)(5+x^2)^{3/2}/15 + C$ **23.** $\ln|\sin x + \sqrt{1+\sin^2 x}| + C$

25. $(1/2)e^x\sqrt{16+e^{2x}} + 8\ln|e^x + \sqrt{16+e^{2x}}| + C$ **27.** $\sqrt{25-x^4} - 5\ln|[5+\sqrt{25-x^4}]/x^2| + C$

29. $(1/6)\cos^2 2x\sin 2x + (1/3)\sin 2x + C$ **31.** $\dfrac{-1}{3}\ln\left|\dfrac{1+2x^3}{x^3}\right| + C$

33. $\dfrac{-1}{3}\sin^2 x\cos x - \dfrac{2}{3}\cos x + C$ **35.** $x^2\ \text{Arcsin}\,x^2 + \sqrt{1-x^4} + C$ **37.** 15 ft^3

39. $v = (1/3)t - (1/9)\ln|1+3t|$ **41.** $i = (25/9)(e^{0.3t})(0.3t-1) + 25/9$

Section 26.8 (page 1090)

Sequential Review: 1. $(-1/6)(1+\cos 2x)^3 + C$ **3.** $(2/3)(\ln x)^{3/2} + C$ **5.** $(1/6)(e^{3x^2-1}) + C$
7. $(1/3)(\text{Arcsin}\,2x)^{3/2} + C$ **9.** $(-1/3)e^{-x^3} + C$ **11.** $(1/2)e^2 - 3/2$ **13.** $x + C$ **15.** 1.91
17. 0.693 **19.** 0 **21.** $-\sec x^{-1} + C$ **23.** $(1/2)\ln|\sec e^{2x} + \tan e^{2x}| + C$ **25.** 0.205
27. $\sqrt{x^2-9} - 3\ \text{Arcsec}\,(x/3) + C$ **29.** $-\sqrt{1-x^2}/x + C$ **31.** $\sqrt{2}/2$ **33.** $-x\cot x + \ln|\sin x| + C$
35. $(x^4/4)\ln x - x^4/16 + C$ **37.** $x\ \text{Arccos}\,x - (1-x^2)^{1/2} + C$ **39.** none of these **41.** power rule
43. $(-1/2)\ln|(2/x+3)| + C$ **45.** $2^x/\ln 2 + C$
Applied Problems: 1. about 6×10^{14} Hz **3.** 0.0377 cm^3 **5.** $V = (-400/3)(1-2t)^{3/2} + 1900/3$
7. $v = 5.32(10^6)e^{3+t^2} - 107(10^6)$ **9.** 0.617 J **11.** 14 cm/s **13.** 4.29 m/s **15.** 4.88×10^{-17} J
17. 0.000033 radians **19.** about 31°
Random Review: 1. $\ln|\sin e^x| + C$ **2.** $(1/5)(1+3e^x)^5 + C$ **3.** 0.881
4. $e^x\ln e^x - e^x + C$ or $xe^x - e^x + C$ **5.** $\ln|4-\cos x| + C$ **6.** $(-x/3)\cos 3x + (1/9)\sin 3x + C$
7. 3.19 **8.** 0 **9.** $e^{\text{Arcsin}\,x} + C$ **10.** $(3/2)e^{2x} - 9e^x + C$
Chapter Test: 1. $2\sqrt{\tan x} + C$ **2.** $(1/3)e^{x^3+1} + C$ **3.** 0.188 **4.** 0.102
5. $\sqrt{x^2-25} - 5\ \text{Sec}^{-1}(x/5) + C$ **6.** $\text{Sin}^{-1}(x/3) + C$ **7.** 5.89 **8.** $\ln|e^x + \sqrt{4+e^{2x}}| + C$
9. 0.02 mA **10.** $s = e^{\sin t} - 1$

CHAPTER 27

Section 27.1 (page 1099)

1. $1 - x^2/2! + x^4/4!$ **3.** $2x - 8x^3/3! + 32x^5/5!$ **5.** $x - x^3/2! + x^5/4!$ **7.** $1 - x + x^2/2!$
9. $x - 4x^2/2! + 12x^3/3!$ **11.** $1 - x + 2x^3/3!$ **13.** $1 + 2x + 3x^2$ **15.** $(2/3) + (2/9)x + (4/27)(x^2/2!)$
17. $4x - 16(x^2/2!) + 128(x^3/3!)$ **19.** $x + 2x^3/3!$ **21.** $x - 2x^3/3!$
23. $0.540 - 0.841\,x - 0.691\,x^2 - 0.270\,x^3$

Section 27.2 (page 1104)

1. $-1 + (x-\pi)^2/2! - (x-\pi)^4/4!$ **3.** $1 - 4(x-\pi/4)^2/2! + 16(x-\pi/4)^4/4!$
5. $e^6 + 2e^6(x-3) + 4e^6(x-3)^2/2!$ **7.** $2 + (1/4)(x-4) - (1/32)(x-4)^2/2!$
9. $(1/3) - (1/9)(x-3) + (2/27)(x-3)^2/2!$ **11.** $\ln 5 - (2/5)(x+2) - (4/25)(x+2)^2/2!$
13. (a) 0.2679492 **(b)** 0.2677805 **(c)** 0.2282844 **15. (a)** 16.44465 **(b)** 3.800000 **(c)** 16.06843
17. (a) 5.196152 **(b)** MacLaurin series does not exist **(c)** 5.200000 **19. (a)** 2.080084
(b) MacLaurin series does not exist **(c)** 2.083333 **21. (a)** 2.924018
(b) MacLaurin series does not exist **(c)** 2.925926 **23.** $c + x^2/(2!c) + x^4/(4!c^3) + \cdots$
25. MacLaurin series: $\alpha = 1 - R^2C/(2L)$ **27.** $v = 1 + t^2/2$

Section 27.3 (page 1109)

1. $1 + x^2 + x^4/2! + x^6/3! + \cdots$ **3.** $x^{1/2} - x^{3/2}/3! + x^{5/2}/5! - \cdots$
5. $1 - 4x^2/2! + 16x^4/4! - \cdots$ **7.** $-x - x^2/2 - x^3/3 - x^4/4 - \cdots$ **9.** $x + x^3/3! + x^5/5! + x^7/7!$
11. $x^2 + x^3 + x^4/2! + x^5/3!$ **13.** $2\sqrt{2}/(4-\pi) + [8\sqrt{2}/(4-\pi)^2 - 2\sqrt{2}/(4-\pi)](x-\pi/4) + [64\sqrt{2}/(4-\pi)^3 - 16\sqrt{2}/(4-\pi)^2 - 2\sqrt{2}/(4-\pi)](x-\pi/4)^2/2!$ **15.** $e^2 \ln 3 + (e^2 \ln 3 + e^2/3)(x-2) + (2e^2/3 - e^2/9 + e^2 \ln 3)(x-2)^2/2!$
17. $x - x^3/(3\cdot 3!) + x^5/(5\cdot 5!) - x^7/(7\cdot 7!) + \cdots$ **19.** $x - x^5/(5\cdot 2!) + x^9/(9\cdot 4!) - x^{13}/(13\cdot 6!) + \cdots$
21. $-x^{-1} + \ln x + x/2! + x^2/(2\cdot 3!) + x^3/(3\cdot 4!) + \cdots$
23. $x^3/3 - x^5/(2\cdot 5) + x^7/(3\cdot 7) - x^9/(4\cdot 9) + \cdots$ **25.** 3100 m
27. $i = 0.125\,t^2 + 0.025\,t^3 + 0.0028\,t^4$ **29.** 0.341 **31. (a)** yes **(b)** no **(c)** they differ by a constant

Section 27.4 (page 1114)

1. 0.017452406, at least ten places **3.** 0.9986292, at least six places **5.** 1.625000 **7.** 0.5
9. 0.692073, zero places **11.** 0.53023 **13.** 22.194518 **15.** 2.1162555
17. Arctan $x = x - x^3/3 + x^5/5$, $\pi = 3.47$ **19. (a)** $y = 25 + 0.00125\,x^2$ **(b)** 25.0003 m **21.** six terms

Section 27.5 (page 1123)

1. $1/2 + (2/\pi)\sin x + 2/(3\pi)\sin 3x$

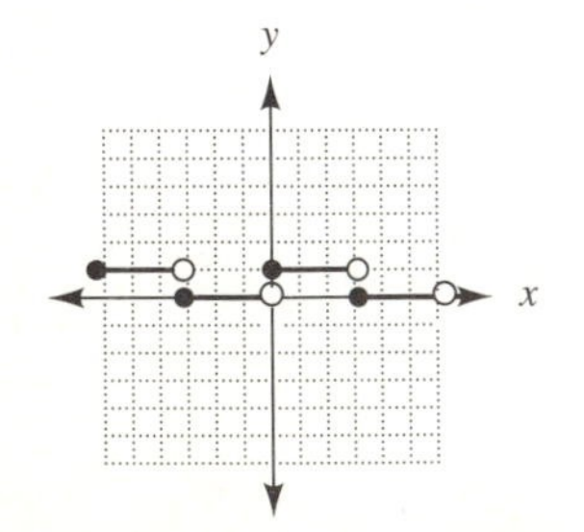

3. $(4/\pi)\sin x + 4/(3\pi)\sin 3x + 4/(5\pi)\sin 5x$

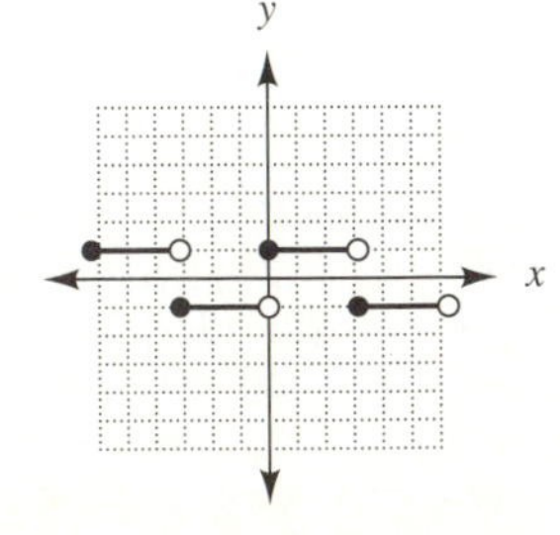

5. $\pi/4 - (2/\pi)\cos x + \sin x - (1/2)\sin 2x$

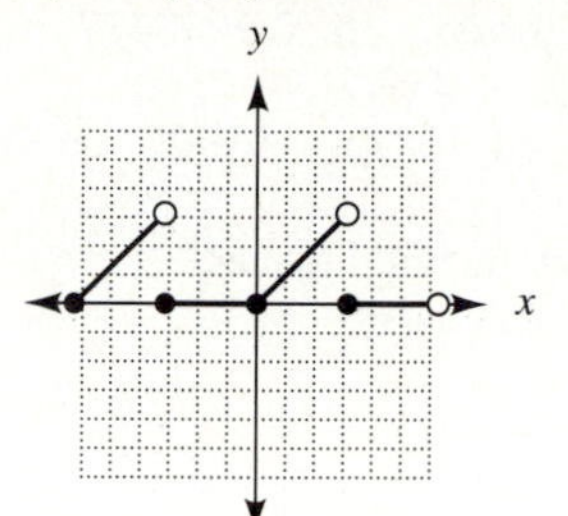

7. $\pi/2 - 4/\pi \cos x - 4/(9\pi)\cos 3x$

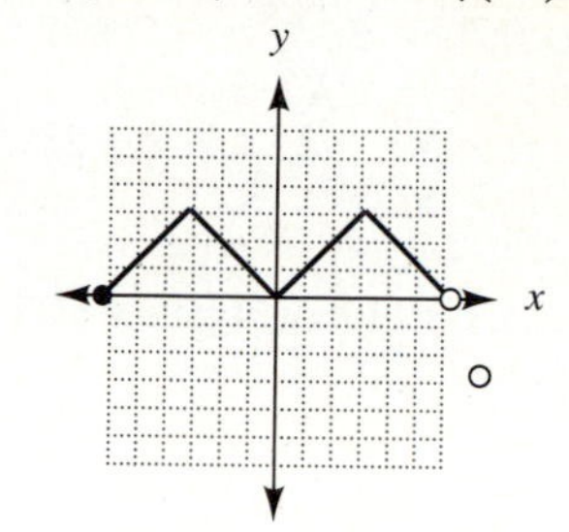

9. $1 + 2\sin x - \sin 2x$

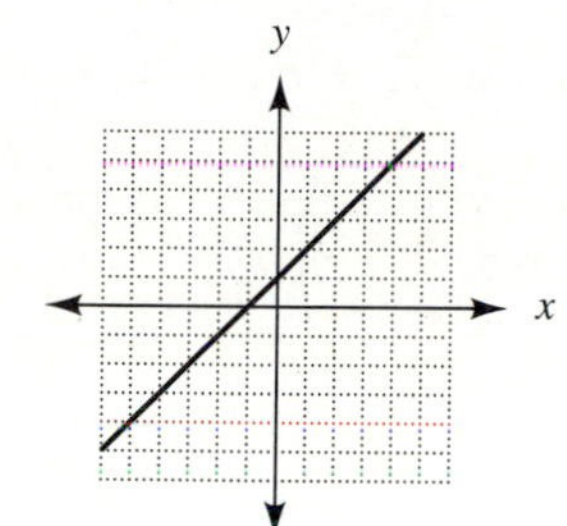

11. 1

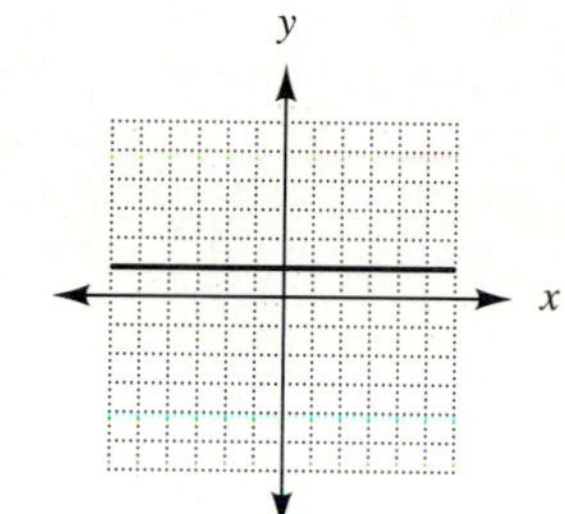

13. $1/\pi - 2/(3\pi)\cos 2x + (1/2)\sin x + \cdots$

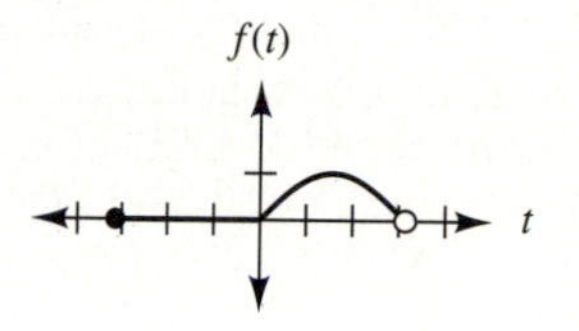

15. $170/\pi$

Section 27.6 (page 1125)

Sequential Review: 1. $(1/2)x - (1/8)x^3/3! + (1/32)x^5/5!$ **3.** $x - 2(x^2/2!) + 3(x^3/3!)$
5. $1 - 2x + 4x^2/2!$ **7.** $\sqrt[3]{3} + (\sqrt[3]{3}/9)x$ **9.** $-3x + 9x^2/2$
11. $-(x - \pi/2) + (x - \pi/2)^3/3! - (x - \pi/2)^5/5!$ **13.** $e + e(x - 2) + e(x - 2)^2/2!$
15. $\ln 3 - (1/3)(x + 2) - (1/9)(x + 2)^2/2!$ **17.** 0.38261 **19.** $x/2 - x^3/(8 \cdot 3!) + x^5/(32 \cdot 5!) - \cdots$
21. $1 + \sqrt{x} + x/2! + x\sqrt{x}/3! + \cdots$ **23.** $-e^{\pi}(x - \pi) - e^{\pi}(x - \pi)^2$
25. $\ln x - x^2/(2 \cdot 2!) + x^4/(4 \cdot 4!) + \cdots$ **27.** $(-2/3)x^{3/2} - (1/2)x^2/2 - (2/5)x^{5/2}/3 - \cdots$ **29.** 6.625
31. at least ten decimal places **33.** at least two decimal places **35.** 0.6427076 **37.** $(1/2 + \pi/4)$
39. $b_n = -1/n$, if n is even; $(-2 + \pi)/(\pi n)$, if n is odd **41.**

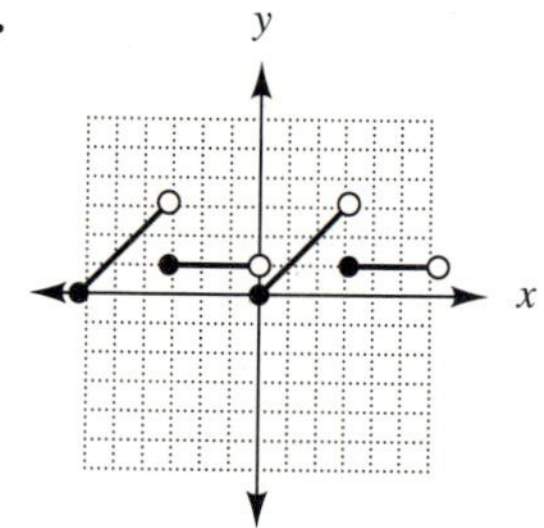

43. $a_n = 0$, if n is even; $4/(n^2\pi)$, if n is odd **45.** $-\pi/2 + (4/\pi)\cos x + 4/(9\pi)\cos 3x$
Applied Problems: 1. at least nine places **3.** $63711.5\,t + 1.5091 \times 10^9\,t^3 + 6.3456 \times 10^9\,t^5$
5. $100.3 \times 10^{-3} + 2.310(v - 0.7) + (53.19/2!)(v - 0.7)^2$ **7.** $5.47 - 10.9\,t + 27.3\,t^2$
9. $4.96 + 9.90(v - 1.3) + 19.81(v - 1.3)^2/2 = 9.906\,v^2 - 15.85\,v + 8.816$
11. $546 \times 10^{-6} + 34.9 \times 10^{-6}(T - 273)$ **13.** $v = 7.2384\,t - 171{,}464t^3$, 723.7 μv
15. $\cos t = 1 - t^2/2$ **17.** $v(t) = 10 \sin \pi/8$ if $0 \le t < 2$ and $v(t) = 0$ if $2 < t \le 16$

Random Review: 1. $1 - x/2! + x^2/4!$ **2.** $1 + 1(x-1) + (x-1)^2/2!$ **3.** 0.424425
4. $2/3 - (2/9)x + (4/27)(x^2/2!)$ **5.** $b_n = -1/n$, if n is even; $1/n + 2/(n\pi)$, if n is odd **6.** 0.9998477
7. $\pi^2/6$ **8.** $x^3/3 - x^5/(5 \cdot 3!) + x^7/(7 \cdot 5!) - \cdots$ **9.** 3.0740741 **10.** $1 - 2x + 4x^2/2!$
Chapter Test: 1. $1 - x/2! + x^2/4!$ **2.** $1 + x - 2x^3/3!$ **3.** $3 + (1/6)(x-9) - (1/108)(x-9)^2/2!$
4. $-1/2 - \sqrt{3}(x - \pi/3) - (x - \pi/3)^2$ **5.** $x^3/3 - x^7/(7 \cdot 3!) + x^{11}/(11 \cdot 5!) - \cdots$
6. $x + (2/3)x^{3/2} + (1/2)x^2/2! + (2/5)x^{5/2}/3! + \cdots$ **7.** 0.9993908, to at least six decimal places
8. $(-\pi/4 + 1/2) + (2/\pi)\cos x + (1 + 2/\pi)\sin x - (1/2)\sin 2x$ **9.** $V = 24.32 + 3.04(P-4)$
10. at least six decimal places

CHAPTER 28

Section 28.1 (page 1134)

1. 1, 1 **3.** 1, 1 **5.** 2, 1 **7.** 2, 1 **9.** 1, 2 **11.** $4 - (1/3)(12 - Be^{-x/3}) = (1/3)Be^{-x/3}$
13. $4Ae^{2x} + 4Be^{-2x} - 4(Ae^{2x} + Be^{-2x}) = 0$ **15.** $-4A\sin 2x - 4B\cos 2x + 4(A\sin 2x + B\cos 2x) = 0$
17. $y' = \dfrac{y}{1/y - x} = \dfrac{y}{(1-xy)/y} = \dfrac{y^2}{1-xy}$
19. $(y\cos y - \sin y + x)[1/(1+\cos y)] = (y\cos y - \sin y + y + \sin y)[1/(1+\cos y)]$
$= y(\cos y + 1)\,[1/(1+\cos y)] = y$
21. $y = C/(2x-1)$ **23.** $y^2(x^2+1) = C$ **25.** $Cy^2 = (x-1)^3$

Section 28.2 (page 1140)

1. $3x^2 + C = y^2$ **3.** $xy + yC = -1$ **5.** $x^2y^2 = 1 + Cx^2$ **7.** $y = Ce^x - 2$
9. $y = 1/[\ln(Ce^x + C)]$ **11.** $e^x = e^y + C$ **13.** $|\sec x + \tan x| = Ce^{-\cos y}$ **15.** $3y^3\cos x + Cy^3 = 1$
17. $4y^3 = -3x^4 + 108$ **19.** $y = e^{x^2/2+1}$ **21.** $\ln y = (1/2)\text{Arctan}\,(x/2)$ **23.** $x + y = 2xy$
25. (a) $y = -10\,e^{-0.1t} + 10$ **(b)** 4 l, to one significant digit **27. (a)** $y = 200\,e^{0.288t}$
(b) 600, to one significant digit **29. (a)** $117\,e^{-2.05t} + 68 = y$ **(b)** 83°

Section 28.3 (page 1146)

1. $ye^{x^2/2} = 3e^{x^2/2} + C$ **3.** $y = 2(x-1) + Ce^{-x}$ **5.** $y = (1/2)e^x + Ce^{-x}$ **7.** $y = 2/3 + Ce^{-3x}$
9. $y = (1/6)x^3 + Cx^{-3}$ **11.** $yx^2 = x^2\sin x - 2\sin x + 2x\cos x + C$ **13.** $y\cos x = x + C$
15. $y = -1 + Ce^{-\cot x}$ **17.** $y = x - 1 + 2e^{-x}$ **19.** $y = (1/4)(2x-1) + (1/4)e^{-2x}$
21. $y = xe^{2x} + 2e^{2x}$ **23.** $y = 1$ **25. (a)** $y = 20e^{-0.223t}$ **(b)** 12.8 g **(c)** 3.11 hours
27. (a) $y = 10{,}000\,e^{0.0002t}$ **(b)** \$10,120.60 **(c)** about 9.5 years **29. (a)** $q = CV - CVe^{-t/RC}$
(b) 0.3 C, to one significant digit **31.** $y^2 - x^2 = C$

Section 28.4 (page 1152)

1. $y = C_1e^{4x} + C_2e^{-3x}$ **3.** $y = C_1e^{(-1/2)x} + C_2e^{2x}$ **5.** $y = C_1 + C_2e^{-x/2}$ **7.** $y = C_1e^{-2x} + C_2xe^{-2x}$
9. $y = C_1e^{-0.5x}(\cos x\sqrt{3}/2) + C_2e^{-0.5x}(\sin x\sqrt{3}/2)$ **11.** $y = 8 - 3e^x$ **13.** $y = 3e^{(-2/3)x} + e^{2x}$
15. $y = 1 - e^{-2x}$ **17.** $y = e^{-x} + xe^{-x}$ **19.** $y = \cos x + \sin x$ **21.** $x = C_1\cos 3t + C_2\sin 3t$
23. $x = C_1e^{-9t} + C_2e^{-t}$ **25.** $z = C_1\cos\theta + C_2\sin\theta$

Section 28.5 (page 1156)

1. $y = C_1e^{2x} + C_2e^{-2x} - (1/2)x$ **3.** $y = C_1e^{(-1/2)x} + C_2e^{-x} + x^2 - 7x + 17$
5. $y = C_1 + C_2e^{4x} - (1/4)e^{2x}$ **7.** $y = C_1e^{(-3/2)x} + C_2e^{-x} + (1/7)e^{2x} + (1/3)x - 5/9$
9. $y = C_1e^{2x} + C_2e^{-3x} + (3/50)\sin x - (21/50)\cos x$

11. $y = C_1 + C_2e^{-x/3} + (1/74)\cos 2x + (3/37)\sin 2x$ **13.** $y = 5\cos 5x - \sin 5x + 0.2x$
15. $y = \cos x - 3\sin x$ **17.** $x = C_1e^{-t} + C_2e^{-3t} - 0.5\cos 2t + 4\sin 2t$
19. transient term: $e^{-\alpha t}(C_1\cos\beta t + C_2\sin\beta t)$, steady state term: $A\cos\omega t + B\sin\omega t$
21. $q = C_1e^{-2t}\cos 4t + C_2e^{-2t}\sin 4t + \cos 4t + 4\sin 4t$

Section 28.6 (page 1162)

All answers assume $s > 0$ **1.** $1/s$ **3.** $1/(s+a)$ **5.** $a/(a^2+s^2)$ **7.** $b/[(a-s)^2+b^2]$, if $s > a$
9. $a^2/[s(s^2+a^2)]$ **11.** $a^3/[s^2(s^2+a^2)]$ **13.** $(a-s)/(s^2+a^2)$ **15.** e^{2x}, $s > 2$ **17.** $\sin x$
19. xe^{-3x} **21.** $2e^{-2x} - 3e^{-3x}$ **23.** $x\cos 2x$

Section 28.7 (page 1167)

1. $y = e^x$ **3.** $y = 2e^{-2x}$ **5.** $y = \sin 2x$ **7.** $y = 2\cos x + \sin x$ **9.** $y = e^{2x} + e^{3x}$
11. $y = x\sin 2x$ **13.** $y = e^x\sin x$ **15.** $y = xe^{-x}$ **17.** 28% **19.** 17 m/s

Section 28.8 (page 1169)

Sequential Review: 1. 2, 1 **3.** 2, 2
5. $-12\sin 2x + 4(1 + 3\sin 2x) = 4$; $-12\sin 2x + 4 + 12\sin 2x = 4$ is an identity **7.** $C(x^2-1) = y^3$
9. $2x^2 + 2x + C = y^2$ **11.** $y = (1/2)e^{-2x} + C$ **13.** $y = x + 3x^2/2 + 1$ **15.** $y = \ln x$
17. $y = (1 - \cos x) + C$ **19.** $y = 2 + Ce^{-x^2/2}$ **21.** $y = -1 + 3e^x$ **23.** $y = x^2 + 3/x$
25. $y = C_1 + C_2e^{(3/2)x}$ **27.** $y = C_1e^{(-1/3)x} + C_2xe^{(-1/3)x}$ **29.** $y = 3 + 2e^{3x}$
31. $y = C_1e^{2x} + C_2e^x + (1/10)\sin x + (3/10)\cos x$ **33.** $y = (3/4)e^{2x} + (1/4)xe^{2x} + 1/4 + (1/4)x$
35. $1/s^2$, if $s > 0$ **37.** $a/[s(s+a)]$, $s > -a$ **39.** $\sin 4x$ **41.** $y = e^{-x}$ **43.** $y = e^{3x} - e^{2x}$
45. $y = e^{-x}\sin 2x$
Applied Problems: 1. $q = (-12/5)e^{25t/13}\cos(5/13)t + 12/5$

3. $6\left[\dfrac{s+1}{(s+1)^2+707^2}\right] + 5.66 \times 10^{-3}\left[\dfrac{707}{(s+1)^2+707^2}\right]$

5. $i' = -28e^{-4t} - 3e^{-t}$, $i'' = 112e^{-4t} + 3e^{-t}$, so $112e^{-4t} + 3e^{-t} + 5(-28e^{-4t} - 3e^{-t}) + 4(7e^{-4t} + 3e^{-t}) = 112e^{-4t} + 3e^{-t} - 140e^{-4t} - 15e^{-t} + 28e^{-4t} + 12e^{-t} = 0$
7. $R = Z\cos\theta$ and $dR/d\theta = -Z\sin\theta = (-R/\cos\theta)\sin\theta = -R\tan\theta$, Thus $dR/d\theta + R\tan\theta = 0$ **9.** $r = e^{4.5313-1/\theta}$ **11.** $i = (3\cos t + \sin t)/10 - 0.3e^{-3t}$

13. $i = 0.376e^{-20.6t} - 0.457e^{-104t} + 0.814$ **15.** $\dfrac{-31+4s}{(s-3)(s-8)}$ **17.** $\dfrac{10s+23.6}{s^3+2s^2+15s}$

Random Review: 1. $y = C_1e^{2x} + C_2e^{-2x} - (1/3)e^x$ **2.** 2, 1 **3.** $y = 4\cos x + 2\sin x$ **4.** xe^x
5. $y^2 = 1/(3 - 2e^x)$ **6.** $2e^{-x} + 8 - 2e^{-x} = 8$ is an identity **7.** $5y^3 = 6x^5 + C$
8. $y = C_1e^{(-5/2)x} + C_2e^x$ **9.** $y = (-1/2)e^x + (7/2)e^{3x}$ **10.** $(s^2 + s - a)/[s^2(s-a)]$, $s > a$
Chapter Test: 1. 2, 1 **2.** $y = Ce^{\sin x}$ **3.** $y = e^x + 1$ **4.** $y = -2 + 4e^{x^2/2}$ **5.** $y = C_1e^{x/3} + C_2e^{-2x}$
6. $y = C_1e^x + C_2xe^x + x^2e^x$ **7.** $1/(s-a) + a/(s^2+a^2)$, $s > a$ **8.** $y = 1 + x$ **9.** 2.93 qt
10. $i = -1.33\cos 377t + C$

INDEX